1 ARMY

2004

GUIDE TO THE EVALUATION OF

Educational Experiences *in the* Armed Services

ACE

American Council on Education • One Dupont Circle NW • Washington, DC 20036-1193

PRODUCTION STAFF FOR THE 2004 GUIDE

Military Programs

James Selbe, Director, Program Evaluations
Cynthia Bruce, Assistant Director, Military Programs
Diana Gross, Assistant Director, Military Registries
Penelope Suritz, Assistant Director, Military Programs
Gwendolyn L. Dozier, Program Coordinator
Jacque McDowell, Program Coordinator
Thia Davis, Program Coordinator
Detra Rouse, Project Coordinator
Karen Montanez-Safadi, Office Assistant

The work reported or presented herein was performed pursuant to a contract with Defense Activity for Non-Traditional Education Support (DANTES) on behalf of the Department of Defense (Contract N00140-03-C- J-093). However, the opinions expressed herein do not necessarily reflect the position or policy of the U.S. Department of Defense, and no official endorsement by the U.S. Department of Defense should be inferred.

Library of Congress Cataloging in Publication Data:
American Council on Education.
Guide to the evaluation of educational experiences in the armed services.
ISBN 1-57356-608-X

Contents

National Association of Independent Colleges & Universities
Mary Brown Bullock, President
Agnes Scott College, GA

National Association of State Universities & Land-Grant Colleges
Kermit L. Hall, President
Utah State University, UT

ELECTED ASSOCIATIONS - FOR ONE-YEAR TERMS

Council for Higher Education Accreditation
Judith Eaton, President
Council for Higher Education Accreditation, DC

National Association of College and University Business Officers
James E. Morley, Jr., President
National Association of College and University Business Officers, DC

DESIGNATED REPRESENTATIVE – FOR ONE-YEAR TERM

Washington Higher Education Secretariat
Carol Schneider, President
Association of American Colleges & Universities, DC

Commission on Lifelong Learning of the
American Council on Education
Washington, DC

Foreword

For more than a half century, the *Guide to the Evaluation of Educational Experiences in the Armed Services* has been the standard reference work for recognizing learning acquired in military life. Since 1942, the American Council on Education has worked cooperatively with the U.S. Department of Defense, the Armed Services, and the U.S. Coast Guard in helping hundreds of thousands of individuals earn academic credit for learning achieved while serving their country.

The ACE *Guide* evaluation system enables students to apply their experience, training, and expertise to their degree work at colleges and universities. Based on sound educational practices, the program maintains academic integrity while providing for an efficient use of financial and educational resources.

Special recognition must be paid to the many faculty members who served as content specialists on ACE evaluation teams and to their institutions for their wholehearted cooperation in this effort. For all this generous support and assistance, we are deeply indebted.

We are pleased to commend this *Guide* to you as you work with servicemembers and veterans in helping them to integrate ACE *Guide*-recommended credit into their degree programs.

David Ward
President
American Council on Education

How to Find and Use Army Course Exhibits

This volume contains recommendations for formal courses offered by the Army, Army Reserve, and Army National Guard with exhibit dates of 1/90 and later.

The instructions that follow provide a step-by-step procedure for finding and using the exhibits and recommendations. Readers unfamiliar with ACE evaluation procedures should read Appendix A.

Step 1

Verify course completion from military records.

It is the responsibility of the institution official to verify course completion. The following military records are used to verify successful completion of military courses.

- *Army/American Council on Education Registry Transcript System* (AARTS) transcript. The AARTS transcript is a convenient, fully-automated transcripting system that carefully tracks all of a servicemember's or veteran's military training, with college credit recommendations for ACE-completed evaluations. It also includes college-level test results. The information contained in an AARTS transcript is presented in a format that is designed to help registrars, counselors, and admissions officers award college credit for these experiences.

  AARTS transcripts are available to enlisted members of the regular Army, Army National Guard, and the Army Reserve. The Army has removed the restriction limiting AARTS transcripts to soldiers with basic active service duty dates on or after October 1, 1981. Therefore, servicemembers with service dates before this date may be eligible to obtain an AARTS transcript. This change may also affect Army National Guard soldiers and veterans on active duty as of January 1, 1993 with basic active service dates on or before October 1, 1981. These individuals are advised to seek further information by contacting the AARTS Operation Center. The toll-free number is 1-866-297-4427; the AARTS web site is http://aarts.army.mil.

If the student has submitted an AARTS transcript, it is not necessary to follow this step-by-step procedure to verify military experiences. See Step 6 if you have questions about the transcript.

If the student is not eligible for a transcript, use the following documentation:

- DD Form 295, *Application for the Evaluation of Learning Experiences During Military Service.* Military education officers can provide this form to active-duty servicemembers, reservists, and National Guard members. The form must be certified by an authorized commissioned officer or his/her designee.
- DD Form 214, *Certificate of Release or Discharge from Active Duty.* If the veteran does not have a copy, one can be obtained, together with other in-service training records, from the National Personnel Records Center (Military Personnel Records), 9700 Page Avenue, St. Louis, MO 63132-5100. The veteran may request service records by submitting U.S. Government Standard Form 180, which is available from a state veterans affairs office, the Veterans Administration, or the National Personnel Records Center, or it can be reproduced from this volume (Appendix B).
- *Course Completion Certificates.* These may be used to complement other records or when service courses are not recorded on official records.

Step 2

Find the course exhibit by identifying the American Council on Education (ACE) ID number in the Course Number Index or the Keyword Index.

- *Course Number Index.* All available military course numbers are listed in the Course Number Index in alphanumeric sequence. If the applicant's military course number cannot be located in the Course Number Index, search for the course title in the Keyword Index.
- *Keyword Index.* Identify all possible keywords within a course title. For example, the keywords in the title "Digital, Analog, and Hybrid Computer Fundamentals" are *Digital, Analog, Hybrid,* and *Computer.* Find one of those keywords in the Keyword Index, and search the listing under the keyword for the course title. If the title cannot be found under one keyword, search all other possible keywords.
- *Identify ACE ID number.* When the title or military course number has been located, note the corresponding ACE ID number. This number refers to the course exhibit's location in the *Guide.* The two-letter prefix refers to the specific service, i.e., AR=Army. ACE ID numbers are presented in numeric sequence.

Step 3

Match the course-identifying information with the corresponding data in the course exhibit.

First determine that the dates of attendance fall within the dates in the exhibit, and select the appropriate version. The version with the highest number is the current version. Other course-identifying information includes the official military title, military course number, length of course, dates of attendance, and location. It is important to match all items.

Step 4

Read the course description.

Consideration should be given not only to the amount of credit and to the subject area, but also to the learning outcomes or course objectives, and instruction cited in the course exhibit. These portions of the exhibit outline the course content and scope and also provide essential information about the nature of the course. This information will help you determine the appropriate placement of credit for each individual student within the requirements and programs at your institution.

Step 5

Award credit as appropriate.

Users may modify the credit recommendations in accordance with institutional policy and the educational goals of each applicant.

Step 6

If assistance is required, contact the American Council on Education.

If assistance is needed, the college can submit a *Request for Course Recommendation* form completed by the student. Contact the Military Evaluations Advisory Service at:

Military Evaluations Program
American Council on Education
One Dupont Circle NW, Suite 250
Washington, DC 20036-1193
(202) 939-9470
FAX: (202) 775-8578
E-mail: mileval@ace.nche.edu

ACE does not make corrections to military transcripts. To request corrections or updates to AARTS transcripts, the student should contact the AARTS Operation Center in writing, stating the specific problem with reference to the in writing, stating the specific problem with reference to the transcript number and dates. Submit the required documentation of course completion or MOSs held. Send to:

AARTS Operation Center transcript number and dates. Submit the required documentation of course completion or MOSs held. Send to:

AARTS Operation Center
415 McPherson Avenue
Ft. Leavenworth, KS 66027-1373
1-866-297-4427 (toll-free)
E-mail: aarts@leavenworth.army.mil

Sample Course Exhibit

AR-1401-0033

FINANCE SPECIALIST

Course Number: *Version 1:* 542-73C10; 542-73C10 (ST). *Version 2:* 542-73C10; 542-73C10 (ST).

Location: *Version 1:* Finance School, Soldier Support Institute, Ft. Benjamin Harrison, IN. *Version 2:* Soldier Support Institute, Ft., Jackson, SC; Finance School, Soldier Support Institute, Ft. Benjamin Harrison, IN.

Length: *Version 1:* 7-8 weeks (244-296 hours). *Version 2:* 7-8 weeks (244-296 hours).

Exhibit Dates: *Version 1: 1/90*-3/91. *Version 2:* 4/91-Present.

Learning Outcomes: *Version 1:* Upon completion of the course, the student will be able to identify and determine the types of incentive pay plans, prepare pay vouchers, complete pay documents, and compute pay allowances. *Version 2:* Upon completion of the course, the student will be able to determine pay status and adjustments, compute payroll and travel allowances, prepare payroll and travel vouchers, process pay inquiries, and access and input data on computer.

Instruction: *Version 1:* Methods of instruction include lectures, role playing, and in-class exercises. Topics covered include financial operations, data entry operations, and payroll processing. *Version 2:* Methods of instruction include lectures, role playing, and in-class exercises. Topics covered include financial operations and payroll processing.

Credit Recommendation: *Version 1.* In the vocational certificate category, 2 semester hours in payroll accounting (4/91); in the lower-division baccalaureate/associate degree category, 1 semester hour in computer applications (4/91). *Version 2:* In the lower-division baccalaureate/associate degree category, 2 semester hours in payroll accounting and 1 in computer applications (1/97).

Related Occupations: 73C

ID Number. An identification number assigned by ACE to identify each course.

Military Course Number. The number assigned to the course by the military. Listed by version, if appropriate.

Length. The length of the course in weeks, with contact hours in parentheses. Listed by version.

Learning Outcomes. Competencies students acquire during the course. Older courses have Objectives.

Credit Recommendation. By version. Recommended in four categories: vocational certificate, lower-division baccalaureate/associate degree, upper division baccalaureate, and graduate. degree. Expressed in semester hours. (See Appendix A for detailed explanation of credit categories.)

Related Occupation. A cross reference to related MOS exhibits. Officials awarding credit for a course and an MOS should compare the exhibit for the course with that for the MOS before awarding credit.

Title. Version 1 (if applicable) is the more recent. If course has only one version, version number is omitted throughout exhibit.

Location. By version. The service school, military installation, state.

Exhibit Dates. Listed by version. Training start date on materials evaluated and, if applicable, the date the training was eliminated. "Present" denotes publication cut-off for this edition of the Guide (5/02). The earliest start date in this edition is 1/90.

Instruction. Description of instruction, including teaching methods, facilities, equipment, major subject areas covered.

Evaluation Date. Date when the credit recommendation was established. Month and year are given in parentheses following each recommendation.

How to Find and Use Army Enlisted MOS Exhibits

This volume contains recommendations for all Army enlisted military occupational specialties (MOSs) with exhibit dates of 1/90 and later.

The instructions that follow provide a step-by-step procedure for finding and using the MOS exhibits and recommendations. Readers unfamiliar with how MOSs are structured and how MOS proficiency is demonstrated should read Appendix A.

Step 1

Have the applicant submit official documentation

Several records may be used for MOS verification. The applicant may submit one or several of the following forms:

- *Army/American Council on Education Registry Transcript System* (AARTS) transcript. The AARTS transcript is a convenient, fully-automated transcripting system that carefully tracks all of a servicemember's or veteran's military training, with college credit recommendations for ACE-completed evaluations. It also includes college-level test results. The information contained in an AARTS transcript is presented in a format that is designed to help registrars, counselors, and admissions officers award college credit for these experiences.

  AARTS transcripts are available to enlisted members of the regular Army, Army National Guard, and the Army Reserve. The Army has removed the restriction limiting AARTS transcripts to soldiers with basic active service duty dates on or after October 1, 1981. Therefore, servicemembers with service dates before this date may be eligible to obtain an AARTS transcript. This change may also affect Army National Guard soldiers and veterans on active duty as of January 1, 1993 with basic active service dates on or before October 1, 1981. These individuals are advised to seek further information by contacting the AARTS Operation Center. The toll-free number is 1-866-297-4427; the AARTS web site is http://aarts.army.mil.

If the student has submitted an AARTS transcript, it is not necessary to follow this step-by-step procedure to verify military experiences. See Step 6 if you have questions about the transcript.

If the student is not eligible for a transcript, use the following documentation:

- DD Form 295, *Application for the Evaluation of Learning Experiences During Military Service.* This form may be used when it contains occupational information. The April 1975 version contains the section "Major Service Jobs or Billets." The June 1979, November 1986, and April 2000 versions have a section titled "Military Occupational History." Check for MOS designation, title, and period during which the MOS was held. The form may also list test scores and dates. The form must be certified by an authorized commissioned officer or his/her designee.
- DD Form 214, *Certificate of Release or Discharge from Active Duty.* This form may be used for verifying an MOS but will not necessarily provide the dates during which the MOS was held. Check Item 11 for MOS designation, title, and years and months in the MOS.
- DA Form 2166-5, *Enlisted Evaluation Report (EER).* This document contains the supervisor's rating. It provides ratings on duty performance traits, demonstrated overall performance of assigned duties, and advancement potential. Because supervisor's ratings are subjective, one EER alone is seldom sufficient to validate proficiency.
- DD Form 2166–5A, *Senior Enlisted Evaluation Report (SEER).* This document provides ratings on performance qualities, leadership skills, demonstrated overall performance, and advancement potential. The maximum score attainable on EERs and SEERs is 125. EERs were formerly given to those in paygrades E-4 and E-5, while E-6 and above received SEERs. Then EERs were discontinued, and those in paygrades E-5 through E-7 were given SEERs. Those in paygrades E-4, E-8, and E-9 were not evaluated.
- DA Form 2166-1, *NCO Evaluation Report (NCOER).* Current supervisor's rating, replacing EERs and SEERs. This document provides description and scope of duties, ratings, and written comments about the NCO's MOS competence, leadership abilities, training responsibilities, and responsibility and accountability. Check Part IV for a success rating (meets standards) or an excellence rating (exceeds standards) in sections b, d, e, and f. In Part V, determine that an overall rating of fully capable or among the best has been achieved.
- USAEREC Form 10A and EREC Form 10, *Enlisted Evaluation Data Report. Enlisted Evaluation Data Report* replaces USAEEC Form 10. This record provides an EER weighted average (EERWA) score. It also lists raw Skill Qualification Score (SQT) scores along with a

percentile score. The percentile compares the soldier with others who have taken the test in that MOS.

- DA Form 2-1, *Personnel Qualifications Record.* This record provides a summary of a soldier's career, including MOSs held, courses taken, and awards received. Item 6 lists each MOS designation and skill level, title, and dates held. Item 35 lists the soldier's assignments, with MOS and skill level designations and the effective date of the assignment.

- DA Form 20, *Enlisted Qualification Record.* Sections 22 and 23 provide the MOS code, title, date awarded, and MOS evaluation scores and dates. Section 38 lists duty positions held and dates.

- *Individual Soldier's Report* (ISR). This form reports SQT results. Check for the MOS designation, skill level, dates, and a score report. After August 1983, a score of 60 or above indicates that the individual has qualified in the MOS for the Army's purposes.

Step 2

Verify each occupational specialty the person has successfully held.

When verifying enlisted MOSs, it is important to identify the skill level. Since the higher skill levels of an MOS encompass the duties of the lower skill levels (i.e., a person with MOS 12B40 is able to perform the duties of 12B30), the credit recommendation for MOS 12B40 would include the recommendation for 12B30. Thus, you need to verify the highest skill level a person held.

To determine an individual's eligibility for the award of MOS credit, determine the dates the MOS was held and follow the appropriate instructions from the following list:

- January 1990--September 1991. Check for a score of 60 or above on the SQT; if score is not available, use NCOERs.

- October 1991-September 1993. The SQT was replaced by the Self-Development Tests (SDT) for skill levels 20-40. The SDT was phased in gradually during this time period and was not reviewed by ACE during its pilot stage. Use NCO Evaluation Reports (NCOERs). There is no longer an Army formal assessment for Skill Level 10 personnel. ACE no longer publishes credit recommendations for Skill Level 10.

- October 1993-February 1995. The Self-Development Test (SDT) received ACE's endorsement for use as an indicator of MOS proficiency. Credit may be awarded for enlisted MOSs at Skill Levels 20-40 when an SDT score of 70 percent or greater is achieved. SDT scores will be reported on the AARTS transcript and on the Individual Soldiers Report (ISR). If score is not available, use NCOERs. For further information, see Appendix A.

- March 1995-Present. The Army discontinued offering the SDT as of February 1995. ACE looked into the possibility of using NCOERs as the sole indicator of MOS proficiency. It was found that while the NCOER is an indicator of competency in management skills, it falls short as an indicator of technical competence.

Further study led to the following policy:

After 3/95, only soldiers in skill levels 30, 40, and 50 will be eligible for management credit based on the NCOER. They will also be eligible to receive the technical credit recommended for the preceeding skill level.

The MOS exhibits in this *Guide* have been modified to reflect this change.

Step 3

Find the appropriate exhibit in the Guide.

Find the information necessary for locating the correct MOS exhibit. Pertinent information includes the MOS designation, the MOS title, the dates each MOS was held, and, when appropriate, the date associated with an MOS evaluation (MOS Evaluation Test score, SQT results, EER/SEER, NCOER). If you cannot verify MOS proficiency through a careful examination of military records, you may wish to conduct your own assessment. *Verification of MOS proficiency should always precede the granting of credit or advanced standing.*

The exhibit(s) for an MOS can easily be found when the MOS designation is known. Each MOS exhibit is assigned an ACE ID number that has three components. The first component, MOS, identifies the exhibit as one that describes an Army military occupational specialty. The second component consists of the MOS designation, e.g., 81C. The third component, a three-digit, sequentially assigned number, e.g., -001, -002, -003, uniquely identifies the exhibit. Enlisted MOS exhibit ID numbers have nine characters, e.g., MOS-81C-001.

If only the title of the MOS is known, the exhibit ID number(s) can be found by referring to the Occupation Title Index.

There may be several exhibits for an MOS. Each time there is a new recommendation for an MOS, there is a new exhibit. For example, three exhibits appear for MOS 12B. The ACE ID numbers for the three exhibits are MOS-12B-001, MOS-12B-002, and MOS-12B-003.

The oldest exhibit for the MOS, the first to be evaluated, is assigned -001 as the last three characters. The last exhibit for an MOS, the exhibit ending in the highest number, is the most recent.

When there is more than one exhibit, select the correct one by referring to the exhibit dates that conform to the time frame in which the individual held the MOS. The exhibit dates for each new exhibit establish continuity with the previous exhibit. If the end date for the first exhibit is 1/91, the start date for the next exhibit is 2/91. In many cases, the date of the soldier's last MOS evaluation will indicate which exhibit is most appropriate.

When the time period that a person held an MOS skill level spans more than one exhibit for the MOS, use your own judgment in deciding which exhibit is more appropriate. If the credit granted in the earlier exhibit does not appear in a later exhibit, the credit that was dropped may be awarded. This is based on the belief that if credit was recommended for a particular skill required during the earlier time frame, it should not be taken away from the individual just because that skill is no longer a requirement of that MOS.

Step 4

Read the entire exhibit.

Although credit is granted for a particular skill level of an MOS, users are urged to read the entire MOS exhibit. Doing so will make possible the most appropriate application of the recommendation to the student's program of study at your institution. Each item in the exhibit has been prepared to help users interpret the credit recommendations. (See sample exhibit.)

The descriptions, which are similar to learning outcome statements of postsecondary courses and programs of study, provide essential information about the learning required for proficiency in the MOS. Comparing the MOS description with a description of the course or program of study that the student will pursue at your institution will help you

- determine how much of the recommended credit applies to the course or program of study at your institution; identify additional areas of possible credit;
- resolve problems with duplication of credit when the applicant has applied for credit for more than one military learning experience; and
- place the student at the appropriate point in the course sequence or program of study.

Step 5

Award credit as appropriate.

The credit recommendations are advisory. They are intended to assist in placing active-duty personnel and veterans in postsecondary programs of study and jobs.

When an applicant has applied for credit for more than one military learning experience, you may have to reduce the total amount of credit recommended to avoid granting duplicate credit.

If the AARTS transcript is being used, all course or MOS credit recommendations will be included as part of the document. As duplication of credit recommendations may exist for more than one learning experience, it is advised that the active transcript be reviewed and adjustments made to avoid awarding duplicate credit.

Step 6

If assistance is required, contact the American Council on Education.

If assistance is needed, the college can submit a *Request for Army Enlisted and Warrant Officer Exhibits* form. Contact the Military Evaluations Advisory Service at:

Military Evaluations
American Council on Education
One Dupont Circle NW, Suite 250
Washington, DC 20036-1193
(202) 939-9470
FAX: (202) 775-8578
E-mail: mileval@ace.nche.edu

Please note that ACE does not make corrections to military transcripts. To request corrections or updates to AARTS transcripts, the student should contact the AARTS Operation Center in writing, stating the specific problem with reference to the transcript number and date. Submit the required documentation of course completion or MOSs held. Send to:

AARTS Operation Center
415 McPherson Avenue
Ft. Leavenworth, KS 66027-1373
1-866-297-4427 (toll-free)
E-mail: aarts@leavenworth.army.mil

Sample Enlisted MOS Exhibit

ID Number. A nine-character code assigned by ACE to identify each enlisted MOS exhibit.

MOS Designation. The five-character codes that identify the MOS and each of its skill levels. An MOS may have as many as five skill levels (e.g., Skill Levels 10, 20, 30, 40 and 50) or as few as one. The enlisted MOS structure is fully described in Appendix A.

Career Management Field. Two digits and a title designating a group of related MOSs; if divided into subfields (three digits and a title), the subfield is also provided. When a person has held more than one MOS in the same career management field or subfield, you should be alert to the possibility of overlapping recommendations

Description. A summary description applying to all skill levels and a specific description for each skill level. Provides information about the duties performed and the qualifications required for proficiency in the MOS. It indicates the rationale behind a given recommendation and provides information that can be used as a starting point should you wish to conduct further assessment of the learning a soldier or veteran has acquired. The description is also useful in making decisions when recommendations from more than one learning experience appear to duplicate each other.

MOS-63S-002

HEAVY WHEEL VEHICLE MECHANIC
63S10
63S20

Exhibit Dates: 8/92–Present.

Career Management Field: 63 (Mechanical Maintenance).

Description

Summary: Performs maintenance on heavy wheel vehicles and material-handling equipment. *Skill Level 10:* Replaces engine components such as fuel pumps, generators, starters, regulators, radiators, universal joints, brake shoes, engine mounts, and lines and fittings; adjusts operating mechanisms; prepares maintenance forms and records. *Skill Level 20:* Able to perform the duties required for Skill Level 10; provides technical guidance and supervision to subordinates; troubleshoots and uses diagnostic equipment to identify malfunctions; interprets complex schematic diagrams; completes maintenance forms and records; conducts on-the-job training.

Recommendation, Skill Level 10

Credit may be granted on the basis of individualized assessment of the student (3/94).

Recommendation, Skill Level 20

In the lower-division baccalaureate/associate degree category, 3 semester hours in diesel fuel systems, 3 in electrical systems, 3 in brake systems, 3 in heavy duty drive trains, 3 in diesel engine fundamentals, 3 in military science, and 3 in personnel supervision NOTE: This recommendation for skill level 20 is valid for the dates 8/92-2/95 only) (8/92).

Title. The official Army title of the MOS during the period of the exhibit dates. If the title changed during that period, the newer title is given first and the older title is given in parentheses on the next line.

Exhibit Dates. Start and end dates by month and year. When an end date is given, the MOS was either discontinued (an explanation is provided) or changed (another exhibit will follow for the MOS, e.g., MOS-13B-004). An end date of "Present" indicates that the MOS still exists and the exhibit is current as of May 2002. Some exhibits will end with the phrase "Pending evaluation." The earliest start date in this edition is 1/90.

Recommendation. By skill level. Only the recommendation for the highest skill level held should be used; skill-level recommendations should not be added.

Educational credit is recommended in four categories: vocational certificate, lower-division baccalaureate/associate degree, upper-division baccalaureate, and graduate degree. (See Appendix A for category definitions.)

Date of Evaluation. By month and year. Follows the recommendation for each skill level.

How to Find and Use Army Warrant Officer MOS Exhibits

This volume contains recommendations for all Army warrant officer military occupational specialties (MOSs) with exhibit dates of 1/90 and later.

The instructions that follow provide a step-by-step procedure for finding and using the MOS exhibits and recommendations. Readers unfamiliar with how the warrant officer MOSs are structured should read Appendix A.

Step 1

Have the applicant submit official documentation.

Several records may be used for warrant officer MOS verification. A servicemember or veteran may submit one or several of the following forms:

- AARTS (Army/American Council on Education Registry Transcript System). The Army is working on adding warrant officers and commissioned officer files to the AARTS database. As entries in these areas may be incomplete, for additional information, please contact the AARTS website at: http://aarts.army.mil or call 866-297-4427.
- DA Form 67-8, *Officer Evaluation Report (OER).* See section on duty description for MOS designation, performance evaluation, and performance and potential evaluation.
- DA Form 67-8-1, *Officer Evaluation Report Support Form.* See written statements by the rated officer and the rater.
- DA Form 2-1, *Personnel Qualifications Record.* See items 3 and 6 for MOS designation, relevant dates, and dates of evaluation.
- DA Form 66, *Officer Qualification Record.* See section I for MOS designations and titles.

Step 2

Verify each occupational specialty the person has successfully held.

Many warrant officers began their careers as enlisted personnel; in such cases, warrant officers should also submit their enlisted records for evaluation. See the section "How to Find and Use Army Enlisted MOS Exhibits" for step-by-step instructions for granting credit for enlisted MOSs.

Army records must be requested by the individual. Records may be obtained as follows:

Active duty. The Army installation to which he or she is assigned.

Army National Guard personnel. The National Guard unit to which he or she is assigned.

Army Reservists or retired personnel. The U.S. Army Reserve Component and Administrative Center (TAGO), 9700 Page Avenue, St. Louis, MO 63132.

Army veteran. National Personnel Records Center (Military Personnel Records), 9700 Page Avenue, St. Louis, MO 63132-5100. The veteran may request service records by submitting U.S. Government Standard Form 180, which may be obtained from a state's veterans affairs office, the Veterans Administration, or the National Personnel Records Center. It can also be reproduced from Appendix B.

Step 3

Find the appropriate exhibit in the Guide.

Find the information necessary for locating the correct MOS exhibit. Pertinent information includes the MOS designation, the MOS title, and the dates each MOS was held. The exhibit(s) for an MOS can easily be found when the MOS designation is known. Each MOS exhibit is assigned an ACE ID number that has three components. The first component, MOS, identifies the exhibit as one that describes an Army warrant officer military occupational specialty. The second component consists of the MOS designation, e.g., 211A, 256A. The third component, a three-digit, sequentially assigned number, for example, -001, -002, -003, uniquely identifies the exhibit. Warrant officer MOS exhibit ID numbers have ten characters, e.g., MOS-271A-001. Exhibits are arranged by ACE ID number.

If only the title of the MOS is known, the exhibit ID number(s) can be found by referring to the Occupation Title Index.

There may be several exhibits for an MOS. Each time there is a new recommendation for an MOS, there is a new exhibit. For example, three exhibits appear for MOS 140B. The ACE ID numbers for the three exhibits are MOS-140B-001, MOS-140B-002, and MOS-140B-003. The oldest exhibit for the MOS, the first to be evaluated, is assigned -001 as the last three characters. The last exhibit for an MOS, the exhibit ending in the highest number, is the most recent.

When there is more than one exhibit, select the correct one by referring to the exhibit dates that conform to the time frame in which the individual held the MOS. The exhibit dates for each new exhibit establish continuity with the previous exhibit. If the end date for the first exhibit is 1/91, the start date for the next exhibit is 2/91. In many cases, the date of the soldier's last MOS evaluation will indicate which exhibit is most appropriate. However, when the time period that a person held an MOS skill level spans more than one exhibit for the MOS, you should exercise your own judgment in deciding which exhibit is more appropriate.

Step 4

Read the entire exhibit.

Although credit is granted for a particular skill level of an MOS, users are urged to read the entire MOS exhibit. Doing so will make possible the most appropriate application of the recommendation to the student's program of study at your institution. Each item in the exhibit has been prepared to help users interpret the credit recommendations. (See sample exhibits.)

The descriptions, which are similar to learning outcome statements of postsecondary courses and programs of study, provide essential information about the learning required for proficiency in the MOS. Comparing the MOS description with a description of the course or program of study that the student will pursue at your institution will help you

- determine how much of the recommended credit applies to the course or program of study at your institution;
- identify additional areas of possible credit;
- resolve problems with duplication of credit when the applicant has applied for credit for more than one military learning experience; and
- place the student at the appropriate point in the course sequence or program of study.

In addition, in warrant officer MOS exhibits, the exhibits contain notes on career progression, other MOSs an individual may have held are listed in the exhibit item Career Pattern.

Step 5

Award credit as appropriate.

The MOS recommendations are advisory. They are intended to assist in placing enlisted soldiers, warrant officers, and veterans in postsecondary programs of study. The recommendations may be modified.

When an applicant has applied for credit for more than one military learning experience, you may have to reduce the total amount of credit recommended to avoid granting duplicate credit.

Step 6

If assistance is required, contact the American Council on Education.

If assistance is needed, the college can submit a *Request for Army Enlisted and Warrant Officer Exhibits* form. Contact the Military Evaluations Advisory Service at:

Military Evaluations
American Council on Education
One Dupont Circle NW, Suite 250
Washington, DC 20036-1193
(202) 939-9470
FAX: (202) 775-8578
E-mail: mileval@ace.nche.edu

Sample Warrant Officer MOS Exhibit

MOS-915A-002

WHEEL VEHICLE MAINTENANCE TECHNICIAN
(LIGHT)
915A0

Exhibit Dates: 8/92–Present.

Career Pattern

May have progressed from any of the following MOSs: 62B, 63B, 63D, 63E, 63F, 63G, 63N, 63S, 63T, 63Y, 63W, or 63Z.

Description

Plans, organizes, and performs unit maintenance on wheel vehicles, light track vehicles (except Bradley), self-propelled artillery systems, fire control, armament, ground support, and power-driven chemical equipment; analyzes unit equipment malfunctions; directs unit work loads; enforces fire and safety programs; manages calibration, oil analysis, and readiness reports; using the Army maintenance management system, manages periodic maintenance, vehicle dispatcher, vehicle repair, and common task training; conducts technical inspections of units; establishes procedures; instructs subordinate personnel; supervises personnel assigned to maintenance sections, motor pools, or similar activities.

Recommendation

In the lower-division baccalaureate/associate degree category, 3 semester hours in personnel supervision, 2 in records and information management, and 2 in computer applications. If rank was CW2 or CW3: 1 additional semester hour in computer applications. If rank was CW4: 2 additional semester hours in computer applications. In the upper-division baccalaureate category, 3 semester hours in organizational management and 3 for field experience in management. If rank was CW2, 3 additional semester hours for field experience in management, 3 in management problems, 3 in operations management, 3 in communication techniques for managers, and 3 in fleet maintenance management. If rank was CW3, 5 additional semester hours for field experience in management, 3 in management problems, 3 in operations management, 3 in communication techniques for managers, and 3 in fleet maintenance management. If rank was CW4, 7 additional semester hours for field experience in management, 3 in management problems, 3 in operations management, 3 in communication techniques for managers, and 3 in fleet maintenance management (8/92).

ID Number. A ten-character code assigned by ACE to identify each warrant officer MOS exhibit.

MOS Designation. The four-character code that identifies the MOS plus the neutral fifth character, zero ("0"). The zero is added as a reminder that the MOS evaluation does not include the evaluation of learning associated with Special Qualifications Identifiers (SQIs). When a person's MOS designation includes an SQI, signified by a letter or number other than zero as the fifth character, additional credit may be granted on the basis of individual assessment (refer to Appendix E).

Description. Provides information about the duties performed and the qualifications required for proficiency in the MOS. It indicates the rationale behind a given credit recommendation and provides information that can be used as a starting point should you wish to conduct further assessment of the learning a warrant officer or veteran has acquired. The description is also useful in making decisions about how much credit to grant when credit recommendations from more than one learning experience appear to duplicate each other.

Title. The official Army title of the MOS during the period of the exhibit dates. If the title changed during that period, the newer title is given first and the older title is given in parentheses on the next line. A title in the exhibit may be different from the title that appears on the person's Army records. The latter title may be a Special Qualifications Identifier.

Exhibit Dates. Start and end dates by month and year. An end date of "Present" indicates that the exhibit is current as of May 2002. The earliest start date in this edition is 1/90. (See questions 1–4 in Questions and Answers.) Some exhibits will end with the phrase "Pending evaluation."

Career Pattern. Enlisted or other warrant officer MOSs that an individual may have held. You should be alert to the possibility of overlapping recommendations when a warrant officer has held one or more of these MOSs.

Recommendation. Educational credit is recommended in four categories: vocational certificate, lower division baccalaureate/associate degree, upper-division baccalaureate, and graduate degree. (See Appendix A for category definitions.)

Date of Evaluation. By month and year.

Questions and Answers

About the *Guide*

The Guide to the Evaluation of Educational Experiences in the Armed Services is the standard reference for recognizing learning acquired during military service. The *Guide* system enables servicemembers and veterans to apply their experience, training, and expertise to their degree programs at colleges and universities worldwide.

The Guide has been published biennially since 1974. Each edition replaces the previous edition, which should be discarded. The 2004 *Guide* contains course and occupation exhibits with exhibit start dates of 1/90 or later. It is a three-volume set: Volume 1 contains Army courses and occupations; Volume 2 contains Navy courses and occupations; and Volume 3 has Air Force, Coast Guard, Department of Defense, and Marine Corps courses; and Coast Guard and Marine Corps occupations.

The *Handbook to the Guide* provides credit recommendations for courses and occupations evaluated since the latest *Guide* was printed. The *Handbook* also includes information on examination evaluations, ACE, forms and instructions, and helpful hints. The *Handbook* is distributed to military education offices and civilian institutions three times between publications of the *Guide*. When you receive a new *Handbook*, discard the previous one.

Formal courses offered by the following military components are eligible to be evaluated: Army, Navy, Air Force, Coast Guard, Marine Corps, and selected Department of Defense schools.

Online *Guide* availability

In addition to the present hardbound copies, the *Guide* is also available online for immediate access at: http://www.militaryguides.acenet.edu. Those users with Internet access are encouraged to view and use the online edition, which is updated daily and contains all course and occupation entries that have been evaluated to date.

In addition to the latest updates of courses and occupations evaluated by ACE since the publication of the hardbound *Guide* and *Handbook to the Guide* editions, the online *Guide* offers immediate search capability for courses and occupations, using the following search criteria: keywords, course number, school/location, dates taken, and ACE identification number. The online version provides the user with steps on how to search for individual courses and occupations, and contains 'Frequently Asked Questions' on the military evaluation process, data currency, browser specification recommendations, and user feedback. Users are also able to print individual exhibits of *Guide* entries, as desired.

1954-89 edition of the *Guide*

The four-volume 1954-89 *Guide*, which contains courses and occupations with exhibit dates of 1/54 through 12/89, has been incorporated into the current online version and is accessible online at: http://www.militaryguides.acenet.edu. Owners of the 1954-89 volume are advised to retain this edition for permanent reference, as it is no longer available in printed format.

The course evaluations conducted by ACE do not include all courses offered by the armed services. Many courses do not meet ACE's criteria and cannot be evaluated. In general, courses evaluated and published in the *Guide* are those offered on a full-time basis. After 1981, ACE began evaluating courses that are 45 academic hours in length. Before that time courses evaluated were at least two weeks long and included a minimum of 60 contact hours of academic instruction. Before 1973, the minimum length requirement was three weeks or 90 contact hours.

Another requirement for evaluation is that a course meet service-wide training requirements and be cited in the formal school listing for the service. This requirement generally excludes locally organized and command-level training programs, as well as courses offered on a one-time basis. Classified courses can only be evaluated if the military school omits classified portions of the course.

In addition, the following categories of military occupational specialties have been evaluated: Army enlisted and warrant officers; Navy general rates, ratings, and selected NECs (Navy Enlisted Classifications); Coast Guard enlisted, warrant officers and aviators; and Marine Corps enlisted.

It is the responsibility of each military school to provide ACE with materials for all new and revised courses. The credit recommendations for occupations are for actual on-the-job experiences.

The *Guide* is constantly being updated. If you cannot locate the exhibit in the *Guide, Handbook*, or online version, contact ACE.

1. Which learning experiences have been evaluated for the Army?

This volume contains credit recommendations for formal courses offered by the Army, Army Reserve, and Army National Guard, and Army enlisted and warrant officer military occupational specialties (MOSs). Courses and occupations with start dates of 1/90 or later are in this volume.

2. Does ACE issue Army transcripts?

No. ACE does not generate military transcripts. The Army/ACE Registry Transcript System AARTS) is a product of the Army and must be ordered from: AARTS Operations Center, 415 McPherson Avenue, Ft. Leavenworth, KS 66027; by phone at 1-866-297-4427 (toll-free); via the AARTS website at: http://aarts.army.mil; or by e-mail at: aarts@leavenworth.army.mil.

3. What information is found on the transcript?

The transcript is divided into separate sections that include:

- Personal servicemember data;
- Military course completions -- all courses that have been evaluated by ACE, with full descriptions and credit recommendations;
- Military occupations – full descriptions, skill levels, and credit recommendations;
- College-level test scores – CLEP, DSSTs, NCPACE, ACT/PEP, and Excelsior Test score data; and
- Other Learning Experiences – additional completed courses and occupations not evaluated by ACE for college credit.

4. Who is eligible to receive an AARTS transcript?

AARTS transcripts are available to enlisted members of the regular Army, Army National Guard, and the Army Reserve. The Army has removed the restriction limiting AARTS transcripts to soldiers with basic active service dates (BASD) on or after October 1, 1981. Regular Army enlisted and soldiers with service dates before this date may therefore be able to obtain an AARTS transcript.

As this change may also affect Army National Guard soldiers and veterans on active duty as of January 1, 1993, with pay entry basic dates/basic active service dates on or before October 1, 1981, these individuals are advised to seek further information by calling AARTS Operations at their toll-free number: 866-297-4427, or by visiting the AARTS website at: http://aarts.army.mil.

5. What if a soldier or veteran is not eligible to receive a transcript?

In such cases, the student should submit a DD Form 295, Application for the Evaluation of Learning Experiences during Military Service, a DD Form 214, Certificate of Release or Discharge from Active Duty, or course completion certificates. The DD Form 295 is available to all active duty servicemembers and may be obtained from Army Education Centers on the onsite duty location. For veterans who do not have these documents, copies may be obtained from the National Personnel Records Center (Military Personnel Records), 9700 Page Avenue, St. Louis, MO 63132-5100 or by visiting their website at www.nara.gov/regional/stlouis.html.

6. If an applicant has served in more than one branch of the Armed Services, can all the information be consolidated into one transcript?

No. However, if servicemembers have served in more than one branch of the military, they may be eligible for more than one transcript. Military transcripts are products of the individual services and must be ordered separately. They are available at no charge to the servicemember or the institution. Please remember that the services maintain their own training and service records, with varying dates of eligibility. Thus, servicemembers are advised to check with their respective Operation Centers to confirm their eligibility to receive a transcript.

7. What is the Community College of the Air Force?

Serrvicemembers who started an Air Force course after April 1972 can obtain a Community College of the Air Force (CCAF) transcript. Servicemembers who take courses after this date should contact CCAF for information on transcript availability. The transcript may be used to request transfer of credit to another institution or to otherwise document college credit. CCAF does not award credit for all Air Force courses, but only those regularly attended by Air Force enlisted personnel and taught by CCAF-affiliated schools. The transcript also has credit for Air Force Specialty codes (occupations).

To obtain a CCAF transcript, send a completed AF Form 2099 (available to active duty personnel from the Education Office) or a brief letter requesting a transcript to CCAF/RRR, 130 West Maxwell Boulevard, Maxwell Air Force Base, AL 36112-6613. A certificate of training must accompany the request. DD Form 214 is acceptable for Basic Training only. A copy of the ID card, front and back, is also needed. Servicemembers requesting transcripts should specify the type of transcript (personal or official) and the address to which the transcript is to be mailed, provide full name and social security number, and sign the request form. The transcripts are free. Official transcripts will be mailed only to institutions.

8. Are Army correspondence courses evaluated?

The Army, with few exceptions, has not established an ongoing proctored end-of-course examination program, one of ACE's criteria for reviewing correspondence courses. Those listed are, for the most part, high-level schools or specialized schools, such as the Army Logistics Management College. They can be found in the Keyword Index under Correspondence for easy identification. The advanced search capability in the online *Guide* (keyword: *correspondence* or *by correspondence*) will also quickly identify and list those courses that ACE has evaluated for college credit for all of the services.

9. Why are most Defense Language Institute (DLI) courses end-dated in 1990? What do I do for foreign language courses taken after that time?

The Defense Language Institute Foreign Language Center has been authorized by Congress, the regional Accrediting Commission for Community and Junior Colleges, and the Western Association of Schools and Colleges to grant an Associate of Arts Degree in Foreign Language to qualified students. Details regarding eligibility, general education requirements, and petition forms can be found on the DLI webpage: http://www.dli.army.mil; click on "AA Degree Program."

For those not eligible for a transcript, credit recommendations are based on successful completion of the Defense Language Proficiency Test (DLPT) end-of-course examination.

ACE's Credit by Examination Program began evaluating DLI's end-of-course examinations in October 1990. Credit recommendations are now based on successful completion of the DLPT III and IV series. These examinations may be taken by servicemembers who complete the DLI courses, as well as by servicemembers who must demonstrate ongoing language proficiency without having taken formal training. Evaluation of the tests is a way to ensure that both groups receive appropriate recognition. See the Defense Language Proficiency Test charts in Volume 3 for the credit recommendations.

An official DLI transcript listing the test results is available to the examinee from Commandant, Defense Language Institute, Foreign Language Center, ATTN: ATFL-DAA-AR, The Presidio of Monterey, CA 93944-5006. Test scores are available by writing to DLI/FLC, /The Presidio of Monterey CA 93944-5006, ATTN: ATFL-EST-M (DLPT Score Reports), or you may call (831) 242-5291 for the Office of the Provost, or (831) 242-5825 or -6459 for the Registrar's Office. For information about the status of a transcript, call (831) 242-5366. The FAX number is (831) 242-5146.

Additional information on these examinations may be obtained from ACE's *Guide to Educational Credit by Examination.*

10. What are USAFI and DANTES? Should I grant credit for those courses and tests listed on an applicant's USAFI or DANTES military test reports?

USAFI was the United States Armed Forces Institute, which offered an extensive educational program to active-duty personnel. USAFI correspondence, seminar, self-study courses, end-of-course tests, and Subject Standardized Tests (SSTs) were made available to servicemembers worldwide until 1974, when USAFI was disestablished. Credit recommendations for USAFI courses and examinations are listed in the *Handbook to the Guide.*

The Defense Activity for Non-Traditional Education Support (DANTES) was established in 1974. DANTES continues the development and administration of Subject Standardized Tests (SSTs) and other educational services. ACE continues to recommend credit for USAFI offerings and DANTES SSTs. Credit recommendations for DANTES SSTs are in the *Handbook to the Guide.* Detailed descriptions are in the *Guide to Educational Credit by Examination,* which is available from the ACE Credit by Examination Program.

In verifying completion of USAFI or DANTES courses or tests, the military test report is not to be considered official. That report is given to all servicemembers who have taken a course or test. To obtain official USAFI or DANTES transcripts, refer to the addresses provided in Appendix A, under "Other Resources."

11. Which military schools grant degrees?

The following list contains the only military schools with degree-granting status at publication time:

- Community College of the Air Force, Maxwell Air Force Base, AL; accredited at the two-year community college level;

- Army Command and General Staff College at Fort Leavenworth, KS; accredited to grant an M.A.;

- Air Force Institute of Technology at Wright-Patterson Air Force Base, OH; accredited through the doctoral level;

- Defense Language Institute Foreign Language Center; accredited to grant Associate of Arts degrees in foreign language;

- Joint Military Intelligence College (formerly Defense Intelligence College), Washington, DC; accredited to grant an M.A.;

- Naval Postgraduate School, Monterey, CA; accredited through the doctoral level;

- Naval War College, Newport, RI; accredited to grant an M.A.;

- Uniformed Services University of the Health Sciences School of Medicine, Bethesda, MD; accredited through the doctoral level;

- The National Defense University's National War College; accredited to grant an M.A.;

- Industrial College of the Armed Forces; accredited to grant an M.A.; and

- The service academies.

12. What is the Defense Acquisition University?

The Defense Acquisition University (DAU) is a consortium of a number of schools offering courses in contracts, logistics, management, and acquisition. A number of years ago, some courses, offered by various sponsors, were evaluated by ACE's CREDIT office, because attendees were primarily civilian. Military Programs evaluated other courses offered by military schools. In the mid-1990s all component schools came under the oversight of the Department of Defense. All courses were standardized, so that now there is consistency in the courses offered by all component schools. CON 101, for example, will have the same content, regardless of the component school offering the course. Because of that standardization, Military Programs recommends credit for the courses when they are offered with DAU as the school of record. Credit recommendations can be found under Department of Defense course exhibits in the *Guide*.

13. What are primary, secondary, and duty MOSs?

All soldiers receive a primary MOS in which they normally work and are evaluated. Some soldiers receive a secondary MOS, which is generally related to their primary MOS. They are evaluated every other year in the secondary MOS.

A soldier works in a duty MOS, which in most cases is the same as the primary MOS. In instances in which the duty MOS is different from the primary MOS, soldiers are evaluated by their supervisor in their duty MOS. Credit may be awarded in either the primary or duty MOS or both. The soldier must maintain proficiency in the primary MOS as well as the duty MOS.

Ordinarily, the primary, secondary, and duty MOSs are in the same or a related career field. If this is the case, there is apt to be some duplication in the credit recommendations.

14. Most MOS exhibits do not carry a specific credit recommendation for Skill Levels 10 and 20. Why is that?

Recent MOS exhibits contain the following statement for Skills Levels 10 and 20: Credit may be granted on the basis on an individualized assessment of the student.

One of ACE's criteria for evaluating an occupational system is that it must provide for the assessment of the individual. Since the Army no longer offers a standardized testing system, ACE recommends credit for Skill Levels 30, 40, and 50. For additional information, see Appendix A.

15. How does a college conduct its own assessment of the student?

The process of conducting an individual assessment to verify MOS proficiency is not very different from making a portfolio assessment of a student's experiences. The *Guide* can simplify the steps in the assessment process.

When a student submits a portfolio, the institution official must (1) determine the thoroughness and preciseness of the documentation of the experience, (2) identify the learning outcomes that have been achieved, (3) judge the equivalence of the learning outcomes to those achieved through the institution's programs of study or courses, and (4) translate the learning outcomes into credit. In considering whether to grant credit for an MOS, keep in mind that ACE evaluation teams have already accomplished three of these steps. They have (1) determined expected learning outcomes of that MOS, (2) judged the compatibility of the MOS training with postsecondary curricula, and (3) translated the learning outcomes into credit recommendations for specific postsecondary courses.

In order to make a recommendation, ACE evaluators first identify the skills, competencies, and knowledge associated with a given MOS. They use various resources, including the official Army MOS regulation, which describes the duties and qualifications for each MOS skill level; technical manuals, soldier's manuals, and other publications used by soldiers in the day-to-day performance of their duties and to prepare for their evaluation tests; and study guides that outline the proficiency requirements for each MOS skill level. During site visits to Army installations, evaluators interview soldiers and may observe them on the job.

Thus, ACE provides a sound initial assessment of MOSs in terms of their equivalency to postsecondary learning. The institution must verify whether the individual actually possesses the skills and knowledge described in the *Guide* entry before awarding the recommended credit. Although the ACE MOS credit recommendations are valid in a general way, they cannot be applied routinely to every individual seeking MOS-related credit. When you can confirm that the individual has attained some or all of the skills and knowledge represented by the MOS, the ACE recommendation can be utilized or appropriately modified.

16. I have looked up several courses and MOSs for one applicant. It appears that a lot of the recommended credit is in the same subject area. How can I avoid granting duplicate credit to this person?

You may grant credit for any combination of learning experiences. In doing so, however, you must be alert to the possibility of overlapping credit recommendations. When the student is applying for credit for more than one learning experience, the recommendations might cover some of the same learning. In such cases, awarding a simple total of the recommended credit could result in the award of more credit than the learning merits.

To determine how much credit should be awarded without duplication, use the following steps:

- Read and compare all the descriptions, and, on the basis of the person's program of study, identify the appropriate recommendation in each exhibit.
- Read and compare all the recommendations. It may be necessary to obtain additional information from the individual through interview or further assessment.
- Make decisions on how much credit might be awarded without duplication. Credit should be awarded as appropriate to the educational goals of the individual and the policies of the institution.

If you cannot determine whether duplication exists, contact ACE.

17. Does the Center for Lifelong Learning still publish a newsletter?

The newsletter, the *Center Update*, has been replaced by an electronic newsletter now available on the web. It is known as *Centerpoint.* It contains new features including: legislative actions affecting adult learning; innovative programs and partnerships; letters to the editor; guest editorials; the latest in adult education; ACE news; and compelling stories of real people who have used the Center's services to build successful futures.

To subscribe to *Centerpoint,* free of charge, simply send an e-mail to centerpoint@listserv.nche.edu and type 'subscribe' in the subject line. With each new issue, you'll receive an e-mail including content highlights and direct links to the stories and information you want.

18. How can I contact ACE if I have questions?

ACE maintains an advisory service, as well as a Military/CREDIT Call Center to assist students and schools with questions about the Military *Guide*, courses and occupations that have been evaluated by its Military Evaluations department, and to provide information on ordering the AARTS and SMART military transcripts. If you have any questions, please contact the ACE Call Center at: **(202) 939-9470 or (202) 939-9434.**

Joint Statement on the Transfer and Award of Credit

The following set of guidelines has been developed by the three national associations whose member institutions are directly involved in the transfer and award of academic credit: the American Association of Collegiate Registrars and Admissions Officers, the American Council on Education, and the Council for Higher Education Accreditation. The need for such a statement came from an awareness of the growing complexity of transfer policies and practices, which have been brought about, in part, by the changing nature of postsecondary education. With increasing frequency, students are pursuing their education in a variety of institutional and extrainstitutional settings. Social equity and the intelligent use of resources require that validated learning be recognized wherever it takes place.

The statement is thus intended to serve as a guide for institutions developing or reviewing policies dealing with transfer, acceptance and award of credit. "Transfer" as used here refers to the movement of students from one college, university or other education provider to another and to the process by which credits representing educational experiences, courses, degrees or credentials that are awarded by an education provider are accepted or not accepted by a receiving institution.

Basic Assumptions

This statement is directed to institutions of postsecondary education and others concerned with the transfer of academic credit among institutions and the award of academic credit for learning that takes place at another institution or education provider. Basic to this statement is the principle that each institution is responsible for determining its own policies and practices with regard to the transfer, acceptance, and award of credit. Institutions are encouraged to review their policies and practices periodically to assure that they accomplish the institutions' objectives and that they function in a manner that is fair and equitable to students. General statements of policy such as this one or others referred to, should be used as guides, not as substitutes, for institutional policies and practices.

Transfer and award of credit is a concept that increasingly involves transfer between dissimilar institutions and curricula and recognition of extra-institutional learning, as well as transfer between institutions and curricula with similar characteristics. As their personal circumstances and educational objectives change, students seek to have their learning, wherever and however attained, recognized by institutions where they enroll for further study. It is important for reasons of social equity and educational effectiveness for all institutions to develop reasonable and definitive policies and procedures for acceptance of such learning experiences, as well as for the transfer of credits earned at another institution. Such policies and procedures should provide maximum consideration for the individual student who has changed institutions or objectives. It is the receiving institution's responsibility to provide reasonable and definitive policies and procedures for determining a student's knowledge in required subject areas. All sending institutions have a responsibility to furnish transcripts and other documents necessary for a receiving institution to judge the quality and quantity of the student's work. Institutions also have a responsibility to advise the student that the work reflected on the transcript may or may not be accepted by a receiving institution as bearing the same (or any) credits as those awarded by the provider institution, or that the credits awarded will be applicable to the academic credential the student is pursuing.

Inter-Institutional Transfer of Credit

Transfer of credit from one institution to another involves at least three considerations:

(1) the educational quality of the learning experience which the student transfers;

(2) the comparability of the nature, content, and level of the learning experience to that offered by the receiving institution; and

(3) the appropriateness and applicability of the learning experience to the programs offered by the receiving institution, in light of the student's educational goals.

Accredited Institutions

Accreditation speaks primarily to the first of these considerations, serving as the basic indicator that an institution meets certain minimum standards. Users of accreditation are urged to give careful attention to the accreditation conferred by accrediting bodies recognized by the Council for Higher Education Accreditation (CHEA). CHEA has a formal process of recognition which requires that all accrediting bodies so recognized must meet the same standards. Under these standards, CHEA has recognized a number of accrediting bodies, including:

(1) regional accrediting commissions (which historically accredited the more traditional colleges and universities but which now accredit proprietary, vocational-technical, distance learning providers, and single-purpose institutions as well);

(2) national accrediting bodies that accredit various kinds of specialized institutions, including distance learning providers and freestanding professional schools; and

(3) professional organizations that accredit programs within multipurpose institutions.

Although accrediting agencies vary in the ways they are organized and in their statements of scope and mission, all accrediting bodies that meet CHEA's standards for recognition function to ensure that the institutions or programs they accredit have met generally accepted minimum standards for accreditation.

Accreditation thus affords reason for confidence in an institution's or a program's purposes, in the appropriateness of its resources and plans for carrying out these purposes, and in its effectiveness in accomplishing its goals, insofar as these things can be judged. Accreditation speaks to the probability, but does not guarantee, that students have met acceptable standards of educational accomplishment.

Comparability and Applicability

Comparability of the nature, content, and level of transfer credit and the appropriateness and applicability of the credit earned to programs offered by the receiving institution are as important in the evaluation process as the accreditation status of the institution at which the transfer credit was awarded. Since accreditation does not address these questions, this information must be obtained from catalogues and other materials and from direct contact between knowledgeable and experienced faculty and staff at both the receiving and sending institutions. When such considerations as comparability and appropriateness of credit are satisfied, however, the receiving institution should have reasonable confidence that students from accredited institutions are qualified to undertake the receiving institution's educational program. In its articulation and transfer policies, the institution should judge courses, programs and other learning experiences on their learning outcomes, and the existence of valid evaluation measures, including third-party expert review, and not on modes of delivery.

Admissions and Degree Purposes

At some institutions there may be differences between the acceptance of credit for admission purposes and the applicability of credit for degree purposes. A receiving institution may accept previous work, place a credit value on it, and enter it on the transcript. However, that previous work, because of its nature and not its inherent quality, may be determined to have no applicability to a specific degree to be pursued by the student. Institutions have a responsibility to make this distinction, and its implications, clear to students before they decide to enroll. This should be a matter of full disclosure, with the best interests of the student in mind. Institutions also should make every reasonable effort to reduce the gap between credits accepted and credits applied toward an educational credential.

Additional Criteria for Transfer Decisions

The following additional criteria are offered to assist institutions, accreditors and higher education associations in future transfer decisions. These criteria are intended to sustain academic quality in an environment of more varied transfer, assure consistency of transfer practice, and encourage appropriate accountability about transfer policy and practice.

Balance in the Use of Accreditation Status in Transfer Decisions. Institutions and accreditors need to assure that transfer decisions are not made solely on the source of accreditation of a sending program or institution. While acknowledging that accreditation is an important factor, receiving institutions ought to make clear their institutional reasons for accepting or not accepting credits that students seek to transfer. Students should have reasonable explanations about how work offered for credit is or is not of sufficient quality when compared with the receiving institution and how work is or is not comparable with curricula and standards to meet degree requirements of the receiving institution.

Consistency. Institutions and accreditors need to reaffirm that the considerations that inform transfer decisions are applied consistently in the context of changing student attendance patterns (students likely to engage in more transfer) and emerging new providers of higher education (new sources of credits and experience to be evaluated). New providers and new attendance patterns increase the number and type of transfer issues that institutions will address—making consistency even more important in the future.

Accountability for Effective Public Communication. Institutions and accreditors need to assure that students and the public are fully and accurately informed about their respective transfer policies and practices. The public has a significant interest in higher education's effective management of transfer, especially in an environment of expanding access and mobility. Public funding is routinely provided to colleges and universities. This funding is accompanied by public expectations that the transfer process is built on a strong commitment to fairness and efficiency.

Commitment to Address Innovation. Institutions and accreditors need to be flexible and open in considering alternative approaches to managing transfer when these approaches will benefit students. Distance learning and other applications of technology generate alternative approaches to many functions of colleges and universities. Transfer is inevitably among these.

Foreign Institutions

In most cases, foreign institutions are chartered and authorized to grant degrees by their national governments, usually through a Ministry of Education or similar appropriate ministerial body. No other nation has a system comparable with voluntary accreditation as it exists in the United States. At an operational level, AACRAO's Office of International Education Services can assist institutions by providing general or specific guidelines on admission and placement of foreign students, or by providing evaluations of foreign educational credentials.

Evaluation of Extra-Institutional and Experiential Learning for Purposes of Transfer and Award of Credit

Transfer and award of credit policies should encompass educational accomplishment attained in extra-institutional settings. In deciding on the award of credit for extra-institutional learning, institutions will find the services of the American Council on Education's Center for Lifelong Learning helpful. One of the Center's functions is to operate and foster programs to determine credit equivalencies for various modes of extra-institutional learning. The Center maintains evaluation programs for formal courses offered by the military and civilian organizations such as business, corporations, government agencies, training providers, institutes, and labor unions. Evaluation services are also available for examination programs, for occupations with validated job proficiency evaluation systems, and for correspondence courses offered by schools accredited by the Distance Education and Training Council. The results are published in a Guide series. Another resource is the General Educational Development (GED) Testing Program, which provides a means for assessing high school equivalency.

For learning that has not been evaluated through the ACE evaluation processes, institutions are encouraged to explore the Council for Adult and Experiential Learning (CAEL) procedures and processes.

Uses of This Statement

Institutions are encouraged to use this statement as a basis for discussions in developing or reviewing institutional policies with regards to the transfer and award of credit. If the statement reflects an institution's policies, that institution may wish to use these guidelines to inform faculty, staff, and students.

It is also recommended that accrediting bodies reflect the essential precepts of this statement in their criteria.

American Association of Collegiate Registrars and Admissions Officers	**9/28/01 (date)**
American Council on Education	**9/28/01 (date)**
Council for Higher Education Accreditation	**9/28/01 (date)**

Bookmark it today!

http://www.militaryguides.acenet.edu

Army Course Exhibits

AR-0101-0002

Department of Defense Pest Management

Course Number: 6H-F12/322-F12.
Location: Academy of Health Sciences, Fort Sam Houston, TX.
Length: 3 weeks (131 hours).
Exhibit Dates: 7/93–Present.
Learning Outcomes: Upon completion of the course, the student will be able to apply local, state, and federal laws to pest management activities; label and apply pesticides; evaluate and maintain various vegetation and turf areas according to accepted pest control principles and techniques; and identify and control arthropod and vertebrate pests.
Instruction: Course includes lecture and practical exercises in pest and vegetation management. Course consists of a 6 day Common Core followed by phases. Phase 1 concentrates on broad areas of vegetation and turf management; phase 2 provides training in areas of biology, identification, and control of arthropods and vertebrates as nuisance pests and vectors of disease.
Credit Recommendation: In the lower-division baccalaureate/associate degree category, for the core, 3 semester hours in pest management; for phase 1, 1 semester hour in agriculture pest management; and for phase 2, 2 semester hours in arthropod and vertebrate pest management (10/95).

AR-0102-0003

Installation Veterinary Services

Course Number: 6G-F3/321-F3; 6G-F3/321-F4; 321-F3.
Location: Academy of Health Sciences, Fort Sam Houston, TX.
Length: 2 weeks (79–84 hours).
Exhibit Dates: 4/85–10/93.
Learning Outcomes: After 10/93 see AR-0102-0008. To provide selected officers and enlisted personnel with training to supervise veterinary activities.
Instruction: Course includes conferences, demonstrations, discussions, and practical exercises in administration, meat and seafood technology and quality assurance, veterinary preventive medicine, and surveillance inspections.
Credit Recommendation: In the lower-division baccalaureate/associate degree category, 1 semester hour in administration, 1 in food technology, and 1 in food inspection procedures (1/95).
Related Occupations: 91R.

AR-0102-0005

Advanced Animal Care Specialist

Course Number: 321-91T30.
Location: Walter Reed Institute of Research, Washington, DC.
Length: 10 weeks (331–348 hours).
Exhibit Dates: 1/83–9/91.
Objectives: After 9/91 see AR-0104-0012. To provide veterinary animal care specialists with training in administration, advanced medical skills, and preventive medicine aspects of community health. Emphasis is given to leadership situations involving practical problems of administrative control in animal disease prevention and control facilities.
Instruction: Course includes lectures, practical exercises, and examinations in administration and public relations, pathogenesis and management of disease, public health and preventive medicine, emergency medicine, surgery and anesthesia, diagnostic procedures, parasitology, laboratory animals in research, and equine medicine and husbandry.
Credit Recommendation: In the lower-division baccalaureate/associate degree category, 2 semester hours in administration and 1 in preventive medicine. In the upper-division baccalaureate category, 6 semester hours in veterinary science/technology, 6 in animal science/technology, or 6 in animal laboratory technology (6/88).
Related Occupations: 91T.

AR-0102-0007

Animal Care Specialist

Course Number: ***Version 1:*** 321-91T10. *Version 2:* 321-91T10. *Version 3:* 321-91T10.
Location: ***Version 1:*** Walter Reed Institute of Research, Washington, DC; Academy of Health Sciences, Fort Sam Houston, TX. *Version 2:* Academy of Health Sciences, Fort Sam Houston, TX. *Version 3:* Academy of Health Sciences, Fort Sam Houston, TX.
Length: ***Version 1:*** 9 weeks (331 hours). *Version 2:* 9 weeks (378–379 hours). *Version 3:* 9 weeks (379 hours).
Exhibit Dates: ***Version 1:*** 9/89–12/97. *Version 2:* 1/98–9/02. *Version 3:* 10/02–Present.
Learning Outcomes: ***Version 1:*** Upon completion of the course, specialists will demonstrate knowledge of the basic principles and techniques of animal care, disease prevention, recognition of disease states, diagnostic laboratory procedures, and treatment methods to enable them to assist a veterinary medical officer. *Version 2:* Upon completion of the course, specialists will demonstrate knowledge of the basic principles and techniques of animal care, disease prevention, recognition of disease states, diagnostic laboratory procedures, and treatment methods to enable them to assist a veterinary medical officer. *Version 3:* Upon completion of the course, specialists will demonstrate knowledge of the basic principles and techniques of animal care, disease prevention, recognition of disease states, diagnostic laboratory procedures, and treatment methods to enable them to assist a veterinary medical officer.
Instruction: ***Version 1:*** Course consists of lectures, laboratory exercises, and demonstrations covering the basic services related to animal structure and function, pathology, elementary pharmacology, animal handling and care, and diagnostic laboratory procedures. Clinical components include radiological techniques, anesthesiology and surgical procedures, and public health education. *Version 2:* Course consists of lectures, laboratory exercises, and demonstrations covering the basic services related to animal structure and function, pathology, elementary pharmacology, veterinary administrative procedures, animal handling and care, and diagnostic laboratory procedures. Clinical components include radiological techniques, anesthesiology and surgical procedures, and public health education. *Version 3:* Course consists of lectures, laboratory exercises, and demonstrations covering the basic services related to animal structure and function, pathology, elementary pharmacology, veterinary administrative procedures, animal handling and care, and diagnostic laboratory procedures. Clinical components include radiological techniques, anesthesiology and surgical procedures, emergency medical care, and public health education.
Credit Recommendation: ***Version 1:*** In the lower-division baccalaureate/associate degree category, 12 semester hours in veterinary science/technology, animal sciences, or laboratory animal science (8/93). *Version 2:* In the lower-division baccalaureate/associate degree category, 9 semester hours in basic animal care concepts, 3 in disease recognition and prevention, and 3 in diagnosis and treatment of disease (5/01). *Version 3:* In the lower-division baccalaureate/associate degree category, 9 semester hours in basic animal care concepts, 9 in clinical practice, 3 in disease recognition and prevention, and 3 in diagnosis and treatment of disease (11/03).
Related Occupations: 91T.

AR-0102-0008

Installation Veterinary Services

Course Number: 6G-F3/321-F3.
Location: Academy of Health Sciences, Fort Sam Houston, TX.
Length: 2 weeks (84 hours).
Exhibit Dates: 11/93–1/98.
Learning Outcomes: Before 11/93 see AR-0102-0003. Upon completion of the course, the student will be expected to identify the mission, organization, force structure policies, issues, and proposed changes which effect installation level veterinary services and discuss current concepts that seek to standardize the performance of food inspection tasks, e.g., inspection, reporting, and management.
Instruction: This course presents such topics as food inspection and reporting and management exercises using the conference and practical exercise format. Classroom instruction is reinforced and evaluated through regular examinations.

Credit Recommendation: In the lower-division baccalaureate/associate degree category, 2 semester hours in food hygiene and quality assurance (1/97).

AR-0102-0009

VETERINARY SERVICES OPERATIONS (RC)

Course Number: 6-8-C9(RC), Phase 2.

Location: Academy of Health Sciences, Fort Sam Houston, TX.

Length: 2 weeks (90 hours).

Exhibit Dates: 7/96–Present.

Learning Outcomes: Upon completion of the course, the student will be able to maintain a sanitary environment for food preservation with an emphasis on diary and meat products and employ disease prevention measures.

Instruction: Phase 1 is conducted through correspondence or executive summaries. This course is taught through the use of lectures, guided practical experiences, and field work. Topics cover food technology/sanitation, food products, food inspection, and animal care.

Credit Recommendation: In the lower-division baccalaureate/associate degree category, 2 semester hours in food technology and sanitation and 1 in meat inspection (10/03).

AR-0102-0010

MARINE MAMMAL VETERINARY MEDICINE

Course Number: 6G-F8/321-F8.

Location: Space and Warfare Systems Center, San Diego, CA.

Length: 2 weeks (79 hours).

Exhibit Dates: 3/99–Present.

Learning Outcomes: Upon completion of the course, the student will apply appropriate laws, facility requirements, water and food quality requirements, and anatomy and physiology to disease identification and care of marine mammals.

Instruction: The methods of instruction include didactic lectures, demonstrations, hands-on-training, quizzes, and examinations on the above topics and how marine mammals fit into Navy programs.

Credit Recommendation: In the graduate degree category, 2 semester hours in marine mammal veterinary care (10/03).

AR-0102-0011

VETERINARY READINESS, MOBILIZATION AND DEPLOYMENT

Course Number: 6G-F9/321-F9.

Location: Academy of Health Sciences, Fort Sam Houston, TX.

Length: 2 weeks (92 hours).

Exhibit Dates: 4/98–Present.

Learning Outcomes: Upon completion of the course, the student will be able to provide leadership and support in establishing services such as food inspection, sanitation, and public health and preventive medicine; procure food and equipment; and perform other activities in support of the veterinary mission.

Instruction: The course is conducted using seminars, demonstrations, field exercises, lectures, and examinations.

Credit Recommendation: In the graduate degree category, 3 semester hours in veterinary supervision and management (7/99).

AR-0104-0001

VETERINARY FOOD INSPECTION SPECIALIST ADVANCED

Course Number: 321-91R20.

Location: Academy of Health Sciences, Fort Sam Houston, TX.

Length: 9 weeks (344–348 hours).

Exhibit Dates: 6/83–9/91.

Objectives: To provide food inspection specialists with a working knowledge of advanced procedures and techniques of food hygiene, safety, and quality assurance inspections so that they may effectively supervise subordinates.

Instruction: Course includes conferences, practical exercises, and demonstrations of the procedures and techniques of food inspection, including principles of sanitation, facility inspections, and quality assurance techniques for produce, dairy, meat, poultry, and seafood products.

Credit Recommendation: In the lower-division baccalaureate/associate degree category, 3 semester hours in food sanitation, 6 in food service and product inspection, and 1 in administration (8/88).

Related Occupations: 91R.

AR-0104-0002

VETERINARY FOOD INSPECTION SPECIALIST (Veterinary Specialist Basic)

Course Number: 321-91R10.

Location: Academy of Health Sciences, Fort Sam Houston, TX.

Length: 8 weeks (302 hours).

Exhibit Dates: 6/76–3/94.

Objectives: After 3/94 see AR-0104-0013. To provide food handlers with the necessary skills to perform food hygiene, safety, and basic quality assurance inspections under supervision.

Instruction: Conferences, demonstrations, and practical exercises cover administration, food technology and inspection, and veterinary preventive medicine. When given applicable documents, inspection equipment, and raw materials for meat, dairy products, poultry, and waterfood, students perform appropriate component, process, or end-item inspection; conduct surveillance inspections for a variety of fresh fruits and vegetables; perform product wholesomeness inspection; and detect unsanitary conditions in food handling facilities.

Credit Recommendation: In the lower-division baccalaureate/associate degree category, 3 semester hours in administration, 4 in meat and waterfood technology, 3 in dairy and poultry science, 3 in food inspection techniques, and 1 in veterinary preventive medicine (3/87).

Related Occupations: 91R.

AR-0104-0006

HOSPITAL FOOD SERVICE SPECIALIST ADVANCED

Course Number: 800-94F30.

Location: Academy of Health Sciences, Fort Sam Houston, TX.

Length: 7 weeks (273–284 hours).

Exhibit Dates: 12/79–8/91.

Objectives: To provide training to selected food service specialists in administrative practices, nutrition, clerical dietetic principles, and food production and supply techniques so that they can effectively manage patient food service. Individuals learn to perform administrative duties, provide nutritional care, and carry out financial management and supply tasks in the operation of a hospital food service operation.

Instruction: Conferences, practical exercises, and examinations cover nutritional care and clinical dietetics principles, personnel management, administration and food production, and supply management.

Credit Recommendation: In the lower-division baccalaureate/associate degree category, 3 semester hours in nutritional care, 3 in food production management, 1 in personnel management, and 1 in administration (8/88).

Related Occupations: 94F.

AR-0104-0010

VETERINARY SERVICES TECHNICIAN (BASIC)

Course Number: 6G-640A (WO).

Location: Academy of Health Sciences, Fort Sam Houston, TX.

Length: 3 weeks (98 hours).

Exhibit Dates: 12/88–4/95.

Learning Outcomes: After 4/95 see AR-0104-0014. Upon completion of the course, the student will be able to provide food inspection services at the organizational level, including coordination and completion of applicable reports and records, assistance to public health officers and veterinarians in preventing and controlling zoonotic and foodborne diseases, and inspection of food in the various categories in accordance with applicable contractual requirements.

Instruction: Prerequisite courses required. Course consists of lectures, practica, and seminars in veterinary science food technology, public health, preventive medicine, and health care administration. Emphasis is placed on leadership problems.

Credit Recommendation: In the upper-division baccalaureate category, 4 semester hours in meat and dairy hygiene and 2 in community health and hygiene (3/90).

Related Occupations: 640A.

AR-0104-0011

VETERINARY FOOD INSPECTION SPECIALIST (BASIC) (RC)

Course Number: 321-91R10 (RC), Phase 2.

Location: Academy of Health Sciences, Fort Sam Houston, TX.

Length: 2 weeks (69–94 hours).

Exhibit Dates: 6/89–Present.

Learning Outcomes: Upon completion of the course, under the direction of a veterinarian medical officer, the specialist is able to inspect food in several categories, including red meats, poultry and eggs, waterfood, dairy products, fresh fruits, and vegetables. Acquired skills enable the specialist to perform these inspections on the basis of food hygiene, safety, and quality assurance.

Instruction: Prerequisite courses are required. This course includes practicum. Topics covered include interpretation of contracts and standards; deteriorative and adulterative changes in foods of various categories; and inspection procedures for red meat, waterfood, poultry and eggs, and dairy products. Sampling techniques for bulk foods, inspection methods for meat byproducts, and methods for decontamination of foods are presented. Facility inspection training is not included in this program.

Credit Recommendation: In the upper-division baccalaureate category, 3 semester hours in food service (10/03).

Related Occupations: 91R.

AR-0104-0012

1. Army Medical Department (AMEDD) Basic Noncommissioned Officer (NCO)
2. Army Medical Department (AMEDD) Noncommissioned Officer (NCO) Basic (NCOES)

Course Number: ***Version 1:*** 6-8-C40 (91T30 Tech Tng); 6-8-C40 (91T). *Version 2:* 6-8-C40 (91T30 Tech Tng).

Location: Academy of Health Sciences, Fort Sam Houston, TX.

Length: ***Version 1:*** 14 weeks (570 hours). *Version 2:* 9 weeks (403 hours).

Exhibit Dates: ***Version 1:*** 10/91–4/98. *Version 2:* 5/98–Present.

Learning Outcomes: ***Version 1:*** Before 10/91 see AR-0102-0005. Upon completion of the course, the student will be able to supervise animal care, including inspection, animal disease prevention, and sanitary inspection of food establishments. *Version 2:* Upon completion of the course, the student will be able to supervise animal care, and disease prevention, recognize animal disease, and inspect food preparation facilities.

Instruction: ***Version 1:*** Course covers military leadership, personnel and physical resource management, and food inspection. Methodology includes lectures, discussions, and practical applications. Course includes a common core of leadership subjects. *Version 2:* Student will have completed 7 week of AMEDD NCO Basic prior to enrolling in technical training phase. This course includes lectures, videos, and laboratories.

Credit Recommendation: ***Version 1:*** In the lower-division baccalaureate/associate degree category, see AR-1406-0090 for common core credit. In the upper-division baccalaureate category, 2 semester hours in military science (8/93). *Version 2:* In the lower-division baccalaureate/associate degree category, 3 semester hours in veterinary assistant and 6 in a veterinary technology practicum (10/03).

Related Occupations: 91T.

AR-0104-0013

1. Veterinary Food Inspection Specialist Basic
2. Veterinary Food Inspection Specialist Basic
3. Veterinary Food Inspection Specialist (Basic)

Course Number: ***Version 1:*** 321-91R10. *Version 2:* 321-91R10. *Version 3:* 321-91R10.

Location: Academy of Health Sciences, Fort Sam Houston, TX.

Length: ***Version 1:*** 8 weeks (302–315 hours). *Version 2:* 8 weeks (315–320 hours). *Version 3:* 8 weeks (336 hours).

Exhibit Dates: ***Version 1:*** 4/94–12/96. *Version 2:* 1/97–3/02. *Version 3:* 4/02–Present.

Learning Outcomes: ***Version 1:*** Before 4/94 see AR-0104-0002. Upon completion of the course, the student will be able to perform food hygiene inspections, food safety inspections, and food quality assurance inspections for the various food categories and discuss the concept of food sampling and preparation for submittal to the laboratory. *Version 2:* Upon completion of the course, the student will be able to perform under supervision food hygiene, safety, and quality assurance inspections. *Version 3:* Upon completion of the course, the student will be able to develop the necessary skills and techniques to perform under direction food hygiene, safety, and quality assurance inspections.

Instruction: ***Version 1:*** Conferences, demonstrations, and practical exercises cover administration, food technology and inspection, and veterinary preventive medicine. When given applicable documents, inspection equipment, and raw materials for meat, dairy products, poultry and waterfood, students can perform appropriate component, process, or end-item inspection; conduct surveillance inspections on a variety of fresh fruits and vegetables; perform product wholesomeness inspection; and detect unsanitary conditions in food-handling facilities. Student instruction is reinforced through testing. *Version 2:* The course includes topics related to the inspection of food products, including meats, fruits, and vegetables; sanitation procedures; food safety; and quality assurance. All students are eligible for the certificate in applied food service sanitation given by the Educational Foundation of the National Restaurant Association. *Version 3:* Audiovisual materials, practical and classroom exercises, discussion, laboratory, case studies, and lecture. Topics include identification, condition, and quantity of food groups; operational rations and semi-perishables; food laboratory analysis; sampling and basic food inspections procedures; and basic food establishment sanitation and safety and quality assurance. Other subjects include animal medicine and selected military subjects. Also included are introduction to chemical, biological, and radiological agents, nuclear devices, and high yield explosives.

Credit Recommendation: ***Version 1:*** In the lower-division baccalaureate/associate degree category, 10 semester hours in food inspection/safety (10/95). *Version 2:* In the lower-division baccalaureate/associate degree category, 4 semester hours in food safety and quality assurance, 4 in introduction to food handling and processing, and 2 in food procurement (5/01). *Version 3:* In the lower-division baccalaureate/associate degree category, 3 semester hours in food safety and quality assurance, 3 in introduction to food handling and processing, 2 in food procurement, and 4 in clinical experience (11/03).

AR-0104-0014

1. Veterinary Services Technician (Basic)
2. Veterinary Services Technician Warrant Officer Basic

Course Number: ***Version 1:*** 6G-640A (WO). *Version 2:* 6G-640A (WO Basic).

Location: Academy of Health Sciences, Fort Sam Houston, TX.

Length: ***Version 1:*** 3 weeks (117 hours). *Version 2:* 5 weeks (197 hours).

Exhibit Dates: ***Version 1:*** 5/95–3/97. *Version 2:* 4/97–Present.

Learning Outcomes: ***Version 1:*** Before 5/95 see AR-0104-0010. Upon completion of the course, the student will be able to assist the veterinarian in preventing and controlling zoonotic and foodborne diseases; apply administrative skills to documents, reports, and forms applicable to food inspection situations; and perform duties relevant to the mission of the Army veterinary services. *Version 2:* Upon completion of the course, the student will be able to assist the veterinarian in preventing and controlling zoonotic and foodborne diseases; apply administrative skills to documents, reports, and forms applicable to food inspection situations; and perform duties relevant to the mission of the Army veterinary services.

Instruction: ***Version 1:*** This course employs the conference type of instruction for topics covering food hygiene, food technology, and health care administration. Animal medicine and combat service support topics are formatted for instruction using demonstrations, seminars, and practical exercises. Reinforcement of instruction is one of the objectives of regular examinations. *Version 2:* This course employs the conference type of instruction for topics, including food hygiene, food technology, and health care administration. Animal medicine and combat service support topics are formatted for instruction using demonstrations, seminars, and practical exercises. Reinforcement of instruction is one of the objectives of regular examinations.

Credit Recommendation: ***Version 1:*** In the lower-division baccalaureate/associate degree category, 6 semester hours in food hygiene and technology and 2 in animal medicine and care (10/95). *Version 2:* In the lower-division baccalaureate/associate degree category, 8 semester hours in introduction to zoonotic and foodborne diseases, 6 in food hygiene and zoonosis, and 2 in animal care (7/99).

Related Occupations: 640A.

AR-0202-0001

1. Graphics Documentation Specialist Basic Noncommissioned Officer (NCO)
2. Graphics Documentation Specialist Basic Noncommissioned Officer (NCO) 25Q30

Course Number: ***Version 1:*** G3AZR23171 000 (25Q30); G3AZR23171 000. *Version 2:* G3AZR23171 000 (25Q30); G3AZR23171 000.

Location: Technical Training Center, Lowry AFB, CO.

Length: ***Version 1:*** 6–7 weeks (216–257 hours). *Version 2:* 7–8 weeks (290 hours).

Exhibit Dates: ***Version 1:*** 10/88–9/91. *Version 2:* 10/91–9/94.

Learning Outcomes: ***Version 1:*** Upon completion of the course, the student will be able to manage a graphics documentation team which provides visual information support services including Diazo overhead transparencies and computer graphics for presentation and publication purposes. Student will plan, supervise, and direct the operation of the graphics documentation team and the maintenance of the equipment. *Version 2:* Upon completion of the course, the student will be capable of managing a graphics documentation team as well as performing individual functions within the organization. Students can plan, supervise, and direct the operation of the graphics documentation team and the maintenance of the equipment.

Instruction: ***Version 1:*** Lectures and practical exercises are reinforced by application in a field operational test where the unit must plan and execute a move to support a military operation requiring visual information support for briefings, documentation, and training. Course also includes a common core of leadership subjects. *Version 2:* Lectures, demonstrations, and practical exercises cover routine visual information documentation, television equipment maintenance, computer applications, photographic facility management, and television production management. Students also receive training in graphics, black-and-white/color theory, fundamentals of lettering and drawing, viewgraph preparation, map overlays, and computer-generated graphics.

Credit Recommendation: ***Version 1:*** In the lower-division baccalaureate/associate degree category, see AR-1406-0090 for common core credit. In the upper-division baccalaureate category, 3 semester hours in graphics management (5/90). *Version 2:* In the upper-division baccalaureate category, 3 semester hours in graphics management (7/94).

Related Occupations: 25M; 25Q.

AR-0202-0002

Still Documentation Specialist Basic Noncommissioned Officer (NCO) (25S30)

Course Number: G3AZR23172 002.

Location: Technical Training Center, Lowry AFB, CO.

Length: 7–8 weeks (290 hours).

Exhibit Dates: 10/91–9/94.

Learning Outcomes: Before 10/91 see AR-1709-0011. Upon completion of the course, the student will be able to manage visual information facilities and organizations including training, planning, supervising, and directing visual information documentation operations and support for unit activities both in the garrison and in the field.

Instruction: Lectures, demonstrations, and practical exercises provide knowledge and skills for the administration, supervision, and operation of a visual documentation team. Topics include management principles and technical information necessary to direct, plan, manage, and review team operations, including instruction in computer graphics, video systems, still photography production, and procedures for historical documentation and unit operations support. Emphasis is on still photography operations and leadership responsibilities.

Credit Recommendation: In the upper-division baccalaureate category, 3 semester hours in audio/visual management (7/94).

Related Occupations: 25S.

AR-0301-0003

Cost Estimating for Engineers

Course Number: ***Version 1:*** ALMC-CC. *Version 2:* ALMC-CC.

Location: Army Logistics Management College, Fort Lee, VA.

Length: ***Version 1:*** 2 weeks (73 hours). *Version 2:* 2 weeks (73 hours).

Exhibit Dates: ***Version 1:*** 8/90–9/92. *Version 2:* 10/92–9/98.

Learning Outcomes: ***Version 1:*** Upon completion of the course, the student will be able to describe the preferred methods of cost estimating, describe the variety of data sources and data adjustments, illustrate the techniques used in developing and analyzing life cycle cost estimates, and illustrate the proper use of cost estimating methods and techniques. *Version 2:* Upon completion of the course, the student will be able to describe the preferred methods of cost estimating; describe the variety of data sources and data adjustments using various techniques, including parametric statistics, uncertainty and validity tests, economic analyses, and regression models; illustrate the techniques used in developing and analyzing life cycle cost estimates; and illustrate the proper use of cost estimating methods and techniques.

Instruction: ***Version 1:*** Lectures, practical exercises, and group workshop methods are used in the qualitative and quantitative techniques of cost estimating and analysis and methods of developing cost estimates and evaluating uncertainties. *Version 2:* Lectures, practical exercises, and group workshop methods are used in the qualitative and quantitative techniques of cost estimating and analysis and methods of developing cost estimates and evaluating uncertainties.

Credit Recommendation: ***Version 1:*** In the upper-division baccalaureate category, 3 semester hours in cost analysis (6/91). *Version 2:* In the upper-division baccalaureate category, 3 semester hours in cost analysis (8/95).

AR-0301-0004

Cost Estimating for Engineers by Correspondence

Course Number: ALMC-CC.

Location: Army Logistics Management College, Fort Lee, VA.

Length: Maximum, 52 weeks.

Exhibit Dates: 8/90–9/91.

Learning Outcomes: Upon completion of the course, the student will be able to describe the preferred methods of cost estimating; describe the variety of data sources and data adjustments; illustrate the techniques used in developing and analyzing life cycle cost estimates; and illustrate the proper use of cost estimating methods and techniques.

Instruction: This is a correspondence, self-instruction program covering methods used in the qualitative and quantitative cost estimating and analysis, developing cost estimates, and evaluating uncertainties.

Credit Recommendation: In the upper-division baccalaureate category, 3 semester hours in cost analysis (6/91).

AR-0306-0001

Reading Improvement

Course Number: 1-19-XQ-100.

Location: Military Police School, Fort McClellan, AL.

Length: Self-paced, 1–2 weeks (50 hours).

Exhibit Dates: 6/81–3/92.

Objectives: To improve the student's ability to read more effectively by increasing both speed and comprehension rates.

Instruction: This is primarily a laboratory course using the Science Research Associates' conceptual approach to reading improvement. Through individualized programmed instruction, the student demonstrates progressive mastery of the multilevel power builder and rate builder reading programs.

Credit Recommendation: In the vocational certificate category, credit in reading improvement on the basis of institutional evaluation (3/82).

AR-0326-0003

Associate Logistics Executive Development

Course Number: 8A-F19.

Location: Army Logistics Management Center, Fort Lee, VA.

Length: 10 weeks.

Exhibit Dates: 7/79–9/88.

Objectives: After 9/88 see AR-1405-0213. To teach officers logistics management.

Instruction: Lectures and practical exercises cover logistics management, including management systems and development; inventory management; personnel management; financial management; statistical methods; maintenance management; logistical organization of the military; procurement; contract law, policies, and definitions; contract development and administration; depot operations management; containerization; distribution management; and maintenance economics. Lecture-conferences, cases, practical exercises, guest speakers, and field trips cover management principles, financial management, the acquisition process, asset management, economic and statistical principles, and logistics management. Note: Course is offered two weeks a year for five years.

Credit Recommendation: In the upper-division baccalaureate category, 4 semester hours in general management, 4 in economic principles and decision making, and 6 in logistics management. In the graduate degree category, 3 semester hours in general management (3/86).

AR-0326-0008

Army Depot Operations Management

Course Number: 8B-F10; ALM-42-0228.

Location: Army Logistics Management College, Fort Lee, VA; Army Logistics Management College, Onsite at various locations.

Length: 4 weeks (156 hours).

Exhibit Dates: 6/80–7/88.

Objectives: After 7/88 see AR-0326-0051. To teach officers and civilian personnel depot directorate-level management.

Instruction: Instructional methods include lectures, case discussions, computer-assisted simulations, an oral project, field trips, written analysis, guided discussion, seminars, workshops, and presentations by guest speakers. Topics include management of the major depot functions of receipt, storage, care, preservation, distribution, and control of materiel.

Credit Recommendation: In the upper-division baccalaureate category, 3 semester hours in management of warehousing operations (5/81).

Related Occupations: 726A; 761A.

AR-0326-0010

Intermediate Pre-Award Contracting
(Management of Defense Acquisition Contracts Advanced)
(Management of Acquisition Contracts Advanced)

Course Number: CON 211; 8D-F12.

Location: Army Logistics Management College, Fort Lee, VA; Army Logistics Management College, Onsite at various locations; Army Logistics Management Center, Fort Lee, VA.

Length: 1–2 weeks (102 hours).

Exhibit Dates: 6/85–5/94.

Objectives: After 5/94 see AR-0326-0057. To train procurement management officers in advanced procurement management.

Instruction: Lectures and practical exercises cover procurement management, including procurement planning, funding, and technical requirement impact: small business, labor surplus, and disaster area relief operations; procurement by formal advertising and negotiation; contract types and purposes; cost and price analysis; contract administration; terminations and remedies; patents and data; and procurement trends. Lectures and practical exercises cover planning for contract management, including acquisition planning, socioeconomic programs, and cost/price analysis and contract management, including bidding, negotiations, modifications, and termination.

Credit Recommendation: In the upper-division baccalaureate category, 2 semester hours in acquisition management. In the graduate degree category, 1 semester hour in contract management (10/88).

AR-0326-0012

Decision Risk Analysis

Course Number: ALMC-DA.

Location: Army Logistics Management College, Fort Lee, VA; Army Logistics Management Center, Fort Lee, VA; Army Logistics Management College, Onsite at various locations.

Length: 2 weeks (73–78 hours).

Exhibit Dates: 7/72–4/04.

Objectives: To provide a basic understanding of the concepts and techniques of decision risk analysis for the design, analysis, and control of large-scale military programs.

Instruction: Lectures and practical exercises cover the theory and application of qualitative and quantitative techniques of decision risk analysis.

Credit Recommendation: In the upper-division baccalaureate category, 2 semester hours in risk analysis. In the graduate degree category, 1 semester hour in risk analysis (8/90).

AR-0326-0013

1. Decision Risk Analysis for Logisticians
2. Decision Analysis for Logisticians

Course Number: ***Version 1:*** None. *Version 2:* ALMC-DC.

Location: ***Version 1:*** Army Logistics Management College, Fort Lee, VA; Army Logistics Management Center, Fort Lee, VA; Army Logistics Management College, Onsite at various locations. *Version 2:* Army Logistics Management College, Fort Lee, VA; Army Logistics Management College, Onsite at various locations.

Length: ***Version 1:*** 2 weeks (74–78 hours). *Version 2:* 2 weeks (74 hours).

Exhibit Dates: ***Version 1:*** 7/72–7/93. *Version 2:* 8/93–Present.

Learning Outcomes: ***Version 1:*** To provide a basic understanding of concepts and techniques of decision risk analysis as applied to logistics problems. *Version 2:* To provide a basic understanding of concepts and techniques of decision analysis as applied to logistics problems.

Instruction: ***Version 1:*** Lectures and practical exercises cover analysis techniques in logistics problems; application of analysis techniques; and case studies of actual problems in such logistics areas as requirements, procurement, distribution, and disposal. *Version 2:* Lectures and practical exercises cover analysis techniques in logistics problems; application of analysis techniques; and case studies of actual problems in such logistics areas as requirements, procurement, distribution, and disposal.

Credit Recommendation: ***Version 1:*** In the upper-division baccalaureate category, 2 semester hours in decision risk analysis. In the graduate degree category, 1 semester hour in decision risk analysis. Credit is to granted in either the upper-division baccalaureate category or graduate degree category, but not in both (8/90). *Version 2:* In the upper-division baccalaureate category, 2 semester hours in decision analysis. In the graduate degree category, 1 semester hour in decision analysis. Credit is to be granted in either the upper-division baccalaureate category or graduate degree category, but not in both (8/95).

AR-0326-0042

1. Army Medical Department (AMEDD) Officer Advanced
2. Army Medical Department (AMEDD) Officer Advanced
3. Army Medical Department Officer (AMEDD) Advanced
4. Army Medical Department Officer (AMEDD) Advanced

Course Number: ***Version 1:*** 6-8-C22. *Version 2:* 6-8-C22. *Version 3:* 6-8-C22. *Version 4:* 6-8-C22.

Location: Academy of Health Sciences, Fort Sam Houston, TX.

Length: ***Version 1:*** 20 weeks (569 hours). *Version 2:* 16–20 weeks (431–738 hours). *Version 3:* 20 weeks (643–923 hours). *Version 4:* 10 weeks (402–450 hours).

Exhibit Dates: ***Version 1:*** 6/87–5/90. *Version 2:* 6/90–12/94. *Version 3:* 1/95–12/96. *Version 4:* 1/97–8/99.

Learning Outcomes: ***Version 1:*** Upon completion of the course, the student will be prepared to take command, leadership, and staff positions of greater responsibility in the Army Medical Corps. *Version 2:* Upon completion of this course, the student will be able to describe the Army family advocacy program, discuss principles of resource and personnel management, discuss military operations and protocols, manage NBC casualties, discuss preventive medicine programs, and discuss the mission of AMEDD divisions in combat. In addition, those in the health care administration track will be able to perform capitation budgetry, discuss the role of health care administration, perform library research, and manage the gateway-to-care program. Those in the command and staff track will be able to discuss command issues, coordinate AMEDD activities in combat, manage a unit awards program, identify current legal issues, discuss unit training programs, conduct combat support services operations, and set up/tear down deployable medical systems. Those in the preventive medicine track will be able to discuss military pest management issues, evaluate sanitary emergency and environmental science issues, and identify environmental hazards. Those in the dental track will be able to identify the role of dental officers in war/mass casualty, discuss dental quality assurance, and describe issues of dental program management. Those in the laboratory science track will be able to discuss current OSHA regulations and laboratory safety, integrate CLIA into laboratory operations, participate in clinical laboratory management activities, operate a field medical laboratory, and manage civilian employees. Those in the medical specialty track will complete an advanced clinical/administration practicum, perform the medical specialist's role in the field, and discuss issues in resource management. Those in the nurse track will be able to implement basic trauma life support; participate in the setup and breakdown of a deployable medical system; identify nursing roles in the Navy and Marine Corps, the Air Force, and the National Disaster Medical System; and complete one of the following outcomes: manage resources, manpower, equipment, and supplies in a clinical environment; integrate roles and responsibilities of the nurse counselor; integrate the roles of community health nursing; or integrate roles of nursing education/staff development. Those in the veterinary track will be able to identify cuts of meat purchased by the Defense Commissary Agency; discuss current issues in veterinary practice, food safety, and sanitary inspections; conduct veterinary field operations; perform a sanitary inspection; and conduct a foodborne disease investigation. Those in the pharmacy track will be able to discuss current issues in pharmacy practice, participate in research activity, and demonstrate knowledge of pharmacokinetics. Those in the audiology track will be able to discuss audiology issues, complete a clinic rotation, complete a research project, and use the hearing evaluation automated registry system manager's module. Those in the optometry track will be able to discuss current trends and issues in optometry, complete a research project, and complete a clinical rotation. Students in the logistics track will be able to discuss current trends and issues, complete a research project, and monitor management indicators using a computer system. Those in the health services human resource manager track will be able to discuss the role and responsibility of the personnel office, discuss current issues, and complete a research paper. Those students in the medical corps track will be able to discuss health care trends, discuss legal-ethical issues, discuss military roles of the physician, and discuss medical treatment of combat situations. Those in the radiation safety track will be able to discuss current issues in radiation safety, discuss current diagnostic and treatment procedures, and discuss military applications and hazards of current technology. Those in the behavioral sciences track will be able to establish a quality assurance program, evaluate clients exhibiting combat stress, organize intervention teams in a disaster/terrorist situation, and discuss specific Army intervention programs. *Version 3:* Upon completion of this course, the student will be able to describe the Army family advocacy program, discuss principles of resource and personnel management, discuss military operations and protocols, manage NBC casualties, discuss preventive medicine programs, and discuss the mission of AMEDD divisions in combat. In addition, those in the health care administration track will be able to perform capitation budgetry, discuss the role of health care administration, perform library research, and manage the gateway-to-care program. Those in the command and staff track will be able to discuss command issues, coordinate AMEDD activities in combat, manage a unit awards program, identify current legal issues, discuss unit training programs, conduct combat support services operations, and set up/tear down deployable medical systems. Those in the preventive medicine track will be able to discuss military pest management issues; evaluate sanitary, emergency, and environmental science issues; and identify environmental hazards. Those in the dental track will be able to identify the role of dental officers in war/mass casualty, discuss dental quality assurance, and describe issues of dental program management. Those in the laboratory science track will be able to discuss current OSHA regulations and laboratory safety, integrate CLIA into laboratory operations, participate in clinical laboratory management activities, operate a field medical laboratory, and manage civilian employees. Those in the medical specialty track will complete an advanced clinical/administration practicum, perform the medical specialist's role in the field, and discuss issues in resource management. Those in the nurse track will be able to implement basic trauma life support; participate in the setup and breakdown of a deployable medical system; identify nursing roles in the Navy and Marine Corps, the Air Force, and the National Disaster Medical System; and complete one of the following outcomes: manage resources, manpower, equipment, or supplies in a clinical environment; integrate roles and responsibilities of the nurse counselor; integrate the roles of community health nursing; or integrate roles of nursing education/staff development. Those in the veterinary track will be able to identify cuts of meat purchased by the Defense Commissary Agency; discuss current issues in veterinary practice, food safety, and sanitary inspections; conduct veterinary field operations; perform a sanitary inspection; and conduct a foodborne disease investigation. Those in the pharmacy track will be able to discuss current issues in pharmacy practice, participate in research activity, and demonstrate knowledge of pharmacokinetics. Those in the audiology track will be able to discuss audiology issues, complete a clinical rotation, complete a research project, and use the hearing evaluation automated registry system manager's module. Those in the optometry track will be able to discuss current trends and issues in optometry, complete a research project, and complete a clinical rotation. Students in the logistics track will be able to discuss current trends and issues, complete a research project, and monitor management indicators using a computer system. Those in the health services human resource manager track will be able to discuss the role and responsibility of personnel offices, discuss current issues, and complete a research paper. Those students in the medical corps track will be able to discuss health care trends, discuss legal-ethical issues, discuss military roles of the physician, and discuss medical treatment of combat situations. Those in the radiation safety track will be able to discuss current issues in radiation safety, discuss current diagnostic and treatment procedures, and discuss military applications and hazards of current technology. Those in the behavioral sciences track will be able to establish a quality assurance program, evaluate clients exhibiting combat stress, organize intervention teams in a disaster/terrorist situation, and discuss specific Army intervention programs. *Version 4:* After 8/99 see AR-0326-0062. Upon completion of the core curriculum of the course, students will be able to analyze the historical perspectives of military medical history; develop organizational team building; employ planning and decision making processes to solve problems and develop courses of action; demonstrate ethical decision-making and leadership; explain military health care delivery systems; supervise military unit operations, including personnel management and training, supply and maintenance operations, risk management and safety, preventive medicine, resource allocation, and accounting; and direct unit logistic support operations under combat and noncombat conditions. Those in the dental track will be able to employ the skills acquired in the core course in the management of a dental clinic and implement quality assurance procedures and hazard communication programs. Those in the medical specialist corps track (AMSC) will be able to employ the skills acquired in the core cores in AMSC management, discuss current developments and missions of each of the AMSC specialists, and maintain continuous quality improvement programs. Those in the nursing track will be able to employ the skills acquired in the core course in nursing management, implement care plans for rapid assessment of trauma patients, and assess current issues affecting the Army nurse corps. Those in the veterinary science track will be able to employ the skills acquired in the core course in veterinary management; supervise the inspection and procurement of meat, dairy, and water products; investigate incidences of disease outbreaks; and implement quality animal training and welfare programs. Those in the medical corps track will be able to discuss current civilian health care trends and future projections, new trends in credentialing, pharmoeconomic developments, ethical considerations, downsizing, and conflict resolution. Those in the medical service corps track will be able to implement total quality management concepts and policies, instill ethical decision making and behavior through the command, and demonstrate in-depth knowledge of managed care programs.

Instruction: *Version 1:* Lectures, seminars, conferences, demonstrations, and practical exercises cover military science topics and health care administration. Students attend a common core, then subject-specific tracks. *Version 2:* Course content covers Army family advocacy program, medical equipment program, facility planning, resource management, health care automation, health care delivery system, quality assurance, military science, preventive medicine and dentistry, and food service operations. In addition, the health care administration track covers capitation budgeting issues in health care administration, roles of health care administrators, marketing, library research, JCAHO accreditation, negotiations, coordinated care, and total quality assurance. The specialty is taught through lectures and practical exercises and does not in include a research paper. The command and staff specialty covers basic concepts of command, inspections, and AMEDD synchronization; awards program; legal issues; and risk management. This specialty does not include a research paper and is conducted through lectures, practical exercises, and field exercises. The preventive medicine track covers pest management, combat preventive medicine, sanitation, epidemiology, environmental hazards, and research. The track is presented through lectures and practical exercises and a research project. The dental specialty covers dental leadership roles in mass casualty, total quality management, personnel management, sterilization and infection, and CPR instructor training. This track is presented through lectures, practical exercises, and advanced individual training. The laboratory science track covers OSHA and laboratory safety, CLIA educational programs in laboratory sciences, budgeting and staffing, quality assurance, total quality management, civilian personnel management, and field exercises. The medical specialty track includes career development, leadership, marketing, clinical practicum, continuous quality improvement, facilities planning, budgeting, health risk appraisal, and the roles of the Army. Methods in this track include lectures, practical exercises, and clinical practicum. The nurse track covers nursing practice standards; quality assurance; infection control; basic trauma life support; altitude stresses; deployable medical systems; Air Force, Navy, Marine Corps, and national disaster medical system; and combat environment. One of the following tracks will be completed: clinical head nurse, recruitment/ROTC nurse corps, preventive medicine/community health nurse, or nursing education staff development. This track is presented through lectures and practical exercises and includes field exercises. This track does not include a research paper. The veterinary track includes use of working dogs, unit training, authority of veterinary inspectors, meat cuts, carcass disposition, antemortem/postmortem findings, veterinary threat analysis, animal rights, quality management, food safety, retail code, and foodborne infections. This track contains no research project. Methodology includes lectures, practical exercises, and field exercises. The pharmacy track covers pharmacy issues, parental nutrition, security of the pharmacy, drug diversion, computer systems, pharmacy management, inspection, specialty pharmacies, quality assurance, budgeting and staffing, and pharmacokinetics. Teaching methods include lectures, practical exercises, and research. The audiology track covers manpower, contracts, audiology issues, personnel management, clinical rotation, research, and hearing evaluation automated registry system manager's module. This track is conducted through lectures, practical exercises, clinical practicum, and research. The optometry track covers manpower, field optometry, quality improvement, profiling, personnel management, facilities planning, current trends, vision conservation, new developments, clinical experiences, and computer networking. Practical exercises and lectures are included along with research. The logistics track covers property management, medical materiel management, biomedical maintenance, central processing, research, and theater medical management information systems. Methods of presentation include lectures, research, and practical exercises. The health services human resources manager track covers the role of health services personnel officers, the role of the adjutant, personnel issues, computer applications, special programs, research, EEOC, civilian personnel management, and AMEDD organizational structure. The medical corps track covers ethics, administration, trends, resource management, quality assurance, leadership, combat injuries and treatment, research, and DRGs. There is minimal research involved in this track. The radiation track covers radiation protection, principles of radiation, JCAHO, quality control, update of specific procedures in diagnosis and treatment, dosimetry, lasers, optical radiation, microwaves, electromagnetic interference, and ultrasound. This track has no research project. The behavioral science track covers quality assurance, role in combat, crisis intervention, alcohol and drug abuse program, case studies, budgeting, and research. There is a research project in this track. *Version 3:* Course content covers Army family advocacy program, medical equipment program, facility planning, resource management, health care automation, health care delivery system, quality assurance, military science, preventive medicine, dentistry, and food service operations. In addition, the health care administration track covers capitation budgeting issues in health care administration, roles of health care administrators, marketing, library research, JCAHO accreditation, negotiations, coordinated care, and total quality assurance. The specialty is taught through lectures and practical exercises and does not include a research paper. There are patient administration, personnel, and human resources subtracks. The command and staff specialty covers basic concepts of command, inspections, and AMEDD synchronization; awards program; legal issues; and risk management. This specialty does not include a research paper and is conducted through lectures, practical exercises, and field exercises. The preventive medicine track covers pest management, combat preventive medicine, sanitation, epidemiology, environmental hazards, and research. The track is presented through lectures and practical exercises and a research project. The dental specialty covers dental leadership roles in mass casualty, total quality management, personnel management, sterilization and infection, and CPR instructor training and is presented through lectures, practical exercises, and advanced individual training. The laboratory science track covers OSHA and laboratory safety, CLIA educational programs in laboratory sciences, budgeting and staffing, quality assurance, total quality management, civilian personnel management, and field exercises. The medical specialty track includes career development, leadership, marketing, clinical practicum, continuous quality improvement, facility planning, budgeting, health risk appraisal, and the roles of the Army. There are lectures, practical exercises, and a clinical practicum. The nurse track covers nursing practice standards; quality assurance; infection control; basic trauma life support; altitude stresses; deployable medical systems; Air Force, Navy, Marine Corps, and national disaster medical system; and combat environment. One of the following tracks will be completed: clinical head nurse, recruitment/ROTC nurse corps, preventive medicine/community health nurse, or nursing education staff development. This track is presented through lectures and practical exercises and includes field exercises. This track does not include a research paper. The veterinary track includes use of working dogs, unit training, authority of veterinary inspectors, meat cuts, carcass disposition, antemortem/postmortem findings, veterinary threat analysis, animal rights, quality management, food safety, retail code, and food-borne infections. This track contains no research project. Methodology includes lectures, practical exercises, and field exercises. The pharmacy track covers pharmacy issues, parental nutrition, security of the pharmacy, drug diversion, computer systems, pharmacy management, inspection, specialty pharmacies, quality assurance, budgeting and staffing, and pharmacokinetics. Teaching methods include lectures, practical exercises, and research. The audiology track covers manpower, contracts, audiology issues, personnel management, clinical rotation, research, and hearing evaluation automated registry system manager's module. This track is conducted through lectures, practical exercises, clinical practicum, and research. The optometry track covers manpower, field optometry, quality improvement, profiling, personnel management, facilities planning, current trends, vision conservation, new developments, clinical experiences, and computer networking. Practical exercises and lectures are included along with research. The logistics track covers property management, medical materiel management, biomedical maintenance, central processing, research, and theater medical management information systems. Methods of presentation include lectures, research, and practical exercises. The health services human resource manager track covers the role of health services personnel officers, the role of the adjutant, personnel issues, computer applications, special programs, research, EEOC, civilian personnel management, and AMEDD organizational structure. The medical corps track covers ethics, administration, trends, resource management, roles, quality assurance, leadership, combat injuries and treatment, research, and DRGs. There is minimal research involved in this specialty. The radiation track covers radiation protection, principles of radiation, JCAHO, quality control, update of specific procedures in diagnosis and treatment, dosimetry, lasers, optical radiation, microwaves, electromagnetic interference, and ultrasound. This track has no research project. The behavioral science track covers quality assurance, role in combat, crisis intervention, alcohol and drug abuse program, case studies, budgeting, and research. This track has a research project. *Version 4:* Phase 1 is by correspondence and not included in the evaluation. Phase 2 consists of a 9 week common core and 6 one-week tracks. Lectures, distinguished guest lecturers, reading assignments, practical exercises, group discussions, research discussions, and field exercises provide instruction in the concepts of overall army medical unit management in peacetime and during combat. Lectures, panel discussions, and briefings provide instruction in dental clinic management for those in that track. Those in the AMSC track are provided with lectures, practical exercises, and field training to provide instruction in the management of medical professional services. The nursing track is presented through lectures and practical exercises with instruction in issues in peacetime and combat nursing management. The veterinary track is presented through lectures, reading assignments, field training, practical exercises, and panel discussions to provide instruction in administration of veterinary functions including food inspection and procurement. The medical corps track is offered through lectures providing instruction in current issues in medical management, and the medical service corps track uses lectures and research study to provide medical service corps officers with executive skills.

Credit Recommendation: ***Version 1:*** In the upper-division baccalaureate category, 9 semester hours in business management and organization (4/88). *Version 2:* In the lower-division baccalaureate/associate degree category, 1 semester hour in CPR instructor training and certification for those in the dental track. In the upper-division baccalaureate category, 1 semester hour in preventive medicine, 3 in personnel management, 5 in military science, and 2 in resource management for all students; for those in the nurse track, 1 semester hour in basic trauma life support. In the graduate degree category, for those in the health care administration track, 4 semester hours in administration and 1 in issues in health care administration; for those in the command and staff track, 5 semester hours in military leadership and operations; for those in the preventive medicine track, 2 in environmental science, 1 in issues in preventive medicine, and 2 for the research project; for those in the dental track, 1 in issues in dental science and 1 in research/advanced skills; for those in the laboratory science track, 1 semester hour in federal and state regulations, 2 in medical laboratory field operations, 1 in issues in laboratory science, and 1 in research/advanced skills; for those in the medical specialty track, 2 semester hours in clinical/administrative practicum and 1 in issues for medical specialists; for those in the nurse track, 2 semester hours in the nursing specialty (clinical head nurse, recruitment/ROTC nurse corps, preventive medicine/community health nurse, or nursing education staff development), and 2 in nursing leadership issues; for those in the veterinary track, 3 semester hours in veterinary field operations and 2 in food safety issues/inspection; for those in the pharmacy track, 1 in the research project, 1 in pharmacy issues, and 3 in pharmacokinetics; for those in the audiology track, 1 in issues in audiology, 2 in the research project, 1 in clinical experience, and 1 in hearing conservation management; for those in the optometry track, 2 in the research project, 1 in clinical experience, and 1 in optometry issues; for those in the logistics track, 2 in the research project, 2 in computer applications, and 1 in current issues; for those in the health services human resources manager track, 2 in the research project, 2 in personnel management, and 1 in current issues; for those in the medical corps track, 1 semester hour in ethical issues in health care, 1 in current trends in health care, and 3 in combat injuries; for those in the radiation safety track, 1 in issues in radiation safety and 4 in advanced radiation safety; and for those in the behavioral science track, 2 semester hours in crisis intervention, 1 in issues in behavioral science, and 2 in clinical research (8/93). *Version 3:* In the graduate degree category, for the core curriculum in module 1 that all students complete, 8 semester hours in health care support service and 5 in military science; for module 2, that all students complete, 3 semester hours in computer applications and an additional 5 in military science; for module 3 credit is based on the track is completed. For those in the health care administration track, 6 semester hours in that area; for those in the patient administration subtrack, 10 semester hours in that area; for those in the personnel subtrack, 10 in that area; and for those in human resources subtrack, 10 in that area. For those in the command and staff track, 10 semester hours in leadership; for those in environmental engineering and science, 6 semester hours in those areas. For those in laboratory track, 9 semester hours in laboratory. For those in the AMSC track, 4 semester hours in management for all, plus 2 in physician assistant for those in that subtrack; 2 semester hours in nutritional care for those in that subtrack; 2 in occupational therapy for those in that subtrack; and 2 in physical therapy for those in that subtrack. For those in the nursing track, 10 semester hours in nursing. For those in the veterinary science track; 10 semester hours in veterinary science. For those in the pharmacy track, 9 semester hours in that area. For those in the audiology track, 6 semester hours in audiology. For those in the optometry track, 7 semester hours in optometry. For those in the logistics track, 9 semester hours in logistics. For those in the nuclear, biological, and chemical track, 10 semester hours in nuclear, biological, and chemical principles. For those in the AMEDD track, 8 in leadership and 2 in military science. And for those in the dental track, 2 semester hours in dental science (10/95). *Version 4:* In the upper-division baccalaureate category, for all students who complete the core curriculum, 3 semester hours in human resource management, 2 in topics in military history, 2 in material management, and 3 in health care administration. In the graduate degree category, for those in the dental track, 1 semester hour in dental clinic administration; for those in the army medical specialist corps track, 1 semester hour in health administration; for those in the nursing track, 1 semester hour in nursing administration; for those in the veterinary science track, 1 semester hour in veterinary administration; for those in the medical corps and medical service corps tracks, 1 semester hour in health administration (9/97).

AR-0326-0044

MASTER PLANNER

Course Number: EH 456.

Location: Combined Arms Training Center, Vilseck, W. Germany.

Length: 2 weeks (80 hours).

Exhibit Dates: 4/81–10/98.

Learning Outcomes: Upon completion of the course, the student will be able to make limited studies in technical areas; coordinate information and approvals required to prepare a master plan; develop documents of the master plan, including environmental impact, land use, general site, utilities, road networks, and technical support services; define functional requirements as Military Construction Army (MCA) programming; provide MCA details regarding cost estimates, construction criteria, energy requirements, and equipment programming; and provide justification paragraphs and site sketches.

Instruction: Demonstrations, conference classes, and practical exercises cover the techniques, policies, and documents necessary for master planning and construction planning. Main topics include development of a master plan, including documents, articulation with a community planning board, construction planning, and cost estimating.

Credit Recommendation: In the upper-division baccalaureate category, 1 semester hour in project management and 1 in community planning (10/88).

AR-0326-0045

1. MANPRINT STAFF OFFICERS
2. MANPRINT ACTION OFFICERS
3. MANPRINT ACTION OFFICERS

Course Number: *Version 1:* 7C-F27/500-F18. *Version 2:* ALMC-MS. *Version 3:* ALMC-MS.

Location: *Version 1:* Humphreys Center, Fort Belvoir, VA; Soldier Support Center, Fort Belvoir, VA; Xerox Center, Leesburg, VA. *Version 2:* Army Logistics Management College, Fort Lee, VA. *Version 3:* Army Logistics Management College, Fort Lee, VA.

Length: *Version 1:* 2–3 weeks (117 hours). *Version 2:* 1–2 weeks (65 hours). *Version 3:* 1–2 weeks (65 hours).

Exhibit Dates: *Version 1:* 10/87–9/90. *Version 2:* 10/90–2/93. *Version 3:* 3/93–Present.

Learning Outcomes: *Version 1:* Upon completion of the course, the student will be able to introduce human factors engineering considerations into the management of the material development and acquisition process in order to increase performance in the total system. *Version 2:* Upon completion of the course, the student will be able to introduce human factors engineering considerations into the management of the material development and acquisition process in order to increase performance in the total system. *Version 3:* Upon completion of the course, the student will be able to introduce human factors engineering considerations into the management of the material development and acquisition process in order to increase performance in the total system.

Instruction: *Version 1:* Course includes classroom instruction, practical exercises, and independent study in the issues of manpower management, personnel, training, safety, human performance and work load analysis and their effect on an acquisition system. *Version 2:* Course includes classroom instruction, practical exercises, and independent study in the issues of manpower management, personnel, training, safety, human performance, and work load analysis and their effect on an acquisition system. *Version 3:* Course includes classroom instruction, practical exercises, and independent study in the issues of manpower management, personnel, training, safety, human performance, and work load analysis and their effect on an acquisition system.

Credit Recommendation: *Version 1:* In the upper-division baccalaureate category, 2 semester hours in organizational behavior (with an emphasis on the nontheoretical approach) (12/88). *Version 2:* In the upper-division baccalaureate category, 2 semester hours in organizational behavior (8/95). *Version 3:* In the upper-division baccalaureate category, 2 semester hours in organizational behavior or human relations (1/02).

AR-0326-0046

ORGANIZATIONAL DEVELOPMENT CONSULTANTS

Course Number: ODCC 1-90.

Location: Professional Education Center, Cp. Joseph Robinson, North Little Rock, AR.

Length: 15 weeks (579 hours).

Exhibit Dates: 9/88–12/97.

Learning Outcomes: Upon completion of the course, the student will be able to directly apply negotiation and consulting skills in accomplishing organizational objectives.

Instruction: Instruction includes lectures, practical experience, structured experience, objective examinations, discussions, reports, student demonstrations, team development, and feedback to individuals.

Credit Recommendation: In the upper-division baccalaureate category, 3 semester hours in human resource management consulting (7/90).

AR-0326-0047

LOGISTICS MANAGEMENT DEVELOPMENT

Course Number: *Version 1:* 8A-F16. *Version 2:* 8A-F16.

Location: Army Logistics Management College, Fort Lee, VA; Army Logistics Management College, Onsite at various locations.

Length: *Version 1:* 4 weeks (152–153 hours). *Version 2:* 4 weeks (153–154 hours).

Exhibit Dates: *Version 1:* 1/89–12/94. *Version 2:* 1/95–Present.

Learning Outcomes: *Version 1:* Upon completion of the course, the student will be able to explain the fundamental concepts of materiel development and apply specific logistics policies and procedures; define and use the life-cycle management model in the contracting, inventory management, distribution, maintenance, and disposal of materiel; use statistical and probability techniques, computer tools, and behavioral theory to solve logistics problems; and integrate financial management theory into logistics operations. *Version 2:* Upon completion of the course, the student will be able to address the management of evolving logistics systems, explain fundamental concepts of material development and apply specific logistic policies and procedures; define and use the life-cycle management model in the contracting, inventory management, distribution, maintenance and disposal of material; use statistical and probability techniques, computer tools, and behavior theory to solve logistics problems; and integrate financial management theory into logistics operations.

Instruction: *Version 1:* Lectures, discussions, and simulation exercises cover life-cycle management model, contracting, material readiness, asset management, financial management, and specific decision-making tools. *Version 2:* Lectures, discussions, and simulation exercises include maintenance planning, distribution, facility needs, contracting, financial management, and specific decision-making tools.

Credit Recommendation: *Version 1:* In the upper-division baccalaureate category, 4 semester hours in logistics management (10/91). *Version 2:* In the graduate degree category, 3 semester hours in logistics management (1/02).

AR-0326-0048

LOGISTICS MANAGEMENT DEVELOPMENT BY CORRESPONDENCE

Course Number: 8A-F16.

Location: Army Logistics Management College, Fort Lee, VA.

Length: Maximum, 52 weeks.

Exhibit Dates: 1/89–12/94.

Learning Outcomes: After 12/94 see AR-0326-0047. Upon completion of the course, the student will be able to explain the fundamental concepts of materiel development and apply specific logistics policies and procedures; define and use the life-cycle management model in the contracting, inventory management, distribution, maintenance, and disposal of materiel; use statistical and probability techniques, computer tools, and behavioral theory to solve logistics problems; and integrate financial management theory into logistics operations.

Instruction: This is a correspondence course. Topics include life-cycle management model, contracting, materiel readiness, asset management, financial management, and specific decision-making tools.

Credit Recommendation: In the upper-division baccalaureate category, 4 semester hours in logistics management (10/91).

AR-0326-0049

Army Maintenance Management

Course Number: *Version 1:* 8A-F3. *Version 2:* 8A-F3.

Location: Army Logistics Management College, Fort Lee, VA; Army Logistics Management College, Onsite at various locations.

Length: *Version 1:* 4 weeks (152 hours). *Version 2:* 2 weeks (77 hours).

Exhibit Dates: *Version 1:* 7/85–12/99. *Version 2:* 1/00–Present.

Learning Outcomes: *Version 1:* Upon completion of the course, the student will be able to increase the readiness and sustainability of the United States and allied forces by applying maintenance doctrine and policies to the planning and managing depot operations and wholesale support of the retail maintenance system. *Version 2:* Upon completion of the course, the student will be able to increase the readiness and sustainability of the United States and allied forces by applying maintenance doctrine and policies to the planning and managing depot operations and wholesale support of the retail maintenance system.

Instruction: *Version 1:* Lectures, case studies, and practical exercises cover maintenance system management, including management development, material requirements, maintenance engineering, and maintenance operations. *Version 2:* Lectures, case studies, and practical exercises cover maintenance system management, including management development, materiel requirements, maintenance engineering, and maintenance operations.

Credit Recommendation: *Version 1:* In the upper-division baccalaureate category, 4 semester hours in materiel management (10/91). *Version 2:* In the upper-division baccalaureate category, 3 semester hours in materiel management (1/02).

AR-0326-0050

Army Maintenance Management by Correspondence

Course Number: 8A-F3.

Location: Army Logistics Management College, Fort Lee, VA.

Length: Maximum, 52 weeks.

Exhibit Dates: 7/85–12/99.

Learning Outcomes: After 12/99 see AR-0326-0049. Upon completion of the course, the student will be able to increase the readiness and sustainability of the United States and allied forces by applying maintenance doctrine and policies and by planning for and exercising management control over depot operations and wholesale support of the retail maintenance system.

Instruction: This is a correspondence course covering topics in maintenance system management, including management development, material requirements, maintenance engineering, and maintenance operations.

Credit Recommendation: In the upper-division baccalaureate category, 4 semester hours in materiel management (10/91).

AR-0326-0051

1. Depot Supply Operations Management
2. Defense Distribution Management

Course Number: *Version 1:* 8B-F10. *Version 2:* 8B-F10.

Location: Army Logistics Management College, Fort Lee, VA; Army Logistics Management College, Satellite Education Program, Fort Lee, VA; Army Logistics Management College, Onsite at various locations.

Length: *Version 1:* 4 weeks (155–157 hours). *Version 2:* 4 weeks (149 hours).

Exhibit Dates: *Version 1:* 8/88–9/93. *Version 2:* 10/93–Present.

Learning Outcomes: *Version 1:* Before 8/88 see AR-0326-0008. Upon completion of the course, the student will be able to perform the major depot functions (receiving, storing, inventorying, issuing, transporting, and controlling of material) and will be able to integrate the major depot into the Army logistics system. *Version 2:* Upon completion of the course, the student will be able to perform the major depot functions (receiving, storing, inventorying, issuing, transporting, and controlling of material) and will be able to integrate the major depot into the Army logistics system.

Instruction: *Version 1:* Instruction is through lecture-conferences, simulations, and workshops and includes general management practices and specific managerial skills in acquisition, maintenance, materiel, and depot supply operations. *Version 2:* Instruction is through lecture-conferences, simulations, and workshops and includes general management practices and specific managerial skills in acquisition, maintenance, materiel, and depot supply operations.

Credit Recommendation: *Version 1:* In the upper-division baccalaureate category, 4 semester hours in warehouse operations management (6/91). *Version 2:* In the upper-division baccalaureate category, 4 semester hours in warehouse/distribution management (7/00).

AR-0326-0052

1. Depot Supply Operations Management by Correspondence
2. Defense Distribution Management by Correspondence

Course Number: *Version 1:* 8B-F10. *Version 2:* 8B-F10.

Location: Army Logistics Management College, Fort Lee, VA.

Length: *Version 1:* Maximum, 52 weeks. *Version 2:* Maximum, 52 weeks.

Exhibit Dates: *Version 1:* 8/88–9/93. *Version 2:* 10/93–Present.

Learning Outcomes: *Version 1:* Upon completion of the course, the student will be able to perform the major depot functions (receiving, storing, inventorying, issuing, transporting, and controlling of material) and will be able to integrate the major depot into the Army logistics system. *Version 2:* Upon completion of the course, the student will be able to perform the major depot functions (receiving, storing, inventorying, issuing, transporting, and controlling of material) and will be able to integrate the major depot into the Army logistics system.

Instruction: *Version 1:* This is a correspondence course covering general management practices and specific managerial skills in acquisition, maintenance, materiel, and depot supply operations. *Version 2:* This is a correspondence course covering general management practices and specific managerial skills in acquisition, maintenance, materiel, and depot supply operations.

Credit Recommendation: *Version 1:* In the upper-division baccalaureate category, 4 semester hours in warehouse operations management (6/91). *Version 2:* In the upper-division baccalaureate category, 4 semester hours in warehouse/distribution management (8/95).

AR-0326-0053

Contracting Fundamentals
(Management of Defense Acquisition Contracts Basic)

Course Number: *Version 1:* CON 101; 8D-4320; 8D-4320 (JT). *Version 2:* CON 101.

Location: *Version 1:* Army Logistics Management College, Fort Lee, VA; Army Logistics Management College, Onsite at various locations; Army Logistics Management College, Satellite Education Program, Fort Lee, VA. *Version 2:* Defense Acquisition University, Various locations in the Continental US; Army Logistics Management College, Fort Lee, VA; Army Logistics Management College, Onsite at various locations; Army Logistics Management College, Satellite Education Program, Fort Lee, VA.

Length: *Version 1:* 4 weeks (156 hours). *Version 2:* 4 weeks (157 hours).

Exhibit Dates: *Version 1:* 1/88–9/93. *Version 2:* 10/93–3/97.

Learning Outcomes: *Version 1:* Upon completion of the course, the student will be able to analyze a given requirement for acquisition and select appropriate methods for satisfying the requirement; employ the various solicitation, evaluation, and award techniques utilized in the sealed bidding and negotiation process; administer contracts in accordance with their terms and conditions including standard and special contract clauses; and apply the statutes and implementing regulations which govern the contracting process. *Version 2:* After 9/97 see DD-1408-0032. Upon completion of the course, the student will be able to analyze a given requirement for acquisition and select appropriate methods for satisfying the requirement; employ the various solicitation, evaluation, and award techniques utilized in the sealed bidding and negotiation process; administer contracts in accordance with their terms and conditions including standard and special contract clauses; apply the statutes and implementing regulations which govern the contracting process; and apply basic procedures as described in the Federal Acquisition Regulation along with other forms of documentation.

Instruction: *Version 1:* Lectures and practical exercises cover contract management, including statutes, regulations, and policies; budgets and funds; specifications and allocations; potential contractors; small business and labor surplus programs; elements of a contract; purchase by sealed bidding; competitive proposals; cost and price analysis; contract types and purposes; clauses and provisions; taxes and other financial considerations; contract administration and modifications; quality control; termination of contracts; labor problems; and contract close-out and appeals. *Version 2:* Lectures and practical exercises cover contract management, including statutes, regulations, and policies; budgets and funds; specifications and allocations; potential contractors; small business and labor surplus programs; elements of a contract; purchase by sealed bidding; competitive proposals; cost and price analysis; contract types and purposes; clauses and provisions; taxes and other financial considerations; contract administration and modifications; quality control; termination of contracts; labor problems; and contract close-out and appeals. Proctored final examinations are administered.

Credit Recommendation: ***Version 1:*** In the upper-division baccalaureate category, 3 semester hours in logistics management (6/91). *Version 2:* In the upper-division baccalaureate category, 3 semester hours in logistics management (9/95).

AR-0326-0054

1. Management of Defense Acquisition Contracts Basic by Correspondence
2. Contracting Fundamentals by Correspondence

Course Number: ***Version 1:*** CON 101; 8D-4320; 8D-4320 (JT). *Version 2:* CON 101.

Location: ***Version 1:*** Army Logistics Management College, Fort Lee, VA. *Version 2:* Defense Acquisition University, Various locations in the Continental US; Army Logistics Management College, Fort Lee, VA.

Length: ***Version 1:*** Maximum, 52 weeks. *Version 2:* Maximum, 52 weeks.

Exhibit Dates: ***Version 1:*** 1/88–9/93. *Version 2:* 10/93–3/97.

Learning Outcomes: ***Version 1:*** Upon completion of the course, the student will be able to analyze a given requirement for acquisition and select appropriate methods for satisfying the requirement; employ the various solicitation, evaluation, and award techniques utilized in the sealed bidding and negotiation process; administer contracts in accordance with their terms and conditions including standard and special contract clauses; and apply the statutes and implementing regulations which govern the contracting process. *Version 2:* Upon completion of the course, the student will be able to analyze a given requirement for acquisition and select appropriate methods for satisfying the requirement; employ the various solicitation, evaluation, and award techniques utilized in the sealed bidding and negotiation process; administer contracts in accordance with their terms and conditions including standard and special contract clauses; apply the statutes and implementing regulations which govern the contracting process; and apply the basic procedures described in the Federal Acquisition Regulations along with other forms of documentation.

Instruction: ***Version 1:*** This is a correspondence course covering contract management, including statutes, regulations, and policies; budgets and funds; specifications and allocations; potential contractors; small business and labor surplus programs; elements of a contract; purchase by sealed bidding; competitive proposals; cost and price analysis; contract types and purposes; clauses and provisions; taxes and other financial considerations; contract administration and modifications; quality control; termination of contracts; labor problems, and contract close-out and appeals. *Version 2:* This is a correspondence course covering contract management, including statutes, regulations, and policies; budgets and funds; specifications and allocations; potential contractors; small business and labor surplus programs; elements of a contract; purchase by sealed bidding; competitive proposals; cost and price analysis; contract types and purposes; clauses and provisions; taxes and other financial considerations; contract administration and modifications; quality control; termination of contracts; labor problems; and contract close-out and appeals. Proctored final examinations are administered.

Credit Recommendation: ***Version 1:*** In the upper-division baccalaureate category, 3 semester hours in logistics management (6/91). *Version 2:* In the upper-division baccalaureate category, 3 semester hours in logistics management (8/95).

AR-0326-0055

1. Integrated Logistics Support Advanced
2. Intermediate Acquisition Logistics

Course Number: ***Version 1:*** ALMC-IT. *Version 2:* LOG 201.

Location: Army Logistics Management College, Fort Lee, VA.

Length: ***Version 1:*** 3 weeks (113 hours). *Version 2:* 3 weeks (115 hours).

Exhibit Dates: ***Version 1:*** 4/88–8/91. *Version 2:* 9/91–5/97.

Learning Outcomes: ***Version 1:*** Upon completion of the course, the student will be able to apply integrated logistics support management techniques in planning materiel acquisitions. *Version 2:* After 2/99 see DD-0326-0009. Upon completion of the course, the student will be able to apply integrated logistics support management techniques in planning materiel acquisitions.

Instruction: ***Version 1:*** Instruction includes lectures, exercises, and computer modeling covering policies, procedures, and concepts of materiel acquisition; integrated logistics support management; and contractual aspects of logistics support management. *Version 2:* Instruction includes lectures, exercises, and computer modeling covering policies, procedures, and concepts of materiel acquisition; integrated logistics support management; and contractual aspects of logistics support management.

Credit Recommendation: ***Version 1:*** In the upper-division baccalaureate category, 2 semester hours in materiel acquisition process and support systems. In the graduate degree category, 1 semester hour in materiel acquisition process and support systems. Credit is to be awarded in either the upper-division or the graduate category but not both (6/91). *Version 2:* In the upper-division baccalaureate category, 2 semester hours in procurement. In the graduate degree category, 1 semester hour in advanced logistics management (8/95).

AR-0326-0056

Logistics Support Analysis
(Defense Basic Logistics Support Analysis)

Course Number: LOG 202; ALMC-LR.

Location: Army Logistics Management College, Fort Lee, VA; Army Logistics Management College, Onsite at various locations; Army Logistics Management College, Satellite Education Program, Fort Lee, VA.

Length: 2 weeks (74 hours).

Exhibit Dates: 8/89–9/96.

Learning Outcomes: Upon completion of the course, the student will be able to use management principles, procedures, and selection techniques to tailor a logistics support analysis to meet the unique needs of a material acquisition program; describe how logistics support analysis supports integrated logistics in a materiel acquisition; perform analysis; and complete data records to build a logistics support analysis record.

Instruction: Lecture-conferences, practical exercises, and computer simulations cover integrated logistics support management. Topics include concepts, principles, and policies of materiel acquisition and logistics management; quantitative techniques in materiel acquisition and integrated logistics support; analytical techniques related to decision making and cost effectiveness; and tradeoff determinations using logistic parameters.

Credit Recommendation: In the upper-division baccalaureate category, 2 semester hours in materiel acquisition process and support systems. In the graduate degree category, 1 semester hour in materiel acquisition process and support systems. Credit is to be granted in either the upper-division or graduate category but not in both (6/91).

AR-0326-0057

Intermediate Contracting

Course Number: CON 211.

Location: Army Logistics Management College, Fort Lee, VA; Army Logistics Management College, Onsite at various locations.

Length: 3 weeks (92–96 hours).

Exhibit Dates: 6/94–4/99.

Learning Outcomes: Before 6/94 see AR-0326-0010. Upon completion of the course, the student will be able to analyze and resolve key elements in the pre-award stages of contract negotiations and recognize the ethical and legal implications involved in the contract decision-making process.

Instruction: Lectures and practical exercises cover procurement management, including procurement planning, funding, and technical requirement impact; small business, labor surplus, and disaster area relief operations; procurement by formal advertising and negotiation; contract types and purposes; cost and price analysis; contract administration; terminations and remedies; patents and data; and procurement trends. Lectures and practical exercises also cover planning for contract management, including acquisition planning, socioeconomic programs, and cost/price analysis and contract management, including bidding, negotiating, modifying, and terminating.

Credit Recommendation: In the upper-division baccalaureate category, 2 semester hours in contract management. In the graduate degree category, 1 semester hour in contract management. Credit is to be awarded in either the upper-division or graduate category but not in both (6/96).

AR-0326-0058

Consultant Development

Course Number: QTC-040.

Location: Professional Education Center, Cp. Joseph Robinson, North Little Rock, AR.

Length: 2 weeks (85 hours).

Exhibit Dates: 10/95–Present.

Learning Outcomes: Upon completion of the course, students will be able to serve as a consultant and change agent in their technical areas of expertise to organizations requiring support from outside the local organization.

Instruction: This by-invitation-only course includes lectures, class discussions, extensive class exercises, presentations and homework assignments in topics, including change theory; systems theory; organizational philosophy and developing intervention techniques, and instrumentation surveys, questionnaires to gain feedback.

Credit Recommendation: In the upper-division baccalaureate category, 3 semester hours in organizational communication or organizational development (6/98).

AR-0326-0059

CONSULTANT CERTIFICATION

Course Number: QTC-050.

Location: Professional Education Center, Cp. Joseph Robinson, North Little Rock, AR.

Length: 2 weeks (80 hours).

Exhibit Dates: 10/95–Present.

Learning Outcomes: Upon completion of the course, students will be able to serve as a consultant facilitator in their technical areas of expertise to organizations requiring support from outside the local organization.

Instruction: This course includes lectures, class discussion, extensive class exercises, presentations and homework assignments in such topics as interviewing techniques, interpersonal communication skills, conflict management, understanding organizations, implementation concepts, obtaining feedback following project, and general administration of a project.

Credit Recommendation: In the lower-division baccalaureate/associate degree category, 3 semester hours in interpersonal skills, human relations in business, or organizational development (6/98).

AR-0326-0062

ARMY MEDICAL DEPARTMENT (AMEDD) OFFICER ADVANCED

Course Number: *Version 1:* 6-8-C22, Phase 2. *Version 2:* 6-8-C22, Phase 2.

Location: Academy of Health Sciences, Fort Sam Houston, TX.

Length: *Version 1:* 10 weeks (465–607 hours). *Version 2:* 9 weeks (349–451 hours).

Exhibit Dates: *Version 1:* 9/99–9/01. *Version 2:* 10/01–Present.

Learning Outcomes: *Version 1:* Before 9/99 see AR-0326-0042. Upon completion of the core curriculum of the course, students will be able to analyze the historical perspectives of military medical history; develop organizational team building; employ planning and decision-making processes to solve problems and develop courses of action; demonstrate ethical decision making and leadership; explain military health care delivery systems; supervise military unit operations, including personnel management and training, supply and maintenance operations, risk management and safety, preventive medicine, resource allocation, and accounting; and direct unit logistic support operations under combat and noncombat conditions. Those in the dental track will be able to employ the skills acquired in the core course in the management of a dental clinic and implement quality assurance procedures and hazard communication programs. Those in the medical specialist corps track (AMSC) will be able to employ the skills acquired in the core course in AMSC management, discuss current developments and missions of each of the AMSC specialists, and maintain continuous quality improvement programs. Those in the nursing track will be able to employ the skills acquired in the core course in nursing management, implement care plans for rapid assessment of trauma patients, and assess current issues affecting the Army nurse corps. Those in the veterinary science track will be able to employ the skills acquired in the core course in veterinary management; supervise the inspection and procurement of meat, dairy, and water products; investigate incidences of disease outbreaks; and implement quality animal training and welfare programs. Those in the medical corps track will be able to discuss current civilian health trends and future projections, new trends in credentialing, pharmo-economic developments, ethical considerations, downsizing, and conflict resolution. Those in the medical service corps track will be able to implement total quality management concepts and policies, instill ethical decision making and behavior throughout the management concepts and policies, instill ethical decision making and behavior through the command, and demonstrate in-depth knowledge of managed-care programs. *Version 2:* Upon completion of the core curriculum of the course, students will be able to analyze the historical perspectives of military medical history; develop organizational team building; employ planning and decision-making processes to solve problems and develop courses of action; demonstrate ethical decision making and leadership; explain military health care delivery systems; supervise military unit operations, including personnel management and training, supply and maintenance operations, risk management and safety, preventive medicine, resource allocation, and accounting; direct unit logistic support operations under combat and noncombatant conditions; identify national security policies impacting Army doctrine; identify concepts of international politics and sovereign states; identify elements of national power, security, and strategy; and analyze lessons learned in Civil War. Those in the dental track will be able to employ the skills acquired in the core course in the management of a dental clinic and implement quality assurance procedures and hazard communication programs. Those in the medical specialist corps track (AMSC) will be able to employ the skills acquired in the core course in AMSC management, discuss current developments and missions of each of the AMSC specialists, and maintain continuous quality improvement programs. Those in the nursing track will be able to employ the skills acquired in the core course in nursing management, implement care plans for rapid assessment of trauma patients, and assess current issues affecting the Army nurse corps. Those in the veterinary science track will be able to employ the skills acquired in the core course in veterinary management; supervise the inspection and procurement of meat, dairy, and water products; investigate incidences of disease outbreaks; and implement quality animal training and welfare programs. Those in the medical corps track will be able to discuss current civilian health trends and future projections, new trends in credentialing, pharmo-economic developments, ethical considerations, downsizing, and conflict resolution. Those in the medical service corps track will be able to implement total quality management concepts and policies, instill ethical decision making and behavior throughout the management concepts and policies, instill ethical decision making and behavior through the command, and demonstrate in-depth knowledge of managed-care programs.

Instruction: *Version 1:* Phase 1 is by correspondence and is not included in the evaluation. Phase 2 consists of a 9-week common core and six one-week tracks. Lectures, distinguished guest lecturers, reading assignments, practical exercises, group discussions, research discussions, and field exercises provide instruction in the concepts of overall army medical unit management in peacetime and during combat. Lectures, panel discussions, and briefings provide instruction in dental clinic management for those in that track. Those in the AMSC track are provided with lectures, practical exercises, and field training to provide instruction in the management of medical professional services. The nursing track is presented through lectures and practical exercises with instruction in issues in peacetime and combat nursing management. The veterinary track is presented through lectures, reading assignments, field training, practical exercises, and panel discussions to provide instruction in administration of veterinary functions including food inspection and procurement. The medical corps track is offered through lectures providing instruction in current issues in medical management, and the medical service corps track uses lectures and research study to provide medical service corps officers with executive skills. *Version 2:* Phase 1 is by correspondence and is not included in the evaluation. Phase 2 consists of a 9-week common core and six one-week tracks. Lectures, distinguished guest lecturers, reading assignments, practical exercises, group discussions, research discussions, and field exercises provide instruction in the concepts of overall army medical unit management in peacetime and during combat. Lectures, panel discussions, and briefings provide instruction in dental clinic management for those in that track. Those in the AMSC track are provided with lectures, practical exercises, and field training to provide instruction in the management of medical professional services. The nursing track is presented through lectures and practical exercises with instruction in issues in peacetime and combat nursing management. The veterinary track is presented through lectures, reading assignments, field training, practical exercises, and panel discussions to provide instruction in administration of veterinary functions including food inspection and procurement. The medical corps track is offered through lectures providing instruction in current issues in medical management, and the medical service corps track uses lectures and research study to provide medical service corps officers with executive skills.

Credit Recommendation: *Version 1:* In the upper-division baccalaureate category, for all students who complete the core curriculum, 3 semester hours in topics in military history, 3 in human resource management, 2 in leadership command, control, and communication, and 2 in advanced military science. In the graduate degree category, for those in the dental track, 1 semester hour in dental clinic administration; for those in the Army medical specialist corps track, 1 semester hour in health administration; for those in the nursing track, 1 semester hour in nursing administration; for those in the veterinary science track, 1 semester hour in veterinary administration; for those in the medical corps and medical service corps tracks, 1 semester hour in health administration (6/02). *Version 2:* In the lower-division baccalaureate/associate degree category, 2 semester hours in topics of military history. In the upper-division baccalaureate category, 3 semester hours in human resource management; 2 in leadership, command, control, and communications; 3 in health care administration; and 2 in advanced military science. In the graduate degree category, for those in the dental track, 1 semester hour in dental administration; for those in the nursing track, 1 semester hour in nursing administration; for those in the veterinary track, 1 semester hour in veterinary administration; and for those in the medical corps and medical services corps track, 1 semester hour in health care administration (11/03).

AR-0418-0001

Travel Clerk

Course Number: 542-F4.

Location: Institute for Personnel and Resource Management, Fort Benjamin Harrison, IN.

Length: 3 weeks (115 hours).

Exhibit Dates: 4/84–1/97.

Objectives: To prepare students to function as travel clerks.

Instruction: Lectures and practical exercises cover travel allowances, travel regulations, and the preparation of travel vouchers.

Credit Recommendation: Credit is not recommended because of the military-specific nature of the course (9/88).

AR-0418-0004

Passenger Travel Specialist

Course Number: 8C-F11/542-F6.

Location: Transportation and Aviation Logistics School, Fort Eustis, VA.

Length: 2 weeks (74 hours).

Exhibit Dates: 5/83–12/90.

Objectives: After 12/90 see AR-0418-0006. To train personnel in basic skills to function as a passenger travel specialist.

Instruction: Lectures and practical exercises cover routing, documentation, entitlements, carrier guides and tariffs, group movements, travel orders, rentals, and electronic reservation and ticketing systems.

Credit Recommendation: In the lower-division baccalaureate/associate degree category, 2 semester hours in travel and tourism (7/85).

AR-0418-0006

Passenger Travel Specialist

Course Number: 8C-F11/542-F6.

Location: Transportation School, Fort Eustis, VA.

Length: 2 weeks (72 hours).

Exhibit Dates: 1/91–Present.

Learning Outcomes: Before 1/91 see AR-0418-0004. Upon completion of the course, the student will be able to demonstrate the skills and knowledge to work with travel publications; interpret travel orders and entitlements; effectively use official airline guides, official rail and bus line guides; and understand associated tariffs.

Instruction: Lectures and practical exercises cover routing documentation, entitlements, carrier guides and tariffs, group movements, travel orders, rentals, and electronic reservation and ticketing systems.

Credit Recommendation: In the lower-division baccalaureate/associate degree category, 2 semester hours in travel agency operations (6/98).

AR-0419-0005

Traffic Management Coordinator
Primary Technical

Course Number: 514-71N20.

Location: Transportation School, Fort Eustis, VA.

Length: 3–5 weeks (90–167 hours).

Exhibit Dates: 9/85–12/93.

Objectives: To provide enlisted personnel with the techniques and procedures necessary to control the movement of supplies and personnel by rail, water, highway, and air.

Instruction: Lectures and practical demonstrations cover modes of transportation; transportation procedures and regulations; personnel and passenger movement; and motor, rail, water, and terminal transportation systems. Topics include the development of transportation tariff and rate structures and regulations of traffic management. Also includes air systems.

Credit Recommendation: In the lower-division baccalaureate/associate degree category, 3 semester hours in transportation or traffic operations (5/86).

Related Occupations: 71N.

AR-0419-0017

Traffic Management Coordinator
Noncommissioned Officer (NCO)
Advanced

Course Number: 5-641-C42; 5-641-C42B.

Location: Transportation and Aviation Logistics School, Fort Eustis, VA.

Length: 7–11 weeks (264–370 hours).

Exhibit Dates: 12/83–9/90.

Objectives: To train noncommissioned officers to handle administrative responsibilities in a transportation unit.

Instruction: Lectures and practical exercises cover leadership styles, supply procedures, traffic management, demurrage, carrier qualifications, claims, freight movements, computerized movement, management, rates and tariffs, freight classification, documentation of freight shipments, loss and damage, and terminal operations.

Credit Recommendation: In the lower-division baccalaureate/associate degree category, 2 semester hours in principles of supervision and 3 in principles of traffic and transportation management (3/88).

Related Occupations: 71N.

AR-0419-0024

Motor Transport Noncommissioned
Officer (NCO) Advanced

Course Number: 8-641-C42.

Location: Transportation and Aviation Logistics School, Fort Eustis, VA; Transportation School, Fort Eustis, VA.

Length: 5–7 weeks (187–238 hours).

Exhibit Dates: 9/77–9/90.

Objectives: To train noncommissioned officers in motor transport operations.

Instruction: Lectures and practical exercises cover the duties of a noncommissioned officer in motor transportation. Topics include military operations; leadership styles and communication; staff functions; truck transport planning; and the organization, operation, mission, and function of specific motor transportation units.

Credit Recommendation: In the lower-division baccalaureate/associate degree category, 2 semester hours in principles of supervision and 2 as a technical elective in transportation (3/85).

Related Occupations: 64C.

AR-0419-0027

Strategic Mobility Planning

Course Number: 8C-ASI3S/822-F10; 8C-F2; 822-F10.

Location: Transportation School, Fort Eustis, VA.

Length: 3–4 weeks (109–137 hours).

Exhibit Dates: 2/81–1/98.

Objectives: To train selected commissioned officers, warrant officers, and enlisted personnel in planning, organizing, and conducting unit movements.

Instruction: Lectures and practical exercises include air movements and planning operations; preparation of equipment; restraint fundamentals; airfield departure and arrival procedures; dangerous cargo; joint combat airlift; helicopter landing procedures; unit movement by rail; railway equipment capabilities; and loading, blocking, and tracing equipment on rail cars.

Credit Recommendation: In the lower-division baccalaureate/associate degree category, 3 semester hours in general transportation and planning (3/88).

AR-0419-0033

Transportation Officer Basic
(Transportation Officer Orientation)

Course Number: 8-55-C20; 8-55-C20-95A/C/D; 8-55-C20-88A/C/D.

Location: Transportation and Aviation Logistics School, Fort Eustis, VA; Transportation School, Fort Eustis, VA.

Length: 12–20 weeks (515–828 hours).

Exhibit Dates: 7/74–12/98.

Objectives: After 3/99 see AR-0419-0070. To train newly commissioned officers for transportation operations management.

Instruction: Lectures and practical exercises cover the organization, planning, and management of air, motor, rail, and water transportation movements, including truck company operations, convoy operations, rail operations, loading and blocking, terminal operations and management, cargo planning and stowing, dangerous cargo, and traffic management and planning.

Credit Recommendation: In the lower-division baccalaureate/associate degree category, 3 semester hours in principles of supervision and 2 in principles of depot operations (3/88).

AR-0419-0035

Defense Advanced Traffic Management
(Advanced Traffic Management)

Course Number: 8C-F3.
Location: Transportation and Aviation Logistics School, Fort Eustis, VA; Transportation School, Fort Eustis, VA.
Length: 3 weeks (108 hours).
Exhibit Dates: 8/85–2/92.
Objectives: After 2/92 see AR-0419-0055. To train commissioned officers with prior schooling in traffic management in advanced managerial procedures for traffic and general transportation.
Instruction: Lectures and practical exercises cover managerial procedures for traffic and general transportation, including economics of transportation; characteristics of various modes and carriers; transportation services and regulations; industrial and military traffic management; transportation in logistics; movement of regular and special cargo (weapons and missiles), household goods, and passengers; claim prevention; and rate negotiation.
Credit Recommendation: In the upper-division baccalaureate category, 2 semester hours in transportation management and 2 in economics of transportation (5/86).

AR-0419-0036

Installation Traffic Management

Course Number: *Version 1:* 8C-F4. *Version 2:* 8C-F4.
Location: Transportation School, Fort Eustis, VA.
Length: *Version 1:* 4 weeks (143–155 hours). *Version 2:* 2 weeks (73 hours).
Exhibit Dates: *Version 1:* 2/61–5/91. *Version 2:* 6/91–Present.
Learning Outcomes: *Version 1:* Upon completion of the course, the student will be able to use knowledge of transportation, including modes of transportation, carrier qualifications, quality control, and traffic management, to plan and control traffic at an installation. *Version 2:* Upon completion of the course, the student will be able to apply transportation principles, including documentation, hazardous materials, security, and traffic management to plan and control transportation at an installation.
Instruction: *Version 1:* Lectures and practical exercises cover installation traffic management, including passenger, freight, and personal property traffic. Topics include carrier qualifications, modes and storage selection, quality control procedures, traffic management, specialized freight movements, small shipment services, organization and field installations, commercial transportation services, procurement, integrated accounting, and movement of units and large groups. *Version 2:* Classroom, practical exercises, and written exams cover freight, military personal, and passenger traffic documentation; hazardous cargo; and security.
Credit Recommendation: *Version 1:* In the upper-division baccalaureate category, 2 semester hours in transportation management (4/84). *Version 2:* In the lower-division baccalaureate/associate degree category, 2 semester hours in traffic management (6/01).

AR-0419-0037

Basic Joint Personal Property
(Joint Personal Property)

Course Number: 8C-SI3V/553-ASIH2; 8C-F5/514-F2.
Location: Transportation and Aviation Logistics School, Fort Eustis, VA; Transportation School, Fort Eustis, VA.
Length: 2 weeks (74–77 hours).
Exhibit Dates: 2/85–9/91.
Objectives: After 9/91 see AR-0419-0063. To train commissioned officers, warrant officers, and civilians in traffic management functions associated with personal property shipments.
Instruction: Lectures, practical exercises, and field experiences cover entitlement and authorization, shipping procedures, carrier contracts, utilization of carrier rate tariffs, establishment of a traffic distribution record, inspection of shipments and facilities of carrier, claim procedures, and traffic and control procedures.
Credit Recommendation: In the lower-division baccalaureate/associate degree category, 2 semester hours in traffic distribution, freight loss, and damage claims (3/88).

AR-0419-0040

Traffic Management Coordinator

Course Number: 514-71N10.
Location: Transportation and Aviation Logistics School, Fort Eustis, VA; Transportation School, Fort Eustis, VA.
Length: 5–6 weeks (202 hours).
Exhibit Dates: 3/85–9/91.
Objectives: To provide enlisted personnel with knowledge of transportation and related activities necessary to perform entry-level duties.
Instruction: Lectures and practical exercises cover performance of critical tasks related to cargo documentation, planning, loss and damage control, unit movements, passenger movements, overview of installation traffic officer functions, and orientation on computerized traffic movement management systems.
Credit Recommendation: In the lower-division baccalaureate/associate degree category, 4 semester hours in introduction to traffic and transportation (5/86).
Related Occupations: 71N.

AR-0419-0041

Traffic Management Coordinator Basic Noncommissioned Officer (NCO)
(Traffic Management Coordinator Basic Technical)

Course Number: 533-88N30; 514-88N30; 514-71N30.
Location: Transportation and Aviation Logistics School, Fort Eustis, VA; Transportation School, Fort Eustis, VA.
Length: 3–8 weeks (92–254 hours).
Exhibit Dates: 10/81–9/91.
Objectives: After 9/91 see AR-0419-0058. To provide enlisted personnel with the techniques and procedures necessary to supervise and control the movement of supplies and personnel by rail, highway, and air.
Instruction: Lectures and practical demonstrations cover modes of transportation; transportation procedures and regulations; management of personnel movement; and supervision of motor, rail, water, and terminal transportation systems. Course includes a common core of leadership subjects.
Credit Recommendation: In the lower-division baccalaureate/associate degree category, 3 semester hours in traffic operations supervision or as a technical elective in transportation and traffic management. See AR-1406-0090 for common core credit (10/91).
Related Occupations: 71N; 88N.

AR-0419-0042

Instructor Training

Course Number: 19-82.
Location: Training Center, Fort Dix, NJ.
Length: 2 weeks (77 hours).
Exhibit Dates: 8/82–6/96.
Objectives: To provide personnel assigned as instructors with the knowledge, skills, and techniques necessary for teaching military subjects.
Instruction: Instruction covers principles of lesson planning, selection of visual aids, instructional strategies and techniques, and teaching skills. Instructional styles including conference classes, demonstrations, performance-oriented instruction, and remedial instruction are emphasized. Students are required to give various types and lengths of presentations which are critiqued by instructional staff.
Credit Recommendation: In the lower-division baccalaureate/associate degree category, 1 semester hour in speech or public speaking. In the upper-division baccalaureate category, 3 semester hours in methods of instruction (9/83).

AR-0419-0043

Transportation Field Grade Officer Refresher

Course Number: 8C-F13.
Location: Transportation and Aviation Logistics School, Fort Eustis, VA.
Length: 2 weeks (72 hours).
Exhibit Dates: 7/86–12/93.
Objectives: To provide a broad knowledge base in the current organization and practices of transportation activities.
Instruction: This course utilizes lectures, demonstrations, and practical exercises to give the individual a broad base of knowledge of current organization and management practices of the Transportation Corps.
Credit Recommendation: In the upper-division baccalaureate category, 1 semester hour in transportation management (5/86).

AR-0419-0044

Basic Freight Traffic
(Defense Basic Traffic Management)

Course Number: 8C-F12/553-F1.
Location: Transportation and Aviation Logistics School, Fort Eustis, VA.
Length: 2 weeks (73 hours).
Exhibit Dates: 3/85–9/92.
Objectives: After 9/92 see AR-0419-0057. To provide selected individuals with a broad base of knowledge in commercial transportation systems as applied to government freight transportation.
Instruction: Lectures and practical exercises cover commercial transportation systems and the procurement of services from commercial sources for Department of Defense freight transportation.

Credit Recommendation: In the lower-division baccalaureate/associate degree category, 3 semester hours in transportation practices (6/91).

AR-0419-0045

WATERCRAFT OPERATOR ADVANCED NONCOMMISSIONED OFFICER (NCO)

Course Number: 062-88K40.

Location: Transportation School, Fort Eustis, VA.

Length: 12–14 weeks (399–503 hours).

Exhibit Dates: 10/87–Present.

Learning Outcomes: Upon completion of the course, the student will be able to perform watercraft supervision and administration, including use of logbooks, presail inspections, and shipboard emergency drills; perform cargo handling operations; obtain weather information; interpret international code flags and flashing light signals; use navigational aids; and serve as a mate on a multiscrew vessel.

Instruction: Lectures and practical exercises cover the operation of multiscrew vessels. Course includes a common core of leadership subjects.

Credit Recommendation: In the lower-division baccalaureate/associate degree category, 2 semester hours in principles of supervision, 3 in marine navigation, and 1 in multiple screw vessel operation. See AR-1404-0035 for common core credit (10/00).

Related Occupations: 88K.

AR-0419-0046

CARGO SPECIALIST

Course Number: *Version 1:* 822-88H10. *Version 2:* 822-88H10.

Location: *Version 1:* Transportation and Aviation Logistics School, Fort Eustis, VA. *Version 2:* Transportation School, Fort Eustis, VA.

Length: *Version 1:* 9–10 weeks (336–338 hours). *Version 2:* 8 weeks (309 hours).

Exhibit Dates: *Version 1:* 1/87–9/98. *Version 2:* 10/98–Present.

Learning Outcomes: *Version 1:* Upon completion of the course, the student will be able to perform cargo handling operations, loading procedures, and transshipment of cargo at air, rail, and marine terminals. *Version 2:* Upon completion of the course, the student will be able to perform cargo handling operations, including rail, air, and marine transportation; operate cargo-handling gear and safety nets, materials handling equipment such as winches and cranes, aboard ship-vehicles for heavy-lift containers, of Hagglund's Crane, and the rough-terrain fork lift; and perform equipment preventive maintenance and servicing.

Instruction: *Version 1:* Course includes lectures, demonstrations, and performance exercises in rigging and positioning ship's cargo-handling gear and safety nets; operating materials-handling equipment, winches, and cranes; lift on/off and roll on/off stevedoring operations aboard ship for vehicles, heavy-lifts, and containers; and general cargo operations of Hagglund's crane and the rough-terrain forklift. Course also includes some preventive maintenance. *Version 2:* Methods of instruction include classroom, practical, simulator, practical skill projects, demonstrations, written and practical exams.

Credit Recommendation: *Version 1:* In the lower-division baccalaureate/associate degree category, 3 semester hours in marine cargo operations (4/92). *Version 2:* In the lower-division baccalaureate/associate degree category, 3 semester hours in marine cargo operations and 1 in heavy equipment maintenance (6/01).

Related Occupations: 88H.

AR-0419-0047

TRANSPORTATION OFFICER ADVANCED RESERVE PHASES 2 AND 4

Course Number: 8-55-C22-RC.

Location: Transportation and Aviation Logistics School, Fort Eustis, VA.

Length: 4 weeks (148 hours).

Exhibit Dates: 5/86–3/92.

Learning Outcomes: Upon completion of the course, the student will be able to manage the movement and handling of cargo, determine and develop marine terminal cargo capacity, account for cargo shipments and receivables, and manage highway movements and rail transportation.

Instruction: Course includes lectures, demonstrations and performance exercises in marine cargo packaging and handling, containerization, ship loading techniques, cargo discharge operations, deployment, cargo accountability, motor transportation planning and control, convoy operations, rail equipment capabilities, rail planning and train operations, yard and terminal operations, railway construction, and rehabilitation. Phases 2 and 4 are resident phases; the others are completed by correspondence.

Credit Recommendation: Credit is not recommended because of the limited, specialized nature of the course (3/88).

AR-0419-0048

MOTOR TRANSPORT OPERATOR BASIC NONCOMMISSIONED OFFICER (NCO)

Course Number: *Version 1:* 811-88M30. *Version 2:* 811-88M30.

Location: Transportation School, Fort Eustis, VA.

Length: *Version 1:* 7 weeks (295 hours). *Version 2:* 8 weeks (459 hours).

Exhibit Dates: *Version 1:* 10/90–12/95. *Version 2:* 1/96–Present.

Learning Outcomes: *Version 1:* Upon completion of the course, the student will be able to supervise motor transport operators in preventive auto/truck/heavy equipment maintenance in deployment, and perform record keeping. *Version 2:* Upon completion of the course, the student will be able to supervise motor transport operators under varying conditions and operations (including preventive auto/truck/heavy equipment maintenance) and perform record keeping.

Instruction: *Version 1:* Lectures and practical applications cover leadership, management, and supervisory functions in the area of motor transport operations. Course includes a common core of leadership subjects. *Version 2:* Methods of instruction include classroom and practical applications covering leadership, management, and supervisory functions in the area of motor transport operations. A common core of leadership subjects is taught throughout the course.

Credit Recommendation: *Version 1:* In the lower-division baccalaureate/associate degree category, 1 semester hour in leadership, 1 in automotive maintenance, and 1 in records and information management. See AR-1406-0090 for common core credit (6/91). *Version 2:* In the lower-division baccalaureate/associate degree category, 1 semester hour in principles of supervision, 1 automotive maintenance, and 1 in records and information management. See AR-1406-0090 for common core credit (6/01).

Related Occupations: 88M.

AR-0419-0049

MOTOR TRANSPORT OPERATOR ADVANCED NONCOMMISSIONED OFFICER (NCO)

Course Number: *Version 1:* 811-88M40; 8-641-C42. *Version 2:* 811-88M40. *Version 3:* 811-88M40.

Location: Transportation School, Fort Eustis, VA.

Length: *Version 1:* 8 weeks (333 hours). *Version 2:* 7–8 weeks (289 hours). *Version 3:* 3 weeks (106 hours).

Exhibit Dates: *Version 1:* 1/90–12/95. *Version 2:* 1/96–12/01. *Version 3:* 1/02–Present.

Learning Outcomes: *Version 1:* Upon completion of the course, the student will be able to provide leadership, effective communication, record keeping, and office management to the motor transport system and plan and execute material procurement. *Version 2:* Upon completion of the course, the student will be able to provide leadership, effective communication, recordkeeping, and office management to the motor transport system and plan and execute materials procurement. *Version 3:* Upon completion of the course, the student will be able to provide leadership, effective communication, recordkeeping, and office management in support of planning and implementation of the motor transport system and materials procurement.

Instruction: *Version 1:* Lectures and practical applications by demonstration and field experience cover planning, execution, and analysis in deployment, group organization, and materials procurement. Course includes a common core of leadership subjects. *Version 2:* Methods of instruction include classroom and practical applications by demonstration and field experience which cover planning, execution, and analysis in deployment, group organization, and material procurement. Course includes a common core of leadership subjects. *Version 3:* Practical and classroom exercises. Topics include advanced leadership and communication training, recordkeeping, analysis, planning, and implementation and procurement training in support of group deployments.

Credit Recommendation: *Version 1:* In the lower-division baccalaureate/associate degree category, 2 semester hours in business communication and 2 in office administration. See AR-1404-0035 for common core credit (6/91). *Version 2:* In the lower-division baccalaureate/associate degree category, 2 semester hours in business communication and 2 in office administration. See AR-1404-0035 for common core credit (6/01). *Version 3:* In the lower-division baccalaureate/associate degree category, 2 semester hours in logistics management. In the upper-division baccalaureate category, 2 semester hours in crew/team resource management (12/03).

Related Occupations: 88M.

AR-0419-0050

Traffic Management Coordinator
Advanced Noncommissioned
Officer (NCO)

Course Number: 5-88-C42B.
Location: Transportation School, Fort Eustis, VA.
Length: 7–8 weeks (176–245 hours).
Exhibit Dates: 10/90–4/04.
Learning Outcomes: Upon completion of the course, the student will be able to perform administrative duties for terminal operations, local transfer, and highway movement for the transportation of vehicles and cargo.
Instruction: Lectures and practical exercises cover leadership styles, supply procedures, traffic management, demurrage, carrier qualifications, claims, freight movements, computerized movements management, rates and tariffs, freight classification, documentation of freight shipments, loss and damage, and terminal operations. Course includes a common core of leadership subjects.
Credit Recommendation: In the lower-division baccalaureate/associate degree category, 1 semester hour in principles of supervision and 2 in introduction to transportation. See AR-1404-0035 for common core credit (10/91).
Related Occupations: 88N.

AR-0419-0051

Cargo Specialist Advanced
Noncommissioned Officer (NCO)

Course Number: *Version 1:* 5-88-C42A. *Version 2:* 822-88H40. *Version 3:* 822-88H40.
Location: *Version 1:* Transportation School, Fort Eustis, VA; Aviation Logistics School, Fort Eustis, VA. *Version 2:* Transportation School, Fort Eustis, VA; Aviation Logistics School, Fort Eustis, VA. *Version 3:* Transportation School, Fort Eustis, VA.
Length: *Version 1:* 7–8 weeks (276 hours). *Version 2:* 9 weeks (223 hours). *Version 3:* 6–7 weeks (174 hours).
Exhibit Dates: *Version 1:* 10/88–9/92. *Version 2:* 10/92–9/94. *Version 3:* 10/94–Present.
Learning Outcomes: *Version 1:* Upon completion of the course, the student will be able to apply the principles of ocean cargo documentation, terminal operations, and vessel loading and discharge. *Version 2:* Upon completion of the course, the student will be able to supervise and perform the duties, functions, operations, and problem solving inherent to water terminal organizations. *Version 3:* Upon completion of the course, the student will be able to supervise and perform duties, functions, operations, and problem solving for water terminal operations.
Instruction: *Version 1:* Lectures and practical exercises cover military operations, leadership and communication, staff functions, water transport operations, and the supervision associated with water terminal operations. Course includes a common core of leadership subjects. *Version 2:* Lectures, practical exercises, and examinations cover water terminal operations, supervision, water transport operations, and military operations. Course includes a common core of leadership subjects. *Version 3:* Methods of instruction include classroom, practical exercises, field exercises and written exams in supervision of water terminal operations. Terminal management, cargo operations (water), operational planning, and computerized deployment system are covered.
Credit Recommendation: *Version 1:* In the lower-division baccalaureate/associate degree category, 2 semester hours in water terminal operations. See AR-1404-0035 for common core credit (6/91). *Version 2:* In the lower-division baccalaureate/associate degree category, 2 semester hours in water terminal operations. See AR-1404-0035 for common core credit (3/94). *Version 3:* In the lower-division baccalaureate/associate degree category, 2 semester hours in water terminal operations and 1 in principles of supervision (10/03).
Related Occupations: 88H.

AR-0419-0052

Air Deployment Planning

Course Number: *Version 1:* 8C-SI3S/553-F4. *Version 2:* 8C-SI3S/553-F4. *Version 3:* 8C-SI3C/553-F4.
Location: *Version 1:* Joint Strategic Deployment Training Center, Fort Eustis, VA. *Version 2:* Transportation School, Fort Eustis, VA. *Version 3:* Transportation School, Fort Eustis, VA.
Length: *Version 1:* 3 weeks (110 hours). *Version 2:* 2 weeks (69 hours). *Version 3:* 2–3 weeks (101 hours).
Exhibit Dates: *Version 1:* 10/88–12/96. *Version 2:* 1/97–12/97. *Version 3:* 1/98–Present.
Learning Outcomes: *Version 1:* Upon completion of the course, the student will be able to plan air movements, determine aircraft limitations, and load and secure cargo. *Version 2:* Upon completion of the course, the student will be able to plan air mobilization movement, including passenger, bulk cargo and vehicle preparation; plan loading; and prepare cargo manifests. *Version 3:* Upon completion of the course, the student will be able to perform duties of loading and securing aircraft cargo and determining weight and balance of aircraft and individual equipment components, based on various safety and cargo handling configurations.
Instruction: *Version 1:* Lectures and limited practical experience cover planning air deployment, aircraft restrictions, and loading and securing cargo. *Version 2:* Methods of instruction include classroom, limited practical experience, and situation planning. *Version 3:* Methods of instruction include classroom, limited practical experiences, and complex scenario-driven problem-solving projects.
Credit Recommendation: *Version 1:* Credit is not recommended because of the limited, specialized nature of the course (6/91). *Version 2:* Credit is not recommended because of the limited, specialized nature of the course (6/01). *Version 3:* In the lower-division baccalaureate/associate degree category, 1 semester hour in aircraft weight, balance, and structural familiarization (6/01).

AR-0419-0053

Unit Movement Officer Deployment
Planning
(Surface Deployment Planning)

Course Number: 8C-F17/553-F5.
Location: Joint Strategic Deployment Training Center, Fort Eustis, VA.
Length: 2 weeks (71–73 hours).
Exhibit Dates: 4/90–Present.
Learning Outcomes: Upon completion of the course, the student will be able to plan surface deployment by highway, rail, or waterway.
Instruction: Lectures and limited practical exercises cover deployment planning, documentation, and security. Course covers deployment by rail, by roadway and by water. After 6/92, course also includes deployment by air.
Credit Recommendation: Credit is not recommended because of the limited, specialized nature of the course (10/00).

AR-0419-0054

Transportation Officer Advanced

Course Number: 8-55-C22.
Location: Transportation School, Fort Eustis, VA.
Length: 20 weeks (809 hours).
Exhibit Dates: 10/89–1/98.
Learning Outcomes: Upon completion of the course, the student will be able to apply the command and staff skills needed to succeed as commanders and staff officers in the transportation field.
Instruction: Lectures and practical exercises cover leadership; management; logistics; data processing; operations research applications; tactics involving nuclear, biological, and chemical operations; air, motor, rail, and water terminals and transport; traffic management; and written and oral communication.
Credit Recommendation: In the lower-division baccalaureate/associate degree category, 3 semester hours in principles of supervision, 2 in introductory data processing, 3 in principles of transportation, 3 in business communication and report writing, and 2 in public speaking. In the upper-division baccalaureate category, 6 semester hours in traffic management (6/91).

AR-0419-0055

Defense Advanced Traffic Management

Course Number: 8C-FC.
Location: Transportation School, Fort Eustis, VA.
Length: 2 weeks (72 hours).
Exhibit Dates: 3/92–Present.
Learning Outcomes: Before 3/92 see AR-0419-0035. Upon completion of the course, the student will be able to perform advanced administrative duties in the transportation field.
Instruction: Guest speakers from the Department of Defense, other government agencies, and commercial industry make presentations. Field trips are included.
Credit Recommendation: Credit is not recommended because of the limited, specialized nature of the course (10/00).

AR-0419-0056

Strategic Deployment Planning

Course Number: 8C-F16/533-F3.
Location: Transportation School, Fort Eustis, VA.
Length: 2 weeks (71 hours).
Exhibit Dates: 3/92–Present.
Learning Outcomes: Upon completion of the course, the student will be able to plan, organize, and execute a deployment plan.
Instruction: Lectures and practical exercises cover deployment planning, plan execution, communications, and organization of deployment and field operations.

Credit Recommendation: Credit is not recommended because the program is a broad-based orientation and contains diverse skill development (10/00).

AR-0419-0057

BASIC FREIGHT TRAFFIC

Course Number: 8C-F12/553-F1.
Location: Transportation School, Fort Eustis, VA.
Length: 2 weeks (28–71 hours).
Exhibit Dates: 10/92–Present.
Learning Outcomes: Before 10/92 see AR-0419-0044. Upon completion of the course, the student will be able to move freight within the US and procure and evaluate commercial transportation services.
Instruction: Lectures and practical exercises cover commercial transportation systems and procurement of services from commercial sources.
Credit Recommendation: In the lower-division baccalaureate/associate degree category, 3 semester hours in transportation practices (10/00).

AR-0419-0058

TRANSPORTATION MANAGEMENT COORDINATOR
(Traffic Management Coordinator Basic Noncommissioned (NCO))

Course Number: 533-88N30, Phase 1; 553-88N30, Phase 4.
Location: Transportation School, Fort Eustis, VA.
Length: 5–7 weeks (257–280 hours).
Exhibit Dates: 10/91–Present.
Learning Outcomes: Before 10/91 see AR-0419-0041. Upon completion of the course, the student will be able to coordinate transportation activities including both personnel and cargo; select modes of transportation; determine handling and packaging requirements; and trace routing.
Instruction: Lectures and practical demonstrations cover modes of transportation; transportation procedures and regulations; management of personnel and passenger movement; and supervision of motor, rail, water, air, and terminal transportation systems. Course includes a common core of leadership subjects.
Credit Recommendation: In the lower-division baccalaureate/associate degree category, 3 semester hours in transportation management supervision and 1 in logistics. See AR-1406-0090 for common core credit (10/00).
Related Occupations: 88N.

AR-0419-0059

INSTALLATION TRAFFIC MANAGEMENT

Course Number: 8C-F4/553-F10.
Location: Transportation School, Fort Eustis, VA.
Length: 2 weeks (73 hours).
Exhibit Dates: 6/91–Present.
Learning Outcomes: Upon completion of the course, the student will be able to perform the duties of the base transportation officer.
Instruction: Lectures and practical exercises cover commercial and military traffic functions. Topics include passenger travel, documentation, shipment of personal property, storage, military transportation and movement procedures, loss and damage detection, demurrage, security, and hazardous cargo.
Credit Recommendation: In the upper-division baccalaureate category, 2 semester hours in transportation management (6/98).

AR-0419-0060

TRAFFIC MANAGEMENT COORDINATOR ADVANCED NONCOMMISSIONED OFFICER (NCO) RESERVE

Course Number: 551-88N40; 551-88N40-RC.
Location: Reserve Component Training Centers, US.
Length: 76 hours.
Exhibit Dates: 10/92–Present.
Learning Outcomes: Upon completion of the course, the student will be able to perform administrative duties for terminal operations, local transfer, and highway movement for the transportation of vehicles and cargo.
Instruction: Lectures and practical exercises cover planning and controlling movement, local transfer, rail and air transportation, and coordination of cargo shipments.
Credit Recommendation: In the lower-division baccalaureate/associate degree category, 2 semester hours in introduction to transportation (3/94).
Related Occupations: 88N.

AR-0419-0061

1. TRAFFIC MANAGEMENT COORDINATOR
2. TRANSPORTATION MANAGEMENT COORDINATOR, PHASE 4

Course Number: ***Version 1:*** 553-88N10. *Version 2:* 553-88N10, Phase 4.
Location: Transportation School, Fort Eustis, VA.
Length: ***Version 1:*** 8–9 weeks (261–317 hours). *Version 2:* 6 weeks (320 hours).
Exhibit Dates: ***Version 1:*** 10/91–9/95. *Version 2:* 10/95–Present.
Learning Outcomes: ***Version 1:*** Upon completion of the course, the student will be able to perform entry-level duties in selection of transportation modes, logistics, routing, personnel and cargo transportation, computerized management systems, and transportation of hazardous materials. *Version 2:* Upon completion of the course, the student will be able to perform the duties in transportation management including modes of transportation, logistics, routing, personnel, hazardous materials, cargo transportation, and use computerized data management system.
Instruction: ***Version 1:*** Lectures and practical exercises cover transportation modes; cargo transport; personnel transport; hazardous chemical classification, handling, and shipping; loss and claims; logistics requirements; computer management systems; and routing. *Version 2:* Classroom, practical, computer-aided instruction, and demonstrations cover documentation; modes of transportation; hazardous materials handling, shipping, and identification; logistics; computer management system; and routing.
Credit Recommendation: ***Version 1:*** In the lower-division baccalaureate/associate degree category, 4 semester hours in transportation management (3/94). *Version 2:* In the lower-division baccalaureate/associate degree category, 4 semester hours in transportation management and 2 in basic computer applications (6/01).
Related Occupations: 88N.

AR-0419-0062

1. TRANSPORTATION OFFICER (BRANCH QUALIFICATION)
2. TRANSPORTATION OFFICER BRANCH QUALIFICATION

Course Number: ***Version 1:*** 8-55-C20-88A/C/D (BQ). *Version 2:* 8-55-020-88A/C/D (BQ).
Location: Transportation School, Fort Eustis, VA.
Length: ***Version 1:*** 4 weeks (144 hours). *Version 2:* 4 weeks (180 hours).
Exhibit Dates: ***Version 1:*** 5/92–9/99. *Version 2:* 10/99–Present.
Learning Outcomes: ***Version 1:*** Upon completion of the course, the student will be able to determine shipping methods, identify/label containers including hazardous materials, and determine adequate packaging. *Version 2:* Upon completion of the course, the student will be able to determine shipping methods, identify and label containers including hazardous materials and determine adequate packaging, loading, and securing of equipment.
Instruction: ***Version 1:*** Course covers a review of fundamentals of transportation, including packaging, paperwork, labeling, shipping method selection, and hazardous materials regulations. *Version 2:* Methods of instruction include classroom, practical, and simulated operations.
Credit Recommendation: ***Version 1:*** In the lower-division baccalaureate/associate degree category, 1 semester hour in transportation management (6/98). *Version 2:* In the lower-division baccalaureate/associate degree category, 1 semester hour in transportation management (6/01).
Related Occupations: 88A; 88C; 88D.

AR-0419-0063

JOINT PERSONAL PROPERTY

Course Number: 8C-SI3V/553-ASIH2.
Location: Transportation School, Fort Eustis, VA.
Length: 2 weeks (73 hours).
Exhibit Dates: 10/91–Present.
Learning Outcomes: Before 10/91 see AR-0419-0037. Upon completion of the course, the student will have a working knowledge of personal property transportation methods and will be able to determine and maintain quality assurance of transported property.
Instruction: Lectures and practical applications cover methods of transportation, temporary storage, long-term storage, records maintenance, quality assurance, basic regulations, and claim procedures.
Credit Recommendation: In the lower-division baccalaureate/associate degree category, 2 semester hours in transportation, freight loss, and claims (10/00).

AR-0419-0064

TRAFFIC MANAGEMENT COORDINATOR BASIC NONCOMMISSIONED OFFICER (NCO) RESERVE

Course Number: 551-88N30-RC.
Location: Reserve Component Training Centers, US.
Length: 91 hours.
Exhibit Dates: 7/92–Present.

Learning Outcomes: Upon completion of the course, the student will be able to supervise and control the movement of personnel and cargo by rail, highway, and air.

Instruction: Lectures and practical exercises cover shipping documents; routing; and air, rail, and highway movement.

Credit Recommendation: Credit is not recommended because of the limited, specialized nature of the course (3/94).

Related Occupations: 88N.

AR-0419-0065

Division Transportation Officer

Course Number: 553-F9; 8C-F25; 8C-F25/553-F9.

Location: Transportation School, Fort Eustis, VA.

Length: 2 weeks (73 hours).

Exhibit Dates: 6/93–Present.

Learning Outcomes: Upon completion of the course, the student will be able to perform transportation and movement control functions within the military structure.

Instruction: Methods of instruction include conferences and performance exercises involving planning, coordinating, controlling, programming, and regulating the allocation and use of transportation resources in order to fulfill movement requirements and the strategic, operational, and tactical levels of war.

Credit Recommendation: In the upper-division baccalaureate category, 2 semester hours in decision making and problem solving (6/01).

AR-0419-0066

Combined Logistics Officer Advanced, Phase 2

Course Number: 8-55-C22 (LOG), Phase 2.

Location: Transportation School, Fort Eustis, VA.

Length: 5 weeks (200 hours).

Exhibit Dates: 12/97–Present.

Learning Outcomes: Upon completion of the course, the student will be able to direct wartime missions in strategic deployment, movement, motor transport, watercraft, and marine operations terminal.

Instruction: Methods of instruction include conferences, performance exercises, role playing, and field exercise in mobilization, joint strategic deployment, movement management, motor transport operations, rail operations, and marine terminal operations.

Credit Recommendation: In the upper-division baccalaureate category, 3 semester hours in transportation management (6/01).

AR-0419-0067

Reserve Component Transportation Officer Advanced (Marine), Phase 2

Course Number: 8-55-C23C, Phase 2.

Location: Transportation School, Fort Eustis, VA.

Length: 2 weeks (74 hours).

Exhibit Dates: 4/00–Present.

Learning Outcomes: Upon completion of the course, the student will be able to manage the movement and handling of marine cargo.

Instruction: Methods of instruction include classroom instruction, demonstrations, problem-solving, and exams.

Credit Recommendation: Credit is not recommended because of the military-specific nature of the course (6/01).

AR-0419-0068

Reserve Component Transportation Officer Advanced (Highway/Rail), Phase 1

Course Number: 8-55-C23D, Phase 1.

Location: Transportation School, Fort Eustis, VA.

Length: 2 weeks (74 hours).

Exhibit Dates: 4/00–Present.

Learning Outcomes: Upon completion of the course, the student will be able to operate and manage highway and rail transportation, including planning, movement, equipment, and recovery.

Instruction: Methods of instruction include classroom conferences and limited practical, written, and situational exams.

Credit Recommendation: Credit is not recommended because of the limited, specialized nature of the course (6/01).

AR-0419-0069

Railway Operations Crewmember

Course Number: 812-88U10-RC.

Location: Transportation School, Fort Eustis, VA.

Length: 5 weeks (176 hours).

Exhibit Dates: 10/00–Present.

Learning Outcomes: Upon completion of the course, the student will be able to perform tasks in the areas of railway operations, engine operation, communications, car coupling and uncoupling, car and cargo inspections, record keeping, minor repairs, and sealing of cars.

Instruction: Methods of instruction include classroom conferences, demonstrations, and practical exercises in engine operations, (hand, flag, fuses, torpedoes, lamps, and engine whistles) communication signaling, cargo tie down, car inspection, coupling and uncoupling, minor repairs, and sealing of cars.

Credit Recommendation: In the lower-division baccalaureate/associate degree category, 2 semester hours in locomotive motorman operations (6/01).

Related Occupations: 88U.

AR-0419-0070

Transportation Officer Basic

Course Number: 8-55-C20-88A/C/D.

Location: Transportation School, Fort Eustis, VA.

Length: 17–18 weeks (766 hours).

Exhibit Dates: 4/99–9/03.

Learning Outcomes: Before 4/99 see AR-0419-0033. Upon completion of the course, the student will be able to lead and plan in the areas of joint strategic deployment and unit movement by rail, air, or sea.

Instruction: Methods of instruction include conferences, computer applications, and field exercises. Topics include leadership, maintenance, fighting, and transporting. The officer is required to participate in practical exercises interacting with trainers, fellow students, host providers, and equipment hardware and software.

Credit Recommendation: In the lower-division baccalaureate/associate degree category, 3 semester hours in principles of supervision, 1 in depot operations, 1 in leadership, and 1 in principles of operating systems (6/01).

Related Occupations: 88A; 88C; 88D.

AR-0419-0071

Transportation Officer Advanced - RC (Rail/Highway)

Course Number: 8-55-C23-88D, Phase 2.

Location: Transportation School, Fort Eustis, VA.

Length: 2 weeks (74 hours).

Exhibit Dates: 4/00–Present.

Learning Outcomes: Upon completion of the course, the student will be able to operate and manage highway and rail transportation, including planning, movement, equipment, and recovery.

Instruction: Practical exercises, discussion, and lecture. Topics include movement management, utilization of a highway network, convoy management, security coordination, operations planning, and rail operations.

Credit Recommendation: In the upper-division baccalaureate category, 2 semester hours in logistics management (12/03).

AR-0505-0001

Equal Opportunity Representative

Course Number: DL 471.

Location: Combined Arms Training Center, Vilseck, W. Germany.

Length: 2 weeks (68 hours).

Exhibit Dates: 3/86–10/98.

Learning Outcomes: Upon completion of the course, the student will be able to gather information on possible equal opportunity violations; lead guided discussions on equal opportunity subjects; identify and discriminate between actual and suspected equal opportunity violations; prepare and present an information briefing; and effectively communicate with all soldiers regardless of race, sex, or religion.

Instruction: Lectures, demonstrations, and seminars are methods used. The student is required to give a three-to-five minute impromptu speech, a ten-minute lecture, and lead a twenty-minute group discussion.

Credit Recommendation: In the lower-division baccalaureate/associate degree category, 1 semester hour in speech communication (10/88).

AR-0602-0007

English 3200: Programmed Course in Grammar and Usage

Course Number: 1-19-XR-100.

Location: Military Police School, Fort McClellan, AL; Chemical School, Fort McClellan, AL.

Length: Self-paced, 2–3 weeks (104 hours).

Exhibit Dates: 10/80–1/92.

Objectives: To provide selected enlisted personnel and noncommissioned and commissioned officers with a working knowledge of English usage.

Instruction: Self-paced instruction uses a programmed workbook. Course covers grammar usage, sentence construction, and punctuation.

Credit Recommendation: In the vocational certificate category, 3 semester hours in grammar or English usage (3/82).

AR-0602-0008

Spanish (QB) Special Operations Forces
Basic Military Language

Course Number: 2E-F117/011-F32(QB).
Location: John F. Kennedy Special Warfare Center and School, Fort Bragg, NC.
Length: 18–19 weeks (713 hours).
Exhibit Dates: 4/97–Present.
Learning Outcomes: Upon completion of the course, the student will have achieved a level in listening, speaking, and reading proficiency equivalent to the ACTFL intermediate-low level and a memorized proficiency in writing.
Instruction: Small-group classroom instruction emphasizes communicative competence in oral/aural skills and task-based reading skills. An integral learning format, directed toward functional language use, incorporates modules that include everyday situations and specialized military tasks as well as cultural and writing components.
Credit Recommendation: In the lower-division baccalaureate/associate degree category, 2 semesters in an intensive beginning-level course in Spanish (4/01).

AR-0602-0009

German (GM) Special Operations Forces
Basic Military Language

Course Number: 2E-F128/011-F43(GM).
Location: John F. Kennedy Special Warfare Center and School, Fort Bragg, NC.
Length: 18–19 weeks (713 hours).
Exhibit Dates: 6/97–Present.
Learning Outcomes: Upon completion of the course, the student will have achieved a level in listening, speaking, and reading proficiency equivalent to the ACTFL intermediate-low level, and a memorized proficiency in writing.
Instruction: Small-group classroom instruction emphasizes communicative competence in oral/aural skills and task-based reading skills. An integral learning format, directed toward functional language use, incorporates modules that include everyday situations and specialized military tasks as well as cultural and writing components.
Credit Recommendation: In the lower-division baccalaureate/associate degree category, 2 semesters in an intensive beginning-level course in German (4/01).

AR-0602-0010

French (FR) Special Operations Forces
Basic Military Language

Course Number: 2E-F120/011-F35 (FR).
Location: John F. Kennedy Special Warfare Center and School, Fort Bragg, NC.
Length: 18–19 weeks (713 hours).
Exhibit Dates: 6/97–Present.
Learning Outcomes: Upon completion of the course, the student will have achieved a level in listening, speaking, and reading proficiency equivalent to the ACTFL intermediate-low level and a memorized proficiency in writing.
Instruction: Small-group classroom instruction emphasizes communicative competence in oral/aural skills and task-based reading skills. An integral learning format, directed toward functional language use, incorporates modules that include everyday situations and specialized military tasks as well as cultural and writing components.
Credit Recommendation: In the lower-division baccalaureate/associate degree category, 2 semesters in an intensive beginning level course in French (4/01).

AR-0602-0011

Arabic (AD) Special Operations Forces
Basic Military Language

Course Number: 2E-F121/011-F36.
Location: John F. Kennedy Special Warfare Center and School, Fort Bragg, NC.
Length: 24–25 weeks (935 hours).
Exhibit Dates: 6/97–Present.
Learning Outcomes: Upon completion of the course, the student will have achieved a level in listening, speaking, and reading proficiency equivalent to the ACTFL intermediate-low level and a memorized proficiency in writing.
Instruction: Small-group classroom instruction emphasizes communicative competence in oral/aural skills and task-based reading skills. An integral learning format, directed toward functional language use, incorporates modules that include everyday situations and specialized military tasks as well as cultural and writing components.
Credit Recommendation: In the lower-division baccalaureate/associate degree category, 2 semesters of an intensive beginning-level course in Arabic (4/01).

AR-0602-0012

Tagalog (TA) Special Operations Forces
Basic Military Language

Course Number: 2E-F125/011-F40 (TA).
Location: John F. Kennedy Special Warfare Center and School, Fort Bragg, NC.
Length: 24–25 weeks (935 hours).
Exhibit Dates: 6/97–Present.
Learning Outcomes: Upon completion of the course, the student will have achieved a level in listening, speaking, and reading proficiency equivalent to the ACTFL intermediate-low level and a memorized proficiency in writing.
Instruction: Small-group classroom instruction emphasizes communicative competence in oral/aural skills and task-based reading skills. An integral learning format, directed toward functional language use, incorporates modules that include everyday situations and specialized military tasks as well as cultural and writing components.
Credit Recommendation: In the lower-division baccalaureate/associate degree category, 2 semesters in an intensive beginning-level course in Tagalog (4/01).

AR-0602-0013

Persian-Farsi (PF) Special Operations
Forces Basic Military Language

Course Number: 2E-F122/011-F37 (PF).
Location: John F. Kennedy Special Warfare Center and School, Fort Bragg, NC.
Length: 24–25 weeks (935 hours).
Exhibit Dates: 6/97–Present.
Learning Outcomes: Upon completion of the course, the student will have achieved a level in listening, speaking, and reading proficiency equivalent to the ACTFL intermediate-low level and a memorized proficiency in writing.
Instruction: Small-group classroom instruction emphasizes communicative competence in oral/aural skills and task-based reading skills. An integral learning format, directed toward functional language use, incorporates modules that include everyday situations and specialized military tasks as well as cultural and writing components.
Credit Recommendation: In the lower-division baccalaureate/associate degree category, 2 semesters in an intensive beginning-level course in Persian-Farsi (4/01).

AR-0602-0014

Czech (CX) Special Operations Forces
Basic Military Language

Course Number: 2E-F114/011-F29 (CX).
Location: John F. Kennedy Special Warfare Center and School, Fort Bragg, NC.
Length: 24–25 weeks (935 hours).
Exhibit Dates: 6/97–Present.
Learning Outcomes: Upon completion of the course, the student will have achieved a level in listening, speaking, and reading-proficiency equivalent to the ACTFL intermediate-low level and a memorized proficiency in writing.
Instruction: Small-group classroom instruction emphasizes communicative competence in oral/aural skills and task-based reading skills. An integral learning format, directed toward functional language use, incorporates modules that include everyday situations and specialized military tasks as well as cultural and writing components.
Credit Recommendation: In the lower-division baccalaureate/associate degree category, 2 semesters in an intensive beginning-level course in Czech (4/01).

AR-0602-0015

Thai (TH) Special Operations Forces
Basic Military Language

Course Number: 2E-F116/011-F31 (TH).
Location: John F. Kennedy Special Warfare Center and School, Fort Bragg, NC.
Length: 24–25 weeks (935 hours).
Exhibit Dates: 6/97–Present.
Learning Outcomes: Upon completion of the course, the student will have achieved a level in listening, speaking, and reading proficiency equivalent to the ACTFL intermediate-low level and a memorized proficiency in writing.
Instruction: Small-group classroom instruction emphasizes communicative competence in oral/aural skills and task-based reading skills. An Integral learning format, directed toward functional language use, incorporates modules that include everyday situations and specialized military tasks as well as cultural and writing components.
Credit Recommendation: In the lower-division baccalaureate/associate degree category, 2 semesters in an intensive beginning-level course in Thai (4/01).

AR-0602-0016

Russian (RU) Special Operations Forces
Basic Military Language

Course Number: 2E-F118/011-F33 (RU).

Location: John F. Kennedy Special Warfare Center and School, Fort Bragg, NC.

Length: 24–25 weeks (935 hours).

Exhibit Dates: 6/97–Present.

Learning Outcomes: Upon completion of the course, the student will have achieved a level in listening, speaking, and reading proficiency equivalent to the ACTFL intermediate-low level and a memorized proficiency in writing.

Instruction: Small-group classroom instruction emphasizes communicative competence in oral/aural skills and task-based reading skills. An integral learning format, directed toward functional language use, incorporates modules that include everyday situations and specialized military tasks as well as cultural and writing components.

Credit Recommendation: In the lower-division baccalaureate/associate degree category, 2 semesters in an intensive beginning-level course in Russian (4/01).

AR-0602-0017

Polish (PL) Special Operations Forces Basic Military Language

Course Number: 2E-F119/011-F34 (PL).

Location: John F. Kennedy Special Warfare Center and School, Fort Bragg, NC.

Length: 24–25 weeks (935 hours).

Exhibit Dates: 6/97–Present.

Learning Outcomes: Upon completion of the course, the student will have achieved a level in listening, speaking, and reading proficiency equivalent to the ACTFL intermediate-low level and a memorized proficiency in writing.

Instruction: Small-group classroom instruction emphasizes communicative competence in oral/aural skills and task-based reading skills. An integral learning format, directed toward functional language use, incorporate modules that include everyday situations and specialized military tasks as well as cultural and writing components.

Credit Recommendation: In the lower-division baccalaureate/associate degree category, 2 semesters in an intensive beginning-level course in Polish (4/01).

AR-0602-0018

Vietnamese (VN) Special Operations Forces Basic Military Language

Course Number: 2E-F126/011-F41 (VN).

Location: John F. Kennedy Special Warfare Center and School, Fort Bragg, NC.

Length: 24–25 weeks (935 hours).

Exhibit Dates: 6/97–Present.

Learning Outcomes: Upon completion of the course, the student will have achieved a level in listening, speaking, and reading-proficiency equivalent to the ACTFL intermediate-low level and a memorized proficiency in writing.

Instruction: Small-group classroom instruction emphasizes communicative competence in oral/aural skills and task-based reading skills. An integral learning format, directed toward functional language use, incorporates modules that include everyday situations and specialized military tasks as well as cultural and writing components.

Credit Recommendation: In the lower-division baccalaureate/associate degree category, 2 semesters in an intensive beginning-level course in Vietnamese (4/01).

AR-0602-0019

Portuguese (PT) Special Operations Forces Basic Military Language

Course Number: 2E-F124/011-F39 (PT).

Location: John F. Kennedy Special Warfare Center and School, Fort Bragg, NC.

Length: 18–19 weeks (713 hours).

Exhibit Dates: 6/97–Present.

Learning Outcomes: Upon completion of the course, the student will have achieved a level in listening, speaking, and reading-proficiency equivalent to the ACTFL intermediate-low level and a memorized proficiency in writing.

Instruction: Small-group classroom instruction emphasizes communicative competence in oral/aural skills and task-based reading skills. An integral learning format, directed toward functional language use, incorporates modules that include everyday situations and specialized military tasks as well as cultural and writing components.

Credit Recommendation: In the lower-division baccalaureate/associate degree category, 2 semesters in an intensive beginning-level course in Portuguese (4/01).

AR-0602-0020

Korean (KP) Special Operations Forces Basic Military Language

Course Number: 2E-F115/011-F30 (KP).

Location: John F. Kennedy Special Warfare Center and School, Fort Bragg, NC.

Length: 24–25 weeks (935 hours).

Exhibit Dates: 6/97–Present.

Learning Outcomes: Upon completion of the course, the student will have achieved a level in listening, speaking, and reading-proficiency equivalent to the ACTFL intermediate-low level and a memorized proficiency in writing.

Instruction: Small-group classroom instruction emphasizes communicative competence in oral/aural skills and task-based reading skills. An integral learning format, directed toward functional language use, incorporates modules that include everyday situations and specialized military tasks as well as cultural and writing components.

Credit Recommendation: In the lower-division baccalaureate/associate degree category, 2 semester hours in an intensive beginning-level course in Korean (4/01).

AR-0701-0002

Dental Hygiene

Course Number: 330-X2.

Location: Academy of Health Sciences, Fort Sam Houston, TX; Medical Field Service School, Fort Sam Houston, TX.

Length: 16 weeks (664 hours).

Exhibit Dates: 12/83–10/90.

Objectives: To provide the student with the knowledge and skills to perform the duties of dental hygienist and an expanded dental assistant under the supervision of a dental corps officer.

Instruction: Instruction includes performing oral hygiene and prophylactic procedures; conducting individual oral disease programs; applying pit and fissure sealants to the teeth; exposing, processing, and mounting dental radiographs; making preliminary dental impressions on patients and fabricating diagnostic casts; placing and finishing dental restorations; and cleaning and sterilizing dental instruments and equipment.

Credit Recommendation: In the lower-division baccalaureate/associate degree category, 3 semester hours in basic dental science, 2 in dental health, 8 in clinical training in dental hygiene, 3 in dental materials, 2 in nutrition, and 2 in dental anatomy and physiology (9/85).

Related Occupations: 91E.

AR-0701-0012

Dental Specialist Reserve Component (Nonresident/Resident) Phase 2

Course Number: 330-91E10-RC.

Location: Academy of Health Sciences, Fort Sam Houston, TX.

Length: 2 weeks (70–82 hours).

Exhibit Dates: 5/83–1/93.

Learning Outcomes: Upon completion of the course, the student will be able to perform the procedures necessary to assist the dental officer in the examination, care, and treatment of the oral region.

Instruction: Phase 2, the resident portion of the course, consists of conferences and practical exercises in basic dental sciences, radiology, materials, and records and also covers the duties of the dental specialist.

Credit Recommendation: In the lower-division baccalaureate/associate degree category, 1 semester hour in dental radiology and 2 in chairside assisting (10/90).

Related Occupations: 91E.

AR-0701-0013

Dental Specialist

Course Number: *Version 1:* 330-91E10. *Version 2:* 330-91E10. *Version 3:* 330-91E10.

Location: Academy of Health Sciences, Fort Sam Houston, TX.

Length: *Version 1:* 6 weeks (221–232 hours). *Version 2:* 7 weeks (266 hours). *Version 3:* 7–8 weeks (324 hours).

Exhibit Dates: *Version 1:* 11/87–9/95. *Version 2:* 10/95–9/01. *Version 3:* 10/01–Present.

Learning Outcomes: *Version 1:* Upon completion of the course, the student will be able to perform the procedures necessary to assist the dental officer in care, examination, and treatment of the oral region. *Version 2:* Upon completion of the course, the student will be able to perform as a dental assistant, including four-handed chairside assisting and taking dental radiographs. *Version 3:* Upon completion of the course, the student will be able to discuss oral anatomy, physiology, pathology, infection control and hygiene; perform and process dental radiographs, chairside assisting, scaling and polishing, and instrument sterilization and set up.

Instruction: *Version 1:* Lectures and practical experience cover the basic principles of performing dental assisting duties: learning to expose, process, and mount dental radiographs for diagnostic value; mixing commonly used dental restoration and impression materials; preparing instrument setups; and sterilizing and disinfecting used dental instruments. *Version 2:* This course includes lectures, discussions, demonstrations, and laboratory practice in biomedical sciences, dental radiology, dental materials, and chairside assisting. *Version 3:* Audio visual materials, practical and classroom exercises, laboratory, and lecture. Topics include oral anatomy, physiology, and pathology; performance and processing of dental radiographs; set-up and sterilization of instruments; infection control and oral hygiene; chairside and surgical assisting; scaling and polishing; patient relations; recordkeeping; and clinical practice.

Credit Recommendation: ***Version 1:*** In the lower-division baccalaureate/associate degree category, 1 semester hour in basic dental science, 2 in dental radiology, and 2 in chairside assisting (8/92). *Version 2:* In the lower-division baccalaureate/associate degree category, 2 semester hours in basic dental science, 2 in dental radiography, and 2 in chairside assisting (5/02). *Version 3:* In the vocational certificate category, 1 semester hour in cardiopulmonary resuscitation. In the lower-division baccalaureate/associate degree category, 2 semester hours in basic dental science, 2 in dental radiography, 1 in infection control, 2 in dental instrumentation, and 4 in clinical internship (11/03).

Related Occupations: 91E.

AR-0701-0014

1. Dental Laboratory Specialist (Basic)
2. Dental Laboratory Specialist

Course Number: ***Version 1:*** 331-42D10. *Version 2:* 331-N5.

Location: Academy of Health Sciences, Fort Sam Houston, TX.

Length: ***Version 1:*** 19 weeks (761–855 hours). *Version 2:* 24 weeks (967 hours).

Exhibit Dates: ***Version 1:*** 10/85–2/95. *Version 2:* 3/95–Present.

Learning Outcomes: ***Version 1:*** Upon completion of the course, the student will be able to perform dental prosthetic laboratory techniques. *Version 2:* Upon completion of the course, the student will be able to fabricate complete dentures, removable partial dentures, sample fixed prosthesis, orthodontic retainers, and occlusal splints and provide dental laboratory support in a field environment.

Instruction: ***Version 1:*** Lectures and practical exercises cover basic dental prosthetic laboratory techniques. Course includes anatomy and tooth morphology, basic dental materials, and complete dental prosthetics. *Version 2:* Course instruction in basic dental science, dental materials and equipment, laboratory skills, and dental devices is provided using conferences and practical exercises. Instruction concerning laboratory equipment and most topics on fabrication procedures are approached using demonstrations. Students are evaluated through regular testing.

Credit Recommendation: ***Version 1:*** In the lower-division baccalaureate/associate degree category, 1 semester hour in tooth anatomy, 1 in dental materials, and 6 in dental laboratory procedures (12/93). *Version 2:* In the lower-division baccalaureate/associate degree category, 1 semester hour in tooth anatomy, 2 in dental materials, 5 in dental laboratory procedures—basic, and 22 in dental laboratory skills—dentures and prosthetics (10/95).

Related Occupations: 42D.

AR-0701-0015

Dental Laboratory Specialist (Senior)

Course Number: 331-42D30.

Location: Academy of Health Sciences, Fort Sam Houston, TX.

Length: 10 weeks (328–393 hours).

Exhibit Dates: 5/88–9/91.

Learning Outcomes: Upon completion of the course, the student will be able to fabricate complex dental appliances to a clinically acceptable standard, including ceramic crowns, porcelain-fused-to-metal crowns, fixed partial dentures, fixed removable semiprecision appliances, and complex orthodontic and surgical appliances.

Instruction: This course includes lectures, discussions, conferences, demonstrations, and practical exercises in coronal morphology, Hallow bulb obturator, maxillary crowns and dentures, acid etched bridge, porcelain jacket crowns, mandibular partial dentures, semiprecision attachments, denture repair, and postsolder techniques.

Credit Recommendation: In the lower-division baccalaureate/associate degree category, 3 semester hours in maxillofacial appliances, 5 in porcelain-fused-to-metal restorations, 2 in semiprecision attachments, and 1 in ceramic alloys (3/90).

Related Occupations: 42D.

AR-0701-0016

Preventive Dentistry Specialty

Course Number: ***Version 1:*** 330-X2. *Version 2:* 330-X2. *Version 3:* 330-X2.

Location: Academy of Health Sciences, Fort Sam Houston, TX.

Length: ***Version 1:*** 16 weeks (645–653 hours). *Version 2:* 16 weeks (645 hours). *Version 3:* 12 weeks (477–478 hours).

Exhibit Dates: ***Version 1:*** 11/90–12/93. *Version 2:* 1/94–3/96. *Version 3:* 4/96–Present.

Learning Outcomes: ***Version 1:*** Upon completion of the course, the student will be able to perform oral hygiene prophylactic procedures, conduct individual oral disease control programs, apply pit and fissure sealants to the teeth, evaluate dental radiographs, make preliminary dental impressions, and place and finish dental restorations. *Version 2:* Upon completion of the course, the student will be able to provide preventive dental services, including performing oral hygiene prophylactic procedures, conducting individual oral disease control programs, applying pit and fissure sealants, exposing and evaluating dental radiographs, placing and finishing dental restorations, cleaning and sterilizing dental equipment and instruments, and performing expanded duties. *Version 3:* Upon completion of the course, the student will be able to treat and prevent dental disease by applying knowledge of dental anatomy and pathology and by performing prophylactic procedures, including manual and ultrasonic sealing, oral inspection, charting, coronal polishing, fluoride application, and patient education.

Instruction: ***Version 1:*** Instruction includes oral hygiene, prophylactic procedures, conducting individual oral disease programs, applying pit and fissure sealants to the teeth, evaluating radiographs, performing expanded duties in dental procedures, making preliminary dental impressions on patients and fabricating diagnostic casts, placing and finishing dental restorations, and cleaning and sterilizing dental instruments and equipment. *Version 2:* This course includes lectures, discussions, demonstrations, and laboratory practice in dental sciences, dental materials, dental anatomy and physiology, expanded duties, and dental pathology. *Version 3:* This course includes lectures, discussions, demonstrations, and laboratory practice in dental sciences, dental materials, dental anatomy and physiology, expanded duties, and dental pathology.

Credit Recommendation: ***Version 1:*** In the lower-division baccalaureate/associate degree category, 3 semester hours in basic dental science, 2 in dental health, 8 in clinical training in dental hygiene, 3 in dental materials, 2 in nutrition, 2 in expanded duties in dental procedures, and 2 in dental anatomy and physiology (1/94). *Version 2:* In the upper-division baccalaureate category, 2 semester hours in dental anatomy, morphology, and physiology; 3 in dental materials; 2 in nutrition; 3 in dental sciences; 2 in dental health; 2 in expanded procedures; and 8 in clinical practice (10/95). *Version 3:* In the upper-division baccalaureate category, 2 semester hours in dental sciences, 2 in dental expanded duties (procedures), 8 in clinical dental practice, and 1 in dental equipment maintenance (5/01).

AR-0701-0017

Army Medical Department (AMEDD) Basic Noncommissioned Officer (NCO) (42D)

Course Number: 6-8-C40 (42D).

Location: Academy of Health Sciences, Fort Sam Houston, TX.

Length: 19 weeks (686 hours).

Exhibit Dates: 10/91–4/04.

Learning Outcomes: Upon completion of the course, the student will be able to supervise and manage a dental laboratory, including personnel, equipment, records, infection control procedures, and X-ray safety. Student will perform all of the procedures expected in the dental laboratory including fabricating dental appliances.

Instruction: The course covers dental laboratory and clinic management, personnel supervision, infection control procedures, dental records, dental anatomy, equipment maintenance, computer use, and the fabrication of dental and orthodontic appliances. Methods of instruction include lectures, laboratories, discussions, small group work, and supervised practice. Course includes a common core of leadership subjects.

Credit Recommendation: In the lower-division baccalaureate/associate degree category, see AR-1406-0090 for common core credit. In the upper-division baccalaureate category, 2 semester hours in dental anatomy, 2 in dental clinic management, 3 in orthodontic appliances, 2 in military science, 3 in management and supervision, and 6 in dental appliances (8/93).

Related Occupations: 42D.

AR-0701-0018

1. Army Medical Department (AMEDD) Noncommissioned Officer (NCO) Basic (91E)
2. Army Medical Department (AMEDD) Noncommissioned Officer (NCO) Basic

Course Number: ***Version 1:*** 6-8-C40 (91E). *Version 2:* 6-8-C40 (Tech Tng) AC; 6-8-C40 (91E30 Tech Tng); 6-8-C40 (91E).

Location: Academy of Health Sciences, Fort Sam Houston, TX.

Length: ***Version 1:*** 7 weeks (252 hours). *Version 2:* 2 weeks (50–80 hours).

Exhibit Dates: ***Version 1:*** 10/91–3/95. *Version 2:* 4/95–Present.

Learning Outcomes: ***Version 1:*** Upon completion of the course, the student will be able to manage and supervise a dental laboratory. *Version 2:* Upon completion of the course, the student (dental technician) will be able to supervise or perform dental, preventive dental, and common dental laboratory procedures in combat and peacetime environments.

Instruction: ***Version 1:*** The course covers personnel supervision and leadership, dental equipment and maintenance, computer use, infection control, occupational safety, patient dental records, and management. Methodologies include laboratories, lectures, discussions, presentations, and practical applications. Course includes a common core of leadership subjects. *Version 2:* Topics including general history and traditions of the dental care system, leadership development, and resource management are presented via conferences and practical exercises. Examinations are used to reinforce instructional materials. The course stresses leadership competencies and dental management.

Credit Recommendation: ***Version 1:*** In the lower-division baccalaureate/associate degree category, see AR-1406-0090 for common core credit. In the upper-division baccalaureate category, 3 semester hours in management and supervision and 2 in military science (8/93). *Version 2:* In the lower-division baccalaureate/associate degree category, 2 semester hours in leadership and management training (5/01).

Related Occupations: 91E.

AR-0701-0019

ADVANCED DENTAL LABORATORY SPECIALIST

Course Number: 331-F1.

Location: Academy of Health Sciences, Fort Sam Houston, TX.

Length: 6 weeks (233 hours).

Exhibit Dates: 5/95–Present.

Learning Outcomes: Upon completion of the course, the student will be expected to wax and carve retainers and pontics to a clinically acceptable level, reproduce missing oral tissues to a clinically acceptable level, discuss principles of occlusion for fixed prostheses, discuss concepts of color and its effects on ceramic and porcelain restorations, apply knowledge of fixed partial dentures, fabricate maxillofacial devices, and fabricate porcelain-fused-to-metal devices.

Instruction: Instruction covers manipulation or fabrication of materials such as porcelain or maxillofacial appliances using the conference style. Instruction that requires student manipulation of materials is usually approached first through demonstration, followed by practical exercises. Students are evaluated through examinations.

Credit Recommendation: In the lower-division baccalaureate/associate degree category, 8 semester hours in dental laboratory (10/95).

Related Occupations: 91E.

AR-0701-0021

TOTAL ARMY TRAINING SYSTEM DENTAL SPECIALIST
(Dental Specialist)

Course Number: 081-91E10 (F); 081-91E10.

Location: Reserve Component Training Centers, US.

Length: 118–120 hours.

Exhibit Dates: 12/98–Present.

Learning Outcomes: Upon completion of the course, the student will be able to perform as a dental assistant including four-handed chairside assisting and taking dental radiographs.

Instruction: This course includes lectures, discussions, demonstrations, and laboratory practice in biomedical sciences, dental radiology, dental materials, and chairside assisting. This recommendation is based on the resident phase only. Students also completed a nonresident phase.

Credit Recommendation: In the lower-division baccalaureate/associate degree category, 2 semester hours clinical experience. Credit recommendation is based on the resident phase only (9/00).

AR-0701-0022

DENTAL MANAGEMENT DEVELOPMENT

Course Number: 6C-F1/331-F2.

Location: Academy of Health Sciences, Fort Sam Houston, TX.

Length: 1–2 weeks (75 hours).

Exhibit Dates: 8/01–Present.

Learning Outcomes: Upon completion of the course, student will be able to supervise a dental treatment facility.

Instruction: Laboratory, classroom exercise, practical exercises, lecture, discussion. Topics include Non- Commissioned Officer In Charge (NCOIC), dental clinic briefing, Total Quality Management (TQM) leadership, Myers Briggs, personnel evaluation/counseling, and sexual harassment/labor relations.

Credit Recommendation: In the upper-division baccalaureate category, 3 semester hours in dental clinic management (6/02).

AR-0702-0005

RENAL DIALYSIS TECHNICIAN
(Dialysis Technician)

Course Number: 300-F2.

Location: Academy of Health Sciences, Fort Sam Houston, TX; Walter Reed Medical Center, Washington, DC.

Length: 20 weeks (734–735 hours).

Exhibit Dates: 1/84–12/97.

Objectives: To provide selected personnel with knowledge and skills required to perform safe and effective hemodialysis and peritoneal dialysis.

Instruction: This course provides the student with a comprehensive introduction to renal dialysis/hemodialysis technician education and consists of clinical and didactic instruction.

Credit Recommendation: In the lower-division baccalaureate/associate degree category, 3 semester hours in introduction to hemo/renal dialysis theory, 2 in anatomy and physiology of cardiovascular system, 2 in anatomy and physiology of renal system, 2 in pathophysiology of cardiac and renal diseases, 8 in instrumental analysis in hemo/renal dialysis, 4 in clinical practicum in hemodialysis, and 4 in clinical practicum in renal dialysis (9/87).

AR-0702-0006

UROLOGY PROCEDURES

Course Number: 300-F12.

Location: Academy of Health Sciences, Fort Sam Houston, TX.

Length: Phase 1, 8 weeks (338–343 hours); Phase 2, 16 weeks (947 hours).

Exhibit Dates: 4/85–4/04.

Objectives: To provide the knowledge and skills necessary to function as a urology technician in both clinics and operating rooms.

Instruction: Didactic, laboratory, and clinical instruction cover all phases of urological care, including, anatomy and physiology, radiology, laboratory procedures, urological examinations, and patient care. The 8-week didactic and laboratory instruction is followed by a 16-week practicum.

Credit Recommendation: In the lower-division baccalaureate/associate degree category, 1 semester hour in laboratory procedures in urology, 1 in anatomy/physiology in urology, 2 in urologic pharmacy/pharmacology, 4 in radiologic procedures in urology, 6 in physical assessment and diagnostic procedures in urology, and 2 in clinical practicum in urology (6/91).

AR-0702-0007

MEDICAL LABORATORY SPECIALIST ADVANCED

Course Number: 6-8-C40 (92B); 311-92B30.

Location: Academy of Health Sciences, Fort Sam Houston, TX.

Length: 50 weeks (2049–2199 hours).

Exhibit Dates: 11/85–12/91.

Learning Outcomes: Upon completion of the course, the student will be able to perform techniques and procedures appropriate to a technical specialist and/or laboratory supervisor.

Instruction: Lectures and practical experience cover medical laboratory supervision, equipment use, clinical chemistry, microbiology, parasitology, hematology, urinalysis, histopathology, and cytotechnology and immunology.

Credit Recommendation: In the lower-division baccalaureate/associate degree category, 30 semester hours in medical technology or 10 in biology or chemistry (11/89).

Related Occupations: 92B.

AR-0702-0011

CYTOLOGY SPECIALIST

Course Number: B-311-0036; 311-M2; 311-92E20.

Location: Brooke Medical Center, Fort Sam Houston, TX; Academy of Health Sciences, Fort Sam Houston, TX.

Length: 50–52 weeks (1829–2060 hours).

Exhibit Dates: 6/88–Present.

Learning Outcomes: Upon completion of the course, the student will be able to process and prepare all types of cytology specimens; identify and characterize the degree of normality or abnormality of the cellular material examined (for all types of specimens); mark abnormal cells for pathologist examination; identify and review histologic specimens for quality control; and demonstrate methods and knowledge of laboratory safety, organization, and management.

Instruction: Course includes lectures, demonstrations, case studies, laboratory, and clinical practice. Topics covered include gynecological and nongynecological body sites, including all organ systems, body cavities, and bone and soft tissue tumors. This is an American Medical Association (CAHEA) accredited program.

Credit Recommendation: In the upper-division baccalaureate category, 4 semester hours for field experience in introduction to cytology (including cytoprep), 3 in female genital tract—benign, 4 in female genital tract—malignant, 2 in respiratory tract, 2 in gastrointestinal tract, 3 in central nervous system and effusion, 2 in urinary tract, 3 in fine needle aspiration cytology (including breast, lymph node, head, and neck bone), 2 in research, and 18 in clinical practicum (5/01).

Related Occupations: 92E.

AR-0702-0012

Cardiovascular Technician

Course Number: 300-Y6, Phase 1.

Location: Academy of Health Sciences, Fort Sam Houston, TX.

Length: 52 weeks (1774 hours).

Exhibit Dates: 11/88–12/94.

Learning Outcomes: After 12/94 see AR-0702-0021. Upon completion of the course, the student will be able to review and/or record patient history and supporting clinical data; perform appropriate procedures and record anatomical, pathological, and/or physiological data for further diagnosis; apply invasive cardiovascular principles of cardiac catheterization, coronary angiography, and angioplasty; apply noninvasive cardiovascular principles of echocardiography, phonocardiography, and electrocardiography; and apply the noninvasive peripheral vascular principles of Doppler ultrasound.

Instruction: Phase 1 is a combination of conference, demonstration, and practical experience in cardiovascular disease and practice. Phase 2 offers clinical practice and internship in cardiovascular diagnosis, treatment, and care in the medical center setting.

Credit Recommendation: In the lower-division baccalaureate/associate degree category, for phase 1, 3 semester hours in anatomy and physiology of coronary and pulmonary system, 8 in principles of radiography, 6 in pathophysiology of cardiopulmonary disease, 2 in pharmacology, 5 in noninvasive cardiology, 9 in invasive cardiology; for phase 2, 19 semester hours in clinical internship in cardiovascular technician (3/90).

AR-0702-0013

Medical Laboratory Specialist Basic

Course Number: B-311-0011; 311-92B10.

Location: Academy of Health Sciences, Fort Sam Houston, TX.

Length: 15 weeks (584–593 hours).

Exhibit Dates: 11/89–12/94.

Learning Outcomes: Upon completion of the course, the student will be able to perform basic laboratory procedures in urinalysis, hematology, immunohematology, microbiology, and clinical chemistry.

Instruction: Lectures and practical exercises cover basic hematology, immunohematology, clinical chemistry, bacteriology, urinalysis, serology, and parasitology procedures.

Credit Recommendation: In the lower-division baccalaureate/associate degree category, 3 semester hours in introduction to medical laboratory technology, 3 in microbiology, 3 in clinical chemistry, 2 in hematology, 2 in immunohematology, and 1 in serology (1/94).

Related Occupations: 92B.

AR-0702-0014

Special Forces Medical Sergeant Advanced Noncommissioned Officer (NCO)

Course Number: ***Version 1:*** 011-18D40. *Version 2:* 011-18D40. *Version 3:* 011-18D40. *Version 4:* 011-18D40.

Location: John F. Kennedy Special Warfare Center and School, Fort Bragg, NC.

Length: ***Version 1:*** 12 weeks (524–543 hours). *Version 2:* 12 weeks (524–543 hours). *Version 3:* 18–19 weeks (1051 hours). *Version 4:* 18 weeks (737 hours).

Exhibit Dates: ***Version 1:*** 6/89–9/93. *Version 2:* 10/93–9/94. *Version 3:* 10/94–10/98. *Version 4:* 11/98–Present.

Learning Outcomes: ***Version 1:*** Upon completion of the course, the student will be able to provide leadership and technical training in areas such as resource management, communications, operations, unconventional warfare planning, intelligence collection and processing, operational planning, administration, and medical-specific tasks including advanced cardiac life support. *Version 2:* Upon completion of the course, the student will provide leadership and technical training in areas such as resource management, communications, operations, unconventional warfare planning, intelligence collection and processing, operational planning, administration, and medical-specific tasks. *Version 3:* Upon completion of the course, the student will be able to manage trauma in a combat situation. *Version 4:* Upon completion of the course, the student will provide leadership and technical training in areas such as resource management, communication, operations, unconventional warfare planning, intelligence collection and processing, operational planning, administration and medical-specific tasks including trauma management.

Instruction: ***Version 1:*** Course includes lectures, demonstrations, practical exercises, and examinations in career management, communication, and medical-specific tasks including advanced cardiac life support. Course includes a common core of leadership subjects. *Version 2:* The course includes lectures, demonstrations, practical exercises, and examinations in this career field, noncommissioned officer development, and medical-specific subjects and tasks. Course includes a common core of leadership subjects. *Version 3:* Course includes lectures, demonstrations, and performance exercises in leadership, military studies, resource management, effective communication, operations, tactics, unconventional warfare planning, intelligence collection and processing, NCO tasks, photography, emergency medical care in a combat situation, and other medical-specific tasks. Course includes a common core of leadership subjects. *Version 4:* Course includes lectures, demonstrations, and performance exercises in leadership, military studies, resource management, effective communication, operations, tactics, unconventional warfare planning, intelligence collection and processing, NCO tasks, photography, emergency medical care in a combat situation, and other medical-specific tasks. Course includes a common core of leadership subjects.

Credit Recommendation: ***Version 1:*** In the lower-division baccalaureate/associate degree category, 1 semester hour in introduction to US foreign policy; 2 in advanced cardiac life support; and 3 in military science, military studies, or leadership. See AR-1404-0035 for common core credit (12/92). *Version 2:* In the lower-division baccalaureate/associate degree category, 3 semester hours in military science, military studies, or leadership and 3 in applied medical sciences. See AR-1404-0035 for common core credit (12/92). *Version 3:* In the lower-division baccalaureate/associate degree category, 1 semester hour in photography, 3 in military science, and 3 in emergency medical technology. See AR-1404-0035 for common core credit (6/96). *Version 4:* In the lower-division baccalaureate/associate degree category, 3 semester hours in military science and 3 in emergency medical technology. See AR-1401-0035 for common core credit (4/01).

Related Occupations: 18D.

AR-0702-0015

Special Forces Medical Sergeant Advanced Noncommissioned Officer (NCO) Reserve Component Phase 2

Course Number: 011-18D40-RC.

Location: John F. Kennedy Special Warfare Center and School, Fort Bragg, NC.

Length: 3 weeks (209 hours).

Exhibit Dates: 9/93–4/04.

Learning Outcomes: Upon completion of the course, the student will be able to serve in a leadership and technical role in areas such as resource management, communications operations, unconventional warfare planning, intelligence collecting and processing, operational planning and techniques, administration, and medical-specific tasks including advanced cardiac life support tasks.

Instruction: Course includes lectures, demonstrations, practical exercises, and examinations in the career field, noncommissioned officer development, and medical-specific tasks.

Credit Recommendation: In the lower-division baccalaureate/associate degree category, 3 semester hours in military science, military studies, or leadership and 3 in applied medical sciences (12/92).

Related Occupations: 18D.

AR-0702-0016

Blood Donor Center Operations

Course Number: 311-M4.

Location: Academy of Health Sciences, Fort Sam Houston, TX.

Length: 3 weeks (108 hours).

Exhibit Dates: 10/90–Present.

Learning Outcomes: Upon completion of the course, the student will be able to prepare blood products for storage and transfusion therapy; determine suitability of donors for donating blood; perform multiple phlebotomies using aseptic techniques; and perform appropriate procedures to produce transfusible units of red blood cells deglycerolized.

Instruction: Topics covered include donor suitability factors; conditions affecting the safety of blood for transfusion; and procedures for providing usable blood components/products and transfusible units of red blood cells, plasma, and transfusible platelet concentrate.

Credit Recommendation: In the lower-division baccalaureate/associate degree category, 4 semester hours in blood banking techniques (1/94).

AR-0702-0017

Army Medical Department (AMEDD) Noncommissioned Officer (NCO) Basic (92B)

Course Number: 6-8-C40 (92B Tech Tng); 6-8-C40 (92B).

Location: Academy of Health Sciences, Fort Sam Houston, TX.

Length: 55 weeks (2253 hours).

Exhibit Dates: 1/92–7/94.

Learning Outcomes: Upon completion of the course, the student will be able to perform as a medical laboratory specialist including assuming such responsibilities as analysis of human specimens and management of the clinical laboratory.

Instruction: Course covers general chemistry, clinical chemistry, hematology, immunology, immunohematology, bacteriology, parasitology, mycology, and clinical laboratory management. Methodologies include lectures, laboratories, group discussions, and significant clinical practice. Course includes a common core of leadership subjects.

Credit Recommendation: In the lower-division baccalaureate/associate degree category, 4 semester hours in general chemistry, 6 in hematology, 4 in immunology, 4 in immunohematology, 6 in clinical laboratory practicum, 6 in clinical chemistry, and 6 in bacteriology. See AR-1406-0090 for common core credit. In the upper-division baccalaureate category, 6 semester hours in clinical laboratory practicum, 4 in personnel supervision, 3 in clinical laboratory management, 2 in military science, 6 in clinical chemistry, 6 in bacteriology, 4 in mycology, and 4 in parasitology (8/93).

Related Occupations: 92B.

AR-0702-0018

Medical Laboratory Specialist (MLT) Initial Entry Training

Course Number: 311-91K10 (MLT) (IET).

Location: Academy of Health Sciences, Fort Sam Houston, TX.

Length: Phase 1, 26 weeks (1170–1176 hours); Phase 2, 26 weeks (1000–1040 hours).

Exhibit Dates: 10/93–12/95.

Learning Outcomes: After 12/95 see AR-0702-0022. Upon completion of the course, the student will be able to perform basic and advanced medical laboratory procedures, process blood for donation, and perform blood banking procedures.

Instruction: Phase 1 of this course includes lectures, discussions, case presentations, demonstrations, and laboratory practice in chemistry, serology, hematology, immunohematology, urinalysis, parasitology, microbiology, and blood banking. Phase 2 is composed of supervised clinical practice at assigned clinical sites such as hospital laboratories. Course is offered at Army Medical Centers at Brooke, Walter Reed, Eisenhower, Tripler, Madigan, William Beaumont, and Womack and Medical Hospitals at Blanchfield, Martin, Moncrief, Darnell, Evans, Leonard Wood, Ireland, Bayne-Jones, Irwin, Reynolds and Winn, and at Navy Hospitals at Great Lakes, Jacksonville, and Pensacola and the School of Health Sciences at Portsmouth.

Credit Recommendation: In the lower-division baccalaureate/associate degree category, for Phase 1, 1 semester hour in urinalysis, 8 in hematology, 5 in general chemistry, 6 in clinical chemistry, 4 in parasitology, 3 in serology, 8 in microbiology, and 8 in blood banking; for Phase 2, 16 semester hours in clinical practice (10/95).

Related Occupations: 91K.

AR-0702-0019

Medical Laboratory Specialist (MLT) Advanced

Course Number: 311-91K30 (MLT) (ADV).

Location: Academy of Health Sciences, Fort Sam Houston, TX.

Length: Phase 1, 26 weeks (1170–1176 hours); Phase 2, 26 weeks (1000–1040 hours).

Exhibit Dates: 10/93–Present.

Learning Outcomes: Upon completion of the course, the student will be able to perform basic and advanced medical laboratory procedures, process blood for donation, and perform blood banking procedures.

Instruction: Phase 1 of this course includes lectures, discussions, case presentations, demonstrations, and laboratory practice in chemistry, serology, hematology, immunohematology, urinalysis, parasitology, microbiology, and blood banking. Phase 2 is composed of supervised clinical practice at assigned clinical sites such as hospital laboratories. Course is offered at Army Medical Centers at Brooke, Walter Reed, Eisenhower, Tripler, Madigan, William Beaumont, and Womack and Medical Hospitals at Blanchfield, Martin, Moncrief, Darnell, Evans, Leonard Wood, Ireland, Bayne-Jones, Irwin, Reynolds and Winn, and at Navy Hospitals at Great Lakes, Jacksonville, and Pensacola and the School of Health Sciences at Portsmouth.

Credit Recommendation: In the lower-division baccalaureate/associate degree category, for Phase 1, 1 semester hour in urinalysis, 8 in hematology, 5 in general chemistry, 6 in clinical chemistry, 4 in parasitology, 3 in serology, 8 in microbiology, and 8 in blood banking; for Phase 2, 16 semester hours in clinical practice (10/95).

Related Occupations: 91K.

AR-0702-0020

1. Cardiovascular Specialist
2. Cardiovascular Specialty

Course Number: *Version 1:* 300-Y6, Phase 1. *Version 2:* 300-Y6, Phase 2.

Location: *Version 1:* Academy of Health Sciences, Fort Sam Houston, TX. *Version 2:* Madigan Medical Center, Tacoma, WA; Brooke Medical Center, Fort Sam Houston, TX.

Length: *Version 1:* 36 weeks (1440 hours). *Version 2:* 36 weeks (1440 hours).

Exhibit Dates: *Version 1:* 1/95–6/97. *Version 2:* 7/97–Present.

Learning Outcomes: *Version 1:* Upon completion of the course, the student will be able to review and/or record patient history and supporting clinical data; perform appropriate procedures and record anatomical, pathological, and/or physiological data for further diagnosis; apply invasive cardiovascular techniques of cardiac catheterization, coronary angiography, and angioplasty; apply noninvasive cardiovascular techniques of echocardiography and electrocardiography; and apply the noninvasive peripheral vascular techniques of Doppler ultrasound. *Version 2:* Upon completion of the course, the student will be able to review and/or record patient history and supporting clinical data; perform appropriate procedures and record anatomical, pathological, and/or physiological data for further diagnosis; apply invasive cardiovascular techniques of cardiac catheterization, coronary angiography, and angioplasty; apply noninvasive cardiovascular techniques of echocardiography and electrocardiography; and apply the noninvasive peripheral vascular techniques of Doppler ultrasound.

Instruction: *Version 1:* Clinical instruction, practice, and internship cover the practice of a cardiovascular technician including diagnosis, treatment, and care in a medical setting. *Version 2:* Clinical instruction, practice, and internship cover the practice of a cardiovascular technician including diagnosis, treatment, and care in a medical setting.

Credit Recommendation: *Version 1:* In the lower-division baccalaureate/associate degree category, 20 in cardiovascular technician (internship) (5/97). *Version 2:* In the lower-division baccalaureate/associate degree category, 20 in cardiovascular technician (internship) (2/99).

AR-0702-0021

1. Cardiovascular Specialist
2. Cardiovascular Specialty

Course Number: *Version 1:* 300-Y6, Phase 1. *Version 2:* 300-Y6, Phase 1.

Location: Academy of Health Sciences, Fort Sam Houston, TX.

Length: *Version 1:* 21 weeks (844 hours). *Version 2:* 21 weeks (840–847 hours).

Exhibit Dates: *Version 1:* 1/95–1/97. *Version 2:* 2/97–Present.

Learning Outcomes: *Version 1:* Before 1/95 see AR-0702-0012. Upon completion of the course, the student will be able to perform basic computer operations; demonstrate understanding of basic anatomy and physiology; discuss cardiovascular pathophysiology and pharmacology; state principles of radiation safety and basic radiography; perform and interpret electrocardiograms; practice sterile technique associated with invasive procedures; discuss and perform simulated echocardiography procedures; discuss and identify the appropriate equipment used in cardiac catheterization procedures, including angiography, electrophysiology studies, angioplasty, and athererectomy; and perform CPR. *Version 2:* Before 1/95 see AR-0702-0012. Upon completion of the course, the student will be able to perform basic computer operations; demonstrate understanding of basic anatomy and physiology; discuss cardiovascular pathophysiology and pharmacology; state principles of radiation safety and basic radiography; perform and interpret electrocardiograms; practice sterile technique associated with invasive procedures; discuss and perform simulated echocardiography procedures; discuss and identify the appropriate equipment used in cardiac catheterization procedures, including angiography, electrophysiology studies, angioplasty, and athererectomy; and perform CPR and advanced cardiac life support.

Instruction: *Version 1:* Phase 1 is a combination of lecture, demonstration, and practical simulation covering basic computer science, anatomy and physiology, cardiovascular pathophysiology and pharmacology, radiation safety, basic radiography procedures, electrocardiogram performance and interpretation, sterile operating room techniques, cardiac catheterization procedures, echocardiography, chemistry, and cardiopulmonary resuscitation. *Version 2:* Phase 1 is a combination of lecture, demonstration, and practical simulation covering basic computer science, anatomy and physiology, cardiovascular pathophysiology and pharmacology, radiation safety, basic radiography procedures, electrocardiogram performance and interpretation, sterile operating room techniques, cardiac catheterization procedures, echocardiography, cardiopulmonary resuscitation, and advanced cardiac life support.

Credit Recommendation: *Version 1:* In the lower-division baccalaureate/associate degree category, 1 semester hour in introduction to computers, 3 in anatomy and physiology, 6 in cardiopathophysiology, 1 in pharmacology, 8 in basic radiography, 1 in chemistry, 1 in EKG, 1 in CPR, 6 in noninvasive procedures, and 10 in invasive procedures (2/99). *Version 2:* In the lower-division baccalaureate/associate degree category, 1 semester hour in introduction to computers, 3 in anatomy and physiology, 6 in cardiopathophysiology, 1 in pharmacology, 8 in basic radiography, 1 in EKG, 1 in CPR, 3 in advanced cardiac life support, 6 in noninvasive procedures, and 10 in invasive procedures (9/00).

AR-0702-0022

MEDICAL LABORATORY SPECIALIST

Course Number: 311-91K10, Phase 1.

Location: Academy of Health Sciences, Fort Sam Houston, TX.

Length: 26 weeks (1191 hours).

Exhibit Dates: 1/96–Present.

Learning Outcomes: Before 1/96 see AR-0702-0018. Upon completion of the course the student will be able to perform basic and advanced medical laboratory procedures, process blood for donation, and perform blood banking procedures.

Instruction: Laboratory, audio visual materials, case studies, discussion, classroom exercise, practical exercises, lecture and learner presentations. Topics include general laboratory, chemical math, specimens, clinical chemistry, hematology, immunohematology, clinical microbiology, and practicum.

Credit Recommendation: In the lower-division baccalaureate/associate degree category, 1 semester hour in introduction to medical laboratory, 5 in general laboratory chemistry, 6 in clinical laboratory chemistry, 5 in hematology, 7 in immunohematology, 10 in clinical microbiology, 1 in urinalysis, 3 in serology, 1 in a medical laboratory practicum, and 4 in parasitology (6/02).

AR-0702-0023

MEDICAL LABORATORY SPECIALIST

Course Number: 311-91K10, Phase 2.

Location: Academy of Health Sciences, Fort Sam Houston, TX.

Length: 26 weeks (1000 hours).

Exhibit Dates: 12/96–Present.

Learning Outcomes: Upon completion of the course, the student will be able to perform basic and advanced medical laboratory procedures, process blood for donation, and perform blood banking procedures.

Instruction: Laboratory, audio visual materials, case studies, discussion, classroom exercise, practical exercises, lecture, learner presentations. Clinical practice including general laboratory procedures & specimen processing, clinical chemistry, hematology/serology, immunohematology, clinical microbiology, immunology/serology, and urinalysis.

Credit Recommendation: In the lower-division baccalaureate/associate degree category, 4 semester hours in clinical chemistry practicum, 4 in hematology practicum, 4 in immunohematology practicum, 3 in microbiology practicum, 1 in immunology practicum, and 1 in urinalysis practicum (6/02).

AR-0703-0010

OPERATING ROOM NURSING

Course Number: 6F-66E.

Location: Academy of Health Sciences, Fort Sam Houston, TX; Brooke Medical Center, Fort Sam Houston, TX; Madigan Medical Center, Tacoma, WA; William Beaumont Medical Center, El Paso, TX.

Length: 16 weeks (585–588 hours).

Exhibit Dates: 3/79–4/04.

Learning Outcomes: Upon completion of the course, junior Army Nurse Corps officers will be able to function as entry-level nurses in the operating room.

Instruction: Classroom instruction and clinical demonstrations cover principles of microbiology, sterilization, disinfection, sterilization and preparation of surgical supplies, operating room nursing, environmental control in hospital operating rooms; and the use of specialized equipment.

Credit Recommendation: In the graduate degree category, 3 semester hours in advanced surgical/operating room nursing and 4 in advanced operating room techniques/practicum (5/91).

Related Occupations: 66E; 66H; 66J.

AR-0703-0015

PRACTICAL NURSE

Course Number: *Version 2:* 300-91C20. *Version 3:* 300-91C10; 300-91C20.

Location: *Version 2:* Academy of Health Sciences, Fort Sam Houston, TX; Phase 2, Brooke Medical Center, Fort Sam Houston, TX; Womack Medical Center, Fort Bragg, NC; Fitzsimmons Medical Center, Aurora, CO; Dwight David Eisenhower Medical Center, Fort Gordon, GA; William Beaumont Medical Center, El Paso, TX. *Version 3:* Academy of Health Sciences, Fort Sam Houston, TX; Phase 2, Brooke Medical Center, Fort Sam Houston, TX; William Beaumont Medical Center, El Paso, TX; Dwight David Eisenhower Medical Center, Fort Gordon, GA; Fitzsimmons Medical Center, Aurora, CO; Madigan Medical Center, Tacoma, WA; Womack Medical Center, Fort Bragg, NC.

Length: *Version 2:* Phase 1, 6 weeks (168–270 hours); Phase 2, 46 weeks (1840 hours). *Version 3:* Phase 1, 6 weeks (140 hours); Phase 2, 46 weeks (1570 hours).

Exhibit Dates: *Version 2:* 7/88–6/92. *Version 3:* 7/92–12/01.

Learning Outcomes: *Version 2:* Upon completion of Phase 1, the student will be able to demonstrate a body of theoretical knowledge by written and oral examinations in preparation for clinical care of clients. In Phase 2, the student will have the opportunity to implement the body of knowledge acquired in Phase 1 by administering basic nursing care to a selected group of patients in a variety of health care settings under the supervision of a registered nurse. Upon completion, the graduate is eligible to take the state board examination for licensure. *Version 3:* After 12/01 see AR-0703-0036. Upon completion of Phase 1, the student will be able to demonstrate theoretical knowledge of basic principles of microbiology, pharmacology, anatomy and physiology, and preventive medicine; calculate medical dosages; and discuss legal/ethical aspects of practical nursing. Upon completion of Phase 2, the student will be able to discuss basic communication skills; identify in transcultural factors nursing; identify principles of infection control; demonstrate basic nursing skills; demonstrate medication administration skills; identify nursing care for the pre-/postoperative patient; develop nursing care plans using the nursing process; demonstrate theoretical knowledge of common diseases and their nursing implications, including gerontology, pediatric, and obstetrical conditions; and, in the clinical portion, administer nursing care to a selected group of patients in a variety of health care settings under the supervision of a registered nurse.

Instruction: *Version 2:* Upon completion of Phase 1, the student will be able to demonstrate a body of theoretical knowledge by written and oral examinations in preparation for the clinical care of clients. In Phase 2, the student will have the opportunity of implementing the body of knowledge acquired in Phase 1 by administering basic nursing care to a selected group of patients in a variety of health care settings under the supervision of a registered nurse. Upon completion, the student is eligible to take the state board examination for licensure. *Version 3:* Phase 1 of this course covers microbiology, anatomy and physiology, nutrition, drug dosage calculations, preventive medicine, pharmacology, and personal/vocational adjustment. Methodology includes lectures and demonstrations. In Phase 1, the course covers nursing fundamentals, common disease processes related to each body system, and nursing care for patients in a variety of settings. Phase 1 also includes documentation of patient care, nursing process, medication administration, and patient assessment. Phase 2 covers clinical experience in providing patient care to a selected group of patients in a variety of health care settings including specialty areas. Clinical practice is done under the supervision of a registered nurse. Methodology includes lectures, demonstrations, and clinical experiences. Graduates are eligible for the National Council Licensure Examination for Practical Nurses, and the program is approved by the Board of Vocational Nurses for the state of Texas.

Credit Recommendation: *Version 2:* In the lower-division baccalaureate/associate degree category, Phase 1, 1 semester hour in medical emergencies/casualties, 2 in dosage calculations and pharmacology, 5 in anatomy and physiology, 2 in microbiology, 2 in nutrition, and 4 in basic nursing care; for Phase 2, 14 semester hours in clinical practicum (3/90). *Version 3:* In the lower-division baccalaureate/associate degree category, for Phase 1, 1 semester hour in first aid/CPR, 1 in dosage calculation, 2 in anatomy and physiology, and 2 in vocational nursing history and legal issues; for Phase 2, 3 semester hours in anatomy and physiology, 4 in fundamentals of nursing, 1 in nursing process, 2 in pharmacology, 4 in medical-surgical nursing theory, 2 in pediatric nursing, 2 in obstetric/newborn nursing theory, 1 in mental health nursing, and 10 in clinical practicum (5/01).

Related Occupations: 91C.

AR-0703-0016

2. Critical Care Nursing
3. Critical Care/Emergency Nursing

Course Number: ***Version 2:*** 6F-F5. *Version 3:* 6F-F5.

Location: ***Version 2:*** Academy of Health Sciences, Fort Sam Houston, TX. *Version 3:* Brooke Medical Center, Fort Sam Houston, TX; Walter Reed Medical Center, Washington, DC; Madigan Medical Center, Tacoma, WA; Academy of Health Sciences, Fort Sam Houston, TX.

Length: ***Version 2:*** 16 weeks (588–594 hours). *Version 3:* 16 weeks (620 hours).

Exhibit Dates: ***Version 2:*** 1/88–6/97. *Version 3:* 7/97–10/97.

Learning Outcomes: ***Version 2:*** Upon completion of the course, the student will be able to amplify the knowledge introduced in a basic baccalaureate program in nursing and demonstrate the knowledge and skills necessary for caring for a critically ill patient. *Version 3:* After 10/97 see AR-0703-0031. Upon completion of the course, the student will be able to amplify the knowledge introduced in a nursing program and demonstrate knowledge and skills in caring for a patient who is critically ill in an emergency room setting. The critical care nurse will care for patients across the life span in a critical care unit. The emergency care nurse will care for patients across the life span in an emergency setting. Both will learn the responsibilities, nursing care, teaching role, and principles and techniques of staff supervision and management in related setting.

Instruction: ***Version 2:*** Course includes theory as well as practical application in caring for patients with medical/surgical, neonatal, and surgical/trauma conditions. *Version 3:* Course includes theory as well as practical application in caring for patients with medical/surgical, neonatal, and surgical trauma conditions.

Credit Recommendation: ***Version 2:*** In the graduate degree category, 16 semester hours in critical care nursing (theory) for the course; 15 semester hours in clinical or practicum for critical care nursing for the practicum (7/93). *Version 3:* In the upper-division baccalaureate category, for critical care track, 9 semester hours in critical care and 4 in clinical. For emergency nursing track, 9 semester hours in emergency nursing and 4 in clinical (7/99).

AR-0703-0017

1. Renal Dialysis Nurse Education, Phase 2 and 3
2. Renal Dialysis Nurse Education

Course Number: ***Version 1:*** 6F-F8. *Version 2:* 6F-F8.

Location: ***Version 1:*** Academy of Health Sciences, Fort Sam Houston, TX. *Version 2:* Walter Reed Medical Center, Washington, DC.

Length: ***Version 1:*** Phase 2, 6 weeks (197 hours); Phase 3, 12 weeks (433 hours). *Version 2:* 6 weeks (254 hours).

Exhibit Dates: ***Version 1:*** 12/87–2/93. *Version 2:* 3/93–Present.

Learning Outcomes: ***Version 1:*** Upon completion of the course, the student will be able to implement the nursing process in caring for patients undergoing renal replacement therapy and plan an educational program for paraprofessional personnel in the dialysis unit. *Version 2:* Upon completion of the course, the student will be able to implement the nursing process in caring for patients undergoing peritoneal dialysis, renal dialysis, or renal transplants; implement patient teaching plans; and supervise paraprofessionals in patient care.

Instruction: ***Version 1:*** Course includes guest speakers, self-paced instruction, care-planning, patient teaching plans, and videotapes. Clinical experience is provided in peritoneal and renal dialysis as well as renal transplants. *Version 2:* Lectures, guest speakers, and self-paced instruction cover the care for patients undergoing renal dialysis. The course includes clinical experience in peritoneal and renal dialysis. Care of the renal transplant patient is included. Course includes a self-study component on nephrology nursing.

Credit Recommendation: ***Version 1:*** In the graduate degree category, for Phase 2, 4 semester hours in practicum (Phase 1 must be completed prior to Phase 2); for Phase 3, 8 in practicum for (Phase 1 and 2 must be completed prior to Phase 3) (4/88). *Version 2:* In the graduate degree category, 3 semester hours in fundamentals of nephrology nursing, 5 in hemodialysis nursing theory, and 3 in clinical renal dialysis/transplant (10/95).

AR-0703-0018

Pediatric Nursing

Course Number: ***Version 1:*** 6F-66D. *Version 2:* 6F-66D.

Location: ***Version 1:*** Fitzsimmons Medical Center, Aurora, CO. *Version 2:* Tripler Medical Center, Honolulu, HI.

Length: ***Version 1:*** 16 weeks (575–582 hours). *Version 2:* 16 weeks (609–612 hours).

Exhibit Dates: ***Version 1:*** 1/88–5/93. *Version 2:* 6/93–Present.

Learning Outcomes: ***Version 1:*** Upon completion of the course, the student will be able to build upon the knowledge acquired in the area of pediatrics for the baccalaureate graduate. *Version 2:* Upon completion of the course, the student will be able to discuss legal/ethical issues of pediatric nursing; discuss growth and development from birth to adolescence; demonstrate pediatric advanced life support; demonstrate pediatric patient assessment; and demonstrate care of the neonatal and pediatric patient in a variety of clinical settings.

Instruction: ***Version 1:*** Course includes lectures and demonstrations as well as practice of additional skills in the practitioner role. Emphasis is on the nursing process as it relates to patient care and family interventions. *Version 2:* Course covers pediatric growth and development, pediatric life support, nursing process, neonatal care, pediatric care, and clinical experiences. Teaching methodologies include lectures, demonstrations, and directed clinical experiences.

Credit Recommendation: ***Version 1:*** In the graduate degree category, 15 semester hours in pediatric nursing for the course; 15 semester hours in clinical practicum for the practicum (4/88). *Version 2:* In the graduate degree category, 2 semester hours in pediatric assessment, 1 in pediatric advanced life support, 5 in clinical practicum, 1 in pediatric growth and development, 3 in neonatal nursing, and 4 in pediatric nursing (8/95).

AR-0703-0019

Operating Room Nursing for Army Nurse Corps Officer (Reserve Component)

Course Number: 6F-66E(RC).

Location: Academy of Health Sciences, Fort Sam Houston, TX; Fitzsimmons Medical Center, Aurora, CO.

Length: Phase I, 8 weeks (278 hours); Phase 2, 6 weeks (224 hours).

Exhibit Dates: 1/88–4/04.

Learning Outcomes: Upon completion of the course, the student will function as entry-level practitioner in the surgical suite.

Instruction: Course includes a review of principles of anatomy, physiology, and microbiology. Clinical practicum is the major component and includes conferences, seminars, and examinations where appropriate.

Credit Recommendation: In the graduate degree category, for Phase 1, 2 semester hours in nursing theory; for Phase 2, 6 semester hours in practicum. NOTE: Must have completed Phase 1 prior to Phase 2 (8/93).

AR-0703-0020

Nurse Practitioner Adult Medical-Surgical Health Care

Course Number: 6F-66H.

Location: Academy of Health Sciences, Fort Sam Houston, TX.

Length: 22 weeks (746 hours).

Exhibit Dates: 1/88–1/93.

Learning Outcomes: Upon completion of the course, the student will serve as a primary health care provider to adults with selected medical conditions, refer patients to other members of the health team to ensure comprehensive and continuous care of the individual, compare and contrast the role of the nurse practitioner in civilian and military environments, and demonstrate knowledge of the law pertaining to the practice of nursing.

Instruction: Teaching methodology includes lectures, conferences, demonstrations, and clinical practice. A review of anatomy, physiology, and pathology is used as a base for preparation of the development of such skills as completing a health history and techniques for screening of various normal/abnormal physical and psychosocial conditions. Clinical experience is provided in acute and chronic illnesses, with delivery of primary health care under the supervision of faculty or assigned medical doctor. The course also includes communication skills, the interview process, problem-oriented medical record, prescription writing, and leadership training.

Credit Recommendation: In the graduate degree category, 10 semester hours in medical-surgical health care theory and 12 in medical-surgical health care clinical practicum (4/88).

AR-0703-0021

1. Army Medical Department Officer Clinical Head Nurse

2. Army Medical Department (AMEDD) Head Nurse Leader Development
(Army Medical Department (AMEDD) Clinical Head Nurse)

Course Number: ***Version 1:*** 6F-F3. *Version 2:* 6F-F3.
Location: Academy of Health Sciences, Fort Sam Houston, TX.
Length: ***Version 1:*** 2 weeks (74–88 hours). *Version 2:* 2 weeks (88–90 hours).
Exhibit Dates: ***Version 1:*** 6/88–8/94. *Version 2:* 9/94–Present.
Learning Outcomes: ***Version 1:*** Upon completion of the course, the student will be able to function as a clinical nurse. *Version 2:* Upon completion of the course, the student will be able to function in a leadership position, with emphases on management of personnel and methods of evaluation.
Instruction: ***Version 1:*** Instruction includes lectures, discussions, and panel participation in managerial principles, communication issues, risk management, the law as it pertains to the practice of nursing, and use of standards of nursing care in developing criteria for monitoring and evaluating activities. *Version 2:* In order to accomplish the outcome, the officer will develop a management plan including current assignment policies; apply theories of adult learning; implement a plan for the decision making process; develop a budget for the areas of personnel, supplies, equipment, and marketing; and apply the law as it pertains to the practice of nursing.
Credit Recommendation: ***Version 1:*** In the upper-division baccalaureate category, 3 semester hours in nursing administration (9/93). *Version 2:* In the upper-division baccalaureate category, 4 semester hours in nursing administration seminar (9/00).

AR-0703-0022

1. Principles of Advanced Nursing Administration for Army Nursing Corps Officers
2. Army Medical Department (AMEDD) Advanced Nurse Leadership
(Principles of Advanced Nursing Administration)

Course Number: ***Version 1:*** 6F-F2. *Version 2:* 6F-F2.
Location: Academy of Health Sciences, Fort Sam Houston, TX.
Length: ***Version 1:*** 2 weeks (74 hours). *Version 2:* 2 weeks (87–89 hours).
Exhibit Dates: ***Version 1:*** 6/88–4/94. *Version 2:* 5/94–Present.
Learning Outcomes: ***Version 1:*** Upon completion of the course, the student will be able to apply management concepts and educational activities in order to function effectively as a charge nurse in a medical facility. *Version 2:* Upon completion of the course, the student will be able to discuss major trends in health care reform, apply principles of financial and personnel management, apply principles of decision making and conflict resolution, and apply principles of organizational management to the health care delivery system.
Instruction: ***Version 1:*** Methods of instruction include lectures, discussions, and role playing on current trends in resource management, health care economics, ethics, and various styles of leadership. *Version 2:* Lectures and participatory learning experiences cover leadership development for nurses in advanced management positions. Course includes personnel and resource management, organizational theories and management, leadership, decision making, and conflict resolution.
Credit Recommendation: ***Version 1:*** In the upper-division baccalaureate category, 2 semester hours in nursing (11/90). *Version 2:* In the graduate degree category, 1 semester hour in current issues/health care reform, 2 in introduction to organizational theories and management, and 3 in leadership/decision making (9/00).

AR-0703-0023

1. Obstetrical and Gynecological Nursing
2. Obstetrical and Gynecological Nursing
3. Obstetrical/Gynecological Nursing

Course Number: ***Version 1:*** 6F-66G. *Version 2:* 6F-66G. *Version 3:* 6F-66G.
Location: Tripler Medical Center, Honolulu, HI; Academy of Health Sciences, Fort Sam Houston, TX.
Length: ***Version 1:*** 16 weeks (472–566 hours). *Version 2:* 16 weeks (582 hours). *Version 3:* 16 weeks (594–616 hours).
Exhibit Dates: ***Version 1:*** 10/88–5/93. *Version 2:* 6/93–4/96. *Version 3:* 5/96–Present.
Learning Outcomes: ***Version 1:*** Upon completion of the course, the student will be able to assume responsibility for the primary health care for women during childbearing and interconceptive periods. Such competency will reflect concern for the physical and psychosocial well-being of the childbearing woman and her family. *Version 2:* Upon completion of the course, the student will be able to discuss legal and ethical issues in OB/GYN; demonstrate knowledge of anatomy and physiology, high-risk factors, and complications of the childbearing woman and the newborn; conduct a prenatal examination including history; perform breast examination; implement a childbirth course; perform duties of a circulating nurse in an operating room; conduct a well-woman gynecological examination; perform physical assessment and CPR on the newborn; and develop nursing care plans. *Version 3:* Upon completion of the course, the student will be able to discuss legal and ethical issues in OB/GYN, risks and complications of pregnancy, gynecological disorders and emergencies, and family planning and dysfunctions; perform prenatal, postnatal, and well-women breast and gynecological exams; implement child birth courses and assist in delivery; perform newborn assessment; and develop nursing care plans.
Instruction: ***Version 1:*** Lectures, discussions, examinations, and clinical demonstrations cover assessment and care of the antepartum and postpartum patient, the bio-psycho-social aspects of childbearing, family planning, and patient/family counseling through the use of the nursing process. *Version 2:* Course includes legal and ethical issues of OB/GYN nursing, anatomy and physiology of the reproductive system, childbearing and delivery, lactation, fetal monitoring, nutrition, high-risk conditions, OB/GYN emergencies, gynecological problems, physical assessment, and clinical experience in inpatient and outpatient settings. Methodologies include lectures, conferences, discussions, practical exercises, and clinical practicum. *Version 3:* Course includes legal and ethical issues of OB/GYN nursing, anatomy and physiology of the reproductive system, childbearing and delivery, lactation, fetal monitoring, nutrition, high-risk conditions, OB/GYN emergencies, gynecological problems, physical assessment, and clinical experience in inpatient and outpatient settings. Methodologies include lectures, conferences, discussions, practical exercises, and clinical practicum.
Credit Recommendation: ***Version 1:*** In the graduate degree category, 2 semester hours in physical assessment, 2 in family planning, 2 in health education and counseling, 5 in normal obstetrical nursing, 1 in newborn care, 2 in high-risk pregnancy, and 9 in clinical practicum (3/90). *Version 2:* In the graduate degree category, 2 semester hours in patient education, 5 in normal obstetrical nursing, 2 in gynecological nursing and family planning, 9 in clinical practicum (OB-GYN), 2 in physical assessment, 1 in neonatal nursing, 2 in high-risk pregnancy (8/95). *Version 3:* In the upper-division baccalaureate category, 10 semester hours in clinical practice. In the graduate degree category, 3 semester hours in patient education, 2 in high-risk pregnancy, 2 in obstetrical nursing, 2 gynecological nursing, and 2 in neonatal nursing (9/00).

AR-0703-0024

Principles of Military Preventive Medicine (MC/VC/DC)

Course Number: 6A-F5.
Location: Academy of Health Sciences, Fort Sam Houston, TX.
Length: 9 weeks (365 hours).
Exhibit Dates: 10/88–5/90.
Learning Outcomes: Upon completion of the course, students will be able to serve as preventive medicine officers, public health veterinarians, or preventive dentists.
Instruction: This course includes lectures, discussions, and clinical practice in preventive medicine, environmental hygiene, infection control, epidemiology, public health, water quality, and communicable diseases.
Credit Recommendation: In the upper-division baccalaureate category, 3 semester hours in preventive medicine, 2 in environmental hygiene, 1 in infection control, 1 in epidemiology, 1 in public health, and 1 in water quality. In the graduate degree category, 2 semester hours in communicable diseases (3/90).

AR-0703-0025

Special Forces Medical Sergeant
(Special Forces Medical Sergeant, Phases 1 and 3)

Course Number: ***Version 1:*** 011-18D30. *Version 2:* 011-18D30.
Location: John F. Kennedy Special Warfare Center and School, Fort Bragg, NC.
Length: ***Version 1:*** 24 weeks (1559 hours). *Version 2:* 24 weeks (1782 hours).
Exhibit Dates: ***Version 1:*** 10/89–9/91. *Version 2:* 10/91–6/96.
Learning Outcomes: ***Version 1:*** Upon completion of the course, the student will be able to perform advanced cardiac life support, basic radiology skills, general anesthesia, endotracheal intubation, and basic laboratory procedures; manage trauma; and perform basic veterinary examinations and treatment. *Version 2:* After 6/96 see AR-0703-0032. Upon completion of the course, the student will be able to perform advanced cardiac life support, apply

basic radiology skills, perform general anesthesia and endotracheal intubation, perform basic laboratory procedures, manage trauma, and perform basic veterinary examinations and treatment.

Instruction: ***Version 1:*** Lectures, laboratories and demonstrations cover advanced cardiac life support, radiology, trauma management, anesthesia, central materials supply, laboratory examinations, veterinary procedures, and records/reports. Course includes common core of special operations and leadership subjects, including air operations, survival, land navigation, combat, special operations, and leadership. *Version 2:* Lectures, laboratories, and demonstrations cover advanced cardiac life support, radiology, trauma management, anesthesia, central materials supply, laboratory examination, veterinary procedures, anatomy, pharmacology, and records/reports. Course includes common core of special operations and leadership subjects, including air operations, survival, land navigation, and combat skills.

Credit Recommendation: ***Version 1:*** In the lower-division baccalaureate/associate degree category, 1 semester hour in advanced cardiac life support, 1 in applied basic radiological science, 5 in emergency medical technology or nursing, 2 in applied veterinary science, 2 in medical laboratory technology (12/92). *Version 2:* In the lower-division baccalaureate/associate degree category, 2 semester hours in advanced cardiac life support, 1 in applied basic radiological science, 6 in emergency medical technology or nursing, 3 in applied veterinary science, and 2 in medical laboratory technology (12/92).

Related Occupations: 18D.

AR-0703-0026

1. Combat Casualty Care
2. Combat Casualty Care (C4)

Course Number: ***Version 1:*** 6AC4. *Version 2:* 6A-C4.

Location: Academy of Health Sciences, Fort Sam Houston, TX.

Length: ***Version 1:*** 1–2 weeks (77–81 hours). *Version 2:* 2 weeks (143 hours).

Exhibit Dates: ***Version 1:*** 1/86–9/01. *Version 2:* 10/01–Present.

Learning Outcomes: ***Version 1:*** Upon completion of the course, the student will be able to assess and manage the trauma victim; demonstrate patient assessment; establish patient management priorities; identify medical support and logistics required for various military operations; provide leadership and identify medical implications for casualty treatment in nuclear, biological, and chemical agent environments; establish treatment priorities for trauma victims based on current military operations; and evacuate trauma victims in accordance with standing protocols. *Version 2:* Upon completion of the course, the student will be able to assess trauma victims utilizing triage methods; medically manage trauma victims; identify logistic support necessary to sustain medical operations; provide effective medical care in a nuclear, biological, and chemical environment; evacuate trauma victims in accordance with standing protocols; and manage mass casualty operation.

Instruction: ***Version 1:*** Covers advanced trauma life support; patient assessment; triage; tri-service medical capabilities; nuclear, biological, and chemical warfare; and environmental injuries. Methodologies include lectures, demonstrations, and field exercises. *Version 2:* Practical and classroom exercises, discussion, and lecture. Topics include advanced trauma life support training; treatment of trauma victims in a contaminated environment; logistics necessary to implement and/or sustain medical support; triage; nuclear, biological, and chemical environments; preventive medicine; medical evacuation policies and procedures, and humanitarian assistance.

Credit Recommendation: ***Version 1:*** In the graduate degree category, 2 semester hours in advanced trauma life support (12/99). *Version 2:* In the graduate degree category, 4 semester hours in emergency health response and 2 in emergency preparedness (11/03).

AR-0703-0027

Clinical Laboratory Officer

Course Number: 6H-71E67; 6H-68F.

Location: Academy of Health Sciences, Fort Sam Houston, TX.

Length: 50–53 weeks (1890–1914 hours).

Exhibit Dates: 2/92–Present.

Learning Outcomes: Upon completion of the course, the student will be able to supervise a clinical laboratory, monitor physical resources, plan and monitor budgets, manage personnel, supervise evaluations, and manage medical laboratory clinical services.

Instruction: Course covers management, microbiology, immunology, blood banking, urinalysis, clinical chemistry, hematology, coagulation, phlebotomy, and anatomical pathology. Methodology includes lectures, laboratory exercises, discussions, and clinical practice.

Credit Recommendation: In the upper-division baccalaureate category, 10 semester hours in microbiology, 3 in management, 6 in immunology, 8 in blood banking, 2 in urinalysis, 7 in clinical chemistry, 8 in hematology, and 2 in phlebotomy (7/95).

AR-0703-0028

Perioperative Nursing

Course Number: ***Version 1:*** 6F-66E. *Version 2:* 6F-66E. *Version 3:* 6F-66E.

Location: Academy of Health Sciences, Fort Sam Houston, TX; Brooke Medical Center, Fort Sam Houston, TX; William Beaumont Medical Center, El Paso, TX; Madigan Medical Center, Tacoma, WA.

Length: ***Version 1:*** 16 weeks (588 hours). *Version 2:* 16 weeks (586–601 hours). *Version 3:* 16 weeks (605 hours).

Exhibit Dates: ***Version 1:*** 5/91–6/93. *Version 2:* 7/93–6/98. *Version 3:* 7/98–Present.

Learning Outcomes: ***Version 1:*** Upon completion of the course, the student will be able to discuss sterilization and disinfection, process surgical supplies for sterilization, develop and implement patient care plans using the nursing process, perform the duties of an operating room nurse in a field medical facility, identify potential safety hazards in the operating room, and coordinate members of the operating room team. *Version 2:* Upon completion of the course, the student will be able to discuss sterilization and disinfection, process surgical supplies for sterilization, develop and implement patient care plans using the nursing process, perform the duties of an operating room nurse in a field medical facility, identify environmental hazards in the operating room, and coordinate members of the operating room team. *Version 3:* Upon completion of the course, the student will be able to discuss sterilization and disinfection, process surgical supplies for sterilization, develop and implement patient care plans using the nursing process, perform the duties of an operating room nurse in a field medical facility, identify environmental hazards in the operating room, and coordinate members of the operating room team.

Instruction: ***Version 1:*** Course covers microbiology, sterilization, surgical supplies, perioperative nursing, environmental hazards, specialized equipment, specialty surgeries, and field nursing. *Version 2:* Course covers microbiology, sterilization, surgical supplies, perioperative nursing, environmental hazards, specialized equipment, specialty surgeries, and field nursing. Methods of instruction include lectures, practical exercises, and clinical experience. *Version 3:* Course covers microbiology, sterilization, surgical supplies, perioperative nursing, environmental hazards, specialized equipment, specialty surgeries, and field nursing. Methods of instruction include lectures, practical exercises, and clinical experience.

Credit Recommendation: ***Version 1:*** In the graduate degree category, 6 semester hours in clinical practicum, 5 in perioperative nursing, and 3 in specialty perioperative nursing (8/93). *Version 2:* In the graduate degree category, 4 semester hours in clinical practicum, 5 in perioperative nursing, and 3 in specialty perioperative nursing (8/96). *Version 3:* In the graduate degree category, 9 semester hours in clinical practicum, 5 in perioperative nursing, and 3 in specialty perioperative nursing (10/03).

AR-0703-0029

Army Medical Department (AMEDD) Noncommissioned Officer (NCO) Basic

Course Number: 6-8-C40 (91C30).

Location: Academy of Health Sciences, Fort Sam Houston, TX.

Length: 7 weeks (300 hours).

Exhibit Dates: 4/95–Present.

Learning Outcomes: Upon completion of the course, the student will be able to lead a military unit; evaluate performance and counsel; supervise and delegate; and perform drills, training, and specific military skills.

Instruction: The course covers military leadership, personal and performance counseling, duties and responsibilities of noncommissioned officers, leadership assessment and development, leadership procedures and combat orders, combat leadership, physical training and marksmanship, property accountability, individual and team training, small-group process, and specific military skills. The methodologies include lectures, discussions, group work, and practical application. This course contains a common core of leadership subjects.

Credit Recommendation: In the lower-division baccalaureate/associate degree category, 3 semester hours in military science and 1 semester hour of elective credit in practical nursing. See AR-1406-0090 for common core credit (11/97).

Related Occupations: 91C.

AR-0703-0030

Emergency Nursing

Course Number: 6F-F6.

Location: Brooke Medical Center, Fort Sam Houston, TX; Academy of Health Sciences, Fort Sam Houston, TX.

Length: 16 weeks (609–620 hours).

Exhibit Dates: 11/97–Present.

Learning Outcomes: Before 11/97 see AR-0703-0016. Upon completion of the course, the student amplify the knowledge introduced in a nursing program and demonstrate knowledge and skills in caring for a patient in an emergency setting.

Instruction: Course includes theory as well as practical application in caring for patients with medical/surgical, neonatal, and surgical/trauma conditions.

Credit Recommendation: In the graduate degree category, 18 semester hours in critical care and 8 in clinical practice (9/00).

AR-0703-0031

Critical Care Nursing

Course Number: 6F-F5.

Location: Walter Reed Medical Center, Washington, DC; Madigan Medical Center, Tacoma, WA; Brooke Medical Center, Fort Sam Houston, TX; Academy of Health Sciences, Fort Sam Houston, TX.

Length: 16 weeks (603 hours).

Exhibit Dates: 11/97–Present.

Learning Outcomes: Before 11/97 see AR-0703-0016. Upon completion of the course, the student will amplify the knowledge introduced in a nursing program and demonstrate knowledge and skills in caring for a patient in a critical care setting. Students will be prepared to care for patients across the life span in a critical care unit, focusing on responsibilities, nursing care, teaching role, principles and techniques of staff supervision, and management of patients.

Instruction: Theory and practical application in caring for patients in a critical care situation. Topics include respiratory; basic and advanced cardiovascular; acute renal failure; hepatic disorders; hematology/oncology; neurological and endocrine; trauma management; pediatrics; OB/GYN and genitourinary; eye, ear, nose, and throat; musculoskeletal; and core emergency care.

Credit Recommendation: In the graduate degree category, 18 semester hours in critical care and 8 in clinical practice (9/00).

AR-0703-0032

Special Forces Medical Sergeant, Phase 2

Course Number: 011-18D30.

Location: John F. Kennedy Special Warfare Center and School, Fort Bragg, NC.

Length: 44 weeks (2242–2243 hours).

Exhibit Dates: 7/96–Present.

Learning Outcomes: Before 7/96 see AR-0703-0025. Upon completion of the course, the student will be able to perform emergency medical technician procedures, critical care procedures, trauma management, basic Ob-gyn procedures, basic veterinary examination and treatment, and general anesthesia procedures; apply radiology skills; and apply basic precontamination and contamination procedures in the biological and chemical areas.

Instruction: Lectures, laboratories, and demonstrations cover advanced cardiac life support, radiology, trauma management, anesthesia, central materials supply, laboratory examination, veterinary procedures, anatomy, pharmacology, and records/reports. Course includes common core of special operations and leadership subjects, including air operations, survival, land navigation, and combat skills.

Credit Recommendation: In the lower-division baccalaureate/associate degree category, 4 semester hours in anatomy and physiology, 3 in critical care, 3 in critical care practicum, 3 in emergency medicine, 3 in emergency medicine practicum, 3 in military science, 3 in basic veterinary science, and 2 in introduction to medical laboratory technology (4/01).

Related Occupations: 18D.

AR-0703-0034

Special Forces Medical Sergeant, Phase 3

Course Number: 011-18D30 (2), Phase 3.

Location: John F. Kennedy Special Warfare Center and School, Fort Bragg, NC.

Length: 6 weeks (590 hours).

Exhibit Dates: 7/96–Present.

Learning Outcomes: Upon completion of the course, the student will be able to conduct air operations; conduct special operations; plan a mission planning exercise; employ small unit tactics, techniques, and battle drills; and plan for an unconventional warfare mission.

Instruction: Lecture, discussion, demonstration, and practical experience, cover learning and applying military operations in the field.

Credit Recommendation: In the lower-division baccalaureate/associate degree category, 2 semester hours in military science (4/01).

Related Occupations: 18D.

AR-0703-0035

Special Forces Medical Sergeant, Phase 1

Course Number: 011-18D30, Phase 1.

Location: John F. Kennedy Special Warfare Center and School, Fort Bragg, NC.

Length: 6–8 weeks.

Exhibit Dates: 4/00–Present.

Learning Outcomes: Upon completion of the course, the student will successfully perform land navigation techniques, utilize small unit tactics, and apply theories of leadership in a variety of circumstances.

Instruction: Lecture, discussion, demonstration, and practical exercises are used to present and practice military science information in land navigation, small unit tactics, and leadership skills.

Credit Recommendation: In the lower-division baccalaureate/associate degree category, 3 semester hours in military science (4/01).

Related Occupations: 18D.

AR-0703-0036

1. Practical Nurse
2. Licensed Practical Nurse

Course Number: ***Version 1:*** 300-M6, Phase 2. *Version 2:* 300-M6, Phase 2.

Location: ***Version 1:*** Brooke Medical Center, Fort Sam Houston, TX; Dwight David Eisenhower Medical Center, Fort Gordon, GA; Walter Reed Medical Center, Washington, DC; Madigan Medical Center, Tacoma, WA; Academy of Health Sciences, Fort Sam Houston, TX. *Version 2:* Academy of Health Sciences, Fort Sam Houston, TX.

Length: ***Version 1:*** 46 weeks (1704 hours). *Version 2:* 38 weeks (1371 hours).

Exhibit Dates: ***Version 1:*** 1/02–10/03. *Version 2:* 11/03–Present.

Learning Outcomes: ***Version 1:*** Before 1/02 see AR-0703-0015. Upon completion of the course the student will be able to discuss basic communication skills; identify transcultural factors in nursing; identify principles of infection control; demonstrate basic nursing skills, demonstrate medication administration skills, identify nursing care for the pre/post-operative patient; develop nursing care plans using the nursing process; demonstrate theoretical knowledge of common diseases and their nursing implications, including gerontology, pediatric, and obstetrical conditions, and in the clinical portion, administer nursing care to a selected group of patients in a variety of health care settings under the supervision of a registered nurse. *Version 2:* Upon completion of the course the student will be able to discuss basic communication skills; identify transcultural factors in nursing; identify principles of infection control; demonstrate basic nursing and medication administration skills, identify nursing care for the pre/post-operative patient; develop nursing care plans using the nursing process; demonstrate theoretical knowledge of common diseases and their nursing implications, including gerontology, pediatric, obstetrical, and mental health conditions; and in the clinical portion, administer nursing care to a selected group of patients in a variety of health care settings under the supervision of a registered nurse.

Instruction: ***Version 1:*** Laboratory, audio visual materials, case studies, discussion, classroom exercise, practical exercises, lecture, and learning presentations. Topics include anatomy and physiology, pharmacology, nutrition, medication administration, nursing process, clinical experience in medical surgical, pediatrics, obstetrics, gerontology, emergency room, and clinical care. *Version 2:* Laboratory, audiovisual materials, case studies, discussion, classroom exercise, practical exercises, lecture, and learning presentations. Topics include anatomy and physiology, pharmacology, medication administration, nursing process, clinical experience in medical surgical, pediatrics, obstetrics, gerontology, mental health, emergency room, and clinical care.

Credit Recommendation: ***Version 1:*** In the lower-division baccalaureate/associate degree category, 12 semester hours in a practical nurse clinical practicum, 2 in human anatomy and physiology (no laboratory), 3 in introduction to nursing science, 3 in foundations of practical nursing, 6 in medical-surgeon nursing for the practical nurse, 2 in obstetrical nursing for the practical nurse, 2 in pediatric nursing for the practical nurse, 1 in medication calculation, 3 in nursing pharmacology, 1 in microbiology (no laboratory), 3 in psychosocial aspects of health and illness, 3 in introduction to nutrition, and 1 in emergency nursing (6/02). *Version 2:* In the lower-division baccalaureate/associate degree category, 9 semester hours in LPN clinical practicum, 3 in nursing science, 3 in foundations of LPN, 6 in medical-surgeon, 2 in obstetrics, 2 in pediatrics, 3 in pharmacology, 1 in microbiology (no lab), and 1 in medical calculation (11/03).

AR-0703-0037

1. Licensed Practical Nurse
2. Practical Nurse

Course Number: *Version 1:* 300-M6, Phase 1. *Version 2:* 300-M6, Phase 1.

Location: Academy of Health Sciences, Fort Sam Houston, TX.

Length: *Version 1:* 6 weeks (218 hours). *Version 2:* 8 weeks (317 hours).

Exhibit Dates: *Version 1:* 1/02–9/03. *Version 2:* 10/03–Present.

Learning Outcomes: *Version 1:* Before 1/02 see AR-0703-0015. Upon completion of the course the student will be able to demonstrate theoretical knowledge of basic principles in microbiology, pharmacology, anatomy and physiology, and preventive medicine; calculate medical dosages; and discuss legal/ethical aspects of practical nursing. *Version 2:* Upon completion of the course the student will be able to discuss basic communication skills; identify transcultural factors in nursing; identify principles of infection control; demonstrate basic nursing skills, demonstrate medication administration skills, identify nursing care for the pre/post-operative patient; develop nursing care plans using the nursing process; demonstrate theoretical knowledge of common diseases and their nursing implications, including gerontology, pediatric, obstetrical and mental health conditions; and in the clinical portion, administer nursing care to a selected group of patients in a variety of health care settings under the supervision of a registered nurse.

Instruction: *Version 1:* Laboratory, audio visual materials, case studies, discussion, classroom exercise, practical exercise, lecture, learner presentations. Topics include basic anatomy and physiology, roles of nursing personnel, medical calculations, basic microbiology, and basic nutrition. *Version 2:* Laboratory, audio visual materials, case studies, discussion, classroom exercise, practical exercises, lecture, and learning presentations. Topics include anatomy and physiology, pharmacology, pharmacology and medication administration, nursing process, clinical experience in medical surgical, pediatrics, obstetrics, gerontology, mental health, emergency room, and clinical care.

Credit Recommendation: *Version 1:* In the lower-division baccalaureate/associate degree category, 2 semester hours in human anatomy and physiology (no lab), 1 in introduction to nursing science, 1 in medication calculations, and 1 in microbiology (no lab) (11/03). *Version 2:* In the lower-division baccalaureate/associate degree category, 3 semester hours in anatomy and physiology, 3 in introduction to nursing science, 1 in medical calculations, 1 in microbiology, and 4 in clinical internship (11/03).

AR-0703-0038

JOINT OPERATIONS MEDICAL MANAGERS

Course Number: 6A-C4A.

Location: Academy of Health Sciences, Fort Sam Houston, TX; Defense Medical Readiness Training Institute, Camp Bullis, TX.

Length: 1–2 weeks (71–77 hours).

Exhibit Dates: 9/92–Present.

Learning Outcomes: Upon completion of the course, the student will be able to function effectively in a combat zone and combat medical treatment facility, assess the wartime environment and health care delivery system, and maximize in-theater returns to duty.

Instruction: Wound management and care are taught using conferences, practical exercises, and demonstrations. The term wound is broadly interpreted and is inclusive of injury from all types of war ordinance, as well as snake bites, psychiatric disorders, dentistry, and insect injury.

Credit Recommendation: In the graduate degree category, 3 semester hours in body wounds and 1 in field exercise (1/03).

AR-0703-0039

LICENSED PRACTICAL NURSE

Course Number: 300-M6, Phase 3.

Location: Madigan Army Medical Center, Fort Lewis, WA; Walter Reed Army Medical Center, Washington, DC; William Beaumont Army Medical Center, Fort Bliss, TX.

Length: 6 weeks (238 hours).

Exhibit Dates: 11/03–Present.

Learning Outcomes: Upon completion of the course, the student will be able to initiate and manage nursing treatments, as well as emergent and supportive medical treatments for casualties and patients with injuries and /or illness in both TOE and TDA settings.

Instruction: Audiovisual materials, practical and classroom exercises, discussion, laboratory, case studies, and lecture. Topics include beginning skills to function in critical care environments, advanced anatomy and physiology, pathophysiology of critical disease processes, integration of pharmacology, and medical and nursing management/treatment modalities for critically ill patients.

Credit Recommendation: In the lower-division baccalaureate/associate degree category, 3 semester hours in advanced anatomy and physiology, 1 in pathophysiology, and 3 in critical care internship (11/03).

AR-0704-0001

PHYSICAL THERAPY SPECIALIST

Course Number: B-303-A051, Phase 1; B-303-0051, Phase 2; 303-91J10.

Location: Academy of Health Sciences, Fort Sam Houston, TX; Selected Hospitals, Continental US; Navy Hospitals, Continental US; Medical Field Service School, Fort Sam Houston, TX.

Length: Phase 1, 17 weeks (646 hours); Phase 2, 8–10 weeks (320–400 hours).

Exhibit Dates: 9/81–2/91.

Objectives: After 2/91 see AR-0704-0012. To provide qualified medical corpsmen with knowledge and skills in physical therapy techniques and to promote continuing education in the profession.

Instruction: Lectures and practical exercises cover functional anatomy and physiology, human growth and development, basic principles and clinical application of physical modalities, ethical standards, proper techniques of handling and positioning patients, therapeutic exercises, and development of patient therapy programs.

Credit Recommendation: In the lower-division baccalaureate/associate degree category, 6 semester hours in functional anatomy and physiology, 3 in applied psychology, 13 in physical therapy principles and procedures, 10 in supervised clinical experience, and 1 in disease pathology (3/91).

Related Occupations: 91A; 91J.

AR-0704-0004

OCCUPATIONAL THERAPY SPECIALIST

Course Number: 303-91L10.

Location: Academy of Health Sciences, Fort Sam Houston, TX; Selected Hospitals, Continental US.

Length: Phase 1, 17 weeks (632 hours); Phase 2, 8 weeks (360 hours).

Exhibit Dates: 6/80–2/90.

Objectives: After 1/90 see AR-0704-0011. To train enlisted personnel in the knowledge of physical and behavioral sciences and principles and concepts related to implementation of remedial, health maintenance, and preventive occupational therapy procedures.

Instruction: Lectures and practical exercises cover anatomy and physiology, occupational therapy processes, psychology, and pathological conditions as related to occupational therapy.

Credit Recommendation: In the lower-division baccalaureate/associate degree category, 4 semester hours in functional anatomy and physiology, 3 in applied psychology, 12 in occupational therapy techniques, 9 in supervised clinical experience, and 1 in disease pathology (3/90).

Related Occupations: 91L.

AR-0704-0008

RESPIRATORY SPECIALIST

Course Number: 300-91V10.

Location: State Chest Hospital, San Antonio, TX; Wilford Hall, Lackland AFB, TX; Phase 2, Academy of Health Sciences, Fort Sam Houston, TX; Phase 1, Academy of Health Sciences, Fort Sam Houston, TX.

Length: Phase 1, 16 weeks (627 hours); Phase 2, 16 weeks (550 hours).

Exhibit Dates: 3/87–1/96.

Learning Outcomes: Upon completion of the course the student will be able to conduct procedures related to aerosol/humidity therapy, ventilation therapy, pulmonary function testing, infection control, cardiopulmonary drug administration and critical patient care performed under the supervision of a physician.

Instruction: Lectures, demonstrations, and practical exercises in selected basic science courses related to respiratory care. Topics include anatomy, respiration, physiology, pharmacology and microbiology, physical therapy, clinical medicine, and medical materiel management.

Credit Recommendation: In the lower-division baccalaureate/associate degree category, for phase 1, 3 semester hours in anatomy and physiology, 3 in general science, 3 in cardiopulmonary physiology, 3 in clinical medicine and 9 in respiratory care techniques; for Phase 2, 9 semester hours in clinical experience (4/88).

AR-0704-0009

ORTHOPEDIC SPECIALIST
(Orthopedic Specialist)

Course Number: *Version 1:* 304-91H10. *Version 2:* 304-P1; 304-91H10.

Location: Academy of Health Sciences, Fort Sam Houston, TX.

Length: *Version 1:* Phase 1, 6 weeks (209 hours); Phase 2, 6 weeks (232 hours). *Version 2:* Phase 1, 6 weeks (227 hours); Phase 2, 6 weeks (232 hours).

Exhibit Dates: *Version 1:* 6/85–2/94. *Version 2:* 3/94–10/00.

Learning Outcomes: *Version 1:* Upon completion of the course, the student will be able to prepare patients for major surgical procedures, assist the physician in minor surgical procedures, instruct patients regarding care, maintain mechanical equipment in an orthopedic clinic, and maintain appropriate medical records. *Version 2:* Upon completion of the course, the student will be able to prepare patients for surgical procedures; assist the physician in minor surgical procedures; apply splints, casts, and traction; instruct patients regarding care; maintain mechanical equipment in an orthopedic clinic; and maintain appropriate medical records.

Instruction: *Version 1:* Lectures and clinical experiences cover anatomy and physiology and the principles of orthopedic surgery, including the preparation of patients for major surgical procedures and assisting the physician in minor surgical procedures; care and treatment of orthopedic patients, including giving proper instruction to patients for care of casts and splints, proper crutch walking techniques, and precautions; fabricating, modifying, and applying plaster casts; applying casts and splints used in surgery; performing dressing changes, sterile scrubs, and suture removal; and observing the principles of asepsis. *Version 2:* Phase 1 includes lectures, discussions, demonstrations, and laboratory practice in musculo-skeletal anatomy and physiology; casting, splinting, traction, and wound care; and related equipment. Phase 2 includes supervised clinical practice at designated hospitals.

Credit Recommendation: *Version 1:* In the lower-division baccalaureate/associate degree category, for Phase 1, 3 semester hours in anatomy, 1 in physiology, 4 in principles of orthopedics and patient care, 4 in orthopedic methods and materials; for Phase 2, 5 semester hours in practicum or clinical application of principles of orthopedics (4/88). *Version 2:* In the lower-division baccalaureate/associate degree category, for Phase 1, 2 semester hours in musculo-skeletal anatomy and physiology, 4 in orthopedics and patient care, and 4 in orthopedic devices, materials, and methods; for Phase 2, 5 semester hours in clinical practice in orthopedics (12/96).

Related Occupations: 91H.

AR-0704-0010

Orthotic Specialist

Course Number: 304-42C10.

Location: Academy of Health Sciences, Fort Sam Houston, TX.

Length: Phase 1, 20 weeks (765 hours); Phase 2, 32 weeks (1272 hours).

Exhibit Dates: 7/88–Present.

Learning Outcomes: Upon completion of the course, the student will be able to take impressions using appropriate techniques and media and measure, design, fabricate and fit orthotics in accordance with prescriptions.

Instruction: Lectures, discussions, demonstrations, conferences, shop, and clinical practice cover the design, manufacture, fitting, and repair of orthotic devices. Course includes medical terminology, ethics, materials, and tools. This course includes extensive instruction in orthopedic anatomy, physiology, and biomechanics.

Credit Recommendation: In the lower-division baccalaureate/associate degree category, for Phase 1, 3 semester hours in anatomy, 2 in physiology, 3 in kinesiology, 8 in orthotics, 3 in orthotic shop techniques, and 10 in clinical practice; for Phase 2, 16 semester hours in clinical practice (3/90).

Related Occupations: 42C.

AR-0704-0011

1. Occupational Therapy Specialist
2. Occupational Therapy Specialist
3. Occupational Therapy Specialty

Course Number: *Version 1:* 303-91L10. *Version 2:* 303-N3; 303-91L10. *Version 3:* 303-N3; B-303-0052, Phase 1; B-303-0053, Phase 2.

Location: *Version 1:* Academy of Health Sciences, Fort Sam Houston, TX. *Version 2:* Air Force Personnel, Wright-Patterson AFB, OH; Phase 1, Academy of Health Sciences, Fort Sam Houston, TX; Phase 2, for Army, Academy of Health Sciences, Fort Sam Houston, TX; Phase 2, for Air Force, Wilford Hall, Lackland AFB, TX. *Version 3:* Phase 1, Academy of Health Sciences, Fort Sam Houston, TX; Phase 2, for Army, Multiple locations; Phase 2, for Navy, Naval Regional Medical Center, Portsmouth, VA; Phase 2, for Air Force, Wilford Hall, Lackland AFB, TX.

Length: *Version 1:* 25 weeks (994 hours). *Version 2:* Phase 1, 17 weeks (741 hours); Phase 2, 12 weeks (480 hours). *Version 3:* Phase 1, 17 weeks (700–701 hours); Phase 2, 12 weeks (480 hours).

Exhibit Dates: *Version 1:* 3/90–10/93. *Version 2:* 11/93–1/96. *Version 3:* 2/96–Present.

Learning Outcomes: *Version 1:* Before 3/90 see AR-0704-0004. Upon completion of the course, the student will be able to select, collect, record, and communicate data that contributes to an occupational therapy treatment program; establish, implement, and modify an occupational therapy treatment plan; maintain effective interpersonal relationships with patients; and manage time, materials, and the treatment environment. *Version 2:* Upon completion of the course, the student will be able to select, collect, record, and communicate data that contributes to an occupational therapy treatment programs; implement an occupational therapy treatment plan; administer occupational therapy treatments to patients; maintain effective interpersonal relationships with patients; and manage time, materials, and the treatment environment. *Version 3:* Before 2/90 see AR-0704-0004. Upon completion of the course, the student will be able to select, collect, record, and communicate data that contributes to an occupational therapy treatment program; implement an occupational therapy treatment plan; administer occupational therapy treatments to patients; maintain effective interpersonal relationships with patients; and manage time, materials, and the treatment environment.

Instruction: *Version 1:* Course covers human anatomy, physiology, and kinesiology; occupational therapy processes; occupational therapy for psychiatric and physical dysfunction; therapeutic application, analysis, and adaptation of activities; and therapeutic recreation and reconditioning sports. Methodology includes lectures, discussions, demonstrations, examinations, field observations, and field experience in mental health and physical dysfunction. *Version 2:* Phase 1 covers human anatomy, physiology, and kinesiology; occupational therapy processes; occupational therapy for psychiatric and physical dysfunction; therapeutic application, analysis, and adaptation of activities; and therapeutic recreation and reconditioning sports. Methodology includes lectures, discussions, demonstrations, examinations, field observations, and field experience in mental health and physical dysfunction. Phase 2 includes supervised clinical practice in active occupational therapy clinical departments. *Version 3:* Phase 1 covers human anatomy, physiology, and kinesiology; occupational therapy processes; occupational therapy for psychiatric and physical dysfunction; therapeutic application, analysis, and adaptation of activities; and therapeutic recreation and reconditioning sports. Methodology includes lectures, discussions, demonstrations, examinations, field observations, and field experience in mental health and physical dysfunction. Phase 2 includes supervised clinical practice in active occupational therapy clinical departments. Note that for the Army, Phase 2 is conducted at the following locations: Bayne-Jones Hospital, Brooke Medical Center, Eisenhower Medical Center, General Leonard Wood Hospital, Irwin Hospital, Walter Reed Medical Center, William Beaumont Medical Center, Blanchfield Hospital, Darnall Hospital, Evans Hospital, Ireland Hospital, Madigan Medical Center, Moncrief Hospital, Reynolds Hospital, Winn Hospital, and Womack Medical Center.

Credit Recommendation: *Version 1:* In the lower-division baccalaureate/associate degree category, for phase 1, 4 semester hours in anatomy and physiology, 12 in occupational therapy techniques, 3 in applied psychology, 1 in disease pathology; for phase 2, 9 semester hours in supervised clinical experience (8/93). *Version 2:* In the lower-division baccalaureate/associate degree category, for Phase 1, 4 semester hours in functional human anatomy and physiology, 2 in pathology, 3 in applied psychology and 12 in occupational therapy techniques; for Phase 2, 10 semester hours in clinical practice in occupational therapy (10/95). *Version 3:* In the lower-division baccalaureate/associate degree category, for Phase 1, 4 semester hours in functional human anatomy and physiology, 2 in pathology, 3 in applied psychology, and 12 in occupational therapy techniques; for Phase 2, 10 semester hours in clinical practice in occupational therapy (6/98).

Related Occupations: 91L.

AR-0704-0012

1. Physical Therapy Specialist
2. Physical Therapy Specialist
3. Physical Therapy Specialty

Course Number: *Version 1:* 303-91J10; B-303-0051; B-303-A051. *Version 2:* 303-N9. *Version 3:* 303-N9.

Location: *Version 1:* Academy of Health Sciences, Fort Sam Houston, TX; Selected Hospitals, Continental US. *Version 2:* Academy of Health Sciences, Fort Sam Houston, TX. *Version 3:* Phase 1, Academy of Health Sciences, Fort Sam Houston, TX; Phase 2, Civilian/Military Medical Facilities, Continental, US; Phase 2, Naval Regional Medical Centers, Continental, US.

Length: *Version 1:* Phase 1, 17 weeks (685 hours); Phase 2, 8–10 weeks (320–400 hours). *Version 2:* Phase 1, 17 weeks (667 hours); Phase 2, 8–10 weeks (400 hours). *Version 3:* Phase 1, 17 weeks (667 hours); Phase 2, 8–10 weeks (400 hours).

Exhibit Dates: *Version 1:* 3/91–12/94. *Version 2:* 1/95–1/96. *Version 3:* 2/96–7/98.

Learning Outcomes: *Version 1:* Before 3/91 see AR-0704-0001. Upon completion of the course, the student will be able to perform the duties of a physical therapy technician (under direction), including applying therapeutic modalities, instructing and guiding patients in specific exercises, and conveying patient care information to other health care providers. *Version 2:* Upon completion of the course, the student will be able to perform the duties of a physical therapy technician (under supervision), including applying therapeutic modalities, instructing and guiding patients in specific exercises, and giving patient care information to other providers. *Version 3:* After 7/98 see AR-0704-0014. Upon completion of the course, the student will be able to perform the duties of a physical therapy technician (under supervision), including applying therapeutic modalities, instructing and guiding patients in specific exercises, and giving patient care information to other providers.

Instruction: *Version 1:* Lectures and practical exercises cover functional anatomy and physiology, human growth and development, basic principles and clinical application of physical modalities, ethical standards, proper techniques in handling and positioning patients, therapeutic exercises and development of patient therapy programs. Army and Coast Guard members attend a 10-week Phase 2; Navy personnel attend an 8-week Phase 2. *Version 2:* Lectures and practical exercises cover functional anatomy and physiology, human growth and development, basic principles and clinical application of physical modalities, ethical standards, proper techniques in handling and positioning patients, therapeutic exercises, and development of patient therapy programs. Army and Coast Guard members attend a 10-week Phase 2; Navy members attend an 8-week Phase 2 at naval facilities. Both are on-the-job in the clinical environment. *Version 3:* Lectures and practical exercises cover functional anatomy and physiology, human growth and development, basic principles and clinical application of physical modalities, ethical standards, proper techniques in handling and positioning patients, therapeutic exercises, development of patient therapy programs. Army and Coast Guard members attend a 10-week Phase 2; Navy members attend an 8-week Phase 2 at naval facilities. Both are on-the-job in the clinical environment.

Credit Recommendation: *Version 1:* In the lower-division baccalaureate/associate degree category, for Phase 1, 6 semester hours in functional anatomy and physiology, 3 in applied psychology, 13 in physical therapy principles and procedures, and 1 in disease pathology; for Phase 2, 10 semester hours in supervised clinical experience (8/93). *Version 2:* In the lower-division baccalaureate/associate degree category, for Phase 1, 6 semester hours in functional anatomy and physiology, 3 in applied psychology, 2 in principles of physical therapy, 10 in rehabilitative techniques, and 2 in pathology. Phase 2, 10 semester hours in clinical practice (10/95). *Version 3:* In the lower-division baccalaureate/associate degree category, for Phase 1, 6 semester hours in functional anatomy and physiology, 3 in applied psychology, 2 in principles of physical therapy, 10 in rehabilitative techniques, and 2 in pathology; for Phase 2, 10 semester hours in clinical practice (9/00).

Related Occupations: 91J.

AR-0704-0013

Army Medical Department (AMEDD) Noncommissioned Officer (NCO) Basic

Course Number: 6-8-C40 (91V30).

Location: Academy of Health Sciences, Fort Sam Houston, TX.

Length: 7 weeks (300 hours).

Exhibit Dates: 4/95–Present.

Learning Outcomes: Upon completion of the course, the student will be able to lead a military unit; evaluate performance and counsel; supervise and delegate; and perform drills, training, and specific military skills.

Instruction: The course covers military leadership, personal and performance counseling, duties and responsibilities of noncommissioned officers, leadership assessment and development, leadership procedures and combat orders, combat leadership, physical training and marksmanship, property accountability, individual and team training, small-group process, and specific military skills. The methodologies include lectures, discussions, group work, and practical application. This course contains a common core of leadership subjects.

Credit Recommendation: In the lower-division baccalaureate/associate degree category, 3 semester hours in military science and 1 in elective credit in a respiratory specialty. See AR-1406-0090 for common core credit (11/97).

AR-0704-0014

Physical Therapy Specialty, Phase 2

Course Number: B-303-0051, Phase 2; 303-N9, Phase 2.

Location: US Coast Guard Academy, New London, CT; Dispensary, USCG TRADEN, Camp May, NJ; Martin ACH, Fort Benning, GA; Reynolds ACH, Fort Sill, OK; Kimbrough ACH, Fort Meade, MD; Keller ACH, West Point, NY; Ireland ACH, Fort Knox, KY; DeWitt ACH, Fort Belvoir, VA; Blanchfield ACH, Fort Campbell, KY; Moncrief ACH, Fort Jackson, SC; Winn ACH, Fort Stewart, GA; Irwin ACH, Fort Riley, KS; Evans ACH, Fort Carson, CO; Naval Hospital, Camp Pendleton, CA; Naval Hospital, Camp Lejeune, NC; Naval Hospital, Bethesda, MD; Naval School of Health Sciences Naval Hospital, San Diego, CA; Madigan Medical Center, Tacoma, WA; Dwight David Eisenhower Medical Center, Fort Gordon, GA; Womack Medical Center, Fort Bragg, NC; Brooke Medical Center, Fort Sam Houston, TX; Walter Reed Medical Center, Washington, DC; William Beaumont Medical Center, El Paso, TX; Tripler Medical Center, Honolulu, HI; Naval Regional Medical Center, Portsmouth, VA; Academy of Health Sciences, Fort Sam Houston, TX.

Length: Army and Coast Guard, 10 weeks (400 hours); Navy, 8 weeks (320 hours).

Exhibit Dates: 8/98–Present.

Learning Outcomes: Before 8/98 see AR-0704-0012. Upon completion of the course, the student will be able to perform in patient and outpatient modalities, outpatient orthopedics, inpatient orthopedics and rehabilitation, hydrotherapy, and other procedures.

Instruction: Clinical instruction is provided by a supervisor in an on-the-job environment. Clinical skills of the physical therapy assistant are assessed. Students complete one case study and in service presentation.

Credit Recommendation: In the lower-division baccalaureate/associate degree category, for Army and Coast Guard 8 semester hours in clinical experience in physical therapy and for the Navy 7 semester hours in clinical experience in physical therapy (7/99).

AR-0704-0015

Physical Therapy Specialty

Course Number: B-303-0050, Phase 1; 303-N9, Phase 1.

Location: Academy of Health Sciences, Fort Sam Houston, TX.

Length: 17 weeks (667 hours).

Exhibit Dates: 8/98–Present.

Learning Outcomes: Before 8/98 see AR-0704-0012. Upon completion of the course, the student will be able to perform the duties of a physical therapy technician under supervision, including applying therapeutic modulates, instructing and guide patients in specific exercises, and giving patient care information to other providers.

Instruction: Lectures and practical exercises cover functional anatomy and physiology, basic principles and clinical application of physical modalities, ethical standards, proper techniques in handling and positioning patients, and therapeutic exercises.

Credit Recommendation: In the lower-division baccalaureate/associate degree category, 6 semester hours in anatomy and physiology, 3 in applied physiology, 2 in principles of physical therapy, 10 in rehabilitative techniques, 2 in pathology, and 2 in psychology (10/03).

AR-0704-0016

Orthopedic Specialist

Course Number: B-300-0141; 304-P1, Phase 1.

Location: Academy of Health Sciences, Fort Sam Houston, TX.

Length: 6 weeks (227 hours).

Exhibit Dates: 11/00–Present.

Learning Outcomes: Upon completion of the course, the student will be able to prepare patients for surgical procedures; assist the physician in minor surgical procedures; apply splints, casts, and traction; instruct patients regarding care; maintain mechanical equipment in an orthopedic clinic; and maintain appropriate medical records.

Instruction: Phase 1 includes lectures, discussions, demonstrations, and laboratory practice in musculo-skeletal anatomy and physiology; casting, splinting, traction, and wound care; and related equipment.

Credit Recommendation: In the lower-division baccalaureate/associate degree category, 2 semester hours in musculo-skeletal anatomy and physiology, 4 in orthopedics and patient care, and 4 in orthopedic devices, materials, and methods (5/01).

Related Occupations: 91H.

AR-0704-0019

Orthopedic Specialist, Phase 2

Course Number: B-300-0141; 304-P1, Phase 2.

Location: Womack Medical Center, Fort Bragg, NC; Martin Community Hospital, Fort

Benning, GA; Darnall Community Hospital, Fort Hood, TX; William Beaumont Medical Center, El Paso, TX; Walter Reed Medical Center, Washington, DC; Madigan Medical Center, Tacoma, WA; Dwight David Eisenhower Medical Center, Fort Gordon, GA; Brooke Medical Center, Fort Sam Houston, TX.

Length: 6 weeks (232 hours).

Exhibit Dates: 11/00–Present.

Learning Outcomes: Upon completion of the course, the student will be able to prepare patients for surgical procedures; assist the physician in minor surgical procedures; apply splints, casts, and traction; instruct patients regarding care; maintain mechanical equipment in an orthopedic clinic; and maintain appropriate medical records. This phase consists of the clinical component.

Instruction: Phase 2 consists of clinical experience.

Credit Recommendation: In the lower-division baccalaureate/associate degree category, 5 semester hours in clinical practice in orthopedics (12/99).

Related Occupations: 91H.

AR-0704-0023

Respiratory Specialist

Course Number: *Version 1:* B-300-1423; 300-91V20, Phase 1. *Version 2:* 300-91V20, Phase 1.

Location: Academy of Health Sciences, Fort Sam Houston, TX.

Length: *Version 1:* 16 weeks (616–627 hours). *Version 2:* 16–20 weeks (616–745 hours).

Exhibit Dates: *Version 1:* 2/96–3/02. *Version 2:* 4/02–Present.

Learning Outcomes: *Version 1:* Upon completion of this course, the students will be able to practice aerosol/humidity therapy, ventilation therapy, pulmonary function testing, infection control, cardiopulmonary drug administration, and critical patient care. *Version 2:* Upon completion of the 16-week course (Navy), the students, under the supervision of a physician, will be able to practice aerosol/humidity therapy, ventilation therapy, pulmonary function testing, infection control, cardiopulmonary drug administration, and critical patient care. Upon completion of the 20-week course (Army), the students, under the supervision of a physician, will be able to practice aerosol/humidity therapy, ventilation therapy, pulmonary function testing, infection control, cardiopulmonary drug administration, critical patient care, cardiopulmonary resuscitation, and basic first aid; take vital signs; and, discuss legal/ethical/professional issues in health care.

Instruction: *Version 1:* Lectures, demonstrations, and practical exercises in selected basic science courses related to respiratory care, including anatomy, respiration, physiology, pharmacology, and microbiology. The course also includes physical therapy, clinical medicine, and medical materiel management. *Version 2:* For the 16-week course, lectures, demonstrations, and practical exercises cover selected basic science courses related to respiratory care, including anatomy, respiration, physiology, pharmacology, and microbiology. The course also includes physical therapy, clinical medicine, and medical materiel management. For the 20-week course, lectures, demonstrations, and practical exercises cover selected basic science courses related to respiratory care, including anatomy, respiration, physiology, pharmacology, and microbiology. The course also includes physical therapy, clinical medicine, medical materiel management, cardiopulmonary resuscitation, first aid, legal and ethical issues, and review of basic anatomy and physiology.

Credit Recommendation: *Version 1:* In the lower-division baccalaureate/associate degree category, 3 semester hours in anatomy and physiology, 3 in general science, 3 in cardiopulmonary physiology, 3 in clinical medicine and 9 in respiratory care techniques (6/98). *Version 2:* In the lower-division baccalaureate/associate degree category, 6 semester hours in cardiopulmonary anatomy and physiology, 4 in respiratory care equipment, 3 in mechanical ventilation, 3 in cardiopulmonary diagnostics, 3 in pulmonary function, and 3 in management of intensive care patients. For those in the Army, add 1 semester hour in basic life support, 1 in basic first aid, 3 in introduction to health sciences, and 1 in basic military science (6/02).

Related Occupations: 91V.

AR-0704-0024

Respiratory Specialist

Course Number: B-300-1424, Phase 2; 300-91V20, Phase 2.

Location: State Chest Hospital, San Antonio, TX; Wilford Hall, Lackland AFB, TX; Academy of Health Sciences, Fort Sam Houston, TX.

Length: 16 weeks (550–600 hours).

Exhibit Dates: 2/96–Present.

Learning Outcomes: Upon completion of this course, the students will be able to practice aerosol/humidity therapy, ventilation therapy, pulmonary function testing, and infection control.

Instruction: Lectures, demonstrations, and practical exercises in selected basic science courses related to respiratory care, including anatomy, respiration, physiology, pharmacology, and microbiology. The course also includes physical therapy, clinical medicine, and medical materiel management.

Credit Recommendation: In the lower-division baccalaureate/associate degree category, 9 semester hours in a respiratory care clinical experience (6/02).

Related Occupations: 91V.

AR-0705-0001

Radiology Specialist
(X-Ray Specialist)
(Radiographic Procedures Basic)
(X-Ray Procedures Basic)

Course Number: 313-91P10.

Location: Academy of Health Sciences, Fort Sam Houston, TX; Selected MEDACS, MEDCENS, Continental US.

Length: Phase 1, 20 weeks (802 hours); Phase 2, 22 weeks (880 hours).

Exhibit Dates: 10/88–1/99.

Objectives: After 1/99 see AR-0705-0013. To provide enlisted personnel with a working knowledge of basic X-ray clinic procedures.

Instruction: Lectures and practical experience cover the fundamentals of X-ray clinic operations and procedures, including applied anatomy and physiology, manual and automatic X-ray film processing, equipment operation and maintenance, sterile techniques, radiation safety measures, and diagnostic radiograph production.

Credit Recommendation: In the lower-division baccalaureate/associate degree category, for Phase 1, 2 semester hours in medical terminology, 4 in anatomy and physiology, 1 in radiographic exposure, 2 in radiographic equipment, 1 in radiographic film processing, 3 in radiation physics, 1 in radiation biology, 1 in radiation protection, 6 in radiographic procedures, 1 in quality assurance, 1 in medical ethics, 1 in computer literacy; for Phase 2, 8 in clinical practicum. Both phases must be completed (12/93).

Related Occupations: 91P.

AR-0705-0003

Radiological Safety

Course Number: 7K-F3.

Location: Ordnance Center and School, Aberdeen Proving Ground, MD; Chemical School, Fort McClellan, AL.

Length: 3 weeks (112–120 hours).

Exhibit Dates: 5/83–12/97.

Objectives: To provide officer, enlisted, and civilian personnel with basic instruction in radiological safety.

Instruction: Conferences, case studies, and practical exercises cover radiological safety principles, including storage, handling, transportation, disposal, reporting, control, and general safety for radioisotopes, neutronic sources, microwave, laser, and machine-produced radiation.

Credit Recommendation: In the upper-division baccalaureate category, 3 semester hours in principles of radiological safety (6/87).

Related Occupations: 214E; 214G; 222B; 223B; 223D; 224B; 241F; 251B; 251D.

AR-0705-0004

Radiology Noncommissioned Officer (NCO) Management

Course Number: 313-F1.

Location: Academy of Health Sciences, Fort Sam Houston, TX.

Length: 1–2 weeks (60 hours).

Exhibit Dates: 4/85–3/94.

Learning Outcomes: After 3/94 see AR-0705-0011. To provide noncommissioned officer management training for personnel assigned to management positions in a department of radiology.

Instruction: Areas of instruction include administrative procedures; supervision and performance of supply and maintenance procedures; implementation of Army quality assurance program; methods of improving radiological health care delivery by application of current techniques; planning, organizing, and operating departments of radiology in the US and overseas; principles of unit and/or patient administration; and personnel management.

Credit Recommendation: In the lower-division baccalaureate/associate degree category, 2 semester hours in radiology management (9/85).

AR-0705-0005

Medical X-Ray Survey Techniques

Course Number: *Version 1:* 6H-F18/322-F18; 6H-F18/323-F18. *Version 2:* 6H-F18/322-F18; 6H-F18/323-F18.

Location: Academy of Health Sciences, Fort Sam Houston, TX.

Length: *Version 1:* 2 weeks (84–90 hours). *Version 2:* 2 weeks (84–90 hours); 2 weeks (85 hours).

Exhibit Dates: *Version 1:* 11/83–3/01. *Version 2:* 4/01–Present.

Objectives: *Version 1:* To provide students with basic knowledge and techniques required to evaluate potential hazards from medical diagnostic radiation exposure according to federal safety standards. *Version 2:* Upon completion of the course, the student will be able to describe radiation hazard precautions; explain the physical principles of x-ray production; discuss the biological and legal hazards of ionizing radiation exposure; meet JCAHO x-ray standards and determine radiation precautions; describe radiation quality assurance; calculate radiation dosage; and meet federal x-ray system performance standards.

Instruction: *Version 1:* Units of instruction include principles of X-ray production, overview of diagnostic procedures, use of survey instruments, biological effects of radiation, and quality assurance. *Version 2:* Practical and classroom exercises, discussion, laboratory, and lecture. Topics include principles of x-ray tubes, generators, transformers, x-ray photon interactions, filtering beams, and scatter radiation; operating principles of the IAW system; quality assurance, x-ray shielding and room design; radiation biological damage; legal aspects; operating principles of specialized equipment; federal performance standards; dose calculation; and JCAHO radiation standards.

Credit Recommendation: *Version 1:* In the lower-division baccalaureate/associate degree category, 3 semester hours in radiation safety (8/95). *Version 2:* In the upper-division baccalaureate category, 3 semester hours in advanced principles of radiation physics and 3 in regulation of x-ray facilities (11/03).

AR-0705-0007

Polyphase X-Ray Systems

Course Number: 4B-F5/198-F5.

Location: Equipment and Optical School, Aurora, CO.

Length: 3 weeks (2–116 hours).

Exhibit Dates: 8/80–8/96.

Objectives: To provide warrant officers and enlisted personnel training on polyphase X-ray systems with emphasis on operation, maintenance, and calibration.

Instruction: This course provides a study of operating procedures and purpose of several X-ray systems and includes the use of manufacturer's literature in operator calibration and maintenance. Practical experience in diagnosis and repair of malfunctioning X-ray equipment is also covered.

Credit Recommendation: In the vocational certificate category, 3 semester hours in X-ray equipment operation (9/85).

AR-0705-0008

BRH Compliance Testing

Course Number: 4B-F6/198-F6.

Location: Equipment and Optical School, Aurora, CO.

Length: 2 weeks (67–76 hours).

Exhibit Dates: 8/80–8/96.

Objectives: To provide warrant officers and enlisted personnel training on the Radiation Control Health and Safety Act of 1968 with particular emphasis on compliance testing procedures.

Instruction: Students learn to perform tests, measurements, and adjustments necessary to keep X-ray equipment within government guidelines for radiation.

Credit Recommendation: In the vocational certificate category, 2 semester hours in X-ray machine calibration (9/85).

AR-0705-0009

1. Radiation Safety Specialist
2. Health Physics Specialist
3. Health Physics Specialty
4. Health Physics Specialty

Course Number: *Version 1:* 322-91X20; 311-91X20. *Version 2:* 322-91X20; 311-91X20; 322-N4. *Version 3:* 322-N4. *Version 4:* 322-N4.

Location: Academy of Health Sciences, Fort Sam Houston, TX.

Length: *Version 1:* 12 weeks (400–506 hours). *Version 2:* 12 weeks (400–506 hours). *Version 3:* 12 weeks (466 hours). *Version 4:* 12 weeks (487 hours).

Exhibit Dates: *Version 1:* 11/86–5/90. *Version 2:* 6/90–4/00. *Version 3:* 5/00–9/02. *Version 4:* 10/02–Present.

Learning Outcomes: *Version 1:* Upon completion of the course, the student will be able to conduct radiation surveys of nuclear medicine clinics, radioisotope laboratories, and medical X-ray machines; inspect incoming radioisotope shipments; dispose of radioactive waste; assist in maintaining acceptable radiation procedures; and assist in evaluating directed energy hazards. *Version 2:* Upon completion of the course, the student will be able to apply knowledge of health physics to assess radiation hazards, perform diagnostic X-ray compliance and radiation safety surveys, process radioactive waste for storage or shipment, select and utilize laboratory equipment, and perform health physics tasks necessary for radionuclide therapy. *Version 3:* Upon completion of the course, the student will be able to apply knowledge of health physics to assess radiation hazards, perform diagnostic X-ray compliance and radiation safety surveys, process radioactive waste for storage or shipment, select and utilize laboratory equipment, and perform health physics tasks necessary for radionuclide therapy. *Version 4:* Upon completion of the course, the student will be able to apply knowledge of health physics to assess radiation hazards; perform diagnostic X-ray compliance and radiation safety surveys; process radioactive waste for storage or shipment; select and utilize laboratory equipment; perform health physics tasks necessary for radionuclide therapy; use dosimeters; and conduct environmental samples for radiology threats and emergency response.

Instruction: *Version 1:* Lectures, conferences, demonstrations, and discussions cover physical science, radiation detection, measurement, dosimetry, regulatory requirements, radiation and the environment, and medical X-ray survey procedures. *Version 2:* Course covers mathematics, basic physics, radiation physics, radiation materials, medical X-ray survey techniques, and dosimetry. Lectures, demonstrations, and laboratory exercises are the methodologies. *Version 3:* Lectures, demonstrations, and laboratory exercises covering mathematics, basic physics, radiation physics, radiation materials, medical X-ray survey techniques, and dosimetry. *Version 4:* Lectures, demonstrations, and laboratory exercises covering mathematics, basic physics, radiation physics, radiation materials, medical X-ray survey techniques, and dosimetry.

Credit Recommendation: *Version 1:* In the lower-division baccalaureate/associate degree category, 3 semester hours in principles of physical sciences, 3 in physical sciences laboratory, and 3 in X-ray survey techniques (4/88). *Version 2:* In the lower-division baccalaureate/associate degree category, 3 semester hours in radiologic techniques, 3 in physical science, and 3 in physical science laboratory (5/01). *Version 3:* In the lower-division baccalaureate/associate degree category, 3 semester hours in radiograph equipment, 3 in radiation physics, 4 in applied health physics, 2 in radiation protection, and 5 in diagnostic procedures of radiographic equipment (6/02). *Version 4:* In the lower-division baccalaureate/associate degree category, 3 semester hours in radiograph equipment, 2 in radiation physics, 6 in radiation processes, 4 in radiation protection, 4 in environmental science (radiation), and 4 in applied health physics (6/02).

Related Occupations: 91X.

AR-0705-0010

Special Operations Medical Sergeant
Phase 2A

Course Number: 011-18D30.

Location: Academy of Health Sciences, Fort Sam Houston, TX.

Length: 21 weeks (854 hours).

Exhibit Dates: 5/88–4/98.

Learning Outcomes: Upon completion of the course, the student will be able to assume medical duties and responsibilities, including assessment and management of specified surgical emergencies in field conditions, patient management, administration of local anesthesia for dental conditions, dosage calculations, and administration of medicinal agents.

Instruction: Course includes lectures, laboratories, and clinical training in the areas of anatomy, physiology, dental sciences, psychiatry, physical therapy, medicine, nursing, and field medical service.

Credit Recommendation: In the lower-division baccalaureate/associate degree category, 2 semester hours in introductory medical laboratory procedures and credit in nursing or physician assisting on the basis of institutional evaluation. In the upper-division baccalaureate category, 3 semester hours in physiology and hygiene and 2 in medical laboratory procedures (4/88).

Related Occupations: 18D.

AR-0705-0011

Radiology Noncommissioned Officer
(NCO) Management

Course Number: *Version 1:* 313-F1. *Version 2:* 313-F1.

Location: Academy of Health Sciences, Fort Sam Houston, TX.

Length: *Version 1:* 1–2 weeks (65 hours). *Version 2:* 1–2 weeks (68 hours).

Exhibit Dates: *Version 1:* 4/94–4/99. *Version 2:* 5/99–Present.

Learning Outcomes: *Version 1:* Before 4/94 see AR-0705-0004. Upon completion of the course, the student will be able to apply basic principles of administration and management to the management of a radiology department. *Version 2:* Upon completion of the course, the student will be able to supervise and manage a radiology department.

Instruction: ***Version 1:*** Areas of instruction include administrative procedures; supervision and performance of supply and maintenance procedures; implementation of Army quality assurance program; methods of improving radiological health care delivery by application of current techniques; planning, organizing, and operating departments of radiology in the US and overseas; principles of unit and/or patient administration; and personnel management. *Version 2:* Lectures and practical assignments address administration procedures, supply and maintenance procedures, and quality assurance.

Credit Recommendation: ***Version 1:*** In the lower-division baccalaureate/associate degree category, 3 semester hours in radiology department administration/management (10/95). *Version 2:* In the lower-division baccalaureate/associate degree category, 3 semester hours in radiology department management (7/99).

AR-0705-0012

Army Medical Department (AMEDD) Noncommissioned Officer (NCO) Basic

Course Number: 6-8-C40 (91P30).

Location: Academy of Health Sciences, Fort Sam Houston, TX.

Length: 7 weeks (300 hours).

Exhibit Dates: 4/95–Present.

Learning Outcomes: Upon completion of the course, the student will be able to lead a military unit; evaluate performance and counsel; supervise and delegate; and perform drills, training, and specific military skills.

Instruction: The course covers military leadership, personal and performance counseling, duties and responsibilities of noncommissioned officers, leadership assessment and development, leadership procedures and combat orders, combat leadership, physical training and marksmanship, property accountability, individual and team training, small-group process, and specific military skills. The methodologies include lectures, discussions, group work, and practical application. This course contains a common core of leadership subjects.

Credit Recommendation: In the lower-division baccalaureate/associate degree category, 3 semester hours in military science and 1 in elective credit in radiology. See AR-1406-0090 for common core credit (11/97).

Related Occupations: 91P.

AR-0705-0013

Radiology Specialist

Course Number: 313-91P10, Phase 2.

Location: Darnall Community Hospital, Fort Hood, TX; Evans Community Hospital, Fort Carson, CO; Ireland Community Hospital, Fort Knox, KY; Lester Community Hospital, Fort Rucker, AL; Winn Community Hospital, Fort Stewart, GA; Martin Community Hospital, Fort Benning, GA; Moncrief Ambulatory Acute Care Center, Fort Jackson, SC; Reynolds Community Hospital, Fort Sill, OK; Kimbrough Ambulatory Acute Care Center, Fort Meade, MD; Blanchfield Community Hospital, Fort Campbell, KY; Womack Medical Center, Fort Bragg, NC; Madigan Medical Center, Tacoma, WA; Walter Reed Medical Center, Washington, DC; William Beaumont Medical Center, El Paso, TX; Dwight David Eisenhower Medical Center, Fort Gordon, GA; Brooke Medical Center, Fort Sam Houston, TX; Academy of Health Sciences, Fort Sam Houston, TX.

Length: 22 weeks (856 hours).

Exhibit Dates: 2/99–Present.

Learning Outcomes: Before 2/99 see AR-0705-0001. Upon completion of the course, the student will be able to perform specific radiologic procedures in a supervised clinical setting. Emphasis is to acquire competency in performing procedures/examinations on actual sick and injured patients.

Instruction: Clinical instruction is provided in a supervised setting at selected medical facilities.

Credit Recommendation: In the lower-division baccalaureate/associate degree category, 14 semester hours in radiographic clinical experience (6/02).

AR-0705-0014

Radiology Specialist

Course Number: ***Version 1:*** 313-91P10, Phase 1. *Version 2:* 313-91P10, Phase 1.

Location: Academy of Health Sciences, Fort Sam Houston, TX.

Length: ***Version 1:*** 20 weeks (806 hours). *Version 2:* 24 weeks (990 hours).

Exhibit Dates: ***Version 1:*** 4/99–9/01. *Version 2:* 10/01–Present.

Learning Outcomes: ***Version 1:*** Before 4/99 see AR-0705-0001. Upon completion of the course, the student will be able to perform radiographic procedures using a foundation in radiation biology, radiation and medical ethics, and law. *Version 2:* Upon completion of the course, the student will be able to perform radiographic procedures, recognize legal/ethical situations, operate radiographic imaging equipment, perform quality control tests, and recognize radiographic pathology.

Instruction: ***Version 1:*** Instruction is provided through lectures and demonstrations. Skills are taught with an emphasis in radiographic principles, radiographic exposure, and imagery equipment and techniques. *Version 2:* Laboratory, audio visual materials, case studies, discussion, classroom exercise, practical exercises, and lecture. Topics include overview of radiology profession, legal/ethical issues, medical terminology, anatomy and physiology for radiology, radiographic images, radiographic equipment, film processing, procedures, quality assurances, radiation physics, radiation protection, radiation biology, radiographic pathology, and positioning.

Credit Recommendation: ***Version 1:*** In the lower-division baccalaureate/associate degree category, 1 semester hour radiology, 1 in radiographic pathology, 2 in medical terminology, 6 in anatomy and physiology, 1 in radiographic exposure, 4 in imaging equipment, 1 in radiographic film processing, 2 in quality assurance, 2 in radiation physics, 1 in radiation protection, 8 in radiographic procedures, 1 in medical ethics, 1 in computer literacy, 3 in radiographic laboratory procedures (6/02). *Version 2:* In the lower-division baccalaureate/associate degree category, 1 semester hour in computer literacy, 3 in introduction to radiology, 1 in medical terminology, 3 in introductory radiographic techniques, 3 in radiation physics, 9 in radiographics procedures, 2 in quality assurance, 3 in radiographic equipment, 6 in radiographic exposure and processing, 6 in human structure, function and pathology, and 1 in practicum in radiology (6/02).

AR-0705-0015

Imaging Procurement/Acceptance Procurement

Course Number: 300-F15.

Location: Academy of Health Sciences, Fort Sam Houston, TX.

Length: 3 weeks (120 hours).

Exhibit Dates: 3/00–Present.

Learning Outcomes: Upon completion of the course, the student will be able to discuss regulation of diagnostic x-ray system procurement; analyze space design and power supplies for diagnostic x-ray systems; validate inspection procedures for diagnostic x-ray procuring systems; and perform post-calibration radiation inspections on radiographic and fluoroscopic x-ray systems.

Instruction: Computer-based training, practical exercises, discussion, and lecture. Topics include regulation of diagnostic x-ray systems procurement; power supplies and space design for diagnostic systems; inspection and acceptance procedures; post-calibration radiation inspection; and fluoroscopic x-ray acceptance inspection.

Credit Recommendation: In the upper-division baccalaureate category, 3 semester hours in diagnostic x-ray system procurement (11/03).

AR-0705-0016

Advanced Digital Imaging Systems Maintenance

Course Number: 300-F14.

Location: Academy of Health Sciences, Fort Sam Houston, TX.

Length: 6 weeks (240 hours).

Exhibit Dates: 3/00–Present.

Learning Outcomes: Upon completion of the course, the student will be able to maintain and repair diagnostic imaging equipment.

Instruction: Practical exercises, discussion, lecture, and computer-based training. Topics include radiographic/fluoroscopic imaging systems, mammography systems, and gamma cameras.

Credit Recommendation: In the lower-division baccalaureate/associate degree category, 3 semester hours in maintenance of diagnostic imaging systems, 3 in repair of diagnostic imaging systems, and 3 in fundamentals of diagnostic imaging systems (11/03).

AR-0705-0017

Advanced Medical Systems Maintenance

Course Number: 300-F13.

Location: Academy of Health Sciences, Fort Sam Houston, TX.

Length: 4 weeks (160 hours).

Exhibit Dates: 2/00–Present.

Learning Outcomes: Upon completion of the course, the student will able to discuss clinical applications and theory of operation of medical laser systems; perform circuit analysis on medical laser systems; and perform practical applications and troubleshooting on medical laser systems.

Instruction: Practical exercises, discussion, lecture, and computer-based training. Topics include basic laser physics and safety; clinical applications, operation, equipment, circuitry, and troubleshooting of medical laser systems, including Nd:YAG, CO2 and Ophthalmic; and evaluation of medical laser systems.

Credit Recommendation: In the upper-division baccalaureate category, 3 semester hours in medical laser systems maintenance (11/03).

AR-0705-0018

Advanced Medical Lab Systems Maintenance

Course Number: 300-F12.

Location: Academy of Health Sciences, Fort Sam Houston, TX.

Length: 4 weeks (160 hours).

Exhibit Dates: 3/00–Present.

Learning Outcomes: Upon completion of the course, the student will be able to discuss clinical applications and operational theory of medical laboratory equipment and instrumentation; perform practical applications and troubleshooting on medical laboratory equipment and instrumentation; and evaluate selected medical laboratory systems.

Instruction: Audio visual materials, learner materials, practical exercises, discussion, lecture, and computer-based training. Topics include clinical and practical applications, operational theory, circuitry, troubleshooting; and evaluation of general clinical laboratory equipment, chemistry and hematology analyzers, cell washers, plasma sterilization, and blood gas analyzers.

Credit Recommendation: In the upper-division baccalaureate category, 3 semester hours in clinical laboratory instrumentation maintenance (11/03).

AR-0706-0002

Eye Specialist

Course Number: 300-P3; 300-91Y10.

Location: Academy of Health Sciences, Fort Sam Houston, TX.

Length: Phase 1, 6 weeks (236–252 hours); Phase 2, 7 weeks (240 hours).

Exhibit Dates: 6/88–Present.

Learning Outcomes: Upon completion of the course, the student will be able to provide full assistance to ophthalmologists and optometrists in military eye clinics, field hospitals, and combat units.

Instruction: Phase 1 of the course is conducted through conferences, demonstrations, and lectures emphasizing principles of ophthalmology and optometry including routine diagnostic medications and minor repairs to spectacles; Phase 2 is a clinical practicum in ophthalmology and optometry.

Credit Recommendation: In the lower-division baccalaureate/associate degree category, for Phase 1, 3 semester hours in principles of ophthalmology, 2 in principles of optometry, and 1 in anatomy and physiology; for Phase 2, 4 semester hours in ophthalmology practicum and 4 in optometry practicum. Both phases must be completed (5/02).

Related Occupations: 91Y.

AR-0707-0010

Army Medical Department (AMEDD) Officer Basic

Course Number: 6-8-C20.

Location: Academy of Health Sciences, Fort Sam Houston, TX.

Length: 8–9 weeks (355 hours).

Exhibit Dates: 1/83–7/90.

Objectives: To train entry level commissioned medical service corps officers and warrant officers in military management skills.

Instruction: Instruction primarily informs the student of military management procedures with a specific health-related unit in the handling of hazardous materiel.

Credit Recommendation: In the vocational certificate category, 1 semester hour in basic EMS first aid. In the upper-division baccalaureate category, 3 semester hours in industrial/occupational hygiene in hazardous materials (9/85).

AR-0707-0011

Basic Environmental Staff
(Basic Environmental Coordinator's)

Course Number: ALMC-BE.

Location: Army Logistics Management College, Fort Lee, VA; Army Logistics Management Center, Fort Lee, VA.

Length: 2 weeks (67–76 hours).

Exhibit Dates: 5/85–Present.

Objectives: To train environmental staff to prepare and evaluate reports on the effects of military actions on environmental pollution and to present an overview of current and relevant statutory and regulatory requirements of the environmental coordinator.

Instruction: Lecture-conferences and case method instruction in the basic concepts of ecology, pollution abatement, environmental regulations, environmental management considerations, and the preparation of environmental documentation.

Credit Recommendation: In the lower-division baccalaureate/associate degree category, 2 semester hours in environmental management (7/00).

AR-0707-0014

Preventive Medicine Program Management

Course Number: ***Version 1:*** 6A-F6. *Version 2:* 6A-F6. *Version 3:* 6A-F6.

Location: Academy of Health Sciences, Fort Sam Houston, TX.

Length: ***Version 1:*** 2 weeks (71 hours). *Version 2:* 2 weeks (71 hours). *Version 3:* 2 weeks (70–74 hours).

Exhibit Dates: ***Version 1:*** 9/87–9/90. *Version 2:* 10/90–10/93. *Version 3:* 11/93–Present.

Learning Outcomes: ***Version 1:*** Upon completion of the course, the student will be able to apply acquired management skills and problem solving in the preventive medicine area. *Version 2:* Upon completion of the course, the student will be able to apply management and problem solving skills in the preventive medicine area. *Version 3:* Upon completion of the course, the student will be able to apply management and problem solving skills in the preventive medicine area.

Instruction: ***Version 1:*** Conferences and practical exercises cover manpower and resource management, current issues in preventive medicine, and leadership. *Version 2:* Course covers topics in management and leadership as applied to functional operation and coordination of preventive medicine services. Methodology includes conferences, case studies, and discussions. *Version 3:* This course covers topics in management and leadership as applied to functional operation and coordination of preventive medicine services. Methodology includes conferences, case studies, and guided discussions. The course includes a discussion of marketing strategies, quality assurance and continuous quality improvement, risk management, and leadership skills.

Credit Recommendation: ***Version 1:*** In the upper-division baccalaureate category, 2 semester hours in business management and organization (4/88). *Version 2:* In the upper-division baccalaureate category, 2 semester hours in organizational management (8/93). *Version 3:* In the upper-division baccalaureate category, 3 semester hours in organizational management (7/99).

AR-0707-0015

1. Sexually Transmitted Diseases Intervention
   (Human Immunodeficiency Virus (HIV)/ Sexually Transmitted Diseases (STD) Intervention)
2. Sexually Transmitted Diseases Intervention
   (Human Immunodeficiency Virus (HIV)/ Sexually Transmitted Diseases (STD) Intervention)
3. Sexually Transmitted and Other Communicable Diseases Intervention

Course Number: ***Version 1:*** 6H-F9/323-F9. *Version 2:* 6H-F9/322-F9. *Version 3:* 6H-F9/322-F9.

Location: Academy of Health Sciences, Fort Sam Houston, TX.

Length: ***Version 1:*** 2 weeks (69–83 hours). *Version 2:* 2 weeks (69–83 hours). *Version 3:* 2 weeks (83 hours).

Exhibit Dates: ***Version 1:*** 6/88–2/93. *Version 2:* 3/93–12/94. *Version 3:* 1/95–Present.

Learning Outcomes: ***Version 1:*** Upon completion of the course, the student will be able to take a major role in the implementation and operation of programs aimed at the prevention and control of sexually transmitted diseases (STDs) including the human immunodeficiency virus (HIV). *Version 2:* Upon completion of the course, the student will be able to conduct a sexual history interview, contact interviews and counseling, contact tracing, and STD education programs. *Version 3:* Upon completion of the course, the student will be able to conduct a sexual history interview, contact interviews and counseling, contact tracing, and sexually transmitted diseases (STD) education programs.

Instruction: ***Version 1:*** This course includes lectures, discussions, and clinical practice. Topics covered include the basic pathology and epidemiology of all of the major STDs, case management, medical and sexual history, interviewing, counseling, reporting requirements, specimen collection, and microscopy techniques. *Version 2:* Course covers case management, STD health education, interviewing, counseling, field investigations, HIV infections, and sexual mores and practices. Methodologies include lectures and audiovisuals. *Version 3:* Course covers case management, STD health education, interviewing, counseling, field investigations, HIV infections, and sexual mores and practices. Methodologies include lectures, discussions, and audiovisuals.

Credit Recommendation: ***Version 1:*** In the upper-division baccalaureate category, 4 semester hours in sexually transmitted disease (3/90). *Version 2:* In the upper-division baccalaureate category, 3 semester hours in sexually transmitted diseases/case management and education (8/93). *Version 3:* In the upper-division baccalaureate category, 3 semester hours in sexually transmitted diseases/case management and education (7/99).

AR-0707-0016

Preventive Medicine Specialist

Course Number: ***Version 1:*** 322-91S10. *Version 2:* 322-91S10. *Version 3:* 322-91S10.

Location: Academy of Health Sciences, Fort Sam Houston, TX.

Length: ***Version 1:*** 15 weeks (624 hours). *Version 2:* 15 weeks (574–630 hours). *Version 3:* 15 weeks (669 hours).

Exhibit Dates: ***Version 1:*** 10/88–9/91. *Version 2:* 10/91–7/00. *Version 3:* 8/00–Present.

Learning Outcomes: ***Version 1:*** Upon completion of the course, the student will be able to apply the principles of sanitary science and disease control, apply the fundamentals of environmental science including occupational safety and health and radiological and chemical protection, and apply advanced concepts of entomological research and practice. Student will be able to apply the principles of environmental health to the control of those environmental factors affecting the health and well being of personnel in military areas. *Version 2:* Upon completion of the course, the student will be able to apply the principles of preventive medicine to the control of environmental factors affecting the health and well being of personnel, assess and maintain sanitary conditions, monitor occupational health and safety, assess community health problems, and monitor entomological conditions. *Version 3:* Upon completion of the course the student will be able to apply principles of preventive medicine to the control of environmental factors affecting the health and well-being of personnel.

Instruction: ***Version 1:*** Conferences, lectures, demonstrations, and practical exercises cover the identification of sanitary deficiencies in water supply and waste water systems; inspection of and discrepancies in food service facilities; identification of hazards relating to the work environment in industrial, technical, and training areas; identification, collection, preservation, and shipment of medically important arthropods and rodents; selection, safe handling, mixing, allocation, and disposal of pesticides including pest control procedures; performance of environmental health surveys; identification of potential health hazards in facilities producing ionizing and non-ionizing radiation; description of disease chain of transmission; and the compilation of medical statistical data. *Version 2:* Course covers sanitary and waste water systems, occupational and industrial hygiene, radiological and chemical protection, community health practices, health promotion and disease prevention, and medical entomology. Methodology includes lectures, demonstrations, and practical exercises. *Version 3:* Laboratory, audio visual materials, discussion, classroom exercise, practical exercises, lecture. Topics include basic epidemiology principles, etology of diseases of major importance, hazard communication, hazardous chemicals, air sampling techniques, personal protective equipment, water sampling techniques, soil sampling techniques, nuclear biological radiological chemical (NBRC), identification and control of arthropods, and pesticides and herbicides.

Credit Recommendation: ***Version 1:*** In the lower-division baccalaureate/associate degree category, 4 semester hours in sanitary science and disease control and 19 semester hours in environmental science or public health. In the upper-division baccalaureate category, 7 semester hours in advanced zoology/entomology (3/90). *Version 2:* In the upper-division baccalaureate category, 4 semester hours in issues in public health, 4 in occupational health and safety, 7 in medical entomology, and 3 in community health practice (5/01). *Version 3:* In the lower-division baccalaureate/associate degree category, 1 semester hour in introduction to community health, 1 in hazardous materials nuclear biological radiological chemical (NBRC), 6 in environmental health, 3 in pesticides and herbicides, 3 in a preventive medicine practicum. In the upper-division baccalaureate category, 4 semester hours in medical entomology (6/02).

Related Occupations: 91S.

AR-0707-0017

1. Preventive Medicine Specialist Reserve Component (Nonresident/Resident) Phase 2
2. Preventive Medicine Specialist Reserve Component

Course Number: ***Version 1:*** 322-91S10-RC. *Version 2:* 322-91S10 (RC), Phase 2.

Location: Academy of Health Sciences, Fort Sam Houston, TX.

Length: ***Version 1:*** 2 weeks (71–85 hours). *Version 2:* 2 weeks (87 hours).

Exhibit Dates: ***Version 1:*** 9/88–6/99. *Version 2:* 7/99–Present.

Learning Outcomes: ***Version 1:*** Upon completion of the course, the student will be able to apply fundamentals of sanitation to food services, water analysis, field waste management, and rodent and pest control. *Version 2:* Upon completion of the course the student will be able to inspect and evaluate sanitation and safety of selected facilities, evaluate waste disposal operations, implement insect and rodent control programs, assist with data collection for communicable disease/occupational disease investigations.

Instruction: ***Version 1:*** This course is a combination of conferences, demonstrations, and practical experience in sanitary and environmental science. Credit is based on Phase 2 only. *Version 2:* Laboratory, audio visual materials, case studies, discussion, classroom exercise, practical exercises, lecture. Water sampling tests, sound levels measure mates, heat/cold/stress management, field sanitation, food and diseases investigation, pesticides (insects/rodents), medical entomology, and environmental health.

Credit Recommendation: ***Version 1:*** In the lower-division baccalaureate/associate degree category, 3 semester hours in community health and sanitary science and 3 in medical entomology (4/94). *Version 2:* In the lower-division baccalaureate/associate degree category, 3 semester hours in medical entomology, and 3 in environmental health (6/02).

Related Occupations: 91S.

AR-0707-0018

Basic Industrial Hygiene Technician

Course Number: 6H-F11/322-F11; 6H-F11/323-F11.

Location: Academy of Health Sciences, Fort Sam Houston, TX.

Length: 2 weeks (79 hours).

Exhibit Dates: 6/88–Present.

Learning Outcomes: Upon completion of the course, the student will be able to teach basic industrial hygiene techniques required for the recognition, evaluation, and control of occupational health hazards.

Instruction: Methods of instruction include lectures, problem solving, calibrations, testing, and measurements. Topics include industrial technology; hazard communication and evaluation; occupational diseases; direct and indirect air sampling instruments; respiratory protection; industrial ventilation; current trends in industrial hygiene; industrial hygiene survey techniques; equipment calibration, utilization, and storage; health hazard information module; hazards associated with chemical reports; and ergonomic stress in the workshop.

Credit Recommendation: In the upper-division baccalaureate category, 4 semester hours in industrial hygiene (7/94).

AR-0707-0019

1. Army Medical Department (AMEDD) Officer Basic, Phase 1 and Phase 2
2. Army Medical Department (AMEDD) Officers Basic, Phases 1 and 2
3. Army Medical Department (AMEDD) Officer Basic (All Corps)

Course Number: ***Version 1:*** 6-8-C20. *Version 2:* 6-8-C20. *Version 3:* 6-8-C20.

Location: Academy of Health Sciences, Fort Sam Houston, TX.

Length: ***Version 1:*** 9–14 weeks (154–347 hours). *Version 2:* Phase 1, 8–9 weeks (336 hours); Phase 2, 1–5 weeks (50–210 hours). *Version 3:* 9–14 weeks (350–578 hours).

Exhibit Dates: ***Version 1:*** 8/90–9/94. *Version 2:* 10/94–8/01. *Version 3:* 9/01–Present.

Learning Outcomes: ***Version 1:*** Upon completion of the core curriculum, Phase 1, the student will be able to discuss the structure and functions of the Army health care system; discuss the basic principles of biomedical supply and Army medical records, personnel management, and medical legal issues; identify the role of Army medical department personnel; discuss the organization of the Army, medical military history, and combat medical issues; and discuss preventive medicine issues, including sexually transmitted diseases, AIDS, NBC, radiation, and entomology. Those students in the Veterinary track will be able to inspect poultry, beef, pork, dairy, fish, fresh fruit and vegetables, eggs, and other food products; inspect food establishments; discuss the principles of animal disease prevention and control; and discuss the role of the veterinary service in the Army. Those in the Dental track will be able to discuss the role of dental science personnel in the military, discuss the organization of the Army dental care system, and discuss the use and management of dental health records. Those in the Pharmacy track will be able to discuss pharmacy service operations in the Army, prepare sterile products, and perform pharmacist duties using military procedures. Those students who are in the Nursing track will be able to perform basic trauma life support, discuss issues of quality assurance, complete Army documentation of the nursing process, and discuss workload issues. Those in the Military Science track will be able to discuss equipment maintenance issues; demonstrate emergency medical care, including heat and cold injuries, chest injuries, head injuries, splinting/bandaging, fractures, shock, hemorrhage, and airway obstruction. Students in the Behavioral Science track will be able to discuss the role of the behavioral science professional in crisis intervention, in disaster, in combat, and in terrorism situations; identify components of Army programs in behavioral science, including alcohol and drug programs, family advocacy program, and exceptional family member program. Those in the Preventive Medicine (68B) track will be able to discuss the principles of radiation protection, measurements, and waste management; discuss biological effects of radiation; and discuss regulatory issues of radiation sources and radiographic equipment. Those in the Preventive Medicine track (68F, 68N, 68P) will be able to inspect a food service facility; discuss sanitation and testing of water supply, swimming pools, and natural bathing areas; identify responsibilities in environmental surveillance; and describe the role of preventive medicine in the Army. Students in the Optometry track will be able to discuss the role of the optometrist in the Army and discuss management responsibilities. Those in the Medicine and Surgery track will be able to discuss the role of the medical officer in the Army, discuss the medical training programs for Army personnel, identify medical implications of combat and of chemical warfare, and discuss management of sick call. Those in the Human Resources Management track will be able to discuss nutrition care in the Army hospital, participate in the medical field feeding exercise, and use the Army's computerized food service system. Those is the Laboratory Science track will be able to discuss the curriculum of medical laboratory specialists and to identify combat laboratory materials. Those students in the Audiology track will be able to demonstrate use of the evaluation automated registry system and describe the role of audiology in the Army. Students will complete only one track. *Version 2:* The course consists of a preparatory phase, a common core, and subject specific tracks. Upon completion of the preparatory phase, the student will be able to discuss the fundamentals of the Uniform Code of Military Justice; describe the mission, function, and structure of the Army and the Department of Defense; discuss the principles of war and military history; discuss basic principles of military leadership and protocol; and implement basic first aid principles and nuclear, biological, and chemical protective measures. Upon completion of the common core, the student will be able to apply basic principles of health care administration, including budgeting, resource management, and the health care delivery systems; discuss legal issues, including military justice, standards of conduct, tort liability, the Federal Tort Claims Act, Geneva Convention of Code and Conduct, the Freedom of Information Act, and the Privacy Act; discuss the medical and dental health care system in the Army; discuss principles of military science, including command and staff functions, combat health support, uniforms, drills and ceremonies; discuss principles of organizational management, riot management/assessment, team building, and EEO; and participate in field experiences. Those in the nursing track will be able to discuss health care concepts and management philosophies, including ethics, resource management, work load management, labor relations, personnel issues, and quality improvement related to nursing and to identify career planning strategies, including empowerment, coping strategies, stress management, team building, and mentorship. Those in the dental corps track will be able to discuss dentistry field operations including the role of dental enlisted personnel and discuss dental care concepts, including dental quality assurance, dental records, and preventive dentistry programs. Those in the medical corps track will be able to discuss the military and medical roles/responsibilities of a brigade-level medical officer and discuss selected diseases of military importance. Those in the medical service corps platoon leader track will be able to discuss the roles and responsibilities of field medical support operations and discuss principles of health services planning, operations, training, and security. Those in the clinical support services laboratory track will be able to discuss the roles and responsibilities of laboratory professional in the military and in warfare situations and discuss issues related to laboratory science, including blood bank/donor center, biological warfare, OSHA, laboratory safety, CLIP/CLIA, laboratory management, and quality assurance. Those in the audiology track will be able to supervise a hearing conservation program perform; patient care in ENT clinics and audiology clinics, including audiometric testing and use of prosthetics; and supervise an audiology clinic. Those in the environmental science/engineering track will be able to identify arthropods, insects, and other animals of military importance; discuss principles of water quality, water supply, food service sanitation, water microbiology, water chemistry, air pollution, industrial hygiene, hazardous waste management and medical water management, foodborne illness, pesticide handling, preventive medicine in disasters, and the medical effects of lasers, microwaves, and directed energy. Those in the medical service corps/NBC will be able to discuss basic principles of radiation of various types, operate various equipment to measure radiation of a number of types, discuss principles of radiation safety including shielding and packaging of radioactive waste and contamination control procedures, and discuss biological effects of radiation exposure. Those in the medical service corps track will be able to discuss combat stress control measures; discuss programs to support the family and individual, including family advocacy, suicide prevention/psychological autopsy, and exceptional family members; and discuss principles of behavioral science administration and staff supervision. Those in the medical service corps/pharmacy track will be able to discuss the role and responsibilities of the pharmacist in the military, discuss principles of sterile product management, and discuss operations of inpatient and outpatient pharmacy service. Those in the optometry track will be able to discuss the roles and responsibility of the optometrist in the military, perform optometry service in an optometry clinic, and serve as a division optometrist. Those in the medical service corps/occupational therapy track will be able to discuss the roles and responsibilities of an occupational therapist in the military. Those in the medical service corps/physical therapy track will be able to discuss the roles and responsibilities of an physical therapist in the military. Those in the medical service corps/dietitian track will be able to discuss the roles and responsibilities of the dietitian in the military. Those in the medical service corps/physician assistant track will be able to discuss the roles and responsibilities of the physician assistant in the military, discuss principles of field sanitation and industrial toxicology, and discuss preventive medical conditions with special significance to the military. And those in the veterinary corps track will be able to discuss the roles and responsibilities of the veterinary corps in the military and discuss current issues and methods of food technology, sanitation, inspection, and animal care. *Version 3:* The course consists of a preparatory phase, a common core, and subject specific tracks. Upon completion of the preparatory phase, the student will be able to discuss the fundamentals of the Uniform Code of Military Justice; describe the mission, function, and structure of the Army and the Department of Defense; discuss the principles of war and military history; discuss basic principles of military leadership and protocol; and implement basic first aid principles and nuclear, biological, and chemical protective measures. Upon completion of the common core, the student will be able to apply basic principles of health care administration, including budgeting, resource management, and the health care delivery systems; discuss legal issues, including military justice, standards of conduct, tort liability, the Federal Tort Claims Act, Geneva Convention of Code and Conduct, the Freedom of Information Act, and the Privacy Act; discuss the medical and dental health care system in the Army; discuss principles of military science, including command and staff functions, combat health support, uniforms, drills and ceremonies; discuss principles of organizational management, riot management/assessment, team building, and EEO; and participate in field experiences. Those in the nursing track will be able to discuss health care concepts and management philosophies, including ethics, resource management, work load management, labor relations, personnel issues, and quality improvement related to nursing and to identify career planning strategies, including empowerment, coping strategies, stress management, team building, and mentorship. Those in the dental corps track will be able to discuss dentistry field operations including the role of dental enlisted personnel and discuss dental care concepts, including dental quality assurance, dental records, and preventive dentistry programs. Those in the medical corps track will be able to discuss the military and medical roles/responsibilities of a brigade-level medical officer and discuss selected diseases of military importance. Those in the medical service corps platoon leader track will be able to discuss the roles and responsibilities of field medical support operations and discuss principles of health services planning, operations, training, and security. Those in the clinical support services laboratory track will be able to discuss the roles and responsibilities of laboratory professional in the military and in warfare situations and discuss issues related to laboratory science, including blood bank/donor center, biological warfare, OSHA, laboratory safety, CLIP/CLIA, laboratory management, and quality assurance. Those in the audiology track will be able to supervise a hearing conservation program perform; patient care in ENT clinics and audiology clinics, including audiometric testing and use of prosthetics; and supervise an audiology clinic. Those in the environmental science/engineering track will be able to identify arthropods, insects, and other animals of military importance; discuss principles of water quality, water supply, food service sanitation, water microbiology, water chemistry, air pollution, industrial hygiene, hazardous waste management and medical water management, foodborne illness, pesticide handling, preventive medicine in disasters, and the medical effects of lasers, microwaves, and directed energy. Those in the medical service corps/NBC will be able to discuss basic principles of radiation of various types, operate various equipment to measure radiation of a number of types, discuss principles of radiation safety including shielding and packaging of radioactive waste and contamination control procedures, and discuss biological effects of radiation exposure. Those in the medical service corps track will be able to discuss combat stress control measures; discuss programs to support the family and individual, including family advocacy, suicide prevention/ psychological autopsy, and exceptional family members; and discuss principles of behavioral science administration and staff supervision. Those in the medical service corps/pharmacy track will be able to discuss the role and responsibilities of the pharmacist in the military, discuss principles of sterile product management, and discuss operations of inpatient and outpatient pharmacy service. Those in the optometry track will be able to discuss the roles and responsibility of the optometrist in the military, perform optometry service in an optometry clinic, and serve as a division optometrist. Those in the medical service corps/occupational therapy track will be able to discuss the roles and responsibilities of an occupational therapist in the military. Those in the medical service corps/physical therapy track will be able to discuss the roles and responsibilities of a physical therapist in the military. Those in the medical service corps/dietitian track will be able to discuss the roles and responsibilities of the dietitian in the military. Those in the medical service corps/physician assistant track will be able to discuss the roles and responsibilities of the physician assistant in the military, discuss principles of field sanitation and industrial toxicology, and discuss preventive medical conditions with special significance to the military. And those in the veterinary corps track will be able to discuss the roles and responsibilities of the veterinary corps in the military and discuss current issues and methods of food technology, sanitation, inspection, and animal care.

Instruction: ***Version 1:*** Phase 1 (core) of this course covers the Army medical-dental care system, biomedical equipment management, manpower management, medical legal issues, personnel management, roles of the Army medical department personnel, Army organization, medical military history, combat medical issues, and preventive medicine. All students are provided with lectures, conferences, and practical exercises in the core (Phase 1), then progress to specialty tracks. The Veterinary track covers inspection of poultry, beef, pork, dairy, and fishery products; animal disease control; mission of veterinary services in the Army; personnel management; quality assurance; and procurement and storage of fresh fruit and vegetables. The Dental track covers dental corps organization and the roles and responsibilities of dental personnel as well as the dental health record. The Pharmacy track covers organization of pharmacy service, pharmacy supply procedures, hyperalimentation, sterile products, personnel management, computerization of the pharmacy, quality assurance, and controlled substances. The Nursing track covers basic trauma life support, mission and organization of the Army nurse corps, documentation of nursing process, workload management, quality assurance, mentoring, and legal issues in nursing. The Military Science track covers maintenance forms/records, preventive maintenance service, and emergency medical care. The Behavioral Science track covers Army mental health services, staff supervision, crisis intervention in combat and disaster, Army family advocacy program, psychological aspects of AIDS and of terrorism, quality assurance, and ethics. The Preventive Medicine track (68B) covers health physics, radiation units, radioactive decay, nuclear statistics, radiation protection, biological effects of radiation, radiation regulatory requirements, radioactive waste management, and safety surveys of radiographic equipment. The Preventive Medicine track (68F, 68N, and 68P) covers water supply, swimming pools, sanitation, industrial hygiene and entomology, and preventive medicine. The Optometry track covers Army personnel management, the role of the optometry service in the military, and equipment maintenance. The Medicine and Surgery track covers the mission of the medical corps, brigade operation and combat, medical training of Army personnel, medical implications of Army combat operations and of chemical warfare, field operations, and sick call. The Human Resources Management track covers nutrition care in Army hospitals, the Army's computerized food service system, field feeding system, thermal injuries, clinical dietetic management, and the dining facility. The Laboratory Sciences track covers the curriculum of medical laboratory specialists and combat laboratory materials. The Audiology track covers audiology clinic management, audiology in the Army, and hearing conservation programs. *Version 2:* This course is divided into three modules. The preparatory and common core are completed by all, who then attend specific tracks. Lectures, practical exercises, and field exercises are included in the program. The preparatory module is for those students who have no prior military experiences. The common core provide initial training in the Army medical and dental health care systems, health care operations, and deployable skills. The track modules provide specialized content related to the health care profession. Only one track is completed by the student. Some track include clinical experience as well as field training exercises. *Version 3:* This course is divided into three modules. The preparatory and common core are completed by all, who then attend specific tracks. Lectures, practical and field exercises are included in the program. The preparatory module is for those students who have no prior military experiences. The common core provides initial training in the Army medical and dental health care systems, health care operations, and deployable skills. The track modules provide specialized content related to the health care profession. Only one track is completed by the student. Some tracks include clinical experience as well as field training exercises.

Credit Recommendation: ***Version 1:*** In the lower-division baccalaureate/associate degree category, 1 semester hours in emergency and preventive medicine for core studies (Phase 1); for those in the Military Science track, 3 semester hours in first aid. In the upper-division baccalaureate category, 3 semester hours in health care administration and 3 in military science for the core studies (Phase 1). For those in the Veterinary track, 1 semester hours in military science, 4 in food and meat inspection, and 1 in animal disease prevention and control. For those in the Pharmacy track, 2 semester hours in sterile production and 1 in clinical practicum. For those in the Nursing track, 1 semester hour in basic trauma life support. For those in the Behavioral Science track, 1 semester hour in crisis intervention. For those in Preventive Medicine (68B) track, 2 semester hours in radiation protection. For those in Preventive Medicine (86F, 68N, 68P) track, 1 semester hour in principles of sanitation. For those in Medicine and Surgery track, 1 semester hour in military science. For those in the Audiology track, 1 semester hour in audiology practicum. For those in the Medicine and Surgery track, 1 semester hour in military science. In the graduate degree category, 2 semester hours in managing chemical injuries for those in the Medicine and Surgery track (8/93). *Version 2:* In the lower-division baccalaureate/associate degree category, 2 semester hours in military science for completion of the preparatory phase. In the upper-division baccalaureate category, 3 semester hours in military science, 3 in basic health care administration, and 3 in basic organizational behavior/management for completion of the common core phase. For those in the nursing track, 1 semester hour in emergency nursing and 1 additional semester hour in military science. For those in the dental track, an additional 2 semester hours in military science. For those in the medical track, 2 semester hours in diseases of military importance and 1 additional in military science. For those in the medical service corps platoon leader track, 2 semester hours in health services planning and operations and 4 additional in military science. For those in the clinical support services track, an additional 2 semester hours in military science and 2 in advanced laboratory practicum. For those in the audiology track, an additional 2 semester hours in military science and 2 in clinical practicum. For those in environmental science/ engineering track, an additional 2 in military science and 2 in preventive medicine/industrial hygiene. For those in the NBC track, 6 semester hours in radiation safety. For those in behavioral science track, an additional 2 in military science and 2 in issues in behavioral science. For those in the pharmacy track, an additional 2 semester hours in military science and 3 in sterile products. For those in the optometry track, an additional 2 in military science and 2 in clinical practicum. For those in the occupational therapy track, the physical therapy track, and the dietician track, an additional 2 semester hours in military science. For those in the physician assistant track, an additional 2 semester hours in military science and 2 in diseases of military importance. For those in the veterinary track, an additional 2 in military science and 3 in current issues in veterinary science (10/95). *Version 3:* In the lower-division baccalaureate/ associate degree category, 2 semester hours in basic military science for completion of the preparatory phase; for completion of the common core, 2 semester hours in basic military science and 2 in wilderness survival; for those in the nursing track, dental track, clinical support track, environmental science/engineering track, behavioral science track, occupational therapy track, physical therapy track, and the dietician track, an additional 2 semester hours in military science; for those in the audiology track, pharmacy track, and optometry track, an additional 1 semester hour in military science. In the upper-division baccalaureate category, for completion of the common core, 3 semester hours in health care administration, 3 in organizational behavior, 2 in leadership command/control and 2 in military law; for those in the nursing track, an additional 1 semester hour in emergency nursing; for those in the medical track, an additional 2 semester hours in military pathophysiology; for those in the clinical support track, an additional 2 semester hours in advanced laboratory practicum; for those in the audiology track, an additional 2 semester hours in audio clinical practicum; for those in the environmental science/engineering track, an additional 2 semester hours in preventive medicine/industrial hygiene; for those in the NBC track, 5 semester hours in radiology; for those in the behavioral science track, an additional 2 semester hours in issues in behavioral science; for those in the pharmacy track, an additional 1 semester hour in a pharmacy clinical practicum; for those in optometry track, an additional 1 semester hour in an optometry clinical practice; for those in the veterinary track, 1 semester hour in sanitation and safety, 3 in food technology, and 3 in veterinary science (6/02).

Related Occupations: 65C; 67B; 68B; 68F; 68L; 68N; 68P; 68R; 68S; 68U.

AR-0707-0020

PRINCIPLES OF MILITARY PREVENTIVE MEDICINE

Course Number: 6A-F5.

Location: Academy of Health Sciences, Fort Sam Houston, TX.

Length: 9 weeks (365 hours).

Exhibit Dates: 6/90–1/93.

Learning Outcomes: Upon completion of the course, the student will be able to apply the principles and concepts of preventive health services, including basic pest control, industrial and occupational considerations, and preventive medicine issues with an emphasis on roles and responsibilities; prevention programs; disease control; and nuclear, radiation, and other safety hazards. Those in the physician assistant/veterinary corps track will be able to apply preventive health service principles and procedures with an emphasis on family health issues, sanitation and water treatment, industrial health problems, and disease prevention and control. Those in the medical service corps (entomology) will be able to apply preventive health service principles with an emphasis on control of pests, sanitation and water treatment, and disease prevention and control. Those in the community health nurse track will have an emphasis on intervention in domestic violence and child abuse, health and nutrition support to families, industrial hygiene issues, and public health strategies in the community. Those in the general medical service corps, excluding entomologists, will have an emphasis on sanitary and environmental hygiene situations.

Instruction: This course covers topics in community health practices; communicable and infectious diseases; epidemiology; statistics; hygiene; health physics; sanitary engineering; and environmental sciences as core topics. Those in the physician assistant/veterinary corps track will cover topics in family case referral, health topics, protective equipment, and disease prevention and control. Those in the medical service corps track (entomologists) cover topics in pest management, waste management, field water supply, protective equipment, and public health issues. Those in the community health nurse track cover topics in wellness, sexual abuse, nutrition and other community issues, public health problems such as pediatric infectious diseases, industrial hygiene, and occupational and behavioral health risks. Those in the medical service corps (except entomologists) cover pest management; field water supply, waste management, protective equipment, and disease prevention and control.

Credit Recommendation: In the upper-division baccalaureate category, for those in the medical service corps (except entomologists) 3 semester hours in principles of industrial hygiene and 3 in community health principles. For those in the veterinary corps/physician assistant track, 3 semester hours in public health and 3 in issues in preventive medicine. For those in the medical corps (entomology) track, 4 in issues in entomology and 2 in public health principles. And for those in the community health nurse track, 3 semester hours in public health principles and 3 in community health nursing (8/93).

AR-0707-0021

INTERMEDIATE INDUSTRIAL HYGIENE TOPICS

Course Number: 6H-F10/322-F10.

Location: Academy of Health Sciences, Fort Sam Houston, TX.

Length: 2 weeks (78 hours).

Exhibit Dates: 3/94–Present.

Learning Outcomes: Upon completion of the course, the student will be able to conduct sampling of airborne contaminants; describe requirements of a respiratory protection program; discuss current trends in industrial hygiene; describe management of asbestos; calibrate industrial hygiene survey equipment; identify elements of ergonomics programs; and discuss other environmental issues, including toxicology, air quality, ventilation, solvents, paint, hearing conservation, and blood-borne pathogens.

Instruction: Lectures, small group discussions, and practical exercises cover industrial hygiene techniques used in the recognition, evaluation, and control of occupational health hazards.

Credit Recommendation: In the lower-division baccalaureate/associate degree category, 4 semester hours in principles of industrial hygiene (10/95).

AR-0707-0022

TROPICAL MEDICINE

Course Number: 6H-F23.

Location: Walter Reed Institute of Research, Washington, DC.

Length: 6 weeks (216 hours).

Exhibit Dates: 1/90–7/95.

Learning Outcomes: Upon completion of the course, the student will be able to discuss the principles of immune response to tropical diseases, diagnose and treat common tropical diseases, prevent tropical diseases in selected populations, and perform appropriate tests to identify tropical diseases.

Instruction: This course provides a broad introduction and overview of tropical medicine. All major diseases, health care problems, and prevention issues in tropical medicine are addressed. Lectures and laboratory experiences are provided.

Credit Recommendation: In the graduate degree category, 10 semester hours in tropical medicine (10/95).

AR-0707-0023

ARMY MEDICAL DEPARTMENT (AMEDD) OFFICER BASIC

Course Number: 6-8-C20(MC/DC/VC).

Location: Academy of Health Sciences, Fort Sam Houston, TX.

Length: 3 weeks (122 hours).

Exhibit Dates: 1/90–9/90.

Learning Outcomes: Upon completion of the core curriculum of the course, the student will be able to discuss roles and responsibilities of the medical, dental, or veterinary officer in the military in peace and wartime. Those in the medical phase will be able to discuss current concepts in the treatment of burns and high-velocity ballistic wounds. Those in the dental phase will be able to discuss the principles of preventive dentistry. Those in the veterinary phase will be able to discuss the principles of human resources.

Instruction: This course includes lectures on the responsibilities of the medical department officer, military customs and traditions, principles of medical support of combat operations, the Uniform Code of Military Justice, NBC protective equipment, and substance abuse.

Credit Recommendation: In the lower-division baccalaureate/associate degree category, 2 semester hours in military science (medical application) for the core curriculum and 1 semester hour in selected track (medical, dental, or veterinary) (10/95).

AR-0707-0024

PRINCIPLES OF MILITARY PREVENTIVE MEDICINE

Course Number: ***Version 1:*** 6A-F5. *Version 2:* 6A-F5.

Location: Academy of Health Sciences, Fort Sam Houston, TX.

Length: ***Version 1:*** 9 weeks (312–354 hours). *Version 2:* 9 weeks (298–379 hours).

Exhibit Dates: ***Version 1:*** 2/96–12/98. *Version 2:* 1/99–Present.

Learning Outcomes: ***Version 1:*** Upon completion of the course, the student will be able to apply entry-level principles of current issues in community and public health. Students in the environmental science and engineering officer track (162 hours) will be able to apply the principles and concepts of sanitation and hygiene, pest management procedures, hazardous materials, water management, and disease prevention. Students in the medical corps, veterinary corps, and physician's assistant tracks (142 hours) will be able to apply the principles and concepts of health promotion and disease prevention and maintenance of occupational and family health. Students in the medical service corps track (entomologists) will be able to apply preventive health service principles with an emphasis on pest management, water treatment, and disease prevention and control. Students in the community health nurse track (122 hours) will be able to apply community health principles with an emphasis on domestic violence and child abuse intervention, health and nutrition support to families, industrial hygiene, and public health intervention in the community. Students in the nuclear medicine science officer track (125 hours) will be able to apply the principles of radiation detection and protection with an emphasis on radiation's effects and management, and current regulatory requirements. Students in the audiologist track (122 hours) will be able to apply the principles of preventive medicine related to audiology with an emphasis on health promotion and disease prevention, occupational health, and public health. *Version 2:* Upon completion of the course, the student will be able to apply entry-level principles of current issues in community and public health. Students in the environmental science and engineering officer track will be able to apply the principles and concepts of sanitation and hygiene, pest management procedures, hazardous materials, and water management, and disease prevention. Students in the medical corps, veterinary corps, and physician's assistant tracks will be able to apply the principles and concepts of health promotion and disease prevention and maintenance of occupational and family health. Students in the medical service corps track (entomologists) will be able to apply preventive health service principles with an emphasis on pest management, water treatment, and disease prevention and control. Students in the community health nurse track will be able to apply community health principles with an emphasis on domestic violence and child abuse intervention, health and nutrition support to families, industrial hygiene, and public health intervention in the community. Students in the nuclear medicine science officer track will be able to apply the principles of radiation detection and protection with an emphasis on radiation's effects and management, and current regulatory requirements. Students in the audiologist track will be able to apply the principles of preventive medicine related to audiology with an emphasis on health promotion and disease prevention, occupational health, and public health.

Instruction: ***Version 1:*** The course is an entry-level core course for several different officer specialty groups and offers common topics in preventive medicine that provide a basis for six specialty track courses. Common topics include sexually transmitted disease, health promotion and disease prevention, industrial hygiene, and occupational medicine. Instruction occurs primarily through lectures, discussions, and guided experiences. For those in the environmental science and engineering officers track, the course covers topics in environmental sciences such as water treatment and management, food service sanitation, and community and industrial hygiene. Instruction is provided through lectures and guided experiences. For those in the medical corps, veterinary corps, and physician's assistant tracks, the course covers topics on the maintenance of optimum health and wellness and occupational health issues. Instruction is through lectures and guided experiences. For those in the medical service corps track (entomologists), the course covers topics in pest control, waste management, and disease control. Instruction is through lectures and guided experiences. For those in the community health nurse track, the course covers topics in optimum health and wellness, family health, sexual abuse, disease prevention, and occupational and community health. Instruction is provided through lectures and guided experiences. For those in the nuclear medical sciences officer track, the course covers topics in radiation detection systems, issues related to radiation protection and control, nuclear waste management, and current regulations and standards. Instruction is provided through lectures and guided experiences. For those in the audiologist track, the course covers topics in maintenance of optimum health and wellness, occupational and environmental health, family health issues such as domestic violence and sexual abuse, and issues related to audiology and preventive medicine. Instruction is provided through lectures and guided experiences. *Version 2:* Instruction is given through lectures, seminars, discussion groups, hands-on demonstrations and examinations. Common topics include sexually transmitted disease, health promotion and disease prevention, industrial hygiene, and occupational medicine.

Credit Recommendation: ***Version 1:*** In the upper-division baccalaureate category, for all students, who will have taken the common core, 3 semester hours in topics in preventive medicine and 2 in issues in public health. For those in the environmental science and engineering officers tracks, 4 semester hours in industrial hygiene; for those in the medical corps, veterinary corps, and physician assistant tracks, 3 semester hours in wellness; for those in the medical service corps track (entomologists), 3 semester hours in environmental and waste control; for those in the community health nurse track, 3 semester hours in issues in community health; for those in the nuclear medical science officer track, 3 semester hours in issues in radiation protection; and for those in the audiologist track, 3 semester hours in current issues in audiology (5/97). *Version 2:* In the upper-division baccalaureate category, 5 semester hours in fundamentals of preventive medicine, 3 in preventive medicine principles of environmental science, 2 in methods in preventive medicine for health professionals, and 3 in topics in nuclear medicine preventive medicine. For audiology track: 2 semester hours in research topics in audiology. For community health nurse track: 3 semester hours in preventive medicine for the community health nurse. For entomology track: 2 semester hours in topics in preventive medicine for entomology (7/99).

AR-0707-0025

Immunization/Allergy Specialty

Course Number: 300-F4.

Location: Walter Reed Medical Center, Washington, DC.

Length: 5 weeks (191 hours).

Exhibit Dates: 10/96–Present.

Learning Outcomes: Upon completion of the course, the student will be able to select appropriate equipment and sites for administering intradermal, subcutaneous, and intramuscular injections; administer adult and pediatric vaccines; discuss the role of the immune system in allergy and immunization; perform and interpret skin tests; prepare and perform allergen immunotherapy; and perform diagnostic procedures for evaluation of patient's immune response.

Instruction: Instruction by lecture, demonstration, and clinical practice includes basic immunology related to hypersensitivity, types of allergens, immunization techniques and schedules, evaluation techniques, and immunotheraphy.

Credit Recommendation: In the lower-division baccalaureate/associate degree category, 2 semester hours in basic nursing skills and 1 in nursing practicum (10/00).

AR-0707-0026

Army Medical Department (AMEDD) Noncommissioned Officer (NCO) Basic

Course Number: 6-8-C40 (91S30).

Location: Academy of Health Sciences, Fort Sam Houston, TX.

Length: 7 weeks (300 hours).

Exhibit Dates: 4/95–Present.

Learning Outcomes: Upon completion of the course, the student will be able to lead a military unit; performance and counsel; supervise and delegate; and perform evaluate drills, training, and specific military skills.

Instruction: The course covers military leadership, personal and performance counseling, duties and responsibilities of noncommissioned officers, leadership assessment and development, leadership procedures and combat orders, combat leadership, physical training and marksmanship, property accountability, individual and team training, small-group process, and specific military skills. The methodologies include lectures, discussions, group work, and practical application. This course contains a common core of leadership subjects.

Credit Recommendation: In the lower-division baccalaureate/associate degree category, 3 semester hours in military science and 1 in elective credit in preventive medicine. See AR-1406-0090 for common core credit (11/97).

AR-0707-0027

Shiploading and Stowage

Course Number: A-8C-0013; 822-F26; 8C-F29; 8C-F29/822-F26.

Location: Transportation School, Fort Eustis, VA.

Length: 2 weeks (80 hours).

Exhibit Dates: 10/94–Present.

Learning Outcomes: Upon completion of the course, the student will be able to preplan and supervise the loading, stowing, and discharge of a cargo ship.

Instruction: Methods of instruction include conferences and performance exercises relating to cargo storage, cargo stability, logistics, and procedures on development of a pre-stow plan.

Credit Recommendation: In the lower-division baccalaureate/associate degree category, 1 semester hour in industrial safety and 1 in basic mathematics (6/01).

AR-0707-0028

Chemical Officer Basic

Course Number: 4-3-C20-74A.

Location: Chemical School, Fort Leonard Wood, MO.

Length: 19–20 weeks (651–652 hours).

Exhibit Dates: 10/00–Present.

Learning Outcomes: Upon completion of the course, the student will be able to detect and identify nuclear, radiation and biological agents, apply basic concepts of meteorology to the dispersal of nuclear, biological, and chemical agents; and communicate orally and in writing.

Instruction: Methods of instruction include classroom lectures, group practical exercises, operational exercises, and written and oral examinations.

Credit Recommendation: In the lower-division baccalaureate/associate degree category, 2 semester hours in introduction to meteorology, 1 in technical writing, 1 in oral communication skills, and 3 in environmental science (1/02).

AR-0707-0029

Basic Industrial Hygiene Techniques

Course Number: 611-F11/322-F11.

Location: Academy of Health Sciences, Fort Sam Houston, TX.

Length: 2 weeks (85 hours).

Exhibit Dates: 1/00–Present.

Learning Outcomes: Upon completion of the course, the student will be able to identify basic principles of industrial hygiene techniques in the recognition, evaluation and control of occupational health hazards.

Instruction: Practical exercises, classroom, laboratory, lecture. Topics include ventilation, survey and sample collection, use of basic industrial hygiene, OSHA regulations, noise pollution abatement, and personal protective equipment hazard evaluation.

Credit Recommendation: In the upper-division baccalaureate category, 4 semester hours in industrial hygiene (11/03).

AR-0708-0001

Psychiatric Specialist
(Psychiatric Procedures Basic)
(Neuropsychiatric Procedures Basic)
(Neuropsychiatric Specialist)

Course Number: B-302-0045; B-302-A045; 302-91F20; 302-91F10.

Location: Phase 1, Academy of Health Sciences, Fort Sam Houston, TX; Phase 2, Civilian/Military Medical Facilities, Continental, US; Phase 2, Naval Regional Medical Centers, Continental, US; Phase 2, Medical Field Service School, Oakland, CA; Phase 2, Naval Regional Medical Center, Portsmouth, VA; Phase 2, Wilford Hall AF Medical Center, Lackland AFB, TX; Phase 2, Audie Murphy Memorial Veterans Hospital, San Antonio, TX; Phase 2, Naval School of Health Sciences Naval Hospital, San Diego, CA.

Length: 8–12 weeks (472–477 hours).

Exhibit Dates: 1/69–6/94.

Learning Outcomes: After 6/94 see AR-0708-0005. To provide nursing personnel with practical knowledge and skills to assist in the care and treatment of psychiatric patients.

Instruction: Lectures and practical exercises cover the performance of managerial, clerical, and security duties of psychiatric attendants; identification, preparation, administration, and observation for side effects of medication as required; observation, recording, and reporting of patient behavior; assistance in psychiatric therapies; evaluation and fulfillment of the physical and psychological nursing needs of psychiatric patients; behavioral recognition; therapeutic intervention procedures; lectures on understanding human development; and etiology, dynamics, and symptomatology of mental illness.

Credit Recommendation: In the lower-division baccalaureate/associate degree category, 3 semester hours in abnormal psychology, 3 in psychiatric nursing, 3 in psychiatric therapies, and 5 in psychiatric or mental health clinical practice. In the upper-division baccalaureate category, credit in psychology and psychiatric nursing on the basis of institutional evaluation (1/90).

Related Occupations: 91F.

AR-0708-0003

Psychiatric Mental Health Nurse

Course Number: 6F-66C.

Location: Dwight David Eisenhower Medical Center, Fort Gordon, GA; Academy of Health Sciences, Fort Sam Houston, TX.

Length: 22 weeks (791–838 hours).

Exhibit Dates: 7/79–6/93.

Objectives: After 6/93 see AR-0708-0004. To prepare Army Nurse Corps officers to provide care and treatment for psychiatric patients in a hospital setting.

Instruction: This course uses lectures, conferences, and practical experience to cover therapeutic psychiatric nursing, psychiatric dysfunction, group therapy, somatic therapy, therapeutic milieu, and nursing unit management. All subject areas include a directed clinical practicum.

Credit Recommendation: In the graduate degree category, 2 semester hours in psychological dysfunction, 3 in group therapy, 3 in therapeutic psychiatric nursing, and 4 in supervised clinical experience in psychiatric nursing (7/93).

AR-0708-0004

1. Psychiatric/Mental Nursing
2. Psychiatric/Mental Health Nursing
3. Psychiatric/Mental Health Nursing
4. Psychiatric/Mental Health Nursing
5. Psychiatric/Mental Health Nursing

Course Number: *Version 1:* 6F-66C. *Version 2:* 6F-66C. *Version 3:* 6F-66C. *Version 4:* 6F-66C. *Version 5:* 6F-66C.

Location: *Version 1:* Academy of Health Sciences, Fort Sam Houston, TX. *Version 2:* Academy of Health Sciences, Fort Sam Houston, TX. *Version 3:* Dwight David Eisenhower Medical Center, Fort Gordon, GA; Academy of Health Sciences, Fort Sam Houston, TX. *Version 4:* Dwight David Eisenhower Medical Center, Fort Gordon, GA. *Version 5:* Academy of Health Sciences, Fort Sam Houston, TX.

Length: *Version 1:* 22 weeks (791 hours). *Version 2:* 22 weeks (792 hours). *Version 3:* 22 weeks (844 hours). *Version 4:* 22 weeks (845 hours). *Version 5:* 16 weeks (645 hours).

Exhibit Dates: *Version 1:* 7/93–12/95. *Version 2:* 1/96–6/98. *Version 3:* 7/98–6/00. *Version 4:* 7/00–12/01. *Version 5:* 1/02–Present.

Learning Outcomes: *Version 1:* Before 7/93 see AR-0708-0003. Upon completion of the course, the student will be able to identify the various roles of the psychiatric treatment team, identify the role of the psychiatric nurse, apply the nursing therapeutic process to the care of emotionally distressed individuals, apply selected theories of personality development, assess patients for psychiatric dysfunction, conduct therapeutic groups, apply somatic therapies to patient care, establish a therapeutic milieu, apply management skills to a psychiatric unit, and apply the nursing process to the care of a broad range of psychiatric patients. *Version 2:* Upon completion of the course, the student will be able to identify the various roles of the psychiatric treatment team, identify the role of the psychiatric nurse, apply the nursing therapeutic process to the care of emotionally distressed individuals, apply selected theories of personality development, assess patients for psychiatric dysfunction, conduct therapeutic groups, apply somatic therapies to patient care, establish a therapeutic milieu, apply management skills to a psychiatric unit, and apply the nursing process to the care of a broad range of psychiatric patients. *Version 3:* Upon completion of the course, the student will be able to identify the various roles of the psychiatric treatment team, identify the role of the psychiatric nurse, apply the nursing therapeutic process to the care of emotionally distressed individuals, apply selected theories of personality development, assess patients for psychiatric dysfunction, conduct therapeutic groups, apply somatic therapies to patient care, establish a therapeutic milieu, apply management skills to a psychiatric unit, and apply the nursing process to the care of a broad range of psychiatric patients. *Version 4:* Upon completion of the course, the student will be able to assess the health care needs of emotionally distressed individuals; provide preventive, supportive, rehabilitative, and restorative intervention for individuals and groups in the hospital setting; establish and maintain a therapeutic environment within the psychiatric nursing care unit; apply basic management skills in a psychiatric care unit. *Version 5:* Upon completion of the course, the student will be able to assess the health care needs of emotionally distressed individuals; provide preventive, supportive, rehabilitative, and restorative intervention for individuals and groups in the hospital setting; establish and maintain a therapeutic environment within the psychiatric nursing care unit; apply basic management skills in a psychiatric care unit.

Instruction: *Version 1:* This course uses lectures, discussions, presentations, clinical experiences, and intervention to assess and provide therapeutic treatment to a group of clients with varied psychiatric illnesses. The course also includes managing a psychiatric nursing unit. *Version 2:* This course uses lectures, discussions, presentations, clinical experiences, and interventions to cover establishing and maintaining a therapeutic environment; providing preventative, rehabilitative, and restorative interventions; determining and applying the appropriate management of various psychiatric dysfunctions; and managing a psychiatric nursing unit. *Version 3:* This course uses lectures, discussions, presentations, clinical experiences, and interventions to cover establishing and maintaining a therapeutic environment; providing preventative, rehabilitative, and restorative interventions; determining and applying the appropriate management of various psychiatric dysfunctions; and managing a psychiatric nursing unit. *Version 4:* Lectures, conferences, discussion, practical exercises, examinations, and clinical practicum. Topics include the fundamentals of psychiatric and mental health nursing; the role of the psychiatric nurse; research; nursing therapeutic processes; group therapy; somatic therapies; nursing unit management; psychiatric dysfunctions; and the role of the psychiatric nurse in the military. *Version 5:* Lectures, conferences, discussion, practical exercises, examinations, and clinical practicum. Topics include the fundamentals of psychiatric and mental health nursing; the role of the psychiatric nurse; research; nursing therapeutic processes; group therapy; somatic therapies; nursing unit management; psychiatric dysfunctions; and the role of the psychiatric nurse in the military.

Credit Recommendation: ***Version 1:*** In the upper-division baccalaureate category, 1 semester hour in orientation to psychiatric nursing, 1 in state-of-the-art psychiatric nursing, 2 in nursing therapeutic process, 1 in developmental theories, 4 in psychiatric dysfunction and group therapy, 1 in therapeutic milieu and nursing unit management, and 14 in clinical practicum (10/95). *Version 2:* In the upper-division baccalaureate category, 14 semester hours in clinical practicum. In the graduate degree category, 3 semester hours in psychiatric dysfunctions, 3 in group therapy, 3 in therapeutic process, and 1 in psychiatric unit management (7/97). *Version 3:* In the upper-division baccalaureate category, 14 semester hours in clinical practicum and 1 in introduction to counseling. In the graduate degree category, 3 semester hours in psychiatric dysfunctions, 3 in group therapy, 3 in therapeutic process, and 1 in psychiatric unit management (7/99). *Version 4:* In the upper-division baccalaureate category, 1 semester hour in introduction to counseling. In the graduate degree category, 14 semester hours in a psychiatric nursing clinical practicum, 3 in nursing therapeutic processes, 2 in psychiatric conditions, 4 in psychiatric therapies, and 1 in psychiatric unit management (6/02). *Version 5:* In the upper-division baccalaureate category, 1 semester hour in introduction to counseling. In the graduate degree category, 3 semester hours in nursing therapeutic processes, 2 in psychiatric conditions, 4 in psychiatric therapies, 1 in psychiatric unit management, and 10 for a psychiatric nursing clinical practicum (6/02).

AR-0708-0005

Psychiatric Specialist

Course Number: 302-91F10.

Location: Academy of Health Sciences, Fort Sam Houston, TX.

Length: 12 weeks (477 hours).

Exhibit Dates: 7/94–9/96.

Learning Outcomes: Before 7/94 see AR-0708-0001. Upon completion of the course, the student will be able to identify common problems in human development, identify common symptoms of psychopathology, demonstrate basic communication techniques, demonstrate appropriate supportive interventions in the care and treatment of psychiatric patients, and demonstrate appropriate protective interventions for the care and treatment of psychiatric patients.

Instruction: Lectures and practical exercises cover the performance of managerial, clerical, and security duties of psychiatric technicians; identification, preparation, administration, and observation for side effects of medication as required; observation, recording, and reporting of patient behavior; assistance in psychiatric therapies; evaluation and fulfillment of the physical and psychological nursing needs of psychiatric patients; behavioral recognition; therapeutic intervention procedures; lectures on understanding human development; and etiology, dynamics, and symptomatology of mental illness.

Credit Recommendation: In the lower-division baccalaureate/associate degree category, 3 semester hours in human development and psychopathology, 2 in psychiatric patient care, 3 in treatment modalities, and 6 in clinical practicum (10/95).

Related Occupations: 91F.

AR-0708-0008

Mental Health Specialist

Course Number: ***Version 1:*** 302-91X10, Phase 1. *Version 2:* 302-91X10, Phase 2.

Location: Academy of Health Sciences, Fort Sam Houston, TX.

Length: ***Version 1:*** 19 weeks (768 hours). *Version 2:* 20 weeks (802 hours).

Exhibit Dates: ***Version 1:*** 10/96–9/01. *Version 2:* 10/01–Present.

Learning Outcomes: ***Version 1:*** Upon completion of the course, the student will be able to recognize, assess, and evaluate individuals and families in need of mental care; assume responsibilities in health/psychiatric facilities; and provide substance abuse intervention. Students will apply an understanding of human development, psychopathological disorders, and appropriate communication techniques to testing, consulting, interviewing, and counseling patients. *Version 2:* Upon completion of the course, the student will be able to recognize, assess, and evaluate individuals and families in need of mental care; assume responsibilities in health/psychiatric facilities; and provide substance abuse intervention. Students will apply an understanding of human development, psychopathological disorders, and appropriate communication techniques to testing, consulting, interviewing, and counseling patients.

Instruction: ***Version 1:*** This course combines didactic classroom instruction, closely supervised practical experiences, and closely supervised clinical practicum. Topics include psychopathological disorders, behavior disorders, psychiatric nursing, and alcohol and drug abuse intervention and control. *Version 2:* Laboratory, audio-visual materials, case studies, discussions, classroom exercises, practical exercises, and lectures. Topics include human development, psychopathological disorders, psychiatric interventions, alcohol and drug abuse, psychological testing, combat stress, interviewing, counseling, and emergency treatment.

Credit Recommendation: ***Version 1:*** In the lower-division baccalaureate/associate degree category, 3 semester hours in psychopathology, 2 in substance abuse, 2 in introduction to psychiatric nursing, 3 in interviewing and testing, 4 in introduction to counseling, and 8 in a mental health clinical practicum (6/02). *Version 2:* In the lower-division baccalaureate/associate degree category, 1 semester hour in developmental psychology, 1 in introduction to mental health careers, 3 in abnormal psychology, 3 in introduction to psychological interventions, 4 in introduction to counseling, 3 in interviewing techniques, 1 in psychological testing, and 6 in a mental health clinical practicum (6/02).

Related Occupations: 91X.

AR-0708-0009

Army Medical Department (AMEDD)
Noncommissioned Officer (NCO)
Basic

Course Number: 6-8-C40 (91X30).

Location: Academy of Health Sciences, Fort Sam Houston, TX.

Length: 7 weeks (300 hours).

Exhibit Dates: 4/95–Present.

Learning Outcomes: Upon completion of the course, the student will be able to lead a military unit; evaluate performance and counsel; supervise and delegate; and perform drills, training, and specific military skills.

Instruction: The course covers military leadership, personal and performance counseling, duties and responsibilities of noncommissioned officers, leadership assessment and development, leadership procedures and combat orders, combat leadership, physical training and marksmanship, property accountability, individual and team training, small-group process, and specific military skills. The methodologies include lectures, discussions, group work, and practical application. This course contains a common core of leadership subjects.

Credit Recommendation: In the lower-division baccalaureate/associate degree category, 3 semester hours in military science and 1 in elective credit in a psychiatric specialty. See AR-1406-0090 for common core credit (11/97).

AR-0709-0002

Ear, Nose, and Throat Specialist
(Eye, Ear, Nose, and Throat Specialist)

Course Number: ***Version 4:*** 300-91U10. *Version 5:* 300-P2.

Location: ***Version 4:*** Academy of Health Sciences, Fort Sam Houston, TX; Selected Medical Centers and Activities, Continental US. *Version 5:* Academy of Health Sciences, Fort Sam Houston, TX.

Length: ***Version 4:*** Phase 1, 4–6 weeks (225 hours); Phase 2, 7–9 weeks (272–528 hours). *Version 5:* Phase 1, 6 weeks (237–238 hours); Phase 2, 7 weeks (264–280 hours).

Exhibit Dates: ***Version 4:*** 6/85–8/92. *Version 5:* 9/92–2/97.

Learning Outcomes: ***Version 4:*** Upon completion of the course, the student will be able to assist physicians and nurses in specialized outpatient services. *Version 5:* After 2/97 see AR-0709-0063. Upon completion of the course, the graduate will be able to identify the structures and functions of the ear, nose, and throat; develop skills in administering various tests and reporting results to appropriate personnel; assist the physician in clinics; perform specific treatments (excluding irrigation of ear) under supervision; identify specific equipment and/or instruments for an ENT clinic; match symptoms, causes, and diseases of the body parts to the correct treatment; and relate the pharmacological aspects to various areas of the clinic.

Instruction: ***Version 4:*** Lectures and practical exercises cover sterile techniques, audiometric and optical examinations, administration of medications, preparation and safeguarding of medical records, ordering and storing medical supplies, and identification and care of specialized instruments and equipment. Optics is not included. *Version 5:* Methods of instruction include didactic and practical experiences in a clinical setting. Topics include anatomy and physiology of ear, nose, throat; pathological conditions; diagnostic phases; sterilization; and assisting in the clinics.

Credit Recommendation: ***Version 4:*** In the vocational certificate category, 6 semester hours in medical assisting. In the lower-division baccalaureate/associate degree category, 1 semester hour in anatomy and physiology, 1 in audiology, and 4 in practicum in otolaryngology (3/90). *Version 5:* In the lower-division baccalaureate/associate degree category, for Phase 1, didactic, 6 semester hours in ENT; for Phase 2, clinical, 6 semester hours in ENT practicum (7/97).

Related Occupations: 91B; 91U.

AR-0709-0005

Pharmacy Specialist

Course Number: 312-91Q10; 8-R-932.2.

Location: Academy of Health Sciences, Fort Sam Houston, TX; Medical Field Service School, Fort Sam Houston, TX.

Length: 17–18 weeks (570–715 hours).

Exhibit Dates: 8/81–9/93.

Objectives: After 9/93 see AR-0709-0052. To provide pharmacy auxiliary personnel with a working knowledge of drugs, their sources, preparation, uses, incompatibilities, and doses; pharmaceutical symbols and terms; and storage, control procedures, and dispensing procedures performed under the supervision of a pharmacist or physician.

Instruction: Lectures and practical experience cover the fundamentals of pharmacy, including manufacture and labeling of pharmaceutical prescriptions, preparation of common drugs and medicines, use of pharmaceutical reference texts, and applied accounting and control procedures.

Credit Recommendation: In the lower-division baccalaureate/associate degree category, 2 semester hours in pharmacy administration and calculation, 1 in pharmaceutical chemistry, 2 in physiology and pathology, 3 in pharmacology, 4 in hospital pharmacy, and 3 in pharmaceutical preparations (3/92).

Related Occupations: 91Q.

AR-0709-0009

Clinical Specialist

Course Number: 300-91C10; 300-91C30.

Location: Hospitals, Continental US.

Length: 40 weeks (1372–1760 hours).

Exhibit Dates: 4/76–1/90.

Objectives: To provide medical specialists with the advanced skills and knowledge necessary to supervise and perform patient care duties appropriate to hospital and field medical assistants.

Instruction: Course includes a study of basic principles of management and administration, including personnel development, professional nursing sciences, pharmacology in patient care, and basic principles of health and environment related to Army medical field service and fundamental clinical experience emphasizing medical and surgical patient care, including surgical follow-through and immediate postoperative care, as well as ambulatory patient, maternal and newborn patient, and pediatric patient care.

Credit Recommendation: In the lower-division baccalaureate/associate degree category, 12 semester hours in clinical nursing experience, 8 in medical-surgical nursing, 4 in anatomy and physiology, 3 in maternal and child nursing, 2 in theory and practice of patient care, 2 in pharmacology, 1 in nutrition, 1 in preventive medicine, 1 in psychiatric medicine, and 3 in environmental health (6/77).

Related Occupations: 91B; 91C; 91Z.

AR-0709-0015

Optical Laboratory Specialist
(Optician/Optical Laboratory Specialist)

Course Number: *Version 2:* 311-42E10. *Version 3:* 311-91H10; B-311-0023; 311-42E10. *Version 4:* 311-91H10; B-311-0023; 311-42E10.

Location: *Version 2:* Medical Equipment and Optical School, Aurora, CO. *Version 3:* Tri-Service Optical School, Yorktown, VA. *Version 4:* Tri-Service Optical School, Yorktown, VA.

Length: *Version 2:* 21 weeks (815–945 hours). *Version 3:* 24 weeks (832–846 hours). *Version 4:* 24 weeks (944 hours).

Exhibit Dates: *Version 2:* 6/82–10/96. *Version 3:* 11/96–3/03. *Version 4:* 4/03–Present.

Learning Outcomes: *Version 2:* To provide students with a theoretical basis and working knowledge of ophthalmic optics, optical laboratory procedures, and optical equipment in order to fabricate prescribed eyewear. *Version 3:* Upon completion of the course, the student will be able to fabricate lenses, maintain and repair surfacing equipment, screen for common diseases and disorders of the eye, and perform selected clinical procedures. *Version 4:* Pending evaluation.

Instruction: *Version 2:* Lectures and practical exercises include prescription analysis, laboratory procedures and production of eye wear, ocular anatomy and physiology, skill development, lens verification, equipment maintenance and repair, and safety precautions. *Version 3:* Methods of instruction include lectures, discussions, and clinical practice in the physics of diseases and disorders of the visual system, administration of ocular pharmaceutical agents, geometrical optics, optical math, ophthalmic lens design, mechanical optics, spectacle fabrication, equipment repair and maintenance, clinical procedures, ophthalmic dispensing, and administration and clinical management. *Version 4:* Pending evaluation.

Credit Recommendation: *Version 2:* In the lower-division baccalaureate/associate degree category, 6 semester hours in ophthalmic optics, 7 in optical laboratory procedures, 7 in optical laboratory production, and 6 in optical laboratory apprenticeship (7/88). *Version 3:* In the lower-division baccalaureate/associate degree category, 1 semester hour in physics, 3 in diseases and disorders of the visual system, 1 in the administration of ophthalmic pharmacological agents, 1 in geometric optics, 2 in optical mathematics, 3 in ophthalmic lens design, 3 in mechanical optics, 3 in spectacle fabrication, 3 in equipment repair and maintenance, 2 in ophthalmic dispensing, 3 in clinical procedures, and 1 in administrative and clinical management (5/98). *Version 4:* Pending evaluation.

Related Occupations: 42E; 91H.

AR-0709-0017

Pharmacy Sterile Products

Course Number: 6H-68H/312-F1.

Location: Academy of Health Sciences, Fort Sam Houston, TX.

Length: Self-paced, maximum, 2 weeks (83 hours).

Exhibit Dates: 3/78–3/92.

Objectives: To provide pharmacy personnel with a working knowledge of the techniques, procedures, and administrative functions involved in preparing sterile products.

Instruction: Lectures, demonstrations, and practical exercises cover the work environment; the preparation, calculation, and administration of sterile products; and the management of a pharmacy sterile product section. Course has two elements: a precourse self-instruction phase requiring student to complete self-paced, exportable modules on pharmaceutical calculations and a two-week resident phase.

Credit Recommendation: In the upper-division baccalaureate category, 3 semester hours in specialized pharmacy or pharmaceutical chemistry (6/89).

Related Occupations: 91Q.

AR-0709-0023

Cardiac Specialist

Course Number: 300-91N10.

Location: Academy of Health Sciences, Fort Sam Houston, TX.

Length: 12 weeks (513 hours).

Exhibit Dates: 7/85–6/96.

Objectives: To provide fundamental techniques necessary to assume duties and responsibilities of the cardiac specialist.

Instruction: Course provides a comprehensive introduction to noninvasive cardiology technology consisting of 159 hours didactic and 354 hours of clinical practicum. Topics include interpretation of EKG patterns, performance of cardiac stress test, and use of quantitative assessment with Holtor monitor and 12 lead EKG for normal and abnormal cardiac conditions.

Credit Recommendation: In the lower-division baccalaureate/associate degree category, 6 semester hours of didactic instruction in noninvasive cardiology technology and 1 in clinical practicum in cardiology technology (9/85).

Related Occupations: 91N.

AR-0709-0027

Graves Registration Specialist Basic Technical

Course Number: 492-57F30.

Location: Quartermaster Center and School, Fort Lee, VA.

Length: 3–9 weeks (120 hours).

Exhibit Dates: 1/85–5/92.

Objectives: To train enlisted personnel in supervision of a graves registration activity.

Instruction: Lectures, conferences, practical exercises, demonstrations, and examinations cover search and recovery missions, mass casualty operations, operating procedures for collecting points, and temporary cemeteries.

Credit Recommendation: In the vocational certificate category, 4 semester hours in mortuary science or forensic sciences (8/88).

Related Occupations: 57F.

AR-0709-0032

Medical Specialist

Course Number: *Version 1:* 300-91A10. *Version 2:* 300-91A10. *Version 3:* 300-91B10. *Version 4:* 300-91B10.

Location: Academy of Health Sciences, Fort Sam Houston, TX.

Length: *Version 1:* 10 weeks (360 hours). *Version 2:* 10 weeks (384 hours). *Version 3:* 10

weeks (394 hours). *Version 4:* 10 weeks (362–451 hours).

Exhibit Dates: ***Version 1:*** 8/87–2/91. *Version 2:* 3/91–3/93. *Version 3:* 4/93–4/96. *Version 4:* 5/96–12/01.

Learning Outcomes: ***Version 1:*** Upon completion of the course, the student will be able to measure and assess vital signs, complete and maintain records, perform CPR, administer oxygen, manage a patient with artificial airways, manage wounds, administer injections and immunizations, treat chemical injuries, initiate measures to prevent spread of communicable diseases, and perform basic field sanitation procedures. *Version 2:* Upon completion of the course, the student will be able to measure and record vital signs, perform CPR and first aid, administer oxygen therapy, perform patient assessment, manage patients with artificial airways, control bleeding, administer immunizations and injections, initiate and maintain intravenous therapy, immobilize orthopedic injuries, perform selected invasive procedures, and complete the National Registry Exam for Emergency Medical Technician-Basic certification. *Version 3:* Upon completion of the course, the student will be able to measure and record vital signs, practice medical-surgical asepsis and infection control, perform CPR and first aid, administer oxygen therapy, perform patient assessment, control bleeding, care for patients with shock, immobilize orthopedic injuries, administer immunizations and injections, initiate and maintain intravenous therapy, and manage patients with artificial airways. Students complete National Registry Exam for Emergency Medical Technician Basic certification during the course. *Version 4:* Upon completion of the course, the student will be able to measure and record vital signs, practice medical-surgical asepsis and infection control, perform CPR and first aid, administer oxygen therapy, perform patient assessment, control bleeding, care for patients with shock, immobilize orthopedic injuries, administer immunizations and injections, initiate and maintain intravenous therapy, and manage patients with artificial airways. Students complete National Registry Exam for Emergency Medical Technician Basic certification during the course.

Instruction: ***Version 1:*** Lectures, laboratories, and demonstrations cover anatomy and physiology, medical records, and nursing care procedures. *Version 2:* Methodology includes lectures, demonstrations, practical exercises, and field training. Course content includes control of bleeding, splinting, medical emergencies, anatomy and physiology, invasive skills, immunizations, blood and intravenous nursing skills, and record keeping. *Version 3:* Course covers bleeding and shock, splinting, medical emergencies, invasive skills, injection, intravenous infusion and blood drawing, nursing skills, anatomy and physiology, acute minor illness, and record keeping. Methodology includes lectures, demonstrations, practical exercises, and field training in CPR. *Version 4:* Course covers bleeding and shock, splinting, medical emergencies, invasive skills, injection, intravenous infusion and blood drawing, nursing skills, anatomy and physiology, acute minor illness, and record keeping. Methodology includes lectures, demonstrations, practical exercises, and field training in CPR.

Credit Recommendation: ***Version 1:*** In the lower-division baccalaureate/associate degree category, 2 semester hours in physiology, 2 in first aid, and 1 in nursing care (4/88). *Version 2:* In the lower-division baccalaureate/associate degree category, 3 semester hours in basic life support and first aid, 3 in emergency medical technology, 2 in anatomy and physiology, and 2 in medical terminology (8/93). *Version 3:* In the lower-division baccalaureate/associate degree category, 3 semester hours in basic life support or first aid, 4 in emergency medical technology, 2 in anatomy and physiology, and 2 in medical terminology (8/93). *Version 4:* In the lower-division baccalaureate/associate degree category, 1 semester hour in anatomy and physiology, 4 in emergency medical technology, 2 in emergency medical technician clinical experience, 2 in basic nursing skills, and 1 in nursing clinical experience (5/01).

Related Occupations: 91A; 91B.

AR-0709-0033

MEDICAL NONCOMMISSIONED OFFICER (NCO)

Course Number: 300-91B20.

Location: Academy of Health Sciences, Fort Sam Houston, TX; Selected US Army Reserve and National Guard Academies, Continental US.

Length: 12 weeks (471 hours).

Exhibit Dates: 9/87–9/91.

Learning Outcomes: After 9/91 see AR-0709-0049. Upon completion of the course, the student will be able to function at the noncommissioned officer level in the battlefield, in garrison prehospital care, or emergency/ambulatory treatment facilities.

Instruction: Lectures and laboratories cover pharmacology, anatomy, physiology, paramedic skills, clinical skills, and infectious diseases.

Credit Recommendation: In the lower-division baccalaureate/associate degree category, 1 semester hour in hygiene and 2 in pharmacology. In the upper-division baccalaureate category, 3 semester hours in physiology. Credit for physician assistant programs as determined by the receiving institution (4/88).

Related Occupations: 91B.

AR-0709-0035

OPERATING ROOM SPECIALIST

Course Number: ***Version 1:*** 301-91D10. *Version 2:* 301-91D10, Phase 1; 301-91D10, Phase 2.

Location: ***Version 1:*** Phase 1, Academy of Health Sciences, Fort Sam Houston, TX; Phase 2, Selected Medical Facilities, various locations. *Version 2:* Phase 1, Academy of Health Sciences, Fort Sam Houston, TX; Phase 2, Selected MEDACS, MEDCENS, Continental US.

Length: ***Version 1:*** Phase 1, 6 weeks (216 hours); Phase 2, 6 weeks (254 hours). *Version 2:* Phase 1, 6 weeks (224 hours); Phase 2, 6 weeks (254 hours).

Exhibit Dates: ***Version 1:*** 12/87–1/95. *Version 2:* 2/95–7/96.

Learning Outcomes: ***Version 1:*** Upon completion of the course, the student will be able to identify, maintain, and sterilize surgical instruments, sutures, needles, blades, and metal ware; carry out the duties of scrub and circulation specialist; transport and position patients; prepare sterile supplies; and handle specimens. *Version 2:* Upon completion of the course, the operating room specialist will be able to apply cognitive and practical skills in the roles of scrub, circulation, and central material supply/service specialists to support surgical procedures.

Instruction: ***Version 1:*** Lectures, laboratories, and clinical practicum cover operating room and sterilization techniques, hygiene, and physiology. Both Phases must be completed. *Version 2:* To accomplish the learning outcome, the topics covered will include principles and methods of sterilization; identification and care of surgical instruments, sutures, needles, and blades; duties of the scrub and circulation specialist; sterile techniques; identification, transporting, and positioning of patients; preparation of sterile supplies; and handling of specimens.

Credit Recommendation: ***Version 1:*** In the vocational certificate category, certification in operating room techniques. In the lower-division baccalaureate/associate degree category, 15 semester hours in operating room techniques and 1 in physiology. Credit is based on successful completion of both phases (4/88). *Version 2:* In the lower-division baccalaureate/associate degree category, for Phase 1, 15 semester hours in didactic operating room technology; for Phase 2, 8 semester hours in clinical operating room technology (10/95).

Related Occupations: 91D.

AR-0709-0036

PATIENT ADMINISTRATION

Course Number: ***Version 1:*** 7M-F3. *Version 2:* 7M-F3. *Version 3:* 7M-F3.

Location: Academy of Health Sciences, Fort Sam Houston, TX.

Length: ***Version 1:*** 6–7 weeks (265 hours). *Version 2:* 6–7 weeks (285 hours). *Version 3:* 5 weeks (224 hours).

Exhibit Dates: ***Version 1:*** 10/87–12/97. *Version 2:* 1/98–12/01. *Version 3:* 1/02–Present.

Learning Outcomes: ***Version 1:*** Upon completion of the course, the student will be able to demonstrate the principles and procedures required for patients administration in Army medical treatment facilities. *Version 2:* Upon completion of the course, the student will demonstrate the supervisory principles and procedures required to perform as a patient administrator. *Version 3:* Upon completion of the course, the student will demonstrate the supervisory principles and procedures required to perform as a patient administrator, health information management and procedures, and a health care administrator.

Instruction: ***Version 1:*** Lectures and practical exercises cover the principles and procedures required for patients administration in Army medical treatment facilities, including patient administration, health care organization and analysis, and human resource management. *Version 2:* Lectures and practical exercises and discussions cover the principles and procedures for patient administration. *Version 3:* Lectures and practical exercises and discussions cover the principles and procedures for patient administration.

Credit Recommendation: ***Version 1:*** In the upper-division baccalaureate category, 6 semester hours in health care organization and management (10/95). *Version 2:* In the upper-division baccalaureate category, 6 semester hours in patient administration (7/99). *Version 3:* In the lower-division baccalaureate/associate degree category, 2 semester hours in principles of health care information management, 2 in management procedures, 3 in health care administration, and 3 in patient care administration. For those in the health care administrator track, 1 semester hour in health information; for those in the patient administrator track, 1 in patient care administration (6/02).

AR-0709-0037

1. Graves Registration Specialist Advanced Noncommissioned Officer (NCO)
2. Mortuary Affairs Specialist Advanced Noncommissioned Officer (NCO) (Graves Registration Specialist Advanced Noncommissioned Officer (NCO))
3. Mortuary Affairs Specialist Advanced Noncommissioned Officer (NCO)
4. Mortuary Affairs Specialist Advanced Noncommissioned Officer (NCO)

Course Number: ***Version 1:*** 492-57F40. *Version 2:* 492-57F40. *Version 3:* 492-92M40. *Version 4:* 101-92M40F, Phase 2; 492-92M40, Phase 2.

Location: Quartermaster Center and School, Fort Lee, VA.

Length: ***Version 1:*** 5–6 weeks (128 hours). *Version 2:* 6–7 weeks (234 hours). *Version 3:* 8–9 weeks (307–313 hours). *Version 4:* 6 weeks (224 hours).

Exhibit Dates: ***Version 1:*** 1/87–5/92. *Version 2:* 6/92–9/95. *Version 3:* 10/95–11/01. *Version 4:* 12/01–Present.

Learning Outcomes: ***Version 1:*** Upon completion of the course, the student will be able to supervise personnel in cemetery operations and mortuary service contracts and coordinate graves registration matters. *Version 2:* Upon completion of the course, the student will be able to supervise personnel in cemetery operations and mortuary service contracts and coordinate graves registration matters. *Version 3:* Upon completion of the course, the student will be able to supervise personnel in mortuary and cemetery operations and coordinate graves registration matters. *Version 4:* Upon completion of the course, the student will be able to supervise personnel in mortuary and cemetery operations and coordinate graves registration matters.

Instruction: ***Version 1:*** Course includes conferences, television, and performance exercises in the review of mortuary service contracts, graves registration operations, cemetery layout requirements, escort services, and arrangement of matters such as temporary cemetery locations, emergency burials, security and disposition of remains and personal effects, and inspection and shipment of remains. *Version 2:* Course includes conferences, television, and performance exercises in the review of mortuary service contracts; graves registration operations, cemetery layout requirements, escort services, and arrangement of matters such as temporary cemetery locations, emergency burials, security and disposition of remains and personal effects, and inspection and shipment of remains. Course includes a common core of leadership subjects. *Version 3:* Methods include conferences, television, and performance exercises in the review of mortuary services, graves registration operations, cemetery layout requirements, escort services, temporary cemetery locations, and emergency burials, security and disposition of remains and personal effects; and inspection and shipment of remains. *Version 4:* Methods include conferences, television, and performance exercises in the review of mortuary services, graves registration operations, cemetery layout requirements, escort services, temporary cemetery locations, and emergency burials, security and disposition of remains and personal effects; and inspection and shipment of remains.

Credit Recommendation: ***Version 1:*** In the lower-division baccalaureate/associate degree category, 3 semester hours in forensic sciences and 1 in management sciences (10/88). *Version 2:* In the lower-division baccalaureate/associate degree category, 3 semester hours in forensic sciences or in mortuary science and 1 in principles of supervision. See AR-1404-0035 for common core credit (2/93). *Version 3:* In the lower-division baccalaureate/associate degree category, 1 semester hour in microcomputer applications, 1 in principles of supervision, and 3 in mortuary science or forensic sciences. See AR-1404-0035 for common core credit (12/96). *Version 4:* In the lower-division baccalaureate/associate degree category, 3 semester hours in principles of supervision, and 3 in forensic science or 3 in mortuary science (6/02).

Related Occupations: 57F; 92M.

AR-0709-0038

Graves Registration Specialist

Course Number: 492-57F10.

Location: Quartermaster Center and School, Fort Lee, VA.

Length: 8 weeks (284–288 hours).

Exhibit Dates: 8/87–4/04.

Learning Outcomes: Upon completion of the course, the student will be responsible for the search, recovery, evacuation, documentation, and proper disposition of remains and associated personal effects.

Instruction: Course includes lectures, demonstrations, television, and performance exercises in mortuary affairs; search, recovery, and collection operations; identification by fingerprinting and by anatomical and skeletal charts; and processing all required forms and records.

Credit Recommendation: In the vocational certificate category, 9 semester hours in mortuary science or forensic studies. In the lower-division baccalaureate/associate degree category, 3 semester hours in forensic sciences (10/90).

Related Occupations: 57E.

AR-0709-0039

Special Forces Diving Medical Technician

Course Number: ***Version 1:*** 011-F22. *Version 2:* 011-ASIQ5. *Version 3:* 011-ASIQ5.

Location: ***Version 1:*** John F. Kennedy Special Warfare Center and School, Fort Bragg, NC. *Version 2:* Naval Air Station, Key West, FL. *Version 3:* John F. Kennedy Special Warfare Center and School, Fort Bragg, NC.

Length: ***Version 1:*** 4 weeks (150 hours). *Version 2:* 4 weeks (208 hours). *Version 3:* 3 weeks (111 hours).

Exhibit Dates: ***Version 1:*** 7/89–9/91. *Version 2:* 10/91–12/95. *Version 3:* 1/96–Present.

Learning Outcomes: ***Version 1:*** Upon completion of the course, the student will be able to understand the effects of pressure on the human body during diving operations, injuries resulting from excessive exposure to pressure, and the diagnostic procedures involved in these injuries; diagnose and treat diving emergencies and injuries peculiar to combat diving operations; and plan diving operations, including the use of recompression chambers for treating diving injuries, neurological examinations, catheterization, intravenous infusions, intubation, and aeromedical evacuation. *Version 2:* Upon completion of the course, the student will be able to understand the effects of pressure on the human body during diving operations, injuries resulting from excessive exposure to pressure, and the diagnostic procedures involved in identifying these injuries; assess and initiate emergency treatment of injuries peculiar to combat diving operations; and plan medical aspects of diving operations, including use of hyperbaric chambers for treating diving injuries, neurological examinations, catheterization, venipuncture, intubations, CPR, and aeromedical evacuation. *Version 3:* Upon completion of the course, the student will be able to understand the effects of pressure on the human body during diving operations, injuries resulting from excessive exposure to pressure, and the diagnostic procedures involved in identifying these injuries; assess and initiate emergency treatment of injuries peculiar to combat diving operations; and plan and initiate medical aspects of diving operations, including use of hyperbaric chambers for treating diving injuries, neurological examination, catheterization, venipuncture, intubations, CPR, and aeromedical evacuation.

Instruction: ***Version 1:*** Course includes lectures, demonstrations, and performance exercises in physiology and diagnostic procedures, diving emergencies and injuries, and diving operations. *Version 2:* Course includes lectures, demonstrations, and practical exercises in cardiopulmonary and neurological anatomy and physiology; physical examination, CPR, diagnostic and treatment procedures, diving emergencies and injuries, and medical aspects of planning diving operations. *Version 3:* Course include lectures, demonstrations, and practical exercises in cardiopulmonary and neurological anatomy and physiology, physical examination, CPR, diagnostic and treatment procedures, diving emergencies and injuries, and medical aspects of planning diving operations.

Credit Recommendation: ***Version 1:*** In the lower-division baccalaureate/associate degree category, 3 semester hours in applied medical science, medical diving technician, or applied science (12/88). *Version 2:* In the lower-division baccalaureate/associate degree category, 1 semester hours in basic life support and 3 in applied medical science, medical diving technician, or applied science (12/92). *Version 3:* In the lower-division baccalaureate/associate degree category, 1 semester hour in basic life support and 3 in applied science (4/01).

AR-0709-0040

1. Flight Medical Aidman
2. Flight Medical Aidman
3. Flight Medic

Course Number: ***Version 1:*** 300-F6. *Version 2:* 300-F6. *Version 3:* 300-F6.

Location: ***Version 1:*** Academy of Health Sciences, School of Aviation Medicine, Fort Rucker, AL. *Version 2:* Aviation School, Fort Rucker, AL. *Version 3:* School of Aviation Medicine, Fort Rucker, AL.

Length: *Version 1:* 4 weeks (120 hours). *Version 2:* 4 weeks (147 hours). *Version 3:* 4 weeks (149 hours).

Exhibit Dates: *Version 1:* 10/89–11/95. *Version 2:* 12/95–12/97. *Version 3:* 1/98–Present.

Learning Outcomes: *Version 1:* Upon completion of the course, the student will be able to train medics in skills needed to be a member of an aeromedical evacuation team, including evaluating wounded from a battlefield; assisting the flight surgeon; installing, inspecting, and maintaining hoist on a helicopter; operating hoist during evacuation; treating unconscious or struggling victims and those with head wounds-; triaging wounded; performing basic life support skills, including patient assessment, bandaging and splinting, emergency childbirth; and transport and evacuation. *Version 2:* Upon completion of the course, the student will be able to monitor and stabilize inflight wounded, perform crash rescue, perform aero-evacuations of sick and wounded individuals, perform high-performance hoist and other complex rescue operations, and conduct pre-Medevac treatment. *Version 3:* Upon completion of the course, the student will be able to perform advanced life support procedures to stabilize and transport critically ill and wounded patients during rescue operations.

Instruction: *Version 1:* Units of instruction include history of Army aviation medicine, aeromedical evacuation, crash rescue, helicopter indoctrination, patient transfer preparation, investigation of aircraft accidents, inflight care and safety, chemistry of fire, aircraft avionics, aircraft maintenance, extraction techniques, aviation physiology, EKG setup and operation, inhalation therapy, psychiatric disorders, toxicology, lifesaving procedures, and examination and evacuation of psychiatric and trauma patients. Also included are underwater survival and egress techniques and training in flight simulator. Course is conducted through a combination of conferences, demonstrations, and practical experience in the fundamentals of basic life support. *Version 2:* Lectures and simulations cover selected conditions related to the treatment, rescue, and evacuation of sick and wounded individuals. *Version 3:* Lectures and simulations are conducted in a practical environment.

Credit Recommendation: *Version 1:* In the lower-division baccalaureate/associate degree category, 4 semester hours in basic emergency medical technology (3/90). *Version 2:* In the lower-division baccalaureate/associate degree category, 6 semester hours in advanced emergency medical service (10/95). *Version 3:* In the lower-division baccalaureate/associate degree category, 6 semester hours in advanced emergency medical services. In the upper-division baccalaureate category, 2 semester hours in aviation physiology (1/03).

AR-0709-0041

Orientation to Army Medical Department (AMEDD) Pharmacy Services

Course Number: 6-8-C26.

Location: Academy of Health Sciences, Fort Sam Houston, TX.

Length: 2–3 weeks (101–108 hours).

Exhibit Dates: 5/88–4/04.

Learning Outcomes: Upon completion of the course, the student will be able to begin working as an Army pharmacist and apply a general knowledge of the organization and operation of an Army pharmacy.

Instruction: Course includes lectures, discussions, and practical experience in the organization of Army hospitals and pharmacies, inventory management, personnel management, pharmacy computerization, staffing, budgeting, and physical security.

Credit Recommendation: In the upper-division baccalaureate category, 2 semester hours in pharmacy management and 1 in personnel management (3/93).

AR-0709-0042

Quality Audit of Subsistence

Course Number: 6G-F5/321-F5.

Location: Academy of Health Sciences, Fort Sam Houston, TX.

Length: 2 weeks (86 hours).

Exhibit Dates: 8/89–Present.

Learning Outcomes: Upon completion of the course, the student will be able to provide quality audits of subsistence using standardized, uniform interpretation of policies and procedures and provide wholesomeness inspection of red meat, poultry, and waterfood.

Instruction: This course consists of lectures, demonstrations, problem solving, and practicum in food science and food hygiene.

Credit Recommendation: In the lower-division baccalaureate/associate degree category, 1 semester hour in food service technology (1/97).

AR-0709-0043

1. Patient Administration Reserve Component (Nonresident/Resident)
2. Patient Administration Reserve Component (RC)

Course Number: *Version 1:* 7M-F3-RC. *Version 2:* 7M-F3 (RC).

Location: Academy of Health Sciences, Fort Sam Houston, TX.

Length: *Version 1:* 2 weeks (56–81 hours). *Version 2:* 2 weeks (84 hours).

Exhibit Dates: *Version 1:* 9/88–3/98. *Version 2:* 4/98–Present.

Learning Outcomes: *Version 1:* Upon completion of the course, the student will be able to identify and list components of quality assurance programs; list and select branches of patient administration, including entitlements—military and civilian, line of duty, admissions, and disposition of trust funds; use medical statistical reports; follow proper procedures for reporting deceased persons; follow procedures relating to release of medical information; manage patients in a combat zone hospital; follow medical accounting practices, follow procedures for evacuation and transfer of patients; understand hospital accreditation procedure; identify and discuss physician/administrator relationship; describe patient-centered/staff-centered factors; describe disposition of medically unfit (percentage of disability), follow procedures for reporting seriously ill/very seriously ill patients; and discuss issues affecting the role of the patient administrator. *Version 2:* Upon completion of the course, the student will be able to plan, organize, direct, and control patient administration in a health care delivery program or patient administration activities a health service facility.

Instruction: *Version 1:* Instructional modes include conference/lecture covering the above topics. *Version 2:* Lectures, conferences, discussions, and examinations cover topics pertinent to patient administration.

Credit Recommendation: *Version 1:* In the lower-division baccalaureate/associate degree category, 2 semester hours in patient care administration (7/94). *Version 2:* In the lower-division baccalaureate/associate degree category, 2 semester hours in patient administration (7/99).

AR-0709-0044

Army Medical Department (TDA) Table of Distribution and Allowances Pre-command

Course Number: 7M-F9.

Location: Academy of Health Sciences, Fort Sam Houston, TX.

Length: 2 weeks (65–83 hours).

Exhibit Dates: 5/88–4/04.

Learning Outcomes: Upon completion of the course, the student will be able to review and discuss current issues in health care administration.

Instruction: Instruction involves a series of guest lectures, covering Army health care administration facilities, dental care systems and family advocacy programs.

Credit Recommendation: Credit is not recommended because of the limited, specialized nature of the course (3/90).

AR-0709-0045

Principles of Military Preventive Medicine (ENTO)

Course Number: 6A-F5.

Location: Academy of Health Sciences, Fort Sam Houston, TX.

Length: 9 weeks (365 hours).

Exhibit Dates: 10/88–5/90.

Learning Outcomes: Upon completion of the course, the student will be able to serve as an entomologist.

Instruction: This course includes lectures, discussions, and practice in preventive medicine, environmental hygiene, infection control, epidemiology, and public health.

Credit Recommendation: In the upper-division baccalaureate category, 3 semester hours in preventive medicine, 2 in environmental hygiene, 1 in infection control, 1 in epidemiology, 3 in entomology, and 1 in public health (3/90).

AR-0709-0046

1. Mortuary Affairs Specialist Basic Noncommissioned (NCO) Officer
2. Mortuary Affairs Specialist Basic Noncommissioned Officer (NCO) (TATS Mortuary Affairs Specialist Basic Noncommissioned Officer (NCO))

Course Number: *Version 1:* 492-57F30. *Version 2:* 101-92M30 (F), Phase 2; 492-92M30 (F), Phase 2; 492-92M30.

Location: Quartermaster Center and School, Fort Lee, VA.

Length: *Version 1:* 7–8 weeks (262 hours). *Version 2:* 3–6 weeks (128–213 hours).

Exhibit Dates: *Version 1:* 6/92–9/95. *Version 2:* 10/95–Present.

Learning Outcomes: *Version 1:* Upon completion of the course, the student will be able to supervise medium-sized mortuary affairs activities or memorial affairs activities in an overseas mortuary. *Version 2:* Upon completion of the course, the student will be able to supervise medium-sized mortuary affairs activities or memorial affairs activities in an overseas mortuary.

Instruction: ***Version 1:*** Lectures, conferences, practical exercises, demonstrations, and examinations cover search and recovery operations, collection point operations, and cemetery/mass casualty and burial/escort operations. Course includes a common core of leadership subjects. *Version 2:* Lectures, conferences, practical exercises, demonstrations, and examinations cover search and recovery operations, collection point operations, and cemetery/mass casualty and burial/escort operations. Course includes a common core of leadership subjects.

Credit Recommendation: ***Version 1:*** In the lower-division baccalaureate/associate degree category, 3 semester hours in mortuary science or forensic sciences. See AR-1406-0090 for common core credit (2/93). *Version 2:* In the lower-division baccalaureate/associate degree category, 3 semester hours in mortuary science. See AR-1406-0090 for common core credit (6/02).

Related Occupations: 57F; 92M.

AR-0709-0047

Mortuary Affairs Officer
(Graves Registration Officer)

Course Number: ***Version 1:*** 8B-S14V. *Version 2:* 8B-SI4V. *Version 3:* 8B-SI4V.

Location: Quartermaster Center and School, Fort Lee, VA.

Length: ***Version 1:*** 2 weeks (72 hours). *Version 2:* 2 weeks (80 hours). *Version 3:* 2 weeks (80 hours).

Exhibit Dates: ***Version 1:*** 10/89–1/93. *Version 2:* 2/93–12/01. *Version 3:* 1/02–Present.

Learning Outcomes: ***Version 1:*** Upon completion of the course, the student will be able to plan, coordinate, and supervise the care and handling of deceased persons and their personal effects in a theater of operations. *Version 2:* Upon completion of the course, the student will be able to plan, coordinate, and supervise the care and handling of deceased persons and their personal effects in a theater of operations. *Version 3:* Upon completion of the course, the student will be able to plan, coordinate, and supervise the care and handling of deceased persons and their personal effects in a theater of operations.

Instruction: ***Version 1:*** Lectures, conferences, films, practical exercises, and examinations cover training, planning, coordination, and supervision of all mortuary affairs matters in a theater of operation in order to meet the need of the total force. *Version 2:* Lectures, conferences, films, practical exercises, and examinations cover training, planning, coordination, and supervision of all mortuary affairs matters in a theater of operation in order to meet the need of the total force. *Version 3:* Lectures, conferences, films, practical exercises, and examinations cover training, planning, coordination, and supervision of all mortuary affairs matters in a theater of operation in order to meet the need of the total force.

Credit Recommendation: ***Version 1:*** In the upper-division baccalaureate category, 3 semester hours in public administration or operations management (2/93). *Version 2:* In the upper-division baccalaureate category, 2 semester hours in public administration or operations management (12/96). *Version 3:* In the upper-division baccalaureate category, 3 semester hours in mortuary management (6/02).

AR-0709-0048

Patient Administration Specialist
Reserve Phase 2

Course Number: 513-71G10-RC, Phase 2.

Location: Academy of Health Sciences, Fort Sam Houston, TX.

Length: 2 weeks (83 hours).

Exhibit Dates: 1/91–4/04.

Learning Outcomes: Upon completion of the course, the student will be able to admit and discharge patients, maintain medical records, and perform basic administrative procedures in a health care facility.

Instruction: Course covers patient management, medical record maintenance, and administrative support processes in a health facility. Methodology includes lectures and practical exercises.

Credit Recommendation: In the lower-division baccalaureate/associate degree category, 3 semester hours in medical record technology (8/93).

Related Occupations: 71G.

AR-0709-0049

1. Army Medical Department (AMEDD) Noncommissioned Officer (NCO) Basic (91B)
2. Army Medical Department (AMEDD) Basic Noncommissioned Officer (NCO)

Course Number: ***Version 1:*** 6-8-C40 (91B Tech Tng). *Version 2:* 6-8-C40 (91B30 Tech Tng).

Location: Academy of Health Sciences, Fort Sam Houston, TX.

Length: ***Version 1:*** 15 weeks (567 hours). *Version 2:* 15 weeks (703 hours).

Exhibit Dates: ***Version 1:*** 10/91–12/93. *Version 2:* 1/94–Present.

Learning Outcomes: ***Version 1:*** Before 10/91 see AR-0709-0033. Upon completion of the course, the student will be able to provide emergency medical services to the sick and injured. *Version 2:* Before 10/91 see AR-0709-0033. Upon completion of the course, the student will be able to provide health service support on the battlefield, perform and document basic medical care, administer IV fluids and blood therapy, discuss the function and structures of the human body, and assess/treat patients with trauma and common pathologies.

Instruction: ***Version 1:*** The course covers emergency medical topics such as pharmacology, intravenous therapy, airway management, physical assessment, medical records, fluid and electrolyte balance, applied anatomy and physiology, and medical emergencies with an emphasis on the treatment of patients with multiple traumatic injuries. Course includes a common core of leadership subjects. *Version 2:* The course covers emergency medical topics such as pharmacology, intravenous therapy, airway management, physical assessment, medical records, fluid and electrolyte balance, applied anatomy and physiology, and medical emergencies with an emphasis on the treatment of patients with multiple traumatic injuries.

Credit Recommendation: ***Version 1:*** In the lower-division baccalaureate/associate degree category, 2 semester hours in anatomy and physiology, 2 in applied pharmacology, 2 in managing medical emergencies, and 6 in managing traumatic injuries. See AR-1406-0090 for common core credit. In the upper-division baccalaureate category, 2 semester hours in military science (8/93). *Version 2:* In the lower-division baccalaureate/associate degree category, 2 semester hours in applied pharmacology, 2 in anatomy and physiology, 6 in assessing/managing injuries, 2 in military science, and 1 in IV therapy (5/01).

Related Occupations: 91B.

AR-0709-0050

1. Special Operations Medical Sergeant Phase 1A
2. Special Operations Medical Sergeant

Course Number: ***Version 1:*** 011-18D30. *Version 2:* 011-18D30.

Location: Academy of Health Sciences, Fort Sam Houston, TX.

Length: ***Version 1:*** 31–33 weeks (1261–1356 hours). *Version 2:* 33 weeks (1355 hours).

Exhibit Dates: ***Version 1:*** 2/91–10/94. *Version 2:* 11/94–Present.

Learning Outcomes: ***Version 1:*** Upon completion of the course, the student will be able to establish and supervise a medical facility in conventional as well as nonconventional settings and assess and manage medical and traumatic pathologies, triage, dental procedures, and related support services. *Version 2:* Upon completion of the course, the student will be able to prepare for a military mission, discuss the function and structure of human body systems, calculate drug dosages, administer medications, and assess/manage medical emergencies.

Instruction: ***Version 1:*** Course content covers anatomy, physiology, medical terminology, pathology, pharmacology, emergency medical technology, dental pathology and procedures, nursing fundamentals, physical assessment, and prevention and patient care with an emphasis on trauma. Methodology includes lectures, discussions, and laboratory and clinical experiences. *Version 2:* Lectures and hands-on experiences cover advanced assessment and treatment skills. Emphases is on management of specific pathophysiological, medical, surgical, and dental conditions.

Credit Recommendation: ***Version 1:*** In the lower-division baccalaureate/associate degree category, 2 semester hours in medical terminology, 5 in anatomy and physiology, 3 in pharmacology, 2 in dental procedures, 5 in emergency medical technology, 4 in nursing fundamentals, 3 in physical assessment, and 3 in pathology (11/93). *Version 2:* In the lower-division baccalaureate/associate degree category, 4 semester hours in basic anatomy and physiology, 3 in basic pharmacology, 10 in emergency medical technician, and 2 in medical terminology (10/95).

Related Occupations: 11D; 18D.

AR-0709-0051

Joint Special Operations Forces Trauma

Course Number: 011-F18.

Location: Academy of Health Sciences, Fort Sam Houston, TX.

Length: 22–23 weeks (901–911 hours).

Exhibit Dates: 6/94–Present.

Learning Outcomes: Upon completion of the course, the student will be able to identify functions and structures of the human body, perform basic drug dosage calculations, discuss basic principles of pharmacology, recognize emergency medical and psychiatric disorders, assess/manage selected medical conditions, and administer medications.

Instruction: Course includes didactic and hands-on instruction in basic patient management, emergency care, basic pharmacology, and physical assessment.

Credit Recommendation: In the lower-division baccalaureate/associate degree category, 4 semester hours in basic anatomy and physiology, 3 in introductory pharmacology, and 10 in emergency medical technology (5/96).

AR-0709-0052

Pharmacy Specialist

Course Number: ***Version 1:*** 312-91Q10. *Version 2:* 312-91Q10.

Location: Academy of Health Sciences, Fort Sam Houston, TX.

Length: ***Version 1:*** 18 weeks (736 hours). *Version 2:* 19 weeks (760 hours).

Exhibit Dates: ***Version 1:*** 10/93–9/99. *Version 2:* 10/99–Present.

Learning Outcomes: ***Version 1:*** Before 10/93 see AR-0709-0005. Upon completion of the course, the pharmacy technician will be able to discuss federal laws and Army regulations controlling pharmacy operation and administration; perform pharmaceutical calculations; discuss basic anatomy, physiology, and pathophysiology as it relates to pharmacology; discuss common medications according to drug classification, including generic name, trade name, uses, and precautions; prepare and evaluate pharmacy sterile products; and perform the duties and functions of a pharmacy technician under the supervision of a registered pharmacist. *Version 2:* Upon completion of the course the student will be able to perform pharmacy procedures for requesting, receiving, controlling, storing, dispensing and compounding drugs, and prepare pharmacy sterile products.

Instruction: ***Version 1:*** Lectures, laboratories, and clinical experience cover the fundamentals of pharmacy, including labeling of prescriptions, pharmaceutical calculations, preparation of common medications, preparation of sterile products, and use of pharmaceutical references. Pharmaceutical preparations and procedures includes dosage forms, drug information, quality assurance, prescription interpretation, pharmacy procedures, and federal regulations. *Version 2:* Laboratory, audio visual materials, case studies, discussion, classroom exercise, practical exercises, learner presentations. Topics include pharmacology, storage of drugs, CBRNE, pharmaceutical preparations, legal issues, and clinical practice.

Credit Recommendation: ***Version 1:*** In the lower-division baccalaureate/associate degree category, 3 semester hours in pharmacology; 2 in hospital pharmacy—clinical; 7 in pharmaceutical preparations and procedures; 2 in pharmaceutical calculations; 1 in pharmaceutical chemistry; 2 in anatomy, physiology, and pathology; and 3 in sterile products (10/95). *Version 2:* In the lower-division baccalaureate/associate degree category, 6 semester hours in pharmacology therapeutics, 3 in pharmacy calculations, 2 in sterile products, 2 in introduction to pharmacy, 3 in institutional pharmacy, 1 in pharmaceutical chemistry, 3 in clinical in-hospital pharmacy, and 2 in human anatomy and physiology (no laboratory) (6/02).

Related Occupations: 91Q.

AR-0709-0053

Pharmacy Sterile Products

Course Number: 6H-67E/321-F1; 6H-68H/312-F1.

Location: Academy of Health Sciences, Fort Sam Houston, TX.

Length: 2 weeks (83 hours).

Exhibit Dates: 3/92–Present.

Learning Outcomes: Upon completion of the course, the student will be able to perform sterile products calculations; explain procedures necessary in the operation of a sterile products work environment; describe special protocols necessary for use of investigational drugs and neoplastic agents; discuss sterile products therapeutics; prepare sterile products; and maintain appropriate standards and records.

Instruction: This course includes self-paced modules, lectures, seminars, and laboratory instruction in the preparation, calculation, and administration of sterile products. The course also includes information related to the work environment and management of a pharmacy sterile product section. Course has a precourse self-study phase, where the student completes self-paced exportable modules, and a two-week resident phase.

Credit Recommendation: In the upper-division baccalaureate category, 3 semester hours in sterile products (10/98).

AR-0709-0054

Mortuary Affairs Specialist

Course Number: ***Version 1:*** 492-92M10. *Version 2:* 492-92M10.

Location: Quartermaster Center and School, Fort Lee, VA.

Length: ***Version 1:*** 6 weeks (240 hours). *Version 2:* 6 weeks (240 hours).

Exhibit Dates: ***Version 1:*** 3/96–12/01. *Version 2:* 1/02–Present.

Learning Outcomes: ***Version 1:*** Upon completion of the course, the student will be able to assist in search and recovery, evacuation of remains, identification and documentation, and conduct proper disposition of remains and personal effects. *Version 2:* Upon completion of the course, the student will be able to assist in search and recovery, evacuation of remains, identification and documentation, and conduct proper disposition of remains and personal effects.

Instruction: ***Version 1:*** Methods include conferences, hands-on exercises, demonstrations, and video tapes. Course topics include procedures used in search, recovery, and evacuation operations; interment/disinterment operations; handling of remains and personal effects; and preparation of forms and records. *Version 2:* Methods include conferences, hands-on exercises, demonstrations, and video tapes. Course topics include procedures used in search, recovery, and evacuation operations; interment/disinterment operations; handling of remains and personal effects; and preparation of forms and records.

Credit Recommendation: ***Version 1:*** In the lower-division baccalaureate/associate degree category, 2 semester hours in mortuary science or forensics (12/01). *Version 2:* In the lower-division baccalaureate/associate degree category, 2 semester hours in mortuary science, and 2 in forensic science (6/02).

Related Occupations: 92M.

AR-0709-0055

Operating Room Specialist

Course Number: ***Version 1:*** 301-91D10, Phase 1. *Version 2:* 301-91D10, Phase 1.

Location: Academy of Health Sciences, Fort Sam Houston, TX.

Length: ***Version 1:*** 6 weeks (224–227 hours). *Version 2:* 9 weeks (382 hours).

Exhibit Dates: ***Version 1:*** 8/96–3/02. *Version 2:* 4/02–Present.

Learning Outcomes: ***Version 1:*** Upon completion of the course, the student will apply appropriate cognitive and psychomotor skills to the role of an operating room technician; practice a variety of sterilization procedures using disinfecting agents, steam, and chemical means; support the operating team in a variety of surgical procedures, including head and neck, genitourinary, orthopedic, thoracic, vascular, and neurological; and demonstrate and differentiate the support roles of scrub versus circulator. *Version 2:* Upon completion of the course, the student will be able to apply appropriate cognitive and psychomotor skills to role of an operating room technician, practice a variety of sterilization procedures using disinfecting agents, steam, and chemical means, support the operating team in a variety of surgical procedures, including head and neck, genitourinary, orthopedic, thoracic, vascular, and neurological, demonstrate and differentiate the support roles of scrub versus circulator; become CPR certified, and apply knowledge of anatomy and physiology to the surgical patient.

Instruction: ***Version 1:*** Instruction is provided by lecture and demonstration in addition to laboratory and prechemical experiences. *Version 2:* Laboratory, audiovisual materials, discussion, classroom exercise, practical exercises, lecture, and learner presentations.

Credit Recommendation: ***Version 1:*** In the lower-division baccalaureate/associate degree category, 3 semester hours in introduction to surgical technology, 3 in sterilization procedures, 5 in surgical procedures, and 3 in circulator and scrub duties (7/99). *Version 2:* In the lower-division baccalaureate/associate degree category, 3 semester hours in introduction to surgical technician, 2 in sterilization process, 5 in surgical procedures, 3 in circulator and scrub, 1 in CPR, 3 in anatomy and physiology, and 1 in an operating room clinical practice (6/02).

Related Occupations: 91D.

AR-0709-0056

Operating Room Specialist

Course Number: ***Version 1:*** 301-91D10, Phase 2. *Version 2:* 301-91D10, Phase 2. *Version 3:* 301-91D10, Phase 2.

Location: ***Version 1:*** Selected MEDACS, MEDCENS, Continental US; Academy of Health Sciences, Fort Sam Houston, TX. *Version 2:* Academy of Health Sciences, Fort Sam Houston, TX. *Version 3:* Academy of Health Sciences, Fort Sam Houston, TX.

Length: ***Version 1:*** 6 weeks (254 hours). *Version 2:* 6 weeks (254 hours). *Version 3:* 9 weeks (382 hours).

Exhibit Dates: *Version 1:* 8/96–12/98. *Version 2:* 1/99–3/02. *Version 3:* 4/02–Present.

Learning Outcomes: *Version 1:* Upon completion of the course, the student will apply appropriate cognitive and psychomotor skills to the role of a scrub and circulator specialist. *Version 2:* Upon completion of the course, the student will apply appropriate cognitive and psychomotor skills in the role of a scrub nurse, circulating specialist, and central material service specialist for direct support of surgical procedures in operating rooms, clinics, and field environments. *Version 3:* Upon completion of the course, the student will be able to apply appropriate cognitive and psychomotor skills in the role of a scrub nurse, circulating specialist, and central material service specialist for direct support of surgical procedures in operating rooms, clinics, and field environments.

Instruction: *Version 1:* Instruction is provided by licensed personnel supervising clinical practice experience. *Version 2:* Instruction is provided by licensed personnel supervising clinical practice experience. *Version 3:* Laboratory. Instruction is provided by licensed personnel supervising clinical practice experience.

Credit Recommendation: *Version 1:* In the lower-division baccalaureate/associate degree category, 6 semester hours in clinical practice in operating room techniques (5/97). *Version 2:* In the upper-division baccalaureate category, 6 semester hours in clinical practice in operating room techniques (7/99). *Version 3:* In the lower-division baccalaureate/associate degree category, 7 semester hours in an operating room clinical practice (6/02).

Related Occupations: 91D.

AR-0709-0057

Army Medical Department (AMEDD) Noncommissioned Officer (NCO) Basic

Course Number: 6-8-C40 (91Q30).

Location: Academy of Health Sciences, Fort Sam Houston, TX.

Length: 7 weeks (300 hours).

Exhibit Dates: 4/95–Present.

Learning Outcomes: Upon completion of the course, the student will be able to lead a military unit; evaluate performance and counsel; supervise and delegate; and perform drills, training, and specific military skills.

Instruction: The course covers military leadership, personal and performance counseling, duties and responsibilities of noncommissioned officers, leadership assessment and development, leadership procedures and combat orders, combat leadership, physical training and marksmanship, property accountability, individual and team training, small-group process, and specific military skills. The methodologies include lectures, discussions, group work, and practical application. This course contains a common core of leadership subjects.

Credit Recommendation: In the lower-division baccalaureate/associate degree category, 3 semester hours in military science and 1 in of elective credit in pharmacy. See AR-1406-0090 for common core credit (11/97).

Related Occupations: 91Q.

AR-0709-0058

Army Medical Department (AMEDD) Noncommissioned Officer (NCO) Basic

Course Number: 6-8-C40 (91D30).

Location: Academy of Health Sciences, Fort Sam Houston, TX.

Length: 7 weeks (300 hours).

Exhibit Dates: 4/95–Present.

Learning Outcomes: Upon completion of the course, the student will be able to lead a military unit; evaluate performance and counsel; supervise and delegate; and perform drills, training, and specific military skills.

Instruction: The course covers military leadership, personal and performance counseling, duties and responsibilities of noncommissioned officers, leadership assessment and development, leadership procedures and combat orders, combat leadership, physical training and marksmanship, property accountability, individual and team training, small-group process, and specific military skills. The methodologies include lectures, discussions, group work, and practical application. This course contains a common core of leadership subjects.

Credit Recommendation: In the lower-division baccalaureate/associate degree category, 3 semester hours in military science and 1 in of elective credit in an operating room specialty. See AR-1406-0090 for common core credit (11/97).

Related Occupations: 91D.

AR-0709-0059

Army Medical Department (AMEDD) Noncommissioned Officer (NCO) Basic

Course Number: 6-8-C40 (71G30).

Location: Academy of Health Sciences, Fort Sam Houston, TX.

Length: 7 weeks (300 hours).

Exhibit Dates: 4/95–Present.

Learning Outcomes: Upon completion of the course, the student will be able to lead a military unit; evaluate performance and counsel; supervise and delegate; and perform drills, training, and specific military skills.

Instruction: The course covers military leadership, personal and performance counseling, duties and responsibilities of noncommissioned officers, leadership assessment and development, leadership procedures and combat orders, combat leadership, physical training and marksmanship, property accountability, individual and team training, small-group process, and specific military skills. The methodologies include lectures, discussions, group work, and practical application. This course contains a common core of leadership subjects.

Credit Recommendation: In the lower-division baccalaureate/associate degree category, 3 semester hours in military science and 1 in elective credit in patient administration. See AR-1406-0090 for common core credit (11/97).

Related Occupations: 71G.

AR-0709-0060

Army Medical Department (AMEDD) Noncommissioned Officer (NCO) Basic

Course Number: 6-8-C40 (42E30).

Location: Academy of Health Sciences, Fort Sam Houston, TX.

Length: 7 weeks (300 hours).

Exhibit Dates: 4/95–Present.

Learning Outcomes: Upon completion of the course, the student will be able to lead a military unit; evaluate performance and counsel; supervise and delegate; and perform drills, training, and specific military skills.

Instruction: The course covers military leadership, personal and performance counseling, duties and responsibilities of noncommissioned officers, leadership assessment and development, leadership procedures and combat orders, combat leadership, physical training and marksmanship, property accountability, individual and team training, small-group process, and specific military skills. The methodologies include lectures, discussions, group work, and practical application. This course contains a common core of leadership subjects.

Credit Recommendation: In the lower-division baccalaureate/associate degree category, 3 semester hours in military science and 1 in elective credit in optical laboratory. See AR-1406-0090 for common core credit (11/97).

Related Occupations: 42E.

AR-0709-0061

Ear, Nose, and Throat (ENT) Specialty

Course Number: 300-P2, Phase 1.

Location: Academy of Health Sciences, Fort Sam Houston, TX.

Length: 6 weeks (237 hours).

Exhibit Dates: 3/97–Present.

Learning Outcomes: Upon completion of the course, the student will be able to identify the structures and function of the ear, nose, and throat; develop skills in administering various tests and reporting results to appropriate personnel; assist otolaryngologists and audiologists in clinics; perform specific treatments (excluding irrigation of ear) under supervision; identify specific equipment and/or instruments for an ENT clinic; match symptoms, causes, and disease of the body parts to the correct treatment; and relate pharmacological aspects to various areas of the clinic.

Instruction: Methods of instruction include didactic and practical experiences in a clinical setting. Topics include anatomy and physiology of ear, nose, and throat; pathological conditions; diagnostic phases; sterilization; and assisting in the clinics.

Credit Recommendation: In the lower-division baccalaureate/associate degree category, 6 semester hours in basic ear, nose, and throat (5/01).

AR-0709-0063

Ear, Nose, and Throat (ENT) Specialty

Course Number: 300-P2, Phase 2.

Location: Blanchfield Community Hospital, Fort Campbell, KY; Evans Community Hospital, Fort Lewis, WA; Brooke Medical Center, Fort Sam Houston, TX; William Beaumont Medical Center, El Paso, TX; Dwight David Eisenhower Medical Center, Fort Gordon, GA; Walter Reed Medical Center, Washington, DC; Madigan Medical Center, Tacoma, WA; Womack Medical Center, Fort Bragg, NC; Academy of Health Sciences, Fort Sam Houston, TX.

Length: 7 weeks (280 hours).

Exhibit Dates: 3/97–Present.

Learning Outcomes: Upon completion of the course, the student will be able to perform basic ear, nose, and throat and audiology procedures and provide clinical support in specialized outpatient services.

Instruction: Clinical instruction is provided at an appropriate military health care center.

Credit Recommendation: In the lower-division baccalaureate/associate degree category, 6 semester hours in ear, nose, and throat practicum (7/99).

AR-0709-0064

TOTAL ARMY TRAINING SYSTEMS (TATS) MEDICAL SPECIALIST

Course Number: 081-91B10 (F), Phase 2; 300-91B10 (RC).
Location: Academy of Health Sciences, Fort Sam Houston, TX.
Length: 79 hours.
Exhibit Dates: 10/01–Present.
Learning Outcomes: Upon completion of the course the student will be able to perform triage, crisis treatment, vaccinations, combat health care and treatment.
Instruction: Laboratory, classroom exercise, practical exercises, and lecture. Topics include military science, patient management, health care, health physics, and hazardous material.
Credit Recommendation: In the lower-division baccalaureate/associate degree category, 1 semester hour in military science, 1 in patient management, 3 in health care specialist, 1 in health physics, 2 in hazardous material (6/02).

AR-0709-0065

HEALTH CARE SPECIALIST

Course Number: ***Version 1:*** 300-91W10. *Version 2:* 300-91W10.
Location: Academy of Health Sciences, Fort Sam Houston, TX.
Length: ***Version 1:*** 16 weeks (910 hours). *Version 2:* 16 weeks (1046 hours).
Exhibit Dates: ***Version 1:*** 1/01–4/03. *Version 2:* 5/03–Present.
Learning Outcomes: ***Version 1:*** Upon completion of the course the student will be able to measure and assess vital signs, complete and maintain records, perform CPR, administer oxygen, manage a patient with artificial airways, manage wounds, administer injections and immunizations, treat chemical injuries, initiate measures to prevent spread of communicable diseases, perform basic field sanitation procedures, and perform emergency and evacuation care. *Version 2:* Upon completion of the course the student will be able to measure and assess vital signs, complete and maintain records, perform CPR, administer oxygen, manage a patient with artificial airways, manage wounds, administer injections and immunizations, treat chemical injuries, initiate measures to prevent spread of communicable diseases, perform basic field sanitation procedures, and perform emergency and evacuation care.
Instruction: ***Version 1:*** Laboratory, audio-visual materials, case studies, discussion, classroom exercise, practical exercises, lecture, learner presentation. Topics include emergency and evacuation care, minor acute care, in-patient basic care procedures, outpatient care, casualty triage and processing, EMT-basic, health protection, supportive care, combat trauma, NBRC, and clinical experiences in ambulance outpatient and hospital settings. *Version 2:* Laboratory, audio visual materials, case studies, discussion, classroom exercise, practical exercises, lecture, learner presentation. Topics include emergency and evacuation care, minor acute care, in-patient basic care procedures, outpatient care, casualty triage and processing, EMT-basic, health protection, supportive care, combat trauma, NBRC, and clinical experiences in ambulance outpatient and hospital settings.
Credit Recommendation: ***Version 1:*** In the lower-division baccalaureate/associate degree category, 1 semester hour in basic life support, 3 in advanced first aid, 4 in human anatomy and physiology (no laboratory), 3 in medical terminology, 7 in basic EMT (theory and clinical), 2 in advanced EMT (theory and clinical), 3 in pharmacology, and 4 in patient care procedures (6/02). *Version 2:* In the lower-division baccalaureate/associate degree category, 1 semester hour in basic life support, 6 in anatomy and physiology (no lab), 4 in terminology, 7 in EMT theory, 9 in advanced EMT, 6 in patient care procedures, and 15 in clinical practicum (11/03).
Related Occupations: 91W.

AR-0709-0066

MEDICAL ASSISTANCE

Course Number: TAC-7, Phase 1.
Location: Western Hemisphere Institute for Security Cooperation (WHINSEC), Fort Benning, GA.
Length: 6 weeks (161 hours).
Exhibit Dates: 4/01–Present.
Learning Outcomes: Upon completion of the course, the student will be able to identify appropriate medical action; administer and perform medical procedures in adverse conditions; utilize preventative medical measures; recognize, evaluate, and treat an injured person in a life-threatening situation.
Instruction: Audio-visual materials, practical and classroom exercises, case studies, learner presentation, laboratory, and lecture. Topics include application of medical treatment, and management of medical situations in emergencies.
Credit Recommendation: In the lower-division baccalaureate/associate degree category, 3 semester hours in medical assistance (10/02).

AR-0709-0067

HEALTH CARE SPECIALIST RECLASSIFICATION

Course Number: 081-91W10, Phase 1.
Location: Academy of Health Sciences, Fort Sam Houston, TX.
Length: 159 hours.
Exhibit Dates: 6/03–Present.
Learning Outcomes: Upon completion of the course, the student will be able to provide emergency medical prescriptions; limited primary care; force health protection; evacuation in a variety of operational and clinical settings, from point of injury or illness through the continuum of military health care.
Instruction: Audiovisual materials, practical and classroom exercises, laboratory, and lecture. Topics include emergency and evacuation care, minor acute care, in-patient basic care procedures, casualty triage and processing, combat trauma, EMT basic, and health protection and supportive care.
Credit Recommendation: In the lower-division baccalaureate/associate degree category, 7 semester hours in basic EMT clinical procedures, 1 in anatomy and physiology (no lab), 1 in medical terminology, and 1 in basic life support (11/03).

AR-0709-0070

HEALTH CARE SPECIALIST RECLASSIFICATION

Course Number: 081-91W10, Phase 3.
Location: Academy of Health Sciences, Fort Sam Houston, TX.
Length: 80 hours.
Exhibit Dates: 10/02–Present.
Learning Outcomes: Upon completion of the course, the student will be able to provide emergency medical treatment and limited primary care, and force health protection and evacuation in a variety of operational and clinical settings, from point of injury through the continuum of medical care, casualty triage, and processing.
Instruction: Practical and classroom exercises, written examinations, laboratory, and lecture. Topics include military operations other than war; casualty care; medical care for patients exposed to weapons of mass destruction; deployable medical systems; aircraft and ground evacuation; and casualty triage and processing. This phase expands to Phase II self-study.
Credit Recommendation: In the lower-division baccalaureate/associate degree category, 3 semester hours in patient care procedures and 3 semester hours in advanced EMT (11/03).

AR-0709-0071

HEALTH CARE SPECIALIST RECLASSIFICATION

Course Number: 081-91W10, Phase 4.
Location: Academy of Health Sciences, Fort Sam Houston, TX.
Length: 204 hours.
Exhibit Dates: 10/01–Present.
Learning Outcomes: Upon completion of the course, the student will be able to provide emergency medical prescriptions, limited primary care, casualty triage and processes, force health protection, and both air and ground evacuation, from point of illness through the continuum of military health care.
Instruction: Audiovisual materials, practical and classroom exercises, lecture, and written exams. Topics include invasive and evacuation procedures, triage, and casualty care.
Credit Recommendation: In the lower-division baccalaureate/associate degree category, 2 semester hours in advanced first aid, 1 in advanced EMT, 2 in patient care procedures, and 3 in clinical practicum (11/03).

AR-0709-0072

MEDICAL NONCOMMISSIONED OFFICER (NCO)

Course Number: 300-91B30.
Location: Academy of Health Sciences, Fort Sam Houston, TX.
Length: 254 hours.
Exhibit Dates: 10/86–12/87.
Learning Outcomes: Upon completion of the course, the student will be able to function at the noncommissioned officer-level on the battlefield or in garrison pre-hospital care and emergency/ambulatory treatment facilities.
Instruction: Lectures, laboratories, and independent-study materials in pharmacology, anatomy, physiology, paramedic skills, clinical skills, and infectious diseases.
Credit Recommendation: In the lower-division baccalaureate/associate degree category, 1 semester hour in hygiene and 2 in phar-

macology. In the upper-division baccalaureate category, 3 semester hours in physiology and credit for physician assistant programs determined by the institution (4/88).

Related Occupations: 91B.

AR-0799-0015

1. AMEDD Reserve Component Officer Advanced (Phase 2)
   (Reserve Component Army Medical Department Officer Advanced, Phase 2)
2. Army Medical Department (AMEDD) Reserve Component Officer Advanced

Course Number: *Version 1:* 6-8-C22 (RC-OAC/P2). *Version 2:* 6-8-C22 (RC-OAC).

Location: *Version 1:* Academy of Health Sciences, Fort Sam Houston, TX; Reserve locations, US. *Version 2:* Reserve Component Training Centers, US; Academy of Health Sciences, Fort Sam Houston, TX.

Length: *Version 1:* 2 weeks (81–85 hours). *Version 2:* Phase 2, 2 weeks (94–95 hours).

Exhibit Dates: *Version 1:* 4/87–2/95. *Version 2:* 3/95–Present.

Learning Outcomes: *Version 1:* Upon completion of the course, the student will be able to apply training in military medical support operations to the duties and responsibilities of AMEDD officers. *Version 2:* Upon completion of the course, the student will be able to apply the concepts of medical unit organization to a combat medical unit.

Instruction: *Version 1:* Lectures, conferences, and practical exercises are presented with primary focus on military science as applied to the medical field. *Version 2:* Lectures and practical exercises are used to teach the fundamentals of combat health support.

Credit Recommendation: *Version 1:* In the upper-division baccalaureate category, 2 semester hours in military science (6/93). *Version 2:* Credit is not recommended because of the military-specific nature of the course (5/97).

AR-0799-0016

Reserve Component Army Medical Department Officer Advanced, Phase 3

Course Number: 6-8-C22 (RC-OAC/P3).

Location: Reserve Component Training Centers, US; Academy of Health Sciences, Fort Sam Houston, TX.

Length: 2 weeks (75 hours).

Exhibit Dates: 4/87–4/98.

Learning Outcomes: Upon completion of the course, the student will be able to apply training in military medical support operations to the duties and responsibilities of AMEDD officers.

Instruction: Lectures, conference, demonstrations, and practical exercise emphasize organization, administration, preventive medicine, and behavioral science.

Credit Recommendation: In the upper-division baccalaureate category, 1 semester hour in military science. In the graduate degree category, 1 semester hour in preventive medicine and 1 in health care administration (4/88).

AR-0799-0017

Army Medical Department Staff Development (Reserve Component) (Nonresident/Resident)

Course Number: 7M-F4-RC/300-F4-RC.

Location: Academy of Health Sciences, Fort Sam Houston, TX.

Length: 2 weeks (80 hours).

Exhibit Dates: 10/87–1/93.

Learning Outcomes: Upon completion of the course, the student will be able to perform in staff assignments at the corps level and above.

Instruction: Lectures, conferences, and practical exercises emphasize military science.

Credit Recommendation: Credit is not recommended because of the limited, specialized nature of the course (4/88).

AR-0799-0018

1. Aeromedical Evacuation Officer
   (Essential Medical Training for Army Medical Department (AMEDD) Aviators)
2. Medical Evacuation Doctrine
   (Aeromedical Evacuation Officer)
   (Essential Medical Training for Army Medical Department (AMEDD) Aviators)

Course Number: *Version 1:* 2C-F7 (MS/WO/RC). *Version 2:* 2C-F7/300-F10; 2C-F7 (MS/WO/RC).

Location: Academy of Health Sciences, Fort Sam Houston, TX; School of Aviation Medicine, Fort Rucker, AL.

Length: *Version 1:* 2 weeks (72–73 hours). *Version 2:* 2 weeks (65 hours).

Exhibit Dates: *Version 1:* 8/89–4/93. *Version 2:* 5/93–Present.

Learning Outcomes: *Version 1:* Upon completion of the course, the student will be able to administer basic life support; assess the sick and wounded, including triage, bandaging, and splinting; administer IV therapy; assess vital signs; perform emergency childbirth; interpret electrocardiographs; and evacuate and transport the sick and injured. *Version 2:* Upon completion of the course, the student will be able to serve as an aeromedical evacuation pilot; evacuate and transport the sick and injured; and assess and treat the sick and wounded, including triage, bandaging, and splinting.

Instruction: *Version 1:* Conferences, lectures, demonstrations, and practical experience cover the administration of basic life support procedures, patient assessment and triage, and technical procedures, including emergency childbirth, IV therapy, and evacuation and transport of the sick and injured. *Version 2:* Methodology includes lectures and discussion on triage, first aid, mass evacuation procedures, emergency communication, emergency planning, and crew coordination.

Credit Recommendation: *Version 1:* In the lower-division baccalaureate/associate degree category, 3 semester hours in EMT first responder and emergency medical technician (3/90). *Version 2:* In the lower-division baccalaureate/associate degree category, 2 semester hours in trauma evaluation (10/03).

AR-0799-0019

1. Health Services Warrant Officer Basic
   (Health Services Maintenance Technician)
2. Health Services Maintenance Warrant Officer Basic
   (Health Services Warrant Officer Basic)
3. Health Services Maintenance Warrant Officer Basic

Course Number: *Version 1:* 4B-670A; 4B-670A/198-670A. *Version 2:* 4B-670A. *Version 3:* 4B-670A.

Location: Academy of Health Sciences, Fort Sam Houston, TX.

Length: *Version 1:* 4 weeks (128–172 hours). *Version 2:* 6 weeks (243–244 hours). *Version 3:* 6 weeks (228 hours).

Exhibit Dates: *Version 1:* 7/88–12/95. *Version 2:* 1/96–3/03. *Version 3:* 4/03–Present.

Learning Outcomes: *Version 1:* Upon completion of the course, the student will be able to manage medical equipment, systems, and maintenance programs; train and supervise personnel performing medical equipment maintenance; determine maintenance and safety requirements; develop recommendations for installation of equipment; maintain records of equipment maintenance; contract for maintenance; inventory repair parts; inspect equipment; determine staffing requirements; and develop procedures for effective scheduling of manpower, time, and finances. *Version 2:* Upon completion of the course, the student will be able to manage medical equipment, systems, and maintenance programs; train and supervise personnel performing medical equipment maintenance; determine maintenance and safety requirements; develop recommendations for installation of equipment; maintain records of equipment maintenance; contract for maintenance; inventory repair parts; inspect equipment; determine staffing requirements; and develop procedures for effective scheduling of manpower, time, and finances. *Version 3:* Upon completion of the course, the student will be able to manage medical equipment, systems, and maintenance programs; train and supervise personnel performing medical equipment maintenance; determine maintenance and safety requirements; develop recommendations for installation of equipment; maintain records of equipment maintenance; contract for maintenance; inventory repair parts; inspect equipment; determine staffing requirements; and develop procedures for effective scheduling of manpower, time, and finances.

Instruction: *Version 1:* Training includes instruction in biomedical maintenance, automated maintenance management systems, logistics planning, manpower management, communication skills, and nonmedical maintenance. Hands-on training is provided on field equipment, electrical safety issues, vehicle maintenance, and deployable medical equipment. *Version 2:* Training includes instruction in biomedical maintenance, automated maintenance management systems, logistics planning, manpower management, communication skills, nonmedical maintenance, quality control, and accreditation documentation. Hands-on training is provided on field equipment, electrical safety issues, vehicle maintenance, and deployable medical equipment. *Version 3:* Training includes instruction in biomedical maintenance, automated maintenance management systems, logistics planning, manpower management, communication skills, nonmedical maintenance, quality control, and accreditation documentation. Hands-on training is provided on field equipment, electrical safety issues, vehicle maintenance, and deployable medical equipment.

Credit Recommendation: *Version 1:* In the upper-division baccalaureate category, 3 semester hours in biomedical equipment maintenance management and 3 for field experience in biomedical equipment maintenance management (8/93). *Version 2:* In the lower-division baccalaureate/associate degree category, 2 semester hours in equipment maintenance management, 2 in personnel supervision, and 2 in materiel management (6/98). *Version 3:* In the lower-division baccalaureate/associate degree category, 2 semester hours in equipment maintenance management, 1 in personnel supervision, and 2 in materiel management (11/03).

Related Occupations: 670A.

AR-0799-0020

PRINCIPLES OF MILITARY PREVENTIVE MEDICINE (CHN)

Course Number: 6A-F5.

Location: Academy of Health Sciences, Fort Sam Houston, TX.

Length: 9 weeks (365 hours).

Exhibit Dates: 10/88–5/90.

Learning Outcomes: Upon completion of the course, the student will be able to serve as a community health nurse.

Instruction: This course includes lectures, discussions, and practice in preventive medicine, environmental hygiene, infection control, epidemiology, and community health nursing.

Credit Recommendation: In the upper-division baccalaureate category, 3 semester hours in preventive medicine, 2 in environmental hygiene, 1 in infection control, 1 in epidemiology, and 4 in community health (3/90).

AR-0799-0021

ARMY MEDICAL DEPARTMENT (AMEDD) AVIATOR ADVANCED

Course Number: 2C-F8.

Location: Academy of Health Sciences, Fort Sam Houston, TX.

Length: 3 weeks (100 hours).

Exhibit Dates: 10/88–1/93.

Learning Outcomes: Upon completion of the course, the student will be able to identify the various command units and list their functions, history, and resources; apply the objectives of planning, programming, and budgeting; list and describe objectives of the manpower management system; describe behavioral objectives (sleep logistics, psychological aspects of warfare); list levels of administration in the Army dental care system; apply principles of triage; and describe the organization and function of the Military Science Division (identify mission, organization, and functions).

Instruction: Instruction consists primarily of lecture/conferences and practical exercises.

Credit Recommendation: In the upper-division baccalaureate category, 2 semester hours in health care administration and 2 in military science (3/90).

AR-0799-0022

1. ARMY FLIGHT SURGEON
2. ARMY FLIGHT SURGEON PRIMARY
3. ARMY FLIGHT SURGEON PRIMARY
4. ARMY FLIGHT SURGEON (PRIMARY)

Course Number: *Version 1:* 6A-61N9D. *Version 2:* 6A-61N9D. *Version 3:* 6A-61N9D. *Version 4:* 6A-61N9D.

Location: Academy of Health Sciences, School of Aviation Medicine, Fort Rucker, AL.

Length: *Version 1:* 7 weeks (302–303 hours). *Version 2:* 6 weeks (224 hours). *Version 3:* 6 weeks (218–224 hours). *Version 4:* 6 weeks (218 hours).

Exhibit Dates: *Version 1:* 6/82–10/91. *Version 2:* 11/91–12/94. *Version 3:* 1/95–3/99. *Version 4:* 4/99–Present.

Learning Outcomes: *Version 1:* Upon completion of the course, the student will be able to supply medical support to Army aviation units, including unit-level health service support, aeromedical staff advice, medical investigation of aircraft accidents, and the application of medical and scientific information and techniques to the conduct of military aviation operations. *Version 2:* Upon completion of the course, the student will be able to supply medical support to Army aviation units, including unit-level health service support, aeromedical staff advice, medical investigation of aircraft accidents, and the application of medical and scientific information and techniques to the conduct of military aviation operations. *Version 3:* Upon completion of the course, the physician or physician's assistant will be able to supply administrative support as an Army flight surgeon, including unit-level health service support, aeromedical staff advice, medical investigation of aircraft accidents, and the application of medical and scientific information and techniques to the conduct of military aviation operations. *Version 4:* Upon completion of the course the student will be able to discuss the basics of rotary wing aviation, aviation medical policies, and medical issues relating to flight.

Instruction: *Version 1:* This course includes lectures, discussions, demonstrations, and practical experiences in aircraft systems, principles of aviation, aviation safety, flight training, atmospheric physics, and clinical and preventive medicine topics as applied to aviation. *Version 2:* This course includes lectures, discussions, demonstrations, and practical experiences in aircraft systems, principles of aviation, aviation safety, atmospheric physics, and clinical and preventive medicine topics as applied to aviation. *Version 3:* This course includes lectures, discussions, demonstrations, and practical experiences in aircraft systems, principles of aviation, aviation safety, atmospheric physics, and clinical and preventive medicine topics as applied to aviation. *Version 4:* Laboratory, audio visual materials, discussion, classroom exercise, practical exercises, and lecture. Topics include basic rotary wing aviation, aviation medicine policies (e.g. flying restrictions, flight physicals, flight status), FAA flight physicals, certifications, aviation safety in-flight emergencies and equipment, accident investigation, forensic pathology, crew coordination, altitude physiology, G forces, spatial disorientation, oxygen systems, altitude chamber flight, aviation personality patterns, night vision, aviation toxicology, and stress fatigue.

Credit Recommendation: *Version 1:* In the upper-division baccalaureate category, 4 semester hours in general aviation and 5 in flight training. In the graduate degree category, 5 semester hours in aviation medicine (3/90). *Version 2:* In the upper-division baccalaureate category, 2 semester hours in general aviation and 2 in aviation administration. In the graduate degree category, 8 semester hours in aviation medicine (12/95). *Version 3:* In the upper-division baccalaureate category, 3 semester hours in aviation medicine logistics. In the graduate degree category, 2 semester hours in aviation medicine administration (9/00). *Version 4:* In the lower-division baccalaureate/associate degree category, 1 semester hour in introduction to aviation. In the graduate degree category, 1 semester hour in administrative issues in flight medicine and 2 in flight physiology (6/02).

Related Occupations: 61N.

AR-0799-0023

REHABILITATION TRAINING INSTRUCTOR

Course Number: 7H-F45 (RTIC).

Location: Military Police School, Fort McClellan, AL.

Length: 2 weeks (98 hours).

Exhibit Dates: 5/92–9/97.

Learning Outcomes: Upon completion of the course, students will teach civilians who are involved in boot camp youthful offender programs to use drill sergeant techniques to maintain a constant level of control and discipline.

Instruction: Lectures and applied exercises cover fitness training, counseling, alcohol and drug abuse, drill techniques, stress management, and behavioral evaluations.

Credit Recommendation: In the lower-division baccalaureate/associate degree category, 3 semester hours in counseling techniques and 3 in physical education/fitness (6/92).

AR-0799-0024

NUCLEAR PHARMACY ORIENTATION

Course Number: 6H-F19.

Location: Academy of Health Sciences, Fort Sam Houston, TX; Fitzsimmons Medical Center, Aurora, CO.

Length: 2 weeks (76 hours).

Exhibit Dates: 5/90–Present.

Learning Outcomes: Upon completion of the course, the student will be able to describe the role of the nuclear pharmacy; discuss trends in nuclear pharmacy practice; prepare radiopharmaceuticals; perform dose calibration and quality control techniques; discuss state, local and federal regulations which pertain to nuclear pharmacy; discuss application of nuclear medicine technologies for evaluation and diagnosis; practice radiation safety; and discuss basic nuclear physics.

Instruction: Course covers nuclear pharmacy management and operations, radiopharmaceutics, radiation safety, nuclear medicine studies, and nuclear physics. Methodology includes lectures, demonstrations, and practical exercises.

Credit Recommendation: In the graduate degree category, 2 semester hours in nuclear pharmacy (7/95).

AR-0799-0025

ARMY MEDICAL DEPARTMENT (AMEDD) TABLE OF DISTRIBUTION (TDA) DEPUTY COMMANDER FOR CLINICAL SERVICES

Course Number: 6A-DCCS.

Location: Academy of Health Sciences, Fort Sam Houston, TX.

Length: 2 weeks (64 hours).

Exhibit Dates: 8/91–4/04.

Learning Outcomes: Upon completion of the course, the student will be able to manage AMEDD resources; manage personnel; apply decision making processes; understand legal issues; practice modern labor relations strategies; and understand risk management issues.

Instruction: Course covers human resource management, patient administration, health care administration, acquisition and management of resources, professional affairs, and legal issues. Methodology includes lectures, discussions, and practical exercises.

Credit Recommendation: In the lower-division baccalaureate/associate degree category, 2 semester hours in management (8/93).

AR-0799-0026

1. Army Medical Department (AMEDD) Basic Noncommissioned Officer (NCO) (76J)
2. Army Medical Department (AMEDD) Noncommissioned Officer (NCO)

Course Number: *Version 1:* 6-8-C40 (76J); 6-8-C40 (76J Tech Tng). *Version 2:* 6-8-C40 (91J30 Tech Tng); 6-8-C40 (76J30 Tech Tng).

Location: Academy of Health Sciences, Fort Sam Houston, TX.

Length: *Version 1:* 7 weeks (252–276 hours). *Version 2:* 2 weeks (80 hours).

Exhibit Dates: *Version 1:* 2/92–9/95. *Version 2:* 10/95–Present.

Learning Outcomes: *Version 1:* Upon completion of the course, the student will be able to supervise a material management department in a health care facility. *Version 2:* Upon completion of the course, the student will be able to support the medical material needs of an organization by supervising material storage, property accounting, inventory management, disposal of hazardous material, material requisitioning, and logistical concerns.

Instruction: *Version 1:* This course covers purchasing, storage, and distribution of a wide variety of materials needed in medical organizations, with significant emphasis on organization, accountability, and record keeping. Methodologies include lectures, discussions, small-group work, presentations, and supervised practical applications. *Version 2:* Concepts of medical material management, including storage and disposal, quality control, logistics of suppliers, and inventory management are covered by lectures and practical exercises.

Credit Recommendation: *Version 1:* In the lower-division baccalaureate/associate degree category, see AR-1406-0090 for common core credit. In the upper-division baccalaureate category, 2 semester hours in military science and 3 in materiel management (8/93). *Version 2:* In the lower-division baccalaureate/associate degree category, 3 in medical material management (10/03).

Related Occupations: 76J.

AR-0799-0027

Army Medical Department (AMEDD) Noncommissioned Officer (NCO) Advanced

Course Number: *Version 1:* 6-8-C42. *Version 2:* 6-8-C42. *Version 3:* 6-8-C42. *Version 4:* 6-8-C42 (Resident).

Location: Academy of Health Sciences, Fort Sam Houston, TX.

Length: *Version 1:* 8 weeks (140–291 hours). *Version 2:* 6 weeks (265 hours). *Version 3:* 6 weeks (185 hours). *Version 4:* 3 weeks (366 hours).

Exhibit Dates: *Version 1:* 10/91–12/94. *Version 2:* 1/95–3/98. *Version 3:* 4/98–4/02. *Version 4:* 5/02–Present.

Learning Outcomes: *Version 1:* Upon completion of the course, the student will be able to apply general management principles to organizational issues such as unit operations, resource management and supply, preventive medicine, and safety and environmental problems. Those students in the tracks for biological sciences assistant; orthotic specialist; optical laboratory specialist; psychiatric specialist; behavioral sciences specialist; orthopedic specialist; physical therapy specialist; occupational therapy specialist; cardiac specialist; ear, nose, and throat specialist; respiratory specialist; nuclear medicine specialist; eye specialist; and cytology specialist will be able to conduct related laboratory procedures in a field environment demonstrating effective training and supervisory skills. Those students in the patient administration specialty will be able to utilize diagnostic-related groups for patient administration. Those students in the dental specialist and dental laboratory specialist tracks will be able to provide preventive maintenance to dental equipment, manage personnel, and handle organizational issues in a military unit. Those who are in the medical specialist track will be able to develop a medical support plan, set up and evacuate a medical field station, and supervise staff in an emergency situation. Students in the veterinary food inspection specialty and the animal care specialty will be able to develop a work load activities report and manage an animal facility. The students in the pharmacy specialty will be able to recognize and prevent drug diversion, apply security procedures to military pharmacies, monitor quality assurance, and manage field operations. Those students in the health physics specialty and the preventive medicine specialty will be able to conduct emergency planning and risk analysis and manage radiological hazards. The students in the X-ray specialty will be able to develop a risk management program, practice in an ethical manner, apply quality control tests, and perform radiographic procedures on pediatric patients. Those in the operating room specialty will be able to monitor operating room equipment and property and manage the operating room suite. Those in the practical nurse specialty will be able to monitor management of medical equipment, apply safety measures to motor vehicle use, and assume responsibility for infection control. Those in the medical supply specialty will be able to monitor the supply of medical equipment; supervise the request, receipt, and storage of medical supplies and equipment; and account for lost, damaged, and destroyed equipment. And those in the medical equipment repairer specialty will be able to develop work schedules and work orders and manage personnel and work centers. *Version 2:* Upon completion of the course, the student will be able to apply general management principles to organizational issues such as unit operation, resource management and supply, preventive medicine, and safety and environmental problems. Those students in the tracks for biological sciences assistant; orthotic specialist; optical laboratory specialist; psychiatric specialist; behavioral sciences specialist; orthopedic specialist; physical therapy specialist; occupational therapy specialist; cardiac specialist; ear, nose, and throat specialist; respiratory specialist; nuclear medicine specialist; eye specialist; and cytology specialist will be able to conduct related laboratory procedures in a field environment demonstrating effective training and supervisory skills. Those students in the patient administration specialty will be able to utilize diagnostic-related groups for patient administration. Those students in the dental specialist and the dental laboratory specialist tracks will be able to perform preventive maintenance on dental equipment, manage personnel, and handle organizational issues in a military unit. Those who are in the medical specialist track will be able to develop a medical support plan, set up and evacuate a medical field station, and supervise staff in an emergency situation. Students in the veterinary food inspection specialty and the animal care specialty will be able to develop a work load activities report and manage an animal facility. The students in the pharmacy specialty will be able to recognize and prevent drug diversion, apply security procedures to military pharmacies, monitor quality assurance, and manage field operations. Those students in the health physics specialty and the preventive medicine specialty will be able to conduct emergency planning and risk analysis and manage radiological hazards. Students in the X-ray specialty will be able to develop a risk management program, practice in an ethical manner, apply quality control tests, and perform radiographic procedures on pediatric patients. Those in the operating room specialty will be able to monitor operating room equipment and property and manage the operating room suite. Those in the practical nurse specialty will be able to monitor management of medical equipment, apply safety measures to motor vehicle use, and assume responsibility for infection control. Those in the medical supply specialty will be able to monitor the supply of medical equipment; supervise the request, receipt, and storage of medical supplies and equipment; and account for lost, damaged, and destroyed equipment. And those in the medical equipment repairer specialty will be able to develop work schedules and work orders and manage personnel and work centers. *Version 3:* Upon completion of the course, the student will be able to apply general management principles to organizational issues such as unit operation, resource management and supply, preventive medicine, and safety and environmental problems. Those students in the tracks for medical specialist include optical laboratory specialist; operating room specialist and respiratory specialist. Students in those specialties will be able to conduct related laboratory procedures in a field environment demonstrating effective training and supervisory skills. Those students in the patient administration specialty will be able to utilize diagnostic-related groups for patient administration. Those students in the dental specialist and the dental laboratory specialist tracks will be able to perform preventive maintenance on dental equipment, manage personnel, and handle organizational issues in a military unit. Those who are in the medical specialist track will be able to develop a medical support plan, set up and evacuate a medical field station, and supervise staff in an emergency situation. Students in the veterinary food inspection specialty and the animal care specialty will be able to develop a work load activities report and manage an animal facility. The students in the pharmacy specialty will be able to recognize and prevent drug diversion, apply security procedures to military pharmacies, monitor quality assurance, and manage field operations. Those students in the health physics specialty and the preventive medicine specialty will be able to conduct emergency planning and risk analysis and manage radiological hazards. Students in the mental health specialty will be able to deal appropriately with psychiatric emergencies, recognize substance users/abusers and apply management measures to minimize the effects of combat stress detachment. Students in the X-ray specialty will be able to develop a risk management program, practice in an ethical manner, apply quality control tests, and perform radiographic procedures on pediatric patients. Those in the operating room specialty will be able to monitor operating room equipment and property and manage the operating room suite. Those in the practical nurse specialty will be able to monitor management of medical equipment, apply safety measures to motor vehicle use, and assume responsibility for infection control. Those in the medical supply specialty will be able to monitor the supply of medical equipment; supervise the request, receipt, and storage of medical supplies and equipment; and account for lost, damaged, and destroyed equipment. And those in the medical equipment repairer specialty will be able to develop work schedules and work orders and manage personnel and work centers. *Version 4:* Pending evaluation.

Instruction: ***Version 1:*** Topics covered include military organization, health service support, elements of military operation and defense, and medical topics related to unit functioning. In addition, topics include laboratory procedures in blood banking, hematology, clinical chemistry, microbiology, and AIDS in the laboratory for students in the following tracks: biological sciences assistant; orthotic, optical laboratory; psychiatric; behavioral sciences; orthopedic; physical therapy; occupational therapy; cardiac; ear, nose, and throat; respiratory; nuclear medicine; eye; and cytology. Covered topics include resource management and diagnostic-related groups for those in the patient administration track. Covered topics include maintaining equipment, managing employees, acquiring and managing equipment, and establishing a unit in the field for those in the dental specialist and in the dental laboratory specialist tracks. Covered topics include medical support planning, clinic operations, medical field equipment, and emergency vehicles for those in the medical specialist track. Covered topics include work load reporting, animal disease prevention and control, and field organization and management for those in the veterinary food inspection and the animal care tracks. Covered topics include drug diversion, pharmacy security, quality assurance, and field operation management for those in the pharmacy specialty track. Covered topics include emergency management, risk analysis, and radiological emergency management for those in the preventive medicine specialist track. Covered topics include risk management, ethics, quality control, and pediatric radiation for those in the X-ray specialty. Covered topics include equipment and property management, personnel scheduling and operations, and operating room maintenance for those in the operating room specialty. Covered topics include work load management, supply and equipment management, military safety, infection control, and AIDS for those in the practical nurse specialty. Covered topics include the management of medical equipment and procedures for acquiring, storing, and dispensing equipment for those in the medical supply specialist track. Covered topics include work load management, maintenance and repair of equipment, and management of the work center for those in the medical equipment repairer track. Methodology includes lectures, conferences, and practical experiences in all tracks. Course includes a common core of leadership subjects. *Version 2:* Topics covered include military organization, health service support, elements of military operation and defense, and medical topics related to unit functioning. In addition, topics include laboratory procedures in blood banking, hematology, clinical chemistry, microbiology, and AIDS in the laboratory for students in the following tracks: biological sciences assistant; orthotic; optical laboratory; psychiatric; behavioral sciences; orthopedic; physical therapy; occupational therapy; cardiac; ear, nose, and throat; respiratory; nuclear medicine; eye; and cytology. Covered topics include resource management and diagnostic-related groups for those in the patient administration track. Covered topics include equipment maintenance, managing employees, acquisition and management of equipment, and establishment of a unit in the field for those in the dental specialist and in the dental laboratory specialist tracks. Covered topics include medical support planning, clinic operations, medical field equipment, and emergency vehicles for those in the medical specialist track. Covered topics include work load reporting, animal disease prevention and control, and field organization and management for those in the veterinary food inspection and the animal care tracks. Covered topics include drug diversion, pharmacy security, quality assurance, and field operation management for those in the pharmacy specialty track. Covered topics include emergency management, risk analysis, and radiological emergency management for those in the preventive medicine specialist track. Covered topics include risk management, ethics, quality control, and pediatric radiation for those in the X-ray specialty. Covered topics include equipment and property management, personnel scheduling and operations, and operating room maintenance for those in the operating room specialty. Covered topics include work load management, supply and equipment management, military safety, infection control, and AIDS for those in the practical nurse specialty. Covered topics include the management of medical equipment, and procedures for acquiring, storing, and dispensing equipment for those in the medical supply specialist track. Covered topics include work load management, maintenance and repair of equipment, and management of the work center for those in the medical equipment repairer track. Methodology includes lectures, conferences, and practical experiences in all tracks. Course includes a common core of leadership subjects. *Version 3:* Topics covered include military organization, health service support, elements of military operation and defense, and medical topics related to unit functioning. In addition, topics include laboratory procedures in blood banking, hematology, clinical chemistry, microbiology, and AIDS in the laboratory for students in the following tracks: biological sciences assistant; orthotic; optical laboratory; psychiatric; orthopedic; physical therapy; occupational therapy; cardiac; ear, nose, and throat; respiratory; nuclear medicine; eye; and cytology. Covered topics include resource management and diagnostic-related groups for those in the patient administration track. Covered topics include equipment maintenance, managing employees, acquisition and management of equipment, and establishment of a unit in the field for those in the dental specialist and in the dental laboratory specialist tracks. Covered topics include medical support planning, clinic operations, medical field equipment, and emergency vehicles for those in the medical specialist track. Covered topics include work load reporting, animal disease prevention and control, and field organization and management for those in the veterinary food inspection and the animal care tracks. Covered topics include combat stress management, recognition and detection of substance users/abusers, and scenarios involving psychiatric emergencies for those in the mental health track. Covered topics include drug diversion, pharmacy security, quality assurance, and field operation management for those in the pharmacy specialty track. Covered topics include emergency management, risk analysis, and radiological emergency management for those in the preventive medicine specialist track. Covered topics include risk management, ethics, quality control, and pediatric radiation for those in the X-ray specialty. Covered topics include equipment and property management, personnel scheduling and operations, and operating room maintenance for those in the operating room specialty. Covered topics include work load management, supply and equipment management, military safety, infection control, and AIDS for those in the practical nurse specialty. Covered topics include the management of medical equipment, and procedures for acquiring, storing, and dispensing equipment for those in the medical supply specialist track. Covered topics include work load management, maintenance and repair of equipment, and management of the work center for those in the medical equipment repairer track. Methodology includes lectures, conferences, and practical experiences in all tracks. Course includes a common core of leadership subjects. *Version 4:* Pending evaluation.

Credit Recommendation: ***Version 1:*** In the lower-division baccalaureate/associate degree category, see AR-1404-0035 for common core credit. In the upper-division baccalaureate category, 3 semester hours in operations management and 1 in preventive medicine for all students. For those in specialties 92B, 01H, 42C, 42E, 91F, 91G, 91H, 91J, 91L, 91N, 91U, 91V, 91W, 91Y, 92E, 71G, 91E, and 42D, 1 semester hour in clinical field experiences. For those in 91R and 91T, 1 semester hour in veterinary field management. For those in 91Q, 1 semester hour in pharmacy field management. For those in the medical noncommissioned officer track, 1 in operations control. For those in 91S and 91X, 1 semester hour in emergency management. For those in the 91P, 1 semester hour in radiographic management. For those in 91D, 1 semester hour in operational management. For those in 91C, 1 semester hour in field management operations. For those in 76J, 1 semester hour in supply management. And for those in 35U, 1 semester hour in organizational management (9/93). *Version 2:* In the lower-division baccalaureate/associate degree category, For completion of the common core, 3 semester hours in military science. Students will complete one track; for the medical equipment repairer track, 1 in medical equipment repair. For the patient administration specialist track, 1 in patient administration. For the medical supply specialist track, 1 in medical supply.

For the medical specialist track, 1 in medical specialist. For the practical nurse track, 1 in practical nursing. For the dental specialist track, 1 in dental. For the medical laboratory specialist track, 1 in medical laboratory. For the radiology specialist track, 1 in radiology specialty. For the pharmacy specialist track, 1 in pharmacy. For the veterinary food inspection track, 1 in veterinary technology. For the preventive medicine specialist track, 1 in preventive medicine. For the hospital food service track, 1 in food service. See AR-1404-0035 for common core credit. Students will complete one track (11/95). *Version 3:* In the lower-division baccalaureate/associate degree category, for completion of the common core, 3 semester hours in military science. Students will complete one track; for the medical equipment repairer track, 1 in medical equipment repair. For the patient administration specialist track, 1 in patient administration. For the medical supply specialist track, 1 in medical supply. For the medical specialist track, 1 in medical specialist. For the practical nurse track, 1 in practical nursing. For the dental specialist track, 1 in dental. For the medical laboratory specialist track, 1 in medical laboratory. For the radiology specialist track, 1 in radiology specialty. For the pharmacy specialist track, 1 in pharmacy. For the veterinary food inspection/animal care track, 1 in veterinary technology. For the preventive medicine specialist track, 1 in preventive medicine. For the hospital food service track, 1 in food service. For the mental health track, 1 in mental health specialty. See AR-1404-0035 for common core credit. Students will complete one track (5/01). *Version 4:* Pending evaluation.

Related Occupations: 01H; 35U; 42C; 42D; 42E; 71G; 76J; 91B; 91C; 91D; 91E; 91F; 91G; 91H; 91J; 91L; 91N; 91P; 91Q; 91R; 91S; 91T; 91U; 91V; 91W; 91X; 91Y; 92B; 92E.

AR-0799-0028

Army Medical Department (AMEDD) Noncommissioned Officer (NCO) Advanced

Course Number: 6-8-C42 (91M40).

Location: Academy of Health Sciences, Fort Sam Houston, TX.

Length: 2 weeks (88 hours).

Exhibit Dates: 1/95–Present.

Learning Outcomes: Upon completion of the course, the student will be able to supervise a medical food service operation, assess patient nutritional status, and present group sessions on nutritional education.

Instruction: Course includes lectures and practical exercises in the areas of nutrition, food production and supply, and personnel management.

Credit Recommendation: In the lower-division baccalaureate/associate degree category, 3 semester hours in hospital food service (4/02).

Related Occupations: 91M.

AR-0799-0029

1. Flight Surgeon (Primary) Reserve Phase 1
2. Army Flight Surgeon (Primary)

Course Number: ***Version 1:*** 6A-61N9D-RC, Phase 1. *Version 2:* 6A-61N9D (RC), Phase 1.

Location: Aviation School, Fort Rucker, AL.

Length: ***Version 1:*** 3 weeks (110–113 hours). *Version 2:* 3 weeks (113 hours).

Exhibit Dates: ***Version 1:*** 11/94–3/99. *Version 2:* 4/99–Present.

Learning Outcomes: ***Version 1:*** Upon completion of the course, the physician or physician's assistant will provide administrative support as a flight surgeon to a military aviation unit including medical advice and assistance with operations and management. *Version 2:* Upon completion of the course the student will be able to discuss the basics of rotary wing aviation, aviation medical policies, and medical issues relating to flight.

Instruction: ***Version 1:*** The course includes topics in aviation medicine administration, aviation principles, and flight physiology. Instruction is provided through lectures, discussions, demonstrations, and guided experiences. *Version 2:* Laboratory, audiovisual materials, discussion, classroom exercise, practical exercises, and lecture. Topics include basic rotary wing aviation, aviation medicine policies (e.g. flying restrictions, flight physicals, flight status), FAA flight physicals, certifications, aviation safety in-flight emergencies and equipment, accident investigation, forensic pathology, crew coordination, altitude physiology, g forces, spatial disorientation, oxygen systems, altitude chamber flight, aviation personality patterns, night vision, aviation toxicology, and stress fatigue.

Credit Recommendation: ***Version 1:*** In the upper-division baccalaureate category, 3 semester hours in aviation medicine logistics (9/00). *Version 2:* In the lower-division baccalaureate/associate degree category, 1 semester hour in introduction to aviation. In the graduate degree category, 1 semester hour in administrative issues in flight medicine and 2 in flight physiology (6/02).

AR-0799-0030

Flight Surgeon (Primary) Reserve

Course Number: 6A-61N9D-RC, Phase 2.

Location: Aviation School, Fort Rucker, AL.

Length: 3 weeks (105–112 hours).

Exhibit Dates: 1/95–3/99.

Learning Outcomes: Upon completion of the course, the physician or physician's assistant will provide administrative support as a flight surgeon to a military aviation unit including medical intervention for health problems in an aviation environment.

Instruction: The course includes topics in aviation medicine related to illness and health problems that can occur in an aviation environment. Instruction is provided through lectures, discussions, and guided experiences.

Credit Recommendation: In the upper-division baccalaureate category, 2 semester hours in aviation medicine administration (11/97).

Related Occupations: 61N.

AR-0799-0031

Brigade Surgeon

Course Number: 7M-F7.

Location: Academy of Health Sciences, Fort Sam Houston, TX.

Length: 2 weeks (80 hours).

Exhibit Dates: 8/96–Present.

Learning Outcomes: Upon completion of the course, the physician will assume the role of brigade surgeon and provide administrative and health services support for a brigade-level command.

Instruction: The course includes instruction through guest lecture and seminars.

Credit Recommendation: Credit is not recommended because of the military-specific nature of the course (10/03).

AR-0799-0032

Aviation Medicine Orientation

Course Number: 6A-F1.

Location: Aviation School, Fort Rucker, AL.

Length: 2 weeks (73–74 hours).

Exhibit Dates: 1/95–Present.

Learning Outcomes: Upon completion of the course, the physician will be able to function as a medical officer and flight surgeon in an aviation unit and provide health services support and medical assistance in the conduct of military aviation operations.

Instruction: The course covers service regulations related to aviation medicine. Instruction is provided primarily through lecture and discussion. Emphasis is on administrative and management procedures and relevant aviation medicine topics appropriate to the flight surgeon in aviation operations.

Credit Recommendation: In the upper-division baccalaureate category, 2 semester hours in introduction to aviation medicine (1/03).

AR-0799-0033

Health Services Human Resources Manager (HRM)

Course Number: 6H-70F67.

Location: Academy of Health Sciences, Fort Sam Houston, TX.

Length: 3 weeks (119 hours).

Exhibit Dates: 9/97–Present.

Learning Outcomes: Upon completion of the course, the student will be able to apply concepts and procedures of personnel management.

Instruction: Lectures, practical exercises, and examinations provide technical training the application of personnel management principles.

Credit Recommendation: In the graduate degree category, 6 semester hours in human resource management (10/03).

AR-0799-0034

Army Medical Department Noncommissioned Officer (NCO) Advanced Reserve

Course Number: ***Version 1:*** 6-8-C42, Phase 2. *Version 2:* 6-8-C42, Phase 2.

Location: Academy of Health Sciences, Fort Sam Houston, TX.

Length: ***Version 1:*** 7–8 weeks (108 hours). *Version 2:* 2 weeks (96–108 hours).

Exhibit Dates: ***Version 1:*** 6/98–9/02. *Version 2:* 10/02–Present.

Learning Outcomes: ***Version 1:*** Upon completion of the course, the student will have the technical, tactical, and leadership trainer qualities necessary to be successful in Army operations as a platoon or operation sergeant. *Version 2:* Upon completion of Annex B, the student will be able to identify organizational structure of medical echelon, develop evacuation plans, identify rear area elements of defense, identify problems and solutions for the command environment, understand the role of the noncommissioned officer, use the manual for crunch material, role of response to threats, identify integrated roles of the active Army and National Guard, identify basic concepts of defense, offense and integrated operations, training, and providing practical exercises in the above areas. For those who follow Medical Equipment Repair are given a maintenance scenario and should deliver the correct action; Patient Administration Specialist will manage patients in a threat of operations; Medical Supply Specialist will provide managerial tasks relative to warehouse management; Health Care Specialist should be able to relocate a battalion aid station, order and monitor supplies, develop a duty roster; Dental Specialist should be able to identify concepts and applications of dental support in a threat of operations; Medical Laboratory Specialist will manage and coordinate lab activities and elements used to order the medical prescription facility; Radiology Specialist will plan a radiology layout to include identification of factors, equipment, personnel; Pharmacy Specialist will manage pharmacy operations referring to the composite health care system; Vet Food Inspectors will perform food audits to include fresh fruits and vegetables, bakery, meat, eggs, focus on animal issues both as food source and military dogs; Preventive Medicine Specialist will identify current trends and issues in preventive medicine; Hospital Food Service Specialist supervise and manage areas of food preparation, supply and personnel management; and the Mental Health Specialist supervise personnel conducting mental health personnel and exit briefing.

Instruction: *Version 1:* Lectures, practical exercises, and examinations provide related technical and leadership training. *Version 2:* Lectures, learner presentations, classroom exercises, and practical exercises.

Credit Recommendation: *Version 1:* In the lower-division baccalaureate/associate degree category, 2 semester hours in career management, 1 in medical technical training, 1 in medical equipment repair for 91A Medical Equipment Repairer, 1 in patient administration for 71G Patient Administration, 1 in medical supply for 76J Medical Supply Specialist, 1 in medical specialist for 91B Medical Specialist, 1 in practical nursing for 91C Practical Nurse, 1 in dental specialist for 91E Dental Specialist, 1 in medical laboratory for 91K Medical Laboratory Specialist, 1 in radiology for 91P Radiology Specialist, 1 in pharmacy for 91Q Pharmacy Specialist, 1 in veterinary food inspection for 91R Vet Food Inspection Specialist, 1 in animal care for 91T Animal Care Specialist, 1 in preventive medicine for 91S Preventive Medicine Specialist, 1 in hospital food service for 91M Hospital Food Service Specialist, and 1 in mental health for 91X Mental Health Specialist (7/99). *Version 2:* In the upper-division baccalaureate category, All tracks receive 2 semester hours in military science and 1 in practicum in military science. For those in the medical equipment repair track, 1 semester hour in equipment repair; for those in the medical logistics specialist track, 1 semester hour in logic; for those in the health care specialist track, 1 semester hour patient administration; for those in the dental specialist track, 1 semester hour in dental management; for those in the medical laboratory specialist, 1 semester hour in laboratory management; for those in the radiology specialist track, 1 semester hour in radiology management; for those in the pharmacy specialist track, 1 semester hour in pharmacy management; for those in veterinary inspection track, 1 semester hour in preventive medicine; for those in the preventative medicine track, 1 semester hour in preventive medicine; for those in the hospital food specialist track, 1 semester hour in food services; and for those in the mental health specialist track, 1 semester hour in mental health (6/02).

AR-0799-0035

Rehabilitation Training Instructor

Course Number: 7H-F45.

Location: Military Police School, Fort Leonard Wood, MO.

Length: 2 weeks (88 hours).

Exhibit Dates: 10/99–Present.

Learning Outcomes: Upon completion of the course, the student will apply principles of counseling, including stress management, leadership counseling and other forms of positive behavior programs in a custodial corrections setting to teach civilian correctional officers working in boot-camp-style youthful offenders programs.

Instruction: Lectures and applied exercises cover fitness training, counseling, alcohol and drug abuse, drill techniques, stress management, and behavioral evaluations.

Credit Recommendation: In the lower-division baccalaureate/associate degree category, 3 semester hours in counseling and 3 in physical education (2/01).

AR-0799-0037

Army Flight Surgeon (Primary)

Course Number: 6A-61N9D, Phase 1.

Location: Aviation Center, Fort Rucker, AL; Academy of Health Sciences, Fort Sam Houston, TX.

Length: 3 weeks (105 hours).

Exhibit Dates: 4/99–Present.

Learning Outcomes: Before 4/99 see AR-0799-0029. Upon completion of the course the student will be able to serve as flight surgeon or aeromedical physician assistant, discuss medical intervention for health problems in an aviation environment, and identify issues of deployment of aviation levels.

Instruction: Laboratory, audio visual materials, case studies, discussion, classroom exercise, practical exercises, lecture. Topics include ENT disorders, ophthalmology, orthopedics, psychiatry, urology, drug and alcohol, sports medicine, laser eye injuries, internal medicine, psychological support, hearing, development planning, flight medical operations, health service support in various behaviors, and nuclear biological radiological chemical (NBRC).

Credit Recommendation: In the graduate degree category, 3 semester hours in aviation medicine (6/02).

AR-0799-0039

Aeromedical Psychology Training

Course Number: 6H-F27.

Location: School of Aviation Medicine, Fort Rucker, AL; Academy of Health Sciences, Fort Sam Houston, TX.

Length: 3 weeks (123 hours).

Exhibit Dates: 11/98–Present.

Learning Outcomes: Upon completion of the course, the student will be able to discuss the basics of wing aviation and aviation medical policies; identify key medical issues in selecting, retaining, and eliminating aviators; relate physiology to all aspects of altitude and flight; and recognize psychological issues in a rotary wing environment.

Instruction: Audio-visual materials, case studies, discussion, classroom exercises, practical exercises, lectures, and learner presentations. Topics include: aviation medicine and policies, role of the flight surgeon, aeromedical evaluation, drugs and alcohol in aviation, flight physiology, aviation psychology, and aviation administration.

Credit Recommendation: In the lower-division baccalaureate/associate degree category, 2 semester hours in introduction to aviation, 1 in flight physiology, and 1 in aviation administration. In the upper-division baccalaureate category, 2 semester hours in aviation psychology (6/02).

AR-0799-0040

Army Medical Department (AMEDD) Noncommissioned Officer Advanced (NCOES)

Course Number: 6-8-C42 (Resident), Phase 2.

Location: Academy of Health Sciences, Fort Sam Houston, TX.

Length: 3 weeks (135–366 hours).

Exhibit Dates: 5/02–Present.

Learning Outcomes: Upon completion of the course, the student will be able to apply general management principles, adapt management principles to support combat health operations, demonstrate technical and tactical leadership attributes, and utilize these skills in a designated health care field.

Instruction: Topics include personnel management, equipment management, professional ethics, Army quality processes, and leadership in military operations.

Credit Recommendation: In the upper-division baccalaureate category, 3 semester hours in human resource management, 2 in medical ethics, and 2 in military science/leadership (11/03).

AR-0799-0041

Army Medical Department (AMEDD) Warrant Officer Advanced (WOAC) (PILOT)

Course Number: 6-8-C32 (640A-670A).

Location: Academy of Health Sciences, Fort Sam Houston, TX.

Length: 6 weeks (394 hours).

Exhibit Dates: 10/02–Present.

Learning Outcomes: Upon completion of the course, the student will be able to perform technical and tactical tasks required (veterinary service/health services maintenance technician.

Instruction: Topics include chemical, biological, and radiological agents; nuclear and high explosive devices; advanced technical skills, policies and procedures, ethical reasoning process; counseling; leadership development; and environmental laws and regulations. For veterinary technicians specifically, topics include food establishment auditing techniques; food safety and inspection; food-borne disease issues in public health; dairy and pasteurization processes; and food vulnerability assessment.

For advanced health services technician specifically, topics include electrical vacuum system, medical gas system preparation for JCAHO visits; environmental; and medical equipment in health care facility training; and OSHA medical of medical maintenance functions.

Credit Recommendation: In the lower-division baccalaureate/associate degree category, for veterinary technician, 3 semester hours in leadership, 3 in food establishment auditing, 3 in food-borne diseases, and 3 in sanitation safety; and for health services technician, 3 semester hours in medical equipment maintenance, 3 in medical logistics, 3 in equipment, and 1 in equipment safety (11/03).

AR-0801-0007

Army Drug and Alcohol Rehabilitation Training Group
(Army Drug and Alcohol Rehabilitation)

Course Number: 5H-F5/302-F5.
Location: Academy of Health Sciences, Fort Sam Houston, TX.
Length: 2 weeks (76–85 hours).
Exhibit Dates: 9/79–Present.
Learning Outcomes: Upon completion of this course, students will have a working knowledge of group process and techniques necessary to effectively counsel clients who are substance abusers.
Instruction: Course includes techniques of group development and process, group phenomena, case notes and reporting, group emotional distress, termination and follow-up, and counselor growth and development.
Credit Recommendation: In the lower-division baccalaureate/associate degree category, 3 semester hours in techniques of group counseling and 2 in practicum in group processing (12/99).

AR-0801-0010

Chemical Operations Specialist Basic Noncommissioned Officer (NCO) Reclassification

Course Number: 494-54B20-T; 494-54B (R).
Location: Chemical School, Fort McClellan, AL.
Length: 11–14 weeks (396–483 hours).
Exhibit Dates: 1/86–2/92.
Objectives: After 2/92 see AR-0801-0027. To train enlisted personnel to operate, maintain, and supervise the use of nuclear, biological, and chemical (NBC) detection and decontamination equipment; smoke generators; and NBC defense measures.
Instruction: Conferences, practical exercises, and examinations cover individual protection skills, radiological and chemical decontamination, smoke, biological operations, tactical training, and new equipment.
Credit Recommendation: In the lower-division baccalaureate/associate degree category, 3 semester hours in radiological safety (1/88).
Related Occupations: 54B.

AR-0801-0012

Chemical Operations Specialist (Basic Noncommissioned Officer (NCO))

Course Number: 494-54B.
Location: Chemical School, Fort McClellan, AL.
Length: 9–12 weeks (325–451 hours).
Exhibit Dates: 1/86–2/92.
Objectives: After 2/92 see AR-0801-0028. To train chemical operations specialists to assist in the establishment, administration, and application of nuclear, biological, and chemical (NBC) defense measures and to operate, maintain, and supervise the use of NBC detection and decontamination equipment and smoke generators.
Instruction: Conferences, practical exercises, and examinations cover individual protection skills, radiological and chemical decontamination, smoke and biological operations, tactical training, and new equipment.
Credit Recommendation: In the lower-division baccalaureate/associate degree category, 3 semester hours in radiological safety (1/88).
Related Occupations: 54B.

AR-0801-0014

Alcohol and Drug Abuse Prevention and Control Program Clinical Director/Clinical Consultant

Course Number: *Version 1:* 5H-F9/302-F9. *Version 2:* 5H-F9/302-F9.
Location: Academy of Health Sciences, Fort Sam Houston, TX.
Length: *Version 1:* 1 week (54–55 hours). *Version 2:* 1 week (41 hours).
Exhibit Dates: *Version 1:* 4/87–2/01. *Version 2:* 3/01–Present.
Learning Outcomes: *Version 1:* Upon completion of the course, the student will be able to implement, administer, and evaluate the clinical aspects of an alcohol and drug abuse prevention and control program. *Version 2:* Upon completion of the course, the student will be able to implement, administer, and evaluate the clinical aspects of an alcohol and drug abuse prevention and control program.
Instruction: *Version 1:* Course presents a review of the role of the Health Services Command Clinical Certification Program, including current initiatives, role of the clinical consultant, and quality assurance. Training includes screening alcohol and other substance abuse referrals, clinical supervision of counselors, current trends in the substance abuse field, and a tour of a drug testing laboratory. Training is also provided in the treatment of substance abuse. *Version 2:* Course presents a review of the role of the Health Services Command Clinical Certification Program, including current initiatives, role of the clinical consultant, and quality assurance. Training includes screening alcohol and other substance abuse referrals, clinical supervision of counselors, current trends in the substance abuse field, and a tour of a drug testing laboratory. Training is also provided in the treatment of substance abuse.
Credit Recommendation: *Version 1:* In the upper-division baccalaureate category, 3 semester hours in substance abuse education (8/91). *Version 2:* In the upper-division baccalaureate category, 3 semester hours in substance abuse education (11/03).

AR-0801-0015

Alcohol and Drug Abuse Prevention and Control Program Alcohol and Drug Control Officer

Course Number: 5H-F6/302-F6.
Location: Academy of Health Sciences, Fort Sam Houston, TX.
Length: 1 week (49 hours).
Exhibit Dates: 1/87–3/91.
Learning Outcomes: Upon completion of the course, the student will be able to supervise an alcohol and drug abuse prevention program including selection and management of program personnel and engage in community resource development.
Instruction: Course content includes understanding policies and regulations relating to the management of substance abuse programs and understanding trends in substance abuse prevention.
Credit Recommendation: In the upper-division baccalaureate category, 3 semester hours in introduction to substance abuse or 3 in health care administration (8/91).

AR-0801-0019

General Medical Orientation

Course Number: 300-F10.
Location: Academy of Health Sciences, Fort Sam Houston, TX.
Length: 3–4 weeks (113–131 hours).
Exhibit Dates: 10/87–4/92.
Learning Outcomes: Upon completion of the course, the student will have a beginning foundation in anatomy and physiology, medical terminology, first aid, medical asepsis, and field sanitation.
Instruction: Lectures, demonstrations, and hands-on experience cover vital signs, CPR techniques, splinting, bleeding management, food and water sanitation, waste disposal, and insect and rodent control.
Credit Recommendation: Credit is not recommended because of the military-specific nature of the course (11/90).

AR-0801-0023

NBC Defense Officer and Noncommissioned Officer (NCO)
(NBC Defense)

Course Number: NBC 54.
Location: Combined Arms Training Center, Vilseck, W. Germany.
Length: 2 weeks (61–80 hours).
Exhibit Dates: 5/79–10/98.
Learning Outcomes: Upon completion of the course, the student will be able to recognize and react to chemical, biological, or radiological hazards; prepare graphic models of hazard zones; survey contaminated zones, using detection kits, radiac instruments, and dosimeters; supervise expeditious measures for decontamination; establish and maintain a library of documents and prepare associated reports; establish a training program for teaching others; and use prediction devices to determine nature of, magnitude of, and hazards due to chemical, biological, and radiological devices.
Instruction: Conferences, lectures, practical exercises, and performance examinations cover toxic and nuclear hazards, detection of chemical agents, radiological monitoring, use of protective clothing, decontamination procedures, hazard predictions, and warning/reporting systems.
Credit Recommendation: Credit is not recommended because of the limited, specialized nature of the course (10/88).

AR-0801-0024

Principles of Military Preventive Medicine (MS)

Course Number: 6A-F5.

Location: Academy of Health Sciences, Fort Sam Houston, TX.

Length: 9 weeks (365 hours).

Exhibit Dates: 10/88–5/90.

Learning Outcomes: Upon completion of the course, the student will be able to serve as nuclear medical science officer, an environmental science officer, or as a sanitary engineer.

Instruction: This course includes lectures, discussions, and clinical practice in preventive medicine, environmental hygiene, infection control, epidemiology, water quality, sanitation, and industrial hygiene.

Credit Recommendation: In the lower-division baccalaureate/associate degree category, 3 semester hours in preventive medicine, 2 in environmental hygiene, 1 in infection control, 1 in epidemiology, 2 in water quality, 1 in sanitation, and 1 in industrial hygiene (3/90).

AR-0801-0025

Drug and Alcohol Rehabilitation Training Individual

Course Number: 5H-F4/302-F4.

Location: Academy of Health Sciences, Fort Sam Houston, TX.

Length: 2 weeks (85 hours).

Exhibit Dates: 1/90–Present.

Learning Outcomes: Upon completion of the course, the student will be able to provide effective counseling to clients with alcohol or drug abuse conditions; assess and screen for substance abuse; apply and discuss appropriate counseling techniques; assess psychiatric emergencies; and understand lapse, relapse, and relapse prevention as applied to rehabilitation.

Instruction: Course covers topics in individual and small group counseling, evaluation and assessment screening for substance abuse, client outreach, and consultation. Methods include conferences, demonstrations, and practical exercises.

Credit Recommendation: In the upper-division baccalaureate category, 3 semester hours in substance abuse rehabilitation counseling (10/03).

AR-0801-0026

Nuclear, Biological, Chemical (NBC) Reconnaissance

Course Number: ***Version 1:*** 494-ASI-L5. *Version 2:* 4K-F10/494-ASIL5.

Location: ***Version 1:*** Chemical School, Fort McClellan, AL. *Version 2:* Chemical School, Fort Leonard Wood, MO.

Length: ***Version 1:*** 5 weeks (237 hours). *Version 2:* 5–6 weeks (243–256 hours).

Exhibit Dates: ***Version 1:*** 7/91–9/98. *Version 2:* 10/98–Present.

Learning Outcomes: ***Version 1:*** Upon completion of the course, the student will be able to describe the organization, mission, and function of the Nuclear, Biological, and Chemical Reconnaissance system (NBCRS); apply the tactics, techniques, and procedures of NBCRS, particularly in areas of movement, surveillance, search, survey, and sampling; practice avoidance of contamination hazards in areas of operation; and detect and identify NBC agents rapidly. *Version 2:* Upon completion of the course, the student will be able to describe the Nuclear, Biological, and Chemical Reconnaissance System (NBCRS); apply general tactics, techniques, and procedures of NBCRS, particularly in the areas of movement, surveillance, search, survey, and sampling; practice contamination hazards in areas of operation; and detect and identify NBC agents rapidly.

Instruction: ***Version 1:*** Instruction includes lectures, group discussions, and practical training in nuclear, biological, and chemical reconnaissance systems. *Version 2:* Instruction includes lectures, group discussions, oral examinations, and practical training exercises in nuclear, biological, and chemical reconnaissance systems.

Credit Recommendation: ***Version 1:*** In the upper-division baccalaureate category, 3 semester hours in environmental science, or environmental studies, or hazardous materials management (9/95). *Version 2:* In the lower-division baccalaureate/associate degree category, 3 semester hours in environmental science, or 3 in environmental studies, or 3 in hazardous materials management (4/04).

AR-0801-0027

Chemical Operations Specialist Basic Noncommissioned Officer (NCO) Reclassification

Course Number: 494-54B20-R.

Location: Chemical School, Fort McClellan, AL.

Length: 15–16 weeks (665–703 hours).

Exhibit Dates: 3/92–12/99.

Learning Outcomes: Before 3/92 see AR-0801-0010. Upon completion of the course, the student will be able to identify and perform critical tasks in nuclear, biological, and chemical operations, including applying biological defense, logistics, and maintenance procedures; detecting and decontaminating equipment and smoke generators; and conducting tactical training.

Instruction: Methods of instruction consist of conferences, practical exercises, and examinations in individual protective skills, radiological and chemical decontamination, smoke and biological operations, and tactical training.

Credit Recommendation: In the lower-division baccalaureate/associate degree category, 3 semester hours in radiological safety, 1 semester hour in technical writing, and 1 semester hour in basic computer literacy (9/95).

AR-0801-0028

Chemical Operations Specialist Basic Noncommissioned Officer (NCO)

Course Number: ***Version 1:*** 494-54B20. *Version 2:* 494-54B20.

Location: ***Version 1:*** Chemical School, Fort McClellan, AL. *Version 2:* Chemical School, Fort Leonard Wood, MO.

Length: ***Version 1:*** 15–16 weeks (647–665 hours). *Version 2:* 15–16 weeks (570–571 hours).

Exhibit Dates: ***Version 1:*** 3/92–9/00. *Version 2:* 10/00–Present.

Learning Outcomes: ***Version 1:*** Before 3/92 see AR-0801-0012. Upon completion of the course, the student will be able to perform critical tasks in nuclear, biological, and chemical applications, including biological defense, logistics, and maintenance procedures; detecting contamination; decontaminating equipment and smoke generators; and conducting tactical training. *Version 2:* Upon completion of the course, the student will be able to identify and perform critical tasks in nuclear, biological, and chemical applications, including biological defense, logistics, and maintenance procedures; detecting decontamination; decontaminating equipment using smoke generators; and conducting tactical training.

Instruction: ***Version 1:*** The course of instruction consists of conferences, practical exercises, and examinations in individual protective skills, radiological and chemical decontamination, smoke and biological operations, and tactical training. This course contains a common core of leadership subjects. *Version 2:* The course of instruction consists of conferences, practical exercises, and examinations in individual protective skills, radiological and chemical decontamination, smoke and biological operations, and tactical training. This course contains a common core of leadership subjects.

Credit Recommendation: ***Version 1:*** In the lower-division baccalaureate/associate degree category, 1 semester hour in technical writing, 3 in radiological safety, and 1 in basic computer literacy. See AR-1406-0090 for common core credit (9/95). *Version 2:* In the lower-division baccalaureate/associate degree category, 1 semester hour in technical writing, 3 in radiological safety, and 1 in introduction to computers. See AR-1406-0090 for common core credit (1/02).

Related Occupations: 54B.

AR-0801-0029

Chemical Operations Specialist Advanced Noncommissioned Officer (NCO)

Course Number: 494-54B40; 4-54-C42.

Location: Chemical School, Fort McClellan, AL.

Length: 13–14 weeks (563–571 hours).

Exhibit Dates: 3/92–12/99.

Learning Outcomes: After 9/00 see AR-1720-0013. Upon completion of the course, the student will be able to perform critical tasks in nuclear and chemical operations, biological defense, logistics, and maintenance management.

Instruction: Instruction includes lectures, small-group discussion, and practical exercises in management practices and concepts, including nuclear and chemical operations, biological defense, and tactical decision making processes. Instruction also includes word processing, data base, spread sheet, and computer graphics, as well as technical writing skills.

Credit Recommendation: In the lower-division baccalaureate/associate degree category, 3 semester hours in nuclear instrumentation, 3 semester hours in radiological safety, 1 semester hour in basic computer literacy, and 1 semester hour in technical writing skills. See AR-1404-0035 for common core credit (9/95).

Related Occupations: 54B.

AR-0801-0030

Chemical Operations Specialist, Phase 1

Course Number: 54B1OST1; 54B10 OSUT (ST1).

Location: Chemical School, Fort McClellan, AL.

Length: 8 weeks (300 hours).

Exhibit Dates: 11/93–12/99.

Learning Outcomes: Upon completion of the course, the student will be a self-disciplined, motivated, and physically fit soldier trained in combat survivability skills, combat techniques, and individual weapons and be prepared to perform NBC detection, identification, and decontamination operations.

Instruction: This 8 week long course includes general military subjects and selected tasks.

Credit Recommendation: Credit is not recommended because of the military-specific nature of the course (9/95).

Related Occupations: 54B.

AR-0801-0031

Chemical Operations Specialist, Phase 2

Course Number: 54B1OST2; 54B10OSUT (ST2).

Location: Chemical School, Fort McClellan, AL.

Length: 10 weeks (539 hours).

Exhibit Dates: 11/93–12/99.

Learning Outcomes: Upon completion of the course, the student will be a well-disciplined, motivated, and physically fit soldier trained in combat survivability skills, combat techniques, and entry-level chemical skills.

Instruction: Instruction in the 10 week long course includes training in tasks necessary to perform map reading/land navigation; M60 and.50 caliber machine gun firing; chemical, biological, and radiological missions; decontamination; and smoke generator operation. This course includes an 8-9 week Basic Training component.

Credit Recommendation: Credit is not recommended because of the military-specific nature of the course (9/95).

Related Occupations: 54B.

AR-0801-0032

Chemical Operations Specialist (AITS Inserts)

Course Number: 494-54B10-OSUT.

Location: Chemical School, Fort McClellan, AL.

Length: 11–12 weeks (441 hours).

Exhibit Dates: 11/93–12/99.

Learning Outcomes: Upon completion of the course, the student will be a self-disciplined, motivated, and physically fit soldier trained in combat survivability skills, combat techniques, and individual weapons and prepared to perform NBC detection, identification, and decontamination operations.

Instruction: Instruction in the 11-12 week long course includes training in tasks necessary to perform map reading/land navigation; M60 and.50 caliber machine gun firing; chemical, biological, and radiological missions; decontamination; and smoke generator operation. This course includes an 8-9 week Basic Training component.

Credit Recommendation: Credit is not recommended because of the military-specific nature of the course (9/95).

Related Occupations: 54B.

AR-0801-0033

Chemical Operations Specialist

Course Number: 54B10-OSUT; 54B10UST; 54B1OSUT.

Location: Chemical School, Fort McClellan, AL.

Length: 18 weeks (867–868 hours).

Exhibit Dates: 11/93–12/99.

Learning Outcomes: Upon completion of the course, the student will be able to perform basic soldier skills, including combat skills and techniques, radiological defense, chemical and biological defense, decontamination, smoke operations, and driver's training.

Instruction: Instruction includes training in combat skills, chemical and biological defense, and chemical specialist training. This course includes an 8-9 week Basic Training component.

Credit Recommendation: Credit is not recommended because of the military-specific nature of the course (9/95).

Related Occupations: 54B.

AR-0801-0034

1. Army Medical Department (AMEDD) Noncommissioned Officer (NCO) Basic
2. Army Medical Department (AMEDD) Noncommissioned Officer Basic (NCOES)

Course Number: ***Version 1:*** 6-8-C40 (91K Tech Tng); 6-8-C40 (91K30 Tech Tng), Phase 1. *Version 2:* 6-8-C40 (91K30 Tech Tng), Phase 2.

Location: Academy of Health Sciences, Fort Sam Houston, TX.

Length: ***Version 1:*** 3 weeks (120 hours). *Version 2:* 6–7 weeks (72 hours).

Exhibit Dates: ***Version 1:*** 4/95–9/03. *Version 2:* 10/03–Present.

Learning Outcomes: ***Version 1:*** Upon completion of the course, the student will be able to manage a laboratory and apply advanced concepts of clinical chemistry, microscopy, and microbiology. *Version 2:* Upon completion of the course, the student will be able to manage a medical laboratory; supervise small groups; apply advanced concepts of chemistry, microscopy and microbiology; assemble a field laboratory; and ensure quality assurance and safety standards.

Instruction: ***Version 1:*** Lectures and practical exercises cover laboratory management, quality assurance, and safety. *Version 2:* Practical exercises, discussion, laboratory, case studies, and lecture. Topics include laboratory management, quality assurance and safety, and assembling a laboratory in a field environment.

Credit Recommendation: ***Version 1:*** In the upper-division baccalaureate category, 2 semester hours in clinical laboratory management and 2 in clinical laboratory procedures (11/97). *Version 2:* In the upper-division baccalaureate category, 3 semester hours in clinical laboratory management (11/03).

Related Occupations: 91K.

AR-0801-0035

Alcohol and Drug Abuse Prevention and Control Program

Course Number: 5H-F6/302-F6.

Location: Academy of Health Sciences, Fort Sam Houston, TX.

Length: 2 weeks (72 hours).

Exhibit Dates: 3/93–3/97.

Learning Outcomes: Upon completion of the course, the student will be able to initiate a program in alcohol and drug abuse prevention as well as provide directors of such programs with training in principles and administration of the clinical aspects.

Instruction: Methods of instruction include discussion of policies and regulations pertaining to management and responsibilities of the drug and alcohol abuse program, with emphasis on current and future trends.

Credit Recommendation: In the upper-division baccalaureate category, 2 semester hours in health care administration seminar (10/95).

AR-0801-0036

Special Operations Combat Medic

Course Number: 300-F8.

Location: John F. Kennedy Special Warfare Center and School, Fort Bragg, NC.

Length: 24 weeks (1091–1092 hours).

Exhibit Dates: 7/96–Present.

Learning Outcomes: Upon completion of the course, the student will be able to manage trauma patients, manage patients prior to medical evacuation, and provide basic medical care to team members.

Instruction: Lectures, laboratories, role playing, simulations, and videos are course methodologies. Students should qualify to sit for the National Registry EMT exam.

Credit Recommendation: In the lower-division baccalaureate/associate degree category, 4 semester hours in anatomy and physiology, 10 in emergency clinical, 12 in emergency care, and 3 in military science (9/00).

AR-0802-0001

Technical Escort

Course Number: ***Version 2:*** 2E-SI5J/494-ASIJ5; 2E-SI5J/494J5; 2E-F21; 494-F3; 2E-ASI5J/494-ASIJ5. *Version 3:* 494-ASIJ5; 2E-SI5J. *Version 4:* 2E-SI5J/494-ASIJ5.

Location: ***Version 2:*** Missile and Munitions School, Redstone Arsenal, AL. *Version 3:* Ordnance, Missile and Munitions School, Redstone Arsenal, AL. *Version 4:* Ordnance, Missile and Munitions School, Redstone Arsenal, AL.

Length: ***Version 2:*** 4–5 weeks (131–168 hours). *Version 3:* 3–4 weeks (126–127 hours). *Version 4:* 3–4 weeks (131 hours).

Exhibit Dates: ***Version 2:*** 1/76–9/92. *Version 3:* 10/92–9/99. *Version 4:* 10/99–Present.

Learning Outcomes: ***Version 2:*** Upon completion of the course, the student will be able to package and store chemical, biological, and radiological agents; describe the effects of these agents on humans; and use instruments to measure radiation. *Version 3:* Upon completion of the course, the student will be able to package and store chemical, biological, and radiological agents; describe the effects of these agents on humans; and use instruments to measure radiation. *Version 4:* Upon completion of the course, the student will be able to perform duties related to identifying, recovering, sampling, transporting and disposing of nuclear, biological, and chemical waste materials; describe the effects of these materials on humans; and use instruments to measure them.

Instruction: ***Version 2:*** Methods include lectures and practical experience on the handling of hazardous materials. *Version 3:* Methods include lectures and practical experience on the handling of hazardous materials. *Version 4:* Methods of instruction include lectures, conferences, practical exercises, and various testing procedures. Topics covered in the course include radiation, chemical agent characteristics and effects on humans, use of protective clothing and respiratory protective devices, biological agent and detection devices, and packaging/sealing/resealing of hazardous materials. Exercises are scheduled to give practical experience on the subjects presented.

Credit Recommendation: ***Version 2:*** In the vocational certificate category, 3 semester hours in hazardous materials handling (10/90). *Version 3:* In the lower-division baccalaureate/associate degree category, 3 semester hours in hazardous materials handling (5/01). *Version 4:* In the lower-division baccalaureate/associate degree category, 3 semester hours in hazardous materials handling (6/01).

Related Occupations: 411A; 4514; 54B; 54D; 54E; 55D.

AR-0802-0002

EXPLOSIVE ORDNANCE DISPOSAL, PHASE 1

Course Number: 4E-F3/431-55D20; 4E-F3; 431-55D10; 431-55D20; 431-413.1.

Location: Missile and Munitions School, Redstone Arsenal, AL; Chemical School, Fort McClellan, AL.

Length: 2–3 weeks (72–76 hours).

Exhibit Dates: 2/65–5/96.

Objectives: To train enlisted personnel and commissioned officers in the disposal of chemical and biological agents found in unexploded ordnance.

Instruction: Lectures and practical exercises cover explosive ordnance disposal. Instruction includes biological agents; chemical and biological defense; recognition of the chemical, physical, and the physiological effects of chemical agents on the body; and treatment procedures for personnel exposed to chemical agents. Upon completion, the student will be able to identify biological materials and chemical munitions, perform leak sealing and packaging procedures, and dispose of chemical and biological materials. Instruction also includes chemical agents and compounds, biological materials, decontaminants, and protective clothing.

Credit Recommendation: Credit is not recommended because of the military-specific nature of the course (5/86).

Related Occupations: 55D.

AR-0802-0019

NUCLEAR WEAPONS TECHNICIAN

Course Number: 4-9-C32-260A0.

Location: Missile and Munitions School, Redstone Arsenal, AL.

Length: 5–6 weeks (199 hours).

Exhibit Dates: 9/85–1/96.

Objectives: To provide warrant officers with the technical training to qualify them in new maintenance techniques and management procedures of special ammunition units.

Instruction: Lectures and practical exercises cover leadership, training management, electronic warfare, and nuclear fundamentals.

Credit Recommendation: Credit is not recommended because of the military-specific nature of the course (4/86).

Related Occupations: 260A.

AR-0802-0020

FIRST SERGEANT

Course Number: ***Version 1:*** 521-SQIM. *Version 2:* 521-SQIM. *Version 3:* 400-FSC, Phase 2; 521-SQIMA, Phase 2; 521-SQM, Phase 2.

Location: ***Version 1:*** McGraw Barracks, Munich, W. Germany; Sergeants Major Academy, Fort Bliss, TX. *Version 2:* Sergeants Major Academy, Fort Bliss, TX. *Version 3:* Reserve Schools, Multiple locations; Sergeants Major Academy, Fort Bliss, TX.

Length: ***Version 1:*** 8 weeks (257 hours). *Version 2:* 4–5 weeks (174–175 hours). *Version 3:* 2–3 weeks (103 hours).

Exhibit Dates: ***Version 1:*** 10/81–9/91. *Version 2:* 10/91–9/00. *Version 3:* 10/00–Present.

Learning Outcomes: ***Version 1:*** Upon completion of the course, the student will be able to supervise units of up to 100 persons. *Version 2:* Upon completion of the course, the student will be able to supervise units of up to 200 persons. *Version 3:* Pending evaluation.

Instruction: ***Version 1:*** Course includes lectures, seminars, demonstrations, and performance exercises in safety, stress management, substance abuse, maintenance and control of equipment, motivation, leadership skills, and written and verbal skills. *Version 2:* Course includes lectures, seminars, demonstrations, and performance exercises in safety, stress management, diversity, substance abuse, maintenance and control of equipment, motivation, leadership skills, and computer and verbal skills. *Version 3:* Pending evaluation.

Credit Recommendation: ***Version 1:*** In the lower-division baccalaureate/associate degree category, 3 semester hours in health and fitness, 1 in supply management, and 1 in fundamentals of communication. In the upper-division baccalaureate category, 1 semester hour in human resource management (9/87). *Version 2:* In the lower-division baccalaureate/associate degree category, 1 semester hour in material management and 2 in fundamentals of communication. In the upper-division baccalaureate category, 1 semester hour in human resource management (6/95). *Version 3:* Pending evaluation.

AR-0802-0021

AMMUNITION TECHNICIAN TECHNICAL AND TACTICAL CERTIFICATION

Course Number: 4E-910A.

Location: Ordnance, Missile and Munitions School, Redstone Arsenal, AL.

Length: 9–10 weeks (331–376 hours).

Exhibit Dates: 10/88–9/94.

Learning Outcomes: After 9/94 see AR-2201-0463. Upon completion of the course, the student will be able to perform duties as an ammunition supply technician.

Instruction: Lectures, discussions, and role playing cover conventional and chemical munitions, the Standard Army Ammunition System, software to control supply of munitions, and munitions safety and storage.

Credit Recommendation: Credit is not recommended because of the military-specific nature of the course (1/91).

Related Occupations: 411A; 910A.

AR-0802-0025

NUCLEAR AND CHEMICAL TARGET ANALYSIS

Course Number: 2E-ASI5H-A.

Location: Field Artillery School, Fort Sill, OK.

Length: 3 weeks (146 hours).

Exhibit Dates: 1/87–10/97.

Learning Outcomes: Upon completion of the course, the student will be able to analyze and provide predictive data on nuclear and chemical detonations, use semi-logarithmic graphs, nomograms, tabular data, and probability statistics; and compute characteristics of overall effects of nuclear and chemical detonations using prescribed techniques and processes.

Instruction: Lectures and practical exercises cover the fundamental effects of nuclear weapons; damage estimation and target analysis; chemical weapons employment and the tactical application of nuclear and chemical weapons; nuclear and chemical trends and developments; foreign nuclear, biological, and chemical capabilities; and math formulas, nomograms, semi-logarithmic graphs, circular map scales, slide rule, and probability statistics.

Credit Recommendation: In the vocational certificate category, 1 semester hour in applied mathematics. In the lower-division baccalaureate/associate degree category, 1 semester hour in nuclear technology and 2 in nuclear safety and damage estimation (6/89).

AR-0802-0026

MUNITIONS/HAZARDOUS MATERIALS FIRE FIGHTING

Course Number: C3AZR57150 001.

Location: Training Center, Chanute AFB, IL.

Length: 1–2 weeks (52 hours).

Exhibit Dates: 3/89–4/04.

Learning Outcomes: Upon completion of the course, the student will be able to identify, classify, and control various hazardous materials emergencies.

Instruction: Lectures and practical exercises cover the above topics.

Credit Recommendation: In the lower-division baccalaureate/associate degree category, 3 semester hours in introduction to hazardous materials emergencies (1/93).

AR-0802-0027

SMALL UNIT LEADER'S FORCE PROTECTION

Course Number: 7K-F19.

Location: Safety Center, Fort Rucker, AL.

Length: 2 weeks (61–62 hours).

Exhibit Dates: 10/95–Present.

Learning Outcomes: Upon completion of the course, the student will be able to develop a unit safety program, manage a reporting system, and institute an active accident prevention program.

Instruction: Group-paced instruction covers safety and accident prevention programs. Course includes identification and evaluation of hazards, as well as developing, implementing, and supervising control measures, force protection requirements, and hazard/accident reporting procedures.

Credit Recommendation: In the lower-division baccalaureate/associate degree category, 2 semester hours in general safety and 1 in record keeping (1/96).

AR-0802-0028

1. EXPLOSIVE ORDNANCE DISPOSAL SPECIALIST ADVANCED NONCOMMISSIONED OFFICER (NCO)
2. EXPLOSIVE ORDNANCE DISPOSAL SPECIALIST ADVANCED

Noncommissioned Officer (NCO), Phase 2

Course Number: *Version 1:* 431-55D40. *Version 2:* 431-55D40, Phase 2.

Location: Ordnance, Missile and Munitions School, Redstone Arsenal, AL.

Length: *Version 1:* 4–5 weeks (165 hours). *Version 2:* 3 weeks (103 hours).

Exhibit Dates: *Version 1:* 5/94–9/00. *Version 2:* 10/00–Present.

Learning Outcomes: *Version 1:* Upon completion of the course, the student will be able to supervise explosive ordnance disposal specialists. *Version 2:* Upon completion of the course, the student will be able to supervise explosive ordnance disposal specialists.

Instruction: *Version 1:* Lectures and group discussions provide training in common military subjects; planning for explosive ordnance disposal support of range clearance operations; planning emergency clearance operations; and handling and disposing of chemical/biological toxins; nuclear; and other munitions. Course includes a common core of leadership subjects. *Version 2:* Lectures and group discussions provide training in common military subjects; planning for explosive ordnance disposal support of range clearance operations; planning emergency clearance operations; and handling and disposing of chemical/biological toxins, nuclear, and other munitions. Course includes a common core of leadership subjects.

Credit Recommendation: *Version 1:* In the lower-division baccalaureate/associate degree category, 3 semester hours in military science. See AR-1404-0035 for common core credit (7/96). *Version 2:* In the lower-division baccalaureate/associate degree category, 3 semester hours in military science. See AR-1404-0035 for common core credit (6/01).

Related Occupations: 55D.

AR-0802-0029

1. Explosive Ordnance Disposal Specialist Phase 3
2. Explosive Ordnance Disposal Specialist, Phase 2

Course Number: *Version 1:* 4E-91E/431-55D10/20. *Version 2:* 4E-91E/431-55D10/20, Phase 2.

Location: Ordnance, Missile and Munitions School, Redstone Arsenal, AL.

Length: *Version 1:* 3–4 weeks (146 hours). *Version 2:* 5 weeks (186 hours).

Exhibit Dates: *Version 1:* 2/94–9/00. *Version 2:* 10/00–Present.

Learning Outcomes: *Version 1:* Upon completion of the course, the student will be able to perform those tasks unique to explosive ordnance disposal. *Version 2:* Upon completion of the course, the student will be able to perform those tasks unique to explosive ordnance disposal.

Instruction: *Version 1:* Methods of instruction are classroom lectures and field demonstrations on safety; equipment; and operation, detection, identification, and recovery of ordnance. *Version 2:* Methods of instruction are classroom lectures and field demonstrations on safety; equipment; and operation, detection, identification, and recovery of ordnance.

Credit Recommendation: *Version 1:* Credit is not recommended because of the military-specific nature of the course (7/96). *Version 2:* Credit is not recommended because of the military-specific nature of the course (6/01).

Related Occupations: 55D.

AR-0802-0030

Warrant Officer Basic 918B

Course Number: 4F 918B, Phase 1.

Location: Ordnance, Missile and Munitions School, Redstone Arsenal, AL.

Length: 6–7 weeks (244 hours).

Exhibit Dates: 5/99–Present.

Learning Outcomes: Upon completion of the course, the student will be able to supervise radiation monitoring procedures, employment of nuclear, biological, and chemical markers, decontamination, platoon compliance with host-nation, federal, state and local environmental laws, implementation of preventive medicine policies; plan, build, develop, and motivate a platoon; perform unit-level Test, Measurement, Diagnostic Equipment (TMDE) management; and act as custodian of radiation calibration operations.

Instruction: Lectures and laboratory exercises cover management and control of nuclear, biological, and chemical material; monitoring of organizational compliance with hazardous material handling.

Credit Recommendation: In the lower-division baccalaureate/associate degree category, 3 semester hours in management of hazardous materials (6/01).

AR-0802-0031

Army Medical Department (AMEDD) Aviation Officer Pre-Command

Course Number: 2C-F11.

Location: Academy of Health Sciences, Fort Sam Houston, TX.

Length: 1–2 weeks (62 hours).

Exhibit Dates: 4/99–Present.

Learning Outcomes: Upon completion of the course the student will be able to identify issues concerning medical evacuation planning; lead, manage, and mentor assigned staff; and serve as an ambulance commander.

Instruction: Audio visual materials, discussion, classroom exercise, and practical exercise. Topics include tactics and doctrine, medical evacuation planning, combat health support, issues of leadership and mentoring, safety, legal issues, risk management, and military briefing.

Credit Recommendation: In the upper-division baccalaureate category, 3 semester hours in disaster management leadership (6/02).

AR-0803-0006

Special Forces Combat Diver Qualification

Course Number: *Version 1:* 2E-SI4W/4Y/011-ASIW7/W9. *Version 2:* 2E-SI4W/011-ASIW7.

Location: *Version 1:* Naval Air Station, Key West, FL. *Version 2:* John F. Kennedy Special Warfare Center and School, Fort Bragg, NC.

Length: *Version 1:* 4–5 weeks (147–215 hours). *Version 2:* 4–5 weeks (195–196 hours).

Exhibit Dates: *Version 1:* 10/88–12/95. *Version 2:* 1/96–Present.

Learning Outcomes: *Version 1:* Upon completion of the course, the student will be able to complete open-water dives in a safe and effective manner using open-circuit and closed-circuit equipment. *Version 2:* Upon completion of the course, the student will be able to complete open-water dives in a safe and effective manner using open-circuit and closed-circuit equipment.

Instruction: *Version 1:* Each successful student completes at least 100 hours of laboratory-based, skill-acquisition activity in scuba diving. *Version 2:* Course includes lectures, demonstrations, and practical experiences. Students complete at least 100 hours of laboratory-based skill-acquisition activity in scuba diving.

Credit Recommendation: *Version 1:* In the lower-division baccalaureate/associate degree category, 1 semester hour in scuba diving (12/92). *Version 2:* In the lower-division baccalaureate/associate degree category, 1 semester hour in scuba diving/leisure recreation (4/01).

AR-0803-0007

Survival, Evasion, Resistance, and Escape (SERE) High Risk (Level C)
(Survival, Evasion, Resistance, and Escape (SERE) High Risk)

Course Number: 3A-F38/012-F27.

Location: John F. Kennedy Special Warfare Center and School, Fort Bragg, NC.

Length: 3–4 weeks (304–310 hours).

Exhibit Dates: 3/88–Present.

Learning Outcomes: Upon completion of the course, the student will demonstrate skills in personal survival techniques, resistance to interrogation and exploitation, and escape and evasion procedures.

Instruction: Lectures and demonstrations cover the subject areas. Course includes 120 hours field training where students execute survival skills and laboratory-based learning activities.

Credit Recommendation: In the lower-division baccalaureate/associate degree category, 1 semester hour in survival skills/outdoor pursuits (11/00).

AR-0803-0008

Special Operations Qualification Reserve Component Phase 2

Course Number: 011-18B30/18C30/18E30/2E-18A.

Location: John F. Kennedy Special Warfare Center and School, Fort Bragg, NC.

Length: 3 weeks (177–178 hours).

Exhibit Dates: 10/88–9/90.

Learning Outcomes: Upon completion of the course, the student will perform map reading, small boat operations, long-range reconnaissance patrolling, fieldcraft, survival, air operations, and the endurance and confidence building requisite to special forces operations.

Instruction: Course material includes map and compass reading, patrolling, fieldcraft, survival, and air operations by means of lectures, demonstrations, and laboratories.

Credit Recommendation: In the lower-division baccalaureate/associate degree category, 1 semester hour in orienteering (12/88).

Related Occupations: 18B; 18C; 18E.

AR-0803-0009

Special Operations Qualification Reserve Component Phase 6

Course Number: 011-18B30-RC/18E30-RC/18C30-RC/2E-18A-RC; 011-18E30-RC; 2E-18A-RC; 011-18C30-RC; 011-18B30-RC.

Location: John F. Kennedy Special Warfare Center and School, Fort Bragg, NC.

Length: 3 weeks (264 hours).

Exhibit Dates: 10/88–9/92.

Learning Outcomes: Upon completion of the course, the student will be able to demonstrate knowledge of landing zone operations, drop zone procedures, conduct an unconventional warfare mission, and conduct detachment pre-mission training.

Instruction: Lectures, demonstrations, and participation in practical field exercises are modes used to ensure acquisition of requisite knowledge and skills.

Credit Recommendation: Credit is not recommended because of the military-specific nature of the course (12/88).

Related Occupations: 18B; 18C; 18E.

AR-0803-0011

Reserve Component Master Fitness Trainer

(Master Fitness Trainer Reserve Component)

Course Number: 7B-F9/562-F4-RC.

Location: Physical Fitness School, Fort Benning, GA; Off-site locations, Worldwide; Soldier Support Institute, Fort Benjamin Harrison, IN.

Length: 2 weeks (97 hours).

Exhibit Dates: 7/88–10/95.

Learning Outcomes: Upon completion of the course, the student will be able to design, set up, and monitor a physical fitness program for a group or an individual.

Instruction: Course includes lectures and practical exercises in the science of human anatomy and physiology, body composition, exercise assessment, nutrition, stress management, exercise techniques and games, and the organization of group and individual fitness and exercise programs.

Credit Recommendation: In the lower-division baccalaureate/associate degree category, 3 semester hours in introduction to health and recreation (7/92).

AR-0803-0012

Master Fitness Trainer

Course Number: 7B-F7/562-F2; 7B-SI6P/562-ASIP5; 7B-FS/562-F2.

Location: Soldier Support Institute, Fort Benjamin Harrison, IN; Physical Fitness School, Fort Benning, GA; Off-site locations, Worldwide.

Length: 2–4 weeks (86–154 hours).

Exhibit Dates: 10/86–Present.

Learning Outcomes: Upon completion of the course, the student will be able to design, set up, and monitor a fitness program for groups or for an individual and perform as unit advisors to their commanders on physical fitness issues.

Instruction: Course includes lectures and practical exercises in the science of human anatomy and physiology, body composition, exercise assessment, nutrition, stress management, exercise techniques and games, and the organization of group and individual fitness and exercise programs.

Credit Recommendation: In the lower-division baccalaureate/associate degree category, 4 semester hours in exercise science of which 2 are specifically in exercise science laboratory, or 3 in introduction to health and recreation (7/99).

AR-0803-0013

Special Operations Training

Course Number: *Version 1:* 2E-F132/011-F45. *Version 2:* 2E-F132/011-F45.

Location: John F. Kennedy Special Warfare Center and School, Fort Bragg, NC.

Length: *Version 1:* 4 weeks (208–209 hours). *Version 2:* 4 weeks (193–194 hours).

Exhibit Dates: *Version 1:* 10/93–9/97. *Version 2:* 10/97–Present.

Learning Outcomes: *Version 1:* Upon completion of the course, the student will be able to provide specialized and comprehensive instruction and training in tactics and techniques and conduct direct action and unilateral special operations by Special Operations Forces. *Version 2:* Upon completion of the course, the student will be able to provide specialized and comprehensive instruction and training in tactics and techniques and conduct direct action and unilateral special operations.

Instruction: *Version 1:* Course includes lectures and practical exercises on survival, evasion, escape, resistance, and aviation-related materials. *Version 2:* Course includes lectures and practical exercises on survival, evasion, escape, resistance, and aviation-related materials.

Credit Recommendation: *Version 1:* In the lower-division baccalaureate/associate degree category, 1 semester hour in survival skills/outdoor pursuits (12/92). *Version 2:* In the lower-division baccalaureate/associate degree category, 1 semester hour in military science (4/01).

AR-0803-0014

Waterborne Infiltration

Course Number: 2E-F138/011-F47.

Location: John F. Kennedy Special Warfare Center and School, Fort Bragg, NC.

Length: 6 weeks (345–348 hours).

Exhibit Dates: 10/89–9/92.

Learning Outcomes: Upon completion of the course, the student will perform individual surface swimming techniques, team surface swimming techniques, and surface swimming planning; small boat operations, kayak operations and maintenance and field repair of small boats, kayaks, and outboard engines; infiltration planning; rubber duck operations; nautical chart reading; and intermediate delivery systems for insertion and extraction of personnel.

Instruction: Lectures and practical exercises cover surface swimming techniques, team surface swimming, and surface swimming planning considerations; small boat and kayak operations; nautical chart reading; and intermediate delivery systems for insertion and extraction.

Credit Recommendation: In the lower-division baccalaureate/associate degree category, 1 semester hour in swimming, 1 in sailing, and 2 in orienteering (12/92).

AR-0803-0015

Survival, Evasion, Resistance, and Escape (SERE) Instructor Qualification

Course Number: O12-F19.

Location: John F. Kennedy Special Warfare Center and School, Fort Bragg, NC.

Length: 5 weeks (338–339 hours).

Exhibit Dates: 1/82–12/00.

Learning Outcomes: Upon completion of the course, the student will be able to train others in the code of conduct, survival fieldcraft with application to worldwide environments, techniques of evasion and resistance to interrogation and exploitation, and escape from captivity. The student will also participate in a survival exercise and in a resistance training laboratory.

Instruction: Lectures and practical exercises cover survival, evasion, resistance, escape, and resistance training laboratory.

Credit Recommendation: In the lower-division baccalaureate/associate degree category, 3 semester hours in survival skills/outdoor pursuits (12/92).

AR-0804-0003

Mountaineering Winter

Course Number: NGB-MTNWAR WINTER.

Location: National Guard Mountain Warfare School, Ethan Allen Firing Range, Jericho, VT.

Length: 2–3 weeks (179 hours).

Exhibit Dates: 7/86–4/04.

Learning Outcomes: Upon completion of the course, the student will be able to navigate during winter conditions in a mountainous environment including areas where substantial snowfall is common as well as in areas with significant glacial formations, using traditional and altimeter navigation techniques. Further, the student will demonstrate proficiency in specialized mountaineering techniques in areas characterized by rugged peaks, step ridges, deep ravines, and valleys during winter conditions. The student will also master survival skills for severe winter conditions.

Instruction: Lectures, demonstrations, and a heavy emphasis on performance experiences and practical exercises cover traditional and altimeter navigation techniques, mountaineering techniques, and severe cold weather survival methods.

Credit Recommendation: In the lower-division baccalaureate/associate degree category, 1 semester hour in orienteering and 3 in mountaineering/winter survival (3/92).

AR-0804-0004

Mountaineering Summer

Course Number: NGB-MTNWAR SUMMER.

Location: National Guard Mountain Warfare School, Ethan Allen Firing Range, Jericho, VT.

Length: 2–3 weeks (172–173 hours).

Exhibit Dates: 7/86–4/04.

Learning Outcomes: Upon completion of the course, the student will be able to navigate during periods of limited visibility in special environments, such as northern regions and deserts as well as mountainous regions, using both traditional land navigation methods and altimeter navigation. Furthermore, the student will demonstrate proficiency in the fundamentals, principles, and techniques of mountaineering that will provide access to otherwise inaccessible rugged mountainous terrain.

Instruction: Lectures, demonstrations, and a heavy emphasis on performance experiences and practical exercises cover skills of traditional and altimeter navigation techniques and military activities in rugged mountainous terrain.

Credit Recommendation: In the lower-division baccalaureate/associate degree category, 1 semester hour in orienteering and 1 in mountaineering (3/92).

AR-1115-0010

1. Basic Statistics for Logistics Management
2. Introduction to Operations Research and Systems Analysis

Course Number: *Version 1:* ALMC-BA; ALM-63-0449. *Version 2:* ALM-64-0046.

Location: *Version 1:* Army Logistics Management Center, Fort Lee, VA; Army Logistics Management College, Onsite at various locations. *Version 2:* Army Logistics Management Center, Fort Lee, VA.

Length: *Version 1:* Self-paced, 2–3 weeks (82 hours). *Version 2:* 2 weeks (80 hours).

Exhibit Dates: *Version 1:* 8/80–9/93. *Version 2:* 8/00–Present.

Learning Outcomes: *Version 1:* To develop and increase knowledge and understanding of statistics as applied to logistics management. *Version 2:* Pending evaluation.

Instruction: *Version 1:* Self-paced individualized instruction covers the areas of descriptive statistics, concepts and methods, inferential statistics, time series analysis, and simple linear regression. This course stresses basic statistical skills; and emphasis is placed on developing the ability to understand, apply, and analyze statistics. *Version 2:* Pending evaluation.

Credit Recommendation: *Version 1:* In the lower-division baccalaureate/associate degree category, 3 semester hours in statistics (12/91). *Version 2:* Pending evaluation.

AR-1115-0012

Resource Management Quantitative

Course Number: 7D-F25; 7D-F25/541-F9.

Location: Finance School, Soldier Support Institute, Fort Benjamin Harrison, IN.

Length: 1–2 weeks (67 hours).

Exhibit Dates: 2/87–12/91.

Learning Outcomes: Upon completion of the course, the student will be able use of statistical measures, including descriptive statistics, inference, regression analysis, PERT, and economic analysis using a microcomputer.

Instruction: Topics covered include descriptive measures, normal curve, statistical sampling, determination of sample size, population inferences, Lotus spreadsheet, regression analysis, goodness of fit, PERT, and present value.

Credit Recommendation: In the lower-division baccalaureate/associate degree category, 2 semester hours in introductory statistics (12/88).

AR-1115-0013

Battlefield Spectrum Management

Course Number: 4C-SI5D/260-ASID9.

Location: Signal School, Fort Gordon, GA.

Length: 4–5 weeks (158–159 hours).

Exhibit Dates: 10/89–5/91.

Learning Outcomes: Upon completion of the course, the student will apply principles of mathematics, wave propagation, transmission lines, and antennas to the coordination, planning, engineering, and controlling of theater radio frequencies.

Instruction: Lectures and practical exercises cover topics in mathematics and electronic communications and their applicability to the battlefield.

Credit Recommendation: In the lower-division baccalaureate/associate degree category, 1 semester hour in technical mathematics and 2 in electronic communications (6/90).

AR-1115-0014

Operations Research and Systems Analysis (ORSA) Familiarization by Correspondence
(Introduction to Operations Research Systems Analysis by Correspondence)

Course Number: ALMC-SC.

Location: Army Logistics Management College, Fort Lee, VA.

Length: Maximum, 52 weeks.

Exhibit Dates: 2/88–9/91.

Learning Outcomes: Upon completion of the course, the student will be able to apply basic operations research and standard techniques and to simple problems.

Instruction: This correspondence instruction consists of reading the assigned chapters in text books and completing assigned homework. Topics include probability and statistics and a range of basic operations research techniques.

Credit Recommendation: In the lower-division baccalaureate/associate degree category, 1 semester hour in statistics. In the upper-division baccalaureate category, 1 semester hour in operations research (6/91).

AR-1205-0006

2. Army Band Advanced Noncommissioned Officer (NCO) Phase 2
3. Army Band Advanced Noncommissioned Officer (NCO)

Course Number: *Version 2:* 450-97-C42. *Version 3:* 4-97-C42.

Location: School of Music, Little Creek, Norfolk, VA.

Length: *Version 2:* 12 weeks (422–435 hours). *Version 3:* 12 weeks (488 hours).

Exhibit Dates: *Version 2:* 10/86–12/96. *Version 3:* 1/97–Present.

Learning Outcomes: *Version 2:* Upon completion of the course, the student will be able to rehearse the band, lead ensembles, and train enlisted Army musicians in technical skills and duties. *Version 3:* Upon completion of the course, the student will be able to rehearse the band; lead ensembles; and conduct ensembles, dance bands, and stage bands; and train enlisted musicians in technical skills and duties.

Instruction: *Version 2:* Topics include leading ensembles, dance bands, and stage bands in rehearsal and performance; analyzing scores for form, style, voicing, instrumentation, and harmony; supervising the band; and administration, operations, and supply. Course includes a common core of leadership subjects. *Version 3:* Topics include conducting ensembles, dance bands, and stage bands in rehearsal and performance; analyzing scores for form, style, voicing, instrumentation, and harmony; supervising the band; and administration, operations, and supply. Course includes a common core of leadership subjects.

Credit Recommendation: *Version 2:* In the lower-division baccalaureate/associate degree category, see AR-1404-0035 for common core credit. In the upper-division baccalaureate category, 1 semester hour in band administration (2/93). *Version 3:* In the lower-division baccalaureate/associate degree category, see AR-1404-0035 for common core credit. In the upper-division baccalaureate category, 1 semester hour in band administration and 2 in conducting (9/98).

Related Occupations: 02B; 02C; 02D; 02E; 02F; 02G; 02H; 02J; 02K; 02L; 02M; 02N; 02T; 02U.

AR-1205-0008

Reserve Component Army Band Advanced Noncommissioned Officer (NCO), Phase 2

Course Number: 4-97-C42-RC, Phase 2; 450-97-C42-RC.

Location: Reserve Component Training Centers, US; School of Music, Little Creek, Norfolk, VA.

Length: 2 weeks (70–72 hours).

Exhibit Dates: 5/89–Present.

Learning Outcomes: Upon completion of the course, the student will be able to apply advanced rehearsal techniques to concert, stage, and marching bands.

Instruction: The course comprises instruction in advanced rehearsal techniques, including concert band rehearsal, stage band rehearsal, and marching band drill. Course includes a common core of leadership subjects.

Credit Recommendation: In the lower-division baccalaureate/associate degree category, see AR-2201-0295 for common core credit. In the upper-division baccalaureate category, 2-3 semester hours in advanced conducting/advanced rehearsal techniques (3/00).

Related Occupations: 02B; 02C; 02D; 02E; 02F; 02G; 02H; 02J; 02K; 02L; 02M; 02N; 02T; 02U.

AR-1205-0009

1. Bandmaster Senior Warrant Officer Training
2. Bandmaster Warrant Officer Advanced

Course Number: *Version 1:* 7-51-C32. *Version 2:* 7-51-C32.

Location: School of Music, Little Creek, Norfolk, VA.

Length: *Version 1:* 2 weeks (72 hours). *Version 2:* 2 weeks (72 hours).

Exhibit Dates: *Version 1:* 3/91–2/96. *Version 2:* 3/96–6/00.

Learning Outcomes: *Version 1:* Upon completion of the course, the student will be able to lead and conduct Army bands with advanced technical proficiency and assume advanced assignments as an Army bandleader. *Version 2:* Upon completion of the course, the student will have strengthened technical proficiency for leading and conducting Army bands by reviewing the fundamentals of musical instruments, audition techniques, sound reinforcement, and advanced Musical Instrument Digital Interface (MIDI) applications.

Instruction: *Version 1:* Lectures, discussions, and practical applications cover problems and solutions of the modern senior bandmaster, equipment modernization for Army bands, show production/choreography, sound reinforcement techniques, advanced concert band conducting, music literature, and band audition techniques (refresher). *Version 2:* The student reviews practical guides for woodwind, brass, percussion, bass, and guitar instruments; rehearsal objectives, including score analysis and fundamentals of conducting; band audition techniques; components of sound reinforcement systems; and the applications, uses, trends, and upgrades of MIDI.

Credit Recommendation: ***Version 1:*** In the lower-division baccalaureate/associate degree category, 1 semester hour in music recording and show production (applicable to collegiate programs such as music theater, music merchandising, etc., or as a music elective in other programs). In the graduate degree category, 2 semester hours in instrumental (band) conducting (2/93). *Version 2:* In the lower-division baccalaureate/associate degree category, 1 semester hour in music education methods and 1 in music show production (9/98).

AR-1205-0010

Bandmaster Warrant Officer Basic
(Warrant Officer Basic)
(Bandmaster/Warrant Officer Technical and Tactical Certification)

Course Number: 7N-420C.

Location: School of Music, Little Creek, Norfolk, VA.

Length: 21 weeks (756 hours).

Exhibit Dates: 1/89–1/00.

Learning Outcomes: Upon completion of the course, the student will be able to rehearse and conduct a military concert band and a stage/dance band; compose, arrange, and transcribe music for a military concert band and a stage/dance band; train a military marching band; administer a military music command; execute administrative, managerial, and supply duties necessary to units and commands; manage technical band operations as a bandmaster/commander; perform as a bandmaster in secondary mission tasks; and assess student progress and the quality and effectiveness of instruction.

Instruction: Training covers instrumental methods (brass, woodwind, and percussion); concert/stage band conducting; style analysis, ear training, sight singing, and keyboard techniques; composition skills (harmony, counterpoint, band scoring, march composition, and arranging for concert/stage band); band performance (drum majoring and marching band instructional techniques and leading the marching band in military ceremonies); Army band management (unit administration, supply procedures, budget development, and technical operations); military leadership as applied to human relations, counseling, drug abuse, military law, military justice, discipline, and physical fitness; administration of band auditions; company-level command principles and procedures; and officer indoctrination.

Credit Recommendation: In the upper-division baccalaureate category, 3 semester hours in instrumental conducting, 3 in instrumental arranging, 1 in performing ensembles (marching band), and 3 in band administration (as an elective) (9/98).

Related Occupations: 420C.

AR-1205-0011

Army Band Officer

Course Number: ***Version 1:*** 7N-42C. *Version 2:* 7N-42C.

Location: School of Music, Little Creek, Norfolk, VA.

Length: ***Version 1:*** 6 weeks (216 hours). *Version 2:* 6 weeks (216 hours).

Exhibit Dates: ***Version 1:*** 1/91–3/97. *Version 2:* 4/97–Present.

Learning Outcomes: ***Version 1:*** Upon completion of the course, the student will be able to function effectively as an Army band officer. *Version 2:* Upon completion of the course, the student will be able to conduct rehearsals, use advanced conducting techniques, plan concerts and presentations, and serve as the principle band administrator.

Instruction: ***Version 1:*** Topics include advanced conducting and rehearsal techniques; stage band, concert band, and ceremonial conducting; jazz style; music arranging, scoring, and composing techniques; piano proficiency; ear training; counterpoint; harmony; score analysis; formulation of stylistic interpretation of selected works; staff band officer duties; technical inspections; review of School of Music courses; administration of band auditions; technical band operations; budget and supplies; and marching band/military ceremonies. *Version 2:* Topics include arranging techniques and scoring projects (for a variety of instrumental ensembles, including marching, concert, and jazz bands); ensemble rehearsal procedures, advanced conducting techniques, and concert planning and presentations (for stage and concert band); keyboard harmony (including score analysis at the keyboard); band administration (including audition procedures, budgets and supplies, technical operations, inspections, and property management); and Musical Instrument Digital Interface (MIDI) training.

Credit Recommendation: ***Version 1:*** In the lower-division baccalaureate/associate degree category, 1 semester hour in keyboard harmony and 1 in ear training/sight singing. In the upper-division baccalaureate category, 1 semester hour in principles of management and 3 in arranging (concert band, stage band, marching band). In the graduate degree category, 3 semester hours in instrumental (band) conducting (2/93). *Version 2:* In the lower-division baccalaureate/associate degree category, 1 semester hour in keyboard harmony, 1 in marching band techniques, and 1 in ear training and sight singing. In the upper-division baccalaureate category, 3 semester hours in instrumental arranging (concert, jazz, and marching band), and 1 in principles of management. In the graduate degree category, 3 semester hours in instrumental (band) conducting (3/00).

AR-1205-0012

Army Band Basic Noncommissioned Officer (NCO)

Course Number: 4-97-C40; 450-02B-02U30.

Location: School of Music, Little Creek, Norfolk, VA.

Length: 7–10 weeks (260–408 hours).

Exhibit Dates: 7/86–Present.

Learning Outcomes: Upon completion of the course, the student will be able to apply the technical skills and duties of a band noncommissioned officer, including rehearsal techniques for a concert/stage band, drum major techniques, squad leader duties in the secondary mission, supply, administration, and operations; describe the historical element of the Army band; identify the NCO's role as it relates to Army bands; and conduct a field training exercise.

Instruction: Block instruction covers elements of music theory/harmony, form and analysis, ear training, conducting, rehearsal techniques, applied music, ensembles (traditional and jazz), marching band drill and mace techniques, establishing and supervising a dismount point, band administration, military branch history, and military field training exercise. Course includes a common core of leadership subjects.

Credit Recommendation: In the lower-division baccalaureate/associate degree category, see AR-1406-0090 for common core credit. In the upper-division baccalaureate category, 2-4 semester hours in advanced conducting/rehearsal techniques (3/00).

Related Occupations: 02B; 02C; 02D; 02E; 02F; 02G; 02H; 02J; 02K; 02L; 02M; 02N; 02S; 02T; 02U.

AR-1205-0013

1. Reserve Component Army Band Basic Noncommissioned Officer (NCO), Phase 2
2. Army Band Noncommissioned Officer (NCO) Reserve Component—RC

Course Number: ***Version 1:*** 450-02B-02U30-RC. *Version 2:* 4-97-C40-RC.

Location: ***Version 1:*** Reserve Centers, US; School of Music, Little Creek, Norfolk, VA. *Version 2:* Reserve Component Training Centers, US; School of Music, Little Creek, Norfolk, VA.

Length: ***Version 1:*** 2 weeks (72 hours). *Version 2:* 2 weeks (72 hours).

Exhibit Dates: ***Version 1:*** 6/88–6/97. *Version 2:* 7/97–Present.

Learning Outcomes: ***Version 1:*** Upon completion Phase 2 of the course, the student will be able to apply the technical skills and duties required of a basic noncommissioned officer, including rehearsal techniques for concert/stage band, mace signals for marching drill, and establishing and supervising a dismount point. *Version 2:* Upon completion of Phase 2 of the course, the student will be able to apply the technical skills and duties required of a basic noncommissioned officer, including rehearsal techniques for concert/stage band, mace signals for marching drill, and establishing and supervising a dismount point.

Instruction: ***Version 1:*** This phase comprises block instruction covering elements of music, voice leading, form/analysis, chord progression, rhythmic analysis, rehearsal conducting techniques, mace techniques for marching band drill, and assisting the military police by establishing and supervising a dismount point. *Version 2:* This phase comprises block instruction covering elements of music, voice leading, form/analysis, chord progression, rhythmic analysis, rehearsal conducting techniques, mace techniques for marching band drill, and assisting the military police by establishing and supervising a dismount point.

Credit Recommendation: ***Version 1:*** In the upper-division baccalaureate category, 2 semester hours in advanced conducting/rehearsal techniques (2/93). *Version 2:* In the upper-division baccalaureate category, 2 semester hours in advanced conducting/rehearsal techniques (3/00).

Related Occupations: 02B; 02C; 02D; 02E; 02F; 02G; 02H; 02J; 02K; 02L; 02M; 02N; 02T; 02U.

AR-1303-0004

Secondary Reference Laboratory Specialist

Course Number: G2ASR32470 016.

Location: Technical Training Center, Lowry AFB, CO.

Length: 16 weeks (622–624 hours).

Exhibit Dates: 8/88–4/04.

Learning Outcomes: Upon completion of the course, the student will be able to perform basic mathematical computations related to error analysis; use secondary standards of length to calibrate gauge blocks, micrometers etc.; apply secondary standards to calibrate torque measuring instrumentation, pressure, vacuum, resistance, and capacitance; apply secondary standards to calibrate frequency counters; and trace errors in secondary standards to primary standards at the National Bureau of Standards.

Instruction: Course includes instruction and practical exercises in the calibration and use of secondary standards of mass, length, time, resistance, voltage, spectral content of lasers, pressure, and allied items.

Credit Recommendation: In the lower-division baccalaureate/associate degree category, 3 semester hours in error analysis and 3 in applied physics (11/91).

AR-1304-0008

Field Artillery Meteorological Crewmember Noncommissioned Officer (NCO) Advanced

Course Number: 420-93F40.

Location: Field Artillery School, Fort Sill, OK.

Length: 5–7 weeks (177–252 hours).

Exhibit Dates: 10/85–9/96.

Learning Outcomes: To provide enlisted personnel with knowledge of the supervisory responsibilities and techniques necessary for the installation, operation, and maintenance of the field artillery meteorology section.

Instruction: Lectures and practical experiences cover positioning the meteorological section, teaching techniques for personnel training, and using test equipment and tools to adjust and maintain section equipment. Course includes a common core of leadership subjects.

Credit Recommendation: Credit is recommended for the common core only. See AR-1404-0035 (1/92).

Related Occupations: 93F.

AR-1304-0009

Marine Artillery Operations Chief

Course Number: 250-0848.

Location: Field Artillery School, Fort Sill, OK.

Length: 13–14 weeks (506 hours).

Exhibit Dates: 7/86–4/04.

Learning Outcomes: Upon completion of the course, the student will be able to perform technical duties in fire direction, artillery meteorology, surveying, targeting, fire support coordination, fire planning, and observed fire.

Instruction: Lectures and practical exercises cover computation of firing data, operating procedures of the backup computer system, firing charts, surveying, ballistics, meteorology, and targeting.

Credit Recommendation: In the lower-division baccalaureate/associate degree category, 1 semester hour in meteorology and 1 in surveying (6/89).

AR-1304-0010

Field Artillery Meteorological Crewmember Basic Noncommissioned Officer (NCO)

Course Number: 420-93F30.

Location: Field Artillery School, Fort Sill, OK.

Length: 5–7 weeks (175–260 hours).

Exhibit Dates: 10/86–9/97.

Learning Outcomes: After 9/00 see AR-1408-0274. Upon completion of the course, the student will be able to serve as meteorological shift supervisor, supporting division artillery firing battalions.

Instruction: Lectures and practical exercises cover the Rawin system for upper-air measurements; computer analysis of Rawin data to produce ballistic temperatures and densities; evaluation and quality control of derived data; and a review of management, maintenance, and leadership techniques. Course includes a common core of leadership subjects.

Credit Recommendation: In the lower-division baccalaureate/associate degree category, 2 semester hours in meteorological measurements. See AR-1406-0090 for common core credit (1/92).

Related Occupations: 93F.

AR-1304-0011

Field Artillery Meteorological Crewmember

Course Number: ***Version 2:*** 420-93F10. *Version 3:* 420-93F10.

Location: Field Artillery School, Fort Sill, OK.

Length: ***Version 2:*** 13 weeks (513 hours). *Version 3:* 9–10 weeks (328–349 hours).

Exhibit Dates: ***Version 2:*** 1/90–9/91. *Version 3:* 10/91–9/97.

Learning Outcomes: ***Version 2:*** Upon completion of the course, the student will be able to install and operate meteorological measuring equipment needed to produce required data for calculation of meteorological effects on artillery ballistics. *Version 3:* After 9/97 see AR-1304-0018. Upon completion of the course, the student will be able to install and operate meteorological measuring equipment needed to produce required data for calculation of meteorological effects on artillery ballistics.

Instruction: ***Version 2:*** Lectures and laboratory exercises use both actual equipment and computer simulations. Topics covered include basic communications; use and setup of radio equipment; basic meteorology; and the operation and use of meteorological equipment, such as anemometers, psychometers, barometers, radiosondes, and meteorology data and measuring sets. *Version 3:* Lectures and laboratory exercises use both actual equipment and computer simulations. Topics covered include basic communications; use and setup of radio equipment; basic meteorology; and the operation and use of meteorological equipment, such as anemometers, psychometers, barometers, radiosondes, and meteorology data and measuring sets.

Credit Recommendation: ***Version 2:*** In the lower-division baccalaureate/associate degree category, 3 semester hours basic meteorology. In the upper-division baccalaureate category, 1 semester hour in meteorological instruments (9/94). *Version 3:* In the lower-division baccalaureate/associate degree category, 2 semester hours in basic meteorology. In the upper-division baccalaureate category, 1 semester hour in meteorological instruments (9/94).

Related Occupations: 93F.

AR-1304-0012

Field Artillery Meteorological Crewmember Reserve

Course Number: 420-93F10-RC.

Location: Field Artillery School, Fort Sill, OK.

Length: 9–10 weeks (328–349 hours).

Exhibit Dates: 10/91–Present.

Learning Outcomes: Upon completion of the course, the student will be able to install and operate meteorological measuring equipment needed to produce required data for calculation of meteorological effects on artillery ballistics.

Instruction: Lectures and laboratory exercises use both actual equipment and computer simulations. Topics covered include basic communications; use and setup of radio equipment; basic meteorology; and the operation and use of meteorological equipment, such as anemometers, psychometers, barometers, radiosondes; and meteorology data and measuring sets.

Credit Recommendation: In the lower-division baccalaureate/associate degree category, 2 semester hours basic meteorology. In the upper-division baccalaureate category, 1 semester hour in meteorological instruments (9/94).

Related Occupations: 93F.

AR-1304-0013

Staff Weather Officer Army Indoctrination (USAF)

Course Number: 5B-F3/420-F2 (USAF).

Location: Intelligence School, Fort Huachuca, AZ.

Length: 2 weeks (53–64 hours).

Exhibit Dates: 10/91–Present.

Learning Outcomes: Upon completion of the course, the student will understand Army organization and force structure, weather resources, and Army requirements.

Instruction: Lectures and practical exercises are the methods of instruction. Topics covered include Army organization, concepts and doctrine, staff procedures, supply and maintenance, communications, security, and weather resources.

Credit Recommendation: Credit is not recommended because of the limited, specialized nature of the course (6/97).

AR-1304-0014

Total Army Training System (TATS) Field Artillery Meteorological Senior Sergeant Advanced Noncommissioned Officer (NCO)

Course Number: 420-93F40 (F).

Location: Field Artillery School, Fort Sill, OK.

Length: 1–4 weeks (96–179 hours).

Exhibit Dates: 2/00–Present.

Learning Outcomes: Upon completion of the course, the student will be able to supervise the installation, operation, and maintenance of the field artillery meteorology section.

Instruction: Lectures and practical experiences cover positioning the meteorological section, teaching techniques for personnel training, and use of tools to adjust and maintain section equipment. This course includes a common core of leadership subjects.

Credit Recommendation: In the lower-division baccalaureate/associate degree category, 3 semester hours in introduction to meteorology. See AR-1404-0035 for common core credit (9/03).

Related Occupations: 93F.

AR-1304-0015

TOTAL ARMY TRAINING SYSTEM (TATS) FIELD ARTILLERY METEOROLOGICAL SENIOR SERGEANT ADVANCED NONCOMMISSIONED OFFICER (NCO)

Course Number: 061-93F40 (F), Phase 2.

Location: Reserve Schools, Multiple locations; Field Artillery School, Fort Sill, OK.

Length: 93–96 hours.

Exhibit Dates: 2/00–Present.

Learning Outcomes: Upon completion of the course, the student will be able to supervise the installation, operation, and maintenance of the field artillery meteorology section.

Instruction: Lectures and practical experiences cover positioning the meteorological section, teaching techniques for personnel training, and use of tools to adjust and maintain section equipment.

Credit Recommendation: In the lower-division baccalaureate/associate degree category, 3 semester hours in introduction to meteorology (9/03).

Related Occupations: 93F.

AR-1304-0016

TOTAL ARMY TRAINING SYSTEM (TATS) FIELD ARTILLERY METEOROLOGICAL SECTION SERGEANT BASIC NONCOMMISSIONED OFFICER (NCO)

Course Number: ***Version 1:*** 420-93F30 (F). *Version 2:* 420-93F30 (F), Phase 2.

Location: Field Artillery School, Fort Sill, OK.

Length: ***Version 1:*** 5–7 weeks (246 hours). *Version 2:* 1–2 weeks (93–100 hours).

Exhibit Dates: ***Version 1:*** 10/97–9/00. *Version 2:* 10/00–Present.

Learning Outcomes: ***Version 1:*** Upon completion of the course, the student will be able to operate and maintain a meteorological data system used to select remote antenna sites and acquire radiosonde signals. Given a description of a weather balloon type, current weather conditions, and any special conditions, the student will be able to select calcium hydride charges for sounding balloon inflation. *Version 2:* Upon completion of the course, the student will be able to lead and supervise personnel in measuring, analyzing, and evaluating meteorological data.

Instruction: ***Version 1:*** Lectures and practical exercises cover air measurement; computer analysis of data; data evaluations; and a review of management, maintenance, and leadership techniques. This course includes a common core of leadership subjects. *Version 2:* Lectures and practical exercises cover Rawin measurements, computer analysis and evaluation of data, management, and leadership subjects.

Credit Recommendation: ***Version 1:*** In the lower-division baccalaureate/associate degree category, 3 semester hours in meteorology. See AR-1406-0090 for common core credit (6/00). *Version 2:* In the lower-division baccalaureate/associate degree category, 2 semester hours for field experience in management. See AR-1406-0090 for common core credit (9/03).

Related Occupations: 93F.

AR-1304-0017

TOTAL ARMY TRAINING SYSTEM (TATS) FIELD ARTILLERY METEOROLOGICAL SECTION SERGEANT BASIC NONCOMMISSIONED OFFICER (NCO)

Course Number: ***Version 1:*** 061-93F30 (F), Phase 2. *Version 2:* 061-93F30 (F), Phase 2.

Location: ***Version 1:*** Reserve Component Training Centers, US; Field Artillery School, Fort Sill, OK. *Version 2:* Reserve Schools, Multiple locations.

Length: ***Version 1:*** 155 hours. *Version 2:* 93–96 hours.

Exhibit Dates: ***Version 1:*** 10/97–9/00. *Version 2:* 10/00–Present.

Learning Outcomes: ***Version 1:*** Upon completion of the course, the student will be able to operate and maintain a meteorological data system used to select remote antenna sites and acquire radiosonde signals. Given a description of a weather balloon type, current weather conditions, and any special conditions, the student will be able to select calcium hydride charges for sounding balloon inflation. *Version 2:* Upon completion of the course, the student will be able to lead and supervise personnel in measuring, analyzing, and evaluating meteorological data.

Instruction: ***Version 1:*** Lectures and practical exercises cover air measurement; computer analysis of data; data evaluations; and a review of management, maintenance, and leadership techniques. *Version 2:* Lectures and practical exercises cover rawin measurements, computer analysis and evaluation of data, management, and leadership subjects.

Credit Recommendation: ***Version 1:*** In the lower-division baccalaureate/associate degree category, 3 semester hours in meteorology (6/00). *Version 2:* In the lower-division baccalaureate/associate degree category, 2 semester hours for field experience in supervision. See AR-2201-0337 for common core credit (9/03).

Related Occupations: 93F.

AR-1304-0018

TOTAL ARMY TRAINING SYSTEM (TATS) FIELD ARTILLERY METEOROLOGICAL CREWMEMBER

Course Number: ***Version 1:*** 420-93F10 (F). *Version 2:* 420-93F10 (F).

Location: Field Artillery School, Fort Sill, OK.

Length: ***Version 1:*** 9–10 weeks (352 hours). *Version 2:* 7–8 weeks (324 hours).

Exhibit Dates: ***Version 1:*** 10/97–9/02. *Version 2:* 10/02–Present.

Learning Outcomes: ***Version 1:*** Before 10/97 see AR-1304-0011. Upon completion of the course, the student will be able to produce meteorological data for the battlefield. *Version 2:* Upon completion of the course, the student will be to produce meteorological data for the modern battlefield.

Instruction: ***Version 1:*** Lectures and practical exercises cover meteorology, communications, balloon operations, and data compilation and analysis. *Version 2:* Lectures and practical exercises cover meteorology, communications, balloon operations, and data compilation and analysis.

Credit Recommendation: ***Version 1:*** In the lower-division baccalaureate/associate degree category, 3 semester hours in basic meteorology and 3 in meteorological laboratory (1/01). *Version 2:* In the lower-division baccalaureate/associate degree category, 3 semester hours in basic meteorology and 3 in meteorological laboratory (10/03).

Related Occupations: 93F.

AR-1304-0019

TOTAL ARMY TRAINING SYSTEM (TATS) FIELD ARTILLERY METEOROLOGICAL CREWMEMBER

Course Number: ***Version 1:*** 061-93F10 (F). *Version 2:* 061-93F10 (F) (DL), Phase 2.

Location: Field Artillery School, Fort Sill, OK.

Length: ***Version 1:*** 115 hours. *Version 2:* 67 hours.

Exhibit Dates: ***Version 1:*** 10/97–9/02. *Version 2:* 10/02–Present.

Learning Outcomes: ***Version 1:*** Before 10/97 see AR-1304-0011. Upon completion of the course, the student will be able to produce meteorological data which is used in command planning. *Version 2:* Upon completion of the course, the student will be able to produce meteorological data for the modern battlefield.

Instruction: ***Version 1:*** Lecture and practical exercises cover meteorology, communications, balloon operations, measurements, and surface observations. *Version 2:* Lecture and practical exercises. Topics include basic meteorology, balloon operations, measurements, and surface observations.

Credit Recommendation: ***Version 1:*** In the lower-division baccalaureate/associate degree category, 3 semester hours in basic meteorology (1/01). *Version 2:* In the lower-division baccalaureate/associate degree category, 3 semester hours in basic meteorology (10/03).

Related Occupations: 93F.

AR-1401-0021

1. DEPARTMENT OF DEFENSE (DOD) COST ACCOUNTING STANDARDS WORKSHOP (Cost Accounting Standards Workshop)
2. COST ACCOUNTING STANDARDS WORKSHOP

Course Number: ***Version 1:*** ALMC-CE; ALM-36-0217. *Version 2:* CON 233.

Location: ***Version 1:*** Army Logistics Management College, Fort Lee, VA; Army Logistics Management College, Onsite at various locations. *Version 2:* Defense Acquisition University, Various locations in the Continental US; Army Logistics Management College, Onsite at various locations; Army Logistics Management College, Fort Lee, VA.

Length: ***Version 1:*** 2 weeks (74 hours). *Version 2:* 2 weeks (75–76 hours).

Exhibit Dates: ***Version 1:*** 5/77–9/93. *Version 2:* 10/93–3/03.

Learning Outcomes: ***Version 1:*** Upon completion of the course, the student will be able to apply cost accounting standards used to defense contracts. *Version 2:* Upon completion of the course, the student will be able to apply cost accounting standards to defense contracts.

Instruction: ***Version 1:*** Lectures/conferences and practical exercises cover the latest regulations of the Cost Accounting Standards Board. *Version 2:* Lectures/conferences and practical exercises cover the latest regulations of the Cost Accounting Standards Board.

Credit Recommendation: ***Version 1:*** In the upper-division baccalaureate category, 2 semester hours in cost accounting standards (10/90). *Version 2:* In the upper-division baccalaureate category, 2 semester hours in cost accounting standards (8/95).

AR-1401-0022

Internal Review

Course Number: RM415.

Location: Combined Arms Training Center, Munich, W. Germany.

Length: 2 weeks (70 hours).

Exhibit Dates: 5/79–6/96.

Objectives: To provide accountants/auditors with an orientation in the requirements for auditing within federal agencies, specifically appropriated/non-appropriated fund accounting systems.

Instruction: This course covers specialized audit standards, terminology, survey of verification techniques, work paper preparation, appropriated/nonappropriated fund control verification, follow-up reviews, and communication.

Credit Recommendation: In the lower-division baccalaureate/associate degree category, 1 semester hour in business report writing and 1 in auditing practices (10/80).

AR-1401-0029

Advanced Noncommissioned Officer (NCO) MOS 73C/D

Course Number: 7-71-C42.

Location: Institute for Personnel and Resource Management, Fort Benjamin Harrison, IN.

Length: 5–17 weeks (168–444 hours).

Exhibit Dates: 3/81–12/91.

Objectives: To provide selected personnel with advanced leadership skills, military subjects, and communication skills for effective supervision of specific MOS tracks.

Instruction: Lectures and practical exercises cover a basic core, including written and oral communication, management principles, unit training, and combat survival. MOS options are recruitment and reenlistment, equal opportunity, club management, legal topics, chaplain assisting, and data processing. Course includes a common core of leadership subjects.

Credit Recommendation: Credit is recommended for the common core only. See AR-1404-0035 (3/81).

Related Occupations: 73C; 73D.

AR-1401-0030

Basic Internal Audit
(Management Audit Level 1 Basic)

Course Number: None.

Location: National Guard Bureau Management Auditer School, Aberdeen, MD.

Length: 2 weeks (65–70 hours).

Exhibit Dates: 1/85–5/99.

Learning Outcomes: Upon completion of the course, the student will be able to identify auditing tools; identify elements in an audit, including auditing techniques, elementary statistical sampling, flow charting, field work, and problem solving; organize findings; and effectively communicate findings.

Instruction: Lectures, discussions, and practical exercises cover auditing tools, auditing standards, auditing techniques, statistical techniques, audit documentation, and presentation of findings and recommendations.

Credit Recommendation: In the lower-division baccalaureate/associate degree category, 1 semester hour in operational auditing (basic) (5/89).

AR-1401-0031

Intermediate Internal Audit
(Management Audit Level 2 Intermediate)

Course Number: None.

Location: National Guard Bureau Management Auditer School, Aberdeen, MD.

Length: 2 weeks (65–74 hours).

Exhibit Dates: 1/85–12/99.

Learning Outcomes: Upon completion of the course, the student will be able to apply human relations principles, expand the use of auditing tools and techniques, understand the role of the audit team leader, and organize and conduct an audit.

Instruction: Lectures, discussions, and practical exercises cover human relations, audit team management, analytical techniques, audit simulation, and presentation of audit findings and recommendations.

Credit Recommendation: In the lower-division baccalaureate/associate degree category, 1 semester hour in operational auditing and 1 in human relations in management (12/89).

AR-1401-0032

Materiel Control and Accounting Specialist Advanced Individual Training

Course Number: 551-76P10.

Location: Quartermaster Center and School, Fort Lee, VA.

Length: 8 weeks (288 hours).

Exhibit Dates: 4/88–10/98.

Learning Outcomes: Upon completion of the course, the student will be able to establish stock record accounts and process documents for the materiel supply function including use of both a manual and an automated standard supply system.

Instruction: Lectures, demonstrations, and practical exercises cover processing supply documents, posting, and maintaining stock record accounts. Topics include use of automated supply procedures, determination of stocking criteria, diskette and tape handling, and diskette processing.

Credit Recommendation: In the vocational certificate category, 3 semester hours in records processing. In the lower-division baccalaureate/associate degree category, 1 semester hour in keyboarding (10/88).

Related Occupations: 76P.

AR-1401-0033

1. Finance Specialist
2. Finance Specialist
3. Financial Management Technician Advanced Individual Training

Course Number: ***Version 1:*** 542-73C10; 542-73C10 (ST). *Version 2:* 542-73C10 (F); 542-73C10; 542-73C10 (ST). *Version 3:* 541-44C10.

Location: ***Version 1:*** Finance School, Soldier Support Institute, Fort Benjamin Harrison, IN. *Version 2:* Soldier Support Institute, Fort Jackson, SC; Finance School, Soldier Support Institute, Fort Benjamin Harrison, IN. *Version 3:* Finance School, Fort Jackson, SC.

Length: ***Version 1:*** 7–8 weeks (244–296 hours). *Version 2:* 7–8 weeks (244–296 hours). *Version 3:* 9 weeks (309 hours).

Exhibit Dates: ***Version 1:*** 10/86–3/91. *Version 2:* 4/91–10/02. *Version 3:* 11/02–Present.

Learning Outcomes: ***Version 1:*** Upon completion of the course, the student will be able to identify and determine the types of incentive pay plans, prepare pay vouchers, complete pay documents, and compute pay allowances. *Version 2:* Upon completion of the course, the student will be able to determine pay status and adjustments, compute payroll and travel allowances, prepare payroll and travel vouchers, process pay inquiries, and access and input data on computer. *Version 3:* Pending evaluation.

Instruction: ***Version 1:*** Methods of instruction include lectures, role playing, and in-class exercises. Topics covered include financial operations, data entry operations, and payroll processing. *Version 2:* Methods of instruction include lectures, role playing, and in-class exercises. Topics covered include financial operations and payroll processing. *Version 3:* Pending evaluation.

Credit Recommendation: ***Version 1:*** In the vocational certificate category, 2 semester hours in payroll accounting. In the lower-division baccalaureate/associate degree category, 1 semester hour in computer applications (4/91). *Version 2:* In the lower-division baccalaureate/associate degree category, 2 semester hours in payroll accounting and 1 in computer applications (1/97). *Version 3:* Pending evaluation.

Related Occupations: 44C.

AR-1401-0034

Military Accounting

Course Number: ***Version 1:*** 7D-F32/541-F4; 7D-44A/541-F4; 7D-44B/541-F4. *Version 2:* 7D-F32/541-F4.

Location: ***Version 1:*** Finance School, Soldier Support Institute, Fort Benjamin Harrison, IN. *Version 2:* Soldier Support Institute, Fort Jackson, SC.

Length: ***Version 1:*** 3–5 weeks (145–165 hours). *Version 2:* 3–4 weeks (115 hours).

Exhibit Dates: ***Version 1:*** 1/86–5/93. *Version 2:* 6/93–Present.

Learning Outcomes: ***Version 1:*** Upon completion of the course, the student will be able to use generally accepted accounting principles in government fund accounting, define basic comptrollership functions, outline the elements of the budgeting process, reconcile cash, use subsidiary ledgers, identify and correct errors, and record year end adjusting transactions. *Version 2:* Upon completion of the course the student will be able to use generally accepted accounting principles in government fund accounting in the areas of expenditure accounting, reimbursement accounting, and miscellaneous accounting, including various reconciliations and end-of period transactions and procedures.

Instruction: ***Version 1:*** Methods of instruction include lectures and class exercises. Topics covered include program planning and budgeting, accounting classifications, funds receipt and disbursement, administration controls, nonappropriated funds, reimbursement, and cash classifications. *Version 2:* Lectures, classroom discussions, examinations, quizzes, and practical exercises uses current PCs, operating systems, and software to analyze the flow of funds and reimbursement activities between various functions, perform cash reconciliations, understand end-of-month and year closing procedures and transactions, and interact with an automated system to support those processes.

Credit Recommendation: ***Version 1:*** In the upper-division baccalaureate category, 3 semester hours in governmental accounting (11/91). *Version 2:* In the upper-division baccalaureate category, 3 semester hours in governmental accounting (9/97).

AR-1401-0035

1. Finance Specialist Basic Noncommissioned Officer (NCO)
2. Finance/Accounting Basic Noncommissioned Officer (NCO)
3. Finance/Accounting Basic Noncommissioned Officer (NCO)

Course Number: ***Version 1:*** 542-73C30; 542-73C/D30. *Version 2:* 542-73C30; 542-73C/D30. *Version 3:* 542-73C/D30.

Location: ***Version 1:*** Finance School, Soldier Support Institute, Fort Benjamin Harrison, IN. *Version 2:* Finance School, Soldier Support Institute, Fort Benjamin Harrison, IN. *Version 3:* Soldier Support Institute, Fort Jackson, SC.

Length: ***Version 1:*** 7–8 weeks (272–280 hours). *Version 2:* 9–10 weeks (348 hours). *Version 3:* 10 weeks (369 hours).

Exhibit Dates: ***Version 1:*** 7/86–9/91. *Version 2:* 10/91–1/96. *Version 3:* 2/96–Present.

Learning Outcomes: ***Version 1:*** Upon completion of the course, the student will be able to review and verify disbursing records, review and compute travel vouchers, develop accountability statements, and prepare financial office reports. *Version 2:* Upon completion of the course, the student will be able to prepare, review, distribute, and maintain payroll records; compute travel vouchers; develop accounting statements; and prepare financial statements. *Version 3:* Upon completion of the course, the student will be able to prepare, maintain, and review dispensing records, books, and reports in payroll systems with special emphasis on travel vouchers, pay vouchers, and commercial service vouchers; prepare cash reconciliation statements and understand an imprest fund; and understand what a budget is and how it is be used.

Instruction: ***Version 1:*** Methods of instruction include lectures, exercises, and class discussions. Topics covered include payroll accounting, accounting classifications, cash reconciliations, and accounting documents. Course includes a common core of leadership subjects. *Version 2:* Lectures and applicable exercises are used in the areas of payroll accounting, payroll disbursement, travel vouchers, cash disbursements, and cash accounting. *Version 3:* Methods of instruction includes lectures, practice problems, and instructor critique of practice problems. Course includes a common core of leadership subjects.

Credit Recommendation: ***Version 1:*** In the vocational certificate category, 2 semester hours in payroll accounting. In the lower-division baccalaureate/associate degree category, see AR-1406-0090 for common core credit. In the upper-division baccalaureate category, 2 semester hours in governmental accounting (12/88). *Version 2:* In the lower-division baccalaureate/associate degree category, 2 semester hours in payroll accounting. In the upper-division baccalaureate category, 2 semester hours in governmental accounting (12/91). *Version 3:* In the lower-division baccalaureate/associate degree category, 2 semester hours in payroll accounting and 1 in introduction to budgeting. See AR-1406-0090 for common core credit (9/97).

Related Occupations: 73C; 73D.

AR-1401-0036

Central Accounting Officer

Course Number: 7D-F15.

Location: Finance School, Soldier Support Institute, Fort Benjamin Harrison, IN.

Length: 2 weeks (77 hours).

Exhibit Dates: 5/86–12/91.

Learning Outcomes: Upon completion of the course, the student will be able to outline the basic concepts and codes of unappropriated grants, work with automated accounting systems, analyze financial statements, and prepare budgets.

Instruction: Methods of instruction include lectures, case studies, and in-class exercises. Topics of discussion include unappropriated funds, financial statements, comparative relationships, budget planning, and internal controls.

Credit Recommendation: In the upper-division baccalaureate category, 1 semester hour in governmental accounting and 1 in financial statement analysis (12/88).

AR-1401-0037

Disbursing Operations

Course Number: ***Version 1:*** 7D-F18/542-F8. *Version 2:* 7D-F18/542-F8.

Location: ***Version 1:*** Finance School, Soldier Support Institute, Fort Benjamin Harrison, IN. *Version 2:* Soldier Support Institute, Fort Jackson, SC.

Length: ***Version 1:*** 2 weeks (74 hours). *Version 2:* 2 weeks (93 hours).

Exhibit Dates: ***Version 1:*** 9/87–5/93. *Version 2:* 6/93–Present.

Learning Outcomes: ***Version 1:*** Upon completion of the course, the student will be able to maintain cash books, review collection and disbursement vouchers, process overages, and process cancelled checks. *Version 2:* Upon completion of the course, the student will understand the processes associated with cash management and control, verification, reconciliation, and the maintenance of cash back and check operations.

Instruction: ***Version 1:*** Lectures, classroom discussions, and practical exercises cover cash and check collection, disbursement vouchers, deposits and reconciliation statements, imprest fund transactions, and accountability statements. *Version 2:* Lectures, classroom discussions, and practical exercises cover cash disbursement and receipt operations, deposits and reconciliation statements, and the use of an automated disbursing operations processing system.

Credit Recommendation: ***Version 1:*** In the vocational certificate category, 1 semester hour in disbursement operations (11/91). *Version 2:* In the vocational certificate category, 1 semester hour in cash operations or cash disbursement and control (9/97).

AR-1401-0038

Resource Management Budget

Course Number: 7D-F23.

Location: Soldier Support Institute, Fort Jackson, SC; Finance School, Soldier Support Institute, Fort Benjamin Harrison, IN.

Length: 2 weeks (71 hours).

Exhibit Dates: 1/87–10/98.

Learning Outcomes: Upon completion of the course, the student will be able to manage fund flow, implement control functions, coordinate the approval process, implement the procurement function, and recommend adjustments.

Instruction: Instruction methods include lectures, class participation, and role playing. Topics covered include budget planning and implementation, funded and nonfunded expenses, and budget review and analysis.

Credit Recommendation: In the upper-division baccalaureate category, 2 semester hours in budget application (9/96).

AR-1401-0041

Accounting Specialist

Course Number: 541-73D10.

Location: Soldier Support Institute, Fort Jackson, SC; Finance School, Soldier Support Institute, Fort Benjamin Harrison, IN.

Length: 8–9 weeks (332–339 hours).

Exhibit Dates: 1/88–Present.

Learning Outcomes: Upon completion of the course, the student will be able to apply generally accepted accounting principles in governmental fund accounting; use a detailed coding procedure; process expenditures, reimbursements, and miscellaneous transactions; describe the basic organizational structure in government accounting; and use the correct forms and format in fund accounting.

Instruction: Methods of instruction consist of lectures and in-class exercises. Topics covered include expenditures and reimbursements, commercial accounts, fiscal codes, and budget execution.

Credit Recommendation: In the upper-division baccalaureate category, 3 semester hours in governmental fund accounting (1/97).

Related Occupations: 73D.

AR-1401-0042

Budget

Course Number: 800.

Location: Reserve Readiness Training Center, Fort McCoy, WI.

Length: 2 weeks (64 hours).

Exhibit Dates: 9/86–5/99.

Learning Outcomes: Upon completion of the course, the student will be able to implement the basic principles of planning, programming, budgeting, and execution system as they apply to a major reserve unit.

Instruction: Topics covered include basic principles and applications of financial management, including the programming, budgeting, and execution system; Army management, manpower, and automated ledger and budget concepts; budget procedures, including interfund transfers, requirements-based budget preparation, and fund distribution; rules of obligation; cost estimating; budget processing including reading and interpreting automated fund control reports; budget execution and tracking; determining status of operating resources; and completion of year-end close.

Credit Recommendation: In the lower-division baccalaureate/associate degree category, 2 semester hours in budgeting or financial management (7/89).

AR-1401-0043

Budget Management

Course Number: *Version 1:* 2300. *Version 2:* 2300; 921-720; 720; 921-2300-LVL2.

Location: Reserve Readiness Training Center, Fort McCoy, WI.

Length: *Version 1:* 2 weeks (50 hours). *Version 2:* 1–2 weeks (64–66 hours).

Exhibit Dates: *Version 1:* 9/87–10/95. *Version 2:* 11/95–Present.

Learning Outcomes: *Version 1:* Upon completion of the course, the student will be able to prepare unit budgets for operations, maintenance, and personnel using proper funds and using Army planning, programming, budgeting, and execution systems (PPBES); identify and interpret accounting classifications and resource codes; identify funds available for budget construction and expenditures; determine unit budget needs for operations, maintenance, and personnel. *Version 2:* Upon completion of the course, the student will be able to prepare unit budgets for operations, maintenance, and personnel using proper funds and using specific planning, programming, budgeting, and execution systems; identify and interpret accounting classifications and resource codes; identify funds available for budget construction and expenditures; and determine unit budget needs for operations, maintenance, and personnel.

Instruction: *Version 1:* Reading and lectures cover PPBES, reserve appropriations, purchase order and contract management, budgetary procedures, fund controls, automated budget system fund distributions, and budget reconciliation reports. Practical exercises cover accounting classifications within regulation guidelines, budgetary expenditures and obligations, identification of budget requirements and funds, distribution of funds, and budget reconciliation reports. *Version 2:* Readings, discussions, projects and lectures cover planning, programming, budgeting, and execution systems; reserve appropriations; purchase order and contract management; budgetary procedures; fund controls; automated budget system fund distributions; and budget reconciliation reports. Practical exercises cover accounting classifications within regulation guidelines, budgetary expenditures and obligations, identification of budget requirements and funds, distribution of funds, and budget reconciliation reports.

Credit Recommendation: *Version 1:* In the lower-division baccalaureate/associate degree category, 1 semester hour in budget management (7/89). *Version 2:* In the lower-division baccalaureate/associate degree category, 1 semester hour in budget management (11/02).

AR-1401-0044

Civilian Payroll Technician

Course Number: 542-F9.

Location: Finance School, Soldier Support Institute, Fort Benjamin Harrison, IN.

Length: 3 weeks (114 hours).

Exhibit Dates: 1/93–1/97.

Learning Outcomes: Upon completion of the course, the student will be able to process civilian pay vouchers, overtime adjustments, tax appropriations, unemployment compensation, sick leave, annual leave, state and federal withholding, and benefits.

Instruction: Lectures and practical exercises cover the subject matter.

Credit Recommendation: In the lower-division baccalaureate/associate degree category, 2 semester hours in payroll accounting (12/91).

AR-1401-0045

Civilian Pay Management

Course Number: 542-F14.

Location: Finance School, Soldier Support Institute, Fort Benjamin Harrison, IN.

Length: 2 weeks (74 hours).

Exhibit Dates: 6/89–1/97.

Learning Outcomes: Upon completion of the course, the student will be able to supervise the processing of civilian pay vouchers, overtime adjustments, tax appropriations, unemployment compensation, sick leave, annual leave, state and federal withholding, and benefits.

Instruction: Lectures and practical applications are used to provide ending competencies.

Credit Recommendation: In the lower-division baccalaureate/associate degree category, 2 semester hours in payroll accounting (12/91).

AR-1401-0046

Finance Officer Advanced Nonresident/ Resident Reserve Component Phases 1 and 3

Course Number: 7-14-C23, Phase 3; 7-14-C23, Phase 1.

Location: Finance School, Soldier Support Institute, Fort Benjamin Harrison, IN.

Length: Phase 1, 2 weeks (73 hours); Phase 3, 2 weeks (77 hours).

Exhibit Dates: 10/90–1/97.

Learning Outcomes: After 9/93 see AR-1401-0054. Upon completion of the course, the student will be able to apply basic accounting principles and procedures, analyze financial statements, and understand accounting systems for specialized situations.

Instruction: Lectures, practical exercises, and case studies cover the subject matter.

Credit Recommendation: In the lower-division baccalaureate/associate degree category, for Phase 3, 2 semester hours in accounting fundamentals; credit is not recommended for Phase 1 because of the military-specific nature of the phase (12/91).

AR-1401-0047

1. Finance Officer Advanced
2. Finance Officer Advanced
3. Finance Captains Career

Course Number: *Version 1:* 7-14-C22. *Version 2:* 7-14-C22. *Version 3:* 7-14-C32, Phase 1.

Location: *Version 1:* Finance School, Soldier Support Institute, Fort Benjamin Harrison, IN. *Version 2:* Soldier Support Institute, Fort Jackson, SC; Finance School, Soldier Support Institute, Fort Benjamin Harrison, IN. *Version 3:* Finance School, Fort Jackson, SC.

Length: *Version 1:* 20 weeks (795 hours). *Version 2:* 20 weeks (762 hours). *Version 3:* 18 weeks (704 hours).

Exhibit Dates: *Version 1:* 7/87–9/90. *Version 2:* 10/90–12/00. *Version 3:* 1/01–Present.

Learning Outcomes: *Version 1:* Upon completion of the course, the student will be able to communicate effectively in oral or written form; provide leadership in a line or staff position; use standard deviation, variance, sampling, and other statistical tools to analyze accounting/financial information; prepare and administer financial control systems; use effective listening, speaking, and writing skills; and supervise a commercial accounts section and a disbursing financial section. *Version 2:* Upon completion of the course, the student will be able to use the microcomputer for accounting and financial applications; prepare and administer financial management and control systems; use standard deviation, variance, sampling, and other statistical analysis tools to analyze accounting/financial information; develop pay procedures and apply management review techniques; understand applications of governmental accounting and financial analysis; provide leadership in a line or staff position; and communicate effectively in written or oral form. *Version 3:* Upon completion of the course, the student will be able to use the microcomputer for accounting and financial applications; and prepare and administer financial management and control systems; use standard deviation variance, sampling, and other statistical analysis tools to analyze accounting and financial information; develop pay procedures and apply management revise techniques; understand applications of governmental accounting and financial analysis; provide leadership in a line or staff position; and communicate effectively in written or oral form.

Instruction: *Version 1:* Lectures and practical exercises cover financial matters. Students choose several electives. Electives have not been evaluated individually. *Version 2:* Lectures, practical exercises, and individual and group oral and written presentations cover financial matters. *Version 3:* Lectures, practical exercises, individual and group discussions, and oral and written presentations cover financial matters.

Credit Recommendation: *Version 1:* In the lower-division baccalaureate/associate degree category, 3 semester hours in leadership, 2 in business communication, and 2 in micro-accounting applications. In the upper-division baccalaureate category, 2 semester hours in business and professional speaking and 2 in statistical methods for management (12/91). *Version 2:* In the lower-division baccalaureate/associate degree category, 2 semester hours in business communication, 1 in technical writing, 2 in microcomputer applications, and 3 in leadership. In the upper-division baccalaureate category, 2 semester hours in statistical methods for management, 2 in business and professional speaking, 2 in governmental accounting, and 2 in financial management (9/96). *Version 3:* In the lower-division baccalaureate/associate degree category, 2 semester hours in business communication, 1 in technical writing, and 2 in computer application. In the upper-division baccalaureate category, 3 semester hours in leadership, 3 in budget control and cost accounting, and 2 in governmental accounting (3/02).

AR-1401-0048

Commercial Accounts Administration

Course Number: *Version 1:* 542-F6. *Version 2:* 542-F6.
Location: *Version 1:* Soldier Support Institute, Fort Jackson, SC; Finance School, Soldier Support Institute, Fort Benjamin Harrison, IN. *Version 2:* Soldier Support Institute, Fort Jackson, SC.
Length: *Version 1:* 1–2 weeks (58 hours). *Version 2:* 1–2 weeks (58 hours).
Exhibit Dates: *Version 1:* 12/89–5/94. *Version 2:* 6/94–Present.
Learning Outcomes: *Version 1:* Upon completion of the course, the student will be able to examine receipts, prepare and process commercial vouchers, and calculate discounts and interest charges. *Version 2:* Upon completion of the course, the student will be able to understand and use manual or computerized accounts payable systems including using a voucher system and taking discounts.
Instruction: *Version 1:* Lectures and practical exercises cover the subject matter. *Version 2:* Methods of instruction include lecture, practice problem solving, and class discussion.
Credit Recommendation: *Version 1:* In the vocational certificate category, 2 semester hours in accounts payable (9/96). *Version 2:* In the vocational certificate category, 1 semester hour in accounts payable clerical procedures (9/97).

AR-1401-0049

Travel Administration and Entitlements

Course Number: 7D-F30/542-F4.
Location: Soldier Support Institute, Fort Jackson, SC; Finance School, Soldier Support Institute, Fort Benjamin Harrison, IN.
Length: 3 weeks (115 hours).
Exhibit Dates: 3/89–Present.
Learning Outcomes: Upon completion of the course, the student will be able to determine travel entitlements, travel allowances, eligibility, and amount of travel payment.
Instruction: Course methods include lectures and practical exercises.
Credit Recommendation: Credit is not recommended because of the limited, technical nature of the course (6/96).

AR-1401-0050

Finance/Accounting Advanced Noncommissioned Officer (NCO)

Course Number: *Version 1:* 542-73C/D40; 542-73C40; 542-73D40. *Version 2:* 542-73C/D40. *Version 3:* 542-73C/D40; 542-73C/D40 (Resident).
Location: *Version 1:* Soldier Support Center, Fort Benjamin Harrison, IN. *Version 2:* Soldier Support Institute, Fort Jackson, SC. *Version 3:* NCO Academy, Fort Jackson, SC.
Length: *Version 1:* 8 weeks (292–302 hours). *Version 2:* 6–7 weeks (241 hours). *Version 3:* 3–4 weeks (146 hours).
Exhibit Dates: *Version 1:* 1/87–9/95. *Version 2:* 10/95–9/01. *Version 3:* 10/01–Present.
Learning Outcomes: *Version 1:* Upon completion of the course, the student will be able to apply leadership skills in a first-line management position, communicate effectively in writing, and perform basic payroll accounting tasks. *Version 2:* Upon completion of the course, the student will be able to perform basic accounting functions related to daily cash reports, the check register, and a payroll system; read and interpret a payroll account; and process and analyze payroll accounts. *Version 3:* Upon completion of the course, the student will be able to perform basic accounting functions related to daily cash reports, the check register, and a payroll system; read and interpret a payroll account; and process and analyze payroll accounts.
Instruction: *Version 1:* Course methods include lectures and practical exercises. *Version 2:* Methods of instruction include lectures, class participation, exercises, and an occasional guest speaker. Course includes a common core of leadership subjects. *Version 3:* Methods of instruction include lectures, class participation, exercises, and an occasional guest speaker. Course includes a common core of leadership subjects.
Credit Recommendation: *Version 1:* In the lower-division baccalaureate/associate degree category, 2 semester hours in leadership, 3 in payroll accounting, and 1 in business communication (12/91). *Version 2:* In the lower-division baccalaureate/associate degree category, 3 semester hours in payroll and cash accounting. See AR-1404-0035 for common core credit (9/97). *Version 3:* In the lower-division baccalaureate/associate degree category, 2 semester hours in payroll and cash accounting. See AR-1404-0035 for common core credit (3/02).
Related Occupations: 73C; 73D.

AR-1401-0051

Advanced Management Accounting Analysis
(Army Management Accounting and Analysis)

Course Number: 7D-F29/541-F12.
Location: Soldier Support Institute, Fort Jackson, SC; Soldier Support Center, Fort Benjamin Harrison, IN.
Length: 2 weeks (75–80 hours).
Exhibit Dates: 1/90–Present.
Learning Outcomes: Upon completion of the course, the student will be able to describe the role of an Army accountant in the management process and apply basic statistical measures in the analysis of accounting functions.
Instruction: Lectures, discussions, and practical exercises give an overview of resource management, fund accounting, obligation management, stock fund management, and cost accounting. Statistical information presented includes mean, median, mode, and standard deviation.
Credit Recommendation: In the upper-division baccalaureate category, 3 semester hours in managerial accounting (1/97).

AR-1401-0052

Unit Pay Administration

Course Number: *Version 1:* 1300; 1300-LVL1. *Version 2:* 921-710.
Location: Reserve Readiness Training Center, Fort McCoy, WI.
Length: *Version 1:* 2 weeks (67–68 hours). *Version 2:* 2 weeks (68 hours).
Exhibit Dates: *Version 1:* 4/96–2/02. *Version 2:* 3/02–Present.
Learning Outcomes: *Version 1:* Upon completion of the course, the student will be able to maintain pay accounts, update personnel records, process travel accounts, process deceased soldiers' records, and process pay adjustments. *Version 2:* Upon completion of the course, the student will be able to summarize the reserve pay process; access key forms and the internet/intranet sites for relevant information; identify criteria, procedures, and substantiating documentation to verify the Master Military Pay File; identify bonus types, eligibility requirements, procedures for payment of the initial and anniversary payment, termination, and recoupment; prepare a transmittal letter; identify drill types, requirements, limitations, usage, and the documentation required to process unit pay; post unit pay for review and payment, including automated incentive pay; complete basic procedures to process individual pay transactions; identify travel entitlements; determine the need for travel vouchers; identify the supporting documentation required for payment; prepare a pay request record for submission; and determine incapacitation pay entitlements and actions necessary to process returned checks, indebtedness transactions, and deceased soldier documents.
Instruction: *Version 1:* Readings, lectures, discussions, and written and hands-on computer exercises are course methods. *Version 2:* Audiovisual materials, practical and classroom exercises, discussion, lecture, and computer-based training. Topics include payroll administration, automated pay process and resources, inactive duty pay, entitlements and allowances, pay inquiry, maintaining the pay account, request for orders, exception to policy for single source data, pay entry basic date/pay date, selected reserve incentive program, transmittal letter, travel, unit commanders pay management report, and miscellaneous pay actions.
Credit Recommendation: *Version 1:* Credit is not recommended because of the military-specific nature of the course (8/02). *Version 2:* In the lower-division baccalaureate/associate degree category, 2 semester hours in compensation (11/02).

AR-1401-0053

Reserve Component Reconfigured Accounting Specialist Phase 2

Course Number: 805A-73D10, Phase 2.
Location: Soldier Support Institute, Fort Jackson, SC.
Length: 3–4 weeks (130 hours).
Exhibit Dates: 10/95–Present.
Learning Outcomes: Upon completion of the course, the student will understand the rules and processes associated with accounting systems supporting the Army, including appropriations, coding of expenditures, budgeting and commercial accounts, and the use of accounting regulations and publications.
Instruction: Lectures, class discussions, homework assignments, examinations, and practical exercises use PCs to understand and chart accounts, the flow of funds, various inputs to the Army standard financial system, disbursement and receipt transactions, cashiering functions, and examination of vouchers.
Credit Recommendation: In the lower-division baccalaureate/associate degree category, 3 semester hours in practical accounting or bookkeeping (9/97).
Related Occupations: 73D.

AR-1401-0054

FINANCE OFFICER ADVANCED RESERVE COMPONENT PHASE 2

Course Number: 7-14-C23, Phase 2.
Location: Soldier Support Institute, Fort Jackson, SC.
Length: 2 weeks (81 hours).
Exhibit Dates: 10/93–Present.
Learning Outcomes: Before 10/93 see AR-1401-0046. Upon completion of the course, the student will have an introduction to accounting and learn to use a voucher system.
Instruction: Methods of instruction are primarily lectures with some practice problems in using the military voucher system.
Credit Recommendation: In the vocational certificate category, 1 semester hour in introduction to accounting (9/97).

AR-1401-0055

NONAPPROPRIATED CHAPLAIN'S FUND CLERK

Course Number: 561-F2.
Location: Chaplain Center and School, Fort Jackson, SC.
Length: 2 weeks (79 hours).
Exhibit Dates: 1/00–Present.
Learning Outcomes: Upon completion of the course, the student will be able to apply the principals of single-entry fund accounting and perform clerical duties.
Instruction: Topics include preparing support documents; assisting in the preparation of annual budgets; and performing bookkeeping functions, including receipts, disbursements, and control of nonappropriated funds.
Credit Recommendation: In the lower-division baccalaureate/associate degree category, 3 semester hours in fund accounting (3/02).

AR-1401-0056

CHAPLAINCY RESOURCES MANAGER

Course Number: 5G-F7/561-F6.
Location: Chaplain Center and School, Fort Jackson, SC.
Length: 2 weeks (79 hours).
Exhibit Dates: 2/02–Present.
Learning Outcomes: Upon completion of the course, the student will be able to manage nonappropriated funds.
Instruction: Methods of instruction include discussion, practical exercises, and case studies. Topics covered include management control, audit trail and sources/uses of funds.
Credit Recommendation: In the lower-division baccalaureate/associate degree category, 1 semester hour in fund accounting. In the upper-division baccalaureate category, 1 semester hour in principles of management (3/02).

AR-1402-0008

PERSINSD AUTOMATIC DATA PROCESSING (ADP) INTERN PROGRAM

Course Number: None.
Location: PERSINSD, Alexandria, VA.
Length: 26 weeks.
Exhibit Dates: 1/75–1/91.
Objectives: To attract, develop, and retain qualified civilian and military personnel designated as management interns to serve as programmers and systems analysts in the Personnel Systems Command, Department of the Army.
Instruction: Through lectures, laboratory periods, video-assisted instruction, practical projects, and directed study, the following areas are covered: introduction to UNIVAC 1108 hardware, principles of data processing, systems analysis and design, COBOL programming, data communication concepts and design, operations research techniques, personnel applications project, and the UNIVAC 1100 executive operating system.
Credit Recommendation: In the lower-division baccalaureate/associate degree category, 4 semester hours in COBOL programming, 3 in data processing principles, and 2 in operating systems. In the upper-division baccalaureate category, 3 semester hours in business systems analysis (5/76).

AR-1402-0026

FIELD ARTILLERY BALLISTIC MISSILE OFFICER

Course Number: 2E-1190.
Location: Artillery and Missile School, Fort Sill, OK.
Length: 9 weeks (319–328 hours).
Exhibit Dates: 3/68–1/90.
Objectives: To train commissioned officers to supervise tactical deployment, system maintenance, and operation of Pershing and Sergeant missile systems.
Instruction: Lectures and practical exercises cover Pershing and Sergeant missile system operations and supervision. Course includes communications and electronics (equipment, procedures, and systems); digital computers; guided missiles, including warheads, launchers, and organizational and preventive maintenance; and tactical subjects.
Credit Recommendation: In the vocational certificate category, 1 semester hour in digital computer familiarization (6/74).

AR-1402-0043

BASIC ASSEMBLER LANGUAGE CODING (ALC)

Course Number: None.
Location: Computer Systems Command, Fort Belvoir, VA; Computer Systems Command, Support Group, Fort Lee, VA; Computer Systems Command, Support Group, Fort McPherson, GA; Computer Systems Command, Support Group, Fort Shafter, HI; Computer Systems Command, Zweibruecken, Germany.
Length: 3 weeks (120 hours).
Exhibit Dates: 1/80–12/90.
Objectives: To provide students with an understanding of basic assembler language for the IBM 360/370.
Instruction: Subject matter includes standard instruction set, decimal instruction set, branching, logic, Boolean instructions, editing, and translation. I/O macros, debugging, indexing, and necessary job control language.
Credit Recommendation: In the lower-division baccalaureate/associate degree category, 3 semester hours in assembler language programming (11/82).

AR-1402-0044

DATA PROCESSING PRINCIPLES

Course Number: None.
Location: Computer Systems Command, Fort Belvoir, VA; Computer Systems Command, Support Group, Fort Lee, VA; Computer Systems Command, Support Group, Fort McPherson, GA; Computer Systems Command, Support Group, Fort Shafter, HI; Computer Systems Command, Zweibruecken, Germany.
Length: 4–5 weeks (176 hours).
Exhibit Dates: 1/80–12/90.
Objectives: To provide students with an understanding of how a computer works and its major components.
Instruction: This course consists of four modules: Data Processing Concepts, Fundamentals of Programming, Structured Programming Technologies Overview, and Introduction to Systems 360/370. Students must complete each module. Subject matter includes programming concepts, problem analysis, decision tables, coding, number systems, and structured design techniques. Additional problem solving techniques, such as program loops, subroutines, counters and switches, and tables are also covered. The architecture and organization of the IBM 360/370 are included.
Credit Recommendation: In the lower-division baccalaureate/associate degree category, 3 semester hours in data processing (11/82).

AR-1402-0045

OPERATING SYSTEMS

Course Number: None.
Location: Computer Systems Command, Fort Belvoir, VA; Computer Systems Command, Support Group, Fort Lee, VA; Computer Systems Command, Support Group, Fort McPherson, GA; Computer Systems Command, Support Group, Fort Shafter, HI; Computer Systems Command, Zweibruecken, Germany.
Length: 4–5 weeks (176–184 hours).
Exhibit Dates: 1/80–12/90.
Objectives: To provide students with an overview of operating systems, with emphasis the IBM OS operating system.
Instruction: Program consists of five modules: OS Concepts and Facilities, OS Job Control Language, OS Advanced Job Control Language Concepts and Techniques, OS Utilities and Workshop, and File Organization and Accessing Methods. Course covers concepts, facilities, vocabulary, and characteristics of the IBM operating systems.
Credit Recommendation: In the lower-division baccalaureate/associate degree category, 3 semester hours in data processing (operating systems) (11/82).

AR-1402-0046

ADVANCED COBOL WORKSHOP

Course Number: None.
Location: Computer Systems Command, Fort Belvoir, VA; Computer Systems Command, Support Group, Fort Lee, VA; Computer Systems Command, Support Group, Fort McPherson, GA; Computer Systems Command, Support Group, Fort Shafter, HI; Computer Systems Command, Zweibruecken, Germany.
Length: 4 weeks (160 hours).
Exhibit Dates: 1/80–12/90.
Objectives: To provide advanced ANS COBOL skills so that students may code complex programs.
Instruction: The lecture/workshop programs involves coding at least three programs in which skills are developed in use of source

language storage system, use of the SORT verb, table handling features, programming techniques, and ISAM applications.

Credit Recommendation: In the lower-division baccalaureate/associate degree category, 3 semester hours in data processing (COBOL) (11/82).

AR-1402-0047

Introduction to ANS COBOL

Course Number: None.

Location: Computer Systems Command, Fort Belvoir, VA; Computer Systems Command, Support Group, Fort Lee, VA; Computer Systems Command, Support Group, Fort McPherson, GA; Computer Systems Command, Support Group, Fort Shafter, HI; Computer Systems Command, Zweibruecken, Germany.

Length: 3 weeks (120 hours).

Exhibit Dates: 1/80–12/90.

Objectives: To provide programmers with an entry-level knowledge of ANS COBOL.

Instruction: COBOL instructions are presented in a structured programming environment with an emphasis on the syntax of the language. Students load at least three programs requiring skills in data manipulation and table processing.

Credit Recommendation: In the lower-division baccalaureate/associate degree category, 3 semester hours in data processing (COBOL) (11/82).

AR-1402-0048

Business Systems Analysis

Course Number: None.

Location: Computer Systems Command, Fort Belvoir, VA; Computer Systems Command, Support Group, Fort Lee, VA; Computer Systems Command, Support Group, Fort McPherson, GA; Computer Systems Command, Support Group, Fort Shafter, HI; Computer Systems Command, Zweibruecken, Germany.

Length: 2 weeks (80 hours).

Exhibit Dates: 1/82–9/91.

Objectives: To provide students with an understanding of systems analysis, design, and implementation for organizational problem solving.

Instruction: The course comprises Systems Analysis Training for Programmers (also called Systems Analysis for Computer Programmers) and Design/Documentation Workshop. Course covers principles and techniques required in the analysis and design of information processing systems. Topics include defining objectives; interviewing and gathering information; documenting; analyzing available information; selecting and designing a solution; types of controls; quantitative techniques; and structured design techniques, including the top-down approach, HIPO, and Program Design Language. Guidelines for distinguishing between good design and bad design are included.

Credit Recommendation: In the lower-division baccalaureate/associate degree category, 3 semester hours in data processing (systems analysis) (11/82).

AR-1402-0051

Security in Automated Systems (SAS)
(Security in Automatic Data Processing Systems (ADPS))

Course Number: ALMC-DX; ALM-51-0297.

Location: Army Logistics Management College, Onsite at various locations; Army Logistics Management College, Fort Lee, VA.

Length: 1–2 weeks (60–78 hours).

Exhibit Dates: 6/76–9/90.

Objectives: To provide students with an awareness of the need for security and the means for achieving a satisfactory level of security in automatic data processing facilities.

Instruction: Lectures and workshop exercises cover the design of security systems for automatic data processing facilities.

Credit Recommendation: In the lower-division baccalaureate/associate degree category, 1 semester hour in ADP systems security (4/89).

AR-1402-0057

Integrated Facilities System

Course Number: ALMC-5D; ALM-59U210; ALM-59U260; ALM-59U310.

Location: Army Logistics Management Center, Fort Lee, VA.

Length: 2–3 weeks (74–95 hours).

Exhibit Dates: 1/75–3/96.

Objectives: To train personnel to convert and update computer files in real property maintenance, asset accounting, and facilities engineering management systems.

Instruction: Programmed instruction texts and practical exercises cover real property maintenance, asset accounting, and facilities engineering management systems.

Credit Recommendation: In the vocational certificate category, 2 semester hours in integrated facilities system application (3/86).

AR-1402-0059

Decentralized Automated Service Support System (DAS3)

Course Number: 8B-F38; 551-ASIU8.

Location: Quartermaster Center and School, Fort Lee, VA.

Length: 4 weeks (141 hours).

Exhibit Dates: 3/80–10/90.

Objectives: To provide selected warrant officers, noncommissioned officers, and enlisted personnel with the comprehensive knowledge and skills required to operate an automated data processing system used to support a logistics management information system and a standard supply system.

Instruction: Types of instruction include equipment demonstrations, practical exercises, lectures, and performance examinations. Topics include introduction to the central processing unit, peripheral devices, the use of application software programs, and the operation of magnetic type subsystems, line printers, and disk drives.

Credit Recommendation: In the lower-division baccalaureate/associate degree category, 2 semester hours in introduction to computer operations (4/81).

AR-1402-0065

Disk Operating Systems (DOS) Concepts and Job Control Language (JCL)

Course Number: ADP-30.

Location: Combined Arms Training Center, Zweibruecken, Germany.

Length: 2 weeks (74 hours).

Exhibit Dates: 1/83–6/96.

Objectives: To train personnel to use disk operating system (DOS) concepts and Job Control Language (JCL).

Instruction: Lectures and hands-on computer application cover DOS supervision, program status words, input/output tables, communication regions, DOS problem program areas, and the nonresident DOS supervisor.

Credit Recommendation: In the lower-division baccalaureate/associate degree category, 3 semester hours in Job Control Language (JCL) (10/82).

Related Occupations: 741A; 74C; 74D; 74F.

AR-1402-0066

Disk Operating System (DOS) Executive Software

Course Number: ADP-31.

Location: Combined Arms Training Center, Zweibruecken, Germany.

Length: 2 weeks (80 hours).

Exhibit Dates: 7/80–6/96.

Objectives: To provide students with the necessary knowledge and skills to use software programs that support a disk operating system.

Instruction: This self-paced independent reading course deals with the application of special military software packages and includes required practical exercises (without hands-on time).

Credit Recommendation: Credit is not recommended because of the limited, specialized nature of the course (10/82).

Related Occupations: 741; 74E.

AR-1402-0067

Structured Coding Workshop
(Structured Coding Using COBOL)

Course Number: None.

Location: Computer Systems Command, Fort Belvoir, VA; Computer Systems Command, Support Group, Fort Lee, VA; Computer Systems Command, Support Group, Fort McPherson, GA; Computer Systems Command, Support Group, Fort Shafter, HI; Computer Systems Command, Zweibruecken, Germany.

Length: 1 week (40–45 hours).

Exhibit Dates: 1/80–9/91.

Objectives: To provide coverage of structured coding techniques using COBOL.

Instruction: The program covers IPO charts, top-down design tools, Nursi-Schneiderman/Chapin charts, and structured flowcharts.

Credit Recommendation: In the lower-division baccalaureate/associate degree category, I semester hours in data processing (structured design) (11/82).

AR-1402-0068

DAS3 Executive Language and Editor

Course Number: None.

Location: Computer Systems Command, Fort Belvoir, VA; Computer Systems Command, Support Group, Fort Lee, VA; Computer Systems Command, Support Group, Fort McPherson, GA; Computer Systems Command, Support Group, Fort Shafter, HI; Computer Systems Command, Zweibruecken, Germany.

Length: 1–2 weeks (59 hours).

Exhibit Dates: 1/80–9/91.

Objectives: To provide programmers with the ability to create, modify, and execute tasks on the Honeywell Level 6/DAS3 computer system.

Instruction: The course consists of two modules: Introduction to DAS3 and DAS3 Executive Language and Editor Function. Course includes detailed instruction and workshop in systems configuration and user interaction with the Level 6 Mod 400 operating system via the command processor and utilities. The course covers commands, data file organization, editor directives, linker, command files and sort/merge.

Credit Recommendation: In the lower-division baccalaureate/associate degree category, 1 semester hour in data processing (executive language JCL) (11/82).

AR-1402-0069

DAS3 COBOL Specifics

Course Number: None.

Location: Computer Systems Command, Fort Belvoir, VA; Computer Systems Command, Support Group, Fort Lee, VA; Computer Systems Command, Support Group, Fort McPherson, GA; Computer Systems Command, Support Group, Fort Shafter, HI; Computer Systems Command, Zweibruecken, Germany.

Length: 1–2 weeks (56 hours).

Exhibit Dates: 1/80–12/90.

Objectives: To provide experienced COBOL programmers with Honeywell Level 6 COBOL differences.

Instruction: This course provides a review of MOD 400 COBOL needed to create and run application programs on the Level 6 computer. Emphasis is placed on communication and file accessing concepts, data representation, and differences between Version 69 and 74 COBOL.

Credit Recommendation: In the lower-division baccalaureate/associate degree category, 1 semester hours in data processing (COBOL) (11/82).

AR-1402-0070

Introduction to Data Base Management Systems

Course Number: None.

Location: Computer Systems Command, Fort Belvoir, VA; Computer Systems Command, Support Group, Fort Lee, VA; Computer Systems Command, Support Group, Fort McPherson, GA; Computer Systems Command, Support Group, Fort Shafter, HI; Computer Systems Command, Zweibruecken, Germany.

Length: 1–2 weeks (64 hours).

Exhibit Dates: 1/80–12/90.

Objectives: To provide personnel with an understanding of data base management systems and how they operate.

Instruction: The course comprises two modules: Introduction to Data Base Concepts and Management Systems and Overview of Data Base Management Systems. Both modules must be completed. Course includes the basic components of a data base management system, various data models such as true structures, and the conversion of logical designs into physical representations. The importance of privacy/security, tuning, and other operational factors are considered. An overview of five data base management systems is presented.

Credit Recommendation: In the upper-division baccalaureate category, 2 semester hours in data processing (11/82).

AR-1402-0071

Tactical Fire (TACFIRE) Support Element Liaison Officer

Course Number: 2G-ASI4D/250-ASIX3; 2G-F52.

Location: Field Artillery School, Fort Sill, OK.

Length: 5–7 weeks (200–255 hours).

Exhibit Dates: 5/81–6/96.

Objectives: To provide selected personnel with the knowledge and skills required to tactically employ, operate, and maintain the Variable Format Message Entry Device.

Instruction: Self-paced instruction covers the use of the tactical fire communications system, the Variable Format Message Entry Device (VFMED), and fire support/liaison procedures using the VFMED.

Credit Recommendation: In the vocational certificate category, 1 semester hour in computer operations (10/83).

AR-1402-0072

Tactical Fire (TACFIRE) Tactical Operations Center

Course Number: 250-F14; 2G-ASI4D/250-F14.

Location: Field Artillery School, Fort Sill, OK.

Length: 4–5 weeks (160 hours).

Exhibit Dates: 1/85–6/96.

Objectives: To provide selected personnel with the knowledge and skills necessary to employ, operate, and maintain TACFIRE remote devices.

Instruction: Instruction covers TACFIRE capabilities, characteristics, organization, applications, and operating procedures.

Credit Recommendation: Credit is not recommended because of the limited, specialized nature of the course (11/85).

AR-1402-0074

Tactical Fire (TACFIRE) Direction System Fire Support
(TACFIRE Fire Direction System Fire Support with Fire Support Element Module)

Course Number: 250-F6; 2G-F30.

Location: Field Artillery School, Fort Sill, OK.

Length: 9–11 weeks (383 hours).

Exhibit Dates: 2/79–6/96.

Objectives: To provide selected fire direction operations and intelligence personnel with the knowledge and skills required to tactically employ, operate, and maintain a tactical fire direction system at the field artillery battalion and division level.

Instruction: Instruction is provided at the systems level in the operation and organizational maintenance of a military tactical fire direction system.

Credit Recommendation: In the lower-division baccalaureate/associate degree category, 3 semester hours in computer operations and 1 in computer maintenance (10/83).

AR-1402-0075

Firefinder Radar Repairer

Course Number: 104-ASIX5.

Location: Field Artillery School, Fort Sill, OK.

Length: 27–28 weeks (1001–1037 hours).

Exhibit Dates: 9/83–6/96.

Objectives: To provide the skills and knowledge required to inspect, test, and perform direct support maintenance at the block-diagram level of field artillery firefinder sets.

Instruction: Lectures and laboratory instruction cover basic electronics, digital fundamentals and logic, operation of radar sets, and direct support maintenance and calibration procedures required to provide mission support maintenance on field artillery firefinder radar equipment.

Credit Recommendation: In the lower-division baccalaureate/associate degree category, 3 semester hours in radar maintenance and operation and 3 in digital theory (11/85).

AR-1402-0078

1. Automatic Message Switching Central AN/TYC-39(V)
   (Automatic Message Switching Central (AN/TYC-39))
2. Automatic Message Switching Center Operations
3. Automatic Message Switching Center Operations

Course Number: ***Version 1:*** 260-ASIZ2. *Version 2:* 260ASIZ2; 260-ASIZ2. *Version 3:* 150-F14(OS)2, Phase 1.

Location: Signal School, Fort Gordon, GA.

Length: ***Version 1:*** 7–8 weeks (263 hours). *Version 2:* 7–8 weeks (266 hours). *Version 3:* 12–13 weeks (488 hours).

Exhibit Dates: ***Version 1:*** 7/82–12/94. *Version 2:* 1/95–9/99. *Version 3:* 10/99–Present.

Learning Outcomes: ***Version 1:*** Upon completion of the course, the student will be able to operate a mobile automatic message switching central. *Version 2:* Upon completion of the course, the student will be able to operate a mobile automatic message switching central. *Version 3:* Upon completion of the course, the student will be able to operate a mobile automatic message switching central control.

Instruction: ***Version 1:*** Course presents non-technical instruction in the operation of the message switching system, including system start-up, management, traffic service, and file control. *Version 2:* Course presents non-technical instruction in the operation of the message switching system, including system start-up, management, traffic service, file control, supervision of on-line and off-line job functions, and operator system analysis. *Version 3:* Practical and classroom exercises, and discussion. Topics include system start-up, system management, traffic service, file control, supervision of the on-line and off-line job function, and operate systems analysis.

Credit Recommendation: ***Version 1:*** Credit is not recommended because of the military-specific nature of the course (5/90). *Version 2:* In the lower-division baccalaureate/associate degree category, 3 semester hours in telecommunications control (2/97). *Version 3:* In the lower-division baccalaureate/associate degree category, 3 semester hours in electronic systems troubleshooting (2/03).

Related Occupations: 72E.

AR-1402-0083

Tactical FIRE (TACFIRE) Battalion Fire Direction Officer

Course Number: 2G-SI4F; 2G-F57.
Location: Field Artillery School, Fort Sill, OK.
Length: 7 weeks (252–266 hours).
Exhibit Dates: 7/85–12/94.
Objectives: To enable selected personnel to perform as tactical fire (TACFIRE) battalion fire direction officers.
Instruction: Lectures and demonstrations cover the TACFIRE system, equipment, organization, operation, operating procedures, and maintenance management.
Credit Recommendation: In the lower-division baccalaureate/associate degree category, 2 semester hours in computer operations and 1 in maintenance management (7/91).

AR-1402-0084

Tactical Fire (TACFIRE) Divarty, Field Artillery Brigade, and Corps Artillery Fire Control Element

Course Number: 250-F33; 2E-ASI4F/250-ASIY1.
Location: Field Artillery School, Fort Sill, OK; Field Artillery School, 7th Army, Grafenwoehr, Germany.
Length: 6–7 weeks (237–280 hours).
Exhibit Dates: 7/85–12/94.
Objectives: After 12/99 see AR-1402-0182. To enable selected personnel to perform as tactical fire (TACFIRE) fire direction control officers and noncommissioned officers.
Instruction: Lectures and demonstrations cover employing, operating, and maintaining the (TACFIRE) direction system, applications of data processing, and fire control elements.
Credit Recommendation: In the lower-division baccalaureate/associate degree category, 2 semester hours in computer operations and 1 in maintenance management (1/92).
Related Occupations: 13C; 13E.

AR-1402-0086

Ammunition Stock Control and Accounting Specialist Basic Noncommissioned Officer (NCO)
(Ammunition Stock Control and Accounting Specialist Basic Technical)

Course Number: 551-55R30.
Location: Missile and Munitions School, Redstone Arsenal, AL.
Length: 7–10 weeks (261–363 hours).
Exhibit Dates: 4/85–6/90.
Objectives: To enable persons to perform ammunition stock control and accounting duties and to operate computer equipment associated with the Standard Army Ammunition System.
Instruction: Personnel are trained to operate the computer equipment associated with the Standard Army Ammunition System. Course includes use of software and routine maintenance of line printer, teleprinter, and magnetic tape units. Course includes a common core of leadership subjects.
Credit Recommendation: In the vocational certificate category, 2 semester hours in introduction to computer operations. In the lower-division baccalaureate/associate degree category, see AR-1406-0090 for common core credit (10/89).
Related Occupations: 55R.

AR-1402-0087

Chaplain Assistant
(Chaplain's Assistant)
(Chapel Activities Specialist)

Course Number: 561-71M10.
Location: Chaplain Center and School, Fort Monmouth, NJ.
Length: 8–9 weeks (306–310 hours).
Exhibit Dates: 6/89–2/99.
Learning Outcomes: After 2/99 see AR-1513-0016. Upon completion of the course, the student will be able to apply interpersonal interaction skills, provide comfort to distressed people, prepare a chapel for various religious services, assist in tactical deployment, operate a motor vehicle safely, and perform various basic clerical skills including typing and filing.
Instruction: Topics include interpersonal skills, chapel preparation, identification of ecclesiastical vestments and items, religious training and worship scheduling, religious practices, motor vehicle operation, tactical deployment equipment, casualty assistance, administrative office practice, typing, and filing.
Credit Recommendation: In the lower-division baccalaureate/associate degree category, 1 semester hour in interpersonal communication, 1 in data entry/keyboarding, and 1 in clerical duties (filing and typing) (2/93).
Related Occupations: 71M.

AR-1402-0088

Pastoral Coordinator

Course Number: 5G-ASI7F; 5G-ASI7F/561-F5.
Location: Chaplain Center and School, Fort Monmouth, NJ.
Length: 2 weeks (80 hours).
Exhibit Dates: 5/84–6/00.
Learning Outcomes: Upon completion of the course, the student will be able to prepare a computerized, comprehensive master plan of religious activities, taking into account budgeting, manpower, and resource considerations; monitor program execution; and account for fund expenditures.
Instruction: Topics include microcomputer applications to planning, programming, budgeting, contracts and property; manpower considerations; and spiritual dimensions.
Credit Recommendation: In the lower-division baccalaureate/associate degree category, 1 semester hour in microcomputer applications. In the upper-division baccalaureate category, 1 semester hour in budget management (6/90).

AR-1402-0089

Technical Control and Analysis Center Operator, AN/TSQ-130 (V)

Course Number: 232-F5.
Location: Intelligence School, Fort Devens, MA.
Length: 3–5 weeks (136–172 hours).
Exhibit Dates: 10/85–3/94.
Learning Outcomes: Upon completion of the course, the student will be able to operate a technical control and analysis center to draft, format, and process messages; manipulate the data base; monitor message flow; and prepare the system for relocation and reinstallation.
Instruction: Instruction includes hands-on training in the operation and reinstallation of a technical control and analysis center including the use of function keys and management of the data base.
Credit Recommendation: In the lower-division baccalaureate/associate degree category, 1 semester hour in computer operations (6/91).

AR-1402-0094

Console Control Operations

Course Number: WHO2.
Location: White House Communications Agency, Washington, DC.
Length: Phase 1, classroom, 10 weeks (96 hours); Phase 2, practicum, 14 weeks (168 hours).
Exhibit Dates: 3/78–12/97.
Learning Outcomes: Upon completion of the course, the student will be able to act as a communications assistant for information; provide communications to the members of the Executive Branch of the government; establish and operate voice transmissions which ensure security, both on telephone lines and radio nets; integrate telephone/radio communications from ground, mobile, and air units; establish telephone communications as requested, including mobile phone calls, matrix calls, conference calls, and special-handling calls between top government officials; and provide emergency communications systems for the Executive Branch as required.
Instruction: Topics include the organization of the Executive Branch; computer training; communications security; installation of communications networks including telephone, radio, and satellite; installation and use of sophisticated telecommunications systems, both commercial and military; maintenance of records and files of daily telephone communications; and operation of switchboard consoles. Instructional methods include lectures, seminars, and practical exercises. Course includes 14 weeks of supervised on-the-job training.
Credit Recommendation: In the lower-division baccalaureate/associate degree category, 1 semester hour in computer literacy, 2 in file management, and 3 in introduction to voice communications networks (9/87).

AR-1402-0095

Standard Port System-Enhanced/Decentralized Automated Service Support System (DASPE-E/DAS3)

Course Number: 531-ASIM6.
Location: Transportation and Aviation Logistics School, Fort Eustis, VA.
Length: 7 weeks (252 hours).
Exhibit Dates: 7/87–1/98.
Learning Outcomes: Upon completion of the course, the student will be able to operate and perform operator maintenance on computer software and hardware.
Instruction: Lectures, practical exercises, and demonstrations cover the use of the Army Standard Post System-Enhanced/Decentralized Automated Service Support System. Course includes identification of major system components, power-up/power-down procedures, entering data, document production, processing data, production of status reports, generating tables and report printouts, and downloading data to microcomputers.
Credit Recommendation: In the lower-division baccalaureate/associate degree cate-

gory, 3 semester hours in computer operation (6/91).

AR-1402-0096

1. Director of Information Management (DOIM)/Echelons Above Corps (EAC) Operations Officer
2. Director of Information Management Operations Officer

Course Number: ***Version 1:*** 7E-F35/531-F10. *Version 2:* 7E-F35/531-F10, Phase 1.

Location: Signal School, Fort Gordon, GA.

Length: ***Version 1:*** 5 weeks (180 hours). *Version 2:* 2 weeks (76 hours).

Exhibit Dates: ***Version 1:*** 4/88–2/92. *Version 2:* 3/97–Present.

Learning Outcomes: ***Version 1:*** Upon completion of the course, the student will be able to disseminate Army information using telecommunications, automation, print media, and visual presentations. *Version 2:* Upon completion of the course, the student will be able to disseminate Army information using telecommunications, automation, print media, and visual presentations.

Instruction: ***Version 1:*** Instruction includes classroom, laboratory exercises, and tours of various media production sites. Topics include terrestrial transmission and satellite communications systems, defense data systems, record management, and print/visual methods of presentation. *Version 2:* Audio visual materials, practical and classroom exercises, discussion, and computer-based training. Topics include information technology and management, printing/publications, records management, automation, networking, and information resource systems.

Credit Recommendation: ***Version 1:*** In the lower-division baccalaureate/associate degree category, 3 semester hours in methods of information dissemination (6/88). *Version 2:* In the lower-division baccalaureate/associate degree category, 3 semester hours in information management and dissemination (2/03).

AR-1402-0097

Advanced Theory and Expert System Building

(Advanced Course in Artificial Intelligence Theory and Expert Systems Building)

Course Number: 7E-F37/150-F11.

Location: Signal School, Fort Gordon, GA.

Length: 3–4 weeks (116–150 hours).

Exhibit Dates: 1/88–12/92.

Learning Outcomes: Upon completion of the course, the student will be able to apply a working knowledge of the theory of artificial intelligence as it relates to expert systems building. The student will also possess a working knowledge of artificial intelligence software.

Instruction: Lectures and practical exercises cover predicate logic, PROLOG, LISP, SEARCH, and knowledge representation. Course includes case studies and hands-on experience with graphics-oriented artificial intelligence software.

Credit Recommendation: In the graduate degree category, 2 semester hours in artificial intelligence. A student completing both this and 7E-F36/150-F10 should receive a total of 3 semester hours in artificial intelligence (6/90).

AR-1402-0098

Expert Systems Architecture and Object-Oriented Programming

Course Number: 7E-F38/150-F12.

Location: Signal School, Fort Gordon, GA.

Length: 3 weeks (116 hours).

Exhibit Dates: 10/89–12/92.

Learning Outcomes: Upon completion of the course, the student will perform advanced programming techniques for expert systems architecture and object-oriented programming environments.

Instruction: Lectures and exercises cover advanced LISP programming, debugging methods, and user/designer interfaces. The course is a logical extension of 7E-F37/150-F11.

Credit Recommendation: In the graduate degree category, 2 semester hours in advanced problem solving in artificial intelligence (6/90).

AR-1402-0099

Knowledge Engineering and Basic Expert Systems Building

Course Number: ***Version 1:*** 7E-F36/150-F10. *Version 2:* 7E-F36/150-F10.

Location: Signal School, Fort Gordon, GA.

Length: ***Version 1:*** 2 weeks (73–78 hours). *Version 2:* 2 weeks (78 hours).

Exhibit Dates: ***Version 1:*** 6/86–9/92. *Version 2:* 10/92–4/95.

Learning Outcomes: ***Version 1:*** Upon completion of the course, the student will demonstrate a working knowledge of problem assessment, design cost analysis, testing techniques, and methodologies for the project management and building of rule-based expert systems. *Version 2:* Upon completion of the course, the student will be able to use a DOS-compatible computer and expert system software to design, implement, and test a rules-based expert program; determine when expert system software may be used to solve a problem; apply expert system software to practice problems; and make oral presentations outlining an expert system design plan.

Instruction: ***Version 1:*** Lectures and practical exercises cover problem solving, construction of knowledge systems, evaluation of knowledge systems, and the construction of expert systems prototypes. *Version 2:* Lectures and practical exercises cover the operation and application of a DOS-based expert system program.

Credit Recommendation: ***Version 1:*** In the graduate degree category, 1 semester hour in artificial intelligence (6/88). *Version 2:* In the upper-division baccalaureate category, 3 semester hours in introduction to application of expert system software (2/94).

AR-1402-0101

Tactical Army Combat Service Support Computer System/Standard Army Maintenance System (TACCS/SAMS)

Course Number: 551-ASIB5.

Location: Quartermaster Center and School, Fort Lee, VA.

Length: 3 weeks (108 hours).

Exhibit Dates: 10/87–Present.

Objectives: Upon completion of the course, the student will be able to describe the operation of a microcomputer-based system and an applications-oriented system designed for this hardware and enter data and commands to the system.

Instruction: Classroom lectures, demonstrations, and practical exercises are used to teach microcomputer hardware, including diskette drives, tape cassette drives, and printer operation and application operation, including all functions associated with data input, processing, reports, and file backups.

Credit Recommendation: In the lower-division baccalaureate/associate degree category, 2 semester hours in microcomputer applications (5/02).

AR-1402-0102

Decentralized Automated Service Support System (DAS3) and The Direct Support Unit Standard Supply System (DS4) Operator

Course Number: 8B-F39/551-ASIM6.

Location: Quartermaster Center and School, Fort Lee, VA.

Length: 6 weeks (212 hours).

Exhibit Dates: 3/88–10/98.

Learning Outcomes: Upon completion of the course, the student will be able to describe the minicomputer environment including peripheral devices and the operating system used to control computer hardware and user interface.

Instruction: Classroom lectures, demonstrations, and practical exercises are used to teach the operation of a minicomputer and associated tape drives, printer, communications gear, and disk drives. The function and operation of the operating system exposes the student to console control of applications, utilities, backup processes, and problem determination.

Credit Recommendation: In the lower-division baccalaureate/associate degree category, 4 semester hours in computer operations (10/88).

AR-1402-0103

Multiple Virtual Storage (MVS) 5 Concepts and Operations

Course Number: MVS 5.

Location: Combined Arms Training Center, Pirmasens, W. Germany.

Length: 2 weeks (80 hours).

Exhibit Dates: 2/86–10/98.

Learning Outcomes: Upon completion of the course, the student will be able to describe the function of IBM 4341 OS/VS2 Multiple Virtual Storage (MVS) Job Control Language (JCL); list the nine JCL statements and code simple JCL statements; identify standard IBM utility programs and be able to code JCL to execute simple utility functions; explain the concept of virtual storage and the functional areas of MVS; describe the path of a job through the system; explain how processing interaction occurs; initialize an OS/VSL MVS system and use system commands; use JES 2 commands; and explain relationships among VTOCS, labels, and control blocks.

Instruction: Methods of instruction include lectures, hands-on experience, examinations, and demonstrations.

Credit Recommendation: In the vocational certificate category, 1 semester hour in computer operations (10/88).

AR-1402-0105

MULTIPLE VIRTUAL STORAGE (MVS) 3
ADVANCED CONCEPTS

Course Number: MVS 3.
Location: Combined Arms Training Center, Pirmasens, W. Germany.
Length: 2 weeks (80 hours).
Exhibit Dates: 2/86–10/98.
Learning Outcomes: Upon completion of the course, the student will be able to code simple skeletal Assembler programs; use macros to invoke the functions of the program, task, and data management systems; and locate and map control blocks and data areas in virtual storage dumps on an IBM 4341 OS/V52 multiple virtual storage system.
Instruction: Methods of instruction include classroom lectures, examinations, practical exercises (including hands-on), and demonstrations.
Credit Recommendation: In the vocational certificate category, 1 semester hour in computer concepts (10/88).

AR-1402-0107

COBOL PROGRAMMING

Course Number: ADP-27.
Location: Combined Arms Training Center, Munich, W. Germany.
Length: 2 weeks (73 hours).
Exhibit Dates: 5/83–10/98.
Learning Outcomes: Upon completion of the course, the student will be able to read, write, and debug simple COBOL programs including program structures (divisions), input/output, and arithmetic and logic operations.
Instruction: Methods of instruction include lectures, demonstrations, program writing, program entry, execution, and analysis.
Credit Recommendation: In the lower-division baccalaureate/associate degree category, 1 semester hour in introduction to COBOL (10/88).

AR-1402-0108

STRUCTURED SYSTEMS ANALYSIS AND DESIGN

Course Number: ADP-50.
Location: Combined Arms Training Center, Munich, W. Germany.
Length: 2 weeks (69 hours).
Exhibit Dates: 8/85–10/98.
Learning Outcomes: Upon completion of the course, the student will be able to analyze data flow and operation of existing systems; design new, structured systems; and prepare specialized forms for analyzing data (grid charts, decision tables, data flow diagrams, data dictionaries, and structure charts).
Instruction: Methods of instruction include lectures, practical exercises (simulations), and study of text material.
Credit Recommendation: In the lower-division baccalaureate/associate degree category, 1 semester hour in introduction to structured systems (10/88).

AR-1402-0109

TACTICAL ARMY COMBAT SERVICE SUPPORT
COMPUTER SYSTEM (TACCS)
PROGRAMMING

Course Number: ADP-30.
Location: Combined Arms Training Center, Munich, W. Germany.
Length: 2 weeks (73 hours).
Exhibit Dates: 1/87–10/98.
Learning Outcomes: Upon completion of the course, the student will be able to describe and explain the computer structure and capabilities of the military computer system (TACCS); unpack, set up, and check out the functioning of the TACCS unit; use floppy diskettes to install software; create and modify text files; use password protection to safeguard files; create an archival volume; access files; describe compilers, interpreters, program development, and the sequence of compile-link-execute; compile and execute programs in COBOL, BASIC, and Pascal; use a systems debugger to locate and modify memory; explain basic communications concepts and use a terminal emulator for initiation of asynchronous terminal emulator communications; build and transmit a file; perform bootstrap ROM diagnostic functions and correct malfunctions noted by monitor error codes; define system errors and status codes; interpret error messages; perform recovery procedures; and perform reboot procedures.
Instruction: Through lecture/conferences and practical exercises, the student is able to perform hands-on exercises in setting up a computer system followed by activation and fault isolation. Instruction includes application of software, using COBOL, BASIC, and Pascal; instruction in procedures for administration functions and security of files; and procedures for performing system manager functions.
Credit Recommendation: In the lower-division baccalaureate/associate degree category, 2 semester hours in fundamentals of computer applications (10/88).

AR-1402-0111

1. TARGET ACQUISITION RADAR TECHNICIAN WARRANT OFFICER ADVANCED
2. TARGET ACQUISITION WARRANT OFFICER ADVANCED

Course Number: ***Version 1:*** 2-6-C32-131A. *Version 2:* 2-6-C32-131A.
Location: Field Artillery School, Fort Sill, OK.
Length: ***Version 1:*** 3–4 weeks (140 hours). *Version 2:* 9–10 weeks (325–341 hours).
Exhibit Dates: ***Version 1:*** 10/88–12/94. *Version 2:* 1/95–Present.
Learning Outcomes: ***Version 1:*** Upon completion of the course, the student will be able to communicate effectively, both orally and in writing; edit military correspondence; provide leadership to technical staff; coordinate with agencies in the military community; use word processing and spreadsheet software; and apply supply and management concepts. *Version 2:* Upon completion of the course, the student will be able to communicate effectively, both orally and in writing; edit military correspondence; provide leadership to the technical staff; and coordinate with agencies in the military community. Students will be able to use word processing, spreadsheet, data base, and management software.
Instruction: ***Version 1:*** Lectures cover the latest technological changes and advanced concepts in occupational fields with emphasis on effective communication skills, leadership ability, supply and management concepts, and hands-on experience in word processing and spreadsheet software. *Version 2:* The course consists of classroom lectures and appropriate hands-on laboratory work on the use of computers, field training on target acquisition devices, and self-paced auto-tutorial instruction.
Credit Recommendation: ***Version 1:*** In the vocational certificate category, 1 semester hour in introduction to application software (Lotus 1-2-3). In the lower-division baccalaureate/associate degree category, 3 semester hours in communication (oral and written) (6/89). *Version 2:* In the lower-division baccalaureate/associate degree category, 3 semester hours in personnel supervision and 3 in communication. In the upper-division baccalaureate category, 3 semester hours in principles of management (9/03).
Related Occupations: 131A.

AR-1402-0112

OH-58D FIELD ARTILLERY AERIAL FIRE
SUPPORT OBSERVER

Course Number: 250-F32.
Location: Field Artillery School, Fort Sill, OK.
Length: 5 weeks (218 hours).
Exhibit Dates: 3/89–12/92.
Learning Outcomes: Upon completion of the course, the student will be able to perform duties as an aerial observer, providing fire support information for artillery organizations at the brigade level; operate specialized, new systems, including a computerized laser-guided system, thermal imagery optics, digital message device, and laser range finder; perform procedures for requesting and directing accurate artillery fire; and provide advice and guidance to commanders on the coordinated use of all fire support assets.
Instruction: Conferences, simulations, and extensive practical exercises are followed by field training exercises which emulate live fire missions by day and night. Course includes flight training and aerial missions to develop skill proficiency, examinations to test competence in such topics as fire support planning, use of communications equipment (voice and digital messages), use of special munitions and their guidance systems, offensive and defensive operations, and procedures for adjusting artillery fire on targets.
Credit Recommendation: In the lower-division baccalaureate/associate degree category, 2 semester hours in communications system operations (6/89).

AR-1402-0113

BRANCH AUTOMATION OFFICER

Course Number: ***Version 1:*** 7E-SI4H. *Version 2:* 7E-SI4H/7E-F40/260-F10.
Location: Signal School, Fort Gordon, GA.
Length: ***Version 1:*** 10–11 weeks (382 hours). *Version 2:* 8 weeks (320 hours).
Exhibit Dates: ***Version 1:*** 1/89–12/92. *Version 2:* 1/93–12/96.
Learning Outcomes: ***Version 1:*** Upon completion of the course, the student will be able to perform the duties of a data processing manager; specify, acquire, and develop data processing systems; use software application packages; demonstrate a working knowledge of operating systems, hardware architecture, local area networks, structured programming, and data base management techniques; and manage implementation and maintenance of data processing systems. *Version 2:* Upon completion of the course, the student will be able to perform the duties of a data processing manager; specify, acquire, and develop data processing

systems; use software application packages; demonstrate a working knowledge of operating systems, hardware architecture, local area networks, structured programming, and data base management techniques; and manage implementation and maintenance of data processing systems.

Instruction: *Version 1:* Lectures, practical exercises, and exams cover computer hardware architecture; operating systems; software application packages; systems analysis and design; local area networks; project management; data base management; structured programming; and system testing, implementation, and management. Both microcomputers and mainframes are covered. *Version 2:* Lectures, practical exercises, and exams cover computer hardware architecture; operating systems; software application packages; systems analysis and design; local area networks; project management; data base management; structured programming; and system testing, implementation, and management. Both microcomputers and mainframes are covered.

Credit Recommendation: *Version 1:* In the lower-division baccalaureate/associate degree category, 3 semester hours in introduction to systems analysis, 3 in computer literacy, 1 in data base management, 1 in operating systems, 1 in computer programming concepts, and 3 in data communications. In the upper-division baccalaureate category, 3 in computer systems management (7/89). *Version 2:* In the lower-division baccalaureate/associate degree category, 4 semester hours in computer literacy, 3 in software engineering, and 2 in data communications (2/94).

AR-1402-0114

Software Analyst
(Programmer/Analyst)

Course Number: *Version 1:* 532-74F10; 532-74F10 (ST). *Version 2:* 532-74F10; 532-74F10 (ST).

Location: Signal School, Fort Gordon, GA; Institute for Administration, Fort Benjamin Harrison, IN.

Length: *Version 1:* 9–11 weeks (322–396 hours). *Version 2:* 10 weeks (360 hours).

Exhibit Dates: *Version 1:* 2/87–7/92. *Version 2:* 8/92–9/95.

Learning Outcomes: *Version 1:* Upon completion of the course, the student will be able to prepare flowcharts for a project; and analyze, write, modify, and debug COBOL programs. *Version 2:* Upon completion of the course, the student will be able to use commonly available personal computer software such as DOS, Lotus, and DBase III. The student will be able to create structured programs in Unix and Ada.

Instruction: *Version 1:* Classroom lectures, exercises, and demonstrations using appropriate computers are course methodologies. *Version 2:* The course uses classroom lectures, exercises, and demonstrations. Topics include computer concepts, MS-DOS, Lotus, DBase III, Ada, JCL, and data communications.

Credit Recommendation: *Version 1:* In the lower-division baccalaureate/associate degree category, 2 semester hours in automated data processing and 5 in COBOL programming (5/90). *Version 2:* In the lower-division baccalaureate/associate degree category, 4 semester hours in computer literacy and 2 in JCL programming (2/94).

Related Occupations: 74F.

AR-1402-0115

Systems Automation

Course Number: *Version 1:* 7E-53A. *Version 2:* 7E-53A. *Version 3:* 7E-53A.

Location: *Version 1:* Signal School, Fort Gordon, GA; Institute for Administration, Fort Benjamin Harrison, IN. *Version 2:* Signal School, Fort Gordon, GA; Institute for Administration, Fort Benjamin Harrison, IN. *Version 3:* Signal School, Fort Gordon, GA.

Length: *Version 1:* 19–20 weeks (716 hours). *Version 2:* 19–20 weeks (716–760 hours). *Version 3:* 19–20 weeks (760 hours).

Exhibit Dates: *Version 1:* 1/89–9/92. *Version 2:* 10/92–9/99. *Version 3:* 10/99–Present.

Learning Outcomes: *Version 1:* Upon completion of the course, the student will be able to perform the duties of an information systems engineer; demonstrate a working knowledge of MS-DOS, UNIX, and a Burroughs operating system; operate a specific commercial microcomputer spreadsheet, word processor, data base, and graphics program; apply security management concepts, acquisition management concepts, information center management concepts, and hardware architecture and data communications concepts; install and troubleshoot an ethernet local area network; demonstrate a working knowledge of two international data communications standards, basic concepts of digital electronics, and concepts of software engineering; design a data base system and program in Ada and DB III; and design, develop, and build expert systems. *Version 2:* Upon completion of the course, the student will be able to perform the duties of an information systems engineer; demonstrate a working knowledge of MS-DOS, UNIX, and a Burroughs operating system; operate a specific commercial microcomputer spreadsheet, word processor, data base, and graphics program; demonstrate security management concepts, acquisition management concepts, information center management concepts, and hardware architecture and data communications concepts; install and troubleshoot an ethernet local area network; demonstrate a working knowledge of two international data communications standards, basic concepts of digital electronics, and concepts of software engineering; design a data base system and program in Ada and DB III; and design, develop, and build expert systems. *Version 3:* Upon completion of the course, the student will encompass a broad range of automation topics to include hardware related fundamentals, operating systems, networking basics, resource planning and management, system analysis, programming and database fundamentals, and network/security fundamentals.

Instruction: *Version 1:* Lectures, practical exercises, computer-aided instruction, and tests cover the course topics. This is a rigorous and intensive course covering the topics in great depth. Only personnel with baccalaureate degrees that support advanced study in computer science and information systems engineering are admitted to the course. *Version 2:* Lectures, practical exercises, computer-aided instruction, and tests cover the course topics. This is a rigorous and intensive course covering the topics in great depth. Only personnel with baccalaureate degrees that support advanced study in computer science and information systems engineering are admitted to the course. *Version 3:* Audio visual materials, laboratory, lectures, practical and classroom exercises, discussion and learner presentations. Topics include computer hardware fundamentals, software applications, and computer networking.

Credit Recommendation: *Version 1:* In the lower-division baccalaureate/associate degree category, 1 semester hour in operating systems, 3 in data communications, 1 in DB III programming, 3 in applications software, and 1 in digital principles. In the upper-division baccalaureate category, 3 semester hours in advanced data communications, 3 in software system design, 3 in Ada programming, and 3 in information systems management. In the graduate degree category, 3 semester hours in expert systems and 2 in system data base (7/89). *Version 2:* In the lower-division baccalaureate/associate degree category, 4 semester hours in computer literacy and 3 in data communications. In the upper-division baccalaureate category, 3 semester hours in software engineering, 3 in information systems management, 6 in Ada programming, 3 in advanced data communications, and 3 in artificial intelligence and expert systems (11/02). *Version 3:* In the upper-division baccalaureate category, 3 semester hours in computer hardware maintenance and repair, 3 in relational databases, 3 in web design/HTML editing, 3 in network infrastructure, 3 in Windows server, 3 in network fundamentals, 3 in software applications, 3 in UNIX server, and 3 in network security (2/03).

Related Occupations: 53A.

AR-1402-0116

Information Systems Operator Basic Noncommissioned Officer (NCO)
(Computer/Machine Operator Basic Noncommissioned Officer (NCO))

Course Number: *Version 1:* 531-74D30. *Version 2:* 531-74D30.

Location: Signal School, Fort Gordon, GA; Institute for Administration, Fort Benjamin Harrison, IN.

Length: *Version 1:* 3–4 weeks (120 hours). *Version 2:* 11 weeks (401 hours).

Exhibit Dates: *Version 1:* 8/86–9/91. *Version 2:* 10/91–8/93.

Learning Outcomes: *Version 1:* Upon completion of the course the student will write operating system control statements; correct a production halt or abnormal termination; and devise a continuity of operation plan for a military data handling computer. *Version 2:* Upon completion of the course, the student will be able to perform the tasks of an information system operator in air-land battle operations; provide leadership needed to perform duties in battle; and perform system administration using the UNIX operating system.

Instruction: *Version 1:* Classroom and practical exercises cover topics using the applicable computer. Course includes a common core of leadership subjects. *Version 2:* Classroom exercises cover automation systems development and deployment and case studies encompassing data communications hardware/software configuration and analysis using Bourne shell programming features and UNIX operating system commands and concepts. Course includes a common core of leadership subjects.

Credit Recommendation: *Version 1:* In the lower-division baccalaureate/associate degree category, 1 semester hour in computer systems

operation. See AR-1406-0090 for common core credit (7/89). *Version 2:* In the lower-division baccalaureate/associate degree category, 2 semester hours in UNIX operating system. See AR-1406-0090 for common core credit. In the upper-division baccalaureate category, 3 semester hours in system design and analyses (2/94).

Related Occupations: 74D.

AR-1402-0117

INFORMATION SYSTEMS OPERATOR
(Computer/Machine Operator)
(Computer Systems Operator)

Course Number: 531-74D10; 531-74D10/531-74D10 (ST).

Location: Signal School, Fort Gordon, GA; Institute for Administration, Fort Benjamin Harrison, IN.

Length: 6–13 weeks (239–446 hours).

Exhibit Dates: 10/87–9/95.

Learning Outcomes: Upon completion of the course, the student will be able to identify the components of a data processing system, supply input and obtain output from the system, use MS-DOS and UNIX operating systems, establish communications between two computers, and perform unit-level maintenance on computer equipment.

Instruction: Classroom and practical exercises in basic data processing, computer operation, MS-DOS and UNIX operating systems, Lotus spreadsheet, DBase III, data communications, and unit-level system maintenance.

Credit Recommendation: In the vocational certificate category, 2 semester hours in computer operation. In the lower-division baccalaureate/associate degree category, 1 semester hour in introduction to data processing, 4 in data processing operating systems, and 2 in data processing applications software (2/94).

Related Occupations: 74D.

AR-1402-0119

DATA PROCESSING ADVANCED
NONCOMMISSIONED OFFICER (NCO)

Course Number: ***Version 2:*** 5-74-C42. *Version 3:* 5-74-C42. *Version 4:* 5-74-C42. *Version 5:* 5-74-C42.

Location: ***Version 2:*** Signal School, Fort Gordon, GA; Institute for Administration, Fort Benjamin Harrison, IN. *Version 3:* Signal School, Fort Gordon, GA; Institute for Administration, Fort Benjamin Harrison, IN. *Version 4:* Signal School, Fort Gordon, GA. *Version 5:* Signal School, Fort Gordon, GA.

Length: ***Version 2:*** 14–15 weeks (518 hours). *Version 3:* 15–16 weeks (567 hours). *Version 4:* 14–15 weeks (551 hours). *Version 5:* 19 weeks (721–722 hours).

Exhibit Dates: ***Version 2:*** 1/89–9/91. *Version 3:* 10/91–9/95. *Version 4:* 10/95–12/97. *Version 5:* 1/98–Present.

Learning Outcomes: ***Version 2:*** Upon completion of the course, the student will be able to use Lotus 1-2-3, Multimate, and DB III; program in DB III and Ada; configure and design data communications networks; manage the acquisition, installation, maintenance, and security of information systems; and design expert systems. *Version 3:* Upon completion of the course, the student will be able to use MS DOS-based word processors, spreadsheets, and data bases; configure data communications networks; manage the installation, maintenance, and security of information systems; and manage the development of an expert system, including problem analysis, system development, prototyping, and testing. *Version 4:* Upon completion of the course, the student will be able to use MS DOS-based word processors, spreadsheets, graphics presentation packages, and data bases; configure data communications networks; manage the installation, maintenance, and security of information systems; program data base management systems; and install, manage, and administer a tactical local area network. *Version 5:* Upon completion of the course, the student will be able to maintain and repair computing systems; install, set up, and maintain network systems using DOS, Windows 95, Windows NT, or Unix; use security measures, fundamental web design, and Access Data base design; and control and manage computer network application systems.

Instruction: ***Version 2:*** Lectures and practical exercises cover the above topics. Course includes a common core of leadership subjects. *Version 3:* Lectures and practical exercises cover the topics above. Course includes a common core of leadership subjects. *Version 4:* Lectures and practical exercises cover the topics above. Course includes a common core of leadership subjects. *Version 5:* Classroom presentation and practice exercises are followed by written and practical tests. This course includes a common core of leadership subjects.

Credit Recommendation: ***Version 2:*** In the lower-division baccalaureate/associate degree category, 3 semester hours in computer applications software, 1 in DB III programming, 2 in Ada programming, and 2 in data communications. See AR-1404-0035 for common core credit. In the upper-division baccalaureate category, 3 in information systems management. In the graduate degree category, 2 semester hours in expert systems (7/89). *Version 3:* In the lower-division baccalaureate/associate degree category, 2 semester hours in introduction to data processing, 1 in applications software, 2 in data communications, and 3 in information systems management. See AR-1404-0035 for common core credit. In the upper-division baccalaureate category, 3 semester hours in expert systems (2/94). *Version 4:* In the lower-division baccalaureate/associate degree category, 2 semester hours in introduction to data processing, 1 in computer applications software, 2 in data communications, 2 in information systems management, and 2 in local area networks. See AR-1404-0035 for common core credit (2/97). *Version 5:* In the lower-division baccalaureate/associate degree category, 2 semester hours in computer systems maintenance and repair and 2 in network systems management and control. See AR-1404-0035 for common core credit. In the upper-division baccalaureate category, 2 semester hours in computer network installation and maintenance (11/02).

AR-1402-0120

PROGRAMMER/ANALYST BASIC
NONCOMMISSIONED OFFICER (NCO)

Course Number: 532-74F30.

Location: Signal School, Fort Gordon, GA; Institute for Administration, Fort Benjamin Harrison, IN.

Length: 10 weeks (368 hours).

Exhibit Dates: 3/87–8/93.

Learning Outcomes: Upon completion of the course, the student will be able to use techniques of systems analysis to design, implement, or modify computer systems; manage such systems, including security, quality assurance, and network considerations; create or manage system data base; write COBOL programs; update and manipulate indexed sequential files, and create screen forms and associated interactive programs.

Instruction: Lectures and case studies cover the introduction to the Honeywell system and the Tactical Army Computer System. Course includes a common core of leadership subjects.

Credit Recommendation: In the lower-division baccalaureate/associate degree category, see AR-1406-0090 for common core credit. In the upper-division baccalaureate category, 6 semester hours in systems analysis and 3 in data base management systems (7/89).

Related Occupations: 74F.

AR-1402-0122

FIRE DIRECTION OFFICER

Course Number: TS111B.

Location: Combined Arms Training Center, Vilseck, W. Germany.

Length: 7–8 weeks (290 hours).

Exhibit Dates: 7/88–6/99.

Learning Outcomes: Upon completion of the course, the student will be able to operate and maintain the computer system known as TACFIRE; power up the system, load data base, compose and transmit message formats, modify the operations system data base, and reconfigure the system to operate under alternative modes; enter, retrieve, and validate files; print plans and lists; and load and perform maintenance procedures to correct a system malfunction.

Instruction: Conferences, practical exercises, simulations of operational missions, exercises in troubleshooting malfunctions, and written examinations cover computer operations, processing messages, malfunctions and corrections, maintenance procedures, and computer systems operation.

Credit Recommendation: In the lower-division baccalaureate/associate degree category, 3 semester hours in computer operations (6/89).

AR-1402-0123

TACTICAL FIRE (TACFIRE) TACTICAL
OPERATIONS CENTER

Course Number: TS112.

Location: Combined Arms Training Center, Vilseck, W. Germany.

Length: 5–6 weeks (211 hours).

Exhibit Dates: 11/88–6/99.

Learning Outcomes: Upon completion of the course, the student will be able to use the TACFIRE computer system to support fire planning in a tactical operations center; operate and maintain the computer system, including the loading the data base, composing and transmitting message formats, modifying data base, configuring the system for alternate modes of operation, and performing tactical fire planning processes; use digital message device to provide data input; retrieve target information from files; and correct for system malfunctions.

Instruction: Conferences, extended practical exercises, simulated fire planning missions, malfunction simulations, and written examinations cover such topics as tactical fire planning, computer operations, message processing, mal-

functions and corrections, maintenance procedures, and computer systems operation.

Credit Recommendation: In the lower-division baccalaureate/associate degree category, 3 semester hours in computer operations (6/89).

AR-1402-0124

Tactical Fire (TACFIRE) Fire Support Element

Course Number: TS113.

Location: Combined Arms Training Center, Vilseck, W. Germany.

Length: 5–6 weeks (211 hours).

Exhibit Dates: 11/88–6/99.

Learning Outcomes: Upon completion of the course, the student will be able to use the TACFIRE computer system to plan artillery fire support and conduct operations in a fire support element; operate and maintain the computer system, including loading the data base, composing and transmitting message formats, modifying the data base, configuring the system for alternate modes of operation, and performing fire planning processes; use special digital message device to provide data input; retrieve target information from files; and correct for system malfunctions.

Instruction: Conferences, extended practical exercises, simulated processes and fire planning missions, malfunction simulations, and written examinations cover such topics as fire planning, computer operations, processing messages, malfunctions and corrections, maintenance procedures, and computer systems operation.

Credit Recommendation: In the lower-division baccalaureate/associate degree category, 3 semester hours in computer operations (6/89).

AR-1402-0125

Tactical Fire (TACFIRE) Operations Specialist

Course Number: TS111A.

Location: Combined Arms Training Center, Vilseck, W. Germany.

Length: 8 weeks (314 hours).

Exhibit Dates: 7/88–6/99.

Learning Outcomes: Upon completion of the course, the student will be able to operate and maintain the computer system known as TACFIRE; power up the system, load data base, compose and transmit message formats, modify the operations system data base, and reconfigure the system to operate under alternative mode operations; enter, retrieve, and validate files; print plans and lists; and load and perform maintenance procedures to correct a system malfunction.

Instruction: Conference, practical exercises, simulations of operational missions, exercises to troubleshoot malfunctions, and written examinations cover computer operations, processing messages, malfunctions and corrections, maintenance procedures, and computer systems operation.

Credit Recommendation: In the lower-division baccalaureate/associate degree category, 3 semester hours in computer operations (6/89).

AR-1402-0126

2. Counter-Signals Intelligence Specialist Basic Noncommissioned Officer (NCO)
3. Multidisciplined Counterintelligence Analyst Basic Noncommissioned Officer (NCO)

Course Number: *Version 2:* 231-97G30. *Version 3:* 231-97G30.

Location: *Version 2:* Intelligence School, Fort Huachuca, AZ. *Version 3:* Intelligence Center and School, Fort Huachuca, AZ.

Length: *Version 2:* 10 weeks (360 hours). *Version 3:* 10 weeks (362 hours).

Exhibit Dates: *Version 2:* 11/89–9/96. *Version 3:* 10/96–Present.

Learning Outcomes: *Version 2:* Upon completion of the course, the student will demonstrate technical, analytical, and managerial skills; make decisions in an intelligence setting; and perform computer simulations. *Version 3:* Upon completion of the course, the student will demonstrate technical, analytical, and supervisory skills; make decisions in an intelligence setting; perform computer simulations; and supervise subordinates in technical operations and general military performance.

Instruction: *Version 2:* Lectures, classroom testing, and computer simulation are methods of instruction. Course includes a common core of leadership subjects. *Version 3:* Lectures, practical exercises, and computer simulation are methods of instruction. Course includes a common core of leadership subjects. Assessment includes performance measures and written exams.

Credit Recommendation: *Version 2:* In the lower-division baccalaureate/associate degree category, 2 semester hours in management problems. See AR-1406-0090 for common core credit (6/90). *Version 3:* In the lower-division baccalaureate/associate degree category, 3 semester hours in data base applications. See AR-1406-0090 for common core credit (6/97).

AR-1402-0127

1. Counter-Signals Intelligence Specialist
2. Multidiscipline Counterintelligence Specialist

Course Number: *Version 1:* 231-97G10. *Version 2:* 231-97G10.

Location: Intelligence School, Fort Huachuca, AZ.

Length: *Version 1:* 7–9 weeks (272–324 hours). *Version 2:* 11 weeks (396 hours).

Exhibit Dates: *Version 1:* 10/87–2/93. *Version 2:* 3/93–Present.

Learning Outcomes: *Version 1:* Upon completion of the course, the student will be able to read maps; configure, set up, and use a data base; safeguard classified information; define signal theory/wave propagation; and explain the intelligence and electronic warfare mission. *Version 2:* Upon completion of the course, the student will be able to read maps; configure, set up, and use a UNIX-based computer application to establish, manipulate, and maintain a data base; safeguard classified information; define signal theory/wave propagation; and explain the intelligence and electronic warfare mission.

Instruction: *Version 1:* Lectures, classroom testing, and computer laboratory work cover the above subject areas. *Version 2:* Lectures, classroom testing, and practical exercises using a UNIX-based computer workstation in collecting, maintaining, and analyzing intelligence information.

Credit Recommendation: *Version 1:* In the lower-division baccalaureate/associate degree category, 1 semester hour in map reading and 3 in data base applications (6/90). *Version 2:* In the lower-division baccalaureate/associate degree category, 3 semester hours in data base applications (6/97).

Related Occupations: 97G.

AR-1402-0128

1. Military Intelligence Warrant Officer Technical/Tactical Certification Common Core, Phase 1
   (Military Intelligence Common Core Phase 1)
2. Military Intelligence Warrant Officer Tactical/Technical Certification Common Core

Course Number: *Version 1:* 3A-WOTTC-CC. *Version 2:* 3A-WOTTC-CC, Phase 1.

Location: Intelligence School, Fort Huachuca, AZ.

Length: *Version 1:* 4–6 weeks (140–144 hours). *Version 2:* 7 weeks (253–254 hours).

Exhibit Dates: *Version 1:* 10/88–4/00. *Version 2:* 5/00–Present.

Learning Outcomes: *Version 1:* Upon completion of the course, the student will be able to apply a basic knowledge of military intelligence; discuss the relationship between the National Security Agency and other intelligence organizations; use the basic terms and concepts associated with data processing; and make decisions on the basis of intelligence data. *Version 2:* Upon completion of the course, the student will be able to apply a basic knowledge of military intelligence; discuss the role of the National Security Agency and other security organizations; use and operate intelligence data processing systems; and analyze intelligence information and brief on conclusions.

Instruction: *Version 1:* Lectures, classroom testing, and case studies are the modes of instruction. Students also participate in a computer simulation of a warfare system in the classroom. Players make decisions as they would in the field involving battle and counterattack. The computer workstation uses all facets of intelligence communication and information. *Version 2:* This course consists of classroom lectures, case studies, and simulations of both war and military operations other than war.

Credit Recommendation: *Version 1:* In the lower-division baccalaureate/associate degree category, 1 semester hour in automated data processing. In the upper-division baccalaureate category, 3 semester hours in simulation exercises in decision making (1/97). *Version 2:* In the lower-division baccalaureate/associate degree category, 3 semester hours in computer simulations. In the upper-division baccalaureate category, 3 semester hours in simulation exercises in decision making (8/01).

AR-1402-0130

1. Tactical Automated Network Technician Warrant Officer Technical and Tactical Certification
2. Tactical Automated Network Technician Warrant Officer Basic

3. Tactical Automated Network Technician Warrant Officer Basic

Course Number: ***Version 1:*** 4C-250B. *Version 2:* 4C-250B. *Version 3:* 4C-250B.

Location: Signal School, Fort Gordon, GA.

Length: ***Version 1:*** 19 weeks (684 hours). *Version 2:* 19 weeks (684 hours). *Version 3:* 19 weeks.

Exhibit Dates: ***Version 1:*** 10/90–3/97. *Version 2:* 4/97–5/98. *Version 3:* 6/98–Present.

Learning Outcomes: ***Version 1:*** Upon completion of the course, the student will be able to present a formal oral report; prepare a written staff study; requisition parts and equipment; maintain property records; make inventory adjustments; maintain equipment status and maintenance records; manage and evaluate preventive maintenance systems; explain the budgeting process; explain basic computer concepts; set up a microcomputer system; use an integrated software system, including word processing, spreadsheet, and data base; explain the principles, capabilities, and limitations of digital multiplexed voice and data communications; explain the operation and management of a message switching network; use diagnostics to isolate faults in network equipment. *Version 2:* Upon completion of the course, the student will display a general knowledge of logistics and inventory management, requisition parts and equipment, manage and maintain equipment status and maintenance records, manage and evaluate preventive maintenance systems, present a formal oral report, prepare an effective written report, and solve financial management problems. The student will also be familiar with DOS and Windows operating systems, word processing, spreadsheets, presentation graphics, database management systems and information systems security, data communications, LANS, packet switching; explain the principles, capabilities, and limitations of digital multiplexed voice and data communications; explain the operation and management of message switching networks; and use diagnostics to isolate faults in network equipment. *Version 3:* Upon completion of the course, the student will be able to apply network essentials to PC hardware, computer operating systems (DOS, Windows NT, and Unix) local area networks (LANS), wide area networks (WANS), TCP/IP, and network security. The student will also be able to present a formal oral report, prepare a written staff study, requisition parts and equipment, maintain equipment status and maintenance records, and manage and evaluate preventive maintenance systems.

Instruction: ***Version 1:*** Classroom lectures, practical exercises, and tours are used to present topics in automated voice and data networks. *Version 2:* Course includes lectures, exams, practical exercises, computer-assisted instruction, field exercises, and interactive multimedia presentations. *Version 3:* Course includes lectures, exams, practical exercises, computer-assisted instruction, field exercises, and interactive multimedia presentations.

Credit Recommendation: ***Version 1:*** In the lower-division baccalaureate/associate degree category, 1 semester hour in technical communication, 3 in computer literacy, 3 in maintenance management, and 2 in data communications (7/92). *Version 2:* In the lower-division baccalaureate/associate degree category, 3 semester hours in materiel management, 3 in computer literacy, 2 in data communications, and 1 in technical communications (5/99). *Version 3:* In the lower-division baccalaureate/associate degree category, 2 semester hours in computer systems maintenance and repair, 2 in computer system applications, and 3 in computer network installation and maintenance. Upon completion of the course the student will be ready to take the Microsoft Certified System Engineer (MCSE) test for networking essentials (5/99).

Related Occupations: 250B.

AR-1402-0131

Signal Systems Maintenance Technician Warrant Officer Technical and Tactical Certification

Course Number: 4C-256A.

Location: Signal School, Fort Gordon, GA.

Length: 19 weeks (684 hours).

Exhibit Dates: 10/89–9/95.

Learning Outcomes: Upon completion of the course, the student will be able to present a formal oral report; prepare a written staff study; requisition parts and equipment; maintain property records; make inventory adjustments; maintain equipment status and maintenance records; manage and evaluate preventive maintenance systems; explain the budgeting process; explain basic computer concepts; set up a microcomputer system; use an integrated software system, including word processing, spreadsheet and data base; explain the principles, capabilities, and limitations of UHF, VHF, and microwave communication systems; troubleshoot a variety of communication systems; explain site security techniques and procedures; and apply communication facilities management procedures.

Instruction: Classroom lectures, practical exercises, and tours are used to present topics in signal systems management.

Credit Recommendation: In the lower-division baccalaureate/associate degree category, 1 semester hour in technical communication, 3 in computer literacy, 3 in maintenance management, and 2 in electronic communications (2/94).

Related Occupations: 256A.

AR-1402-0132

Communications Security (COMSEC) Technician Warrant Officer Technical and Tactical Certification

Course Number: ***Version 1:*** 4C-250A. *Version 2:* 4C-250A.

Location: Signal School, Fort Gordon, GA.

Length: ***Version 1:*** 19 weeks (760 hours). *Version 2:* 19 weeks (684 hours).

Exhibit Dates: ***Version 1:*** 10/89–1/92. *Version 2:* 2/92–Present.

Learning Outcomes: ***Version 1:*** Upon completion of the course, the student will be able to present a formal oral report; prepare a written staff study; requisition parts and equipment; maintain property records; make inventory adjustments; maintain equipment status and maintenance records; manage and evaluate preventive maintenance systems; explain the budgeting process; explain basic computer concepts; set up a microcomputer system; use an integrated software system, including word processing, spreadsheet and data base; apply site security techniques and procedures; and explain communication facilities management procedures. *Version 2:* Upon completion of the course, the student will be able to manage large radio-linked telephone networks and apply communications security concepts.

Instruction: ***Version 1:*** Classroom lectures, practical exercises, and tours are used to present topics in communications site management. *Version 2:* Lectures and practical exercises cover effective oral and written communication, logistics management, automatic data processing, communications security, electronic warfare, training management, communications system planning, and radio communications systems.

Credit Recommendation: ***Version 1:*** In the lower-division baccalaureate/associate degree category, 1 semester hour in technical communication, 3 in computer literacy, and 3 in maintenance management (6/90). *Version 2:* In the lower-division baccalaureate/associate degree category, 1 semester hour in technical communication, 3 in computer literacy, 3 in secure communications management, and 3 in communications system maintenance management (2/94).

Related Occupations: 250A.

AR-1402-0133

Test, Measurement, and Diagnostic Equipment (TMDE) Maintenance Support Specialist Advanced Noncommissioned Officer (NCO)

Course Number: G2ASR32470 017.

Location: Technical Training Center, Lowry AFB, CO.

Length: 13 weeks (504 hours).

Exhibit Dates: 10/89–4/04.

Learning Outcomes: Upon completion of the course, the student will be able to apply microcomputer concepts; use an integrated software system for word processing, spreadsheet, and data transmission via modem; archive files; and supervise and train subordinates involved in maintenance and calibration of test and measurement equipment.

Instruction: Classroom lectures and practical exercises are used to present topics in supervision and management of a test and measurement equipment maintenance unit. Course includes a common core of leadership subjects.

Credit Recommendation: In the lower-division baccalaureate/associate degree category, 2 semester hours in computer literacy and 3 in maintenance management. See AR-1404-0035 for common core credit (6/90).

AR-1402-0135

Introduction to Personal Computing

Course Number: PEC-C-IPC.

Location: Professional Education Center, Cp. Joseph Robinson, North Little Rock, AR.

Length: 1 week (45 hours).

Exhibit Dates: 5/88–12/97.

Learning Outcomes: Upon completion of the course, the student will be able to apply concepts of hardware and software, boot MS-DOS computers and manipulate applications and files in DOS, run Enable software, and use word processing, spreadsheet, and data base modules effectively.

Instruction: Hands-on performance-based instruction with specific exercises, examples, and problems are the modes of instruction. The

course includes a mandatory laboratory component.

Credit Recommendation: In the lower-division baccalaureate/associate degree category, 2 semester hours in personal computers (7/90).

AR-1402-0136

SUPPLY ACCOUNTING MANAGEMENT INFORMATION SYSTEM

Course Number: SAMIS 90.

Location: Professional Education Center, Cp. Joseph Robinson, North Little Rock, AR.

Length: 1–2 weeks (80 hours).

Exhibit Dates: 8/89–12/97.

Learning Outcomes: Upon completion of the course, the student will be able to input data into the supply accounting management information system on Burroughs/UNIS45 equipment; organize data for input; generate specific reports using the SAMIS system; and input, output, and review data and reports generated by the SAMIS system.

Instruction: Course is performance-based, with students working with the SAMIS system, inputting data and generating reports. Evaluation is continuous and covers 21 critical tasks.

Credit Recommendation: In the vocational certificate category, 1 semester hour in data entry operator (7/90).

AR-1402-0137

INTELLIGENT TECHNOLOGIES FOR OPERATIONS RESEARCH

Course Number: ***Version 1:*** ALMC-IJ. *Version 2:* ALMC-IJ.

Location: ***Version 1:*** Army Logistics Management College, Fort Lee, VA; Army Logistics Management College, Onsite at various locations. *Version 2:* Army Logistics Management College, Fort Lee, VA.

Length: ***Version 1:*** 5 weeks (196 hours). *Version 2:* 5 weeks (194 hours).

Exhibit Dates: ***Version 1:*** 3/91–2/93. *Version 2:* 3/93–Present.

Learning Outcomes: ***Version 1:*** Upon completion of the course, the student will be able to apply expert system technologies to solve problems. *Version 2:* Upon completion of the course, the student will be able to apply expert system technology to solving problems.

Instruction: ***Version 1:*** Lecture-conferences, practical exercises, and case studies include applications of artificial intelligence and the expert system. *Version 2:* Lecture-conferences, practical exercises, and case studies include such topics as applications of artificial intelligence and the expert system.

Credit Recommendation: ***Version 1:*** In the graduate degree category, 6 semester hours in computer science or operations research (6/91). *Version 2:* In the graduate degree category, 6 semester hours in computer science or operations research (7/00).

AR-1402-0138

OPERATIONS RESEARCH SYSTEMS ANALYSIS (ORSA) MILITARY SKILLS DEVELOPMENT

Course Number: ALMC-OM.

Location: Army Logistics Management College, Fort Lee, VA.

Length: 3 weeks (115–117 hours).

Exhibit Dates: 12/89–12/92.

Learning Outcomes: Upon completion of the course, the student will be able to analyze and critique studies using ORSA techniques and interpret study results to decision makers.

Instruction: Instruction consists of lecture-conferences, practical exercises, and case studies. Topics include ORSA principles and methods, probability and statistics, computer programming, and quantitative analysis applications.

Credit Recommendation: In the lower-division baccalaureate/associate degree category, 1 semester hour in computer programming (DBase or Lotus 1-2-3). In the upper-division baccalaureate category, 3 semester hours in quantitative methods (6/91).

AR-1402-0139

OPERATIONS RESEARCH SYSTEMS ANALYSIS MILITARY APPLICATIONS 1 (ORSA MAC 1)

Course Number: ALMC-SB.

Location: Army Logistics Management College, Fort Lee, VA.

Length: 13 weeks (500 hours).

Exhibit Dates: 2/89–12/92.

Learning Outcomes: After 12/92 see AR-1402-0164. Upon completion of the course, the student will be able to perform operations research systems analysis (ORSA) studies, evaluate ORSA studies, and interpret the results of ORSA studies to Department of Defense decision makers in operational terms.

Instruction: Lecture-conferences, practical exercises, and case studies are used to cover probability and statistics, mathematical modeling, decision analysis, computer programming and simulation, and artificial intelligence.

Credit Recommendation: In the lower-division baccalaureate/associate degree category, 3 semester hours in computer programming (DGS and Lotus) and 3 in probability and statistics. In the upper-division baccalaureate category, 3 semester hours in quantitative methods. In the graduate degree category, 3 semester hours in advanced probability and statistics, 6 in advanced management mathematics, and 6 in operations research (6/91).

AR-1402-0140

OPERATIONS RESEARCH SYSTEMS ANALYSIS MILITARY APPLICATIONS 2 (ORSA MAC 2)

Course Number: ALMC-SK.

Location: Army Logistics Management College, Fort Lee, VA; Army Logistics Management College, Onsite at various locations.

Length: 2 weeks (74–77 hours).

Exhibit Dates: 6/89–8/00.

Learning Outcomes: Upon completion of the course, the student will be able to conduct ORSA studies using the latest techniques in addressing military problems.

Instruction: The instruction consists of lectures, conferences, and practical exercises and includes the probability and statistics, mathematical programming, modeling, simulations, and both decision analysis and cost/effectiveness analysis.

Credit Recommendation: In the upper-division baccalaureate category, 3 semester hours in quantitative methods (6/91).

AR-1402-0141

ELECTRONIC PROCESSING AND DISSEMINATION SYSTEM OPERATOR/ANALYST

Course Number: 233-F13/3B-ASI9C.

Location: Intelligence School, Fort Devens, MA.

Length: 8–9 weeks (318–319 hours).

Exhibit Dates: 1/87–3/94.

Learning Outcomes: Upon completion of the course, the student will be able to operate electronic processing and dissemination systems and perform operational analysis on the electronic order of battle resident data base.

Instruction: Instruction includes classroom lectures, discussions, and practical exercises in the description and operation of the electronic processing and dissemination systems.

Credit Recommendation: In the lower-division baccalaureate/associate degree category, 2 semester hours in computer operations (6/91).

AR-1402-0142

ENHANCED TACTICAL USERS TERMINAL OPERATOR/ANALYST

Course Number: ***Version 1:*** 3A-SI3E/ASI9C/233-ASIT4; 3B-F14/233-ASIT4; 233-F14/3B-F14. *Version 2:* 3A-SI3E/ASI9C/233-ASIT4; 3B-F14/233-ASIT4; 233-F14/3B-F14.

Location: ***Version 1:*** Intelligence School, Fort Devens, MA. *Version 2:* Intelligence Center and School, Fort Huachuca, AZ.

Length: ***Version 1:*** 6–9 weeks (232–324 hours). *Version 2:* 8–9 weeks (314 hours).

Exhibit Dates: ***Version 1:*** 7/89–4/92. *Version 2:* 5/92–Present.

Learning Outcomes: ***Version 1:*** Upon completion of the course, the student will be able to operate an enhanced users terminal and perform associated computer operations. *Version 2:* Upon completion of the course the student will be able to install and operate a complex computer application program on a computer workstation.

Instruction: ***Version 1:*** Lectures, presentations, and practical exercises cover the operation of the enhanced tactical users terminal. Course includes computer files, messages, program operation, and data entry. *Version 2:* Instruction consists of predominately hands-on exercises with limited lectures and demonstrations. Evaluations consists of performance-based competency assessments from part task to a final four-day simulation exercise under actual operating scenarios. Only 25 percent of grade is based on written exams. Remedial instruction outside normal class hours is provided for each of the five phases. Knowledge and skills include network communications setup and operation; installation and operation of a specialized computer applications program, involving text and graphic file creation, manipulation, and integration of sensor, text, and visual data inputs; preparation of routine and special analysis programs; and the editing and electronic distribution of the resulting reports. Operations are performed under time constraints and within rule-based procedures. Telecommunications skills include setup and operation of radar and radio inputs through analog/digital conversion system for integration with computer analysis application program. Basic systems troubleshooting and system maintenance are included.

Credit Recommendation: *Version 1:* In the lower-division baccalaureate/associate degree category, 2 semester hours in computer operation (6/91). *Version 2:* In the vocational certificate category, 3 semester hours in computer workstation application for text and graphic files. In the lower-division baccalaureate/associate degree category, 3 semester hours in computer system operations for manipulating text and graphic data bases (6/97).

AR-1402-0143

Ammunition Stock Control and Accounting Specialist Basic Noncommissioned Officer (NCO)

Course Number: 551-55R30.
Location: Missile and Munitions School, Redstone Arsenal, AL.
Length: 9–10 weeks (347 hours).
Exhibit Dates: 7/90–9/92.
Learning Outcomes: Upon completion of the course, the student will be able to manage the instruction of small groups in a computer environment, maintain ammunition storage drawings, maintain safety data, manage ammunition assets, manage a data processing installation, supervise subordinates in a simulated combat environment, and interpret round reports and create hard-copy reports from disk and tape files.
Instruction: Lectures and practical exercises cover complaint procedures, site selection, and ammunition destruction. Course includes a common core of leadership subjects.
Credit Recommendation: In the lower-division baccalaureate/associate degree category, 2 semester hours in introduction to data processing. See AR-1406-0090 for common core credit (8/91).
Related Occupations: 55R.

AR-1402-0144

Software Engineering and Ada

Course Number: None.
Location: Information Systems Software Center, Fort Belvoir, VA.
Length: 1–2 weeks (48 hours).
Exhibit Dates: 1/91–4/04.
Learning Outcomes: Upon completion of the course, the student will be able to explain the major elements of the Ada programming language, apply concepts and problems in software engineering, and use techniques developed in the Ada language to solve software problems.
Instruction: Lectures and written exercises cover features of the Ada language and software engineering concepts. Course consists of three modules: Ada Overview, Software Engineering and Ada methodology, and Ada Programming Structures Overview.
Credit Recommendation: In the lower-division baccalaureate/associate degree category, 2 semester hours in introduction to Ada and 1 in introduction to software engineering (2/92).

AR-1402-0145

DBASE III Plus

Course Number: None.
Location: Information Systems Software Center, Fort Belvoir, VA.
Length: 1–2 weeks (56 hours).
Exhibit Dates: 1/91–4/04.
Learning Outcomes: Upon completion of the course, the student will be able to create, edit, modify, sort, enter, and retrieve data from DBASE III+ data bases; create simple report and label forms; use memory variables to summarize data; formulate complex queries in the DBASE III+ language; use DBASE III+ functions; change system memory variables and operating environment parameters; convert data sets between DBASE III+ and other applications such as Lotus 1-2-3; design, develop, code, execute, test, and debug DBASE III+ programs that manipulate data files; format the screen; and generate printed output.
Instruction: Lectures and practical exercises cover creating, maintaining, modifying, entering, and retrieving data from DBASE III+ data bases; formulating queries; using relational operations on multiple files; and creating programs in the DBASE III+ language. Course consists of three modules: Introduction to DBASE III+, Advanced DBASE III+, and Programming in DBASE III+.
Credit Recommendation: In the lower-division baccalaureate/associate degree category, 3 semester hours in introduction to data base management and programming using DBASE III+ (2/92).

AR-1402-0146

Introduction to Database Management

Course Number: None.
Location: Information Systems Software Center, Fort Belvoir, VA.
Length: 1–2 weeks (56 hours).
Exhibit Dates: 1/91–4/04.
Learning Outcomes: Upon completion of the course, the student will be able to define the principles underlying a data base management system (DBMS) and the components of a DBMS as compared to other systems; define data models, including tree, network, and relational structures; describe logical as well as the physical organization of data and normalization of data to 3NF; create and use a data dictionary and data definition language; query via SQL; describe the need for data integrity, backup, and recovery; use the facilities of the XDB software package for microcomputers to create data in tables and fields, edit input data, and create indexes; and use commands to conduct basic queries, perform data manipulation, record updating, and report writer functions.
Instruction: Class presentations, examination, and hands-on projects are included. Course consists of two modules: Introduction to Database Management Systems, and Introduction to XDB.
Credit Recommendation: In the lower-division baccalaureate/associate degree category, 2 semester hours in introduction to data base management systems and 2 in introduction to the XDB data base software package (2/92).

AR-1402-0147

Ada Programming Language

Course Number: None.
Location: Information Systems Software Center, Fort Belvoir, VA.
Length: 2 weeks (80 hours).
Exhibit Dates: 1/91–4/04.
Learning Outcomes: Upon completion of the course, the student will be able to design, code, compile, link, execute, test, and debug programs in the Ada programming language using knowledge of data types and classes, declarations, objects, packages, the lexical elements of the Ada language, control statements, and composite data types.
Instruction: Lectures and practical exercises cover program design and development in the Ada programming language, including program units, lexical elements, scalar and composite data classes, and control structures.
Credit Recommendation: In the lower-division baccalaureate/associate degree category, 4 semester hours in programming in Ada (2/92).

AR-1402-0148

C Programming Language

Course Number: None.
Location: Information Systems Software Center, Fort Belvoir, VA.
Length: 2 weeks (80 hours).
Exhibit Dates: 1/91–4/04.
Learning Outcomes: Upon completion of the course, the student will be able to design, code, compile, link, execute, test, and debug functions and computer programs in the ANSI C programming language including computation, control logic, I/O, scalar, aggregate and complex data types, proprocessor directives and macro, pointers and argument passing.
Instruction: Lectures and practical exercises in program design and development in ANSI C including statement syntax, control flows, scalar and aggregate data types such as arrays, structures, and unions, pointers, argument passing, proprocessor syntax including macros, character and block file I/O functions, buffers and complex aggregates such as arrays of structures and C++ objects. Course consists of two modules: Introduction to the C Programming Language, and Advanced C Programming.
Credit Recommendation: In the lower-division baccalaureate/associate degree category, 4 semester hours in programming in C (2/92).

AR-1402-0149

Introduction to Personal Computers

Course Number: None.
Location: Information Systems Software Center, Fort Belvoir, VA.
Length: 2 weeks (80 hours).
Exhibit Dates: 1/91–4/04.
Learning Outcomes: Upon completion of the course, the student will be able use data processing terminology; identify the components associated with a PC; explain the purpose and the proper handling of computer hardware; load and start DOS on a PC; execute DOS commands to manage files and subdirectories; create, edit, and revise text files; manipulate PC keyboard and printer data flow; create batch files and turnkey systems; create, edit, and save Lotus 1-2-3 spreadsheets; perform mathematical computations; create graphs; query Lotus 1-2-3 for individual records; use Wordstar or Word Perfect to create, revise, and print documents; create a dBASE III+ data base and input data; edit, sort, and retrieve data; create and execute simple programs with reports; and summarize data using dBASE III+ statistical commands.
Instruction: Lectures, practical exercises introduce students to microcomputer hardware and software, DOS and some of its more advanced commands, Lotus 1-2-3, Wordstar or

Word Perfect, and dBASE III+ data base software. Course consists of six modules: Microcomputer Overview, Introduction to MS DOS, Advanced MS DOS, Introduction to Lotus 1-2-3, Introduction to dBASE III+ and Introduction to Wordstar or Word Perfect.

Credit Recommendation: In the lower-division baccalaureate/associate degree category, 3 semester hours in introduction to personal computers (2/92).

AR-1402-0150

INTRODUCTION TO COBOL
(Basic ANS COBOL)

Course Number: None.
Location: Information Systems Software Center, Fort Belvoir, VA.
Length: 4–5 weeks (180–190 hours).
Exhibit Dates: 1/91–4/04.
Learning Outcomes: Upon completion of the course, the student will be able to develop, code, compile, and test ANS COBOL programs; create and access sequential files in an IBM environment; perform data manipulation and arithmetic operations employing complex logic; write programs employing the concepts of sorting and table processing; write Job Control Language to support COBOL programming; locate erroneous instructions and data using system abnormal termination codes; and use binary and hexadecimal number formats to interpret and respond to system dumps.
Instruction: Lectures, practical exercises, and examinations support comprehensive coverage of the COBOL programming language in a structured programming environment with an emphasis on table processing. Course also includes familiarization with the tools and methods or determining the cause of abnormal termination of COBOL programs. Course consists of two modules: ANS COBOL and OS/MVS COBOL Debugging.
Credit Recommendation: In the lower-division baccalaureate/associate degree category, 4 semester hours in data processing (COBOL) (2/92).

AR-1402-0151

DATA PROCESSING PRINCIPLES

Course Number: None.
Location: Information Systems Software Center, Fort Belvoir, VA.
Length: 3 weeks (120 hours).
Exhibit Dates: 1/91–4/04.
Learning Outcomes: Upon completion of the course, the student will understand the basics of computer systems, including the components of systems and computers, software overview, presentation of data by a computer, and concepts of programming. The student will be able to solve problems; use charts, tables, flow charts, and logic patterns; and manipulate data. Student will be able to use the BASIC language.
Instruction: Lectures, quizzes, and examinations are used with a major component being hands-on experience. Course consists of three modules: Data Processing Concepts, Fundamentals of Programming, and Introduction to the BASIC Programming Language. Structured concepts are stressed and proficiency is developed in editing programs using BASIC commands (LET, if...then, for...next, etc.), formatting output, subroutines, and file processing.
Credit Recommendation: In the lower-division baccalaureate/associate degree category, 3 semester hours in data processing principles, introduction to computers, or fundamentals of programming and 3 in computer programming in BASIC (2/92).

AR-1402-0152

UNIX OPERATING SYSTEM PRINCIPLES

Course Number: None.
Location: Information Systems Software Center, Fort Belvoir, VA.
Length: 2 weeks (80 hours).
Exhibit Dates: 1/91–4/04.
Learning Outcomes: Upon completion of the course, the student will be able to describe the UNIX file and directory structure; use file, directory, and process commands to create, edit, maintain, and remove files; use redirection to create output files and read input files; build compound processes; create background processes; use editors to search and process data files; and create, execute, and debug simple shell scripts using variables, options, parameters, arguments, single and multiple decision structures, and counted and logical loop structures. The student will also be able to design, code, compile, link, execute, test, and debug computer programs in the ANSI I programming language, including computation, control logic, simple I/O, scalar and array data types, and pointers.
Instruction: Lectures and practical exercises cover the UNIX file structure, common commands, file creation and editing, redirection and pipes, foreground and background processes, editors, shell programming, and program design and development in ANSI C, including statement syntax, control of flow, and scalar and array data types, and pointers. Course consists of three modules: Fundamentals of UNIX, UNIX Shell, and Introduction to the C Programming Language.
Credit Recommendation: In the lower-division baccalaureate/associate degree category, 2 semester hours in introduction to programming in C and 2 in introduction to UNIX (2/92).

AR-1402-0153

OPERATING SYSTEMS

Course Number: None.
Location: Information Systems Software Center, Fort Belvoir, VA.
Length: 3 weeks (120 hours).
Exhibit Dates: 1/91–4/04.
Learning Outcomes: Upon completion of the course, the student will be able to describe the rationale behind the development, function, and use of operating systems and apply principles of comparative operating systems primarily in the IBM environment, including data representation, storage and addressing, multiprogramming, job initiation, and registers. Student will be able to code Job Control Language (JCL) job streams including the use of proper syntax for job, route, print, execute, data definition, and other statements and describe program stages and catalogued and in-stream procedures. Student will be able to code JCL needed to involve other OS utilities including IEBUPOTE, IEBGENER, IEHLIST, and others.
Instruction: Lectures, quizzes, examinations, and hand-on laboratory experiences are provided. Course consists of three modules: OS Concepts and Facilities, OS Job Control language, and OS Utilities and Workshop.
Credit Recommendation: In the lower-division baccalaureate/associate degree category, 2 semester hours in OS utilities, 3 in OS JCL, and 2 in operating systems (2/92).

AR-1402-0154

NETWORK MANAGEMENT/INFORMATION
SYSTEMS STAFF OFFICER

Course Number: 4C-25E.
Location: Signal School, Fort Gordon, GA.
Length: 20 weeks (779 hours).
Exhibit Dates: 4/91–Present.
Learning Outcomes: Upon completion of the course, the student will be able to manage the integration and interconnection of diverse types of automation and communications systems into interoperable local area and long-haul information networks. Students apply techniques to determine the overall network architecture to support battlefield user requirements, exercise network control, and conduct recurring network analysis. The student will apply practical systems engineering principles to plan, design, and install tactical and nontactical information networks.
Instruction: The course uses classroom lectures and exercises. Topics include automated network management; automatic switching; mobile subscriber equipment planning; probability and statistics; reliability, maintainability, and testing; quantitative testing techniques; acquisition and life cycle management; communications systems; operating system concepts; data communications; local area networks; and wide area networks.
Credit Recommendation: In the lower-division baccalaureate/associate degree category, 3 semester hours in data communications, 6 in communications systems, 3 in computer literacy, and 6 in data communications and local area networks. In the upper-division baccalaureate category, 5 semester hours in statistical mathematics and process control (2/94).

AR-1402-0155

COMMUNICATIONS-ELECTRONICS
MAINTENANCE CHIEF ADVANCED
NONCOMMISSIONED OFFICER (NCO)

Course Number: 1-29-C42.
Location: Signal School, Fort Gordon, GA.
Length: 14 weeks (529 hours).
Exhibit Dates: 3/92–9/95.
Learning Outcomes: Upon completion of the course, the student will be able to use basic data base and spreadsheet functions on MS DOS-based computers; plan and supervise maintenance of on-site radio and data communications; and review military files and correspondence for compliance with approved formats.
Instruction: Lectures and practical exercises cover elementary MS DOS operating system and applications programs, installation and operation of mobile and wire communications systems, and maintenance of military records. Course includes a common core of leadership subjects.
Credit Recommendation: In the vocational certificate category, 2 semester hours in communications systems operation and maintenance. In the lower-division baccalaureate/associate degree category, 3 semester hours in

introduction to data processing. See AR-1404-0035 for common core credit (2/94).

AR-1402-0156

Telecommunications Operations Chief Advanced Noncommissioned Officer (NCO)

Course Number: *Version 1:* 101-31W40. *Version 2:* 260-31W40; 101-31W40.

Location: Signal School, Fort Gordon, GA.

Length: *Version 1:* 19–20 weeks (733 hours). *Version 2:* 16–17 weeks (635–648 hours).

Exhibit Dates: *Version 1:* 10/93–9/94. *Version 2:* 10/94–Present.

Learning Outcomes: *Version 1:* Before 10/93 see AR-1406-0109. Upon completion of the course, the student will be able to plan, supervise, coordinate, operate, and manage large area radio/telephone communications systems. *Version 2:* Upon completion of the course, the student will be able to plan, supervise, coordinate, operate, and manage large area radio/telephone communications systems; install, manage, and administer a tactical area network; and use MS DOS-based word processors, spreadsheets, graphics presentation package, and databases.

Instruction: *Version 1:* Lectures and practical exercises cover computer literacy, telephone subscriber systems, digital communications, communications system planning, and radio communications systems. Course includes a common core of leadership subjects. *Version 2:* Lectures and practical exercises cover computer literacy, local area networks, telephone subscriber systems, digital communications, communications system planning, and radio communications systems. Course includes a common core of leadership subjects.

Credit Recommendation: *Version 1:* In the lower-division baccalaureate/associate degree category, 1 semester hour in computer literacy and 6 in telecommunications system management. See AR-1404-0035 for common core credit (2/94). *Version 2:* In the lower-division baccalaureate/associate degree category, 1 semester hour in computer applications software, 1 in local area networks, and 5 in telecommunications system management. See AR-1404-0035 for common core credit (11/02).

Related Occupations: 31W.

AR-1402-0157

1. Data Processing Technician Warrant Officer Technical and Tactical Certification
2. Information Systems Warrant Officer Basic Technical and Tactical Certification
3. Information Systems Technician Warrant Officer Basic

Course Number: *Version 1:* 7E-251A. *Version 2:* 7E-251A. *Version 3:* 7E-251A, Phase 2.

Location: Signal School, Fort Gordon, GA.

Length: *Version 1:* 19 weeks (684 hours). *Version 2:* 19 weeks (758–759 hours). *Version 3:* 22 weeks (808–832 hours).

Exhibit Dates: *Version 1:* 10/91–12/97. *Version 2:* 1/98–9/01. *Version 3:* 10/01–Present.

Learning Outcomes: *Version 1:* Upon completion of the course, the student will be able to present a formal oral report; prepare a written staff study; requisition parts and equipment; maintain records and inventory; evaluate preventive maintenance systems; explain budgetary process; explain basic computer concepts for a UNIX-based system; use integrated software system, including word processing, spreadsheet, data base, and graphics; and explain principles of high-frequency data communications systems and local area networks and management. *Version 2:* Upon completion of the course, the student will be able to maintain and repair computing systems; install, setup, and maintain network systems using DOS, Windows 95, Windows NT or Unix; manage and control the security of networked systems; program in visual basic; and develop and maintain data bases in network systems. *Version 3:* Upon completion of the course, the student will be able to demonstrate an understanding of CISCO router networks, manage and control the security of networked systems, and develop and maintain data bases in network systems.

Instruction: *Version 1:* Classroom lectures and practical exercises cover topics in data processing. *Version 2:* Classroom presentations and practice laboratory exercises are followed by written and practical tests. *Version 3:* Practical, classroom exercises and computer-based training. Topics include CISCO router networks, system administration, data base management, and object oriented programming.

Credit Recommendation: *Version 1:* In the lower-division baccalaureate/associate degree category, 1 semester hour in technical communication, 3 in computer literacy, 3 in maintenance management, 2 in data communications, and 2 in local area networks (2/94). *Version 2:* In the lower-division baccalaureate/associate degree category, 2 semester hours in computer systems maintenance and repair, 1 in network security management and control, 2 in network systems management and control, and 2 in materials management. In the upper-division baccalaureate category, 3 semester hours in network installation and maintenance and 3 in programming in Visual Basic (5/99). *Version 3:* In the lower-division baccalaureate/associate degree category, 12 semester hours in CISCO router network, 2 in network system management, and 2 in data base management (2/03).

Related Occupations: 251A.

AR-1402-0158

Joint Tactical Communications System Management

Course Number: 4C-F33.

Location: Signal School, Fort Gordon, GA.

Length: 3 weeks (108 hours).

Exhibit Dates: 10/90–6/95.

Learning Outcomes: Upon completion of the course, the student will be able to plan and operate joint tactical communications networks.

Instruction: The course uses classroom lectures and laboratory exercises to teach network architecture, transmission systems, digital group multiplexing, tactical automated message switching, and automated network planning and management.

Credit Recommendation: Credit is recommended for the common core only. See AR-1406-0090 (2/94).

AR-1402-0159

Communications Computer System Repairer Basic Noncommissioned Officer (NCO)

Course Number: 150-39G30.

Location: Signal School, Fort Gordon, GA.

Length: 7 weeks (252 hours).

Exhibit Dates: 10/91–9/94.

Learning Outcomes: After 9/94 see AR-1402-0166. Upon completion of the course, the student will be able to supervise subordinates in the establishment of data communications between data processing systems.

Instruction: The course uses classroom lectures and practical exercises to teach communication between personal computers and the DAS3 system, between DAS3 and TACCS, and between DAS3 to DAS3.

Credit Recommendation: Credit is not recommended because of the military-specific nature of the course (2/94).

Related Occupations: 39G; 74G.

AR-1402-0160

Cannon Fire Direction Specialist Basic Noncommissioned Officer (NCO)

Course Number: 250-8-13E30.

Location: Field Artillery School, Fort Sill, OK.

Length: 7 weeks (289 hours).

Exhibit Dates: 10/92–9/97.

Learning Outcomes: After 9/00 see AR-2201-0519. Upon completion of the course, the student will be able to supervise a battery fire direction center equipped with a battery computer system.

Instruction: Lectures and practical exercises cover leadership, tactics, communications, land navigation, weapons, fire data, and maintenance. Course includes a common core of leadership subjects.

Credit Recommendation: In the lower-division baccalaureate/associate degree category, 2 semester hours in computer operations. See AR-1406-0090 for common core credit (9/94).

Related Occupations: 13E.

AR-1402-0161

Field Artillery Tactical Fire Direction System Repairer

Course Number: 160-ASIA7.

Location: Field Artillery School, Fort Sill, OK.

Length: 7–8 weeks (277 hours).

Exhibit Dates: 11/93–12/94.

Learning Outcomes: Upon completion of the course, the student will be able to maintain, repair, and troubleshoot, down to the component level, a computer-driven fire control system.

Instruction: Lectures and laboratory exercises cover the major subsystems (computer group and central processor, magnetic tape unit, line printer, digital data terminal, digital plotter, power systems, and communications equipment) and their component parts, including circuit boards, wiring diagrams, and computer driven diagnostics/fault isolation. Students will be able to use diagnostics, diagnostic codings, schematics, and other technical documentation as well as previously acquired knowledge of electronic theory and practice to identify, isolate, and repair faults and perform standard operation and preventive maintenance actions.

Credit Recommendation: In the lower-division baccalaureate/associate degree category, 2 semester hours in computer system architecture and 3 in computer maintenance and troubleshooting (9/94).

AR-1402-0162

Tactical Fire (TACFIRE) Fire Support Leader

Course Number: 2E-514D.
Location: Field Artillery School, Fort Sill, OK.
Length: 7 weeks (263 hours).
Exhibit Dates: 1/91–10/93.
Learning Outcomes: Upon completion of the course, the student will be able to perform and supervise fire support operations in sections equipped with the digital message device, variable format message entry device, electronic tactical display, and upgraded counterfire equipment.
Instruction: Lectures and practical exercises cover the TACFIRE system, communications, ammunition data, mission control, supervision of a fire plan, and target analysis.
Credit Recommendation: In the lower-division baccalaureate/associate degree category, 2 semester hours in computer operations (9/94).

AR-1402-0163

Light Tactical Fire (LTACFIRE) Operator

Course Number: 250-F39.
Location: Field Artillery School, Fort Sill, OK.
Length: 3 weeks (120 hours).
Exhibit Dates: 6/92–12/94.
Learning Outcomes: Upon completion of the course, the student will be able to operate, employ, and maintain the light tactical fire computer targeting system.
Instruction: Lectures and practical exercises cover communications, ammunition and fire unit data, battlefield geometry, meteorological data, target intelligence, fire planning, and fire control.
Credit Recommendation: In the lower-division baccalaureate/associate degree category, 1 semester hour in map reading and 1 in data analysis (9/94).

AR-1402-0164

Operations Research Systems Analysis Military Applications (ORSA MAC), Phases 1 and 2

Course Number: ALMC-SA, Phase 1; ALMC-SB, Phase 2.
Location: Army Logistics Management College, Fort Lee, VA.
Length: Phase 1, 5 weeks (140–178 hours); Phase 2, 9 weeks (326 hours).
Exhibit Dates: 1/93–Present.
Learning Outcomes: Before 1/93 see AR-1402-0139. Upon completion of the course, the student will be able to perform operations research analysis and studies and to interpret the results of that research in operational terms in reports to decision makers.
Instruction: Phase 1 provides a comprehensive study of probability and statistics. Phase 1 also includes computer literacy, simple linear regression, calculus, and linear algebra. Phase 2 presents analysis of variance; artificial intelligence; case studies; combat models; cost, economic, and operation effectiveness analysis; decision making; design and analysis of experiments; forecasting; inventory; mathematical programming; multiple linear regression; networks; queuing theory; and operations research military applications process. Instruction is provided through lectures and guest speakers and is accompanied by typical applications of the analytical techniques. Exercises, case studies, and examinations are graded to determine mastery, and practical exercises outside of the classroom are scheduled. Students must complete requirements for both phases to qualify for credit recommendation.
Credit Recommendation: In the lower-division baccalaureate/associate degree category, 3 semester hours in computer programming (DGS and Lotus) and 3 in probability and statistics. In the upper-division baccalaureate category, 3 in quantitative methods. In the graduate degree category, 3 in advanced probability and statistics, 6 in advanced management mathematics, and 6 in operations research. Credit is to be granted only upon completion of both phases (7/00).

AR-1402-0165

Information Systems Operator Analyst Basic Noncommissioned Officer (BNCOC)

Course Number: 531-74B30; 531-74B30, Phase 2; 531-74B30, Phase 7.
Location: Signal School, Fort Gordon, GA.
Length: 13–19 weeks (510–695 hours).
Exhibit Dates: 6/96–Present.
Learning Outcomes: Upon completion of the course, the student will be able to manage an automated information system and supervise personnel who are computer systems operators, software analysts, installers, and maintainers.
Instruction: Formal classroom instruction and practical exercises focus on teaching microcomputer software, operating systems, software utilities, security, assembling computers, unit level maintenance, data communications, local area networks, system administration, data base programming, and global positioning systems. Course includes a common core of leadership subjects.
Credit Recommendation: In the lower-division baccalaureate/associate degree category, 3 semester hours in computer applications software, 3 in data base programming, and 3 in systems management. See AR-1406-0090 for common core credit (2/03).
Related Occupations: 74B.

AR-1402-0166

Telecommunications Computer Operator-Maintainer Basic Noncommissioned Officer (NCO)

Course Number: 150-74G30.
Location: Signal School, Fort Gordon, GA.
Length: 8–9 weeks (342 hours).
Exhibit Dates: 10/94–Present.
Learning Outcomes: Before 10/94 see AR-1402-0159. Upon completion of the course, the student will be able to supervise the installation, operation, troubleshooting, and maintenance of telecommunications computer systems.
Instruction: This course uses classroom lectures and practical exercises in network planning and calibration and managing the installation, operation, troubleshooting, and maintenance of telecommunications computer systems.
Credit Recommendation: In the lower-division baccalaureate/associate degree category, 3 semester hours in telecommunications/computer operations (2/97).
Related Occupations: 74G.

AR-1402-0167

1. Data Communications and Local Area Networks
2. Local Area Networks-Concepts and Configuration

Course Number: *Version 1:* 7E-F46/150-F18. *Version 2:* 7E-F46/150-F18.
Location: Signal School, Fort Gordon, GA.
Length: *Version 1:* 2 weeks (80 hours). *Version 2:* 2 weeks (80 hours).
Exhibit Dates: *Version 1:* 11/96–9/00. *Version 2:* 10/00–Present.
Learning Outcomes: *Version 1:* Upon completion of the course, the student will demonstrate an understanding of local area network (LAN) concepts and install, maintain, and troubleshoot LANs. *Version 2:* Upon completion of the course, the student will demonstrate an understanding of local area network (LAN) concepts and install, maintain, and troubleshoot LANs.
Instruction: *Version 1:* Classroom lectures and laboratory exercises cover LAN definitions and background, topologies and transmission, LAN access control methods, IEEE 802.3 Ethernet, IEEE 802.5 token ring, ISO open systems reference model, ISO link layer, LAN operating systems and applications, extended LANs and internet working, LAN implementation and design, network management, network administrator responsibilities, and troubleshooting networks. *Version 2:* Topics include install or do an unattended installation of Windows 2000 server; Windows 2000 server file systems; NTFS security and configuration for shared folders; create DFS; install and administer Active Directory; create and administer user accounts, groups, and group policy objects; manage and administer printers; Windows 2000 protocols and services; configure TCP/IP, DHCP, WINS, and DNS; configure certificate services and EFS; manage hardware devices and drivers; back-up and restore data; manage storage disk systems; monitor system performance; install and configure terminal services and licenses.
Credit Recommendation: *Version 1:* In the lower-division baccalaureate/associate degree category, 2 semester hours in local area networks (2/97). *Version 2:* In the lower-division baccalaureate/associate degree category, 3 semester hours in network fundamentals (2/03).

AR-1402-0168

Information Systems Operator-Analyst

Course Number: *Version 1:* 531-74B10, Phase 5; 531-74B10. *Version 2:* 531-74B10.
Location: Signal School, Fort Gordon, GA.
Length: *Version 1:* 11–18 weeks (453–632 hours). *Version 2:* 15 weeks (552 hours).
Exhibit Dates: *Version 1:* 10/95–12/01. *Version 2:* 1/02–Present.
Learning Outcomes: *Version 1:* Upon completion of the course, the student will have a basic working knowledge and understanding of microcomputer software, computer operating systems, software utilities, assembly/disassembly of microcomputers, data communications, local area networks, problem solving using structural design techniques, data base system design and development, and error recovery

and security. Students will be able to use personal computer hardware and basic computer applications software. *Version 2:* Upon completion of the course, the student will be able to demonstrate a basic working knowledge and understanding of microcomputer software, computer operating systems, software utilities, assembly/disassembly of microcomputers, data communications, local area networks, problem solving using structural design techniques, data base system design, develop and error recovery and security. Students will be able to use personal computer hardware and basic computer applications software.

Instruction: ***Version 1:*** Knowledge skills and understanding are provided through formal classroom lectures and demonstrations, as well as intensive actual laboratory experiences with computers and computer software. *Version 2:* Topics include computer hardware repair in an A+ environment, windows desktop workstation administration, windows server deployment, networking fundamentals, CISCO router fundamentals, UNIX computer integration, and Microsoft office user application.

Credit Recommendation: ***Version 1:*** In the lower-division baccalaureate/associate degree category, 3 semester hours in introduction to personal computers, 3 in computer applications software, 1 in computers and communications, and 2 in local area networks (2/03). *Version 2:* In the lower-division baccalaureate/associate degree category, 3 semester hours in CISCO router fundamentals, 3 in computer hardware repair, 3 in network fundamentals, 3 in windows desktop operating system administration, 3 in windows server administration, 3 in computer applications, and 1 in computers and communication (2/03).

Related Occupations: 74B.

AR-1402-0170

LAND COMBAT SUPPORT SYSTEM TEST SPECIALIST (TRANSITION)

Course Number: 121-35B10 (T).

Location: Ordnance, Missile and Munitions School, Redstone Arsenal, AL.

Length: 10 weeks (365 hours).

Exhibit Dates: 10/93–9/98.

Learning Outcomes: Upon completion of the course, the student will be able to repair and maintain land combat support systems and associated test equipment.

Instruction: Lectures and practical exercises cover maintaining computerized electronic and mechanical test equipment, including logic circuits, programming, and operating systems.

Credit Recommendation: In the lower-division baccalaureate/associate degree category, 2 semester hours in digital logic fundamentals (5/97).

Related Occupations: 35B.

AR-1402-0171

TELECOMMUNICATIONS TERMINAL DEVICE REPAIRER BASIC NONCOMMISSIONED OFFICER (NCO)

Course Number: 160-35J30.

Location: Signal School, Fort Gordon, GA; Ordnance, Missile and Munitions School, Redstone Arsenal, AL.

Length: 10–11 weeks (436 hours).

Exhibit Dates: 10/94–6/00.

Learning Outcomes: Upon completion of the course, the student will be able to supervise repair and maintenance of battery-operated telecommunications equipment, including computer controlled systems, IBM compatible personal computers, and peripheral devices and perform necessary maintenance management tasks.

Instruction: Course instruction includes lectures, practical exercises, demonstration, and small group instruction covering topics of maintenance and testing of the AN/TYQ-33 TACCS computer systems, IBM compatible personal computer and peripherals, and supervision of a maintenance facility. Course includes of a common core of leadership subjects.

Credit Recommendation: In the lower-division baccalaureate/associate degree category, 3 semester hours in maintenance management. See AR-1406-0090 for common core credit (5/97).

Related Occupations: 35J.

AR-1402-0172

AIR TRAFFIC CONTROL EQUIPMENT REPAIRER ADVANCED NONCOMMISSIONED OFFICER (NCO)

Course Number: 102-35D40.

Location: Combined Arms Support Command, Fort Lee, VA; Signal School, Fort Gordon, GA; Ordnance, Missile and Munitions School, Redstone Arsenal, AL.

Length: 7–8 weeks (284 hours).

Exhibit Dates: 3/96–6/00.

Learning Outcomes: Upon completion of the course, the student will be able to supervise air traffic control equipment repair.

Instruction: Lectures and practical exercises cover leadership, communications, training, management, computer literacy, and computer networking. Also, including are current electronic technology as used in air traffic control, and microcomputer use including components, terminology, DOS, programming languages, word processing, files, spreadsheets, data base programs, and graphic software. Course includes a common core of leadership subjects.

Credit Recommendation: In the lower-division baccalaureate/associate degree category, 2 semester hours in introduction to computer applications and 3 in maintenance management. See AR-1404-0035 for common core credit (5/97).

Related Occupations: 35D.

AR-1402-0173

ELECTRONIC MAINTENANCE CHIEF ADVANCED NONCOMMISSIONED OFFICER (NCO)

Course Number: 1-35-C42.

Location: Signal School, Fort Gordon, GA; Ordnance, Missile and Munitions School, Redstone Arsenal, AL.

Length: 13–14 weeks (533 hours).

Exhibit Dates: 10/94–6/00.

Learning Outcomes: Upon completion of the course, the student will be able to plan, supervise, and train maintenance personnel for ordnance maintenance facilities.

Instruction: Lectures and practical exercises in leadership, communications, training management; marksmanship; combat leadership; computer technology an introduction to digital computers, DOS operating systems, software, programming languages, spreadsheets, graphics, and presentation programs; and ordnance electronic maintenance and repair. Course contains a common core of leadership subjects.

Credit Recommendation: In the lower-division baccalaureate/associate degree category, 3 semester hours in introduction to computer applications and 3 in maintenance management. See AR-1404-0035 for common core credit (5/97).

AR-1402-0174

ELECTRONIC SYSTEMS MAINTENANCE TECHNICIAN WARRANT OFFICER

Course Number: 4C-11-C32-918B.

Location: Ordnance, Missile and Munitions School, Fort Gordon, GA; Ordnance, Missile and Munitions School, Redstone Arsenal, AL.

Length: 14–15 weeks (276 hours).

Exhibit Dates: 10/94–10/99.

Learning Outcomes: Upon completion of the course, the student will be able to perform in a senior level position in managing an electronic systems maintenance unit.

Instruction: Lectures and practical exercises cover electronic warfare, logistics, automated data processing, and current technology in electronic systems. Microcomputer technology, topics include software, spreadsheets, graphics, data base management, and E-mail.

Credit Recommendation: In the lower-division baccalaureate/associate degree category, 3 semester hours in introduction to computer applications (5/97).

Related Occupations: 918B.

AR-1402-0175

1. ALL SOURCE ANALYSIS SYSTEM (ASAS) MASTER ANALYST
2. ALL SOURCE ANALYSIS SYSTEM MASTER ANALYST

Course Number: ***Version 1:*** 3A-F71/232-F11. *Version 2:* 3A-F71/232-ASI1F.

Location: Intelligence School, Fort Huachuca, AZ.

Length: ***Version 1:*** 8 weeks (283 hours). *Version 2:* 8–9 weeks (317 hours).

Exhibit Dates: ***Version 1:*** 2/98–8/98. *Version 2:* 9/98–Present.

Learning Outcomes: ***Version 1:*** Upon completion of the course, the student will be qualified to function as the unit's master of ASAS operations, the military intelligence commander's mentor, and the commander's advisor. *Version 2:* Upon completion of the course, the student will be qualified to function as the unit's master of the ASAS operations, the military intelligence commander's mentor, and the commander's advisor.

Instruction: ***Version 1:*** Lectures and practical exercises cover maintenance responsibilities, software problem reports, software engineering, data base structure, and system-level troubleshooting. *Version 2:* Lectures and practical exercises cover maintenance responsibilities, software problem reports, software engineering, data base structure, and system-level troubleshooting.

Credit Recommendation: ***Version 1:*** In the lower-division baccalaureate/associate degree category, 3 semester hours in computer system diagnostics (6/97). *Version 2:* In the lower-division baccalaureate/associate degree category, 3 semester hours in computer system diagnostics (8/01).

AR-1402-0176

Constant Course Operator Terminal Maintenance

Course Number: 150-F28.
Location: Intelligence School, Fort Huachuca, AZ.
Length: 2–3 weeks (104 hours).
Exhibit Dates: 1/96–Present.
Learning Outcomes: Upon completion of the course, the student is qualified to repair and/or maintain the constant source operator terminal.
Instruction: Lectures and practical exercises cover SUN workstation maintenance and constant source terminal maintenance.
Credit Recommendation: In the lower-division baccalaureate/associate degree category, 3 semester hours in computer peripheral maintenance (6/97).

AR-1402-0177

Non-Morse Interceptor/Analyst Basic Noncommissioned Officer (NCO)

Course Number: 231-98K30.
Location: Intelligence School, Fort Huachuca, AZ.
Length: 7–8 weeks (281 hours).
Exhibit Dates: 10/96–Present.
Learning Outcomes: Before 10/96 see AR-1717-0078. Upon completion of the course, the student will be able to analyze electronic signals and generate computerized reports.
Instruction: Lectures, presentations, and practical exercises cover analysis of modulation techniques and report generation. This course contains a common core of leadership subjects.
Credit Recommendation: In the lower-division baccalaureate/associate degree category, 2 semester hours in data base applications. See AR-1406-0090 for common core credit (6/97).
Related Occupations: 98K.

AR-1402-0178

Tactical Exploitation of Natural Capabilities (TENCAP) Data Analyst

Course Number: 3A-S13E/ASI9C/233-ASI2T.
Location: Intelligence School, Fort Huachuca, AZ.
Length: 10 weeks.
Exhibit Dates: 1/97–Present.
Learning Outcomes: Upon completion of the course, the student will be able to perform operational analysis on intelligence and to operate the TENCAP UNIX-based system. Student will maintain data base files, perform global positioning, and manipulate software on a UNIX-based system; create, edit, and correlate disposition message data base files; transmit, receive, and report imagery through the communications interface unit; and create, correlate, and maintain exploitation of signal parametrics data base files. Students will demonstrate total system use in a real world, scenario-driven example.
Instruction: Course consists of applied instruction and hands-on exercises for a secure communication TENCAP system consisting of three units.
Credit Recommendation: In the lower-division baccalaureate/associate degree category, 3 semester hours in data base applications (6/97).
Related Occupations: 513E.

AR-1402-0179

Electronic Switching System Technical Manager

Course Number: *Version 1:* 4C-250B (T). *Version 2:* 4C-250B (T).
Location: Signal School, Fort Gordon, GA.
Length: *Version 1:* 11 weeks (396 hours). *Version 2:* 11 weeks (396 hours).
Exhibit Dates: *Version 1:* 10/96–5/98. *Version 2:* 6/98–Present.
Learning Outcomes: *Version 1:* Upon completion of the course, the student will display a general knowledge of logistics and inventory management and be able to manage and maintain equipment status and maintenance records. They student will be also familiar with DOS and windows operating systems, word processing, spreadsheets, presentation graphics, data base management systems. Student will be able to explain the principles, capabilities and limitations of digital multiplexed voice and data communications, explain the operation and management of message switching networks; and provide computer maintenance and repair. *Version 2:* Upon completion of the course, the student will be able to operate data communications and local area networks; plan a voice switch network; develop data bases; develop key matrix plans perform system fault isolation procedures; operate tactical switched networks; configure computer network systems; and manage the integration, transmission, reception, and control of voice and data over computer network systems.
Instruction: *Version 1:* Course includes lectures, demonstrations, and practical exercises. *Version 2:* Conferences and practical exercises cover local area network topologies, transmission techniques, services, standards, and interfaces. Applications include multiplexing of voice and data signals over computer network systems.
Credit Recommendation: *Version 1:* In the lower-division baccalaureate/associate degree category, 2 semester hours in materials management, 3 in computer literacy, and 2 in data communications (5/99). *Version 2:* In the lower-division baccalaureate/associate degree category, 2 semester hours in computer systems maintenance and repair and 2 in computer system applications. In the upper-division baccalaureate category, 3 semester hours in computer network installation and maintenance. Upon completion of the course the student will be ready to take Microsoft Computer System Engineer certification test for networking essentials (5/99).

AR-1402-0180

Enhanced Position Location Reporting System (EPLRS) Net Control Station Operator/Repair

Course Number: 201-ASIT1.
Location: Signal School, Fort Gordon, GA.
Length: 6–10 weeks (240–346 hours).
Exhibit Dates: 10/98–Present.
Learning Outcomes: Upon completion of the course, the student will be able to troubleshoot system and correct faults in the Enhanced Position Locator Reporting System net control station.
Instruction: Lectures and practical exercises, and demonstrations by instructor, cover basics in the enhanced position location reporting system.
Credit Recommendation: In the lower-division baccalaureate/associate degree category, 3 semester hours in computer systems maintenance and repair (2/03).

AR-1402-0181

Information Systems Technician Warrant Officer Advanced
(Information Systems Senior Warrant Officer)

Course Number: 4-11-C32-251A, Phase 1; 4-11-C32-251A.
Location: Signal School, Fort Gordon, GA.
Length: 18–20 weeks (736–746 hours).
Exhibit Dates: 1/98–Present.
Learning Outcomes: Upon completion of the course, the student will be able to maintain and repair computing systems; install, setup, and maintain network systems using Unix and Windows NT; and manage and control security of networked systems.
Instruction: Classroom presentations and practice laboratory exercises are followed by written and practical tests.
Credit Recommendation: In the lower-division baccalaureate/associate degree category, 1 semester hour in computer systems maintenance and repair, 1 in network security management and control, 2 in network systems management and control, and 1 in principles of supervision. In the upper-division baccalaureate category, 3 semester hours in computer network installation and maintenance (11/02).

AR-1402-0182

Fire Support Automated Fire Control Element

Course Number: 250-F33.
Location: Field Artillery School, Fort Sill, OK.
Length: 2–3 weeks (96 hours).
Exhibit Dates: 1/00–Present.
Learning Outcomes: Before 1/00 see AR-1402-0084. Upon completion of the course, the student will use supervisory and computer skills in a fire support and combined arms operation department.
Instruction: Classroom instruction is coupled with hands-on training in initial fire support automation equipment. Computing support covers training in basic computing including use of and access to data base systems.
Credit Recommendation: In the lower-division baccalaureate/associate degree category, 2 semester hours in basic computer operations (1/01).

AR-1402-0183

Electronic Maintenance and Calibration Advanced Noncommissioned Officer (NCO)

Course Number: *Version 1:* 1-35-C42. *Version 2:* 1-35-C42.
Location: Ordnance, Missile and Munitions School, Redstone Arsenal, AL.
Length: *Version 1:* 8 weeks (283 hours). *Version 2:* 5 weeks (180 hours).
Exhibit Dates: *Version 1:* 5/99–9/00. *Version 2:* 10/00–Present.

Learning Outcomes: *Version 1:* Upon completion of the course, the student will be able to plan, supervise, and train maintenance operations in ordnance maintenance facilities. *Version 2:* Upon completion of the course, the student will be able to plan, supervise, and train maintenance operations in ordnance maintenance facilities.

Instruction: *Version 1:* Lectures and practical exercises cover leadership, communication, training management, computer applications, computer networking, and field training exercises. Course contains a common core of leadership subjects. *Version 2:* Audiovisual materials, learner presentations, computer-based training, practical exercises, discussion, case studies, and lecture. Topics include Microsoft Office, Windows, IE, and leadership and management specific to electronic maintenance operations.

Credit Recommendation: *Version 1:* In the lower-division baccalaureate/associate degree category, 3 semester hours in computer applications and 3 in maintenance management. See AR-1404-0035 for common core credit (6/01). *Version 2:* In the lower-division baccalaureate/associate degree category, 2 semester hours in computer applications and 2 in maintenance management. Credit no to be awarded to students who have successfully completed 1-35-C40 (2/04).

AR-1402-0184

Warrant Officer Basic 918B

Course Number: 4F-918B, Phase 2.

Location: Ordnance Electronic Maintenance Training, Fort Gordon, GA.

Length: 12 weeks (424 hours).

Exhibit Dates: 3/00–Present.

Learning Outcomes: Before 3/00 see AR-1408-0240. Upon completion of the course, the student will be able to diagnose and repair common microcomputer systems; apply knowledge of UNIX, DOS, and Windows NT operating systems; install, set up and maintain a local area network in client-server and peer-to-peer configurations; and apply principles of multichannel electronic communications.

Instruction: This course provides lectures and laboratory exercises in computer repair, operating systems, local area network, installation and management, and principles of electronic communications.

Credit Recommendation: In the lower-division baccalaureate/associate degree category, 3 semester hours in computer repair, 3 in local area network installation and management, and 3 in digital communications (6/01).

Related Occupations: 918B.

AR-1402-0185

Warrant Officer Advanced, Phase 2

Course Number: 4-9-C32-918B (Consol), Phase 2.

Location: Ordnance, Missile and Munitions School, Redstone Arsenal, AL.

Length: 6–7 weeks (242 hours).

Exhibit Dates: 6/00–Present.

Learning Outcomes: Upon completion of the course, the student will be able to use UNIX, DOS, Windows, and Windows NT operating systems; use local area networks; and wide area networks; identify typical commercial off-the-shelf computer-related hardware; and inspect and supervise personnel employed in maintenance and repair of avionics communications equipment.

Instruction: Conferences and practical exercises cover computer operating systems and networking related to avionics equipment and maintenance procedures.

Credit Recommendation: In the lower-division baccalaureate/associate degree category, 3 semester hours in computer applications, 2 in networking, and 3 in maintenance management (6/01).

AR-1402-0186

Integrated Computerized Deployment System

Course Number: 8C-F26/822-F25; 822-F25; 8C-F26.

Location: Transportation School, Fort Eustis, VA.

Length: 2 weeks (73 hours).

Exhibit Dates: 4/98–Present.

Learning Outcomes: Upon completion of the course, the student will be able to operate the integrated computerized deployment system and provide load planning on the microcomputer system.

Instruction: Methods of instruction include conferences and performances exercises to create and deliver load plans, including reports and diagrams imposed by the Military Traffic Management Command utilizing a microcomputer system.

Credit Recommendation: In the lower-division baccalaureate/associate degree category, 2 semester hours in principles of computer operations systems (6/01).

AR-1402-0187

Worldwide Port System

Course Number: 533-F12; 8C-F31; 8C-F31/553-F12.

Location: Transportation School, Fort Eustis, VA.

Length: 3 weeks (108 hours).

Exhibit Dates: 10/96–Present.

Learning Outcomes: Upon completion of the course, the student will be able to operate worldwide port system (WPS) computer software and hardware.

Instruction: Methods of instruction include classroom conferences, computer aided instruction, practical exercises, and written exams in the use of WPS software and hardware.

Credit Recommendation: In the lower-division baccalaureate/associate degree category, 2 semester hours in principles of computer operation systems (6/01).

AR-1402-0188

1. Mobility Warrant Officer Basic
2. Mobility Officer Warrant Officer Basic

Course Number: *Version 1:* 8C-882A. *Version 2:* 8C-882A.

Location: Transportation School, Fort Eustis, VA.

Length: *Version 1:* 13 weeks (513 hours). *Version 2:* 18 weeks (710 hours).

Exhibit Dates: *Version 1:* 1/01–9/02. *Version 2:* 10/02–Present.

Learning Outcomes: *Version 1:* Upon completion of the course, the student will be able to conduct, manage, and supervise movement operations by air, highway, rail, or water. *Version 2:* Upon completion of the course, the student will be able to conduct, manage, and supervise resource movement operations by air, highway, rail, or water.

Instruction: *Version 1:* Methods of instruction include classroom, practical exercises, computer-aided instruction, and written exams. Extensive field exercises provide experiences in supervision, conducting and managing personnel, and cargo and vehicle transportation operations. *Version 2:* Computer-based training, practical and classroom exercises, case studies, and lecture. Topics include plan, implement, supervise resource movement operations by air, highway, rail, or water to include personnel, cargo, and vehicles.

Credit Recommendation: *Version 1:* In the lower-division baccalaureate/associate degree category, 3 semester hours in principles of computer operations systems. In the upper-division baccalaureate category, 3 semester hours in transportation management (6/01). *Version 2:* In the lower-division baccalaureate/associate degree category, 4 semester hours in information systems. In the upper-division baccalaureate category, 3 semester hours in intermodel resource management (12/03).

AR-1402-0189

Chemical Operations Specialist Basic Noncommissioned Officer (NCO) Total Army Training System (TATS)

Course Number: 494-54B20/30-R, Phase 3.

Location: Chemical School, Fort Leonard Wood, MO.

Length: 134 hours.

Exhibit Dates: 1/01–Present.

Learning Outcomes: Upon completion of the course, the student will be able to perform critical tasks in nuclear, biological, and chemical procedures using appropriate writing and computer skills.

Instruction: Lectures and demonstrations cover computers and technical writing skills. The bulk of the time is spent on an extensive field exercise.

Credit Recommendation: In the lower-division baccalaureate/associate degree category, 1 semester hour in technical writing and 1 in introduction to computing (1/02).

Related Occupations: 54B.

AR-1402-0190

Chemical Operations Specialist Basic Noncommissioned Officer (NCO) Total Army Training System (TATS), Phase 3

Course Number: 494-54B20/30, Phase 3.

Location: Chemical School, Fort Leonard Wood, MO.

Length: 134 hours.

Exhibit Dates: 1/01–Present.

Learning Outcomes: Upon completion of the course, the student will be able to perform nuclear, biological, and chemical procedures using appropriate writing and computer skills.

Instruction: Lectures and demonstrations cover computers and technical writing skills. The bulk of the time is spent on an extensive field exercise.

Credit Recommendation: In the upper-division baccalaureate category, 1 semester hour in technical writing and 1 in introduction to computers (1/02).

Related Occupations: 54B.

AR-1402-0191

Total Army Training System (TATS) Chaplain Assistant Advanced Noncommissioned Officer (NCO), Phase 2

Course Number: 561-56M40 (Option 1), Phase 2.
Location: Chaplain Center and School, Fort Jackson, SC.
Length: 6–7 weeks (233–237 hours).
Exhibit Dates: 10/01–Present.
Learning Outcomes: Upon completion of the course, the student will be able to lead a medium-sized organization.
Instruction: Methods of instruction include classroom exercises, practical exercises, lectures, and small group discussions. Topics include leadership, supervision, and military operations.
Credit Recommendation: In the lower-division baccalaureate/associate degree category, 3 semester hours in principles of supervision (3/02).
Related Occupations: 56M.

AR-1402-0192

Computer Literacy Training

Course Number: DEV-1, Phase 1.
Location: Western Hemisphere Institute for Security Cooperation (WHINSEC), Fort Benning, GA.
Length: 2 weeks (68 hours).
Exhibit Dates: 2/01–Present.
Learning Outcomes: Upon completion of the course, the student will be able to perform basic DOS and windows-based applications using word-processing graphics, database, and spreadsheet software.
Instruction: Audio-visual materials, classroom and practical exercises, and lecture. Topics include the use of standard computer software programs for administrative, analytical, and instructional tasks.
Credit Recommendation: In the lower-division baccalaureate/associate degree category, 3 semester hours in computer literacy (8/02).

AR-1402-0193

Computer Network Defense

Course Number: 921-640, Phase 3.
Location: Reserve Readiness Training Center, Fort McCoy, WI.
Length: 2–3 weeks (94 hours).
Exhibit Dates: 4/01–Present.
Learning Outcomes: Upon completion of the course, the student will be able to demonstrate an understanding of the legal and ethical issues of intrusion defense, demonstrate an understanding of common computer security flaws, implement encryption in the network, execute post-intrusion actions, demonstrate an understanding of the methodologies of computer intrusions, and utilize network perimeter defense tools.
Instruction: Audiovisual materials, practical and classroom exercises, discussion, laboratory, lecture, final examination, and web-employed training. Topics include legal and ethical issues of intrusion defense, common security flaws, implementation encryption, post-intrusion actions, methodologies of computer intrusions, and network perimeter defense tools.
Credit Recommendation: In the lower-division baccalaureate/associate degree category, 3 semester hours in computer security (11/02).

AR-1402-0194

Information Systems Security Monitoring Certification

Course Number: 3B-F17/232ASI2G.
Location: Intelligence Center and School, Fort Huachuca, AZ.
Length: 3–4 weeks (120 hours).
Exhibit Dates: 3/02–Present.
Learning Outcomes: Upon completion of the course, the student will be able to supervise information systems, analyze and interpret information system vulnerabilities, and recommend corrective measures to be taken when system vulnerabilities are present.
Instruction: Practical exercises and case studies. Topics include analyzing telecommunications, pager, cellular, e-mail, telephone, radio, information systems, and legal requirements of information systems.
Credit Recommendation: In the lower-division baccalaureate/associate degree category, 1 semester hour in information systems analysis or information systems supervision (12/02).

AR-1402-0195

EW/Intercept Systems Repairer Basic Noncommissioned Officer (BNCOC)

Course Number: 102-33W30, Phase 2.
Location: Intelligence Center and School, Fort Huachuca, AZ.
Length: 10 weeks (361 hours).
Exhibit Dates: 10/02–Present.
Learning Outcomes: Upon completion of the course, the student will be able to maintain fiber optic cables and connectors, Solaris and Windows NT operating systems, configure and troubleshoot LAN's consisting of Solaris and Windows NT.
Instruction: Audiovisual materials, practical and classroom exercises, discussion, laboratory, and computer-based training. Topics include fiber optic cable troubleshooting, SUN workstation operation and repair, configuration of SUN and Solaris workstations, computer network components and procedures, and security protocols.
Credit Recommendation: In the lower-division baccalaureate/associate degree category, 2 semester hours in maintenance and troubleshooting of computer workstations and networks (12/02).

AR-1402-0196

IBM Workstation Maintenance

Course Number: 102-F119.
Location: Intelligence Center and School, Fort Huachuca, AZ.
Length: 5 weeks (196 hours).
Exhibit Dates: 10/99–Present.
Learning Outcomes: Upon completion of the course, the student will be able to identify IBM workstation configuration; perform power up/down procedures; operate, configure, and troubleshoot IBM workstations in stand alone and networked environments; and distinguish between software and hardware faults.
Instruction: Audiovisual materials, practical and classroom exercises, laboratory, and lecture. Topics include operation, configuration, and troubleshooting IBM workstations; repair workstations; operation of workstations in stand alone and networked systems.
Credit Recommendation: In the lower-division baccalaureate/associate degree category, 3 semester hours computer repair and maintenance (12/02).

AR-1402-0197

SUN Server and Workstation Maintenance

Course Number: 102-F121.
Location: Intelligence Center and School, Fort Huachuca, AZ.
Length: 5 weeks (196 hours).
Exhibit Dates: 1/02–Present.
Learning Outcomes: Upon completion of the course, the student will be able to identify SUN workstation configurations; perform power up/down procedures; operate, configure, and troubleshoot SUN RISC/UNIX workstations in stand alone and networked environments.
Instruction: Audiovisual materials, practical and classroom exercises, laboratory, and lecture. Topics include operation, configuration, and troubleshooting SUN RISC/UNIX servers; repair workstations; and prepare SUN services for operation in stand-alone mode.
Credit Recommendation: In the lower-division baccalaureate/associate degree category, 3 semester hours in computer maintenance (12/02).

AR-1402-0198

Network Management Technician Warrant Officer Basic (WOBC)

Course Number: 4C-250N.
Location: Signal School, Fort Gordon, GA.
Length: 19 weeks (690 hours).
Exhibit Dates: 1/02–Present.
Learning Outcomes: Upon completion of the course, the student will be able to manage LAN and WAN networks and manage satellite communication systems.
Instruction: Audio visual materials, practical exercises, discussion, lecture, and computer-based training. Topics include CISCO routers, network security, network control, and satellite communications.
Credit Recommendation: In the lower-division baccalaureate/associate degree category, 12 semester hours in CISCO router networks, 3 in network security, and 3 in communication systems (2/03).

AR-1402-0199

System Administrator/Network Manager Security

Course Number: 7E-F66/531-F21.
Location: Signal School, Fort Gordon, GA.
Length: 2 weeks (80 hours).
Exhibit Dates: 10/01–Present.
Learning Outcomes: Upon completion of the course, the student will be able to recognize and defeat security threats to a computer network made up of individual computers running under UNIX or M52K, hosted by a web server

running under MS Internet software, and operate and maintain the web server.

Instruction: Audio visual materials, practical exercises, discussion, lecture, and computer-based training. Topics include UNIX and MS2K operating system security, system security administration, network security management, network management, network security software, and encryption.

Credit Recommendation: In the lower-division baccalaureate/associate degree category, 3 semester hours in network security (2/03).

AR-1402-0200

TATS Network Management Technician-WOAC

Course Number: 4-11-C32-250N.

Location: Signal School, Fort Gordon, GA.

Length: 16 weeks (616 hours).

Exhibit Dates: 3/03–Present.

Learning Outcomes: Upon completion of the course, the student will be able to configure, operate and manage a wide area network (WAN) system using CISCO routers and Windows 2000/UNIX based workstations.

Instruction: Audio visual materials, practical exercises, discussion, lecture, and computer-based training. Topics include CISCO routers, Windows 2000/UNIX, network security, network management, and network control.

Credit Recommendation: In the lower-division baccalaureate/associate degree category, 12 semester hours in CISCO router networks, 2 in personal computer operations systems, 3 in network management, and 3 in network security (2/03).

AR-1402-0201

Tactical Automated Network Technician Warrant Officer Advanced

Course Number: 4-11-C32-250B.

Location: Signal School, Fort Gordon, GA.

Length: 14 weeks (504 hours).

Exhibit Dates: 6/98–Present.

Learning Outcomes: Upon completion of the course, the student will be able to demonstrate an understanding of a computer operating system, basic networking fundamentals, local area networks, routers, network security issues, military communication, transmission, switching systems, and network management procedures.

Instruction: Practical and classroom exercises. Topics include personal computer hardware, operating systems, networking fundamentals, local area networks, routers, network security, military communication, transmission, switching systems, and network management.

Credit Recommendation: In the lower-division baccalaureate/associate degree category, 2 semester hours in computer fundamentals, 2 in networking fundamentals, 2 in local area networks, and 1 in network security (2/03).

AR-1402-0202

Signal Systems Support Technician Warrant Officer Basic (WOB)

Course Number: 7E-254A, Phase 2.

Location: Signal School, Fort Gordon, GA.

Length: 22 weeks (1532 hours).

Exhibit Dates: 10/02–Present.

Learning Outcomes: Upon completion of the course, the student will be able to demonstrate an understanding of CISCO router networks, and manage and control the security of networked systems.

Instruction: Practical, classroom exercises, and computer-based training. Topics include CISCO router networks, operating systems, network systems security, and network management.

Credit Recommendation: In the lower-division baccalaureate/associate degree category, 12 semester hours in CISCO router networks, 2 in network systems management, and 2 in operating systems (2/03).

AR-1402-0203

Joint Revised Battlefield Electronic Communications (RBECS)

Course Number: 4C-F58/201-F16.

Location: Signal School, Fort Gordon, GA.

Length: 2 weeks (78 hours).

Exhibit Dates: 10/96–Present.

Learning Outcomes: Upon completion of the course, the student will be able to operate and manage a computer based military communication system and train others to do the same.

Instruction: Audio visual materials, practical exercises, discussion, lecture, and computer-based training. Topics include single channel radio system and personal computer hardware and software.

Credit Recommendation: In the lower-division baccalaureate/associate degree category, 1 semester hour in introduction to computers (2/03).

AR-1402-0205

Joint Automated COMMS Engineering System

Course Number: 4C-F66/260-F21, Phase 1.

Location: Signal School, Fort Gordon, GA.

Length: 2 weeks (80 hours).

Exhibit Dates: 10/02–Present.

Learning Outcomes: Upon completion of the course, the student will be able to operate and manage a Windows NT-based program used to manage a communications network.

Instruction: Topics include automated communications engineering software, communications network planning, and radio networks.

Credit Recommendation: In the lower-division baccalaureate/associate degree category, 2 semester hours in introduction to computers (2/03).

AR-1402-0207

Information Systems Operations Leveler

Course Number: 7E-F70, Phase 1.

Location: Signal School, Fort Gordon, GA.

Length: 10 weeks (385 hours).

Exhibit Dates: 9/00–Present.

Learning Outcomes: Upon completion of the course, the student will be able to perform signal organizations and hierarchy, telecommunications and electronics fundamentals, Army and Joint information systems, planning, implementation and management of data networks (computer and computer networks).

Instruction: Topics include signal hierarchy, electronic fundamentals, data communications, automation, and communications network employment.

Credit Recommendation: In the lower-division baccalaureate/associate degree category, 12 semester hours in CISCO certified network associate academy, 3 in Windows workstation, 3 in Windows server, 3 in data communications, and 1 in electronics fundamentals (2/03).

AR-1402-0208

Mounted Leader Digital Training

Course Number: 2E-F195/020-F27.

Location: Armor School, Fort Knox, KY.

Length: 3 weeks (116 hours).

Exhibit Dates: 6/03–Present.

Learning Outcomes: Upon completion of the course, the student will be able to utilize digital information and decision making tools.

Instruction: Practical and classroom exercises, lecture, and computer-based training. Topics include division XXI equipment usage, digitized command posts, tactical internets, and battle control systems.

Credit Recommendation: In the lower-division baccalaureate/associate degree category, 3 semester hours in introduction to computers (6/03).

AR-1402-0209

Advanced Field Artillery Tactical Data System Command and Staff

Course Number: 2E-F110/041-F7.

Location: Field Artillery School, Fort Sill, OK.

Length: 3 weeks (108 hours).

Exhibit Dates: 10/01–Present.

Learning Outcomes: Upon completion of the course, the student will be able to operate the Advanced Field Artillery Tactical Data Systems (AFATDS).

Instruction: Practical and classroom exercises, lecture, and computer-based training. Topics include database operations, computer operation fundamentals, AFATDS operations, and AFATDS communications.

Credit Recommendation: In the lower-division baccalaureate/associate degree category, 2 semester hours in computer fundamentals and 1 in database systems (10/03).

AR-1402-0210

TATS Field Artillery Tactical Data System Specialist Basic Noncommissioned Officer

Course Number: 250-13D30 (F).

Location: Field Artillery School, Fort Sill, OK.

Length: 5–6 weeks (221 hours).

Exhibit Dates: 10/01–Present.

Learning Outcomes: Upon completion of the course, the student will be able to lead, train, and direct an Advanced Field Artillery Data Systems (AFATDS)-equipped section in executing the delivery of artillery fire.

Instruction: Practical exercises, lecture, and computer-based training. Topics include military branch history, manual computation of muzzle velocities, safety procedures, AFATDS setup and operations, and mission planning.

Credit Recommendation: In the lower-division baccalaureate/associate degree category, 1 semester hour in basic military science and 2 in computer fundamentals (10/03).

AR-1402-0211

TATS Field Artillery Tactical Data Systems Specialist Basic Noncommissioned Officer

Course Number: 061-13D30 (F).

Location: Field Artillery School, Fort Sill, OK.

Length: 166 hours.

Exhibit Dates: 10/01–Present.

Learning Outcomes: Upon completion of the course, the student will be able to lead, train, and direct an Advanced Field Artillery Data System (AFTADS)-equipped section in executing the delivery of artillery fire.

Instruction: Practical exercises, lecture, and computer-based training. Topics include, military branch history, manual computation of muzzle velocities, safety procedures, AFTADS setup and operations, and mission planning.

Credit Recommendation: In the lower-division baccalaureate/associate degree category, 1 semester hour in basic military science and 1 in computer fundamentals (10/03).

AR-1402-0212

Light Tactical Fire (LTACFIRE) Operator

Course Number: 2E-F180/250-F39.

Location: Field Artillery School, Fort Sill, OK.

Length: 2 weeks (76–77 hours).

Exhibit Dates: 12/00–Present.

Learning Outcomes: Upon completion of the course, the student will be able to operate the Light Tactical Fire (LTACFIRE) and (IFAS) systems.

Instruction: Audiovisual materials, practical exercises, discussion, laboratory, and computer-based training. Topics include computer communications, fire planning parameters, battlefield processing of data, and conducting tactical fire control.

Credit Recommendation: In the vocational certificate category, 2 semester hours in computer operations. In the lower-division baccalaureate/associate degree category, 1 semester hour in surveying (10/03).

AR-1402-0213

TATS Field Artillery Tactical Data Systems Specialist ANCOC

Course Number: 250-13D40 (F).

Location: Field Artillery School, Fort Sill, OK.

Length: 3 weeks (144 hours).

Exhibit Dates: 10/01–Present.

Learning Outcomes: Upon completion of the course, the student will be able to supervise the operations of a fire control elements equipped with Advanced Field Artillery Data System (AFATDS) equipment.

Instruction: Practical exercises, lectures, and computer-based training. Topics include unit history, muzzle velocity computations, AFATDS, fire support, and operational moves planning.

Credit Recommendation: In the lower-division baccalaureate/associate degree category, 2 semester hours in basic military science and 2 in computer fundamentals (10/03).

AR-1402-0214

TATS Field Artillery Tactical Data Systems Specialist ANCOC

Course Number: 061-13D40 (F).

Location: Reserve Schools, Multiple locations.

Length: 3 weeks (144 hours).

Exhibit Dates: 10/01–Present.

Learning Outcomes: Upon completion of the course, the student will be able to supervise the operations of a fire control elements-equipped with Advanced Field Artillery Tactical Data System (AFATDS) equipment.

Instruction: Practical exercises, lectures, and computer-based training. Topics include unit history, muzzle velocity computations, Advanced Field Artillery Tactical Data System (AFATDS) systems administration, fire support, and operational moves planning.

Credit Recommendation: In the lower-division baccalaureate/associate degree category, 2 semester hours in basic military science and 2 in computer fundamentals (10/03).

AR-1402-0215

TATS Advanced Field Artillery Tactical Data System (AFATDS)

Course Number: 2E-F182/250-ASIY1/F9 (F), Phase 1.

Location: Field Artillery School, Fort Sill, OK.

Length: 3 weeks (120 hours).

Exhibit Dates: 8/03–Present.

Learning Outcomes: Upon completion of the course, the student (MOS 13C Soldier) will be able to transition to MOS 13D to perform manual gunnery and safety procedures.

Instruction: Topics include establishing of communications, building and maintaining a database, tactical and technical fire control, establishing plans, and planning movement operations, and performing various administrative functions, including the interoperability between DOD systems this phase specifically designed to train MOS 13C on manual gunnery and safety computation.

Credit Recommendation: In the vocational certificate category, 1 semester hour in radar operation and 1 in computer operation. In the lower-division baccalaureate/associate degree category, 3 semester hours in surveying and 2 in astronomy (10/03).

AR-1402-0216

TATS Advanced Field Artillery Tactical Data System (AFATDS)

Course Number: 2E-F182/250-ASIY1/F9 (F), Phase 2.

Location: Field Artillery School, Fort Sill, OK.

Length: 2 weeks (92 hours).

Exhibit Dates: 8/03–Present.

Learning Outcomes: Upon completion of the course, the student will be able to conduct manual gunnery and safety procedures, operate equipment used by field artillery tactical data system specialist, and conduct fire support planning and execution tasks. MOS 13C personnel can be transitioned to MOS 13D upon completion of this course.

Instruction: Audiovisual materials, practical and classroom exercises, discussions, laboratory, and computer-based training. Topics include how to establish communications, how to build and maintain a database, tactical and technical fire control, and how to plan movement operations and administrative functions, including the interoperability between DOD systems.

Credit Recommendation: In the vocational certificate category, 1 semester hour in radar operation and 1 in computer operation. In the lower-division baccalaureate/associate degree category, 3 semester hours in surveying and 1 in astronomy (10/03).

AR-1402-0217

TATS Advanced Field Artillery Tactical Data System (AFATDS)

Course Number: 2E-F182/250-ASIY1/F9 (F).

Location: Field Artillery School, Fort Sill, OK.

Length: 1–2 weeks (64 hours).

Exhibit Dates: 8/03–Present.

Learning Outcomes: Upon completion of the course, the student will be able to conduct manual gunnery and safety procedures, operate equipment used by field artillery tactical data system specialist, and conduct fire support planning and execution tasks. MOS 13C personnel can be transitioned to MOS 13D upon completion of this course.

Instruction: Audiovisual materials, practical and classroom exercises, discussions, laboratory, and computer-based training. Topics include how to establish communications, how to build and maintain a database, tactical and technical fire control, how to plan movement operations and administrative functions, including the interoperability between DOD systems.

Credit Recommendation: In the vocational certificate category, 1 semester hour in radar operation and 1 in computer operation. In the lower-division baccalaureate/associate degree category, 3 semester hours in surveying and in astronomy (10/03).

AR-1402-0218

Telemedicine Equipment Maintenance Course

Course Number: 300-F16.

Location: Academy of Health Sciences, Fort Sam Houston, TX.

Length: 3 weeks (120 hours).

Exhibit Dates: 5/00–Present.

Learning Outcomes: Upon completion of the course, the student will be able to discuss the functions, applications and benefits of picture archiving and communication system; demonstrate ability to interact/network among MDIS, UNIX, Solaris, and MacIntosh Operating Systems; and perform MDIS backup and recovery, calibration, and malfunction troubleshooting.

Instruction: Practical and classroom exercises, discussion, and lecture. Topics include overview of picture archiving and communications system, MDIS systems with HUB and spoke configuration; computer networking, Solaris Operating Systems and MacIntosh Interface, client/server database concepts and UNIX interaction; and MDIS troubleshooting, system backup and calibration.

Credit Recommendation: In the lower-division baccalaureate/associate degree category, 3 semester hours in computer networking. In the upper-division baccalaureate category, 3 semester hours in teleradiation equipment maintenance (11/03).

AR-1402-0219

Total Army Training Systems Field Artillery Tactical Data System Specialist

Course Number: 13D10-OSUT.

Location: Field Artillery School, Fort Sill, OK.

Length: 14 weeks (287 hours).

Exhibit Dates: 1/03–Present.

Learning Outcomes: Upon completion of the course, the student will be able to operate a field artillery tactical data system and communications system.

Instruction: Practical and classroom exercises, lectures, and computer-based training. Topics include military tactics, data communications, Advanced Field Artillery Tactical Data System (AFATADS) operation, and manual fire direction.

Credit Recommendation: In the lower-division baccalaureate/associate degree category, 1 semester hour in basic military science and 3 in computer fundamentals (10/03).

AR-1403-0002

Legal Specialist
(Legal Clerk)

Course Number: 512-71D10; 512-71D10 (ST).

Location: Institute for Personnel and Resource Management, Fort Benjamin Harrison, IN.

Length: 10 weeks (345–395 hours).

Exhibit Dates: 8/82–6/90.

Objectives: After 6/90 see AR-1403-0017. To provide enlisted personnel with introductory concepts of military administrative law and Army claims systems with emphasis on practical processing skills and procedures.

Instruction: Lectures cover Army claims system and board proceedings, including investigative boards, Article 15, nonjudicial punishment, identification and preparation of charges and specifications for summary court martial and special court martial proceedings, and post-trial actions. Practical exercises cover the use of basic forms in the above areas, including drafting documents, applying rules of law to facts, digesting testimony, and performing preliminary legal research.

Credit Recommendation: In the lower-division baccalaureate/associate degree category, 3 semester hours in typing and 4 as an elective in paralegal training (1/89).

Related Occupations: 71B; 71D.

AR-1403-0007

Administrative Specialist

Course Number: 510-71L10.

Location: Administration School, Fort Jackson, SC.

Length: Self-paced, 4–10 weeks (284–327 hours).

Exhibit Dates: 12/79–2/96.

Objectives: After 2/96 see AR-1403-0021. To prepare selected enlisted personnel to function as administrative clerks.

Instruction: Lectures and practical exercise include typing and assembling various types of correspondence, establishing and maintaining files, keeping records, and processing classified material. Top students will enter fast track, an additional 36 hours in advanced, in-depth study.

Credit Recommendation: In the lower-division baccalaureate/associate degree category, 3 semester hours in typing, 1 in filing, and 2 in business communications (9/89).

Related Occupations: 71L.

AR-1403-0010

Patient Administration Specialist

Course Number: ***Version 1:*** 513-71G10. *Version 2:* 513-91G10.

Location: Academy of Health Sciences, Fort Sam Houston, TX.

Length: ***Version 1:*** 7–8 weeks (264–284 hours). *Version 2:* 6–7 weeks (291 hours).

Exhibit Dates: ***Version 1:*** 9/87–9/99. *Version 2:* 10/99–Present.

Learning Outcomes: ***Version 1:*** Upon completion of the course, the student will be able to provide a wide range of patient support services in a health facility, including patient management, maintenance of patient records, and automated hospital support systems; use medical terminology; and prepare medical statistical reports. *Version 2:* Upon completion of the course, the student will be able to provide a wide range of patient support services in a health facility, including patient management, maintenance of patient records, and automated hospital support systems; use medical terminology; and prepare medical statistical reports.

Instruction: ***Version 1:*** Lectures and classroom exercises cover preparation, processing, review of medical records and reports, and general administrative procedures. *Version 2:* Discussion, classroom exercise, practical exercises, and lectures. Topics will cover preparation processing, review of medical records and reports, and general administrative procedures.

Credit Recommendation: ***Version 1:*** In the lower-division baccalaureate/associate degree category, 8 semester hours in medical records technology (8/93). *Version 2:* In the lower-division baccalaureate/associate degree category, 3 semester hours in health care management, 2 in health care coding, 2 in patient administration clinical practicum, and 2 in medical terminology (6/02).

Related Occupations: 71G.

AR-1403-0011

Unit/Organization Supply Procedures

Course Number: LOG 73.

Location: Combined Arms Training Center, Vilseck, W. Germany.

Length: 2 weeks (71 hours).

Exhibit Dates: 8/87–10/98.

Learning Outcomes: Upon completion of the course, the student will be able to review Army supply publications, supply records, forms, and inventory control records; recognize when an inventory should be conducted; conduct an inventory; apply the inventory concepts of control, balance, and centralization/decentralization; and apply inventory concepts to ammunition, clothing, equipment, petroleum, oil and lubricants, meals (ready to eat), and arms.

Instruction: Lectures and practical exercises use actual forms and records.

Credit Recommendation: In the lower-division baccalaureate/associate degree category, 1 semester hour in supply management and 1 in clerical procedures (10/88).

AR-1403-0013

Personnel Information System Management Specialist

Course Number: ***Version 1:*** 500-75F10 (ST); 500-75F10; 500-75F10/75F10 (ST). *Version 2:* 500-71F10.

Location: ***Version 1:*** Soldier Support Institute, Fort Jackson, SC; Soldier Support Institute, Fort Benjamin Harrison, IN. *Version 2:* Adjutant General School, Fort Jackson, SC.

Length: ***Version 1:*** 8–11 weeks (240–370 hours). *Version 2:* 8 weeks (308 hours).

Exhibit Dates: ***Version 1:*** 10/85–4/98. *Version 2:* 5/98–Present.

Learning Outcomes: ***Version 1:*** Upon completion of the course, the student will be able to identify the components of a personnel information system, identify errors in the system and take corrective action, enter data, and review personnel system output reports. *Version 2:* Upon completion of the course, the student will be able to enter, retrieve, and organize personnel data.

Instruction: ***Version 1:*** Classroom and practical exercises cover the use and operation of a microcomputer and entering, processing, and retrieving data in a personnel information system. *Version 2:* Classroom and practical exercises cover the use and operation of a microcomputer and entering, processing, and retrieving data in a personnel information system.

Credit Recommendation: ***Version 1:*** In the vocational certificate category, 1 semester hour in personnel record keeping. In the lower-division baccalaureate/associate degree category, 1 semester hour in microcomputer applications and 1 in keyboarding (9/96). *Version 2:* In the lower-division baccalaureate/associate degree category, 3 semester hours in records management, or database management, and 3 in introduction to computer (3/02).

Related Occupations: 71F; 75F.

AR-1403-0014

Personnel Service Center Basic Noncommissioned Officer (NCO)

Course Number: 500-75C/D/E30.

Location: Soldier Support Institute, Fort Benjamin Harrison, IN.

Length: 7 weeks (263 hours).

Exhibit Dates: 1/89–9/91.

Learning Outcomes: Upon completion of the course, the student will be able to review manning reports; prepare requisitions; determine eligibility for promotion; perform appropriate actions regarding pay and benefits and changes in personnel classification; and execute specific applications in word processing, graphics, and spreadsheets.

Instruction: Instruction includes extensive classroom lectures on personnel record supervision, review of personnel benefits, training and leadership skills. Course also includes experience in word processing, graphics, and spreadsheets as applied to personnel activities. Brief attention is given to counseling.

Credit Recommendation: In the lower-division baccalaureate/associate degree category, 3 semester hours in office administration and 1 in filing and records control (12/88).

Related Occupations: 75C; 75D; 75E.

AR-1403-0016

Unit Clerk

Course Number: PEC-P-UC1.
Location: Professional Education Center, Cp. Joseph Robinson, North Little Rock, AR.
Length: 2–3 weeks (97 hours).
Exhibit Dates: 3/90–5/95.
Learning Outcomes: Upon completion of the course, the student will be able to perform basic personnel and administrative actions with emphasis on the preparation of personnel forms.
Instruction: Lecture and in-class exercises are designed to prepare the student to complete forms addressing administrative, financial, and information management actions.
Credit Recommendation: In the vocational certificate category, 2 semester hours in business systems and procedures (7/90).

AR-1403-0017

1. Legal Specialist
2. Legal Specialist
3. Paralegal Specialist
(71D Advanced Individual Training)

Course Number: ***Version 1:*** 512-71D10. *Version 2:* 512-71D10. *Version 3:* 512-27D10.
Location: ***Version 1:*** Adjutant General School, Fort Benjamin Harrison, IN. *Version 2:* Soldier Support Center, Fort Jackson, SC. *Version 3:* Soldier Support Center, Fort Jackson, SC.
Length: ***Version 1:*** 9 weeks (365 hours). *Version 2:* 8–9 weeks (310 hours). *Version 3:* 10–11 weeks (328–383 hours).
Exhibit Dates: ***Version 1:*** 7/90–9/95. *Version 2:* 10/95–4/01. *Version 3:* 5/01–Present.
Learning Outcomes: ***Version 1:*** Before 7/90 see AR-1403-0002. Upon completion of the course, the student will be able to apply rules of law within the military justice system, prepare legal documents using word processing software, and use military justice terminology in proper military format. *Version 2:* Upon completion of the course, the student will be able to apply rules of law within the military justice system, summarize testimony, prepare legal documents using word processing software, and use military terminology in proper military law format. *Version 3:* Upon completion of the course, the student will be able to apply rules of law within the military justice system, summarize testimony, prepare legal documents using word processing software, and use military terminology in proper military law format.
Instruction: ***Version 1:*** Lectures and discussions cover concepts of Army administrative law, Army claims systems, board proceedings, nonjudicial proceedings, and military separations. Practical exercises include keyboarding, using word processing software and hardware, drafting documents, and applying rules of law. *Version 2:* Lectures and discussions cover concepts of Army administrative law, Army claims systems, board proceedings, nonjudicial proceedings, and military separations. Practical exercises include keyboarding, using word processing software and hardware, drafting documents, and applying rules of law. *Version 3:* Lectures and discussions cover concepts of Army administrative law, Army claims systems, board proceedings, nonjudicial proceedings, and military separations. Practical exercises include keyboarding, using word processing software and hardware, drafting documents, and applying rules of law.
Credit Recommendation: ***Version 1:*** In the lower-division baccalaureate/associate degree category, 1 semester hour in keyboarding, 2 in word processing, 2 in microcomputer applications, and 2 in legal writing (12/91). *Version 2:* In the lower-division baccalaureate/associate degree category, 1 semester hour in keyboarding, 1 in word processing, 2 in law office computer applications, and 3 in legal writing (9/97). *Version 3:* In the lower-division baccalaureate/associate degree category, 3 semester hours in word processing, 2 in law office computer application, and 3 in legal writing (5/02).
Related Occupations: 27D; 71D.

AR-1403-0018

Reserve Component Reconfigured (RC3)
Personnel Services Specialist
Phase 2

Course Number: 805C-75H10, Phase 2.
Location: Soldier Support Institute, Fort Jackson, SC.
Length: 3 weeks (93 hours).
Exhibit Dates: 3/97–Present.
Learning Outcomes: Upon completion of the course, the student will be able to perform selected clerical operations in a personnel office related to classification/reclassification actions, transfers, personnel records, and office administration.
Instruction: Lectures, class discussions, and examinations relates to processing, promotion rosters, promotion work sheets, evaluation reports, personnel records, casualty records, and selected functions of an PC-based personnel records system.
Credit Recommendation: In the lower-division baccalaureate/associate degree category, 3 semester hours in office procedures, clerical procedures, or recording keeping (9/97).
Related Occupations: 75H.

AR-1403-0019

Reserve Component Reconfigured (RC3)
Personnel Services Specialist
Phase 1

Course Number: 805C-75H10, Phase 1.
Location: Soldier Support Institute, Fort Jackson, SC.
Length: 2 weeks (72 hours).
Exhibit Dates: 3/97–Present.
Learning Outcomes: Upon completion of the course, the student will be able to perform selected clerical operations in a personnel office related to classification/reclassification actions, transfers, personnel records, and office administration.
Instruction: Lectures, class discussions, and examinations relate to processing voluntary and involuntary reclassifications, reassignments, military records, actions related to personnel affairs, preparation of dependent ID cards, and discharge actions.
Credit Recommendation: In the lower-division baccalaureate/associate degree category, 2 semester hours in office procedures, clerical procedures, or record keeping (9/97).
Related Occupations: 75H.

AR-1403-0020

Reserve Component Reconfigured
Personnel Administration
Specialist

Course Number: 805C-75B10.
Location: Soldier Support Institute, Fort Jackson, SC.
Length: 3 weeks (97 hours).
Exhibit Dates: 1/97–Present.
Learning Outcomes: Upon completion of the course, the student will be able to perform selected clerical operations in a personnel office related to classification/reclassification actions, transfers, personnel records, and office administration.
Instruction: Lectures, class discussions, and examinations relate to processing personnel records, file organization and maintenance, preparing staff changes and setup, and performing selected functions of PC-based personnel records system.
Credit Recommendation: In the lower-division baccalaureate/associate degree category, 3 semester hours in office clerical procedures or record keeping (9/97).
Related Occupations: 75B.

AR-1403-0021

Administrative Specialist

Course Number: ***Version 1:*** 510-71L10. *Version 2:* 510-71L10(F). *Version 3:* 510-42L10.
Location: ***Version 1:*** Soldier Support Institute, Fort Jackson, SC. *Version 2:* Soldier Support Center, Fort Jackson, SC. *Version 3:* Soldier Support Center, Fort Jackson, SC.
Length: ***Version 1:*** 5 weeks (224 hours). *Version 2:* 5 weeks (148 hours). *Version 3:* 5 weeks (224 hours).
Exhibit Dates: ***Version 1:*** 3/96–9/00. *Version 2:* 10/00–9/02. *Version 3:* 10/02–Present.
Learning Outcomes: ***Version 1:*** Before 3/96 see AR-1403-0007. Upon completion of the course, the student will be able to keyboard and perform basic clerical procedures dealing with correspondence and files. *Version 2:* Upon completion of the course, the student will be able to perform basic word processing and basic records management. *Version 3:* Pending evaluation.
Instruction: ***Version 1:*** Methods of instruction include lecture and hands-on training. *Version 2:* Topics covered include database management and information management. *Version 3:* Pending evaluation.
Credit Recommendation: ***Version 1:*** In the lower-division baccalaureate/associate degree category, 2 semester hours in keyboarding and 2 in clerical procedures (9/97). *Version 2:* In the lower-division baccalaureate/associate degree category, 3 semester hours in information management (3/02). *Version 3:* Pending evaluation.
Related Occupations: 42L.

AR-1403-0022

Unit Clerk

Course Number: FT-010.
Location: Professional Education Center, Cp. Joseph Robinson, North Little Rock, AR.
Length: 2–3 weeks (98–99 hours).
Exhibit Dates: 6/95–Present.
Learning Outcomes: Upon completion of the course, the student will be able to prepare various personnel reports and correspondence; prepare and process personnel action documents including medical care, direct deposit, and federal and state tax forms; compute/verify pay rates; and prepare sworn statements and wellness/fitness forms.

Instruction: Course methods include class presentations, discussions, and a final examination. Course emphasizes completing and checking various documents for accuracy and correctness and maintaining files and records.

Credit Recommendation: In the lower-division baccalaureate/associate degree category, 3 semester hours in personnel record keeping (6/98).

AR-1403-0023

Personnel Sergeant

Course Number: FT-011.

Location: Professional Education Center, Cp. Joseph Robinson, North Little Rock, AR.

Length: 2 weeks (62 hours).

Exhibit Dates: 11/96–Present.

Learning Outcomes: Upon completion of the course, the student will be able to supervise the preparation and correctness of various personnel documents including wellness/fitness reports, correspondence, employee performance evaluations, re-enlistment processing, and line of data investigations.

Instruction: Class presentations, discussions, and final examination provide an emphasis on the preparation of various personnel action documents for completeness and accuracy.

Credit Recommendation: In the lower-division baccalaureate/associate degree category, 2 semester hours in office procedures/supervision (6/98).

AR-1403-0024

Total Army Training System (TATS) Postal Operations, Phase 2

Course Number: 510-AS1F5 (71L10) (F), Phase 2.

Location: Adjutant General School, Fort Jackson, SC.

Length: 2 weeks (77–78 hours).

Exhibit Dates: 10/98–Present.

Learning Outcomes: Upon completion if the course, the student will be able to prepare money order transactions, operate an integrated retail terminal, and process registered mail.

Instruction: Methods of instruction include conferences and hardware/software practical applications.

Credit Recommendation: In the lower-division baccalaureate/associate degree category, 1 semester hour in office procedures (3/02).

Related Occupations: 71L.

AR-1403-0025

Patient Administration Specialist

Course Number: 513-91G10 (RC), Phase 2.

Location: Academy of Health Sciences, Fort Sam Houston, TX.

Length: 2 weeks (75 hours).

Exhibit Dates: 1/02–Present.

Learning Outcomes: Upon completion of the course, the student will be able to provide a wide range of patient support services in a health facility, including patient management, maintenance of patient records, and automated hospital support systems; use medical terminology; and prepare medical statistical reports.

Instruction: Practical and classroom exercises, discussion, laboratory, and lecture. Topics include fundamentals of computers in health information administration, automated patient administration support systems, medical statistical reporting, vital statistics processing, and patient dispositions.

Credit Recommendation: In the upper-division baccalaureate category, 3 semester hours in health information systems directed practice (11/03).

AR-1403-0026

Patient Administration Specialist

Course Number: 513-91G10 (RC), Phase 1.

Location: Academy of Health Sciences, Fort Sam Houston, TX.

Length: 3–4 weeks (138 hours).

Exhibit Dates: 6/03–Present.

Learning Outcomes: Upon completion of the course, the student will be able to provide administrative services that support health care delivery, and initiate, maintain, archive, and retrieve medical records.

Instruction: Audiovisual materials, computer-based training, practical and classroom exercises, discussion, and lecture. Topics include ICD-9-CM coding, patient accountability, medical records administration, medical terminology, and storage and retrieval systems.

Credit Recommendation: In the lower-division baccalaureate/associate degree category, 3 semester hours in introduction to health information systems, 3 in health information systems (coding and classification), and 2 in medical terminology (11/03).

AR-1404-0001

3. Postal Operations
4. Total Army Training System (TATS) Postal Operations (Postal Operations)

Course Number: *Version 3:* 510-ASIF5. *Version 4:* 510-ASIF5 (F); 510-ASIF5.

Location: *Version 3:* Institute for Administration, Fort Benjamin Harrison, IN; Adjutant General School, Fort Benjamin Harrison, IN. *Version 4:* Adjutant General School, Fort Jackson, SC; Soldier Support Institute, Fort Jackson, SC.

Length: *Version 3:* 4–5 weeks (155–181 hours). *Version 4:* Self paced, 5 weeks (187 hours).

Exhibit Dates: *Version 3:* 10/88–7/94. *Version 4:* 8/94–9/97.

Objectives: *Version 3:* To provide officers and enlisted personnel with the fundamental techniques and procedures of military post office management. *Version 4:* After 9/97 see AR-1405-0279. Upon completion of the course, the student should be able to process domestic and international mail; requisition, safeguard, and issue stamps, money orders, supplies and equipment, maintain and audit postal accounts, and operate a postal service center and integrated retail terminals.

Instruction: *Version 3:* Lectures cover postal organization, facilities, and operations and practical experience cover mail handling. *Version 4:* Lectures cover postal organization, facilities, operations, and practical experience cover mail handling.

Credit Recommendation: *Version 3:* In the vocational certificate category, 3 semester hours in postal operations (6/92). *Version 4:* In the vocational certificate category, 3 semester hours in postal operations (9/98).

Related Occupations: 71F.

AR-1404-0006

Signal Officer Basic

Course Number: 4-11-C20.

Location: Signal School, Fort Gordon, GA.

Length: 19–20 weeks (725–814 hours).

Exhibit Dates: 2/85–9/92.

Objectives: After 9/92 see AR-1408-0209. To provide newly commissioned officers with the basic understanding of principles of different methods of communications at a tactical division communications center.

Instruction: Instruction covers leadership, personnel and operations management, technician electronics, supply operations, and military subjects.

Credit Recommendation: In the vocational certificate category, 6 semester hours in basic electricity and electronics. In the lower-division baccalaureate/associate degree category, 3 semester hours in business organization and management (6/90).

AR-1404-0009

1. Single Channel Radio Operator
2. Radio Operator-Maintainer
3. TATS Radio Operator-Maintainer (Radio Operator - Maintainer)
4. Radio Operator-Maintainer

Course Number: *Version 1:* 201-31C10. *Version 2:* 201-31C10. *Version 3:* 201-31C10 (F). *Version 4:* 201-31C10 (F).

Location: *Version 1:* Southeastern Signal School, Fort Gordon, GA; Signal School, Fort Gordon, GA. *Version 2:* Signal School, Fort Gordon, GA. *Version 3:* Signal School, Fort Gordon, GA. *Version 4:* Signal School, Fort Gordon, GA.

Length: *Version 1:* 12–14 weeks (437–477 hours). *Version 2:* 9–10 weeks (358 hours). *Version 3:* 9–10 weeks (358 hours). *Version 4:* 11–12 weeks (414 hours).

Exhibit Dates: *Version 1:* 6/83–9/95. *Version 2:* 10/95–1/98. *Version 3:* 2/98–9/02. *Version 4:* 10/02–Present.

Learning Outcomes: *Version 1:* Upon completion of the course, the student will be trained to install, operate, and maintain field radio teletypewriter sets and related equipment. *Version 2:* Upon completion of the course, the student will be trained to install, operate, and maintain field radio teletypewriter sets and related equipment. *Version 3:* Upon completion of the course, the student will be trained to install, operate, and maintain field radioteletype and radiotelephone communications systems. *Version 4:* Upon completion of the course, the student will be able to install, operate, and maintain field radioteletype and radio telephone communication systems.

Instruction: *Version 1:* Lectures and practical exercises cover teletypewriter set installation, operation, and maintenance, including keyboard operation, radiotelegraph procedures, and equipment familiarization. *Version 2:* Lectures, demonstrations, and practical exercises cover installation, operation, and maintenance of computerized teletypewriter set, including keyboard operation, data communications, radiotelegraph procedures, and equipment familiarization. *Version 3:* Lectures, demonstrations, and practical exercises cover installation, operation, and maintenance of computerized radiotelephone systems, including data communications and equipment familiarization. *Version 4:* Lectures, demonstrations, and practical exercises cover installation, operation, and maintenance of computerized radiotelephone systems, including data communications and equipment familiarization.

Credit Recommendation: *Version 1:* In the lower-division baccalaureate/associate degree category, 1 semester hour in typing (5/90). *Version 2:* In the lower-division baccalaureate/associate degree category, 1 semester hour in basic electronics, 6 in radio operation and maintenance, and 1 in data communications (2/97). *Version 3:* In the lower-division baccalaureate/associate degree category, 1 semester hour in basic electronics, 3 in radio operation and maintenance, and 1 in data communications systems (5/99). *Version 4:* In the lower-division baccalaureate/associate degree category, 1 semester hour in basic electronics, 3 in radio communications, 3 in troubleshoot/maintenance of radio equipment, and 1 in data communications (2/03).

Related Occupations: 05C; 31C.

AR-1404-0011

International Morse Code

Course Number: 231-ASIA4.

Location: Signal School, Fort Gordon, GA.

Length: 4–8 weeks (165–288 hours).

Exhibit Dates: 11/78–2/92.

Objectives: Provides student with knowledge and capability to send and receive international Morse code at a minimum speed of 15 words per minute.

Instruction: Course includes practice in sending and receiving international Morse code, military lettering, phonetic alphabet, and radio telegraph procedures.

Credit Recommendation: In the vocational certificate category, 3 semester hours in Morse code radio operator (1/94).

Related Occupations: 05B; 05C.

AR-1404-0029

AN/UGC-74 Operator

Course Number: UGC-3.

Location: Combined Arms Training Center, Flint Kaserne, Bad Toelz, Germany.

Length: 1–2 weeks (60–74 hours).

Exhibit Dates: 11/81–6/96.

Objectives: To train radio teletypewriter operators and combat telecommunications center operators in the operation of a new teletypewriter set.

Instruction: Topics include the purpose, use, and components of a teletypewriter set and editing, transmitting, and receiving messages using the teletypewriter set.

Credit Recommendation: Credit is not recommended because of the military-specific nature of the course (10/82).

Related Occupations: 05C; 72E.

AR-1404-0035

3. Advanced Noncommissioned Officer (NCO) Common Core
4. Advanced Noncommissioned Officer (NCO) Common Leader (Common Core)
5. Advanced Noncommissioned Officer (NCO) Common Core
6. Advanced Noncommissioned Officer (NCO) Common Core

Course Number: *Version 3:* ANCOC; 400-ANCOC. *Version 4:* ANCOC; 400-ANCOC; 400-ANCOC-MTA. *Version 5:* 600-ANCOC, Phase 1. *Version 6:* 600-ANCOC, Phase 1.

Location: *Version 3:* Installations, Continental US and overseas. *Version 4:* NCO Academies, US. *Version 5:* NCO Academy, Worldwide. *Version 6:* NCO Academy, Worldwide.

Length: *Version 3:* 1–2 weeks (57 hours). *Version 4:* 2 weeks (83 hours). *Version 5:* 2–3 weeks (78 hours). *Version 6:* 2 weeks (78 hours).

Exhibit Dates: *Version 3:* 2/89–3/94. *Version 4:* 4/94–9/00. *Version 5:* 10/00–9/01. *Version 6:* 10/01–Present.

Learning Outcomes: *Version 3:* Upon completion of the course the student will be able to provide mid-level supervision and leadership in units of up to 400 persons. *Version 4:* Upon completion of the course the student will be able to provide mid-level supervision and leadership at the operational level in units of up to 50 persons. *Version 5:* Upon completion of the course the student will be able to provide mid-level supervision and leadership at the operational level to platoon size units. *Version 6:* Upon completion of the course the student will be able to provide mid-level supervision and performance-oriented training at the operational level to groups up to platoon-sized groups into combat.

Instruction: *Version 3:* Course includes lectures, demonstrations, and performance exercises in motivation, training, morale, supervision, effective writing, oral communications, and listening skills. *Version 4:* Topics include development of mid-level skills in leadership and training and motivating personnel, building morale, supervising, using effective written and oral communications, using appropriate listening skills, and managing logistics and material. *Version 5:* Topics to include the development and application of the following skills: leadership, written and oral communication, health, professional, training, and warfighter documented by both written and practical examinations. *Version 6:* Audiovisual materials, Case studies, practical and classroom exercises, discussions, learner presentations, laboratory, lectures, and conferences. Topics to include the development and application of the following skills: leadership, written and oral communication, research, health, professional, training, and war fighter documented by both written and practical examinations.

Credit Recommendation: *Version 3:* In the lower-division baccalaureate/associate degree category, 1 semester hour in military science, and 1 in communication (4/90). *Version 4:* In the lower-division baccalaureate/associate degree category, 1 semester hour in military science and 1 in fundamentals of communication (6/95). *Version 5:* In the lower-division baccalaureate/associate degree category, 2 semester hours in military science, 1 in fundamentals of communication, and 1 in introduction to personnel management (7/02). *Version 6:* In the lower-division baccalaureate/associate degree category, 2 semester hours in leadership, 1 in fundamentals of communications, and 1 in introduction to personnel management (8/03).

AR-1404-0037

1. Single Channel Radio Operator Basic Noncommissioned Officer (NCO)
2. Radio Operator-Maintainer Basic Noncommissioned Officer (NCO)
3. Radio Operator-Maintainer Basic Noncommissioned Officer (NCO)

Course Number: *Version 1:* 201-31C20. *Version 2:* 201-31C30. *Version 3:* 201-31C30, Phase 2.

Location: Signal School, Fort Gordon, GA.

Length: *Version 1:* 6 weeks (216 hours). *Version 2:* 8 weeks (290 hours). *Version 3:* 4 weeks (180 hours).

Exhibit Dates: *Version 1:* 3/86–5/96. *Version 2:* 6/96–9/01. *Version 3:* 10/01–Present.

Learning Outcomes: *Version 1:* Upon completion of the course, the student will be able to supervise personnel in the installation, operation, and maintenance of radio and radioteletype equipment. *Version 2:* Upon completion of the course, the student will be able to supervise personnel in the installation, operation, and maintenance of radio and associated equipment. *Version 3:* Upon completion of the course, the student will be able to train and supervise subordinate in the installation, operation, and maintenance of military radio communication systems and LANS.

Instruction: *Version 1:* Lectures and practical exercises cover supervising the installation and maintenance of AM/single sideband radio and radioteletype systems. Course includes a common core of leadership subjects. *Version 2:* Lectures and practical exercises cover basic electronics and safety as well as supervision and inspection of AM/single sideband radios, generators, and related equipment. This course includes a common core of leadership subjects. *Version 3:* Audio visual materials, discussion, laboratory, lectures and practical exercises. Topics include tactical radio systems, tactical antenna systems, digital telephone networks, and local area networks.

Credit Recommendation: *Version 1:* In the vocational certificate category, 1 semester hour in AM and single sideband radio communications. In the lower-division baccalaureate/associate degree category, see AR-1406-0090 for common core credit (6/88). *Version 2:* In the lower-division baccalaureate/associate degree category, 3 semester hours in radio communications. See AR-1406-0090 for common core credit (2/03). *Version 3:* In the lower-division baccalaureate/associate degree category, 3 semester hours in radio communications and 3 in network fundamentals. See AR-1406-0090 for common core credit (2/03).

Related Occupations: 31C.

AR-1404-0038

Single Channel Radio Team Chief

Course Number: SIG 7.

Location: Combined Arms Training Center, Bad Toelz, W. Germany.

Length: 3 weeks (102 hours).

Exhibit Dates: 2/87–4/04.

Learning Outcomes: Upon completion of the course, the student will be able to set up, operate, and maintain a radio network for secure voice and teletypewriter communications; determine appropriate antenna lengths for various types of applications; identify components of a radio teletype message, a radio teletypewriter, and encryption equipment; and troubleshoot and isolate faults in DC loop circuits.

Instruction: Lectures, practical experience (hands-on and simulation), and examinations (written and performance) cover the duties of a single channel radio team chief.

Credit Recommendation: In the lower-division baccalaureate/associate degree category, 2 semester hours in radio and radio telegraphy and security (10/88).

AR-1404-0040

Postal Supervisor

Course Number: ***Version 1:*** 7A-F59/510-ASIF5. *Version 2:* 7A-F59/510-ASIF5.

Location: ***Version 1:*** Soldier Support Center, Fort Benjamin Harrison, IN. *Version 2:* Soldier Support Institute, Fort Jackson, SC.

Length: ***Version 1:*** 5–6 weeks (224 hours). *Version 2:* 2–3 weeks (93 hours).

Exhibit Dates: ***Version 1:*** 4/88–9/95. *Version 2:* 10/95–Present.

Learning Outcomes: ***Version 1:*** Upon completion of the course, the student will be able to process domestic and international mail; requisition, safeguard, and stamps, money orders, supplies, and equipment; maintain and audit a variety of postal accounts; operate/supervise postal service center and integrated retail terminals; and understand supervise overall postal operations. *Version 2:* Upon completion of the course, the student will be able to supervise employees in processing domestic and international mail, requisition, issue, safeguard and issue stamps, money orders, supplies, and equipment; maintain and audit postal accounts; supervise operation of postal service center and integrated retail terminals; and overall postal operations.

Instruction: ***Version 1:*** Lectures, practical exercises, and programmed instruction are the methods of instruction. *Version 2:* Lectures, practical exercises, and programmed instruction are the methods of instruction.

Credit Recommendation: ***Version 1:*** In the lower-division baccalaureate/associate degree category, 4 semester hours in postal services management (6/92). *Version 2:* In the lower-division baccalaureate/associate degree category, 3 semester hours in postal services management (9/97).

AR-1404-0041

Special Forces Communications Sergeant Advanced Noncommissioned Officer (NCO)

Course Number: ***Version 1:*** 011-18E40. *Version 2:* 011-18E40. *Version 3:* 011-18E40.

Location: John F. Kennedy Special Warfare Center and School, Fort Bragg, NC.

Length: ***Version 1:*** 12 weeks (523–524 hours). *Version 2:* 18–19 weeks (1051 hours). *Version 3:* 18 weeks (737 hours).

Exhibit Dates: ***Version 1:*** 6/89–9/94. *Version 2:* 10/94–10/98. *Version 3:* 11/98–Present.

Learning Outcomes: ***Version 1:*** Upon completion of the course, the student will be able to determine communications needs, identify appropriate communications equipment, develop an extensive communications system, and supervise communications systems. *Version 2:* Upon completion of the course, the student will be able to apply advanced leadership and technical training techniques. *Version 3:* Upon completion of the course, the student will be able to lead and train Special Forces units in communications and intelligence.

Instruction: ***Version 1:*** Methods of instruction include lectures, practical exercises, and simulation in communications planning and applications. Course includes a common core of leadership subjects. *Version 2:* Course includes lectures, discussions, small group instruction, peer-to-peer interaction, and practical experiences. Topics covered include group dynamics, military battle tactics, historical studies, military etiquette, and Special Forces photography. Course includes a common core of leadership subjects. *Version 3:* Lectures and practical exercises cover leadership, management, communications, tactics, planning, and intelligence. This course includes common core leadership subjects.

Credit Recommendation: ***Version 1:*** In the lower-division baccalaureate/associate degree category, 2 semester hours in communications theory and 1 in communications practicum. See AR-1404-0035 for common core credit (12/92). *Version 2:* In the lower-division baccalaureate/associate degree category, 3 semester hours in military science and 1 in photography. See AR-1404-0035 for common core credit (6/96). *Version 3:* In the lower-division baccalaureate/associate degree category, 1 semester hour in leadership, 2 in intelligence techniques, 1 in photography, and 1 in electronic communications. See AR-1404-0035 for common core credit (4/01).

Related Occupations: 18E.

AR-1404-0042

Special Forces Communications Sergeant Advanced Noncommissioned Officer (NCO) Reserve Component Phase 2

Course Number: 011-18E40-RC.

Location: John F. Kennedy Special Warfare Center and School, Fort Bragg, NC.

Length: 3 weeks (209 hours).

Exhibit Dates: 8/91–4/04.

Learning Outcomes: Upon completion of the course, the student will be able to determine communications needs, identify appropriate communications systems, develop an extensive communications system, supervise communications systems.

Instruction: Methods of instruction include lectures, practical exercises, and simulation in communications planning and application.

Credit Recommendation: In the lower-division baccalaureate/associate degree category, 2 semester hours in communications theory and 1 in communications practicum (12/92).

Related Occupations: 18E.

AR-1404-0043

Traffic Analysis Technician Warrant Officer Technical/Tactical Certification (Reserve Component) Phases 2 and 3

Course Number: 3B-352C-RC.

Location: Intelligence School, Fort Huachuca, AZ.

Length: 4 weeks (135 hours).

Exhibit Dates: 10/88–4/04.

Learning Outcomes: Upon completion of the course, the student will be able to manage intelligence collection, using collection and processing equipment, and operate, analyze, and evaluate intelligence traffic.

Instruction: Phase 1 offers nonresident instruction. Phase 2 (72 hours) and Phase 3 (58 hours) require active duty training time at the Intelligence Center. Lectures, field trips, and laboratory-based exercises develop skills in the analysis and reporting of intelligence traffic. Credit is based on the resident phases only.

Credit Recommendation: In the upper-division baccalaureate category, 3 semester hours in operations management (10/93).

Related Occupations: 352C.

AR-1404-0044

1. Mobile Subscriber Equipment (MSE) Network Switching System Operator Basic Noncommissioned Officer (NCO)
2. Network Switching Systems Operator/Maintainer

Course Number: ***Version 1:*** 260-31F20. *Version 2:* 260-31F30.

Location: Signal School, Fort Gordon, GA.

Length: ***Version 1:*** 11–12 weeks (446 hours). *Version 2:* 12 weeks (462–463 hours).

Exhibit Dates: ***Version 1:*** 10/91–9/94. *Version 2:* 8/96–Present.

Learning Outcomes: ***Version 1:*** Upon completion of the course, the student will be able to provide leadership and technical assistance in the deployment, operation, and maintenance of mobile subscriber equipment communications systems. *Version 2:* Upon completion of the course, the student will be able to provide leadership and technical assistance in the deployment, operation, and maintenance of mobile subscriber equipment communications systems and be able to use MS DOS-based word processors, spreadsheets, graphics presentation package, and data bases.

Instruction: ***Version 1:*** Lectures and practical exercises cover communications equipment deployment, operation, and maintenance, including voice, switching, digital, and microwave equipment. Course includes a common core of leadership subjects. *Version 2:* Lectures and practical exercises cover computer literacy, and communications equipment deployment, operation, and maintenance, including voice, switching, digital, and microwave equipment. Course includes a common core of leadership subjects.

Credit Recommendation: ***Version 1:*** In the vocational certificate category, 3 semester hours in communications systems operation. In the lower-division baccalaureate/associate degree category, see AR-1406-0090 for common core credit (2/94). *Version 2:* In the vocational certificate category, 3 semester hours in communications systems operations. In the lower-division baccalaureate/associate degree category, 1 semester hour in computer applications software. See AR-1406-0090 for common core credit (2/97).

Related Occupations: 31F.

AR-1404-0045

Systems Automation 2

Course Number: 7E-F41.

Location: Signal School, Fort Gordon, GA.

Length: 4–5 weeks (190 hours).

Exhibit Dates: 10/92–10/95.

Learning Outcomes: Upon completion of the course, the student will be able to select and evaluate various software packages, evaluate and implement computer security systems, evaluate and establish data management plans, and provide leadership for large-scale software projects.

Instruction: Classroom instruction and laboratory exercises are used. Topics include computer hardware, application software, information security, and automatic data processing equipment.

Credit Recommendation: In the lower-division baccalaureate/associate degree category, 3 semester hours in software applications. In the upper-division baccalaureate category, 4 semester hours in system analysis and design (2/94).

AR-1404-0046

1. Record Telecommunications Operator/Maintainer
(Record Telecommunications Center Operator)
2. Telecommunications Operator/Maintainer

Course Number: *Version 1:* 260-74C10. *Version 2:* 260-74C10.

Location: Signal School, Fort Gordon, GA.

Length: *Version 1:* 10–12 weeks (374–405 hours). *Version 2:* 9 weeks (324 hours).

Exhibit Dates: *Version 1:* 10/92–9/95. *Version 2:* 10/95–Present.

Learning Outcomes: *Version 1:* Upon completion of the course, the student will be able to install, operate, and perform unit level maintenance on tactical telecommunications centers and type messages at a minimum speed of 25 wpm with no less than 95 percent accuracy. *Version 2:* Upon completion of the course, the student will be able to install, operate, and perform maintenance on telecommunications equipment.

Instruction: *Version 1:* Course includes computer operations and associated CAI typing using MS DOS 2.1 and Typing Tutor IV programs. Course contains exercises in installation, operation, and unit level maintenance on telecommunications shelters, associated communications security devices, and power generators; procedural functions in message formatting by typing; and incoming/outgoing message processing within an automated switching center. *Version 2:* Lectures and practical exercises cover telecommunications operations, including message formatting, processing I/O messages within an automated switching system, preventive maintenance for computer terminal devices, and installation and operation of microcomputers and software.

Credit Recommendation: *Version 1:* In the vocational certificate category, 3 semester hours in communications system operation and maintenance. In the lower-division baccalaureate/associate degree category, 2 semester hours in keyboarding (3/95). *Version 2:* In the lower-division baccalaureate/associate degree category, 3 semester hours in communications systems operation/maintenance and 2 in computer applications software (2/97).

Related Occupations: 74C.

AR-1404-0047

1. Record Telecommunications Center Operator
(Record Telecommunications Center Operator Basic Noncommissioned Officer (NCO))
2. Telecommunications Operator-Maintainer (BNCOC)

Course Number: *Version 1:* 260-74C30; 260-74C20. *Version 2:* 260-74C30.

Location: Signal School, Fort Gordon, GA.

Length: *Version 1:* 9 weeks (335 hours). *Version 2:* 9–10 weeks (335–349 hours).

Exhibit Dates: *Version 1:* 10/92–9/93. *Version 2:* 10/93–Present.

Learning Outcomes: *Version 1:* Upon completion of the course, the student will be able to perform as a combat first-line supervisor in tactical and strategic telecommunications centers and perform mail server functions in an automated message switching center. *Version 2:* Upon completion of the course, the student will be able to manage a tactical and strategic telecommunications center and perform mail server functions in an automated message switching center.

Instruction: *Version 1:* This course includes supervisory and operational exercises in a strategic telecommunication center, including message processing, service actions, unit-level maintenance of the center, preparation of reports, and automated message switching using DINAH Software for an AUTODIN Mail Server. Course includes a common core of leadership subjects. *Version 2:* Course includes supervisory and operational exercises in a strategic telecommunications center, including message processing and service actions, communication-security duties, processing incoming and outgoing messages, performing inventories, safeguarding materials, performing unit-level maintenance of the center, preparing reports, automated message switching, overseeing distribution functions, managing inquiries, and compiling annual reports. Course includes a common core of leadership subjects.

Credit Recommendation: *Version 1:* In the lower-division baccalaureate/associate degree category, 1 semester hour in computer literacy. See AR-1406-0090 for common core credit (7/95). *Version 2:* In the lower-division baccalaureate/associate degree category, 1 semester hour in computer literacy. See AR-1406-0090 for common core credit (2/03).

Related Occupations: 74C.

AR-1404-0048

Morse Interceptor

Course Number: *Version 1:* 231-98H10. *Version 2:* CVS; CSW; A-231-0044; 1N2X1; 231-98H10.

Location: Intelligence School, Fort Huachuca, AZ.

Length: *Version 1:* 11 weeks (440 hours). *Version 2:* 21–22 weeks (840 hours).

Exhibit Dates: *Version 1:* 10/95–9/98. *Version 2:* 10/98–Present.

Learning Outcomes: *Version 1:* Upon completion of the course, the student will be able to intercept and process radio transmission keyed in Morse code. *Version 2:* Upon completion of the course, the student will be able to intercept and process radio transmission keyed in Morse code.

Instruction: *Version 1:* Lectures and practical exercises cover Morse code, message processing, and message forwarding equipment. *Version 2:* Lectures and practical exercises cover Morse code, message processing, and message forwarding equipment.

Credit Recommendation: *Version 1:* Credit is not recommended because of the limited, specialized nature of the course (6/97). *Version 2:* Credit is not recommended because of the military-specific nature of the course (8/01).

Related Occupations: 98H.

AR-1404-0049

Noncommunications Instructor/Analyst Basic Noncommissioned Officer (NCO)

Course Number: 233-98J30.

Location: Intelligence School, Fort Huachuca, AZ.

Length: 6–7 weeks (235 hours).

Exhibit Dates: 10/95–9/98.

Learning Outcomes: After 9/98 see AR-1404-0050. Upon completion of the course, the student will be able to supervise and manage personnel in the interception of electronic signals and the preparation of reports and analysis of technical data.

Instruction: Instruction is presented through lectures, demonstrations, and practical exercises in personnel supervision, technical file management, data analysis, and operation and maintenance of technical equipment. Course includes a common core of leadership subjects.

Credit Recommendation: In the lower-division baccalaureate/associate degree category, 3 semester hours in data base applications. See AR-1406-0090 for common core credit (6/97).

AR-1404-0050

Total Army Training System (TATS) Electronic Intelligence Interceptor/Analyst Basic Noncommissioned Officer (NCO)

Course Number: 233-98J30, Phase 2; 233-98J30.

Location: Intelligence School, Fort Huachuca, AZ.

Length: 4–7 weeks (144–235 hours).

Exhibit Dates: 10/98–Present.

Learning Outcomes: Before 10/00 see AR-1404-0051. Upon completion of the course, the student will be able to perform common leadership tasks and demonstrate skills associated with electronic intelligence interception.

Instruction: This course contains lectures, discussions, and exercises dealing with common leadership skills, military writing, tactical leadership, and assorted topics dealing with electronic intelligence interception and analysis.

Credit Recommendation: Credit is recommended for the common core only. See AR-1406-0090 (12/02).

Related Occupations: 98J.

AR-1404-0051

Total Army Training System (TATS) Electronic Intelligence Interceptor/Analyst Basic Noncommissioned Officer (NCO)

Course Number: 233-98J30.

Location: Intelligence School, Fort Huachuca, AZ; Reserve Schools, Multiple locations.

Length: 90 hours.

Exhibit Dates: 10/98–9/00.

Learning Outcomes: After 9/00 see AR-1404-0050. Upon completion of the course, the student will be able to perform common leadership tasks and demonstrate skills associated with electronic intelligence interception.

Instruction: This course contains lectures, discussions and exercises dealing with common leadership skills, military writing, tactical leadership, and assorted topics dealing with electronic intelligence interception and analysis.

Credit Recommendation: Credit is not recommended because of the military-specific nature of the course (8/01).

AR-1404-0052

Satellite/Microwave Systems Chief, Advance Noncommissioned Officer (ANCOC)

Course Number: 101-31P40, Phase 2.

Location: Signal School, Fort Gordon, GA.

Length: 9–10 weeks (364 hours).

Exhibit Dates: 10/01–Present.

Learning Outcomes: Upon completion of the course, the student will be able to perform and manage common signal training, computer literacy, and mobile subscriber equipment. This training will provide the ANCOC graduate with the skills and knowledge required to be a platoon sergeant of a tactical signal company or a site chief of a satellite at a microwave communications facility.

Instruction: Audio visual materials, practical and classroom exercises, laboratory, lecture, and computer-based training. Topics include computer literacy, site administration, strategic signal operations, satellite communication systems, site operations, tactical signal operations, mobile subscriber equipment, and plans/operations.

Credit Recommendation: In the lower-division baccalaureate/associate degree category, 3 semester hours in network fundamentals, and 3 in radio communications (2/03).

AR-1405-0024

Medical Supply Specialist
(Medical Supply Procedures)
(Medical Material Procedures)

Course Number: 551-76J10; 551-76J20; 551-767.1; 8-E-44.

Location: Academy of Health Sciences, Fort Sam Houston, TX; Medical Field Service School, Fort Sam Houston, TX.

Length: 5–7 weeks (182–242 hours).

Exhibit Dates: 12/76–10/93.

Objectives: After 10/93 see AR-1405-0241. To train enlisted personnel in medical supply procedures.

Instruction: This course provides lectures in publications, property control and stock-accounting procedures, and supply and services procedures at the installation level and a survey of management methods.

Credit Recommendation: In the lower-division baccalaureate/associate degree category, 3 semester hours in stock control accounting and inventory management (7/87).

Related Occupations: 098; 76A; 76J.

AR-1405-0061

Unit Supply Specialist Advanced Individual Training
(Unit Supply Specialist)

Course Number: 552-76Y10.

Location: Quartermaster Center and School, Fort Lee, VA; Quartermaster School, Fort Jackson, SC.

Length: 7–8 weeks (254–280 hours).

Exhibit Dates: 1/83–9/90.

Objectives: After 9/90 see AR-1405-0235. To train selected enlisted personnel to perform supply duties and to maintain small arms.

Instruction: Lectures and practical exercises cover supply operations and the maintenance of small arms. Course includes basic mathematics, typewriting, use of technical publications, unit and organization supply procedures, and organizational maintenance of small arms.

Credit Recommendation: In the vocational certificate category, 1 semester hour in records maintenance and 1 in supply supervision and maintenance (10/90).

Related Occupations: 76Y.

AR-1405-0065

Defense Reutilization and Marketing Operations Advanced
(Defense Advanced Disposal Management)

Course Number: *Version 3:* 8B-F17. *Version 4:* 8B-F17.

Location: *Version 3:* Army Logistics Management Center, Fort Lee, VA; Army Logistics Management College, Onsite at various locations. *Version 4:* Army Logistics Management College, Fort Lee, VA; Army Logistics Management College, Onsite at various locations.

Length: *Version 3:* 3 weeks (115 hours). *Version 4:* 3 weeks (114–115 hours).

Exhibit Dates: *Version 3:* 9/83–8/93. *Version 4:* 9/93–9/99.

Learning Outcomes: *Version 3:* Upon completion of the course, the student will be able to manage property disposal activities. *Version 4:* Upon completion of the course, the student will be able to manage property disposal activities.

Instruction: *Version 3:* Lecture-conferences, practical exercises, and cases cover property disposal management. Topics include management principles and techniques, the Department of Defense disposal program, management of disposal operations marketing, property control, materiel processing, and contracting procedures. *Version 4:* Methods of instruction include lecture-conferences, case studies, guest speakers, student presentations, and practical exercises. Topics include management principles and techniques, the Department of Defense disposal program, management of disposal operations marketing, property control, materiel processing, contracting procedures, and management communications.

Credit Recommendation: *Version 3:* In the lower-division baccalaureate/associate degree category, 1 semester hours in general management (10/89). *Version 4:* In the upper-division baccalaureate category, 3 semester hours in principles of management, 1 in management communication, and 1 in contract law (7/00).

Related Occupations: 411A; 621A; 630A; 631A; 632A.

AR-1405-0069

Quartermaster Officer Basic

Course Number: 8-10-C20-81A/82A/92A.

Location: Quartermaster Center and School, Fort Lee, VA.

Length: 19 weeks (794 hours).

Exhibit Dates: 5/82–9/92.

Objectives: After 9/92 see AR-1405-0236. To train newly commissioned quartermaster officers for their first duty assignment.

Instruction: Lectures and practical exercises cover basic quartermaster officer training, including officer orientation, skill functions, unit functions, combined arms, and skill application. Practical exercises, conferences, demonstrations, films, television, seminars, case studies, and examinations all present materials designed to train newly commissioned quartermaster officers as either petroleum, subsistence supply, or supply and service management officers. Course covers officer orientation and professional development, unit logistics and functions, unit supply and maintenance, materiel management, automatic data processing, subsistence management, food service management, petroleum operations, and water supply operations.

Credit Recommendation: In the lower-division baccalaureate/associate degree category, 2 semester hours in introduction to management and 2 in supply management. In the upper-division baccalaureate category, 4 semester hours in organizational behavior (8/85).

Related Occupations: 81A; 82A; 92A.

AR-1405-0074

Ordnance Maintenance Management Officer Basic

Course Number: 4-9-C20-91B; 4-9-C20-91A.

Location: Ordnance Center and School, Aberdeen Proving Ground, MD.

Length: 19–20 weeks (669–759 hours).

Exhibit Dates: 1/86–8/91.

Learning Outcomes: After 8/91 see AR-1408-0244. Upon completion of the course, the student will be able to perform the duties of a vehicle maintenance administrator, including logistics, instruction, communication, and skills and knowledge required for vehicle maintenance.

Instruction: Lectures, demonstrations, and practical exercises in unit administration, logistics, supply, effective communication, and the technical skills required for vehicle maintenance.

Credit Recommendation: In the lower-division baccalaureate/associate degree category, 1 semester hour in communication skills and 2 in automotive/heavy equipment fundamentals. In the upper-division baccalaureate category, 3 semester hours in principles of supervision, 3 in vehicle maintenance management, 2 in fundamentals of instruction, and 4 in logistics (3/87).

Related Occupations: 91A; 91B.

AR-1405-0103

Major Item Management (ALMC-MH)

Course Number: ALMC-MH; ALM-41-0339.

Location: Army Logistics Management Center, Fort Lee, VA; Army Logistics Management College, Onsite at various locations.

Length: 2–3 weeks (98–114 hours).

Exhibit Dates: 7/79–12/97.

Objectives: To develop and increase knowledge and understanding of materiel managers in the management of major items.

Instruction: Methods of instruction include lecture-conferences, practical exercises, workshops, and computer-assisted simulations within decision making exercises. The course covers the entire life cycle of major items from the entry of new equipment to the inventory to the disposal of the equipment.

Credit Recommendation: In the upper-division baccalaureate category, 2 semester hours in management of high-value equipment (7/87).

AR-1405-0110

STORAGE SUPPLYMAN NONCOMMISSIONED OFFICER (NCO) ADVANCED

Course Number: None.
Location: Quartermaster Center and School, Fort Lee, VA.
Length: 4–5 weeks (134 hours).
Exhibit Dates: 3/84–1/90.
Objectives: To provide enlisted personnel and skills required to perform as noncommissioned officers.
Instruction: Types of instruction include lectures, conferences, case studies, demonstrations, and practical exercises. Topics include management of supplies, storage operations management, organizational effectiveness, and inventory management.
Credit Recommendation: In the lower-division baccalaureate/associate degree category, 1 semester hour in organizational behavior and 3 in supply management (8/85).

AR-1405-0114

LOGISTICS MANAGEMENT FOR AUDITORS

Course Number: ALM-41-0251.
Location: Army Logistics Management Center, Fort Lee, VA; Army Logistics Management College, Onsite at various locations.
Length: 2 weeks (64 hours).
Exhibit Dates: 8/80–6/96.
Objectives: To inform audit personnel of Army logistics functions, systems, facilities, and activities.
Instruction: Methods include lectures, guest speakers, practical exercises, and the use of simulated models of the logistics system throughout the Army. Course includes differentiation of management systems comprising the Department of Army wholesale and retail systems and their relationship to planning, programming, and controlling functions.
Credit Recommendation: In the upper-division baccalaureate category, 2 semester hours in auditing of the materiel management function (5/81).

AR-1405-0116

DEFENSE REUTILIZATION AND MARKETING PROPERTY ACCOUNTING
(Defense Integrated Disposal Management System)

Course Number: ALMC-IC; ALM-44-0282.
Location: Army Logistics Management College, Fort Lee, VA; Army Logistics Management Center, Fort Lee, VA; Army Logistics Management College, Onsite at various locations.
Length: 2–3 weeks (77–114 hours).
Exhibit Dates: 11/79–7/93.
Objectives: After 7/93 see AR-1405-0243. To improve the proficiency of personnel engaged in record keeping functions so that they can initiate and maintain accurate computerized property records and be able to use management output listings for the control of property at the Defense Reutilization and Marketing Office.
Instruction: Instructional methods include the use of lecture-conferences, practical exercises, study group workshops, guest speakers, examinations, and critiques. Topics include automated record keeping procedures related to property disposal, transfers, sales, adjustments, and reversals.
Credit Recommendation: In the vocational certificate category, 3 semester hours in data entry and automated systems output (10/89).

AR-1405-0118

COMMODITY COMMAND STANDARD SYSTEM (CCSS) FUNCTIONAL

Course Number: ALMC-3L; ALM-56-M115.
Location: Army Logistics Management College, Fort Lee, VA; Army Logistics Management Center, Fort Lee, VA; Army Logistics Management College, Onsite at various locations.
Length: 2 weeks (69–80 hours).
Exhibit Dates: 8/80–8/92.
Objectives: After 8/92 see AR-1405-0244. To enhance the ability of those working at the Army Material Development and Readiness Command to perform their duties more efficiently and effectively through the use of a particular automated system.
Instruction: Methods of instruction include lecture-conferences, practical exercises, workshops, videotaped lectures, and computer-assisted simulation. Topics include logistics management concepts and computerized data systems and how they interface with the initial provisioning and replenishment cycles.
Credit Recommendation: In the upper-division baccalaureate category, 2 semester hours in automated systems in logistics management (8/90).

AR-1405-0123

1. AUTOMATED INVENTORY MANAGEMENT
2. AUTOMATED INVENTORY MANAGEMENT
3. MEDICAL LOGISTICS SYSTEMS MANAGEMENT
(Automated Inventory Management)

Course Number: ***Version 1:*** 8B-F22/551-F22; 8B-F22. *Version 2:* 8B-F22/551-F22. *Version 3:* 8B-F22/551-F22.
Location: Academy of Health Sciences, Fort Sam Houston, TX.
Length: ***Version 1:*** 2 weeks (83–84 hours). *Version 2:* 2 weeks (82 hours). *Version 3:* 2 weeks (82 hours).
Exhibit Dates: ***Version 1:*** 1/80–2/96. *Version 2:* 3/96–9/02. *Version 3:* 10/02–Present.
Learning Outcomes: ***Version 1:*** Upon completion of the course, the student will have a functional knowledge of automated inventory management procedures, including concepts and interfaces with other automated and manual supply management areas. *Version 2:* Upon completion of the course, the student will have a functional knowledge of automated inventory management procedures, including concepts and interfaces with other automated and manual supply management areas. *Version 3:* Upon completion of the course, the student will be able to discuss automated inventory management systems; process customer reports, receipts, turn-ins, and destructions and stock management reports; and perform inventory counts, management functions, and medical logistics management functions during a disaster.
Instruction: ***Version 1:*** Conferences, self-instruction, and practical exercises cover inventory management, data processing and storage, and purchase procedures. *Version 2:* Conferences, self-instruction, and practical exercises cover inventory management, data processing and storage, and purchase procedures. *Version 3:* Practical exercises, discussion, and lecture. Topics include customer support and requests; stock management; receipts, turn-ins; destruction inventory processes and counts; managerial functions; and disaster logistics management of material.
Credit Recommendation: ***Version 1:*** In the lower-division baccalaureate/associate degree category, 3 semester hours in management science (5/87). *Version 2:* In the lower-division baccalaureate/associate degree category, 2 semester hours in inventory management (5/97). *Version 3:* In the lower-division baccalaureate/associate degree category, 3 semester hours in inventory management and 3 in introduction to logistics (11/03).

AR-1405-0140

COMBAT SERVICES SUPPORT PRE-COMMAND

Course Number: 2G-F46.
Location: Quartermaster Center and School, Fort Lee, VA.
Length: 2 weeks (77 hours).
Exhibit Dates: 5/85–9/93.
Objectives: To assist individuals in preparation for large organizational management responsibilities and to ensure a common understanding of training, personnel, logistics management, and tactical doctrine. This is a refresher course for selected management-level persons.
Instruction: Discussions, lectures, tours, and reviews cover management concepts, including supply, maintenance, and training management. Logistics management is also covered.
Credit Recommendation: In the upper-division baccalaureate category, 2 semester hours in senior seminar in logistics management (8/88).

AR-1405-0145

FACILITIES ENGINEERING SUPPLY SYSTEM MANAGEMENT

Course Number: ALMC-SC.
Location: Army Logistics Management Center, Fort Lee, VA.
Length: 2 weeks (79 hours).
Exhibit Dates: 5/85–3/96.
Objectives: To develop the operational and management skills required for positions in the automated Facilities Engineering Supply System (FESS).
Instruction: Course covers the daily operation of the FESS system, systems interface, data input, management and system reports, and procedures.
Credit Recommendation: In the lower-division baccalaureate/associate degree category, 2 semester hours in automated supply systems (3/86).

AR-1405-0146

ORDNANCE OFFICER ADVANCED MUNITIONS MATERIEL MANAGEMENT FIELD OPERATIONS

Course Number: 4E-F0A-F10.
Location: Missile and Munitions School, Redstone Arsenal, AL.
Length: 4 weeks (152 hours).
Exhibit Dates: 5/85–4/96.
Objectives: To prepare the company grade officer to function as a company commander in support of munitions materiel.
Instruction: Students receive training in the areas of war planning strategies, battalion materiel operations, standard army ammunition systems reports, and decision making.
Credit Recommendation: Credit is not recommended because of the military-specific nature of the course (4/86).

AR-1405-0147

Ordnance Officer Advanced Missile Materiel Management Field Operations

Course Number: FE-F0A-F19.
Location: Missile and Munitions School, Redstone Arsenal, AL.
Length: 3 weeks (110 hours).
Exhibit Dates: 5/85–1/96.
Objectives: To prepare company grade ordnance officers to function effectively as commander or officer in support of missile systems.
Instruction: Students receive training in the areas of equipment updates, logistics updates, and management concepts and policy at the brigade level and below for air defense, land combat, and ballistic weapons systems.
Credit Recommendation: Credit is not recommended because of the military-specific nature of the course (4/86).

AR-1405-0148

Ordnance Officer Advanced Wholesale Missile/Munitions Materiel Management

Course Number: 4E-F0A-F11; 4F-F0A-F20.
Location: Missile and Munitions School, Redstone Arsenal, AL.
Length: 4 weeks (152 hours).
Exhibit Dates: 5/85–1/96.
Objectives: To prepare the company grade ordnance officer to function as a munitions materiel manager and ammunition plant executive officer assigned to any one of the munitions commodity commands or depots.
Instruction: Students receive training in the areas of logistics management, budget system overview, and mission area analysis. Students receive an overview of ammunition industrial base management and an understanding of the Army Ammunition Management System.
Credit Recommendation: Credit is not recommended because of the military-specific nature of the course (4/86).

AR-1405-0150

Ordnance Officer Advanced Maintenance Management

Course Number: 4-9-C22-91A.
Location: Ordnance Center and School, Aberdeen Proving Ground, MD.
Length: 20 weeks (720 hours).
Exhibit Dates: 1/85–12/97.
Learning Outcomes: Upon completion of the course, the student will be able to lead and train personnel; provide personnel with management concepts as they apply to designated systems and equipment; and test and evaluate production, procurement, supply, and storage.
Instruction: Instruction includes logistics, procurement, management (finances, personnel and equipment) law, and communication skills.
Credit Recommendation: In the lower-division baccalaureate/associate degree category, 1 semester hour in financial management. In the upper-division baccalaureate category, 3 semester hours in maintenance management, 3 in personnel management, and 1 in logistics (3/87).
Related Occupations: 91A.

AR-1405-0151

Ordnance Maintenance Management Officer Advanced

Course Number: 4-9-C22-91A-RC.
Location: Ordnance Center and School, Aberdeen Proving Ground, MD.
Length: Phase 2, 2 weeks (66 hours); Phase 4, 2 weeks (66 hours).
Exhibit Dates: 3/86–12/97.
Learning Outcomes: Upon completion of the course, the student will be able to carry out management duties required by the student's individual specialty.
Instruction: Instruction includes maintenance management procedures and inventory management. Credit is recommended based on the two resident phases. Phases 1 and 3 are taken by correspondence and are not prerequisite to the resident phases.
Credit Recommendation: In the lower-division baccalaureate/associate degree category, 1 semester hour in maintenance management and 1 in inventory management (3/87).

AR-1405-0152

Repair Shop Technician Warrant Officer Advanced

Course Number: 4-9-C32-441A.
Location: Ordnance Center and School, Aberdeen Proving Ground, MD.
Length: 12 weeks (429–445 hours).
Exhibit Dates: 3/85–12/97.
Learning Outcomes: Upon completion of the course, the student will be able to manage a repair shop, operate and maintain machine shop equipment, operate and maintain welding equipment, and keep maintenance records on equipment coming into the repair shop.
Instruction: Instruction includes lectures and hands-on experience covering management and communication skills, general shop operations, body repair, and welding. Emphasis on machine shop operation and maintenance management.
Credit Recommendation: In the lower-division baccalaureate/associate degree category, 1 semester hour in communication skills, 1 in management, 1 in heavy equipment maintenance management, 6 in general machine shop, 2 in welding, and 1 in body repair (3/87).
Related Occupations: 441A.

AR-1405-0153

Light Systems Maintenance Technician Warrant Officer Advanced

Course Number: 4-9-C32-630B.
Location: Ordnance Center and School, Aberdeen Proving Ground, MD.
Length: 14 weeks (610–620 hours).
Exhibit Dates: 1/85–9/90.
Learning Outcomes: Upon completion of the course, the student will be able to maintain, manage, and keep maintenance records on light and medium duty equipment and perform and supervise repairs on light and medium duty equipment.
Instruction: Instruction includes lectures and hands-on experience with emphasis on management, maintenance management, and communication skills. Considerable emphasis is placed on troubleshooting and repair of light and medium duty equipment.
Credit Recommendation: In the lower-division baccalaureate/associate degree category, 1 semester hour in communication skills, 1 in management, 3 in heavy equipment maintenance management, 1 in automotive electrical systems, and 3 in diesel and heavy equipment mechanics (3/87).
Related Occupations: 630B.

AR-1405-0154

Ordnance Officer Advanced Maintenance Additional Specialty

Course Number: 4N-91A.
Location: Ordnance Center and School, Aberdeen Proving Ground, MD.
Length: 11 weeks (406–407 hours).
Exhibit Dates: 7/85–9/90.
Learning Outcomes: Upon completion of the course, the student will be able to manage facilities, personnel, and maintenance and logistics management and prepare status reports directed to the upper management and administrative levels.
Instruction: Instruction includes management in the areas of organization, finance, material maintenance, contract management, and logistics.
Credit Recommendation: In the upper-division baccalaureate category, 2 semester hours in maintenance management, 3 in material management, and 1 in logistics (3/87).

AR-1405-0155

Ammunition Stock Control and Accounting Specialist

Course Number: 551-55R10/551-55R10-RC.
Location: Ordnance, Missile and Munitions School, Redstone Arsenal, AL.
Length: 8–9 weeks (300 hours).
Exhibit Dates: 8/86–1/96.
Learning Outcomes: Upon completion of the course, the student will be able to identify demolition materials, pyrotechnics, and explosives; keep proper records and stock and inventory control of ammunition; understand computer printouts, registers, and lot location records from computerized subsystems; and perform basic duties as a supply clerk.
Instruction: Lectures and hands-on training cover data processing equipment. Discussions, role playing, and testing cover data processing, inventory and stock control, and record keeping.
Credit Recommendation: In the lower-division baccalaureate/associate degree category, 3 semester hours in stock control and 2 in introduction to data processing (8/91).
Related Occupations: 55R.

AR-1405-0156

Ordnance Missile Materiel Management Officer-Reserve

Course Number: None.

Location: Ordnance, Missile and Munitions School, Redstone Arsenal, AL.

Length: 4 weeks (144 hours).

Exhibit Dates: 7/86–1/96.

Learning Outcomes: Upon completion of the course, the student will be able to practice techniques of good management; supervise and train maintenance personnel; and practice proper storage procedures, inventory control, and logistics.

Instruction: Lectures, discussions, tests, and role-playing cover management skills and responsibilities, materiel management, and evaluation of shop supply operations. Evaluation included only resident phases.

Credit Recommendation: In the lower-division baccalaureate/associate degree category, 3 semester hours in supply management (4/88).

AR-1405-0157

Ordnance Munitions Materiel Management Officer Reserve

Course Number: None.

Location: Ordnance, Missile and Munitions School, Redstone Arsenal, AL.

Length: 4 weeks (144 hours).

Exhibit Dates: 7/86–9/90.

Learning Outcomes: Upon completion of the course, the student will apply a working knowledge of the Standard Army Ammunition System and its distribution channels from manufacturer to user to the maintenance and upkeep of ammunition.

Instruction: Discussions, lectures, and field training cover munitions fundamentals, organization, and management. Only the two resident phases (Phase 2 and Phase 4) were evaluated.

Credit Recommendation: Credit is not recommended because of the military-specific nature of the course (4/88).

AR-1405-0158

Ordnance Munitions Materiel Management Officer Basic

Course Number: ***Version 1:*** 4-9-C20-91D. *Version 2:* 4-9-C20-91D.

Location: Ordnance, Missile and Munitions School, Redstone Arsenal, AL.

Length: ***Version 1:*** 18–19 weeks (750 hours). *Version 2:* 17–18 weeks (713 hours).

Exhibit Dates: ***Version 1:*** 12/87–9/92. *Version 2:* 10/92–10/00.

Learning Outcomes: ***Version 1:*** Upon completion of the course, the student will be able to write, interpret, and follow training plans through performance objectives; use proper communication techniques for letter and report writing; develop leadership and decision making techniques; interpret financial management and budgeting plans; and analyze problems in budgeting. *Version 2:* Upon completion of the course, the student will be trained as a newly commissioned officer with a background for their first duty assignment in ordnance materiel management.

Instruction: ***Version 1:*** Course includes logistics subjects, including support shop operations, direct support supply/maintenance exercises, and logistics assistance. Course also includes unit and organizational management, combined arms, and tactical field training. *Version 2:* This course consists of instructor led delivery in common military subjects, small unit field operations, conventional munitions, and leadership principles.

Credit Recommendation: ***Version 1:*** In the lower-division baccalaureate/associate degree category, 3 semester hours in principles of finance. In the upper-division baccalaureate category, 3 semester hours in principles of management (4/88). *Version 2:* In the lower-division baccalaureate/associate degree category, 4 semester hours in military science. In the upper-division baccalaureate category, 3 semester hours in principles of leadership (7/96).

Related Occupations: 91D.

AR-1405-0159

Ordnance Munitions Materiel Management Officer Basic

Course Number: 4-9-C20-91C.

Location: Ordnance, Missile and Munitions School, Redstone Arsenal, AL.

Length: 18–19 weeks (750 hours).

Exhibit Dates: 12/87–1/96.

Learning Outcomes: Upon completion of the course, the student will be able to write, interpret, and follow training plans through performance objectives; use proper communication techniques in letter and report writing; develop leadership and decision making techniques; interpret financial management and budgeting plans; and analyze problems in budgeting.

Instruction: Course includes logistics subjects, including support shop operations, direct support supply/maintenance exercises, and logistics assistance. Course also includes unit and organizational management, combined arms, and tactical field training.

Credit Recommendation: In the lower-division baccalaureate/associate degree category, 3 semester hours in principles of finance. In the upper-division baccalaureate category, 3 semester hours in principles of management (4/88).

Related Occupations: 91C.

AR-1405-0160

Standard Army Ammunition System (SAAS) Manager Training

Course Number: 8B-F44/551-F16.

Location: Ordnance, Missile and Munitions School, Redstone Arsenal, AL.

Length: 2 weeks (75 hours).

Exhibit Dates: 11/85–9/90.

Learning Outcomes: Upon completion of the course, the student will be able to understand the internal operation and responsibilities of the SAAS, perform minor tasks in the data processing area of supply, and understand the requisition and reporting procedures for the SAAS.

Instruction: Topics include stock control/stock accounting and documentation of data processing procedures and structures. Mode is lectures, discussions, and the use of manuals. The course includes hands-on training with computer equipment.

Credit Recommendation: In the vocational certificate category, 2 semester hours in data input. In the lower-division baccalaureate/associate degree category, 3 semester hours in stock management and control (4/88).

AR-1405-0161

Materiel Control and Accounting Specialist Basic Noncommissioned Officer (NCO)

Course Number: 551-76P30.

Location: Quartermaster Center and School, Fort Lee, VA.

Length: 7 weeks (252 hours).

Exhibit Dates: 6/88–10/98.

Learning Outcomes: Upon completion of the course, the student will be able to supervise the operation of both manual and automated supply systems and supervise inventory procedures.

Instruction: Lectures, demonstrations, practical exercises, and examinations cover supervising a manual and an automated supply system including procedures for conducting a physical inventory. Course includes a common core of leadership subjects.

Credit Recommendation: In the vocational certificate category, 3 semester hours in supply supervision and 1 in computerized record keeping. In the lower-division baccalaureate/associate degree category, see AR-1406-0090 for common core credit (10/88).

Related Occupations: 76P.

AR-1405-0162

Unit Supply Specialist Advanced Noncommissioned Officer (NCO)

Course Number: ***Version 2:*** 552-76Y40. *Version 3:* 552-76Y40. *Version 4:* 551-92Y40, Phase 2; 552-92Y40.

Location: Quartermaster Center and School, Fort Lee, VA.

Length: ***Version 2:*** 8 weeks (283 hours). *Version 3:* 8–9 weeks (313–316 hours). *Version 4:* 8–9 weeks (317–324 hours).

Exhibit Dates: ***Version 2:*** 4/89–9/92. *Version 3:* 10/92–9/95. *Version 4:* 10/95–Present.

Learning Outcomes: ***Version 2:*** Upon completion of the course, the student will be able to supervise the requisition and receipt of supply items and the property accounting function and apply computerized record keeping procedures. *Version 3:* Upon completion of the course, the student will be able to supervise the requisition and receipt of supply items and the property accounting function. *Version 4:* Upon completion of the course, the student will be able to supervise the requisition and receipt of supply items and the property accounting function.

Instruction: ***Version 2:*** Lectures, class discussions, practical exercises, and examinations cover computerized record keeping, supply management, including supply of bulk petroleum, and transportation management. Course includes a common core of leadership subjects. *Version 3:* Lectures, discussions, practical exercises, and examinations cover supply and transportation management. Course includes a common core of leadership subjects. *Version 4:* Lectures, discussions, practical exercises, and examinations cover supply management. The course includes a common core of leadership subjects.

Credit Recommendation: ***Version 2:*** In the vocational certificate category, 1 semester hour in computerized record keeping. In the lower-division baccalaureate/associate degree category, 6 semester hours in materiel management or supply management. See AR-1404-0035 for common core credit (2/93). *Version 3:* In the lower-division baccalaureate/associate degree category, 3 semester hours in material management or supply management. See AR-1404-0035 for common core credit (2/93). *Version 4:* In the lower-division baccalaureate/associate degree category, 1 semester hour in microcomputer applications, 1 in administrative management skills, and 2 in supply management. See AR-1404-0035 for common core credit (5/02).

Related Occupations: 76Y; 92Y.

AR-1405-0163

MATERIEL STORAGE AND HANDLING SPECIALIST ADVANCED NONCOMMISSIONED OFFICER (NCO)

Course Number: 551-76V40.

Location: Quartermaster Center and School, Fort Lee, VA.

Length: 7–8 weeks (257–259 hours).

Exhibit Dates: 7/87–10/98.

Learning Outcomes: Upon completion of the course, the student will be able to supervise warehouse storage and materiel-handling operations.

Instruction: Lectures, demonstrations, practical exercises, and examinations cover storage operations. Topics include selection of materiel-handling equipment, quality assurance for supplies in storage, preservation and repacking, and maintenance of storage records. Course includes a common core of leadership subjects.

Credit Recommendation: In the lower-division baccalaureate/associate degree category, 3 semester hours in material management. See AR-1404-0035 for common core credit (10/88).

Related Occupations: 76V.

AR-1405-0164

MATERIEL STORAGE AND HANDLING SPECIALIST BASIC NONCOMMISSIONED OFFICER (NCO)

Course Number: 551-76V30.

Location: Quartermaster Center and School, Fort Lee, VA.

Length: 5–6 weeks (211–213 hours).

Exhibit Dates: 6/87–10/98.

Learning Outcomes: Upon completion of the course, the student will be able to supervise storage operations and the taking of inventories; apply techniques for preservation and packing of materials; and perform computerized record keeping.

Instruction: Lectures, demonstrations, practical exercises, and examinations cover storage operations, inventory control, preservation of supplies, packing operations, and the processing of transactions in a computerized supply system. Course includes a common core of leadership subjects.

Credit Recommendation: In the vocational certificate category, 3 semester hours in inventory control and 1 in record keeping. In the lower-division baccalaureate/associate degree category, see AR-1406-0090 for common core credit (10/88).

Related Occupations: 76V.

AR-1405-0165

STANDARD PROPERTY BOOK SYSTEM FUNCTIONAL

Course Number: 8A-F32/552-F2.

Location: Quartermaster Center and School, Fort Lee, VA.

Length: 3 weeks (108–119 hours).

Exhibit Dates: 4/85–10/98.

Learning Outcomes: Upon completion of the course, the student will be able to prepare input transactions, understand process transactions, and read and interpret the reports of a computerized inventory control system.

Instruction: Course includes classroom lectures and practical exercise (hands-on) in preparing input transactions, reading reports, identifying system codes, and other clerical tasks involved in the use of a computerized inventory control system.

Credit Recommendation: In the vocational certificate category, 1 semester hour in inventory control and 1 in computerized record keeping (10/88).

AR-1405-0166

1. STANDARD PROPERTY BOOK SYSTEM REDESIGNED
2. STANDARD PROPERTY BOOK SYSTEM-REDESIGNED INSTALLATION/TABLE OF DISTRIBUTION AND ALLOWANCES

Course Number: ***Version 1:*** 8B-F46/551-F23. *Version 2:* 8B-F46/551-ASIG3; 8B-F46/551-F23.

Location: Quartermaster Center and School, Fort Lee, VA.

Length: ***Version 1:*** 4 weeks (152 hours). *Version 2:* 3 weeks (108 hours).

Exhibit Dates: ***Version 1:*** 3/87–9/95. *Version 2:* 10/95–Present.

Learning Outcomes: ***Version 1:*** Upon completion of the course, the student will understand the operating procedures, prepare input transactions, process transactions, and read and interpret reports of a microcomputer-based inventory control system. *Version 2:* Upon completion of the course, the student will be able to use the military standard property book system computerized software systems.

Instruction: ***Version 1:*** Course includes classroom instruction, demonstrations, and hands-on exercises in the basic operations of a microcomputer, preparation of input transactions, processing transactions, and reading reports of an automated inventory control system. *Version 2:* Classroom instruction and practical exercises cover the operating procedures, transaction processing, and report production capabilities of the computerized inventory control system.

Credit Recommendation: ***Version 1:*** In the vocational certificate category, 1 semester hour in inventory control and 1 in computerized record keeping (10/88). *Version 2:* In the lower-division baccalaureate/associate degree category, 1 semester hour in inventory control and 1 in computerized records and information systems (5/02).

Related Occupations: 761A; 76Y; 76Z; 92A; 92B.

AR-1405-0167

NONCOMMISSIONED OFFICER (NCO) LOGISTICS PROGRAM

Course Number: 551-SQIK; 551-F5.

Location: Quartermaster Center and School, Fort Lee, VA.

Length: 6 weeks (209 hours).

Exhibit Dates: 10/88–1/91.

Learning Outcomes: Upon completion of this course, the student will understand the processes involved in logistics management and will have the knowledge to function in all levels of an organization.

Instruction: Classroom lectures and practical exercises enable the student to understand the requirements of logistics operation as it applies to many levels of an organization, including supply, ammunition, transportation, and maintenance. Automated systems are introduced in support of the logistics function.

Credit Recommendation: In the lower-division baccalaureate/associate degree category, 3 semester hours in introduction to logistics (5/02).

AR-1405-0169

STANDARD ARMY INTERMEDIATE LEVEL SUPPLY (SAIL)
(Standard Army Intermediate Level Supply (SAILS) ABX)

Course Number: 551-ASIF3.

Location: Quartermaster Center and School, Fort Lee, VA.

Length: 4 weeks (144–148 hours).

Exhibit Dates: 10/86–10/96.

Learning Outcomes: Upon completion of the course, the student will be able to operate an automated supply system.

Instruction: A significant number of classroom exercises and some lectures introduce the student to the processes involved in operating an automated supply system to support all aspects of an inventory control operation.

Credit Recommendation: In the vocational certificate category, 2 semester hours in inventory control and 1 in computerized record keeping (5/02).

Related Occupations: 76J; 76P; 76V; 76W; 76Z.

AR-1405-0170

GENERAL SUPPLY TECHNICIAN WARRANT OFFICER TECHNICAL CERTIFICATION

Course Number: 8B-761A.

Location: Quartermaster Center and School, Fort Lee, VA.

Length: 9 weeks (324–343 hours).

Exhibit Dates: 1/86–10/90.

Learning Outcomes: Upon completion of the course, the student will be able to manage a property accounting function, apply the fundamentals of stock fund accounting, apply the basic concepts of microcomputer applications, and solve problems in the use of an integrated supply management system.

Instruction: Lectures, class discussion, demonstrations, case studies, and practical exercises cover property accounting, maintenance management, governmental accounting (stock funds, microcomputer applications), and integrated supply management systems.

Credit Recommendation: In the lower-division baccalaureate/associate degree category, 1 semester hour in microcomputer applications, 1 in governmental accounting, and 3 in material management (10/88).

Related Occupations: 761A.

AR-1405-0173

SENIOR SUPPLY/SERVICES SERGEANT

Course Number: 551-76Z50.

Location: Quartermaster Center and School, Fort Lee, VA.

Length: 10 weeks (360 hours).

Exhibit Dates: 1/87–10/98.

Learning Outcomes: Upon completion of the course, the student will be able to manage supply and service operations in large organizations including planning and coordinating personnel and equipment for large stationary as well as mobile operations.

Instruction: Methods of instruction include lectures, group discussions, practical exercises, and written examinations. Topics covered include supply and service operations; leadership; automated supply systems; supervision techniques; staffing procedures; and the relationship of supply to other logistical functions, including maintenance, transportation, and management in a large organization. Course also includes procurement and types of physical inventories.

Credit Recommendation: In the upper-division baccalaureate category, 3 semester hours in supply and service management (10/88).

Related Occupations: 76Z.

AR-1405-0174

Subsistence Supply Specialist Basic Noncommissioned Officer (NCO)

Course Number: 551-76X30.

Location: Quartermaster Center and School, Fort Lee, VA.

Length: 3–4 weeks (180 hours).

Exhibit Dates: 2/89–10/98.

Learning Outcomes: Upon completion of the course, the student will be able to supervise, motivate, and manage subordinates; supervise the inventory control process; use an inventory system; and calculate supply requisitions.

Instruction: Classroom lecture, seminars, and practical exercises cover the inventory control processes of requisitions, receipts, storage, security, and issue of supplies; use of an inventory system to control the above; and motivation and supervision of subordinates to meet the objectives of an organization responsible for an inventory control function. Course includes a common core of leadership subjects.

Credit Recommendation: In the lower-division baccalaureate/associate degree category, 1 semester hours in principles of supervision and 2 in inventory control/materials management. See AR-1406-0090 for common core credit (10/88).

Related Occupations: 76X.

AR-1405-0175

Equipment Records and Parts Specialist Advanced Individual Training

Course Number: 551-76C10.

Location: Quartermaster Center and School, Fort Lee, VA.

Length: 10–11 weeks (370–375 hours).

Exhibit Dates: 10/87–10/98.

Learning Outcomes: Upon completion of the course, the student will be able to function as a clerk in manual as well as automated situations.

Instruction: Practical exercises, demonstrations, and some classroom lectures provide the student with the skills necessary to be a stock clerk, including parts inventory maintenance, use of catalogs and reference materials, completion of forms, processing repair parts, and preventive maintenance record keeping.

Credit Recommendation: In the vocational certificate category, 4 semester hours in stock keeping or parts inventory control (10/88).

Related Occupations: 76C.

AR-1405-0176

1. Direct Support Unit Standard Supply System (DS4)
2. Direct Support Unit Standard Supply System

Course Number: ***Version 1:*** 8B-F37/551-ASIT8. *Version 2:* 8B-F37/551-ASIT8.

Location: Quartermaster Center and School, Fort Lee, VA.

Length: ***Version 1:*** 4 weeks (144 hours). *Version 2:* 3 weeks (109 hours).

Exhibit Dates: ***Version 1:*** 1/88–6/93. *Version 2:* 7/93–Present.

Learning Outcomes: ***Version 1:*** Upon completion of the course, the student will be able to manage the inventory control function using a standard supply system. *Version 2:* Upon completion of the course, the student will be able to operate and supervise DS4 software applications on a desktop computer.

Instruction: ***Version 1:*** Lectures, simulations, and practical exercises cover managing the inventory control function using a standard supply system. Topics include processing supply requisitions and receipts, producing demand analysis listings, and conducting inventory using the standard system. *Version 2:* Course includes lectures and substantial use of applications on a desktop computer.

Credit Recommendation: ***Version 1:*** In the vocational certificate category, 2 semester hours in inventory control, warehousing, and/or storekeeping, 1 in computerized record keeping. In the lower-division baccalaureate/associate degree category, 2 semester hours in inventory management (10/88). *Version 2:* In the lower-division baccalaureate/associate degree category, 2 semester hours in computerized records and information management (12/96).

AR-1405-0177

Supply and Services Management Officer

Course Number: ***Version 2:*** 8B-92A/B/8B-920A/920B. *Version 3:* 8B-92A/B/8B-920A/920B, Phase 1.

Location: Quartermaster Center and School, Fort Lee, VA.

Length: ***Version 2:*** 8–9 weeks (313–331 hours). *Version 3:* 5–6 weeks (197–210 hours).

Exhibit Dates: ***Version 2:*** 10/88–12/94. *Version 3:* 1/95–Present.

Learning Outcomes: ***Version 2:*** Upon completion of the course, the student will be able to apply logistics management techniques to all classes of supply and field services. *Version 3:* Upon completion of the course, the student will be able to apply the principles of logistics management to various classes of supply and field service operations.

Instruction: ***Version 2:*** Classroom lectures and practical exercises enable the student to understand the management of a logistics operation, including facets of stock ordering, receipt, control, and movement. *Version 3:* Primarily classroom lectures and practical exercises are used to enable the student to learn the management of a logistics operation, including facets of stock ordering, receipt, control, and movement.

Credit Recommendation: ***Version 2:*** In the lower-division baccalaureate/associate degree category, 3 semester hours in inventory control or material management and 1 in microcomputer applications (5/90). *Version 3:* In the lower-division baccalaureate/associate degree category, 3 semester hours in supply management and 1 in computerized record keeping (5/02).

Related Occupations: 920A; 920B; 92A.

AR-1405-0178

Materiel Control and Accounting Specialist Advanced Noncommissioned Officer (NCO)

Course Number: 5-761-C42.

Location: Quartermaster Center and School, Fort Lee, VA.

Length: 4 weeks (145 hours).

Exhibit Dates: 8/87–10/98.

Learning Outcomes: Upon completion of the course, the student will be able to supervise the use of manual and automated inventory control procedures including data base construction and describe the flow of requests from customers to source of supply.

Instruction: Lectures, demonstration, and practical exercises cover inventory control procedures including use of an automated supply system. Topics include preparing support parameter cards, constructing a data base, processing and editing supply requests, and understanding the flow of requests from customer to the source of supply. Course includes a common core of leadership subjects.

Credit Recommendation: In the lower-division baccalaureate/associate degree category, 3 semester hours in supply management. See AR-1404-0035 for common core credit (10/88).

AR-1405-0179

Materiel Storage and Handling Specialist Advanced Individual Training

Course Number: 551-76V10.

Location: Quartermaster Center and School, Fort Lee, VA.

Length: 6–7 weeks (246 hours).

Exhibit Dates: 7/87–1/98.

Learning Outcomes: Upon completion of the course, the student will be able to serve as a materials handling and packaging storage operator using an automated supply system.

Instruction: Methodologies include lectures, demonstrations, and operation of equipment. Topics include packing and crating, warehousing, materials handling, and inventory of physical items and petroleum products. Also included is familiarization with use of an automated supply system.

Credit Recommendation: In the vocational certificate category, 3 semester hours in materials handling and storage (10/88).

Related Occupations: 76V.

AR-1405-0180

Troop Subsistence Activity Management

Course Number: 8E-F3/551-F15.

Location: Quartermaster Center and School, Fort Lee, VA.

Length: 4 weeks (144 hours).

Exhibit Dates: 4/89–10/90.

Learning Outcomes: Upon completion of the course, the student will be able to use basic management skills and administrative and warehousing practices and procedures to effectively conduct troop issue subsistence activities. Student will be able to use a basic food management information system in the troop issue subsistence activity.

Instruction: Methods of instruction include lectures, a comprehensive warehouse practical exercise, review, and written tests. Course content includes management functions; supervisory skills; subsistence record keeping (receiving, processing, and issuing); and inventory maintenance, sanitation, and control. Course also includes a computer applications component.

Credit Recommendation: In the lower-division baccalaureate/associate degree category, 1 semester hour in computer applications, 2 in personnel management, and 1 in storeroom operations (10/88).

AR-1405-0181

Repair Parts Technician Senior Warrant Officer Reserve Components

Course Number: 8-10-C32-920C-RC.

Location: Quartermaster Center and School, Fort Lee, VA.

Length: 2 weeks (76 hours).

Exhibit Dates: 10/88–10/91.

Learning Outcomes: Upon completion of the course, the student will apply the basic techniques in material acquisition and control.

Instruction: Classroom lectures and some exercises develop an awareness of requirements, distribution, financial management, and data processing support.

Credit Recommendation: In the lower-division baccalaureate/associate degree category, 1 semester hour in logistics management (10/88).

Related Occupations: 920C.

AR-1405-0183

Subsistence Supply Specialist

Course Number: 822-76X10.

Location: Quartermaster Center and School, Fort Lee, VA.

Length: 7 weeks (253 hours).

Exhibit Dates: 10/88–10/98.

Learning Outcomes: Upon completion of the course, the student will be able to use prescribed methods of warehousing, including receiving, storing, and issuing substance items; maintain sanitation standards; and use preventive maintenance scenarios for materials and/or warehousing equipment.

Instruction: Methods of instruction include lectures by instructors and students, practical exercises by participants, written examinations, and a comprehensive test. Topics covered include warehouse sanitation; requisitioning of food and supplies; receiving, storing and shipping semiperishable subsistence; materials equipment handling; and preventive maintenance.

Credit Recommendation: In the vocational certificate category, 3 semester hours in material handling (10/88).

Related Occupations: 76X.

AR-1405-0184

Property Accounting Technician Warrant Officer Advanced
(Property Accounting Technician Senior Warrant Officer Reserve Component)
(Property Book Technician Senior Warrant Officer Training Reserve Component)

Course Number: 8-10-C32-920A-RC.

Location: Quartermaster Center and School, Fort Lee, VA.

Length: 2 weeks (69–71 hours).

Exhibit Dates: 10/88–Present.

Learning Outcomes: Upon completion of the course, the student will be able to provide technical leadership in a major/large scale property management activity.

Instruction: Course includes lectures, demonstrations, and practical exercises in basic data processing concepts, regulations, logistics concepts, and a computerized inventory control system.

Credit Recommendation: In the vocational certificate category, 1 semester hour in computerized record keeping and 1 in inventory control (5/02).

Related Occupations: 920A.

AR-1405-0185

Property Accounting Technician Warrant Officer Basic
(Property Accounting Technician Quartermaster Warrant Officer Technical/Tactical Certification)
(Property Book Technician Quartermaster Warrant Officer Technical/Tactical Certification)

Course Number: ***Version 1:*** 8B-920A. *Version 2:* 8B-920A.

Location: Quartermaster Center and School, Fort Lee, VA.

Length: ***Version 1:*** 9 weeks (324–325 hours). *Version 2:* 9 weeks (347 hours).

Exhibit Dates: ***Version 1:*** 10/88–12/98. *Version 2:* 1/99–Present.

Learning Outcomes: ***Version 1:*** Upon completion of the course, the student will be able to apply principles of property management, including accountability, audit trails, work flow, and physical considerations such as safety, storage, and security; describe the operational requirements of accounting for various types of property; use a microcomputer-based property management/inventory control system; and provide technical leadership in a small to mid-sized property management activity. *Version 2:* Upon completion of the course, the student will be able to apply principles of property management, including accountability, audit trails and physical considerations; use the major Army information systems for inventory and property management; produce management reports for planning and control.

Instruction: ***Version 1:*** Course includes mostly lectures on the principles of property management and the specific procedures of accounting for various types of property. Lectures and practical hands-on exercises cover the operations, transaction requirements, and capabilities of a property control system. After 5/92, the course includes materials management and wholesale logistics. *Version 2:* Classroom exercise and lecture. Topics include property management principles, specific procedures for property accounting, hands-on exercises with Army property and inventory, materials management, and information systems.

Credit Recommendation: ***Version 1:*** In the vocational certificate category, 1 semester hour in computerized record keeping. In the lower-division baccalaureate/associate degree category, 3 semester hours in inventory control (5/96). *Version 2:* In the lower-division baccalaureate/associate degree category, 2 semester hours in computer applications, 3 in inventory control and materials management, and 1 in leadership (6/02).

Related Occupations: 920A.

AR-1405-0186

Property Accounting Technician Warrant Officer Technical/Tactical Certification Reserve
(Property Book Technician Warrant Officer Technical/Tactical Certification Reserve)

Course Number: ***Version 1:*** 8B-920A-RC. *Version 2:* 8B-920A-RC.

Location: Quartermaster Center and School, Fort Lee, VA.

Length: ***Version 1:*** 2 weeks (76 hours). *Version 2:* 2 weeks (76 hours).

Exhibit Dates: ***Version 1:*** 10/88–6/92. *Version 2:* 7/92–8/92.

Learning Outcomes: ***Version 1:*** Upon completion of the course, the student will be able to provide technical leadership in a small to mid-sized property management activity. *Version 2:* After 6/92 see AR-1405-0247. Upon completion of the course, the student will be able to provide technical leadership in a small to mid-sized property management activity.

Instruction: ***Version 1:*** Course includes classroom instruction in the principles of property management and the specific procedures of accounting for various types of property. Classroom instruction, demonstrations, and practical exercise cover the use and operation of a computerized inventory control system. *Version 2:* Lectures, hands-on demonstrations, and use of the automated system emphasize concepts and processes. The principles of property management and the specific procedures of accounting for various types of property through the use of a computerized record keeping system are presented.

Credit Recommendation: ***Version 1:*** In the vocational certificate category, 1 semester hour in computerized record keeping and 1 in inventory control (10/88). *Version 2:* In the lower-division baccalaureate/associate degree category, 2 semester hours in inventory control, or records and information management, or supply management (2/93).

Related Occupations: 920A.

AR-1405-0187

1. Property Accounting Technician Senior Warrant Officer
   (Property Book Technician Senior Warrant Officer Training)
2. Property Accounting Technician Warrant Officer Advanced
   (Property Accounting Technician Senior Warrant Officer)
   (Property Book Technician Senior Warrant Officer Training)
3. Property Accounting Technician Warrant Officer Advanced

Course Number: *Version 1:* 8-10-C32-920A. *Version 2:* 8-10-C32-920A. *Version 3:* 8-10-C32-920A.

Location: Quartermaster Center and School, Fort Lee, VA.

Length: *Version 1:* 11 weeks (394 hours). *Version 2:* 10 weeks (355–376 hours). *Version 3:* 10 weeks (360 hours).

Exhibit Dates: *Version 1:* 10/88–6/91. *Version 2:* 7/91–8/99. *Version 3:* 9/99–Present.

Learning Outcomes: *Version 1:* Upon completion of the course, the student will be able to apply skills in all phases of property management, regulations, procedures, and computerized inventory control systems to the management of a major large-scale property control activity. *Version 2:* Upon completion of the course, the student will apply communication and leadership skills to the management of a major large-scale property control activity. *Version 3:* Upon completion of the course, the student will receive advanced training to communicate and lead from a broader perspective, expand their knowledge and skill in property accounting and contracting, and make extensive use of Army information systems in all phases of property management at multi-decision levels.

Instruction: *Version 1:* Course includes classroom instruction, demonstrations, and practical exercises in supervision, communication skills, general military topics, supply management, property management, inventory control, elementary accounting, and the use of a computerized property management system. *Version 2:* Lectures, demonstrations, classroom presentations, and considerable hands-on experience with an automated inventory control system are used to introduce the student to the principles of inventory and logistics management. All phases of property management; inventory control; storage, receipt, and issue of supplies and materials; and logistics planning for food services, bulk petroleum, and water supply are presented. *Version 3:* Audio-visual materials, practical exercise, lecture, in-class problem solving, in-class presentations, and active interaction with several key supply information systems.

Credit Recommendation: *Version 1:* In the vocational certificate category, 2 semester hours in computerized record keeping. In the lower-division baccalaureate/associate degree category, 2 semester hours in business communication, 1 in supply management, 1 in inventory control, and 1 in principles of supervision (10/88). *Version 2:* In the lower-division baccalaureate/associate degree category, 2 semester hours in business communication; 1 in principles of leadership or supervision; 3 in inventory management, inventory control, or supplies management; and 2 in introduction to computers or records and information management (6/95). *Version 3:* In the lower-division baccalaureate/associate degree category, 2 semester hours in principles of communication; 1 in principles of leadership; 3 in inventory, material, and supply management; and 2 in computer applications (6/02).

Related Occupations: 920A.

AR-1405-0188

1. Supply Systems Technician Quartermaster Warrant Officer Technical/Tactical Certification (Repair Parts Technician Quartermaster Warrant Officer Technical/Tactical Certification)
2. Supply Systems Technician Warrant Officer Basic
3. Supply Systems Technician Warrant Officer Basic

Course Number: *Version 1:* 8B-920B. *Version 2:* 8B-920B. *Version 3:* 8B-920B.

Location: Quartermaster Center and School, Fort Lee, VA.

Length: *Version 1:* 9 weeks (321–326 hours). *Version 2:* 9 weeks (321–349 hours). *Version 3:* 9 weeks (353 hours).

Exhibit Dates: *Version 1:* 10/88–3/94. *Version 2:* 4/94–12/98. *Version 3:* 1/99–Present.

Learning Outcomes: *Version 1:* Upon completion of the course, the student will apply appropriate logistics support techniques to the repair parts supply activity. *Version 2:* Upon completion of the course, the student will be able to perform management tasks in the areas of maintenance, finance, petroleum and water operations, and materiel. *Version 3:* Upon completion of the course, the student will be able to perform management tasks in the areas of maintenance, finance, petroleum and water resources and materials; understand the management of the supply support function; use information systems for transactions, record keeping, material release, and order control.

Instruction: *Version 1:* Classroom lectures and a significant amount of practical exercises are designed to teach the student the inventory processes of requisitioning, receiving, stocking, and issuing various types of supplies. Automated support of inventory systems is also introduced. *Version 2:* Course includes classroom lectures and experiential exercises. Substantial course content is in automated supply systems. *Version 3:* Practical exercise and lecture. Topics include classroom lectures, detailed exercises and briefings, and field trips to major supply facilities.

Credit Recommendation: *Version 1:* In the lower-division baccalaureate/associate degree category, 3 semester hours in inventory control warehousing and/or stockkeeping and 1 in microcomputer applications (12/92). *Version 2:* In the lower-division baccalaureate/associate degree category, 2 semester hours in supply management and 2 in computerized record and information management (12/96). *Version 3:* In the lower-division baccalaureate/associate degree category, 3 semester hours in inventory supply and materials management, 2 in computer applications, and 1 in principles of leadership (6/02).

Related Occupations: 920A; 920B.

AR-1405-0190

Supply Systems Technician Warrant Officer Basic - Reserve
(Repair Parts Technician Quartermaster Warrant Officer Technical/Tactical Certification Reserve Component)
(Supply Systems Technician Quartermaster Warrant Officer Technical/Tactical Certification Reserve)

Course Number: 8B-920B-RC.

Location: Quartermaster Center and School, Fort Lee, VA.

Length: 2 weeks (76–78 hours).

Exhibit Dates: 10/88–4/04.

Learning Outcomes: Upon completion of the course, the student will be able to describe the logistics operations of materiel acquisition and control.

Instruction: Classroom lectures and some exercises develop an awareness of the logistics functions of requirements, distribution support, financial management, and data processing support.

Credit Recommendation: In the lower-division baccalaureate/associate degree category, 1 semester hour in logistics management (12/92).

Related Occupations: 920B.

AR-1405-0191

Advanced Logistics Management

Course Number: 8A-F37.

Location: Quartermaster Center and School, Fort Lee, VA.

Length: 4 weeks (144 hours).

Exhibit Dates: 1/89–10/98.

Learning Outcomes: Upon completion of the course, the student will be able to manage the activities associated with procurement, storage, issue, and supply of all items necessary to maintain the operating level of an organization.

Instruction: Instruction is primarily a classroom activity through seminars and lectures to develop an understanding of the logistics process, including usage requirements, contract principles, negotiation, contract administration, project management, depot operations, and automated systems support.

Credit Recommendation: In the upper-division baccalaureate category, 3 semester hours in logistics management (10/88).

AR-1405-0194

Supply Support Activity (SSA) Management

Course Number: LOG 71.

Location: Combined Arms Training Center, Vilseck, W. Germany.

Length: 3 weeks (120 hours).

Exhibit Dates: 10/87–10/98.

Learning Outcomes: Upon completion of the course, the student will be able to describe the basic functions of supply support activities and storage operations; implement a system of issuing, receiving, and shipping procedures; build and maintain a supply system, including constructing a data base, managing parameters, updating data files, processing supply status documents, follow-ups, and cancellation requests, and controlling inventory based on demand analysis; use management reports and supply performance measures; and analyze future supply systems.

Instruction: Lectures and practical exercises using actual documentation and publications.

Credit Recommendation: In the lower-division baccalaureate/associate degree category, 3 semester hour in supply management (10/88).

AR-1405-0195

The Army Maintenance Management System (TAMMS) and Prescribed Loading List (PLL)

Course Number: LOG 74.

Location: Combined Arms Training Center, Vilseck, W. Germany.

Length: 2 weeks (72 hours).

Exhibit Dates: 8/87–10/98.

Learning Outcomes: Upon completion of the course, the student will be able to accurately prepare, maintain, and dispose of necessary maintenance and supply forms and records; maintain a publications library for unit equipment, supply, and maintenance operations; prepare, maintain, and dispose of historical records; use microfiche and records to identify and locate specific requirements for maintenance and supply activities and; demonstrate a basic understanding of the prescribed load list policies and procedures.

Instruction: Lectures, demonstrations, and seminars cover identifying, preparing, and stating the disposition of various maintenance and supply forms.

Credit Recommendation: In the lower-division baccalaureate/associate degree category, 1 semester hour in clerical procedures (10/88).

AR-1405-0196

1. Personnel Information Systems Management Specialist Basic Noncommissioned Officer (NCO)
2. Personnel Information Systems Management Specialist Basic Noncommissioned Officer (NCO)
3. Personnel Information System Management Specialist

Course Number: ***Version 1:*** 500-75F30. *Version 2:* 500-75F30. *Version 3:* 500-75F30.

Location: ***Version 1:*** Soldier Support Institute, Fort Benjamin Harrison, IN. *Version 2:* Adjutant General School, Fort Benjamin Harrison, IN. *Version 3:* Soldier Support Institute, Fort Jackson, SC.

Length: ***Version 1:*** 7 weeks (285 hours). *Version 2:* 8 weeks (296 hours). *Version 3:* 8–9 weeks (287 hours).

Exhibit Dates: ***Version 1:*** 10/86–9/91. *Version 2:* 10/91–12/95. *Version 3:* 1/96–Present.

Learning Outcomes: ***Version 1:*** Upon completion of the course, the student will be able to make corrections in a personnel information system and review data input. *Version 2:* Upon completion of the course, the student will be able to make corrections in a personnel information system and review data input. *Version 3:* Upon completion of the course the student will be able to process personnel data via an automated personnel information system, including input to the system, review of output, and correction of errors.

Instruction: ***Version 1:*** Classroom instruction and practical experience cover the use and operation of a microcomputer, including data entry, processing, and data output. *Version 2:* Classroom instruction and practical experience cover the use and operation of a microcomputer, including data entry, processing, and data output. *Version 3:* Lecture, class discussions, and personal computer exercises cover the setup of a personal computer system and operation of the system, including data input, processing, output error correction, and report generation.

Credit Recommendation: ***Version 1:*** In the vocational certificate category, 1 semester hour in personnel record keeping. In the lower-division baccalaureate/associate degree category, 1 semester hour in computer applications and 1 in keyboard operations (12/88). *Version 2:* In the lower-division baccalaureate/associate degree category, 3 semester hours in clerical procedures (12/91). *Version 3:* In the lower-division baccalaureate/associate degree category, 1 semester hour in personnel record keeping and 2 in records and information management (9/97).

Related Occupations: 75F.

AR-1405-0197

Logistics Management

Course Number: ***Version 1:*** 600. *Version 2:* 600. *Version 3:* 921-430.

Location: Reserve Readiness Training Center, Fort McCoy, WI.

Length: ***Version 1:*** 2 weeks (100 hours). *Version 2:* 2 weeks (82 hours). *Version 3:* 2 weeks (74 hours).

Exhibit Dates: ***Version 1:*** 11/87–5/93. *Version 2:* 6/93–2/99. *Version 3:* 3/99–Present.

Learning Outcomes: ***Version 1:*** Upon completion of the course, the student will be able to administer and maintain systems for transportation, delivery, and movement of supplies, including how to use documents for automation, facility management, publications, safety, and food service; how to prepare for mobilization using such systems as unit mobilization, force integration, inventory control, and load planning; and hazardous cargo planning. *Version 2:* Upon completion of the course, the student will be able to administer and maintain systems for transportation, delivery, and movement of supplies and use documents for automation, facility management, safety, and food service. *Version 3:* Pending evaluation.

Instruction: ***Version 1:*** Course is taught using lectures and practical applications, quizzes, handouts, take-home materials, and a final examination. *Version 2:* Course is taught using lectures and practical applications, quizzes, handouts, take-home materials, and a final examination. *Version 3:* Pending evaluation.

Credit Recommendation: ***Version 1:*** In the lower-division baccalaureate/associate degree category, 2 semester hours in material or supply management. In the upper-division baccalaureate category, 2 semester hours in logistics management (7/89). *Version 2:* In the lower-division baccalaureate/associate degree category, 2 semester hours in material or supply management and 1 in transportation management. In the upper-division baccalaureate category, 2 semester hours in logistics management (3/97). *Version 3:* Pending evaluation.

AR-1405-0198

1. Unit Mobilization Planners
2. RC Unit Mobilization Planners

Course Number: ***Version 1:*** 2000; 2000-LVL2. *Version 2:* 921-340.

Location: Reserve Readiness Training Center, Fort McCoy, WI.

Length: ***Version 1:*** 2 weeks (76–92 hours). *Version 2:* 2–3 weeks (92 hours).

Exhibit Dates: ***Version 1:*** 5/87–8/00. *Version 2:* 9/00–Present.

Learning Outcomes: ***Version 1:*** Upon completion of the course, the student will plan unit mobilization for Army reserve units, including transportation operating agency actions and relationships, mobilization and development planning and execution, and installation activities during post-mobilization actions. *Version 2:* Upon completion of the course, the student will be able to plan and execute mobilization actions for Army reserve units, including transportation, communications, and integration with active Army units, and perform appropriate post-mobilization actions.

Instruction: ***Version 1:*** Classroom lectures and discussions, workbook exercises, quizzes, and final examination cover fundamentals of Department of Defense and Department of the Army mobilization planning process; coordination of transportation and ADP systems mobilization requirements; mobilization planning responsibilities in the areas of security, personnel, and administration; logistics and movement planning; and completion of a detailed unit mobilization plan. *Version 2:* Audiovisual materials, practical and classroom exercises, discussion, case studies, lecture, and quizzes. Topics include foundations of mobilization planning, unit level mobilization planning, personnel and administration mobilization planning, operations and training mobilization planning, logistics mobilization planning, mobilization movement planning, mobilization file review and evaluation, mobilization exercise, and demobilization.

Credit Recommendation: ***Version 1:*** In the upper-division baccalaureate category, 2 semester hours in logistics management (7/96). *Version 2:* In the upper-division baccalaureate category, 2 semester hours in workforce management or logistics management (11/02).

AR-1405-0199

Training Management

Course Number: ***Version 1:*** 500. *Version 2:* 500. *Version 3:* 921-310.

Location: Reserve Readiness Training Center, Fort McCoy, WI.

Length: ***Version 1:*** 2 weeks (75 hours). *Version 2:* 2 weeks (70–71 hours). *Version 3:* 2 weeks (71 hours).

Exhibit Dates: ***Version 1:*** 7/87–8/94. *Version 2:* 9/94–4/01. *Version 3:* 5/01–Present.

Learning Outcomes: ***Version 1:*** Upon completion of the course, the student will be able to identify Army reserve training needs and priorities, develop short and long-range training strategies, and plan integrated budget and administrative support to satisfy various mobilization scenarios. *Version 2:* Upon completion of the course, the student will be able to establish long-range training goals, develop training objectives, develop a training strategy, write a yearly training guidance plan, plan a mission essential task-list-based training event, complete a post mobilization training plan, and present a yearly training briefing. *Version 3:* Upon completion of the course, the student will be able to establish long-range training goals, develop training objectives, develop a training strategy, write a yearly training guidance plan, plan a mission essential task-list-based training event, complete a post mobilization training plan, and present a yearly training briefing.

Instruction: ***Version 1:*** Methods of instruction include lectures, discussions, projects, quizzes, and final examinations. Topics include an overview of Army force structure, fundamentals of the Army Training Management System, short and long-range training strategies, reconciliation of training needs with requisite budget and administrative resources, and mobilization levels and phases and their impact on training requirements and readiness. *Version 2:* Methods of instruction include lectures, discussions, projects, quizzes, and final examinations. *Version 3:* Methods of instruction include lectures, discussions, projects, quizzes, and final examinations. Topics include an overview of Army force structure, fundamentals of the Army Training Management System, short and long-range training strategies, reconciliation of training needs with requisite budget and administrative resources, and mobilization levels and phases and their impact on training requirements and readiness.

Credit Recommendation: ***Version 1:*** In the upper-division baccalaureate category, 2 semester hours in training management (7/89). *Version 2:* In the upper-division baccalaureate category, 2 semester hours in training management (3/97). *Version 3:* In the upper-division baccalaureate category, 2 semester hours in training management (11/02).

AR-1405-0200

1. Reserve Supply
2. Company Supply/ULLS-S4 (Reserve Supply)

Course Number: ***Version 1:*** 1000. *Version 2:* 921-440.

Location: Reserve Readiness Training Center, Fort McCoy, WI.

Length: ***Version 1:*** 2 weeks (80 hours). *Version 2:* 2 weeks (76 hours).

Exhibit Dates: ***Version 1:*** 4/87–6/96. *Version 2:* 10/00–Present.

Learning Outcomes: ***Version 1:*** Upon completion of the course, the student will be able to use control inventory and work flow principles as they relate to requisition procedures and forms, hand-receipt procedures, property disposition, physical security, and material condition status reports. *Version 2:* Upon completion of the course, the student will be able to perform inventory control, perform requisition procedures (automated and manual), perform turn in procedures, use computer software packages, and use supply audit trains.

Instruction: ***Version 1:*** Lectures, discussions, quizzes, final examination, structured exercises are course methods. *Version 2:* Practical exercises, discussion, lecture, and computer-based training. Topics include modern Army Record Keeping System, supply publications for the Army Authorization Document Redesign manual/automated document registers, requisitions, receipt of supplies, equipment responsibilities, inventories, soldier issue files, turn-in supplies, and Unit Level Logistics Systems.

Credit Recommendation: ***Version 1:*** In the lower-division baccalaureate/associate degree category, 1 semester hour in office procedures (7/89). *Version 2:* In the lower-division baccalaureate/associate degree category, 2 semester hours in supply management (11/02).

AR-1405-0201

1. Movement Officer
2. Unit Movement Officer
3. Unit Movement Officer

Course Number: ***Version 1:*** 3400. *Version 2:* 3400; 921-3400-LVL1. *Version 3:* 921-410.

Location: Reserve Readiness Training Center, Fort McCoy, WI.

Length: ***Version 1:*** 2 weeks (80 hours). *Version 2:* 2 weeks (78–79 hours). *Version 3:* 2–3 weeks (104 hours).

Exhibit Dates: ***Version 1:*** 12/88–8/93. *Version 2:* 9/93–9/00. *Version 3:* 10/00–Present.

Learning Outcomes: ***Version 1:*** Upon completion of the course, the student will be able to complete and analyze a unit's load plan for hazardous cargo, including proper marking and documentation requirements; develop a convoy plan and record its elements; develop a rail movement plan for a given unit; and develop an air movement plan, including a load plan for a specific aircraft and air lift support elements, teams, and team responsibilities. *Version 2:* Upon completion of the course, the student will be able to complete and analyze a unit's load plan for hazardous cargo, including proper marking and documentation requirements; develop a rail, water, and motor movement plan for a given unit; and develop an air movement plan, including a load plan for a specific aircraft and air lift support elements, teams, and team responsibilities. *Version 3:* Upon completion of the course, the student will be able to identify phases and levels of mobilization and deployment; prepare and conduct a redeployment; develop the components of a mobilization movement plan to ensure validity and workability; supervise the movement of hazardous materials; identify the characteristics and purpose of the Intransit Visibility System; prepare for and conduct a redeployment and demobilization; demonstrate the ability to plan, organize, and execute a movement by motor, rail, and air; and demonstrate an understanding of water movement operations.

Instruction: ***Version 1:*** Course includes homework and a series of practical exercises for each unit. Lectures, discussions, quizzes, and final examinations are part of the total point requirement. *Version 2:* Course includes homework and a series of practical exercises for each unit. Lectures, discussions, quizzes, and final examinations are part of the total point requirement. *Version 3:* Audiovisual materials, practical and classroom exercises, discussion, learner presentation, lecture, and computer-based training. Topics include mobilization and movement management; movement of hazardous materials; reception, staging, onward movement and integration; redeployment; demobilization; validation; requesting commercial transportation; Intransit Visibility System; deployment movement plan; planning for motor, rail, and air movement; and water movement operations.

Credit Recommendation: ***Version 1:*** In the upper-division baccalaureate category, 3 semester hours in logistics management (7/89). *Version 2:* In the upper-division baccalaureate category, 2 semester hours in traffic management or 3 in industrial supervision (3/97). *Version 3:* In the upper-division baccalaureate category, 2 semester hours in logistics management (11/02).

AR-1405-0202

Medical Supply Specialist Reserve
(Medical Supply Specialist (Basic) Reserve Component, Phase 2)

Course Number: 551-76J10-RC.

Location: Academy of Health Sciences, Fort Sam Houston, TX.

Length: 2 weeks (82–90 hours).

Exhibit Dates: 11/89–Present.

Learning Outcomes: Upon completion of the course, the student will be able to manage the supply and equipment inventory system of a medical unit using the federal supply catalog, DoD Medical Catalog, microfiche viewer, DLA customer supply assistance booklet, and an automated system containing the Theater Army Medical Information System.

Instruction: This is a two phase course. Phase 1 is conducted as a correspondence course. Phase 2 is the resident portion. This second phase includes lectures, discussions, and practical exercises in the federal supply catalog, military standard requisitioning issue procedures, general administrative procedures, manual and automated property and inventory control, stock accounting, and the Theater Army Medical Management Information System.

Credit Recommendation: In the lower-division baccalaureate/associate degree category, 3 semester hours in material management (supply management) (2/96).

Related Occupations: 76J.

AR-1405-0203

Medical Logistics Noncommissioned Officer (NCO)

Course Number: 551-F1.

Location: Academy of Health Sciences, Fort Sam Houston, TX.

Length: 4 weeks (157 hours).

Exhibit Dates: 7/88–4/04.

Learning Outcomes: Upon completion of the course, the student will be able to interpret and implement policies and procedures using computers in inventory management, financial management, and management information system.

Instruction: This course is a combination of conferences and practical experiences covering computer literacy and management applications to finance, inventory, and a management information system.

Credit Recommendation: In the lower-division baccalaureate/associate degree category, 3 semester hours in systems management (3/90).

AR-1405-0204

Supply Noncommissioned Officer (NCO)

Course Number: SNCO 90.

Location: Professional Education Center, Cp. Joseph Robinson, North Little Rock, AR.

Length: 1 week (54–56 hours).

Exhibit Dates: 10/89–12/97.

Learning Outcomes: Upon completion of the course, the student will be able to address supply issues and personnel issues related to subsistence, maintenance of equipment, and security.

Instruction: Instruction strategies include lectures, discussions, structural exercises, and practical exercises. Content includes key security, clothing control, maintenance requests, meal accounting, subsections, cash vouchers, and other supply issues.

Credit Recommendation: In the vocational certificate category, 1 semester hour in office procedures (7/90).

AR-1405-0205

Senior Level Logistics

Course Number: SLL 90.

Location: Professional Education Center, Cp. Joseph Robinson, North Little Rock, AR.

Length: 1–2 weeks (56 hours).

Exhibit Dates: 10/89–12/97.

Learning Outcomes: Upon completion of the course, the student will be able to supervise the accountability of medical, financial, environmental, food service, and property resources.

Instruction: Instructional strategies include practical experience, structural exercises, lectures, and discussions. Course content includes accountability management of medical, financial, maintenance, environmental, food service, and property resources.

Credit Recommendation: In the lower-division baccalaureate/associate degree category, 1 semester hour in supply management (7/90).

AR-1405-0206

ORGANIZATION DEVELOPMENT FACILITATORS

Course Number: PEC-O ODF.

Location: Professional Education Center, Cp. Joseph Robinson, North Little Rock, AR.

Length: 3 weeks (114 hours).

Exhibit Dates: 6/90–12/97.

Learning Outcomes: Upon completion of the course, the student will be able to facilitate group discussion on issues critical to organizational group process.

Instruction: Instruction includes lectures, discussions, practical exercises, demonstrations, reports, and objective evaluations. Content covered includes learning theory; group concepts; group process; and briefing/reporting skills, including research, organization, delivery, and audiovisual aids.

Credit Recommendation: In the lower-division baccalaureate/associate degree category, 2 semester hours in oral group communication skills (7/90).

AR-1405-0208

SENIOR EXECUTIVE TRANSITION

Course Number: PEC-T-O-SET91; PEC-T-O-SET.

Location: Professional Education Center, Cp. Joseph Robinson, North Little Rock, AR.

Length: 1–2 weeks (51–66 hours).

Exhibit Dates: 4/88–12/97.

Learning Outcomes: Upon completion of the course, the student will be able to identify various human relations, self-analysis, and communication skills use those skills to achieve strategic objectives in an organization.

Instruction: Lectures, group discussions, and in-class exercises cover areas in interpersonal communication, team building, situational leadership, values, assessment, and stress management.

Credit Recommendation: Credit is not recommended because there is no provision for formal student evaluation (7/90).

AR-1405-0209

ORGANIZATIONAL MAINTENANCE SHOP (OMS)/UNIT TRAINING EQUIPMENT SITE (UTES)

Course Number: 90 OMS/UTES.

Location: Professional Education Center, Cp. Joseph Robinson, North Little Rock, AR.

Length: 1 week (63 hours).

Exhibit Dates: 3/90–12/97.

Learning Outcomes: Upon completion of the course, the student will be able to supervise the safety of hazardous materials, supervise security of materials in a unit accounting inventory, and maintain supply records.

Instruction: Lectures, discussions, and exercises are the primary instructional methods. The content includes management of an office in the areas of safety, hazardous material safety, inventory, and accounting.

Credit Recommendation: In the lower-division baccalaureate/associate degree category, 1 semester hour in supply supervision (7/90).

AR-1405-0210

ORGANIZATIONAL MAINTENANCE SHOP (OMS)/UNIT TRAINING EQUIPMENT SITE (UTES) TOOL AND PARTS ATTENDANT

Course Number: 90 OMS/UTES.

Location: Professional Education Center, Cp. Joseph Robinson, North Little Rock, AR.

Length: 1 week (49 hours).

Exhibit Dates: 3/90–12/97.

Learning Outcomes: Upon completion of the course, the student will be able to supervise safety, hazardous materials, security, and accounting of tools and parts.

Instruction: Course topics include correspondence, securities, accounting, safety, records, inventory, and hazardous materials.

Credit Recommendation: Credit is not recommended because of the military-specific nature of the course (7/90).

AR-1405-0211

LOGISTICS EXECUTIVE DEVELOPMENT

Course Number: ***Version 1:*** 8A-F17. *Version 2:* 8A-F17. *Version 3:* 8A-F17.

Location: Army Logistics Management College, Fort Lee, VA.

Length: ***Version 1:*** 19–20 weeks (708–755 hours). *Version 2:* 19–20 weeks (730 hours). *Version 3:* 16 weeks (594 hours).

Exhibit Dates: ***Version 1:*** 12/89–2/93. *Version 2:* 3/93–12/97. *Version 3:* 1/98–Present.

Learning Outcomes: ***Version 1:*** Upon completion of the course, the student will be able to serve in an executive position in logistics management. *Version 2:* Upon completion of the course, the student will be able to serve in an executive position in logistics management. *Version 3:* Upon completion of the course, the student will be able to serve in an executive position in logistics management.

Instruction: ***Version 1:*** Lectures, discussions, case studies, computer-assisted simulation, practical exercises, guest lecturers, and applied research are course methodologies. Topics covered include development of strategy and logistics policies, acquisition management, materiel readiness, tactical logistics, historical importance of logistics, managerial communication, resource management, managerial economics, analytical techniques, and current computer applications. *Version 2:* Lectures, discussions, case studies, computer-assisted simulation, practical exercises, guest lecturers, and applied research are course methodologies. Topics covered include development of strategies and logistics policies, acquisition management, materiel readiness, tactical logistics, historical importance of logistics, managerial communication, resource management (including inventory control systems), managerial economics, macroeconomic principles, quantitative methods and analytical techniques, and current computer applications. *Version 3:* Lectures, discussions, case studies, computer-assisted simulation, practical exercises, guest lecturers, and applied research are course methodologies. Topics covered include development of strategies and logistics policies, acquisition management, materiel readiness, tactical logistics, historical importance of logistics, managerial communication, resource management (including inventory control systems), managerial economics, macroeconomic principles, quantitative methods and analytical techniques, and current computer applications.

Credit Recommendation: ***Version 1:*** In the upper-division baccalaureate category, 3 semester hours in general management, 2 in managerial economics, 1 in strategic management, 3 in procurement management, 6 in logistics management, 3 in quantitative analysis. In the graduate degree category, 3 semester hours in managerial economics and 6 in management science (10/91). *Version 2:* In the lower-division baccalaureate/associate degree category, 3 semester hours in macroeconomic principles. In the upper-division baccalaureate category, 3 semester hours in principles of management, 3 in strategy and policy, 3 in procurement management, 6 in logistics management, and 3 in quantitative analysis. In the graduate degree category, 3 semester hours in managerial economics, 3 in systems management, and 3 in management science (7/00). *Version 3:* In the upper-division baccalaureate category, 3 semester hours in introduction to management, 2 in managerial economics, 1 in strategic management, 3 in procurement management, 6 in logistics management and 2 in quantitative analysis. In the graduate degree category, 6 semester hours in general management and 3 in managerial economics (8/02).

AR-1405-0212

LOGISTICS EXECUTIVE DEVELOPMENT BY CORRESPONDENCE

Course Number: ***Version 1:*** 8A-F17. *Version 2:* 8A-F17.

Location: Army Logistics Management College, Fort Lee, VA.

Length: ***Version 1:*** Maximum, 52 weeks. *Version 2:* Maximum, 52 weeks.

Exhibit Dates: ***Version 1:*** 12/89–2/93. *Version 2:* 3/93–8/94.

Learning Outcomes: ***Version 1:*** Upon completion of the course, the student will be able to serve in an executive position in logistics management. *Version 2:* Upon completion of the course, the student will be able to serve in an executive position in logistics management.

Instruction: ***Version 1:*** This is a correspondence course. Topics covered include development of strategy and logistics policies, acquisition management, materiel readiness, tactical logistics, historical importance of logistics, managerial communication, resource management, managerial economics, analytical techniques, and current computer applications. *Version 2:* This is a correspondence course. Topics covered include development of strategies and logistics policies, acquisition management, materiel readiness, tactical logistics, historical importance of logistics, managerial communication, resource management (including inventory control systems), managerial economics and macroeconomic principles, quantitative methods and analytical techniques, and current computer applications.

Credit Recommendation: ***Version 1:*** In the upper-division baccalaureate category, 3 semester hours in general management, 2 in managerial economics, 1 in strategic management, 3 in procurement management, 6 in logistics management, 3 in quantitative analysis. In the graduate degree category, 3 semester hours in managerial economics and 6 in management science (6/91). *Version 2:* In the lower-division baccalaureate/associate degree category, 3 semester hours in macroeconomic principles. In the upper-division baccalaureate category, 3 semester hours principles of management, 3 in strategy and policy, 3 in procurement management, 6 in logistics management, and 3 in quantitative analysis. In the graduate degree category, 3 semester hours in managerial economics, 3 in systems management, and 3 in management science (7/00).

AR-1405-0213

1. Associate Logistics Executive Development Reserve Component
2. Associate Logistics Executive Development

Course Number: ***Version 1:*** 8A-F19. *Version 2:* 8A-F19 (ALMC-AL, -AC, -AD, -AE, -AM).

Location: Army Logistics Management College, Fort Lee, VA; Army Logistics Management College, Onsite at various locations.

Length: ***Version 1:*** 10 weeks (365 hours). *Version 2:* 10 weeks (360 hours).

Exhibit Dates: ***Version 1:*** 10/88–2/93. *Version 2:* 3/93–Present.

Learning Outcomes: ***Version 1:*** Before 10/88 see AR-0326-0003. Upon completion of the course, the student will be able to assume executive responsibilities in logistics management. *Version 2:* Upon completion of the course, the student will be able to assume executive responsibilities in logistics management.

Instruction: ***Version 1:*** Lectures, discussions, exercises, computer-assisted simulations, and applied research are the delivery modes. Topics covered include organizational, financial, and personnel management in logistics; acquisition management; distribution maintenance and materiel management; managerial economics, statistics, and operations research; and logistics support concepts. The course is offered as resident, correspondence, or a combination of both delivery systems. Proctored final examinations are required in all delivery systems. *Version 2:* Lectures, discussions, exercises, computer-assisted simulations, and applied research cover organizational, financial, and personnel management in logistics; acquisition management; distribution maintenance, including facilities, equipment, and transportation; materiel management including inventory control systems; managerial economics, statistics, and operations research; and logistics support concepts. The course is offered as resident, correspondence, or a combination of both delivery systems. Proctored final examinations are required for all delivery systems. This course consists of five phases, all of which must be completed before credit it granted. The phases are Combat Logistics Part 1 (ALMC-AL), Acquisition Process (ALMC-AC), Materiel Readiness (ALMC-AD), Decision Sciences (ALMC-AE), and Combat Logistics Part 2 (ALMC-AM).

Credit Recommendation: ***Version 1:*** In the upper-division baccalaureate category, 3 semester hours in economic principles and decision making, 3 in logistics management, 3 in materiel management, and 3 in principles of management. In the graduate degree category, 3 semester hours in advanced logistics management (6/91). *Version 2:* In the lower-division baccalaureate/associate degree category, 1 semester hour in computer applications in logistics management. In the upper-division baccalaureate category, 3 semester hours in economic analysis for decision making, 3 in principles of management, 3 in logistics management, 3 in materiel management, and 2 in inventory control principles. In the graduate degree category, 3 semester hours in advanced logistics management (7/00).

AR-1405-0214

1. Associate Logistics Executive Development Reserve Component by Correspondence
2. Associate Logistics Executive Development by Correspondence

Course Number: ***Version 1:*** 8A-F19. *Version 2:* 8A-F19.

Location: Army Logistics Management College, Fort Lee, VA.

Length: ***Version 1:*** Maximum, 52 weeks. *Version 2:* Maximum, 52 weeks.

Exhibit Dates: ***Version 1:*** 10/88–2/93. *Version 2:* 3/93–Present.

Learning Outcomes: ***Version 1:*** Upon completion of the course, the student will be able to assume executive responsibilities in logistics management. *Version 2:* Upon completion of the course, the student will be able to assume executive responsibilities in logistics management.

Instruction: ***Version 1:*** This is a correspondence course. Topics covered include organizational, financial, and personnel management in logistics; acquisition management; distribution maintenance and materiel management; managerial economics, statistics, and operations research; and logistics support concepts. The course is offered as resident, correspondence, or a combination of both delivery systems. Proctored final examinations are required in all delivery systems. *Version 2:* Topics covered include organizational, financial, and personnel management in logistics; acquisition management; distribution maintenance, including facilities, equipment, and transportation; materiel management including inventory control systems; managerial economics, statistics, and operations research; and logistics support concepts. The course is offered as resident, correspondence, or a combination of both delivery systems. Proctored final examinations are required for all systems.

Credit Recommendation: ***Version 1:*** In the upper-division baccalaureate category, 3 semester hours in economic principles and decision making, 3 in logistics management, 3 in materiel management, and 3 in principles of management. In the graduate degree category, 3 semester hours in advanced logistics management (6/91). *Version 2:* In the upper-division baccalaureate category, 3 semester hours in economic analysis for decision making, 3 in principles of management, 3 in logistics management, 3 in materiel management, and 2 in inventory control principles. In the graduate degree category, 3 semester hours in advanced logistics management. Credit is to be awarded upon completion of entire program (8/95).

AR-1405-0216

Defense Reutilization and Marketing Operations Basic

Course Number: ***Version 1:*** 8G-F1. *Version 2:* 8G-F1.

Location: Army Logistics Management College, Fort Lee, VA.

Length: ***Version 1:*** 3 weeks (112 hours). *Version 2:* 3 weeks (81 hours).

Exhibit Dates: ***Version 1:*** 2/89–6/95. *Version 2:* 7/95–Present.

Learning Outcomes: ***Version 1:*** Upon completion of the course, the student will be able to communicate with appropriate organizational elements of the reutilization and marketing service, other governmental agencies, and the public; promote the programs and perform duties associated with reutilization, transfer, and donation of property; carry out sales functions in order to optimize the monetary return to the government, including sales planning, promotion, item description writing, and contracting functions of sales; and process property requiring special handling into and through the reutilization and marketing cycle. *Version 2:* Upon completion of the course, the student will be able to communicate with appropriate organizational elements of the reutilization and marketing service, other governmental agencies, and the public; promote programs and perform duties associated with property accounting, reutilization, transfer, and donation of property; carry out sales functions in order to optimize the monetary return to the government, including sales planning, promotion, item description writing, and contracting functions of sales; and process property requiring special handling including hazardous property.

Instruction: ***Version 1:*** Instructional methods include workshops, practical exercises, and lecture-conferences using guest speakers as well as resident faculty. Topics covered include basic disposal operations, including objectives, policies, and procedures involved in the reutilization, donation, sale, abandonment, or destruction and demolition of DoD excess and surplus property. *Version 2:* Instructional methods include lecture-conferences, practical exercises, workshops, and guest speakers. Topics covered include basic disposal operations, including objectives, policies, and procedures involved in the reutilization, donation, sale, abandonment, or destruction of property; property accounting; demilitarization of DoD excess and surplus property; and property accounting.

Credit Recommendation: ***Version 1:*** In the upper-division baccalaureate category, 3 semester hours in applied marketing (6/91). *Version 2:* In the upper-division baccalaureate category, 3 semester hours in marketing distribution systems (8/95).

AR-1405-0217

Defense Reutilization and Marketing Operations Basic by Correspondence

Course Number: 8G-F1.

Location: Army Logistics Management College, Fort Lee, VA.

Length: Maximum, 52 weeks.

Exhibit Dates: 2/89–4/93.

Learning Outcomes: Upon completion of the course, the student will be able to communicate with appropriate organizational elements of the reutilization and marketing service, other governmental agencies, and the public; promote the programs and perform duties associated with reutilization, transfer, and donation of property; carry out sales functions in order to optimize the monetary return to the government, including sales planning, promotion, item description writing, and contracting functions of sales; and process property requiring special handling into and through the reutilization and marketing cycle.

Instruction: This is a correspondence course with a proctored final exam. Topics covered include basic disposal operations, including objectives, policies and procedures involved in the reutilization, donation, sale, abandonment, or destruction and demolition of DoD excess and surplus property.

Credit Recommendation: In the upper-division baccalaureate category, 3 semester hours in applied marketing (6/91).

AR-1405-0218

COMMODITY COMMAND STANDARD SYSTEM PHYSICAL INVENTORY MANAGEMENT

Course Number: ALMC-6C.

Location: Army Logistics Management College, Fort Lee, VA.

Length: 1–2 weeks (50 hours).

Exhibit Dates: 1/88–10/92.

Learning Outcomes: Upon completion of the course, the student will be able to identify and prepare inventory application input documents, analyze and interpret inventory application output products, and describe the major processes and procedures of inventory applications.

Instruction: Programmed instruction and practical exercises cover requesting/control procedures, materiel release denials, zero balance flasher, and quality control. Reinforcement of knowledge gained in the location record audit/match is depicted in an audio tape case. Videotaped lectures are used to introduce each block of instruction. Initial, intermediate, and final surveys are used to examine the students throughout the course.

Credit Recommendation: In the lower-division baccalaureate/associate degree category, 1 semester hour in inventory management (6/91).

AR-1405-0219

STANDARD DEPOT SYSTEM DEPOT PHYSICAL INVENTORY MANAGEMENT

Course Number: ALMC-6D.

Location: Army Logistics Management College, Fort Lee, VA; Army Logistics Management College, Onsite at various locations.

Length: 60 hours.

Exhibit Dates: 6/89–10/91.

Learning Outcomes: Upon completion of the course, the student will be able to manage an overall physical inventory management system, correct discrepancies between depot records and material at the location, and coordinate procedures for performing quarterly and annual audits.

Instruction: The course is presented through the use of multimedia individualized techniques. Topics include physical inventory, material release denials, quality control checks, and report preparation. Course brings together several teaching techniques such as lesson books, programmed instruction, videotape lectures, and examinations in a self-paced course of instruction.

Credit Recommendation: In the lower-division baccalaureate/associate degree category, 3 semester hours in inventory management (6/91).

AR-1405-0220

STANDARD DEPOT SYSTEM DEPOT MAINTENANCE WORKLOADING

Course Number: ALMC-7X.

Location: Army Logistics Management College, Fort Lee, VA; Army Logistics Management College, Onsite at various locations.

Length: 80 hours.

Exhibit Dates: 6/89–10/91.

Learning Outcomes: Upon completion of the course, the student will be able to determine material requirements; describe the interlinking process used to accomplish depot maintenance workloading; prepare input to individual job responsibilities applicable to work areas; and identify output products from maintenance workloading, including frequencies, data elements, and managerial control information.

Instruction: Course is presented through multimedia, individualized instruction using lesson books, reference books, and practical exercises. Course content covers maintenance quality function and managerial responsibilities for material requirements determination and maintenance workloading. Input formats and output products are also covered in some detail. Course brings together several teaching techniques such as lesson books, programmed instruction, videotape lectures, and examinations in a self-paced course of instruction.

Credit Recommendation: In the lower-division baccalaureate/associate degree category, 1 semester hour in material management (6/91).

AR-1405-0221

PROVISIONING
(Army Provisioning Process)

Course Number: *Version 1:* LOG 303; ALMC-AH. *Version 2:* ALMC-AH; LOG 205.

Location: *Version 1:* Army Logistics Management College, Fort Lee, VA; Army Logistics Management College, Onsite at various locations; Army Logistics Management College, Satellite Education Network. *Version 2:* Army Logistics Management College, Fort Lee, VA; Army Logistics Management College, Onsite at various locations; Army Logistics Management College, Satellite Education Program, Fort Lee, VA.

Length: *Version 1:* 2 weeks (72–74 hours). *Version 2:* 2 weeks (75 hours).

Exhibit Dates: *Version 1:* 1/87–12/92. *Version 2:* 1/93–9/94.

Learning Outcomes: *Version 1:* Upon completion of the course, the student will be able to describe the provisioning process; select, code, catalogue, and compute requirements; and distribute support items. *Version 2:* Upon completion of the course, the student will be able to describe the provisioning process; select, code, catalogue, and compute requirements; and distribute support items.

Instruction: *Version 1:* Lecture-conferences, practical exercises, workshops, and cases include topics such as planning for functional interrelationships in the provisioning process (procurement, maintenance, and supply functions), provisioning management controls and techniques, and logistics support engineering and the provisioning process. *Version 2:* Instructional methods used are lecture-conferences, practical exercises, workshops, and case studies. Topics include functional interrelationships and planning in the provisioning process (procurement, maintenance, and supply functions), provisioning management controls and techniques, and logistics support engineering and the provisioning process.

Credit Recommendation: *Version 1:* In the lower-division baccalaureate/associate degree category, 3 semester hours in supply management (6/91). *Version 2:* In the lower-division baccalaureate/associate degree category, 3 semester hours in supply management (8/95).

AR-1405-0222

LOGISTICS ASSISTANCE PROGRAM OPERATIONS
(Logistics Assistance Program Workshop I)

Course Number: ALMC-LV.

Location: Army Logistics Management College, Fort Lee, VA.

Length: 2 weeks (76–77 hours).

Exhibit Dates: 1/88–12/93.

Learning Outcomes: Upon completion of the course, the student will be able to undertake logistics duties in a military environment.

Instruction: Instructional methods include workshops, lecture-conferences, and practical exercises on the defense structure, personal requirements for the job, administration, and career management.

Credit Recommendation: Credit is not recommended because of the limited, specialized content of the course (6/91).

AR-1405-0223

LOGISTICS ASSISTANCE PROGRAM SUPPLY SUPPORT

Course Number: ALMC-LW.

Location: Army Logistics Management College, Fort Lee, VA.

Length: 2 weeks.

Exhibit Dates: 7/90–9/91.

Learning Outcomes: Upon completion of the course, the student will be able to share information and experiences that will enhance Army readiness.

Instruction: A combination of lecture-conferences, guest lecturers, and workshops are be used. The course stresses the pooling of experiences related to the logistics assistance program. Specific topics related to this field are introduced to stimulate thought and discussion.

Credit Recommendation: Credit is not recommended because of the military-specific nature of the course (6/91).

AR-1405-0224

MAINTENANCE PROVISIONING PROCEDURES

Course Number: ALMC-MP.

Location: Army Logistics Management College, Fort Lee, VA; On-site, Material Command Major, Subordinate Commands, US.

Length: 2–3 weeks (77 hours).

Exhibit Dates: 7/89–9/94.

Learning Outcomes: Upon completion of the course, the student will be able to use contract data requirement documents for various acquisition strategies; identify maintenance support items; validate and assign failure factors/maintenance replacement rates; develop and validate maintenance allocation charts; and determine maintenance codes.

Instruction: Instruction includes lectures and exercises on system acquisition, documentation, and supportability; data generation description and acquisition; maintenance allocation chart development; maintenance coding; and maintenance provisioning review.

Credit Recommendation: In the lower-division baccalaureate/associate degree category, 3 semester hours in material management (6/91).

AR-1405-0225

New Equipment Training Management

Course Number: ALMC-NE.

Location: Army Logistics Management College, Fort Lee, VA.

Length: 2 weeks (73 hours).

Exhibit Dates: 10/90–9/95.

Learning Outcomes: Upon completion of the course, the student will be able to describe the Army modernization process and its impact on the Army; identify and integrate documentation for new equipment training throughout the acquisition process; and show relationships among the Army modernization process, Integrated logistic support, and new equipment training.

Instruction: Methods of instruction include lecture-conferences, practical exercises, and guest speakers. Topics covered are an introduction to Army modernization training processes; new equipment training, planning, development, and evaluation; the life-cycle system management model, and the Army modernization training automation system.

Credit Recommendation: Credit is not recommended because of the military-specific nature of the course (6/91).

AR-1405-0226

1. Materiel Acquisition Management for Reserve Components, Phases 2 and 4
2. Materiel Acquisition Management, Phases 1, 2, 3, 4, for Reserve Components

Course Number: *Version 1:* ALM-31-0238. *Version 2:* ALMC-MN; ALMC-38; ALMC-MM; ALMC-37.

Location: Army Logistics Management College, Fort Lee, VA.

Length: *Version 1:* 4 weeks (156 hours). *Version 2:* Phase 2, 2 weeks (69 hours); Phase 4, 2 weeks (67 hours).

Exhibit Dates: *Version 1:* 9/86–7/93. *Version 2:* 8/93–2/97.

Learning Outcomes: *Version 1:* Upon completion of the course, the student will be able to describe the environment in which material is conceived, developed, acquired, and operated; participate in the research and development process; evaluate and apply the mechanisms and disciplines of systems engineering and integrated logistics support; use managerial tools pertaining to financial management and cost controls; manage the transition of material from development to production and practice the basic functions of production management; and apply the principles of contract formulation, negotiation, and administration. *Version 2:* Upon completion of the course, the student will be able to describe the environment in which material is conceived, developed, acquired, and operated; participate in the research and development process; evaluate and apply the mechanisms and disciplines of systems engineering and integrated logistics support; use managerial tools pertaining to financial management and cost controls; manage the transition of material from development to production and practice the basic functions of production management; apply the principles of contract formulation, negotiation, and administration; prepare, review, and evaluate requests for proposals; and compare and contrast risk analysis and decision risk analysis. Phases 1 and 3 are accomplished through correspondence, with proctored examinations required to progress.

Instruction: *Version 1:* Lecture-conferences, guest speakers, and correspondence instruction cover materials acquisition management, including financial and cost management, acquisition concepts and policies, research and development, testing and evaluation, integrated logistics support, force modernization, and contract management. *Version 2:* Completion of the four phases encompasses lecture-instruction and its ancillary forms of classroom techniques for Phases 2 and 4. Phases 1 and 3 are offered by correspondence and together a series of topic are covered including financial and cost management, acquisition concepts and polices, research and development, testing and evaluation, integrated logistics support, force modernization, and contract management.

Credit Recommendation: *Version 1:* In the upper-division baccalaureate category, 3 semester hours in logistics management. In the graduate degree category, 3 semester hours in contract management (6/91). *Version 2:* In the upper-division baccalaureate category, 4 semester hours in contract management. In the graduate degree category, 4 semester hours in contract management. Credit is to be granted in either the upper-division or in the graduate category, but not in both. All four phases must be completed before credit is to be granted (8/95).

AR-1405-0227

Quicklook 2 Maintenance Training

Course Number: 102-F92.

Location: Intelligence School, Fort Devens, MA.

Length: 12 weeks (359 hours).

Exhibit Dates: 7/88–3/94.

Learning Outcomes: Upon completion of this course, the student will be able to manage maintenance and repair operations for tactical equipment, determine repairs necessary, and secure needed replacement parts.

Instruction: Lectures, demonstrations, and practical exercises cover combat logistics, quality management, supply operations, maintenance, and field operations.

Credit Recommendation: In the lower-division baccalaureate/associate degree category, 3 semester hours in industrial management or maintenance technology (6/91).

Related Occupations: 33R.

AR-1405-0229

Ammunition Stock Control and Accounting Specialist Reserve

Course Number: 093-55R10-RC.

Location: Missile and Munitions School, Redstone Arsenal, AL.

Length: 2 weeks (92–122 hours).

Exhibit Dates: 8/88–1/96.

Learning Outcomes: Upon completion of the course, the student will be able to identify ammunition, complete inventory procedures and shipping documents, process receipts and transactions, and conduct reconciliation procedures.

Instruction: Topics include standard Army ammunition system operations, both automated and manual.

Credit Recommendation: Credit is not recommended because of the limited, specialized nature of the course (8/91).

Related Occupations: 55R.

AR-1405-0230

Commodity Command Standard System Physical Inventory Management by Correspondence

Course Number: ALMC-6C.

Location: Army Logistics Management College, Fort Lee, VA.

Length: Maximum, 52 weeks.

Exhibit Dates: 2/89–3/92.

Learning Outcomes: Upon completion of the course, the student will be able to identify and prepare inventory application input documents, analyze and interpret inventory application output products, and describe the major process and procedures of inventory applications.

Instruction: This is a correspondence course and covers requesting/controlling procedures, materiel release denials, zero balance flasher, and quality control. Review of knowledge gained in the location record audit/match is required. Initial, intermediate, and final surveys are used to examine students throughout the course.

Credit Recommendation: In the lower-division baccalaureate/associate degree category, 1 semester hour in inventory management (6/91).

AR-1405-0231

Commodity Command Standard System (CCSS) Functional by Correspondence

Course Number: *Version 1:* ALMC-3L; ALM-56-M115. *Version 2:* ALMC-3L.

Location: Army Logistics Management College, Fort Lee, VA.

Length: *Version 1:* Maximum, 52 weeks. *Version 2:* Maximum, 52 weeks.

Exhibit Dates: *Version 1:* 8/80–9/92. *Version 2:* 10/92–6/96.

Learning Outcomes: *Version 1:* Upon completion of the course, the student will manage and supervise persons using a particular automated system. *Version 2:* Upon completion of the course, the student will be able to use data processing and total logistics support systems.

Instruction: *Version 1:* This is a correspondence course. Topics include logistics management concepts, computerized data systems, and how the above interface with the initial provisioning and replenishment cycles. *Version 2:* This is a correspondence course. Topics include logistics management concepts, computerized data systems, and how the above interface with the initial provisioning and replenishment cycles.

Credit Recommendation: *Version 1:* In the upper-division baccalaureate category, 2 semester hours in automated systems in logistics management (8/90). *Version 2:* In the upper-division baccalaureate category, 2 semester hours in logistics management (8/95).

AR-1405-0232

Utilities Operation and Maintenance Technician

Course Number: 4-5-C32-210A.

Location: Engineer Housing Support Center, Fort Belvoir, VA; Engineer School, Fort Leonard Wood, MO.

Length: 8 weeks (311 hours).

Exhibit Dates: 1/91–9/94.

Learning Outcomes: After 12/93 see AR-1405-0272. Upon completion of the course, the student will be able to manage personnel; manage the operation of utilities; and plan, program, and budget facility operations.

Instruction: Lectures and practical exercises cover management, including staff studies, schedule preparation, employee relations, and report preparation; planning, operation, and maintenance in the installation of utility systems; quality control in construction projects; facility engineering, including real property, fire protection, and energy conservation; and planning, programming, and budgeting for the operation of a utility facility.

Credit Recommendation: In the lower-division baccalaureate/associate degree category, 3 semester hours in facility management and 1 in personnel supervision (1/93).

Related Occupations: 210A.

AR-1405-0233

Unit Supply Specialist Basic Noncommissioned Officer (NCO)

Course Number: ***Version 1:*** 552-76Y30. *Version 2:* 552-92Y30; 552-76Y30. *Version 3:* 551-92Y30, Phase 2.

Location: Quartermaster Center and School, Fort Lee, VA.

Length: ***Version 1:*** 8 weeks (243 hours). *Version 2:* 9–10 weeks (338–367 hours). *Version 3:* 8 weeks (295 hours).

Exhibit Dates: ***Version 1:*** 2/87–6/91. *Version 2:* 7/91–11/01. *Version 3:* 12/01–Present.

Learning Outcomes: ***Version 1:*** Upon completion of the course, the student will be able to perform as the supervisor of a supply organization; teach, evaluate, and counsel junior leaders; and conduct collective training. *Version 2:* Upon completion of the course, the student will be able to perform as a supervisor of a supply organization; teach, evaluate, and counsel junior leaders; and conduct collective training. *Version 3:* Upon completion of the course, the student will be able to perform as a supervisor of a supply organization; teach, evaluate and counsel junior leaders; and conduct collective training.

Instruction: ***Version 1:*** Conferences, practical exercises, and seminars review the nature of supply organizations and the role of the supply supervisor in records management, using manual and automated systems and procedures. There is emphasis on inventory management, preventive maintenance, order control, accounting, and distribution control of large quantities in diversified inventories. *Version 2:* Conferences, practical exercises, and seminars review the nature of supply organizations and the role of the supply supervisor in records management, using manual and automated systems and procedures. There is emphasis on inventory management, preventive maintenance, order control, accounting, and distribution control of large quantities in diversified inventories. Course includes a common core of leadership subjects. *Version 3:* Discussion, classroom exercise, practical exercises, lecture, examination, conferences, and seminars review the nature of supply organizations and the role of the supply supervisor in using manual and automated supply records, management systems, and procedures. Emphasis is placed on the transport of supplies, forecasting supply needs and special topics including small arms inventory, security of the installation, use of appropriated funds, and property book functions.

Credit Recommendation: ***Version 1:*** In the lower-division baccalaureate/associate degree category, 3 semester hours in records management and 3 in inventory management (2/93). *Version 2:* In the lower-division baccalaureate/associate degree category, 3 semester hours in records management and 3 in inventory management. See AR-1406-0090 for common core credit (5/96). *Version 3:* In the lower-division baccalaureate/associate degree category, 2 semester hours in records management, 3 in inventory management, and 3 in principles of supervision (6/02).

Related Occupations: 76Y; 92Y.

AR-1405-0234

Logistics Officer (S4)

Course Number: 8A-F33.

Location: Quartermaster Center and School, Fort Lee, VA.

Length: 4–5 weeks (154 hours).

Exhibit Dates: 5/88–10/94.

Learning Outcomes: Upon completion of the course, the student will apply the principles of logistics from a management perspective, including processing forms and documentation; setting priorities for supply; preparing logistics estimates, plans, and orders; identifying contingency procedures; and using an automated system to support logistics functions.

Instruction: Classroom instruction and examinations test the student's knowledge and application of skills for planning, coordinating, and controlling a logistics supply operation.

Credit Recommendation: In the lower-division baccalaureate/associate degree category, 3 semester hours in logistics management or physical distribution management (5/02).

AR-1405-0235

Unit Supply Specialist

Course Number: ***Version 1:*** 552-92Y10; 552-76Y10. *Version 2:* 552-92Y10. *Version 3:* 552-92Y10.

Location: ***Version 1:*** Quartermaster Center and School, Fort Lee, VA; Training Center, Fort Jackson, SC. *Version 2:* Quartermaster Center and School, Fort Lee, VA. *Version 3:* Quartermaster Center and School, Fort Lee, VA.

Length: ***Version 1:*** 7–8 weeks (217–288 hours). *Version 2:* 7–8 weeks (277 hours). *Version 3:* 7–8 weeks (277 hours).

Exhibit Dates: ***Version 1:*** 10/90–9/97. *Version 2:* 10/97–12/99. *Version 3:* 1/00–Present.

Learning Outcomes: ***Version 1:*** Before 10/90 see AR-1405-0061. Upon completion of the course, the student will be able to perform unit and organization supply procedures and organizational maintenance on small arms. *Version 2:* Upon completion of the course, the student will have an understanding of unit and organizational supply procedures and organizational maintenance of small arms. *Version 3:* To provide the student with the skills necessary to use an automated and manual system to request, receive, store, issue and maintain records associated with expendable supplies, equipment and small arms at a unit level.

Instruction: ***Version 1:*** Lectures and practical exercises cover supply operations and the maintenance of small arms. Course includes basic mathematics, typewriting, use of technical publications, unit and organization supply procedures, and organizational maintenance of small arms. *Version 2:* Discussion, classroom exercise, practical exercise, lecture and examination. Topics include requesting, receiving, storing, issuing, maintaining, and establishing accountability of individual, installation, organizational, and expandable/durable supplies and equipment; security and administration of a unit arms room; and organizational maintenance of small arms. *Version 3:* Discussion, classroom exercise, practical exercise, lecture and examination. Topics include establishing manual and automated procedures, setting up and operating a computer system that supports the process, preparing unit supply files and signature cards, establishing security records, and requesting, receiving, storing and issuing supplies and materials.

Credit Recommendation: ***Version 1:*** In the vocational certificate category, 3 semester hours in records maintenance and 3 in supply supervision and maintenance (5/96). *Version 2:* In the lower-division baccalaureate/associate degree category, 3 semester hours in inventory management and control (6/02). *Version 3:* In the lower-division baccalaureate/associate degree category, 3 semester hours in inventory management and control, and 3 in computer applications (6/02).

Related Occupations: 76Y; 92Y.

AR-1405-0236

Quartermaster Officer Basic

Course Number: ***Version 1:*** 8-10-C20-92A. *Version 2:* 8-10-C20; 8-10-C20-92A. *Version 3:* 8-10-C20.

Location: Quartermaster Center and School, Fort Lee, VA.

Length: ***Version 1:*** 17 weeks (660 hours). *Version 2:* 14–15 weeks (525–526 hours). *Version 3:* 14 weeks (526 hours).

Exhibit Dates: ***Version 1:*** 10/92–12/95. *Version 2:* 1/96–8/98. *Version 3:* 9/98–Present.

Learning Outcomes: ***Version 1:*** Before 10/92 see AR-1405-0069. Upon completion of the course, the student will be able to manage a unit responsible for providing support in the areas of supply management and operations, petroleum and water operations, mortuary services, airborne and field services, and subsistence and food services. *Version 2:* Upon completion of the course, the student will be able to manage a unit responsible for providing support in the areas of supply management and logistics operations, petroleum and water operations, mortuary services, field services, and subsistence and food services. *Version 3:* The course will provide student with the knowledge and skill development to lead a platoon including: interpersonal communication, leadership, technical quartermaster functions, specialized training in material management, subsistence management, petroleum and water operations, and aerial delivery.

Instruction: *Version 1:* Instruction is via classroom lectures, presentations, demonstrations, practice, and examinations. Topics covered include professional development, unit organization and leadership, automatic data processing support, and unit logistics. *Version 2:* Instruction is conducted through classroom lectures, presentations, demonstrations, practice, and examinations. Topics covered include professional development, unit organization and leadership, automatic data processing support, and unit logistics. *Version 3:* Audio visual materials, discussion, practical exercise, lecture, and learner presentations, field trips. Instruction is conducted through classroom lectures, presentations, demonstrations, practice, and examinations. Topics covered include professional development, unit organization and leadership, automatic data processing support, and unit logistics.

Credit Recommendation: *Version 1:* In the lower-division baccalaureate/associate degree category, 2 semester hours in introduction to management and 2 in supply management. In the upper-division baccalaureate category, 2 semester hours in personnel supervision (2/93). *Version 2:* In the lower-division baccalaureate/associate degree category, 2 semester hours in personnel supervision and 1 in supply management. In the upper-division baccalaureate category, 2 semester hours in logistics management and 1 in organizational management (5/02). *Version 3:* In the lower-division baccalaureate/associate degree category, 1 semester hour in interpersonal communication, and 3 in human resource management. In the upper-division baccalaureate category, 3 semester hours in logistics and supply system management (6/02).

Related Occupations: 92A.

AR-1405-0237

1. Combined Logistics Officer Advanced Quartermaster Branch Specific Track
2. Combined Logistics Captains Career

Course Number: *Version 1:* 8-10-C22 (LOG). *Version 2:* 8-10-C22 (LOG), Phase 2.

Location: Quartermaster Center and School, Fort Lee, VA.

Length: *Version 1:* 5 weeks (196 hours). *Version 2:* 5 weeks (200 hours).

Exhibit Dates: *Version 1:* 8/92–12/96. *Version 2:* 1/97–Present.

Learning Outcomes: *Version 1:* Upon completion the course, the student will apply a working knowledge of logistics management and the functions of inventory control to the supervision of a unit that issues, receives, and stores supplies. *Version 2:* Upon completion of the course, the student will be prepared for command and staff assignments through knowledge, acquisitions and skill development in material management, automated logistics systems, and operations support functions.

Instruction: *Version 1:* Classroom instruction and hands-on projects using an automated system to support inventory control and logistics management are included. Skills are developed in food service, supply, water, and petroleum. Automated systems skills are developed, and the principles of ADP systems in general are presented. Use of the automated system is a major topic of study. *Version 2:* Laboratory, audio visual materials, classroom exercise, practical exercise, and lecture. Topics include detailed skills and knowledge of the Army information and management systems for material and logistics.

Credit Recommendation: *Version 1:* In the lower-division baccalaureate/associate degree category, 3 semester hours in inventory control or inventory management and 3 in computer literacy, computers and computing, or introduction to computers (2/93). *Version 2:* In the lower-division baccalaureate/associate degree category, 3 semester hours in computer applications in inventory and logistics management. In the upper-division baccalaureate category, 3 semester hours in inventory management (6/02).

AR-1405-0238

1. Automated Logistics Specialist
2. Automated Logistical Specialist
3. Automated Logistical Specialist
4. Automated Logistical Specialist

Course Number: *Version 1:* 551-92A10. *Version 2:* 551-92A10. *Version 3:* 551-92A10. *Version 4:* 551-92A10.

Location: Quartermaster Center and School, Fort Lee, VA.

Length: *Version 1:* 13 weeks (467–474 hours). *Version 2:* 11 weeks (391 hours). *Version 3:* 12 weeks (424 hours). *Version 4:* 12 weeks (424 hours).

Exhibit Dates: *Version 1:* 5/92–6/95. *Version 2:* 7/95–9/98. *Version 3:* 10/98–4/99. *Version 4:* 5/99–Present.

Learning Outcomes: *Version 1:* Upon completion of the course, the student will be able to use manual and automated processes to control inventory and apply proper procedures to receive, store, and protect supplies and equipment. *Version 2:* Upon completion of the course, the student will be able to use manual and computer-based processes to control inventory and apply proper procedures to control, store, and protect supplies and equipment. *Version 3:* Upon completion of the course, the student will be able to use manual and computer-based processes to control inventory and apply proper procedures to control, store, and protect supplies and equipment. *Version 4:* Upon completion of the course, the student will be able to use manual and computer-based processes to receive, issue, store, turn-in, and reconcile supplies and equipment.

Instruction: *Version 1:* Lectures, class presentations, demonstrations, practice, and examinations cover manual and automated processes; inventory receipt, issue, and reconciliation; material handling; packing; movement; and storage. *Version 2:* Lectures, class presentations, demonstrations, practice, and examinations cover manual and automated processes; inventory receipt, issue, and reconciliation; material handling; packing; movement; and storage. *Version 3:* Lectures, class presentations, demonstrations, practice, and examinations cover manual and automated processes; inventory receipt, issue, and reconciliation; material handling; packing; movement; and storage. *Version 4:* Audio visual materials, discussion, examination, classroom and practical exercise, and lecture. Topics include primarily using automated system to control inventory processes associated with supplies and equipment, repair parts, subsistence items, retail supplies, oil and petroleum products and hazardous materials and includes system access for generating reports, data inquiry and backup and recovery.

Credit Recommendation: *Version 1:* In the lower-division baccalaureate/associate degree category, 2 semester hours in introduction to material handling, 2 in record keeping or clerical procedures, and 2 in introduction to computers (2/93). *Version 2:* In the lower-division baccalaureate/associate degree category, 2 semester hours in record keeping, 2 in computerized records and information management, and 1 in supply management (12/96). *Version 3:* In the lower-division baccalaureate/associate degree category, 3 semester hours in inventory management and control and 3 in computer applications (6/02). *Version 4:* In the lower-division baccalaureate/associate degree category, 2 semester hours in computer applications, 2 in inventory management, and 2 in information/record management (6/02).

Related Occupations: 92A.

AR-1405-0239

1. Automated Logistics Management Specialist Basic Noncommissioned Officer (NCO)
2. Automated Logistical Specialist Basic Noncommissioned Officer (NCO)
3. Automated Logistical Specialist Basic Noncommissioned Officer (NCO)

Course Number: *Version 1:* 551-92A30. *Version 2:* 551-92A30. *Version 3:* 551-92A30, Phase 1.

Location: Quartermaster Center and School, Fort Lee, VA.

Length: *Version 1:* 9 weeks (325 hours). *Version 2:* 10–11 weeks (388 hours). *Version 3:* 8 weeks (309 hours).

Exhibit Dates: *Version 1:* 1/93–9/95. *Version 2:* 10/95–11/01. *Version 3:* 12/01–Present.

Learning Outcomes: *Version 1:* Upon completion of the course, the student will be able to lead small groups in performing inventory and issue tasks using an automated system. *Version 2:* Upon completion of the course, the student will be able to lead small groups in performing inventory and issue tasks using standard automated systems. *Version 3:* Upon completion of the course, the student will be able to lead small groups in performing control, receipt and issue of inventory using an automated inventory control system.

Instruction: *Version 1:* Lectures, class presentations, demonstrations, practice, and examinations develop improved skills, understanding, and knowledge in the areas of material handling, warehouse layout planning, and issue and receipt processing. Course includes a common core of leadership subjects. *Version 2:* Lectures, class presentations, demonstrations, practice, and examinations develop improved skills, understanding, and knowledge in the areas of material handling, warehouse layout planning, and issue and receipt processing. Course includes a common core of leadership subjects. *Version 3:* Discussion, classroom exercise, practical exercise, lectures and examinations. Topics include supervision of a small group and hands-on ability to perform all functions of an automated inventory control system including system reporting, system administrator functions, storage space planning, supervising physical inventories, supervising loading teams and subsistence supply planning.

Credit Recommendation: *Version 1:* In the lower-division baccalaureate/associate degree category, 2 semester hours in advanced materials handling and 3 in records and information management. See AR-1406-0090 for common core credit (2/93). *Version 2:* In the lower-division baccalaureate/associate degree category, 1 semester hour in supply management, 2 in computerized records and information management, and 2 in advanced computerized records and information management. See AR-1406-0090 for common core credit (12/96). *Version 3:* In the lower-division baccalaureate/associate degree category, 2 semester hours in inventory management and control, 2 in computer applications, and 2 in supply chain management. See AR-1406-0090 for common core credit (6/02).

Related Occupations: 92A.

AR-1405-0240

1. Automated Logistics Management Advanced Noncommissioned Officer
2. Automated Logistical Specialist Advanced Noncommissioned Officer (NCO)
3. Automated Logistical Specialist Advanced Noncommissioned Officer (NCO)

Course Number: *Version 1:* 551-92A40. *Version 2:* 551-92A40. *Version 3:* 551-92A40, Phase 2.

Location: Quartermaster Center and School, Fort Lee, VA.

Length: *Version 1:* 9–10 weeks (351 hours). *Version 2:* 12 weeks (437 hours). *Version 3:* 10 weeks (367 hours).

Exhibit Dates: *Version 1:* 5/92–9/95. *Version 2:* 10/95–11/01. *Version 3:* 12/01–Present.

Learning Outcomes: *Version 1:* Upon completion of the course, the student will be able to supervise and manage a group using an automated supply/inventory system, including inventory control, warehouse/storage processes, material and supplies movement, and security and protection. *Version 2:* Upon completion of the course, the student will be able to supervise and manage a group using an automated supply/inventory system, including inventory control, warehouse/storage processes, material and supplies movement, and security and protection. *Version 3:* Upon completion of the course, the student will be able to supervise and manage a group using an automated inventory system.

Instruction: *Version 1:* Lectures, class presentations, demonstrations, practice, and examinations build skills. Course includes a common core of leadership subjects. *Version 2:* Lectures, class presentations, demonstrations, practice, and examinations build skills. Course includes a common core of leadership subjects. *Version 3:* Discussion, classroom exercise, practical exercise, lecture, and examination. Topics include all facets of inventory control; receipt and disbursement of goods and supplies, warehouse and storage processes materials; and supplies movement; and security and protection.

Credit Recommendation: *Version 1:* In the lower-division baccalaureate/associate degree category, 3 semester hours in personnel supervision or introduction to supervision. See AR-1404-0035 for common core credit. In the upper-division baccalaureate category, 3 semester hours in logistics management (2/93). *Version 2:* In the lower-division baccalaureate/associate degree category, 1 semester hour in administrative management skills, 2 in computerized records and information management, and 1 in microcomputer applications. See AR-1404-0035 for common core credit. In the upper-division baccalaureate category, 3 semester hours in logistics management (12/96). *Version 3:* In the lower-division baccalaureate/associate degree category, 2 semester hours in principles of supervision, 2 in information management, and 2 in computer applications. In the upper-division baccalaureate category, 3 semester hours in logistics management (6/02).

Related Occupations: 92A.

AR-1405-0241

Medical Supply Specialist

Course Number: 551-76J10.

Location: Academy of Health Sciences, Fort Sam Houston, TX.

Length: 5–6 weeks (216 hours).

Exhibit Dates: 11/93–Present.

Learning Outcomes: After 12/01 see AR-1405-0280. Upon completion of the course, the student will be able to manage the medical materiel requirements of a health care facility; manage medical equipment; and establish and monitor quality control.

Instruction: Course covers medical supply publications, property control, and stock accounting procedures. Methodology includes lectures and practical exercises.

Credit Recommendation: In the lower-division baccalaureate/associate degree category, 4 semester hours in materials management (1/96).

Related Occupations: 76J.

AR-1405-0243

Defense Reutilization and Marketing Property Accounting

Course Number: *Version 1:* ALMC-IC. *Version 2:* ALMC-IC.

Location: *Version 1:* Army Logistics Management College, Fort Lee, VA; Army Logistics Management College, Onsite at various locations. *Version 2:* Army Logistics Management College, Fort Lee, VA.

Length: *Version 1:* 2 weeks (72–73 hours). *Version 2:* 2 weeks (75 hours).

Exhibit Dates: *Version 1:* 8/93–3/98. *Version 2:* 4/98–Present.

Learning Outcomes: *Version 1:* Before 8/93 see AR-1405-0116. Upon completion of the course, the student will be able to input and maintain accurate computerized property records and interpret and analyze management data for control and accountability property. *Version 2:* The student will be able to interpret and analyze management data for control and accountability purposes and use computerized accounting packages to update and store those records.

Instruction: *Version 1:* Instructional methods include the use of lecture-conferences, practical exercises, study-group workshops, guest speakers, and computer-based practica. Topics include automated record keeping procedures related to property disposal, transfers, sales, adjustments, and reversals and interpretation and analysis of management data. *Version 2:* Instructional methods include the use of lecture-conferences, practical exercises, study-group workshops, guest speakers, and computer-based practice. Topics include automated record keeping procedures related to property disposal, transfers, sales, adjustments, and reversals and interpretation and analysis of management data.

Credit Recommendation: *Version 1:* In the lower-division baccalaureate/associate degree category, 3 semester hours in computer operations and 1 in materiel management accounting and control (8/95). *Version 2:* In the upper-division baccalaureate category, 3 semester hours in managerial accounting (1/02).

AR-1405-0244

Commodity Command Standard System (CCSS) Functional

Course Number: ALMC-3L.

Location: Army Logistics Management College, Fort Lee, VA.

Length: 1–2 weeks (73 hours).

Exhibit Dates: 9/92–Present.

Learning Outcomes: Before 9/92 see AR-1405-0118. Upon completion of the course, the students will be able to use data processing and total logistics support systems.

Instruction: Methods of instruction include lectures, work group exercises, and computer simulations. Topics include logistics management concepts, and how computerized data systems interface with the initial provisioning and replenishing cycles.

Credit Recommendation: In the upper-division baccalaureate category, 2 semester hours in logistics management (8/95).

AR-1405-0245

Combined Logistics Officer Advanced Phase 2

Course Number: 4-9-C22-91D (LOG).

Location: Ordnance, Missile and Munitions School, Redstone Arsenal, AL.

Length: 4–5 weeks (193 hours).

Exhibit Dates: 10/92–10/01.

Learning Outcomes: Upon completion of the course, the student will have an understanding of ammunition concept, doctrine from the theater through the ammunition transfer point, field operations, and wholesale logistics.

Instruction: Lecture and self-study prepare the soldier in military logistics, formulation of budgets basic tactical army combat, ammunition storage, and principles of operation.

Credit Recommendation: In the lower-division baccalaureate/associate degree category, 3 semester hours in military science and logistics (7/96).

AR-1405-0246

Supply Systems Technician Warrant Officer Advanced Reserve Component

Course Number: 8-10-C32-920B-RC.

Location: Quartermaster Center and School, Fort Lee, VA.

Length: 2 weeks (72 hours).

Exhibit Dates: 7/92–Present.

Learning Outcomes: Upon completion of the course, the student will be able to operate the standard Army retail supply system military computerized software system.

Instruction: Classroom instruction and practical exercises cover military stocking procedures and operation of the computerized system supporting this function.

Credit Recommendation: In the lower-division baccalaureate/associate degree cate-

gory, 1 semester hour in computerized record and information management (5/02).

AR-1405-0247

Property Accounting Technician Warrant Officer Basic Reserve Component

Course Number: 8B-920A-RC.

Location: Quartermaster Center and School, Fort Lee, VA.

Length: 2 weeks (76 hours).

Exhibit Dates: 7/95–Present.

Learning Outcomes: Before 7/92 see AR-1405-0186. Upon completion of the course, the student will be able to provide technical leadership in a small to mid-sized property management activity.

Instruction: Course includes classroom lectures and exercises.

Credit Recommendation: In the lower-division baccalaureate/associate degree category, 1 semester hour in record keeping (5/02).

AR-1405-0248

Standard Army Retail Supply System (SARSS) 2AD/2AC/2B

Course Number: 8B-F53/551-ASIG2; 8B-F53/551-F27.

Location: Quartermaster Center and School, Fort Lee, VA.

Length: 3 weeks (98 hours).

Exhibit Dates: 10/95–Present.

Learning Outcomes: Upon completion of the course, the student will be able to operate computer equipment and manage or supervise standard Army retail supply system software applications.

Instruction: Methods of instruction include lectures and practical exercises. Topics include concepts, principles, computer operations, and functional procedures associated with the standard Army retail supply system software.

Credit Recommendation: In the lower-division baccalaureate/associate degree category, 2 semester hours in computerized records and information management (5/02).

AR-1405-0249

1. Automated Property Book (Automated Property Book Course II (APBC II))
2. Standard Property Book System Redesign (SPBS-R)

Course Number: *Version 1:* 2200; 2600; 942-442; 921-2600. *Version 2:* 2600; 2200; 921-442.

Location: Reserve Readiness Training Center, Fort McCoy, WI.

Length: *Version 1:* 2 weeks (71–80 hours). *Version 2:* 2 weeks (80 hours).

Exhibit Dates: *Version 1:* 5/95–6/01. *Version 2:* 7/01–Present.

Learning Outcomes: *Version 1:* Upon completion of the course, the student will be able to identify the applications, capabilities, and files of the standard property and book system-redesign and use a computer system. *Version 2:* Upon completion of the course, the student will be able to use the SDBS-R computer system to establish system parameters, maintain property accountability, complete military standard requisition and issue procedures, produce systems generated management report transactions, and recognize and fix mistakes.

Instruction: *Version 1:* Lectures, demonstrations, and practical exercises cover code table files, catalog file maintenance, unit header update, property accountability, automated hand receipts, automated requisitions, transfer of units, record and data management, and systems maintenance. *Version 2:* Lectures, practical and classroom exercises, discussion, and computer-based training. Topics include collecting information, automated requisitions, analyzing and documenting information, using information for accurate data entry, recognize and fix mistakes, property book transactions, and code tables.

Credit Recommendation: *Version 1:* Credit is not recommended because of the limited, specialized nature of the course (11/02). *Version 2:* In the lower-division baccalaureate/associate degree category, 2 semester hours in computer applications (11/02).

AR-1405-0250

Facility Manager

Course Number: 2500.

Location: Reserve Readiness Training Center, Fort McCoy, WI.

Length: 71 hours.

Exhibit Dates: 3/96–6/01.

Learning Outcomes: Upon completion of the courses, the student will be able to order materials; maintain appropriate property books; develop contracts; develop accident-prevention and loss-control methods, procedures, and programs in industrial establishments; apply codes; apply safety-engineering and management principles; develop an awareness and comprehension of disaster response program; and transport, store, and handle hazardous materials.

Instruction: Course includes homework, extensive practical exercises, lectures, discussions, and quizzes.

Credit Recommendation: In the lower-division baccalaureate/associate degree category, 2 semester hours in management and enforcement of accident prevention. In the upper-division baccalaureate category, 1 semester hour in industrial supervision and 1 in hazardous materials and emergency planning (3/97).

AR-1405-0251

Strength Management

Course Number: 3500.

Location: Reserve Readiness Training Center, Fort McCoy, WI.

Length: 2 weeks (80 hours).

Exhibit Dates: 11/95–Present.

Learning Outcomes: Upon completion of the course, the student will be able to identify the basic fundamentals of personnel management actions as they apply to retention; identify the purposes, procedures, and management of key retention programs as they apply to the strength management office; and identify and develop various techniques, plans, and coordination methods to enhance the implementation of effective retention efforts.

Instruction: Methods include lectures, discussions, practical exercises and quizzes. Topics covered include personnel management systems, retention programs, and retention operations.

Credit Recommendation: Credit is not recommended because of the military-specific nature of the course (3/97).

AR-1405-0252

Armor Crewman Noncommissioned Officer (NCO) Advanced

Course Number: 020-19E40/19K40.

Location: Armor Center and School, Fort Knox, KY.

Length: 12 weeks (467 hours).

Exhibit Dates: 1/91–Present.

Learning Outcomes: Upon completion of the course, the student will be able to perform the leadership duties of a platoon sergeant.

Instruction: Small group instruction using lecture-demonstration and practical exercises is employed to cover the topics of tactical training, gunnery preparation, maintenance management, training management, leadership skills, and staff support functions. Course includes a common core of leadership subjects.

Credit Recommendation: In the lower-division baccalaureate/associate degree category, 1 semester hour in technical writing, 2 in principles of management, and 1 in training management. See AR-1404-0035 for common core credit (3/97).

Related Occupations: 19E; 19K.

AR-1405-0253

Total Army Training System (TATS) Armor Crewman Noncommissioned Officer (NCO) Advanced Phase 1

Course Number: 020-19K40 (F).

Location: Armor Center and School, Fort Knox, KY.

Length: 28 weeks (253–254 hours).

Exhibit Dates: 12/94–Present.

Learning Outcomes: Upon completion of the course, the student will be able to demonstrate leadership, organizational and systems management, and other skills necessary to perform the duties of platoon sergeant in armor units.

Instruction: Methods to provide this instruction include two weeks of resident classroom training and a second 28-week component which includes distance learning materials consisting of videos, computer-assisted instruction, interactive video disks, and print media. Topics include principles of management, principles of supervision, leadership, and other military-specific topics.

Credit Recommendation: In the lower-division baccalaureate/associate degree category, 3 semester hours in principles of management, 3 in personnel supervision, and 3 in leadership (2/01).

AR-1405-0254

1. Total Army Training System (TATS) Armor Crewman Noncommissioned Officer (NCO) Advanced, Phase 2
2. Armor Crewman Advanced Noncommissioned Officer (NCO)

Course Number: *Version 1:* 020-19K40 (F), Phase 2. *Version 2:* 020-19K40, Phase 2.

Location: *Version 1:* Armor Center and School, Fort Knox, KY. *Version 2:* Armor School, Fort Knox, KY.

Length: *Version 1:* 3 weeks (125 hours). *Version 2:* 11–12 weeks (614 hours).

Exhibit Dates: ***Version 1:*** 4/94–9/00. *Version 2:* 10/00–Present.

Learning Outcomes: ***Version 1:*** Upon completion of the course, the student will be able to function as a leader in the areas of military tactics, weapons, and maintenance supervision. *Version 2:* Upon completion of the course, the student will be able to function as a leader in the areas of military tactics, weapons, and maintenance supervision. Students will apply property accountability concepts and be able to use train-the-trainer concepts.

Instruction: ***Version 1:*** Lecture-discussion and practical experience are used to cover the topics of leadership techniques, management, electronic communications, obstacle preparation, military justice, battle positioning, gunnery techniques, platoon rearming/resupplying, hasty river crossing, and physical fitness. *Version 2:* Lecture-discussion and practical experience are used to cover leadership techniques, management, electronic communications, obstacle preparation, military justice, battle positioning, gunnery techniques, platoon rearming/resupplying, hasty river crossing, physical fitness, supervising and evaluating a maintenance program, maintenance status reports, forms and records, and technical and tactical information.

Credit Recommendation: ***Version 1:*** Credit is not recommended because of the military-specific nature of the course (3/97). *Version 2:* In the lower-division baccalaureate/associate degree category, 6 semesters hours in military tactics and 3 in maintenance supervision. Students will have completed Phase 1, Advanced Noncommissioned Officer (NCO) Common Core (6/01).

Related Occupations: 19K.

AR-1405-0255

1. Standard Army Ammunition System Reserve Component
2. Standard Army Ammunition System Modernization Functional

(Standard Army Ammunition System Modernization Functional, Reserve Component)

Course Number: ***Version 1:*** 645-F2-RC. *Version 2:* 645-F2-AC/RC; 645-F2-AR/RC.

Location: Ordnance, Missile and Munitions School, Redstone Arsenal, AL.

Length: ***Version 1:*** 2 weeks (78 hours). *Version 2:* 2 weeks (70 hours).

Exhibit Dates: ***Version 1:*** 10/93–9/00. *Version 2:* 10/00–Present.

Learning Outcomes: ***Version 1:*** Upon completion of the course, the student will be able to perform administrative duties, including automated shipping and receiving and explosive safety. *Version 2:* Upon completion of the course, the student will be able to perform administrative duties, including automated computer-based shipping and receiving and explosive safety.

Instruction: ***Version 1:*** Lecture and practical exercises in automated ammunition stores control and accountability using software and computer terminals. This includes access, master files, security, placing orders, quantity adjust, transportation, and safety reports. *Version 2:* Lectures and practical exercises cover automated ammunition stores control and accountability using software and computer terminals. This includes access, master files, security, placing orders, quantity adjustment, transportation, and safety reports.

Credit Recommendation: ***Version 1:*** In the lower-division baccalaureate/associate degree category, 2 semester hours in computer-based materiel management (5/97). *Version 2:* In the lower-division baccalaureate/associate degree category, 2 semester hours in computer-based materiel management (1/04).

AR-1405-0256

1. Standard Army Ammunition System (SAAS) 1/3 Reserve Component
2. Standard Army Ammunition System Modernization Functional, Reserve Component

Course Number: ***Version 1:*** 645-F3-RC. *Version 2:* 645-F3-AC/RC.

Location: Ordnance, Missile and Munitions School, Redstone Arsenal, AL.

Length: ***Version 1:*** 2 weeks (71 hours). *Version 2:* 2 weeks (71 hours).

Exhibit Dates: ***Version 1:*** 10/93–9/00. *Version 2:* 10/00–Present.

Learning Outcomes: ***Version 1:*** Upon completion of the course, the student will be able to conduct transactions and management of ammunition, including stock control and accounting. *Version 2:* Upon completion of the course, the student will be able to conduct computer-based transactions and management of ammunition, including inventory control, distribution, and accounting.

Instruction: ***Version 1:*** Lecture and practical exercises cover the Army ammunition system and methods of inventory control, distribution, and accounting. *Version 2:* Lectures and practical exercises on the Army ammunition system and methods of inventory control, distribution, and accounting.

Credit Recommendation: ***Version 1:*** In the lower-division baccalaureate/associate degree category, 1 semester hour in material management (5/97). *Version 2:* In the lower-division baccalaureate/associate degree category, 1 semester hour in material management (6/01).

AR-1405-0257

Ordnance Munitions Materiel Management Officer Advanced Reserve Component, Phase 2

Course Number: 4-9-C23-91D-RC, Phase 2.

Location: Ordnance, Missile and Munitions School, Redstone Arsenal, AL.

Length: 2–3 weeks (108 hours).

Exhibit Dates: 6/94–10/00.

Learning Outcomes: Upon completion of the course, the student will be able to perform logistics management of ammunition supply in a battlefield or general support environment.

Instruction: Lecture and practical exercises on ammunition shipment, transport, routes, delivery, storage, security, and issuance to battlefield units.

Credit Recommendation: In the lower-division baccalaureate/associate degree category, 1 semester hour in materiel management (5/97).

Related Occupations: 91D.

AR-1405-0258

Basic Property Accounting

Course Number: LTC-001.

Location: Professional Education Center, Cp. Joseph Robinson, North Little Rock, AR.

Length: 1 week (45 hours).

Exhibit Dates: 6/93–Present.

Learning Outcomes: Upon completion of the course, the student will be able to possess the skills necessary to install and operate a PC workstation and software applications in support of an automated inventory and property control system, including update of control tables, maintenance of property files, transfer processing, requisitions and item receipts, and preparation of routine and ad hoc reports.

Instruction: Class presentations, discussions, demonstrations, and hands-on exercises train the student in PC hardware subsystems, connecting the PC subsystems, and operating the PC in all aspects of property control and inventory maintenance. This includes updating control tables; adding, changing, and deleting property records; processing transfers; requisitioning and receiving items; and printing reports.

Credit Recommendation: In the lower-division baccalaureate/associate degree category, 2 semester hours in automated inventory control or inventory management (10/03).

AR-1405-0259

Standard Army Maintenance System 2 Sustainment (Basic)

Course Number: LTC-003.

Location: Professional Education Center, Cp. Joseph Robinson, North Little Rock, AR.

Length: 1 week (47 hours).

Exhibit Dates: 7/91–Present.

Learning Outcomes: Upon completion of the course, the student will be able to install and operate a PC workstation and software applications in support of an automated maintenance system at the division or headquarters level. Student will be able to monitor, create, and maintain master records; schedule work and monitor status; monitor stock inventory and replacement; and prepare routine and ad hoc reports.

Instruction: Class presentations, discussions, demonstrations, hands-on exercises, and an examination train the student in PC components, connecting components, PC operations, and in all aspects of a maintenance system. This includes add/modify/delete master records, register and modify work orders, work order status, stock replenishment, and printing of ad hoc and production reports.

Credit Recommendation: In the lower-division baccalaureate/associate degree category, 2 semester hours in automated maintenance systems or maintenance systems (6/98).

AR-1405-0260

Standard Army Maintenance System 1 Sustainment (Basic)

Course Number: LTC-002.

Location: Professional Education Center, Cp. Joseph Robinson, North Little Rock, AR.

Length: 2 weeks (56 hours).

Exhibit Dates: 10/91–Present.

Learning Outcomes: Upon completion of the course, the student will be able to install and operate a PC workstation and software applications in support of an automated unit maintenance system, including master records, work orders, parts supply, task scheduling and status, and stock replacement. The student will be able to generate routine and ad hoc production reports from the system.

Instruction: Class presentations, discussions, demonstrations, hands-on exercises and an examination train the student in PC components, connecting components, and PC operations in all aspects of a maintenance system. This includes add/modify/delete master records, register and modify work orders, and tasks, parts and status, stock replenishment, and printing of adhoc and production reports.

Credit Recommendation: In the lower-division baccalaureate/associate degree category, 2 semester hours in automated maintenance systems (6/98).

AR-1405-0261

OMS TAMMS and Readiness

Course Number: LTC-018.

Location: Professional Education Center, Cp. Joseph Robinson, North Little Rock, AR.

Length: 2 weeks (67 hours).

Exhibit Dates: 12/96–Present.

Learning Outcomes: Upon completion of the course, the student will be able to operate a maintenance shop, including supervision and training in safety and accident reporting, operation of an automated maintenance system, hazardous material handling, and supply and property accountability.

Instruction: Class presentations and discussions introduce the student to maintenance shop operations. This includes safety and accident reporting, introduction to an automated maintenance system, supply receipt and distribution, petroleum and lubricant control, hazardous material management, and property accounting.

Credit Recommendation: In the lower-division baccalaureate/associate degree category, 2 semester hours maintenance and inventory systems or operations (10/03).

AR-1405-0262

OMS Basic Course

Course Number: 922-LTC-017 (R), Phase 2; LTC-017.

Location: Professional Education Center, Cp. Joseph Robinson, North Little Rock, AR.

Length: 2 weeks (67 hours).

Exhibit Dates: 12/96–Present.

Learning Outcomes: Upon completion of the course, the student will be able to supervise the staff operations of a maintenance management system to inventory, control, and dispense maintenance supplies.

Instruction: Class presentations and discussions introduce the student to maintenance management and shop operations. This includes safety and accident reporting, introduction to an automated mainframe system, personnel management, petroleum and lubricant control, hazardous material management, request and receipt of supplies, and property accounting.

Credit Recommendation: In the lower-division baccalaureate/associate degree category, 3 semester hours in maintenance and inventory system supervision (10/03).

AR-1405-0263

1. Battalion Logistics Noncommissioned Officer (NCO)
2. Battalion Supply

Course Number: *Version 1:* LTC-019. *Version 2:* LTC-019.

Location: Professional Education Center, Cp. Joseph Robinson, North Little Rock, AR.

Length: *Version 1:* 2 weeks (69 hours). *Version 2:* 1 week (59 hours).

Exhibit Dates: *Version 1:* 1/95–7/02. *Version 2:* 8/02–Present.

Learning Outcomes: *Version 1:* Upon completion of the course, the student will be able to supervise staff to determine needs, order supplies, receive inventory, and disburse of all manner of items. *Version 2:* Pending evaluation.

Instruction: *Version 1:* Class presentations and discussion develop the student's expertise in logistics management. This includes the logistics management of petroleum and lubricants, food service, ammunition, physical security, status reporting, disbursement, and receipt of items. *Version 2:* Pending evaluation.

Credit Recommendation: *Version 1:* In the upper-division baccalaureate category, 3 semester hours in logistics management, materials management, or supply management (6/98). *Version 2:* Pending evaluation.

AR-1405-0264

Unit Supply Noncommissioned Officer (NCO)

Course Number: LTC-27, Phase 1; LTC-027, Phase 1.

Location: Professional Education Center, Cp. Joseph Robinson, North Little Rock, AR.

Length: 2 weeks (71 hours).

Exhibit Dates: 6/95–Present.

Learning Outcomes: Upon completion of the course, the student will be able to requisition, receive, place in stock, and disburse various items of equipment and supplies.

Instruction: Class presentations and discussion develop the student's expertise in logistics supply. This includes the preparation of requisitions, receipt of supplies, disposition of property, physical security, personal clothing disbursement, and ammunition management.

Credit Recommendation: In the lower-division baccalaureate/associate degree category, 3 semester hours in inventory management, supply management, or materials management (6/98).

AR-1405-0265

Personnel Officer

Course Number: FT-012.

Location: Professional Education Center, Cp. Joseph Robinson, North Little Rock, AR.

Length: 2 weeks (62 hours).

Exhibit Dates: 11/96–Present.

Learning Outcomes: Upon completion of the course, the student will be able to manage a personnel section responsible for the preparation, and accuracy of personnel documents, including survivor benefits, pay adjustments, entry pay rates, retirement points, and employee ratings, for NCOs and officers, promotion verification, and change of specialty.

Instruction: The course includes class presentations, discussions, and a final examination. Emphasis is placed on the accuracy and timeliness of various personnel action documents.

Credit Recommendation: In the lower-division baccalaureate/associate degree category, 2 semester hours in records and information management (6/98).

AR-1405-0266

Standard Army Retail Supply System Level 3 (SARSS2 A/B) Supervision

Course Number: LTC-054.

Location: Professional Education Center, Cp. Joseph Robinson, North Little Rock, AR.

Length: 2 weeks (118 hours).

Exhibit Dates: 1/96–Present.

Learning Outcomes: Upon completion of the course, the student will be able to manage the Army's automated retail supply system operation, including operators training maintaining records, and generating reports.

Instruction: Computer-based simulations, hands-on practice, and problem solving scenarios are course methods. Assessment is based on actual exercises using the system.

Credit Recommendation: In the lower-division baccalaureate/associate degree category, 3 semester hours in automated supply systems operation. In the upper-division baccalaureate category, 2 semester hours in managing an automated supply operation (6/98).

AR-1405-0267

Standard Army Retail Supply System - Level 3 (SSARSS-2) Operator

Course Number: LTC-053.

Location: Professional Education Center, Cp. Joseph Robinson, North Little Rock, AR.

Length: 2 weeks (94 hours).

Exhibit Dates: 1/96–Present.

Learning Outcomes: Upon completion of the course, the student will be able to operate the Army's automated retail supply system operation including file creation, maintenance, standard report procedures, and ad hoc queries.

Instruction: Computer-based instruction, simulations for hands-on practices, and problem solving scenarios are course methods. Assessment is based on actual exercises using the system.

Credit Recommendation: In the lower-division baccalaureate/associate degree category, 3 semester hours in automated supply systems operations (6/98).

AR-1405-0268

Aviation Precommand

Course Number: 2G-F42.

Location: Aviation School, Fort Rucker, AL.

Length: 2–3 weeks (113–114 hours).

Exhibit Dates: 10/96–Present.

Learning Outcomes: Before 10/96 see AR-1606 0173. Upon completion of the course, students will be able to apply the principals and techniques of training management to their command and demonstrate a familiarization with systems and equipment applicable to their branch.

Instruction: Course includes briefings to field grade officers on tactics; training management; maintenance and logistics; safety; aviation medicine; fire support; communications-electronics; nuclear, biological, and chemical trends; and general military subjects.

Credit Recommendation: Credit is not recommended because of the limited, specialized nature of the course (11/98).

AR-1405-0269

Aviation Warrant Officer Advanced

Course Number: ***Version 1:*** 2-1-C32. *Version 2:* 2-1-C32, Phase 1;.

Location: ***Version 1:*** Aviation School, Fort Rucker, AL. *Version 2:* Aviation Center, Fort Rucker, AL.

Length: ***Version 1:*** 11 weeks (411 hours). *Version 2:* 8 weeks (343 hours).

Exhibit Dates: ***Version 1:*** 10/95–9/01. *Version 2:* 10/01–Present.

Learning Outcomes: ***Version 1:*** Before 10/95 see AR-1408-0063. Upon completion of the course, the student will apply principles of leadership, risk management process, and aviation deployment to logistical operations. *Version 2:* Upon completion of the course, the student will apply principles of leadership, risk management process, and aviation deployment to logistical operations.

Instruction: ***Version 1:*** Lectures and practical exercises in management, including general instruction in principles, philosophies, and concepts; general instruction in aviation safety including both accident prevention and investigation; microcomputer terminology; programming with hands-on operating experience; logistics management; communication skills; and interpersonal relations. *Version 2:* Lectures and practical exercises in management, including general instruction in principles, philosophies, and concepts; general instruction in aviation safety, including both accident prevention and investigation; microcomputer terminology; programming with hands-on operating experience; logistics management; communication skills; and interpersonal relations.

Credit Recommendation: ***Version 1:*** In the lower-division baccalaureate/associate degree category, 2 semester hours in communication skills and 3 in basic microcomputers. In the upper-division baccalaureate category, 3 semester hours in principles of management, 3 in interpersonal relations, and 1 in logistics management (11/98). *Version 2:* In the lower-division baccalaureate/associate degree category, 2 semester hours in communication skills and 1 in basic microcomputer applications. In the upper-division baccalaureate category, 3 semester hours in principles of management, 3 in interpersonal relationships, and 1 in logistics management (4/03).

AR-1405-0270

Aviation Captains Career-Reserve Component

Course Number: 2-1-C23, Phase 2; 2-1-C23.

Location: Aviation Center, Fort Rucker, AL.

Length: 2 weeks (143–167 hours).

Exhibit Dates: 5/94–Present.

Learning Outcomes: Upon completion of the course, the student will have a general understanding of combined arms operations, be familiar with organizational structures, will be able to provide briefing, and be able to plan and execute a variety of logistical operations.

Instruction: Small group instruction, briefings, and exercises cover combined arms operations, aviation tactical operations, principles of management, training/training management, threat update, and intelligence preparation of the battlefield.

Credit Recommendation: In the upper-division baccalaureate category, 3 semester hours in principles of management (4/03).

AR-1405-0271

Aviation Captains Career

Course Number: 2-1-C22, Phase 1; 2-1-C22.

Location: Aviation Center, Fort Rucker, AL.

Length: 18–20 weeks (647–767 hours).

Exhibit Dates: 4/96–Present.

Learning Outcomes: Upon completion of the course, the student will have a general understanding of combined arms operations, will be familiar with organizational structures. The student will also be able to provide briefing plan and execute a variety of logistical operations. The student will understand training management junctions, organizational concepts and applications, characteristics of offensive operations, military history as it relates to future operations, and be able to use effective listening, appropriate grammar, and effective speaking techniques.

Instruction: Lectures and practical exercises cover military subjects, leadership, principles of management, logistics management, technical writing skills, and combat planning.

Credit Recommendation: In the lower-division baccalaureate/associate degree category, 2 semester hours in technical writing. In the upper-division baccalaureate category, 3 semester hours in logistics management and 3 in maintenance management (4/03).

AR-1405-0272

Utilities Operations and Maintenance Technician Warrant Officer Advanced

Course Number: 4-5-C32-210A.

Location: Engineer School, Fort Leonard Wood, MO.

Length: 7 weeks (268–285 hours).

Exhibit Dates: 1/94–Present.

Learning Outcomes: Before 1/94 see AR-1405-0232. Upon completion of the course, the student will be able to manage personnel; manage the operation of utilities; and plan, program, and budget facility operations.

Instruction: Lectures and practical exercises cover management, including staff studies, schedule preparation, employee relations, and report preparation; planning, operation, and maintenance in the installation of utility systems; quality control in construction projects; facility engineering, including real property, fire protection, and energy conservation; and planning, programming, and budgeting for the operation of a utility facility.

Credit Recommendation: In the lower-division baccalaureate/associate degree category, 3 semester hours in facility management and 1 in personnel supervision (6/99).

AR-1405-0273

Transportation Management Coordinator

Course Number: 553-88N40.

Location: Transportation School, Fort Eustis, VA.

Length: 7 weeks (252 hours).

Exhibit Dates: 10/00–Present.

Learning Outcomes: Upon completion of the course, the student will be able to perform management duties in cargo and passenger movement organizations.

Instruction: Methods of instruction include classroom, conferences, and practical and field training exercises in planning, coordinating, and managing a cargo and passenger movement operation. This course includes a common core of leadership subjects.

Credit Recommendation: In the lower-division baccalaureate/associate degree category, see AR-1404-0035 for common core credit. In the upper-division baccalaureate category, 2 semester hours in transportation management (6/01).

Related Occupations: 88N.

AR-1405-0274

Total Army Training System (TATS) OH-58A/C Scout Helicopter Repairer Supervisor Basic Noncommissioned Officer (NCO), Phase 4

Course Number: 552-67V30-RC (F), Phase 4.

Location: Aviation Logistics School, Fort Eustis, VA.

Length: 166 hours.

Exhibit Dates: 10/01–Present.

Learning Outcomes: Upon completion of the course, the student will be able to troubleshoot and inspect the documents and maintenance for the OH-58 observation/scout helicopter.

Instruction: Methods of instruction include conferences, video, small groups, and practical exercises in the classroom and in the shop environment. Technical inspection and supervisory skills are learned. Students are evaluated through written and practical examinations.

Credit Recommendation: In the lower-division baccalaureate/associate degree category, 2 semester hours in aircraft maintenance supervision (6/01).

Related Occupations: 67V.

AR-1405-0275

Total Army Training System (TATS) UH-60 Helicopter Repairer Supervisor Basic Noncommissioned Officer (NCO), Phase 2

(Total Army Training System (TATS) CMF67 Basic Noncommissioned Officer (NCO) CAM)

Course Number: 552-CMF67-CAM-PH2, Phase 2.

Location: Aviation Logistics School, Fort Eustis, VA.

Length: 3 weeks (129 hours).

Exhibit Dates: 10/00–Present.

Learning Outcomes: Upon completion of the course, the student will be able to use appropriate standards and safety and production control procedures to supervises personnel and perform the duties of a technical inspector.

Instruction: Methods of instruction include student conferences, practical exercises and small group experience. Evaluation is based on written examinations.

Credit Recommendation: In the lower-division baccalaureate/associate degree category, 2 semester hours in aircraft maintenance supervision (6/01).

AR-1405-0276

Total Army Training System (TATS) UH-60 Helicopter Repairer Supervisor Basic Noncommissioned Officer (NCO), Phase 4

Course Number: 600-67T30 (F), Phase 4.
Location: Aviation Logistics School, Fort Eustis, VA.
Length: 3–4 weeks (116 hours).
Exhibit Dates: 10/00–Present.
Learning Outcomes: Upon completion of the course, the student will be able to supervise and inspect maintenance performed on the UH-60 aircraft.
Instruction: Methods of instruction include student conferences, demonstrations, and practical exercises in classroom and shop. Evaluation is accomplished by written and practical examinations.
Credit Recommendation: In the lower-division baccalaureate/associate degree category, 2 semester hours in aircraft maintenance supervision (6/01).
Related Occupations: 67T.

AR-1405-0277

Total Army Training System UH-60 Helicopter Repairer Supervisor Basic Noncommissioned Officer (NCO), Phase 2

Course Number: 552-CMF-67-CAM, Phase 2.
Location: Aviation Logistics School, Fort Eustis, VA.
Length: 129 hours.
Exhibit Dates: 10/00–Present.
Learning Outcomes: Upon completion of the course, the student will be able to supervise repairs, inspections, maintenance, and troubleshooting of aircraft.
Instruction: Methods of instruction include classroom presentations and small-group exercises.
Credit Recommendation: In the lower-division baccalaureate/associate degree category, 2 semester hours in aircraft maintenance supervision (6/01).

AR-1405-0278

OH-58D Armament/Electrical/Avionic Systems Repairer Supervisor Basic Noncommissioned Officer (NCO)

Course Number: 102-68J30 (OH-58D); 102-68S30, Phase 2.
Location: Aviation Logistics School, Fort Eustis, VA.
Length: 74–560 hours.
Exhibit Dates: 10/97–Present.
Learning Outcomes: This course consist of four modules. Upon completion of module A, the PH-2 TATS common aircraft maintenance (CAM), the student will be able to function as a supervisory and technical inspector, enforce policy, and follow maintenance, repair, and safety procedures. Upon completion of module B, the OH-58D enabling skills and electrical systems (AAD), the student will be able to perform maintenance inspections on control display, electrical, and lighting systems; use technical manuals; and follow safety procedures. Upon completion of module C, the OH-58D avionics systems, the student will be able to supervise, inspect, troubleshoot, and maintain avionic systems and subsystems including intercommunications, navigation/control, the Mast Mounted Sight Subsystem (MMSS), and the Aircraft Survivability Equipment (ASE). Upon completion of module D, the OH-58 armament/missile systems, the student will be able to supervise the troubleshooting, inspection and maintenance of OH-58D armament missile systems including armaments, rocket, machine guns, and air systems; and follow air transport procedures.
Instruction: Methods of instruction include computer-assisted instruction, video, conferences, practical exercises, and small-group interaction. Evaluation is based on written and practical examinations.
Credit Recommendation: In the lower-division baccalaureate/associate degree category, 2 semester hours in aircraft maintenance supervision for completion of module A; 1 in aircraft maintenance management and inspection for completion of module B; 1 in aircraft maintenance management and inspection for completion of module C; and 2 in aircraft maintenance management and inspection for completion of module D. If all four modules are completed, 4 semester hours in aircraft maintenance management and inspection and 2 in aircraft maintenance supervision (6/01).
Related Occupations: 68J; 68S.

AR-1405-0279

Total Army Training System (TATS) Postal Operations

Course Number: 7A-F67/510-ASIF5.
Location: Adjutant General School, Fort Jackson, SC.
Length: 5 weeks (187 hours).
Exhibit Dates: 10/97–Present.
Learning Outcomes: Before 10/97 see AR-1404-0001. Upon completion of this course, students will be able to process domestic and international mail; requisition, safeguard, and issue stamps, money orders, supplies and equipment; maintain and audit postal accounts; and operate a postal service center and integrated retail terminals.
Instruction: Topics include postal organization, facilities, and operation.
Credit Recommendation: In the lower-division baccalaureate/associate degree category, 3 credit hours in records management or office procedures (3/02).

AR-1405-0280

Medical Logistics Specialist

Course Number: 551-91J10.
Location: Academy of Health Sciences, Fort Sam Houston, TX.
Length: 5–6 weeks (251 hours).
Exhibit Dates: 1/02–Present.
Learning Outcomes: Before 1/02 see AR-1405-0241. Upon completion of the course, the student will be able to read and use medical supply publications and resources to search, locate and record medical supply information; perform property control procedures for medical supplies and equipment; and perform inventory management.
Instruction: Laboratory, audio visual materials, case studies, discussion, classroom exercise, practical exercises, and lectures. Topics include introduction to supply, material publications, property control, inventory management, special storage and handling of medical material, and field training exercise.
Credit Recommendation: In the lower-division baccalaureate/associate degree category, 2 semester hours in military science, 3 in property control and inventory, and 2 in medical material management (6/02).
Related Occupations: 91J.

AR-1405-0281

Supply System Technician Warrant Officer Advanced

Course Number: 8-10-C32-92B.
Location: Quartermaster Center and School, Fort Lee, VA.
Length: 10 weeks (358 hours).
Exhibit Dates: 9/98–Present.
Learning Outcomes: Upon completion of the course, the student will be able to communicate and lead from a broader perspective; perform management functions across the range of quartermaster specialties; utilize the multi-levels of the supply information systems; generate and analyze supply system reports; and track transactions, and plan and control supply levels and activities.
Instruction: Classroom and practical exercises, lecture, learner presentations in-class problem solving, in-class presentations, active interaction with several key supply information systems, and field trips to defense and commercial supply/warehouse operations.
Credit Recommendation: In the lower-division baccalaureate/associate degree category, 2 semester hours in principles of communication; 1 in principles of leadership; 3 in inventory, material, and supply management; and 2 in computer applications (6/02).

AR-1405-0282

Combined Logistics Captains Career

Course Number: 2-1-C22 (LOG), Phase 2.
Location: Aviation Center, Fort Rucker, AL.
Length: 5 weeks (162 hours).
Exhibit Dates: 10/01–Present.
Learning Outcomes: Upon completion of the course, the student will be able to conduct aviation operation coordination for battlefield combat and attack operations, serve as a company commander or battalion/brigade staff officer, and execute tactical operations.
Instruction: Audio visual materials, practical and classroom exercises, discussion, lecture, and computer-based training. Topics include doctrinal Army aviation roles, missions and capabilities for a combined logistics operation at a brigade level, military decision making, combat and attack operations.
Credit Recommendation: In the upper-division baccalaureate category, 3 semester hours in logistics management (4/03).

AR-1405-0283

Aviation Warrant Officer Advanced (RC)

Course Number: 2-1-C23-RC, Phase 2.
Location: Aviation Center, Fort Rucker, AL.
Length: 2–3 weeks (140 hours).
Exhibit Dates: 10/04–Present.
Learning Outcomes: Upon completion of the course, the student will be able to apply principles of leadership, risk management process, and aviation deployment to logistical operations.
Instruction: Audio visual materials, practical and classroom exercises, discussion, laboratory, lecture, and computer-based training. Topics include management, general instruction in the principles, philosophies, concepts, and scope of management; general instruction in aviation safety including both accident prevention and investigation; micro-computer terminology; programming with hands-on

experience; logistics management; communication skills; and interpersonal relations.

Credit Recommendation: In the lower-division baccalaureate/associate degree category, 2 semester hours in communication skills, and 1 in basic computer applications. In the upper-division baccalaureate category, 3 semester hours in principles of management, 3 in interpersonal relationships, and 1 in logistics management (4/03).

AR-1405-0284

Aviation Captains Career DL

Course Number: 2-1-C23 (DL), Phase 1.
Location: Aviation Center, Fort Rucker, AL.
Length: 192 hours.
Exhibit Dates: 1/01–Present.
Learning Outcomes: Upon completion of the course, the student will have a general understanding of combined arms operations, be familiar with organizational structures, be able to provide briefing, and be able to plan and execute a variety of logistical operations.
Instruction: Small group instruction, briefings, and exercises cover combined arms operations, aviation tactical operations, principles of management, training/training management, threat update, and intelligence preparation of the battlefield.
Credit Recommendation: In the lower-division baccalaureate/associate degree category, 3 semester hours in principles of management (4/03).

AR-1405-0285

Armor Captains Career

Course Number: 2-17-C23 (DL), Phase 1.
Location: Armor School, Fort Knox, KY.
Length: 9–10 weeks (373 hours).
Exhibit Dates: 10/00–Present.
Learning Outcomes: Upon completion of the course, the student will be able to apply skills and knowledge in professional ethics, leadership values and conduct training; demonstrate ability to effectively communicate with subordinates, superiors, and peers; and display knowledge of Navy and Marine Corps roles in supporting national defense.
Instruction: Audiovisual materials, practical exercises, discussion, learner presentation, lecture, and computer-based training. Topics include tactical operations, troop leading procedures, integration of electronic warfare, fundamentals of Army operations, integration of engineer assets, and United States Marine Corps and Navy roles and missions.
Credit Recommendation: In the lower-division baccalaureate/associate degree category, 3 semester hours in leadership, 1 in ethics, 3 in organizational communication, and 1 in logistics and planning (6/03).

AR-1405-0286

Armor Captains Career

Course Number: 2-17-C23, Phase 2.
Location: Armor School, Fort Knox, KY.
Length: 2 weeks (125 hours).
Exhibit Dates: 10/02–Present.
Learning Outcomes: Upon completion of the course, the student will be able to demonstrate knowledge and skills in technical training, logistical and maintenance procedures, and perform leadership and staff functions at the battalion brigade levels.
Instruction: Audiovisual materials, discussion, learner presentation, lecture, and classroom and practical exercises. Topics include mission analysis, logistics and maintenance procedures, course-of-action analysis, and decision making leadership, task force operations.
Credit Recommendation: In the upper-division baccalaureate category, 3 semester hours in logistics and operations management and 3 in managerial planning. In the graduate degree category, 1 semester hour in decision making and problem solving (6/03).

AR-1405-0287

Ordnance Maintenance Management Officer Advanced Reserve Component

Course Number: 4-9-C23-91B/C RC.
Location: Ordnance Center and School, Aberdeen Proving Ground, MD.
Length: 90 hours.
Exhibit Dates: 8/93–Present.
Learning Outcomes: Upon completion of the course, the student will be able to demonstrate the ability to lead train, fight, and maintain units as maintenance company commanders and battalion/brigade staff officers.
Instruction: Audio visual materials, learner presentations, computer-based training, practical exercises, classroom exercises, discussion, laboratory, and lecture. Topics include overview of the Army logistics system, NBC logistics, maintenance management, depot maintenance/operations, accounting for property, company operations, direct support maintenance operations, and standard Army retail supply system.
Credit Recommendation: In the upper-division baccalaureate category, 3 semester hours in management planning and supervision (8/03).

AR-1405-0288

Health Services Materiel Officer (RC)

Course Number: 8B-70K67 (RC), Phase 2.
Location: Academy of Health Sciences, Fort Sam Houston, TX.
Length: 2 weeks (81 hours).
Exhibit Dates: 4/03–Present.
Learning Outcomes: Upon completion of the course, the student will be able to apply principles and techniques used in management of logistics functions in Army hospitals and field medical units and identify the mission of combat support services.
Instruction: Audiovisual materials, learner presentations, computer-based training, practical and classroom exercises, discussion, laboratory, case studies, and lecture. Topics include supply terms, combat health logic systems in a theatre of operations, combat service support, and organization of medical supply office.
Credit Recommendation: In the lower-division baccalaureate/associate degree category, 3 semester hours in introduction to logistics. In the upper-division baccalaureate category, 3 semester hours in advanced logistics (11/03).

AR-1405-0289

Health Services Materiel Officer

Course Number: 8B-70K67.
Location: Academy of Health Sciences, Fort Sam Houston, TX.
Length: 10 weeks (382 hours).
Exhibit Dates: 2/02–Present.
Learning Outcomes: Upon completion of the course, the student will be able to apply the principles and techniques used in health care management of logistical functions in Army hospitals and field medical units; utilize the hazard communication program and OSHA; and understand the concepts of theatre blood management upon implementation of medical force.
Instruction: Audiovisual materials, computer-based training, practical and classroom exercises, discussion, and lecture. Topics include logistics systems and organizations; essential elements of Army financial budget formulation, structure and organization; maintenance of audit trail covering life cycle process for property from the time it enters the system to when it leaves; quality control; combat service support; capital expenditure; hazard communication program; OSHA; theatre blood management; and chemical, biological, radiological, nuclear and high exposures.
Credit Recommendation: In the lower-division baccalaureate/associate degree category, 3 semester hours in introduction to logistics. In the upper-division baccalaureate category, 3 semester hours in advanced logistics and 9 in clinical internship (logistics) (11/03).

AR-1406-0011

Personnel Records Specialist

Course Number: 500-75D10.
Location: Training Center, Fort Jackson, SC.
Length: 7 weeks (211–244 hours).
Exhibit Dates: 9/82–9/98.
Objectives: To train individuals to maintain personnel records.
Instruction: Conferences and practical exercises cover personnel records maintenance through such source documents as court-martial orders efficiency reports, suspension of favorable personnel actions, requested and completed personnel actions; in- and out-processing of personnel records and the initiation, use, and updating of standard installation/division personnel system documentation.
Credit Recommendation: In the lower-division baccalaureate/associate degree category, 2 semester hours in typewriting, 3 in clerical bookkeeping, and 2 in office procedures (8/88).
Related Occupations: 09B; 09E; 75B; 75D.

AR-1406-0020

Retention Noncommissioned Officer (NCO)

Course Number: 501-79D30; 501-00R30.
Location: Institute for Administration, Fort Benjamin Harrison, IN; Adjutant General School, Fort Benjamin Harrison, IN.
Length: 6–9 weeks (223–317 hours).
Exhibit Dates: 1/85–9/90.
Objectives: To train enlisted personnel to manage a recruiting and reenlistment office.
Instruction: Lectures and practical exercises cover the management of a recruiting and reenlistment office. Course includes principles and techniques of recruiting and reenlistment and effective speaking. Conference instruction and practical exercises cover service computation,

bonuses, interviewing and counseling techniques, and maintenance of enlistment program.

Credit Recommendation: Credit is not recommended because of the military-specific nature of the course (11/91).

Related Occupations: 00E; 00R; 79D.

AR-1406-0026

Personnel Actions Specialist

Course Number: 500-75E10; 500-75E10 (ST).

Location: Training Center, Fort Jackson, SC.

Length: 7–8 weeks (277 hours).

Exhibit Dates: 9/82–8/98.

Objectives: To train students to prepare, process, and review personnel records.

Instruction: Lectures, field training experiences, and practical exercises cover the preparation, processing, and review of records of personnel actions.

Credit Recommendation: In the lower-division baccalaureate/associate degree category, 2 semester hours in typing and 3 in clerical record keeping (8/88).

Related Occupations: 09B; 09E; 75B; 75E.

AR-1406-0032

Recruiting Station Commanders

Course Number: 501-F2.

Location: Institute for Administration, Fort Benjamin Harrison, IN.

Length: 2 weeks (76 hours).

Exhibit Dates: 7/83–9/90.

Objectives: To provide selected enlisted personnel with the skills necessary to perform as a recruiting station supervisor.

Instruction: Conferences and practical exercises cover responsibility, leadership, management, personnel training, sales promotion, and administration of recruiting stations.

Credit Recommendation: In the upper-division baccalaureate category, 2 semester hours in principles of supervision (3/85).

Related Occupations: 00E.

AR-1406-0034

Leadership and Management Development Trainers Course

Course Number: None.

Location: Organizational Effectiveness Center and School, Fort Ord, CA; Selected Army locations.

Length: 4 weeks (148–149 hours).

Exhibit Dates: 5/79–6/96.

Objectives: To provide instruction in training to officer and senior enlisted personnel selected to instruct the Leadership and Management Development Trainers Course. Selected training techniques for leadership methods and related advanced management development and behavioral science skills are incorporated.

Instruction: This course is a basic introduction to leadership/interpersonal relations training. Focus of the course is the application of training and instructional techniques which provide students with skills in conducting a basic course in leadership and management development. Topics include group dynamics, communication modes, interpersonal relationship models, and performance and personal counseling.

Credit Recommendation: In the lower-division baccalaureate/associate degree category, 2 semester hours in leadership/interpersonal relations. In the upper-division baccalaureate category, 1 semester hour in training and development (4/81).

AR-1406-0035

Air Defense Artillery (ADA) Advanced Noncommissioned Officer (NCO)

Course Number: 0-16-C42; 1-23-C42; 1-27-C42B; 1-27-C42.

Location: Air Defense Artillery School, Fort Bliss, TX.

Length: 8–9 weeks (312 hours).

Exhibit Dates: 3/84–2/92.

Objectives: To provide senior enlisted personnel with a working knowledge of duties, instructional methods, and general knowledge necessary to perform as senior noncommissioned officers in air defense management.

Instruction: The conferences and practical exercises include small unit self-defense, pilot tactics, and instruction in HIMAD and SHORAD system deployment. Course includes a common core of leadership subjects.

Credit Recommendation: Credit is recommended for the common core only. See AR-1404-0035 (11/88).

AR-1406-0036

A Systems Approach to Training (Instructional Systems Development)

Course Number: 5K061FD90; 5K061FD92.

Location: Field Artillery School, Fort Sill, OK.

Length: 3 weeks (120 hours).

Exhibit Dates: 1/80–6/96.

Learning Outcomes: The student prepares instructional materials and learns task analysis and teaching by objectives.

Instruction: Through lectures, demonstrations, and projects, the student learns to recognize tasks and objectives and develops several types of instructional materials including self-paced instructional materials and validation techniques. The material covered emphasizes using student performance objectives and criterion testing.

Credit Recommendation: In the upper-division baccalaureate category, 3 semester hours in development of instructional materials (10/84).

AR-1406-0037

Instructor Training

Course Number: 5K061FD90.

Location: Field Artillery School, Fort Sill, OK.

Length: 2 weeks (83 hours).

Exhibit Dates: 8/79–11/95.

Objectives: To provide faculty with the necessary skills and techniques needed for classroom instruction.

Instruction: The course includes instruction in the principles of assessing needs, selecting proper audiovisual and demonstration equipment, classroom management, criterion testing, and editing instructional materials. The student is required to give various types and lengths of presentations that are critiqued by the instructional staff.

Credit Recommendation: In the lower-division baccalaureate/associate degree category, 1 semester hour in speech or public speaking. In the upper-division baccalaureate category, 3 semester hours in teaching methods (11/85).

AR-1406-0038

Technical Services Officer (TSO) Leadership Workshop

Course Number: None.

Location: Deputy Chief of Staff for Personnel, Atlanta, GA; Deputy Chief of Staff for Personnel, Baltimore, MD; Deputy Chief of Staff for Personnel, Dallas, TX; Deputy Chief of Staff for Personnel, San Francisco, CA.

Length: 1–2 weeks (64 hours).

Exhibit Dates: 1/80–1/90.

Objectives: To provide an overview of certain aspects of the personnel management function and its interface with other areas of the organization.

Instruction: Lectures and practical exercises cover the fundamentals of personnel management, including human resource planning, the regulatory base, wage and salary administration, records management, and labor relations. No formal evaluation of students is conducted.

Credit Recommendation: In the lower-division baccalaureate/associate degree category, 1 semester hour in personnel management based on departmental evaluation (assessment/interview) (8/81).

AR-1406-0039

Basic Training and Development

Course Number: None.

Location: Deputy Chief of Staff for Personnel, Atlanta, GA; Deputy Chief of Staff for Personnel, Baltimore, MD; Deputy Chief of Staff for Personnel, Dallas, TX; Deputy Chief of Staff for Personnel, San Francisco, CA.

Length: 1–2 weeks (64 hours).

Exhibit Dates: 1/80–6/96.

Objectives: To provide an overview of training and development policies, administrative practices, management procedures, and program constraints within the civilian personnel force of the Department of the Army.

Instruction: The course uses a variety of instructional methods, including precourse text readings, lectures, handouts, case study, audiovisual materials, and practical exercises.

Credit Recommendation: In the lower-division baccalaureate/associate degree category, 1 semester hour in training and development within the Department of the Army, subject to departmental interview (assessment/evaluation) (8/81).

AR-1406-0052

Principles of Counseling

Course Number: 1-19-XL-100.

Location: Military Police School, Fort McClellan, AL; Chemical School, Fort McClellan, AL.

Length: Self-paced, 1–2 weeks (50 hours).

Exhibit Dates: 10/80–3/92.

Objectives: To train instructors/trainers in the basic skills of counseling in an academic environment.

Instruction: This course focuses on developing the basic skills of counseling required to improve instructor/trainer-student communication. The course is conducted in a self-paced workshop and provides learning materials, multimedia instruction, and resource personnel for students.

Credit Recommendation: In the upper-division baccalaureate category, 2 semester hours in guidance and counseling (3/82).

AR-1406-0054

BASIC INSTRUCTOR TRAINING COURSE

Course Number: 1-19-XB-100; 19H-BIT.
Location: Military Police School, Fort McClellan, AL.
Length: Self-paced, 3–4 weeks (94–110 hours).
Exhibit Dates: 1/77–1/92.
Objectives: To provide students with knowledge and skills needed to become effective basic instructors.
Instruction: Cognitive and practical experiences are used to teach a systems approach to instructional needs analysis, design, presentation, and evaluation. Instructional strategies and platform skills are emphasized.
Credit Recommendation: In the upper-division baccalaureate category, 3 semester hours in methods of instruction (3/82).

AR-1406-0055

TECHNIQUES AND PROCEDURES FOR FIELD MANUAL DEVELOPMENT

Course Number: 1-19-XT-100.
Location: Military Police School, Fort McClellan, AL.
Length: Self-paced, 1–2 weeks (58 hours).
Exhibit Dates: 6/81–12/97.
Objectives: To provide student with techniques and strategies for designing and developing various instructional materials.
Instruction: This course is an introduction to manual writing techniques and procedures. Performance evaluation of the actual writing project is coordinated with a course manager, editor, and graphics specialist.
Credit Recommendation: In the lower-division baccalaureate/associate degree category, 2 semester hours in instructional materials design (3/82).

AR-1406-0056

INSTRUCTOR TRAINING COURSE

Course Number: None.
Location: Chemical School, Fort McClellan, AL.
Length: 2–3 weeks (102 hours).
Exhibit Dates: 1/82–1/92.
Objectives: To provide selected officers, noncommissioned officers, enlisted personnel, and civilians with the knowledge and skills needed to be competent instructors.
Instruction: Course includes a systems approach to instructor training that emphasizes analysis, design, and appropriate instructional strategies. The course focuses on developing behavioral learning objectives and generating a variety of platform skills.
Credit Recommendation: In the upper-division baccalaureate category, 3 semester hours in methods of instruction (3/82).

AR-1406-0057

JOB AND TASK ANALYSIS (ISD PHASE 1)

Course Number: 1-19XC-100.
Location: Military Police School, Fort McClellan, AL.
Length: Self-paced, 3–4 weeks (120 hours).
Exhibit Dates: 10/80–1/94.
Objectives: To provide selected enlisted personnel, noncommissioned, and commissioned officers with general knowledge of the instructional system design process.
Instruction: Self-paced instruction and practical exercises cover job analysis, task identification, objectives, and job performance measures as defined by criterion tests. Participants receive instruction in selecting instructional materials, instructional settings, and delivery systems. Management decisions relevant to the job and task analysis process are covered, along with internal and external evaluations for instructional system revision.
Credit Recommendation: In the lower-division baccalaureate/associate degree category, 4 semester hours in job and task analysis (3/82).

AR-1406-0058

DESIGN (ISD PHASE 2); DEVELOP (ISD PHASE 3); CONTROL/EVALUATION (ISD PHASE 5); VALIDATION PROCEDURES FOR COURSE MATERIALS AND TESTS

Course Number: 1-19-XD-100; 1-19-XE-100; 1-19-XG-100; 1-19-XM-100.
Location: Military Police School, Fort McClellan, AL.
Length: Self-paced, 3 weeks (89–90 hours).
Exhibit Dates: 6/81–1/94.
Objectives: To train participants in designing instruction using job analysis information, developing courses and course material, and validating testing.
Instruction: These four short courses emphasize objectives, mastery, test design, sequencing of instruction, learning activities, material and media selection, and evaluation/control.
Credit Recommendation: In the upper-division baccalaureate category, 3 semester hours in instructional materials and methods of teaching, for the completion of all four courses (3/82).

AR-1406-0061

ARMY REENLISTMENT OFFICER

Course Number: 7C-F17.
Location: Recruiting and Retention School, Fort Benjamin Harrison, IN.
Length: 1 week (50 hours).
Exhibit Dates: 7/84–12/91.
Objectives: To prepare officers for responsibilities as unit/installation reenlistment officers.
Instruction: Instruction and practical exercises include policies and responsibilities in an reenlistment program, including separation, retention, reenlistment ceremony, awards and incentives, and computation of service time.
Credit Recommendation: Credit is not recommended because of the military-specific nature of the course (3/85).

AR-1406-0068

BASIC NONCOMMISSIONED OFFICER (NCO) 11B TRACK

Course Number: *Version 1:* 010-1-11B30; 010-2-11B30; 010-3-11B30; 010-4-11B30; 010-5-11B30; 010-6-11B30; 010-7-11B30; 010-9-11B30; 010-10-11B30; 010-12-11B30; 010-13-11B30; 010-14-11B30; 010-15-11B30; 010-16-11B30; 010-18-11B30. *Version 2:* 010-1-11B30; 010-2-11B30; 010-3-11B30; 010-4-11B30; 010-5-11B30; 010-6-11B30; 010-10-11B30; 010-12-11B30; 010-13-11B30; 010-14-11B30; 010-15-11B30; 010-16-11B30; 010-18-11B30.
Location: *Version 1:* Infantry School, Fort Benning, GA; NCO Academy, Fort Knox, KY; NCO Academy, Fort Bragg, NC; NCO Academy, Fort Benning, GA; NCO Academy, Fort Stewart, GA; NCO Academy, Fort Campbell, KY; NCO Academy, Fort Polk, LA; NCO Academy, Fort Riley, KS; NCO Academy, Fort Hood, TX; NCO Academy, Fort Carson, CO; NCO Academy, Fort Ord, CA; NCO Academy, Fort Lewis, WA; NCO Academy, AK; NCO Academy, HI; NCO Academy, Panama; NCO Academy, Europe. *Version 2:* NCO Academy, Europe, Panama, Hawaii, Alaska, and Federal R; Infantry School, Fort Benning, GA; NCO Academy, Fort Drum, NY; NCO Academy, Fort Bragg, NC; NCO Academy, Fort Benning, GA; NCO Academy, Fort Stewart, GA; NCO Academy, Fort Campbell, KY; NCO Academy, Fort Polk, LA; NCO Academy, Fort Carson, CO; NCO Academy, Fort Ord, CA; NCO Academy, Fort Lewis, WA.
Length: *Version 1:* 5 weeks (222 hours). *Version 2:* 7 weeks (328 hours).
Exhibit Dates: *Version 1:* 8/83–9/90. *Version 2:* 10/90–9/92.
Objectives: *Version 1:* To train infantry leaders to manage, lead, train, and direct subordinates to maintain, operate, and employ weapons and equipment. *Version 2:* To train infantry leaders to manage, lead, train, and direct subordinates to maintain, operate, and employ weapons and equipment.
Instruction: *Version 1:* Performance-oriented training techniques cover leadership and management skills using lectures and demonstrations. This course includes leadership, maintenance, tactics, communications, navigation, weapons, and management. *Version 2:* Performance-oriented training techniques cover leadership and management skills using lectures and demonstrations. This course includes leadership, maintenance, tactics, communications, navigation, weapons, and management. Course includes a common core of leadership subjects.
Credit Recommendation: *Version 1:* In the lower-division baccalaureate/associate degree category, 1 semester hour in personnel management (9/85). *Version 2:* In the lower-division baccalaureate/associate degree category, 1 semester hour in personnel management. See AR-1406-0090 for common core credit (10/90).
Related Occupations: 11B.

AR-1406-0069

BASIC NONCOMMISSIONED OFFICER (NCO) 11C TRACK

Course Number: *Version 1:* 010-1-11C30; 010-2-11C30; 010-3-11C30; 010-4-11C30; 010-5-11C30; 010-6-11C30; 010-7-11C30; 010-9-11C30; 010-10-11C30; 010-12-11C30; 010-13-11C30; 010-14-11C30; 010-15-11C30; 010-16-11C30; 010-18-11C30. *Version 2:* 010-3-11C30; 010-5-11C30; 010-9-11C30; 010-13-11C30; 010-18-11C30.
Location: *Version 1:* NCO Academy, Europe, Panama, Hawaii, Alaska, and Federal R; Infantry School, Fort Benning, GA; NCO Academy, Fort Knox, KY; NCO Academy, Fort Bragg, NC; NCO Academy, Fort Benning, GA; NCO Academy, Fort Stewart, GA; NCO Academy, Fort Campbell, KY; NCO Academy, Fort

Polk, LA; NCO Academy, Fort Riley, KS; NCO Academy, Fort Hood, TX; NCO Academy, Fort Carson, CO; NCO Academy, Fort Ord, CA; NCO Academy, Fort Lewis, WA. *Version 2:* NCO Academy, Europe, Panama, Hawaii, Alaska, and Federal R; NCO Academy, Fort Benning, GA; NCO Academy, Fort Campbell, KY; NCO Academy, Fort Hood, TX; NCO Academy, Fort Lewis, WA.

Length: *Version 1:* 5 weeks (205 hours). *Version 2:* 7 weeks (315 hours).

Exhibit Dates: *Version 1:* 6/83–9/90. *Version 2:* 10/90–5/93.

Objectives: *Version 1:* To train infantry leaders to lead, train, and direct subordinates to maintain, operate, and employ weapons and equipment. *Version 2:* To train infantry leaders to lead, train, and direct subordinates to maintain, operate, and employ weapons and equipment.

Instruction: *Version 1:* Performance-oriented training techniques cover management and leadership skills using lectures and demonstrations. The course includes maintenance, tactics, communications, navigation, and training management. *Version 2:* Performance oriented training techniques cover management and leadership skills using lectures and demonstrations. The course includes maintenance, tactics, communications, navigation, and training management. Course includes a common core of leadership subjects.

Credit Recommendation: *Version 1:* In the lower-division baccalaureate/associate degree category, 1 semester hour in personnel management (9/85). *Version 2:* In the lower-division baccalaureate/associate degree category, 1 semester hour in personnel management. See AR-1406-0090 for common core credit (9/85).

AR-1406-0070

Basic Noncommissioned Officer (NCO) 11M Track

Course Number: *Version 1:* 010-1-11M30; 010-2-11M30; 010-3-11M30; 010-4-11M30; 010-5-11M30; 010-6-11M30; 010-7-11M30; 010-9-11M30; 010-10-11M30; 010-12-11M30; 010-13-11M30; 010-14-11M30; 010-15-11M30; 010-16-11M30; 010-18-11M30. *Version 2:* 010-3-11M30; 010-4-11M30; 010-6-11M30; 010-7-11M30; 010-9-11M30; 010-10-11M30.

Location: *Version 1:* NCO Academy, Europe, Panama, Hawaii, Alaska, and Federal R; Infantry School, Fort Benning, GA; NCO Academy, Fort Knox, KY; NCO Academy, Fort Bragg, NC; NCO Academy, Fort Benning, GA; NCO Academy, Fort Stewart, GA; NCO Academy, Fort Campbell, KY; NCO Academy, Fort Polk, LA; NCO Academy, Fort Riley, KS; NCO Academy, Fort Hood, TX; NCO Academy, Fort Carson, CO; NCO Academy, Fort Ord, CA; NCO Academy, Fort Lewis, WA. *Version 2:* NCO Academy, Europe, Panama, Hawaii, Alaska, and Federal R; NCO Academy, Fort Benning, GA; NCO Academy, Fort Stewart, GA; NCO Academy, Fort Riley, KS; NCO Academy, Fort Polk, LA; NCO Academy, Fort Hood, TX.

Length: *Version 1:* 5–6 weeks (263 hours). *Version 2:* 7 weeks (315 hours).

Exhibit Dates: *Version 1:* 6/83–9/90. *Version 2:* 10/90–12/92.

Objectives: *Version 1:* To train personnel in management, leadership, and critical job tasks. *Version 2:* To train personnel in management, leadership, and critical job tasks.

Instruction: *Version 1:* Performance-oriented training techniques cover management and leadership skill development using lectures and demonstrations. Students are rotated through leadership positions during the course. *Version 2:* Performance-oriented training techniques cover management and leadership skill development using lectures and demonstrations. Students are rotated through leadership positions during the course. Course includes a common core of leadership subjects.

Credit Recommendation: *Version 1:* In the lower-division baccalaureate/associate degree category, 2 semester hours in personnel management (9/85). *Version 2:* In the lower-division baccalaureate/associate degree category, 2 semester hours in personnel management. See AR-1406-0090 for common core credit (10/90).

AR-1406-0071

Basic Noncommissioned Officer (NCO) 11H Track

Course Number: *Version 1:* 010-1-11H30; 010-2-11H30; 010-3-11H30; 010-4-11H30; 010-5-11H30; 010-6-11H30; 010-7-11H30; 010-9-11H30; 010-10-11H30; 010-12-11H30; 010-13-11H30; 010-14-11H30; 010-15-11H30; 010-16-11H30; 010-18-11H30. *Version 2:* 010-2-11H30; 010-3-11H30; 010-6-11H30; 010-18-11H30.

Location: *Version 1:* Infantry School, Fort Benning, GA; NCO Academy, Fort Knox, KY; NCO Academy, Fort Bragg, NC; NCO Academy, Fort Benning, GA; NCO Academy, Fort Stewart, GA; NCO Academy, Fort Campbell, KY; NCO Academy, Fort Polk, LA; NCO Academy, Fort Riley, KS; NCO Academy, Fort Hood, TX; NCO Academy, Fort Carson, CO; NCO Academy, Fort Ord, CA; NCO Academy, Fort Lewis, WA; NCO Academy, Fort Greeley, AK; NCO Academy, Hawaii; NCO Academy, Panama, CZ; NCO Academy, Europe. *Version 2:* NCO Academy, Fort Benning, GA; NCO Academy, Fort Bragg, NC; NCO Academy, Fort Polk, LA; NCO Academy, Europe.

Length: *Version 1:* 5–6 weeks (187 hours). *Version 2:* 7 weeks (318 hours).

Exhibit Dates: *Version 1:* 6/83–9/90. *Version 2:* 10/90–5/93.

Objectives: *Version 1:* To train infantry section leaders to lead, train, and direct subordinates to maintain, operate, and employ weapons. *Version 2:* To train infantry section leaders to lead, train, and direct subordinates to maintain, operate, and employ weapons.

Instruction: *Version 1:* Performance-oriented training covers management and leadership skills and techniques using lectures and demonstrations. *Version 2:* Performance-oriented training covers management and leadership skills and techniques using lectures and demonstrations. Course includes a common core of leadership subjects.

Credit Recommendation: *Version 1:* In the lower-division baccalaureate/associate degree category, 1 semester hour in personnel management (9/85). *Version 2:* In the lower-division baccalaureate/associate degree category, 1 semester hour in personnel management. See AR-1406-0090 for common core credit (10/90).

AR-1406-0072

T-42 Instructor Pilot (IP) Methods of Instruction

Course Number: None.

Location: Aviation Center, Fort Rucker, AL.

Length: Self-paced, 4 weeks (157 hours).

Exhibit Dates: 5/79–1/90.

Objectives: To train aviators to conduct transition training for T-42 instructor pilots.

Instruction: This is a self-paced proficiency advancement course employing completely individualized instruction in T-42 aircraft systems, flight training, and methods of instruction.

Credit Recommendation: In the upper-division baccalaureate category, 3 semester hours in principles and methods of instruction (11/86).

AR-1406-0075

TH-55 Instructor Pilot Methods of Instruction

Course Number: None.

Location: Aviation Center, Fort Rucker, AL.

Length: 6 weeks (189 hours).

Exhibit Dates: 8/79–1/90.

Objectives: To qualify rotary-wing pilots as instructor pilots for primary rotary-wing flight training.

Instruction: Course provides academic and flight instruction, including regulations and procedures, aircraft systems, forms and records, instructing fundamentals, flight safety, aerodynamics, transition flight training, and human relations.

Credit Recommendation: In the upper-division baccalaureate category, 3 semester hours in principles and methods of instruction (11/86).

Related Occupations: 100B; 15A.

AR-1406-0076

UH-1 Contact Instructor Pilot (Interim)

Course Number: None.

Location: Aviation Center, Fort Rucker, AL.

Length: 4–6 weeks (153–225 hours).

Exhibit Dates: 1/82–1/90.

Objectives: To train selected commissioned and warrant officer aviators as unit instructor pilots for transition/standardization training in the UH-1 helicopter.

Instruction: Lectures and practical training cover day/night visual flight training; before-flight, basic-flight, approach/landing, and emergency tasks; principles and techniques of flight instruction; and aircraft systems and maintenance.

Credit Recommendation: In the upper-division baccalaureate category, 3 semester hours in principles and methods of instruction (11/86).

AR-1406-0077

UH-1 Contact Instructor Pilot Methods of Instruction

Course Number: None.

Location: Aviation Center, Fort Rucker, AL.

Length: 6 weeks (225 hours).

Exhibit Dates: 7/86–11/96.

Objectives: To train selected commissioned and warrant officer aviators as unit instructor pilots in transition/standardization training in the UH-1 helicopter.

Instruction: Lectures and practical training cover day/night visual flight rules; before-flight, basic-flight, approach/landing, and emergency tasks; principles and techniques of flight instruction; and aircraft systems and maintenance.

Credit Recommendation: In the upper-division baccalaureate category, 3 semester hours in principles and methods of instruction (11/86).

AR-1406-0079

UH-60 Instructor Pilot

Course Number: 2C-F33.

Location: Aviation Center, Fort Rucker, AL.

Length: 4–5 weeks (175–187 hours).

Exhibit Dates: 2/80–11/96.

Objectives: To train UH-60 aviators as instructor pilots capable of teaching and evaluating UH-60 flight training.

Instruction: Flight and academic instruction covers the UH-60 helicopter and in methods of instruction.

Credit Recommendation: In the upper-division baccalaureate category, 3 semester hours in basic aerodynamics and theory of flight, 3 in basic helicopter and flight systems, 3 in principles and methods of instruction, and credit for flight training on the basis of institutional evaluation (11/86).

AR-1406-0080

UH-1 Combat Skills Day/Night/Night Vision Goggles Methods of Instruction

Course Number: None.

Location: Aviation Center, Fort Rucker, AL.

Length: 8 weeks (209 hours).

Exhibit Dates: 7/83–12/97.

Objectives: To qualify selected helicopter instructor pilots as UH-1 Combat Skills Day/Night/Night Vision Goggles instructor pilots.

Instruction: Flight and academic training covers combat skills, night/day night vision goggle, UH-1 aircrew training, and manual maneuver tasks under combat conditions.

Credit Recommendation: Credit is not recommended because of the military-specific nature of the course (11/86).

AR-1406-0081

Office Machine Repairer Noncommissioned Officer (NCO) Advanced

(Mechanical Maintenance Noncommissioned Officer (NCO) Advanced, MOS 41J)

Course Number: 670-41J40; 6-63-C42 (41J).

Location: Ordnance Center and School, Aberdeen Proving Ground, MD.

Length: 6–8 weeks (217–355 hours).

Exhibit Dates: 10/85–1/98.

Learning Outcomes: Upon completion of the course, the student will be able to apply skills and technical knowledge to serve as a supervisor.

Instruction: Lectures, demonstrations, and practical exercises cover leadership, communication skills, fundamentals of instruction, logistics management, and refresher training in the appropriate military occupational specialty. Emphasis in on leadership and on verbal and written communication. Course includes a common core of leadership subjects.

Credit Recommendation: Credit is recommended for the common core only. See AR-1404-0035 (1/88).

Related Occupations: 41J.

AR-1406-0082

Chaplain Candidate

Course Number: 5G-F5.

Location: Chaplain Center and School, Fort Monmouth, NJ.

Length: 5–6 weeks (204 hours).

Exhibit Dates: 10/83–12/97.

Learning Outcomes: Upon completion of the course, the student will be able to perform basic soldier survival skills; apply knowledge of military protocol, customs, and organization; describe the role of the chaplain in the Army; and plan lessons and training.

Instruction: Topics include military organization, customs, courtesies, and ceremonies; physical fitness, military survival skills, and health and first aid; military administration and training; and pastoral ministry to minorities, hospital patients, and other chaplains.

Credit Recommendation: In the lower-division baccalaureate/associate degree category, 1 semester hour in technical communication and 2 in health and first aid. In the upper-division baccalaureate category, 1 semester hour in human relations (4/87).

AR-1406-0083

Division Chaplain

Course Number: *Version 1:* 5G-F4. *Version 2:* 5G-F4.

Location: Chaplain Center and School, Fort Monmouth, NJ.

Length: *Version 1:* 2 weeks (84 hours). *Version 2:* 2 weeks (70 hours).

Exhibit Dates: *Version 1:* 4/85–12/03. *Version 2:* 1/04–Present.

Learning Outcomes: *Version 1:* Upon completion of the course, the student will be able to serve as the senior supervising chaplain in an Army division (10,000 persons) using management skills (personnel, fiscal, logistics, and training) in the tactical garrison environment and in mobilization and deployment conditions. *Version 2:* Pending evaluation.

Instruction: *Version 1:* Topics include command and staff relationships, pastoral responsibilities, personnel management, the master religious program, training programs, fund management, logistical support, spiritual dimensions, and civilian/military religious liaison. *Version 2:* Pending evaluation.

Credit Recommendation: *Version 1:* Credit is not recommended because of the limited, specialized nature of the course (6/90). *Version 2:* Pending evaluation.

AR-1406-0084

Installation Chaplain

Course Number: *Version 2:* 5G-F3. *Version 3:* 5G-F3.

Location: *Version 2:* Chaplain Center and School, Fort Monmouth, NJ. *Version 3:* Chaplain Center and School, Fort Jackson, SC.

Length: *Version 2:* 2 weeks (81 hours). *Version 3:* 2 weeks (72 hours).

Exhibit Dates: *Version 2:* 4/88–11/01. *Version 3:* 12/01–Present.

Learning Outcomes: *Version 2:* Upon completion of the course, the student will be able to serve as supervising chaplain at an Army installation, including managing chapel facilities and fiscal resources; supervising personnel; planning and programming activities; and advising senior command and staff personnel on the religious, ethical, and moral climate of the installation. *Version 3:* Pending evaluation.

Instruction: *Version 2:* Topics include command and staff relationships, pastoral responsibilities, supervision and management of religious ministry resources, personnel programs and training, deployment responsibilities, and transitioning issues. *Version 3:* Pending evaluation.

Credit Recommendation: *Version 2:* In the lower-division baccalaureate/associate degree category, 2 semester hours in principles of management (6/90). *Version 3:* Pending evaluation.

AR-1406-0085

Staff and Parish Development Consultant

Course Number: 5G-ASI7D.

Location: Chaplain Center and School, Fort Monmouth, NJ.

Length: 3 weeks (132 hours).

Exhibit Dates: 5/84–12/97.

Learning Outcomes: Upon completion of the course, the student will be able to plan and implement interventions, apply staff/parish development concepts to organizational scenarios, perform data analysis, and evaluate interventions.

Instruction: Topics include entry into a consultant/client contractual relationship, contracting, data gathering skills, concepts in intervention planning, intervention strategies, group and organizational issues, feedback and analysis concepts, evaluation of interventions, and termination of a consultant/client contractual relationship. Course includes small-group scenarios.

Credit Recommendation: In the upper-division baccalaureate category, 3 semester hours in organizational group dynamics (4/87).

AR-1406-0087

Advanced Noncommissioned Officer (NCO) MOS 00U

Course Number: 7-71-C42.

Location: Institute for Personnel and Resource Management, Fort Benjamin Harrison, IN.

Length: 5–17 weeks (168–444 hours).

Exhibit Dates: 3/81–12/91.

Objectives: To provide selected personnel with advanced leadership skills, military subjects, and communication skills for effective supervision of specific MOS tracks.

Instruction: Lectures and practical exercises cover a basic core, including written and oral communication, management principles, unit training, and combat survival. MOS options are recruitment and reenlistment, equal opportunity, club management, legal topics, chaplain assisting, and data processing.

Credit Recommendation: In the lower-division baccalaureate/associate degree category, see AR-1404-0035 for common core credit. In the upper-division baccalaureate category, 1 semester hour in personnel management (3/85).

Related Occupations: 00U.

AR-1406-0088

Advanced Noncommissioned Officer (NCO) MOS 00E

Course Number: 7-71-C42.

Location: Institute for Personnel and Resource Management, Fort Benjamin Harrison, IN.

Length: 5–17 weeks (168–444 hours).

Exhibit Dates: 3/81–9/91.

Objectives: To provide selected personnel with advanced leadership skills, military subjects, and communication skills for effective supervision of specific MOS tracks.

Instruction: Lectures and practical exercises cover a basic core including written and oral communication, management principles, unit training, and combat survival. MOS options are recruitment and reenlistment, equal opportunity, club management, legal topics, chaplain assisting, and data processing.

Credit Recommendation: In the lower-division baccalaureate/associate degree category, 2 semester hours in principles of supervision. See AR-1404-0035 for common core credit (3/85).

Related Occupations: 00E.

AR-1406-0090

1. Basic Noncommissioned Officer (NCO) Common Leader Training
   (Basic Noncommissioned Officer (NCO) Common Core)
2. Basic Noncommissioned Officer (NCO) Common Leader Training
3. Basic Noncommissioned Officer (NCO)

Course Number: ***Version 1:*** None. *Version 2:* 400-BNCOC-MTA. *Version 3:* 600-BNCOC, Phase 1.

Location: Training Centers, US and Overseas.

Length: ***Version 1:*** 1–2 weeks (45–49 hours). *Version 2:* 2 weeks (85 hours). *Version 3:* 2 weeks (74 hours).

Exhibit Dates: ***Version 1:*** 3/86–3/94. *Version 2:* 4/94–9/01. *Version 3:* 10/01–Present.

Learning Outcomes: ***Version 1:*** Upon completion of the course, the student will be able to lead a squad or section, train soldiers to fight and survive on a modern battlefield, develop and maintain discipline, supervise maintenance activities and property accountability, and use a systems approach to training. *Version 2:* Upon completion of the course, the student will be able to provide entry-level leadership to a group of up to 12 subordinates, train these individuals to successfully perform their jobs, instill proper behavioral habits in subordinates, supervise organizational maintenance activities and property accountability, and employ the systems approach to personnel training. *Version 3:* Upon completion of the course, the selected student will be able to perform the duties and execute the responsibilities of noncommissioned officers, using the small group instruction process to teach the theory and principles of battle-focused common core training and war-fighting skills required to lead a squad/section-sized element into combat.

Instruction: ***Version 1:*** Topics include leadership, personal and performance counseling, training, management, resource management, and military skills. *Version 2:* Topics include entry-level leadership skills, personal and performance counseling, training techniques, basic management skills, and introductory resource management skills. *Version 3:* Laboratory, practical exercises, discussion, lecture, and classroom exercises. Topics include lecture and practice exercises to develop communications, leadership, resource, professional training, and warfighting skills in noncommissioned officers to effectively lead a squad/section-sized element into combat.

Credit Recommendation: ***Version 1:*** In the lower-division baccalaureate/associate degree category, 1 semester hour in personnel supervision (10/88). *Version 2:* In the lower-division baccalaureate/associate degree category, 2 semester hours in supervision or in leadership (6/95). *Version 3:* In the lower-division baccalaureate/associate degree category, 3 semester hours in military science or basic supervision (8/03).

AR-1406-0091

Skill Qualification Test (SQT) Developer's Workshop

Course Number: None.

Location: Various locations, Continental US.

Length: 80 hours.

Exhibit Dates: 1/87–6/96.

Learning Outcomes: Upon completion of the workshop, the student will be able to develop, validate, prepare for printing, and maintain quality control/assurance for skill qualification tests.

Instruction: Self-paced instruction in criterion-referenced measurement is presented. The workshop modules cover task selection; item writing; item tryout and analysis; training standards, time limits, and minimum passing score; preparing final test materials; creating scoring templates; and monitoring test performance and quality control/assurance.

Credit Recommendation: In the upper-division baccalaureate category, 2 semester hours in an introductory course in criterion-referenced measurement (6/87).

AR-1406-0092

Staff and Faculty Development
(Faculty Development)

Course Number: 5K-F3/520-F3.

Location: Academy of Health Sciences, Fort Sam Houston, TX.

Length: 3 weeks (117–121 hours).

Exhibit Dates: 4/87–4/96.

Learning Outcomes: Upon completion of the course, the student will be assigned as an instructor at the Academy of Health Sciences.

Instruction: Lectures, demonstrations, and practical exercises include instructional objectives, test construction, student counseling, lesson plan preparation, teaching aids, school organization and regulations, and management of the learning environment.

Credit Recommendation: In the upper-division baccalaureate category, 4 semester hours in instructional methods (8/94).

AR-1406-0093

Telephone Central Office Repair Basic Noncommissioned Officer (NCO)

Course Number: 622-29N30.

Location: Signal School, Fort Gordon, GA.

Length: 6–7 weeks (230–232 hours).

Exhibit Dates: 10/86–9/95.

Learning Outcomes: Upon completion of the course, the student will be able to supervise subordinates who perform maintenance, troubleshooting, and repair on dial/manual central office telephone exchange equipment and perform technical inspections.

Instruction: Classes, demonstrations, and lectures cover operating, testing, maintaining, and repairing specific telephone exchange equipment. Course includes a common core of leadership subjects.

Credit Recommendation: In the vocational certificate category, 3 semester hours in telephone central exchange supervision. In the lower-division baccalaureate/associate degree category, 2 semester hours in electronic systems troubleshooting and maintenance. See AR-1406-0090 for common core credit (5/90).

Related Occupations: 29N.

AR-1406-0094

Instructor Training Workshop

Course Number: ***Version 1:*** IT24. *Version 2:* ITW 24.

Location: Combined Arms Training Center, Vilseck, W. Germany.

Length: ***Version 1:*** 2 weeks (68 hours). *Version 2:* 2 weeks (80 hours).

Exhibit Dates: ***Version 1:*** 9/81–1/98. *Version 2:* 2/98–Present.

Learning Outcomes: ***Version 1:*** Upon completion of the course, the student will be able to write and edit instructional objectives, performance tests, and lesson plans; select and develop training aids, student handouts, and practical exercises; counsel students with educational problems; and prepare and present classes using a practical exercise, performance test, and self-paced/small group presentation. *Version 2:* Upon completion of the course, the student will be able to plan and write instructional objectives, develop lesson plans, deliver small group instruction, conduct formal and informal after action reviews, use training aids, and present lessons.

Instruction: ***Version 1:*** Lectures, demonstrations, and seminars are modes of instruction. The student is required to demonstrate mastery of instructional skills and methods by means of student presentations. *Version 2:* Lectures, demonstrations, and seminars are modes of instruction. The student is required to demonstrate mastery of communication and instructional skills and methods by means of student presentations.

Credit Recommendation: ***Version 1:*** In the lower-division baccalaureate/associate degree category, 2 semester hours in instructional methods (10/88). *Version 2:* In the lower-division baccalaureate/associate degree category, 2 semester hours in instructional methods and 1 in speech communication (9/03).

AR-1406-0095

Antiterrorism Instructor Qualification

Course Number: ***Version 1:*** 5K-F5/012-F30. *Version 2:* 5K-F5/012-F30.

Location: John F. Kennedy Special Warfare Center and School, Fort Bragg, NC.

Length: ***Version 1:*** 1–2 weeks (67–70 hours). *Version 2:* 1–2 weeks (70 hours).

Exhibit Dates: ***Version 1:*** 10/88–2/97. *Version 2:* 3/97–Present.

Learning Outcomes: ***Version 1:*** Upon completion of the course, the student will be able to teach antiterrorism to individuals and units deploying overseas in order to minimize vulnerability to enemy attack. *Version 2:* Upon completion of the course, the student will be able to teach antiterrorism to individuals and units deploying overseas in order to minimize vulnerability to enemy attack.

Instruction: ***Version 1:*** Lectures and examinations cover introduction to terrorism, terrorism operations, individual protective measures, and hostage survival techniques. *Version 2:* Lectures and examinations cover introduction to terrorism, terrorism operations, individual protective measures, and hostage survival techniques.

Credit Recommendation: ***Version 1:*** In the upper-division baccalaureate category, 1 semester hour in test construction (7/91). *Version 2:* In the upper-division baccalaureate category, 1 semester hour in assessment and evaluation and 1 in instructional technology (4/01).

AR-1406-0096

Special Forces Technician Warrant Officer Basic
(Special Forces Technician Warrant Officer Technical/Tactical Certification)
(Special Operations Technician Warrant Officer Technical Certification)

Course Number: ***Version 1:*** 2E-180A. *Version 2:* 2E-180A.

Location: John F. Kennedy Special Warfare Center and School, Fort Bragg, NC.

Length: ***Version 1:*** 13–14 weeks (618 hours). *Version 2:* 18 weeks (942–968 hours).

Exhibit Dates: ***Version 1:*** 2/88–9/93. *Version 2:* 10/93–Present.

Learning Outcomes: ***Version 1:*** Upon completion of the course, the student will be able to implement leadership and management duties to conduct special operations in the areas of security, intelligence, psychological operations, civil affairs, operations command, deployment, and reconnaissance. *Version 2:* Upon completion of the course, the student will be able to implement leadership and management duties to conduct special operations in the areas of security, intelligence, psychological operations, civil affairs, operations command, deployment, and reconnaissance and employ effective writing and computer word processing skills.

Instruction: ***Version 1:*** Lectures, discussion, demonstrations, and field-based exercises are used to instruct students. Included are topics in supervision and management. *Version 2:* Lectures, discussions, demonstrations, and field-based exercises are used to instruct students. Included are topics in supervision and management, role playing, and computer use.

Credit Recommendation: ***Version 1:*** In the lower-division baccalaureate/associate degree category, 3 semester hours in business organization and management and 2 in principles of supervision (12/88). *Version 2:* In the lower-division baccalaureate/associate degree category, 3 semester hours in business organization and management, 2 in principles of supervision, and 2 in word processing (9/00).

Related Occupations: 180A.

AR-1406-0100

Personnel Administration Specialist Basic Noncommissioned Officer (NCO)

Course Number: ***Version 1:*** 500-75B30. *Version 2:* 500-75B30. *Version 3:* 500-75B30.

Location: ***Version 1:*** Adjutant General School, Fort Benjamin Harrison, IN; Soldier Support Institute, Fort Benjamin Harrison, IN. *Version 2:* Adjutant General School, Fort Benjamin Harrison, IN; Soldier Support Institute, Fort Benjamin Harrison, IN. *Version 3:* Soldier Support Institute, Fort Jackson, SC.

Length: ***Version 1:*** 5–6 weeks (209 hours). *Version 2:* 6–7 weeks (266 hours). *Version 3:* 7–8 weeks (260 hours).

Exhibit Dates: ***Version 1:*** 1/89–9/91. *Version 2:* 10/91–2/96. *Version 3:* 3/96–Present.

Learning Outcomes: ***Version 1:*** Upon completion of the course, the student will be able to perform appropriate functions in a personnel office, including review of personnel evaluations, personnel classification, and assignment of personnel based on qualifications; describe promotion policies and procedures; and use a computer to access, create, edit, file, and print documents. *Version 2:* Upon completion of the course, the student will be able to perform appropriate functions in a personnel office, including review of personnel evaluations, personnel classification, and assignment of personnel based on qualifications; describe promotion policies and procedures; and use a computer to access, create, edit, file, and print documents. *Version 3:* Upon completion of the course, the student will be able to perform appropriate functions in a personnel office to ensure accuracy of evaluations, classifications, personnel assignments based on qualifications, use of promotion policies and procedures, and access to an automated personnel reporting system.

Instruction: ***Version 1:*** Instruction involves lectures, demonstrations, and in-class exercises in principles of supervision, computer operations, and personnel management. *Version 2:* Instruction involves lectures, demonstrations, and in-class exercises in principles of supervision, computer operations, and personnel management. *Version 3:* Lecture, class discussions, PC exercises, and examinations to supervise a personnel section and utilize automated personnel reporting systems to access, create, and edit data and print reports.

Credit Recommendation: ***Version 1:*** In the lower-division baccalaureate/associate degree category, 3 semester hours in principles of supervision and 1 in introduction to computer applications. In the upper-division baccalaureate category, 1 semester hour in personnel management (12/88). *Version 2:* In the lower-division baccalaureate/associate degree category, 3 semester hours in personnel supervision and 1 in computer applications (12/91). *Version 3:* In the lower-division baccalaureate/associate degree category, 2 semester hours in human resource management and 1 in computer applications systems. See AR-1406-0090 for common core credit (9/97).

Related Occupations: 75B.

AR-1406-0101

1. National Guard Recruiting Advanced
2. National Guard Recruiting Field Sales Manager

Course Number: ***Version 1:*** 501-F16. *Version 2:* 501-F31 (ARNG).

Location: ***Version 1:*** Recruiting and Retention School, Soldier Support Institute, Fort Benjamin Harrison, IN. *Version 2:* Professional Education Center, Cp. Joseph Robinson, North Little Rock, AR.

Length: ***Version 1:*** 1–2 weeks (56 hours). *Version 2:* 2 weeks (80 hours).

Exhibit Dates: ***Version 1:*** 10/85–9/90. *Version 2:* 10/90–12/97.

Learning Outcomes: ***Version 1:*** Upon completion of the course, the student will be able to develop a sales presentation, demonstrate interpersonal skills, and analyze demographic data. *Version 2:* Upon completion of the course, the student will be able to evaluate selling skills, provide recruiter training, interpret and analyze market trends, and identify the developmental needs of a sales team.

Instruction: ***Version 1:*** Lectures and classroom discussion cover interpersonal communication, selling skills, positive reinforcement, performance evaluation techniques, and the use of demographic data. *Version 2:* Lectures, discussions, practical exercises and supervised self-study are methods used.

Credit Recommendation: ***Version 1:*** In the vocational certificate category, 1 semester hour in sales fundamentals and 1 in interpersonal communication (12/88). *Version 2:* In the lower-division baccalaureate/associate degree category, 2 semester hours in sales management and 1 in interpersonal communication (12/91).

AR-1406-0102

Personnel Sergeant Advanced Noncommissioned Officer (NCO)

Course Number: ***Version 1:*** 500-75Z40. *Version 2:* 5-71-C42. *Version 3:* 5-71-C42.

Location: ***Version 1:*** Soldier Support Institute, Fort Benjamin Harrison, IN. *Version 2:* Adjutant General School, Fort Benjamin Harrison, IN. *Version 3:* Soldier Support Institute, Fort Jackson, SC.

Length: ***Version 1:*** 10 weeks (360–369 hours). *Version 2:* 12 weeks (477 hours). *Version 3:* 11–12 weeks (403–404 hours).

Exhibit Dates: ***Version 1:*** 10/86–12/92. *Version 2:* 1/93–12/95. *Version 3:* 1/96–Present.

Learning Outcomes: ***Version 1:*** Upon completion of the course, the student will be able to manage a medium-sized personnel activity. *Version 2:* Upon completion of the course, the student will be able to prepare clear, well-organized correspondence; supervise general personnel activities at the first-line level; and supervise development and processing of computerized personnel records. *Version 3:* Upon completion of the course, the student will be able to supervise a personnel unit and be familiar with selected aspects of personnel information processing including assigning staff, preparing readiness reports, reviewing transition processing, auditing personnel records, reviewing evaluations, and supervising the use of automated personnel reporting systems.

Instruction: ***Version 1:*** Course includes classroom instruction and practical exercises in leadership, communication skills, training management, personnel records management, evaluation and classification systems and programs, and military-specific subjects. *Version 2:* Lectures and practical exercises cover the above subject matter. *Version 3:* Lectures, class discussions and practical exercises ensure the proper functioning of a personnel office. Topics include personnel records management, classification and evaluation processes, and use of computer-based systems to support the functioning of the office.

Credit Recommendation: ***Version 1:*** In the upper-division baccalaureate category, 3 semester hours in personnel management (12/88). *Version 2:* In the lower-division baccalaureate/associate degree category, 1 semester hour in business communication, 2 in personnel supervision, and 3 in office administration (12/91). *Version 3:* In the lower-division baccalaureate/associate degree category, 2 semester hours in office administration, 2 in records and information management, and 2 in human resource management (9/97).

Related Occupations: 75Z.

AR-1406-0103

1. Recruiter
2. Army Recruiter
3. Army Recruiter

Course Number: ***Version 1:*** 501-SQI4. *Version 2:* 501-SQI4. *Version 3:* 501-SQI4.

Location: ***Version 1:*** Recruiting and Retention School, Soldier Support Institute, Fort Benjamin Harrison, IN. *Version 2:* Soldier Support Institute, Fort Jackson, SC. *Version 3:* Recruiting and Retention School, Fort Jackson, SC.

Length: ***Version 1:*** 6 weeks (238–249 hours). *Version 2:* 5–6 weeks (244 hours). *Version 3:* 8 weeks (390 hours).

Exhibit Dates: ***Version 1:*** 10/86–9/95. *Version 2:* 10/95–12/00. *Version 3:* 1/01–Present.

Learning Outcomes: ***Version 1:*** Upon completion of the course, the student will be able to set goals and establish practices, develop a sales presentation, use competitive strategies, evaluate market data, use proper interviewing skills, and make promotional presentations. *Version 2:* Upon completion of the course, the student will be able to set goals, establish procedures for prospecting using current marketing analysis, develop sales presentations, use competitive strategies, use proper interviewing skills, and make promotional presentations. *Version 3:* Upon completion of the course, the student will be able to prospect, develop sales presentations, interview prospective recruits, and make a sales presentation.

Instruction: ***Version 1:*** Methods of instruction include lectures, role playing, and classroom discussions. Topics covered include sales prospecting, market analysis, effective communication, public speaking, interviewing skills and techniques, and time management. *Version 2:* Methods of instruction include lectures, role playing, and classroom discussions. Topics covered include sales prospecting, market analysis, effective communication, public speaking, interviewing skills and techniques, and time management. *Version 3:* Methods of instruction include lectures, role playing, and classroom discussions. Topics covered include sales prospecting, market analysis, effective communication, public speaking, interviewing skills and techniques, and time management.

Credit Recommendation: ***Version 1:*** In the vocational certificate category, 2 semester hours in sales fundamentals and 2 in interpersonal communication (5/92). *Version 2:* In the lower-division baccalaureate/associate degree category, 3 semester hours in salesmanship and 1 in interpersonal communication (9/97). *Version 3:* In the upper-division baccalaureate category, 3 semester hours in selling, or marketing, 3 in human resources management, and 3 in business interpersonal communication (3/02).

AR-1406-0104

Battalion S1/Adjutant Staff Officer (Battalion S1)

Course Number: 7C-F14.

Location: Soldier Support Institute, Fort Benjamin Harrison, IN.

Length: 5–6 weeks (214–244 hours).

Exhibit Dates: 5/86–12/91.

Learning Outcomes: Upon completion of the course, the student will be able to manage the personnel resource activity of a medium to large organization.

Instruction: Course includes classroom and practical exercises in business communication, safety programs, training, leadership, motivation, performance management, classification systems, evaluation techniques, and personnel information systems and record keeping.

Credit Recommendation: In the lower-division baccalaureate/associate degree category, 1 semester hour in business communication. In the upper-division baccalaureate category, 3 semester hours in human resource management (12/88).

AR-1406-0105

Personnel Administration Specialist

Course Number: 500-75B10; 500-75B10/75B10 (ST).

Location: Soldier Support Institute, Fort Benjamin Harrison, IN.

Length: 7–9 weeks (239–306 hours).

Exhibit Dates: 2/86–12/91.

Learning Outcomes: After 12/91 see AR-1406-0153. Upon completion of the course, the student will be able to perform clerical duties, including typewriting, filing, office procedures, and computer operations.

Instruction: Classroom instruction and practical exercises include basic typewriting, computer operations, unit files, publications, reports, and correspondence.

Credit Recommendation: In the lower-division baccalaureate/associate degree category, 1 semester hour in keyboarding, 2 in office procedures, and 1 in clerical record keeping (11/91).

Related Occupations: 75B.

AR-1406-0107

Electronics Equipment Maintenance Advanced Noncommissioned Officer (NCO)

Course Number: 1-29-C42 (39X).

Location: Signal School, Fort Gordon, GA.

Length: 14–15 weeks (553 hours).

Exhibit Dates: 10/88–3/92.

Learning Outcomes: Upon completion of the course, the student will be able to use microcomputers, including using DOS, word processors, data bases, and spreadsheets; receive and send Army messages; prepare Army forms related to shop management; and describe at the overview-level a numbers of pieces of communications equipment.

Instruction: Lectures, practical exercises, and exams cover the subject material. Course includes a common core of leadership subjects.

Credit Recommendation: In the lower-division baccalaureate/associate degree category, 3 semester hours in computer literacy. See AR-1404-0035 for common core credit (7/89).

Related Occupations: 39X.

AR-1406-0108

Test, Measurement, and Diagnostic Equipment (TMDE) Advanced Noncommissioned Officer (NCO)

Course Number: 1-29-C42 (35H).

Location: Signal School, Fort Gordon, GA.

Length: 11–12 weeks (417 hours).

Exhibit Dates: 10/88–2/91.

Learning Outcomes: Upon completion of the course, the student will be able to use microcomputers, including using DOS, word processors, data bases, and spreadsheets; receive and send Army messages; prepare Army forms related to shop management; and describe at the overview-level a number of pieces of communications equipment.

Instruction: Lectures, practical exercises, and exams cover the subject material. Course includes a common core of leadership subjects.

Credit Recommendation: In the lower-division baccalaureate/associate degree category, 3 semester hours in computer literacy. See AR-1404-0035 for common core credit (7/89).

Related Occupations: 35H.

AR-1406-0109

Signal Operations Advanced Noncommissioned Officer (NCO)

Course Number: 1-31-C42.

Location: Signal School, Fort Gordon, GA.

Length: 18–19 weeks (713–749 hours).

Exhibit Dates: 10/88–9/93.

Learning Outcomes: After 9/93 see AR-1402-0156. Upon completion of the course, the student will supervise, coordinate, and provide technical assistance in installing, operating, and managing single and multichannel communications facilities.

Instruction: Lectures and practical experience cover microcomputer concepts and applications, including components and functions, operating systems, programming languages, and types of software. The course also includes operation and management of communications facilities. There is also a common core of leadership subjects.

Credit Recommendation: In the lower-division baccalaureate/associate degree category, 3 semester hours in computer literacy. See AR-1404-0035 for common core credit (2/94).

AR-1406-0110

Personnel Management

Course Number: 300; 300-LVL2.

Location: Reserve Readiness Training Center, Fort McCoy, WI.

Length: 2 weeks (80–85 hours).

Exhibit Dates: 9/86–2/01.

Learning Outcomes: Upon completion of the course, the student will be able to identify personnel and administrative requirements of the preparatory and alert phase of mobilization; determine the unit's personnel requirements and recommend a personnel management plan; determine requirements for conducting military boards, investigations, and review actions; interpret reenlistment eligibility criteria; and evaluate officer and enlisted promotion criteria and officer appointment criteria.

Instruction: Lectures and discussions cover the personnel phase of organizational planning, including mobilization, authorization documents, personnel reporting requirements, and manning reports; personnel managements systems including officer and enlisted personnel systems; personnel actions, including promotions/appointments, reenlistments, qualitative management, casualty reporting, and awards and decorations; and strength management, including retention, strength accounting, and benefits for Army reserve members and families. Course also includes practical activities in determining personnel and position requirements, recommending personnel management plans, determining requirements for conducting military boards and investigations, interpreting reenlistment eligibility criteria, and evaluating officer and enlisted promotion criteria and officer appointment criteria.

Credit Recommendation: In the lower-division baccalaureate/associate degree category, 2 semester hours in administrative management (7/96).

AR-1406-0111

1. Instructor
2. Instructor Training
3. Total Army Instructor Training (TAITC)

Course Number: *Version 1:* 5000. *Version 2:* 5000. *Version 3:* 921-350.

Location: Reserve Readiness Training Center, Fort McCoy, WI.

Length: *Version 1:* 2 weeks (48 hours). *Version 2:* 2–3 weeks (94 hours). *Version 3:* 2 weeks (80 hours).

Exhibit Dates: *Version 1:* 7/87–8/93. *Version 2:* 9/93–5/99. *Version 3:* 6/99–Present.

Learning Outcomes: *Version 1:* Upon completion of the course, the student will be able to analyze, develop, and design course and training materials using various learning strategies; make presentations to adult participants using audience participation, illustrations, and humor; prepare and sequence objectives and competencies for future classes; prepare examination questions for quizzes and pre- and post-tests; develop graphics, audiovisual, and written instructional materials; gather and validate instructional materials; and develop lesson plans, class outlines, and materials list needed for classroom presentations. *Version 2:* Upon completion of the course, the student will be able to develop training materials, present a block of instruction, develop lesson plans, make instructional decisions, design instructional aids, and deliver instruction using a variety of methods. *Version 3:* Upon completion of the course, the student will be able to identify course requirements, prepare learning objectives, develop a training outline and training session, conduct presentation evaluations, communicate real time evaluations, conduct a counseling session, and determine intervention techniques.

Instruction: *Version 1:* Readings, lectures, and discussions cover effective presentations, teaching methods, audiovisual materials, testing and evaluation, and ordering of equipment and materials needed for classroom use. Practical exercises cover preparing exam questions, preparing sequence of objectives/learning elements, developing graphics, using audiovisual equipment, preparing stock and equipment orders for classroom materials, and visiting classes to observe teaching techniques and classroom presentations. *Version 2:* Readings, lectures, and discussions cover effective presentations, teaching methods, audiovisual materials, testing and evaluation, and ordering of equipment and materials needed for classroom use. Practical exercises cover preparing exam questions, preparing sequence of objectives/learning elements, developing graphics, using audiovisual equipment, preparing stock and equipment orders for classroom materials, and visiting classes to observe teaching techniques and classroom presentations. *Version 3:* Audiovisual materials, practical and classroom exercises, discussion, learner presentations, and lecture. Topics include identification of course requirements, preparing learning objectives, developing a training session, instructional techniques, communicate real time evaluations, intervention techniques, conducting a counseling session, guidelines for preparing presentations, and practice presentations.

Credit Recommendation: *Version 1:* In the upper-division baccalaureate category, 3 semester hours in instructional methods (7/89). *Version 2:* In the upper-division baccalaureate category, 3 semester hours in methods of teaching or instructional strategies (3/97). *Version 3:* In the lower-division baccalaureate/associate degree category, 1 semester hour in speech communication. In the upper-division baccalaureate category, 3 semester hours in instructional methods (9/03).

AR-1406-0112

Warrant Officer Professional Development

Course Number: 2600; 8500.

Location: Reserve Readiness Training Center, Fort McCoy, WI.

Length: 2 weeks (48 hours).

Exhibit Dates: 4/87–10/92.

Learning Outcomes: Upon completion of the course, the student will be able to solve problems and make decisions, apply leadership and planning techniques, delegate authority, manage resources, develop an information feedback system, and communicate with peers and subordinates using effective communication skills, both oral and written.

Instruction: Readings, lectures, and discussions cover leadership orientation, including professional ethics within the warrant officer system, supervisory skills, principles and techniques in directing, delegating authority, and managing resources and leadership topics, including arms on the air-land battlefield, soldier team development, stress and battlefield leadership, and other leadership skills. The course addresses the skills needed to lead a unit; communication skills; unit assessment; and the effects of fear, panic, and stress on unit leadership.

Credit Recommendation: In the lower-division baccalaureate/associate degree category, 2 semester hours in principles of supervision (7/89).

AR-1406-0113

UH-1 Contact, Tactics, Night/Night Vision Goggle (N/NVG) Instructor Pilot Methods of Instruction

Course Number: 2C-SI5K/SQIC (UH-1).

Location: Aviation Center, Fort Rucker, AL.

Length: 14 weeks (461 hours).

Exhibit Dates: 10/88–9/95.

Learning Outcomes: Upon completion of the course, pilots will be qualified as instructors on the UH-1 helicopter.

Instruction: Course includes principles of instruction, aerodynamics, aircraft systems, night tactics, aeromedical factors, weapon systems, combat skills, and appropriate flight training.

Credit Recommendation: In the upper-division baccalaureate category, 3 semester hours in flight instructor academic preparation and 3 in flight training for flight instructor (4/90).

AR-1406-0114

AH-1 Instructor Pilot Methods of Instruction (MOI) (Contact, Tactics, and Gunnery) (Night Vision Goggle)

Course Number: 000-SI5K/SQIC (AH-1).

Location: Aviation Center, Fort Rucker, AL.

Length: 10–11 weeks (368 hours).

Exhibit Dates: 10/88–7/03.

Learning Outcomes: Upon completion of the course, pilots will be qualified as instructors on the AH-1 helicopters.

Instruction: Course includes academic instruction in principles of instruction, aerodynamics, aircraft systems, aeromedical factors, night flight, gunnery, weapon systems, and appropriate flight training.

Credit Recommendation: In the upper-division baccalaureate category, 3 semester hours in flight instructor academic preparation and 3 in flight training for flight instructor (4/90).

AR-1406-0115

1. UH-60 Method of Instruction
2. UH-60 Instructor Pilot Methods of Instruction

Course Number: *Version 1:* 2C-SI5K/SQIC (UH-60). *Version 2:* 2C-SIG2/SQIC (UH-60 GRAD).

Location: Aviation Center, Fort Rucker, AL.

Length: *Version 1:* 8 weeks (278 hours). *Version 2:* 9 weeks (339–346 hours).

Exhibit Dates: *Version 1:* 10/88–6/96. *Version 2:* 7/96–Present.

Learning Outcomes: *Version 1:* Upon completion of the course, pilots will be qualified as instructors on the UH-60 helicopter. *Version 2:* Upon completion of the course, the student will be able to teach and evaluate academic and flight instruction.

Instruction: *Version 1:* Course includes fundamentals of instructing, communications, aerodynamics, aeromedical factors, night operations, aircraft systems of the UH-60, and appropriate flight training. *Version 2:* This course consists of theory of flight, instructing fundamentals, aircrew training, regulations, safety, automatic flight control systems, power train and hydraulics and electrical systems, malfunction analysis, night academic, methods of instruction for contact maneuvers, dual and single engine operations, emergency procedures, and Command Instrument System operations. Sufficient depth of knowledge will be gained to enable the graduate to teach academic and flight instruction in the UH-60 helicopter.

Credit Recommendation: *Version 1:* In the upper-division baccalaureate category, 3 semester hours in flight instructor academic preparation and 3 in flight training for flight instructor (4/90). *Version 2:* In the upper-division baccalaureate category, 3 semester hours flight instruction (helicopter), 2 in helicopter systems, and 1 in methods of instruction (1/03).

AR-1406-0116

U-21 Instructor Pilot Methods of Instruction

(U-21 Instructor Pilot)

Course Number: 2B-SIG2/SQIC (U-21 MOI) (CT); 2B-SI5K/2B-SQIC (U-21).
Location: Aviation Center, Fort Rucker, AL.
Length: 6–7 weeks (239 hours).
Exhibit Dates: 5/89–Present.
Learning Outcomes: Upon completion of the course, pilots will be qualified as instructors on the U-21 aircraft.
Instruction: Course includes fundamentals of instruction, aerodynamics, navigation, air traffic control, air crew communications, aircraft systems, aeromedical and physiological factors, and appropriate flight training.
Credit Recommendation: In the upper-division baccalaureate category, 3 semester hours in flight instructor academic preparation and 3 in flight training for flight instructor (11/98).

AR-1406-0117

AH-1S Instructor Pilot

Course Number: 2C-SI5K/2C-SQIC.
Location: Aviation Center, Fort Rucker, AL.
Length: 10–11 weeks (373 hours).
Exhibit Dates: 10/87–7/03.
Learning Outcomes: Upon completion of the course, pilots will be qualified as instructors in the AH-1S helicopter.
Instruction: Course includes academic instruction in principles of instruction, aerodynamics, aircraft systems, aeromedical factors, night flight, weapon systems, combat skills, and appropriate flight training.
Credit Recommendation: In the upper-division baccalaureate category, 3 semester hours in flight instructor academic preparation and 3 in flight training for flight instructor (4/90).

AR-1406-0118

1. OH-58D Instructor Pilot Methods of Instruction (MOI)
2. OH-58D Warrior Instructor Pilot Methods of Instruction (MOI)

Course Number: *Version 1:* 2C/SI5K/SQIC (OH-58D). *Version 2:* 2C-SIG2/SQIC (OH58D MOI), Phase 2; 2C-SIG2/SQIC (OH58D MO1); 2C-SIG2/SQIC (OH-58D).
Location: Aviation Center, Fort Rucker, AL.
Length: *Version 1:* 10–11 weeks (431 hours). *Version 2:* 11 weeks (479–480 hours).
Exhibit Dates: *Version 1:* 10/88–9/96. *Version 2:* 10/96–Present.
Learning Outcomes: *Version 1:* Upon completion of the course, pilots will be qualified as instructors on the OH-58D helicopter. *Version 2:* Upon completion of the course, the student will be able to perform as an instructor pilot in the OH-58D helicopter.
Instruction: *Version 1:* Course consists of mental and physical skills required for accomplishing instructor pilot duties (day and night tasks), including aircraft systems, airway communications, principles of instruction, navigation, mast-mounted sight, airborne target, handover systems, cockpit procedures trainer, mission planning, safety, and aeromedical factors. *Version 2:* Lectures and computer-assisted instruction are used to provide training in OH-58D aircraft systems, mast-mounted sight and laser range finder/designator systems, night vision device operations, aeromedical factors for pilots, aerodynamics, fundamentals of instruction, crew co-ordination, OH-58D communications/navigation systems, and OH-58D weapons systems.
Credit Recommendation: *Version 1:* In the upper-division baccalaureate category, 3 semester hours in flight instructor academic preparation and 3 in flight training for flight instructor (4/90). *Version 2:* In the upper-division baccalaureate category, 3 semester hours in flight instructor academic preparation and 3 in training for flight instructor (1/03).

AR-1406-0119

1. OH-58D Instructor Pilot
2. OH-58D Warrior Instructor Pilot

Course Number: *Version 1:* 2C-SI 5K/2C-SQIC (OH-58D). *Version 2:* 2C-SIG2/2C-SQIC (OH-58D).
Location: Aviation Center, Fort Rucker, AL.
Length: *Version 1:* 8–9 weeks (349 hours). *Version 2:* 10 weeks (449 hours).
Exhibit Dates: *Version 1:* 10/89–6/96. *Version 2:* 7/96–Present.
Learning Outcomes: *Version 1:* Upon completion of the course, pilots will be qualified as instructors on the OH-58D helicopter. *Version 2:* Upon completion of the course, the student will be qualified as instructor pilot for the OH-58D helicopter.
Instruction: *Version 1:* Course consists of mental and physical skills required for accomplishing instructor pilot duties (day and night tasks), including aircraft systems airway communications, principles of instruction, navigation, mast-mounted sight, airborne target, handover systems, cockpit procedures trainer, mission planning, safety, and aeromedical factors. *Version 2:* Classroom lectures, cockpit procedures trainer, practical exercises, and advanced flight instructions are used in this course to teach advanced helicopter aerodynamics, aerospace, principles of instruction, cockpit management, night vision devices, communications, electronic navigation, and aeromedial factors.
Credit Recommendation: *Version 1:* In the upper-division baccalaureate category, 3 semester hours in flight instructor academic preparation and 3 in flight training for flight instructor (4/90). *Version 2:* In the upper-division baccalaureate category, 3 semester hours in advanced helicopter flight instructor academic preparation and 3 in advanced helicopter flight training (11/98).

AR-1406-0120

CH-47D Instructor Pilot

Course Number: 2C-SI5K/2C-SQIC (CH-47D).
Location: Aviation Center, Fort Rucker, AL.
Length: 10 weeks (386 hours).
Exhibit Dates: 10/88–9/96.
Learning Outcomes: Upon completion of the course, pilots will be qualified as instructors for CH-47D helicopters.
Instruction: Course includes fundamentals of instruction, aviation medical factors, night flying orientation, CH-47D aircraft systems, and practical flight training.
Credit Recommendation: In the upper-division baccalaureate category, 3 semester hours in flight instructor academic preparation and 3 in flight training for flight instructor (4/90).

AR-1406-0121

AH-64 Instructor Pilot Methods of Instruction

Course Number: 2C-SI5K/SQIC (AH-64).
Location: Aviation Center, Fort Rucker, AL.
Length: 9–10 weeks (361 hours).
Exhibit Dates: 10/88–12/97.
Learning Outcomes: After 12/97 see AR-1406-0196. Upon completion of the course, pilots will be qualified as instructors of the AH-64 helicopter.
Instruction: Course includes aviation subjects, both academic and flight, including regulations, principles of instruction, aerodynamics, effective communication, night vision systems, aeromedical factors, weapon systems, aircraft systems, and flight training in AH-64 helicopter.
Credit Recommendation: In the upper-division baccalaureate category, 3 semester hours in flight instructor academic preparation and 3 in flight training for flight instructor (4/90).

AR-1406-0122

OH-58A/C Instructor Pilot

Course Number: *Version 1:* 2C-SI5K/2C-SQIC. *Version 2:* 2C-SIG2/SQIC (OH-58).
Location: *Version 1:* Aviation Center, Fort Rucker, AL. *Version 2:* Aviation School, Fort Rucker, AL.
Length: *Version 1:* 8 weeks (292 hours). *Version 2:* 8 weeks (302 hours).
Exhibit Dates: *Version 1:* 10/88–6/97. *Version 2:* 7/97–Present.
Learning Outcomes: *Version 1:* Upon completion of the course, pilots will be qualified as instructors in OH-58 A/C helicopters. *Version 2:* Upon completion of the course, the student will be qualified as instructor for the OH-58A/C helicopter.
Instruction: *Version 1:* Course consists of mental and physical skills required for accomplishing instructor pilot duties (day and night tasks), including aircraft systems, airway communications, principles of instruction, navigation, mast-mounted sight, airborne target handover systems, cockpit procedures trainer, mission planning, safety, and aeromedical factors. *Version 2:* Classroom lectures, cockpit procedure trainer, practical exercises, and advanced flight instructions are utilized in this course to teach advanced helicopter aerodynamics, aerospace, principles of instruction, cockpit management, night vision devices, communications, electronic navigation, and aeromedical factors.
Credit Recommendation: *Version 1:* In the upper-division baccalaureate category, 3 semester hours in flight instructor academic preparation and 3 in flight training for flight instructor (4/90). *Version 2:* In the upper-division baccalaureate category, 3 semester hours in advanced helicopter flight instructor methods and 3 in advanced helicopter flight training (11/98).

AR-1406-0123

Personnel and Logistics Staff Noncommissioned Officer (NCO)

Course Number: 500-F22.
Location: Sergeants Major Academy, Fort Bliss, TX.
Length: 4–5 weeks (123 hours).
Exhibit Dates: 1/88–4/92.

Learning Outcomes: Upon completion of the course, the student will be able to supervise those providing logistical and personnel services.

Instruction: This course includes lectures, small-group instruction, and performance exercises in leadership and resource management.

Credit Recommendation: In the lower-division baccalaureate/associate degree category, 2 semester hours in military science (4/90).

AR-1406-0125

Military Intelligence Advanced Noncommissioned Officer (NCO)

Course Number: 2-96-C42.

Location: Intelligence School, Fort Huachuca, AZ.

Length: 9–10 weeks (407 hours).

Exhibit Dates: 1/87–9/99.

Learning Outcomes: Upon completion of the course, the student will be able to lead, manage, supervise, and train intelligence personnel in performing the tactical and strategic intelligence tasks of a military intelligence unit and prepare and communicate plans and reports in written and oral form.

Instruction: Instruction includes classroom and practical exercises in leadership, management, administration, and communication skills. Course includes a common core of leadership subjects.

Credit Recommendation: In the lower-division baccalaureate/associate degree category, 3 semester hours in technical writing. See AR-1404-0035 for common core credit. In the upper-division baccalaureate category, 1 semester hour in leadership and 2 in human resource management (5/01).

AR-1406-0126

Electronic Warfare (EW)/Cryptographic Advanced Noncommissioned Officer (NCO)

Course Number: 2-98-C42.

Location: Intelligence School, Fort Huachuca, AZ.

Length: 9–10 weeks (407 hours).

Exhibit Dates: 1/87–9/98.

Learning Outcomes: Upon completion of the course, the student will be able to lead, manage, supervise, and train intelligence personnel in performing the tactical and strategic intelligence tasks of a military intelligence unit and prepare and communicate plans and reports in written and oral form.

Instruction: Instruction includes classroom and practical exercises in leadership, management, administration, and communication skills. Course includes a common core of leadership subjects.

Credit Recommendation: In the lower-division baccalaureate/associate degree category, 3 semester hours in technical writing. See AR-1404-0035 for common core credit. In the upper-division baccalaureate category, 1 semester hour in leadership and 2 in human resource management (5/01).

AR-1406-0127

Electronic Warfare (EW)/Intercept System Maintenance Analyst Advanced Noncommissioned Officer (NCO)

Course Number: 2-33-C42.

Location: Intelligence School, Fort Huachuca, AZ.

Length: 12–13 weeks (510 hours).

Exhibit Dates: 1/87–5/01.

Learning Outcomes: Upon completion of the course, the student will be able to lead, manage, supervise, and train intelligence personnel in performing the tactical and strategic intelligence tasks of a military intelligence unit; prepare and communicate plans and reports in written and oral form; plan and supervise maintenance operations on mechanical and electronic equipment; and manage supply procedures for a military intelligence unit.

Instruction: Instruction includes classroom and practical exercises in leadership, management, administration, communication skills, maintenance concepts and plans, and supervision. Course includes a common core of leadership subjects.

Credit Recommendation: In the lower-division baccalaureate/associate degree category, 3 semester hours in technical writing and 3 in maintenance management. See AR-1404-0035 for common core credit. In the upper-division baccalaureate category, 1 semester hour in leadership and 2 in human resource management (6/90).

AR-1406-0128

1. Military Intelligence Officer Advanced (Signals Intelligence/Electronic Warfare (SIGINT/EW))
2. Military Intelligence Officer Advanced
3. Military Intelligence Officer Advanced

Course Number: ***Version 1:*** 3-30-C22-35G. *Version 2:* 3-30-C22-35G. *Version 3:* 3-30-C22.

Location: ***Version 1:*** Intelligence School, Fort Huachuca, AZ. *Version 2:* Intelligence School, Fort Huachuca, AZ. *Version 3:* Intelligence Center and School, Fort Huachuca, AZ.

Length: ***Version 1:*** 20 weeks (767–795 hours). *Version 2:* 20 weeks (798–799 hours). *Version 3:* 20 weeks (871 hours).

Exhibit Dates: ***Version 1:*** 6/87–1/92. *Version 2:* 2/92–5/94. *Version 3:* 6/94–9/99.

Learning Outcomes: ***Version 1:*** Upon completion of the course, the student will perform military intelligence operations, including threat analysis and staff and management functions, make decisions based on intelligence data, write effectively, and apply basic principles of electronic communications. *Version 2:* Upon completion of the course, the student will be able to perform responsibilities and tasks for military intelligence unit commander, including management, technical writing, investigation techniques, counterintelligence operations, surveillance operations and counterterrorism intelligence. *Version 3:* Upon completion of the course, the student will be able to perform responsibilities and tasks for military intelligence unit commander, including management, technical writing, investigation techniques, counterintelligence operations, surveillance operations, and counterterrorism intelligence.

Instruction: ***Version 1:*** Topics include management, leadership, writing, military history, and geopolitical studies of conflict areas in Central and South America and the USSR/Eastern Europe. Students also participate in computer simulation of a warfare system. Players make decisions as they would in the field involving battle and counterattack. The workstation uses all facets of intelligence information and communications. Electronics topics include radio wave propagation, modulation, antennas, transmitters, and receivers. *Version 2:* Topics include management, leadership, technical writing, investigative techniques, counterterrorism, and human and electronic surveillance techniques. Students participate in computer simulation of a warfare system, and the workstation uses all facets of intelligence information and communications. Electronics topics include radio wave propagation, modulation, antennas, transmitters, and receivers. *Version 3:* Topics include management, leadership, technical writing, investigative techniques, counterterrorism, and human and electronic surveillance techniques. Students participate in computer simulation of a warfare system, and the workstation uses all facets of intelligence information and communications. Electronics topics include radio wave propagation, modulation, antennas, transmitters, and receivers.

Credit Recommendation: ***Version 1:*** In the lower-division baccalaureate/associate degree category, 3 semester hours in technical writing and 1 in electronic communications. In the upper-division baccalaureate category, 3 semester hours in simulation exercise in decision making, 1 in principles of personnel and organizational management, and 2 in international studies in areas of conflict (10/91). *Version 2:* In the lower-division baccalaureate/associate degree category, 2 semester hours in technical writing, 1 in counterterrorism procedures, 1 in investigative techniques, and 1 in surveillance techniques. In the upper-division baccalaureate category, 3 semester hours in principles of management (6/97). *Version 3:* In the lower-division baccalaureate/associate degree category, 2 semester hours in technical writing, 1 in investigative techniques, 1 in surveillance techniques, and 1 in counterterrorism procedures. In the upper-division baccalaureate category, 3 semester hours in principles of management (5/01).

Related Occupations: 35G.

AR-1406-0129

Tactical Telecommunications Center Operator/Automatic Data Telecommunications Center Operator

Course Number: 260-72E10/260-72G10.

Location: Signal School, Fort Gordon, GA.

Length: 11–12 weeks (405 hours).

Exhibit Dates: 1/90–1/91.

Learning Outcomes: Upon completion of the course, the student will be able to operate a telecommunications center and prepare and type messages.

Instruction: Lectures and practical exercises focus on learning to type, operating the communications equipment, and preparing messages.

Credit Recommendation: In the lower-division baccalaureate/associate degree category, 2 semester hours in keyboarding (6/90).

Related Occupations: 72E; 72G.

AR-1406-0130

Chaplain Assistant Basic Noncommissioned Officer (NCO) Reserve Component

Course Number: 161-71M20-RC.

Location: Reserve Component Training Centers, US; Chaplain Center and School, Fort Monmouth, NJ.

Length: 117 hours.

Exhibit Dates: 4/90–4/04.

Learning Outcomes: Upon completion of the course, the student will be able to lead and organize small group activities involving land navigation, maintenance, training, tactical deployment, and equipment operation; account for appropriated funds; communicate effectively; and care for battle fatigue casualties.

Instruction: Topics include leadership, communication, battle fatigue casualties, religious program management, volunteer management, tactical deployment, equipment operation, tactical operations, and leadership skills.

Credit Recommendation: Credit is not recommended because of the military-specific nature of the course (6/90).

Related Occupations: 71M.

AR-1406-0132

Personnel Staff Noncommissioned Officer (NCO) Military Personnel Technician

Course Number: PEC-P-PS2.

Location: Professional Education Center, Cp. Joseph Robinson, North Little Rock, AR.

Length: 2 weeks (64 hours).

Exhibit Dates: 3/90–4/04.

Learning Outcomes: Upon completion of the course, the student will be able to perform basic administrative and personnel actions with emphasis on the preparation and review of military personnel forms.

Instruction: Lectures and in-class exercises are designed to prepare the student to complete forms addressing administrative, financial, and information management actions as well as retention and fitness concerns.

Credit Recommendation: In the lower-division baccalaureate/associate degree category, 1 semester hour in office management (7/90).

AR-1406-0133

Battalion Training Officer

Course Number: PEC-T-BTO-X-XX.

Location: Professional Education Center, Cp. Joseph Robinson, North Little Rock, AR.

Length: 1–2 weeks (46 hours).

Exhibit Dates: 3/90–12/97.

Learning Outcomes: Upon completion of the course, the student will be able to prepare training outlines, prepare training and evaluation plans, ensure career-path coordination with training objectives, and evaluate training calendars.

Instruction: This course is performance-based, with exercises performed by students individually and in groups. Evaluation is accomplished through exercises and testing.

Credit Recommendation: In the lower-division baccalaureate/associate degree category, 1 semester hour in training and development (7/90).

AR-1406-0134

1. National Guard (ARNG) Advanced Recruiter and Sales
2. National Guard (ARNG) Recruiting and Retention Management

Course Number: ***Version 1:*** 501-F16-ARNG; 501-F16 (ARNG). *Version 2:* 7C-F21-ARNG.

Location: Professional Education Center, Cp. Joseph Robinson, North Little Rock, AR.

Length: ***Version 1:*** 2 weeks (79–83 hours). *Version 2:* 2 weeks (64 hours).

Exhibit Dates: ***Version 1:*** 9/89–12/91. *Version 2:* 1/92–Present.

Learning Outcomes: ***Version 1:*** Upon completion of the course, the student will be able to function effectively as a recruiter by using sales management techniques. *Version 2:* Upon completion of the course, the student will be able to manage a recruiting program, including market and trend analysis, program planning and evaluation, budget management, report preparation and presentation, and production management of recruiting and retention program.

Instruction: ***Version 1:*** Lectures and in-class exercises cover advanced sales management, human effectiveness, and goal setting. *Version 2:* Course is performance based with numerous team and individual exercises. Extensive use is made of scenario materials and commercial educational materials on advanced sales management, increasing human effectiveness, and goals setting.

Credit Recommendation: ***Version 1:*** In the lower-division baccalaureate/associate degree category, 2 semester hours in sales management (11/91). *Version 2:* In the upper-division baccalaureate category, 3 semester hours in sales management (6/98).

AR-1406-0135

Noncommissioned Officer (NCO) in Charge

Course Number: PEC-T-NCOIC-X-XX.

Location: Professional Education Center, Cp. Joseph Robinson, North Little Rock, AR.

Length: 2 weeks (55–56 hours).

Exhibit Dates: 3/90–12/97.

Learning Outcomes: Upon completion of the course, the student will be able to practice good management skills in the areas of listening, stress management, dynamic subordination, personnel evaluation, delegation, and the basic concepts of situational leadership.

Instruction: This course is performance-based, with individual and group exercises performed by the students. Evaluation is accomplished through exercises and testing.

Credit Recommendation: In the lower-division baccalaureate/associate degree category, 1 semester hour in personnel supervision (7/90).

AR-1406-0136

Training Noncommissioned Officer (NCO)

Course Number: PEC-T-TNCO-X-XX.

Location: Professional Education Center, Cp. Joseph Robinson, North Little Rock, AR.

Length: 3 weeks (72 hours).

Exhibit Dates: 3/90–4/04.

Learning Outcomes: Upon completion of the course, the student will be able to prepare training outlines, prepare training and evaluation plans, ensure career-path coordination with training objectives, identify proper resources needed to implement a training program, and set up physical requirements to accommodate training.

Instruction: Course is performance-based, with exercises performed by students individually and in groups. Evaluation is accomplished through exercises and testing.

Credit Recommendation: In the lower-division baccalaureate/associate degree category, 2 semester hours in training and development (7/90).

AR-1406-0137

Battalion Supply

Course Number: 90-BSC; BSC-90.

Location: Professional Education Center, Cp. Joseph Robinson, North Little Rock, AR.

Length: 1–2 weeks (54 hours).

Exhibit Dates: 10/89–12/97.

Learning Outcomes: Upon completion of the course, the student will be able to write and process inventory, property, equipment, and financial reports.

Instruction: Instructional strategies include proctored exercises, structured exercises, discussions, and lectures. Content includes forms and procedures related to property, equipment, and finance.

Credit Recommendation: In the vocational certificate category, 1 semester hour in office personnel (7/90).

AR-1406-0138

Battalion Training Enlisted

Course Number: PEC-T-BTE-X-XX.

Location: Professional Education Center, Cp. Joseph Robinson, North Little Rock, AR.

Length: 2 weeks (46 hours).

Exhibit Dates: 3/90–12/97.

Learning Outcomes: Upon completion of the course, the student will be able to prepare training outlines and training and evaluation plans, ensure career-path coordination with training objectives, and evaluate training calendars.

Instruction: This course is performance based, with exercises performed by students individually and in groups. Evaluation is accomplished through exercises and testing.

Credit Recommendation: In the lower-division baccalaureate/associate degree category, 1 semester hour in training and development (7/90).

AR-1406-0139

National Guard (ARNG) Recruiter and Sales

Course Number: 501-00E20 (ARNG).

Location: Professional Education Center, Cp. Joseph Robinson, North Little Rock, AR.

Length: 4 weeks (149 hours).

Exhibit Dates: 10/88–4/04.

Learning Outcomes: After 9/96 see AR-1406-0188. Upon completion of the course, the student will be able to function effectively as a recruiter by using sales management skills and techniques.

Instruction: Lectures and in-class exercises cover stress management, interpersonal communication, market analysis, interviewing, and sales techniques.

Credit Recommendation: In the lower-division baccalaureate/associate degree category, 3 semester hours in sales management (11/91).

Related Occupations: 00E.

AR-1406-0140

Turbine Engine Driven Generator Repairer Basic Noncommissioned Officer (NCO) Common Core

Course Number: 690-52F30, Phase 1.
Location: Ordnance Center and School, Aberdeen Proving Ground, MD.
Length: 3–4 weeks (126–127 hours).
Exhibit Dates: 10/87–4/04.
Learning Outcomes: Upon completion of the course, the student will be able to manage, supervise, and provide leadership.
Instruction: Academic lectures provide concepts and processes in leadership doctrine to further improve leadership abilities, communication and management skills, and the ability to instill discipline. Course includes a common core of leadership subjects.
Credit Recommendation: In the lower-division baccalaureate/associate degree category, 1 semester hour in the principles of supervision. See AR-1406-0090 for common core credit (12/90).
Related Occupations: 52F.

AR-1406-0141

Power Generation Equipment Repairer Basic Noncommissioned Officer (NCO) Common Core

Course Number: 662-52D30, Phase 1.
Location: Ordnance Center and School, Aberdeen Proving Ground, MD.
Length: 3–5 weeks (126–127 hours).
Exhibit Dates: 10/87–9/90.
Learning Outcomes: After 9/96 see AR-1732-0019. Upon completion of the course, the student will be able to manage, supervise, and provide leadership.
Instruction: Academic lectures provide concepts and processes in leadership doctrine to further improve leadership abilities, communication and management skills, and the ability to instill discipline. Course includes a common core of leadership subjects.
Credit Recommendation: In the lower-division baccalaureate/associate degree category, 1 semester hour in the principles of supervision. See AR-1406-0090 for common core credit (10/91).
Related Occupations: 52D.

AR-1406-0142

Senior Utilities, Quartermaster/Chemical Equipment Repairer Basic Noncommissioned Officer (NCO) Common Core

Course Number: 690-52C30, Phase 1.
Location: Ordnance Center and School, Aberdeen Proving Ground, MD.
Length: 3–4 weeks (126–127 hours).
Exhibit Dates: 10/87–4/04.
Learning Outcomes: Upon completion of the course, the student will be able to manage, supervise, and provide leadership.
Instruction: Academic lectures provide concepts and processes in leadership doctrine to further improve leadership abilities, communication and management skills, and the ability to instill discipline. Course includes a common core of leadership subjects.
Credit Recommendation: In the lower-division baccalaureate/associate degree category, 1 semester hour in the principles of supervision. See AR-1406-0090 for common core credit (12/90).
Related Occupations: 52C.

AR-1406-0143

Utilities Equipment Repairer Basic Noncommissioned Officer (NCO)

Course Number: 662-53C30, Phase 1.
Location: Ordnance Center and School, Aberdeen Proving Ground, MD.
Length: 4–5 weeks (170 hours).
Exhibit Dates: 10/91–9/96.
Learning Outcomes: After 9/96 see AR-1701-0004. Upon completion of the course, the student will be able to manage, supervise, and provide leadership.
Instruction: Academic lectures provide concepts and processes in leadership doctrine to further improve leadership abilities, communication and management skills, and the ability to instill discipline. Course includes a common core of leadership subjects.
Credit Recommendation: In the lower-division baccalaureate/associate degree category, 1 semester hour in principles of supervision. See AR-1406-0090 for common core credit (4/91).
Related Occupations: 53C.

AR-1406-0144

Manpower and Force Management

Course Number: *Version 1:* ALMC-MG. *Version 2:* ALMC-MG.
Location: *Version 1:* Army Logistics Management College, Fort Lee, VA; Army Logistics Management College, Satellite Education Program, Fort Lee, VA. *Version 2:* Army Logistics Management College, Fort Lee, VA.
Length: *Version 1:* 2 weeks (72–75 hours). *Version 2:* 2 weeks (71–74 hours).
Exhibit Dates: *Version 1:* 5/87–1/95. *Version 2:* 2/95–Present.
Learning Outcomes: *Version 1:* Upon completion of the course, the student will be able to explain manpower and force functions; explain the relationship between manpower management and other organizational systems; and apply manpower management techniques, principles, and procedures. *Version 2:* Upon completion of the course, the student will be able to explain manpower and force functions; explain the relationship between manpower management and other organizational systems; and apply manpower management techniques, principles, and procedures.
Instruction: *Version 1:* Instruction includes lectures, discussions, exercises, and case studies. Topics include manpower requirement determinations; planning, programming, and budgeting; allocations; documentation; and analysis and evaluation. *Version 2:* Instruction includes lectures, discussions, exercises, and case studies. Topics include manpower requirement determinations; planning, programming, and budgeting; allocations; documentation; and analysis and evaluation.
Credit Recommendation: *Version 1:* In the upper-division baccalaureate category, 2 semester hours in manpower management (6/91). *Version 2:* In the upper-division baccalaureate category, 3 semester hours in logistics management (1/02).

AR-1406-0147

Personnel Management Specialist

Course Number: *Version 1:* 500-75C10. *Version 2:* 500-75C10.
Location: Soldier Support Center, Fort Benjamin Harrison, IN.
Length: *Version 1:* 10–11 weeks (371 hours). *Version 2:* 9–10 weeks (350–354 hours).
Exhibit Dates: *Version 1:* 10/88–1/91. *Version 2:* 2/91–1/97.
Learning Outcomes: *Version 1:* Upon completion of the course, the student will be able to keyboard at the rate of 20 net words per minute on a five minute test and use accepted personnel procedures in the handling of records including security and EEOC guidelines. *Version 2:* Upon completion of the course, the student will be able to keyboard at the rate of 20 net words per minute on a five minute test; perform basic word processing functions in personnel correspondence and documentation; and use accepted personnel procedures in the handling of records including security and EEOC guidelines.
Instruction: *Version 1:* Lectures and practical exercises cover keyboarding skills, record keeping, personnel procedures, and business correspondence. *Version 2:* Lectures and practical exercises cover keyboarding skills, basic word processing functions, record keeping, personnel procedures, and business correspondence.
Credit Recommendation: *Version 1:* In the lower-division baccalaureate/associate degree category, 2 semester hours in keyboarding and 1 in record keeping (12/91). *Version 2:* In the lower-division baccalaureate/associate degree category, 2 semester hours in keyboarding, 1 in record keeping, and 2 in basic word processing (12/91).
Related Occupations: 75C.

AR-1406-0148

Personnel Management Specialist Basic Noncommissioned Officer (NCO)

Course Number: 500-75C30.
Location: Adjutant General School, Fort Benjamin Harrison, IN.
Length: 8–9 weeks (310 hours).
Exhibit Dates: 10/91–9/95.
Learning Outcomes: After 9/95 see AR-1406-0181. Upon completion of the course, the student will be able to supervise and maintain personnel files and correspondence; process orders; operate computer equipment in order to edit, create directories, compose, and format personnel correspondence; and evaluate personnel reports (files) and make appropriate recommendations.
Instruction: Lectures and practical exercises cover establishing and maintaining personnel procedures, including record keeping, record security, information processing, supervising, and setting priorities.
Credit Recommendation: In the upper-division baccalaureate category, 3 semester hours in human resource management (12/91).
Related Occupations: 75C.

AR-1406-0149

Personnel Records Specialist Basic Noncommissioned Officer (NCO)

Course Number: 500-75D30.
Location: Adjutant General School, Fort Benjamin Harrison, IN.
Length: 8–9 weeks (310 hours).

Exhibit Dates: 10/91–9/95.

Learning Outcomes: Upon completion of the course, the student will be able to supervise and maintain personnel files and correspondence; process orders; operate computer equipment in order to edit, create directories, compose, and format personnel correspondence; and evaluate personnel reports (files) and make appropriate recommendations.

Instruction: Lectures and practical exercises cover establishing and maintaining personnel procedures, including record keeping, record security, information processing, supervising, and setting priorities.

Credit Recommendation: In the upper-division baccalaureate category, 3 semester hours in human resource management (12/91).

Related Occupations: 75D.

AR-1406-0150

Personnel Actions Specialist Basic Noncommissioned Officer (NCO)

Course Number: 500-75E30.

Location: Adjutant General School, Fort Benjamin Harrison, IN.

Length: 8–9 weeks (310 hours).

Exhibit Dates: 10/91–9/95.

Learning Outcomes: Upon completion of the course, the student will be able to supervise and maintain personnel files and correspondence; process orders; operate computer equipment in order to edit, create directories, compose, and format personnel correspondence; and evaluate personnel reports (files) and make appropriate recommendations.

Instruction: Lectures and practical exercises cover establishing and maintaining personnel procedures, including record keeping, record security, information processing, supervising, and setting priorities.

Credit Recommendation: In the upper-division baccalaureate category, 3 semester hours in human resource management (12/91).

Related Occupations: 75E.

AR-1406-0151

Military Personnel Technician Warrant Officer Technical/Tactical Certification

Course Number: 7C-420A.

Location: Soldier Support Center, Fort Benjamin Harrison, IN.

Length: 4–5 weeks (167 hours).

Exhibit Dates: 3/89–1/97.

Learning Outcomes: Upon completion of the course, the student will be able to apply military personnel principles, practices, and procedures.

Instruction: Lectures, discussions, videotapes, and seminar groups are the modes of instruction.

Credit Recommendation: Credit is not recommended because of the military-specific nature of the course (12/91).

Related Occupations: 420A.

AR-1406-0152

Military Personnel Technician Warrant Officer Technical and Tactical Certification Reserve Component Phase 2

Course Number: 7C-420A-RC.

Location: Soldier Support Center, Fort Benjamin Harrison, IN.

Length: 2 weeks (79 hours).

Exhibit Dates: 3/90–1/97.

Learning Outcomes: Upon completion of the course, the student will be able to apply military personnel principles, practices, and procedures.

Instruction: Lectures and practical exercises on the subject matter.

Credit Recommendation: Credit is not recommended because of the limited, specialized nature of the course (12/91).

Related Occupations: 420A.

AR-1406-0153

Personnel Administration Specialist

Course Number: *Version 1:* 500-75B10. *Version 2:* 500-75B10 (F); 500-75B10.

Location: *Version 1:* Soldier Support Institute, Fort Jackson, SC; Soldier Support Center, Fort Benjamin Harrison, IN. *Version 2:* Adjutant General School, Fort Jackson, SC.

Length: *Version 1:* 7–8 weeks (280–305 hours). *Version 2:* 8–9 weeks (329–330 hours).

Exhibit Dates: *Version 1:* 1/92–4/98. *Version 2:* 5/98–Present.

Learning Outcomes: *Version 1:* Before 1/92 see AR-1406-0105. Upon completion of the course, the student will be able to perform clerical duties, including clerical procedures and computer applications; keyboard at a minimum of 20 net words per minute on a five minute test; and perform basic computer operations in the area of office communication. *Version 2:* Upon completion of the course, the student will be able to perform clerical duties, including clerical procedures and computer applications; keyboard at minimum of 20 net words per minute on a five minute test; and perform basic computer operations in the area of office communication.

Instruction: *Version 1:* Lectures and practical exercises on clerical procedures, keyboarding, and computer operations. *Version 2:* Lectures and practical exercises on clerical procedures, keyboarding, and computer operations.

Credit Recommendation: *Version 1:* In the lower-division baccalaureate/associate degree category, 1 semester hour in keyboarding and 3 in clerical procedures (9/96). *Version 2:* In the lower-division baccalaureate/associate degree category, 2 semester hours in records management and 3 in clerical procedures (3/02).

Related Occupations: 75B.

AR-1406-0154

Administrative Senior Warrant Officer Reserve Component

Course Number: 7-12-C32-RC.

Location: Adjutant General School, Fort Benjamin Harrison, IN.

Length: 2 weeks (91 hours).

Exhibit Dates: 3/91–1/97.

Learning Outcomes: Upon completion of the course, the student will be able to apply supervisory and leadership skills at the mid-management level.

Instruction: Course includes lectures and practical exercises.

Credit Recommendation: In the lower-division baccalaureate/associate degree category, 1 semester hour in leadership (12/91).

AR-1406-0156

1. Recruiting and Retention Basic Noncommissioned Officer (NCO)
2. Recruiting Basic Noncommissioned Officer (NCO)

Course Number: *Version 1:* 501-00R/00E30. *Version 2:* 501-79R30.

Location: *Version 1:* Recruiting and Retention School, Fort Benjamin Harrison, IN. *Version 2:* Soldier Support Institute, Fort Jackson, SC.

Length: *Version 1:* 3 weeks (112 hours). *Version 2:* 3–4 weeks (133 hours).

Exhibit Dates: *Version 1:* 10/91–9/95. *Version 2:* 10/95–Present.

Learning Outcomes: *Version 1:* Upon completion of the course, the student will be able to apply leadership skills in a first-line position. *Version 2:* Upon completion of the course, the student will be able to apply leadership skills in a first-line position.

Instruction: *Version 1:* Lectures and practical exercises cover leadership and management. *Version 2:* Lectures and practical exercises cover leadership and management; writing, and personal and performance counseling.

Credit Recommendation: *Version 1:* In the lower-division baccalaureate/associate degree category, 1 semester hour in leadership (12/91). *Version 2:* Credit is recommended for the common core only. See AR-1406-0090 (9/96).

Related Occupations: 00E; 00R.

AR-1406-0157

Recruiting Operations
(Recruiting Operations (S3))

Course Number: 7C-F36/501-F18.

Location: Soldier Support Institute, Fort Jackson, SC; Recruiting and Retention School, Fort Benjamin Harrison, IN.

Length: 2 weeks (71–75 hours).

Exhibit Dates: 10/90–Present.

Learning Outcomes: Upon completion of the course, the student will be able to complete all operational forms in a recruiting battalion and determine eligibility for enlistment.

Instruction: Lectures and case studies cover the subject matter.

Credit Recommendation: Credit is not recommended because of the military-specific nature of the course (9/96).

AR-1406-0158

Recruiting and Retention Advanced Noncommissioned Officer (NCO)

Course Number: 5-79-C42.

Location: Recruiting and Retention School, Fort Benjamin Harrison, IN.

Length: 7–8 weeks (440 hours).

Exhibit Dates: 10/91–6/96.

Learning Outcomes: After 6/96 see AR-1406-0185. Upon completion of the course, the student will be able to evaluate written communication for correctness, implement leadership/management strategies, evaluate recruiting action plans and make recommendations, apply appropriate sales techniques to recruiting, and employ effective interpersonal communication skills.

Instruction: Lectures, small-group role play and discussions cover the topics above.

Credit Recommendation: In the lower-division baccalaureate/associate degree category, 2 semester hours in salesmanship, 1 in interpersonal communication, and 1 in human relations (12/91).

AR-1406-0159

Retention Noncommissioned Officer
(NCO)

Course Number: 501-00R30.
Location: Soldier Support Institute, Fort Jackson, SC; Recruiting and Retention School, Fort Benjamin Harrison, IN.
Length: 8–9 weeks (317 hours).
Exhibit Dates: 10/90–4/98.
Learning Outcomes: After 4/98 see AR-1406-0228. Upon completion of the course, the student will be able to interpret personnel records, assist personnel in the reenlistment process, determine reenlistment or extension eligibility, prepare and present retention training, and conduct retention interview/counseling sessions.
Instruction: Methods include lectures, discussions, guest speakers, and practical exercises.
Credit Recommendation: In the lower-division baccalaureate/associate degree category, 1 semester hour in interpersonal communication, 2 in salesmanship, and 1 in human relations (9/96).
Related Occupations: 00R.

AR-1406-0160

Guidance Counselor

Course Number: 501-ASIV7.
Location: Recruiting and Retention School, Fort Benjamin Harrison, IN.
Length: 3 weeks (112–115 hours).
Exhibit Dates: 1/89–1/97.
Learning Outcomes: After 6/97 see AR-1406-0229. Upon completion of the course, the student will be able to interview prospective enlistees, communicate effectively, determine eligibility, and prepare enlistment documents.
Instruction: Methods include lectures, discussions, and practical exercises.
Credit Recommendation: In the lower-division baccalaureate/associate degree category, 1 semester hour in salesmanship and 1 in counseling techniques (12/91).

AR-1406-0161

Health Care Recruiting
(Nurse Recruiting)

Course Number: ***Version 1:*** 6E-F1/501-F17; 7C-F33/501-F17. *Version 2:* 6E-F1/501-F17.
Location: ***Version 1:*** Soldier Support Institute, Fort Jackson, SC; Recruiting and Retention School, Fort Benjamin Harrison, IN. *Version 2:* Recruiting and Retention School, Fort Jackson, SC; Soldier Support Institute, Fort Jackson, SC.
Length: ***Version 1:*** 2 weeks (77–88 hours). *Version 2:* 3 weeks (110–113 hours).
Exhibit Dates: ***Version 1:*** 4/90–9/97. *Version 2:* 10/97–Present.
Learning Outcomes: ***Version 1:*** Upon completion of the course, the student will be able to obtain leads in recruiting nurses, conduct interviews, determine eligibility, and process applicants. *Version 2:* Upon completion of the course, the student will be able to use sales techniques, including prospecting and market analysis; conduct interviews to determine eligibility; and process applicant documents.
Instruction: ***Version 1:*** Methods of instructions include lectures, role play, practical exercises, and guest speakers. *Version 2:* Methods of instruction include guest lectures, practical exercises, lectures, and role playing. Topics covered include recruitment market analysis, sales techniques, prospecting, interviewing techniques, and processing of enlistment documents.
Credit Recommendation: ***Version 1:*** In the lower-division baccalaureate/associate degree category, 1 semester hour in salesmanship (9/96). *Version 2:* In the lower-division baccalaureate/associate degree category, 2 semester hours in salesmanship (2/02).

AR-1406-0162

Recruiting Company Commander
(Recruiting Commander)

Course Number: 7C-F13.
Location: Soldier Support Institute, Fort Jackson, SC; Recruiting and Retention School, Fort Benjamin Harrison, IN.
Length: 2–3 weeks (91–130 hours).
Exhibit Dates: 10/88–1/95.
Learning Outcomes: Upon completion of the course, the student will be able to provide leadership in the recruitment mission, including market analysis, sales techniques, and small and large group presentations and evaluate recruiting results.
Instruction: Lectures, discussions, case studies, role play, videotapes and guest speakers are modes of instruction.
Credit Recommendation: In the lower-division baccalaureate/associate degree category, 1 semester hour in leadership, 2 in sales management, and 1 in interpersonal communication (9/98).

AR-1406-0163

Station Commander

Course Number: ***Version 1:*** 501-F2. *Version 2:* 501-F2. *Version 3:* 501-F2.
Location: ***Version 1:*** Recruiting and Retention School, Fort Benjamin Harrison, IN. *Version 2:* Soldier Support Institute, Fort Jackson, SC. *Version 3:* Soldier Support Center, Fort Jackson, SC.
Length: ***Version 1:*** 3 weeks (114 hours). *Version 2:* 3–4 weeks (154 hours). *Version 3:* 5 weeks (197 hours).
Exhibit Dates: ***Version 1:*** 10/90–6/97. *Version 2:* 7/97–9/00. *Version 3:* 10/00–Present.
Learning Outcomes: ***Version 1:*** Upon completion of the course, the student will be able to apply leadership skills in a managerial position and supervise individuals responsible for conducting market analysis, sales interviews, counseling sessions, and other sales-related activities. *Version 2:* Upon completion of the course, the student will be able to apply leadership skills in a managerial position and supervise individuals responsible for conducting market analysis, sales interviews, counseling sessions, and other sales-related activities. *Version 3:* Upon completion of the course, the student will be able to apply leadership skills in a managerial position and supervise individuals responsible for conducting market analysis, sales interviews, counseling sessions, and other sales-related activities.
Instruction: ***Version 1:*** Methods include role play and group exercises. *Version 2:* Methods used include conferences, discussion, practical examinations, role playing, practical applications, and guest speakers. Topics covered include leadership, management systems, training assessment, planning, execution, and evaluation. *Version 3:* Methods of instruction include conferences, discussion, practical examinations, role playing, practical applications, and guest speakers. Topics covered include leadership, management systems, training assessment, planning, execution, and evaluation.
Credit Recommendation: ***Version 1:*** In the lower-division baccalaureate/associate degree category, 3 semester hours in sales management (12/91). *Version 2:* In the lower-division baccalaureate/associate degree category, 3 semester hours in sales management (5/01). *Version 3:* In the upper-division baccalaureate category, 3 semester hours in sales management (3/02).

AR-1406-0164

Reserve Retention Noncommissioned
Officer (NCO)

Course Number: 501-F13.
Location: Reserve Readiness Training Center, Fort McCoy, WI.
Length: 2 weeks (70 hours).
Exhibit Dates: 1/90–4/04.
Learning Outcomes: Upon completion of the course, the student will be able to evaluate raters/senior raters, conduct training of sales team, evaluate sales presentations, evaluate sales team, and conduct seminar training.
Instruction: Methods of instruction include lectures, discussions, videotapes, and practical exercises.
Credit Recommendation: In the lower-division baccalaureate/associate degree category, 2 semester hours in personnel supervision (12/91).

AR-1406-0165

Reserve Component Transition
(Transition Noncommissioned Officer (NCO))
(In-Service Recruiter/Retention Noncommissioned Officer (NCO))

Course Number: 501-F14.
Location: Recruiting and Retention School, Fort Jackson, SC; Soldier Support Institute, Fort Jackson, SC; Recruiting and Retention School, Fort Benjamin Harrison, IN.
Length: 2 weeks (73–81 hours).
Exhibit Dates: 3/89–Present.
Learning Outcomes: Upon completion of the course, the student will be able to process transition personnel using a particular software system.
Instruction: Methods of instruction include small-group discussions, lectures, and practical exercises.
Credit Recommendation: Credit is not recommended because of the military-specific nature of the course (5/01).

AR-1406-0166

Recruiting First Sergeant

Course Number: 501-F7.
Location: Recruiting and Retention School, Fort Benjamin Harrison, IN.
Length: 2 weeks (79–92 hours).
Exhibit Dates: 10/88–1/97.
Learning Outcomes: Upon completion of the course, the student will be able to advise the recruiting company commander on recruiting, conduct interviews, determine eligibility of candidates, and process applications.

Instruction: Lectures, role playing, practical exercises, and guest speakers comprise the teaching methods.

Credit Recommendation: In the lower-division baccalaureate/associate degree category, 1 semester hour in salesmanship (12/91).

AR-1406-0167

Faculty Development

Course Number: ALMC-FF.

Location: Army Logistics Management College, Fort Lee, VA.

Length: 1–2 weeks (61–62 hours).

Exhibit Dates: 10/92–4/97.

Learning Outcomes: Upon completion of the course, the student will be able to develop lesson plans, select appropriate evaluation and measurement instruments, present units of instruction using varied learning strategies and accompanying instructional aids, and apply appropriate classroom management procedures.

Instruction: Topics include development of learning objectives, methods of instruction, use of instructional aids, evaluation and measurement procedures, platform techniques, and design of lesson plans. Methodology includes lectures, panel and group discussions, individual and group presentations, and written exercises.

Credit Recommendation: In the lower-division baccalaureate/associate degree category, 3 semester hours in educational methods (8/95).

AR-1406-0168

Air Defense Artillery Advanced Noncommissioned Officer (NCO)

Course Number: *Version 1:* 0-16-C42. *Version 2:* 0-16-C42 (TATS).

Location: Air Defense Artillery School, Fort Bliss, TX.

Length: *Version 1:* 7–8 weeks (311–323 hours). *Version 2:* 7–8 weeks (251 hours).

Exhibit Dates: *Version 1:* 3/92–3/99. *Version 2:* 4/99–12/00.

Learning Outcomes: *Version 1:* Upon completion of the course, the student will be able to lead, train, and direct subordinates to maintain, operate, and employ weapons on the battlefield and to employ fighting techniques. *Version 2:* Upon completion of the course, the student will be able to lead, train, and direct subordinates to maintain, operate, and employ weapons on the battlefield and to employ fighting techniques.

Instruction: *Version 1:* Lectures and practical exercises cover maintenance procedures, weapon systems, firing positions, and tactical maneuvering. Course includes a common core of leadership subjects. *Version 2:* Lectures and practical exercises cover maintenance procedures, weapon systems, firing positions, and tactical maneuvering. Course includes a common core of leadership subjects.

Credit Recommendation: *Version 1:* Credit is recommended for the common core only. See AR-1404-0035 (8/95). *Version 2:* Credit is recommended for the common core only. See AR-1404-0035 (4/01).

AR-1406-0169

Orientation to Systems Approach to Training

Course Number: 5K-F4/520-F4.

Location: Academy of Health Sciences, Fort Sam Houston, TX.

Length: 1 week (45 hours).

Exhibit Dates: 6/94–Present.

Learning Outcomes: Upon completion of the course, the student will be able to write a measurable behavioral objective; identify the characteristics of the adult learner; analyze the elements of a task; write examinations based upon cognitive learning objectives; identify common methods for group instruction; design a performance test based upon standards, behaviors, and conditions of the task; and develop a lesson plan.

Instruction: This course uses lectures and presentations to teach lesson plan development. Adult learning theories are discussed. Strategies for writing behavioral objectives, examination items, and a lesson plan are described. Group teaching techniques and performance-based evaluation are covered.

Credit Recommendation: In the graduate degree category, 2 semester hours in introduction to teaching/learning process (10/95).

AR-1406-0170

Instructor
(Interservice Instructor)

Course Number: 4A-SI5K/SQI8/710-SQIH; FL-ITC-SF.

Location: Engineer School, Fort Leonard Wood, MO.

Length: 3 weeks (120 hours).

Exhibit Dates: 6/95–Present.

Learning Outcomes: Upon completion of the course, the student will be able to perform the basic techniques and fundamentals of instruction, including preparation, classroom management, and implementation; describe the responsibilities of an effective instructor and the essentials of effective communication; master the essential conditions for learning; create lesson plans; make instructional decisions; develop instructional aids; and deliver a variety of instructional methods.

Instruction: Course delivery is through lectures, discussions, demonstrations, practice, and critiques. After exposure to fundamentals and techniques of planning, management, and implementation of instruction, participants engage in extensive supervised practice and cover the development of instructional materials.

Credit Recommendation: In the upper-division baccalaureate category, 3 semester hours in design and delivery of instruction (9/98).

AR-1406-0171

Instructor Training

Course Number: 2C-SI5K/SQI8/600-SQIH.

Location: Aviation School, Fort Rucker, AL.

Length: 3 weeks (112–113 hours).

Exhibit Dates: 6/91–Present.

Learning Outcomes: Upon completion of the course, students will be able to describe theory and practice relating to concepts, methods, techniques, and technology of performance-oriented training; describe the principles of the learning process; develop training objectives and aids; perform the fundamentals of counseling; perform testing and learning analysis; and practice the essentials of effective communication.

Instruction: Course delivery is through lectures, discussions, demonstrations, practice, and critiques. After exposure to techniques of planning, management, and implementation of instruction, participants engage in extensive supervised practice of training delivery.

Credit Recommendation: In the lower-division baccalaureate/associate degree category, 1 semester hour in speech communication. In the upper-division baccalaureate category, 3 semester hours in design and delivery of instruction (9/03).

AR-1406-0172

Instructor Training

Course Number: 4C-SI5K/SQI8/260-SQIH.

Location: Signal School, Fort Gordon, GA.

Length: 2 weeks (80 hours).

Exhibit Dates: 3/91–Present.

Learning Outcomes: Upon completion of the course, the student will be able to describe the basic techniques and fundamentals of instruction including preparation and implementation, describe the responsibilities of an effective instructor and the essentials of effective communication, describe the general conditions for learning, develop lesson plans, make instructional decisions, design instructional aids, and deliver instruction in a variety of methods.

Instruction: Course delivery is through lectures, discussions, demonstrations, practice, and critiques. After exposure to fundamentals and techniques of planning, management, and implementation of instruction, participants engage in supervised practice and critique.

Credit Recommendation: In the lower-division baccalaureate/associate degree category, 1 semester hour in speech communication. In the upper-division baccalaureate category, 2 semester hours in techniques of instruction (9/03).

AR-1406-0173

Faculty Instructor

Course Number: OE-7; SOA-515K/SQI8/SQIH.

Location: School of the Americas, Fort Benning, GA.

Length: 3 weeks (119–126 hours).

Exhibit Dates: 1/92–Present.

Learning Outcomes: Upon completion of the course, the student will be able to design, develop, and implement instructional programs focusing on military academic subjects and use the systems approach to training to develop learning objectives, lesson plans, criterion-referenced test items, training aids, instructional materials, and techniques for classroom management.

Instruction: Course delivery is through lectures, discussions, small-group instruction, and practical application. After exposure to techniques of planning, management, and implementation of instruction, participants engage in extensive material development and supervised practice.

Credit Recommendation: In the upper-division baccalaureate category, 3 semester hours in design and delivery of instruction (7/96).

AR-1406-0174

Instructor Training

Course Number: 2E-SI5K/SQI8/494-SQIH.

Location: Ordnance, Missile and Munitions School, Redstone Arsenal, AL.

Length: 2 weeks (80 hours).

Exhibit Dates: 10/91–Present.

Learning Outcomes: Upon completion of the course, the student will be able to describe the basic techniques of instruction, including preparation, classroom management, and implementation; demonstrate academic counseling and communication skills; describe the responsibilities of an effective instructor, the essential conditions for learning, and the principles of development for effective lesson plans; make instructional decisions; develop instructional aids; and deliver instruction in a variety of methods.

Instruction: Course delivery is through lectures, discussions, demonstrations, role playing, counseling situations and practice, and the presentation of instruction followed by critique. After exposure to fundamentals and techniques of planning, management, communication skills, and implementation of instruction, participants apply these through supervised practice.

Credit Recommendation: In the lower-division baccalaureate/associate degree category, 2 semester hours in speech communication. In the upper-division baccalaureate category, 2 semester hours in techniques of instruction (9/03).

AR-1406-0175

Staff and Faculty Development

Course Number: 2F-SI5K/SQI8/202-SQIH.

Location: Soldier Support Center, Fort Jackson, SC; Soldier Support Institute, Fort Benjamin Harrison, IN.

Length: 2–3 weeks (79–96 hours).

Exhibit Dates: 10/91–1/96.

Learning Outcomes: Upon completion of the course, the student will be able to perform the five phases of the systems approach to training, including analysis, design, development, implementation, and evaluation; engage in extensive practice in the implementation phase including the operation of audiovisual equipment; develop training aids; conduct demonstrations; deliver various types of presentations; assess the target audience and the learning environment; select appropriate methods; and design evaluation based on the learning objectives.

Instruction: Course delivery is through lectures, discussions, demonstrations, role playing, practice, and critiques. After exposure to the five-phase systems approach to training, participants engage in extensive supervised practice and the development of instructional materials.

Credit Recommendation: In the upper-division baccalaureate category, 3 semester hours in design and delivery of instruction (9/96).

AR-1406-0176

Instructor Training

Course Number: 8C-SI5K/SQI8/811-SQIH.

Location: Transportation School, Fort Eustis, VA; Aviation Logistics School, Fort Eustis, VA.

Length: 3 weeks (120 hours).

Exhibit Dates: 5/90–Present.

Learning Outcomes: Upon completion of the course, the student will demonstrate familiarity with the basic techniques of instruction including preparation, classroom management, and implementation; describe the responsibilities of an effective instructor and the essentials of effective communication; describe the essential conditions for learning; create lesson plans; make instructional decisions; develop instructional aids; deliver a variety of instructional methods; and test and evaluate.

Instruction: Course delivery is through lectures, discussions, demonstrations, practice, and critique. After exposure to techniques of planning, management, communication skills, and implementation of instruction, participants engage in extensive supervised practice and the development of lesson outlines, unit plans, and presentations.

Credit Recommendation: In the lower-division baccalaureate/associate degree category, 1 semester hour in speech communication. In the upper-division baccalaureate category, 3 semester hours in design and delivery of instruction (9/03).

AR-1406-0177

Company Trainers

Course Number: *Version 1:* 0700. *Version 2:* 921-320.

Location: Reserve Readiness Training Center, Fort McCoy, WI.

Length: *Version 1:* 69 hours. *Version 2:* 2 weeks (77 hours).

Exhibit Dates: *Version 1:* 2/95–1/02. *Version 2:* 2/02–Present.

Learning Outcomes: *Version 1:* Upon completion of the course, the student will be able to plan, assess, and deliver training and perform administrative requirements necessary to support the training. *Version 2:* Upon completion of the course, the student will be able to explain reserve component role in the force structure; develop the training program mission essential task list; forecast training needs; identify and request training areas/ranges; identify training resources; determine training support and individual training requirements; explain the types of training, assemblies, and duties; identify tasks for implementing a structured exercise; plan a structured exercise; utilize automation to support unit training activities; conduct unit training assessment; and evaluate unit training events and outcomes.

Instruction: *Version 1:* Lectures, discussions, and practical exercises cover training. *Version 2:* Audiovisual materials, practical and classroom exercises, discussion, case studies, lecture, and computer-based training. Topics include training process, training resources, standardized and structured training exercises, individual training management, and automation.

Credit Recommendation: *Version 1:* Credit is not recommended because of the limited, specialized nature of the course (3/97). *Version 2:* In the upper-division baccalaureate category, 2 semester hours in training management (11/02).

AR-1406-0178

Advanced Retention Noncommissioned Officer (NCO)

Course Number: 8700.

Location: Reserve Readiness Training Center, Fort McCoy, WI.

Length: 83 hours.

Exhibit Dates: 4/93–Present.

Learning Outcomes: Upon completion of the course, the student will be able to apply theories of human behavior in organized systems, motivate individuals, plan and conduct training, relieve stress in a working environment, and manage time.

Instruction: Course includes homework and a series of performance exams, lectures, and discussions.

Credit Recommendation: In the upper-division baccalaureate category, 1 semester hour in techniques of oral reporting and 1 instructional strategies (3/97).

AR-1406-0179

1. Unit Conduct of Fire Trainer (UCOFT) Senior Instructor/Operator (I/O)
2. Unit Conduct of Fire Trainer Senior Instructor/Operator

Course Number: *Version 1:* 020-F13. *Version 2:* 020-F13.

Location: Armor Center and School, Fort Knox, KY.

Length: *Version 1:* 3 weeks (115–120 hours). *Version 2:* 3 weeks (120 hours).

Exhibit Dates: *Version 1:* 4/91–4/94. *Version 2:* 5/94–Present.

Learning Outcomes: *Version 1:* Upon completion of the course, the student will be able to operate, perform maintenance checks, and brief others on the unit conduct of fire trainer (weaponry). *Version 2:* Upon completion of the course, the student will be able to supervise, instructor/operators, enforce performance standards, perform as a training manager, integrate training into the gunnery program, serve as an instructor for new instructor/operators, and certify and recertify instructor/operators.

Instruction: *Version 1:* Methods of instruction include lectures, demonstrations, and practical exercises and cover the topics of operation, maintenance checks, and briefing of others on the fire trainer (weaponry). *Version 2:* Lectures, discussions and practical experiences are used to cover the topics of power up/power down of fire trainer (weaponry), developing training programs, instructional supervision, implementation of exportable training, emergency procedures, and crew record keeping.

Credit Recommendation: *Version 1:* In the lower-division baccalaureate/associate degree category, 1 semester hour in teaching methods and 1 in training management (3/97). *Version 2:* In the lower-division baccalaureate/associate degree category, 1 semester hour in teaching methods and 1 in training management (9/00).

AR-1406-0180

Faculty Development

Course Number: 5K-F3/520-F3.

Location: Academy of Health Sciences, Fort Sam Houston, TX.

Length: 2 weeks (101 hours).

Exhibit Dates: 5/95–Present.

Learning Outcomes: Upon completion of the course, the student will be able to develop lesson plans, select appropriate evaluation and measurement instruments, present units of instruction using varied learning strategies and instructional aids, and apply appropriate classroom management procedures.

Instruction: Topics include development of learning objectives, methods of instruction, use of instructional aids, evaluation and measurement procedures, instructional delivery, and design of lesson plans. Methodology includes lectures, demonstrations, individual instructional presentations, and written exercises.

Credit Recommendation: In the upper-division baccalaureate category, 3 semester hours in educational methods (5/97).

AR-1406-0181

Personnel Service Company Basic Noncommissioned Officer

Course Number: 500-75C30; 500-75D30; 500-75E30.

Location: Soldier Support Institute, Fort Jackson, SC.

Length: 9–10 weeks (343 hours).

Exhibit Dates: 10/95–Present.

Learning Outcomes: Before 10/95 see AR-1406-0148. Upon completion of the course, the student will be able to supervise and maintain personnel system, including maintenance of personnel files, correspondence, and the processing of orders; operate computer equipment in order to edit, create directories, compose, and format personnel correspondence; evaluate personnel records and make appropriate recommendations about separations and awards; and conduct personal and performance counseling.

Instruction: Lectures and practical exercises cover establishing and maintaining personnel procedures, including record keeping, record security, information processing, supervising, and setting priorities.

Credit Recommendation: In the upper-division baccalaureate category, 4 semester hours in human resource management (9/97).

Related Occupations: 75C; 75D; 75E.

AR-1406-0182

Personnel Services Specialist

Course Number: *Version 1:* 500-75H10. *Version 2:* 500-75H10.

Location: *Version 1:* Soldier Support Institute, Fort Jackson, SC. *Version 2:* Soldier Support Center, Fort Jackson, SC.

Length: *Version 1:* 8 weeks (308 hours). *Version 2:* 8–9 weeks (305–308 hours).

Exhibit Dates: *Version 1:* 10/96–12/99. *Version 2:* 1/00–Present.

Learning Outcomes: *Version 1:* Upon completion of the course, the student will be able to process personnel information, including maintaining records, processing promotions, documenting personnel actions, and accessing and updating an automated personnel information system. *Version 2:* Upon completion of the course, the student will be able to process personnel information, including maintaining records, processing promotions, documenting personnel actions, and accessing and updating an automated personnel information system.

Instruction: *Version 1:* Methods include lectures, class discussions, examinations, and personal computer exercises. Topics include classifications, assigning and reassigning personnel, establishing promotion rosters, maintaining records, processing separations, and updating automated personnel system. *Version 2:* Methods of instruction include lectures, class discussions, examinations, and personal computer exercises. Topics include classifications, assigning and reassigning personnel, establishing promotion rosters, maintaining records, processing separations, and updating automated personnel system.

Credit Recommendation: *Version 1:* In the lower-division baccalaureate/associate degree category, 3 semester hours in human resource management and 2 in records and information management (9/97). *Version 2:* In the lower-division baccalaureate/associate degree category, 3 semester hours in records management and 3 in human resource management (3/02).

Related Occupations: 75H.

AR-1406-0183

Personnel Services Sergeant Basic Noncommissioned Officer (NCO)

Course Number: *Version 1:* 500-75H30. *Version 2:* 500-75H30, Phase 2.

Location: *Version 1:* Soldier Support Institute, Fort Jackson, SC. *Version 2:* Soldier Support Institute, Fort Jackson, SC; Adjutant General School, Fort Jackson, SC.

Length: *Version 1:* 12 weeks (480 hours). *Version 2:* 6–7 weeks (290 hours).

Exhibit Dates: *Version 1:* 10/96–9/01. *Version 2:* 10/01–Present.

Learning Outcomes: *Version 1:* Upon completion of the course, the student will be able to organize and manage a personnel section; validate and process personnel action documents to receive, transfer, promote, and evaluate staff; and operate via PC all functions of an automated personnel information system. *Version 2:* Upon completion of the course, the student will be able to organize and manage a personnel section; validate and process personnel action documents to receive, transfer, promote, and evaluate staff; and operate via PC all functions of an automated personnel information system.

Instruction: *Version 1:* Lectures, class discussions, examinations, and practical exercises use a PC to complete and validate personnel documents, including classification, evaluation, promotion, and transitioning documents and to enter, edit, update, and generate reports from several automated systems supporting personnel processing. *Version 2:* Lectures, class discussions, examinations, and practical exercises use a PC to complete and validate personnel documents, including classification, evaluation, promotion, and transitioning documents and to enter, edit, update, and generate reports from several automated systems supporting personnel processing.

Credit Recommendation: *Version 1:* In the lower-division baccalaureate/associate degree category, 2 semester hours in records and information management and 2 in human resource management (9/97). *Version 2:* In the lower-division baccalaureate/associate degree category, 3 semester hours in information management (3/02).

Related Occupations: 75H.

AR-1406-0184

Recruiting Officer (Nurse Counseling Training) Phase 3

Course Number: 7C-F13, Phase 3.

Location: Soldier Support Institute, Fort Jackson, SC.

Length: 1 week (40 hours).

Exhibit Dates: 10/95–Present.

Learning Outcomes: Upon completion of the course, the student will be able to be a recruiting officer counseling trainer.

Instruction: Methods of instruction include practical exercises, conferences, and guest speakers.

Credit Recommendation: Credit is not recommended because of the military-specific nature of the course (9/97).

AR-1406-0185

Recruiting Advanced Noncommissioned Officer (NCO) Phase 2

Course Number: 5-79-C42 (REC); 5-79-C42, Phase 2.

Location: Soldier Support Institute, Fort Jackson, SC.

Length: 9 weeks (322 hours).

Exhibit Dates: 7/96–Present.

Learning Outcomes: Before 7/96 see AR-1406-0158. Upon completion of the course, the student will be able to prepare and/or evaluate written communications employ interpersonal communication skills and leadership/management strategies, develop and evaluate recruiting action plans and make recommendations, and apply appropriate sales and marketing techniques to recruiting.

Instruction: Methods of instruction used include lectures, discussions, small group instruction, and role playing production management, written communication, interpersonal communication, in management strategies, development of action plans, and proper use of sales and marketing techniques in recruiting. Course includes a common core of leadership subjects.

Credit Recommendation: In the lower-division baccalaureate/associate degree category, 3 semester hours in salesmanship, 1 in interpersonal communication, and 1 in human relations. See AR-1404-0035 for common core credit (9/97).

AR-1406-0186

Army National Guard 79T Advanced Noncommissioned Officer

Course Number: 501-79T4-ARNG.

Location: Professional Education Center, Cp. Joseph Robinson, North Little Rock, AR; Soldier Support Institute, Fort Jackson, SC.

Length: 5 weeks (163 hours).

Exhibit Dates: 1/96–Present.

Learning Outcomes: Upon completion of the course, the student will be able to use goal setting and sales techniques as a recruiter.

Instruction: Methods used include conference, demonstrations, small group exercises, and independent study covering the use of goal setting and sales techniques in recruiting. Course includes a common core of leadership subjects.

Credit Recommendation: In the lower-division baccalaureate/associate degree category, 2 semester hours in sales techniques. See AR-2201-0295 for common core credit (9/97).

Related Occupations: 79T.

AR-1406-0187

Army National Guard 79D Advanced Noncommissioned Officer

Course Number: 501-79D40-ARNG.

Location: Professional Education Center, Cp. Joseph Robinson, North Little Rock, AR; Soldier Support Institute, Fort Jackson, SC.

Length: 5 weeks (266 hours).

Exhibit Dates: 10/94–Present.

Learning Outcomes: Upon completion of the course, the student will be knowledgeable in common leadership and war fighting skills.

Instruction: Instruction emphasizes leadership, communication skills, training management, and common military skills. Course includes self-awareness and personality types, personal and professional growth techniques, reinforcement training analysis, interpersonal management skills, and discussion of attrition management programs and concerns. Course contains a common core of leadership subjects.

Credit Recommendation: Credit is recommended for the common core only. See AR-2201-0295 (9/97).

Related Occupations: 79D.

AR-1406-0188

79T Basic Course

Course Number: 501-79T20.

Location: Professional Education Center, Cp. Joseph Robinson, North Little Rock, AR.

Length: 6 weeks (238 hours).

Exhibit Dates: 10/96–Present.

Learning Outcomes: After 1/01 see AR-1406-0224. Upon completion of the course, the student will be able to perform recruiting and retention duties and responsibilities for a large organization including administration of recruiting and retention program, planning and executing recruiting programs, interviewing and classifying prospects, preparing and presenting recruiting/retention briefings, and ethics of recruiting and retention.

Instruction: Performance-based instruction significant homework assignments, lectures and laboratory exercises culminate in individual presentations. Extensive classwork and practical exercises in interviewing, stress management, staffing analysis, prospecting, market analysis and sales techniques are presented.

Credit Recommendation: In the lower-division baccalaureate/associate degree category, 3 semester hours in salesmanship, 3 in human resource management, 2 in interpersonal communication, and 1 in presentation skills (6/98).

AR-1406-0189

Unit Retention

Course Number: 501-U-ARNG.

Location: Professional Education Center, Cp. Joseph Robinson, North Little Rock, AR.

Length: 2 weeks (73 hours).

Exhibit Dates: 1/94–Present.

Learning Outcomes: Upon completion of the course, the student will be able to compile and analyze retention data prepare organizational attrition reports apply use incentives, and, interviewing and counseling techniques.

Instruction: Course is performance based with lectures and practical exercises on locating and quantifying retention figures, attrition cycles, incentive programs and preparing and conducting retention interviews and counseling.

Credit Recommendation: In the upper-division baccalaureate category, 3 semester hours in human resource management (6/98).

AR-1406-0190

Trainer Development

Course Number: QTC-020.

Location: Professional Education Center, Cp. Joseph Robinson, North Little Rock, AR.

Length: 2 weeks (80 hours).

Exhibit Dates: 10/95–Present.

Learning Outcomes: Upon completion of the course, the student will be able to serve as assistant instructor for quality awareness, meeting management, and process action team courses. Students will apply instructor/trainer skills in instructional methods, instructional design, and practical knowledge of process acting team concepts and statistical process control techniques.

Instruction: Performance-based instruction is conducted in small group seminars with assessment of individual performance in scenario-based exercises.

Credit Recommendation: In the lower-division baccalaureate/associate degree category, 3 semester hours in training development (6/98).

AR-1406-0191

Trainer Certification

Course Number: QTC-030.

Location: Professional Education Center, Cp. Joseph Robinson, North Little Rock, AR.

Length: 2 weeks (81 hours).

Exhibit Dates: 10/95–Present.

Learning Outcomes: Upon completion of the course, the student will be qualified to conduct quality training courses. Given a quality training course assignment, students will design, develop, and present the program of instruction and lesson plans for the course.

Instruction: Students in small facilitator/instructor teams design and instruct on quality training subjects selected from the meeting management, quality awareness, or process action team course. Evaluation is a performance assessment that, when successfully accomplished, results in a trainer certification.

Credit Recommendation: In the lower-division baccalaureate/associate degree category, 3 semester hours in instructional design and presentation skills (6/98).

AR-1406-0192

Readiness Noncommissioned Officer (NCO)/Noncommissioned Officer In-Charge (NCOIC)

Course Number: FT-020.

Location: Professional Education Center, Cp. Joseph Robinson, North Little Rock, AR.

Length: 2 weeks (69 hours).

Exhibit Dates: 6/95–Present.

Learning Outcomes: Upon completion of the course, the student will be able to plan training requirements for long-and-short term organizational staffing requirements.

Instruction: Performance-based instruction consists of classroom instruction and scenario-based practical exercises in long-and short-range training plans.

Credit Recommendation: In the upper-division baccalaureate category, 2 semester hours in training management (6/98).

AR-1406-0193

Facilitator

Course Number: QTC-010.

Location: Professional Education Center, Cp. Joseph Robinson, North Little Rock, AR.

Length: 2 weeks (80 hours).

Exhibit Dates: 6/95–Present.

Learning Outcomes: Upon completion of the course, the student will be able to facilitate meetings. Knowledge and skills include meeting management, team building, group dynamics, adult and experiential learning, communication, and conflict management.

Instruction: The course is taught in small group seminars with emphasis on student individual assignments and classroom discussion. Instructor serves to guide and facilitate the learning environment. Exams are performance-based evaluations of student exercises.

Credit Recommendation: In the lower-division baccalaureate/associate degree category, 3 semester hours in small group communication (6/98).

AR-1406-0194

OH-58 Aeroscout Instructor Pilot Methods of Instruction (Graduate)

Course Number: 2C-SIG2/SQIC (OH-58GRAD), Phase 3; 2C-SIG2/SQIC (OH-58GRAD), Phase 2.

Location: Aviation School, Fort Rucker, AL.

Length: 7–8 weeks (282–283 hours).

Exhibit Dates: 7/97–Present.

Learning Outcomes: Upon completion of the course, the student will be qualified to teach and evaluate flight instruction in the OH-58.

Instruction: Course includes flight and academic training and practical application of principles and techniques of flight instruction, academic instruction and practical application of the OH-58 aircraft systems, combat skills, and aeromedical factors. Sufficient depth of knowledge is gained to enable the graduate to teach both academic and flight instruction.

Credit Recommendation: In the upper-division baccalaureate category, 3 semester hours in flight training, 1 in methods of instruction, and 1 in helicopter systems (3/03).

AR-1406-0195

Fixed Wing Multi-Engine Methods of Instruction

Course Number: 2B-SIG2/SQIC (C-12 MOI) CT.

Location: Aviation School, Fort Rucker, AL.

Length: 5–6 weeks (214 hours).

Exhibit Dates: 12/97–Present.

Learning Outcomes: Upon completion of the course, the student will be qualified as an instructor for the Beech King Air twin turbine engine airplane.

Instruction: Classroom lectures, practical exercises, flight simulation instruction, and advanced flight instruction are used to teach electronic navigation, advanced airplane aerodynamics, principle of instruction, airplane systems, weight and balance, aircraft performance charts, and aeromedical factors.

Credit Recommendation: In the upper-division baccalaureate category, 3 semester hours in advanced airplane flight instructor methods and 3 in advanced twin turbine airplane flight training (11/98).

AR-1406-0196

AH-64A Instructor Pilot Methods of Instruction

Course Number: 2C-SIG2/SQIC (AH-64A MOI), Phase 2.
Location: Aviation School, Fort Rucker, AL.
Length: 9 weeks (367 hours).
Exhibit Dates: 1/98–Present.
Learning Outcomes: Before 1/98 see AR-1406-0121. Upon completion of the course, the student will be qualified as instructor in the AH-64A helicopter.
Instruction: Classroom lecture, practical exercise, and advanced flight instruction are used in this course to teach advanced helicopter aerodynamics, aerospace, principles of instruction, cockpit management, night vision devices, communications, electronic navigation, and aeromedical factors.
Credit Recommendation: In the upper-division baccalaureate category, 3 semester hours in advanced helicopter flight instructor methods and 3 in advanced helicopter flight training (11/98).

AR-1406-0197

CH-47D Instructor Pilot Methods of Instruction

Course Number: 2C-SIG2/SQIC (CH-47D), Phase 2; 2C-SIG2/SQIC (CH-47D).
Location: Aviation School, Fort Rucker, AL.
Length: 10 weeks (367 hours).
Exhibit Dates: 10/96–Present.
Learning Outcomes: Upon completion of the course, the student will be qualified as instructor pilot capable of teaching and evaluating academic and flight instruction in the CH-47D helicopter.
Instruction: The course consists of instructor pilot training, night vision goggle training, and flight and academic training in day and night visual flight rules tasks. The tasks include before flight, basic flight, approach/landings, and emergency tasks and academic instruction and practical application of CH-47D aircraft system subjects.
Credit Recommendation: In the upper-division baccalaureate category, 3 semester hours in advanced helicopter systems, 1 in methods of instruction, and 3 in advanced flight instruction (9/00).

AR-1406-0198

AH-1F Instructor Pilot Methods of Instruction (Graduate)

Course Number: 2C-SIG2/SQIC (AH-1GRAD).
Location: Aviation School, Fort Rucker, AL.
Length: 10–11 weeks (374 hours).
Exhibit Dates: 10/95–Present.
Learning Outcomes: Upon completion of the course, the student will be capable of conducting contact and combat skills flight instruction in the AH-1F.
Instruction: This course consists of flight and academic training in contact, tactics, night vision, and gunnery tasks. The course also includes academic instruction and practical application of principles and techniques of flight instruction and academic instruction and practical application of AH-1 aircraft systems, weapon systems, combat skills, and aeromedical factors. Sufficient depth of knowledge is gained to enable the graduate to present both academic and flight instruction in flight, weapons, and maintenance areas.
Credit Recommendation: In the upper-division baccalaureate category, 3 semester hours in flight training, 1 in methods of instruction, and 2 in helicopter systems (11/98).

AR-1406-0199

OH-58 Aeroscout Night Vision Device Instructor Pilot Methods of Instruction Initial Entry Rotary Wing

Course Number: 2C-SIG2/SQIC (OH58NVDIERW).
Location: Aviation School, Fort Rucker, AL.
Length: 3–4 weeks (104 hours).
Exhibit Dates: 3/98–Present.
Learning Outcomes: Upon completion of the course, the student will be able to provide flight and academic instruction on the aircraft night vision device systems.
Instruction: Classroom instruction and practical exercises cover night vision devices on the OH-58 helicopter. Course covers leadership and professional and military skills and maintenance management. Supply and logistic and tactical operations are stressed.
Credit Recommendation: Credit is not recommended because of the military-specific nature of the course (11/98).

AR-1406-0200

UH-1 Instructor Pilot Methods of Instruction

Course Number: 2C-SIG2-SQIC (UH-1), Phase 1; 2C-SIG2/SQIC (UH-1).
Location: Aviation School, Fort Rucker, AL.
Length: Self-paced, 7–8 weeks (271–274 hours).
Exhibit Dates: 7/95–Present.
Learning Outcomes: Upon completion of the course, the student will be able to conduct contact and combat skills flight instruction in the UH-1.
Instruction: The course consists of flight and academic training in day visual flight rules and maneuver tasks, including preflight, basic flight, approach/landing, emergency tasks, tactical operations, and post flight operations. Included also is academic instruction and practical application of principles and techniques of flight instruction, academic instruction, and practical application of UH-1 aircraft systems and maintenance subjects. Sufficient depth of knowledge is gained to enable graduates to teach/evaluate basic and tactical flight and academic subjects.
Credit Recommendation: In the upper-division baccalaureate category, 3 semester hours in flight training, 2 in methods of instruction, 1 in helicopter systems (9/00).

AR-1406-0201

AH-1 Night Vision Goggles Instructor Pilot Methods of Instruction (Graduate) Phase 2

Course Number: 2C-SIG2/SQIC (AH-1NVGGRAD).
Location: Aviation School, Fort Rucker, AL.
Length: 1–2 weeks (57 hours).
Exhibit Dates: 10/92–Present.
Learning Outcomes: Upon completion of the course, the student will have an understanding of the basics of night vision physiology, night flight operations, and night vision equipment. The student will also be qualified to provide night vision flight instruction.
Instruction: Classroom and practical exercises cover night vision flight training and operations, night vision physiological characteristics, mission planning, and night vision devices equipment.
Credit Recommendation: Credit is not recommended because of the military-specific nature of the course (11/98).

AR-1406-0202

OH-58 Aeroscout Night Vision Device Instructor Pilot Methods of Instruction Graduate

Course Number: 2C-SIG2/SQIC (OH58NVDGRAD).
Location: Aviation School, Fort Rucker, AL.
Length: 3–4 weeks (104 hours).
Exhibit Dates: 10/92–Present.
Learning Outcomes: Upon completion of the course, the student will be qualified to evaluate night vision flight instruction and provide night vision-related flight instruction in the OH-58 helicopter. The student will also be able to describe the basics of night vision physiology, night flight mission planning, and night vision device requirement.
Instruction: Classroom and practical exercises cover night vision flight training and operations, flight evaluation and critique, night terrain interpretation, mission planning, and night vision devices equipment.
Credit Recommendation: Credit is not recommended because of the military-specific nature of the course (11/98).

AR-1406-0203

Rotary Wing Instrument Instructor Pilot Methods of Instruction

Course Number: 2C-F54 (RWIIP MOI) (CT), Phase 1; 2C-F54(RWIIP MOI) (CT).
Location: Aviation School, Fort Rucker, AL.
Length: 4–5 weeks (174 hours).
Exhibit Dates: 1/96–Present.
Learning Outcomes: Upon completion of the course, the student will be qualified as instrument instructor of military students. Graduates are able to teach basic and advanced instrument procedures to non rated rotary wing student pilots under simulated and actual conditions.
Instruction: Course consists of academic, simulator, and TH-67 flight training in day instrument flight rules, flight procedures, instrument instruction techniques, use of the TH-67 instrument package, and student management techniques.
Credit Recommendation: In the upper-division baccalaureate category, 2 semester hours in advanced flight instruction, 2 in methods of instruction, and 2 in advanced helicopter systems (2/03).

AR-1406-0204

UH-1 Night/Night Vision Goggle Instructor Pilot Methods of Instruction

Course Number: 2C-SIG2/SQIC (UH-1 N/NVG), Phase 3; 2C-SIG2/SQIC (UH-1 N/NVG), Phase 2; 2C-SIG2/SQIC (UH-1 N/NVG).

Location: Aviation School, Fort Rucker, AL.

Length: 4 weeks (114–115 hours).

Exhibit Dates: 10/95–Present.

Learning Outcomes: Upon completion of the course, the student will use basics of night vision physiology, night flight operations, and night vision device equipment to provide night vision flight instruction.

Instruction: This course includes principles of instruction, aerodynamics, aircraft systems, night tactics, aeromedical factors, weapon systems, combat skills, and appropriate flight training.

Credit Recommendation: Credit is not recommended because of the military-specific nature of the course (3/03).

AR-1406-0205

OH-58 Aeroscout Instructor Pilot Methods of Instruction (Initial Entry Rotary Wing)

Course Number: 2C-SIG2/SQIC (OH-58IERW), Phase 2; 2C-SIG2/SQIC (OH58IERW).

Location: Aviation School, Fort Rucker, AL.

Length: 7–8 weeks (281–284 hours).

Exhibit Dates: 10/96–Present.

Learning Outcomes: Upon completion of the course, the student will be able to evaluate flight instruction and provide instruction in helicopter flight, aircraft systems, aeromedical factors, and night flight in the OH-58 helicopter.

Instruction: Classroom and practical training cover day/night flying, principles and techniques of flight instruction, aircraft systems, combat skills, aeromedical factors, and night vision goggles.

Credit Recommendation: In the upper-division baccalaureate category, 4 semester hours in advanced aerodynamics and theory of flight, 4 in advanced helicopter flight and systems, and 3 in principles and methods of instruction (1/03).

AR-1406-0206

CH-47D Instructor Pilot

Course Number: 2C-SIG2/2C-SQIC (CH-47D), Phase 3; 2C-SIG2/2C-SQIC (CH-47D).

Location: Aviation School, Fort Rucker, AL.

Length: 8 weeks (360–367 hours).

Exhibit Dates: 10/96–Present.

Learning Outcomes: Upon completion of the course, the student will be capable of teaching and evaluating academic and flight instruction in the CH-47D helicopter.

Instruction: The course consists of CH-47D instructor pilot training and night vision goggles training, flight and academic training in day/night visual flight rules tasks. The tasks include before-flight, basic flight, approach/landings, and emergency tasks; academic instruction; and practical application of CH-47D aircraft system subjects.

Credit Recommendation: In the upper-division baccalaureate category, 3 semester hours in helicopter systems, 3 in methods of instruction, and 3 in advanced flight training (10/00).

AR-1406-0207

AH-64A Instructor Pilot

Course Number: 2C-SIG2/2C-SQIC (AH-64A).

Location: Aviation School, Fort Rucker, AL.

Length: 7 weeks (296–297 hours).

Exhibit Dates: 1/98–Present.

Learning Outcomes: Upon completion of the course, the student will be capable of teaching and evaluating flight training in the AH-64A.

Instruction: The course consists of flight and academic instruction to qualify rotary-wing aviators who have completed qualification training in the AH-64 in methods of instruction to teach and evaluate AH-64A contact flight, night vision systems, weapon systems, and the combat mission simulator. Applied aerodynamics, regulations, flight planning, and arrival phases of an instrument flight procedure are stressed.

Credit Recommendation: In the upper-division baccalaureate category, 3 semester hours in helicopter systems and 3 in methods of instruction (11/98).

AR-1406-0208

AH-1F Instructor Pilot

Course Number: 2C-SIG2/2C-SQIC (AH-1F).

Location: Aviation School, Fort Rucker, AL.

Length: 10 weeks (367 hours).

Exhibit Dates: 1/90–Present.

Learning Outcomes: Upon completion of the course, the student will be qualified as an instructor for the AH-1F helicopter.

Instruction: Classroom lectures, practical exercise, and advanced flight instruction are used in this course to teach advanced helicopter aerodynamics, aerospace, principles of instruction, crew coordination, communications, electronic navigation, and aeromedical factors.

Credit Recommendation: In the upper-division baccalaureate category, 3 semester hours in advanced helicopter flight instructor methods and 3 in advanced helicopter flight training (11/98).

AR-1406-0209

Fixed Wing Multi-engine Instructor Pilot

Course Number: 2B-SIG2/2B-SQIC (CT), Phase 2; 2B-SIG2/SQIC (CT).

Location: Aviation School, Fort Rucker, AL.

Length: 4–5 weeks (214 hours).

Exhibit Dates: 10/96–Present.

Learning Outcomes: Upon completion of the course, the student will be able to apply theory of flight; aerodynamics; regulations; instructing fundamentals; safety; navigation systems; aircraft systems; and methods of instruction for contact, instrument, and emergency procedures to multi-engine pilot training.

Instruction: This course qualifies pilots to conduct transition and instructor pilot training in the multi-engine airplane.

Credit Recommendation: In the upper-division baccalaureate category, 3 semester hours in multi-engine instructor pilot (1/03).

AR-1406-0210

UH-60 Instructor Pilot

Course Number: 2C-SIG2/2C-SQIC (UH-60), Phase 1; 2C-SIG2/2C-SQIC (UH-60).

Location: Aviation School, Fort Rucker, AL.

Length: 8 weeks (309 hours).

Exhibit Dates: 10/98–Present.

Learning Outcomes: Upon completion of the course, the student will be able to give flight instruction in a twin-engine helicopter such as the UH-60. The student will be able to give ground and/or flight instruction in theory of flight, fundamentals of instruction, safety, helicopter systems, and helicopter flight procedures.

Instruction: The course uses lecture, conference, practical exercise, and flight instruction. The course covers advanced aviation ground and flight topics, methods of instructions, and advanced flight training.

Credit Recommendation: In the lower-division baccalaureate/associate degree category, 2 semester hours in helicopter systems. In the upper-division baccalaureate category, 3 semester hours in advanced theory of flight and 3 in advanced helicopter flight training (10/00).

AR-1406-0211

UH-1 Instructor Pilot

Course Number: 2C-SIG2/2C-SQIC (UH-1).

Location: Aviation School, Fort Rucker, AL.

Length: 8 weeks (278–284 hours).

Exhibit Dates: 10/97–Present.

Learning Outcomes: Upon completion of the course, the student will be qualified to teach transition and standardization training in the UH-1 helicopter. The student will also be able to teach day/night visual flight rules including before-flight, basic flight, approach/landing, and emergency tasks; academic and practical application of UH-1; and systems and maintenance subjects.

Instruction: Lectures and practical training cover UH-1 aircraft systems, aviation subjects, aeromedical subjects, tactical academics, flight training, night academics, and course critique.

Credit Recommendation: In the upper-division baccalaureate category, 3 semester hours in principles and methods of instruction (5/01).

AR-1406-0212

Night (N) Night Vision Goggles Instructor

Course Number: 2C-F37, Phase 1; 2C-F37.

Location: Aviation School, Fort Rucker, AL.

Length: 4 weeks (116 hours).

Exhibit Dates: 1/95–Present.

Learning Outcomes: Upon completion of the course, the student will apply to night vision flight instruction, the basics of night vision physiology, night flight mission planning, and night vision device equipment.

Instruction: Classroom and practical exercises cover night vision flight training and operations, night vision physiological characteristics, mission planning, and night vision devices equipment.

Credit Recommendation: Credit is not recommended because of the limited, specialized nature of the course (2/03).

AR-1406-0213

Total Army Training System (TATS) Fire Support Sergeant Basic Noncommissioned Officer (NCO)

Course Number: 061-13F30 (F); 061-13F30 (F), Phase 3.

Location: Reserve Schools, Multiple locations.

Length: 108–143 hours.

Exhibit Dates: 10/97–Present.

Learning Outcomes: Upon completion of the course, the student will be able to lead, train, and direct a fire support team in requesting and coordinating the delivery of artillery fire and supervise the maintenance of associated equipment.

Instruction: Lectures and practical training relate to leadership, maintenance, tactics, land navigation, weapons, fire support, and coordination.

Credit Recommendation: In the lower-division baccalaureate/associate degree category, 3 semester hours for field experience in management and 1 in map reading (1/01).

AR-1406-0215

Total Army Training System (TATS) Fire Support Sergeant Basic Noncommissioned Officer (NCO)

Course Number: *Version 1:* 250-13F30 (F). *Version 2:* 250-13F30.

Location: Field Artillery School, Fort Sill, OK.

Length: *Version 1:* 5–6 weeks (244 hours). *Version 2:* 2–3 weeks (143 hours).

Exhibit Dates: *Version 1:* 10/97–9/00. *Version 2:* 10/00–Present.

Learning Outcomes: *Version 1:* Before 10/97 see AR-2201-0256. Upon completion of the course, the student will be able to lead, train, and direct a fire support team in requesting the delivery of artillery fire and supervise the maintenance of associated equipment. *Version 2:* Upon completion of the course, the student will be able to lead, train, and direct a fire support team in requesting the delivery of artillery fire and supervise the maintenance of associated equipment.

Instruction: *Version 1:* Lectures and practical training relate to leadership, maintenance, tactics, land navigation, weapons, fire support, and coordination. This course includes a common core of leadership subjects. *Version 2:* Lectures and practical training relate to leadership, maintenance, tactics, land navigation, weapons, fire support, and coordination. This course includes a common core of leadership subjects.

Credit Recommendation: *Version 1:* In the lower-division baccalaureate/associate degree category, 3 semester hours for field experience in management and 1 in map reading. See AR-1406-0090 for common core credit (1/01). *Version 2:* In the lower-division baccalaureate/associate degree category, 3 semester hours for field experience in management and 1 in map reading. See AR-2201-0337 for common core credit (1/01).

Related Occupations: 13F.

AR-1406-0217

Instructor Training

Course Number: 2E-SI5K/SQI8/011-SQIH.

Location: John F. Kennedy Special Warfare Center and School, Fort Bragg, NC.

Length: 2 weeks (77 hours).

Exhibit Dates: 6/96–Present.

Learning Outcomes: Upon completion of the course, the student will be able to describe the basic techniques of instruction, including preparation, classroom management, and implementation; demonstrate academic counseling and communication skills; describe the responsibilities of an effective instructor, the essential conditions for learning, and the principles of development for effective lesson plans; make instructional decisions; develop instructional aids; and deliver instruction in a variety of methods.

Instruction: Course delivery is through lectures, discussions, demonstrations, role playing, counseling situations, and practice, and the presentation of instruction is followed by critique. After exposure to fundamentals and techniques of planning, management, communication skills, and implementation of instruction, participants apply these through supervised practice.

Credit Recommendation: In the lower-division baccalaureate/associate degree category, 2 semester hours in instructional strategies and 1 in speech communication (9/03).

AR-1406-0218

Air Defense Artillery Advanced Noncommissioned Officer (NCO)

Course Number: 0-14-C42 (TATS), Phase 2; 0-16-C42 (TATS).

Location: Air Defense Artillery School, Fort Bliss, TX.

Length: 152–171 hours.

Exhibit Dates: 4/99–Present.

Learning Outcomes: Upon completion of the course, the student will be able to lead, train, and direct subordinates to maintain, operate, and employ weapons on the battlefield and to employ fighting techniques.

Instruction: Lectures and practical exercises cover maintenance procedures, weapon systems, firing positions, and tactical maneuvering.

Credit Recommendation: Credit is not recommended because of military-specific nature of the course (4/01).

AR-1406-0219

Total Army Training System (TATS) 98 CMF Military Intelligence Advanced Noncommissioned Officer (NCO)

Course Number: 2-98-C42.

Location: Intelligence School, Fort Huachuca, AZ.

Length: 9–10 weeks (354 hours).

Exhibit Dates: 10/98–Present.

Learning Outcomes: Upon completion of the course, the student will be able to lead, manage, supervise, and train intelligence personnel to perform the tactical and strategic intelligence tasks in a military intelligence unit and prepare and communicate plans and reports in written and oral form.

Instruction: Instruction includes classroom and practical exercises in leadership, management, administration, and communication skills. Course includes a common core of leadership subjects.

Credit Recommendation: In the lower-division baccalaureate/associate degree category, 3 semester hours in technical writing and 1 in human resource management. See AR-1404-0035 for common core credit (8/01).

AR-1406-0220

Total Army Training System (TATS) 98 CMF Military Intelligence Advanced Noncommissioned Officer (NCO)

Course Number: 2-98-C42.

Location: Reserve Schools, Multiple locations; Intelligence School, Fort Huachuca, AZ.

Length: 114–115 hours.

Exhibit Dates: 10/98–Present.

Learning Outcomes: Upon completion of the course, the student will be able to lead, manage, supervise, and train intelligence personnel to perform the tactical and strategic intelligence tasks in a military intelligence unit and prepare and communicate plans and reports in written and oral form.

Instruction: Instruction includes classroom and practical exercises in leadership, management, administration, and communication skills.

Credit Recommendation: In the lower-division baccalaureate/associate degree category, 3 semester hours in technical writing and 1 in human resource management (8/01).

AR-1406-0221

Total Army Training System (TATS) 33 CMF Military Intelligence Advanced Noncommissioned Officer (NCO)

Course Number: 1-33-C42.

Location: Intelligence School, Fort Huachuca, AZ.

Length: 6–12 weeks (245–411 hours).

Exhibit Dates: 6/01–Present.

Learning Outcomes: Before 10/01 see AR-1406-0222. Upon completion of the course, the student will be able to use a computer in an office environment, supervise personnel, assign duties, and plan and organize intelligence activities.

Instruction: This course consists of classroom instruction and hands-on experience in basic computer usage and testing procedures to validate the students skills. This course contains a common core of leadership subjects.

Credit Recommendation: In the lower-division baccalaureate/associate degree category, 3 semester hours in basic management skills. See AR-1404-0035 for common core credit (12/02).

AR-1406-0222

Total Army Training System (TATS) CMF 33 Military Intelligence Advanced Noncommissioned Officer (NCO)

Course Number: 1-33-C42.

Location: Reserve Schools, Multiple locations; Intelligence School, Fort Huachuca, AZ.

Length: 240–241 hours.
Exhibit Dates: 6/01–9/01.
Learning Outcomes: After 9/01 see AR-1406-0221. Upon completion of the course, the student will be able to use a computer in an office environment, supervise personnel, assign duties, and plan and organize intelligence activities.
Instruction: This course consists of classroom instruction and hands-on experience in basic computer use and testing procedures to validate the students skills.
Credit Recommendation: In the lower-division baccalaureate/associate degree category, 3 semester hours in basic management skills (8/01).

AR-1406-0223

Total Army Training System (TATS) 96 CMF Military Intelligence Noncommissioned Officer (NCO) Advanced

Course Number: 2-96-C42.
Location: Reserve Schools, Multiple locations.
Length: 114–115 hours.
Exhibit Dates: 10/99–Present.
Learning Outcomes: Upon completion of the course, the student will be able to lead, manage, supervise, and train intelligence personnel to perform the tactical and strategic intelligence tasks in a military intelligence unit and prepare and communicate plans and reports in written and oral form.
Instruction: Instruction includes classroom and practical exercises in leadership, management, administration, and communication skills.
Credit Recommendation: In the lower-division baccalaureate/associate degree category, 3 semester hours in technical writing. In the upper-division baccalaureate category, 1 semester hour in leadership and 2 in human resource management (8/01).

AR-1406-0224

Recruiting Advanced Noncommissioned Officer (NCO), Phase 2

Course Number: 5-79R-C42 (REC), Phase 2.
Location: Recruiting and Retention School, Fort Jackson, SC.
Length: 2 weeks (76–77 hours).
Exhibit Dates: 2/01–Present.
Learning Outcomes: Before 2/01 see AR-1406-0188. Upon completion of the course, the student will be able to prepare and/or evaluate written communication, employ interpersonal communication skills and leadership/management strategies, develop and evaluate recruiting action plans and make recommendations, and apply appropriate sales and marketing techniques to recruiting.
Instruction: Methods of instruction used include lectures, discussions, small group instruction and role playing, production management, written communication, interpersonal communication in management strategies, development of action plans, and proper use of sales and marketing techniques in recruiting.
Credit Recommendation: In the upper-division baccalaureate category, 2 semester hours in personal selling, or human resources management, or interpersonal communication (3/02).

AR-1406-0225

Military Personnel Technician Warrant Officer Basic

Course Number: 7C-420A.
Location: Adjutant General School, Fort Jackson, SC.
Length: 6–7 weeks (262 hours).
Exhibit Dates: 10/00–Present.
Learning Outcomes: Upon completion of the course, the student will be able to apply personnel principles, practices, and procedures.
Instruction: Methods of instruction include lectures, discussions, videotapes, and group discussions. Topics covered include personnel activities such as recruiting, selecting, staffing, promotion, and termination, as well as personnel policies and procedures.
Credit Recommendation: In the upper-division baccalaureate category, 3 semester hours in military science and 3 in supervision, or management (3/02).
Related Occupations: 420A.

AR-1406-0226

Recruiter Trainer Noncommissioned Officer (NCO)

Course Number: 501-F19.
Location: Recruiting and Retention School, Fort Jackson, SC.
Length: 2 weeks (75–76 hours).
Exhibit Dates: 7/02–Present.
Learning Outcomes: Upon completion of the course, the student will be able to prospect, identify, and recruit Army personnel using accepted selling techniques.
Instruction: Methods of instruction include classroom exercises, practical exercises, and lectures.
Credit Recommendation: In the upper-division baccalaureate category, 2 semester hours in personal selling, or 2 in human resources management, or training development (3/02).

AR-1406-0227

Recruiting Company Commander

Course Number: 7C-F40.
Location: Recruiting and Retention School, Fort Jackson, SC.
Length: 3 weeks (117 hours).
Exhibit Dates: 8/01–Present.
Learning Outcomes: Upon completion of the course, the student will be able to provide leadership in the recruitment mission, including sales, time management, and personnel recruitment.
Instruction: Methods of instruction include classroom exercises, practical exercises, discussions, and lectures.
Credit Recommendation: In the lower-division baccalaureate/associate degree category, 2 semester hours in leadership and 2 in sales management (3/02).

AR-1406-0228

Total Army Training System (TATS) Career Counselor

Course Number: 501-79S30.
Location: Recruiting and Retention School, Fort Jackson, SC.
Length: 8 weeks (361 hours).
Exhibit Dates: 5/98–Present.
Learning Outcomes: Before 5/98 see AR-1406-0159. Upon completion of the course, the student will be able to interpret personnel records, assist personnel in the reenlistment process, determine reenlistment or extension eligibility, prepare and present retention training, and conduct retention interview/counseling sessions.
Instruction: Methods of instruction include lectures, discussions, guest speakers, practical exercises. Topics covered include evaluation of personnel files, retention processing, interviewing techniques, retention identification, counseling, and follow-up.
Credit Recommendation: In the upper-division baccalaureate category, 3 semester hours in counseling or career counseling and 3 in human resources management (3/02).
Related Occupations: 79S.

AR-1406-0229

Guidance Counselor Operations

Course Number: 501-ASIV7.
Location: Recruiting and Retention School, Fort Jackson, SC.
Length: 4 weeks (167 hours).
Exhibit Dates: 7/97–Present.
Learning Outcomes: Before 7/97 see AR-1406-0160. Upon completion of the e course, the student will be able to interview prospective enlistees, communicate effectively, determine eligibility, and prepare enlistment documents.
Instruction: Topics covered include counseling, record keeping, personal selling techniques, and marketing concepts.
Credit Recommendation: In the lower-division baccalaureate/associate degree category, 3 semester hours in records management and 3 in counseling. In the upper-division baccalaureate category, 1 semester hour in personal selling/sales (3/02).

AR-1406-0230

Total Army Instructor Training

Course Number: 8E-SI5K/SQI8/551-SQIH.
Location: Quartermaster Center and School, Fort Lee, VA.
Length: 2–3 weeks (96 hours).
Exhibit Dates: 6/99–Present.
Learning Outcomes: Upon completion of the course, the student will be able to describe the basic techniques of instruction including preparation, classroom management and implementation, demonstrate academic counseling and communication skills, describe the responsibility of an effective instructor, the essential conditions for learning and development of lesson plans, make instructional decisions, develop instructional aids, and deliver instruction in a variety of methods.
Instruction: Audio visual materials, classroom and practical exercises, lecture and learner presentations. Topics include foundation of instruction, counseling sessions, power point, five minute, thirty minute and fifty minute teaching exercises followed by peer critique, and lesson plan preparation.
Credit Recommendation: In the lower-division baccalaureate/associate degree category, 3 semester hours in oral communications or 3 in instructional techniques (6/02).

AR-1406-0231

Instructor Training

Course Number: DEV-2, Phase 1.
Location: Western Hemisphere Institute for Security Cooperation (WHINSEC), Fort Benning, GA.
Length: 2–3 weeks (85 hours).
Exhibit Dates: 9/02–Present.
Learning Outcomes: Upon completion of the course, the student will be able to provide theory and practice in the concepts, methods, techniques, and technology of performance-oriented training.
Instruction: Topics include a systems approach to instructor training that emphasizes analysis, design, and appropriate instructional strategies. The course focuses on developing plans and tests to measure learning.
Credit Recommendation: In the upper-division baccalaureate category, 3 semester hours in methods of instruction (8/02).

AR-1406-0232

Aviation Tactical Operations Officer

Course Number: 2C-SQII.
Location: Aviation Center, Fort Rucker, AL.
Length: 3 weeks (115 hours).
Exhibit Dates: 3/03–Present.
Learning Outcomes: Upon completion of the course, the student will be able to perform the duties of an aviation tactical operations officer involved in training and management, joint operations and supervising battlefield operations.
Instruction: Audio visual materials, practical exercises, discussion, lecture, and computer-based training. Topics include use of digital equipment and maps to obtain information from multi-service electronic combat bulletin boards, air tasking, current threat information; management of Army battle command system; and joint operations and training to supervise battlefield operations.
Credit Recommendation: In the upper-division baccalaureate category, 2 semester hours in logistics management (4/03).

AR-1406-0233

AH-64D Instructor Pilot MOI

Course Number: 2C-SIG2/SQIC (AH-64D MOI), Phase 1.
Location: Aviation Center, Fort Rucker, AL.
Length: 10 weeks (440 hours).
Exhibit Dates: 10/02–Present.
Learning Outcomes: Upon completion of the course, the student will be able to qualify as an instructor pilot and be capable of teaching and evaluating advanced helicopter flight training for AH-64D.
Instruction: Audio visual materials, practical and classroom exercises, laboratory, discussion, lecture, and computer-based training. Topics include instructional methods, flight training, and aircraft systems operation and management for AH-64D helicopter.
Credit Recommendation: In the upper-division baccalaureate category, 3 semester hours in advanced aircraft systems operations and management and 6 in instructional methods (4/03).

AR-1406-0234

AH-64D Instructor Pilot MOI

Course Number: 2C-SIG2/2C-SQIC (MOI), Phase 1.
Location: Aviation Center, Fort Rucker, AL.
Length: 10 weeks (439 hours).
Exhibit Dates: 10/02–Present.
Learning Outcomes: Upon completion of the course, the student will be able to teach and evaluate flight training in AH-64D helicopter.
Instruction: Audio visual materials, practical and classroom exercises, discussion, laboratory, lecture, computer-based training, and simulator. Topics include fundamentals of instruction of effective communication; aviation subjects such as regulations aerodynamics in helicopter AH-64D, aircraft electrical, fuel, pressurized air, hydraulic, flight controls, data and radio communication, navigation, drive train and utility systems; aeromedical subjects to cover disorientation, stress and fatigue, night vision and noise in aviation systems; weapons and sight/sensor systems; and flight training.
Credit Recommendation: In the upper-division baccalaureate category, 3 semester hours in helicopter systems and 6 in methods of instruction (4/03).

AR-1406-0235

AH-64D Instructor Pilot

Course Number: 2C-SIG2/2C-SQIC (AH-64D).
Location: Aviation Center, Fort Rucker, AL.
Length: 10 weeks (423 hours).
Exhibit Dates: 2/00–Present.
Learning Outcomes: Upon completion of the course, the student will be able to qualify as an instructor pilot and be capable of teaching and evaluating advanced helicopter flight training for AH-64D.
Instruction: Audio visual materials, practical, and classroom exercises, discussion, laboratory, lecture, and computer-based training. Topics include instructional methods, flight training, and aircraft systems operations and management for AH-64D helicopter.
Credit Recommendation: In the upper-division baccalaureate category, 3 semester hours in advanced aircraft systems operations and management, and 6 in instructional methods (4/03).

AR-1406-0236

Fixed Wing IFE Qualification

Course Number: 011-2C-15A/155A, Phase 1.
Location: Aviation Center, Fort Rucker, AL.
Length: 1–2 weeks (61 hours).
Exhibit Dates: 6/99–Present.
Learning Outcomes: Upon completion of the course, the student will be knowledgeable in subjects common to all fixed wing instrument flights.
Instruction: Audio visual materials, discussion, classroom exercises, lecture, and computer-based training. Topics include regulations for Army aircraft, air traffic control, radio navigation, instrument approach criteria, weather, flight information publications, flight planning, and TERPs.
Credit Recommendation: In the upper-division baccalaureate category, 1 semester hour in flight instructor ground school (4/03).

AR-1406-0237

Fixed Wing Multi-Engine IP (USAF)

Course Number: 2C-F79 (OS) (CT).
Location: Aviation Center, Fort Rucker, AL.
Length: 2 weeks (57 hours).
Exhibit Dates: 10/99–Present.
Learning Outcomes: Upon completion of the course, the student will be able to perform the duties as an instructor pilot for the C-12 fixed wing multi-engine aircraft.
Instruction: Audiovisual materials, practical exercises, laboratory, lecture, and computer-based training. Topics include methods of instruction, flight training, and C-12 aircraft systems.
Credit Recommendation: In the upper-division baccalaureate category, 1 semester hour methods of instructions (4/03).

AR-1406-0238

Primary Instructor Pilot MOI (CIV Contractor)

Course Number: 2C-SIG2/SQIC (TH-67) (CT), Phase 1.
Location: Aviation Center, Fort Rucker, AL.
Length: 5–6 weeks (199 hours).
Exhibit Dates: 6/96–Present.
Learning Outcomes: Upon completion of the course, the student will be able to qualify as an instructor pilot and be capable of teaching and evaluating primary VFR helicopter flight training.
Instruction: Audio visual training, practical and classroom exercises, discussion, laboratory, lecture, and computer-based training. Topics include instructional methods for VFR operations including pre-flight, enroute flight, approach and landings, and emergency procedures; instruction and techniques of flight instruction; and aircraft systems and maintenance fundamentals.
Credit Recommendation: In the upper-division baccalaureate category, 3 semester hours in methods of instruction, and 1 in advanced aircraft systems operation and management (4/03).

AR-1406-0239

OH-58D Warrior Instructor Pilot

Course Number: 2C-SIG2/2C-SQIC (OH58D), Phase 2.
Location: Aviation Center, Fort Rucker, AL.
Length: 10 weeks (450 hours).
Exhibit Dates: 10/99–Present.
Learning Outcomes: Upon completion of the course, the student will be able to qualify as instructor pilot and be capable of teaching and evaluating advanced helicopter flight training.
Instruction: Audio visual materials, practical and classroom exercises, discussion, laboratory, lecture, computer-based training. Topics include instructional methods, aircraft systems operation and management, and flight training.
Credit Recommendation: In the upper-division baccalaureate category, 3 semester hours in advanced aircraft systems operation and management, and 6 in instructional methods (4/03).

AR-1406-0240

OH-58D (R) Warrior Instructor Pilot

Course Number: 2C-SIG2/SQIC (OH-58D) (R).
Location: Aviation Center, Fort Rucker, AL.
Length: 11 weeks (482 hours).
Exhibit Dates: 12/99–Present.
Learning Outcomes: Upon completion of the course, the student will be able to qualify as an instructor pilot and be capable of teaching

and evaluating advanced helicopter flight training.

Instruction: Audio visual training, practical and classroom exercises, discussion, laboratory, lecture, and computer-based training. Topics include instructional methods, aircraft systems operation and management, and flight training.

Credit Recommendation: In the upper-division baccalaureate category, 3 semester hours in advanced aircraft systems operation and management, and 6 in instructional methods (4/03).

AR-1406-0241

OH-58D (R) Warrior Instructor Pilot MOI

Course Number: 2C-SIG2/SQIC (OH58D/R MOI).

Location: Aviation Center, Fort Rucker, AL.

Length: 12 weeks (512 hours).

Exhibit Dates: 12/99–Present.

Learning Outcomes: Upon completion of the course, the student will be able to qualify as an instructor pilot and be capable of teaching and evaluating advanced helicopter flight training.

Instruction: Audio visual materials, practical and classroom exercises, discussion, laboratory, lecture, and computer-based training. Topics include instructional methods, aircraft systems operation and management, and flight training.

Credit Recommendation: In the upper-division baccalaureate category, 3 semester hours in advanced aircraft systems operation and management, and 6 in instructional methods (4/03).

AR-1406-0243

Systems Approach to Training

Course Number: 131-SAT-1.

Location: Ordnance Center and School, Aberdeen Proving Ground, MD.

Length: 2 weeks (80 hours).

Exhibit Dates: 10/01–Present.

Learning Outcomes: Upon completion of the course, the student will be able to prepare instructional materials, learn task analysis, and teach by objectives with an emphasis on test construction.

Instruction: Lectures, demonstrations, and projects. Topics include recognizing instructional tasks and objectives and developing appropriate validated education tests to promote the Army's System Approach to Training (SAT).

Credit Recommendation: In the lower-division baccalaureate/associate degree category, 3 semester hours in educational test construction (8/03).

AR-1406-0245

Total Army Training Systems (TATS) Automated Fire Support Systems Senior Sergeant ANCOC

Course Number: 061-13C40 (F), Phase 3.

Location: Field Artillery School, Fort Sill, OK.

Length: 104 hours.

Exhibit Dates: 10/00–Present.

Learning Outcomes: Upon completion of the course, the student will be able to perform automated fire support systems senior sergeant tasks in a field artillery technical data systems equipped unit.

Instruction: Audiovisual materials, practical exercises, lectures, and individualized self paced. Topics include technical writing, map drawing, leadership, resource management, military skills, operations, intelligence and war fighting skills, basic supervision, and oral communications.

Credit Recommendation: In the lower-division baccalaureate/associate degree category, 1 semester hour in military science, 1 in map preparation, and 3 in personnel supervision (10/03).

AR-1407-0003

Advanced Noncommissioned Officer (NCO) MOS 71D/E

Course Number: 7-71-C42.

Location: Institute for Personnel and Resource Management, Fort Benjamin Harrison, IN.

Length: 5–17 weeks (168–444 hours).

Exhibit Dates: 3/81–12/91.

Objectives: To provide selected personnel with advanced leadership skills, military subjects, and communication skills for effective supervision of specific MOS tracks.

Instruction: Lectures and practical exercises include a basic core, including written and oral communication, management principles, unit training, and combat survival. MOS options are recruitment and reenlistment, equal opportunity, club management, legal topics, chaplain assisting, and data processing.

Credit Recommendation: In the lower-division baccalaureate/associate degree category, 1 semester hour in paralegal training. See AR-1404-0035 for common core credit (3/85).

Related Occupations: 71D; 71E.

AR-1407-0004

Advanced Noncommissioned Officer (NCO) MOS 71C

Course Number: 7-71-C42.

Location: Institute for Personnel and Resource Management, Fort Benjamin Harrison, IN.

Length: 5–17 weeks (168–444 hours).

Exhibit Dates: 3/81–9/91.

Objectives: To provide selected personnel with advanced leadership skills, military subjects, and communication skills for effective supervision of specific MOS tracks.

Instruction: Lectures and practical exercises include a basic core, including written and oral communication, management principles, unit training, and combat survival. MOS options are recruitment and reenlistment, equal opportunity, club management, legal topics, chaplain assisting, and data processing.

Credit Recommendation: Credit is not recommended because of the limited content of the course (3/85).

Related Occupations: 71C.

AR-1407-0005

Advanced International Morse Code (IMC)

Course Number: 011-ASIA4 (CMF 18).

Location: John F. Kennedy Special Warfare Center and School, Fort Bragg, NC.

Length: 8 weeks (296 hours).

Exhibit Dates: 10/86–2/98.

Learning Outcomes: Upon completion of the course, the student will be able to copy and send international Morse code at a speed of 13 five-letter code groups per minute.

Instruction: Instruction includes practice in receiving and sending international Morse code.

Credit Recommendation: In the vocational certificate category, 3 semester hours in Morse code techniques (12/88).

AR-1407-0006

Court Reporter Noncommissioned Officer (NCO) Advanced

Course Number: 512-71E40.

Location: Soldier Support Institute, Fort Benjamin Harrison, IN.

Length: 5 weeks (180 hours).

Exhibit Dates: 10/87–12/91.

Learning Outcomes: Upon completion of the course, the student will be able to describe various styles of leadership, use feedback to modify activities, evaluate progress, and apply counseling techniques.

Instruction: Methods of instruction include lectures and group discussions. Topics include leadership, motivation, counseling, communication, and discipline. This course contains a common core of leadership subjects.

Credit Recommendation: Credit is recommended for the common core only. See AR-1404-0035 (12/88).

Related Occupations: 71D; 71E.

AR-1407-0007

1. Legal Specialist Noncommissioned Officer (NCO) Advanced
2. Legal Specialist Advanced Noncommissioned Officer (NCO)
3. Paralegal Specialist Advanced Noncommissioned Officer (NCO)

(71D Advanced Noncommissioned Officer (NCO))

Course Number: *Version 1:* 512-71D/40/50; 512-71D/E/40/50; 512-71D40. *Version 2:* 512-71D/40/50; 512-71D40. *Version 3:* 512-27D40; 512-71D40; 512-27D40, Phase 2.

Location: *Version 1:* Soldier Support Institute, Fort Benjamin Harrison, IN. *Version 2:* Soldier Support Institute, Fort Jackson, SC. *Version 3:* NCO Academy, Fort Jackson, SC.

Length: *Version 1:* 5 weeks (180 hours). *Version 2:* 6–7 weeks (222–223 hours). *Version 3:* 5–6 weeks (80 hours).

Exhibit Dates: *Version 1:* 10/87–3/96. *Version 2:* 4/96–9/00. *Version 3:* 10/00–Present.

Learning Outcomes: *Version 1:* Upon completion of the course, the student will be able to describe various styles of leadership, use feedback to modify activities, evaluate progress, and apply counseling techniques. *Version 2:* Upon completion of the course, the student will be able to describe various styles of leadership, use feedback to modify activities, evaluate progress, and apply counseling techniques. *Version 3:* Upon completion of the course, the student will be able to describe various styles of leadership, use feedback to modify activities, evaluate progress, and apply counseling techniques.

Instruction: *Version 1:* Methods of instruction include lectures and group discussions. Topics include leadership, motivation, counseling, communication, and discipline. This course contains a common core of leadership subjects.

Version 2: Methods of instruction include lectures and group discussions. Topics include leadership, motivation, counseling, communication, and discipline. This course contains a common core of leadership subjects. *Version 3:* Methods of instruction include lectures and group discussions. Topics include leadership, motivation, counseling, communication, and discipline. This course contains a common core of leadership subjects.

Credit Recommendation: ***Version 1:*** Credit is recommended for the common core only. See AR-1404-0035 (12/88). *Version 2:* Credit is recommended for the common core only. See AR-1404-0035 (9/97). *Version 3:* In the lower-division baccalaureate/associate degree category, 1 semester hour in military science and 1 in business communication (5/02).

Related Occupations: 27D; 71D.

AR-1407-0008

Executive Administrative Assistant

Course Number: ***Version 1:*** 511-71C10/20; 511-71C10. *Version 2:* 510-ASIE3.

Location: ***Version 1:*** Adjutant General School, Fort Benjamin Harrison, IN. *Version 2:* Soldier Support Institute, Fort Jackson, SC.

Length: ***Version 1:*** 12 weeks (426–440 hours). *Version 2:* 5 weeks (180 hours).

Exhibit Dates: ***Version 1:*** 10/86–4/96. *Version 2:* 5/96–Present.

Learning Outcomes: ***Version 1:*** Upon completion of the course, the student will be able to type, file, write business correspondence using good English grammar, transcribe, and use proper social protocol. *Version 2:* Upon completion of the course, the student will be able to provide executive administrative support duties, including typing, filing, maintaining calendar, writing business correspondence, and understanding and using proper social protocol.

Instruction: ***Version 1:*** Lectures and practical exercises cover office administration, keyboarding, word processing, filing, correspondence, English grammar and usage, transcription, and social protocol. *Version 2:* Lectures, discussions, practical exercises, examinations in keyboarding, cover using a PC as a word processor, preparing correspondence, transcribing from dictating machines and using proper protocol for attending social functions, declining/accepting invitations, and using acceptable English grammar.

Credit Recommendation: ***Version 1:*** In the lower-division baccalaureate/associate degree category, 2 semester hours in English composition, 3 in typing, 1 in word processing concepts, 1 in office procedures, and 3 in machine transcription (11/91). *Version 2:* In the lower-division baccalaureate/associate degree category, 1 semester hour in keyboarding, 1 in word processing, and 2 in English composition (9/97).

Related Occupations: 71C.

AR-1407-0009

1. Legal Specialist/Court Reporter Basic Noncommissioned Officer (NCO)
2. Legal Specialist Basic Noncommissioned Officer (NCO)
3. Paralegal Specialist Basic Noncommissioned Officer (NCO) (71D Basic Noncommissioned Officers (NCO))

Course Number: ***Version 1:*** 512-71D/20/30; 512-71D/E/20/30; 512-71D20/30; 512-71D/E30. *Version 2:* 512-71D/20/30; 512-71D30. *Version 3:* 512-27D30; 512-71D30; 512-27D30, Phase 2.

Location: ***Version 1:*** Soldier Support Institute, Fort Benjamin Harrison, IN; Adjutant General School, Fort Benjamin Harrison, IN. *Version 2:* Soldier Support Institute, Fort Jackson, SC. *Version 3:* NCO Academy, Fort Jackson, SC.

Length: ***Version 1:*** 5–7 weeks (193–286 hours). *Version 2:* 7–8 weeks (277 hours). *Version 3:* 2–3 weeks (152 hours).

Exhibit Dates: ***Version 1:*** 10/86–3/96. *Version 2:* 4/96–9/00. *Version 3:* 10/00–Present.

Learning Outcomes: ***Version 1:*** Upon completion of the course, the student will be able to review administrative separations, evaluate personal property claims, review courts-martial, conduct legal research, and prepare files for disposal. *Version 2:* Upon completion of the course, the student will be able to review administrative separations, evaluate personal property claims, review courts-martial, conduct legal research, and prepare files for disposal. *Version 3:* Upon completion of the course, the student will be able to review administrative separations, evaluate personal property claims, review courts-martial, conduct legal research, and prepare files for disposal.

Instruction: ***Version 1:*** Instruction includes lectures and class discussions on file preparation and disposal, legal research, administrative separations, court-martial documents, and personal property claims. *Version 2:* Instruction includes lectures, class discussions, and computer applications on file preparation and disposal, legal research, administrative separations, court-martial documents, and personal property claims. Course includes a common core of leadership subjects. *Version 3:* Instruction includes lectures, class discussions, and computer applications on file preparation and disposal, legal research, administrative separations, court-martial documents, and personal property claims. Course includes a common core of leadership subjects.

Credit Recommendation: ***Version 1:*** In the lower-division baccalaureate/associate degree category, 2 semester hours in legal research and 2 in a criminal justice elective (11/91). *Version 2:* In the lower-division baccalaureate/associate degree category, 2 semester hours in criminal justice and 2 in legal research. See AR-1406-0090 for common core credit (9/97). *Version 3:* In the lower-division baccalaureate/associate degree category, 1 semester hour in criminal justice and 1 in legal research (5/02).

Related Occupations: 27D; 71D; 71E.

AR-1408-0019

Air Defense Artillery (ADA) Officer Advanced

Course Number: 2-44-C22.

Location: Air Defense Artillery School, Fort Bliss, TX.

Length: 20–26 weeks (731–986 hours).

Exhibit Dates: 12/83–7/91.

Objectives: To train air defense artillery officers for command and staff duties in air defense battalions and brigades.

Instruction: Lectures and practical experience cover introduction to automatic data processing; flow charting; BASIC language programming; financial management systems; maintenance systems; quality assurance and control; nuclear, biological, and chemical weapon employment; Nike and Hawk air defense system characteristics and capabilities; air defense tactics; and electives in two of the following: automatic data processing, communicative arts, logistics management, electronics, or insurgent warfare.

Credit Recommendation: Credit is not recommended because of the military-specific nature of the course (10/89).

AR-1408-0028

Field Artillery Officer Advanced

Course Number: 2-6-C22.

Location: Artillery and Missile School, Fort Sill, OK; Air Defense Artillery School, Fort Bliss, TX.

Length: 20 weeks (710–720 hours).

Exhibit Dates: 1/85–3/94.

Objectives: After 3/94 see AR-1408-0213. To train field artillery officers for command and staff duties.

Instruction: Lectures and practical exercises cover battle doctrine, artillery fire support, target acquisition, tactics, combined arms, technical writing, organizational communication, supervision, and data processing.

Credit Recommendation: In the lower-division baccalaureate/associate degree category, 3 semester hours in data processing/word processing and 1 in technical writing. In the upper-division baccalaureate category, 2 semester hours in supervision (11/85).

AR-1408-0035

Defense Specification Management

Course Number: ***Version 2:*** 8D-F1. *Version 3:* PQM 103; ALM-34-0235.

Location: ***Version 2:*** Army Logistics Management College, Fort Lee, VA; Army Logistics Management College, Onsite at various locations. *Version 3:* Army Logistics Management College, Fort Lee, VA.

Length: ***Version 2:*** 2 weeks (69–71 hours). *Version 3:* 2 weeks (74 hours).

Exhibit Dates: ***Version 2:*** 8/73–4/93. *Version 3:* 5/93–3/98.

Learning Outcomes: ***Version 2:*** Upon completion of the course, military and civilian personnel will have become familiar with the development, preparation, and use of contract specifications. *Version 3:* After 3/98 see DD-1408-0029. Upon completion of the course, the student will be able to apply procurement improvement tools to government and commercial procedures and use market research techniques to identify existing or create new item descriptions or documents to support acquisition.

Instruction: ***Version 2:*** Lecture-conferences and practical exercises cover fundamentals of specifications, requirements of specifications, specifications and procurement, quality assurance, preparation for delivery, and management practices. *Version 3:* Lecture-conferences and practical exercises cover fundamentals of specifications, requirements of specifications, specifications and procurement, quality assurance, preparation for delivery, and management practices.

Credit Recommendation: ***Version 2:*** In the lower-division baccalaureate/associate degree category, 1 semester hour in procurement (9/90). *Version 3:* In the lower-division baccalaureate/associate degree category, 1 semester hour in procurement (5/96).

AR-1408-0056

Cost Analysis for Decision Making

Course Number: ALMC-CB.
Location: Army Logistics Management Center, Fort Lee, VA; Army Logistics Management College, Fort Lee, VA.
Length: 4 weeks (153–157 hours).
Exhibit Dates: 5/73–9/98.
Objectives: To train commissioned officers and civilian personnel in the theory and application of cost and economic analysis as applied to the development, acquisition, and operation of Army weapons systems and facilities.
Instruction: Lectures, practical exercises and workshops cover cost analysis theory and application. The course also includes regression analysis, learning curve theory, nonstatistical estimating methods, uncertainty analysis, inflation theory and practice, design-to-cost relationships, and development of cost estimating relationships.
Credit Recommendation: In the upper-division baccalaureate category, 3 semester hours in cost analysis. In the graduate degree category, 1 semester hour in cost analysis (8/90).

AR-1408-0063

Aviation Senior Warrant Officer Training
(Aviation Warrant Officer Advanced)

Course Number: 2-1-C32.
Location: Aviation Center, Fort Rucker, AL.
Length: 12 weeks (415 hours).
Exhibit Dates: 10/88–9/95.
Learning Outcomes: After 9/95 see AR-1405-0269. To prepare aviation officers to perform in a variety of unit aviation functions.
Instruction: Lectures and practical exercises cover management, including general instruction in the principles, philosophies, concepts, and scope of management; general instruction in aviation safety in both accident prevention and investigation; and microcomputer terminology and programming with hands-on operating experience.
Credit Recommendation: In the lower-division baccalaureate/associate degree category, 2 semester hours in communication skills and 3 in basic microcomputers. In the upper-division baccalaureate category, 3 semester hours in principles of management, 2 in aircraft safety, and 3 in interpersonal relations (4/90).
Related Occupations: 100C; 100D; 100E; 100F; 100G; 100P; 100Q; 100R; 100Z; 101B; 101C; 102A; 103A; 152B; 152C; 152D; 152F; 152G; 153A; 153B; 153C; 153D; 154A; 154B; 154C; 155A; 155D; 155E; 156A; 671B; 671C; 671D; 671E.

AR-1408-0067

Medical Logistics Management

Course Number: 8B-F20.
Location: Academy of Health Sciences, Fort Sam Houston, TX.
Length: 10 weeks (356–431 hours).
Exhibit Dates: 2/75–1/92.
Objectives: To provide officers with knowledge of concepts and tools of management logistics as related to Army hospitals and field medical units.
Instruction: Lectures and practical exercises cover medical material management; planning, programming, and budgeting; basic computer concepts; and logistics.
Credit Recommendation: In the upper-division baccalaureate category, 3 semester hours in material management (5/92).

AR-1408-0068

Army Medical Department (AMEDD) Noncommissioned Officer (NCO) Basic

Course Number: 6-8-C40.
Location: Academy of Health Sciences, Fort Sam Houston, TX.
Length: 8–9 weeks (262 hours).
Exhibit Dates: 5/78–9/91.
Objectives: To provide enlisted personnel with the knowledge and skills required to perform administrative duties in the Army Medical Department.
Instruction: Conferences, demonstrations, and practical exercises cover military correspondence, oral communication, supervisory management, and general military topics. Instruction also includes resource management and problem solving techniques.
Credit Recommendation: In the lower-division baccalaureate/associate degree category, 6 semester hours in administration, 1 in interpersonal communication, and 1 in military science (8/79).
Related Occupations: 35G; 71G; 76J; 94F.

AR-1408-0072

3. Master Warrant Officer
4. Warrant Officer Staff

Course Number: ***Version 3:*** 1-250-C8. *Version 4:* 1-250-C8.
Location: Aviation Center, Fort Rucker, AL.
Length: ***Version 3:*** 8 weeks (288 hours). *Version 4:* 4 weeks (171 hours).
Exhibit Dates: ***Version 3:*** 9/88–9/95. *Version 4:* 10/95–Present.
Learning Outcomes: ***Version 3:*** To provide senior warrant officers with a working knowledge of subject matter essential to their careers in various duty assignments and to provide a background required for the progression of senior warrant officers into technical staff positions. *Version 4:* To provide senior warrant officers with a working knowledge of subject matter essential to their careers in various duty assignments and to provide a background required for the progression of senior warrant officers into technical staff positions.
Instruction: ***Version 3:*** Lectures and practical exercises cover management principles, philosophies, concepts, and scope and microcomputers, operating systems, programming, and hands-on computer operation. This course includes 100 hours of nonresident training. *Version 4:* Lectures, practical and classroom exercises, audio visual materials, discussion, case studies, learner presentations, and computer-based training. Topics include management principles, philosophies and concepts of genuine issues and problems, including environmental awareness, risk and training management, and total quality management.
Credit Recommendation: ***Version 3:*** In the lower-division baccalaureate/associate degree category, 3 semester hours in technical writing, 2 in interpersonal relations, and 1 in basic microcomputers. In the upper-division baccalaureate category, 3 semester hours in human resource management and 3 in management problems (4/90). *Version 4:* In the lower-division baccalaureate/associate degree category, 3 semester hours in technical writing or 3 in communications. In the upper-division baccalaureate category, 3 semester hours in human resource management (4/03).

AR-1408-0075

Directorate of Logistics
(Directorate of Industrial Operations)

Course Number: ALMC-DT; ALM-61-0235.
Location: Army Logistics Management College, Fort Lee, VA.
Length: 3–4 weeks (100–143 hours).
Exhibit Dates: 6/76–9/95.
Objectives: To enable officer and civilian graduates to improve their management performance in various installation support activities.
Instruction: Lectures, practical exercises, and group discussions cover general management principles; financial management; and management of such support activities as transportation, housing, supply, and facility maintenance.
Credit Recommendation: In the upper-division baccalaureate category, 3 semester hours in general management (10/90).

AR-1408-0085

Advanced Controllership Symposium

Course Number: None.
Location: Institute for Administration, Fort Benjamin Harrison, IN.
Length: 1–2 weeks (65–67 hours).
Exhibit Dates: 5/79–12/91.
Objectives: To broaden the professional development of commissioned officer and civilian comptrollers.
Instruction: Lectures and small group discussions cover military comptrollership policies and procedures, including current issues, manpower management, budget, financial management, and cost analysis.
Credit Recommendation: In the upper-division baccalaureate category, 1 semester hour in comptrollership (3/85).

AR-1408-0102

Current Issues in the Veterinary Service

Course Number: 6-8-C8.
Location: Academy of Health Sciences, Fort Sam Houston, TX.
Length: 2 weeks (79 hours).
Exhibit Dates: 6/80–12/93.
Objectives: After 12/93 see AR-1408-0218. To provide Veterinary Corps officers with a review of policies, procedures, and programs with emphasis on problem areas and solutions.
Instruction: Conferences, demonstrations, and practical exercises cover administration, food technology/science, surveillance inspection, and veterinary preventive medicine.
Credit Recommendation: In the lower-division baccalaureate/associate degree category, 3 semester hours in management science (7/90).

AR-1408-0104

Army Medical Department Officer (AMEDD) Basic Uniformed

Services University of Health Sciences

Course Number: 6-8-C20 (USUHS).

Location: Academy of Health Sciences, Fort Sam Houston, TX.

Length: 4 weeks (143–171 hours).

Exhibit Dates: 4/79–6/96.

Objectives: To provide basic orientation and training to students of the Uniformed Services University of Health Sciences including the responsibilities and duties of an officer.

Instruction: Lectures, conferences, and practical exercises cover health care administration, preventive medicine, and military science.

Credit Recommendation: In the lower-division baccalaureate/associate degree category, 6 semester hours in military science, 1 in administration, and 1 in preventive medicine (5/82).

AR-1408-0105

TOW/Dragon Repairer Basic Technical (BT)

Course Number: 121-27E30.

Location: Missile and Munitions School, Redstone Arsenal, AL.

Length: 7–10 weeks (276–374 hours).

Exhibit Dates: 7/82–4/98.

Learning Outcomes: Upon completion of the course, the student will be able to prepare a material condition status report, schedule activities which support maintenance actions, identify facts about alcohol and other drugs, solve problems associated with personnel or performance, and provide technical assistance to subordinates in equipment maintenance.

Instruction: Classroom and laboratory exercises cover the activities and procedures that support unit maintenance actions.

Credit Recommendation: In the lower-division baccalaureate/associate degree category, 3 semester hours in organizational management (4/88).

Related Occupations: 27E.

AR-1408-0107

Maintenance Management Additional Specialty

Course Number: XX91.

Location: Ordnance Center and School, Aberdeen Proving Ground, MD.

Length: 10–11 weeks (362 hours).

Exhibit Dates: 1/84–9/90.

Objectives: To train officers to command maintenance units and to perform duties of a field maintenance staff officer.

Instruction: Course includes instruction in organizational supply and maintenance, automotive and support equipment fundamentals, supply operations, material acquisition, research and development, and wholesale logistics.

Credit Recommendation: In the lower-division baccalaureate/associate degree category, 1 semester hours in automotive and related equipment fundamentals. In the upper-division baccalaureate category, 6 semester hours in maintenance management (2/84).

AR-1408-0108

Ordnance Officer Advanced

Course Number: 4-9-C22.

Location: Ordnance Center and School, Aberdeen Proving Ground, MD.

Length: 26 weeks (847 hours).

Exhibit Dates: 7/83–2/94.

Objectives: To prepare officers to command maintenance units and to perform duties of a field maintenance staff officer.

Instruction: Lectures and practical exercises cover leadership development and supervision, personnel management, professional ethics and values, training management, logistics and supply operations, financial management, budgets and records, stock accounting, and procurement. This course also covers maintenance management including material and manpower utilization and introduction to military tactics, operations, and training.

Credit Recommendation: In the upper-division baccalaureate category, 3 semester hours in personnel management, 3 in supervision, 3 in records management, 6 in maintenance management, and 1 in communication—writing (2/84).

AR-1408-0114

Logistics Pre-Command

Course Number: ***Version 1:*** ALMC-PD. *Version 2:* ALMC-PD, Phase 3.

Location: ***Version 1:*** Army Logistics Management Center, Fort Lee, VA. *Version 2:* Army Logistics Management College, Fort Lee, VA.

Length: ***Version 1:*** 3 weeks (109–115 hours). *Version 2:* 2 weeks (125 hours).

Exhibit Dates: ***Version 1:*** 4/85–6/92. *Version 2:* 7/92–Present.

Learning Outcomes: ***Version 1:*** Upon completion of the course the student will be able to use problem indicators to recognize types and causes of management problems typical to specific types of organizations and recognize analytical concepts applicable to the management of the organization's missions and resources. *Version 2:* Upon completion of the course, the student will be able to use problem indicators to recognize types and causes of management problems typical to specific types of organizations and recognize analytical concepts applicable to the management of the organization's missions and resources.

Instruction: ***Version 1:*** Lectures, seminars, field trips, and guest speakers cover manpower management, civilian personnel management and labor relations, and management systems organization. *Version 2:* Lecture format, with senior management guest speakers, cover key management concepts in the governmental chain of command.

Credit Recommendation: ***Version 1:*** In the lower-division baccalaureate/associate degree category, 3 semester hours in administrative management (11/89). *Version 2:* Credit is not recommended because of the limited, specialized nature of the course (1/02).

AR-1408-0116

Installation Logistics Management

Course Number: ALMC-IN.

Location: Army Logistics Management College, Onsite at various locations; Army Logistics Management Center, Fort Lee, VA.

Length: 2 weeks (66–67 hours).

Exhibit Dates: 11/85–9/90.

Objectives: To provide knowledge of the policies and procedures governing management of the logistics function at Army installations, including supply, maintenance, transportation, and contracting.

Instruction: Lecture-conferences and case method instruction cover management required for installation logistics management, including supply, transportation, maintenance, and procurement. NOTE: There are no quizzes, written exams, or research papers required in this course. Students must present their solutions to a budgeting case orally. The course is graded pass-fail.

Credit Recommendation: In the upper-division baccalaureate category, 1 semester hour in logistics management (10/90).

AR-1408-0118

Force Modernization and Sustaining (Equipping the Force)
(Force Modernization Management (Equipping the Force))

Course Number: ALMC-FG.

Location: Army Logistics Management College, Fort Lee, VA.

Length: 2–3 weeks (69–73 hours).

Exhibit Dates: 5/83–1/95.

Objectives: To provide an understanding of the processes involved in the organizational interpretation of supportable equipment systems within the Department of the Army.

Instruction: Lectures and exercises cover introduction to the modernization process, the resource management process, the materiel distribution and acquisition processes, and force development.

Credit Recommendation: In the upper-division baccalaureate category, 1 semester hour in resource planning (11/90).

AR-1408-0120

Automated Information Systems Contracting
(Defense Contracting for Information Resources)

Course Number: CON 241; ALMC-ZX.

Location: Army Logistics Management College, Onsite at various locations; Army Logistics Management College, Fort Lee, VA.

Length: 2 weeks (57–74 hours).

Exhibit Dates: 8/85–3/93.

Learning Outcomes: After 3/93 see DD-1402-0006. To understand the impact of legislation and other congressional action on the acquisition of information resources and to apply correct decision making techniques for making such acquisitions.

Instruction: Lectures and practical exercises provide detailed instructions in all significant aspects of contracting for information resources. It is designed to provide a working familiarity with pertinent government regulations, policies, and procedures.

Credit Recommendation: In the upper-division baccalaureate category, 2 semester hours in contracting management (information resources) (10/88).

AR-1408-0121

Commercial Activities Process
(Commercial Activities Review Process)

Course Number: ALMC-CM.

Location: Army Logistics Management College, Onsite at various locations; Army Logistics Management College, Fort Lee, VA.

Length: 1–2 weeks (56–74 hours).

Exhibit Dates: 8/85–10/92.

Objectives: To provide instruction in preparing performance work statements and cost analyses of commercial contracting and comparable military activities that provide services to military installations.

Instruction: Lecture-conferences and case method instruction cover work measurement, writing of performance work statements, and comparative cost analyses of governmental versus commercial contracting activities.

Credit Recommendation: In the upper-division baccalaureate category, 2 semester hours in managerial analysis (3/92).

AR-1408-0122

Materiel Acquisition Management

Course Number: *Version 1:* ALMC-ML. *Version 2:* ALMC-ML. *Version 3:* ALMC-ML.

Location: *Version 1:* Army Logistics Management Center, Fort Lee, VA; Army Logistics Management College, Fort Lee, VA. *Version 2:* Army Logistics Management College, Fort Lee, VA. *Version 3:* Army Logistics Management College, Fort Lee, VA.

Length: *Version 1:* 8–9 weeks (307–343 hours). *Version 2:* 8 weeks (307 hours). *Version 3:* 7 weeks (279 hours).

Exhibit Dates: *Version 1:* 10/84–5/93. *Version 2:* 6/93–7/99. *Version 3:* 8/99–9/02.

Learning Outcomes: *Version 1:* Upon completion of the course, the student will be able to perform entry-level duties in the Army's materiel acquisition process. *Version 2:* Upon completion of the course, the student will be able to identify needs, manage the procurement process, determine requirements for disposing of materiel, and apply cost and budgeting techniques to contract negotiations. *Version 3:* Upon completion of the course, the student will be able to identify needs, manage projects using scheduling software, apply cost and budgeting techniques to contract negotiations, and create an acquisition strategy using MS Project 2000.

Instruction: *Version 1:* Lectures, case studies, exercises, and discussions cover materiel acquisition management, including financial and cost management, integrated logistics support, force modernization, production management, and contract management. *Version 2:* Lectures, case studies, exercises, and discussions cover materiel acquisition management, including financial and cost management, integrated logistics support, force modernization, production management, and contract management. *Version 3:* This course uses lectures, case studies, exercises and projects to measure learning outcome. The course covers materiel acquisition, government contracting, software acquisition, and budgeting.

Credit Recommendation: *Version 1:* In the upper-division baccalaureate category, 1 semester hour in logistics management. In the graduate degree category, 2 semester hours in contract management, 2 in financial/cost management, and 1 in research and development management (6/93). *Version 2:* In the upper-division baccalaureate category, 1 semester hour in logistics management. In the graduate degree category, 1 semester hour in research and development, 2 in financial and cost management, and 2 in contract management (8/95). *Version 3:* In the graduate degree category, 3 semester hours in acquisition management and 3 in contract management (1/02).

AR-1408-0123

Pre-Command Phase 3 Combat Service Support Missile and Munitions

Course Number: 9-26-F45.

Location: Missile and Munitions School, Redstone Arsenal, AL.

Length: 2 weeks (77 hours).

Exhibit Dates: 9/85–4/96.

Objectives: To provide missile and munitions command designees with update training on administration, supply, maintenance, and training.

Instruction: Lectures and practical exercises cover force development, supply procedures, readiness reporting, planographs, special weapons, materiel management, and ammunition transportation.

Credit Recommendation: In the lower-division baccalaureate/associate degree category, 2 semester hours in material management (4/86).

AR-1408-0126

Inspector General

Course Number: None.

Location: Inspector General School, Fort Belvoir, VA.

Length: 2–6 weeks (76–188 hours).

Exhibit Dates: 4/83–12/90.

Learning Outcomes: Upon completion of the course, the student will be able to apply fundamental inspector general policies and procedures and the functional life-cycle model of the Army to operational inspections.

Instruction: Course covers policies and processes in performing inspections and general functions of administrative investigations. Students learn the functional life cycle of the Army as it applies to inspector general policies and processes. Student is provided with knowledge of the functions, operations, systems, policies, and command structure of the Army in order to analyze potential problems and to conduct systemic inspections.

Credit Recommendation: In the lower-division baccalaureate/associate degree category, 3 semester hours in introduction to internal organizational auditing. Credit is not recommended for those students attending only the reserve portion (the first two weeks) of the course (2/87).

AR-1408-0127

Chaplain Assistant Advanced Noncommissioned Officer (NCO)

Course Number: 661-71M40.

Location: Chaplain Center and School, Fort Monmouth, NJ.

Length: 7 weeks (254 hours).

Exhibit Dates: 5/89–4/04.

Learning Outcomes: Upon completion of the course, the student will be able to lead, supervise, manage, and communicate effectively; prepare a religious activities program and budget; conduct training and maintenance activities; and manage the budget.

Instruction: Topics include advanced leadership and management skills, general military subjects, communication skills, training management, logistics management, and the unit ministry team in combat. This course contains a common core of leadership subjects.

Credit Recommendation: In the lower-division baccalaureate/associate degree category, 1 semester hour in principles of supervision. See AR-1404-0035 for common core credit (6/90).

Related Occupations: 71M.

AR-1408-0128

Chaplain Reserve Component General Staff College

Course Number: 5G-F2.

Location: Chaplain Center and School, Fort Monmouth, NJ.

Length: 6 weeks (216 hours).

Exhibit Dates: 10/87–4/04.

Learning Outcomes: Upon completion of the course, the student will be able to assess management and leadership styles and principles, demonstrate an understanding of appropriated and nonappropriated fund support, recommend procedures to be used during mobilization, and develop training plans for a religious ministries team.

Instruction: Topics include writing skills, management, budgeting, tactical logistics, mobilization issues, strategic studies, and the ministry in combat. Course includes a simulated combat scenario, case studies, an integration paper, and a seminar. This course also includes a nonresident portion.

Credit Recommendation: In the lower-division baccalaureate/associate degree category, 3 semester hours in report writing. In the upper-division baccalaureate category, 3 semester hours in principles of management and 3 in business organization and management (6/90).

AR-1408-0129

Chaplain Training Manager

Course Number: 5G-ASI7E.

Location: Chaplain Center and School, Fort Monmouth, NJ.

Length: 2 weeks (77 hours).

Exhibit Dates: 4/84–1/96.

Learning Outcomes: Upon completion of the course, the student will be able to manage a chaplain's training program at an Army installation; apply instructional methods; and demonstrate knowledge of training resources, staff coordination, supervision, evaluation, and management of training programs.

Instruction: Topics include short- and long-range planning for training programs, identifying training resources and personnel, maintaining records, developing training goals and objectives, and monitoring/evaluating training proficiency.

Credit Recommendation: In the upper-division baccalaureate category, 1 semester hour in training programs management (6/90).

AR-1408-0130

Nonappropriated Chaplains' Fund Clerk

Course Number: 561-F2.

Location: Chaplain Center and School, Fort Monmouth, NJ.

Length: 2 weeks (72 hours).

Exhibit Dates: 10/87–6/00.

Learning Outcomes: Upon completion of the course, the student will be able to apply the principles of single-entry fund accounting and perform clerical duties.

Instruction: Topics include preparing support documents; assisting in the preparation of annual budgets; and performing bookkeeping functions, including receipts, disbursements, and control of nonappropriated funds.

Credit Recommendation: In the lower-division baccalaureate/associate degree category, 1 semester hour in fund accounting (6/90).

AR-1408-0131

Nonappropriated Chaplains' Fund Custodian

Course Number: 5G-F6/561-F1.

Location: Chaplain Center and School, Fort Monmouth, NJ.

Length: 2 weeks (74 hours).

Exhibit Dates: 4/89–6/00.

Learning Outcomes: Upon completion of the course, the student will be able to use management by objectives and accounting control techniques to perform the functions of custodian, supervisor, and accountant for the nonappropriated chaplain's fund.

Instruction: Topics include single-entry accounting, organizing and managing documents, controlling nonappropriated funds, managing the receipt of funds, preparing and certifying fund reports, and supervising clerical activities associated with nonappropriated funds.

Credit Recommendation: In the lower-division baccalaureate/associate degree category, 1 semester hour in fund accounting and 1 in principles of management (6/90).

AR-1408-0132

Electronic Warfare (EW)/Intercept Equipment Repair Technician Warrant Officer Technical Certification

Course Number: 4C-285A.

Location: Intelligence School, Fort Devens, MA.

Length: 12–13 weeks (482 hours).

Exhibit Dates: 12/84–9/90.

Learning Outcomes: Upon completion of the course, the student will be able to manage the operation, maintenance, and repair of computer-controlled equipment; operate, maintain, and repair computer-controlled equipment; program in BASIC and Assembler language; write diagnostic programs and use them to troubleshoot CPUs and peripherals; and employ the techniques of systems analysis to troubleshoot and repair computer-controlled systems.

Instruction: Lectures, tapes, demonstrations, written exams, and performance exercises cover managing the operation, maintenance, and repair of computer-controlled systems; programming in BASIC and Assembler language; writing diagnostic programs and using them to troubleshoot and repair computer-controlled systems; and developing and using systems analysis techniques.

Credit Recommendation: In the lower-division baccalaureate/associate degree category, 2 semester hours in maintenance management; 4 in maintenance and repair of peripherals; 2 in computer programming in BASIC; 2 in Assembler language programming; and 4 in microcomputer maintenance and repair, 4 in systems analysis and maintenance, or 4 in computer systems maintenance and repair (5/87).

Related Occupations: 285A.

AR-1408-0133

Morse Intercept Technician Warrant Officer Technical/Tactical Certification Phase 2
(Morse Intercept Technician Warrant Officer Technical Certification)

Course Number: 3B-352H; 3B-984A.

Location: Intelligence Center and School, Fort Huachuca, AZ; Intelligence School, Fort Devens, MA.

Length: 3–4 weeks (127–128 hours).

Exhibit Dates: 12/84–Present.

Learning Outcomes: Upon completion of the course, the student will be able to manage Morse intercept and collection systems, operate and coordinate systems, and evaluate and analyze Morse intercept data.

Instruction: Instruction includes lectures, field trips, and performance exercises in the management of the operation of Morse intercept systems including evaluation and analysis of Morse intercept data.

Credit Recommendation: In the lower-division baccalaureate/associate degree category, 3 semester hours in operations management (1/97).

Related Occupations: 352H; 984A.

AR-1408-0134

Non-Morse Intercept Technician Warrant Officer Technical Certification

Course Number: 3B-985A; 3B-352K.

Location: Intelligence School, Fort Devens, MA.

Length: 3–4 weeks (130 hours).

Exhibit Dates: 12/84–3/94.

Learning Outcomes: Upon completion of the course, the student will be able to manage non-Morse intercept and collection systems, operate and coordinate systems, and evaluate and analyze non-Morse intercept data.

Instruction: Instruction includes lectures, field trips, and performance exercises in the management of non-Morse intercept systems including evaluation and analysis of non-Morse intercept data.

Credit Recommendation: In the lower-division baccalaureate/associate degree category, 3 semester hours in operations management (5/87).

Related Occupations: 352K; 985A.

AR-1408-0135

Emitter Location/Identification Technician Warrant Officer Technical Certification

Course Number: 3B-352D; 3B-986A.

Location: Intelligence Center and School, Fort Huachuca, AZ; Intelligence School, Fort Devens, MA.

Length: 3–5 weeks (112–179 hours).

Exhibit Dates: 7/89–4/04.

Learning Outcomes: Upon completion of the course, the student will be able to manage the siting, employment, control, and operation of direction-finding equipment and set up office procedures to process data from these activities.

Instruction: Instruction includes lectures, field trips, and performance exercises in the management of the location, employment, control, and operation of direction-finding systems.

Credit Recommendation: In the lower-division baccalaureate/associate degree category, 3 semester hours in operations management (6/91).

Related Occupations: 352D; 986A.

AR-1408-0136

2. Emanations Analysis Technician Warrant Officer Technical/Tactical Certification
3. Emanations Analysis Technician Warrant Officer Basic

Course Number: *Version 2:* 3B-352J. *Version 3:* 3B-352J.

Location: *Version 2:* Intelligence Center and School, Fort Huachuca, AZ; Intelligence School, Fort Devens, MA. *Version 3:* Intelligence Center and School, Fort Huachuca, AZ.

Length: *Version 2:* 3–4 weeks (134 hours). *Version 3:* 4–5 weeks (181–182 hours).

Exhibit Dates: *Version 2:* 5/89–7/96. *Version 3:* 8/96–12/97.

Learning Outcomes: *Version 2:* Upon completion of the course, the student will be able to supervise and conduct training programs for persons assigned to electronic intelligence gathering units. *Version 3:* Upon completion of the course, the student will be able to supervise and conduct training programs for persons assigned to electronic intelligence gathering units.

Instruction: *Version 2:* Lectures, group interaction, and exercises cover the management and operation of intelligence units. *Version 3:* Lectures, group interaction, and exercises cover the management and operation of intelligence units.

Credit Recommendation: *Version 2:* In the lower-division baccalaureate/associate degree category, 3 semester hours in operations management (6/91). *Version 3:* Credit is not recommended because of the limited, specialized nature of the course (6/97).

Related Occupations: 983A.

AR-1408-0137

1. Traffic Analysis Technician Warrant Officer Technical/Tactical Certification Phase 2
(Traffic Analysis Technician Warrant Officer Technical Certification)
2. Traffic Analysis Technician Warrant Officer Basic Phase 2

Course Number: *Version 1:* 3B-352C, Phase 2; 3B-982A. *Version 2:* 3B-352C.

Location: *Version 1:* Intelligence Center and School, Fort Huachuca, AZ; Intelligence School, Fort Devens, MA. *Version 2:* Intelligence Center and School, Fort Huachuca, AZ.

Length: *Version 1:* 5–6 weeks (197–211 hours). *Version 2:* 5–6 weeks (213 hours).

Exhibit Dates: *Version 1:* 12/84–7/96. *Version 2:* 8/96–Present.

Learning Outcomes: *Version 1:* Upon completion of the course, the student will be able to manage intelligence collection/processing equipment, including operating, collecting, processing, analyzing, and evaluating intelligence traffic. *Version 2:* Upon completion of the course, the student will be able to supervise intelligence collection/processing equipment, including operating, collecting, processing, analyzing, and evaluating intelligence traffic.

Instruction: *Version 1:* Lectures, field trips, and performance exercises cover the management of intelligence collection and processing equipment, including analysis and evaluation of intelligence traffic. *Version 2:* Lectures, field trips, and performance exercises cover the installation and supervision of intelligence collection and processing equipment, including analysis and evaluation of intelligence traffic.

Credit Recommendation: ***Version 1:*** In the lower-division baccalaureate/associate degree category, 3 semester hours in operations management (6/91). *Version 2:* In the lower-division baccalaureate/associate degree category, 2 semester hours in introduction to local area network (11/02).

Related Occupations: 352C; 982A.

AR-1408-0138

1. Voice Intercept Technician Warrant Officer Technical/Tactical Certification Phase 2
   (Voice Intercept Technician Warrant Officer Technical Certification)
2. Voice Intercept Technician Warrant Officer Basic Phase 2
3. Voice Intercept Technician Warrant Officer Basic

Course Number: ***Version 1:*** 3B-352G, Phase 2; 3B-988A. *Version 2:* 3B-352G. *Version 3:* 3B-352G, Phase 2.

Location: ***Version 1:*** Intelligence Center and School, Fort Huachuca, AZ; Intelligence School, Fort Devens, MA. *Version 2:* Intelligence Center and School, Fort Huachuca, AZ. *Version 3:* Intelligence Center and School, Fort Huachuca, AZ.

Length: ***Version 1:*** 3–6 weeks (134–222 hours). *Version 2:* 3–4 weeks (131–148 hours). *Version 3:* 3–4 weeks (130–148 hours).

Exhibit Dates: ***Version 1:*** 12/84–7/96. *Version 2:* 8/96–9/00. *Version 3:* 10/00–Present.

Learning Outcomes: ***Version 1:*** Upon completion of the course, the student will be able to manage voice intercept and collection systems, including operating, coordinating, analyzing, and evaluating the systems. *Version 2:* Upon completion of the course, the student will be able to manage voice intercept and collection systems, including operating, coordinating, analyzing, and evaluating the systems. *Version 3:* Upon completion of the course, the student will be able to manage voice intercept and collection systems, including operating, coordinating, analyzing, and evaluating the systems.

Instruction: ***Version 1:*** Instruction includes lectures, field trips, and performance exercises in the management of voice intercept systems, including analysis and evaluation of voice intercept data. *Version 2:* Instruction includes lectures, field trips, and performance exercises in the management of voice intercept systems, including analysis and evaluation of voice intercept data. *Version 3:* Instruction includes lectures, field trips, and performance exercises in the management of voice intercept systems, including analysis and evaluation of voice intercept data.

Credit Recommendation: ***Version 1:*** In the lower-division baccalaureate/associate degree category, 3 semester hours in operations management (6/91). *Version 2:* Credit is not recommended because of the limited, specialized nature of the course (5/01). *Version 3:* Credit is not recommended because of the military-specific nature of the course (8/02).

Related Occupations: 352G; 988A.

AR-1408-0139

Chaplain Assistant Basic Noncommissioned Officer (NCO)
(Chapel Activities Specialist Basic Noncommissioned Officer (NCO))

Course Number: 561-71M20.

Location: Chaplain Center and School, Fort Monmouth, NJ.

Length: 5 weeks (177 hours).

Exhibit Dates: 5/89–4/04.

Learning Outcomes: Upon completion of the course, the student will be able to lead and organize small group activities involving maintenance, land navigation, training, tactical deployment, equipment operation, and computer hardware and software operations; account for appropriated funds; communicate effectively; and care for battle fatigue casualties.

Instruction: Topics include leadership, communications, battle fatigue casualty care, religious program management, volunteer management, computer hardware and software operations, tactical equipment operation, and leadership skills. This course contains a common core of leadership subjects.

Credit Recommendation: In the lower-division baccalaureate/associate degree category, 1 semester hour in microcomputer applications. See AR-1406-0090 for common core credit (6/90).

Related Occupations: 71M.

AR-1408-0140

Chaplain Officer Advanced

Course Number: 5-16-C22.

Location: Chaplain Center and School, Fort Monmouth, NJ.

Length: 20 weeks (710–820 hours).

Exhibit Dates: 5/89–6/00.

Learning Outcomes: After 11/01 see AR-1513-0014. Upon completion of the course, the student will be able to perform administrative and ministerial duties in a large (up to 500 persons) military organization, including pastoral counseling, resolving ethical and theological issues, managing fiscal affairs, and performing chaplain's combat functions.

Instruction: Topics include leadership, ethics, management, pastoral counseling, fiscal management, combat and tactical functions, appropriated and nonappropriated funds, marriage and the family, theological issues, and homily preparation. The final two weeks of the course are dedicated to a study of a directed elective in one of the following fields: homiletics/liturgics, marriage and family pastoral counseling, and automated religious support.

Credit Recommendation: In the upper-division baccalaureate category, 3 semester hours in ethics and theological issues and 3 in organizational management; if the individual has completed the automated religious support elective, 1 additional semester hour in microcomputer applications. In the graduate degree category, 3 semester hours in homiletics and 3 in marriage and family counseling; if the individual has completed the homiletics/liturgics elective, 1 additional semester hour in homiletics; if the individual has completed the marriage and family pastoral counseling elective, 1 additional semester hour in marriage and family counseling (6/90).

AR-1408-0141

Chaplain Officer Basic

Course Number: 5-16-C20.

Location: Chaplain Center and School, Fort Monmouth, NJ.

Length: 11–12 weeks (408 hours).

Exhibit Dates: 1/90–5/96.

Learning Outcomes: Upon completion of this course, the student will be able to perform pastoral counseling, administrative duties, and pastoral ministry duties in a military setting; demonstrate leadership, management, supervision, and ethics in conducting a battalion chaplain's duties; resolve pastoral problems; and demonstrate knowledge of internal controls in fund management.

Instruction: Topics include military and administrative skills, supervision skills, management, ethics, pastoral ministry in the military environment, pastoral counseling, internal controls, component training, and chaplain support activities.

Credit Recommendation: In the lower-division baccalaureate/associate degree category, 2 semester hours in principles of supervision. In the upper-division baccalaureate category, 1 semester hour in human relations, 2 in counseling, and 3 in principles of management (6/90).

Related Occupations: 56A.

AR-1408-0142

Advanced Noncommissioned Officer (NCO) MOS 71L

Course Number: 7-71-C42.

Location: Institute for Personnel and Resource Management, Fort Benjamin Harrison, IN.

Length: 5–17 weeks (168–444 hours).

Exhibit Dates: 3/81–9/91.

Objectives: To provide selected personnel with advanced leadership skills, military subjects, and communication skills for effective supervision of specific MOS tracks.

Instruction: Lectures and practical exercises include a basic core of written and oral communication, management principles, unit training, and combat survival. MOS options are recruitment and reenlistment, equal opportunity, club management, legal topics, chaplain assisting, and data processing.

Credit Recommendation: Credit is not recommended because of the limited content of the course (3/85).

Related Occupations: 71L.

AR-1408-0144

Armor Noncommissioned Officer (NCO) Advanced 19D
(19D Armor Noncommissioned Officer (NCO) Advanced)

Course Number: O-19-C42.

Location: Armor Center and School, Fort Knox, KY.

Length: 14–15 weeks (614 hours).

Exhibit Dates: 9/83–12/97.

Learning Outcomes: Upon completion of the course, the student will be able to provide leadership in a specialized field.

Instruction: Course includes tactical training, especially in maintenance procedures, training methods and techniques, leadership skills, and staff support.

Credit Recommendation: In the lower-division baccalaureate/associate degree category, 2 semester hours in basic surveying, 2 in communication skills, 3 in maintenance management, and 3 in personnel management (7/87).

Related Occupations: 19D.

AR-1408-0148

Sergeants Major

Course Number: ***Version 2:*** 1-250-C5. *Version 3:* 1-250-C5. *Version 4:* 1-250-C5.

Location: Sergeants Major Academy, Fort Bliss, TX.

Length: ***Version 2:*** 22 weeks (680–729 hours). *Version 3:* 22 weeks (729–730 hours). *Version 4:* 38–40 weeks (895–1440 hours).

Exhibit Dates: ***Version 2:*** 1/88–12/90. *Version 3:* 1/91–8/95. *Version 4:* 9/95–Present.

Learning Outcomes: ***Version 2:*** Upon completion of the course, the student will be able to provide leadership and management to the upper-level national defense structure. *Version 3:* Upon completion of the course, the student will be able to provide leadership and management to the upper-level national defense structure. *Version 4:* Upon completion of the course, the student will be able to provide leadership and management to the upper-level national defense structure.

Instruction: ***Version 2:*** Course includes lectures, demonstrations, and performance exercises in planning, group interaction, human relations, leadership, idea synthesis, oral and written communication, public speaking, listening, community understanding, history, resource management, military training, geopolitics, international studies, ideologies, US foreign policy, conflict resolution, ethics, human motivation, small-group communication, leadership theories, and management skills. *Version 3:* Course includes lectures, demonstrations, and performance exercises in planning, group interaction, human relations, leadership, idea synthesis, oral and written communication, introduction to research, public speaking, listening, community understanding, history, resource management, military training, geopolitics, international studies, ideologies, US foreign policy, conflict resolution, ethics, human motivation, small-group communication, leadership theories, and management skills. *Version 4:* Course includes lectures, demonstrations, and performance exercises in planning, group interaction, human relations, leadership, idea synthesis, oral and written communication, introduction to research, public speaking, listening, introduction to health and physiology, community understanding, history, resource management, military training, geopolitics, international studies, ideologies, US foreign policy, conflict resolution, ethics, human motivation, small-group communication, leadership theories, and management skills.

Credit Recommendation: ***Version 2:*** In the lower-division baccalaureate/associate degree category, 2 semester hours in principles of supervision, 3 in group dynamics, and 3 in communication. In the upper-division baccalaureate category, 4 semester hours in military science, 4 in international relations, 1 in human resource management, and 1 in ethics (1/91). *Version 3:* In the lower-division baccalaureate/associate degree category, 1 semester hour in written communication, 3 in interpersonal communication, 2 in public speaking, 3 in principles of supervision, and 2 in introduction to research. In the upper-division baccalaureate category, 1 semester hour in ethics, 3 in military science, and 3 in international relations (6/95). *Version 4:* In the lower-division baccalaureate/associate degree category, 3 semester hours in interpersonal communication, 2 in introduction to research, 1 in written communication, 1 in principles of supervision, 1 in public speaking, and 3 in health/physiology. In the upper-division baccalaureate category, 1 semester hour in ethics, 3 in human resource management, 6 in military science, and 1 in international relations (5/01).

AR-1408-0149

2. Sergeants Major Corresponding Studies
(Sergeants Major Correspondence Studies)
3. Sergeants Major Corresponding Studies
(Sergeants Major Correspondence Studies)
4. Sergeants Major Nonresident

Course Number: ***Version 2:*** 1-250-C5 (ACCP). *Version 3:* 1-250-C5 (ACCP). *Version 4:* 1-250-C5 (ACCP).

Location: Sergeants Major Academy, Fort Bliss, TX.

Length: ***Version 2:*** Last 2 weeks in residence, maximum, 104 weeks. *Version 3:* Last 2 weeks in residence, maximum, 104 weeks. *Version 4:* In residence, 2 weeks, maximum, 104 weeks.

Exhibit Dates: ***Version 2:*** 7/89–6/90. *Version 3:* 7/90–11/94. *Version 4:* 12/94–Present.

Learning Outcomes: ***Version 2:*** Upon completion of the course, the student will be able to provide leadership and management to the upper-level national defense structure. *Version 3:* Upon completion of the course, the student will be able to provide leadership and management to the upper-level national defense structure. *Version 4:* Upon completion of the course, the student will be able to provide leadership and management to the upper-level national defense structure.

Instruction: ***Version 2:*** The instructional mode is primarily correspondence study plus two weeks resident training in planning, group interaction, human relations, oral and written communication, effective listening, history, resource management, military training, international studies, US foreign policy, conflict resolution, human motivation, small group dynamics, leadership theories, and management skills. *Version 3:* The instructional mode is primarily correspondence study plus two weeks resident training in planning, group interaction, human relations, oral and written communication, effective listening, history, resource management, military training, international studies, US foreign policy, conflict resolution, human motivation, small-group dynamics, leadership theories, and management skills. *Version 4:* The instructional mode is primarily correspondence study plus two weeks resident training in planning, group interaction, human relations, oral and written communication, effective listening, history, resource management, military training, international studies, US foreign policy, conflict resolution, human motivation, small-group dynamics, leadership theories, and management skills.

Credit Recommendation: ***Version 2:*** In the lower-division baccalaureate/associate degree category, 2 semester hours in group dynamics. In the upper-division baccalaureate category, 4 semester hours in military science, 3 in international relations, and 1 in human resource management (4/90). *Version 3:* In the lower-division baccalaureate/associate degree category, 2 semester hours in communication and 2 in group dynamics. In the upper-division baccalaureate category, 4 semester hours in military science, 3 in international relations, and 1 in human resource management (4/90). *Version 4:* In the lower-division baccalaureate/associate degree category, 1 semester hour in public speaking, 1 in interpersonal communication, 1 in written communication, and 1 in principles of supervision. In the upper-division baccalaureate category, 4 semester hours in military science and 3 in international relations (6/95).

AR-1408-0150

Senior Officer Preventive Logistics
(Senior Officer Logistics Management)

Course Number: 8A-F23.

Location: Armor Center and School, Fort Knox, KY.

Length: Total of 2 weeks; Track 1, 1–2 weeks (80 hours); Track 2, 38 hours.

Exhibit Dates: 11/86–12/90.

Objectives: Upon completion of Track 1, the student will be able to perform management functions in the area of preventive logistics with emphasis on supply; transportation; maintenance procedures; and Army concepts, publications, and records. Upon completion of Track 2, the student will be able to perform management functions in the area of preventive logistics with an emphasis on supply, transportation, and maintenance procedures.

Instruction: Track 1 is conducted through lectures, group discussions, and practical exercises and includes topics in policy review, records, accountability, supply, transportation, maintenance procedures, and inspection techniques. Track 1 is offered for Army officers. Track 2 is conducted through group discussions and practical exercises and includes topics in supply, transportation, maintenance procedures, and inspection techniques. This track is for Marine Corps, Air Force, Navy, and allied officers only.

Credit Recommendation: In the graduate degree category, for students completing Track 1, 3 semester hours in management systems. For students completing Track 2, 2 semester hours in management systems (7/87).

AR-1408-0152

Ordnance Officer Missile/Munitions Additional Specialty (ADSPEC)

Course Number: 4D-91C/D.

Location: Ordnance, Missile and Munitions School, Redstone Arsenal, AL.

Length: 11–12 weeks (408 hours).

Exhibit Dates: 1/88–9/90.

Learning Outcomes: Upon completion of the course, the student will be able to apply leadership techniques in the areas of logistics, warehousing, and storage of munitions and supplies; organize subjects for briefings, conferences, and departmental meetings; organize and design a conference; and lead the conference using competent management techniques.

Instruction: Lectures, role playing, discussions, films, and testing cover materiel management, leadership, administration, and personnel activities.

Credit Recommendation: In the lower-division baccalaureate/associate degree category, 3 semester hours in materiel management (4/88).

Related Occupations: 91C; 91D.

AR-1408-0153

ORDNANCE OFFICER ADVANCED MISSILE/ MUNITIONS MATERIEL MANAGEMENT

Course Number: 4-9-C22-91C/D.

Location: Ordnance, Missile and Munitions School, Redstone Arsenal, AL.

Length: 20 weeks (742 hours).

Exhibit Dates: 1/88–1/96.

Learning Outcomes: Upon completion of the course, the student will be able to apply leadership techniques in logistics, warehousing, and storage of munitions and supplies; organize subjects for briefings, conferences, and departmental meetings; organize and design a conference and lead the conference using competent management techniques; and manage materiel and munitions in mobilization, as well as in peacetime.

Instruction: Lectures, role playing, discussions, films, and testing cover materiel management, leadership, administration, and personnel activities. The course also includes US/Soviet army organization, Warsaw Pact, ammunition issue and disposal, nuclear weapons, and missile support systems.

Credit Recommendation: In the lower-division baccalaureate/associate degree category, 3 semester hours in materiel management (8/91).

Related Occupations: 91C; 91D.

AR-1408-0154

MEDICAL LOGISTICS MANAGEMENT RESERVE

Course Number: *Version 1:* 8B-F20-RC. *Version 2:* 8B-F20-RC.

Location: Academy of Health Sciences, Fort Sam Houston, TX.

Length: *Version 1:* 2 weeks (78 hours). *Version 2:* 2 weeks (88 hours).

Exhibit Dates: *Version 1:* 2/86–2/93. *Version 2:* 3/93–Present.

Learning Outcomes: *Version 1:* Upon completion of the course, the student will be able to apply concepts and tools of management logistics as they relate to Army hospitals and field medical units. *Version 2:* Upon completion of the course, the student will be able to administer the admission and disposition of patients in health care facilities; maintain medical records; and process medical information.

Instruction: *Version 1:* Lectures and practical exercises cover medical material management; planning, programming, and budgeting; basic computer concepts; and logistics. *Version 2:* Course covers maintenance of medical records, medical statistical reporting, administrative support of patient movements and transfers, disability processing, and medical evaluation.

Credit Recommendation: *Version 1:* In the lower-division baccalaureate/associate degree category, 1 semester hour in material management (4/88). *Version 2:* In the upper-division baccalaureate category, 1 semester hour in material management (5/96).

AR-1408-0155

QUARTERMASTER OFFICER ADVANCED

Course Number: 8-10-C22.

Location: Quartermaster Center and School, Fort Lee, VA.

Length: 20 weeks (709–765 hours).

Exhibit Dates: 9/87–10/94.

Learning Outcomes: Upon completion of the course, the student will be able to manage the logistics, military justice, administrative, and organizational functions in materiel, services, subsistence, and petroleum units.

Instruction: Lectures, conferences, practical exercises, and examinations cover general management subjects, including organizational supply, maintenance management, financial and acquisition management, materiel management, transportation, automated systems, and systems management. Other topics include military justice, military history, graves registration, and combined arms. The course includes petroleum, materiel, and food tracks which are completed by all students.

Credit Recommendation: In the lower-division baccalaureate/associate degree category, 3 semester hours in introduction to management science/quantitative methods. In the upper-division baccalaureate category, 3 semester hours in logistics management (3/93).

AR-1408-0156

AIRDROP SYSTEMS TECHNICIAN WARRANT OFFICER ADVANCED

(Airdrop Systems Technician Senior Warrant Officer Training (Technical/Tactical Certification))

Course Number: 8-10-C32-921A.

Location: Quartermaster Center and School, Fort Lee, VA.

Length: 10–11 weeks (358–360 hours).

Exhibit Dates: 10/88–Present.

Learning Outcomes: Upon completion of the course, the student will master general military subjects and supervise the technical aspects of an airdrop operation, including packing, rigging, maintaining equipment, and keeping records.

Instruction: Course includes classroom lectures and practical exercises in supervision; communication skills; general military subjects; and packing, rigging, planning, and supervising an airdrop operation.

Credit Recommendation: In the vocational certificate category, 2 semester hours in advanced cargo rigging. In the lower-division baccalaureate/associate degree category, 2 semester hours in business communication and 1 in principles of supervision (5/02).

Related Occupations: 921A.

AR-1408-0157

1. FOOD SERVICE TECHNICIAN SENIOR WARRANT OFFICER
   (Quartermaster Senior Warrant Officer)
2. FOOD SERVICE TECHNICIAN WARRANT OFFICER ADVANCED
   (Food Service Technician Senior Warrant Officer)
   (Quartermaster Senior Warrant Officer)

Course Number: *Version 1:* 8-10-C32-922A. *Version 2:* 8-10-C32-922A.

Location: Quartermaster Center and School, Fort Lee, VA.

Length: *Version 1:* 11 weeks (396 hours). *Version 2:* 10–11 weeks (359–360 hours).

Exhibit Dates: *Version 1:* 10/88–9/92. *Version 2:* 10/92–Present.

Learning Outcomes: *Version 1:* Upon completion of the course, the student will be able to give an effective speech, operate a microcomputer, describe behavioral relationships within an organization, apply principles of government financial accounting (stock funds), and manage the food service function of a large organization. *Version 2:* Upon completion of the course, the student will be able to communicate effectively in orally and in writing, describe behavioral relationships within an organization, apply principles of government financial accounting, and manage the food service function of a large organization.

Instruction: *Version 1:* Lectures, class discussions, and practical exercises cover effective speaking, computer applications, organizational behavior, leadership, government financial accounting (including stock funds), and food service management and nutrition. *Version 2:* Lectures, class discussions, and practical exercises cover effective speaking, computer applications, organizational behavior, leadership, government financial accounting (including stock funds), and food service management and nutrition.

Credit Recommendation: *Version 1:* In the vocational certificate category, 1 semester hour in analysis of food service operations. In the lower-division baccalaureate/associate degree category, 1 semester hour in microcomputer applications, 2 in business communications, 1 in organizational behavior, and 1 in governmental financial accounting (10/88). *Version 2:* In the lower-division baccalaureate/associate degree category, 2 semester hours in business communication, 1 in organizational behavior (leadership), 1 in governmental financial management, and 2 in food service management (5/02).

Related Occupations: 922A.

AR-1408-0158

SUPPLY SYSTEMS TECHNICIAN WARRANT OFFICER ADVANCED

(Repair Parts Technician Senior Warrant Officer)

Course Number: 8-10-C32-920B.

Location: Quartermaster Center and School, Fort Lee, VA.

Length: 10–11 weeks (359–397 hours).

Exhibit Dates: 10/88–10/98.

Learning Outcomes: Upon completion of the course, the student will understand many general military subjects and supervise the technical aspects of logistics management including inventory control and reporting.

Instruction: Classroom lectures and some practical exercises prepare the student to function as a mid-level manager of a supply operation. Topics include stock control, receipt and issue, and appropriate automated and manual systems to monitor the operation.

Credit Recommendation: In the lower-division baccalaureate/associate degree category, 2 semester hours in business communication, 1 in principles of supervision, and 3 in material management and/or logistics management (5/02).

Related Occupations: 920B.

AR-1408-0159

Command Sergeants Major

Course Number: CSM 206.

Location: Combined Arms Training Center, Vilseck, W. Germany.

Length: 1 week (40–75 hours).

Exhibit Dates: 4/86–10/98.

Learning Outcomes: Upon completion of the course, the student will be able to interpret regulations in the context of existing policies; assist others in the management of drug and alcohol abuse programs; make judgments relating to family advocacy programs, evaluation of enlisted personnel, equal opportunity considerations, and the noncommissioned officer education system; and demonstrate leadership through advising and consulting on matters of mission, policies, and available support services.

Instruction: A sequence of guest lectures, seminars, and conferences provides an orientation to the role and functions of the command sergeant major or sergeant major. Topics include military justice, drug and alcohol abuse prevention, personnel matters, support services, maintenance management, and training management.

Credit Recommendation: Credit is not recommended because the nature of the program is a broad-based orientation and briefing (10/88).

AR-1408-0160

Company Commander and First Sergeant
(Company Commander)

Course Number: CC/FS25; CC205/FS202; CC205.

Location: Combined Arms Training Center, Vilseck, W. Germany.

Length: 1–2 weeks (46–86 hours).

Exhibit Dates: 4/78–10/98.

Learning Outcomes: Upon completion of the course, the student will be able to make command/leadership decisions based on subjects treated in the course; establish programs in the areas of maintenance management, administration, training management, and supply management; provide guidance in matters of pay and allowances, military law, personnel reliability, and safety; and implement equal opportunities policies.

Instruction: Conference classes and guest speakers cover selected military subjects, particularly in those programs of interest to commanders and leaders. Topics of concentration include maintenance management, training management, military law, safety, personnel and administration matters, and equal opportunity policies.

Credit Recommendation: Credit is not recommended because the course is a broad-based orientation and briefing (10/88).

AR-1408-0162

Special Forces Technician Senior Warrant Officer
(Special Operations Technician Senior Warrant Officer)

Course Number: 2-33-C32.

Location: John F. Kennedy Special Warfare Center and School, Fort Bragg, NC.

Length: 9–10 weeks (331–379 hours).

Exhibit Dates: 10/87–Present.

Learning Outcomes: Upon completion of the course, the student will be able to perform common tasks and special operation-specific tasks and demonstrate leadership and management skills in Special Operations units and on Army and joint staffs.

Instruction: Lectures, discussions, and written projects cover warrant officer common subjects (communication arts, military history, leadership and ethics, program management, force integration, employment and tactics of US forces, and logistics) and advanced special operations subjects (low-intensity conflicts, civil affairs, psychological operations, threats, and S-2/S-3 duties).

Credit Recommendation: In the lower-division baccalaureate/associate degree category, 3 semester hours in business organization and management, 1 in psychology, and 1 in leadership studies (9/00).

AR-1408-0163

Advanced Special Operations Techniques

Course Number: 011-F27; 2E-F141/011-F27.

Location: John F. Kennedy Special Warfare Center and School, Fort Bragg, NC.

Length: 12–13 weeks (738–798 hours).

Exhibit Dates: 7/88–1/91.

Learning Outcomes: Upon completion of the course, the student will be able to employ advanced skills in defensive source nets in support of unconventional warfare, foreign internal defense, terrorism counteraction, and strategic reconnaissance.

Instruction: Course includes lectures, demonstrations, and performance exercises in instructional techniques necessary to train US and foreign personnel in advanced special operations intelligence activities. Course also includes the tactics and techniques employed in advanced areas of intelligence as they relate to special operations mission. The course places strong emphasis on practical application in both rural and urban environments.

Credit Recommendation: In the lower-division baccalaureate/associate degree category, 3 semester hours in business organization and management and 3 in principles of supervision (12/92).

AR-1408-0164

1. Administrative Specialist Basic Noncommissioned Officer (NCO) (71L30)
2. Executive Administrative Assistant/ Administrative Specialist Basic Noncommissioned Officer (NCO)
3. Administrative Specialist Basic Noncommissioned Officer (NCO)

Course Number: *Version 1:* 510-71L30. *Version 2:* 510-71C/L30. *Version 3:* 510-71L30.

Location: *Version 1:* Noncommissioned Officer Academy, Fort Benjamin Harrison, IN; Soldier Support Institute, Fort Benjamin Harrison, IN. *Version 2:* Adjutant General School, Fort Benjamin Harrison, IN. *Version 3:* Soldier Support Institute, Fort Jackson, SC.

Length: *Version 1:* 3–4 weeks (129 hours). *Version 2:* 5 weeks (186 hours). *Version 3:* 5–6 weeks (179–180 hours).

Exhibit Dates: *Version 1:* 10/86–9/91. *Version 2:* 10/91–12/95. *Version 3:* 1/96–Present.

Learning Outcomes: *Version 1:* Upon completion of the course, the student will be able to lead a squad or section; train others to fight and survive on a battlefield; and develop, maintain, and supervise maintenance activities and accountability. *Version 2:* Upon completion of the course, the student will be able to prepare business correspondence; maintain records; review publications for corrections; and set up the computer, activate the computer system, and prepare various documents. *Version 3:* Upon completion of the course, the student will be able to prepare business correspondence; maintain records; review publications for corrections; and set up the computer, activate the computer system, and prepare various documents.

Instruction: *Version 1:* Topics include leadership subjects, personal and performance counseling, training, and document preparation and verification. This course contains a common core of leadership subjects. *Version 2:* Lectures and appropriate applications cover the above subjects. *Version 3:* Lectures and appropriate applications cover the above subjects. This course contains a common core of leadership subjects.

Credit Recommendation: *Version 1:* Credit is recommended for the common core only. See AR-1406-0090 (12/88). *Version 2:* In the lower-division baccalaureate/associate degree category, 2 semester hours in clerical procedures and 1 in computer applications (12/91). *Version 3:* In the lower-division baccalaureate/associate degree category, 2 semester hours in clerical procedures and 1 in computer applications. See AR-1406-0090 for common core credit (9/97).

Related Occupations: 71C; 71L.

AR-1408-0165

Planning, Programming, Budgeting and Execution Systems

Course Number: *Version 1:* 7D-45A/541-F13; 7D-45A/B. *Version 2:* 7D-45A/541-F13, Phase 1.

Location: *Version 1:* Soldier Support Institute, Fort Jackson, SC; Finance School, Soldier Support Institute, Fort Benjamin Harrison, IN. *Version 2:* Finance School, Fort Jackson, SC.

Length: *Version 1:* 1–2 weeks (60–66 hours). *Version 2:* 2 weeks (75 hours).

Exhibit Dates: *Version 1:* 8/87–12/02. *Version 2:* 1/03–Present.

Learning Outcomes: *Version 1:* Upon completion of the course, the student will be able to explain the process of fund flow, discuss control functions, identify approval levels, describe the procurement process, and examine the program and budget cycle for security assistance. *Version 2:* Pending evaluation.

Instruction: *Version 1:* Methods of instruction include lectures, class discussions, and in-class exercises. Topics covered include budget planning process, acquisition of funds, cost analysis, internal controls, organization efficiency, and audit compliance. *Version 2:* Pending evaluation.

Credit Recommendation: *Version 1:* In the upper-division baccalaureate category, 2 semester hours in budgeting theory (9/96). *Version 2:* Pending evaluation.

Related Occupations: 45A.

AR-1408-0166

Finance Officer Basic

Course Number: *Version 1:* 7-14-C20-44C; 7-14-C20; 7-14-C20-44A. *Version 2:* 7-14-C20-44A, Phase 1.

Location: *Version 1:* Soldier Support Institute, Fort Jackson, SC; Finance School, Soldier Support Institute, Fort Benjamin Harrison, IN. *Version 2:* Training Center, Fort Jackson, SC.

Length: *Version 1:* 15–16 weeks (567–601 hours). *Version 2:* 15–16 weeks (615 hours).

Exhibit Dates: *Version 1:* 10/86–7/01. *Version 2:* 8/01–Present.

Learning Outcomes: *Version 1:* Upon completion of the course, the student will be able to supervise and lead, communicate effectively, and perform supervisory accounting functions. *Version 2:* Pending evaluation.

Instruction: *Version 1:* Lectures and structured exercises replicating practical situations cover supervision, leadership, and oral and written communication. Financial aspects include budgeting, disbursement, documentation, and fund accounting. *Version 2:* Pending evaluation.

Credit Recommendation: *Version 1:* In the lower-division baccalaureate/associate degree category, 1 semester hour in principles of supervision, 1 as a communication elective, and 2 in governmental accounting (9/97). *Version 2:* Pending evaluation.

Related Occupations: 44A.

AR-1408-0167

Administrative Senior Warrant Officer Advanced
(Administrative Warrant Officer Advanced)

Course Number: 7-12-C32.

Location: Soldier Support Institute, Fort Benjamin Harrison, IN.

Length: 5–6 weeks (202 hours).

Exhibit Dates: 6/89–1/97.

Learning Outcomes: After 9/00 see AR-1408-0289. Upon completion of the course, the student will be able to manage a medium-sized organization including personnel supervision, computer equipment, and software utilization and provide leadership for military and civilian personnel.

Instruction: Lectures, practical exercises and discussion cover the above topics.

Credit Recommendation: In the lower-division baccalaureate/associate degree category, 1 semester hour in leadership, 1 in introduction to microcomputers, and 2 in personnel supervision (12/91).

AR-1408-0168

1. Adjutant General Officer Advanced
2. Adjutant General Officer Advanced
3. Adjutant General Captains Career, Phase 1

Course Number: *Version 1:* 7-12-C22. *Version 2:* 7-12-C22. *Version 3:* 7-12-C22, Phase 1.

Location: *Version 1:* Adjutant General School, Fort Benjamin Harrison, IN. *Version 2:* Soldier Support Institute, Fort Jackson, SC; Soldier Support Institute, Fort Benjamin Harrison, IN. *Version 3:* Adjutant General School, Fort Jackson, SC.

Length: *Version 1:* 20 weeks (710–756 hours). *Version 2:* 20 weeks (785 hours). *Version 3:* 18 weeks (629 hours).

Exhibit Dates: *Version 1:* 4/88–6/94. *Version 2:* 7/94–9/00. *Version 3:* 10/00–Present.

Learning Outcomes: *Version 1:* Upon completion of the course, the student will be able to manage human resources in a large organization. *Version 2:* Upon completion of the course, the student will be able to supervise subordinates and manage human resources in a large organization. The student will also apply written and oral communication skills and describe selected topics in military science relating to staffing, census, and resource planning. *Version 3:* Upon completion of the course, the student will be able to manage human resources in a medium-sized organization.

Instruction: *Version 1:* Lectures, practical exercises, and self-guided instruction cover business communication, safety programs, training programs, leadership, motivation, performance management, classification systems, evaluation techniques, goal setting, budgeting, fund accounting, PPB systems, ethics, and American military history. *Version 2:* Lectures, practical exercises, class discussions, and examinations cover business communication, supervision of individuals and groups, leadership principles, performance reviews, safety programs, motivation, and classification systems. *Version 3:* Lectures, practical exercises, class discussions, and examinations cover business communication, leadership principles, performance reviews, automation, human resource management activities, and military science.

Credit Recommendation: *Version 1:* In the lower-division baccalaureate/associate degree category, 1 semester hour in business communication and 1 in American military history. In the upper-division baccalaureate category, 3 semester hours in principles of supervision, 3 in human resource management, and 1 in governmental accounting (11/91). *Version 2:* In the lower-division baccalaureate/associate degree category, 2 semester hours in business communication, 2 in military science, and 3 in personnel supervision. In the upper-division baccalaureate category, 3 semester hours in human resource management (1/97). *Version 3:* In the lower-division baccalaureate/associate degree category, 3 semester hours in personnel supervision. In the upper-division baccalaureate category, 3 semester hours in military science and 3 in human resource management (3/02).

AR-1408-0169

Administrative Specialist Advanced Noncommissioned Officer (NCO)

Course Number: *Version 1:* 510-71L40. *Version 2:* 510-71L40. *Version 3:* 510-71L40, Phase 2.

Location: *Version 1:* Soldier Support Institute, Fort Benjamin Harrison, IN. *Version 2:* Soldier Support Institute, Fort Jackson, SC. *Version 3:* Soldier Support Center, Fort Jackson, SC.

Length: *Version 1:* 7 weeks (251–260 hours). *Version 2:* 5–6 weeks (191–192 hours). *Version 3:* 2–3 weeks (156 hours).

Exhibit Dates: *Version 1:* 10/87–9/95. *Version 2:* 10/95–9/00. *Version 3:* 10/00–Present.

Learning Outcomes: *Version 1:* Upon completion of the course, the student will be able to supervise and lead units of up to 400 persons. *Version 2:* Upon completion of the course, the student will be able to review correspondence, files, forms, and publications. *Version 3:* Upon completion of the course, the student will be able to teach, evaluate, and counsel personnel; conduct collective training; and perform as a supervisor.

Instruction: *Version 1:* Instruction includes lectures and practical exercises in a common core,. including written and oral communication, management principles, unit training, and combat survival. *Version 2:* Methods used are lecture and Army-specific forms. Course includes a common core of leadership subjects. *Version 3:* Methods of instruction include classroom exercises, lectures, computer laboratories, practical exercises, and discussion. Topics include military field training, supervision, and administrative operations.

Credit Recommendation: *Version 1:* Credit is recommended for the common core only. See AR-1404-0035 (11/91). *Version 2:* In the lower-division baccalaureate/associate degree category, 1 semester hour in clerical procedures. See AR-1404-0035 for common core credit (9/97). *Version 3:* In the lower-division baccalaureate/associate degree category, 2 semester hours in military science and 3 in principles of supervision (3/02).

Related Occupations: 71L.

AR-1408-0170

Target Acquisition/Surveillance Radar Repairer Basic Noncommissioned Officer (NCO)

Course Number: *Version 1:* 104-39C30. *Version 2:* 104-39C30.

Location: Field Artillery School, Fort Sill, OK.

Length: *Version 1:* 6–8 weeks (218–355 hours). *Version 2:* 8 weeks (355 hours).

Exhibit Dates: *Version 1:* 1/87–12/91. *Version 2:* 1/92–12/96.

Learning Outcomes: *Version 1:* Upon completion of the course, the student will be able to supervise and provide technical assistance in the areas of maintenance management, logistics support, and quality control for a target acquisition/surveillance radar system. *Version 2:* Upon completion of the course, the student will be able to supervise and provide technical assistance in the areas of maintenance management, logistics support, and quality control for a target acquisition/surveillance radar system.

Instruction: *Version 1:* This course covers the administration and operation of a communications-electronics repair facility, including maintenance management, supply procedures, quality control procedures, diagnostic examinations, and future trends. This course contains a common core of leadership subjects. *Version 2:* Lectures and practical exercises cover general maintenance management, supply logistics, and quality control procedures and practices. The student also studies the operation and maintenance of the target acquisition command/control system and its associated computer, radar, rawinsonde, and meteorological data system.

Credit Recommendation: *Version 1:* In the lower-division baccalaureate/associate degree category, 2 semester hours in material management. See AR-1406-0090 for common core credit (1/92). *Version 2:* In the lower-division baccalaureate/associate degree category, 3 semester hours in maintenance management. See AR-1406-0090 for common core credit (9/94).

Related Occupations: 39C.

AR-1408-0171

Unit Administration

(Unit Administration Basic)

Course Number: 921-110; 0100; 100; 100-LVL1.

Location: Reserve Readiness Training Center, Fort McCoy, WI.

Length: 2 weeks (80–120 hours).

Exhibit Dates: 7/87–Present.

Learning Outcomes: Upon completion of the course, the student will be able to prepare and administer financial procedures; describe the organizational structure and chain of command; track personnel from selection through education, training, assignment, evaluation, and promotion using management systems; and administer retention programs, including reenlistment, service obligations, participation requirements, and enforcement procedures.

Instruction: Lectures, discussions, and simulations follow a precourse survey. A comprehensive performance examination is administered for each major unit.

Credit Recommendation: In the lower-division baccalaureate/associate degree category, 2 semester hours in office administration or records and information management (11/02).

AR-1408-0172

Unit Manager

Course Number: 400.

Location: Reserve Readiness Training Center, Fort McCoy, WI.

Length: 2 weeks (85 hours).

Exhibit Dates: 4/87–10/92.

Learning Outcomes: Upon completion of the course, the student will be able to supervise civilian employees; write or change job descriptions for civilian employees; administer alcohol and drug abuse prevention programs; manage work scheduling and leave administration including overtime and compensatory time within the Fair Labor Standards Act; recruit, interview, and select personnel using Equal Employment Opportunity guidelines; write standards to be used in performance appraisals; manage personnel using proper time management procedures, group decision making processes, stress management techniques, and problems solving techniques; use automated systems for unit management; determine the validity of data recorded for reserve training; oversee training administration and entitlements; use Army National Guard and Reserve (AGR) assets; use Fulltime Unit Status model; fill vacancies and generate job descriptions for AGR soldiers; evaluate personnel; use the promotion, assignment/reassignment system; record and use proper fund control in budget expenditures; and use the requirements-based budgeting concept as applied to unit management, force integration, and manpower needs.

Instruction: Reading, lectures, and discussions cover the role of the supervisor of civilian and military personnel, including job classification, work scheduling, leave administration, recruitment and selection of personnel, discipline, grievances, recognition of employees, Equal Employment Opportunity Standards, safety and occupational health, performance appraisal systems, resolution of work place conflicts, interpersonal relationships, motivation techniques, and the use of AGR assets. Practical exercises and case studies cover the use of AGR assets, group decision making in problem solving, leadership styles in problem solving, evaluation plans and alternatives, validity of unit records of reserve training, applicability of automated systems to unit management, readiness, grievances, employee recognition, discipline, Department of the Army guidelines, and union contracts.

Credit Recommendation: In the lower-division baccalaureate/associate degree category, 3 semester hours in principles of supervision (7/89).

AR-1408-0173

1. Unit Records Administration
2. Unit Records Administration/ Standard Installation Division Personnel System (SIDPERS)
3. Unit Records Administration Automation

Course Number: *Version 1:* 1200. *Version 2:* 1200; 1200-LVL1. *Version 3:* 921-120.

Location: Reserve Readiness Training Center, Fort McCoy, WI.

Length: *Version 1:* 2 weeks (62 hours). *Version 2:* 2 weeks (71 hours). *Version 3:* 2 weeks (80 hours).

Exhibit Dates: *Version 1:* 6/89–3/94. *Version 2:* 4/94–9/00. *Version 3:* 10/00–Present.

Learning Outcomes: *Version 1:* Upon completion of the course, the student will be able to establish and maintain the various records required by a unit in the Army. *Version 2:* Upon completion of the course, the student will be able to establish and maintain personnel records, unit files, and publications accounts and perform basic filing. *Version 3:* Upon completion of the course, the student will be able to establish and maintain a publications account, establish unit files, prepare file labels and establish unit files, establish military personnel records, define family care plan requirements, and update personnel qualification records.

Instruction: *Version 1:* Classroom lectures and discussions, workbook exercises, quizzes, final examination are the modes of instruction. Topics covered are application of the Freedom of Information act and Privacy act; establishing a records system for a Army unit, including health and dental records, automated personnel records, updates to the automated personnel data base, and authorization documents; and personnel management activities. *Version 2:* Classroom lectures and discussions, workbook exercises, quizzes, and final examination are the modes of instruction. Topics covered are application of the Freedom of Information act and Privacy act; establishing a records system for a Army unit, including health and dental records, automated personnel records, updates to the automated personnel data base, and authorization documents; and personnel management activities. *Version 3:* Practical and classroom exercises, and computer-based training. Topics include freedom of information act, publications, files management, health records, promotions, and family care plan.

Credit Recommendation: *Version 1:* In the lower-division baccalaureate/associate degree category, 2 semester hours in records management (7/89). *Version 2:* In the lower-division baccalaureate/associate degree category, 2 semester hours in records management (3/97). *Version 3:* In the lower-division baccalaureate/associate degree category, 2 semester hours in computer applications (11/02).

AR-1408-0174

Field Artillery Officer Advanced Reserve Component

Course Number: 2-6-C26.

Location: Field Artillery School, Fort Sill, OK.

Length: 12 weeks (481 hours).

Exhibit Dates: 2/86–12/96.

Learning Outcomes: Upon completion of the course, the student will be able to serve as a fire support officer, battery commander, and field artillery staff officer.

Instruction: Lectures, discussions, and practical exercises cover battle doctrine, effective communication, tactics, fire support, targeting, weapons, and security. Other topics include leadership, administration, report writing, procurement, supply accounting, map reading, and training.

Credit Recommendation: In the lower-division baccalaureate/associate degree category, 2 semester hours in leadership, 2 in personnel supervision, and 1 in technical writing (6/89).

AR-1408-0176

Collection Management/Tencaps Operations Assignment Specific Module

Course Number: 3A-S13E/ASI9C/243-ASIT4.

Location: Intelligence School, Fort Huachuca, AZ.

Length: 3 weeks (116 hours).

Exhibit Dates: 10/88–4/04.

Learning Outcomes: Upon completion of the course, the student will perform intelligence collection duties and manage an operation for tactical exploitation of national capabilities.

Instruction: Lectures and practical exercises cover analyzing all sources of intelligence collection, reconnaissance, and surveillance; determining the availability and capability of collection resources; implementing collection procedures; evaluating intelligence reports; and updating collection plans including space intelligence collection activities.

Credit Recommendation: In the upper-division baccalaureate category, 3 semester hours in planning and operations management (6/90).

AR-1408-0177

Military Intelligence Officer Basic

Course Number: *Version 1:* 3-30-C20-35D. *Version 2:* 3-30-C20-35D. *Version 3:* 3-30-C20-35D.

Location: *Version 1:* Intelligence School, Fort Huachuca, AZ. *Version 2:* Intelligence School, Fort Huachuca, AZ. *Version 3:* Intelligence Center and School, Fort Huachuca, AZ.

Length: *Version 1:* 23–24 weeks (915–922 hours). *Version 2:* 18–23 weeks (818–950 hours). *Version 3:* 18 weeks (650 hours).

Exhibit Dates: *Version 1:* 10/86–12/92. *Version 2:* 1/93–2/98. *Version 3:* 3/98–Present.

Learning Outcomes: *Version 1:* Upon completion of the course, the student will have the skills in leadership, personnel management, security, and communication necessary to command a military intelligence platoon. *Version 2:* Upon completion of the course, the student will be prepared to perform leadership and management duties in the intelligence arena. *Version 3:*

Upon completion of the course, the student will be prepared to accept the duties and responsibilities of a tactical all-source intelligence officer.

Instruction: *Version 1:* Instruction includes lectures and practical exercises in writing effectiveness; organization and planning; leadership and personnel management; maintenance; security; supply; signal interpretation; signal theory; and Soviet, North Korean, Warsaw Pact, and NATO capabilities. *Version 2:* Lectures and practical exercises cover leadership and ethics, effective communication skills, tactics, counterintelligence, combat communications, signal theory, battle-focused training, and military law. *Version 3:* Lectures and practical exercises cover leadership and ethics, effective communication skills, tactics, counterintelligence, combat communications, signal theory, battle-focused training, and military law.

Credit Recommendation: *Version 1:* In the lower-division baccalaureate/associate degree category, 3 semester hours in principles of physical security and 1 in business writing. In the upper-division baccalaureate category, 1 semester hour in political science (areas of geopolitical struggle with emphasis on Eastern/Western Europe), 3 in human resource management, and 3 in principles of organization and management (1/92). *Version 2:* In the lower-division baccalaureate/associate degree category, 1 semester hour in communication and 2 in principles of supervision (6/97). *Version 3:* In the lower-division baccalaureate/associate degree category, 1 semester hour in communication and 2 in principles of supervision (8/01).

Related Occupations: 35D.

AR-1408-0180

Chaplain Officer Advanced Reserve Component

Course Number: 5-16-C23.

Location: Chaplain Center and School, Fort Monmouth, NJ.

Length: 18 weeks (681 hours).

Exhibit Dates: 1/89–4/04.

Learning Outcomes: Upon completion of the course, the student will be able to perform administrative and ministerial duties in a large (up to 500 persons) military organization.

Instruction: Topics include leadership, ethics, management, pastoral counseling, fiscal management, combat and tactical functions, appropriated and nonappropriated funds, marriage and the family, theological issues, and homily preparation.

Credit Recommendation: In the upper-division baccalaureate category, 3 semester hours in ethics and theological issues and 3 in organizational management. In the graduate degree category, 3 semester hours in homiletics and 3 in marriage and family counseling (6/90).

AR-1408-0182

Administrative Officer

Course Number: PEC-P-A03.

Location: Professional Education Center, Cp. Joseph Robinson, North Little Rock, AR.

Length: 2 weeks (64 hours).

Exhibit Dates: 3/90–4/04.

Learning Outcomes: Upon completion of the course, the student will be able to perform and supervise others in basic administration and personnel activities including the review of military personnel forms.

Instruction: Lectures and in class exercises are designed to prepare the student to supervise the preparation of forms addressing administrative, financial, and information management actions, as well as retention and fitness concerns.

Credit Recommendation: In the lower-division baccalaureate/associate degree category, 1 semester hour in administrative management (7/90).

AR-1408-0183

National Guard (ARNG) Recruiting and Retention Detachment Commander

Course Number: FC-F21 (ARNG).

Location: Professional Education Center, Cp. Joseph Robinson, North Little Rock, AR.

Length: 2 weeks (77 hours).

Exhibit Dates: 4/86–4/04.

Learning Outcomes: Upon completion of the course, the student will be able to evaluate markets, develop a sales team, and conduct recruiter training programs.

Instruction: Lectures and in-class exercises cover the areas of time management, team building, human and material resource management, sales management, and personality measurement techniques. Course provides the student with technical expertise and interactive skills which effectively direct and influence the performance of individual recruiters.

Credit Recommendation: In the lower-division baccalaureate/associate degree category, 2 semester hours in general management (7/90).

AR-1408-0184

Guidance Counselor

Course Number: 501 ASIV7 (ARNG).

Location: Professional Education Center, Cp. Joseph Robinson, North Little Rock, AR.

Length: 2 weeks (75 hours).

Exhibit Dates: 5/89–12/97.

Learning Outcomes: Upon completion of the course, the student will be able to provide guidance to personnel on retention opportunities.

Instruction: Structural exercises, lectures, and discussions are the instructional strategies. Topics are recruitment, retention, and processing information related to personnel.

Credit Recommendation: Credit is not recommended because of the military-specific nature of the course (10/90).

AR-1408-0185

1. National Guard (ARNG) Recruiting Field Sales Manager
2. ARNG Recruiting Noncommissioned Officer (NCO) In-Charge

Course Number: *Version 1:* 501-F31 (ARNG). *Version 2:* 501-F31-ARNG.

Location: Professional Education Center, Cp. Joseph Robinson, North Little Rock, AR.

Length: *Version 1:* 2 weeks (80 hours). *Version 2:* 2 weeks (65 hours).

Exhibit Dates: *Version 1:* 7/89–9/93. *Version 2:* 10/93–Present.

Learning Outcomes: *Version 1:* Upon completion of the course, the student will be able to effectively use skills in time management, sales team development, interpersonal management, stress management, and self-awareness studies (Meyers-Briggs) to improve communication and motivation. *Version 2:* Upon completion of the course, the student will be able to develop a recruiting team, and including training and evaluation of recruiter performance use self-awareness and personality- type instruments, time management, stress identification, and team management and training.

Instruction: *Version 1:* This course is performance-based with many exercises, usually performed on an individual basis. Extensive use is made of materials developed privately as well as in the Army National Guard. Evaluation of exercises and material occurs throughout the course. *Version 2:* This course is performance-based with many exercises, usually performed on an individual basis. Extensive use is made of materials developed privately as well as in the Army National Guard. Evaluation of exercises and material occurs throughout the course.

Credit Recommendation: *Version 1:* In the upper-division baccalaureate category, 2 semester hours in sales management (7/90). *Version 2:* In the upper-division baccalaureate category, 3 semester hours in sales management (6/98).

AR-1408-0186

Army Installation Management

Course Number: 1B-F1.

Location: Army Logistics Management College, Fort Lee, VA.

Length: 3 weeks (107–122 hours).

Exhibit Dates: 4/87–9/95.

Learning Outcomes: Upon completion of the course, the student will be able to manage a staff, using new techniques affecting support operations, and avoid or solve common problems. Graduate will serve in a key position, managing the garrison as a community, and will formulate strategies for putting action plans into operation.

Instruction: Instruction includes guest speakers, small-group instruction, and lectures with topics such as garrison operations, resource management, budget and finance, labor relations, and other installation management issues.

Credit Recommendation: In the upper-division baccalaureate category, 3 semester hours in organization and management (8/95).

AR-1408-0187

Purchasing Fundamentals
(Small Purchase Fundamentals)
(Defense Small Purchase Basic)

Course Number: PUR 101; ALMC-B3.

Location: Army Logistics Management College, Fort Lee, VA; Army Logistics Management College, Satellite Education Program, Fort Lee, VA; Army Logistics Management College, Onsite at various locations.

Length: 2 weeks (76–79 hours).

Exhibit Dates: 10/89–9/97.

Learning Outcomes: After 9/97 see DD-0326-0008. Upon completion of the course, the student will be able to define and discuss policies and procedures for standards of conduct, explain regulatory guidance for small purchases within the DoD, determine the authority of the contracting officer and buyer to make awards, use mandatory sources of supply or services, maintain small purchase files, determine that elements required for a valid purchase request are present, list requirements for maintaining source lists, explain competition and market research, employ the most appropriate small purchase method, identify applicable clauses and/or provisions, employ the appropriate solicitation method and complete the solicitation phase, handle quotations, amend or cancel RFQs, use price analysis techniques, demonstrate the procedures used in placing blanket purchase agreements, identify small purchase categories, and explain post-award functions including contractor's failure to perform.

Instruction: Methods of instruction include critiques, demonstrations, lectures, practical exercises, readings, and examinations. Topics include small purchase policies and procedures, presolicitation phase, solicitation phase, evaluation phase, award methods, special small purchase categories, and administration phase.

Credit Recommendation: In the upper-division baccalaureate category, 2 semester hours in basic purchasing (11/93).

AR-1408-0188

Executive Small Purchase
(Defense Small Purchase Advanced)

Course Number: PUR 201; ALMC-B4.

Location: Army Logistics Management College, Fort Lee, VA; Army Logistics Management College, Onsite at various locations.

Length: 1–2 weeks (62–63 hours).

Exhibit Dates: 9/90–11/97.

Learning Outcomes: After 9/97 see DD-1408-0031. Upon completion of the course, the student will be able to perform as a procurement agent within DoD, handle complex purchase requests (under $25 thousand), evaluate small purchase quotations, conduct negotiations for small purchases, and handle all aspects of small purchases (not to exceed $25,000).

Instruction: Using lectures and practical exercise, the course covers all phases of small purchase procurement, including analyzing bid quotations, conducting negotiations, and determining the criteria for technical evaluation.

Credit Recommendation: In the upper-division baccalaureate category, 2 semester hours in purchasing (6/95).

AR-1408-0189

Contracting Officer's Representative

Course Number: ALMC-CL.

Location: Army Logistics Management College, Fort Lee, VA; Army Logistics Management College, Satellite Education Program, Fort Lee, VA; Army Logistics Management College, Onsite at various locations.

Length: 2 weeks (65 hours).

Exhibit Dates: 2/89–6/90.

Learning Outcomes: Upon completion of the course, the student will be able to describe the process and the requirements of planning, awarding, and administering service contracts; apply managerial tools and techniques to monitoring contract performance; and exercise the authority, assume the responsibility, and carry out the activities of a contracting officer's representative.

Instruction: Instruction includes lectures, cases, and exercises on pre-award and award considerations for service contracts, writing and interpreting work statements, and post-award functions.

Credit Recommendation: In the lower-division baccalaureate/associate degree category, 2 semester hours in business communication/contract writing (6/91).

AR-1408-0190

Management of Installation Level Contracts

Course Number: ALMC-IB.

Location: Army Logistics Management College, Fort Lee, VA; Army Logistics Management College, Onsite at various locations.

Length: 2 weeks (66–73 hours).

Exhibit Dates: 3/88–9/93.

Learning Outcomes: Upon completion of the course, the student will be able to explain the team concept of contract management; make sound pre-award decisions in terms of organizational needs, governmental regulations, and ethical considerations; and identify DoD policies, contract law provisions, and legislative acts that directly affect the contracting officer/team.

Instruction: Lectures, practical exercises, and case studies cover contract negotiations and administration of services on installations.

Credit Recommendation: In the upper-division baccalaureate category, 1 semester hour in contract law and 1 in contract administration (6/91).

AR-1408-0191

Integrated Item Manager's

Course Number: ALMC-IF.

Location: Army Logistics Management College, Fort Lee, VA.

Length: 4 weeks (154–155 hours).

Exhibit Dates: 4/89–9/96.

Learning Outcomes: Upon completion of the course, the student will be able to apply catalogue data (inventory management codes and numbers) to daily material management functions; forecast requirements based on data analysis, projection of need, and allocation of funds; use available systems in making sound, logical materiel decisions; and calculate and project materiel depot maintenance (repair/overhaul) programs.

Instruction: Lectures and practical exercises provide instruction on the integrated material management functions as they relate to the management of major and secondary items. Subjects covered include the life cycle of the materiel system with emphasis placed on wholesale requirement computation, forecasting, and financial management.

Credit Recommendation: In the lower-division baccalaureate/associate degree category, 3 semester hours in materiel management (10/92).

AR-1408-0192

Defense Contracting for Information Resource

Course Number: ALMC-ZX.

Location: Army Logistics Management College, Fort Lee, VA.

Length: 1–2 weeks (70–71 hours).

Exhibit Dates: 4/90–9/98.

Learning Outcomes: Upon completion of the course, the student will be able to interpret the impact of DoD directives, federal regulations, and legislation on the acquisition of information resources; identify the difference in contracting for information resources compared with other resources/services; and implement all stages of contracting for information resources.

Instruction: Instruction includes lectures, discussions, exercises, and case studies covering pre-acquisition planning, acquisition plan development, presolicitation, solicitation development, evaluation, award, and performance evaluation of information resources contracts.

Credit Recommendation: In the upper-division baccalaureate category, 2 semester hours in contract management (6/91).

AR-1408-0193

Marine Warrant Officer Advanced
(Marine Warrant Officer Senior Warrant Officer)

Course Number: 8-55-C32-880A/881A.

Location: Transportation School, Fort Eustis, VA.

Length: 6 weeks (216–274 hours).

Exhibit Dates: 10/90–Present.

Learning Outcomes: Upon completion of the course, the student will be able to describe the laws governing weapons, treatment of captives, and rights of prisoners of war; report criminal acts; develop spreadsheets and data bases; prepare written course assignments; prepare personnel evaluations and recommendations for awards; manage and counsel civilian personnel in federal employment; perform electronic navigation; identify points of entry for Army vessels; and act as liaison with harbormaster and marine engineering officer.

Instruction: Lectures, examinations, and practical exercises cover laws, drugs, weapons, Warsaw Pact, electronic warfare, computer literacy, military and civilian personnel standards, customs, reports, management, marine safety, and military writing.

Credit Recommendation: In the lower-division baccalaureate/associate degree category, 3 semester hours in personnel supervision and 1 in computer literacy (10/00).

Related Occupations: 880A; 881A.

AR-1408-0194

Intelligence/Electronic Warfare (EW) Equipment Technician Warrant Officer Technical/Tactical Certification, Phase 2

Course Number: 4C-353A.

Location: Intelligence Center and School, Fort Huachuca, AZ; Intelligence School, Fort Devens, MA.

Length: 10 weeks (357 hours).

Exhibit Dates: 10/90–2/99.

Learning Outcomes: Upon completion of the course, the student will be able to initiate, operate, and manage an industrial maintenance, safety, and quality assurance program and a logistics and supply organization that supports the operation and maintenance of computer-controlled equipment and software.

Instruction: Lectures, demonstrations, and practical exercises cover operations, supply, maintenance, safety, and quality systems that support computer-based equipment and software.

Credit Recommendation: In the lower-division baccalaureate/associate degree category, 3 semester hours in industrial management, safety, and maintenance (5/01).

Related Occupations: 353A.

AR-1408-0196

Ordnance Missile Material Management (Phase 1 and 3)

Course Number: 093-AOC-91C-RC.

Location: Missile and Munitions School, Redstone Arsenal, AL.

Length: Phase 1, 2 weeks (88 hours); Phase 2, 2 weeks (87 hours).

Exhibit Dates: 6/90–1/96.

Learning Outcomes: Upon completion of Phase 1, the student will be able to plan, support, conduct, and evaluate training in material-handling assignments. Upon completion of Phase 3, the student will be able to manage directing movements; provide in-transit visibility, authorization levels, and requisition assets; and perform automated accounting and material management (Phase 3).

Instruction: Phase 1 includes lectures, role playing, discussions, and practical exercises in missile material management. Phase 3 includes lectures and practical exercises in material management.

Credit Recommendation: In the lower-division baccalaureate/associate degree category, 1 semester hour in principles of management for Phase 1 and 1 semester hour in material management for Phase 3 (8/91).

Related Occupations: 91C.

AR-1408-0197

Ordnance Munitions Material Management (Phase 1 and 3)

Course Number: 093-AOC-91D-RC.

Location: Missile and Munitions School, Redstone Arsenal, AL.

Length: Phase 1, 2 weeks (106 hours); Phase 3, 2 weeks (76 hours).

Exhibit Dates: 11/88–1/96.

Learning Outcomes: Upon completion of Phase 1, the student will be able to plan, support, conduct, and evaluate training in material-handling assignments. Upon completion of Phase 3, the student will be able to manage directing movements; provide in-transit visibility, authorization levels, and requisition assets; and perform automated accounting and material management.

Instruction: Phase 1 includes lectures, role playing, discussions, and practical exercises in munitions material management. Phase 3 includes lectures and practical exercises in material management.

Credit Recommendation: In the lower-division baccalaureate/associate degree category, 1 semester hour in principles of management for Phase 1 and 1 semester hour in material management for Phase 3 (8/91).

Related Occupations: 91D.

AR-1408-0198

Ammunition Technician Senior Warrant Officer Reserve (Phase 1 and 2)

Course Number: 4-9-C32-910A-RC.

Location: Missile and Munitions School, Redstone Arsenal, AL.

Length: Phase 1, 2 weeks (100 hours); Phase 2, 2 weeks (80 hours).

Exhibit Dates: 10/90–Present.

Learning Outcomes: Upon completion of the course, the student will be able to provide effective leadership and management; describe the fundamentals of electronic warfare; describe maintenance reporting procedures; and process missile and rocket ammunition reports.

Instruction: Lectures, practical exercises, and field experience in communication skills, effective leadership, military justice, battle doctrine, electronic warfare, maintenance management, history file manipulation, system administrator procedures, and new developments in ammunition.

Credit Recommendation: In the upper-division baccalaureate category, 2 semester hours for field experience in management (5/01).

Related Occupations: 910A.

AR-1408-0199

Defense Small Purchase Basic by Correspondence

Course Number: ALMC-B3.

Location: Army Logistics Management College, Fort Lee, VA.

Length: Maximum, 52 weeks.

Exhibit Dates: 10/89–1/00.

Learning Outcomes: Upon completion of the course, the student will be able to define and discuss policies and procedures for standards of conduct and examples of improper business practices, explain regulatory guidance for small purchase within the DoD, determine the authority of the contracting officer and buyer to make awards, identify and use mandatory sources of supply or services; maintain small purchase files, determine that elements required for a valid purchase request are present, list requirements for maintaining source lists, explain competition and market research, employ the most appropriate small purchase method, identify the applicable clauses and/or provisions, employ the appropriate solicitation method and complete the solicitation phase, handle quotations, amend or cancel RFQs, use price analysis techniques, apply the procedures used in placing blanket purchase agreements, identify small purchase categories, and explain post-award functions including contractor's failure to perform.

Instruction: Topics include small purchase policies and procedures, presolicitation phase, solicitation phase, evaluation phase, award methods, special small purchase categories, and administration phase.

Credit Recommendation: In the upper-division baccalaureate category, 2 semester hours in basic purchasing (6/91).

AR-1408-0200

Adjutant General Officer Basic

Course Number: 7-12-C20-42A.

Location: Soldier Support Institute, Fort Jackson, SC; Adjutant General School, Fort Benjamin Harrison, IN.

Length: 13 weeks (564 hours).

Exhibit Dates: 8/88–Present.

Learning Outcomes: Upon completion of the course, the student will be able to write memos, letters, and reports; supervise personnel activities, including classification, record keeping, career development, awards, and transitions; and describe operational and supervisory aspects of personnel administration systems.

Instruction: Lectures and practical exercises cover the subjects above.

Credit Recommendation: In the lower-division baccalaureate/associate degree category, 1 semester hour in records management, 2 in office administration, 3 in business communication, and 3 in personnel supervision (9/96).

Related Occupations: 42A.

AR-1408-0201

1. Adjutant General Officer Advanced (Company Command Module) Reserve Component Phase 1
2. Adjutant General Officer Advanced Reserve Component Phase 2

Course Number: *Version 1:* 7-12-C23. *Version 2:* 7-12-C23.

Location: *Version 1:* Adjutant General School, Fort Benjamin Harrison, IN. *Version 2:* Soldier Support Institute, Fort Jackson, SC.

Length: *Version 1:* 2 weeks (89 hours). *Version 2:* 2 weeks (101 hours).

Exhibit Dates: *Version 1:* 10/90–3/95. *Version 2:* 4/95–Present.

Learning Outcomes: *Version 1:* Upon completion of the course, the student will be able to serve as a company commander and perform in the unique environment of the battlefield. *Version 2:* Upon completion of the course, the student will apply basic supervisory skills.

Instruction: *Version 1:* Lectures and practical exercises give a broad overview of the military environment. *Version 2:* Methods include class presentation by students and simulations.

Credit Recommendation: *Version 1:* Credit is not recommended because of the military-specific nature of the course (12/91). *Version 2:* In the lower-division baccalaureate/associate degree category, 2 semester hours in personnel supervision (9/97).

AR-1408-0202

Engineer Pre-Command

Course Number: 2G-F27.

Location: Engineer School, Fort Leonard Wood, MO.

Length: 1–2 weeks (94 hours).

Exhibit Dates: 8/92–4/04.

Learning Outcomes: Upon completion of the course, the student will be able to lead an engineering unit, organize and deliver training programs, manage maintenance policies and procedures, and manage personnel.

Instruction: Lectures and practical exercises cover the above topics.

Credit Recommendation: In the upper-division baccalaureate category, 2 semester hours in organizational management (1/93).

AR-1408-0203

Army Medical Department (AMEDD) Noncommissioned Officer (NCOES) Basic

Course Number: *Version 1:* 6-8-C40. *Version 2:* 6-8-C40 (Mock Entry); 6-8-C40. *Version 3:* 6-8-C40. *Version 4:* 6-8-C40 (Resident).

Location: Academy of Health Sciences, Fort Sam Houston, TX.

Length: *Version 1:* 5 weeks (188 hours). *Version 2:* 7 weeks (300 hours). *Version 3:* 7 weeks (301 hours). *Version 4:* 4–5 weeks (207 hours).

Exhibit Dates: *Version 1:* 10/91–3/95. *Version 2:* 4/95–5/98. *Version 3:* 6/98–4/02. *Version 4:* 5/02–Present.

Learning Outcomes: *Version 1:* Upon completion of the course, the student will be able to lead a military unit; perform performance review and counseling; supervise and delegate; and perform drills, training, and specific military skills. *Version 2:* Upon completion of the course, the student will be able to lead a military unit; perform performance review and counseling; supervise and delegate; and perform drills, training, and specific military skills. *Version 3:* Upon completion of the course, the student will be able to lead a military unit; perform performance review and counseling; supervise and delegate; and perform drills, training, and specific military skills. Students will apply relevant principles, concepts, and techniques of resource management. *Version 4:* Upon completion of the course, the student will be able to lead a military unit; perform performance review and counseling; supervise and delegate; and perform drills, training, and specific military skills. Students will apply relevant principles, concepts, and techniques of resource management.

Instruction: ***Version 1:*** The course covers military leadership, personal and performance counseling, duties and responsibilities of noncommissioned officers, leadership assessment and development, leadership procedures and combat orders, combat leadership, physical training and marksmanship, property accountability, individual and team training, small-group process, and specific military skills. The methodologies include lectures, discussions, group work, and practical application. This course contains a common core of leadership subjects. *Version 2:* The course covers military leadership, personal and performance counseling, duties and responsibilities of noncommissioned officers, leadership assessment and development, leadership procedures and combat orders, combat leadership, physical training and marksmanship, property accountability, individual and team training, small-group process, and specific military skills. The methodologies include lectures, discussions, group work, and practical application. This course contains a common core of leadership subjects. Tracks relate to military occupational specialty and are completed by only those students in that designation. *Version 3:* The course covers military leadership, personal and performance counseling, duties and responsibilities of noncommissioned officers, leadership assessment and development, leadership procedures and combat orders, combat leadership, physical training and marksmanship, property accountability, individual and team training, small-group process, and specific military skills. The methodologies include lectures, discussions, group work, and practical application. This course contains a common core of leadership subjects. *Version 4:* The course covers military leadership, personal and performance counseling, duties and responsibilities of noncommissioned officers, leadership assessment and development, leadership procedures and combat orders, combat leadership, physical training and marksmanship, property accountability, individual and team training, small-group process, and specific military skills. The methodologies include lectures, discussions, group work, and practical application. This course contains a common core of leadership subjects.

Credit Recommendation: ***Version 1:*** In the lower-division baccalaureate/associate degree category, see AR-1406-0090 for common core credit. In the upper-division baccalaureate category, 2 semester hours in military science (8/93). *Version 2:* In the lower-division baccalaureate/associate degree category, 3 semester hours in military science. For those in the Behavioral Science track, 1 semester hour of elective credit in behavioral science. For those in the Medical Supply Specialist basic track, 1 semester hour of elective credit in medical supply. For those in the Medicine Specialist track, 1 semester hour of elective credit in medical specialty. For those in the Dental Specialist track, 1 semester hour of elective credit in basic medical equipment repair. For those in the Medical Laboratory specialist track, 1 semester hour of elective credit in medical laboratory. For those in the Hospital Food Service Specialist track, 1 semester hour of elective credit in hospital food service. For those in the Veterinary Food Inspector specialist track, 1 semester hour of elective credit in veterinary food inspection. And for those in the Animal Care Specialist track, 1 semester hour of elective credit in animal care specialty. See AR-1406-0090 for common core credit (11/97). *Version 3:* In the lower-division baccalaureate/associate degree category, 3 semester hours in military science. See AR-1406-0090 for common core credit (5/01). *Version 4:* In the lower-division baccalaureate/associate degree category, 3 semester hours in military science, 3 in human resource management, and 3 in leadership (11/03).

AR-1408-0204

Medical Logistics Management

Course Number: 8B-F20.

Location: Academy of Health Sciences, Fort Sam Houston, TX.

Length: 10 weeks (378–380 hours).

Exhibit Dates: 6/92–Present.

Learning Outcomes: Upon completion of the course, the student will be able to manage the logistics of patient services in a health care facility; process patients from admission to discharge; maintain records; process medical information; conduct medical care evaluations; and prepare medical statistical reports.

Instruction: Lectures and practical exercises cover medical material management; planning, programming, and budgeting; basic computer concepts; and logistics.

Credit Recommendation: In the upper-division baccalaureate category, 4 semester hours in material management (5/96).

AR-1408-0205

Emanations Analysis Technician Warrant Officer Technical/Tactical Certification (WOTTC)—RC, Phases 2 and 3

Course Number: 3B-352J-RC.

Location: Intelligence School, Fort Devens, MA.

Length: 3–4 weeks (134 hours).

Exhibit Dates: 1/90–3/94.

Learning Outcomes: Upon completion of the course, the student will be able to supervise subordinates and conduct training programs for those assigned to electronic intelligence gathering units.

Instruction: Lectures, group interaction, and exercises cover management and operation of intelligence units.

Credit Recommendation: In the lower-division baccalaureate/associate degree category, 3 semester hours in operations management (6/91).

Related Occupations: 352J; 983A.

AR-1408-0206

Voice Intercept Technician Warrant Officer Technical/Tactical Certification (WOTTC)—RC Phases 2 and 3

Course Number: 3B-352G-RC.

Location: Intelligence School, Fort Devens, MA.

Length: 3–6 weeks (134–222 hours).

Exhibit Dates: 1/90–3/94.

Learning Outcomes: Upon completion of the course, the student will be able to manage voice intercept and collection systems, including operation, coordination, analysis, and evaluation of the systems.

Instruction: Instruction includes lectures, field trips, and performance exercises in the management of the operation of voice intercept systems including analysis and evaluation of voice intercept data.

Credit Recommendation: In the lower-division baccalaureate/associate degree category, 3 semester hours in operations management (6/91).

Related Occupations: 352G.

AR-1408-0207

Intelligence/Electronic Warfare (EW) Equipment Technician Warrant Officer Technical/Tactical Certification (Reserve Component), Phases 2 and 3

Course Number: 4C-353A.

Location: Intelligence School, Fort Huachuca, AZ.

Length: 4 weeks (202 hours).

Exhibit Dates: 10/90–2/99.

Learning Outcomes: Upon completion of the course, the student will be able to initiate, manage, and operate electronic warfare intercept systems, including diagnostic troubleshooting, system management, and logistic support.

Instruction: Lectures, demonstrations, and classroom computer-driven simulations develop capabilities in the management and operations of service/repair centers. Student will be familiar with logistics and supply policies and with the operation and repair of computer-controlled equipment.

Credit Recommendation: In the lower-division baccalaureate/associate degree category, 1 semester hour in service center management. In the upper-division baccalaureate category, 3 semester hours in operations management (5/01).

Related Occupations: 353A.

AR-1408-0208

Collection Management Operations

Course Number: 3A-SI3E/ASI9C/243-ASIT4.

Location: Intelligence School, Fort Huachuca, AZ.

Length: 3 weeks (110 hours).

Exhibit Dates: 6/92–4/04.

Learning Outcomes: Upon completion of the course, the student will be able to describe intelligence-gathering means available to the armed forces, including tactical and strategic sources, and formulate management plans for organizing the collection of intelligence to meet military objectives.

Instruction: Lectures and practical exercises cover the collection and management of intelligence from a variety of sources to meet various military conflict situations.

Credit Recommendation: Credit is not recommended because of the military-specific nature of the course (10/93).

AR-1408-0209

SIGNAL OFFICER BASIC

Course Number: ***Version 1:*** 4-11-C20. *Version 2:* 4-11-C20. *Version 3:* 4-11-C20.

Location: Signal School, Fort Gordon, GA.

Length: ***Version 1:*** 23–24 weeks (983 hours). *Version 2:* 18 weeks (705 hours). *Version 3:* 18 weeks (726 hours).

Exhibit Dates: ***Version 1:*** 10/92–4/95. *Version 2:* 5/95–9/00. *Version 3:* 10/00–Present.

Learning Outcomes: ***Version 1:*** Before 10/92 see AR-1404-0006. Upon completion of the course, the student will be able to plan and manage the communications interface for microwave-based and troposcatter-based combat net radio. *Version 2:* Upon completion of the course, the student will be able to plan and manage the communications requirements of a signal unit to include computing support, data communications support, leadership of staff, tactics, logistics and the technology of military communications. *Version 3:* Upon completion of the course, the student will be able to act as a Nodal Platoon Leader of a mobile subscriber division, or a Battalion Signal Staff Officer.

Instruction: ***Version 1:*** Instruction covers leadership functions, personnel and operations management, technician electronics, computer use, supply operations, and military subjects. *Version 2:* Lectures, classroom discussion, projects, and examinations cover introduction to computing, data communications and networking, leadership functions, personnel and operations management, supply operations, electrical circuits, fundamentals of electronics, and the communications technologies. Practical exercises are used to simulate military operations and train in the use of equipment and technology. *Version 3:* Audio visual materials, practical and classroom exercises, discussion, lecture, and computer-based training. Topics include computer hardware, Microsoft Windows NT, LAN networks, webpage design, radio telecommunications, and communication systems management.

Credit Recommendation: ***Version 1:*** In the vocational certificate category, 3 semester hours in basic electricity and 3 in basic electronics. In the lower-division baccalaureate/associate degree category, 1 semester hour in computer literacy and 3 in business organization and management (2/94). *Version 2:* In the lower-division baccalaureate/associate degree category, 2 semester hours in introduction to computing, 2 in introduction to data communications and networking, 2 in supervision or introduction to management, 1 in maintenance operations, and 3 in communications systems (5/00). *Version 3:* In the lower-division baccalaureate/associate degree category, 2 semester hours in computer hardware and software, 3 in computer networks, 2 in webpage design, 2 in introduction to management, 2 in maintenance management, and 3 in communications systems (2/03).

AR-1408-0210

SIGNAL OFFICER (BRANCH QUALIFICATION)

Course Number: ***Version 1:*** 4-11-C20 (BQ). *Version 2:* 4-11-C20 (BQ).

Location: Signal School, Fort Gordon, GA.

Length: ***Version 1:*** 10 weeks (380 hours). *Version 2:* 10 weeks (393 hours).

Exhibit Dates: ***Version 1:*** 10/92–1/96. *Version 2:* 2/96–Present.

Learning Outcomes: ***Version 1:*** Upon completion of the course, the student will be able to manage communications equipment operators and plan appropriate tactics and doctrine. *Version 2:* Upon completion of the course, the student will be able to manage communications requirements in a signal unit, including computing support, data communications support, and military communications.

Instruction: ***Version 1:*** The course employs lectures, demonstrations, and practical laboratory exercises. The main topics covered include information technology, communications security manager functions, combat net radio, area common user systems, and signal officer indoctrination. *Version 2:* Lectures, classroom discussion, projects, and examination cover introduction to computing, data communications and networking, electrical circuits, fundamentals of electronics and the communications technologies. Practical exercises are used to simulate military operations and train in the use of equipment and technology.

Credit Recommendation: ***Version 1:*** In the lower-division baccalaureate/associate degree category, 2 semester hours in field service management (2/94). *Version 2:* In the lower-division baccalaureate/associate degree category, 2 semester hours in introduction to computing, 2 in introduction to data communications and networking, and 3 in communications systems (5/00).

AR-1408-0211

AVIATION LIFE SUPPORT EQUIPMENT TECHNICIAN (ALSET), RESERVE COMPONENT

Course Number: 2C-F51/4D-ASIH2/552-ASIQ2-RC.

Location: Reserve Component Training Centers, US; Reserve training centers, US.

Length: 92 hours.

Exhibit Dates: 6/92–Present.

Learning Outcomes: Upon completion of the course, the student will be able to provide logistics, preventive maintenance, and repair support to aviation life support equipment.

Instruction: Lectures and practical application cover preventive maintenance, repair, and logistics support.

Credit Recommendation: In the lower-division baccalaureate/associate degree category, 2 semester hours in logistics and 1 in medical equipment repair technician (3/94).

AR-1408-0212

AVIATION LIFE SUPPORT EQUIPMENT TECHNICIAN

Course Number: ***Version 1:*** 2C-F51/4D-ASIH2/600-ASIQ2. *Version 2:* 2C-F51/4D-ASIH2/600-ASIQ2, Phase 1; 2C-F51/4D-ASIH2/600-ASIQ2.

Location: ***Version 1:*** Transportation School, Fort Eustis, VA. *Version 2:* Aviation School, Fort Rucker, AL.

Length: ***Version 1:*** 5–6 weeks (196 hours). *Version 2:* 5 weeks (200 hours).

Exhibit Dates: ***Version 1:*** 10/91–9/96. *Version 2:* 10/96–Present.

Learning Outcomes: ***Version 1:*** Before 10/91 see AR-1704-0008. Upon completion of the course, the student will be able to maintain and repair aviation life support equipment including survival equipment. *Version 2:* Upon completion of the course, the student will be able to inspect, maintain, and repair aviation life support equipment and survival equipment, including flotation equipment, anti-exposure suits, first aid kits, survival vests, radios, aircrew body armor and restraint harnesses, flight helmets, chemical and biological protective masks, oxygen masks and systems and conduct aircrew survival training.

Instruction: ***Version 1:*** Lectures and practical applications cover logistics and publications and testing, maintenance, and repair of life support and survival equipment. *Version 2:* Lectures and practical exercises cover logistics, publication, testing and maintenance and repair of life support and survival equipment.

Credit Recommendation: ***Version 1:*** In the lower-division baccalaureate/associate degree category, 2 semester hours in logistics and 2 in medical equipment repair technician (3/94). *Version 2:* Credit is not recommended because of the limited, specialized nature of the course (1/03).

AR-1408-0213

FIELD ARTILLERY OFFICER ADVANCED

Course Number: ***Version 1:*** 2-6-C22. *Version 2:* 2-6-C22.

Location: Field Artillery School, Fort Sill, OK.

Length: ***Version 1:*** 20 weeks (766 hours). *Version 2:* 15–20 weeks (630–850 hours).

Exhibit Dates: ***Version 1:*** 4/94–9/95. *Version 2:* 10/95–Present.

Learning Outcomes: ***Version 1:*** Before 4/94 see AR-1408-0028. Upon completion of the course, the student will be able to make decisions or recommend courses of action relating to combat support, threat forces, and doctrine; use fire support systems in combined arms operations; and use concepts, doctrine, and tactics in making decisions, performing missions, and in supervising subordinates. *Version 2:* Upon completion of the course, the student will be able to use appropriate field artillery hardware, perform data analysis, and perform combat service support. Student will use effective communication techniques to lead and supervise personnel.

Instruction: ***Version 1:*** Lectures, conferences, practical exercises, and field exercises cover battle doctrine, tactics, combined arms, fire support for battle, and automated and manual fire support systems; concepts and procedures for survey, target acquisition (radar), ballistics, and battle orders; and maintenance, supply, communication, safety procedures, and technical writing. *Version 2:* This course consists of classroom lectures, conferences with practical exercises, computer laboratory, and use of appropriate field artillery hardware. Skills learned include data analysis for Army operation doctrine, fire support system; combined arms operations; threat forces; combat service support; leadership; and command control.

Credit Recommendation: *Version 1:* In the lower-division baccalaureate/associate degree category, 1 semester hour in data analysis, 2 in technical writing, and 1 word processing. In the upper-division baccalaureate category, 3 semester hours in management problems (9/94). *Version 2:* In the upper-division baccalaureate category, 3 semester hours in principles of management and 3 in management communications (6/00).

AR-1408-0215

Combined Logistics Officer Advanced, Phases 1 and 3

Course Number: 8-10-C22(LOG)P3; 8-10-C22 (LOG)P1; 8-10-C22 (LOG)P1, Phase 1; 8-10-C22 (LOG)P3, Phase 3.

Location: Army Logistics Management College, Fort Lee, VA.

Length: Phase 1, 7 weeks (255–286 hours); Phase 3, 8 weeks (276–288 hours).

Exhibit Dates: 6/92–9/99.

Learning Outcomes: Upon completion of Phase 1, the student will be able to plan the phases of a military organization with emphasis on preparation for staff positions and apply newly developed communications skills, personnel management techniques, training skills, and professional administrative skills to improve leadership. Upon completion of Phase 3, students will be able to identify and define logistics functions and plan functions of logistics management for large-scale activities.

Instruction: Lectures include leadership and personnel functions, unit maintenance and supply, military organizations, and history. In Phase 1, emphasis is placed on communication skills. In Phase 3, the instruction is of a capstone mode, where training for other phases is integrated. (Phase 2 is branch specific, conducted at the designated school.) A concluding model exercise is developed and presented, using the techniques, principles, and procedures acquired in the course.

Credit Recommendation: In the graduate degree category, 3 semester hours in management for Phase 1 and 3 semester hours in management for Phase 3 (1/98).

AR-1408-0216

Installation Logistics Management

Course Number: *Version 1:* ALMC-IN. *Version 2:* ALMC-IN.

Location: *Version 1:* Army Logistics Management College, Onsite at various locations. *Version 2:* Army Logistics Management College, Fort Lee, VA.

Length: *Version 1:* 2 weeks (67–68 hours). *Version 2:* 2 weeks (83 hours).

Exhibit Dates: *Version 1:* 10/90–5/97. *Version 2:* 6/97–Present.

Learning Outcomes: *Version 1:* Upon completion of the course, the student will be able to improve the efficiency and effectiveness of installation procedures for materiel and equipment. *Version 2:* Upon completion of the course, the student will apply general knowledge of the duties, responsibilities, and difficulties associated with installation-level logistics.

Instruction: *Version 1:* Lecture-conferences and case studies cover the various facets of management required for installation logistics, including supply, transportation, maintenance, and procurement. NOTE: There are no quizzes, written exams, or research papers required in this course. Students must present their solutions to a budgeting case orally. The course is graded pass-fail. *Version 2:* Course contains one case study/practical exercise with group collaboration and presentation.

Credit Recommendation: *Version 1:* In the upper-division baccalaureate category, 1 semester hour in logistics management (8/95). *Version 2:* In the upper-division baccalaureate category, 1 semester hour in introduction to logistics management (2/02).

AR-1408-0217

Air Defense Artillery Officer Advanced

Course Number: 2-44-C22.

Location: Air Defense Artillery School, Fort Bliss, TX.

Length: 20 weeks (744–800 hours).

Exhibit Dates: 8/91–9/03.

Learning Outcomes: Upon completion of the course, the student will be able to manage and administer army air defense operational units in the areas of supply, logistics, personnel/human factors, and Army air defense weapon systems and tactics.

Instruction: Lectures and practical exercises cover written and oral communication, leadership training management, staff organization and procedures, supply and logistic systems, personnel problems and management, and specific Army air defense weapon systems and tactics.

Credit Recommendation: In the lower-division baccalaureate/associate degree category, 1 semester hour in written and oral communication (8/95).

AR-1408-0218

Current Issues in the Veterinary Service

Course Number: 6-8-C8.

Location: Academy of Health Sciences, Fort Sam Houston, TX.

Length: 2 weeks (86 hours).

Exhibit Dates: 1/94–Present.

Learning Outcomes: Before 1/94 see AR-1408-0102. Upon completion of the course, the student will be expected to describe the mission of the Veterinary Corps; discuss current issues and policies in animal care and treatment; discuss current issues and policies in food safety, quality, and inspection; and evaluate all aspects of a veterinary service program.

Instruction: Conferences, demonstrations, and practical exercises cover administration, food technology/science, surveillance inspection, and veterinary preventive medicine.

Credit Recommendation: In the lower-division baccalaureate/associate degree category, 5 semester hours in current issues in veterinary science (10/03).

AR-1408-0219

Health Services Human Resources Manager

Course Number: 6H-70F67.

Location: Academy of Health Sciences, Fort Sam Houston, TX.

Length: 3–4 weeks (124–173 hours).

Exhibit Dates: 11/94–Present.

Learning Outcomes: Upon completion of the course, the student will be able to effectively function as a health service personnel manager in any Army medical unit.

Instruction: Methods of instruction include classroom lectures, role playing, special projects, and discussions. Topics include evaluation, organization of military officers, record keeping, promotion and discharge procedures, personnel actions and correspondence, principles of management as applied to civilian/military personnel, and medical organizations.

Credit Recommendation: In the lower-division baccalaureate/associate degree category, 2 semester hours in health science management (11/97).

AR-1408-0220

1. Combined Logistics Officer Advanced, Phase 2
2. Combined Logistics Officer Advanced Phase 2

Course Number: *Version 1:* 6-8-C22 (CLOAC). *Version 2:* 6-8-C22 (CLOAC).

Location: Academy of Health Sciences, Fort Sam Houston, TX.

Length: *Version 1:* 5 weeks (188 hours). *Version 2:* 5 weeks (191 hours).

Exhibit Dates: *Version 1:* 10/94–8/96. *Version 2:* 9/96–Present.

Learning Outcomes: *Version 1:* Upon completion of the course, the student will be able to identify appropriate preventive medicine strategies based on specific situations, discuss alternative systems in the delivery of health care, apply concepts and principles of combat health support, provide instruction to veterinary TOE teams, provide instruction on hearing conservation and nutrition, discuss multidisciplinary, health rolls and programs. *Version 2:* Upon completion of the course, the student will be able to apply principles of materials management and procurement; and health services planning, delivery, and support.

Instruction: *Version 1:* Lectures and simulations are used to assist the learner in applying principles of preventive medicine and combat health support. Alternative health care delivery systems are discussed as are multidisciplinary roles and programs. *Version 2:* Lecture and practical application provide the student with the skills to plan, support, and deliver health services and to maintain those services, employing materials procurement and management techniques.

Credit Recommendation: *Version 1:* In the graduate degree category, 4 semester hours in health services administration (10/95). *Version 2:* In the graduate degree category, 3 semester hours in medical services management (10/03).

AR-1408-0221

Medical Management of Chemical and Biological Casualties

Course Number: 6H-F26.

Location: Army Medical Research Institute of Infectious Disease, Fort Detrick, MD; Army Research Institute for Chemical Defense, Aberdeen Proving Ground, MD.

Length: 1–2 weeks (51 hours).

Exhibit Dates: 3/94–Present.

Learning Outcomes: Upon completion of the course, students will be able to recognize patients who have been exposed to chemical or biological agents, decontaminate and/or treat patients who have been exposed to chemical or biological agents, and protect themselves from contamination by chemical or biological agents.

Instruction: Methods include lectures, discussions, and practical experience. Topics include biological and chemical warfare agents and diagnosis, treatment, decontamination, and protection from biological and chemical agents.

Credit Recommendation: In the upper-division baccalaureate category, 3 semester hours in medical management of chemical and biological casualties (10/95).

AR-1408-0222

Maneuver Forward Air Defense Technician Warrant Officer Technical Tactical Certification

Course Number: 4F-917A.

Location: Ordnance, Missile and Munitions School, Redstone Arsenal, AL.

Length: 6–7 weeks (276 hours).

Exhibit Dates: 3/95–6/00.

Learning Outcomes: Upon completion of the course, the student will understand concepts of professional development, perform shop management tasks, training in the Integrated Family of Test Equipment, and develop supervision skills for maintenance operations of a missile weapons system.

Instruction: Lectures and limited practical exercises in professional development, shop management, and maintenance operations supervision.

Credit Recommendation: In the lower-division baccalaureate/associate degree category, 1 semester hour in military science and 3 in maintenance management (7/96).

Related Occupations: 917A.

AR-1408-0223

Missile Maintenance And Auto Test Equipment Basic Noncommissioned Officer (NCO)

Course Number: 1-27-C40.

Location: Ordnance, Missile and Munitions School, Redstone Arsenal, AL.

Length: 4–5 weeks (171 hours).

Exhibit Dates: 12/95–6/00.

Learning Outcomes: Upon completion of the course, the student will be able to manage missile maintenance and field ops.

Instruction: Lectures and limited practical exercises in professional development, leadership, maintenance management, and field operations. Course contains a common core of leadership subjects.

Credit Recommendation: In the lower-division baccalaureate/associate degree category, 1 semester hour in maintenance management. See AR-1406-0090 for common core credit (2/01).

AR-1408-0224

Test Measurement and Diagnostic Equipment (TMDE) Maintenance Support Specialist Advanced Noncommissioned Officer (NCO)

Course Number: 198-35H40.

Location: Ordnance, Missile and Munitions School, Redstone Arsenal, AL.

Length: 4–5 weeks (197 hours).

Exhibit Dates: 2/95–9/00.

Learning Outcomes: Upon completion of the course, the student will demonstrate the skills necessary to supervise, train, and lead subordinates in electronic maintenance support units, teams, and detachments under tactical conditions.

Instruction: Lectures, practical exercises, videos, and small group instruction covers standard operating procedures, automated data processing, quality assurance inspections, maintenance operations, and leadership. Course contains a common core of leadership subjects.

Credit Recommendation: In the lower-division baccalaureate/associate degree category, 2 semester hours in military science and 1 in maintenance management. See AR-1404-0035 for common core credit (5/01).

Related Occupations: 35H.

AR-1408-0225

Wire Systems Equipment Repairer

Course Number: 622-35N30.

Location: Ordnance, Missile and Munitions School, Redstone Arsenal, AL.

Length: 8 weeks (313 hours).

Exhibit Dates: 10/94–9/98.

Learning Outcomes: Upon completion of the course, the student will demonstrate the skills necessary to perform final inspection of manual, semiautomatic and automatic tactical electronic switching systems and associated equipment.

Instruction: Lectures, practical exercises, and small group instruction cover basic supervision, safety, switchboard assembly inspection, power supply inspection, installation of switching systems, and maintenance management and quality control procedures.

Credit Recommendation: In the lower-division baccalaureate/associate degree category, 2 semester hours in maintenance management (7/96).

Related Occupations: 35N.

AR-1408-0226

Combined Logistics Officer Advanced Phase 2

Course Number: 8-10-C22 (LOG).

Location: Quartermaster Center and School, Fort Lee, VA.

Length: 5 weeks (198 hours).

Exhibit Dates: 3/96–10/98.

Learning Outcomes: Upon completion of the course, the student will manage a supply support system from the wholesale to the direct support level, including logistics automation, field feeding, petroleum and water, mortuary affairs, and airborne field services.

Instruction: Lectures, data interpretations, and professional development project are methods of presentation of this course which covers the above topics.

Credit Recommendation: In the upper-division baccalaureate category, 2 semester hours in logistics management (5/02).

AR-1408-0227

Basic Noncommissioned Officer (NCO) Cavalry Scout 19D

Course Number: 250-1-19D30 (ITV); 250-7-19D30 (ITV); 250-9-19D30 (ITV); 250-10-19D30 (ITV); 250-18-19D30 (ITV).

Location: NCO Academy, Fort Knox, KY; NCO Academy, Fort Riley, KS; NCO Academy, Fort Hood, TX; NCO Academy, Fort Carson, CO; NCO Academy, 7th Army, Europe.

Length: 8–9 weeks (339–340 hours).

Exhibit Dates: 1/90–Present.

Learning Outcomes: Upon completion of the course, the student will be able to supervise and train personnel in the maintenance, operation, and employment of weapons.

Instruction: The course includes lectures, demonstrations, and practice in leadership and supervision of personnel in the maintenance, operation, and use of weapons. Topics include leadership techniques; land navigation; nuclear, biological, chemical defense; communication; tactics; maintenance; mine warfare; and demolitions. Course contains a common core of leadership subjects.

Credit Recommendation: Credit is recommended for the common core only. See AR-1406-0090 (2/01).

Related Occupations: 19D.

AR-1408-0228

1. Armor Officer Basic (CLI)
2. Armor Officer Basic (M1A1)
3. Armor Officer Basic (M1A1)

Course Number: *Version 1:* 2-17-C20. *Version 2:* 2-17-C20 (M1A1). *Version 3:* 2-17-C20 (M1A1).

Location: Armor Center and School, Fort Knox, KY.

Length: *Version 1:* 15–16 weeks (737 hours). *Version 2:* 17 weeks (915 hours). *Version 3:* 17 weeks (984–1106 hours).

Exhibit Dates: *Version 1:* 5/92–5/98. *Version 2:* 6/98–9/01. *Version 3:* 10/01–Present.

Learning Outcomes: *Version 1:* Upon completion of the course, the student will be able to demonstrates the basic administrative, executive, and supervisory management skills. *Version 2:* Upon completion of the course, the student will be able to demonstrate the fundamentals of tactics, gunnery, maintenance, and leadership. *Version 3:* Upon completion of the course, the student will be able to demonstrate the fundamentals of tactics, gunnery, maintenance, and leadership.

Instruction: *Version 1:* Instruction is delivered in real or simulated settings, primarily through demonstrations, practices, and hands-on activities. Some instruction is given in the classroom with traditional media or lectures. Topics include military training, supervisory management, ethics, decision making, communication, and administration. *Version 2:* Lectures, demonstrations, and practical exercises in tactics, maintenance, leadership, and communications. This is a two track course with students in either M1A1 specific or M1A1 and M1A1 and M1A2 combined equipment and tactics. Credit recommended is applicable to either Track 1 or Track 2. *Version 3:* Audiovisual materials, practical exercises, discussion, lecture, computer-based training, and lecture. Topics include military problem solving, risk management, values and ethics, military justice, law of war, administrative and personnel actions, maintenance management, property accountability, equal opportunity, and sexual harassment.

Credit Recommendation: *Version 1:* In the lower-division baccalaureate/associate degree category, 3 semester hours in personnel supervision (3/97). *Version 2:* In the lower-division baccalaureate/associate degree category, 3 semester hours in military tactics and 3 in leadership (6/01). *Version 3:* In the lower-division baccalaureate/associate degree category, 3 semester hours in military tactics and 3 in leadership. In the upper-division baccalaureate category, 2 in human resource management, 1 in human relations, and 3 in managerial planning (6/03).

AR-1408-0229

1. Armor Officer Advanced
2. Armor Captains Career
3. Armor Captains Career

Course Number: *Version 1:* 2-17-C22. *Version 2:* 2-17-C22 (F). *Version 3:* 2-17-C22.

Location: *Version 1:* Armor Center and School, Fort Knox, KY. *Version 2:* Armor School, Fort Knox, KY. *Version 3:* Armor Center, Fort Knox, KY.

Length: *Version 1:* 19–20 weeks (764 hours). *Version 2:* 18 weeks (705 hours). *Version 3:* 18 weeks (720 hours).

Exhibit Dates: *Version 1:* 10/93–9/98. *Version 2:* 10/98–10/01. *Version 3:* 11/01–Present.

Learning Outcomes: *Version 1:* Upon completion of the course, the student will be able to apply skills in professional ethics, leadership, values, and the conduct of training. *Version 2:* Upon completion of the course, the student will be able to apply skills in professional communication, ethics, leadership, values, and the conduct of training. *Version 3:* Upon completion of the course, the student will be able to apply skills, knowledge and training in professional leadership, values and conduct of training; and serve as a staff officer.

Instruction: *Version 1:* Instruction includes lectures and practical exercises in preventive maintenance, values, ethics, stress, leadership, principles of defense, team building, and combat orders. *Version 2:* Lectures and practical exercises in maintenance/supply, leadership, ethics, and cavalry tactics simulation. *Version 3:* Practical and classroom exercises, discussions, lectures, and computer based training. Topics include management strategy, quantitative/qualitative analysis and decision-making, maintenance/supply, operational planning, and teambuilding.

Credit Recommendation: *Version 1:* In the lower-division baccalaureate/associate degree category, 3 semester hours in oral communication and 2 in leadership (3/97). *Version 2:* In the lower-division baccalaureate/associate degree category, 3 semester hours in leadership and 1 in report writing. In the upper-division baccalaureate category, 3 semester hours in principles of management (6/01). *Version 3:* In the lower-division baccalaureate/associate degree category, 3 semester hours in technical report writing and 3 in material management. In the upper-division baccalaureate category, 3 semester hours in training and development (6/03).

AR-1408-0230

1. Cavalry Scout Basic Noncommissioned Officer (NCO)
2. Basic Noncommissioned Officer (NCO) Cavalry Scout 19D
3. Cavalry Scout Basic Noncommissioned Officer (NCO)

Course Number: *Version 1:* 250-9-19D30; 250-10-19D30; 250-18-19D30; 250-1-19D30. *Version 2:* 250-1-19D30. *Version 3:* 250-19D30.

Location: *Version 1:* Armor School, Fort Carson, CO; Armor School, Fort Hood, TX; Armor Center and School, Fort Knox, KY. *Version 2:* Armor Center and School, Fort Knox, KY. *Version 3:* Armor School, Fort Knox, KY.

Length: *Version 1:* 8 weeks (405–406 hours). *Version 2:* 8 weeks (477 hours). *Version 3:* 8 weeks (453 hours).

Exhibit Dates: *Version 1:* 11/91–12/96. *Version 2:* 1/97–9/97. *Version 3:* 10/97–Present.

Learning Outcomes: *Version 1:* Upon completion of the course, the student will be able to lead, train, and direct section/squad leaders in mine warfare communications, land navigation, safety, marksmanship, tactics, armor, and combat skills. *Version 2:* Upon completion of the course, the student will be able to fight, maintain, train, and sustain the section and perform as a scout section sergeant. *Version 3:* Upon completion of the course, the student will be able to lead, train, and direct section squad leaders in mine warfare communications, land navigation, safety, marksmanship, tactics, armor, weapons, security, and maintenance.

Instruction: *Version 1:* Lectures and practical exercises cover mine warfare, demolitions, tactics, maintenance, and situation training. *Version 2:* Seminars and practical exercises cover mine warfare, communication, tactics, demolition, maintenance, safety, leadership, and combat exercises. *Version 3:* The small-group instructional model is used for the course with leadership practice in a scout platoon environment and tactical seminars. Topics include leadership procedures, training management, communication, and military subjects.

Credit Recommendation: *Version 1:* In the lower-division baccalaureate/associate degree category, 1 semester hour in principles of supervision (3/97). *Version 2:* In the lower-division baccalaureate/associate degree category, 1 semester hour principles of supervision (3/97). *Version 3:* In the lower-division baccalaureate/associate degree category, 1 semester hour in personnel supervision (6/01).

Related Occupations: 19D.

AR-1408-0232

Armor Pre-Command Branch Phase

Course Number: *Version 1:* 2G-F24. *Version 2:* 2G-F24.

Location: *Version 1:* Armor Center and School, Fort Knox, KY. *Version 2:* Armor School, Fort Knox, KY.

Length: *Version 1:* 3–4 weeks (126–127 hours). *Version 2:* 2 weeks (145 hours).

Exhibit Dates: *Version 1:* 1/90–11/99. *Version 2:* 12/99–Present.

Learning Outcomes: *Version 1:* Upon completion of the course, the student will be able to apply logistics, supply material, and operations management skills to command in the armor environment. *Version 2:* Upon completion of the course, the student will be able to apply logistics, supply material, and operations management skills to command in the armor environment.

Instruction: *Version 1:* Instruction is provided through traditional classroom lectures, discussions, and demonstrations with military-specific systems and devices. Topics include communications technology, leadership, and principles of supervision. *Version 2:* Lectures and practical exercises in military tactics, communications, leadership, and supervision.

Credit Recommendation: *Version 1:* In the lower-division baccalaureate/associate degree category, 2 semester hours in personnel supervision (3/97). *Version 2:* In the lower-division baccalaureate/associate degree category, 3 semester hours in leadership (4/03).

AR-1408-0233

Cavalry Leader

Course Number: *Version 1:* 2E-FOA-F134. *Version 2:* 2E-FOA-F134/250-F46.

Location: *Version 1:* Armor Center and School, Fort Knox, KY. *Version 2:* Armor School, Fort Knox, KY.

Length: *Version 1:* 3–4 weeks (116–122 hours). *Version 2:* 3 weeks (111 hours).

Exhibit Dates: *Version 1:* 6/92–9/00. *Version 2:* 10/00–Present.

Learning Outcomes: *Version 1:* Upon completion of the course, the student will demonstrate an ability to lead and manage human resources. *Version 2:* Upon completion of the course, the student will be able to plan a successful operation strategy; develop strategy resulting in combat support and service; and plan reconnaissance operations.

Instruction: *Version 1:* Instruction is provided through lectures, demonstrations, and simulations. Topics include specific military training and the management of human resources. *Version 2:* Practical exercises and discussion. Topics include tactical employment of calvary troops, economy of force and reconnaissance and security missions, integration and synchronization of combat power and combat service support.

Credit Recommendation: *Version 1:* In the lower-division baccalaureate/associate degree category, 1 semester hour in personnel supervision (9/00). *Version 2:* In the lower-division baccalaureate/associate degree category, 3 semester hours in military science and 3 in communication. In the upper-division baccalaureate category, 3 semester hours in operations management and 3 in organizational planning and management (6/03).

AR-1408-0234

1. Scout Platoon Leader
2. Scout Leader

Course Number: *Version 1:* 2E-F137/521-F2. *Version 2:* 2E-F137/521-F2.

Location: *Version 1:* Armor Center and School, Fort Knox, KY. *Version 2:* Armor Center, Fort Knox, KY.

Length: *Version 1:* 2–3 weeks (184–202 hours). *Version 2:* 3–4 weeks (218 hours).

Exhibit Dates: *Version 1:* 5/92–9/02. *Version 2:* 10/02–Present.

Learning Outcomes: *Version 1:* Upon completion of the course, the student will demonstrate planning, organizing, directing, and controlling skills. *Version 2:* Upon completion of the course, the student will demonstrate planning, organizing, directing, and controlling skills.

Instruction: *Version 1:* Instruction is provided through classroom lectures, demonstrations, and exercises in the field. Topics include specific military training in leading reconnaissance and security missions, as well as supervision, logistics, supply materials, operations, and human resource management. *Version 2:* Audiovisual materials, lectures, practical and classroom exercises. Topics include specific military training in leading reconnaissance and security missions, as well as supervision, logistics, supply materials, operations, and human resource management.

Credit Recommendation: *Version 1:* In the lower-division baccalaureate/associate degree category, 1 semester hour in supervisory management (2/01). *Version 2:* In the lower-division baccalaureate/associate degree category, 2 semester hours in supervisory management (6/03).

AR-1408-0236

M1 Armor Crewman Basic Noncommissioned Officer (NCO) Reserve Component

Course Number: RC-171-19K30.

Location: Armor Center and School, Fort Knox, KY.

Length: 2 weeks (47 hours).

Exhibit Dates: 10/92–Present.

Learning Outcomes: Upon completion of the course, the students will be able to lead, manage, train, and direct subordinates to maintain, operate, and employ tank weapon systems.

Instruction: Instruction is provided through lectures, demonstrations, laboratories and use of self-paced instructional materials. Topics include supply, material, operations, human resource management, as well as leadership and other military-specific topics.

Credit Recommendation: In the lower-division baccalaureate/associate degree category, 1 semester hour in principles of management (2/01).

AR-1408-0237

Senior Officer Logistics Management

Course Number: *Version 1:* 8A-F23. *Version 2:* 8A-F23. *Version 3:* 8A-F23.

Location: *Version 1:* Armor Center and School, Fort Knox, KY. *Version 2:* Armor Center and School, Fort Knox, KY. *Version 3:* Armor School, Fort Knox, KY.

Length: *Version 1:* 1 week (46 hours). *Version 2:* 1 week (46–61 hours). *Version 3:* 1 week (61–63 hours).

Exhibit Dates: *Version 1:* 6/92–3/95. *Version 2:* 4/95–9/96. *Version 3:* 10/96–Present.

Learning Outcomes: *Version 1:* Upon completion of the course, the student will be able to manage preventive maintenance and logistics, particularly supply transportation and maintenance at the user level. *Version 2:* Upon completion of the course, the student will be able to perform unit-level logistics management in the areas of regulations, maintenance, supply, transportation, and medical systems; and preventive maintenance and logistics, particularly supply transportation and maintenance at the user level. *Version 3:* Upon completion of the course, the student will be prepared to manage activities related to maintenance and logistics.

Instruction: *Version 1:* Methods used for the course include lectures, tours, demonstrations, and discussions. Topics include logistics, supply, material, operations, and human resource and maintenance management in areas of preventive maintenance and logistics. *Version 2:* Lecture, demonstrations, discussions, and practical exercises cover supply, maintenance, safety, tactics, and combat techniques. Topics include logistics, supply material, operations, and human resource and maintenance management in areas of preventive maintenance and logistics. *Version 3:* Lectures, demonstrations, and practical exercises in logistics, supply, material, personnel, and management.

Credit Recommendation: *Version 1:* In the lower-division baccalaureate/associate degree category, 2 semester hours in maintenance management (3/97). *Version 2:* In the lower-division baccalaureate/associate degree category, 2 semester hours in maintenance management (3/97). *Version 3:* In the lower-division baccalaureate/associate degree category, 3 semester hours in maintenance management (6/01).

AR-1408-0238

Air Traffic Control Equipment Repair Basic Noncommissioned Officer (NCO), Phase 1

Course Number: 102-35D30.

Location: Ordnance, Missile and Munitions School, Redstone Arsenal, AL.

Length: 6 weeks (236 hours).

Exhibit Dates: 10/95–6/00.

Learning Outcomes: Upon completion of the course, the student will be able to apply leadership and advanced technical skills. In the areas of supervision, administration, operations, resource management, and technical assistance.

Instruction: Lectures and practical exercises in leadership, maintenance management, supervision, and advanced technical skills in air traffic control equipment. Course contains a common core of leadership subjects.

Credit Recommendation: In the lower-division baccalaureate/associate degree category, 3 semester hours in maintenance management. See AR-1406-0090 for common core credit (5/97).

Related Occupations: 35D.

AR-1408-0239

Ordnance Munitions Material Management (Branch Qualification)

Course Number: 4-9-C20-91D (BQ).

Location: Ordnance, Missile and Munitions School, Redstone Arsenal, AL.

Length: 5–6 weeks (216 hours).

Exhibit Dates: 4/92–9/00.

Learning Outcomes: Upon completion of the course, the student will be able to supervise a ammunition logistics system, including safety, storage, supply procedures, and accountability.

Instruction: Lectures and practical exercises cover ammunition marking, supply, storage, inventory, transport, and disposal.

Credit Recommendation: In the lower-division baccalaureate/associate degree category, 2 semester hours in materiel management (5/97).

Related Occupations: 91D.

AR-1408-0240

Electronic Systems Maintenance Technician Warrant Officer (WOBC)

Course Number: 4C-918B.

Location: Ordnance, Missile and Munitions School, Fort Gordon, GA; Ordnance, Missile and Munitions School, Redstone Arsenal, AL.

Length: 19 weeks (682 hours).

Exhibit Dates: 10/94–6/00.

Learning Outcomes: After 2/00 see AR-1402-0184. Upon completion of the course, the student will be able to manage logistics, ordnance missions, electronic warfare, site defense, and communications devices.

Instruction: Lectures and practical exercises in communications, maintenance management, ordnance functions and missions, electronic warfare site defense, avionics/navigation equipment, ground satellite equipment, and transmission communications equipment.

Credit Recommendation: In the lower-division baccalaureate/associate degree category, 3 semester hours in maintenance management. In the upper-division baccalaureate category, 3 semester hours in logistics management (5/97).

Related Occupations: 918B.

AR-1408-0241

Maneuver Forces Air Defense (MFADS) Technician Warrant Officers

Course Number: 4-9-C32-917A.

Location: Ordnance, Missile and Munitions School, Redstone Arsenal, AL.

Length: 10–11 weeks (313 hours).

Exhibit Dates: 3/95–6/00.

Learning Outcomes: Upon completion of the course, the student will be able to perform technical duties and manage maintenance operations.

Instruction: Lectures and practical exercises cover leadership, communications, management, maintenance operations, and technical proficiency.

Credit Recommendation: In the lower-division baccalaureate/associate degree category, 3 semester hours in maintenance management (5/97).

Related Occupations: 917A.

AR-1408-0242

Land Combat Missile Systems Technician Warrant Officer Advanced

Course Number: 4-9-C32-912A.

Location: Ordnance, Missile and Munitions School, Redstone Arsenal, AL.

Length: 5 weeks (178 hours).

Exhibit Dates: 11/94–5/00.

Learning Outcomes: Upon completion of the course, the student will be able to manage a land combat missile unit, including missile maintenance management, new maintenance techniques, and training updates.

Instruction: Lectures and practical exercises cover missile maintenance, maintenance management, and update training on technical knowledge.

Credit Recommendation: In the lower-division baccalaureate/associate degree category, 3 semester hours in maintenance management (5/97).

Related Occupations: 912A.

AR-1408-0243

Wire Systems Equipment Repairer Basic Noncommissioned Officer (NCO)

Course Number: 622-35N30.

Location: Signal School, Fort Gordon, GA; Ordnance, Missile and Munitions School, Redstone Arsenal, AL.

Length: 8 weeks (313 hours).

Exhibit Dates: 10/96–6/00.

Learning Outcomes: Upon completion of the course, the student will be able to provide supervision and technical assistance in an electronic maintenance facility.

Instruction: Lectures, practical exercises, and small-group instruction cover installation, programming, operation, repair, and facial inspection of manual, semiautomatic, and automatic electronic switching systems and associated equipment. Course includes a common core of leadership subjects.

Credit Recommendation: In the lower-division baccalaureate/associate degree category, 3 semester hours in maintenance management. See AR-1406-0090 for common core credit (5/97).

Related Occupations: 35N.

AR-1408-0244

Ordnance Maintenance Management Officer Branch Qualification

Course Number: 4-9-C20-91B/C (BQ).

Location: Ordnance Center and School, Aberdeen Proving Ground, MD.

Length: 4 weeks (136 hours).

Exhibit Dates: 10/91–Present.

Learning Outcomes: Before 9/91 see AR-1405-0074. Upon completion of the course, the student will be able to complete, organize, and manage maintenance control forms, their follow through, and completion.

Instruction: Instruction begins with the history and organization of ordnance maintenance management. Safety, operations, procedures, and computer skills are involved.

Credit Recommendation: In the lower-division baccalaureate/associate degree category, 1 semester hour in business management (6/97).

AR-1408-0245

Basic Noncommissioned Officer (NCO) 19D CFV

Course Number: 250-1-19D30 (CFV); 250-4-19D30 (CFV); 250-9-19D30 (CFV); 250-18-19D30 (CFV).

Location: NCO Academy, Fort Knox, KY; NCO Academy, Fort Stewart, GA; NCO Academy, Fort Hood, TX; NCO Academy, 7th Army, Europe.

Length: 6 weeks (342–343 hours).

Exhibit Dates: 1/90–Present.

Learning Outcomes: Upon completion of the course, the student will be able to supervise and train personnel in the maintenance, operation, and employment of weapons and equipment.

Instruction: The course includes lectures, demonstration, and practice in leadership and supervision of personnel relative to the maintenance, operation, and employment of weapons. Topics include leadership techniques, land navigation, nuclear/biological/chemical defense, communication, tactics, maintenance, mine warfare, and demolitions.

Credit Recommendation: Credit is recommended for the common core only. See AR-1406-0090 (3/97).

Related Occupations: 19D.

AR-1408-0246

Finance Officer Branch Qualification

Course Number: 7-14-C20-44A (BQ).

Location: Soldier Support Institute, Fort Jackson, SC.

Length: 4 weeks (146 hours).

Exhibit Dates: 2/92–Present.

Learning Outcomes: Upon completion of the course, the student will be able to differentiate between appropriated and nonappropriated funds, certify vouchers for payment, describe the functions of the cashier and cash control, and compute military pay and military travel.

Instruction: Methods of instruction include lectures, practice problems, simulations, and use of specialized software.

Credit Recommendation: In the lower-division baccalaureate/associate degree category, 3 semester hours in cash disbursements (accounting) (9/97).

AR-1408-0247

1. Retention Advanced Noncommissioned Officer
2. Retention Advanced Noncommissioned Officer (NCO), Phase 2

Course Number: ***Version 1:*** 5-79-C42 (RET). *Version 2:* 5-79-C42 (RET), Phase 2.

Location: ***Version 1:*** Soldier Support Institute, Fort Jackson, SC. *Version 2:* NCO Academy, Fort Jackson, SC.

Length: ***Version 1:*** 6–7 weeks (244 hours). *Version 2:* 3–4 weeks (132 hours).

Exhibit Dates: ***Version 1:*** 3/96–9/01. *Version 2:* 10/01–Present.

Learning Outcomes: ***Version 1:*** Before 3/96 see AR-1406-0158. Upon completion of the course, the student will be able to teach, evaluate, and counsel personnel; conduct collective training; and perform as a supervisor. *Version 2:* Upon completion of the course, the student will be able to teach, evaluate, and counsel personnel; conduct collective training; and perform as a supervisor.

Instruction: ***Version 1:*** Methods used include conferences, small groups instruction, practical exercises, and lectures covering management, leadership, training, and supervision of subordinates. This course includes a common core of leadership subjects. *Version 2:* Methods used include conferences, small group instruction, practical exercises, and lectures covering management, leadership, and supervision of subordinates.

Credit Recommendation: ***Version 1:*** In the lower-division baccalaureate/associate degree category, 3 semester hours in supervisory management. See AR-1404-0035 for common core credit (9/97). *Version 2:* In the lower-division baccalaureate/associate degree category, 1 semester hour in military science and 3 in principles of supervision (3/02).

AR-1408-0248

Reserve Component Administrative Specialist

Course Number: 805C-71L10.

Location: Soldier Support Institute, Fort Jackson, SC.

Length: 3 weeks (104 hours).

Exhibit Dates: 1/97–Present.

Learning Outcomes: Upon completion of the course, the student will be able to handle basic correspondence, use a computer (word processing, graphics, data base, and spreadsheet applications), and apply keyboarding skills.

Instruction: Methods include lecture and hands-on practice.

Credit Recommendation: In the lower-division baccalaureate/associate degree category, 2 semester hours in clerical procedures (9/97).

Related Occupations: 71L.

AR-1408-0249

Reserve Component Reconfigured Administrative Specialist, Advanced Noncommissioned Officer Phase 2

Course Number: 805C-71L40, Phase 2.

Location: Soldier Support Institute, Fort Jackson, SC.

Length: 2 weeks (60 hours).

Exhibit Dates: 6/97–Present.

Learning Outcomes: Upon completion of the course, the student will be able to review correspondence, manage files, manage publications, and/or blank forms, inventory classified documents, and destroy classified documents.

Instruction: Lectures, classroom discussions, exercises, and examinations are course methods.

Credit Recommendation: Credit is not recommended because of the military-specific nature of the course (9/97).

Related Occupations: 71L.

AR-1408-0250

Master Marksmanship Trainer

Course Number: MTC-003.

Location: Professional Education Center, Cp. Joseph Robinson, North Little Rock, AR.

Length: 2 weeks (94 hours).

Exhibit Dates: 4/96–Present.

Learning Outcomes: Upon completion of the course, the student will be able to conduct range operations and training for small arms, squad assault weapons, grenade launchers, and light and heavy machine guns, including qualification firing, nomenclature, assembly, and disassembly.

Instruction: Performance-based small group instruction includes demonstration and hands-on exercises. Assessment is performance observation and written exams.

Credit Recommendation: In the lower-division baccalaureate/associate degree category, 3 semester hours in small arms range operations (6/98).

AR-1408-0251

Training and Mobilization Management

Course Number: FT-022.

Location: Professional Education Center, Cp. Joseph Robinson, North Little Rock, AR.

Length: 2 weeks (70 hours).

Exhibit Dates: 11/96–Present.

Learning Outcomes: Upon completion of the course, the student will be able to apply skills in organizational management, administration, and logistics to supervise unit readiness and mobilization of a National Guard Unit.

Instruction: Course is performance based with lectures and practical exercises on administrative, training, and logistics requirements for organizational readiness.

Credit Recommendation: In the lower-division baccalaureate/associate degree category, 3 semester hours in supervision and planning (6/98).

AR-1408-0252

Small Arms Instructor Range Operations

Course Number: MTC-001.

Location: Professional Education Center, Cp. Joseph Robinson, North Little Rock, AR.

Length: 2 weeks (134 hours).
Exhibit Dates: 7/95–Present.
Learning Outcomes: Upon completion of the course, the student will be able to instruct small arms maintenance, and marksmanship and conduct marksmanship range operations.
Instruction: Performance-based instruction consists of lecture, demonstration, and practice. The preponderance of training occurs on the marksmanship ranges. Evaluation is based on observation of performance and written exam.
Credit Recommendation: In the lower-division baccalaureate/associate degree category, 1 semester hour in small arms range operations (6/98).

AR-1408-0253

STANDARD ARMY RETAIL SUPPLY SYSTEM - LEVEL 1 - SUPERVISOR (SARSS1)

Course Number: LTC-051.
Location: Professional Education Center, Cp. Joseph Robinson, North Little Rock, AR.
Length: 2 weeks (61 hours).
Exhibit Dates: 1/96–Present.
Learning Outcomes: Upon completion of the course, the student will be able to supervise and operate a computer application program to conduct retail stock operations, including receiving, stocking, and issuing transactions suspense files; and audit and tracking functions.
Instruction: This performance-based course uses computer simulations for hands-on training of an automated retail supply system.
Credit Recommendation: In the upper-division baccalaureate category, 3 semester hours in retail supply supervision (6/98).

AR-1408-0254

STANDARD ARMY RETAIL SUPPLY SYSTEM - LEVEL 1 - OPERATOR (SARSS-1)

Course Number: *Version 1:* LTC-050. *Version 2:* LTC-050.
Location: Professional Education Center, Cp. Joseph Robinson, North Little Rock, AR.
Length: *Version 1:* 2 weeks (54 hours). *Version 2:* 2 weeks (63 hours).
Exhibit Dates: *Version 1:* 1/96–9/02. *Version 2:* 10/02–Present.
Learning Outcomes: *Version 1:* Upon completion of the course, the student will be able to operate a computer application program to conduct retail stock operations, including ordering, stocking, and issuing transactions; suspense files; and audit and tracking functions. *Version 2:* Pending evaluation.
Instruction: *Version 1:* This performance-based course uses computer simulations for hands-on training of an automated supply system functions. *Version 2:* Pending evaluation.
Credit Recommendation: *Version 1:* In the lower-division baccalaureate/associate degree category, 3 semester hours in retail supply operations (6/98). *Version 2:* Pending evaluation.

AR-1408-0255

INITIAL ENTRY ROTARY WING AVIATOR COMMON CORE PROFESSIONAL DEVELOPMENT

Course Number: 2C-IERW (Common Core) (Off).
Location: Aviation School, Fort Rucker, AL.
Length: 2 weeks (64 hours).
Exhibit Dates: 11/94–Present.
Learning Outcomes: Upon completion of the course, the student will be able to describe personnel and general management techniques and procedures.
Instruction: The course uses lectures, conferences, briefings, and practical exercise methods of instruction. The topics covered include leadership, personnel, general management, and crisis analysis.
Credit Recommendation: In the lower-division baccalaureate/associate degree category, 2 semester hours in principles of supervision (11/98).

AR-1408-0256

INITIAL ENTRY/ROTARY WING (COMM CORE) WARRANT OFFICER PROFESSIONAL DEVELOPMENT

Course Number: 2C-IERW (Common Core) (WO).
Location: Aviation School, Fort Rucker, AL.
Length: 4 weeks (189 hours).
Exhibit Dates: 10/94–Present.
Learning Outcomes: Upon completion of the course, the student will be qualified to function as a pilot-in-command and will be technically and tactically proficient in Army aviation operations and combined arms warfare.
Instruction: Course includes classroom instruction, seminars, field training, and practical exercises in support and management fundamentals, including logistics, maintenance, personnel, leadership, ethics, and problem solving; combat operations; survival and liaison; tactical doctrine; mission planning; and survival skills.
Credit Recommendation: In the lower-division baccalaureate/associate degree category, 2 semester hours in principles of supervision (11/98).

AR-1408-0257

ARMY MEDICAL DEPARTMENT (AMEDD) NONCOMMISSIONED OFFICER NCO BASIC RESERVE PHASE 2

Course Number: 081-6-8-C40.
Location: Academy of Health Sciences, Fort Sam Houston, TX.
Length: 7 weeks (136 hours).
Exhibit Dates: 9/98–Present.
Learning Outcomes: Upon completion of the course, the student will have the be able to apply relevant principles, concepts and techniques of resource management to resolve problems to participate in planning, organizing, and conducting mission-specific operations.
Instruction: Lectures, practical exercises and examinations provide technical and leadership training.
Credit Recommendation: In the lower-division baccalaureate/associate degree category, 3 semester hours in career management. An additional 1 semester hour for only one of the following 17 MOS Specialty options: 91A, Medical Equipment Repairer, Basic; 42E, Optical Laboratory Specialist; 71G, Patient Administration Specialist; 91C, Practical Nurse; 91D, Operating Room Specialist; 91X, Mental Health Specialist; 91P, Radiology Specialist; 91Q, Pharmacy Specialist; 91S, Preventive Medicine Specialist; 91V, Respiratory Specialist; 91T, Animal Care Specialist; 76J, Medical Supply Specialist; 91B, Medical Specialist; 91E, Dental Specialist; 91K, Medical Laboratory Specialist; 91M, Hospital Food Service Specialist; or 91R, Veterinary Food Inspector Specialist (7/99).

AR-1408-0258

PUBLIC AFFAIRS SUPERVISOR BASIC NONCOMMISSIONED OFFICER (NCO)

Course Number: 570-46Q/R30.
Location: Signal Corps Regimental NCO Det., Fort Meade, MD.
Length: 5–6 weeks (225 hours).
Exhibit Dates: 4/99–Present.
Learning Outcomes: Upon completion of the course, the student will be able to plan and conduct training, supervise and coordinate the activities of a public affairs office, and manage media production.
Instruction: Instruction is provided in a mix of lecture, demonstration, and performance. Evaluation includes written examination and practical experience analysis and review. Topics include media training, print journalism, electronic journalism, and techniques of supervisor.
Credit Recommendation: In the upper-division baccalaureate category, 3 semester hours in media management (3/01).

AR-1408-0259

PUBLIC AFFAIRS SUPERVISOR ADVANCED NONCOMMISSIONED OFFICER (NCO)

Course Number: 570-46Q/R40.
Location: Signal Corps Regimental NCO Academy Det., Fort Meade, MD.
Length: 5–6 weeks (212 hours).
Exhibit Dates: 8/98–Present.
Learning Outcomes: Upon completion of the course, the student will be able to supervise the operation of a public affairs office and plan, supervise, and implement public relations activities.
Instruction: The instruction is delivered using a mix of lecture, demonstration, and project performance. Evaluations include written evaluation and project review. Topics include management of resources, print and electronic journalism resources, and risk management and assessment.
Credit Recommendation: In the upper-division baccalaureate category, 3 semester hours public relations management (2/01).

AR-1408-0260

PUBLIC AFFAIRS SUPERVISOR RESERVE COMPONENT BASIC NONCOMMISSIONED OFFICER (NCO)

Course Number: 570-46Q/R30-RC.
Location: Signal Corps Regimental NCO Academy Det., Fort Meade, MD.
Length: 2 weeks (116 hours).
Exhibit Dates: 6/98–Present.
Learning Outcomes: Upon completion of the course, the student will be able to plan and direct the activities of a public affairs office.
Instruction: Instruction is provided in a mix of lecture, demonstration, and performance. Evaluation includes written examinations, practical experience analysis, and review. Topics include media training in print, electronic journalism, and techniques for supervision.
Credit Recommendation: In the upper-division baccalaureate category, 2 semester hours in media management (3/01).

AR-1408-0261

Public Affairs Supervisor Reserve Component Advanced Noncommissioned Officer (NCO)

Course Number: 570-46Q/R40-RC.
Location: Signal Corps Regimental NCO Academy Det., Fort Meade, MD.
Length: 2 weeks (112 hours).
Exhibit Dates: 7/98–Present.
Learning Outcomes: Upon completion of the course, the student will be able to supervise the planning and implementation of public relations activities in a public affairs office.
Instruction: The course is taught using a combination of lecture, demonstration, and projects. Evaluation includes written examinations, project review, and analysis. Topics include management of resources, print and electronic journalism resources, risk management, and assessment.
Credit Recommendation: In the upper-division baccalaureate category, 2 semester hours in public relations management (3/01).

AR-1408-0262

Director of Information Management Operations Officer

Course Number: 7E-F35/531-F10.
Location: Signal School, Fort Gordon, GA.
Length: 2 weeks (76 hours).
Exhibit Dates: 3/97–Present.
Learning Outcomes: Upon completion of the course, the student will be able to manage information at a post, camp, or station.
Instruction: This course includes lectures, classroom discussions, and examinations covering an introduction to information management and management of the functional areas within information management, including visual information, library, printing, telecommunications, and records management. Students are introduced to contracting, security, information requirement studies, planning for information management, and systems configuration management.
Credit Recommendation: In the lower-division baccalaureate/associate degree category, 2 semester hours in information services and 1 in information management (5/00).

AR-1408-0263

Joint Task Force Systems

Course Number: 4C-F55/260-F15.
Location: Signal School, Fort Gordon, GA.
Length: 9 weeks (357 hours).
Exhibit Dates: 10/95–Present.
Learning Outcomes: Upon completion of the course, the student will be able to provide, plan, and manage communications using military and commercial off-the-shelf systems.
Instruction: This course includes lectures, classroom discussions, projects, and examinations to introduce the student to the planning and management necessary to integrate and interconnect diverse types of automation and communications systems into local-area and wide-area networks. Student will exercise network control and analysis to determine the status of the system.
Credit Recommendation: In the lower-division baccalaureate/associate degree category, 2 semester hours in communications systems and 3 in data communications and networking (5/00).

AR-1408-0264

Field Artillery Officer Advanced Reserve, Phase 2

Course Number: 2-6-C23, Phase 2.
Location: Field Artillery School, Fort Sill, OK.
Length: 2 weeks (110 hours).
Exhibit Dates: 4/94–Present.
Learning Outcomes: Upon completion of the course, the student will be prepared to train and supervise field artillery reserve officers.
Instruction: The course consists of classroom lectures and conferences; appropriate hands-on training in gunnery, fire support, and combined arms operations; and laboratory experiences using computing equipment.
Credit Recommendation: In the lower-division baccalaureate/associate degree category, 3 semester hours in principles of management and 1 in management communications (6/00).

AR-1408-0265

Total Army Training System (TATS) Automated Fire Support Systems Senior Sergeant Advanced Noncommissioned Officer (NCO)

Course Number: 061-13C40 (F), Phase 2.
Location: Reserve Schools, Multiple locations.
Length: 162 hours.
Exhibit Dates: 10/00–Present.
Learning Outcomes: Before 10/00 see AR-2201-0369. Upon completion of the course, the student will be able to provide leadership and training to personnel in the automated fire support unit.
Instruction: Lectures and field exercises cover tactical operations, situation planning, fire support, equipment maintenance, and leadership.
Credit Recommendation: In the upper-division baccalaureate category, 3 semester hours for field experience in management (9/03).
Related Occupations: 13C.

AR-1408-0266

Total Army Training System (TATS) Automated Fire Support Systems Senior Sergeant Advanced Noncommissioned Officer (NCO)

Course Number: 250-13C40 (F), Phase 2.
Location: Field Artillery School, Fort Sill, OK.
Length: 3 weeks (162 hours).
Exhibit Dates: 10/00–Present.
Learning Outcomes: Before 10/00 see AR-2201-0369. Upon completion of the course, the student will be able to provide leadership and training to personnel in an automated fire support unit.
Instruction: Lectures and field exercises cover tactical operations, situation planning, fire support, equipment maintenance, and leadership.
Credit Recommendation: In the upper-division baccalaureate category, 3 semester hours in field experience in management (1/01).
Related Occupations: 13C.

AR-1408-0267

Total Army Training System (TATS) Automated Fire Control Section Sergeant Basic Noncommissioned Officer (NCO)

Course Number: 250-13C30 (F), Phase 2.
Location: Field Artillery School, Fort Sill, OK.
Length: 2–3 weeks (149 hours).
Exhibit Dates: 10/00–Present.
Learning Outcomes: Upon completion of the course, the student will be able to use leadership, training, resource management, and data analysis techniques.
Instruction: Lectures and practical exercises cover leadership, training, management, military skills, and data analysis.
Credit Recommendation: In the upper-division baccalaureate category, 3 semester hours in field experience in management (9/03).
Related Occupations: 13C.

AR-1408-0268

Total Army Training System (TATS) Automated Fire Control Section Sergeant Basic Noncommissioned Officer (NCO)

Course Number: 061-13C30 (F), Phase 2.
Location: Field Artillery School, Fort Sill, OK.
Length: 149–152 hours.
Exhibit Dates: 10/00–Present.
Learning Outcomes: Upon completion of the course, the student will be able to apply skills in leadership, training, resource management, and data analysis.
Instruction: Lectures and practical exercises in leadership, training, management, military skills, and data analysis.
Credit Recommendation: In the upper-division baccalaureate category, 3 semester hours in field experience in management (9/03).

AR-1408-0269

Total Army Training System (TATS) Automated Fire Control Section Sergeant Basic Noncommissioned Officer (NCO)

Course Number: 250-13C30 (F).
Location: Field Artillery School, Fort Sill, OK.
Length: 5 weeks (228 hours).
Exhibit Dates: 10/97–Present.
Learning Outcomes: Before 10/97 see AR-1715-0638. Upon completion of the course, the student will be able to lead and manage data analysis and fire planning activities.
Instruction: Lectures and practical exercises cover employing, operating, and maintaining the limited fire support automation system; leadership; training management; and resource management.
Credit Recommendation: In the lower-division baccalaureate/associate degree category, 3 semester hours in field experience in management (6/00).
Related Occupations: 13C.

AR-1408-0270

Total Army Training System (TATS) Automated Fire Control Section Sergeant Basic Noncommissioned Officer (NCO)

Course Number: 061-13C30 (F), Phase 2.
Location: Field Artillery School, Fort Sill, OK.
Length: 143–152 hours.

Exhibit Dates: 10/97–Present.

Learning Outcomes: Before 10/97 see AR-1715-0638. Upon completion of the course, the student will be able to lead and manage data analysis and fire planning activities.

Instruction: Lectures and practical exercises cover employing, operating, and maintaining the limited fire support automation system; leadership; training management; and resource management.

Credit Recommendation: In the lower-division baccalaureate/associate degree category, 3 semester hours in field experience in management (9/03).

AR-1408-0271

TOTAL ARMY TRAINING SYSTEM (TATS)
MULTIPLE LAUNCH ROCKET SYSTEM
SECTION CHIEF, BASIC
NONCOMMISSIONED OFFICER (NCO)

Course Number: 042-13M30 (F), Phase 2.

Location: Field Artillery School, Fort Sill, OK.

Length: 2 weeks (118–121 hours).

Exhibit Dates: 10/00–Present.

Learning Outcomes: Before 10/00 see AR-2201-0509. Upon completion of the course, the student will be able to perform supervision tasks.

Instruction: Lectures and practical exercises cover map reading battle exercises, leadership concepts, military procedures for safety, and reload operations.

Credit Recommendation: In the lower-division baccalaureate/associate degree category, 3 semester hours in principles of supervision (9/03).

Related Occupations: 13M.

AR-1408-0272

TOTAL ARMY TRAINING SYSTEM (TATS)
MULTIPLE LAUNCH ROCKET SYSTEM
SECTION CHIEF, BASIC
NONCOMMISSIONED OFFICER (NCO)

Course Number: 061-13M30 (F), Phase 2.

Location: Reserve Schools, Multiple locations.

Length: 118–121 hours.

Exhibit Dates: 10/00–Present.

Learning Outcomes: Before 10/00 see AR-2201-0510. Upon completion of the course, the student will be able to perform supervision tasks.

Instruction: Lectures and practical exercises cover map reading, battle exercises, leadership concepts, military procedures for safety, and reload operations.

Credit Recommendation: In the lower-division baccalaureate/associate degree category, 3 semester hours in principles of supervision (9/03).

Related Occupations: 13M.

AR-1408-0275

TOTAL ARMY TRAINING SYSTEM (TATS)
MULTIPLE LAUNCHER ROCKET SYSTEM
(MLRS) OPERATIONS/FIRE DIRECTION
SENIOR SERGEANT ADVANCED
NONCOMMISSIONED OFFICER (NCO)

Course Number: 043-13P40, Phase 2.

Location: Reserve Schools, Multiple locations.

Length: 1–2 weeks (104 hours).

Exhibit Dates: 10/00–Present.

Learning Outcomes: Upon completion of the course, the student will be able to perform personnel management, equipment maintenance, and fire control tasks.

Instruction: Lectures and field exercises cover firing data computation, personnel management, equipment maintenance, and fire control.

Credit Recommendation: In the lower-division baccalaureate/associate degree category, 3 semester hours in field experience in management (1/01).

Related Occupations: 13P.

AR-1408-0276

TOTAL ARMY TRAINING SYSTEM (TATS)
MULTIPLE LAUNCHER ROCKET SYSTEM
(MLRS) OPERATIONS/FIRE DIRECTION
SENIOR SERGEANT ADVANCED
NONCOMMISSIONED OFFICER (NCO)

Course Number: 061-13P40, Phase 2.

Location: Field Artillery School, Fort Sill, OK.

Length: 104 hours.

Exhibit Dates: 10/00–Present.

Learning Outcomes: Upon completion of the course, the student will be able to operate and maintain fire direction control equipment and apply personnel management concepts.

Instruction: Lecture and field exercises cover computing fire data, personnel management, equipment maintenance, and fire control.

Credit Recommendation: In the lower-division baccalaureate/associate degree category, 3 semester hours in field experience in management (1/01).

AR-1408-0277

AIR DEFENSE ARTILLERY PHASE, CAPTAIN

Course Number: 2-44-C22, Phase 1.

Location: Air Defense Artillery School, Fort Bliss, TX.

Length: 18 weeks (648 hours).

Exhibit Dates: 10/03–Present.

Learning Outcomes: Upon completion of the course, the student will be able to supervise the operations of an air defense unit and function as liaison to higher commands.

Instruction: Lectures and practical exercises cover both written (operational orders) and oral communication, strategic and tactical air defense operations with associated briefing techniques, leadership training, and other military-specific subjects. Extensive training and experience are provided in overall air defense operations and integrated military decision making.

Credit Recommendation: In the lower-division baccalaureate/associate degree category, 1 semester hour in written and oral communications (4/01).

AR-1408-0278

AIR DEFENSE ARTILLERY PHASE, CAPTAIN'S
CAREER

Course Number: 2-44-C22, Phase 1.

Location: Air Defense Artillery School, Fort Bliss, TX.

Length: 2 weeks (110 hours).

Exhibit Dates: 10/03–Present.

Learning Outcomes: Upon completion of the course, student will be able to prepare a field exercise and to generate the appropriate operational orders.

Instruction: The primary means of instruction in this course is practical exercises coupled with limited hours devoted to lectures and testing. Specific testing in the areas of threat, air defense systems, combined arms, and battery command is included.

Credit Recommendation: Credit is not recommended because of the military-specific nature of the course (4/01).

AR-1408-0279

UNMANNED AERIAL VEHICLE OPERATIONS
SUPERVISOR BASIC
NONCOMMISSIONED OFFICER (NCO)

Course Number: 243-96U30, Phase 2; 243-96U30.

Location: Intelligence School, Fort Huachuca, AZ.

Length: 5–6 weeks (217–584 hours).

Exhibit Dates: 10/98–Present.

Learning Outcomes: Upon completion of the course, the student will be able to lead teams of soldiers in unmanned aerial vehicles operations.

Instruction: This course includes lectures and exercises covering common leadership skills, physical fitness training, effective military writing skills, and applied leadership theory.

Credit Recommendation: In the lower-division baccalaureate/associate degree category, 2 semester hours in management. See AR-1406-0090 for common core credit (8/02).

Related Occupations: 96U.

AR-1408-0280

TOTAL ARMY TRAINING SYSTEM (TATS)
INTELLIGENCE ELECTRONIC WARFARE
EQUIPMENT TECHNICIAN WARRANT
OFFICER BASIC

Course Number: 4C-353A.

Location: Intelligence School, Fort Huachuca, AZ.

Length: 357 hours.

Exhibit Dates: 3/99–Present.

Learning Outcomes: Upon completion of the course, the student will be able to initiate, operate, and manage an industrial maintenance, safety, and quality assurance programs and a logistic and supply organization that supports the operation and maintenance of computer-controlled equipment and software.

Instruction: Lectures, demonstrations, and practical exercises cover operations, supply, maintenance, safety, and quality systems that support computer-based equipment and software.

Credit Recommendation: In the lower-division baccalaureate/associate degree category, 1 semester hour in quality control and 3 in maintenance operations (8/01).

Related Occupations: 353A.

AR-1408-0281

RESERVE COMPONENT MULTIFUNCTIONAL
COMBAT SERVICE SUPPORT

Course Number: ALMC-RC.

Location: Army Logistics Management College, Fort Lee, VA.

Length: 2 weeks (81 hours).

Exhibit Dates: 6/01–Present.

Learning Outcomes: Upon completion of the course, the student will perform duties in multifunctional combat service support organizations.

Instruction: Topics include supply and services, ordnance support, combat health support, transportation support, and battlefield sustainment exercises.

Credit Recommendation: In the lower-division baccalaureate/associate degree category, 3 semester hours in logistics management (1/02).

AR-1408-0282

ARMY SECONDARY ITEM MANAGEMENT

Course Number: ALMC-SI.

Location: Army Logistics Management College, Fort Lee, VA.

Length: 2 weeks (73–74 hours).

Exhibit Dates: 10/96–Present.

Learning Outcomes: Upon completion of the course, the student will understand the Army Materiel Command (AMC) integrated items management of secondary items and apply logistics management tools to support the weapons system acquisition process.

Instruction: Lectures, discussions, and simulation exercises cover the item management plan. The focus of this course is on the materiel management process and the relationship of secondary items to weapon systems. Special emphasis is placed on wholesale requirements computation and the distribution processes.

Credit Recommendation: In the graduate degree category, 3 semester hours in logistics management (1/02).

AR-1408-0283

COMBINED LOGISTICS CAPTAINS CAREER, PHASE 1

Course Number: 8-10-C22 (LOG) P1, Phase 1.

Location: Army Logistics Management College, Fort Lee, VA.

Length: 6–7 weeks (231 hours).

Exhibit Dates: 10/99–Present.

Learning Outcomes: Upon completion of the course, the student will be able to apply communication skills, personnel management techniques, training skills, and professional administrative skills to improve leadership.

Instruction: Lectures include leadership and personnel functions, unit maintenance and supply, military organization, and history. Emphasis is placed on communication skills.

Credit Recommendation: In the graduate degree category, 3 semester hours in logistics management. This credit does not duplicate the credit in Phase 3 (1/02).

AR-1408-0284

COMBINED LOGISTICS CAPTAINS CAREER, PHASE 3

Course Number: 8-10-C22 (LOG) P3, Phase 3.

Location: Army Logistics Management College, Fort Lee, VA.

Length: 6–7 weeks (239 hours).

Exhibit Dates: 10/99–Present.

Learning Outcomes: Upon completion of the course, students will be able to identify and define logistics functions and manage logistics for large-scale activities.

Instruction: Lectures include leadership and personnel functions, unit maintenance and supply, military organizations, and history. The instruction is of capstone mode, where training for other phases is integrated. (Phase 2 is branch specific, conducted at the designated school.) A concluding model exercise is developed and presented, using the techniques, principles, and procedures acquired in the course.

Credit Recommendation: In the graduate degree category, 3 semester hours in logistics management. This credit does not duplicate that in Phase 1 (1/02).

AR-1408-0285

TOTAL ARMY TRAINING SYSTEM (TATS) CHAPLAIN ASSISTANT BASIC NONCOMMISSIONED OFFICER (NCO)

Course Number: 561-56M30 (Option 1).

Location: Chaplain Center and School, Fort Jackson, SC.

Length: 6 weeks (163 hours).

Exhibit Dates: 10/97–Present.

Learning Outcomes: Upon completion of the course, the student will be able to assist chaplains in identifying personnel in need of ministering, assist chaplain in preparing for spiritual guidance for personnel, and perform officer management duties.

Instruction: Methods of instruction include group discussions, examinations, practical exercises on computers and in classroom settings. Topics covered include spiritual fitness, preparation for worship and spiritual renewal and enrichment, and office procedures.

Credit Recommendation: In the lower-division baccalaureate/associate degree category, 1 semester hour in office procedures. In the upper-division baccalaureate category, 1 semester hour in ethics or religion and 1 in organizational psychology (3/02).

Related Occupations: 56M.

AR-1408-0286

INITIAL MILITARY TRAINING FOR CHAPLAINS AND CHAPLAIN CANDIDATES (Chaplain Officer Basic Common Core)

Course Number: 5-16-C20 (Common Core); 5-16-C20/C25; 5-16-C20.

Location: Chaplain Center and School, Fort Jackson, SC.

Length: 2–4 weeks (65–195 hours).

Exhibit Dates: 6/96–Present.

Learning Outcomes: Upon completion of the course, the student will have been exposed to general military topics.

Instruction: Methods of instruction include conference/seminar and student exercises.

Credit Recommendation: In the lower-division baccalaureate/associate degree category, 1 semester hour in military science (3/04).

AR-1408-0287

SENIOR PERSONNEL SERVICES SERGEANT

Course Number: 500-75H40.

Location: NCO Academy, Fort Jackson, SC.

Length: 5 weeks (240 hours).

Exhibit Dates: 10/01–Present.

Learning Outcomes: Upon completion of the course, the student will be able to teach, evaluate, lead, and counsel personnel; conduct collective training, and perform as a supervisor.

Instruction: Method of instruction includes practical exercises, discussions, lectures, and classroom exercises. Topics include leadership, communication skills, training management, common military skills, and technical services.

Credit Recommendation: In the lower-division baccalaureate/associate degree category, 3 semester hours in military science and 3 in principles of supervision (3/02).

Related Occupations: 75H.

AR-1408-0288

STANDARD INSTALLATION/DIVISION PERSONNEL SYSTEM (SIDPERS) 3

Course Number: 7C-F39/500-F29.

Location: Adjutant General School, Fort Jackson, SC.

Length: 2 weeks (72–73 hours).

Exhibit Dates: 1/98–9/02.

Learning Outcomes: Upon completion of the course, the student will be able to utilize a personal computer for information management.

Instruction: This course is offered completely as hands-on/performance based utilizing personal computers.

Credit Recommendation: In the lower-division baccalaureate/associate degree category, 3 semester hours in information management (3/02).

AR-1408-0289

TOTAL ARMY TRAINING SYSTEM (TATS) ADMINISTRATIVE WARRANT OFFICER ADVANCED

Course Number: 7-12-C32.

Location: Adjutant General School, Fort Jackson, SC.

Length: 7–8 weeks (302 hours).

Exhibit Dates: 10/00–Present.

Learning Outcomes: Before 10/00 see AR-1408-0167. Upon completion of the course, the student will be able to manage a medium-sized organization including personnel supervision, computer equipment, and software utilization and provide leadership for military and civilian personnel.

Instruction: Methods of instruction include lectures, practical exercises, examinations, and discussions. Topics covered include personnel management and information technology.

Credit Recommendation: In the upper-division baccalaureate category, 3 semester hours in leadership or supervision and 3 in information management (3/02).

AR-1408-0290

ADMINISTRATIVE WARRANT OFFICER ADVANCED

Course Number: 7-12-C33, Phase 1.

Location: Adjutant General School, Fort Jackson, SC.

Length: 2 weeks (121–122 hours).

Exhibit Dates: 10/01–Present.

Learning Outcomes: Upon completion of the course, the student will be able to perform in mid-level personnel management activities including leadership, information management, and office administration.

Instruction: Methods of instruction include classroom exercises, practical exercises, discussions, lectures, and computer laboratories.

Credit Recommendation: In the upper-division baccalaureate category, 3 semester hours in decision making and problem solving (3/02).

AR-1408-0291

ADMINISTRATIVE WARRANT OFFICER ADVANCED

Course Number: 7-12-C33, Phase 2.

Location: Adjutant General School, Fort Jackson, SC.

Length: 2 weeks (120 hours).

Exhibit Dates: 10/01–Present.

Learning Outcomes: Upon completion of the course, the student will be able to perform in mid-level managerial personnel management activities including leadership, information management, and office administration.

Instruction: Methods of instruction include classroom exercises, practical exercises, discussions, lectures, and computer laboratories.

Credit Recommendation: In the upper-division baccalaureate category, 3 semester hours in management information systems (3/02).

AR-1408-0292

JOINT MORTUARY AFFAIRS SENIOR NONCOMMISSIONED OFFICER (NCO)

Course Number: 492-F1, Phase 1.

Location: Quartermaster Center and School, Fort Lee, VA.

Length: 2 weeks (80 hours).

Exhibit Dates: 10/96–Present.

Learning Outcomes: Upon completion of the course, the student will be able to plan, coordinate, and supervise the care and handling of deceased personnel and their personal effects; supervise search, recovery, identification, evacuation, collection point operations, interment, and disinterment operations.

Instruction: Practical exercises, lecture, and discussion. Topics include search, recovery, evacuation, and collection point operations; interment and disinterment operations; mortuary affairs operations; and the development of mass fatality plans.

Credit Recommendation: In the lower-division baccalaureate/associate degree category, 3 semester hours in principles of supervision (6/02).

AR-1408-0293

DEPARTMENTAL RESOURCE MANAGEMENT

Course Number: CMS-3, Phase 1.

Location: Western Hemisphere Institute for Security Cooperation (WHINSEC), Fort Benning, GA.

Length: 7 weeks (226 hours).

Exhibit Dates: 2/01–Present.

Learning Outcomes: Upon completion of the course, the student will be able to use statistical measures, including descriptive statistics, cost analysis, comparison, and effect on decision-making; plan a budget, apply decision-making tools, and implement a budget within economic constraints of an individual country; apply the basic concepts of force structure planning, acquisition, and materiel readiness.

Instruction: Audio-visual materials, practical and classroom exercises, learner presentations, discussion, laboratory, case studies, and lecture. Topics include operational research, quantitative analysis, statistics, scientific management, decision-making, microeconomics, cyclical budget process, microeconomics in southern hemisphere countries, logistics, budgeting, planning, human resources, force structure planning, constructing, and materiel readiness.

Credit Recommendation: In the upper-division baccalaureate category, 3 semester hours in operations research, quantitative analysis or quantitative materials in decision-making, and 3 in logistics management. In the graduate degree category, 1 semester hour in regional microeconomics (10/02).

AR-1408-0294

JOINT OPERATIONS

Course Number: OPME-5, Phase 2.

Location: Western Hemisphere Institute for Security Cooperation (WHINSEC), Fort Benning, GA.

Length: 5 weeks (162 hours).

Exhibit Dates: 2/01–Present.

Learning Outcomes: Upon completion of the course, the student will be able to participate in joint multinational and interagency strategic coalition organizational planning, execution, and review; design and apply strategic and organizational concepts; synthesize and analyze complex strategic planning processes; integrate multinational logistics and transportation management; information processing, and operations planning concepts on a regional basis; understand the application of ends, ways, and means of a joint planning strategy.

Instruction: Audio-visual materials, practical and classroom exercises, discussion, laboratory, lecture, learner presentation and case studies. Topics include strategic and operational planning, contingency and deliberate planning, joint force simulation and modeling, operations (other than war), peacekeeping, humanitarian assistance, disaster relief, civil/military operations, future concepts, human rights awareness, and management of physical training.

Credit Recommendation: In the graduate degree category, 3 semester hours in organizational planning and 3 in operations analysis and management (10/02).

AR-1408-0295

COMPANY MAINTENANCE/ULLS-G

Course Number: 921-420.

Location: Reserve Readiness Training Center, Fort McCoy, WI.

Length: 2 weeks (73 hours).

Exhibit Dates: 4/99–Present.

Learning Outcomes: Upon completion of the course, the student will be able to order publications, utilize the modern Army record keeping system, operate the unit level logistics system-ground, conduct all aspects of the automated unit level maintenance processes, complete the requirements of the Army analysis oil program certification, request and receive supplies, monitor equipment calibration, utilize safety procedures, and identify physical security requirements.

Instruction: Practical and classroom exercises, lecture, and computer-based training. Topics include maintenance operations (automated), Army oil analysis program certification, supply operations (automated), and final performance exercises (automated).

Credit Recommendation: In the lower-division baccalaureate/associate degree category, 2 semester hours in automotive maintenance management (11/02).

AR-1408-0296

RETENTION AND TRANSITION MANAGERS

Course Number: ***Version 1:*** 921-181. *Version 2:* 921-181.

Location: Reserve Readiness Training Center, Fort McCoy, WI.

Length: ***Version 1:*** 2 weeks (80 hours). *Version 2:* 2 weeks (80 hours).

Exhibit Dates: ***Version 1:*** 9/00–2/02. *Version 2:* 3/02–Present.

Learning Outcomes: ***Version 1:*** Upon completion of the course, the student will be able to demonstrate knowledge of eligibility for re-enlistment extension, demonstrate knowledge of eligibility for incentives/entitlements, develop a budget based upon analysis of the yearly retention and transition training plan, analyze and interpret retention/transition performance data, set-up and conduct a meeting, explain the Warrant Officer selection process and mentorship program, demonstrate the steps necessary to order retention publicity items, maintain a retention bulletin board, prepare a request for an antedated contract, prepare a request for an exception to policy, determine the most appropriate ways to combat stress in the workplace, and identify the key aspects of the family support program. *Version 2:* Upon completion of the course, the student will be able to develop a training outline, conduct a training session, conduct an effective training assessment program, evaluate a retention interview, prepare a time management work plan, determine proper procedures, maintenance requirements, reporting requirements, procedures for accounting for and maintaining all assigned equipment, analyze the use of applicable reports for retention purposes, perform quality control inspections on sample packets and identify submission and filing requirements for Warrant Officer packets, analyze data for retention purposes, and identify and analyze various reports for retention purposes.

Instruction: ***Version 1:*** Audiovisual materials, practical and classroom exercises, discussion, and lecture. Topics include eligibility for re-enlistment/extension, incentives/entitlements, roles and responsibilities of the Retention and Transition Manager, elements of the yearly retention and transition training plan, unit climate profile, time management work plan, time transition process, Warrant Officer transition process, automated systems, antedated contract, request for an exception to policy for incentives, and family support program. *Version 2:* Audiovisual materials, practical and classroom exercises, discussion, and lecture. Topics include roles and responsibilities of the area manager, recruiting partnership council, leadership, retention management, management systems, retention missions, retention processing systems, training outline, session and assessment, interview evaluation, time management work plan, property and maintenance, and quality control.

Credit Recommendation: ***Version 1:*** In the upper-division baccalaureate category, 2 semester hours in workforce management (11/02). *Version 2:* In the upper-division baccalaureate category, 2 semester hours in workforce management (11/02).

AR-1408-0297

IMAGERY INTELLIGENCE (IMINT) OFFICER

Course Number: 3A-35C.

Location: Intelligence Center and School, Fort Huachuca, AZ.

Length: 5 weeks (192 hours).

Exhibit Dates: 5/99–Present.

Learning Outcomes: Upon completion of the course, the student will be able to take responsibility for the conduct of imagery intelligence operations, and manage imagery intelligence assets.

Instruction: Practical and classroom exercises, discussion, lecture, and computer-based training. Topics include study of US imagery systems; knowledge, skills, and techniques used by imagery analysts; training of officers to manage imagery intelligence collection, exploitation, processing, production, systems, and organization.

Credit Recommendation: In the lower-division baccalaureate/associate degree category, 2 semester hours in basic management skills (12/02).

AR-1408-0298

Signal Basic Officer Leader

Course Number: 4-11-C20 (P), Phase 2.

Location: Signal School, Fort Gordon, GA.

Length: 12 weeks (540 hours).

Exhibit Dates: 6/03–Present.

Learning Outcomes: Upon completion of the course, the student will be able to lead a mobile subscriber equipment (MSE) division or Corps Signal Battalion or be assigned as a Signal Staff Officer (S-6) in a maneuver battalion.

Instruction: Practical and classroom exercises, and lecture. Topics include computer basics, computer hardware, operating systems, network essentials, local area networks, network/system security, leadership and military subjects, signal support planning and operations.

Credit Recommendation: In the lower-division baccalaureate/associate degree category, 3 semester hours in network fundamentals (2/03).

AR-1408-0299

Signal Staff Officer (S-6)

Course Number: 4C-F40, Phase 1.

Location: Signal School, Fort Gordon, GA.

Length: 4 weeks (160 hours).

Exhibit Dates: 6/02–Present.

Learning Outcomes: Upon completion of the course, the student will be able to plan, execute, manage and allocate resources in tactical communications; provide communications/automation interface unique to non-signal warfighting Battalion/Brigade; telecommunications; installation, operations, and maintenance of data communications; user owned systems; network encryption devices; teleprocessing and distributed database systems; tactical network management; and systems administration at the S-6 level.

Instruction: Audiovisual materials, practical and classroom exercises, discussion, laboratory, lecture, learner presentation, and computer-based training. Topics include Windows server, radio communications, and network fundamentals.

Credit Recommendation: In the lower-division baccalaureate/associate degree category, 3 semester hours in Window server, 3 in radio communications, and 3 in network fundamentals (2/03).

AR-1408-0300

Warrant Officer Candidate School

Course Number: 911-09W-RC, Phase 2; 911-09W-RC, Phase 1.

Location: Aviation Center, Fort Rucker, AL.

Length: 11 weeks (440 hours).

Exhibit Dates: 10/94–Present.

Learning Outcomes: Upon completion of the course, the student will be able to train Warrant Officer candidates in the skills of all warrant officers.

Instruction: Audio visual materials, practical and classroom exercises, discussion, laboratory, lecture, learner presentations, and computer-based training. Topics include leadership, ethics, professional development, communication skills, personnel management, military history, deployment, and tactics.

Credit Recommendation: In the lower-division baccalaureate/associate degree category, 6 semester hours in leadership and supervision (4/03).

AR-1408-0301

Armor Basic Officer Leader (M1A2)

Course Number: 2-17-C20 (M1A2).

Location: Armor School, Fort Knox, KY.

Length: 17 weeks (984–1106 hours).

Exhibit Dates: 10/01–Present.

Learning Outcomes: Upon completion of the course, the student will be able to demonstrate the fundamentals of tactics, gunnery, maintenance, and leadership.

Instruction: Audiovisual materials, practical exercises, discussion, lecture, computer-based training, and lecture. Topics include military problem solving, risk management, values and ethics, military justice, law of war, administrative and personnel actions, maintenance management, property accountability, equal opportunity, and sexual harassment.

Credit Recommendation: In the lower-division baccalaureate/associate degree category, 3 semester hours in military tactics and 3 in leadership. In the upper-division baccalaureate category, 2 semester hours in human resource management, 1 in human relations, and 2 in managerial planning (6/03).

AR-1408-0302

Maintenance Leader (DL)

Course Number: 8C-F21/020-F23 (DL), Phase 1.

Location: Armor School, Fort Knox, KY.

Length: 1–2 weeks (65 hours).

Exhibit Dates: 5/01–Present.

Learning Outcomes: Upon completion of the course, the student will be able to manage and supervise in the areas of maintenance at the battalion level and below; and evaluate and manage class procedures and unit level logistics systems/ground reports.

Instruction: Practical exercises and computer-based training. Topics include managing unit records, planning and organizing unit maintenance operations in the field, developing maintenance management reports, conducting inspections of unit equipment, and directing vehicle recovery operations using battalion-level assets.

Credit Recommendation: In the lower-division baccalaureate/associate degree category, 1 semester hour in technical report writing, 1 in records management, and 1 in maintenance management (6/03).

AR-1408-0303

Ordnance Maintenance Management Officer (BQ)

Course Number: 4-9-C20-91B/C (BQ), Phase 1.

Location: Ordnance Center and School, Aberdeen Proving Ground, MD.

Length: 4 weeks (211 hours).

Exhibit Dates: 1/92–Present.

Learning Outcomes: Upon completion of the course, the student will be able to demonstrate a knowledge of support supply, support maintenance, and technical maintenance.

Instruction: Audio visual materials, learner presentations, computer-based training, practical exercises, classroom exercises, discussion, laboratory, and lecture. Topics include military-specific items on logistics and reports military publications supply system, and overviews on motor pool operations, metalworking, and weapons department.

Credit Recommendation: In the lower-division baccalaureate/associate degree category, 2 semester hours in business management (8/03).

AR-1408-0304

Ordnance Officer Basic

Course Number: 4-9-C20-91A, Phase 1.

Location: Ordnance Center and School, Aberdeen Proving Ground, MD.

Length: 18–19 weeks (700 hours).

Exhibit Dates: 7/01–Present.

Learning Outcomes: Upon completion of the course, the student will be able to function at the unit level as a platoon leader, maintenance control officer, munitions operations officer or as a munitions operations officer at the division level. This course is designed to train newly commissioned officers for their subsequent assignment.

Instruction: Audio visual materials, learner presentations, computer-based training, practical exercises, classroom exercises, discussion, laboratory, and lecture. Topics include leadership, tactics, communications skills, maintenance and supply, conventional munitions, metalworking, ground support equipment maintenance, wheel vehicle maintenance, recovery principles, automotive/armament maintenance, and an overview of electronic and missile maintenance.

Credit Recommendation: In the lower-division baccalaureate/associate degree category, 3 semester hours in leadership, 3 in speech, and 3 in personnel supervision. In the upper-division baccalaureate category, 3 semester hours maintenance management (8/03).

AR-1408-0305

Senior Automotive Maintenance Officer Warrant Officer Advanced

Course Number: 4-9-C32-915E, Phase 1.

Location: Ordnance Center and School, Aberdeen Proving Ground, MD.

Length: 14–15 weeks (515 hours).

Exhibit Dates: 10/02–Present.

Learning Outcomes: Upon completion of the course, the student will be able to manage and supervise various maintenance operations.

Instruction: Practical exercises, classroom exercises, discussion, laboratory, and lecture. Topics include management and supervision of direct/general maintenance, recovery operations, fuel and electrical component repairs, and wheeled vehicles and tanks.

Credit Recommendation: In the upper-division baccalaureate category, 3 semester hours in maintenance management (8/03).

AR-1408-0306

BASIC OFFICER LEADERSHIP COURSE (BOLC)

Course Number: 4-9-C20-91A, Phase 2.

Location: Ordnance Center and School, Aberdeen Proving Ground, MD.

Length: 12–13 weeks (469 hours).

Exhibit Dates: 3/02–Present.

Learning Outcomes: Upon completion of the course, the student will be able to demonstrate knowledge of leadership, communication skills, supply management, material and munitions management.

Instruction: Audio visual materials, learner presentations, computer-based training, practical exercises, classroom exercises, discussion, laboratory, and lecture. Topics include leadership communications skills, maintenance and supply, conventional munitions, metalworking, GSE, recovery operations, and automotive/armament maintenance.

Credit Recommendation: In the lower-division baccalaureate/associate degree category, 1 semester hours in leadership, 1 in speech, and 3 in personnel supervision. In the upper-division baccalaureate category, 3 semester hours in maintenance management (8/03).

AR-1409-0011

EMITTER LOCATOR/IDENTIFIER
(Electronic Warfare (EW)/Signal Intelligence (SIGINT) Emitter Identifier/Locator)

Course Number: *Version 1:* 231-98D10; 231-05D10. *Version 2:* 231-98D10. *Version 3:* 231-98D10.

Location: *Version 1:* Intelligence School, Fort Devens, MA. *Version 2:* Intelligence School, Fort Huachuca, AZ. *Version 3:* Intelligence Center and School, Fort Huachuca, AZ.

Length: *Version 1:* 18–28 weeks (644–859 hours). *Version 2:* 7–8 weeks (259–290 hours). *Version 3:* 18 weeks (659 hours).

Exhibit Dates: *Version 1:* 2/84–12/91. *Version 2:* 1/92–9/97. *Version 3:* 10/97–10/98.

Learning Outcomes: *Version 1:* Upon completion of the course, the student will be able to send and receive international Morse code; install, operate, and troubleshoot radio teletype equipment; and type at 20 words per minute. *Version 2:* Upon completion of the course, the student will be able to send and receive international Morse code at 18 groups per minute, operate and troubleshoot radio teletype equipment, and type at 20 words per minute without error. *Version 3:* Upon completion of the course, the student will be able to send and receive international Morse code at 18 groups per minute and operate radio teletype equipment.

Instruction: *Version 1:* Instruction includes lectures, computer-aided instruction, and performance-based exercises in code copying and transcription; typing; deployment, installation, and operation of a direction-finding system, and equipment inspection and maintenance procedures. *Version 2:* Instruction includes lectures, computer-aided instruction, and performance-based exercises in code copying and transcription; typing proficiency; deployment, installation, and operation of a direction-finding system; equipment inspection and maintenance procedures; fix analysis; and performance reports. *Version 3:* Instruction includes lectures, computer-aided instruction, and performance-based exercises in code copying and transcription; typing proficiency; deployment, installation, and operation of a direction finding system; equipment inspection and maintenance procedures; fix analysis; and performance reports.

Credit Recommendation: *Version 1:* In the vocational certificate category, 3 semester hours in Morse code. In the lower-division baccalaureate/associate degree category, 2 hours in typing (10/90). *Version 2:* In the vocational certificate category, 3 semester hours in Morse code. In the lower-division baccalaureate/associate degree category, 2 hours in keyboarding (5/97). *Version 3:* In the vocational certificate category, 6 semester hours in Morse code. In the lower-division baccalaureate/associate degree category, 2 semester hours in keyboarding (1/00).

Related Occupations: 05D; 98D.

AR-1409-0012

ELECTRONIC WARFARE (EW)/SIGNAL INTELLIGENCE (SIGINT) MORSE INTERCEPTOR

Course Number: 231-05H10.

Location: Intelligence School, Fort Devens, MA.

Length: 21–23 weeks (678–829 hours).

Exhibit Dates: 10/85–3/94.

Learning Outcomes: Upon completion of the course, the student will be able to copy Morse code; detect, acquire, identify, and record foreign communications; type 25 words per minute; perform basic word processing; and evaluate incoming signals.

Instruction: Instruction includes lectures and performance exercises in typing proficiency, copying Morse code, operating equipment, and generating required data.

Credit Recommendation: In the vocational certificate category, 3 semester hours in Morse code. In the lower-division baccalaureate/associate degree category, 2 semester hours in beginning typing (1/88).

Related Occupations: 05H.

AR-1511-0012

CIVIL AFFAIRS ENLISTED

Course Number: 500-SQID.

Location: Institute for Military Assistance, Fort Bragg, NC.

Length: 2 weeks (70–80 hours).

Exhibit Dates: 3/81–4/04.

Learning Outcomes: Upon completion of the course, the student will be able to carry out a wide range of civilian-related practical activities, including demonstrating awareness of psychological effects of military action; preparing documents affecting public health, evacuees, and refugees; and becoming sensitive to effects of cultural differences and the difficulties of cross-cultural communication.

Instruction: Course consists of exercises and lectures involving role playing sessions in a wide range of civil affairs subject matter, including public health, refugees, and psychological and cross-cultural problems. Student may also prepare an area study.

Credit Recommendation: In the upper-division baccalaureate category, 2 semester hours in public administration (12/92).

AR-1511-0013

SENIOR NONCOMMISSIONED OFFICER (NCO) OPERATIONS AND INTELLIGENCE

Course Number: 243-F11.

Location: McGraw Barracks, Munich, W. Germany; Sergeants Major Academy, Fort Bliss, TX.

Length: 10 weeks (347 hours).

Exhibit Dates: 1/86–4/92.

Learning Outcomes: Upon completion of the course, the student will be able to supervise operations and intelligence sections.

Instruction: Course includes lectures, seminars, demonstrations, and practical exercises in battlefield tactics, military procedures, military doctrine, world situations, and military potential.

Credit Recommendation: In the lower-division baccalaureate/associate degree category, 2 semester hours in military science and 1 in international relations (9/87).

AR-1511-0014

1. CIVIL AFFAIRS OFFICER ADVANCED NONRESIDENT/RESIDENT RESERVE COMPONENT
2. CIVIL AFFAIRS OFFICER ADVANCED NONRESIDENT/RESIDENT RESERVE COMPONENT PHASE 2

(Civil Affairs Officer Advanced Nonresident/Resident Reserve Component)

Course Number: *Version 1:* 5-41-C23. *Version 2:* 5-41-C23.

Location: John F. Kennedy Special Warfare Center and School, Fort Bragg, NC.

Length: *Version 1:* Phase 2, 2 weeks (66–69 hours); Phase 4, 2 weeks (69–70 hours). *Version 2:* Phase 2, 2 weeks (70 hours).

Exhibit Dates: *Version 1:* 7/88–9/93. *Version 2:* 10/93–Present.

Learning Outcomes: *Version 1:* Upon completion of the course, the student will serve as civil affairs officer; advise the commander as to the establishment/administration of local civilian government; communicate the psychological and cultural effects of unintended offending actions; analyze the cost of proper programs; cite the danger of terrorist activities; control civilian-refugee problems; and preserve art, documents, and cultural treasures under conditions of military conflict. *Version 2:* Upon completion of the course, the student will be able to serve as a civil affairs officer; conduct an assessment of civil affairs needs; advise the commander of local economic, social, and political factors affecting a military occupied community; plan and recommend actions for control of dislocated civilians; and serve as a liaison between the local community and military command.

Instruction: *Version 1:* Lectures, seminars, students presentations, individual research papers, and team area study papers cover planning of civil affairs activities, psychological and cross-cultural analysis, economic considerations, civil-military operations, sociology, low-intensity conflict and terrorism, civil affairs, and legal considerations. *Version 2:*

Lectures, seminars, student presentations, individual research papers, and team area study papers cover planning of civil affairs activities and psychological and cross-cultural analysis, economic considerations, civil-military operations, sociology, low-intensity conflict and terrorism, civil affairs, and legal considerations.

Credit Recommendation: *Version 1:* In the upper-division baccalaureate category, For Phase 2, 3 semester hours in international relations or public administration. For Phase 4, 3 additional semester hours in international relations or public administration (10/90). *Version 2:* In the upper-division baccalaureate category, 1 semester hour in international relations and 3 in public administration planning (9/00).

AR-1511-0016

Civil Affairs

Course Number: 5D-F6/500-F4.
Location: John F. Kennedy Special Warfare Center and School, Fort Bragg, NC.
Length: 5–6 weeks (205 hours).
Exhibit Dates: 7/88–5/96.
Learning Outcomes: After 5/96 see AR-1511-0029. Upon completion of the course, the student will be able to interpret Communist strategy, analyze effects of military operations on civilian and political affairs, describe how the Army structure deals with these questions, handle effects of civilian dislocation, and recognize and deal with low-intensity conflict.
Instruction: Lectures, seminars, a country study paper, and simulations of field situations are modes of instruction. Topics include civil-military history, psychological operations, air/land battle, economic considerations, cultural and religious impact, sociology, and legal considerations.
Credit Recommendation: In the upper-division baccalaureate category, 6 semester hours in public administration and 3 in sociology (12/88).

AR-1511-0018

Special Operations Reconnaissance
(Strategic Reconnaissance)

Course Number: 2E-F66/011-F25.
Location: John F. Kennedy Special Warfare Center and School, Fort Bragg, NC.
Length: 8 weeks (469–503 hours).
Exhibit Dates: 6/86–2/92.
Learning Outcomes: Upon completion of the course, the student will be able to construct antennas, perform elementary maintenance, and operate special field portable radio sets; identify and treat various medical ailments with little or no supplies or medication; recognize and treat effects of combat stress; recognize and report on Soviet Order of Battle Intelligence; identify Soviet military equipment and weapon systems; demonstrate proper evasive techniques; and use field photographic equipment.
Instruction: Course includes lectures and hands-on training in basic and field radio setup and operation, Warsaw Pact order of battle, weapon systems, minor medical emergencies, use of laser acquisition and beacon devices, evasive techniques for humans and attack/sentry dogs, and proper use of field photographic equipment.
Credit Recommendation: In the upper-division baccalaureate category, 2 semester hours in Soviet studies (12/88).

AR-1511-0019

Special Forces Operations and Intelligence Resident/ Nonresident Phase 4

Course Number: 244-ASIF1-RC.
Location: John F. Kennedy Special Warfare Center and School, Fort Bragg, NC.
Length: 2 weeks (135–149 hours).
Exhibit Dates: 6/87–9/98.
Learning Outcomes: Upon completion of the course, the student will be able to perform intelligence functions in support of Special Forces missions, analyze Soviet military organization, and conduct clandestine operations.
Instruction: Lectures, seminars, special conferences on intelligence, written reports, and oral presentations cover aspects of Soviet military organization and the intelligence operations of US forces.
Credit Recommendation: In the lower-division baccalaureate/associate degree category, 1 semester hour in communication. In the upper-division baccalaureate category, 2 semester hours in Soviet studies, 1 in psychology, and 1 in political science (12/88).

AR-1511-0020

Special Forces Operations and Intelligence Resident/ Nonresident Phase 2

Course Number: 244-ASIF1-RC.
Location: John F. Kennedy Special Warfare Center and School, Fort Bragg, NC.
Length: 2 weeks (128–137 hours).
Exhibit Dates: 4/87–9/98.
Learning Outcomes: Upon completion of the course, the student will be able to support Special Forces missions, coordinate air/ground/nautical missions, analyze maps, and support air/land battle.
Instruction: Lectures, seminars, simulation, and discussion cover Special Forces operations, map interpretation, role of Special Forces in air/land battle, aspects of Army organization, and terrorism.
Credit Recommendation: In the lower-division baccalaureate/associate degree category, 1 semester hour in map interpretation. In the upper-division baccalaureate category, 1 semester hour in political science, 1 in psychology, and 1 in social science (12/88).

AR-1511-0021

Army War College

Course Number: None.
Location: Army War College, Carlisle Barracks, PA.
Length: 44 weeks.
Exhibit Dates: 8/88–Present.
Learning Outcomes: Upon completion of the course, the student will have an understanding of the science of warfare in the area of national security studies, strategy, force planning, and international relations in order to prepare officers for higher command positions. Students will develop an understanding of the process by which national strategy is formulated, including meeting changing threats to national security. Students will demonstrate heightened awareness of personal values, strengths, and leadership styles and develop an ethical framework for reasoning and value judgments; creative skills for leadership in a complex organization; understanding of historical perspectives as they relate to operational and organizational perspectives; and understanding of budgeting, human resource management, and program planning.
Instruction: This course is presented through lectures, seminars, reading and writing, simulation exercises, oral presentations, field trips, case studies, self-assessment techniques, audiovisual presentations, and computer exercises. Individual group work is supported by extensive library access. There is a large resident faculty supplemented by distinguished guest speakers and by field trips. There is a good faculty-student ratio, and there are several group projects. Topics include ethics, leadership, joint service operations, science of military warfare including implementation at the regional and global levels and its application to future warfare, human resource techniques, budgeting, professionalism, self-assessment techniques, American military history, theory of war, international relations, and American foreign policy. Students will take up to eight individually-selected advanced courses in national security policy and strategy, regional studies, operations, and command and leadership.
Credit Recommendation: In the upper-division baccalaureate category, 3 semester hours in international economics, 3 in political science, 3 in human resource management, and 3 in geography. In the graduate degree category, 6 semester hours in management; 3-6 for the research paper, based on the receiving institution's review of the paper; 3 in national security studies; 2 in American military history; 1 in international politics; and 3 in American foreign relations. Additionally, up to 4 semester hours may be granted in national security policy and strategy, regional studies, operations, and command and leadership based on the receiving institution's review of materials submitted by the student (11/97).

AR-1511-0022

Army War College Corresponding (by Correspondence)

Course Number: None.
Location: Army War College, Carlisle Barracks, PA.
Length: Maximum, 104 weeks.
Exhibit Dates: 8/88–Present.
Learning Outcomes: Upon completion of the course, the student will have an understanding of the science of warfare in the areas of national security studies, strategy, force planning, and international relations. Students will have an understanding of the process by which national strategy is formulated including meeting changing threats to national security. Students will demonstrate heightened awareness of personal values, strengths, and leadership styles; develop an ethical framework for reasoning and value judgment; develop creative skills for leadership in a complex organization; understand historical perspectives as they relate to operational and organizational perspectives; and understand budgeting, human resources management, and program planning.
Instruction: This correspondence course, offered over two years, is designed to train senior military officers in leadership and management of a large, complex organization, through readings and writing in military history and various strategic questions. There is a two-week seminar at the War College at the end of each year of correspondence study. Students are

required to complete substantial writing assignments which are critiqued. Topics include ethics, leadership, joint service operations, the science of military warfare including implementation at the regional and global levels and its application to future warfare, human resource techniques, budgeting techniques, professionalism, self-assessment techniques, American military history, theory of war, international relations, and American foreign policy.

Credit Recommendation: In the upper-division baccalaureate category, 3 semester hours in international economics, 3 in political science, 3 in human resource management, and 3 in geography. In the graduate degree category, 6 semester hours in management, 3-6 for the research paper, based on the receiving institution's review of the paper, 3 in national security studies, 2 in American military history, 1 in international politics, and 3 in American foreign relations. It is recommended that the receiving institution delay awarding of credit until the student has completed 3-6 graduate hours (11/97).

AR-1511-0023

Defense Strategy by Correspondence

Course Number: None.
Location: Army War College, Carlisle Barracks, PA.
Length: 18–26 weeks.
Exhibit Dates: 1/89–1/90.
Learning Outcomes: Upon completion of the course, the student will be able to describe the domestic and international environment as it impacts on the formulation of national security policy.
Instruction: This is a correspondence course presenting national security strategy (domestic and international environments) and the development of a strategy to serve it. The writing component consists of several very short papers as well as one 1500-word paper on national strategy.
Credit Recommendation: In the upper-division baccalaureate category, 3 semester hours in national security. There is no proctored final examination (1/92).

AR-1511-0024

Intelligence in Combatting Terrorism
(Intelligence in Terrorism Counteraction)

Course Number: 3C-F14/244-F8.
Location: Intelligence School, Fort Huachuca, AZ.
Length: 2 weeks (72–83 hours).
Exhibit Dates: 10/87–Present.
Learning Outcomes: Upon completion of the course, the student will be able to describe terrorist movements, tactics, organizations, and ideologies and apply terrorism threat assessment techniques.
Instruction: Lectures, discussions, and classroom testing are the methods of instruction.
Credit Recommendation: In the lower-division baccalaureate/associate degree category, 3 semester hours in international relations and terrorism (8/01).

AR-1511-0025

Foreign Internal Defense/Internal Defense and Development

Course Number: 3A-F59.
Location: John F. Kennedy Special Warfare Center and School, Fort Bragg, NC.
Length: 5 weeks (196 hours).
Exhibit Dates: 5/91–9/92.
Learning Outcomes: Upon completion of the course, the student will be able to analyze and assess both insurgency and counterinsurgency warfare, apply intelligence and psychological operations to various kinds of warfare, and apply knowledge of the role of the United States in international affairs.
Instruction: Lectures, seminars, and oral presentations are used in conjunction with readings in political science and economics case studies.
Credit Recommendation: In the upper-division baccalaureate category, 3 semester hours in political science (12/92).

AR-1511-0026

Civil Affairs Specialist Reserve

Course Number: *Version 1:* 500-38A10-RC. *Version 2:* 500-38A10-RC.
Location: *Version 1:* John F. Kennedy Special Warfare Center and School, Fort Bragg, NC. *Version 2:* Reserve Schools, Multiple locations.
Length: *Version 1:* 11 weeks (436 hours). *Version 2:* 9–10 weeks (246 hours).
Exhibit Dates: *Version 1:* 10/93–3/97. *Version 2:* 4/97–Present.
Learning Outcomes: *Version 1:* Upon completion of the course, the student will be able to interpret US and foreign maps; apply urban and area research techniques on documents; conduct civil, governmental, humanitarian, and defense assistance; conduct research and updates of area studies; and learn and apply organizational and leadership skills required in field operations. *Version 2:* Upon completion of the course, the student will be able to interpret US and foreign maps; conduct civil, governmental, humanitarian, and defense assistance; apply organizational and leadership skills required in field operations; and conduct research on documents and other aspects of urban and regional studies.
Instruction: *Version 1:* Methods of instruction include lectures, guided discussions, seminars, and practical exercises. Topics covered are map reading; land navigation; administration; communications; civil affairs; management; disaster aid; and domestic security, defense, and control. *Version 2:* Lecture, discussion, and practical exercises include map reading; land navigation; administration; communications; civil affairs; disaster aid; and domestic security, defense, and control.
Credit Recommendation: *Version 1:* In the lower-division baccalaureate/associate degree category, 1 semester hour in orienteering. In the upper-division baccalaureate category, 3 semester hours in public administration (12/92). *Version 2:* In the lower-division baccalaureate/associate degree category, 1 semester hour in map reading. In the upper-division baccalaureate category, 3 semester hours in public administration (4/01).
Related Occupations: 38A.

AR-1511-0027

Regional Studies

Course Number: *Version 1:* 5D-F5. *Version 2:* 5D-F5. *Version 3:* 5D-F5.
Location: John F. Kennedy Special Warfare Center and School, Fort Bragg, NC.
Length: *Version 1:* 14–15 weeks (557 hours). *Version 2:* 16–17 weeks (620 hours). *Version 3:* 15 weeks (568 hours).
Exhibit Dates: *Version 1:* 2/90–9/91. *Version 2:* 10/91–6/96. *Version 3:* 7/96–Present.
Learning Outcomes: *Version 1:* Upon completion of the course, the student will apply the fundamentals of area studies in a military context, demonstrate an in-depth knowledge of Africa, Asia, Europe, Latin America, or the Middle East; and demonstrate proficiency in cross-cultural issues, geography, history, economics, political science, and foreign policy. *Version 2:* Upon completion of the course, the student will apply the fundamentals of area studies in a military context; demonstrate an in-depth knowledge of Africa, Asia, Europe, Latin America, or the Middle East; and demonstrate proficiency in cross-cultural issues, geography, history, economics, political science, and foreign policy. *Version 3:* Upon completion of the course, the student will apply the fundamentals of area studies in a military context; demonstrate an in-depth knowledge of Africa, Asia, Europe, Latin America, or the Middle East; and demonstrate proficiency in cross-cultural issues, geography, history, economics, political science, and foreign policy.
Instruction: *Version 1:* Methods of instruction include lectures, discussions, seminars, texts, handbooks, and secondary and primary documents. Topics covered include a general study of Africa, Asia, Europe, Latin America, and the Middle East and an in-depth analysis of one of these five regions. Cross-cultural communication, geography, history, economics, political science, and foreign policy are studied at the local, national, regional, and world levels. *Version 2:* Methods of instruction include lectures, discussions, seminars, texts, handbooks, and secondary and primary documents. Topics covered include a general study of Africa, Asia, Europe, Latin American, and the Middle East and an in-depth analysis of one of those five regions. Cross-cultural communication, geography, history, economics, political science, and foreign policy are studied at the local, national, regional, and world levels. *Version 3:* Methods of instruction include lectures, discussions, seminars, research papers, oral presentations, and analyses. Topics covered include a general study of Africa, Asia, Europe, Latin America, and the Middle East and in-depth analysis of one of those five regions. Cross-cultural communication, geography, history, economics, political science, and foreign policy are studied at the local, national, regional, and world levels.
Credit Recommendation: *Version 1:* In the upper-division baccalaureate category, 3 semester hours in national power, 3 in geopolitics of world regions, 3 in regional geography (of one of the following: Africa, Asia, Latin America, Europe, or the Middle East), and 3 in comparative politics. In the graduate degree category, up to 6 semester hours may be awarded based on institutional evaluation (12/92). *Version 2:* In the upper-division baccalaureate category, 4 semester hours in national power, 3 in geopolitics of world regions, 3 in regional geography (in one of the following areas: Africa, Asia, Latin America, Middle East, Europe), and 3 in comparative politics. In the graduate degree category, up to 6 semester hours may be awarded based on institutional evaluation (12/93). *Version 3:* In the upper-division baccalau-

reate category, 3 semester hours in international relations, 3 in US foreign policy, 3 in comparative politics, and 3 in world regional geography. In the graduate degree category, 3 semester hours in international relations and 3 in comparative politics (4/01).

AR-1511-0028

EURASIAN FOREIGN AREA OFFICER

Course Number: None.
Location: George C. Marshall, European Center for Security Studies, Europe, Germany.
Length: 78 weeks.
Exhibit Dates: 1/90–Present.
Learning Outcomes: Upon successful completion of the entire 18-month program, including material incorporated into the Political-Military Studies, the Regional Studies, Military Studies, and language courses, the student will understand the basic historical, ethnic, and geographic make-up of the former Soviet Union region; apply this knowledge in interpreting current political military events and considering specific problems in that region; conduct oral presentations or participate in discussions of individual topics concerning the region in reasonable depth; work in positions requiring familiarity with Russian/Eurasian security issues; work in positions requiring command of advanced terminology in Russian; function effectively in the Defense Attache Office (DAO) and embassy environment, be familiar with US government political-military goals and programs with the host nation; interact with military and civilian counterparts in a foreign country in the language of that country; be able to discuss current political, military, economic, cultural, and religious trends in the host nation and assess how they relate to US goals and policies; and apply their Foreign Area Officer (FAO) skills, including language, security studies, and policy analysis, among others. Students are evaluated on the basis of how well they performed the tasks and met the objectives and on the demonstrated improvement in their language skills (pre- and post-test) following the completion of the internship.
Instruction: This is an 18-month long program designed to prepare graduates for key assignments in or concerned with the countries of the former Soviet Union (FSU) and Baltic states. Graduates will have been provided with practical hands-on knowledge of living, working, and traveling in the territory of the former Soviet Union. Phase one of the program is devoted to intensive language training, normally Russian. Training may also be provided in Ukrainian or other former Soviet Union languages. The courses in the academic phase are all conducted in Russian. Readings are primarily in Russian supplemented by English readings. For credit for the language portion of this program, refer to the Defense Language Proficiency Test (DLPT) recommendations. The second phase of the course is divided into the resident core curriculum in political-military and regional studies and the internships or field studies portion. Course content includes a general overview of Russian military history, with special emphasis on the Soviet era; a selective review of security issues and policies of the various FSU states; and a highlighting of the historic importance of military/security concerns in Russian society. Students are introduced to historic and contemporary political-military events as viewed from the Russian/FSU states perspective. There will be one substantial project on a specialized topic of the student's choice related to FSU history or policies. This is normally, though not always, research-oriented. Course provides for an introduction to some social science approaches to policy analysis and an examination of Russian-language sources. The course also includes advanced military studies which would apply in an undergraduate advanced military science program. The internship program is designed to provide FAO Senior Fellows with practical, hands-on experience to supplement and extend the resident training and classroom instruction they receive during their course of study. Students are required to complete a minimum of one three-month internship working in support of a US agency in Eastern Europe or Eurasia. The internships expose students to policy issues confronting the countries and regions in which they are working; enhances their language and interpretation skills through daily application and use; increases their understanding of contemporary culture, customs, and socio-political-economic trends; and exposes them to the day-to-day workings of the office of the Defense Attache. Students typically support the Defense Attache through a variety of assignments and tasks including preparation for and execution of cooperative military exercises, Security Assistance Organization, International Military and Training Programs, Partnership for Peace programs, as well as other DAO functions and embassy activities. Upon completing the internship, students must submit a military "After Action Review" (AAR) in which they summarize the activities and duties they performed, identify critical issues and problems they encountered, describe the policy-related measures taken to address the problems, and make recommendations as to how the internship experience could be enhanced and improved for future interns. In many instances students are directly involved in policy advising and coordination roles in high-level activities. The internship is comparable to a Master's level internship at US colleges and universities in degree programs such as International Affairs, Technology and International Development, and Public Policy (as illustrative examples) where students are typically required to complete an internship with a governmental or private agency or organization conducting policy-oriented work in an area of interest to the student and related to their field of study. Such internships also require the application of skills knowledge acquired in the classroom. Students usually must submit some written assignment summarizing and evaluating their internship experience, generally relating it to the work they have done in the classroom.
Credit Recommendation: In the graduate degree category, 3 semester hours in Russian history/Russian military history with an emphasis on the history of the Soviet and immediate post-Soviet era, 3 in political science (comparative politics/international relations), and 3 in internship credit/field experience at those institutions which normally offer credit for internships (7/98).

AR-1511-0029

CIVIL AFFAIRS OFFICER

Course Number: 5D-39C.
Location: John F. Kennedy Special Warfare Center and School, Fort Bragg, NC.
Length: 4–5 weeks (183 hours).
Exhibit Dates: 6/96–Present.
Learning Outcomes: Before 6/96 see AR-1511-0016. Upon completion of the course, the student will be able to plan and conduct civil affairs operations; procure resources for civil affairs operations; recognize the impact of religious and cultural beliefs and customs; assess the purpose of psychological operations; and demonstrate knowledge of societal and culture change, foreign service functions and organization, and US security assistance programs.
Instruction: Lectures, seminars, simulations, and a country study paper are instructional modes. Topics include civil-military history; psychological operations; cultural change and value systems; procurement of goods, labor, and services; group structures; religious impact on society; foreign service functions and organization; and the US security assistance program.
Credit Recommendation: In the lower-division baccalaureate/associate degree category, 1 semester hour in communication studies. In the upper-division baccalaureate category, 6 semester hours in public administration and 3 in sociology (4/01).

AR-1511-0030

CIVIL AFFAIRS SPECIALIST NONCOMMISSIONED OFFICER (NCO) ADVANCED-RESERVE COMPONENT

Course Number: 331-38A40.
Location: Reserve Schools, Multiple locations.
Length: 92 hours.
Exhibit Dates: 3/96–Present.
Learning Outcomes: Upon completion of the course, the student will be able to perform tasks in the areas of civil affairs, including command and control, legal affairs, resource control, and civil administration.
Instruction: Lectures and practical exercises cover civil affairs, administration, host nations, civic actions, combined forces operations, cultural awareness, legal affairs, terrorism, and health/sanitation issues.
Credit Recommendation: In the lower-division baccalaureate/associate degree category, 2 semester hours in introduction to public administration (4/01).

AR-1511-0031

CIVIL AFFAIRS SPECIALIST RESERVE COMPONENT, PHASE 2

Course Number: 331-38A10-RC, Phase 2.
Location: Reserve Schools, Multiple locations.
Length: 93 hours.
Exhibit Dates: 1/98–Present.
Learning Outcomes: Upon completion of the course, the student will be able to perform tasks in the area of civil administration and affairs, information collection, data processing, and basic analysis.
Instruction: Lecture, discussion, and field exercises cover automatic data processing and information collection and analyses.
Credit Recommendation: In the lower-division baccalaureate/associate degree category, 2 semester hours in introduction to public administration (4/01).
Related Occupations: 38A.

AR-1511-0032

Human Rights Instructor

Course Number: CMS-5, Phase 1.
Location: Western Hemisphere Institute for Security Cooperation (WHINSEC), Fort Benning, GA.
Length: 3 weeks (94 hours).
Exhibit Dates: 6/02–Present.
Learning Outcomes: Upon completion of the course, the student will be able to apply concepts, methods, and techniques for effective teaching and integration of human rights topics into other instructional environments.
Instruction: Lectures, discussions, case studies, practical exercises, learner presentations, and seminars. The course focuses on a human rights core of instruction including ethics, doctrine of just war, the historical development of human rights, and human rights law. Topics include a systems approach to instructor training that emphasizes performance-oriented instruction, learning analysis, and developing plans tests to measure learning.
Credit Recommendation: In the upper-division baccalaureate category, 1 semester hour in methods of instruction and 3 in political science (9/02).

AR-1512-0010

Interpersonal Communications

Course Number: 1-19-XB-135.
Location: Military Police School, Fort McClellan, AL.
Length: Self-paced, 1–2 weeks (48 hours).
Exhibit Dates: 2/82–6/96.
Objectives: To acquaint instructors and trainers with the skills necessary for effective interpersonal communication.
Instruction: Lectures, seminars, demonstrations, and practical exercises (role-playing) cover the art of helping as related to effective interpersonal communication. Course includes basic principles of interpersonal communication, including attending skills, listening/responding skills, summarizing helpee expressions, problem solving, and goal development.
Credit Recommendation: In the upper-division baccalaureate category, 2 semester hours in human relations or as an elective in guidance and counseling (3/82).

AR-1512-0011

Family Advocacy Staff Training

Course Number: *Version 1:* 5H-F20/302-F20. *Version 2:* 5H-F20/302-F20.
Location: Academy of Health Sciences, Fort Sam Houston, TX.
Length: *Version 1:* 3 weeks (128 hours). *Version 2:* 2 weeks (78 hours).
Exhibit Dates: *Version 1:* 4/87–6/92. *Version 2:* 7/92–Present.
Learning Outcomes: *Version 1:* Upon completion of the course, the student will be able to provide specialized training to personnel working directly or indirectly in the area of family advocacy. *Version 2:* Upon completion of the course, the student will be able to describe the dynamics of child abuse, including child sexual abuse, family advocacy planning, and implementation; implement a family program plan, emphasizing prevention, legal issues, and resource management; and present information to others through a briefing or other instructional experience.
Instruction: *Version 1:* Course includes lectures, demonstrations, and experiential learning in the areas of management, needs assessment, evaluation, and counseling of child/spouse abuse. *Version 2:* Course covers topics in family advocacy programs in the Department of Defense, family advocacy planning and evaluation, records management, resource management, substance abuse, spouse and child abuse, continuous quality improvement, legal issues related to the family, and prevention programs. Methods include lectures and practical exercises. Students who do not successfully complete one or more critical portions of the course will receive a certificate of attendance but not a certificate of completion. Students do not receive an academic evaluation report.
Credit Recommendation: *Version 1:* In the upper-division baccalaureate category, 3 semester hours in family counseling (1/90). *Version 2:* In the upper-division baccalaureate category, 3 semester hours in family advocacy and support intervention (10/03).

AR-1512-0012

Alcohol and Drug Abuse Prevention and Control Program Advanced Counseling Skills

Course Number: DA 484.
Location: Combined Arms Training Center, Munich, W. Germany.
Length: 1–2 weeks (54 hours).
Exhibit Dates: 7/82–12/97.
Learning Outcomes: Upon completion of the course, the student is able to provide advanced counseling to substance abusers.
Instruction: Topics include a review of the disease concept of alcoholism, counseling the alcoholic client, and family dynamics. Included also are the characteristics of drinking problems in women and current cultural/racial counseling issues with alcoholism. Instruction is provided on the recovery process.
Credit Recommendation: In the lower-division baccalaureate/associate degree category, 3 semester hours in the counseling of alcoholics (7/87).

AR-1512-0013

Alcohol and Drug Abuse Prevention and Control Program Family Counseling

Course Number: *Version 1:* DA 486. *Version 2:* 5H-F7/302-F7.
Location: *Version 1:* Combined Arms Training Center, Munich, W. Germany. *Version 2:* Academy of Health Sciences, Fort Sam Houston, TX.
Length: *Version 1:* 1–2 weeks (54 hours). *Version 2:* 2 weeks (73–85 hours).
Exhibit Dates: *Version 1:* 8/86–7/91. *Version 2:* 8/91–Present.
Learning Outcomes: *Version 1:* Upon completion of the course, the student will be able to assess and counsel families and function as a clinical director/clinical supervisor in community and residential facilities. *Version 2:* Upon completion of the course, the student will be able to assess communities and provide families with counseling services, including intake interviews, family assessment, and effective substance abuse counseling. Student will be able to deal ethically and legally with family-related issues.
Instruction: *Version 1:* The methods of instruction include lectures, demonstrations, and practice. Topics include family systems theory, alcohol abuse, ethical practice, community resources, patient assessment techniques, therapeutic techniques, and family counseling applications. *Version 2:* The course covers topics in family systems theory, patient assessment techniques, ethical issues, legal issues, legal aspects of drug abuse, and community assessment and intervention. Methods of instruction include seminars, conferences, and practical exercises.
Credit Recommendation: *Version 1:* In the upper-division baccalaureate category, 3 semester hours in family counseling (7/87). *Version 2:* In the upper-division baccalaureate category, 3 semester hours in community intervention techniques and the family (6/02).

AR-1512-0014

Alcohol and Drug Abuse Prevention and Control Program Group Counseling Skills

Course Number: DA 483.
Location: Combined Arms Training Center, Munich, W. Germany.
Length: 2 weeks (69 hours).
Exhibit Dates: 1/80–12/97.
Learning Outcomes: Upon completion of the course, the student will have beginning skills in the organizing and conducting counseling groups.
Instruction: This course includes the design and development of groups, including establishment of group rules, norms, and goals. Course includes both theory and practice related to leadership styles, negotiation of leader-member roles, and related facilitative skills.
Credit Recommendation: In the upper-division baccalaureate category, 3 semester hours in group dynamics (7/87).

AR-1512-0015

Alcohol and Drug Prevention and Control Program Individual Counseling Skills

Course Number: DA 482.
Location: Combined Arms Training Center, Munich, W. Germany.
Length: 1–2 weeks (53–54 hours).
Exhibit Dates: 9/82–12/97.
Learning Outcomes: Upon completion of the course, the student will be able to provide individual counseling services in a substance abuse agency.
Instruction: The course provides students with knowledge, skills, and practice in the basic concepts and applications of reality therapy, communication, drug and alcohol assessment, and counseling practice.
Credit Recommendation: In the lower-division baccalaureate/associate degree category, 3 semester hours in introductory counseling (7/87).

AR-1512-0016

Behavioral Science Specialist

Course Number: *Version 1:* 302-91G10. *Version 2:* 302-91G10.
Location: Academy of Health Sciences, Fort Sam Houston, TX.

Length: *Version 1:* 15 weeks (571–576 hours). *Version 2:* 14–15 weeks (530–588 hours).

Exhibit Dates: *Version 1:* 12/85–2/93. *Version 2:* 3/93–9/96.

Learning Outcomes: *Version 1:* Upon completion of the course, the student will be able, under professional supervision, to interview clients in several settings; administer and score psychological tests; and assist in determining client need for referral. *Version 2:* Upon completion of the course, the student, under professional supervision, will be able to administer and score psychological tests; interview clients to identify psychopathological disorders; counsel patients in drug abuse situations; identify, treat, and refer soldiers exhibiting combat stress; conduct and record client interviews; and determine the need for referral of clients with psychopathological disorders.

Instruction: *Version 1:* Lectures and practical exercises cover social, psychological, and psychiatric principles; human development; group concepts; factors influencing behavior; behavioral problems; substance abuse disorders; psychological testing; and counseling. *Version 2:* Lectures and practical exercises cover social, psychological, and psychiatric principles; human development; group concepts; factors influencing behavior; behavioral problems; substance abuse disorders; psychological testing; and counseling.

Credit Recommendation: *Version 1:* In the lower-division baccalaureate/associate degree category, 3 semester hours in human development, 3 in counseling and testing, 3 in psychology, and 1 in sociology. In the upper-division baccalaureate category, 3 semester hours in abnormal psychology (12/91). *Version 2:* In the lower-division baccalaureate/associate degree category, 2 semester hours in substance abuse, 3 in psychology, 2 in sociology, and 4 in human development. In the upper-division baccalaureate category, 3 semester hours in abnormal psychology, 3 in psychological testing, and 3 in psychological counseling (10/95).

Related Occupations: 91G.

AR-1512-0017

Reserve Component Psychological Operations Officer

Course Number: 5E-39B-RC; 3A-SI5E-RC.

Location: John F. Kennedy Special Warfare Center and School, Fort Bragg, NC.

Length: Phase 1, 2 weeks (79 hours); Phase 2, 2 weeks (80 hours); Phase 3, 3 weeks (65 hours).

Exhibit Dates: 6/87–10/93.

Learning Outcomes: After 3/96 see AR-1512-0028. Upon completion of the course, the student will be able to communicate the role and structure of psychological operations (psyops) intelligence in support of national objectives; plan, organize, direct, and supervise the use of psyops in all military operations across the spectrum of conflict; train personnel in psyops doctrines and techniques; demonstrate knowledge of social science theory and procedures in relation to psyops; analyze cross-cultural communication; design propaganda programs for target audiences; and coordinate, supervise, and disseminate psyops products using media, leadership, and decision-making skills.

Instruction: Lectures, discussions, seminars, simulations, and practical exercises are on developing a scenario in a mythical region of the world. Topics include psychological operations; doctrine and organization of US psyops policy formulation and dissemination; and social science theory and methodology, including cross-cultural communication, anthropological context in the execution of psyops, target analysis, propaganda development, and use of media. Credit recommended is for the resident portions of the course.

Credit Recommendation: In the upper-division baccalaureate category, For Phase 1, 1 semester hour in psychology, 1 in area studies, 1 in marketing and advertising, and 1 in communication. For Phase 2, 1 semester hour in psychology, 1 in area studies, and 1 in cross-cultural communication. For Phase 3, 1 semester hour in media production and 1 in leadership studies. For those completing more than one phase, the credit should be added (10/91).

AR-1512-0018

Psychological Operations Officer

Course Number: *Version 1:* 5E-39B; 3A-SI5E. *Version 2:* 5E-39B.

Location: John F. Kennedy Special Warfare Center and School, Fort Bragg, NC.

Length: *Version 1:* 6 weeks (218–224 hours). *Version 2:* 4–5 weeks (184 hours).

Exhibit Dates: *Version 1:* 6/87–5/96. *Version 2:* 6/96–Present.

Learning Outcomes: *Version 1:* Upon completion of the course, the student will communicate the role and structure of psychological operations (psyops) intelligence in support of national objectives; plan, organize, direct, and supervise the use of psyops in all military operations across the spectrum of conflict; train personnel in psyops doctrine and techniques; demonstrate knowledge of social science theory and procedures in relation to psyops; analyze cross-cultural communication; design propaganda programs for target audiences; and coordinate, supervise, and disseminate psyops products using media, leadership, and decision-making skills. *Version 2:* Upon completion of the course, the student will communicate the role and structure of psychological operations and intelligence information in support of national objectives; plan, organize, direct, and supervise the use of psychological operations (psyops) in military operations, train personnel in psyops doctrine and techniques; analyze cross-cultural and international communication; design propaganda programs for target audiences; and coordinate and apply psychological operations products using media, leadership, and decision-making skills.

Instruction: *Version 1:* Lectures, discussions, seminars, simulations, and practical exercises are based on developing a scenario in a mythical region of the world. Topic include psychological operations; doctrine and organization of US psyops policy formulation and dissemination; and social science theory and methodology, including cross-cultural communication, anthropological context in the execution of psyops, target analysis, propaganda development, and use of media. *Version 2:* Lectures, discussions, seminars, simulations, and practical exercises cover psychological operations, US psyops policy formulation and dissemination, social science concepts of cross-cultural communication, anthropological context for psychological operations, analysis of target audience, analysis of mass media and public opinion, persuasive techniques, economic conditions, cultural influence, and counterpropaganda.

Credit Recommendation: *Version 1:* In the upper-division baccalaureate category, 2 semester hours in psychology, 2 in communication theory, 2 in area studies, 1 in media production, 1 in marketing and advertising, and 1 in leadership studies (10/91). *Version 2:* In the upper-division baccalaureate category, 3 semester hours in psychology and 2 in communication and media studies (4/01).

AR-1512-0019

Special Operations Staff Officer

Course Number: 2E-F112.

Location: John F. Kennedy Special Warfare Center and School, Fort Bragg, NC.

Length: 4–7 weeks (153–275 hours).

Exhibit Dates: 10/88–9/96.

Learning Outcomes: Upon completion of the course, the student will exhibit skills in planning and managing resources needed to implement military operations, including managing military forces, equipment, and supplies and advising foreign policy developers in Special Operations activities with emphasis on peacetime operations.

Instruction: Lectures and practical exercises cover planning and employment of Special Operations forces in low- and mid-intensity spectrums. Instruction includes Special Operations organization, mission, employment, and command and control structure; US foreign policy; regional studies; campaign planning; crisis action systems; joint deployment system; and threat forces.

Credit Recommendation: In the upper-division baccalaureate category, 3 semester hours in principles of management, 3 in logistics management, and 3 in international relations (12/92).

AR-1512-0020

Psychological Operations Basic Noncommissioned Officer (NCO)

Course Number: *Version 1:* 243-96F30. *Version 2:* 243-96F30. *Version 3:* 243-37F30. *Version 4:* 243-37F30. *Version 5:* 243-37F30.

Location: John F. Kennedy Special Warfare Center and School, Fort Bragg, NC.

Length: *Version 1:* 4 weeks (151 hours). *Version 2:* 5–6 weeks (225 hours). *Version 3:* 6–7 weeks (285 hours). *Version 4:* 6–7 weeks (266 hours). *Version 5:* 4–5 weeks (283 hours).

Exhibit Dates: *Version 1:* 4/88–5/90. *Version 2:* 6/90–9/92. *Version 3:* 10/92–9/96. *Version 4:* 10/96–9/00. *Version 5:* 10/00–Present.

Learning Outcomes: *Version 1:* Upon completion of the course, the student will be able to classify and document intelligence materials; brief superiors; work with social science research methodology; describe and critique target analysis process; identify and use counterpropaganda techniques; analyze Soviet psychological awareness programs; select, employ, and evaluate communication media in developing a psychological operations campaign; coordinate supporting units; use automated data system; and use communications technology and audiovisual equipment. *Version 2:* Upon completion of the course, the student will be able to classify and document intelligence materials, brief superiors, work with social science research methodology, describe and critique target analysis process, use counterpropaganda techniques, employ and evaluate communications media in developing a psychological operations campaign, coordinate supporting units, use automated data system, and use communications technology and audiovisual equipment. *Version 3:* Upon completion of the course, the student will be able to classify and document primary materials; apply social science research techniques; and select, employ, deliver, and evaluate verbal, print, and telecommunications information. *Version 4:* Upon completion of the course, the student will demonstrate knowledge of small-group dynamics, military science, historical studies, military etiquette, and map reading. *Version 5:* Upon completion of the course, the student will be able to categorize and document intelligence materials, identify and analyze counterpropaganda techniques, use communication media for psychological operations campaigns, demonstrate a knowledge of small-group dynamics, and use maps for planning land navigation.

Instruction: ***Version 1:*** Lectures, student discussions, and exercises cover topics such as map analysis and land navigation, intelligence, psychological operations, social science research methodology, propaganda analysis, target analysis and media selection, communications, automated data systems, and counterpropaganda techniques. This course contains a common core of leadership subjects. *Version 2:* Lectures, student discussions, and exercises cover map analysis and land navigation, intelligence, psychological operations, social science research methodology, propaganda analysis, target analysis, media selection, communication, automated data systems, and counterpropaganda techniques. This course contains a common core of leadership subjects. *Version 3:* Lectures, directed discussions, and practical exercises cover map reading and land navigation; social science research techniques; and media selection, analysis, and distribution. This course contains a common core of leadership subjects. *Version 4:* Students engage in lectures, discussions, small-group instruction, and practical exercises in map reading, counterpropaganda techniques, and land navigation. Course contains a common core of leadership subjects. *Version 5:* Lectures, student discussion, and exercises cover topics such as map analysis and land navigation, intelligence, psychological operations, social science research methodology, propaganda analysis, target audience analysis, media selection communications, small-group instruction, and counterpropaganda techniques.

Credit Recommendation: ***Version 1:*** In the lower-division baccalaureate/associate degree category, 1 semester hour in data processing, 1 in map interpretation, and 1 in audiovisual techniques. See AR-1406-0090 for common core credit. In the upper-division baccalaureate category, 1 semester hour in social science, 1 in marketing or advertising, and 1 in communications production (12/88). *Version 2:* In the lower-division baccalaureate/associate degree category, 1 semester hour in map interpretation, 3 in communications, and 1 in audiovisual techniques. See AR-1406-0090 for common core credit (12/92). *Version 3:* In the lower-division baccalaureate/associate degree category, 1 semester hour in map interpretation, 1 in audiovisual techniques, and 3 in communications. See AR-1406-0090 for common core credit (12/92). *Version 4:* In the lower-division baccalaureate/associate degree category, 1 semester hour in map reading and 2 in military science. See AR-1406-0090 for common core credit (6/96). *Version 5:* In the lower-division baccalaureate/associate degree category, 1 semester hour in map reading, 1 in psychology, and 2 in military science (4/01).

Related Occupations: 37F; 96F.

AR-1512-0021

PSYCHOLOGICAL OPERATIONS SPECIALIST

Course Number: ***Version 1:*** 243-37F10; 243-96F10. *Version 2:* 243-37F10; 243-96F10. *Version 3:* 243-37F10.

Location: John F. Kennedy Special Warfare Center and School, Fort Bragg, NC.

Length: ***Version 1:*** 9–11 weeks (333–443 hours). *Version 2:* 9–11 weeks (333–443 hours). *Version 3:* 11 weeks (421 hours).

Exhibit Dates: ***Version 1:*** 10/87–9/91. *Version 2:* 10/91–2/97. *Version 3:* 3/97–Present.

Learning Outcomes: ***Version 1:*** Upon completion of the course, the student will be able to use US and foreign maps in support of tactical psychological operations (psyops) missions, process and use relevant intelligence to support tactical operations, identify and conceptualize psyops, analyze and respond to foreign propaganda, target an audience and market a propaganda product, use media equipment, identify Soviet psychological warfare techniques and capabilities, and apply persuasive techniques in conducting counterpropaganda campaigns. *Version 2:* Upon completion of the course, the student will be able to use US and foreign maps in support of tactical psychological operations (psyops) missions, process and use relevant intelligence to support tactical operations, identify and conceptualize psyops, analyze and respond to foreign propaganda, target an audience and market a propaganda product, use media equipment, identify Soviet psychological warfare techniques and capabilities, and apply persuasive techniques in conducting counterpropaganda campaigns. *Version 3:* Upon completion of the course, the student will be able to use US and foreign maps in support of tactical psychological operations (psyops) missions, process and use relevant intelligence to support tactical operations, identify and conceptualize psyops, analyze and respond to foreign propaganda, target an audience and market a propaganda product, use media equipment, identify psychological warfare techniques and capabilities, and apply persuasive techniques in conducting counterpropaganda campaigns.

Instruction: ***Version 1:*** Lectures, small-group discussions, and seminars cover map reading and land navigation, audience analysis for propaganda campaigns, use of media/audiovisual equipment, use of printed media and its production, examination of intelligence sources and reports, US policy formulation, dynamics of interpersonal and cross-cultural communication, cultural biases, and Soviet psyops. Course includes a simulation exercise dealing with the application of intelligence procedures in the field. *Version 2:* Lectures, small-group discussions, and seminars cover map reading and land navigation, audience analysis for propaganda campaigns, use of media/audiovisual equipment, use of printed media and its production, examination of intelligence sources and reports, US policy formulation, dynamics of interpersonal and cross-cultural communication, cultural biases, and Soviet psyops. Course includes a simulation exercise dealing with the application of intelligence procedures in the field. *Version 3:* Lectures, small-group discussions, and seminars cover map reading and land navigation, audience analysis for propaganda campaigns, use of media/audiovisual equipment, use of printed media and its production, examination of intelligence sources and reports, US policy formulation, dynamics of interpersonal and cross-cultural communication, cultural biases, psychological operations, and application of intelligence procedures in the field.

Credit Recommendation: ***Version 1:*** In the vocational certificate category, 2 semester hours in communications technology. In the lower-division baccalaureate/associate degree category, 1 semester hour in map interpretation, 1 in marketing or advertising, 1 in record keeping, 1 in communications theory, 1 in principles of psychology, 1 in area studies, and 2 in media production (10/91). *Version 2:* In the vocational certificate category, 2 semester hours in communications technology. In the lower-division baccalaureate/associate degree category, 2 semester hours in map interpretation, 1 in marketing or advertising, 2 in communication theory, 1 in principles of psychology, and 2 in media production (12/92). *Version 3:* In the lower-division baccalaureate/associate degree category, 2 semester hours in communications technology and 1 in psychology (4/01).

Related Occupations: 37F; 96F.

AR-1512-0022

1. RESERVE COMPONENT PSYCHOLOGICAL OPERATIONS SPECIALIST
2. RESERVE COMPONENT PSYCHOLOGICAL OPERATIONS SPECIALIST
3. PSYCHOLOGICAL OPERATIONS SPECIALIST RESERVE

(Psychological Operations Specialist (Reclass)-RC)

Course Number: ***Version 1:*** 243-96F10-RC IDT/ADT. *Version 2:* 331-37F10-RC IDT/ADT; 243-96F10-RC IDT/ADT. *Version 3:* 331-37F10-RC, Phase 2.

Location: ***Version 1:*** John F. Kennedy Special Warfare Center and School, Fort Bragg, NC. *Version 2:* John F. Kennedy Special Warfare Center and School, Fort Bragg, NC. *Version 3:* Reserve Component Training Centers, US.

Length: ***Version 1:*** Phase 1, 3 weeks (118 hours); Phase 2, 3 weeks (113 hours). *Version 2:* Phase 2, 98 hours. *Version 3:* 126 hours.

Exhibit Dates: ***Version 1:*** 10/88–9/90. *Version 2:* 10/90–5/01. *Version 3:* 6/01–Present.

Learning Outcomes: ***Version 1:*** Upon completion of the course, the student will be able to use US and foreign maps in support of tactical psychological operations (psyops) missions, process and use relevant intelligence to support tactical operations, identify and conceptualize psyops, analyze and respond to foreign propaganda, target an audience and market a propaganda product, use media equipment, identify Soviet psychological warfare techniques and capabilities, and apply persuasive techniques in conducting counterpropaganda campaigns. *Version 2:* Upon completion of the course, the student will be able to operate and maintain media/audiovisual equipment and prepare and distribute written persuasive material.

Version 3: Upon completion of the course, the student will be able to use US and foreign maps in support of tactical psychological operations, process and use relevant intelligence information to support tactical operations, analyze and respond to foreign propaganda, identify and analyze target audiences, and develop audiovisual materials.

Instruction: ***Version 1:*** Lectures, small-group discussions, and seminars cover map reading and land navigation, audience analysis for propaganda campaigns, use of media/audiovisual equipment, use of printed media and its production, examination of intelligence sources and reports, US policy formulation, dynamics of interpersonal and cross-cultural communication, cultural biases, and Soviet psyops. Course includes a simulation exercise dealing with the application of intelligence procedures in the field. *Version 2:* Lectures, applied exercises, discussions, and simulations cover the selection, operation, and application of media/audiovisual equipment to meet specific needs. *Version 3:* Lectures, small-group discussions, and seminars cover map reading, land navigation, target audience analysis, audiovisual products, media development, intelligence sources and reports, and propaganda message analysis.

Credit Recommendation: ***Version 1:*** In the vocational certificate category, 2 semester hours in communications technology. In the lower-division baccalaureate/associate degree category, 2 semester hours in media production (12/88). *Version 2:* In the vocational certificate category, 2 semester hours in communications technology. In the lower-division baccalaureate/associate degree category, 2 semester hours in media production. Credit is for Phase 2 only (12/93). *Version 3:* In the lower-division baccalaureate/associate degree category, 2 semester hours in communications studies and 2 in communications technology (4/01).

Related Occupations: 37F; 96F.

AR-1512-0023

JOINT PSYCHOLOGICAL OPERATIONS STAFF PLANNING
(Joint Operations Staff Planning)

Course Number: 3A-F53/243-F12.

Location: John F. Kennedy Special Warfare Center and School, Fort Bragg, NC.

Length: 2 weeks (76–77 hours).

Exhibit Dates: 11/87–10/93.

Learning Outcomes: After 3/98 see AR-1512-0030. Upon completion of the course, the student will be able to use planning skills for the preparation and support of joint operations; understand role, mission, and functions of psychological operations; and analyze foreign psychological operations, doctrines, and practices.

Instruction: Lectures and exercises cover the above topics.

Credit Recommendation: Credit is not recommended because of the military-specific nature of the course (10/89).

AR-1512-0024

ALCOHOL AND DRUG ABUSE PREVENTION AND CONTROL (ADAPCP) FAMILY COUNSELING

Course Number: 5H-F8/302-F8.

Location: Academy of Health Sciences, Fort Sam Houston, TX.

Length: 2 weeks (72–83 hours).

Exhibit Dates: 8/91–Present.

Learning Outcomes: Upon completion of the course, the student will be able to determine the most appropriate strategies and intervention for treatment of a family when a substance abusing member is present and implement a plan taking the family structures, specific outcomes of the intervention, and eventual termination of the relationship into consideration.

Instruction: Course covers topics in theories of family treatment, family structures, intervention techniques and applications, and termination of the counseling relationship. Methods include conferences, demonstrations, and extensive practical exercises.

Credit Recommendation: In the upper-division baccalaureate category, 3 semester hours in family counseling (4/94).

AR-1512-0025

PSYCHOLOGICAL OPERATIONS ADVANCED NONCOMMISSIONED OFFICER (NCO)

Course Number: ***Version 1:*** 243-37F40. *Version 2:* 243-37F40.

Location: John F. Kennedy Special Warfare Center and School, Fort Bragg, NC.

Length: ***Version 1:*** 6–7 weeks (254 hours). *Version 2:* 4–5 weeks (185 hours).

Exhibit Dates: ***Version 1:*** 10/93–9/00. *Version 2:* 10/00–Present.

Learning Outcomes: ***Version 1:*** Upon completion of the course, the student will be able to describe and identify unconventional warfare techniques, identify personnel and resource needs for psychological operations, evaluate specific psychological operations environments, discuss the purpose and mission of the psychological operations program, and apply basic leadership procedures. *Version 2:* Upon completion of the course, the student will be able to describe and identify unconventional warfare techniques, identify personnel and resource needs for psychological operations (psyops), evaluate specific psyops environment, discuss the purpose and mission of the psychological operations program, and apply basic leadership procedures.

Instruction: ***Version 1:*** Instruction includes lectures, tours, demonstrations, exercises, and written and oral exams in leadership skills and unconventional warfare planning. This course contains a common core of leadership subjects. *Version 2:* Instruction includes lectures, demonstrations, exercises, and examinations in practicing group dynamics, planning and directing psychological operations programs, responding to civil disturbance, planning countermeasures, and directing audio and audiovisual psyops programs.

Credit Recommendation: ***Version 1:*** Credit is recommended for the common core only. See AR-1404-0035 (3/96). *Version 2:* In the lower-division baccalaureate/associate degree category, 1 semester hour in psychology and 2 in military science (4/01).

Related Occupations: 37F.

AR-1512-0026

PSYCHOLOGICAL OPERATIONS ADVANCED NONCOMMISSIONED OFFICER (NCO) RESERVE COMPONENT PHASE 2

Course Number: 243-37R40-RC.

Location: John F. Kennedy Special Warfare Center and School, Fort Bragg, NC.

Length: 2–3 weeks (120 hours).

Exhibit Dates: 10/93–10/00.

Learning Outcomes: Upon completion of the course, the student will be able to evaluate the battlefield's psychological environment and apply propaganda techniques for dissemination of information, compare and contrast intelligence information, and provide choices of action for the field commander.

Instruction: Instruction consists of practical exercises; review of literature, manuals, and text materials; group discussions; exercise critiques; and completion of oral exams.

Credit Recommendation: Credit is not recommended because of the limited, specialized nature of the course (3/96).

Related Occupations: 37R.

AR-1512-0027

FAMILY ADVO STAFF (ADV) (FASTA)
(Family Advocacy Staff)

Course Number: 5H-F21/302-F21.

Location: Academy of Health Sciences, Fort Sam Houston, TX.

Length: 5 weeks (190 hours).

Exhibit Dates: 11/97–Present.

Learning Outcomes: Upon completion of the course, the student will build on basic skills and be trained to prevent, identify, investigate, and treat family violence cases and respond to acts of family violence.

Instruction: Lectures, conferences, discussions, and examinations cover family advocacy staff training. Course consists of 5 modules, each 1 week in length. All segments must be completed before credit is to be granted.

Credit Recommendation: In the upper-division baccalaureate category, 3 semester hours in social problems and 1 in group therapy. NOTE. Credit is to be granted only upon presentation of certificate of completion for all five segments (10/03).

AR-1512-0028

PSYCHOLOGICAL OPERATIONS OFFICER RESERVE COMPONENT

Course Number: None.

Location: Reserve Schools, Multiple locations.

Length: 2 weeks.

Exhibit Dates: 4/96–Present.

Learning Outcomes: Before 4/96 see AR-1512-0017. Upon completion of the course, the student will be able to communicate the role and structure of psychological operations (psyops) intelligence in support of national objectives; plan, organize, direct, and supervise the use of psyops in all military operations across the spectrum of conflict; train personnel in psyops doctrines and techniques; demonstrate knowledge of social science theory and procedures in relation to psyops; analyze cross-cultural communication; design propaganda programs for target audiences; and coordinate, supervise, and disseminate psyops products using media, leadership, and decision-making skills.

Instruction: Lectures, discussions, seminars, simulations, and practical exercises are on developing a scenario in a mythical region of the world. Topics include psychological operations; doctrine and organization of US psyops policy formulation and dissemination; and social science theory and methodology, including cross-cultural communication, anthropological context in the execution of psyops, target analysis, propaganda development, and use of media. Credit recommended is for the resident portion of the course.

Credit Recommendation: In the lower-division baccalaureate/associate degree category, 2 semester hours in communications technology. In the upper-division baccalaureate category, 3 semester hours in psychology (4/01).

AR-1512-0029

PSYCHOLOGICAL OPERATIONS ADVANCED NONCOMMISSIONED OFFICER (NCO) RESERVE COMPONENT

Course Number: 243-37-F40-RC, Phase 2.

Location: Reserve Schools, Multiple locations.

Length: 2–3 weeks (80–110 hours).

Exhibit Dates: 10/00–Present.

Learning Outcomes: Upon completion of the course, the student will be able to evaluate the battlefield's psychological environment, apply propaganda techniques for dissemination of information, compare and contrast intelligence information, and provide choices of action for the field commander.

Instruction: Instruction consists of practical exercises; review of literature, manuals, and text materials; group discussions; exercise critiques; and completion of oral exams.

Credit Recommendation: Credit is not recommended because of the limited, specialized nature of the course (4/01).

AR-1512-0030

JOINT PSYCHOLOGICAL OPERATIONS STAFF PLANNING

Course Number: 3A-F53/243-F12.

Location: John F. Kennedy Special Warfare Center and School, Fort Bragg, NC.

Length: 2 weeks (77 hours).

Exhibit Dates: 4/98–Present.

Learning Outcomes: Before 4/98 see AR-1512-0023. Upon completion of the course, the student will be able to use planning skills for the preparation and support of joint operations; describe the role, mission, and functions of psychological operations; and analyze foreign psychological operations, doctrines, and practices.

Instruction: Lectures and exercises cover the above topics.

Credit Recommendation: Credit is not recommended because of the military-specific nature of the course (4/01).

AR-1512-0031

PSYCHOLOGICAL OPERATIONS BASIC NONCOMMISSIONED OFFICER, PHASE 2

Course Number: 331-37F30 (F), Phase 2.

Location: John F. Kennedy Special Warfare Center and School, Fort Bragg, NC.

Length: 156 hours.

Exhibit Dates: 4/00–Present.

Learning Outcomes: Upon completion of the course, the student will be able to analyze messages, categorize relevant psychological operations information, conduct intelligence preparation for battle space, identify resource needs for psychological operations, and provide guidance to surbordinates in collecting psychological operations information.

Instruction: Instruction includes lectures, demonstrations, exercises, and examinations in psychological operations and information analysis, interviewing of prisoners, intelligence preparation for battle space, and collection of information for psychological operations.

Credit Recommendation: Credit is not recommended because of the limited, specialized nature of the course (4/01).

Related Occupations: 37F.

AR-1512-0032

INFORMATION OPERATIONS

Course Number: TAC-4, Phase 2.

Location: Western Hemisphere Institute for Security Cooperation (WHINSEC), Fort Benning, GA.

Length: 8 weeks (226 hours).

Exhibit Dates: 1/02–Present.

Learning Outcomes: Upon completion of the course, the student will be able to communicate the role and structure of psychological operations intelligence in support of national objectives; become a liaison to civilian authorities in the implementation of civil/military activities, such as humanitarian relief operations and peacekeeping; analyze and manage demographic, cross-cultural, and international communications models in support of national strategies.

Instruction: Audio-visual materials, practical and classroom exercises, discussion, case studies, learner presentations, lecture, and laboratory. Topics include information management, integration of all media resources in product development and public affairs, diffusion techniques, demographics and the nature of target audiences, interpersonal and intrapersonal communications, and computer and electronics security systems.

Credit Recommendation: In the upper-division baccalaureate category, 3 semester hours in social psychology, 2 in communication or public administration, and 1 in computer and electronic systems security (10/02).

AR-1513-0007

ARMY COMMUNITY SERVICE

Course Number: None.

Location: Institute for Personnel and Resource Management, Fort Benjamin Harrison, IN.

Length: 1 week (49 hours).

Exhibit Dates: 10/83–6/96.

Objectives: To train managerial personnel to manage community service centers.

Instruction: Conferences and practical applications include publicity, funding sources, private organizations, foster care, referrals and follow-ups, and community relations.

Credit Recommendation: Credit is not recommended because of the military-specific nature of the course (3/85).

AR-1513-0008

ARMY COMMUNITY SERVICE MANAGEMENT

Course Number: None.

Location: Institute for Personnel and Resource Management, Fort Benjamin Harrison, IN.

Length: 3 weeks (114 hours).

Exhibit Dates: 10/83–6/96.

Objectives: To provide managerial personnel within the Army community service area with the skills necessary for successful management of community service programs.

Instruction: Lectures and practice in needs assessment, programs, resource management, budgeting, staffing and administration prepares students to manage Army community service centers.

Credit Recommendation: In the lower-division baccalaureate/associate degree category, 2 semester hours in resource management and budgeting (3/85).

AR-1513-0009

INDIVIDUAL TERRORISM AWARENESS

Course Number: ***Version 1:*** 3A-F40/011-F21. *Version 2:* 3A-F40/011-F21.

Location: John F. Kennedy Special Warfare Center and School, Fort Bragg, NC.

Length: ***Version 1:*** 1 week (58–59 hours). *Version 2:* 1 week (59 hours).

Exhibit Dates: ***Version 1:*** 10/88–3/97. *Version 2:* 4/97–Present.

Learning Outcomes: ***Version 1:*** Upon completion of the course, the student will be able to describe the nature of terrorism; evaluate motivation, organization, and tactics of terrorist groups; describe terrorist operations including planning and execution of terrorist attacks; reduce vulnerability to terrorist attack; demonstrate counterterrorist surveillance, survival, and individual protective measures; use hostage survival techniques and clandestine communication skills; describe the relationship between terrorism and the media and minimize terrorist use of media; teach others how to limit their vulnerability to terrorist selection and attack; and teach techniques to survive a terrorist assault and maximize survival in a hostage situation. *Version 2:* Upon completion of the course, the student will have an understanding of the nature of terrorism, be able to minimize personal vulnerability to terrorist attack, and practice self-defense, and how to react and survive in a hostage situation.

Instruction: ***Version 1:*** Lectures and discussions include introduction to terrorism, terrorist operations, conducting terrorist surveillance, individual protective measures, survival, shooting techniques, hostage survival techniques including resistance to interrogation, clandestine communication techniques, and terrorism and the media. *Version 2:* Lectures, demonstrations and experiential exercises cover terrorism, terrorist operations, and surveillance; individual protective measures such as defensive driving; and hostage survival techniques.

Credit Recommendation: ***Version 1:*** In the upper-division baccalaureate category, 3 semester hours in social science (3/92). *Version 2:* In the lower-division baccalaureate/associate degree category, 2 semester hours in self-defense (4/01).

AR-1513-0010

DIRECTOR OF PERSONNEL AND COMMUNITY ACTIVITIES (DPCA)

Course Number: 7C-F15/500-F17.

Location: Soldier Support Institute, Fort Benjamin Harrison, IN.

Length: 3–4 weeks (154 hours).

Exhibit Dates: 10/87–12/91.

Learning Outcomes: Upon completion of the course, the student will be able to describe the duties and responsibilities of a human services manager, perform physical training exercises, manage budgets, develop internal controls, and staff and manage a human services operation.

Instruction: Methods of instruction include lectures and classroom discussions. Topics covered are human resource operations, personnel counseling, personal counseling, and family support services.

Credit Recommendation: In the upper-division baccalaureate category, 3 semester hours in human services management (12/88).

AR-1513-0011

Clinical Pastoral Education

Course Number: 5G-CPE-7R.

Location: Madigan Medical Center, Tacoma, WA; Dwight David Eisenhower Medical Center, Fort Gordon, GA; Walter Reed Medical Center, Washington, DC; Brooke Medical Center, Fort Sam Houston, TX.

Length: 49 weeks (2004 hours).

Exhibit Dates: 3/97–Present.

Learning Outcomes: Upon completion of the course, the student will be able to function effectively in institutional ministries at medical centers, medical departments, medical brigades, and combat support hospitals.

Instruction: Lectures, practical exercises and examinations are structured to produce competent chaplains.

Credit Recommendation: In the graduate degree category, 12 semester hours in pastoral education and 3 in clinical pastoral education (10/03).

AR-1513-0012

Chaplain Officer Basic, Phase 2

Course Number: 5-16-C25, Phase 2.

Location: Chaplain Center and School, Fort Jackson, SC.

Length: 3–4 weeks (98–100 hours).

Exhibit Dates: 6/96–Present.

Learning Outcomes: Upon completion of the course, the student will be able to perform many aspects of pastoral ministry, including worship service, with homily/sermon, pastoral care and counseling in specialized situations, and religious education.

Instruction: Topics include planning worship services with homily/sermon, ministry to HIV victims, ministry in a multicultural setting, major schools of pastoral counseling, marriage and family counseling, counseling to the mentally ill, and personnel pastoral issues.

Credit Recommendation: In the graduate degree category, 2 semester hours in pastoral ministry (3/02).

AR-1513-0013

Chaplain Officer Basic, Phase 3

Course Number: 5-16-C20/25, Phase 3.

Location: Chaplain Center and School, Fort Jackson, SC.

Length: 3 weeks (115 hours).

Exhibit Dates: 6/96–Present.

Learning Outcomes: Upon completion of the course, the student will be able to plan and conduct worship and preaching in military and multireligious settings.

Instruction: Topics include worship, preaching, observances for honoring the dead, sensitivity to multifaith situations for religious services.

Credit Recommendation: In the graduate degree category, 1 semester hour in worship (3/02).

AR-1513-0014

Chaplain Career, Phase 1

Course Number: 5-16-C22, Phase 1.

Location: Chaplain Center and School, Fort Jackson, SC.

Length: 6 weeks (264 hours).

Exhibit Dates: 12/01–Present.

Learning Outcomes: Before 12/01 see AR-1408-0140. Upon completion of the course, the student will be able to conduct pastoral ministry in settings of cultural and religious diversity, including leadership in worship and consideration of the ethical dimensions of a variety of cultural, theological, and social issues.

Instruction: Topics include ministry in multicultural and multireligious settings; characteristics of spiritual leadership; homiletics and supervision of preaching; ethical dimensions of sexuality; multiculturalism and pluralism; liberation theology; just war and pacifism; nationalism and international politics.

Credit Recommendation: In the upper-division baccalaureate category, 3 semester hours in decision making and problem solving. In the graduate degree category, 3 semester hours in pastoral ministry (3/02).

AR-1513-0015

Chaplain Career, Phase 2

Course Number: 5-16-C23, Phase 2.

Location: Chaplain Center and School, Fort Jackson, SC.

Length: 2 weeks (62 hours).

Exhibit Dates: 12/01–Present.

Learning Outcomes: Upon completion of the course, the student will be able to provide pastoral care and counseling to individuals and in small group settings.

Instruction: Topics include pastoral counseling techniques, including appropriate procedures in critical event, spiritual formation in small groups, and pastoral care to individuals and small groups. Attention is also given to supervision of worship and religious education in small groups.

Credit Recommendation: In the graduate degree category, 2 semester hours in pastoral care and counseling (3/02).

AR-1513-0016

1. Chaplain Assistant 56M10 (AIT)
2. Chaplain Assistant

Course Number: *Version 1:* 561-56M10. *Version 2:* 561-56M10.

Location: Chaplain Center and School, Fort Jackson, SC.

Length: *Version 1:* 7 weeks (237 hours). *Version 2:* 7 weeks (253 hours).

Exhibit Dates: *Version 1:* 3/99–12/03. *Version 2:* 1/04–Present.

Learning Outcomes: *Version 1:* Before 3/99 see AR-1402-0087. Upon completion of the course, the student will be able to apply interpersonal skills, provide comfort to distressed people, prepare a facility for various religious services, assist in tactical deployment and perform various office procedures. *Version 2:* Pending evaluation.

Instruction: *Version 1:* Methods of instruction include lectures, practical exercises, examinations, and demonstrations. Topics covered include word processing, computer applications, religious rituals, ethics and grief counseling. *Version 2:* Pending evaluation.

Credit Recommendation: *Version 1:* In the lower-division baccalaureate/associate degree category, 1 semester hour in computer application and 1 in word processing. In the upper-division baccalaureate category, 1 semester hour in religion, 1 in ethics, and 1 in grief counseling (3/02). *Version 2:* Pending evaluation.

AR-1601-0030

4. Petroleum Laboratory Specialist
5. Petroleum Laboratory Specialist AIT

Course Number: *Version 4:* 491-77L10. *Version 5:* 491-77L10; 491-92L10.

Location: Quartermaster Center and School, Fort Lee, VA.

Length: *Version 4:* 11 weeks (408–428 hours). *Version 5:* 9–10 weeks (396–397 hours).

Exhibit Dates: *Version 4:* 3/87–9/96. *Version 5:* 10/96–Present.

Learning Outcomes: *Version 4:* Upon completion of the course, the student will be able to conduct laboratory testing and analysis of petroleum products and operate and maintain CFR test engines. *Version 5:* Upon completion of the course, the student will apply knowledge of the chemistry and properties of petroleum products; conduct laboratory analysis and testing of products; practice proper laboratory protocol in testing, handling, storage, and disposal of petroleum products; and use, care for, and maintain laboratory equipment.

Instruction: *Version 4:* Lectures, demonstrations, and practical exercises cover basic petroleum laboratory subjects; tests required for surveillance of petroleum products; procedures and techniques for the identification, evaluation, and disposition of petroleum products; and the organization and operation of the quality surveillance program in military petroleum facilities. *Version 5:* Lectures, demonstrations, and practical exercises, including extensive laboratory experience and equipment use, are used throughout the course.

Credit Recommendation: *Version 4:* In the vocational certificate category, 9 semester hours in testing and analysis of petroleum products. In the lower-division baccalaureate/associate degree category, 3 semester hours in physical science (4/91). *Version 5:* In the vocational certificate category, 2 semester hours in chemical laboratory procedures and 9 in testing and analysis of petroleum products (4/04).

Related Occupations: 77L; 92C.

AR-1601-0054

Field Artillery Target Acquisition and Field Artillery Survey Officer

Course Number: 2E-13D.

Location: Field Artillery School, Fort Sill, OK.

Length: 7–8 weeks (267 hours).

Exhibit Dates: 8/85–6/98.

Objectives: To train field artillery officers in the field of target acquisition/reconnaissance and survey.

Instruction: Lectures, demonstrations, and field exercises cover target acquisition systems, including planning, employment, supervision, and training in survey, targeting, and radar.

Credit Recommendation: In the vocational certificate category, 1 semester hour in radar operations. In the lower-division baccalaureate/associate degree category, 6 semester hours in surveying and 2 in astronomy (6/88).

AR-1601-0055

Marine Corps Field Artillery Fire Controlman
(Marine Field Artillery Fire Controlman)

Course Number: 250-0844; 041-0844.
Location: Field Artillery School, Fort Sill, OK.
Length: 7 weeks (267 hours).
Exhibit Dates: 2/82–10/96.
Objectives: To train Marine Corps enlisted personnel in map reading, artillery surveying, and cannon fire direction.
Instruction: Lectures, demonstrations, and field exercises cover map reading and plotting, cannon fire direction, artillery survey, and gunnery. Student is taught basic mathematics including logarithms and right triangle trigonometry.
Credit Recommendation: In the lower-division baccalaureate/associate degree category, 1 semester hour in map reading, 1 in surveying, 1 in technical mathematics (10/86).

AR-1601-0056

Multiple Launch Rocket Systems (MLRS)/Lance Operations
(Lance Operations/Fire Direction Assistant)

Course Number: 250-15J10.
Location: Field Artillery School, Fort Sill, OK.
Length: 11 weeks (392–411 hours).
Exhibit Dates: 4/84–6/96.
Objectives: To qualify enlisted personnel to become fire direction assistants in a Lance missile unit or multiple launch rocket system unit.
Instruction: Lectures and practical exercises cover the operation and maintenance of Lance operation/MLRS fire direction system platoon leader digital message device, hand-held calculator, and the use of fire direction system message formats.
Credit Recommendation: In the vocational certificate category, 1 semester hour in map reading. In the lower-division baccalaureate/associate degree category, 2 semester hours in computer operations (11/85).
Related Occupations: 15J.

AR-1601-0065

Advanced Petroleum Management

Course Number: 8-10-OA-81A.
Location: Quartermaster Center and School, Fort Lee, VA.
Length: 8 weeks (293 hours).
Exhibit Dates: 5/85–10/91.
Objectives: To prepare commissioned officers to perform duties in fuel and energy management.
Instruction: Lectures, conferences, laboratories, practical exercises, inspections, tours, and demonstrations all present the materials covered in this course. Topics include advanced petroleum operations; basic petroleum mathematics and hydraulics; petroleum equipment and field operations; petroleum operations for each military service; the support relationships for joint operations, organization, and operation of a quality surveillance program; and evaluation and disposition of petroleum products.
Credit Recommendation: In the lower-division baccalaureate/associate degree category, 3 semester hours in introduction to petroleum management (8/85).

AR-1601-0068

Air Defense Artillery (ADA) Operations and Intelligence Assistant

Course Number: 16H10-OSUT, Phase 1; 16H10-OSUT (ST), Phase 1; 16H10-OSUT (ST), Phase 2; 16H10-OSUT; 250-16H10-OSUT; 250-16H10; 250-16H10-AIT.
Location: Air Defense Artillery School, Fort Bliss, TX.
Length: 5–29 weeks (209–1259 hours).
Exhibit Dates: 3/83–6/94.
Objectives: To train enlisted personnel to become Air Defense Artillery operators and intelligence assistants.
Instruction: This course consists of conferences and practical exercises in basic soldier skills and military occupational specialty skills. Topics include map reading, communication skills, tactical operations center operations, and the AN/TSQ-73 system command/control console operator tasks. This course includes an 8-9 week Basic Training component.
Credit Recommendation: Credit is not recommended because of the military-specific nature of the course (11/88).
Related Occupations: 16H.

AR-1601-0070

Combat Engineer

Course Number: 12B10-OSUT (ST); 030-12B10; 12B10-OSUT.
Location: Training Center Engineer, Fort Leonard Wood, MO.
Length: 7–13 weeks (281–567 hours).
Exhibit Dates: 10/81–12/91.
Objectives: To provide personnel with specific military skills and knowledge required of a combat engineer.
Instruction: Lectures and practical experiences cover specific military areas of demolition, mine/countermine operations, field fortifications, land navigation, camouflage operations, and offensive and defensive tactical operations. Course may include basic training.
Credit Recommendation: Credit is not recommended because of the military-specific nature of the course (12/83).
Related Occupations: 12B.

AR-1601-0071

Petroleum Laboratory Specialist Basic Noncommissioned Officer (NCO)

Course Number: 491-92C30.
Location: Quartermaster Center and School, Fort Lee, VA.
Length: 4 weeks (147 hours).
Exhibit Dates: 10/88–9/91.
Objectives: To educate and train enlisted personnel in the supervision of a petroleum laboratory test activity.
Instruction: Practical exercises, conferences and lectures, demonstrations, television, and examinations are the methods of presenting laboratory safety, sampling and gauging procedures, petroleum laboratory testing procedures, petroleum laboratory maintenance program, and quality surveillance operations. This course contains a common core of leadership subjects.
Credit Recommendation: In the vocational certificate category, 3 semester hours in petroleum laboratory test activity. In the lower-division baccalaureate/associate degree category, see AR-1406-0090 for common core credit (8/88).
Related Occupations: 92C.

AR-1601-0072

Petroleum Supply Specialist Basic Noncommissioned Officer (NCO)

Course Number: *Version 2:* 821-76W30. *Version 3:* 821-77F30. *Version 4:* 821-77F30. *Version 5:* 821-92F30.
Location: Quartermaster Center and School, Fort Lee, VA.
Length: *Version 2:* 5 weeks (180 hours). *Version 3:* 9 weeks (335 hours). *Version 4:* 5–11 weeks (288–374 hours). *Version 5:* 9 weeks (325 hours).
Exhibit Dates: *Version 2:* 3/86–9/91. *Version 3:* 10/91–9/95. *Version 4:* 10/95–9/02. *Version 5:* 10/02–Present.
Learning Outcomes: *Version 2:* Upon completion of the course, the student will be able to function as a petroleum supervisor. *Version 3:* Upon completion of the course, the student will be able to supervise pipeline or pump station operations, petroleum supply storage facilities, and water supply and distribution systems. *Version 4:* Upon completion of the course, the student apply knowledge and skills to supervise integrated petroleum supply operations, storage facilities, and quality surveillance. *Version 5:* Pending evaluation.
Instruction: *Version 2:* The course uses practical exercises, demonstrations, inspections, field exercises, and examinations in general petroleum subjects, supply point and terminal operations, pipeline systems, water supply operations, and laboratory tests. *Version 3:* The course uses practical exercises, demonstrations, inspections, field exercises, and examinations in general petroleum subjects, supply point and terminal operations, pipeline systems, water supply operations, and laboratory tests. This course contains a common core of leadership subjects. *Version 4:* Lectures, demonstrations, and classroom and field exercises are course methods. The course includes common leader and leader combat tasks. Civilians may be enrolled in the five-week component only. Course contains a common core of leadership subjects for military personnel. *Version 5:* Pending evaluation.
Credit Recommendation: *Version 2:* In the vocational certificate category, 3 semester hours in petroleum supply (10/88). *Version 3:* In the lower-division baccalaureate/associate degree category, 3 semester hours in petroleum supply specialist, petroleum supply technology, or in applied science. See AR-1406-0090 for common core credit (2/93). *Version 4:* In the lower-division baccalaureate/associate degree category, 3 in petroleum supply operations. See AR-1406-0090 for common core credit. Civilian students will not attend the common core (4/02). *Version 5:* Pending evaluation.
Related Occupations: 76W; 77F.

AR-1601-0075

Petroleum Supply Specialist
(Petroleum Supply Specialist Advanced Individual Training)

Course Number: *Version 1:* 821-77F10. *Version 2:* 821-77F10. *Version 3:* 821-92F10.
Location: Quartermaster Center and School, Fort Lee, VA.
Length: *Version 1:* 9–10 weeks (338–362 hours). *Version 2:* 8–10 weeks (332–360 hours). *Version 3:* 10 weeks (392 hours).

Exhibit Dates: *Version 1:* 5/87–5/93. *Version 2:* 6/93–3/04. *Version 3:* 4/04–Present.

Learning Outcomes: *Version 1:* Upon completion of the course, the student will be able to receive, store, issue, dispense, and ship petroleum products. *Version 2:* Upon completion of the course, the student apply knowledge and skills to perform operations tasks in the receiving, storing, issuing, dispensing, and shipping of petroleum products. *Version 3:* Pending evaluation.

Instruction: *Version 1:* Lectures and practical exercises cover packing, supply point, terminal, and pipeline operations. *Version 2:* Lectures, demonstrations, and practical exercises cover phases of petroleum supply. *Version 3:* Pending evaluation.

Credit Recommendation: *Version 1:* In the vocational certificate category, 3 semester hours in petroleum storage procedures (10/90). *Version 2:* In the vocational certificate category, 3 semester hours in petroleum systems operation (5/02). *Version 3:* Pending evaluation.

Related Occupations: 77F; 92F.

AR-1601-0076

PETROLEUM SUPPLY RESERVE SPECIALIST RESERVE COMPONENT

Course Number: 821-77F10-RC.

Location: Reserve Component Training Centers, US; Quartermaster Center and School, Fort Lee, VA.

Length: 2 weeks (88 hours).

Exhibit Dates: 10/87–10/91.

Learning Outcomes: Upon completion of the course, the student will be able to perform skills in direct and general support of all areas involving petroleum product and water supply operations.

Instruction: Lectures, demonstrations, and practical exercises cover general petroleum subjects, including fire and safety precautions, health and handling hazards, class III supply paint operations, terminal and pipeline operations, and water supply operations.

Credit Recommendation: In the vocational certificate category, 6 semester hours in petroleum products (10/88).

Related Occupations: 77F.

AR-1601-0078

CANNON CREWMEMBER ADVANCED NONCOMMISSIONED OFFICER (NCO)

Course Number: 041-13B40.

Location: Field Artillery School, Fort Sill, OK.

Length: 8–9 weeks (288–349 hours).

Exhibit Dates: 3/89–12/97.

Learning Outcomes: After 12/96 see AR-1601-0114. Upon completion of the course, the student will be able to use optical instruments, surveying instruments, communications equipment, and computer equipment; provide leadership and skill in planning, organizing, supervising, and coordinating effective functioning of the unit; demonstrate skills needed for reconnaissance and emplacement of an artillery battery; supervise all details for occupation of the selected location; and coordinate the surveying procedures needed for accurate firing.

Instruction: Methods of instruction include lectures, conferences, activities, and field exercises covering topics in leadership, decision making, supervision, tactics, surveying procedures, and communications systems. This course contains a common core of leadership subjects. Through performance exercises and examinations, the student demonstrates competence in the topics taught.

Credit Recommendation: In the lower-division baccalaureate/associate degree category, 1 semester hour in surveying, 1 in technical communication, and 1 in principles of management. See AR-1404-0035 for common core credit (1/92).

Related Occupations: 13B.

AR-1601-0079

FIELD ARTILLERY SURVEYOR

Course Number: 412-82C10/412-82C10 (ST); 412-82C10.

Location: Field Artillery School, Fort Sill, OK.

Length: 9–11 weeks (373–424 hours).

Exhibit Dates: 10/86–9/97.

Learning Outcomes: After 9/97 see AR-1601-0108. Upon completion of the course, the student will be able to employ proper map reading techniques, operate secure radio and wire communications equipment, and establish distance and location of points by mechanical, electronic, and astronomical methods.

Instruction: Lectures and practice cover various map displays; astronomical observations and computations; use of aiming circle; horizontal taping; theodolite triangulation and electronic distance-measuring equipment; gyroscopic azimuthal determinations; closure errors; methods of encoding and decoding messages; setup and operation of various field radios; and computation of grid locations, distances, directions, heights, and astronomic positions from field data with the aid of a computer and/or calculator.

Credit Recommendation: In the lower-division baccalaureate/associate degree category, 3 semester hours in general surveying and 1 for basic field practice in surveying. In the upper-division baccalaureate category, 3 semester hours in advanced surveying methods and 1 for field practice in surveying (10/89).

Related Occupations: 82C.

AR-1601-0080

SENIOR UTILITIES QUARTERMASTER/CHEMICAL EQUIPMENT REPAIRER BASIC NONCOMMISSIONED OFFICER (NCO) TECHNICAL PORTION

Course Number: 690-52C30, Phase 2.

Location: Ordnance Center and School, Aberdeen Proving Ground, MD.

Length: 6 weeks (212 hours).

Exhibit Dates: 10/87–9/96.

Learning Outcomes: After 9/96 see AR-1601-0102. Upon completion of the course, the student will be able to repair, maintain, and troubleshoot pumps and laundry, clothing repair, chemical decontamination, and water purification equipment.

Instruction: Course consists of lectures and practical exercises on maintenance, repair, and troubleshooting. The course also includes basic electricity, schematic reading, pumps, laundry equipment including washers and dryers, water heaters, light and medium sewing machines, filter units, air compressors, smoke generators, decontamination apparatus, erdlators, and steam cleaners.

Credit Recommendation: In the lower-division baccalaureate/associate degree category, 4 semester hours in equipment repair (12/90).

Related Occupations: 52C.

AR-1601-0081

COMBAT ENGINEER BASIC NONCOMMISSIONED OFFICER (NCO)

Course Number: 030-20-12B30; 030-18-12B30; 030-13-12B30; 030-12-12B30; 030-9-12B30; 030-6-12B30; 030-5-12B30; 030-4-12B30; 030-2-12B30.

Location: Engineer Training Army School Systems (TASS) Battalions, Worldwide; Engineer School, Fort Leonard Wood, MO; Training Center, Fort Leonard Wood, MO; Training Center, Grafenwoehr, Germany; Training Center, Fort Ord, CA; Training Center, Fort Hood, TX; Training Center, Ft. Campbell, KY; Training Center, Fort Stewart, GA; Training Center, Fort Bragg, NC.

Length: 7–9 weeks (298–335 hours).

Exhibit Dates: 10/91–Present.

Learning Outcomes: Upon completion of the course, the student will be able to supervise demolition operations, prepare reconnaissance reports, supervise minefield operations, and supervise military bridge construction.

Instruction: Lectures, demonstrations, and practical experiences cover demolitions; reconnaissance reports; minefields; and Bailey bridge, fixed bridge, and floating bridge construction. This course contains a common core of leadership subjects.

Credit Recommendation: Credit is recommended for the common core only. See AR-1406-0090 (6/99).

AR-1601-0082

1. COMBAT ENGINEERING ADVANCED NONCOMMISSIONED OFFICER (NCO)
2. TATS COMBAT ENGINEERING ADVANCED NONCOMMISSIONED OFFICER (NCO)

Course Number: *Version 1:* 0-12-C42. *Version 2:* 0-12-C42.

Location: *Version 1:* Engineer School, Fort Leonard Wood, MO. *Version 2:* Engineer Training Army School Systems (TASS) Battalions, Worldwide; Engineer School, Fort Leonard Wood, MO.

Length: *Version 1:* 7–10 weeks (310–369 hours). *Version 2:* 9–10 weeks (342–347 hours).

Exhibit Dates: *Version 1:* 3/92–9/97. *Version 2:* 10/97–Present.

Learning Outcomes: *Version 1:* Upon completion of the course, the student will be able to support combat operations by being proficient in demolitions; by practicing survival skills; by understanding mine/countermine warfare; by determining military load classifications for timber trestle, concrete slab, concrete T-beam, and masonry arch bridges; by understanding the use, design, and construction of military fixed/float bridges; and by constructing wire obstacles and antivehicle ditches. *Version 2:* Upon completion of the course, the student will be able to support combat operations by being proficient in demolitions; by practicing survival skills; by understanding mine/countermine warfare; by determining military load classifications for timber trestle, concrete slab, concrete T-beam, and masonry arch bridges; by understanding the use, design, and construction of military fixed/float bridges; and by constructing wire obstacles and antivehicle ditches.

Instruction: *Version 1:* Lectures, demonstrations, and practical exercises cover the above topics. This course contains a common core of leadership subjects. *Version 2:* Lectures, demonstrations, and practical exercises cover the above topics. This course contains a common core of leadership subjects.

Credit Recommendation: *Version 1:* In the lower-division baccalaureate/associate degree category, 2 semester hours in general construction. See AR-1404-0035 for common core credit (4/87). *Version 2:* In the lower-division baccalaureate/associate degree category, 2 semester hours in general construction. Common Core credit for active duty only, see AR-1404-0035 (6/99).

AR-1601-0083

Construction Equipment Supervisor Basic Noncommissioned Officer (NCO)

Course Number: 713-62N30.

Location: Engineer School, Fort Leonard Wood, MO.

Length: 7–10 weeks (260–399 hours).

Exhibit Dates: 6/90–9/01.

Learning Outcomes: After 9/01 see AR-1601-0122. Upon completion of the course, the student will be able to supervise, inspect, and plan heavy/highway construction projects; operate heavy construction equipment; and prepare reconnaissance reports.

Instruction: Lectures, demonstrations, and practical exercises cover the above topics. This course contains a common core of leadership subjects.

Credit Recommendation: In the lower-division baccalaureate/associate degree category, 2 semester hours in project supervision. See AR-1406-0090 for common core credit (1/93).

Related Occupations: 62N.

AR-1601-0084

1. Bridge Crewmember Basic Noncommissioned Officer (NCO)
2. TATS Bridge Crewmember Basic Noncommissioned Officer (NCO)

Course Number: *Version 1:* 030-12C30. *Version 2:* 030-12C30.

Location: *Version 1:* Engineer School, Fort Leonard Wood, MO. *Version 2:* Engineer Training Army School Systems (TASS) Battalions, Worldwide; Engineer School, Fort Leonard Wood, MO.

Length: *Version 1:* 8–9 weeks (310–337 hours). *Version 2:* 7–8 weeks (292–300 hours).

Exhibit Dates: *Version 1:* 10/91–9/97. *Version 2:* 3/98–Present.

Learning Outcomes: *Version 1:* Upon completion of the course, the student will be able to supervise the installation of fixed bridging and river crossing operations and perform rigging, demolition, and reconnaissance missions. *Version 2:* Upon completion of the course, the student will be able to supervise the installation of fixed bridging and river crossing operations and perform rigging, demolition, and reconnaissance missions.

Instruction: *Version 1:* Lectures, demonstrations, and practical exercises cover various military bridges, river crossings, demolition, rigging, and site selection. This course contains a common core of leadership subjects. *Version 2:* Lectures, demonstrations, and practical exercises cover various military bridges, river crossing, demolition, rigging, and site selection. This course contains a common core of leadership subjects. Reserves and National Guard will take a year to complete the same training.

Credit Recommendation: *Version 1:* Credit is recommended for the common core only. See AR-1406-0090 (1/93). *Version 2:* Credit is recommended for the common core only. See AR-1406-0090 (6/99).

Related Occupations: 12C.

AR-1601-0085

Carpentry/Masonry Specialist Reserve Component

Course Number: 052-51B10-RC.

Location: Engineer School, Fort Leonard Wood, MO.

Length: 3–4 weeks (120 hours).

Exhibit Dates: 2/92–4/04.

Learning Outcomes: Upon completion of the course, the student will be able to identify materials and tools of construction; construct wood frame floor, wall, and roof systems; perform building layout; mix and place concrete; and erect masonry walls.

Instruction: Lectures and practical exercises cover tools and materials, wood frame construction practices, building layout, and construction of masonry and concrete components.

Credit Recommendation: In the lower-division baccalaureate/associate degree category, 2 semester hours in carpentry and 1 in masonry (1/93).

Related Occupations: 51B.

AR-1601-0086

Combat Engineer

Course Number: 030-12B10-OSUT; 030-12B10 ST; 030-12B10-TATS AIT; 030-12B10 AIT.

Location: Engineer School, Fort Leonard Wood, MO.

Length: 6–14 weeks (239–682 hours).

Exhibit Dates: 1/92–9/01.

Learning Outcomes: Upon completion of the course, the student will be able to describe mine operations, demolition activities, construction operations, river crossings, and rigging. The 13 and 14 week versions include basic training.

Instruction: Lectures, demonstrations, and exercises cover engineer tools, mine/countermine operations, basic combat construction, rigging, demolition, fixed bridging, river crossing operations, and vehicle operations. This course includes an 8-9 week long Basic Training component. NOTE: The Reserves and National Guard can take one year to complete this course (ADT & IDT).

Credit Recommendation: Credit is not recommended because of the military-specific nature of the course (6/99).

Related Occupations: 12B.

AR-1601-0087

Construction Surveyor

Course Number: 412-82B10.

Location: Engineer School, Fort Leonard Wood, MO.

Length: 10–14 weeks (413–523 hours).

Exhibit Dates: 10/91–4/95.

Learning Outcomes: Upon completion of the course, the student will be able to perform basic surveying measurements of distances, angles, and elevations by use of the steel tape, engineer's level, one-minute theodolite, electronic distance measuring devices, basic trigonometric functions, and the electronic hand calculator; collect topographic data by transit stadia and plane table methods and prepare a map from data collected; perform construction layout of buildings and utilities; perform centerline profile and cross-section surveys; determine curve data and lay out horizontal curves; perform basic road construction computation; and interpret site plans and construction drawings.

Instruction: Lectures, demonstrations, and practical exercises cover the use of standard surveying equipment, topographic mapping techniques, construction layout, reading of plans and drawings, and computation techniques.

Credit Recommendation: In the lower-division baccalaureate/associate degree category, 3 semester hours in basic surveying and 3 in construction surveying (3/93).

Related Occupations: 82B.

AR-1601-0088

Materials Quality Specialist

Course Number: 491-51G10.

Location: Engineer School, Fort Leonard Wood, MO.

Length: 11–12 weeks (439 hours).

Exhibit Dates: 10/89–4/95.

Learning Outcomes: Upon completion of the course, the student will be able to test aggregates, soils, concrete, and bituminous materials; interpret material test results; assist in site evaluation; assist the military engineer in design and construction; and perform basic quality control operations.

Instruction: Lectures, demonstrations, and practical experiences cover soil types and classifications; soil testing procedures and equipment ranging from Speedy Moisture Kits to nuclear densimeters; concrete batten design and testing, including slump, air, and compression tests and analysis; identification of bituminous materials such as asphalt/aggregate mixtures; testing of hot mixes; and design of bituminous mix by Marshall Stability method.

Credit Recommendation: In the lower-division baccalaureate/associate degree category, 3 semester hours in soil mechanics and testing, 3 in concrete mix design and testing, and 3 in asphalt mix design and testing (1/93).

Related Occupations: 51G.

AR-1601-0089

1. Bridge Specialist
2. TATS 12C10 Bridge Crewmember

Course Number: *Version 1:* 12C10-OSUT/030-12C10/12C10-OSUT (ST). *Version 2:* 030-12C10-OSUT; 030-12C10 (AIT); 030-12C10 ST.

Location: *Version 1:* Engineer School, Fort Leonard Wood, MO. *Version 2:* Engineer Training Army School Systems (TASS) Battalions, Worldwide; Engineer School, Fort Leonard Wood, MO.

Length: *Version 1:* 6–13 weeks (272–589 hours). *Version 2:* 6–14 weeks (265–685 hours).

Exhibit Dates: *Version 1:* 10/86–9/97. *Version 2:* 10/97–Present.

Learning Outcomes: *Version 1:* Upon completion of the course, the student will be able to assist in the installation of fixed bridging, in rigging, in river crossing operations, and in vehicle operation and maintenance. *Version 2:* Upon completion of the course, the student will be able to assist in the installation of fixed bridging, in rigging, in river crossing operations, and in vehicle operation and maintenance. The reserves and National Guard people can take one year to complete this course.

Instruction: *Version 1:* Lectures and practical exercises cover the above topics. The 13-week version includes basic training. *Version 2:* Lectures and practical exercises cover the above topics. The 13-14 week version includes basic training. This course includes an 8-9 week long Basic Training component.

Credit Recommendation: *Version 1:* Credit is not recommended because of the limited, specialized nature of the course (1/93). *Version 2:* Credit is not recommended because of the limited, specialized nature of the course (6/99).

Related Occupations: 12C.

AR-1601-0090

Technical Engineering Supervisor Basic Noncommissioned Officer (NCO)

Course Number: 413-51T30.

Location: Engineer School, Fort Leonard Wood, MO.

Length: 11–15 weeks (445–527 hours).

Exhibit Dates: 2/92–9/01.

Learning Outcomes: After 9/01 see AR-1601-0127. Upon completion of the course, the student will be able to perform basic soil tests, prepare architectural drawings, perform basic surveying operations, and provide on-site scheduling and supervision.

Instruction: Lectures, demonstrations, and practical exercises cover the above topics. This course contains a common core of leadership subjects.

Credit Recommendation: In the lower-division baccalaureate/associate degree category, 2 semester hours in basic soil testing, 2 in architectural drafting, 2 in basic surveying, and 2 in project supervision. See AR-1406-0090 for common core credit (6/99).

Related Occupations: 51T.

AR-1601-0091

1. Engineer Officer Basic
2. Engineer Officer Basic
3. Engineer Officer Basic (00)

Course Number: *Version 1:* 4-5-C20. *Version 2:* 4-5-C20. *Version 3:* 4-5-C20, Phase 1.

Location: Engineer School, Fort Leonard Wood, MO.

Length: *Version 1:* 17 weeks (957–958 hours). *Version 2:* 17 weeks (819 hours). *Version 3:* 17 weeks (804 hours).

Exhibit Dates: *Version 1:* 6/89–9/97. *Version 2:* 10/97–8/00. *Version 3:* 9/00–Present.

Learning Outcomes: *Version 1:* Upon completion of the course, the student will be able to design and build Bailey and floating bridges, design and inspect concrete construction, and plan and manage construction projects. *Version 2:* Upon completion of the course, the student will be able to apply structural theory in the design and construction of bridges and buildings and plan and manage horizontal and vertical construction projects. *Version 3:* Upon completion of the course, the student will be able to interpret CADD-supported construction, plumbing and electrical utility drawings; perform site and soil investigations, and earthwork estimates; design wearing surfaces for roads and airfields.

Instruction: *Version 1:* Lectures, demonstrations, and practical exercises cover leadership and ethics, maintenance and supply, training, tactical doctrine, construction operations, and general engineering support. *Version 2:* Lectures, demonstrations, and practical exercises cover leadership and ethics, maintenance and supply, training, tactical doctrine, construction operations, and general engineering support. *Version 3:* Lectures, demonstrations, and practical exercises cover leadership and ethics, maintenance and supply, training, tactical doctrine, construction operations, and general engineering support.

Credit Recommendation: *Version 1:* In the lower-division baccalaureate/associate degree category, 3 semester hours in materials and methods of construction (1/93). *Version 2:* In the upper-division baccalaureate category, 3 semester hours in materials and methods of construction (6/99). *Version 3:* In the upper-division baccalaureate category, 3 semester hours in construction project supervision (1/02).

AR-1601-0092

1. Engineer Officer Advanced
2. Engineer Officer Advance
   (Maneuver Support Center (MANSCEN) Engineer Officer Advanced)
   (Officer Advanced)

Course Number: *Version 1:* 4-5-C22. *Version 2:* 4-5-C22.

Location: Engineer School, Fort Leonard Wood, MO.

Length: *Version 1:* 20 weeks (866 hours). *Version 2:* 18–20 weeks (757–767 hours).

Exhibit Dates: *Version 1:* 3/92–9/96. *Version 2:* 10/96–Present.

Learning Outcomes: *Version 1:* Upon completion of the course, the student will be able to lead and develop subordinates' physical and psychological capabilities; implement team-building concepts; manage and coordinate various units within an organization, including production, equipment, and supplies; design, plan, and manage construction projects, including soil mechanics, drainage, concrete and asphalt paving, bridges, utilities, and structures; and inspect and ensure quality in concrete, wood, and metal structures. *Version 2:* Upon completion of the course, the student will develop awareness of leadership and ethical qualities; make effective use of human resources through understanding human motivation, behavior patterns, conflict management, group functioning and process; plan the design, materials, procurement, and construction of roads, airfields, and supporting military facilities, including soil mechanics, drainage, concrete and asphalt pavements, bridges, utilities, and structures; and inspect and ensure quality in concrete, wood, and metal structures.

Instruction: *Version 1:* Lectures, demonstrations, and practical exercises cover the management and control of an organization in the areas of production, equipment, and supply. *Version 2:* Lectures, demonstrations, and practical exercises cover the management and control of an organization in the areas of production, equipment, and supply.

Credit Recommendation: *Version 1:* In the lower-division baccalaureate/associate degree category, 4 semester hours in methods/materials of heavy construction; 3 in design, analysis, and construction of structures; and 3 in heavy construction/highway administration (1/93). *Version 2:* In the upper-division baccalaureate category, 3 semester hours in human relations, 3 in basic structural engineering, 3 in engineering materials, and 3 in construction project supervision (2/01).

AR-1601-0093

Engineer Equipment Repair Technician Warrant Officer Technical Tactical Certification Reserve

Course Number: 4L-213A-RC.

Location: Engineer School, Fort Leonard Wood, MO.

Length: Phase 2, 2 weeks (114 hours); Phase 4, 2 weeks (117 hours).

Exhibit Dates: 5/91–4/04.

Learning Outcomes: Upon completion of Phase 2, the student will be able to manage a repair facility; direct shop personnel; and diagnose, repair, and maintain power generation equipment. Upon completion of Phase 4, the student will be able to diagnose; repair; and maintain construction equipment, including clutches, brakes, tracks, air compressor systems, and hydraulics.

Instruction: Lectures and practical exercises cover the above topics.

Credit Recommendation: In the lower-division baccalaureate/associate degree category, for Phase 2, 3 semester hours in repair facility management. For Phase 4, 3 semester hours in construction equipment repair (1/93).

Related Occupations: 213A.

AR-1601-0094

Utilities Operator and Maintenance Technician Senior Warrant Reserve Component

Course Number: 4-5-C32-210A-RC.

Location: Engineering Housing Support Center, Fort Belvoir, VA; Engineer School, Fort Leonard Wood, MO.

Length: 4 weeks (146 hours).

Exhibit Dates: 10/90–4/04.

Learning Outcomes: Upon completion of the course, the student will be able to manage utilities, construction, and maintenance activities.

Instruction: Lectures and practical exercises cover management, including staff studies, schedule preparation, employee relations, and report preparation; planning, operation, and maintenance in the installation of utility systems; quality control in construction projects; and operation of a utility facility.

Credit Recommendation: In the lower-division baccalaureate/associate degree category, 3 semester hours in facility management and 1 in personnel supervision (1/93).

Related Occupations: 210A.

AR-1601-0095

Petroleum Laboratory Specialist Basic Noncommissioned Officer (NCO)

Course Number: *Version 1:* 491-77L30. *Version 2:* 491-77L30; 491-92L30.

Location: Quartermaster Center and School, Fort Lee, VA.

Length: *Version 1:* 6 weeks (213–215 hours). *Version 2:* 4–6 weeks (128 hours).

Exhibit Dates: *Version 1:* 10/91–9/99. *Version 2:* 10/99–Present.

Learning Outcomes: *Version 1:* Upon completion of the course, the student will be able to supervise laboratory safety, sampling and gauging procedures, and maintenance and quality surveillance operations in a medium-sized petroleum laboratory test activity. *Version 2:* Upon completion of the course, the student will be able to assume management responsibility for a medium petroleum test activity, including supervision of sampling, gauging, safety, quality, and maintenance.

Instruction: *Version 1:* Practical exercises, conferences and lectures, demonstrations, television, and examinations are the methods of presenting laboratory safety, sampling and gauging procedures, petroleum laboratory maintenance, and quality surveillance operations. This course contains a common core of leadership subjects. *Version 2:* Laboratory, audio-visual materials, discussions, practical exercises, and lectures, in classroom and laboratory facilities. Topics include all components of petroleum laboratory operations.

Credit Recommendation: *Version 1:* In the lower-division baccalaureate/associate degree category, 3 semester hours in petroleum laboratory specialist, petroleum laboratory science, or applied science. See AR-1406-0090 for common core credit (5/96). *Version 2:* In the lower-division baccalaureate/associate degree category, 3 semester hours in petroleum lab science or 3 in applied science (4/04).

Related Occupations: 77L.

AR-1601-0097

Petroleum Officer

Course Number: *Version 1:* 8B-92F. *Version 2:* 31H; 8B-92F.

Location: Quartermaster Center and School, Fort Lee, VA.

Length: *Version 1:* 8 weeks (289–304 hours). *Version 2:* 8 weeks (307 hours).

Exhibit Dates: *Version 1:* 2/90–9/95. *Version 2:* 10/95–Present.

Learning Outcomes: *Version 1:* Upon completion of the course, the student will be able to perform petroleum and water management duties in staff and operations assignments and supervise supply/distribution point and quality surveillance/control operations. *Version 2:* Upon completion of the course, the student will apply skills and knowledge to perform petroleum management duties, supervise supply/distribution point operations and quality surveillance, and perform water management functions.

Instruction: *Version 1:* Lectures and practical exercises cover general petroleum subjects, basic laboratory subjects and tests, petroleum supply operations, and terminal and pipeline operations. *Version 2:* The course is conducted through demonstrations, lectures, exercises, and field trips involving the above topics. Navy, Air Force and Marine officers may attend only the first 80 hours of the course and will receive a certificate of completion for that portion.

Credit Recommendation: *Version 1:* In the upper-division baccalaureate category, 3 semester hours in petroleum supply management or materials management and 2 in petroleum laboratory techniques (2/93). *Version 2:* In the upper-division baccalaureate category, 2 semester hours in petroleum supply management and 1 in water supply management for those completing the entire program (4/04).

AR-1601-0098

Petroleum Supply Specialist Advanced Noncommissioned Officer (NCO)
(Petroleum and Water Specialist Advanced NCO)

Course Number: 8-77-C42(1).

Location: Quartermaster Center and School, Fort Lee, VA.

Length: 11–12 weeks (415 hours).

Exhibit Dates: 10/95–9/00.

Learning Outcomes: Before 10/95 see AR-1717-0092. Upon completion of the course, the student will be able to perform tasks in petroleum supply, petroleum laboratory, and water treatment operations.

Instruction: This course is presented through lectures, demonstrations, exercises, and problem solving components. The course includes common leader training and combat training.

Credit Recommendation: In the lower-division baccalaureate/associate degree category, 1 semester hour in microcomputer applications, 1 in material management, 1 in applied science, 1 in water supply management, and 2 in petroleum supply operations and management. See AR-1404-0035 for common core credit (12/96).

Related Occupations: 77F.

AR-1601-0099

Petroleum Laboratory Specialist Advanced Noncommissioned Officer (NCO)
(Petroleum and Water Specialist Advanced NCO)

Course Number: 8-77-C42(2).

Location: Quartermaster Center and School, Fort Lee, VA.

Length: 11–12 weeks (415 hours).

Exhibit Dates: 10/95–9/00.

Learning Outcomes: Before 10/95 see AR-1717-0092. Upon completion of the course, the student will be able to perform and manage petroleum supply, storage, and laboratory operations.

Instruction: This course is conducted through small group instruction, practical exercises in management and occasional field trips. Topics include petroleum supply ad storage. Students receive limited coverage of supervision of field water operations. Course contains a common core of leadership subjects.

Credit Recommendation: In the lower-division baccalaureate/associate degree category, 1 semester hour in microcomputer applications, 1 in materiel management, 1 in water supply management, and 3 in petroleum supply operations and management. See AR-1404-0035 for common core credit (12/96).

AR-1601-0102

1. Senior Quartermaster/Chemical Equipment Repairer Basic Noncommissioned Officer (NCO)
2. Senior Quartermaster/Chemical Equipment Repairer Basic Noncommissioned Officer (BNCO)

Course Number: *Version 1:* 662-52C30. *Version 2:* 662-52C30.

Location: Ordnance Center and School, Aberdeen Proving Ground, MD.

Length: *Version 1:* 11 weeks (415 hours). *Version 2:* 11–12 weeks (415 hours).

Exhibit Dates: *Version 1:* 10/96–12/97. *Version 2:* 1/98–Present.

Learning Outcomes: *Version 1:* Before 10/96 see AR-1601-0080. Upon completion of the course laundry equipment, the student will be able to repair clothing, maintain, and troubleshoot pumps, heaters, water purification laundry equipment, and chemical decontamination. *Version 2:* Upon completion of the course laundry equipment, the student will be able to repair clothing, maintain, and troubleshoot pumps, heaters, water purification laundry equipment, and chemical decontamination.

Instruction: *Version 1:* This course consists of lectures and practical exercises on maintenance, repair, and troubleshooting. The course also includes basic electricity; schematic reading; pumps; laundry equipment, including washers and dryers; water heaters; light and medium sewing machines; filter units; air compressors; smoke generators; decontamination apparatus; erdlators; and steam cleaners. Course contains a common core of leadership subjects. *Version 2:* Audio visual materials, practical exercises, classroom exercises, discussion, laboratory, and lecture. Topics include electrical and mechanical maintenance, repair, and troubleshooting of pumps, heaters, laundry, bath, chemical decontamination, and water purification equipment.

Credit Recommendation: *Version 1:* In the lower-division baccalaureate/associate degree category, 4 semester hours in mechanical maintenance. See AR-1406-0090 for common core credit (1/98). *Version 2:* In the lower-division baccalaureate/associate degree category, 3 semester hours in mechanical systems maintenance and 3 in electrical systems maintenance. See AR-1406-0090 for common core credit (8/03).

Related Occupations: 52C.

AR-1601-0103

General Engineer Advanced Noncommissioned Officer

Course Number: 7-51-C42.

Location: Engineer School, Fort Leonard Wood, MO.

Length: 6–7 weeks (358 hours).

Exhibit Dates: 10/94–Present.

Learning Outcomes: Upon completion of the course, the student will be able to supervise and inspect horizontal or vertical construction projects, including site selection, layout, installation of utilities, and quality control.

Instruction: Lectures, demonstrations, and practical exercises cover construction planning, operations, and supervision. Course includes a common core of leadership subjects.

Credit Recommendation: In the lower-division baccalaureate/associate degree category, 3 semester hours in construction project supervision. See AR-1404-0035 for common core credit (6/99).

AR-1601-0104

Construction and Engineering Supervisor Basic Noncommissioned Officer

Course Number: 710-51H30.

Location: Engineer School, Fort Leonard Wood, MO.

Length: 9–10 weeks (396–397 hours).

Exhibit Dates: 10/94–9/01.

Learning Outcomes: Upon completion of the course, the student will be able to interpret construction plans, design and construct concrete forms, lay out a wood frame building, supervise the installation of interior electrical and plumbing systems, prepare a bill of materials, and schedule construction work.

Instruction: Lectures, demonstrations, and practical exercises cover light vertical construction and construction scheduling using the critical path method. Course contains a common core of leadership subjects.

Credit Recommendation: In the lower-division baccalaureate/associate degree category, 2 semester hours in construction plan reading and 3 in construction project supervision. See AR-1406-0090 for common core credit (6/99).

Related Occupations: 51H.

AR-1601-0105

Total Army Training System (TATS) Field Artillery Surveyor Section Chief Basic Noncommissioned Officer (NCO)

Course Number: 412-82C30 (F).

Location: Field Artillery School, Fort Sill, OK.

Length: 5–6 weeks (246 hours).

Exhibit Dates: 10/97–9/00.

Learning Outcomes: After 9/00 see AR-1601-0116. Upon completion of the course, the student will be able to plan and conduct field artillery surveys, maintain survey equipment systems, and check and supervise the surveyors.

Instruction: Lectures and practical exercises cover astronomical observation, field artillery survey, solution of triangulation, location of traverse errors, triangulation and resection, leadership, training management, resource management, and military skills. This course contains common core of leadership subjects.

Credit Recommendation: In the lower-division baccalaureate/associate degree category, 3 semester hours in general surveying. See AR-1406-0090 for common core credit (6/00).

Related Occupations: 82C.

AR-1601-0106

Total Army Training System (TATS) Field Artillery Surveyor Section Chief Basic Noncommissioned Officer (NCO)

Course Number: 061-82C30 (F).

Location: Reserve Component Training Centers, US; Field Artillery School, Fort Sill, OK.

Length: 158 hours.

Exhibit Dates: 10/97–9/00.

Learning Outcomes: After 9/00 see AR-1601-0115. Upon completion of the course, the student will be able to plan and conduct field artillery surveys, maintain survey equipment systems, and check and supervise the duties of surveyors.

Instruction: Lectures and practical exercises cover astronomical observation, field artillery survey, solution of triangulation, location of traverse errors, triangulation and resection, leadership, training management, resource management, and military skills.

Credit Recommendation: In the lower-division baccalaureate/associate degree category, 3 semester hours in general surveying (6/00).

Related Occupations: 82C.

AR-1601-0107

Total Army Training System (TATS) Field Artillery Surveyor

Course Number: 061-82C10, Phase 2.

Location: Reserve Schools, Multiple locations.

Length: 128–139 hours.

Exhibit Dates: 10/97–Present.

Learning Outcomes: Upon completion of the course, the student will be able to employ proper map reading techniques; determine the distance between survey stations by mechanical and electronic means; determine direction by astronomic observation and gyroscopic means; operate angular measuring instruments and automated survey systems; prepare schematic sketches of field surveys; and use field data to compute direction distance, grid coordinates, height, and astronomic azimuth.

Instruction: Lecture, and field exercises cover field artillery survey operations; communications; map reading; field data; use of aiming circle; gyroscope azimuthal determination; computation of grid location, distance, heights, and astronomical position from field data; and fire coordination.

Credit Recommendation: In the lower-division baccalaureate/associate degree category, 2 semester hours in map reading and 2 in astronomic surveying (9/03).

Related Occupations: 82C.

AR-1601-0108

Total Army Training System (TATS) Field Artillery Surveyor

Course Number: 412-82C10 (F); 412-82C10.

Location: Field Artillery School, Fort Sill, OK.

Length: 7–9 weeks (256–306 hours).

Exhibit Dates: 10/97–Present.

Learning Outcomes: Before 10/97 see AR-1601-0079. Upon completion of the course, the student will be able to employ proper map reading techniques; determine the distance between survey stations by mechanical and electronic means; determine direction by astronomic observation and gyroscopic means; operate angular measuring instruments and automated survey systems; prepare schematic sketches of field surveys; and use field data to compute direction, distance, grid coordinates, height, and astronomic azimuth.

Instruction: Lectures and field exercises cover field artillery survey operations; communications; map reading; field data; use of aiming circle; gyroscope azimuthal determination; computation of grid location, distance, heights, and astronomical position from field data; and fire coordination.

Credit Recommendation: In the lower-division baccalaureate/associate degree category, 1 semester hour in land navigation and 3 in astronomic surveying (9/03).

AR-1601-0109

Total Army Training System (TATS) Fire Support Sergeant Advanced Noncommissioned Officer (NCO)

Course Number: 250-13F40 (F).

Location: Field Artillery School, Fort Sill, OK.

Length: 4–9 weeks (173–370 hours).

Exhibit Dates: 10/95–Present.

Learning Outcomes: Before 10/95 see AR-2201-0363. Upon completion of the course, the student will be able to apply leadership skills in combat operations, use equipment needed for field operations, including communications gear and guidance and targeting equipment; analyze enemy forces and their capabilities; and determine support needs to carry out offensive maneuvers.

Instruction: This course consists of classroom lectures, field exercises, and appropriate field equipment laboratories. A final section includes training in military methods, including writing style, Army family team building, leadership concepts as they relate to warfare and military justice, risk management, and leading troop operations.

Credit Recommendation: In the lower-division baccalaureate/associate degree category, 3 semester hours in principles of supervision and 2 in map reading and terrain analysis (9/03).

Related Occupations: 13F.

AR-1601-0110

Total Army Training System (TATS) Fire Support Sergeant Advanced Noncommissioned Officer (NCO)

Course Number: 061-13F40 (F), Phase 3.

Location: Reserve Schools, Multiple locations.

Length: 129–142 hours.

Exhibit Dates: 10/95–Present.

Learning Outcomes: Before 10/95 see AR-2201-0363. Upon completion of the course, the student will be able to apply leadership skills in combat operations, use equipment needed for field operations, including communications gear and guidance and targeting equipment; analyze enemy forces and their capabilities; and determine support needs to carry out offensive maneuvers.

Instruction: This course consists of classroom lectures, field exercises, and appropriate field equipment laboratories. A final section includes training in military methods, including writing style, Army family team building, leadership concepts as they relate to warfare and military justice, risk management, and leading troop operations.

Credit Recommendation: In the lower-division baccalaureate/associate degree category, 3 semester hours in principles of supervision and 2 in map reading and terrain analysis (9/03).

Related Occupations: 13F.

AR-1601-0111

Total Army Training System (TATS) Multiple Launch Rocket System (MLRS) Crewmember

Course Number: 042-13M10 (F).
Location: Field Artillery School, Fort Sill, OK.
Length: 4 weeks (191–205 hours).
Exhibit Dates: 10/97–Present.
Learning Outcomes: Before 10/97 see AR-2201-0282. Upon completion of the course, the student will be able to drive the launcher, perform land navigation, perform fire control maintenance, and use communications equipment.
Instruction: Lectures and exercises cover driving the launcher, loading, fire control, communications, and land navigation.
Credit Recommendation: In the lower-division baccalaureate/associate degree category, 1 semester hour in land navigation and 1 in communications systems (9/03).
Related Occupations: 13M.

AR-1601-0112

Total Army Training System (TATS) Multiple Launch Rocket System (MLRS) Crewmember

Course Number: 061-13M10 (F).
Location: Reserve Schools, Multiple locations.
Length: 128–152 hours.
Exhibit Dates: 10/97–Present.
Learning Outcomes: Upon completion of the course, the student will be able to drive the launcher, perform land navigation, perform fire control maintenance, and use communications equipment.
Instruction: Lectures and exercises cover driving the launcher, loading, fire control, communications, and land navigation.
Credit Recommendation: In the lower-division baccalaureate/associate degree category, 1 semester hour in land navigation and 1 in communications systems (9/03).
Related Occupations: 13M.

AR-1601-0113

Total Army Training System (TATS) Field Artillery Platoon Sergeant Advanced Noncommissioned Officer (NCO)

Course Number: *Version 1:* 041-13B40 (F). *Version 2:* 041-13B40 (F), Phase 2.
Location: Field Artillery School, Fort Sill, OK.
Length: *Version 1:* 7 weeks (283–284 hours). *Version 2:* 178–183 hours.
Exhibit Dates: *Version 1:* 1/97–9/00. *Version 2:* 10/00–Present.
Learning Outcomes: *Version 1:* Upon completion of the course, the student will be able to place weapons; perform maintenance; and apply knowledge of fire control, communications, and navigation. *Version 2:* Upon completion of the course, the student will be able to place weapons; perform maintenance; and apply knowledge of fire control, communications, and navigation.
Instruction: *Version 1:* Lectures and practical exercises cover weapons emplacements, ammunition, communications, maintenance, fire control, surveying, and navigation. This course contains a common core of leadership subjects. *Version 2:* Lectures and practical exercises cover weapons emplacements, ammunition, communications, maintenance, fire control, surveying, and navigation. This course contains a common core of leadership subjects.
Credit Recommendation: *Version 1:* In the lower-division baccalaureate/associate degree category, 1 semester hour in map reading. See AR-1404-0035 for common core credit (1/01). *Version 2:* In the lower-division baccalaureate/associate degree category, 2 semester hours in applied surveying and 2 in radar systems. See AR-1404-0035 for common core credit (9/03).
Related Occupations: 13B.

AR-1601-0114

1. Total Army Training System (TATS) Field Artillery Platoon Sergeant Advanced Noncommissioned Officer (NCO), Phase 3
2. Total Army Training System (TATS) Field Artillery Platoon Sergeant Advanced Noncommissioned Officer (NCO)

Course Number: *Version 1:* 061-13B40 (F), Phase 3. *Version 2:* 061-13B40 (F), Phase 3.
Location: *Version 1:* Field Artillery School, Fort Sill, OK. *Version 2:* Reserve Schools, Multiple locations; Field Artillery School, Fort Sill, OK.
Length: *Version 1:* 142 hours. *Version 2:* 158–178 hours.
Exhibit Dates: *Version 1:* 1/97–9/00. *Version 2:* 10/00–Present.
Learning Outcomes: *Version 1:* Before 1/97 see AR-1601-0078. Upon completion of the course, the student will be able to place weapons; perform maintenance; and apply knowledge of fire control, communications, and navigation. *Version 2:* Upon completion of the course, the student will be able to apply skills in the areas of field artillery operations, radar, surveying, intelligence data, leadership, and management.
Instruction: *Version 1:* Lectures, and practical exercises cover weapons emplacement, ammunition, communications, maintenance, and land navigation. *Version 2:* Lectures and practical exercises cover leadership, training, management, intelligence data, operations, and tactical fire control.
Credit Recommendation: *Version 1:* In the lower-division baccalaureate/associate degree category, 1 semester hour in map reading (1/01). *Version 2:* In the lower-division baccalaureate/associate degree category, 2 semester hours in applied surveying and 2 in radar systems (9/03).
Related Occupations: 13B.

AR-1601-0115

Total Army Training System (TATS) Field Artillery Surveyor Section Chief Basic Noncommissioned Officer (NCO)

Course Number: 061-82C30 (F), Phase 2.
Location: Reserve Schools, Multiple locations.
Length: 146 hours.
Exhibit Dates: 10/00–Present.
Learning Outcomes: Before 10/00 see AR-1601-0106. Upon completion of the course, the student will be able to lead and supervise battery surveys and perform land navigation.
Instruction: Lectures and practical exercises cover astronomic observation, field artillery survey, solution of triangles, location of traverse errors, triangulation and resection, leadership, training management, resource management, and military skills.
Credit Recommendation: In the lower-division baccalaureate/associate degree category, 3 semester hours in land navigation (1/01).
Related Occupations: 82C.

AR-1601-0116

Total Army Training System (TATS) Field Artillery Surveyor Section Chief Basic Noncommissioned Officer (NCO)

Course Number: 412-82C30 (F), Phase 2.
Location: Field Artillery School, Fort Sill, OK.
Length: 2–3 weeks (148 hours).
Exhibit Dates: 10/00–Present.
Learning Outcomes: Before 10/00 see AR-1601-0105. Upon completion of the course, the student will be able to make astronomic observations, conduct field artillery surveys, and perform triangulation and resection.
Instruction: Lectures and practical exercises cover astronomic observation, field artillery survey, solution of triangles, location of traverse errors, triangulation and resection, leadership, training management, resource management, and military skills.
Credit Recommendation: In the lower-division baccalaureate/associate degree category, 3 semester hours in land navigation. See AR-1406-0090 for common core credit (1/01).
Related Occupations: 82C.

AR-1601-0117

Construction Engineer Supervisor Advanced Noncommissioned Officer (NCO)

Course Number: 713-51H40.
Location: Engineer School, Fort Leonard Wood, MO.
Length: 5–6 weeks (161 hours).
Exhibit Dates: 10/01–Present.
Learning Outcomes: Upon completion of the course, the student will teach senior engineers supervision skills that evaluate project management, utility plan design and timber bridge design, and supervises quality assurance and control of projects.
Instruction: Specific training and instruction in managing construction projects, designing utilities plans, and designing timber trestle bridges.
Credit Recommendation: In the lower-division baccalaureate/associate degree category, 3 semester hours in project management (1/02).

AR-1601-0118

Construction Engineer Supervisor Advanced Noncommissioned Officer (NCO), Phase 2

Course Number: 713-51H40, Phase 2.
Location: Engineer School, Fort Leonard Wood, MO.
Length: 75 hours.
Exhibit Dates: 10/01–Present.
Learning Outcomes: Upon completion of the course, the student will be able to manage a construction project.
Instruction: Classroom lectures to evaluate plans and specifications and perform construction calculations for material quantities.

Credit Recommendation: In the lower-division baccalaureate/associate degree category, 2 semester hours in project management (1/02).

AR-1601-0119

Construction Engineer Supervisor Basic Noncommissioned Officer (NCO)

Course Number: 713-51H30.
Location: Engineer School, Fort Leonard Wood, MO.
Length: 5–6 weeks (250 hours).
Exhibit Dates: 10/01–Present.
Learning Outcomes: Upon completion of the course, the student will be able to supervise, inspect, and plan vertical construction projects, and calculate material quantities.
Instruction: Lectures, demonstrations and practical exercises to cover vertical construction projects.
Credit Recommendation: In the lower-division baccalaureate/associate degree category, 3 semester hours in construction supervision (1/02).

AR-1601-0120

Construction Engineer Supervisor Basic Noncommissioned Officer (NCO)

Course Number: 713-51H30.
Location: Engineer School, Fort Leonard Wood, MO.
Length: 115–139 hours.
Exhibit Dates: 10/01–Present.
Learning Outcomes: Upon completion of the course, the student will be able to supervise, inspect, and plan vertical construction projects.
Instruction: Lectures, demonstrations and practical exercises to cover vertical construction projects.
Credit Recommendation: In the lower-division baccalaureate/associate degree category, 2 semester hours in construction supervision (1/02).

AR-1601-0121

Construction Equipment Supervisor Basic Noncommissioned Officer (NCO)

Course Number: 713-62N30.
Location: Engineer School, Fort Leonard Wood, MO.
Length: 7 weeks (311 hours).
Exhibit Dates: 10/01–Present.
Learning Outcomes: Before 10/01 see AR-1601-0083. Upon completion of the course, the student will be able to supervise, inspect, and plan heavy/highway construction projects; supervise the operation of heavy construction equipment; and prepare reconnaissance reports.
Instruction: Lectures, demonstrations, and practical exercises cover the above topics.
Credit Recommendation: In the lower-division baccalaureate/associate degree category, 2 semester hours in project supervision (1/02).
Related Occupations: 62N.

AR-1601-0122

Construction Equipment Supervisor Basic Noncommissioned Officer (NCO), Phase 2

Course Number: 713-62N30, Phase 2.
Location: Engineer School, Fort Leonard Wood, MO.
Length: 130 hours.
Exhibit Dates: 10/01–Present.
Learning Outcomes: Before 10/01 see AR-1601-0083. Upon completion of the course, the student will be able to supervise, inspect, and plan heavy/highway construction projects; operation of heavy construction equipment; and prepare reconnaissance reports.
Instruction: Lectures, demonstrations, and practical exercises cover the above topics.
Credit Recommendation: In the lower-division baccalaureate/associate degree category, 2 semester hours in project supervision (1/02).
Related Occupations: 62N.

AR-1601-0123

Construction Equipment Supervisor Basic Noncommissioned Officer (NCO), Phase 3

Course Number: 713-62N30, Phase 3.
Location: Engineer School, Fort Leonard Wood, MO.
Length: 166 hours.
Exhibit Dates: 10/01–Present.
Learning Outcomes: Before 10/01 see AR-1601-0083. Upon completion of the course, the student will be able to supervise, inspect, and plan heavy/highway construction projects; supervise the operation of heavy construction equipment; and prepare reconnaissance reports.
Instruction: Lectures, demonstrations, and practical exercises cover the above topics.
Credit Recommendation: In the lower-division baccalaureate/associate degree category, 2 semester hours in project supervision (1/02).
Related Occupations: 62N.

AR-1601-0124

Combat Engineer

Course Number: 030-12B10 OSUT.
Location: Engineer School, Fort Leonard Wood, MO.
Length: 14 weeks (700 hours).
Exhibit Dates: 10/01–Present.
Learning Outcomes: Upon completion of the course, the student will be able to describe mine operations, demolition activities, construction operations, river crossings, and rigging.
Instruction: Lectures, demonstrations, and exercises cover engineer tools, mine/countermine operations, basic combat construction, rigging, demolition, fixed bridging, river crossing operations, and vehicle operations. This course includes an 8-9 week long Basic Training component.
Credit Recommendation: Credit is not recommended because of the military-specific nature of the course (1/02).
Related Occupations: 12B.

AR-1601-0125

12B10 Combat Engineer

Course Number: 030-12B10 AIT.
Location: Engineer School, Fort Leonard Wood, MO.
Length: 120 hours.
Exhibit Dates: 10/01–Present.
Learning Outcomes: Upon completion of the course, the student will be able to describe mine operations, demolition activities, construction operations, river crossings, and rigging.
Instruction: Lectures, demonstrations, and exercises cover engineer tools, mine/countermine operations, basic combat construction, rigging, demolition, fixed bridging, river crossing operations, and vehicle operations.
Credit Recommendation: Credit is not recommended because of the military-specific nature of the course (1/02).
Related Occupations: 12B.

AR-1601-0126

12B10 Combat Engineer

Course Number: 030-12B10 ST.
Location: Engineer School, Fort Leonard Wood, MO.
Length: 14 weeks (700 hours).
Exhibit Dates: 10/01–Present.
Learning Outcomes: Upon completion of the course, the student will be able to describe mine operations, demolition activities, construction operations, river crossings, and rigging.
Instruction: Lectures, demonstrations, and exercises cover engineer tools, mine/countermine operations, basic combat construction, rigging, demolition, fixed bridging, river crossing operations, and vehicle operations. This course includes an 8-9 week long Basic Training Component.
Credit Recommendation: Credit is not recommended because of the military-specific nature of the course (1/02).
Related Occupations: 12B.

AR-1601-0127

Technical Engineer Supervisor Basic Noncommissioned Officer (NCO)

Course Number: 413-51T30.
Location: Engineer School, Fort Leonard Wood, MO.
Length: 9–10 weeks (385 hours).
Exhibit Dates: 10/01–Present.
Learning Outcomes: Before 10/01 see AR-1601-0090. Upon completion of the course, the student will be able to supervise, inspect, and plan technical projects on paper and at a computer workstation, including bill of materials for the construction mission, construction drawings and sketches, soil test results, perform engineering tasks to prepare horizontal construction plans and computations. The student undergoes combat engineering training and an extensive field exercise putting into practice the lessons taught.
Instruction: Instruction and practical exercises cover supervising field engineering operations as described above.
Credit Recommendation: In the lower-division baccalaureate/associate degree category, 3 semester hours in construction management (1/02).

AR-1601-0128

Technical Engineer Supervisor

Course Number: 413-51T30.
Location: Engineer School, Fort Leonard Wood, MO.
Length: 235 hours.
Exhibit Dates: 10/01–Present.
Learning Outcomes: Upon completion of the course, the student will be able to supervise, inspect, and plan technical projects on paper

and at a computer workstation, including bill of materials for the construction mission, construction drawings and sketches; and soil test results, perform engineering construction reconnaissance and site analysis; and supervise surveying tasks to prepare horizontal construction plans and computations.

Instruction: Instruction and practical exercises cover supervising field engineering operations as described above.

Credit Recommendation: In the lower-division baccalaureate/associate degree category, 3 semester hours in construction management (1/02).

Related Occupations: 51T.

AR-1601-0129

Total Army Training System (TATS)
Combat Engineering Advanced
Noncommissioned Officer
(ANCOC)

Course Number: 0-12-C42.

Location: Engineer School, Fort Leonard Wood, MO.

Length: 150 hours.

Exhibit Dates: 10/97–Present.

Learning Outcomes: Upon completion of the course, the student will be able to support combat operations by being proficient in demolitions; by practicing survival skills; by understanding mine/countermine warfare; by determining military load classifications for timber trestle, concrete slab, concrete T-beam; and by understanding the use, design, and construction of military fixed/float bridges; and by constructing wire obstacles and antivehicle ditches.

Instruction: Lectures, demonstrations, and practical exercises cover the above topics.

Credit Recommendation: In the lower-division baccalaureate/associate degree category, 2 semester hours in general construction (1/02).

AR-1601-0130

Total Army Training System (TATS)
Petroleum Supply Specialist
(reclass)

Course Number: 101-77F10 (R) (F).

Location: Quartermaster Center and School, Fort Lee, VA.

Length: 6 weeks (214 hours).

Exhibit Dates: 10/00–Present.

Learning Outcomes: Upon completion of the course, the student will have the knowledge and skills necessary to receive, store, issue, dispense, and ship petroleum products.

Instruction: Discussion, classroom and practical exercises, and lecture. Topics included in the course are safety, considerations; examination for contamination; gaging and sampling; pump theory; filtration and monitoring; water front operations; manifolds, pipes, valves, and fittings; railcar depot and truck stand transfer, and aviation fuel quality assurance.

Credit Recommendation: In the lower-division baccalaureate/associate degree category, 3 semester hours in petroleum operations (6/02).

AR-1601-0131

Total Army Training System (TATS)
Petroleum Supply Specialist
(reclass)

Course Number: 101-77F10 (R) (F), Phase 1; 101-77F10 (R) (F), Phase 2.

Location: Quartermaster Center and School, Fort Lee, VA.

Length: 120 hours.

Exhibit Dates: 10/00–Present.

Learning Outcomes: Upon completion of the course, the student will have the necessary skills and knowledge to receive, store, issue, dispense, and ship petroleum products.

Instruction: Discussion, classroom exercises, practical exercises, lecture and examination. Topics include safety considerations; examination for contamination; gaging and sampling; pump theory; filtration and monitor theory; waterfront operations; manifolds, pipes, valves, and fittings; railcar, depot and truck transfer; and aviation fuel quality assurance.

Credit Recommendation: In the lower-division baccalaureate/associate degree category, 2 semester hours in petroleum operations (6/02).

AR-1601-0132

Quartermaster/Chemical Equipment
Repairer

Course Number: 091-63J10 TATSC, Phase 2.

Location: Reserve Schools, Multiple locations.

Length: 118 hours.

Exhibit Dates: 10/99–Present.

Learning Outcomes: Upon completion of the course, the student will be able to repair, maintain, and trouble shoot pumps, heaters, water purification, laundry, and chemical decontamination equipment.

Instruction: Audio visual materials, practical exercises, classroom exercises, discussion, laboratory, and lecture. Topics include lectures and practical exercises on the electrical and mechanical maintenance, repair, and troubleshooting of pumps, heaters, laundry and bath, chemical decontamination, and water purification equipment.

Credit Recommendation: In the lower-division baccalaureate/associate degree category, 2 semester hours in mechanical systems maintenance and 2 in electrical systems maintenance (8/03).

AR-1606-0006

Flight Simulator (UH1FS) Specialist
(Flight Simulator (2B24) Specialist)

Course Number: 191-F1; 191-ASIF8.

Location: Aviation Center, Fort Rucker, AL.

Length: 2–3 weeks (91 hours).

Exhibit Dates: 6/75–11/96.

Objectives: To enable enlisted personnel to perform as flight simulator system (2B24 or UH1FS) operators in order to provide instruction to rotary-wing rated Army aviators.

Instruction: Practical exercises and programmed instruction cover the synthetic flight training system, UH1FS or 2B24.

Credit Recommendation: Credit is not recommended because of the military-specific nature of the course (11/98).

Related Occupations: 93D; 93J.

AR-1606-0047

OH-58 Instructor Pilot
(OH-58 Transition/Gunnery Instructor Pilot
(IP) Qualification)

Course Number: 2C-F23.

Location: Aviation Center, Fort Rucker, AL.

Length: 4–5 weeks (150 hours).

Exhibit Dates: 7/85–11/96.

Objectives: To train selected pilots as instructor pilots in the OH-58 helicopter.

Instruction: Course includes flight and academic training in day/night visual flight rules; before-flight, basic flight, approach/landing, and emergency tasks; principles and techniques of flight instruction; aircraft systems and maintenance; and principles and methods of instruction.

Credit Recommendation: In the upper-division baccalaureate category, 3 semester hours in basic aerodynamics and theory of flight, 3 in basic helicopter flight and systems, 3 in principles and methods of instruction, and credit for flight training on the basis of institutional evaluation (11/86).

Related Occupations: 100B; 100C; 100D; 100E; 102A.

AR-1606-0053

UH-1 Contact Instructor Pilot
(UH-1 Instructor Pilot)

Course Number: 2C-ASI5K/2C SQIC.

Location: Aviation Center, Fort Rucker, AL.

Length: 6 weeks (225 hours).

Exhibit Dates: 10/86–11/96.

Objectives: To train selected pilots as instructor pilots in the UH-1 helicopter.

Instruction: Lectures and practical training cover day/night visual; flight rules for the UH-1, before-flight, basic flight, approach/landing, and emergency tasks; principles and techniques of flight instruction, and aircraft systems and maintenance.

Credit Recommendation: In the upper-division baccalaureate category, 3 semester hours in principles and methods of instruction (11/86).

Related Occupations: 100B; 100C; 100D; 100E; 102A.

AR-1606-0058

AH-1S Aviator Qualification
(AH-1G Aviator Qualification)

Course Number: 2C-ASI1M/2C-100E.

Location: Aviation Center, Fort Rucker, AL.

Length: 6–7 weeks (251–254 hours).

Exhibit Dates: 2/82–4/95.

Objectives: After 4/95 see AR-1606-0220. To provide rated pilots with transition training on AH-1G (Cobra) or AH-1S helicopters.

Instruction: Lectures and practical exercises cover transition flight training on AH-1S helicopters, including preflight inspections, ground procedures, night operations, instrument flying, regulations, flight planning, maintenance procedures, and principles of aerodynamics.

Credit Recommendation: In the upper-division baccalaureate category, 3 semester hours in basic aerodynamics and theory of flight, 3 in basic helicopter flight and systems, and credit for flight training on the basis of institutional evaluation (6/87).

Related Occupations: 100E.

AR-1606-0064

CH-47B/C Aviator Qualification

Course Number: 2C-ASI 1G-D/2C-100C-D; 2C-ASI 1G-B/2C-100C-B; 2C-SI1G/2C-154B.

Location: Aviation Center, Fort Rucker, AL.
Length: 2–7 weeks (108–315 hours).
Exhibit Dates: 7/84–11/96.

Objectives: To provide rated pilots with transition training on CH-47 (Chinook) helicopters.

Instruction: Lectures and practical exercises cover transition flight training on CH-47 aircraft, including preflight inspection, ground procedures, night operations, instrument flying, regulations, flight planning, and maintenance procedures.

Credit Recommendation: In the upper-division baccalaureate category, 3 semester hours in basic aerodynamics and theory of flight, 3 in basic helicopter flight and systems, and credit for flight training on the basis of institutional evaluation (11/86).

Related Occupations: 100B; 100C; 102A.

AR-1606-0074

Officer/Warrant Officer Rotary Wing Aviator
(Warrant Officer Candidate Rotary Wing Aviator)

Course Number: 1-N-1981C; 1-R-1981C; 1-A-1981B; 1-B-1981B; 1-R-062C-C; 1-R-062B-C; 1-H-062C; 1-H-062B; 1-H-1981B; 2C-062B-B; 2C-1981-B; 2C-100B-C; 2C-100G; 2C-15A; 2C-100B-B; 2C-15ABX; 2C-100-B; 2C-15B; 2C-15B/2C-100B-B.
Location: Primary Helicopter School, Fort Wolters, TX; Aviation School, Fort Rucker, AL.
Length: 28–36 weeks (904–1472 hours).
Exhibit Dates: 3/59–11/96.

Objectives: To train selected officers and warrant officers in primary rotary-wing flying techniques and helicopter use.

Instruction: Flight experience and lectures cover aerodynamics, meteorology, navigation, aircraft maintenance, instrument flight theory, and aviation physiology. This course is equivalent to private and commercial pilot requirements.

Credit Recommendation: In the lower-division baccalaureate/associate degree category, 3 semester hours in primary helicopter flight training subjects, 3 for flight experience, 3 in aerodynamics, 3 in meteorology, 3 in navigation, 3 in aircraft maintenance, and 3 in aviation physiology (11/86).

Related Occupations: 100B.

AR-1606-0133

UH-60 Aviator Qualification

Course Number: 2C-ASI IN.
Location: Aviation Center, Fort Rucker, AL.
Length: 6 weeks (241 hours).
Exhibit Dates: 10/85–9/95.

Objectives: After 9/95 see AR-1606-0222. To provide rated pilots with training on UH-60 helicopters.

Instruction: Flight and academic instruction covers the UH-60 helicopter, maneuvers, combat skills, and night vision goggles.

Credit Recommendation: In the upper-division baccalaureate category, 3 semester hours in basic aerodynamics and theory of flight, 3 in basic helicopter flight and systems, and credit for flight training on the basis of institutional evaluation (11/86).

Related Occupations: 100B.

AR-1606-0139

Night Vision Goggle (NVG) AN/PVS5 Methods of Instruction

Course Number: None.
Location: Aviation Center, Fort Rucker, AL.
Length: 2–3 weeks (77–78 hours).
Exhibit Dates: 4/77–11/96.

Objectives: To train selected rotary-wing instructor pilots as night vision goggle instructor pilots.

Instruction: Topics include night visual goggle flight training and night academics.

Credit Recommendation: Credit is not recommended because of the military-specific nature of the course (11/86).

AR-1606-0144

Special Forces Diving Supervisor

Course Number: None.
Location: Institute for Military Assistance, Fort Bragg, NC.
Length: 3–4 weeks (122 hours).
Exhibit Dates: 3/80–4/92.

Objectives: To train Special Forces commissioned and noncommissioned officers as diving supervisors.

Instruction: Course concentrates on diving plans, preparation, and supervision.

Credit Recommendation: Credit is not recommended because of the limited, specialized nature of the course (4/83).

AR-1606-0161

OH-58D Aviator Qualification

Course Number: 2C-ASI1A/2C-152D; 2C SI 1A/2C-152D.
Location: Aviation Center, Fort Rucker, AL.
Length: 7–8 weeks (293–302 hours).
Exhibit Dates: 10/86–11/96.

Objectives: After 11/96 see AR-1606-0224. To train commissioned and warrant officer rotary-wing aviators as OH-58D pilots.

Instruction: Classroom and practical instruction cover aircraft systems, flight training, communications, navigation, mast-mounted sight, airborne target handover systems, mission planning, and safety.

Credit Recommendation: In the upper-division baccalaureate category, 3 semester hours in basic aerodynamics and theory of flight, 3 in basic helicopter flight and systems, and credit for flight training on the basis of institutional evaluation (11/86).

Related Occupations: 152D.

AR-1606-0163

Aerospace Medicine Residency Flight Training

Course Number: None.
Location: Aviation Center, Fort Rucker, AL.
Length: 4–5 weeks (165 hours).
Exhibit Dates: 7/87–11/96.

Objectives: To familiarize resident physicians with turbine-powered aircraft systems and rotary-wing flight principles and operation.

Instruction: Course includes training in electrical system, power plant and related systems, fuel systems, rotor systems, power train systems, flight control system, weight and balance, and flight training on the UH-1 helicopter.

Credit Recommendation: In the lower-division baccalaureate/associate degree category, 3 semester hours in aircraft systems. In the upper-division baccalaureate category, 3 semester hours in basic aerodynamics and theory of flight, 3 in basic helicopter flight and systems, and credit for flight training on the basis of institutional evaluation (11/86).

AR-1606-0165

OH-58 Aeroscout Instructor Pilot Methods of Instruction

Course Number: None.
Location: Aviation Center, Fort Rucker, AL.
Length: 10–12 weeks (397–420 hours).
Exhibit Dates: 1/78–11/96.

Objectives: To train commissioned and warrant officer aviators as OH-58 instructor pilots for contact, night, night vision goggles, and aeroscout flying.

Instruction: Classroom and practical training covers day/night flying, principles and techniques of flight instruction, aircraft systems, combat skills, aeromedical factors, and night vision goggles.

Credit Recommendation: In the upper-division baccalaureate category, 3 semester hours in basic aerodynamics and theory of flight, 3 in basic helicopter flight and systems, 3 in principles and methods of instruction, and credit for flight training on the basis of institutional evaluation (11/86).

AR-1606-0166

Rotary Wing Aviator Refresher Training

Course Number: 2C-F31.
Location: Aviation Center, Fort Rucker, AL.
Length: 4–6 weeks (151–221 hours).
Exhibit Dates: 10/80–3/99.

Learning Outcomes: To train, refresh, and requalify rotary-wing aviators returning to flying status.

Instruction: Classroom and practical exercises cover UH-1 helicopter systems and flight training sequences.

Credit Recommendation: Credit is not recommended because of the limited, specialized nature of the course (3/89).

AR-1606-0168

AH-64 Instructor Pilot

Course Number: 2C-F36.
Location: Aviation Center, Fort Rucker, AL.
Length: 16–17 weeks (659 hours).
Exhibit Dates: 9/84–11/96.

Objectives: To provide training in flight and academic instruction for the AH-64 helicopter and in teaching and evaluating AH-64 flight training.

Instruction: Topics include flight and academic instruction in AH-64 helicopter contact flying skills, gunnery, target acquisition and designation sight, combat skills, pilot night vision sensor system, and methods of instructing and evaluating AH-64 flight training.

Credit Recommendation: In the upper-division baccalaureate category, 3 semester hours in basic aerodynamics and theory of flight, 3 in basic helicopter flight systems, 3 in principles and methods of instruction, and credit for flight training on the basis of institutional evaluation (11/86).

AR-1606-0169

OH-58D Field Artillery Aerial Observer

Course Number: 2E-F109.
Location: Aviation Center, Fort Rucker, AL.

Length: 5–6 weeks (203–207 hours).
Exhibit Dates: 12/86–12/96.
Objectives: To train personnel to perform field artillery aerial observer duties on the OH-58D helicopter.
Instruction: Lectures and practical training cover aerial observation, including aircraft systems, avionics, mast-mounted sight, airborne handover systems, aircraft handling, mission planning, and safety.
Credit Recommendation: Credit is not recommended because of the limited, specialized nature of the course (12/86).

AR-1606-0170

Aviation Officer Basic

Course Number: 2-1-C20.
Location: Aviation Center, Fort Rucker, AL.
Length: 9–10 weeks (366–367 hours).
Exhibit Dates: 10/85–12/94.
Objectives: After 12/94 see AR-1606-0225. To provide newly commissioned aviation officers with common military and combat arms training.
Instruction: Course consists of lectures and practical exercises in common task skills, leadership, and arms.
Credit Recommendation: In the lower-division baccalaureate/associate degree category, 3 semester hours in personnel supervision and 1 in technical writing (6/88).

AR-1606-0171

Aviation Officer Advanced

Course Number: 2-1-C22.
Location: Aviation Center, Fort Rucker, AL.
Length: 20 weeks (716 hours).
Exhibit Dates: 5/89–3/96.
Learning Outcomes: To provide commissioned officers with advanced training in military subjects.
Instruction: Lectures and practical exercises cover military subjects, leadership, management, technical writing skills, and combat planning.
Credit Recommendation: In the lower-division baccalaureate/associate degree category, 2 semester hours in technical writing. In the upper-division baccalaureate category, 3 semester hours in principles of management (4/90).

AR-1606-0172

Night/Night Vision Goggle Instructor Pilot

Course Number: 2C-F37.
Location: Aviation Center, Fort Rucker, AL.
Length: 4–5 weeks (115 hours).
Exhibit Dates: 10/83–12/94.
Objectives: To qualify selected rotary-wing instructor pilots as night vision goggle instructor pilots.
Instruction: Topics include night visual goggle flight training and night academics.
Credit Recommendation: Credit is not recommended because of the military-specific nature of the course (3/90).

AR-1606-0173

Aviation Pre-Command

Course Number: 2G-F42.
Location: Aviation Center, Fort Rucker, AL.
Length: 2–3 weeks (78–110 hours).
Exhibit Dates: 8/79–9/96.
Objectives: After 9/96 see AR-1405-0268. To provide update and refresher training in Army aviation subjects.
Instruction: Training includes tactics; training management; maintenance and logistics; safety; aviation medicine; fire support; communications-electronics; nuclear, biological, and chemical trends; and general military subjects.
Credit Recommendation: Credit is not recommended because of the military-specific nature of the course (11/88).

AR-1606-0174

Aviation Safety Officer

Course Number: ***Version 1:*** 7K-F-12. *Version 2:* 7K-F-12. *Version 3:* 7K-F-12.
Location: Safety Center, Fort Rucker, AL.
Length: ***Version 1:*** 6 weeks (183–186 hours). *Version 2:* 6 weeks (221 hours). *Version 3:* 6 weeks (298 hours).
Exhibit Dates: ***Version 1:*** 8/86–9/93. *Version 2:* 10/93–1/03. *Version 3:* 2/03–Present.
Learning Outcomes: ***Version 1:*** Upon completion of this course, the student will understand the safety management system and the safety professional's role in accident prevention. The student will also be able to establish countermeasures and control measures for system defects using evaluations and survey data, initiate a safety and accident prevention program and implement all safety program requirements, demonstrate a working knowledge of OSHA Safety and Health Standards, and identify risks associated with particular operations and weigh these risks against overall training value to be gained. *Version 2:* Upon completion of this course, the student will understand the safety management system and the safety professional's role in accident prevention. The student will also be able to establish countermeasures and control measures for system defects using evaluations and survey data, initiate a safety and accident prevention program and implement all safety program requirements, demonstrate a working knowledge of OSHA Safety and Health Standards, and identify risks associated with particular operations and weigh these risks against overall training value to be gained. *Version 3:* Upon completion of this course, the student will understand the safety management system and the safety professional's role in accident prevention. The student will also be able to establish countermeasures and control measures for system defects using evaluations and survey data, initiate a safety and accident prevention program and implement all safety program requirements, demonstrate a working knowledge of OSHA Safety and Health Standards, and identify risks associated with particular operations and weigh these risks against overall training value to be gained.
Instruction: ***Version 1:*** Lectures, demonstrations, and practical exercises cover industrial safety and accident prevention. *Version 2:* Lectures, demonstrations, and practical exercises cover industrial safety and accident prevention. *Version 3:* Audio visual materials, practical and classroom exercises, discussion, laboratory, case studies, lecture, learner presentation and computer-based training. Topics include safety program management, accident investigation and analysis, including certification; Occupational Safety and Health (OSHA); ergonomics, hazardous materials/communication, transportation safety, risk management, accident causation, environmental law; and hearing conservation, weapons safety, aviation safety, research methods, and legal aspects of safety.
Credit Recommendation: ***Version 1:*** In the upper-division baccalaureate category, 3 semester hours in principles of industrial safety, 3 in safety engineering, and 3 in risk management (3/88). *Version 2:* In the upper-division baccalaureate category, 3 semester hours in principles of industrial safety, 3 in safety engineering, and 1 in risk management (6/95). *Version 3:* In the upper-division baccalaureate category, 3 semester hours in accident investigation and analysis. In the graduate degree category, 3 semester hours in risk management and applications, and 3 in industrial safety (4/03).

AR-1606-0175

Combat Intelligence

Course Number: INT 3-OE; INT 3.
Location: Combined Arms Training Center, Vilseck, W. Germany.
Length: 3 weeks (108–114 hours).
Exhibit Dates: 10/74–10/98.
Learning Outcomes: Upon completion of the course, the student will be able to perform the duties of an intelligence officer/noncommissioned officer in a battalion or a brigade; prepare reports based on the process of planning, collecting, processing, and disseminating data; analyze and collate information from multiple sources; write periodic reports and summary reports based on information collected; read maps and determine influence of terrain on military operations; maintain journal and workbook of activities; develop a situation map overlay; coordinate surveillance activities; prepare and present oral briefings; and develop and implement security measures, with emphasis on personnel security and security of documents.
Instruction: Lectures, conferences, demonstrations, seminar classes, practical exercises, and a comprehensive simulation exercise cover intelligence topics. The student functions in an intelligence role within the simulated situation.
Credit Recommendation: Credit is not recommended because of the military-specific nature of the course (10/88).

AR-1606-0176

Special Forces Operations and Intelligence

Course Number: 244-ASIF1.
Location: John F. Kennedy Special Warfare Center and School, Fort Bragg, NC.
Length: 16 weeks (666 hours).
Exhibit Dates: 3/87–9/98.
Learning Outcomes: Upon completion of the course, the student will perform operations and intelligence missions, advise and train in military operations, coordinate air/ground and nautical missions, develop and exploit intelligence information, apply principles of security and threats, and work with Special Forces reporting systems.
Instruction: Lectures, discussions, and command post exercises cover psychological operations in air/land battle, support of unconventional warfare, cross-cultural communication, impact of terrorism, US foreign policy, legal aspects of unconventional warfare, photography and map assessment, area studies, intelligence publications, mission planning process, clandestine operations, Soviet military

structure and capabilities, intelligence collection and interpretation, and foreign internal defense.

Credit Recommendation: In the lower-division baccalaureate/associate degree category, 1 semester hour in photography, 1 in map interpretation, and 1 in communications. In the upper-division baccalaureate category, 1 semester hour in cross-cultural communication, 2 in psychology, 2 in Soviet studies, 1 in social science, 2 in public administration, 1 in political science, and 2 in police administration (12/88).

AR-1606-0177

Special Forces Combat Diving Supervisor

Course Number: 2E-F65/011-F23; 2E-F65/011-ASIS6.

Location: John F. Kennedy Special Warfare Center and School, Fort Bragg, NC; Naval Air Station, Key West, FL.

Length: 2–4 weeks (79–120 hours).

Exhibit Dates: 6/86–Present.

Learning Outcomes: Upon completion of the course, the student will be able to plan, prepare, and conduct dives; conduct inspections and postdiving checks; teach use of closed- and open-circuit apparatus; teach emergency medical treatment; teach field repair of equipment and handling of air/oxygen systems; teach use of dive tables and tide and current charts; and teach submarine and chamber operations.

Instruction: Lectures, demonstrations, and laboratory participation cover the supervision of combat diving.

Credit Recommendation: In the lower-division baccalaureate/associate degree category, 1 semester in advanced scuba diving and 1 in scuba instruction (1/01).

AR-1606-0178

CH-47D Aviator Qualification

Course Number: *Version 1:* 2C-SI 1B/2C-154C. *Version 2:* 2C-SIC2/2C-154C.

Location: Aviation Center, Fort Rucker, AL.

Length: *Version 1:* 10 weeks (371 hours). *Version 2:* 10 weeks (344–360 hours).

Exhibit Dates: *Version 1:* 10/87–9/94. *Version 2:* 10/94–Present.

Learning Outcomes: *Version 1:* Upon completion of the course, the student will be qualified as pilot of a CH-47D helicopter. *Version 2:* Upon completion of the course, the student will be qualified as pilot of a CH-47D helicopter.

Instruction: *Version 1:* Lectures and practical exercise cover flight training on the CH-47D helicopter, including preflight, night operation, emergency procedures, engine and power train systems, and fuel and flight minor maintenance procedures. *Version 2:* Lectures, briefings, and block flight instruction covers basic helicopter systems, including, engines, power train, fuel, hydraulics, flight controls, malfunction analysis, emergency procedures, and necessary skills to achieve pilot qualification.

Credit Recommendation: *Version 1:* In the lower-division baccalaureate/associate degree category, 3 semester hours in aircraft systems and 3 in helicopter flight training (4/90). *Version 2:* In the lower-division baccalaureate/associate degree category, 3 semester hours in aircraft systems and 3 in helicopter flight training (1/03).

AR-1606-0179

AH-64 Aviator Qualification

Course Number: *Version 1:* 2C-SI1L/2C-152F. *Version 2:* 2C-SID5/152F.

Location: Aviation Center, Fort Rucker, AL.

Length: *Version 1:* 10 weeks (430 hours). *Version 2:* 12–13 weeks (480 hours).

Exhibit Dates: *Version 1:* 2/88–9/95. *Version 2:* 10/95–Present.

Learning Outcomes: *Version 1:* Upon completion of the course, the student will be qualified as a pilot of an AH-64 helicopter. *Version 2:* Upon completion of the course, the student will be qualified as a pilot of an AH-64 helicopter.

Instruction: *Version 1:* Instruction covers the basic AH-64 helicopter systems, including weapon; airframe, electrical, fuel, hydraulic, caution/warning, and pressurized air systems; digital stabilization; fault location; the modular forward looking infrared radar; the integrated helmet and sighting system; navigation systems; flight instrumentation; and flight instruction. *Version 2:* Lectures, briefings, and block flight instruction cover the basic helicopter systems, including weapon, airframe, electrical, fuel, hydraulic, caution/warning, and pressurized air systems; digital stabilization; fault location; modular forward looking infrared radar; integrated helmet and sighting system; navigation systems; and flight instruments.

Credit Recommendation: *Version 1:* In the lower-division baccalaureate/associate degree category, 3 semester hours in aircraft systems and 3 in helicopter flight training (4/90). *Version 2:* In the lower-division baccalaureate/associate degree category, 3 semester hours in aircraft systems and 3 in helicopter flight training (1/03).

AR-1606-0180

1. Rotary Wing Qualification
2. Rotary Wing Qualification, Phase 2
3. Rotary Wing Qualification, Phase 3

Course Number: *Version 1:* 2C-ASI1Q/2C-100B-D. *Version 2:* 2C-SIB1/2C-153B. *Version 3:* 2C-SIB1/2C-153B, Phase 3.

Location: *Version 1:* Aviation Center, Fort Rucker, AL. *Version 2:* Aviation School, Fort Rucker, AL. *Version 3:* Aviation Center, Fort Rucker, AL.

Length: *Version 1:* 11–12 weeks (372 hours). *Version 2:* 11 weeks (325 hours). *Version 3:* 11–12 weeks (372 hours).

Exhibit Dates: *Version 1:* 2/86–9/97. *Version 2:* 10/97–12/99. *Version 3:* 1/00–Present.

Learning Outcomes: *Version 1:* Upon completion of the course, the student will transition from an accomplished fixed-wing aircraft pilot to a UH-1 rotary-wing aircraft pilot. *Version 2:* Upon completion of the course, the fixed-wing pilot will be able to fly the UH-1 helicopter. *Version 3:* Upon completion of the course, the fixed-wing pilot will be able to fly the UH-1 helicopter.

Instruction: *Version 1:* Lectures and practical exercises cover officer/warrant officer transition to the UH-1 rotary-wing aircraft, including flight training, basic flight subjects, maintenance procedures, tactical flight procedures, navigation, instrument flight, and recognition of hostile action. *Version 2:* Lectures and practical exercises cover airplane pilot transition to the UH-1 helicopter, including flight training, basic flight subjects, maintenance procedures, tactical flight procedures, navigation, instrument flight, and operational environment. *Version 3:* Lectures and practical exercises cover airplane pilot transition to the UH-1 helicopter, including flight training, basic flight subjects, maintenance procedures, tactical flight procedures, navigation, instrument flight, and operational environment.

Credit Recommendation: *Version 1:* In the lower-division baccalaureate/associate degree category, 3 semester hours in helicopter flight training and 3 in helicopter systems (4/90). *Version 2:* In the lower-division baccalaureate/associate degree category, 3 semester hours in aerodynamics and theory of flight, 1 in helicopter systems, and 3 in helicopter flight training (11/98). *Version 3:* In the lower-division baccalaureate/associate degree category, 3 semester hours in aerodynamics and theory of flight, 1 in helicopter systems, and 3 in helicopter flight training (2/01).

Related Occupations: 100B; 153B.

AR-1606-0181

1. Initial Entry Rotary-Wing Aviator Common Core
2. Initial Entry Rotary Wing Aviator Common Core

Course Number: *Version 1:* 2C-OH-58/UH-1/AH-1/UH-60. *Version 2:* 2C-IERW (Common Core).

Location: *Version 1:* Aviation Center, Fort Rucker, AL. *Version 2:* Aviation School, Fort Rucker, AL.

Length: *Version 1:* 20 weeks (764 hours). *Version 2:* 20 weeks (722 hours).

Exhibit Dates: *Version 1:* 4/88–6/96. *Version 2:* 7/96–Present.

Learning Outcomes: *Version 1:* Upon completion of the course, the student will be able to perform as a private pilot on rotorcraft helicopters. *Version 2:* Upon completion of the course, the student will be able to perform as an instrument-rated private pilot on rotorcraft helicopters.

Instruction: *Version 1:* Lectures and practical exercises cover basic rotary-wing flight maneuvers, normal and emergency procedures, flight planning, basic instruments, and associated safety considerations. *Version 2:* Lectures, practical exercises, and flight training covering basic rotary-wing flight, normal and emergency procedures, flight planning, instrument flight, human factors, physiology of flight, aviation meteorology, aircraft systems, and federal aviation regulations.

Credit Recommendation: *Version 1:* In the lower-division baccalaureate/associate degree category, 3 semester hours in flight regulations, 3 in aircraft navigation (visual), 3 in aviation meteorology, 3 in aircraft engines and systems, 3 in physiology of flight, 3 in aircraft performance, and 6 in private pilot flight training (helicopter) (4/90). *Version 2:* In the lower-division baccalaureate/associate degree category, 1 semester hour in aircraft navigation (visual), 2 in aviation meteorology, 1 in aircraft engines and systems, 1 in physiology of flight, 3 in private pilot ground, 3 in private pilot flight (helicopter), 3 in instrument rating ground, 3 in instrument rating flight, and 1 in human factors (11/98).

AR-1606-0182

Initial Entry Rotary-Wing Aviator (UH-60)

Course Number: 2C-SI 1N/2C-153D (UH-60).
Location: Aviation Center, Fort Rucker, AL.
Length: 18–19 weeks (665 hours).
Exhibit Dates: 4/88–9/01.
Learning Outcomes: After 9/01 see AR-1606-0246. Upon completion of the course, the student will be able to perform at the level of private pilot on rotorcraft helicopters.
Instruction: Lectures and practical exercises cover advanced rotary-wing flight maneuvers, normal and emergency procedures, instrument flight planning, advanced instruments, and associated safety considerations.
Credit Recommendation: In the lower-division baccalaureate/associate degree category, 3 semester hours in aircraft navigation (instrument), 3 in cockpit procedures training (helicopter), 3 in instrument rating (helicopter), and 6 in commercial pilot flight training (helicopter) (4/90).
Related Occupations: 153D.

AR-1606-0183

Initial Entry Rotary Wing Aviator (UH-1)

Course Number: *Version 1:* 2C-SI 1E/2C-153B (UH-1). *Version 2:* 2C-15A/67J/SIB1/153B (UH-1), Phase 4.
Location: Aviation Center, Fort Rucker, AL.
Length: *Version 1:* 16–17 weeks (572 hours). *Version 2:* 12 weeks (514 hours).
Exhibit Dates: *Version 1:* 4/88–7/01. *Version 2:* 8/01–Present.
Learning Outcomes: *Version 1:* Upon completion of the course, the student will be able to perform at the level of private pilot on rotorcraft helicopters. *Version 2:* Upon completion of the course, the student will be able to perform at the level of commercial pilot on rotorcraft helicopters.
Instruction: *Version 1:* Lectures and practical exercises cover advanced rotary-wing flight maneuvers, normal and emergency procedures, instrument flight planning, advanced instruments, and associated safety considerations. *Version 2:* Lectures and practical exercises cover advanced rotary-wing flight maneuvers, normal and emergency procedures, instrument flight planning, advanced instruments, and associated safety considerations.
Credit Recommendation: *Version 1:* In the lower-division baccalaureate/associate degree category, 3 semester hours in aircraft navigation (instrument), 3 in cockpit procedures training (helicopter), 3 in instrument rating (helicopter), and 6 in commercial pilot flight training (helicopter) (4/90). *Version 2:* In the lower-division baccalaureate/associate degree category, 3 semester hours in aircraft navigation (instrument) and 3 in instrument rating (helicopter). In the upper-division baccalaureate category, 3 semester hours in cockpit procedures training (helicopter) and 6 in commercial pilot advanced flight training (4/03).
Related Occupations: 153B.

AR-1606-0184

Initial Entry Rotary-Wing Aviator (OH-58)

Course Number: 2C-SI 1D/2C-152B (OH-58).
Location: Aviation Center, Fort Rucker, AL.
Length: 16–17 weeks (592 hours).
Exhibit Dates: 4/88–9/96.
Learning Outcomes: After 9/96 see AR-1606-0217. Upon completion of the course, the student will be able to perform at the level of private pilot on rotorcraft helicopters.
Instruction: Lectures and practical exercises cover advanced rotary-wing flight maneuvers, normal and emergency procedures, instrument flight planning, advanced instruments, and associated safety considerations.
Credit Recommendation: In the lower-division baccalaureate/associate degree category, 3 semester hours in aircraft navigation (instrument), 3 in cockpit procedures training (helicopter), 3 in instrument rating (helicopter), and 6 in commercial pilot flight training (helicopter) (4/90).
Related Occupations: 152B.

AR-1606-0185

Initial Entry Rotary-Wing Aviator (AH-1)

Course Number: 2C-SI 1M/2C-152G (AH-1).
Location: Aviation Center, Fort Rucker, AL.
Length: 20–21 weeks (772 hours).
Exhibit Dates: 4/88–7/03.
Learning Outcomes: Upon completion of the course, the student will be able to perform at the level of private pilot on rotorcraft helicopters.
Instruction: Lectures and practical exercises cover advanced rotary-wing flight maneuvers, normal and emergency procedures, instrument flight planning, advanced instruments, and associated safety considerations.
Credit Recommendation: In the lower-division baccalaureate/associate degree category, 3 semester hours in aircraft navigation (instrument), 3 in cockpit procedures training (helicopter), 3 in instrument rating (helicopter), and 6 in commercial pilot flight training (helicopter) (4/90).
Related Occupations: 152G.

AR-1606-0186

Aeroscout Observer

Course Number: 600-93B10.
Location: Aviation Center, Fort Rucker, AL.
Length: 14–15 weeks (556 hours).
Exhibit Dates: 10/86–7/03.
Learning Outcomes: Upon completion of the course, the student will be able to perform the duties of aeroscout observer in the OH-58A/C helicopter.
Instruction: Course includes lectures and practical sessions in map reading; nontactical navigation; combat operations; target identification; aerial adjustment of artillery; nuclear, biological, and chemical threats; night vision devices; fundamentals of aviation medicine; communication procedures; extensive aeroscout observer flight training including very low-level and high-altitude operations and emergency aircraft piloting instruction; aircraft systems; and survival, evasion, resistance, and escape practices.
Credit Recommendation: In the lower-division baccalaureate/associate degree category, 2 semester hours in aerial navigation and 1 in helicopter flight instruction (4/90).
Related Occupations: 93B.

AR-1606-0187

Aeroscout Observer Basic Noncommissioned Officer (NCO)

Course Number: 600-93B30.
Location: Aviation Center, Fort Rucker, AL.
Length: 5 weeks (209 hours).
Exhibit Dates: 10/90–7/03.
Learning Outcomes: Upon completion of the course, the student will be able to employ tactical operations consistent with aeroscout procedures.
Instruction: Lectures prepare the noncommissioned officer for tactical operations, field exercises, aviation-related tasks, aeroscout procedures, and operations procedures. This course contains a common core of leadership subjects.
Credit Recommendation: Credit is recommended for the common core only. See AR-1406-0090 (4/90).
Related Occupations: 93B.

AR-1606-0188

OH-58A/C Field Artillery Aerial Observer

Course Number: 2E-F108.
Location: Aviation Center, Fort Rucker, AL.
Length: 7 weeks (274 hours).
Exhibit Dates: 7/86–7/03.
Learning Outcomes: Upon completion of the course, the student will be able to serve as a field artillery officer in OH-58A/C helicopter systems operation, target identification and location, and aerial navigation.
Instruction: Topics include aerial navigation, radio procedures, practices and safety considerations in terrain-following flights, combat and target identification, night operations, aviation medicine, and practical emergency pilot training.
Credit Recommendation: In the lower-division baccalaureate/associate degree category, 1 semester hour in helicopter flight instruction (4/90).

AR-1606-0189

CH-47D Flight Engineer Instructor

Course Number: 600-ASIN1, Phase 2; 600-ASIN1.
Location: Aviation Center, Fort Rucker, AL.
Length: 6 weeks (227–253 hours).
Exhibit Dates: 3/86–Present.
Learning Outcomes: Upon completion of the course, the student will be able to serve as flight engineer instructor for the CH-47D helicopter.
Instruction: Lectures, practical exercises, and in-flight instruction cover preflight, inflight, and postflight tasks. In-depth coverage of aircraft systems, tactical flight training tasks, physiology of flight, and aviation safety are included. Student receives training in methods of instruction for both academics and flight.
Credit Recommendation: In the lower-division baccalaureate/associate degree category, 2 semester hours in methods of instruction, 2 in advanced aircraft systems, and 2 in inflight procedure training (1/03).

AR-1606-0190

Rotary Wing Instrument Flight Examiner

Course Number: *Version 1:* 2C-SI1V/2C-SQIF. *Version 2:* 2C-SIG1/2C-SQIF, Phase 1; 2C-SIG1/2C-SQIF.
Location: *Version 1:* Aviation Center, Fort Rucker, AL. *Version 2:* Aviation School, Fort Rucker, AL.

Length: *Version 1:* 7–8 weeks (303 hours). *Version 2:* 7–8 weeks (303 hours).

Exhibit Dates: *Version 1:* 10/88–9/98. *Version 2:* 10/98–Present.

Learning Outcomes: *Version 1:* Upon completion of the course, the student will be able to train selected flight instructors as instrument flight examiners. *Version 2:* Upon completion of the course, the student will be able to apply basic and advanced instruments, emergency tasks, regulations, flight plans, air traffic control, instrument approach criteria, cockpit communications, and fundamentals of instruction skills to the areas of instrument instructor and flight examiner (instrument).

Instruction: *Version 1:* Lectures and practical exercises cover helicopter basic and advanced instrument flight and emergency tasks and instructor/examiner techniques. Course includes regulations, flight planning, air traffic control, approach criteria, cockpit communications, and fundamentals of instruction. *Version 2:* This course consists of instrument flight training, instrument flight subjects, and critiques. This course is designed to qualify and standardize instrument pilots and prepare them to evaluate instrument-rated aviation in simulated or actual instrument conditions.

Credit Recommendation: *Version 1:* In the upper-division baccalaureate category, 3 semester hours in airline transport pilot (helicopter) and 3 in advanced instruction techniques (helicopter) (4/90). *Version 2:* In the graduate degree category, 3 semester hours in transport category aircraft operations (helicopter) and 3 in advanced instruction techniques (5/01).

AR-1606-0191

Imagery Analyst Basic Noncommissioned Officer (NCO)

Course Number: *Version 1:* 242-96D30. *Version 2:* 242-96D30.

Location: Intelligence School, Fort Huachuca, AZ.

Length: *Version 1:* 6–7 weeks (236 hours). *Version 2:* 7–8 weeks (292 hours).

Exhibit Dates: *Version 1:* 10/87–8/92. *Version 2:* 12/94–5/01.

Learning Outcomes: *Version 1:* Upon completion of the course, the student will be able to identify, plot, measure, and analyze remotely-sensed data and images, including conventional photographs and infrared, radar, and electro-optical sensors; input data into computer-based files; use the computer to record, store, sort, transmit, and display aerial data; maintain a data base; and generate displays of computer graphics. *Version 2:* After 5/01 see AR-1606-0231. Upon completion of the course, the student will be able to identify, plot, measure, and analyze remotely-sensed data and images, including conventional photographs and infrared, radar, and electro-optical sensors; input data into computer-based files; use the computer to record, store, sort, transmit, and display aerial data; maintain a data base; and generate displays of computer graphics.

Instruction: *Version 1:* Instruction is achieved through lectures, laboratories, and practical exercises. Topics include basic map reading, photo and image interpretation, computer-assisted cartography, and the creation of geographic information systems. This course contains a common core of leadership subjects. *Version 2:* Instruction is achieved through lectures, laboratories, and practical exercises. Topics include basic map reading, photo and image interpretation, computer-assisted cartography, and the creation of geographic information systems. This course contains a common core of leadership subjects.

Credit Recommendation: *Version 1:* In the lower-division baccalaureate/associate degree category, 3 semester hours in displays of geographic information. See AR-1406-0090 for common core credit. In the upper-division baccalaureate category, 3 semester hours in geographic information systems (6/90). *Version 2:* In the lower-division baccalaureate/associate degree category, 2 semester hours in display of geographic information and 2 in geographic information systems. See AR-1406-0090 for common core credit (5/01).

Related Occupations: 96D.

AR-1606-0192

Intelligence Analyst Basic Noncommissioned Officer (NCO)

Course Number: *Version 1:* 243-96B30. *Version 2:* 243-96B30.

Location: Intelligence School, Fort Huachuca, AZ.

Length: *Version 1:* 6–7 weeks (180–270 hours). *Version 2:* 7–8 weeks (265–266 hours).

Exhibit Dates: *Version 1:* 10/86–9/95. *Version 2:* 10/95–5/00.

Learning Outcomes: *Version 1:* Upon completion of the course, the student will be able to identify types of intelligence reports and their uses; use appropriate training techniques; and supervise selected field maintenance tasks. *Version 2:* Upon completion of the course, the student will possess the technical, analytical and managerial skills required for successful performance as an intelligence analyst.

Instruction: *Version 1:* Lectures, practical exercises and simulations cover the review of intelligence reports. *Version 2:* Lectures and practical exercises cover intelligence preparation, analysis of requirements, message parsing, and collection management. Course contains a common core of leadership subjects.

Credit Recommendation: *Version 1:* Credit is not recommended because of the military-specific nature of the course (9/92). *Version 2:* Credit is recommended for the common core only. See AR-1406-0090 (5/01).

Related Occupations: 96B.

AR-1606-0193

Imagery Analyst

Course Number: *Version 1:* 242-96D10. *Version 2:* 242-96D10.

Location: Intelligence School, Fort Huachuca, AZ.

Length: *Version 1:* 16–18 weeks (583–648 hours). *Version 2:* 14–16 weeks (607–664 hours).

Exhibit Dates: *Version 1:* 1/87–7/92. *Version 2:* 10/96–Present.

Learning Outcomes: *Version 1:* Upon completion of the course, the student will be able to read and interpret large-scale maps, aerial photographs, and radar and infrared images; write reports; and identify transportation and communication routes. *Version 2:* Upon completion of the course, the student will be able to read and interpret large-scale maps, aerial photographs, and radar and infrared images; write reports; and identify transportation and communication routes.

Instruction: *Version 1:* Instruction is achieved through classroom and laboratory exercises. *Version 2:* Instruction is achieved through classroom and laboratory exercises.

Credit Recommendation: *Version 1:* In the lower-division baccalaureate/associate degree category, 3 semester hours in displays of geographic information. In the upper-division baccalaureate category, 3 semester hours in aerial photography interpretation and 3 in remote sensing (6/90). *Version 2:* In the lower-division baccalaureate/associate degree category, 3 semester hours in displays of geographic information, 3 aerial photography interpretation, and 3 in remote sensing (6/97).

Related Occupations: 96D.

AR-1606-0194

Military Intelligence Officer Advanced (Imagery Exploitation)

Course Number: 3-30-C22-35C.

Location: Intelligence School, Fort Huachuca, AZ.

Length: 6–10 weeks (240–360 hours).

Exhibit Dates: 6/86–9/99.

Learning Outcomes: After 9/99 see AR-1606-0236. Upon completion of the course, the student will be able to apply the techniques and skills of aerial imagery analysis, including measuring features in aerial photographs, using the aerial slide rule and/or calculator, transferring features between maps and aerial photographs, identifying communication and transportation arteries, using image keys, and analyzing radar and multispectral images.

Instruction: Instruction is achieved through small classes with an emphasis on laboratories and practical exercises.

Credit Recommendation: In the upper-division baccalaureate category, 3 semester hours in aerial photograph interpretation and 3 in remote sensing (5/01).

Related Occupations: 35C.

AR-1606-0195

Senior Warrant Officer Training Reserve Component
(Military Intelligence Warrant Officer Reserve Component)

Course Number: 3-30-C32-RC.

Location: Intelligence School, Fort Huachuca, AZ.

Length: 1–2 weeks (88–96 hours).

Exhibit Dates: 6/89–9/99.

Learning Outcomes: Upon completion of the course, the student will be able to identify specific computer hardware, identify Warsaw pact nations, identify other world threats, and apply collection management techniques.

Instruction: Lectures cover world threats and computer awareness.

Credit Recommendation: Credit is not recommended because of the military-specific nature of the course (5/01).

AR-1606-0196

Military Intelligence Officer Advanced Common Core

Course Number: C-30-C22-35.

Location: Intelligence School, Fort Huachuca, AZ.

Length: 9 weeks (411 hours).

Exhibit Dates: 4/89–9/91.

Learning Outcomes: Upon completion of the course, the student will be able to perform military intelligence operations including advanced threat analysis and staff and management functions, make decisions on the basis of intelligence data, and write effectively.

Instruction: Instruction includes leadership and management, effective writing, military history, and geopolitical studies of conflict areas. Students participate in computer simulations of a warfare system. Players make decisions as they would in the field, involving attack and counterattack. Data analysis and resulting decisions are the major skills developed in this intensive 60-hour exercise.

Credit Recommendation: In the lower-division baccalaureate/associate degree category, 3 semester hours in technical writing. In the upper-division baccalaureate category, 3 semester hours in simulation exercise in decision making, 1 in principles of personnel and organizational management, and 2 in international studies in areas of conflict (6/90).

AR-1606-0197

Military Intelligence Pre-Command

Course Number: 2G-F41.

Location: Intelligence School, Fort Huachuca, AZ.

Length: 2 weeks (79 hours).

Exhibit Dates: 4/89–4/04.

Learning Outcomes: Upon completion of the course, the student will be familiar with general military subjects, recognize the role of various officers and command titles, will be familiar with military intelligence.

Instruction: Lectures, discussions and forums cover the subject matter.

Credit Recommendation: Credit is not recommended because of the limited, specialized nature of the course (6/90).

AR-1606-0198

Interrogator Basic Noncommissioned Officer (NCO)

Course Number: 241-97E30.

Location: Intelligence School, Fort Huachuca, AZ.

Length: 5–7 weeks (197–259 hours).

Exhibit Dates: 1/89–9/99.

Learning Outcomes: After 9/99 see AR-1606-0235. Upon completion of the course, the student will use military vocabulary and job-specific language skills and prepare battlefield intelligence.

Instruction: Lectures and practical exercises cover the above subject matter. This course contains a common core of leadership subjects.

Credit Recommendation: Credit is recommended for the common core only. See AR-1406-0090 (5/01).

Related Occupations: 97E.

AR-1606-0199

1. Human Intelligence Collection Technician Warrant Officer Basic Phase 2
   (Interrogation Technical Warrant Officer Technical/Tactical Certification)
   (Warrant Officer Technical/Tactical Certification Phase 2 MOS 351E Interrogation Technician)
2. Human Intelligence Collection Technician Warrant Officer Basic, Phase 2

Course Number: *Version 1:* 3A-351E. *Version 2:* 3A-351E.

Location: Intelligence School, Fort Huachuca, AZ.

Length: *Version 1:* 2–3 weeks (89–111 hours). *Version 2:* 2–3 weeks (101 hours).

Exhibit Dates: *Version 1:* 11/88–12/99. *Version 2:* 1/00–Present.

Learning Outcomes: *Version 1:* Upon completion of the course, the student will be able to perform all phases of interrogation and write interrogation reports. *Version 2:* Upon completion of the course, the student will be able to perform all phases of interrogation and write interrogation reports.

Instruction: *Version 1:* Lectures and simulation exercises comprise the methods of instruction. *Version 2:* Lectures and simulation exercises comprise the methods of instruction.

Credit Recommendation: *Version 1:* Credit is not recommended because of the military-specific nature of the course (6/97). *Version 2:* Credit is not recommended because of the military-specific nature of the course (8/01).

Related Occupations: 351E.

AR-1606-0200

1. Imagery Intelligence Technician Warrant Officer Technical/ Tactical Certification
   (Warrant Officer Technical/Tactical Certification Phase 2 350D Imagery Intelligence Technician)
2. Imagery Intelligence Technician Warrant Officer Technical Tactical Certification Common Core

Course Number: *Version 1:* 3A-350D. *Version 2:* 3A-350D, Phase 2.

Location: Intelligence School, Fort Huachuca, AZ.

Length: *Version 1:* 2–3 weeks (93 hours). *Version 2:* 2–3 weeks (86 hours).

Exhibit Dates: *Version 1:* 11/88–9/99. *Version 2:* 10/99–Present.

Learning Outcomes: *Version 1:* Upon completion of the course, the student will be updated on technology and new equipment, describe imagery intelligence materials, and perform selected types of imagery interpretation. *Version 2:* Upon completion of the course, the student will be updated on technology and new equipment, describe imagery intelligence materials, and perform selected types of imagery.

Instruction: *Version 1:* Lectures and classroom exercises comprise the methods of instruction. *Version 2:* Lectures and classroom exercises comprise the methods of instruction.

Credit Recommendation: *Version 1:* Credit is not recommended because of the military-specific nature of the course (1/97). *Version 2:* Credit is not recommended because of the military-specific nature of the course (8/01).

Related Occupations: 350D.

AR-1606-0201

Counterintelligence Technician Warrant Officer Basic, Phase 2
(Warrant Officer Technical/Tactical Certification Phase 2 MOS 351B Counterintelligence Technician)

Course Number: 3C-351B.

Location: Intelligence School, Fort Huachuca, AZ.

Length: 2–3 weeks (86–97 hours).

Exhibit Dates: 11/88–Present.

Learning Outcomes: Upon completion of the course, the student will be able to perform basic field office management and recognize basic counterintelligence functional areas.

Instruction: Lectures and simulation exercises are the methods of instruction.

Credit Recommendation: Credit is not recommended because of the limited, specialized nature of the course (5/01).

Related Occupations: 351B.

AR-1606-0202

Military Intelligence Senior Warrant Officer

Course Number: 3-30-C32.

Location: Intelligence School, Fort Huachuca, AZ.

Length: 4–5 weeks (142–175 hours).

Exhibit Dates: 3/89–9/99.

Learning Outcomes: After 9/99 see AR-1606-0230. Upon completion of the course, the student will be aware of technological changes and advanced concepts within the intelligence disciplines; be able to describe world threats; and apply data processing principles.

Instruction: Lecture is the mode of instruction.

Credit Recommendation: Credit is not recommended because of the military-specific nature of the course (5/01).

AR-1606-0203

Common Ground Station Operator Basic Noncommissioned Officer (NCO)
(Imagery Ground Station Supervisor)
(Aerial Intelligence Specialist Basic Noncommissioned Officer (NCO))

Course Number: 233-96H30, Phase 2; 233-96H30.

Location: Intelligence School, Fort Huachuca, AZ.

Length: 3–5 weeks (116–200 hours).

Exhibit Dates: 10/88–Present.

Learning Outcomes: Upon completion of the course, the student will supervise the planning, employment, and training of aerial surveillance operations including visual/photo reconnaissance missions; survival, evasion, resistance, and escape equipment use; and flight-line operations.

Instruction: Instruction includes classroom and practical exercises in visual/photo reconnaissance, technical operations, and supervision. This course contains a common core of leadership subjects.

Credit Recommendation: Credit is recommended for the common core only. See AR-1406-0090 (8/02).

Related Occupations: 96H.

AR-1606-0204

Quickfix 2 System Qualification

Course Number: 2C-15C/2B-ASIB5.

Location: Intelligence School, Fort Huachuca, AZ.

Length: 4–9 weeks (164–173 hours).

Exhibit Dates: 4/90–Present.

Learning Outcomes: Upon completion of the course, the student will perform flight mission profiles, identify the major characteristics of low-intensity conflict, define terrorism, describe aerial exploitation, and define radar terms and operational concepts.

Instruction: Lectures and practical exercises cover the above subjects.

Credit Recommendation: Credit is not recommended because of the military-specific nature of the course (8/98).

AR-1606-0205

Quickfix 2 Crew Certification

Course Number: 231-F35; 233-F15.

Location: Intelligence School, Fort Huachuca, AZ.

Length: 3–4 weeks (111–130 hours).

Exhibit Dates: 4/90–Present.

Learning Outcomes: Upon completion of the course, the student will be able to operate the Quickfix 2 system during airborne exercises and perform preflight and postflight procedures on the system.

Instruction: Lectures, demonstrations, and performance exercises covering inflight procedures are the methods of instruction.

Credit Recommendation: Credit is not recommended because of the military-specific nature of the course (8/98).

AR-1606-0207

1. Ground Surveillance System Supervisor
 (Ground Surveillance System Operator Basic Noncommissioned Officer (NCO))
2. Ground Surveillance System Supervisor Basic Noncommissioned Officer (BNCOC)

Course Number: ***Version 1:*** 243-96R30; 231-96R30. *Version 2:* 243-96R30, Phase 2.

Location: Intelligence School, Fort Huachuca, AZ.

Length: ***Version 1:*** 5–6 weeks (182–234 hours). *Version 2:* 3 weeks (115 hours).

Exhibit Dates: ***Version 1:*** 10/89–9/00. *Version 2:* 10/00–Present.

Learning Outcomes: ***Version 1:*** Upon completion of the course, the student will be able to select a site, perform ground surveillance mission planning, and program the sensor signal simulator. *Version 2:* Upon completion of the course, the student will be able to describe ground surveillance systems and their capabilities, program and operate a sensor signal simulator, select a site, and perform ground surveillance mission planning.

Instruction: ***Version 1:*** Lectures and practical exercises cover the use of ground surveillance systems and the supervision of system operators. This course contains a common core of leadership subjects. *Version 2:* Lectures and practical exercises cover the use of ground surveillance systems and the supervision of system operators. This course contains a common core of leadership subjects.

Credit Recommendation: ***Version 1:*** Credit is recommended for the common core only. See AR-1406-0090 (6/97). *Version 2:* Credit is recommended for the common core only. See AR-1406-0090 (12/02).

Related Occupations: 96R.

AR-1606-0208

Aerial Intelligence Specialist

Course Number: 233-96H10.

Location: Intelligence School, Fort Huachuca, AZ.

Length: 8–9 weeks (298–313 hours).

Exhibit Dates: 3/87–5/97.

Learning Outcomes: Upon completion of the course, the student will read maps and plot information on them, operate and perform operator maintenance on airborne photographic and radar surveillance systems; and troubleshoot and make minor repairs to aerial surveillance systems.

Instruction: Small classes are held in hangars, classrooms, flight line and inflight.

Credit Recommendation: In the lower-division baccalaureate/associate degree category, 3 semester hours in introduction to map reading (10/91).

Related Occupations: 96H.

AR-1606-0209

Diver Pre-Qualification
 (Diver Transition)

Course Number: 433-F2.

Location: Transportation School, Fort Eustis, VA.

Length: 1–2 weeks (60 hours).

Exhibit Dates: 10/87–4/04.

Learning Outcomes: Upon completion of the course, the student will be able to dive under controlled conditions not exceeding 30 feet.

Instruction: Lectures and practical experience are conducted within a controlled environment. Course includes signals, equipment, and safety procedures.

Credit Recommendation: In the lower-division baccalaureate/associate degree category, 1 semester hour in basic diver (PADI) certification (6/91).

AR-1606-0210

Special Forces Assistant Operations and Intelligence Sergeant

Course Number: ***Version 1:*** 011-18F40. *Version 2:* 011-18F40.

Location: John F. Kennedy Special Warfare Center and School, Fort Bragg, NC.

Length: ***Version 1:*** 13 weeks (585 hours). *Version 2:* 16 weeks (699 hours).

Exhibit Dates: ***Version 1:*** 10/92–9/93. *Version 2:* 10/93–12/94.

Learning Outcomes: ***Version 1:*** Upon completion of the course, the student will perform operations and intelligence missions, advise and train in military operations, coordinate air/ground and nautical missions, develop and exploit intelligence information, and demonstrate principles of security. *Version 2:* After 12/94 see AR-2201-0459. Upon completion of the course, the student will perform operations and intelligence missions, advise and train in military operations, coordinate air/ground and nautical missions, develop and exploit intelligence information, and demonstrate the principles of security.

Instruction: ***Version 1:*** Lectures, discussions, and command post exercises cover psychological operations in air/land battle, support of unconventional warfare, cross-cultural communication, impact of terrorism, area studies, and US foreign policy. *Version 2:* Lectures, discussions, and command post exercises cover psychological operations in air/land battle, support of unconventional warfare, cross-cultural communication, impact of terrorism, area studies, and US foreign policy.

Credit Recommendation: ***Version 1:*** In the lower-division baccalaureate/associate degree category, 1 semester hour in photography, 1 in communications, and 1 in map interpretation. In the upper-division baccalaureate category, 1 semester hour in cross-cultural communication, 2 in psychology, 3 in social science, 2 in political science, and 2 in public administration (12/92). *Version 2:* In the lower-division baccalaureate/associate degree category, 1 semester hour in photography, 1 in communications, and 1 in map interpretation. In the upper-division baccalaureate category, 2 semester hours in cross-cultural communication, 2 in psychology, 2 in social science, 2 in political science, and 2 in public administration (12/92).

Related Occupations: 18F.

AR-1606-0211

Special Forces Assistant Operations and Intelligence Sergeant Reserve Component Phase 2

Course Number: 011-18F40-RC.

Location: John F. Kennedy Special Warfare Center and School, Fort Bragg, NC.

Length: 2–3 weeks (121 hours).

Exhibit Dates: 10/93–9/98.

Learning Outcomes: Upon completion of the Phase 2 of this reserve course, the student will be able to support Special Forces missions, coordinate air/ground/nautical missions, and analyze maps and support air/land battle.

Instruction: Lectures, seminars, simulations, and discussions in Phase 2 of this reserve course cover Special Forces operations, map interpretation, and the role of Special Forces in air/land battle.

Credit Recommendation: In the upper-division baccalaureate category, 1 semester hour in map interpretation, 1 in political science, 1 in psychology, and 1 in social science. Credit recommended is for Phase 2 (12/92).

Related Occupations: 18F.

AR-1606-0212

Special Forces Assistant Operations and Intelligence Sergeant Reserve Component Phase 4

Course Number: 011-18F40-RC.

Location: John F. Kennedy Special Warfare Center and School, Fort Bragg, NC.

Length: 2–3 weeks (158 hours).

Exhibit Dates: 10/93–9/98.

Learning Outcomes: Upon completion of Phase 4 of this reserve course, the student will be able to perform intelligence functions in support of Special Forces mission, analyze an opponent's military organization, and conduct clandestine operations.

Instruction: Lectures, seminars, and special conferences in Phase 4 of this Reserve course cover intelligence operations. Written reports and oral presentation are other methods of instruction and evaluation.

Credit Recommendation: In the lower-division baccalaureate/associate degree category, 1 semester hour in photographic interpretation and 1 in communication. In the upper-division baccalaureate category, 2 semester hours in social science (elective). Credit recommended is for completion of Phase 4 (12/92).

Related Occupations: 18F.

AR-1606-0213

1. Guardrail Common Sensor/Pilot Qualification
2. Guardrail/Common Sensor Pilot Qualification

Course Number: *Version 1:* 2C-15C/2B-ASIF4 (PQ). *Version 2:* 2C-15C/2B-ASIF4 (PQ).

Location: Intelligence School, Fort Huachuca, AZ.

Length: *Version 1:* 9–10 weeks (352–353 hours). *Version 2:* 9–10 weeks (348–352 hours).

Exhibit Dates: *Version 1:* 8/96–12/99. *Version 2:* 1/00–Present.

Learning Outcomes: *Version 1:* Upon completion of the course, the student will be qualified to operate and pilot special electronic mission aircraft (RC-12K/N) for military intelligence missions. *Version 2:* Upon completion of the course, the student will be qualified to operate and pilot special electronic mission aircraft (RC-12K/NN) for military intelligence missions.

Instruction: *Version 1:* Instruction covers Guardrail common sensor training, overview of onboard equipment, and flight operational procedures related to the aircraft. Instruction is provided in the classroom and inflight. *Version 2:* Instruction covers Guardrail common sensor training, overview of onboard equipment, and flight operational procedures related to the aircraft. Instruction is provided in the classroom and inflight.

Credit Recommendation: *Version 1:* Credit is not recommended because of the military-specific nature of the course (6/97). *Version 2:* Credit is not recommended because of the military-specific nature of the course (8/02).

AR-1606-0214

1. Telemetry Collection Operations
2. Telemetry Collection Operations
3. Basic Foreign Instrumentation Signals Intelligence (FISINT)

Course Number: *Version 1:* 233-ASIJ1. *Version 2:* 233-ASIJ1. *Version 3:* 233-ASIJ1.

Location: *Version 1:* Intelligence School, Fort Huachuca, AZ. *Version 2:* Intelligence Center and School, Fort Huachuca, AZ. *Version 3:* Intelligence Center and School, Fort Huachuca, AZ.

Length: *Version 1:* 11–12 weeks (410 hours). *Version 2:* 8 weeks (295–296 hours). *Version 3:* 10 weeks (384 hours).

Exhibit Dates: *Version 1:* 1/91–10/96. *Version 2:* 11/96–9/02. *Version 3:* 10/02–Present.

Learning Outcomes: *Version 1:* Upon completion of the course, the student will be able to analyze foreign instrumentation signals using special analysis equipment, identify missile and earth satellite vehicle designs, and use various reporting techniques. *Version 2:* Upon completion of the course, the student will be able to analyze foreign instrumentation signals by using analysis and telemetry equipment and be able to report the results. *Version 3:* Upon completion of the course, the student will be able to analyze foreign instrumentation signals using analog and digital techniques and equipment.

Instruction: *Version 1:* Lectures cover electronic instruments and audio signals; familiarization with target vehicles, ranges, and facilities; relationships of orbits to the function of a satellite; composition of foreign instrumentation signals; analysis techniques; real-time signal reports; and postmission analysis to determine the identity of an unknown target. *Version 2:* Laboratory, audio visual materials, practical exercises, lecture and learner presentations. *Version 3:* Practical and classroom exercises, discussion, and lecture. Topics include analog and digital analysis equipment usage identification of telemetry systems, signal analysis, software processing of telemetry signals, and reporting of analytical results.

Credit Recommendation: *Version 1:* Credit is not recommended because of the limited, specialized nature of the course (6/97). *Version 2:* In the vocational certificate category, 1 semester hour in electronic equipment operation. In the lower-division baccalaureate/associate degree category, 1 semester hour in electronic equipment operation (8/01). *Version 3:* In the vocational certificate category, 2 semester hours in electronic equipment operation (12/02).

AR-1606-0215

Military Intelligence Officer Transition Reserve
(Reserve Component Military Intelligence Officer Transition Phase 2)

Course Number: 3A-35D (T) RC.

Location: Intelligence School, Fort Huachuca, AZ.

Length: 85 hours.

Exhibit Dates: 6/95–Present.

Learning Outcomes: Upon completion of the course, the student will be provided with the necessary basic military intelligence basic skills to advance to the reserve component military intelligence officers advanced course.

Instruction: Lectures are related to threat evaluation, intelligence estimation, and ground surveillance. Phase 1 was offered by correspondence and was not included in this evaluation.

Credit Recommendation: Credit is not recommended because of the military-specific nature of the course (6/97).

Related Occupations: 35D.

AR-1606-0216

Military Intelligence Officer Advanced Reserve Component, Phase 2

Course Number: 3-30-C23.

Location: Intelligence School, Fort Huachuca, AZ.

Length: 86 hours.

Exhibit Dates: 8/93–9/99.

Learning Outcomes: After 9/99 see AR-1606-0238. Upon completion of the course, the student will be able to lead and manage intelligence and electronic warfare systems.

Instruction: Course includes lectures in threat evaluation, electronic warfare, counterintelligence support, terrorists, and terrorist groups. Phase 1 is by correspondence and not included in this evaluation.

Credit Recommendation: Credit is not recommended because of the military-specific nature of the course (5/01).

AR-1606-0217

Initial Entry Rotary Wing OH-58C Track

Course Number: 2C-15A/SIA2/152B (TH-67) (C).

Location: Aviation School, Fort Rucker, AL.

Length: 12 weeks (512 hours).

Exhibit Dates: 10/96–Present.

Learning Outcomes: Before 10/96 see AR-1606-0184. Upon completion of the course, the student will be qualified as an OH-58 helicopter pilot.

Instruction: This course uses lectures, conferences, and practical exercises as methods of instruction. Topics include flight maneuvers, emergency and safety procedures, flight planning, instrument flight tasks, and basic helicopter systems.

Credit Recommendation: In the lower-division baccalaureate/associate degree category, 3 semester hours in aviation theory of flight and systems, 1 in helicopter systems, and 3 in helicopter flight training (11/98).

AR-1606-0218

C-12 Aviator Qualification (Flight) Phase 2

Course Number: 2B-SIE5/2B-155E.

Location: Aviation School, Fort Rucker, AL.

Length: 2–3 weeks (69 hours).

Exhibit Dates: 6/92–9/99.

Learning Outcomes: Upon completion of the course, the student will be qualified as pilot in the Beech King Air twin turbine airplane.

Instruction: This course uses inflight instruction in the Beech King Air airplane. The topics include flight maneuver, instrument flight procedures, and emergency procedures.

Credit Recommendation: In the upper-division baccalaureate category, 2 semester hours in advanced multi-engine turbine aircraft flight (11/98).

AR-1606-0219

C-12 Aviator Qualification (Simulator) Phase 1

Course Number: 2B-SIE5/2B-155E (CT).

Location: Aviation School, Fort Rucker, AL.

Length: 2–3 weeks (88 hours).

Exhibit Dates: 12/97–9/99.

Learning Outcomes: Upon completion of the course, the student will be prepared to enter the flight training phase of Beech King Air twin turbine for pilot qualification.

Instruction: Classroom instruction and flight simulation instruction are used to teach flight procedures, instrument flight, emergency procedures, systems, performance charts, and weight and balance.

Credit Recommendation: In the upper-division baccalaureate category, 2 semester hours in advanced flight theory and aircraft systems (11/98).

AR-1606-0220

AH-1 Aviator Qualification

Course Number: 2C-SID2/152G.

Location: Aviation School, Fort Rucker, AL.

Length: 6 weeks (248 hours).

Exhibit Dates: 5/95–Present.

Learning Outcomes: Before 5/95 see AR-1606-0058. Upon completion of the course, the student will be able to fly the AH-1 Cobra helicopter and operate the weapon systems.

Instruction: Lectures and practical exercises cover transition flight training on AH-1S helicopters, including preflight inspections, ground procedures, night operations, instrument flying, regulations, flight planning, maintenance procedures, and principles of aerodynamics.

Credit Recommendation: In the upper-division baccalaureate category, 3 semester hours in advanced aerodynamics and theory of flight, 3 in advanced helicopter procedures and systems, and 3 in helicopter flight training (11/98).

AR-1606-0221

Fixed Wing Multi-Engine Qualification

Course Number: 2C-15A/67J/SIE4/2B-155A/D.

Location: Aviation School, Fort Rucker, AL.

Length: 14 weeks (480 hours).

Exhibit Dates: 5/98–Present.

Learning Outcomes: Upon completion of the course, the student will qualify as a multi-engine airplane pilot.

Instruction: The course uses classroom, practical exercise, and inflight instruction. The topics include aerodynamics, multi-engine airplane performance, weight and balance, federal airspace, instrument flight procedures, systems, and environmental flight factors.

Credit Recommendation: In the lower-division baccalaureate/associate degree category, 3 semester hours in airplane multi-engine theory of flight and systems, 2 in aircraft systems, and 3 in airplane multi-engine flight training (11/98).

AR-1606-0222

UH-60 Aviator Qualification

Course Number: 2C-SIB2/2C-153D, Phase 2; 2C-SIB2/2C-153D.

Location: Aviation School, Fort Rucker, AL.

Length: 6 weeks (237–243 hours).

Exhibit Dates: 10/95–Present.

Learning Outcomes: Before 10/95 see AR-1606-0133. Upon completion of the course, the helicopter pilot will be qualified as a pilot in the UH-60 twin turbine helicopter.

Instruction: Flight and academic instruction covers the UH-60 helicopter, maneuvers, combat skills, and night vision goggles.

Credit Recommendation: In the upper-division baccalaureate category, 3 semester hours in advanced theory of flight and systems and 3 in advanced helicopter flight (1/03).

AR-1606-0223

UH-1H Undergraduate Pilot Qualification (Air Force)

Course Number: 2C-F76 (OS).

Location: Aviation School, Fort Rucker, AL.

Length: 12 weeks (377–378 hours).

Exhibit Dates: 11/97–Present.

Learning Outcomes: Upon completion of the course, the student will qualify as a pilot for the UH-1 helicopter.

Instruction: The course used conference, computer-based instruction, practical exercises and flight training as methods of instruction. The topics include aerodynamics, navigation, communications, night vision systems, and flight maneuvers.

Credit Recommendation: In the lower-division baccalaureate/associate degree category, 3 semester hours in aerodynamics and theory of flight. In the upper-division baccalaureate category, 3 semester hours in advanced helicopter flight training (11/98).

AR-1606-0224

OH-58D Warrior Aviator Qualification

Course Number: ***Version 1:*** 2C-SIA4/152D. *Version 2:* 2C-SIA4/152D, Phase 3.

Location: ***Version 1:*** Aviation School, Fort Rucker, AL. *Version 2:* Aviation Center, Fort Rucker, AL.

Length: ***Version 1:*** 11 weeks (471 hours). *Version 2:* 11 weeks (458 hours).

Exhibit Dates: ***Version 1:*** 12/96–4/99. *Version 2:* 5/99–Present.

Learning Outcomes: ***Version 1:*** Before 12/96 see AR-1606-0161. Upon completion of the course, the student will qualify as a pilot for the OH-58D helicopter and operate its electronic systems. *Version 2:* Upon completion of the course, the student will qualify as a pilot for the OH-58D helicopter and operate its electronic systems.

Instruction: ***Version 1:*** Classroom and practical instruction cover aircraft systems, flight training, communications, navigation, mast-mounted sight, airborne target handover systems, mission planning, and safety. *Version 2:* Topics include classroom and practical instruction covering aircraft systems, flight training, communications, cockpit/crew resource management, navigation, mast-mounted sight, airborne target handover systems, mission planning, and safety.

Credit Recommendation: ***Version 1:*** In the upper-division baccalaureate category, 3 semester hours in advanced helicopter procedures and systems and 3 in advanced helicopter flight training (11/98). *Version 2:* In the upper-division baccalaureate category, 3 semester hours in advanced aircraft systems operation and management, and 3 in crew/cockpit resource management (4/03).

Related Occupations: 152D.

AR-1606-0225

Aviation Officer Basic, Phase 1

Course Number: 2-1-C20.

Location: Aviation School, Fort Rucker, AL.

Length: 4 weeks (160 hours).

Exhibit Dates: 1/95–3/96.

Learning Outcomes: Before 1/95 see AR-1606-0170. Upon completion of the course, the student will have the knowledge and skills to be an entry-level supervisor.

Instruction: This course consists of lectures and practical exercises in communication skills, leadership, and arms.

Credit Recommendation: In the lower-division baccalaureate/associate degree category, 1 semester hour in principles of supervision and 1 in technical writing (11/98).

AR-1606-0226

Aviation Officer Basic

Course Number: 2-1-C20.

Location: Aviation School, Fort Rucker, AL.

Length: 6 weeks (287 hours).

Exhibit Dates: 4/96–Present.

Learning Outcomes: Upon completion of the course, the student will have the knowledge and skills to be an entry-level supervisor and manager.

Instruction: This course consists of lectures and practical exercises in communication skills, leadership, and arms.

Credit Recommendation: In the lower-division baccalaureate/associate degree category, 2 semester hours in principles of supervision (11/98).

AR-1606-0227

Strategic Intelligence Officer Transition (FA34)

Course Number: 3A-34; 3A-34 (T).

Location: Intelligence School, Fort Huachuca, AZ.

Length: 7–10 weeks (264–360 hours).

Exhibit Dates: 5/99–Present.

Learning Outcomes: Upon completion of the course, the student will be able to analyze intelligence reports, raw data, maps and charts, photos, and write reports and brief superiors in intelligence matters.

Instruction: Lectures, practical exercises, student briefings, and presentations are course methodologies.

Credit Recommendation: In the lower-division baccalaureate/associate degree category, 1 semester hour in intelligence methods and 2 in critical analysis (12/02).

AR-1606-0228

Intelligence Analyst Basic Noncommissioned Officer (NCO)

Course Number: 243-96B30.

Location: Reserve Schools, Multiple locations; Intelligence School, Fort Huachuca, AZ.

Length: 182–183 hours.

Exhibit Dates: 6/00–Present.

Learning Outcomes: Upon completion of the course, the student will be able to identify types of intelligence reports and their uses, use appropriate training techniques, and supervise selected field maintenance tasks.

Instruction: Lectures, practical exercises, and simulations cover the review of intelligence reports.

Credit Recommendation: Credit is not recommended because of the military-specific nature of the course (8/01).

Related Occupations: 96B.

AR-1606-0229

Intelligence Analyst Basic Noncommissioned Officer (NCO)

Course Number: 243-96B30.

Location: Intelligence School, Fort Huachuca, AZ.

Length: 9 weeks (323–324 hours).

Exhibit Dates: 6/00–Present.

Learning Outcomes: Upon completion of the course, the student will possess the technical, analytical, and managerial skills required for successful performance as an intelligence analyst.

Instruction: Lectures and practical exercises cover intelligence preparation, analysis of requirements, message parsing, and collection management. Course contains a common core of leadership subjects.

Credit Recommendation: Credit is recommended for the common core only. See AR-1406-0090 (8/01).

Related Occupations: 96B.

AR-1606-0230

Total Army Training System (TATS) Military Intelligence Warrant Officer Advanced

Course Number: 3-30-C32-RC, Phase 2; 3-30-C32-RC, Phase 1; 3-30-C32.

Location: Intelligence Center and School, Fort Huachuca, AZ.

Length: 4 weeks (172 hours).

Exhibit Dates: 10/99–Present.

Learning Outcomes: Before 10/99 see AR-1606-0202. Upon completion of the course, the student will be able to develop a persuasive presentation that is logical, cohesive, convincing, clearly written, and concise reports.

Instruction: Lectures and practical exercises cover fundamentals of speech, and effective writing.

Credit Recommendation: In the lower-division baccalaureate/associate degree category, 3 semester hours in persuasive speech and 3 in technical writing (12/02).

AR-1606-0231

Total Army Training System (TATS) Imagery Analyst Basic Noncommissioned Officer (NCO), Phase 2

Course Number: 242-96D30 (F), Phase 2.

Location: Intelligence School, Fort Huachuca, AZ.

Length: 8–9 weeks (504 hours).

Exhibit Dates: 6/01–Present.

Learning Outcomes: Before 6/01 see AR-1606-0191. Upon completion of the course, the student will be able to identify, plot, measure, and analyze remotely-sensed data and images, including conventional photographs and infrared, radar, and electro-optical sensors; input data into computer-based files; use the computer to record, store, sort, transmit, and display aerial data; maintain a data base; and generate displays of computer graphics.

Instruction: Instruction is achieved through lectures, laboratories, and practical exercises. Topics include basic map reading, photo and image interpretation, computer-assisted cartography, and the creation of geographic information systems. This course contains a common core of leadership subjects.

Credit Recommendation: In the lower-division baccalaureate/associate degree category, 2 semester hours in display of geographic information and 2 in geographical information. See AR-1406-0090 for common core credit (8/01).

Related Occupations: 96D.

AR-1606-0232

Total Army Training System (TATS) Imagery Analyst Basic Noncommissioned Officer (NCO), Phase 2

Course Number: 242-96D30 (F), Phase 2.

Location: Reserve Schools, Multiple locations; Intelligence School, Fort Huachuca, AZ.

Length: 279 hours.

Exhibit Dates: 6/01–Present.

Learning Outcomes: Upon completion of the course, the student will be able to identify, plot, measure, and analyze remotely-sensed data and images, including conventional photographs and infrared, radar, and electro-optical sensors; input data into computer-based files; use the computer to record, store, sort, transmit, and display aerial data; maintain a data base; and generate displays of computer graphics.

Instruction: Instruction is achieved through lectures, laboratories, and practical exercises. Topics include basic map reading, photo and image interpretation, computer-assisted cartography, and the creation of geographic information systems.

Credit Recommendation: In the lower-division baccalaureate/associate degree category, 2 semester hours in display of geographic information and 2 in geographical information systems (8/01).

AR-1606-0233

Interim Total Army Training System (TATS) Imagery Analyst Reserve Component

Course Number: 301-96D10-RC.

Location: Reserve Schools, Multiple locations; Intelligence School, Fort Huachuca, AZ.

Length: 210 hours.

Exhibit Dates: 10/99–Present.

Learning Outcomes: Upon completion of the course, the student will be able to read and interpret imagery and recommend exploitation of such data.

Instruction: This course utilizes classroom lectures, hands-on practical exercises, and testing.

Credit Recommendation: Credit is not recommended because of the military-specific nature of the course (8/01).

Related Occupations: 96D.

AR-1606-0234

Total Army Training System (TATS-C) Interrogator Basic Noncommissioned Officer (NCO)

Course Number: 241-97E30 (F), Phase 2; 241-97E30.

Location: Intelligence School, Fort Huachuca, AZ.

Length: 6–7 weeks (219–528 hours).

Exhibit Dates: 10/99–Present.

Learning Outcomes: Upon completion of the course, the student will be able to perform the tasks of a common leader, use military vocabulary, perform job specific language skills, and prepare battle field intelligence.

Instruction: Lectures and practical exercises covering the above subject matter. This course contains a common core of leadership subjects.

Credit Recommendation: Credit is recommended for the common core only. See AR-1406-0090 (12/02).

Related Occupations: 97E.

AR-1606-0235

Total Army Training System (TATS-C) Interrogator Basic Noncommissioned Officer (NCO)

Course Number: 241-97E30.

Location: Reserve Schools, Multiple locations; Intelligence School, Fort Huachuca, AZ.

Length: 150 hours.

Exhibit Dates: 10/99–4/02.

Learning Outcomes: After 4/02 see AR-1606-0234. Upon completion of the course, the student will use military vocabulary and job-specific language skills and prepare battlefield intelligence.

Instruction: Lectures and practical exercises cover the above subject matter.

Credit Recommendation: Credit is not recommended because of the military-specific nature of the course (8/01).

AR-1606-0236

Military Intelligence Captains Career Total Army Training System (TATS)

Course Number: 3-30-C22.

Location: Intelligence School, Fort Huachuca, AZ.

Length: 18 weeks (696 hours).

Exhibit Dates: 10/99–Present.

Learning Outcomes: Upon completion of the course, the student will be able to lead and manage intelligence and electronic warfare systems.

Instruction: This course includes lectures in military decision making, the identification of military intelligence systems, intelligence briefings, intelligence plans, and military threat determination, and all source intelligence analysis.

Credit Recommendation: Credit is not recommended because of the military-specific nature of the course (8/01).

AR-1606-0237

Total Army Training System (TATS) Military Intelligence Captains Career Reserve Component, Phase 1

Course Number: 3-30-C23, Phase 1.

Location: Reserve Schools, Multiple locations; Intelligence School, Fort Huachuca, AZ.

Length: 240 hours.

Exhibit Dates: 10/99–Present.

Learning Outcomes: Upon completion of the course, the student will be able to lead and manage intelligence and electronic warfare systems.

Instruction: This course includes lectures in military decision making, the identification of military intelligence systems, intelligence briefings, intelligence plans, military threat determination of all source intelligence analysis.

Credit Recommendation: Credit is not recommended because of military-specific nature of the course (8/01).

AR-1606-0238

Total Army Training System (TATS) Military Intelligence Captains Career Reserve Component, Phase 2

Course Number: 3-30-C23, Phase 2.

Location: Reserve Schools, Multiple locations; Intelligence School, Fort Huachuca, AZ.

Length: 240 hours.

Exhibit Dates: 10/99–Present.

Learning Outcomes: Upon completion of this course, the student will be able to lead and manage intelligence and electronic warfare systems.

Instruction: This course includes lectures in military decision-making, the identification of military intelligence systems, intelligence briefings, intelligence plans, military threat determination of all source intelligence analysis.

Credit Recommendation: Credit is not recommended because of the military-specific nature of the course (8/01).

AR-1606-0239

Counterintelligence/Force Protection Source Operations

Course Number: 3C-F17/244-F9.
Location: Intelligence School, Fort Huachuca, AZ.
Length: 5 weeks (219 hours).
Exhibit Dates: 10/98–Present.
Learning Outcomes: Upon completion of the course, the student will be able to analyze mission requirements, plan team composition, and develop input to the collection plan.
Instruction: Lectures, preparation of reports, practical exercises, and course examinations are course methods.
Credit Recommendation: Credit is not recommended because of the military-specific nature of the course (8/01).

AR-1606-0242

OH-58D (R) Warrior Aviator Qualification

Course Number: 2C-SIA4/152D (OH-58D) (R).
Location: Aviation Center, Fort Rucker, AL.
Length: 12–13 weeks (502 hours).
Exhibit Dates: 12/99–Present.
Learning Outcomes: Upon completion of the course, the student will be able to qualify as a pilot for the OH-58D helicopter.
Instruction: Practical and classroom exercises, laboratory, lecture, and computer-based training. Topics include aircraft electrical controls and displays, power train, power plant, main rotor, tail rotor, fuel, hydraulics, flight controls, communications, navigation, sensor, weapon systems, malfunctions and emergency procedures, and flight planning and training.
Credit Recommendation: In the upper-division baccalaureate category, 3 semester hours in advanced helicopter flight training, and 3 in advanced aircraft systems and operation (4/03).

AR-1606-0243

Oh-58A/C Aviator Qualification

Course Number: 2C-SIA2/152B, Phase 1.
Location: Aviation Center, Fort Rucker, AL.
Length: 3 weeks (113 hours).
Exhibit Dates: 10/02–Present.
Learning Outcomes: Upon completion of the course, the student will be able to qualify as pilot for the OH-58A/C helicopter.
Instruction: Practical and classroom exercises, laboratory, and lecture. Topics include aircraft familiarization, weight and balance, electrical, fuel, power plant, hydraulics, rotor, power train systems; malfunctions and emergency procedures; and flight training.
Credit Recommendation: In the upper-division baccalaureate category, 1 semester hour in advanced helicopter flight training, and 2 in advanced aircraft systems operation and management (4/03).

AR-1606-0244

OH-58C Aviator Qualification

Course Number: 2C-SIA2/152B.
Location: Aviation Center, Fort Rucker, AL.
Length: 3 weeks (111 hours).
Exhibit Dates: 10/99–Present.
Learning Outcomes: Upon completion of the course, the student will be able to qualify as a pilot for the OH-58C helicopter.
Instruction: Practical and classroom exercises, laboratory, and lecture. Topics include aircraft familiarization, weight and balance, electrical, fuel power plant, hydraulics, rotor, and power train systems; malfunctions and emergency procedures; and flight training.
Credit Recommendation: In the upper-division baccalaureate category, 1 semester hour in advanced helicopter flight training, and 2 in advanced aircraft systems operation and management (4/03).

AR-1606-0245

Initial Entry Rotary Wing CH-47D Track

Course Number: 2C-15A/SIC2/154C.
Location: Aviation Center, Fort Rucker, AL.
Length: 14 weeks (548 hours).
Exhibit Dates: 10/02–Present.
Learning Outcomes: Upon completion of the course, the student will be able to achieve pilot qualifications and designation as an Army combat aviator in the CH-47D aircraft system.
Instruction: Audio visual materials, practical and classroom exercises, discussion, laboratory, lecture, and computer-based training. Topics include initial entry rotary wing CH-47D aircraft flight training; aviator tasks such as hovering, takeoff, basic flight, approach and landing, emergency, instrument, tactical and special mission after landing; combat skills and night academics.
Credit Recommendation: In the upper-division baccalaureate category, 3 semester hours in advanced aircraft systems operation and management, and 3 in advanced helicopter flight training (4/03).

AR-1606-0246

Initial Entry Rotary Wing (IERW) UH-60 Track

Course Number: 2C-15A/SIB2/153D.
Location: Aviation Center, Fort Rucker, AL.
Length: 12 weeks (510 hours).
Exhibit Dates: 10/01–Present.
Learning Outcomes: Before 10/01 see AR-1606-0182. Upon completion of the course, the student will be able to achieve pilot qualification and designation as an Army combat aviator in the UH-60 aircraft systems.
Instruction: Audio visual materials, practical and classroom exercises, discussion, laboratory, lecture, and computer-based training. Topics include initial entry rotary wing UH-60 aircraft flight training, aircraft and navigation systems including automatic flight control system, GPS, navigation and command, instrument system, electrical, fuel, power trains, rotor systems, malfunction analysis, combat skills, and night academics.
Credit Recommendation: In the upper-division baccalaureate category, 3 semester hours in advanced aircraft systems operation and management, and 3 in advanced helicopter flight training (4/03).

AR-1606-0247

AH-64D Aviator Qualification (Supplemental)

Course Number: 2C-SID7/152H, Phase 1.
Location: Aviation Center, Fort Rucker, AL.
Length: 6–7 weeks (249 hours).
Exhibit Dates: 10/01–Present.
Learning Outcomes: Upon completion of the course, the helicopter pilot will be qualified to fly AH-64D attack helicopter.
Instruction: Audio visual materials, practical and classroom exercises, discussion, laboratory, computer-based training, and helicopter stimulator. Topics include AH-64D aircraft survivability equipment, navigation and communication systems, weapons subsystem, sight sensors, night vision, aviation mission planning, and AH 64-D flight training.
Credit Recommendation: In the upper-division baccalaureate category, 3 semester hours in advanced aircraft systems operation and management (4/03).

AR-1606-0248

AH-64D Instructor Pilot (Supplemental)

Course Number: 2C-SIG2/2C-SQIC (SUP-64D).
Location: Aviation Center, Fort Rucker, AL.
Length: 3 weeks (97 hours).
Exhibit Dates: 10/99–Present.
Learning Outcomes: Upon completion of the course, the student will be able to update the AH-64D pilot to the improved and new additions to the AH-64 helicopter.
Instruction: Audio visual materials, practical and classroom exercises, discussion, laboratory, learner presentations, lecture, computer-based training. Topics include updates for AH-64D attack helicopter systems.
Credit Recommendation: Credit is not recommended because of the limited, specialized nature of the course (4/03).

AR-1606-0249

AH-64D Aviator Qualification

Course Number: 2C-SID7/152H (AH-64D).
Location: Aviation Center, Fort Rucker, AL.
Length: 14 weeks (579 hours).
Exhibit Dates: 10/02–Present.
Learning Outcomes: Upon completion of the course, the student will be able to perform flight tasks for AH-64D fire control radar/target acquisition; combat using gunnery, crew coordination; and pilot an AH-64D attack helicopter.
Instruction: Audio visual materials, practical and classroom exercises, laboratory, lecture, computer-based training and helicopter simulator. Topics include flight tasks and training in a AH-64D helicopter, night vision systems, gunnery, combat skills, flight evaluation, flight controls, electrical, fuel, engine, pressurized air, drive train, rotor, utility, hydraulic, navigation, tactical and radio communication, weapons systems, and combat skills.
Credit Recommendation: In the upper-division baccalaureate category, 3 semester hours in advanced aircraft systems operation and management, 3 semester hours in advanced helicopter flight (4/03).

AR-1606-0250

AH-64D Aviator Qualification (Supplemental)

Course Number: 2C-SID7/152F.
Location: Aviation Center, Fort Rucker, AL.

Length: 6–7 weeks (299 hours).
Exhibit Dates: 10/99–Present.
Learning Outcomes: Upon completion of the course, the helicopter pilot will be qualified to fly AH-64D attack helicopter.
Instruction: Audio visual materials, practical and classroom exercises, discussion, laboratory, lecture, and helicopter simulator. Topics include AH-64D aircraft survivability equipment, navigation and communication systems, weapons subsystems, sight sensors and night vision, aviation mission planning, and AH-64D flight training.
Credit Recommendation: In the upper-division baccalaureate category, 3 semester hours in advanced aircraft systems operation and management (4/03).

AR-1606-0251

C-23 Aircraft Qualification

Course Number: NGB-FAQC-F080-6.
Location: Aviation Center, Fort Rucker, AL.
Length: 5–6 weeks (204 hours).
Exhibit Dates: 10/00–Present.
Learning Outcomes: Upon completion of the course, the student will be qualified as a pilot in the C-23 airplane under VMC or IMC conditions.
Instruction: Audio visual materials, practical and classroom exercises, discussion, laboratory, lecture, and computer-based training. Topics include classroom and flight instruction in addition to in-flight instruction in the C-23 airplane; flight procedures, instrument flight, emergency procedures, systems, performance charts, weight and balance, and multi-engine flight maneuvers.
Credit Recommendation: In the upper-division baccalaureate category, 2 semester hours in advanced flight theory and aircraft systems, and 2 in advanced multi-engine turbine aircraft flight (4/03).

AR-1606-0252

C-12 IFE Qualification

Course Number: 011-2C-15A/155A (C12), Phase 2.
Location: Aviation Center, Fort Rucker, AL.
Length: 3 weeks (132 hours).
Exhibit Dates: 6/99–Present.
Learning Outcomes: Upon completion of the course, the student will be able to ensure adequate emphasis is placed in training selected aviators on aircraft specific cockpit instrumentation and equipment. The student should also be able to fully integrate all the cockpit instrumentation and equipment into the instrument flight learning and evaluation process.
Instruction: Audio visual materials, practical and classroom exercises, discussion, laboratory, learner presentations, and computer-based training. Topics include basic and advanced instrument procedures, and flight instructor techniques, academic subjects specific to the C-12 airplane to include airborne weather radar, and advanced avionics.
Credit Recommendation: In the upper-division baccalaureate category, 1 semester hour in advanced aircraft systems and management, and 1 in advanced fundamentals of instruction (4/03).

AR-1606-0254

C-12 Aviator Qualification (CT)

Course Number: 2B-SIE5/155E (CT).
Location: Aviation Center, Fort Rucker, AL.
Length: 4 weeks (142 hours).
Exhibit Dates: 10/99–Present.
Learning Outcomes: Upon completion of the course, the student will be qualified as a pilot in the Beech King Air twin turbine airplane under VMC or IMC conditions.
Instruction: Audiovisual materials, practical and classroom exercises, discussion, laboratory, lecture, and computer-based training. Topics include classroom instruction and flight simulation in addition to in-flight instruction in the Beech King Air (C-12) airplane, flight procedures, instrument flight, emergency procedures, systems, performance charts, weight and balance, and multi-engine flight maneuvers.
Credit Recommendation: In the upper-division baccalaureate category, 2 semester hours in advanced flight theory and aircraft systems, and 2 in advanced multi-engine turbine aircraft flight (4/03).

AR-1606-0255

C-26 IFE Qualification

Course Number: 011-2C-15A/155A (C26), Phase 2.
Location: Aviation Center, Fort Rucker, AL.
Length: 3 weeks (151 hours).
Exhibit Dates: 6/99–Present.
Learning Outcomes: Upon completion of the course, the student will be able to ensure adequate emphasis is placed in training selected aviators on aircraft-specific cockpit instrumentation and equipment and to fully integrate all cockpit instrumentation and equipment into the instrument flight learning and evaluation process.
Instruction: Audio visual materials, practical and classroom exercises, discussion, laboratory, lecture, and computer-based training. Topics include basic and advanced instrument procedures, basic and advanced instrument flight instructor techniques, academic subjects specific to the C-26 airplane, airborne weather radar, advanced avionics, and flight management system.
Credit Recommendation: In the upper-division baccalaureate category, 1 semester hour in advanced aircraft systems and management, and 1 in advanced fundamentals of instruction (4/03).

AR-1606-0256

Fixed Wing Multi-Engine Qualification

Course Number: 2C-15A/67J/SIE4/2B-155A, Phase 4.
Location: Aviation Center, Fort Rucker, AL.
Length: 12 weeks (397 hours).
Exhibit Dates: 10/99–Present.
Learning Outcomes: Upon completion of the course, the student will be able to qualify as a multi-engine fixed wing aircraft pilot.
Instruction: Practical and classroom exercises, laboratory, lecture and flight simulator. Topics include multi-engine turboprop flight training, aircraft systems, performance, and aerodynamics.
Credit Recommendation: In the upper-division baccalaureate category, 3 semester hours in advanced aircraft systems operation and management, and 3 in multi-engine ground school and flight laboratory (fixed wing) (4/03).

AR-1606-0257

OH-58D Warrior Aviator Refresher Training

Course Number: 2C-A3 (OH-58D)-IRR-REFR.
Location: Aviation Center, Fort Rucker, AL.
Length: 3–4 weeks (132 hours).
Exhibit Dates: 10/01–Present.
Learning Outcomes: Upon completion of the course the student will be able to provide aviators, returning to flying assignments, with refresher training to successfully complete a OH-58D currency flight.
Instruction: Audio visual materials, practical and classroom exercises, discussion, laboratory, and lecture. Topics include refresher training in the OH-58D helicopter, OH-58D academics, weapons review, and navigational review.
Credit Recommendation: Credit is not recommended because of the limited, specialized nature of the course (4/03).

AR-1701-0003

Utilities Equipment Repairer

Course Number: ***Version 1:*** 662-52C10. *Version 2:* 662-52C10 ITRO.
Location: Ordnance Center and School, Aberdeen Proving Ground, MD.
Length: ***Version 1:*** 15–16 weeks (562 hours). *Version 2:* 11 weeks (423 hours).
Exhibit Dates: ***Version 1:*** 10/90–9/96. *Version 2:* 10/96–Present.
Learning Outcomes: ***Version 1:*** Before 10/90 see AR-1702-0003. Upon completion of the course, the student will be able to apply the theories of basic electricity and refrigeration to power equipment, perform service procedures, diagnose, and troubleshoot. *Version 2:* Upon completion of the course, the student will be able to apply the basic concepts of electricity and refrigeration to service, repair, and maintain air conditioning, ice making, halon, heating, and refrigeration equipment; and meet certification requirements for sections 608 and 609 of the Clean Air act of 1990.
Instruction: ***Version 1:*** Course includes lectures on electrical and refrigeration theory and demonstrations using tools and test equipment on appropriate components. Extensive hands-on experience on service manual repair procedures and troubleshooting techniques is provided. *Version 2:* This course includes lectures on electrical and refrigeration theory and demonstrations using tools and test equipment on appropriate components. Extensive hands-on experience on service manual repair procedures and troubleshooting techniques is provided.
Credit Recommendation: ***Version 1:*** In the lower-division baccalaureate/associate degree category, 1 semester hour in general shop practice, 2 hours in basic electricity, 3 in basic air conditioning, and 3 in air conditioning and refrigeration service (12/90). *Version 2:* In the lower-division baccalaureate/associate degree category, 2 semester hours in basic electricity, 3 in basic refrigeration and air conditioning, and 3 in service and repair of refrigeration and air conditioning equipment (8/03).
Related Occupations: 52C.

AR-1701-0004

Senior Utilities Equipment Repairer

Course Number: 662-52C30 (63J).

Location: Ordnance Center and School, Aberdeen Proving Ground, MD.

Length: 13–14 weeks (482 hours).

Exhibit Dates: 10/96–Present.

Learning Outcomes: Before 10/96 see AR-1406-0143. Upon completion of the course, the student will be able to service, repair, and maintain air conditioning and refrigeration equipment and meet certification requirements for sections 608 and 609 of the Clean Air act of 1990 as well as provide basic supervision of personnel.

Instruction: This course includes lectures and practical demonstrations on electrical and refrigeration theory using tools and test equipment. Extensive hands-on work covers repair procedures and troubleshooting techniques. Basic supervision and management skills are also covered. Course contains a common core of leadership subjects.

Credit Recommendation: In the lower-division baccalaureate/associate degree category, 1 semester hour in supervision/management, 3 in basic refrigeration and air conditioning, and 3 in service and repair of refrigeration and a/c equipment. See AR-1406-0090 for common core credit (4/01).

Related Occupations: 52C; 63J.

AR-1701-0005

Utilities Equipment Repairer

Course Number: 091-52C10 (TATS), Phase 3.

Location: Reserve Schools, Multiple locations.

Length: 100 hours.

Exhibit Dates: 10/99–Present.

Learning Outcomes: Upon completion of the course, the student will be able to apply the basic concepts of electricity and air conditioning to service, repair, and maintain air conditioning, ice making, halon, heating, and refrigeration equipment and meet certification requirements for sections 608 and 609 of the Clean Air Act of 1990.

Instruction: Audio visual materials, practical exercises, classroom exercises, discussion, laboratory, and lecture. Topics include lectures on electrical and refrigeration theory and demonstrations using tools and test equipment on appropriate components. Extensive hands-on experience on service manual repair procedures and troubleshooting techniques is provided.

Credit Recommendation: In the lower-division baccalaureate/associate degree category, 2 semester hours in basic electricity, 2 in basic refrigeration and air conditioning, and 2 in service and repair of refrigeration and air conditioning equipment (8/03).

AR-1702-0003

Utilities Equipment Repairer

Course Number: 662-52C10.

Location: Engineer School, Fort Belvoir, VA.

Length: 11 weeks (393 hours).

Exhibit Dates: 2/79–9/90.

Objectives: After 9/90 see AR-1701-0003. To train enlisted personnel to operate and maintain gas turbine engines and multi-output package equipment.

Instruction: Conferences and practical exercises cover diagnostic procedures and maintenance and repair of gas turbine engines and refrigeration, air conditioning, and heating equipment.

Credit Recommendation: In the lower-division baccalaureate/associate degree category, 1 semester hour in introduction to electricity and 2 in air conditioning and refrigeration (9/79).

Related Occupations: 52C.

AR-1703-0026

Fuel and Electrical Systems Repairer

Course Number: *Version 2:* 610-63G10/ 610-3524. *Version 3:* 610-63G10. *Version 4:* 610-63G10.

Location: Ordnance Center and School, Aberdeen Proving Ground, MD.

Length: *Version 2:* 4–14 weeks (158–508 hours). *Version 3:* 12–13 weeks (465 hours). *Version 4:* 9 weeks (324 hours).

Exhibit Dates: *Version 2:* 1/88–9/95. *Version 3:* 10/95–9/99. *Version 4:* 10/99–Present.

Learning Outcomes: *Version 2:* Upon completion of the course, the student will describe fundamental operating principles of automotive fuel and electrical systems, use electrical and electronic automotive test equipment, and perform basic maintenance on fuel and electrical systems on gasoline and diesel engines in automotive or stationary applications. *Version 3:* Upon completion of the course, the student will describe fundamental operating principles of automotive fuel and electrical systems, use electrical and electronic automotive test equipment, and perform basic maintenance on fuel and electrical systems on gasoline and diesel engines in automotive and stationary applications. *Version 4:* Upon completion of the course, the student will be able to describe the operating principles of automatic brakes, transmission/drive systems, engines, and electrical systems; use shop hand tools and basic electrical management equipment; and operate and repair (as appropriate) engines, electrical systems, and the 500 amp test stand.

Instruction: *Version 2:* Instruction includes applied electricity, ignition, starting, and charging systems; gasoline and diesel fuel systems; and hydraulic brake system component repair. Course also includes carburetor repair, two-barrel overhaul, injector pumps and nozzle, and power unit injector pump. *Version 3:* Instruction includes applied electricity, ignition, starting and charging systems, and gasoline and diesel fuel systems. Course also includes air induction and fuel systems, along with diesel pumps and injectors. *Version 4:* Audio visual materials, learner presentations, computer-based training, practical exercises, classroom exercises, discussion, laboratory, and lecture. Topics include engine construction, transmission/drive lines/suspension, principles of electricity, automotive electrical systems, brakes, safety, publications, preventative maintenance, test equipment, and troubleshooting practice.

Credit Recommendation: *Version 2:* In the lower-division baccalaureate/associate degree category, 5 semester hours in automotive electrical systems, 3 in diesel fuel systems, and 3 in gasoline engine fuel systems. In the upper-division baccalaureate category, 5 semester hours in automotive engine diagnosis (12/90). *Version 3:* In the lower-division baccalaureate/associate degree category, 5 semester hours in automotive electrical systems, 3 in diesel fuel systems, 1 in gasoline engine fuel systems, and 5 in automotive engine diagnosis (1/98). *Version 4:* In the lower-division baccalaureate/associate degree category, 3 semester hours in automotive engines, 2 in automotive hydraulic systems, 5 in automotive electrical systems, and 1 in shop safety (8/03).

Related Occupations: 63G.

AR-1703-0027

Wheel Vehicle Repairer

Course Number: *Version 2:* 610-63W10 (CT). *Version 3:* 610-63W10. *Version 4:* 610-63W10.

Location: Ordnance Center and School, Aberdeen Proving Ground, MD.

Length: *Version 2:* 13–16 weeks (481–582 hours). *Version 3:* 13–14 weeks (520 hours). *Version 4:* 11–12 weeks (425 hours).

Exhibit Dates: *Version 2:* 1/87–6/95. *Version 3:* 7/95–9/99. *Version 4:* 10/99–Present.

Learning Outcomes: *Version 2:* Upon completion of the course, the student will be able to perform maintenance on wheeled heavy equipment vehicles, including troubleshooting, removing, replacing, and adjusting transmissions, and repairing front and rear differentials, suspensions, steering, brakes, and engines. *Version 3:* Upon completion of the course, the student will be able to perform maintenance on wheeled heavy-duty equipment vehicles, including power train adjustment and replacement, diesel injection service, servicing of front and rear axles, suspension, steering, and brakes. *Version 4:* Upon completion of the course, the student will be able to perform maintenance and troubleshooting on wheeled heavy-duty equipment vehicles utilizing hydraulic, electrical, mechanical systems knowledge of power train, steering, suspension, and transmission systems.

Instruction: *Version 2:* Lectures and practical exercises cover shop practice using hand tools and measurement devices, introduction to the engine and its systems, troubleshooting and repairing suspension and steering systems, replacing and adjusting transmissions and front and rear differentials, replacing fuel injection pump, troubleshooting and repairing air and hydraulic brakes, troubleshooting and replacing charging and starting system components, testing and repairing engines, and repairing heavy-duty hydraulic systems. *Version 3:* Lectures and practical exercises cover shop practice using hand tools and measurement devices, introduction to the engine and its systems, troubleshooting and repairing suspension and steering systems, replacing and adjusting transmissions and front and rear differentials, replacing fuel injection pump, troubleshooting and repairing air and hydraulic brakes, troubleshooting and replacing charging and starting system components, testing and repairing engines, and repairing heavy-duty hydraulic systems. *Version 4:* Audio visual materials, computer-based training, practical exercises, classroom exercises, discussion, laboratory, and lecture. Topics include shop practice using hand tools and measurement devices for hydraulic, electrical, and mechanical systems in the maintenance and troubleshooting support of suspension and steering systems, transmission and power train systems, and diesel engines.

Credit Recommendation: *Version 2:* In the lower-division baccalaureate/associate degree category, 3 semester hours in heavy duty diesel/gas engine repair, 1 in heavy-duty brakes, 1 in automotive electrical, 1 in heavy-duty suspension and steering, 2 in heavy-duty power trains, and 2 in heavy-duty hydraulics (12/90). *Version 3:* In the lower-division baccalaureate/associate degree category, 3 semester hours in introduction to diesel/gas engine repair, 1 in basic brakes, 1 in basic automotive electrical DC theory, 2 in heavy-duty power trains, and 1 in basic hydraulics (1/98). *Version 4:* In the lower-division baccalaureate/associate degree category, 3 semester hours in diesel/gas engine repair, 2 in basic electrical systems, 2 in basic hydraulic systems, and 1 in basic mechanical systems (8/03).

Related Occupations: 63W.

AR-1703-0028

Light Wheel Vehicle Mechanic

Course Number: 610-63B10 (ST); 610-63B10.

Location: Training Center, Aberdeen Proving Ground, MD; Training Center, Fort Leonard Wood, MO; Training Center, Fort Dix, NJ; Training Center, Fort Jackson, SC.

Length: 14–15 weeks (534–535 hours).

Exhibit Dates: 10/88–9/95.

Learning Outcomes: Upon completion of the course, the student will be able to repair automotive wheel vehicles, including engine troubleshooting and overhaul, electrical system repair and troubleshooting, brake system repair and troubleshooting, and steering system repair and troubleshooting.

Instruction: Lectures and practical exercises cover troubleshooting and repairing of engines; basic fuel, electrical, and cooling systems; suspension and steering; brakes; preventive maintenance; and recovery operations.

Credit Recommendation: In the lower-division baccalaureate/associate degree category, 2 semester hours in automotive engine servicing, 3 in basic automotive electrical, 3 in automotive brakes, and 2 in automotive steering and suspension (12/90).

Related Occupations: 63B.

AR-1703-0029

Light Wheel Vehicle Mechanic Basic Noncommissioned Officer (NCO)

Course Number: *Version 2:* 610-63B30; 610-63B30 (CT). *Version 3:* 610-63B30.

Location: Ordnance Center and School, Aberdeen Proving Ground, MD.

Length: *Version 2:* 12 weeks (427–428 hours). *Version 3:* 12 weeks (435 hours).

Exhibit Dates: *Version 2:* 4/87–9/96. *Version 3:* 10/96–Present.

Learning Outcomes: *Version 2:* Upon completion of the course, the student will be able to use diagnostic equipment to troubleshoot and maintain wheeled and track vehicles and power generation equipment. *Version 3:* Upon completion of the course, the student will be able to use diagnostic equipment to troubleshoot and maintain wheeled and track vehicles and power generation equipment.

Instruction: *Version 2:* Lectures and practical exercises cover troubleshooting, repairing, and correcting disorders in engines, fuel (gas and diesel) systems, electrical cooling systems, power trains, chassis and suspension systems, components of power generation equipment, and wheeled and track equipment. This course contains a common core of leadership subjects. *Version 3:* Lectures and practical exercises cover troubleshooting and repairing vehicle systems (wheel and track), fuel (gas and diesel), lubrication, hydraulic, chassis and suspension, vehicle electrical and cooling, and portable power generation equipment. This course contains a common core of leadership subjects.

Credit Recommendation: *Version 2:* In the lower-division baccalaureate/associate degree category, 3 semester hours in heavy equipment repair, 3 in power generation equipment repair and maintenance, and 7 in heavy equipment maintenance and repair of which 3 should be specifically in truck repair and maintenance. See AR-1406-0090 for common core credit (12/90). *Version 3:* In the lower-division baccalaureate/associate degree category, 7 semester hours in vehicle equipment repair and 1 in power generation equipment repair. See AR-1406-0090 for common core credit (5/01).

Related Occupations: 63B.

AR-1703-0030

Metal Worker
(Vehicle Body Mechanic)
(Vehicle Body Repairman)
(Vehicle Body Mechanic (USAF))
(Vehicle Body Repairman (USMC))

Course Number: *Version 2:* 704-47233; 704-44B10; 704-44B10 (USA). *Version 3:* 704-44B10; 704-44B10 (USA); 704-3513 (OS); 704-47233 (OS); 704-47233.

Location: Ordnance Center and School, Aberdeen Proving Ground, MD.

Length: *Version 2:* 9–15 weeks (325–486 hours). *Version 3:* 9–15 weeks (325–486 hours).

Exhibit Dates: *Version 2:* 1/89–9/91. *Version 3:* 10/91–12/97.

Learning Outcomes: *Version 2:* Upon completion of the course, the student will be able to set up and use oxyacetylene, electric arc, and MIG and TIG welding equipment to perform auto body repair. *Version 3:* After 12/97 see AR-1723-0012. Upon completion of the course, the student will be able to repair and install metal body components, radiators, and fuel tanks; weld components; and perform common maintenance tasks.

Instruction: *Version 2:* Lectures and practical exercises cover oxyacetylene and inert gas welding; use of metal body repair tool and equipment; after-operations checks; and trim, hardware, and glasswork. Course also covers electric arc and MIG and TIG welding; repairing, repainting, installing, and aligning metal body components; and care and use of tools and equipment. *Version 3:* Lectures and laboratory exercises cover safety, oxyacetylene welding, MIG and TIG welding, metal body repair, arc welding, radiator repair, and glass replacement.

Credit Recommendation: *Version 2:* In the lower-division baccalaureate/associate degree category, for Army personnel, 2 semester hours in oxyacetylene, 2 in electric arc, 2 in MIG and TIG, and 3 in automotive body repair. For Air Force personnel, 1 semester hour in oxyacetylene, 1 in electric arc, and 3 in automotive body repair. For Marine Corps personnel, 2 semester hours in oxyacetylene, 1 in MIG and TIG, and 3 in automotive body repair (7/90). *Version 3:* In the lower-division baccalaureate/associate degree category, 3 semester hours in automotive body repair and painting, 2 in shop safety fundamentals, and 3 in oxyacetylene welding. For Army personnel, 3 additional semester hours in arc welding and 3 in MIG/TIG welding. For Marine Corps personnel, 1 additional semester hour in arc welding and 2 in MIG/TIG welding. For Air Force personnel, 2 additional semester hours in arc welding (11/91).

Related Occupations: 44B.

AR-1703-0031

Track Vehicle Repairer

Course Number: *Version 1:* 611-63H10 (CT); 611-63H10. *Version 2:* 611-63H10. *Version 3:* 611-63H10.

Location: Ordnance Center and School, Aberdeen Proving Ground, MD.

Length: *Version 1:* 16–17 weeks (593–603 hours). *Version 2:* 15 weeks (548 hours). *Version 3:* 14 weeks (503 hours).

Exhibit Dates: *Version 1:* 9/85–6/95. *Version 2:* 7/95–9/99. *Version 3:* 10/99–Present.

Learning Outcomes: *Version 1:* Upon completion of the course, the student will be able to disassemble, inspect, assemble, and test various components and systems on track vehicles. *Version 2:* Upon completion of the course, the student will be able to troubleshoot and repair heavy vehicles (M1/M1A1 Tank), disassemble, inspect, and test various heavy vehicle components. *Version 3:* Upon completion of the course, the student will be able to troubleshoot and repair heavy vehicles (M1/M1A1 Tank), disassemble, inspect, and test various heavy vehicle components.

Instruction: *Version 1:* Lectures, demonstrations, and practical exercises cover electrical systems, diesel and gasoline engines, fuel delivery systems, driveline components, and hydraulic systems. Principles of operation, troubleshooting, and diagnostic test equipment are stressed. *Version 2:* Lectures, demonstrations, and practical exercises cover hydraulics, transmissions, driveline components, diesel engines, fuel delivery systems, and steering/suspension systems for track vehicles. Principles of operation, troubleshooting, and the use of diagnostic test equipment are emphasized in this course. *Version 3:* Lectures, demonstrations, and practical exercises. Topics include hydraulics, transmissions, driveline components, diesel engines, fuel delivery systems, and steering/suspension systems for track vehicles. Principles of operation, troubleshooting, and the use of diagnostic test equipment are stressed.

Credit Recommendation: *Version 1:* In the lower-division baccalaureate/associate degree category, 3 semester hours in diesel engine systems and 2 in automotive electrical systems (1/88). *Version 2:* In the lower-division baccalaureate/associate degree category, 3 semester hours in diesel engine systems, 2 in introduction to automotive systems, and 2 in automotive electrical systems (1/98). *Version 3:* In the lower-division baccalaureate/associate degree category, 3 semester hours in automotive fundamentals, 3 in heavy equipment service, and 2 in diesel engine systems (8/03).

Related Occupations: 63H.

AR-1703-0032

Field Artillery Vehicle Maintenance Technician Warrant Officer Advanced

Course Number: 4-9-C32-630C.

Location: Ordnance Center and School, Aberdeen Proving Ground, MD.

Length: 17 weeks (623 hours).

Exhibit Dates: 1/85–9/90.

Learning Outcomes: Upon completion of the course, the student will be able to troubleshoot, repair, and manage field artillery vehicles.

Instruction: Instruction includes lecture and hands-on with emphasis on management, maintenance management, and communication skills. Considerable emphasis is placed on troubleshooting and repair of field artillery vehicles.

Credit Recommendation: In the lower-division baccalaureate/associate degree category, 1 semester hour in communication skills, 1 in management, 3 in heavy equipment maintenance management, 1 in automotive electrical systems, 1 in hydraulics, and 5 in diesel and heavy equipment mechanics (3/87).

Related Occupations: 630C.

AR-1703-0033

ARMOR/CAVALRY MAINTENANCE TECHNICIAN WARRANT OFFICER ADVANCED

Course Number: 4-9-C32-630D.

Location: Ordnance Center and School, Aberdeen Proving Ground, MD.

Length: 19 weeks (700–745 hours).

Exhibit Dates: 1/85–9/90.

Learning Outcomes: Upon completion of the course, the student will be able to troubleshoot, repair, and manage, armor/cavalry equipment.

Instruction: Instruction includes lectures and hands-on with emphasis on management, maintenance management, and communication skills. Considerable emphasis is placed on troubleshooting and repair of armor/cavalry equipment.

Credit Recommendation: In the lower-division baccalaureate/associate degree category, 1 semester hour in communication skills, 1 in management, 1 in heavy equipment maintenance management, 1 in hydraulics, 1 in automotive electrical, and 5 in diesel and heavy equipment mechanics (3/87).

Related Occupations: 630D.

AR-1703-0034

ORDNANCE OFFICER ADVANCED FIELD MAINTENANCE MANAGEMENT

Course Number: 4N-FOA-F6.

Location: Ordnance Center and School, Aberdeen Proving Ground, MD.

Length: 6 weeks (212 hours).

Exhibit Dates: 6/85–9/90.

Learning Outcomes: Upon completion of the course, the student will be able to serve as an automotive maintenance manager.

Instruction: Lectures, demonstrations, and practical exercises cover maintenance procedures, principles of weapon systems, logistic support of land battles, and Army supply systems.

Credit Recommendation: In the lower-division baccalaureate/associate degree category, 2 semester hours in fundamentals of automotive maintenance (3/87).

AR-1703-0035

ORDNANCE OFFICER ADVANCED WHOLESALE MAINTENANCE MANAGEMENT

Course Number: 4N-FOA-F7.

Location: Ordnance Center and School, Aberdeen Proving Ground, MD.

Length: 6 weeks (212 hours).

Exhibit Dates: 6/85–9/90.

Learning Outcomes: Upon completion of the course, the student will be able to serve as an automotive maintenance material manager.

Instruction: Instruction includes lectures, demonstrations, and practical exercises in maintenance procedures for various military vehicles, principles of weapons systems, supply procedures, budgeting, and distribution and procurement procedures using computer assistance. Role playing and student-developed program techniques are widely used to support classroom instruction.

Credit Recommendation: In the lower-division baccalaureate/associate degree category, 2 semester hours in fundamentals of automotive maintenance and 2 in automotive parts management (3/87).

AR-1703-0036

REFRESHER COURSE, HEAVY WHEEL VEHICLE MECHANIC INDIVIDUAL READY RESERVE

Course Number: 610-F4 (IRR).

Location: Ordnance Center and School, Aberdeen Proving Ground, MD.

Length: 2 weeks (90 hours).

Exhibit Dates: 10/86–9/90.

Learning Outcomes: Upon completion of the course, the student will be able to use tools and automotive test equipment to troubleshoot malfunctions in various automotive systems.

Instruction: This is a refresher course consisting of lectures, demonstrations, and practical exercises in diagnostic testing and troubleshooting of automotive engines, fuel systems, engine electrical systems, brake systems, cooling systems, and related systems.

Credit Recommendation: In the lower-division baccalaureate/associate degree category, 1 semester hour in automotive electrical systems and 1 in automotive diagnostics and testing (3/87).

Related Occupations: 63S.

AR-1703-0037

HEAVY WHEEL VEHICLE MECHANIC

Course Number: *Version 2:* 610-63S10. *Version 3:* 610-63S10. *Version 4:* 610-63S10, Phase 1; 610-63S10.

Location: Training Center, Fort Jackson, SC; Ordnance Center and School, Aberdeen Proving Ground, MD.

Length: *Version 2:* 13 weeks (476 hours). *Version 3:* 11–12 weeks (427 hours). *Version 4:* 9 weeks (325 hours).

Exhibit Dates: *Version 2:* 4/87–2/90. *Version 3:* 3/90–9/95. *Version 4:* 10/95–Present.

Learning Outcomes: *Version 2:* Upon completion of the course, the student will be able to perform repairs on heavy automotive wheel vehicles and material-handling equipment, including gas/diesel engine troubleshooting and overhaul, electrical systems, basic brake systems, basic steering and suspension system, power train maintenance, and hydraulic system maintenance. *Version 3:* Upon completion of the course, the student will be able to perform repairs on heavy automotive wheel vehicles and material-handling equipment, including gas/diesel engine troubleshooting and overhaul, electrical systems, basic brake systems, basic steering and suspension systems, power train maintenance, and hydraulic system maintenance. *Version 4:* Upon completion of this course, the student will be able to perform repairs on heavy automotive-wheeled vehicles including basic repair of the engine, suspension, brake, electrical and exhaust systems, and perform maintenance on the power train and hydraulic system.

Instruction: *Version 2:* Lectures and practical exercises cover troubleshooting and repairing engine, fuel, air, and exhaust systems; automotive electrical systems; brake systems; steering and suspension systems; and hydraulic systems. *Version 3:* Lectures and practical exercises cover troubleshooting and repairing engine, fuel, air, and exhaust systems; automotive electrical systems; brake systems; steering and suspension systems; and hydraulic systems. *Version 4:* Practical and classroom exercises, laboratory, and lecture. Topics include engine and engine systems, automotive electricity, heavy equipment steering and suspension, power train, and hydraulic maintenance.

Credit Recommendation: *Version 2:* In the lower-division baccalaureate/associate degree category, 3 semester hours in heavy duty gasoline engine systems, 1 in automotive electrical, 1 in heavy duty brakes hydraulic/air, and 1 in heavy duty steering and suspension servicing (12/90). *Version 3:* In the lower-division baccalaureate/associate degree category, 2 semester hours in automotive engine servicing, 2 in basic automotive electrical, 1 in basic automotive brakes, 1 in basic steering and suspension systems, and 2 in basic hydraulic systems (12/90). *Version 4:* In the lower-division baccalaureate/associate degree category, 2 semester hours in diesel engines, 2 in hydraulics systems, 2 in heavy equipment suspension and steering, and 2 in automotive electrical systems (8/03).

Related Occupations: 63S.

AR-1703-0038

TRACK VEHICLE REPAIRER INDIVIDUAL READY RESERVE (63H30)

Course Number: 611-F6 (IRR).

Location: Ordnance Center and School, Aberdeen Proving Ground, MD.

Length: 2 weeks (77 hours).

Exhibit Dates: 10/86–9/90.

Learning Outcomes: Upon completion of the course, the student will be able to use tools and automotive test equipment to troubleshoot malfunctions in various automotive systems.

Instruction: This is a refresher course consisting of lectures, demonstrations, and practical exercises in diagnostic testing and troubleshooting automotive engines, fuel systems, engine electrical systems, cooling systems, and related systems.

Credit Recommendation: In the lower-division baccalaureate/associate degree category, 1 semester hour in automotive electrical systems (3/87).

Related Occupations: 63H.

AR-1703-0040

M1 ABRAMS TANK SYSTEM MECHANIC BASIC NONCOMMISSIONED OFFICER (NCO)

Course Number: 611-63E30.

Location: Armor Center and School, Fort Knox, KY.

Length: For the 63E Track, 12–13 weeks (406 hours); For the 45E Track, 11 weeks (400 hours).

Exhibit Dates: 10/86–12/97.

Learning Outcomes: Upon completion of the course, the student is able to supervise and perform unit maintenance on the hull and tank turret of special track vehicles and wheel vehicles.

Instruction: Course includes maintenance management; recovery operations; diagnostics; supervision; training management; and inspecting, testing, and repairing systems and subsystems on track and wheel vehicles.

Credit Recommendation: In the lower-division baccalaureate/associate degree category, 2 semester hours in basic automotive electricity laboratory and 2 in maintenance management for either the 63E or the 45E Tracks (3/87).

Related Occupations: 45E; 63E.

AR-1703-0042

M60A1/A3 Tank Systems Mechanic

Course Number: 611-63N10.

Location: Armor Center and School, Fort Knox, KY.

Length: 8–9 weeks (304–332 hours).

Exhibit Dates: 4/83–12/97.

Learning Outcomes: Upon completion of the course, the student is able to perform organizational maintenance on the chassis system and components of track and wheel vehicles and conduct recovery operations using light and medium recovery vehicles.

Instruction: Instruction includes servicing, lubricating, removing and replacing, adjusting, testing, and troubleshooting systems, components, and assemblies of engine, fuel, exhaust, cooling, and electrical systems; engine power trains; tracks; suspensions; steering controls; and hulls of specialized military track and wheel vehicles.

Credit Recommendation: In the lower-division baccalaureate/associate degree category, 1 semester hour in basic automotive electricity laboratory (3/87).

Related Occupations: 63N.

AR-1703-0043

2. Track Vehicle Mechanic Initial Entry Training
3. Track Vehicle Mechanic
4. Track Vehicle Mechanic

Course Number: ***Version 2:*** 611-63Y10. *Version 3:* 611-63Y10. *Version 4:* 611-63Y10.

Location: ***Version 2:*** Armor Center, Fort Knox, KY; Ordnance Center and School, Aberdeen Proving Ground, MD. *Version 3:* Ordnance Center and School, Aberdeen Proving Ground, MD. *Version 4:* Ordnance Center and School, Aberdeen Proving Ground, MD.

Length: ***Version 2:*** 11–13 weeks (468 hours). *Version 3:* 11–12 weeks (407–433 hours). *Version 4:* 12–13 weeks (434 hours).

Exhibit Dates: ***Version 2:*** 10/87–6/95. *Version 3:* 7/95–9/99. *Version 4:* 10/99–Present.

Learning Outcomes: ***Version 2:*** Upon completion of the course, the student will apply basic automotive vehicle power train and chassis theory to the maintenance and routine servicing of the common vehicle systems. *Version 3:* Upon completion of the course, the student will be able to perform unit-level maintenance on automotive support vehicles. *Version 4:* Upon completion of the course, the student will be able to perform unit-level maintenance on automotive vehicles and understand basic theory of operation and repair of automotive vehicles.

Instruction: *Version 2:* Lectures, laboratory experience, projects, and demonstrations cover automotive vehicles power train, and chassis components, using appropriate tools, service manuals, and test equipment. *Version 3:* Lectures and practical exercises cover maintenance procedures, tools, diagnostic equipment, safety, troubleshooting, and vehicle performance characteristics. *Version 4:* Lectures and practical exercises. Topics include maintenance procedures, tools, diagnostic equipment, safety, troubleshooting, and vehicle performance characteristics.

Credit Recommendation: ***Version 2:*** In the lower-division baccalaureate/associate degree category, 2 semester hours in automotive fundamentals, 2 in automotive electrical systems, and 2 in basic automotive laboratory (12/90). *Version 3:* In the lower-division baccalaureate/associate degree category, 1 semester hour in automotive fuel systems, 2 in automotive electrical systems, 2 in gasoline and diesel engines, and 2 in basic hydraulic systems (3/03). *Version 4:* In the lower-division baccalaureate/associate degree category, 3 semester hours in automotive power plants and 3 in general automotive service laboratory (8/03).

Related Occupations: 63Y.

AR-1703-0044

M60A3 Tank System Mechanic Individual Ready Reserve Refresher

Course Number: 63N20; 611-F11 (IRR).

Location: Armor Center and School, Fort Knox, KY; Ordnance Center and School, Aberdeen Proving Ground, MD.

Length: 2 weeks (116–120 hours).

Exhibit Dates: 6/86–12/97.

Learning Outcomes: Upon completion of the course, the student will be able to use appropriate tool and diagnose equipment faults for the MOS in which the student was trained.

Instruction: Lectures and practical exercises cover automotive systems, vehicle recovery operations, and tank turret systems.

Credit Recommendation: Credit is not recommended because of the limited, specialized nature of the course (3/87).

Related Occupations: 63N.

AR-1703-0045

M2/3 Bradley Fighting Vehicle System Turret Mechanic
(Bradley Fighting Vehicle System Turret Mechanic)

Course Number: 643-45T10.

Location: Armor Center and School, Fort Knox, KY.

Length: 9–11 weeks (322–393 hours).

Exhibit Dates: 6/83–12/94.

Learning Outcomes: After 12/94 see AR-1710-0184. Upon completion of the course, the student will be able to troubleshoot, remove, repair, replace, and adjust components and assemblies on the Bradley fighting vehicle turrets and perform preventive checks and services as required using manuals and maintenance forms.

Instruction: Instruction includes lectures, and performance exercises on electrical and hydraulic systems of the Bradley fighting vehicle system turret. Emphasis is placed on troubleshooting, repair, replacement, and adjustment of turret components and assemblies.

Credit Recommendation: Credit is not recommended because of the military-specific nature of the course (7/91).

Related Occupations: 45T.

AR-1703-0046

Advanced Machinist
(Machinist)

Course Number: 702-42750.

Location: Ordnance Center and School, Aberdeen Proving Ground, MD.

Length: 4–5 weeks (188–192 hours).

Exhibit Dates: 6/87–9/90.

Learning Outcomes: Upon completion of the course, the student will be able to perform measuring and gauging operations, advanced milling and grinding, and layout and heat treatment of tool materials.

Instruction: Lectures and practical exercises cover precision measuring and gauging, milling, tool grinding, cutting threads, reading shop drawings, fabrication of jigs and fixtures, tool and die making and heat treatment of materials.

Credit Recommendation: In the lower-division baccalaureate/associate degree category, 1 semester hour in precision measuring and gauging, 1 in advanced machine shop, and 2 in tool and die making (12/90).

AR-1703-0047

ITV/IFV/CFV Systems Mechanic Recovery Specialist

Course Number: 611-ASIH8 (63T).

Location: Armor Center and School, Fort Knox, KY.

Length: 4–5 weeks (163 hours).

Exhibit Dates: 10/84–12/97.

Learning Outcomes: Upon completion of the course, the student will be able to operate light and medium recovery vehicles and associated equipment to recover track vehicles.

Instruction: Instruction includes operation, servicing, and use of light and medium recovery vehicles and equipment and procedures for recovery and transportation of disabled vehicles.

Credit Recommendation: In the vocational certificate category, 1 semester hour in rigging and winch operation, 1 in vehicle operation, and 1 in servicing and maintenance (3/87).

Related Occupations: 63T.

AR-1703-0049

Metalworker Individual Ready Reserve (44B20)

Course Number: 704-F1-IRR.

Location: Ordnance Center and School, Aberdeen Proving Ground, MD.

Length: 2 weeks (90 hours).

Exhibit Dates: 10/86–9/90.

Learning Outcomes: Upon completion of the course, the student will be able to use welding and auto body repair tools and equipment for the MOS in which the student was trained.

Instruction: This is a refresher course in the operation and use of oxyacetylene, arc, and inert gas welding equipment and auto body repair techniques.

Credit Recommendation: Credit is not recommended because of the limited, specialized nature of the course (3/87).

Related Occupations: 44B.

AR-1703-0050

M1 Abrams Tank Turret Mechanic

Course Number: 643-45E10.
Location: Armor Center and School, Fort Knox, KY.
Length: 9 weeks (316–350 hours).
Exhibit Dates: 6/83–6/93.
Learning Outcomes: After 6/93 see AR-2201-0490. Upon completion of the course, the student will be able to perform maintenance on the fire control and hydraulic systems of a tank turret.
Instruction: Lectures, demonstrations, and practical exercises cover basic electricity and electrical systems, principles and practices of hydraulics, special tools, shop safety, and the tank's fire control and the armament systems.
Credit Recommendation: In the lower-division baccalaureate/associate degree category, 1 semester hour in fundamentals of hydraulic systems and 1 in electrical systems shop (3/87).
Related Occupations: 45E.

AR-1703-0051

M60A1/A3 Tank Recovery Specialist

Course Number: 611-ASIH8 (63N).
Location: Armor Center and School, Fort Knox, KY.
Length: 4–5 weeks (140 hours).
Exhibit Dates: 10/84–3/97.
Learning Outcomes: Upon completion of the course, the student will be able to operate wheel and tracked recovery vehicles.
Instruction: Lectures, demonstrations, and practical exercises cover the operation and preventive of wheeled and tracked recovery vehicles. Other topics include driving instruction, boom operation, and oxyacetylene cutting.
Credit Recommendation: In the lower-division baccalaureate/associate degree category, 1 semester hour in heavy equipment operation (3/87).
Related Occupations: 63N.

AR-1703-0052

Light Wheel Vehicle Mechanic (Recovery Specialist)

Course Number: 610-63B10 (ASIH8); 610-63B10 (ASI-H8) (ST); 610-63B10 (ASI-H8).
Location: Training Center, Fort Leonard Wood, MO; Training Center, Fort Dix, NJ; Training Center, Fort Jackson, SC.
Length: 2 weeks (79 hours).
Exhibit Dates: 10/84–4/04.
Learning Outcomes: Upon completion of the course, the student will be able to operate a wrecker truck for the recovery of light vehicles.
Instruction: Lectures, demonstration, and practical experience cover the operation of a wheeled wrecker truck including the operation of recovery boom, rigging, and safety procedures.
Credit Recommendation: Credit is not recommended because of the limited, specialized nature of the course (1/90).
Related Occupations: 63B.

AR-1703-0053

Track Vehicle Repairer Individual Ready Reserve Refresher

Course Number: 611-F4 (IRR).
Location: Ordnance Center and School, Aberdeen Proving Ground, MD.
Length: 2 weeks (79 hours).
Exhibit Dates: 10/86–9/90.
Learning Outcomes: Upon completion of the course, the student will be able to disassemble, identify, inspect, reassemble, and test charging system components in diesel fuel systems and troubleshoot and repair the electrical and fuel systems on a diesel engine, turbine engine, 6U53T Detroit Diesel engine, and Allison X200-4 automatic transmission.
Instruction: Course includes lectures and hand-on training, stressing diagnosing and troubleshooting malfunctions, on Detroit Diesel 6U53T engines, turbine engines, Allison X200-4 automatic transmissions, electrical systems, and fuel systems. Repair of charging and fuel system components and repair of failed systems after diagnosis are also covered.
Credit Recommendation: In the lower-division baccalaureate/associate degree category, 2 semester hours in heavy equipment mechanics (3/87).
Related Occupations: 63H.

AR-1703-0055

Wheel Vehicle Repairer Individual Ready Reserve (IRR) Refresher (63W20)

Course Number: 610-F3 (IRR).
Location: Ordnance Center and School, Aberdeen Proving Ground, MD.
Length: 2 weeks (76 hours).
Exhibit Dates: 10/86–9/90.
Learning Outcomes: Upon completion of the course, the student will be able to use tools and automotive test equipment to troubleshoot malfunctions in various automotive systems.
Instruction: This is a refresher course consisting of lectures, demonstrations, and practical exercises in troubleshooting automotive engines, fuel systems, electrical systems, brake systems, and steering and suspension systems.
Credit Recommendation: In the lower-division baccalaureate/associate degree category, 1 semester hour in automotive steering and suspension systems (3/87).
Related Occupations: 63W.

AR-1703-0057

Junior Officer Maintenance

Course Number: 8C-77D; 8C-F21.
Location: Armor Center, Fort Knox, KY.
Length: 6 weeks (167–214 hours).
Exhibit Dates: 1/86–10/99.
Learning Outcomes: After 9/94 see AR-1703-0064. Upon completion of the course, the student will be able to supervise the maintenance of wheeled and tracked vehicles and associated armament equipment.
Instruction: Instruction includes lectures and practical exercises in maintenance and management of wheeled and tracked vehicles and associated armament equipment. Topics include records; maintenance forms; administrative control of licensing and dispatch; use and control of tools, test equipment, and repair parts; and maintenance management. Emphasis is also placed on vehicle recovery and power plant maintenance and troubleshooting.
Credit Recommendation: In the upper-division baccalaureate category, 4 semester hours in equipment maintenance management (10/89).

AR-1703-0058

Track Vehicle Repairer Individual Ready Reserve Refresher

Course Number: 63H30 (IRR).
Location: Ordnance Center and School, Aberdeen Proving Ground, MD.
Length: 2 weeks (77 hours).
Exhibit Dates: 10/86–9/90.
Learning Outcomes: Upon completion of the course, the student will be able to disassemble, identify, inspect, reassemble, and test charging system components and diesel fuel system components and troubleshoot and repair the electrical and fuel systems on Cummins diesel engines, turbine engines, and power generation systems.
Instruction: Course includes lectures and hands-on training, stressing diagnosing, troubleshooting, and repairing malfunctions, on Cummins diesel engines, turbine engines, power generation units, electrical and fuel systems. Course also includes new equipment updates.
Credit Recommendation: In the lower-division baccalaureate/associate degree category, 2 semester hours in heavy equipment mechanic (3/87).
Related Occupations: 63H.

AR-1703-0059

Motor Transport Operator
(Motor Vehicle Operator)

Course Number: *Version 1:* 811-64C10; 811-3521 (OS); 811-88M10. *Version 2:* 811-88M10.
Location: *Version 1:* Air Defense Artillery School, Fort Bliss, TX; Training Center, Fort Leonard Wood, MO. *Version 2:* ITRO Motor Transportation School, Fort Leonard Wood, MO.
Length: *Version 1:* 5–8 weeks (182–302 hours). *Version 2:* 6 weeks (222 hours).
Exhibit Dates: *Version 1:* 4/85–9/97. *Version 2:* 10/97–Present.
Learning Outcomes: *Version 1:* Upon completion of this interservice training course, the student will be able to operate light and medium tactical wheeled vehicles under varying conditions on/off road, over a commitment route, and in convoy operations; perform vehicle recovery; perform operator maintenance checks on the vehicle; perform vehicle inspection; and prepare operator, maintenance, and accident forms. *Version 2:* Upon completion of this interservice training course, the student will be able to operate light and medium tactical wheeled vehicles under varying conditions on and off the road, over a commitment route, and in convoy operations; perform vehicle recovery; perform operator maintenance; perform vehicle inspection; and prepare operator, maintenance, and accident forms.
Instruction: *Version 1:* Lectures and practical exercises cover the operation of five-ton cargo trucks and teaches motor skills such as controlling speed, space management, night driving, skid control, accident procedures and fires, winter driving, and hazard perception. Also covered are backing while driving on roadway or through streams, brush, mud, sand, and ice; executing downhill braking procedures; identifying flags; communicating by radio; defending against attack; and driving through contaminated areas. Course covers use of strip map in transport operations; personal

health and hazardous materials rules; and basic principles of inspecting, balancing, weighing, and securing cargo. This is an interservice course. *Version 2:* Methods of instruction include classroom and practical exercises covering the operation of five-ton cargo trucks and teaches motor skills such as controlling speed, space management, night driving, skid control, accident procedures and fires, winter driving, and hazard perception. Also covered are braking while driving on roadway or through streams, brush, mud sand and ice; executing downhill braking procedures; identifying flags; communicating by radio; defending against attack; and driving through contaminated areas. Course covers use of strip map in transport operations; personal health and hazardous materials rules; and basic principles of inspecting, balancing, weighing, and securing cargo. This is an interservice course.

Credit Recommendation: ***Version 1:*** In the vocational certificate category, 2 semester hours in truck driving (10/00). *Version 2:* In the lower-division baccalaureate/associate degree category, 2 semester hours in truck driving or operation of vehicles in difficult terrain (10/03).

Related Occupations: 64C; 88M.

AR-1703-0061

Fuel and Electrical Systems Repairer

Course Number: 610-3524.

Location: Ordnance Center and School, Aberdeen Proving Ground, MD.

Length: 4–5 weeks (158 hours).

Exhibit Dates: 1/88–4/04.

Learning Outcomes: Upon completion of the course, the Marine student will be able to perform basic gasoline and diesel engine electrical and fuel system maintenance and repair, use diagnostic equipment, and troubleshoot these system.

Instruction: Instruction for Marines includes applied electricity, ignition, starting, and charging systems; gasoline and diesel fuel systems; carburetor repair; two-barrel overhaul; injector pumps and injector nozzle; and power and injector pump.

Credit Recommendation: In the lower-division baccalaureate/associate degree category, 2 semester hours in diesel engine laboratory and 2 in automotive electrical systems laboratory (12/90).

AR-1703-0062

Engineer Tracked Vehicle Crewman Basic Noncommissioned Officer (NCO)

Course Number: 030-12F30; 030-18-12F30/030-20-12F30.

Location: Training Centers, Europe; Engineer School, Fort Leonard Wood, MO.

Length: 7–8 weeks (265–295 hours).

Exhibit Dates: 10/91–9/96.

Learning Outcomes: Upon completion of the course the student will be able to operate a tracked vehicle and prepare the vehicle for firing.

Instruction: Lectures and practical exercises cover the above topics. This course contains a common core of leadership subjects.

Credit Recommendation: Credit is recommended for the common core only. See AR-1406-0090 (1/93).

Related Occupations: 12F.

AR-1703-0063

Maintenance, Reserve

Course Number: 1100; 921-1100-LVL1.

Location: Reserve Readiness Training Center, Fort McCoy, WI.

Length: 2 weeks (68 hours).

Exhibit Dates: 4/93–Present.

Learning Outcomes: Upon completion of the course, the student will be able to perform basic unit maintenance administration.

Instruction: Course includes practical exercises, readings, lectures and discussions, audio-visuals, quizzes, and tests. Topics include establishing and maintaining records and files needed to administer maintenance operations, motor pool layout, and safety and physical security of the motor pool.

Credit Recommendation: Credit is not recommended because of the military-specific nature of the course (3/97).

AR-1703-0064

Battalion Maintenance Officer

Course Number: 8C-F21.

Location: Armor Center and School, Fort Knox, KY.

Length: 4–5 weeks (176 hours).

Exhibit Dates: 10/94–9/00.

Learning Outcomes: Before 10/94 see AR-1703-0057. Upon completion of the course, the student will be able to manage and supervise vehicle maintenance.

Instruction: Lectures, demonstrations, and practical exercises are used. Topics include maintenance forms and records; licensing and dispatch; use and control of tools and test equipment; parts and materials; supervisors responsibility for planning, coordinating, and controlling maintenance services; and familiarization with components and functions of vehicle systems and power generation equipment.

Credit Recommendation: In the lower-division baccalaureate/associate degree category, 2 semester hours in service management, 1 in records management, and 1 in supply management (3/97).

AR-1703-0065

Self-Propelled Field Artillery System Mechanic

Course Number: ***Version 1:*** 611-63D10. *Version 2:* 611-63D10.

Location: Ordnance Center and School, Aberdeen Proving Ground, MD.

Length: ***Version 1:*** 9–10 weeks (348 hours). *Version 2:* 8–9 weeks (298 hours).

Exhibit Dates: ***Version 1:*** 10/95–9/99. *Version 2:* 10/99–Present.

Learning Outcomes: ***Version 1:*** Upon completion of the course, the student will be able to perform maintenance on automotive systems and components of self-propelled artillery and cargo carriers. *Version 2:* Upon completion of the course, the student will be able to perform maintenance on automotive systems and components of self-propelled artillery and tracked cargo carriers.

Instruction: ***Version 1:*** Lectures and practical exercises cover service, maintenance, parts replacement, and troubleshooting of electrical systems and power trains using test/diagnostic equipment. *Version 2:* Practical exercises and lecture. Topics include service, maintenance, parts replacement, and troubleshooting of electrical, mechanical, and hydraulic systems, using test/diagnostic equipment.

Credit Recommendation: ***Version 1:*** In the lower-division baccalaureate/associate degree category, 2 semester hours in basic hydraulic systems, 2 in automotive electrical systems, 2 in gasoline and diesel engines, and 1 in automotive fuel systems (1/98). *Version 2:* In the lower-division baccalaureate/associate degree category, 2 semester hours in basic electrical systems, 2 in basic hydraulic systems, 2 in basic mechanical systems, and 2 in gasoline/diesel engines (8/03).

Related Occupations: 63D.

AR-1703-0066

Light Wheel Vehicle Mechanic Basic Noncommissioned Officer

Course Number: 091-63B30, Phase 3.

Location: Reserve Schools, Multiple locations.

Length: 3–4 weeks (120 hours).

Exhibit Dates: 10/00–Present.

Learning Outcomes: Upon completion of the course, the student will be able to use diagnostic equipment to troubleshoot and maintain wheeled and track vehicles, as well as understand basic vehicle recovery procedures.

Instruction: Audio visual, practical exercises, classroom exercises, discussion, laboratory, and lecture. Topics include lectures and practical exercises covering troubleshooting and repairing vehicle systems (wheeled and track), fuel (gas and diesel), lubrication, hydraulic, chassis and suspension, vehicle electrical and cooling, and basic vehicle recovery procedures. This course contains a common core of leadership subjects.

Credit Recommendation: In the lower-division baccalaureate/associate degree category, 3 semester hours in vehicle equipment repair (8/03).

AR-1703-0067

Light Wheel Vehicle Mechanic Basic Noncommissioned Officer

Course Number: 610-63B30.

Location: Ordnance Center and School, Aberdeen Proving Ground, MD.

Length: 3–4 weeks (120 hours).

Exhibit Dates: 10/00–Present.

Learning Outcomes: Upon completion of the course, the student will be able to use diagnostic equipment to troubleshoot and maintain wheeled and track vehicles, as well as understand basic vehicle recovery procedures.

Instruction: Audio visual, practical exercises, classroom exercises, discussion, laboratory, and lecture. Topics include lectures and practical exercises covering troubleshooting and repairing vehicle systems (wheeled and track), fuel (gas and diesel), lubrication, hydraulic, chassis and suspension, vehicle electrical and cooling, and basic vehicle recovery procedures. This course contains a common core of leadership subjects.

Credit Recommendation: In the lower-division baccalaureate/associate degree category, 3 semester hours in vehicle equipment repair. See AR-1406-0090 for common core credit (8/03).

AR-1703-0068

TATS Fuel and Electrical Systems Repairer

Course Number: 091-63G10.

Location: Ordnance Center and School, Aberdeen Proving Ground, MD.

Length: 104 hours.

Exhibit Dates: 10/99–Present.

Learning Outcomes: Upon completion of the course, the student will be able to use appropriate technical manuals, precision test measurement and diagnostic equipment, and troubleshoot electrical equipment, using schematics.

Instruction: Audiovisual materials, learner presentations, computer-based training, practical exercises, classroom exercises, discussion, and laboratory. Topics include fundamental shop skills, electrical repair (wiring harnesses, etc.), and repair/troubleshooting of electrical components.

Credit Recommendation: In the lower-division baccalaureate/associate degree category, 1 semester hour in shop safety and 4 in automotive electrical systems (8/03).

AR-1703-0071

TATS Track Vehicle Repairer

Course Number: 091-63H10, Phase 2.

Location: Reserve Schools, Multiple locations.

Length: 111 hours.

Exhibit Dates: 1/99–Present.

Learning Outcomes: Upon completion of the course, the student will be able to perform maintenance and troubleshooting on tracked heavy-duty equipment vehicles utilizing electrical and mechanical systems knowledge of multi-cylinder engines, transmissions, and turbine systems.

Instruction: Audiovisual materials, computer-based training, practical exercises, classroom exercises, discussion, laboratory, and lecture. Topics include shop practice using hand tools and measurement devices for electrical and mechanical systems in the maintenance and troubleshooting support of multi-cylinder, transmissions and turbines.

Credit Recommendation: In the lower-division baccalaureate/associate degree category, 2 semester hours in basic electrical systems, 2 in basic mechanical systems, and 1 in diesel/gas engine repair (8/03).

AR-1703-0072

Track Vehicle Mechanic (TATS)

Course Number: 091-63Y10, Phase 2.

Location: Ordnance Center and School, Aberdeen Proving Ground, MD.

Length: 115 hours.

Exhibit Dates: 10/98–Present.

Learning Outcomes: Upon completion of the course, the student will be able to perform routine maintenance on automotive heavy equipment systems and components, troubleshoot, diagnose and repair automotive heavy equipment systems, practice automotive laboratory safety, and effectively use service and repair manuals.

Instruction: Audio visual materials, practical exercises, classroom exercises, discussion, laboratory, and lecture. Topics include automotive laboratory safety, automotive vehicle preventative maintenance, diagnosis and repair of automotive power plant systems and driveline components, effective use of equipment service manuals, and use of electrical test equipment.

Credit Recommendation: In the lower-division baccalaureate/associate degree category, 2 semester hours in automotive heavy equipment service (8/03).

AR-1703-0073

Track Vehicle Mechanic (Reclassification)

Course Number: 091-63Y10-TATS.

Location: Reserve Schools, Multiple locations.

Length: 3–4 weeks (197 hours).

Exhibit Dates: 10/01–Present.

Learning Outcomes: Upon completion of the course, the student will be able to use automotive service tools and measuring instruments, understand basic automotive vehicle and operating principles, perform routine preventative maintenance checks, basic troubleshooting, and diagnosis of automotive systems, and understand and practice safety rules and regulations of automotive service.

Instruction: Audio visual materials, learner presentations, practical exercises, classroom exercises, discussion, laboratory, and lecture. Topics include use of automotive service tools and equipment, safety in the automotive/heavy equipment laboratory, basic systems in automotive troubleshooting and diagnosis, and basic automotive engine design.

Credit Recommendation: In the lower-division baccalaureate/associate degree category, 3 semester hours in automotive fundamentals and 2 in general automotive service (8/03).

AR-1703-0074

Wheel Recovery Vehicle Specialist

Course Number: 091-ASIH8 (63B/S/W).

Location: Reserve Schools, Multiple locations.

Length: 1–2 weeks (85 hours).

Exhibit Dates: 12/01–Present.

Learning Outcomes: Upon completion of the course, the student will be able to operate and maintain recovery vehicles for recovering and transporting wheeled vehicles.

Instruction: Audio visual materials, practical exercises, classroom exercises, discussion, laboratory, and lecture. Topics include extensive practical exercises with appropriate field demonstrations of the safe recovery and transport of wheeled vehicles.

Credit Recommendation: In the vocational certificate category, 2 semester hours in rigging systems. In the lower-division baccalaureate/associate degree category, 1 semester hour in mechanical safety systems laboratory (8/03).

AR-1703-0075

Wheel Recovery Vehicle Specialist

Course Number: 611-ASIH8 (63W).

Location: Ordnance Center and School, Aberdeen Proving Ground, MD.

Length: 2–3 weeks (85 hours).

Exhibit Dates: 10/99–Present.

Learning Outcomes: Upon completion of the course, the student will be able to operate and maintain recovery vehicles for recovering and transporting wheeled vehicles.

Instruction: Audio visual materials, practical exercises, classroom exercises, discussion, laboratory, and lecture. Topics include extensive practical exercises with appropriate field demonstrations of the safe recovery and transport of wheeled vehicles.

Credit Recommendation: In the vocational certificate category, 2 semester hours in rigging systems. In the lower-division baccalaureate/associate degree category, 1 semester hour in mechanical safety systems laboratory (8/03).

AR-1703-0076

Wheel Vehicle Repairer

Course Number: 091-63W10.

Location: Reserve Schools, Multiple locations.

Length: 109 hours.

Exhibit Dates: 10/99–Present.

Learning Outcomes: Upon completion of the course, the student will be able to perform maintenance and troubleshooting on wheeled automotive vehicles utilizing electrical, hydraulic, and mechanical systems knowledge of automotive engines, transmissions, suspension, and brake systems.

Instruction: Audio visual materials, computer-based training, practical exercises, classroom exercises, discussion, laboratory, and lecture. Topics include shop practice using hand tools and measurements devices for electrical, hydraulic, and mechanical systems on automotive engines, transmissions, suspensions, and brakes.

Credit Recommendation: In the lower-division baccalaureate/associate degree category, 2 semester hours in automotive systems repair, 1 in basic electrical systems, 1 in basic hydraulic systems, and 1 in basic mechanical systems (8/03).

AR-1703-0077

Light Wheel Vehicle Mechanic

Course Number: 091-63B10-TATSC, Phase 1.

Location: Reserve Schools, Multiple locations.

Length: 109 hours.

Exhibit Dates: 10/98–Present.

Learning Outcomes: Upon completion of the course, the student will be able to understand the basic operating principles of automotive vehicles, use tools and equipment to complete basic service on the engine and drive line, and do basic, routine repair of automotive brake and suspension systems.

Instruction: Audio visual materials, practical exercises, classroom exercises, discussion, laboratory, and lecture. Topics include automotive power plants and power trains, engine support systems, basic electricity and automotive electrical systems, and overview of brakes and suspension systems on light automotive vehicles.

Credit Recommendation: In the lower-division baccalaureate/associate degree category, 3 semester hours in general automotive laboratory (8/03).

AR-1703-0078

Heavy Wheel Vehicle Mechanic

Course Number: 091-63S10 (TATS).

Location: Reserve Schools, Multiple locations.

Length: 116 hours.

Exhibit Dates: 10/97–Present.

Learning Outcomes: Upon completion of the course, the student will be able to apply basic operating principles of automotive heavy equipment, and perform basic maintenance and service heavy automotive vehicles.

Instruction: Audio visual materials, practical exercises, discussion, laboratory, and lecture. Topics include overview of basic operating principles of internal combustion engines and heavy equipment drive trains, basic electricity and automotive electrical service, overview of brakes and suspension systems on heavy automotive vehicles, and basic hydraulics and hydraulic system service.

Credit Recommendation: In the lower-division baccalaureate/associate degree category, 3 semester hours in heavy automotive equipment basic service (8/03).

AR-1704-0003

Air Traffic Control Radar Controller
(Air Traffic Control Ground Control Approach Specialist)

Course Number: 222-93J20; 222-93J10.

Location: Aviation Center, Fort Rucker, AL.

Length: 13–17 weeks (482–553 hours).

Exhibit Dates: 5/73–1/90.

Objectives: To train enlisted personnel in air traffic control UFR/IFR operations.

Instruction: Course includes practical exercises and lectures in aviation weather, navigational aids, Federal Aviation Administration regulations, and air traffic control procedures.

Credit Recommendation: In the lower-division baccalaureate/associate degree category, 6 semester hours in air traffic control (11/86).

Related Occupations: 93J.

AR-1704-0008

Aviation Life Support Equipment Technician

Course Number: 2C-F51/4D-ASI1F/600-ASIQ2; 2C-F51/4D-ASIH2/600-ASIQ2.

Location: Transportation School, Fort Eustis, VA.

Length: 5–6 weeks (196–204 hours).

Exhibit Dates: 3/88–9/91.

Objectives: After 9/91 see AR-1408-0212. To train personnel to maintain and repair life support survival equipment.

Instruction: Lectures and practical exercises cover the testing and maintenance of life support systems, including life preservers, life rafts, armor vests, helmets, oxygen masks, and first aid kits. The course also includes publications, forms, and record control procedures and shop, supply, and safety procedures.

Credit Recommendation: Credit is not recommended because of the military-specific nature of the course (10/91).

AR-1704-0012

Aircraft Power Train Repairer
(Aircraft Power Train Repair)

Course Number: 602-68D10.

Location: Transportation School, Fort Eustis, VA; Transportation and Aviation Logistics School, Fort Eustis, VA.

Length: 13–14 weeks (463–493 hours).

Exhibit Dates: 1/86–9/94.

Objectives: After 9/94 see AR-1704-0222. To train enlisted personnel to maintain and repair helicopter power trains and associated equipment.

Instruction: Practical experience covers helicopter power train maintenance and repair, structural hardware, equipment record systems; bearings, gears, and component preservation; and penetrant, magnetic particle, and radiographic and ultrasonic inspection. The course also includes nondestructive testing.

Credit Recommendation: In the vocational certificate category, 12 semester hours in helicopter power train maintenance and repair. In the lower-division baccalaureate/associate degree category, 6 semester hours in theory, operation, maintenance, and repair procedures on helicopter power train systems (10/89).

Related Occupations: 68D.

AR-1704-0018

UH-60 Helicopter Repairer Transition

Course Number: 600-67T20-T.

Location: Transportation School, Fort Eustis, VA.

Length: 8–9 weeks (252 hours).

Exhibit Dates: 5/78–12/93.

Objectives: To train enlisted personnel to maintain and repair the UH-60 series helicopter.

Instruction: Lectures and practical exercises cover aircraft maintenance and repair procedures; helicopter equipment; auxiliary equipment; and power plant and power train maintenance, repair, and inspections including rotor and blade systems.

Credit Recommendation: In the lower-division baccalaureate/associate degree category, 3 semester hours in helicopter maintenance (3/83).

Related Occupations: 67T.

AR-1704-0033

Utility Helicopter Repairer (UH-1)

Course Number: 600-67N20; 600-67N10 (ST); 600-67N10.

Location: Transportation School, Fort Eustis, VA; Aviation School, Fort Rucker, AL.

Length: 10–11 weeks (388–396 hours).

Exhibit Dates: 7/84–3/92.

Objectives: After 3/92 see AR-1704-0205. To provide personnel with the training necessary to perform organizational, direct, and general support maintenance tasks on the UH-1 utility helicopter.

Instruction: Lectures and practical exercises cover helicopter system fundamentals, including airframe, fuel, lubrication, hydraulic, rotor, and electrical maintenance and repair; inspection and troubleshooting of helicopter systems; flight control; and ground-handling equipment maintenance.

Credit Recommendation: In the vocational certificate category, 7 semester hours in helicopter maintenance. In the lower-division baccalaureate/associate degree category, 3 semester hours in helicopter maintenance (4/92).

Related Occupations: 002; 09B; 67N; 67W; 67Z.

AR-1704-0038

Observation Airplane Repairer
(OV-1 Airplane Repair)

Course Number: 600-67H10.

Location: Transportation and Aviation Logistics School, Fort Eustis, VA.

Length: 9–12 weeks (341–498 hours).

Exhibit Dates: 7/83–12/93.

Objectives: To train enlisted personnel in entry-level maintenance of multi-engine, turbine-powered aircraft.

Instruction: Lectures and practical exercises cover the fundamentals of multi-engine, turbine-powered aircraft, including operation, maintenance, troubleshooting, and repair of hydraulic, electrical, instrument, fuel, power plant, propeller, landing gear, and flight control systems.

Credit Recommendation: In the lower-division baccalaureate/associate degree category, 5 semester hours in aviation maintenance (7/85).

Related Occupations: 67F; 67H; 67Z.

AR-1704-0044

Smoke Operation Specialist

Course Number: 494-54C10.

Location: Chemical School, Fort McClellan, AL.

Length: 6–7 weeks (265–290 hours).

Exhibit Dates: 6/80–3/92.

Objectives: To provide enlisted personnel with training in smoke operations and fuel handling.

Instruction: Lectures, demonstrations, and practical experience cover chemical, smoke generator, fuel-handling, and tactical operations.

Credit Recommendation: Credit is not recommended because of the military-specific nature of the course (3/82).

Related Occupations: 54C.

AR-1704-0046

Smoke Operations Specialist (Transition)

Course Number: 494-54C20-T; 494-54C20T.

Location: Chemical School, Fort McClellan, AL.

Length: 6 weeks (193 hours).

Exhibit Dates: 8/81–1/92.

Objectives: To train reclassified enlisted personnel for jobs as smoke and fuel supply team/squad leaders.

Instruction: Conferences and practical exercises cover decontamination and operation and maintenance of smoke generators, wind measuring sets, tank and pump units, and hand-driven dispensing pumps.

Credit Recommendation: Credit is not recommended because of the military-specific nature of the course (3/82).

Related Occupations: 54C.

AR-1704-0062

Aircraft Armament Maintenance Technician

Course Number: 4D-SQIE.

Location: Transportation and Aviation Logistics School, Fort Eustis, VA.

Length: 12–13 weeks (444 hours).

Exhibit Dates: 1/84–9/90.

Objectives: To teach officers to supervise aircraft armament maintenance personnel in weapons system and fire control system repair.

Instruction: Lectures and practical exercises cover helicopter armament systems and subsystems, basic electricity and electronics, and missile systems and subsystems.

Credit Recommendation: In the lower-division baccalaureate/associate degree category,

gory, 3 semester hours in electricity and electronics theory (7/85).

Related Occupations: 100B; 100E.

AR-1704-0066

3. Observation/Scout Helicopter Repairer
(OH-58 Helicopter Repairer)
4. Observations/Scout Helicopter Repairer

Course Number: *Version 3:* 600-67V10; 600-67V10 (ST) (OH-58); 600-67V10 (OH-58). *Version 4:* 600-67V10 (OH-58), Phase 1.

Location: *Version 3:* Aviation School, Fort Rucker, AL. *Version 4:* Aviation Center, Fort Rucker, AL.

Length: *Version 3:* 8–10 weeks (214–356 hours). *Version 4:* 9 weeks (368 hours).

Exhibit Dates: *Version 3:* 1/80–9/98. *Version 4:* 10/98–Present.

Learning Outcomes: *Version 3:* To train enlisted personnel in organizational, direct, and general support maintenance of the OH-58 helicopter. *Version 4:* Upon completion of the course, the student will be able to inspect, maintain, troubleshoot, replace, and repair various helicopter systems and components.

Instruction: *Version 3:* Lectures and practical exercises cover maintenance of OH-58 helicopters, including power plant, rotor gear train (main and tail rotor), and hydraulics. Course also includes inspection procedures, electrical system inspection and troubleshooting, and fundamentals of alternating current and direct current. *Version 4:* Practical exercises, and lecture. Topics include airframe inspection and repair; maintenance, inspection, and repair of helicopter hydraulic, electrical, landing gear, tail rotor, drive train, flight control, power plant, and main rotor systems.

Credit Recommendation: *Version 3:* In the vocational certificate category, 6 semester hours in helicopter repair. In the lower-division baccalaureate/associate degree category, 3 semester hours in helicopter repair (4/92). *Version 4:* In the lower-division baccalaureate/associate degree category, 6 semester hours in aviation maintenance or 3 in aviation maintenance, and 3 in basic aircraft systems (4/03).

Related Occupations: 00Z; 67V; 67W; 67Z.

AR-1704-0075

Aircraft Electrician

Course Number: 602-68F10.

Location: Transportation School, Fort Eustis, VA; Transportation and Aviation Logistics School, Fort Eustis, VA.

Length: 21–24 weeks (738–768 hours).

Exhibit Dates: 10/84–9/94.

Objectives: After 9/94 see AR-1704-0223. To train enlisted personnel to perform direct and general support maintenance on aircraft electrical systems and components.

Instruction: Lectures and practical exercises cover basic electrical and electronic theory, soldering techniques, AC and DC circuits, semiconductors, wiring diagrams, and test equipment. Reinforcement of these skills is accomplished by operating, testing, troubleshooting, and repairing aircraft electrical systems.

Credit Recommendation: In the lower-division baccalaureate/associate degree category, 3 semester hours in fundamentals of electricity, 3 in basic electronics, 6 in electrical/electronics laboratory, 3 in aircraft electrical systems, and 3 in aircraft instrument repair (4/92).

Related Occupations: 68F.

AR-1704-0087

Ejection Seat Repairer
(Martin-Baker J-5 Ejection Seat Specialist)

Course Number: 600-F6; 600-ASIB7; 600-ASIB7 (67H).

Location: Transportation and Aviation Logistics School, Fort Eustis, VA; Transportation School, Fort Eustis, VA.

Length: 3–4 weeks (108–128 hours).

Exhibit Dates: 7/67–12/93.

Objectives: To train personnel with previous technical experience to perform as J-5 ejection seat specialists.

Instruction: Lectures and practical exercises cover the inspection, repair, and organizational and support maintenance of Martin-Baker J-5 ejection seats, aircraft survival seat kits, and 9E2A ejection seat simulators; ejection seat removal, disassembly, inspection, maintenance, and reassembly; personnel parachutes; OV-1 aircraft survival seat kits; SEEK-2 life vest; fitting and adjusting personal equipment; introduction to 9E2A ejection seat simulators; function of related systems; and physiological effects.

Credit Recommendation: Credit is not recommended because of the limited, specialized nature of the course (6/91).

Related Occupations: 67H; 67K.

AR-1704-0093

Observation/Scout Helicopter Repairer (OH-6)

Course Number: 600-67V10 (OH-6).

Location: Transportation School, Fort Eustis, VA.

Length: 7 weeks (214 hours).

Exhibit Dates: 7/87–1/92.

Objectives: To train personnel to provide direct and general support maintenance for the scout helicopter.

Instruction: Lectures and practical exercises cover general aircraft maintenance and repair procedures on helicopter equipment, auxiliary equipment, and power plant.

Credit Recommendation: In the vocational certificate category, 6 semester hours in helicopter repair. In the lower-division baccalaureate/associate degree category, 3 semester hours in helicopter repair (3/83).

Related Occupations: 67V.

AR-1704-0095

AH-1 Attack Helicopter Repairer
(Attack Helicopter Repairer)
(Attack Helicopter Repairman)

Course Number: 600-67Y10.

Location: Transportation and Aviation Logistics School, Fort Eustis, VA.

Length: 11–13 weeks (418–455 hours).

Exhibit Dates: 7/84–9/95.

Objectives: After 9/95 see AR-1704-0221. To train enlisted personnel in entry-level mechanical tasks related to the repair of helicopters.

Instruction: Lectures, programmed instruction, and practical exercises cover general aircraft maintenance procedures, helicopter equipment, auxiliary equipment, and power plant maintenance and inspection.

Credit Recommendation: In the lower-division baccalaureate/associate degree category, 3 semester hours in helicopter maintenance and 2 in helicopter systems inspection and repair (4/92).

Related Occupations: 67Y.

AR-1704-0100

Aircraft Structure Repair

Course Number: 603-68G10.

Location: Transportation School, Fort Eustis, VA; Transportation and Aviation Logistics School, Fort Eustis, VA.

Length: 15 weeks (547 hours).

Exhibit Dates: 7/84–9/92.

Objectives: After 9/92 see AR-1704-0207. To train enlisted personnel in entry-level tasks in the maintenance of aircraft structural members, sheet metal surfaces, fiberglass, and rotor blade repair.

Instruction: Course involves practical exercises including aircraft riveting; repair of aircraft skin, structural members, and rotor blades; maintenance and equipment record systems; forming of aircraft parts; and shop and flight-line safety.

Credit Recommendation: In the lower-division baccalaureate/associate degree category, 3 semester hours in airframe sheet metal shop and 3 in airframe structural repairs (1/88).

Related Occupations: 68G.

AR-1704-0106

Aircraft Power Plant Repairer Supervisor Basic Noncommissioned Officer (NCO)
(Aircraft Power Plant Repairer Basic Noncommissioned Officer (NCO))
(Aircraft Power Plant Repairer Basic Technical)

Course Number: *Version 2:* 601-68B30. *Version 3:* 601-68B30.

Location: Transportation and Aviation Logistics School, Fort Eustis, VA.

Length: *Version 2:* 4–9 weeks (129–221 hours). *Version 3:* 10 weeks (368 hours).

Exhibit Dates: *Version 2:* 10/85–9/91. *Version 3:* 10/91–9/96.

Objectives: *Version 2:* To train selected enlisted personnel in aviation maintenance management principles and procedures. *Version 3:* After 9/96 see AR-1704-0218. To train selected enlisted personnel in aviation maintenance management principles and procedures.

Instruction: *Version 2:* The course consists of lectures and practical experiences in applying the policies and procedures associated with approved aviation maintenance standards. *Version 3:* The course consists of lectures and practical experiences in applying the policies and procedures associated with approved aviation maintenance standards. This course contains a common core of leadership subjects.

Credit Recommendation: *Version 2:* In the lower-division baccalaureate/associate degree category, 2 semester hours in aviation maintenance management and 3 in principles of supervision (3/88). *Version 3:* In the lower-division baccalaureate/associate degree category, 2 semester hours in aviation maintenance management and 3 in principles of supervision. See AR-1406-0090 for common core credit (3/88).

Related Occupations: 68B.

AR-1704-0107

Aircraft Power Train Repairer Basic Noncommissioned Officer (NCO)
(Aircraft Power Train Repairer Basic Technical)

Course Number: 602-68D30.
Location: Transportation and Aviation Logistics School, Fort Eustis, VA.
Length: 3–5 weeks (117–175 hours).
Exhibit Dates: 10/85–3/90.
Objectives: To train selected enlisted personnel in aviation maintenance management principles and procedures.
Instruction: Lectures and practical experiences cover the supervision of helicopter power train systems maintenance and repair. This course contains a common core of leadership subjects.
Credit Recommendation: In the lower-division baccalaureate/associate degree category, 3 semester hours in supervision of aircraft power train system maintenance. See AR-1406-0090 for common core credit (7/85).
Related Occupations: 68D.

AR-1704-0108

Aircraft Electrician Repairer Supervisor Basic Noncommissioned Officer (NCO)
(Aircraft Electrician Repairer Basic Noncommissioned Officer (NCO))
(Aircraft Electrician Repairer Basic Technical)

Course Number: 602-68F30.
Location: Transportation and Aviation Logistics School, Fort Eustis, VA.
Length: 9–12 weeks (354–505 hours).
Exhibit Dates: 10/85–3/90.
Objectives: To train selected enlisted personnel in aviation maintenance management principles and procedures.
Instruction: Lectures and practical exercises cover conducting thorough inspections on aircraft including electrical subsystems and components and applying approved aviation maintenance standards. This course contains a common core of leadership subjects.
Credit Recommendation: In the lower-division baccalaureate/associate degree category, 2 semester hours in basic electronics and 2 in aviation maintenance management. See AR-1406-0090 for common core credit (3/88).
Related Occupations: 68F.

AR-1704-0109

Aircraft Structural Repairer Basic Noncommissioned Officer (NCO)
(Aircraft Structural Repairer Basic Technical)

Course Number: 603-68G30.
Location: Transportation and Aviation Logistics School, Fort Eustis, VA.
Length: 3–7 weeks (105–200 hours).
Exhibit Dates: 10/84–3/90.
Objectives: After 3/90 see AR-1704-0184. To train selected enlisted personnel in aviation maintenance management principles and procedures.
Instruction: The course consists of lectures and practical experiences in applying approved aviation maintenance standards. This course contains a common core of leadership subjects.
Credit Recommendation: In the lower-division baccalaureate/associate degree category, 2 semester hours in aviation maintenance management. See AR-1406-0090 for common core credit (7/85).
Related Occupations: 68G.

AR-1704-0114

Aircraft Pneudraulic Repairer Basic Noncommissioned Officer (NCO)

Course Number: 603-68H30.
Location: Transportation and Aviation Logistics School, Fort Eustis, VA.
Length: 5–6 weeks (100 hours).
Exhibit Dates: 10/85–3/90.
Objectives: After 3/90 see AR-1704-0187. To train selected enlisted personnel in aviation maintenance management principles and procedures.
Instruction: The course consists of lectures and practical experiences in applying approved aviation maintenance standards to aircraft pneudraulic systems. This course contains a common core of leadership subjects.
Credit Recommendation: In the lower-division baccalaureate/associate degree category, 2 semester hours in aviation maintenance management. See AR-1406-0090 for common core credit (5/86).
Related Occupations: 68H.

AR-1704-0115

Utility/Cargo Airplane Repairer Basic Technical

Course Number: 600-67G30.
Location: Transportation and Aviation Logistics School, Fort Eustis, VA.
Length: 5–6 weeks (188 hours).
Exhibit Dates: 10/85–12/93.
Objectives: To train selected enlisted personnel in aviation maintenance management principles and procedures.
Instruction: The course consists of lectures and practical experiences in applying approved aviation maintenance standards to utility/cargo aircraft.
Credit Recommendation: In the lower-division baccalaureate/associate degree category, 2 semester hours in aircraft maintenance management and 3 in principles of supervision (5/86).
Related Occupations: 67G.

AR-1704-0118

Heavy Lift Helicopter Repairer Basic Technical

Course Number: 600-67X30.
Location: Transportation and Aviation Logistics School, Fort Eustis, VA.
Length: 3 weeks (105 hours).
Exhibit Dates: 10/84–12/93.
Objectives: To train selected enlisted personnel in aviation maintenance management principles and procedures.
Instruction: The course consists of lectures and practical experiences in applying approved aviation maintenance standards to helicopter repair.
Credit Recommendation: In the lower-division baccalaureate/associate degree category, 2 semester hours in aviation maintenance management (7/85).
Related Occupations: 67X.

AR-1704-0119

Aircraft Fire Control Repairer Basic Technical

Course Number: 600-68J30.
Location: Transportation and Aviation Logistics School, Fort Eustis, VA.
Length: 15 weeks (538 hours).
Exhibit Dates: 3/83–7/95.
Objectives: To train enlisted personnel to perform and supervise the maintenance and repair of aircraft fire control systems and associated subsystems.
Instruction: Lectures and practical experience cover supervising and repairing aircraft fire control systems and subsystems. The course also includes weapon systems and armament electrical/electronic training.
Credit Recommendation: In the lower-division baccalaureate/associate degree category, 3 semester hours in electrical/electronics, 2 in electrical laboratory, and 3 in supervision and inspection (7/85).
Related Occupations: 68J.

AR-1704-0123

UH-1 Utility Helicopter Technical Inspector
(Utility Helicopter Technical Inspector)

Course Number: 600-66N20.
Location: Transportation and Aviation Logistics School, Fort Eustis, VA.
Length: 12–13 weeks (446–459 hours).
Exhibit Dates: 7/85–3/90.
Objectives: To train enlisted personnel to inspect maintenance performed by helicopter repairers and allied trades personnel.
Instruction: Lectures and practical exercises cover inspections on helicopter systems, subsystems, and components before, during, and after maintenance activities to ensure compliance with proper helicopter configuration control. Course includes Army oil analysis, test equipment calibration, directives, technical manuals, work orders, standard operating procedure, and other policies and procedures.
Credit Recommendation: In the lower-division baccalaureate/associate degree category, 3 semester hours in helicopter inspections and 3 in power plant inspections (3/88).
Related Occupations: 66N.

AR-1704-0126

Observation/Scout Helicopter Technical Inspector

Course Number: 600-66V20.
Location: Transportation and Aviation Logistics School, Fort Eustis, VA.
Length: 10–14 weeks (372–488 hours).
Exhibit Dates: 10/84–12/93.
Objectives: To train enlisted personnel to inspect maintenance performed by helicopter repairers and allied trades personnel.
Instruction: Lecture and practical exercises cover inspections on helicopter systems, subsystems, and components before, during, and after maintenance activities to ensure compliance with proper helicopter configuration control. Course includes Army oil analysis, test equipment calibration, directives, technical manuals, work orders, unit standard operating procedure, and other policies and procedures.
Credit Recommendation: In the lower-division baccalaureate/associate degree category, 3 semester hours in helicopter inspections

and 3 semester hours in power plant inspections (3/88).

Related Occupations: 66V.

AR-1704-0127

ATTACK HELICOPTER TECHNICAL INSPECTOR

Course Number: 600-66Y20.

Location: Transportation and Aviation Logistics School, Fort Eustis, VA.

Length: 12–13 weeks (439–461 hours).

Exhibit Dates: 7/85–12/93.

Objectives: To train enlisted personnel to inspect maintenance performed by helicopter repairers and allied trades personnel.

Instruction: Lectures and practical exercises cover inspections on helicopter systems, subsystems, and components before, during, and after maintenance activities to ensure compliance with proper helicopter configuration control. Course also includes Army oil analysis, test equipment calibration, directives, technical manuals, work orders, unit standard operating procedure, and other policies and procedures.

Credit Recommendation: In the lower-division baccalaureate/associate degree category, 3 semester hours in helicopter inspections and 3 in power plant inspections (3/88).

Related Occupations: 66Y.

AR-1704-0130

CH-54 HELICOPTER REPAIRER

Course Number: 600-67X10.

Location: Transportation and Aviation Logistics School, Fort Eustis, VA.

Length: 13–14 weeks (481 hours).

Exhibit Dates: 10/84–12/93.

Objectives: To train personnel in entry-level mechanic tasks in helicopter repair.

Instruction: Lectures, programmed instruction, and practical exercises cover general aircraft maintenance procedures and helicopter equipment, power plant, and power train maintenance, repair, and inspection.

Credit Recommendation: In the lower-division baccalaureate/associate degree category, 3 semester hours in helicopter maintenance and 2 in helicopter systems inspections and repair (7/85).

Related Occupations: 67X.

AR-1704-0133

LIGHTER AIR-CUSHION VEHICLE OPERATOR

Course Number: 062-ASIE6 (61B).

Location: Transportation and Aviation Logistics School, Fort Eustis, VA.

Length: 9 weeks (324 hours).

Exhibit Dates: 7/86–9/93.

Objectives: After 9/93 see AR-1704-0210. To train selected personnel in the operation of lighter air-cushion vehicles.

Instruction: Lectures and practical exercises cover craft operation, use of emergency equipment and procedures, logistics procedures, and operational safety.

Credit Recommendation: Credit is not recommended because of the limited, specialized nature of the course (5/86).

Related Occupations: 61B.

AR-1704-0141

ARMAMENT/FIRE CONTROL MAINTENANCE SUPERVISOR NONCOMMISSIONED OFFICER (NCO) ADVANCED
(Mechanical Maintenance Noncommissioned Officer (NCO) Advanced, MOS 45Z)

Course Number: 6-63-C42 (45Z); 640-45Z40.

Location: Ordnance Center and School, Aberdeen Proving Ground, MD.

Length: 10–12 weeks (376–547 hours).

Exhibit Dates: 10/85–9/96.

Learning Outcomes: Upon completion of the course, students will be able to serve as supervisors in the military occupational specialty for which they are trained.

Instruction: Lectures, demonstrations, and practical exercises cover leadership, communication skills, fundamentals of instruction, logistics management, and refresher training in the appropriate military occupational specialty. Emphasis in on leadership and on verbal and written communication.

Credit Recommendation: In the lower-division baccalaureate/associate degree category, 1 semester hour in basic electricity. See AR-1404-0035 for common core credit (1/88).

Related Occupations: 45Z.

AR-1704-0142

M1 ABRAMS TANK SYSTEMS MECHANIC

Course Number: ***Version 1:*** 611-63E10. *Version 2:* 611-63E10.

Location: Armor Center, Fort Knox, KY.

Length: ***Version 1:*** 9–10 weeks (328–329 hours). *Version 2:* 13–14 weeks (479–502 hours).

Exhibit Dates: ***Version 1:*** 6/83–5/93. *Version 2:* 6/93–Present.

Learning Outcomes: ***Version 1:*** Upon completion of the course, the student will be able to test, troubleshoot, replace, repair, and adjust selected components and systems on various vehicles. *Version 2:* Upon completion of the course, the student will be able to supervise and perform unit maintenance on the hull and turret of the Abrams tank and perform unit maintenance on armored personnel carriers, recovery vehicles, and personnel vehicles.

Instruction: ***Version 1:*** Lectures, demonstrations, and practical exercises cover the use of special test equipment to troubleshoot and service electrical, suspension, driveline, and braking systems. *Version 2:* This course includes lectures, demonstrations, conferences and performance exercises in testing and troubleshooting systems; inspecting, servicing, lubricating, replacing, and adjusting components; and using publications and special tools.

Credit Recommendation: ***Version 1:*** In the lower-division baccalaureate/associate degree category, 1 semester hour in automotive electricity and 2 in heavy equipment repair (7/87). *Version 2:* In the lower-division baccalaureate/associate degree category, 3 semester hours in electrical, 2 in brakes, 3 in engine fundamentals, 2 in power train, 2 in suspension, and 1 in diesel engines (10/00).

Related Occupations: 63E.

AR-1704-0144

LIGHTER AIR-CUSHION VEHICLE MAINTENANCE TECHNICIAN

Course Number: 601-ASIE6 (68B).

Location: Transportation and Aviation Logistics School, Fort Eustis, VA; Transportation School, Fort Story, VA.

Length: 3–4 weeks (123 hours).

Exhibit Dates: 7/87–1/98.

Learning Outcomes: Upon completion of the course, the student will be able to maintain the lighter air-cushion vehicle in safe and efficient operating condition; operate and maintain the many aircraft-related components of the vehicle, including fans and propellers, gas turbine engine, auxiliary and ground power units, and associated controls; operate and maintain air conditioning and heating systems and conduct pre- and postoperative checks; repair and replace the skirts on the vehicle; and observe and enforce flight-line safety and hearing conservation.

Instruction: Lectures, practical exercises, and examinations cover power plants and their operation and controls, air conditioning and heating, and skirt repair and replacement. The noise pollution and flight-line safety are also covered.

Credit Recommendation: In the lower-division baccalaureate/associate degree category, 2 semester hours in gas turbine theory and operation (3/88).

Related Occupations: 68B.

AR-1704-0146

AIRCRAFT ARMAMENT TECHNICAL INSPECTOR

Course Number: 600-66J30.

Location: Transportation and Aviation Logistics School, Fort Eustis, VA.

Length: 10–11 weeks (395 hours).

Exhibit Dates: 10/87–3/90.

Learning Outcomes: Upon completion of the course, the student will be able to plan maintenance procedures and aviation inventories; use data communications hardware and software; use ground support equipment; establish policies and procedures for hangar safety; supervise the inspection and repair of boresight assembly, ground support systems, and turret and fire control systems; and read electrical wiring diagrams and compute and measure electrical values of DC and AC circuits using oscilloscopes and digital multimeters.

Instruction: Lectures and practical exercise cover the above subjects.

Credit Recommendation: In the lower-division baccalaureate/associate degree category, 2 semester hours in aviation maintenance management and 2 in principles of supervision (3/88).

Related Occupations: 66J.

AR-1704-0147

AIRCRAFT/ARMAMENT/MISSILE SYSTEMS SUPERVISOR BASIC NONCOMMISSIONED OFFICER (NCO)
(Aircraft Fire Control Systems Supervisor Basic Noncommissioned Officer (NCO) Basic)

Course Number: 646-68J30.

Location: Transportation and Aviation Logistics School, Fort Eustis, VA.

Length: 16–17 weeks (597 hours).

Exhibit Dates: 10/87–3/90.

Learning Outcomes: Upon completion of the course, the student will be able to plan maintenance procedures and aviation inventories; use data communication hardware and software; use ground support equipment; establish policies and procedures for shop and hangar safety; supervise the inspection and repair of boresight assembly, ground support systems, and turret and fire control systems; read electrical wiring diagrams and compute and measure electrical values of DC and AC circuits using oscilloscopes and digital multimeters; and supervise inspection and repair of transistor/amplifier, function generator/interface, relay, and integrated circuits.

Instruction: Lectures, demonstrations, and practical exercises cover the above subjects. This course contains a common core of leadership subjects.

Credit Recommendation: In the lower-division baccalaureate/associate degree category, 2 semester hours in aviation maintenance management and 3 in principles of supervision. See AR-1406-0090 for common core credit (3/88).

Related Occupations: 68J.

AR-1704-0148

CH-54 Heavy Lift Helicopter Repairer Reserve Component

Course Number: 600-67X10-RC.

Location: Transportation and Aviation Logistics School, Fort Eustis, VA.

Length: 10–11 weeks (403 hours).

Exhibit Dates: 10/87–12/93.

Learning Outcomes: Upon completion of the course, the student will be able to perform maintenance, excluding repair of system components, on the CH-54 heavy lift helicopter under direct supervision; identify, remove, and replace subsystem and system components in electrical, hydraulic, auxiliary power plant, utility, power plant, and rotor and power train systems; identify, remove, and replace subsystem components; assist in operational checks of the flight control, automated flight control, and fuel systems; identify and control types of corrosion; identify, remove, and replace fuselage subsystem and system components; and assist in maintenance and operational checks.

Instruction: Lectures, demonstrations, examinations, and practical work are methods of instruction. Subjects include aircraft familiarization and operation, maintenance publications and tools, foreign object damage and its prevention, shop and flight-line safety, corrosion control and prevention, and aircraft systems.

Credit Recommendation: In the lower-division baccalaureate/associate degree category, 8 semester hours in helicopter maintenance (3/88).

Related Occupations: 67X.

AR-1704-0149

OV-1 Observation Airplane Repairer Supervisor Basic Noncommissioned Officer (NCO)
(OV-1 Observation Airplane Repairer Basic Noncommissioned Officer (NCO))

Course Number: ***Version 1:*** 600-67H30. *Version 2:* 600-67H30. *Version 3:* 600-67H30.

Location: Transportation and Aviation Logistics School, Fort Eustis, VA.

Length: ***Version 1:*** 9–10 weeks (315 hours). *Version 2:* 15 weeks (451 hours). *Version 3:* 16–17 weeks (621 hours).

Exhibit Dates: ***Version 1:*** 10/87–3/90. *Version 2:* 4/90–9/91. *Version 3:* 10/91–4/04.

Learning Outcomes: ***Version 1:*** Upon completion of the course, the student will be able to plan maintenance procedures and aviation inventories, use data communication hardware and software for oil analysis; use ground support equipment and precision tools; establish policies and procedures for shop and hangar safety; and supervise maintenance on fuel systems, pneudraulic systems, electrical systems, utility systems, power plants, rotor and drivetrain systems, aircraft structures, landing gear, aircraft weighing procedures, and engine control and flight control systems. *Version 2:* Upon completion of the course, the student will be able to plan maintenance procedures and aviation inventories; use data communication hardware and software for oil analysis; use ground support equipment and precision tools; establish policies and procedures for shop and hangar safety; and supervise maintenance on fuel systems, pneudraulic systems, electrical systems, utility systems, power plants, rotor and drivetrain systems, aircraft structures, landing gear, aircraft weighing procedures, and engine control and flight control systems. *Version 3:* Upon completion of the course, the student will be able to plan maintenance procedures and aviation inventories; use data communication hardware and software for oil analysis; use ground support equipment and precision tools; establish policies and procedures for shop and hangar safety; and supervise maintenance on fuel systems, pneudraulic systems, electrical systems, utility systems, power plants, rotor and drivetrain systems, aircraft structures, landing gear, aircraft weighing procedures, and engine control and flight control systems.

Instruction: ***Version 1:*** Lectures, demonstrations, and practical exercises cover supervision, aircraft maintenance, and maintenance management. This course contains a common core of leadership subjects. *Version 2:* Lectures, demonstrations, and practical exercises cover supervision, aircraft maintenance, and maintenance management. Course no longer contains common core. *Version 3:* Lectures, demonstrations, and practical exercises cover supervision, aircraft maintenance, and maintenance management. This course contains a common core of leadership subjects.

Credit Recommendation: ***Version 1:*** In the lower-division baccalaureate/associate degree category, 2 semester hours in aviation maintenance management and 3 in principles of supervision. See AR-1406-0090 for common core credit (3/88). *Version 2:* In the lower-division baccalaureate/associate degree category, 3 semester hours in aviation maintenance management and 3 in principles of supervision (6/91). *Version 3:* In the lower-division baccalaureate/associate degree category, 3 semester hours in aviation maintenance management and 3 in principles of supervision. See AR-1406-0090 for common core credit (6/91).

Related Occupations: 67H.

AR-1704-0150

UH-60 Helicopter Repairer Supervisor Basic Noncommissioned Officer (NCO)
(UH-60 Tactical Transport Helicopter Basic Noncommissioned Officer (NCO))

Course Number: ***Version 1:*** 600-67T30. *Version 2:* 600-67T30. *Version 3:* 600-67T30. *Version 4:* 600-67T30.

Location: ***Version 1:*** Transportation and Aviation Logistics School, Fort Eustis, VA. *Version 2:* Transportation and Aviation Logistics School, Fort Eustis, VA. *Version 3:* Transportation and Aviation Logistics School, Fort Eustis, VA. *Version 4:* Aviation Logistics School, Fort Eustis, VA.

Length: ***Version 1:*** 8–9 weeks (277 hours). *Version 2:* 12–13 weeks (386 hours). *Version 3:* 14 weeks (502 hours). *Version 4:* 14 weeks (534 hours).

Exhibit Dates: ***Version 1:*** 10/87–3/90. *Version 2:* 4/90–9/91. *Version 3:* 10/91–9/98. *Version 4:* 10/98–9/00.

Learning Outcomes: ***Version 1:*** Upon completion of the course, the student will be able to plan maintenance procedures and aviation inventories; use data communication hardware and software for oil analysis; use ground support equipment and precision tools; establish policies and procedures for shop and hangar safety; and supervise maintenance on fuel systems, pneudraulic systems, electrical systems, utility systems, power plants, rotor and drivetrain systems, aircraft structures, landing gear, aircraft weighing procedures, and engine control and flight control systems. *Version 2:* Upon completion of the course, the student will be able to plan maintenance procedures and aviation inventories; use data communication hardware and software for oil analysis; use ground support equipment and precision tools; establish policies and procedures for shop and hangar safety; and supervise maintenance on fuel systems, pneudraulic systems, electrical systems, utility systems, power plants, rotor and drivetrain systems, aircraft structures, landing gear, aircraft weighing procedures, and engine control and flight control systems. *Version 3:* Upon completion of the course, the student will be able to plan maintenance procedures and aviation inventories; use data communication hardware and software for oil analysis programs; use ground support equipment and precision tools; establish policies and procedures for shop and hangar safety; and supervise maintenance on fuel systems, pneudraulic systems, electrical systems, utility systems, power plants, rotor and drivetrain systems, aircraft structures, landing gear, aircraft weighing procedures, and engine control and flight control systems. *Version 4:* After 9/00 see AR-1717-0185. Upon completion of the course, the student will be able to plan maintenance procedures and aviation inventories; use data communication hardware and software for oil analysis programs; use ground support equipment and precision tools; establish policies and procedures for shop and hangar safety; and supervise maintenance on fuel systems, pneudraulic systems, electrical systems, utility systems, power plants, rotor and drivetrain systems, aircraft structures, landing gear, aircraft weighing procedures, and engine control and flight control systems.

Instruction: ***Version 1:*** Lectures, demonstrations, and practical exercises cover supervision, aircraft maintenance, and maintenance management. This course contains a common core of leadership subjects. *Version 2:* Lectures, demonstrations, and practical exercises cover supervision, aircraft maintenance, and maintenance management. *Version 3:* Lectures, demonstrations, and practical exercises cover supervision, aircraft maintenance, and maintenance management. This course contains a common core of leadership subjects. *Version 4:* Lectures, demonstrations, and practical exercises cover supervision, aircraft maintenance, and maintenance management. This course contains a common core of leadership subjects.

Credit Recommendation: ***Version 1:*** In the lower-division baccalaureate/associate degree category, 2 semester hours in aviation maintenance management and 3 in principles of supervision. See AR-1406-0090 for common core credit (3/88). *Version 2:* In the lower-division baccalaureate/associate degree category, 2 semester hours in principles of supervision and 3 in aviation maintenance management (6/91). *Version 3:* In the lower-division baccalaureate/associate degree category, 2 semester hours in principles of supervision and 3 in aviation maintenance management. See AR-1406-0090 for common core credit (6/91). *Version 4:* In the lower-division baccalaureate/associate degree category, 3 semester hours in aviation maintenance management and 2 in principles of supervision. See AR-1406-0090 for common core credit (8/99).

Related Occupations: 67T.

AR-1704-0151

CH-47 Helicopter Repairer Supervisor Basic Noncommissioned Officer (NCO)
(CH-47 Helicopter Repairer Basic Noncommissioned Officer (NCO))
(CH-47 Medium Helicopter Repairer Basic Noncommissioned Officer (NCO))

Course Number: ***Version 2:*** 600-67U30. *Version 3:* 600-67U30. *Version 4:* 600-67U30.

Location: ***Version 2:*** Transportation and Aviation Logistics School, Fort Eustis, VA. *Version 3:* Transportation and Aviation Logistics School, Fort Eustis, VA. *Version 4:* Aviation Logistics School, Fort Eustis, VA.

Length: ***Version 2:*** 12–13 weeks (355 hours). *Version 3:* 14 weeks (513 hours). *Version 4:* 13–14 weeks (522–525 hours).

Exhibit Dates: ***Version 2:*** 10/89–9/91. *Version 3:* 10/91–9/94. *Version 4:* 10/94–9/00.

Learning Outcomes: ***Version 2:*** Upon completion of the course, the student will be able to plan maintenance procedures and aviation inventories; use data communications hardware and software for oil analysis; use ground support equipment and precision tools; establish policies and procedures for shop and hangar safety; and supervise maintenance on fuel systems, pneudraulic systems, electrical systems, utility systems, power plants, rotor and drivetrain systems, aircraft structures, landing gear, aircraft weighing procedures, and engine control and flight control systems. *Version 3:* Upon completion of the course, the student will be able to plan maintenance procedures and aviation inventories; use data communications hardware and software for oil analysis; use ground support equipment and precision tools; establish policies and procedures for shop and hangar safety; and supervise maintenance on fuel systems, pneudraulic systems, electrical systems, utility systems, power plants, rotor and drivetrain systems, aircraft structures, landing gear, aircraft weighing procedures, and engine control and flight control systems. *Version 4:* After 9/00 see AR-1717-0179. Upon completion of the course, the student will be able to plan maintenance procedures and aviation inventories; use data communications hardware and software for oil analysis; use ground support equipment and precision tools; establish policies and procedures for shop and hangar safety; and supervise maintenance on fuel systems, pneudraulic systems, electrical systems, utility systems, power plants, rotor and drivetrain systems, aircraft structures, landing gear, aircraft weighing procedures, and engine control and flight control systems.

Instruction: ***Version 2:*** Lectures, demonstrations, and practical exercises cover supervision, aircraft maintenance, and maintenance management. *Version 3:* Lectures, demonstrations, and practical exercises cover supervision, aircraft maintenance, and maintenance management. This course contains a common core of leadership subjects. *Version 4:* Lectures, demonstrations, and practical exercises cover supervision, aircraft maintenance, and maintenance management. This course contains a common core of leadership subjects.

Credit Recommendation: ***Version 2:*** In the lower-division baccalaureate/associate degree category, 3 semester hours in aviation maintenance management and 2 in principles of supervision (6/91). *Version 3:* In the lower-division baccalaureate/associate degree category, 3 semester hours in aviation maintenance management and 2 in principles of supervision. See AR-1406-0090 for common core credit (6/91). *Version 4:* In the lower-division baccalaureate/associate degree category, 3 semester hours in aviation maintenance management, 2 in principles of supervision, and 2 in aircraft inspection. See AR-1406-0090 for common core credit (8/99).

Related Occupations: 67U.

AR-1704-0152

1. OH-58 Observation/Scout Helicopter Repairer Basic Noncommissioned Officer (NCO)
2. OH-58 Observation/Scout Helicopter Repairer Basic Noncommissioned Officer (NCO)
3. OH-58 Observation/Scout Helicopter Repairer Basic Noncommissioned Officer (NCO)
4. OH-58 Observation/Scout Helicopter Repairer Supervisor Basic Noncommissioned Officer (NCO)

Course Number: ***Version 1:*** 600-67V30. *Version 2:* 600-67V30. *Version 3:* 600-67V30. *Version 4:* 600-67V30.

Location: ***Version 1:*** Transportation and Aviation Logistics School, Fort Eustis, VA. *Version 2:* Transportation and Aviation Logistics School, Fort Eustis, VA. *Version 3:* Transportation and Aviation Logistics School, Fort Eustis, VA. *Version 4:* Aviation Logistics School, Fort Eustis, VA.

Length: ***Version 1:*** 6 weeks (217 hours). *Version 2:* 11–12 weeks (336 hours). *Version 3:* 13–14 weeks (506 hours). *Version 4:* 13–14 weeks (522 hours).

Exhibit Dates: ***Version 1:*** 10/87–3/90. *Version 2:* 4/90–9/91. *Version 3:* 10/91–9/94. *Version 4:* 10/94–Present.

Learning Outcomes: ***Version 1:*** Upon completion of the course, the student will be able to plan maintenance procedures and aviation inventories; use data communication hardware and software for oil analysis; use ground support equipment and precision tools; establish policies and procedures for shop and hangar safety; and supervise maintenance on fuel systems, pneudraulic systems, electrical systems, utility systems, power plants, rotor and drivetrain systems, aircraft structures, landing gear, aircraft weighing procedures, and engine control and flight control systems. *Version 2:* Upon completion of the course, the student will be able to plan maintenance procedures and aviation inventories; use data communication hardware and software for oil analysis; use ground support equipment and precision tools; establish policies and procedures for shop and hangar safety; and supervise maintenance on fuel systems, pneudraulic systems, electrical systems, utility systems, power plants, rotor and drivetrain systems, aircraft structures, landing gear, aircraft weighing procedures, and engine control and flight control systems. *Version 3:* Upon completion of the course, the student will be able to plan maintenance procedures and aviation inventories; use data communication hardware and software for oil analysis; use ground support equipment and precision tools; establish policies and procedures for shop and hangar safety; and supervise maintenance on fuel systems, pneudraulic systems, electrical systems, utility systems, power plants, rotor and drivetrain systems, aircraft structures, landing gear, aircraft weighing procedures, and engine control and flight control systems. *Version 4:* Upon completion of the course, the student will be able to plan maintenance procedures and aviation inventories; use data communication hardware and software for oil analysis; use ground support equipment and precision tools; establish policies and procedures for shop and hangar safety; and supervise maintenance on fuel systems, pneudraulic systems, electrical systems, utility systems, power plants, rotor and drivetrain systems, aircraft structures, landing gear, aircraft weighing procedures, and engine control and flight control systems.

Instruction: ***Version 1:*** Lectures, demonstrations, and practical exercises cover supervision, aircraft maintenance, and maintenance management. This course contains a common core of leadership subjects. *Version 2:* Lectures, demonstrations, and practical exercises cover supervision, aircraft maintenance, and maintenance management. *Version 3:* Lectures, demonstrations, and practical exercises cover supervision, aircraft maintenance, and maintenance management. This course contains a common core of leadership subjects. *Version 4:* Lectures, demonstrations, and practical exercises cover supervision, aircraft maintenance, and maintenance management. This course contains a common core of leadership subjects.

Credit Recommendation: ***Version 1:*** In the lower-division baccalaureate/associate degree category, 2 semester hours in aviation maintenance management and 1 in principles of supervision. See AR-1406-0090 for common core credit (3/88). *Version 2:* In the lower-division baccalaureate/associate degree category, 3 semester hours in aviation maintenance management and 2 in principles of supervision (6/91). *Version 3:* In the lower-division baccalaureate/associate degree category, 3 semester hours in aviation maintenance management and 2 in principles of supervision. See AR-1406-0090 for common core credit (6/91). *Version 4:* In the lower-division baccalaureate/associate degree category, 3 semester hours in aviation maintenance management, 3 in principles of supervision, and 3 in aircraft inspections. See AR-1406-0090 for common core credit (6/98).

Related Occupations: 67V.

AR-1704-0153

AH-1 Attack Helicopter Repairer Supervisor Basic Noncommissioned Officer (NCO)

(AH-1 Attack Helicopter Repairer Basic Noncommissioned Officer (NCO))

Course Number: *Version 1:* 600-67Y30. *Version 2:* 600-67Y30. *Version 3:* 600-67Y30. *Version 4:* 600-67Y30.

Location: *Version 1:* Transportation and Aviation Logistics School, Fort Eustis, VA. *Version 2:* Transportation and Aviation Logistics School, Fort Eustis, VA. *Version 3:* Transportation and Aviation Logistics School, Fort Eustis, VA. *Version 4:* Aviation Logistics School, Fort Eustis, VA.

Length: *Version 1:* 7–8 weeks (210 hours). *Version 2:* 13–14 weeks (401 hours). *Version 3:* 15–16 weeks (570 hours). *Version 4:* 15–16 weeks (583 hours).

Exhibit Dates: *Version 1:* 10/87–3/90. *Version 2:* 4/90–9/91. *Version 3:* 10/91–9/94. *Version 4:* 10/94–Present.

Learning Outcomes: *Version 1:* Upon completion of the course, the student will be able to plan maintenance procedures and aviation inventories; use data communication hardware and software for oil analysis; use ground support equipment and precision tools; establish policies and procedures for shop and hangar safety; and supervise maintenance on fuel systems, pneudraulic systems, electrical systems, utility systems, power plants, rotor and drivetrain systems, aircraft structures, landing gear, aircraft weighing procedures, and engine control and flight control systems. *Version 2:* Upon completion of the course, the student will be able to plan maintenance procedures and aviation inventories; use data communication hardware and software for oil analysis; use ground support equipment and precision tools; establish policies and procedures for shop and hangar safety; and supervise maintenance on fuel systems, pneudraulic systems, electrical systems, utility systems, power plants, rotor and drivetrain systems, aircraft structures, landing gear, aircraft weighing procedures, and engine control and flight control systems. *Version 3:* Upon completion of the course, the student will be able to plan maintenance procedures and aviation inventories; use data communication hardware and software for oil analysis; use ground support equipment and precision tools; establish policies and procedures for shop and hangar safety; and supervise maintenance on fuel systems, pneudraulic systems, electrical systems, utility systems, power plants, rotor and drivetrain systems, aircraft structures, landing gear, aircraft weighing procedures, and engine control and flight control systems. *Version 4:* Upon completion of the course, the student will be able to plan maintenance procedures and aviation inventories; use data communication hardware and software for oil analysis; use ground support equipment and precision tools; establish policies and procedures for shop and hangar safety; and supervise maintenance on fuel systems, pneudraulic systems, electrical systems, utility systems, power plants, rotor and drivetrain systems, aircraft structures, landing gear, aircraft weighing procedures, and engine control and flight control systems.

Instruction: *Version 1:* Lectures, demonstrations, and practical exercises cover supervision, aircraft maintenance, and maintenance management. This course contains a common core of leadership subjects. *Version 2:* Lectures, demonstrations, and practical exercises cover supervision, aircraft maintenance, and maintenance management. *Version 3:* Lectures, demonstrations, and practical exercises in supervision, aircraft maintenance, and maintenance management. This course contains a common core of leadership subjects. *Version 4:* Lectures, demonstrations, and practical exercises cover supervision, aircraft maintenance, and maintenance management. This course contains a common core of leadership subjects.

Credit Recommendation: *Version 1:* In the lower-division baccalaureate/associate degree category, 2 semester hours in aviation maintenance management and 1 in principles of supervision. See AR-1406-0090 for common core credit (3/88). *Version 2:* In the lower-division baccalaureate/associate degree category, 3 semester hours in aviation maintenance management and 1 in principles of supervision (6/91). *Version 3:* In the lower-division baccalaureate/associate degree category, 3 semester hours in aviation maintenance management and 1 in principles of supervision. See AR-1406-0090 for common core credit (6/91). *Version 4:* In the lower-division baccalaureate/associate degree category, 3 semester hours in aircraft inspections, 5 in aviation maintenance management, and 1 in principles of supervision. See AR-1406-0090 for common core credit (6/98).

Related Occupations: 67Y.

AR-1704-0154

CH-47 Helicopter Repairer
(Medium Helicopter Repairer (CH-47))

Course Number: 600-67U10.

Location: Aviation Logistics School, Fort Eustis, VA.

Length: 15–18 weeks (569–667 hours).

Exhibit Dates: 10/86–Present.

Learning Outcomes: Upon completion of the course, the student will be able to use publications, forms, and records related to maintenance tasks; practice shop and flight-line safety procedures; use common hand tools, aircraft hardware, and ground support equipment; locate, identify, remove, and replace faulty parts and components in the airframe and landing gear systems, utility and hydraulic systems, fuel, electrical, power plant, rotor and power train systems; and assist in performing operational checks and inspections.

Instruction: Lectures, demonstrations and practical exercises cover the skills needed to perform maintenance tasks and develop the knowledge to replace and check helicopter system components.

Credit Recommendation: In the lower-division baccalaureate/associate degree category, 6 semester hours in helicopter maintenance (2/01).

Related Occupations: 67U.

AR-1704-0155

OH-6-RC Observation/Scout Helicopter Repairer Reserve Component

Course Number: 600-67V10 (OH-6).

Location: Transportation and Aviation Logistics School, Fort Eustis, VA.

Length: 7–8 weeks (253 hours).

Exhibit Dates: 6/87–12/93.

Learning Outcomes: Upon completion of the course, the student will be able to use publications, forms, and records related to maintenance tasks; practice shop and flight-line safety procedures; use common hand tools, aircraft hardware, and ground support equipment; locate, identify, remove, and replace parts and components in the airframe and landing gear systems, utility and hydraulic systems, fuel, electrical, power plant, rotor and power train systems; and assist in performing operational checks and inspections.

Instruction: Lectures, demonstrations, and practical exercises cover the skills needed to perform maintenance tasks and develop the knowledge to replace and check helicopter system components.

Credit Recommendation: In the lower-division baccalaureate/associate degree category, 3 semester hours in helicopter maintenance (3/88).

Related Occupations: 67V.

AR-1704-0156

1. AH-64 Attack Helicopter Repairer
2. AH-64A Attack Helicopter Repairer
3. AH-64A Attack Helicopter Repairer

Course Number: *Version 1:* 600-67R10. *Version 2:* 600-67R10; 600-67R10 (AH-64A). *Version 3:* 600-67R10 (AH-64A); 600-15R10-RC (AH-64A).

Location: *Version 1:* Transportation and Aviation Logistics School, Fort Eustis, VA. *Version 2:* Aviation Logistics School, Fort Eustis, VA. *Version 3:* Aviation Logistics School, Fort Eustis, VA.

Length: *Version 1:* 15–18 weeks (573–662 hours). *Version 2:* 14–15 weeks (516–585 hours). *Version 3:* 14–15 weeks (590 hours).

Exhibit Dates: *Version 1:* 10/87–2/96. *Version 2:* 3/96–9/01. *Version 3:* 10/01–Present.

Learning Outcomes: *Version 1:* Upon completion of the course, the student will be able to use publications, forms, and records related to maintenance tasks; practice shop and flight-line safety procedures; use common hand tools, aircraft hardware, and ground support equipment; locate, identify, remove, and replace parts and components in the airframe and landing gear systems, utility and hydraulic systems, fuel, electrical, power plant, rotor and power train systems; and assist in performing operational checks and inspections. *Version 2:* Upon completion of the course, the student will be able to use publications, forms, and records related to maintenance tasks; practice shop and flight-line safety procedures; use common hand tools, aircraft hardware, and ground support equipment; locate, identify, remove, and replace parts and components in the airframe and landing gear systems, utility and hydraulic systems, fuel, electrical, power plant, rotor and power train systems; and assist in performing operational checks and inspections. *Version 3:* Upon completion of the course, the student will be able to use publications, forms, and records related to maintenance tasks; practice shop and flight-line safety procedures; use common hand tools, aircraft hardware, and ground support equipment; locate, identify, remove, and replace parts and components in the airframe and landing gear systems, utility and hydraulic systems, fuel, electrical, power plant, rotor and power train systems; and assist in performing operational checks and inspections.

Instruction: *Version 1:* Lectures, demonstrations and practical exercises cover the skills needed to perform maintenance tasks and develop the knowledge to replace and check helicopter system components. *Version 2:* Lectures, demonstrations and practical exercises cover the skills needed to perform maintenance tasks and develop the knowledge to replace and check helicopter system components. *Version 3:* Audiovisual materials, practical exercises, computer-based training, and lecture. Topics include basic aircraft hydraulics, airframe, landing gear system, pneumatics, fuel and utility systems, crew chief duties, and environmental control systems.

Credit Recommendation: ***Version 1:*** In the lower-division baccalaureate/associate degree category, 6 semester hours in helicopter maintenance (6/91). *Version 2:* In the lower-division baccalaureate/associate degree category, 6 semester hours in helicopter maintenance (2/01). *Version 3:* In the lower-division baccalaureate/associate degree category, 3 semester hours in aircraft hydraulic systems fundamentals, 2 in aircraft pneumatic systems fundamentals and 1 in maintenance management (4/04).

Related Occupations: 67R.

AR-1704-0157

Airborne Laser Tracker (ALT) Maintenance Repairer

Course Number: 4D-F4/600-F11.

Location: Transportation and Aviation Logistics School, Fort Eustis, VA.

Length: 2 weeks (63 hours).

Exhibit Dates: 10/87–12/93.

Learning Outcomes: Upon completion of the course, the student will be able to maintain airborne laser tracking systems and operate supporting test equipment.

Instruction: Lectures and practical exercises cover the principles of operation, testing, and maintenance of airborne laser tracking systems.

Credit Recommendation: Credit is not recommended because of the limited-specialized nature of the course (3/88).

AR-1704-0158

1. AH-64 Attack Helicopter Repairer Transition
2. AH-64 Attack Helicopter Repairer Transition
3. AH-64A Attack Helicopter Repairer (Transition) TATS (AH-64A Attack Helicopter Repairer (Transition))
4. AH-64D Attack Helicopter Repairer Transition

Course Number: ***Version 1:*** 600-67R20/30-T. *Version 2:* 600-67R20/30-T. *Version 3:* 600-67R2/30 (AH-64A) (T)(F); 600-67R2/30 (T). *Version 4:* 600-67R2/30 (AH-64A) (T) (F); 600-ASIY1 (67R).

Location: ***Version 1:*** Transportation and Aviation Logistics School, Fort Eustis, VA. *Version 2:* Aviation Logistics School, Fort Eustis, VA. *Version 3:* Aviation Logistics School, Fort Eustis, VA. *Version 4:* Aviation Logistics School, Fort Eustis, VA.

Length: ***Version 1:*** 8–9 weeks (310 hours). *Version 2:* 6 weeks (217 hours). *Version 3:* 6 weeks (236–240 hours). *Version 4:* 5–6 weeks (182–240 hours).

Exhibit Dates: ***Version 1:*** 10/87–3/91. *Version 2:* 4/91–12/95. *Version 3:* 1/96–3/01. *Version 4:* 4/01–Present.

Learning Outcomes: ***Version 1:*** Upon completion of the course, the student will be able to perform safety check procedures on the AH-64 including weapons, using proper safety procedures; distinguish materials and methods of construction in assemblies and structures; identify major airframe sections; assemble the helicopter-mounted maintenance crane; maintain and replace airframe components; perform aviation unit maintenance and intermediate maintenance, including removing components and installing, adjusting, and requisitioning repair parts; perform limited visual inspections; identify and maintain common and special tools; select correct fuels and lubricants; and use proper shop and flight-line safety procedures. *Version 2:* Upon completion of the course, the student will be able to perform safety check procedures on the AH-64 including weapons; distinguish materials and methods of construction in assemblies and structures; identify major airframe sections; assemble the helicopter-mounted maintenance crane; maintain and replace airframe components; perform aviation unit maintenance and intermediate maintenance, including removing components and installing, adjusting, and requisitioning repair parts; perform limited visual inspections; identify and maintain common and special tools; select correct fuels and lubricants; and use proper shop and flight-line safety procedures. *Version 3:* Upon completion of the course, the student will be able to perform safety check procedures on the AH-64 including weapons, using proper safety procedures; distinguish materials and methods of construction in assemblies and structures; identify major airframe sections; assemble the helicopter-mounted maintenance crane; maintain and replace airframe components; perform aviation unit maintenance and intermediate maintenance, including removing components and installing, adjusting, and requisitioning repair parts; perform limited visual inspections; identify and maintain common and special tools; select correct fuels and lubricants; and use proper shop and flight-line safety procedures. *Version 4:* Upon completion of the course, the student will be able to test, remove, and replace components of the pressure air, flight control, environmental control, auxiliary power unit, and fire control radar systems on the AH-64D helicopter, and use the portable on-system repair tool along with the interactive electronic technical manual for system maintenance. All instruction is accomplished on the AH-64D helicopter.

Instruction: ***Version 1:*** Lectures, practical exercises, and examinations cover tools; repair manuals; airframe, utility, landing gear, fuel, drive, flight control, rotor, hydraulic, electrical, instrument, avionic, armament, and fire control systems; plus power plant and auxiliary and ground power units. *Version 2:* Lectures, practical exercises, and examinations cover tools; repair manuals; and airframe, utility, landing gear, fuel, drive, flight control, rotor, hydraulic, electrical, instrument, avionic, armament, and fire control systems; plus power plant and auxiliary and ground power units. *Version 3:* Lectures, practical exercises, and examinations cover tools; repair manuals; and airframe, utility, landing gear, fuel, drive, flight control, rotor, hydraulic, electrical, instrument, avionic, armament, and fire control systems; plus power plant and auxiliary and ground power units. *Version 4:* Methods of instruction include lecture, laboratory, and practical exercises. Student evaluation is based on written and practical testing.

Credit Recommendation: ***Version 1:*** In the lower-division baccalaureate/associate degree category, 4 semester hours in helicopter maintenance (3/88). *Version 2:* In the lower-division baccalaureate/associate degree category, 5 semester hours in helicopter maintenance (3/94). *Version 3:* In the lower-division baccalaureate/associate degree category, 5 semester hours in helicopter maintenance (2/01). *Version 4:* In the lower-division baccalaureate/associate degree category, 4 semester hours in helicopter maintenance (10/03).

Related Occupations: 67R.

AR-1704-0159

Aircraft Armament Missile Systems Repairer

Course Number: 646-ASIX1 (68J).

Location: Transportation and Aviation Logistics School, Fort Eustis, VA.

Length: 11–12 weeks (387 hours).

Exhibit Dates: 1/88–9/92.

Learning Outcomes: Upon completion of the course, the student will be able to perform maintenance tasks on AH-64 systems, including target acquisition and designation, night vision sensor, integrated helmet and display sight, area weapon, aerial rocket control, point target weapons, boresight, and chaff.

Instruction: Course includes lectures, demonstrations and performance exercises in principles of operation, troubleshooting, testing, component replacement, operational checks, and specialized maintenance procedures.

Credit Recommendation: Credit is not recommended because of the military-specific nature of the course (6/91).

Related Occupations: 68J.

AR-1704-0160

Utility/Cargo and Observation Airplane Supervisor Advanced Noncommissioned Officer (NCO)

Course Number: 600-67G40.

Location: Transportation and Aviation Logistics School, Fort Eustis, VA.

Length: 8–9 weeks (229 hours).

Exhibit Dates: 10/87–12/93.

Learning Outcomes: Upon completion of the course, the student will be able to apply policies and standards of aviation maintenance management, determine training requirements, counsel personnel, perform production and quality control procedures, introduce unit safety programs, conduct maintenance shop performance inspections, comply with aviation logistics and supply procedures, maintain and confirm the accuracy of aviation forms and records, prepare readiness reports, and use data communication hardware and software.

Instruction: Lectures and practical exercises cover the administration of a maintenance management program. This course contains a common core of leadership subjects.

Credit Recommendation: In the lower-division baccalaureate/associate degree category, 4 semester hours in administrative management. See AR-1404-0035 for common core credit (3/88).

Related Occupations: 67G.

AR-1704-0161

OV-1 Observation Airplane Technical Inspector

Course Number: 600-66H20.

Location: Transportation and Aviation Logistics School, Fort Eustis, VA.

Length: 14–15 weeks (547 hours).

Exhibit Dates: 10/87–3/98.

Learning Outcomes: Upon completion of the course, the student will be able to perform nondestructive inspections; inspect power plant hot section, cold section, and accessory modules; operate and inspect pneudraulic systems including landing gear; inspect structural repairs; analyze and troubleshoot electrical/electronic systems; and compute weight and balance data.

Instruction: Lectures, demonstrations, and practical exercises cover inspection techniques, performance evaluation, and operational checks on the power plant, structure, and systems of aircraft.

Credit Recommendation: In the lower-division baccalaureate/associate degree category, 3 semester hours in aircraft structural and system inspections, 3 in aircraft power plant inspections, and 1 in nondestructive testing (3/88).

Related Occupations: 66H.

AR-1704-0162

TACTICAL TRANSPORT HELICOPTER TECHNICAL INSPECTOR

Course Number: 600-66T20.

Location: Transportation and Aviation Logistics School, Fort Eustis, VA.

Length: 13 weeks (475 hours).

Exhibit Dates: 10/87–3/90.

Learning Outcomes: Upon completion of the course, the student will be able to perform nondestructive inspections; inspect power plant hot section, cold section, and accessory modules; operate and inspect pneudraulic systems including landing gear; inspect helicopter structural repairs; analyze and troubleshoot electrical/electronic systems; and compute weight and balance data.

Instruction: Lectures, demonstrations, and practical exercises cover inspection techniques, performance evaluation, and operational checks on the power plant, structure, and systems of helicopters.

Credit Recommendation: In the lower-division baccalaureate/associate degree category, 3 semester hours in helicopter structural and systems inspections, 3 in helicopter power plant inspections, and 1 in nondestructive testing (3/88).

Related Occupations: 66T.

AR-1704-0163

2. AH-64 ATTACK/OH-58D SCOUT HELICOPTER REPAIRER SUPERVISOR ADVANCED NONCOMMISSIONED OFFICER (NCO)
3. AH-64 ATTACK/OH-58D SCOUT HELICOPTER REPAIRER SUPERVISOR ADVANCED NONCOMMISSIONED OFFICER (NCO)
4. AH-64 ATTACK HELICOPTER REPAIRER SUPERVISOR ADVANCED NONCOMMISSIONED OFFICER (NCO)
5. AH-64 ATTACK HELICOPTER REPAIRER SUPERVISOR ADVANCED NONCOMMISSIONED OFFICER (NCO)

Course Number: ***Version 2:*** 600-67R40. *Version 3:* 600-67R40. *Version 4:* 600-67R40. *Version 5:* 600-67R40, Phase 2; 600-15R40, Phase 2.

Location: ***Version 2:*** Transportation and Aviation Logistics School, Fort Eustis, VA. *Version 3:* Transportation and Aviation Logistics School, Fort Eustis, VA. *Version 4:* Aviation School, Fort Rucker, AL. *Version 5:* Aviation Center, Fort Rucker, AL.

Length: ***Version 2:*** 9–10 weeks (173–174 hours). *Version 3:* 8–9 weeks (303 hours). *Version 4:* 8–9 weeks (345 hours). *Version 5:* 6 weeks (262 hours).

Exhibit Dates: ***Version 2:*** 10/89–9/91. *Version 3:* 10/91–9/97. *Version 4:* 10/97–9/00. *Version 5:* 10/00–Present.

Learning Outcomes: ***Version 2:*** Upon completion of the course, the student will be able to apply policies and standards of aviation maintenance management, determine training requirements, counsel personnel, apply production and quality control procedures, introduce unit safety programs, conduct maintenance shop performance inspections, comply with aviation logistics and supply procedures, maintain and confirm the accuracy of aviation forms and records, prepare readiness reports, and use data communication hardware and software. *Version 3:* Upon completion of the course, the student will be able to apply policies and standards of aviation maintenance management, determine training requirements, counsel personnel, apply production and quality control procedures, introduce unit safety programs, conduct maintenance shop performance inspections, comply with aviation logistics and supply procedures, maintain and confirm the accuracy of aviation forms and records, prepare readiness reports, and use data communication hardware and software. *Version 4:* Upon completion of the course, the student will be able to apply policies and standards of aviation maintenance management, determine training requirements, counsel personnel, apply production and quality control procedures, introduce unit safety programs, conduct maintenance shop performance inspections, comply with aviation logistics and supply procedures, maintain and confirm the accuracy of aviation forms and records, prepare readiness reports, and use data communication hardware and software. *Version 5:* Upon completion of the course, the student will be able to demonstrate basic management, supervisory, and leadership skills in an aircraft maintenance environment.

Instruction: ***Version 2:*** Lectures and practical exercises cover the administration of a maintenance management program. *Version 3:* Lectures and practical exercises cover the administration of a maintenance management program. This course contains a common core of leadership subjects. *Version 4:* Lectures and practical exercises cover the administration of a maintenance management program. This course contains a common core of leadership subjects. *Version 5:* Lecture, practical and classroom exercises. Topics include fundamentals of aviation maintenance management, logistics management, maintenance duties and responsibilities, safety, training, and maintenance supervision.

Credit Recommendation: ***Version 2:*** In the lower-division baccalaureate/associate degree category, 3 semester hours in administrative management (6/91). *Version 3:* In the lower-division baccalaureate/associate degree category, 3 semester hours in administrative management. See AR-1404-0035 for common core credit (6/91). *Version 4:* In the lower-division baccalaureate/associate degree category, 3 semester hours in maintenance management. See AR-1404-0035 for common core credit (11/98). *Version 5:* In the lower-division baccalaureate/associate degree category, 3 semester hours in maintenance management. See AR-1404-0035 for common core credit (4/03).

Related Occupations: 67R.

AR-1704-0164

UH-60 HELICOPTER REPAIRER SUPERVISOR ADVANCED NONCOMMISSIONED OFFICER (NCO)
(Utility/Tactical Transport Helicopter Repairer Supervisor Advanced Noncommissioned Officer (NCO))

Course Number: ***Version 2:*** 600-67T40. *Version 3:* 600-67T40. *Version 4:* 600-67T40. *Version 5:* 600-67T40, Phase 2.

Location: ***Version 2:*** Transportation and Aviation Logistics School, Fort Eustis, VA. *Version 3:* Transportation and Aviation Logistics School, Fort Eustis, VA. *Version 4:* Aviation School, Fort Rucker, AL. *Version 5:* Aviation Center, Fort Rucker, AL.

Length: ***Version 2:*** 9–10 weeks (146 hours). *Version 3:* 8–9 weeks (303 hours). *Version 4:* 8–9 weeks (345 hours). *Version 5:* 6–7 weeks (262 hours).

Exhibit Dates: ***Version 2:*** 10/89–9/91. *Version 3:* 10/91–9/97. *Version 4:* 10/97–9/00. *Version 5:* 10/00–12/03.

Learning Outcomes: ***Version 2:*** Upon completion of the course, the student will be able to apply policies and standards of aviation maintenance management, determine training requirements, counsel personnel, apply production and quality control procedures, introduce unit safety programs, conduct maintenance shop performance inspections, comply with aviation logistics and supply procedures, maintain and confirm the accuracy of aviation forms and records, prepare readiness reports, and use data communication hardware and software. *Version 3:* Upon completion of the course, the student will be able to apply policies and standards of aviation maintenance management, determine training requirements, counsel personnel, apply production and quality control procedures, introduce unit safety programs, conduct maintenance shop performance inspections, comply with aviation logistics and supply procedures, maintain and confirm the accuracy of aviation forms and records, prepare readiness reports, and use data communication hardware and software. *Version 4:* Upon completion of the course, the student will be able to apply policies and standards of aviation maintenance management, determine training requirements, counsel personnel, apply production and quality control procedures, introduce unit safety programs, conduct maintenance shop performance inspections, comply with aviation logistics and supply procedures, maintain and confirm the accuracy of aviation forms and records, prepare readiness reports, and use data communication hardware and software. *Version 5:* Upon completion of the course, the student will be able to conduct convoy and tactical operations, employ protective techniques in field operations, train personnel to manage aviation maintenance, prepare technical reports, manage Army Safety Program and production control, conduct aviation deployment and field training exercises, and operate basic computers.

Instruction: ***Version 2:*** Lectures and practical exercises cover the administration of a maintenance management program. *Version 3:* Lectures and practical exercises cover the administration of a maintenance management program. This course contains a common core of leadership subjects. *Version 4:* Lectures and practical exercises cover the administration of a maintenance management program. This course contains a common core of leadership subjects. *Version 5:* Topics include preventive maintenance checks and services; unit status report preparation; military history and convoy operations; single channel ground/airborne radio systems; operations and order process; aviation maintenance and logistics management; identify maintenance duties and responsibilities; administer Army Safety Program such as aviation deployment procedures, MS DOS, MS Office, and field training exercises.

Credit Recommendation: ***Version 2:*** In the lower-division baccalaureate/associate degree category, 3 semester hours in administrative management (6/91). *Version 3:* In the lower-division baccalaureate/associate degree category, 3 semester hours in administrative management. See AR-1404-0035 for common core credit (6/91). *Version 4:* In the lower-division baccalaureate/associate degree category, 3 semester hours in maintenance management. See AR-1404-0035 for common core credit (11/98). *Version 5:* In the lower-division baccalaureate/associate degree category, 3 semester hours in maintenance management. See AR-1404-0035 for common core credit (4/03).

Related Occupations: 67T.

AR-1704-0165

CH-47 Helicopter Repairer Supervisor Advanced Noncommissioned Officer (NCO)
(CH-47 Medium/CH-54 Heavy Lift Helicopter Repairer Supervisor Advanced Noncommissioned Officer (NCO))

Course Number: ***Version 2:*** 600-67U40. *Version 3:* 600-67U40. *Version 4:* 600-67U40. *Version 5:* 600-67U40, Phase 2; 600-15U40, Phase 2.

Location: ***Version 2:*** Transportation and Aviation Logistics School, Fort Eustis, VA. *Version 3:* Transportation and Aviation Logistics School, Fort Eustis, VA. *Version 4:* Aviation School, Fort Rucker, AL. *Version 5:* Aviation Center, Fort Rucker, AL.

Length: ***Version 2:*** 9–10 weeks (146 hours). *Version 3:* 8–9 weeks (303 hours). *Version 4:* 8–9 weeks (345 hours). *Version 5:* 6 weeks (262 hours).

Exhibit Dates: ***Version 2:*** 10/89–9/91. *Version 3:* 10/91–9/97. *Version 4:* 10/97–1/01. *Version 5:* 2/01–Present.

Learning Outcomes: ***Version 2:*** Upon completion of the course, the student will be able to apply policies and standards of aviation maintenance management, determine training requirements, counsel personnel, apply production and quality control procedures, introduce unit safety programs, conduct maintenance shop performance inspections, comply with aviation logistics and supply procedures, maintain and confirm the accuracy of aviation forms and records, prepare readiness reports, and use data communication hardware and software. *Version 3:* Upon completion of the course, the student will be able to apply policies and standards of aviation maintenance management, determine training requirements, counsel personnel, apply production and quality control procedures, introduce unit safety programs, conduct maintenance shop performance inspections, comply with aviation logistics and supply procedures, maintain and confirm the accuracy of aviation forms and records, prepare readiness reports, and use data communication hardware and software. *Version 4:* Upon completion of the course, the student will be able to apply policies and standards of aviation maintenance management, determine training requirements, counsel personnel, apply production and quality control procedures, introduce unit safety programs, conduct maintenance shop performance inspections, comply with aviation logistics and supply procedures, maintain and confirm the accuracy of aviation forms and records, prepare readiness reports, and use data communication hardware and software. *Version 5:* Upon completion of the course, the student will be able to demonstrate basic management, supervisory and leadership skills in an aircraft maintenance environment.

Instruction: ***Version 2:*** Lectures and practical exercises cover the administration of a maintenance management program. *Version 3:* Lectures and practical exercises cover the administration of a maintenance management program. This course contains a common core of leadership subjects. *Version 4:* Lectures and practical exercises cover the administration of a maintenance management program. This course contains a common core of leadership subjects. *Version 5:* Lectures, practical and classroom exercises. Topics include fundamentals of aviation maintenance management, logistics management, maintenance duties and responsibilities, safety, training, and maintenance supervision.

Credit Recommendation: ***Version 2:*** In the lower-division baccalaureate/associate degree category, 4 semester hours in administrative management (6/91). *Version 3:* In the lower-division baccalaureate/associate degree category, 4 semester hours in administrative management. See AR-1404-0035 for common core credit (6/91). *Version 4:* In the lower-division baccalaureate/associate degree category, 3 semester hours in maintenance management. See AR-1404-0035 for common core credit (11/98). *Version 5:* In the lower-division baccalaureate/associate degree category, 3 semester hours in maintenance management. See AR-1404-0035 for common core credit (4/04).

Related Occupations: 67U.

AR-1704-0166

1. Aircraft Component Repairer Supervisor Advanced Noncommissioned Officer (NCO)
2. Aircraft Component Repairer Supervisor Advanced Noncommissioned Officer (NCO)
3. Aircraft Component Repairer Supervisor Advanced Noncommissioned Officer (NCO)
4. Aircraft Component Repairer Supervisor Advanced Noncommissioned Officer (NCO)
5. Aircraft Component Repairer Supervisor Advanced Noncommissioned Officer (ANCO)

Course Number: ***Version 1:*** 600-67K40. *Version 2:* 600-67K40. *Version 3:* 602-68K40. *Version 4:* 602-68K40. *Version 5:* 602-68K40.

Location: ***Version 1:*** Transportation and Aviation Logistics School, Fort Eustis, VA. *Version 2:* Transportation and Aviation Logistics School, Fort Eustis, VA. *Version 3:* Transportation and Aviation Logistics School, Fort Eustis, VA. *Version 4:* Aviation Logistics School, Fort Eustis, VA. *Version 5:* Aviation Logistics School, Fort Eustis, VA.

Length: ***Version 1:*** 9–10 weeks (342 hours). *Version 2:* 12–13 weeks (249 hours). *Version 3:* 10–11 weeks (393 hours). *Version 4:* 11–12 weeks (425 hours). *Version 5:* 13–14 weeks (490 hours).

Exhibit Dates: ***Version 1:*** 10/87–3/90. *Version 2:* 4/90–9/91. *Version 3:* 10/91–9/95. *Version 4:* 10/95–9/98. *Version 5:* 10/98–Present.

Learning Outcomes: ***Version 1:*** Upon completion of the course, the student will be able to apply policies and standards of aviation maintenance management, determine training requirements, counsel personnel, apply production and quality control procedures, introduce unit safety programs, conduct maintenance shop performance inspections, comply with aviation logistics and supply procedures, maintain and confirm the accuracy of aviation forms and records, prepare readiness reports, and use data communications hardware and software. *Version 2:* Upon completion of the course, the student will be able to apply policies and standards of aviation maintenance management, determine training requirements, counsel personnel, apply production and quality control procedures, introduce unit safety programs, conduct maintenance shop performance inspections, comply with aviation logistics and supply procedures, maintain and confirm the accuracy of aviation forms and records, prepare readiness reports, and use data communications hardware and software. *Version 3:* Upon completion of the course, the student will be able to apply policies and standards of aviation maintenance management, determine training requirements, counsel personnel, apply production and quality control procedures, introduce unit safety programs, conduct maintenance shop performance inspections, comply with aviation logistics and supply procedures, maintain and confirm the accuracy of aviation forms and records, prepare readiness reports, and use data communications hardware and software. *Version 4:* Upon completion of the course, the student will be able to apply policies and standards of aviation maintenance management, determine training requirements, counsel personnel, apply production and quality control procedures, introduce unit safety programs, conduct maintenance shop performance inspections, comply with aviation logistics and supply procedures, maintain and confirm the accuracy of aviation forms and records, prepare readiness reports, and use data communications hardware and software. *Version 5:* Upon completion of the course, the student will be able to apply leadership supervision, military resource management, and combat technical procedures; inspect, evaluate, and calibrate aircraft components; ad supervise maintenance and technical publications.

Instruction: ***Version 1:*** Lectures and practical exercises cover the administration of a maintenance management program. This course contains a common core of leadership subjects. *Version 2:* Lectures and practical exercises cover the administration of a maintenance management program. *Version 3:* Lectures and practical exercises cover the administration of a maintenance management program. This course contains a common core of leadership subjects. *Version 4:* Lectures and practical exercises cover the administration of a maintenance management program. This course contains a common core of leadership subjects. *Version 5:*

Methods of instruction include programmed instruction with small-group activities, conferences, demonstrations, videos, and practical exercises. Student evaluation is based on written examinations. Credit recommended is based on completion of the entire 13-14 week long course at Ft. Eustis.

Credit Recommendation: ***Version 1:*** In the lower-division baccalaureate/associate degree category, 4 semester hours in administrative management. See AR-1404-0035 for common core credit (3/88). *Version 2:* In the lower-division baccalaureate/associate degree category, 3 semester hours in administrative management (6/91). *Version 3:* In the lower-division baccalaureate/associate degree category, 3 semester hours in administrative management. See AR-1404-0035 for common core credit (6/91). *Version 4:* In the lower-division baccalaureate/associate degree category, 3 semester hours in aviation maintenance management. See AR-1404-0035 for common core credit (6/98). *Version 5:* In the lower-division baccalaureate/associate degree category, 2 semester hours in aircraft maintenance supervision and 2 in office administration. See AR-1404-0035 for common core credit (7/01).

Related Occupations: 67K; 68K.

AR-1704-0167

2. Aircraft Armament/Missile Systems Repairer Supervisor Advanced Noncommissioned Officer (NCO)
   (Aircraft Fire Control Systems Supervisor Advanced Noncommissioned Officer (NCO))
3. Aircraft Armament/Missile Systems Repairer Supervisor Advanced Noncommissioned Officer (NCO)
   (Aircraft Fire Control Systems Supervisor Advanced Noncommissioned Officer (NCO))
4. Aircraft Armament/Missile Systems Repairer Supervisor Advanced Noncommissioned Officer (NCO)
5. Aircraft Armanent/Missile Systems Repairer Supervisor Advanced Noncommissioned Officer (NCO)

Course Number: ***Version 2:*** 646-68J40. *Version 3:* 646-68J40. *Version 4:* 602-68J40. *Version 5:* 602-68J40, Phase 2.

Location: ***Version 2:*** Transportation and Aviation Logistics School, Fort Eustis, VA. *Version 3:* Transportation and Aviation Logistics School, Fort Eustis, VA. *Version 4:* Aviation School, Fort Rucker, AL. *Version 5:* Aviation Center, Fort Rucker, AL.

Length: ***Version 2:*** 9–10 weeks (146 hours). *Version 3:* 8–9 weeks (303 hours). *Version 4:* 8–9 weeks (345 hours). *Version 5:* 6–7 weeks (262 hours).

Exhibit Dates: ***Version 2:*** 10/89–9/91. *Version 3:* 10/91–12/96. *Version 4:* 1/97–9/00. *Version 5:* 10/00–Present.

Learning Outcomes: ***Version 2:*** Upon completion of the course, the student will be able to apply policies and standards of aviation maintenance management, determine training requirements, counsel personnel, apply production and quality control procedures, introduce unit safety programs, conduct maintenance shop performance inspections, comply with aviation logistics and supply procedures, maintain and confirm the accuracy of aviation forms and records, prepare readiness reports, and use data communication hardware and software. *Version 3:* Upon completion of the course, the student will be able to apply policies and standards of aviation maintenance management, determine training requirements, counsel personnel, apply production and quality control procedures, introduce unit safety programs, conduct maintenance shop performance inspections, comply with aviation logistics and supply procedures, maintain and confirm the accuracy of aviation forms and records, prepare readiness reports, and use data communication hardware and software. *Version 4:* Upon completion of the course, the student will be able to apply policies and standards of aviation maintenance management, determine training requirements, counsel personnel, apply production and quality control procedures, introduce unit safety programs, conduct maintenance shop performance inspections, comply with aviation logistics and supply procedures, maintain and confirm the accuracy of aviation forms and records, prepare readiness reports, and use data communication hardware and software. *Version 5:* Upon completion of the course, the student will be able to demonstrate basic management, supervisory, and leadership skills in an aircraft maintenance environment.

Instruction: ***Version 2:*** Lectures and practical exercises in the administration of a maintenance management program. *Version 3:* Lectures and practical exercises cover the administration of a maintenance management program. This course contains a common core of leadership subjects. *Version 4:* Lectures and practical exercises cover the administration of a maintenance management program. This course contains a common core of leadership subjects. *Version 5:* Lectures, practical and classroom exercises. Topics include fundamentals of aviation maintenance management, logistics management, maintenance duties and responsibilities, safety, training, and maintenance supervision.

Credit Recommendation: ***Version 2:*** In the lower-division baccalaureate/associate degree category, 3 semester hours in administrative management (6/91). *Version 3:* In the lower-division baccalaureate/associate degree category, 3 semester hours in administrative management. See AR-1404-0035 for common core credit (6/91). *Version 4:* In the lower-division baccalaureate/associate degree category, 3 semester hours in maintenance management. See AR-1404-0035 for common core credit (11/98). *Version 5:* In the lower-division baccalaureate/associate degree category, 3 semester hours in maintenance management. See AR-1404-0035 for common core credit (4/03).

Related Occupations: 68J.

AR-1704-0169

Organizational Maintenance Supervisors

Course Number: LOG 213 (OMS-33).

Location: Combined Arms Training Center, Vilseck, W. Germany.

Length: 2 weeks (65 hours).

Exhibit Dates: 10/76–10/98.

Learning Outcomes: Upon completion of the course, the student will be able to select and interpret required regulations, manuals, pamphlets, and forms; determine the materials and requirements for establishing on-the-job training, driver training, safety, operations, and readiness programs; select proper test equipment; and perform several common maintenance tests.

Instruction: Lectures and seminars cover assembling data and completing various forms and reports necessary to supervise an organizational maintenance facility.

Credit Recommendation: Credit is not recommended because of the military-specific nature of the course (10/88).

AR-1704-0170

M60-M113A1 Series Unit Maintenance Training

Course Number: TVM 218.

Location: Combined Arms Training Center, Vilseck, W. Germany.

Length: 3 weeks (102 hours).

Exhibit Dates: 2/86–10/98.

Learning Outcomes: Upon completion of the course, the student will be able to use repair manuals to perform track vehicle maintenance in the areas of suspension systems, fuel systems, exhaust systems, air intake systems, band adjustment on the transmission, and removal and installation of diesel engines and troubleshoot the starting, charging, and electrical systems.

Instruction: Lectures and laboratory exercises cover troubleshooting the electrical systems using proper test equipment and performing track vehicle maintenance procedures.

Credit Recommendation: In the lower-division baccalaureate/associate degree category, 2 semester hours in maintenance procedures (10/88).

AR-1704-0171

Wheel and Power Generation Unit Maintenance

Course Number: WPG-222.

Location: Combined Arms Training Center, Vilseck, W. Germany.

Length: 3 weeks (112 hours).

Exhibit Dates: 8/87–10/98.

Learning Outcomes: Upon completion of the course, the student will be able to isolate and repair faults in the cooling, air, fuel, and electrical systems of light and heavy trucks and small to large portable power generators using test equipment and repair/replace components in the mechanical and electrical systems.

Instruction: Conference and practical exercises cover safety and basic electrical and mechanical fundamentals for automotive and power generating equipment.

Credit Recommendation: In the lower-division baccalaureate/associate degree category, 3 semester hours in maintenance procedures (10/88).

AR-1704-0172

Air Traffic Control Systems, Subsystems, and Equipment Repairer Supervisor Advanced Noncommissioned Officer (NCO)

Course Number: 102-93D40.

Location: Aviation Center, Fort Rucker, AL.

Length: 8–9 weeks (326–332 hours).

Exhibit Dates: 1/89–7/03.

Learning Outcomes: Upon completion of the course, the student will be able to perform as a noncommissioned officer in MOS 93D40.

Instruction: Topics include aviation subjects in general, field site training, and related skills. There is an emphasis on maintenance supervision. This course contains a common core of leadership subjects.

Credit Recommendation: In the lower-division baccalaureate/associate degree category, 3 semester hour in maintenance management. See AR-1404-0035 for common core credit (4/90).

Related Occupations: 93D.

AR-1704-0173

1. Flight Operations Basic Noncommissioned Officer (NCO)
2. Aviation Operations Specialist Basic Noncommissioned Officer (NCO)
3. Aviation Operations Specialist Basic Noncommissioned Officer (NCO)

Course Number: ***Version 1:*** 556-93P30. *Version 2:* 556-93P30. *Version 3:* 556-93P30, Phase 2.

Location: ***Version 1:*** Aviation Center, Fort Rucker, AL. *Version 2:* Aviation School, Fort Rucker, AL. *Version 3:* Aviation Center, Fort Rucker, AL.

Length: ***Version 1:*** 4 weeks (160 hours). *Version 2:* 5–6 weeks (224 hours). *Version 3:* 3 weeks (150 hours).

Exhibit Dates: ***Version 1:*** 10/87–9/97. *Version 2:* 10/97–1/01. *Version 3:* 2/01–Present.

Learning Outcomes: ***Version 1:*** Upon completion of the course, the student will be able to provide leadership at a flight operations facility. *Version 2:* Upon completion of the course, the student will be able to provide leadership at a flight operations facility. *Version 3:* Upon completion of the course, the student will be able to demonstrate knowledge of basic airport/airfield operations, management, and leadership skills.

Instruction: ***Version 1:*** Studies cover leadership, military skills, professional skills, resource management, training management, airfield procedures, tactical operations, flight planning procedures, vehicle loading, marshalling procedures, field management training, supervision of dispatch and flight planning area, and publications. This course contains a common core of leadership subjects. *Version 2:* Studies cover leadership, military skills, professional skills, resource management, training management, airfield procedures, tactical operations, flight planning procedures, vehicle loading, marshalling procedures, field management training, supervision of dispatch and flight planning area, and publications. This course contains a common core of leadership subjects. *Version 3:* Topics include basic leadership, resource management, training management skills, airfield procedures, airfield operations, dispatch, flight planning, and aviation publications.

Credit Recommendation: ***Version 1:*** In the lower-division baccalaureate/associate degree category, 3 semester hours in airfield/airport operations management. See AR-1406-0090 for common core credit (4/90). *Version 2:* In the lower-division baccalaureate/associate degree category, 2 semester hours in airfield/airport operations management. See AR-1406-0090 for common core credit (11/98). *Version 3:* In the lower-division baccalaureate/associate degree category, 1 semester hour in airport operations or 1 in flight/dispatch operations and 1 in airport management. See AR-1406-0090 for common core credit (4/03).

AR-1704-0174

1. Flight Operations Advanced Noncommissioned Officer (NCO)
2. Aviation Operations Specialist Advanced Noncommissioned Officer (NCO)
3. Aviation Operations Specialist Advanced Noncommissioned Officer (NCO)

Course Number: ***Version 1:*** 556-93P40. *Version 2:* 556-93P40. *Version 3:* 556-93P40, Phase 2.

Location: ***Version 1:*** Aviation Center, Fort Rucker, AL. *Version 2:* Aviation School, Fort Rucker, AL. *Version 3:* Aviation Center, Fort Rucker, AL.

Length: ***Version 1:*** 8–9 weeks (324–333 hours). *Version 2:* 5–6 weeks (240 hours). *Version 3:* 4–5 weeks (190 hours).

Exhibit Dates: ***Version 1:*** 10/88–9/94. *Version 2:* 10/94–9/02. *Version 3:* 10/02–Present.

Learning Outcomes: ***Version 1:*** Upon completion of the course, the student will be able to provide supervision, management, and training in a flight operations facility. *Version 2:* Upon completion of the course, the student will be able to provide supervision, management, and training in a flight operations facility. *Version 3:* Upon completion of the course, the student will be able to provide supervision, management, and training in a flight operations facility.

Instruction: ***Version 1:*** Topics include aviation safety, administrative procedures, flight programs, field training, tactical operations, and reports. This course contains a common core of leadership subjects. *Version 2:* Topics include aviation safety, administrative procedures, flight programs, field training, tactical operations, and reports. This course contains a common core of leadership subjects. *Version 3:* Topics include aviation safety, administrative procedures, flight programs, field training, tactical operations, and reports. This course contains a common core of leadership subjects.

Credit Recommendation: ***Version 1:*** In the lower-division baccalaureate/associate degree category, 3 semester hours in air traffic management. See AR-1404-0035 for common core credit. In the upper-division baccalaureate category, 3 semester hours in advanced airfield/airport operations management (4/90). *Version 2:* In the lower-division baccalaureate/associate degree category, 2 semester hours in aviation flight operations. See AR-1404-0035 for common core credit. In the upper-division baccalaureate category, 2 semester hours in airfield/airport operations (11/98). *Version 3:* In the lower-division baccalaureate/associate degree category, 2 in flight/dispatch operations. In the upper-division baccalaureate category, 2 semester hours in airport operations or 2 in airport management (4/03).

Related Occupations: 93P.

AR-1704-0175

Air Traffic Control Operator Basic Noncommissioned Officer (NCO)

Course Number: ***Version 1:*** 222-93C30. *Version 2:* 222-93C30. *Version 3:* 222-93C30. *Version 4:* 222-93C30, Phase 2.

Location: ***Version 1:*** Aviation Center, Fort Rucker, AL. *Version 2:* Aviation School, Fort Rucker, AL. *Version 3:* Aviation School, Fort Rucker, AL. *Version 4:* Aviation Center, Fort Rucker, AL.

Length: ***Version 1:*** 4 weeks (153 hours). *Version 2:* 5–6 weeks (243 hours). *Version 3:* 7 weeks (294 hours). *Version 4:* 4–5 weeks (213 hours).

Exhibit Dates: ***Version 1:*** 10/87–9/94. *Version 2:* 10/94–12/97. *Version 3:* 1/98–1/01. *Version 4:* 2/01–Present.

Learning Outcomes: ***Version 1:*** Upon completion of the course, the student will be able to provide basic leadership in an air traffic control facility. *Version 2:* Upon completion of the course, the student will be able to perform basic air traffic controller functions as well as support the operation of an ATC facility through the application of basic leadership skills. *Version 3:* Upon completion of the course, the student will be able to perform basic air traffic controller functions as well as support the operation of an ATC facility through the application of basic leadership skills. *Version 4:* Upon completion of the course, the student will be able to perform basic air traffic controller functions as well as support the operation of an ATC facility through the application of basic leadership skills.

Instruction: ***Version 1:*** Training covers leadership, professional skills, resource management, training management, air traffic control facility operations, administration, training programs, and supervisory tasks for fixed-base and tactical operations. This course contains a common core of leadership subjects. *Version 2:* Training covers leadership; professional skills; resource management; training management; administrative functions; supervisory tasks for fixed-base and tactical operations; and air traffic control facility operations, including helicopter GCA approaches, airspace control, and helicopter nonprecision approaches. *Version 3:* Training covers leadership; professional skills; resource management; training management; administrative functions; supervisory tasks for fixed-base and tactical operations; and air traffic control facility operations, including helicopter and fixed-wing ground controller approaches, airspace control, and fixed/rotary-wing nonprecision approaches. *Version 4:* Training covers leadership; professional skills; resource management; training management; administrative functions; supervisory tasks for fixed-base and tactical operations; and air traffic control facility operations, including helicopter and fixed-wing ground controller approaches, airspace control, and fixed/rotary-wing nonprecision approaches.

Credit Recommendation: ***Version 1:*** Credit is recommended for the common core only. See AR-1406-0090 (4/90). *Version 2:* In the lower-division baccalaureate/associate degree category, see AR-1406-0090 for common core credit. In the upper-division baccalaureate category, 2 semester hours in air traffic administration (11/98). *Version 3:* In the lower-division baccalaureate/associate degree category, see AR-1406-0090 for common core credit. In the upper-division baccalaureate category, 3 semester hours in air traffic administration (11/98). *Version 4:* In the lower-division baccalaureate/associate degree category, 3 semester hours in air traffic control operations. See AR-1406-0090 for common core credit (4/03).

Related Occupations: 93C.

AR-1704-0176

Air Traffic Control Operator Advanced Noncommissioned Officer (NCO)

Course Number: ***Version 1:*** 222-93C40. *Version 2:* 222-93C40. *Version 3:* 222-93C40. *Version 4:* 222-93C40, Phase 2.

Location: ***Version 1:*** Aviation Center, Fort Rucker, AL. *Version 2:* Aviation School, Fort Rucker, AL. *Version 3:* Aviation School, Fort Rucker, AL. *Version 4:* Aviation Center, Fort Rucker, AL.

Length: ***Version 1:*** 9 weeks (326–345 hours). *Version 2:* 7–8 weeks (317 hours). *Version 3:* 7 weeks (289 hours). *Version 4:* 4–5 weeks (212 hours).

Exhibit Dates: ***Version 1:*** 10/88–9/94. *Version 2:* 10/94–12/97. *Version 3:* 1/98–1/01. *Version 4:* 2/01–Present.

Learning Outcomes: ***Version 1:*** Upon completion of the course, the student will be able to manage site and field training for air traffic control facilities. *Version 2:* Upon completion of the course, the student will be able to manage site and field training for air traffic control facilities; establish terminal instrument approaches; construct standard instrument approach procedures for different categories of aircraft; and develop nondirectional beacon approach and precision radar approaches, special use airspace, precision and nonprecision criteria, and criteria for determining holding pattern airspace. *Version 3:* Upon completion of the course, the student will be able to manage site and field training for air traffic control facilities, supervise the establishment of terminal instrument approach procedures, determine airplane command and control procedures, interpret and implement airspace management procedures, and prepare an operations overlay. *Version 4:* Upon completion of the course, the student will be able to manage site and field training for air traffic control facilities, supervise the establishment of terminal instrument approach procedures, determine airplane command and control procedures, interpret and implement airspace management procedures, and prepare an operations overlay.

Instruction: ***Version 1:*** Lectures and practical exercises cover unit training and administration, supply, communications and security management, preventive medicine and survival, civil operations, air traffic control procedures, personnel management and training, facility management, advanced air traffic control procedures, site selection, and field training. This course contains a common core of leadership subjects. *Version 2:* Classroom instruction, small-group instruction, and practical exercises cover training and administration, communications, airspace, air traffic control operations and procedures, approaches, holding patterns, personnel management and training, site selection, and field training. *Version 3:* Classroom instruction, small-group instruction, practical exercises, and field experience. Topics include training and administration, communications management, air traffic control procedures, personnel management and training, facility management, advanced air traffic control operations and procedures, site selection, and field training. *Version 4:* Classroom instruction, small-group instruction, practical exercises, and field experience. Topics include training and administration, communications management, air traffic control procedures, personnel management and training, facility management, advanced air traffic control operations and procedures, site selection, and field training.

Credit Recommendation: ***Version 1:*** In the lower-division baccalaureate/associate degree category, 3 semester hours in air traffic management. See AR-1404-0035 for common core credit. In the upper-division baccalaureate category, 3 semester hours for field experience in air traffic control (4/90). *Version 2:* In the lower-division baccalaureate/associate degree category, see AR-1404-0035 for common core credit. In the upper-division baccalaureate category, 3 semester hours in advanced air traffic control operations and 3 in air traffic management (11/98). *Version 3:* In the lower-division baccalaureate/associate degree category, see AR-1404-0035 for common core credit. In the upper-division baccalaureate category, 2 semester hours in advanced air traffic control operations and 3 in air traffic management (11/98). *Version 4:* In the upper-division baccalaureate category, 3 semester hours in advanced air traffic control operations (4/03).

Related Occupations: 93C.

AR-1704-0177

Air Traffic Control Operator

Course Number: 222-93C10, Phase 1; 222-93C10.

Location: Aviation Center, Fort Rucker, AL.

Length: 11–15 weeks (457–563 hours).

Exhibit Dates: 10/87–Present.

Learning Outcomes: Upon completion of the course, the student will be able to perform air traffic control duties with apprentice-level knowledge of visual flight rules (VFR); instrument flight rules (IFR); and air traffic control regulations, concepts, and procedures.

Instruction: Course includes audiovisual materials, discussion, lecture, laboratory, computer -based training, learner presentations, lectures and practical experience in basic air traffic control (ATC), VFR, IFR, and control tower operations, including aviation weather, Federal Aviation Administration regulations, air traffic control procedures in radar, and GCA and tactical ATC equipment.

Credit Recommendation: In the lower-division baccalaureate/associate degree category, 6 semester hours in air traffic control (4/03).

Related Occupations: 93C.

AR-1704-0178

Aviation Operations Specialist (Flight Operations Coordinator)

Course Number: 556-93P10, Phase 1; 556-93P10.

Location: Aviation Center, Fort Rucker, AL.

Length: 6–8 weeks (247–303 hours).

Exhibit Dates: 1/87–Present.

Learning Outcomes: Upon completion of the course, the student will be able to schedule and coordinate aircraft flight operations and perform related support functions.

Instruction: Lectures and practical exercises include land navigation; aeronautical charts; flight information publications; notices to airmen; and airfield operations, administration, and communications.

Credit Recommendation: In the lower-division baccalaureate/associate degree category, 3 semester hours in aviation flight operations (4/03).

Related Occupations: 93P.

AR-1704-0179

Avionics Equipment Maintenance Supervisor Advanced Noncommissioned Officer (NCO)

Course Number: 102-35P40.

Location: Aviation Center, Fort Rucker, AL.

Length: 8–9 weeks (313–331 hours).

Exhibit Dates: 10/88–7/03.

Learning Outcomes: Upon completion of the course, the student will be able to implement effective group communication and perform resource management, quality control, and leadership functions.

Instruction: Topics include leadership, communication skills, training management, and supervision of aviation/avionics communications and electrical/electronic systems equipment. This course contains a common core of leadership subjects.

Credit Recommendation: Credit is recommended for the common core only. See AR-1404-0035 (4/90).

Related Occupations: 35P.

AR-1704-0180

AH-64 Aircraft Power Train Repairer

Course Number: 602-ASIX1 (68D).

Location: Aviation Center, Fort Rucker, AL.

Length: 3–4 weeks (115 hours).

Exhibit Dates: 10/89–7/03.

Learning Outcomes: Upon completion of the course, the student will be able to perform aviation maintenance on the AH-64 attack helicopter power train systems and associated equipment.

Instruction: Course consists of lectures, demonstrations, and hands-on disassembly, repair, and assembly of all gear boxes and transmission of the AH-64 helicopter.

Credit Recommendation: In the lower-division baccalaureate/associate degree category, 1 semester hour in power train repair (aircraft) (4/90).

Related Occupations: 68D.

AR-1704-0181

1. Turbine Engine Driver Generator Repairer
2. Turbine Engine Driven Generator Repairer

Course Number: ***Version 1:*** 662-52F10. *Version 2:* 662-52F10.

Location: Ordnance Center and School, Aberdeen Proving Ground, MD.

Length: ***Version 1:*** 12–13 weeks (464 hours). *Version 2:* 10 weeks (360 hours).

Exhibit Dates: ***Version 1:*** 10/90–9/96. *Version 2:* 10/96–Present.

Learning Outcomes: ***Version 1:*** Upon completion of the course, the student will be able to troubleshoot, diagnose, and repair the turbine and generator by applying the principles of gas turbine driven AC/DC generator sets. *Version 2:* Upon completion of the course, the student will be able to perform maintenance on turbine driven generators and associated equipment.

Instruction: ***Version 1:*** Lectures and demonstrations cover basic electricity and gas turbine design and servicing procedures.

Extensive laboratory and field experience are provided in diagnosis and repair of failed units. *Version 2:* Lectures and practical exercises cover maintenance functions, electrical fundamentals, turbine engine maintenance, and generators.

Credit Recommendation: ***Version 1:*** In the lower-division baccalaureate/associate degree category, 3 semester hours in gas turbine engine service, 3 in basic electricity, and 2 in AC/DC generators (12/90). *Version 2:* In the lower-division baccalaureate/associate degree category, 3 semester hours in fundamentals of electricity and 3 in turbine engine AC/DC generators (1/98).

Related Occupations: 52F.

AR-1704-0182

1. Aircraft Electrician Repairer Supervisor Basic Noncommissioned Officer (NCO)
2. Aircraft Electrician Repairer Supervisor Basic Noncommissioned Officer (NCO)
3. Aircraft Electrician Supervisor Basic Noncommissioned Officer (NCO)

Course Number: ***Version 1:*** 602-68F30. *Version 2:* 602-68F30. *Version 3:* 602-68F30.

Location: Aviation Logistics School, Fort Eustis, VA.

Length: ***Version 1:*** 13–14 weeks (399 hours). *Version 2:* 12–13 weeks (452 hours). *Version 3:* 10–11 weeks (373 hours).

Exhibit Dates: ***Version 1:*** 4/90–9/91. *Version 2:* 10/91–4/94. *Version 3:* 5/94–Present.

Learning Outcomes: ***Version 1:*** Upon completion of the course, the student will be able to supervise repair and maintenance of electrical and electronic systems on helicopters. *Version 2:* Upon completion of the course, the student will be able to supervise repair and maintenance of electrical and electronic systems on helicopters. *Version 3:* Upon completion of the course, the student will be able to supervise repair and maintenance of electrical and electronic systems on helicopters.

Instruction: ***Version 1:*** Lectures and practical exercises cover the materials required to conduct inspections on aircraft including maintenance of electrical subsystems and components and apply the policies and procedures of approved aviation maintenance standards. *Version 2:* Lectures and practical exercises cover the materials required to conduct inspections on aircraft including maintenance of electrical subsystems and components and apply the policies and procedures of approved aviation maintenance standards. This course contains a common core of leadership subjects. *Version 3:* Lectures and practical exercises cover the materials required to conduct inspections on aircraft including maintenance of electrical subsystems and components and apply the policies and procedures of approved aviation maintenance standards. This course contains a common core of leadership subjects.

Credit Recommendation: ***Version 1:*** In the lower-division baccalaureate/associate degree category, 3 semester hours in aviation maintenance management (6/91). *Version 2:* In the lower-division baccalaureate/associate degree category, 3 semester hours in aviation maintenance management. See AR-1406-0090 for common core credit (6/91). *Version 3:* In the lower-division baccalaureate/associate degree category, 3 semester hours in aviation maintenance management and 2 in aircraft electrical systems. See AR-1406-0090 for common core credit (6/98).

Related Occupations: 68F.

AR-1704-0183

1. AH-64 Aircraft Armament Maintenance Technician
2. AH-64 Armament Maintenance Technician

Course Number: ***Version 1:*** 4D-SQIE (AH-64). *Version 2:* 4D-SQIE (152F).

Location: Aviation Logistics School, Fort Eustis, VA.

Length: ***Version 1:*** 13 weeks (470 hours). *Version 2:* 12–13 weeks (479 hours).

Exhibit Dates: ***Version 1:*** 10/90–12/98. *Version 2:* 1/99–Present.

Learning Outcomes: ***Version 1:*** Upon completion of the course, the student will be able to troubleshoot and repair aircraft armament systems used on AH-64 helicopters, including TADS-night sensor assembly, pilot's night vision sensor, helmet, video recorder, weapons systems, external stores, aerial rocket and control system, point target weapon system, chaff dispenser, and aircraft interface subsystems. *Version 2:* Upon completion of the course, the student will be able to troubleshoot and repair aircraft electrical systems on AH-64D Helicopters including TADS-night sensor assembly, pilot's night vision sensor, helmet, video recorder, weapons systems, external stores, aerial rocket and control system, point target system, embedded global positioning systems and area weapons system.

Instruction: ***Version 1:*** Lectures and practical exercises cover helicopter armament systems and subsystems, basic electricity and electronics, and missile systems and subsystems. *Version 2:* Methods of instruction include classroom and practical exercises.

Credit Recommendation: ***Version 1:*** In the lower-division baccalaureate/associate degree category, 3 semester hours in electronic systems troubleshooting and maintenance (6/91). *Version 2:* In the lower-division baccalaureate/associate degree category, 3 semester hours in aircraft electrical systems (6/01).

Related Occupations: 152F.

AR-1704-0184

1. Aircraft Structural Repairer Supervisor Basic Noncommissioned Officer (NCO)
2. Aircraft Structural Repairer Supervisor Basic Noncommissioned Officer (NCO)
3. TATS Aircraft Structural Repairer Supervisor Basic Noncommissioned Officer (NCO)

Course Number: ***Version 1:*** 603-68G30. *Version 2:* 603-68G30. *Version 3:* 603-68G30, Phase 2.

Location: Aviation Logistics School, Fort Eustis, VA.

Length: ***Version 1:*** 9–10 weeks (272 hours). *Version 2:* 9–11 weeks (366–376 hours). *Version 3:* 6 weeks (266 hours).

Exhibit Dates: ***Version 1:*** 4/90–9/91. *Version 2:* 10/91–9/00. *Version 3:* 10/00–Present.

Learning Outcomes: ***Version 1:*** Before 4/90 see AR-1704-0109. Upon completion of the course, the student will be able to inspect, evaluate, and supervise aircraft structural repair. *Version 2:* Upon completion of the course, the student will be able to inspect, evaluate, and supervise aircraft structural repair. *Version 3:* Upon completion of the course, the student will be able to inspect, evaluate, and supervise aircraft structured repairs, IAW technical publications, and operational standards.

Instruction: ***Version 1:*** Lectures and practical exercises cover the policies and procedures associated with approved maintenance standards. *Version 2:* Lectures and practical exercises cover the approved procedures for the maintenance, inspection, and repair of aircraft structural components. This course contains a common core of leadership subjects. *Version 3:* Practical and classroom exercises and lecture. Topics include approved procedures for maintenance, inspection, and repair of structural components; and common core leadership subjects.

Credit Recommendation: ***Version 1:*** In the lower-division baccalaureate/associate degree category, 3 semester hours in aviation maintenance management (6/91). *Version 2:* In the lower-division baccalaureate/associate degree category, 3 semester hours in aviation maintenance management. See AR-1406-0090 for common core credit (6/98). *Version 3:* In the lower-division baccalaureate/associate degree category, 3 semester hours in aviation maintenance management (12/03).

Related Occupations: 68G.

AR-1704-0185

Aircraft Power Train Repairer Supervisor Basic Noncommissioned Officer (NCO)

Course Number: ***Version 1:*** 602-68D30. *Version 2:* 602-68D30. *Version 3:* 602-68D30.

Location: Aviation Logistics School, Fort Eustis, VA.

Length: ***Version 1:*** 12 weeks (357 hours). *Version 2:* 12 weeks (434 hours). *Version 3:* 12 weeks (452 hours).

Exhibit Dates: ***Version 1:*** 4/90–9/91. *Version 2:* 10/91–9/96. *Version 3:* 10/96–9/00.

Learning Outcomes: ***Version 1:*** Upon completion of the course, the student will be able to supervise maintenance and repair performed on helicopter power trains, main and tail rotors, transmissions, gear boxes, and hubs. *Version 2:* Upon completion of the course, the student will be able to supervise maintenance and repair performed on helicopter power trains, main and tail rotors, transmissions, gear boxes, and hubs. *Version 3:* After 9/00 see AR-1704-0241. Upon completion of the course, the student will be able to supervise maintenance and repair performed on helicopter power trains, main and tail rotors, transmissions, gear boxes, and hubs.

Instruction: ***Version 1:*** Lectures and practical exercises cover the supervision of helicopter power train systems maintenance and repair. *Version 2:* Lectures and practical exercises cover the supervision of helicopter power train systems maintenance and repair. This course contains a common core of leadership subjects. *Version 3:* Lectures and practical exercises cover the supervision of helicopter power train systems maintenance and repair. This course contains a common core of leadership subjects.

Credit Recommendation: ***Version 1:*** In the lower-division baccalaureate/associate degree category, 3 semester hours in aviation maintenance management (6/91). *Version 2:* In the lower-division baccalaureate/associate degree category, 3 semester hours in aviation maintenance management. See AR-1406-0090 for common core credit (6/91). *Version 3:* In the lower-division baccalaureate/associate degree category, 4 semester hours in aviation maintenance management and 3 in power plant airworthiness inspections. See AR-1406-0090 for common core credit (6/98).

Related Occupations: 68D.

AR-1704-0186

Aircraft Powerplant Repairer

Course Number: ***Version 1:*** 601-68B10. *Version 2:* 601-68B10. *Version 3:* 601-68B10.

Location: Aviation Logistics School, Fort Eustis, VA.

Length: ***Version 1:*** 15–17 weeks (576–619 hours). *Version 2:* 17 weeks (643 hours). *Version 3:* 17 weeks (703 hours).

Exhibit Dates: ***Version 1:*** 10/89–12/98. *Version 2:* 1/99–10/02. *Version 3:* 11/02–Present.

Learning Outcomes: ***Version 1:*** Upon completion of the course, the student will be able to troubleshoot, maintain, repair, and replace aircraft turboshaft engines, including the T53, T55, and T-700 series in various helicopter airframes. *Version 2:* Upon completion of the course, the student will be able to troubleshoot, maintain, repair, and replace aircraft turboshaft engines and supporting equipment. *Version 3:* Upon completion of the course, the student will be able to troubleshoot, maintain, repair, and replace aircraft turboshaft engines and supporting equipment.

Instruction: ***Version 1:*** Lectures and practical experience cover the theory, operation, maintenance, troubleshooting, and repair of various types of turboshaft turbine engines, components, and subassemblies. *Version 2:* Practical exercises, discussion, classroom exercises, laboratory, and lecture. Topics include theory, operation, maintenance, troubleshooting, and repair of various types of turboshaft engines, components, and subassemblies. *Version 3:* Practical exercises, discussion, classroom exercises, laboratory, and lecture. Topics include theory, operation, maintenance, troubleshooting, and repair of various types of turboshaft engines, components, and subassemblies.

Credit Recommendation: ***Version 1:*** In the lower-division baccalaureate/associate degree category, 4 semester hours in turbine engine maintenance, repair, and troubleshooting (6/98). *Version 2:* In the lower-division baccalaureate/associate degree category, 3 semester hours in aircraft powerplant systems and components and 3 in aircraft powerplant maintenance and inspection (12/03). *Version 3:* In the lower-division baccalaureate/associate degree category, 3 semester hours in aircraft powerplant systems and components and 4 in aircraft powerplant maintenance and inspection (12/03).

Related Occupations: 68B.

AR-1704-0187

1. Aircraft Pneudraulic Repairer Supervisor Basic Noncommissioned Officer (NCO)
2. Aircraft Pneudraulics Repairer Supervisor

(Aircraft Pneudraulic Repairer Supervisor Basic Noncommissioned Officer (NCO))

Course Number: ***Version 1:*** 603-68H30. *Version 2:* 603-68H30.

Location: Aviation Logistics School, Fort Eustis, VA.

Length: ***Version 1:*** 7–8 weeks (200 hours). *Version 2:* 8–9 weeks (314–331 hours).

Exhibit Dates: ***Version 1:*** 4/90–9/91. *Version 2:* 10/91–Present.

Learning Outcomes: ***Version 1:*** Before 4/90 see AR-1704-0114. Upon completion of the course, the student will be able to supervise the maintenance of hydraulic systems and components. *Version 2:* Upon completion of the course, the student will be able to supervise the maintenance, repair, and inspection of aircraft hydraulic/pneumatic system components.

Instruction: ***Version 1:*** Lectures and practical exercises cover maintenance standards of aircraft pneudraulic systems. *Version 2:* Lectures and practical exercises cover maintenance, inspection, and repair of aircraft hydraulic/pneumatic systems and components. This course contains a common core of leadership subjects.

Credit Recommendation: ***Version 1:*** In the lower-division baccalaureate/associate degree category, 3 semester hours in aviation maintenance management (6/91). *Version 2:* In the lower-division baccalaureate/associate degree category, 3 semester hours in aviation maintenance management. See AR-1406-0090 for common core credit (6/98).

Related Occupations: 68H.

AR-1704-0188

1. AH-1 Attack/OH-58A/C Observation/ Scout Helicopter Repairer Supervisor Advanced Noncommissioned Officer (NCO)
2. AH-1 Attack/OH-58A/C Observation/ Scout Helicopter Repairer Supervisor Advanced Noncommissioned Officer (NCO)
3. AH-1 Attack/Observation/Scout Helicopter Repairer Supervisor Advanced Noncommissioned Officer (NCO)
4. AH-1 Attack/Observation/Scout Helicopter Repairer Supervisor Advanced Noncommissioned Officer (NCO)

Course Number: ***Version 1:*** 600-67Y40. *Version 2:* 600-67Y40. *Version 3:* 600-67Y40. *Version 4:* 600-67Y40, Phase 2.

Location: ***Version 1:*** Aviation Logistics School, Fort Eustis, VA. *Version 2:* Aviation Logistics School, Fort Eustis, VA. *Version 3:* Aviation School, Fort Rucker, AL. *Version 4:* Aviation Center, Fort Rucker, AL.

Length: ***Version 1:*** 9–10 weeks (260 hours). *Version 2:* 8 weeks (303 hours). *Version 3:* 8–9 weeks (345 hours). *Version 4:* 6 weeks (262 hours).

Exhibit Dates: ***Version 1:*** 10/89–9/91. *Version 2:* 10/91–9/97. *Version 3:* 10/97–9/00. *Version 4:* 10/00–Present.

Learning Outcomes: ***Version 1:*** Upon completion of the course, the student will be able to apply policies and standards of aviation maintenance management, determine training requirements, counsel personnel, apply production and quality control procedures, introduce unit safety programs, conduct maintenance shop performance inspections, comply with logistic and supply procedures, prepare reports, and use data communications hardware and software. *Version 2:* Upon completion of the course, the student will be able to apply policies and standards of aviation maintenance management, determine training requirements, counsel personnel, apply production and quality control procedures, introduce unit safety programs, conduct maintenance shop performance inspections, comply with logistic and supply procedures, prepare reports, and use data communications hardware and software. *Version 3:* Upon completion of the course, the student will be able to apply policies and standards of aviation maintenance management, determine training requirements, counsel personnel, apply production and quality control procedures, introduce unit safety programs, conduct maintenance shop performance inspections, comply with logistic and supply procedures, prepare reports, and use data communications hardware and software. *Version 4:* Upon completion of the course, the student will be able to demonstrate basic management, supervisory, and leadership skills in an aircraft maintenance environment.

Instruction: ***Version 1:*** Lectures and practical exercises cover the administration of a maintenance management program. *Version 2:* Lectures and practical exercises cover the administration of a maintenance management program. This course contains a common core of leadership subjects. *Version 3:* Lectures and practical exercises cover the administration of a maintenance management program. This course contains a common core of leadership subjects. *Version 4:* Practical, classroom exercises, and lecture. Topics include fundamentals of aviation maintenance management; logistics management, maintenance duties and responsibilities, safety, training, and maintenance supervision.

Credit Recommendation: ***Version 1:*** In the lower-division baccalaureate/associate degree category, 3 semester hours in administrative management (6/91). *Version 2:* In the lower-division baccalaureate/associate degree category, 3 semester hours in administrative management. See AR-1404-0035 for common core credit (6/91). *Version 3:* In the lower-division baccalaureate/associate degree category, 3 semester hours in maintenance management. See AR-1404-0035 for common core credit (11/98). *Version 4:* In the lower-division baccalaureate/associate degree category, 3 semester hours in maintenance management. See AR-1404-0035 for common core credit (4/03).

Related Occupations: 67Y.

AR-1704-0189

UH-1 Helicopter Repairer Basic Noncommissioned Officer (NCO) Reserve Component

(Utility Helicopter Repairer Basic Noncommissioned Officer (NCO) Reserve Component)

Course Number: 600-67N30-RC.

Location: Reserve Component Training Centers, US; Aviation Logistics School, Fort Eustis, VA.

Length: 5 weeks (189 hours).

Exhibit Dates: 10/90–4/04.

Learning Outcomes: Upon completion of the course, the student will be able to supervise helicopter maintenance, repair, systems troubleshooting, and inspection.

Instruction: Lectures and practical exercises cover the utility helicopter. This course contains a common core of leadership subjects.

Credit Recommendation: In the lower-division baccalaureate/associate degree category, 3 semester hours in aircraft maintenance management. See AR-2201-0337 for common core credit (6/91).

Related Occupations: 67N.

AR-1704-0190

APPRENTICE AIRCRAFT MECHANIC (AIT)

Course Number: 600-67A10.

Location: Aviation Logistics School, Fort Eustis, VA.

Length: 9–10 weeks (336–376 hours).

Exhibit Dates: 10/90–12/93.

Learning Outcomes: Upon completion of the course, the student will be able to assist maintenance technicians in routine helicopter maintenance and minor repairs on two types of aircraft.

Instruction: Lectures and practical experience cover basic aircraft maintenance and repair, tools and equipment, hardware, safety devices, structural repair, corrosion control and painting, power plant maintenance, and pneudraulic systems.

Credit Recommendation: In the lower-division baccalaureate/associate degree category, 2 semester hours in introduction to aircraft maintenance (6/91).

Related Occupations: 67A.

AR-1704-0191

1. AH-64 ARMAMENT/ELECTRICAL SYSTEM REPAIRER
2. AH-64 ARMAMENT/ELECTRICAL SYSTEMS REPAIRER
3. AH-64A ARMAMENT/ELECTRICAL SYSTEMS REPAIRER

Course Number: *Version 1:* 646-68X10. *Version 2:* 646-68X10. *Version 3:* 646-68X10 (AH-64A).

Location: Aviation Logistics School, Fort Eustis, VA.

Length: *Version 1:* 20–23 weeks (774–858 hours). *Version 2:* 18–19 weeks (736 hours). *Version 3:* 20–21 weeks (775 hours).

Exhibit Dates: *Version 1:* 1/90–9/95. *Version 2:* 10/95–9/97. *Version 3:* 10/97–9/00.

Learning Outcomes: *Version 1:* Upon completion of the course, the student will be able to maintain electrical and electronic systems on aircraft. *Version 2:* Upon completion of the course, the student will be able to troubleshoot, maintain, and repair aircraft electrical and electronic systems. *Version 3:* Upon completion of the course, the student will be able to perform maintenance, preventive maintenance, and repair on aircraft electrical and electronic systems on AH-64A helicopters.

Instruction: *Version 1:* Lectures, demonstrations, and practical experience cover all phases of basic electricity, troubleshooting, repair, and operational checks. *Version 2:* Lectures, demonstrations, and practical experience cover basic electricity and electronics and the troubleshooting, repair, and maintenance of aircraft electrical/armament systems. *Version 3:* Methods of instruction include classroom conferences and practical exercises.

Credit Recommendation: *Version 1:* In the lower-division baccalaureate/associate degree category, 3 semester hours in basic electricity, 3 in electronics, 1 in introduction to aircraft maintenance, and 3 in aircraft electrical systems (3/94). *Version 2:* In the lower-division baccalaureate/associate degree category, 3 semester hours in basic electricity, 3 in electronics, 1 in aircraft maintenance, and 3 in aircraft electrical systems or electronic systems troubleshooting and maintenance (6/98). *Version 3:* In the lower-division baccalaureate/associate degree category, 3 semester hours in aircraft basic electricity, 3 in aircraft electrical systems, 3 in electronic systems, and 3 in fundamentals of aircraft maintenance (6/01).

Related Occupations: 68X.

AR-1704-0192

1. AH-64 AIRCRAFT ARMAMENT/MISSILE SYSTEMS SUPERVISOR BASIC NONCOMMISSIONED OFFICER (NCO)
2. AH-64 ARMAMENT/ELECTRICAL SYSTEMS REPAIRER SUPERVISOR BASIC NONCOMMISSIONED OFFICER (NCO)
3. AH-64 ARMAMENT/ELECTRICAL SYSTEMS REPAIRER BASIC NONCOMMISSIONED OFFICER (NCO)

Course Number: *Version 1:* 646-68J30 (AH-64). *Version 2:* 646-68X30. *Version 3:* 646-68X30.

Location: Aviation Logistics School, Fort Eustis, VA.

Length: *Version 1:* 15–16 weeks (477 hours). *Version 2:* 15–16 weeks (477 hours). *Version 3:* 15 weeks (576 hours).

Exhibit Dates: *Version 1:* 4/90–9/91. *Version 2:* 10/91–9/94. *Version 3:* 10/94–9/00.

Learning Outcomes: *Version 1:* Upon completion of the course, the student will be able to repair, maintain, and supervise the electronic systems on missile guidance systems and weapon control systems. *Version 2:* Upon completion of the course, the student will be able to repair, maintain, and supervise the electronic systems on missile guidance systems and weapon control systems. *Version 3:* After 9/00 see AR-1704-0236. Upon completion of the course, the student will be able to troubleshoot, maintain, and repair aircraft missile guidance systems and weapon control systems. The student will also be able to supervise the maintenance of aircraft armament/electrical systems.

Instruction: *Version 1:* Lectures and practical exercises cover AC/DC circuits, electricity and electronics, and supervision and management. *Version 2:* Lectures and practical exercises cover AC/DC circuits, electricity and electronics, and supervision and management. This course contains a common core of leadership subjects. *Version 3:* Lectures and practical exercises cover AC/DC circuits, electricity and electronics, and supervision and management. This course contains a common core of leadership subjects.

Credit Recommendation: *Version 1:* In the lower-division baccalaureate/associate degree category, 3 semester hours in aviation maintenance management and 2 in electronic systems (6/91). *Version 2:* In the lower-division baccalaureate/associate degree category, 3 semester hours in aviation maintenance management and 2 in electronic systems. See AR-1406-0090 for common core credit (9/92). *Version 3:* In the lower-division baccalaureate/associate degree category, 3 semester hours in aviation maintenance management and 6 in electronic systems troubleshooting and maintenance. See AR-1406-0090 for common core credit (6/98).

Related Occupations: 68J; 68X.

AR-1704-0193

1. AIRCRAFT ARMAMENT/MISSILE SYSTEMS SUPERVISOR BASIC NONCOMMISSIONED OFFICER (NCO)
2. AIRCRAFT ARMAMENT/MISSILE SYSTEMS SUPERVISOR BASIC NONCOMMISSIONED OFFICER (NCO)
3. AH-1F ARMAMENT/MISSILE SYSTEMS SUPERVISOR BASIC NONCOMMISSIONED OFFICER (NCO)

Course Number: *Version 1:* 646-68J30 (AH-1). *Version 2:* 646-68J30 (AH-1). *Version 3:* 602-68J30.

Location: Aviation Logistics School, Fort Eustis, VA.

Length: *Version 1:* 19 weeks (609 hours). *Version 2:* 17 weeks (643 hours). *Version 3:* 14–15 weeks (555 hours).

Exhibit Dates: *Version 1:* 4/90–9/91. *Version 2:* 10/91–9/95. *Version 3:* 10/95–Present.

Learning Outcomes: *Version 1:* Upon completion of the course, the student will be able to supervise and maintain aircraft electrical and electronic systems. *Version 2:* Upon completion of the course, the student will be able to supervise the maintenance of aircraft electrical and electronic systems and perform maintenance on these systems. *Version 3:* Upon completion of the course, the student will be able to supervise the maintenance of aircraft electrical and electronic systems and perform maintenance on these systems.

Instruction: *Version 1:* Course consists of lectures and practical exercises in aircraft electrical and electronic system troubleshooting and maintenance. *Version 2:* Lectures and practical exercises cover aircraft electrical and electronic system troubleshooting and maintenance and aircraft maintenance supervision. This course contains a common core of leadership subjects. *Version 3:* Lectures and practical exercises cover aircraft electrical system troubleshooting and maintenance and aircraft maintenance supervision. This course contains a common core of leadership subjects.

Credit Recommendation: *Version 1:* In the lower-division baccalaureate/associate degree category, 3 semester hours in aviation maintenance management and 3 in aviation electronic systems troubleshooting and maintenance (6/91). *Version 2:* In the lower-division baccalaureate/associate degree category, 3 semester hours in aviation maintenance management and 3 in avionic electronic systems troubleshooting and maintenance. See AR-1406-0090 for common core credit (3/94). *Version 3:* In the lower-division baccalaureate/associate degree category, 3 semester hours in aviation maintenance management and 6 in electronic systems troubleshooting and maintenance. See AR-1406-0090 for common core credit (6/98).

Related Occupations: 68J.

AR-1704-0194

OH-58D HELICOPTER REPAIRER
(OH-58D Advanced Scout Helicopter Repairer)

Course Number: *Version 1:* 600-67S10. *Version 2:* 600-67S10. *Version 3:* 600-67S10.

Location: Aviation Logistics School, Fort Eustis, VA.

Length: *Version 1:* 9–11 weeks (343–399 hours). *Version 2:* 12–13 weeks (478–490 hours). *Version 3:* 13–14 weeks (522 hours).

Exhibit Dates: ***Version 1:*** 1/88–9/97. *Version 2:* 10/97–9/02. *Version 3:* 10/02–Present.

Learning Outcomes: ***Version 1:*** Upon completion of the course, the student will be able to perform routine maintenance, inspection, and repair on helicopters. *Version 2:* Upon completion of the course, the student will be able to inspect and maintain the OH-58D helicopter engine, rotor, fuel, hydraulic, electrical, and flight control systems. *Version 3:* Upon completion of the course, the student will be able to perform routine inspection, maintenance and repair to powerplant, powertrain, rotor, fuel, and electrical and flight control systems and IAW OH-580 technical publications.

Instruction: ***Version 1:*** Lectures and practical applications develop skills in record keeping, inspection, routine maintenance, and repairs. *Version 2:* Methods of instruction include lecture, laboratory, and practical experience on OH-58D helicopter. Written and practical examinations are used for student evaluation. *Version 3:* Audiovisual materials, practical and classroom exercises, discussion, and lecture. Topics include inspection, troubleshooting, maintenance and repair of aircraft powertrain; powerplant, rotor, fuel, electrical and flight control systems; and IAW technical publications.

Credit Recommendation: ***Version 1:*** In the lower-division baccalaureate/associate degree category, 3 semester hours in aircraft maintenance/inspection (4/92). *Version 2:* In the lower-division baccalaureate/associate degree category, 3 semester hours in helicopter inspection and maintenance (6/01). *Version 3:* In the lower-division baccalaureate/associate degree category, 3 semester hours in aircraft maintenance and repair and 3 in aircraft systems troubleshooting and maintenance (12/03).

Related Occupations: 67S.

AR-1704-0195

UH-60 Helicopter Repairer Supervisor Basic Noncommissioned Officer (NCO) Reserve Component
(Tactical Transport Helicopter Repairer Basic Noncommissioned Officer (NCO) Reserve Component)

Course Number: 600-67T30-RC.

Location: Aviation Logistics School, Fort Eustis, VA.

Length: 5 weeks (196 hours).

Exhibit Dates: 10/90–4/04.

Learning Outcomes: Upon completion of the course, the student will be able to supervise the maintenance, repair, and system troubleshooting of the UH-60 aircraft.

Instruction: Lectures, demonstrations and practical exercises cover the UH-60. This course contains a common core of leadership subjects.

Credit Recommendation: In the lower-division baccalaureate/associate degree category, 3 semester hours in aircraft maintenance management. See AR-2201-0337 for common core credit (6/91).

Related Occupations: 67N; 67T.

AR-1704-0196

Aircraft Maintenance (CMF 67) Advanced Noncommissioned Officer (NCO) Reserve

Course Number: 552-CMF 67 ANCOC-RC.

Location: Aviation Logistics School, Fort Eustis, VA.

Length: 1–2 weeks (87 hours).

Exhibit Dates: 2/88–4/04.

Learning Outcomes: Upon completion of the course, the student will be able to supervise aircraft record keeping, inventory control, tool calibration, safety program, and preventive maintenance on ground support equipment.

Instruction: Classroom lectures and simulated paperwork cover supply control, maintenance files, production controls, unit moves, tool calibration, and preventive maintenance checks on ground support equipment. This course contains a common core of leadership subjects.

Credit Recommendation: In the lower-division baccalaureate/associate degree category, 1 semester hour in principles of supervision. See AR-2201-0295 for common core credit (6/91).

AR-1704-0197

AH-64 Aircraft Pneudraulics Repairer

Course Number: 602-ASIX1 (68H).

Location: Aviation Logistics School, Fort Eustis, VA.

Length: 5–6 weeks (191 hours).

Exhibit Dates: 10/90–4/04.

Learning Outcomes: Upon completion of the course, the student will be able to maintain and repair aircraft pneudraulic systems.

Instruction: Lectures and practical skill development using training aids and practical aircraft projects are the methods of instruction.

Credit Recommendation: In the lower-division baccalaureate/associate degree category, 3 semester hours in aircraft pneudraulics repair and maintenance (6/91).

Related Occupations: 68H.

AR-1704-0198

AH-64 Aircraft Power Train Repairer

Course Number: 602-ASIX1 (68D).

Location: Aviation Logistics School, Fort Eustis, VA.

Length: 3–4 weeks (115 hours).

Exhibit Dates: 1/89–4/04.

Learning Outcomes: Upon completion of the course, the student will be able to perform unit and routine maintenance on aircraft power trains.

Instruction: Lectures and practical exercises cover maintenance, troubleshooting, inspection, and replacement of various aircraft power train components.

Credit Recommendation: Credit is not recommended because of the limited, specialized nature of the course (6/91).

Related Occupations: 68D.

AR-1704-0199

1. Aviation Maintenance Warrant Officer Technical/Tactical Certification
2. Aviation Maintenance Technician Warrant Officer Basic
3. Aviation Maintenance Technician

Course Number: ***Version 1:*** 4D-151A. *Version 2:* 4D-151A. *Version 3:* 4D-151A.

Location: Aviation Logistics School, Fort Eustis, VA.

Length: ***Version 1:*** 18 weeks (613–661 hours). *Version 2:* 8 weeks (273 hours). *Version 3:* 9 weeks (320 hours).

Exhibit Dates: ***Version 1:*** 10/89–8/95. *Version 2:* 9/95–9/00. *Version 3:* 10/00–Present.

Learning Outcomes: ***Version 1:*** Upon completion of the course, the student will be able to manage aircraft maintenance and logistics. *Version 2:* Upon completion of the course, the student will be able to manage aircraft maintenance and logistics. *Version 3:* Upon completion of the course, the student will be able to manage aircraft maintenance and logistics of powertrain components, hydraulic and utility systems, structural repairs to panels and rotors, and avionics/electrical systems.

Instruction: ***Version 1:*** Classroom instruction covers logistics; aviation maintenance management; aeronautical technology subjects; power train; propulsion; and structural, pneudraulic, and electrical systems. *Version 2:* Classroom instruction covers logistics; aviation maintenance management; aeronautical technology subjects; power train; propulsion; and structural, pneudraulic, and electrical systems. *Version 3:* Practical exercises, discussion, laboratory, and lecture. Topics include, monitor maintenance and logistics of powertrain components, structural repairs to panels and rotors, and aircraft primary systems.

Credit Recommendation: ***Version 1:*** In the lower-division baccalaureate/associate degree category, 3 semester hours in administrative management (10/92). *Version 2:* In the lower-division baccalaureate/associate degree category, 3 semester hours in aviation administration and management (6/98). *Version 3:* In the lower-division baccalaureate/associate degree category, 3 semester hours in aviation maintenance management (12/03).

Related Occupations: 151A.

AR-1704-0200

Observation Airplane Repairer Supervisor Advanced Noncommissioned Officer (NCO)

Course Number: ***Version 1:*** 600-67H40. *Version 2:* 600-67H40.

Location: Aviation Logistics School, Fort Eustis, VA.

Length: ***Version 1:*** 9–10 weeks (346 hours). *Version 2:* 8–9 weeks (303 hours).

Exhibit Dates: ***Version 1:*** 10/89–9/91. *Version 2:* 10/91–4/04.

Learning Outcomes: ***Version 1:*** Upon completion of the course, the student will be able to determine training requirements, counsel personnel, apply production and quality control procedures, introduce unit safety programs, conduct maintenance shop performance inspections, comply with aviation logistics and supply procedures, maintain and confirm accuracy of aviation forms and records, prepare readiness reports, and use data communications hardware and software. *Version 2:* Upon completion of the course, the student will be able to determine training requirements, counsel personnel, apply production and quality control procedures, introduce unit safety programs, conduct maintenance shop performance inspections, comply with aviation logistics and supply procedures, maintain and confirm accuracy of aviation forms and records, prepare readiness reports, and use data communications hardware and software.

Instruction: *Version 1:* Lectures and practical exercises cover the administration of a maintenance management program. *Version 2:* Lectures and practical exercises cover the administration of a maintenance management program. This course contains a common core of leadership subjects.

Credit Recommendation: *Version 1:* In the lower-division baccalaureate/associate degree category, 3 semester hours in administrative management (6/91). *Version 2:* In the lower-division baccalaureate/associate degree category, 3 semester hours in administrative management. See AR-1404-0035 for common core credit (6/91).

Related Occupations: 67H.

AR-1704-0201

OH-58D Power Train Repairer

Course Number: 602-ASIW5 (68D).

Location: Aviation Logistics School, Fort Eustis, VA.

Length: 3 weeks (97 hours).

Exhibit Dates: 10/88–12/93.

Learning Outcomes: Upon completion of the course, the student will be able to maintain and repair main and tail rotor components and parts associated with the Bell helicopter.

Instruction: Lectures and practical exercises cover the Bell helicopter.

Credit Recommendation: In the lower-division baccalaureate/associate degree category, 1 semester in helicopter rotor maintenance and repair (6/91).

Related Occupations: 68D.

AR-1704-0202

OH-58D Helicopter Repairer Supervisor Basic Noncommissioned Officer (NCO)
(OH-58D Scout Helicopter Repairer Supervisor Basic Noncommissioned Officer (NCO))

Course Number: *Version 1:* 600-67S30. *Version 2:* 600-67S30. *Version 3:* 600-67S30. *Version 4:* 600-67S30.

Location: Aviation Logistics School, Fort Eustis, VA.

Length: *Version 1:* 7–8 weeks (189 hours). *Version 2:* 14 weeks (424 hours). *Version 3:* 15–16 weeks (579 hours). *Version 4:* 15–16 weeks (591 hours).

Exhibit Dates: *Version 1:* 10/88–3/90. *Version 2:* 4/90–9/91. *Version 3:* 10/91–9/94. *Version 4:* 10/94–Present.

Learning Outcomes: *Version 1:* Upon completion of the course, the student will be able to supervise maintenance personnel and organize maintenance procedures on helicopters. *Version 2:* Upon completion of the course, the student will be able to maintain, repair, and supervise maintenance and repairs on helicopters. *Version 3:* Upon completion of the course, the student will be able to maintain, repair, and supervise maintenance and repairs on helicopters. *Version 4:* Upon completion of the course, the student will be able to plan maintenance procedures and aviation inventories; use data communication hardware and software for oil analysis; use ground support equipment and precision tools; establish policies and procedures for shop and hangar safety; and supervise maintenance on fuel systems, pneudraulic systems, electrical systems, utility systems, power plants, rotor and drivetrain systems, aircraft structures, landing gears, aircraft weighing procedures, and engine and flight control systems.

Instruction: *Version 1:* Lectures and practical exercises cover supervision of maintenance personnel and maintenance procedures. *Version 2:* Lectures and practical exercises cover the maintenance and repair of helicopters and supervision of personnel. *Version 3:* Lectures and practical exercises cover the maintenance and repair of helicopters and supervision of personnel. This course contains a common core of leadership subjects. *Version 4:* Lectures, demonstrations, and practical exercises cover supervision, aircraft maintenance, and management. This course contains a common core of leadership subjects.

Credit Recommendation: *Version 1:* In the lower-division baccalaureate/associate degree category, 2 semester hours in maintenance management and 1 in principles of supervision (6/91). *Version 2:* In the lower-division baccalaureate/associate degree category, 2 semester hours in aviation maintenance and 2 in maintenance management (6/91). *Version 3:* In the lower-division baccalaureate/associate degree category, 2 semester hours in aviation maintenance and 2 in maintenance management. See AR-1406-0090 for common core credit (6/91). *Version 4:* In the lower-division baccalaureate/associate degree category, 3 semester hours in aircraft inspections, 2 in aviation maintenance management, and 2 in helicopter maintenance. See AR-1406-0090 for common core credit (6/98).

Related Occupations: 67S.

AR-1704-0203

OH-58D Power Plant Repairer

Course Number: 601-ASIW5 (68B).

Location: Aviation Logistics School, Fort Eustis, VA.

Length: 3 weeks (97 hours).

Exhibit Dates: 10/88–12/93.

Learning Outcomes: Upon completion of the course, the student will be able to perform unit and routine maintenance on aviation power plants.

Instruction: Lectures and practical exercises cover maintenance, troubleshooting, inspection, and replacement of various aircraft power plant components.

Credit Recommendation: Credit is not recommended because of the limited, specialized nature of the course (6/91).

Related Occupations: 68B.

AR-1704-0204

UH-60 Helicopter Repairer
(UH-60 Tactical Transport Helicopter Repairer)

Course Number: 600-67T10.

Location: Aviation Logistics School, Fort Eustis, VA.

Length: 13–14 weeks (483–641 hours).

Exhibit Dates: 10/89–Present.

Learning Outcomes: Upon completion of the course, the student will be able to repair rotary-wing aircraft at the entry level.

Instruction: Lectures, programmed instruction, and practical exercises cover aircraft; general maintenance procedures of rotary-wing aircraft; helicopter systems, including fuel, electrical, hydraulic, flight control, power plant, main and tail rotor, and utility systems; and troubleshooting and aircraft inspection.

Credit Recommendation: In the lower-division baccalaureate/associate degree category, 3 semester hours in helicopter maintenance and 2 in helicopter systems inspection and repair (10/03).

Related Occupations: 67T.

AR-1704-0205

UH-1 Helicopter Repairer

Course Number: *Version 1:* 600-67N10. *Version 2:* 600-67N10 (UH-1).

Location: *Version 1:* Aviation Center, Fort Rucker, AL. *Version 2:* Aviation School, Fort Rucker, AL.

Length: *Version 1:* 10 weeks (396 hours). *Version 2:* 9–10 weeks (360–400 hours).

Exhibit Dates: *Version 1:* 4/92–9/95. *Version 2:* 10/95–Present.

Learning Outcomes: *Version 1:* Before 4/92 see AR-1704-0033. Upon completion of the course, the student will be able to perform organizational, direct, and support maintenance tasks on the UH-1 utility helicopter. *Version 2:* Before 4/92 see AR-1704-0033. Upon completion of the course, the student will be able to perform organizational, direct, and support maintenance tasks on the UH-1 utility helicopter.

Instruction: *Version 1:* Practical exercises and examinations cover helicopter system fundamentals, including airframe, fuel, lubrication, hydraulic, rotor, and electrical system maintenance procedures; inspecting and troubleshooting helicopter systems and subsystems; and maintaining and troubleshooting ground handling equipment. *Version 2:* Practical exercises and examinations cover UH-1 helicopter system fundamentals, including flight control, power plant, drivetrain, airframe, fuel, lubrication, hydraulic, rotor, and electrical system maintenance procedures and inspection, troubleshooting, and removal and replacement of helicopter systems and subsystems.

Credit Recommendation: *Version 1:* In the vocational certificate category, 7 semester hours in helicopter maintenance. In the lower-division baccalaureate/associate degree category, 3 semester hours in helicopter maintenance (3/94). *Version 2:* In the lower-division baccalaureate/associate degree category, 9 semester hours in helicopter maintenance (1/03).

Related Occupations: 67N; 76N.

AR-1704-0206

1. AH-64 Armament/Electrical System Repairer (Transition)
2. AH-64A Armament/Electrical Systems Repairer Transition

Course Number: *Version 1:* 646-68X2/30-T. *Version 2:* 646-68X2/30 (T).

Location: Aviation Logistics School, Fort Eustis, VA.

Length: *Version 1:* 12–13 weeks (484 hours). *Version 2:* 11–12 weeks (449 hours).

Exhibit Dates: *Version 1:* 10/92–9/01. *Version 2:* 10/01–Present.

Learning Outcomes: *Version 1:* Upon completion of the course, the student will be able to maintain and troubleshoot helicopter systems, including electrical, area weapon, aerial rocket control, boresight, night vision sensor, target acquisition and designation, integrated helmet and display sight, and chaff. *Version 2:* Upon completion of the course, the student will be able to inspect electrical sections of AH-64A helicopter, including lighting, caution and warning, instrument, environmental control engine, and auxiliary power and supervise maintenance record keeping, quality control, and spare parts supply.

Instruction: *Version 1:* Lectures, practical exercises, and examinations cover principles of operation, troubleshooting, testing, component replacement, operational checks, and specialized maintenance procedures. Fault detection in AC/DC electrical systems is also included. *Version 2:* Methods of instruction include lecture, laboratory, and practical application. Student evaluation is based on written and practical testing methods.

Credit Recommendation: *Version 1:* In the lower-division baccalaureate/associate degree category, 3 semester hours in AC/DC aircraft electrical system maintenance (3/94). *Version 2:* In the lower-division baccalaureate/associate degree category, 3 in AC/DC aircraft electrical system maintenance (7/01).

Related Occupations: 68X.

AR-1704-0207

AIRCRAFT STRUCTURAL REPAIRER

Course Number: *Version 1:* 603-68G10. *Version 2:* 603-68G10.

Location: Aviation Logistics School, Fort Eustis, VA.

Length: *Version 1:* 13–15 weeks (482–539 hours). *Version 2:* 14 weeks (551 hours).

Exhibit Dates: *Version 1:* 10/92–7/00. *Version 2:* 8/00–Present.

Learning Outcomes: *Version 1:* Before 10/92 see AR-1704-0100. Upon completion of the course, the student will be able to perform entry-level tasks in maintenance of aircraft structural components, sheet metal surfaces, fiberglass, and helicopter rotor blade and will be able to fabricate aircraft parts. *Version 2:* Upon completion of the course, the student will be able to fabricate, maintain and repair aircraft structural members, sheet metal surfaces, and fiberglass and composite rotor blade components.

Instruction: *Version 1:* Course involves practical exercises including aircraft riveting; repair of aircraft skin, structural members, and rotor blade; maintenance and equipment record systems; forming aircraft parts; and shop and flight-line safety. *Version 2:* Practical exercises and lecture. Topics include basic aircraft maintenance, corrosion control; fabrication, repair and replacement of aircraft structural components and sheet metal parts; repair of fiberglass and composite components; and rotor blade repair and painting.

Credit Recommendation: *Version 1:* In the lower-division baccalaureate/associate degree category, 3 semester hours in airframe sheet metal shop and 3 in airframe structural repair (6/98). *Version 2:* In the lower-division baccalaureate/associate degree category, 3 semester hours in aircraft structure fundamentals and 3 in aircraft systems and structure maintenance and inspection (12/03).

Related Occupations: 68G.

AR-1704-0208

AVIATION LOGISTICS OFFICER ADVANCED

Course Number: 2-55-C22.

Location: Aviation Logistics School, Fort Eustis, VA.

Length: 5 weeks (185 hours).

Exhibit Dates: 10/91–2/93.

Learning Outcomes: Upon completion of the course, the student will be able to manage material systems and serve in higher positions in aviation and Army logistics.

Instruction: Lectures, practical exercises, and examinations cover maintenance management, logistics management, advanced aviation logistics, aviation life support equipment, common technical systems, and propulsion systems. Course includes record keeping, parts procedures, aircraft safety, and life-cycle analysis.

Credit Recommendation: In the lower-division baccalaureate/associate degree category, 2 semester hours in aviation maintenance management. In the upper-division baccalaureate category, 2 semester hours in advanced logistics management (3/94).

AR-1704-0209

LIGHTER AIR-CUSHION VEHICLE SUPERVISOR

Course Number: 8C-F20/652-F7.

Location: Transportation School, Fort Eustis, VA.

Length: 3 weeks (108 hours).

Exhibit Dates: 4/91–1/98.

Learning Outcomes: Upon completion of the course, the student will be able to perform supervisory duties in the operation of the 30-ton lighter air-cushion vehicle.

Instruction: Lectures and practical exercises cover location, function, and operation of the systems and components of the 30-ton lighter air-cushion vehicle.

Credit Recommendation: Credit is not recommended because of the limited, specialized nature of the course (3/94).

AR-1704-0210

LIGHTER AIR-CUSHION VEHICLE 30 TON OPERATOR

Course Number: 062-ASIE6 (88K).

Location: Transportation School, Fort Eustis, VA.

Length: 9 weeks (324 hours).

Exhibit Dates: 10/93–1/98.

Learning Outcomes: Before 10/93 see AR-1704-0133. Upon completion of the course, the student will be able to operate a 30-Ton lighter air-cushion vehicle.

Instruction: Lectures and practical exercises cover craft operation, use of emergency equipment and procedures, logistics procedures, and operational safety.

Credit Recommendation: Credit is not recommended because of the limited, specialized nature of the course (3/94).

Related Occupations: 88K.

AR-1704-0211

LIGHTER AIR-CUSHION 30 TON SUBSYSTEM

Course Number: 062-F7.

Location: Transportation School, Fort Eustis, VA.

Length: 3–4 weeks (130 hours).

Exhibit Dates: 10/92–1/98.

Learning Outcomes: Upon completion of the course, the student will be able to perform preventive maintenance on the various subsystems of the lighter air-cushion vehicle.

Instruction: Lectures and practical exercises cover the maintenance of various subsystems of the vehicle.

Credit Recommendation: Credit is not recommended because of the limited, specialized nature of the course (3/94).

AR-1704-0213

LIGHTER AIR-CUSHION VEHICLE 30 TON INTRODUCTION AND THEORY

Course Number: 062-ASIE6.

Location: Transportation School, Fort Eustis, VA.

Length: 3 weeks (112 hours).

Exhibit Dates: 10/92–1/98.

Learning Outcomes: Upon completion of the course, the student will be able to perform preventive maintenance on the 30-Ton lighter air-cushion vehicle.

Instruction: Lectures and practical exercises cover maintaining the vehicle.

Credit Recommendation: Credit is not recommended because of the limited, specialized nature of the course (3/94).

AR-1704-0214

AVIONIC COMMUNICATION SYSTEM REPAIRER BASIC NONCOMMISSIONED OFFICER (NCO)

Course Number: 102-35L30.

Location: Signal School, Fort Gordon, GA; Ordnance, Missile and Munitions School, Redstone Arsenal, AL.

Length: 6 weeks (252 hours).

Exhibit Dates: 10/96–6/00.

Learning Outcomes: Upon completion of the course, the student will be able to supervise and provide technical assistance in an electronic communication maintenance facility.

Instruction: Lectures, practical exercises, and small group instruction cover maintenance management, fundamental computer applications, and military leadership. Course contains a common core leadership subjects.

Credit Recommendation: In the lower-division baccalaureate/associate degree category, 2 semester hours in maintenance management. See AR-1406-0090 for common core credit (5/97).

Related Occupations: 35L.

AR-1704-0215

AVIONIC FLIGHT SYSTEM REPAIRER BASIC NONCOMMISSIONED OFFICER (NCO)

Course Number: 102-35Q30.

Location: Combined Arms Support Command, Fort Lee, VA; Ordnance, Missile and Munitions School, Redstone Arsenal, AL.

Length: 5–6 weeks (208 hours).

Exhibit Dates: 10/95–6/00.

Learning Outcomes: Upon completion of the course, the student will be able to provide leadership and avionic flight systems.

Instruction: Lectures and practical exercises cover leadership, resource management, communications, training management, advanced electronics, avionics systems, and maintenance management. Course contains a common core of leadership subjects.

Credit Recommendation: In the lower-division baccalaureate/associate degree category, 3 semester hours in maintenance management. See AR-1406-0090 for common core credit (5/97).

AR-1704-0216

AVIONICS RADAR REPAIRER BASIC NONCOMMISSIONED OFFICER (NCO)

Course Number: 102-35R30.

Location: Ordnance, Missile and Munitions School, Redstone Arsenal, AL.

Length: 6–7 weeks (272 hours).

Exhibit Dates: 10/95–6/00.

Learning Outcomes: Upon completion of the course, the student will be able to provide leadership and will possess the advanced technical skills necessary to function as an avionics radar repair technician.

Instruction: Lectures and practical exercises cover current avionic radar equipment, supervision, maintenance operations, and the management of maintenance operations. Course contains a common core of leadership subjects.

Credit Recommendation: In the lower-division baccalaureate/associate degree category, 3 semester hours in maintenance management. See AR-1406-0090 for common core credit (5/97).

Related Occupations: 35R.

AR-1704-0217

TURBINE ENGINE DRIVEN GENERATOR REPAIRER

Course Number: 662-52F30.

Location: Ordnance Center and School, Aberdeen Proving Ground, MD.

Length: 9–10 weeks (353 hours).

Exhibit Dates: 10/96–Present.

Learning Outcomes: Upon completion of the course, the student will be able to provide technical, logistical, operational, and support for repair personnel.

Instruction: Lectures and practical exercises cover logistics, battalion maintenance, turbine engine power generators, and the aviation ground power unit. Course contains a common core leadership subjects.

Credit Recommendation: In the lower-division baccalaureate/associate degree category, 3 semester hours in turbine engine AC/DC generators. See AR-1406-0090 for common core credit (1/98).

Related Occupations: 52F.

AR-1704-0218

AIRCRAFT POWER PLANT REPAIRER SUPERVISOR BASIC NONCOMMISSIONED OFFICER (NCO)

Course Number: 601-68B30.

Location: Aviation Logistics School, Fort Eustis, VA.

Length: 10–11 weeks (399 hours).

Exhibit Dates: 10/96–9/00.

Learning Outcomes: After 9/00 see AR-1717-0174. Upon completion of the course, the student will be able to apply the principles and procedures of aviation maintenance management as they apply to various turbine engines.

Instruction: The course consists of lectures and practical experiences in applying the policies and procedures associated with approved aviation maintenance standards. This course contains a common core of leadership subjects.

Credit Recommendation: In the lower-division baccalaureate/associate degree category, 2 semester hours in aviation maintenance management and 3 in principles of supervision. See AR-1406-0090 for common core credit (6/98).

Related Occupations: 68B.

AR-1704-0219

1. OH-58D HELICOPTER REPAIRER (TRANSITION)
2. OH-58D HELICOPTER REPAIRER (TRANSITION) TATS

Course Number: *Version 1:* 552-67S2/30-RC (T); 600-67S2/30 (T). *Version 2:* 552-67S2/30-RC (T); 600-67S2/30 (T); 552-67S2/30-RC (T) (F).

Location: Aviation Logistics School, Fort Eustis, VA.

Length: *Version 1:* 7–8 weeks (284 hours). *Version 2:* 7–8 weeks (284 hours).

Exhibit Dates: *Version 1:* 5/94–9/03. *Version 2:* 10/03–Present.

Learning Outcomes: *Version 1:* Upon completion of the course, the student will be able to repair, service, and maintain rotary-wing aircraft. *Version 2:* Upon completion of the course, the student will be able to repair, service, and maintain rotary-wing aircraft.

Instruction: *Version 1:* Lectures, programmed instruction, and practical exercises cover helicopter maintenance and repair procedures, including fuel, electrical, hydraulic, flight control, power plant, main and tail rotor, transmission, and utility systems; troubleshooting; and aircraft inspection. *Version 2:* Computer-based training, practical exercises, and lecture. Topics include helicopter maintenance and repair procedures, including fuel, electrical, hydraulic, flight control, power plant, main and tail rotor, transmission, and utility systems; troubleshooting; and aircraft inspection.

Credit Recommendation: *Version 1:* In the lower-division baccalaureate/associate degree category, 3 semester hours in helicopter maintenance (6/03). *Version 2:* In the lower-division baccalaureate/associate degree category, 2 semester hours in aircraft primary systems and 1 in maintenance management (12/03).

AR-1704-0220

UH-60 HELICOPTER REPAIRER - TRANSITION

Course Number: 600-6772/30 (T) (F).

Location: Aviation Logistics School, Fort Eustis, VA.

Length: 8–9 weeks.

Exhibit Dates: 10/96–Present.

Learning Outcomes: Upon completion of the course, the student will be able to repair, service, and maintain rotary-wing aircraft.

Instruction: Lectures, programmed instruction, and practical exercises cover helicopter maintenance and repair procedures, including fuel, electrical, hydraulic, flight control, power plant, main and tail rotor, transmission, and utility systems; troubleshooting; and aircraft inspection.

Credit Recommendation: In the lower-division baccalaureate/associate degree category, 3 semester hours in helicopter maintenance (6/98).

AR-1704-0221

AH-1 ATTACK HELICOPTER REPAIRER

Course Number: 600-67Y10.

Location: Aviation Logistics School, Fort Eustis, VA.

Length: 10–12 weeks (424–460 hours).

Exhibit Dates: 10/95–9/03.

Learning Outcomes: Before 10/95 see AR-1704-0095. Upon completion of the course, the student will be able to use publications, forms, and records related to maintenance tasks; practice shop and flight-line safety procedures; use common hand tools, aircraft hardware, and ground support equipment; locate, identify, remove, and replace parts and components in the airframe and landing gear systems, utility and hydraulic systems, fuel, electrical, power plant, rotor, and power train systems; and assist in performing operational checks and inspections.

Instruction: Lectures, demonstrations, and practical exercises cover the skills needed to perform general aircraft maintenance procedures on helicopter equipment, auxiliary equipment, and power plant and to remove, inspect, repair, and replace helicopter system components.

Credit Recommendation: In the lower-division baccalaureate/associate degree category, 3 semester hours in helicopter maintenance and 5 in helicopter systems inspection and repair (8/99).

Related Occupations: 67Y.

AR-1704-0222

AIRCRAFT POWER TRAIN REPAIRER

Course Number: 602-68D10.

Location: Aviation Logistics School, Fort Eustis, VA.

Length: 17–19 weeks (612–699 hours).

Exhibit Dates: 10/94–Present.

Learning Outcomes: Before 10/94 see AR-1704-0012. Upon completion of the course, the student will be able to repair, maintain, and inspect rotary-wing aircraft power train systems.

Instruction: Lectures and practical exercises cover rotary-wing aircraft power train components maintenance, inspection, and repair; structural hardware; equipment record systems; bearing, gear, and component preservation; and penetrant, magnetic particle, radiographic, and ultrasonic nondestructive testing.

Credit Recommendation: In the lower-division baccalaureate/associate degree category, 6 semester hours in theory, operation, maintenance, repair, and inspection of helicopter power train components (12/03).

Related Occupations: 68D.

AR-1704-0223

AIRCRAFT ELECTRICIAN

Course Number: *Version 1:* 602-68F10. *Version 2:* 602-68F10.

Location: Aviation Logistics School, Fort Eustis, VA.

Length: *Version 1:* 20 weeks (739 hours). *Version 2:* 18–19 weeks (672 hours).

Exhibit Dates: *Version 1:* 10/94–9/98. *Version 2:* 10/98–Present.

Learning Outcomes: *Version 1:* Before 10/94 see AR-1704-0075. Upon completion of the course, the student will be able to troubleshoot, maintain, and repair aircraft electrical systems and instrumentation. *Version 2:* Upon completion of the course, the student will be able to troubleshoot, maintain, and repair electrical systems on AH-1F, CH-47D and UH-60 helicopters.

Instruction: *Version 1:* Lectures and practical exercises cover basic electrical and electronic theory, soldering techniques, AC and DC circuits, semiconductors, wiring diagrams, and test equipment. Reinforcement of these skills is accomplished by operating, testing, troubleshooting, and repairing aircraft electrical systems. *Version 2:* Methods of instruction include lecture, laboratory, and practical experience on aircraft electrical systems. Written and practical tests are used for student evaluation.

Credit Recommendation: ***Version 1:*** In the lower-division baccalaureate/associate degree category, 3 semester hours in basic electricity, 3 in basic electronics, 6 in electronic systems troubleshooting and maintenance, 3 in aircraft electrical systems, and 3 in aircraft instrument systems (6/98). *Version 2:* In the lower-division baccalaureate/associate degree category, 3 semester hours in basic electricity, 3 in basic electronics, and 4 in aircraft electrical systems (7/01).

Related Occupations: 68F.

AR-1704-0224

OH-58D Aircraft Electrical Repairer

Course Number: 602-ASIW (68F).

Location: Aviation Logistics School, Fort Eustis, VA.

Length: 3 weeks (95 hours).

Exhibit Dates: 10/96–Present.

Learning Outcomes: Before 10/96 see AR-1714-0033. Upon completion of the course, the student will be able to maintain, repair, and troubleshoot aircraft electrical systems and components.

Instruction: Lectures and practical exercises cover aircraft electrical/electronic systems, including components, aircraft instrument systems, lighting systems, and utility electrical systems. Reinforcement of these skills is accomplished by operating, testing, troubleshooting, and repairing aircraft electrical systems.

Credit Recommendation: In the lower-division baccalaureate/associate degree category, 3 semester hours in electronic systems troubleshooting and maintenance (6/98).

AR-1704-0225

Guardrail Common Sensor/Instructor Pilot (1P)

Course Number: 2C-15C/2B-ASIF4 (1P).

Location: Intelligence Center and School, Fort Huachuca, AZ.

Length: 9–10 weeks (352–353 hours).

Exhibit Dates: 8/96–Present.

Learning Outcomes: Upon completion of the course, the student will be able to function as an instructor pilot in the RC-12K/N/P aircraft equipped with the guardrail common sensor intelligence equipment.

Instruction: Instruction covers military intelligence mission training, aircraft performance and weight and balance, calculations, multi-engine turbine engine flight training, and turbine aircraft systems operation, including electrical, fuel, pneumatic, environmental, pressurization, anti-ice, flight control, landing gear, and radar systems. Methods of instruction include lecture and flight training.

Credit Recommendation: In the upper-division baccalaureate category, 3 semester hours in theory of flight and systems and 3 in advanced multi-engine flight (11/98).

AR-1704-0226

1. Aircraft Armament/Electrical Systems Repairer Supervisor Advanced Noncommissioned Officer (NCO)
2. AH-64A Armament/Electrical Systems Repairer Supervisor Advanced Noncommissioned Officer Advanced Noncommissioned Officer (NCO)

Course Number: ***Version 1:*** 602-68X40. *Version 2:* 602-68X40, Phase 2.

Location: ***Version 1:*** Aviation School, Fort Rucker, AL. *Version 2:* Aviation Center, Fort Rucker, AL.

Length: ***Version 1:*** 8–9 weeks (345 hours). *Version 2:* 5–6 weeks (262 hours).

Exhibit Dates: ***Version 1:*** 10/97–9/02. *Version 2:* 10/02–Present.

Learning Outcomes: ***Version 1:*** Upon completion of the course, the student will provide leadership and management in aircraft armament/electrical systems maintenance and repair. In addition the student will be able to demonstrate leadership and supervisory skills in the training environment, prepare various required forms, operate communications equipment, and demonstrate basic knowledge of computer systems and software. *Version 2:* Upon completion of the course, the student will be able to conduct personnel training and resource management for electrical systems repairer; perform supply and logistic management; perform brigade and battalion simulation exercises; conduct a situational training exercise; and use DOS, MS Office for word processing, and basic computer operations.

Instruction: ***Version 1:*** Classroom instruction and practical exercises cover leadership, military skills, professional skills, resource management, communication skills, training management systems, maintenance management, supply and logistics, and tactical operations. *Version 2:* Classroom instruction and practical exercises cover leadership, military skills, professional skills, resource management, communication skills, training management systems, maintenance management, supply and logistics, and tactical operations.

Credit Recommendation: ***Version 1:*** In the lower-division baccalaureate/associate degree category, 3 semester hours in maintenance management. See AR-1404-0035 for common core credit (11/98). *Version 2:* In the lower-division baccalaureate/associate degree category, 3 semester hours in maintenance management. See AR-1404-0035 for common core credit (4/03).

AR-1704-0227

1. OH-58D Scout Helicopter Repairer Supervisor Advanced Noncommissioned Officer (NCO)
2. OH-58D Helicopter Repairer Supervisor ANCOC
   (OH-58D Scout Helicopter Repairer Supervisor Advanced Noncommissioned Officer (NCO))

Course Number: ***Version 1:*** 600-67S40. *Version 2:* 600-67S40, Phase 2; 600-15S40, Phase 2.

Location: ***Version 1:*** Aviation School, Fort Rucker, AL. *Version 2:* Aviation Center, Fort Rucker, AL.

Length: ***Version 1:*** 8–9 weeks (345 hours). *Version 2:* 6 weeks (262 hours).

Exhibit Dates: ***Version 1:*** 10/97–9/00. *Version 2:* 10/00–Present.

Learning Outcomes: ***Version 1:*** Upon completion of the course, the student will be able to apply policies and standards of aviation maintenance management, determine training requirements, counsel personnel, apply production and quality control procedures, introduce unit safety programs, conduct maintenance shop performance inspections, comply with aviation logistics and supply procedures, maintain and confirm the accuracy of aviation forms and records, prepare readiness reports, and use data communication hardware and software. *Version 2:* Upon completion of the course, the student will be able to demonstrate basic management, supervisory, and leadership skills in an aircraft maintenance environment.

Instruction: ***Version 1:*** Lectures and practical exercises cover the administration of a maintenance management program. This course contains a common core of leadership subjects. *Version 2:* Practical, classroom exercises and lecture. Topics include fundamentals of aviation maintenance management, logistics management, maintenance duties and responsibilities, safety, training, and maintenance supervision.

Credit Recommendation: ***Version 1:*** In the lower-division baccalaureate/associate degree category, 3 semester hours in maintenance management. See AR-1404-0035 for common core credit (11/98). *Version 2:* In the lower-division baccalaureate/associate degree category, 3 semester hours in maintenance management. See AR-1404-0035 for common core credit (4/04).

AR-1704-0230

Railway Section Repairer

Course Number: 850-88T10-RC.

Location: Transportation School, Fort Eustis, VA.

Length: 5 weeks (176 hours).

Exhibit Dates: 10/00–Present.

Learning Outcomes: Upon completion of the course, the student will be able to inspect, assess, maintain, and repair track locomotive.

Instruction: Methods of instruction include conferences, performance exercises, and demonstrations on inspecting, repairing, and maintaining railway tracks, roadbeds, signals, and switches.

Credit Recommendation: In the lower-division baccalaureate/associate degree category, 2 semester hours in industrial maintenance technology (6/01).

Related Occupations: 88T.

AR-1704-0231

Total Army Training System (TATS) OH-58A/C Observation/Scout Helicopter Repairer Supervisor Basic Noncommissioned Officer (NCO)

Course Number: 600-67V30-RC (F), Phase 3.

Location: Aviation Logistics School, Fort Eustis, VA.

Length: 3 weeks (129 hours).

Exhibit Dates: 10/03–Present.

Learning Outcomes: Upon completion of the course, the student will be able to perform basic nondestructive testing and inspections on

OH-58A/C helicopter and components including structural, power train, and rotor systems; T63-A700 power plants; AC/DC power systems; and instrument pilot-static systems inspections.

Instruction: Methods of instruction include classroom and practical exercises.

Credit Recommendation: In the lower-division baccalaureate/associate degree category, 3 semester hours in aircraft inspection (6/01).

Related Occupations: 67V.

AR-1704-0232

1. OH-58D Armament/Electrical Systems Repairer
2. OH-58D Armament/Electrical/Avionic Systems Repairer

Course Number: ***Version 1:*** 102-68J10 (OH-58D). *Version 2:* 102-68S10.

Location: Aviation Logistics School, Fort Eustis, VA.

Length: ***Version 1:*** 19–20 weeks (732 hours). *Version 2:* 29 weeks (1152 hours).

Exhibit Dates: ***Version 1:*** 1/99–9/00. *Version 2:* 10/00–Present.

Learning Outcomes: ***Version 1:*** Upon completion of the course, the student will be able to troubleshoot, maintain and repair OH-58D helicopter electrical and avionics systems. *Version 2:* Upon completion of the course, the student will be able to troubleshoot, maintain, and repair various aircraft electrical and avionics systems.

Instruction: ***Version 1:*** Lecture and laboratory experience with written and practical examinations are used for student instruction and evaluation. *Version 2:* Audiovisual materials, practical and classroom exercises, discussion, laboratory, and lecture. Topics include electronics fundamentals; operation, troubleshooting, and maintenance of aircraft electrical, avionics, and weapon systems.

Credit Recommendation: ***Version 1:*** In the lower-division baccalaureate/associate degree category, 4 semester hours in aircraft electrical system maintenance and repair and 3 in avionics system maintenance and repair (6/01). *Version 2:* In the lower-division baccalaureate/associate degree category, 3 semester hours in basic electricity, 4 in aircraft electrical system maintenance, and 4 in avionics systems troubleshooting and maintenance (12/03).

AR-1704-0233

AH-64D Armament Maintenance Technician

Course Number: 4D-SQIE (152H).

Location: Aviation Logistics School, Fort Eustis, VA.

Length: 18–19 weeks (714 hours).

Exhibit Dates: 6/00–Present.

Learning Outcomes: Upon completion of the course, the student will be able to perform maintenance, preventive maintenance, troubleshooting, and repair on AH-64D electrical and avionic systems.

Instruction: Methods of instruction include classroom, practical, and in-shop training.

Credit Recommendation: In the lower-division baccalaureate/associate degree category, 4 semester hours in aircraft electrical systems and 3 in avionics systems (6/01).

Related Occupations: 152H.

AR-1704-0234

OH-58D Armament Maintenance Technician

Course Number: ***Version 1:*** 4D-SQIE (152D). *Version 2:* 4D-SQIE (OH-58D).

Location: Aviation Logistics School, Fort Eustis, VA.

Length: ***Version 1:*** 8 weeks (297 hours). *Version 2:* 4 weeks (146 hours).

Exhibit Dates: ***Version 1:*** 12/98–9/02. *Version 2:* 10/02–Present.

Learning Outcomes: ***Version 1:*** Upon completion of the course, the student will be able to perform maintenance, inspection, and repair of electrical systems on OH-58D helicopter. *Version 2:* Upon completion of the course, the student will be able to inspect, test, troubleshoot and repair helicopter weapon systems; supervise and manage helicopter weapon systems troubleshooting and maintenance.

Instruction: ***Version 1:*** Methods of instruction include classroom conferences and practical application of skills. *Version 2:* Practical and classroom exercises, discussion, laboratory, and lecture. Topics include inspection, maintenance, and troubleshooting of helicopter weapon systems and components; supervision; and management of aviation troubleshooting and maintenance.

Credit Recommendation: ***Version 1:*** In the lower-division baccalaureate/associate degree category, 4 semester hours in aircraft electrical systems (6/01). *Version 2:* In the lower-division baccalaureate/associate degree category, 2 semester hours in aircraft electrical systems troubleshooting and maintenance and 2 in aircraft maintenance management (12/03).

Related Occupations: 152D.

AR-1704-0235

Total Army Training System (TATS) AH-64A Armament/Electrical Systems Supervisor Basic Noncommissioned Officer (NCO)

Course Number: 646-68X30 (F), Phase 2.

Location: Aviation Logistics School, Fort Eustis, VA.

Length: 12 weeks (499 hours).

Exhibit Dates: 10/00–Present.

Learning Outcomes: Before 10/00 see AR-1704-0192. Upon completion of the course, the student will be able to supervise maintenance, inspect, troubleshoot, and repair electrical and electronic systems on the AH-64A helicopter.

Instruction: Lectures and practical exercises include AC/DC circuits, electrical and avionics systems, supervision, and management.

Credit Recommendation: In the lower-division baccalaureate/associate degree category, 2 semester hours in aircraft maintenance supervision, 3 in electronic systems troubleshooting and maintenance, and 3 in electrical systems inspection (6/01).

Related Occupations: 68X.

AR-1704-0236

Total Army Training System (TATS) AH-64A Armament/Electrical Systems Repairer Supervisor Basic Noncommissioned Officer (NCO)

Course Number: 646-68X30 (F), Phase 2.

Location: Aviation Logistics School, Fort Eustis, VA.

Length: 129 hours.

Exhibit Dates: 10/00–Present.

Learning Outcomes: Before 10/00 see AR-1704-0192. Upon completion of the course, the student will be able to supervise repairs, maintenance, and troubleshooting on AH-64A armament electrical systems.

Instruction: Methods of instruction include practical and classroom exercise covering AC/DC circuits, electrical and avionic systems, supervision, and management.

Credit Recommendation: In the lower-division baccalaureate/associate degree category, 2 semester hours in aircraft maintenance supervision (6/01).

Related Occupations: 68X.

AR-1704-0237

Total Army Training System (TATS) AH-64A Armament/Electrical Systems Repairer Supervisor Basic Noncommissioned Officer (NCO)

Course Number: 646-68X30 (F), Phase 3.

Location: Aviation Logistics School, Fort Eustis, VA.

Length: 160 hours.

Exhibit Dates: 10/00–Present.

Learning Outcomes: Upon completion of the course, the student will be able to perform inspections on AH-64A helicopter electrical systems.

Instruction: Methods of instruction include classroom and practical exercises covering AH-64A electrical/electronic systems.

Credit Recommendation: In the lower-division baccalaureate/associate degree category, 3 semester hours in electrical systems inspection (6/01).

Related Occupations: 68X.

AR-1704-0238

Total Army Training System (TATS) AH-64A Armament/Electrical Systems Repairer Supervisor Basic Noncommissioned Officer (NCO)

Course Number: 646-68X30 (F), Phase 4.

Location: Aviation Logistics School, Fort Eustis, VA.

Length: 210 hours.

Exhibit Dates: 10/00–Present.

Learning Outcomes: Upon completion of the course, the student will be able to perform and supervise maintenance, and inspect and repair AH-64A helicopter electronic systems.

Instruction: Methods of instruction include classroom and practical and situational exercises on AH-64A armament electronic systems.

Credit Recommendation: In the lower-division baccalaureate/associate degree category, 3 semester hours in aircraft maintenance supervision and 3 in electronic systems troubleshooting and maintenance (6/01).

Related Occupations: 68X.

AR-1704-0239

Total Army Training System Aircraft Powertrain Repairer Supervisor Basic Noncommissioned Officer (NCO), Phase 3

Course Number: 552-68D30 (F), Phase 3.

Location: Aviation Logistics School, Fort Eustis, VA.

Length: 225 hours.

Exhibit Dates: 10/00–Present.

Learning Outcomes: Upon completion of the course, the student will be able to supervise and conduct inspections and repairs on helicopter powertrains.

Instruction: Methods of instruction include classroom and practical exercises, including powertrain components and supervisory skill development.

Credit Recommendation: In the lower-division baccalaureate/associate degree category, 3 semester hours in aircraft maintenance supervision and 3 in aircraft maintenance management and inspection (6/01).

Related Occupations: 68D.

AR-1704-0240

Total Army Training System (TATS) Aircraft Powertrain Repairer Supervisor Basic Noncommissioned Officer (NCO), Phase 2

Course Number: 552-CMF67-CAM PH-2, Phase 2.

Location: Aviation Logistics School, Fort Eustis, VA.

Length: 129 hours.

Exhibit Dates: 10/00–Present.

Learning Outcomes: Upon completion of the course, the student will be able to supervise repairs, inspections, maintenance, and troubleshooting on aircraft powertrains.

Instruction: Methods of instruction include practical and classroom exercises in maintenance supervision.

Credit Recommendation: In the lower-division baccalaureate/associate degree category, 2 semester hours in aircraft maintenance supervision (6/01).

AR-1704-0241

Total Army Training System (TATS) Aircraft Powertrain Repairer Supervisor Basic Noncommissioned Officer (NCO)

Course Number: 602-68D30 (F), Phase 2; 552-68D30 (F), Phase 2.

Location: Aviation Logistics School, Fort Eustis, VA.

Length: 9 weeks (354 hours).

Exhibit Dates: 10/00–Present.

Learning Outcomes: Before 10/00 see AR-1704-0185. Upon completion of the course, the student will be able to supervise and conduct repairs, inspection, and troubleshooting of helicopter powertrains.

Instruction: Methods of instruction include classroom and practical exercises including powertrain components and supervisory skill development.

Credit Recommendation: In the lower-division baccalaureate/associate degree category, 5 semester hours in aircraft maintenance supervision and 3 in aircraft maintenance management and inspection (10/03).

Related Occupations: 68D.

AR-1704-0242

Total Army Training System (TATS) UH-1 Helicopter Repairer Supervisor Basic Noncommissioned Officer (NCO), Phase 3

Course Number: 552-67N30 (F), Phase 3.

Location: Aviation Logistics School, Fort Eustis, VA.

Length: 3 weeks (128 hours).

Exhibit Dates: 10/00–Present.

Learning Outcomes: Upon completion of the course, the student will be able to inspect the powertrain; structure; and pneudraulic propulsion, and electrical systems on the UH-1 helicopter.

Instruction: Methods of instruction include conferences, demonstrations, and practical exercises. Students are evaluated based on written and practical examinations.

Credit Recommendation: In the lower-division baccalaureate/associate degree category, 3 semester hours in aircraft maintenance management and inspection (6/01).

Related Occupations: 67N.

AR-1704-0246

Unmanned Aerial Vehicle Operator

Course Number: 243-96U10, Phase 1.

Location: Intelligence Center and School, Fort Huachuca, AZ.

Length: 23–24 weeks (851 hours).

Exhibit Dates: 10/00–Present.

Learning Outcomes: Upon completion of the course, the student will be able to plan missions, prepare Unmanned Aerial Vehicles (UAV) for launch and flight, perform maintenance on vehicles, and recognize ground vehicles and equipment.

Instruction: Audiovisual materials, practical and classroom exercises, laboratory, and lecture. Topics include simulation training, and flight training.

Credit Recommendation: In the lower-division baccalaureate/associate degree category, 2 semester hours in introduction aviation, and 1 in aircraft maintenance (12/02).

AR-1704-0247

Hunter UAV Mechanical Maintenance Specialist

Course Number: 662-F5.

Location: Intelligence Center and School, Fort Huachuca, AZ.

Length: 10 weeks (360 hours).

Exhibit Dates: 1/02–Present.

Learning Outcomes: Upon completion of the course, the student will be able to troubleshoot and repair power plants and airframes of unmanned aerial vehicles (UAV).

Instruction: Audiovisual materials, practical exercises, discussion, and lecture. Topics include examination of systems, repair of composite materials, fueling and de-fueling procedures, and repair of airframes and powerplants.

Credit Recommendation: In the vocational certificate category, 1 semester hour in airframe maintenance, and 1 in aircraft power plant maintenance (12/02).

AR-1704-0248

Armament Repairer

Course Number: 091-45K10, Phase 3.

Location: Reserve Schools, Multiple locations.

Length: 108 hours.

Exhibit Dates: 10/00–Present.

Learning Outcomes: Upon completion of the course, the student will be able to repair and maintain electrical, hydraulic and mechanical systems on the Abrams military tank turret.

Instruction: Audio visual materials, practical exercises, discussion, laboratory, and lecture. Topics include inspection of cannon tube, cannon turret control systems, and turret troubleshooting, diagnosis, and repair.

Credit Recommendation: In the lower-division baccalaureate/associate degree category, 1 semester hour in mechanical systems maintenance and 1 in electrical and hydraulic systems repair (8/03).

AR-1704-0249

Armament Repairer

Course Number: 091-45K10, Phase 4.

Location: Reserve Schools, Multiple locations.

Length: 110 hours.

Exhibit Dates: 10/00–Present.

Learning Outcomes: Upon completion of the course, the student will be able to maintain, troubleshoot, and repair turret on the Bradley Fighting Vehicle (BFV); maintain Howitzer cab system on the BFV; and maintain the microclimate air conditioning system.

Instruction: Topics include troubleshooting and repair of the Bradley Fighting Vehicle (BFV) tank turret; troubleshooting, diagnosis, and repair of the Howitzer cab systems; and maintenance of microclimate conditioning system (self-contained air conditioning unit).

Credit Recommendation: In the lower-division baccalaureate/associate degree category, 1 semester hour in mechanical systems maintenance and 1 in electrical and hydraulic systems repair (8/03).

AR-1704-0250

AH-64D Armament/Electrical/Avionic Systems Repairer Transition

Course Number: 646-68Y2/30 (T).

Location: Aviation Logistics School, Fort Eustis, VA.

Length: 11 weeks (408 hours).

Exhibit Dates: 1/98–Present.

Learning Outcomes: Upon completion of the course, the student will be able to test, troubleshoot, maintain, and repair aircraft electrical, communications, navigation, data management, weapons, flight control, and survivability systems.

Instruction: Practical and classroom exercises, discussion, laboratory, and lecture. Topics include operation, troubleshooting and maintenance of aircraft data management, electrical power, communications, weapons, navigation, flight control, and survivability systems.

Credit Recommendation: In the lower-division baccalaureate/associate degree category, 2 semester hours in aircraft electrical systems troubleshooting and maintenance and 3 in avionics systems troubleshooting and maintenance (12/03).

AR-1704-0251

AH 64D Armament Maintenance Technician

Course Number: 4D-SQIE (AH-64D), Phase 2.

Location: Aviation Logistics School, Fort Eustis, VA.

Length: 14–15 weeks (563 hours).

Exhibit Dates: 10/02–Present.

Learning Outcomes: Upon completion of the course, the student will be able to supervise and manage the testing, repair, troubleshooting, and maintenance of various aircraft systems.

Instruction: Computer-based training, practical and classroom exercises, discussion, laboratory, and lecture. Topics include inspection, troubleshooting, and maintenance of aircraft data management systems, electrical systems, flight control systems, weapons systems, and aircraft survivability equipment.

Credit Recommendation: In the lower-division baccalaureate/associate degree category, 3 semester hours in avionics systems troubleshooting and maintenance and 3 in aircraft systems troubleshooting and maintenance (12/03).

AR-1704-0252

AH-64D Armament/Electrical/Avionic Systems Repairer

Course Number: 646-68Y10.

Location: Aviation Logistics School, Fort Eustis, VA.

Length: 24 weeks (960 hours).

Exhibit Dates: 10/99–Present.

Learning Outcomes: Upon completion of the course, the student will be able to maintain, test, repair, and troubleshoot aircraft weapons, avionics, and electrical systems and their associated test equipment.

Instruction: Audiovisual materials, practical and classroom exercises, discussion, laboratory, and lecture. Topics include basic electronics; troubleshooting and maintenance of aircraft data management; electrical power; communications; navigation flight control; weapons; and survivability systems and support equipment.

Credit Recommendation: In the lower-division baccalaureate/associate degree category, 3 semester hours in basic electricity, 4 in aircraft electrical systems troubleshooting and maintenance, and 4 in avionics systems troubleshooting and maintenance (12/03).

AR-1704-0253

AH-64D Attack Helicopter Repairer

Course Number: 600-67R10 (AH-64D).

Location: Aviation Logistics School, Fort Eustis, VA.

Length: 16–17 weeks (675 hours).

Exhibit Dates: 10/02–Present.

Learning Outcomes: Upon completion of the course, the student will be able to use publications, forms, and records related to maintenance tasks; practice shop and flight-line safety procedures; use common hand tools, aircraft hardware, and ground support equipment; locate, identify, remove, and replace parts and components in the airframe and landing gear systems, utility and hydraulic systems, fuel, electrical, power plant, rotor and power train systems; and assist in performing operational checks and inspections.

Instruction: Audiovisual materials, computer-based training, practical exercises, and lecture. Topics include basic aircraft hydraulics, airframe, landing gear system, pneumatics, fuel and utility systems, crew chief duties, and environment control systems.

Credit Recommendation: In the lower-division baccalaureate/associate degree category, 3 semester hours in aircraft hydraulic systems fundamentals, 2 in aircraft pnuematic systems fundamentals, and 1 in maintenance management (12/03).

AR-1704-0254

Apprentice Aircraft Pneudraulic System Specialist

Course Number: J3ABR2A635 002 (68H10); J3ABR2A635 002 (15H10).

Location: Air Education and Training Command, Sheppard Air Force Base, Texas.

Length: 7–10 weeks (200–267 hours).

Exhibit Dates: 10/03–Present.

Learning Outcomes: Upon completion of the course, the student will be able to remove, repair, and replace aircraft pneudraulic systems and sub-systems.

Instruction: Practical exercises, discussion, and lecture. Topics include fabrication of aircraft tube and hose assemblies; hydraulic component repair and maintenance; and aircraft ground power unit, repair maintenance.

Credit Recommendation: In the lower-division baccalaureate/associate degree category, 2 semester hours in aircraft hydraulic systems and 1 in aircraft primary systems troubleshooting and maintenance (12/03).

AR-1704-0255

TATS AH-64D Armament/Electrical/Avionic Systems Supervisor Basic Noncommissioned Officer (NCO)

Course Number: 646-68Y30, Phase 2.

Location: Aviation Logistics School, Fort Eustis, VA.

Length: 14–15 weeks (556 hours).

Exhibit Dates: 10/00–Present.

Learning Outcomes: Upon completion of the course, the student will be able to troubleshoot, supervise, and inspect maintenance performed on various aircraft electrical and avionics systems.

Instruction: Practical and classroom exercises, discussion, laboratory, and lecture. Topics include troubleshooting and the inspection and supervision of maintenance for aircraft data management, electrical, navigation, flight control, sensor, and weapons systems.

Credit Recommendation: In the lower-division baccalaureate/associate degree category, 2 semester hours in aircraft electrical systems troubleshooting and maintenance and 3 in avionics systems troubleshooting and maintenance, or 5 in aircraft maintenance management (12/03).

Related Occupations: 68Y.

AR-1704-0256

TATS AH-64D Attack Helicopter Repairer Supervisor Basic Noncommissioned Officer (BNCO)

Course Number: 600-67R30 (AH-64D) (F).

Location: Aviation Logistics School, Fort Eustis, VA.

Length: 14–15 weeks (581 hours).

Exhibit Dates: 10/00–Present.

Learning Outcomes: Upon completion of the course, the student will be able to plan maintenance procedures and aviation inventories; use data communications hardware and software including oil analysis programs; use ground support equipment and precision tools; establish policies and procedures for shop and hangar safety; and supervise maintenance of fuel systems, pneudraulic systems, electrical systems, utility systems, power plants, rotor and drivetrain systems, aircraft structures, landing gear, aircraft weight and balance procedures, engine control, and flight control systems.

Instruction: Lectures, demonstrations, and practical exercises cover supervision, aircraft maintenance, and maintenance management. This course contains a common core of leadership subjects.

Credit Recommendation: In the lower-division baccalaureate/associate degree category, 3 semester hours in aircraft systems troubleshooting and maintenance. In the upper-division baccalaureate category, 3 semester hours in aviation maintenance management (12/03).

AR-1704-0258

TATS Aircraft Electrician Supervisor Basic Noncommissioned Officer

Course Number: 602-68F30 (F), Phase 2.

Location: Aviation Logistics School, Fort Eustis, VA.

Length: 6–7 weeks (264 hours).

Exhibit Dates: 10/00–Present.

Learning Outcomes: Upon completion of the course, the student will be able to inspect the repair and supervise the maintenance of various aircraft electrical system components.

Instruction: Audiovisual materials, practical and classroom exercises, discussion, laboratory, and lecture. Topics include inspection, repair, and maintenance supervision of aircraft electrical and electronic systems and their components.

Credit Recommendation: In the lower-division baccalaureate/associate degree category, 1 semester hour in aircraft electrical systems, 2 in aircraft systems troubleshooting and maintenance, or 3 in aircraft maintenance management (12/03).

AR-1704-0259

TATS Aircraft Electrician Supervisor Basic Noncommissioned Officer (BNCO)

Course Number: 603-68F30 (F), Phase 3.

Location: Aviation Logistics School, Fort Eustis, VA.

Length: 3–4 weeks (135 hours).

Exhibit Dates: 10/00–Present.

Learning Outcomes: Upon completion of the course, the student will be able to inspect, repair, and perform maintenance on electrical and electronic systems on helicopters and supervise performance of electrical systems maintenance on helicopters.

Instruction: Computer-based training, practical exercises, discussion, and lecture. Topics include inspection and repair of aircraft wirings, corrosion in electrical systems, electrical maintenance, fuel and utility system components on helicopters and supervision of performance of electrical maintenance on helicopters.

Credit Recommendation: In the lower-division baccalaureate/associate degree category, 1 semester hour in electrical systems fundamentals and 1 in aircraft maintenance management (12/03).

AR-1704-0260

TATS Aircraft Pneudraulics Repair Supervisor

Course Number: 602-68H30 (F), Phase 2.

Location: Aviation Logistics School, Fort Eustis, VA.

Length: 6 weeks (236 hours).
Exhibit Dates: 10/00–Present.
Learning Outcomes: Upon completion of the course, the student will be able to supervise and provide technical guidance for maintenance of aircraft pneudraulics systems and evaluate maintenance operations for compliance with technical and operational publications and standards.
Instruction: Practical exercises, discussion, and lecture. Topics include interpretation and application of technical, operational, and quality control publications relating to supervision and technical guidance for performing aircraft pneudraulic systems maintenance.
Credit Recommendation: In the lower-division baccalaureate/associate degree category, 3 semester hours in aircraft primary systems (12/03).

AR-1704-0261

TATS Aircraft Pneudraulics Repair Supervisor

Course Number: 602-68H30 (F), Phase 3.
Location: Aviation Logistics School, Fort Eustis, VA.
Length: 3 weeks (107 hours).
Exhibit Dates: 10/02–Present.
Learning Outcomes: Upon completion of the course, the student will be able to supervise and provide technical guidance to aircraft pneudraulics maintenance personnel.
Instruction: Practical and classroom exercises, discussion, laboratory, and lecture. Topics include pneudraulics maintenance, inspection, and supervision; use of test equipment; and pneudraulics component repair.
Credit Recommendation: In the lower-division baccalaureate/associate degree category, 2 semester hours in aircraft maintenance management (12/03).

AR-1704-0262

TATS OH-58 Helicopter Repair Supervisor Basic Noncommissioned Officer

Course Number: 600-67S30 (F), Phase 2.
Location: Aviation Logistics School, Fort Eustis, VA.
Length: 12 weeks (484 hours).
Exhibit Dates: 10/00–Present.
Learning Outcomes: Upon completion of the course, the student will be able to maintain, repair, and supervise maintenance and repairs on helicopters.
Instruction: Lectures and practical exercises cover the maintenance and repair of helicopters and supervision of personnel. This course contains a common core of leadership subjects.
Credit Recommendation: In the lower-division baccalaureate/associate degree category, 2 semester hours in aviation maintenance and 2 in maintenance management (12/03).

AR-1704-0263

TATS OH-58D Helicopter Repair Supervisor Basic Noncommissioned Officer

Course Number: 552-67S30 (F), Phase 4.
Location: Aviation Logistics School, Fort Eustis, VA.
Length: 7 weeks (273 hours).
Exhibit Dates: 10/02–Present.
Learning Outcomes: Upon completion of the course, the student will be able to inspect and supervise the maintenance on aircraft components and systems.
Instruction: Audiovisual materials, computer-based training, practical exercises, discussion, and lecture. Topics include basic aircraft maintenance; electrical, hydraulic, environmental, control, fuel, and powertrain systems; flight controls; and supervision of aircraft maintenance.
Credit Recommendation: In the lower-division baccalaureate/associate degree category, 2 semester hours in aircraft systems troubleshooting and maintenance and 2 in aircraft maintenance management (12/03).

AR-1704-0265

TATS Aircraft Component Repairer Supervisor Advanced Noncommissioned Officer

Course Number: 552-68K40 (F), Phase 3; 602-68K40 (F), Phase 3.
Location: Aviation Logistics School, Fort Eustis, VA.
Length: 5–6 weeks (202–238 hours).
Exhibit Dates: 10/01–Present.
Learning Outcomes: Upon completion of the course, the student will be able to troubleshoot, supervise, and inspect the maintenance of various aircraft systems and components.
Instruction: Practical and classroom exercises, discussion, laboratory, and lecture. Topics include troubleshooting, inspection, supervision, planning, and management of aircraft maintenance.
Credit Recommendation: In the lower-division baccalaureate/associate degree category, 3 semester hours in aviation maintenance management (12/03).

AR-1704-0266

TATS AH-64D Armament/Electrical/ Avionic Systems Repairer Supervisor Basic Noncommissioned Officer (NCO)

Course Number: 646-68Y30-RC (F), Phase 3.
Location: Aviation Logistics School, Fort Eustis, VA.
Length: 11 weeks (556 hours).
Exhibit Dates: 10/03–Present.
Learning Outcomes: Upon completion of the course, the student will be able to troubleshoot, supervise, and inspect maintenance performed on various aircraft electrical and avionics systems.
Instruction: Practical and classroom exercises, discussion, laboratory, and lecture. Topics include troubleshooting and the inspection and supervision of maintenance for aircraft data management, electrical, navigation, flight control, sensor, weapons, and survivability systems.
Credit Recommendation: In the lower-division baccalaureate/associate degree category, 2 semester hours in aircraft electrical systems troubleshooting and maintenance and 2 in avionics systems troubleshooting and maintenance, or 4 in aircraft maintenance management (12/03).

AR-1704-0267

TATS AH-64D Armament/Electrical/ Avionic Systems Repairer Supervisor Basic Noncommissioned Officer (NCO)

Course Number: 552-68Y30 (F), Phase 3.
Location: Aviation Logistics School, Fort Eustis, VA.
Length: 11 weeks (556 hours).
Exhibit Dates: 10/03–Present.
Learning Outcomes: Upon completion of the course, the student will be able to troubleshoot, supervise, and inspect maintenance performed on various aircraft electrical and avionics systems.
Instruction: Practical and classroom exercises, discussion, laboratory, and lecture. Topics include troubleshooting and the inspection and supervision of maintenance for aircraft data management, electrical, navigation, flight control, sensor, weapons, and survivability systems.
Credit Recommendation: In the lower-division baccalaureate/associate degree category, 2 semester hours in aircraft electrical systems troubleshooting and maintenance and 3 in avionics systems troubleshooting and maintenance (12/03).
Related Occupations: 68Y.

AR-1709-0010

Aerial Sensor Specialist (OV-1D)

Course Number: 221-96H10.
Location: Intelligence School, Fort Huachuca, AZ.
Length: 10 weeks (305–372 hours).
Exhibit Dates: 5/80–9/96.
Objectives: To train enlisted military personnel in the operation and maintenance of an aerial sensing system.
Instruction: Lectures and practical application cover the operation and maintenance of an aerial sensing system.
Credit Recommendation: Credit is not recommended because of the military-specific nature of the course (5/82).
Related Occupations: 96H.

AR-1709-0011

Still Documentation Specialist Basic Noncommissioned Officer (NCO)

Course Number: G3AZR23172 002.
Location: Technical Training Center, Lowry AFB, CO.
Length: 6–7 weeks (216–257 hours).
Exhibit Dates: 10/88–9/91.
Learning Outcomes: After 9/91 see AR-0202-0002. Upon completion of the course, the student will be able to manage a still photography team in performing visual information support services, including darkroom development and printing; plan, supervise, and direct the operation of the team and the maintenance of photography equipment and portable facilities at a fixed installation or under combat conditions; and plan mobile support operations for training, documentation, and briefing requirements.
Instruction: Lectures and practical exercises are reinforced by application in a field operational test where the unit must plan and execute a move to support a military operation requiring visual information.
Credit Recommendation: In the lower-division baccalaureate/associate degree category, see AR-1406-0090 for common core credit. In the upper-division baccalaureate category, 3 semester hours in the management of a photographic service (5/90).

AR-1710-0062

BRADLEY INFANTRY FIGHTING VEHICLE COMMANDER

Course Number: 2E-F58/010-F5/7-11M30/40; 2E-F58/010-F5.
Location: Infantry School, Fort Benning, GA.
Length: 6 weeks (242–285 hours).
Exhibit Dates: 7/82–10/95.
Objectives: To Train the soldier on the capabilities, characteristics, and components of the Bradley infantry fighting vehicle (BIFV).
Instruction: Lectures and practical exercises cover BIFV tactical operations, driving skills, vehicle safety support weapons, and daily preventive maintenance and service.
Credit Recommendation: In the lower-division baccalaureate/associate degree category, 3 semester hours in heavy equipment operations (10/85).
Related Occupations: 11M.

AR-1710-0063

BRADLEY INFANTRY FIGHTING VEHICLE GUNNER

Course Number: 010-F6; 010-11M20.
Location: Infantry School, Fort Benning, GA.
Length: 4–5 weeks (141–195 hours).
Exhibit Dates: 11/82–3/90.
Objectives: To train the soldier on the capabilities, characteristics, and components of the Bradley infantry fighting vehicle (BIFV).
Instruction: Lectures and practical exercises cover BIFV operations and driving skills, vehicle safety support weapons, and daily preventive maintenance and service.
Credit Recommendation: In the vocational certificate category, 3 semester hours in heavy equipment operations (9/85).
Related Occupations: 11M.

AR-1710-0065

SYSTEM MECHANIC M60 SERIES VEHICLE

Course Number: CNC20.
Location: Combined Arms Training Center, Vilseck, W. Germany.
Length: 4–5 weeks (165 hours).
Exhibit Dates: 3/81–1/94.
Objectives: To troubleshoot and maintain hydraulic, mechanical, and electrical systems in tanks, light trucks, and power generating equipment up to 60kW.
Instruction: Lectures and practical exercises cover tank turret systems, operation, and maintenance, including vehicle recovery and evacuation, live engine maintenance, basic electricity, fuel systems, and power generation. Included are related safety procedures.
Credit Recommendation: In the lower-division baccalaureate/associate degree category, 1 semester hour in electrical laboratory or electrical shop (1/82).
Related Occupations: 63N.

AR-1710-0066

SYSTEM MECHANIC SELF-PROPELLED (SP) ARTILLERY M109/M110 SERIES VEHICLES

Course Number: CNC22.
Location: Combined Arms Training Center, Vilseck, W. Germany.
Length: 6–7 weeks (234 hours).
Exhibit Dates: 3/81–1/94.
Objectives: To troubleshoot and maintain hydraulic, mechanical, and electrical systems in tanks, light trucks, and power generating equipment up to 60kW.
Instruction: Lectures and practical exercises cover tank turret systems, operation, and maintenance, including vehicle recovery and evacuation, live engine maintenance, basic electricity, fuel systems, and power generation. Included are related safety procedures.
Credit Recommendation: In the lower-division baccalaureate/associate degree category, 2 semester hours in electrical laboratory or electrical shop (1/82).
Related Occupations: 63D.

AR-1710-0067

SYSTEM MECHANIC SELF-PROPELLED (SP) ARTILLERY M109/M110 SERIES VEHICLES

Course Number: CNC25.
Location: Combined Arms Training Center, Vilseck, W. Germany.
Length: 5–6 weeks (198 hours).
Exhibit Dates: 3/81–1/94.
Objectives: To troubleshoot and maintain hydraulic, mechanical, and electrical systems in tanks, light trucks, and generating equipment up to 60kW.
Instruction: Lectures and practical exercises cover tank turret systems, operation, and maintenance, including vehicle recovery and evacuation, live engine maintenance, basic electricity, fuel systems, and power generation. Included are related safety procedures.
Credit Recommendation: In the lower-division baccalaureate/associate degree category, 2 semester hours in electrical laboratory or electrical shop (1/82).
Related Occupations: 63D.

AR-1710-0068

SYSTEM MECHANIC M60 SERIES VEHICLE

Course Number: CNC24.
Location: Combined Arms Training Center, Vilseck, W. Germany.
Length: 5–6 weeks (198 hours).
Exhibit Dates: 3/81–5/92.
Objectives: To troubleshoot and maintain hydraulic, mechanical, and electrical systems in tanks, light trucks, and generating equipment up to 60kW.
Instruction: Lectures and practical exercises cover tank turret systems, operation, and maintenance, including vehicle recovery and evacuation, live engine maintenance, basic electricity, fuel systems, and power generation. Included are related safety procedures.
Credit Recommendation: In the lower-division baccalaureate/associate degree category, 2 semester hours in electrical laboratory or electrical shop (1/82).
Related Occupations: 63N.

AR-1710-0069

SYSTEM MECHANIC FOR M113 SERIES VEHICLES

Course Number: CNC26.
Location: Combined Arms Training Center, Vilseck, W. Germany.
Length: 3–4 weeks (129 hours).
Exhibit Dates: 3/81–1/94.
Objectives: To troubleshoot and maintain hydraulic, mechanical, and electrical systems in tanks, light trucks, and generating equipment up to 60kW.
Instruction: Lectures and practical exercises in tank turret systems, operation, and maintenance, including vehicle recovery and evacuation, live engine maintenance, basic electricity, fuel systems, and power generation. Included are related safety procedures.
Credit Recommendation: Credit is not recommended because of the limited, specialized nature of the course (1/82).
Related Occupations: 63T.

AR-1710-0070

SYSTEM MECHANIC M113 SERIES VEHICLE

Course Number: CNC21.
Location: Combined Arms Training Center, Vilseck, W. Germany.
Length: 4–5 weeks (165 hours).
Exhibit Dates: 3/81–1/94.
Objectives: To troubleshoot and maintain hydraulic, mechanical, and electrical systems in tanks, light trucks, and generating equipment up to 60kW.
Instruction: Lectures and practical exercises in tank turret systems, operation, and maintenance, including vehicle recovery and evacuation, live engine maintenance, basic electricity, fuel systems, and power generation. Included are related safety procedures.
Credit Recommendation: In the lower-division baccalaureate/associate degree category, 1 semester hour in electrical laboratory or electrical shop (1/82).
Related Occupations: 63T.

AR-1710-0073

SYSTEM MECHANIC M60 SERIES VEHICLE

Course Number: CNC23.
Location: Combined Arms Training Center, Vilseck, W. Germany.
Length: 6–7 weeks (234 hours).
Exhibit Dates: 3/81–1/94.
Objectives: To troubleshoot and maintain hydraulic, mechanical, and electrical systems in tanks, light trucks, and generating equipment up to 60kW.
Instruction: Lectures and practical exercises cover tank turret systems, operation, and maintenance, including vehicle recovery and evacuation, live engine maintenance, basic electricity, fuel systems, and power generation. Included are related safety procedures.
Credit Recommendation: In the lower-division baccalaureate/associate degree category, 2 semester hours in electrical laboratory or electrical shop (1/82).
Related Occupations: 63N.

AR-1710-0080

M109/M110 ORGANIZATIONAL MAINTENANCE TRAINING

Course Number: TVM 220.
Location: Combined Arms Training Center, Vilseck, W. Germany.
Length: 3 weeks (108 hours).
Exhibit Dates: 11/81–1/94.
Objectives: To train selected supervisory personnel in troubleshooting and maintaining self-propelled howitzers.
Instruction: Lectures and practical exercises cover troubleshooting and maintaining self-pro-

pelled howitzers with emphasis on electrical and suspension systems.

Credit Recommendation: Credit is not recommended because of the limited, specialized nature of the course (1/82).

Related Occupations: 63D.

AR-1710-0082

Organizational Mechanic Conversion

Course Number: CNC28.

Location: Combined Arms Training Center, Vilseck, W. Germany.

Length: 6 weeks (216–217 hours).

Exhibit Dates: 11/81–1/94.

Objectives: To upgrade organizational mechanics in the field of heavy equipment, including light and heavy vehicles, light track vehicles, and power generation equipment.

Instruction: This course consists of lectures and demonstrations with great emphasis on practical exercises, including troubleshooting light and heavy vehicles and light tracked vehicles. Emphasis is placed on the use and interpretation of diagnostic test instruments. Material-handling equipment also forms a significant point of the course, with performance examinations being the major method for evaluation.

Credit Recommendation: In the lower-division baccalaureate/associate degree category, 2 semester hours in heavy equipment ignition, starting, charging, and electrical systems; 2 in heavy equipment brakes, suspension, and steering systems laboratory; and 1 in heavy equipment general maintenance (1/82).

Related Occupations: 63B.

AR-1710-0084

Heavy Construction Equipment Operator

Course Number: 62E10-OSUT (ST); 62E10-OSUT; 730-62E10.

Location: Training Center Engineer, Fort Leonard Wood, MO.

Length: 8–16 weeks (224–639 hours).

Exhibit Dates: 5/78–2/92.

Objectives: To train personnel as heavy construction equipment operators.

Instruction: Lectures and practical exercises cover the operation of heavy equipment, including crawler tractor, front-end loader, motor grader, and wheeled tractor. Course may include basic training.

Credit Recommendation: In the vocational certificate category, 3 semester hours in heavy construction equipment operation (12/83).

Related Occupations: 62E.

AR-1710-0086

General Construction Equipment Operator

Course Number: 62J10-OSUT (ST); 62J10-OSUT; 730-62J10.

Location: Training Center Engineer, Fort Leonard Wood, MO.

Length: 8–15 weeks (248–598 hours).

Exhibit Dates: 5/78–7/92.

Objectives: To train personnel in general construction equipment other than heavy construction equipment.

Instruction: Lectures and practical exercises cover the operation of construction equipment, including sweepers, rollers, air compressors, rotary tillers, earth augers, ditching machines, and tractors. This course includes an 8-9 week long Basic Training component.

Credit Recommendation: In the vocational certificate category, 3 semester hours in operation and maintenance of light construction equipment (12/83).

Related Occupations: 62J.

AR-1710-0087

Construction Equipment Repairer Primary Technical
(Construction Equipment Repair)

Course Number: 612-62B20.

Location: Training Center Engineer, Fort Leonard Wood, MO.

Length: 6 weeks (190–216 hours).

Exhibit Dates: 11/79–12/93.

Objectives: To provide advanced training in engineering construction equipment repair.

Instruction: Lectures and practical exercises cover the repair and maintenance of engineering construction equipment, include cutting and welding; air induction systems; engine fuel systems; engine removal, disassembly, repair, and assembly; power trains; and pneumatic equipment. Course may include basic training.

Credit Recommendation: In the vocational certificate category, 5 semester hours in advanced maintenance and repair of engineer construction equipment (12/83).

Related Occupations: 62B.

AR-1710-0089

Construction Equipment Operator

Course Number: 730-55131.

Location: Training Center Engineer, Fort Leonard Wood, MO.

Length: 11 weeks (382 hours).

Exhibit Dates: 3/82–10/92.

Objectives: To provide training in operating and maintaining common construction equipment.

Instruction: This course for Air Force personnel is presented through lectures and practical experiences in basic operational skills for common construction equipment, including mechanical and hydraulic cranes, clamshell and draglines, forklifts, crawler tractors, front-end loaders, street sweepers, steel-wheeled and pneumatic-tired rollers, and water distributors. Maintenance and safety procedures are also covered.

Credit Recommendation: In the vocational certificate category, 7 semester hours in basic construction equipment operation and maintenance (12/83).

AR-1710-0090

Construction Equipment Operator

Course Number: 730-1345.

Location: Training Center Engineer, Fort Leonard Wood, MO.

Length: 9 weeks (317 hours).

Exhibit Dates: 9/81–1/90.

Objectives: To provide Marine personnel to operate and maintain common construction equipment.

Instruction: Lectures and practical experiences cover basic operational skills for common construction equipment, including crawler tractors, front-end loaders, motor graders, and forklifts. Maintenance safety procedures are also covered.

Credit Recommendation: In the vocational certificate category, 6 semester hours in basic construction equipment operation and maintenance (12/83).

AR-1710-0091

Concrete and Asphalt Equipment Operator

Course Number: 62H10-OSUT (ST); 730-62H10; 62H10-OSUT.

Location: Training Center Engineer, Fort Leonard Wood, MO.

Length: 8–15 weeks (335–512 hours).

Exhibit Dates: 10/81–4/90.

Objectives: To provide training in operation and maintenance of concrete and asphalt equipment.

Instruction: Lectures and practical experiences cover operation, maintenance, and safety procedures required for proper use of concrete and asphalt equipment, including concrete mobile mixers, bituminous distributors, asphalt kettles, 15 kW generators, oil heaters, asphalt melters, dryer-mixers, and aggregate spreaders. Maintenance of flexible pavement is included. Course may include basic training.

Credit Recommendation: In the vocational certificate category, 3 semester hours in operation and maintenance of basic concrete and asphalt equipment (12/83).

Related Occupations: 62H.

AR-1710-0092

Lifting and Loading Equipment Operator

Course Number: 62F10-OSUT (ST); 730-62F10; 62F10-OSUT.

Location: Training Center Engineer, Fort Leonard Wood, MO.

Length: 9–16 weeks (276–611 hours).

Exhibit Dates: 3/78–3/92.

Objectives: To train personnel in the proper use and maintenance of lifting and loading equipment.

Instruction: Lectures and practical experiences cover operating and maintaining equipment, including forklifts, mechanical cranes, clamshell and draglines, and hydraulic cranes. This course includes an 8-9 week long Basic Training component.

Credit Recommendation: In the vocational certificate category, 3 semester hours in operation and maintenance of lifting and loading equipment (12/83).

Related Occupations: 62F.

AR-1710-0095

Quartermaster/Chemical Equipment Repairer Basic Technical
(Quartermaster/Chemical Equipment Repairer)

Course Number: 690-63J30.

Location: Ordnance Center and School, Aberdeen Proving Ground, MD.

Length: 7–8 weeks (274–279 hours).

Exhibit Dates: 5/82–9/90.

Objectives: To train enlisted personnel in the management and supervision of mechanical equipment.

Instruction: Lectures and practical exercises cover shop management and maintenance procedures; leadership and training skills; counseling and personnel evaluation; and inspecting, testing, and supervising the maintenance and repair of various items of mechanical equipment.

Credit Recommendation: In the lower-division baccalaureate/associate degree category, 3 semester hours in supervision and 3 in basic management techniques (11/86).

Related Occupations: 63J.

AR-1710-0101

M1 Abrams Tank Fire Control Systems Repairer

Course Number: 113-ASIL8 (45G).

Location: Ordnance Center and School, Aberdeen Proving Ground, MD.

Length: 3–4 weeks (126 hours).

Exhibit Dates: 10/82–9/90.

Objectives: To teach enlisted personnel to perform maintenance on fire control system and related test equipment of the M1 Abrams tank.

Instruction: This course instructs maintenance personnel in testing and repairing gunner's primary sight, direct support electrical system test set, laser range finder, computer control panel, electronic units, line of sight electronic unit, thermal imagine system test set, thermal mock-up system, power control unit, image control unit, and thermal receiver unit.

Credit Recommendation: Credit is not recommended because of the military-specific nature of the training (2/84).

Related Occupations: 45G; 63G.

AR-1710-0104

M1 Abrams Tank Turret Repair

Course Number: 643-ASIL8 (45K).

Location: Ordnance Center and School, Aberdeen Proving Ground, MD.

Length: 5–7 weeks (180–210 hours).

Exhibit Dates: 10/82–9/90.

Objectives: To train enlisted personnel to perform general maintenance and repair on the armament installed on specific armored vehicles.

Instruction: Lectures and practical exercises cover the testing, maintenance, and repair of various mechanical and electrical systems and components on the armaments installed on specific armored vehicles.

Credit Recommendation: Credit is not recommended because of the military-specific nature of the course (2/84).

Related Occupations: 45K.

AR-1710-0105

M2/M3 Bradley Fighting Vehicle Turret Repair

Course Number: 643-ASIM5 (45K).

Location: Ordnance Center and School, Aberdeen Proving Ground, MD.

Length: 4 weeks (144–157 hours).

Exhibit Dates: 3/83–9/90.

Objectives: To train enlisted personnel to perform general maintenance and repair on the armament installed on specific armored vehicles.

Instruction: Lectures and practical exercises cover the testing, maintenance, and repair of various mechanical and electrical systems and components on the armament installed on specific armored vehicles.

Credit Recommendation: Credit is not recommended because of the limited, specialized nature of the course (2/87).

Related Occupations: 45K.

AR-1710-0108

Bradley Fighting Vehicle Systems TOW/Dragon Repairer

Course Number: 4F-F3/121-ASID3.

Location: Missile and Munitions School, Redstone Arsenal, AL.

Length: 4–6 weeks (134–215 hours).

Exhibit Dates: 9/85–10/98.

Objectives: To provide active and reserve enlisted and warrant officer with training to perform and supervise intermediate maintenance on TOW subsystems and the Bradley fighting vehicle.

Instruction: Personnel are trained to troubleshoot, replace, and align the components and interface of TOW subsystems of the Bradley fighting vehicle.

Credit Recommendation: Credit is not recommended because of the military-specific nature of the course (10/88).

AR-1710-0109

1. Tank Turret Repairer Basic Noncommissioned Officer (NCO)
2. Armament Repairer
3. Armament Repairer

Course Number: *Version 1:* 643-45K30. *Version 2:* 643-45K30. *Version 3:* 643-45K30.

Location: Ordnance Center and School, Aberdeen Proving Ground, MD.

Length: *Version 1:* 19–20 weeks (686–711 hours). *Version 2:* 19–20 weeks (711–712 hours). *Version 3:* 19–20 weeks (556 hours).

Exhibit Dates: *Version 1:* 3/88–9/96. *Version 2:* 10/96–9/98. *Version 3:* 10/98–Present.

Learning Outcomes: *Version 1:* Upon completion of the course, the student will be able to perform maintenance on electrical/electronic and hydraulic components and systems of tank turrets and self-propelled and towed artillery. *Version 2:* Upon completion of the course, the student will be able to perform maintenance on small arms, and mechanical, electronic, and hydraulic components/systems of tank turrets and howitzers. *Version 3:* Upon completion of the course, the student will be able to perform maintenance on small arms, and mechanical, electronic, and hydraulic components/systems of tank turrets and howitzers.

Instruction: *Version 1:* Lectures and practical exercises cover inspection and basic knowledge and skills in electrical/electronic and hydraulic components of tank turrets and self-propelled and towed vehicles. This course contains a common core of leadership subjects. *Version 2:* Lectures and practical exercises cover the maintenance and inspection of small arms, tank turrets, and howitzers. This course contains a common core of leadership subjects. *Version 3:* Lectures and practical exercises cover the maintenance and inspection of small arms, tank turrets, and howitzers. This course contains a common core of leadership subjects.

Credit Recommendation: *Version 1:* In the lower-division baccalaureate/associate degree category, 2 semester hours in basic automotive electricity/electronics and 1 in basic hydraulics. See AR-1406-0090 for common core credit (10/91). *Version 2:* In the lower-division baccalaureate/associate degree category, 1 semester hour in hydraulic systems and 2 in electrical systems. See AR-1406-0090 for common core credit (4/01). *Version 3:* In the lower-division baccalaureate/associate degree category, 1 semester hour in hydraulic systems and 2 in electrical systems. See AR-1406-0090 for common core credit (8/03).

Related Occupations: 45K.

AR-1710-0110

Bradley Fighting Vehicle System Mechanic Basic Noncommissioned Officer (NCO)

Course Number: 611-63T30.

Location: Armor Center and School, Fort Knox, KY.

Length: 17–18 weeks (638–639 hours).

Exhibit Dates: 6/88–12/95.

Learning Outcomes: After 12/95 see AR-1710-0185. Upon completion of the course, the student will be able to supervise enlisted personnel in the maintenance of vehicles, perform basic maintenance procedures, and teach basic procedures to others.

Instruction: Lectures cover leadership skills and service facility management. Vehicle system maintenance and service techniques and inspection and testing of electrical, fuel, and accessory systems on heavy vehicles are included. The course also includes practice in the use of service manuals and appropriate records. This course contains a common core of leadership subjects.

Credit Recommendation: In the lower-division baccalaureate/associate degree category, 2 semester hours in automotive fundamentals, 2 in equipment maintenance, and 3 in heavy equipment service. See AR-1406-0090 for common core credit. In the upper-division baccalaureate category, 2 semester hours in service management (12/90).

Related Occupations: 45K; 63T.

AR-1710-0111

2. Track Vehicle Repairer Basic Noncommissioned Officer (NCO)
3. Track Vehicle Repairer Basic Noncommissioned Officer (NCO)
4. Track Vehicle Repairer Basic Noncommissioned Officer (BNCOC)

Course Number: *Version 2:* 611-63H30. *Version 3:* 611-63H30. *Version 4:* 611-63H30.

Location: Ordnance Center and School, Aberdeen Proving Ground, MD.

Length: *Version 2:* 18–19 weeks (648–674 hours). *Version 3:* 12–13 weeks (665 hours). *Version 4:* 16 weeks (578 hours).

Exhibit Dates: *Version 2:* 3/88–9/96. *Version 3:* 10/96–12/97. *Version 4:* 1/98–Present.

Learning Outcomes: *Version 2:* Upon completion of the course, the student will be able to supervise maintenance procedures and perform general maintenance, repair, troubleshooting, and inspection on wheel and tracked vehicles. *Version 3:* Upon completion of the course, the student will be able to supervise maintenance procedures and perform general maintenance, repair, troubleshooting, and inspection on wheel and track vehicles. *Version 4:* Upon completion of the course, the student will be able to supervise maintenance procedures and perform general maintenance, repair, troubleshooting, and inspection on wheel and tracked vehicles.

Instruction: *Version 2:* Lectures, demonstrations, and practical exercises cover internal combustion engines, fuel and electrical systems, starters, regulators, alternators, brake system components, hydraulic principles, steering and suspension systems, and engine tune-up and measuring tools. Construction, operation, and diagnostic testing procedures are stressed. This course contains a common core of leadership subjects. *Version 3:* Lectures, demonstrations, and practical exercises cover internal combustion engines, fuel and electrical systems, starters, regulators, alternators, brake system components, hydraulic principles, steering and suspension systems, and engine tune-up and measuring tools. Construction, operation, and diagnostic testing procedures are stressed. This course contains a common core of leadership subjects. *Version 4:* Lectures, demonstrations, and practical exercises. Topics include internal combustion engines, fuel and electrical systems, starters, regulators, alternators, brake system components, hydraulic principles, steering and suspension systems, and engine tune-up and measuring tools. Construction, operation, and diagnostic testing procedures are stressed. This course contains a common core of leadership subjects.

Credit Recommendation: ***Version 2:*** In the lower-division baccalaureate/associate degree category, 1 semester hour in heavy equipment repair, 2 in diesel engine systems and 3 in basic automotive electrical systems. See AR-1406-0090 for common core credit (3/92). *Version 3:* In the lower-division baccalaureate/associate degree category, 2 semester hours in heavy equipment repair, 2 in diesel engine systems, and 3 in basic automotive electrical systems. See AR-1406-0090 for common core credit (1/98). *Version 4:* In the lower-division baccalaureate/associate degree category, 2 semester hours in automotive power plants, 3 in automotive electrical systems, and 3 in heavy equipment maintenance and repair (8/03).

Related Occupations: 63H.

AR-1710-0112

Self-Propelled Field Artillery System Basic Noncommissioned Officer (NCO)

Course Number: 611-63D30; 611-63D30 (08T2).

Location: Ordnance Center and School, Aberdeen Proving Ground, MD.

Length: 8 weeks (320 hours).

Exhibit Dates: 5/88–9/96.

Learning Outcomes: After 9/96 see AR-1710-0186. Upon completion of the course, the student will be able to train enlisted personnel to perform organizational maintenance and repair on self-propelled field artillery cannon weapon systems and light and medium recovery vehicles.

Instruction: Topics include training of a maintenance noncommissioned officer in a tactical environment, general military training subjects, logistics training, and recovery operations. Those in MOS 63D learn armament maintenance. Those in MOS 45D have instruction in replacing, removing/installing, inspecting, testing, and troubleshooting components/systems of tracked vehicles. This course contains a common core of leadership subjects.

Credit Recommendation: In the lower-division baccalaureate/associate degree category, 2 semester hours in heavy equipment maintenance/repair management and 2 in automotive and heavy equipment repair. See AR-1406-0090 for common core credit (9/90).

Related Occupations: 45D; 63D.

AR-1710-0113

M1 Abrams Tank Vehicle Repairer

Course Number: 610-ASIL8 (63H).

Location: Ordnance Center and School, Aberdeen Proving Ground, MD.

Length: 3–4 weeks (132 hours).

Exhibit Dates: 1/85–9/90.

Learning Outcomes: Upon completion of the course, the student will be able to perform maintenance on light automotive components on specialized track vehicles.

Instruction: Course includes turbine engine construction and operation; removal and replacement of engine; troubleshooting of engine; performance testing and troubleshooting of hydraulic systems; track link repair and adjustment; and transmission removal, disassembly, inspection, repair, and reassembly.

Credit Recommendation: In the vocational certificate category, 1 semester hour in transmission mechanic for heavy track vehicle. In the lower-division baccalaureate/associate degree category, 2 semester hours in turbine engine laboratory (3/87).

Related Occupations: 63H.

AR-1710-0114

Tank Turret Repairer (USA/USA-ST)
(Tank Turret Repairman (USMC))

Course Number: 643-2146 (45K); 643-45K10 (CT).

Location: Ordnance Center and School, Aberdeen Proving Ground, MD.

Length: Army, 20 weeks; Marine Corps, 13 weeks (468 hours).

Exhibit Dates: 10/86–12/95.

Learning Outcomes: Upon completion of the course, the student will be able to maintain mechanical, electrical, and hydraulic components on tank turrets.

Instruction: Lectures and practical exercises cover testing, maintenance, and repair of various mechanical, electrical, and electronic components of tank turrets.

Credit Recommendation: In the lower-division baccalaureate/associate degree category, 1 semester hour in automotive electrical or 1 in maintenance technology (12/90).

Related Occupations: 45K.

AR-1710-0115

2. Warrant Officer Technical/Tactical Certification
(Allied Trades Technician Warrant Officer Technical/Tactical Certification)
3. Allied Trades Repair Technician Warrant Officer Basic

Course Number: ***Version 2:*** 4L-914A. *Version 3:* 4L-914A.

Location: Ordnance Center and School, Aberdeen Proving Ground, MD.

Length: ***Version 2:*** 17–18 weeks (634–635 hours). *Version 3:* 12–13 weeks (460 hours).

Exhibit Dates: ***Version 2:*** 3/89–12/99. *Version 3:* 1/00–Present.

Learning Outcomes: ***Version 2:*** Upon completion of the course, the student will be able to manage track vehicle repair facilities, machine shops, and welding facilities. *Version 3:* Upon completion of the course, the student will be able to manage and supervise various shop operations.

Instruction: ***Version 2:*** Methods of instruction include lectures, demonstrations, shop experience, and field trials. Topics include communication skills, inventory control, machining, welding, and body repair. *Version 3:* Audio visual materials, practical exercises, classroom exercises, discussion, laboratory, and lecture. Topics include management and supervision of direct support maintenance, vehicle recovery, machine shop, welding operation, and material repair.

Credit Recommendation: ***Version 2:*** In the lower-division baccalaureate/associate degree category, 1 semester hour in automotive body repair, 2 in communication skills, 2 in automotive electricity, 2 in arc welding, 1 in inert gas welding, 5 in basic machine shop (lathes), and 5 in advanced machine shop. In the upper-division baccalaureate category, 3 semester hours in maintenance management (12/90). *Version 3:* In the lower-division baccalaureate/associate degree category, 3 semester hours in maintenance management or 3 in supervision (8/03).

Related Occupations: 914A.

AR-1710-0116

3. Warrant Officer Technical/Tactical Certification
(Unit Maintenance Technician (Light) Warrant Officer Technical/Tactical Certification)
4. Unit Maintenance Technician (Light)
5. Unit Maintenance Officer

Course Number: ***Version 3:*** 4L-915A. *Version 4:* 4L-915A. *Version 5:* 4L-915A.

Location: Ordnance Center and School, Aberdeen Proving Ground, MD.

Length: ***Version 3:*** 17–18 weeks (626 hours). *Version 4:* 12–13 weeks (412 hours). *Version 5:* 12–13 weeks (444 hours).

Exhibit Dates: ***Version 3:*** 3/89–1/94. *Version 4:* 2/94–9/02. *Version 5:* 10/02–Present.

Learning Outcomes: ***Version 3:*** Upon completion of the course, the student will be able to assume the duties and responsibilities of managing an automotive facility with tracked vehicles and truck/trailer wheeled vehicles. *Version 4:* Upon completion of the course, the student will be able to demonstrate a knowledge of common technical and ordnance subjects which apply to unit maintenance, wheel vehicles, power generation, light systems and track vehicles, and field artillery. This course is designed to train unit maintenance officers to supervise and perform organizational maintenance. *Version 5:* Upon completion of the course, the student will be able to demonstrate a knowledge of common technical and ordnance subjects which apply to unit maintenance, wheel vehicles, power generation, light systems and track vehicles, and field artillery. This course is designed to train unit maintenance officers to supervise and perform organizational maintenance.

Instruction: ***Version 3:*** Course content includes professional development subjects, maintenance management, wheel vehicle maintenance, power generation, and building maintenance. *Version 4:* Audio visual materials, learner presentations, computer-based training, practical exercises, classroom exercises, discussion, laboratory, and lecture. Topics include supply and maintenance management; general military subjects; power generation, recovery, and support equipment; automotive systems; artillery and artillery carrier systems; small arms; and tow vehicles. *Version 5:* Audio visual materials, learner presentations, computer-based training, practical exercises, classroom exercises, discussion, laboratory, and lecture. Topics include small arms, recovery operations, maintenance of wheel vehicles, automotive systems, power generation equipment, supply management, motor pool operations, unit level maintenance management, maintenance of towed artillery, and general military subjects.

Credit Recommendation: *Version 3:* In the lower-division baccalaureate/associate degree category, 3 semester hours in communication skills, 3 in basic electrical laboratory, and 3 in automotive electricity. In the upper-division baccalaureate category, 5 semester hours in maintenance management (12/90). *Version 4:* In the lower-division baccalaureate/associate degree category, 3 semester hours in technical problem solving, 2 in automotive hydraulics, 2 in automotive electricity, and 3 in personnel supervision (8/03). *Version 5:* In the lower-division baccalaureate/associate degree category, 3 semester hours in technical problem solving, 3 in personnel supervision, 2 in automotive hydraulics, and 2 in automotive electricity. In the upper-division baccalaureate category, 3 semester hours in maintenance and supervision (8/03).

Related Occupations: 915A.

AR-1710-0117

Heavy Wheel Vehicle Recovery Specialist

Course Number: 610-ASIH8 (63S).

Location: Training Center, Fort Jackson, SC.

Length: 3 weeks (106 hours).

Exhibit Dates: 10/86–12/97.

Learning Outcomes: Upon completion of the course, the student will be able to operate wheeled vehicle recovery vehicles (heavy-duty wrecker) and equipment to recover and transport wheeled vehicles.

Instruction: Course includes operating and servicing recovery vehicles (heavy duty wreckers) and equipment and using proper procedures in recovering and transporting heavy wheeled vehicles.

Credit Recommendation: In the vocational certificate category, 3 semester hours in truck driving (3/87).

Related Occupations: 63S.

AR-1710-0118

Wheel Vehicle Recovery Specialist

Course Number: 610-ASIH8 (63W).

Location: Ordnance Center and School, Aberdeen Proving Ground, MD.

Length: 2 weeks (80–81 hours).

Exhibit Dates: 6/84–12/97.

Learning Outcomes: Upon completion of the course, the student will be able to operate wheeled recovery vehicles and associated equipment to recover and transport wheeled vehicles.

Instruction: Instruction includes operating and servicing recovery vehicles and equipment and procedures used in recovery and transport.

Credit Recommendation: In the vocational certificate category, 2 semester hours in truck driving (3/87).

Related Occupations: 63W.

AR-1710-0119

Wheel Vehicle Repairer Individual Ready Reserve Refresher

Course Number: 610-F3 (IRR).

Location: Ordnance Center and School, Aberdeen Proving Ground, MD.

Length: 2 weeks (76 hours).

Exhibit Dates: 10/86–9/90.

Learning Outcomes: Upon completion of the course, the student will be ale to describe basic operating principles of diesel fuel systems, including Cummins, Detroit Diesel, Stanadyne DB-2, and Bosch distributor pumps, engines, brake systems, and electrical systems and use diagnostic equipment to troubleshoot malfunctions in the fuel, engine, electrical, transmission, brake, engine retarder, power train, suspension, steering, and hydraulic systems of a heavy truck.

Instruction: Course includes lectures on the basic principles of operation of the internal combustion engine, fuel system, electrical system, and associated maintenance. Hands-on instruction includes using diagnostic test equipment to troubleshoot malfunctions in the fuel, engine, electrical, transmission, brake, engine retarder, power train, suspension, steering, and hydraulic system of a heavy highway truck/ tractor.

Credit Recommendation: In the vocational certificate category, 2 semester hours in heavy truck mechanic (3/87).

Related Occupations: 63W.

AR-1710-0121

Light Wheel Vehicle Mechanic Individual Ready Reserve Refresher

Course Number: 610-63B20-IRR.

Location: Ordnance Center and School, Aberdeen Proving Ground, MD.

Length: 2 weeks (75 hours).

Exhibit Dates: 6/85–9/90.

Learning Outcomes: Upon completion of the course, students will be able perform unit maintenance on wheel vehicles.

Instruction: This is a refresher course in vehicle maintenance and includes such topics as maintenance management; principles of fuel systems, brake systems, and electrical systems; and troubleshooting.

Credit Recommendation: In the lower-division baccalaureate/associate degree category, 1 semester hour in automotive maintenance management (3/87).

Related Occupations: 63B.

AR-1710-0122

Light Wheel Vehicle Mechanic Individual Ready Reserve Refresher

Course Number: 610-63B30 (IRR).

Location: Ordnance Center and School, Aberdeen Proving Ground, MD.

Length: 2 weeks (75 hours).

Exhibit Dates: 6/85–9/90.

Learning Outcomes: Upon completion of the course, the student will be able to supervise maintenance and perform preventive maintenance and troubleshoot light and medium duty wheel and track vehicles and power generation equipment.

Instruction: Lectures cover management, preventive maintenance, and troubleshooting of gasoline engines, diesel engines, electrical systems, brakes, hydraulic systems, and power generation equipment. Some hands-on experience is included.

Credit Recommendation: In the lower-division baccalaureate/associate degree category, 2 semester hours in preventive maintenance supervision (3/87).

Related Occupations: 63B.

AR-1710-0123

Armament Repair Technician Warrant Officer Advanced

Course Number: 4-9-C32-421A.

Location: Ordnance Center and School, Aberdeen Proving Ground, MD.

Length: 15 weeks (546–549 hours).

Exhibit Dates: 3/85–12/97.

Learning Outcomes: Upon completion of the course, the student will be able to manage, maintain, and repair armament equipment as needed; keep records on armament equipment repair; and maintain the electrical and laser systems on armament equipment.

Instruction: Instruction includes lectures and hands-on practice with emphasis on management, maintenance management, and communication skills. Considerable emphasis is placed on the electrical and laser systems on armament equipment.

Credit Recommendation: In the lower-division baccalaureate/associate degree category, 1 semester hour in communication skills, 1 in management, 1 in heavy equipment maintenance management, 4 in basic electronics, and 2 in automotive electrical (3/87).

Related Occupations: 421A.

AR-1710-0124

Track Vehicle Recovery Specialist

Course Number: 611-ASI-H8 (63H) (63Y) (63D).

Location: Ordnance Center and School, Aberdeen Proving Ground, MD.

Length: 4–5 weeks (166–167 hours).

Exhibit Dates: 10/88–1/90.

Learning Outcomes: Upon completion of the course, the student will be able to operate vehicles and equipment used to recover vehicles in various situations.

Instruction: Instruction includes lectures and operation and servicing of recovery vehicles and equipment. Procedures used in the recovery of disabled track vehicles, oxyacetylene cutting, selection of rigging equipment, and rigging are also covered.

Credit Recommendation: In the vocational certificate category, 4 semester hours in heavy equipment operation (10/88).

Related Occupations: 63D; 63H; 63Y.

AR-1710-0126

Track Vehicle Mechanic Individual Ready Reserve Refresher

Course Number: 611-F5 (IRR).

Location: Ordnance Center and School, Aberdeen Proving Ground, MD.

Length: 2 weeks (94–95 hours).

Exhibit Dates: 10/86–9/90.

Learning Outcomes: Upon completion of the course, the student will be able to describe the basic principles of operation of the internal combustion engine, fuel system, electrical system (includes starting and charging), hydraulic system, and tracked vehicle power train and brakes and use diagnostic test equipment, references, and basic knowledge to troubleshoot and record malfunctions of track vehicle engines and fuel, electrical, hydraulic, and power train systems.

Instruction: Instruction includes lectures on the basic principles of operation of the internal combustion engine, fuel system, electrical system (including starting and charging), hydraulic system, and track vehicle power train and brakes; care and use of test equipment; hands-on troubleshooting of induced malfunctions on engine, fuel system, electrical system, hydraulic system, and power train and brakes; and recording diagnosis on correct forms.

Credit Recommendation: In the vocational certificate category, 3 semester hours in construction equipment mechanic (3/87).

AR-1710-0127

Track Vehicle Repairer Noncommissioned Officer (NCO) Advanced
(Mechanical Maintenance Noncommissioned Officer (NCO) Advanced, MOS 63H)

Course Number: 6-63-C42 (63H); 610-63H40.
Location: Ordnance Center and School, Aberdeen Proving Ground, MD.
Length: 9–11 weeks (336–504 hours).
Exhibit Dates: 10/85–7/95.
Learning Outcomes: Upon completion of the course, the student will work as a maintenance supervisor.
Instruction: Lectures, demonstrations, and practical exercises cover leadership, communication skills, fundamentals of instruction, logistics management, and refresher training in track vehicle repair. Emphasis in on leadership and on verbal and written communication. This course contains a common core of leadership subjects.
Credit Recommendation: Credit is recommended for the common core only. See AR-1404-0035 (1/88).
Related Occupations: 63H.

AR-1710-0128

Construction Equipment Repairer Noncommissioned Officer (NCO) Advanced
(Mechanical Maintenance Noncommissioned Officer (NCO) Advanced, MOS 62B)

Course Number: 6-63-C42 (62B); 612-62B40.
Location: Ordnance Center and School, Aberdeen Proving Ground, MD.
Length: 8–10 weeks (280–459 hours).
Exhibit Dates: 10/85–7/95.
Learning Outcomes: Upon completion of the course, the student will serve as a maintenance supervisor.
Instruction: Lectures, demonstrations, and practical exercises cover leadership, communication skills, fundamentals of instruction, logistics management, and refresher training in construction equipment repair. Emphasis in on leadership and on verbal and written communication. This course contains a common core of leadership subjects.
Credit Recommendation: Credit is recommended for the common core only. See AR-1404-0035 (1/88).
Related Occupations: 62B.

AR-1710-0129

Bradley Fighting Vehicle System Mechanic Noncommissioned Officer (NCO)
(Mechanical Maintenance Noncommissioned Officer (NCO) Advanced, MOS 63B, D, E, N, T)

Course Number: 6-63-C42 (63B, D, E, N, T); 611-63T40.
Location: Ordnance Center and School, Aberdeen Proving Ground, MD.
Length: 8–10 weeks (295–445 hours).
Exhibit Dates: 10/85–7/95.
Learning Outcomes: Upon completion of the course, the student will be able to supervise maintenance.
Instruction: Lectures, demonstrations, and practical exercises cover leadership, communication skills, fundamentals of instruction, logistics management, and refresher training in the Bradley fighting vehicle. Emphasis in on leadership and on verbal and written communication. This course contains a common core of leadership subjects.
Credit Recommendation: Credit is recommended for the common core only. See AR-1404-0035 (1/88).
Related Occupations: 63B; 63D; 63E; 63N; 63T.

AR-1710-0130

Tank M60A1/A3 Tank System Mechanic Noncommissioned Officer (NCO) Advanced
(Mechanical Maintenance Noncommissioned Officer (NCO) Advanced, MOS 63B, D, E, N, T)

Course Number: 6-63-C42 (63B, D, E, N, T); 611-63N40.
Location: Ordnance Center and School, Aberdeen Proving Ground, MD.
Length: 8–10 weeks (295–445 hours).
Exhibit Dates: 10/85–9/92.
Learning Outcomes: Upon completion of the course, the student will be able to serve as a supervisor of M60A1/A3 tank maintenance.
Instruction: Lectures, demonstrations, and practical exercises cover leadership, communication skills, fundamentals of instruction, logistics management, and refresher training in the supervision of tank maintenance. Emphasis in on leadership and on verbal and written communication. This course contains a common core of leadership subjects.
Credit Recommendation: Credit is recommended for the common core only. See AR-1404-0035 (1/88).
Related Occupations: 63B; 63D; 63E; 63N; 63T.

AR-1710-0131

M1 Abrams Tank System Mechanic Noncommissioned Officer (NCO) Advanced
(Mechanical Maintenance Noncommissioned Officer (NCO) Advanced, MOS 63B, D, E, N, T)

Course Number: 6-63-C42 (63B, D, E, N, T); 611-63E40.
Location: Ordnance Center and School, Aberdeen Proving Ground, MD.
Length: 8–10 weeks (295–445 hours).
Exhibit Dates: 10/85–7/95.
Learning Outcomes: Upon completion of the course, the student will be able to work as a maintenance supervisor.
Instruction: Lectures, demonstrations, and practical exercises cover leadership, communication skills, fundamentals of instruction, logistics management, and refresher training in maintenance supervision. Emphasis in on leadership and on verbal and written communication. This course contains a common core of leadership subjects.
Credit Recommendation: Credit is recommended for the common core only. See AR-1404-0035 (1/88).
Related Occupations: 63B; 63D; 63E; 63N; 63T.

AR-1710-0132

Light Wheel Vehicle Mechanic Noncommissioned Officer (NCO) Advanced
(Mechanical Maintenance Noncommissioned Officer (NCO) Advanced, MOS 63B, D, E, N, T)

Course Number: 6-63-C42 (63B, D, E, N, T); 610-63B40.
Location: Ordnance Center and School, Aberdeen Proving Ground, MD.
Length: 8–10 weeks (295–445 hours).
Exhibit Dates: 10/85–7/95.
Learning Outcomes: Upon completion of the course, the student will be able to serve as a maintenance supervisor.
Instruction: Lectures, demonstrations, and practical exercises cover leadership, communication skills, fundamentals of instruction, logistics management, and refresher training in light wheel vehicle repair. Emphasis in on leadership and on verbal and written communication. This course contains a common core of leadership subjects.
Credit Recommendation: Credit is recommended for the common core only. See AR-1404-0035 (1/88).
Related Occupations: 63B; 63D; 63E; 63N; 63T.

AR-1710-0133

Special Purpose Equipment Repairer Noncommissioned Officer (NCO) Advanced

Course Number: 662-52X40.
Location: Ordnance Center and School, Aberdeen Proving Ground, MD.
Length: 7 weeks (259 hours).
Exhibit Dates: 1/87–7/95.
Learning Outcomes: Upon completion of the course, the student will be able to supervise maintenance.
Instruction: Lectures, demonstrations, and practical exercises cover leadership, communication skills, fundamentals of instruction, logistics management, and refresher training in equipment maintenance. Emphasis in on leadership and on verbal and written communication. This course contains a common core of leadership subjects.
Credit Recommendation: Credit is recommended for the common core only. See AR-1404-0035 (1/88).
Related Occupations: 52X.

AR-1710-0135

M1 Abrams Tank Recovery Specialist

Course Number: 611-ASIH8 (63E).
Location: Armor Center, Fort Knox, KY.
Length: 3–4 weeks (131 hours).
Exhibit Dates: 10/84–12/97.

Learning Outcomes: Upon completion of the course, the student will be able to operate and maintain recovery vehicles and wreckers.

Instruction: Instruction includes lectures, practical exercises, and hands-on operation and maintenance of recovery vehicles and wreckers, including minor repairs and adjustments, boom and winch operation, on and off the road driving skills, and recovery of disabled vehicles.

Credit Recommendation: In the lower-division baccalaureate/associate degree category, 1 semester hour in heavy equipment operation (7/87).

Related Occupations: 63E.

AR-1710-0136

Armor Officer Basic National Guard/ Army Reserve
(Armor Officer Basic Reserve Component)

Course Number: 2-17-C25.

Location: Armor Center, Fort Knox, KY.

Length: 8 weeks (510–580 hours).

Exhibit Dates: 4/84–Present.

Learning Outcomes: Upon completion of the course, the student will be able to perform the duties of an armor platoon leader.

Instruction: Instruction includes lectures and practical exercises in operation, maintenance, and tactical employment of platoon weapons, equipment, and vehicles; communication; tactical movement leadership; management; and navigation. Emphasis on field exercises.

Credit Recommendation: In the lower-division baccalaureate/associate degree category, 3 semester hours in management skills, 2 in basic surveying, and 4 in military science (7/92).

AR-1710-0138

M60A1/A3 Tank Systems Mechanic

Course Number: 611-63N30.

Location: Armor Center, Fort Knox, KY.

Length: 8–10 weeks (299–338 hours).

Exhibit Dates: 8/86–12/97.

Learning Outcomes: Upon completion of the course, the student will be able to manage maintenance and recovery operations and supervise diagnostic procedures, inspection, and testing of maintenance functions on engines; power trains; and electrical, suspension, and hydraulic systems of the hull turret.

Instruction: Instruction includes lectures, demonstrations, shop, and performance exercises on maintenance management, recovery, diagnostics, supervision, and inspection and testing of maintenance functions on power trains, electrical systems, and hydraulic systems of the automotive and turret systems. Training management and basic leadership are also emphasized. Track A (MOS 45N) emphasizes M113, M88A1, and M151 series vehicle maintenance while Track B (MOS 63N) emphasizes armament maintenance on M60A3 vehicles.

Credit Recommendation: In the lower-division baccalaureate/associate degree category, For track 1, 2 semester hours in maintenance management, 1 in automotive electricity, and 1 in automotive engines. For Track 2, 3 semester hours in maintenance management, 1 in automotive electricity, and 1 in automotive engines (7/87).

Related Occupations: 63N.

AR-1710-0139

Recovery Specialist-ASIH8/Track Recovery Operations (M88A1/ M578)
(Recovery Operations (Track))

Course Number: 221T; RCY 221T.

Location: Combined Arms Training Center, Vilseck, W. Germany.

Length: 2 weeks (80 hours).

Exhibit Dates: 12/87–10/98.

Learning Outcomes: Upon completion of the course, the student will be able to determine correct procedures to use in the recovery of disabled vehicles; state basic principles of rigging including mechanical advantage and lift capabilities; select and use ropes, cables, mechanical anchors, and blocks; operate boom, hoist, and winch from hand signals; use acetylene torch to cut through metal; and recover disabled or overturned tracked vehicles.

Instruction: Conference classes and practical exercises cover determining correct procedures, operating equipment, and using load computations to recover vehicles. Students learn to recover, hoist, and tow tracked vehicles and operate and maintain recovery equipment.

Credit Recommendation: In the lower-division baccalaureate/associate degree category, 1 semester hour in recovery equipment operation (10/88).

AR-1710-0142

Maintenance Supervisor

Course Number: LOG 215.

Location: Combined Arms Training Center, Vilseck, W. Germany.

Length: 2 weeks (74 hours).

Exhibit Dates: 1/88–10/98.

Learning Outcomes: Upon completion of the course, the student will be able to use maintenance publications in the supervising of maintenance procedures, record data on maintenance records, prepare status reports of equipment, maintain records of equipment status, prepare document register for supply actions, verify completion of maintenance services, evaluate a maintenance facility, and implement a quality assurance program.

Instruction: Conferences, lectures, and practical exercises are the course methods. Students demonstrate knowledge of procedures and maintenance system records.

Credit Recommendation: In the lower-division baccalaureate/associate degree category, 2 semester hours in automotive maintenance systems and administration (10/88).

AR-1710-0143

M2/3 Bradley Fighting Vehicle (BFV) Unit Turret Maintenance

Course Number: TRT 316.

Location: Combined Arms Training Center, Vilseck, W. Germany.

Length: 2 weeks (74 hours).

Exhibit Dates: 8/87–10/98.

Learning Outcomes: Upon completion of the course, the student will be able to operate, troubleshoot, and repair the mechanical systems of the 25mm automatic gun using test equipment and repair procedures and operate, troubleshoot, and repair the electrical, mechanical, and hydraulic systems of the M2/M3 Bradley fighting vehicle turret system.

Instruction: Lectures and conferences cover electrical and mechanical system fundamentals and safety. Practical exercises cover component location, system operation, and troubleshooting using test equipment and isolating and repairing/replacing components on both weapon systems and electrical and mechanical components of the turret systems.

Credit Recommendation: In the lower-division baccalaureate/associate degree category, 1 semester hour in maintenance procedures and 1 in electrical system troubleshooting (10/88).

AR-1710-0144

M1 Unit Turret Mechanic

Course Number: TRT 317.

Location: Combined Arms Training Center, Vilseck, W. Germany.

Length: 2 weeks (74 hours).

Exhibit Dates: 9/87–10/98.

Learning Outcomes: Upon completion of the course, the student will be able to operate and troubleshoot all systems using test equipment and repair/replace components of the M1 tank turret.

Instruction: Lectures cover safety and electronic, electrical, and mechanical fundamentals. Practical exercises cover locating and identifying components, operating and troubleshooting all systems of the turret, using test equipment, isolating faults, and repairing and replacing turret.

Credit Recommendation: In the lower-division baccalaureate/associate degree category, 1 semester hour in electrical systems troubleshooting and 1 in maintenance procedures (10/88).

AR-1710-0145

M60A3 Unit Turret Maintenance

Course Number: TRT 318.

Location: Combined Arms Training Center, Vilseck, W. Germany.

Length: 3 weeks (109 hours).

Exhibit Dates: 11/86–10/98.

Learning Outcomes: Upon completion of the course, the student will be able to locate, identify, and troubleshoot all components of the tank turret; isolate faults; and repair/replace components on the weapon, fire control, electrical, and hydraulic systems.

Instruction: Lectures cover safety and basic electrical and hydraulic system fundamentals. Practical exercises cover operation, use of test equipment, isolation of faults, and repair of components.

Credit Recommendation: In the lower-division baccalaureate/associate degree category, 1 semester hour in basic electrical maintenance, 1 in hydraulic maintenance, and 1 in maintenance procedures (10/88).

AR-1710-0146

M113 Series Unit Maintenance Training

Course Number: TVM 219.

Location: Combined Arms Training Center, Vilseck, W. Germany.

Length: 2 weeks (74 hours).

Exhibit Dates: 9/87–10/98.

Learning Outcomes: Upon completion of the course, the student will be able to operate and troubleshoot automotive and electrical systems and use test equipment and wiring diagrams to isolate faults and repair/replace components.

Instruction: Lectures cover safety and basic mechanical and electrical fundamentals. Practical exercises cover troubleshooting, using test equipment and wiring diagrams, to isolate and repair/replace components on the diesel fuel, air, cooling, and electrical systems on the M113 personnel carrier; performing steering and brake adjustments; and removing and replacing power plant.

Credit Recommendation: In the lower-division baccalaureate/associate degree category, 1 semester hour in maintenance procedures (10/88).

AR-1710-0147

M2/3 Bradley Fighting Vehicle (BFV) Unit Maintenance Training

Course Number: TVM 216.

Location: Combined Arms Training Center, Vilseck, W. Germany.

Length: 2–3 weeks (74–111 hours).

Exhibit Dates: 9/87–10/98.

Learning Outcomes: Upon completion of the course, the student will be able to perform basic maintenance on track vehicles using service manuals, test equipment, and special tools.

Instruction: Lectures and laboratory exercises cover safety, repair parts manuals, electricity, test equipment, brake systems, special test equipment, transmission and steering linkage adjustments, electrical systems, and general maintenance procedures. The three-week course also includes cooling systems, lubricating systems, and fuel systems.

Credit Recommendation: In the lower-division baccalaureate/associate degree category, 1 semester hour in heavy equipment procedures (10/88).

AR-1710-0148

M1 Unit Automotive Maintenance

Course Number: TVM 217.

Location: Combined Arms Training Center, Vilseck, W. Germany.

Length: 2 weeks (74 hours).

Exhibit Dates: 11/87–10/98.

Learning Outcomes: Upon completion of the course, the student will be able to troubleshoot and diagnose vehicle malfunctions on the M1 Abrams tank, using various test, measurement, and diagnostic equipment, and special tools; identify Ohm's law, terms and symbols, wiring schematics, and types of electrical circuits; and troubleshoot charging system, fuel system, starting system, transmission, and power distribution systems.

Instruction: Lectures and practical experience cover basic electricity, special tools and test equipment, testing and diagnosing components and circuits, and maintenance safety.

Credit Recommendation: In the lower-division baccalaureate/associate degree category, 2 semester hours in electrical systems troubleshooting (10/88).

AR-1710-0149

M109-M110, M548 Unit Maintenance Training

Course Number: TVM 220.

Location: Combined Arms Training Center, Vilseck, W. Germany.

Length: 3 weeks (102 hours).

Exhibit Dates: 11/86–10/98.

Learning Outcomes: Upon completion of the course, the student will be able to use test equipment to troubleshoot critical automotive and electrical systems on the M109/M110/M548 track vehicles; use vehicle service manuals and wiring diagrams; isolate faults in the automotive systems, including fuel, air, electrical, and cooling systems; and perform repair/replacement.

Instruction: Conferences cover safety and basic automotive and electrical fundamentals. Practical exercises cover troubleshooting the automotive and electrical system components, using test equipment, isolating faults, and determining repair/replacement needs.

Credit Recommendation: In the lower-division baccalaureate/associate degree category, 2 semester hours in maintenance procedures (10/88).

AR-1710-0150

Cargo Specialist Basic Noncommissioned Officer (NCO)

Course Number: ***Version 1:*** 822-88H20/30; 822-88H20. *Version 2:* 822-88H20/30; 882-88H30. *Version 3:* 882-88H30.

Location: ***Version 1:*** Transportation and Aviation Logistics School, Fort Eustis, VA; Transportation School, Fort Eustis, VA. *Version 2:* Transportation School, Fort Eustis, VA. *Version 3:* Transportation School, Fort Eustis, VA.

Length: ***Version 1:*** 4–8 weeks (138–203 hours). *Version 2:* 5–6 weeks (204–251 hours). *Version 3:* 5–6 weeks (204 hours).

Exhibit Dates: ***Version 1:*** 4/87–9/00. *Version 2:* 10/00–6/03. *Version 3:* 7/03–Present.

Learning Outcomes: ***Version 1:*** Upon completion of the course, the student will be able to supervise cargo and hatch operations, loading procedures, and transshipment at air, rail, motor, inland barge, and marine terminals. *Version 2:* Upon completion of the course, the student will be able to perform, supervise, and train personnel in cargo handling and loading in air, motor, and marine operations, including equipment preventive maintenance, hazardous material handling, and equipment operations. *Version 3:* Upon completion of the course, the student will be able to perform, supervise, and train personnel in cargo handling and loading in air, motor, and marine operations, including equipment preventive maintenance, hazardous material handling, and equipment operations.

Instruction: ***Version 1:*** Lectures, demonstrations, and performance exercises cover leadership, cargo operations (rail), terminal and water transport, hazardous cargo, operation of Hagglund's crane, maintenance and property accountability, supervision and planning, and conducting and evaluating performance-oriented training. This course contains a common core of leadership subjects. *Version 2:* Audiovisual materials, practical exercises, discussion, laboratory, lecture. Topics include cargo, terminal, and water transport operations; operation of Hagglunds crane; hazardous cargo; estimation of terminal capacity; and operation of rough terrain container crane. *Version 3:* Audiovisual materials, practical exercises, discussion, laboratory, lecture. Topics include cargo, terminal, and water transport operations; operation of Hagglunds crane; hazardous cargo; estimation of terminal capacity; and operation of rough terrain container crane.

Credit Recommendation: *Version 1:* In the lower-division baccalaureate/associate degree category, 1 semester hour in principles of supervision. See AR-1406-0090 for common core credit (3/94). *Version 2:* In the lower-division baccalaureate/associate degree category, 1 semester hour in principles of supervision and 3 in principles of heavy equipment operations (12/03). *Version 3:* In the lower-division baccalaureate/associate degree category, 3 semester hours in heavy equipment operations and 1 in inter-model transportation operations (12/03).

Related Occupations: 88H.

AR-1710-0151

Special Forces Engineer Sergeant

Course Number: ***Version 1:*** 011-18C30. *Version 2:* 011-18C30.

Location: John F. Kennedy Special Warfare Center and School, Fort Bragg, NC.

Length: ***Version 1:*** 23–24 weeks (1423–1618 hours). *Version 2:* 26 weeks (1979 hours).

Exhibit Dates: ***Version 1:*** 10/89–6/00. *Version 2:* 7/00–Present.

Learning Outcomes: ***Version 1:*** Upon completion of the course, the student will be able to use aerial photos and maps to determine distances, routes, and point locations; practice survival techniques; read construction plans; determine tool and material use; prepare a site and lay out a building; construct small bridges; understand basic concrete construction; use explosives in demolition; and practice basic military techniques and operations. *Version 2:* Upon completion of the course, the student will be able to perform duties in the areas of land navigation, construction, bridging, demolitions, tactics, and mission planning.

Instruction: ***Version 1:*** Lectures, demonstrations, and practical exercises cover the use of maps and aerial photos, survival techniques, use of tools and equipment, reading plans, basic construction techniques, and use of explosives. This course contains a common core of leadership subjects. *Version 2:* Lectures and practical exercises cover navigation, tactics, construction, bridging, demolitions, and mission planning.

Credit Recommendation: ***Version 1:*** In the lower-division baccalaureate/associate degree category, 1 semester hour in orienteering, 2 in basic construction techniques, and 1 in blueprint reading. See AR-1406-0090 for common core credit (1/93). *Version 2:* In the lower-division baccalaureate/associate degree category, 3 semester hours in land navigation, 3 in construction technology, and 3 in explosives and demolition. See AR-1406-0090 for common core credit (4/01).

Related Occupations: 18C.

AR-1710-0152

Special Operations Engineer Sergeant Reserve Component Phase 4

Course Number: 011-18C30-RC.

Location: John F. Kennedy Special Warfare Center and School, Fort Bragg, NC.

Length: 3 weeks (169 hours).

Exhibit Dates: 10/88–9/91.

Learning Outcomes: After 9/01 see AR-1710-0207. Upon completion of the course, the student will be able to describe military and commercial explosives, their limitations and proper use; prepare plan for use of explosives; and demonstrate, through field exercise, an aerial approach and demolition of a specified target.

Instruction: Instruction includes lectures on foreign mine warfare, military and commercial explosives, care and destruction of unserviceable explosives, safety procedures, proper placement and type of explosives to obtain desired results, and extensive field exercises.

Credit Recommendation: In the vocational certificate category, 3 semester hours in explosive techniques (12/88).

Related Occupations: 18C.

AR-1710-0153

Recovery Specialist-ASIH8/Wheel Recovery Operations (M816/936/984E1)
(Recovery Operations (Wheel))

Course Number: 221W; RCY/221W.

Location: Combined Arms Training Center, Vilseck, W. Germany.

Length: 2 weeks (80 hours).

Exhibit Dates: 12/87–10/98.

Learning Outcomes: Upon completion of the course, the student will be able to determine correct procedures to use in recovery of disabled vehicles; state basic principles of rigging including mechanical advantage and lift capabilities; select cables, ropes, mechanical anchors, and blocks; operate boom, hoist, and winch from hand signals; use acetylene torch to cut through metal; and recover disabled and overturned wheeled vehicles through proper operations and with proper equipment.

Instruction: Conference classes and practical exercises cover determining correct procedures, operating equipment, and using load computations to recover vehicles. The student learns to recover, hoist, and tow wheeled vehicles; operate recovery equipment; and perform maintenance on recovery equipment.

Credit Recommendation: In the lower-division baccalaureate/associate degree category, 1 semester hour in recovery equipment operation (10/88).

AR-1710-0156

Field Artillery Weapons Maintenance

Course Number: ***Version 1:*** 041-ASIU6. *Version 2:* 041-ASIU6.

Location: Field Artillery School, Fort Sill, OK.

Length: ***Version 1:*** 3 weeks (109 hours). *Version 2:* 3 weeks (109 hours).

Exhibit Dates: ***Version 1:*** 1/90–12/97. *Version 2:* 1/98–Present.

Learning Outcomes: ***Version 1:*** Upon completion of the course, the student will be able to perform basic maintenance and tests on the M198 and M119 towed howitzer hydraulic and electrical systems. *Version 2:* Upon completion of the course, the student will be able to perform unit maintenance on canon weapons armament systems in towed units.

Instruction: ***Version 1:*** Lectures and performance exercises cover organizational maintenance procedures employed on the howitzers, including care and use of tools; shop safety; and inspection, repair, and lubrication. *Version 2:* The course consists of lectures and performance exercises related to shop safety; Army maintenance management system; and maintenance of firing components, fire control equipment, and subsystems of M198 howitzers.

Credit Recommendation: ***Version 1:*** In the lower-division baccalaureate/associate degree category, 2 semester hours in heavy equipment maintenance and 1 in maintenance management (6/89). *Version 2:* In the lower-division baccalaureate/associate degree category, 2 semester hours in heavy equipment maintenance and 1 in maintenance management (6/00).

AR-1710-0157

1. Self-Propelled Field Artillery Turret Mechanic
2. Self-Propelled Field Artillery Turret Mechanic
3. Field Artillery Turret Mechanic

Course Number: ***Version 1:*** 643-45D10. *Version 2:* 642-45D10. *Version 3:* 642-45D10, Phase 2.

Location: ***Version 1:*** Field Artillery School, Fort Sill, OK. *Version 2:* Ordnance Center and School, Aberdeen Proving Ground, MD. *Version 3:* Ordnance Center and School, Aberdeen Proving Ground, MD.

Length: ***Version 1:*** 8 weeks (284–315 hours). *Version 2:* 8–9 weeks (303–312 hours). *Version 3:* 8–9 weeks (302 hours).

Exhibit Dates: ***Version 1:*** 1/88–9/95. *Version 2:* 10/95–9/01. *Version 3:* 10/01–Present.

Learning Outcomes: ***Version 1:*** Upon completion of the course, the student will be able to perform unit maintenance on the electrical, cab, hydraulic, and fire control systems of the M992 tracked vehicle. *Version 2:* Upon completion of the course, the student will be able to perform maintenance on turret systems and components, including armament, fire control, electrical, and hydraulic systems. *Version 3:* Upon completion of the course, the student will be able to perform maintenance on turret systems and components, including armament, fire control, electrical, and hydraulic systems.

Instruction: ***Version 1:*** Lectures and practical exercises cover lubricating, inspecting, removing, replacing, testing, and troubleshooting the turret system. Course also includes publications, tools, maintenance forms, and safety. *Version 2:* Lectures and practical exercises cover the repair and maintenance of self-propelled artillery units, including electrical and hydraulic systems. *Version 3:* Audio visual materials, practical exercises, classroom exercises, discussion, laboratory, and lecture. Lectures and practical exercises cover the repair and maintenance of self-propelled artillery units, including electrical and hydraulic systems.

Credit Recommendation: ***Version 1:*** In the vocational certificate category, 1 semester hour in basic mechanics, 1 in basic hydraulics, and 1 in basic electricity (6/89). *Version 2:* In the lower-division baccalaureate/associate degree category, 3 semester hours in mechanical, electrical, and hydraulic fundamentals (3/03). *Version 3:* In the lower-division baccalaureate/associate degree category, 1 semester hour in basic mechanics, 1 in basic electricity, 1 in basic hydraulics, and 2 in vehicle equipment repair (8/03).

Related Occupations: 45D.

AR-1710-0158

Abrams/Bradley Supervisor Diagnostics

Course Number: SD-315.

Location: Combined Arms Training Center, Vilseck, W. Germany.

Length: 3–4 weeks (144–160 hours).

Exhibit Dates: 1/89–10/99.

Learning Outcomes: Upon completion of the course, the student will be able to maintain and troubleshoot the M1A1 Abrams tank and Bradley fighting vehicle turret and automotive systems.

Instruction: The course is divided into four tracks. In Track A (MOS's 45G, 45K, 45Z, and 913A), topics include basic electrical fundamentals and components, repair of wiring harnesses, overview of hydraulic components and operation, turret electrical and hydraulic systems, diagnostic procedures, use of specialized test equipment and service manuals, and basic electrical troubleshooting procedures to isolate malfunctions. In Track B (MOS's 63G, 63H, 915B, 915D, and 915E), topics include basic electrical components, repair of wiring harnesses and overview of hydraulic components and operation, turbine engines, operation and troubleshooting procedures to determine malfunctions in the hydraulic transmission system, system functions, electrical circuits and troubleshooting, power plant functioning and fault isolation, and specialized test equipment. In Track C (MOS's 45E and 63E), topics include basic electrical components, repair of wiring harnesses, overview of hydraulic components and operation, turret electrical and hydraulic systems, diagnostic procedures, specialized test equipment, turbine engines, and operation and troubleshooting procedures to determine malfunctions in the hydraulic transmission system. In Track D (MOS's 45T and 63T), topics include basic electrical fundamentals and components, repair of wiring harnesses, overview of hydraulic components and operation, use of specialized test equipment and manuals, basic electrical circuits and troubleshooting, procedures to isolate malfunctions, systems function, power plant functioning, and fault isolation.

Credit Recommendation: In the lower-division baccalaureate/associate degree category, for Track A, 1 semester hour in basic electricity, 2 in electrical systems troubleshooting, and 1 in mechanical systems troubleshooting. For Track B, 1 semester hour in basic electricity, 1 in hydraulic systems troubleshooting, and 1 in mechanical systems troubleshooting. For Track C, 1 semester hour in basic electricity, 1 in electrical systems troubleshooting, 1 in mechanical systems troubleshooting, and 1 in hydraulic systems troubleshooting. For Track D, 1 semester hour in basic electricity, 1 in electrical systems troubleshooting, and 1 in mechanical systems troubleshooting (10/89).

AR-1710-0159

Quartermaster/Chemical Equipment Repairer

Course Number: ***Version 1:*** 690-63J10. *Version 2:* 690-63J10.

Location: Ordnance Center and School, Aberdeen Proving Ground, MD.

Length: ***Version 1:*** 11–13 weeks (417–449 hours). *Version 2:* 10–16 weeks (387–550 hours).

Exhibit Dates: ***Version 1:*** 6/88–9/95. *Version 2:* 10/95–Present.

Learning Outcomes: ***Version 1:*** Upon completion of the course, the student will be able to troubleshoot, maintain, and repair small engines, pumps, and laundry/bath, water purification, steam cleaning, textile, chemical, and burner equipment. *Version 2:* Upon completion of the course, the student will be able to troubleshoot, maintain, and repair small engines, pumps, and laundry/bath, water purification, steam cleaning, textile, chemical, and burner equipment.

Instruction: *Version 1:* Lectures and practical exercises cover troubleshooting, maintaining, and repairing basic fuel, air, and exhaust systems, test and measurement equipment, principles of basic automotive electrical and hydraulic systems, small engines, pumps, laundry and bath equipment, water purification, steam cleaning, textile, chemical, and burner equipment. *Version 2:* Lectures and practical exercises cover troubleshooting, maintaining, and repairing basic fuel, air and exhaust systems, test and measurement equipment, principles of basic automotive electrical and hydraulic systems, small engines, pumps laundry and bath equipment, water purification, steam cleaning, textile, chemical, and burner equipment.

Credit Recommendation: *Version 1:* In the lower-division baccalaureate/associate degree category, 1 semester hour in basic automotive and 5 in mechanical maintenance (12/90). *Version 2:* In the lower-division baccalaureate/associate degree category, 1 semester hour in small engine repair and 5 in mechanical maintenance (8/03).

Related Occupations: 63J.

AR-1710-0160

Quartermaster/Chemical Equipment Repairer Transition

Course Number: 690-63J30-T.

Location: Ordnance Center and School, Aberdeen Proving Ground, MD.

Length: 3–4 weeks (116–117 hours).

Exhibit Dates: 4/87–4/04.

Learning Outcomes: Upon completion of the course, the student will be able to troubleshoot and repair pumps; laundry equipment; portable heaters; and clothing repair, water purification, and chemical handling equipment.

Instruction: Lectures and practical exercises cover troubleshooting and repair of pumps, water heaters, air compressors, washers, dryers, heaters, sewing machines, filter units, water purification units, decontamination units, smoke generators, and shelter systems.

Credit Recommendation: In the lower-division baccalaureate/associate degree category, 4 semester hours in general machine maintenance (12/90).

Related Occupations: 63J.

AR-1710-0161

Transition Machinist

Course Number: 702-45850.

Location: Ordnance Center and School, Aberdeen Proving Ground, MD.

Length: 3–4 weeks (152 hours).

Exhibit Dates: 10/89–4/04.

Learning Outcomes: Upon completion of the course, the student will be able to safely use small metalworking hand tools, metal lathes, and milling machines.

Instruction: Lectures and demonstrations cover the safe use of metalworking hand tools and machine lathes and mills. Students design, plan, and construct projects in a laboratory situation.

Credit Recommendation: In the lower-division baccalaureate/associate degree category, 1 semester hour in basic metalworking and 2 in machine tools laboratory (12/90).

AR-1710-0164

Watercraft Engineer

Course Number: *Version 1:* 652-88L10. *Version 2:* 652-88L10.

Location: Transportation School, Fort Eustis, VA.

Length: *Version 1:* 9–10 weeks (360–372 hours). *Version 2:* 8 weeks (355 hours).

Exhibit Dates: *Version 1:* 10/88–9/95. *Version 2:* 10/95–Present.

Learning Outcomes: *Version 1:* Upon completion of the course, the student will be able to perform engine room maintenance and repair including marine electrical systems. *Version 2:* Upon completion of the course, the student will be able to operate and maintain vessel engines and machinery, including hydraulic and electrical systems.

Instruction: *Version 1:* Lectures and practical exercises cover marine engineering systems on a small craft, including damage control, watchstanding duties, engine room operation, and marine electrical systems. Classroom instruction is followed by a 72-hour performance exercise. *Version 2:* Methods of instruction include conferences, performance exercises, demonstrations and televised information on small watercraft, including damage control, watch standing duties, diesel engine operations and maintenance, and marine electrical and hydraulic systems maintenance.

Credit Recommendation: *Version 1:* In the lower-division baccalaureate/associate degree category, 3 semester hours in marine engine room operation (10/92). *Version 2:* In the lower-division baccalaureate/associate degree category, 1 semester hour in small engine repair, 1 in basic hydraulics, and 1 in basic electricity (6/01).

Related Occupations: 88L.

AR-1710-0165

1. Ammunition Specialist Reserve Phase 2
2. Ammunition Specialist Reserve Component Phase 2

Course Number: *Version 1:* 645-55B10-RC. *Version 2:* 645-55B10-RC, Phase 2.

Location: Missile and Munitions School, Redstone Arsenal, AL.

Length: *Version 1:* 2 weeks (75–165 hours). *Version 2:* 2–3 weeks (105–106 hours).

Exhibit Dates: *Version 1:* 3/90–9/92. *Version 2:* 10/92–6/00.

Learning Outcomes: *Version 1:* Upon completion of the course, the student will be able to operate a rough terrain forklift, perform operator maintenance on a forklift, and perform ammunition supply point operations. *Version 2:* Upon completion of the course, the student will be able to inspect and classify ammunition, handle ammunition, issue receipts, prepare automated documents, and operate and maintain forklifts.

Instruction: *Version 1:* Lectures and practical exercises cover forklift operation, ammunition loading and storage, ammunition inspection, fire fighting, and inert ammunition preparation. *Version 2:* Lectures and practical exercises cover ammunition safety, ammunition storage, ammunition defect determination, shipping and storage container inspection, ammunition disposal, material management, and safety.

Credit Recommendation: *Version 1:* Credit is not recommended because of the military-specific nature of the course (10/92). *Version 2:* Credit is not recommended because of the military-specific nature of the course (7/96).

Related Occupations: 55B.

AR-1710-0166

Construction Equipment Repairer Supervisor Basic Noncommissioned Officer (NCO)

Course Number: 612-62B30.

Location: Engineer School, Fort Leonard Wood, MO.

Length: 9–11 weeks (357–432 hours).

Exhibit Dates: 10/91–10/01.

Learning Outcomes: After 9/01 see AR-1710-0200. Upon completion of the course, the student will have the diagnostic abilities to troubleshoot faults in diesel engine hydraulic, electrical, and power train systems; supervise and inspect an engineer equipment maintenance operation; and instruct subordinates in maintenance operations.

Instruction: Lectures, demonstrations, and practical experiences cover troubleshooting, maintenance shop operations, and management. This course contains a common core of leadership subjects.

Credit Recommendation: In the lower-division baccalaureate/associate degree category, 3 semester hours in advanced diesel mechanics, 2 in construction equipment troubleshooting, and 2 in diesel repair shop operations. See AR-1406-0090 for common core credit (6/99).

Related Occupations: 62B.

AR-1710-0167

Engineer Equipment Repairer (Construction Equipment Repairer)

Course Number: 612-62B10.

Location: Engineer School, Fort Leonard Wood, MO.

Length: 7–9 weeks (282–352 hours).

Exhibit Dates: 10/89–9/01.

Learning Outcomes: After 9/01 see AR-1710-0205. Upon completion of the course, the student will be able to use and maintain tools necessary for heavy equipment repair; identify technical data and specifications; diagnose and service automotive electrical systems on diesel engines, including air, cooling, lubrication, and fuel systems; tune diesel engines; diagnose and repair hydraulic systems, power trains, clutches, and bearings on various types of heavy construction equipment; maintain records; and perform preventive maintenance.

Instruction: Lectures, demonstrations, and practical exercises cover heavy equipment repair. Army personnel will have a unit on welding.

Credit Recommendation: In the lower-division baccalaureate/associate degree category, 3 semester hours in heavy equipment repair, 3 in diesel equipment repair, and 1 in automotive electrical system repair (6/99).

Related Occupations: 62B.

AR-1710-0168

Crane Operator

Course Number: *Version 1:* 713-62F10. *Version 2:* 713-62F10.

Location: Engineer School, Fort Leonard Wood, MO.

Length: *Version 1:* 7 weeks (288–289 hours). *Version 2:* 6–7 weeks (242–243 hours).

Exhibit Dates: *Version 1:* 4/92–9/95. *Version 2:* 10/95–9/01.

Learning Outcomes: *Version 1:* Upon completion of the course, the student will be able to perform basic preventive maintenance checks and operate 20-ton mechanical cranes, 12-ton mechanical cranes and draglines, 25-ton hydraulic cranes, pile drivers, and crane carriers. *Version 2:* After 9/01 see AR-1710-0203. Upon completion of the course, the student will be able to operate heavy equipment, including a five-ton dump truck, semitractor trailer, 25-ton hydraulic crane, and pile-driver; basic clamshell operations; and perform preventive maintenance.

Instruction: *Version 1:* Lectures, demonstrations, and practical exercises cover maintenance, maneuverability, and operation of the cranes. *Version 2:* Instruction includes lectures, demonstrations, and practical exercises in preventive maintenance and operating techniques on the above listed equipment.

Credit Recommendation: *Version 1:* Credit is not recommended because of the limited, specialized nature of the course (1/93). *Version 2:* In the lower-division baccalaureate/associate degree category, 3 semester hours in crane operations (6/99).

Related Occupations: 62F.

AR-1710-0169

General Construction Equipment Operator

Course Number: *Version 1:* 713-62J10; 713-55131. *Version 2:* 713-62J10; 3E2X1 (OS). *Version 3:* 713-62J10.

Location: Engineer School, Fort Leonard Wood, MO.

Length: *Version 1:* Army, 7 weeks (276 hours); Air Force, 4–5 weeks (152 hours). *Version 2:* 6–7 weeks (250–251 hours). *Version 3:* 6–7 weeks (241–242 hours).

Exhibit Dates: *Version 1:* 8/92–7/95. *Version 2:* 8/95–9/01. *Version 3:* 10/01–Present.

Learning Outcomes: *Version 1:* Upon completion of the course, Army students will be able to perform basic maintenance checks and service and operate dump trucks, rollers, water distributors, air compressors, and small emplacement excavators with backhoe and front bucket. Air Force students will be able to perform basic preventive maintenance checks and services and maneuver and operate forklifts, rollers, utility tractors, water distributors, dump trucks, and rotary sweepers. *Version 2:* Upon completion of the course, the student will operate and on a five-ton dump truck; semitractor trailer, small emplacement excavator and backhoe (SEE tractor) air compressor; water distributor; and compaction equipment, including nine-wheel rollers, tamping foot rollers, and vibratory roller. Air Force students will be able to perform preventive maintenance and operate forklifts and rotary sweepers but not air compressors. *Version 3:* Upon completion of the course, the student will operate and perform basic maintenance on a five-ton dump truck, semitractor trailer, small emplacement excavator and backhoe (SEE tractor) air compressor; water distributor; and compaction equipment, including nine-wheel rollers, tamping foot rollers, and vibratory roller. Air Force students will be able to perform preventive maintenance and operate five ton dump trucks, semitractor trailer, and small emplacement excavator and backhoe (SEE tractor), forklifts and rotary sweepers. Navy students will be able to operate and perform basic maintenance on five ton dump trucks, semitractor trailer, air compressor and compaction equipment.

Instruction: *Version 1:* Lectures, demonstrations, and practical exercises cover basic preventive maintenance and operation of the above pieces of equipment. *Version 2:* Instruction includes lectures, demonstration and practical exercises in operator preventive maintenance and operating techniques on the above listed equipment. *Version 3:* Instruction includes lectures, demonstration and practical exercises cover basic maintenance and operation of the above equipment. Army, Navy and Air Force personnel attend portions of this course. Credit recommended is for completion of the entire course only.

Credit Recommendation: *Version 1:* Credit is not recommended because of the limited, specialized nature of the course (1/93). *Version 2:* In the lower-division baccalaureate/associate degree category, 3 semester hours in heavy equipment operation (6/99). *Version 3:* In the lower-division baccalaureate/associate degree category, 3 semester hours for Army student in heavy equipment operation. Credit is not recommended for Air Force or Navy students, who attend only portions of the course (1/02).

Related Occupations: 62J.

AR-1710-0170

1. Heavy Construction Equipment Operator
   (USAF Apprentice Construction Equipment Operator)
2. Heavy Construction Equipment Operator (USA)

Course Number: *Version 1:* 713-55131-05; 713-62E10. *Version 2:* 713-62E10 (ITRO).

Location: Engineer School, Fort Leonard Wood, MO.

Length: *Version 1:* Army, 7–8 weeks (315 hours); Air Force, 4–5 weeks (164 hours). *Version 2:* 7–8 weeks (322–323 hours).

Exhibit Dates: *Version 1:* 3/92–9/95. *Version 2:* 10/95–9/01.

Learning Outcomes: *Version 1:* Upon completion of the course, the student will be able to perform maintenance checks and operate five-ton semitractor trailer, dozer, scooploader, and grader. The Army version also includes a scraper. *Version 2:* Upon completion of this course, the student will be able to perform operator preventive maintenance on a five-ton dump truck, semitractor trailer, dozer, grader, scraper, and scooploader.

Instruction: *Version 1:* Lectures and practical exercises cover the operation and operator maintenance of the above equipment. *Version 2:* Instruction includes lectures, demonstrations, and practical exercises in preventive maintenance and operating techniques on the above listed equipment. NOTE: The Air Force curriculum does not includes scraper operations.

Credit Recommendation: *Version 1:* Credit is not recommended because of the limited, specialized nature of the course (1/93). *Version 2:* In the lower-division baccalaureate/associate degree category, 3 semester hours in heavy equipment operation (6/99).

Related Occupations: 62E.

AR-1710-0171

Quarrying Specialist

Course Number: *Version 1:* 730-62G10. *Version 2:* 713-62G10.

Location: Engineer School, Fort Leonard Wood, MO.

Length: *Version 1:* 6–7 weeks (259 hours). *Version 2:* 6 weeks (248 hours).

Exhibit Dates: *Version 1:* 7/89–2/96. *Version 2:* 3/96–3/99.

Learning Outcomes: *Version 1:* Upon completion of the course, the student will be able to erect, operate, and perform operational maintenance on a rock crushing plant. *Version 2:* Upon completion of the course, the student will be able to erect, operate, and perform operational maintenance on a rock crushing plant.

Instruction: *Version 1:* Lectures and practical exercises cover procedures needed to assemble a rock crushing plant, the operation of the plant, the drilling operation of the plant, the demolition operation of the plant, and the maintenance of a rock crushing plant. *Version 2:* Lectures and practical exercises cover procedures needed to assemble a rock crushing plant, the operation of the plant, the drilling operation of the plant, the demolition operation of the plant, and the maintenance of a rock crushing plant.

Credit Recommendation: *Version 1:* Credit is not recommended because there is no comparable course of instruction offered in postsecondary institutions (1/93). *Version 2:* Credit is not recommended because there is no comparable course of instruction offered in postsecondary institutions (6/99).

Related Occupations: 62G.

AR-1710-0172

Plumber

Course Number: 720-51K10.

Location: Engineer School, Fort Leonard Wood, MO.

Length: 7–8 weeks (288–289 hours).

Exhibit Dates: 1/89–4/95.

Learning Outcomes: Upon completion of the course, the student will be able to read construction plans; cut and join steel, cast iron, copper, and plastic pipes; install and test basic plumbing systems; determine proper tool and material use; repair valve and control devices; and construct, repair, and replace pipelines, plumbing components, and steel tank assemblies.

Instruction: Lectures, demonstrations and practical exercises cover tool and material selection, sewer and drain system installation, water supply lines, installation of water closets and lavatories, construction of pipeline systems, and erection of steel storage tanks.

Credit Recommendation: In the lower-division baccalaureate/associate degree category, 3 semester hours in basic plumbing (1/93).

Related Occupations: 51K.

AR-1710-0173

Carpentry/Masonry Specialist

Course Number: 712-51B10.

Location: Engineer School, Fort Leonard Wood, MO.

Length: 7–8 weeks (288–289 hours).

Exhibit Dates: 10/87–3/95.

Learning Outcomes: After 7/00 see AR-1710-0209. Upon completion of the course, the student will be able to maintain and use carpentry and masonry tools; identify wood and masonry material; construct wood frame floors, walls, and roofing systems including door and window openings and wall and ceiling coverings; and perform masonry and concrete construction, including layout, form construction, reinforcement, wall erection, and forming, mixing, pouring, and finishing concrete.

Instruction: Conferences and practical exercises cover identification, maintenance, and use of hand and power tools; building layout of wood frame, floor, wall, and roofing systems; and interior walls and ceiling. Also includes masonry and concrete foundations, walls, formwork, and reinforcement.

Credit Recommendation: In the lower-division baccalaureate/associate degree category, 2 semester hours in general building construction, 4 semester hours in carpentry, and 2 in masonry and concrete construction (1/93).

Related Occupations: 51B.

AR-1710-0174

1. Concrete and Asphalt Equipment Operator
2. Asphalt and Concrete Equipment Operator

Course Number: *Version 1:* 713-62H10 (VALID); 730-62H10. *Version 2:* 713-62H10.

Location: Engineer School, Fort Leonard Wood, MO.

Length: *Version 1:* 5–6 weeks (211–235 hours). *Version 2:* 5 weeks (200 hours).

Exhibit Dates: *Version 1:* 5/90–9/01. *Version 2:* 10/01–Present.

Learning Outcomes: *Version 1:* Upon completion of the course, the student will be able to maintain and operate asphalt and concrete batching equipment and operate asphalt paving equipment. *Version 2:* Upon completion of the course, the student will be able to maintain and operate asphalt and concrete batching equipment and operate asphalt paving equipment.

Instruction: *Version 1:* Lectures and laboratories cover the above topics. *Version 2:* Lectures and laboratories cover the above topics.

Credit Recommendation: *Version 1:* Credit is not recommended because of the limited, specialized nature of the course (6/99). *Version 2:* Credit is not recommended because of the limited, specialized nature of the course (1/02).

Related Occupations: 62H.

AR-1710-0175

Engineer Tracked Vehicle Crewman

Course Number: 030-12F10.

Location: Engineer School, Fort Leonard Wood, MO.

Length: 8–9 weeks (373 hours).

Exhibit Dates: 10/89–1/94.

Learning Outcomes: Upon completion of the course, the student will be able to inspect and perform basic maintenance on various track vehicles including checking and adjusting track; lubricate and maintain air, cooling, and hydraulic systems; operate various vehicles including dozer and grader; install and operate radio communications devices; and guide vehicle movement using hand signals.

Instruction: Lectures, demonstrations and practical exercises cover the above topics.

Credit Recommendation: In the lower-division baccalaureate/associate degree category, 2 semester hours in heavy equipment maintenance (1/93).

Related Occupations: 12F.

AR-1710-0176

Utilities Operation and Maintenance Technician

Course Number: 4A-210A-RC.

Location: Engineer School, Fort Leonard Wood, MO.

Length: Phase 2, 2 weeks (114 hours); Phase 4, 2 weeks (103 hours).

Exhibit Dates: 6/91–4/04.

Learning Outcomes: Upon completion of the course, the student will be able to install, operate, and repair electrical power generation systems; use sound construction management practices; and supervise utility systems.

Instruction: Lectures and practical exercises cover the installation, operation, troubleshooting, and repair of electrical power generation equipment. Construction management includes planning, preparing, and supervising both light and heavy construction.

Credit Recommendation: In the lower-division baccalaureate/associate degree category, 1 semester hour in electrical generator and motor repair, 2 in construction management, and 1 in construction blueprint reading (1/93).

Related Occupations: 210A.

AR-1710-0177

Utilities Operations and Maintenance Technician Warrant Officer

Course Number: 4A-210A.

Location: Engineer School, Fort Leonard Wood, MO.

Length: 9–12 weeks (325–421 hours).

Exhibit Dates: 1/91–Present.

Learning Outcomes: Upon completion of the course, the student will be able to install, operate, and repair electrical power generation systems; use sound construction management practices; and supervise utility systems.

Instruction: Lectures and practical exercises cover the installation, operation, troubleshooting, and repair of electrical power generation equipment. Construction management includes planning, preparing, and supervising both light and heavy construction.

Credit Recommendation: In the lower-division baccalaureate/associate degree category, 1 semester hour in AC/DC circuits, 1 in electrical generator and motor repair, 3 in construction management, and 3 in construction blueprint reading (6/99).

Related Occupations: 210A.

AR-1710-0178

Lighter Air-Cushion Vehicle 30 Ton Propulsion System

Course Number: 062-F6.

Location: Transportation School, Fort Eustis, VA.

Length: 7 weeks (258 hours).

Exhibit Dates: 7/92–1/98.

Learning Outcomes: Upon completion of the course, the student will be able to troubleshoot and maintain a gas turbine engine.

Instruction: Lectures and practical exercises cover the servicing requirements of a gas turbine engine and related systems.

Credit Recommendation: In the lower-division baccalaureate/associate degree category, 1 semester hour in gas turbine theory (3/94).

AR-1710-0179

Track Vehicle Recovery Specialist

Course Number: 611-ASIH8 (63E).

Location: Armor Center and School, Fort Knox, KY.

Length: 2–3 weeks (109 hours).

Exhibit Dates: 1/97–Present.

Learning Outcomes: Upon completion of the course, the student will be able to conduct recovery duties for tracked utilities in field locations.

Instruction: Lectures and practical exercises cover topography, map reading, radio operations, preventive maintenance, rigging, winches, and transport of recovered vehicles.

Credit Recommendation: In the vocational certificate category, 3 semester hours in tow truck operations (3/97).

AR-1710-0180

Track Vehicle Recovery Specialist

Course Number: 611-ASIH8 (63T).

Location: Armor Center and School, Fort Knox, KY.

Length: 2–3 weeks (109 hours).

Exhibit Dates: 1/97–Present.

Learning Outcomes: Upon completion of the course, the student will be able to operate tracked recovery vehicles in the recovery of tracked vehicles in the field.

Instruction: Lecture and practical exercises cover tracked recovery vehicles and associated recovery equipment.

Credit Recommendation: In the vocational certificate category, 3 semester hours in tow truck operations (3/97).

AR-1710-0182

1. M1A1 Abrams Tank System Mechanic
2. M1A1 Abrams Tank System Maintainer

Course Number: *Version 1:* 611-63E30. *Version 2:* 611-63A30.

Location: Armor Center and School, Fort Knox, KY.

Length: *Version 1:* 11 weeks (655 hours). *Version 2:* 15 weeks (538 hours).

Exhibit Dates: *Version 1:* 1/96–9/00. *Version 2:* 10/00–Present.

Learning Outcomes: *Version 1:* Upon completion of the course, the student will be able to supervise and perform unit maintenance on the hull and turret of the Abrams tank, armored personnel vehicles, and recovery vehicles. *Version 2:* Upon completion of the course, the student will be able to supervise and perform unit maintenance on the hull and turret of the Abrams tank, armored personnel vehicles, and recovery vehicles.

Instruction: *Version 1:* This course includes lectures, demonstrations, videos, and performance exercises in maintenance management, recovery operations, diagnostics, supervision, training management, theory of automotive materials, theory of operation, turret inspection and testing, and repairing of systems and subsystems on vehicles. Course contains a common core of leadership subjects. *Version 2:* This course includes lectures, demonstrations, videos, and performance exercises in maintenance management, recovery operations, diagnostics, supervision, training management, theory of automotive materials, theory of operation, turret inspection and testing, and repairing of systems and subsystems on vehicles.

Credit Recommendation: ***Version 1:*** In the lower-division baccalaureate/associate degree category, 3 semester hours in electrical systems, 3 in diesel engine fundamentals, 3 in hydraulics, and 1 in maintenance management. See AR-1404-0035 for common core credit (3/97). *Version 2:* In the lower-division baccalaureate/associate degree category, 3 semester hours in electrical systems (DC), 3 in diesel engine fundamentals, 3 in basic hydraulics, 1 in maintenance management, and 3 in military science (5/03).

Related Occupations: 63A; 63E.

AR-1710-0183

1. M2/3 Bradley Fighting Vehicle Systems Mechanic, Phase 1
2. M2/3 Bradley Fighting Vehicle System Maintainer

Course Number: ***Version 1:*** 611-63T10. *Version 2:* 611-63M10.

Location: ***Version 1:*** Armor Center and School, Fort Knox, KY. *Version 2:* Armor Center, Fort Knox, KY.

Length: ***Version 1:*** 12 weeks (444 hours). *Version 2:* 14 weeks (520 hours).

Exhibit Dates: ***Version 1:*** 10/95–6/00. *Version 2:* 7/00–Present.

Learning Outcomes: ***Version 1:*** Upon completion of the course, the student will be able to perform unit-level maintenance on Bradley fighting vehicles, recovery vehicles, armored personnel carriers, and trucks. *Version 2:* Upon completion of the course, the student will be able to perform unit-level maintenance on M2/3 Bradley fighting vehicle.

Instruction: ***Version 1:*** This course includes lectures, demonstrations, conferences, videos, and performance exercises in engine systems, electrical systems, hydraulic systems, power trains, and vehicle recovery. *Version 2:* Learning outcomes are accomplished by demonstrations, classroom activities and practical experience. Topics include engine systems and operation; fuel, electrical (starting and charging), mechanical brakes, suspension systems, hull, turret operation and repair, fire control and weapon systems; vehicle recovery; and safety.

Credit Recommendation: ***Version 1:*** In the lower-division baccalaureate/associate degree category, 2 semester hours in introduction to automotive technology, 2 in automotive electrical, and 1 in diesel engine (3/97). *Version 2:* In the vocational certificate category, 1 semester hour in tow truck operation. In the lower-division baccalaureate/associate degree category, 3 semester hours in electrical systems (DC), 3 in diesel engine fundamentals, 3 in basic hydraulics, and 1 in maintenance management (6/01).

Related Occupations: 63M; 63T.

AR-1710-0184

M2/M3A2 Bradley Fighting Vehicle System Turret

Course Number: 643-45T10.

Location: Armor Center and School, Fort Knox, KY.

Length: 9–10 weeks (370 hours).

Exhibit Dates: 1/95–9/95.

Learning Outcomes: Before 1/95 see AR-1703-0045. Upon completion of the course, the student will be able to perform unit maintenance on mounted turret systems and other components of the vehicle.

Instruction: The instructional methods used include lecture, demonstration, videotape, print material, and practical experiences. The topics covered include inspection, service, troubleshooting, basic hydraulics, basic mechanical, and basic electrical. Special tools, diagnostic equipment, and testing techniques are also covered.

Credit Recommendation: In the lower-division baccalaureate/associate degree category, 1 semester hour in basic hydraulics and 1 in preventive equipment maintenance (3/97).

Related Occupations: 45T.

AR-1710-0185

1. Bradley Fighting Vehicle Systems Mechanic
2. M2/3 Bradley Fighting Vehicle System Maintainer
3. M2/3 Bradley Fighting Vehicle System Maintainer

Course Number: ***Version 1:*** 611-63T30, Phase 2. *Version 2:* 611-63M30; 611-63M30, Phase 2. *Version 3:* 611-63M30.

Location: ***Version 1:*** Armor Center and School, Fort Knox, KY. *Version 2:* Armor Center, Fort Knox, KY. *Version 3:* Armor School, Fort Knox, KY.

Length: ***Version 1:*** 17 weeks (638–639 hours). *Version 2:* 12–13 weeks (453 hours). *Version 3:* 12–13 weeks (454 hours).

Exhibit Dates: ***Version 1:*** 1/96–9/00. *Version 2:* 10/00–9/01. *Version 3:* 10/01–Present.

Learning Outcomes: ***Version 1:*** Before 1/96 see AR-1710-0110. Upon completion of the course, the student will be able to supervise and perform maintenance on vehicle and turret systems. *Version 2:* Upon completion of the course, the student will be able to supervise and perform maintenance on M2/3 Bradley Fighting Vehicle Systems. *Version 3:* Upon completion of the course, the student will be able to supervise and perform maintenance on M2/3 Bradley Fighting Vehicle Systems.

Instruction: ***Version 1:*** Instruction includes demonstrations and practical experiences. Topics covered are safety, parts, wiring diagrams, electrical systems, fuel systems, internal combustion engine, power train and suspension, brakes, starting and charging systems, turret systems, basic hydraulics, weapon feed systems, automotive weapon firing systems, disabled vehicle recovery, and leadership. Course includes a common core of leadership subjects. *Version 2:* Learning outcomes are accomplished by demonstrations and practical experiences. Topics include engine systems and operations, fuel, electrical, mechanical (starting and changing), and brake; suspension systems; turret operation and repair; weapon systems; vehicle recovery; safety; and leadership. *Version 3:* Discussion, learner presentations, and classroom and practical exercises. Topics include engine systems and operations; fuel; electrical; mechanical (starting and changing) and brake; suspension systems; turret operation and repair; vehicle recovery; safety; computer operating systems and associated software usage; basic electricity; tool and equipment skills for maintenance and operation of M2/3 Bradley Fighting Vehicle.

Credit Recommendation: ***Version 1:*** In the vocational certificate category, 1 semester hour in tow truck operations. In the lower-division baccalaureate/associate degree category, 3 semester hours in introduction to automotive technology. See AR-1404-0035 for common core credit (3/97). *Version 2:* In the vocational certificate category, 1 semester hour in tow truck operation. In the lower-division baccalaureate/associate degree category, 3 semester hours in introduction to automotive technology (3/03). *Version 3:* In the lower-division baccalaureate/associate degree category, 3 semester hours in introduction to computers, 1 in basic electronics, 2 in industrial technology shop skills, and 2 in industrial safety (6/03).

Related Occupations: 63M; 63T.

AR-1710-0186

Self-Propelled Field Artillery System Mechanics Basic Noncommissioned Officer (NCO)

Course Number: 611-63D30 (45D).

Location: Ordnance Center and School, Aberdeen Proving Ground, MD.

Length: 9–12 weeks (324–406 hours).

Exhibit Dates: 10/96–Present.

Learning Outcomes: Before 10/96 see AR-1710-0112. Upon completion of the course, the student will be able to perform maintenance on self-propelled field artillery systems, recovery vehicles, and ammunition vehicles.

Instruction: Lectures and practical exercises cover maintenance management, supervision, shop operation, troubleshooting, hydraulics, training, and theory of maintenance. Course contains a common core of leadership subjects.

Credit Recommendation: In the lower-division baccalaureate/associate degree category, 3 semester hours in heavy equipment maintenance. See AR-1406-0090 for common core credit (5/03).

Related Occupations: 45D; 63D.

AR-1710-0187

1. Self-Propelled Field Artillery Systems Mechanics Basic Noncommissioned Officer (NCO)
2. Self-Propelled Field Artillery System Mechanics (BNCOC)

Course Number: ***Version 1:*** 611-63D30. *Version 2:* 611–63D30.

Location: Ordnance Center and School, Aberdeen Proving Ground, MD.

Length: ***Version 1:*** 12 weeks (438–439 hours). *Version 2:* 9–10 weeks (357 hours).

Exhibit Dates: ***Version 1:*** 10/96–4/00. *Version 2:* 5/00–Present.

Learning Outcomes: ***Version 1:*** Upon completion of the course, the student will be able to perform, and supervise and train others in unit maintenance of self-propelled field artillery weapon systems, light and medium recovery vehicles, and carrier ammunition-tracked vehicles. *Version 2:* Upon completion of the course, the student will be able to perform unit maintenance on self-propelled field artillery cannon weapon systems, light and medium recovery vehicles, and carrier ammunition-tracked vehicles.

Instruction: ***Version 1:*** Lectures, practical demonstrations, and hands-on experience cover shop operations, recovery operations, diagnostic and troubleshooting, inspection and testing, hydraulic and electrical systems, as well as basic leadership and supervision skills. *Version 2:* Practical exercises and lecture. Topics include logistics, heavy-wheeled vehicle maintenance carrier, ammunition, and tracked vehicle maintenance, recovery/evacuation management, and self-propelled Howitzer.

Credit Recommendation: ***Version 1:*** In the lower-division baccalaureate/associate degree category, 2 semester hours in maintenance and shop management and 2 in automotive and heavy equipment repair. See AR-1406-0090 for common core credit (1/98). *Version 2:* In the lower-division baccalaureate/associate degree category, 2 semester hours in maintenance and shop management and 2 in automotive and heavy equipment repair. See AR-1406-0090 for common core credit (8/03).

Related Occupations: 63D.

AR-1710-0188

ENGINEER EQUIPMENT REPAIR TECHNICIAN

Course Number: 4L-919A.

Location: Engineer School, Fort Leonard Wood, MO.

Length: 17 weeks (612 hours).

Exhibit Dates: 8/97–Present.

Learning Outcomes: Upon completion of the course, the student will be able to supervise all aspects of engineer equipment shop maintenance, including power generation, heating, ventilation, and air conditioning equipment; electrical systems; engines; hydraulics; brake systems; and power trains.

Instruction: Lectures, demonstrations, and practical exercises cover maintenance management and specific, hands-on demonstration of skills in each functional area listed in the learning outcomes.

Credit Recommendation: In the lower-division baccalaureate/associate degree category, 3 semester hours in automotive maintenance and 3 in heavy equipment maintenance (6/99).

AR-1710-0189

TECHNICAL ENGINEERING (USA)

Course Number: 491-51T10.

Location: Engineer School, Fort Leonard Wood, MO.

Length: 18–19 weeks (729 hours).

Exhibit Dates: 1/95–9/01.

Learning Outcomes: After 9/01 see AR-1710-0196. Upon completion of the course, the student will be able to perform quality control testing on concrete such as slump, air content, and compressive and flexural strength; perform quality tests on soils such as identification, moisture content, specific gravity, density, and bearing ratio; prepare basic architectural, civil, and mechanical drawings by both conventional and computer aided (CAD) methods; and perform basic and construction surveying including horizontal and vertical control, with both conventional and automated instruments.

Instruction: Instruction includes lectures, demonstrations, and practical exercises covering concrete and soil testing procedures, drafting (conventional and CAD), basic and construction surveying procedures and instruments, and automated surveying techniques. NOTE: Air Force personnel do not take concrete and soils testing portion of this course.

Credit Recommendation: In the lower-division baccalaureate/associate degree category, 3 semester hours in construction drafting, 3 in basic surveying, and 3 in construction surveying. For all but Air Force, 3 semester hours in soil testing and 2 in concrete testing. For Army personnel, 2 semester hours in automated surveying (6/99).

AR-1710-0190

ABRAMS TANK SYSTEM MAINTAINER

Course Number: ***Version 1:*** 611-63A10. *Version 2:* 611-63A10.

Location: Armor Center and School, Fort Knox, KY.

Length: ***Version 1:*** 16 weeks (577 hours). *Version 2:* 16 weeks (574 hours).

Exhibit Dates: ***Version 1:*** 6/00–12/00. *Version 2:* 1/01–Present.

Learning Outcomes: ***Version 1:*** Upon completion of the course, the student will be able to supervise and perform unit maintenance on the hull and turret of the Abrams tank, armored personnel vehicles, and recovery vehicles. *Version 2:* Upon completion of the course, the student will be able to supervise and perform unit maintenance on the hull and turret of the Abrams tank, armored personnel vehicles, and recovery vehicles.

Instruction: ***Version 1:*** Learning outcomes are accomplished by demonstrations, classroom activities, and practical experience. Topics include engine systems and operation; theory of automotive materials; theory of operation; fuel, electrical (starting and charging), mechanical, brake, and suspension systems; hull turret operation and repair; fire control and weapon system; vehicle recovery operations; safety and testing; and repair of systems and subsystems. *Version 2:* Learning outcomes are accomplished by demonstrations, classroom activities, and practical experience. Topics include engine systems and operation; theory of automotive materials; theory of operation; fuel, electrical (starting and charging), mechanical, brake, and suspension systems; hull, turret operation and repair; fire control and weapon system; vehicle recovery operations; safety and testing; and repair of systems and subsystems.

Credit Recommendation: ***Version 1:*** In the vocational certificate category, 1 semester hour in tow truck operation. In the lower-division baccalaureate/associate degree category, 3 semester hours in electrical (basic DC), 3 in diesel engine fundamentals, 3 in military science, 3 in basic hydraulics, and 1 in maintenance management (6/01). *Version 2:* In the lower-division baccalaureate/associate degree category, 3 semester hours in electrical (DC), 3 in diesel engine fundamentals, 3 in basic hydraulics, 1 in maintenance management, and 3 in military science (6/01).

Related Occupations: 63A.

AR-1710-0192

ASPHALT AND CONCRETE EQUIPMENT OPERATOR

Course Number: 713-62H10.

Location: Engineer School, Fort Leonard Wood, MO.

Length: 121 hours.

Exhibit Dates: 10/01–Present.

Learning Outcomes: Upon completion of the course, the student will be able to maintain and operate asphalt and concrete batching equipment and operate asphalt paving equipment.

Instruction: Lectures and laboratories cover the above topics.

Credit Recommendation: Credit is not recommended because of the limited, specialized nature of the course (1/02).

Related Occupations: 62H.

AR-1710-0193

QUARRY SPECIALIST

Course Number: 713-62G10.

Location: Engineer School, Fort Leonard Wood, MO.

Length: 6 weeks (231 hours).

Exhibit Dates: 10/01–Present.

Learning Outcomes: Upon completion of the course, the student will be able to erect, operate, and perform operational maintenance on a rock crushing plant.

Instruction: Lectures and practical exercises cover procedures needed to assemble a rock crushing plant, the operation of the plant, the drilling operation of the plant, the demolition operation of the plant, and the maintenance of a rock crushing plant.

Credit Recommendation: Credit is not recommended because of the limited, specialized nature of the course (1/02).

Related Occupations: 62G.

AR-1710-0194

QUARRY SPECIALIST

Course Number: 713-62G10.

Location: Engineer School, Fort Leonard Wood, MO.

Length: 125 hours.

Exhibit Dates: 10/01–Present.

Learning Outcomes: Upon completion of the course, the student will be able to erect, operate, and perform operational maintenance on a rock crushing plant.

Instruction: Lectures and practical exercises cover procedures needed to assemble a rock crushing plant, the operation of the plant, the drilling operation of the plant, the demolition operation of the plant, and the maintenance of a rock crushing plant.

Credit Recommendation: Credit is not recommended because of the limited, specialized nature of the course (1/02).

Related Occupations: 62G.

AR-1710-0195

QUARRY SPECIALIST

Course Number: 713-62G10.

Location: Engineer School, Fort Leonard Wood, MO.

Length: 66 hours.

Exhibit Dates: 10/01–Present.

Learning Outcomes: Upon completion of the course, the student will be able to erect, operate, and perform operational maintenance on a rock crushing plant.

Instruction: Lectures and practical exercises cover procedures needed to assemble a rock crushing plant, the operation of the plant, the drilling operation of the plant, the demolition operation of the plant, and the maintenance of a rock crushing plant.

Credit Recommendation: Credit is not recommended because of the limited, specialized nature of the course (1/02).

AR-1710-0196

TECHNICAL ENGINEER

Course Number: 413-51T10.
Location: Engineer School, Fort Leonard Wood, MO.
Length: 115–404 hours.
Exhibit Dates: 10/01–Present.
Learning Outcomes: Before 10/01 see AR-1710-0189. Course includes operation of a CAD system to prepare mechanical drawings and architectural working drawings of a small structure. The student learns the theory and math needed to perform differential leveling, a traverse survey of at least three points within an allowed error, and a site survey depicting all terrain and manmade object. The student performs quality testing on concrete and performs soil tests for: identification, moisture, specific gravity, density, and bearing ratio.
Instruction: Lectures, demonstrations, and practical exercises. Topics include operation of a CAD system to prepare mechanical drawings and architectural working drawings of a small structure. The student learns the theory and math needed to perform differential leveling, a traverse survey of at least three points within an allowed error, and a site survey depicting all terrain and manmade object. The student performs quality testing on concrete and performs soil tests for: identification, moisture, specific gravity, density, and bearing ratio.
Credit Recommendation: In the lower-division baccalaureate/associate degree category, for completion of phase one, 3 semester hours in basic AutoCAD and 1 in basic surveying; for completion of phase two, 2 semester hours in basic surveying and 3 in construction surveying; for completion of phase three, 3 semester hours in soil testing and 2 in concrete testing; and for completion of all three phases, 3 semester hours in basic AutoCAD, 3 in soil testing, 3 in basic surveying, 2 in concrete testing and 3 in construction surveying (1/02).
Related Occupations: 51T.

AR-1710-0200

CONSTRUCTION EQUIPMENT REPAIRER
SUPERVISOR BASIC
NONCOMMISSIONED OFFICER (NCO)

Course Number: 612-62B30.
Location: Engineer School, Fort Leonard Wood, MO.
Length: 10–11 weeks (329 hours).
Exhibit Dates: 10/01–Present.
Learning Outcomes: Before 10/01 see AR-1710-0166. Upon completion of the course, the student will be able to troubleshoot faults in diesel engines and construction equipment systems, including electrical, hydraulic, brake, and power train systems, and have knowledge of repair procedures in order to supervise others in construction equipment repair.
Instruction: Course includes lectures, demonstrations, and practical exercises in diesel engine and construction equipment systems troubleshooting, repair operations, and repair supervision.
Credit Recommendation: In the lower-division baccalaureate/associate degree category, 2 semester hours in hydraulic and electrical systems troubleshooting, 3 in diesel engine troubleshooting, 1 in construction equipment systems troubleshooting, and 2 in personnel supervision (1/02).
Related Occupations: 62B.

AR-1710-0201

CONSTRUCTION EQUIPMENT REPAIRER
SUPERVISOR BASIC
NONCOMMISSIONED OFFICER (NCO),
PHASE 4

Course Number: 612-62B30, Phase 4.
Location: Engineer School, Fort Leonard Wood, MO.
Length: 114–161 hours.
Exhibit Dates: 10/01–Present.
Learning Outcomes: Upon completion of the course, the student will be able to troubleshoot faults in the brake and power train systems of construction equipment and will be able to supervise repair procedures.
Instruction: Course covers lectures, demonstrations, and practical exercises in troubleshooting the brake and power train systems of construction equipment repair procedures, and repair supervision.
Credit Recommendation: In the lower-division baccalaureate/associate degree category, 2 semester hours in troubleshooting of construction equipment brake and power train systems (1/02).
Related Occupations: 62B.

AR-1710-0202

CONSTRUCTION EQUIPMENT REPAIRER
SUPERVISOR BASIC
NONCOMMISSIONED OFFICER (NCO),
PHASE 3

Course Number: 612-62B30, Phase 3.
Location: Engineer School, Fort Leonard Wood, MO.
Length: 125 hours.
Exhibit Dates: 10/01–Present.
Learning Outcomes: Upon completion of the course, the student will be able to troubleshoot faults in diesel engines and the electrical and hydraulic systems of construction equipment and will be able to supervise repair procedures.
Instruction: Course includes lectures, demonstrations, and practical exercises in the troubleshooting of diesel engines, the electrical and hydraulic systems and repair procedures.
Credit Recommendation: In the lower-division baccalaureate/associate degree category, 3 semester hours in diesel engine troubleshooting and 2 in hydraulic and electrical systems troubleshooting (1/02).
Related Occupations: 62B.

AR-1710-0203

CRANE OPERATOR

Course Number: 713-62F10.
Location: Engineer School, Fort Leonard Wood, MO.
Length: 6–7 weeks (109–234 hours).
Exhibit Dates: 10/01–Present.
Learning Outcomes: Before 10/01 see AR-1710-0168. Upon completion of the course, the student will be able to operate heavy equipment, including a five-ton dump truck, semi-tractor trailer, 25-ton hydraulic crane, and pile driver; clamshell operations; and perform preventive maintenance.
Instruction: Instruction includes lectures, demonstrations, and practical exercises in preventive maintenance and operating techniques on the above listed equipment.
Credit Recommendation: In the lower-division baccalaureate/associate degree category, 3 semester hours in crane operations (1/02).

AR-1710-0205

CONSTRUCTION EQUIPMENT REPAIRER

Course Number: 612-62B10.
Location: Engineer School, Fort Leonard Wood, MO.
Length: 8 weeks (298 hours).
Exhibit Dates: 10/01–Present.
Learning Outcomes: Before 10/01 see AR-1710-0167. Upon completion of the course, the student will be able to correct electrical system malfunctions in internal combustion engines and electrical systems, hydraulic systems, power train systems, and brake systems in construction equipment, including two-and four-cycle diesel engines.
Instruction: Lectures and practical exercises cover repair of construction equipment following current maintenance management procedures, locating repair parts, and maintaining records. The student follows required shop safety practice, proper use of tools, and diagnostic equipment to locate, repair, and maintain the various systems.
Credit Recommendation: In the lower-division baccalaureate/associate degree category, 3 semester hours in diesel engine repair and maintenance and 3 in construction equipment repair and maintenance (1/02).
Related Occupations: 62B.

AR-1710-0206

CONSTRUCTION EQUIPMENT REPAIRER

Course Number: 612-62B10.
Location: Engineer School, Fort Leonard Wood, MO.
Length: 241 hours.
Exhibit Dates: 10/01–Present.
Learning Outcomes: Upon completion of the course, the student will be able to correct electrical system malfunctions in internal combustion engines and electrical systems, hydraulic systems, power train systems, and brake systems in construction equipment, including two- and four-cycle diesel engines.
Instruction: Lectures and practical exercises cover repair of construction equipment following current maintenance management procedures, locating repair parts, and maintaining records. The student follows required shop safety practice, proper use of tools, and diagnostic equipment to locate, repair, and maintain the various systems.
Credit Recommendation: In the lower-division baccalaureate/associate degree category, 3 semester hours in diesel engine repair and maintenance and 3 in construction equipment repair and maintenance (1/02).
Related Occupations: 62B.

AR-1710-0207

PLUMBER

Course Number: 720-51K10.
Location: Engineer School, Fort Leonard Wood, MO.
Length: 4–5 weeks (242 hours).
Exhibit Dates: 10/01–Present.
Learning Outcomes: Before 10/01 see AR-1710-0152. Upon completion of the course, the student will be able to read construction plans; cut and join steel, cast iron, copper, and plastic pipes; install and test basic plumbing systems;

determine proper tool and material use; repair valve and control devices; and construct, repair, and replace pipelines, plumbing components, and steel tank assemblies.

Instruction: Lectures, demonstrations and practical exercises cover tool and material selection, sewer and drain system installation, water supply lines, installation of water closets and lavatories, construction of pipeline systems, and erection of steel storage tanks.

Credit Recommendation: In the lower-division baccalaureate/associate degree category, 3 semester hours in basic plumbing (1/02).

AR-1710-0208

PLUMBER

Course Number: 720-51K10.

Location: Engineer School, Fort Leonard Wood, MO.

Length: 136 hours.

Exhibit Dates: 8/00–Present.

Learning Outcomes: Upon completion of the course, the student will be able to read construction plans; cut and join steel, cast iron, copper, and plastic pipes; install and test basic plumbing systems; determine proper tool and material use; repair valve and control devices; and construct, repair, and replace pipelines, plumbing components, and steel tank assemblies.

Instruction: Lectures, demonstrations and practical exercises cover tool and material selection, sewer and drain system installation, water supply lines, installation of water closets and lavatories, construction of pipeline systems, and erection of steel storage tanks.

Credit Recommendation: In the lower-division baccalaureate/associate degree category, 3 semester hours in basic plumbing (1/02).

AR-1710-0209

CARPENTRY AND MASONRY SPECIALIST

Course Number: 720-51B10.

Location: Engineer School, Fort Leonard Wood, MO.

Length: 7–8 weeks (266 hours).

Exhibit Dates: 8/00–Present.

Learning Outcomes: Before 8/00 see AR-1710-0173. Upon completion of the course, the student will be able to perform construction mathematics, print reading, and materials take-off listing; maintain and use carpentry and masonry tools; identify wood and masonry material; construct wood frame floors, walls, and roofing systems, including door and window openings and wall and ceiling coverings; and perform masonry and concrete construction, including layout, form construction, reinforcement, wall erection, and forming, mixing, pouring, and finishing concrete.

Instruction: Conferences and practical exercises cover identification, maintenance, and use of hand and power tools; building layout of wood frame, floor, wall, and roofing systems; and interior walls and ceiling. Also includes masonry and concrete foundations, walls, formwork, and reinforcement.

Credit Recommendation: In the lower-division baccalaureate/associate degree category, 2 semester hours in construction fundamentals, 3 in concrete and masonry, and 3 in framing and roofing (1/02).

Related Occupations: 51B.

AR-1710-0210

CARPENTRY AND MASONRY SPECIALIST

Course Number: 720-51B10.

Location: Engineer School, Fort Leonard Wood, MO.

Length: 120 hours.

Exhibit Dates: 8/00–Present.

Learning Outcomes: Upon completion of the course, the student will be able to maintain and use carpentry and masonry tools; identify wood and masonry material; construct wood frame floors, walls, and roofing systems, including door and window openings and wall and ceiling coverings; and perform masonry and concrete construction, including layout, form construction, reinforcement, wall erection, and forming, mixing, pouring, and finishing concrete.

Instruction: Conferences and practical exercises cover identification, maintenance, and use of hand and power tools; building layout of wood frame, floor, wall, and roofing systems; and interior walls and ceiling. Also includes masonry and concrete foundations, walls, formwork, and reinforcement.

Credit Recommendation: In the lower-division baccalaureate/associate degree category, 2 semester hours in concrete and masonry and 3 in framing and roofing (1/02).

Related Occupations: 51B.

AR-1710-0212

M2/3 BRADLEY FIGHTING VEHICLE SYSTEM MAINTAIN

Course Number: 63M10-OSUT.

Location: Armor School, Fort Knox, KY.

Length: 21 weeks (867 hours).

Exhibit Dates: 10/02–Present.

Learning Outcomes: Upon completion of the course, the student will be able to perform maintenance management and basic military duties.

Instruction: Practical exercises, discussions, laboratory, lectures, and computer-based training. Topics include general military subjects; inspection, service, and maintenance of the Bradley Vehicle; principles of machinery, and mechanical devices.

Credit Recommendation: In the lower-division baccalaureate/associate degree category, 3 semester hours in military science and 3 in maintenance management (6/03).

AR-1710-0215

M1A1 ABRAMS TANK SYSTEM MAINTAINER

Course Number: 63A10-OSUT.

Location: Armor School, Fort Knox, KY.

Length: 23 weeks (926 hours).

Exhibit Dates: 10/02–Present.

Learning Outcomes: Upon completion of the course, the student will be able to perform maintenance management and basic military duties.

Instruction: Practical exercises, discussion, laboratory, lecture, and computer-based training. Topics include general military subjects, inspection, service, maintenance of Abrams Tank, and principles of machinery.

Credit Recommendation: In the lower-division baccalaureate/associate degree category, 3 semester hours in military science and 3 in maintenance management (6/03).

AR-1710-0216

UNIT MAINTENANCE TECHNICIAN (LIGHT)

Course Number: 4L-915A RC, Phase 2.

Location: Ordnance Center and School, Aberdeen Proving Ground, MD.

Length: 2 weeks (103 hours).

Exhibit Dates: 1/94–Present.

Learning Outcomes: Upon completion of the course, the student will be able to demonstrate knowledge in the technical subjects that apply to unit maintenance, logistics management, wheel vehicle maintenance, power generation, and light systems/field artillery.

Instruction: Topics include standard Army maintenance systems, retail parts operations, and vehicle recovery.

Credit Recommendation: In the lower-division baccalaureate/associate degree category, 2 semester hours in maintenance shop management (8/03).

AR-1710-0217

UNIT MAINTENANCE TECHNICIAN (LIGHT)

Course Number: 4L-915A RC, Phase 4.

Location: Ordnance Center and School, Aberdeen Proving Ground, MD.

Length: 2 weeks (107 hours).

Exhibit Dates: 1/94–Present.

Learning Outcomes: Upon completion of the course, the student will be able to demonstrate knowledge of technical subjects that apply to unit maintenance, logistics maintenance, wheel vehicle maintenance, power generation, and light systems/field artillery. This course is designed to train the warrant officer in basic skills needed to become a technician.

Instruction: Practical exercises, classroom exercises, laboratory, and lecture. Topics include maintenance terms and maintenance management in the following areas: logistics, chemical equipment and storage, heavy-duty transmission, and power generation equipment.

Credit Recommendation: In the lower-division baccalaureate/associate degree category, 2 semester hours in maintenance shop management (8/03).

AR-1710-0218

ABRAMS TANK SYSTEM MAINTAINER

Course Number: 091-63A10-TATS, Phase 3.

Location: Armor School, Fort Knox, KY.

Length: 177 hours.

Exhibit Dates: 10/02–Present.

Learning Outcomes: Upon completion of this course, the student will be able to perform corrective and preventative maintenance on the Abrams Tank systems.

Instruction: Audiovisual materials, learner presentations, practical and classroom exercises, discussion, laboratory, and lecture. Topics include turret hydraulic, electrical, and M1A1 fire control systems.

Credit Recommendation: In the lower-division baccalaureate/associate degree category, 3 semester hours in electrical systems troubleshooting and maintenance and 3 in hydraulic systems (8/03).

AR-1710-0219

ABRAMS TANK SYSTEM MAINTAINER

Course Number: 091-63A10-TATS, Phase 2.

Location: Armor School, Fort Knox, KY.
Length: 158 hours.
Exhibit Dates: 10/02–Present.
Learning Outcomes: Upon completion of the course, the student will be able to perform unit level maintenance on the M1A1 Abrams Tank systems.
Instruction: Audiovisual materials, learner presentations, practical and classroom exercises, discussion, laboratory, and lecture. Topics include chassis systems, turret systems, and direct support tasks for the M1A1 systems.
Credit Recommendation: In the lower-division baccalaureate/associate degree category, 1 semester hour hydraulic systems and 3 in electronic troubleshooting and maintenance (8/03).

AR-1710-0220

M1A1 Tank System Mechanic

Course Number: 091-63E10-TATS, Phase 2.
Location: Reserve Schools, Multiple locations.
Length: 120 hours.
Exhibit Dates: 10/99–Present.
Learning Outcomes: Upon completion of the course, the student will be able to demonstrate knowledge of the M1A1 tank and be able to troubleshoot electrical and drivetrain systems in the tank.
Instruction: Topics include tank familiarization, M1A1/M998 electrical systems, M1A1/M998/M113 fuel systems, M1A1/M998 brake and drivetrain systems, and M1A1 suspension systems.
Credit Recommendation: In the lower-division baccalaureate/associate degree category, 2 semester hours in electrical systems maintenance, 1 in fuel systems, and 1 in hydraulics (8/03).

AR-1710-0221

M1A1 ASI

Course Number: 113-ASIK4.
Location: Ordnance Center and School, Aberdeen Proving Ground, MD.
Length: 2–3 weeks (100 hours).
Exhibit Dates: 10/96–Present.
Learning Outcomes: Upon completion of the course, the student will be able to perform maintenance and repair on advancing laser and thermal energy-based fire control systems of the M1A2 tank.
Instruction: Audio visual materials, practical exercises, classroom exercises, discussion, laboratory, and lecture. Topics include lecture and extensive practical exercises on the maintenance and repair of advanced laser and thermal energy-based fire control systems including range finders, imagers, viewers, displays, and sights.
Credit Recommendation: In the lower-division baccalaureate/associate degree category, 2 semester hours in fundamentals of thermal energy systems and 2 in thermal energy systems laboratory (8/03).

AR-1710-0222

Track Recovery Vehicle Specialist

Course Number: 611-ASIH8 (63D,H,Y).
Location: Ordnance Center and School, Aberdeen Proving Ground, MD.
Length: 3–4 weeks (118 hours).
Exhibit Dates: 10/99–Present.
Learning Outcomes: Upon completion of the course, the student will be able to use automotive recovery vehicles (common lay term-wreckers) to recover stranded vehicles; understand safety concepts and policies related to rigging, hoisting and load transport; and safely contain areas contaminated with automotive lubricants, coolants, and fuels.
Instruction: Practical exercises, classroom exercises, laboratory, and lecture. Topics include basic maintenance of wheeled- and tracked-vehicles, heavy rigging and hoisting techniques, oxyacetylene welding and cutting techniques, and safety and contamination issues concerning automotive vehicle liquid spills.
Credit Recommendation: In the vocational certificate category, 2 semester hours in rigging systems. In the lower-division baccalaureate/associate degree category, 1 semester hour in industrial safety (8/03).

AR-1710-0223

Track Recovery Vehicle Specialist

Course Number: 091-ASIH8 (63D/E/H/T/Y).
Location: Ordnance Center and School, Aberdeen Proving Ground, MD.
Length: 118 hours.
Exhibit Dates: 10/99–Present.
Learning Outcomes: Upon completion of the course, the student will be able to use automotive recovery vehicles (common lay term-wreckers) to recover stranded vehicles; understand safety concepts and policies related to rigging, hoisting and load transport; and safely contain areas contaminated with automotive lubricants, coolants, and fuels.
Instruction: Audio visual materials, learner presentations, practical exercises, classroom exercises, discussion, laboratory, and lecture. Topics include basic maintenance of wheeled- and tracked-vehicles, oxyacetylene cutting techniques, and safety and contamination issues concerning automotive vehicle liquid spills.
Credit Recommendation: In the vocational certificate category, 2 semester hours in rigging systems. In the lower-division baccalaureate/associate degree category, 1 in mechanical safety systems laboratory (8/03).

AR-1710-0224

Self-Propelled Field Artillery System Mechanic (BNCOC)

Course Number: 091-63D30, Phase 4.
Location: Reserve Schools, Multiple locations.
Length: 67 hours.
Exhibit Dates: 10/00–Present.
Learning Outcomes: Upon completion of the course, the student will be able to perform unit level maintenance on the automotive and armament systems of self-propelled field artillery.
Instruction: Audio visual materials, learner presentations, practical exercises, classroom exercises, discussion, laboratory, and lecture. Topics include recovery operations, training exercises, maintenance deployment procedures, unit level logistics, and shop operations.
Credit Recommendation: In the lower-division baccalaureate/associate degree category, 1 semester hour in personnel supervision (8/03).

AR-1710-0225

Self-Propelled Field Artillery System Mechanic (BNCOC)

Course Number: 091-63D30, Phase 3.
Location: Reserve Schools, Multiple locations.
Length: 120 hours.
Exhibit Dates: 10/00–Present.
Learning Outcomes: Upon completion of the course, the student will be able to perform unit-level maintenance on the automotive and armament systems of the self-propelled field artillery, medium recovery vehicles, and carrier, ammunition-tracked vehicles.
Instruction: Audio visual materials, learner presentations, practical exercises, classroom exercises, discussion, laboratory, and lecture. Topics include general maintenance.
Credit Recommendation: Credit is not recommended because of the military-specific nature of the course (8/03).

AR-1710-0226

TATS Self-Propelled Field Artillery System Mechanic

Course Number: 091-63D10, Phase 2.
Location: Ordnance Center and School, Aberdeen Proving Ground, MD.
Length: 117 hours.
Exhibit Dates: 10/99–Present.
Learning Outcomes: Upon completion of the course, the student will be able to perform basic maintenance and troubleshooting on electrical, mechanical, and automotive systems, as well as components of self-propelled field artillery, tracked carriers, and recovery equipment.
Instruction: Audio visual materials, practical exercises, classroom exercises, discussion, laboratory, and lecture. Topics include the service, maintenance, parts replacement, and troubleshooting of electrical and mechanical systems using test/diagnostic equipment.
Credit Recommendation: In the lower-division baccalaureate/associate degree category, 2 semester hours in maintenance and troubleshooting of electrical systems and 2 in maintenance and troubleshooting of mechanical systems (8/03).

AR-1710-0227

TATS Self Propelled Field Artillery Turret Mechanic

Course Number: 091-45D10.
Location: Ordnance Center and School, Aberdeen Proving Ground, MD.
Length: 120 hours.
Exhibit Dates: 10/99–Present.
Learning Outcomes: Upon completion of the course, the student will be able to perform maintenance on turret systems and components, including armament, fire control, electrical, and hydraulic systems.
Instruction: Audio visual materials, practical exercises, classroom exercises, discussion, laboratory, and lecture. Topics include the repair and maintenance of self-propelled Army units, electrical, and hydraulic systems.
Credit Recommendation: In the lower-division baccalaureate/associate degree category, 2 semester hours in vehicle equipment repair (8/03).

AR-1710-0228

M1A2 ASI (45K)

Course Number: 643-ASIK4 (45K).
Location: Ordnance Center and School, Aberdeen Proving Ground, MD.
Length: 2–3 weeks (103 hours).
Exhibit Dates: 10/96–Present.
Learning Outcomes: Upon completion of the course, the student will be able to perform maintenance and repair on advanced electrical and electronics-based turret systems of the M1A2 tank.
Instruction: Audiovisual materials, practical exercises, classroom exercises, discussion, laboratory, and lecture. Topics include lecture and extensive practice exercises on the maintenance and repair of advanced electrical and electronics-based turret systems, including radio and navigational units, hull, switching, digital and analog displays, and software test units.
Credit Recommendation: In the lower-division baccalaureate/associate degree category, 3 semester hours in electronic systems II laboratory (8/03).

AR-1710-0229

M2/3 Bradley Fighting Vehicle System Maintain

Course Number: 091-63M10-TATS, Phase 3.
Location: Armor Center and School, Fort Knox, KY.
Length: 167 hours.
Exhibit Dates: 10/02–Present.
Learning Outcomes: Upon completion of the course, the student will be able to maintain turret and gun operation on the M2/3 Bradley fighting vehicle system.
Instruction: Practical and classroom exercises, discussion, laboratory, and lecture. Topics include operation and maintenance of turret and guns on the M2/3 Bradley fighting vehicle systems.
Credit Recommendation: Credit is not recommended because of the military-specific nature of the course (8/03).

AR-1710-0230

M2/3 Bradley Fighting Vehicle System Maintain

Course Number: 091-63M10-TATS, Phase 2.
Location: Armor Center and School, Fort Knox, KY.
Length: 168 hours.
Exhibit Dates: 10/02–Present.
Learning Outcomes: Upon completion of the course, the student will be able to maintain all aspects of a M2/3 Bradley Fighting Vehicle system.
Instruction: Practical and classroom exercises, discussion, laboratory, and lecture. Topics include operation of the M2/3 Bradley engine including electrical systems, starter, charging and fuel system maintenance of power unit, cooling, steering, brake, suspension, fire suppression, chemical protective system, and turret operation of the M2/3 Bradley Fighting Vehicle system.
Credit Recommendation: In the lower-division baccalaureate/associate degree category, 3 semester hours in engine maintenance or electromechanical systems (8/03).

AR-1710-0231

M2/3 Bradley Fighting Vehicle System Maintain

Course Number: 091-63M30, Phase 3.
Location: Reserve Schools, Multiple locations.
Length: 178 hours.
Exhibit Dates: 10/02–Present.
Learning Outcomes: Upon the completion of the course, the student will be able to maintain the gun, turret, firing systems and fire control on the M2/3 Bradley Fighting Vehicle System; perform diagnostics on the transmission, power distribution, starter, charging, and hydraulics of M2/3 Bradley Fighting Vehicle System.
Instruction: Practical exercises, classroom exercises, discussion, laboratory, and lecture. Topics include maintenance and diagnostics of transmissions, power distribution, fire controls, and hydraulics.
Credit Recommendation: In the lower-division baccalaureate/associate degree category, 3 semester hours in maintenance on electromechanical systems (8/03).

AR-1710-0232

M2/3 Bradley Fighting Vehicle System Maintain

Course Number: 091-63M30, Phase 4.
Location: Reserve Schools, Multiple locations.
Length: 140 hours.
Exhibit Dates: 10/02–Present.
Learning Outcomes: Upon completion of the course, the student will be able to maintain the turret and fire control systems on M2/3 Bradley Fighting Vehicle Systems.
Instruction: Practical exercises, classroom exercises, discussion, laboratory, and lecture. Topics include testing and maintenance of turret and fire control systems.
Credit Recommendation: In the lower-division baccalaureate/associate degree category, 2 semester hours in electromechanical systems maintenance (8/03).

AR-1710-0233

M2/3 Bradley Fighting Vehicle System Mechanic

Course Number: 091-63T10-TATS, Phase 1.
Location: Reserve Schools, Multiple locations.
Length: 111 hours.
Exhibit Dates: 10/98–Present.
Learning Outcomes: Upon completion of the course, the student will be able to maintain electrical, hydraulic systems, the engine, power train, and chassis of the M2/3 Bradley Fighting Vehicle System.
Instruction: Practical exercises, classroom exercises, discussion, laboratory, and lecture. Topics include maintenance of electrical and hydraulic systems, and the engine and power train of the M2/3 Bradley Fighting Vehicle System.
Credit Recommendation: In the lower-division baccalaureate/associate degree category, 1 semester hour in maintenance on electrical systems, 1 in mechanical systems, and 1 in hydraulic systems (8/03).

AR-1710-0234

TATS M1A1 Abrams Tank Turret Mechanic (Reclass)

Course Number: 091-45E10, Phase 2.
Location: Reserve Schools, Multiple locations.
Length: 120 hours.
Exhibit Dates: 10/00–Present.
Learning Outcomes: Upon completion of the course, the student will be able to perform the duties of a M1A1 Tank turret mechanic. This course is designed to train personnel transferring from another MOS.
Instruction: Topics include forms and records, M1A1 tank familiarization, M1A1 fire control system, M1A1 azimuth/elevation system, and test and diagnostic equipment.
Credit Recommendation: In the lower-division baccalaureate/associate degree category, 2 semester hours in electronic systems maintenance and 2 in hydraulics (8/03).

AR-1710-0235

Abrams Tank System Maintainer (ADT)

Course Number: 091-63A30, Phase 3.
Location: Reserve Schools, Multiple locations.
Length: 144 hours.
Exhibit Dates: 10/02–Present.
Learning Outcomes: Upon completion of the course, the student will be able to perform unit maintenance on the hull and turret of the M1A1 Abrams tank and the M88 series tracked recovery vehicle at an advanced skill level.
Instruction: Audiovisual materials, learner presentations, computer-based training, practical exercises, classroom exercises, discussion, laboratory, and lecture. Topics include M1A1 tank system orientation, M1A1 hull maintenance (ADT), and M1A1 hull maintenance.
Credit Recommendation: In the lower-division baccalaureate/associate degree category, 2 semester hours in electrical system maintenance, 2 in automotive hydraulics, and 1 electrical machinery (8/03).

AR-1710-0236

Abrams Tank System Maintainer (ADT)

Course Number: 091-63A30, Phase 4.
Location: Reserve Schools, Multiple locations.
Length: 177 hours.
Exhibit Dates: 10/02–Present.
Learning Outcomes: Upon completion of the course, the student will be able to supervise and perform unit maintenance on the hull and turret of the M1A1 Abrams Tank Systems.
Instruction: Audio visual materials, learner presentations, computer-based training, practical exercises, classroom exercises, discussion, laboratory, and lecture. Topics include M1A1 turret maintenance, M1A1 turret and fire control systems, M1A1 direct support, field exercise to reinforce the skill training concepts provided.
Credit Recommendation: In the lower-division baccalaureate/associate degree category, 2 semester hours in hydraulic systems, 3 in electronic troubleshooting and maintenance (8/03).

AR-1710-0237

Aviation Generator/Aviation Ground Power Unit

Course Number: 662-ASID4.
Location: Ordnance Center and School, Aberdeen Proving Ground, MD.
Length: 4–5 weeks (160 hours).
Exhibit Dates: 1/98–Present.
Learning Outcomes: Upon completion of the course, the student will be able to understand the theory of turbine engine operation and maintain and troubleshoot the electrical systems of generators and ground power units.
Instruction: Practical exercises, classroom exercises, discussion, laboratory, and lecture. Topics include theory of turbine engines, maintenance and troubleshooting of DC systems and electrical controls of generators, and maintenance and troubleshooting of AC/DC of ground power units.
Credit Recommendation: In the lower-division baccalaureate/associate degree category, 3 semester hours in electrical power systems (8/03).

AR-1712-0002

WATERCRAFT ENGINEER PRIMARY TECHNICAL

Course Number: 652-61C20.
Location: Transportation School, Fort Eustis, VA.
Length: 11 weeks (408 hours).
Exhibit Dates: 5/86–5/96.
Objectives: To provide enlisted personnel with technical training in diesel engine repair and maintenance for marine engineering certification.
Instruction: Lectures and practical exercises cover maintenance and repair of diesel engines and associated systems, including hydraulic, electrical, drive train, and pneumatic systems. Practical training includes generation and maintenance of refrigeration systems, fire fighting systems, and associated shipboard equipment.
Credit Recommendation: In the lower-division baccalaureate/associate degree category, 2 semester hours in diesel engine operation and maintenance and 1 in electricity (5/86).
Related Occupations: 61A; 61C.

AR-1712-0016

M2/M3 BRADLEY FIGHTING VEHICLE REPAIRER

Course Number: 611-ASIM5 (63H).
Location: Ordnance Center and School, Aberdeen Proving Ground, MD.
Length: 3 weeks (108 hours).
Exhibit Dates: 6/83–9/90.
Objectives: To train enlisted personnel in the general maintenance and servicing of diesel engines.
Instruction: Lectures, demonstrations, and practical exercises cover troubleshooting, adjusting, testing, and repairing various systems and components of the diesel engine. The course includes adjusting and repairing valve trains, troubleshooting, replacing and servicing transmissions and gear boxes, and adjusting fuel injectors.
Credit Recommendation: In the lower-division baccalaureate/associate degree category, 4 semester hours in diesel engine technology (2/84).
Related Occupations: 63H.

AR-1712-0018

WATERCRAFT ENGINEER BASIC NONCOMMISSIONED OFFICER (NCO)

Course Number: ***Version 1:*** 652-88L30. *Version 2:* 652-88L30.
Location: ***Version 1:*** Transportation and Aviation Logistics School, Fort Eustis, VA. *Version 2:* Transportation School, Fort Eustis, VA; Transportation and Aviation Logistics School, Fort Eustis, VA.
Length: ***Version 1:*** 7 weeks (238 hours). *Version 2:* 14–15 weeks (471–505 hours).
Exhibit Dates: ***Version 1:*** 1/87–9/91. *Version 2:* 10/91–Present.
Learning Outcomes: ***Version 1:*** Upon completion of the course, the student will be able to supervise and train enlisted personnel in fire fighting, damage control, emergency and administrative duties, pollution control, and shipboard sanitation and operate, service, troubleshoot, and repair marine hydraulic systems, marine electrical systems, and marine air conditioning and refrigeration systems. *Version 2:* Upon completion of the course, the student will be able to supervise and train enlisted personnel in fire fighting, damage control, emergency and administrative duties, pollution control, and shipboard sanitation and operate, service, troubleshoot, and repair marine hydraulic systems, marine electrical systems, diesel engines, and marine air conditioning and refrigeration systems.
Instruction: ***Version 1:*** Lectures, exercises, and examinations cover common marine technical tasks, marine hydraulics, electrical systems, air conditioning, and refrigeration. This course contains a common core of leadership subjects. *Version 2:* Lectures, exercises, and examinations cover common marine technical tasks, marine hydraulics, electrical systems, diesel engines, air conditioning, and refrigeration. This course contains a common core of leadership subjects.
Credit Recommendation: ***Version 1:*** In the lower-division baccalaureate/associate degree category, 1 semester hour in marine engineering supervision, 1 in marine hydraulics, 1 in marine electricity, and 1 in marine air conditioning and refrigeration. See AR-1406-0090 for common core credit (3/88). *Version 2:* In the lower-division baccalaureate/associate degree category, 1 semester hour in marine engineering, 1 in hydraulic maintenance and repair, 2 in marine electrical system maintenance and repair, 1 in refrigeration and air conditioning, and 3 in diesel engine maintenance and overhaul. See AR-1406-0090 for common core credit (10/00).
Related Occupations: 88L.

AR-1712-0019

WATERCRAFT ENGINEER ADVANCED NONCOMMISSIONED OFFICER (NCO)

Course Number: ***Version 2:*** 652-88L40. *Version 3:* 652-88L40. *Version 4:* 652-88L40.
Location: ***Version 2:*** Transportation and Aviation Logistics School, Fort Eustis, VA. *Version 3:* Transportation School, Fort Eustis, VA. *Version 4:* Transportation School, Fort Eustis, VA.
Length: ***Version 2:*** 13–14 weeks (497 hours). *Version 3:* 14–15 weeks (458–561 hours). *Version 4:* 12–15 weeks (463 hours).
Exhibit Dates: ***Version 2:*** 10/88–3/92. *Version 3:* 4/92–9/96. *Version 4:* 10/96–Present.
Learning Outcomes: ***Version 2:*** Upon completion of the course, the student will be able to supervise the operation and maintenance of marine diesel engines; diesel fuel injector systems; electrical systems; marine refrigeration and air conditioning systems; air compressors and controls; pumps, gears, and drive couplings; and line shafts, bearings and propellers.; perform technical inspections of marine engine components and welds; and assume shipboard duties as a marine engineer including management and administrative duties. *Version 3:* Upon completion of the course, the student will be able to supervise the performance of watercraft operation, maintenance, and repair, including marine diesel engines, fuel injection systems, refrigeration and air conditioning systems, hydraulics, and electrical components and perform nondestructive testing on welds. *Version 4:* Upon completion of the course, the student will be able to perform as a maintenance supervisor and successfully complete the Marine Technical Examination (MTE). The student will be able to supervise the operation and maintenance of marine diesel engines; common engineering tasks; technical inspections; fuel injection; electrical systems and refrigeration systems; hydraulics; water purification systems, cardiopulmonary resuscitation, familiarization of the EMD, cummins, and caterpillar engines; marine welding, inspection, and testing of marine engine components; assume shipboard duties as a marine engineer including management and administrative duties.
Instruction: ***Version 2:*** Lectures, demonstrations, and practical exercises cover marine engineering, administration and management. *Version 3:* Lectures, demonstrations, and practical exercises cover marine engineering, administration, and management. This course contains a common core of leadership subjects. *Version 4:* Methods of instruction include conference, performance exercises, demonstrations, televised information on marine engineering administration and management.
Credit Recommendation: ***Version 2:*** In the lower-division baccalaureate/associate degree category, 3 semester hours in marine engineering supervision, 3 in marine engineering operation, 2 in marine electrical systems, and 1 in marine diesel engines (6/91). *Version 3:* In the vocational certificate category, student receives the equivalent of certificates in CPR and air conditioning (freon). In the lower-division baccalaureate/associate degree category, 3 semester hours in supervision, 2 in electrical component repair, 1 in diesel engine theory and maintenance, 1 in refrigeration and air conditioning, and 1 in hydraulic system theory and maintenance. See AR-1404-0035 for common core credit (3/94). *Version 4:* In the lower-division baccalaureate/associate degree category, 3 semester hours in marine engineering supervision, 3 in marine engineering operations, 2 in electrical systems maintenance, 1 in small engine repair, and 1 in hydraulic system maintenance (6/01).
Related Occupations: 88L.

AR-1712-0020

2. MARINE ENGINEERING OFFICER WARRANT OFFICER TECHNICAL/TACTICAL CERTIFICATION
(Marine Engineering Officer Warrant Officer Technical Certification)
3. MARINE ENGINEERING OFFICER-WARRANT OFFICER BASIC

Course Number: ***Version 2:*** 4H-881A. *Version 3:* 4H-881A.

Location: ***Version 2:*** Transportation and Aviation Logistics School, Fort Eustis, VA. *Version 3:* Transportation School, Fort Eustis, VA.

Length: ***Version 2:*** 39 weeks (1410–1412 hours). *Version 3:* 39 weeks (1501–1527 hours).

Exhibit Dates: ***Version 2:*** 10/88–2/93. *Version 3:* 3/93–Present.

Learning Outcomes: ***Version 2:*** Upon completion of the course, the student will be able to direct the operation, maintenance, repair, and overhaul of engineering machinery and equipment installed in or on Army watercraft; meet the academic and vessel-specific requirements for the Army Marine License Annotated Chief Engineer of Class A-1 Motor Vessel and Assistant Engineer of Class A-2 Unlimited Motor Vessel; operate and maintain auxiliary shipboard systems, including fresh and raw water pumps and systems, oil-water separator, centrifuges, and low-pressure boiler and heating systems; operate, maintain, and repair shipboard electrical apparatus and instruments, including AC and DC motors, generators and switchboards, batteries, refrigeration systems and controls, and windlass, winch, and capstan; operate and maintain diesel engines and associated equipment; perform propeller, shaft, rudder, and plastic repairs and inspect, test, repair, and replace marine piping systems; and maintain, inspect, and replace rudders, propellers, propeller shafts, bearings, and stuffing boxes. *Version 3:* Upon completion of the course, the student will be able to direct the operations of Army watercraft and to successfully meet and complete the academic and vessel-specific requirements for the Army Marine License, Chief Engineer of Class A-1 Motor Vessels or Assistant Engineer of Class A-2 Unlimited Motor Vessels.

Instruction: ***Version 2:*** Lectures, examinations, and practical exercises cover shipboard nomenclature, drills, emergencies, damage control, fire fighting, drownproofing, survival, sanitation, plus customs, reports, and courtesies. Topics also include marine electricity; refrigeration; diesel engine theory, operation, and maintenance; and hull, propulsion, and steering maintenance. *Version 3:* Methods of instruction include conferences, performance exercises, demonstrations, televised information, and field training with marine technical examination covering supply procedures, maintenance management, marine engineering principles and practices; marine refrigeration (HVAC), marine sanitation, and marine auxiliary equipment maintenance and repair; high speed and main two- and four-stroke cycle marine engines and driven reduction gear; alternating and direct current marine electrical generation and distribution systems; hydraulic systems; and transfer control systems for academic certification and vessel-specific licensing.

Credit Recommendation: ***Version 2:*** In the lower-division baccalaureate/associate degree category, 3 semester hours in diesel engineering, 2 in refrigeration, 2 in marine electricity, and 4 in practicum in open water vessel operations. In the upper-division baccalaureate category, 2 semester hours in field experience in management (7/92). *Version 3:* In the lower-division baccalaureate/associate degree category, 1 in industrial maintenance technology, 3 in diesel engineering, 2 in refrigeration (HVAC) systems principles, 2 in electrical systems maintenance, 4 for a practicum in open-water vessel operations, 2 in welding (oxyacetylene and arc) and 1 in nondestructive testing (6/01).

Related Occupations: 510A; 881A.

AR-1712-0021

MARINE ENGINEERING OFFICER, A2 CERTIFICATION

Course Number: ***Version 2:*** 4H-SQI2 (881A). *Version 3:* 4H-SQI2 (881A).

Location: ***Version 2:*** Transportation and Aviation Logistics School, Fort Eustis, VA. *Version 3:* Transportation School, Fort Eustis, VA.

Length: ***Version 2:*** 12 weeks (436–437 hours). *Version 3:* 12 weeks (436 hours).

Exhibit Dates: ***Version 2:*** 10/89–9/96. *Version 3:* 10/96–Present.

Learning Outcomes: ***Version 2:*** Upon completion of the course, the student will be able to perform the duties of a chief engineer in operating, maintaining, and repairing marine engineering systems on ships in open water, under tow, and in salvage and resupply operations. *Version 3:* Upon completion of the course, the student will be able to perform the duties of a chief engineer, in operating, maintaining, and repairing engineering systems aboard Army watercraft in open ocean waters for towing, salvage, and resupply operations; and successfully complete the academic requirements for Army Marine Certification as Chief Engineer of Class A2 Unlimited Motor Vessels Upon Oceans (SQI 2).

Instruction: ***Version 2:*** Lectures and practical exercises cover the troubleshooting and repair of high- and low-speed diesel engines, fuel injection systems, advanced marine electrical systems, marine auxiliary systems, and air conditioning and refrigeration systems. Course is followed by a 16-hour exam for the Army Marine Certification as Chief Engineer of Class A Unlimited Motor Vessels upon Oceans. *Version 3:* Methods of instruction include conferences, performance exercises, demonstrations, computer- assisted instruction, and televised information on marine inspection, piping systems, fuel and waste oil systems, fire fighting systems, nondestructive testing, steering systems, boilers, air systems, propeller shafts and rudder systems, diesel engines, barges, lifting gear, and electrical systems. Upon successful completion of the final examination Army Certification as Chief Engineer of Class A2 Unlimited Motor Vessels Upon Oceans (SQI 2).

Credit Recommendation: ***Version 2:*** In the lower-division baccalaureate/associate degree category, 2 semester hours in advanced marine electrical systems, 1 in advanced diesel engine maintenance, and 2 in principles of supervision (10/92). *Version 3:* In the lower-division baccalaureate/associate degree category, 2 semester hours in advanced electrical system maintenance, 1 in advanced diesel engine maintenance, and 2 in principles of supervision (6/01).

Related Occupations: 510A; 881A.

AR-1712-0022

ENGINEER EQUIPMENT REPAIR TECHNICIAN SENIOR WARRANT OFFICER

Course Number: 4-5-C32-213A.

Location: Engineer School, Fort Leonard Wood, MO.

Length: 10–11 weeks (375 hours).

Exhibit Dates: 10/89–4/04.

Learning Outcomes: Upon completion of the course, the student will be able to manage equipment maintenance and repair facilities and diagnose, repair, and maintain power generation equipment and diesel and gas driven equipment.

Instruction: Lectures and practical exercises cover the above topics.

Credit Recommendation: In the lower-division baccalaureate/associate degree category, 2 semester hours in facility and personnel management and 3 in diesel and gas equipment repair (1/93).

Related Occupations: 213A.

AR-1712-0023

ENGINEER EQUIPMENT REPAIR TECHNICIAN SENIOR WARRANT OFFICER RESERVE

Course Number: 4-5-C32-213A-RC (SWOT).

Location: Engineer School, Fort Leonard Wood, MO.

Length: Phase 2, 2 weeks (87 hours); Phase 4, 2 weeks (83 hours).

Exhibit Dates: 10/90–4/04.

Learning Outcomes: Upon completion of Phase 2, the student will be able to diagnose, repair, and maintain power generation equipment and diesel and gas engines. Upon completion of Phase 4, the student will be able to diagnose, repair, and maintain power trains, hydraulics, and pneumatics.

Instruction: Phase 2 includes conferences, discussion, and practical exercises on diagnosing, maintaining, and repairing power generation equipment and gas and diesel engines. Phase 4 includes conferences, discussions, and practical exercises on diagnosing, maintaining, and repairing power train, hydraulic, and pneumatic systems.

Credit Recommendation: In the lower-division baccalaureate/associate degree category, for Phase 2, 2 semester hours in diesel and gas engine repair. For Phase 4, 2 semester hours in power train, pneumatic, and hydraulic system repair (1/93).

Related Occupations: 213A.

AR-1713-0009

TECHNICAL DRAFTING SPECIALIST

Course Number: 413-81B10.

Location: Engineer School, Fort Leonard Wood, MO.

Length: 11–12 weeks (340–455 hours).

Exhibit Dates: 10/89–4/04.

Learning Outcomes: Upon completion of the course, the student will be able to perform tasks in mechanical drafting, architectural drawing, structural drafting, civil engineering drafting, and technical illustrating and perform construction estimating.

Instruction: Lectures, demonstrations, and practical exercises cover the use of drafting equipment; mechanical drawing, including single view, multiple view, pictorial drawings, and sketching; architectural drawing, including floor plans, function plans, building section details and elevations, and electrical, plumbing, and HVAC plans; structural drafting, including wood, steel, and concrete; civil engineering drafting, including topographic maps and road plans/profiles; and technical illustrating including charts and overlays using ink and color shading materials.

Credit Recommendation: In the lower-division baccalaureate/associate degree category, 3 semester hours in mechanical drawing, 4 in architectural drawing, 1 in structural drafting, 1 in civil engineering drafting, 1 in technical illustrating, and 1 in construction estimating (1/93).

Related Occupations: 81B.

AR-1714-0015

3. Wire Systems Installer
 (Wire System Installer/Operator)
4. Cable Systems Installer-Maintainer
5. Cable Systems Installer-Maintainer

Course Number: ***Version 3:*** 621-31L10; 621-36C10-OSUT; 621-36C10. *Version 4:* 621-31L10. *Version 5:* 621-31L10; 621-31L10 TATS.

Location: Signal School, Fort Gordon, GA.

Length: ***Version 3:*** 5–18 weeks (296–653 hours). *Version 4:* 7–8 weeks (252–288 hours). *Version 5:* 7 weeks (252 hours).

Exhibit Dates: ***Version 3:*** 10/78–6/94. *Version 4:* 7/94–4/98. *Version 5:* 5/98–Present.

Learning Outcomes: ***Version 3:*** Upon completion of the course, the student will be able to install and maintain field wire and cable networks and telephones. *Version 4:* Upon completion of the course, the student will be able to install and maintain field wire and cable networks and telephones. *Version 5:* Upon completion of the course, the student will be able to install operate, and perform unit-level maintenance on cable fiber, and wire systems; state procedures and practices pertaining to electrical safety; calculate voltage, current, and resistance using Ohm's law formula; troubleshoot fiber and cable assembly, and install tactical telephone employing preinstalled multiplexers.

Instruction: ***Version 3:*** Lectures and practical exercises cover construction, maintenance, and recovery of open wire, aerial cable, field wire, and field cable system; and installation of telephones, unattended repeaters, and field switchboards. Course includes fixed-cable construction, fixed-telephone installation, field cable transmission lines, and field training exercises. *Version 4:* Lectures, demonstrations, and practical exercises cover basic electricity, field wire construction and maintenance operations, maintenance of reel units, field cable construction and maintenance; telephone installation; and use of test sets. *Version 5:* Lectures, demonstrations, and practical exercises cover basic electricity, field wire construction and maintenance operations, maintenance of reel units, field cable construction and maintenance; telephone installation; and use of test sets. This course includes an 8-9 week long Basic Training component.

Credit Recommendation: ***Version 3:*** In the vocational certificate category, 4 semester hours in telephone lineman/installer (5/90). *Version 4:* In the vocational certificate category, 9 semester hours in cable system installation and maintenance (2/97). *Version 5:* In the lower-division baccalaureate/associate degree category, 2 semester hours in telecommunications cable system installation and maintenance (2/03).

Related Occupations: 31L; 36C.

AR-1714-0027

Switching Systems Operator
 (Wire Systems Operator)

Course Number: 621-36M10.

Location: Signal School, Fort Gordon, GA.

Length: 7–9 weeks (227–295 hours).

Exhibit Dates: 8/82–10/92.

Objectives: To provide personnel with the knowledge to install, maintain, and operate a tactical manual and automatic switchboard.

Instruction: Student learns proper operation procedures for a manual switchboard. Extensive practical exercises on cable interconnections and equipment setup in the field are provided. Preventive maintenance, equipment checkout, and cable fault troubleshooting procedures are also covered.

Credit Recommendation: Credit is not recommended because of the military-specific nature of the course (6/88).

Related Occupations: 36M.

AR-1714-0028

Switching Systems Operator Basic Noncommissioned Officer (NCO)
 (Wire Systems Operator Primary Technical)

Course Number: 621-36M20.

Location: Signal School, Fort Gordon, GA.

Length: 5–10 weeks (205–352 hours).

Exhibit Dates: 1/84–10/92.

Objectives: To provide experienced personnel with the supervisory skills in the installation, maintenance, and operation of a tactical manual and automatic switchboard.

Instruction: Student learns specific supervisory skills to direct and inspect field installation of cables and equipment setup. Preventive maintenance, equipment checkout, and cable fault troubleshooting procedures are also covered.

Credit Recommendation: Credit is not recommended because of the military-specific nature of the course (2/94).

Related Occupations: 36M.

AR-1714-0029

M1 Abrams Fuel and Electrical Systems Repairer

Course Number: 610-ASIL8 (63G).

Location: Ordnance Center and School, Aberdeen Proving Ground, MD.

Length: 3 weeks (108 hours).

Exhibit Dates: 6/85–9/90.

Learning Outcomes: Upon completion of the course, the student will be able to troubleshoot electrical systems using wiring diagrams and diagnostic equipment; remove and replace failed electrical components and wiring connectors; repair wiring harnesses; solder and unsolder circuit boards and electronic components; and troubleshoot fuel pump and fuel transfer pump.

Instruction: Course includes troubleshooting procedures; replacing or repairing electrical system parts, test driver alert panel, driver master panel, power distribution box, electronic control box, and wire harness; using publications, tools, and test equipment; soldering; using wiring diagrams; operating solid state circuits and safety precautions; and troubleshooting fuel pump and fuel transfer pump.

Credit Recommendation: In the lower-division baccalaureate/associate degree category, 3 semester hours in automotive electrical systems and 2 in basic electronics (3/87).

Related Occupations: 63G.

AR-1714-0030

Fire Control System Repairer
 (Electro-Optical Ordnance Repairer USMC)

Course Number: 113-45G10/670-2172.

Location: Ordnance Center and School, Aberdeen Proving Ground, MD.

Length: Army personnel, 26 weeks (936 hours); Marine Corps personnel, 25–26 weeks (910 hours).

Exhibit Dates: 10/88–9/96.

Learning Outcomes: Upon completion of the course, the student will be able to operate, maintain, and repair laser infrared observation sets.

Instruction: Course includes maintenance forms and publications; safety procedures; basic AC and DC electronics, including concepts of electricity, safety, voltage, current and resistance, conversion units, Ohm's law, color codes, and analyzing and troubleshooting series, parallel, and series parallel circuits. Course also covers batteries, multimeters, introduction to oscilloscopes, magnetism, inductance, generation of AC, capacitance, resonance, transistor fundamentals (NPN-PNP), power supplies, and rectifiers. Basic digital circuits are introduced, along with the binary system; Boolean algebra; diode logic; soldering practices; maintaining, troubleshooting, and adjusting infrared observation set; and troubleshooting and adjusting xenon search light fire control system, laser range finder, and ballistics computer.

Credit Recommendation: In the lower-division baccalaureate/associate degree category, 5 semester hours in basic electricity and electronics and 2 in basic computer (12/90).

Related Occupations: 45G.

AR-1714-0031

1. Aircraft Armament/Missile Systems Repairer
2. AH-1F Armament/Missile Systems Repairer
3. AH-1F Armament/Missile Systems Repairer

Course Number: ***Version 1:*** 646-68J10. *Version 2:* 602-68J20. *Version 3:* 102-68J10.

Location: Aviation Logistics School, Fort Eustis, VA.

Length: ***Version 1:*** 21–23 weeks (757–814 hours). *Version 2:* 20–21 weeks (767 hours). *Version 3:* 18–19 weeks (689 hours).

Exhibit Dates: ***Version 1:*** 10/89–9/94. *Version 2:* 10/94–9/96. *Version 3:* 10/96–Present.

Learning Outcomes: ***Version 1:*** Upon completion of the course, the student will be able to maintain and troubleshoot all electrical and electronic systems of the aircraft's fire control system and its components. *Version 2:* Upon completion of the course, the student will be able to maintain and troubleshoot all electrical and electronic systems of the aircraft's fire control system and its components. *Version 3:* Upon completion of the course, the student will be able to maintain and troubleshoot all electrical and electronic systems of the aircraft's fire control system and its components.

Instruction: ***Version 1:*** Lectures and practical experience cover repairing aircraft fire control systems and subsystems, including wing stores/ejector racks, turret systems, boresights, and armament electrical/electronic systems. *Version 2:* Lectures and practical experience cover repairing aircraft fire control systems and subsystems, including wing stores/ejector racks, turret systems, boresights, and armament electrical/electronic systems. *Version 3:* Lectures and practical experience cover the maintenance and repair of aircraft fire control systems and subsystems, including wing stores/ejector racks, turret systems, air data systems, and armament electrical/electronic systems.

Credit Recommendation: ***Version 1:*** In the lower-division baccalaureate/associate degree category, 3 semester hours in basic electricity, 3 in electronics, 1 in aircraft familiarization, and 3 in aircraft electrical (4/92). *Version 2:* In the lower-division baccalaureate/associate degree category, 3 semester hours in basic electricity, 3 in solid state electronics, and 6 in electronic systems troubleshooting and maintenance (6/98). *Version 3:* In the lower-division baccalaureate/associate degree category, 3 semester hours in basic electricity, 3 in solid state electronics, and 5 in electronic systems troubleshooting and maintenance (6/98).

Related Occupations: 68J.

AR-1714-0032

AH-64 Aircraft Electrical Repairer

Course Number: 602-ASIX1 (68F).

Location: Aviation Logistics School, Fort Eustis, VA.

Length: 4 weeks (138 hours).

Exhibit Dates: 10/90–9/91.

Learning Outcomes: Upon completion of the course, the student will be able to operate, troubleshoot, and repair AC/DC electrical systems including airframe and engine.

Instruction: Lectures and practical exercises cover repair, operation, and troubleshooting of aircraft and engine electrical systems.

Credit Recommendation: In the lower-division baccalaureate/associate degree category, 3 semester hours in AC/DC aircraft/engine electrical systems maintenance and repair (6/91).

Related Occupations: 68F.

AR-1714-0033

OH-58D Aircraft Electrical Repairer

Course Number: 610-ASIW5 (68F).

Location: Aviation Logistics School, Fort Eustis, VA.

Length: 3 weeks (94 hours).

Exhibit Dates: 10/88–9/96.

Learning Outcomes: After 9/96 see AR-1704-0224. Upon completion of the course, the student will be able to operate, troubleshoot, and repair aircraft and engine AC/DC electrical systems.

Instruction: Lectures and practical experience cover the repair, troubleshooting, and operational testing of aircraft and engine electrical systems.

Credit Recommendation: In the lower-division baccalaureate/associate degree category, 3 semester hours in AC/DC aircraft/engine electrical systems maintenance and repair (10/91).

Related Occupations: 68F.

AR-1714-0034

Basic Communications-Electronics Installation
(Communications-Electronics Basic Installer)

Course Number: ***Version 1:*** 829-F2. *Version 2:* 829-F2.

Location: ***Version 1:*** 1199th Signal Battalion, Fort Huachuca, AZ. *Version 2:* 504th Signal Battalion, Fort Huachuca, AZ.

Length: ***Version 1:*** 7 weeks (280 hours). *Version 2:* 5 weeks (192 hours).

Exhibit Dates: ***Version 1:*** 1/91–9/93. *Version 2:* 10/93–4/04.

Learning Outcomes: ***Version 1:*** Upon completion of the course, the student will be able to install a communications system by splicing cables, running conduit, aligning cabinets, identifying cable color, and connecting to terminals. *Version 2:* Upon completion of the course, the student will be able to install a variety of telephone, fiber-optic, and RF communications systems by splicing cables, running conduit, aligning cabinets, identifying cable color, and connecting to terminals.

Instruction: ***Version 1:*** Lectures and practical exercises cover installing communications systems. *Version 2:* Lectures and practical exercises cover installing communications systems.

Credit Recommendation: ***Version 1:*** In the vocational certificate category, 6 semester hours in cable installation (10/91). *Version 2:* In the vocational certificate category, 6 semester hours in telephone, fiber-optic, and RF cable installation (10/93).

AR-1714-0035

Interior Electrician

Course Number: ITRO-51R20; 721-51R10.

Location: Engineer School, Fort Leonard Wood, MO.

Length: 6–7 weeks (232–254 hours).

Exhibit Dates: 10/86–Present.

Learning Outcomes: Upon completion of the course, the student will be able to read construction plans; install service entrances using Romex conduit and armored cable; determine proper tool and material use; test basic electrical circuits; install branch circuits; and determine color code and terminals for installation of switches, outlets, and light fixtures.

Instruction: Lectures and practical exercises cover the above topics.

Credit Recommendation: In the lower-division baccalaureate/associate degree category, 3 semester hours in basic electrical installation (6/99).

Related Occupations: 51R.

AR-1714-0036

Direct Current and Low Frequency Reference Measurement and Calibration

Course Number: E3AZRP051 012.

Location: Ordnance, Missile and Munitions School, Redstone Arsenal, AL.

Length: 5–6 weeks (216 hours).

Exhibit Dates: 10/95–9/00.

Learning Outcomes: Upon completion of the course, the student will be able to operate, maintain, calibrate, troubleshoot, and repair direct current and low-frequency secondary reference laboratory and measurement equipment.

Instruction: Lectures, practical exercises, and demonstrations provide instruction in the operation, calibration, and application of such test equipment as oscilloscopes, multimeters, power amplifiers, and test circuits such as voltage dividers, resistance, capacitance bridge, inductors, and current shunts.

Credit Recommendation: In the lower-division baccalaureate/associate degree category, 3 semester hours in electrical maintenance and troubleshooting (7/96).

AR-1714-0037

Fire Control Systems Repairer

Course Number: ***Version 1:*** 113-45G10. *Version 2:* 113-45G10.

Location: Ordnance Center and School, Aberdeen Proving Ground, MD.

Length: ***Version 1:*** 25–26 weeks (988 hours). *Version 2:* 22–23 weeks (908 hours).

Exhibit Dates: ***Version 1:*** 10/96–9/00. *Version 2:* 10/00–Present.

Learning Outcomes: ***Version 1:*** Upon completion of the course, the student will be able to diagnose, maintain, operate, and repair laser infrared observations sets. *Version 2:* Upon completion of the course, the student will be able to perform both direct and general support for the control devices used in weapons systems (fire control).

Instruction: ***Version 1:*** Course includes maintenance forms and publications; safety procedures; basic AC and DC electronics, including concepts of electricity, safety, voltage, current and resistance, conversion units, Ohm's law, color codes; and analyzing and troubleshooting series, parallel, and series-parallel circuits. Course also covers batteries, multi-meters, introduction to oscilloscopes, magnetism, inductance, generation of AC, capacitance, resonance, transistor fundamentals (NPN-PNP), power supplies, and rectifiers. Basic digital circuits are introduced along with the binary system; Boolean algebra; diode logic; soldering practices; maintaining, troubleshooting, and adjusting infrared observation set; and troubleshooting and adjusting xenon search light fire control system, laser range finder, and ballistics computer. *Version 2:* Audio visual materials, practical exercises, classroom exercises, laboratory, and lecture. Topics include AC/DC electronics, digital electronics, Abram fire control system, laser devices, artillery fire control, digitized repair of towed artillery, Bradley tow electro-optic missile guidance systems, technical publications, hand tools/test equipment, and military-specific training.

Credit Recommendation: ***Version 1:*** In the lower-division baccalaureate/associate degree category, 5 semester hours in basic electricity and electronics and 2 in basic computer operations (1/98). *Version 2:* In the lower-division baccalaureate/associate degree category, 2 semester hours in machine shop orientation, 2 in digital circuits, 3 in basic electricity, 2 in basic electronics, 2 in safety, 3 in electronic troubleshooting and maintenance, and 2 in technical mathematics (8/03).

Related Occupations: 45G.

AR-1714-0038

Deployable Power Generation and Distribution System

Course Number: 661-DPGDS-OP.

Location: Prime Power Production School, Fort Belvoir, VA.

Length: 2 weeks (80 hours).

Exhibit Dates: 4/02–Present.

Learning Outcomes: Upon completion of the course, the student will be able to safely perform siting and instillation of the MEP-PU-

810A1B and DPGDS systems; perform preventative maintenance checks and services, including startup and shutdown.

Instruction: Practical exercises, laboratory, and lecture. Topics include site selection and installation, primary and secondary power distribution, air intake and exhaust; cooling fuel, hydraulics lubrication systems, housing and trailer maintenance, remote operator terminal, and startup and shutdown procedures.

Credit Recommendation: In the lower-division baccalaureate/associate degree category, 2 semester hours in power plant operation (8/03).

AR-1714-0039

Deployable Power Generation and Distribution System (DPGDS) Support Maintenance

Course Number: 661-DBGDS-SM.

Location: Prime Power Production School, Fort Belvoir, VA.

Length: 2 weeks (83 hours).

Exhibit Dates: 4/02–Present.

Learning Outcomes: Upon completion of the course, the student will be able to maintain the MED-PU-810 A/B's electrical systems; perform field removal and replacement of the primary distribution center; construct medium voltage power cables; use associated troubleshooting procedures; and properly observe safety precautions associated with electrical system maintenance.

Instruction: Practical exercises, laboratory, and lecture. Topics include electrical systems maintenance, and MEP-PU-810 A/B prime mover maintenance and safety.

Credit Recommendation: In the lower-division baccalaureate/associate degree category, 3 semester hours in power plant maintenance practicum (8/03).

AR-1714-0040

MEP-PU-810A/B DPGDS New Equipment

Course Number: 661-DPGDS-NET.

Location: Prime Power Production School, Fort Belvoir, VA.

Length: 2–3 weeks (84 hours).

Exhibit Dates: 9/02–Present.

Learning Outcomes: Upon completion of the course, the student will be able to safely perform siting and installation of the NET-DPGDS-01 system; perform preventative maintenance checks and services; and startup and shutdown single and multi-unit parallel operations.

Instruction: Laboratory and lecture. Topics include primary switching, secondary distribution power panels; and coverage of the following systems: primary/secondary cable, remote area lighting, air intake, exhaust, cooling, hydraulic, and lubrication.

Credit Recommendation: In the lower-division baccalaureate/associate degree category, 2 semester hours in power plant operation theory (8/03).

AR-1714-0041

Utility Operation & Maintenance Technical Advance Warrant Officer

Course Number: 4-5-C32-210A-RC, Phase 4.

Location: Prime Power Production School, Fort Belvoir, VA.

Length: 2 weeks (80 hours).

Exhibit Dates: 4/02–Present.

Learning Outcomes: Upon completion of the course, the student will be able to demonstrate through fundamental knowledge and applications of infrared survey, transformer operation and maintenance, HV cable splicing, testing and termination, protective relay maintenance, voltage regulator operations, applicable electrical codes; and correctly install welded fuel systems; and use power plant safety rules and regulations.

Instruction: Practical and classroom exercises and lecture. Topics include infrared survey, transformer and circuit breaker maintenance, switch gear, cable splicing and testing, grounding concepts, welding safety, and plant deployment.

Credit Recommendation: In the lower-division baccalaureate/associate degree category, 2 semester hours in power distribution maintenance practicum, and 2 in industrial safety (flammable/hazardous materials) (8/03).

AR-1714-0042

Utility Operation & Maintenance Technical Advance Warrant Officer

Course Number: 4-5-C32-210A-AC, Phase 2.

Location: Prime Power Production School, Fort Belvoir, VA.

Length: 2 weeks (80 hours).

Exhibit Dates: 4/02–Present.

Learning Outcomes: Upon completion of the course, the student will be able to demonstrate through fundamental knowledge the applications of infrared survey, transformer operation and maintenance, HV cable splicing, testing and termination, protective relay maintenance, voltage regulator operations, and applicable electrical codes; correctly install welded fuel systems; and use power plant safety rules and regulations.

Instruction: Practical and classroom exercises and lecture. Topics include infrared survey; transformer, maintenance switch gear, circuit breaker maintenance, cable splicing and testing, grounding concepts, welding, safety, and deployment.

Credit Recommendation: In the lower-division baccalaureate/associate degree category, 2 semester hours in power distribution maintenance practicum, and 2 in industrial safety (flammable/hazardous materials) (8/03).

AR-1714-0043

Mast and Electric Power Plant Maintenance

Course Number: 662-ASIC9.

Location: Ordnance Center and School, Aberdeen Proving Ground, MD.

Length: 4–5 weeks (155 hours).

Exhibit Dates: 1/98–Present.

Learning Outcomes: Upon completion of the course, the student will be able to perform maintenance on the hydraulic and pneumatic mast group, and on an electric power plant and its controls.

Instruction: Practical exercises, classroom exercises, discussion, laboratory, and lecture. Topics include maintenance of hydraulic and pneumatic systems of the mast group, an electric power plant, and its AC/DC controls.

Credit Recommendation: In the lower-division baccalaureate/associate degree category, 2 semester hours in hydraulic-pneumatic systems maintenance and 2 in electrical systems maintenance (8/03).

AR-1715-0047

Satellite Communications (SATCOM) Terminal AN/TSC-54 Repair

Course Number: 102-F6.

Location: Signal School, Fort Gordon, GA; Signal School, Fort Monmouth, NJ.

Length: 8–10 weeks (298–366 hours).

Exhibit Dates: 8/68–1/92.

Objectives: To train enlisted personnel to inspect, test, and repair a satellite communications terminal.

Instruction: Lectures and practical exercises cover satellite communications terminal inspection, testing, and repair, including system familiarization, transmitter and receiver operation and repair, and antenna system inspection and repair.

Credit Recommendation: Credit is not recommended because of the limited, specialized nature of the course (8/79).

Related Occupations: 26Y.

AR-1715-0053

Multichannel Communications Systems Operator
(Multichannel Communications Equipment Operator)

Course Number: 202-31M10; 31M10-CELT; 202-31M20; 202-293.1; 11-R-293.1; 11-E-46.

Location: Signal School, Fort Gordon, GA.

Length: 8–16 weeks (280–714 hours).

Exhibit Dates: 4/57–9/95.

Objectives: To train personnel to install, operate, and maintain field radio relay and carrier systems and associated equipment.

Instruction: Lectures and practical exercises cover installation, operation, and maintenance of radio sets, telephone systems, telegraph systems, antennas and generators, FDM systems, and PMC sets; area communications system equipment maintenance; electronic warfare; and field exercises.

Credit Recommendation: Credit is not recommended because of the military-specific nature of the course (6/90).

Related Occupations: 31L; 31M.

AR-1715-0060

1. Special Forces Base Communications Operator
(Special Forces Base Communications Systems)
2. Special Operations Communications System Operator

Course Number: *Version 1:* 101-F12. *Version 2:* 101-F12.

Location: *Version 1:* Institute for Military Assistance, Fort Bragg, NC. *Version 2:* John F. Kennedy Special Warfare Center and School, Fort Bragg, NC.

Length: *Version 1:* 6–7 weeks (187–264 hours). *Version 2:* 8 weeks (320 hours).

Exhibit Dates: *Version 1:* 8/72–5/96. *Version 2:* 6/96–Present.

Learning Outcomes: *Version 1:* Upon completion of the course, the student will be able to install, operate, and perform preventive maintenance on Special Forces base communications systems. *Version 2:* Upon completion of the course, the student will be able to install and operate special operation communications systems.

Instruction: *Version 1:* Lectures and practical exercises cover the operation and components of communications systems, including radio teletypewriter and radio set, tuning and cording, communications central equipment, and message center group. Technical training is limited to that necessary to perform preventive maintenance and to interconnect components for optimum performance. Military considerations are emphasized throughout. *Version 2:* Lecture and practical exercises cover the installation, operations, troubleshooting, and maintenance of communications systems used in special operations.

Credit Recommendation: *Version 1:* In the vocational certificate category, credit in electrical laboratory on the basis of institutional evaluation (5/93). *Version 2:* In the lower-division baccalaureate/associate degree category, 3 semester hours in electronic communications systems (4/01).

Related Occupations: 31E; 31Z; 32A; 72B.

AR-1715-0084

HERCULES ELECTRONIC MECHANIC
(Nike Hercules Electronic Maintenance)

Course Number: 121-24U10.

Location: Air Defense Artillery School, Fort Bliss, TX.

Length: 8–9 weeks (316 hours).

Exhibit Dates: 12/85–6/94.

Objectives: To train warrant officers and enlisted personnel to assemble, install, maintain, calibrate, and repair Nike Hercules guided missiles, associated testing and handling equipment, and launch equipment.

Instruction: Lectures and laboratories cover basic electricity; basic electronics; radio circuits; warhead familiarization, assembly, and servicing; and system malfunction analysis. Course includes a study of the fundamentals of AC/DC circuits and a study of the Hercules guided missile system.

Credit Recommendation: In the lower-division baccalaureate/associate degree category, 2 semester hours in AC/DC circuits (11/86).

Related Occupations: 221B; 221C; 22F; 24U.

AR-1715-0091

UNIT LEVEL COMMUNICATIONS MAINTAINER
(Tactical Communications Systems Operator/Mechanic)

Course Number: 101-31V10.

Location: Field Artillery School, Fort Sill, OK.

Length: 13–14 weeks (490 hours).

Exhibit Dates: 10/88–12/93.

Objectives: To train enlisted personnel to install, operate, and maintain radio transmitters and receivers.

Instruction: Lectures and practical exercises cover radio transmitter, receiver, and antenna fundamentals; operation and maintenance of AM, FM, and single-sideband radio sets; and radio equipment security procedures. Instruction includes Ohm's law, multimeters to check circuit components and continuity, soldering skills, and considerable emphasis on hands-on training in troubleshooting techniques and maintenance procedures.

Credit Recommendation: In the vocational certificate category, 3 semester hours in radio repair. In the lower-division baccalaureate/associate degree category, 1 semester hour in electronics laboratory based on institutional evaluation (6/89).

Related Occupations: 31B; 31E; 31G; 31V.

AR-1715-0120

CHAPARRAL SYSTEM MECHANIC

Course Number: 121-24N10.

Location: Air Defense Artillery School, Fort Bliss, TX.

Length: 13–17 weeks (478–608 hours).

Exhibit Dates: 5/85–8/93.

Objectives: After 8/93 see AR-1715-0956. To train enlisted personnel to maintain Chaparral weapon systems and associated equipment.

Instruction: Lectures and demonstrations cover the organizational maintenance of equipment assemblies and subassemblies of the Chaparral missile system.

Credit Recommendation: In the vocational certificate category, 4 semester hours in maintenance of electromechanical equipment (2/90).

Related Occupations: 09B; 24N.

AR-1715-0121

BALLISTIC METEOROLOGY CREWMAN, SPECIALIST CANDIDATE
(Ballistic Meteorology Crewman, Skill Development Base)

Course Number: 420-93F20-1.

Location: Field Artillery School, Fort Sill, OK.

Length: 11–12 weeks (394–455 hours).

Exhibit Dates: 2/69–1/92.

Objectives: To train ballistic meteorology crewmen as supervisors or technicians on ballistic meteorology crews.

Instruction: Lectures and practical exercises cover meteorology, meteorological equipment, maintenance procedures, and tactical employment of meteorology sections. The course also includes leadership topics.

Credit Recommendation: Credit is not recommended because of the limited, specialized nature of the course (3/74).

Related Occupations: 93F.

AR-1715-0122

PERSHING ELECTRICAL-MECHANICAL REPAIR
(Ballistic Missile Electrical Mechanical Repair (Pershing))

Course Number: 631-46N10.

Location: Missile and Munitions School, Redstone Arsenal, AL.

Length: 13–18 weeks (488–639 hours).

Exhibit Dates: 9/78–9/90.

Objectives: To train enlisted personnel to inspect, test, and repair Pershing missile electrical, mechanical, and hydraulic systems and associated test and ground support equipment.

Instruction: Lectures and laboratory exercises include basic electricity; concepts of work, energy, and power; DC and AC circuit theory; series, parallel, and series-parallel circuits; inductive, capacitive, and LCR circuits; soldering techniques and practices; basic manual skills; missile hydraulic systems; and associated test and ground support equipment.

Credit Recommendation: In the vocational certificate category, 5 semester hours in electronic repair techniques. In the lower-division baccalaureate/associate degree category, 1 semester hour in basic electricity/electronic fundamentals (10/87).

Related Occupations: 46N.

AR-1715-0157

MARINE RADAR OBSERVER

Course Number: 8C-F10/062-F3; 8C-F10; 062-F3; 8C-F1; 813-F1.

Location: Transportation School, Fort Eustis, VA.

Length: 2 weeks (72–84 hours).

Exhibit Dates: 10/72–9/91.

Objectives: After 9/91 see AR-1715-0919. To provide watercraft operators with a working knowledge of the fundamentals in the operation, use, interpretation, and analysis of radar.

Instruction: Training includes principles of radar, radar navigation and plotting, and troubleshooting.

Credit Recommendation: In the vocational certificate category, 1 semester hour in radar operation (3/88).

Related Occupations: 61B.

AR-1715-0169

FORWARD AREA ALERTING RADAR (FAAR) REPAIRER
(Forward Area Alerting Radar (FAAR) Repair)

Course Number: 104-27N10.

Location: Missile and Munitions School, Redstone Arsenal, AL.

Length: 25 weeks (933–951 hours).

Exhibit Dates: 3/81–1/96.

Objectives: To train enlisted personnel to perform direct and general support maintenance on specific radar systems and associated equipment.

Instruction: Lectures and practical exercises cover maintenance procedures for the FAAR, including system logic; power distribution system; receiving, display, and data link systems; circuit analysis; various testing systems; stimulus generator; and system troubleshooting. Topics also include DC/AC electricity and microwave and digital circuit fundamentals.

Credit Recommendation: In the lower-division baccalaureate/associate degree category, 3 semester hours in electronic equipment maintenance, 3 in basic electricity/electronics, 1 in microwave fundamentals, and 2 in fundamentals of computer circuits (10/88).

Related Occupations: 26C; 26W; 27N.

AR-1715-0178

GROUND SURVEILLANCE SYSTEMS OPERATOR
(GROUND SURVEILLANCE RADAR CREWMAN)

Course Number: 221-17K10; 243-96R10.

Location: Intelligence School, Fort Huachuca, AZ.

Length: 5–7 weeks (168–278 hours).

Exhibit Dates: 2/80–9/99.

Learning Outcomes: After 9/99 see AR-1715-1069. Upon completion of the course, the student will be able to operate ground surveillance radar equipment.

Instruction: Lectures, practical exercises, and demonstrations cover specific ground surveillance radar equipment operation and maintenance procedures.

Credit Recommendation: In the vocational certificate category, 1 semester hour in logical troubleshooting techniques. In the lower-division baccalaureate/associate degree category, 1 semester hour in electromechanical maintenance and 1 in map interpretation (5/01).

Related Occupations: 17K; 96R.

AR-1715-0195

Secure Voice Access System Repair

Course Number: 160-ASIK8; 160-F37.

Location: Signal School, Fort Gordon, GA; Signal School, Fort Monmouth, NJ.

Length: 11–12 weeks (424 hours).

Exhibit Dates: 9/84–11/91.

Objectives: To train dial central office repairmen to maintain the automatic dial central office and the secure voice access console.

Instruction: Lectures and practical exercises cover automatic dial control office equipment operation, logic gating and correlation, simulators, controllers, maintenance panel operation, and troubleshooting procedures; secure voice access console, basic circuits and logic functions, and power distribution; and maintenance, alignment, and troubleshooting procedures.

Credit Recommendation: In the lower-division baccalaureate/associate degree category, 3 semester hours in troubleshooting techniques (10/90).

Related Occupations: 36H.

AR-1715-0199

Ground Control Radar Repair

Course Number: 104-26D10.

Location: Signal School, Fort Gordon, GA.

Length: 32 weeks (1200 hours).

Exhibit Dates: 11/79–6/96.

Objectives: To train enlisted personnel to maintain ground control approach radars and associated IFF equipment.

Instruction: Course includes lectures and practical experience in the general operation, malfunction analysis, and repair procedures of a ground control radar system. Course includes instruction on solid state power supplies, audio amplifiers, oscillators, pulse circuitry, standard digital logic, microwave transmitters, modulators, and receivers. This basic background knowledge is used to study, in particular, specific ground control radar sets, interrogators to identify specific aircraft, navigational beacons, and moving target circuitry. Instruction and practice in the repair of printed circuit boards is a part of this course.

Credit Recommendation: In the lower-division baccalaureate/associate degree category, 3 semester hours in basic electronics and electricity, 4 in theory of microwave transmission, and 6 in microwave system analysis (11/81).

Related Occupations: 09B; 26D; 26W.

AR-1715-0204

Nuclear Weapons Specialist
(Nuclear Weapons Maintenance Specialist)
(Nuclear Weapons Maintenance)

Course Number: 644-55G10 (ST); 644-55G10; 644-55G20.

Location: Missile and Munitions School, Redstone Arsenal, AL.

Length: 6–13 weeks (204–512 hours).

Exhibit Dates: 6/72–1/96.

Objectives: To train enlisted personnel to assemble, disassemble, maintain, test, and inspect nuclear weapons and to follow proper procedures in nuclear emergencies.

Instruction: Lectures and practical exercises cover the assembly, disassembly, maintenance, testing, and repair of nuclear weapons and procedures to follow in nuclear emergency situations. Course includes nuclear fundamentals, principles and components of nuclear weapons, electrical systems, electrical test equipment, care and handling of nuclear weapons, common shop operations, various examining procedures, emergency team operations, effects of nuclear weapons, and emergency destruction.

Credit Recommendation: Credit is not recommended because of the limited, specialized nature of the course (4/91).

Related Occupations: 09B; 55G.

AR-1715-0226

Communications and Electronics Staff Officer
(Communication/Electronic Staff Officer)

Course Number: 4C-ASI6B; 4C-25A.

Location: Field Artillery School, Fort Sill, OK; Artillery and Missile School, Fort Sill, OK.

Length: 9–10 weeks (281–364 hours).

Exhibit Dates: 12/79–12/94.

Objectives: To train signal officers to supervise and coordinate the installation, operation, and maintenance of communications systems and selected electronic equipment.

Instruction: Lectures and practical exercises cover supervision of the installation, operation, and maintenance of communications systems and selected electronic equipment, including AM and FM radio fundamentals and equipment, applied communications, radiotelephony, automatic data processing equipment, tactics, ground surveillance equipment, and management procedures.

Credit Recommendation: In the vocational certificate category, 1 semester hour in communications system installation. In the lower-division baccalaureate/associate degree category, 2 semester hours in communications equipment maintenance management, 1 in communications systems planning, and 1 in technical writing (6/89).

AR-1715-0250

Field Artillery Radar Crew Member
(Field Artillery Radar Operator)
(Field Artillery Radar Crewman)

Course Number: 221-17B10; 221-17B10 (ST); 221-17B20; 6-R-156.1; 6-R-211.1; 6-E-22.

Location: Field Artillery School, Fort Sill, OK.

Length: 7–8 weeks (262 hours).

Exhibit Dates: 10/84–1/96.

Objectives: To train enlisted personnel to operate and perform preventive maintenance on field artillery radar.

Instruction: Lectures and practical exercises cover field artillery radar operation and preventive maintenance, including radar site evaluation; equipment emplacement, adjustment, calibration, and orientation; radar gunnery exercises; maintenance procedures; and map reading and plotting routines.

Credit Recommendation: In the vocational certificate category, 1 semester hour in map reading and 1 in radar operations (11/85).

Related Occupations: 09B; 17B.

AR-1715-0253

Digital Subscriber Terminal Equipment (DSTE) Repairer

Course Number: 150-29G10; 150-34F20.

Location: Signal School, Fort Gordon, GA.

Length: 23–24 weeks (849–853 hours).

Exhibit Dates: 12/77–1/90.

Objectives: To train senior technicians to operate, maintain, and repair subscriber terminals of digital communications systems.

Instruction: Practical exercises cover the troubleshooting and repair of digital subscriber terminal equipment.

Credit Recommendation: In the vocational certificate category, 6 semester hours in computer peripheral equipment repair (10/86).

Related Occupations: 09B; 29G; 34D; 34F.

AR-1715-0299

1. TOW/Dragon Repairer
2. Land Combat Electronic Missile System Repairer
3. Land Combat Electronic Missile System Repairer

Course Number: ***Version 1:*** 121-27E10/121-27E10 (ST). *Version 2:* 121-27E10. *Version 3:* 121-27E10.

Location: ***Version 1:*** Missile and Munitions School, Redstone Arsenal, AL. *Version 2:* Ordnance, Missile and Munitions School, Redstone Arsenal, AL. *Version 3:* Ordnance, Missile and Munitions School, Redstone Arsenal, AL.

Length: ***Version 1:*** 18–20 weeks (672–714 hours). *Version 2:* 16 weeks (627 hours). *Version 3:* 24 weeks (864 hours).

Exhibit Dates: ***Version 1:*** 9/83–9/95. *Version 2:* 10/95–6/01. *Version 3:* 7/01–Present.

Learning Outcomes: ***Version 1:*** Upon completion of the course, the student will be able to repair wire guided missile systems. *Version 2:* Upon completion of the course, the student will be able to diagnose, repair, and maintain TOW and Dragon guided missile systems, Bradley fighting vehicle, TOW/TOW 2 subsystem, related night sights, and ancillary test equipment. *Version 3:* Upon completion of the course, the student will be able to perform maintenance on the TOW/TOW 2 and Dragon guided missile systems, the Bradley fighting vehicle, the Improved Target Aquisition System, the Javelin system, and ancillary test equipment. The student will demonstrate competence in AC/DC theory, transistor circuit theory, logic circuits, mechanical hand tools, and the use of various electronic test equipment.

Instruction: ***Version 1:*** Lectures and practical exercises cover sinusoidal AC and DC circuits, multimeter operation, transients in RLC circuits, power supplies, basic transistors, transistor amplifiers, FETs, and oscillators. Digital electronics includes number systems, arithmetic, gates, and flip-flops. This course includes a section on precision soldering. Maintenance, troubleshooting, and repair of the TOW system are also covered. *Version 2:* This course provides instruction through lectures, practical exercises, and demonstrations on DC circuit theory; AC circuit analysis; basic electronics; diodes; transistors; power supplies; soldering; and maintenance, troubleshooting, and repair of land combat electronic missile system launchers. *Version 3:* Instruction is provided using lectures, practical exercises, examinations, and computer-assisted instruction modules. Practical exercises include a hands-on laboratory using various electronic meters and mechanical hand tools. The class course is a field exercise where the student performs authorized levels of maintenance and repair.

Credit Recommendation: *Version 1:* In the lower-division baccalaureate/associate degree category, 2 semester hours in precision soldering techniques, 3 in introduction to AC/DC circuits, 1 in introduction to digital circuits, 3 in solid state circuitry, and 3 in electronic systems maintenance (12/88). *Version 2:* In the lower-division baccalaureate/associate degree category, 3 semester hours in introduction to AC/DC circuits, 1 in soldering techniques, 1 in basic electronics, and 3 in electronic system maintenance (7/96). *Version 3:* In the lower-division baccalaureate/associate degree category, 5 semester hours in AC/DC circuits, 2 in electronic circuits, 3 in logic circuit theory, and 3 in electronic systems maintenance (6/01).

Related Occupations: 27E.

AR-1715-0300

Radar Engagement Simulator AN/TPQ-29

Course Number: 121-ASIG6.

Location: Air Defense Artillery School, Fort Bliss, TX.

Length: 9 weeks (329 hours).

Exhibit Dates: 2/83–1/93.

Objectives: To train enlisted personnel to operate, adjust, and maintain the Hawk simulator system.

Instruction: Lectures and practical exercises cover the maintenance of the Hawk simulator system. Topics include AC and DC control systems, electronic test equipment, basic computer logic, radar fundamentals, and circuit analysis. Course also includes theory of operation, operational checks, alignment, and organizational maintenance of the Hawk missile simulator system.

Credit Recommendation: In the lower-division baccalaureate/associate degree category, 2 semester hours in electronic systems troubleshooting and maintenance (11/88).

Related Occupations: 24E.

AR-1715-0303

Shillelagh Repairer

(Shillelagh Missile System Repair)

Course Number: 121-ASIL9.

Location: Missile and Munitions School, Redstone Arsenal, AL.

Length: 6 weeks (216–217 hours).

Exhibit Dates: 1/86–1/96.

Objectives: To train enlisted personnel to inspect, test, and maintain Shillelagh missile systems and associated support equipment.

Instruction: Lectures and practical exercises cover Shillelagh fundamentals, control circuits, fault locater, range selector, transmitter, and modulator troubleshooting and repair.

Credit Recommendation: In the lower-division baccalaureate/associate degree category, 3 semester hours in electronic equipment maintenance (4/86).

Related Occupations: 27H; 27Z.

AR-1715-0324

Pershing Officer

(Pershing II Officer)

Course Number: 2F-13C; 2F-1190P.

Location: Field Artillery School, Fort Sill, OK.

Length: 5–8 weeks (206–297 hours).

Exhibit Dates: 2/72–12/90.

Objectives: To train officers to supervise the maintenance and operation of the Pershing missile system.

Instruction: Lectures and practical exercises cover supervision of the maintenance and operation of the Pershing missile system and its associated communications equipment. Topics include missile assembly and disassembly and firing section operations.

Credit Recommendation: Credit is not recommended because of the limited, technical nature of the course (2/88).

AR-1715-0382

Hawk Officer

(Improved Hawk Officer)

Course Number: 2F-14F (PIP II); 2F-14D; 2F-1180C.

Location: Air Defense Artillery School, Fort Bliss, TX.

Length: 6–19 weeks (304–688 hours).

Exhibit Dates: 6/73–9/91.

Objectives: After 9/91 see AR-1715-0959. To provide commissioned officers with knowledge of the characteristics, capabilities, functions, and maintenance of a radar-controlled missile system.

Instruction: Instruction includes electronic warfare, missile system operation, and system components.

Credit Recommendation: Credit is not recommended because of the military-specific nature of the course (10/91).

AR-1715-0397

Hawk Radar Signal Simulator Station Repairer

(Improved Hawk Radar Signal Simulator Station Repairer)

Course Number: 104-ASIW2; 104-F15.

Location: Missile and Munitions School, Redstone Arsenal, AL.

Length: 9 weeks (221–356 hours).

Exhibit Dates: 4/84–9/90.

Objectives: To provide the knowledge required for performing support maintenance on the electronic systems of the Hawk simulator station and associated test equipment.

Instruction: Lectures and practical exercises cover the inspection, testing, and repair of electronic circuits including power distribution circuits, target simulation circuits, oscillators, radar trigger and video synchronizers, and pulse signal generation.

Credit Recommendation: In the lower-division baccalaureate/associate degree category, 3 semester hours in electronic equipment repair (2/89).

Related Occupations: 24H.

AR-1715-0434

Combat Signaler

(Tactical Wire Operations Specialist)

Course Number: 621-31K10; 621-36K10-OSUT.

Location: Signal School, Fort Gordon, GA.

Length: 8–12 weeks (254–496 hours).

Exhibit Dates: 9/78–12/92.

Objectives: Provide enlisted personnel with a general working knowledge of the installation and maintenance of field wave lines, tactical switchboards, and radio systems.

Instruction: Course includes installation, maintenance, and operation of field wire switchboards and telephones. Student installs and repairs line systems in the field. Some individuals will complete basic training through One Station Unit Training (OSUT). This course includes an 8-9 week long Basic Training component.

Credit Recommendation: In the vocational certificate category, 4 semester hours in lineman-installer and 2 in switchboard operator (10/89).

Related Occupations: 31K; 36K.

AR-1715-0438

Satellite Communications (SATCOM) Operator, Navy

Course Number: 102-F43; 102-F43 (26Y); 102-F43 (26Y)/102-F43.

Location: Signal School, Fort Gordon, GA.

Length: 8–11 weeks (320–370 hours).

Exhibit Dates: 1/85–5/98.

Objectives: After 12/98 see AR-1715-1015. To provide personnel with skills required to operate a satellite communications terminal and a digital communications subsystem.

Instruction: Course includes principles of satellite communications including digital communications and terminal subsystems.

Credit Recommendation: In the vocational certificate category, 3 semester hours in satellite communications operation (2/94).

Related Occupations: 26Y.

AR-1715-0486

Pershing II Missile Crewmember

(Pershing Missile Crewmember)

Course Number: 043-15E10; 15E10-OSUT.

Location: Field Artillery School, Fort Sill, OK.

Length: 5–14 weeks (186–223 hours).

Exhibit Dates: 3/84–12/90.

Objectives: To provide the student with the skills, knowledge, and experience required to perform the duties of a Pershing missile crewmember.

Instruction: The student is instructed in the procedures required to set up a Pershing missile system and perform daily maintenance. These tasks include assembling the missile sections and mounting them to the launcher, operating the power generator, and laying cables. This course may include 8 weeks of basic training.

Credit Recommendation: Credit is not recommended because of the military-specific nature of the course (6/89).

Related Occupations: 15E.

AR-1715-0488

Pershing System Maintenance

Course Number: 4F-214EB; 121-21G20.

Location: Field Artillery School, Fort Sill, OK.

Length: 7 weeks (255 hours).

Exhibit Dates: 8/77–1/92.

Objectives: To train advanced students in troubleshooting and repair techniques on both the manual and electrical components of the Pershing missile ground support equipment.

Instruction: Students are instructed in procedures for tracing AC and DC voltage levels and system signals to isolate faults and repair electrical subsystems. Approximately 25 percent of the course is devoted to working knowledge of electronic devices and circuitry, including DC waves and parallel circuits, AC and DC motor functional characteristics, solid state diode applications, transistor biasing and functional operation, and octal number conversions.

Credit Recommendation: In the vocational certificate category, 2 semester hours in electric and electronic circuits. In the lower-division baccalaureate/associate degree category, 2 semester hours as a technical elective for students in a nontechnical program (1/80).

Related Occupations: 21G.

AR-1715-0492

Tactical Communications Chief Basic Noncommissioned Officer (NCO)
(Tactical Communications Chief Basic Technical)
(Tactical Communications Chief)

Course Number: 101-31V30; 103-31G30.

Location: Field Artillery School, Fort Sill, OK.

Length: 9–11 weeks (331–371 hours).

Exhibit Dates: 11/82–9/91.

Objectives: After 9/91 see AR-1715-0915. To provide selected enlisted personnel with a working knowledge of those duties required to perform as tactical communications systems operators/mechanics.

Instruction: Topics include wire communications, FM and AM radio systems, communications procedures, electronic warfare, and a very basic overview of radio fundamentals. Most instruction centers around the planning of field exercises. This course contains a common core of leadership subjects.

Credit Recommendation: In the vocational certificate category, 2 semester hours in basic electronics. In the lower-division baccalaureate/associate degree category, see AR-1406-0090 for common core credit (5/88).

Related Occupations: 31G; 31V.

AR-1715-0493

Field Artillery Moving Target Locating Radar/Sensor Operator
(Field Artillery (AN/TPS-58B) Radar Operator)

Course Number: 221-ASIL3; 221-F6.

Location: Field Artillery School, Fort Sill, OK.

Length: 2–4 weeks (80–133 hours).

Exhibit Dates: 1/85–6/99.

Objectives: To train enlisted personnel to operate and maintain field artillery radar.

Instruction: Self-paced instruction and practical exercises cover field artillery radar operation and preventive maintenance and includes radar site evaluation and equipment emplacement, adjustment, calibration, and orientation.

Credit Recommendation: In the vocational certificate category, 2 semester hours in radar operations (6/89).

AR-1715-0498

OV/RV-1D (Mohawk) Systems Qualification

Course Number: 2C-15C/2B-ASIF1; 3A-F42.

Location: Intelligence School, Fort Huachuca, AZ.

Length: 7–8 weeks (216–268 hours).

Exhibit Dates: 10/84–5/91.

Objectives: After 5/91 see AR-1715-0912. To familiarize selected commissioned and warrant officers with all special electronic mission aircraft and to train rated aviation officers in two specific military aircraft (OV/RV-1D).

Instruction: Instruction covers the use of special airborne electronic equipment used in military intelligence and flight procedures related to the aircraft. Instruction is provided in the classroom and in flight.

Credit Recommendation: Credit is not recommended because of the military-specific nature of the course (6/90).

AR-1715-0499

Improved Hawk Information and Coordination Central Maintenance Transition (PIP)

Course Number: 104-24G1T/2T/4T; 104-24G1T/2T/3T/4T.

Location: Air Defense Artillery School, Fort Bliss, TX.

Length: 4–5 weeks (149–159 hours).

Exhibit Dates: 10/79–8/91.

Objectives: To provide personnel with knowledge of organizational maintenance for the product improvement program of the Improved Continuous Wave Acquisition Radar.

Instruction: This course provides lectures and laboratory instruction in theory of operation of equipment, use of test equipment and procedures, and troubleshooting at the block-diagram level.

Credit Recommendation: In the vocational certificate category, 3 semester hours in electronics laboratory (11/88).

Related Occupations: 24G.

AR-1715-0500

Aerial Electronic Warning/Defense Equipment Repairer

Course Number: 4C-F18; 102-ASIW6 (35K).

Location: Signal School, Fort Gordon, GA; Intelligence School, Fort Huachuca, AZ.

Length: 2–5 weeks (91–144 hours).

Exhibit Dates: 7/82–9/95.

Objectives: To train personnel in the maintenance of survivability equipment for military aircraft at the organizational or aviation unit maintenance level.

Instruction: This course teaches the necessary maintenance procedures to maintain infrared jammers and radar warning receivers at the block-diagram level.

Credit Recommendation: Credit is not recommended because of the military-specific nature of the course (7/92).

Related Occupations: 35K.

AR-1715-0501

Operation of the Automated Multimedia Exchange (AMME) System

Course Number: 260-ASIU3.

Location: Communications Electronics Engineering Installation Agency, Fort Huachuca, AZ.

Length: 2–6 weeks (80–200 hours).

Exhibit Dates: 12/79–4/04.

Objectives: To provide specialized operator training on the Automated Multimedia Exchange (AMME) system.

Instruction: This program includes lectures and practical exercises in the general operation of the Automatic Multimedia Exchange system. The course includes traffic flow, system configuration, equipment identification and operation, system initialization, message entry, and traffic management.

Credit Recommendation: Credit is not recommended because of the limited, specialized nature of the course (6/90).

AR-1715-0511

Hawk Missile Crewmember

Course Number: 16D10-OSUT, Phase 1; 16D10-OSUT; 043-16D10; 043-16D10-0SUT.

Location: Air Defense Artillery School, Fort Bliss, TX.

Length: 5–8 weeks (206–276 hours).

Exhibit Dates: 10/87–1/93.

Objectives: To train enlisted personnel as Hawk Missile crewmembers.

Instruction: Lectures and practical exercises cover the procedures and practices used in operating specific Hawk missile systems and equipment. This course includes an 8-9 week Basic Training component.

Credit Recommendation: Credit is not recommended because of the military-specific nature of the course (9/91).

Related Occupations: 16D.

AR-1715-0521

Improved Hawk Master Mechanic

Course Number: 121-24R50-B.

Location: Air Defense Artillery School, Fort Bliss, TX.

Length: 30–31 weeks (1064–1084 hours).

Exhibit Dates: 5/83–12/91.

Objectives: To train experienced enlisted personnel to perform operational and organizational checks and maintenance on the Improved Hawk Missile System.

Instruction: Lectures and practical exercises cover digital logic and computer fundamentals, including logic gates; computer flow charts; computer maintenance; operational checks; and fault isolation involving power distribution systems, computer control systems, continuous wave radar sets, pulsed wave radar sets, range-only radar sets, and other equipment relating to the Hawk missile support system.

Credit Recommendation: In the vocational certificate category, 10 semester hours in electronic equipment maintenance. In the lower-division baccalaureate/associate degree category, 3 semester hours in electronic troubleshooting and maintenance and 3 in radar systems maintenance and operations (11/84).

Related Occupations: 24R.

AR-1715-0529

Chaparral Redeye Repairer

Course Number: 121-27G10.

Location: Missile and Munitions School, Redstone Arsenal, AL.

Length: 28 weeks (1024–1026 hours).

Exhibit Dates: 1/84–1/96.

Objectives: To provide enlisted personnel with the skills and knowledge required to perform maintenance on the Chaparral and Redeye guided missile systems and ancillary test equipment.

Instruction: Lectures and practical experiences cover basic electricity, AC/DC circuits, transistors, linear and nonlinear devices, power supplies, amplifiers, sinusoidal and nonsinusoidal oscillators, logic circuits, and computer fundamentals. Practical experiences cover the troubleshooting and repair of Chaparral and Redeye systems. This course includes precision soldering.

Credit Recommendation: In the lower-division baccalaureate/associate degree category, 3 semester hours in introduction to AC/DC circuits, 2 in precision soldering, 1 in introduction to digital circuits, and 3 in solid state circuits (10/88).

Related Occupations: 27G.

AR-1715-0530

Ballistic/Land Combat Missile and Light Air Defense Weapons Systems Maintenance Noncommissioned Officer (NCO) Advanced

Course Number: 1-27-C42A.

Location: Missile and Munitions School, Redstone Arsenal, AL.

Length: 8–20 weeks (279–713 hours).

Exhibit Dates: 8/83–4/90.

Objectives: To provide selected enlisted personnel with career development training in general military subjects, specialty-related subjects, and logistics management.

Instruction: Course includes specific tracks in certified soldering, missile materiel management, Pershing electronics and ground support equipment, and general support and handling equipment for the LCSS Lance, TOW Dragon, Vulcan, Chaparral Redeye, Shillelagh, and FAAR systems.

Credit Recommendation: In the lower-division baccalaureate/associate degree category, 1 semester hour in electronics systems maintenance (4/86).

AR-1715-0549

Medium Capacity Multichannel Communication Systems
(Medium Capacity Multichannel Communication Systems Team Chief)

Course Number: SIG 2.

Location: Combined Arms Training Center, Flint Kaserne, Bad Toelz, Germany.

Length: 2 weeks (74 hours).

Exhibit Dates: 1/80–10/96.

Objectives: To provide students with the skills and knowledge necessary to supervise operators of a specific Army radio communications system.

Instruction: Instruction covers installing, operating, aligning, and performing preventive maintenance on the components of radio communications systems, including radio receivers, radio transmitters, and antennas.

Credit Recommendation: Credit is not recommended because of the military-specific nature of the course (10/82).

Related Occupations: 25A; 25D; 31M; 31Z.

AR-1715-0554

FM Radio Alignment
(Radio and Repair Alignment)

Course Number: SIG 35.

Location: Combined Arms Training Center, Flint Kaserne, Bad Toelz, Germany.

Length: 2 weeks (67–73 hours).

Exhibit Dates: 11/81–7/96.

Objectives: To provide radio repairers with the skills and knowledge required to align FM radio sets.

Instruction: Topics include the use of test equipment used to align FM radios, block diagrams of the radios, technical characteristics of the radios, and alignment procedures.

Credit Recommendation: In the vocational certificate category, 2 semester hours in FM radio alignment. In the lower-division baccalaureate/associate degree category, 1 semester hour in communications equipment alignment (10/82).

Related Occupations: 31E.

AR-1715-0555

AN/UGC-74 Teletypewriter General Support Maintenance

Course Number: UGC-1.

Location: Combined Arms Training Center, Flint Kaserne, Bad Toelz, Germany.

Length: 2 weeks (119 hours).

Exhibit Dates: 11/81–6/96.

Objectives: To provide teletypewriter maintenance personnel with the skills and knowledge necessary to troubleshoot and repair teletype communications terminal equipment.

Instruction: Topics include installing, operating, and maintaining teletypewriter systems. Maintenance procedures include module replacement, disassembly and reassembly, testing and fault isolation, alignment procedures, and the sending and receiving of test messages between two terminals.

Credit Recommendation: Credit is not recommended because of the military-specific nature of the course (10/82).

Related Occupations: 31J.

AR-1715-0556

AN/UGC-74 Teletypewriter Direct Support Maintenance

Course Number: UGC-2.

Location: Combined Arms Training Center, Flint Kaserne, Bad Toelz, Germany.

Length: 2 weeks (80 hours).

Exhibit Dates: 11/81–6/96.

Objectives: To provide teletypewriter maintenance personnel with the skills and knowledge necessary to troubleshoot and repair teletype communications terminal equipment AN/UGC-74(V)3.

Instruction: Topics include installing, operating, and maintaining UGC-1 teletypewriter systems. Maintenance procedures include module replacement, disassembly and reassembly, testing and fault isolation, alignment procedures, and the sending and receiving of test messages between two terminals.

Credit Recommendation: Credit is not recommended because of the military-specific nature of the course (10/82).

Related Occupations: 31J.

AR-1715-0557

AN/MYQ

Course Number: ADP-15.

Location: Combined Arms Training Center, Zweibruecken, Germany.

Length: 3 weeks.

Exhibit Dates: 10/82–9/96.

Objectives: To train personnel to operate and perform routine operator maintenance on a minicomputer system.

Instruction: Lectures, videotapes, and hands-on experience cover operation and routine operator maintenance of the AN/MYQ-4 computer system. Topics include printers, tape drives, card readers, disk drives, video terminals, system power-up, operation control panel, peripheral equipment, error recovery, performance examination, and save and restore procedures.

Credit Recommendation: In the vocational certificate category, 4 semester hours in minicomputer system operations. In the lower-division baccalaureate/associate degree category, 1 semester hour in survey of computer hardware (10/82).

AR-1715-0558

Target Acquisition Radar Technician-Warrant Officer Technical Certification
(Target Acquisition Radar Technician)

Course Number: 4C-211A.

Location: Field Artillery School, Fort Sill, OK.

Length: 27 weeks (1061 hours).

Exhibit Dates: 7/85–6/96.

Objectives: To provide radar warrant officers with tactical and technical skills to locate, operate, and repair radar sets.

Instruction: Instruction includes theory of radar operations, weapon location, radar set operation, electronic circuits, radar maintenance and repair, and electronic systems diagnosis and repair.

Credit Recommendation: In the lower-division baccalaureate/associate degree category, 6 semester hours in basic electricity and electronics, 3 in electronic devices, 3 in digital circuit theory, 3 in digital computer fundamentals, 3 in radar theory, and 3 in system maintenance and troubleshooting (11/85).

Related Occupations: 211A.

AR-1715-0562

Lance System Repairer
(Lance Repairer)

Course Number: 121-27L10.

Location: Missile and Munitions School, Redstone Arsenal, AL.

Length: 16–18 weeks (622 hours).

Exhibit Dates: 9/82–1/96.

Objectives: To train enlisted personnel to perform direct and general support maintenance on Lance Missile systems (less carrier and warhead) and on associated test equipment.

Instruction: Lectures and practical experience cover basic electricity, DC/AC circuits, series/parallel circuits, linear and nonlinear devices, transistors, amplifiers, power supplies, and use of test equipment, including multimeters, oscilloscopes, and function generators. Included are practical experiences in testing, troubleshooting, and repairing electronic subsystems.

Credit Recommendation: In the lower-division baccalaureate/associate degree category, 3 semester hours in basic electricity/electronics, 1 in electronic shop practices, and 3 in electronic equipment repair (10/88).

Related Occupations: 27L.

AR-1715-0564

1. Multiple Launch Rocket System (MLRS) Repairer
2. Multiple Launch Rocket System (MLRS) Repairer
3. Multiple Launch Rocket System (MLRS) Repairer
4. Multiple Launch Rocket System Repairer

Course Number: ***Version 1:*** 611-27M10; 121-27M10; 632-27M10. *Version 2:* 121-27M10. *Version 3:* 121-27M10. *Version 4:* 121-27M10.

Location: ***Version 1:*** Missile and Munitions School, Redstone Arsenal, AL. *Version 2:* Ordnance, Missile and Munitions School, Redstone Arsenal, AL. *Version 3:* Ordnance, Missile and Munitions School, Redstone Arsenal, AL. *Version 4:* Ordnance, Missile and Munitions School, Redstone Arsenal, AL.

Length: ***Version 1:*** 18–19 weeks (682–684 hours). *Version 2:* 14–15 weeks (597 hours). *Version 3:* 23–24 weeks (823 hours). *Version 4:* 18 weeks (680 hours).

Exhibit Dates: ***Version 1:*** 12/82–9/95. *Version 2:* 10/95–9/00. *Version 3:* 10/00–12/01. *Version 4:* 1/02–Present.

Learning Outcomes: ***Version 1:*** Upon completion of the course, the student will be able to perform direct and general support maintenance on the multiple launch rocket system. *Version 2:* Upon completion of the course, the student will be able to perform maintenance and repair on the multiple launch rocket system. *Version 3:* Upon completion of the course, the student will be able to perform direct and general support maintenance on the multiple launch rocket system. *Version 4:* Upon completion of the course, the student will be able to inspect, test, and adjust system components to specified tolerance for a rocket launch system.

Instruction: ***Version 1:*** Lectures and practical exercises cover basic electricity, circuits, devices, test equipment, and maintenance practices. Course also includes extensive training in analysis of circuit diagrams, schematics, and electronic equipment using multimeters, oscilloscopes, and function generators. Topics include series and parallel resonance circuits, power supplies, transistor amplifiers, relays, and switches. *Version 2:* Lectures and practical exercises ensure comprehension of the subject matter. Topics include DC circuits, basic soldering, hydraulics and gear train systems, test equipment, and maintenance practices. *Version 3:* Lectures and practical exercises basic electricity, soldering techniques, through-hole soldering cover and wire wrap. *Version 4:* Computer-based training, practical and classroom exercises, and laboratory. Topics include fundamentals of DC/AC circuit characteristics, operation and troubleshooting procedures; introduction to RL and RC transients and filters practical exercises to measure time constants using multimeter, function generator; oscilloscope, and soldering techniques.

Credit Recommendation: ***Version 1:*** In the lower-division baccalaureate/associate degree category, 3 semester hours in basic electricity/electronics, 1 in electronic shop practices, and 3 in electronic equipment maintenance (2/88). *Version 2:* In the lower-division baccalaureate/associate degree category, 1 semester hour in soldering, 2 in DC circuits, and 3 in electromechanical service and repair (10/00). *Version 3:* In the lower-division baccalaureate/associate degree category, 2 semester hours in AC/DC circuits, 1 in electronic shop practice, 3 in electronic systems troubleshooting and maintenance, and 1 in soldering techniques (6/01). *Version 4:* In the lower-division baccalaureate/associate degree category, 2 semester hours in DC/AC circuits, 1 in electronic shop practice, and 1 in electronic system troubleshooting and maintenance (2/04).

Related Occupations: 27M.

AR-1715-0565

Equipment Operation/Maintenance Test Set Guided Missile Set (TOW Ground)
(TOW Field Test Set Repairer)

Course Number: 121-ASIJ9; 121-F28.

Location: Missile and Munitions School, Redstone Arsenal, AL.

Length: Self-paced, 3–4 weeks (114–128 hours).

Exhibit Dates: 7/81–9/95.

Objectives: After 9/95 see AR-1715-1000. To provide students with the skills and knowledge required to operate, troubleshoot, and repair the TOW field test set.

Instruction: This self-paced course covers the theory and operation of the TOW field test set and ancillary equipment. Tools, test equipment, and service manuals used in the repair of power supplies and controller are also incorporated into this course.

Credit Recommendation: In the lower-division baccalaureate/associate degree category, 1 semester hour in electronic laboratory and 2 in electronic equipment maintenance (11/83).

Related Occupations: 27E.

AR-1715-0566

Field Artillery Computer Repairer
(Field Artillery Digital Automatic Computer Radar Technician)

Course Number: 041-34Y10, Phase 1; 104-2885.

Location: Ordnance Center and School, Aberdeen Proving Ground, MD.

Length: 9–18 weeks (330–623 hours).

Exhibit Dates: 1/84–6/96.

Objectives: To train enlisted personnel to perform basic electronic maintenance and direct general support maintenance of field artillery digital automatic equipment and related test equipment.

Instruction: The course provides instruction to enlisted Army personnel in basic electronics and includes AC/DC circuits; fundamentals of electricity; color coding; series, parallel, and series-parallel circuits; batteries; multimeters; introduction to oscilloscopes; magnetism; inductance; generating AC; capacitance; resonance; transistor fundamentals, including transistors (NPN-PNP), power supplies, and rectifiers; basic digital circuits, including binary system, Boolean algebra, diode logic, and logic trainer; and precision soldering practices, including wire stripping, soldering various types of connections, and removing and replacing defective circuit components. Army and Marine personnel are instructed in the maintenance of signal data reproducers, including inspection, troubleshooting, and repair/replacement of frame-mounted components, main power circuit breakers, test sets, and computer logic units. Army and Marine Corps personnel also learn to test, troubleshoot, and repair computer gun direction, perform computer operational inspections, test and repair power circuits, replace memory units, and perform reliability tests.

Credit Recommendation: In the lower-division baccalaureate/associate degree category, for Army personnel, 5 semester hours in basic electricity, 3 in basic electronics, and 2 in basic computers; for Marine Corps personnel, 2 semester hours in basic computers (2/84).

Related Occupations: 34Y.

AR-1715-0568

RC-12D Improved Guardrail Systems Qualification (RC-12D (Improved Guardrail V) Systems Qualification)

Course Number: 2C-15C/2B-ASIF3; 3A-F45.

Location: Intelligence School, Fort Huachuca, AZ.

Length: 4–9 weeks (200–321 hours).

Exhibit Dates: 10/84–5/91.

Objectives: After 5/91 see AR-1715-0913. To familiarize selected commissioned and warrant officers with special electronic mission aircraft and train rated aviation officers in a specific military reconnaissance airplane (RC-12D).

Instruction: Instruction covers the use of special airborne electronic equipment used in military intelligence and flight operational procedures related to the aircraft. Instruction is provided in the classroom and in flight.

Credit Recommendation: Credit is not recommended because of the military-specific nature of the course (6/90).

AR-1715-0569

Guardrail V System Qualification (RU-21 (Guardrail V) (2R) (RU-21 (Guardrail)) Systems Qualification)

Course Number: 2C-15C/2B-ASIF2; 2B-ASI2K; 3A-F43.

Location: Intelligence School, Fort Huachuca, AZ.

Length: 5–7 weeks (200–243 hours).

Exhibit Dates: 10/84–4/04.

Objectives: To familiarize selected commissioned and warrant officers with special electronic mission aircraft and to train rated aviation officers in a specific reconnaissance airplane (RU-21H).

Instruction: Instruction covers the use of special airborne electronic equipment used in military intelligence and flight operational procedures related to the aircraft. Instruction is provided in the classroom and in flight.

Credit Recommendation: Credit is not recommended because of the military-specific nature of the course (6/90).

AR-1715-0574

Improved Hawk Organizational Maintenance Supervisor/Master Mechanic (PIP 2) Transition

Course Number: 4F-F15/121-24R50-T.

Location: Air Defense Artillery School, Fort Bliss, TX.
Length: 6–8 weeks (219–263 hours).
Exhibit Dates: 8/85–12/91.
Objectives: To provide warrant officers and senior enlisted personnel with the knowledge of organizational maintenance procedures utilized in the product improvement program.
Instruction: Course uses conferences and practical exercises to provide instruction in the maintenance of the HPI/RAM transmitters, receiver signal processor, and TAS Improved Hawk Missile System at the management level. Training is superficial in the area of circuitry.
Credit Recommendation: Credit is not recommended because of the military-specific nature of the course (11/88).
Related Occupations: 24R.

AR-1715-0576

STINGER GUNNER OPERATOR (USMC)
(Redeye and Stinger Gunner/Operator)

Course Number: 121-7212.
Location: Air Defense Artillery School, Fort Bliss, TX.
Length: 6 weeks (188–189 hours).
Exhibit Dates: 1/80–10/99.
Objectives: To provide training in the operation and employment of the Redeye and Stinger missile system in the Fleet Marine Forces.
Instruction: Instruction covers the organization of Stinger/Redeye platoons, aircraft identification, principles of operation of missile systems, operating procedures, employment of the system, communications procedures, and mapping procedures.
Credit Recommendation: Credit is not recommended because of the military-specific nature of the course (10/89).

AR-1715-0578

PATRIOT MISSILE CREWMEMBER

Course Number: 16T10-OSUT.
Location: Air Defense Artillery School, Fort Bliss, TX.
Length: 18–20 weeks (732–890 hours).
Exhibit Dates: 8/83–6/94.
Objectives: To train enlisted personnel to perform duties of a crewmember on the Patriot missile system.
Instruction: Instruction covers march order and emplacement of Patriot missile system-related equipment, survey functions, missile reload procedures, preventive maintenance procedures, launcher tests, logistics functions, communications procedures, and vehicle operations. This course includes an 8-9 week Basic Training component.
Credit Recommendation: In the vocational certificate category, 3 semester hours in electromechanical maintenance procedures (11/85).
Related Occupations: 16T.

AR-1715-0579

PATRIOT MISSILE CREWMEMBER
(Patriot Missile Crewmember Transition)

Course Number: 043-16T10; 043-16T1/2/3/4-T.
Location: Air Defense Artillery School, Fort Bliss, TX.
Length: 11–13 weeks (377–470 hours).
Exhibit Dates: 8/83–7/93.
Objectives: After 7/93 see AR-1715-0941. To train enlisted personnel to perform the duties of a crewmember on the Patriot missile system.
Instruction: Instruction covers march order and emplacement of all Patriot missile system-related equipment, survey functions, missile reload procedures, preventive maintenance procedures, launcher tests, logistic functions, communications procedures, and vehicle operations.
Credit Recommendation: In the vocational certificate category, 3 semester hours in electromechanical maintenance procedures (4/91).
Related Occupations: 16T.

AR-1715-0580

PATRIOT MISSILE CREWMEMBER

Course Number: 16T10-OSUT (24T), Phase 1; 16T10-OSUT, Phase 1.
Location: Air Defense Artillery School, Fort Bliss, TX.
Length: 14–15 weeks (548 hours).
Exhibit Dates: 8/83–6/94.
Objectives: To train entry-level personnel to perform duties of a crewmember on the Patriot missile system.
Instruction: Instruction covers march order and emplacement of Patriot missile system-related equipment, survey functions, missile reload procedures, preventive maintenance procedures, communications procedures, and light and heavy vehicle operation. This course includes an 8-9 week Basic Training component.
Credit Recommendation: In the vocational certificate category, 3 semester hours in electromechanical maintenance procedures (11/86).
Related Occupations: 16T; 24T.

AR-1715-0581

PATRIOT MISSILE CREWMEMBER TRANSITION

Course Number: 043-16T1/2/3/4-T (24T), Phase 1; 043-16T10 (24T), Phase 1.
Location: Air Defense Artillery School, Fort Bliss, TX.
Length: 7–8 weeks (286 hours).
Exhibit Dates: 8/83–6/96.
Objectives: To train enlisted personnel to perform the duties of a crewmember on the Patriot missile system.
Instruction: Instruction covers march order and emplacement of all Patriot missile system-related equipment, survey functions, missile reload procedures, preventive maintenance procedures, launcher tests, logistic functions, and communications procedures.
Credit Recommendation: In the vocational certificate category, 3 semester hours in electromechanical maintenance procedures (12/84).
Related Occupations: 16T; 24T.

AR-1715-0582

PATRIOT OPERATOR AND SYSTEM MECHANIC

Course Number: 632-24T10.
Location: Air Defense Artillery School, Fort Bliss, TX.
Length: 32–34 weeks (1150–1220 hours).
Exhibit Dates: 1/90–Present.
Learning Outcomes: Upon completion of the course, the student will be able to employ, maintain, and repair the Patriot missile system.
Instruction: Lectures and practical exercises cover operation, repair, and maintenance management of the Patriot missile system. Course includes operator repair procedures on, checks and adjustments, and fault isolation.
Credit Recommendation: In the lower-division baccalaureate/associate degree category, 3 semester hours in radar systems, 2 in AC/DC circuits, 2 in solid state electronics, and 3 in electronic systems troubleshooting and maintenance (6/95).
Related Occupations: 24T.

AR-1715-0585

FORWARD AREA ALERTING RADAR (FAAR)
ORGANIZATIONAL MAINTENANCE

Course Number: 121-ASIX7.
Location: Air Defense Artillery School, Fort Bliss, TX.
Length: 5–7 weeks (193–226 hours).
Exhibit Dates: 10/82–12/90.
Objectives: To provide personnel with the knowledge and skills needed to perform organizational maintenance on radar and associated equipment.
Instruction: This course, presented through lecture and laboratory training, teaches the students to energize radar and to troubleshoot and repair all radar components.
Credit Recommendation: In the vocational certificate category, 3 semester hours in radar system—repair and maintenance. In the lower-division baccalaureate/associate degree category, 3 semester hours in electronic troubleshooting and repair (9/89).

AR-1715-0594

COMMUNICATIONS SECURITY (COMSEC)
EQUIPMENT TSEC/KG-81 REPAIR

Course Number: 160-F20.
Location: Signal School, Fort Gordon, GA.
Length: 6–8 weeks (246–280 hours).
Exhibit Dates: 6/81–9/95.
Objectives: To provide enlisted personnel with the skills and knowledge to repair one specific piece of military electronics equipment.
Instruction: Student practices disassembly, reassembly, testing, and troubleshooting of a particular piece of communications security equipment.
Credit Recommendation: In the lower-division baccalaureate/associate degree category, 3 semester hours in systems maintenance (5/90).

AR-1715-0596

DS/GS TEST AND REPAIR STATION
MAINTENANCE (AH-64A)

Course Number: 198-ASIX1.
Location: Signal School, Fort Gordon, GA.
Length: 6–8 weeks (216–288 hours).
Exhibit Dates: 1/85–2/90.
Objectives: After 2/90 see AR-1715-0878. To provide enlisted personnel with skills necessary to perform maintenance and repair on a specific piece of military electronic equipment.
Instruction: Lectures and practicum cover troubleshooting the AH-64A using programmed diagnostic procedures for fault isolation and making appropriate repairs on all components of the system, such as test console, interface, video, pneumatic, and power subsystems.
Credit Recommendation: In the lower-division baccalaureate/associate degree category, 3 semester hours in diagnostic procedures (6/89).

AR-1715-0598

ELECTRONIC SWITCHING SYSTEMS TECHNICAL MANAGER

Course Number: 4C-F39; 4C-ASI3T; 4C-F20.
Location: Signal School, Fort Gordon, GA.
Length: 7–9 weeks (275–324 hours).
Exhibit Dates: 1/83–9/94.
Learning Outcomes: Upon completion of the course, students will be able to maintain an automatic telephone central office and automatic message switching control.
Instruction: This course provides a general familiarization of certain telephone central office equipment. Technical documentation and procedures to maintain this equipment are presented to provide a supervisor with sufficient knowledge to prepare, initiate, and direct support maintenance.
Credit Recommendation: Credit is not recommended because of the limited, specialized nature of the course (6/90).
Related Occupations: 290A.

AR-1715-0605

JOINT TACTICAL AUTOMATED SWITCHING NETWORK SUPERVISOR
(Joint Tactical Communications (TRI-TAC) Operations)

Course Number: 260-ASIK7.
Location: Signal School, Fort Gordon, GA.
Length: 6–9 weeks (244–324 hours).
Exhibit Dates: 12/82–6/95.
Objectives: To train enlisted personnel to plan and supervise the installation and operation of joint tactical communications equipment.
Instruction: Lectures and practical exercises cover preparing the diagrams for the installation and initialization of joint tactical communications equipment.
Credit Recommendation: Credit is not recommended because of the limited, specialized nature of the course (6/90).

AR-1715-0606

1. MULTICHANNEL COMMUNICATIONS SYSTEMS OPERATOR BASIC NONCOMMISSIONED OFFICER (NCO)
   (Multichannel Communications Equipment Operator Basic Noncommissioned Officer (NCO))
2. MULTICHANNEL TRANSMISSION SYSTEMS OPERATOR/MAINTAINER BASIC NONCOMMISSIONED OFFICER (NCO)

Course Number: ***Version 1:*** 202-31M20. *Version 2:* 202-31R30.
Location: Signal School, Fort Gordon, GA.
Length: ***Version 1:*** 5–7 weeks (191–234 hours). *Version 2:* 12 weeks (455–456 hours).
Exhibit Dates: ***Version 1:*** 3/85–9/94. *Version 2:* 1/96–Present.
Learning Outcomes: ***Version 1:*** Upon completion of the course, the student will be able to supervise, direct operation, and check maintenance of multichannel communications equipment. *Version 2:* Upon completion of the course, the student will be able to supervise, and direct operation and organizational maintenance of multichannel communications equipment.
Instruction: ***Version 1:*** Training includes supervision of installation, operation, and operator's/organizational maintenance of low- and medium-capacity multichannel communications equipment. This course contains a common core of leadership subjects. *Version 2:* Training includes supervision of installation, operation, and operator's/organizational maintenance of multichannel communications equipment. Specific content includes basic electronics, the multimeter, digital communications theory, switching equipment, system control, deployment of transmission systems, global positioning systems, communications security, logistics support, electronic warfare, encryption/decryption, node switching, digital group multiplexers and antennas. This course contains a common core of leadership subjects.
Credit Recommendation: ***Version 1:*** Credit is recommended for the common core only. See AR-1406-0090 (5/88). *Version 2:* Credit is recommended for the common core only. See AR-1406-0090 (2/97).
Related Occupations: 31M; 31R.

AR-1715-0607

SATELLITE COMMUNICATIONS (SATCOM) TERMINAL OPERATOR/MAINTAINER
(Satellite Communications (SATCOM) Terminal AN/MSC-64 General Support Maintenance)

Course Number: 101-ASIV9 (29E); 101-ASIV9.
Location: Signal School, Fort Gordon, GA.
Length: 2–7 weeks (101–244 hours).
Exhibit Dates: 3/82–9/95.
Objectives: To train enlisted personnel to perform ground support maintenance on a satellite communications terminal and associated equipment.
Instruction: Lectures and practical exercises cover maintenance procedures of the satellite communications terminal and various receivers and digital modems.
Credit Recommendation: Credit is not recommended because of the limited, specialized nature of the course (7/89).
Related Occupations: 29E.

AR-1715-0608

SATELLITE COMMUNICATIONS (SATCOM) TERMINAL OPERATOR/MAINTAINER

Course Number: 201-ASIV9 (31C); 201-ASIV9 (05C).
Location: Signal School, Fort Gordon, GA.
Length: 4–11 weeks (160–394 hours).
Exhibit Dates: 7/82–6/93.
Objectives: To train enlisted personnel to install, operate, and perform organizational maintenance on a satellite terminal.
Instruction: Lectures and practical exercises cover the installation, operation, and maintenance of a satellite communications terminal; installation of equipment as a system; and installation/deinstallation of components of the system including the satellite tracking antenna.
Credit Recommendation: Credit is not recommended because of the limited, specialized nature of the course (7/89).
Related Occupations: 05C; 31C.

AR-1715-0614

STANDARDIZED COMMUNICATIONS SECURITY (COMSEC) CUSTODIAN

Course Number: 4C-F22/160-F23 (F); SIG 34; 4C-F22/160-F23.
Location: Training Center, Yongsan, Korea; Training Centers, US; Training Centers, Europe; Signal School, Fort Gordon, GA.
Length: 1–2 weeks (64–75 hours).
Exhibit Dates: 12/83–Present.
Objectives: To provide enlisted and commissioned personnel who are COMSEC custodians with proper COMSEC accounting procedures.
Instruction: Instruction includes security measures, COMSEC accounts, emergency and maintenance procedures, and logistics for COMSEC.
Credit Recommendation: Credit is not recommended because of the military-specific nature of the course (2/03).

AR-1715-0617

TRI-TAC COMMUNICATIONS SECURITY (COMSEC) EQUIPMENT LIMITED MAINTENANCE (DIRECT SUPPORT)

Course Number: 160-F21.
Location: Signal School, Fort Gordon, GA.
Length: 7 weeks (254 hours).
Exhibit Dates: 1/84–12/92.
Objectives: To train students to repair equipment by substitution of pluggable assemblies.
Instruction: Students receive practice exercises in the troubleshooting of several pieces of military electronic equipment.
Credit Recommendation: Credit is not recommended because of the limited, specialized nature of the course (5/90).

AR-1715-0618

2. WIRE SYSTEMS INSTALLER BASIC NONCOMMISSIONED OFFICER (NCO)
   (Wire Systems Installer Primary Technical)
3. CABLE SYSTEMS INSTALLER-MAINTAINER BASIC NONCOMMISSIONED OFFICER (NCO)
4. CABLE SYSTEMS INSTALLER-MAINTAINER

Course Number: ***Version 2:*** 621-31L20; 621-36C20. *Version 3:* 621-31L30. *Version 4:* 621-31L30, Phase 2.
Location: Signal School, Fort Gordon, GA.
Length: ***Version 2:*** 4–7 weeks (149–221 hours). *Version 3:* 8 weeks (296 hours). *Version 4:* 5 weeks (153 hours).
Exhibit Dates: ***Version 2:*** 5/85–5/96. *Version 3:* 6/96–9/01. *Version 4:* 10/01–Present.
Learning Outcomes: ***Version 2:*** Upon completion of the course, the student will be able to lead a wire team in the installation, maintenance, and recovery of wire and cable systems. *Version 3:* Upon completion of the course, the student will be able to lead a wire team in the installation, maintenance, and recovery of wire and cable systems. *Version 4:* Upon completion of the course, the student will be able to lead a wire team in the installation, maintenance, and recovery of wire and cable systems.
Instruction: ***Version 2:*** Practical exercises cover determining material needs, personnel requirements, and work schedules to install, replace, and recover telephone cable networks. Course includes field exercises and procedures to check proper installation. This course contains a common core of leadership subjects. *Version 3:* Practical and classroom exercises, and computer-based training. Topics include determining material needs, personnel requirements, and work schedules to install, replace, and recover telephone cable networks. Course

includes field exercises and procedures to check proper installation. This course contains a common core of leadership subjects. *Version 4:* Practical and classroom exercises, and computer-based training. Topics include determining material needs, personnel requirements and work schedules, installation, replacing and recovering telephone cable networks, and Network Essential Plus. Course includes field exercises and procedures to check two proper installations. This course contains a common core of leadership subjects.

Credit Recommendation: *Version 2:* In the vocational certificate category, 2 semester hours in telephone line field installation supervision. In the lower-division baccalaureate/associate degree category, see AR-1406-0090 for common core credit (6/88). *Version 3:* In the vocational certificate category, 2 semester hours in telephone line field installation supervision. In the lower-division baccalaureate/associate degree category, see AR-1406-0090 for common core credit (2/97). *Version 4:* In the lower-division baccalaureate/associate degree category, 3 semester hours in networking fundamentals. See AR-1406-0090 for common core credit (2/03).

Related Occupations: 31L; 36C.

AR-1715-0619

BIOMEDICAL EQUIPMENT MAINTENANCE ORIENTATION

Course Number: 4B-F20.

Location: Equipment and Optical School, Aurora, CO.

Length: 16 weeks (646 hours).

Exhibit Dates: 8/82–6/96.

Objectives: To train commissioned officers in electronic theory and to provide orientation in practical skills essential for calibration, repair, and maintenance of biomedical equipment.

Instruction: This course covers study of AC and DC circuits including series and parallel circuits, inductive and capacitive reactance, resonance, and time constants. Also included are transistor theory and circuitry; digital electronics, computer theory, and microprocessors; and a study of the circuitry of several types of biomedical equipment.

Credit Recommendation: In the lower-division baccalaureate/associate degree category, 3 semester hours in AC/DC circuit theory, 3 in semiconductor devices and circuits, 3 in introduction to microprocessors, and 3 in biomedical equipment repair (9/85).

AR-1715-0620

MICROPROCESSORS

Course Number: 4B-F7/198-F7; 198-F7.

Location: Equipment and Optical School, Aurora, CO.

Length: 3 weeks (120 hours).

Exhibit Dates: 8/80–8/95.

Learning Outcomes: Upon completion of the course, the student will be able to calibrate equipment and analyze and diagnose minor malfunctions on certain microprocessors and microprocessor-based medical equipment.

Instruction: Topics include basic and advanced digital computer architecture, MMD-1 systems, and block-diagram descriptions of the digital computer and 8080A microprocessor and its instruction set. The course also includes interfacing, I/O techniques, and flags and interrupts.

Credit Recommendation: In the lower-division baccalaureate/associate degree category, 3 semester hours in introduction to microprocessors (9/85).

AR-1715-0621

ADVANCED DIGITAL THEORY

Course Number: 4B-F4/198-F4; 198-F4.

Location: Equipment and Optical School, Aurora, CO.

Length: 1 week (40 hours).

Exhibit Dates: 8/80–1/93.

Objectives: To provide warrant officers and enlisted personnel with a familiarization of digital circuit operations required for repair of microprocessor-based medical equipment.

Instruction: This course familiarizes personnel with binary numbers, logic gates, arithmetic circuits, flip-flops, registers, and memories.

Credit Recommendation: In the lower-division baccalaureate/associate degree category, 1 semester hour in introduction to digital concepts (9/85).

AR-1715-0622

CLOSED CIRCUIT TELEVISION

Course Number: 4B-F8/198-F8; 198-F8.

Location: Equipment and Optical School, Aurora, CO.

Length: 2 weeks (67–77 hours).

Exhibit Dates: 8/80–6/96.

Objectives: To train warrant officers and enlisted personnel in the repair and calibration of specific closed circuit television equipment.

Instruction: Lectures and practical exercises cover the maintenance and repair of specific monitors and cameras.

Credit Recommendation: In the vocational certificate category, 2 semester hours in closed circuit television/camera/monitor repair (9/85).

AR-1715-0624

PERSHING ELECTRONIC MATERIAL SPECIALIST BASIC NONCOMMISSIONED OFFICER (NCO)

Course Number: 121-21G30.

Location: Field Artillery School, Fort Sill, OK.

Length: 3–5 weeks (115–191 hours).

Exhibit Dates: 10/87–12/91.

Objectives: To train enlisted personnel to perform the tasks required of a Pershing materiel supervisor.

Instruction: Topics include Pershing missile maintenance operation, countdown operations, system electronics, maintenance procedures, troubleshooting, system tactics, and communications. This course contains a common core of leadership subjects.

Credit Recommendation: Credit is recommended for the common core only. See AR-1406-0090 (3/88).

Related Occupations: 21G.

AR-1715-0629

METEOROLOGICAL TECHNICIAN WARRANT OFFICER TECHNICAL CERTIFICATION

Course Number: 5B-201A.

Location: Field Artillery School, Fort Sill, OK.

Length: 16–17 weeks (594 hours).

Exhibit Dates: 10/84–6/96.

Objectives: To train meteorological warrant officer candidates in the technical aspects of meteorological instruments and equipment.

Instruction: Lectures and practical experience cover the operation of a meteorological section and organizational maintenance of electronic, optical, and mechanical meteorological instruments.

Credit Recommendation: In the vocational certificate category, 6 semester hours in troubleshooting and maintenance of electronic systems. In the lower-division baccalaureate/associate degree category, 3 semester hours in meteorology and 2 in basic electricity and electronics (11/85).

Related Occupations: 201A.

AR-1715-0630

FIREFINDER MAINTENANCE SUPERVISOR

Course Number: 4C-F24.

Location: Field Artillery School, Fort Sill, OK.

Length: 19–20 weeks (744–774 hours).

Exhibit Dates: 11/84–12/96.

Objectives: To provide technicians with the knowledge and skills required to supervise the maintenance and repair of the Firefinder radar.

Instruction: Lectures and demonstrations cover AC/DC circuits, solid state devices, integrated circuits, digital electronic logic circuits, and computers.

Credit Recommendation: In the lower-division baccalaureate/associate degree category, 6 semester hours in fundamentals of electricity and electronics, 3 in digital circuit theory, 3 in microcomputer fundamentals, 3 in electronic devices, 3 in radar theory, and 3 in radar/computer systems maintenance and troubleshooting (6/89).

AR-1715-0631

FIELD ARTILLERY DIGITAL SYSTEM REPAIRER

Course Number: 113-34L10.

Location: Field Artillery School, Fort Sill, OK.

Length: 16 weeks (583 hours).

Exhibit Dates: 11/84–6/96.

Objectives: To train enlisted personnel in basic and digital electronic maintenance, mortar fire control systems, message devices, battery computer systems, and related test equipment.

Instruction: Lectures and practical exercises cover AC/DC circuits, solid state devices, special solid state devices, integrated circuits, digital electronic logic circuits, maintaining and troubleshooting digital computer system, and performing basic soldering.

Credit Recommendation: In the lower-division baccalaureate/associate degree category, 6 semester hours in basic electricity and electronics, 3 in digital circuit theory, 3 in electronic devices, and 3 in digital computer theory (11/85).

Related Occupations: 34L.

AR-1715-0633

FIELD ARTILLERY CANNON WEAPON SYSTEMS QUALIFICATION

Course Number: 2E-FOA-F77.

Location: Field Artillery School, Fort Sill, OK.

Length: 2–3 weeks (52–80 hours).

Exhibit Dates: 6/85–6/96.

Objectives: To provide training to commissioned officers in the operation, maintenance, and employment of cannon systems.

Instruction: Lectures, demonstrations, and field instruction cover artillery ammunition, battery operations, and maintenance systems.

Credit Recommendation: Credit is not recommended because of the military-specific nature of the course (9/86).

AR-1715-0635

Field Artillery Radar Basic Noncommissioned Officer (NCO)
(Field Artillery Radar Basic Technical)

Course Number: 221-13R/17B30; 221-13R30/221-17B30.

Location: Field Artillery School, Fort Sill, OK.

Length: 5–6 weeks (201–245 hours).

Exhibit Dates: 10/86–6/96.

Objectives: After 9/00 see AR-1715-1038. To provide the technical knowledge and skills required to support field artillery firing units.

Instruction: Topics include operations and site selection for specific fire control radars, radar gunnery and survey, and preventive maintenance on the radar and support equipment. This course contains a common core of leadership subjects.

Credit Recommendation: In the vocational certificate category, 3 semester hours in radar operations. In the lower-division baccalaureate/associate degree category, see AR-1406-0090 for common core credit (12/91).

Related Occupations: 13R; 17B.

AR-1715-0636

Pershing II Missile Crewmember Basic Noncommissioned Officer (NCO)

Course Number: 042-15E30.

Location: Field Artillery School, Fort Sill, OK.

Length: 5 weeks (179 hours).

Exhibit Dates: 10/86–10/96.

Objectives: To train selected enlisted personnel to perform tasks required of a Pershing missile crewmember.

Instruction: Lectures and practical experience cover Pershing missile firing operations, missile mating, tactics, communications, maintenance, and special weapons. This course contains a common core of leadership subjects.

Credit Recommendation: Credit is recommended for the common core only. See AR-1406-0090 (1/85).

Related Occupations: 15E.

AR-1715-0638

Tactical Fire (TACFIRE) Operations Specialist Basic Noncommissioned Officer (NCO)
(Tactical Fire (TACFIRE) Operations Specialist Basic Technical)

Course Number: 250-13C30.

Location: Field Artillery School, Fort Sill, OK.

Length: 7–11 weeks (249–420 hours).

Exhibit Dates: 1/82–9/97.

Learning Outcomes: After 9/97 see AR-1408-0270. To provide TACFIRE operation specialists with additional training on the tactical fire direction center system.

Instruction: Lectures and practical experience cover employing, operating, and maintaining the tactical fire direction computer system.

Credit Recommendation: Credit is not recommended because of the military-specific nature of the course (1/92).

Related Occupations: 13C.

AR-1715-0639

Medical Equipment Repairer Advanced

Course Number: 4B-670A/198-91A30; 4B-670A/198-35U30; 4B-670A/198-35U20; B-326-0011.

Location: Medical Equipment and Optical School, Aurora, CO.

Length: 30 weeks (1144–1150 hours).

Exhibit Dates: 10/90–Present.

Learning Outcomes: Upon completion of the course, the personnel will be able to perform as a medical equipment repair technician and or medical equipment repairmen. This is an interservice training course and available to active Army, Army reserve, Navy, Coast Guard, and Department of Defense civilian personnel.

Instruction: Lectures and practical exercises cover installation, inspection, maintenance, repair, calibration, and adjustment of selected electrically- and electronically-controlled medical equipment. Topics include advanced circuit analysis of linear, digital, and microprocessor theory; application required to troubleshoot equipment down to the component level; electricity and electronics; maintenance management; fundamentals of AC and DC circuits; solid state devices and circuits; digital devices and circuits; and principles of equipment operation.

Credit Recommendation: In the lower-division baccalaureate/associate degree category, 3 semester hours in DC/AC fundamentals, 3 in solid state devices and circuits, 3 in digital/microprocessor electronics, and 6 in biomedical equipment maintenance (2/98).

Related Occupations: 35U; 670A; 91A.

AR-1715-0641

Chaparral/Redeye Repairer Basic Noncommissioned Officer (NCO)
(Chaparral/Redeye Repairer Basic Technical (BT))

Course Number: 121-27G30.

Location: Missile and Munitions School, Redstone Arsenal, AL.

Length: 7–8 weeks (243–278 hours).

Exhibit Dates: 6/85–1/96.

Objectives: To provide Army personnel with the skills and knowledge required to instruct, assist, and supervise subordinates in the maintenance of the Chaparral and Redeye weapon systems and ancillary equipment.

Instruction: Personnel are trained in system knowledge as well as in skills required for training and supervising an intermediate-level missile maintenance unit. Course includes training in software fundamentals and such hardware basics as address decoding, memories, peripherals, and control circuits. Also included is troubleshooting using several electronic modules and test sets.

Credit Recommendation: In the lower-division baccalaureate/associate degree category, 2 semester hours in introduction to microprocessor/computer hardware and 1 in electronic system maintenance (4/88).

Related Occupations: 27G.

AR-1715-0642

Forward Area Alerting Radar (FAAR) Repairer Basic Noncommissioned Officer (NCO)

Course Number: 104-27N30.

Location: Missile and Munitions School, Redstone Arsenal, AL.

Length: 10 weeks (359 hours).

Exhibit Dates: 1/86–1/96.

Objectives: To provide Army personnel with skills and knowledge required to instruct, assist, and supervise subordinates in the maintenance of Forward Area Alerting Radar and ancillary equipment.

Instruction: Personnel are trained in systems knowledge and in skills required for training and supervision of an intermediate-level missile maintenance unit. Course includes training in software fundamentals and such hardware basics as address decoding, memories, peripherals, and control circuits. Also includes troubleshooting on several electronic modules and test sets.

Credit Recommendation: In the lower-division baccalaureate/associate degree category, 2 semester hours in introduction to microprocessor/computer hardware and 2 in electronic systems maintenance (10/87).

Related Occupations: 27N.

AR-1715-0644

Pershing II Support Repair Technician (DS/GS)

Course Number: 4F-F21.

Location: Missile and Munitions School, Redstone Arsenal, AL.

Length: 12–13 weeks (468 hours).

Exhibit Dates: 5/85–9/90.

Objectives: To provide selected warrant officers with the technical training to qualify them on new maintenance techniques and additional equipment.

Instruction: Lectures and practical exercises cover quality control, inventory control, portable test equipment, and electronic troubleshooting and repair.

Credit Recommendation: In the lower-division baccalaureate/associate degree category, 3 semester hours in electronic systems maintenance (4/86).

AR-1715-0647

Hawk Missile Support Repair Technician (DS/GS)

Course Number: 4F-F22.

Location: Missile and Munitions School, Redstone Arsenal, AL.

Length: 9–10 weeks (347 hours).

Exhibit Dates: 4/85–9/91.

Objectives: To train selected warrant officers to perform and supervise intermediate maintenance on the Hawk missile system.

Instruction: Lectures and practical exercises cover automated support, soldering, computer logic and logic circuits, and computer and microprocessor fundamentals.

Credit Recommendation: In the lower-division baccalaureate/associate degree category, 2 semester hours in introduction to digital circuitry and 3 in introduction to computer/microprocessor hardware (4/86).

AR-1715-0649

Pershing Electrical-Mechanical Repairer Basic Noncommissioned Officer (NCO)

Course Number: 631-46N30.
Location: Missile and Munitions School, Redstone Arsenal, AL.
Length: 5 weeks (179 hours).
Exhibit Dates: 6/85–9/90.
Objectives: To train enlisted personnel to instruct, assist, and supervise subordinates in the maintenance of Pershing missile system equipment.
Instruction: This course includes a study of shop operation and soldering and Pershing missile system equipment.
Credit Recommendation: Credit is not recommended because of the limited, specialized nature of the course (10/87).
Related Occupations: 46N.

AR-1715-0652

Land Combat Support System Test Specialist Basic Noncommissioned Officer (NCO)
(Land Combat Support System Test Specialist)

Course Number: 121-27B30.
Location: Missile and Munitions School, Redstone Arsenal, AL.
Length: 13 weeks (474–475 hours).
Exhibit Dates: 3/85–1/96.
Objectives: To provide personnel with the skills and knowledge required to instruct, assist, and supervise subordinates in maintenance on the land combat support system and ancillary equipment.
Instruction: Personnel are trained in management, system knowledge, and skills required for the training and supervision of a missile maintenance unit. The course includes training in microprocessor hardware.
Credit Recommendation: In the lower-division baccalaureate/associate degree category, 1 semester hour in introduction to microprocessor hardware (4/86).
Related Occupations: 27B.

AR-1715-0654

Patriot Communications Crewmember

Course Number: 202-ASIX4; 202-F4.
Location: Air Defense Artillery School, Fort Bliss, TX.
Length: 4–5 weeks (144–152 hours).
Exhibit Dates: 10/85–4/04.
Objectives: To provide personnel with the knowledge and skills necessary to operate, emplace, and perform organizational maintenance on missile system equipment.
Instruction: Course provides training on equipment, techniques of operation, traffic and data flow, and organizational maintenance.
Credit Recommendation: Credit is not recommended because of the military-specific nature of the course (1/92).

AR-1715-0655

Air Defense Artillery (ADA) Officer Advanced Reserve Component (RC)

Course Number: None.
Location: Air Defense Artillery School, Fort Bliss, TX.
Length: 4 weeks (151–161 hours).
Exhibit Dates: 6/85–6/94.
Objectives: To prepare reserve officers for command assignments in Air Defense Artillery.
Instruction: Instruction includes topics in air defense command and control, tactics, electronic warfare, weapons, and combined arms.
Credit Recommendation: Credit is not recommended because of the military-specific nature of the course (11/86).

AR-1715-0656

Warrant Officer Technical Certification (WOTC) Chaparral/Vulcan Systems Technician

Course Number: 4F-140B; 4F-224B.
Location: Air Defense Artillery School, Fort Bliss, TX.
Length: 18–20 weeks (669–694 hours).
Exhibit Dates: 2/85–9/90.
Objectives: To train personnel in the operation and maintenance of a radar air defense system.
Instruction: Study includes principles of operation and maintenance of a number of subsystems, including radar, fire control, hydraulic drive, power distribution, transmitting, receiving, display, and antenna. A study of basic digital concepts is included.
Credit Recommendation: In the lower-division baccalaureate/associate degree category, 2 semester hours in digital fundamentals and 3 in electronic systems troubleshooting and maintenance (11/88).
Related Occupations: 140B; 224B.

AR-1715-0658

Avionic Mechanic

Course Number: *Version 1:* 102-68N10; 102-35K10. *Version 2:* 102-68N10; 102-35K10. *Version 3:* 102-68N10. *Version 4:* 102-68N10.
Location: *Version 1:* Signal School, Fort Gordon, GA. *Version 2:* Signal School, Fort Gordon, GA. *Version 3:* Aviation Logistics School, Fort Eustis, VA. *Version 4:* Aviation Logistics School, Fort Eustis, VA.
Length: *Version 1:* 23–24 weeks (883–930 hours). *Version 2:* 20–23 weeks (752–808 hours). *Version 3:* 21–22 weeks (781–791 hours). *Version 4:* 25 weeks (924 hours).
Exhibit Dates: *Version 1:* 10/86–9/90. *Version 2:* 10/90–9/98. *Version 3:* 10/98–9/02. *Version 4:* 10/02–Present.
Learning Outcomes: *Version 1:* Upon completion of the course, the student will be able to maintain and repair avionic communications, navigation, stabilization, and radar systems. Given defective avionic equipment, the student will test, operate, identify, and replace faulty unit or component. *Version 2:* Upon completion of the course, the student will be able to identify series, parallel, and series/parallel resistance circuits and determine values of resistance, voltage, and current in the circuit and maintain and repair avionic communications, navigation, stabilization, and radar systems. Given defective avionic equipment, the student will test, operate, identify, and replace faulty unit or component. *Version 3:* Upon completion of the course, the student will be able to operate, test, repair, replace, maintain, and troubleshoot various aircraft communications, navigation, flight control, stabilization, counter-measures, and radar systems. *Version 4:* Upon completion of the course, the student will be able to operate, test, repair, replace, maintain, and troubleshoot various aircraft communications, navigation, flight control, stabilization, counter-measures, and radar systems.
Instruction: *Version 1:* Training includes topics in DC/AC circuits, reading schematic and wiring diagrams, filters, transformers, synchros, and servos. Also included are extensive theory and laboratory exercises in avionic systems, including navigation systems, radio receiving systems, and absolute altimeter systems. *Version 2:* Training includes topics in DC/AC circuits, reading schematic and wiring diagrams, filters, transformers, synchros, and servos. Also included are extensive theory and laboratory exercises in avionic systems, including navigation systems, radio receiving systems, and absolute altimeter systems. *Version 3:* Audiovisual materials, practical and classroom exercises, discussion, laboratory, and lecture. Topics include basic AC and DC fundamentals; operation, troubleshooting, and repair of aircraft communications systems, navigation systems, radar systems and special equipment, avionics aircraft wiring, flight control systems, and aircraft survivability equipment. *Version 4:* Audiovisual materials, practical and classroom exercises, discussion, laboratory, and lecture. Topics include basic AC and DC fundamentals; operation, troubleshooting, and repair of aircraft communications systems, navigation systems, radar systems and special equipment, avionics aircraft wiring, flight control systems, and aircraft survivability equipment.
Credit Recommendation: *Version 1:* In the lower-division baccalaureate/associate degree category, 3 semester hours in electronic system troubleshooting and 2 in DC/AC circuits (6/90). *Version 2:* In the lower-division baccalaureate/associate degree category, 3 semester hours in electronic system troubleshooting, 3 in AC/DC circuits, and 1 in radar systems (2/94). *Version 3:* In the lower-division baccalaureate/associate degree category, 2 semester hours in avionics systems fundamentals, 2 in avionics troubleshooting and maintenance, 3 in basic electronics, and 1 radar systems (12/03). *Version 4:* In the lower-division baccalaureate/associate degree category, 2 semester hours in avionics systems fundamentals, 3 in avionics troubleshooting and maintenance, 3 in basic electronics, and 1 in radar systems (12/03).
Related Occupations: 35K; 68N.

AR-1715-0660

2. Avionic Radar Repairer
   (Avionics Special Equipment Repair)
3. Avionic Radar Repairer
4. Avionic Systems Repairer

Course Number: *Version 2:* 102-35R10; 102-68R10. *Version 3:* 102-35R10. *Version 4:* 102-35R10.
Location: *Version 2:* Aviation School, Fort Rucker, AL; Signal School, Fort Gordon, GA. *Version 3:* Signal School, Fort Gordon, GA; Ordnance, Missile and Munitions School, Redstone Arsenal, AL. *Version 4:* Signal School, Fort Gordon, GA.
Length: *Version 2:* 24–27 weeks (928–1026 hours). *Version 3:* 23–24 weeks (852 hours). *Version 4:* 28 weeks (1012 hours).
Exhibit Dates: *Version 2:* 10/88–9/95. *Version 3:* 10/95–9/99. *Version 4:* 10/99–Present.

Learning Outcomes: *Version 2:* Upon completion of the course, the student will be able to identify series, parallel, and series-parallel circuits and determine values of resistance, voltage, and current in the circuits; describe concepts of alternating current, including frequency, phase, and reactance; explain concepts of series and parallel resonance and the basics of transformer operation; explain basic operation of semiconductors, junction diodes, and bipolar transistors and identify their use in rectifier circuits and voltage regulators; troubleshoot and explain the operation of basic transistor amplifiers, power amplifiers, and operational amplifiers; explain the operation of FETs, SCRs, and UJTs and troubleshoot and explain the operation of sine-wave oscillators, multivibrator circuits, and Schmitt trigger circuits; describe single-sideband communications techniques including signal tracing and alignment of equipment; explain principles of communications using amplitude modulation techniques; trace signal path, troubleshoot, and align receivers and transmitters; describe frequency modulation communications techniques, including signal tracing and alignment of receiver and transmitter; convert between and perform arithmetic in the decimal, binary, octal, and hexadecimal number systems; analyze logic circuits, flip-flop circuits, registers, and counters; explain the theory and operation of microprocessor-based computers, including the concept of buses, memory, control signals, and addressing; and solder and desolder wire to various types of electronic terminals. *Version 3:* Upon completion of the course, the student will be able to provide maintenance on test, diagnostic, and measurement equipment used to diagnose problems in avionic radar equipment. *Version 4:* Upon completion of the course, the student will be able to inspect, test, and perform maintenance, troubleshoot, and repair avionic equipment.

Instruction: *Version 2:* This course includes electrical, electronic, digital, and communications. Practical exercises cover maintenance procedures for complex electronic systems using specialized test equipment. *Version 3:* Lectures and practical exercises covering DC/AC circuits, power supply circuits, amplifier and oscillator circuits, AM/FM fundamentals, digital and microprocessor fundamentals, and soldering. *Version 4:* Lectures, practical exercises, and classroom exercises covering practical theory of AC/DC circuits including power supplies, amplifiers, oscillators and transceivers with emphasis on troubleshooting techniques. A method of troubleshooting logic circuits and an introduction to microcomputers is presented. Lectures and practical exercises using common and specialized test equipment required to maintain and repair avionic radar equipment, aviator's night vision imaging systems, transponders, radar signal detecting systems, laser detecting systems, jammers, infrared countermeasure sets, missile approach detectors, navigation systems, and flight control/stabilization systems.

Credit Recommendation: *Version 2:* In the vocational certificate category, 2 semester hours in soldering. In the lower-division baccalaureate/associate degree category, 1 semester hour in AC/DC circuits, 3 in solid state electronics, 2 in electronic communications, 1 in digital fundamentals, 1 in microprocessors, 3 in electronic systems troubleshooting and maintenance, and 3 in radar systems (7/92). *Version 3:* In the vocational certificate category, 2 semester hours in AC/DC circuits and 1 in soldering. In the lower-division baccalaureate/associate degree category, 3 semester hours in basic electronic systems and 3 in introduction to digital circuits and microcomputers (5/97). *Version 4:* In the lower-division baccalaureate/associate degree category, 2 semester hours in AC/DC circuits, 3 in digital circuits and microcomputers, and 2 in troubleshooting electronic systems (2/04).

Related Occupations: 35R; 68R.

AR-1715-0661

Field Communications Security (COMSEC) Equipment Repair

Course Number: 160-29S10.
Location: Signal School, Fort Gordon, GA.
Length: 18–21 weeks (675–756 hours).
Exhibit Dates: 10/88–9/95.

Learning Outcomes: Upon completion of the course, the student will be able to perform simple circuit analysis of electronic and transistor circuits, connect and test these circuits, and perform limited maintenance on various types of communications security equipment.

Instruction: Lectures and performance exercises cover operation and maintenance of COMSEC equipment. Strong emphasis is placed on limited maintenance of secure communications equipment.

Credit Recommendation: In the lower-division baccalaureate/associate degree category, 2 semester hours in basic electronics laboratory, 1 in DC circuits, 1 in AC circuits, 1 in solid state electronics, and 3 in electronic systems troubleshooting and maintenance (7/92).

Related Occupations: 29S.

AR-1715-0663

Radio Repairer

Course Number: 101-29E10.
Location: Signal School, Fort Gordon, GA.
Length: 25–27 weeks (948–1023 hours).
Exhibit Dates: 3/88–9/95.

Learning Outcomes: Upon completion of the course, the student will be able to identify series, parallel, and series-parallel circuits and determine values of resistance, voltage, and current in the circuits; explain the basic operation of semiconductors, junction diodes, and bipolar transistors and identify their use in rectifier circuits and voltage regulators; explain concepts of alternating current, including frequency, phase, and reactance; explain concepts of series and parallel resonance and the basics of transformer operation; troubleshoot and explain the operation of basic transistor amplifiers, power amplifiers, and operational amplifiers; explain the operation of FETs, SCRs, and UJTs and troubleshoot and explain the operation of sine-wave, oscillator, multivibrator, and Schmitt trigger circuits; convert between and perform arithmetic in the decimal, binary, octal, and hexadecimal number systems; analyze logic circuits, flip-flop circuits, registers, and counters; explain principles of communications using amplitude modulation techniques; trace signal path, troubleshoot, and align receivers and transmitters; explain frequency modulation communications techniques including signal tracing and alignment of receiver and transmitter; comprehend single-sideband communications techniques including signal tracing and alignment of equipment; solder and desolder electronic components to printed circuit boards; align, troubleshoot, and repair radio receivers, transmitters, and associated equipment.

Instruction: Lectures and practical exercises provide the necessary skills to perform general electronic repair of AM, FM, and single-sideband communications equipment. Instruction includes basic electronic circuits, electron tubes, solid state devices, receiver and transmitter alignment, power supplies, digital logic circuits, and radio teletype equipment.

Credit Recommendation: In the vocational certificate category, 2 semester hours in soldering techniques. In the lower-division baccalaureate/associate degree category, 1 semester hour in DC circuits, 1 in AC circuits, 1 in digital principles, 3 in solid state electronics, 3 in electronic communications, 2 in basic electronics laboratory, and 4 in electronic systems troubleshooting and maintenance (6/89).

Related Occupations: 29E.

AR-1715-0665

Digital Radio and Multiplexer Acquisition (DRAMA) Communications Systems Repair
(Digital Radio Multiplexer Acquisition (DRAMA))

Course Number: 101-F22.
Location: Signal School, Fort Gordon, GA.
Length: 4 weeks (144 hours).
Exhibit Dates: 10/85–12/91.

Learning Outcomes: Upon completion of the course, the student will be able to operate, test, troubleshoot, and repair digital multiplexers and radios in a digital radio and multiplexer acquisition system.

Instruction: Course includes lectures and performance exercises in the operation and maintenance of the communications system.

Credit Recommendation: Credit is not recommended because of the limited, specialized nature of the course (10/87).

AR-1715-0666

Digital European Backbone (DEB) Communications System Repair

Course Number: 101-F16.
Location: Signal School, Fort Gordon, GA.
Length: 7 weeks (252 hours).
Exhibit Dates: 10/85–8/91.

Learning Outcomes: Upon completion of the course, the student will be able to operate, align, test, troubleshoot, and repair digital multiplexers and radios used in a digital communications system.

Instruction: Course includes lectures and performance exercises on the operation and maintenance of digital communications systems.

Credit Recommendation: In the lower-division baccalaureate/associate degree category, 1 semester hour in electronic systems troubleshooting (3/88).

AR-1715-0667

Digital Communications Satellite Subsystems
(Digital Communications Subsystems)

Course Number: 102-F40(2); 102-F40; 102-F40/102-F40 (26Y).
Location: Signal School, Fort Gordon, GA.
Length: 5–8 weeks (216–288 hours).

Exhibit Dates: 2/85–9/95.

Learning Outcomes: Upon completion of the course, the student will be able to operate, maintain, configure, and troubleshoot digital communications subsystems.

Instruction: Lectures and practical exercises cover the operation and maintenance of digital communications subsystems. Topics include multiplexers, modems, encoders, decoders, and communications subsystems. Laboratory experience is given in configuration, alignment, and troubleshooting of digital communications systems.

Credit Recommendation: In the lower-division baccalaureate/associate degree category, 2 semester hours in digital communications (1/01).

Related Occupations: 26Y.

AR-1715-0668

TRANSPORTABLE AUTOMATIC SWITCHING SYSTEMS OPERATOR/MAINTAINER

Course Number: ***Version 1:*** 622-36L10. *Version 2:* 622-36L10.

Location: Signal School, Fort Gordon, GA.

Length: ***Version 1:*** 26–28 weeks (951–1052 hours). *Version 2:* 21–22 weeks (785 hours).

Exhibit Dates: ***Version 1:*** 9/89–5/93. *Version 2:* 6/93–6/95.

Learning Outcomes: ***Version 1:*** Upon completion of the course, the student will be able to demonstrate a basic knowledge of DC/AC circuit theory, various solid state devices such as multiplexers and gates, digital numbering systems, and organized troubleshooting techniques and a detailed knowledge of the design, operation, and maintenance of automatic telephone central switching systems. *Version 2:* Upon completion of the course, the student will be able to describe basic DC/AC circuits, diodes, power supplies, transistors, logic gates, counters, binary and hexadecimal numbering systems, basic soldering techniques and operate install, maintain, and troubleshoot specified automatic telephone central switching systems.

Instruction: ***Version 1:*** Training includes topics in DC/AC fundamentals, circuit theory, basic electronics, solid state technology, digital numbering systems, and digital devices such as multiplexers and gates and design, operation, and maintenance procedures for the automatic telephone switching system. *Version 2:* Lectures and practical exercises cover DC/AC fundamentals, diodes, power supplies, transistors, digital logic gates and counters, basic soldering practices and the operation, installation, and troubleshooting of specified automatic telephone switching systems.

Credit Recommendation: ***Version 1:*** In the vocational certificate category, 3 semester hours in automatic telephone switching systems. In the lower-division baccalaureate/associate degree category, 3 semester hours in basic electronics laboratory, 1 in DC circuits, 2 in AC/DC circuits, 2 in digital principles, 1 in solid state electronics, and 3 in electronic systems troubleshooting and maintenance (6/90). *Version 2:* In the vocational certificate category, 1 semester hour in soldering techniques. In the lower-division baccalaureate/associate degree category, 3 semester hours in AC/DC circuits, 1 in solid state electronics, and 1 in digital principles (2/94).

Related Occupations: 36L.

AR-1715-0669

FIXED COMMUNICATIONS SECURITY (COMSEC) EQUIPMENT REPAIR

Course Number: 160-29F10.

Location: Signal School, Fort Gordon, GA.

Length: 13–16 weeks (491–562 hours).

Exhibit Dates: 1/89–7/90.

Learning Outcomes: Upon completion of the course, the student will identify series, parallel, and series-parallel DC circuits and determine values of resistance, voltage, and current in the circuits; explain AC circuit concepts, including frequency, phase, and reactance; solder and desolder wire to various types of electronic terminals; and perform limited maintenance on communications security equipment.

Instruction: Classroom and practical exercises cover topics in DC and AC circuits, soldering techniques, and on the specific equipment.

Credit Recommendation: In the vocational certificate category, 2 semester hours in soldering techniques. In the lower-division baccalaureate/associate degree category, 1 semester hour in DC circuits, 1 in AC circuits, 1 in basic electronics laboratory, and 1 in electronic systems troubleshooting and maintenance (3/90).

Related Occupations: 29F.

AR-1715-0670

3. SWITCHING CENTRAL REPAIRER (Telephone Central Office Repairer)
4. WIRE SYSTEMS EQUIPMENT REPAIRER
5. WIRE SYSTEMS EQUIPMENT REPAIRER
6. WIRE SYSTEMS EQUIPMENT REPAIRER

Course Number: ***Version 3:*** 622-29N10. *Version 4:* 622-35N10. *Version 5:* 622-35N10. *Version 6:* 622-35N10.

Location: ***Version 3:*** Signal School, Fort Gordon, GA. *Version 4:* Signal School, Fort Gordon, GA; Ordnance, Missile and Munitions School, Redstone Arsenal, AL. *Version 5:* Ordnance, Missile and Munitions School, Redstone Arsenal, AL. *Version 6:* Ordnance, Missile and Munitions School, Redstone Arsenal, AL.

Length: ***Version 3:*** 18–19 weeks (700 hours). *Version 4:* 16–17 weeks (598 hours). *Version 5:* 16 weeks (598 hours). *Version 6:* 17 weeks (622 hours).

Exhibit Dates: ***Version 3:*** 10/89–9/95. *Version 4:* 10/95–12/97. *Version 5:* 1/98–9/03. *Version 6:* 10/03–Present.

Learning Outcomes: ***Version 3:*** Upon completion of the course, the student will be able to identify series, parallel, and series-parallel resistive circuits and determine values of resistance, voltage, and current in the circuits; recognize the meaning of frequency, period, sine wave, phase shift, inductance, capacitance, reactance, and impedance; calculate voltage and current in transformers; calculate quality, bandwidth, and cutoff frequencies of filters; calculate impedances, voltages, and current in capacitive and inductive AC circuits, power supplies, and digital circuits; diagnose defective automatic telephone switchboards; and perform repairs. *Version 4:* Upon completion of the course, the student will apply knowledge of basic DC/AC circuits, transformers, power supply circuit, including voltage regulators and basic digital logic circuits, to the troubleshooting and repairing of manual, semiautomatic, and automated telephone systems. *Version 5:* Upon completion of the course, the student will apply knowledge of basic DC/AC circuits, transformers, power supply circuits, including voltage regulators and basic digital logic circuit to the troubleshooting and repairing of manual, semiautomatic, and automated telephone systems. *Version 6:* Upon completion of the course, the student will be able to apply knowledge of basic DC/AC circuits, power supply circuits, including voltage regulators and basic digital logic circuits to the troubleshooting and repair of manual, semi-automatic, and automated telephone systems.

Instruction: *Version 3:* Lectures and practical exercises cover manual, semiautomatic, and automatic telephone switching systems and associated equipment. *Version 4:* The course includes lectures and practical exercises covering DC/AC circuits; power supply circuits; digital fundamentals; and installation, operation, troubleshooting, and repair of telephone switchboards. *Version 5:* The course includes lectures and practical exercises covering DC/AC circuits; power supply circuits; digital fundamentals; and installation, operation, troubleshooting, and repair of telephone switchboards. *Version 6:* Practical exercises and lecture. Topics include DC/AC circuits; power supply circuits; digital fundamentals; installation, operation, troubleshooting, and repair for telephone switchboards.

Credit Recommendation: ***Version 3:*** In the vocational certificate category, 2 semester hours in telephone switching equipment repair. In the lower-division baccalaureate/associate degree category, 2 semester hours in basic electronics laboratory, 2 in AC/DC circuits, 1 in digital principles, and 1 in electronic systems troubleshooting and maintenance (6/90). *Version 4:* In the vocational certificate category, 2 semester hours in telephone switching repair and 1 in electronic systems troubleshooting and maintenance. In the lower-division baccalaureate/associate degree category, 2 semester hours in AC/DC circuits (10/00). *Version 5:* In the vocational certificate category, 2 semester hours in telephone switching repair and 1 in electronic systems troubleshooting. In the lower-division baccalaureate/associate degree category, 2 semester hours in AC/DC circuits (6/01). *Version 6:* In the lower-division baccalaureate/associate degree category, 2 semester hours in DC/AC circuits; 1 in electronic systems, troubleshooting, and maintenance; and 1 in digital circuits and troubleshooting logic circuits (2/04).

Related Occupations: 29N; 35N.

AR-1715-0671

TACTICAL SATELLITE/MICROWAVE REPAIRER

Course Number: 101-26L10; 101-29M10.

Location: Signal School, Fort Gordon, GA.

Length: 28–30 weeks (1022–1140 hours).

Exhibit Dates: 7/88–6/92.

Learning Outcomes: Upon completion of the course, the student will be able to identify series, parallel, and series-parallel DC circuits and determine values of resistance, voltage, and current in the circuits; describe AC circuit concepts, including frequency, phase, and reactance; solder and desolder wire to various types of electronic terminals; troubleshoot and explain operation of basic transistor amplifiers, power amplifiers, and operational amplifiers; explain operation of FETs, SCRs, and UJTs; troubleshoot and explain the operation of sine-wave oscillators, multivibrator circuits, and

Schmitt trigger circuits; describe FM communications systems, digital communications (including multiplexing) and satellite communications; trace signals and align receivers and transmitters; convert between and perform arithmetic in decimal, binary, octal, and hexadecimal number systems; analyze logic circuits, flip-flop circuits, registers, and counters; and perform limited maintenance on military-specific equipment.

Instruction: Lectures and practical exercises emphasize troubleshooting of the circuits and systems studied.

Credit Recommendation: In the vocational certificate category, 2 semester hours in soldering techniques. In the lower-division baccalaureate/associate degree category, 1 semester hour in DC circuits, 1 in AC circuits, 3 in solid state electronics, 1 in digital principles, 3 in basic electronics laboratory, 3 in electronic systems troubleshooting and maintenance, and 2 in electronic communications (6/90).

Related Occupations: 26L; 29M.

AR-1715-0672

2. Microwave Systems Operator-Maintainer
(Strategic Microwave Systems Repairer)
3. Microwave Systems Operator-Maintainer
(Strategic Microwave Systems Repairer)
4. Microwave Systems Operator-Maintainer
(Strategic Microwave Systems Repairer)
5. Microwave Systems Operator-Maintainer
6. Microwave Systems Operator Maintainer
7. Microwave Systems Operator-Maintainer

Course Number: *Version 2:* 101-29V10; 101-26V10. *Version 3:* 101-29V10. *Version 4:* 101-31P10. *Version 5:* 101-31P10. *Version 6:* 101-31P10-ITRO; 101-31P10. *Version 7:* 101-31P10.

Location: Signal School, Fort Gordon, GA.

Length: *Version 2:* 30–31 weeks (1106 hours). *Version 3:* 30–32 weeks (1121–1287 hours). *Version 4:* 31–32 weeks (1145 hours). *Version 5:* 24–25 weeks (972 hours). *Version 6:* 27 weeks (1080 hours). *Version 7:* 25 weeks (904 hours).

Exhibit Dates: *Version 2:* 10/89–6/90. *Version 3:* 7/90–9/93. *Version 4:* 10/93–9/95. *Version 5:* 10/95–9/97. *Version 6:* 10/97–12/01. *Version 7:* 1/02–Present.

Learning Outcomes: *Version 2:* Upon completion of the course, the student will be able to identify series, parallel, and series-parallel resistive circuits and determine values of resistance, voltage, and current; recognize the meaning of frequency, period, sine wave, phase shift, inductance, reactance, capacitive reactance, and impedance; calculate voltage and current in transformers; explain the basic operation of semiconductors, junction diodes, and transistors and identify their uses in rectifier circuits and voltage regulators; troubleshoot and explain the operation of basic transistor amplifiers and power amplifiers; identify operational amplifiers; explain the operation of FETs, SCRs and UJTs; troubleshoot and explain the operation of sine-wave oscillators, multivibrator circuits, and Schmitt trigger circuits; describe frequency modulation communications techniques; trace signal paths, troubleshoot, and align FM receivers and transmitters; convert between and perform arithmetic in the decimal, binary, octal, and hexadecimal number systems; analyze logic circuits, flip-flop circuits, registers, and counters; explain the theory and operation of microprocessor-based computers, including the concepts of buses, memory, control signals, and addressing; explain how a microcomputer performs arithmetic and logic operations; use machine language instructions; solder and desolder wire to various types of electronic terminals; and troubleshoot and repair microwave radio systems. *Version 3:* Upon completion of the course, the student will be able to identify series, parallel, and series-parallel resistance circuits and determine values of resistance, voltage, and current in the circuits; explain the basic operation of semiconductors, junction diodes, and transistors and identify their uses in rectifier circuits and voltage regulators; explain frequency modulation communications techniques; trace signal path, troubleshoot, and align FM receivers and transmitters; convert between and perform arithmetic in the decimal, binary, octal, and hexadecimal number systems; analyze logic circuits, flip-flop circuits, registers, and counters; explain the theory and operation of microprocessor-based computers, including the concepts of buses, memory, control signals, and addressing; solder and desolder wire to various types of electronic terminals; operate, maintain, troubleshoot, and repair digital communications systems; and troubleshoot and repair fiber-optic cable assemblies. *Version 4:* Upon completion of the course, the student will be able to identify series, parallel, and series-parallel resistance circuits and determine values of resistance, voltage, and current in the circuits; explain the basic operation of semiconductors, junction diodes, and transistors and identify their uses in rectifier circuits and voltage regulators; explain frequency modulation communications techniques; trace signal path, troubleshoot, and align FM receivers and transmitters; convert between and perform arithmetic in the decimal, binary, octal, and hexadecimal number systems; analyze logic circuits, flip-flop circuits, registers, and counters; solder and desolder wire to various types of electronic terminals; and perform maintenance on microwave FM and digital radio communications systems. *Version 5:* Upon completing of the course, the student will be able to identify series, parallel, and series-parallel resistance circuits and determine values of resistance, voltage, and current in the circuits; explain the basic operation of semiconductors, junction diodes, and transistors and identify their uses in rectifier circuits and voltage regulators; explain frequency modulation communications techniques; trace signal paths, troubleshoot and align FM receivers and transmitters; convert between and perform arithmetic in the decimal, binary, octal, and hexadecimal number systems; and analyze logic circuits, flip-flop circuits, registers, and counters. *Version 6:* Upon completion of the course, the student will be able to identify series, parallel, and series-parallel resistance circuits; explain basic application of semiconductor devices as used in rectifier and regulator circuits; explain frequency modulation in communications techniques; trace signal paths and troubleshoot and align FM receivers and transmitters; identify digital logic circuits; analyze circuit function; and perform decimal, binary, octal, and hexadecimal conversions. *Version 7:* Upon completion of the course, the student will be able to demonstrate basic knowledge of AC/DC and basic digital circuits; explain the use of rectifier and regulator circuits in power supplies; explain modulation techniques in microwave communication radios; and apply troubleshooting principles on microwave radio systems.

Instruction: *Version 2:* Lectures and practical exercises cover electronics topics. *Version 3:* Lectures and practical exercises cover DC circuits, AC circuits, regulated DC power supplies, FM receivers, digital electronics, microprocessors, digital communications systems, and fiber-optic cable fabrication and repair. Special emphasis is given to the use of test and measurement equipment at microwave frequencies. *Version 4:* Classroom and laboratory exercises cover basic DC/AC circuits, diodes, transistors, power supplies, FM transmitters and receivers, digital logic gates and counters, and basic soldering. Specific instruction and exercises in aligning and maintaining microwave communications equipment and related devices are also included. *Version 5:* Classroom and laboratory exercises cover basic DC/AC circuits, diodes, transistors, power supplies, FM transmitters and receivers, digital logic gates and counters, and basic soldering. Specific instruction and exercises in aligning and maintaining microwave communications equipment and related devices are also included. *Version 6:* Classroom and laboratory exercises cover basic DC/AC circuits, diodes, transistors, power supplies, FM transmitters and receivers, digital logic gates and counters, and basic soldering. Specific instruction and exercises in aligning and maintaining microwave communications equipment and related devices are also included. *Version 7:* Classroom and laboratory exercises cover basic DC/AC circuits, diodes, transistors, power supplies, FM transmitters and receivers, digital logic gates and counters, and basic soldering. Specific instruction and exercises in aligning and maintaining microwave communications equipment and related devices are also included.

Credit Recommendation: *Version 2:* In the vocational certificate category, 2 semester hours in soldering techniques. In the lower-division baccalaureate/associate degree category, 3 semester hours in solid state electronics, 2 in microprocessors, 1 in DC circuits, 2 in electronic communications, 2 in AC circuits, and 3 in electronic systems troubleshooting and maintenance (6/90). *Version 3:* In the vocational certificate category, 1 semester hour in soldering techniques. In the lower-division baccalaureate/associate degree category, 3 semester hours in AC/DC circuits, 3 in solid state electronics, 2 in microprocessors, 2 in digital principles, 2 in electronic communications, 2 in digital communications, and 3 in electronic systems troubleshooting and maintenance (2/94). *Version 4:* In the vocational certificate category, 1 semester hour in soldering techniques. In the lower-division baccalaureate/associate degree category, 3 semester hours in AC/DC circuits, 1 in solid state electronics, 1 in digital principles, 1 in electronic communications, 3 in basic electronics laboratory, and 3 in electronic systems troubleshooting and maintenance (2/94). *Version 5:* In the lower-division baccalaureate/associate degree category, 3 semester hours in AC/DC circuits, 1 in solid state electronics, 1 in digital principles, 1 in electronic communications, 3 in basic electronics laboratory, and 3 in electronic systems troubleshooting and maintenance (2/97). *Version 6:* In the lower-division baccalaureate/associate degree category, 3 semester hours in electronic systems troubleshooting and maintenance, 3 in basic electronics laboratory, 3 in AC/DC circuits, 1 in digital principles and 1 in electronic communications (5/99). *Version 7:* In the lower-division baccalaureate/associate degree category, 3 semester hours in AC/DC electronics; 3 in radio systems troubleshooting; 3 in radio communications; and 1 in digital circuits (2/03).

Related Occupations: 26V; 29V; 31P.

AR-1715-0673

USAF Organizational/Intermediate Maintenance of the AN/MSC-64 Satellite Communications (SATCOM) Terminal

Course Number: 101-F20.
Location: Signal School, Fort Gordon, GA.
Length: 3–5 weeks (138–168 hours).
Exhibit Dates: 4/86–6/93.

Learning Outcomes: Upon completion of the course, student will be able to operate, test, troubleshoot, and repair a satellite communications terminal.

Instruction: Course includes lectures and performance exercises on the operation and maintenance of a SATCOM terminal.

Credit Recommendation: Credit is not recommended because of the limited, specialized nature of the course (6/88).

AR-1715-0677

Field Communications Security (COMSEC) Equipment Full Maintenance

Course Number: 160-ASIG7.
Location: Signal School, Fort Gordon, GA.
Length: 18–22 weeks (658–785 hours).
Exhibit Dates: 10/88–9/95.

Learning Outcomes: Upon completion of the course, the student will be able to explain the basic operation of semiconductors, junction diodes, and bipolar transistors and identify their use in rectifier circuits and voltage regulators; troubleshoot and explain the operation of basic transistor amplifiers, power amplifiers, and operational amplifiers; explain the operation of FETs, SCRs, and UJTs; troubleshoot and explain the operation of sine-wave oscillators, multivibrator circuits, and Schmitt trigger circuits; convert between and perform arithmetic in the decimal, binary, octal, and hexadecimal number systems; analyze logic circuits, flip-flop circuits, registers, and counters; solder and desolder wire to various types of electronic terminals; solder and desolder electronic components to printed circuit boards; and troubleshoot and repair a number of military communications systems.

Instruction: Course includes lectures and performance exercises on solid state electronics, electronic circuits, and digital circuits and on various types of COMSEC equipment. Strong emphasis is placed on troubleshooting and repair of circuit boards.

Credit Recommendation: In the vocational certificate category, 4 semester hours in soldering techniques. In the lower-division baccalaureate/associate degree category, 3 semester hours in solid state electronics, 1 in digital principles, 1 in basic electronics laboratory, and 3 in electronic systems troubleshooting and maintenance (7/89).

AR-1715-0678

Strategic Microwave Systems Repairer Basic Noncommissioned Officer (NCO)

Course Number: 101-29V30.
Location: Signal School, Fort Gordon, GA.
Length: 6–16 weeks (239–609 hours).
Exhibit Dates: 4/86–6/94.

Learning Outcomes: Upon completion of the course, the student will be able to provide technical assistance and supervision to subordinates in the maintenance and operation of microwave multichannel radio and multiplex equipment and systems.

Instruction: This course contains a review of fundamentals of digital communications, including transistor filters, microwave propagation, digital techniques, nodal architecture, digital group multiplexer, and emergency power sources. Also included are maintenance administration, communications security, and site supervision. This course contains a common core of leadership subjects.

Credit Recommendation: In the lower-division baccalaureate/associate degree category, 2 semester hours in microwave communications. See AR-1406-0090 for common core credit (2/94).

Related Occupations: 29V.

AR-1715-0679

Satellite Communication (SATCOM) Fundamentals

Course Number: 102-F39.
Location: Signal School, Fort Gordon, GA.
Length: 12 weeks (432 hours).
Exhibit Dates: 10/88–9/95.

Learning Outcomes: After 12/97 see AR-1715-1016. Upon completion of the course, the student will be able to operate and maintain digital communications subsystems and perform functional analysis, operational checks, signal tracing, and troubleshooting of digital receivers and transmitters.

Instruction: Topics include fundamentals of defense satellite communications systems; digital electronics techniques; and operation, maintenance, and troubleshooting digital communications subsystems. Course methods include readings, lectures, discussions, and evaluation by test and performance.

Credit Recommendation: In the lower-division baccalaureate/associate degree category, 3 semester hours in digital communications and 3 in electronic systems troubleshooting and maintenance (7/92).

AR-1715-0680

Dial Central Office Repairer

Course Number: 101-ASIV2.
Location: Signal School, Fort Gordon, GA.
Length: 13 weeks (494 hours).
Exhibit Dates: 4/86–9/95.

Learning Outcomes: Upon completion of the course, the student will be able to perform direct and general support as well as depot-level maintenance on dial central office telephone exchange equipment.

Instruction: Course includes lectures, demonstrations, and performance exercises in the operation and maintenance of dial central office systems.

Credit Recommendation: In the vocational certificate category, 3 semester hours in dial central office equipment maintenance (2/87).

AR-1715-0681

Fixed Communications Security (COMSEC) Equipment Repairer (Transition 32G)
(Fixed Cryptographic Equipment Repair Transition 32F)

Course Number: 160-29F10/20/30-T (32F); 160-29F10/20/30-T.
Location: Signal School, Fort Gordon, GA.
Length: 3–8 weeks (144–296 hours).
Exhibit Dates: 3/86–10/90.

Learning Outcomes: Upon completion of the course, the student will be able to restore fixed cryptographic equipment to operation by substituting pluggable assemblies, subassemblies, elements, and printed wiring assemblies and perform limited preventive maintenance and troubleshooting of communications terminals.

Instruction: Topics in limited maintenance and troubleshooting of secure communications terminals are covered using readings, lectures, discussions, exercises, and evaluation by tests and performance.

Credit Recommendation: In the vocational certificate category, 2 semester hours in electronic systems troubleshooting (7/89).

Related Occupations: 29F; 32F; 32G.

AR-1715-0682

Tactical Satellite/Microwave Supervisor Basic Noncommissioned Officer (NCO)

Course Number: 101-29M30.
Location: Signal School, Fort Gordon, GA.
Length: 7–8 weeks (280 hours).
Exhibit Dates: 3/89–11/91.

Learning Outcomes: Upon completion of the course, the student will be able to establish and maintain a parts inventory, requisition parts, maintain records, explain concepts of digital multiplexing for voice and data transmission, explain the operation of modems and group multiplexing equipment, and troubleshoot data communications equipment.

Instruction: Classroom lectures and practical exercises are used to present topics on supply management and digital communications equipment. This course contains a common core of leadership subjects.

Credit Recommendation: In the lower-division baccalaureate/associate degree category, 1 semester hour in supply management, 2 in data communications, and 1 in electronic systems troubleshooting and maintenance. See AR-1406-0090 for common core credit (6/90).

Related Occupations: 29M.

AR-1715-0685

Radio Repairer Basic Noncommissioned Officer (NCO)
(Communication Electronics Radio Repairer Basic Noncommissioned Officer (NCO))

Course Number: ***Version 1:*** 101-29E30. *Version 2:* 101-29E30.
Location: Signal School, Fort Gordon, GA.

Length: *Version 1:* 6–9 weeks (216–319 hours). *Version 2:* 10–11 weeks (367 hours).

Exhibit Dates: *Version 1:* 3/86–9/91. *Version 2:* 10/91–9/95.

Learning Outcomes: *Version 1:* Upon completion of the course, the student will be able to provide technical assistance and supervision to subordinates in the maintenance of radio communications equipment. *Version 2:* Upon completion of the course, the student will be able to set up and manage a repair facility and supervise repair personnel.

Instruction: *Version 1:* The course contains a review of general Army communications systems, maintenance management, supply procedures, and inventory control. This course contains a common core of leadership subjects. *Version 2:* The course includes a review of general Army communications systems, maintenance management, supply procedures, and inventory control. This course contains a common core of leadership subjects.

Credit Recommendation: *Version 1:* In the lower-division baccalaureate/associate degree category, 2 semester hours in maintenance management. See AR-1406-0090 for common core credit (6/90). *Version 2:* In the lower-division baccalaureate/associate degree category, 3 semester hours in maintenance management. See AR-1406-0090 for common core credit (2/94).

Related Occupations: 29E.

AR-1715-0687

Fixed Ciphony Repairer
(Fixed Ciphony Repairer (Transition) (32G))

Course Number: 160-29F10/20/30-T (32G).

Location: Signal School, Fort Gordon, GA.

Length: 3–7 weeks (130–243 hours).

Exhibit Dates: 3/86–10/90.

Learning Outcomes: Upon completion of the course, the student will be able to restore fixed ciphony equipment to operation by substitution of pluggable assemblies, subassemblies, elements, and printed wiring assemblies and perform limited maintenance, operational testing, and troubleshooting of the equipment.

Instruction: Topics in limited maintenance, operational testing, and troubleshooting secure communications terminals are covered using readings, lectures, discussions, exercises, and evaluation by testing and performance.

Credit Recommendation: In the vocational certificate category, 2 semester hours in electronic systems troubleshooting (7/89).

Related Occupations: 29F.

AR-1715-0688

Inertial Navigation Set AN/ASN-86 Repair

Course Number: 102-ASIR5.

Location: Signal School, Fort Gordon, GA.

Length: 10 weeks (362 hours).

Exhibit Dates: 6/87–7/91.

Learning Outcomes: Upon completion of the course, the student will be able to inspect, test, and perform organizational, direct, and general support maintenance on the inertial navigation set.

Instruction: Course includes lectures and performance exercises on the operation and maintenance of the navigation set.

Credit Recommendation: In the vocational certificate category, 3 semester hours in electronic systems troubleshooting (2/87).

AR-1715-0689

1. Teletypewriter Equipment Repairer Basic Noncommissioned Officer (NCO)
2. Telecommunications Terminal Device Repairer Basic Noncommissioned Officer (NCO)

Course Number: *Version 1:* 160-29J30. *Version 2:* 160-29J30.

Location: Signal School, Fort Gordon, GA.

Length: *Version 1:* 10–11 weeks (389 hours). *Version 2:* 13–14 weeks (505 hours).

Exhibit Dates: *Version 1:* 1/89–9/91. *Version 2:* 10/91–9/95.

Learning Outcomes: *Version 1:* Upon completion of the course, the student will be able to convert between and perform arithmetic in the decimal, binary, octal, and hexadecimal number systems; analyze logic circuits, flip-flop circuits, registers, and counters; explain the theory and operation of microprocessor-based computers, including the concepts of buses, memory, control signals, and addressing; and supervise repair facilities for teletype equipment. *Version 2:* Upon completion of the course, the student will be able to explain the theory and operation of microprocessor-based computers, including the concepts of buses, memory, control signals, addressing, machine implementation of assembly language instructions, interfacing, and interrupt processing and supervise repair facilities for telecommunications equipment.

Instruction: *Version 1:* Lectures cover supervisory practices for repair facilities, record keeping, supply procedures, and quality control. This course contains a common core of leadership subjects. *Version 2:* Lectures and practical exercises cover microprocessor fundamentals, instruction execution, and installation and maintenance of military-specific computer and communications systems. This course contains a common core of leadership subjects.

Credit Recommendation: *Version 1:* In the lower-division baccalaureate/associate degree category, 1 semester hour in digital principles, 1 in microprocessors, and 3 in electronic systems troubleshooting and maintenance. See AR-1406-0090 for common core credit (7/89). *Version 2:* In the lower-division baccalaureate/associate degree category, 2 semester hours in microprocessors and 3 in electronic systems troubleshooting and maintenance. See AR-1406-0090 for common core credit (2/94).

Related Occupations: 29J.

AR-1715-0690

Communications Security (COMSEC) Equipment TSEC/KI-1A Repair

Course Number: 160-F30 (29S).

Location: Signal School, Fort Gordon, GA.

Length: 7–8 weeks (274 hours).

Exhibit Dates: 10/86–9/95.

Learning Outcomes: Upon completion of the course, the student will be able to troubleshoot and repair specific communications security equipment to the component level.

Instruction: Lectures and practical exercises emphasize troubleshooting. Student uses oscilloscope, multimeter, and hand tools to troubleshoot and repair.

Credit Recommendation: In the lower-division baccalaureate/associate degree category, 3 semester hours in electronic systems troubleshooting (2/87).

Related Occupations: 29S.

AR-1715-0691

Defense Satellite Communications System (DSCS) Ground Mobile Forces (GMF) Controller

Course Number: 102-ASIQ1.

Location: Signal School, Fort Gordon, GA.

Length: 14 weeks (504 hours).

Exhibit Dates: 10/88–9/91.

Learning Outcomes: Upon completion of the course, the student will be able to operate, troubleshoot, and repair the satellite communications ground mobile forces control equipment.

Instruction: Course includes lectures, demonstrations, and performance exercises on the operation and maintenance of GMF equipment.

Credit Recommendation: In the lower-division baccalaureate/associate degree category, 3 semester hours in electronics systems troubleshooting and maintenance (7/89).

Related Occupations: 26Y.

AR-1715-0693

AN/TYC-39(V) Operator (USAF)

Course Number: 260-F1 (72E).

Location: Signal School, Fort Gordon, GA.

Length: 3–5 weeks (136–196 hours).

Exhibit Dates: 5/83–6/92.

Learning Outcomes: Upon completion of the course, the student will be able to install, operate, and perform operator maintenance on a tactical automatic message switching central and load paper, prepare equipment for operation, interpret service messages, make log entries, and enter commands.

Instruction: Message processing, major components within the message processing shelter, power initialization/shutdown procedures are covered using readings, lectures, discussions, exercises, and evaluation by test and performance.

Credit Recommendation: Credit is not recommended because of the limited, specialized nature of the course (2/88).

Related Occupations: 72E.

AR-1715-0694

Satellite Communications (SATCOM) Controller

Course Number: 4C-F19/102-ASIY1; 4C-F19/102-ASIA1.

Location: Signal School, Fort Gordon, GA.

Length: 10–15 weeks (360–540 hours).

Exhibit Dates: 10/85–9/91.

Learning Outcomes: Upon completion of the course, the student will be able to operate a satellite communications system.

Instruction: Lectures and practical exercises include the basics of satellite communications, signal flow in satellites and ground stations, satellite control, and system operation.

Credit Recommendation: Credit is not recommended because of the limited, specialized nature of the course (7/89).

Related Occupations: 29Y.

AR-1715-0695

GROUND MOBILE FORCES (GMF) INTERFACE SUBSYSTEM
(Data Processing Group OL-230/MSC-66(V))

Course Number: *Version 1:* 102-F55; 102-F55 (26Y). *Version 2:* 102-F55; 102-F55 (26Y).
Location: Signal School, Fort Gordon, GA.
Length: *Version 1:* 2–3 weeks (72–88 hours). *Version 2:* 4 weeks (144 hours).
Exhibit Dates: *Version 1:* 5/85–5/92. *Version 2:* 6/92–6/94.
Learning Outcomes: *Version 1:* Upon completion of the course, the student will be able to operate and perform operator maintenance on the data processing group OL-230/MSC-66(V). *Version 2:* Upon completion of the course, the student will be able to operate, troubleshoot, align, and repair military-specific data communications equipment.
Instruction: *Version 1:* Course includes lectures, demonstrations, and performance exercises on maintaining the equipment. *Version 2:* Lectures and practical exercises cover the operation, maintenance, and alignment of a multiplexer system and a modem system.
Credit Recommendation: *Version 1:* Credit is not recommended because of the limited, specialized nature of the course (6/89). *Version 2:* In the vocational certificate category, 3 semester hours in electronic systems operation and maintenance (2/94).
Related Occupations: 26Y.

AR-1715-0696

AH-64A AVIONIC MECHANIC

Course Number: 102-ASIX1 (35K); 102-ASIX1.
Location: Signal School, Fort Gordon, GA.
Length: 3–5 weeks (104–154 hours).
Exhibit Dates: 10/85–9/95.
Learning Outcomes: Upon completion of the course, the student will be able to test, troubleshoot, and repair AH-64A avionic communications and navigation systems.
Instruction: Lectures and performance exercises cover troubleshooting communications and navigation systems on a helicopter.
Credit Recommendation: In the lower-division baccalaureate/associate degree category, 1 semester hour in communications systems troubleshooting and maintenance (2/94).
Related Occupations: 35K.

AR-1715-0699

TACTICAL SATELLITE/MICROWAVE SYSTEMS OPERATOR BASIC NONCOMMISSIONED OFFICER (NCO)
(Tactical Satellite/Microwave Systems Operator Primary Technical)

Course Number: 201-31Q20; 201-26Q20.
Location: Signal School, Fort Gordon, GA.
Length: 6–8 weeks (216–290 hours).
Exhibit Dates: 4/85–6/93.
Learning Outcomes: Upon completion of the course, the student will be able to act as team chief in the installation, operator maintenance, and troubleshooting of satellite and microwave communications equipment.
Instruction: Lectures and practical exercises cover directing installation and maintenance of power generator sets and a radio set. This course contains a common core of leadership subjects.
Credit Recommendation: Credit is recommended for the common core only. See AR-1406-0090 (2/94).
Related Occupations: 26Q; 31Q.

AR-1715-0700

TACTICAL SIGNAL STAFF OFFICER
(Corps/Division Communications Operations Officer)

Course Number: 4C-FOA-F26.
Location: Signal School, Fort Gordon, GA.
Length: 5–6 weeks (180–216 hours).
Exhibit Dates: 6/85–6/95.
Learning Outcomes: Upon completion of the course, the student will be able to function in a corps/division as a communications specialist.
Instruction: Lectures and practical exercises cover subjects related to procedures and equipment used in division/corps communications.
Credit Recommendation: Credit is not recommended because of the limited, specialized nature of the course (6/90).

AR-1715-0701

SATELLITE COMMUNICATIONS (SATCOM) TERMINAL AN/GSC-49

Course Number: *Version 1:* 102-ASIP4 (CT). *Version 2:* 102-F113; 102-ASIP4.
Location: Signal School, Fort Gordon, GA.
Length: *Version 1:* 18 weeks (648 hours). *Version 2:* 4 weeks (144 hours).
Exhibit Dates: *Version 1:* 10/88–11/91. *Version 2:* 12/91–10/95.
Learning Outcomes: *Version 1:* Upon completion of the course, the student will be able to operate and maintain a military-specific communications terminal. *Version 2:* Upon completion of the course, the student will be able to operate and maintain a military satellite communications terminal, including VHF, digital, frequency conversion, amplifiers, and antenna subsystems.
Instruction: *Version 1:* Classroom and practical exercises cover the particular equipment. *Version 2:* Lectures and practical exercises cover satellite communications systems and antennas and equipment failure and repair.
Credit Recommendation: *Version 1:* In the lower-division baccalaureate/associate degree category, 2 semester hours in satellite communications systems and 3 in electronic systems troubleshooting and maintenance (6/89). *Version 2:* In the lower-division baccalaureate/associate degree category, 3 semester hours in communications systems troubleshooting and maintenance (8/96).
Related Occupations: 29Y.

AR-1715-0702

SATELLITE COMMUNICATIONS (SATCOM) TERMINAL AN/FSC-78/79 AND AN/GSC-39

Course Number: 102-F42 (31S) (P); 102-F42.
Location: Signal School, Fort Gordon, GA.
Length: 8 weeks (288 hours).
Exhibit Dates: 12/86–7/98.
Learning Outcomes: Upon completion of the course, the student will be able to operate and maintain two SATCOM terminals and operate and maintain up/down converters and communications, antenna positioning, frequency generating, and control/monitor equipment.
Instruction: Lectures and practical exercises include system overview, frequency conversion and synthesis, SATCOM subsystems, control and monitoring equipment, and troubleshooting to the lowest replacement unit. Methods include readings, lectures, discussions, exercises, and evaluation by test and performance.
Credit Recommendation: In the lower-division baccalaureate/associate degree category, 3 semester hours in electronic systems troubleshooting (2/94).
Related Occupations: 31S.

AR-1715-0704

TELECOMMUNICATIONS TECHNICIAN WARRANT OFFICER ADVANCED

Course Number: 4-11-C32-250A; 4-11-C32-290A.
Location: Signal School, Fort Gordon, GA.
Length: 14 weeks (527 hours).
Exhibit Dates: 10/85–10/99.
Learning Outcomes: Upon completion of the course, the student will have the necessary managerial and leadership abilities for supervisory and staff positions and have a thorough understanding of electronic warfare, NBC operations, air/land battle operations, automatic data processing fundamentals, and logistics.
Instruction: This course includes lectures, demonstrations, and performance exercises in all of the skill areas required for a communications electronics warrant officer including the technical aspects of managing a telecommunication system and the general aspects of personnel supervision and logistics.
Credit Recommendation: In the lower-division baccalaureate/associate degree category, 2 semester hours in introduction to data processing and 1 in leadership (5/90).
Related Occupations: 250A; 290A.

AR-1715-0707

CALIBRATION AND REPAIR TECHNICIAN WARRANT OFFICER ADVANCED

Course Number: 4-11-C32-252A.
Location: Signal School, Fort Gordon, GA.
Length: 14 weeks (527 hours).
Exhibit Dates: 10/85–6/92.
Learning Outcomes: Upon completion of the course, the student will have the necessary managerial and leadership abilities for supervisory and staff positions and have a thorough understanding of electronic warfare, NBC operations, air/land battle operations, automatic data processing fundamentals, and logistics.
Instruction: This course includes lectures, demonstrations, and performance exercises in all of the skill areas required of a calibration and repair warrant officer including the technical aspects of managing a maintenance system and the general aspects of personnel supervision and logistics.
Credit Recommendation: In the lower-division baccalaureate/associate degree category, 2 semester hours in introduction to data processing and 1 in leadership (2/87).
Related Occupations: 252A.

AR-1715-0708

AN/TTC-39(V) OPERATOR (USAF)

Course Number: 260-F2.

Location: Signal School, Fort Gordon, GA.

Length: 5–6 weeks (200 hours).

Exhibit Dates: 5/83–12/97.

Learning Outcomes: Upon completion of the course, the student will be able to install, operate, and perform operator maintenance on a specific piece of tactical automatic telephone central office equipment. Given appropriate equipment, the student will initialize the system, manually load the display unit, send messages, and identify operator-level faults.

Instruction: The course includes topics in power initialization, automatic data processing start-up procedures, initiation of communications security equipment, message handling, operator fault identification and isolation. Methods of instruction include readings, lectures, discussions, exercises, and evaluation by test and performance.

Credit Recommendation: Credit is not recommended because of the limited, specialized nature of the course (8/87).

AR-1715-0710

2. Noncommunications Interceptor/ Analyst
 (Electronic Warfare (EW) Signal Intelligence (SIGINT) Noncommunications Interceptor)
3. Noncommunications Interceptor/ Analyst
 (Electronic Warfare (EW) Signal Intelligence (SIGINT) Noncommunications Interceptor)
4. Electronic Intelligence Interceptor/ Analyst
 (Noncommunications Interceptor/Analyst)
 (Electronic Warfare (EW) Signal Intelligence (SIGINT) Noncommunications Interceptor)

Course Number: ***Version 2:*** 233-98J10. *Version 3:* 233-98J10. *Version 4:* 233-98J10.

Location: ***Version 2:*** Intelligence School, Fort Devens, MA. *Version 3:* Intelligence Center and School, Fort Huachuca, AZ. *Version 4:* Intelligence Center and School, Fort Huachuca, AZ.

Length: ***Version 2:*** 16–19 weeks (510–610 hours). *Version 3:* 10–11 weeks (336–394 hours). *Version 4:* 12 weeks (432 hours).

Exhibit Dates: ***Version 2:*** 10/88–9/92. *Version 3:* 10/92–9/00. *Version 4:* 10/00–Present.

Learning Outcomes: ***Version 2:*** Upon completion of the course, the student will be able to use technical references and equipment to search for selected classes of intercepted noncommunication electro-optic signals and record, analyze, and identify these signals. *Version 3:* Upon completion of the course, the student will be able to describe the characteristics and theory of electromagnetic signals, analyze and identify radar and radio waveform emissions using electronic intercept information, establish and maintain digital files, process data for analysis, and prepare intelligence reports. *Version 4:* Upon completion of the course, the student will be able to perform entry level duties of an electronic intelligence intercept analyst.

Instruction: ***Version 2:*** Lectures, demonstrations, and performance exercises cover map reading; data and signal analysis; basic electrical and communications theory; report writing; operation and maintenance of radio communications equipment; Soviet and North Korean defense systems; and the identification, collection, and reporting of intercepted noncommunications signals. *Version 3:* Lectures, demonstrations, and practical exercises cover signal theory, radio and radar signal analysis, intelligence processing and reporting, operation of electronic intercept equipment, and maintenance of digital files on electronic intercept information. *Version 4:* Practical exercises, discussion, and lecture. Topics cover analysis, correlation and reporting of noncommunication signals, waveform and radar basics, and intelligence processing and reporting.

Credit Recommendation: ***Version 2:*** In the lower-division baccalaureate/associate degree category, 2 semester hours in introduction to radar systems and 2 in personnel management (6/91). *Version 3:* In the lower-division baccalaureate/associate degree category, 3 semester hours in telecommunications signal theory and waveform analysis and 3 in signal analysis and security (6/97). *Version 4:* In the lower-division baccalaureate/associate degree category, 2 semester hours in introduction to radar systems and basic telecommunications (12/02).

Related Occupations: 98J.

AR-1715-0711

Technical Surveillance Countermeasures, Phase 1
 (Defense Against Sound Equipment (DASE), Phase 1)

Course Number: 3C-SQIW/244-ASIG9; 3C-ASI9L/244-ASIG9.

Location: Intelligence School, Fort Devens, MA.

Length: 15–16 weeks (561–565 hours).

Exhibit Dates: 1/86–3/94.

Learning Outcomes: Upon completion of the course, the student will be conversant in fundamentals of electronics, communications, and digital circuits.

Instruction: Lectures and practical exercises cover DC circuits, AC circuits, solid state circuit analysis, communications system signal analysis, and introduction to digital principles.

Credit Recommendation: In the lower-division baccalaureate/associate degree category, 3 semester hours in DC circuits, 3 in AC circuits, 4 in solid state circuits, 3 in communications systems, and 2 in digital fundamentals (10/90).

AR-1715-0713

Electronic Warfare (EW)/Intercept Aerial Sensor Repairer

Course Number: ***Version 1:*** 102-33V10. *Version 2:* 102-33V10.

Location: Intelligence School, Fort Devens, MA.

Length: ***Version 1:*** Phase 1, 24 weeks (849 hours). *Version 2:* 18 weeks (656 hours).

Exhibit Dates: ***Version 1:*** 10/86–2/91. *Version 2:* 3/91–9/91.

Learning Outcomes: ***Version 1:*** Upon completion of the course, the student be able to identify and correct problems in solid state, digital, and microprocessor-based communications subsystems, including A/D-D/A converters, UHF/VHF radio receiver, and video displays. *Version 2:* Upon completion of the course, the student will be able to describe the theory, system concepts, and circuit design of solid state electronics; conduct laboratory experimentation with linear and digital solid state electronics, analog communications, and 8-bit microcomputer hardware using Assembly language programming; and describe microcomputer peripherals including A/D and D/A converters.

Instruction: ***Version 1:*** Lectures, laboratories, demonstrations, and performance exercises cover electronic warfare/intercept aerial sensor systems. *Version 2:* Lectures, demonstrations, computer-assisted instruction, and laboratory exercises cover solid state devices and circuits, communications systems, analog and digital electronics, and 8-bit microcomputer hardware and software including I/O devices.

Credit Recommendation: ***Version 1:*** In the vocational certificate category, 4 semester hours in communications laboratory and 1 in PACE solder laboratory. In the lower-division baccalaureate/associate degree category, 4 semester hours in electricity/electronics (DC/AC), 4 in solid state circuit analysis, 4 in digital logic, and 3 in microprocessor fundamentals (5/87). *Version 2:* In the lower-division baccalaureate/associate degree category, 3 semester hours in AC/DC circuits, 3 in digital logic, and 3 in analog communications. In the upper-division baccalaureate category, 3 in operational amplifiers and applications and 3 in introduction to microcomputers (7/91).

Related Occupations: 33V.

AR-1715-0714

Electronic Warfare (EW)/Intercept Aviation Systems Basic Noncommissioned Officer (NCO)
 (Electronic Warfare (EW)/Intercept Aviation Equipment Repairer Basic Noncommissioned Officer (NCO))

Course Number: ***Version 1:*** 102-33R30. *Version 2:* 102-33R30.

Location: Intelligence School, Fort Devens, MA.

Length: ***Version 1:*** 16–17 weeks (595–596 hours). *Version 2:* 19–20 weeks (713 hours).

Exhibit Dates: ***Version 1:*** 3/89–12/90. *Version 2:* 1/91–3/94.

Learning Outcomes: ***Version 1:*** Upon completion of the course, the student will be able to teach and supervise enlisted personnel in state-of-the-art troubleshooting and alignment techniques associated with electronic warfare/intercept aviation systems. *Version 2:* After 11/94 see AR-1715-1006. Upon completion of the course, the student will be able to teach and supervise enlisted personnel in state-of-the-art troubleshooting and alignment techniques associated with electronic warfare/intercept aviation systems.

Instruction: ***Version 1:*** Lectures, demonstrations, and practical exercises cover counseling, personnel supervision, operations management, modulation techniques, time and frequency domain analysis, and electronic equipment maintenance and troubleshooting. This course contains a common core of leadership subjects. *Version 2:* Lectures, demonstrations, and practical exercises cover counseling, personnel supervision, operations management, modulation techniques, computer organization and maintenance, and electronic equipment maintenance and troubleshooting. This course contains a common core of leadership subjects.

Credit Recommendation: ***Version 1:*** In the vocational certificate category, 3 semester hours in electronic systems troubleshooting and maintenance. In the lower-division baccalaureate/associate degree category, 4 semester hours in electronic communications and 1 in personnel management. See AR-1406-0090 for common core credit (7/91). *Version 2:* In the vocational certificate category, 3 semester hours in electronic systems troubleshooting and maintenance and 3 in computer system troubleshooting and maintenance. In the lower-division baccalaureate/associate degree category, 2 semester hours in electronic communications, 2 in computer systems and organization, and 2 in personnel management. See AR-1406-0090 for common core credit (11/91).

Related Occupations: 33R.

AR-1715-0715

ELECTRONIC WARFARE (EW)/INTERCEPT AVIATION SYSTEMS REPAIRER

Course Number: ***Version 1:*** 102-33R10. *Version 2:* 102-33R10.

Location: ***Version 1:*** Intelligence School, Fort Devens, MA. *Version 2:* Intelligence Center and School, Fort Huachuca, AZ.

Length: ***Version 1:*** 35–36 weeks (1273 hours). *Version 2:* 35–39 weeks (1274–1534 hours).

Exhibit Dates: ***Version 1:*** 10/87–9/93. *Version 2:* 10/93–12/01.

Learning Outcomes: ***Version 1:*** Upon completion of the course, the student will be able to use common laboratory instruments and perform electronic laboratory work; analyze DC series, parallel, and series-parallel circuits; analyze RL and RC circuits; analyze filters and resonant circuits; explain the operation of solid state devices; analyze analog circuits including class A, B, and C power amplifiers, power supplies, and oscillators; explain operational amplifier operation and circuits; describe the operation of vacuum tubes and vacuum tube circuits; perform conversions between decimal, binary, octal, and hexadecimal numbers; add and subtract binary and twos-complement numbers; explain the operation of digital circuit elements, including logic gates, flip-flops, counters, registers, decoders, and encoders; explain microprocessor system architecture and register structure; write Assembly and machine language microprocessor programs; explain electronic communications concepts and devices, including wave propagation, transmitters, AM, FM, pulse code modulation, receivers, transmission lines, antennas, and microwaves; troubleshoot a wide variety of electronic circuits and systems; and perform electronic soldering. *Version 2:* Upon completion of the course, the student will be able to use common laboratory instruments and perform electronic laboratory work; analyze DC series, parallel, and series-parallel circuits; analyze RL and RC circuits; analyze filters and resonant circuits; explain the operation of solid state devices; analyze analog circuits including class A, B, and C power amplifiers, power supplies, and oscillators; explain operational amplifier operation and circuits; describe the operation of vacuum tubes and vacuum tube circuits; perform conversions between decimal, binary, octal, and hexadecimal numbers; add and subtract binary and twos-complement numbers; explain the operation of digital circuit elements, including logic gates, flip-flops, counters, registers, decoders, and encoders; explain microprocessor system architecture and register structure; write Assembly and machine language microprocessor programs; explain electronic communications concepts and devices, including wave propagation, transmitters, AM, FM, pulse code modulation, receivers, transmission lines, antennas, and microwaves; troubleshoot a wide variety of electronic circuits and systems; and perform electronic soldering.

Instruction: ***Version 1:*** Lectures, laboratories, demonstrations, and performance exercises cover electronic warfare/intercept aviation systems. *Version 2:* Lectures, laboratories, demonstrations, and performance exercises cover electronic warfare/intercept aviation systems.

Credit Recommendation: ***Version 1:*** In the vocational certificate category, 2 semester hours in soldering techniques. In the lower-division baccalaureate/associate degree category, 3 semester hours in basic electronics laboratory, 1 in DC circuits, 3 in AC circuits, 4 in digital principles, 4 in solid state electronics, 3 in microprocessors, 4 in electronic systems troubleshooting and maintenance, and 4 in electronic communications (12/88). *Version 2:* In the vocational certificate category, 2 semester hours in soldering techniques. In the lower-division baccalaureate/associate degree category, 3 semester hours in basic electronics laboratory, 1 in DC circuits, 3 in AC circuits, 4 in digital principles, 4 in solid state electronics, 3 in microprocessors, 4 in electronic systems troubleshooting and maintenance, and 4 in electronic communications (6/97).

Related Occupations: 33R.

AR-1715-0717

ELECTRONIC WARFARE (EW)/INTERCEPT STRATEGIC RECEIVING SUBSYSTEM REPAIRER BASIC NONCOMMISSIONED OFFICER (NCO)

Course Number: 102-33P30.

Location: Intelligence School, Fort Devens, MA.

Length: 19 weeks (683–689 hours).

Exhibit Dates: 6/89–12/90.

Learning Outcomes: Upon completion of the course, the student will be able to teach and supervise enlisted personnel in state-of-the-art troubleshooting and alignment techniques associated with electronic warfare/intercept systems.

Instruction: Lectures, demonstrations, and practical exercises cover counseling, personnel supervision, operations management, modulation techniques, computer organization and maintenance, and electronic equipment maintenance and repair. This course contains a common core of leadership subjects.

Credit Recommendation: In the vocational certificate category, 3 semester hours in electronic systems troubleshooting and maintenance and 3 in computer systems troubleshooting and maintenance. In the lower-division baccalaureate/associate degree category, 3 semester hours in electronic communications, 2 in computer systems and organization, and 2 in personnel management. See AR-1406-0090 for common core credit (7/91).

Related Occupations: 33P.

AR-1715-0718

ELECTRONIC WARFARE (EW)/INTERCEPT STRATEGIC SIGNAL PROCESSING/STORAGE SUBSYSTEM REPAIRER

Course Number: 102-33Q10.

Location: Intelligence School, Fort Devens, MA.

Length: 46–47 weeks (1715 hours).

Exhibit Dates: 10/84–3/94.

Learning Outcomes: Upon completion of the course, the student will be able to identify and correct problems in digital, analog, solid state, and microprocessor-based signal processing/storage subsystems, including modulators/demodulators, A/D-D/A converters, power supplies, demultiplexers, and digital recorders.

Instruction: Classroom study and practical experience through the use of laboratory work are the methods of instruction.

Credit Recommendation: In the vocational certificate category, 4 semester hours in communications laboratory and 1 in PACE solder laboratory. In the lower-division baccalaureate/associate degree category, 4 semester hours in electricity/electronics (AC/DC) theory, 4 in solid state circuit analysis, 4 in digital logic, and 3 in microprocessor fundamentals (5/87).

Related Occupations: 33Q.

AR-1715-0719

ELECTRONIC WARFARE (EW)/INTERCEPT STRATEGIC SIGNAL PROCESSING AND STORAGE SUBSYSTEM REPAIRER BASIC NONCOMMISSIONED OFFICER (NCO)

Course Number: 102-33Q30.

Location: Intelligence School, Fort Devens, MA.

Length: 18–19 weeks (672–678 hours).

Exhibit Dates: 1/85–3/94.

Learning Outcomes: Upon completion of the course, the student will be able to teach and supervise selected enlisted personnel in troubleshooting and alignment techniques associated with electronic warfare intercept systems.

Instruction: Topics include military safety, map reading, first aid, counseling, personnel supervision, operations management, report writing, electronic troubleshooting, communications systems, and electronic maintenance. This course contains a common core of leadership subjects.

Credit Recommendation: In the vocational certificate category, 2 semester hours in map reading and 3 in advanced electronic maintenance. In the lower-division baccalaureate/associate degree category, 4 semester hours in communications systems, 5 in electronic troubleshooting, and 2 in personnel supervision. See AR-1406-0090 for common core credit (6/89).

Related Occupations: 33Q.

AR-1715-0720

ELECTRONIC WARFARE (EW)/INTERCEPT STRATEGIC SYSTEMS ANALYST AND COMMAND AND CONTROL SUBSYSTEMS REPAIRER

Course Number: 102-33M20.

Location: Intelligence School, Fort Devens, MA.

Length: 34 weeks (1236 hours).

Exhibit Dates: 10/86–9/90.

Learning Outcomes: Upon completion of the course, the student will be able to perform maintenance on high-speed serial RS-232 interface bus, IEE-488 parallel interface bus, disk and tape drive controllers, line printers, and minicomputers and write and run diagnostic Assembly language programs.

Instruction: Lectures, laboratories, demonstrations, and performance exercises cover electronic warfare command and control subsystems.

Credit Recommendation: In the lower-division baccalaureate/associate degree category, 4 semester hours in computer architecture, 3 in computer programming, 4 in minicomputer programming, 3 in minicomputer hardware, and 6 in computer maintenance (5/87).

Related Occupations: 33M.

AR-1715-0721

Electronic Warfare (EW)/Intercept Strategic Systems Analyst and Command and Control Subsystems Repairer Basic Noncommissioned Officer (NCO)

Course Number: 102-33M30.

Location: Intelligence School, Fort Devens, MA.

Length: 18–19 weeks (663 hours).

Exhibit Dates: 7/89–12/90.

Learning Outcomes: Upon completion of the course, the student will be able to analyze analog and digital communications system signals and modulation and signal processing circuits and maintain a minicomputer system.

Instruction: Lectures and laboratory exercises describe the theory and practice of electronic communications system and minicomputer system repair. This course contains a common core of leadership subjects.

Credit Recommendation: In the lower-division baccalaureate/associate degree category, 2 semester hours in computer maintenance and 2 in personnel management. See AR-1406-0090 for common core credit. In the upper-division baccalaureate category, 3 semester hours in communications systems (7/91).

Related Occupations: 33M.

AR-1715-0722

Electronic Warfare (EW)/Intercept Tactical Systems Repairer

Course Number: ***Version 1:*** 102-33T10. *Version 2:* 102-33T10.

Location: Intelligence School, Fort Devens, MA.

Length: ***Version 1:*** 39–40 weeks (1413–1424 hours). *Version 2:* 32 weeks (1192 hours).

Exhibit Dates: ***Version 1:*** 10/84–3/91. *Version 2:* 4/91–9/93.

Learning Outcomes: ***Version 1:*** Upon completion of the course, the student will be able to identify and correct problems in solid state, analog, digital, and microprocessor-based communications subsystems, including A/D-D/A converters, modulators-demodulators, UHF/VHF radio receivers, and video displays. *Version 2:* Upon completion of the course, the student will describe the theory, system concepts, and circuit design of linear and digital solid state electronics, analog communications, and 8-bit microcomputer hardware and organization; write Assembly-language code; and use microcomputer peripherals including A/D and D/A converters.

Instruction: ***Version 1:*** Lectures, laboratories, demonstrations, and performance exercises cover electronic warfare/intercept tactical systems repair. *Version 2:* Lectures, demonstrations, computer-assisted instruction, and laboratory exercises cover solid state devices and circuits, communications systems, analog and digital electronics, and 8-bit microcomputer hardware and software including I/O devices.

Credit Recommendation: ***Version 1:*** In the vocational certificate category, 4 semester hours in communications laboratory and 1 in PACE solder laboratory. In the lower-division baccalaureate/associate degree category, 4 semester hours in electricity/electronics (AC/DC theory), 4 in solid state circuit analysis, 4 in digital logic, and 3 in microprocessor fundamentals (5/87). *Version 2:* In the lower-division baccalaureate/associate degree category, 3 semester hours in AC/DC circuits, 3 in digital logic, and 3 in analog communications. In the upper-division baccalaureate category, 3 in operational amplifiers and applications and 3 in introduction to microcomputers (7/91).

Related Occupations: 33T.

AR-1715-0723

1. Electronic Warfare (EW)/Intercept Tactical Systems Repairer Basic Noncommissioned Officer (NCO)
2. Tactical Systems Repairer Basic Noncommissioned Officer (NCO)

Course Number: ***Version 1:*** 102-33T30. *Version 2:* 102-33T30.

Location: Intelligence School, Fort Devens, MA.

Length: ***Version 1:*** 17 weeks (617 hours). *Version 2:* 18–19 weeks (668–671 hours).

Exhibit Dates: ***Version 1:*** 10/86–12/90. *Version 2:* 1/91–3/94.

Learning Outcomes: ***Version 1:*** Upon completion of the course, the student will be able to teach and supervise selected enlisted personnel in troubleshooting and alignment techniques associated with electronic warfare intercept systems. *Version 2:* After 11/94 see AR-1715-1005. Upon completion of the course, the student will be able to teach and supervise selected enlisted personnel in troubleshooting and alignment techniques associated with electronic warfare intercept systems.

Instruction: ***Version 1:*** Topics include military safety, map reading, first aid practices, counseling, personnel supervision, operations management, report writing, electronic troubleshooting, communications systems, and electronic maintenance practices. This course contains a common core of leadership subjects. *Version 2:* Lectures, demonstrations, and practical exercises cover counseling, personnel supervision, operations management, modulation techniques, and electronic troubleshooting. This course contains a common core of leadership subjects.

Credit Recommendation: ***Version 1:*** In the vocational certificate category, 2 semester hours in map reading and 3 in advanced electronic maintenance. In the lower-division baccalaureate/associate degree category, 4 semester hours in communications systems, 5 in electronic troubleshooting, and 2 in personnel supervision. See AR-1406-0090 for common core credit (5/87). *Version 2:* In the vocational certificate category, 3 semester hours in electronic systems troubleshooting and maintenance. In the lower-division baccalaureate/associate degree category, 2 semester hours in electronic communications and 2 in personnel management. See AR-1406-0090 for common core credit (6/91).

Related Occupations: 33T.

AR-1715-0726

1. Electronic Warfare (EW)/Signal Intelligence (SIGINT) Noncommunications Interceptor Analyst Basic Noncommissioned Officer (NCO)
2. Noncommunications Interceptor/Analyst Basic Noncommissioned Officer (NCO)
3. Noncommunications Interceptor/Analyst Basic Noncommissioned Officer (NCO)

Course Number: ***Version 1:*** 233-98J30. *Version 2:* 233-98J30. *Version 3:* 233-98J30.

Location: Intelligence School, Fort Devens, MA.

Length: ***Version 1:*** 12 weeks (448 hours). *Version 2:* 9 weeks (328 hours). *Version 3:* 8–9 weeks.

Exhibit Dates: ***Version 1:*** 10/85–12/90. *Version 2:* 1/91–12/92. *Version 3:* 1/93–1/94.

Learning Outcomes: ***Version 1:*** Upon completion of the course, the student will be able to supervise and manage personnel in the interception of electro-optic signals, including the preparation of reports, selection of operations sites, and the interpretation and analysis of technical data and reports and describe basic radar operating principles and radio wave propagation and reception. *Version 2:* Upon completion of the course, the student will be able to supervise and manage personnel in the interpretation of electronic signals and the preparation of reports and analysis of technical data. *Version 3:* After 9/95 see AR-1404-0049. Upon completion of the course, the student will be able to supervise and manage personnel in the interception of electronic signals and the preparation of reports and analysis of technical data.

Instruction: ***Version 1:*** Instruction includes lectures, demonstrations, and performance exercises in office practice, personnel supervision, report preparation, map reading, radar systems and theory, and radar signal analysis. *Version 2:* Lectures, demonstrations, and performance exercises cover office practices, personnel supervision, report generation, and equipment operation. This course contains a common core of leadership subjects. *Version 3:* Instruction is presented through lectures, demonstrations, and performance exercises in personnel supervision, technical file management, data analysis, and operation and maintenance of technical equipment. This course contains a common core of leadership subjects.

Credit Recommendation: ***Version 1:*** In the lower-division baccalaureate/associate degree category, 1 semester hour in office management procedures and 2 in introduction to radar systems (5/87). *Version 2:* In the lower-division baccalaureate/associate degree category, 2 semester hours in personnel management. See AR-1406-0090 for common core credit (6/91). *Version 3:* In the lower-division baccalaureate/associate degree category, 3 semester hours in data base applications. See AR-1406-0090 for common core credit (6/97).

Related Occupations: 98J.

AR-1715-0728

1. Electronic Warfare (EW) Signals Intelligence (SIGINT) Voice Interceptor Basic Noncommissioned Officer (NCO)

2. Voice Interceptor Basic Noncommissioned Officer (NCO)

Course Number: *Version 1:* 232-98G3LXX. *Version 2:* 232-98G3LXX.

Location: Intelligence School, Fort Devens, MA.

Length: *Version 1:* 13–14 weeks (490 hours). *Version 2:* 9 weeks (327 hours).

Exhibit Dates: *Version 1:* 4/86–9/91. *Version 2:* 10/91–6/93.

Learning Outcomes: *Version 1:* Upon completion of the course, the student will be able to manage voice signal collection and processing activities, perform analysis of nonclear and nonvoice signals in a specific foreign language, operate equipment to collect and record specific radio transmissions, and assume operations management and supervisory responsibilities in voice signal interception. *Version 2:* Upon completion of the course, the student will be able to supervise and manage voice intercept operation and processing activities for cryptologic and electronic warfare operations.

Instruction: *Version 1:* Lectures, audiovisual presentations, case studies, seminars, and performance exercises cover personnel management, office management procedures, supervision techniques in voice signal collection, and foreign language transcription. *Version 2:* Lectures and experiential exercises cover supervision, planning, and management techniques. This course contains a common core of leadership subjects.

Credit Recommendation: *Version 1:* In the lower-division baccalaureate/associate degree category, 2 semester hours in operations management (supervision of maintenance management personnel) and 1 in office management procedures (5/87). *Version 2:* In the lower-division baccalaureate/associate degree category, 2 semester hours in personnel management. See AR-1406-0090 for common core credit (6/91).

Related Occupations: 98G.

AR-1715-0729

Telemetry Collection Operations
(Foreign Instrumentation Signals Externals Analysis)

Course Number: 233-ASIJ1.

Location: Intelligence School, Fort Devens, MA.

Length: 8–12 weeks (299–410 hours).

Exhibit Dates: 2/87–3/94.

Learning Outcomes: Upon completion of the course, the student will be able to identify missile and earth satellite vehicle designs, components, functions, and capabilities by the analysis of telemetry signals and use assorted reporting techniques.

Instruction: Lectures, audiovisual presentations, and performance exercises cover missile identification, knowledge of earth satellite vehicle characteristics, and reporting procedures involving signal analysis.

Credit Recommendation: Credit is not recommended because of the limited, specialized nature of the course (10/93).

AR-1715-0730

Quicklook 2 Operator

Course Number: 233-F8.

Location: Intelligence School, Fort Devens, MA.

Length: 4 weeks (135–136 hours).

Exhibit Dates: 2/83–3/94.

Learning Outcomes: Upon completion of the course, the student will be able to operate an automated data processing terminal to build files and generate reports, initialize system, process and edit data, and prepare reports.

Instruction: Lectures, demonstrations, and hands-on training cover file preparation, file management, and preparation of mission files and reports.

Credit Recommendation: In the lower-division baccalaureate/associate degree category, 2 semester hours in computer familiarization (3/88).

AR-1715-0731

Telemetry Internals Analysis

Course Number: 233-ASIZ7.

Location: Intelligence School, Fort Devens, MA.

Length: 5–6 weeks (200–208 hours).

Exhibit Dates: 10/84–6/91.

Learning Outcomes: Upon completion of the course, the student will be able to identify missile and earth satellite vehicle design, components, function, and capabilities by the analysis of telemetry signals and use assorted reporting techniques.

Instruction: Lectures, audiovisual presentations, and performance exercises cover missile identification, earth satellite vehicles, and reporting procedures involving signal analysis.

Credit Recommendation: Credit is not recommended because of the limited, specialized nature of the course (5/87).

AR-1715-0733

AN/TSQ-133 Special Purpose Receiving System (Tracechain) Maintenance
(Tracechain AN/FSQ-133 Special Purpose Receiving System Maintenance)

Course Number: 102-F61.

Location: Intelligence School, Fort Devens, MA; Field Station, Camp Humphries, Korea.

Length: 14–19 weeks (514–712 hours).

Exhibit Dates: 10/87–10/01.

Learning Outcomes: Upon completion of the course, the student will be able to maintain, test, and troubleshoot the AN/FSQ-133 special purpose receiving system (Tracechain) using standard and specialized software routines to locate faults and identify lowest replaceable units.

Instruction: Instruction includes the operation, maintenance, troubleshooting, and repair of the PDP11/73 with VCO receivers, spectrum-controlled 64KB command link, and Tracechain switch matrix using RTOSS with application/diagnostic software.

Credit Recommendation: In the vocational certificate category, 6 semester hours in computer systems troubleshooting and repair. In the lower-division baccalaureate/associate degree category, 3 semester hours in computer systems maintenance (10/91).

AR-1715-0734

1. Electronic Warfare (EW)/Signals Intelligence (SIGINT) Analyst Basic Noncommissioned Officer (NCO)
2. Signals Intelligence (SIGINT) Analyst Basic Noncommissioned Officer (NCO)

Course Number: *Version 1:* 232-98C30. *Version 2:* 232-98C30.

Location: Security School, Fort Devens, MA.

Length: *Version 1:* 10–14 weeks (379–502 hours). *Version 2:* 9 weeks (326 hours).

Exhibit Dates: *Version 1:* 3/86–12/90. *Version 2:* 1/91–3/94.

Learning Outcomes: *Version 1:* Upon completion of the course, the student will be able to read, construct, and use maps; prepare technical reports; recover encryption systems; perform critical mode analysis; perform quality control on product reports; and manage data bases. *Version 2:* Upon completion of the course, the student will be able to read, construct, and use maps; prepare technical reports; recover encryption systems; perform critical mode analysis; perform quality control on product reports; and manage data bases.

Instruction: *Version 1:* Lectures and exercises cover map reading, technical report preparation, quality control in product reports, encryption system relocation and recovery, and data base management. *Version 2:* Lectures and exercises cover map reading, technical report preparation, quality control in product reports, encryption system relocation and recovery, and data base management. This course contains a common core of leadership subjects.

Credit Recommendation: *Version 1:* In the lower-division baccalaureate/associate degree category, 2 semester hours in office management procedures and 1 in microcomputer applications (10/88). *Version 2:* In the lower-division baccalaureate/associate degree category, 2 semester hours in personnel management. See AR-1406-0090 for common core credit (6/91).

Related Occupations: 98C.

AR-1715-0737

Equipment Operation/Maintenance Test Set, Guided Missile System (TOW-Ground) Reserve

Course Number: 121-ASIJ9(RC).

Location: Ordnance, Missile and Munitions School, Redstone Arsenal, AL.

Length: 2 weeks (72 hours).

Exhibit Dates: 6/85–1/96.

Learning Outcomes: Upon completion of the course, the student will be able to operate, test, troubleshoot, and repair a guided missile test set.

Instruction: Lectures and practical exercises cover the test set.

Credit Recommendation: Credit is not recommended because of the military-specific nature of the course (4/88).

AR-1715-0738

Hawk Continuous Wave (CW) Radar Repairer

Course Number: 104-24K10.

Location: Ordnance, Missile and Munitions School, Redstone Arsenal, AL.

Length: 29–33 weeks (1152–1180 hours).

Exhibit Dates: 10/87–1/96.

Learning Outcomes: Upon completion of the course, the student will be able to solve series/parallel resistive networks and use test equipment to measure these quantities; apply the concepts of AC electricity to capacitors and resistors; analyze the operation of transistors, amplifiers, power supplies, oscillators, voltage regulators, and nonlinear transistor circuits; use binary, hexadecimal, and octal numbers; analyze circuits containing logic gates, counters, registers, adders, encoders, decoders, multiplexers, flip-flops, and analog/digital converters; apply the principles of computer organization and microprocessors, including register structure, memory, I/O, and addressing; write simple Assembly language programs; use standard laboratory equipment to measure and test analog and digital circuits and systems; measure and operate basic microwave devices; and troubleshoot and repair a complex military radar system.

Instruction: Theory and practical exercises cover the fundamentals of AC/DC circuits, transistor circuits, digital electronics, computers, microprocessors, and microwaves. The course includes soldering techniques and detailed troubleshooting and repair of a complex military radar system.

Credit Recommendation: In the vocational certificate category, 2 semester hours in soldering techniques. In the lower-division baccalaureate/associate degree category, 2 semester hours in DC circuits, 1 in AC circuits, 5 in solid state circuits, 1 in vacuum tube electronics, 3 in digital principles, 3 in microprocessors, 3 in computer systems and organization, 1 in radar systems, and 1 in electronic systems troubleshooting and maintenance (10/89).

Related Occupations: 24K.

AR-1715-0739

Hawk Fire Control Repairer

Course Number: 104-24H10.

Location: Ordnance, Missile and Munitions School, Redstone Arsenal, AL.

Length: 51 weeks (1854 hours).

Exhibit Dates: 10/87–1/96.

Learning Outcomes: Upon completion of the course, the student will be able to describe principles of DC/AC circuits including Ohm's law, series/parallel circuits, voltage dividers, and resonance; describe basic behavior of solid state devices such as transistors and zener diodes and circuits such as RC-coupled amplifiers, oscillators, multivibrators, and power supplies; use typical test equipment in troubleshooting; describe the basics of digital computers, including binary, octal, and hex arithmetic, AND/OR gates, flip-flops, counters, registers, arithmetic units, and A/D converters; explain I/O principles and use of typical memory devices; apply principles of stored-program computer operations; explain operation of sweep generators, limiters, video amplifiers, and voltage regulators using both solid state devices and vacuum tubes; repair electronic equipment, using approved soldering tools and techniques, on chassis, printed circuit board, and cables; operate microprocessor-based equipment using Assembly language programming; use logic probes and signature analyzer to troubleshoot; and troubleshoot and repair complex military equipment.

Instruction: Lectures and laboratory exercises include computer-directed lessons and tests. Detailed studies cover maintenance of complex military radar equipment.

Credit Recommendation: In the vocational certificate category, 2 semester hours in soldering techniques. In the lower-division baccalaureate/associate degree category, 2 semester hours in DC circuits, 1 in AC circuits, 4 in solid state devices, 1 in vacuum tube principles, 3 in digital principles, 3 in computer systems and organization, 3 in electronic systems troubleshooting and maintenance, and 3 in microprocessors (4/88).

Related Occupations: 24H.

AR-1715-0740

Hawk Launcher and Mechanical Systems Repairer Basic Noncommissioned Officer (NCO)

Course Number: 121-24L30.

Location: Ordnance, Missile and Munitions School, Redstone Arsenal, AL.

Length: 17–18 weeks (630–700 hours).

Exhibit Dates: 7/88–3/90.

Learning Outcomes: Upon completion of the course, the student will describe the operation of digital computers and computer-based signal processing systems and be able to troubleshoot computer and electronic systems to the component and module level.

Instruction: Course includes lectures, demonstrations, and performance exercises in digital logic devices, logic circuits, computer organization, computer programming, and troubleshooting of computers and computer-based signal processing systems.

Credit Recommendation: In the lower-division baccalaureate/associate degree category, 3 semester hours in computer organization and operation, 3 in electronic systems troubleshooting, and 3 in computer systems troubleshooting (8/91).

Related Occupations: 24L.

AR-1715-0741

Hawk Fire Control Repairer Basic Noncommissioned Officer (NCO)

Course Number: *Version 1:* 104-24H30. *Version 2:* 104-24H30.

Location: Ordnance, Missile and Munitions School, Redstone Arsenal, AL.

Length: *Version 1:* 11–12 weeks (430–455 hours). *Version 2:* 12–13 weeks (463 hours).

Exhibit Dates: *Version 1:* 4/86–3/90. *Version 2:* 4/90–1/96.

Learning Outcomes: *Version 1:* Upon completion of the course, the student will be able to troubleshoot and repair a complex military fire control system. *Version 2:* Upon completion of the course, the student will describe the operation of a radar-based electronic signal processing system and troubleshoot linear electronic systems to the module and component levels.

Instruction: *Version 1:* Theory and practical exercises cover the fundamentals of microprocessors and microprocessor-based computer systems, including address decoding, memory, instruction sets, software, and troubleshooting. A detailed study of a complex military missile control system is included. This course contains a common core of leadership subjects. *Version 2:* Course includes lectures, demonstrations, and practical exercises in electronic signal processing systems, continuous wave radar systems, and electronic signal processing systems. This course contains a common core of leadership subjects.

Credit Recommendation: *Version 1:* In the lower-division baccalaureate/associate degree category, 3 semester hours in microprocessors and 3 in electronic systems troubleshooting and maintenance. See AR-1406-0090 for common core credit (4/88). *Version 2:* In the lower-division baccalaureate/associate degree category, 3 semester hours in electronic system troubleshooting and 3 in continuous wave radar system organization. See AR-1406-0090 for common core credit (8/91).

Related Occupations: 24H.

AR-1715-0742

1. Hawk Pulse Radar Repairer Basic Noncommissioned Officer (NCO)
2. Hawk Field Maintenance Equipment/Pulse Acquisition Radar Repairer Basic Noncommissioned Officer (NCO)

Course Number: *Version 1:* 104-24J30. *Version 2:* 104-24J30.

Location: Ordnance, Missile and Munitions School, Redstone Arsenal, AL.

Length: *Version 1:* 19 weeks (696–725 hours). *Version 2:* 17–18 weeks (670 hours).

Exhibit Dates: *Version 1:* 4/86–3/90. *Version 2:* 4/90–1/96.

Learning Outcomes: *Version 1:* Upon completion of the course, the student will be able to troubleshoot and repair a complex military radar system. *Version 2:* Upon completion of the course, the student will be able to describe logic devices, circuits, and the organization and operation of digital computer CPUs and systems; write, execute, and debug Assembly language programs; troubleshoot a computer-based system to the module and component levels; and operate and troubleshoot a continuous wave radar-based signal processing system.

Instruction: *Version 1:* Theory and practical exercises cover the fundamentals of microprocessors and microprocessor-based computer systems, including address decoding, memory, instruction sets, software, and troubleshooting. A detailed study of a complex military missile control system is included. This course contains a common core of leadership subjects. *Version 2:* This course includes lectures, demonstrations, and performance exercises in digital logic devices; logic circuits; computer organization; computer programming; and troubleshooting computers, computer-based signal processing systems, and continuous wave radar-based systems. This course contains a common core of leadership subjects.

Credit Recommendation: *Version 1:* In the lower-division baccalaureate/associate degree category, 3 semester hours in microprocessors and 3 in electronic systems troubleshooting and maintenance. See AR-1406-0090 for common core credit (9/88). *Version 2:* In the lower-division baccalaureate/associate degree category, 3 semester hours in computer organization and operation, 3 in computer system troubleshooting, 3 in electronic systems troubleshooting, and 3 in continuous wave radar system organization. See AR-1406-0090 for common core credit (8/91).

Related Occupations: 24J.

AR-1715-0743

Hawk Continuous Wave (CW) Radar Repairer Basic Noncommissioned Officer (NCO)

Course Number: *Version 1:* 104-24K30. *Version 2:* 104-24K30.

Location: Ordnance, Missile and Munitions School, Redstone Arsenal, AL.

Length: *Version 1:* 15 weeks (551–580 hours). *Version 2:* 15–16 weeks (593 hours).

Exhibit Dates: *Version 1:* 2/89–3/90. *Version 2:* 4/90–1/96.

Learning Outcomes: *Version 1:* Upon completion of the course, the student will be able to troubleshoot and repair a complex military missile control system. *Version 2:* Upon completion of the course, the student will be able to troubleshoot and repair a complex military missile control system.

Instruction: *Version 1:* Theory and practical exercises cover the fundamentals of microprocessors and microprocessor-based computer systems, including address decoding, memory, instruction sets, software, and troubleshooting. A detailed study of a complex military missile control system is included. This course contains a common core of leadership subjects. *Version 2:* Theory and practical exercises cover the fundamentals of microprocessors and microprocessor-based computer systems, including address decoding, memory, instruction sets, software, and troubleshooting. A detailed study of a complex military missile control system is included. This course contains a common core of leadership subjects.

Credit Recommendation: *Version 1:* In the lower-division baccalaureate/associate degree category, 3 semester hours in microprocessors and 3 in electronic systems troubleshooting and maintenance. See AR-1406-0090 for common core credit (9/88). *Version 2:* In the lower-division baccalaureate/associate degree category, 3 semester hours in electronic systems troubleshooting and 3 in computer system troubleshooting. See AR-1406-0090 for common core credit (8/91).

Related Occupations: 24K.

AR-1715-0744

Hawk Pulse Radar Repairer Transition

Course Number: 104-24J10/20/30-T.

Location: Ordnance, Missile and Munitions School, Redstone Arsenal, AL.

Length: 8–9 weeks (308 hours).

Exhibit Dates: 11/87–1/96.

Learning Outcomes: Upon completion of the course, the student will be able to identify components and major functional blocks by use of schematics and manuals and perform service checks and repairs on complex military electronic equipment using specialized test equipment.

Instruction: Lectures and practical exercises cover the specific military equipment.

Credit Recommendation: In the lower-division baccalaureate/associate degree category, 2 semester hours in electronic systems troubleshooting and maintenance (4/88).

Related Occupations: 24J.

AR-1715-0745

Hawk Field Maintenance Equipment/ Pulse Acquisition Radar Repairer

Course Number: 104-24J10.

Location: Ordnance, Missile and Munitions School, Redstone Arsenal, AL.

Length: 46 weeks (1669 hours).

Exhibit Dates: 1/89–1/96.

Learning Outcomes: Upon completion of the course, the student will be able to describe the principles of DC/AC circuits, including Ohm's law, series/parallel circuits, voltage dividers, and resonance; describe basic behavior of solid state devices such as bipolar transistors, field-effect transistors, and zener diodes and circuits such as RC-coupled amplifier, oscillators, multivibrators, and power supplies; use typical test equipment in troubleshooting; describe the basics of digital computers, including binary, octal, and hexadecimal arithmetic units and analog/digital converters; explain I/O principles and use of typical memory devices; apply principles of stored-program computer operation; explain operation of sweep generators, limiters, video amplifiers, and voltage regulators using both solid state and vacuum tubes; repair electronic equipment, using approved soldering tools and techniques, on chassis, printed boards, and cables; and operate, maintain, and troubleshoot electronic systems and radar-based missile fire control systems.

Instruction: Lectures and laboratory instruction cover basic electricity, solid state devices and circuits, soldering techniques, basic computer operation, radar systems, and signal processing equipment.

Credit Recommendation: In the vocational certificate category, 1 semester hour in soldering techniques. In the lower-division baccalaureate/associate degree category, 3 semester hours in introduction to computer operations, 3 in basic electricity, 3 in solid state devices and circuits, 3 in troubleshooting electronic systems, and 3 in troubleshooting radar systems (8/91).

Related Occupations: 24J.

AR-1715-0746

1. Intermediate Maintenance Patriot Missile System Phases 1 and 2
2. Patriot System Repairer Phases 1 and 2

Course Number: *Version 1:* 632-ASIT5. *Version 2:* 632-27X2/3/4.

Location: *Version 1:* Air Defense Artillery School, Fort Bliss, TX; Ordnance, Missile and Munitions School, Redstone Arsenal, AL. *Version 2:* Phase 1, Ordnance, Missile and Munitions School, Redstone Arsenal, AL; Phase 2, Missile and Munitions School, Fort Bliss, TX.

Length: *Version 1:* Phase 1, 19–20 weeks (694 hours); Phase 2, 40 weeks (1440 hours). *Version 2:* Phase 1, 19–20 weeks (698 hours); Phase 2, 38–39 weeks (1378 hours).

Exhibit Dates: *Version 1:* 1/87–5/90. *Version 2:* 6/90–9/00.

Learning Outcomes: *Version 1:* Upon completion of the course, the student will be able to apply Ohm's law to solve series and parallel resistive circuits; use test equipment to measure DC circuit quantities; apply the concepts of AC electricity to the circuit behavior of inductors, capacitors, and combinations of resistors, inductors, and capacitors; use test equipment to generate and measure AC quantities; describe the behavior of BJT and FET transistor devices and the operation of amplifiers, power supplies, oscillators, and nonlinear circuits containing these devices; perform standard soldering and wire wrap repair techniques; apply the concepts of binary, octal, and hexadecimal numbers, Boolean AND, OR, NOR, XOR operators, and flip-flops; analyze the operation of digital circuits including decoders, encoders, registers, counters, and D/A and A/D converters; describe the operation of digital computers to the ALU and control register level; describe the operation of microprocessor-based computers to the subsystem and component level; write and debug Assembly language programs and apply standard diagnostic software and hardware techniques to find faults at the system and component level; describe the basic operation of microwave devices and systems; maintain, operate and troubleshoot a computer-based weapon system and associated radar systems. *Version 2:* Upon completion of the course, the student will be able to apply Ohm's law to solve series and parallel resistive circuits; use test equipment to measure DC circuit quantities; apply the concepts of AC electricity to the circuit behavior of inductors, capacitors, and combinations of resistors, inductors, and capacitors; use test equipment to generate and measure AC quantities; describe the behavior of BJT and FET transistor devices and the operation of amplifiers, power supplies, oscillators, and nonlinear circuits containing these devices; apply standard soldering and wire wrap repair techniques; apply the concepts of binary, octal, and hexadecimal numbers, Boolean AND, OR, NOR, XOR operators, and flip-flops; analyze the operation of digital circuits, including decoders, encoders, registers, counters, and D/A and A/D converters; describe the operation of digital computers to the ALU and control register level; describe the operation of microprocessor-based computers to the subsystem and component level; write and debug Assembly language programs and apply standard diagnostic software and hardware techniques to find faults at the system and component level; describe the basic operation of microwave devices and systems; and maintain, operate, and troubleshoot a computer-based weapon system and associated radar systems.

Instruction: *Version 1:* Lectures and laboratory exercises cover the above subject matter. Field exercises use simulated problems on complex, military computer- and radar-based systems. Course is for enlisted personnel. *Version 2:* Lectures, demonstrations, and practical exercises cover troubleshooting and repair of computer, radar, signal processing, and associated interface systems to the module level. Field exercises use simulated problems on complex military, computer-based and radar-based systems.

Credit Recommendation: *Version 1:* In the vocational certificate category, for Phase 1, 2 semester hours in solder and wire wrap techniques. In the lower-division baccalaureate/associate degree category, for Phase 1, 2 semester hours in DC circuits, 1 in AC circuits, 4 in solid state electronics, 3 in digital principles, 3 in computer systems and organization, and 4 in microprocessors. For Phase 2, 9 semester hours in computer systems troubleshooting and maintenance and 2 in radar systems (4/88). *Version 2:* In the vocational certificate category, for Phase 1, 1 semester hour in soldering techniques and 1 in wire wrap techniques. In the lower-division baccalaureate/associate degree category, for Phase 1, 3 semester hours in basic electricity, 5 in solid state devices and circuits, and 3 in computer organization and operation. For Phase 2, 3 semester hours in radar systems troubleshooting, 3 in computer system troubleshooting, and 3 in electronic system troubleshooting (8/91).

Related Occupations: 27X.

AR-1715-0747

Lance Repairer Basic Noncommissioned Officer (NCO)

Course Number: 121-27L30.

Location: Ordnance, Missile and Munitions School, Redstone Arsenal, AL.

Length: 7–8 weeks (277 hours).

Exhibit Dates: 8/88–1/96.

Learning Outcomes: Upon completion of the course, the student will be able to use binary, hexadecimal, and octal numbers; analyze circuits containing logic gates, counters, registers, adders, encoders, decoders, multiplexers, flip-flops, and analog/digital converters; perform precision soldering; and operate a missile maintenance unit.

Instruction: Lectures and practical exercises cover digital electronics, soldering, and operation of a missile maintenance unit. This course contains a common core of leadership subjects.

Credit Recommendation: In the vocational certificate category, 1 semester hour in soldering techniques. In the lower-division baccalaureate/associate degree category, 1 semester hour in electronic systems troubleshooting and maintenance and 3 in digital principles. See AR-1406-0090 for common core credit (10/88).

Related Occupations: 27L.

AR-1715-0748

LAND COMBAT SUPPORT SYSTEM TEST SPECIALIST

Course Number: ***Version 1:*** 121-27B10. *Version 2:* 121-35B10.

Location: Ordnance, Missile and Munitions School, Redstone Arsenal, AL.

Length: ***Version 1:*** 31 weeks (1154 hours). *Version 2:* 19–20 weeks (727 hours).

Exhibit Dates: ***Version 1:*** 9/87–9/95. *Version 2:* 10/95–9/00.

Learning Outcomes: ***Version 1:*** Upon completion of the course, the student will be able to apply Ohm's law to the solution of series and parallel resistive circuits; use test equipment to measure DC circuit quantities; apply the concepts of AC electricity to the circuit behavior of inductors, capacitors, and combinations of resistors, inductors, and capacitors; use test equipment to generate and measure AC quantities; describe the behavior of BJT and FET transistor devices and the operation of amplifiers, power supplies, oscillators, and nonlinear circuits containing these devices; apply the concepts of binary, octal, and hexadecimal numbers, Boolean algebra, and AND, OR, NOR, and XOR operators and flip-flops; describe the operation of counters, encoders, decoders, shift registers, and A/D and D/A circuits; and test, maintain, and repair electronic test sets, assemblies, and subassemblies that support various weapon systems. *Version 2:* Upon completion of the course, the student will be able to apply DC/AC fundamentals, basic electronic principles, and digital logic theory to test, maintain, and repair electronic test sets, assemblies, and subassemblies that support various weapons systems.

Instruction: ***Version 1:*** Lectures and practical exercises cover DC/AC electricity, semiconductor devices, and digital circuit fundamentals. Course includes exercises in Dragon, Lance, Shillelagh, and TOW missile systems. This course also includes soldering. *Version 2:* Lectures and practical exercises cover DC/AC electricity, semiconductor devices, and digital circuit fundamentals. Course includes exercises related to missile systems.

Credit Recommendation: ***Version 1:*** In the vocational certificate category, 1 semester hour in soldering techniques. In the lower-division baccalaureate/associate degree category, 2 semester hours in DC circuits, 1 in AC circuits, 3 in digital principles, 4 in solid state electronics, and 3 in electronic systems troubleshooting and maintenance (4/88). *Version 2:* In the lower-division baccalaureate/associate degree category, 3 semester hours in basic electricity, 3 in solid state electronics, and 3 in electronic systems troubleshooting and maintenance (5/01).

Related Occupations: 27B; 35B.

AR-1715-0749

MULTIPLE LAUNCH ROCKET SYSTEM (MLRS) REPAIRER BASIC NONCOMMISSIONED OFFICER (NCO)

Course Number: 121-27M30.

Location: Ordnance, Missile and Munitions School, Redstone Arsenal, AL.

Length: 7–8 weeks (273–277 hours).

Exhibit Dates: 7/86–1/96.

Learning Outcomes: Upon completion of the course, the student will be able to apply the concepts of binary, octal, and hexadecimal numbers, and Boolean functions including gates and flip-flops; analyze the operation of encoders, decoders, registers, arithmetic units, and A/D and D/A converters; and apply standard soldering procedures.

Instruction: Lectures and practical exercises cover digital logic devices and circuits. Practice covers basic soldering techniques. This course contains a common core of leadership subjects.

Credit Recommendation: In the vocational certificate category, 1 semester hour in soldering techniques. In the lower-division baccalaureate/associate degree category, 3 semester hours in digital principles. See AR-1406-0090 for common core credit (4/88).

Related Occupations: 27M.

AR-1715-0750

1. NUCLEAR WEAPONS TECHNICIAN WARRANT OFFICER TECHNICAL AND TACTICAL CERTIFICATION
2. NUCLEAR WEAPONS TECHNICIAN WARRANT OFFICER TECHNICAL/TACTICAL CERTIFICATION

Course Number: ***Version 1:*** 4E-911A. *Version 2:* 4E-911A.

Location: Ordnance, Missile and Munitions School, Redstone Arsenal, AL.

Length: ***Version 1:*** 19 weeks (688 hours). *Version 2:* 6–7 weeks (228 hours).

Exhibit Dates: ***Version 1:*** 10/88–5/91. *Version 2:* 6/91–1/96.

Learning Outcomes: ***Version 1:*** Upon completion of the course, the student will be able to identify prefire and cancelled fire procedures pertaining to nuclear weapons; weld; calibrate torque wrenches; calibrate, troubleshoot, and repair nuclear weapons test equipment; and conduct battle damage assessment and repair. *Version 2:* Upon completion of the course, the student will be able to identify prefire and cancelled fire procedures pertaining to nuclear weapons; weld; calibrate torque wrenches; calibrate, troubleshoot, and repair nuclear weapons test equipment; and conduct battle damage assessment and repair.

Instruction: ***Version 1:*** Lectures and practical exercises cover nuclear weapons operations. *Version 2:* This course covers nuclear weapons topics including calibration and inspection of nuclear weapons and associated equipment.

Credit Recommendation: ***Version 1:*** In the vocational certificate category, 1 semester hour in soldering techniques. In the lower-division baccalaureate/associate degree category, 2 semester hours in DC circuits and 1 in AC circuits (8/91). *Version 2:* Credit is not recommended because of the specialized nature of the course (8/91).

Related Occupations: 260A; 911A.

AR-1715-0751

PERSHING ELECTRONICS REPAIRER BASIC NONCOMMISSIONED OFFICER (NCO)

Course Number: 121-21L30.

Location: Ordnance, Missile and Munitions School, Redstone Arsenal, AL.

Length: 10 weeks (375 hours).

Exhibit Dates: 6/87–1/96.

Learning Outcomes: Upon completion of the course, the student will be able to apply the concepts of binary, octal, and hexadecimal numbers, Boolean operators, and gates and flip-flops; analyze encoders, decoders, registers, and A/D and D/A converters; describe the operation of digital computers to the ALU register level; write and debug machine language programs for computers and microprocessors; and analyze malfunctions of computer circuits and systems.

Instruction: Lectures, demonstrations, and practical exercises cover the subject matter. This course contains a common core of leadership subjects.

Credit Recommendation: In the vocational certificate category, 1 semester hour in soldering techniques. In the lower-division baccalaureate/associate degree category, 3 semester hours in digital principles, 3 in computer systems and organization, and 3 in microprocessors. See AR-1406-0090 for common core credit (4/88).

Related Occupations: 21L.

AR-1715-0752

PERSHING ELECTRONICS REPAIRER

Course Number: 121-21L10.

Location: Ordnance, Missile and Munitions School, Redstone Arsenal, AL.

Length: 32–33 weeks (1185 hours).

Exhibit Dates: 10/87–1/96.

Learning Outcomes: Upon completion of the course, the student will be able to apply Ohm's law to the solution of series and parallel resistive networks; use test equipment to measure DC circuit quantities; apply the concepts of AC electricity to the circuit behavior of inductors, capacitors, and combinations of resistors, inductors, and capacitors; use test equipment to generate and measure AC quantities; describe the behavior of BJT and FET transistor devices and the operation of amplifiers, power supplies, oscillators, and nonlinear circuits containing these devices; use binary, octal, and hexadecimal numbers to analyze circuits containing combinational gates and flip-flops, including decoders, encoders, registers, counters, and A/D and D/A converters; and test, maintain, and repair electronic test sets, assemblies, and subassemblies which support the Pershing missile system.

Instruction: Lectures and practical exercises cover Pershing missile electronic assembly operation, maintenance, and repair, including electronic fundamentals, missile electronics, electronic equipment maintenance, safety, portable test equipment, system analysis, and shop operation. Course also includes soldering.

Credit Recommendation: In the vocational certificate category, 2 semester hours in soldering techniques. In the lower-division baccalaureate/associate degree category, 2 semester hours in DC circuits, 1 in AC circuits, 4 in solid state electronics, 3 in digital principles, and 3 in electronic systems troubleshooting and maintenance (4/88).

Related Occupations: 21L.

AR-1715-0753

PHASE 1 AND PHASE 2 PATRIOT SYSTEM INTERMEDIATE MAINTENANCE TECHNICIAN

Course Number: 4F-F17.

Location: Air Defense Artillery School, Fort Bliss, TX; Ordnance, Missile and Munitions School, Redstone Arsenal, AL.

Length: Phase 1, 13–14 weeks (496 hours); Phase 2, 37–38 weeks (1357 hours).

Exhibit Dates: 10/87–9/91.

Learning Outcomes: Upon completion of the course, the student will be able to apply digital principles including number systems, logic gates, registers, counters, and logic analysis; describe the behavior of complex electronic devices and techniques such as digital signal processor, fiber optics, and time domain reflectometry; repair electronic equipment using approved soldering techniques and wire wrap procedures; describe computer system organization including flow charts, instruction sets, machine language, and peripherals; troubleshoot and repair faults in complex digital computer systems; operate complex test equipment, including logic analyzer, signature analyzer, and spectrum analyzer; and operate, troubleshoot, and repair microwave communications and radar equipment.

Instruction: Lectures, laboratory exercises, and field exercise use simulated problems on complex military computer equipment. Course is for warrant officers.

Credit Recommendation: In the vocational certificate category, for Phase 1, 2 semester hours in soldering techniques. In the lower-division baccalaureate/associate degree category, for Phase 1, 3 semester hours in digital principles, 3 in computer systems and organization, and 4 in microprocessors. For Phase 2, 9 semester hours in computer systems troubleshooting and maintenance and 2 in radar systems (4/88).

AR-1715-0754

TOW2 REPAIRER TRANSITION

Course Number: 4F-F14/121-F29.

Location: Ordnance, Missile and Munitions School, Redstone Arsenal, AL.

Length: 2–3 weeks (86–87 hours).

Exhibit Dates: 1/85–1/96.

Learning Outcomes: Upon completion of the course, the student will be able to test, troubleshoot, and repair a military weapon system.

Instruction: Classroom and practical exercises cover a specific weapon system.

Credit Recommendation: Credit is not recommended because of the limited, technical, nature of the course (4/88).

AR-1715-0755

TOW/DRAGON REPAIRER

Course Number: 121-27E10-RC.

Location: Ordnance, Missile and Munitions School, Redstone Arsenal, AL.

Length: 5 weeks (380 hours).

Exhibit Dates: 8/86–Present.

Learning Outcomes: Upon completion of the course, the student will be able to describe basic behavior of electricity; perform safety procedures; describe characteristics of components, including resistors, capacitors, inductors, diodes, and transformers; and troubleshoot and repair military electronic equipment.

Instruction: Home study includes electronics topics. Resident study includes military-specific equipment.

Credit Recommendation: In the vocational certificate category, 1 semester hour in fundamentals of electricity (5/01).

Related Occupations: 27E.

AR-1715-0756

VULCAN REPAIRER BASIC NONCOMMISSIONED OFFICER (NCO)

Course Number: 121-27F30.

Location: Ordnance, Missile and Munitions School, Redstone Arsenal, AL.

Length: 11 weeks (399–421 hours).

Exhibit Dates: 6/86–1/96.

Learning Outcomes: Upon completion of the course, the student will to apply concepts of microprocessor software and hardware, including number systems, address decoding, and use of memory and peripherals and perform troubleshooting and maintenance on complex military electronic equipment.

Instruction: Lectures and practical exercises cover typical microprocessor equipment and soldering. This course contains a common core of leadership subjects.

Credit Recommendation: In the vocational certificate category, 1 semester hour in soldering techniques. In the lower-division baccalaureate/associate degree category, 2 semester hours in microprocessors and 3 in electronic systems troubleshooting and maintenance. See AR-1406-0090 for common core credit (4/88).

Related Occupations: 27F.

AR-1715-0757

VULCAN REPAIRER TRANSITION

Course Number: 4F-F16/121-F30.

Location: Ordnance, Missile and Munitions School, Redstone Arsenal, AL.

Length: 6 weeks (216 hours).

Exhibit Dates: 3/88–1/96.

Learning Outcomes: Upon completion of the course, the student will be able to operate and maintain a military gun system.

Instruction: Lectures and practical exercises cover the operation and maintenance of a military gun system.

Credit Recommendation: Credit is not recommended because of the military specific-nature of the course (4/88).

AR-1715-0758

VULCAN REPAIRER

Course Number: 121-27F10.

Location: Ordnance, Missile and Munitions School, Redstone Arsenal, AL.

Length: 32–33 weeks (1233 hours).

Exhibit Dates: 10/87–1/96.

Learning Outcomes: Upon completion of the course, the student will be able to solve series/parallel resistive circuits; apply the concepts of AC electricity to the circuit behavior of inductors, capacitors, and resistors; analyze the operation of transistors, amplifiers, power supplies, oscillators, voltage regulators, and nonlinear transistor circuits; use binary, octal, and hexadecimal numbers; analyze circuits containing logic gates, counters, registers, adders, encoders, decoders, multiplexers, flip-flops, and analog/digital converters; measure and operate basic microwave devices; use standard laboratory equipment to measure and test analog and digital circuits and systems; and troubleshoot and repair a complex military fire control and radar system.

Instruction: Theory and practical exercises cover the fundamentals of AC/DC circuits, transistor circuitry, and digital electronics. The course also includes soldering techniques and detailed troubleshooting and repair of complex military radar and fire control equipment.

Credit Recommendation: In the vocational certificate category, 2 semester hours in soldering techniques. In the lower-division baccalaureate/associate degree category, 2 semester hours in DC circuits, 1 in AC circuits, 4 in solid state electronics, 2 in digital principles, 1 in radar systems, and 3 in electronic systems troubleshooting and maintenance (4/88).

Related Occupations: 27F.

AR-1715-0759

MEDICAL EQUIPMENT REPAIRER (UNIT LEVEL)

Course Number: *Version 1:* 4B-F2/198-35G10. *Version 2:* 4B-F2/198-35G10-Y-1. *Version 3:* 4B-670A/198-91A10; B-326-0010; 4B-F2/198-91A10.

Location: Medical Equipment and Optical School, Aurora, CO.

Length: *Version 1:* 38 weeks (1524 hours). *Version 2:* 38 weeks (1480–1567 hours). *Version 3:* 38 weeks (1480–1567 hours).

Exhibit Dates: *Version 1:* 9/87–9/90. *Version 2:* 10/90–11/95. *Version 3:* 11/95–Present.

Learning Outcomes: *Version 1:* Upon completion of the course the student will be able to solve resistive series/parallel DC circuits; analyze circuits containing capacitors and inductors; solve series and parallel resonance circuits; analyze diode and transistor circuits, including power amplifiers, operational amplifiers, and oscillators; analyze digital circuits, including gates, flip-flops, and counters; and troubleshoot and maintain a number of specific pieces of biomedical equipment, including special refrigerator, centrifuge, ultrasonic cleaner, sterilizer, blood cell counter, analyzer, blood gas apparatus, coagulation analyzer, spectrophotometer, defibrillator, electrocardiograph, X-ray system, respirator, and microscope. Student will also be able to identify major human anatomical features and physiological systems. *Version 2:* Upon completion of the course the student will be able to solder and desolder circuit components; solve resistive series/parallel DC circuits; analyze circuits containing capacitors and inductors; solve series and parallel resonance circuits; analyze diode and transistor circuits, including power amplifiers, operational amplifiers, and oscillators; analyze digital circuits, including gates, flip-flops, and counters; and troubleshoot and maintain a number of specific pieces of biomedical equipment, including special refrigerator, centrifuge, ultrasonic cleaner, sterilizer, blood cell counter, analyzer, blood gas apparatus, coagulation analyzer, spectrophotometer, defibrillator, electrocardiograph, X-ray system, respirator, and microscope. Student will also be able to identify major human anatomical features and physiological systems. *Version 3:* Upon completion of the course, the student will be able to replace modules and component parts; repair printed circuit boards; repair and adjust medical equipment using test, measurement, and diagnostic equipment; perform scheduled preventive maintenance checks and servicing; and calibrate, verify, and certify medical equipment. This is an interservice course open to active Army, Army reserve, Navy, and Department of Defense civilian personnel.

Instruction: ***Version 1:*** Theory and practical exercises cover AC/DC circuits, solid state electronics, digital electronics, and a wide variety of pieces of biomedical equipment. *Version 2:* Theory and practical exercises cover AC/DC circuits, solid state electronics, digital electronics, and a wide variety of biomedical equipment. *Version 3:* Theory and practical exercises cover AC/DC circuits, solid state electronics, digital electronics, and a wide variety of biomedical equipment.

Credit Recommendation: ***Version 1:*** In the lower-division baccalaureate/associate degree category, 2 semester hours in DC circuits, 2 in AC circuits, 4 in solid state electronics, 1 in digital principles, 9 in biomedical equipment troubleshooting and maintenance, and 1 in basic anatomy (1/90). *Version 2:* In the lower-division baccalaureate/associate degree category, 1 semester hour in soldering techniques, 3 in DC circuits, 3 in AC circuits, 4 in solid state electronics, 2 in digital principles, 9 in biomedical equipment troubleshooting and maintenance, and 1 in basic anatomy (7/95). *Version 3:* In the lower-division baccalaureate/associate degree category, 2 semester hours in solid state electronics; 6 in electronic theory; and 10 in basic medical equipment maintenance, troubleshooting, and repair. In the upper-division baccalaureate category, 6 semester hours in advanced medical equipment maintenance, troubleshooting, and repair (2/98).

Related Occupations: 35G; 91A.

AR-1715-0761

Intermediate Level Maintenance of AN/MSM-105(V)1

Course Number: 198-ASIW3.

Location: Signal School, Fort Gordon, GA.

Length: 18 weeks (686 hours).

Exhibit Dates: 10/86–9/90.

Learning Outcomes: Upon completion of the course, the student will be able to troubleshoot and repair automatic test system, including printed circuit boards, wiring harnesses, and cables with the aid of manual fault isolation procedures, if the internal self-diagnostic program fails.

Instruction: Classroom instruction on various number systems is included. Laboratory simulations are done using trainers for various computer systems. Training includes instruction in the procedures necessary to ensure quality.

Credit Recommendation: In the lower-division baccalaureate/associate degree category, 1 semester hour in DC circuits, 3 in digital principles, 2 in computer systems and organization, 4 in computer troubleshooting and maintenance, and 1 in electronic systems troubleshooting and maintenance (6/88).

AR-1715-0762

Radio Terminal/Repeater Set Operations Maintenance

Course Number: 201-ASIR6.

Location: Signal School, Fort Gordon, GA.

Length: 4–5 weeks (168 hours).

Exhibit Dates: 6/88–12/91.

Learning Outcomes: Upon completion of the course, the student will be able to install, operate, troubleshoot, and perform organizational maintenance on line-of-sight radio terminal, repeater, and the quick erect expandable mast.

Instruction: Classroom and practical exercises cover radio terminal/repeater sets.

Credit Recommendation: Credit is not recommended because of the military-specific nature of the course (6/88).

AR-1715-0763

Communications Security (COMSEC) TSEC/ST-58 Repair

Course Number: 160-F32.

Location: Signal School, Fort Gordon, GA.

Length: 10–11 weeks (432 hours).

Exhibit Dates: 3/88–9/95.

Learning Outcomes: Upon completion of the course, the student will be able to analyze and troubleshoot digital and microprocessor circuits and a specific military computer.

Instruction: Classroom and practical exercises cover topics in digital electronics, microprocessors, and the military computer.

Credit Recommendation: In the lower-division baccalaureate/associate degree category, 2 semester hours in digital principles, 2 in microprocessors, and 3 in computer systems troubleshooting and maintenance (6/88).

AR-1715-0764

Fixed Communications Security (COMSEC) Equipment Full Maintenance

Course Number: 160-ASIG8.

Location: Signal School, Fort Gordon, GA.

Length: 6–7 weeks (231 hours).

Exhibit Dates: 1/89–9/95.

Learning Outcomes: Upon completion of the course, the student will be to explain the basic operation of semiconductors, junction diodes, and bipolar transistors and identify their use in rectifier circuits and voltage regulators; troubleshoot and explain the operation of basic transistor amplifiers, power amplifiers, and operational amplifiers; solder and desolder electronic components to printed circuits boards; and maintain, troubleshoot, and repair a specific military communications system.

Instruction: Instruction is provided by classroom presentations, laboratory exercises, and troubleshooting and testing communications security equipment.

Credit Recommendation: In the vocational certificate category, 2 semester hours in soldering techniques. In the lower-division baccalaureate/associate degree category, 2 semester hours in solid state electronics, 1 in basic electronics laboratory, and 2 in electronic systems troubleshooting and maintenance (7/89).

AR-1715-0767

AN/TYC-39(V) I/O Maintenance (USAF)

Course Number: 150-F14.

Location: Signal School, Fort Gordon, GA.

Length: 8–9 weeks (320 hours).

Exhibit Dates: 10/87–12/97.

Learning Outcomes: Upon completion of the course, the student will be able to troubleshoot and repair a specific piece of telephone automatic message switching equipment; perform line installation procedures, patch procedures, and data entry; and troubleshoot associated peripheral equipment.

Instruction: Classroom and practical exercises are focused on equipment maintenance.

Credit Recommendation: In the lower-division baccalaureate/associate degree category, 3 semester hours in electronic systems troubleshooting and maintenance (6/88).

AR-1715-0768

AN/TTC-39A O/I Maintenance

Course Number: 150-F13.

Location: Signal School, Fort Gordon, GA.

Length: 9–10 weeks (360 hours).

Exhibit Dates: 10/87–4/04.

Learning Outcomes: Upon completion of the course, the student will be able to maintain and repair a piece of automatic central office telephone equipment.

Instruction: Instruction is presented by classroom and by practical experience in troubleshooting, testing, and data base entry using the above equipment.

Credit Recommendation: In the lower-division baccalaureate/associate degree category, 3 semester hours in electronic systems troubleshooting and maintenance (6/88).

AR-1715-0770

Signal Officer Advanced

Course Number: ***Version 1:*** 4-11-C22-25C. *Version 2:* 4-11-C22-25C.

Location: Signal School, Fort Gordon, GA.

Length: ***Version 1:*** 20 weeks (715–724 hours). *Version 2:* 20 weeks (756 hours).

Exhibit Dates: ***Version 1:*** 10/86–5/94. *Version 2:* 6/94–Present.

Learning Outcomes: ***Version 1:*** Upon completion of the course, the student will be able to supervise communications-electronics operations. *Version 2:* Upon completion of the course, the student will be able to supervise signal and communications operations.

Instruction: ***Version 1:*** Lectures cover combined arms subjects, including air/land battle doctrine; communications systems planning, management, and control, digital and analog engineering operations; and communications interfaces. Military subjects include electronic warfare, NBC defense, leadership, personnel administration, property accounting, training management, force integration, military justice, and tactics and doctrine. *Version 2:* Lectures, classroom discussion, projects and examinations cover the practical aspects of supporting a large military unit with necessary communications functions, including systems planning, management, and control; communications systems and interfaces; leadership; personnel administration; property accounting; and signal systems tactics and doctrine.

Credit Recommendation: ***Version 1:*** Credit is not recommended because of the military-specific nature of the course (7/89). *Version 2:* In the lower-division baccalaureate/associate degree category, 2 semester hours in introduction to computing, 2 in principles of supervision or introduction to management, 1 in human resource management, 3 in communications systems, and 2 in communications systems planning (5/00).

Related Occupations: 25C.

AR-1715-0771

Fixed Communications Security (COMSEC) Supervisor Basic Noncommissioned Officer (NCO)

Course Number: 160-29F30.

Location: Signal School, Fort Gordon, GA.

Length: 8 weeks (288 hours).

Exhibit Dates: 3/87–12/90.

Learning Outcomes: Upon completion of the course, the student will be able to lead and supervise personnel engaged in the maintenance of communications security equipment.

Instruction: Lectures and practical exercise cover topics in leadership, map reading, air/land battle, maintenance management, and shop practice. This course contains a common core of leadership subjects.

Credit Recommendation: Credit is recommended for the common core only. See AR-1406-0090 (6/89).

Related Occupations: 29F.

AR-1715-0772

Field Communication Security (COMSEC) Maintenance Supervisor Basic Noncommissioned Officer (NCO)

Course Number: 160-29S30.

Location: Signal School, Fort Gordon, GA.

Length: 6–8 weeks (230–288 hours).

Exhibit Dates: 3/87–9/95.

Learning Outcomes: Upon completion of the course, the student will be able to lead and supervise personnel engaged in the maintenance of communications security equipment.

Instruction: Lectures and practical exercises cover topics in leadership, map reading, air/land battle, maintenance management, and shop practice. This course contains a common core of leadership subjects.

Credit Recommendation: Credit is recommended for the common core only. See AR-1406-0090 (2/94).

Related Occupations: 29S.

AR-1715-0774

Satellite Communications (SATCOM) Terminal AN/GSC-52(V)

Course Number: ***Version 1:*** 102-ASIQ7 (P); 102-ASIQ7; 102-ASIQ7 (CT). *Version 2:* 102-ASIQ7 (P); 102-ASIQ7; 102-ASIQ7 (CT).

Location: Signal School, Fort Gordon, GA.

Length: ***Version 1:*** 7–8 weeks (273–281 hours). *Version 2:* 7–8 weeks (280 hours).

Exhibit Dates: ***Version 1:*** 10/86–9/93. *Version 2:* 10/93–4/95.

Learning Outcomes: ***Version 1:*** Upon completion of the course, the student will have an overall knowledge of the satellite communications terminal, its operational characteristics, and necessary preventive maintenance. *Version 2:* Upon completion of the course, the student will be able to operate and maintain a SATCOM communications terminal, including transmitting, receiving, frequency conversion, computer control, computer control subsystems, and fiber-optic assemblies.

Instruction: ***Version 1:*** Lectures and practical exercises cover the overall design, operation, and preventive maintenance of the SATCOM terminal. *Version 2:* Lectures and practical exercises cover the overall design, operation, and preventive maintenance of the SATCOM terminal and fiber-optic assemblies.

Credit Recommendation: ***Version 1:*** In the lower-division baccalaureate/associate degree category, 2 semester hours in principles of satellite communications systems (6/89). *Version 2:* In the lower-division baccalaureate/associate degree category, 1 semester hour in communications system troubleshooting and maintenance and 2 in principles of satellite communications systems (2/94).

Related Occupations: 29Y.

AR-1715-0775

Radio Terminal/Repeater Set Operator

Course Number: 202-ASIY1 (31M).

Location: Signal School, Fort Gordon, GA.

Length: 4 weeks (144 hours).

Exhibit Dates: 6/88–12/91.

Learning Outcomes: Upon completion of the course, the student will be able to install, operate, troubleshoot, and perform unit level maintenance on a radio set.

Instruction: Classroom and practical exercises are centered around the specific radio equipment.

Credit Recommendation: In the lower-division baccalaureate/associate degree category, 1 semester hour in electronic systems troubleshooting and maintenance (5/90).

Related Occupations: 31M.

AR-1715-0778

Automatic Telephone Central AN/TCC-38(V) Repair
(Automatic Telephone Central AN/TCC-38 Repair)

Course Number: 622-F29; 622-F22 (36L); 622-ASIP2.

Location: Signal School, Fort Gordon, GA.

Length: 12 weeks (437–459 hours).

Exhibit Dates: 8/83–6/95.

Learning Outcomes: Upon completion of the course, the student will be able to install and perform general support maintenance on the automatic telephone central office AN/TCC-38(V).

Instruction: Instruction includes classroom and laboratory experience necessary to service the equipment.

Credit Recommendation: In the lower-division baccalaureate/associate degree category, 3 semester hours in electronic systems troubleshooting and maintenance (11/89).

Related Occupations: 36L.

AR-1715-0779

Special Electronic Devices Repair

Course Number: 198-39E10.

Location: Signal School, Fort Gordon, GA.

Length: 25–28 weeks (933–1008 hours).

Exhibit Dates: 2/87–9/95.

Learning Outcomes: Upon completion of the course, the student will be able to solder, apply principles of basic DC and AC circuits and components, basic solid state electronics, and digital logic and numbering systems; and troubleshoot and repair special electronic devices, including night vision equipment, mine detection equipment, and thermal viewers through general support maintenance.

Instruction: Training includes topics in DC/AC circuits; basic electronics; precision soldering; solid state technology; and use of binary, octal, and hexadecimal numbering systems. Laboratory experience on troubleshooting techniques is included.

Credit Recommendation: In the vocational certificate category, 2 semester hours in soldering techniques. In the lower-division baccalaureate/associate degree category, 3 semester hours in basic electronics laboratory, 1 in DC circuits, 2 in AC circuits, 2 in digital principles, 3 in solid state electronics, and 3 in electronic systems troubleshooting and maintenance (8/92).

Related Occupations: 39E.

AR-1715-0781

Satellite Communications (SATCOM) Systems Repairer
(Satellite Communications (SATCOM) Equipment Repairer)

Course Number: ***Version 1:*** 102-29Y10. *Version 2:* 102-29Y10.

Location: Signal School, Fort Gordon, GA.

Length: ***Version 1:*** 18–41 weeks (642–1481 hours). *Version 2:* 37 weeks (1356 hours).

Exhibit Dates: ***Version 1:*** 3/88–9/91. *Version 2:* 10/91–9/94.

Learning Outcomes: ***Version 1:*** Upon completion of the course, the student will be able to solve DC series/parallel resistive circuits; calculate RLC AC circuits; calculate transients in RL and RC circuits; analyze transistor circuits, including power supplies, voltage regulators, amplifiers, oscillators, and multivibrators; use binary, octal, and hexadecimal numbering systems; analyze digital circuits containing gates, flip-flops, counters, and registers; analyze the circuitry of FM transmitters and receivers; troubleshoot and align FM transmitters and receivers; solder; use common test equipment and read schematics; and troubleshoot and repair several pieces of military satellite communications equipment. *Version 2:* Upon completion of the course, the student will be able to identify series, parallel, and series/parallel resistance circuits and determine values of resistance, voltage, and current in the circuit; recognize the meaning of frequency, period, sine wave, phase shift, inductive reactance, capacitive reactance, and impedance; calculate voltage and current in transformers; calculate quality factor, bandwidth, and cutoff frequencies of resonant circuits; calculate impedances, voltages, and current in capacitive and inductive AC circuits; explain the basic operation of semiconductors, junction diodes, and transistors and identify their uses in rectifier circuits and voltage regulators; describe communication techniques; trace signal path, troubleshoot, and align mobile FM transceivers; convert between and perform arithmetic in the decimal, binary, octal, and hexadecimal number systems; analyze logic circuits, flip-flop circuits, registers, and counters; operate, maintain, troubleshoot, and repair satellite earth terminal equipment including digital communications systems, using standard electronic test equipment and spectrum analyzer (using spread spectrum techniques).

Instruction: ***Version 1:*** Lectures and laboratory exercises cover topics in DC and AC circuits, solid state electronics, digital electronics, FM communications, satellite communications, soldering techniques, and theory of operation and troubleshooting methods for satellite communications equipment and devices. *Version 2:* Lectures and laboratory exercises cover topics in DC and AC circuits, solid state electronics, digital electronics, FM communications, satellite communications, soldering techniques, and theory of operation and troubleshooting methods for satellite communications equipment and devices.

Credit Recommendation: ***Version 1:*** In the vocational certificate category, 2 semester hours in soldering techniques. In the lower-division baccalaureate/associate degree category, 1 semester hour in DC circuits, 2 in AC circuits, 2 in digital principles, 3 in solid state electronics, 1 in electronic communications, 3 in basic electronics laboratory, and 3 in electronic systems troubleshooting and maintenance (6/89). *Version 2:* In the vocational certificate category, 2 semester hours in soldering techniques. In the lower-division baccalaureate/associate degree category, 3 semester hours in DC circuits, 1 in AC circuits, 1 in digital principles, 1 in solid state electronics, 2 in electronic communications, 2 in digital communications, 3 in basic electronics laboratory, and 3 in electronic system troubleshooting and maintenance (2/94).

Related Occupations: 29Y; 31S.

AR-1715-0782

Tactical Satellite/Microwave Systems Operator

Course Number: 201-31Q10.

Location: Signal School, Fort Gordon, GA.

Length: 15–16 weeks (580 hours).

Exhibit Dates: 12/87–6/94.

Learning Outcomes: Upon completion of the course, the student will be able to install, line up, operate, troubleshoot, and perform first-level maintenance and preventive maintenance on tactical, line-of-sight, and tropospheric scatter systems.

Instruction: Lectures, discussions, and demonstrations are employed. Evaluation is by test and by operational procedure and maintenance procedure checks.

Credit Recommendation: In the vocational certificate category, 3 semester hours in electronic system operation and maintenance (2/94).

Related Occupations: 31Q.

AR-1715-0783

1. OH-58D Aviation Intermediate Maintenance Avionic Repairer
2. OH-58D Special Electronic
3. OH-58D Special Electronics Equipment Repairer

Course Number: ***Version 1:*** 102-ASIW5 (35R). *Version 2:* 102-ASIW5 (68R). *Version 3:* 102-ASIW5 (35R).

Location: ***Version 1:*** Signal School, Fort Gordon, GA. *Version 2:* Signal School, Fort Gordon, GA. *Version 3:* Ordnance, Missile and Munitions School, Redstone Arsenal, AL.

Length: ***Version 1:*** 3–4 weeks (120 hours). *Version 2:* 6 weeks (216 hours). *Version 3:* 6 weeks (226 hours).

Exhibit Dates: ***Version 1:*** 10/87–2/93. *Version 2:* 3/93–9/95. *Version 3:* 10/95–Present.

Learning Outcomes: ***Version 1:*** Upon completion of the course, the student will be able to use test support system assemblies and perform diagnostic testing, troubleshooting, and repair of OH-58D mast-mounted sight and control display systems. *Version 2:* Upon completion of the course, the student will be able to use test support system assemblies; perform diagnostic testing, troubleshooting, and repair of a computer-controlled helicopter gun system; and maintain associated test equipment. *Version 3:* Upon completion of the course, the student will be able to maintain and repair avionic subsystems and components of the Kiowa Warrior aircraft.

Instruction: ***Version 1:*** Instruction is accomplished by lectures and laboratory experiences on the system. *Version 2:* Lectures and practical exercises cover the mechanical and electronic systems associated with the gun sight. *Version 3:* Lectures and practical exercises cover the test support system, control display subsystem avionics, and support equipment for the Kiowa Warrior Aircraft.

Credit Recommendation: ***Version 1:*** In the lower-division baccalaureate/associate degree category, 1 semester hour in electronic systems troubleshooting and maintenance (6/88). *Version 2:* In the lower-division baccalaureate/associate degree category, 3 semester hours in electronic systems troubleshooting and maintenance (3/94). *Version 3:* Credit is not recommended because of the military-specific nature of the course (5/01).

Related Occupations: 68R.

AR-1715-0784

AHIP Avionic Mechanic (OH-58D)

Course Number: ***Version 1:*** 102-ASIW5 (35K). *Version 2:* 102-ASIW5 (68N).

Location: Signal School, Fort Gordon, GA.

Length: ***Version 1:*** 3–4 weeks (128 hours). *Version 2:* 5–6 weeks (206 hours).

Exhibit Dates: ***Version 1:*** 10/87–3/91. *Version 2:* 4/91–9/95.

Learning Outcomes: ***Version 1:*** Upon completion of the course, the student will be able to maintain and repair avionic systems, including control/display subsystems, communications systems, Doppler/gyrocompass systems, radar altimeters, automatic target systems, altitude/heading systems, mast-mounted sight systems, and data bus interface. *Version 2:* Upon completion of the course, the student will be able to maintain and repair avionic systems, including control/display subsystems, communications systems, Doppler/gyrocompass systems, radar altimeters, automatic target systems, altitude/heading systems, mast-mounted sight systems, aircraft survivability equipment, and data bus interface.

Instruction: ***Version 1:*** Instruction is accomplished through lectures, computer-aided instruction, unit level maintenance tasks on avionic systems, and practical exercises. *Version 2:* Instruction is accomplished through lectures, computer-aided instruction, unit level maintenance tasks on avionic systems, and practical exercises.

Credit Recommendation: ***Version 1:*** In the vocational certificate category, 2 semester hours in aviation electronics systems (6/88). *Version 2:* In the vocational certificate category, 3 semester hours in avionic system troubleshooting and maintenance (2/94).

Related Occupations: 35K; 68N.

AR-1715-0787

Air Traffic Control Systems, Subsystems, and Equipment Repairer

(Tactical Air Traffic Control Systems, Subsystems, and Equipment Repairer)

Course Number: ***Version 1:*** 103-93D10; 102-F1. *Version 2:* 102-35D10; 102-93D10.

Location: ***Version 1:*** Signal School, Fort Gordon, GA. *Version 2:* Ordnance, Missile and Munitions School, Redstone Arsenal, AL; Signal School, Fort Gordon, GA.

Length: ***Version 1:*** 27–28 weeks (1077 hours). *Version 2:* 28–32 weeks (1035–1152 hours).

Exhibit Dates: ***Version 1:*** 10/88–9/90. *Version 2:* 10/90–9/95.

Learning Outcomes: ***Version 1:*** Upon completion of the course, the student will be able to identify series, parallel, and series/parallel circuits and determine values of resistance, voltage, and current in the circuits; describe concepts of alternating current, including frequency, phase, and reactance; explain concepts of series and parallel resonance and the basics of transformer operation; explain the basic operation of semiconductors, junction diodes, and bipolar transistors and identify their use in rectifier circuits and voltage regulators; troubleshoot and explain the operation of basic transistor amplifiers and operational amplifiers; explain the operation of FETs, SCRs, and UJTs and troubleshoot and explain the operation of sine wave oscillators, multivibrator circuits, and Schmitt trigger circuits; convert between and perform arithmetic in the decimal, binary, octal, and hexadecimal number systems; analyze logic circuits, flip-flop circuits, registers, and counters; explain theory and operation of microprocessor-based computers, including the concepts of buses, memory, control signals, and addressing; explain how a microcomputer performs arithmetic and logic operations; use machine language instructions; solder and desolder wire to various types of electronic terminals; solder and desolder electrical wires, cables, and terminals; and align, adjust, troubleshoot, and repair specific air traffic control systems. *Version 2:* After 9/95 see AR-1715-0993. Upon completion of the course, the student will be able to identify series, parallel, and series/parallel circuits and determine values of resistance, voltage, and current in the circuits; describe concepts of alternating current, including frequency, phase, and reactance; explain concepts of series and parallel resonance and the basics of transformer operation; explain the basic operation of semiconductors, junction diodes, and bipolar transistors and identify their use in rectifier circuits and voltage regulators; troubleshoot and explain the operation of basic transistor amplifiers and operational amplifiers; explain the operation of FETs, SCRs, and UJTs; troubleshoot and explain the operation of sine wave oscillator, multivibrator circuits, and Schmitt trigger circuits; convert between and perform arithmetic in the decimal, binary, octal, and hexadecimal number systems; analyze logic circuits, flip-flop circuits, registers, and counters; explain theory and operation of microprocessor-based computers, including the concepts of buses, memory, control signals, and addressing; explain how a microcomputer performs arithmetic and logic operations; use machine language instructions; solder and desolder wire to various types of electronic terminals; solder and desolder electrical wires, cables, and terminals; align, adjust, troubleshoot, and repair specific air traffic control systems.

Instruction: *Version 1:* Lectures and practical exercises cover the air traffic control system and its repair. *Version 2:* Lectures and practical exercises cover the air traffic control system and its repair.

Credit Recommendation: *Version 1:* In the vocational certificate category, 4 semester hours in soldering techniques. In the lower-division baccalaureate/associate degree category, 1 semester hour in DC circuits, 1 in AC circuits, 3 in solid state electronics, 3 in basic electronics laboratory, 1 in digital principles, 2 in microprocessors, and 3 in electronic systems troubleshooting and maintenance (6/89). *Version 2:* In the vocational certificate category, 1 semester hour in soldering techniques. In the lower-division baccalaureate/associate degree category, 2 semester hours in DC circuits, 1 in AC circuits, 4 in solid state electronics, 3 in basic electronics laboratory, 1 in digital principles, 2 in microprocessors, 3 in electronic systems troubleshooting and maintenance, and 3 in radar systems (8/96).

Related Occupations: 35D; 93D.

AR-1715-0788

WATCHMATE EQUIPMENT MAINTENANCE TRAINING

Course Number: 102-F59.

Location: Intelligence School, Fort Devens, MA.

Length: 3–4 weeks (116 hours).

Exhibit Dates: 2/87–3/94.

Learning Outcomes: Upon completion of the course, the student will be able to troubleshoot selected signal processing subsystem equipment to the line replacement unit level.

Instruction: Lectures and practical exercises in troubleshooting and maintenance of the Watchmate system.

Credit Recommendation: In the vocational certificate category, 2 semester hours in electronics laboratory (9/88).

AR-1715-0789

AN/TSQ-138 TRAILBLAZER OPERATOR

Course Number: 231-F20.

Location: Intelligence School, Fort Devens, MA.

Length: 2 weeks (70 hours).

Exhibit Dates: 7/88–3/94.

Learning Outcomes: Upon completion of the course, the student will be able to operate and perform preventive maintenance on the AN/TSQ-138 Trailblazer.

Instruction: Lectures and practical exercises cover to system setup, searching techniques, message handling, and antenna testing.

Credit Recommendation: Credit is not recommended due to the limited, specialized content of the course (9/88).

AR-1715-0790

ELECTRONIC PROCESSING AND DISSEMINATION SYSTEM OPERATOR/ANALYST

Course Number: *Version 1:* 3B-ASI9C/233-ASIT4; 233-F13/3B-ASI9C. *Version 2:* 3B-ASI9C/233-ASIT4; 233-F13/3B-ASI9C.

Location: *Version 1:* Intelligence School, Fort Devens, MA. *Version 2:* Intelligence Center and School, Fort Huachuca, AZ.

Length: *Version 1:* 8–9 weeks (293–326 hours). *Version 2:* 9–10 weeks (335–336 hours).

Exhibit Dates: *Version 1:* 1/87–9/92. *Version 2:* 10/95–12/96.

Learning Outcomes: *Version 1:* Upon completion of the course, the student will be able to operate an electronic processing and dissemination system and perform operational analysis on the electronic order-of-battle resident data base. *Version 2:* Upon completion of the course, the student will be qualified to operate the AN/TSQ-134(V) electronic processing and dissemination system and perform operational analysis on intelligence data.

Instruction: *Version 1:* Lectures and practical exercises cover the electronic processing and dissemination system, preventive maintenance, tactics, operations security, and message formats. *Version 2:* Lectures and practical exercises cover electronic processing and dissemination system and subsystems, data base maintenance, multichannel radio system, and use of specialized utility software.

Credit Recommendation: *Version 1:* In the lower-division baccalaureate/associate degree category, 2 semester hours in computer literacy (10/90). *Version 2:* In the lower-division baccalaureate/associate degree category, 3 semester hours in data base applications (6/97).

AR-1715-0791

ORGANIZATIONAL COMMUNICATIONS-ELECTRONICS (C-E) OPERATOR/MECHANICS

Course Number: SIG 9.

Location: Combined Arms Training Center, Bad Toelz, W. Germany.

Length: 3–4 weeks (108–137 hours).

Exhibit Dates: 10/85–11/98.

Learning Outcomes: Upon completion of the course, the student will be able to install and operate communications-electronics systems at the company-sized unit level; maintain the systems; identify capabilities and characteristics of FM radio sets, teletype equipment, switchboards, and security equipment; describe antenna types and radiation patterns; operate radioteletype equipment; install wire lines and telephone equipment; install remoting (radio/telephone) devices; explain maintenance programs and repair part stocking; evaluate the operation of repaired equipment; and troubleshoot faulty radios and teletype equipment using multimeters and test sets.

Instruction: Conferences and practical exercises emphasize diagnosis and fault isolation. Major topics include radio fundamentals and equipment, cable/wire systems and equipment, operational techniques and procedures, and maintenance/diagnosis.

Credit Recommendation: In the vocational certificate category, 1 semester hour in electronics circuitry. In the lower-division baccalaureate/associate degree category, 1 semester hour in radio and telephone communications systems (11/88).

Related Occupations: 31G; 31V.

AR-1715-0792

INTERMEDIATE LEVEL MAINTENANCE OF THE AN/GRC-106 AND THE MD-522

Course Number: SIG 35 AM.

Location: Combined Arms Training Center, Bad Toelz, W. Germany.

Length: 3 weeks (102 hours).

Exhibit Dates: 9/86–10/98.

Learning Outcomes: Upon completion of the course, the student will be able to operate test sets, multimeters, and counters; set up, operate, troubleshoot, and repair radios and modems; explain modes of operation of communications equipment (e.g., duplex vs. one-way reversible); and adjust test equipment and radio and modem components.

Instruction: Lectures and practical exercises cover analyzing, locating, troubleshooting, and correcting malfunctions in radio sets and modems.

Credit Recommendation: In the vocational certificate category, 2 semester hour in FM radio alignment. In the lower-division baccalaureate/associate degree category, 1 semester hour in communications equipment alignment (10/88).

AR-1715-0793

MULTICHANNEL COMMUNICATIONS SYSTEMS

Course Number: SIG 1.

Location: Combined Arms Training Center, Bad Toelz, W. Germany.

Length: 3 weeks (102 hours).

Exhibit Dates: 6/85–10/98.

Learning Outcomes: Upon completion of the course, the student will be able to plot simple radiopath profiles; install and orient antennas used with multichannel radio equipment; operate multichannel equipment in an electronic warfare environment; install and operate multichannel equipment configurations and components; and install and operate specific equipment configurations in 12/24-channel radio, 12/24-channel cable, integrated cable and radio, all in secure mode, for a specific military radio system.

Instruction: Lectures and hands-on instruction cover installing, operating, aligning, and performing preventive maintenance on radio communications systems, including receivers, transmitters, and antennas and their components.

Credit Recommendation: In the vocational certificate category, 2 semester hours in multichannel communications systems (10/88).

AR-1715-0794

TACTICAL TELECOMMUNICATIONS CENTER SUPERVISOR

Course Number: SIG 33.

Location: Combined Arms Training Center, Bad Toelz, W. Germany.

Length: 3 weeks (102 hours).

Exhibit Dates: 1/88–10/98.

Learning Outcomes: Upon completion of the course, the student will be able to establish, operate, and maintain radio/teletype communications; supervise the installation and operation of telecommunications center systems; supervise the processing and formatting of messages; install encrypting devices for message security; identify circuit capabilities for telegraph and teletype; troubleshoot and locate faults in circuits; and plan the installation of telecommunications centers.

Instruction: Conference classes and practical exercises cover the concepts and technology of telecommunications centers. Major subjects include communications center systems and communications center management.

Credit Recommendation: In the lower-division baccalaureate/associate degree cate-

gory, 2 semester hours in radio and radiotelegraphy and security (10/88).

AR-1715-0795

TACTICAL COMMUNICATIONS OFFICER AND CHIEF

Course Number: SIG 29.

Location: Combined Arms Training Center, Bad Toelz, W. Germany.

Length: 3–4 weeks (106–160 hours).

Exhibit Dates: 12/87–10/98.

Learning Outcomes: Upon completion of the course, the student will be able to plan and supervise the installation of wire, radio, or combined communications systems; supervise the operation and maintenance of communications systems; identify the capability and characteristics of AM/FM radio sets and telephone sets used in tactical communications centers; supervise the establishment of integrated wire/radio nets, including use of switchboards, remoting devices, and cryptographic devices; describe telegraphic principles, telegraphic equipment, and operations; describe multichannel/multiplex systems; operate radioteletype equipment; troubleshoot and debug communications systems; describe wave propagation, waveforms, wave fields, and antennas; and compute required antenna length.

Instruction: Conference classes, practical exercises, and other activities cover wire communications, radio communications, multiplexing, integrated wire/radio systems, radiotelegraphing, and radioteletyping.

Credit Recommendation: In the lower-division baccalaureate/associate degree category, 2 semester hours in radio and telephone communications systems (10/88).

AR-1715-0796

AIR DEFENSE ARTILLERY (ADA) OFFICER ADVANCED PATRIOT FOLLOW-ON
(Air Defense Artillery (ADA) Officer Advanced Follow-On Training Patriot Track)

Course Number: 2F-FOA-F15.

Location: Air Defense Artillery School, Fort Bliss, TX.

Length: 4–5 weeks (164–184 hours).

Exhibit Dates: 3/87–Present.

Learning Outcomes: Upon completion of the course, the student will be able to perform as an officer in a Patriot missile system unit.

Instruction: Lectures and practical exercises cover topics on the Patriot missile system, including preventive maintenance, deployment, emplacement, communications procedures, and battle tactics.

Credit Recommendation: Credit is not recommended because of the military-specific nature of the course (9/00).

AR-1715-0797

1. PATRIOT SYSTEM TECHNICIAN WARRANT OFFICER TECHNICAL/TACTICAL CERTIFICATION
(Patriot Missile System Technician Warrant Officer Technical Certification (WOTC), Phases 1 and 2)
2. PATRIOT SYSTEM TECHNICIAN WARRANT OFFICER BASIC
(Patriot System Technician Warrant Officer Technical/Tactical Certification)
(Patriot Missile System Technician Warrant Officer Technical Certification (WOTC) Phases 1 and 2)
3. PATRIOT SYSTEM TECHNICIAN WARRANT OFFICER BASIC
4. PATRIOT SYSTEM TECHNICIAN WARRANT OFFICER BASIC
5. PATRIOT SYSTEM TECHNICIAN WARRANT OFFICER BASIC
6. PATRIOT SYSTEM TECHNICIAN WARRANT OFFICER BASIC

Course Number: ***Version 1:*** 4B-140E. *Version 2:* 4F-140E. *Version 3:* 4F-140E. *Version 4:* 4F-140E. *Version 5:* 4F-140E, Phase 1. *Version 6:* 4F-140E.

Location: Air Defense Artillery School, Fort Bliss, TX.

Length: ***Version 1:*** 10–32 weeks (377–1152 hours). *Version 2:* 28–29 weeks (1016–1037 hours). *Version 3:* 27–28 weeks (984 hours). *Version 4:* 27–28 weeks (984 hours). *Version 5:* 31–32 weeks (1136 hours). *Version 6:* 34 weeks (1238 hours).

Exhibit Dates: ***Version 1:*** 1/89–10/91. *Version 2:* 11/91–1/97. *Version 3:* 2/97–12/97. *Version 4:* 1/98–5/01. *Version 5:* 6/01–10/02. *Version 6:* 11/02–Present.

Learning Outcomes: ***Version 1:*** Upon completion of Phase 1, the student will be able to perform operational checks and isolate faults in digital circuits and describe the functional operation and perform organizational maintenance on the Patriot missile system. Upon completion of Phase 2, the student will be able to describe the organization and operations of a weapons central computer and use diagnostic software to detect and isolate faults in the weapons system and associated radar system. *Version 2:* Upon completion of the course, the student will be able to perform operational checks and isolate faults in digital circuits, describe the functional operation and perform organizational maintenance on the Patriot missile system, describe the organization and operation of a weapons central computer, and use diagnostic software to detect and isolate faults in the weapons system and associated radar system. *Version 3:* Upon completion of the course, the graduate will be able to troubleshoot, operate, maintain, and repair radar, communications, and computer peripheral systems and equipment and describe the Patriot missile system and its components. *Version 4:* Upon completion of the course, the graduate will be able to troubleshoot, operate, maintain, and repair radar, communications, and computer peripheral systems and equipment and describe the Patriot missile system and its components. *Version 5:* Upon completion of the course, the graduate will be able to troubleshoot, operate, maintain, and repair radar, communications, and computer peripheral systems and equipment and describe the Patriot missile system and its components. *Version 6:* Pending evaluation.

Instruction: ***Version 1:*** Lectures, demonstrations, and practical experiences (Phase 1) cover basic digital circuits and the Patriot missile system. In Phase 2 the weapons control computer and radar system are covered. *Version 2:* Lectures, demonstrations, and practical experiences cover basic digital circuits and the Patriot missile system. Weapons control computer and radar system are also covered. *Version 3:* Classroom presentations, demonstrations, practical exercises, and examinations are used to create a knowledge base and repair and maintenance skills. Topics include radar function, operation, maintenance, and repair; electronic and digital signal processing; weapons systems; electronic communications; and computer peripheral function, operation, maintenance, and repair. *Version 4:* Classroom presentations, demonstrations, practical exercises, and examinations are used to create a knowledge base and repair and maintenance skills. Topics include radar function, operation, maintenance, and repair; electronic and digital signal processing; weapons systems; electronic communications; and computer peripheral function, operation, maintenance, and repair. *Version 5:* Classroom presentations, demonstrations, practical exercises, and examinations are used to create a knowledge base and repair and maintenance skills. Topics include radar function, operation, maintenance and repair; electronic and digital signal processing; weapons systems; electronic communications; and computer peripheral function, operation, maintenance and repair. *Version 6:* Pending evaluation.

Credit Recommendation: ***Version 1:*** In the lower-division baccalaureate/associate degree category, for Phase 1, 4 semester hours in digital principles, 1 in computer systems troubleshooting and maintenance, 2 in radar systems, and 1 in electronic systems troubleshooting and maintenance. For Phase 2, 3 semester hours in computer systems troubleshooting and maintenance, 1 in computer systems and organization, 2 in radar systems, and 2 in electronic systems troubleshooting and maintenance. If both phases are taken, credit should be added (11/88). *Version 2:* In the lower-division baccalaureate/associate degree category, 3 semester hours in digital circuits, 4 in computer systems troubleshooting and maintenance, 3 in radar systems, and 3 in electronic troubleshooting (6/95). *Version 3:* In the lower-division baccalaureate/associate degree category, 3 semester hours in electronics system troubleshooting, maintenance, and repair; 3 in radar system troubleshooting, maintenance, and repair; and 2 in computer system troubleshooting and repair (4/01). *Version 4:* In the lower-division baccalaureate/associate degree category, 2 semester hours in electronic system troubleshooting, maintenance, and repair and 3 in radar system troubleshooting, maintenance, and repair (4/01). *Version 5:* In the lower-division baccalaureate/associate degree category, 2 semester hours in electronics systems troubleshooting, maintenance, and repair and 5 in radar system troubleshooting, maintenance, and repair (4/01). *Version 6:* Pending evaluation.

Related Occupations: 140E.

AR-1715-0798

PATRIOT OPERATOR AND SYSTEM MECHANIC

Course Number: 632-24T1/2/3/4-T.

Location: Air Defense Artillery School, Fort Bliss, TX.

Length: 35–36 weeks (1267 hours).

Exhibit Dates: 4/88–11/98.

Learning Outcomes: Upon completion of the course, the student will be able to use common laboratory instruments and perform electronic laboratory work; explain the operation of solid state devices, including bipolar and FET transistors, diodes, UJTs, and SCRs; analyze solid state circuits, including amplifiers, voltage regulators, and operational amplifiers; perform conversions between decimal, binary, octal, and hexadecimal number systems; explain the operation of digital circuit elements, including logic gates, flip-flops, counters, registers, adders, and subtracters; and follow systematic troubleshooting procedures.

Instruction: Lectures and practical exercises cover the fundamentals of electrical measurements, solid state electronics, power supplies and power distribution, digital and computer fundamentals, electronic equipment operation and maintenance, and radar operation and maintenance.

Credit Recommendation: In the lower-division baccalaureate/associate degree category, 4 semester hours in solid state electronics, 3 in digital principles, 1 in basic electronics laboratory, 1 in industrial electronics, and 3 in electronic systems troubleshooting and maintenance (11/88).

Related Occupations: 24T.

AR-1715-0799

HAWK MASTER MECHANIC

Course Number: *Version 1:* 121-24R40-A. *Version 2:* 121-24R40-A.

Location: Air Defense Artillery School, Fort Bliss, TX.

Length: *Version 1:* 19–20 weeks (730 hours). *Version 2:* 13–14 weeks (500 hours).

Exhibit Dates: *Version 1:* 5/88–2/91. *Version 2:* 3/91–1/93.

Learning Outcomes: *Version 1:* Upon completion of the course, the student will be able to perform conversions between decimal, binary, octal, and hexadecimal numbers; describe the logic functions and circuit operation of logic gates, flip-flops, counters, registers, and adders; and troubleshoot and repair military radar systems. *Version 2:* Upon completion of the course, the student will be able to perform conversions between decimal, binary, octal, and hexadecimal numbers; describe the logic functions and circuit operation of logic gates, flip-flops, counters, registers, and adders; and troubleshoot and repair military radar systems.

Instruction: *Version 1:* Lectures and practical exercises cover topics in digital electronics and in the troubleshooting and repair of missile equipment. *Version 2:* Instruction in this course includes both lecture and hands-on laboratory experiences. Some subjects include conversions of various number systems; logic gates, counters, registers, adders, flip-flops, and various timing considerations; serial data stream analysis; radar transmitter/receiver circuits; microprocessor component analysis; and IFF transponder operations.

Credit Recommendation: *Version 1:* In the lower-division baccalaureate/associate degree category, 2 semester hours in digital principles, 2 in radar systems, and 3 in electronic systems troubleshooting and maintenance (11/88). *Version 2:* In the lower-division baccalaureate/associate degree category, 2 semester hours in digital principles, 2 in radar systems, and 2 in electronic systems troubleshooting and maintenance (6/95).

Related Occupations: 24R.

AR-1715-0800

MAN PORTABLE AIR DEFENSE (MANPAD) CREWMEMBER (STINGER) RESERVE COMPONENT

Course Number: 043-16S10-RC.

Location: Reserve Component Training Centers, US.

Length: 4 weeks (164 hours).

Exhibit Dates: 10/87–12/98.

Learning Outcomes: Upon completion of the course, the student will be able to visually identify projected aircraft images, maintain and operate weapon/training devices, program an IFF interrogator, track a radio-controlled aerial target, operate and maintain man portable air defense communications equipment, select a firing position and fire, and destroy weapons to prevent enemy use.

Instruction: Lectures and practical experience cover the Stinger weapon system and its maintenance and communications systems.

Credit Recommendation: Credit is not recommended because of the military-specific nature of the course (11/88).

Related Occupations: 16S.

AR-1715-0801

MAN PORTABLE AIR DEFENSE (MANPAD) CREW MEMBER (REDEYE) RESERVE COMPONENT

Course Number: 043-16S10-RC.

Location: Reserve Component Training Centers, US.

Length: 4 weeks (157 hours).

Exhibit Dates: 10/87–11/98.

Learning Outcomes: Upon completion of the course, the student will be able to visually identify projected aircraft images, maintain and operate weapon/training devices, charge M-49 batteries, track a radio-controlled aerial target, operate and maintain man portable air defense communications equipment, select a firing position and fire, and destroy weapons to prevent enemy use.

Instruction: Lectures and practical experience cover the skills required to prepare and fire the Redeye weapon system and maintain the system at the operator level.

Credit Recommendation: Credit is not recommended because of the military-specific nature of the course (11/88).

Related Occupations: 16S.

AR-1715-0802

HAWK MASTER MECHANIC

Course Number: 121-24R40-B.

Location: Air Defense Artillery School, Fort Bliss, TX.

Length: 7–9 weeks (281–300 hours).

Exhibit Dates: 6/88–1/92.

Learning Outcomes: Upon completion of the course, the student will be able to troubleshoot and repair the loader-transporter, missile launcher, and radar system of the Hawk missile.

Instruction: Lectures and practical exercises focus on troubleshooting and repairing the equipment.

Credit Recommendation: In the lower-division baccalaureate/associate degree category, 3 semester hours in electronic systems troubleshooting and maintenance (3/91).

Related Occupations: 24R.

AR-1715-0803

AIR DEFENSE ARTILLERY OFFICER ADVANCED HAWK FOLLOW-ON
(Air Defense Artillery (ADA) Officer Advanced Follow-On Training Hawk Track)

Course Number: 2F-FOA-F14.

Location: Air Defense Artillery School, Fort Bliss, TX.

Length: 4–5 weeks (156 hours).

Exhibit Dates: 3/88–6/96.

Learning Outcomes: Upon completion of the course, the student will be able to describe the operation and deployment of each item of equipment in the Hawk missile system.

Instruction: Lectures and practical exercises cover the Hawk missile system, its components, operation, and deployment.

Credit Recommendation: Credit is not recommended because of the military-specific nature of the course (10/90).

AR-1715-0807

AIR DEFENSE ARTILLERY (ADA) COMMAND AND CONTROL SYSTEMS TECHNICIAN WARRANT OFFICER TECHNICAL CERTIFICATION

Course Number: 4C-140A; 4C-225B.

Location: Air Defense Artillery School, Fort Bliss, TX.

Length: 26–27 weeks (942 hours).

Exhibit Dates: 1/88–6/94.

Learning Outcomes: Upon completion of the course, the student will be able to use common laboratory instruments and perform electronic laboratory work; describe the operation of vacuum tubes, motors, generators, relays, synchro systems, and servo systems; describe the basic principles of radar; and troubleshoot and repair radar and computer equipment.

Instruction: Lectures and practical exercises focus on operation and maintenance of the command and control system which includes radar, communications, and data processing equipment.

Credit Recommendation: In the lower-division baccalaureate/associate degree category, 3 semester hours in basic electronics laboratory, 3 in electronic systems troubleshooting and maintenance, 2 in radar systems, 2 in computer systems troubleshooting and maintenance, and 1 in industrial electronics (11/88).

Related Occupations: 140A; 225B.

AR-1715-0808

FORWARD AREA ALERTING RADAR (FAAR) OPERATOR
(Defense Acquisition Radar Operator)

Course Number: 043-16J10.

Location: Air Defense Artillery School, Fort Bliss, TX.

Length: 5–8 weeks (200–276 hours).

Exhibit Dates: 10/87–6/94.

Learning Outcomes: Upon completion of the course, the student will be able to explain the general operation of a Forward Area Alerting Radar system, operate the system, and perform operational checks.

Instruction: Lectures, demonstrations, and practical exercises, including field training, cover the operation of the radar system.

Credit Recommendation: Credit is not recommended because of the military-specific nature of the course (10/90).

Related Occupations: 16J.

AR-1715-0809

BASIC NONCOMMISSIONED OFFICER (NCO) 24N TRACK
(Chaparral System Mechanic Basic Noncommissioned Officer (NCO))

Course Number: 121-24N30.
Location: Air Defense Artillery School, Fort Bliss, TX.
Length: 5 weeks (172–220 hours).
Exhibit Dates: 10/87–6/94.
Learning Outcomes: Upon completion of the course, the student will be able to demonstrate common leadership skills, a working knowledge of maintenance management, and supervisory skills in the maintenance of the diesel power unit of a Chaparral weapon system.
Instruction: Lectures and practical experience cover leadership skills, maintenance and supply management, and maintenance of the diesel power unit of a Chaparral weapon system. This course contains a common core of leadership subjects.
Credit Recommendation: Credit is recommended for the common core only. See AR-1406-0090 (3/89).
Related Occupations: 24N.

AR-1715-0810

Vulcan System Mechanic Basic Noncommissioned Officer (NCO)

Course Number: 121-24M30.
Location: Air Defense Artillery School, Fort Bliss, TX.
Length: 5 weeks (168–220 hours).
Exhibit Dates: 10/87–7/94.
Learning Outcomes: Upon completion of the course, the student will plan, implement, supervise, and evaluate Vulcan air defense system maintenance.
Instruction: Lectures and practical exercises focus on military procedures and the specific equipment. This course contains a common core of leadership subjects.
Credit Recommendation: Credit is recommended for the common core only. See AR-1406-0090 (3/89).
Related Occupations: 24M.

AR-1715-0812

1. Air Defense Artillery (ADA) Crew Member
2. Chaparral Crewmember

Course Number: *Version 1:* 043-16P10. *Version 2:* 043-16P10.
Location: Air Defense Artillery School, Fort Bliss, TX.
Length: *Version 1:* 8–9 weeks (296 hours). *Version 2:* 8 weeks (286 hours).
Exhibit Dates: *Version 1:* 6/87–7/93. *Version 2:* 8/93–10/95.
Learning Outcomes: *Version 1:* Upon completion of the course, the student will be able to identify aircraft and operate communications equipment, missile equipment, and a missile carrier. *Version 2:* Upon completion of the course, the student will be able to operate and perform corrective maintenance on Chaparral communications equipment and vehicle.
Instruction: *Version 1:* Lectures and practical exercises cover the above topics. *Version 2:* Conferences, demonstrations, and practical exercises cover the basic operator skills of the Chaparral system, including operation, corrective maintenance, and crew/gunner functions.
Credit Recommendation: *Version 1:* Credit is not recommended because of the military-specific nature of the course (11/88). *Version 2:* In the vocational certificate category, 1 semester hour in automotive maintenance (6/95).
Related Occupations: 16P.

AR-1715-0813

Basic Noncommissioned Officer (NCO) 24T Track
(Patriot Operator and System Mechanic Basic Noncommissioned Officer (NCO))

Course Number: 632-24T30.
Location: Air Defense Artillery School, Fort Bliss, TX.
Length: 5–7 weeks (185–278 hours).
Exhibit Dates: 1/87–6/94.
Learning Outcomes: Upon completion of the course, the student will plan, implement, supervise, and evaluate tactical operations of the Patriot missile system and maintain system equipment.
Instruction: Lectures and practical exercises focus on the operation and maintenance on the missile system. This course contains a common core of leadership subjects.
Credit Recommendation: Credit is recommended for the common core only. See AR-1406-0090 (3/89).
Related Occupations: 24T.

AR-1715-0814

Hawk Firing Section Mechanic (PIP II)

Course Number: *Version 1:* 121-24C10. *Version 2:* 121-24C10.
Location: Air Defense Artillery School, Fort Bliss, TX.
Length: *Version 1:* 26 weeks (934 hours). *Version 2:* 16 weeks (590 hours).
Exhibit Dates: *Version 1:* 10/87–9/90. *Version 2:* 10/90–12/92.
Learning Outcomes: *Version 1:* Upon completion of the course, the student will be able to use common laboratory instruments to perform electronic laboratory work; analyze DC series, parallel, and series/parallel circuits; analyze series and parallel RL and RC circuits; make time constant calculations; make transformer calculations; explain the operation of solid state devices, including bipolar and FET transistors, diodes, UJTs, and SCRs; analyze solid state circuits, including class A, B, and C amplifiers, voltage regulators, operational amplifiers, oscillators, power supplies, multivibrators, and CRT circuits; describe the operation of vacuum tubes; perform conversions among decimal, binary, octal, and hexadecimal number systems; explain the operation of digital circuit elements, including logic gates, flip-flops, counters, registers, adders, subtracters, multiplexers, and demultiplexers; describe digital-to-analog and analog-to-digital converters; analyze superheterodyne receiver circuits; follow systematic troubleshooting procedures; and perform operational checks, fault isolation, and repair of Hawk missile and radar equipment. *Version 2:* Upon completion of the course, the student will be able to operate and maintain the launcher, the loader-transporter, and the illuminator radar of the Hawk system and integrate overall system control signals.
Instruction: *Version 1:* Lectures and practical exercises cover topics in basic electronics and maintenance of the missile system. *Version 2:* This course is taught using lectures coupled with extensive hands-on laboratory experiences. Topics include AC/DC circuits, transistors/FETs, thyristors, voltage regulation, power supplies, filters, resonance, and oscillators; digital concepts and devices, including counters, registers, A/D and D/A conversion, flip-flops, and logic gates; and basic hydraulic theory and power distribution.
Credit Recommendation: *Version 1:* In the lower-division baccalaureate/associate degree category, 2 semester hour in AC circuits, 1 in DC circuits, 4 in solid state electronics, 4 in digital principles, 1 in radar systems, 3 in basic electronics laboratory, and 3 in electronic systems troubleshooting and maintenance (11/88). *Version 2:* In the lower-division baccalaureate/associate degree category, 2 semester hours in DC circuits, 1 in AC circuits, 1 in solid state electronics, 2 in digital circuits, 2 in radar, and 2 in electronic systems troubleshooting and maintenance (6/95).
Related Occupations: 24C.

AR-1715-0815

Hawk Information Coordination Central Mechanic (PIP II)

Course Number: *Version 1:* 104-24G10 (PIP II). *Version 2:* 104-24G10 (PIP II).
Location: Air Defense Artillery School, Fort Bliss, TX.
Length: *Version 1:* 45 weeks (1618 hours). *Version 2:* 17 weeks (613–619 hours).
Exhibit Dates: *Version 1:* 4/88–10/90. *Version 2:* 11/90–12/92.
Learning Outcomes: *Version 1:* Upon completion of the course, the student will be able to use common laboratory instruments and perform electronic laboratory work; analyze DC series, parallel, and series/parallel circuits; analyze series and parallel RL and RC circuits; make time constant calculations; make transformer calculations; explain the operation of solid state devices, including bipolar and FET transistors, diodes, UJTs, and SCRs; analyze solid state circuits, including class A, B, and C amplifiers, voltage regulators, operational amplifiers, oscillators, power supplies, multivibrators, and CRT circuits; describe the operation of vacuum tubes and analyze various vacuum tube circuits; perform conversions between decimal, binary, octal, and hexadecimal number systems; explain the operation and truth table for digital circuit elements, including logic gates, flip-flops, counters, registers, adders, subtracters, multiplexers, and demultiplexers; describe digital-to-analog and analog-to-digital converters; analyze superheterodyne receiver circuits; and follow systematic troubleshooting procedures. *Version 2:* Upon completion of the course, the student will be able to perform unit maintenance on the radar, interrogator set, data processor, and control console systems; use common laboratory instruments and perform electronic laboratory work; analyze DC series, parallel, and series/parallel circuits; analyze series and parallel RL and RC circuits; make time constant calculations; make transformer calculations; explain the operation of solid state devices, including bipolar and FET transistors, diodes, UJTs, and SCRs; analyze solid state circuits, including class A, B, and C amplifiers, voltage regulators, operational amplifiers, oscillators, power supplies, multivibrators, and CRT circuits; describe the operation of vacuum tubes and analyze various vacuum tube circuits; perform conversions between decimal, binary, octal, and hexadecimal number systems; explain the operation and truth table for digital circuit elements, including logic gates, flip-flops, counters, registers,

adders, subtracters, multiplexers, and demultiplexers; describe digital-to-analog and analog-to-digital converters; analyze superheterodyne receiver circuits; and follow systematic troubleshooting procedures.

Instruction: *Version 1:* Lectures, demonstrations, and practical experiences cover AC and DC circuits, solid state electronics, digital circuits, automated data processing equipment, and radar systems. *Version 2:* Lectures and practical exercises cover the Hawk missile system and includes operational checks, adjustments, fault isolation, and unit level maintenance of the complete system. Topics include AC and DC circuits, solid state electronics, digital circuits, automated data processing equipment, and radar systems.

Credit Recommendation: *Version 1:* In the lower-division baccalaureate/associate degree category, 2 semester hours in AC circuits, 1 in DC circuits, 3 in solid state electronics, 4 in digital principles, 2 in radar systems, 3 in basic electronics laboratory, 1 in vacuum tube electronics, 3 in electronic systems troubleshooting and maintenance, and 3 in computer systems troubleshooting and maintenance (11/88). *Version 2:* In the lower-division baccalaureate/associate degree category, 2 semester hours in radar systems, 3 in AC/DC circuits, and 3 in electronic systems troubleshooting and maintenance (6/95).

Related Occupations: 24G.

AR-1715-0817

MAN PORTABLE AIR DEFENSE (MANPAD) CREW MEMBER

Course Number: 043-16S10.

Location: Air Defense Artillery School, Fort Bliss, TX.

Length: 7–8 weeks (253–261 hours).

Exhibit Dates: 12/86–10/01.

Learning Outcomes: Upon completion of the course, the student will be able to visually identify at least 30 aircraft types; operate and maintain Stinger and Redeye weapons systems and training devices; and operate and maintain associated transportation and communications equipment, such as 1/4 ton truck and trailer and radio.

Instruction: Lectures and practical exercises cover the operation and maintenance of weapon system components.

Credit Recommendation: Credit is not recommended because of the military-specific nature of the course (10/91).

Related Occupations: 16S.

AR-1715-0818

HAWK FIRE CONTROL CREW MEMBER

Course Number: 043-16E10 (PIP II).

Location: Air Defense Artillery School, Fort Bliss, TX.

Length: 7–9 weeks (278–313 hours).

Exhibit Dates: 11/88–1/93.

Learning Outcomes: Upon completion of the course, the student will be able to perform checks and procedures on a 60kW generator and a 2 1/2 ton truck; perform checks and procedures on radar; prepare Hawk for travel and emplacement; perform Hawk system orientation, alignment, and maintenance; and read, write, transmit, and receive coded tactical messages.

Instruction: Lectures and demonstrations cover the installation, operation, and preventive maintenance of equipment in the system.

Credit Recommendation: Credit is not recommended because of the military-specific nature of the course (6/91).

Related Occupations: 16E.

AR-1715-0819

VULCAN CREW MEMBER

Course Number: 041-16R10.

Location: Air Defense Artillery School, Fort Bliss, TX.

Length: 9–10 weeks (345 hours).

Exhibit Dates: 1/88–6/94.

Learning Outcomes: Upon completion of the course, the student will be able to visually recognize types of aircraft, perform maintenance on Vulcan systems including radar, and prepare Vulcan systems for operation in a tactical environment.

Instruction: Lectures and practical exercises cover the procedures for operation and maintenance of air defense weapons systems.

Credit Recommendation: Credit is not recommended because of the military-specific nature of the course (11/88).

Related Occupations: 16R.

AR-1715-0820

1. AN/TSQ-73 AIR DEFENSE ARTILLERY (ADA) COMMAND AND CONTROL SYSTEM OPERATOR/REPAIRER
2. AN/TSQ-73 AIR DEFENSE ARTILLERY (ADA) COMMAND AND CONTROL SYSTEM OPERATOR/MAINTAINER
   (AN/TSQ-73 Air Defense Artillery (ADA) Command and Control System Operator/Repairer)
3. AN-TSQ-73 AIR DEFENSE ARTILLERY COMMAND AND CONTROL SYSTEM OPERATOR/MAINTAINER

Course Number: *Version 1:* 150-25L10. *Version 2:* 150-25L10. *Version 3:* 150-25L10.

Location: Air Defense Artillery School, Fort Bliss, TX.

Length: *Version 1:* 31 weeks (1112 hours). *Version 2:* 24 weeks (878 hours). *Version 3:* 19–20 weeks (692 hours).

Exhibit Dates: *Version 1:* 11/87–7/90. *Version 2:* 8/90–12/95. *Version 3:* 1/96–12/98.

Learning Outcomes: *Version 1:* Upon completion of this course, the student will be able to use common test instruments and perform electronic tests; analyze DC series-parallel, and series/parallel circuits; analyze series and parallel RL and RC circuits; make time constant calculations; make transformer calculations; explain the operation of solid state devices, including bipolar and FET transistors, diodes, UJTs, and SCRs; analyze solid state circuits, including class A, B, and C amplifiers, voltage regulators, operational amplifiers, oscillators, power supplies, multivibrators, and CRT circuits; describe the operation of vacuum tubes and analyze various vacuum tube circuits; perform conversions among decimal, binary, octal, and hexadecimal number systems; explain the operation and truth table for digital circuit elements, including logic gates, flip-flops, counters, registers, adders, subtracters, multiplexers, and demultiplexers; describe digital-to-analog and analog-to-digital converters; analyze superheterodyne receiver circuits; and follow systematic troubleshooting procedures. *Version 2:* Upon completion of this course, the student will be able to use system console switches, auxiliary readout data fields, technical manuals, and test instruments to operate, maintain, and isolate system problems; identify console switch functions; hook up and operate the system in tactical modes; establish transmission zones, data communications, voice nets, and intercom functions; and perform signal tracing, system alignment, and safety procedures. *Version 3:* Upon completion of the course, the student will be able to use test equipment and troubleshooting techniques to maintain the analog and digital circuitry of communications and radar equipment.

Instruction: *Version 1:* Lectures and practical exercises cover basic electronics and the maintenance of a radar system and an interrogation system. *Version 2:* Conferences, demonstrations, and practical exercises cover basic system operations, maintenance, and fault isolation of an air defense artillery command and control system, including pulse acquisition radar and interrogator set. *Version 3:* The instruction is provided using lectures, written tests, and practical exercises. Much of the course is devoted to hands-on learning through the use of the practical exercises. Initialization and maintenance of computers and computer peripherals are taught in the course, and extensive practical exercise time is devoted to this area.

Credit Recommendation: *Version 1:* In the lower-division baccalaureate/associate degree category, 1 semester hour in AC circuits, 1 in DC circuits, 3 in solid state electronics, 4 in digital principles, 2 in radar systems, 3 in basic electronics laboratory, and 3 in electronic systems troubleshooting and maintenance (11/88). *Version 2:* In the lower-division baccalaureate/associate degree category, 3 semester hours in radar systems, 3 in electronic systems troubleshooting and maintenance, 1 in DC circuits, 1 in AC circuits, 1 in solid state electronics, and 1 in digital principles (5/95). *Version 3:* In the lower-division baccalaureate/associate degree category, 3 semester hours in radar systems, 3 in electronic systems troubleshooting and repair, and 3 in microcomputer system troubleshooting and repair (4/01).

Related Occupations: 25L.

AR-1715-0821

HAWK FIRE CONTROL MECHANIC (PIP II)
(Hawk Fire Control Maintenance)

Course Number: 121-24E10.

Location: Air Defense Artillery School, Fort Bliss, TX.

Length: 36–37 weeks (1321 hours).

Exhibit Dates: 2/88–2/90.

Learning Outcomes: Upon completion of the course, the student will be able to use common laboratory instruments and perform electronic laboratory work; analyze DC series, parallel, and series/parallel circuits; analyze series and parallel RL and RC circuits; make time constant calculations; make transformer calculations; explain the operation of solid state devices, including bipolar and FET transistors, diodes, UJTs, and SCRs; analyze solid state circuits including class A, B, and C amplifiers, voltage regulators, operational amplifiers, oscillators, power supplies, multivibrators, and CRT circuits; describe the operation of vacuum tubes and analyze various vacuum tube circuits; perform conversions among decimal, binary, octal, and hexadecimal number systems; explain the operation and truth table for digital circuit elements, including logic gates, flip-flops, counters, registers, adders, subtracters, multiplexers, and demultiplexers; describe digital-to-analog and analog-to-digital converters; analyze superheterodyne receiver circuits; follow systematic troubleshooting procedures; perform operational checks, fault isolation, and repair of the Hawk missile fire control system including radar.

Instruction: Lectures and practical exercises cover basic electronics and maintenance of the missile system.

Credit Recommendation: In the lower-division baccalaureate/associate degree category, 2 semester hours in AC circuits, 1 in DC circuits, 4 in solid state electronics, 4 in digital principles, 1 in radar systems, 3 in basic electronics laboratory, 1 in vacuum tube electronics, and 3 in electronic systems troubleshooting and maintenance (11/88).

Related Occupations: 24E.

AR-1715-0822

1. COMMUNICATIONS ELECTRONIC WARFARE (EW) OPERATIONS
   (Communications Electronic Warfare (EW) Equipment Operations)
   (Electronic Warfare (EW) Operations, Phase I)
2. COMMUNICATIONS ELECTRONIC WARFARE EQUIPMENT OPERATOR

Course Number: ***Version 1:*** 231-F32; 231-ASIK3. *Version 2:* 231-F32.

Location: ***Version 1:*** Intelligence School, Fort Huachuca, AZ; Intelligence School, Fort Devens, MA. *Version 2:* Intelligence School, Fort Huachuca, AZ.

Length: ***Version 1:*** 5–9 weeks (216–355 hours). *Version 2:* 6 weeks (231 hours).

Exhibit Dates: ***Version 1:*** 10/89–4/00. *Version 2:* 5/00–Present.

Learning Outcomes: ***Version 1:*** Upon completion of the course, the student will be able to operate and maintain electronic countermeasures equipment and radio receiver sets for purposes of communications jamming, data collection, and data reporting. *Version 2:* Upon completion of the course, the student will be able to operate and maintain electronic countermeasures equipment and radio receiver sets for purposes of communications jamming, data collection, and data reporting.

Instruction: ***Version 1:*** Lectures, demonstrations, and performance exercises cover basic radio communications theory, the operation and maintenance of electronic countermeasures equipment and radio receiver sets, and technical documentation and reporting procedures. *Version 2:* Lectures, demonstrations, and performance exercises cover basic radio communications theory, the operation and maintenance of electronic countermeasures equipment and radio receiver sets, and technical documentation and reporting procedures.

Credit Recommendation: ***Version 1:*** Credit is not recommended because of the limited, specialized nature of the course (6/97). *Version 2:* Credit is not recommended because of the limited, specialized nature of the course (8/01).

AR-1715-0823

PERSHING II ELECTRONICS MATERIEL SPECIALIST

Course Number: 121-21G10.

Location: Field Artillery School, Fort Sill, OK.

Length: 10 weeks (361 hours).

Exhibit Dates: 6/89–12/90.

Learning Outcomes: Upon completion of the course, the student will be able to operate and perform supervised preventive and corrective maintenance on the Pershing missile system.

Instruction: Lectures and laboratory exercises cover basic electricity/electronics theory and troubleshooting, generator operation, basic hydraulic systems, inertial measurement systems, and general organizational-level corrective maintenance.

Credit Recommendation: In the vocational certificate category, 3 semester hours in electrical laboratory (6/89).

Related Occupations: 21G.

AR-1715-0824

SPECIAL FORCES COMMUNICATIONS SERGEANT
(Special Operations Communications Sergeant)

Course Number: ***Version 1:*** 011-18E30. *Version 2:* 011-18E30, Phase 1.

Location: John F. Kennedy Special Warfare Center and School, Fort Bragg, NC.

Length: ***Version 1:*** 23–24 weeks (1568–1645 hours). *Version 2:* 28 weeks (1907 hours).

Exhibit Dates: ***Version 1:*** 10/87–3/00. *Version 2:* 4/00–Present.

Learning Outcomes: ***Version 1:*** Upon completion of the course, the student will be able to determine, by use of aerial photographs and military maps, proper locations, distances, elevations, and desirable routes; locate points on maps by use of azimuth distance and intersection techniques; care for and launch small boats; navigate to a selected landing site; prepare military orders for special forces operations; install antennas and operate specialized radio equipment to encrypt and decrypt messages; send and receive international Morse code at a speed of 15 five-letter code groups per minute; and operate field communications systems under adverse conditions. *Version 2:* Upon completion of the course, the student will be able to perform land navigation, tactics, communications, and marksmanship.

Instruction: ***Version 1:*** Instruction includes lectures and practical exercises covering maps, photographs, specialized radio sets, short- and long-range reconnaissance techniques, security systems, infiltration and exfiltration techniques, special survival techniques, airborne operations, cross-cultural communication, extensive unconventional warfare field experience, basic radio communications, antenna design and installation, international Morse code, and field operations in extreme environmental conditions. This course contains a common core of leadership subjects. *Version 2:* Lectures and practical exercises cover land navigation, patrol tactics, communications systems, Morse code, and marksmanship. This course contains a common core of leadership subjects.

Credit Recommendation: ***Version 1:*** In the vocational certificate category, 3 semester hours in Morse code and 3 in basic radio communications. In the lower-division baccalaureate/associate degree category, 1 semester hour in orienteering, 1 in survival/outdoor pursuits, 3 in electronic communications, and 1 in water safety. See AR-1406-0090 for common core credit (12/92). *Version 2:* In the lower-division baccalaureate/associate degree category, 3 semester hours in land navigation, 1 in map reading, and 3 in electronic communications. See AR-1406-0090 for common core credit (4/01).

Related Occupations: 18E.

AR-1715-0825

SPECIAL FORCES BURST COMMUNICATIONS SYSTEM MAINTENANCE

Course Number: 101-F23.

Location: John F. Kennedy Special Warfare Center and School, Fort Bragg, NC.

Length: 8 weeks (279–306 hours).

Exhibit Dates: 3/88–9/94.

Learning Outcomes: After 5/98 see AR-1715-1045. Upon completion of the course, the student will be able to plan, conduct, and support installation and unit-level maintenance of a burst communications system.

Instruction: Instruction includes lectures and hands-on training in general theory of burst communications systems, alignment and initialization procedures, high-frequency transmitters and subsystems, receiver and communications security subsystems, and field radio and power supplies for field communications systems.

Credit Recommendation: In the vocational certificate category, 3 semester hours in electronic troubleshooting. In the lower-division baccalaureate/associate degree category, 3 semester hours in electronic communications (1/91).

AR-1715-0826

SPECIAL OPERATIONS COMMUNICATIONS SERGEANT RESERVE COMPONENT PHASE 4

Course Number: 011-18E30-RC.

Location: John F. Kennedy Special Warfare Center and School, Fort Bragg, NC.

Length: 3 weeks (193 hours).

Exhibit Dates: 10/88–9/91.

Learning Outcomes: Upon completion of the course, the student will be able to send and receive international Morse code at a speed of ten five-letter code groups per minute; construct antennas; and install, tune, and operate special forces radio equipment.

Instruction: Instruction includes lectures in radio theory, differences in specialized radio sets, international Morse code, cryptographic systems, and installation and operation of specialized radio equipment.

Credit Recommendation: In the vocational certificate category, 3 semester hours in electronic communications (12/88).

Related Occupations: 18E.

AR-1715-0827

ELECTRONIC WARFARE (EW)/INTERCEPT STRATEGIC RECEIVING SUBSYSTEM REPAIRER

Course Number: 102-33P10.

Location: Intelligence School, Fort Devens, MA.

Length: 42–43 weeks (1562–1563 hours).

Exhibit Dates: 10/84–3/94.

Learning Outcomes: Upon completion of the course, the student will be able to identify and correct problems in solid state, analog, digital and microprocessor communication subsystems, including modulators/demodulators,

UHF/VHF radio receivers, video displays, and servo mechanisms.

Instruction: Lectures, laboratories, demonstrations, and performance exercises cover electronic warfare/intercept strategic receiving subsystems.

Credit Recommendation: In the vocational certificate category, 4 semester hours in communications laboratory and 2 in PACE solder laboratory. In the lower-division baccalaureate/associate degree category, 4 semester hours in electricity/electronics (AC/DC) theory, 4 in solid state circuit analysis, 4 in digital logic, and 3 in microprocessor fundamentals (5/87).

Related Occupations: 33P.

AR-1715-0829

FIELD ARTILLERY FIREFINDER RADAR OPERATOR

Course Number: 221-13R10.

Location: Field Artillery School, Fort Sill, OK.

Length: 5–8 weeks (190–277 hours).

Exhibit Dates: 10/88–9/97.

Learning Outcomes: Upon completion of the course, the student will be able to select a site for an artillery sighting radar and emplace, orient, check, adjust, and perform preventive maintenance on radar systems.

Instruction: Lectures and practical exercises cover the deployment and operator maintenance of radar sets that support field artillery; the use of maps for fire control; analysis of target data to differentiate between friendly and hostile fire; and emplacement, operation, and preventive maintenance of fire control radars.

Credit Recommendation: In the vocational certificate category, 3 semester hours in radar operations. In the lower-division baccalaureate/associate degree category, 3 semester hours in map reading (10/91).

Related Occupations: 13R.

AR-1715-0831

FIELD ARTILLERY DIGITAL SYSTEMS REPAIRER

Course Number: 113-39L10.

Location: Field Artillery School, Fort Sill, OK.

Length: 16–17 weeks (601 hours).

Exhibit Dates: 12/89–6/99.

Learning Outcomes: Upon completion of the course, the student will be able to use basic test equipment; apply basic principles of AC/DC circuits, analog and digital systems, computers, and radar systems; test, troubleshoot, and repair field artillery systems; and solder.

Instruction: Lectures and laboratory exercises cover the fundamentals of AC/DC circuits, solid state devices, digital circuits and systems and troubleshooting and repair of field digital systems and battery computer units.

Credit Recommendation: In the vocational certificate category, 2 semester hours in soldering. In the lower-division baccalaureate/associate degree category, 4 semester hours in AC/DC circuits, 3 in solid state electronics, 3 in digital principles, 3 in computer fundamentals, and 3 in electronic systems troubleshooting and repair (6/89).

Related Occupations: 39L.

AR-1715-0832

TACTICAL FIRE (TACFIRE) OPERATIONS SPECIALIST

Course Number: 250-13C10.

Location: Field Artillery School, Fort Sill, OK.

Length: 5 weeks (181 hours).

Exhibit Dates: 3/89–9/98.

Learning Outcomes: After 9/98 see AR-1715-1023. Upon completion of the course, the student will be able to tactically employ, operate, and maintain the Tactical Fire (TACFIRE) Direction System.

Instruction: Lectures and demonstrations cover the employment, operation, and maintenance of TACFIRE equipment; operation of the backup computer system and variable format message entry device; field artillery operations; and the duties, responsibilities, and mission of individuals working in a TACFIRE operations center.

Credit Recommendation: In the vocational certificate category, 1 semester hour in radio operations and 1 in operational maintenance. In the lower-division baccalaureate/associate degree category, 1 semester hour in map reading and 1 in computer operations (6/89).

Related Occupations: 13C.

AR-1715-0835

1. METEOROLOGICAL EQUIPMENT MAINTENANCE METEOROLOGICAL DATA SYSTEM (Meteorological Equipment Maintenance)
2. METEOROLOGICAL EQUIPMENT MAINTENANCE METEOROLOGICAL DATA SYSTEMS (Meteorological Equipment Maintenance)

Course Number: *Version 1:* 420-ASIH1. *Version 2:* 420-ASIH1.

Location: Field Artillery School, Fort Sill, OK.

Length: *Version 1:* 19 weeks (693 hours). *Version 2:* 10 weeks (360 hours).

Exhibit Dates: *Version 1:* 10/88–9/91. *Version 2:* 10/91–9/96.

Learning Outcomes: *Version 1:* Upon completion of the course, the student will be able to perform complete maintenance on the GMD/1 Rawin and MDS upper air measurement systems. *Version 2:* After 9/96 see AR-1715-1041. Upon completion of the course, the student will be able to perform organizational maintenance on a meteorological data system and a meteorological measuring system.

Instruction: *Version 1:* Lectures and practical exercises cover the individual sections of the GMD/1 and MDS Rawin systems, troubleshooting procedures for all portions of system, calibration and performance evaluations, repair of mechanical systems, standard test instruments, and review of all technical manuals. *Version 2:* Lectures and laboratory exercises cover the maintenance, troubleshooting, and repair of all parts of the meteorological data system and the meteorological measuring system. The initial part of the course covers basic DC/AC circuits and solid state theory.

Credit Recommendation: *Version 1:* In the vocational certificate category, 3 semester hours in introduction to electricity. In the lower-division baccalaureate/associate degree category, 6 semester hours in electronic systems troubleshooting and maintenance (6/89). *Version 2:* In the vocational certificate category, 3 semester hours in DC/AC circuit theory and solid state electronics. In the lower-division baccalaureate/associate degree category, 3 semester hours in electronic systems troubleshooting and maintenance (9/94).

AR-1715-0838

FIELD ARTILLERY FIREFINDER UNIT MAINTENANCE

Course Number: *Version 1:* 221-ASIX5. *Version 2:* 221-ASIX5.

Location: Field Artillery School, Fort Sill, OK.

Length: *Version 1:* 11–16 weeks (389–576 hours). *Version 2:* 8–12 weeks (246–428 hours).

Exhibit Dates: *Version 1:* 10/85–2/91. *Version 2:* 3/91–3/95.

Learning Outcomes: *Version 1:* Upon completion of the course, the student will be able to use the theory and specialized test equipment required for the calibration, alignment, and corrective maintenance on the field artillery firefinder radar set. *Version 2:* Upon completion of the course, the student will be able to use the theory and specialized test equipment required for the calibration, alignment, and corrective maintenance of a field artillery firefinder radar set.

Instruction: *Version 1:* Lectures and laboratory exercises cover basic electronics, digital circuits, number systems, solid state theory, and radar theory; use of oscilloscope and diagnostic test procedures to troubleshoot the radar computer, magnetic drive, and line printer; and corrective maintenance procedures for the radar buffer, synchronizer, and video processor. *Version 2:* Lectures and laboratory exercises cover the operation and maintenance of a radar system and its subsystems, including power distribution, signal processor, magnetic drive, receiver/exciter, beam steering unit, antenna positioning, line printer, etc. Topics covered include functional parts and their interrelationships; purpose and use of controls; the use of calibration, alignment, and troubleshooting procedures and systems; and an overview of DC/AC circuits and basic electronics.

Credit Recommendation: *Version 1:* In the vocational certificate category, 3 semester hours in basic electronics and 3 in radar laboratory. In the lower-division baccalaureate/associate degree category, 3 semester hours in electronic systems troubleshooting and maintenance (6/89). *Version 2:* In the vocational certificate category, 3 semester hours in basic DC/AC/digital/solid state circuits and 3 in radar laboratory (9/94).

AR-1715-0839

TARGET ACQUISITION/SURVEILLANCE RADAR REPAIRER

Course Number: 104-39C10.

Location: Field Artillery School, Fort Sill, OK.

Length: 23–26 weeks (889–932 hours).

Exhibit Dates: 10/88–4/04.

Learning Outcomes: Upon completion of the course, the student will be able to use basic test equipment; apply basic principles of AC/DC circuits, analog and digital systems, computers, and radar systems; and test, troubleshoot, and repair target acquisition/surveillance radars.

Instruction: Lectures and laboratory practice cover DC/AC/solid state/digital circuits and radar systems, computer fundamentals, and reliable soldering and troubleshooting techniques.

Credit Recommendation: In the vocational certificate category, 2 semester hours in soldering. In the lower-division baccalaureate/associate degree category, 4 semester hours in AC/DC circuits, 3 in solid state electronics, 3 in digital principles, 3 in computer fundamentals, 2 in radar systems, and 2 in electronic troubleshooting and repair (6/89).

Related Occupations: 39C.

AR-1715-0840

1. Target Acquisition Radar Technician Warrant Officer Technical Certification
2. Total Army Training System (TATS) Field Artillery Targeting Technician Warrant Officer Basic

Course Number: *Version 1:* 4C-131A; 4C-211A. *Version 2:* 4C-131A (F); 4C-131A.

Location: Field Artillery School, Fort Sill, OK.

Length: *Version 1:* 29–30 weeks (1088 hours). *Version 2:* 28–32 weeks (1011–1155 hours).

Exhibit Dates: *Version 1:* 12/87–9/98. *Version 2:* 10/98–Present.

Learning Outcomes: *Version 1:* Upon completion of the course, the student will be able to supervise and provide technical, operational, and corrective maintenance support for the target acquisition radar system. *Version 2:* Upon completion of the course, the student will be able to manage personnel and provide technical, operational, and corrective maintenance support for the target acquisition radar system.

Instruction: *Version 1:* Lectures and extensive laboratory exercises cover radar site selection; digital, solid state and basic electronics; radar operations and maintenance; antenna selection; test equipment; and system troubleshooting. *Version 2:* Lectures and laboratory exercises cover radar site selection; digital, solid state, and basic electronics; radar operations, systems, and maintenance; antenna selection; test equipment; and system troubleshooting to prepare the student to manage, operate, and maintain a field artillery radar control targeting system. Lectures and practical exercises to practice methodologies for various targeting scenarios are also included.

Credit Recommendation: *Version 1:* In the lower-division baccalaureate/associate degree category, 3 semester hours in basic electronics/electricity, 3 in electronics laboratory, 6 in electronic systems troubleshooting and repair, 3 in digital principles, 2 in microprocessors, and 3 in radar systems operation. In the upper-division baccalaureate category, 3 semester hours in radar systems (6/89). *Version 2:* In the lower-division baccalaureate/associate degree category, 3 semester hours in radar systems operations, troubleshooting, and repair; and 2 in digital principles. In the upper-division baccalaureate category, 3 semester hours in management (10/03).

Related Occupations: 131A; 211A.

AR-1715-0841

Target Acquisition Radar Technician Warrant Officer Technical Certification (Reserve Component)

Course Number: 4C-131A-RC.

Location: Field Artillery School, Fort Sill, OK.

Length: For Phases 2 and 4, 14–15 weeks (534 hours).

Exhibit Dates: 1/89–4/95.

Learning Outcomes: After 7/95 see AR-1715-1025. Upon completion of the course, the student will be able to supervise and provide technical, operational, and corrective maintenance support for the target acquisition radar system.

Instruction: Lectures and extensive laboratory exercises cover radar site selection; digital, solid state, and basic electronics; radar maintenance and operation; antenna selection; test equipment; and system troubleshooting.

Credit Recommendation: In the lower-division baccalaureate/associate degree category, 3 semester hours in basic electronics/electricity, 3 in electronics laboratory, 6 in electronic systems troubleshooting and repair, 3 in digital principles, 2 in microprocessors, and 3 in radar systems operation. In the upper-division baccalaureate category, 3 semester hours in radar systems (6/89).

Related Occupations: 131A.

AR-1715-0842

Pershing Missile Systems Technician Warrant Officer Technical Certification

Course Number: 4F-214E; 4F-130A.

Location: Field Artillery School, Fort Sill, OK.

Length: 18–25 weeks (668–901 hours).

Exhibit Dates: 11/84–12/90.

Learning Outcomes: Upon completion of the course, the student will be able to secure nuclear weapons and perform emergency nuclear weapons operations.

Instruction: Lectures and practical exercises cover AC/DC circuits, transistors and transistor circuits, logic circuits, microprocessors, Pershing missile system, and message authentication.

Credit Recommendation: In the lower-division baccalaureate/associate degree category, 2 semester hours in electronic systems troubleshooting and repair, 2 in DC circuits, 2 in AC circuits, 2 in solid state electronics, and 2 in microprocessors (6/89).

Related Occupations: 130A; 214E.

AR-1715-0843

Lance Missile System Technician Warrant Officer Technical Certification

Course Number: 4F-130B.

Location: Field Artillery School, Fort Sill, OK.

Length: 7–8 weeks (266 hours).

Exhibit Dates: 5/89–12/92.

Learning Outcomes: Upon completion of the course, the student will be able to identify the correct procedures for missile inspections, handling, and assembly; inspect, detect, and correct unit level maintenance problems; and update forms and records associated with the equipment.

Instruction: Lectures and laboratories cover the organization and mission of the Lance missile system; tactical employment and operation of the system; assembly, storage, and inspection procedures; characteristics and nomenclature of missile components; and the preparation and submission of missile material readiness reports.

Credit Recommendation: In the lower-division baccalaureate/associate degree category, 2 semester hours in DC circuits (6/89).

Related Occupations: 130B.

AR-1715-0844

Cannon Fire Direction Specialist

Course Number: 250-13E10.

Location: Field Artillery School, Fort Sill, OK.

Length: 7–8 weeks (250–280 hours).

Exhibit Dates: 3/89–9/98.

Learning Outcomes: After 9/98 see AR-1715-1028. Upon completion of the course, the student will be able to construct firing charts, plot reference points, target locations, make corrections for subsequent firing, input necessary meteorological data to the backup computer system, use various maps, and prepare and operate specialized radio sets.

Instruction: Topics include review of basic mathematics, lectures and practice in the construction and use of firing charts, use of meteorological data for input to the backup computer system, and specialized radio systems.

Credit Recommendation: In the vocational certificate category, 1 semester hour in radio communications. In the lower-division baccalaureate/associate degree category, 1 semester hour in computer operations and 1 in map reading (1/92).

Related Occupations: 13E.

AR-1715-0846

Fire Support Specialist
(Field Artillery Fire Support Specialist)

Course Number: *Version 1:* 250-13F10. *Version 2:* 250-13F10.

Location: Field Artillery School, Fort Sill, OK.

Length: *Version 1:* 6–7 weeks (216–243 hours). *Version 2:* 6–7 weeks (243–255 hours).

Exhibit Dates: *Version 1:* 10/88–12/91. *Version 2:* 1/92–9/97.

Learning Outcomes: *Version 1:* Upon completion of the course, the student will be able to prepare digital messages and operate message devices, determine target locations by any of several methods, operate laser and thermal sighting devices, determine and convert map coordinates, and use specialized radio systems. *Version 2:* Upon completion of the course, the student will be able to locate, request, and adjust indirect fire support using lasers and digital communications equipment.

Instruction: *Version 1:* Instruction includes lectures and practical exercises in map reading, land navigation, resection, polar plotting, installation and operation of radio systems, thermal and visual sighting devices, and transmission of artillery sighting information. *Version 2:* Lectures and practical exercises cover target acquisition, map reading, land navigation, terrain analysis, radio communications, and firing procedures.

Credit Recommendation: *Version 1:* In the vocational certificate category, 1 semester hour in radio operation. In the lower-division baccalaureate/associate degree category, 1 semester hour in map reading (1/92). *Version 2:* In the lower-division baccalaureate/associate degree category, 2 semester hours in map reading and terrain analysis (6/95).

Related Occupations: 13F.

AR-1715-0847

Multiple Launch Rocket System (MLRS)/ Lance Operations Fire Specialist

Course Number: 043-13P10.

Location: Field Artillery School, Fort Sill, OK.

Length: 10–12 weeks (384–452 hours).

Exhibit Dates: 2/89–9/97.

Learning Outcomes: After 9/97 see AR-1715-1044. Upon completion of the course, the student will be able to read and interpret maps; plot locations; compute distance, direction, and altitude; use radio communications equipment; use digital message devices; operate a computer system; and install and operate telephones with security equipment.

Instruction: Lectures, group activities, and performance exercises cover map interpretation, communications equipment and maintenance, and computer use.

Credit Recommendation: In the vocational certificate category, 1 semester hour in radio operations. In the lower-division baccalaureate/associate degree category, 1 semester hour in map reading (6/91).

Related Occupations: 13P.

AR-1715-0848

1. Telecommunications Terminal Device Repairer (Teletypewriter Equipment Repairer)
2. Telecommunications Terminal Device Repairer (Teletypewriter Equipment Repairer)
3. Telecommunications Terminal Device Repairer
4. Computer/Automation Systems Repairer

Course Number: ***Version 1:*** 160-29J10. *Version 2:* 160-29J10. *Version 3:* 160-35J10. *Version 4:* 160-35J10.

Location: ***Version 1:*** Signal School, Fort Gordon, GA. *Version 2:* Signal School, Fort Gordon, GA. *Version 3:* Signal School, Fort Gordon, GA; Ordnance, Missile and Munitions School, Redstone Arsenal, AL. *Version 4:* Ordnance Missile and Munitions School, Fort Gordon, GA.

Length: ***Version 1:*** 17–18 weeks (712 hours). *Version 2:* 20–26 weeks (740–920 hours). *Version 3:* 20 weeks (733 hours). *Version 4:* 13–15 weeks (528–540 hours).

Exhibit Dates: ***Version 1:*** 1/89–9/91. *Version 2:* 10/91–9/95. *Version 3:* 10/95–12/98. *Version 4:* 10/99–Present.

Learning Outcomes: ***Version 1:*** Upon completion of the course, the student will be able to identify series, parallel, and series-parallel circuits and determine values of resistance, voltage, and current in the circuits; apply concepts of alternating current, including frequency, phase, and reactance; explain the basic operation of semiconductors, junction diodes, and bipolar transistors and identify their use in rectifier circuits and voltage regulators; convert between and perform arithmetic in the decimal, binary, octal, and hexadecimal number systems; analyze logic circuits, flip-flop circuits, registers, and counters; explain the theory and operation of microprocessor-based computers, including the concepts of buses, memory, control signals, and addressing; and install, troubleshoot, and repair teletypewriters, facsimile equipment, bar code scanners, and console panels. *Version 2:* Upon completion of the course, the student will be able to identify series, parallel, and series-parallel resistance circuits and determine values of resistance, voltage, and current in the circuit; explain the basic operation of semiconductors, junction diodes, and transistors and identify their uses in rectifier circuits and voltage regulators; convert between and perform arithmetic in the decimal, binary, octal, and hexadecimal number systems; analyze logic circuits, flip-flop circuits, registers, and counters; explain the theory and operation of microprocessor-based computers, including the concepts of buses, memory, control signals, and addressing; explain how a microcomputer performs arithmetic and logic operations; use machine language instructions; solder and desolder wire to various types of electronic terminals; and solder and desolder electronic components to printed circuit boards. *Version 3:* Upon completion of the course, the student will be able to perform maintenance on various computerized telecommunications devices. *Version 4:* Upon completion of the course, the student will be able to use basic electric fundamentals to perform maintenance on various computerized telecommunications devices.

Instruction: ***Version 1:*** Lectures, practical exercises, and exams cover the subject matter. *Version 2:* Lectures, practical exercises, and exams cover the subject matter. *Version 3:* Lectures and practical exercises cover DC/AC circuits; power supply circuits; computer application and repair; soldering; and such communication terminal devices as printers, CRT monitors and facsimile units. *Version 4:* Lectures and practical exercises cover basic electronics and repair of such terminal devices as printers, monitors, and facsimile units.

Credit Recommendation: ***Version 1:*** In the lower-division baccalaureate/associate degree category, 1 semester hour in AC circuits, 1 in DC circuits, 1 in solid state electronics, 1 in digital principles, 1 in microprocessors, 3 in electronic systems troubleshooting and maintenance, and 3 in basic electronics laboratory (7/89). *Version 2:* In the vocational certificate category, 1 semester hour in soldering techniques. In the lower-division baccalaureate/associate degree category, 1 semester hour in AC circuits, 2 in DC circuits, 1 in solid state electronics, 1 in digital principles, 1 in microprocessors, 3 in electronic systems troubleshooting and maintenance, and 3 in basic electronics laboratory (2/94). *Version 3:* In the vocational certificate category, 1 semester hour in soldering and 2 in AC/DC circuits. In the lower-division baccalaureate/associate degree category, 4 semester hours in microcomputer systems and repair (5/97). *Version 4:* In the lower-division baccalaureate/associate degree category, 2 semester hours in AC/DC circuits, 4 in microcomputer systems and repair, 2 in digital fundamentals, and 1 in soldering (11/02).

Related Occupations: 29J; 35J.

AR-1715-0849

Communications Systems/Circuit Controller

Course Number: 202-31N10.

Location: Signal School, Fort Gordon, GA.

Length: 21–24 weeks (778–864 hours).

Exhibit Dates: 10/88–4/04.

Learning Outcomes: Upon completion of the course, the student will be able to identify series, parallel, and series/parallel DC circuits and determine values of resistance, voltage, and current in the circuits; apply AC circuit concepts, including frequency, phase, and reactance; explain basic operation of semiconductors, junction diodes, and bipolar transistors and identify their use in rectifier circuits and voltage regulators; troubleshoot and explain the operation of basic transistor amplifiers, power amplifiers, and operational amplifiers; and install, operate, and maintain military communications equipment.

Instruction: Classroom and laboratory exercises involve DC and AC circuits, typical amplifier circuits, and the specific equipment.

Credit Recommendation: In the lower-division baccalaureate/associate degree category, 1 semester hour in AC circuits, 1 in DC circuits, 2 in basic electronics laboratory, 2 in solid state electronics, and 2 in electronic systems troubleshooting and maintenance (6/90).

Related Occupations: 31N.

AR-1715-0850

Automatic Test Equipment Operator/ Maintainer

Course Number: 198-39B10.

Location: Signal School, Fort Gordon, GA.

Length: 18–23 weeks (653–820 hours).

Exhibit Dates: 1/89–9/90.

Learning Outcomes: After 8/90 see AR-1715-1001. Upon completion of the course, the student will be able to perform high reliability precision soldering; inspect, test, and repair printed circuit boards at the component level; and operate, troubleshoot, and repair automatic test equipment.

Instruction: Lectures and practical experience cover certified precision soldering; testing and inspecting printed circuit boards; and the use, operation, and maintenance of automatic test equipment.

Credit Recommendation: In the vocational certificate category, 4 semester hours in soldering techniques. In the lower-division baccalaureate/associate degree category, 3 semester hours in electronic systems troubleshooting and maintenance (5/90).

Related Occupations: 39B.

AR-1715-0851

Communications-Electronics Maintenance Advanced Noncommissioned Officer (NCO)

Course Number: 1-29-C42 (29X).

Location: Signal School, Fort Gordon, GA.

Length: 14–15 weeks (543 hours).

Exhibit Dates: 10/88–9/95.

Learning Outcomes: Upon completion of the course, the student will be able to use microcomputers, including using DOS, word processors, data bases, and spreadsheets; receive and send Army messages; prepare Army forms related to shop management; and describe at the overview level a number of pieces of communications equipment.

Instruction: Lectures, practical exercises, and exams cover the subject material. This course contains a common core of leadership subjects.

Credit Recommendation: In the lower-division baccalaureate/associate degree category, 3 semester hours in computer literacy. See AR-1404-0035 for common core credit (7/89).

Related Occupations: 29X.

AR-1715-0852

TRI-TAC Communications Security (COMSEC) Equipment Full Maintenance

Course Number: *Version 1:* 160-F27. *Version 2:* 160-F27.

Location: *Version 1:* Signal School, Fort Gordon, GA. *Version 2:* Signal School, Fort Gordon, GA; Ordnance, Missile and Munitions School, Redstone Arsenal, AL.

Length: *Version 1:* 4–5 weeks (151 hours). *Version 2:* 3 weeks (108 hours).

Exhibit Dates: *Version 1:* 11/87–9/94. *Version 2:* 10/94–9/95.

Learning Outcomes: *Version 1:* Upon completion of the course, the student will be able to troubleshoot and repair communications security equipment to the component level using automatic test equipment as well as conventional test instruments. *Version 2:* Upon completion of the course, the student will be able to troubleshoot and repair communications security equipment to the component level using automatic test equipment as well as conventional test instruments.

Instruction: *Version 1:* Lectures and practical experience cover fault isolation and maintenance of communications security equipment including power supplies and analog subassemblies using automatic test equipment as well as conventional test sets. *Version 2:* Lectures and practical exercises cover fault isolation and maintenance of communications security equipment including power supplies and analog subassemblies using automatic test equipment as well as conventional test sets.

Credit Recommendation: *Version 1:* In the lower-division baccalaureate/associate degree category, 3 semester hours in electronic systems troubleshooting and maintenance (7/89). *Version 2:* In the lower-division baccalaureate/associate degree category, 2 semester hours in electronic systems troubleshooting and maintenance (5/97).

AR-1715-0853

Satellite Communications (SATCOM) AN/USC-28(V) Set

Course Number: 102-F60.

Location: Signal School, Fort Gordon, GA.

Length: 8 weeks (288 hours).

Exhibit Dates: 12/88–9/95.

Learning Outcomes: Upon completion of the course, the student will be able to troubleshoot and repair a satellite communications set.

Instruction: Lectures and laboratory exercises cover the operation and maintenance of a satellite ground terminal.

Credit Recommendation: In the lower-division baccalaureate/associate degree category, 3 semester hour in electronic systems troubleshooting and maintenance (7/92).

AR-1715-0854

1. Supervisor Special Electronic Devices Repairer Basic Noncommissioned Officer (NCO)
2. Special Electronic Devices Repairer Basic Noncommissioned Officer (NCO)

Course Number: *Version 1:* 198-39E30. *Version 2:* 198-35F30.

Location: *Version 1:* Signal School, Fort Gordon, GA. *Version 2:* Ordnance, Missile and Munitions School, Redstone Arsenal, AL.

Length: *Version 1:* 8–11 weeks (318–398 hours). *Version 2:* 10–11 weeks (403–404 hours).

Exhibit Dates: *Version 1:* 7/89–9/94. *Version 2:* 10/94–6/00.

Learning Outcomes: *Version 1:* Upon completion of the course, the student will be able to explain the theory and operation of a microprocessor-controlled computer and use these and other concepts to assist subordinates in the maintenance of electronic equipment including night vision devices and chemical agent detection systems. *Version 2:* Upon completion of the course, the student will be able to supervise and provide technical assistance to maintenance personnel.

Instruction: *Version 1:* Course includes lectures and practical exercises in microprocessors, diagnostic testing, application of device theory, and schematic diagrams in order to troubleshoot and repair night vision devices and chemical agent detection devices. This course contains a common core of leadership subjects. *Version 2:* Lectures and practical exercises cover new technology, maintenance techniques, and equipment in the repair of mine detectors, night vision devices, battlefield illumination equipment, position and azimuth equipment, and test equipment. This course contains a common core of leadership subjects.

Credit Recommendation: *Version 1:* In the lower-division baccalaureate/associate degree category, 1 semester hour in microprocessors and 1 in electronic systems troubleshooting and maintenance. See AR-1406-0090 for common core credit (2/94). *Version 2:* In the lower-division baccalaureate/associate degree category, 3 semester hours in maintenance management. See AR-1406-0090 for common core credit (5/97).

Related Occupations: 35F; 39E.

AR-1715-0855

Transportable Automatic Switching Systems Operator/Maintainer Basic Noncommissioned Officer (NCO)

Course Number: 622-36L30.

Location: Signal School, Fort Gordon, GA.

Length: 9–10 weeks (351 hours).

Exhibit Dates: 1/89–10/95.

Learning Outcomes: Upon completion of the course, the student will be able to supervise personnel involved in switching network operations and maintenance troubleshoot and repair specific electronic equipment.

Instruction: Classroom and practical exercises cover military specific automatic message switching equipment. This course contains a common core of leadership subjects.

Credit Recommendation: In the lower-division baccalaureate/associate degree category, 2 semester hours in electronic systems troubleshooting and maintenance. See AR-1406-0090 for common core credit (7/89).

Related Occupations: 36L.

AR-1715-0856

Calibration Specialist Basic Noncommissioned Officer (NCO) (35H30)

Course Number: G2ASR32470 218 (35H30).

Location: Training Center, Lowry AFB, CO.

Length: 16–17 weeks (633–644 hours).

Exhibit Dates: 10/88–4/04.

Learning Outcomes: Upon completion of the course, the student will be able to operate, maintain, calibrate, troubleshoot, and repair selected calibration standards and supervise, train, and lead subordinates in these duties.

Instruction: Classroom and laboratory exercises cover the calibration equipment. This course contains a common core of leadership subjects.

Credit Recommendation: In the vocational certificate category, 3 semester hours in test equipment maintenance and calibration. In the lower-division baccalaureate/associate degree category, 3 semester hours in electronic systems troubleshooting and maintenance, 1 in digital principles, and 1 in microprocessors. See AR-1406-0090 for common core credit (5/90).

Related Occupations: 35H.

AR-1715-0858

Communications-Electronics Technician Warrant Officer Technical Certification Reserve Component Phases 2 and 4

Course Number: 4C-286A-RC; 4C-256A-RC.

Location: Signal School, Fort Gordon, GA.

Length: Phase 2, 2 weeks (96 hours); Phase 4, 2 weeks (96 hours).

Exhibit Dates: 5/88–9/95.

Learning Outcomes: Upon completion of the course, the student will perform in senior-level supervisory positions using managerial and leadership concepts and will be able to manage the maintenance of Army communications systems.

Instruction: Lectures, practical exercises, and tests cover topics in logistics, record keeping, shop planning, and communications systems.

Credit Recommendation: In the lower-division baccalaureate/associate degree category, 1 semester hour in electronic communications (7/89).

Related Occupations: 256A; 286A.

AR-1715-0859

Telecommunications Technician Warrant Officer Technical Certification Reserve Component Phases 2 and 4

Course Number: 4C-290A-RC, Phase 4; 4C-290A-RC, Phase 2; 4C-250A-RC, Phase 4; 4C-250A-RC, Phase 2.

Location: Signal School, Fort Gordon, GA.

Length: Phase 2, 2 weeks (96 hours); Phase 4, 2 weeks (96 hours).

Exhibit Dates: 5/88–10/94.

Learning Outcomes: Upon completion of the course, the student will be able to perform in senior-level supervisory positions using managerial and leadership concepts and will be able to manage a telecommunications center.

Instruction: Lectures, practical exercises, and exams cover topics in logistics, record keeping, planning, and telecommunications systems.

Credit Recommendation: Credit is not recommended because of the military-specific nature of the course (7/89).

Related Occupations: 250A; 290A.

AR-1715-0860

Communications-Electronics Repair Technician Senior Warrant Officer

Course Number: 4-11-C32-256A.
Location: Signal School, Fort Gordon, GA.
Length: 14 weeks (527 hours).
Exhibit Dates: 10/86–9/95.

Learning Outcomes: Upon completion of the course, the student will perform in senior-level supervisory positions using managerial concepts and leadership concepts; will be able to use computer hardware and software; and be able to troubleshoot and repair several specific Army communications systems.

Instruction: Lectures, practical exercises, and exams cover the subjects above. A core of leadership subjects is included.

Credit Recommendation: In the lower-division baccalaureate/associate degree category, 1 semester hour in computer literacy, 3 in electronic systems troubleshooting and maintenance, and 1 in leadership. For those students who complete annex N, satellite communications equipment, 2 additional semester hours in satellite communications systems (7/92).

Related Occupations: 256A.

AR-1715-0867

Avionic Flight Systems Repairer
(Avionic Navigation and Flight Control Equipment Repairer)

Course Number: ***Version 1:*** 102-35M10; 102-68Q10. *Version 2:* 102-35Q10.

Location: ***Version 1:*** Signal School, Fort Gordon, GA. *Version 2:* Signal School, Fort Gordon, GA; Ordnance, Missile and Munitions School, Redstone Arsenal, AL.

Length: ***Version 1:*** 21–24 weeks (779–912 hours). *Version 2:* 20 weeks (695–696 hours).

Exhibit Dates: ***Version 1:*** 10/88–9/95. *Version 2:* 10/95–9/99.

Learning Outcomes: ***Version 1:*** Upon completion of the course, the student will be able to identify series, parallel, and series-parallel circuits and determine values of resistance, voltage, and current in the circuits; apply concepts of alternating current, including frequency, phase, and reactance; explain concepts of series and parallel resonance and basics of transformer operation; explain the basic operation of semiconductors, junction diodes, and bipolar transistors and identify their use in rectifier circuits and voltage regulators; troubleshoot and explain the operation of basic transistor amplifiers, power amplifiers, and operational amplifiers; explain the operation of FETs, SCRs, and UJTs and troubleshoot and explain the operation of sine-wave oscillators, multivibrator circuits, and Schmitt trigger circuits; explain the principles of communications using amplitude modulation techniques; trace signal paths, troubleshoot, and align receivers and transmitters; convert between and perform arithmetic in the decimal, binary, octal, and hexadecimal number systems; analyze logic circuits, flip-flop circuits, registers, and counters; and solder and desolder wire to various types of electronic terminals. *Version 2:* Upon completion of the course, the student will understand basic DC/AC circuits, transformers, solid state devices, including FETs, SCRs, and UJTs, transistor and operational amplifiers, oscillator circuits, power supply circuits including voltage regulation, and basic digital logic circuits; troubleshoot and align AM/FM transmitters and receivers; and use appropriate soldering techniques in repair.

Instruction: ***Version 1:*** This course includes electrical, electronic, digital, and communications non-calculus-based theory and hands-on evaluation and maintenance procedures on complex electronic systems using specialized test equipment. *Version 2:* Course includes lectures and practical exercises in topics, including DC/AC circuits, power supply circuits, amplifier circuits, AM/FM fundamentals, digital fundamentals, soldering techniques, and repair and maintenance procedures on complex electronic systems using specialized test equipment.

Credit Recommendation: ***Version 1:*** In the vocational certificate category, 2 semester hours in soldering. In the lower-division baccalaureate/associate degree category, 1 semester hour in AC/DC circuits, 1 in DC circuits, 3 in solid state electronics, 2 in electronic communications, 3 in electronic systems troubleshooting and maintenance, and 3 in aircraft control augmentation (7/92). *Version 2:* In the vocational certificate category, 1 semester hour in soldering for electricity/electronics. In the lower-division baccalaureate/associate degree category, 2 semester hours in AC/DC circuits, 3 in solid state electronics, 2 in electronic communications, 3 in electronic systems troubleshooting and maintenance, and 3 in aircraft flight systems (5/97).

Related Occupations: 35M; 35Q; 68Q.

AR-1715-0868

Avionic Communications Equipment Repairer

Course Number: ***Version 1:*** 102-35L10; 102-68L10. *Version 2:* 102-35L10. *Version 3:* 102-35L10. *Version 4:* 102-35L10.

Location: ***Version 1:*** Signal School, Fort Gordon, GA. *Version 2:* Signal School, Fort Gordon, GA; Ordnance, Missile and Munitions School, Redstone Arsenal, AL. *Version 3:* Ordnance Electronic Maintenance Training, Fort Gordon, GA. *Version 4:* Signal School, Fort Gordon, GA.

Length: ***Version 1:*** 23–26 weeks (875–980 hours). *Version 2:* 22 weeks (1582–1583 hours). *Version 3:* 23 weeks (827 hours). *Version 4:* 19 weeks (699 hours).

Exhibit Dates: ***Version 1:*** 10/88–9/91. *Version 2:* 10/95–9/99. *Version 3:* 10/99–9/02. *Version 4:* 10/02–Present.

Learning Outcomes: ***Version 1:*** Upon completion of the course, the student will be able to identify series, parallel, and series-parallel circuits and determine values of resistance, voltage, and current in the circuits; apply concepts of alternating current, including frequency, phase, and reactance; explain concepts of series and parallel resonance and basics of transformer operation; explain the basic operation of semiconductors, junction diodes, and bipolar transistors and identify their use in rectifier circuits and voltage regulators; troubleshoot and explain the operation of basic transistor amplifier, power amplifier, and operational amplifier; explain the operation of FETs, SCRs, and UJTs; troubleshoot and explain the operations of sine-wave oscillators, multivibrator circuits, and Schmitt trigger circuits; apply single-sideband communications techniques, including signal tracing and alignment of receiver and transmitter; explain the principles of communications using amplitude modulation techniques; trace signal paths, troubleshoot, and align receivers and transmitters; convert between and perform arithmetic in the decimal, binary, octal, and hexadecimal number systems; analyze logic circuits, flip-flop circuits, registers, and counters; explain the theory and operation of microprocessor-based computers, including the concepts of buses, memory, control signals, and addressing; and solder and desolder wire to various types of electronic terminals. *Version 2:* Upon completion of the course, the student will be able to perform troubleshooting procedures on solid state power supplies, audio amplifiers, audio signal generators, pulse generators, and digital circuitry as related to avionics communications equipment. *Version 3:* Upon completion of the course, the student will be able to identify series, parallel, series-parallel DC/AC circuits; determine values for resistance, voltage, and current; apply concepts of frequency, phase, and reactance; explain basic operations of amplifiers, oscillators and power supplies; and perform soldering techniques. *Version 4:* Upon completion of the course, the student will be able to perform troubleshooting and maintenance for solid-state power supplies, audio amplifiers, audio signal generators, pulse generators, and digital circuitry as related to avionic communications equipment.

Instruction: ***Version 1:*** This course includes electrical, electronic, and communications. Hands-on evaluation and maintenance procedures for VHF and UHF communications equipment, standard lightweight avionic installations, and frequency-modulated communications equipment are also included. *Version 2:* Topics covered are series, parallel, and series-parallel circuits and determining values of resistance, voltage, and current in the circuits; concepts of alternating current, including frequency, phase, and reactance; concepts of series and parallel resonance and basics of transformer operation; basic operation of semiconductors, junction diodes, and bipolar transistors and their use in rectifier circuits and voltage regulators; the operation of basic transistor amplifier, power amplifier, and operational amplifier; the operation of FETs, SCRs, and UJTs; troubleshooting transistor amplifier circuits, sine-wave oscillators, multivibrator circuits, and Schmitt trigger circuits; single-sideband communications techniques including signal amplitude modulation techniques; tracing signal paths; troubleshooting receivers and transmitters; number systems; logic circuits, flip-flop circuits, registers, and counters; theory and operation of microprocessor-based computers, including the concepts of buses, memory, control signals, and addressing; and soldering and desoldering wire to various types of electronic terminals. *Version 3:* Topics include electrical, electronic and communications theory; laboratory experiments in VHF/UHF communications; standard avionic installations; digital logic, and soldering methods. *Version 4:* Instruction is provided using practical exercises, discussion and lecture. Topics include AC/DC circuits, power supply circuits, amplifier circuits, oscillator circuits, digital and microprocessor fundamentals, AM and FM transceiver fundamentals, and troubleshooting and repair techniques.

Credit Recommendation: ***Version 1:*** In the vocational certificate category, 2 semester hours in soldering. In the lower-division baccalaureate/associate degree category, 1 semester hour in AC circuits, 1 in DC circuits, 3 in solid state electronics, 3 in electronic communications, 1 in digital fundamentals, 1 in microprocessors, and 3 in electronic systems troubleshooting and maintenance (4/90). *Version 2:* In the vocational certificate category, 1 semester hour in soldering. In the lower-division baccalaureate/associate degree category, 1 semester hour in AC circuits, 1 in DC circuits, 3 in solid state electronics, 2 in electronic communications, 1 in digital fundamentals, 1 in microprocessors, and 3 in electronic systems troubleshooting and maintenance (5/97). *Version 3:* In the lower-division baccalaureate/associate degree category, 1 semester hour in DC circuits, 1 in AC circuits, 3 in solid state electronics, 3 in electronic communications, 1 in digital electronics, 1 in microprocessors, 3 in electronic systems troubleshooting and repair, and 1 in soldering (6/01). *Version 4:* In the lower-division baccalaureate/associate degree category, 3 semester hours in AC/DC circuits, 1 in solid state electronics, 1 in solid state electronics lab, 2 in digital logic and microprocessor fundamentals, 2 in electronic systems, 1 in troubleshooting and maintenance, and 1 in AC/DC circuits lab (2/04).

Related Occupations: 35L; 68L.

AR-1715-0869

Electronic Warfare (EW)/Intercept Aerial Sensor Repairer Phase 2

Course Number: 102-33V10; 104-33V10.

Location: Intelligence School, Fort Huachuca, AZ.

Length: 19–22 weeks (608–687 hours).

Exhibit Dates: 10/86–4/04.

Learning Outcomes: Upon completion of the course, the student will be able to maintain, troubleshoot, and repair a radar surveillance set, a radar data receiving set, a data system, and an infrared countermeasures set.

Instruction: Lectures and practical exercises emphasize troubleshooting and repair of the equipment above.

Credit Recommendation: In the lower-division baccalaureate/associate degree category, 3 semester hours in electronic systems troubleshooting and maintenance (6/90).

Related Occupations: 33V.

AR-1715-0870

Joint Surveillance Target Attack Radar System (JStars) Ground Station Module (GSM) Operator/Supervisor

Course Number: 242-F5.

Location: Intelligence School, Fort Huachuca, AZ.

Length: 5–12 weeks (180 hours).

Exhibit Dates: 10/89–12/90.

Learning Outcomes: Upon completion of the course, the student will be able to operate a specific military surveillance radar.

Instruction: Lectures and practical exercises focus on the use and operation of the surveillance radar. The first seven-weeks of this course is contractor-taught and was not evaluated.

Credit Recommendation: Credit is not recommended because of the military-specific nature of the course (6/90).

AR-1715-0871

Special Electronic Mission Aircraft (SEMA) Common Core

Course Number: 2C-15C/2B.

Location: Intelligence School, Fort Huachuca, AZ.

Length: 2 weeks (70–71 hours).

Exhibit Dates: 10/89–4/04.

Learning Outcomes: Upon completion of the course, the student will perform as an aviator in special aircraft used in military intelligence.

Instruction: Lectures and practical exercises cover topics in military intelligence, surveillance radar, and threats to aircraft.

Credit Recommendation: Credit is not recommended because of the military-specific nature of the course (6/90).

AR-1715-0872

Advanced Mobile Subscriber Equipment (MSE) Node Switch/Large Extension Node Switch Operator

Course Number: 160-ASIV4 (CT).

Location: Signal School, Fort Gordon, GA.

Length: 3 weeks (117 hours).

Exhibit Dates: 2/90–6/94.

Learning Outcomes: Upon completion of the course, the student will be able to initialize, modify, and troubleshoot links in a communications system.

Instruction: Lectures and practical exercises cover the operating and troubleshooting of switch links.

Credit Recommendation: In the lower-division baccalaureate/associate degree category, 3 semester hours in electronic systems troubleshooting and maintenance (6/90).

AR-1715-0873

Mobile Subscriber Equipment (MSE) Systems Maintainer

Course Number: 101-ASIV8 (29M) (CT).

Location: Signal School, Fort Gordon, GA.

Length: 3–4 weeks (121 hours).

Exhibit Dates: 7/89–12/91.

Learning Outcomes: Upon completion of the course, the student will be able to troubleshoot, align, and repair radio and associated communications equipment.

Instruction: Lectures and practical exercises cover the functional characteristics and maintenance of radio access unit and line-of-sight and down-the-hill super high frequency electronic communications equipment.

Credit Recommendation: In the lower-division baccalaureate/associate degree category, 3 semester hours in systems troubleshooting and maintenance (6/90).

Related Occupations: 29M.

AR-1715-0874

1. Automated Communications Computer Systems Repairer
2. Automated Communications Computer Systems Repairer
3. Telecommunications Computer Operator-Maintainer
4. Telecommunications Computer Operator-Maintainer

Course Number: *Version 1:* 150-39G10. *Version 2:* 150-39G10. *Version 3:* 150-74G10 ITRO; 150-74G10. *Version 4:* 150-74G10.

Location: Signal School, Fort Gordon, GA.

Length: *Version 1:* 28–29 weeks (1036 hours). *Version 2:* 28–29 weeks (1036 hours). *Version 3:* 22–25 weeks (914 hours). *Version 4:* 19 weeks (748 hours).

Exhibit Dates: *Version 1:* 2/90–2/93. *Version 2:* 3/93–9/94. *Version 3:* 10/94–9/98. *Version 4:* 10/98–Present.

Learning Outcomes: *Version 1:* Upon completion of the course, the student will be able to identify series, parallel, and series-parallel resistive circuits and determine values of resistance, voltage, and current in the circuits; calculate values of voltage and current in reactive AC circuits; explain the basic operation of semiconductors, junction diodes, and transistors and identify their use in rectifier circuits and voltage regulators; troubleshoot and explain the operation of basic transistor amplifiers and power amplifiers; identify operational amplifiers; convert between and perform arithmetic in the decimal, binary, octal, and hexadecimal number systems; analyze logic circuits, flip-flop circuits, registers, and counters; explain the theory and operation of microprocessor-based computers, including the concepts of buses, memory, control signals, and addressing; explain how a microcomputer performs arithmetic and logic operations; use machine language instructions; solder and desolder electronic components to printed circuit boards; and troubleshoot and repair video display terminals, printers, card punch readers, magnetic tape units, and hard disk drives. *Version 2:* Upon completion of the course, the student will be able to identify series, parallel, and series-parallel resistive circuits and determine values of resistance, voltage, and current in the circuit; calculate values of voltage and current in reactive AC circuits; explain the basic operation of semiconductors, junction diodes, and transistors and identify their use in rectifier circuits and voltage regulators; troubleshoot and explain the operation of basic transistor amplifiers and power amplifiers; identify operational amplifiers; convert between and perform arithmetic in the decimal, binary, octal, and hexadecimal number systems; analyze logic circuits, flip-flop circuits, registers, and counters; explain the theory and operation of microprocessor-based computers, including the concepts of buses, memory, control signals, and addressing; explain how a microcomputer performs arithmetic and logic operations; use machine language instructions; solder and desolder wires, cables, and terminals; and troubleshoot and repair video display terminals, printers, card punch readers, magnetic tape units, and hard disk drives. *Version 3:* Upon completion of the course, the student will be able to identify series, parallel, and series-parallel resistive circuits and determine values of resistance, voltage, and current in the circuit; calculate values of voltage and current in reactive AC circuits; explain the basic operation of semiconductors, junction diodes, and transistors and identify their use in rectifier circuits and voltage regulators; troubleshoot and explain the operation of basic transistor amplifiers and power amplifiers; identify operational amplifiers; convert between and perform arithmetic in the decimal, binary, octal, and hexadecimal number systems; analyze logic circuits, flip-flop circuits, registers, and counters; explain the theory and operation of microprocessor-based computers, including the concepts of buses, memory, control signals, and addressing; explain how a microcomputer performs arithmetic and logic operations; use machine language instructions; solder and desolder wires, cables, and terminals; and troubleshoot and repair video display terminals, printers, card punch readers, magnetic tape units, and hard disk drives. *Version 4:* Upon completion of the course, the student will be able to identify series, parallel, and series-parallel resistive circuits and determine values of resistance, voltage, and current in the circuit; measure values of voltage and current in reactive AC circuits; explain the basic operation of semiconductors, junction diodes, and transistors and identify their use in rectifier circuits and voltage regulators; identify operational amplifiers; convert between and perform arithmetic in the decimal, binary, octal, and hexadecimal number systems; analyze logic circuits, flip-flop circuits, registers, and counters; explain the theory and operation of microprocessor-based computers, including the concepts of buses, memory, control signals, and addressing; explain how a microcomputer performs arithmetic and logic operations; use machine language instructions; solder and desolder wires, cables, and terminals; and troubleshoot and repair video display terminals, printers, card punch readers, magnetic tape units, and hard disk drives.

Instruction: *Version 1:* Lectures and practical exercises cover topics in basic electronics and the troubleshooting and repair of a computer system. *Version 2:* Lectures and practical exercises cover topics in basic electronics and the troubleshooting and repair of a computer system. *Version 3:* Lectures and practical exercises cover topics in basic electronics and the troubleshooting and repair of a computer system. *Version 4:* Lectures and practical exercises cover topics in basic electronics and the troubleshooting and repair of a computer system.

Credit Recommendation: *Version 1:* In the vocational certificate category, 2 semester hours in soldering techniques. In the lower-division baccalaureate/associate degree category, 1 semester hour in AC circuits, 1 in DC circuits, 2 in solid state electronics, 1 in digital fundamentals, 2 in microprocessors, 3 in basic electronics laboratory, and 3 in computer systems troubleshooting and maintenance (6/90). *Version 2:* In the vocational certificate category, 1 semester hour in soldering techniques. In the lower-division baccalaureate/associate degree category, 3 semester hours in AC/DC circuits, 2 in solid state electronics, 1 in digital fundamentals, 2 in microprocessors, 3 in basic electronics laboratory, and 3 in computer systems troubleshooting and maintenance (2/94). *Version 3:* In the lower-division baccalaureate/associate degree category, 1 semester hour in soldering techniques, 3 in AC/DC circuits, 2 in solid state electronics techniques, 1 in digital fundamentals, 2 in microprocessors, 3 in basic electronics laboratory, and 3 in computer systems troubleshooting and maintenance (12/97). *Version 4:* In the lower-division baccalaureate/associate degree category, 3 semester hours in digital fundamentals, 3 in AC/DC circuits, 3 in basic electronics laboratory, and 3 in computer systems troubleshooting and maintenance (5/99).

Related Occupations: 39G; 74G.

AR-1715-0875

MOBILE SUBSCRIBER EQUIPMENT (MSE) SYSTEM PLANNER/SYSTEM CONTROL CENTER OPERATOR

Course Number: 260-ASIY3 (CT).
Location: Signal School, Fort Gordon, GA.
Length: 11–12 weeks (410 hours).
Exhibit Dates: 8/89–6/94.

Learning Outcomes: Upon completion of the course, the student will be able to operate, troubleshoot, and maintain the control center of an electronic communications system.

Instruction: Lectures and practical exercises cover the specific equipment.

Credit Recommendation: In the lower-division baccalaureate/associate degree category, 3 semester hours in systems troubleshooting and maintenance (6/90).

AR-1715-0876

MOBILE SUBSCRIBER EQUIPMENT (MSE) CENTRAL OFFICE REPAIRER

Course Number: 622-ASIV8 (29N) (CT).
Location: Signal School, Fort Gordon, GA.
Length: 7 weeks (276 hours).
Exhibit Dates: 1/88–4/04.

Learning Outcomes: Upon completion of the course, the student will be able to locate and repair faults in telephone central office switching equipment.

Instruction: Lectures and practical exercises cover the functional operation, technical characteristics, and maintenance of network switch, controllers, timing equipment, power equipment, and processing equipment.

Credit Recommendation: In the lower-division baccalaureate/associate degree category, 3 semester hours in electronic systems troubleshooting and maintenance (6/90).

Related Occupations: 29N.

AR-1715-0877

TACTICAL SATELLITE/MICROWAVE REPAIRER TRANSITION

Course Number: 113-29M10(T)-HTRTS-M.

Location: High Tech Regional Training Site, Sacramento, CA; High Tech Regional Training Site, Tobyhanna, PA.
Length: 2 weeks (80 hours).
Exhibit Dates: 10/90–4/04.

Learning Outcomes: Upon completion of the course, the student will be able to troubleshoot, repair, and maintain a satellite/microwave system.

Instruction: Lectures and practical exercises cover troubleshooting.

Credit Recommendation: In the lower-division baccalaureate/associate degree category, 3 semester hours in electronic systems troubleshooting and maintenance (6/90).

Related Occupations: 29M.

AR-1715-0878

DS/G S TEST AND REPAIR STATION MAINTENANCE (AH-64)

Course Number: 201-ASIX1; 198-ASIX1.
Location: Signal School, Fort Gordon, GA.
Length: 13–15 weeks (468–545 hours).
Exhibit Dates: 3/90–9/95.

Learning Outcomes: Before 3/90 see AR-1715-0596. Upon completion of the course, the student will be able to troubleshoot, repair, and maintain specific electronic test equipment.

Instruction: Lectures and practical exercises cover troubleshooting and repairing specific equipment.

Credit Recommendation: In the lower-division baccalaureate/associate degree category, 3 semester hours in electronic systems troubleshooting and maintenance (8/96).

AR-1715-0879

COMMUNICATIONS SYSTEMS/CIRCUIT CONTROLLER (RC)

Course Number: 113-31N10-RC.
Location: Various locations, US and overseas.
Length: 2–3 weeks (72–91 hours).
Exhibit Dates: 10/90–4/04.

Learning Outcomes: Upon completion of the course, the student will be able to install and test cabled communications systems using basic electronic test instruments.

Instruction: Lectures and practical exercises cover the specific communications system.

Credit Recommendation: In the lower-division baccalaureate/associate degree category, 2 semester hours in basic electronics laboratory (6/90).

Related Occupations: 31N.

AR-1715-0880

MASTER OPERATOR MANEUVER CONTROL SYSTEM

Course Number: 4C-F41/101-F29.
Location: Signal School, Fort Gordon, GA.
Length: 3 weeks (108 hours).
Exhibit Dates: 4/90–6/94.

Learning Outcomes: Upon completion of the course, the student will be able to supervise the installation, operation, and maintenance of the maneuver control system.

Instruction: Lectures and practical exercises cover the maneuver control system.

Credit Recommendation: Credit is not recommended because of the military-specific nature of the course (7/92).

AR-1715-0881

TACTICAL TELECOMMUNICATIONS CENTER OPERATOR BASIC NONCOMMISSIONED OFFICER (NCO) AUTOMATIC DATA
(Telecommunications Center Operator Basic Noncommissioned Officer (NCO))

Course Number: 260-72E20; 260-72G20.
Location: Signal School, Fort Gordon, GA.
Length: 7–8 weeks (276 hours).
Exhibit Dates: 1/90–12/90.

Learning Outcomes: Upon completion of the course, the student will be able to supervise the installation, operation, and maintenance of tactical telecommunications centers.

Instruction: Lectures and practical exercises focus on operating a tactical telecommunications system. This course contains a common core of leadership subjects.

Credit Recommendation: Credit is recommended for the common core only. See AR-1406-0090 (6/90).

Related Occupations: 72E; 72G.

AR-1715-0882

TELECOMMUNICATIONS TERMINAL DEVICE REPAIRER SUSTAINMENT

Course Number: 113-29J10-HTRTS-M.
Location: High Tech Regional Training Site, Sacramento, CA; High Tech Regional Training Site, Tobyhanna, PA.
Length: 2 weeks (80 hours).
Exhibit Dates: 10/90–4/04.

Learning Outcomes: Upon completion of the course, the student will be able to install, troubleshoot, and maintain a digital telecommunications system.

Instruction: Lectures and practical exercises cover troubleshooting and maintaining digital terminal devices.

Credit Recommendation: In the lower-division baccalaureate/associate degree category, 3 semester hours in electronic systems troubleshooting and maintenance (6/90).

Related Occupations: 29J.

AR-1715-0883

DAS3 Computer Systems Repairer

Course Number: 150-39D10.

Location: Signal School, Fort Gordon, GA.

Length: 22–26 weeks (819–914 hours).

Exhibit Dates: 10/88–12/92.

Learning Outcomes: Upon completion of the course, the student will be able to identify series, parallel, and series/parallel resistive circuits and determine values of resistance, voltage, and current in the circuits; identify the meaning of frequency, period, sine wave, phase shift, inductance, reactance, capacitive reactance, and impedance; calculate voltages and currents in transformers and in reactive circuits; convert between and perform arithmetic in the decimal, binary, octal, and hexadecimal number systems; analyze logic circuits, flip-flop circuits, registers, and counters; explain how a microcomputer performs arithmetic and logic operations; use machine language instructions; explain the theory and operation of microprocessor-based computers, including the concepts of buses, memory, control signals, and addressing; use basic electronic laboratory equipment; and troubleshoot, maintain, and repair computer peripherals, including printers, card readers, magnetic tape units, and hard disks.

Instruction: Lectures and practical exercises cover electronics and computer peripherals.

Credit Recommendation: In the lower-division baccalaureate/associate degree category, 3 semester hours in basic electronics laboratory, 1 in DC circuits, 1 in AC circuits, 1 in digital principles, 2 in microprocessors, and 3 in electronic systems troubleshooting and maintenance (6/90).

Related Occupations: 39D.

AR-1715-0884

Mobile Subscriber Equipment (MSE) Transmission System Operator

Course Number: 260-31D10 (CT).

Location: Signal School, Fort Gordon, GA.

Length: 9–10 weeks (348 hours).

Exhibit Dates: 5/89–4/04.

Learning Outcomes: Upon completion of the course, the student will be able to install, operate, and maintain transmission equipment.

Instruction: Lectures and exercises cover the operation and maintenance of transmission equipment.

Credit Recommendation: Credit is not recommended because of the military-specific nature of the course (6/90).

Related Occupations: 31D.

AR-1715-0885

Mobile Subscriber Equipment (MSE) Systems Maintainer

Course Number: 201-ASIV8 (29E) (CT).

Location: Signal School, Fort Gordon, GA.

Length: 2–3 weeks (80 hours).

Exhibit Dates: 8/89–9/95.

Learning Outcomes: Upon completion of the course, the student will be able to maintain and repair single-channel radios and associated electronic equipment.

Instruction: Lectures and practical exercises cover the direct support maintenance and repair of the logic unit, radio control unit, radio subassemblies, and power unit of single-channel radios.

Credit Recommendation: In the lower-division baccalaureate/associate degree category, 2 semester hours in systems troubleshooting and maintenance (6/90).

Related Occupations: 29E.

AR-1715-0886

1. Electronic Switching System Operator (Mobile Subscriber Equipment (MSE) Network Switching System Operator)
2. Network Switching System Operator-Maintainer
3. TATS Network Switching Systems Operator-Maintainer (Network Switching Systems Operator-Maintainer)

Course Number: *Version 1:* 260-31F10 (CT); 260-31F10. *Version 2:* 260-31F10 (CT). *Version 3:* 260-31F10 (CT) (F); 260-31F10 (CT).

Location: Signal School, Fort Gordon, GA.

Length: *Version 1:* 10–15 weeks (375–508 hours). *Version 2:* 16–17 weeks (644 hours). *Version 3:* 17–21 weeks (644 hours).

Exhibit Dates: *Version 1:* 3/89–9/95. *Version 2:* 10/95–9/97. *Version 3:* 10/97–Present.

Learning Outcomes: *Version 1:* Upon completion of the course, the student will be able to install, operate, troubleshoot, and maintain the mobile subscriber equipment network switching system. *Version 2:* Upon completion of the course, the student will be able to install, operate, troubleshoot, and perform maintenance on electronic switching systems and related equipment as well as maintenance on the system control center. *Version 3:* Upon completion of the course, the student will be able to install, operate, troubleshoot and maintain electronic switching systems; practice safety precautions when working with electronics systems; and solder and repair units.

Instruction: *Version 1:* Lectures and demonstrations cover the skills necessary to perform operator/organizational maintenance on MSE switching equipment. *Version 2:* The course includes lectures and practical exercises to install, operate, troubleshoot, and perform maintenance on node switches, large and small extension switches, automatic telephone central offices, electronic switchboards, and systems control centers. *Version 3:* The course includes lectures and practical exercises to install, operate, troubleshoot, and perform maintenance on node switches, large and small extension switches, automatic telephone central offices, electronic switchboards, and systems control centers.

Credit Recommendation: *Version 1:* Credit is not recommended because of the military-specific nature of the course (2/94). *Version 2:* In the vocational certificate category, 9 semester hours in electronic switching systems. In the lower-division baccalaureate/associate degree category, 1 semester hour in basic electronics (2/97). *Version 3:* In the lower-division baccalaureate/associate degree category, 3 semester hours in electronic troubleshooting and maintenance and 1 in basic electronics (11/02).

Related Occupations: 31F.

AR-1715-0887

1. AH-1 Aircraft Armament Maintenance Technician
2. AH-1F Aircraft Armament Maintenance Technician

Course Number: *Version 1:* 4D-SQIE (AH-1). *Version 2:* 4D-SQIE (AH-1F).

Location: Aviation Logistics School, Fort Eustis, VA.

Length: *Version 1:* 10 weeks (366 hours). *Version 2:* 10 weeks (368 hours).

Exhibit Dates: *Version 1:* 10/90–12/95. *Version 2:* 1/96–9/03.

Learning Outcomes: *Version 1:* Upon completion of the course, the student will be able to supervise maintenance and repair of armament systems and troubleshoot and repair electronic systems. *Version 2:* Upon completion of the course, the student will be able to supervise maintenance and repair of armament systems and troubleshoot and repair electronic systems.

Instruction: *Version 1:* Lectures and practical experience cover repair, maintenance, and troubleshooting of electronic armament systems. *Version 2:* Lectures and practical experience cover repair, maintenance, and troubleshooting of electronic armament systems.

Credit Recommendation: *Version 1:* In the lower-division baccalaureate/associate degree category, 2 semester hours in administrative management and 3 in electronic systems troubleshooting and maintenance (6/91). *Version 2:* In the lower-division baccalaureate/associate degree category, 2 semester hours in aviation maintenance management and 3 in electronic systems troubleshooting and maintenance (6/98).

AR-1715-0888

Strategic Systems Repairer Basic Noncommissioned Officer (NCO)

Course Number: 102-33Y30.

Location: Intelligence School, Fort Devens, MA.

Length: 17–18 weeks (641 hours).

Exhibit Dates: 1/91–3/94.

Learning Outcomes: After 11/94 see AR-1715-1004. Upon completion of the course, the student will be able to perform tests, alignment, calibration, and analysis of communications receivers, detectors, and signal processing circuits and test and maintain minicomputers and computer systems.

Instruction: Laboratory exercises use test instruments, diagnostic algorithms, and experimental analysis methods for communications circuit evaluation and computer systems repair. This course contains a common core of leadership subjects.

Credit Recommendation: In the lower-division baccalaureate/associate degree category, 3 semester hours in computer systems maintenance and 2 in personnel management. See AR-1406-0090 for common core credit. In the upper-division baccalaureate category, 2 semester hours in electronic communications laboratory (6/91).

Related Occupations: 33Y.

AR-1715-0889

Strategic Systems Repairer

Course Number: *Version 1:* 102-33Y10. *Version 2:* 102-33Y10.

Location: *Version 1:* Intelligence School, Fort Devens, MA. *Version 2:* Intelligence Center and School, Fort Huachuca, AZ.

Length: *Version 1:* 40 weeks (1497 hours). *Version 2:* 38–39 weeks (1379 hours).

Exhibit Dates: *Version 1:* 10/91–9/93. *Version 2:* 10/93–12/01.

Learning Outcomes: *Version 1:* Upon completion of the course, the student will apply electronic theory, system concepts, and circuit design to laboratory experimentation with linear and digital solid state electronics, analog communications, eight-bit microcomputer hardware, assembly-language programming, and microcomputer peripherals including A/D and D/A converters. *Version 2:* Upon completion of the course, the student will be able to analyze and troubleshoot electric circuits, solid-state and digital circuits, and microcomputers and maintain and troubleshoot SUN computer workstations.

Instruction: *Version 1:* Lectures, demonstrations, computer-assisted instruction, and laboratory exercises cover solid state devices and circuits, communications systems, analog and digital electronics, and eight-bit microcomputer hardware and software including I/O devices. *Version 2:* Lectures, demonstrations, computer-assisted instruction, and laboratory exercises cover DC and AC circuits, solid-state devices, digital electronics, operational amplifiers, microprocessors, communications receiver fundamentals, soldering techniques, electronic troubleshooting, computer system fundamentals, and testing and troubleshooting of SUN workstations in a distributed process network configuration.

Credit Recommendation: *Version 1:* In the vocational certificate category, 3 semester hours in computer system maintenance. In the lower-division baccalaureate/associate degree category, 3 semester hours in digital logic, 3 in AC/DC circuits, and 3 in analog communications. In the upper-division baccalaureate category, 3 in operational amplifiers and applications and 3 in introduction to microcomputers (7/91). *Version 2:* In the vocational certificate category, 1 semester hour in soldering techniques. In the lower-division baccalaureate/associate degree category, 3 semester hours in DC/AC circuits, 3 in solid state electronics, 3 in digital logic, 3 in electronic communications, 3 in microprocessors, and 3 in computer systems troubleshooting and maintenance (6/97).

Related Occupations: 33Y.

AR-1715-0890

Electronic Warfare (EW)/Intercept Aerial Sensor Repairer Basic Noncommissioned Officer (NCO) Phase 2

Course Number: *Version 1:* 102-33V30. *Version 2:* 102-33V30.

Location: Intelligence School, Fort Devens, MA.

Length: *Version 1:* 10 weeks (359 hours). *Version 2:* 6 weeks (223–224 hours).

Exhibit Dates: *Version 1:* 2/89–12/90. *Version 2:* 1/91–3/94.

Learning Outcomes: *Version 1:* Upon completion of the course, the student will perform experiments on communications circuits to determine modulated waveform types, signal-to-noise ratio, and bit error rates. *Version 2:* Upon completion of the course, the student will be able to supervise personnel in the operation of sophisticated electronic equipment.

Instruction: *Version 1:* Laboratory exercises use test instruments and experimental methods for the evaluation of digital communication systems. This course contains a common core of leadership subjects. *Version 2:* Lectures, demonstrations, and practical exercises cover counseling, personnel supervision, and operations management. This course contains a common core of leadership subjects.

Credit Recommendation: *Version 1:* In the lower-division baccalaureate/associate degree category, 1 semester hour in operations management. See AR-1406-0090 for common core credit. In the upper-division baccalaureate category, 2 semester hours in electronic communications laboratory (7/91). *Version 2:* In the lower-division baccalaureate/associate degree category, 2 semester hours in personnel management. See AR-1406-0090 for common core credit (7/91).

Related Occupations: 33V.

AR-1715-0891

Wideband Collection System Maintenance

Course Number: 102-F63.

Location: Intelligence School, Fort Devens, MA.

Length: 5 weeks (180 hours).

Exhibit Dates: 1/91–3/94.

Learning Outcomes: Upon completion of the course, the student will be able to maintain and repair the wide band collection system down to the faulty component.

Instruction: Lectures, discussions, and practical exercises cover troubleshooting wide band collection system, RF collection groups, signal monitor and quality groups, recording and tape quality control groups, and system test and maintenance groups.

Credit Recommendation: In the vocational certificate category, 3 semester hours in electronics maintenance and repair (6/91).

AR-1715-0892

Special Purpose Countermeasures System AN/ALQ-151(V)2 Operator

Course Number: 231-F26.

Location: Intelligence School, Fort Devens, MA.

Length: 2 weeks (75–76 hours).

Exhibit Dates: 10/90–12/90.

Learning Outcomes: Upon completion of the course, the student will be able to check and operate special electronic countermeasures equipment.

Instruction: Lectures, audiovisual presentations, demonstrations, and performance exercises cover the pre-operation, testing, and operation of specialized electronic equipment.

Credit Recommendation: Credit is not recommended because of the limited, specialized nature of the course, which is operational and neither theory nor laboratory oriented (6/91).

AR-1715-0893

Improved Guardrail V Operator

Course Number: 231-F24.

Location: Intelligence School, Fort Devens, MA.

Length: 5 weeks (184 hours).

Exhibit Dates: 10/88–3/94.

Learning Outcomes: After 9/96 see AR-1715-1002. Upon completion of the course, the student will be able to operate special electronic equipment for the acquisition and analysis of electronic signals.

Instruction: Lectures, presentations, demonstrations, computer-assisted instruction, and performance exercises cover the operation of specialized electronic equipment.

Credit Recommendation: Credit is not recommended because of the limited, specialized nature of the course which is operational and neither theory nor laboratory oriented (6/91).

AR-1715-0894

Land Combat Missile Systems Technician Warrant Officer Technical and Tactical Certification Reserve (Phases 2 and 4)

Course Number: 4F-912A-RC.

Location: Missile and Munitions School, Redstone Arsenal, AL.

Length: Phase 2, 3–4 weeks (167 hours); Phase 4, 3–4 weeks (164 hours).

Exhibit Dates: 10/88–6/00.

Learning Outcomes: Upon completion of the course, the student will be able to demonstrate a knowledge of electronics systems maintenance and electronics shop management skills.

Instruction: This course includes lectures and practical exercises in the maintenance, troubleshooting, repair, and calibration associated with night sight, TOW 1, TOW 2, Dragon, and multiple launch rocket systems.

Credit Recommendation: In the lower-division baccalaureate/associate degree category, 4 semester hours in electronic systems maintenance and 1 in electronic shop management (8/96).

Related Occupations: 912A.

AR-1715-0896

1. Ammunition Specialist (Phase 2) Basic Noncommissioned Officer (NCO) Reserve
2. Ammunition Specialist (Phase 3) Basic Noncommissioned Officer (NCO) Reserve

Course Number: *Version 1:* 645-55B30-RC. *Version 2:* 645-55B30-RC.

Location: Missile and Munitions School, Redstone Arsenal, AL.

Length: *Version 1:* 1–2 weeks (116–120 hours). *Version 2:* 2 weeks (83–88 hours).

Exhibit Dates: *Version 1:* 1/90–9/92. *Version 2:* 10/92–8/00.

Learning Outcomes: *Version 1:* Upon completion of the course, the student will be able to identify, receive, store, issue, and destroy ammunition; use a computer for ammunition inventory; and direct fire fighting and rewarehousing operations. *Version 2:* Upon completion of the course, the student will be able to identify, receive, store, issue, and destroy ammunition; use a computer for ammunition inventory; and direct fire fighting and rewarehousing operations.

Instruction: *Version 1:* Lectures and practical exercises cover ammunition identification, emergency destruction techniques, automated inventory procedures, and supervision of ammunition handling and rewarehousing operations. *Version 2:* Lectures and practical exercises cover ammunition identification, emergency destruction techniques, automated inventory procedures, and supervision of ammunition handling and rewarehousing operations.

Credit Recommendation: *Version 1:* Credit is not recommended because of the military-specific nature of the course (8/91). *Version 2:* Credit is not recommended because of the military-specific nature of the course (8/91).

Related Occupations: 55B.

AR-1715-0897

SHORADS Direct Support/General Support (DS/GS) Maintenance Repair Technician

Course Number: 4F-SQIV.

Location: Missile and Munitions School, Redstone Arsenal, AL.

Length: 19–20 weeks (721 hours).

Exhibit Dates: 10/89–9/91.

Learning Outcomes: Upon completion of the course, the student will be able to demonstrate shop management skills, troubleshoot electronic circuits and logic circuits, demonstrate computer architecture skills, operate equipment, describe the operation of microprocessors, and demonstrate soldering techniques.

Instruction: Lectures and practical exercises cover soldering techniques, digital logic circuits, computer operation and architecture, microprocessors. Theoretical and practical exercises cover shop management skills.

Credit Recommendation: In the vocational certificate category, 2 semester hours in soldering techniques. In the lower-division baccalaureate/associate degree category, 3 semester hours in digital logic, 2 in computer operation and architecture, and 2 in microprocessors (8/91).

AR-1715-0899

1. Land Combat Missile Systems Technician Warrant Officer Technical and Tactical Certification
2. Land Combat Missile Systems Technician Warrant Officer Technical/Tactical Certification

Course Number: *Version 1:* 4F-912A. *Version 2:* 4F-912A.

Location: *Version 1:* Missile and Munitions School, Redstone Arsenal, AL. *Version 2:* Ordnance, Missile and Munitions School, Redstone Arsenal, AL.

Length: *Version 1:* 18–20 weeks (683–710 hours). *Version 2:* 17–18 weeks (627 hours).

Exhibit Dates: *Version 1:* 10/88–9/92. *Version 2:* 10/94–6/00.

Learning Outcomes: *Version 1:* Upon completion of the course, the student will be able to troubleshoot, test, and repair electronic and logic circuits, computers, and microprocessors; test and repair Bradley TOW 1/TOW 2, Dragon, night sight, and multiple launch rocket system electronic subsystems; and perform shop management tasks such as planning, supervising, and reporting. *Version 2:* After 5/00 see AR-1715-1063. Upon completion of the course, the student will be able to perform as a land combat missile systems technician.

Instruction: *Version 1:* This course includes a balanced program of lectures and practical experiences with standard training equipment, including computers, microprocessors, power supply trainers, specific subsystems, and associated test equipment. *Version 2:* Lectures and practical exercises cover inspection, testing, and troubleshooting electronic, electrical, mechanical, hydraulic, and cryogenic assemblies, subassemblies, modules, and circuit elements of weapon systems.

Credit Recommendation: *Version 1:* In the lower-division baccalaureate/associate degree category, 3 semester hours in microprocessors, 3 in computer operation and architecture, 3 in shop management, and 3 in electronic system maintenance (8/91). *Version 2:* In the lower-division baccalaureate/associate degree category, 1 semester hour in military science, 2 in maintenance management, and 3 in electronic systems troubleshooting and maintenance (7/96).

Related Occupations: 912A.

AR-1715-0900

Land Combat Missile Systems Repair Technician Senior Warrant Officer

Course Number: 4-9-C32-912A.

Location: Missile and Munitions School, Redstone Arsenal, AL.

Length: 8–9 weeks (319 hours).

Exhibit Dates: 10/88–6/00.

Learning Outcomes: Upon completion of the course, the student will be able to check, inspect, and evaluate maintenance on land combat missile systems.

Instruction: The course contains lectures and practical exercises in the management of the maintenance of the Dragon, Tow 2, multiple launch rocket system, land combat support systems, and associated test equipment.

Credit Recommendation: Credit is not recommended because of the military-specific nature of the course (8/91).

Related Occupations: 912A.

AR-1715-0901

Hawk Advanced Noncommissioned Officer (NCO)

Course Number: 1-27-C42C.

Location: Missile and Munitions School, Redstone Arsenal, AL.

Length: 6 weeks (217 hours).

Exhibit Dates: 4/90–1/96.

Learning Outcomes: Upon completion of the course, the student will be able to manage a missile maintenance support organization.

Instruction: Lectures and practical experience cover organizing, planning, and directing a missile maintenance organization. This course contains a common core of leadership subjects.

Credit Recommendation: Credit is recommended for the common core only. See AR-1404-0035 (8/91).

AR-1715-0902

Land Combat Advanced Noncommissioned Officer (NCO)

Course Number: 1-27-C42A.

Location: Missile and Munitions School, Redstone Arsenal, AL.

Length: 15 weeks (551 hours).

Exhibit Dates: 4/90–1/96.

Learning Outcomes: Upon completion of the course, the student will be able to demonstrate leadership skills; complete forms; operate a computer; direct maintenance on land combat support systems, such as TOW, DRAGON, Lance, and multiple launch rocket system; and prepare and execute a field training exercise.

Instruction: This course contains a balance of lectures and practical exercises in leadership skills in the teaching and direction of field training in the land combat support systems. This course contains a common core of leadership subjects.

Credit Recommendation: Credit is recommended for the common core only. See AR-1404-0035 (8/91).

AR-1715-0903

Air Defense (SHORADS) Advanced Noncommissioned Officer (NCO)

Course Number: 1-27-C42B.

Location: Missile and Munitions School, Redstone Arsenal, AL.

Length: 13–14 weeks (484–487 hours).

Exhibit Dates: 4/90–1/96.

Learning Outcomes: Upon completion of the course, the student will be able to direct the maintenance of various missile and radar systems and auxiliary equipment.

Instruction: The course includes lectures and practical exercises in material management, testing, troubleshooting, and repairing the Vulcan, Chaparral/Redeye, and FAAR systems and auxiliary equipment. This course contains a common core of leadership subjects.

Credit Recommendation: In the lower-division baccalaureate/associate degree category, 4 semester hours in electronic systems maintenance and 1 in shop management. See AR-1404-0035 for common core credit (8/91).

AR-1715-0904

Hawk Fire Control/Continuous Wave (CW) Radar Repairer Transition

Course Number: 104-27KIT/2T/3T/4T.

Location: Missile and Munitions School, Redstone Arsenal, AL.

Length: 19–20 weeks (711 hours).

Exhibit Dates: 3/90–1/96.

Learning Outcomes: Upon completion of the course, the student will be able to describe the operation of eight-bit microprocessors to the CPU register level, including memory organization, address decoding, I/O circuits, and common I/O peripherals; program at the Assembly code level and execute and debug assembly language programs; troubleshoot and repair microprocessor-based systems to the PC board and component level; describe the principal components of a fire control radar system including radar transmitter and fire control servo systems; and operate and maintain a radar-based fire control system.

Instruction: Lectures cover binary math, logic devices, computer circuits, Assembly language programming, radar systems, radar signal processing, and antenna servomechanism components. Practical exercises cover maintaining, troubleshooting, and repairing radar systems; radar signal processing systems; and antenna positioning systems.

Credit Recommendation: In the lower-division baccalaureate/associate degree category, 3 semester hours in electronic systems troubleshooting, 3 in introduction to radar systems and signal processing, 4 in computer organization, 2 in microprocessor laboratory, 3 in computer system troubleshooting, and 1 in servomechanism laboratory (8/91).

Related Occupations: 27K.

AR-1715-0905

HAWK FIRE CONTROL/CONTINUOUS WAVE (CW) RADAR REPAIRER

Course Number: 104-27K10 (PIP 111).

Location: Missile and Munitions School, Redstone Arsenal, AL.

Length: 45 weeks (1800 hours).

Exhibit Dates: 10/89–9/90.

Learning Outcomes: Upon completion of the course, the student will be able to apply basic DC and AC circuit theory to the troubleshooting and repair of computer- and radar-based electronic systems; describe the operation of linear, digital, and computer circuits; maintain, troubleshoot, and repair computer- and radar-based electronic systems to the module and component level; and apply standard soldering and printed circuit board repair techniques to the repair of electronic assemblies.

Instruction: Lectures cover the fundamental operation of passive and active electronic components, digital logic gates and computer circuits, and computer programming concepts. Practical exercises cover soldering, troubleshooting, computer-based systems, and troubleshooting CW radar systems.

Credit Recommendation: In the vocational certificate category, 1 semester hour in soldering techniques. In the lower-division baccalaureate/associate degree category, 3 semester hours in computer system troubleshooting, 3 in radar systems troubleshooting, 3 in basic electricity, 3 in solid state devices, 6 in computer organization and operation, and 3 in electronic systems troubleshooting (8/91).

Related Occupations: 27K.

AR-1715-0906

HAWK MAINTENANCE CHIEF (PIP 3)

Course Number: 4F-F12/121-F27 (PIP 3).

Location: Missile and Munitions School, Redstone Arsenal, AL.

Length: 34–35 weeks (1253 hours).

Exhibit Dates: 2/90–1/96.

Learning Outcomes: Upon completion of the course, the student will be able to operate, program, and troubleshoot digital computer systems; describe the operation of continuous wave (CW), pulse, and phased array radar systems and associated signal processing systems; and troubleshoot and repair pulse, phased array, and CW radar systems and associated signal processing, power generation/distribution, and mechanical hydraulic systems.

Instruction: Lectures, demonstrations, and practical exercises cover digital computer organization, operation, programming, and troubleshooting; pulse, CW, and phased array radar receivers/transmitters and associated signal processing systems; and missile launcher electromechanical and hydraulic systems.

Credit Recommendation: In the lower-division baccalaureate/associate degree category, 3 semester hours in computer organization and operation; 4 in continuous wave, pulse, and phased array radar systems; 3 in computer system troubleshooting; and 3 in electronic system troubleshooting (8/91).

AR-1715-0907

HAWK FIRING SECTION REPAIRER BASIC NONCOMMISSIONED OFFICER (NCO)

Course Number: 121-27H30.

Location: Missile and Munitions School, Redstone Arsenal, AL.

Length: 19 weeks (741 hours).

Exhibit Dates: 4/90–1/96.

Learning Outcomes: Upon completion of the course, the student will be able to describe logic devices, circuits, and the organization of digital computer systems; write, debug, and execute Assembly language programs; demonstrate troubleshooting skills on computer-based systems; troubleshoot electronic systems to the module and component level; troubleshoot computer-based signal processing systems to the module and component level; and troubleshoot continuous-wave-radar-based electronic systems to the component level.

Instruction: Instruction includes lectures, demonstrations, and performance exercises in digital logic devices, logic circuits, computer organization, computer programming, and troubleshooting of computers and computer-based signal processing systems. Troubleshooting continuous-wave-based radar systems and electronic signal processing systems is included. This course contains a common core of leadership subjects.

Credit Recommendation: In the lower-division baccalaureate/associate degree category, 3 semester hours in computer organization and operation, 3 in computer system troubleshooting, 3 in electronic system troubleshooting, and 3 semester hours in continuous wave radar system organization. See AR-1406-0090 for common core credit (8/91).

Related Occupations: 27H.

AR-1715-0908

NUCLEAR WEAPONS SPECIALIST BASIC NONCOMMISSIONED OFFICER (NCO)

Course Number: 644-55F30.

Location: Missile and Munitions School, Redstone Arsenal, AL.

Length: 7–8 weeks (261 hours).

Exhibit Dates: 10/91–1/96.

Learning Outcomes: Upon completion of the course, the student will be able to direct and train personnel performing the assembly, disassembly, maintenance, calibration, testing, and inspection of test equipment and nuclear weapons.

Instruction: Lectures and practical experiences cover inspection and maintenance of nuclear weapons. This course contains a common core of leadership subjects.

Credit Recommendation: Credit is recommended for the common core only. See AR-1406-0090 (8/91).

Related Occupations: 55F.

AR-1715-0909

HAWK FIRING SECTION REPAIRER

Course Number: 121-27H10.

Location: Missile and Munitions School, Redstone Arsenal, AL.

Length: 28 weeks (1064 hours).

Exhibit Dates: 10/89–9/90.

Learning Outcomes: Upon completion of the course, the student will apply basic electrical DC and AC circuit theory to the operation of solid state active devices and circuits and troubleshoot and repair electrical, electronic, and hydraulic systems to the module and component level.

Instruction: Lectures and practical exercises are geared toward a thorough working knowledge of DC and AC circuits and parallel resistive circuits; characteristics of inductors and capacitors; principles of transformers; RL, RC, and RLC circuits and vacuum tube; and semiconductor devices and their applications. Emphasis is on electrical, mechanical, and hydraulic system maintenance and inspecting, testing, and repairing the Improved Hawk and associated test equipment.

Credit Recommendation: In the vocational certificate category, 1 semester hour in soldering techniques. In the lower-division baccalaureate/associate degree category, 3 semester hours in basic electricity, 3 in solid state devices, 3 in electronic systems troubleshooting, and 3 in hydraulic systems maintenance and repair (8/91).

Related Occupations: 27H.

AR-1715-0911

SIGNAL INTELLIGENCE/ELECTRONIC WARFARE (SIGINT/EW) OFFICER (AST-35G)

Course Number: 3B-F16 (35G).

Location: Intelligence School, Fort Huachuca, AZ.

Length: 5 weeks (176 hours).

Exhibit Dates: 7/92–4/04.

Learning Outcomes: Upon completion of the course, the student will be able to conduct intelligence operations involving electronic signals and electronic warfare data sources.

Instruction: Lectures and practical exercises cover signal types and identification and the means to collect, organize, and apply these signals to military intelligence operations.

Credit Recommendation: Credit is not recommended because of the military-specific nature of the course (10/93).

Related Occupations: 35G.

AR-1715-0912

OV/RV-1D (MOHAWK) SYSTEM QUALIFICATION

Course Number: 2C-15C/2B-ASIF1.

Location: Intelligence Center and School, Fort Huachuca, AZ.

Length: 8 weeks (280 hours).

Exhibit Dates: 6/91–4/04.

Learning Outcomes: Before 6/91 see AR-1715-0498. Upon completion of the course, the student will be able to work as special electronics mission aircraft (SEMA) Mohawk system aviator and military intelligence officer; develop flight mission profiles and evasive maneuvers; operate aircraft survivability equipment; and describe other SEMA-related issues.

Instruction: Instruction covers the use of special airborne electronic equipment used in military intelligence and flight operational procedures related to the aircraft. Instruction is provided in the classroom and in flight.

Credit Recommendation: Credit is not recommended because of the military-specific nature of the course (10/93).

AR-1715-0913

Guardrail Pilot Qualification (PQ)
(Guardrail Systems Qualification)
(RC-12D Guardrail Systems Qualification (IGV))

Course Number: 2C-15C/2B-ASIF3; 2C-15C/2B-ASIF3 (IGV).

Location: Intelligence School, Fort Huachuca, AZ.

Length: 5–8 weeks (207–268 hours).

Exhibit Dates: 6/91–Present.

Learning Outcomes: Before 6/91 see AR-1715-0568. Upon completion of the course, the student will be able to identify all special electronic mission aircraft and operate a specific military reconnaissance airplane (RC-12D) and associated electronic intelligence-gathering systems.

Instruction: Instruction covers the use of special airborne electronic equipment used in military intelligence and flight operational procedures related to the aircraft. Instruction is provided in the classroom and inflight.

Credit Recommendation: Credit is not recommended because of the military-specific nature of the course (8/02).

AR-1715-0914

Emitter Locater/Identifier Basic Noncommissioned Officer (NCO)

Course Number: 231-98D30.

Location: Intelligence School, Fort Huachuca, AZ.

Length: 10–11 weeks (385 hours).

Exhibit Dates: 1/91–Present.

Learning Outcomes: Upon completion of the course, the student will be able to select a site; install radio direction finding equipment; and collect, interpret, and report data; train, assign duties, and supervise the section; provide for logistical requirements; and establish the necessary maintenance program.

Instruction: Instruction includes lectures/conferences, small-group practical exercises, and demonstrations of equipment installation, use, and maintenance. A full-scale major training exercise synthesizes learning at the end of the course. This course contains a common core of leadership subjects.

Credit Recommendation: In the lower-division baccalaureate/associate degree category, 3 semester hours in leadership, 2 in materiel management, 1 in electronic communications, and 1 in keyboarding. See AR-1406-0090 for common core credit (1/97).

Related Occupations: 98D.

AR-1715-0915

Tactical Communications Chief Basic Noncommissioned Officer (NCO)

Course Number: 101-31G30.

Location: Field Artillery School, Fort Sill, OK.

Length: 16–17 weeks (592 hours).

Exhibit Dates: 10/91–12/94.

Learning Outcomes: Before 10/91 see AR-1715-0492. Upon completion of the course, the student will be able to operate, maintain, and organize an FM radio and land wire telephone communications system.

Instruction: Lectures and practical exercises cover the areas of FM radio communication, telephone network systems, and maintenance/supply procedures. This course contains a common core of leadership subjects.

Credit Recommendation: In the vocational certificate category, 1 semester hour in electrical fundamentals laboratory and 1 in telephone system installation. In the lower-division baccalaureate/associate degree category, 2 semester hours in communications systems troubleshooting and maintenance. See AR-1406-0090 for common core credit (2/94).

Related Occupations: 31G.

AR-1715-0916

Transportable Automatic Switching Systems Operator/Maintainer Reserve Component

Course Number: 622-36L10-RC.

Location: Signal School, Fort Gordon, GA.

Length: 17–18 weeks (619 hours).

Exhibit Dates: 10/93–10/95.

Learning Outcomes: Upon completion of the course, the student will be able to operate and maintain an automatic telephone central office electronic switching system; identify series, parallel, and series/parallel resistive circuits and determine values of resistance, voltage, and current in the circuit; explain the basic operation of semiconductors, junction diodes, and transistors and identify their uses in rectifier circuits and voltage regulators; convert between and perform arithmetic in the decimal, binary, octal, and hexadecimal number systems; analyze logic circuits, flip-flop circuits, registers, and counters; and solder and desolder wire to various types of electronics terminals.

Instruction: Lectures and practical exercises cover the basic electronics and computer operations that support the electronic switch.

Credit Recommendation: In the vocational certificate category, 1 semester hour in soldering techniques and 3 in telephone system troubleshooting and maintenance. In the lower-division baccalaureate/associate degree category, 3 semester hours in DC/AC circuits, 1 in solid state electronics, 3 in electronic systems troubleshooting and maintenance, and 1 in digital principles (2/94).

Related Occupations: 36L.

AR-1715-0917

Signal Support Systems Specialist Basic Noncommissioned Officer (NCO)

Course Number: *Version 1:* 101-31U30. *Version 2:* 101-31U30.

Location: Signal School, Fort Gordon, GA.

Length: *Version 1:* 15–16 weeks (601 hours). *Version 2:* 10–11 weeks (417 hours).

Exhibit Dates: *Version 1:* 10/92–9/94. *Version 2:* 10/94–Present.

Learning Outcomes: *Version 1:* Upon completion of the course, the student will be able to supervise, install, integrate, troubleshoot, and maintain battlefield manual and automated signal support systems. *Version 2:* Upon completion of the course, the student will be able to supervise, install, integrate, troubleshoot, and maintain battlefield manual and automated signal support systems.

Instruction: *Version 1:* Lectures and laboratory exercises cover DC and AC circuits, system troubleshooting techniques, MS DOS operating system, setup and troubleshooting of local area and wide area networks, and installation and integration of digital communications system. This course contains a common core of leadership subjects. *Version 2:* Lectures and practical exercises cover the installation, operation, maintenance, and management of automated signal support systems. This course also includes training in leadership as well as system troubleshooting techniques of local and wide area networks and installation/integration of digital communications systems.

Credit Recommendation: *Version 1:* In the lower-division baccalaureate/associate degree category, 2 semester hours in AC/DC circuits, 2 in communications system troubleshooting and maintenance, 2 in computer literacy, 3 in local and wide area networks, and 3 in digital systems integration. See AR-1406-0090 for common core credit (2/94). *Version 2:* In the lower-division baccalaureate/associate degree category, 2 semester hours in computer applications software, 3 in local area networks, and 3 in data communications. See AR-1406-0090 for common core credit (2/97).

Related Occupations: 31U.

AR-1715-0918

1. Satellite Communications (SATCOM) Systems Repairer Basic Noncommissioned Officer (NCO)
2. Satellite Communications (SATCOM) Systems Operator/Maintainer Basic Noncommissioned Officer (NCO)

(Satellite Communications (SATCOM) Systems Repairer Basic Noncommissioned Officer (NCO))

Course Number: *Version 1:* 102-29Y30. *Version 2:* 102-31S30.

Location: Signal School, Fort Gordon, GA.

Length: *Version 1:* 10 weeks (384 hours). *Version 2:* 10 weeks (384–385 hours).

Exhibit Dates: *Version 1:* 10/91–9/93. *Version 2:* 10/93–Present.

Learning Outcomes: *Version 1:* Upon completion of the course, the student will be able to supervise, manage, and perform maintenance on a satellite communications system. *Version 2:* Upon completion of the course, the student will be able to supervise, manage, and perform maintenance on a satellite communications system.

Instruction: *Version 1:* Lectures and practical exercises cover satellite communications system equipment, operation, and maintenance management. This course contains a common core of leadership subjects. *Version 2:* Lectures and practical exercises cover satellite communications system equipment, operation, and maintenance management. This course contains a common core of leadership subjects.

Credit Recommendation: *Version 1:* In the lower-division baccalaureate/associate degree category, 4 semester hours in communications systems maintenance management and 1 in computer literacy. See AR-1406-0090 for common core credit (2/94). *Version 2:* In the lower-division baccalaureate/associate degree category, 4 semester hours in communications systems maintenance management and 1 in computer applications software. See AR-1406-0090 for common core credit (2/97).

Related Occupations: 29Y; 31S.

AR-1715-0919

Marine Radar Observer

Course Number: 8C-F10/062-F3, Phase 1; 8C-F10/062-F3.

Location: Transportation School, Fort Eustis, VA.

Length: 1–2 weeks (74 hours).

Exhibit Dates: 10/91–Present.

Learning Outcomes: Before 10/91 see AR-1715-0157. Upon completion of the course, the student will be able to operate marine radar and interpret radar plots.

Instruction: This course is conducted through general classroom and applied simulator operations.

Credit Recommendation: In the lower-division baccalaureate/associate degree category, 1 semester hour in radar operation and plotting (10/03).

AR-1715-0920

SIGNAL SUPPORT SYSTEMS SUPERVISOR ADVANCED NONCOMMISSIONED OFFICER (NCO)

Course Number: *Version 1:* 101-31U40. *Version 2:* 101-31U40.

Location: Signal School, Fort Gordon, GA.

Length: *Version 1:* 14–15 weeks (576 hours). *Version 2:* 11–12 weeks (455 hours).

Exhibit Dates: *Version 1:* 10/93–9/94. *Version 2:* 10/94–Present.

Learning Outcomes: *Version 1:* Upon completion of the course, the student will be able to supervise, coordinate, and provide technical assistance in the management, operation, and installation of communications systems at both the strategic and tactical levels. *Version 2:* Upon completion of the course, the student will be able to supervise, coordinate, and provide technical assistance in the management, operation, and installation of communications systems at both the strategic and tactical levels.

Instruction: *Version 1:* Lectures and practical exercises cover the installation, operation, maintenance, and management of a communications system. This course also includes training in leadership, communication skills, computer operations, and total quality management. This course contains a common core of leadership subjects. *Version 2:* Lectures and practical exercises cover the installation, operation, maintenance, and management of a communications system. Course also includes training in leadership, communication skills and computer operations.

Credit Recommendation: *Version 1:* In the lower-division baccalaureate/associate degree category, 2 semester hours in computer literacy and 4 in communications system troubleshooting and maintenance. See AR-1404-0035 for common core credit (2/94). *Version 2:* In the lower-division baccalaureate/associate degree category, 2 semester hours in computer applications software, 4 in communications systems troubleshooting/repair, and 2 in supervisory management. See AR-1404-0035 for common core credit (2/97).

Related Occupations: 31U.

AR-1715-0922

SATELLITE/MICROWAVE SYSTEMS CHIEF ADVANCED NONCOMMISSIONED OFFICER (NCO)

Course Number: *Version 1:* 101-31S/31P40. *Version 2:* 101-31S/31P40.

Location: Signal School, Fort Gordon, GA.

Length: *Version 1:* 13–14 weeks (544 hours). *Version 2:* 11–12 weeks (468 hours).

Exhibit Dates: *Version 1:* 1/94–9/95. *Version 2:* 10/95–Present.

Learning Outcomes: *Version 1:* Upon completion of the course, the student will be able to direct and supervise the use of computer systems in the maintenance, operation, and administration of a satellite or microwave communications station. *Version 2:* Upon completion of the course, the student will be able to plan, supervise, coordinate, and provide technical assistance in managing, operating, and installing satellite/microwave communications systems.

Instruction: *Version 1:* The course consists of classroom exercises and practical exercises. Topics include computer literacy, signal security, single-channel ground and airborne radio systems, mobile subscriber systems, digital communications system, microwave communications system, and satellite communications system. This course contains a common core of leadership subjects. *Version 2:* The course consists of classroom exercises and practical exercises. Topics include computer literacy, signal security, single-channel ground and airborne radio systems, mobile subscriber systems, digital communications system, microwave communications system, and satellite communications systems. This course contains a common core of leadership subjects.

Credit Recommendation: *Version 1:* In the lower-division baccalaureate/associate degree category, 2 semester hours in computer literacy, 3 in digital and microwave communications system operation and maintenance, and 3 in satellite system applications and operation. See AR-1404-0035 for common core credit (2/94). *Version 2:* In the lower-division baccalaureate/associate degree category, 2 semester hours in computer applications software, 2 in digital and microwave communications system operation and maintenance, and 2 in satellite systems applications. See AR-1404-0035 for common core credit (2/97).

Related Occupations: 31P; 31S.

AR-1715-0924

COMMUNICATIONS SYSTEMS/CIRCUIT CONTROLLER BASIC NONCOMMISSIONED OFFICER (NCO)

Course Number: 101-31N30.

Location: Signal School, Fort Gordon, GA.

Length: 10–11 weeks (396 hours).

Exhibit Dates: 10/91–1/94.

Learning Outcomes: Upon completion of the course, the student will be able to supervise the operation of digital communications systems and tactical technical control facilities.

Instruction: The course consists of classroom lectures, demonstrations, and practical exercises. Topics include digital communications system circuits, digital patch and access system, and tactical technical control facilities. This course contains a common core of leadership subjects.

Credit Recommendation: In the lower-division baccalaureate/associate degree category, 3 semester hours in operating, maintaining, and troubleshooting communications equipment. See AR-1406-0090 for common core credit (2/94).

Related Occupations: 31N.

AR-1715-0925

SATELLITE SYSTEMS/NETWORK COORDINATOR

Course Number: 4C-F19/102-ASI1C (2); 4C-F19/102-ASI1C.

Location: Signal School, Fort Gordon, GA.

Length: 16–20 weeks (576–711 hours).

Exhibit Dates: 10/91–Present.

Learning Outcomes: Upon completion of the course, the student will be able to operate, maintain, and manage a network of satellite communications ground stations and apply electronic countermeasures to ensure reliable communications.

Instruction: Lectures and practical exercises cover the operation of ground station communications equipment and the application of electronic countermeasures using high-frequency test instruments including spectrum analyzers.

Credit Recommendation: In the vocational certificate category, 2 semester hours in computer systems maintenance. In the lower-division baccalaureate/associate degree category, 3 semester hours in satellite communications system operation and management (2/03).

Related Occupations: 31S.

AR-1715-0926

SATELLITE COMMUNICATIONS TERMINAL AN/TSC-85/TSC-93 REPAIRER

Course Number: 101-ASI7A.

Location: Signal School, Fort Gordon, GA.

Length: 5–6 weeks (183 hours).

Exhibit Dates: 10/91–4/95.

Learning Outcomes: Upon completion of the course, the student will be able to operate, troubleshoot, and maintain a military satellite communications terminal.

Instruction: Classroom lectures and practical exercises cover FM tactical satellite terminals using test sets and spectrum analyzers.

Credit Recommendation: In the vocational certificate category, 3 semester hours in satellite communications systems (2/94).

AR-1715-0928

1. SIGNAL SUPPORT SYSTEMS SPECIALIST
2. TATS SIGNAL SUPPORT SYSTEMS SPECIALIST

Course Number: *Version 1:* 101-31U10. *Version 2:* 101-31U10 (F); 101-31U10.

Location: Signal School, Fort Gordon, GA.

Length: *Version 1:* 17 weeks (612 hours). *Version 2:* 17 weeks (612–617 hours).

Exhibit Dates: *Version 1:* 10/92–9/95. *Version 2:* 10/95–Present.

Learning Outcomes: *Version 1:* Upon completion of the course, the student will be able to install, maintain, and troubleshoot battlefield manual and automated signal support combat net radio systems and terminal equipment; perform basic functions on MS DOS-based computers; provide technical assistance to users of signal equipment; and deploy and operate dedicated retransmission stations. *Version 2:* Upon completion of the course, the student will be able to install, troubleshoot, and perform unit level maintenance on manual and automated signal support systems and terminal equipment; provide technical assistance in implementing information systems; and deploy and operate dedicated computer systems.

Instruction: *Version 1:* Lectures and laboratory exercises cover installation, operation, maintenance, and troubleshooting of manual and automated combat net radio signal support systems and terminal equipment. The course includes minor repairs on cables using basic soldering techniques. *Version 2:* Lectures and laboratory experiences cover installing, operating, and maintaining manual and automated signal support systems, terminal equipment, PC communications, local area networks and computer testing utilities and test software.

Credit Recommendation: ***Version 1:*** In the lower-division baccalaureate/associate degree category, 3 semester hours in electronic systems troubleshooting and maintenance and 1 in data processing and operating systems (8/96). *Version 2:* In the lower-division baccalaureate/associate degree category, 3 semester hours in introduction to computers, 1 in computers and communications, and 1 in local area networks (2/03).

Related Occupations: 31U.

AR-1715-0929

TRANSPORTABLE MULTICHANNEL SATELLITE SYSTEMS OPERATOR

Course Number: 202-ASI7A.

Location: Signal School, Fort Gordon, GA.

Length: 4–5 weeks (152 hours).

Exhibit Dates: 6/92–4/95.

Learning Outcomes: Upon completion of the course, the student will be able to install, operate, and maintain a satellite communications ground terminal.

Instruction: Lectures and practical exercises cover the installation, operation, and maintenance of a military-specific satellite communications ground terminal.

Credit Recommendation: In the lower-division baccalaureate/associate degree category, 3 semester hours in electronic systems troubleshooting and maintenance (2/94).

AR-1715-0930

OFFICER REBRANCHED SIGNAL

Course Number: 4C-F37.

Location: Signal School, Fort Gordon, GA.

Length: 2–3 weeks (79 hours).

Exhibit Dates: 1/90–6/94.

Learning Outcomes: Upon completion of the course, the student will be able to demonstrate basic tactical communications management principles including the design of a tactical communications system.

Instruction: The course uses classroom lectures and laboratory exercises. Topics include the history and background of the Signal Corps, electronic communications, tactical communications, and communications equipment.

Credit Recommendation: In the lower-division baccalaureate/associate degree category, 3 semester hours in electronic communications (2/94).

AR-1715-0932

1. MICROWAVE SYSTEMS OPERATOR MAINTENANCE BASIC NONCOMMISSIONED OFFICER (NCO)
2. MICROWAVE SYSTEMS OPERATOR BASIC NONCOMMISSIONED OFFICER (NCO)
3. MICROWAVE SYSTEMS OPERATOR-MAINTAINER (BNCOC)

Course Number: ***Version 1:*** 101-31P30. *Version 2:* 101-31P30. *Version 3:* 101-31P30.

Location: Signal School, Fort Gordon, GA.

Length: ***Version 1:*** 16 weeks (608 hours). *Version 2:* 13–14 weeks (517 hours). *Version 3:* 11 weeks (428 hours).

Exhibit Dates: ***Version 1:*** 2/94–9/94. *Version 2:* 10/94–9/01. *Version 3:* 10/01–Present.

Learning Outcomes: ***Version 1:*** Upon completion of the course, the student will be able to implement telecommunications service orders and supervise the operation and maintenance of microwave communications and control systems. *Version 2:* Upon completion of the course, the student will be able to implement telecommunications service orders and supervise the operation and maintenance of microwave communications and control systems. *Version 3:* Upon completion of the course, the student will be able to implement telecommunications service orders and supervise the operation and maintenance of microwave communications and control systems.

Instruction: ***Version 1:*** Lectures and practical exercises cover signal administration, system procedures, and digital microwave systems operation. This course contains a common core of leadership subjects. *Version 2:* Lectures and practical exercises cover signal administration, system procedures, and digital microwave systems operation. This course contains a common core of leadership subjects. *Version 3:* Lectures and practical exercises. Topics cover signal administration, system procedures, and digital microwave systems operation. This course contains a common core of leadership subjects.

Credit Recommendation: ***Version 1:*** In the lower-division baccalaureate/associate degree category, 1 semester hour in computer literacy, 3 in microwave communications systems management, and 3 in electronic systems troubleshooting and maintenance. See AR-1406-0090 for common core credit (2/94). *Version 2:* In the lower-division baccalaureate/associate degree category, 1 semester hour in computer software applications, 3 in microwave communications systems management, and 3 in electronic systems troubleshooting and maintenance. See AR-1406-0090 for common core credit (2/97). *Version 3:* In the lower-division baccalaureate/associate degree category, 1 semester hour in computer literacy, 3 in microwave communications system management, 3 in electronic systems troubleshooting and maintenance, and 3 in network fundamentals (2/03).

Related Occupations: 31P.

AR-1715-0933

COMMUNICATIONS SYSTEM CONTROL ELEMENT OPERATIONS

Course Number: 4C-F43/260-F11.

Location: Signal School, Fort Gordon, GA.

Length: 5–6 weeks (208 hours).

Exhibit Dates: 1/93–6/95.

Learning Outcomes: Upon completion of the course, the student will be able to install, operate, and maintain a communications system controller based on a DEC Microvax workstation.

Instruction: Lectures and practical exercises cover the installation, operation, and maintenance of military-specific communications system control software running on a commercial workstation.

Credit Recommendation: In the lower-division baccalaureate/associate degree category, 1 semester hour in computer systems troubleshooting and maintenance and 1 in computer literacy (2/94).

AR-1715-0934

TELEVISION EQUIPMENT SPECIALIST

Course Number: G3ABR30435.

Location: Technical Training Center, Lowry AFB, CO.

Length: 12–13 weeks (484 hours).

Exhibit Dates: 1/90–2/97.

Learning Outcomes: Upon completion of the course, the student will be able to apply the theory of television and video systems, use television test instruments, perform routine maintenance, and install and troubleshoot video systems.

Instruction: Training includes the theory of television, television test instruments, video camera, distribution systems, color TV principles, studio audio and video equipment, television receiver and transmitter principles, maintenance, videotape recording, alignment, and troubleshooting procedures. Methods include readings, lectures, exercises, and laboratory experience with television equipment. Evaluation is by test and performance checks. Seventy percent accuracy is required.

Credit Recommendation: In the lower-division baccalaureate/associate degree category, 3 semester hours in electronic test instruments and 3 in television and video theory (2/87).

AR-1715-0935

MULTIPLE LAUNCH ROCKET SYSTEM (MLRS) CADRE

Course Number: 2F-13B/042-13M20/30/40-T.

Location: Field Artillery School, Fort Sill, OK.

Length: 4–6 weeks (119–208 hours).

Exhibit Dates: 1/90–12/99.

Learning Outcomes: Upon completion of the course, the student will be able to position, employ, and supervise a multiple launch rocket system (MLRS) ammunition and firing section.

Instruction: Lectures and practical exercises cover communications, targeting, weapons, navigation, map reading, and system maintenance.

Credit Recommendation: In the vocational certificate category, 1 semester hour in radio system operation. In the lower-division baccalaureate/associate degree category, 1 semester hour in map reading (9/94).

Related Occupations: 13B; 13M.

AR-1715-0936

FIELD ARTILLERY TACTICAL FIRE DIRECTION SYSTEM REPAIRER

Course Number: 113-39Y10.

Location: Field Artillery School, Fort Sill, OK.

Length: 21 weeks (787 hours).

Exhibit Dates: 1/90–10/93.

Learning Outcomes: Upon completion of the course, the student will be able to maintain, repair, and troubleshoot a computer-driven fire control system to the component level.

Instruction: Lectures and laboratory exercises cover the major subsystems, including computer group and central processor, magnetic and core memory units, line printer, digital data terminal, digital plotter, power systems, and communications equipment and their component parts, including circuit boards, wiring diagrams, and computer-driven diagnostics/fault isolation. Students initiate diagnostics, interpret diagnostic coding, and use schematics and other technical documentation in conjunction with previously acquired knowledge of electronic theory and practice, to identify, isolate, and repair faults and perform standard operation and preventive maintenance actions.

Credit Recommendation: In the vocational certificate category, 2 semester hours in soldering. In the lower-division baccalaureate/associate degree category, 2 semester hours in computer operations, 3 in computer maintenance and troubleshooting, 4 in AC/DC circuits, 3 in solid state electronics, 3 in digital principles, and 3 in computer fundamentals (9/94).

Related Occupations: 39Y.

AR-1715-0937

AVENGER SYSTEM CREWMEMBER

Course Number: 043-ASI Y1 (16S); 043-14S10.

Location: Air Defense Artillery School, Fort Bliss, TX.

Length: 5–10 weeks (180–385 hours).

Exhibit Dates: 6/90–7/98.

Learning Outcomes: After 7/98 see AR-1715-0953. Upon completion of the course, the student will be able to operate and maintain the Avenger system.

Instruction: Lectures and practical exercises cover the missile system, machine gun emplacement, and field applications.

Credit Recommendation: Credit is not recommended because of the military-specific nature of the course (8/95).

Related Occupations: 16S.

AR-1715-0938

AVENGER CREWMEMBER (TRANSITION)
(Avenger System Crewmember (Transition))

Course Number: 043-12S2/3/40 (T); 043-ASIY1 (16S20/30/40).

Location: Air Defense Artillery School, Fort Bliss, TX.

Length: 3 weeks (112 hours).

Exhibit Dates: 7/90–10/00.

Learning Outcomes: Upon completion of the course, the student will be able to operate and maintain the Avenger air defense system.

Instruction: Lectures and practical exercises cover the missile system, machine gun emplacement, and field applications.

Credit Recommendation: Credit is not recommended because of the military-specific nature of the course (6/95).

Related Occupations: 14S; 16S.

AR-1715-0939

HAWK MISSILE SYSTEM CREWMEMBER

Course Number: ***Version 1:*** 043-14D10. *Version 2:* 043-14D10.

Location: Air Defense Artillery School, Fort Bliss, TX.

Length: ***Version 1:*** 8–9 weeks (320 hours). *Version 2:* 10 weeks (372–374 hours).

Exhibit Dates: ***Version 1:*** 1/91–7/92. *Version 2:* 8/92–6/96.

Learning Outcomes: ***Version 1:*** Upon completion of the course, the student will be able to energize/deenergize the system; prepare the system for travel and emplace the system; and train personnel on system orientations, alignment, integration, systems checks, system operations, and maintenance. *Version 2:* Upon completion of the course, the student will perform maintenance checks and services; emplacement; system handling procedures; orientation, alignment, and integration checks; system operations; and crew training.

Instruction: ***Version 1:*** Conferences, demonstrations, and practical exercises cover basic system-level operation, emplacement, daily checks, and crew drill. *Version 2:* Conferences, demonstrations, and practical exercises cover basic system-level operation, emplacement, daily checks, and crew drill.

Credit Recommendation: ***Version 1:*** In the lower-division baccalaureate/associate degree category, 3 semester hours in radar systems (6/95). *Version 2:* In the lower-division baccalaureate/associate degree category, 3 semester hours in radar systems and 1 in electronic system maintenance (6/95).

Related Occupations: 14D.

AR-1715-0940

HAWK INFORMATION COORDINATION CENTRAL MECHANIC TRANSITION

Course Number: 104-24G1/2/30 (PIP 3)-T.

Location: Air Defense Artillery School, Fort Bliss, TX.

Length: 13 weeks (464 hours).

Exhibit Dates: 6/91–6/96.

Learning Outcomes: Upon completion of the course, the student will be able to perform operational checks, adjustments, fault isolation, and repair of the Hawk missile system with PIP III changes.

Instruction: Conferences, demonstrations, and practical exercises cover system operation, adjustments, maintenance, fault isolation, and repair.

Credit Recommendation: In the lower-division baccalaureate/associate degree category, 3 semester hours in radar systems and 3 in electronic system maintenance (6/95).

Related Occupations: 24G.

AR-1715-0941

PATRIOT MISSILE CREWMEMBER

Course Number: 043-16T10.

Location: Air Defense Artillery School, Fort Bliss, TX.

Length: 9 weeks (325 hours).

Exhibit Dates: 8/93–Present.

Learning Outcomes: Before 8/93 see AR-1715-0579. Upon completion of the course, the student will be able to operate and maintain the missile truck delivery system, the launching station, and, to a limited extent, the radar system.

Instruction: Instruction is provided using lectures and laboratory experience. Detailed instruction on the truck, preventive maintenance for the truck, and the obtaining of proper driver license is a part of the course. Missile reload procedures, preventive maintenance, launcher tests, logistics functions, and communications procedures are contained in the course.

Credit Recommendation: In the vocational certificate category, 3 semester hours in electromechanical maintenance procedures (6/95).

Related Occupations: 16T.

AR-1715-0942

1. LINE OF SIGHT-FORWARD-HEAVY CREWMEMBER
2. LINE OF SIGHT-FORWARD-HEAVY CREWMEMBER (BRADLEY LINEBACKER CREWMEMBER)

Course Number: *Version 1:* 043-14R10. *Version 2:* 043-14R10, 043-14R10 (TATS).

Location: Air Defense Artillery School, Fort Bliss, TX.

Length: ***Version 1:*** 9–10 weeks (352 hours). *Version 2:* 12–13 weeks (457–461 hours).

Exhibit Dates: ***Version 1:*** 3/93–12/95. *Version 2:* 1/96–Present.

Learning Outcomes: ***Version 1:*** Upon completion of the course, this student will be able to operate the Bradley fighting vehicle (BSFV), to fire on-board weapons systems, and to perform preventive maintenance on both the BSFV and its weapons. *Version 2:* Upon completion of he course, the student will be able to visually identify numerous aircraft and armored vehicles and operate and perform operator maintenance on special armored vehicles and their weapons.

Instruction: ***Version 1:*** This course is taught using both lectures and associated labs. It was designed to teach operation and maintenance of the Bradley system. It includes instruction in visual aircraft recognition, armored vehicle hull and turret system operation, gun operation, driving skills, and associated preventive maintenance. *Version 2:* Lectures and practical exercises include the use of radios, guns, missiles, and an armored vehicle.

Credit Recommendation: ***Version 1:*** In the vocational certificate category, 2 semester hours in diesel equipment maintenance (6/95). *Version 2:* Credit is not recommended because of the military-specific nature of the course (4/01).

Related Occupations: 14R.

AR-1715-0943

HAWK MISSILE SYSTEM MECHANIC (TRANSITION)

Course Number: 121-23R10-B-T.

Location: Air Defense Artillery School, Fort Bliss, TX.

Length: 17–18 weeks (638 hours).

Exhibit Dates: 10/90–6/96.

Learning Outcomes: Upon completion of the course, the student will be able to describe the functional operations of the Hawk missile system and radar.

Instruction: Lectures and practical demonstrations cover operations, adjustments, fault isolation, and maintenance management.

Credit Recommendation: In the lower-division baccalaureate/associate degree category, 3 semester hours in radar systems (6/95).

Related Occupations: 23R.

AR-1715-0944

HAWK MISSILE SYSTEM MECHANIC (TRANSITION)

Course Number: 121-23R10-A-T.

Location: Air Defense Artillery School, Fort Bliss, TX.

Length: 14 weeks (511 hours).

Exhibit Dates: 10/90–5/91.

Learning Outcomes: Upon completion of the course, the student will be able to describe the functional operation of the Hawk missile system and radar.

Instruction: Lectures and practical demonstrations cover operations, adjustments, fault isolation, and maintenance management.

Credit Recommendation: In the lower-division baccalaureate/associate degree category, 3 semester hours in radar systems (6/95).

Related Occupations: 23R.

AR-1715-0945

HAWK FIRING SECTION MECHANIC TRANSITION

Course Number: 121-24C1/2/30 (PIP 3)-T.

Location: Air Defense Artillery School, Fort Bliss, TX.

Length: 10–11 weeks (386 hours).

Exhibit Dates: 6/91–6/96.

Learning Outcomes: Upon completion of the course, the student will be able to describe the functional operation of the Hawk missile system and radar.

Instruction: Lectures and practical demonstrations cover operations, adjustments, fault isolation, and maintenance management.

Credit Recommendation: In the lower-division baccalaureate/associate degree category, 3 semester hours in radar systems (6/95).

Related Occupations: 24C.

AR-1715-0946

HAWK MISSILE SYSTEM MECHANIC

Course Number: ***Version 1:*** 121-23R10. *Version 2:* 121-23R10.

Location: Air Defense Artillery School, Fort Bliss, TX.

Length: ***Version 1:*** 34–35 weeks (1256 hours). *Version 2:* 28–31 weeks (996–1095 hours).

Exhibit Dates: ***Version 1:*** 3/90–6/91. *Version 2:* 7/91–6/96.

Learning Outcomes: ***Version 1:*** Upon completion of the course, the student will be able to perform organizational maintenance, fault isolation, and repair of the Hawk missile system. *Version 2:* Upon completion of the course, the student will be able to perform unit maintenance, fault isolation, and corrective maintenance on the major end items of the Hawk missile system.

Instruction: ***Version 1:*** Conferences, demonstrations, and practical exercises cover missile system operation, maintenance, fault isolation, and repair on major end items of the Hawk missile system. *Version 2:* Conferences, demonstrations, and practical exercises cover missile system operation, maintenance, fault isolation, and repair on major end items of the Hawk missile system.

Credit Recommendation: ***Version 1:*** In the lower-division baccalaureate/associate degree category, 3 semester hours in radar systems, 3 in electronic systems maintenance, 1 in hydraulic systems, 2 in DC circuits, 1 in AC circuits, and 2 in electronic circuit analysis (6/95). *Version 2:* In the lower-division baccalaureate/associate degree category, 3 semester hours in radar systems, 3 in electronic systems maintenance, 1 in hydraulic systems, 2 in DC circuits, 1 in AC circuits, and 2 in electronic circuit analysis (6/95).

Related Occupations: 23R.

AR-1715-0947

HAWK MISSILE SYSTEM TECHNICIAN WARRANT OFFICER TECHNICAL/TACTICAL CERTIFICATION

Course Number: 4F-140D; 4F-140D (PIP 3).

Location: Air Defense Artillery School, Fort Bliss, TX.

Length: 26–27 weeks (935–945 hours).

Exhibit Dates: 1/90–6/96.

Learning Outcomes: Upon completion of the course, the student will be able to operate and maintain the Hawk air defense missile system and support equipment.

Instruction: Lectures and practical exercises cover hydraulics, high voltage, radar, computer controls, and power generation equipment.

Credit Recommendation: In the lower-division baccalaureate/associate degree category, 3 semester hours in radar systems, 3 in electronic systems troubleshooting and maintenance, and 1 in hydraulics (6/95).

Related Occupations: 140D.

AR-1715-0948

SHORADS SYSTEMS TECHNICIAN WARRANT OFFICER TECHNICAL/TACTICAL CERTIFICATION, PHASE 2

Course Number: 4F-140B.

Location: Air Defense Artillery School, Fort Bliss, TX.

Length: 10–11 weeks (370–371 hours).

Exhibit Dates: 6/94–10/00.

Learning Outcomes: Upon completion of the course, the student will be able to operate and maintain air defense weapons and support equipment.

Instruction: Lectures and practical exercises cover the theory, maintenance, operation, repair, and support equipment of the air defense systems. This includes Chaparral, Bradley Stinger, Avenger, and early warning systems.

Credit Recommendation: Credit is not recommended because of the military-specific nature of the course (6/95).

Related Occupations: 140B.

AR-1715-0949

HAWK MISSILE AIR DEFENSE ARTILLERY OFFICER RECLASSIFICATION RESERVE

Course Number: 2F-14D (PIP 2)-RC.

Location: Air Defense Artillery School, Fort Bliss, TX.

Length: Phase 2, 2 weeks (91 hours); Phase 4, 2 weeks (80 hours).

Exhibit Dates: 1/92–6/96.

Learning Outcomes: Upon completion of the course the reserve officer will be able to perform in an initial assignment within a Hawk air defense battery or battalion.

Instruction: This course is organized into 4 separate segments (two of which are correspondence) and is designed to prepare a reserve officer for an initial assignment within a Hawk air defense battery or battalion. The four segments are nonresident instruction provided as read-ahead materials; resident hands-on Hawk system; command and control read-ahead materials; and resident hands-on command and control. Instruction in phases 2 and 4 at Ft. Bliss is a combination of lecture and laboratory experiences. Material in the courses (2 and 4) are military specific.

Credit Recommendation: Credit is not recommended because of the military-specific nature of the course (6/95).

AR-1715-0950

HAWK MISSILE SYSTEM WITH COMPUTER UPGRADE (PHASE 2A) (MAINTAINER)

Course Number: 4F-F29/121-F35.

Location: Air Defense Artillery School, Fort Bliss, TX.

Length: 4–5 weeks (163 hours).

Exhibit Dates: 1/91–1/95.

Learning Outcomes: Upon completion of the course, the student will maintain the Hawk missile system with computer upgrade, perform maintenance on built-in test equipment, and perform block-level system checks using integrated system checks and equipment test programs.

Instruction: Conferences, demonstrations, and practical exercises cover basic system operations, maintenance, and fault isolation.

Credit Recommendation: In the lower-division baccalaureate/associate degree category, 1 semester hour in radar systems, 1 in electronic systems, and 1 in electronic system troubleshooting (6/95).

AR-1715-0951

FIELD ARTILLERY AIR DEFENSE OFFICER RECLASSIFICATION RESERVE

Course Number: 2E-14B-RC.

Location: Air Defense Artillery School, Fort Bliss, TX.

Length: 2–3 weeks (99 hours).

Exhibit Dates: 11/93–Present.

Learning Outcomes: Upon completion of the course, the student will be able to describe the characteristics, limitations, and tactics of air defense artillery systems.

Instruction: Lectures and practical demonstrations cover air defense artillery, including the Chaparral, Avenger, Stinger, and Bradley Stinger systems.

Credit Recommendation: Credit is not recommended because of the military-specific nature of the course (6/95).

AR-1715-0952

HAWK MISSILE SYSTEM WITH COMPUTER UPGRADE (PHASE 2A) (OPERATOR)

Course Number: 2F-F17/043-F10.

Location: Air Defense Artillery School, Fort Bliss, TX.

Length: 4–5 weeks (163 hours).

Exhibit Dates: 1/91–1/95.

Learning Outcomes: Upon completion of the course, the student will be able to emplace, energize/deenergize, orient, align, select sites, test systems, and operate the system.

Instruction: Through classroom instruction, practical demonstrations, and examinations the student will be able to use built-in test equipment and line of sight orientation and perform launch alignment and system interface tests.

Credit Recommendation: In the lower-division baccalaureate/associate degree category, 1 semester hour in radar systems and 2 in electronic equipment maintenance (6/95).

AR-1715-0953

EARLY WARNING SYSTEM OPERATOR

Course Number: ***Version 1:*** 221-14J10. *Version 2:* 221-14J10.

Location: Air Defense Artillery School, Fort Bliss, TX.

Length: ***Version 1:*** 10 weeks (360 hours). *Version 2:* 15 weeks (540 hours).

Exhibit Dates: ***Version 1:*** 10/92–3/95. *Version 2:* 4/95–8/00.

Learning Outcomes: *Version 1:* Upon completion of the course, the student will be able to operate and perform limited maintenance on the early warning system. *Version 2:* After 8/00 see AR-1715-1048. Upon completion of the course, the student will be able to operate and perform limited maintenance on all the equipment of the early warning system, including radar, radio system, and truck delivery system.

Instruction: *Version 1:* Instruction is provided, using lectures and practical exercises in the operation, preventive maintenance, and limited corrective maintenance on the radar, communications, and truck delivery system. In the area of electronics/electrical, instruction is to the block level with power flow included for the onboard generator. Concepts relevant to the system operations, including NBC, communications concepts and mapping, are included in the course. *Version 2:* Instruction is provided, using lectures and practical exercises, in all aspects of the system. Mapping and global positioning, operations in an NBC environment, various communications concepts, and preventive/corrective maintenance operations are included in this course. In general, however, all of the instruction is geared toward operational concepts, and instruction in maintenance is to the block-diagram level with very limited component and test equipment use.

Credit Recommendation: *Version 1:* In the lower-division baccalaureate/associate degree category, 2 semester hours in introduction to technology (6/95). *Version 2:* In the lower-division baccalaureate/associate degree category, 2 semester hours in introduction to technology (6/95).

Related Occupations: 14J.

AR-1715-0954

Hawk Missile System Crewmember (Transition)

Course Number: 043-14D10-T.

Location: Air Defense Artillery School, Fort Bliss, TX.

Length: 5–6 weeks (204 hours).

Exhibit Dates: 10/90–6/96.

Learning Outcomes: Upon completion of the course, the student will be able to identify the electromagnetic threat, prepare for travel, and emplace the Hawk launcher, loader-transporter, and launch section control box; perform before, during, and after operation checks; and perform as a crew chief/member. The 16D transition student will be able to perform operational checks and energize/deenergize and prepare the continuous wave acquisition radar, high powered illuminator radar with video tracking group, platoon command post, identify friend or foe, and pulse acquisition radar for transport. The 16D/16E overall Hawk PIP 3 operator will initialize, orient, and align the Hawk system, including integrated systems checks, launcher crew chief/crew member duties, and arming/disarming procedures.

Instruction: Conferences, demonstrations, and practical exercises cover basic system operation, performance checks, and system deployment.

Credit Recommendation: In the lower-division baccalaureate/associate degree category, 2 semester hours in radar systems, 1 in electronic equipment maintenance for all students; for MOS 16E transition, an additional 3 semester hours in radar systems; and for MOS 16D transition, an additional 1 semester hour in radar systems, and 2 in electronic equipment maintenance (6/95).

Related Occupations: 14D.

AR-1715-0955

1. Hawk Missile System Technician/ Master Mechanic (PIP 3) Transition
2. Hawk System Technician/Master Mechanic PIP 3 Transition

Course Number: *Version 1:* 4F-140D/121-24R40/50-T. *Version 2:* 4F-140D/632-24R40/50-T.

Location: Air Defense Artillery School, Fort Bliss, TX.

Length: *Version 1:* 8–9 weeks (307 hours). *Version 2:* 5–6 weeks (202 hours).

Exhibit Dates: *Version 1:* 10/90–9/92. *Version 2:* 10/92–1/95.

Learning Outcomes: *Version 1:* Upon completion of the course, the student will be able to perform operational checks, maintenance services, and fault isolation procedures on the high-powered illumination radar (HIPIR), continuous wave acquisition radar (CWAR), platoon command post (PCP) and present instruction on operational checks, fault isolation, and maintenance services on the Hawk missile system. *Version 2:* Upon completion of the course, the student will be able to perform operational checks, maintenance services, and fault isolation procedures on high-powered illumination radar (HIPIR), continuous wave acquisition radar (CWAR), and platoon command post (PCP) and present instruction on the operational checks, fault isolation, and maintenance services on the Hawk missile system.

Instruction: *Version 1:* Conferences, demonstrations, and practical exercises cover basic system-level operations, performance checks, and fault isolation on the HIPIR, CWAR, and PCP systems. *Version 2:* Conferences, demonstrations, and practical exercises cover basic system-level operations, performance checks, and fault isolation on the HIPIR, CWAR, and PCP systems.

Credit Recommendation: *Version 1:* In the lower-division baccalaureate/associate degree category, 3 semester hours in radar systems and 3 in electronic systems maintenance (6/95). *Version 2:* In the lower-division baccalaureate/associate degree category, 3 semester hours in radar systems and 2 in electronic systems maintenance (6/95).

Related Occupations: 24R.

AR-1715-0956

Chaparral System Mechanic

Course Number: 632-24N10.

Location: Air Defense Artillery School, Fort Bliss, TX.

Length: 13 weeks (472 hours).

Exhibit Dates: 9/93–10/95.

Learning Outcomes: Before 9/93 see AR-1715-0120. Upon completion of the course, the student will be able to operate, perform preventive maintenance, and troubleshoot the Chaparral air defense system and all support equipment.

Instruction: Conferences, demonstrations, and practical exercises cover basic operation, maintenance, and troubleshooting of the Chaparral air defense system and all support equipment.

Credit Recommendation: In the lower-division baccalaureate/associate degree category, 1 semester hour in radar systems, 2 in electronic systems maintenance, and 1 in DC circuits (6/95).

Related Occupations: 24N.

AR-1715-0957

Hawk Missile System Technician Warrant Officer Technical/ Tactical Certification (PIP 2)

Course Number: 4F-140D (PIP 2).

Location: Air Defense Artillery School, Fort Bliss, TX.

Length: 25–26 weeks (910 hours).

Exhibit Dates: 1/90–1/95.

Learning Outcomes: Upon completion of the course, the student will be able to operate and maintain the radar units associated with the Hawk missile system.

Instruction: Instruction is provided using both lecture and laboratory experiences. The primary focus of the course is on the radar systems (high power, continuous wave, and pulse acquisition radars). Included also is instruction on the IFF system.

Credit Recommendation: In the lower-division baccalaureate/associate degree category, 3 semester hours in radar systems, 2 in power distribution, 3 in electronic troubleshooting, and 2 in microprocessor troubleshooting (6/95).

Related Occupations: 140D.

AR-1715-0958

Hawk Missile System with Computer Upgrade (Phase 3A) Operator

Course Number: 2F-F18/043-F11.

Location: Air Defense Artillery School, Fort Bliss, TX.

Length: 4–5 weeks (163 hours).

Exhibit Dates: 1/91–1/95.

Learning Outcomes: Upon completion of the course, the student will be able to operate, emplace, energize/deenergize, orient, and align the Hawk missile system with computer upgrade and use built-in test equipment to perform line of sight orientation, launcher alignment checks, and system interface tests.

Instruction: Conferences, demonstrations, and practical exercises cover basic missile system operation, maintenance, and fault isolation to the organizational level.

Credit Recommendation: In the lower-division baccalaureate/associate degree category, 1 semester hour in radar systems and 2 in electronic equipment maintenance (6/95).

AR-1715-0959

Hawk Officer

Course Number: 2F-14D (PIP 3).

Location: Air Defense Artillery School, Fort Bliss, TX.

Length: 13–14 weeks (497 hours).

Exhibit Dates: 10/91–6/96.

Learning Outcomes: Before 10/91 see AR-1715-0382. Upon completion of the course, the officer will be able to describe the characteristics, capabilities, and functions of the Hawk battery and the command and control system.

Instruction: Lectures and demonstrations cover safety, technical procedures, equipment requirements, site selection, field procedures, and simulated raids.

Credit Recommendation: Credit is not recommended because of the military-specific nature of the course (5/95).

AR-1715-0960

HAWK MISSILE SYSTEM WITH COMPUTER UPGRADE (PHASE 3A) (MAINTAINER)

Course Number: 4F-F30/121-F36.

Location: Air Defense Artillery School, Fort Bliss, TX.

Length: 4–5 weeks (163 hours).

Exhibit Dates: 1/91–1/95.

Learning Outcomes: Upon completion of the course, the student will be able to use built-in test equipment to perform maintenance, block-level system checks, integrated systems checks, and equipment test programs.

Instruction: Conferences, demonstrations, and practical exercises cover basic system operation, maintenance, and fault isolation.

Credit Recommendation: In the lower-division baccalaureate/associate degree category, 1 semester hour in radar systems, 1 in electronic systems, and 1 in electronic systems troubleshooting (5/95).

AR-1715-0961

HAWK PULSE ACQUISITION RADAR MECHANIC

Course Number: 4F-F31/121-F37.

Location: Air Defense Artillery School, Fort Bliss, TX.

Length: 4–5 weeks (158 hours).

Exhibit Dates: 2/92–6/96.

Learning Outcomes: Upon completion of the course, the student will be able to perform unit maintenance on the pulse acquisition radar.

Instruction: Lectures and practical exercises cover radar operation, system functions, and fault isolation procedures.

Credit Recommendation: In the lower-division baccalaureate/associate degree category, 1 semester hour in radar systems (5/95).

AR-1715-0962

SHORADS SYSTEMS TECHNICIAN WARRANT OFFICER TECHNICAL/TACTICAL CERTIFICATION

Course Number: 4F-140B.

Location: Air Defense Artillery School, Fort Bliss, TX.

Length: 16 weeks (584 hours).

Exhibit Dates: 1/90–9/95.

Learning Outcomes: Upon completion of the course, the student will be able to describe the logistical and tactical abilities of the Vulcan and Chaparral systems used in a field environment.

Instruction: Lectures and practical demonstrations cover the theory, operation, and maintenance of the Vulcan and Chaparral missile systems. Topics include radar, electrical circuits, power generation, positioning, fire control, range computers, and ordnance.

Credit Recommendation: In the lower-division baccalaureate/associate degree category, 2 semester hours in AC/DC circuits (6/95).

Related Occupations: 140B.

AR-1715-0963

AIR DEFENSE ARTILLERY OFFICER ADVANCED RESERVE

Course Number: 2-44-C23.

Location: Air Defense Artillery School, Fort Bliss, TX.

Length: Phase 1, 86–106 hours; Phase 2, 100–121 hours.

Exhibit Dates: 7/93–Present.

Learning Outcomes: Upon completion of the course, the student will be able to load and direct an air defense battery.

Instruction: Lectures and practical exercises cover the use of defensive missile systems in battle.

Credit Recommendation: Credit is not recommended because of the military-specific nature of the course (8/95).

AR-1715-0964

AIR DEFENSE ARTILLERY WARRANT OFFICER ADVANCED

Course Number: 2-44-C32.

Location: Air Defense Artillery School, Fort Bliss, TX.

Length: 6–7 weeks (238–252 hours).

Exhibit Dates: 1/90–10/92.

Learning Outcomes: After 5/00 see AR-1715-1050. Upon completion of the course, the student will be able to use effective written and oral communication techniques; employ leadership principles; describe missile systems; integrate personal computer components; use MS DOS; use word processing software, and use a personal computer graphics program.

Instruction: Lectures and practical exercises include writing and editing techniques, oral presentation skills, leadership skills, specific missile systems, and personal computer hardware and software.

Credit Recommendation: In the lower-division baccalaureate/associate degree category, 1 semester hour in written and oral communication, 1 in personal computer hardware, 1 in DOS, and 1 in word processing (8/95).

AR-1715-0966

AVIONIC MAINTENANCE SUPERVISOR ADVANCED NONCOMMISSIONED OFFICER (NCO) RESERVE PHASE 2

Course Number: RC-011-68P40.

Location: Reserve Component Training Centers, US; Reserve Centers, US; Aviation Center, Fort Rucker, AL.

Length: 56 hours.

Exhibit Dates: 8/91–Present.

Learning Outcomes: Upon completion of the course, the student will have the knowledge and skills required to function effectively as an intermediate avionic equipment maintenance supervisor.

Instruction: Classroom lectures cover verbal and written communication skills, training management, common military skills, logistic and maintenance management, and supervision of avionic maintenance facilities. This course contains a common core of leadership subjects.

Credit Recommendation: Credit is recommended for the common core only. See AR-2201-0295 (1/96).

Related Occupations: 68P.

AR-1715-0967

AVIONIC COMMUNICATIONS EQUIPMENT REPAIRER BASIC NONCOMMISSIONED OFFICER (NCO) RESERVE

Course Number: RC-011-68L30.

Location: Reserve Component Training Centers, US; Aviation Center, Fort Rucker, AL; Reserve Centers, US.

Length: 61 hours.

Exhibit Dates: 7/92–Present.

Learning Outcomes: Upon completion of the course, the student will be able to function effectively as a first-line supervisor of avionics communications equipment repairers in an avionic maintenance facility.

Instruction: Course include providing technical guidance to subordinates on avionics equipment so that maintenance is performed in accordance with applicable technical documentation. This course contains a common core of leadership subjects.

Credit Recommendation: Credit is recommended for the common core only. See AR-2201-0337 (1/96).

Related Occupations: 68L.

AR-1715-0968

TOW/DRAGON MISSILE REPAIRER RESERVE PHASE 4

Course Number: 121-27E10-RC.

Location: Ordnance, Missile and Munitions School, Redstone Arsenal, AL.

Length: 4 weeks (119 hours).

Exhibit Dates: 10/91–9/96.

Learning Outcomes: Upon completion of the course, the student will be able perform maintenance and repair of TOW/TOW 2 missile system.

Instruction: Self-paced instruction provides an introduction to DC/AC circuits, basic electronics, and maintenance and repair electronic launcher systems.

Credit Recommendation: In the lower-division baccalaureate/associate degree category, 1 semester hour in introduction to AC/DC circuits, 1 in basic electronics, and 1 in electronic system maintenance (7/96).

Related Occupations: 27E.

AR-1715-0969

BRADLEY FIGHTING VEHICLE SYSTEM TRANSITION

Course Number: 121-ASIY2 (27E).

Location: Ordnance, Missile and Munitions School, Redstone Arsenal, AL.

Length: 5–6 weeks (209 hours).

Exhibit Dates: 10/93–9/98.

Learning Outcomes: Upon completion of the course, the student will be able to troubleshoot, repair, and test Bradley fighting vehicle TOW/TOW 2 subsystems and subsystem support equipment.

Instruction: Lectures and practical exercises cover the TOW/TOW 2 subsystems and support systems and include safety hazards, guidance set, missile launcher, integrated sight unit, and support equipment.

Credit Recommendation: In the lower-division baccalaureate/associate degree category, 2 semester hours in electronic systems troubleshooting and maintenance (7/96).

Related Occupations: 27E.

AR-1715-0970

HAWK FIRE CONTROL/CONTINUOUS WAVE (CW) RADAR REPAIRER

Course Number: 104-27K10 (PIP 3).

Location: Ordnance, Missile and Munitions School, Redstone Arsenal, AL.

Length: 65–66 weeks (2636 hours).

Exhibit Dates: 10/94–9/97.

Learning Outcomes: Upon completion of the course, the student will be able to apply DC and AC circuit theory, linear and digital circuit theory, and computer systems theory to the troubleshooting and repair of computer- and radar-based electronic systems to the module and component level and apply standard soldering and printed circuit board repair techniques to the repair of electronic assemblies.

Instruction: Lectures cover fundamental operation of passive and active electronic components, digital logic gates, computer circuits, and computer programming concepts. Practical exercises cover soldering, troubleshooting computer-based systems, and troubleshooting continuous wave radar systems.

Credit Recommendation: In the lower-division baccalaureate/associate degree category, 1 semester hour in soldering techniques, 3 in AC/DC circuits, 3 in solid state electronics, 2 in digital principles, 2 in microprocessor fundamentals, and 3 in radar systems troubleshooting (7/96).

Related Occupations: 27K.

AR-1715-0971

1. Integrated Family Test Equipment Operator/Maintainer
2. Integrated Family of Test Equipment

Course Number: *Version 1:* 198-35Y10. *Version 2:* 198-35Y10.

Location: Ordnance, Missile and Munitions School, Redstone Arsenal, AL.

Length: *Version 1:* 23–24 weeks (958 hours). *Version 2:* 33 weeks (1195 hours).

Exhibit Dates: *Version 1:* 10/93–9/98. *Version 2:* 10/98–Present.

Learning Outcomes: *Version 1:* Upon completion of the course, the student will be able to operate and perform basic repair and adjustment to missile system line-replaceable units and ship-replaceable units and perform direct support/general support maintenance on test equipment subsystems. *Version 2:* Upon completion of the course, the student will be able to operate and perform basic repair and adjustment to missile system line-replaceable units and perform direct support/general support level maintenance on test equipment subsystems.

Instruction: *Version 1:* Instruction includes lectures and practical exercise on electricity, electronics, digital logic, soldering techniques, and system operation and maintenance. *Version 2:* Instruction include lectures and practical exercises in electricity, electronics, digital logic, soldering techniques, multilayer circuit repair, microprocessors, and electronic systems operation and maintenance.

Credit Recommendation: *Version 1:* In the lower-division baccalaureate/associate degree category, 2 semester hours in basic electricity, 3 in basic electronics, 2 in solid state electronics, 3 in electronic systems troubleshooting and maintenance, and 1 in soldering (7/96). *Version 2:* In the lower-division baccalaureate/associate degree category, 2 semester hours in AC/DC circuits, 3 in basic electronics, 3 in electronic systems troubleshooting, 2 in digital electronics, 2 in microprocessors, and 1 in soldering (6/01).

Related Occupations: 35Y.

AR-1715-0972

High-to-Medium Altitude Air Defense (HIMAD) Direct Support/General Support (DS/GS) Maintenance Technician Warrant Officer Technical and Tactical Certification

Course Number: 4F-916A.

Location: Ordnance, Missile and Munitions School, Redstone Arsenal, AL.

Length: 18–19 weeks (724 hours).

Exhibit Dates: 10/91–6/00.

Learning Outcomes: Upon completion of the course, the student will be able to perform and support maintenance of the Hawk missile system.

Instruction: Lectures and practical laboratory exercises cover computer equipment; associated test equipment; radar, launcher, and loader systems; and missile shop management procedures.

Credit Recommendation: In the lower-division baccalaureate/associate degree category, 1 semester hour in military science and 3 in electronic systems troubleshooting and maintenance (7/96).

Related Occupations: 916A.

AR-1715-0973

Ammunition Specialist Basic Noncommissioned Officer (NCO) Reserve Component

Course Number: 645-55B30-RC.

Location: Ordnance, Missile and Munitions School, Redstone Arsenal, AL.

Length: 2–3 weeks (83 hours).

Exhibit Dates: 6/94–9/00.

Learning Outcomes: Upon completion of the course, the student will be able to identify, receive, store, issue, and destroy ammunition; use a computer for ammunition inventory; and direct fire fighting and rewarehousing operations.

Instruction: Lectures and practical exercises cover ammunition identification, emergency destruction techniques, automated inventory procedures, and supervision of ammunition handling and rewarehousing operations.

Credit Recommendation: Credit is not recommended because of the military-specific nature of the course (7/96).

Related Occupations: 55B.

AR-1715-0974

1. Test Measurement and Diagnostic Equipment (TMDE) Maintenance Support Specialist Basic Noncommissioned Officer (NCO)
2. Test Measurement and Diagnostic Equipment (TMDE) Maintenance Support Specialist Basic Noncommissioned Officer (NCO)
3. Test Measurement and Diagnostic Equipment Maintenance Support Specialist Basic Noncommissioned Officer (NCO)

Course Number: *Version 1:* 198-35H30. *Version 2:* 198-35H30. *Version 3:* 198-35H30.

Location: Ordnance, Missile and Munitions School, Redstone Arsenal, AL.

Length: *Version 1:* 18–19 weeks (677 hours). *Version 2:* 9 weeks (328 hours). *Version 3:* 9 weeks (328 hours).

Exhibit Dates: *Version 1:* 10/95–9/98. *Version 2:* 10/98–9/00. *Version 3:* 10/00–Present.

Learning Outcomes: *Version 1:* Upon completion of the course, the student will be able to supervise, train, and lead subordinates in TMDE support and operate, maintain, calibrate, troubleshoot, and repair selected calibration standards. *Version 2:* Upon completion of the course, the student will be able to supervise, train, and lead subordinates in Test Measurement and Diagnostic Equipment support and operate, maintain, calibrate, troubleshoot, and repair selected calibration standards. *Version 3:* Upon completion of the course, the student will be able to supervise, train, and lead subordinates in Test Measurement and Diagnostic Equipment support and operate, maintain, calibrate, troubleshoot, and repair selected calibration standards.

Instruction: *Version 1:* This course includes lectures, demonstrations, and laboratory exercises in optics, microprocessor technology, maintenance and repair of electronic equipment, and automated supply operations/data base maintenance. This course contains a common core of leadership subjects. *Version 2:* This course includes lectures, demonstrations, and laboratory exercises in maintenance and repair of electronic equipment and automated supply operations/database maintenance. This course contains a common core of leadership subjects. *Version 3:* This course includes lectures, demonstrations, and laboratory exercises in maintenance and repair of electronic equipment and automated supply operations/database maintenance. This course contains a common core of leadership subjects.

Credit Recommendation: *Version 1:* In the lower-division baccalaureate/associate degree category, 3 semester hours in electronic systems troubleshooting and maintenance, 2 in military science/leadership, 2 in maintenance management, and 3 in microprocessor technology. See AR-1406-0090 for common core credit (7/96). *Version 2:* In the lower-division baccalaureate/associate degree category, 3 semester hours in electronic troubleshooting and maintenance, 2 in military science/leadership, 2 in maintenance management, and 3 in microcomputer applications (6/01). *Version 3:* In the lower-division baccalaureate/associate degree category, 3 semester hours in electronic systems troubleshooting and maintenance, 2 in military science/leadership, 2 in maintenance management, 3 in microcomputer applications (6/01).

Related Occupations: 35H.

AR-1715-0975

Microwave Measurement and Calibration

Course Number: E3AZR2P051 013.

Location: Ordnance, Missile and Munitions School, Redstone Arsenal, AL.

Length: 3–6 weeks (200 hours).

Exhibit Dates: 10/95–9/01.

Learning Outcomes: Upon completion of the course, the student will be able to operate a microwave secondary reference laboratory and calibrate, troubleshoot, and repair test measurement and diagnostic equipment.

Instruction: Lectures, practical exercises, and demonstrations cover transmission line theory, traveling and standing wave theory, cable termination, microwave attenuator calibration, power supply standard calibration, and signal generation workstation calibration and repair.

Credit Recommendation: In the lower-division baccalaureate/associate degree category, 3 semester hours in microwave system maintenance (7/96).

AR-1715-0976

Patriot System Repairer Phase 1

Course Number: 121-27X2/3/4.
Location: Ordnance, Missile and Munitions School, Redstone Arsenal, AL.
Length: 9 weeks (352 hours).
Exhibit Dates: 10/91–9/03.
Learning Outcomes: Upon completion of the course, the student will be able to diagnose, troubleshoot, repair, and maintain the electronic board assemblies of the Patriot missile system.
Instruction: Lectures, laboratory exercises, and performance evaluation cover solid state pulse circuits, receiver and transmitter circuits, timing and display circuits, three-phase power circuits, switching power supplier, phase modulation, digital logic fundamentals, computer technology, microprocessors, logic analyzer operation, and microwave technology.
Credit Recommendation: In the lower-division baccalaureate/associate degree category, 1 semester hour in digital logic fundamentals, 3 in microprocessors, 1 in microwave fundamentals, and 3 in electronic systems troubleshooting and maintenance (7/96).
Related Occupations: 27X.

AR-1715-0977

TOW/Dragon Repairer Reserve Phase 2

Course Number: 121-27E10-RC.
Location: Ordnance, Missile and Munitions School, Redstone Arsenal, AL.
Length: 3–4 weeks (137 hours).
Exhibit Dates: 10/91–9/96.
Learning Outcomes: Upon completion of the course, the student will be able to troubleshoot and maintain the TOW/TOW 2 and Dragon weapon systems.
Instruction: This course provides instruction through lectures and practical exercises in DC circuit theory, AC circuit theory, and basic electronics and maintaining, troubleshooting, and replacing of electronic circuit boards for the TOW and Dragon missiles.
Credit Recommendation: In the lower-division baccalaureate/associate degree category, 2 semester hours in introduction to AC/DC circuits and 1 in basic electronics (7/96).
Related Occupations: 27E.

AR-1715-0978

Integrated Family Test Equipment DS/GS Transition

Course Number: 198-35Y10-T.
Location: Ordnance, Missile and Munitions School, Redstone Arsenal, AL.
Length: 7 weeks (252 hours).
Exhibit Dates: 6/95–9/00.
Learning Outcomes: Upon completion of the course, the student will be able to operate, test, repair, and perform adjustments to the test equipment line-replaceable unit and ship-replaceable unit and perform general support maintenance test equipment subsystems.
Instruction: Instruction includes lectures and practical exercises in single- and multi-layer printed circuit board repair, surface mount component replacement, and computer-based diagnostic testing procedures.
Credit Recommendation: In the lower-division baccalaureate/associate degree category, 3 semester hours in electronic systems troubleshooting and maintenance (7/96).
Related Occupations: 35Y.

AR-1715-0979

1. Patriot System Repairer Phase 2
2. Patriot System Repairer, Phase 2

Course Number: *Version 1:* 121-27X2/3/4. *Version 2:* 121-27X2/3/4, Phase 2.
Location: Air Defense Artillery School, Fort Bliss, TX.
Length: *Version 1:* 31 weeks (628 hours). *Version 2:* 37–38 weeks (1349 hours).
Exhibit Dates: *Version 1:* 10/92–9/99. *Version 2:* 10/99–Present.
Learning Outcomes: *Version 1:* Upon completion of the course, the student will be able to operate the missile support system, subsystems, launcher station, test set, and radar subsystems for the Patriot missile system. *Version 2:* Upon completion of the course, the student will be able to troubleshoot and repair the radar subsystem of a missile system, launch station, test set, and missile support system.
Instruction: *Version 1:* Lectures and practical exercises cover safety, documentation, operational function, test and diagnostic equipment, power distribution system, weapons control computer, launcher station, test set, and radar set. *Version 2:* Lectures and practical exercises cover safety, documentation, operational function, test and diagnostic equipment, power distribution system, weapons control computer, launcher station, test set, and radar set.
Credit Recommendation: *Version 1:* In the lower-division baccalaureate/associate degree category, 3 semester hours in electronic systems maintenance and troubleshooting (7/96). *Version 2:* In the lower-division baccalaureate/associate degree category, 3 semester hours in electronic systems troubleshooting and maintenance and 3 in radar systems (4/01).
Related Occupations: 27X.

AR-1715-0980

High-to-Medium Altitude Air Defense (HIMAD) Direct Support/General Support (DS/GS)

Course Number: 4-9-C32-916A.
Location: Ordnance, Missile and Munitions School, Redstone Arsenal, AL.
Length: 10–11 weeks (422 hours).
Exhibit Dates: 10/92–6/00.
Learning Outcomes: Upon completion of the course, the student will be able to describe military career paths and professional development; the responsibilities of missile maintenance personnel; and newly developed systems, equipment, and techniques and operate, repair, and manage Integrated Family Test Equipment (IFTE) operations.
Instruction: Lectures and limited practical exercises cover leadership; military subjects; overview of missile maintenance support systems; newly developed systems, equipment, and techniques; and current training on the IFTE system.
Credit Recommendation: In the lower-division baccalaureate/associate degree category, 2 semester hours in military science, 1 in maintenance management, and 3 in electronic systems troubleshooting and maintenance (7/96).
Related Occupations: 916A.

AR-1715-0981

1. Signal Support Systems Specialist Reserve Component
2. TATS Signal Support Systems Specialist

Course Number: *Version 1:* 113-31U10-RC. *Version 2:* 113-31U10.
Location: Signal School, Fort Gordon, GA.
Length: *Version 1:* 3 weeks (114 hours). *Version 2:* 17 weeks (617 hours).
Exhibit Dates: *Version 1:* 10/93–9/02. *Version 2:* 10/02–Present.
Learning Outcomes: *Version 1:* Upon completion of the course, the student will be able to install, troubleshoot, and maintain battlefield manual and automotive signal support systems and terminal equipment. *Version 2:* Upon completion of the course, the student will be able to install, troubleshoot, and perform unit level maintenance on manual and automated signal support systems and terminal equipment; provide technical assistance in implementing information systems; and deploy and operate dedicated computer systems.
Instruction: *Version 1:* Lectures and practical experience cover installation, operation, maintenance, and troubleshooting manual and automated combat net radio terminal devices, battlefield automated systems, and dedicated FM radio retransmission stations. *Version 2:* Lectures and laboratory experiences cover installing, operating, and maintaining manual and automated signal support systems, terminal equipment, PC communications, local area networks and computer testing utilities and test software.
Credit Recommendation: *Version 1:* In the lower-division baccalaureate/associate degree category, 2 semester hours in electronic systems troubleshooting and maintenance (2/97). *Version 2:* In the lower-division baccalaureate/associate degree category, 3 semester hours in introduction to computers, 1 in computers and communications, and 1 in local area networks (2/03).
Related Occupations: 31U.

AR-1715-0982

Enhanced Position Location Reporting System (EPLRS) New Control Station Operator

Course Number: 201-F8.
Location: Signal School, Fort Gordon, GA.
Length: 6 weeks (238 hours).
Exhibit Dates: 4/95–Present.
Learning Outcomes: Upon completion of the course, each student will be able to operate and perform preventive maintenance checks and service on the enhanced positioning locator reporting system.
Instruction: Instruction is provided via classroom presentations by instructors and actual operation of equipment. Instruction includes equipment characteristics, theory of operation, and preventive maintenance.
Credit Recommendation: In the lower-division baccalaureate/associate degree category, 3 semester hours in radio operation and maintenance (2/97).

AR-1715-0983

TATS Multichannel Transmissions System Operator-Maintainer

(Multichannel Transmissions System Operator-Maintainer)

Course Number: 202-31R10 (CT) (F); 202-31R10 (CT).

Location: Signal School, Fort Gordon, GA.

Length: 13–14 weeks (492 hours).

Exhibit Dates: 10/95–Present.

Learning Outcomes: Upon completion of the course, the student will be able to install, operate, and perform unit level maintenance on mobile subscriber equipment (MSE) and tri-services tactical communications multichannel transmission systems in a field environment.

Instruction: Classroom and laboratory exercises cover tactical communications, subscriber equipment security, and communications theory; AC/DC principles; antenna principles; radio telephone; mobile subscriber equipment; radio access unit (RAU, AN/TRC-191) installation, operation, and troubleshooting; line-of-sight (LOS) equipment operation; tactical considerations in LOS net operations; radio telephone procedures; and installation and operation of digital group multiplexing equipment with shelters.

Credit Recommendation: In the vocational certificate category, 5 semester hours in radiotelephone communications (11/02).

AR-1715-0984

Electronic Switching System Operator/Maintainer-Reserve Component

Course Number: 260-31F10-RC.

Location: Signal School, Fort Gordon, GA.

Length: 4 weeks (158 hours).

Exhibit Dates: 10/94–Present.

Learning Outcomes: Upon completion of the course, the student will be able to install, operate, troubleshoot, and perform direct support level maintenance on electronic switching systems and related equipment.

Instruction: Course includes lectures and practical experiences to install, operate, troubleshoot, and perform maintenance on electronic switching systems, including central office, telephone, and other associated equipment.

Credit Recommendation: In the vocational certificate category, 3 semester hours in electronic switching systems (2/97).

Related Occupations: 31F.

AR-1715-0985

Single Channel Radio Operator Reserve Component

Course Number: 113-31C10-RC.

Location: Signal School, Fort Gordon, GA.

Length: 2 weeks (80 hours).

Exhibit Dates: 10/93–Present.

Learning Outcomes: Upon completion of the course, the student will be able to install, troubleshoot, and maintain tactical radio teletypewriter set and associated equipment.

Instruction: Lectures and practical experience cover installing, operating, and maintaining AM radio teletypewriter sets and related equipment including power generating equipment. The student also transposes, processes, transmits, receives, and maintains record communications.

Credit Recommendation: In the lower-division baccalaureate/associate degree category, 1 semester hour in electronic systems troubleshooting and maintenance (2/97).

Related Occupations: 31C.

AR-1715-0986

M1A2 Abrams Master Gunner Transition

Course Number: 020-ASIK8 (T).

Location: Armor Center and School, Fort Knox, KY.

Length: 4 weeks (144 hours).

Exhibit Dates: 11/95–Present.

Learning Outcomes: Upon completion of the course, the student will be able to perform advanced gunnery methods, turret weapons system maintenance, and use diagnostic test equipment.

Instruction: Lectures and practical exercises cover tank safety, weapons firing, turret systems troubleshooting, and turret maintenance.

Credit Recommendation: In the lower-division baccalaureate/associate degree category, 1 semester hour in hydraulic systems and 3 in electrical/electronic system troubleshooting and repair (4/03).

AR-1715-0987

Target Acquisition/Surveillance Radar Repairer Phase 1

Course Number: 121-35C10, Phase 1.

Location: Field Artillery School, Fort Sill, OK; Ordnance, Missile and Munitions School, Redstone Arsenal, AL.

Length: 9 weeks (343 hours).

Exhibit Dates: 10/95–9/96.

Learning Outcomes: Upon completion of the course, the student will demonstrate proficiency in basic DC/AC circuit theory, operation of circuit application of solid state devices, microwave systems and antennas, and basic logic circuits.

Instruction: Lectures and practical exercises cover series, parallel, and series-parallel DC and AC resistive circuits; AC capacitive circuits; RL and RC transients; transformers; power supplies; and circuit components. Practical exercises include the use of multimeters, oscilloscopes, function generators, and trainer modules. Also covered are characteristics of diodes, transistors, and FETs and their operation in power supplies and amplifiers; troubleshooting voltage measurements to localize a fault; microwave equipment including transmission lines, waveguide antennas, and amplifiers with minimum theoretical discussion; and survey of computer arithmetic, logic functions, flip-flops, and analog and digital conversion.

Credit Recommendation: In the vocational certificate category, 1 semester hour in microwave laboratory and 1 digital fundamentals. In the lower-division baccalaureate/associate degree category, 4 semester hours in introduction to DC/AC circuit analysis theory, 1 in introduction to DC/AC circuit analysis laboratory, 3 in solid state devices and amplifiers theory, and 2 in solid state devices and amplifiers laboratory (10/00).

AR-1715-0988

Special Electronic Devices Repairer

Course Number: *Version 1:* 198-35F10. *Version 2:* 198-35F10.

Location: Ordnance, Missile and Munitions School, Redstone Arsenal, AL.

Length: *Version 1:* 24–25 weeks (883 hours). *Version 2:* 22–23 weeks (816 hours).

Exhibit Dates: *Version 1:* 10/95–9/99. *Version 2:* 10/99–Present.

Learning Outcomes: *Version 1:* Upon completion of the course, the student will be able to repair night vision devices, battlefield illumination equipment, position and azimuth determining systems, detection and warning systems, and test measurement and diagnostic equipment. *Version 2:* Upon completion of the course, the student will be able to repair mine detectors, night vision devices, battlefield illumination equipment, position and azimuth determining systems, detection and warning systems, and test measurement and diagnostic equipment.

Instruction: *Version 1:* Lectures and practical exercises cover DC/AC circuits, power supply circuits, amplifier and oscillator circuits, digital and computer fundamentals, soldering, and special electronic devices. *Version 2:* Lectures and practical exercises cover DC/AC circuits, power supply circuits, amplifier and oscillator circuits, digital and computer fundamentals, soldering, and special electronic devices.

Credit Recommendation: *Version 1:* In the vocational certificate category, 1 semester hour in soldering. In the lower-division baccalaureate/associate degree category, 2 semester hours AC/DC circuits, 3 in basic electronics, and 3 in digital and microcomputer fundamentals (5/97). *Version 2:* In the lower-division baccalaureate/associate degree category, 3 semester hours AC/DC circuits, 3 in digital fundamentals, 3 in electronic systems troubleshooting, and 1 in soldering (6/01).

Related Occupations: 35F.

AR-1715-0989

Avenger System Repairer

Course Number: *Version 1:* 121-27T10. *Version 2:* 121-27T10.

Location: Ordnance, Missile and Munitions School, Redstone Arsenal, AL.

Length: *Version 1:* 15–16 weeks (562 hours). *Version 2:* 19–20 weeks (706 hours).

Exhibit Dates: *Version 1:* 1/95–9/97. *Version 2:* 10/97–Present.

Learning Outcomes: *Version 1:* Upon completion of the course, the student will be able to perform maintenance on the Avenger missile system and ancillary test equipment. *Version 2:* Upon completion of the course, the student will be able to perform maintenance on the Avenger missile system and ancillary test equipment.

Instruction: *Version 1:* Lectures and practical exercises cover AC/DC and resonance circuits, transistors, soldering, logic, and microprocessors. Repair and service of electronic, mechanical, hydraulic, and cryogenic assemblies are also taught. *Version 2:* Lectures and practical exercises cover AD/DC and resonance circuits, transistors, soldering, logic, and microprocessors. Repair and service of electronic, mechanical, hydraulic, and cryogenic assemblies are also taught.

Credit Recommendation: *Version 1:* In the vocational certificate category, 1 semester hour in basic soldering and 1 in introduction to logic circuits. In the lower-division baccalaureate/associate degree category, 4 semester hours in AC/DC circuits and 2 in basic electronics (5/97). *Version 2:* In the lower-division baccalaureate/associate degree category, 4 semester hours in AC/DC circuits, 1 in electronic devices, 2 in digital fundamentals, and 3 in troubleshooting and repair of electromechanical systems (6/01).

Related Occupations: 27T.

AR-1715-0990

Hawk Field Maintenance Equipment Firing Section Repairer

Course Number: 121-27H10.

Location: Ordnance, Missile and Munitions School, Redstone Arsenal, AL.

Length: 36 weeks (1308 hours).

Exhibit Dates: 10/94–9/98.

Learning Outcomes: Upon completion of the course, the student will apply basic DC/AC circuit theory; operate solid state devices, microwave systems, and antennas; analyze basic logic circuits; apply soldering and fabrication skills; understand all systems and subsystems of a Hawk pulse acquisition radar; and perform field and shop maintenance of this equipment.

Instruction: Lectures and practical exercises cover series, parallel, and series-parallel DC and AC resistive circuits, AC inductive and capacitive circuits, RL and RC transients, transformers, power supplies, and circuit components. Practical exercises include the use of multimeters, oscilloscopes, function generators, pulse generators, shop equipment, and trainer modules. The course includes characteristic of diodes, transistors, and FETs and their operation in power supplies and amplifiers. Practical exercises include troubleshooting using voltage measurement to localize a fault; microwave equipment including transmission lines, waveguides, antennas, and amplifiers with minimum theoretical discussion; survey of computer arithmetic, logic functions, flip-flops and analog and digital conversion; advanced soldering skills, cable, and metal fabrication. And hydraulic systems. This training is directed towards inspecting, testing, and repairing the electro-mechanical, hydraulic, and hydropneumatic portions of the Hawk missile systems.

Credit Recommendation: In the vocational certificate category, 3 semester hours in soldering and fabrication and 1 in microwave laboratory. In the lower-division baccalaureate/associate degree category, 4 semester hours in introduction to DC/AC circuit analysis theory, 1 in introduction to DC/AC circuit analysis laboratory, 3 in solid state devices and amplifiers theory, 1 in solid state devices and amplifiers laboratory, 2 in digital logic systems, and 2 in troubleshooting and repairing electromechanical systems (5/97).

Related Occupations: 27H.

AR-1715-0991

1. Radio/COMSEC Repairer
2. Radio/Communication Security (COMSEC) Repairer (Radio/COMSEC Repairer)
3. Radio/Communication Security (COMSEC) Repairer

Course Number: *Version 1:* 101-35E10. *Version 2:* 101-35E10. *Version 3:* 101-35E10.

Location: Signal School, Fort Gordon, GA.

Length: *Version 1:* 25 weeks (897–898 hours). *Version 2:* 26–27 weeks (956 hours). *Version 3:* 24 weeks (855 hours).

Exhibit Dates: *Version 1:* 10/95–9/00. *Version 2:* 10/00–9/04. *Version 3:* 10/04–Present.

Learning Outcomes: *Version 1:* Upon completion of the course, the student will be able to repair and maintain FM receiver-transmitters, to the assembly level using test sets and test equipment. *Version 2:* Upon completion of the course, the student will be able to utilize basic electronic and electrical training to repair and maintain radio receivers, transmitters, communication secure systems, and associated equipment. *Version 3:* Upon completion of the course, the student will be able to utilize basic electronic and electrical training to repair and maintain radio receiver, transmitters, communication secure systems, and associated equipment.

Instruction: *Version 1:* Lectures and practical exercises cover basic DC/AC circuits; power supply circuits; amplifier circuits; soldering; fundamentals of FM communications; and fault isolation, repair, and alignment of FM receiver-transmitters. *Version 2:* Lectures and practical exercises. Topics include basic electronics, AC/DC circuits, power supplies, amplifiers, soldering, FM communications, communications secure equipment, and the repair/maintenance of specific Army radio-based communications systems. *Version 3:* Lectures and practical exercises. Topics include basic electronics, AC/DC circuits, power supplies, amplifiers, oscillators, soldering, FM communications; communications secure equipment, and the repair/maintenance of specific systems to the component level.

Credit Recommendation: *Version 1:* In the vocational certificate category, 2 semester hours in electronic soldering and 2 in AC/DC circuit analysis. In the lower-division baccalaureate/associate degree category, 3 semester hours in electronic communications systems and 3 in electronic systems maintenance and repair (5/97). *Version 2:* In the vocational certificate category, 2 semester hours in soldering. In the lower-division baccalaureate/associate degree category, 3 semester hours in basic electricity, 3 in basic electronics, 2 in electronic communications systems, and 2 in electronic communications maintenance (8/03). *Version 3:* In the lower-division baccalaureate/associate degree category, 3 semester hours in AC/DC circuit theory, 1 in AC/DC circuit lab, 1 in basic electronic theory, 1 in basic electronic lab, 1 in digital logic fundamentals, and 2 in electronic communication maintenance lab (2/04).

Related Occupations: 35E.

AR-1715-0992

Enhanced Position Location Reporting System (EPLRS) Maintenance

Course Number: 101-F28.

Location: Ordnance, Missile and Munitions School, Fort Gordon, GA; Signal School, Fort Gordon, GA; Ordnance, Missile and Munitions School, Redstone Arsenal, AL.

Length: 3 weeks (108 hours).

Exhibit Dates: 10/95–Present.

Learning Outcomes: Upon completion of the course, the student will be able to operate, maintain, and troubleshoot a receiver/transmitter set to the subassembly level.

Instruction: Instruction includes lectures and practical exercises covering control function, block diagrams, signal flow, fault isolation, alignment procedures, and replacement of selected assembler, subassembler, and circuit cards.

Credit Recommendation: In the vocational certificate category, 2 semester hours in radio communications maintenance (5/01).

AR-1715-0993

1. Air Traffic Control Systems, Subsystems, and Equipment Repairer
2. Air Traffic Control Equipment Repairer

Course Number: *Version 1:* 102-35D10. *Version 2:* 102-35D10.

Location: *Version 1:* Signal School, Fort Gordon, GA; Ordnance, Missile and Munitions School, Redstone Arsenal, AL. *Version 2:* Signal School, Fort Gordon, GA.

Length: *Version 1:* 28–29 weeks (1035 hours). *Version 2:* 33 weeks (1195 hours).

Exhibit Dates: *Version 1:* 10/95–12/98. *Version 2:* 1/99–Present.

Learning Outcomes: *Version 1:* Before 10/95 see AR-1715-0787. Upon completion of the course, the student will be able to troubleshoot and repair, to the lowest replaceable module/component level, radar systems, communications equipment, fixed base radios, communications switching systems, and associated test equipment. *Version 2:* Upon completion of the course, the student will be able to repair tactical radar, IFF/SIF, nondirectional beacon and communications equipment, fixed-base radio, communications switching systems, recording/reproducing equipment, and a fixed-base nondirectional beacon system.

Instruction: *Version 1:* Lectures and practical exercises cover DC/AC circuits; amplifier and oscillator circuits; digital fundamentals; computer applications; soldering; and operation, alignment, troubleshooting, and repair of radar systems and communications equipment, receivers and transmitters, and associated test equipment. *Version 2:* Lectures and practical exercises cover basic electronics and air traffic control system equipment and repair.

Credit Recommendation: *Version 1:* In the vocational certificate category, 1 semester hour in soldering. In the lower-division baccalaureate/associate degree category, 2 semester hours in AC/DC circuits, 3 in solid state electronics, 3 in radar systems maintenance, and 3 in electronic systems troubleshooting and maintenance (5/97). *Version 2:* In the lower-division baccalaureate/associate degree category, 3 semester hours in DC circuits, 3 in AC circuits, 2 in digital fundamentals, 1 in soldering, 3 in radar systems, and 3 in electronic systems troubleshooting and repair (6/01).

Related Occupations: 35D.

AR-1715-0994

Equipment Operator/Maintenance Test Set, Guided Missile Set (TOW-Ground)

Course Number: 121-ASIJ9-RC; 4F-F24.

Location: Ordnance, Missile and Munitions School, Redstone Arsenal, AL.

Length: 2 weeks (82 hours).

Exhibit Dates: 10/95–Present.

Learning Outcomes: Upon completion of the course, the student will be able to operate, test, troubleshoot, and repair the TOW field test set and ancillary equipment.

Instruction: Lectures and practical exercises cover the operation and use of universal counter, function generator, and digital multimeter, and alignment and troubleshooting procedures on hardware, and repairing faults.

Credit Recommendation: In the lower-division baccalaureate/associate degree cate-

gory, 2 semester hours in electronic equipment maintenance and repair (5/01).

AR-1715-0995

Identification Friend/Foe (IFF) Repairer

Course Number: 121-ASI8A(27X); 121-ASI8A.

Location: Air Defense Artillery School, Fort Bliss, TX; Computer Systems Command, Support Group, Fort Lee, VA; Ordnance, Missile and Munitions School, Redstone Arsenal, AL.

Length: 4–7 weeks (144–243 hours).

Exhibit Dates: 7/95–9/98.

Learning Outcomes: Upon completion of the course, the student will be able to inspect, test, and adjust components and repair malfunctions in electronic assemblies, modules, and circuits of the Patriot friend or foe systems.

Instruction: Lecture and practical exercises cover components, assemblies, modules, and circuits, including test equipment, of the Patriot friend or foe system.

Credit Recommendation: Credit is not recommended because of the military-specific nature of the course (5/01).

AR-1715-0996

Radio Repairer (Transition)

Course Number: 160-F38 (T).

Location: Ordnance, Missile and Munitions School, Redstone Arsenal, AL.

Length: 5 weeks (180 hours).

Exhibit Dates: 10/94–9/99.

Learning Outcomes: Upon completion of the course, the student will be able to operate, troubleshoot, and repair an FM radio set and supporting test instrumentation.

Instruction: Lectures and practical exercises cover amplifiers, oscillators, FM transmitters and receivers, and an introduction to single-side band communications.

Credit Recommendation: In the lower-division baccalaureate/associate degree category, 3 semester hours in AM/FM communications systems (5/97).

AR-1715-0997

1. Firefinder/Ground Base Sensor Radar Repairer Phase 2
2. Radar Repairer

Course Number: *Version 1:* 121-35M10, Phase 2. *Version 2:* 121-35M10, Phase 2.

Location: *Version 1:* Field Artillery School, Fort Sill, OK; Ordnance, Missile and Munitions School, Redstone Arsenal, AL. *Version 2:* Field Artillery School, Fort Sill, OK.

Length: *Version 1:* 22 weeks (794 hours). *Version 2:* 26–27 weeks (958 hours).

Exhibit Dates: *Version 1:* 10/95–3/98. *Version 2:* 4/98–12/99.

Learning Outcomes: *Version 1:* Upon completion of the course, the student will be able to test, troubleshoot, and isolate defective components in a fire control radar system and perform maintenance, repair, and alignment tasks. *Version 2:* Upon completion of the course, the student will be able to test, troubleshoot, and isolate defective components in a fire control radar system and perform maintenance, repair, and alignment tasks.

Instruction: *Version 1:* This course includes application of prior training in DC and AC circuits, digital and microprocessor systems, and microwave system to the alignment, adjustment, and troubleshooting of a computer fire control radar system. *Version 2:* This course includes application of prior training in DC and AC circuits, digital and microprocessor systems, and microwave system to the alignment, adjustment, and troubleshooting of a computer fire control radar system. The student learns the operation of a field radar fire control system at the block-diagram level including the power distribution system, directional antenna, and computer/peripherals, and interfaces. Use of a power meter, spectrum analyzer, and basic test equipment are included.

Credit Recommendation: *Version 1:* In the lower-division baccalaureate/associate degree category, 2 semester hours in troubleshooting digital and analog systems (5/97). *Version 2:* In the lower-division baccalaureate/associate degree category, 2 semester hours in troubleshooting digital and analog systems (1/01).

Related Occupations: 35M.

AR-1715-0998

1. Firefinder/Ground Base Sensor Radar Repairer
2. Radar Repairer (Firefinder/Ground Base Sensor Radar Repairer)

Course Number: *Version 1:* 121-35M10, Phase 1. *Version 2:* 121-35M10, Phase 1.

Location: Ordnance, Missile and Munitions School, Redstone Arsenal, AL.

Length: *Version 1:* 11 weeks (406 hours). *Version 2:* 11 weeks (404 hours).

Exhibit Dates: *Version 1:* 10/95–9/02. *Version 2:* 10/02–Present.

Learning Outcomes: *Version 1:* Upon completion of the course, the student will demonstrate an understanding of basic DC/AC circuit theory, operation of circuit application of solid-state devices, microwave systems and antennas, and logic circuits and microcomputers. *Version 2:* Upon completion of the course, the student will be able to analyze, build, and test electrical and electronic systems, both analog and digital.

Instruction: *Version 1:* Lectures and practical exercises cover series, parallel, and series-parallel DC and AC resistive circuits; AC inductive and capacitive circuits; RL and RC transients; transformers; power supplies; and circuit components. Practical exercises include the use of multimeters, oscilloscopes, function generators, and trainer modules. The course includes characteristics of diodes, transistors, and FETs and their operation in power supplies and amplifiers. Practical exercises include troubleshooting using voltage measurements to determine proper operation and exercises with microwave equipment, including transmission lines, waveguides, antennas, and amplifiers with minimum discussion. Also included are survey of computer arithmetic, logic functions, flip-flops, and analog and digital conversion; basic digital systems; lectures and laboratory exercises on the 8080 microcomputer architecture; and programming, timing, interrupts, and troubleshooting. *Version 2:* Computer-based training, practical and classroom exercises, and lecture. Topics include electrical safety, DC and AC circuits, transistor and amplifier circuits, microwave circuits, logic circuits, microprocessors, and soldering.

Credit Recommendation: *Version 1:* In the vocational certificate category, 1 semester hour in microwave laboratory. In the lower-division baccalaureate/associate degree category, 3 semester hours in solid state devices and amplifiers theory, 2 in solid state devices and amplifiers laboratory, 4 in introduction to DC/AC circuit analysis theory, 1 in introduction to DC/AC circuit analysis laboratory, 2 in survey of digital fundamentals and microcomputers for non-electronics technology majors theory, and 1 in survey of digital fundamentals and microcomputers for non-electronics technology majors laboratory (6/01). *Version 2:* In the lower-division baccalaureate/associate degree category, 3 semester hours in basic circuit analysis, 1 in circuit lab, 3 in solid state electronics, 1 in electronics lab, and 2 in digital logic and microprocessors (2/04).

Related Occupations: 35M.

AR-1715-0999

Target Acquisition/Surveillance Radar Repairer Phase 2

Course Number: 121-35C10, Phase 2.

Location: Field Artillery School, Fort Sill, OK; Ordnance, Missile and Munitions School, Redstone Arsenal, AL.

Length: 10 weeks (370 hours).

Exhibit Dates: 10/95–9/00.

Learning Outcomes: Upon completion of the course, the student will be able to maintain, test, troubleshoot, and isolate fault in a target acquisition and surveillance radar system.

Instruction: The course includes application of prior training in DC and AC circuits, basic digital fundamentals, and microwave systems to the alignment, adjustment, troubleshooting, and repair of a radar system that includes a transmitter, receiver, and their subsystems to the subsystem level.

Credit Recommendation: In the lower-division baccalaureate/associate degree category, 2 semester hours in troubleshooting electronic systems laboratory (10/00).

Related Occupations: 35C.

AR-1715-1000

Equipment Operator/Maintenance Test Set, Guided Missile Set (GMS)

Course Number: 121-ASIJ9.

Location: Ordnance, Missile and Munitions School, Redstone Arsenal, AL.

Length: 3–4 weeks (125–126 hours).

Exhibit Dates: 10/95–Present.

Learning Outcomes: Before 10/95 see AR-1715-0565. Upon completion of the course, the student will be able to operate, repair, test, and troubleshoot the TOW field test set and ancillary equipment.

Instruction: Lectures and practical exercises cover the operation and use of the universal counter function generator; and digital multimeter, alignment, troubleshooting and repair procedures on hardware faults, and written and performance tests.

Credit Recommendation: In the lower-division baccalaureate/associate degree category, 2 semester hours in electronic equipment maintenance and repair (5/01).

AR-1715-1001

1. Automatic Test Equipment Operator/Maintainer
2. Apache Attack Helicopter Systems Repairer

Course Number: *Version 1:* 201-39B10. *Version 2:* 198-39B10.

Location: Ordnance, Missile and Munitions School, Redstone Arsenal, AL.

Length: *Version 1:* 29 weeks (1037 hours). *Version 2:* 38 weeks (1334–1384 hours).

Exhibit Dates: *Version 1:* 10/95–12/99. *Version 2:* 1/00–Present.

Learning Outcomes: *Version 1:* Before 9/90 see AR-1715-0850. Upon completion of the course, the student will apply knowledge of basic DC/AC circuit theory, diodes and transistor circuits, digital logic circuits, microcomputers and subsystems, and microprocessors to perform high-reliability soldering; inspect, test, and repair multilevel printed circuit boards to the component level, and operate, troubleshoot, and repair automatic test equipment. *Version 2:* Before 9/90 see AR-1715-0850. Upon completion of the course, the student will apply knowledge of basic DC/AC circuit theory, diodes and transistor circuits, digital logic circuits, microcomputers and subsystems and microprocessors to perform high-reliability soldering; inspect test, and repair multilevel printed circuit boards to the component level, and operate, troubleshoot, and repair automatic test equipment.

Instruction: *Version 1:* Lectures and practical exercises cover series-parallel, and series/parallel DC and AC resistive circuits, AC inductive and capacitive circuits, RL and RC transients, transformer, power supplies, and circuit components. Practical exercises include the use of multimeters, oscilloscopes, and function generators; characteristics of diodes; transistor and basic amplifier circuits, and troubleshooting techniques; and practical skills and knowledge of electrical connections, soldering station operation and maintenance, terminal and pin connections, PC board mounting of components, coaxial cable connection desoldering techniques, cable repair, and multilayer printed wire board repair. *Version 2:* Lectures and practical exercises cover series-parallel, and series/parallel DC and AC resistive circuits, AC inductive and capacitive circuits, RL and RC transients, transformer, power supplies, and circuit components. Practical exercises include the use of multimeters, oscilloscopes, and function generators; characteristics of diodes; transistor and basic amplifier circuits, and troubleshooting techniques; and practical skills and knowledge of electrical connections, soldering station operation and maintenance, terminal and pin connections, PC board mounting of components, coaxial cable connection, soldering techniques, cable repair, multilayer printed wire board repair, and the AH 64 electronic equipment test facility.

Credit Recommendation: *Version 1:* In the vocational certificate category, 4 semester hours in precision soldering techniques and 2 in solid state devices and systems. In the lower-division baccalaureate/associate degree category, 4 semester hours in introduction to DC/AC circuit analysis, 2 in digital logic systems, 2 in microprocessor operation and troubleshooting, and 1 in digital and microcomputer laboratory (5/97). *Version 2:* In the lower-division baccalaureate/associate degree category, 1 semester hour in precision soldering techniques, 3 in AC/DC circuits, 3 in electronic circuits, 2 in digital electronics, 2 in microprocessors, and 3 in electronic system troubleshooting (6/01).

Related Occupations: 39B.

AR-1715-1002

Guardrail Systems Operator

Course Number: 231-F34.

Location: Intelligence Center and School, Fort Huachuca, AZ.

Length: 3–4 weeks (131 hours).

Exhibit Dates: 10/96–Present.

Learning Outcomes: Before 10/96 see AR-1715-0893. Upon completion of the course, the student will be able to operate special electronic equipment for the acquisition and analysis of electronic signals.

Instruction: Lectures, presentations, demonstrations, computer-assisted instruction, and performance exercises cover the operation of specialized electronic equipment.

Credit Recommendation: Credit is not recommended because of the limited, specialized nature of the course (6/97).

AR-1715-1003

Voice Interceptor Basic Noncommissioned Officer (NCO)

Course Number: *Version 1:* 232-98G3LXX. *Version 2:* 231-98G3LXX.

Location: *Version 1:* Intelligence School, Fort Devens, MA. *Version 2:* Intelligence School, Fort Huachuca, AZ.

Length: *Version 1:* 6–7 weeks (238–239 hours). *Version 2:* 6 weeks (214 hours).

Exhibit Dates: *Version 1:* 7/93–12/96. *Version 2:* 1/97–Present.

Learning Outcomes: *Version 1:* Upon completion of the course, the student will be able to supervise a voice intercept and jamming team including preparing, operating, reporting, and maintaining of electronic equipment. *Version 2:* Upon completion of the course, the student will be able to supervise a voice intercept and jamming team including preparing, operating, reporting, and maintaining of electronic equipment.

Instruction: *Version 1:* Lectures, demonstrations, and practical exercises cover supervision, planning, operation and maintenance procedures for voice signal intercept and interference. This course contains a common core of leadership subjects. *Version 2:* Lectures, demonstrations, and practical exercises cover supervision, planning, operation and maintenance procedures for voice signal intercept and interference. This course contains a common core of leadership subjects.

Credit Recommendation: *Version 1:* In the lower-division baccalaureate/associate degree category, 1 semester hour in oral reports. See AR-1406-0090 for common core credit (6/97). *Version 2:* Credit is recommended for the common core only. See AR-1406-0090 (6/97).

AR-1715-1004

Electronic Warfare/Intercept Strategic Systems Repairer Basic Noncommissioned Officer (NCO)

Course Number: 102-33Y30.

Location: Intelligence School, Fort Huachuca, AZ.

Length: 17–18 weeks (638–639 hours).

Exhibit Dates: 12/94–Present.

Learning Outcomes: Before 12/94 see AR-1715-0888. Upon completion of the course, the student will be able to troubleshoot and align strategic electronic warfare/intercept systems and manage a maintenance shop.

Instruction: Laboratory exercises test and measurement equipment to troubleshoot, repair, and align receivers, magnetic tape recorders, multiplexers, and demultiplexers. The course also includes 32-bit multiprocessor architecture, VAX computers, and network fault isolation. Managing a maintenance shop, risk management, squad tactical operations, team building, maintenance management and advanced applications are also included. This course includes a common core of leadership subjects.

Credit Recommendation: In the lower-division baccalaureate/associate degree category, 2 semester hours in electronic measurement techniques, 2 in introduction to digital communications, 3 in computer system maintenance, and 3 in maintenance management. See AR-1406-0090 for common core credit (6/97).

Related Occupations: 33Y.

AR-1715-1005

Electronic Warfare/Intercept Tactical Systems Repairer Basic Noncommissioned Officer (NCO)

Course Number: 102-33T30.

Location: Intelligence School, Fort Huachuca, AZ.

Length: 17–18 weeks (637–638 hours).

Exhibit Dates: 12/94–Present.

Learning Outcomes: Before 12/94 see AR-1715-0723. Upon completion of the course, the student will be able to perform repair functions, supervise subordinates performing the fault isolation and alignment techniques necessary to maintain of electronic warfare/intercept tactical systems, and supervise shop operations.

Instruction: Laboratory exercises use advanced test procedures for receivers, multiplexers, demultiplexers and advanced troubleshooting and alignment techniques for recorders. The course also includes 32-bit multiprocessor architecture, VAX computers, and electronic warfare/intercept tactical systems. Course content includes managing a maintenance shop and supervising personnel with an emphasis on risk management, quality control, team building, squad tactical operations, maintenance management, and property accountability. This course contains a common core of leadership subjects.

Credit Recommendation: In the lower-division baccalaureate/associate degree category, 2 semester hours in electronic measurement techniques, 2 in introduction to digital communications, 3 in computer system maintenance, and 3 in maintenance management. See AR-1406-0090 for common core credit (6/97).

Related Occupations: 33T.

AR-1715-1006

Electronic Warfare/Intercept Aviation Systems Repairer Basic Noncommissioned Officer (NCO)

Course Number: 102-33R30.

Location: Intelligence School, Fort Huachuca, AZ.

Length: 18–19 weeks (687–688 hours).

Exhibit Dates: 12/94–Present.

Learning Outcomes: Before 12/94 see AR-1715-0714. Upon completion of the course, the student will be able to teach and supervise enlisted personnel in the alignment, troubleshooting, and repair of state-of-the-art communications equipment and personal and mainframe computers associated with electronic warfare/intercept aviation systems.

Instruction: Lectures, demonstrations, and practical exercises cover personnel training, supervision, and operations management; advanced high frequency receiver alignment and test procedures. The course includes 32-bit architecture personal and mainframe computer troubleshooting to the faulty circuit card/module; tape and disk drive test and replacement using basic electronic test equipment, including frequency synthesizer, spectrum analyzer, sweep generator, and noise and distortion analyzer; and complete maintenance shop operations. The course contains a common core of leadership subjects.

Credit Recommendation: In the lower-division baccalaureate/associate degree category, 3 semester hours in electronic troubleshooting and repair, 3 in PC troubleshooting and maintenance, 2 in electronic communications, 2 in computer systems and organization, and 2 in maintenance management. See AR-1406-0090 for common core credit (6/97).

Related Occupations: 33R.

AR-1715-1007

ARMAMENT REPAIRER

Course Number: 643-45K10.

Location: Ordnance Center and School, Aberdeen Proving Ground, MD.

Length: 18 weeks (714–720 hours).

Exhibit Dates: 10/96–Present.

Learning Outcomes: Upon completion of the course, the student will be able to maintain tank turrets, components, and firing systems.

Instruction: Lectures and practical exercises cover the inspection, diagnosis, maintenance, and repair of tank and artillery weapon systems including self-propelled artillery.

Credit Recommendation: In the lower-division baccalaureate/associate degree category, 3 semester hours in fundamentals of AC/DC circuits (3/03).

Related Occupations: 45K.

AR-1715-1008

MEDICAL LASER SYSTEMS

Course Number: 4B-F20/198-F20.

Location: Medical Equipment and Optical School, Aurora, CO.

Length: 2 weeks (91 hours).

Exhibit Dates: 1/96–Present.

Learning Outcomes: Upon completion of the course, the student will be able to operate, maintain, calibrate, and perform optical alignment on medical laser systems.

Instruction: The course provides instruction on laser physics, safety, inspection, advanced circuit analysis, repair, preventive maintenance, calibration, and alignment of medical laser systems. Laboratory exercises include repairing circuit faults to the component level and performing calibration and preventive maintenance on advanced medical laser systems including CO2 and YAG surgical lasers.

Credit Recommendation: In the lower-division baccalaureate/associate degree category, 3 semester hours in biomedical application of lasers (4/98).

AR-1715-1009

1. AVIONICS MECHANIC BASIC NONCOMMISSIONED OFFICER (NCO)
2. AVIONICS MECHANIC SUPERVISOR BASIC NONCOMMISSIONED OFFICER (NCO)

Course Number: ***Version 1:*** 102-68N30. *Version 2:* 106-68N30, Phase 2.

Location: ***Version 1:*** Aviation School, Fort Rucker, AL. *Version 2:* Aviation Center, Fort Rucker, AL.

Length: ***Version 1:*** 6 weeks (254 hours). *Version 2:* 3–4 weeks (169 hours).

Exhibit Dates: ***Version 1:*** 1/97–1/01. *Version 2:* 2/01–Present.

Learning Outcomes: ***Version 1:*** Upon completion of the course, the student will be able to provide technical guidance and perform technical inspections on avionic communications, navigation, radar, and flight systems. *Version 2:* Upon completion of the course, the student will be able to train enlisted personnel in leadership and aviation related-tasks, conduct field training exercises, and provide technical guidance on avionics communication equipment.

Instruction: ***Version 1:*** Lectures and practical exercises cover leadership, military skills, professional skills, resource management, communicative skills, supervision training management systems, advanced electronic principles, theory for avionic communications equipment, maintenance management, and technical inspection criteria. *Version 2:* Topics include explanation of operation of single channel ground/airborne radio system using substation frequency hopping and secure modes; field training exercises including vehicle loading, emplacement, nighttime moves and after action review; supervisory, and technical skills in performance of duties as intermediate maintenance supervisor.

Credit Recommendation: ***Version 1:*** Credit is recommended for the common core only. See AR-1406-0090 (11/98). *Version 2:* Credit is recommended for the common core only. See AR-1406-0090 (4/03).

Related Occupations: 68N.

AR-1715-1010

1. AIRCRAFT SURVIVABILITY EQUIPMENT/ ELECTRONIC WARFARE OFFICER
2. ELECTRONIC WARFARE OFFICER

Course Number: ***Version 1:*** 2C-ASIH3. *Version 2:* 2C-ASIH3.

Location: ***Version 1:*** Aviation School, Fort Rucker, AL. *Version 2:* Aviation Center, Fort Rucker, AL.

Length: ***Version 1:*** 2 weeks (77 hours). *Version 2:* 3 weeks (111 hours).

Exhibit Dates: ***Version 1:*** 10/94–9/02. *Version 2:* 10/02–Present.

Learning Outcomes: ***Version 1:*** Upon completion of the course, the student will be able to successfully supervise and manage aircraft survivability equipment training for operators and evaluate and provide assistance for maintenance, training, and logistics. *Version 2:* Upon completion of the course, the student will be able to supervise and manage electronic warfare training for operators; apply fundamentals of lasers, infrared devices and radars; continuous wave (CW) radars detection and jamming techniques; evaluate radar systems pre and post-repair and maintenance; and perform mission planning and management.

Instruction: ***Version 1:*** Classroom instruction and practical exercises cover radar theory; infrared/laser theory; intervisibility; analysis of threat systems; electronic warfare theory, physics, and principles; various detection and jamming principles; expendable countermeasures systems; aircraft survivability equipment operation; and training and logistics analysis and procedures. *Version 2:* Topics include duties and responsibilities of electronic warfare officer (EW); inspection of EW systems and components; diagnostic and maintenance principles of radar, laser, and infrared systems; detection and jamming of radar systems; training coordination; EW tactics, techniques, and procedures; and mission planning and digital operations.

Credit Recommendation: ***Version 1:*** In the lower-division baccalaureate/associate degree category, 1 semester hour in radar principles, 1 in electronic systems troubleshooting and maintenance, and 1 in maintenance management (11/98). *Version 2:* In the lower-division baccalaureate/associate degree category, 1 semester hour in radar principles, 2 in maintenance management, and 1 in electronic systems troubleshooting and maintenance (4/03).

AR-1715-1011

ARMY MEDICAL DEPARTMENT (AMEDD) NONCOMMISSIONED OFFICER (NCO) BASIC

Course Number: 6-8-C40 (91A30).

Location: Academy of Health Sciences, Fort Sam Houston, TX.

Length: 7 weeks (300 hours).

Exhibit Dates: 4/95–Present.

Learning Outcomes: Upon completion of the course, the student will be able to lead a military unit; perform performance review and counseling; supervise and delegate; and perform drills, training, and specific military skills.

Instruction: The course covers military leadership, personal and performance counseling, duties and responsibilities of noncommissioned officers, leadership assessment and development, leadership procedures and combat orders, combat leadership, physical training and marksmanship, property accountability, individual and team training, small-group process, and specific military skills. The methodologies include lectures, discussions, group work, and practical application. This course contains a common core of leadership subjects.

Credit Recommendation: In the lower-division baccalaureate/associate degree category, 3 semester hours in military science and 1 semester hour of elective credit in basic medical equipment repair. See AR-1406-0090 for common core credit (11/97).

Related Occupations: 91A.

AR-1715-1012

SATELLITE COMMUNICATIONS SET AN/USC-28 (V)

Course Number: 102-F60 (2); E5AZA2E151-039; A-102-0276; 102-F60 (2), Phase 1.

Location: Signal School, Fort Gordon, GA.

Length: 5–6 weeks (200 hours).

Exhibit Dates: 1/98–Present.

Learning Outcomes: Upon completion of the course, the student will be able to troubleshoot and repair a satellite communications set.

Instruction: Lectures and laboratory exercises cover the operation and maintenance of a satellite ground terminal.

Credit Recommendation: In the lower-division baccalaureate/associate degree cate-

gory, 3 semester hours in electronic systems troubleshooting and maintenance (11/02).

AR-1715-1013

Satellite Communications (SATCOM) Systems Operator/Maintainer

Course Number: 102-31S10.
Location: Signal School, Fort Gordon, GA.
Length: 35 weeks (1392 hours).
Exhibit Dates: 1/98–4/02.
Learning Outcomes: Upon completion of the course, the student will be able to identify series, parallel, and series/parallel resistance circuits and determine values of resistance, voltage, and current in the circuit; recognize the meaning of frequency, period, sine wave, phase shift, inductive reactance, capacitive reactance, and impedance; calculate voltage and current in transformers; calculate quality factor, bandwidth, and cutoff frequencies of resonant circuits; calculate impedances, voltages, and current in capacitive and inductive AC circuits; explain the basic operation of semiconductors, junction diodes, and transistors and identify their uses in rectifier circuits and voltage regulators; describe communications techniques; trace signal path, troubleshoot, and align mobile FM transceivers; convert between and perform arithmetic in the decimal, binary, octal, and hexadecimal number systems; analyze logic circuits, flip-flop circuits, registers, and counters; operate, maintain, troubleshoot, and repair satellite earth terminal equipment including digital communications systems, using standard electronic test equipment and spectrum analyzer (using spread spectrum techniques).
Instruction: Lectures and laboratory exercises cover topics in DC and AC circuits, solid state electronics, digital electronics, FM communications, satellite communications, soldering techniques, and theory of operation and troubleshooting methods for satellite communications equipment and devices. Course methods include computer-assisted instruction modules.
Credit Recommendation: In the lower-division baccalaureate/associate degree category, 3 semester hours in basic DC circuits, 1 in AC circuits, 1 in digital circuits, 1 in solid state electronics, 1 in electronic communications, 3 in basic electronics laboratory, and 3 in electronic system troubleshooting and maintenance (2/03).

AR-1715-1014

1. Digital Communications Subsystems (Digital Communications Satellite Subsystems)
2. Digital Communications Satellite Subsystems

Course Number: *Version 1:* 102-F40 (2). *Version 2:* 102-F40 (2).
Location: Signal School, Fort Gordon, GA.
Length: *Version 1:* 5–6 weeks (216 hours). *Version 2:* 9–10 weeks (376 hours).
Exhibit Dates: *Version 1:* 1/99–9/01. *Version 2:* 10/01–Present.
Learning Outcomes: *Version 1:* Upon completion of the course, the student will be able to operate, maintain, configure, and troubleshoot digital communications subsystems. *Version 2:* Upon completion of the course, the student will be able to operate program, maintain, troubleshoot, and repair a digital communications satellite and associated control subsystems.
Instruction: *Version 1:* Lectures and practical exercises cover the operation and maintenance of digital communications subsystems. Topics include multiplexers, modems, encoders, decoders, and communications subsystems. Laboratory experience is given in configuration, alignment, and troubleshooting of digital communications systems. *Version 2:* Lectures, laboratory, and practical exercises. Topics include digital communication system, tactical communication system, communication automation system, system control console modem, and digital communication switch.
Credit Recommendation: *Version 1:* In the lower-division baccalaureate/associate degree category, 3 semester hours in digital communications systems (5/99). *Version 2:* In the lower-division baccalaureate/associate degree category, 3 semester hours in digital communication system operation maintenance and troubleshooting (2/03).

AR-1715-1015

Satellite Communication (SATCOM) Operator, Navy

Course Number: A-260-0034; 102-F43 (OS).
Location: Signal School, Fort Gordon, GA.
Length: 7–10 weeks (288–368 hours).
Exhibit Dates: 1/99–Present.
Learning Outcomes: Before 1/99 see AR-1715-0438. Upon completion of the course, the student will be able to operate a satellite communications terminal and a digital communications subsystem.
Instruction: Course includes principles of satellite communications, including digital communications and terminal subsystems.
Credit Recommendation: In the lower-division baccalaureate/associate degree category, 3 semester hours in electronic systems troubleshooting and maintenance (2/03).

AR-1715-1016

Strategic SHF Satellite Communications (SATCOM) Systems Maintenance

Course Number: A-101-0729; 102-F39 (OS).
Location: Navy Supply Corps School Detachment, Fort Gordon, GA; Signal School, Fort Gordon, GA.
Length: 23–27 weeks (880–1053 hours).
Exhibit Dates: 1/98–Present.
Learning Outcomes: Before 1/98 see AR-1715-0679. Upon completion of the course, the student will be able to operate and maintain digital communications subsystems and perform functional analysis, operational checks, signal tracing, and troubleshooting of digital receivers and transmitters.
Instruction: Topics include fundamentals of defense satellite communications systems; digital electronics techniques; and operation, maintenance, and troubleshooting digital communications subsystems. Course methods include readings, lectures, discussions, and evaluation by test and performance.
Credit Recommendation: In the lower-division baccalaureate/associate degree category, 3 semester hours in digital communications systems and 3 in electronic systems troubleshooting and maintenance (2/03).

AR-1715-1017

Visual Information Equipment Operator Maintainer

Course Number: 400-25R30.
Location: Signal Regimental NCO Academy Det., Fort Meade, MD.
Length: 8–9 weeks (243 hours).
Exhibit Dates: 4/96–Present.
Learning Outcomes: Upon completion of the course, the student will apply the basic concepts required to perform as a highly trained technician and maintenance supervisor of the components within a visual information facility.
Instruction: Lecture, demonstration, and laboratory exercise introduce the student to such topics as rack-mounted systems and color television camera systems, including electronic field production systems, multimedia systems, and electronic editing computer systems.
Credit Recommendation: In the lower-division baccalaureate/associate degree category, 3 semester hours in electronic systems troubleshooting and maintenance (11/99).

AR-1715-1018

Visual Information Operations Chief Advanced Noncommissioned Officer (NCO)

Course Number: 4-25-C42.
Location: Signal Corps Regimental NCO Academy Det., Fort Meade, MD.
Length: 11–12 weeks (433 hours).
Exhibit Dates: 7/99–Present.
Learning Outcomes: Upon completion of the courses the student will be able to direct the operation of a visual information and maintenance facility.
Instruction: Through lectures and laboratory exercises, the student is presented with the necessary managerial and leadership skills to supervise a visual information and repair facility. Computer programs dealing with data base management systems serve as a keystone to the standard professional training.
Credit Recommendation: In the lower-division baccalaureate/associate degree category, 3 semester hours in principles of management. In the upper-division baccalaureate category, 3 semester hours in supervisory management (11/99).

AR-1715-1019

Spectrum RISC/UNIX Workstation Maintenance

Course Number: A-102-0406; 102-F112.
Location: Intelligence Center and School, Fort Huachuca, AZ.
Length: 5 weeks (180 hours).
Exhibit Dates: 10/94–Present.
Learning Outcomes: Upon completion of the course, the student will understand system architecture and perform maintenance on the SUN, IBM, Hewitt Packard, Silicon Graphics, or Digital Equipment corporation workstations.
Instruction: Lectures and practical exercises related to a SUN workstation. Topics include operation, both in stand-alone mode and in a distributed processing network; troubleshoot the SUN workstation in the stand-alone modes to the circuit card/module level; troubleshoot a distributed processing network to the component/node/circuit card/ module level.
Credit Recommendation: In the lower-division baccalaureate/associate degree cate-

gory, 3 semester hours in introduction to computers and 3 in computer systems troubleshooting (1/00).

AR-1715-1021

Battlefield Spectrum Management

Course Number: 4C-SI5D/260-ASID9.

Location: Signal School, Fort Gordon, GA.

Length: 9 weeks (355 hours).

Exhibit Dates: 10/94–Present.

Learning Outcomes: Upon completion of the course, the student will be able to plan and coordinate communications.

Instruction: This course includes lectures, classroom discussion, projects, and examinations covering introduction to communications electronics principles; radio wave propagation theory; systems engineering covering HF, VHF/UHF, line-of-sight, satellite, and tropospheric systems; electromagnetic compatibility; and interference resolution.

Credit Recommendation: In the upper-division baccalaureate category, 1 semester hour in principles of electronics and 3 in communications systems (5/00).

AR-1715-1022

Total Army Training System (TATS) Automated Fire Support Systems Specialist

Course Number: 061-13C10 (F).

Location: Reserve Component Training Centers, US; Field Artillery School, Fort Sill, OK.

Length: 99–119 hours.

Exhibit Dates: 10/98–Present.

Learning Outcomes: Upon completion of the course, the student will be able to employ, operate, and maintain a fire support system.

Instruction: Lectures and demonstrations cover the employment, operation, and maintenance of the computer system and communications and artillery operations.

Credit Recommendation: In the lower-division baccalaureate/associate degree category, 1 semester hour in communications systems and 1 in computer operations (6/00).

Related Occupations: 13C.

AR-1715-1023

Total Army Training System (TATS) Automated Fire Support Systems Specialist

Course Number: 250-13C10 (F).

Location: Field Artillery School, Fort Sill, OK.

Length: 3–4 weeks (120–121 hours).

Exhibit Dates: 10/98–Present.

Learning Outcomes: Before 10/98 see AR-1715-0832. Upon completion of the course, the student will be able to employ, operate, and maintain a fire support system.

Instruction: Lectures and demonstrations cover the employment, operation, and maintenance of the computer system and communications and artillery operations.

Credit Recommendation: In the lower-division baccalaureate/associate degree category, 1 semester hour in communications systems and 1 in computer operations (6/00).

Related Occupations: 13C.

AR-1715-1024

Target Acquisition Warrant Officer Basic Reserve Phase 4

Course Number: 4C-131A-RC, Phase 4.

Location: Field Artillery School, Fort Sill, OK.

Length: 2 weeks (106 hours).

Exhibit Dates: 5/95–Present.

Learning Outcomes: Upon completion of this set of courses, the student will be able to supervise and provide unit corrective maintenance on a field artillery target acquisition radar system and field artillery firefinder radar system.

Instruction: The course consists of classroom lectures and laboratory experiences on signal processors and power distribution systems and appropriate field training on how to troubleshoot radar and antenna positioning systems.

Credit Recommendation: In the lower-division baccalaureate/associate degree category, 3 semester hours in radar systems and 2 in introduction to electronics (6/00).

Related Occupations: 131A.

AR-1715-1025

Target Acquisition Warrant Officer Basic Reserve, Phase 2

Course Number: 4C-131A-RC.

Location: Field Artillery School, Fort Sill, OK.

Length: 2 weeks (80 hours).

Exhibit Dates: 8/95–Present.

Learning Outcomes: Before 8/95 see AR-1715-0841. Upon completion of the course, the student will be able to conduct site selection and operate a field artillery target acquisition radar system.

Instruction: The course consists of classroom lectures and laboratory field experiences in the use of power generators, radar equipment, and digital mapping systems.

Credit Recommendation: In the lower-division baccalaureate/associate degree category, 3 semester hours in radar systems (6/00).

Related Occupations: 131A.

AR-1715-1026

Total Army Training System (TATS) M109A6 (Paladin) Commander

Course Number: 041-F9 (F).

Location: Field Artillery School, Fort Sill, OK.

Length: 2 weeks (72–73 hours).

Exhibit Dates: 1/97–Present.

Learning Outcomes: Upon completion of the course, the student will be able to lead a howitzer crew, perform the technical processes used in computer-based firing on designated target coordinates, and determine coordinate location of the howitzer using advanced global positioning system.

Instruction: Demonstrations, lectures, hands-on practice, and performance exercises cover the use of computer-driven digital data for delivering artillery fire on designated targets; use of menus and screens for data processing; use of NAVSTAR advanced global positioning system for coordinate and locations; and introduction to FM radio communications, radio operation, and the transmitting of data. Emphasis is also placed on computer maintenance programs for the vehicle and its onboard equipment.

Credit Recommendation: In the lower-division baccalaureate/associate degree category, 1 semester hour in satellite communications systems, 1 in applied global positioning system, and 3 in personnel supervision (6/00).

AR-1715-1027

Total Army Training System (TATS) M109A6 (Paladin) Commander

Course Number: 061-2E-F165, Phase 3.

Location: Reserve Component Training Centers, US; Field Artillery School, Fort Sill, OK.

Length: 73 hours.

Exhibit Dates: 1/97–Present.

Learning Outcomes: Upon completion of the course, the student will be able to lead a howitzer crew, perform the technical processes used in computer-based firing on designated target coordinates, and determine coordinate location of the howitzer using advanced global positioning system.

Instruction: Demonstrations, lectures, hands-on practice, and performance exercises cover the use of computer-driven digital data for delivering artillery fire on designated targets; use of menus and screens for data processing; use of NAVSTAR advanced global positioning system for coordinate locations; and introduction to FM radio communications, radio operation, and the transmitting of data. Emphasis is also placed on computer maintenance programs for the vehicle and its on-board equipment.

Credit Recommendation: In the lower-division baccalaureate/associate degree category, 1 semester hour in satellite communications systems, 1 in applied global positioning system, and 3 in personnel supervision (6/00).

AR-1715-1028

Total Army Training System (TATS) Cannon Fire Direction Specialist

Course Number: ***Version 1:*** 250-13E10 (F). *Version 2:* 250-13E10 (F).

Location: Field Artillery School, Fort Sill, OK.

Length: ***Version 1:*** 6–7 weeks (281–320 hours). *Version 2:* 6–7 weeks (281 hours).

Exhibit Dates: ***Version 1:*** 10/98–9/01. *Version 2:* 10/01–Present.

Learning Outcomes: ***Version 1:*** Before 10/98 see AR-1715-1029. Upon completion of the course, the student will be able to determine the direction of fire when aiming a cannon using special equipment. *Version 2:* Upon completion of the course, the student will be able to determine the direction of fire when aiming a cannon using special equipment.

Instruction: ***Version 1:*** Lectures and practical exercises cover vehicle maintenance, communications equipment, tactics, and fire control systems. Course includes maintenance of communications equipment. *Version 2:* Lectures and practical exercises cover vehicle maintenance, communications equipment, tactics, and fire control systems. Course includes maintenance of communications equipment.

Credit Recommendation: ***Version 1:*** In the lower-division baccalaureate/associate degree category, 1 semester hour in topography and 1 in radio operation and maintenance (1/01). *Version 2:* In the lower-division baccalaureate/

associate degree category, 1 semester hour in cartography and 1 in electronic communications (10/03).
Related Occupations: 13E.

AR-1715-1029

TOTAL ARMY TRAINING SYSTEM (TATS) CANNON FIRE DIRECTION SPECIALIST

Course Number: ***Version 1:*** 061-13E10 (F). *Version 2:* 061-13E10 (F).
Location: Reserve Schools, Multiple locations.
Length: ***Version 1:*** 134–152 hours. *Version 2:* 134 hours.
Exhibit Dates: ***Version 1:*** 10/98–9/01. *Version 2:* 10/01–Present.
Learning Outcomes: ***Version 1:*** Before 10/98 see AR-1715-1028. Upon completion of the course, the student using special equipment, will be able to determine the direction of fire when aiming a cannon. *Version 2:* Upon completion of the course, the student using special equipment, will be able to determine the direction of fire when aiming a cannon.
Instruction: ***Version 1:*** Lectures and practical exercises cover vehicle and communications equipment maintenance, tactics, and fire control systems. *Version 2:* Lectures and practical exercises cover vehicle and communications equipment maintenance, tactics, and fire control systems.
Credit Recommendation: ***Version 1:*** In the lower-division baccalaureate/associate degree category, 1 semester hour in cartography and 1 in radio operation and maintenance (1/01). *Version 2:* In the lower-division baccalaureate/associate degree category, 1 semester hour in cartography and 1 in radio operation and maintenance (10/03).
Related Occupations: 13E.

AR-1715-1031

TOTAL ARMY TRAINING SYSTEM (TATS) FIELD ARTILLERY FIREFINDER RADAR OPERATOR

Course Number: 221-13R10 (F).
Location: Field Artillery School, Fort Sill, OK.
Length: 8–11 weeks (345–375 hours).
Exhibit Dates: 10/97–Present.
Learning Outcomes: Upon completion of the course, the student will be able to emplace, operate, and maintain radar and support equipment.
Instruction: Lectures and practical exercises cover radar operations, communications, survey requirements, and preventive maintenance on radar equipment.
Credit Recommendation: In the lower-division baccalaureate/associate degree category, 3 semester hours in radar operations (10/03).
Related Occupations: 13R.

AR-1715-1032

TOTAL ARMY TRAINING SYSTEM (TATS) FIRE SUPPORT SPECIALIST

Course Number: 250-13F10 (F).
Location: Field Artillery School, Fort Sill, OK.
Length: 7 weeks (259 hours).
Exhibit Dates: 10/97–9/99.
Learning Outcomes: Upon completion of the course, the student will be able to operate fire support vehicles, operate lasers and digital communications equipment, prepare and install support equipment, and perform land navigation.
Instruction: This course consists of classroom lectures and conferences and hands-on training in communications and fire support devices and navigation.
Credit Recommendation: In the lower-division baccalaureate/associate degree category, 1 semester hour in radio operation and 3 in land navigation (1/01).
Related Occupations: 143F.

AR-1715-1033

TOTAL ARMY TRAINING SYSTEM (TATS) FIRE SUPPORT SPECIALIST

Course Number: 106-13F10 (F), Phase 2.
Location: Field Artillery School, Fort Sill, OK.
Length: 106–134 hours.
Exhibit Dates: 10/97–9/99.
Learning Outcomes: Upon completion of the course, the student will be able to operate fire support vehicles, operate lasers and digital communications equipment, prepare and install support equipment, and perform land navigation.
Instruction: This course consists of classroom lectures and conferences and appropriate hands-on training in communications and fire support devices and navigation.
Credit Recommendation: In the lower-division baccalaureate/associate degree category, 1 in radio operation and 1 in land navigation (1/01).
Related Occupations: 13F.

AR-1715-1034

TOTAL ARMY TRAINING SYSTEM (TATS) FIRE SUPPORT SPECIALIST

Course Number: 250-13F10 (F).
Location: Field Artillery School, Fort Sill, OK.
Length: 7 weeks (255–263 hours).
Exhibit Dates: 10/00–Present.
Learning Outcomes: Upon completion of the course, the student will be able to operate fire support vehicles, operate lasers and digital communications equipment, prepare and install support equipment, and perform land navigation.
Instruction: Class consists of lectures/conferences and hands-on training in communications, fire support devices, and navigation.
Credit Recommendation: In the lower-division baccalaureate/associate degree category, 3 semester hours in land navigation (9/03).
Related Occupations: 13F.

AR-1715-1035

TOTAL ARMY TRAINING SYSTEM (TATS) FIRE SUPPORT SPECIALIST

Course Number: 061-13F10 (F), Phase 2.
Location: Reserve Schools, Multiple locations.
Length: 128–263 hours.
Exhibit Dates: 10/00–Present.
Learning Outcomes: Upon completion of the course, the student will be able to perform navigation procedures, operate lasers and digital communication equipment, and prepare and operate related equipment.
Instruction: Class consists of lectures/conferences and hands-on training in navigation.
Credit Recommendation: In the lower-division baccalaureate/associate degree category, 3 semester hours in land navigation (9/03).
Related Occupations: 13F.

AR-1715-1036

TOTAL ARMY TRAINING SYSTEM (TATS) FIELD ARTILLERY RADAR SECTION CHIEF BASIC NONCOMMISSIONED OFFICER (NCO)

Course Number: 221-13R30 (F).
Location: Field Artillery School, Fort Sill, OK.
Length: 5–6 weeks (233 hours).
Exhibit Dates: 10/97–9/00.
Learning Outcomes: After 9/97 see AR-1715-1037. Upon completion of the course, the student will have a working knowledge of an artillery firing unit, including general radar operation and survey procedures.
Instruction: Lectures and practical experiences cover radar systems, survey procedures, and preventive maintenance on radar and support equipment. Course includes a common core of leadership subjects.
Credit Recommendation: In the lower-division baccalaureate/associate degree category, 3 semester hours in radar operations and 1 in map reading. See AR-1406-0090 for common core credit (1/01).
Related Occupations: 13R.

AR-1715-1037

TOTAL ARMY TRAINING SYSTEM (TATS) FIELD ARTILLERY RADAR SECTION CHIEF BASIC NONCOMMISSIONED OFFICER (NCO)

Course Number: 061-13R30 (F), Phase 2.
Location: Reserve Schools, Multiple locations.
Length: 146 hours.
Exhibit Dates: 10/97–9/00.
Learning Outcomes: Before 10/97 see AR-1715-1036. Upon completion of the course, the student will have a working knowledge of an artillery firing unit, including general radar operation and survey procedures.
Instruction: Lectures and practical experiences cover radar systems, survey procedures, and preventive maintenance on radar and support equipment. Course includes a common core of leadership subjects.
Credit Recommendation: In the lower-division baccalaureate/associate degree category, 3 semester hours in radar operations and 1 in map reading. See AR-1406-0090 for common core credit (1/01).
Related Occupations: 13R.

AR-1715-1038

TOTAL ARMY TRAINING SYSTEM (TATS) FIELD ARTILLERY RADAR SECTION CHIEF BASIC NONCOMMISSIONED OFFICER (NCO)

Course Number: 221-13R30 (F), Phase 2.
Location: Field Artillery School, Fort Sill, OK.
Length: 3 weeks (163–166 hours).
Exhibit Dates: 10/00–Present.
Learning Outcomes: After 9/00 see AR-1715-1039. Upon completion of the course, the

student will be able to conduct site surveys, perform maintenance, provide leadership, and supervise fire control operations.

Instruction: Lectures and practical exercises cover radar setup, operations, and maintenance; survey techniques for radar sites; and personnel supervision.

Credit Recommendation: In the lower-division baccalaureate/associate degree category, 1 semester hour in radar operations and maintenance (9/03).

Related Occupations: 13R.

AR-1715-1039

Total Army Training System (TATS) Field Artillery Radar Section Chief Basic Noncommissioned Officer (NCO)

Course Number: 061-13R30 (F), Phase 2.

Location: Reserve Schools, Multiple locations.

Length: 152–163 hours.

Exhibit Dates: 10/00–Present.

Learning Outcomes: Before 10/00 see AR-1715-1038. Upon completion of the course, the student will be able to conduct site surveys, perform maintenance, provide leadership, and supervise fire control operations.

Instruction: Lectures and practical exercises cover radar setup, operations, and maintenance; survey techniques for radar sites; and personnel supervision.

Credit Recommendation: In the lower-division baccalaureate/associate degree category, 1 semester hours in radar operations and maintenance (9/03).

Related Occupations: 13R.

AR-1715-1040

Total Army Training System (TATS) Meteorological Equipment Maintenance - Meteorological Data System (MDS)

Course Number: 420-ASIH1 (F).

Location: Field Artillery School, Fort Sill, OK.

Length: 6–10 weeks (120–360 hours).

Exhibit Dates: 10/96–Present.

Learning Outcomes: Before 10/96 see AR-1715-0835. Upon completion of the course, the student will apply basic concepts of electricity, magnetism, and semiconductors to the maintenance of field artillery equipment, meteorological data systems, and related components, including antennas and direction finders.

Instruction: This course consists of classroom lectures and examinations, hands-on laboratory experiences, use of appropriate field artillery meteorological devices, and troubleshooting and repairing meteorological equipment.

Credit Recommendation: In the lower-division baccalaureate/associate degree category, 3 semester hours in AC/DC electricity fundamentals and 1 in electronic systems troubleshooting and maintenance (1/01).

AR-1715-1041

Total Army training System (TATS) Meteorological Equipment Maintenance - Meteorological Data System

Course Number: 061-ASIH1 (F).

Location: Field Artillery School, Fort Sill, OK.

Length: 97–121 hours.

Exhibit Dates: 10/96–Present.

Learning Outcomes: Before 10/96 see AR-1715-0835. Upon completion of the course, the student will be able to perform maintenance on meteorological data systems.

Instruction: This course includes classroom lectures and laboratory on basic AC and DC electricity, magnetism, multimeters, basic electronics, and repair techniques; classroom lectures and laboratory on fundamentals of solid state electronics; and troubleshooting and repair of meteorological data systems.

Credit Recommendation: In the lower-division baccalaureate/associate degree category, 1 semester hour in maintenance and repair of electronic systems (1/01).

AR-1715-1042

Total Army Training System (TATS) Multiple Launch Rocket System (MLRS) Fire Direction Specialist

Course Number: ***Version 1:*** 043-13P10 (F). *Version 2:* 043-13P10 (F).

Location: Field Artillery School, Fort Sill, OK.

Length: ***Version 1:*** 12–13 weeks (454 hours). *Version 2:* 7 weeks (304 hours).

Exhibit Dates: ***Version 1:*** 10/97–9/02. *Version 2:* 10/02–Present.

Learning Outcomes: ***Version 1:*** Before 10/97 see AR-1715-0847. Upon completion of the course, the student will be able to read and interpret maps; plot locations; compute distance, direction, and altitude; use radio communications equipment; use digital message devices; operate a computer system; and install and operate telephones with security equipment. *Version 2:* Upon completion of the course, the student will be able to read and interpret maps; plot locations; compute distance, direction, and altitude; use radio communications equipment; use digital message devices; operate a computer system; and install and operate telephones with security equipment.

Instruction: ***Version 1:*** Lectures, group activities, and performance exercises cover map interpretation, communications equipment and maintenance, and computer use. *Version 2:* Lectures, group activities, and performance exercises cover map interpretation, communications equipment and maintenance, computer use, and the operation and maintenance of the Fire Detection System (FDS).

Credit Recommendation: ***Version 1:*** In the vocational certificate category, 1 semester hour in radio operations. In the lower-division baccalaureate/associate degree category, 1 semester hour in topography and 1 in safety (1/01). *Version 2:* In the vocational certificate category, 1 semester hour in radio operations and 1 in computer operations. In the lower-division baccalaureate/associate degree category, 1 semester hour in topography and 1 in safety (10/03).

AR-1715-1043

Total Army Training System (TATS) Multiple Launch Rocket System (MLRS) Fire Direction Senior Sergeant Advanced Noncommissioned (NCO)

Course Number: ***Version 1:*** 061-13P10 (F), Phase 1. *Version 2:* 061-13P10 (F), Phase 1.

Location: Reserve Schools, Multiple locations.

Length: ***Version 1:*** 130 hours. *Version 2:* 152 hours.

Exhibit Dates: ***Version 1:*** 10/97–9/02. *Version 2:* 10/02–Present.

Learning Outcomes: ***Version 1:*** Before 10/97 see AR-1715-0847. Upon completion of the course, the student will be able to read and interpret maps; plot locations; compute distance, direction, and altitude; operate a computer system; and use digital message devices. *Version 2:* Upon completion of the course, the student will be able to read and interpret maps; plot locations; compute distance, direction, and altitude; operate a computer system; and use digital message devices.

Instruction: ***Version 1:*** Lectures, group activities, and performance exercises cover map interpretation, communications equipment and maintenance, and computer use. *Version 2:* Lectures, group activities, and performance exercises cover map interpretation, communications equipment and maintenance, computer use, and the operation and maintenance of the Fire Direction System (FDS).

Credit Recommendation: ***Version 1:*** In the lower-division baccalaureate/associate degree category, 1 semester hour in topography and 1 in safety (1/01). *Version 2:* In the lower-division baccalaureate/associate degree category, 1 semester hour in cartography and 1 in safety (10/03).

Related Occupations: 13P.

AR-1715-1044

Total Army Training System (TATS) Multiple Launch Rocket System (MLRS) Fire Direction Senior Sergeant Advanced Noncommissioned Officer (NCO)

Course Number: 061-13P10 (F), Phase 3.

Location: Field Artillery School, Fort Sill, OK.

Length: 122 hours.

Exhibit Dates: 10/97–Present.

Learning Outcomes: Before 10/97 see AR-1715-0847. Upon completion of the course, the student will be able to read and interpret maps; plot locations; compute distance, direction, and altitude; operate a computer system; and install and operate telephones with security equipment.

Instruction: Lectures, group activities, and performance exercises cover map interpretation, communications equipment and maintenance, and computer use.

Credit Recommendation: In the lower-division baccalaureate/associate degree category, 1 semester hour in topography and 1 in safety (1/01).

Related Occupations: 13P.

AR-1715-1045

Special Operations Communications System Maintenance

Course Number: 101-F23.

Location: John F. Kennedy Special Warfare Center and School, Fort Bragg, NC.

Length: 5–6 weeks (210 hours).

Exhibit Dates: 6/98–Present.

Learning Outcomes: Before 6/98 see AR-1715-0825. Upon completion of the course, the student will be able to plan, conduct, and support unit-level installation and maintenance of communications systems.

Instruction: Lectures and practical exercises cover installation, troubleshooting, repair, and maintenance of communications systems.

Credit Recommendation: In the lower-division baccalaureate/associate degree category, 3 semester hours in high-frequency electronic communications (4/01).

AR-1715-1046

PATRIOT FIRE CONTROL ENHANCED OPERATOR/MAINTAINER

Course Number: 043-14E10 CONFIG3; 043-14E10 (TATS).

Location: Air Defense Artillery School, Fort Bliss, TX.

Length: 27 weeks (934–975 hours).

Exhibit Dates: 10/96–Present.

Learning Outcomes: Upon completion of the course the student will be able to operate, maintain, align, and troubleshoot the various electronic components of Patriot radar and communications equipment.

Instruction: The course uses lectures, written examinations, and practical exercises to provide instruction. Experience is emphasized, focusing on communications, electronics, and computer and radar circuits and systems. Computer initialization/maintenance, software installation, and modifications are included. Extensive practical troubleshooting skills are developed in both analog and digital electronics.

Credit Recommendation: In the lower-division baccalaureate/associate degree category, 3 semester hours in microcomputer systems troubleshooting and repair, 3 in radar systems, and 3 in electronic system troubleshooting and repair (4/01).

Related Occupations: 14E.

AR-1715-1047

AIR DEFENSE ARTILLERY COMMAND, CONTROL, COMMUNICATIONS, COMPUTER, AND INTELLIGENCE ENHANCED SYSTEM OPERATOR/MAINTAINER

Course Number: 2.

Location: Air Defense Artillery School, Fort Bliss, TX.

Length: 19 weeks (694 hours).

Exhibit Dates: 4/97–8/97.

Learning Outcomes: Upon completion of the course, the graduate will be able to set up, initialize, maintain, and operate a tactical operation center; establish local area networks; use computers and operating systems; and set up and perform basic electronic testing.

Instruction: Methods include classroom instruction, demonstration, practical exercises, and examinations. Topics include system test equipment, operating systems (UNIX, DOS), networking fundamentals, local area networks, safety procedures, network diagrams, and media.

Credit Recommendation: In the lower-division baccalaureate/associate degree category, 2 semester hours in introduction to networking and 1 in electronic systems troubleshooting and maintenance (4/01).

Related Occupations: 14J.

AR-1715-1048

1. AIR DEFENSE ARTILLERY COMMAND, CONTROL, COMMUNICATIONS, COMPUTERS, AND INTELLIGENCE (C4I) TACTICAL OPERATIONS CENTER ENHANCED OPERATOR/MAINTAINER
2. AIR DEFENSE ARTILLERY HIGH TO MEDIUM AIR DEFENSE (C4I) TACTICAL OPERATIONS CENTER ENHANCED OPERATOR/MAINTAINER-HIMAD
   (Air Defense Artillery Command, Control, Communications, Computers, and Intelligence (C4I) Tactical Operations Center Enhanced Operator/Maintainer)

Course Number: ***Version 1:*** 043-14J10 (TATS). *Version 2:* 043-ASIQ3.

Location: Air Defense Artillery School, Fort Bliss, TX.

Length: ***Version 1:*** 19 weeks (684 hours). *Version 2:* 9 weeks (324 hours).

Exhibit Dates: ***Version 1:*** 9/00–9/02. *Version 2:* 10/02–Present.

Learning Outcomes: ***Version 1:*** Before 9/00 see AR-1715-0953. Upon completion of the course, the graduate will be able to set up, initialize, maintain, and operate a tactical operations center; establish local area networks; use computers and operating systems including UNIX; and troubleshoot complex systems. *Version 2:* Pending evaluation.

Instruction: ***Version 1:*** Methods include classroom instruction, demonstration, practical exercises, and examinations. Topics include system test equipment, UNIX operating systems, networking fundamentals, local area networks, safety procedures, network diagrams, media, and troubleshooting procedures. *Version 2:* Pending evaluation.

Credit Recommendation: ***Version 1:*** In the lower-division baccalaureate/associate degree category, 1 semester hour in introduction to networking, 1 in electronics system troubleshooting and maintenance, and 1 in UNIX system operations (4/01). *Version 2:* Pending evaluation.

Related Occupations: 14J.

AR-1715-1049

AIR DEFENSE ARTILLERY COMMAND, CONTROL, COMMUNICATIONS, COMPUTERS, AND INTELLIGENCE TACTICAL OPERATIONS CENTER ENHANCED OPERATOR/MAINTAINER

Course Number: 043-14J10 (TATS).

Location: Air Defense Artillery School, Fort Bliss, TX.

Length: 272 hours.

Exhibit Dates: 9/00–9/02.

Learning Outcomes: Upon completion of the course, the graduate will be able to operate an air defense tactical operations center.

Instruction: Methods include lectures, practical exercises, and examinations.

Credit Recommendation: Credit is not recommended because of the military-specific nature of the course (4/01).

Related Occupations: 14J.

AR-1715-1050

AIR DEFENSE ARTILLERY WARRANT OFFICER ADVANCED, PHASE 2

Course Number: 2-44-C32, Phase 2.

Location: Air Defense Artillery School, Fort Bliss, TX.

Length: 6–7 weeks (240 hours).

Exhibit Dates: 6/00–Present.

Learning Outcomes: Before 6/00 see AR-1715-0964. Upon completion of the course, the student will be able to use effective written and oral communication techniques, employ leadership principles, describe missile systems, integrate personal computer components, use word processing software, and use a personal computer graphics program.

Instruction: Lectures and practical exercises include writing and editing techniques, oral presentation skills, leadership skills, specific missile systems, and personal computer hardware and software.

Credit Recommendation: In the lower-division baccalaureate/associate degree category, 1 semester hour in written and oral communication, 1 in introduction to computers, and 1 in personal computer hardware (4/01).

AR-1715-1052

BRADLEY LINEBACKER CREWMEMBER

Course Number: 043-14R10 (TATS).

Location: Air Defense Artillery School, Fort Bliss, TX.

Length: 114–118 hours.

Exhibit Dates: 10/00–Present.

Learning Outcomes: Upon completion of the course, the student will be able to visually identify numerous aircraft and armored vehicles and operate and perform operator maintenance on a specific armored vehicle and its weapons.

Instruction: Lectures and practical exercises include the use of radios, guns, missiles, and an armored vehicle.

Credit Recommendation: Credit is not recommended because of the military-specific nature of the course (4/04).

Related Occupations: 14R.

AR-1715-1053

1. COMMAND AND CONTROL SYSTEM INTEGRATOR
2. AIR DEFENSE COMMAND AND CONTROL SYSTEMS INTEGRATOR WARRANT OFFICER TECHNICAL/TACTICAL CERTIFICATION

Course Number: ***Version 1:*** 4F-140A. *Version 2:* 4F-140A (TATS).

Location: Reserve Schools, Multiple locations; Air Defense Artillery School, Fort Bliss, TX.

Length: ***Version 1:*** 19–20 weeks (697 hours). *Version 2:* 19–20 weeks (698 hours).

Exhibit Dates: ***Version 1:*** 4/97–6/00. *Version 2:* 7/00–Present.

Learning Outcomes: ***Version 1:*** Upon completion of the course, student will be able to operate, maintain, and troubleshoot the equipment associated with an air defense operation center and operate and maintain a local area network. *Version 2:* Upon completion of the course, student will be able to operate, maintain, and troubleshoot the equipment associated with an air defense operation center and operate and maintain a local area network.

Instruction: ***Version 1:*** Lectures and practical exercises cover computer networks, communications equipment, military tactics, and troubleshooting and maintaining techniques for the center's equipment. *Version 2:* Lectures and practical exercises cover computer networks, communications equipment, military tactics, and troubleshooting and maintaining techniques for the center's equipment.

Credit Recommendation: *Version 1:* In the lower-division baccalaureate/associate degree category, 3 semester hours in electronic systems troubleshooting and repair and 1 in computer networking (4/01). *Version 2:* In the lower-division baccalaureate/associate degree category, 3 semester hours in electronic systems troubleshooting and repair and 1 in computer networking. Recommendation for credit is based on completion of 19-20 week long course (4/01).

Related Occupations: 140A.

AR-1715-1054

Patriot Launching Station Enhanced Operator/Maintainer

Course Number: 043-14T10 CONFIG3.

Location: Air Defense Artillery School, Fort Bliss, TX.

Length: 11–12 weeks (406 hours).

Exhibit Dates: 6/01–Present.

Learning Outcomes: Upon completion of the course, the student is able to operate, troubleshoot, and maintain a Patriot missile system launcher.

Instruction: Lectures and practical exercises are provided on driving skills, missile reload, launch station operation, preventive maintenance, and troubleshooting.

Credit Recommendation: In the lower-division baccalaureate/associate degree category, 1 semester hour in electronic systems troubleshooting and repair (4/01).

Related Occupations: 14T.

AR-1715-1055

Patriot Launching Station Enhanced Operator/Maintainer

Course Number: 043-14T10 CONFIG 3.

Location: Air Defense Artillery School, Fort Bliss, TX.

Length: 133 hours.

Exhibit Dates: 6/01–Present.

Learning Outcomes: Upon completion of the course, the student is able to operate, troubleshoot, and maintain a Patriot missile system launcher.

Instruction: Lectures and practical exercises are provided on driving skills, missile reload, launch station operation, preventive maintenance, and troubleshooting.

Credit Recommendation: In the lower-division baccalaureate/associate degree category, 1 semester hour in electronic systems troubleshooting and repair (4/01).

Related Occupations: 14T.

AR-1715-1056

Patriot Launching Station Enhanced Operator/Maintainer

Course Number: 043-14T10 (TATS).

Location: Air Defense Artillery School, Fort Bliss, TX.

Length: 120 hours.

Exhibit Dates: 10/00–Present.

Learning Outcomes: Upon completion of the course, student is able to operate, troubleshoot, and maintain a Patriot missile system launcher.

Instruction: Lectures and practical exercises are provided on driving skills, missile reload, launch station operation, preventive maintenance and troubleshooting.

Credit Recommendation: In the lower-division baccalaureate/associate degree category, 1 semester hour in electronic systems troubleshooting and repair (4/01).

Related Occupations: 14T.

AR-1715-1057

1. Patriot Launcher Station Operator and Patriot Launching Station Enhanced Operator/Maintainer
2. Patriot Launching Station ENH Operator/Maintainer

(Patriot Launcher Station Operator and Patriot Launching Station Enhanced Operator/Maintainer)

Course Number: *Version 1:* 043-14T10. *Version 2:* 043-14T10.

Location: Air Defense Artillery School, Fort Bliss, TX.

Length: *Version 1:* 11 weeks (397 hours). *Version 2:* 12 weeks (446 hours).

Exhibit Dates: *Version 1:* 10/96–9/02. *Version 2:* 10/02–Present.

Learning Outcomes: *Version 1:* Upon completion of the course, student is able to operate, troubleshoot, and maintain a Patriot missile system launcher. *Version 2:* Pending evaluation.

Instruction: *Version 1:* Lectures and practical exercises are provided on driving skills, missile reload, launch station operation, preventive maintenance, and troubleshooting. *Version 2:* Pending evaluation.

Credit Recommendation: *Version 1:* In the lower-division baccalaureate/associate degree category, 1 semester hour in electronic systems troubleshooting and repair (4/01). *Version 2:* Pending evaluation.

Related Occupations: 14T.

AR-1715-1059

Manportable Air Defense System Crewmember

Course Number: 043-14M10.

Location: Air Defense Artillery School, Fort Bliss, TX.

Length: 6–7 weeks (227 hours).

Exhibit Dates: 8/98–Present.

Learning Outcomes: Upon completion of the course, the student will be able to prepare and fire the Stinger weapon system.

Instruction: Lectures and practical exercises cover recognizing aircraft, performing preventive maintenance checks on the weapon, using a GPS receiver, and firing the weapon.

Credit Recommendation: Credit is not recommended because of the military-specific nature of the course (4/01).

Related Occupations: 14M.

AR-1715-1060

Manportable Air Defense System Crewmember

Course Number: 043-14M10 (TATS).

Location: Reserve Schools, Multiple locations; Air Defense Artillery School, Fort Bliss, TX.

Length: 111–119 hours.

Exhibit Dates: 7/00–Present.

Learning Outcomes: Upon completion of the course, the student will be able to prepare and fire the Stinger weapon system.

Instruction: Lectures and practical exercises cover recognizing aircraft, performing preventive maintenance checks on the weapon, using a GPS receiver, and firing the weapon.

Credit Recommendation: Credit is not recommended because of the military-specific nature of the course (4/01).

Related Occupations: 14M.

AR-1715-1061

Avenger Crewmember

Course Number: 043-14S10 (TATS); 043-14S10.

Location: Air Defense Artillery School, Fort Bliss, TX.

Length: 10 weeks (359–367 hours).

Exhibit Dates: 8/98–Present.

Learning Outcomes: Upon completion of the course, the student will be able to operate equipment associated with an air defense missile system.

Instruction: Classroom instruction includes demonstrations, practical exercises, and examinations. Topics include operation and maintenance procedures for an air missile system, including radio equipment and global positioning equipment.

Credit Recommendation: Credit is not recommended because of the military-specific nature of the course (4/01).

Related Occupations: 14S.

AR-1715-1062

Warrant Officer Basic 918D

Course Number: 4F-918D (Consol).

Location: Ordnance, Missile and Munitions School, Redstone Arsenal, AL.

Length: 19–20 weeks (705 hours).

Exhibit Dates: 1/00–Present.

Learning Outcomes: Before 1/00 see AR-1715-0899. Upon completion of the course, the student will be able to supervise electronic systems maintenance and manage the TOW/TOW2, Dragon, and Javelin Weapons systems and the Avenger air defense system.

Instruction: Instruction is provided using lectures, conferences, practical exercises, and computer-aided instruction. Topics covered in this course include AC circuits, transistors, microwave fundamentals, logic fundamentals, microprocessors, and overall systems for the TOW2/Dragon/Javelin, the Bradley Fighting Vehicle TOW2/Linebacker systems, and the Avenger air defense system.

Credit Recommendation: In the lower-division baccalaureate/associate degree category, 2 semester hours in AC circuits, 3 in electronic circuits, 2 in digital electronics, 3 in electronic systems troubleshooting, and 2 in maintenance management (6/01).

Related Occupations: 918D.

AR-1715-1063

Warrant Officer Advanced 918D, Phase 2

Course Number: 4-9-C32-918D (Consol), Phase 2.

Location: Ordnance, Missile and Munitions School, Redstone Arsenal, AL.

Length: 5 weeks (180 hours).

Exhibit Dates: 6/00–Present.

Learning Outcomes: Before 6/00 see AR-1715-0899. Upon completion of the course, the student will be able to inspect, test, and perform intermediate maintenance on radar systems and forward area air defense ground-based sensors.

Instruction: Lectures and practical exercises cover leadership, communications, management, maintenance operations, and technical proficiency.

Credit Recommendation: In the lower-division baccalaureate/associate degree category, 3 semester hours in maintenance management and 3 in electronic systems troubleshooting and maintenance (6/01).

Related Occupations: 918D.

AR-1715-1064

Test, Measurement, and Diagnostic Equipment (TMDE) Maintenance Support Specialist

Course Number: 198-35H10.

Location: Ordnance, Missile and Munitions School, Redstone Arsenal, AL.

Length: 33 weeks (1219 hours).

Exhibit Dates: 9/99–Present.

Learning Outcomes: Upon completion of the course, the student will be able to operate, calibrate and troubleshoot equipment. To achieve this proficiency, the student must demonstrate competence in AC/DC circuits, transistor circuits, microwave devices, logic electronics, microprocessors, and various test equipment.

Instruction: Instruction will be provided using lectures, conferences, practical exercises, tests, demonstrations, and computer-assisted instruction.

Credit Recommendation: In the lower-division baccalaureate/associate degree category, 1 semester hour in soldering, 3 in DC circuits, 3 in AC circuits, 5 in electronic circuits, 1 in microwave, 3 in digital electronics, and 3 in microprocessors (6/01).

Related Occupations: 35H.

AR-1715-1065

Electronic Maintenance Basic Noncommissioned Officer (NCO)

Course Number: ***Version 1:*** 1-35-C40. *Version 2:* 1-35-C40.

Location: Ordnance, Missile and Munitions School, Redstone Arsenal, AL.

Length: ***Version 1:*** 8 weeks (288 hours). *Version 2:* 5 weeks (203 hours).

Exhibit Dates: ***Version 1:*** 10/98–9/00. *Version 2:* 10/00–Present.

Learning Outcomes: ***Version 1:*** Upon completion of the course, the student will be able to plan and supervise the daily operations of an electronics maintenance shop. The student will demonstrate competence in shop operations, supply management, computer library, and common leader training. The student will successfully complete a field training exercise encompassing various aspects of the training. *Version 2:* Upon completion of the course, the student will be able to use Microsoft Office Program Suite within Windows operating systems. The student will be able to plan and supervise the daily operations of an electronics shop. The student will demonstrate competence in shop operations, supply management, computer library, and common leader training. The student will successfully complete a field training exercise encompassing various aspects of the training.

Instruction: ***Version 1:*** Instruction is provided using lectures, conferences, practical exercises, and a field training exercise. Topics include leadership, supply management and operations, and computer applications. *Version 2:* Audiovisual materials, practical exercises, and computer-based training. Topics include Microsoft Office Suite usage, review of basic theories of electronics, development of the supervisory skills and knowledge required to manage supply and shop operations, and practical field exercises using realistic mission scenarios.

Credit Recommendation: ***Version 1:*** In the lower-division baccalaureate/associate degree category, 1 semester hour in computer applications, 2 in leadership, and 2 in maintenance management (6/01). *Version 2:* In the upper-division baccalaureate category, 2 semester hours in introduction to Microsoft Office and 2 in maintenance management (2/04).

AR-1715-1066

Radar Repairer

Course Number: 121-35M10, Phase 2.

Location: Ordnance, Missile and Munitions School, Redstone Arsenal, AL.

Length: 33 weeks (1199 hours).

Exhibit Dates: 1/00–Present.

Learning Outcomes: Upon completion of the course, the student will be able to perform organizational, direct, and general support maintenance on a series of electronic systems, including maintenance, troubleshooting, repair, and alignment.

Instruction: Lectures and exercises cover basic digital, computer, and radar fundamentals and testing, troubleshooting, and defective components.

Credit Recommendation: In the lower-division baccalaureate/associate degree category, 3 semester hours in radar testing and troubleshooting and 3 in digital and analog troubleshooting (1/04).

Related Occupations: 35M.

AR-1715-1068

Aircraft Armament Maintenance Technician

Course Number: 4D-SQIE (151A).

Location: Aviation Logistics School, Fort Eustis, VA.

Length: 23–24 weeks (908 hours).

Exhibit Dates: 10/00–9/03.

Learning Outcomes: Upon completion of the course, the student will be able to perform maintenance, preventive maintenance, troubleshooting, and repairs on electrical and avionics systems for both OH-58D and AH-64 series helicopters.

Instruction: Methods of instruction include classroom, practical, and shop exercises.

Credit Recommendation: In the lower-division baccalaureate/associate degree category, 4 semester hours in aircraft electrical systems and 3 in avionics systems (6/01).

Related Occupations: 151A.

AR-1715-1069

Ground Surveillance Systems Operator

Course Number: 243-96R10.

Location: Intelligence School, Fort Huachuca, AZ.

Length: 5–6 weeks (208–209 hours).

Exhibit Dates: 10/99–Present.

Learning Outcomes: Before 10/99 see AR-1715-0178. Upon completion of the course, the student will be able to operate ground surveillance radar equipment.

Instruction: Lectures, practical exercises, and demonstrations cover specific ground surveillance radar equipment operation and maintenance procedures.

Credit Recommendation: In the vocational certificate category, 1 semester hour in troubleshooting techniques. In the lower-division baccalaureate/associate degree category, 1 semester hour in electromechanical maintenance and 1 in map interpretation (8/02).

Related Occupations: 96R.

AR-1715-1070

Ground Surveillance Systems Operator, Phase 3

Course Number: 243-96R10, Phase 3.

Location: Intelligence School, Fort Huachuca, AZ.

Length: 119 hours.

Exhibit Dates: 10/00–Present.

Learning Outcomes: Upon completion of the course, the student will be able to operate ground surveillance systems.

Instruction: This course includes tactical radio operations, operation and maintenance of specific radar sets, employment of remote battlefield sensors, and basic preventive maintenance checks.

Credit Recommendation: Credit is not recommended because of the military-specific nature of the course (8/01).

Related Occupations: 96R.

AR-1715-1071

Imagery Ground Station Operator

Course Number: 233-96H10.

Location: Intelligence School, Fort Huachuca, AZ.

Length: 19–20 weeks (713 hours).

Exhibit Dates: 8/99–Present.

Learning Outcomes: Upon completion of the course, the student will read maps, plot information on them, and brief responsible authorities on information obtained; set up and operate various information producing sources; and perform basic maintenance on aerial intelligence devices.

Instruction: Small classes are held in hangars, classrooms on the flight line, and inflight.

Credit Recommendation: In the lower-division baccalaureate/associate degree category, 1 semester hour in electronic surveillance systems and 3 in introduction to map reading (8/01).

Related Occupations: 96H.

AR-1715-1073

Telecommunications Systems Engineer

Course Number: 4C-24A.

Location: Signal School, Fort Gordon, GA.

Length: 20 weeks (998 hours).

Exhibit Dates: 5/01–Present.

Learning Outcomes: Upon completion of the course, the student will be able to design, engineer, troubleshoot and perform systems acceptance tests on telecommunications systems and networks both at the conceptual and physical hardware levels; assess network security; understand various system protocols, routing algorithms, encryption methods, and the methods of monitoring traffic flow; plan network evolution; manage various combinations of voice and data networks; master appropriate math skills related to the theoretical understanding of the telecommunications systems addressed in the course; and have a basic understanding of both civilian and military communications systems.

Instruction: Discussion, lectures, homework projects, small group projects, class presentations, midterm and final examinations by module, an in-depth research project, and group and laboratory exercises, including hands-on utilization of all types of data communications and networking equipment including routers, switchers and hubs. These techniques vary over the various components of the instructional programs. Topics include mathematics fundamentals relevant to telecommunications including number theory, exponential and logarithmic functions, series and probability, an understanding of channel activity and noise, the use of simulations, queuing theory, and the analysis of mathematical concepts as related to communications network behavior; C programming including running programs, the use of variables, constants, expressions and functions, program flow control and data storage using strings and arrays; telecommunications system models in the private and public sectors, traffic engineering, switching transmission and signaling; transport services and systems and system control, access, and termination; in-depth knowledge of the OSI model layer 1-3 and a working familiarity of level 4-7 including media (build, terminate and test copper, single and multi-mode fiber cables), encoding, data link control, routing protocols, common data network protocols, and architectures; switching for data and voice over a wide area; ethernet switching and VLAN's ATM and VOIP; circuit and store forward; switching protocols including photonic, optical, wireless and cellular; network security including cryptography, PKI, SSL and various other private and public key algorithms, IP security and host-to-host security protocols; network operations including and understanding of MIB's, SNMP, RMON, ASN, service level agreements and QOS; structured top-down network design in the commercial and DII environments, functions of access; distributed and core networks; performance requirements, analysis of traffic flow and behavior; flat, hierarchical and mesh topologies; network redundancy; integration of new and legacy networks; linking to commercial networks and integration of various commercially available equipment with existing networks; a research project related to the above subjects including writing, modeling, testing, and defending a proposal. (This research project, however, would not be appropriate for transferred credit to another graduate program.).

Credit Recommendation: In the lower-division baccalaureate/associate degree category, 3 semester hours in C programming. In the upper-division baccalaureate category, 3 semester hours in applied mathematics or mathematics for data communication and networking. In the graduate degree category, no more than a total of 9 semester hours from the following credit: Up to 3 semester hours in telecommunications systems; up to 6 in data communications I and II; up to 3 in digital transmission or methods of telecommunications; up to 3 in network switching and switching systems; up to 3 in network security; up to 3 in network management or network operations and management; up to 6 in network design; and up to 3 in future developments in telecommunications systems (8/02).

Related Occupations: 24A.

AR-1715-1074

1. IEW Systems Repairer
2. MI Systems Maintainer/Integrator

Course Number: ***Version 1:*** 102-33W10. *Version 2:* 102-33W10.

Location: Intelligence Center and School, Fort Huachuca, AZ.

Length: ***Version 1:*** 41 weeks (1483 hours). *Version 2:* 43–44 weeks (1577 hours).

Exhibit Dates: ***Version 1:*** 1/00–12/01. *Version 2:* 1/02–Present.

Learning Outcomes: ***Version 1:*** Upon completion of the course, the student will be able to demonstrate mastery electronic theory, AC/DC fundamentals, and digital and analog circuits; troubleshoot mainframe, high performance workstations, and computer networks; and perform entry level maintenance on intelligence electronic warfare equipment. *Version 2:* Upon completion of the course, the student will be able to use common laboratory instruments and perform electronic laboratory work; analyze DC series, parallel, and series-parallel circuits; analyze RL and RC circuits; analyze filters and resonant circuits; explain the operation of solid state devices; analyze analog circuits including class A, B, and C power amplifiers, power supplies, and oscillators; explain operational amplifier operation and circuits; describe the operation of vacuum tube circuits; perform conversions between decimal, binary, octal, and hexadecimal numbers; add and subtract binary and twos-complement numbers; explain the operation of digital circuit elements, including logic gates, flip-flops, counters, registers, decoders, and encoders; explain microprocessor system architecture and register structure; write Assembly and machine language microprocessor programs; explain electronic communications concepts and devices, including wave propagation, transmitters, AM, FM, pulse code modulation, receivers, transmission lines, antennas, and microwaves; troubleshoot a wide variety of electronic circuits and systems; perform electronic soldering; and troubleshoot and maintain Windows NT and UNIX workstations and their networks.

Instruction: ***Version 1:*** Lectures, laboratory, discussion, and practical and classroom exercises. Topics include AC/DC fundamentals, analog, digital circuits, communications, and computer architecture; and troubleshooting NT workstations, RISK and UNIX systems, imagery analysis, multi-sub systems, and direction finding equipment. *Version 2:* Lectures, laboratories, demonstrations, and performance exercises cover electronics warfare/intercept aviation systems.

Credit Recommendation: ***Version 1:*** In the lower-division baccalaureate/associate degree category, 3 semester hours in AC/DC fundamentals, 2 in computer maintenance and repair, and 2 in digital circuits (12/02). *Version 2:* In the vocational certificate category, 2 semester hours in soldering techniques. In the lower-division baccalaureate/associate degree category, 3 semester hours in basic electronics laboratory, 1 in DC circuits, 3 in AC circuits, 4 in digital principles, 4 in solid state electronics, 3 in microprocessors, 4 in electronic systems troubleshooting and maintenance, and 4 in electronic communications (8/02).

AR-1715-1075

Satellite Communication (SATCOM) Strategic Terminals

Course Number: A-102-0402; 102-F104, Phase 4.

Location: Signal School, Fort Gordon, GA.

Length: 7–9 weeks (288–340 hours).

Exhibit Dates: 10/95–Present.

Learning Outcomes: Upon completion of the course, the student will have an over all knowledge of the satellite communications terminal, its operational characteristics, and necessary preventive maintenance.

Instruction: Lectures and practical exercises cover the overall design, operation, and preventive maintenance of the SATCOM terminal.

Credit Recommendation: In the lower-division baccalaureate/associate degree category, 2 semester hours in principles of satellite communications systems (2/03).

AR-1715-1076

SATCOM Systems Operator/Maintainer (Strategic P-AOT)

Course Number: 102-31S10 Strategic P-AOT.

Location: Signal School, Fort Gordon, GA.

Length: 9 weeks (360 hours).

Exhibit Dates: 5/02–Present.

Learning Outcomes: Upon completion of the course, the student will be able to operate and maintain a satellite communications terminal.

Instruction: Audio visual materials, practical and classroom exercises, discussion, and computer-based training. Topics include satellite communication terminals, satellite communication antennas, and associated satellite communication control systems.

Credit Recommendation: In the lower-division baccalaureate/associate degree category, 1 semester hour in communication system (2/03).

AR-1715-1077

SATCOM Systems Operator/Maintainer (Tactical P-AOT)

Course Number: 102-31S10 Tactical P-AOT.

Location: Signal School, Fort Gordon, GA.

Length: 6 weeks (240 hours).

Exhibit Dates: 5/02–Present.

Learning Outcomes: Upon completion of the course, the student will be able to configure, operate, maintain, troubleshoot, and repair satellite communication (SATCOM) terminals.

Instruction: Audio visual materials, practical and classroom exercises, discussion, and computer-based training. Topics include installing SATCOM terminals, operating SATCOM terminals, and maintaining and repairing SATCOM terminals.

Credit Recommendation: In the lower-division baccalaureate/associate degree category, 1 semester hour in communication systems (2/03).

AR-1715-1078

SATCOM Systems Operator/Maintainer (Common Core P-AOT)

Course Number: 102-31S10 CC P-AOT.

Location: Signal School, Fort Gordon, GA.
Length: 16–17 weeks (656 hours).
Exhibit Dates: 2/02–Present.

Learning Outcomes: Upon completion of the course, the student will be able to identify series, parallel, and series/parallel resistance circuits and determine values of resistance, voltage, and current in the circuit; recognize the meaning of frequency, period, sine wave, phase shift, inductive reactance, capacitive reactance, and impedance; calculate voltage and current in transformers; calculate quality factor, bandwidth, and cutoff frequencies of resonant circuits; calculate impedances, voltages, and current in capacitive and inductive AC circuits; explain the basic operation of semiconductors, junction diodes, and transistors and identify their uses in rectifier circuits and voltage regulators; describe communications techniques; trace signal path, troubleshoot, and align mobile FM transceivers; convert between and perform arithmetic in the decimal, binary, octal, and hexadecimal number systems; analyze logic circuits, flip-flop circuits, registers, and counters; operate, maintain, troubleshoot, and repair satellite earth terminal equipment including digital communications systems, using standard electronic test equipment and spectrum analyzer (using spread spectrum techniques).

Instruction: Lectures and laboratory exercises cover topics in DC and AC circuits, solid state electronics, digital electronics, FM communications, satellite communications, soldering techniques, and theory of operation and troubleshooting methods for satellite communications equipment and devices. Course methods include computer-assisted instruction modules.

Credit Recommendation: In the lower-division baccalaureate/associate degree category, 3 semester hours in basic DC circuits, 1 in AC circuits, 1 in digital circuits, 1 in solid state electronics, 1 in electronic communications, 3 in basic electronics laboratory, and 3 in electronic system troubleshooting and maintenance (2/03).

AR-1715-1079

Satellite Wideband-TASCAT Terminals AN/TSC-85B and AN/TSC-93B
(TACSAT Terminals AN/TSC-85B and AN/TSC-93B)

Course Number: 101-F35 (OS), Phase 6.
Location: Signal School, Fort Gordon, GA.
Length: 4 weeks (160 hours).
Exhibit Dates: 3/98–Present.

Learning Outcomes: Upon completion of the course, the student will be able to configure, program and operate satellite communications baseband equipment.

Instruction: Topics include baseband equipment operation and maintenance, control terminal operation and maintenance, network MODEM, and multiplexer operation and maintenance.

Credit Recommendation: In the lower-division baccalaureate/associate degree category, 1 semester hour in electronic systems operation and maintenance (2/03).

AR-1715-1080

Ground Mobile Forces SATCOM Technician

Course Number: 102-F44 (31S) (OS), Phase 4.
Location: Signal School, Fort Gordon, GA.
Length: 6 weeks (223 hours).
Exhibit Dates: 1/95–Present.

Learning Outcomes: Upon completion of the course, the student will be able to operate, troubleshoot, align and perform maintenance on wideband frequency modulated tactical satellite terminals and associated equipment.

Instruction: Practical and classroom exercises. Topics include covering operation alignment, troubleshooting of radio frequency equipment, antennas server control unit, high voltage power supply, and auxiliary equipment.

Credit Recommendation: In the lower-division baccalaureate/associate degree category, 3 semester hours in electronic systems troubleshooting (2/03).

AR-1715-1081

M1A2 Unit Level Maintenance
(M1A2/SEP Unit Level Maintenance)

Course Number: *Version 1:* 643-ASIK4 (63A). *Version 2:* 643-ASIK4 (63A).
Location: Armor School, Fort Knox, KY.
Length: *Version 1:* 3 weeks (120 hours). *Version 2:* 4 weeks (160 hours).
Exhibit Dates: *Version 1:* 6/00–9/01. *Version 2:* 10/01–Present.

Learning Outcomes: *Version 1:* Upon completion of the course, the student will be able to maintain the 1553B Data Bus, power management system, and fire control system; troubleshoot using BIT and FIT; remove and install selected components; perform download/verification with CTS III or SPORT; and perform Boresight/Plumb and Sync. *Version 2:* Upon completion of the course, the student will be able to maintain the 1553B Data Bus, power management system, fire control system, and fire suppression system; troubleshoot using BIT and FIT; remove and install selected components; perform download/verification with CTS III or SPORT; and perform Boresight/Plumb and Sync.

Instruction: *Version 1:* Audiovisual materials, practical and classroom exercises, learner presentations, and laboratory. Topics include M1A2 Vehicle/HOT familiarization, maintenance of 1553B Data Bus, fire control system, power management and performance of boresighting/plumb, and synchronization and maintenance of fire suppression system. *Version 2:* Audiovisual materials, learner presentations, practical and classroom exercises, and laboratory. Topics include M1A2 Vehicle/HOT familiarization, maintenance of 1553B Data Bus, fire control system, power management and performance of boresighting/plumb, and synchronization and maintenance of fire suppression system.

Credit Recommendation: *Version 1:* In the lower-division baccalaureate/associate degree category, 3 semester hours in electronic systems troubleshooting and repair (6/03). *Version 2:* In the lower-division baccalaureate/associate degree category, 4 semester hours in electronic systems troubleshooting and repair (6/03).

AR-1715-1083

Bradley A3 System maintainer

Course Number: 643-ASIB9.
Location: Armor School, Fort Knox, KY.
Length: 4 weeks (144 hours).
Exhibit Dates: 10/01–Present.

Learning Outcomes: Upon completion of the course, the student will be able to perform unit level maintenance on the M2/M3A3 Bradley Fighting Vehicle system.

Instruction: Audiovisual materials, practical exercises, discussion, learner presentations, classroom exercises. Topics include familiarization and maintenance of the M2/M3A3 tank; skill, knowledge, and task summarization of tank components.

Credit Recommendation: In the lower-division baccalaureate/associate degree category, 4 semester hours in electronic systems troubleshooting and repair (6/03).

AR-1715-1084

M3A3 Cavalry Fighting Vehicle Operator (19D)

Course Number: 250-ASIB9.
Location: Armor School, Fort Knox, KY.
Length: 1 week (42 hours).
Exhibit Dates: 6/02–Present.

Learning Outcomes: Upon completion of the course, the student will be able to operate and maintain the M3A3 Bradley Fighting Vehicle, and define the obligations of a M3A3 Cavalry Fighting Vehicle Operator.

Instruction: Practical exercises, discussion, learner presentations, and lecture. Topics include operation, maintenance, orientation and safety features of a Bradley tank; intercom system; field position; and M3A3 Bradley navigation system maintenance and operation.

Credit Recommendation: In the lower-division baccalaureate/associate degree category, 3 semester hours in introduction to electronics and 1 in occupational health and safety (6/03).

AR-1715-1085

M1A2 SEP and Bradley A3 DS Maintenance

Course Number: 091-ASI U1 913A/45K30.
Location: Training Center, Fort Hood, TX.
Length: 5–6 weeks (198 hours).
Exhibit Dates: 2/03–Present.

Learning Outcomes: Upon completion of this course, the student will be able to provide direct support and general maintenance support for the M1A2 and M3/A3 electronic systems.

Instruction: Discussion, lecture, practical exercises, and laboratory. Topics include M1A2 tank system introduction and maintenance and M3A3 Bradley Fighting Vehicle system.

Credit Recommendation: In the lower-division baccalaureate/associate degree category, 2 semester hours in electronic systems maintenance (8/03).

AR-1715-1086

M1A2 SEP and Bradley A3 DS Maintenance

Course Number: 091-ASI U1 45K.
Location: Training Center, Fort Hood, TX.
Length: 4–5 weeks (158 hours).
Exhibit Dates: 2/03–Present.

Learning Outcomes: Upon completion of this course, the student will be able to provide direct support and general maintenance support for the M1A2 and M3/A3 electronic systems.

Instruction: Discussion, lecture, practical exercises, and laboratory. Topics include M1A2 tank system introduction and maintenance and M3A3 Bradley Fighting Vehicle system.

Credit Recommendation: In the lower-division baccalaureate/associate degree category-gory, 2 semester hours in electronic systems maintenance (8/03).

AR-1715-1087

M1A2 SEP AND BRADLEY A3 DS MAINTENANCE

Course Number: 091-ASI U1 45G.
Location: Training Center, Fort Hood, TX.
Length: 4–5 weeks (160 hours).
Exhibit Dates: 2/03–Present.
Learning Outcomes: Upon completion of this course, the student will be able to provide direct support and general maintenance support for the M1A2 and M3/A3 electronic systems.
Instruction: Discussion, lecture, practical exercises, and laboratory. Topics include M1A2 tank system introduction and maintenance and M3A3 Bradley Fighting Vehicle system.
Credit Recommendation: In the lower-division baccalaureate/associate degree category, 2 semester hours in electronic systems maintenance (8/03).

AR-1715-1088

ARMAMENT REPAIR TECHNICIAN

Course Number: 4-9-C32-913A RC, Phase 2.
Location: Ordnance Center and School, Aberdeen Proving Ground, MD.
Length: 2–3 weeks (106 hours).
Exhibit Dates: 1/94–Present.
Learning Outcomes: Upon completion of the course, the student will be able to troubleshoot and maintain various pieces of equipment using test measurement and diagnostic equipment.
Instruction: Practical exercises, classroom exercises, discussion, laboratory, and lecture. Topics include the use of test measurement and diagnostic equipment in troubleshooting.
Credit Recommendation: In the lower-division baccalaureate/associate degree category, 2 semester hours in electromechanical maintenance (8/03).

AR-1715-1089

ARMAMENT REPAIR TECHNICIAN WARRANT OFFICER ADVANCED

Course Number: 4-9-C32-913A, Phase 2.
Location: Ordnance Center and School, Aberdeen Proving Ground, MD.
Length: 3–4 weeks (144 hours).
Exhibit Dates: 10/99–Present.
Learning Outcomes: Upon completion of the course, the student will be able to perform troubleshooting on the M1A2 turret and fire control and similar systems.
Instruction: Practical exercises, classroom exercises, discussion, laboratory, and lecture. Topics include troubleshooting of tank turret and fire control systems and the M109 series Howitzer.
Credit Recommendation: In the lower-division baccalaureate/associate degree category, 3 semester hours in electromechanical maintenance (8/03).

AR-1715-1090

ARMAMENT REPAIR TECHNICIAN WARRANT OFFICER ADVANCED

Course Number: 091-4-9-C32-913A, Phase 2.
Location: Reserve Schools, Multiple locations.
Length: 98 hours.
Exhibit Dates: 10/99–Present.
Learning Outcomes: Upon completion of the course, the student will be able to perform troubleshooting on the M1A2 turret and fire control systems.
Instruction: Topics include troubleshooting of tank turret and fire control systems.
Credit Recommendation: In the lower-division baccalaureate/associate degree category, 2 semester hours in electromechanical maintenance (8/03).

AR-1715-1092

TOTAL ARMY TRAINING SYSTEM (TATS) M1096 (PALADIN) COMMANDER

Course Number: 061-2E-F165/041-F9, Phase 2.
Location: Field Artillery School, Fort Sill, OK.
Length: 73 hours.
Exhibit Dates: 1/97–Present.
Learning Outcomes: Upon completion of the course, the student will be able to provide technical and tactical knowledge to maintain, operate, and deploy the M109A6 Paladin Howitzer, perform crew-level maintenance on the Howitzer, and supervise maintenance and operation of the automatic fire control system equipment associated with the Howitzer.
Instruction: Practical exercises, lecture, and demonstration. Topics include communications, electronics, FM radio operations, safety, troubleshooting weapon systems, and gunnery technology.
Credit Recommendation: In the lower-division baccalaureate/associate degree category, 1 semester hour in basic electronics and 1 in military science (10/03).

AR-1715-1094

COMPUTER-BASED MEDICAL SYSTEM MAINTENANCE

Course Number: 300-F11.
Location: Academy of Health Sciences, Fort Sam Houston, TX.
Length: 4 weeks (160 hours).
Exhibit Dates: 4/00–Present.
Learning Outcomes: Upon completion of the course, the student will be able to perform basic and advanced computer and fiber-optic troubleshooting and repair, and install local area networks.
Instruction: Practical exercises, discussion, lecture, and computer-based training. Topics include personal computer components and installation; testing equipment and operational verification; configuration of operating systems; fiber-optic principles; cord patching and performance measurement; and network fundamentals, components, installation, software, and building.
Credit Recommendation: In the lower-division baccalaureate/associate degree category, 4 semester hours in computer repair, 2 in fiber optics, and 2 in basic networking (11/03).

AR-1715-1095

COMMON ARMAMENT MAINTENANCE CORE

Course Number: 4D-SQIE-CORE, Phase 1.
Location: Aviation Logistics School, Fort Eustis, VA.
Length: 4 weeks (151 hours).
Exhibit Dates: 10/02–Present.
Learning Outcomes: Upon completion of the course, the student will be able to demonstrate knowledge of basic electronic and integrated circuit inspection and troubleshooting procedures.
Instruction: Audiovisual materials, practical and classroom exercises, discussion, laboratory, and lecture. Topics include shop operations; logistics; and inspection and troubleshooting of AC circuits, electrical components, integrated circuits, and electrical systems.
Credit Recommendation: In the lower-division baccalaureate/associate degree category, 3 semester hours in electrical systems troubleshooting and maintenance (12/03).

AR-1715-1096

WIRE SYSTEMS EQUIPMENT REPAIRER

Course Number: 622-35N10.
Location: Missile and Munitions School, Redstone Arsenal, AL.
Length: 17 weeks (622 hours).
Exhibit Dates: 10/03–Present.
Learning Outcomes: Pending evaluation.
Instruction: Pending evaluation.
Credit Recommendation: Pending evaluation.

AR-1715-1098

EXPLOSIVE ORDNANCE DISPOSAL SPECIALIST

Course Number: 4E-91E/645-55D10/20, Phase 1; 4E-89E/645-89D10/20, Phase 1.
Location: Ordnance, Missile and Munitions School, Redstone Arsenal, AL.
Length: 10 weeks (355 hours).
Exhibit Dates: 4/03–Present.
Learning Outcomes: Upon completion of the course, the student will be able to perform those tasks unique to explosive ordnance disposal.
Instruction: Instruction is provided using practical exercises and lectures. Topics include explosive ordnance disposal procedures and fundamentals, safety procedures, ordnance identification, specialty tools, field training exercises, explosive and explosive effects, basic demolition procedures, modern demolition procedures, and initiators.
Credit Recommendation: In the lower-division baccalaureate/associate degree category, 2 semester hours in basic DC/AC electricity and 1 in electricity lab (2/04).
Related Occupations: 55D.

AR-1715-1099

1. PATRIOT FIRE CONTROL OPERATOR
2. PATRIOT FIRE CONTROL ENH OPERATOR/MAINTAINER

Course Number: *Version 1:* 043-14E10. *Version 2:* 043-14E10; 043-14E10/20/3Q (T).
Location: Air Defense Artillery School, Fort Bliss, TX.
Length: *Version 1:* 26 weeks (934 hours). *Version 2:* 18–26 weeks (641–934 hours).
Exhibit Dates: *Version 1:* 10/96–9/02. *Version 2:* 10/02–Present.
Learning Outcomes: *Version 1:* Upon completion of the course the student will be able to operate, maintain, align, and troubleshoot the various electronic components of Patriot radar and communications equipment. *Version 2:* Pending evaluation.

Instruction: *Version 1:* The course uses lectures, written examinations, and practical exercises to provide instruction. Experience is emphasized, focusing on communications, electronics, and computer and radar circuits and systems. Computer initialization/maintenance, software installation, and modifications are included. Extensive practical troubleshooting skills are developed in both analog and digital electronics. *Version 2:* Pending evaluation.

Credit Recommendation: *Version 1:* In the lower-division baccalaureate/associate degree category, 3 semester hours in microcomputer systems troubleshooting and repair, 3 in radar systems, and 3 in electronic system troubleshooting and repair (4/01). *Version 2:* Pending evaluation.

Related Occupations: 14E.

AR-1716-0009

Laundry and Bath Specialist, Reserve

Course Number: 840-57E10-RC.

Location: Quartermaster Center and School, Fort Lee, VA.

Length: 3–4 weeks (133 hours).

Exhibit Dates: 1/81–12/91.

Learning Outcomes: Upon completion of the course, the student will be able to operate and maintain laundry, bath, and fumigation equipment.

Instruction: Lectures and practical exercises cover the operation and maintenance of field bath, heater, and delousing equipment; exchange procedures; operation and maintenance of field laundry equipment; and laundry processing procedures.

Credit Recommendation: In the vocational certificate category, 3 semester hours in laundry equipment maintenance (10/88).

Related Occupations: 57E.

AR-1716-0010

1. Laundry and Bath/Fabric Repair Specialist Basic Noncommissioned Officer (NCO)
2. Laundry and Bath/Fabric Repair Specialist Basic Noncommissioned Officer (NCO)
3. Laundry and Shower/Fabric Repair Specialist Basic Noncommissioned Officer (NCO)

Course Number: *Version 1:* 840-57E/43M30. *Version 2:* 840-57E/43M30. *Version 3:* 840-57E/43M30.

Location: Quartermaster Center and School, Fort Lee, VA.

Length: *Version 1:* 6–7 weeks (241 hours). *Version 2:* 8–9 weeks (301 hours). *Version 3:* 5–6 weeks (198 hours).

Exhibit Dates: *Version 1:* 1/89–5/92. *Version 2:* 6/92–9/95. *Version 3:* 10/95–1/02.

Learning Outcomes: *Version 1:* Upon completion of the course, the student will be able to supervise small and medium-sized textile and repair facilities. *Version 2:* Upon completion of the course, the student will be able to supervise small and medium-sized textile, laundry, and fabric repair facilities. *Version 3:* Upon completion of the course, the student will to supervise small and medium-sized textile, laundry, shower, and fabric repair facilities.

Instruction: *Version 1:* Conferences, practical exercises, visual aids, and demonstrations are used to cover fabric repair, shop safety, preventive maintenance, work flow, supervision and quality control, shop operations, and records maintenance in a fabric repair/laundry and bath operation. This course contains a common core of leadership subjects. *Version 2:* Conferences, practical exercises, visual aids, and demonstrations are used to cover fabric repair, shop safety, preventive maintenance, work flow, supervision and quality control, shop operations, and records maintenance in a fabric repair/laundry and bath operation. This course contains a common core of leadership subjects. *Version 3:* Course includes lectures, demonstrations, practical exercises, and a leadership common core.

Credit Recommendation: *Version 1:* In the lower-division baccalaureate/associate degree category, 3 semester hours in supervisory management. See AR-1406-0090 for common core credit (10/88). *Version 2:* In the lower-division baccalaureate/associate degree category, 3 semester hours in textile care and maintenance. See AR-1406-0090 for common core credit (2/93). *Version 3:* In the lower-division baccalaureate/associate degree category, 1 semester hour in facilities management. See AR-1406-0090 for common core credit (5/02).

Related Occupations: 43M; 57E.

AR-1716-0011

1. Fabric Repair Specialist
2. Fabric Repair Specialist Advanced Individual Training

Course Number: *Version 1:* 760-43M10. *Version 2:* 101-43M10 (F) Phase 1 & 2; 760-43M10.

Location: Quartermaster Center and School, Fort Lee, VA.

Length: *Version 1:* 8–9 weeks (303 hours). *Version 2:* 6 weeks (228–229 hours).

Exhibit Dates: *Version 1:* 10/88–9/95. *Version 2:* 10/95–Present.

Learning Outcomes: *Version 1:* Upon completion of the course, the student will be able to alter and repair organizational clothing, textiles, canvas, and webbed items. *Version 2:* Upon completion of the course, the student will be able to alter and repair organizational clothing and textiles and canvas and webbed equipment items.

Instruction: *Version 1:* Demonstrations and practical exercise are used to develop the students' ability to alter and repair organizational clothing, textiles, canvas, and webbed items. Appropriate sewing machines and other equipment are used and procedures to perform preventive maintenance, troubleshooting, and adjustment of the equipment are covered. *Version 2:* Lectures, demonstration, and practical exercises cover clothing, textiles, canvas, and webbed items. Appropriate machinery is used.

Credit Recommendation: *Version 1:* In the vocational certificate category, 7 semester hours in textile repair and machine operations (2/93). *Version 2:* In the vocational certificate category, 4 semester hours in clothing and textile repair and machine operations (5/02).

Related Occupations: 43M.

AR-1716-0012

1. Laundry and Bath/Fabric Repair Specialists Advanced Noncommissioned Officer (NCO)
2. Laundry and Shower/Fabric Repair Specialist Advanced Noncommissioned Officer (NCO)

Course Number: *Version 1:* 840-57E/43M40. *Version 2:* 840-57E/43M40.

Location: Quartermaster Center and School, Fort Lee, VA.

Length: *Version 1:* 9–10 weeks (344 hours). *Version 2:* 4–7 weeks (174–257 hours).

Exhibit Dates: *Version 1:* 6/92–9/95. *Version 2:* 10/95–Present.

Learning Outcomes: *Version 1:* Upon completion of the course, the student will be able to supervise the identification, care, and maintenance of textiles including the management of a commercial laundry and fabric repair shop. *Version 2:* Upon completion of the course, the student will be able to supervise shower, laundry and fabric repair facilities.

Instruction: *Version 1:* Conference and small group instruction coupled with practical exercises apply the principles of fabric care to the operation of a commercial laundry and fabric repair shop. Topics also include laundry and repair equipment, laundry exchange procedures, and the preparation of cost and administrative reports. This course contains a common core of leadership subjects. *Version 2:* Lectures cover supply topics, common core leader training, and combined arms support.

Credit Recommendation: *Version 1:* In the lower-division baccalaureate/associate degree category, 4 semester hours in textile care and maintenance. See AR-1404-0035 for common core credit (2/93). *Version 2:* In the lower-division baccalaureate/associate degree category, 1 semester hour in supply management and 1 in microcomputer applications. See AR-1404-0035 for common core credit (5/02).

Related Occupations: 43M; 57E.

AR-1716-0013

Laundry and Shower Specialist

Course Number: 101-57E10R; 840-57E10.

Location: Quartermaster Center and School, Fort Lee, VA.

Length: 4 weeks (144 hours).

Exhibit Dates: 10/95–Present.

Learning Outcomes: Upon completion of the course, the soldier will be able to operate and maintain shower and laundry units.

Instruction: Lectures, demonstrations, and exercises cover personnel facilities including shower and laundry units.

Credit Recommendation: In the vocational certificate category, 2 semester hours in personnel facilities operations and maintenance (5/02).

AR-1716-0014

Laundry/Textile Specialist

Course Number: 840-92S40.

Location: Quartermaster Center and School, Fort Lee, VA.

Length: 4–5 weeks (180 hours).

Exhibit Dates: 10/01–Present.

Learning Outcomes: Upon completion of the course, the student will be able to use computer for word processing, develop graphic programs, supervise laundry advanced systems, supervise shower operations, inspect shop equipment, and supervise laundry operations.

Instruction: Classroom exercise, practical exercise, lecture, discussion. Topics include computer software applications, supervision of laundry advanced systems, supervision and shower operations, preventative maintenance of laundry and textile equipment, and shop operations.

Credit Recommendation: In the lower-division baccalaureate/associate degree category, 3 semester hours in principles of supervision (6/02).

AR-1716-0015

Laundry/Textile Specialist

Course Number: 840-92S10.

Location: Quartermaster Center and School, Fort Lee, VA.

Length: 7 weeks (265 hours).

Exhibit Dates: 10/01–Present.

Learning Outcomes: Upon completion of the course, the student will be able to setup, maintain and dismantle several types of sewing machines, mobile laundries, field showers, and perform a range of sewing operations on clothing and other light textiles.

Instruction: Audio visual materials, practical exercise, and lectures. Topics include video procedures and exercises for students in all phases of content.

Credit Recommendation: In the vocational certificate category, 3 semester hours in installation and maintenance, plumbing and sewing, and 1 semester hour in clothing/textile repair (6/02).

AR-1716-0016

Laundry/Textile Specialist

Course Number: 840-92S30, Phase 1.

Location: Quartermaster Center and School, Fort Lee, VA.

Length: 3 weeks (113 hours).

Exhibit Dates: 10/01–Present.

Learning Outcomes: Upon completion of the course, the student will be able to supervise personnel in laundry and textile operations, supervise operations including: setting up, maintaining and dismantling facilities, develop schedules, and quality control.

Instruction: Laboratory, audio visual materials, classroom exercise, practical exercise, and lecture. Topics include lectures in all compounds, group planning exercises, and other exercises relating to facility operation.

Credit Recommendation: In the lower-division baccalaureate/associate degree category, 3 semester hours in principles of supervision (6/02).

AR-1717-0054

Signal Security (SIGSEC) Specialist Basic Technical

Course Number: 231-05G30.

Location: Intelligence School, Fort Devens, MA.

Length: 7–8 weeks (266–335 hours).

Exhibit Dates: 4/77–3/94.

Objectives: To prepare personnel who have been trained as signal security specialists for supervision and management positions.

Instruction: Instructions and practical exercises cover map reading, maintenance management procedures, supervision of maintenance personnel, communications/telecommunications operations, communications security, and military science.

Credit Recommendation: In the lower-division baccalaureate/associate degree category, 3 semester hours in maintenance management procedures and supervision of personnel (2/83).

Related Occupations: 05G.

AR-1717-0067

Lance Missile Mechanic

Course Number: 043-ASIZ3; 043-F4.

Location: Field Artillery School, Fort Sill, OK.

Length: 2 weeks (72 hours).

Exhibit Dates: 8/82–6/96.

Objectives: To train selected personnel to perform as Lance missile mechanics. The student must have completed Lance Missile Crewmember course 043-15D10.

Instruction: Instruction covers the assembly, maintenance, and transportation of the Lance missile.

Credit Recommendation: In the vocational certificate category, 1 semester hour in the use and care of hand tools (10/83).

AR-1717-0075

Aviation Maintenance Officer and Repair Technician Phase 2 CH-47D
(Aviation Maintenance Officer Phase 2 CH-47D)

Course Number: 4D-15T/153A/154CG1G; 4D-71A/160A/100CGH (CH-47D).

Location: Transportation and Aviation Logistics School, Fort Eustis, VA.

Length: 2–4 weeks (75–135 hours).

Exhibit Dates: 10/85–3/98.

Objectives: To provide the aviation commissioned officer and warrant officer with the skills and knowledge required to perform test flights and maintain the CH-47D aircraft.

Instruction: Instruction includes maintenance concepts, management of aircraft maintenance resources, inspection and diagnostic procedures, subsystems and support equipment, and operational checks to determine the airworthiness of the CH-47D aircraft.

Credit Recommendation: In the lower-division baccalaureate/associate degree category, 1 semester hour in aviation maintenance management and 1 in advanced helicopter flight techniques (3/88).

Related Occupations: 100C; 153A; 154C; 160A.

AR-1717-0077

Warrant Officer Technical Certification
(Armament Repair Technician)

Course Number: 4E-421A.

Location: Ordnance Center and School, Aberdeen Proving Ground, MD.

Length: 18 weeks (658 hours).

Exhibit Dates: 3/85–12/97.

Learning Outcomes: Upon completion of the course, the student will be able to assume duties and responsibilities of an upgrade technician in armament repair.

Instruction: Topics include management practices and procedures; basic electricity and electronics; computers; and inspection and repair of artillery, small arms, and fire control equipment.

Credit Recommendation: In the lower-division baccalaureate/associate degree category, 3 semester hours in maintenance management, 2 in communication skills, 2 in basic electricity, 2 in basic electronics, and 1 in basic computer (3/87).

Related Occupations: 421A.

AR-1717-0078

Non-Morse Interceptor/Analyst Basic Noncommissioned Officer (NCO)
(Electronic Warfare (EW) Signals Intelligence (SIGINT) Non-Morse Interceptor Basic Noncommissioned Officer (NCO))

Course Number: *Version 1:* 231-98K30; 231-05K30. *Version 2:* 231-98K30; 231-05K30.

Location: Intelligence School, Fort Devens, MA.

Length: *Version 1:* 11–12 weeks (380–381 hours). *Version 2:* 9 weeks (328–329 hours).

Exhibit Dates: *Version 1:* 4/88–12/90. *Version 2:* 1/91–3/94.

Learning Outcomes: *Version 1:* Upon completion of the course, the student will be able to analyze electronic signals and generate computerized reports. *Version 2:* After 9/96 see AR-1402-0177. Upon completion of the course, the student will be able to analyze electronic signals and generate computerized reports.

Instruction: *Version 1:* Lectures, presentations, and practical exercises cover analysis of modulation techniques and report generation. *Version 2:* Lectures, presentations, and practical exercises cover analysis of modulation techniques and report generation. This course contains a common core of leadership subjects.

Credit Recommendation: *Version 1:* In the vocational certificate category, 3 semester hours in communications theory. In the lower-division baccalaureate/associate degree category, 2 semester hours in personnel management (7/91). *Version 2:* In the vocational certificate category, 3 semester hours in communications theory. In the lower-division baccalaureate/associate degree category, 2 semester hours in personnel management. See AR-1406-0090 for common core credit (7/91).

Related Occupations: 05K; 98K.

AR-1717-0079

1. Electronic Warfare (EW)/Signals Intelligence (SIGINT) Morse Interceptor Basic Noncommissioned Officer (NCO)
2. Morse Interceptor Basic Noncommissioned Officer (NCO)

Course Number: *Version 1:* 231-05H30. *Version 2:* 231-98H30.

Location: Intelligence School, Fort Devens, MA.

Length: *Version 1:* 6–8 weeks (234–281 hours). *Version 2:* 6 weeks (218–220 hours).

Exhibit Dates: *Version 1:* 10/85–12/90. *Version 2:* 1/91–3/94.

Learning Outcomes: *Version 1:* Upon completion of the course, the student will be able to manage and supervise personnel in the analysis of data for target identification and location including work assignments and record keeping. *Version 2:* Upon completion of the course, the student will be able to manage and supervise personnel in the analysis of data for target identification and location including work assignments and record keeping.

Instruction: *Version 1:* Instruction includes lectures, demonstrations, and performance exercises in office practice, personnel supervision, report preparation, map reading, interpreting signal intelligence, and target identification and location. *Version 2:* Instruction includes lectures, demonstrations, and performance exercises in office practice, personnel supervision, report preparation, map reading, interpreting signal intelligence, and target identification and location. This course contains a common core of leadership subjects.

Credit Recommendation: ***Version 1:*** In the lower-division baccalaureate/associate degree category, 1 semester hour in office management procedures (7/88). *Version 2:* In the lower-division baccalaureate/associate degree category, 2 semester hours in personnel management. See AR-1406-0090 for common core credit (6/91).

Related Occupations: 05H; 98H.

AR-1717-0082

1. UH-1 Maintenance Manager/Test Pilot
2. UH-1 Maintenance Test Pilot

Course Number: ***Version 1:*** 4D-SIB1/2C-SQIG; 4D-SI1E/2C-SQIG. *Version 2:* 4D-SIG6/SQIG (UH-1), Phase 1; 4D-SIG6/SQIG (UH-1).

Location: ***Version 1:*** Transportation and Aviation Logistics School, Fort Eustis, VA. *Version 2:* Aviation Center, Fort Rucker, AL.

Length: ***Version 1:*** 8–14 weeks (483–488 hours). *Version 2:* 3–4 weeks (135–150 hours).

Exhibit Dates: ***Version 1:*** 10/87–9/94. *Version 2:* 10/94–Present.

Learning Outcomes: ***Version 1:*** Upon completion of the course, the student will be able to perform all the duties of a manager/maintenance test pilot, including repair and supply procedures; use of publications and logistics performance indicators; development of estimates and reports; use of forms and records; inspection of fuels, lubricants, hydraulic and electrical systems, flight controls, and power plants; performance analysis; rotor and power train systems; maintenance test flight; and evaluations. *Version 2:* Upon completion of the course, the student will have been provided with information and training on UH-1 maintenance troubleshooting and test flight procedures.

Instruction: ***Version 1:*** This course includes lectures, demonstrations, and performance exercises in maintenance officer responsibilities, including logistics, aviation maintenance management, aeronautical technology, maintenance test flight, and quality assurance. *Version 2:* This course includes lectures, demonstrations, and performance exercises in maintenance officer responsibilities for the UH-1. Students should have completed Maintenance Manager, 2C-15D/4D-F6.

Credit Recommendation: ***Version 1:*** In the lower-division baccalaureate/associate degree category, 3 semester hours in aviation maintenance management (10/91). *Version 2:* Credit is not recommended because of the limited, specialized nature of the course (4/03).

AR-1717-0083

1. AH-1S Maintenance Manager/Test Pilot
2. AH-1 Maintenance Test Pilot

Course Number: ***Version 1:*** 4D-SIG6/SQIG (AH-1); 4D-SI1M/2C-SQIG. *Version 2:* 4D-SIG6/SQIG (AH-1).

Location: ***Version 1:*** Transportation and Aviation Logistics School, Fort Eustis, VA. *Version 2:* Aviation Center, Fort Rucker, AL.

Length: ***Version 1:*** 13–14 weeks (482–492 hours). *Version 2:* 3–4 weeks (147 hours).

Exhibit Dates: ***Version 1:*** 10/87–9/94. *Version 2:* 10/94–Present.

Learning Outcomes: ***Version 1:*** Upon completion of the course, the student will be able to perform all the duties of a manager/maintenance test pilot, including repair and supply procedures; use of publications and logistics performance indicators; development of estimates and reports; use of forms and records; inspection of fuels, lubricants, hydraulic and electrical systems, flight controls, and power plants; performance analysis; rotor and power train systems; maintenance test flight; and evaluations. *Version 2:* Upon completion of the course, the student will have been provided with information and training on AH-1 maintenance troubleshooting and test flight procedures.

Instruction: ***Version 1:*** This course includes lectures, demonstrations, and performance exercises in maintenance officer responsibilities, including logistics, aviation maintenance management, aeronautical technology, maintenance test flight, and quality assurance. *Version 2:* This course includes lectures, demonstrations, and performance exercises in maintenance officer responsibilities for the AH-1. Students should have completed Maintenance Manager, 2C-15D/4D-F6.

Credit Recommendation: ***Version 1:*** In the lower-division baccalaureate/associate degree category, 3 semester hours in aviation maintenance management (10/92). *Version 2:* Credit is not recommended because of the limited, specialized nature of the course (11/96).

AR-1717-0084

1. CH-47D Maintenance Manager/Test Pilot
   (CH-47A/B/C Maintenance Manager/Test Pilot)
2. CH-47D Maintenance Test Pilot

Course Number: ***Version 1:*** 4D-SIG6/SQIG; 4D-SIC2/2C-SQIG; 4D-SI1B/2C-SQIG; 4D-SI1G/2C-SQIG. *Version 2:* 4D-SIG6/SQIG (CH-47D), Phase 2; 4D-SIG6/SQIG (CH-47D), Phase 1; 4D-SIG6/SQIG (CH-47D).

Location: ***Version 1:*** Transportation and Aviation Logistics School, Fort Eustis, VA. *Version 2:* Aviation Center, Fort Rucker, AL.

Length: ***Version 1:*** 13–14 weeks (485 hours). *Version 2:* 3–4 weeks (142–186 hours).

Exhibit Dates: ***Version 1:*** 10/87–9/94. *Version 2:* 10/94–Present.

Learning Outcomes: ***Version 1:*** Upon completion of the course, the student will be able to perform all the duties of a manager/maintenance test pilot, including repair and supply procedures; use of publications and logistics performance indicators; development of estimates and reports; use of forms and records; inspection of fuels, lubricants, hydraulic and electrical systems, flight controls, and power plants; performance analysis; rotor and power train systems; maintenance test flight; and evaluations. *Version 2:* Upon completion of the course, the student will have been provided with information and training on CH-47D maintenance troubleshooting and test flight procedures.

Instruction: ***Version 1:*** This course includes lectures, demonstrations, and performance exercises in maintenance officer responsibilities, including logistics, aviation maintenance management, aeronautical technology, maintenance test flight, and quality assurance. *Version 2:* This course includes lectures, demonstrations, and performance exercises in maintenance officer responsibilities for the CH-47D. Students should have completed Maintenance Manager, 2C-15D/4D-F6.

Credit Recommendation: ***Version 1:*** In the lower-division baccalaureate/associate degree category, 3 semester hours in aviation maintenance management (6/91). *Version 2:* Credit is not recommended because of the limited, specialized nature of the course (4/03).

AR-1717-0085

1. OH-58A/C Maintenance Manager/Test Pilot
2. OH-58A/C Maintenance Test Pilot

Course Number: ***Version 1:*** 4D-SIG6/SQIG (OH-58A/C); 4D-SI1P/2C-SQIG. *Version 2:* 4D-SIG6/SQIG (OH-58A/C).

Location: ***Version 1:*** Transportation and Aviation Logistics School, Fort Eustis, VA. *Version 2:* Aviation Center, Fort Rucker, AL.

Length: ***Version 1:*** 13 weeks (475 hours). *Version 2:* 3–4 weeks (125–130 hours).

Exhibit Dates: ***Version 1:*** 10/87–9/94. *Version 2:* 10/94–Present.

Learning Outcomes: ***Version 1:*** Upon completion of the course, the student will be able to perform all the duties of a manager/maintenance test pilot, including repair and supply procedures; use of publications and logistics performance indicators; development of estimates and reports; use of forms and records; inspection of fuels, lubricants, hydraulic and electrical systems, flight controls, and power plants; performance analysis; rotor and power train systems; maintenance test flight; and evaluations. *Version 2:* Upon completion of the course, the student will have information and training on OH-58A/C troubleshooting and test flight procedures.

Instruction: ***Version 1:*** This course includes lectures, demonstrations, and performance exercises in maintenance officer responsibilities, including logistics, aviation maintenance management, aeronautical technology, maintenance test flight, and quality assurance. *Version 2:* This course includes lectures, demonstrations, and performance exercises in maintenance officer responsibilities for the OH-58A/C. Students should have completed Maintenance Manager, 2C-15D/4D-F6.

Credit Recommendation: ***Version 1:*** In the lower-division baccalaureate/associate degree category, 3 semester hours in aviation maintenance management (10/92). *Version 2:* Credit is not recommended because of the limited, specialized nature of the course (11/96).

AR-1717-0086

1. OH-58D Maintenance Manager/Test Pilot
2. OH-58D Warrior Maintenance Test Pilot
   (OH-58D (Armed) Maintenance Manager/Maintenance Test Pilot)

Course Number: ***Version 1:*** 4D-SI1A/2C-SQIG. *Version 2:* 4D-SIG6/SQI (OH-58D); 4D-SIG6/SQIG (OH-58D).

Location: ***Version 1:*** Transportation and Aviation Logistics School, Fort Eustis, VA. *Version 2:* Aviation Center, Fort Rucker, AL.

Length: *Version 1:* 13–14 weeks (485 hours). *Version 2:* 3–5 weeks (130–236 hours).

Exhibit Dates: *Version 1:* 10/87–9/94. *Version 2:* 10/94–Present.

Learning Outcomes: *Version 1:* Upon completion of the course, the student will be able to perform all the duties of a manager/maintenance test pilot, including repair and supply procedures; use of publications and logistics performance indicators; development of estimates and reports; use of forms and records; inspection of fuels, lubricants, hydraulic and electrical systems, flight controls, and power plants; performance analysis; rotor and power train systems; maintenance test flight; and evaluations. *Version 2:* Upon completion of the course, the student will be able to describe OH-58D maintenance troubleshooting and test flight procedures.

Instruction: *Version 1:* This course includes lectures, demonstrations, and performance exercises in maintenance officer responsibilities, including logistics, aviation maintenance management, aeronautical technology, maintenance test flight, and quality assurance. *Version 2:* This course includes lectures, demonstrations, and performance exercises in maintenance officer responsibilities for the OH-58D.

Credit Recommendation: *Version 1:* In the lower-division baccalaureate/associate degree category, 3 semester hours in aviation maintenance management (3/88). *Version 2:* Credit is not recommended because of the limited, specialized nature of the course (11/98).

AR-1717-0087

1. UH-60 Maintenance Manager/Test Pilot
2. UH-60 Maintenance Test Pilot

Course Number: *Version 1:* 4D-SI1N/2C-SQIG. *Version 2:* 4D-SIG6/SQIG (UH-60), Phase 1; 4D-SIG6/SQIG (UH-60); 4D-SIG6/SQIG.

Location: *Version 1:* Transportation and Aviation Logistics School, Fort Eustis, VA. *Version 2:* Aviation Center, Fort Rucker, AL.

Length: *Version 1:* 13–14 weeks (504 hours). *Version 2:* 4 weeks (159 hours).

Exhibit Dates: *Version 1:* 10/87–9/94. *Version 2:* 10/94–Present.

Learning Outcomes: *Version 1:* Upon completion of the course, the student will be able to perform all the duties of a manager/maintenance test pilot, including repair and supply procedures; use of publications and logistics performance indicators; development of estimates and reports; use of forms and records; inspection of fuels, lubricants, hydraulic and electrical systems, flight controls, and power plants; performance analysis; rotor and power train systems; maintenance test flight; and evaluations. *Version 2:* Upon completion of the course, the student will have been provided with information and training on UH-60 maintenance troubleshooting and test flight procedures.

Instruction: *Version 1:* This course includes lectures, demonstrations, and performance exercises in maintenance officer responsibilities, including logistics, aviation maintenance management, aeronautical technology, maintenance test flight, and quality assurance. *Version 2:* This course includes lectures, demonstrations, and performance exercises in maintenance officer responsibilities for the UH-60. Students should have completed Maintenance Manager, 2C-15D/4D-F6.

Credit Recommendation: *Version 1:* In the lower-division baccalaureate/associate degree category, 3 semester hours in aviation maintenance management (3/88). *Version 2:* Credit is not recommended because of the limited, specialized nature of the course (8/98).

AR-1717-0089

1. UH-60 Maintenance Manager/Test Pilot
   (4UH-1 Utility Helicopter Repairer Supervisor Basic Noncommissioned Officer (NCO))
   (4UH-1 Utility Helicopter Basic Noncommissioned Officer (NCO))
2. UH-1 Utility Helicopter Repairer Basic Noncommissioned Officer (NCO)
3. UH-1 Utility Helicopter Repairer Supervisor Basic Noncommissioned Officer (NCO)
4. UH-1 Helicopter Repairer Supervisor Basic Noncommissioned Officer (NCO)

Course Number: *Version 1:* 600-67N30. *Version 2:* 600-67N30. *Version 3:* 600-67N30. *Version 4:* 600-67N30.

Location: *Version 1:* Transportation and Aviation Logistics School, Fort Eustis, VA. *Version 2:* Transportation and Aviation Logistics School, Fort Eustis, VA. *Version 3:* Transportation and Aviation Logistics School, Fort Eustis, VA. *Version 4:* Aviation Logistics School, Fort Eustis, VA.

Length: *Version 1:* 7–8 weeks (231 hours). *Version 2:* 12 weeks (352 hours). *Version 3:* 13–14 weeks (493 hours). *Version 4:* 6–7 weeks (509–512 hours).

Exhibit Dates: *Version 1:* 10/87–3/90. *Version 2:* 4/90–9/91. *Version 3:* 10/91–9/94. *Version 4:* 10/94–9/00.

Learning Outcomes: *Version 1:* Upon completion of the course, the student will be able to plan maintenance procedures and aviation inventories; use data communications hardware and software including oil analysis programs; use ground support equipment and precision tools; establish policies and procedures for shop and hanger safety; and supervise maintenance of fuel systems, pneudraulic systems, electrical systems, utility systems, power plants, rotor and drivetrain systems, aircraft structures, landing gear, aircraft weight and balance procedures, engine control, and flight control systems. *Version 2:* Upon completion of the course, the student will be able to plan maintenance procedures and aviation inventories; use data communications hardware and software including oil analysis programs; use ground support equipment and precision tools; establish policies and procedures for shop and hanger safety; and supervise maintenance of fuel systems, pneudraulic systems, electrical systems, utility systems, power plants, rotor and drivetrain systems, aircraft structures, landing gear, aircraft weight and balance procedures, engine control, and flight control systems. *Version 3:* Upon completion of the course, the student will be able to plan maintenance procedures and aviation inventories; use data communications hardware and software including oil analysis programs; use ground support equipment and precision tools; establish policies and procedures for shop and hanger safety; and supervise maintenance of fuel systems, pneudraulic systems, electrical systems, utility systems, power plants, rotor and drivetrain systems, aircraft structures, landing gear, aircraft weight and balance procedures, engine control, and flight control systems. *Version 4:* Upon completion of the course, the student will be able to plan maintenance procedures and aviation inventories; use data communications hardware and software including oil analysis programs; use ground support equipment and precision tools; establish policies and procedures for shop and hanger safety; and supervise maintenance of fuel systems, pneudraulic systems, electrical systems, utility systems, power plants, rotor and drivetrain systems, aircraft structures, landing gear, aircraft weight and balance procedures, engine control, and flight control systems.

Instruction: *Version 1:* Lectures, demonstrations, and practical exercises cover supervision, aircraft maintenance, and maintenance management. This course contains a common core of leadership subjects. *Version 2:* Lectures, demonstrations, and practical exercises cover supervision, aircraft maintenance, and maintenance management. *Version 3:* Lectures, demonstrations, and practical exercises cover supervision, aircraft maintenance, and maintenance management. This course contains a common core of leadership subjects. *Version 4:* Lectures, demonstrations, and practical exercises cover supervision, aircraft maintenance, and maintenance management. This course contains a common core of leadership subjects.

Credit Recommendation: *Version 1:* In the lower-division baccalaureate/associate degree category, 2 semester hours in aviation maintenance management and 2 in principles of supervision. See AR-1406-0090 for common core credit (3/88). *Version 2:* In the lower-division baccalaureate/associate degree category, 3 semester hours in aviation maintenance management and 2 in principles of supervision (6/91). *Version 3:* In the lower-division baccalaureate/associate degree category, 3 semester hours in aviation maintenance management and 2 in principles of supervision. See AR-1406-0090 for common core credit (6/91). *Version 4:* In the lower-division baccalaureate/associate degree category, 3 semester hours in aviation maintenance management and 2 in principles of supervision. See AR-1406-0090 for common core credit (6/98).

Related Occupations: 67N.

AR-1717-0090

AH-64 Attack Helicopter Repairer Supervisor Basic Noncommissioned Officer (NCO)

Course Number: *Version 1:* 600-67R30. *Version 2:* 600-67R30. *Version 3:* 600-67R30. *Version 4:* 600-67R30. *Version 5:* 600-67R30.

Location: *Version 1:* Transportation and Aviation Logistics School, Fort Eustis, VA. *Version 2:* Transportation and Aviation Logistics School, Fort Eustis, VA. *Version 3:* Transportation and Aviation Logistics School, Fort Eustis, VA. *Version 4:* Aviation Logistics School, Fort Eustis, VA. *Version 5:* Aviation Logistics School, Fort Eustis, VA.

Length: *Version 1:* 7–8 weeks (272 hours). *Version 2:* 14–15 weeks (430 hours). *Version 3:* 14 weeks (534 hours). *Version 4:* 14–15 weeks (546 hours). *Version 5:* 14–15 weeks (566 hours).

Exhibit Dates: *Version 1:* 10/87–3/90. *Version 2:* 4/90–9/91. *Version 3:* 10/91–9/94. *Version 4:* 10/94–9/98. *Version 5:* 10/98–9/00.

Learning Outcomes: *Version 1:* Upon completion of the course, the student will be able to plan maintenance procedures and aviation inventories; use data communications hardware and software including oil analysis programs; use ground support equipment and precision tools; establish policies and procedures for shop and hangar safety; and supervise maintenance on fuel systems, pneudraulic systems, electrical systems, utility systems, power plants, rotor and drivetrain systems, aircraft structures, landing gear, aircraft weight and balance procedures, engine control, and flight control systems. *Version 2:* Upon completion of the course, the student will be able to plan maintenance procedures and aviation inventories; use data communications hardware and software including oil analysis programs; use ground support equipment and precision tools; establish policies and procedures for shop and hangar safety; and supervise maintenance on fuel systems, pneudraulic systems, electrical systems, utility systems, power plants, rotor and drivetrain systems, aircraft structures, landing gear, aircraft weight and balance procedures, engine control, and flight control systems. *Version 3:* Upon completion of the course, the student will be able to plan maintenance procedures and aviation inventories; use data communications hardware and software including oil analysis programs; use ground support equipment and precision tools; establish policies and procedures for shop and hangar safety; and supervise maintenance of fuel systems, pneudraulic systems, electrical systems, utility systems, power plants, rotor and drivetrain systems, aircraft structures, landing gear, aircraft weight and balance procedures, engine control, and flight control systems. *Version 4:* Upon completion of the course, the student will be able to plan maintenance procedures and aviation inventories; use data communications hardware and software including oil analysis programs; use ground support equipment and precision tools; establish policies and procedures for shop and hangar safety; and supervise maintenance of fuel systems, pneudraulic systems, electrical systems, utility systems, power plants, rotor and drivetrain systems, aircraft structures, landing gear, aircraft weight and balance procedures, engine control, and flight control systems. *Version 5:* After 9/00 see AR-1717-0181. Upon completion of the course, students will be able to supervise helicopter inspection and maintenance using computer-based information production control methods. Students will be able to inspect airframe, power plant, rotors, and all associated systems on the AH-64A helicopter using non-destructive and computer-aided inspection methods.

Instruction: *Version 1:* Lectures, demonstrations, and practical exercises cover supervision, aircraft maintenance, and maintenance management. This course contains a common core of leadership subjects. *Version 2:* Lectures, demonstrations, and practical exercises cover supervision, aircraft maintenance, and maintenance management. *Version 3:* Lectures, demonstrations, and practical exercises cover supervision, aircraft maintenance, and maintenance management. This course contains a common core of leadership subjects. *Version 4:* Lectures, demonstrations, and practical exercises cover supervision, aircraft maintenance, and maintenance management. This course contains a common core of leadership subjects. *Version 5:* Methods of instruction include lecture and practical experience. Student evaluation is based on written and practical testing methods.

Credit Recommendation: *Version 1:* In the lower-division baccalaureate/associate degree category, 2 semester hours in aviation maintenance management and 3 in principles of supervision. See AR-1406-0090 for common core credit (3/88). *Version 2:* In the lower-division baccalaureate/associate degree category, 3 semester hours in aviation maintenance management and 2 in principles of supervision (6/91). *Version 3:* In the lower-division baccalaureate/associate degree category, 3 semester hours in aviation maintenance management and 2 in principles of supervision. See AR-1406-0090 for common core credit (6/91). *Version 4:* In the lower-division baccalaureate/associate degree category, 2 semester hours in helicopter systems troubleshooting, 3 in aviation maintenance management, and 3 in principles of supervision. See AR-1406-0090 for common core credit (6/98). *Version 5:* In the lower-division baccalaureate/associate degree category, 5 semester hours in aircraft maintenance management and 4 in aircraft inspection methods. See AR-1406-0090 for common core credit (6/01).

Related Occupations: 67R.

AR-1717-0092

PETROLEUM AND WATER SPECIALIST ADVANCED NONCOMMISSIONED OFFICER (NCO)

Course Number: *Version 1:* 8-77-C42. *Version 2:* 8-77-C42.

Location: Quartermaster Center and School, Fort Lee, VA.

Length: *Version 1:* 12 weeks (445 hours). *Version 2:* 12–13 weeks (453–460 hours).

Exhibit Dates: *Version 1:* 1/88–9/91. *Version 2:* 10/91–9/95.

Learning Outcomes: *Version 1:* Upon completion of the course, the student will be able to train noncommissioned officers in advancement, management, petroleum storage and supply operations, and water operations. *Version 2:* After 9/95 see AR-1732-0016. Upon completion of the course, the student will be able to perform petroleum supply, petroleum laboratory, and water treatment operations.

Instruction: *Version 1:* Lectures and practical exercises cover supply and distribution operations, waterfront and terminal operations, pipeline operations, aircraft refueling and defueling operations, administration and planning, and petroleum laboratory and water treatment operations. This course contains a common core of leadership subjects. *Version 2:* Lectures and practical exercises cover supply and distribution operations, waterfront and terminal operations, pipeline operations, aircraft refueling and defueling operations, administration and planning, and petroleum laboratory and water treatment operations. This course contains a common core of leadership subjects.

Credit Recommendation: *Version 1:* In the lower-division baccalaureate/associate degree category, 1 semester hour in supervisory management, 3 in petroleum operations management, and 2 in water operations management. See AR-1404-0035 for common core credit (10/88). *Version 2:* In the lower-division baccalaureate/associate degree category, 1 semester hour in supervisory management, 3 in petroleum operations management, 2 in water operations management, and 1 in laboratory science. See AR-1404-0035 for common core credit (2/93).

Related Occupations: 77F; 77L; 77W.

AR-1717-0093

1. AH-64 MAINTENANCE MANAGER/TEST PILOT
2. AH-64A MAINTENANCE MANAGER/ MAINTENANCE TEST PILOT

Course Number: *Version 1:* 4D-SI1L/2C-SQIG. *Version 2:* 4D-SIG6/SQIG (AH-64A).

Location: *Version 1:* Transportation and Aviation Logistics School, Fort Eustis, VA. *Version 2:* Aviation Center, Fort Rucker, AL.

Length: *Version 1:* 15 weeks (568 hours). *Version 2:* 5–6 weeks (219 hours).

Exhibit Dates: *Version 1:* 10/88–9/94. *Version 2:* 10/94–Present.

Learning Outcomes: *Version 1:* Upon completion of the course, the student will be able to perform the duties of a manager/maintenance test pilot, including repair and supply procedures; use of publications and logistics performance indicators; development of estimates and reports; use of forms and records; inspection of fuels, lubricants, hydraulic and electrical systems, flight controls, and power plants; performance analysis; rotor and power train systems; maintenance test flight; and evaluations. *Version 2:* Upon completion of the course, the student will be able to describe AH-64A maintenance troubleshooting and test flight procedures.

Instruction: *Version 1:* This course includes lectures, demonstrations, and performance exercises in maintenance officer responsibilities, logistics, aviation maintenance management, aeronautical technology, maintenance test flight, and quality assurance. *Version 2:* This course includes lectures, demonstrations, and performance exercises in maintenance officer responsibilities for the AH-64A.

Credit Recommendation: *Version 1:* In the lower-division baccalaureate/associate degree category, 3 semester hours in aviation maintenance management (3/88). *Version 2:* Credit is not recommended because of the limited, specialized nature of the course (11/98).

AR-1717-0094

MAINTENANCE MANAGEMENT

Course Number: 700.

Location: Reserve Readiness Training Center, Fort McCoy, WI.

Length: 2 weeks (64 hours).

Exhibit Dates: 9/86–12/92.

Learning Outcomes: Upon completion of the course, the student will be able to supervise and manage personnel responsible for customer service, requisition, issue and turn in, property records, and bulk petroleum accounts within proper authorization levels; supervise maintenance management systems, including motor pool operations, material management, production/quality control, and shop layout and design; manage and support personnel involved in records management, safety and security programs, physical safety of facilities, arms, ammunition, and key controls; and provide assistance in budget preparation, command maintenance policy directives, and force integration actions and requirements.

Instruction: Lectures and discussions cover rapport and cooperation with customers, requisition procedures, property records, authorization documents, petroleum accounting, maintenance management systems, motor pool operations, maintenance channels, production/quality control, records management procedures, time management, and effective communications skills. Practical exercises require the student to identify and correct a report of a survey and statement of charges; prepare a consolidated material condition status report; design a shop layout involving the flow of materials, efficient movement of personnel, and safety; and develop input into the budget system.

Credit Recommendation: In the lower-division baccalaureate/associate degree category, 3 semester hours in maintenance management (7/89).

AR-1717-0095

1. Staff/Support Maintenance Technician Senior Warrant Officer
2. Support/Staff Maintenance Technician

Course Number: *Version 1:* 4-9-C32-915E (CT). *Version 2:* 4-9-C32-915E.

Location: Ordnance Center and School, Aberdeen Proving Ground, MD.

Length: *Version 1:* 11 weeks (400 hours). *Version 2:* 12–13 weeks (393 hours).

Exhibit Dates: *Version 1:* 5/90–12/93. *Version 2:* 1/94–Present.

Learning Outcomes: *Version 1:* Upon completion of the course, the student will be able to manage support-level supply and equipment maintenance. *Version 2:* Upon completion of the course, the student will be able to manage support-level supply and equipment maintenance.

Instruction: *Version 1:* Lectures and practical exercises cover general military topics, communication, and professional development skills. Logistics topics include shop operations and maintenance management. Other subjects include the skills and knowledge required to properly use a service section, including machine shop and welding shop operations. Supervision and performance of maintenance on selected fuel and electrical system components and major assemblies are included, as is maintenance on selected heavy track vehicles and the supervision, adjustment, repair, and service of specific fighting vehicles. *Version 2:* Lectures and practical exercises cover general military topics, communication, and professional development skills. Logistics topics include shop operations and maintenance management. Other subjects include the skills and knowledge required to properly use a service section, including machine shop and welding shop operations. Supervision and performance of maintenance on selected fuel and electrical system components and major assemblies are included, as is maintenance on selected heavy track vehicles and the supervision, adjustment, repair, and service of specific fighting vehicles.

Credit Recommendation: *Version 1:* In the upper-division baccalaureate category, 2 semester hours in maintenance management (12/90). *Version 2:* In the upper-division baccalaureate category, 3 semester hours in maintenance management (8/03).

Related Occupations: 915E.

AR-1717-0096

1. Warrant Officer Technical/Tactical Certification (Armament Repair Technician Warrant Officer Technical/Tactical Certification)
2. Armament Repair Technician Warrant Officer Basic

Course Number: *Version 1:* 4E-913A. *Version 2:* 4E-913A.

Location: Ordnance Center and School, Aberdeen Proving Ground, MD.

Length: *Version 1:* 19 weeks (690 hours). *Version 2:* 14–15 weeks (530 hours).

Exhibit Dates: *Version 1:* 3/89–12/99. *Version 2:* 1/00–Present.

Learning Outcomes: *Version 1:* Upon completion of the course, the student will be able to assume the duties and responsibilities of managing armament repair facilities and mobile power generation facilities. *Version 2:* Upon completion of the course, the student will be able to assume the duties and responsibilities of managing armament repair facilities and mobile power generation facilities.

Instruction: *Version 1:* Instruction includes power generation, maintenance management, professional development subjects, communication skills, and hydraulics and electrical troubleshooting. *Version 2:* Topics include management and supervision of direct support maintenance, tank turrets, fire control, artillery, tow systems, and small arms materials.

Credit Recommendation: *Version 1:* In the lower-division baccalaureate/associate degree category, 2 semester hours in communication skills and 3 semester hours in electrical laboratory in industrial electricity. In the upper-division baccalaureate category, 3 semester hours in maintenance management (12/90). *Version 2:* In the upper-division baccalaureate category, 3 semester hours in maintenance management (8/03).

Related Occupations: 913A.

AR-1717-0097

Unit Maintenance Technician (Light) Warrant Officer Technical/Tactical Certification Reserve Component Phases 2 and 4

Course Number: 4L-915A-RC.

Location: Ordnance Center and School, Aberdeen Proving Ground, MD.

Length: 4 weeks (227 hours).

Exhibit Dates: 10/88–12/93.

Objectives: Upon completion of the course, the student will be able to perform the duties and responsibilities of maintenance management in wheeled vehicle repair facilities.

Instruction: Instruction includes a review of management principles, automotive electricity theory and application, power train theory, and basic welding. Phases 1 and 3 are by correspondence.

Credit Recommendation: In the upper-division baccalaureate category, for Phase 2, 1 semester hour in maintenance management. For Phase 4, 1 in maintenance management (12/90).

Related Occupations: 915A.

AR-1717-0098

Unit Maintenance Technician (Heavy) Senior Warrant Officer Reserve Component, Phase 2 and 4

Course Number: 4-9-C32-915D-RC.

Location: Ordnance Center and School, Aberdeen Proving Ground, MD.

Length: 4 weeks (222 hours).

Exhibit Dates: 6/90–12/93.

Learning Outcomes: Upon completion of the course, the student will be able to supervise maintenance operations on tanks under field conditions.

Instruction: Instruction methods include correspondence, lectures, shop repair procedures, and examinations. Topics covered include communication techniques; inventory control; and troubleshooting and repairing of heavy electrical, fuel, and hydraulic systems.

Credit Recommendation: In the upper-division baccalaureate category, for Phase 2, 1 semester hour maintenance management. For Phase 4, 1 in maintenance management (12/90).

Related Occupations: 915D.

AR-1717-0099

Allied Trades Technician Senior Warrant Officer

Course Number: 4-9-C32-914A (CT).

Location: Ordnance Center and School, Aberdeen Proving Ground, MD.

Length: 4–5 weeks (167 hours).

Exhibit Dates: 5/90–12/93.

Learning Outcomes: Upon completion of the course, the student will be able to manage allied trade specialties, supply management, and maintenance management.

Instruction: Lectures and practical exercises cover general military topics, communication skills, and professional development skills. Logistics topics include shop operations and maintenance management. Students are provided with information on maintenance trends in the metalworking field.

Credit Recommendation: In the upper-division baccalaureate category, 2 semester hours in maintenance management (12/90).

Related Occupations: 914A.

AR-1717-0100

Armament Repair Technician Senior Warrant Officer

Course Number: 4-9-C32-913A (CT).

Location: Ordnance Center and School, Aberdeen Proving Ground, MD.

Length: 5 weeks (185 hours).

Exhibit Dates: 5/90–4/04.

Learning Outcomes: Upon completion of the course, the student will be able to manage supply operations and armament maintenance including weapons systems.

Instruction: Lectures and practical exercises cover general military topics, communication skills, and professional development skills. Logistics topics include shop operations and maintenance management. Students are provided with information on the future of armament maintenance and equipment changes.

Credit Recommendation: In the upper-division baccalaureate category, 2 semester hours in vehicle maintenance management (12/90).

Related Occupations: 913A.

AR-1717-0101

Unit Maintenance Technician (Heavy) Senior Warrant Officer

Course Number: 4-9-C32-915D (CT).

Location: Ordnance Center and School, Aberdeen Proving Ground, MD.

Length: 10–11 weeks (369–370 hours).

Exhibit Dates: 5/90–12/93.

Learning Outcomes: Upon completion of the course, the student will be able to manage unit maintenance (heavy), logistics, and unit maintenance in mechanized infantry, armor battalions, and in cavalry squadrons.

Instruction: Lectures and practical exercises cover general military topics, communication skills, and professional development skills. Logistics topics include shop operations and maintenance management. Other subjects include skills required to supervise, troubleshoot, adjust, repair, and maintain specific combat tanks and fighting vehicles and combat vehicle small arms systems.

Credit Recommendation: In the upper-division baccalaureate category, 2 semester hours in maintenance management (12/90).

Related Occupations: 915D.

AR-1717-0102

Repair Shop Technician, Warrant Officer Technical/Tactical Certification, Phases 2 and 4

Course Number: 4L-914A-RC.

Location: Ordnance Center and School, Aberdeen Proving Ground, MD.

Length: 4 weeks (217 hours).

Exhibit Dates: 10/88–12/93.

Learning Outcomes: Upon completion of the course, the student will be able to manage maintenance on wheeled and tracked vehicles and armament.

Instruction: Instruction includes maintenance management, welding, and machining with lathes and milling machines.

Credit Recommendation: In the lower-division baccalaureate/associate degree category, for Phase 2, 2 semester hours in general machine tools. For Phase 4, 2 in general welding. In the upper-division baccalaureate category, 3 semester hours in maintenance management (12/90).

Related Occupations: 914A.

AR-1717-0103

Armament Repair Technician
(Armament Repair Technician Warrant Officer Technical/Tactical Certification, Phase 2 and 4)

Course Number: 4E-913A RC, Phase 4; 4E-913A-RC, Phase 2.

Location: Ordnance Center and School, Aberdeen Proving Ground, MD.

Length: Phase 2, 2 weeks (99–119 hours); Phase 4, 2 weeks (104–114 hours).

Exhibit Dates: 10/88–Present.

Learning Outcomes: Upon completion of the course, the student will be able to manage armament repair facilities.

Instruction: Instruction includes power supplies and rectifiers, basic digital circuits, and trouble shooting with oscilloscope.

Credit Recommendation: In the lower-division baccalaureate/associate degree category, for Phase 2,1 semester hour in DC circuits. Credit is not recommended for Phase 4 (3/03).

Related Occupations: 913A.

AR-1717-0104

Support Maintenance Technician Senior Warrant Officer Reserve Component, Phases 2 and 4

Course Number: 4-9-C32-915E-RC.

Location: Ordnance Center and School, Aberdeen Proving Ground, MD.

Length: Resident, 4 weeks (227 hours).

Exhibit Dates: 7/90–12/93.

Learning Outcomes: Upon completion of the course, the student will be able to supervise vehicle maintenance on tracked and wheeled vehicles.

Instruction: Instruction includes lectures, correspondence, shop practice, and written exams. Topics include military history, management and inventory control methods, and troubleshooting diesel and hydraulic systems.

Credit Recommendation: In the lower-division baccalaureate/associate degree category, for Phase 2, 1 semester hour in vehicle maintenance management. For Phase 4, 1 in vehicle maintenance management (12/90).

Related Occupations: 915E.

AR-1717-0105

Armament Repair Technician Senior Warrant Officer Reserve Component Phase 2

Course Number: 4-9-C32-913A.

Location: Ordnance Center and School, Aberdeen Proving Ground, MD.

Length: 2 weeks (114 hours).

Exhibit Dates: 11/89–12/93.

Learning Outcomes: Upon completion of the course, the student will be able to supervise the activities of armament repair technicians.

Instruction: Methods include lectures, demonstrations, and practical exercises. Topics covered include inventory control, logistics management, and communication skills.

Credit Recommendation: In the upper-division baccalaureate category, 1 semester hour in maintenance management (12/90).

Related Occupations: 913A.

AR-1717-0106

Repair Shop Technician Senior Warrant Officer Reserve Component

Course Number: 4-9-C32-914A-RC.

Location: Ordnance Center and School, Aberdeen Proving Ground, MD.

Length: 2 weeks (114 hours).

Exhibit Dates: 8/89–12/92.

Learning Outcomes: Upon completion of the course, the student will be able to supervise the activities of allied trades technicians.

Instruction: Methods of instruction include lectures, demonstrations, and practical exercise. Topics include training management, inventory control, and communication skills.

Credit Recommendation: In the upper-division baccalaureate category, 1 semester hour in maintenance management (12/90).

Related Occupations: 914A.

AR-1717-0107

Ordnance Officer Advanced Maintenance Management

Course Number: 4-9-C22-91B.

Location: Ordnance Center and School, Aberdeen Proving Ground, MD.

Length: 20 weeks (728 hours).

Exhibit Dates: 10/89–4/04.

Learning Outcomes: Upon completion of the course, the student will be able to manage maintenance companies in the areas of material development, testing, and procurement.

Instruction: Methods of instruction include lectures, demonstrations, field exercises, guest lectures, field trips, and graded examinations. Topics include material-handling flow charts, procurement systems from design through completion, purchasing systems, computer applications, and wholesale supply.

Credit Recommendation: In the upper-division baccalaureate category, 6 semester hours in maintenance management and 3 in logistics management. In the graduate degree category, 3 semester hours in management practicum (12/90).

Related Occupations: 91B.

AR-1717-0108

1. Test Measurement and Diagnostic Equipment (TMDE) Maintenance Support Technician Warrant Officer Technical and Tactical Certification
2. Testing and Measurement of Diagnostic Equipment (TMDE) Maintenance Support Technician Warrant Officer Technical and Tactical Certification

Course Number: *Version 1:* 4B-918A. *Version 2:* 4B-918A.

Location: *Version 1:* Missile and Munitions School, Redstone Arsenal, AL. *Version 2:* Ordnance, Missile and Munitions School, Redstone Arsenal, AL.

Length: *Version 1:* 15–16 weeks (560 hours). *Version 2:* 14–15 weeks (563 hours).

Exhibit Dates: *Version 1:* 10/90–1/92. *Version 2:* 2/92–6/00.

Learning Outcomes: *Version 1:* Upon completion of the course, the student will be able to demonstrate shop management skills and a general knowledge of tactical operations, logic, computers, and microprocessor concepts. *Version 2:* Upon completion of the course, the student will be able to demonstrate an understanding of digital logic circuits, microcomputer subsystems, microprocessors, microcomputer troubleshooting, use of test equipment, shop management skills, and general knowledge of tactical operations.

Instruction: *Version 1:* This course provides instruction in tactical certification subjects, maintenance management, logic, computers, and microprocessors. *Version 2:* Lectures and laboratory exercises cover binary arithmetic, logic functions, Boolean algebra, logic families, flip-flops, AD and DA converters, arithmetic circuits, decoders, encoders, multiplexers, and demultiplexers. Lectures and laboratory exercises also cover microcomputer architecture, programming, timing, interrupts, and troubleshooting. Practical training in shop management of test measurement and diagnostic equipment laboratory includes an understanding of calibration procedures, customer relations, and safety practice.

Credit Recommendation: *Version 1:* In the lower-division baccalaureate/associate degree category, 3 semester hours in maintenance (shop) management, 3 in digital logic, 2 in introduction to data processing, and 2 in microprocessors (8/91). *Version 2:* In the lower-division baccalaureate/associate degree category, 2 semester hours in digital logic, 2 in microcomputer operation and troubleshooting, 1 in digital and microcomputer laboratory, and 3 in maintenance management of electronic equipment (5/97).

Related Occupations: 918A.

AR-1717-0109

Test Measurement and Diagnostic Equipment (TMDE) Maintenance Support Technician Senior Warrant Officer

Course Number: 4-9-C32-918A.

Location: Missile and Munitions School, Redstone Arsenal, AL.

Length: 8–9 weeks (308–354 hours).

Exhibit Dates: 10/90–6/00.

Learning Outcomes: After 4/00 see AR-1717-0171. Upon completion of the course, the student will be able to make policy for company mission/activities; write calibration/maintenance directives; manage technology training programs; and perform tasks such as directing, evaluating, and reporting quality assurance/control, maintenance support, and supply systems.

Instruction: This course offers a balanced program of lectures and practical exercises in management duties and responsibilities, production, and control.

Credit Recommendation: In the lower-division baccalaureate/associate degree category, 2 semester hours in maintenance management (8/96).

Related Occupations: 918A.

AR-1717-0162

Apprentice Aircraft Metals Technology Specialist

Course Number: 702-45830.

Location: Ordnance Center and School, Aberdeen Proving Ground, MD.

Length: 7 weeks (288 hours).

Exhibit Dates: 9/89–4/04.

Learning Outcomes: Upon completion of the course, the student will be able to identify and use small metalworking hands tools; operate lathes, milling, and boring machines; use band saws and other specialized machine tools; design cutting tools; and use proper sharpening techniques.

Instruction: Lecture-demonstrations cover the safe use of metalworking hand tools and large machine tools. Students have extensive laboratory practice and guidance in machine tool use and cutting tool sharpening techniques.

Credit Recommendation: In the lower-division baccalaureate/associate degree category, 1 semester hour in basic metalworking, 6 in metal machine tools, and 2 in cutting tool design (12/90).

AR-1717-0163

Maintenance Manager

Course Number: *Version 1:* 2C-15D/4D-F6. *Version 2:* 2C-15D/4D-F6.

Location: *Version 1:* Aviation School, Fort Rucker, AL; Aviation Center, Fort Rucker, AL. *Version 2:* Aviation School, Fort Rucker, AL.

Length: *Version 1:* 9–10 weeks (366 hours). *Version 2:* 9–10 weeks (376 hours).

Exhibit Dates: *Version 1:* 10/94–12/96. *Version 2:* 1/97–Present.

Learning Outcomes: *Version 1:* Upon completion of the course, the student will be able to perform all the duties of a manager/maintenance test pilot, including repair and supply procedures; use of publications and logistics performance indicators; development of estimates and reports; use of forms and records; inspection of fuels, lubricants, hydraulic and electrical systems, flight controls, and power plants; performance analysis; rotor and power train systems; maintenance test flight; and evaluations. *Version 2:* Upon completion of the course, the student will be able to perform all the duties of a maintenance manager, including repair and supply procedures; use of publications and logistics performance indicators; development of estimates and reports; use of forms and records; inspection of fuels, lubricants, hydraulic and electrical systems, flight controls, and power plants; performance analysis; rotor and power train systems; maintenance test flight; and evaluations.

Instruction: *Version 1:* This course includes lectures, demonstrations, and performance exercises in maintenance officer responsibilities, including logistics, aviation maintenance management, aeronautical technology, maintenance test flight, and quality assurance. *Version 2:* This course includes lectures, demonstrations, and performance exercises in maintenance officer responsibilities, including logistics, aviation maintenance management, aeronautical technology, maintenance test flight, and quality assurance.

Credit Recommendation: *Version 1:* In the lower-division baccalaureate/associate degree category, 3 semester hours in aviation maintenance management (9/98). *Version 2:* In the lower-division baccalaureate/associate degree category, 3 semester hours in aviation maintenance management. In the upper-division baccalaureate category, 3 semester hours in logistics management (11/98).

AR-1717-0164

Shop Operations

Course Number: 1900.

Location: Reserve Readiness Training Center, Fort McCoy, WI.

Length: 2 weeks (71 hours).

Exhibit Dates: 11/96–Present.

Learning Outcomes: Upon completion of the course, the student will be able to perform basic administrative functions in Army shop operations and will perform safety and health functions in an Army shop.

Instruction: Methods include lectures, discussions, and a comprehensive in-basket exercise. Topics include maintenance administration, supply operations, physical security, safety, environmental issues, and maintenance programs.

Credit Recommendation: Credit is not recommended because of the military-specific nature of the course (3/97).

AR-1717-0165

Radio/COMSEC Repairer Basic Noncommissioned Officer (NCO)

Course Number: 101-35E30.

Location: Ordnance, Missile and Munitions School, Redstone Arsenal, AL.

Length: 10 weeks (390 hours).

Exhibit Dates: 10/94–6/00.

Learning Outcomes: Upon completion of the course, the student will be able to provide supervision and technical assistance in an electronic maintenance facility.

Instruction: Lectures, practical exercises, and small group instruction cover communications security, facility administration, maintenance management, and maintenance procedures. This course includes a common core of leadership subjects.

Credit Recommendation: In the lower-division baccalaureate/associate degree category, 3 semester hours in maintenance management. See AR-1406-0090 for common core credit (5/97).

Related Occupations: 35E.

AR-1717-0166

Tank/Automotive Materiel Management Officer Basic

Course Number: 4-9-C20-91B.

Location: Ordnance Center and School, Aberdeen Proving Ground, MD.

Length: 18–19 weeks (737 hours).

Exhibit Dates: 8/91–Present.

Learning Outcomes: Upon completion of the course, the student will be able to perform duties of a vehicle maintenance administrator including logistics, communication, and supervision and will have an understanding of automotive fundamentals.

Instruction: Lectures and performances are used in this course. Topics include wheel vehicle maintenance, metalworking procedures, supply management, and operation of automotive engines.

Credit Recommendation: In the lower-division baccalaureate/associate degree category, 1 semester hour in communication skills, 2 in automotive fundamentals, 3 in principles of supervision, and 3 in shop management (6/97).

AR-1717-0167

Combined Logistics Officer Advanced, Phase 2

Course Number: 4-9-C22-91B/C, Phase 2.

Location: Ordnance Center and School, Aberdeen Proving Ground, MD.

Length: 5 weeks (193 hours).

Exhibit Dates: 1/97–Present.

Learning Outcomes: Upon completion of the course, the student will be able to manage maintenance functions with emphasis on maintenance and ground/missile equipment.

Instruction: Lectures, student presentations, and group discussions cover maintenance procedures and the management of maintenance companies.

Credit Recommendation: In the upper-division baccalaureate category, 3 semester hours in maintenance management (1/98).

AR-1717-0168

Total Army Training System (TATS) Multiple Launch Rocket System (MLRS) Fire Direction Senior Sergeant Advanced Noncommissioned Officer (NCO), Phase 2

Course Number: 061-13P40 (F), Phase 2.

Location: Reserve Component Training Centers, US; Field Artillery School, Fort Sill, OK.

Length: 96–104 hours.

Exhibit Dates: 1/98–Present.

Learning Outcomes: Upon completion of the course, the student will be able to compute firing data, select communications equipment, write target/mission reports, and lead a section.

Instruction: Lectures and practical exercises cover the supervision, operation, and maintenance of the fire direction system.

Credit Recommendation: In the lower-division baccalaureate/associate degree category, 1 semester hour in maintenance management and 1 in safety (9/03).

Related Occupations: 13P.

AR-1717-0169

Total Army Training System (TATS) Multiple Launch Rocket System (MLRS) Fire Direction Senior Sergeant Advanced Noncommissioned Officer (NCO)

Course Number: 043-13P40 (F), Phase 2.

Location: Field Artillery School, Fort Sill, OK.

Length: 3–4 weeks (104–179 hours).

Exhibit Dates: 1/98–Present.

Learning Outcomes: Upon completion of the course, the student will be able to compute firing data, select communications equipment, write target/mission reports, and lead a section.

Instruction: Lectures and practical exercises cover the supervision, operation, and maintenance of the fire direction system. Course includes a common core of leadership subjects.

Credit Recommendation: In the lower-division baccalaureate/associate degree category, 1 semester hour in maintenance management and 1 in safety. See AR-1404-0035 for common core credit (9/03).

Related Occupations: 13P.

AR-1717-0170

Warrant Officer Advanced 918D, Phase 1

Course Number: 4-9-C32-918D, Phase 1.

Location: Ordnance, Missile and Munitions School, Redstone Arsenal, AL.

Length: 4–5 weeks (160 hours).

Exhibit Dates: 5/00–Present.

Learning Outcomes: Upon completion of the course, the student will be able to manage missile maintenance.

Instruction: Instruction is provided by using lectures, conferences, and practical exercises. Topics covered in this course include equal opportunity/sexual harassment, critical thinking skills, decision-making skills, the Army's maintenance information system, logistics management, and a weapons system update.

Credit Recommendation: In the lower-division baccalaureate/associate degree category, 3 semester hours in maintenance management (6/01).

Related Occupations: 918D.

AR-1717-0171

Warrant Officer Advanced 918B, Phase 1

Course Number: 4-9-C32-918B, Phase 1.

Location: Ordnance, Missile and Munitions School, Redstone Arsenal, AL.

Length: 6 weeks (223 hours).

Exhibit Dates: 5/00–Present.

Learning Outcomes: Before 5/00 see AR-1717-0109. Upon completion of the course, the student will be able to make policy for company missions/activities; write calibration/maintenance directives; manage technology training programs; and perform tasks such as directing, evaluating, and reporting quality assurance/control, maintenance support, and supply systems.

Instruction: This course offers a balanced program of lectures and practical exercises cover management duties and responsibilities, production, and control.

Credit Recommendation: In the lower-division baccalaureate/associate degree category, 2 semester hours in maintenance management (6/01).

Related Occupations: 918B.

AR-1717-0172

M1A2/M1A2 Sep Abrams Armor Crewman Basic Noncommissioned Officer (NCO), Phase 2

Course Number: 020-19K30 (M1A2), Phase 2.

Location: Armor Center and School, Fort Knox, KY.

Length: 6–7 weeks (362 hours).

Exhibit Dates: 4/00–Present.

Learning Outcomes: Upon completion of the course, the student will be technically proficient in the oversight of all aspects of tank crew maintenance, training, fighting, and sustaining performance.

Instruction: Lectures and combat-simulated tactical exercises and instruction in armor tactics, secure communications, maintenance, tank gunnery, mine warfare, tank weapons, safety, physical fitness, and situational training. The small-group instruction model is chiefly utilized.

Credit Recommendation: In the lower-division baccalaureate/associate degree category, 3 semester hours in maintenance management and 3 in military tactics (6/01).

AR-1717-0173

Railway Equipment Repairer

Course Number: 690-88P10-RC.

Location: Transportation School, Fort Eustis, VA.

Length: 5 weeks (176 hours).

Exhibit Dates: 10/00–Present.

Learning Outcomes: Upon completion of the course, the student will be able to perform maintenance and preventive maintenance on locomotives, including diesel engines and electrical systems.

Instruction: Methods of instruction include classroom conferences, practical activities, and written exams.

Credit Recommendation: In the lower-division baccalaureate/associate degree category, 1 semester hour in electrical systems maintenance and 3 in diesel engine maintenance (6/01).

Related Occupations: 88P.

AR-1717-0174

Total Army Training System (TATS) Aircraft Powerplant Repairer Supervisor Basic Noncommissioned Officer (NCO)

Course Number: 601-68B30, Phase 2.

Location: Aviation Logistics School, Fort Eustis, VA.

Length: 7–8 weeks (301 hours).

Exhibit Dates: 10/00–Present.

Learning Outcomes: Before 10/00 see AR-1704-0218. Upon completion of the course, the student will be able to apply principles of supervision to the management of an aircraft turbine engine overhaul facility.

Instruction: Methods of instruction include lectures interactive and practical application situations. Student evaluation is based on written and practical testing methods.

Credit Recommendation: In the lower-division baccalaureate/associate degree category, 2 semester hours in aircraft maintenance supervision and 3 in aircraft maintenance management and inspection (6/01).

Related Occupations: 68B.

AR-1717-0175

Total Army Training System Aircraft Powerplant Repairer Supervisor Basic Noncommissioned Officer (NCO), Phase 2

Course Number: 552-CMF67-(AM-Ph2), Phase 2.

Location: Aviation Logistics School, Fort Eustis, VA.

Length: 7 weeks (129 hours).

Exhibit Dates: 10/00–Present.

Learning Outcomes: Before 10/00 see AR-1704-0218. Upon completion of the course, the student will be able to apply principles of supervision in the management of a aircraft turbine engine overhaul facility.

Instruction: Methods of instruction include lecture, interactive and practical application situations. Student evaluation is based on written and practical testing methods. This includes only Module A.

Credit Recommendation: In the lower-division baccalaureate/associate degree category, 2 semester hours in aircraft maintenance supervision (6/01).

AR-1717-0176

Total Army Training System Aircraft Powerplant Repairer Supervisor Basic Noncommissioned Officer (NCO)

Course Number: 552-68B30, Phase 3.

Location: Aviation Logistics School, Fort Eustis, VA.

Length: 172 hours.

Exhibit Dates: 10/00–Present.

Learning Outcomes: Upon completion of the course, the student will be able to supervise shop maintenance and analyze engine performance on the T-5s and T 703-AD helicopter turbine engines. Student will also supervise maintenance on auxiliary power units.

Instruction: Methods of instruction include lecture and practical instruction on turbine engine shop supervision. Student evaluation is based on written and practical testing methods. This includes only Module B.

Credit Recommendation: In the lower-division baccalaureate/associate degree category, 3 semester hours in aircraft maintenance management and inspection (6/01).

Related Occupations: 68B.

AR-1717-0177

Total Army Training System CH-47 Helicopter Repairer Supervisor Basic Noncommissioned Officer (NCO), Phase 2

Course Number: 600-67U30 (F), Phase 2.
Location: Aviation Logistics School, Fort Eustis, VA.
Length: 11 weeks (424 hours).
Exhibit Dates: 10/00–Present.
Learning Outcomes: Before 10/00 see AR-1704-0151. Upon completion of the course, the student will be able to inspect and supervise the repair of CH-47 systems, including powerplant and powertrain, structural components, hydraulic systems, electrical systems, fuel systems, and the auxiliary power unit.
Instruction: Methods of instruction include student conferences, small group activities, and practical exercises directed at leadership development. Evaluation is based written examinations.
Credit Recommendation: In the lower-division baccalaureate/associate degree category, 2 semester hours in aircraft maintenance supervision and 4 in aircraft maintenance management and inspection (6/01).
Related Occupations: 67U.

AR-1717-0178

Total Army Training System (TATS) CH-47 Helicopter Repairer Supervisor Basic Noncommissioned Officer (NCO), Phase 3

Course Number: 552-67U30, Phase 3.
Location: Aviation Logistics School, Fort Eustis, VA.
Length: 115 hours.
Exhibit Dates: 10/00–Present.
Learning Outcomes: Before 10/00 see AR-1704-0151. Upon completion of the course, the student will be able to inspect the powerplant, powertrain, structures, hydraulic, flight, and fuel systems of the CH-47 helicopter.
Instruction: Methods of instruction include student conferences and practical exercises. Evaluation is based written examinations.
Credit Recommendation: In the lower-division baccalaureate/associate degree category, 2 semester hours in aircraft maintenance management and inspection (6/01).
Related Occupations: 67U.

AR-1717-0179

Total Army Training System (TATS) CH-47 Helicopter Repairer Supervisor Basic Noncommissioned Officer (NCO), Phase 4

Course Number: 552-67U30, Phase 4.
Location: Aviation Logistics School, Fort Eustis, VA.
Length: 120 hours.
Exhibit Dates: 10/00–Present.
Learning Outcomes: Before 10/00 see AR-1704-0151. Upon completion of the course, the student will be able to supervise and inspect aircraft maintenance of the engine, landing gear, hydraulic, electrical, fuel, flight control, and auxiliary power systems of the CH-47 helicopter.
Instruction: Methods of instruction include student conferences and practical exercises in the classroom and maintenance shop. Evaluation is based written examinations.
Credit Recommendation: In the lower-division baccalaureate/associate degree category, 2 semester hours in aircraft maintenance management and inspection (6/01).
Related Occupations: 67U.

AR-1717-0180

Total Army Training System (TATS) AH-64A Attack Helicopter Repairer Supervisor Basic Noncommissioned Officer (NCO), Phase 2

Course Number: 600-67R30 (AH-64A) (F), Phase 2.
Location: Aviation Logistics School, Fort Eustis, VA.
Length: 11–12 weeks (453 hours).
Exhibit Dates: 10/00–Present.
Learning Outcomes: Before 10/00 see AR-1717-0090. Upon completion of the course, the student will be able to supervise helicopter inspection and maintenance using computer-based information and data recording for quality and production control. The student will be able to inspect airframe, powerplant, rotors, and all associated systems. The student will be able to use nondestructive and computer-aided inspection methods to determine the condition of an AH-64A helicopter.
Instruction: Methods of instruction include lecture and practical experience. Student evaluation is based on written and practical testing methods.
Credit Recommendation: In the lower-division baccalaureate/associate degree category, 4 semester hours in aviation maintenance management and inspection and 4 in aircraft maintenance supervision (6/01).
Related Occupations: 67R.

AR-1717-0181

Total Army Training System AH-64A Attack Helicopter Repairer Supervisor Basic Noncommissioned Officer (NCO), Phase 2

Course Number: 552-CMF67-CAM PH2, Phase 2.
Location: Aviation Logistics School, Fort Eustis, VA.
Length: 129 hours.
Exhibit Dates: 10/00–Present.
Learning Outcomes: Before 10/00 see AR-1717-0090. Upon completion of the course, the student will be able to perform supervisory and technical inspector duties on AH-64A aircraft including maintenance, inspection, quality control, supply acquisition, and product control documentation (forms and records).
Instruction: Methods of instruction include classroom instruction and classroom exercises.
Credit Recommendation: In the lower-division baccalaureate/associate degree category, 2 semester hours in aircraft maintenance supervision (6/01).
Related Occupations: 67R.

AR-1717-0182

Total Army Training System (TATS) AH-64A Attack Helicopter Repairer Supervisor Basic Noncommissioned Officer (NCO), Phase 4

Course Number: 552-67R30 (AH-64A), Phase 4.
Location: Aviation Logistics School, Fort Eustis, VA.
Length: 236 hours.
Exhibit Dates: 10/00–Present.
Learning Outcomes: Before 10/00 see AR-1717-0090. Upon completion of the course, the student will be able to supervise, troubleshoot and inspect the landing gear, powerplant, rotors, flight control, hydraulic, pneumatic, environmental control, fuel, electrical, and avionics systems on the AH-64A helicopter.
Instruction: Methods of instruction include lecture and practical laboratory experience. Written examinations are used for student evaluation.
Credit Recommendation: In the lower-division baccalaureate/associate degree category, 2 semester hours in aviation maintenance management and inspection and 2 in aircraft maintenance supervision (6/01).
Related Occupations: 67R.

AR-1717-0183

Total Army Training System (TATS) AH-64A Attack Helicopter Repairer Supervisor Basic Noncommissioned Officer (NCO), Phase 3

Course Number: 552-67R30 (AH-64A) (F), Phase 3.
Location: Aviation Logistics School, Fort Eustis, VA.
Length: 88 hours.
Exhibit Dates: 10/00–Present.
Learning Outcomes: Before 10/00 see AR-1717-0090. Upon completion of the course, the student will be able to inspect the engine, structure, bonded panels, rotors, electric wiring, and battery of the AH-64 helicopter. Students will utilize non-destructive testing methods for inspection. Students will troubleshoot the T700-GE-701 engine.
Instruction: Methods of instruction include lecture and practical laboratory exercises. Student evaluation is based on written and practical testing methods.
Credit Recommendation: In the lower-division baccalaureate/associate degree category, 2 semester hours in aviation maintenance management and inspection (6/01).
Related Occupations: 67R.

AR-1717-0184

Total Army Training System (TATS) UH-60 Helicopter Repairer Supervisor Basic Noncommissioned Officer (NCO), Phase 2

Course Number: 600-67T30 (F), Phase 2.
Location: Aviation Logistics School, Fort Eustis, VA.
Length: 11–12 weeks (433 hours).
Exhibit Dates: 10/00–Present.
Learning Outcomes: Before 10/00 see AR-1704-0150. Upon completion of the course, the student will be able to provide supervision and technical inspections of Army aircraft and maintenance on the UH-50 aircraft.
Instruction: Methods of instruction include classroom conferences, small groups, and practical exercises.
Credit Recommendation: In the lower-division baccalaureate/associate degree category, 2 semester hours in aircraft maintenance management and inspection and 4 in aircraft maintenance supervision (6/01).
Related Occupations: 67T.

AR-1717-0185

Total Army Training System (TATS) UH-60 Helicopter Repairer Supervisor Basic Noncommissioned Officer (NCO), Phase 3

Course Number: 600-67T30 (F), Phase 3.
Location: Aviation Logistics School, Fort Eustis, VA.
Length: 3 weeks (128 hours).
Exhibit Dates: 10/00–Present.
Learning Outcomes: Before 10/00 see AR-1704-0150. Upon completion of the course, the student will be able to supervise the inspection of UH-60 systems, including powertrain, structure, pneudraulic, propulsion, and electrical systems.
Instruction: Methods of instruction include student conferences, demonstrations, and practical exercises in the classroom and shop. Evaluation is accomplished by written and practical examinations.
Credit Recommendation: In the lower-division baccalaureate/associate degree category, 2 semester hours in aircraft maintenance management and inspection (6/01).
Related Occupations: 67T.

AR-1717-0191

Total Army Training System (TATS) UH-1 Helicopter Repairer Supervisor Basic Noncommissioned Officer (NCO), Phase 2

Course Number: 600-67N30 (F), Phase 2.
Location: Aviation Logistics School, Fort Eustis, VA.
Length: 11 weeks (411 hours).
Exhibit Dates: 10/00–9/03.
Learning Outcomes: Upon completion of the course, the student will be able to perform supervisory and technical inspections of UH-1 helicopter repairs; apply basic principles of aircraft maintenance management, including Army-specific procedures, safety programs, and quality control; and perform troubleshooting and inspection procedures on airframe and powertrain components.
Instruction: Methods of instruction include conferences, demonstrations, and practical exercises. Computer presentations, slide presentations, overhead projection, and video are used to provide instruction. Written and practical examinations are given throughout the course.
Credit Recommendation: In the lower-division baccalaureate/associate degree category, 4 semester hours in aircraft maintenance management and inspection and 2 in aircraft maintenance supervision (6/01).
Related Occupations: 67N.

AR-1717-0192

Total Army Training System (TATS) UH-1 Helicopter Repairer Supervisor Basic Noncommissioned Officer (NCO), Phase 2

Course Number: 552-CMF67-CAM-PH2, Phase 2.
Location: Aviation Logistics School, Fort Eustis, VA.
Length: 3 weeks (129 hours).
Exhibit Dates: 10/00–Present.
Learning Outcomes: Upon completion of the course, the student will be able to function as a supervisor and technical inspector on the UH-1 helicopter, including policies, procedures, safety, and quality control.
Instruction: Methods of instruction include conferences, practical exercises, and small-group activities. Evaluation is based on written examinations.
Credit Recommendation: In the lower-division baccalaureate/associate degree category, 2 semester hours in aircraft maintenance supervision (6/01).
Related Occupations: 67N.

AR-1717-0193

Total Army Training System (TATS) UH-1 Helicopter Repairer Supervisor Basic Noncommissioned Officer (NCO), Phase 4

Course Number: 552-67N30 (F), Phase 4.
Location: Aviation Logistics School, Fort Eustis, VA.
Length: 3 weeks (94 hours).
Exhibit Dates: 10/00–Present.
Learning Outcomes: Upon completion of the course, the student will be able to supervise and inspect UH-1 aircraft maintenance, including drivetrain, flight control, fuel, hydraulic and tail rotor system.
Instruction: Methods of instruction include conferences, demonstrations, and practical exercises. Evaluation is based on written examinations. Topics also include principles of weight and balance.
Credit Recommendation: In the lower-division baccalaureate/associate degree category, 1 semester hour in aircraft maintenance management and inspection (6/01).
Related Occupations: 67N.

AR-1717-0194

Morse Interceptor Basic Noncommissioned Officer (NCO)

Course Number: 231-98H30.
Location: Intelligence School, Fort Huachuca, AZ.
Length: 7–8 weeks (278–279 hours).
Exhibit Dates: 10/98–Present.
Learning Outcomes: Upon completion of the course, the student will be able to provide the leadership required to provide electronic warfare and signal intelligence to users.
Instruction: Lectures, practical exercises, demonstrations, map reading, and target identification are course methods. Course includes a common core of leadership subjects.
Credit Recommendation: Credit is recommended for the common core only. See AR-1406-0090 (8/01).
Related Occupations: 98H.

AR-1717-0195

OH-58D (R) Warrior Maintenance Test Pilot

Course Number: 4D-SIG6/SQIG (OH-58D/R).
Location: Aviation Center, Fort Rucker, AL.
Length: 6–7 weeks (292 hours).
Exhibit Dates: 10/02–Present.
Learning Outcomes: Upon completion of the course, the student will be able to qualify as a maintenance test pilot for the OH-58D/R helicopter, use proper procedures to analyze various aircraft maintenance problems, and prescribe proper corrective actions.
Instruction: Practical and classroom exercises, laboratory, and lecture. Topics include aircraft systems familiarization, and maintenance test flight training.
Credit Recommendation: Credit is not recommended because of the limited, specialized nature of the course (4/03).

AR-1717-0196

AH-64 Maintenance Test Pilot

Course Number: 4D-SIG6/SQIG (AH-64), Phase 1.
Location: Aviation Center, Fort Rucker, AL.
Length: 5–6 weeks (232 hours).
Exhibit Dates: 10/99–Present.
Learning Outcomes: Upon completion of the course, the student will be able to qualify as a maintenance test pilot for the AH-64 helicopter, use proper procedures to analyze various aircraft maintenance problems, and prescribe proper corrective actions.
Instruction: Practical and classroom exercises, laboratory, and lecture. Topics include aircraft systems familiarization, and maintenance test flight training.
Credit Recommendation: Credit is not recommended because of the limited, specialized nature of the course (4/03).

AR-1717-0197

AH-64D Maintenance Test Pilot (Supplemental)

Course Number: 4D-SIG6/SQIG (AH-64D), Phase 2.
Location: Aviation Center, Fort Rucker, AL.
Length: 2 weeks (68 hours).
Exhibit Dates: 10/99–Present.
Learning Outcomes: Upon completion of the course, the student will be able to qualify as a maintenance test pilot for the AH-64D helicopter, use proper procedures to analyze various aircraft maintenance problems, and prescribe proper corrective actions.
Instruction: Practical and classroom exercises, laboratory, and lecture. Topics include aircraft systems familiarization, and maintenance test flight training.
Credit Recommendation: Credit is not recommended because of the limited, specialized nature of the course (4/03).

AR-1717-0198

AH-64D Maintenance Test Pilot

Course Number: 4D-SIG6/SQIC (AH-64D).
Location: Aviation Center, Fort Rucker, AL.
Length: 6 weeks (236 hours).
Exhibit Dates: 4/00–Present.
Learning Outcomes: Upon completion of the course, the student will be able to qualify as a maintenance test pilot for the AH-64D helicopter, use proper procedures to analyze various aircraft maintenance problems, and prescribe proper corrective actions.
Instruction: Practical and classroom exercises, laboratory, and lecture. Topics include aircraft systems familiarization, and maintenance test flight training.
Credit Recommendation: Credit is not recommended because of the limited, specialized nature of the course (4/03).

AR-1717-0199

COMBINED LOGISTICS CAPTAINS CAREER COURSE

Course Number: 4-9-C22-91A, Phase 2.
Location: Ordnance Center and School, Aberdeen Proving Ground, MD.
Length: 5 weeks (187 hours).
Exhibit Dates: 6/03–Present.
Learning Outcomes: Upon completion of the course, the student will be able to effectively command maintenance and ammunition companies, maintain inventory control systems, and train subordinates in organization and leadership techniques.
Instruction: Computer-based training, practical exercises, classroom exercises, discussion, and lecture. Topics include management techniques, inventory control, problem-solving and assessment techniques, environmental compliance agency rules and regulations regarding refrigerants and other substances.
Credit Recommendation: In the upper-division baccalaureate category, 2 semester hours in maintenance management and 2 in materials handling (8/03).

AR-1717-0200

SUPPORT/STAFF MAINTENANCE TECHNICIAN (RESERVE COMPONENT)

Course Number: 4-9-C32-915E RC, Phase 2.
Location: Ordnance Center and School, Aberdeen Proving Ground, MD.
Length: 2 weeks (110 hours).
Exhibit Dates: 1/94–Present.
Learning Outcomes: Upon completion of the course, the student will be able to troubleshoot and repair various systems on Army field-tracked and wheeled vehicles.
Instruction: Practical exercises, classroom exercises, discussion, laboratory, and lecture. Topics include troubleshooting and repair of electrical systems, hydraulics, fire suppression, and turbine engines.
Credit Recommendation: In the lower-division baccalaureate/associate degree category, 3 semester hours in vehicle maintenance (8/03).

AR-1717-0201

ALLIED TRADES TECHNICIAN

Course Number: 4L-914A RC, Phase 2.
Location: Ordnance Center and School, Aberdeen Proving Ground, MD.
Length: 2–3 weeks (98 hours).
Exhibit Dates: 1/94–Present.
Learning Outcomes: Upon completion of the course, the student will be able to manage maintenance on wheeled and tracked vehicles and armament.
Instruction: Instruction includes maintenance management, welding, and machining with lathes and milling machines.
Credit Recommendation: In the lower-division baccalaureate/associate degree category, 2 semester hours in general machine tools (8/03).

AR-1717-0202

ALLIED TRADES TECHNICIAN

Course Number: 4L-914A RC, Phase 4.
Location: Ordnance Center and School, Aberdeen Proving Ground, MD.
Length: 2 weeks (98 hours).
Exhibit Dates: 1/94–Present.
Learning Outcomes: Upon completion of the course, the student will be able to manage maintenance on wheeled and tracked vehicles and armament.
Instruction: Instruction includes maintenance management, welding, and machining with lathes and milling machines.
Credit Recommendation: In the lower-division baccalaureate/associate degree category, 2 semester hours in general welding. In the upper-division baccalaureate category, 3 semester hours in maintenance management (8/03).

AR-1717-0203

ARMAMENT REPAIR TECHNICIAN WARRANT OFFICER ADVANCED

Course Number: 4-9-C32-913A, Phase 1.
Location: Ordnance Center and School, Aberdeen Proving Ground, MD.
Length: 4–5 weeks (146 hours).
Exhibit Dates: 10/99–Present.
Learning Outcomes: Upon completion of the course, the student will be able to develop and implement a plan for Army family team building and manage a direct support maintenance forward support mission.
Instruction: Practical exercises, classroom exercises, discussion, laboratory, and lecture. Topics include training plan development, management of a company, and management of a site and mission.
Credit Recommendation: In the upper-division baccalaureate category, 3 semester hours in organizational management (8/03).

AR-1717-0204

ARMAMENT REPAIR TECHNICIAN WARRANT OFFICER ADVANCED

Course Number: 091-4-9-C32-913A, Phase 1.
Location: Reserve Schools, Multiple locations.
Length: 94 hours.
Exhibit Dates: 10/99–Present.
Learning Outcomes: Upon completion of the course, the student will be able to develop and implement a plan for Army team building.
Instruction: Practical exercises, classroom exercises, discussion, laboratory, and lecture. Topics include training plan development and management of a company.
Credit Recommendation: In the upper-division baccalaureate category, 3 semester hours in organizational management (8/03).

AR-1717-0205

ARMAMENT REPAIR TECHNICIAN WARRANT OFFICER ADVANCED

Course Number: 091-4-9-C32-913A, Phase 3.
Location: Reserve Schools, Multiple locations.
Length: 99 hours.
Exhibit Dates: 10/99–Present.
Learning Outcomes: Upon completion of the course, the student will be able to maintain and supervise various standard Army maintenance systems.
Instruction: Practical exercises, classroom exercises, discussion, laboratory, and lecture. Topics include maintenance and supervision of several maintenance operations.
Credit Recommendation: In the upper-division baccalaureate category, 2 semester hours in electromechanical maintenance supervision (8/03).

AR-1717-0206

ALLIED TRADES REPAIR TECHNICIAN WARRANT OFFICER ADVANCED

Course Number: 091-4-9-C32-914A, Phase 1.
Location: Reserve Schools, Multiple locations.
Length: 94 hours.
Exhibit Dates: 10/99–Present.
Learning Outcomes: Upon completion of the course, the student will be able to develop and implement a plan for Army team building.
Instruction: Practical exercises, classroom exercises, discussion, laboratory, and lecture. Topics include training plan development and management of a company.
Credit Recommendation: In the upper-division baccalaureate category, 3 semester hours in organizational management (8/03).

AR-1717-0207

ALLIED TRADES REPAIR TECHNICIAN WARRANT OFFICER ADVANCED

Course Number: 091-4-9-C32-914A, Phase 2.
Location: Reserve Schools, Multiple locations.
Length: 117 hours.
Exhibit Dates: 10/99–Present.
Learning Outcomes: Upon completion of the course, the student will be able to understand conditions and procedures for vehicle recovery, understand CNC programming and its applications in recovery operations, and become a senior warrant officer.
Instruction: Practical exercises, classroom exercises, discussion, laboratory, and lecture. Topics include vehicle recovery, CNC programming, and logistics management.
Credit Recommendation: In the lower-division baccalaureate/associate degree category, 2 semester hours in CNC programming. In the upper-division baccalaureate category, 2 semester hours in supply management (8/03).

AR-1717-0208

ALLIED TRADES REPAIR TECHNICIAN WARRANT OFFICER ADVANCED

Course Number: 4-9-C32-914A, Phase 2.
Location: Ordnance Center and School, Aberdeen Proving Ground, MD.
Length: 3–4 weeks (117 hours).
Exhibit Dates: 10/99–Present.
Learning Outcomes: Upon completion of the course, the student will be able to understand conditions and procedures for vehicle recovery, understand CNC programming and its application in recovery operations, and become a senior warrant officer.
Instruction: Practical exercises, classroom exercises, discussion, laboratory, and lecture. Topics include vehicle recovery, CNC programming, and logistics management.
Credit Recommendation: In the lower-division baccalaureate/associate degree category, 2 semester hours in CNC programming.

In the upper-division baccalaureate category, 2 semester hours in supply management (8/03).

AR-1717-0209

Allied Trade Technician Warrant Advanced

Course Number: 4-9-C32-914A RC, Phase 2.

Location: Ordnance Center and School, Aberdeen Proving Ground, MD.

Length: 2 weeks (105 hours).

Exhibit Dates: 1/93–Present.

Learning Outcomes: Upon completion of the course, the student will be able to supervise and direct retail repair parts operations and perform various machining operations.

Instruction: Practical exercises, classroom exercises, discussion, laboratory, and lecture. Topics include various machining operations utilizing different machine tools, supervision of retail repair parts operation, and packaging and crating.

Credit Recommendation: In the lower-division baccalaureate/associate degree category, 3 semester hours in basic machine shop operation (8/03).

AR-1717-0210

Allied Trades Repair Technician Warrant Officer Advanced

Course Number: 4-9-C32-914A, Phase 1.

Location: Ordnance Center and School, Aberdeen Proving Ground, MD.

Length: 2–3 weeks (94 hours).

Exhibit Dates: 10/99–Present.

Learning Outcomes: Upon completion of the course, the student will be able to develop and implement a plan for Army family team building.

Instruction: Practical and classroom exercises, discussion, laboratory, and lecture. Topics include training plan development and management of a company.

Credit Recommendation: In the upper-division baccalaureate category, 3 semester hours in organizational management (8/03).

AR-1717-0211

Allied Trades Technician Warrant Advanced

Course Number: 4-9-C32-914A RC, Phase 2.

Location: Ordnance Center and School, Aberdeen Proving Ground, MD.

Length: 2 weeks (105 hours).

Exhibit Dates: 1/93–Present.

Learning Outcomes: Upon completion of the course, the student will be able to supervise and direct retail repair parts operations and perform various machining operations.

Instruction: Practical and classroom exercises, discussion, laboratory, and lecture. Topics include various machining operations utilizing different machine tools, supervision of retail repair parts operation, and packaging and crating.

Credit Recommendation: In the lower-division baccalaureate/associate degree category, 3 semester hours in basic machine shop operation (8/03).

AR-1717-0212

TATS Aircraft Powerplant Repair Supervisor Basic Noncommissioned Officer

Course Number: 601-68B30-RC (F), Phase 3.

Location: Reserve Schools, Multiple locations.

Length: 172 hours.

Exhibit Dates: 10/00–Present.

Learning Outcomes: Upon completion of the course, the student will be able to supervise and manage turbine engine maintenance at the unit and intermediate maintenance levels.

Instruction: Practical exercises and lecture. Topics include supervising powerplant shop operations and turbine engine maintenance.

Credit Recommendation: In the lower-division baccalaureate/associate degree category, 3 semester hours in aviation maintenance management (12/03).

AR-1719-0004

Visual Information/Audio Documentation Systems Specialist Basic Noncommissioned Officer (NCO)

Course Number: ***Version 1:*** G3AZR23270 001 (25P30); G3AZR23270 001. *Version 2:* G3AZR23270 001 (25P30); G3AZR23270 001.

Location: Technical Training Center, Lowry AFB, CO.

Length: ***Version 1:*** 6–7 weeks (216–257 hours). *Version 2:* 7–8 weeks (290 hours).

Exhibit Dates: ***Version 1:*** 10/88–9/91. *Version 2:* 10/91–9/94.

Learning Outcomes: ***Version 1:*** Upon completion of the course, the student will be able to manage a media system or a team involved in film editing, writing, and other critical graphic arts; photographic skills; and maintenance skills. *Version 2:* Upon completion of the course, the student will be able to manage media systems or teams involved in film editing, writing, and other critical graphic arts; photographic skills; and maintenance skills.

Instruction: ***Version 1:*** Lectures and practical exercises cover management of television production, motion picture production, and other audiovisual operations. Course also includes a common core of leadership subjects. *Version 2:* Lectures, demonstrations, and practical exercises cover management of television production, graphic facilities, television maintenance, computer technology, photographic facilities, and other audiovisual operations.

Credit Recommendation: ***Version 1:*** In the lower-division baccalaureate/associate degree category, see AR-1406-0090 for common core credit. In the upper-division baccalaureate category, 3 semester hours in audiovisual management (5/90). *Version 2:* In the upper-division baccalaureate category, 3 semester hours in audiovisual management (10/94).

Related Occupations: 25P.

AR-1719-0005

Visual Information Advanced Noncommissioned Officer (NCO)

Course Number: ***Version 1:*** G3AAR23199 000. *Version 2:* 4-84-C42/G3AAR23199 000.

Location: Technical Training Center, Lowry AFB, CO.

Length: ***Version 1:*** 9–10 weeks (358 hours). *Version 2:* 11 weeks (417 hours).

Exhibit Dates: ***Version 1:*** 4/89–9/91. *Version 2:* 10/91–Present.

Learning Outcomes: ***Version 1:*** Upon completion of the course, the student will be able to lead and train others in planning, supervising, and directing the operation of visual information facilities and organizations, including photographic, television, documentation, and graphic production operations. *Version 2:* Upon completion of the course, the student will be able to lead and manage visual information facilities and organizations, including training, planning, supervising, and directing visual information documentation operations.

Instruction: ***Version 1:*** Lectures and practical exercises cover the management of motion picture production, visual aids preparation, computer graphics, photographic procedures, still video camera systems, historical documentation procedures, and maintenance facilities. This course also includes a common core of leadership subjects. *Version 2:* Lectures, demonstrations, and practical exercises are the primary forms of instruction. Topics covered include computer training basic management of graphics, still documentation, motion media production, maintenance, and combat camera operations. The course also includes management of production functions, including visual aids preparation, computer graphics, traditional photographic procedures, still video camera systems, and historical documentation procedures.

Credit Recommendation: ***Version 1:*** In the lower-division baccalaureate/associate degree category, see AR-1404-0035 for common core credit. In the upper-division baccalaureate category, 3 semester hours in audiovisual management (5/90). *Version 2:* In the upper-division baccalaureate category, 3 semester hours in audiovisual management (7/94).

AR-1720-0001

Marine Corps Nuclear, Biological, Chemical (NBC) Defense

Course Number: 2G-5702; 494-5711.

Location: Chemical School, Fort McClellan, AL.

Length: Phase 1, 5 weeks (180 hours); Phase 2, 3 weeks (288 hours).

Exhibit Dates: 1/86–6/95.

Objectives: To train selected Marine Corps officers, warrant officers, noncommissioned officers, and enlisted personnel in nuclear, biological, and chemical defense.

Instruction: Phase 2 consists of conferences, demonstrations, practical exercises, and examinations in chemical/biological defense and operations, including nuclear warfare, radiological defense, and decontamination. Phase 2 consists of conferences, demonstrations, practical exercises, and examinations in nuclear warfare and radiological defense, chemical and biological defense operations, decontamination, and nuclear and chemical target analysis.

Credit Recommendation: In the lower-division baccalaureate/associate degree category, for Phase 1, 3 semester hours in introduction to physical sciences. For Phase 2, 3 semester hours in introduction to physical sciences (11/86).

AR-1720-0002

Prime Power Production Specialist
(Health Physics and Plant Chemistry Specialty)

Course Number: None.
Location: Prime Power Production School, Fort Belvoir, VA; Engineer Reactors Group, Fort Belvoir, VA.
Length: 41–43 weeks (1477 hours).
Exhibit Dates: 9/81–3/92.
Objectives: To provide enlisted personnel with the skills and knowledge necessary to operate and maintain nuclear and conventional power plants and to perform as specialists in all health/biological/chemical/radiological monitoring systems of nuclear and conventional power plants.
Instruction: Students complete academic phase training which covers algebra, trigonometry and general mathematics, general physics, chemistry, A/C circuit analysis, D/C circuit analysis, and mechanical engineering (thermodynamics) prior to specialty training. Specialty concentration provides enlisted personnel with a working knowledge of water chemistry and health physics. Upon completion of program, student will be able to perform tests in health monitoring, chemical analysis, and water treatment and analysis. Lectures and practical experiences include fundamentals of electrical and mechanical systems. Operator phase is equivalent to cooperative work experience in professional practice. This phase may be considered for credit depending on the policies of receiving institution with regard to cooperative work experience.
Credit Recommendation: In the lower-division baccalaureate/associate degree category, 6 semester hours in mathematics, 6 in physics, 2 in chemistry, 3 in thermodynamics, 3 in mechanical engineering, 3 in DC circuit analysis, 3 in AC circuit analysis, 3 in motors and generators, 6 in general chemistry, 3 in water analysis and treatment, and 0—6 in cooperative work experience. In the upper-division baccalaureate category, 6 semester hours in health physics (1/81).
Related Occupations: 52E; 52L; 52M.

AR-1720-0003

Prime Power Production Specialist
(Electrical Specialty)

Course Number: None.
Location: Prime Power Production School, Fort Belvoir, VA; Engineer Reactors Group, Fort Belvoir, VA.
Length: 48–50 weeks (1729 hours).
Exhibit Dates: 9/81–3/92.
Objectives: After 3/92 see AR-1720-0009. To provide enlisted personnel with the skills and knowledge necessary to operate and maintain nuclear and conventional power plants and to perform as specialists in the maintenance of all electrical systems of nuclear and conventional power plants.
Instruction: Students complete an academic phase which covers algebra, trigonometry, general mathematics, general physics, chemistry, A/C circuits, D/C circuit, and mechanical engineering including thermodynamics. Specialty phase includes analysis of power plant electrical systems, including motors and generators, transformers, instrumentation, and protection apparatus. This course includes control circuits, regulations, and preventive maintenance procedures. The operator phase is equivalent to cooperative work experience in professional practice. This phase may be considered for credit depending on the policies of the receiving institution on cooperative work experience.
Credit Recommendation: In the vocational certificate category, 2 semester hours in automatic controls, 3 in electrical wiring, 5 in power distribution, and 2 in air conditioning. In the lower-division baccalaureate/associate degree category, 3 semester hours in industrial electronics, 1 in active devices, 2 in instrumentation, 6 in mathematics, 6 in physics, 2 in chemistry, 3 in DC circuits, 3 in AC circuits, 3 in thermodynamics, 3 in mechanical engineering, 3 in motors and generators, and 0—6 in work experience or practicum (8/81).
Related Occupations: 52E; 52J; 52M.

AR-1720-0004

Prime Power Production Specialist
(Mechanical Specialty)

Course Number: None.
Location: Prime Power Production School, Fort Belvoir, VA; Engineer Reactors Group, Fort Belvoir, VA.
Length: 49–52 weeks (1745 hours).
Exhibit Dates: 9/81–3/92.
Objectives: After 3/92 see AR-1720-0011. To provide enlisted personnel with the skills and knowledge necessary to operate and maintain nuclear and conventional power plants and to perform as specialists in all mechanical systems of nuclear and conventional power plants.
Instruction: Students complete an academic phase which covers algebra, trigonometry and general mathematics, general physics, chemistry, A/C circuits, D/C circuits, and mechanical engineering. Course includes lectures and practical exercises covering electrical and mechanical components of power plants and the operation and maintenance of its auxiliary, standby, and related equipment. Course includes electronic components, cable splicing, boilers, vibration analysis, and steam engines. Operation phase is equivalent to cooperative work experience in professional practice. This phase may be considered for credit depending on the policies of the receiving institution on cooperative work experience.
Credit Recommendation: In the vocational certificate category, 2 semester hours in welding, 4 in basic machine shop, 6 in engines, 2 in industrial electronics, 4 in boilers and heat exchangers, and 2 in air conditioning. In the lower-division baccalaureate/associate degree category, 6 semester hours in mathematics, 6 in physics, 2 in chemistry, 3 in thermodynamics, 3 in mechanical engineering, 3 in DC circuit analysis, 3 in AC circuit analysis, 3 in motors and generators, and 0—6 in work experience or practicum (8/81).
Related Occupations: 52E; 52H; 52M.

AR-1720-0005

Prime Power Production Specialist
(Instrumentation Specialty)

Course Number: None.
Location: Prime Power Production School, Fort Belvoir, VA; Engineer Reactors Group, Fort Belvoir, VA.
Length: 48–50 weeks (1777 hours).
Exhibit Dates: 9/81–3/92.
Objectives: After 3/92 see AR-1720-0010. To provide enlisted personnel with the skills and knowledge necessary to operate and maintain nuclear and conventional power plants and to perform as specialists in all electrical and mechanical systems of nuclear and conventional power plants.
Instruction: The academic phase covers algebra, trigonometry and general mathematics, general physics, chemistry, A/C circuit analysis, D/C circuit analysis, and mechanical engineering including thermodynamics. Sensors and measurement technology, electronics for instrumentation, signal processing, elements of automatic control including controllers and their tuning are included. Course also includes amplifiers, digital systems, process instruments, active devices, and automatic control and piping systems. The operator phase is equivalent to cooperative work experience in professional practice. This phase may be considered for credit depending on the policies of the receiving institution on cooperative work experience.
Credit Recommendation: In the vocational certificate category, 3 semester hours in air conditioning and 3 in pipe fitting. In the lower-division baccalaureate/associate degree category, 8 semester hours in active devices, 7 in mathematics, 3 in digital fundamentals, 1 in microprocessors, 5 in instrumentation/ process control, 3 in physics, 2 in chemistry, 3 in thermodynamics, 3 in mechanical engineering, 3 in DC circuits, 3 in AC circuits, 3 in motors and generators, and 0—6 work experience or practicum (8/81).
Related Occupations: 52E; 52K; 52M.

AR-1720-0007

Nuclear, Biological, Chemical (NBC) Defense
(Nuclear, Biological, Chemical (NBC) Defense Officer Noncommissioned Officer (NCO))

Course Number: 2E-ASI3R/494-SQIC.
Location: Reserve Component Training Centers, US; Chemical School, Fort McClellan, AL; National Guard Academies, Continental USA.
Length: Self-paced, 1–3 weeks (50–80 hours).
Exhibit Dates: 1/80–6/96.
Objectives: To provide active Army and reserve components of all MOS's with those duties required of NBC defense officers or NCOs at the battalion and company level.
Instruction: Through the use of films and practical experiences such topics as survival in a chemical, biological and nuclear environment are taught. Other areas covered include chemical and nuclear equipment and maintenance, logistics, radiological monitoring, smoke operations, and decontamination.
Credit Recommendation: Credit is not recommended because of the military-specific nature of the course (11/86).

AR-1720-0008

Ordnance Officer Advanced Nuclear Materiel Management
(Ordnance Officer Advanced Nuclear Weapons Materiel Management)

Course Number: 4E-F0A-F9.
Location: Missile and Munitions School, Redstone Arsenal, AL.
Length: 2 weeks (72 hours).
Exhibit Dates: 10/84–1/96.

Objectives: To prepare the company-grade ordnance officer to function effectively while assigned to an ordnance nuclear weapons unit.

Instruction: Training includes logistics management, budget system overview, and mission area analysis. The course also includes ammunition industrial base management.

Credit Recommendation: Credit is not recommended because of the military-specific nature of the course (1/88).

Related Occupations: 91D.

AR-1720-0009

1. Prime Power Production Specialist, Electrical Specialty
2. Prime Power Production Electrical Specialist

Course Number: ***Version 1:*** 661-52E-ASIS3. *Version 2:* 661-52E/ASI-S3.

Location: Prime Power Production School, Fort Belvoir, VA.

Length: ***Version 1:*** 17–18 weeks (556 hours). *Version 2:* 18 weeks (556 hours).

Exhibit Dates: ***Version 1:*** 4/92–3/02. *Version 2:* 4/02–Present.

Learning Outcomes: ***Version 1:*** Before 4/92 see AR-1720-0003. Upon completion of the course, the student will be able to operate, maintain, test, repair, and troubleshoot electrical and electronic components found in military electrical power and distribution equipment. *Version 2:* Upon completion of the course, the student will be able to operate, maintain, test, repair, and troubleshoot electrical and electronic components found in military electrical power and distribution equipment.

Instruction: ***Version 1:*** Lectures, practical exercises, and conferences cover AC/DC fundamentals, transistors, amplifiers, power supplies, electrical safety, transformers, motors and generators, motor controllers, cable splicing, power distribution, and circuit breakers. *Version 2:* Lectures, practical exercises, and conferences cover AC/DC fundamentals, transistors, amplifiers, power supplies, electrical safety, transformers, motors and generators, motor controllers, cable splicing, power distribution, and circuit breakers.

Credit Recommendation: ***Version 1:*** In the vocational certificate category, 1 semester hour in maintenance and repair of electrical safety control circuits; 1 in application of solid state devices and circuits; 2 in maintenance and repair of transformers, rotating machines, and auxiliary controls; 1 in power cable repair and maintenance; and 1 in power system maintenance. In the lower-division baccalaureate/associate degree category, 2 semester hours in solid state electronics, 2 in industrial safety, 3 in DC/AC machines and transformers, and 4 in power distribution system maintenance (9/94). *Version 2:* In the lower-division baccalaureate/associate degree category, 1 semester hour in maintenance and repair of electrical safety control circuits; 1 in application of solid state devices and circuits; 2 in maintenance and repair of transformers, rotating machines, and auxiliary controls; 1 in power cable repair and maintenance; 1 in power system maintenance; 2 in solid state electronics, 2 in industrial safety, 3 in DC/AC machines and transformers, 4 in power distribution system maintenance, 2 in industrial safety, and 3 in materiels engineering (8/03).

AR-1720-0010

Prime Power Production Instrumentation Specialist

Course Number: ***Version 1:*** 661-52E/ASIE5. *Version 2:* 661-52E/ASI-E5.

Location: Prime Power Production School, Fort Belvoir, VA.

Length: ***Version 1:*** 16–18 weeks (556–595 hours). *Version 2:* 16 weeks (554 hours).

Exhibit Dates: ***Version 1:*** 4/92–3/02. *Version 2:* 4/02–Present.

Learning Outcomes: ***Version 1:*** Before 4/92 see AR-1720-0005. Upon completion of the course, the student will be able to maintain, troubleshoot, and repair instrumentation equipment associated with prime power generation equipment using proper safety procedures. *Version 2:* Upon completion of the course, the student will be able to maintain, troubleshoot, and repair instrumentation equipment associated with prime power generation equipment using proper safety procedures.

Instruction: ***Version 1:*** Lectures and practical exercises cover electrical and electronic test equipment, review of DC/AC circuits, solid state devices, basic transistor amplifier circuits, differential amplifiers, operational amplifiers, digital logic, D/A converters, relay theory and maintenance process instrumentation, and automatic control systems. *Version 2:* Lectures and practical exercises cover electrical and electronic test equipment, review of DC/AC circuits, solid state devices, basic transistor amplifier circuits, differential amplifiers, operational amplifiers, digital logic, D/A converters, relay theory and maintenance process instrumentation, and automatic control systems.

Credit Recommendation: ***Version 1:*** In the vocational certificate category, 3 semester hours in maintenance and troubleshooting of solid state components and electronic circuits; 1 in maintenance and troubleshooting of power system protective relays; and 2 in calibration, maintenance, and process instruments. In the lower-division baccalaureate/associate degree category, 3 semester hours in industrial electronics, 5 in electronic devices and circuits, and 3 in instrumentation/process control (9/94). *Version 2:* In the vocational certificate category, 3 semester hours in maintenance and troubleshooting of solid state components and electronic circuits; 1 in maintenance and troubleshooting of power system protective relays; and 2 in calibration, maintenance, and process instruments. In the lower-division baccalaureate/associate degree category, 4 semester hours in industrial electronics, 6 in electronic devices and circuits, and 4 in instrumentation/process control (8/03).

Related Occupations: 52E.

AR-1720-0011

Prime Power Production Specialist, Mechanical Specialty

Course Number: 661-52E20/S2; 661-52E/ASI-S2.

Location: Prime Power Production School, Fort Belvoir, VA.

Length: 16–17 weeks (556–582 hours).

Exhibit Dates: 4/92–Present.

Learning Outcomes: Before 4/92 see AR-1720-0004. Upon completion of the course, the student will be able to troubleshoot, maintain, test, and repair mechanical systems and components found in mobile electrical power generation plants.

Instruction: Lectures and practical exercises cover machine shop practices, gas welding, piping systems, valves, pumps, heat exchangers, speed control, diesel engines, gas turbines, steam turbines, boilers, and vibration analysis.

Credit Recommendation: In the vocational certificate category, 3 semester hours in basic machine shop, hand tools, and necessary equipment; 1 in gas welding and soldering technology; 3 in piping system construction, operation, and maintenance; 1 in heat exchanger operation and maintenance; 1 in mechanical speed control systems; 6 in diesel operations and maintenance; 3 in gas turbine operation and maintenance; and 1 in vibration measurement. In the lower-division baccalaureate/associate degree category, 3 semester hours in fluid flow and heat transfer systems; 4 in internal combustion engine technology (diesel and gas turbine); and 3 in steam plant systems, turbines, boilers, heat exchangers, and condensers (8/03).

AR-1720-0012

1. Prime Power Production Specialist, Phases 1 and 2
2. Prime Power Production Specialist

Course Number: ***Version 1:*** 661-52E20. *Version 2:* 661-52E20.

Location: Prime Power Production School, Fort Belvoir, VA.

Length: ***Version 1:*** 28–29 weeks (982 hours). *Version 2:* 20–21 weeks (808 hours).

Exhibit Dates: ***Version 1:*** 4/92–9/95. *Version 2:* 10/95–Present.

Learning Outcomes: ***Version 1:*** Upon completion of the academic phase of the course, the student will be able to apply basic AC/DC concepts and various electrical measurement techniques, operate AC/DC machinery, describe three-phase electrical systems, and use protective devices and proper electrical safety procedures. Upon completion of the operator training phase, students will be able to operate, maintain, and troubleshoot special power generation equipment (diesel and gas turbine); practice proper safety procedures; and interpret electrical and mechanical schematics and wiring diagrams. *Version 2:* Upon completion of the academic phase of the course, the student will be able to apply basic AC/DC concepts and various electrical measurement techniques, operate AC/DC machinery, describe three-phase electrical systems, and use protective devices and proper electrical safety procedures. Upon completion of the operator training phase, students will be able to operate, maintain, and troubleshoot special power generation equipment (diesel); practice proper safety procedures; and interpret electrical and mechanical schematics and wiring diagrams.

Instruction: ***Version 1:*** The academic training phase includes lectures, practical exercises, and conferences about basic algebra and trigonometry, physics of motion, fundamentals of thermodynamics, fundamentals of DC and AC circuits, reading circuit diagrams, using relays and circuit breakers, and proper lubrication procedures. The operator training phase includes lectures and practical exercises in power plant technology, diesel and gas turbine generators, reading and interpretation of electrical and mechanical schematics and diagrams, maintenance and troubleshooting of electrical generators, electrical protection devices, and proper safety procedures. *Version 2:* The academic training phase includes lectures, practical exercises, and conferences about basic algebra and trigonometry, physics of motion, fundamentals of thermodynamics, fundamentals of DC and AC circuits, reading of circuit diagrams, use of relays and circuit breakers, and proper lubrication procedures. The operator training phase includes lectures and practical exercises in power plant technology, diesel generators, reading and interpretation of electrical and mechanical schematics and diagrams, maintenance and troubleshooting of electrical generators, electrical protection devices, and proper safety procedures.

Credit Recommendation: ***Version 1:*** In the vocational certificate category, 3 semester hours in electrical/mechanical blueprint (schematics and diagrams) reading; 3 in electrical power generation, distribution, and control; 4 in diesel engine operation and maintenance; and 3 in gas turbine operation and maintenance. In the lower-division baccalaureate/associate degree category, 2 semester hours in electrical power distribution, 2 in electrical power generation, 2 in internal combustion engine technology, 0-6 in cooperative education, 3 in technical mathematics, 3 in physics of mechanics, 3 in DC circuits, 3 in AC circuits, 3 in thermodynamics, 3 in AC/DC machinery, and 1 in electrical safety (9/94). *Version 2:* In the lower-division baccalaureate/associate degree category, 3 semester hours in blue print reading and schematics, 9 in power engineering practicum, 4 in internal combustion engine theory, 2 in internal combustion engine practicum, 3 in technical mathematics, 6 in AC/DC circuitry, 3 in AC/DC practicum, 3 in thermodynamics, 2 in industrial safety, and 3 in power engineering (8/03).

Related Occupations: 52E.

AR-1720-0013

(MOS 54B) Advanced Noncommissioned Officer (NCO)

Course Number: 4-54-C24; 4-54-C42ANCOC, Phase 2.

Location: Chemical School, Fort Leonard Wood, MO.

Length: 6–7 weeks (222–223 hours).

Exhibit Dates: 10/00–Present.

Learning Outcomes: Before 10/00 see AR-0801-0029. Upon completion of the course, the student will be able to perform critical tasks in nuclear and chemical operations, biological defense, logistics, and maintenance management.

Instruction: Lectures, small-group discussion, oral and written examinations, and practical training exercises cover basic management practices and concepts, including nuclear and chemical operations, biological defense, and tactical decision making.

Credit Recommendation: In the lower-division baccalaureate/associate degree category, 2 semester hours in radiation detection, 1 in radiation detection laboratory, 3 in radiological safety, 1 in technical writing, and 1 in oral communication skills (4/04).

AR-1720-0014

(MOS 54B) Advanced Noncommissioned Officer (NCO) Chemical Operations Specialist

Course Number: 4-54-C42TATS09.

Location: Chemical School, Fort Leonard Wood, MO.

Length: 134 hours.

Exhibit Dates: 10/00–Present.

Learning Outcomes: Upon completion of the course, the student will be able to perform critical tasks in nuclear and chemical operations, biological defense, logistics, and maintenance management.

Instruction: Lectures, small-group discussion, oral and written examinations, and practical training exercises cover basic management practices and concepts, including nuclear and chemical operations, biological defense, and tactical decision making.

Credit Recommendation: In the lower-division baccalaureate/associate degree category, 2 semester hours in radiation detection, 1 in radiation detection laboratory, 1 in oral communication skills, and 1 in technical writing (1/02).

AR-1720-0015

Chemical Operations Specialist Basic Noncommissioned Officer (NCO) Total Army Training System (TATS)

Course Number: 494-54B20-TATS-RES.

Location: Chemical School, Fort Leonard Wood, MO.

Length: 9–10 weeks (336–337 hours).

Exhibit Dates: 1/01–Present.

Learning Outcomes: Upon completion of the course, the student will be able to perform critical tasks in nuclear, biological, and chemical applications, including applying biological defense, logistics, and maintenance procedures; detecting contamination; decontaminating equipment and smoke generators; and conducting tactical training.

Instruction: The course of instruction consists of conferences, practical exercises, and examinations in individual protective skills, radiological and chemical decontamination, smoke and biological operations, and tactical training. This course contains a common core of leadership subjects.

Credit Recommendation: In the lower-division baccalaureate/associate degree category, 1 semester hour in technical writing, 3 in radiological safety, and 1 in introduction to computers. See AR-1406-0090 for common core credit (1/02).

Related Occupations: 54B.

AR-1720-0016

Chemical Operations Specialist Basic Noncommissioned Officer (NCO) Total Army Training System

Course Number: 494-54B20-TATS-RECLAS.

Location: Chemical School, Fort Leonard Wood, MO.

Length: 10–11 weeks (369 hours).

Exhibit Dates: 1/01–Present.

Learning Outcomes: Upon completion of the course, the student will be able to perform critical tasks in nuclear, biological, and chemical applications, including applying biological defense, logistics, and maintenance procedures; detecting contamination; decontaminating equipment and smoke generators; and conducting tactical training.

Instruction: The course instructions consists of conferences, practical exercises, and examinations in individual protective skills, radiological and chemical decontamination, smoke and biological operations, and tactical training.

Credit Recommendation: In the lower-division baccalaureate/associate degree category, 3 semester hours in radiological safety, 1 in technical writing, and 1 in introduction to computers (1/02).

Related Occupations: 54B.

AR-1720-0017

Chemical Operations Specialist Basic Noncommissioned Officer (BNCOC) Total Army Training System

Course Number: 494-54B20/30-R, Phase 1.

Location: Chemical School, Fort Leonard Wood, MO.

Length: 157 hours.

Exhibit Dates: 1/01–Present.

Learning Outcomes: Upon completion of the course, the student will be able to perform critical tasks in nuclear, biological, and chemical applications, including applying biological defense, logistics, and maintenance procedures; detecting contamination; decontaminating equipment and smoke generators; and conducting tactical training.

Instruction: The course of instruction consists of conferences, practical exercises, and examinations in individual protective skills, radiological and chemical decontamination, smoke and biological operations, and tactical training.

Credit Recommendation: In the lower-division baccalaureate/associate degree category, 3 semester hours in radiological safety (1/02).

Related Occupations: 54B.

AR-1720-0018

Chemical Operations Specialist Basic Noncommissioned Officer (NCO) Total Army Training System

Course Number: 494-54B20/30, Phase 1.

Location: Chemical School, Fort Leonard Wood, MO.

Length: 110 hours.

Exhibit Dates: 1/01–Present.

Learning Outcomes: Upon completion of the course, the student will be able to perform critical tasks in nuclear, biological, and chemical applications, including applying biological defense, logistics, and maintenance procedures; detecting decontamination; decontaminating equipment and smoke generators; and conducting tactical training.

Instruction: The course of instruction consists of conferences, practical exercises, and examinations in individual protective skills, radiological and chemical decontamination, smoke and biological operations, and tactical training.

Credit Recommendation: In the lower-division baccalaureate/associate degree category, 3 semester hours in radiological safety (1/02).

Related Occupations: 54B.

AR-1720-0019

Chemical Captain Career

Course Number: 4-3-C22-74A.

Location: Chemical School, Fort Leonard Wood, MO.

Length: 18 weeks (674 hours).

Exhibit Dates: 10/00–Present.

Learning Outcomes: Upon completion the course, the student will be able to apply skills in leadership, personnel management, written and oral communication, management, and training of support and staff personnel; be able to identify nuclear radiation and the effects of radiation; to detect radiation; and to implement radiation protection measures.

Instruction: Course instruction includes classroom lecture, discussion, practical exercises, and written and oral examinations.

Credit Recommendation: In the lower-division baccalaureate/associate degree category, 1 semester hour in oral communication skills, 1 in technical writing, 3 in introduction to health physics, and 3 in leadership or 3 in student personnel services (1/02).

AR-1720-0020

Chemical Operations Specialist

Course Number: 54B10-OSUT.

Location: Chemical School, Fort Leonard Wood, MO.

Length: 19–20 weeks (857–880 hours).

Exhibit Dates: 10/00–Present.

Learning Outcomes: Upon completion of the course, the student will be able to perform basic soldier skills, including combat skills and techniques, radiological defense, chemical and biological defense, decontamination, smoke operations, and driving.

Instruction: Instruction includes training in combat skills, chemical and biological defense, and chemical specialist training. This course includes an 8-9 week long basic training component.

Credit Recommendation: Credit is not recommended because of the military specific nature of the course (4/04).

Related Occupations: 54B.

AR-1720-0021

Prime Power Production Specialist Rapid Reserve Supervisor

Course Number: 661-52E30/40-1RR-RTUP.

Location: Prime Power Production School, Fort Belvoir, VA.

Length: 120 hours.

Exhibit Dates: 10/01–Present.

Learning Outcomes: Upon completion of the course, the student will be able to safely perform critical tasks in inspecting, testing, troubleshooting, maintenance, and repair of related systems, equipment, and components used in prime power generation plants.

Instruction: Audiovisual materials and discussion. Topics include cardiopulmonary resuscitation, electrical safety hazards, diesel engine theory and components, AC/DC power theory, engine lubricants, site selection, installation and startup, and distribution system connection.

Credit Recommendation: In the lower-division baccalaureate/associate degree category, 2 semester hours in industrial safety, and 2 in portable power generation (8/03).

AR-1720-0022

Prime Power Production Supervisor

Course Number: 662-52E30 (BNCOC), Phase 2.

Location: Prime Power Production School, Fort Belvoir, VA.

Length: 4–5 weeks (180 hours).

Exhibit Dates: 4/02–Present.

Learning Outcomes: Upon completion of the course, the student will be able to plan the safe deployment, operation, and retrieval of portable prime power; control and distribute both high and low voltage power; schedule maintenance; inspect mechanical repairs; identify parts and systems of diesel engines; and maintain substations.

Instruction: Practical exercises and lecture. Topics include maintenance management, industrial/occupational health and safety, diesel engine systems, supply systems management, welding processes, hazardous flammable fluids handling, high voltage cabling, power voltage and quality properties, low voltage power generation, transformers and protective relays, switchgear and substation maintenance, deployment and retrieval of power plant, and distribution systems equipment.

Credit Recommendation: In the lower-division baccalaureate/associate degree category, 1 semester hour in industrial safety; 1 in production planning; 3 in power plant systems, components, and deployments; 1 in substation maintenance; and 2 in mechanical systems theorem (8/03).

AR-1721-0002

Fire Control Instrument Repair
(Optical Instrument Repairman (USMC))

Course Number: 670-41C10 (CT).

Location: Ordnance Center and School, Aberdeen Proving Ground, MD.

Length: 15–16 weeks (564–565 hours).

Exhibit Dates: 10/88–4/04.

Learning Outcomes: Upon completion of the course, the student will be able to inspect, adjust, and repair optical, nonoptical, and electrical precision instruments and related equipment.

Instruction: Lectures and practical exercises cover optics fundamentals and repair techniques; care and use of common hand tools and measuring instruments; inspection and troubleshooting of binoculars, periscopes, telescopes, aiming circles, and tank fire control systems; and diagnosis and repair of electromechanical computers and gun aiming systems. Training includes telescope mount, periscope (passive) and mount, and panel assemblies. The course includes basic concepts of electricity, including testing, inspecting, and troubleshooting electrical circuits and inspection and repair of binoculars, periscopes, telescopes, aiming circles, mounts and artillery, and tank fire control mechanisms.

Credit Recommendation: In the lower-division baccalaureate/associate degree category, 5 semester hours in optical instrument repair, 3 in basic automotive electricity, and 5 in basic electricity and electronics (9/90).

Related Occupations: 09B; 41C.

AR-1722-0002

Watercraft Operator
(Watercraft Operation)
(Crewman)

Course Number: 062-88K10; 062-76B10.

Location: Transportation and Aviation Logistics School, Fort Eustis, VA.

Length: 6–7 weeks (230–282 hours).

Exhibit Dates: 4/85–12/91.

Objectives: After 12/91 see AR-1722-0008. To provide entry-level enlisted personnel with a working knowledge of the duties and responsibilities of a watercraft operator on Army watercraft and amphibians.

Instruction: Lectures and practical exercises cover ship radio communications and code flags; rigging; piloting, navigation, and charts; and watercraft operations.

Credit Recommendation: In the lower-division baccalaureate/associate degree category, 3 semester hours in seamanship (6/91).

Related Occupations: 61A; 61B; 76B; 88K.

AR-1722-0004

Watercraft Operator Certification
(Watercraft Operator Primary Technical)

Course Number: 062-F5; 062-61B20.

Location: Transportation School, Fort Eustis, VA.

Length: 5–6 weeks (187–190 hours).

Exhibit Dates: 1/82–Present.

Objectives: To train enlisted personnel to operate watercraft and serve in senior positions aboard vessels with two to four crewmen.

Instruction: Lectures and practical exercises cover marlinspike seamanship, meteorology, vessel communications, piloting and coastwise navigation, and shipboard first aid.

Credit Recommendation: In the lower-division baccalaureate/associate degree category, 3 semester hours in navigation, 1 in safety, and 1 in seamanship (10/03).

Related Occupations: 61B.

AR-1722-0005

Marine Deck Officer A2 Certification
(SQI2)

Course Number: ***Version 2:*** 8C-SQI2 (880A). *Version 3:* 8C-SQI2 (880A).

Location: ***Version 2:*** Transportation and Aviation Logistics School, Fort Eustis, VA. *Version 3:* Transportation School, Fort Eustis, VA.

Length: ***Version 2:*** 17 weeks (613 hours). *Version 3:* 17 weeks (680 hours).

Exhibit Dates: ***Version 2:*** 10/89–9/00. *Version 3:* 10/00–Present.

Learning Outcomes: ***Version 2:*** Upon completion of the course, the student will be able to command, operate, and maintain watercraft in open ocean waters for towing, salvage, and resupply operations and be responsible for ship's business; beach and port evaluation; damage control; ship repair; first aid, lifesaving, and fire fighting equipment; cargo operation; meteorology; electronic and celestial navigation; piloting; advanced ship handling; and communications. Student will command the docking and undocking of large vessels and also convoy and towing operations. *Version 3:* Upon completion of the course, the student will be able to command, operate, and maintain watercraft in open waters for towing, salvage and resupply operations, and be responsible for ship's business, beach and port evaluation, damage control, ship repair, first aid and lifesaving, fire fighting equipment, cargo operation, meteorology, electronic and celestial navigation, piloting, advanced ship handling, and communications. Student will command the docking and undocking of large vessels and also convoy and towing operations.

Instruction: *Version 2:* Lectures and practical exercises cover general marine subjects, including piloting and sailing, electronic navigation, advanced ship handling, and celestial navigation. The course includes a 192-hour shipboard training exercise, of this 80 hours take place at night. Navigation systems covered are LORAN, OMEGA, NAVSAT, RDF, and marine radar. Student will complete the comprehensive examination for Master of Class A2 Unlimited Motor Vessels Upon Oceans and the Coast Guard Radar Observer Certifications. *Version 3:* Methods of instruction include classroom demonstrations, examinations, and practical exercises in general marine subjects, including piloting and sailing, electronic navigation, advanced ship handling and celestial navigation. The course includes a 200-hour shipboard training exercise of with 80 hours occuring at night. Navigation systems covered are LORAN, OMEGA, NAVSAT/GPS, RDF and marine radar. Student will complete the comprehensive examination for Master of Class A2 Unlimited Motor Vessels Upon Oceans and the Coast Guard Radar Observer Certification.

Credit Recommendation: *Version 2:* In the lower-division baccalaureate/associate degree category, 3 semester hours in marine piloting, 3 in celestial navigation, 3 in electronic navigation, 3 in advanced ship handling, and 4 for a practicum in open-water ship operation (6/91). *Version 3:* In the lower-division baccalaureate/associate degree category, 3 semester hours in marine piloting, 3 in celestial navigation, 3 in electronic navigation, 3 in advanced ship handling, 1 in principles of operating systems, and 4 for a practicum in open-water ship operation (6/01).

Related Occupations: 500A; 880A.

AR-1722-0006

MARINE DECK OFFICER WARRANT OFFICER BASIC

(Marine Deck Officer Warrant Officer Technical Certification)

Course Number: *Version 1:* 8C-880A. *Version 2:* 8C-880A.

Location: *Version 1:* Transportation and Aviation Logistics School, Fort Eustis, VA. *Version 2:* Transportation School, Fort Eustis, VA.

Length: *Version 1:* 31–32 weeks (1153–1216 hours). *Version 2:* 31–32 weeks (1225–1251 hours).

Exhibit Dates: *Version 1:* 10/87–12/97. *Version 2:* 1/98–Present.

Learning Outcomes: *Version 1:* Upon completion of the course, the student will be able to serve as a licensed Master of Class A1 Motor vessels, as a Mate of A2 motor Vessels, or as a Marine Radar Observer; supervise deck and engine students in common marine subjects, including ship fire fighting, damage control, general deck seamanship, and cargo handling; use visual and radio communications, navigation rules, radar vector analysis, charts, publications, tide tables, and weather information; and use marine sextant, magnetic compass, marine radar, and celestial and electronic navigation. *Version 2:* Upon completion of the course, the student will be able to serve as a license Master of Class A1 Motor Vessels, as a Mate of A2 Motor Vessels, or as a Marine Radar Observer; supervise deck and engine students in marine subjects, including ship fire fighting, damage control general deck seamanship, and cargo handling; use visual and radio communications, navigation rules, radar vector analysis, charts, publications, tide tables, and weather information; and use marine sextant, magnetic compass, marine radar, and celestial and electronic navigation.

Instruction: *Version 1:* Lectures, demonstrations, practical exercises, and hands-on training covers the skills necessary to be certified as a Marine Deck Officer. The student will be issued the Coast Guard Marine Radar Observer's Certificate. *Version 2:* Methods of instruction include classroom, demonstrations, written examinations, practical exercises, and hands-on training covering the skills necessary to be certified as a marine deck officer. This course includes 304 hours of shipboard training exercises with both day and night operations. The global maritime distress and safety system is covered in addition to the regular electronic navigation systems. The student will be issued the Coast Guard Marine Radar Observer's Certificate.

Credit Recommendation: *Version 1:* In the lower-division baccalaureate/associate degree category, 3 semester hours in principles of supervision, 3 in deck seamanship, 3 in radar navigation, 3 in celestial navigation, 3 in piloting, 3 in ship handling, and 6 in practicum in inland and open water ship operation (10/92). *Version 2:* In the lower-division baccalaureate/associate degree category, 3 semester hours in maritime safety, 3 in basic navigation, 3 in deck seamanship, 3 in ship piloting, 3 in ship handling, 1 in computer applications, and 4 in practicum in inland and open-water ship operation. In the upper-division baccalaureate category, 3 semester hours in principles of supervision, 2 in crew/team resource management, and 3 in advanced navigation (12/03).

Related Occupations: 880A.

AR-1722-0007

1. WATERCRAFT ENGINEER CERTIFICATION (88L20)
2. WATERCRAFT ENGINEER CERTIFICATION

Course Number: *Version 1:* 652-F2. *Version 2:* 652-F2.

Location: *Version 1:* Transportation School, Fort Eustis, VA; Transportation and Aviation Logistics School, Fort Eustis, VA. *Version 2:* Transportation School, Fort Eustis, VA.

Length: *Version 1:* 14–15 weeks (531–557 hours). *Version 2:* 14–15 weeks (556 hours).

Exhibit Dates: *Version 1:* 10/88–9/96. *Version 2:* 10/96–Present.

Learning Outcomes: *Version 1:* Upon completion of the course, the student will be able to perform marine engineering tasks and successfully complete a marine certification examination. *Version 2:* Upon completion of the course, the student will be able to perform marine engineering tasks and successfully complete the Marine Technical Examination.

Instruction: *Version 1:* Lectures and practical exercises cover the identification, inspection, operation, maintenance, and overhaul of marine electrical equipment, refrigeration systems, diesel fuel injection equipment, high-speed diesel engine, and oxyacetylene and arc welding. *Version 2:* Methods of instruction include conferences, performance exercises, demonstrations, and televised information covering high speed watercraft, diesel engines, marine electrical systems, hydraulic and pneumatic systems, heating and refrigeration systems, propellers, fuel systems, and utility and environmental controls.

Credit Recommendation: *Version 1:* In the lower-division baccalaureate/associate degree category, 3 semester hours in marine electrical systems, 3 in diesel fuel injection systems, 3 in diesel engine overhaul, and 2 in oxyacetylene and arc welding (3/94). *Version 2:* In the lower-division baccalaureate/associate degree category, 1 semester hour in diesel engine maintenance and repair, 2 in diesel engine overhaul, 2 in oxyacetylene and arc welding, 2 in AC/DC electrical systems principles, 1 in electrical system maintenance, 1 in hydraulic system principles, and 1 in heating and refrigeration (HVAC) principles (6/01).

Related Occupations: 88L.

AR-1722-0008

WATERCRAFT OPERATOR

Course Number: *Version 1:* 062-88K10. *Version 2:* 062-88K10.

Location: *Version 1:* Transportation School, Fort Eustis, VA. *Version 2:* Transportation School, Fort Eustis, VA; Training DIrectorate, Fort Lee, VA.

Length: *Version 1:* 6 weeks (217 hours). *Version 2:* 6 weeks (255 hours).

Exhibit Dates: *Version 1:* 1/92–3/98. *Version 2:* 4/98–Present.

Learning Outcomes: *Version 1:* Before 1/92 see AR-1722-0002. Upon completion of the course, the student will be able to apply fire fighting techniques; demonstrate water survival, watercraft safety, and seamanship skills; and operate watercraft. *Version 2:* Upon completion of the course, the student will be able to operate watercraft; demonstrate water survival, watercraft safety, and seamanship skills; and apply fire-fighting techniques.

Instruction: *Version 1:* Lectures and physical application cover fire fighting, water survival/rescue, and watercraft safety and operation. *Version 2:* Methods of instruction include classroom conferences and physical applications that cover fire fighting, water survival, water rescue, and watercraft safety and operation.

Credit Recommendation: *Version 1:* In the lower-division baccalaureate/associate degree category, 3 semester hours in seamanship, 1 in water safety, and 1 in fire fighting (3/94). *Version 2:* In the lower-division baccalaureate/associate degree category, 3 semester hours in seamanship, 1 in water safety, and 1 in firefighting (6/01).

Related Occupations: 88K.

AR-1723-0008

MACHINIST

(Apprentice Machinist)

Course Number: *Version 1:* 702-42730; 702-44E10. *Version 2:* 702-42730; 702-44E10. *Version 3:* 702-44E10.

Location: Ordnance Center and School, Aberdeen Proving Ground, MD.

Length: *Version 1:* 15–16 weeks (544–550 hours). *Version 2:* 13–16 weeks (486–540 hours). *Version 3:* 13–14 weeks (528 hours).

Exhibit Dates: *Version 1:* 2/85–6/87. *Version 2:* 7/87–12/90. *Version 3:* 1/91–Present.

Learning Outcomes: *Version 1:* Upon completion of the course, the student will be able to fabricate metal parts by using metal working tools and equipment such as lathes, milling machines, shapers, grinders, and heat treating techniques. *Version 2:* Upon completion of the course, the student will be able to fabricate metal parts by using metal working tools and equipment such as lathes, milling machines, shapers, grinders, and heat treating techniques. *Version 3:* Upon completion of the course, the student will be able to lay out and measure fabric metal parts by using tools and equipment such as metal lathes, milling machines, shapers, and the metal cutting band saw. Precision measuring tools, print reading, and heat-treating are also covered.

Instruction: ***Version 1:*** Practical exercises in the use of the metal working tools and equipment are stressed. *Version 2:* Practical exercises in the use of the metal working tools and equipment are stressed. *Version 3:* Lectures, shop work, and performance testing are employed in this course.

Credit Recommendation: ***Version 1:*** In the lower-division baccalaureate/associate degree category, for the Army, 6 semester hours in lathe operation, 2 in milling machine operation, and 3 in machine trades or technology; for the Air Force, 4 semester hours in machine trades or technology, 5 in lathe operation, 1 in shaper operation, and 1 in precision grinding (3/87). *Version 2:* In the lower-division baccalaureate/associate degree category, 10 semester hours in machine tool technology (7/97). *Version 3:* In the lower-division baccalaureate/associate degree category, 3 semester hours in milling machine operations, 2 in lathe operations, 3 in precision measurement and bench work, and 3 in machine shop theory (8/03).

Related Occupations: 44E.

AR-1723-0009

MACHINIST BASIC NONCOMMISSIONED OFFICER (NCO)

Course Number: ***Version 2:*** 702-44E30 (44B). *Version 3:* 702-44E30 (44B). *Version 4:* 702-44E30 (44B).

Location: Ordnance Center and School, Aberdeen Proving Ground, MD.

Length: ***Version 2:*** 17–18 weeks (632 hours). *Version 3:* 14–15 weeks (522 hours). *Version 4:* 19 weeks (678 hours).

Exhibit Dates: ***Version 2:*** 10/86–11/91. *Version 3:* 12/91–9/96. *Version 4:* 10/96–Present.

Learning Outcomes: ***Version 2:*** To train enlisted personnel with MOS 44E20 or 44B20 to supervise and perform the duties of metalworker and machinist at skill level 30. *Version 3:* Upon completion of the course, the student will be able to supervise subordinate personnel and perform the duties of a metalworker and machinist at an advanced skill level. *Version 4:* Upon completion of the course, the student will be able to perform as metalworkers and machinists in the areas of welding, machine shop operations, fuel tank repair, training personnel, and maintenance/supply management.

Instruction: ***Version 2:*** Lectures and practical exercises cover leadership training, the noncommissioned officer in a tactical environment, general military as well as logistical training, and advanced training in metalworking and the machine trades. Those in MOS 44B will also receive basic training in the machine trades, including lectures and practical exercises in shop fundamentals, vertical band saw, lathe, milling machine, and lathe attachments. This course contains a common core of leadership subjects. *Version 3:* Lectures and practical exercises cover machine shop theory, lathe operations, milling machine operations, oxyacetylene and arc welding, and advanced machining. This course contains a common core of leadership subjects. *Version 4:* Lectures and practical exercises cover welding, machine shop operations, training, and maintenance/supply management. This course contains a common core of leadership subjects.

Credit Recommendation: ***Version 2:*** In the lower-division baccalaureate/associate degree category, for those completing the 44B track, 1 semester hour in supervision/management, 4 in machine lathe operation, 1 in milling machine operation, 5 in advanced machine shop, and 2 in advanced metalwork. See AR-1406-0090 for common core credit (3/87). *Version 3:* In the lower-division baccalaureate/associate degree category, 3 semester hours in lathe operations, 3 in milling machine operations, 3 in advanced machining, 3 in machine shop theory. See AR-1406-0090 for common core credit (1/98). *Version 4:* In the lower-division baccalaureate/associate degree category, 3 semester hours in milling machine operations, 3 in lathe operations, 3 in advanced machining, and 3 in machine shop theory. See AR-1406-0090 for common core credit (10/00).

Related Occupations: 44B; 44E.

AR-1723-0010

MACHINIST NONCOMMISSIONED OFFICER (NCO) ADVANCED
(Mechanical Maintenance Noncommissioned Officer (NCO) Advanced, MOS 44E)

Course Number: 702-44E40; 6-63-C42 (44E).

Location: Ordnance Center and School, Aberdeen Proving Ground, MD.

Length: 7–9 weeks (251–398 hours).

Exhibit Dates: 10/85–7/95.

Learning Outcomes: Upon completion of the course, students will be able to serve as a supervisor in the selected military occupational specialty for which they are trained.

Instruction: Lectures, demonstrations, and practical exercises cover leadership, communication skills, fundamentals of instruction, logistics management, and refresher training in the appropriate military occupational specialty. Emphasis in placed on leadership and on verbal and written communication.

Credit Recommendation: Credit is recommended for the common core only. See AR-1404-0035 (1/88).

Related Occupations: 44E.

AR-1723-0011

METALWORKER BASIC NONCOMMISSIONED OFFICER (NCO)
(Machinist Basic Noncommissioned Officer (NCO))

Course Number: ***Version 2:*** 702-44E30. *Version 3:* 702-44E30. *Version 4:* 702-44E30.

Location: Ordnance Center and School, Aberdeen Proving Ground, MD.

Length: ***Version 2:*** 14–15 weeks (538 hours). *Version 3:* 12 weeks (427–428 hours). *Version 4:* 17 weeks (615 hours).

Exhibit Dates: ***Version 2:*** 10/86–11/91. *Version 3:* 12/91–9/96. *Version 4:* 10/96–Present.

Learning Outcomes: ***Version 2:*** To train enlisted personnel with MOS 44E20 or 44B20 to supervise and perform the duties of metalworker and machinist at skill level 30. *Version 3:* Upon completion of the course, the student will be able to perform as a machinist in the areas of machine shop operation, training personnel, and maintenance/supply management. *Version 4:* Upon completion of the course, the student will be able to weld; repair radiators, metal body, and fuel tanks; and perform supply and maintenance management.

Instruction: ***Version 2:*** Lectures and practical exercises cover leadership training, the noncommissioned officer in a tactical environment, general military as well as logistical training, and advanced training in metalworking and the machine trades. Those in MOS 44E receive lectures and practical exercises in shop fundamentals; oxyacetylene, electric arc, and inert gas welding; metal body, radiator, glass working, and fuel tank repair; and metal body painting. *Version 3:* Lectures and practical exercises cover welding, machine shop operations, training, and maintenance/supply management. *Version 4:* Lectures and practical exercises cover oxyacetylene, shielded metal arc and inert gas welding; radiator and fuel tank repair; and machining. This course contains common core of leadership subjects.

Credit Recommendation: ***Version 2:*** In the lower-division baccalaureate/associate degree category, for those completing the 44E track, 1 semester hours in supervision/management, 2 in oxyacetylene welding, 1 in electric arc welding, 1 in MIG and TIG welding, 1 in advanced welding, 3 in automotive body repair, and 2 in advanced metalwork. See AR-1406-0090 for common core credit (3/87). *Version 3:* In the lower-division baccalaureate/associate degree category, 3 semester hours in welding operations and 3 in advanced machining (1/98). *Version 4:* In the lower-division baccalaureate/associate degree category, 3 semester hours in oxyacetylene welding, 3 in shielded metal arc welding, 3 in inert gas welding, and 3 in machine shop operations. See AR-1406-0090 for common core credit (5/01).

Related Occupations: 44E.

AR-1723-0012

METALWORKER

Course Number: ***Version 1:*** 704-44B10. *Version 2:* 704-44B10.

Location: Ordnance Center and School, Aberdeen Proving Ground, MD.

Length: ***Version 1:*** 12–13 weeks (486 hours). *Version 2:* 12–14 weeks (488–536 hours).

Exhibit Dates: ***Version 1:*** 1/98–9/00. *Version 2:* 10/00–Present.

Learning Outcomes: ***Version 1:*** Before 1/98 see AR-1703-0030. Upon completion of the course, the student will be able to apply oxyacetylene, electric arc, MIG, and TIG welding procedures to maintain, repair, and install metal body components, radiators, and fuel tanks, and perform basic vehicle body repair. *Version 2:* Upon completion of the course, the student will be able to understand metal fabrication and bonding techniques, be competent in completing bonding tasks, perform automotive body and glass repair, and repair automotive radiators and fuel tanks.

Instruction: *Version 1:* This course includes lectures, audiovisual presentations, practical demonstrations, and laboratory experiences in oxyacetylene, electric arc, MIG, TIG, shielded metal arc, gas metal arc, and flux core welding techniques and practices. Also covered are radiator, fuel tank, and metal body repair. *Version 2:* Audiovisual materials, practical exercises, laboratory, and lecture. Topics include metal fabrication tools and equipment soldering and oxy-level gas welding and cutting; shielded metal arc and gas metal arc welding techniques and practice; and automotive body, glass, radiator, and fuel tank repair.

Credit Recommendation: *Version 1:* In the lower-division baccalaureate/associate degree category, 3 semester hours in oxyacetylene and electric arc welding, 3 in MIG and TIG welding, and 2 in vehicle body repair (1/98). *Version 2:* In the lower-division baccalaureate/associate degree category, 3 semester hours in basic metals processing, 3 in metal bonding techniques and practice, and 2 in automotive body repair (8/03).

Related Occupations: 44B.

AR-1723-0013

1. Ordnance Advanced Noncommissioned Officer (ANCOC)
2. Advanced Noncommissioned Officer

Course Number: *Version 1:* 6-63-C42. *Version 2:* 6-63-C42, Phase 3.

Location: Ordnance Center and School, Aberdeen Proving Ground, MD.

Length: *Version 1:* 9–10 weeks (348 hours). *Version 2:* 3–4 weeks (123 hours).

Exhibit Dates: *Version 1:* 8/95–9/00. *Version 2:* 10/00–Present.

Learning Outcomes: *Version 1:* Upon completion of the course, the student will be able to serve as a supervisor in the selected military occupational specialty for which they are trained. *Version 2:* Upon completion of the course, the student will be able to demonstrate enhanced skills in leadership, communication and logistics management.

Instruction: *Version 1:* Lectures, demonstrations, and practical exercises cover leadership, communication skills, fundamentals of instruction, logistics management, and refresher training in the appropriate military occupational specialty. Emphasis is placed on leadership and on verbal and written communication. The course contains a common core of leadership subjects. *Version 2:* Practical and classroom exercises, discussions, learner presentations, lectures. Topics include ordnance history, stress management, platoon administration, automated logistics system, maintenance operations management, and field training exercises.

Credit Recommendation: *Version 1:* Credit is recommended for the common core only. See AR-1404-0035 (1/98). *Version 2:* In the lower-division baccalaureate/associate degree category, 2 semester hours in maintenance management (8/03).

AR-1723-0016

Basic Metal Worker

Course Number: 091-44B10, Phase 2.

Location: Reserve Schools, Multiple locations.

Length: 146 hours.

Exhibit Dates: 10/02–Present.

Learning Outcomes: Upon completion of the course, the student will be able to understand various techniques used in bonding (welding, soldering, etc.) metals, and fabricate and/or repair components or products by using various bonding techniques.

Instruction: Practical exercises, discussion, laboratory, and lecture. Topics include brief overview of metal fabrication and bonding techniques; laboratory practice in welding; soldering and other metal bonding processes; and safety in the metal fabricating and bonding laboratory.

Credit Recommendation: In the lower-division baccalaureate/associate degree category, 3 semester hours in basic metal bonding techniques and practice (8/03).

AR-1723-0017

Machinist

Course Number: 091-44E10, Phase 2.

Location: Ordnance Center and School, Aberdeen Proving Ground, MD.

Length: 93 hours.

Exhibit Dates: 10/02–Present.

Learning Outcomes: Upon completion of the course, the student will be able to function with the necessary technical, operational, and logistic skills as a machinist (lathe operator), level 1.

Instruction: Audiovisual materials, practical exercises, classroom exercises, discussion, laboratory, and lecture. Topics include lectures and numerous practical exercises in level 1 lathe operation, including turning, parting, grooving, filing, polishing, knurling, drilling, boring, reaming, threading, grinding, and tapping.

Credit Recommendation: In the lower-division baccalaureate/associate degree category, 3 semester hours in basic machine shop operations (8/03).

AR-1723-0018

Machinist

Course Number: 091-44E10, Phase 3.

Location: Ordnance Center and School, Aberdeen Proving Ground, MD.

Length: 86 hours.

Exhibit Dates: 10/02–Present.

Learning Outcomes: Upon completion of the course, the student will be able to function with the necessary technical, operational, and logistical skills as a machinist (milling operator), level 1.

Instruction: Audio visual materials, practical exercises, classroom exercises, discussion, laboratory, and lecture. Topics include lectures and numerous practical exercises in level 1 milling operation, including sawing, gear cutting, keywaying, versa-mil operating, drilling, grinding, boring, shaping, and indexing.

Credit Recommendation: In the lower-division baccalaureate/associate degree category, 3 semester hours in intermediate machine shop operations (8/03).

AR-1728-0007

Correctional Specialist
(Basic Corrections)
(Basic Correctional Specialist)

Course Number: 830-95C10; 95C10-OSUT; 831-95C10.

Location: Military Police School, Fort McClellan, AL.

Length: Self-paced, 6–14 weeks (224–671 hours).

Exhibit Dates: 2/79–4/92.

Objectives: To provide enlisted personnel with training necessary to perform duties as an entry-level correctional specialist.

Instruction: Seminars and self-paced instruction cover correctional security, custody, and control, basic marksmanship, and treatment of prisoners. This course may include up to eight weeks of basic training.

Credit Recommendation: In the lower-division baccalaureate/associate degree category, 3 semester hours in correctional security (3/82).

Related Occupations: 95C.

AR-1728-0008

Polygraph Examiner Training
(Lie Detection)
(Lie Detector Operation)

Course Number: 7H-F11.

Location: Military Police School, Fort McClellan, AL.

Length: 12–14 weeks (446–532 hours).

Exhibit Dates: 11/82–9/90.

Objectives: After 9/90 see DD-1728-0003. To qualify military and federal civilian investigative/intelligence personnel as polygraph examiners.

Instruction: This course covers lectures in polygraph theory, maintenance management, mental and physical evaluation of examinee, polygraph instrumentation and examination, post-test procedures and practical exercises in zone comparison, peak of tension, and general question techniques.

Credit Recommendation: In the upper-division baccalaureate category, 3 semester hours in interview and interrogation and 3 in applied psychology. In the graduate degree category, 6 semester hours in forensic science or criminal justice and 1—3 in research (based upon institutional evaluation of individual research project) (11/86).

Related Occupations: 951A; 951C; 951D; 951E; 951F; 951G.

AR-1728-0010

Criminal Investigation

Course Number: 7H-951A; 19-OE-12; 7H-951A/832-95D10; 832-95D10.

Location: Military Police School, Fort McClellan, AL.

Length: 15 weeks (559–582 hours).

Exhibit Dates: 11/80–12/97.

Objectives: To qualify enlisted personnel and warrant officers for criminal investigation duty in field units of the Army Criminal Investigation Command.

Instruction: This course covers operation and administration of criminal investigation units; methods and techniques of investigation; crime scene investigation; collection, evaluation, and preservation of evidence; death investigation; investigation of crimes against people; investigation of crimes against property; capabilities and limitations of scientific analysis of evidence; investigative photography; fingerprinting; report writing; and testifying in court.

Credit Recommendation: In the lower-division baccalaureate/associate degree category, 3 semester hours in criminal investigation,

3 in criminal evidence/court procedures, 3 in introduction to criminalistics, and 3 in commercial security. In the upper-division baccalaureate category, 3 semester hours in controlled substances (drugs) (11/86).

Related Occupations: 951A; 951C; 951D; 951E; 951F; 951G; 95D.

AR-1728-0023

BASIC MILITARY POLICE
(Law Enforcement)
(Military Policeman)

Course Number: 830-95B10; 95B10-OSUT; 830-95B10-OST; 830-95B10OST.

Location: Military Police School, Fort McClellan, AL; Military Police School, Fort Gordon, GA.

Length: 7–16 weeks (342–776 hours).

Exhibit Dates: 1/81–7/92.

Objectives: To train the soldier to become disciplined, highly motivated, and capable of performing the duties of an entry-level military policeman including law enforcement and nuclear physical security skills.

Instruction: This course includes military law, civil disturbance, interpersonal communications, and patrol incidents. May include up to eight weeks of basic training.

Credit Recommendation: In the lower-division baccalaureate/associate degree category, 3 semester hours in law enforcement technology (11/88).

Related Occupations: 95B.

AR-1728-0026

MILITARY POLICE OFFICER BASIC
(Military Police Officer Orientation)

Course Number: 7-19-C20-31A; 7-19-C20-31.

Location: Military Police School, Fort McClellan, AL; Provost Marshal General's School, Fort Gordon, GA.

Length: 15–16 weeks (649–782 hours).

Exhibit Dates: 1/84–6/91.

Objectives: After 6/91 see AR-1728-0090. To prepare newly commissioned military police corps officers to supervise police operations, traffic, investigations, and administrative units.

Instruction: Course includes leadership, unit operations, law, supervision of tactical police platoon, and personnel management.

Credit Recommendation: In the lower-division baccalaureate/associate degree category, 3 semester hours in personnel leadership and techniques of management and 3 in police operations. In the upper-division baccalaureate category, 3 semester hours in advanced military science (12/88).

AR-1728-0027

CID CERTIFICATION
(CID Orientation)
(Advanced Investigative Management)
(Criminal Investigation Supervision)

Course Number: 7H-F22/830-F15.

Location: Military Police School, Fort McClellan, AL.

Length: 2 weeks (73–76 hours).

Exhibit Dates: 8/85–12/98.

Objectives: To provide officers, warrant officers, and senior NCOs with management skills needed for selected management positions.

Instruction: This course includes operations and management; evidence, CID reports, administration, records and files, liaison, authority and jurisdiction, evidence, criminal intelligence, crime prevention, surveillance operations, raids, polygraph, protective service, release of information, search and apprehension, notes and sketches, quality control of photos, collection and preservation of evidence, evidence accountability, evidence depository, confessions and admissions, and criminal investigation procedures and reports. Emphasis is on management techniques through CID case studies.

Credit Recommendation: In the upper-division baccalaureate category, 3 semester hours in principles of management (12/88).

Related Occupations: 951A; 951C; 951D; 951E; 951F; 951G.

AR-1728-0028

MILITARY POLICE INVESTIGATOR
(Military Police Investigation)

Course Number: 830-ASIV5; 830-F8.

Location: Military Police School, Vilseck, Germany; Military Police School, Fort McClellan, AL; Military Police School, Fort Gordon, GA.

Length: 8 weeks (307 hours).

Exhibit Dates: 9/85–5/90.

Objectives: After 5/90 see AR-1728-0101. To qualify selected military policemen for duty as military police investigators.

Instruction: Topics include crime scene processing; testimonial evidence; management of sources of information; investigative planning; evaluating evidence; investigating drug, arson, and child abuse offenses; examination of questioned documents; and covert investigations.

Credit Recommendation: In the lower-division baccalaureate/associate degree category, 3 semester hours in criminal investigative techniques and 3 in a criminal justice elective (6/88).

Related Occupations: 95B.

AR-1728-0032

POLYGRAPH EXAMINER ADVANCED

Course Number: 7H-F10.

Location: Military Police School, Fort McClellan, AL.

Length: 3 weeks (110–118 hours).

Exhibit Dates: 10/76–6/96.

Objectives: To provide polygraph examiners with advanced training for certification or requalification.

Instruction: This course provides an update on technique advancements, proper equipment use, question and test structure, question formulation, chart interpretation, pre-test and post-test procedures, and advanced practical exercises.

Credit Recommendation: Credit is not recommended because of the refresher nature of the course and because the new material is insufficient to justify additional credit beyond that recommended for Polygraph Examiner Training (11/86).

Related Occupations: 951A; 951C; 951D; 951E; 951F; 951G.

AR-1728-0035

MILITARY POLICE OFFICER ADVANCED
(Military Police Officer Career)

Course Number: 7-19-C22.

Location: Military Police School, Fort McClellan, AL.

Length: 20 weeks (712 hours).

Exhibit Dates: 1/85–2/91.

Objectives: To prepare military police officers for command of mid-level Army units and to provide refresher training on basic officer skills for managerial duty.

Instruction: This course covers military leadership; logistics; chemical and biological operations; administration at the battalion level; oral and written communication; military organization, operations, and provost marshal functions; planning and conducting military police operations during domestic emergencies; military law; criminal investigation activities, records, and procedures; management, operation, and supervision of confinement facilities; traffic planning and control; methods of handling prisoners of war; command and staff procedures; military police communications; and command post and field exercises. The course also includes command and staff responsibilities, law enforcement operations, investigative supervision, computer literacy, and terrorism and security.

Credit Recommendation: In the lower-division baccalaureate/associate degree category, 3 semester hours in personnel supervision and management, 3 in police operations, and 2 in computer literacy. In the upper-division baccalaureate category, 3 semester hours in human relations in management, 3 in administrative case study, and additional credit based on institutional evaluation of student case study reports (11/86).

AR-1728-0056

DEPARTMENT OF DEFENSE STRATEGIC DEBRIEFING (DEPARTMENT OF DEFENSE STRATEGIC DEBRIEFING AND INTERROGATION TRAINING)

Course Number: 3A-SI3Q/3C-ASI9N/241-ASIN7; 3A-F37/241-F1.

Location: Intelligence School, Fort Huachuca, AZ.

Length: 4–6 weeks (223–262 hours).

Exhibit Dates: 6/83–4/04.

Objectives: To train personnel in strategic briefing and debriefing procedures and advanced interrogation techniques.

Instruction: Instruction includes report writing, security procedures, strategic intelligence, and interrogation including briefing and debriefing.

Credit Recommendation: In the lower-division baccalaureate/associate degree category, 3 semester hours in criminal justice (interviews and interrogations) (4/91).

AR-1728-0063

ADVANCED FRAUD INVESTIGATION

Course Number: 7H-ASI9D/832-ASIU7; 7H-F21/832-ASIU7.

Location: Military Police School, Fort McClellan, AL.

Length: 4 weeks (152 hours).

Exhibit Dates: 4/85–12/97.

Objectives: To qualify accredited Army Criminal Investigation Command special agents, warrant officers, and enlisted personnel in the investigation of fraud within Army logistics systems.

Instruction: Lectures and practical exercises cover interviews and interrogations, sources of information, economic crime investigation, covert operations, use of special investigation funds, evidence processing, investigative accounting and auditing, acquisitions fraud, computer-related crime, computers as investigative aids, presentation of cases to prosecutors, and testifying.

Credit Recommendation: In the upper-division baccalaureate category, 3 semester hours in special topics in criminal justice (4/87).

AR-1728-0064

SPECIAL AMMUNITION SECURITY
(Advanced Security)
(Nuclear Physical Security)

Course Number: 7H-F14/830-F10.
Location: Military Police School, Fort McClellan, AL.
Length: 2 weeks (74–77 hours).
Exhibit Dates: 11/81–8/99.
Objectives: To provide training in planning and applying advanced security measures to protect special weapons storage facilities and firing batteries against theft, espionage, sabotage, and other hazards.
Instruction: Lectures and practical exercises cover applying physical security techniques to special weapons facilities and supervising access control, lock and key control, ID operators, use of military police dogs, and employment of barriers and lighting. The course also includes evaluation of vulnerabilities, leadership and problem solving, and civil disobedience operations.
Credit Recommendation: Credit is not recommended because of the military-specific nature of the course (8/89).

AR-1728-0065

CONVENTIONAL SECURITY/CRIME PREVENTION
(Conventional Physical Security)

Course Number: 7H-31D/830-ASIH3.
Location: Military Police School, Fort McClellan, AL.
Length: 2 weeks (72–74 hours).
Exhibit Dates: 6/82–1/90.
Objectives: To provide enlisted, officer, warrant officer, and civilian personnel with a working knowledge of physical security measures to control pilferage, espionage, sabotage, and other hazards.
Instruction: Lectures and practical exercises cover legal aspects of security, physical security planning, protective lighting, ID systems, key control, bomb threat procedures, and security services.
Credit Recommendation: Credit is not recommended because of the limited, specialized nature of the course (2/90).

AR-1728-0067

HOSTAGE NEGOTIATIONS

Course Number: 7H-F19/830-F14 (CT); 7H-F19/830-F14.
Location: Military Police School, Fort McClellan, AL.
Length: 2 weeks (76 hours).
Exhibit Dates: 12/86–1/96.
Objectives: After 9/99 see AR-1728-0146. To provide training in the basic principles and tactics of hostage negotiations.
Instruction: Lectures and practical exercises are presented on managing a hostage situation. Topics include negotiations, lessons learned, bargaining techniques, communication, profiling the offender, and collecting data to resolve such activities.
Credit Recommendation: In the lower-division baccalaureate/associate degree category, 1 semester hour in special patrol tactics or special topics in law enforcement (11/86).

AR-1728-0068

INSTALLATION PROVOST MARSHAL

Course Number: 7H-F16.
Location: Military Police School, Fort McClellan, AL.
Length: 2 weeks (72 hours).
Exhibit Dates: 8/85–4/92.
Objectives: To provide newly-appointed installation provost marshals and deputy provost marshals with current doctrine pertaining to military police law enforcement operations, organizational concepts, and policies.
Instruction: Lectures cover current military police doctrine and procedures relating to management of military police operations, provost marshal responsibilities, traffic management, crime prevention, and budgeting.
Credit Recommendation: In the upper-division baccalaureate category, 1 semester hour in a criminal justice seminar (11/86).

AR-1728-0070

MILITARY POLICE PRE-COMMAND

Course Number: 2G-F39.
Location: Military Police School, Fort McClellan, AL.
Length: 2 weeks (78 hours).
Exhibit Dates: 4/85–7/91.
Objectives: To provide refresher training in logistics management, tactical operations, corrections, criminal investigations, and prisoner of war doctrine.
Instruction: Instruction includes logistics, training management, law, and military police operations.
Credit Recommendation: Credit is not recommended because of the limited, specialized nature of the course (11/86).

AR-1728-0071

PROTECTIVE SERVICES TRAINING

Course Number: 7H-F18/830-F13.
Location: Military Police School, Fort McClellan, AL.
Length: 3 weeks (116–135 hours).
Exhibit Dates: 10/86–9/97.
Objectives: After 1/00 see AR-1728-0141. To train personnel in the planning and execution of personnel (dignitary) protection.
Instruction: This course includes instruction in methods of motorcade security, advance operations and threat assessment, building and vehicle searches, and emergency medical procedures.
Credit Recommendation: In the lower-division baccalaureate/associate degree category, 2 semester hours in personnel security (6/90).

AR-1728-0072

SPECIAL REACTION TEAM TRAINING

Course Number: 7H-F17/830-F12.
Location: Military Police School, Fort McClellan, AL.
Length: 2–3 weeks (99–139 hours).
Exhibit Dates: 10/84–8/91.
Objectives: After 12/99 see AR-1728-0144. To provide training in high-risk and specialized patrol operations to special reaction teams.
Instruction: Lectures and practical exercises cover special reaction response to terrorism activities, hostage situations, and the barricaded offender. The use of special antiterrorist equipment is included.
Credit Recommendation: In the lower-division baccalaureate/associate degree category, 1 semester hour in special patrol operations (3/89).

AR-1728-0074

CORRECTIONS ADMINISTRATION

Course Number: 7H-F20.
Location: Military Police School, Fort McClellan, AL.
Length: 2 weeks (80 hours).
Exhibit Dates: 8/85–4/92.
Objectives: To provide commissioned officers with a working knowledge of the correctional principles, theories, and practices which are required for administering, operating, and supervising Army corrections facilities at posts, camps, or stations.
Instruction: This course includes management, operation, and administration of installation confinement facilities; history of penal treatment; stockade organization; legal considerations of confinement; prisoner accountability; stockade logistics; confessions; report writing; disciplinary measures; prisoner counseling; problem prisoners; sentence computation.
Credit Recommendation: Credit is not recommended because of the limited, specialized nature of the course (11/86).

AR-1728-0076

1. EXPLOSIVE ORDNANCE DISPOSAL SPECIALIST BASIC NONCOMMISSIONED OFFICER (NCO)
2. EXPLOSIVE ORDNANCE DISPOSAL SPECIALIST BASIC NONCOMMISSIONED OFFICER (NCO)
3. EXPLOSIVE ORDNANCE DISPOSAL SPECIALIST BASIC NONCOMMISSIONED OFFICER (NCO), PHASE 2

Course Number: ***Version 1:*** 431-55D30. *Version 2:* 431-55D30. *Version 3:* 431-55D30, Phase 2.
Location: Ordnance, Missile and Munitions School, Redstone Arsenal, AL.
Length: ***Version 1:*** 6–7 weeks (239–244 hours). *Version 2:* 6–7 weeks (259 hours). *Version 3:* 6–7 weeks (237 hours).
Exhibit Dates: ***Version 1:*** 5/88–12/94. *Version 2:* 1/95–9/00. *Version 3:* 10/00–Present.
Learning Outcomes: ***Version 1:*** Upon completion of the course, the student will be able to identify explosive materials, direct disposal activities, and deactivate conventional and nonconventional explosive devices. *Version 2:* Upon completion of the course, the student will be able to perform procedures on unexploded ordnance (conventional, nuclear, and/or chemical) during incidents/accidents that involve either known or unknown munitions and devices. *Version 3:* Upon completion of the course, the student will be able to provide leadership in identifying explosive materials, directing disposal activities, reporting technical intelligence reporting, in developing render-

safe procedures on unknown ordnance and improvised explosive devices (IEDs), and supervising of emergency personnel decontamination operations.

Instruction: *Version 1:* Lectures and field exercises use typical military ordnance in simulated situations. Course contains a core of leadership subjects. *Version 2:* Lectures and field exercises use typical military ordnance in simulated situations. End-of-course evaluation is by a comprehensive written examination on selected tasks. The course contains a common core of leadership subjects. *Version 3:* Conferences and practical exercises using typical military ordnance in simulated situations.

Credit Recommendation: *Version 1:* In the lower-division baccalaureate/associate degree category, 2 semester hours in explosives handling. See AR-1406-0090 for common core credit (10/91). *Version 2:* In the lower-division baccalaureate/associate degree category, 2 semester hours in explosive disposal and handling. See AR-1406-0090 for common core credit (7/96). *Version 3:* In the lower-division baccalaureate/associate degree category, 2 semester hours in explosive disposal and handling (6/01).

Related Occupations: 55D.

AR-1728-0077

INSTALLATION SECURITY

Course Number: MP-30.

Location: Combined Arms Training Center, Vilseck, W. Germany.

Length: 1–2 weeks (56 hours).

Exhibit Dates: 4/80–10/98.

Learning Outcomes: Upon completion of the course, the student will be able to apply the principles and characteristics of physical security to the protection of an installation, motor pool, vehicles, medical supplies, aircraft, weapons, arms and ammunition, unit mail rooms, automatic data processing equipment, supply storage areas, and other locations of a sensitive nature; conduct physical security surveys and inspections; conduct internal and external vulnerability assessment; conduct physical security education and training; assist in the drafting of a physical security plan for disaster incidents; determine the proper use of security devices, including perimeter barriers, intrusion detection systems, protective lighting, locks and keys, guard force and dogs, and identification and pass control of individuals; and identify the characteristics of terrorism and methods of countering terrorism and handling bomb threats.

Instruction: Lectures cover military regulations and field training manuals.

Credit Recommendation: In the vocational certificate category, 1 semester hour in physical security measures or 1 semester hour in crime prevention (10/88).

AR-1728-0078

TRAFFIC ACCIDENT INVESTIGATION

Course Number: MP-23.

Location: Combined Arms Training Center, Vilseck, W. Germany.

Length: 3 weeks (104 hours).

Exhibit Dates: 6/85–10/98.

Learning Outcomes: Upon completion of the course, the student will be able to function as a traffic accident investigator with specific expertise in countries having a status of forces agreement with the United States; prepare written reports reflecting the cause of traffic accidents, the military and civilian laws involved, specific evidence collected and processed; prepare traffic accident (crime scene) sketches and obtain photographs; effect legal detentions, apprehensions, search, seizures, inspections, and frisks of suspected individuals; conduct interviews and interrogations of suspects and witnesses; obtain written statements, confessions, admissions, dying declarations, and spontaneous exclamations; determine types of traffic enforcement techniques needed in a particular location; and apply the methods for determining speed violations and the proper procedures for the apprehension/detention of intoxicated drivers.

Instruction: Lectures cover military law, apprehension, search and seizure, confessions, admissions and statements, elements of proof, interviews and interrogations, traffic accidents and their causes, traffic accident investigations, examination of the vehicle, collection and preservation of evidence, identification of tire skid marks and scuffs, photographs in traffic accidents, investigation records and forms, traffic enforcement techniques, intoxicated drivers, and hit and run investigation. Also included are lectures and practical exercises in the use of mathematical formulas in traffic investigations, the use of a traffic accident template, measurements at a traffic accident, and preparation of sketches and diagrams.

Credit Recommendation: In the vocational certificate category, 3 semester hours in traffic accident investigation. In the lower-division baccalaureate/associate degree category, 3 semester hours elective credit in traffic accident investigation (10/88).

AR-1728-0080

COUNTERINTELLIGENCE AGENT TRANSITION

Course Number: 244-97B20-T.

Location: Intelligence School, Fort Huachuca, AZ.

Length: 10–11 weeks (386–393 hours).

Exhibit Dates: 4/87–4/04.

Learning Outcomes: Upon completion of the course, the student will be able to perform counterintelligence operations, including conducting counterintelligence investigations, collecting intelligence, planning and conducting surveillance, applying legal principles to counterintelligence operations, and conducting counterintelligence interviews.

Instruction: Lectures, simulated field environment exercises, and classroom testing cover counterintelligence activities.

Credit Recommendation: In the lower-division baccalaureate/associate degree category, 3 semester hours in criminal investigation (investigation and interviews) (6/90).

Related Occupations: 97B.

AR-1728-0081

COUNTERINTELLIGENCE AGENT BASIC NONCOMMISSIONED OFFICER (NCO)

Course Number: 244-97B30.

Location: Intelligence School, Fort Huachuca, AZ.

Length: 7–8 weeks (271–278 hours).

Exhibit Dates: 1/90–9/00.

Learning Outcomes: After 9/00 see AR-1728-0155. Upon completion of the course, the student will be able to apply legal principles to counterintelligence operations, conduct counterintelligence investigations of espionage and sabotage activities, and prepare an investigation.

Instruction: Lectures, classroom testing, and practical exercises cover counterintelligence activities. This course contains a common core of leadership subjects.

Credit Recommendation: In the lower-division baccalaureate/associate degree category, 2 semester hours in investigative techniques. See AR-1406-0090 for common core credit (5/01).

Related Occupations: 97B.

AR-1728-0082

COUNTERINTELLIGENCE AGENT

Course Number: 244-97B20.

Location: Intelligence School, Fort Huachuca, AZ.

Length: 19–20 weeks (830 hours).

Exhibit Dates: 1/88–4/04.

Learning Outcomes: Upon completion of the course, the student will be able to perform counterintelligence operations (conflict analysis), conduct counterintelligence investigations (terrorism, surveillance), conduct security inspections (interviews, planning), and apply legal principles to counterintelligence operations (constitutional law, US codes).

Instruction: Lectures and simulation exercises cover counterintelligence operations and investigations.

Credit Recommendation: In the lower-division baccalaureate/associate degree category, 3 semester hours in investigative principles, 1 in legal principles, and 3 in an investigative/security practicum (6/90).

Related Occupations: 97B.

AR-1728-0083

INTELLIGENCE ANALYST

Course Number: *Version 1:* 243-96B10. *Version 2:* 243-96B10.

Location: Intelligence School, Fort Huachuca, AZ.

Length: *Version 1:* 13–15 weeks (515–587 hours). *Version 2:* 16 weeks (535–536 hours).

Exhibit Dates: *Version 1:* 10/86–9/96. *Version 2:* 10/96–9/00.

Learning Outcomes: *Version 1:* Upon completion of the course, the student will be able interpret topographical maps, describe threat forces in the world, describe personnel security systems, process intelligence reports, and compile data for information systems. *Version 2:* After 9/00 see AR-1728-0151. Upon completion of the course, the student will be able interpret topographical maps, describe threat forces in the world, describe personnel security systems, process intelligence reports, and compile data for information systems.

Instruction: *Version 1:* Lectures, classroom testing, and field evaluation exercises cover intelligence activities. *Version 2:* Instruction consists of lectures, classroom exercises, and field application exercises in topographical analysis; information collection, processing, and analysis using data base applications; and security procedures for information and personnel.

Credit Recommendation: *Version 1:* In the lower-division baccalaureate/associate degree category, 3 semester hours in map reading, 3 in principles of information management, and 1 in security of information/personnel (1/92). *Version 2:* In the vocational certificate category, 1 semester hour in information/personnel security. In the lower-division baccalaureate/associate degree category, 3 semester hours in map analysis and 3 in data base applications (5/01).

Related Occupations: 96B.

AR-1728-0084

INTERROGATOR

Course Number: ***Version 1:*** 241-96C10; 241-97E10. *Version 2:* 241-97E10. *Version 3:* 241-97E10. *Version 4:* 241-97E10.

Location: Intelligence School, Fort Huachuca, AZ.

Length: ***Version 1:*** 9–14 weeks (337–561 hours). *Version 2:* 9–10 weeks (354 hours). *Version 3:* 8–9 weeks (338 hours). *Version 4:* 15–16 weeks (512–632 hours).

Exhibit Dates: ***Version 1:*** 3/85–12/91. *Version 2:* 1/92–9/95. *Version 3:* 10/95–9/00. *Version 4:* 10/00–Present.

Learning Outcomes: ***Version 1:*** Upon completion of the course the student will be able to read maps, coordinate the phases of an interrogation, assess and control the interrogation during questioning, interrogate using an interpreter, and use basic military terminology. *Version 2:* Upon completion of the course, the student will be able to coordinate the phases of an interrogation, assess and control the interrogation during questioning, and interrogate using an interpreter. *Version 3:* Upon completion of the course, the student will be able to coordinate the phases of an interrogation, assess and control the interrogation during questioning, and interrogate using an interpreter. *Version 4:* Upon completion of the course, the student will be able to read maps, coordinate the phases of an interrogation, assess and control the interrogation during questioning, interrogate using an interpreter, and use basic military terminology.

Instruction: ***Version 1:*** Small class lectures, simulated interrogation in a classroom environment, and field training exercises are instructional methodologies. *Version 2:* Small class lectures, simulated interrogation in a classroom environment, and field training exercises are instructional methodologies applied in presenting interrogation principles; document analysis; and planning, conducting, and terminating an interrogation. *Version 3:* Small class lectures, simulated interrogation in a classroom environment, and field training exercises cover interrogation principles, interrogation planning and preparation, approaches and termination, and in screening applications. *Version 4:* Audiovisual materials, practical and classroom exercises, discussion, case studies, laboratory. Topics include small class lectures, simulated interrogation in a classroom environment, field training exercises that cover interrogation principles, interrogation planning and preparation, approaches and termination, and in screening applications.

Credit Recommendation: ***Version 1:*** In the lower-division baccalaureate/associate degree category, 1 semester hour in map reading and 6 in criminal investigation (interrogation/interviewing) (6/90). *Version 2:* In the lower-division baccalaureate/associate degree category, 1 semester hour in map analysis and 6 in interrogation techniques (6/97). *Version 3:* In the lower-division baccalaureate/associate degree category, 6 semester hours in criminal investigation (interrogation/interviewing) (5/01). *Version 4:* In the lower-division baccalaureate/associate degree category, 3 semester hours in interviewing and interrogation (12/02).

Related Occupations: 96C; 97E.

AR-1728-0085

MILITARY INTELLIGENCE OFFICER ADVANCED COUNTERINTELLIGENCE

Course Number: 3-30-C22-35E.

Location: Intelligence School, Fort Huachuca, AZ.

Length: 10 weeks (354 hours).

Exhibit Dates: 3/88–9/99.

Learning Outcomes: Upon completion of the course, the student will conduct investigations and surveillance, interview subjects and suspects, and apply security measures to intelligence operations.

Instruction: Instruction includes lectures and practical exercises in theory and legal principles of investigation, physical security, sabotage and terrorism, surveillance planning and procedures, interviewing, interrogation, personnel security, and planning and conducting human intelligence operations.

Credit Recommendation: In the lower-division baccalaureate/associate degree category, 3 semester hours in criminal investigation and 2 in physical security (5/01).

Related Occupations: 35E.

AR-1728-0086

1. COUNTERINTELLIGENCE ASSISTANT
2. COUNTERINTELLIGENCE AGENT

Course Number: ***Version 1:*** 244-97B10. *Version 2:* 244-97B10.

Location: Intelligence School, Fort Huachuca, AZ.

Length: ***Version 1:*** 10–11 weeks (377 hours). *Version 2:* 17 weeks (589–652 hours).

Exhibit Dates: ***Version 1:*** 1/88–9/90. *Version 2:* 10/90–9/00.

Learning Outcomes: ***Version 1:*** Upon completion of the course, the student will be able to perform personnel security, identify the objectives of a security program, perform a security manager's inspection, identify and describe automation sensitivity levels and security operating modes, and apply legal principles. *Version 2:* Upon completion of the course, the student will be able to perform personnel security, identify the objectives of a security program, perform a security manager's inspection, identify and describe automation sensitivity levels and security operating modes, and apply legal principles.

Instruction: ***Version 1:*** Lectures, testing, and field training cover the subject matter. *Version 2:* Lectures, testing, and field training cover the subject matter.

Credit Recommendation: ***Version 1:*** In the lower-division baccalaureate/associate degree category, 3 semester hours in security techniques and 1 in legal principles (6/90). *Version 2:* In the lower-division baccalaureate/associate degree category, 3 semester hours in security techniques and 1 in legal principles (5/01).

Related Occupations: 97B.

AR-1728-0087

MILITARY INTELLIGENCE OFFICER TRANSITION (Tactical All-Source Intelligence Officer Force Alignment Plan 3)

Course Number: 3A-35D; 3A-35D (T).

Location: Intelligence School, Fort Huachuca, AZ.

Length: 7–9 weeks (258–310 hours).

Exhibit Dates: 12/86–9/99.

Learning Outcomes: After 9/99 see AR-1728-0159. Upon completion of the course, the student will be able to implement physical security measures, perform information and personnel security, analyze threat forces, and define functions of battlefield preparation and tactical intelligence.

Instruction: Lectures and field exercises cover tactical intelligence operations.

Credit Recommendation: In the lower-division baccalaureate/associate degree category, 3 semester hours in introduction to security (personnel/information) (5/01).

Related Occupations: 35D.

AR-1728-0088

LEGAL SPECIALIST/COURT REPORTER ADVANCED NONCOMMISSIONED OFFICER (NCO)

Course Number: 512-71D/E40.

Location: Adjutant General School, Fort Benjamin Harrison, IN.

Length: 8–9 weeks (330 hours).

Exhibit Dates: 1/92–1/97.

Learning Outcomes: Upon completion of the course, the student will be able to prepare wills/powers of attorney, apply appropriate doctrine in establishing, maintaining, and sustaining legal support; provide necessary skills to manage/supervise a legal office; and manage a legal training program.

Instruction: Lectures and practical exercises cover the preparation of wills/powers of attorney and practical exercises in training management.

Credit Recommendation: In the lower-division baccalaureate/associate degree category, 1 semester hour in wills, trusts, and probate administration; 3 in law office management; and 1 in leadership (12/91).

Related Occupations: 71D; 71E.

AR-1728-0089

LEGAL SPECIALIST PHASE 2

Course Number: 512-71D10-RC.

Location: Adjutant General School, Fort Benjamin Harrison, IN.

Length: 2 weeks (113 hours).

Exhibit Dates: 1/90–1/97.

Learning Outcomes: Upon completion of the course, the student will be able to research and prepare court martial charge sheets, research and prepare court martial convening orders, research and prepare referrals to trial, conduct legal research, draft documents, identify and prepare charges, and use word processing equipment and appropriate software.

Instruction: Lectures and practical exercises cover word processing, researching documents, and legal writing.

Credit Recommendation: In the lower-division baccalaureate/associate degree category, 1 semester hour in word processing and 4 in legal research and writing (12/91).

Related Occupations: 71D.

AR-1728-0090

MILITARY POLICE OFFICER BASIC

Course Number: *Version 1:* 7-19-C20-31A; 7-19-C20-31. *Version 2:* 7-19-C20-31A; 7-19-C20-31.

Location: *Version 1:* Military Police School, Fort McClellan, AL. *Version 2:* Military Police School, Fort Leonard Wood, MO.

Length: *Version 1:* 15 weeks (628 hours). *Version 2:* 15 weeks (700–715 hours).

Exhibit Dates: *Version 1:* 7/91–9/97. *Version 2:* 10/97–12/00.

Learning Outcomes: *Version 1:* Before 7/91 see AR-1728-0026. Upon completion of the course, the student will be able to train and supervise military police in police operations, traffic investigations, and police administration; maintain the material readiness of the platoon; and lead a military police platoon in combat. *Version 2:* Upon completion of the course, the student will be able to train, supervise, and lead a military police platoon in the application of military police operations in a military environment.

Instruction: *Version 1:* Leadership, management, and supervision of a police unit are the main topics. Course also covers introduction to law enforcement, patrol operations, investigations, traffic, crime scene processing, use of force, records, interviewing, terrorism, military science, and military law. *Version 2:* Lectures and practical exercises cover leadership, management, and supervision of a police unit. Course also covers introduction to law enforcement, patrol operations, investigations, traffic, crime scene processing, use of force, records, interviewing, terrorism, military science, and military law. Student will be familiar with basic platoon offensive and defensive operations.

Credit Recommendation: *Version 1:* In the lower-division baccalaureate/associate degree category, 3 semester hours in police supervision and 3 in police operations. In the upper-division baccalaureate category, 3 semester hours in advanced military science (6/92). *Version 2:* In the upper-division baccalaureate category, 3 semester hours in advanced military science and 1 in police supervision (2/01).

AR-1728-0091

Military Police Officer Advanced

Course Number: 7-19-C22.

Location: Military Police School, Fort McClellan, AL.

Length: 20 weeks (739 hours).

Exhibit Dates: 3/91–9/97.

Learning Outcomes: After 9/98 see AR-1728-0147. Upon completion of the course, the student will be able to manage mid-level Army units.

Instruction: The course is battle focused and stresses technical, tactical, and leadership skills; military ethics; and military customs. All training is conducted in the five-phase leader excellence scenario.

Credit Recommendation: In the lower-division baccalaureate/associate degree category, 3 semester hours in personnel supervision, 3 in physical security, and 3 in computer applications. In the upper-division baccalaureate category, 3 semester hours in operational planning and control and 2 in case studies in management (5/92).

AR-1728-0092

Criminal Investigation Division (CID) Senior Warrant Officer

Course Number: 7-19-C32.

Location: Military Police School, Fort McClellan, AL.

Length: 4 weeks (144 hours).

Exhibit Dates: 10/88–2/98.

Learning Outcomes: After 2/00 see AR-1728-0138. Upon completion of the course, the student will be able to serve as a criminal investigation division district or field office operations officer.

Instruction: Course topics covered include legal issues, office management, personnel management, resource management, principles of organization, supervision, inspection systems, criminal profiling, and managing hostage/terrorist negotiations.

Credit Recommendation: In the lower-division baccalaureate/associate degree category, 3 semester hours in principles of supervision. In the upper-division baccalaureate category, 3 semester hours in principles of management (6/92).

AR-1728-0093

Special Reaction Team Training Phase 1 and Special Reaction Training Marksman/Observer Phase 2

Course Number: 7H-F17/830-F12.

Location: Military Police School, Fort McClellan, AL.

Length: Phase 1, 2 weeks (99 hours); Phase 2, 1 week (51 hours).

Exhibit Dates: 8/91–11/97.

Learning Outcomes: After 1/00 see AR-1728-0129. Upon completion of Phase 1, the student will be able to combat terrorism including hostage situation management; use firearms instinctively; and report surveillance/intelligence activities. Upon completion of Phase 2 (Marksman/Observer), the student will be able to perform those specialized physical and technical tasks needed to serve as a member of a special reaction team.

Instruction: Phase 1 provides lectures, demonstrations, and performance exercises in special reaction team responses to terrorism activities, hostage situations, barricaded subjects, and drug raids. Phase 2 includes shooting positions, examinations, and qualification on special reaction team firearms.

Credit Recommendation: In the lower-division baccalaureate/associate degree category, 2 semester hours in patrol procedures. Credit is to be awarded only upon completion of both phases (6/92).

AR-1728-0094

Conventional Physical Security/Crime Prevention

Course Number: 7H-31D/830-ASIH3.

Location: Military Police School, Fort McClellan, AL.

Length: 2 weeks (72 hours).

Exhibit Dates: 2/90–1/98.

Learning Outcomes: After 9/00 see AR-1728-0137. Upon completion of the course, the student will be able to plan and implement physical security/crime prevention measures to control pilferage, espionage, sabotage, and other physical security threats.

Instruction: Lectures, tests, and performance exercises cover legal aspects of security, crime prevention, physical security, planning, intrusion detection, crime analysis, and security services.

Credit Recommendation: In the lower-division baccalaureate/associate degree category, 2 semester hours in crime prevention and control (6/92).

AR-1728-0095

Counter Drug Special Weapons and Tactics

Course Number: 7H-F38.

Location: Military Police School, Fort McClellan, AL.

Length: 1 week (56–57 hours).

Exhibit Dates: 11/91–10/97.

Learning Outcomes: After 12/99 see AR-1728-0127. Upon completion of the course, the student will be able to use special weapons and tactics to conduct raids on houses, woodland areas, and convoy areas used to conceal and transport narcotics.

Instruction: Lectures, demonstrations, and performance exercises cover special weapons and tactics, raid preplanning, building entry techniques, and explosive identification.

Credit Recommendation: Credit is not recommended because of the military-specific nature of the course content (5/92).

AR-1728-0096

Counter Drug Field Tactical Police Operations

Course Number: 7H-F36.

Location: Military Police School, Fort McClellan, AL.

Length: 1 week (48 hours).

Exhibit Dates: 11/91–10/97.

Learning Outcomes: Upon completion of the course, the student will be able to perform field tactical police operations.

Instruction: Lectures, demonstrations, and performance exercises cover criminal/tactical intelligence, drug identification, risk management, map reading, land navigation; and raid planning.

Credit Recommendation: Credit is not recommended because of the military-specific nature of the course (6/92).

AR-1728-0097

Criminal Investigation Division (CID) Warrant Officer Technical and Tactical Certification

Course Number: 7H-311A.

Location: Military Police School, Fort McClellan, AL.

Length: 5–6 weeks (186 hours).

Exhibit Dates: 12/90–2/00.

Learning Outcomes: After 2/00 see AR-1728-0142. Upon completion of the course, the student will be able to perform as a team chief and special agent in charge of smaller criminal investigation field elements.

Instruction: Lectures, demonstration, and performance exercises cover leadership situation training exercises and tactical skills in the field.

Credit Recommendation: In the lower-division baccalaureate/associate degree category, 3 semester hours in police supervision (6/92).

Related Occupations: 311A.

AR-1728-0098

Criminal Investigation Division (CID) Warrant Officer Technical and

Tactical Certification Reserve Phase 2

Course Number: 7H-311A-RC.

Location: Military Police School, Fort McClellan, AL.

Length: 2 weeks (72 hours).

Exhibit Dates: 10/88–4/04.

Learning Outcomes: Upon completion of the resident phase (Phase 2) of the course, the student will be able to supervise and manage investigative systems, programs, and operations.

Instruction: Lectures and performance exercises in Phase 2, the resident phase, cover management of a major crime, management of criminal intelligence operations, and supervision of a criminal investigation division source program.

Credit Recommendation: Credit is not recommended because of the military-specific nature of the content. This is based on Phase 2, the resident phase (6/92).

Related Occupations: 311A.

AR-1728-0100

Military Police Advanced Noncommissioned Officer (NCO)

Course Number: 8-95-C42.

Location: Military Police School, Fort McClellan, AL.

Length: 11–12 weeks (485–519 hours).

Exhibit Dates: 3/89–3/00.

Learning Outcomes: After 3/00 see AR-1728-0143. Upon completion of the course, the student will be able to perform the technical duties of platoon sergeant, operations sergeant, confinement supervisor, or chief of investigative support, depending on the military occupational specialty in which the student concentrates.

Instruction: Lectures and applied exercises cover leadership, police equipment, police tactics, battlefield control, military police operations, and advanced noncommissioned officer duties. Separate tracks beyond the common core provide specific concentration in law enforcement operations (95B); corrections/confinement (95C); or criminal investigation (95D).

Credit Recommendation: In the lower-division baccalaureate/associate degree category, for those in the 95B track, 2 semester hours in law enforcement operations. For those in the 95C track, 3 semester hours in correctional supervision. For those in the 95D track, 3 semester hours in administration (6/92).

Related Occupations: 95B; 95C; 95D.

AR-1728-0101

Military Police Investigator

Course Number: 830-ASIV5.

Location: Military Police School, Fort McClellan, AL; Military Police School, Vilseck, Germany.

Length: 8 weeks (316 hours).

Exhibit Dates: 6/90–9/99.

Learning Outcomes: Before 6/90 see AR-1728-0028. Upon completion of the course, the student will be able to perform duties as a public investigator in the fields of criminal investigation, confession and admission, fingerprint evidence, forensic science, crime scene processing, and processing of questioned documents.

Instruction: Lectures, demonstrations, and performance exercises cover law, forensic evidence, testimonial evidence, crime scene processing, investigation of crimes, and special operations.

Credit Recommendation: In the lower-division baccalaureate/associate degree category, 3 semester hours in criminal investigation and 3 in crime scene processing (6/92).

AR-1728-0102

Military Police Basic Noncommissioned Officer (NCO)

Course Number: 830-95B/C/D30.

Location: Military Police School, Fort McClellan, AL.

Length: 8–9 weeks (347–352 hours).

Exhibit Dates: 3/89–9/98.

Learning Outcomes: After 3/00 see AR-1728-0128. Upon completion of the course, the student will be able to serve as a team leader or squad leader in support of combat services.

Instruction: Lectures and practical applications cover combat support, tactics, police missions, equipment systems, leadership, weapons, and security.

Credit Recommendation: In the lower-division baccalaureate/associate degree category, 3 semester hours in law enforcement operations and 1 in personnel supervision (5/92).

Related Occupations: 95B; 95C; 95D.

AR-1728-0103

Corrections Noncommissioned Officer (NCO)

Course Number: 830-95C20.

Location: Military Police School, Fort McClellan, AL.

Length: 4 weeks (144–145 hours).

Exhibit Dates: 4/89–4/04.

Learning Outcomes: Upon completion of the course, the student will describe the legal aspects of confinement, practice physical control of prisoners, and administer correctional services.

Instruction: Lectures, demonstrations, and performance exercises cover corrections, corrections administration, prisoner supervision, interpersonal communication with offenders, records management, and contemporary issues in corrections.

Credit Recommendation: In the lower-division baccalaureate/associate degree category, 2 semester hours in correctional administration/correctional elective (5/92).

Related Occupations: 95C.

AR-1728-0104

Apprentice Criminal Investigation Division (CID) Special Agent

Course Number: 832-95D20/30 (CT).

Location: Military Police School, Fort McClellan, AL.

Length: 15 weeks (575 hours).

Exhibit Dates: 4/88–10/99.

Learning Outcomes: After 10/99 see AR-1728-0136. Upon completion of the course, the student will be able to conduct criminal investigations, process crime scenes, collect and preserve physical evidence, write investigative reports, conduct covert operations, and plan and conduct raids.

Instruction: Lectures, demonstrations, and practical exercises cover criminal law, fundamentals of criminal investigation, crime scene processing, crimes against persons and property, physical evidence, drug investigations, fraud and fraud investigations, reports, covert operations, raids, surveillance, interviews and interrogations, and protective services.

Credit Recommendation: In the lower-division baccalaureate/associate degree category, 3 semester hours in criminal investigation, 3 in criminalistics, and 3 in interview and interrogation (6/92).

Related Occupations: 95D.

AR-1728-0105

Apprentice Criminal Investigation Division (CID) Special Agent Reserve Component Phase 2 Resident

Course Number: 832-95D20/30-RC, Phase 2.

Location: Military Police School, Fort Leonard Wood, MO; Military Police School, Fort McClellan, AL.

Length: 2 weeks (83–84 hours).

Exhibit Dates: 10/87–Present.

Learning Outcomes: Upon completion of the course, the student will be able to process crime scenes, collect and preserve physical evidence, and conduct crime preventive services.

Instruction: Lectures and practical exercises cover crime scene processing, physical evidence, and fraud and waste.

Credit Recommendation: In the lower-division baccalaureate/associate degree category, 3 semester hours in evidence and criminal procedures (5/00).

Related Occupations: 95D.

AR-1728-0106

Apprentice Criminal Investigation Division (CID) Special Agent Reserve Component Phase 4 Resident

Course Number: 832-95D20/30 (T), Phase 4.

Location: Military Police School, Fort Leonard Wood, MO; Military Police School, Fort McClellan, AL.

Length: 2 weeks (87–88 hours).

Exhibit Dates: 2/87–Present.

Learning Outcomes: Upon completion of the course, the student will be able to conduct specialized criminal investigations, participate in protective services operations, write investigative reports, and conduct interviews and interrogations.

Instruction: Lectures and practical exercises cover special investigative techniques in protective service operations, interviewing and interrogation, and in investigative reports. This credit recommendation is based on phase 4, the resident phase.

Credit Recommendation: In the lower-division baccalaureate/associate degree category, 3 semester hours in criminal investigation. This credit recommendation is based on phase 4, the resident phase (5/00).

Related Occupations: 95D.

AR-1728-0107

Basic Military Police/Basic Correction Specialist Phases 1—4, Phase 5 (95B) and Phase 5 (95C)

Course Number: 95B10-OSUT (ST); 95B10-OSUT; 95C10-OSUT; 831-95C10; 830-95B10/95C10; 830-95B10; 830-95C10.

Location: Military Police School, Fort McClellan, AL.

Length: Phases 1-4, 11 weeks (458–459 hours); Phase 5 (95B), 5 weeks (180–181 hours); Phase 5 (95C), 5 weeks (204 hours).

Exhibit Dates: 8/92–8/00.

Learning Outcomes: Upon completion of the course, the student will be able to demonstrate discipline, motivation, physical readiness, and military skills for combat survivability.

Instruction: Lectures and practical exercises include first aid, physical education, physical security, military knowledge and skills, and field training. Student will complete this course (Phase 1), then go to career-specific training in either the military police or corrections specialists fields. The basic military police track, 95B, includes lectures and practical exercises in human behavior, criminal law, patrol procedures, interviewing, and report writing. The corrections specialist track, 95C, includes classroom discussions and lectures in introduction to corrections, administration of facilities, prisoner administration and services, customs and control, investigations, and human behavior. This track also includes practical exercises using a mock confinement facility.

Credit Recommendation: In the lower-division baccalaureate/associate degree category, for Phase 5, 95B track, 3 semester hours in human behavior, 1 in basic criminal law, and 3 in patrol procedures. For Phase 5, 95C track, 3 semester hours in corrections procedures, and 3 in corrections supervision (6/92).

Related Occupations: 95B; 95C.

AR-1728-0111

Fire Protection Technology

Course Number: C3AZR57170 001.

Location: Training Center, Chanute AFB, IL.

Length: 2 weeks (72 hours).

Exhibit Dates: 3/89–4/04.

Learning Outcomes: Upon completion of the course, the student will be able to employ tactics and strategy to establish command and control of fire situations, combat an aircraft/flight-line incident, develop a contingency plan for specific types of operations, and develop a prefire plan for a structural fire situation.

Instruction: Informal lectures, practical exercises, and demonstrations in fire operations include command and control, tactics and strategy, hazardous materials, aircraft fire suppression, contingency operations, and structural fire suppression.

Credit Recommendation: In the lower-division baccalaureate/associate degree category, 3 semester hours in fire fighting command (1/93).

AR-1728-0112

Fire Protection Management Principles
(Army 51M20/30)

Course Number: C3AZR57170 009.

Location: Training Center, Chanute AFB, IL.

Length: 3 weeks (120 hours).

Exhibit Dates: 7/87–4/04.

Learning Outcomes: Upon completion of the course, the student will be able to perform duties of assistant chief, station chief, fire team chief within a fire department, including personnel qualifications, fire protection program staffing requirements, training, communication, records, mobilization, maintenance functions, fire investigations, occupational safety, rescue concepts, and fire fighting principles.

Instruction: Lectures, demonstrations, and practical exercises cover the above topics.

Credit Recommendation: In the lower-division baccalaureate/associate degree category, 4 semester hours in principles of fire service management (1/93).

Related Occupations: 51M.

AR-1728-0113

Apprentice Fire Protection Specialist
(Army 51M10)

Course Number: C3ABR57130 002.

Location: Training Center, Chanute AFB, IL.

Length: 7 weeks (282 hours).

Exhibit Dates: 8/89–4/04.

Learning Outcomes: Upon completion of the course, the student will be able to identify causes and control of fires involving various combustible materials; assess and control aircraft and crash fires; assess and control structural fires; and administer emergency first aid.

Instruction: Lectures, discussions, and practical exercises teach students control and suppression of aircraft and structural fires. The course also includes emergency first aid.

Credit Recommendation: In the lower-division baccalaureate/associate degree category, 3 semester hours in emergency first aid, 3 in aircraft fire suppression, and 3 in structural fire suppression (1/93).

Related Occupations: 51M.

AR-1728-0114

1. Judge Advocate General Officer Basic Phase 1
2. Judge Advocate Officer Basic
(Judge Advocate General Officer Basic Phase 1)

Course Number: *Version 1:* 5-27-C20. *Version 2:* 5-27-C20.

Location: Quartermaster Center and School, Fort Lee, VA.

Length: *Version 1:* 2 weeks (55 hours). *Version 2:* 4 weeks (213 hours).

Exhibit Dates: *Version 1:* 10/92–10/96. *Version 2:* 4/04–Present.

Learning Outcomes: *Version 1:* Upon completion of the course, the student will have a basic familiarity with the Army environment including uniform and dress, courtesy, inspection, drill, fitness, map reading, first aid, rifle training, and code of conduct. *Version 2:* Pending evaluation.

Instruction: *Version 1:* Lectures and practical exercises cover military leadership and management, law enforcement, and tactical operations. *Version 2:* Pending evaluation.

Credit Recommendation: *Version 1:* Credit is not recommended because of the military-specific nature of the course (5/02). *Version 2:* Pending evaluation.

AR-1728-0115

Firearms and Toolmarks Examiner

Course Number: None.

Location: Army Criminal Investigations Laboratory, Fort Gillem, GA.

Length: 104 weeks (3422 hours).

Exhibit Dates: 1/90–Present.

Learning Outcomes: Upon completion of the course, the student will possess the credentials, background, and training to be qualified as an expert witness in firearms identification, toolmark comparison, and serial number restorations before federal, state, municipal, and military courts including grand juries and hearings within the criminal justice system. The student will examine firearms as to function, compare bullets and other projectiles and shell cases for class and individual characteristics, and give expert witness as to match or nonmatch between questioned and known items; use Association of Firearms and Toolmarks Examiners consensus firearms and toolmarks terminology; write expert reports giving results of examinations; testify as an expert witness in court on all aspects of firearm and toolmark identification and comparison; describe firearm and ammunition manufacturing process and ammunition function; restore obliterated serial numbers on metal surfaces; and conduct appropriate tests to estimate muzzle-to-target distances in shooting cases.

Instruction: Topics of instruction include the history of firearms identification; firearm development; cartridge development; handling of evidence; test firing and specimen recovery; laboratory instrumentation; principles of comparative evidence; distance determination; serial number restoration; report writing and court preparation; and advanced criminal investigation, including chemical, photographic, and physical examinations. Course is presented through a combination of lectures, directed readings, practical exercises, comprehensive examinations throughout the course of study, academic research, supervised work on actual criminal cases, and independent project work. The special project/independent study consists of an independent, original investigation of a problem or method under supervision of a qualified examiner. Data resulting from these projects may be presented at appropriate professional meetings. The research project may be result in publication in peer-review forensic sciences journal. Internship experience provides an opportunity for practice on real-world casework exhibits under supervision of a qualified examiner. Proficiency testing constitutes part of this practical training.

Credit Recommendation: In the vocational certificate category, 30 semester hours in firearm and toolmark examination. In the lower-division baccalaureate/associate degree category, 9 semester hours in basic firearm/toolmark examination. In the upper-division baccalaureate category, 9 semester hours in advanced firearm/toolmark examination, 3 in crime scene or pattern evidence, 3 in expert witness testimony, 9 in internship or practicum in firearm/toolmark examination, and 3 in special project/independent study. In the graduate degree category, 3 semester hours in specialized firearm/toolmark examination and 0—6 in based on the receiving institution's review of the student's research work (6/94).

AR-1728-0116

Questioned Documents Examination

Course Number: None.

Location: Army Criminal Investigations Laboratory, Fort Gillem, GA.

Length: 104 weeks (3280 hours).

Exhibit Dates: 1/90–Present.

Learning Outcomes: Upon completion of the course, the student will possess the credentials, background, and training to be qualified as an expert witness in questioned documents examination before federal, state, municipal, and military courts, including grand juries and hearings within the criminal justice system. Students will be able to perform forensic examination of documentary evidence with emphasis on solving various questioned document problems both conceptually and through interpretation and analysis using professional papers and instrumentation.

Instruction: Course is presented through a combination of lectures, directed readings, practical exercises, comprehensive examinations, academic research, supervised work on actual criminal cases, and independent project work. The special project/independent study consists of an independent, original investigation of a problem or method, under supervision of a qualified examiner. Data resulting from these projects may be presented at appropriate professional meetings. Research project may be result in publication in peer-review forensic science journal. Internship experience provides an opportunity for practice on real-world casework exhibits, under supervision of a qualified examiner. Proficiency testing constitutes part of this practical training. Topics include the examination of paper, ink, typewriters, printout devices, and writing instruments. Also covered is presenting expert testimony, internship, counterfeiting, commercial printing, and document photography.

Credit Recommendation: In the vocational certificate category, 30 semester hours (certificate) in questioned document examination. In the lower-division baccalaureate/associate degree category, 6 semester hours in questioned documents and 6 in typewriters and printout devices (photocopiers, fax machines, and printers). In the upper-division baccalaureate category, 3 semester hours in expert witness testimony, 9 in internship or practicum in questioned document examination, 3 in special project/independent study, and 9 in advanced questioned document examination. In the graduate degree category, 3 semester hours in a specialized questioned document examination and 0—6 based on the receiving institution's review of the student's research work (6/94).

AR-1728-0117

Latent Print Examiner

Course Number: None.

Location: Army Criminal Investigations Laboratory, Fort Gillem, GA.

Length: 104 weeks (4160 hours).

Exhibit Dates: 1/90–Present.

Learning Outcomes: Upon completion of the course, the student will possess the credentials, background, and training to be qualified as an expert witness in questioned document examination before federal, state, municipal, and military courts including grand juries and hearings within the criminal justice system. The student will be able to apply the principles of preservation and packaging of physical evidence; classify, develop, and enhance latent fingerprints and other friction ridge impressions using physical, chemical, and instrumental techniques; evaluate latent fingerprints and other friction ridge impressions as to value for identification; compare fingerprints and other friction ridge impressions and provide expert opinion of results of comparison; input data to and extract information from the Automated Fingerprint Identification System; describe the history and development of fingerprint comparison and methods for development of latent friction ridge skin impression; provide written reports and testify as an expert witness in court on fingerprint and other friction ridge skin impression comparisons; develop prints on various surfaces using appropriate techniques; use laser and/or alternate light sources to enhance prints; provide crime scene support for latent prints; and describe other types of prints, such as footwear, lip, ear, palm, tire, and fabric.

Instruction: Course is presented through a combination of lectures, directed readings, practical exercises, comprehensive examinations, academic research, supervised work on actual criminal cases, and independent project work. The special project/independent study consists of an independent, original investigation of a problem or method, under supervision of a qualified examiner. Data resulting from these projects may be presented at appropriate professional meetings. The research project may be result in publication in peer-review forensic sciences journal. Internship experience provides an opportunity for practice on real-world casework exhibits, under supervision of a qualified examiner. Proficiency testing constitutes part of this practical training. There is a requirement to complete various courses external to this school. Topics include history of latent prints; friction ridge biology and physiology; fingerprint pattern recognition and classification; preservation of evidence; latent print development on various surfaces using both chemical and lighting techniques; identification and development of other pattern evidence such as tire, footwear, palm, lip, ear, and fabric impressions; and legal aspects of prints including moot court experience.

Credit Recommendation: In the vocational certificate category, 30 semester hours (certificate) in fingerprint identification. In the lower-division baccalaureate/associate degree category, 9 semester hours in basic fingerprints. In the upper-division baccalaureate category, 9 semester hours in advanced fingerprints, 3 in crime scene/pattern evidence, 3 in expert witness testimony, 9 in internship or practicum in fingerprint examination, and 3 in special project/independent study. In the graduate degree category, 3 semester hours in specialized fingerprint examination and 0—6 based on the receiving institution's review of the student's research work (6/94).

AR-1728-0118

Forensic Photography

Course Number: None.

Location: Army Criminal Investigations Laboratory, Fort Gillem, GA.

Length: 104 weeks (4160 hours).

Exhibit Dates: 1/90–Present.

Learning Outcomes: Upon completion of the course, the student will possess the credentials, background, and training to be qualified as an expert witness in questioned documents examination before federal, state, municipal, and military courts including grand juries and hearings within the criminal justice system. The student will be able to produce negatives and prints using black-and-white or color; set up and photograph evidence such as latent prints, palm prints, footwear, and basic impressions; provide photographic support for all laboratory divisions; provide crime scene photography as required for any investigation; use correct and appropriate close-up macrophotography and microphotography; use videotape and image enhancement techniques; describe legal issues for forensic photography; and describe the production of visual and training aids using color and black-and-white photography.

Instruction: This course is presented through a combination of lectures, directed readings, practical exercises, comprehensive examinations, academic research, supervised work on actual criminal cases, and independent project work. The special project/independent study consists of an independent, original investigation of a problem or method, under supervision of a qualified examiner. Data resulting from these projects may be presented at appropriate professional meetings. Research project may be result in publication in peer-review forensic sciences journal. Internship experience provides an opportunity for practice on real-world casework exhibits, under supervision of a qualified examiner. Proficiency testing constitutes part of this practical training. Topics include principles of photography, evidence photography, photographic optics, principles of light and its management, principles of the camera and its operation, photographic theory, photographic densitometry, photographic chemistry, color photography, infrared and ultraviolet photography, photomicrography, photomacrography, photographic techniques in support of all division laboratory divisions, and investigative activities.

Credit Recommendation: In the vocational certificate category, 30 semester hours (certificate) in photography (forensic). In the lower-division baccalaureate/associate degree category, 3 semester hours in basic photography, 6 in forensic evidence photography, and 3 in color photography. In the upper-division baccalaureate category, 6 semester hours in specialized photographic techniques (micro/macro), 6 in advanced photography, 3 in expert witness testimony, 9 in internship or practicum in photography, and 3 in special project/independent study. In the graduate degree category, 3 semester hours in specialized forensic or evidence photography, 0—6 based on the receiving institution's review of the student's research work (6/94).

AR-1728-0120

Basic Military Police, Phase 4—5 (ITRO/USMC)

Course Number: 830-95B10 (ITRO/USMC).

Location: Military Police School, Fort McClellan, AL.

Length: 9 weeks (356 hours).

Exhibit Dates: 1/93–2/98.

Learning Outcomes: Upon completion of the course, the student will be able to perform police and law enforcement procedures.

Instruction: The course includes lectures, demonstrations, and practical exercises in military law, apprehension and arrest techniques, accident investigation, patrol procedures, report writing, and case preparation and presentation. Course is conducted for Marine Corps personnel.

Credit Recommendation: In the vocational certificate category, 3 semester hours in apprehension and arrest techniques and 2 in accident investigation. In the lower-division baccalaureate/associate degree category, 3 semester hours in military/criminal law and procedures and 2 in patrol operations (3/96).

AR-1728-0121

Warrant Officer Advanced Reserve

Course Number: 8-10-C32-922A-RC.
Location: Quartermaster Center and School, Fort Lee, VA.
Length: 2 weeks (82 hours).
Exhibit Dates: 11/93–Present.
Learning Outcomes: Upon completion of the course, the student will be able to plan and organize food service operations, menu planning, and field bakery operations.
Instruction: Lectures and practical exercises cover field food service operations, menu planning, and field bakery operations.
Credit Recommendation: In the lower-division baccalaureate/associate degree category, 1 semester hour in food service management (4/02).

AR-1728-0122

Counterintelligence Officer (AST-35E)

Course Number: 3C-F16.
Location: Intelligence School, Fort Huachuca, AZ.
Length: 7 weeks (341 hours).
Exhibit Dates: 7/92–9/99.
Learning Outcomes: After 9/99 see AR-1728-0157. Upon completion of the course, the student will be perform duties in counterintelligence and counterespionage operations.
Instruction: Lectures and practical exercises cover evidence procedures, credential control, planning for surveillance, interview techniques, investigation of espionage and sabotage, investigation reports, and special operations.
Credit Recommendation: In the lower-division baccalaureate/associate degree category, 3 semester hours in investigative procedures (5/01).

AR-1728-0123

Internment/Resettlement Specialist

Course Number: 95C10-OSUT; 95C-OSUT.
Location: Military Police School, Fort Leonard Wood, MO.
Length: 17–18 weeks (859 hours).
Exhibit Dates: 9/00–Present.
Learning Outcomes: Upon completion of the course, the student will perform prisoner administration and services, exercise security and control, move prisoners, and discuss contemporary issues in confinement facilities.
Instruction: This course includes basic training as the first 8-9 weeks. Lectures, and practical exercises use a mock confinement facility to introduce corrections subjects.
Credit Recommendation: In the lower-division baccalaureate/associate degree category, 3 semester hours in introduction to corrections (2/01).
Related Occupations: 95C.

AR-1728-0124

Internment/Resettlement Specialist Reclassification

Course Number: 831-95C10.
Location: Military Police School, Fort Leonard Wood, MO.
Length: 8–9 weeks (250 hours).
Exhibit Dates: 9/00–Present.
Learning Outcomes: Upon completion of the course, the student will perform prisoner administration and services, exercise security and control, move prisoners, and discuss contemporary issues in confinement facilities.
Instruction: Lectures and practical exercises use a mock confinement facility to introduce corrections subjects.
Credit Recommendation: In the lower-division baccalaureate/associate degree category, 3 semester hours in introduction to corrections (2/01).
Related Occupations: 95C.

AR-1728-0125

Total Army Training System (TATS) Military Police Captain's Career, Phase 3

Course Number: 7-19-C23, Phase 3.
Location: Military Police School, Fort Leonard Wood, MO.
Length: 2 weeks (122 hours).
Exhibit Dates: 10/00–Present.
Learning Outcomes: Upon completion of the course, the student will be able to describe the military critical thinking process.
Instruction: Course covers ethics, leadership skills, and professional attitude.
Credit Recommendation: Credit is not recommended because of the military-specific nature of the course (2/01).

AR-1728-0126

Total Army Training System (TATS) Military Police Captain's Career

Course Number: 7-19-C23, Phase 1.
Location: Military Police School, Fort Leonard Wood, MO.
Length: 2 weeks (122–123 hours).
Exhibit Dates: 10/00–Present.
Learning Outcomes: Upon completion of the course, the student will be able to describe leadership subjects as well as military ethics, customs, and traditions.
Instruction: Course topics include leadership and military ethics.
Credit Recommendation: Credit is not recommended because of the military-specific nature of the course (2/01).

AR-1728-0127

Counterdrug Special Reaction Team

Course Number: 7H-F38.
Location: Military Police School, Fort Leonard Wood, MO.
Length: 1 week (47–48 hours).
Exhibit Dates: 1/00–Present.
Learning Outcomes: Before 1/00 see AR-1728-0095. Upon completion of the course, the student will be able to use special weapons and tactics to conduct raids on houses, woodland areas, and convoy areas used to conceal and transport narcotics.
Instruction: Lectures, demonstrations, and performance exercises cover special weapons and tactics, raid preplanning, building entry techniques, and explosive identification.
Credit Recommendation: In the vocational certificate category, 1 semester hour in special tactics (2/01).

AR-1728-0128

Military Police Basic Noncommissioned Officer (NCO)

Course Number: 830-95B/C/D30 (F).
Location: Military Police School, Fort Leonard Wood, MO.
Length: 9–10 weeks (466–467 hours).
Exhibit Dates: 4/00–Present.
Learning Outcomes: Before 4/00 see AR-1728-0102. Upon completion of the course, the student will be able to lead in the area of combat services.
Instruction: Lectures and practical applications cover combat support, tactics, police missions, equipment systems, leadership, weapons, and security.
Credit Recommendation: In the lower-division baccalaureate/associate degree category, 3 semester hours in law enforcement operations and 1 in personnel supervision (2/01).
Related Occupations: 95B; 95C; 95D.

AR-1728-0129

Special Reaction Team Marksman Observer

Course Number: 7H-F17/830-F12, Phase 2.
Location: Military Police School, Fort Leonard Wood, MO.
Length: 1 week (52 hours).
Exhibit Dates: 2/00–Present.
Learning Outcomes: Before 2/00 see AR-1728-0093. Upon completion of the course, the student will be able to demonstrate effective measures to neutralize hostile targets during special threat situations through the use of psychological aspects of marksmanship, marksmanship fundamentals, and practical exercises.
Instruction: Lectures and practical exercise cover marksmanship techniques and applications.
Credit Recommendation: In the vocational certificate category, 1 semester hour in police special operations (2/01).

AR-1728-0130

Child Abuse Prevention and Investigative Technician

Course Number: 7H-F43.
Location: Military Police School, Fort Leonard Wood, MO.
Length: 1–2 weeks (64–65 hours).
Exhibit Dates: 10/99–Present.
Learning Outcomes: Upon completion of the course, the student will be able to define child abuse, perform investigative techniques in the operational planning of child abuse investigations, investigate childhood deaths, and describe forensic and psychological techniques in child abuse investigations.
Instruction: Lectures and practical exercises cover child abuse investigations, crime scene processing, and interview/interrogations.
Credit Recommendation: In the lower-division baccalaureate/associate degree category, 3 semester hours in special topics in law enforcement (2/01).

AR-1728-0131

Total Army Training System (TATS) Basic Military Police Reclassification, Phase 2

Course Number: 191-95B10(F), Phase 2.

Location: Military Police School, Fort Leonard Wood, MO.
Length: 122 hours.
Exhibit Dates: 10/00–Present.
Learning Outcomes: Upon completion of the course, the student will be able to perform preventive maintenance checks, operate tactical communications equipment, operate crew-served weapons, read maps, deal with NBC situations, conduct military operations in urban terrain, and carry out tactical vehicle operations.
Instruction: Lectures and practical exercises cover weapons, communications, and map reading.
Credit Recommendation: Credit is not recommended because of the military-specific nature of the course (2/01).
Related Occupations: 95B.

AR-1728-0132

COUNTERDRUG MARKSMAN OBSERVER

Course Number: 7H-F42.
Location: Military Police School, Fort Leonard Wood, MO.
Length: 1 week (50 hours).
Exhibit Dates: 2/00–Present.
Learning Outcomes: Upon completion of the course, the student will be able to demonstrate effective measures to neutralize hostile targets during special threat situations through the use of psychological aspects of marksmanship, marksmanship fundamentals, and practical exercises.
Instruction: Lectures and practical exercises on marksmanship techniques and application.
Credit Recommendation: In the vocational certificate category, 1 semester hour in police special operations (2/01).

AR-1728-0133

BASIC MILITARY POLICE

Course Number: 95B10-OSUT.
Location: Military Police School, Fort Leonard Wood, MO.
Length: 17–18 weeks (609 hours).
Exhibit Dates: 9/00–Present.
Learning Outcomes: Upon completion of the course, the student will apply knowledge of law enforcement operations to patrol procedures, traffic enforcement, first responder, investigations, and special operations, to conform to constitutional standards.
Instruction: The course includes basic training as the first 8-9 weeks. Lectures and practical exercises cover introductory methods of law enforcement operations and procedures.
Credit Recommendation: In the lower-division baccalaureate/associate degree category, 3 semester hours in introduction to law enforcement (2/01).
Related Occupations: 95B.

AR-1728-0134

MILITARY POLICE RECLASSIFICATION

Course Number: 830-95B10.
Location: Military Police School, Fort Leonard Wood, MO.
Length: 8–9 weeks (220 hours).
Exhibit Dates: 9/00–Present.
Learning Outcomes: Upon completion of the course, the student will apply knowledge of law enforcement operations, to patrol procedures, traffic enforcement, first responder investigations, and special operations to conform to constitutional standards.
Instruction: Lectures and practical exercises cover introductory methods of law enforcement operations and procedures.
Credit Recommendation: In the lower-division baccalaureate/associate degree category, 3 semester hours in introduction to law enforcement (2/01).
Related Occupations: 95B.

AR-1728-0135

BASIC MILITARY POLICE, SPLIT OPTION, PHASE 2

Course Number: 95B10-OSUT, Phase 2.
Location: Military Police School, Fort Leonard Wood, MO.
Length: 8–9 weeks (220 hours).
Exhibit Dates: 9/00–Present.
Learning Outcomes: Upon completion of the course, the student will apply knowledge of law enforcement operations, to patrol procedures, traffic enforcement, first responder, investigations and special operations, to conform to constitutional standards.
Instruction: Lectures and practical exercises cover introductory methods of law enforcement operations and procedures. This course includes an 8-9 week long Basic Training component.
Credit Recommendation: In the lower-division baccalaureate/associate degree category, 3 semester hours in introduction to law enforcement (2/01).
Related Occupations: 95B.

AR-1728-0136

APPRENTICE CRIMINAL INVESTIGATION DIVISION (CID) SPECIAL AGENT

Course Number: 832-95D20/30 (CT).
Location: Military Police School, Fort Leonard Wood, MO.
Length: 15 weeks (581 hours).
Exhibit Dates: 11/99–Present.
Learning Outcomes: Before 11/99 see AR-1728-0104. Upon completion of the course, the student will be able to conduct criminal investigations, process crime scenes, collect and preserve physical evidence, write investigative reports, conduct covert operations, plan and conduct raids, and describe concepts of procedural and criminal law.
Instruction: Lectures, demonstrations, and practical exercises cover criminal law, fundamentals of criminal investigation, crime scene processing, crimes against persons and property, physical evidence, drug investigations, fraud and fraud investigations, reports, covert operations, raids, surveillance, interviews and interrogations, and protective services.
Credit Recommendation: In the lower-division baccalaureate/associate degree category, 3 semester hours in criminal investigation, 3 in criminal law, 3 in criminalistics/forensic science, and 3 in interview and interrogation (2/01).
Related Occupations: 95D.

AR-1728-0137

CONVENTIONAL PHYSICAL SECURITY/CRIME PREVENTION

Course Number: 7H-31D/830-ASI H3.
Location: Military Police School, Fort Leonard Wood, MO.
Length: 2 weeks (76 hours).
Exhibit Dates: 10/00–Present.
Learning Outcomes: Before 10/00 see AR-1728-0094. Upon completion of the course, student will be able to plan and implement physical security/crime prevention measures to control pilferage, espionage, sabotage, and other physical security threats.
Instruction: Lectures, tests, and performance exercises cover legal aspects of security, crime prevention, physical security, planning, intrusion detection, crime analysis, and security services.
Credit Recommendation: In the lower-division baccalaureate/associate degree category, 3 semester hours in introduction to security (2/01).

AR-1728-0138

CRIMINAL INVESTIGATION DIVISION (CID) WARRANT OFFICER ADVANCED, PHASE 1

Course Number: 7-19-C32, Phase 1.
Location: Military Police School, Fort Leonard Wood, MO.
Length: 2 weeks (80 hours).
Exhibit Dates: 3/00–Present.
Learning Outcomes: Before 3/00 see AR-1728-0092. Upon completion of the course, the student will be able to perform advanced criminal investigation operations and police administration.
Instruction: Course topics covered include legal issues, office management, personnel management, resource management, principles of organization, supervision, inspection systems, criminal profiling, and managing hostage/terrorist negotiations.
Credit Recommendation: In the upper-division baccalaureate category, 3 semester hours in criminal investigation and police administration. This credit does not duplicate that recommended for 7-19-C32, Phase 2 (2/01).

AR-1728-0139

CRIMINAL INVESTIGATION DIVISION (CID) SPECIAL AGENT WARRANT OFFICER BASIC, PHASE 2

Course Number: 7H-311A, Phase 2.
Location: Military Police School, Fort Leonard Wood, MO.
Length: 2 weeks (65–66 hours).
Exhibit Dates: 3/00–Present.
Learning Outcomes: Before 3/00 see AR-1728-0097. Upon completion of the course, the student will be able to supervise investigations in the areas of economic crimes, environmental crimes, criminal intelligence, crisis negotiations, protective services, personal security, violations, investigative task forces, and major crime investigations, such as homicide and sexual assault.
Instruction: Lectures, demonstrations and performance exercises cover leadership training exercises and tactical skills in the field.
Credit Recommendation: In the lower-division baccalaureate/associate degree category, 3 semester hours in police administration (2/01).
Related Occupations: 311A.

AR-1728-0140

CRIMINAL INVESTIGATION DIVISION (CID) WARRANT OFFICER ADVANCED

Course Number: 7-19-C32, Phase 2.
Location: Military Police School, Fort Leonard Wood, MO.
Length: 2 weeks (67 hours).
Exhibit Dates: 3/00–Present.
Learning Outcomes: Before 3/00 see AR-1728-0092. Upon completion of the course, the student will be able to apply techniques and skills of advanced criminal investigation operations and police administration.
Instruction: Course topics covered include legal issues, office management, personnel management, resource management, principles of organization, supervision, inspection systems, criminal profiling, and managing hostage/terrorist negotiations.
Credit Recommendation: In the upper-division baccalaureate category, 3 semester hours in criminal investigation and police administration. This credit does not duplicate credit recommended for 7-19-C32, Phase 1 (2/01).

AR-1728-0141

Protective Services Training

Course Number: 7H-F18/830-F13.
Location: Military Police School, Fort Leonard Wood, MO.
Length: 3 weeks (125 hours).
Exhibit Dates: 2/00–Present.
Learning Outcomes: Before 2/00 see AR-1728-0071. Upon completion of the course, the student will be able to plan and execute procedures to protect dignitaries.
Instruction: This course includes instruction in methods of motorcade security, advance operations and threat assessment, building and vehicle searches, and emergency medical procedures.
Credit Recommendation: In the lower-division baccalaureate/associate degree category, 3 semester hours in personnel security (2/01).

AR-1728-0142

Criminal Investigation Division (CID) Special Agent Warrant Officer Basic

Course Number: 7H-311A, Phase 1.
Location: Military Police School, Fort Leonard Wood, MO.
Length: 2 weeks (78–79 hours).
Exhibit Dates: 3/00–Present.
Learning Outcomes: Before 3/00 see AR-1728-0097. Upon completion of the course, the student will be able to use technical investigation equipment and supervise and review cases involving risk assessment, hate/bias crimes, drug suppression/special operations, gang crimes, and domestic violence.
Instruction: Lectures, demonstrations, and performance exercises cover leadership training exercises and tactical skills in the field.
Credit Recommendation: In the lower-division baccalaureate/associate degree category, 3 semester hours in management of a criminal justice organization (2/01).
Related Occupations: 311A.

AR-1728-0143

Military Police Advanced Noncommissioned Officer

Course Number: 8-95-C42.
Location: Military Police School, Fort Leonard Wood, MO.
Length: 11 weeks (472–480 hours).
Exhibit Dates: 4/00–Present.
Learning Outcomes: Before 4/00 see AR-1728-0100. Upon completion of the course, the student will be able to perform advanced supervisory, administrative, and managerial tasks in the areas of law enforcement operations, corrections, or criminal investigations.
Instruction: Lectures and applied exercises cover leadership, police equipment, police tactics, battlefield control, military police operations, and advanced noncommissioned officer duties. Separate tracks beyond the leadership common core provide specific concentration in law enforcement operations (95B), corrections/confinement (95C), or criminal investigation (95D).
Credit Recommendation: In the lower-division baccalaureate/associate degree category, 3 semester hours in law enforcement operations (95B), 3 in corrections (95C), or 3 in criminal investigation operations (95D). See AR-1404-0035 for common core credit (2/01).

AR-1728-0144

Special Reaction Team, Phase 1

Course Number: 7H-F17/830-F12, Phase 1.
Location: Military Police School, Fort Leonard Wood, MO.
Length: 2 weeks (105 hours).
Exhibit Dates: 1/00–Present.
Learning Outcomes: Before 1/00 see AR-1728-0072. Upon completion of the course, the student will be able to react to barricaded subject and sniper incidents, apprehend dangerous subjects, rescue hostages, and conduct counterterrorism operations and drug raids.
Instruction: Lectures and practical exercise cover counterterrorism activities, hostage situations, and barricaded offenders.
Credit Recommendation: In the vocational certificate category, 1 semester hour in special patrol operations (2/01).

AR-1728-0145

Military Police Investigation

Course Number: 830-ASIV5.
Location: Military Police School, Fort Leonard Wood, MO.
Length: 8 weeks (311 hours).
Exhibit Dates: 10/99–Present.
Learning Outcomes: Upon completion of the course, the student will be able to perform duties as a public investigator in the fields of criminal investigation, confession and admission, fingerprint evidence, forensic science, crime scene processing, including processing of questioned documents and contemporary investigations in gang crimes, economic crimes, and domestic disturbances and child abuse.
Instruction: Lectures, demonstrations, and performance exercises cover law, forensic evidence, testimonial evidence, crime scene processing, investigation of crimes, and special operations.
Credit Recommendation: In the lower-division baccalaureate/associate degree category, 3 semester hours in criminal investigation and 3 in criminalistics and forensic science (2/01).

AR-1728-0146

Hostage/Crisis Negotiation

Course Number: 7H-F19/830-F14(CT).
Location: Military Police School, Fort Leonard Wood, MO.
Length: 2 weeks (75–76 hours).
Exhibit Dates: 10/99–Present.
Learning Outcomes: Before 10/99 see AR-1728-0067. Upon completion of the course, student will apply techniques and operational planning in hostage/crisis negotiations situations.
Instruction: Lectures and practical exercises are presented on managing a hostage situation. Topics include negotiations, lessons learned, bargaining techniques, communication, profiling the offender, and collecting data to resolve such activities.
Credit Recommendation: In the lower-division baccalaureate/associate degree category, 2 semester hours in special topics in law enforcement and criminal justice (2/01).

AR-1728-0147

Military Police Captains' Career

Course Number: 7-19-C22.
Location: Military Police School, Fort Leonard Wood, MO.
Length: 18 weeks (657–693 hours).
Exhibit Dates: 10/98–Present.
Learning Outcomes: Before 10/98 see AR-1728-0091. Upon completion of the course, the student will supervise personnel and manage police functions, including investigation, security, traffic management, and protective services.
Instruction: Lectures and practical exercises cover police management subjects.
Credit Recommendation: In the upper-division baccalaureate category, 3 semester hours in police management (2/01).

AR-1728-0148

Apprentice Criminal Investigation Division Special Agent, Reserve Component, Phase 2

Course Number: 832-95D20/30, Phase 2.
Location: Military Police School, Fort Leonard Wood, MO.
Length: 2 weeks (83 hours).
Exhibit Dates: 11/99–Present.
Learning Outcomes: Upon completion of the course, the student will apply knowledge of basic criminal law, to the processing of crime scenes, the handling of physical evidence, the investigation of fraud and waste, and brig management.
Instruction: Lectures and practical exercises cover criminal investigation.
Credit Recommendation: In the lower-division baccalaureate/associate degree category, 3 semester hours in criminal investigation. This credit does not duplicate the credit in 832-95D20/30, Phase 4 (2/01).

AR-1728-0149

Apprentice Criminal Investigation Division Special Agent, Reserve Component, Phase 4

Course Number: 832-95D20/30, Phase 4.
Location: Military Police School, Fort Leonard Wood, MO.
Length: 2 weeks (87 hours).
Exhibit Dates: 11/99–Present.
Learning Outcomes: Upon completion of the course, the student will apply knowledge of special investigative techniques to protective

services, interviews and interrogation, and investigation report writing.

Instruction: Lectures and practical exercises cover investigative techniques, interviews, and interrogation.

Credit Recommendation: In the lower-division baccalaureate/associate degree category, 3 semester hours in investigation. This credit does not duplicate the credit in 832-95D20/30, Phase 2 (2/01).

AR-1728-0150

TOTAL ARMY TRAINING SYSTEM (TATS-C) INTERROGATOR

Course Number: 241-97E10.

Location: Reserve Schools, Multiple locations; Intelligence School, Fort Huachuca, AZ.

Length: 90–91 hours.

Exhibit Dates: 10/01–Present.

Learning Outcomes: Upon completion of the course, the student will be able to read maps, coordinate the phases of an interrogation, assess and control the interrogation during questioning, interrogate using an interpreter, and use basic military terminology.

Instruction: Small class lectures, simulated interrogation in a classroom environment, and field training exercises cover interrogation principles, interrogation planning and preparation, approach and termination, and in-screening applications. Reserve students have up to 2 years to complete the course.

Credit Recommendation: In the lower-division baccalaureate/associate degree category, 2 semester hours in interrogation (8/01).

AR-1728-0151

INTELLIGENCE ANALYST

Course Number: 243-96B10.

Location: Intelligence School, Fort Huachuca, AZ.

Length: 16–17 weeks (583–597 hours).

Exhibit Dates: 10/00–Present.

Learning Outcomes: Before 10/00 see AR-1728-0083. Upon completion of the course, the student will be able to read maps, plot symbols on maps, use a specific military intelligence data base, analyze battlefield conditions, and provide intelligence reports.

Instruction: The course uses lectures, exercises, and research activities. Topics covered include map reading, plotting symbols on maps, battlefield analysis, remote workstation operation, providing briefings based on maps.

Credit Recommendation: In the lower-division baccalaureate/associate degree category, 1 semester hour in map reading and use (11/02).

AR-1728-0152

INTELLIGENCE ANALYST, PHASE 2 AND PHASE 4

Course Number: 243-96B10, Phase 4; 243-96B10, Phase 2.

Location: Reserve Schools, Multiple locations; Intelligence School, Fort Huachuca, AZ.

Length: 235 hours.

Exhibit Dates: 10/00–Present.

Learning Outcomes: Before 10/00 see AR-1728-0083. Upon completion of the course, the student will be able to operate a remote workstation, analyze battlefield conditions, and provide intelligence reports.

Instruction: The course uses lectures, exercises, and research activities. Topics covered include battlefield analysis, remote workstation operation, and providing briefings based on maps.

Credit Recommendation: Credit is not recommended because of the military-specific nature of the course (11/02).

AR-1728-0153

TOTAL ARMY TRAINING SYSTEM (TATS) COUNTERINTELLIGENCE AGENT

Course Number: 244-97B10 (F); 244-97B10.

Location: Intelligence School, Fort Huachuca, AZ.

Length: 15–16 weeks (605–606 hours).

Exhibit Dates: 11/99–Present.

Learning Outcomes: Upon completion of the course, the student will be able to perform personnel security, identify the objectives of a security program, perform a security manager's inspection, identify and describe automation sensitivity levels and security operating modes, and apply legal principles.

Instruction: Lectures, practical exercises, small group activities, computer software and hardware training.

Credit Recommendation: In the lower-division baccalaureate/associate degree category, 3 semester hours in investigative techniques (8/01).

AR-1728-0154

TOTAL ARMY TRAINING SYSTEM (TATS) COUNTERINTELLIGENCE AGENT

Course Number: 244-97B10.

Location: Intelligence School, Fort Huachuca, AZ; Reserve Schools, Multiple locations.

Length: 232 hours.

Exhibit Dates: 10/00–Present.

Learning Outcomes: Upon completion of the course, the student will be able to perform personnel security, identify the objectives of a security program, perform a security manager's inspection, identify and describe automation sensitivity levels and security operating modes, and apply legal principles.

Instruction: Lectures, practical exercises, small group activities, computer software and hardware training.

Credit Recommendation: In the lower-division baccalaureate/associate degree category, 3 semester hours in investigative techniques (8/01).

AR-1728-0155

TOTAL ARMY TRAINING SYSTEM (TATS) COUNTERINTELLIGENCE AGENT BASIC NONCOMMISSIONED OFFICER (NCO)

Course Number: 244-97B30.

Location: Intelligence School, Fort Huachuca, AZ.

Length: 7–8 weeks (246 hours).

Exhibit Dates: 10/00–Present.

Learning Outcomes: Before 10/00 see AR-1728-0081. Upon completion of the course, the student will be able to apply legal principles to counterintelligence operations, prepare and supervise counterintelligence investigations, and conduct investigations of allegations of espionage, sabotage, and subversive activities.

Instruction: Lectures, classroom testing, small group activities, and practical exercises covering counterintelligence activities. This course covers a common core of leadership subjects.

Credit Recommendation: In the lower-division baccalaureate/associate degree category, 2 semester hours in investigative techniques. See AR-1406-0090 for common core credit (8/01).

AR-1728-0156

TOTAL ARMY TRAINING SYSTEM (TATS) COUNTERINTELLIGENCE AGENT BASIC NONCOMMISSIONED OFFICER (NCO)

Course Number: 244-97B30.

Location: Intelligence School, Fort Huachuca, AZ; Reserve Schools, Multiple locations.

Length: 123 hours.

Exhibit Dates: 10/00–Present.

Learning Outcomes: Upon completion of the course, the student will be able to apply legal principles to counterintelligence operations and supervise and conduct counterintelligence investigations of espionage and sabotage activities.

Instruction: Lectures, classroom testing, small-group activities, and practical exercises cover counterintelligence activities.

Credit Recommendation: In the lower-division baccalaureate/associate degree category, 2 semester hours in investigative techniques (8/01).

Related Occupations: 97B.

AR-1728-0157

COUNTERINTELLIGENCE OFFICER

(Total Army Training System (TATS) Counterintelligence Officer)

Course Number: 3C-35E.

Location: Intelligence School, Fort Huachuca, AZ.

Length: 7–8 weeks (318–331 hours).

Exhibit Dates: 10/99–Present.

Learning Outcomes: Before 10/99 see AR-1728-0122. Upon completion of the course, the student will be able to perform duties in counter intelligence and counterespionage operations.

Instruction: Lectures and practical exercises cover evidence procedures, credential control planning for surveillance, interview techniques, investigation of espionage and sabotage, role playing, investigation reports, and special operations.

Credit Recommendation: In the lower-division baccalaureate/associate degree category, 3 semester hours in investigative techniques (8/02).

AR-1728-0158

TOTAL ARMY TRAINING SYSTEM (TATS) MILITARY INTELLIGENCE OFFICER TRANSITION

Course Number: 3A-35D (T).

Location: Intelligence School, Fort Huachuca, AZ.

Length: 4 weeks (150 hours).

Exhibit Dates: 10/99–Present.

Learning Outcomes: Before 10/99 see AR-1728-0087. Upon completion of the course, the student will be able to describe threat weapons and tactics, use military intelligence terminology, employ intelligence assets, and process information which turns into intelligence.

Instruction: Lectures, practical exercises, and practice briefings are course methods.

Credit Recommendation: Credit is not recommended because of the military-specific nature of the course (8/01).

AR-1728-0159

Total Army Training System (TATS) Military Intelligence Officer Transition
(Military Intelligence Officer Tactician)

Course Number: 3A-35D; 3A-35D (T).

Location: Intelligence Center and School, Fort Huachuca, AZ; Reserve Schools, Multiple locations.

Length: 4 weeks (106–150 hours).

Exhibit Dates: 10/99–Present.

Learning Outcomes: Before 10/99 see AR-1728-0087. Upon completion of the course, the student will be able to implement physical security measures, perform information and personnel security, analyze threat forces, and define functions of battlefield preparation and tactical intelligence.

Instruction: Lectures, and field exercises cover tactical intelligence operations.

Credit Recommendation: In the lower-division baccalaureate/associate degree category, 3 semester hours in introduction to security (personnel/information) (12/02).

AR-1728-0160

Peace Operations

Course Number: CMS-6, Phase 1.

Location: Western Hemisphere Institute for Security Cooperation (WHINSEC), Fort Benning, GA.

Length: 7 weeks (187 hours).

Exhibit Dates: 2/01–Present.

Learning Outcomes: Upon completion of the course, the student will be familiar with US peace operation doctrine, logistics support, command structures, psychological operations and intelligence capabilities and assets.

Instruction: Audio-visual materials, classroom and practical exercises, and lecture. Participants are military officers (lieutenant to major) and civilian officials.

Credit Recommendation: In the upper-division baccalaureate category, 3 semester hours in political science or public administration or public policy or international relations (8/02).

AR-1728-0161

Counterdrug Operations

Course Number: TAC-6, Phase 2.

Location: Western Hemisphere Institute for Security Cooperation (WHINSEC), Fort Benning, GA.

Length: 12 weeks (755 hours).

Exhibit Dates: 2/01–Present.

Learning Outcomes: Upon completion of the course, students will have a familiarity with staff planning, leadership development, and counterdrug interdiction. Police officers will learn how to apply military techniques in counterdrug activities. Military personnel will be trained in operations in a peacetime environment.

Instruction: Audio-visual materials classroom and practical exercises, and lecture. Participants must be either a non-commissioned or commissioned officer with a national or military police unit dedicated to counterdrug interdiction operations.

Credit Recommendation: In the vocational certificate category, 6 semester hours in law enforcement. In the lower-division baccalaureate/associate degree category, 3 semester hours in criminal justice or law enforcement or security (8/02).

AR-1728-0162

Intelligence Officer

Course Number: TAC-2, Phase 2.

Location: Western Hemisphere Institute for Security Cooperation (WHINSEC), Fort Benning, GA.

Length: 8 weeks (277 hours).

Exhibit Dates: 6/01–Present.

Learning Outcomes: Upon completion of the course, the student will be familiar with the intelligence cycle, counterintelligence, information security, threat analysis, facility integrity, and information processing.

Instruction: Audio-visual materials, classroom and practical exercises, and lecture. Topics include the intelligence cycle, threat analysis, information collection and dissemination. Students must be military intelligence officers in the grade of lieutenant to captain.

Credit Recommendation: In the upper-division baccalaureate category, 3 semester hours in political science (8/02).

AR-1728-0163

Captains Career

Course Number: OPME-3, Phase 1.

Location: Western Hemisphere Institute for Security Cooperation (WHINSEC), Fort Benning, GA.

Length: 18 weeks (665 hours).

Exhibit Dates: 2/01–Present.

Learning Outcomes: Upon completion of the course, the student will be able to demonstrate leadership, decision making and communication skills; analyze problem-solving techniques; synthesize, conduct, and evaluate a simulated operational plan; and demonstrate and evaluate an understanding of the United States Government structure.

Instruction: Audio visual materials, practical and classroom exercises, discussion, and case studies. Topics include leadership doctrine, discussion and analysis of management strategies, and supervision skills.

Credit Recommendation: In the lower-division baccalaureate/associate degree category, 3 semester hours in principles of supervision. In the upper-division baccalaureate category, 3 semester hours in police administration or criminal justice and 3 in correctional administration and management (10/02).

AR-1729-0004

Hospital Food Service Specialist

Course Number: 800-94F10.

Location: Academy of Health Sciences, Fort Sam Houston, TX.

Length: 5 weeks (203–225 hours).

Exhibit Dates: 7/82–9/94.

Learning Outcomes: After 9/94 see AR-1729-0050. To train personnel to perform food preparation and distribution activities and apply diet therapy principles to the preparation of food required in patient meal service.

Instruction: Conferences, self-paced examinations, and practical exercises cover diet preparation, clinical dietetics, and hospital food service operations.

Credit Recommendation: In the vocational certificate category, 2 semester hours in hospital food service and 3 in modified diet preparation. In the lower-division baccalaureate/associate degree category, 3 semester hours in nutritional care (7/88).

Related Occupations: 94F.

AR-1729-0021

Advanced Food Management

Course Number: 8-10-C22-82A.

Location: Quartermaster Center and School, Fort Lee, VA.

Length: 8 weeks (280 hours).

Exhibit Dates: 5/85–6/96.

Objectives: To train military and civilian personnel to manage food service operations.

Instruction: Topics include basic receiving and storeroom operations, food sanitation, food products and technology, and various aspects of food service management. Classroom instruction is complemented by visits to various food service operations.

Credit Recommendation: In the lower-division baccalaureate/associate degree category, 7 semester hours in food service management (8/85).

Related Occupations: 82A.

AR-1729-0028

1. Veterinary Service in Theater of Operations
2. Department of Defense (DOD) Red Meats Certification

Course Number: ***Version 1:*** 6G-F2/321-F2. *Version 2:* 6G-F2/321-F2.

Location: Academy of Health Sciences, Fort Sam Houston, TX.

Length: ***Version 1:*** 3 weeks (126 hours). *Version 2:* 2 weeks (84 hours).

Exhibit Dates: ***Version 1:*** 1/88–7/96. *Version 2:* 8/96–Present.

Learning Outcomes: ***Version 1:*** Upon completion of the course, the student will be able to identify problem areas in food inspection and make recommendations for their resolution and supervise and train personnel in food inspection with emphasis on optimum performance standards and sanitary conditions. *Version 2:* Upon completion of this course, the student will be able to effectively provide standardized inspections and train personnel in meat inspections.

Instruction: ***Version 1:*** This course contains a review of anatomy, physiology, and pathological conditions related to various animals in their normal and/or abnormal status. Lectures, demonstrations, and practical exercises cover procurement and surveillance inspection of various dairy products, meats, fresh fruits, and vegetables. *Version 2:* Lectures, demonstrations, and practical exercises cover procurement and surveillance inspection of meat products.

Credit Recommendation: ***Version 1:*** In the lower-division baccalaureate/associate degree category, 1 semester hour in administration and 2 in food technology (11/89). *Version 2:* In the lower-division baccalaureate/associate degree category, 2 semester hours in meat inspection (6/99).

AR-1729-0029

Commissary Management

Course Number: 8G-82D/551-ASIU5.
Location: Quartermaster Center and School, Fort Lee, VA.
Length: 6 weeks (218 hours).
Exhibit Dates: 6/86–10/98.
Learning Outcomes: Upon completion of the course, the student will be able to manage a retail supply and food market facility.
Instruction: This course includes classroom and practical exercises in principles of supervision, record keeping, basic accounting, facility design and layout, merchandising and promotional techniques, security issues, food storage and display, sanitation, warehousing techniques, and inventory control.
Credit Recommendation: In the lower-division baccalaureate/associate degree category, 2 semester hours in principles of supervision, 2 in food sanitation, and 1 in retail management or 2 semester hours in food sanitation and 3 in retail management of which 2 are specifically in principles of supervision (10/88).

AR-1729-0030

Field Bread Baking Operations

Course Number: 800-ASID1.
Location: Quartermaster Center and School, Fort Lee, VA.
Length: 2 weeks (72 hours).
Exhibit Dates: 7/87–10/98.
Learning Outcomes: Upon completion of the course, the student will be able to manage and operate a mobile field baking plant.
Instruction: This course includes lectures and performance exercises in principles and procedures of mobile bread baking.
Credit Recommendation: In the vocational certificate category, 3 semester hours in bread baking (10/88).
Related Occupations: 94B.

AR-1729-0031

Subsistence Officer

Course Number: *Version 1:* 8E-F4. *Version 2:* 8E-F4.
Location: Quartermaster Center and School, Fort Lee, VA.
Length: *Version 1:* 6 weeks (211–216 hours). *Version 2:* 6 weeks (211–216 hours).
Exhibit Dates: *Version 1:* 1/88–9/93. *Version 2:* 10/93–2/94.
Learning Outcomes: *Version 1:* Upon completion of the course, the student will be able to manage a food supply operation, exercise technical supervision over food service activities, and contract for dining facility operations. *Version 2:* Upon completion of the course, the student will be able to manage a food supply operation, supervise food service sanitation and operations, and oversee dining facility contracts.
Instruction: *Version 1:* Lectures, small group instruction, field trips, and practical exercises cover food service supply management, food service operations, contracting for dining facility operations, and commercial and government food preparation and food procurement operations. *Version 2:* Lectures, small group instruction, field trips, and practical exercises cover food service supply management, food service operations, contracting for dining facility operations, and commercial and government food preparation and food procurement operations.
Credit Recommendation: *Version 1:* In the upper-division baccalaureate category, 6 semester hours in food service management (10/89). *Version 2:* In the lower-division baccalaureate/associate degree category, 2 semester hours in food service sanitation. In the upper-division baccalaureate category, 4 semester hours in food service management (5/02).
Related Occupations: 92G.

AR-1729-0032

Food Service Management

Course Number: *Version 2:* 8E-92G/8E-922A/800-F8; 8E-82G/8E-041A/800-F8. *Version 3:* 8E-92G/8E-922A-F8. *Version 4:* 8E-F6/800-F8.
Location: Quartermaster Center and School, Fort Lee, VA.
Length: *Version 2:* 3 weeks (108 hours). *Version 3:* 4 weeks (144 hours). *Version 4:* 4 weeks (144 hours).
Exhibit Dates: *Version 2:* 6/89–12/95. *Version 3:* 1/96–9/01. *Version 4:* 10/01–Present.
Learning Outcomes: *Version 2:* Upon completion of the course, students will be able to apply the principles and procedures of food service management and accept the responsibilities of food service management. *Version 3:* Upon completion of the course, the student will be able to apply principles and procedures of food service management in accordance with sanitary standards and accept the responsibilities of food service management. *Version 4:* Upon completion of the course, the student will be able to apply principles and procedures of food service management in accordance with sanitary standards and accept the responsibilities of food service management.
Instruction: *Version 2:* This course is conducted through demonstrations, lectures, and performance exercises. Major topics covered in the course include record keeping, administrative procedures, the basics of food preparation, sanitation, nutrition, evaluation of foods, and selected computer applications. *Version 3:* Methods include demonstrations, lectures, and performance exercises. The major topics covered include record keeping, administrative procedures, contract management, basic food preparation, nutrition, sanitation, evaluation of foods, and computer applications. *Version 4:* Methods include demonstrations, lectures, and performance exercises. The major topics covered include record keeping, administrative procedures, contract management, basic food preparation, nutrition, sanitation, evaluation of foods, and computer applications.
Credit Recommendation: *Version 2:* In the lower-division baccalaureate/associate degree category, 3 semester hours in food service management (2/93). *Version 3:* In the lower-division baccalaureate/associate degree category, 2 semester hours in food service management and 1 in sanitation (12/96). *Version 4:* In the lower-division baccalaureate/associate degree category, 2 semester hours in food service management and 1 in food service sanitation (6/02).
Related Occupations: 041A; 82C; 922A; 92G.

AR-1729-0034

1. Food Service Technician Warrant Officer Technical/Tactical Certification
2. Food Service Technician Warrant Officer Technical/Tactical Certification
3. Food Service Technician Warrant Officer Basic

(Warrant Officer Basic)

Course Number: *Version 1:* 8E-922A. *Version 2:* 8E-922A. *Version 3:* 8E-922A.
Location: Quartermaster Center and School, Fort Lee, VA.
Length: *Version 1:* 9 weeks (330 hours). *Version 2:* 8–9 weeks (313 hours). *Version 3:* 9 weeks (338–369 hours).
Exhibit Dates: *Version 1:* 1/88–9/92. *Version 2:* 10/92–11/95. *Version 3:* 12/95–Present.
Learning Outcomes: *Version 1:* Upon completion of the course, the student will be able to apply a knowledge of sanitation and nutrition to the supervision of food service operations. *Version 2:* Upon completion of the course, the student will be able to manage food service facilities and contracts in accordance with nutrition and sanitation standards. *Version 3:* Upon completion of the course, the student will be able to manage food service facilities and contracts in accordance with nutrition and sanitation standards.
Instruction: *Version 1:* Lectures, practical exercises, and demonstrations cover the topics of management theory, effective writing, nutrition, sanitation, and supervision of food service operations. *Version 2:* Lectures, practical exercises, and demonstrations cover nutrition, food preparation and presentation, record keeping, sanitation, and supervision of a food service operation. *Version 3:* Lectures, practical exercises, and demonstrations cover nutrition, food preparation and presentation, record keeping, sanitation, supply management, and management of a food service operation.
Credit Recommendation: *Version 1:* In the lower-division baccalaureate/associate degree category, 1 semester hour in principles of management, 2 in principles of nutrition, 2 in food service supervision, and 1 in food service sanitation (10/88). *Version 2:* In the lower-division baccalaureate/associate degree category, 2 semester hours in food service sanitation, 2 in basic nutrition, and 2 in food service management (2/93). *Version 3:* In the lower-division baccalaureate/associate degree category, 3 semester hours in food service management, 1 in food service sanitation, and 3 in supply management (6/02).
Related Occupations: 22A; 922A.

AR-1729-0036

Food Service Technician Senior Warrant Officer, Reserve

Course Number: 8-10-C32-922A-RC.
Location: Quartermaster Center and School, Fort Lee, VA.
Length: 2 weeks (84 hours).
Exhibit Dates: 10/88–10/93.
Learning Outcomes: Upon completion of the course, the student will be able to plan and organize food service technical functions for field operations.
Instruction: Lectures, seminars, demonstrations, and practical exercises cover field food

service operations, menu planning, and field bakery operations.

Credit Recommendation: In the lower-division baccalaureate/associate degree category, 1 semester hour in food service management (10/88).

Related Occupations: 922A.

AR-1729-0037

1. Food Service Technician Warrant Officer Technical/Tactical Certification Reserve Component (Reserve Component Warrant Officer)
2. Food Service Technician Warrant Officer Basic-RC (RC Warrant Officer)

Course Number: ***Version 1:*** 8E-922A-RC. *Version 2:* 8E-922A-RC.

Location: Quartermaster Center and School, Fort Lee, VA.

Length: ***Version 1:*** 2 weeks (82 hours). *Version 2:* 2 weeks (82 hours).

Exhibit Dates: ***Version 1:*** 10/88–10/93. *Version 2:* 11/93–Present.

Learning Outcomes: ***Version 1:*** Upon completion of the course, the student will be able to perform the organizational management functions of a food advisor and plan, coordinate, and direct sanitation and food service mobile operations. *Version 2:* Upon completion of the course, the student will be able to perform the organizational management functions of a food advisor and plan, coordinate, and direct sanitation and food service mobile operations.

Instruction: ***Version 1:*** This course includes lectures and performance exercises in sanitation, the mobile operations of a food service activity, and management. *Version 2:* This course includes lectures and performance exercises in sanitation, the mobile operations of a food service activity, and management.

Credit Recommendation: ***Version 1:*** In the lower-division baccalaureate/associate degree category, 3 semester hours in food service management (3/00). *Version 2:* In the lower-division baccalaureate/associate degree category, 3 semester hours in food service management (4/04).

Related Occupations: 922A.

AR-1729-0039

Food Service Specialist Basic Noncommissioned Officer (NCO)

Course Number: ***Version 1:*** 800-94B30. *Version 2:* 800-94B30. *Version 3:* 800-92G30. *Version 4:* 800-92G30, Phase 2.

Location: Quartermaster Center and School, Fort Lee, VA.

Length: ***Version 1:*** 8 weeks (297 hours). *Version 2:* 11 weeks (400 hours). *Version 3:* 10 weeks (364–365 hours). *Version 4:* 8 weeks (299 hours).

Exhibit Dates: ***Version 1:*** 10/86–9/92. *Version 2:* 10/92–9/95. *Version 3:* 10/95–10/00. *Version 4:* 10/01–Present.

Learning Outcomes: ***Version 1:*** Upon completion of the course, the student will be able to use correct preparation and serving procedure in cafeteria/dining operations, follow industry-acceptable sanitation and hygiene practices, and supervise personnel in the food production and service areas. *Version 2:* Upon completion of the course, the student will be able to prepare and serve food in accordance with quality assurance and basic nutrition guidelines in a safe and sanitary environment. *Version 3:* Upon completion of the course, the student will be able to supervise the preparation and serving of food in accordance with quality assurance and basic nutritional guidelines and a safe and sanitary environment. *Version 4:* Upon completion of the course, the student will be able to perform all production manager duties field feeding operations; food service sanitation; quantity food preparation and production; food inspection and quality assurance; supervision of food service personnel; and operation of the Army Food Information Management System.

Instruction: ***Version 1:*** Topics include meat processing, sanitation, food preparation, dining facility management, and field kitchen operations. This course contains a common core of leadership subjects. *Version 2:* Conferences, small group instruction, and practical exercises are used to cover nutrition, food preparation and serving techniques, food service sanitation, equipment maintenance, and food service management. Other topics include security, basic computer applications, and record keeping. This course contains a common core of leadership subjects. *Version 3:* Conferences, small group instruction, and practical exercises are used to cover nutrition, food preparation and serving techniques, food service sanitation, equipment maintenance, and food service management. Other topics include security, basic computer applications, and record keeping. This course contains a common core of leadership subjects. *Version 4:* Laboratory, discussion, classroom exercise, lecture, learner presentations, conferences, and small-group instruction, are used to cover nutrition; food preparation and serving techniques; food service sanitation; equipment maintenance; food service management; security; basic computer applications; and record keeping.

Credit Recommendation: ***Version 1:*** In the lower-division baccalaureate/associate degree category, 1 semester hour in sanitation and 2 in dining facility management. See AR-1406-0090 for common core credit (12/88). *Version 2:* In the lower-division baccalaureate/associate degree category, 3 semester hours in food service management and 1 in food service sanitation. See AR-1406-0090 for common core credit (2/93). *Version 3:* In the lower-division baccalaureate/associate degree category, 3 semester hours in food service management, 1 in computer applications, and 1 in sanitation. See AR-1406-0090 for common core credit (12/96). *Version 4:* In the lower-division baccalaureate/associate degree category, 3 semester hours in food management, 3 in intermediate food preparation, and 1 in food service sanitation (6/02).

Related Occupations: 92G; 94B.

AR-1729-0040

Food Service Specialist Advanced Noncommissioned Officer (NCO)

Course Number: ***Version 1:*** 800-94B40. *Version 2:* 800-94B40. *Version 3:* 800-92G40. *Version 4:* 800-92G40, Phase 2.

Location: Quartermaster Center and School, Fort Lee, VA.

Length: ***Version 1:*** 10 weeks (362 hours). *Version 2:* 11–12 weeks (416 hours). *Version 3:* 10 weeks (363 hours). *Version 4:* 8–9 weeks (320 hours).

Exhibit Dates: ***Version 1:*** 3/89–9/92. *Version 2:* 10/92–9/95. *Version 3:* 10/95–9/01. *Version 4:* 10/01–Present.

Learning Outcomes: ***Version 1:*** Upon completion of the course, the student will be able to apply sanitation practices in stationary and field settings, apply general management and human resource theory, solve logistical problems in food service/dining/field facilities, evaluate contract food service operations, and apply nutritional concepts in food preparation and service. *Version 2:* Upon completion of the course, the student will be able to supervise the preparation and serving of foods in accordance with nutritional and sanitary standards. *Version 3:* Upon completion of the course, the student will be able to supervise the preparation and serving of foods in accordance with nutritional and sanitary standards. *Version 4:* Upon completion of the course, the student will be able to perform advanced food specialist duties related to the following: army field feeding system, food service sanitation and safety, dining facility management, automated management systems, and supervise food service personnel.

Instruction: ***Version 1:*** Conference instruction and guest speakers present course content, including sanitation practices, management principles and procedures, dining facility accounting, dining security procedures, staffing, and strategic planning in dining facility equipment replacement. This course contains a common core of leadership subjects. *Version 2:* Conference and small group instruction present course content in the areas of nutrition, sanitation, facility management, equipment replacement, contract administration, security, record keeping, and computer applications. This course contains a common core of leadership subjects. *Version 3:* Conference and small group instruction present course content in the areas of nutrition, sanitation, facility management, equipment replacement, contract administration, security, record keeping, and computer applications. This course contains a common core of leadership subjects. *Version 4:* Conference and small group instruction present course content in the areas of nutrition, sanitation, facility management, equipment replacement, contract administration, security, record keeping, and computer applications. This course contains a common core of leadership subjects.

Credit Recommendation: ***Version 1:*** In the lower-division baccalaureate/associate degree category, 2 semester hours in dining facility management, 2 in human resource management, and 3 in basic nutrition. See AR-1404-0035 for common core credit (10/88). *Version 2:* In the lower-division baccalaureate/associate degree category, 1 semester hour in food service sanitation, 1 in basic nutrition, and 3 in food service management. See AR-1404-0035 for common core credit (2/93). *Version 3:* In the lower-division baccalaureate/associate degree category, 2 semester hours in microcomputer applications, 3 in food service management, and 1 in food service sanitation. See AR-1404-0035 for common core credit (12/01). *Version 4:* In the lower-division baccalaureate/associate degree category, 3 semester hours in food service management, 1 in food service sanitation, and 3 in principles of supervision. See AR-1404-0035 for common core credit (6/02).

Related Occupations: 92G; 94B.

AR-1729-0041

1. Food Service Specialist Advanced Individual Training
2. Food Service Specialist Advanced Individual Training
3. Food Service Specialist

Course Number: *Version 1:* 800-94B10. *Version 2:* 800-92G10. *Version 3:* 800-92G10, Phase 1.

Location: *Version 1:* Quartermaster Center and School, Fort Lee, VA; Training Center, Fort Dix, NJ; Training Center, Fort Jackson, SC. *Version 2:* Quartermaster Center and School, Fort Lee, VA. *Version 3:* Quartermaster Center and School, Fort Lee, VA.

Length: *Version 1:* 9 weeks (324 hours). *Version 2:* 8 weeks (320 hours). *Version 3:* 8 weeks (299 hours).

Exhibit Dates: *Version 1:* 1/87–9/95. *Version 2:* 10/95–10/01. *Version 3:* 11/01–Present.

Learning Outcomes: *Version 1:* Upon completion of the course, the student will be able to use standard food service terminology; prepare and serve food in dining, kitchen, and field settings; and maintain sanitation standards and food preparation equipment. *Version 2:* Upon completion of the course, the student will be able to use standard food service terminology; prepare and serve food in dining, kitchen, and field settings; and maintain sanitation standards and food preparation equipment. *Version 3:* Upon completion of the course, the student will be able to perform food service duties relative to basic food preparation for a garrison dining facility; bake breads and pastries; maintain and operate field feeding equipment; and practice food service sanitation and safety procedures.

Instruction: *Version 1:* Instruction includes small- and large-quantity cooking, baking, and meat cutting; basic food theory; care and operation of equipment under field conditions; and duties and responsibilities of food service personnel. *Version 2:* Instructional methods include lecture, discussion, demonstration, classroom practical applications, and conferences. Topics covered include baking, small- and large-quantity cooking, basic food theory, nutrition, care and operation of equipment, and duties and responsibilities of food service personnel. *Version 3:* Laboratory, discussion, classroom exercise, practical exercise, lecture, and learner presentations. Topics include baking; small and large-quantity cooking; basic food theory; nutrition; care and operation of equipment, and duties and responsibilities of food service personnel.

Credit Recommendation: *Version 1:* In the lower-division baccalaureate/associate degree category, 2 semester hours in food preparation—small quantity and 2 in institutional cooking—large quantity (8/90). *Version 2:* In the lower-division baccalaureate/associate degree category, 3 semester hours in basic food preparation and 2 in fundamentals of baking (12/96). *Version 3:* In the lower-division baccalaureate/associate degree category, 3 semester hours in basic food preparation, 3 semester hours in baking, and 1 semester hour in food service sanitation (6/02).

Related Occupations: 92G; 94B.

AR-1729-0042

Dining Facility Management

Course Number: LOG 75.

Location: Combined Arms Training Center, Vilseck, W. Germany.

Length: 3 weeks (107 hours).

Exhibit Dates: 4/80–10/98.

Learning Outcomes: Upon completion of the course, the student will be able to manage a military dining facility; establish dining facility files; prepare and maintain dining facility forms, such as cash collection sheets, production schedules, subsistence forms, ration requests, master menus, and food preparation charts; supervise food sanitation policies; identify operational requirements, responsibilities, and procedures; recommend equipment replacement.

Instruction: Conference classes, practical exercises, and videotapes are designed to upgrade technical skills required for management of a dining facility. Topics include policies of sanitation, records and reports, files, managing personnel, menu development, and the receipt and storage of rations.

Credit Recommendation: In the lower-division baccalaureate/associate degree category, 2 semester hours in food service management (10/88).

AR-1729-0043

Hospital Food Service Management

Course Number: 800-F3.

Location: Academy of Health Sciences, Fort Sam Houston, TX.

Length: 3 weeks (103 hours).

Exhibit Dates: 8/88–4/04.

Learning Outcomes: Upon completion of the course, the student will apply fundamentals of nutrition management to clinical dietetic nutrition care; principles of food service management, including kitchen layout, design, and equipment; fundamentals of oral and written communication; and principles of motivation and ethical decision making.

Instruction: Conferences and practical exercises cover clinical dietetic nutrition, food service management, and communications.

Credit Recommendation: In the lower-division baccalaureate/associate degree category, 2 semester hours in introduction to nutritional care/commercial dietetics, 2 in fundamentals of food service management/institutional management, and 1 in written and or oral communication (3/90).

AR-1729-0044

Hospital Food Service Specialist (Basic) Reserve Component

Course Number: 800-91M10-RC; 800-94F10-RC.

Location: Academy of Health Sciences, Fort Sam Houston, TX.

Length: 2 weeks (141–160 hours).

Exhibit Dates: 7/88–12/96.

Learning Outcomes: Upon completion of the course, the student will apply fundamentals of nutrition care, including taking a dietary history and evaluating patient menus, and apply food service management practices.

Instruction: Course consists of conferences and practical exercises in clinical dietetics including preparation of specialized diets and quantity food preparation in mobile facilities. Phase 1 is by correspondence; phase 2 is in residence, and phase 3 at a training unit.

Credit Recommendation: In the lower-division baccalaureate/associate degree category, for Phase 2, 3 semester hours in clinical dietetics. For Phase 3,1 in practice of clinical dietetics (4/94).

Related Occupations: 91M; 94F.

AR-1729-0045

Food and Beverage Management

Course Number: 800-F15.

Location: Soldier Support Center, Fort Benjamin Harrison, IN.

Length: 3 weeks (137 hours).

Exhibit Dates: 2/91–1/97.

Learning Outcomes: Upon completion of the course, the student will be able to apply proper sanitation and food presentation procedures; establish and maintain beverage control; receive, store, and issue food items; grill, fry, and broil food; and prepare salads and basic sauces.

Instruction: Lectures, demonstrations, practicum, and evaluation are course methods.

Credit Recommendation: In the lower-division baccalaureate/associate degree category, 3 semester hours in basic food preparation and 1 in food safety and sanitation (12/91).

AR-1729-0046

Food Service Specialist Reserve

Course Number: 101-94B10-RC.

Location: Reserve Component Training Centers, US.

Length: 2 weeks (96 hours).

Exhibit Dates: 1/92–4/04.

Learning Outcomes: Upon completion of the course, the student will be able to apply food service principles to small- and large-quantity cooking and baking.

Instruction: Conferences, demonstrations, and practical exercises cover basic knowledge of small quantity cooking along with some applications to large quantity food operations.

Credit Recommendation: In the vocational certificate category, 4 semester hours in food preparation (2/93).

Related Occupations: 94B.

AR-1729-0047

Advanced Culinary Skills

Course Number: 8E-F5/800-F17, Phase 1.

Location: Quartermaster Center and School, Fort Lee, VA.

Length: 3–4 weeks (122–144 hours).

Exhibit Dates: 1/92–Present.

Learning Outcomes: Upon completion of the course, the student will be able to apply food preparation principles; use proper techniques in the preparation of a variety of food items, including stocks, baked goods, sauces, vegetables, potatoes, meats and fish, salads, hors d'oeuvres and pate, savories and canapes, and desserts; and apply principles of menu planning.

Instruction: Conferences, demonstrations, and practical laboratory exercises are used to present the principles and techniques of food preparation and basic menu planning.

Credit Recommendation: In the lower-division baccalaureate/associate degree category, 4 semester hours in principles of food preparation (4/02).

AR-1729-0048

1. Army Medical Department (AMEDD) Basic Noncommissioned Officer (NCO) (91R)
2. Army Medical Department (AMEDD) Basic Noncommissioned Officer (NCO)

Course Number: ***Version 1:*** 6-8-C40 (91R). *Version 2:* 6-8-C40 (91R30 Tech Tng).

Location: Academy of Health Sciences, Fort Sam Houston, TX.

Length: ***Version 1:*** 12 weeks (487–491 hours). *Version 2:* 6–7 weeks (288–297 hours).

Exhibit Dates: ***Version 1:*** 10/91–12/97. *Version 2:* 1/98–Present.

Learning Outcomes: ***Version 1:*** Upon completion of the course, the student will be able to inspect food, food supply, and storage and perform food preparation services in compliance with approved standards of wholesomeness and quality. *Version 2:* Upon completion of the course, the student will be able to inspect foods, food supply, and storage and food preparation services for compliance with approved standards of wholesomeness and quality.

Instruction: ***Version 1:*** Course covers the inspection of poultry, eggs, dairy, beef, pork, lamb, fish, and other foods; the inspection of preparation and storage food service facilities; animal food supplies; and contract administration. Methodologies include lectures, laboratories, discussions, and practical exercises. *Version 2:* Course covers the inspection of poultry, eggs, dairy, beef, pork, lamb, fish, and other foods; the inspection of preparation and storage food service facilities; animal food supplies; and contract administration. Methodologies include lectures, laboratories, discussions, and practical exercises.

Credit Recommendation: ***Version 1:*** In the lower-division baccalaureate/associate degree category, 4 semester hours in food sanitation and 4 in food inspection. In the upper-division baccalaureate category, 2 semester hours in military science (6/94). *Version 2:* In the lower-division baccalaureate/associate degree category, 3 semester hours in fundamentals of food technology, 3 in food sanitation and inspection, and 1 in military science (5/01).

Related Occupations: 91R.

AR-1729-0049

1. Army Medical Department (AMEDD) Noncommissioned Officer (NCO) (91M)
2. Army Medical Department (AMEDD) Noncommissioned Officer (NCO) Basic

Course Number: ***Version 1:*** 6-8-C40 (91M30 Tech Tng); 6-8-C40 (91M Tech Tng); 6-8-C40 (91M). *Version 2:* 6-8-C40 (91M30 Tech Tng).

Location: Academy of Health Sciences, Fort Sam Houston, TX.

Length: ***Version 1:*** 5–10 weeks (232–420 hours). *Version 2:* 5–6 weeks (232 hours).

Exhibit Dates: ***Version 1:*** 9/91–2/95. *Version 2:* 3/95–Present.

Learning Outcomes: ***Version 1:*** Upon completion of the course, the student will be able to manage a medical food preparation department, including ordering, inventory control, quality assurance, equipment operation and maintenance, planning, budgeting, cost accounting, evaluation, and serve special diets. *Version 2:* Upon completion of the course, the student will be able to manage a medical food preparation department, including ordering, inventory control, quality assurance, equipment operation and maintenance, planning, budgeting, cost accounting, evaluation, and serve special diets.

Instruction: ***Version 1:*** This course covers communication and interviewing, equipment selection, planning and supervising food production, applied nutrition and dietetics, sanitation, budgeting and cost accounting, and evaluation and quality assurance. Methodologies include lectures, laboratories, discussions, and supervised practice. This course contains a common core of leadership subjects. *Version 2:* Laboratory, audio visual materials, discussion, classroom exercise, practical exercise, lecture, and learner presentations. Topics include communication and interviewing, equipment selection, planning and supervising food production, applied nutrition and dietetics, sanitation, budgeting and cost accounting, evaluation, and quality assurance.

Credit Recommendation: ***Version 1:*** In the lower-division baccalaureate/associate degree category, 3 semester hours in food production management and 4 in applied nutrition and dietetics. See AR-1406-0090 for common core credit. In the upper-division baccalaureate category, 2 semester hours in military science (5/01). *Version 2:* In the lower-division baccalaureate/associate degree category, 3 semester hours in clinical management, 3 in food production, and 4 in applied nutrition. In the upper-division baccalaureate category, 2 in military science (6/02).

Related Occupations: 91M.

AR-1729-0050

1. Hospital Food Service Specialist Basic (Hospital Food Service Specialist)
2. Hospital Food Service Specialist (Basic)

Course Number: ***Version 1:*** 800-91M10 (Basic); 800-91M10. *Version 2:* 800-91M10.

Location: Academy of Health Sciences, Fort Sam Houston, TX.

Length: ***Version 1:*** 6–7 weeks (308–310 hours). *Version 2:* 7 weeks (311–319 hours).

Exhibit Dates: ***Version 1:*** 10/94–2/02. *Version 2:* 3/02–Present.

Learning Outcomes: ***Version 1:*** Before 10/94 see AR-1729-0004. Upon completion of the course, the student will be able to discuss basic nutrition and diet therapy, interview patients to determine basic dietary history, prepare food according to modified diet recipes, assemble patient trays, and prepare nourishment for tube feedings. *Version 2:* Upon completion of the course, the student will be able to discuss basic nutrition and diet therapy, interview patients to determine basic dietary history, prepare food according to modified diet recipes, assemble patient trays, prepare nourishment for tube feedings, and perform CPR when given a simulated casualty.

Instruction: ***Version 1:*** Lectures and practical exercises cover basic nutrition, dietary history, and modified diets. *Version 2:* Laboratory, and classroom exercises cover basic nutrition, dietary history, and modified diets.

Credit Recommendation: ***Version 1:*** In the lower-division baccalaureate/associate degree category, 3 semester hours in hospital food service and preparation and 3 in diet therapy (5/01). *Version 2:* In the lower-division baccalaureate/associate degree category, 1 semester hour in CPR, 3 in introduction to nutrition, 3 in hospital food service clinical practice, and 2 in diet therapy (6/02).

Related Occupations: 91M.

AR-1729-0052

Food Product Subject Matter Expert (Pilot)

Course Number: 6G-F10/321-F10.

Location: Academy of Health Sciences, Fort Sam Houston, TX.

Length: 2 weeks (85 hours).

Exhibit Dates: 3/99–Present.

Learning Outcomes: Upon completion of the course, the student will be able to procure, receive, survey, and inspect food and apply proper techniques of food handling.

Instruction: Lecture, videos, computer applications, and laboratory are course methods.

Credit Recommendation: In the lower-division baccalaureate/associate degree category, 3 semester hours in food science (7/99).

AR-1729-0053

Hospital Food Service Specialist (Basic) (RC), Phase 3

Course Number: 800-91M10 (RC).

Location: Civilian/Military Medical Facilities, Continental, US.

Length: 2 weeks (104–105 hours).

Exhibit Dates: 1/97–3/98.

Learning Outcomes: Upon completion of the course, the student will apply fundamentals of nutrition care, including taking dietary history and evaluating patient menus and apply food management practices.

Instruction: Course consists of conferences and practical exercises in clinical dietetics, preparation of special diets, and quality food preparation in mobile facilities. Courses conducted at a training unit. Student will have completed Phase 1 by correspondence. Phase 3 is on-the-job training.

Credit Recommendation: In the lower-division baccalaureate/associate degree category, 1 semester hour in clinical dietetics (7/99).

AR-1729-0054

Hospital Food Service Specialist Basic Reserve, Phase 2

Course Number: ***Version 1:*** 800-91M10 (RC). *Version 2:* 800-91M10 (RC), Phase 2; 800-91M10 (RC).

Location: Academy of Health Sciences, Fort Sam Houston, TX.

Length: ***Version 1:*** 2 weeks (109 hours). *Version 2:* 2 weeks (134 hours).

Exhibit Dates: ***Version 1:*** 1/97–3/98. *Version 2:* 4/98–Present.

Learning Outcomes: ***Version 1:*** Upon completion of the course, the student will apply fundamentals of nutrition care, including taking a dietary history and evaluating patient menus and apply food service management practices. *Version 2:* Upon completion of the course, the student will apply fundamentals of nutrition care, including taking a dietary history and evaluating patient menus and apply food service management practices.

Instruction: *Version 1:* Course consists of conferences and practical exercises in clinical dietetics, including preparation of specialized diets and quantity food preparation in mobile facilities. Phase 1 is by correspondence; phase 2 is in residence, and phase 3 is at a training unit. *Version 2:* Course consists of conferences and practical exercises in clinical dietetics, including preparation of specialized diets and quantity food preparation in mobile facilities. Phase 1 is by correspondence; phase 2 is in residence.

Credit Recommendation: *Version 1:* In the lower-division baccalaureate/associate degree category, for Phase 2, 3 semester hours in clinical dietetics (7/99). *Version 2:* In the lower-division baccalaureate/associate degree category, for Phase 2, 3 semester hours in clinical dietetics (7/99).

Related Occupations: 91M.

AR-1732-0009

WATER TREATMENT SPECIALIST BASIC NONCOMMISSIONED OFFICER (NCO)
(Water Treatment Specialist Basic Technical)

Course Number: 720-51N30.

Location: Quartermaster Center and School, Fort Lee, VA.

Length: 6 weeks (237 hours).

Exhibit Dates: 10/87–9/91.

Objectives: After 9/91 see AR-1732-0015. To provide the student with an overview of general water treatment operations.

Instruction: Through conferences, demonstrations, visual aids and hands-on experiences, the course covers analysis; inspection; and the nuclear, biological, and chemical functions of a water treatment specialist. This course contains a common core of leadership subjects.

Credit Recommendation: In the vocational certificate category, 5 semester hours in water treatment specialist. In the lower-division baccalaureate/associate degree category, see AR-1406-0090 for common core credit (5/88).

Related Occupations: 51N.

AR-1732-0010

1. WATER TREATMENT SPECIALIST ADVANCED INDIVIDUAL TRAINING
(Water Treatment Specialist)
2. WATER TREATMENT SPECIALIST ADVANCED INDIVIDUAL TRAINING
(Water Treatment Specialist)
3. WATER TREATMENT SPECIALIST
4. WATER TREATMENT SPECIALIST AIT
(Water Treatment Specialist)

Course Number: *Version 1:* 720-77W10. *Version 2:* 720-77W10. *Version 3:* 720-77W10. *Version 4:* 720-92W10.

Location: Quartermaster Center and School, Fort Lee, VA.

Length: *Version 1:* 8–9 weeks (296–359 hours). *Version 2:* 11 weeks (396 hours). *Version 3:* 9–11 weeks (379–380 hours). *Version 4:* 10–11 weeks (411 hours).

Exhibit Dates: *Version 1:* 11/87–9/91. *Version 2:* 10/91–9/95. *Version 3:* 10/95–6/03. *Version 4:* 7/03–Present.

Learning Outcomes: *Version 1:* Upon completion of the course, the student will be able to install, operate, and maintain equipment used to provide potable water and perform water quality analysis tests. *Version 2:* Upon completion of the course, the student will be able to install, operate, and maintain equipment used to provide potable water and perform water quality analysis tests. *Version 3:* Upon completion of the course, the student will be able to install, operate, and maintain equipment used to provide potable water and perform water quality analysis tests. *Version 4:* Pending evaluation.

Instruction: *Version 1:* Lectures, demonstration, and practical exercises cover both laboratory and field water purification and treatment; water quality analysis; generator and pump operations; and operating in a nuclear, biological, or chemically contaminated environment. *Version 2:* Lectures, demonstration, and practical exercises cover both laboratory and field water purification and treatment; water quality analysis; generator and pump operations; and operating in a nuclear, biological, or chemically contaminated environment. *Version 3:* Lectures, demonstrations, and practical exercises cover water purification and treatment, quality analysis, and equipment operation and maintenance. *Version 4:* Pending evaluation.

Credit Recommendation: *Version 1:* In the vocational certificate category, 6 semester hours in water treatment plant operations (10/90). *Version 2:* In the lower-division baccalaureate/associate degree category, 6 semester hours in basic water treatment technology, water treatment specialist, or in applied science (12/92). *Version 3:* In the vocational certificate category, 6 semester hours in basic water treatment technology (6/02). *Version 4:* Pending evaluation.

Related Occupations: 77W.

AR-1732-0011

WATER TREATMENT SPECIALIST RESERVE COMPONENT

Course Number: 720-77W10-RC.

Location: Reserve Component Training Centers, US; Quartermaster Center and School, Fort Lee, VA.

Length: 100–120 hours.

Exhibit Dates: 10/87–12/91.

Learning Outcomes: Upon completion of the course, the student will be able to purify, store, and distribute potable and palatable water to field troops.

Instruction: Lectures, demonstrations, and practical exercises cover both laboratory and field work in water purification, water quality analysis, and related equipment. Included are operations with water contaminated with nuclear, biological, and chemical materials.

Credit Recommendation: In the vocational certificate category, 3 semester hours in water treatment plant operation (10/88).

Related Occupations: 77W.

AR-1732-0013

POWER GENERATION EQUIPMENT REPAIRER

Course Number: *Version 1:* 662-52D10. *Version 2:* 662-52D10. *Version 3:* 662-52D10.

Location: *Version 1:* Ordnance School, Fort Belvoir, VA. *Version 2:* Ordnance Center and School, Aberdeen Proving Ground, MD. *Version 3:* Ordnance Center and School, Aberdeen Proving Ground, MD.

Length: *Version 1:* 11 weeks (397 hours). *Version 2:* 11–13 weeks (425–469 hours). *Version 3:* 10–11 weeks (387 hours).

Exhibit Dates: *Version 1:* 10/90–9/95. *Version 2:* 10/95–9/02. *Version 3:* 10/02–Present.

Learning Outcomes: *Version 1:* Upon completion of the course, the student will apply operating principles of gasoline and diesel engines used to drive electrical generators and apply basic electricity, portable generator design theory, and service and diagnostic techniques. *Version 2:* Upon completion of the course, the student will apply operating principles of gasoline and diesel engines used to drive electrical generator sets and will have an introduction to AC and DC circuits, portable generator theory, and diagnosis and repair of this same equipment. *Version 3:* Upon completion of the course, the student will test, repair and maintain various types of power generation equipment.

Instruction: *Version 1:* Lectures and demonstrations cover small gasoline and diesel-engine theory and AC/DC generators. Extensive laboratory and field practice cover diagnosis and troubleshooting of gasoline and diesel engine-driven generating sets. *Version 2:* Lectures, demonstrations and practical exercises cover DC circuit fundamentals, portable generators, and small engines theory and repair. Extensive laboratory and field practice is provided in troubleshooting gasoline- and diesel engine-driven generating sets. *Version 3:* Topics include maintenance management, basic math, soldering skills, fundamentals of electricity, basic engine design, support of 2kw generator set, and support of tactical generator sets.

Credit Recommendation: *Version 1:* In the lower-division baccalaureate/associate degree category, 1 semester hour in automotive shop practice, 2 in small engine service, 1 in diesel engine fundamentals, 3 in basic electricity, and 1 in AC/DC motors and generators (12/90). *Version 2:* In the lower-division baccalaureate/associate degree category, 1 semester hour in automotive shop practice, 2 in small engine service, 3 in AC/DC circuits, and 1 in AC/DC motors and generators (4/01). *Version 3:* In the lower-division baccalaureate/associate degree category, 2 semester hours in basic electricity, 2 of basic electricity laboratory, and 2 in small engine service (8/03).

Related Occupations: 52D.

AR-1732-0014

WATER TREATMENT SPECIALIST RESERVE COMPONENT

Course Number: 101-77W10.

Location: Reserve Component Training Centers, US.

Length: 2–3 weeks (67–89 hours).

Exhibit Dates: 3/92–4/04.

Objectives: Upon completion of the course, the student will be able to operate and maintain a potable water treatment, storage, and distribution system using the 600 GPH or the 3000 GPH reverse osmosis water purification system.

Instruction: Lectures, demonstrations, and practical exercises cover distributing and storing water; analyzing water quality; erecting, operating, and dismantling a tactical water distribution system; operating a generator and pump; and erecting, operating, and dismantling a specific reverse osmosis water purification system.

Credit Recommendation: In the lower-division baccalaureate/associate degree category, 3 semester hours in water treatment operations or 3 in applied science (2/93).

Related Occupations: 77W.

AR-1732-0015

1. Water Treatment Specialist Basic Noncommissioned Officer (NCO)
2. Water Treatment Specialist Basic Noncommissioned Officer

Course Number: ***Version 1:*** 720-77W30. *Version 2:* 720-77W30; 720-92W30.

Location: Quartermaster Center and School, Fort Lee, VA.

Length: ***Version 1:*** 8–10 weeks (271–378 hours). *Version 2:* 8–10 weeks (271–378 hours).

Exhibit Dates: ***Version 1:*** 10/91–9/99. *Version 2:* 10/99–Present.

Learning Outcomes: ***Version 1:*** Before 10/91 see AR-1732-0009. Upon completion of the course, the student will be able to manage and supervise water purification operations and distribution under field operations, including nuclear, biological, and chemical environments. *Version 2:* Upon completion of the course, the student will be able to manage and supervise water purification operations and distribution under field operations, including nuclear, biological, and chemical environments.

Instruction: ***Version 1:*** Through conferences, demonstrations, visual aids, and hands-on experiences, the course covers analysis; inspection; and the nuclear, biological, and chemical functions of a water treatment specialist. This course contains a common core of leadership subjects. *Version 2:* Through conferences, demonstrations, visual aids, and hands-on experiences, the course covers analysis; inspection; and the nuclear, biological, and chemical functions of a water treatment specialist. This course contains a common core of leadership subjects.

Credit Recommendation: ***Version 1:*** In the lower-division baccalaureate/associate degree category, 6 semester hours in advanced water treatment technology or applied science. See AR-1406-0090 for common core credit (5/96). *Version 2:* In the lower-division baccalaureate/associate degree category, 6 semester hours in advanced water treatment technology or applied science (4/04).

Related Occupations: 77W.

AR-1732-0016

Water Treatment Specialist Advanced Noncommissioned Officer (NCO)

Course Number: 8-77-C42(3).

Location: Quartermaster Center and School, Fort Lee, VA.

Length: 11–12 weeks (415 hours).

Exhibit Dates: 10/95–9/00.

Learning Outcomes: Before 10/95 see AR-1717-0092. Upon completion of the course, the student will be able to apply logistics principles in petroleum supply and storage and perform and manage petroleum supply, storage, and laboratory operations.

Instruction: Small-group instruction and practical exercises cover the subject matter. The course contains a common core of leadership subjects. This course is for Water Treatment Specialists (NCO) who have already received all of the available courses in water treatment.

Credit Recommendation: In the lower-division baccalaureate/associate degree category, 1 semester hour in microcomputer applications, 1 in materiel management, 1 in applied science, and 3 in petroleum supply operations and management. See AR-1404-0035 for common core credit (12/96).

AR-1732-0017

Water Treatment Specialist Reserve Component Basic Noncommissioned Officer (NCO)

Course Number: RC-101-77W30.

Location: Quartermaster Center and School, Fort Lee, VA.

Length: 92 hours.

Exhibit Dates: 10/92–Present.

Learning Outcomes: Upon completion of the course, the student will be able to analyze water quality and monitor water treatment operations and distribution.

Instruction: Lectures, demonstrations, and practical exercises cover water treatment subjects.

Credit Recommendation: In the lower-division baccalaureate/associate degree category, 1 semester hour in water treatment technology (12/96).

AR-1732-0019

Senior Power Generation Equipment Repairer

Course Number: 662-52D30.

Location: Ordnance Center and School, Aberdeen Proving Ground, MD.

Length: 17 weeks (610 hours).

Exhibit Dates: 10/96–Present.

Learning Outcomes: Before 10/96 see AR-1406-0141. Upon completion of the course, the student will be able to manage, supervise, and provide leadership to power generation equipment repair.

Instruction: Lectures and practical exercises are used to provide concepts in leadership doctrine to develop management skills. General subjects such as electrical theory and troubleshooting are presented in a hands-on laboratory format. This course contains a common core of leadership subjects.

Credit Recommendation: In the lower-division baccalaureate/associate degree category, 3 semester hours in AC/DC electrical generator troubleshooting. See AR-1406-0090 for common core credit (5/01).

Related Occupations: 52D.

AR-1732-0020

Petroleum and Water Specialist Advanced Non Commissioned Officer

Course Number: 101-77F/L/W40, Phase 3.

Location: Quartermaster Center and School, Fort Lee, VA.

Length: 5 weeks (312 hours).

Exhibit Dates: 10/00–Present.

Learning Outcomes: Upon completion of the course the student will be able to apply logistics principles in petroleum supply and storage and perform and manage petroleum supply, storage, and laboratory operations.

Instruction: Small group instruction and practical exercises cover the subject matter. The course contains a common core of leadership subjects. This course is for water treatment specialists (NCO) who have already received all of the available courses in water treatment. Discussion, classroom exercise, practical exercise, lecture, and examination. Topics include in the course puts emphasize on management of small groups to support water supply operations, laboratory test operations, petroleum terminal operation, pipe line systems and storage and distribution of petroleum at a supply point.

Credit Recommendation: In the lower-division baccalaureate/associate degree category, 3 semester hours in personnel supervision, 3 in petroleum supply operations and management, and 2 in water treatment facility operations (6/02).

AR-1732-0021

Power Generation Equipment Repairer

Course Number: 091-52D10-TATSC, Phase 2.

Location: Reserve Schools, Multiple locations.

Length: 120 hours.

Exhibit Dates: 8/99–Present.

Learning Outcomes: Upon completion of the course, the student will be able to perform maintenance on diesel engine drive generators.

Instruction: Practical exercises, classroom exercises, discussion, laboratory, and lecture. Topics include testing, troubleshooting and maintenance of various aspects of diesel engine-driven generators.

Credit Recommendation: In the lower-division baccalaureate/associate degree category, 2 semester hours in generator repair (8/03).

AR-1732-0022

Petroleum and Water Specialist Advanced Noncommissioned Officer (NCO)

Course Number: ***Version 1:*** 8-77-C42, Phase 2. *Version 2:* 8-77-C42.

Location: Quartermaster Center and School, Fort Lee, VA.

Length: ***Version 1:*** 9–10 weeks (359 hours). *Version 2:* 10 weeks (392 hours).

Exhibit Dates: ***Version 1:*** 10/00–12/02. *Version 2:* 1/03–Present.

Learning Outcomes: ***Version 1:*** Upon completion of the course, the student will be able to apply logistics principles in petroleum supply and storage and perform and manage petroleum supply, storage, and laboratory operations. *Version 2:* Pending evaluation.

Instruction: ***Version 1:*** Discussion, classroom exercise, practical exercise, lecture, and examination. Topics include the management of small groups to support water supply operations, laboratory test operations, and petroleum terminal operation; pipe line systems; and storage and distribution of petroleum at a supply point. *Version 2:* Pending evaluation.

Credit Recommendation: ***Version 1:*** In the lower-division baccalaureate/associate degree category, 3 semester hours in personnel supervision, 3 in petroleum supply operations and management, and 2 in water treatment facility operations (6/02). *Version 2:* Pending evaluation.

AR-1733-0002

Aerial Delivery and Materiel Officer

Course Number: ***Version 1:*** 8B-92D. *Version 2:* 8B-92D.

Location: Quartermaster Center and School, Fort Lee, VA.

Length: ***Version 1:*** 11–13 weeks (394–442 hours). *Version 2:* 5–6 weeks (200–223 hours).

Exhibit Dates: ***Version 1:*** 10/84–9/90. *Version 2:* 10/90–Present.

Learning Outcomes: ***Version 1:*** Upon completion of the course, the student will be able to perform as an aerial and materiel officer. *Version 2:* Upon completion of the course, the student will be able to perform as an aerial and materiel officer.

Instruction: ***Version 1:*** This course includes the management and technical skills required for the inspection, packing, delivery, rigging, recovery, storing, and maintenance of airdrop equipment. *Version 2:* Conferences, demonstrations, and practice exercises cover management and technical skills required for the inspection, packing, delivery, rigging, recovery, storing, and maintenance of airdrop equipment.

Credit Recommendation: ***Version 1:*** In the vocational certificate category, 2 semester hours in aerial delivery and materiel. In the upper-division baccalaureate category, 1 semester hour in air logistics (8/88). *Version 2:* Credit is not recommended because of the military-specific nature of the course (4/04).

AR-1733-0004

Parachute Rigger Advanced Noncommissioned Officer (NCO)
(Service Noncommissioned Officer (NCO) Advanced (Parachute Rigger))

Course Number: ***Version 2:*** 860-43E40. *Version 3:* 860-92R40.

Location: Quartermaster Center and School, Fort Lee, VA.

Length: ***Version 2:*** 6 weeks (191 hours). *Version 3:* 7 weeks (257–293 hours).

Exhibit Dates: ***Version 2:*** 4/86–9/95. *Version 3:* 10/95–Present.

Learning Outcomes: ***Version 2:*** To provide the student with the ability to perform parachute rigging. *Version 3:* Upon completion of the course, the student will apply leadership skills in airdrops planning and battle-focused operations and use computers as a management tool through the use of word processing and graphics software.

Instruction: ***Version 2:*** Conferences, demonstrations, visual aids, and practical exercises cover airdrop rigging and equipment repair, personnel, and small cargo parachutes. This course contains a common core of leadership subjects. *Version 3:* Instruction cover military subjects, including military history, military map symbols, radio operations, combat leadership skills, and field operations. This course contains a common core of leadership subjects.

Credit Recommendation: ***Version 2:*** In the vocational certificate category, 3 semester hours in airdrop and parachute rigging and repair. In the lower-division baccalaureate/associate degree category, see AR-1404-0035 for common core credit (8/85). *Version 3:* In the lower-division baccalaureate/associate degree category, 1 semester hour in microcomputer applications. See AR-1404-0035 for common core credit (6/02).

Related Occupations: 43E; 92R.

AR-1733-0007

Parachute Rigger Basic Noncommissioned Officer (NCO)

Course Number: ***Version 1:*** 860-43E30. *Version 2:* 860-43E3P. *Version 3:* 860-92R30.

Location: Quartermaster Center and School, Fort Lee, VA.

Length: ***Version 1:*** 5 weeks (208 hours). *Version 2:* 7 weeks (250 hours). *Version 3:* 5 weeks (150 hours).

Exhibit Dates: ***Version 1:*** 5/86–12/91. *Version 2:* 1/92–9/95. *Version 3:* 10/95–Present.

Learning Outcomes: ***Version 1:*** Upon completion of the course, the student will be able to supervise the packing of personal and small cargo parachutes, supervise the operation of an airdrop rigging activity, supervise the recovery of cargo parachutes and related airdrop equipment, and perform jumpmaster duties during an airdrop operation. *Version 2:* Upon completion of the course, the student will be able to supervise the packing of personal and small cargo parachutes, supervise the operation of an airdrop rigging activity, supervise the recovery of cargo parachutes and related airdrop equipment, and perform jumpmaster duties during an airdrop operation. *Version 3:* Upon completion of the course, students will apply advanced training in parachute packing and cargo rigging.

Instruction: ***Version 1:*** This course includes classroom demonstrations, lectures, and practical exercises in packing parachutes, rigging cargo for an airdrop, planning and coordinating airdrop operations, and jumpmaster duties and responsibilities during an airdrop. This course contains a common core of leadership subjects. *Version 2:* This course includes classroom demonstrations, lectures, and practical exercises in packing parachutes, rigging cargo for an airdrop, planning and coordinating airdrop operations, and jumpmaster duties and responsibilities during an airdrop. This course contains a common core of leadership subjects. *Version 3:* This course includes small group instruction, demonstrations, and practical exercises in packing parachutes, cargo rigging, and military-specific leadership subjects.

Credit Recommendation: ***Version 1:*** In the vocational certificate category, 3 semester hours in advanced cargo rigging. In the lower-division baccalaureate/associate degree category, see AR-1406-0090 for common core credit (10/88). *Version 2:* In the vocational certificate category, 3 semester hours in cargo rigging. In the lower-division baccalaureate/associate degree category, 2 semester hours in personnel supervision or leadership. See AR-1406-0090 for common core credit (2/93). *Version 3:* In the vocational certificate category, 3 semester hours in cargo rigging. In the lower-division baccalaureate/associate degree category, see AR-1406-0090 for common core credit (6/02).

Related Occupations: 43E; 92R.

AR-1733-0008

1. Parachute Rigger Advanced Individual Training (Parachute Rigger)
2. Parachute Rigger Advanced Individual Training (Parachute Rigger)
3. Parachute Rigger (Parachute Rigger Advanced Individual Training)

Course Number: ***Version 1:*** 860-43E10 (ST); 860-43E10. *Version 2:* 860-43E1P; 860-43E10. *Version 3:* 860-92R1P.

Location: Quartermaster Center and School, Fort Lee, VA.

Length: ***Version 1:*** 12–13 weeks (442–445 hours). *Version 2:* 11 weeks (396 hours). *Version 3:* 10 weeks (355–356 hours).

Exhibit Dates: ***Version 1:*** 2/85–7/92. *Version 2:* 8/92–9/95. *Version 3:* 10/95–Present.

Learning Outcomes: ***Version 1:*** Upon completion of the course, the student will be able to pack parachutes; rig cargo for airdrop; and repair, store, and recover airdrop equipment. *Version 2:* Upon completion of the course, the student will be able to pack parachutes; rig cargo for airdrop; and repair, store, and recover airdrop equipment. *Version 3:* Upon completion of the course, the student will be able to pack different types of parachutes; rig cargo for airdrop; and repair, store, and recover equipment.

Instruction: ***Version 1:*** This course includes some lectures but is presented largely through demonstrations and practical exercises in parachute packing, cargo rigging and recovering, and storing and maintaining airdrop equipment. *Version 2:* This course includes some lectures but is presented largely through demonstrations and practical exercises in parachute packing, cargo rigging and recovering, and storing and maintaining airdrop equipment. *Version 3:* The course includes some lectures, but is largely demonstration and practical exercises, in parachute packing, cargo rigging and recovering, and storing and maintaining airdrop equipment.

Credit Recommendation: ***Version 1:*** In the vocational certificate category, 4 semester hours in cargo rigging (10/88). *Version 2:* In the vocational certificate category, 6 semester hours in cargo rigging. In the lower-division baccalaureate/associate degree category, 1 semester hour in aerial delivery and repair (2/93). *Version 3:* In the vocational certificate category, 6 semester hours in cargo rigging and 1 in airdrop equipment repair (5/02).

Related Occupations: 43E; 92R.

AR-1733-0009

Fabrication of Aerial Delivery Loads

Course Number: ***Version 1:*** 860-F3. *Version 2:* 860-F3 (ITRO).

Location: Quartermaster Center and School, Fort Lee, VA.

Length: ***Version 1:*** 4–5 weeks (161–162 hours). *Version 2:* 3 weeks (273–274 hours).

Exhibit Dates: ***Version 1:*** 6/87–11/92. *Version 2:* 12/92–Present.

Learning Outcomes: ***Version 1:*** Upon completion of the course, the student will be able to pack parachutes, rig cargo, load and secure cargo, and recover equipment used in airdrops. *Version 2:* Upon completion of the course, the student will be able to pack personal and cargo parachutes, rig; load, and secure cargo; and recover equipment used in airdrops.

Instruction: ***Version 1:*** This course includes demonstrations and practical exercises in packing parachutes; rigging, loading, and securing cargo; and recovering equipment in airdrops. *Version 2:* This course includes demonstrations and practical exercises in packing parachutes; rigging, loading, and securing cargo; and recovering equipment in airdrops.

Credit Recommendation: ***Version 1:*** In the vocational certificate category, 2 semester hours in cargo rigging (10/88). *Version 2:* In the vocational certificate category, 1 semester hour in cargo rigging (4/02).

AR-1733-0010

1. Airdrop Systems Technician Warrant Officer Technical/Tactical Certification
2. Airdrop Systems Warrant Officer Basic
(Airdrop Systems Technician Warrant Officer Basic)

Course Number: ***Version 1:*** 4N-921A. *Version 2:* 4N-921A.
Location: Quartermaster Center and School, Fort Lee, VA.
Length: ***Version 1:*** 8–9 weeks (341 hours). *Version 2:* 9 weeks (339–340 hours).
Exhibit Dates: ***Version 1:*** 10/88–3/94. *Version 2:* 4/94–Present.
Learning Outcomes: ***Version 1:*** Upon completion of the course, the student will be able to plan all the operational requirements of an airdrop operation, including logistical support, drop site layouts, staff needs, work flow, and costs and recovery and manage the technical functions, including packing the parachutes, rigging the cargo, repairing airdrop equipment, maintaining records, enforcing safety regulations, and inspecting equipment. *Version 2:* Upon completion of the course, the student will be able to plan all the operational requirements of an airdrop operation to include logistical support; drop site layout; staffing needs; work flow; recovery; recordkeeping and financial and property management; and the management of requisite technical functions such as parachute packing, cargo rigging, airdrop equipment repair, and equipment inspection.
Instruction: ***Version 1:*** This course includes classroom lectures and practical exercises in supervising the packing of parachutes, rigging cargo, using and repairing equipment, developing plans, and identifying needs of airdrop operations. *Version 2:* Classroom instruction in the non-technical course subjects and demonstrations and practical exercises in the technical components.
Credit Recommendation: ***Version 1:*** In the vocational certificate category, 3 semester hours in advanced cargo rigging. In the lower-division baccalaureate/associate degree category, 1 semester hour in cargo transportation (10/88). *Version 2:* In the vocational certificate category, 3 semester hours in advanced cargo rigging and equipment repair. In the lower-division baccalaureate/associate degree category, 2 semester hours in records and information management (4/02).
Related Occupations: 921A.

AR-1733-0011

Explosive Ordnance Disposal Parachute Rigging

Course Number: 431-F3 (Navy); 431-F3.
Location: Quartermaster Center and School, Fort Lee, VA.
Length: 4 weeks (145 hours).
Exhibit Dates: 10/88–Present.
Learning Outcomes: Upon completion of the course, the student will be certified for membership in the Navy Explosive Ordnance Disposal team and will be able to pack parachutes, select airdrop rigging, and maintain equipment for aerial delivery.
Instruction: Demonstrations, practical exercises, and examinations cover parachute packing, airdrop rigging, and airdrop equipment repair.
Credit Recommendation: In the lower-division baccalaureate/associate degree category, 3 semester hours in aerial delivery and material or in air drop equipment and repair (4/02).

AR-1733-0012

Airdrop Systems Technician Warrant Officer Advanced

Course Number: 8-10-C32-921A (IRR) (RT).
Location: Quartermaster Center and School, Fort Lee, VA.
Length: 3 weeks (180 hours).
Exhibit Dates: 10/01–Present.
Learning Outcomes: Upon completion of this course, the mid-career warrant officer will be re-introduced to the knowledge required to perform successfully as an airdrop technician at company and battalion levels.
Instruction: Discussion, classroom exercise, lecture. Topics include refining communication skills, and briefing techniques; budgeting and financial management; Army fundamental operations; military leadership; quartermaster functions; materials management; supply functions; airdrop concepts; drop zone selection; new developments in airdrop technology; airdrop parachute techniques; rigger techniques; and pre-jump planning.
Credit Recommendation: In the upper-division baccalaureate category, 3 semester hours in transportation management (6/02).

AR-2101-0002

Rough Terrain Container Handling Operator
(Rough Terrain Container Handler Operator)

Course Number: 822-F24; 822-ASIB1.
Location: Transportation School, Fort Story, VA; Transportation and Aviation Logistics School, Fort Eustis, VA.
Length: 2–3 weeks (78 hours).
Exhibit Dates: 6/83–Present.
Objectives: To train qualified enlisted personnel in the operation of a heavy-duty forklift.
Instruction: Lectures and practical experiences in the familiarization, operation, lifting procedures, and preventive maintenance of heavy-duty forklifts.
Credit Recommendation: In the lower-division baccalaureate/associate degree category, 1 semester hour in forklift operation (10/00).

AR-2101-0003

Petroleum Vehicle Operator

Course Number: 811-F3; 821-ASIH7.
Location: Training Center, Fort Dix, NJ.
Length: 5 weeks (180 hours).
Exhibit Dates: 10/86–9/98.
Learning Outcomes: After 9/00 see AR-2101-0005. Upon completion of the course, the student will be able to operate light- and heavy-cargo trucks, tractors, and tanker vehicles under all road conditions and perform operator maintenance on the vehicles.
Instruction: Lectures and practical exercises cover the operation of quarter-ton and five-ton cargo trucks and the five-ton, 5,000-gallon tanker.
Credit Recommendation: In the vocational certificate category, 2 semester hours in truck driving (6/91).

AR-2101-0004

Rough Terrain Container Handler Operator Reserve

Course Number: 551-ASIB1-RC.
Location: Reserve Component Training Centers, US.
Length: 73 hours.
Exhibit Dates: 9/92–Present.
Learning Outcomes: Upon completion of the course, the student will be able to perform the duties of a heavy-duty forklift operator.
Instruction: Lectures and practical exercises cover familiarization, operation, lifting procedures, and preventive maintenance of heavy-duty forklifts.
Credit Recommendation: In the lower-division baccalaureate/associate degree category, 1 semester hour in forklift operation (3/94).

AR-2101-0005

Petroleum Vehicle Operator

Course Number: 821-ASIH7.
Location: Transportation School, Fort Eustis, VA.
Length: 5 weeks (177–180 hours).
Exhibit Dates: 10/00–Present.
Learning Outcomes: Before 10/00 see AR-2101-0003. Upon completion of the course, the student will be able to operate light/heavy cargo trucks, tractors, and tanker vehicles including 10-ton HEMTT tanker, 5-ton tractor, and 5000-gallon tanker under all road conditions.
Instruction: Methods of instruction include classroom instruction, practical exercises, demonstrations, extensive driving and (CDL) License Testing (Class B).
Credit Recommendation: In the lower-division baccalaureate/associate degree category, 3 semester hours in truck driver training/certification (10/03).

AR-2201-0022

Ammunition Specialist
(Ammunition Storage)

Course Number: 645-55B20; 645-55B10 (ST); 645-55B10.
Location: Missile and Munitions School, Redstone Arsenal, AL.
Length: 4–8 weeks (145–260 hours).
Exhibit Dates: 2/74–9/90.
Learning Outcomes: Upon completion of the course students will be able to identify and handle various ammunition material; prepare, store, and issue ammunition; perform explosive decontamination; and perform emergency destruction of ammunition.
Instruction: Conferences and practical exercises cover ammunition material, service procedures, storage, handling, movement, inspection, maintenance, and destruction.
Credit Recommendation: In the vocational certificate category, 3 semester hours in explosives handling (10/90).
Related Occupations: 55A; 55B.

AR-2201-0029

Chemical Operations Specialist Noncommissioned Officer (NCO)
(Chemical (54-CMF) Noncommissioned Officer (NCO) Advanced)
(Chemical (CL-CMF) Noncommissioned Officer (NCO) Advanced)

Course Number: 4-CL-C42; 4-54-C42.
Location: Chemical School, Fort McClellan, AL; Ordnance and Chemical School, Aberdeen Proving Ground, MD.
Length: 12 weeks (435 hours).
Exhibit Dates: 10/86–2/92.
Objectives: After 2/92 see AR-0801-0029. To provide enlisted personnel with a basic understanding of nuclear, chemical, and biological defense and to qualify them as master sergeants.
Instruction: Lectures cover Army management practices and concepts, including logistical and maintenance management; chemical, biological, and nuclear warfare; and radiological defense.
Credit Recommendation: In the lower-division baccalaureate/associate degree category, 3 semester hours in radiological safety, 3 in nuclear instrumentation, and 3 in introduction to management (11/86).
Related Occupations: 54B; 54C; 54D; 54E; 54F; 92D.

AR-2201-0037

Electronic Warfare (EW)/Cryptologic Advanced Noncommissioned Officer (NCO)
(Military Intelligence Advanced (98 CMF) Noncommissioned Officer (NCO))

Course Number: 230-F1.
Location: Security Agency School, Fort Devens, MA.
Length: 7–9 weeks (231–294 hours).
Exhibit Dates: 5/77–12/90.
Objectives: To train senior noncommissioned officers to manage electronic warfare/cryptologic operations of the Army Security Agency.
Instruction: Training in management and supervision techniques pertains to military systems, military science, and intelligence and security techniques.
Credit Recommendation: In the lower-division baccalaureate/associate degree category, 3 semester hours in maintenance management procedures and supervision of maintenance personnel (2/83).

AR-2201-0038

Pathfinder

Course Number: 011-F3; 001-SQIY; 2E-F3; 2E-F3/011-SQIY.
Location: Infantry School, Fort Benning, GA.
Length: 3–5 weeks (129–247 hours).
Exhibit Dates: 1/66–7/90.
Objectives: To train officer and enlisted parachutists to provide navigational assistance to drop zone aircraft and deploying troops.
Instruction: Lectures and practical exercises cover the organization, mission, and training of pathfinder platoons; electronics; communications; and map reading.
Credit Recommendation: Credit is not recommended because of the military-specific nature of the course (10/87).

AR-2201-0041

Ranger

Course Number: 011-F2; 2E-F2; 7-D-F4; 7-OE-15; 2E-F2/011-F2.
Location: Infantry School, Fort Benning, GA.
Length: 7–9 weeks (499–1052 hours).
Exhibit Dates: 7/54–8/92.
Objectives: After 8/92 see AR-2201-0434. To develop leadership qualities in selected officers and enlisted soldiers by providing training in self-discipline and obedience.
Instruction: Lectures and practical exercises cover map reading and land navigation, tactical training in guerilla operations, demolition, patrolling, leadership, intelligence, endurance, confidence, and physical training. The course includes approximately 780 hours non duty hours.
Credit Recommendation: In the lower-division baccalaureate/associate degree category, 6 semester hours in physical education (9/85).

AR-2201-0042

Marine Artillery Operations Chief

Course Number: 250-F5; 250-848.
Location: Field Artillery School, Fort Sill, OK.
Length: 13–14 weeks (472–502 hours).
Exhibit Dates: 8/73–11/96.
Objectives: To train noncommissioned officers to supervise field artillery operations.
Instruction: Lectures and practical exercises cover gunnery, target acquisition, tactics and combined arms, artillery meteorology, surveying, fire support coordination, fire planning, and observed fire.
Credit Recommendation: Credit is not recommended because of the military-specific nature of the course (11/86).

AR-2201-0047

Field Artillery Officer Advanced Preparatory

Course Number: 2G-F50.
Location: Field Artillery School, Fort Sill, OK.
Length: 3 weeks (110 hours).
Exhibit Dates: 5/81–6/96.
Objectives: To prepare selected officers for the field artillery officer advanced course.
Instruction: This course includes firing procedures, target location, firing charts, fire direction control, and related artillery fundamentals.
Credit Recommendation: In the vocational certificate category, 1 semester hour in technical mathematics (10/83).

AR-2201-0056

Ammunition Inspector
(Military Ammunition Inspector)
(Ammunition Inspector)

Course Number: 645-55X40; 645-55X30; 645-55X30 (AC).
Location: Missile and Munitions School, Redstone Arsenal, AL.
Length: 7–12 weeks (233–420 hours).
Exhibit Dates: 2/71–9/90.
Objectives: To train enlisted personnel to inspect ammunition.
Instruction: Conferences and practical exercises cover ammunition material; rocket and guided missile inspection and maintenance; storage, handling, and transportation of ammunition; surveillance and maintenance of ammunition; and demilitarization and destruction of ammunition.
Credit Recommendation: In the vocational certificate category, 3 semester hours in explosives handling (10/86).
Related Occupations: 55B; 55C; 55D; 55G; 55X; 55Z.

AR-2201-0063

Chaparral/Vulcan Officer Qualification
(Chaparral/Vulcan Officer and Noncommissioned Officer (NCO) Qualification)
(Chaparral/Vulcan Noncommissioned Officer (NCO) Qualification)

Course Number: 121-F17; 2F-F8; 2E-1174; 2E-14B.
Location: Air Defense Artillery School, Fort Bliss, TX.
Length: 4–8 weeks (171–294 hours).
Exhibit Dates: 7/68–11/98.
Objectives: To provide training in Chaparral/Vulcan air defense system characteristics, capabilities, limitations, and air defense tactics with emphasis on battery or battalion operations.
Instruction: Lectures and practical exercises cover the organization, maintenance, system control, and test and firing procedures of the Vulcan and Chaparral air defense systems.
Credit Recommendation: Credit is not recommended because of the military-specific nature of the course (11/88).

AR-2201-0085

Infantry Mortar Leader
(Infantry Mortar Platoon)

Course Number: *Version 1:* 010-F1; 2E-1543; 2E-ASI32; 2E-11B/C; 2E-ASI3Z/010-F1; 2E-SI3Z/010-F1. *Version 2:* 2E-SI3Z/010-F1.
Location: Infantry School, Fort Benning, GA.
Length: *Version 1:* 5–7 weeks (216–274 hours). *Version 2:* 5–6 weeks (216 hours).
Exhibit Dates: *Version 1:* 3/69–10/94. *Version 2:* 4/99–Present.
Learning Outcomes: *Version 1:* To train officers and noncommissioned officers to command infantry heavy mortar units. *Version 2:* Upon completion of the course, the student will be able to supervise an infantry mortar team using a plotting board and ballistic tables to calculate a target acquisition, use a mortar ballistics computer to acquire a target, set up a mortar firing position within 90 seconds, maintain heavy mortar equipment, and field support planning under live fire exercises.
Instruction: *Version 1:* Lectures and practical exercises cover the supervision of a heavy mortar platoon in support of infantry combat operations, including tactical employment, methods of employment, tactical considerations, tactical training, communications, weapons, mechanical training, crew drill, forward observation procedures, fire direction center procedures, and heavy mortar platoon communications. *Version 2:* Practical and classroom exercises, and discussion. Topics include fire support planning, live fire exercises, target acquisition using a plotting board, and a ballistics computer heavy mortar maintenance.
Credit Recommendation: *Version 1:* Credit is not recommended because of the military-specific nature of the course (5/92). *Version 2:* In the lower-division baccalaureate/associate degree category, 1 semester hour in computer operations (10/02).

Related Occupations: 11B; 11C.

AR-2201-0095

NIKE HERCULES OFFICER

Course Number: 2F-14C.

Location: Air Defense Artillery School, Fort Bliss, TX.

Length: 4–8 weeks (151–288 hours).

Exhibit Dates: 9/63–6/94.

Objectives: To train commissioned officers in the essential materiel and tactics of the Nike Hercules system.

Instruction: Lectures and practical exercises cover the essential materiel and tactics of the Nike Hercules system. Course includes a general description of the Nike Hercules system, block diagrams, operational procedures, general maintenance requirements, handling of nuclear weapons, and air defense tactics.

Credit Recommendation: Credit is not recommended because of the limited, technical nature of the course (1/88).

AR-2201-0096

1. SMALL ARMS REPAIR
2. SMALL ARMS REPAIRER
3. SMALL ARMS REPAIRER

Course Number: ***Version 1:*** 9-E-11; 641-2111 (USMC); 641-45B20; 641-45B10/641-2111. *Version 2:* A0121M1; 21M; 641-2111. *Version 3:* 641-2111.

Location: ***Version 1:*** Ordnance Center and School, Aberdeen Proving Ground, MD. *Version 2:* Marine Corps Detachment, Ordnance Center and School, Aberdeen Proving Ground, MD. *Version 3:* Ordnance Center and School, Aberdeen Proving Ground, MD.

Length: ***Version 1:*** 5–12 weeks (210–543 hours). *Version 2:* 9–10 weeks (363 hours). *Version 3:* 9 weeks (368 hours).

Exhibit Dates: ***Version 1:*** 5/56–7/96. *Version 2:* 8/96–9/01. *Version 3:* 10/01–Present.

Learning Outcomes: ***Version 1:*** To train enlisted personnel to maintain small arms and materiel. *Version 2:* Upon completion of the course, the student will be able to disassemble, inspect, identify malfunction, repair, and reassemble small arms, including pistols, rifles, shotguns, light and heavy machine guns, grenade launchers, and mortars. Students will also be able to perform preventive maintenance on small arms. *Version 3:* Pending evaluation.

Instruction: ***Version 1:*** Lectures and practical exercises cover small arms and materiel maintenance, including rifles, machine guns, submachine guns, mortars, rocket launchers, recoilless weapons, pistols, and grenade launchers. *Version 2:* Lectures, demonstrations, and practical experiences cover safety precautions, parts identification, inspection, determination of malfunction, repair, disassembly and reassembly techniques, and test firing of small arms including pistols, rifles, shotguns, light and heavy machine guns, grenade launchers, and mortars. *Version 3:* Pending evaluation.

Credit Recommendation: ***Version 1:*** Credit is not recommended because of the military-specific nature of the course (11/91). *Version 2:* In the lower-division baccalaureate/associate degree category, 3 semester hours in gunsmithing/small arms repair (1/03). *Version 3:* Pending evaluation.

Related Occupations: 45B; 45Z.

AR-2201-0145

MARINE ARTILLERY SCOUT OBSERVER

Course Number: 250-0846; 250-0841.

Location: Field Artillery School, Fort Sill, OK.

Length: 4–5 weeks (152–162 hours).

Exhibit Dates: 3/75–4/04.

Objectives: To train enlisted personnel in the technical aspects of planning, acquiring, and controlling artillery fire.

Instruction: Lectures and practical exercises cover communications and electronics, gunnery, map and aerial photograph reading, and the development of artillery fire support plans.

Credit Recommendation: Credit is not recommended because of the limited, technical nature of the course (1/90).

AR-2201-0162

SPECIAL FORCES OFFICER
(Special Forces Officer (Modified))

Course Number: 33-G-F3; 2E-F8.

Location: Institute for Military Assistance, Fort Bragg, NC; John F. Kennedy Special Warfare Center and School, Fort Bragg, NC.

Length: 12–13 weeks (574–873 hours).

Exhibit Dates: 8/75–4/92.

Objectives: To provide students with a general knowledge of the latest doctrine and techniques of Special Forces operations.

Instruction: Lectures and practical exercises cover development and employment of resistance forces, infiltration and exfiltration, unconventional warfare organization and capabilities, and psychological operations.

Credit Recommendation: In the lower-division baccalaureate/associate degree category, 3 semester hours in physical education. In the upper-division baccalaureate category, 3 semester hours in military science and tactics (2/80).

AR-2201-0167

OFFICER CANDIDATE

Course Number: 7-E-14; 7-N-F1; 2-7-F1.

Location: Infantry School, Fort Benning, GA.

Length: 14 weeks (500–1056 hours).

Exhibit Dates: 1/84–2/92.

Objectives: To prepare selected personnel to become officers in the reserve component of the Army.

Instruction: Lectures, conferences, demonstrations, and practical exercises cover tactical doctrine, armed forces operations, training management, military leadership, land navigation, physical fitness, and drill and command.

Credit Recommendation: In the lower-division baccalaureate/associate degree category, 3 semester hours in principles of management, 3 in personnel management, 3 in physical education, and 1 in map reading (10/88).

AR-2201-0168

FIELD ARTILLERY OFFICER BASIC (RESERVE COMPONENT)
(Field Artillery Officer Candidate (Reserve Component))

Course Number: 2-N-F2; 2-6-F2; 2-6-C25.

Location: Artillery and Missile School, Fort Sill, OK.

Length: 8–11 weeks (293–493 hours).

Exhibit Dates: 1/58–6/96.

Objectives: To prepare personnel to qualify as second lieutenants in National Guard or reserve field artillery units.

Instruction: Lectures and practical exercises cover maintenance management, communications electronics, gunnery, tactics, map reading, military law, target acquisition, drill and ceremonies, physical training, and leadership.

Credit Recommendation: In the upper-division baccalaureate category, 3 semester hours in military science (11/85).

AR-2201-0181

MANEUVER COMBAT ARMS INFANTRY NONCOMMISSIONED OFFICER (NCO) ADVANCED

Course Number: O-111-C42.

Location: Infantry School, Fort Benning, GA.

Length: 10 weeks (400 hours).

Exhibit Dates: 7/85–6/96.

Objectives: To provide middle-grade enlisted personnel with sufficient knowledge to perform infantry armor duties.

Instruction: Lectures and practical exercises include tactical doctrine, fundamentals of combat in built-up areas, stability operations in insurgent areas, engineer operations, intelligence and logistics, personnel management, map and air photo reading, military leadership, physical training and combat, communications electronics, crew-served weapons, and written and oral communication.

Credit Recommendation: In the lower-division baccalaureate/associate degree category, 3 semester hours in personnel supervision and management and 3 in communication (9/85).

AR-2201-0192

INFANTRY OFFICER BASIC
(Basic Infantry Officer)

Course Number: 7-A-C1; 2-7-C20.

Location: Infantry School, Fort Benning, GA.

Length: 17 weeks (964 hours).

Exhibit Dates: 7/85–5/92.

Objectives: After 5/92 see AR-2201-0428. To train newly commissioned officers in the duties and responsibilities of infantry platoon leaders in rifle and weapons platoons.

Instruction: Lectures, computer-assisted instruction, and practical exercises cover armed forces operations; nuclear, biological, and chemical operations; military leadership; land navigation; communications; maintenance management; individual weapons; crew-served weapons; and antiarmor weapons systems.

Credit Recommendation: In the lower-division baccalaureate/associate degree category, 3 semester hours in map reading and 3 in principles of management. In the upper-division baccalaureate category, 3 semester hours in maintenance management (9/85).

AR-2201-0193

FIELD ARTILLERY OFFICER BASIC
(Regular Army Field Artillery Officer Basic)
(Field Artillery Officer Orientation)

Course Number: 6-O-A; 6-A-C1; 6-A-C20; 2-6-C20 (RA); 2-6-C20.

Location: Field Artillery School, Fort Sill, OK; Artillery and Missile School, Fort Sill, OK.

Length: 18–19 weeks (716 hours).
Exhibit Dates: 10/85–12/90.
Objectives: After 12/90 see AR-2201-0411. To prepare Field Artillery officers to function in the areas of training and maintenance management, leadership, fire direction techniques, and coordination of artillery firing.
Instruction: Instruction topics include maintenance management, leadership, communications equipment, precision techniques in gunnery, manual and FADAC computer, meteorological effects and computations, map and terrain association, survey procedures, and artillery fire coordination and tactics.
Credit Recommendation: In the lower-division baccalaureate/associate degree category, 2 semester hours in leadership, 1 in electronic communications systems, and 2 in maintenance management (11/85).
Related Occupations: 11B; 11C; 13E.

AR-2201-0203

SPACE COLLECTION OPERATIONS
(Space Collection (SPACOL) Operations)
(Telemetry Collection Operations)

Course Number: 233-F7; 233-ASIG4.
Location: Intelligence School, Fort Devens, MA.
Length: 5–7 weeks (168–248 hours).
Exhibit Dates: 7/75–3/94.
Learning Outcomes: Upon completion of the course, the student will be able to intercept, recognize, and record electromagnetic emissions emanating from airborne equipment, missiles, and satellites.
Instruction: Lectures and practical experience cover the maintenance and operation of space collector equipment, including the types, parameters, and identification of space signals, intercept equipment, satellite plotting, and report format. Students will be trained to intercept, log, record, and identify noncommunication electronic emissions that emanate from missiles, earth satellite, and associated equipment.
Credit Recommendation: Credit is not recommended because of the military-specific nature of the course (5/87).

AR-2201-0212

WARRANT OFFICER CANDIDATE MILITARY DEVELOPMENT

Course Number: 2C-F32.
Location: Aviation Center, Fort Rucker, AL.
Length: 6 weeks (212–220 hours).
Exhibit Dates: 8/77–1/90.
Objectives: To train warrant officer candidates in leadership, physical conditioning, and military subjects.
Instruction: Military subjects covered include communicative arts, military personnel management, maintenance management, strategy, and warrant officer development.
Credit Recommendation: Credit is not recommended because of the military-specific nature of the course (11/86).

AR-2201-0213

FIELD ARTILLERY CANNON (FA-CMF) NONCOMMISSIONED OFFICER (NCO) ADVANCED

Course Number: 0-13-C42.
Location: Field Artillery School, Fort Sill, OK.
Length: 8–11 weeks (277–402 hours).
Exhibit Dates: 8/78–6/96.
Objectives: To train the noncommissioned officer to teach modern battlefield tactics, techniques, and procedures.
Instruction: Subjects include target acquisition, communications electronics, tactics and combined arms, weapons, gunnery, and field artillery. Leadership training and supervision skills are emphasized for all students. Map reading and plotting to determine azimuth and elevation are emphasized for those students holding MOS's 13B and 13E.
Credit Recommendation: In the lower-division baccalaureate/associate degree category, for all students completing the course, 2 semester hours in leadership fundamentals. For those students holding MOS's 13B and 13E, 1 additional semester hour in map reading (10/83).
Related Occupations: 13B; 13E; 13Z.

AR-2201-0219

FIELD ARTILLERY CANNON BATTERY OFFICER

Course Number: 2E-13A.
Location: Field Artillery School, Fort Sill, OK.
Length: 8 weeks (320 hours).
Exhibit Dates: 10/75–1/92.
Objectives: To train officers in the techniques, processes, and procedures to function in a cannon-equipped unit.
Instruction: Lectures and practical exercises cover communications equipment, weapon characteristics and capabilities, tactics, nuclear projectiles and safety, surveying concepts, and FADAC computer use. Major emphasis is in gunnery fire techniques, including ballistic corrections, high-angle trajectory, and special employment techniques.
Credit Recommendation: Credit is granted on the basis of institutional evaluation (1/80).

AR-2201-0252

LIGHT AIR DEFENSE ARTILLERY (ADA) CREWMEMBER

Course Number: 043-16F10; 043-16F10-OSUT; 16F10-OSUT.
Location: Air Defense Artillery School, Fort Bliss, TX.
Length: 6–13 weeks (239–508 hours).
Exhibit Dates: 2/79–6/94.
Objectives: To train enlisted personnel to become light air defense artillery crewmembers.
Instruction: Training covers basic military equipment and procedures and practices in the use, operation, and assembly of a specific piece of artillery. This course includes an 8-9 week long Basic Training component.
Credit Recommendation: Credit is not recommended because of the military-specific nature of the course (11/88).
Related Occupations: 16F.

AR-2201-0253

1. PRIMARY LEADERSHIP DEVELOPMENT
2. TOTAL ARMY TRAINING SYSTEM (TATS) PRIMARY LEADERSHIP DEVELOPMENT
(Primary Leadership Development)

Course Number: *Version 1:* 687-21-PLDC; NGB-PLDC (AC); 665-20-PLDC; 605-19-PLDC; 695-18-PLDC; 693-17-PLDC; 694-16-PLDC; 692-15-PLDC; 696-14-PLDC; 675-13-PLDC; 672-12-PLDC; 620-11-PLDC; 640-10-PLDC; 690-09-PLDC; 635-08-PLDC; 645-07-PLDC; 662-06-PLDC; 685-05-PLDC; 682-04-PLDC; 698-03-PLDC; 680-02-PLDC; 612-01-PLDC. *Version 2:* 665-20-PLDC; 694-16-PLDC; 693-17-PLDC; 692-15-PLDC; 687-21-PLDC; 682-4-PLDC; 680-2-PLDC; 675-13-PLDC; 672-12-PLDC; 662-6-PLDC; 645-7-PLDC; 635-8-PLDC; 620-11-PLDC; 612-1-PLDC; 698-3-PLDC; 690-P-PLDC; 685-5-PLDC; 600-00-PLDC.
Location: *Version 1:* NCO Academy, 10th Mountain Division (L), Fort Drum, NY; National Guard Bureau NCO Academy, Camp Beauregard, LA; NCO Academy, Fort Leonard Wood, MO; NCO Academy, Fort Dix, NJ; NCO Academies, Europe; NCO Academy, EUSA, Korea; NCO Academy, Fort Sherman, Panama; NCO Academy, Schofield Barracks, HI; NCO Academy, Fort Richardson, AK; NCO Academy, Fort Lewis, WA; NCO Academy, Fort Ord, CA; NCO Academy, Fort Bliss, TX; NCO Academy, Fort Carson, CO; NCO Academy, Fort Hood, TX; NCO Academy, Fort Sill, OK; NCO Academy, Fort Riley, KS; NCO Academy, Fort Polk, LA; NCO Academy, Fort Campbell, KY; NCO Academy, Fort Stewart, GA; NCO Academy, Fort Benning, GA; NCO Academy, Fort Bragg, NC; NCO Academy, Fort Knox, KY. *Version 2:* NCO Academies, US; NCO Academies, Korea; NCO Academies, Germany.
Length: *Version 1:* 4–6 weeks (292–307 hours). *Version 2:* 4–5 weeks (362–390 hours).
Exhibit Dates: *Version 1:* 3/89–9/93. *Version 2:* 10/93–Present.
Learning Outcomes: *Version 1:* Upon completion of the course, the student will be able to perform all basic tasks related to noncommissioned officer leadership responsibilities. *Version 2:* Upon completion of the course, the student will be able to perform all basic tasks related to noncommissioned officer leadership responsibilities.
Instruction: *Version 1:* Lectures and practical exercises cover leadership, communication, resource management, training management, and professional skills, including introduction to leadership, principles of leadership, human behavior, character of leaders, ethics, problem solving, leadership styles, principles of motivation, counseling, and the responsibility of authority. Emphasis is on teaching to teach and to lead soldiers who will work and fight under the student's leadership. Course content includes defensive/offensive operations and field training exercises in which lessons are applied. *Version 2:* This course includes introductory material on personal leadership, communication, resource management, and training management. The leadership block includes material on leader characteristics, ethics, problem solving, styles of leadership, human motivation, personal counseling, and the exercise of authority. Emphasis is placed on teaching to teach and on leading personnel in a small-unit environment. Substantial attention is given to practical application in a field-based environment.
Credit Recommendation: *Version 1:* In the lower-division baccalaureate/associate degree category, 1 semester hour in principles of supervision and 2 in military science (12/91). *Version 2:* In the lower-division baccalaureate/associate degree category, 2 semester hours in principles of supervision and 2 in military science (8/02).

AR-2201-0254

Effective Military Writing

Course Number: 1-19-XP-100.
Location: Military Police School, Fort McClellan, AL.
Length: Self-paced, 1–2 weeks (48 hours).
Exhibit Dates: 8/81–6/96.
Objectives: To teach selected enlisted, noncommissioned, and commissioned officers to write effectively.
Instruction: This course includes instruction in organization, mechanics, punctuation, and sentence construction. It concentrates on Army written communication and the staff study.
Credit Recommendation: Credit is not recommended because of the military-specific nature of the course (3/82).

AR-2201-0255

Initial Entry Training (IET) Cadre Course (Active Army)

Course Number: 1B-F10/012-F18; 1B-F6/012-F14; 1B-F9/012-F17; 1B-F7/012-F15; 1B-F4/012-F12; 1B-F5/012-F13; 1B-F3/012-F11; 1B-F8/012-F16.
Location: Training Centers, Continental US.
Length: Self-paced, 2–3 weeks (80–120 hours).
Exhibit Dates: 7/80–10/97.
Objectives: To provide training for selected initial entry training for the drill sergeant and the trainee.
Instruction: Instruction is individually and group-paced and consists of classroom, outdoor, range, and other field environments. Emphasis is on selecting techniques to manage initial entry training. Instruction includes skills such as how to discriminate levels of willingness and ability in soldiers. Counseling techniques and managing stress are included.
Credit Recommendation: Credit is not recommended because of the military-specific nature of the course (10/87).

AR-2201-0256

Field Artillery Cannon Fire Support Specialist Basic Noncommissioned Officer (NCO)

Course Number: 250-18-13F30; 250-13-13F30; 250-10-13F30; 250-9-13F30; 250-8-13F30; 250-5-13F30; 250-2-13F30.
Location: NCO Academy, Fort Bragg, NC; NCO Academy, Fort Lewis, WA; NCO Academy, Fort Carson, CO; NCO Academy, Fort Hood, TX; NCO Academy, Fort Sill, OK; NCO Academy, Fort Campbell, KY; NCO Academy, 7th Army, Europe.
Length: 5 weeks (189–237 hours).
Exhibit Dates: 10/86–9/97.
Learning Outcomes: After 9/97 see AR-1406-0215. Upon completion of the course the student will be able to supervise and manage a fire support team in an artillery firing battery.
Instruction: This course emphasizes map reading, leadership, counseling, charting and plotting, the Army supply and training management system, and first aid. This course contains a common core of leadership subjects.
Credit Recommendation: In the lower-division baccalaureate/associate degree category, 1 semester hour in map reading, 1 in supply and maintenance systems, and additional credit on the basis of institutional evaluation. See AR-1406-0090 for common core credit (10/90).
Related Occupations: 13F.

AR-2201-0257

Cannon Fire Direction Specialist Basic Noncommissioned Officer (NCO)
(Field Artillery Cannon Fire Direction Specialist Basic Noncommissioned Officer (NCO))

Course Number: 250-18-13E30; 250-13-13E30; 250-10-13E30; 250-9-13E30; 250-8-13E30; 250-5-13E30; 250-2-13E30.
Location: NCO Academy, 7th Army, Europe; NCO Academy, Fort Lewis, WA; NCO Academy, Fort Carson, CO; NCO Academy, Fort Hood, TX; NCO Academy, Fort Sill, OK; NCO Academy, Fort Campbell, KY; NCO Academy, Fort Bragg, NC.
Length: 5–7 weeks (222–289 hours).
Exhibit Dates: 10/86–4/04.
Objectives: To provide skills and knowledge required by a battery chief in a field artillery firing battery and practical experience in directing and managing a fire direction section.
Instruction: This course provides instruction in nuclear, biological, and chemical warfare; laws of warfare; alcohol and drug abuse; communications equipment; battalion training management system; land navigation; and map reading. This course contains a common core of leadership subjects.
Credit Recommendation: In the lower-division baccalaureate/associate degree category, 1 semester hour in map reading, 1 in supply and maintenance systems, and additional credit on the basis of institutional evaluation. See AR-1406-0090 for common core credit (1/92).
Related Occupations: 13E.

AR-2201-0259

Field Artillery Pre-Command

Course Number: 2G-F23.
Location: Field Artillery School, Fort Sill, OK.
Length: 1–3 weeks (64–120 hours).
Exhibit Dates: 8/83–Present.
Objectives: To provide an understanding of training and maintenance doctrine.
Instruction: This course is an open forum on the organization and mission of the field artillery center, including discussions on updating techniques, fire planning, coordination, doctrine changes, and general army maintenance procedures.
Credit Recommendation: Credit is not recommended because of the military-specific nature of the course (5/99).

AR-2201-0260

Nuclear Cannon Assembly
(Atomic Cannon Eight Inch)

Course Number: 041-ASIM5; 4F-F5; 041-F5.
Location: Field Artillery School, Fort Sill, OK.
Length: 1–2 weeks (55–70 hours).
Exhibit Dates: 10/81–6/96.
Objectives: To provide a working knowledge of the inspection, setup, and prefire operations of the M753 and M422 atomic projectiles.
Instruction: Instruction covers prefire procedures; safety procedures; and the assembly, disassembly, storage, and security of the eight-inch atomic projectile.
Credit Recommendation: Credit is not recommended because of the military-specific nature of the course (10/83).

AR-2201-0261

Patriot Air Defense Officer

Course Number: 2F-14E; 2F-14EX.
Location: Air Defense Artillery School, Fort Bliss, TX.
Length: 8–10 weeks (294–306 hours).
Exhibit Dates: 7/83–11/93.
Objectives: After 11/93 see AR-2201-0421. To teach commissioned officers the operational requirements, tactical employment, and operational procedures of the Patriot air defense system.
Instruction: Lectures and laboratory assignments are designed to provide the student with the basic knowledge of the Patriot air defense weapon, including system operation, crew drill, maintenance, and command and control.
Credit Recommendation: Credit is not recommended because of the military-specific nature of the course (7/91).

AR-2201-0262

Basic Noncommissioned Officer (NCO) (16R30)

Course Number: 043-11-16R30.
Location: NCO Academy, 7th Army, Germany; NCO Academy, Fort Campbell, KY; NCO Academy, Fort Bragg, NC; NCO Academy, Fort Lewis, WA; NCO Academy, Fort Carson, CO; NCO Academy, Fort Bliss, TX.
Length: 5 weeks (176–226 hours).
Exhibit Dates: 9/88–6/94.
Objectives: To train qualified personnel to lead, train, and direct subordinates to maintain, operate, and employ weapons and equipment.
Instruction: This course includes conferences and practical exercises as well as workshops to train a Vulcan squad and to instruct in the deployment and supervision of a Vulcan squad. This course contains a common core of leadership subjects.
Credit Recommendation: Credit is recommended for the common core only. See AR-1406-0090 (3/89).
Related Occupations: 16R.

AR-2201-0263

Basic Noncommissioned Officer (NCO) (16P30)

Course Number: 043-11-16P30.
Location: NCO Academy, 7th Army, Germany; NCO Academy, Fort Bliss, TX.
Length: 5 weeks (192–234 hours).
Exhibit Dates: 9/88–6/94.
Objectives: To train air defense artillery short range missile squad leaders to lead, train, and direct subordinates to maintain, operate, and employ weapons and equipment.
Instruction: This course consists of conferences and practical exercises in training fundamentals, training a Chaparral squad, and deployment and supervision of a Chaparral squad. This course contains a common core of leadership subjects.
Credit Recommendation: Credit is recommended for the common core only. See AR-1406-0090 (3/89).
Related Occupations: 16P.

AR-2201-0264

BASIC NONCOMMISSIONED OFFICER (NCO) (16S30)

Course Number: 043-5-16S30; 043-2-16S30; 043-18-16S30; 043-13-16S30; 043-15-16S30; 043-9-16S30; 043-09-16S30; 043-10-16S30; 043-11-16S30.

Location: NCO Academy, Worldwide; NCO Academy, Fort Bragg, NC; NCO Academy, Fort Campbell, KY; NCO Academy, 7th Army, Europe; NCO Academy, Fort Lewis, WA; NCO Academy, Fort Hood, TX; NCO Academy, Fort Carson, CO; NCO Academy, Fort Bliss, TX.

Length: 4–5 weeks (176–236 hours).

Exhibit Dates: 3/86–6/94.

Objectives: To teach selected personnel to lead, train, and direct subordinates to maintain, operate, and employ weapons and equipment.

Instruction: Conferences, practical exercises, and workshops help teach the soldiers to prepare and conduct training and to train, deploy, and supervise a Redeye/Stinger section. This course contains a common core of leadership subjects.

Credit Recommendation: Credit is recommended for the common core only. See AR-1406-0090 (3/89).

Related Occupations: 16S.

AR-2201-0265

AIR DEFENSE ARTILLERY (ADA) PRE-COMMAND

Course Number: 2G-F25 (TATS); 2G-F25.

Location: Air Defense Artillery School, Fort Bliss, TX.

Length: 2–3 weeks (68–103 hours).

Exhibit Dates: 8/81–Present.

Objectives: To provide selected battalion, group, and brigade command designees with a refresher course on logistics and maintenance management, training management, tactics, command perceptions, and professional development electives.

Instruction: This course includes the latest available information on maintenance management and logistics, training management, tactics, general theory of operation, and evaluation of readiness of air defense weapon systems.

Credit Recommendation: Credit is not recommended because of the military-specific nature of the course (1/01).

AR-2201-0266

MAN PORTABLE AIR DEFENSE SYSTEM (MANPADS) FORSCOM CADRE NET

Course Number: 043-F5.

Location: Air Defense Artillery School, Fort Bliss, TX.

Length: 3 weeks (118 hours).

Exhibit Dates: 1/83–6/96.

Objectives: To provide Redeye supervisory personnel with the training to qualify them to operate and maintain the Stinger weapon system and to prepare them as instructors for the Stinger weapon system.

Instruction: Using conferences and practical exercises, introduction to the Stinger system, its operation and maintenance, moving target simulation, and live aircraft tracking are presented.

Credit Recommendation: Credit is not recommended because of the military-specific nature of the course (12/84).

Related Occupations: 16S.

AR-2201-0267

BASIC NONCOMMISSIONED OFFICER (NCO) (16J30)

Course Number: 221-11-16J30.

Location: Air Defense Artillery School, Fort Bliss, TX.

Length: 5 weeks (185–227 hours).

Exhibit Dates: 3/86–6/94.

Objectives: To teach air defense artillery forward area alerting radar (FAAR) section chiefs to lead, train, and direct subordinates to maintain, operate, and deploy equipment.

Instruction: This course consists of practical exercises and lectures in training and personnel supervision in a FAAR equipped section. This course contains a common core of leadership subjects.

Credit Recommendation: Credit is recommended for the common core only. See AR-1406-0090 (3/89).

Related Occupations: 16J.

AR-2201-0268

BASIC NONCOMMISSIONED OFFICER (NCO) (16D30)

Course Number: 043-11-16D30.

Location: Air Defense Artillery School, Fort Bliss, TX.

Length: 5 weeks (180–227 hours).

Exhibit Dates: 3/86–6/94.

Objectives: To prepare Hawk Missile Firing Section chiefs to lead, train, and direct subordinates to maintain, operate, and deploy weapons and equipment.

Instruction: This course employs conferences, practical exercises, and workshops to develop and conduct section training, train a Hawk section in common tasks, train a Hawk launcher section, and deploy and supervise a Hawk section. This course contains a common core of leadership subjects.

Credit Recommendation: Credit is recommended for the common core only. See AR-1406-0090 (3/89).

Related Occupations: 16D.

AR-2201-0269

BASIC NONCOMMISSIONED OFFICER (NCO) (16E30)

Course Number: 043-11-16E30.

Location: Air Defense Artillery School, Fort Bliss, TX.

Length: 5 weeks (197–248 hours).

Exhibit Dates: 3/86–6/94.

Objectives: To prepare a Hawk Missile Fire Control Section chief to lead, train, and direct subordinates to maintain, operate, and deploy weapons and equipment.

Instruction: This course consists of conferences, workshops, and practical exercises to develop and conduct section training; to train a Hawk section in common tasks; and to train, deploy, and supervise a Hawk section. This course contains a common core of leadership subjects.

Credit Recommendation: Credit is recommended for the common core only. See AR-1406-0090 (3/89).

Related Occupations: 16E.

AR-2201-0270

SIGNAL PRE-COMMAND (Battalion/Brigade Pre-Command)

Course Number: 2G-F40.

Location: Signal School, Fort Gordon, GA.

Length: 2–3 weeks (81–114 hours).

Exhibit Dates: 4/84–4/04.

Objectives: To prepare student for battalion- and brigade-level command.

Instruction: This course includes a review of policies and procedures inherent in signal organizations including personnel and maintenance procedures and training reports.

Credit Recommendation: Credit is not recommended because of the limited, specialized nature of the course (6/90).

AR-2201-0271

ARMY MEDICAL DEPARTMENT (AMEDD) BATTALION/BRIGADE PRE-COMMAND

Course Number: 7M-F2.

Location: Academy of Health Sciences, Fort Sam Houston, TX.

Length: 2 weeks (77–84 hours).

Exhibit Dates: 8/81–Present.

Objectives: To prepare individuals for battalion- and brigade-level command and to ensure an understanding of current Army training, personnel, logistics, and tactical doctrine.

Instruction: Areas of instruction include military science as related to medical personnel, health care administration, preventive medicine, and combat stress and battle fatigue preparedness.

Credit Recommendation: Credit is not recommended because of the military-specific nature of the course (5/96).

AR-2201-0272

INFANTRY PRE-COMMAND

Course Number: 2G-F26.

Location: Infantry School, Fort Benning, GA.

Length: 1–3 weeks (87–167 hours).

Exhibit Dates: 6/85–9/92.

Objectives: To train the student in the overall management of unit operations in preparation for the assumption of command responsibility.

Instruction: Lectures and demonstrations cover offensive and defensive tactical doctrine, field artillery, and field engineering.

Credit Recommendation: Credit is not recommended because of the military-specific nature of the course (10/92).

AR-2201-0273

JUMPMASTER

Course Number: 2E-F60/011-F16; 2E-F3/011-F16.

Location: Infantry School, Fort Benning, GA.

Length: 1–2 weeks (77–80 hours).

Exhibit Dates: 3/82–9/90.

Objectives: After 9/90 see AR-2201-0449. To train students (qualified parachutists) in jumpmaster techniques.

Instruction: Lectures, demonstrations, and practical exercises cover parachuting techniques.

Credit Recommendation: In the vocational certificate category, 2 semester hours in parachuting techniques (9/85).

AR-2201-0274

PRIMARY NONCOMMISSIONED OFFICER (NCO) FOR COMBAT ARMS

Course Number: None.
Location: NCO Academy, Worldwide.
Length: 4 weeks (212 hours).
Exhibit Dates: 11/80–6/96.
Objectives: To develop leadership skills and techniques in a combat arms soldier who has been selected as having potential to become an NCO.
Instruction: Lectures and practical exercises provide a thorough understanding of the responsibilities of NCOs, leadership techniques that apply to their positions, and training and managing their soldiers.
Credit Recommendation: Credit is not recommended because of the limited, specialized nature of the course (9/85).

AR-2201-0275

LIGHT LEADER

Course Number: 2E-F68/010-F8.
Location: Infantry School, Fort Benning, GA.
Length: 4 weeks (160–389 hours).
Exhibit Dates: 2/85–6/96.
Objectives: To train company grade officers and enlisted personnel to command and fill leadership positions in light infantry divisions.
Instruction: Lectures and practical exercises cover collective and individual combat skills, including land navigation, marksmanship, survival, and small unit tactics to develop a cohesive command structure.
Credit Recommendation: Credit is not recommended because of the limited, specialized nature of the course (9/85).

AR-2201-0276

BRADLEY FIGHTING VEHICLE INFANTRYMAN
(Bradley Fighting Vehicle Infantryman Phase 1, Phase 2, Prior Service)

Course Number: 010-11M10; 11M10-OSUT.
Location: Infantry School, Fort Benning, GA.
Length: 8–15 weeks.
Exhibit Dates: 4/84–10/94.
Objectives: To train soldiers to serve as infantry squad members.
Instruction: This course provides performance-oriented hands-on training in Bradley fighting vehicle operation, TOE missile, communications, and weapons. This course includes an 8-9 week Basic Training component.
Credit Recommendation: Credit is not recommended because of the military-specific nature of the course (9/85).
Related Occupations: 11M.

AR-2201-0277

IMPROVED TOE VEHICLE (ITV) TRAINER

Course Number: 2E-F57/010-ASIE9.
Location: Infantry School, Fort Benning, GA.
Length: 3 weeks (120 hours).
Exhibit Dates: 5/85–9/90.
Objectives: To provide the student with instruction on the tactical employment of the improved TOE Vehicle (ITV).
Instruction: This course contains practical exercises, demonstrations, and conferences/lectures on all aspects of the ITV.
Credit Recommendation: Credit is not recommended because of the military-specific nature of the course (9/85).

AR-2201-0278

RESERVE OFFICER TRAINING COURSE (ROTC) RANGER

Course Number: 011-F14.
Location: Infantry School, Fort Benning, GA.
Length: 9 weeks (321 hours).
Exhibit Dates: 10/75–6/96.
Objectives: To develop leadership skills in selected ROTC cadets.
Instruction: Lectures and practical exercises cover parachuting, communications, reconnaissance, survival, weapons, first aid, simulated combat, physical conditioning, and marksmanship.
Credit Recommendation: In the lower-division baccalaureate/associate degree category, 6 semester hours in physical education (9/85).

AR-2201-0279

HEAVY ANTIARMOR WEAPONS INFANTRYMAN
(Heavy Antiarmor Weapons Infantryman Phase 1, Phase 2, Prior Service)

Course Number: 010-11H10-E9; 11H10-E9-OSUT; 010-11H10; 11H10-OSUT.
Location: Infantry School, Fort Benning, GA.
Length: 5–13 weeks (255–598 hours).
Exhibit Dates: 1/79–10/94.
Objectives: To prepare personnel for infantry units.
Instruction: Lectures, demonstrations, and practical exercises cover discipline, motivation, and physical fitness. This course includes an 8-9 week long Basic Training component.
Credit Recommendation: Credit is not recommended because of the military-specific nature of the course (1/92).
Related Occupations: 11H.

AR-2201-0280

MULTIPLE LAUNCH ROCKET SYSTEM (MLRS) MECHANIC

Course Number: 042-ASIS8; 042-ASIS8 (CT).
Location: Field Artillery School, Fort Sill, OK.
Length: 2 weeks (63–71 hours).
Exhibit Dates: 2/83–4/98.
Objectives: To train selected personnel to perform organizational maintenance on the missile launch rocket system.
Instruction: Lectures and demonstrations cover preventive maintenance, servicing, and system troubleshooting.
Credit Recommendation: Credit is not recommended because of the limited, specialized nature of the course (4/88).

AR-2201-0282

MULTIPLE LAUNCH ROCKET SYSTEM (MLRS) CREWMAN

Course Number: 042-13M10.
Location: Field Artillery School, Fort Sill, OK.
Length: 6–7 weeks (222–250 hours).
Exhibit Dates: 10/87–9/97.
Learning Outcomes: After 9/97 see AR-1601-0111. Upon completion of the course, the student will be able to perform the duties of a multiple launch rocket system crewman.
Instruction: Lectures and demonstrations cover electronic communications, weapon systems, target acquisition, fire control system operations, supply procedures, and preventive maintenance checks and services.
Credit Recommendation: In the vocational certificate category, 2 semester hours in map reading (10/91).
Related Occupations: 13M.

AR-2201-0284

NUCLEAR WARHEAD DETACHMENT

Course Number: 2E-FOA-F76; 2E-F96/041-F8.
Location: Field Artillery School, Fort Sill, OK.
Length: 2–3 weeks (79–108 hours).
Exhibit Dates: 8/85–12/93.
Objectives: To prepare selected commissioned officers to command nuclear warhead detachments.
Instruction: Lectures cover the mission, duties, functions, and responsibilities of nuclear warhead detachment personnel.
Credit Recommendation: Credit is not recommended because of the military-specific nature of the course (10/90).

AR-2201-0285

WARRANT OFFICER CANDIDATE
(Warrant Officer Entry)

Course Number: 911-09W.
Location: Aviation Center, Fort Rucker, AL; Reserve Readiness Training Center, Fort McCoy, WI.
Length: 4–7 weeks (217–410 hours).
Exhibit Dates: 1/87–Present.
Learning Outcomes: To train warrant officer candidates in the skills required of all warrant officers.
Instruction: Lectures, demonstrations, and practical exercises cover leadership, ethics, professional development, communication skills, personnel management, military history, deployment, and tactics.
Credit Recommendation: In the lower-division baccalaureate/associate degree category, 6 semester hours in leadership and supervision (4/03).

AR-2201-0288

PERSHING II MISSILE CREWMEMBER ADVANCED NONCOMMISSIONED OFFICER (NCO)

Course Number: 043-15E40.
Location: Field Artillery School, Fort Sill, OK.
Length: 3–9 weeks (124–263 hours).
Exhibit Dates: 8/85–12/91.
Objectives: To train noncommissioned officers to teach modern field tactics.
Instruction: This course is taught in two parts. The first part, the common core, is the Sergeant Major Academy's advanced noncommissioned officers course and includes leadership, ethics, personnel management, doctrine, and tactics. The second part provides instruction in the use of weapons systems; inspection, testing, and security of the Pershing II missile system; and techniques to lead, evaluate, train, and apply management training techniques.

Credit Recommendation: Credit is recommended for the common core only. See AR-1404-0035 (6/89).
Related Occupations: 15E.

AR-2201-0289

Multiple Launch Rocket System (MLRS)/ Lance Fire Direction Supervisor Advanced Noncommissioned Officer (NCO)

Course Number: 250-15J40.
Location: Field Artillery School, Fort Sill, OK.
Length: 8–9 weeks (297 hours).
Exhibit Dates: 8/85–6/96.
Objectives: To train the noncommissioned officer to teach modern battlefield tactics and to supervise MLRS/Lance fire direction systems.
Instruction: Lectures and practical experiences cover leadership, ethics, military doctrine and tactics, training and evaluating fire direction center soldiers, and the application of training management techniques. This course contains a common core of leadership subjects.
Credit Recommendation: Credit is recommended for the common core only. See AR-1404-0035 (11/85).
Related Occupations: 15J.

AR-2201-0290

Multiple Launch Rocket System (MLRS)/ Lance Advanced Noncommissioned Officer (NCO)

Course Number: 043-15D40.
Location: Field Artillery School, Fort Sill, OK.
Length: 3–8 weeks (132–262 hours).
Exhibit Dates: 8/85–10/98.
Objectives: To train noncommissioned officers to teach modern battlefield tactics and to supervise the Lance and MLRS missile system.
Instruction: Lectures and practical exercises cover leadership, ethics, personnel management, military doctrine, and tactics; emplacement, inspection, testing, and security of the multiple launch rocket system; and techniques necessary to lead, evaluate, and train Lance and MLRS system soldiers. This course contains a common core of leadership subjects.
Credit Recommendation: Credit is recommended for the common core only. See AR-1404-0035 (10/88).
Related Occupations: 15D.

AR-2201-0291

Pershing Electronics Materiel Specialist Advanced Noncommissioned Officer (NCO)
(Pershing II Electronics Materiel Specialist Advanced Noncommissioned Officer (NCO))

Course Number: 121-21G40.
Location: Field Artillery School, Fort Sill, OK.
Length: 4–8 weeks (157–267 hours).
Exhibit Dates: 9/85–12/91.
Objectives: To train missile section chiefs and noncommissioned officers in battlefield tactics, leadership, technical areas, and supervision.
Instruction: Lectures and demonstrations cover leadership; ethics; personnel management; doctrine; tactics; and inspections, maintenance, and testing of Pershing missile systems. This course contains a common core of leadership subjects.
Credit Recommendation: Credit is recommended for the common core only. See AR-1404-0035 (6/89).
Related Occupations: 21G.

AR-2201-0292

Field Artillery Radar Crewmember Noncommissioned Officer (NCO) Advanced

Course Number: 221-13R/17B40; 221-17B/13R40.
Location: Field Artillery School, Fort Sill, OK.
Length: 5–7 weeks (185–248 hours).
Exhibit Dates: 10/85–6/96.
Objectives: To train selected personnel in the technical and supervisory skills necessary to perform as a field artillery crewmember noncommissioned officer.
Instruction: Lectures and demonstrations cover field artillery tactics, communications, electronics, weapons, and target acquisition. This course contains a common core of leadership subjects.
Credit Recommendation: Credit is recommended for the common core only. See AR-1404-0035 (10/90).
Related Occupations: 13R; 17B.

AR-2201-0293

Field Artillery Surveyor Advanced Noncommissioned Officer (NCO)
(Field Artillery Surveyor Noncommissioned Officer (NCO) Advanced)

Course Number: 411-82C40.
Location: Field Artillery School, Fort Sill, OK.
Length: 3–7 weeks (134–249 hours).
Exhibit Dates: 10/85–9/98.
Learning Outcomes: After 9/00 see AR-2201-0521. Upon completion of the course, the student will be able to plan and supervise conventional survey operations.
Instruction: Lectures and practical exercise cover training other soldiers in planning, field work, special survey techniques, organizational maintenance of survey equipment, vehicle maintenance, and supply procedures.
Credit Recommendation: Credit is not recommended because of the limited, specialized nature of the course (1/92).
Related Occupations: 82C.

AR-2201-0294

1. Primary Leadership Development Reserve Component
2. Primary Leadership Development Reserve Component
3. Primary Leadership Development

Course Number: ***Version 1:*** PLDC; NGB-PLDC (RC); PLDC-RC; 400-PLDC. *Version 2:* NGB-PLDC (RC); 400-PLDC; PLDC-RC. *Version 3:* 400-PLDC.
Location: ***Version 1:*** Reserve Schools, US; National Guard Academies, US. *Version 2:* Reserve Component Training Centers, US; Reserve Centers, US; National Guard Academies, US. *Version 3:* Reserve Schools, Multiple locations; Various Training Locations, US.
Length: ***Version 1:*** 123 hours. *Version 2:* 150 hours. *Version 3:* 4–5 weeks (184 hours).
Exhibit Dates: ***Version 1:*** 4/85–12/94. *Version 2:* 1/95–9/03. *Version 3:* 10/03–Present.
Learning Outcomes: ***Version 1:*** Upon completion of the course, the student will have been provided with the skills and knowledge necessary to assuming first line supervisors responsibilities including the fundamentals and techniques of leadership, group behavior, and resource management in a military organization. *Version 2:* Upon completion of the course, the student will be able to assume first-line supervision responsibilities and apply the fundamentals and techniques of leadership, group behavior, and resource management in a military organization. *Version 3:* Pending evaluation.
Instruction: ***Version 1:*** The course combines lectures, small-group seminars, and practical application in the areas of leadership, communication, resource and training management, military studies, and professional skills in drill and ceremony, map reading/land navigation, and combat operations for small units. Course is conducted over 15 days or on four weekends with an eight-day resident component. *Version 2:* The course combines lectures, small-group seminars, and practical application in the areas of leadership, communication, resource and training management, military studies, and professional skills. Subjects covered in the course include counseling, sexual harassment, team building, motivation, styles of leadership, and problem solving. *Version 3:* Pending evaluation.
Credit Recommendation: ***Version 1:*** In the lower-division baccalaureate/associate degree category, 3 semester hours in principles of supervision and leadership (5/87). *Version 2:* In the lower-division baccalaureate/associate degree category, 1 semester hour in principles of supervision and 1 in military science (6/95). *Version 3:* Pending evaluation.

AR-2201-0295

1. Advanced Noncommissioned Officer (NCO) Reserve Component
2. Advanced Noncommissioned Officer (NCO) Common Leader Reserve, Phase 1

Course Number: ***Version 1:*** 400-ANCOC-RC; ANCOC. *Version 2:* ANCOC; 400-ANCOC-RC, Phase 1.
Location: Reserve Component Training Centers, US; National Guard Academies, US.
Length: ***Version 1:*** 6 weekends and, 2 weeks (109 hours). *Version 2:* 1 weekend a month for 6 consecutive months or, 2 weeks (83 hours).
Exhibit Dates: ***Version 1:*** 7/80–9/94. *Version 2:* 10/94–9/00.
Learning Outcomes: ***Version 1:*** Upon completion of the course, the student will be able to provide mid-level supervision and leadership at the operational level in units of up to 400 persons. *Version 2:* Upon completion of the course, the student will be able to provide mid-level supervision and leadership at the operational level in units of up to 50 persons.
Instruction: ***Version 1:*** This course includes lectures, demonstrations, and performance exercises in training, motivation, morale, supervision, effective writing, verbal communication, and listening skills. *Version 2:* Topics include development of mid-level skills in training and motivation of personnel, logistics and materiel management, building morale,

effective supervision of personnel, quality written and oral communications, implementation of appropriate listening skills, and mid-level leadership.

Credit Recommendation: *Version 1:* In the lower-division baccalaureate/associate degree category, 1 semester hour in leadership and 1 in fundamentals of communication (2/92). *Version 2:* In the lower-division baccalaureate/associate degree category, 1 semester hour in military science and 1 in fundamentals of communication (6/95).

AR-2201-0296

Senior Course for Reserve Component Noncommissioned Officer (NCO)

Course Number: None.
Location: National Guard Academies, US.
Length: 50 hours.
Exhibit Dates: 7/82–6/96.
Objectives: To develop leadership skills and techniques at the upper operational levels.
Instruction: Lectures and practical exercises, and small-group study cover training, effective writing, public speaking, counseling, and maintenance management. Course is conducted over eight days or four weekends.
Credit Recommendation: In the lower-division baccalaureate/associate degree category, for those students completing the Advanced Course for Reserve Noncommissioned Officer (NCO) (AR-2201-0295), 1 semester hour in group dynamics and training. For all other students, 2 semester hours in group dynamics and training (8/86).

AR-2201-0299

1. First Sergeant Course for Reserve Component
2. First Sergeant for Reserve Components

Course Number: *Version 1:* None. *Version 2:* 400-SQIM.
Location: Reserve Component Training Centers, US; Reserve Centers, US; National Guard Academies, US.
Length: *Version 1:* 80 hours. *Version 2:* Phase 2, in residence, 2 weeks (110 hours).
Exhibit Dates: *Version 1:* 9/84–9/95. *Version 2:* 10/95–Present.
Learning Outcomes: *Version 1:* Upon completion of the course, the student will have developed skills in administrative procedures and processes. *Version 2:* Upon completion of the course, the student will be able to supervise up to 200 persons.
Instruction: *Version 1:* Through lectures, practical exercises, seminars, and problem solving the student develops skills and knowledge in records administration and organizational processes and procedures. *Version 2:* This course includes lectures, seminars, demonstrations, and performance exercises in safety, stress management, diversity, substance abuse, maintenance and control of equipment, motivation, leadership skills, and computer and verbal skills.
Credit Recommendation: *Version 1:* In the lower-division baccalaureate/associate degree category, 2 semester hours in administrative procedures and processes (8/86). *Version 2:* In the lower-division baccalaureate/associate degree category, 1 semester hour in material management and 2 in fundamentals of communication. In the upper-division baccalaureate category. 1 semester hour in human resource management (10/95).

AR-2201-0300

National Guard Officer Candidate

Course Number: OCS.
Location: National Guard Academies, US.
Length: 265–272 hours.
Exhibit Dates: 2/82–8/94.
Objectives: Upon completion of the course, the student will demonstrate leadership and management skills in training, motivation, counseling, disciplining, and evaluating personnel; managing and maintaining systems; planning, organizing, and directing organization processes and procedures; reading and interpreting complex maps; making navigational calculations; and translating from two to three dimensions.
Instruction: Topics include human resource management, leadership, writing, records, maintenance, training, direction, and motivation and course methods include readings, lectures, discussions, exercises, and evaluation by tests and performance. There are two 15 day components plus twelve months of two-day sessions.
Credit Recommendation: In the lower-division baccalaureate/associate degree category, 5 semester hours in leadership and management and 2 in geography/topography (8/86).

AR-2201-0301

Light Air Defense Artillery (ADA) Crewman

Course Number: 16F10-OSUT.
Location: Air Defense Artillery School, Fort Bliss, TX.
Length: 14 weeks (186 hours).
Exhibit Dates: 1/83–6/94.
Objectives: To provide National Guard enlisted personnel with the skills necessary to perform as light air defense artillery crewmen.
Instruction: Course includes training in basic soldier skills, 40mm gun operation, tracked vehicle operation, and firing. This course includes an 8-9 week Basic Training component.
Credit Recommendation: In the vocational certificate category, 1 semester hour in first aid (11/86).
Related Occupations: 16F.

AR-2201-0303

Cadet Military Secondary Training/ ROTC

Course Number: 2C-F34.
Location: Aviation Center, Fort Rucker, AL.
Length: 4 weeks (124 hours).
Exhibit Dates: 6/85–2/90.
Objectives: To introduce military academy and ROTC cadets to the operation and uses of rotary-wing aircraft in a military environment.
Instruction: Topics include TH-55 helicopter maintenance, aerodynamics, tactical subjects, aeromedical subjects, and flight training.
Credit Recommendation: In the upper-division baccalaureate category, 3 semester hours in basic aerodynamics and theory of flight and 3 in basic helicopter flight and systems (11/86).

AR-2201-0306

Signal Officer Advanced Reserve Component

Course Number: 4-11-C26-RC.
Location: Signal School, Fort Gordon, GA.
Length: Resident phase, 12 weeks (432–442 hours).
Exhibit Dates: 6/86–Present.
Learning Outcomes: Upon completion of the course, the student will have the knowledge to be assigned a signal command position at company, battalion, brigade, or division/corps level.
Instruction: This course includes instruction in combined arms, battle doctrine, communications planning, technical characteristics of various tactical radios and wire equipment, review of basic electricity, electromagnetic propagation, and procedures necessary for secure communications.
Credit Recommendation: Credit is not recommended because of the military-specific nature of the course (2/94).

AR-2201-0307

Signal Officer Basic Reserve Components

Course Number: 4-11-C25-RC.
Location: Signal School, Fort Gordon, GA.
Length: Resident phase, 8 weeks (293 hours).
Exhibit Dates: 2/85–4/04.
Learning Outcomes: Upon completion of the course, the student will be able to demonstrate a basic knowledge of communications systems in use by tactical units of the Signal Corps, manage personnel, and manage a tactical communications center.
Instruction: Lectures and hands-on experience cover the various tactical wire and radio communications systems, and lectures cover combat communications concepts, doctrine, and procedures used in tactical communications systems. Only the resident phase was evaluated.
Credit Recommendation: In the lower-division baccalaureate/associate degree category, 3 semester hours in fundamentals of communications systems (6/90).

AR-2201-0308

Artillery Repairer

Course Number: 642-45L10.
Location: Ordnance Center and School, Aberdeen Proving Ground, MD.
Length: 13–14 weeks (491 hours).
Exhibit Dates: 6/86–4/92.
Learning Outcomes: Upon completion of the course, the student will be able to perform general support maintenance on towed and self-propelled artillery.
Instruction: Instruction includes lectures, demonstrations, and practical exercises in the use and care of hand tools and hardware, basic hydraulic principles, electrical fundamentals, and the maintenance and troubleshooting of artillery weapon systems.
Credit Recommendation: In the lower-division baccalaureate/associate degree category, 1 semester hour in basic electricity (10/88).
Related Occupations: 45L.

AR-2201-0309

Towed Artillery Repairman (USMC) (Self-Propelled Artillery System Technician)

(Artillery Repairer)

Course Number: 641-2131; 642-2143 (45L); 642-3121 (45L).
Location: Ordnance Center and School, Aberdeen Proving Ground, MD.
Length: 6–13 weeks (240–468 hours).
Exhibit Dates: 6/86–2/96.
Learning Outcomes: Upon completion of the course, the student will be able to perform general support maintenance on towed and self-propelled artillery.
Instruction: Instruction includes lectures, demonstrations, and practical exercises in the use and care of hand tools and hardware, basic hydraulic principles, electrical fundamentals, and the maintenance and troubleshooting of artillery weapon systems.
Credit Recommendation: In the lower-division baccalaureate/associate degree category, 1 semester hour in basic electricity (10/88).
Related Occupations: 45L.

AR-2201-0310

Light Weapons Infantryman
(Light Weapons Infantryman Phase 1, Phase 2, Prior Service)

Course Number: 010-11B10-C2; 11B10-C2-OSUT; 010-11B10; 11B10-OSUT.
Location: Infantry School, Fort Benning, GA.
Length: 6–13 weeks (255–591 hours).
Exhibit Dates: 1/79–10/94.
Objectives: To prepare personnel for infantry units.
Instruction: Lectures, demonstrations, and practical exercises cover discipline, motivation, and physical fitness. This course is offered in a 12-13 week long presentation or in split training where course is conducted as two 6-7 week long phases. This course includes an 8-9 week long Basic Training component.
Credit Recommendation: Credit is not recommended because of the military-specific nature of the course (9/85).
Related Occupations: 11B.

AR-2201-0311

Indirect Fire Infantryman
(Indirect Fire Infantryman Phase 1, Phase 2, Prior Service)

Course Number: 010-11C10; 11C10-OSUT.
Location: Infantry School, Fort Benning, GA.
Length: 6–13 weeks (255–519 hours).
Exhibit Dates: 1/79–10/94.
Objectives: To prepare personnel for infantry units.
Instruction: Lectures, demonstrations, and practical exercises cover discipline, motivation, and physical fitness. This course is offered in a 12-13 week long presentation or in split training where course is conducted as two 6-7 week long phases. This course includes an 8-9 week long Basic Training component.
Credit Recommendation: Credit is not recommended because of the military-specific nature of the course (1/92).
Related Occupations: 11C.

AR-2201-0312

Basic Noncommissioned Officer (NCO)
16T Track

Course Number: 043-11-16T30.
Location: NCO Academy, Fort Bliss, TX.
Length: 5 weeks (200–209 hours).
Exhibit Dates: 7/87–6/94.
Learning Outcomes: Upon completion of the course, students will be able to lead, train, and direct subordinates to maintain, operate, and employ weapons and equipment.
Instruction: Training is in leadership; maintenance; tactics; communications; land navigation; weapons; nuclear, biological, and chemical warfare; and training management. This course contains a common core of leadership subjects.
Credit Recommendation: Credit is recommended for the common core only. See AR-1406-0090 (3/89).
Related Occupations: 16T.

AR-2201-0313

Electronic Warfare (EW) Analyst

Course Number: *Version 1:* 232-F8. *Version 2:* 232-F8.
Location: *Version 1:* Intelligence School, Fort Devens, MA. *Version 2:* Intelligence Center and School, Fort Huachuca, AZ.
Length: *Version 1:* 7–9 weeks (293–357 hours). *Version 2:* 7 weeks (252–253 hours).
Exhibit Dates: *Version 1:* 10/86–9/92. *Version 2:* 10/92–Present.
Learning Outcomes: *Version 1:* Upon completion of the course, the student will be able to isolate, identify, and report tactical signal intelligence; develop data bases; use tactical communications equipment and vehicles in support of tactical operations, and operate a microfix computer. *Version 2:* Upon completion of the course, the student will be able to process, analyze, and report tactical signal intelligence; set up, operate, and perform preventive maintenance checks on a tactical communication equipment; develop data bases; and operate a tactical automated data processing computer system.
Instruction: *Version 1:* Lectures, demonstrations, and performance exercises cover report writing, radio communications, data base development, operation and maintenance of tactical communications equipment and vehicles, and operation of the microfix computer. *Version 2:* Lectures, demonstrations, and performance exercises cover report writing, tactical communications equipment, and data processing equipment.
Credit Recommendation: *Version 1:* Credit is not recommended because of the military-specific nature of the course (6/89). *Version 2:* Credit is not recommended because of the military-specific nature of the course (6/97).

AR-2201-0315

Pre-Command Phase 3, Systems Proficiency

Course Number: 3B-F11.
Location: Intelligence School, Fort Devens, MA.
Length: 1 week (45 hours).
Exhibit Dates: 4/85–12/90.
Learning Outcomes: Upon completion of the course, the student will be proficient in electronic warfare/signal intelligence collection and processing systems.
Instruction: Instruction includes orientation to basic electronic warfare topics and to missions and procedures related to tactical command sites.
Credit Recommendation: Credit is not recommended because of the limited, specialized nature of the course (5/87).

AR-2201-0316

Electronic Warfare (EW)/Signals Intelligence (SIGINT) Emitter Locater/Identifier Basic Noncommissioned Officer (NCO)

Course Number: 231-05D30.
Location: Security School, Fort Devens, MA; Intelligence School, Fort Devens, MA.
Length: 7–8 weeks (266–267 hours).
Exhibit Dates: 1/89–12/90.
Learning Outcomes: Upon completion of the course, the student will assume the position of squad leader, select an direction finding site, determine logistical requirements to deploy to the site, establish and maintain an advanced technique identification library, and provide quality control in data and records.
Instruction: Instruction includes classroom lectures and practical exercises in leadership training, direction finding, deployment techniques, quality control for records and reports, and establishing and maintaining library systems. This course contains a common core of leadership subjects.
Credit Recommendation: In the lower-division baccalaureate/associate degree category, 2 semester hours in personnel management. See AR-1406-0090 for common core credit (7/91).
Related Occupations: 05D.

AR-2201-0317

Emitter Identification Techniques
(Advanced Identification Techniques Analyst)

Course Number: 231-F7; 231-F29.
Location: Intelligence School, Fort Devens, MA; Intelligence Center and School, Fort Huachuca, AZ.
Length: 3–4 weeks (107–120 hours).
Exhibit Dates: 4/86–Present.
Learning Outcomes: Upon completion of the course, the student will be able to use advanced identification techniques; calibrate and operate specific equipment; and analyze techniques, records, and reports.
Instruction: Lectures, audiovisual presentations, demonstrations, and performance exercises cover the calibration and operation of equipment used in analyzing, recording, and reporting signals.
Credit Recommendation: Credit is not recommended because of the limited, specialized nature of the course (2/96).

AR-2201-0318

Ordnance Pre-Command

Course Number: 2G-F44.
Location: Ordnance Center and School, Aberdeen Proving Ground, MD.
Length: 2 weeks (72 hours).
Exhibit Dates: 1/85–9/90.
Learning Outcomes: Upon completion of the course, the student will be able to assume battalion- and brigade-level commands.
Instruction: Instruction includes a series of lectures, demonstrations, and practical exercises of short duration on military topics

designed to aid the ordnance officer in assuming command of a military unit.

Credit Recommendation: Credit is not recommended because of the military-specific nature of the course (3/87).

AR-2201-0320

Self-Propelled Field Artillery System Mechanic Noncommissioned Officer (NCO) Advanced
(Mechanical Maintenance Noncommissioned Officer (NCO) Advanced, MOS 63B, D, E, N, T)

Course Number: 6-63-C42 (63B, D, E, N, T); 611-63D40.

Location: Ordnance Center and School, Aberdeen Proving Ground, MD.

Length: 8–10 weeks (295–445 hours).

Exhibit Dates: 10/85–7/95.

Learning Outcomes: Upon completion of the course, students will be able to serve as a supervisor in the selected military occupational specialty for which they are trained.

Instruction: Lectures, demonstrations, and practical exercises cover leadership, communication skills, fundamentals of instruction, logistics management, and refresher training in the appropriate military occupational specialty. Emphasis in on leadership and on verbal and written communication. This course contains a common core of leadership subjects.

Credit Recommendation: Credit is recommended for the common core only. See AR-1404-0035 (1/88).

Related Occupations: 63B; 63D; 63E; 63N; 63T.

AR-2201-0321

Armor Officer Advanced Reserve Component

Course Number: 2-17-C26.

Location: Armor Center, Fort Knox, KY.

Length: 12–13 weeks (465–504 hours).

Exhibit Dates: 8/85–10/99.

Learning Outcomes: Upon completion of the course, the student will be able to perform management skills.

Instruction: Instruction includes lectures and practical exercises in military science, organizational support and functions, leadership, communication skills, maintenance and inspection procedures, and logistics.

Credit Recommendation: In the lower-division baccalaureate/associate degree category, 2 semester hours in communication skills. In the upper-division baccalaureate category, 2 semester hours in management systems, 3 in business management, and 2 in military science (10/89).

AR-2201-0325

Basic Noncommissioned Officer (NCO) M60A1/M60A3 19K M1 Abrams

Course Number: 020-18-19E30-A3; 020-18-19K30-M1; 020-10-19E30-A3; 020-9-19E30-A1; 020-9-19K30-M1; 020-7-19E30-A3; 020-6-19E30-A1; 020-4-19E30-A3; 020-1-19E30-A3; 020-1-19K30-M1; 020-1-19E30/19K30.

Location: Armor School, USAREUR; Armor School, Fort Carson, CO; Armor School, Fort Hood, TX; Armor School, Fort Riley, KS; Armor School, Fort Polk, LA; Armor School, Fort Stewart, GA; Armor Center and School, Fort Knox, KY.

Length: 6 weeks (312–314 hours).

Exhibit Dates: 10/86–5/91.

Learning Outcomes: Upon completion of the course, the student will be able to train, direct, and lead personnel in the maintenance, operation, and employment of specialized tank equipment.

Instruction: Course includes lectures and training in leadership, maintenance, tactics, communications, land navigation, and training management.

Credit Recommendation: In the lower-division baccalaureate/associate degree category, 3 semester hours in personnel management, 4 in military science, and 2 in basic surveying (4/97).

Related Occupations: 19E; 19K.

AR-2201-0326

Bradley Fighting Vehicle System Organizational Maintenance

Course Number: 611-ASID3 (63T).

Location: Armor Center, Fort Knox, KY.

Length: 2–3 weeks (92 hours).

Exhibit Dates: 8/83–12/97.

Learning Outcomes: Upon completion of the course, the student will be able to perform minor adjustments and repairs on the Bradley Fighting vehicle.

Instruction: Instruction includes lectures and practical exercises in repairs and adjustments of the engine, steering, braking, cooling, and lubrication systems of the Bradley Fighting vehicle.

Credit Recommendation: Credit is not recommended because of the limited, specialized nature of the course (7/87).

Related Occupations: 63T.

AR-2201-0329

M1/M1A1 Master Gunner Transition

Course Number: 020-ASIA8-T.

Location: Armor Center, Fort Knox, KY.

Length: 4–5 weeks.

Exhibit Dates: 3/86–9/94.

Learning Outcomes: Upon completion of the course, the student will be able to perform the duties of a M1/M1A1 tank master gunner.

Instruction: Instruction includes lectures and practical exercises in the differences between the M60A/A3 and the M1/M1A1 tank weapons systems.

Credit Recommendation: Credit is not recommended because of the military-specific nature of the course (10/88).

AR-2201-0330

M1/M1A1 Abrams Armor Crewman

Course Number: 19K10-OSUT.

Location: Armor Center, Fort Knox, KY.

Length: 14–15 weeks (562–619 hours).

Exhibit Dates: 9/85–1/95.

Learning Outcomes: After 1/95 see AR-2201-0479. Upon completion of the course, the student will be able to perform the duties of an armor crewman.

Instruction: Instruction includes lectures and practical exercises in basic training and skills required of a tank crewman, including weapons training, land navigation, first aid, physical conditioning, tank field exercises, and basic military subjects. This course includes an 8-9 week Basic Training component.

Credit Recommendation: In the lower-division baccalaureate/associate degree category, 5 semester hours in military science (1/96).

Related Occupations: 19K.

AR-2201-0331

M3 Bradley/CFV Cavalry Scout

Course Number: *Version 1:* 19D10-OSUT (M3). *Version 2:* 19D10-OSUT (M3). *Version 3:* 19D10-OSUT (M3), Phase 2; 19D10-OSUT (M3), Phase 1; 19D10-OSUT (M3).

Location: *Version 1:* Armor Center, Fort Knox, KY. *Version 2:* Armor Center and School, Fort Knox, KY. *Version 3:* Armor School, Fort Knox, KY.

Length: *Version 1:* 14–16 weeks (666–792 hours). *Version 2:* 17–18 weeks (716 hours). *Version 3:* 16 weeks (743 hours).

Exhibit Dates: *Version 1:* 4/85–9/94. *Version 2:* 10/94–9/02. *Version 3:* 10/02–Present.

Learning Outcomes: *Version 1:* Upon completion of the course, the student will be able to perform the duties of an armored reconnaissance vehicle crewman. *Version 2:* Upon completion of the course, the student will be able to perform basic soldier tasks. *Version 3:* Upon completion of the course, the student will be able to perform basic cavalry scout duties.

Instruction: *Version 1:* Instruction includes lectures and practical exercises in basic training and the skills required of an armored reconnaissance vehicle crewman, including weapons training, land navigation, first aid, physical conditioning, vehicle field exercises, and basic military subjects. *Version 2:* Lecture, discussion, and practical experience used to cover the topics of combat, physical fitness, first aid, nuclear/biological/chemical contamination, communications, land navigation, weapons, tactical training, intelligence, and tank operation/maintenance. This course includes a 8-9 week Basic Training component. *Version 3:* Lecture, discussion, and practical exercises. Topics include combat, physical fitness, first aid, nuclear/biological/chemical contamination, communications, land navigation, weapons, tactical training, intelligence, and tank operation/maintenance. This course includes an 8-9 week Basic Training component.

Credit Recommendation: *Version 1:* In the lower-division baccalaureate/associate degree category, 5 semester hours in military science (1/96). *Version 2:* In the lower-division baccalaureate/associate degree category, 1 semester hour in basic first aid (3/97). *Version 3:* In the lower-division baccalaureate/associate degree category, 1 semester hour in first aid and 3 in military science (6/03).

Related Occupations: 19D.

AR-2201-0332

M113 Cavalry Scout

Course Number: 19D10-OSUT (M113).

Location: Armor Center, Fort Knox, KY.

Length: 13–14 weeks (561–607 hours).

Exhibit Dates: 1/85–10/95.

Learning Outcomes: Upon completion of the course, the student will be able to perform the duties of an armored reconnaissance vehicle crewman.

Instruction: Instruction includes lectures and practical exercises in basic training and the skills required of an armored reconnaissance vehicle crewman, including weapons training, land navigation, first aid, physical conditioning, vehicle field exercises, and basic military subjects. This course includes an 8-9 week Basic Training component.

Credit Recommendation: In the lower-division baccalaureate/associate degree category, 5 semester hours in military science (11/88).

Related Occupations: 19D.

AR-2201-0333

M60A3 Armor Crewman

Course Number: 19E10-OSUT (M60A3).
Location: Armor Center, Fort Knox, KY.
Length: 14–15 weeks (532–690 hours).
Exhibit Dates: 7/87–10/95.

Learning Outcomes: Upon completion of the course, the student will be able to perform the duties of an armor crewmember.

Instruction: The instruction includes lectures and practical exercises in basic training subjects and the skills required of a tank crewman, including weapons training, land navigation, first aid, physical conditioning, tank field exercises, and basic military subjects. This course includes an 8-9 week Basic Training component.

Credit Recommendation: In the lower-division baccalaureate/associate degree category, 5 semester hours in military science (11/88).

Related Occupations: 19E.

AR-2201-0334

Master Gunner M60A1

Course Number: 020-ASIC5 (M60A1).
Location: Armor Center, Fort Knox, KY.
Length: 11 weeks.
Exhibit Dates: 1/86–1/90.

Learning Outcomes: Upon completion of the course, the student will be able to assist commanders in the planning and implementation of gunnery training programs.

Instruction: Instruction includes lectures and practical exercises in tank weapon systems.

Credit Recommendation: Credit is not recommended because of the military-specific nature of the course (7/87).

AR-2201-0336

Master Gunner M60A3 Transition

Course Number: 020-ASID8.
Location: Armor Center, Fort Knox, KY.
Length: 2 weeks (66–72 hours).
Exhibit Dates: 1/86–10/91.

Learning Outcomes: Upon completion of the course, the student will be able to perform duties as a M60A3 tank master gunner.

Instruction: Instruction includes lectures and practical exercises in the differences between the M60A1 and M60A3 tank weapon systems.

Credit Recommendation: Credit is not recommended because of the military-specific nature of the course (10/88).

AR-2201-0337

1. Basic Noncommissioned Officer (NCO) Reserve Component Phase I Common Leader Training
2. Basic Noncommissioned Officer (NCO) Common Leader Reserve Training

Course Number: *Version 1:* 400-BNCOC-RC; 400-BNCOC; BNCOC. *Version 2:* 400-BNCOC 400-BNCOC-RC.

Location: *Version 1:* Training Centers, US. *Version 2:* Training Centers, US and Overseas.

Length: *Version 1:* 6 days or three weekends, 1 week. *Version 2:* 1 weekend a month for 6 consecutive months or, 2 weeks (85 hours).

Exhibit Dates: *Version 1:* 10/86–9/94. *Version 2:* 10/94–9/00.

Learning Outcomes: *Version 1:* Upon completion of the course, the student will be able to provide lower-level leadership to units of up to 35 persons. *Version 2:* Upon completion of the course, the student will be able to provide entry-level leadership to a group of up to 12 subordinates, train these individuals to perform their jobs, instill proper behavior, supervise organizational maintenance activities and property accountability, and employ the systems approach to personnel training.

Instruction: *Version 1:* The course includes lectures, demonstrations, and performance exercises in supervision, training, and counseling. *Version 2:* Topics include entry-level leadership skills, personal and performance counseling, training techniques, basic management skills, and introductory resource management skills.

Credit Recommendation: *Version 1:* In the lower-division baccalaureate/associate degree category, 1 semester hour in leadership (2/92). *Version 2:* In the lower-division baccalaureate/associate degree category, 2 semester hours in supervision or leadership (6/95).

AR-2201-0339

1. Drill Sergeant
2. Total Army Training System Drill Sergeant
   (Drill Sergeant)
3. Drill Sergeant
   (Total Army Training System (TATS) Drill Sergeant)

Course Number: *Version 1:* 012-SQIX; 012-F4. *Version 2:* 012-SQIX; 012-SQIX (F); 012-F4. *Version 3:* 012-SQIX; 012-SQIX (F).

Location: *Version 1:* Training Center, Fort Knox, KY; Training Center, Fort Sill, OK; Training Center, Fort McClellan, AL; Training Center, Fort Leonard Wood, MO; Training Center, Fort Jackson, SC; Training Center, Fort Dix, NJ; Training Center, Fort Benning, GA. *Version 2:* Training Center, Fort Jackson, SC; Training Center, Fort Leonard Wood, MO; Training Center, Fort Benning, GA. *Version 3:* Drill Sergeant School, Fort Jackson, SC; Training Center Engineer, Fort Leonard Wood, MO; Infantry School, Fort Benning, GA; Training Center, Fort Benning, GA; Training Center, Fort Jackson, SC.

Length: *Version 1:* 9 weeks (318–319 hours). *Version 2:* 8–9 weeks (311 hours). *Version 3:* 9 weeks (372 hours).

Exhibit Dates: *Version 1:* 7/86–2/97. *Version 2:* 3/97–7/00. *Version 3:* 8/00–Present.

Learning Outcomes: *Version 1:* Upon completion of the course, the student will be able to supervise, train, motivate, and evaluate basic trainees. *Version 2:* Upon completion of the course, the student will be able to supervise, train, motivate, and evaluate basic trainees. *Version 3:* Upon completion of the course, the student will be able to supervise, instruct, motivate and evaluate trainees.

Instruction: *Version 1:* This course includes lectures, demonstrations, and performance exercises in personal counseling, individual human behavior, interviewing, interpersonal communication, diet and nutrition, personal health and fitness, physical training techniques, weapon proficiency and safety, instructional techniques, and training aids. *Version 2:* This course includes lectures, demonstrations, and performance exercises in leadership, personal counseling, individual human behavior, interviewing, interpersonal communication, diet and nutrition, personal health and fitness, physical training techniques, weapon proficiency and safety, instructional techniques, and training aids. *Version 3:* Methods of instruction include audiovisual materials, discussion, classroom exercises, practical exercises, and lectures.

Credit Recommendation: *Version 1:* In the lower-division baccalaureate/associate degree category, 2 semester hours in counseling, 1 in physical fitness, and 1 in marksmanship. In the upper-division baccalaureate category, 1 semester hour in principles of teaching (9/87). *Version 2:* In the lower-division baccalaureate/associate degree category, 1 semester hour in introduction to supervision or leadership and 1 in physical fitness. In the upper-division baccalaureate category, 1 semester hour in counseling techniques and 1 in principles of teaching (9/98). *Version 3:* In the lower-division baccalaureate/associate degree category, 3 semester hours in leadership, 1 in methods of instruction, and 1 in marksmanship (3/02).

AR-2201-0340

Watercraft Operator Basic Noncommissioned Officer (NCO)

Course Number: *Version 1:* 062-88K30. *Version 2:* 062-88K30. *Version 3:* 062-88K30.

Location: Transportation School, Fort Eustis, VA.

Length: *Version 1:* 6–7 weeks (192 hours). *Version 2:* 8 weeks (302 hours). *Version 3:* 9–10 weeks (307 hours).

Exhibit Dates: *Version 1:* 1/87–9/90. *Version 2:* 10/90–9/91. *Version 3:* 10/91–Present.

Learning Outcomes: *Version 1:* Upon completion of the course, the student will be able to supervise personnel in watercraft operation in such areas as navigation, ship rigging, cargo handling, damage control, shipboard fire fighting, vessel operation, and marlinspike seamanship. The student will be able to successfully complete the Marine Certification Examination. *Version 2:* Upon completion of the course, the student will be able to supervise personnel in watercraft operation in such areas as navigation, ship rigging, cargo handling, damage control, shipboard fire fighting, vessel operation, and marlinspike seamanship. The student will be able to successfully complete the Marine Certification Examination. *Version 3:* Upon completion of the course, the student will be able to effectively supervise personnel in watercraft operation, including navigation, ship rigging, cargo handling, damage control, shipboard fire fighting, vessel operation, and marlinspike seamanship. The student will be able to successfully complete the Marine Certification Examination.

Instruction: *Version 1:* This course includes lectures, demonstrations, and practical exercises in leadership skills, marine deck operations, quartermaster and deck operations, navigation, and marine regulations. *Version 2:* This course includes lectures, demonstrations, and practical exercises in leadership skills, marine deck operations, quartermaster and deck operations, navigation, and marine regulations. This course contains a common core of leadership subjects. *Version 3:* This course includes lectures, demonstrations, and practical exercises in leadership skills, marine deck operations, quartermaster and deck operations, navigation, and marine regulations. This course contains a common core of leadership subjects.

Credit Recommendation: ***Version 1:*** In the lower-division baccalaureate/associate degree category, 3 semester hours in seamanship and 1 in harbor craft operation (3/88). *Version 2:* In the lower-division baccalaureate/associate degree category, 2 semester hours in principles of supervision. See AR-1406-0090 for common core credit. In the upper-division baccalaureate category, 2 semester hours for field experience in management (6/90). *Version 3:* In the lower-division baccalaureate/associate degree category, 1 semester hour in seamanship and 2 in land navigation. See AR-1406-0090 for common core credit (10/03).

Related Occupations: 88K.

AR-2201-0341

Small Arms Maintenance for Unit Armorers

Course Number: SAM 31.

Location: Combined Arms Training Center, Vilseck, W. Germany.

Length: 2 weeks (68–74 hours).

Exhibit Dates: 4/78–10/98.

Learning Outcomes: Upon completion of the course, the student will be able to establish a small arms room which meets security requirements; identify technical publications to be used for regulations and procedures for security and maintenance of an arms room; determine proper inventory, accounting, and issue/turn-in procedures for weapons; initiate a preventive maintenance schedule; and provide preventive maintenance for all weapons stored in the arms room.

Instruction: Conferences, lectures, and practical exercises on security regulations, small arms maintenance, and accountability of weapons located within the arms room. Special emphasis is given to the disassembly, cleaning, troubleshooting, and safety testing of all weapons.

Credit Recommendation: Credit is not recommended because of the military-specific nature of the course (10/88).

AR-2201-0342

Air Defense Artillery (ADA) Officer Basic

Course Number: 2-44-C20.

Location: Air Defense Artillery School, Fort Bliss, TX.

Length: 10 weeks (365 hours).

Exhibit Dates: 10/87–9/90.

Learning Outcomes: Upon completion of the course, the student will be able to perform duties of an officer in the areas of leadership, management, logistics, and personnel administration.

Instruction: Lectures and practical exercises cover the basics of air defense artillery, map reading, leadership, logistics management, tactics, communication, and military writing.

Credit Recommendation: In the lower-division baccalaureate/associate degree category, 3 semester hours in principles of supervision and 2 in supply management (12/88).

AR-2201-0343

Civil Military Operations

Course Number: 4N-48F/500-F3.

Location: John F. Kennedy Special Warfare Center and School, Fort Bragg, NC.

Length: 2 weeks (70–71 hours).

Exhibit Dates: 2/87–9/97.

Learning Outcomes: Upon completion of the course, the student will perform civil-military operations in the field including civil affairs and psychological operations, apply public health considerations, and implement procedures for dealing with refugees and dislocated civilians.

Instruction: Lectures, seminars, and oral presentations cover civil-military operations, psychological operations, several aspects of civil affairs, and cross-cultural communication.

Credit Recommendation: In the lower-division baccalaureate/associate degree category, 2 semester hours in public administration (12/92).

AR-2201-0344

Military Freefall Jumpmaster
(Special Forces Military Freefall Jumpmaster)

Course Number: 2E-F56/011-F15.

Location: John F. Kennedy Special Warfare Center and School, Fort Bragg, NC.

Length: 3 weeks (143–153 hours).

Exhibit Dates: 1/86–Present.

Objectives: Upon completion of the course, the student will be able to use altimeters, automatic rip cord release devices, and canopy control; demonstrate emergency procedures; calculate wind drift; demonstrate proper spotting techniques; supervise packing and rigging; and plan procedures for day and night jumps.

Instruction: This course covers basic principles in free fall parachuting through lectures, discussions, and parachuting.

Credit Recommendation: In the lower-division baccalaureate/associate degree category, 1 semester hour in parachuting/skydiving (11/00).

AR-2201-0345

Military Freefall Parachutist
(Special Forces Military Freefall Parachutist)

Course Number: 2E-SI4X/ASI4X/011-ASIW8.

Location: John F. Kennedy Special Warfare Center and School, Fort Bragg, NC; Wright-Patterson AFB, OH; Shaw AFB, SC.

Length: 4–6 weeks (240–257 hours).

Exhibit Dates: 6/88–Present.

Learning Outcomes: Upon completion of the course, the student will be able to complete free-fall parachute jumps from selected altitudes and under selected conditions.

Instruction: Each successful candidate completes at least 200 hours of laboratory-based, skill acquisition activity in free-fall parachuting. Starting in January 1991, Phase 2 was 3 weeks in length (158 hours), Phase 2 was 2—3 weeks in length (100 hours).

Credit Recommendation: In the lower-division baccalaureate/associate degree category, 1 semester hour in parachuting/sky diving (11/00).

AR-2201-0346

1. Special Forces Detachment Officer
2. Special Forces Detachment Officer Qualification

Course Number: ***Version 1:*** 2E-18A. *Version 2:* 2E-18A.

Location: John F. Kennedy Special Warfare Center and School, Fort Bragg, NC.

Length: ***Version 1:*** 23–24 weeks (1560–1690 hours). *Version 2:* 26 weeks (2201 hours).

Exhibit Dates: ***Version 1:*** 10/87–3/00. *Version 2:* 4/00–Present.

Learning Outcomes: ***Version 1:*** Upon completion of the course, the student will be able to determine, by use of aerial photographs and maps, locations, elevations, desirable routes of travel and location of specific points; practice proper procedures for small boat operation; navigate to designated points; prepare military orders; practice survival techniques; practice proper climbing procedures; practice psychological warfare operations; practice proper cultural interaction with selected populations; conduct orienteering over large areas; use special photographic equipment; specify and use explosive devices for specific targets; use mortars and small arms; and plan estuarine and ocean landing procedures as well as inshore navigation. *Version 2:* Upon completion of the course, the student will be able to determine, by use of aerial photographs and maps, locations, elevations, desirable routes of travel and location of specific points; practice proper procedures for small boat operation; navigate to designated points; prepare military orders; practice survival techniques; practice proper climbing procedures; practice psychological warfare operations; practice proper cultural interaction with selected populations; conduct orienteering over large areas; use special photographic equipment; specify and use explosive devices for specific targets; use mortars and small arms; and plan estuarine and ocean landing procedures as well as inshore navigation.

Instruction: ***Version 1:*** Lectures and exercises cover the use of aerial photographs and maps; small boat handling; air/ground operations; intelligence gathering; climbing; cultural differences and practices; use of explosives, mortars, and small arms; radio and written message procedures; and navigation and landing procedures. *Version 2:* Lectures and exercises cover the use of aerial photographs and maps; small boat handling; air/ground operations; intelligence gathering; climbing; cultural differences and practices; use of explosives, mortars, and small arms; radio and written message procedures; and navigation and landing procedures.

Credit Recommendation: ***Version 1:*** In the vocational certificate category, 1 semester hour in explosive techniques. In the lower-division baccalaureate/associate degree category, 1 semester hour in orienteering (12/92). *Version 2:* In the lower-division baccalaureate/associate degree category, 2 semester hours in military science (4/01).

AR-2201-0347

Special Forces Detachment Officer Qualification Reserve Component Phase 4

Course Number: 2E-18A-RC.
Location: John F. Kennedy Special Warfare Center and School, Fort Bragg, NC.
Length: 3 weeks (184 hours).
Exhibit Dates: 10/88–9/91.
Learning Outcomes: Upon completion of the course, the student will demonstrate skills in the areas of weapons, engineering, communication, medicine, and training management.
Instruction: Lectures, demonstrations, and written and performance exercises cover duty in operations detachments.
Credit Recommendation: Credit is not recommended because of the military-specific nature of the course (12/88).

AR-2201-0348

Special Operations Weapons Sergeant Reserve Component Phase 4

Course Number: 011-18B30-RC.
Location: John F. Kennedy Special Warfare Center and School, Fort Bragg, NC.
Length: 3 weeks (156 hours).
Exhibit Dates: 10/88–9/91.
Learning Outcomes: Upon completion of the course, the student will be able to install, plot firing patterns, and operate mortars, small arms, and antitank weapons.
Instruction: Instruction includes lectures and exercises in mortars, fire control, selected US and foreign small arms, and antitank weapons.
Credit Recommendation: Credit is not recommended because of the military specific nature of the course (12/88).
Related Occupations: 18B.

AR-2201-0349

Special Operations Target Interdiction

Course Number: ***Version 1:*** 2E-F67/011-F19. *Version 2:* 2E-F67/011-ASIW3.
Location: John F. Kennedy Special Warfare Center and School, Fort Bragg, NC.
Length: ***Version 1:*** 6 weeks (274–288 hours). *Version 2:* 6 weeks (290–291 hours).
Exhibit Dates: ***Version 1:*** 6/86–9/99. *Version 2:* 10/99–Present.
Learning Outcomes: ***Version 1:*** Upon completion of the course, the student will be able to demonstrate advanced rifle marksmanship, target selection, and stalking and interdiction techniques. *Version 2:* Upon completion of the course, the student will be able to function as a sniper in field conditions.
Instruction: ***Version 1:*** Lectures, demonstrations, and field experience are used to develop marksmanship skills. *Version 2:* Lectures and field exercises cover mission planning, land navigation, concealment, marksmanship, equipment maintenance and repair, and evasion/escape.
Credit Recommendation: ***Version 1:*** In the lower-division baccalaureate/associate degree category, 1 semester hour in marksmanship/riflery (10/90). *Version 2:* In the lower-division baccalaureate/associate degree category, 3 semester hours in marksmanship and 1 in map reading and land navigation (4/01).

AR-2201-0350

Special Forces Weapons Sergeant
(Special Operations Weapons Sergeant)

Course Number: ***Version 1:*** 011-18B30. *Version 2:* 011-18B30. *Version 3:* 011-18B30.
Location: John F. Kennedy Special Warfare Center and School, Fort Bragg, NC.

Length: ***Version 1:*** 23–24 weeks (1387–1572 hours). *Version 2:* 23–24 weeks (1387–1572 hours). *Version 3:* 26 weeks (1884 hours).

Exhibit Dates: ***Version 1:*** 10/89–9/91. *Version 2:* 10/91–5/99. *Version 3:* 6/99–Present.

Learning Outcomes: ***Version 1:*** Upon completion of the course, the student will be able to determine, by use of aerial photographs and maps, distances, elevations, desirable routes, and location of specific points; practice proper techniques for small boat operation; navigate to designated points; prepare proper military orders; practice survival techniques; plan air drop-zone operations; practice proper procedures for acquisition of intelligence data; direct and fire selected field mortars, small arms, and antitank weapons; and select proper offensive weapons for Special Forces operations. *Version 2:* Upon completion of the course, the student will be able to determine, by use of aerial photographs and maps, distances, elevations, desirable routes, and location of specific points; practice proper techniques for small boat operation; navigate to designated points; prepare proper military orders; practice survival techniques; plan air drop-zone operations; practice proper procedures for acquisition of intelligence data; direct and fire selected field mortars, small arms, and antitank weapons; and select proper offensive weapons for Special Forces operations. *Version 3:* Upon completion of the course, the student will be able to perform land navigation, tactics, weapons, mission planning, and training.

Instruction: ***Version 1:*** Instruction includes lectures and exercises in the use of aerial photographs, military maps, and associated orienteering equipment; navigation and the use of small boats; survival techniques; air/ground operations; and weapon selection. This course contains a common core of leadership subjects. *Version 2:* Instruction includes lectures and exercises in the use of aerial photographs, military maps, and associated orienteering equipment; navigation and the use of small boats; survival techniques; air/ground operations; and weapon selection. This course contains a common core of leadership subjects. *Version 3:* Lectures and practical exercises cover land navigation, weapon systems, small arms, mortars, planning, and training. This course contains a common core of leadership subjects. This course contains a common core of leadership subjects.

Credit Recommendation: ***Version 1:*** In the vocational certificate category, 1 semester hour in explosive techniques. In the lower-division baccalaureate/associate degree category, 1 semester hour in orienteering, 1 in water safety, and 1 in physical training. See AR-1406-0090 for common core credit (12/93). *Version 2:* In the vocational certificate category, 1 semester hour in explosive techniques. In the lower-division baccalaureate/associate degree category, 1 semester hour in survival skills, 1 in orienteering, 1 in water safety, and 1 in physical training. See AR-1406-0090 for common core credit (12/93). *Version 3:* In the lower-division baccalaureate/associate degree category, 1 semester hour in map reading and 3 in land navigation. See AR-1406-0090 for common core credit (4/01).

Related Occupations: 18B.

AR-2201-0352

Military Enlistment Processing Station (MEPS) Guidance Counselor National Guard (ARNG)

Course Number: 501-ASIV7-ARNG.
Location: Soldier Support Institute, Fort Benjamin Harrison, IN; Professional Education Center, Cp. Joseph Robinson, North Little Rock, AR.
Length: 1–2 weeks (46–75 hours).
Exhibit Dates: 10/87–4/04.
Objectives: Upon completion of the course, the student will be able to function as a career guidance counselor for Army National Guard personnel.
Instruction: Lectures and classroom exercises cover enlistment programs, eligibility requirements, and the duties and responsibilities of a guidance counselor.
Credit Recommendation: Credit is not recommended because of the limited, specialized nature of the course (12/91).

AR-2201-0356

Reserve Retention Noncommissioned Officer (NCO) Advanced

Course Number: 501-F8.
Location: Recruiting and Retention School, Soldier Support Institute, Fort Benjamin Harrison, IN.
Length: 2 weeks (74 hours).
Exhibit Dates: 10/84–12/91.
Learning Outcomes: Upon completion of the course, the student will be able to manage the retention function in an Army reserve unit, train subordinate noncommissioned officers in all aspects of the Army reserve retention program, and evaluate some features of an established retention program.
Instruction: Presentation includes classroom instruction and practical exercises in training and evaluation techniques related to an existing retention program.
Credit Recommendation: Credit is not recommended as the topics included are military-specific and lack depth (12/88).

AR-2201-0357

1. Army Reserve Retention Noncommissioned Officer (NCO)
2. Basic Retention and Transition NCO

Course Number: ***Version 1:*** 8600; 501-79D30-USAR. *Version 2:* 921-180.
Location: ***Version 1:*** Reserve Readiness Training Center, Fort McCoy, WI; Recruiting and Retention School, Soldier Support Institute, Fort Benjamin Harrison, IN. *Version 2:* Reserve Readiness Training Center, Fort McCoy, WI.
Length: ***Version 1:*** 2 weeks (72–84 hours). *Version 2:* 3 weeks (120 hours).
Exhibit Dates: ***Version 1:*** 5/86–1/01. *Version 2:* 2/01–Present.
Learning Outcomes: ***Version 1:*** Upon completion of the course, the student will have a good working knowledge of the Army's retention program and be able to function as the retention noncommissioned officer of an Army Reserve unit. *Version 2:* Upon completion of the course, the student will be able to develop a time management work plan; identify the qualifications, duties, roles, and responsibilities of the retention NCO; determine eligibility for immediate re-enlistment; utilize automated systems for retention activities; evaluate retention

data; explain the selected reserve incentive program; conduct a retention and re-enlistment interview; develop a sponsorship program; conduct a briefing; counsel on separations and transfers; conduct career planning; conduct a unit climate profile; conduct retention training; explain the warrant officer program; and conduct a transition interview.

Instruction: *Version 1:* This course includes classroom instruction and practical exercises in retention techniques, publicity efforts, interview skills, career development counseling, and other topics related to retention programs. *Version 2:* Audiovisual materials, practical and classroom exercises, discussion, lecture, and computer-based training. Topics include time management, retention activities, re-enlistment management, automated systems, retention training, transition and retention training plan, warrant officer program, non-regular service retirement, and transition administration and sales.

Credit Recommendation: *Version 1:* Credit is not recommended due to the military-specific nature of the course (8/96). *Version 2:* In the upper-division baccalaureate category, 3 semester hours in workforce management (11/02).

Related Occupations: 79V.

AR-2201-0359

Multiple Launch Rocket System (MLRS)/ Lance Fire Direction Supervisor Advanced Noncommissioned Officer (NCO)

Course Number: *Version 1:* 043-13P40. *Version 2:* 043-13P40.

Location: Field Artillery School, Fort Sill, OK.

Length: *Version 1:* 4–6 weeks (154–244 hours). *Version 2:* 4–6 weeks (154–244 hours).

Exhibit Dates: *Version 1:* 1/89–9/90. *Version 2:* 10/90–12/97.

Learning Outcomes: *Version 1:* Upon completion of the course, the student will be able to lead, evaluate, and train Lance missile fire direction center soldiers in gunnery techniques, fire direction, management, and control. *Version 2:* Upon completion of the course, the student will be able to lead, evaluate, and train Lance missile fire direction center soldiers in gunnery techniques, fire direction, management, and control.

Instruction: *Version 1:* Lectures and group discussions cover leadership, ethics, personnel management, doctrine, and tactics. *Version 2:* Lectures and group discussions cover leadership, ethics, personnel management, doctrine, and tactics. This course contains a common core of leadership subjects.

Credit Recommendation: *Version 1:* Credit is recommended for the common core only. See AR-1404-0035 (6/89). *Version 2:* Credit is recommended for the common core only. See AR-2201-0295 (1/92).

Related Occupations: 13P.

AR-2201-0360

Lance Crewmember Advanced Noncommissioned Officer (NCO)

Course Number: 042-15D40; 042-13N40.

Location: Field Artillery School, Fort Sill, OK.

Length: 4–5 weeks (143–189 hours).

Exhibit Dates: 10/87–12/93.

Learning Outcomes: Upon completion of the course, the student will be able to perform tasks required of a firing platoon sergeant or an assembly and transport platoon sergeant in a Lance battalion.

Instruction: Lectures and practical exercises cover leadership, ethics, personnel management, tactics, maintenance management, Lance technical training, and maintenance of system-specific forms and records. This course contains a common core of leadership subjects.

Credit Recommendation: Credit is recommended for the common core only. See AR-1404-0035 (5/92).

Related Occupations: 13N; 15D.

AR-2201-0362

Cannon Crewmember Basic Noncommissioned Officer (NCO) (Field Artillery Cannon Basic Noncommissioned Officer (NCO))

Course Number: 041-4-13B30; 041-18-13B30; 041-15-13B30; 041-14-13B30; 041-13-13B30; 041-12-13B30; 041-10-13B30; 041-9-13B30; 041-7-13B30; 041-5-13B30; 041-2-13B30; 041-8-13B30.

Location: NCO Academy, Fort Stewart, GA; NCO Academy, 7th Army, Germany; NCO Academy, Schofield Barracks, HI; NCO Academy, Fort Richardson, AK; NCO Academy, Fort Lewis, WA; NCO Academy, Fort Ord, CA; NCO Academy, Fort Carson, CO; NCO Academy, Fort Hood, TX; NCO Academy, Fort Riley, KS; NCO Academy, Fort Campbell, KY; NCO Academy, Fort Bragg, NC; Field Artillery School, Fort Sill, OK.

Length: 4–6 weeks (153–259 hours).

Exhibit Dates: 1/87–12/96.

Learning Outcomes: After 9/00 see AR-2201-0524. Upon completion of the course, the student will be able to lead, train, and direct subordinates in the maintenance, operation, and deployment of weapons and equipment.

Instruction: Lectures and practical exercises cover leadership, maintenance, tactics, communication, land navigation, weapons, and training management. This course contains a common core of leadership subjects.

Credit Recommendation: Credit is recommended for the common core only. See AR-1406-0090 (4/92).

Related Occupations: 13B; 14B.

AR-2201-0363

Fire Support Specialist Advanced Noncommissioned Officer (NCO) (Field Support Advanced Noncommissioned Officer (NCO))

Course Number: 250-13F40.

Location: Field Artillery School, Fort Sill, OK.

Length: 8–9 weeks (295–329 hours).

Exhibit Dates: 3/88–9/94.

Learning Outcomes: After 9/95 see AR-1601-0110. Upon completion of the course, the student will be able to train others in modern battlefield tactics, techniques, and procedures.

Instruction: Instruction covers the capability and organization of the fire support team, including the duties, responsibilities, and missions of fire support team personnel and techniques necessary to perform, lead, train, and evaluate personnel. This course contains a common core of leadership subjects.

Credit Recommendation: Credit is recommended for the common core only. See AR-1404-0035 (1/92).

Related Occupations: 13F.

AR-2201-0364

Lance Crewmember Basic Noncommissioned Officer (NCO)

Course Number: 042-15D30; 042-13N30 (CT); 042-13N30.

Location: Field Artillery School, Fort Sill, OK.

Length: 5–7 weeks (178–270 hours).

Exhibit Dates: 1/87–12/93.

Learning Outcomes: Upon completion of the course, the student will be able to perform the tasks that are required of a firing section chief or an assembly and transport section chief on a Lance Missile firing battery.

Instruction: Lectures and practical exercises cover Lance Missile firing operations, assembly and transport operations, system tactics, special weapons, operator and crew maintenance, organizational maintenance, and maintenance of system-specific forms and records. This course contains a common core of leadership subjects.

Credit Recommendation: Credit is recommended for the common core only. See AR-1406-0090 (1/92).

Related Occupations: 13N; 15D.

AR-2201-0365

Multiple Launch Rocket System (MLRS)/ Lance Operations Fire Direction Specialist Basic Noncommissioned Officer (NCO)

Course Number: 043-13P30.

Location: Field Artillery School, Fort Sill, OK.

Length: 5–8 weeks (209–312 hours).

Exhibit Dates: 10/88–9/97.

Learning Outcomes: Upon completion of the course, the student will be able to perform the tasks required of a liaison sergeant, fire direction operator, or assistant chief fire direction specialist.

Instruction: Lectures and practical exercises cover common noncommissioned officer skills (leadership, tactics, etc.) and specific tasks required in Multiple Launch Rocket System/ Lance operations, digital fire direction, and communications. This course contains a common core of leadership subjects.

Credit Recommendation: In the lower-division baccalaureate/associate degree category, 1 semester hour in maintenance management. See AR-1406-0090 for common core credit (1/92).

Related Occupations: 13P.

AR-2201-0366

Multiple Launch Rocket System (MLRS) Basic Noncommissioned Officer (NCO)

Course Number: 042-13M30.

Location: Field Artillery School, Fort Sill, OK.

Length: 2–4 weeks (76–191 hours).

Exhibit Dates: 10/86–9/97.

Learning Outcomes: Upon completion of the course, the student will be able to perform the tasks that are required of a section chief in a Multiple Launch Rocket System firing and ammunition section.

Instruction: Lectures and practical exercises cover map reading, air/land battle, leadership, military inspections, launcher destruction, safety procedures, and reload operations. This course contains a common core of leadership subjects.

Credit Recommendation: Credit is recommended for the common core only. See AR-1406-0090 (1/92).

Related Occupations: 13M.

AR-2201-0367

Multiple Launch Rocket System (MLRS) Crewmember Advanced Noncommissioned Officer (NCO)

Course Number: 042-13M40.

Location: Field Artillery School, Fort Sill, OK.

Length: 3–4 weeks (132–179 hours).

Exhibit Dates: 1/87–9/98.

Learning Outcomes: Upon completion of the course, the student will be able to perform tasks that are required of a firing platoon sergeant or an ammunition platoon sergeant in a Multiple Launch Rocket System firing battery.

Instruction: Lectures and practical exercises cover message device operation; ammunition resupply; platoon defense; combat service support; and preparation of messages, maps, and charts. This course contains a common core of leadership subjects.

Credit Recommendation: Credit is recommended for the common core only. See AR-1404-0035 (1/92).

Related Occupations: 13M.

AR-2201-0368

Field Artillery Cannon Battery Officer

Course Number: 2E-13E.

Location: Field Artillery School, Fort Sill, OK.

Length: 6 weeks (280 hours).

Exhibit Dates: 8/88–12/92.

Learning Outcomes: Upon completion of the course, the student will be able to lead in cannon battery supply operations and maintenance and fire direction of a cannon battery field unit.

Instruction: Lectures and practical exercises cover cannon systems, fire direction, battery operations, and maintenance and supply.

Credit Recommendation: In the lower-division baccalaureate/associate degree category, 2 semester hours in maintenance management (9/94).

AR-2201-0369

Tactical Fire (TACFIRE) Operation Specialist Advanced Noncommissioned Officer (NCO) Advanced
(Field Artillery Tactical Fire (TACFIRE) Noncommissioned Officer (NCO) Advanced)

Course Number: 250-13C40.

Location: Field Artillery School, Fort Sill, OK.

Length: 3–7 weeks (137–256 hours).

Exhibit Dates: 11/87–9/98.

Learning Outcomes: After 9/00 see AR-1408-0265. Upon completion of the course, the student will be able to train selected personnel in modern battlefield tactics, techniques, and procedures.

Instruction: Practical applications include automatic data processing and the duties, responsibilities, and mission involved in a computer-controlled fire direction center. This course contains a common core of leadership subjects.

Credit Recommendation: Credit is recommended for the common core only. See AR-1404-0035 (3/92).

Related Occupations: 13C.

AR-2201-0370

Pershing Missile Crewmember Basic Noncommissioned Officer (NCO)

Course Number: 043-15E30.

Location: Field Artillery School, Fort Sill, OK.

Length: 3 weeks (104 hours).

Exhibit Dates: 7/88–12/91.

Learning Outcomes: Upon completion of the course, the student will be able to prepare missile sections for assembly, inspect and install initiators and detonators, and direct countdown operations in the event of a countdown malfunction.

Instruction: Lectures and practical exercises cover launcher operations, missile assembly/disassembly, handling countdown malfunctions, and countdown operations. This course contains a common core of leadership subjects.

Credit Recommendation: Credit is recommended for the common core only. See AR-1406-0090 (6/89).

Related Occupations: 15E.

AR-2201-0371

Light Missile Field Artillery (Lance)

Course Number: 2F-13B.

Location: Field Artillery School, Fort Sill, OK.

Length: 5–6 weeks (211 hours).

Exhibit Dates: 9/88–12/92.

Learning Outcomes: Upon completion of the course, the student will be able to operate, maintain, and deploy the Lance missile system, including component identification, system capabilities, preventive maintenance, safety, transport, targeting, and firing.

Instruction: Lectures and field exercises cover the capabilities of the missile system, system components, safety, transport, targeting, and firing.

Credit Recommendation: Credit is not recommended because of the military-specific nature of the course (6/89).

Related Occupations: 13B.

AR-2201-0372

Cannon Crewman

Course Number: 13B10-OSUT; 041-13B10.

Location: Field Artillery School, Fort Sill, OK.

Length: 5–13 weeks (182 hours).

Exhibit Dates: 7/88–9/94.

Learning Outcomes: Upon completion of the course, the student will be able to perform basic combat survival skills, load and fire howitzers, and perform preventive and other maintenance on artillery equipment.

Instruction: Lectures and field exercises cover the maintenance and use of howitzers including loading and firing. This course includes an 8-9 week Basic Training component.

Credit Recommendation: Credit is not recommended because of the military-specific nature of the course (6/89).

Related Occupations: 13B.

AR-2201-0374

Marine Corps Battery Commander

Course Number: 2E-FOA-F90 (USMC).

Location: Field Artillery School, Fort Sill, OK.

Length: 2 weeks (76 hours).

Exhibit Dates: 1/87–Present.

Learning Outcomes: Upon completion of the course, the student will be able to conduct fire support planning and coordination for amphibious operations.

Instruction: This is a refresher course with lectures, discussions, and exercises on target acquisition, surveying, communications, fire mission planning, and command responsibilities.

Credit Recommendation: Credit is not recommended because of the military-specific nature of the course (8/95).

AR-2201-0375

Lance Crewmember
(Lance Missile Crewmember)

Course Number: 042-15D10; 042-13N10.

Location: Field Artillery School, Fort Sill, OK.

Length: 6–7 weeks (232–254 hours).

Exhibit Dates: 1/87–12/92.

Learning Outcomes: Upon completion of the course, the student will be able to perform preventive maintenance, position and set up equipment, fire missiles, perform safety procedures, and operate the radio set on the Lance missile system.

Instruction: Lectures and performance exercises cover the operation and maintenance of Lance missile system equipment, including assembly, transportation, and daily preventive maintenance.

Credit Recommendation: Credit is not recommended because of the military-specific nature of the course (6/89).

Related Occupations: 13N; 15D.

AR-2201-0376

Field Artillery Surveyor Basic Noncommissioned Officer (NCO)

Course Number: 412-82C30.

Location: Field Artillery School, Fort Sill, OK.

Length: 6–8 weeks (205–314 hours).

Exhibit Dates: 1/86–9/97.

Learning Outcomes: After 9/00 see AR-1601-0116. Upon completion of the course, the student will be able to perform as section chief in artillery and target acquisition battery survey sections.

Instruction: Lectures and practical exercises cover astronomic observation, field artillery survey, solution of triangles, location of traverse errors, triangulation and resection, leadership, training management, resource management, and military skills. This course contains a common core of leadership subjects.

Credit Recommendation: In the lower-division baccalaureate/associate degree category, 3 semester hours in general surveying and

1 for basic field practice in surveying. See AR-1406-0090 for common core credit (1/92).
Related Occupations: 82C.

AR-2201-0377

OH-58D Field Artillery Instructor Pilot Fire Support

Course Number: 2E-F130.
Location: Field Artillery School, Fort Sill, OK.
Length: 2 weeks (69–77 hours).
Exhibit Dates: 1/87–12/92.
Learning Outcomes: Upon completion of the course, the student will be able to serve as an instructor pilot at division, field, and brigade artillery levels and provide instruction and perform evaluations of aerial fire support observers.
Instruction: Lectures and practical exercises cover tactical missions, artillery fire planning, message procedures, targeting, illumination, and ground/air coordination.
Credit Recommendation: Credit is not recommended because of the military-specific nature of the course (10/89).

AR-2201-0378

Professional Artillery Refresher
(Special Career Level Artillery Advanced Gunnery (USMC))

Course Number: 2E-F49 (USMC); 2E-F49.
Location: Field Artillery School, Fort Sill, OK.
Length: 3–6 weeks (115–195 hours).
Exhibit Dates: 6/88–Present.
Learning Outcomes: Upon completion of the course, the student will be able to perform the supervisory duties of a battalion fire direction officer or a battery commander.
Instruction: This is a refresher course on artillery fire planning, coordination, and control. Lectures and class discussions cover the management of a field artillery unit using manual and computer-assisted techniques.
Credit Recommendation: Credit is not recommended because of the military-specific nature of the course (8/95).

AR-2201-0379

Nuclear Cannon Assembly

Course Number: 4F-F5/041-ASIJ4.
Location: Field Artillery School, Fort Sill, OK.
Length: 2 weeks (68 hours).
Exhibit Dates: 3/88–12/92.
Learning Outcomes: Upon completion of the course, the student will be able to perform inspection, unpackaging, and prefire procedures of the M454, M753, and M422 atomic projectiles.
Instruction: Lectures and practical exercises cover the procedures for assembly, prefire, and disassembly of the projectile; weapon maintenance; safety; accident/incident response; and tie-down procedures.
Credit Recommendation: Credit is not recommended because of the military-specific nature of the course (6/89).

AR-2201-0380

Automated Fire Support Leader

Course Number: 2E-F145.
Location: Field Artillery School, Fort Sill, OK.
Length: 6 weeks (205 hours).
Exhibit Dates: 2/89–6/99.
Learning Outcomes: Upon completion of the course, the student will be able to supervise fire support sections in the use of automated equipment that enhances fire support coordination, planning, and execution.
Instruction: Lectures and practical exercises cover meteorological data, message devices, the tactical fire system, fire support coordination, intelligence data, targeting, fire planning, and fire support execution.
Credit Recommendation: Credit is not recommended because of the military-specific nature of the course (6/89).

AR-2201-0381

Nuclear Warhead Detachment

Course Number: 2E-F96; 2E-F96/041-F8.
Location: Field Artillery School, Fort Sill, OK.
Length: 3 weeks (108 hours).
Exhibit Dates: 1/86–12/93.
Learning Outcomes: Upon completion of the course, the student will be able to operate Lance and Pershing missiles using proper security, response to terrorist threats, warhead inspection, and safety considerations.
Instruction: Lectures and practical exercises cover security, courier duties, publications, threats, Lance operations, Pershing operations, weapon storage, and handling and safety rules.
Credit Recommendation: Credit is not recommended because of the military-specific nature of the course (6/89).

AR-2201-0383

Battery Computer System Cadre Reserve Component

Course Number: 2E-F71/250-F16.
Location: Field Artillery School, Fort Sill, OK.
Length: 2–3 weeks (72–108 hours).
Exhibit Dates: 1/85–6/99.
Learning Outcomes: Upon completion of the course, the student will be able to connect the various components of a mobile unit computer system, perform proper procedure to power-up and load computer programs, initialize and construct data bases, perform procedures necessary to process fire missions, and transmit gun commands in an autonomous mode of operation.
Instruction: This is primarily a hands-on course providing training in use and troubleshooting techniques for the battery computer system.
Credit Recommendation: Credit is not recommended because of the military-specific nature of the course (6/89).

AR-2201-0384

1. All-Source Intelligence Technician Warrant Officer Technical/ Tactical Certification, Phase 2
   (Warrant Officer Technical/Tactical Certification, Phase 2 350B All-Source Intelligence Technician)
   (Warrant Officer Technical/Tactical Certification, Phase 2 MOS 350B Order of Battle Technician)
2. All-Source Intelligence Technician-Warrant Officer Technical Tactical Certification (WOTTC), Phase 2

Course Number: *Version 1:* 3A-350B. *Version 2:* 3B-350B.
Location: *Version 1:* Intelligence School, Fort Huachuca, AZ. *Version 2:* Intelligence Center and School, Fort Huachuca, AZ.
Length: *Version 1:* 2–3 weeks (92 hours). *Version 2:* 5 weeks (185 hours).
Exhibit Dates: *Version 1:* 11/88–9/99. *Version 2:* 10/99–Present.
Learning Outcomes: *Version 1:* Upon completion of the course, the student will be updated on new equipment and technology, identify intelligence indicators, prepare an order of battle briefing, and recognize various threats. *Version 2:* Upon completion of the course, the student will be updated on new equipment and technology, identify intelligence indicators, prepare an order of battle briefing, and recognize various threats.
Instruction: *Version 1:* Lectures and classroom testing cover the subject matter. *Version 2:* Lectures and classroom testing cover the subject matter.
Credit Recommendation: *Version 1:* Credit is not recommended because of the military-specific nature of the course (1/97). *Version 2:* Credit is not recommended because of the military-specific nature of the course (8/01).
Related Occupations: 350B.

AR-2201-0385

Battlefield Deception

Course Number: 3A-F56/230-ASID4.
Location: Intelligence School, Fort Huachuca, AZ.
Length: 6–7 weeks (235–246 hours).
Exhibit Dates: 10/88–4/04.
Learning Outcomes: Upon completion of the course, the student will be able to perform deception planning, identify electronic deception, identify and explain the need for sonic deception, perform and evaluate a deception operation, and read maps.
Instruction: Lectures, classroom testing, and field tests cover the above material.
Credit Recommendation: Credit is not recommended because of the military-specific nature of the course (6/90).

AR-2201-0386

Battalion/Brigade Signal Officer

Course Number: 4C-SI6B.
Location: Signal School, Fort Gordon, GA.
Length: 9 weeks (324 hours).
Exhibit Dates: 8/89–10/94.
Learning Outcomes: Upon completion of the course, the student will be able to supervise the installation, operation, and maintenance of communications systems in nonsignal units.
Instruction: Lectures and practical exercises cover communications procedures, AM and FM radio equipment, radio fundamentals, and troubleshooting.
Credit Recommendation: Credit is not recommended because of the military-specific nature of the course (6/90).

AR-2201-0387

Leadership Development

Course Number: PEC-T-O-LDC.
Location: Professional Education Center, Cp. Joseph Robinson, North Little Rock, AR.
Length: 2 weeks (76 hours).
Exhibit Dates: 1/90–12/97.

Learning Outcomes: Upon completion of the course, the student will be able to provide information on direct- and indirect-level leadership based upon human relations and conceptual skills.

Instruction: Lectures, discussions, group exercises, and practical experience are the methods of instruction. The topics covered include brain dominance, conflict management, counseling, delegating, communication, stress management, team building, time management, visioning, negotiating, and problem solving.

Credit Recommendation: Credit is not recommended because there is no provision for formal student evaluation (7/90).

AR-2201-0388

NATIONAL GUARD (ARNG) RETENTION NONCOMMISSIONED OFFICER (NCO)

Course Number: 501-79D30 (ARNG).

Location: Professional Education Center, Cp. Joseph Robinson, North Little Rock, AR.

Length: 2 weeks (72 hours).

Exhibit Dates: 10/88–12/97.

Learning Outcomes: After 9/96 see AR-1406-0188. Upon completion of the course, the student will be able to effectively use presentation, communication, conferencing, and interviewing skills to deliver and receive appropriate information dealing with employee satisfaction levels, desires, and opportunities.

Instruction: Performance-based instruction is used with significant homework assignments, individual presentations, and small-group interviewing and conferencing. Evaluation is accomplished through exercises and examinations.

Credit Recommendation: In the lower-division baccalaureate/associate degree category, 1 semester hour in personnel management (11/91).

Related Occupations: 79D.

AR-2201-0389

AMMUNITION SPECIALIST

Course Number: 645-55B10.

Location: Missile and Munitions School, Redstone Arsenal, AL.

Length: 6 weeks (212 hours).

Exhibit Dates: 10/90–2/94.

Learning Outcomes: After 2/94 see AR-2201-0465. Upon completion of the course students will be able to identify and handle various ammunition material; prepare, store, and issue ammunition; perform explosive decontamination; and perform emergency destruction of ammunition.

Instruction: Conferences and practical exercises cover ammunition material, service procedures, storage, handling, movement, inspection, maintenance, and destruction.

Credit Recommendation: In the lower-division baccalaureate/associate degree category, 3 semester hours in safety risk management (8/91).

Related Occupations: 55B.

AR-2201-0390

AMMUNITION INSPECTOR

Course Number: 645-55X30.

Location: Missile and Munitions School, Redstone Arsenal, AL.

Length: 7–8 weeks (277 hours).

Exhibit Dates: 10/90–9/92.

Learning Outcomes: Upon completion of the course, the student will be able to identify ammunition; inspect field storage; prepare automated system documents; inspect handling operations, artillery ammunition, land mines, and guided missiles; prepare samples for shipment; and perform malfunction investigations.

Instruction: Conferences and practical exercises cover ammunition material; rocket and guided missile inspection and maintenance; storage, handling, and transportation of ammunition; surveillance and maintenance of ammunition; and demilitarization and destruction of ammunition.

Credit Recommendation: In the lower-division baccalaureate/associate degree category, 3 semester hours in safety risk management (8/91).

Related Occupations: 55X.

AR-2201-0391

NUCLEAR WEAPONS TECHNICIAN SENIOR WARRANT OFFICER

Course Number: 4-9-C32-911A.

Location: Missile and Munitions School, Redstone Arsenal, AL.

Length: 3–7 weeks (109–234 hours).

Exhibit Dates: 10/88–1/96.

Learning Outcomes: Upon completion of the course, the student will be able to account for a nuclear stock pile.

Instruction: Annex A includes lectures and practical exercises in military leadership, communications, command climate, rules and codes of war, military justice, and force integration. Annex B includes lectures and discussions of duties and responsibilities, nuclear systems updates, DoE-DoD interrelationships, nuclear weapons support, ionizing radiation protection, and nuclear safety.

Credit Recommendation: Credit is not recommended because of the military-specific nature of the course (8/91).

Related Occupations: 911A.

AR-2201-0392

AMMUNITION SPECIALIST BASIC NONCOMMISSIONED OFFICER (NCO)

Course Number: 645-55B30.

Location: Missile and Munitions School, Redstone Arsenal, AL.

Length: 5–7 weeks (203–243 hours).

Exhibit Dates: 10/87–9/92.

Learning Outcomes: After 9/92 see AR-2201-0467. Upon completion of the course, the student will be able to identify, receive, store, issue, and destroy ammunition; use a computer for ammunition inventory control; and operate an ammunition supply point.

Instruction: Lectures and practical exercises cover threat recognition; ammunition identification; ammunition storage, planning and handling procedures; inventory control; and fire fighting. This course contains a common core of leadership subjects.

Credit Recommendation: Credit is recommended for the common core only. See AR-1406-0090 (10/91).

Related Occupations: 55B.

AR-2201-0393

AMMUNITION INSPECTOR ADVANCED NONCOMMISSIONED OFFICER (NCO)

Course Number: 645-55X40.

Location: Missile and Munitions School, Redstone Arsenal, AL.

Length: 8–9 weeks (284 hours).

Exhibit Dates: 4/90–1/96.

Learning Outcomes: Upon completion of the course, the student will be able to supervise subordinates in the processing of ammunition records and reports and in the storage, inventory, warehousing, disposal, and maintenance of ammunition.

Instruction: This course includes lectures, practical exercises, and small-group instruction in marksmanship, ammunition support, personnel utilization, and document preparation. This course contains a common core of leadership subjects.

Credit Recommendation: Credit is recommended for the common core only. See AR-1404-0035 (8/91).

Related Occupations: 55X.

AR-2201-0394

AMMUNITION TECHNICIAN WARRANT OFFICER TECHNICAL AND TACTICAL CERTIFICATION RESERVE (PHASE 2 AND 4)

Course Number: 4E-910A-RC.

Location: Missile and Munitions School, Redstone Arsenal, AL.

Length: 2 weeks (206 hours).

Exhibit Dates: 10/88–Present.

Learning Outcomes: Upon completion of the course, the student will be able to perform ammunition support, develop ammunition operating procedures, prepare malfunction reports, identify characteristics of toxic and nontoxic chemical agents, and operate a decontamination station.

Instruction: Lectures and practical exercises cover ammunition support and operating procedures, writing malfunction reports, leak sealing for defective chemical weapons, and decontamination center operation. Phase 4 includes the use of a laptop computer system and basic peripherals and software management.

Credit Recommendation: Credit is not recommended because of the military-specific nature of the course (5/01).

Related Occupations: 910A.

AR-2201-0395

AMMUNITION SPECIALIST ADVANCED NONCOMMISSIONED OFFICER (NCO) RESERVE

Course Number: 093-55B40-RC.

Location: Reserve Component Training Centers, US; Missile and Munitions School, Redstone Arsenal, AL.

Length: 2 weeks (98 hours).

Exhibit Dates: 4/89–6/00.

Learning Outcomes: Upon completion of the course, the student will be able to develop a security plan, locate documentation errors, understand error codes, use correct log-in procedures, and describe deployment procedures for vans.

Instruction: Lectures and practical exercises cover site plan preparations, standard Army ammunition system, recovery of data after a power failure, system log-in, and password techniques. This course contains a common core of leadership subjects.

Credit Recommendation: Credit is recommended for the common core only. See AR-2201-0295 (8/91).

Related Occupations: 55B.

AR-2201-0396

AMMUNITION SPECIALIST RESERVE COMPONENT PHASE 2

Course Number: 645-55X30-RC.

Location: Missile and Munitions School, Redstone Arsenal, AL.

Length: 2 weeks (97 hours).

Exhibit Dates: 10/90–6/00.

Learning Outcomes: Upon completion of the course, the student will be able to inspect ammunition, handle ammunition and issue receipts, prepare automated documents, classify grenades and mines and record defects.

Instruction: Lectures and practical exercises in ammunition safety, microfiche readers and data extraction, ammunition storage ammunition defect determination and shipping and storage container inspection.

Credit Recommendation: Credit is not recommended because of the military-specific nature of the course (8/91).

Related Occupations: 55X.

AR-2201-0397

AMMUNITION STOCK CONTROL AND ACCOUNTING SPECIALIST BASIC NONCOMMISSIONED OFFICER (NCO) RESERVE

Course Number: 093-55R30.

Location: Reserve Component Training Centers, US; Missile and Munitions School, Redstone Arsenal, AL.

Length: 2 weeks (82 hours).

Exhibit Dates: 4/89–1/96.

Learning Outcomes: Upon completion of the course, the student will be able to compose reports normally encountered on the job at the Corps Material Management Center.

Instruction: Practical exercises cover operating an automatic data processing system, requisitioning ammunition, producing management reports, preparing stock records, and producing inventory control forms. This course contains a common core of leadership subjects.

Credit Recommendation: Credit is recommended for the common core only. See AR-2201-0337 (8/91).

Related Occupations: 55R.

AR-2201-0398

1. AMMUNITION ADVANCED NONCOMMISSIONED OFFICER (NCO)
2. NUCLEAR WEAPONS SPECIALIST ADVANCED NONCOMMISSIONED OFFICER (NCO)

Course Number: *Version 1:* 6-55-C42A; 645-55B/B40. *Version 2:* 645-55B/R40.

Location: Missile and Munitions School, Redstone Arsenal, AL.

Length: *Version 1:* 13–14 weeks (489 hours). *Version 2:* 7 weeks (248 hours).

Exhibit Dates: *Version 1:* 5/90–3/91. *Version 2:* 4/91–1/96.

Learning Outcomes: *Version 1:* Upon completion of the course, the student will be able to recognize stress in prisoners of war; conduct nuclear, biological, and chemical defense; describe storage standards; plan disposal operations; report malfunctions; review documents; use computer-based inventory control; and maintain stock levels. *Version 2:* Upon completion of the course, the student will be able to recognize stress in prisoners of war; conduct nuclear, biological, and chemical defense; describe storage standards; plan disposal operations; report malfunctions; review documents; use computer-based inventory control; and maintain stock levels.

Instruction: *Version 1:* Lectures and extensive practical exercises cover POW survival, ammunition storage standards, disposal operations, maintenance operations, ammunition logistics, leadership characteristics, and battlefield operations. This course contains a common core of leadership subjects. *Version 2:* Lectures and extensive practical exercises cover POW survival, ammunition storage standards, disposal operations, maintenance operations, ammunition logistics, leadership characteristics, and battlefield operations. This course contains a common core of leadership subjects.

Credit Recommendation: *Version 1:* In the lower-division baccalaureate/associate degree category, 3 semester hours in principles of management and 1 in communications. See AR-1404-0035 for common core credit (8/91). *Version 2:* Credit is recommended for the common core only. See AR-1404-0035 (8/91).

Related Occupations: 55B; 55R.

AR-2201-0399

BASIC COMBAT TRAINING
(Basic Training)
(Recruit Training)

Course Number: *Version 1:* 750-BT; 21-114, Phase 1. *Version 2:* 21-114, Phase 1; 750-BT.

Location: *Version 1:* Training Center, Fort Sill, OK; Training Center, Fort McClellan, AL; Training Center, Fort Leonard Wood, MO; Training Center, Fort Knox, KY; Training Center, Fort Jackson, SC; Training Center, Fort Dix, NJ; Training Center, Fort Bliss, TX; Training Center, Fort Benning, GA. *Version 2:* Training Center, Fort Jackson, SC; Training Center, Fort Knox, KY; Training Center, Fort Leonard Wood, MO; Training Center, Fort Sill, OK; Infantry School, Fort Benning, GA.

Length: *Version 1:* 8–9 weeks (370–425 hours). *Version 2:* 9 weeks (464 hours).

Exhibit Dates: *Version 1:* 7/85–2/00. *Version 2:* 3/00–Present.

Learning Outcomes: *Version 1:* Upon completion of the course, the recruit will be able to demonstrate general knowledge of military organization and culture, mastery of individual and group combat skills including marksmanship and first aid, achievement of minimal physical conditioning standards, and application of basic safety and living skills in an outdoor environment. *Version 2:* Upon completion of the course, the student will be able to demonstrate the skills necessary for survival in a combat environment including marksmanship, physical conditioning, navigation, and combat techniques.

Instruction: *Version 1:* Instruction includes lectures, demonstrations, and performance exercises in basic military culture/subjects, including marksmanship, physical conditioning, first aid, and outdoor adaptation/living skills. *Version 2:* Audio-visual materials, practical and classroom exercises, discussion, and lecture. Topics include physical fitness, marksmanship, communication skills, first aid, combat skills, tactics, navigation, N.B.C. weapons, and testing.

Credit Recommendation: *Version 1:* In the lower-division baccalaureate/associate degree category, 1 semester hour in personal physical conditioning, 1 in outdoor skills practicum, 1 in marksmanship, and 1 in first aid (10/00). *Version 2:* In the lower-division baccalaureate/associate degree category, 2 semester hours in physical conditioning, 2 in marksmanship, and 1 in first aid (10/02).

AR-2201-0401

SPECIAL FORCES WEAPONS SERGEANT ADVANCED NONCOMMISSIONED OFFICER (NCO)

Course Number: *Version 1:* 011-18B40. *Version 2:* 011-18B40. *Version 3:* 011-18B40.

Location: John F. Kennedy Special Warfare Center and School, Fort Bragg, NC.

Length: *Version 1:* 12 weeks (524–543 hours). *Version 2:* 18–19 weeks (1051 hours). *Version 3:* 18–19 weeks (737 hours).

Exhibit Dates: *Version 1:* 6/89–9/94. *Version 2:* 10/94–10/98. *Version 3:* 11/98–Present.

Learning Outcomes: *Version 1:* Upon completion of the course, the student will be able to serve in selected leadership and staff positions. *Version 2:* Upon completion of the course, the student will be able to serve in selected leadership and staff positions in a Special Forces detachment. *Version 3:* Upon completion of the course, the student will be able to demonstrate advanced leadership skills in the areas of Special Forces planning, operations, intelligence, military tactics, and photography.

Instruction: *Version 1:* Lectures and practical exercises cover leadership, military studies, professional skills, resource management, effective communication, operations, tactics, psychological operations, unconventional warfare planning, and operational planning and techniques. *Version 2:* Lectures and practical exercises cover leadership, military studies, professional skills, communications operations, tactics, photography, unconventional warfare planning, operational planning and techniques, and intelligence collection and processing. This includes a common core of leadership subjects. *Version 3:* Lectures and practical exercises cover small-group leadership, effective communication, operations, tactics, psychological operations, planning, and intelligence collection. This includes a common core of leadership subjects.

Credit Recommendation: *Version 1:* Credit is not recommended because of the limited, specialized nature of the course (12/92). *Version 2:* In the lower-division baccalaureate/associate degree category, 3 semester hours in military science and 1 in photography. See AR-1404-0035 for common core credit (6/96). *Version 3:* In the lower-division baccalaureate/associate degree category, 1 semester hour in photography and 3 in military intelligence. See AR-1404-0035 for common core credit (4/01).

Related Occupations: 18B.

AR-2201-0402

SPECIAL FORCES WEAPONS SERGEANT ADVANCED NONCOMMISSIONED OFFICER (NCO) RESERVE COMPONENT PHASE 2

Course Number: 011-18B40-RC.

Location: John F. Kennedy Special Warfare Center and School, Fort Bragg, NC.
Length: 3 weeks (209 hours).
Exhibit Dates: 8/91–4/04.
Learning Outcomes: Upon completion of the course, the student will be able to serve in selected leadership and staff positions.
Instruction: Lectures and practical exercises cover operations, planning, training, intelligence, and organization.
Credit Recommendation: Credit is not recommended because of the limited, specialized nature of the training (12/92).
Related Occupations: 18B.

AR-2201-0403

SPECIAL FORCES ENGINEER SERGEANT ADVANCED NONCOMMISSIONED OFFICER (NCO) RESERVE COMPONENT PHASE 2

Course Number: 011-18C40-RC.
Location: John F. Kennedy Special Warfare Center and School, Fort Bragg, NC.
Length: 3 weeks (209 hours).
Exhibit Dates: 8/91–4/04.
Learning Outcomes: Upon completion of the course, the student will be able to demonstrate basic leadership and professional skills in resource management, communications, tactics, psychological operations, unconventional warfare planning, operational planning, intelligence collection, and basic administration.
Instruction: Lectures, demonstrations, and practical exercises cover leadership skills, communications, psychological operations, unconventional warfare planning, and administration.
Credit Recommendation: Credit is not recommended because of the military-specific course content (1/93).
Related Occupations: 18C.

AR-2201-0404

SPECIAL FORCES ENGINEER SERGEANT ADVANCED NONCOMMISSIONED OFFICER (NCO)

Course Number: ***Version 1:*** 011-18C40. *Version 2:* 011-18C40. *Version 3:* 011-18C40.
Location: John F. Kennedy Special Warfare Center and School, Fort Bragg, NC.
Length: ***Version 1:*** 12 weeks (524–543 hours). *Version 2:* 18–19 weeks (1075 hours). *Version 3:* 18–19 weeks (737 hours).
Exhibit Dates: ***Version 1:*** 6/89–9/94. *Version 2:* 10/94–10/98. *Version 3:* 11/98–Present.
Learning Outcomes: ***Version 1:*** Upon completion of the course, the student will be able to demonstrate basic leadership and professional skills in the areas of resource management, communication, tactics, psychological operations, unconventional warfare planning, operational planning, intelligence collection, and basic administration. *Version 2:* Upon completion of the course, the student will be able to function at a advanced level in leadership and technical training. *Version 3:* Upon completion of the course, the student will demonstrate advanced leadership and professional skills in the area of Special Forces planning, operations, intelligence, field engineering, construction, and demolition techniques; basic photography; and military tactics.
Instruction: ***Version 1:*** Lectures, demonstrations, and practical exercises cover leadership skills, communication, psychological operations, unconventional warfare planning, and administration. This course contains a common core of leadership subjects. *Version 2:* Lectures, discussions, and practical exercises cover group dynamics, military tactics, historical battles, photography, and military etiquette. This course contains a common core of leadership subjects. *Version 3:* Lecture, demonstration, and experiential exercises include a common core of group dynamics and leadership skills, construction and demolition techniques, and photography.
Credit Recommendation: ***Version 1:*** Credit is recommended for the common core only. See AR-1404-0035 (1/93). *Version 2:* In the lower-division baccalaureate/associate degree category, 3 semester hours in military science and 1 in photography. See AR-1404-0035 for common core credit (6/96). *Version 3:* In the lower-division baccalaureate/associate degree category, 3 semester hours in field engineering technology and 1 in basic photography. See AR-1404-0035 for common core credit (4/01).
Related Occupations: 18C.

AR-2201-0405

SAPPER LEADER

Course Number: 2E-F73/030-F3.
Location: Engineer School, Fort Leonard Wood, MO.
Length: 4 weeks (149 hours).
Exhibit Dates: 3/92–4/04.
Learning Outcomes: Upon completion of the course the student will be proficient in demolitions, mountaineering, water and aerial operations, troop leading procedures, and specialized battle techniques.
Instruction: Lectures and practical exercises cover demolitions (conventional and expedient); aerial, water, and mountaineering operations; troop leadership; and battle drills.
Credit Recommendation: Credit is not recommended because of the military-specific nature of the course (1/93).

AR-2201-0406

QUARTERMASTER OFFICER BRANCH QUALIFICATION

Course Number: 8-10-C20-BQ.
Location: Quartermaster Center and School, Fort Lee, VA.
Length: 4 weeks (146 hours).
Exhibit Dates: 10/91–8/94.
Learning Outcomes: Upon completion of the course, the student will be able to use automated systems to manage food service, mortuary affairs, procurement and contracting, petroleum and water supply, and other supply operations.
Instruction: Classroom lectures, practical problems, demonstrations, and examinations are course methods.
Credit Recommendation: Credit is not recommended because of the military-specific nature of the course (5/02).

AR-2201-0407

QUARTERMASTER OFFICER ADVANCED RESERVE, PHASE 2

Course Number: 8-10-C23, Phase 2.
Location: Quartermaster Center and School, Fort Lee, VA.
Length: 2 weeks (108–114 hours).
Exhibit Dates: 10/92–Present.
Learning Outcomes: Upon completion of the course, the student will be able to describe the principles of war, offensive/defensive operations, supply function, and types of logistics; apply leadership skills: and use automated systems.
Instruction: Classroom lectures, practical problems, and examinations are course methods.
Credit Recommendation: Credit is not recommended because of the military-specific nature of the course (5/02).

AR-2201-0408

MULTIPLE LAUNCH ROCKET SYSTEM (MLRS) FAMILY OF MUNITIONS

Course Number: 2F-F16/042-ASIY1 (13M).
Location: Field Artillery School, Fort Sill, OK.
Length: 1–2 weeks (64 hours).
Exhibit Dates: 10/90–12/93.
Learning Outcomes: Upon completion of the course, the student will be able to operate and supervise the operation of the MLRS.
Instruction: Lectures and practical exercises cover fire control systems, fire missions, position calibration, and system troubleshooting.
Credit Recommendation: Credit is not recommended because of the military-specific nature of the course (9/94).
Related Occupations: 13M.

AR-2201-0409

MARINE CORPS FIELD ARTILLERY CHIEF

Course Number: AQD.
Location: Field Artillery School, Fort Sill, OK.
Length: 3 weeks (147 hours).
Exhibit Dates: 10/90–Present.
Learning Outcomes: Upon completion of the course, the student will be able to emplace, operate, test, and maintain the artillery weapon system; apply survey techniques; and select and control ammunition velocity.
Instruction: Lectures, field applications, written exams, and performance exams cover map reading, hasty surveys, fire alignment, and position defense.
Credit Recommendation: Credit is not recommended because of the military-specific nature of the course (9/94).

AR-2201-0410

MARINE CORPS CANNON CREWMAN

Course Number: O82; 041-0881-OS.
Location: Field Artillery School, Fort Sill, OK.
Length: 5 weeks (184–188 hours).
Exhibit Dates: 11/92–Present.
Learning Outcomes: Upon completion of the course, the student will be able to load and fire howitzers, perform preventive maintenance, determine cannon positioning, and use the gun display unit.
Instruction: Lectures and field exercises cover the use and maintenance of howitzers, including loading, firing, and positioning.
Credit Recommendation: Credit is not recommended because of the military-specific nature of the course (1/97).

AR-2201-0412

Battle Staff Noncommissioned Officer (NCO)

Course Number: 250-ASI2S.
Location: Sergeants Major Academy, Fort Bliss, TX.
Length: 6 weeks (259–260 hours).
Exhibit Dates: 1/91–5/98.
Learning Outcomes: Upon completion of the course, the student will be able to apply management principles; support senior officers in a headquarters or a field environment by providing information and recommendations on battlefield operations; supervise people, operations and logistics; and employ all current automated battlefield communications systems.
Instruction: This course includes performance exercises in management, including planning, organization, leadership, human relations, motivation, oral and written communication, public speaking, small-group communication and interaction, resource management, and the use of automated information systems.
Credit Recommendation: In the lower-division baccalaureate/associate degree category, 1 semester hour in health and fitness, 1 in resource management, and 1 in introduction to information systems. In the upper-division baccalaureate category, 1 semester hour in operations management (6/95).

AR-2201-0413

Reserve Component Multifunctional Combat Service Support

Course Number: ALMC-RC.
Location: Army Logistics Management College, Fort Lee, VA.
Length: 2 weeks (87–88 hours).
Exhibit Dates: 1/95–5/01.
Learning Outcomes: Upon completion of the course, the student will be able to assume duties in battlefield combat sustainment while retaining combat readiness.
Instruction: Instruction is primarily a study of field manuals.
Credit Recommendation: Credit is not recommended because of the military-specific nature of the course (8/95).

AR-2201-0414

Support Operations, Phase 2

Course Number: ALMC-SO.
Location: Army Logistics Management College, Fort Lee, VA.
Length: 2 weeks (75 hours).
Exhibit Dates: 4/92–1/00.
Learning Outcomes: Upon completion of the course, the student will be able to prepare and organize support functions and manage combat service support during peace and wartime.
Instruction: Lectures and demonstrations cover arming, moving, sustaining, and fueling the force.
Credit Recommendation: Credit is not recommended because of the military-specific nature of the course (8/95).

AR-2201-0415

Forward Area Air Defense Officer Reclassification, Phase 2

Course Number: 2E-14B-RC.
Location: Air Defense Artillery School, Fort Bliss, TX.
Length: 59 hours.
Exhibit Dates: 6/95–10/00.
Learning Outcomes: Upon completion of the course, the student will be able to direct the use of specific missile systems.
Instruction: Lectures and practical exercises involve the battlefield tactics necessary for several specific missile systems.
Credit Recommendation: Credit is not recommended because of the military-specific nature of the course (8/95).

AR-2201-0416

SHORADS Officer Reclassification Reserve

Course Number: 2E-14B-RC.
Location: Air Defense Artillery School, Fort Bliss, TX.
Length: 2 weeks (109 hours).
Exhibit Dates: 7/90–Present.
Learning Outcomes: Upon completion of the course, the student will be able to direct the use of Stinger, Chaparral, or Vulcan missile systems in battle.
Instruction: Lectures and practical exercises include the characteristics, capabilities, limitations, and tactics of specific air defense systems.
Credit Recommendation: Credit is not recommended because of the military-specific nature of the course (8/95).
Related Occupations: 14B.

AR-2201-0417

Air Defense Artillery Patriot Officer Reclassification Reserve Phase 2 and 4

Course Number: 2E-14E-RC.
Location: Air Defense Artillery School, Fort Bliss, TX.
Length: 101 hours.
Exhibit Dates: 1/95–10/00.
Learning Outcomes: Upon completion of the course, the student will be able to describe the operation and use of a specific army missile.
Instruction: Phases 1 and 3 directed reading (nonresident). Phases 2 and 4 are resident portions which provide lectures and practical exercises on the Patriot missile system.
Credit Recommendation: Credit is not recommended because of the military-specific nature of the course (8/95).
Related Occupations: 14E.

AR-2201-0418

Air Defense Artillery Warrant Officer Advanced Reserve, Phase 2

Course Number: 2-44-C32-RC.
Location: Air Defense Artillery School, Fort Bliss, TX.
Length: 2 weeks (94 hours).
Exhibit Dates: 1/94–Present.
Learning Outcomes: Upon completion of the course, the student will be able to describe military doctrine and missile systems and use effective written and oral communication techniques.
Instruction: Lectures and practical exercises cover military doctrine and history, writing and speaking, and a specific missile system.
Credit Recommendation: In the lower-division baccalaureate/associate degree category, 1 semester hour in written and oral communication (8/95).

AR-2201-0419

Hawk Missile Air Defense Artillery Officer Reclassification Reserve

Course Number: 2F-14D (PIP 2)-RC.
Location: Air Defense Artillery School, Fort Bliss, TX.
Length: Phase 2, 2 weeks (70–110 hours); Phase 4, 4 weeks (80–104 hours).
Exhibit Dates: 1/90–6/96.
Learning Outcomes: Upon completion of the course, the student will be able to direct the employment, checkout, and firing of Hawk and Stinger missiles and associated command and control systems.
Instruction: Lectures and practical exercises include the Hawk, Stinger, and associated command and control equipment.
Credit Recommendation: Credit is not recommended because of the military-specific nature of the course (8/95).
Related Occupations: 14D.

AR-2201-0420

Air Defense Artillery Officer Advanced SHORADS Follow-On

Course Number: 2E-FOA-F142.
Location: Air Defense Artillery School, Fort Bliss, TX.
Length: 3 weeks (87–109 hours).
Exhibit Dates: 1/90–10/97.
Learning Outcomes: Upon completion of the course, the student will be able to describe the characteristics, capabilities, function, and operation of three missile and radar systems.
Instruction: Lectures and practical exercises focus on the specific military systems.
Credit Recommendation: Credit is not recommended because of the military-specific nature of the course (8/95).
Related Occupations: 14B.

AR-2201-0421

Patriot Air Defense Officer

Course Number: 2F-14EX.
Location: Air Defense Artillery School, Fort Bliss, TX.
Length: 8 weeks (291 hours).
Exhibit Dates: 12/93–Present.
Learning Outcomes: Before 12/93 see AR-2201-0261. Upon completion of the course, the student will be able to lead a platoon and direct the use of a Patriot Army air defense missile system.
Instruction: Lectures and practical exercises include the deployment, emplacement, and tactical use of the Patriot air defense missile system and associated battlefield command and control systems.
Credit Recommendation: Credit is not recommended because of the limited, specialized nature of the course (8/95).

AR-2201-0422

Air Defense Artillery Basic Noncommissioned Officer (NCO)

Course Number: 0-14-C40.
Location: Reserve Component Training Centers, US; Air Defense Artillery School, Fort Bliss, TX.
Length: 7 weeks (311–344 hours).
Exhibit Dates: 1/94–12/00.

Learning Outcomes: After 12/00 see AR-2201-0530. Upon completion of the course, the student will be able to lead, train, and direct subordinates in the skills required to participate in combat operations.

Instruction: Lectures and practical exercises are conducted in a 24-hour a day environment. Topics include leadership, tactics, communications, and weapons. This course contains a common core of leadership subjects.

Credit Recommendation: Credit is not recommended because of the military-specific nature of the course (4/01).

AR-2201-0423

Air Defense Artillery Officer Advanced Reserve

Course Number: 2-44-C26.

Location: Air Defense Artillery School, Fort Bliss, TX.

Length: Phase 1, 2 weeks (90 hours); Phase 3, 2 weeks (93 hours).

Exhibit Dates: 1/90–Present.

Learning Outcomes: Upon completion of the course, the student will be able to describe several missile systems and their use in battle.

Instruction: Lectures and practical exercises cover air defense missile systems.

Credit Recommendation: Credit is not recommended because of the military-specific nature of the course (8/95).

AR-2201-0424

1. Air Defense Artillery Officer Basic Common
2. Air Defense Artillery Officer Basic

Course Number: *Version 1:* 2-44-C20. *Version 2:* 2-44-C20 (F); 2-44-C20; 2-44-C20 (TATS).

Location: Air Defense Artillery School, Fort Bliss, TX.

Length: *Version 1:* 9–10 weeks (378 hours). *Version 2:* 10 weeks (360 hours).

Exhibit Dates: *Version 1:* 10/90–12/98. *Version 2:* 1/99–Present.

Learning Outcomes: *Version 1:* Upon completion of the course, the student will be able to manage the training, supply, and logistics operations of an Army air defense weapons platoon; lead the platoon; and direct the combat operations of an air defense weapon system. *Version 2:* Upon completion of the course, the student will be able to lead and manage; supervise logistics and maintenance operations; and make oral presentations, including persuasive speech, to brief military and media.

Instruction: *Version 1:* Lectures and practical exercises cover communication skills, leadership, training, logistics, personnel administration, and military-specific subjects dealing with Army air defense weapons and tactical fighting doctrine. *Version 2:* Lectures and practical exercises cover communication skills, leadership, logistics, maintenance procedures, NBC warfare, personnel supervision, and other military specific subjects.

Credit Recommendation: *Version 1:* In the lower-division baccalaureate/associate degree category, 3 semester hours in principles of supervision (8/95). *Version 2:* In the lower-division baccalaureate/associate degree category, 3 semester hours in speech, 1 in first aid, and 3 in principles of supervision (4/01).

AR-2201-0425

Air Defense Artillery Officer Basic Hawk Track (PIP 2) Track
(Hawk Weapon System Qualification (Phase 3) Air Defense Artillery Officer Basic)

Course Number: 2-44-C20 (14D).

Location: Air Defense Artillery School, Fort Bliss, TX.

Length: 9–10 weeks (379–384 hours).

Exhibit Dates: 10/90–6/96.

Learning Outcomes: Upon completion of the course, the student will be able to supervise a platoon of personnel and manage the maintenance, operation, and tactical use of the Hawk Army air defense missile system.

Instruction: Lectures and practical exercises include the operation, maintenance, deployment, and battlefield tactics of the Hawk air defense missile system and associated communication and central systems.

Credit Recommendation: Credit is not recommended because of the military-specific nature of the course (8/95).

Related Occupations: 14D.

AR-2201-0426

Air Defense Artillery Officer Basic (Patriot Track)
(Patriot Weapon System Qualification Air Defense Artillery Officer Basic)
(Air Defense Artillery Patriot Weapon System)

Course Number: 2-44-C20 (14E).

Location: Air Defense Artillery School, Fort Bliss, TX.

Length: 9–10 weeks (348–368 hours).

Exhibit Dates: 10/90–Present.

Learning Outcomes: Upon completion of the course, the student will be able to supervise a platoon and manage the maintenance, operation, and tactical deployment of a Patriot Army air defense missile system.

Instruction: Lectures and practical exercises cover the operation, maintenance, deployment, and battlefield tactics of the Patriot air defense missile system and associated communications and control systems.

Credit Recommendation: Credit is not recommended because of the limited, specialized nature of the course (4/04).

Related Occupations: 14E.

AR-2201-0427

1. Air Defense Artillery Officer Basic (Forward Area Air Defense Officer Track)
   (Air Defense Artillery Officer Basic SHORADS Track)
2. Air Defense Artillery Officer Basic (SHORAD Track)
   (Air Defense Artillery Officer Basic (SHORAD Track) Officer Basic)

Course Number: *Version 1:* 2-44-C20 (14B). *Version 2:* 2-44-C20 (14B) SHORAD Track (TATS); 2-44-C20 (14B) SHORAD Track.

Location: Air Defense Artillery School, Fort Bliss, TX.

Length: *Version 1:* 7–10 weeks (266–370 hours). *Version 2:* 9–10 weeks (356 hours).

Exhibit Dates: *Version 1:* 10/90–7/98. *Version 2:* 8/98–Present.

Learning Outcomes: *Version 1:* Upon completion of the course, the student will be able to operate and direct the deployment of the air defense weapon systems. *Version 2:* Upon completion of the course, the student will be able to describe the components of an air defense artillery system, including communications, computer, and intelligence systems; integrate these components into the overall operational system; and operate and supervise the deployment of the system.

Instruction: *Version 1:* Lectures and practical exercises cover the operation, maintenance, and deployment of the Chaparral, Avenger, and Vulcan Army air defense systems. *Version 2:* Instruction is provided using lectures, demonstrations, computer simulations, and practical exercises. The components of the course include Stinger/Avenger early warning systems, Bradley Stinger fighting vehicle, tactics, and various associated equipment such as global positioning system.

Credit Recommendation: *Version 1:* Credit is not recommended because of the military-specific nature of the course (8/95). *Version 2:* Credit is not recommended because of the military-specific nature of the course (4/01).

Related Occupations: 14B.

AR-2201-0428

Infantry Officer Basic

Course Number: *Version 1:* 2-7-C20. *Version 2:* 2-7-C20.

Location: Infantry School, Fort Benning, GA.

Length: *Version 1:* 16 weeks (949–1080 hours). *Version 2:* 16 weeks (1050 hours).

Exhibit Dates: *Version 1:* 6/92–12/99. *Version 2:* 1/00–Present.

Learning Outcomes: *Version 1:* Before 6/92 see AR-2201-0192. Upon completion of the course, the student will be prepared to serve as an infantry platoon leader and as a tactical leader in combat. *Version 2:* Upon completion of the course, the student will be able to demonstrate leadership skills in training subordinates in weapon systems, equipment, and combat tactics.

Instruction: *Version 1:* Lectures, field exercises, and practical applications cover Army doctrine, drill, weapons/tactics, map reading, equipment control, logistics, personnel supervision, combat leadership, and conducting training exercises. *Version 2:* Audio-visual materials, practical and classroom exercises, discussion, learner presentation, laboratory, lecture, and computer-based training. Topics include military orientation, equipment operation and control, land navigation, fitness assessment, physical fitness, marksmanship, legal and administrative practices, processes, and communications.

Credit Recommendation: *Version 1:* In the lower-division baccalaureate/associate degree category, 3 semester hours in map reading, 3 in personnel supervision, and 3 in physical training. In the upper-division baccalaureate category, 3 semester hours in materiel management and 3 in leadership (9/00). *Version 2:* In the lower-division baccalaureate/associate degree category, 3 semester hours in land navigation, 2 in fitness assessment, 2 in leadership, 1 in physical fitness, and 1 in marksmanship (10/02).

AR-2201-0429

1. Officer Candidate
2. Officer Candidate
3. Officer Candidate School

Course Number: *Version 1:* 2-7-F1. *Version 2:* 2-7-F1. *Version 3:* 2-7-F1.

Location: Infantry School, Fort Benning, GA.

Length: *Version 1:* 14 weeks (918–919 hours). *Version 2:* 14 weeks (995 hours). *Version 3:* 14 weeks (995 hours).

Exhibit Dates: *Version 1:* 10/92–5/98. *Version 2:* 6/98–5/02. *Version 3:* 6/02–Present.

Learning Outcomes: *Version 1:* Upon completion of the course, the student will be able to describe arms tactics, Army doctrine, and drill and command; apply leadership and supervision techniques, read maps, and use effective communication techniques. *Version 2:* Upon completion of the course, the student will be able to apply combined arms tactics, leadership, personnel, logistics, training, management, communications, weapons systems, and land navigation in a field environment. *Version 3:* Upon completion of the course, the student will be able to perform leadership as a second lieutenant and apply arms tactics in a field environment.

Instruction: *Version 1:* Methods include lectures and field exercises in arms tactics, Army doctrine, field operations, weaponry on range, military drill and command, planning, leadership techniques, and physical training. *Version 2:* Topics include tactics, leadership, personnel, management, communications, weapons, and navigation. *Version 3:* Topics include tactics, ethics, leadership, land navigation, heritage, and history.

Credit Recommendation: *Version 1:* In the lower-division baccalaureate/associate degree category, 3 semester hours in leadership, 3 in physical education, 3 in map reading, 3 in communication skills, and 3 in personnel supervision (12/95). *Version 2:* In the lower-division baccalaureate/associate degree category, 3 semester hours in land navigation and 3 in physical fitness. In the upper-division baccalaureate category, 3 in field experience in management (10/02). *Version 3:* In the lower-division baccalaureate/associate degree category, 3 semester hours in land navigation, 3 in military ethics, 3 leadership principles and practice and 1 in oral communication (10/02).

AR-2201-0430

Indirect Fire Infantryman Advanced Noncommissioned Officer (NCO)

Course Number: *Version 1:* 010-11C40. *Version 2:* 010-11C40, Phase 2.

Location: Infantry School, Fort Benning, GA.

Length: *Version 1:* 10–11 weeks (402–413 hours). *Version 2:* 8 weeks (276 hours).

Exhibit Dates: *Version 1:* 1/96–8/01. *Version 2:* 9/01–Present.

Learning Outcomes: *Version 1:* Upon completion of the course, the student will be able to serve as a combat leader in technical/tactical battle missions and lead a platoon. *Version 2:* Upon completion of the course, the student will be able to train infantry fire section leaders to lead, train, and direct subordinates in maintaining, operating, and employing combat weapons and equipment.

Instruction: *Version 1:* Lectures and practical exercises cover plotting; computer data; ordering fire missions; determining firing positions; applying battle-focused leadership; and maintaining weapons, equipment, and supplies. *Version 2:* Audio-visual materials, practical and classroom exercises, laboratory, discussion, lectures, and computer based-training. Topics include advanced skill development in application of combat tactics and operations, use and employment of mortar ballistic computer and plot board, and platoon-level offensive and defensive tactics.

Credit Recommendation: *Version 1:* In the lower-division baccalaureate/associate degree category, 1 semester hour in map reading and 1 in materiel management. See AR-1404-0035 for common core credit (9/00). *Version 2:* In the lower-division baccalaureate/associate degree category, 1 semester hour in computer operations. See AR-1404-0035 for common core credit (10/02).

Related Occupations: 11C.

AR-2201-0431

1. Bradley Infantry Fighting Vehicle System Master Gunner
2. Bradley Fighting Vehicle (BFV) System Master Gunner

Course Number: *Version 1:* 010-ASIJ3. *Version 2:* 010-ASIJ3.

Location: Infantry School, Fort Benning, GA.

Length: *Version 1:* 12 weeks (515–516 hours). *Version 2:* 13 weeks (560 hours).

Exhibit Dates: *Version 1:* 6/92–5/99. *Version 2:* 6/99–Present.

Learning Outcomes: *Version 1:* Upon completion of the course, the student will be able to describe types ammunition, fire weapon, operate Bradley Infantry Fighting Vehicle (BIFV) vehicle in combat, apply tactical procedures, operate a range facility, and perform maintenance. *Version 2:* Upon completion of the course, the student will be able to perform the operation of all systems in the bradley fighting vehicle, including weapon systems, targeting, and general maintenance.

Instruction: *Version 1:* This course includes lectures, live exercises, and demonstrations on gunnery standards and tactical weaponry. *Version 2:* Practical exercises, discussion, and lecture. This course covers lectures, live exercises, and demonstrations on gunnery standards and tactical weaponry.

Credit Recommendation: *Version 1:* In the lower-division baccalaureate/associate degree category, 2 semester hours in physical education (12/95). *Version 2:* Credit is not recommended because of the military-specific nature of the course (10/02).

AR-2201-0432

Fighting Vehicle Infantryman Advanced Noncommissioned Officer (NCO)

Course Number: 010-11M40.

Location: Infantry School, Fort Benning, GA.

Length: 10–11 weeks (391–400 hours).

Exhibit Dates: 4/95–Present.

Learning Outcomes: Upon completion of the course, the student will be able to serve as a Bradley platoon sergeant.

Instruction: Lectures and practical exercises cover combat skills, physical fitness, logistics, maintenance, assault operations, and communications.

Credit Recommendation: In the lower-division baccalaureate/associate degree category, 1 semester hour in map reading and 1 in materiel management. See AR-1404-0035 for common core credit (10/00).

Related Occupations: 11M.

AR-2201-0433

Heavy Antiarmor Weapons Infantryman Basic Noncommissioned Officer (NCO)

(Basic Noncommissioned Officer (NCO) 11H Heavy Antiarmor Weapons Infantryman)

Course Number: 010-3-11H30; 010-11H30.

Location: Infantry School, Fort Benning, GA.

Length: 7–8 weeks (307–317 hours).

Exhibit Dates: 6/93–Present.

Learning Outcomes: Upon completion of the course, the student will be able to lead, train, and direct subordinates to maintain and employ weapons and equipment.

Instruction: Lectures and practical exercises cover tactics, weapons, and vehicles.

Credit Recommendation: Credit is recommended for the common core only. See AR-1406-0090 (10/00).

Related Occupations: 11H.

AR-2201-0434

Ranger

Course Number: *Version 1:* 2E-S15S-5R/-11-SQIV-G, Phase 1; 2E-SI5S-5R/011-SQIV-G. *Version 2:* 2E-SI5S-5R/011-SQIV-G.

Location: Infantry School, Fort Benning, GA.

Length: *Version 1:* 8–9 weeks (1112–1727 hours). *Version 2:* 8–9 weeks (1144 hours).

Exhibit Dates: *Version 1:* 3/00–Present. *Version 2:* 9/92–2/00.

Learning Outcomes: *Version 1:* Upon completion of the course, the student will be able to complete rigorous physical fitness training, swim 15 miles carrying combat equipment, prepare combat planning operations, use survival skills under harsh conditions; conduct combat patrols; ambush defensive operations in mountainous and swampy terrain; and conduct extended platoon operations using helicopters, inflatable watercraft, and vehicles. *Version 2:* Before 9/92 see AR-2201-0041. Upon completion of the course, the student will be able to provide combat leadership in assault operations.

Instruction: *Version 1:* Practical and classroom exercises, and discussion. Topics include rigorous physical training including swimming and prevention measures for exhaustion, heat stroke, dehydration, and cold weather injuries; performing combat field exercises following various performance task using military mountaineering equipment to climb, repel, and operate in mountainous terrain; and performing waterborne operations under various tactical tasks. *Version 2:* Lectures and field exercises cover combat engagement and survival skills under harsh conditions and includes land navigation, intelligence, and physical training.

Credit Recommendation: *Version 1:* In the lower-division baccalaureate/associate degree category, 6 semester hours in physical education (10/02). *Version 2:* In the lower-division baccalaureate/associate degree category, 6 semester hours in physical education (12/95).

AR-2201-0435

1. Infantry Officer Advanced
2. Infantry Captains Career

Course Number: *Version 1:* 2-7-C22. *Version 2:* 2-7-C22, Phase 1.
Location: Infantry School, Fort Benning, GA.
Length: *Version 1:* 20 weeks (705–706 hours). *Version 2:* 18 weeks (673 hours).
Exhibit Dates: *Version 1:* 7/92–12/98. *Version 2:* 1/99–Present.
Learning Outcomes: *Version 1:* Upon completion of the course, the student will be able to provide leadership and service support. *Version 2:* Upon completion of the course, the student will be able to train officers in leadership warfighting, and service support skills required to serve as company commanders and battalion/brigade staff officers.
Instruction: *Version 1:* Lectures and practical exercises cover tactics, personnel, intelligence, logistics, management, training, land navigation, legal aspects, communications, and weapons. *Version 2:* Topics include combined arms tactics, staff subjects, military leadership, land navigation, legal issues, medical care, communications, and maintenance management.
Credit Recommendation: *Version 1:* In the lower-division baccalaureate/associate degree category, 2 semester hours in leadership, 2 in records and information management, 2 in maintenance management, and 1 in physical fitness (9/98). *Version 2:* In the lower-division baccalaureate/associate degree category, 3 semester hours in leadership (10/02).

AR-2201-0436

Heavy Antiarmor Weapons Infantryman Advanced Noncommissioned Officer (NCO)

Course Number: 010-11H40.
Location: Infantry School, Fort Benning, GA.
Length: 10 weeks (393–442 hours).
Exhibit Dates: 1/96–Present.
Learning Outcomes: Upon completion of the course, the student will be able to lead a heavy antiarmor weapons platoon.
Instruction: Lectures, demonstrations, and field exercises cover computer literacy, world threats, tactical doctrine/battle skills, map reading, battle logistics, equipment maintenance, and gunnery overview.
Credit Recommendation: In the lower-division baccalaureate/associate degree category, 1 semester hour in map reading and 1 in materiel management. See AR-1404-0035 for common core credit (9/00).
Related Occupations: 11H.

AR-2201-0437

Infantry Officer Advanced Reserve Component

Course Number: 2-7-C26.
Location: Infantry School, Fort Benning, GA.
Length: 12 weeks (472–473 hours).
Exhibit Dates: 4/93–Present.
Learning Outcomes: Upon completion of the course, the student will be able to demonstrate proper use of special purpose weapons, identify the major components of health threats to field forces, describe advantages and disadvantages of aerial photographs compared to maps, and describe decontamination operations.
Instruction: This course includes conferences, lectures, demonstrations, seminars, television, and field training/practical exercise.
Credit Recommendation: In the lower-division baccalaureate/associate degree category, 2 semester hours in physical education, 2 in maintenance management, 2 in communication skills, and 2 in leadership (12/95).

AR-2201-0438

Infantry Officer Basic Reserve Component

Course Number: 2-7-C25.
Location: Infantry School, Fort Benning, GA.
Length: 8 weeks (604 hours).
Exhibit Dates: 6/93–Present.
Learning Outcomes: Upon completion of the course, the student will be able to serve as an infantry platoon leader in combat conditions.
Instruction: Lectures and practical exercises cover weapon systems, tactics, leadership, intelligence, and land navigation.
Credit Recommendation: In the lower-division baccalaureate/associate degree category, 2 semester hours in applied geography (12/95).

AR-2201-0439

Infantryman Basic Noncommissioned Officer (NCO)
(Basic Noncommissioned Officer (NCO) 11B Infantryman)

Course Number: *Version 1:* 010-11B30. *Version 2:* 010-11B30, Phase 2.
Location: Infantry School, Fort Benning, GA.
Length: *Version 1:* 7–8 weeks (315–339 hours). *Version 2:* 5 weeks (264 hours).
Exhibit Dates: *Version 1:* 10/92–8/01. *Version 2:* 9/01–Present.
Learning Outcomes: *Version 1:* Upon completion of the course, the student will be able to lead and direct subordinates and maintain, operate, and employ weapons and equipment. *Version 2:* Upon completion of the course, the student will be able to train infantry squad leaders to lead, train, and direct subordinates to maintain, operate, and employ weapons and equipment.
Instruction: *Version 1:* Lectures and practical exercises cover leadership, combat skills, logistics, navigation, communications, and assault operations. *Version 2:* Audio visual materials, practical and classroom exercises, discussion, laboratory, lecture, and computer-based training. Topics include planning and conducting squad (group) exercises, and demonstrating leadership skills and initiative.
Credit Recommendation: *Version 1:* Credit is recommended for the common core only. See AR-1406-0090 (9/00). *Version 2:* In the lower-division baccalaureate/associate degree category, 2 semester hours in principles of supervision. See AR-1406-0090 (10/02).
Related Occupations: 11B.

AR-2201-0440

1. Basic Noncommissioned Officer (NCO) Indirect Fire Infantryman
2. Indirect Fire Infantryman Basic Noncommissioned Officer

Course Number: *Version 1:* 010-3-11C30; 010-11C30. *Version 2:* 010-11C30, Phase 2.
Location: Infantry School, Fort Benning, GA.
Length: *Version 1:* 7–8 weeks (315–318 hours). *Version 2:* 5 weeks (271 hours).
Exhibit Dates: *Version 1:* 6/93–8/01. *Version 2:* 9/01–Present.
Learning Outcomes: *Version 1:* Upon completion of the course, the student will be able to lead, train, and direct subordinates in the maintenance and employment of weapons and equipment. *Version 2:* Upon completion of the course, the student will be able to lead, train, and direct subordinates in the maintenance and employment of weapons and equipment.
Instruction: *Version 1:* Lectures and practical exercises in leadership, combat skills, physical fitness, safety, communications, and assault operations. *Version 2:* Audio-visual materials, practical and classroom exercises, discussion, laboratory, lecture, learner presentations, and computer-based training. Topics include basic orientation, use of desktop computer with word processing, and skill development in use of motor ballistic computer and plot board.
Credit Recommendation: *Version 1:* In the lower-division baccalaureate/associate degree category, 1 semester hour in computer operations. See AR-1406-0090 for common core credit (9/00). *Version 2:* In the lower-division baccalaureate/associate degree category, 1 semester hour in computer operations. See AR-1406-0090 for common core credit (10/02).
Related Occupations: 11C.

AR-2201-0441

1. Bradley Fighting Vehicle Infantryman Basic Noncommissioned Officer (NCO)
2. Fighting Vehicle Infantryman Basic Noncommissioned Officer (NCO)

Course Number: *Version 1:* 010-11M30; 010-18-11M30; 010-10-11M30; 010-9-11M30; 010-6-11M30; 010-4-11M30. *Version 2:* 010-3-11M30.
Location: *Version 1:* NCO Academy, Fort Stewart, GA; NCO Academy, Fort Riley, KS; NCO Academy, Fort Hood, TX; NCO Academy, Fort Carson, CO; NCO Academy, 7th Army, Europe; Infantry School, Fort Benning, GA. *Version 2:* Infantry School, Fort Benning, GA.
Length: *Version 1:* 7–8 weeks (309 hours). *Version 2:* 7–8 weeks (334 hours).
Exhibit Dates: *Version 1:* 1/93–9/97. *Version 2:* 10/97–Present.
Learning Outcomes: *Version 1:* Upon completion of the course, the student will be prepared to supervise the maintenance, employment, and operation of weapons and equipment. *Version 2:* Upon completion of the course, the student will be prepared to supervise the maintenance, employment, and operation of weapons and equipment.
Instruction: *Version 1:* Lecture and practical exercises cover maintenance of equipment, combat operations, and weapons systems. *Version 2:* Lecture and practical exercises cover maintenance of equipment, combat operations, and weapons systems.
Credit Recommendation: *Version 1:* In the lower-division baccalaureate/associate degree category, 1 semester hour physical fitness. See AR-1406-0090 for common core credit (12/95). *Version 2:* In the lower-division baccalaureate/

associate degree category, 1 physical fitness (12/95). See AR-1406-0090 for common core credit (9/00).

Related Occupations: 11M.

AR-2201-0442

SNIPER

Course Number: 010-ASIB4.
Location: Infantry School, Fort Benning, GA.
Length: 5 weeks (310–376 hours).
Exhibit Dates: 5/92–Present.
Learning Outcomes: Upon completion of the course, the student will be able to function as a sniper at the unit level in field combat conditions.
Instruction: Lectures and field exercises cover sniper tactics, intelligence, command and control, stalking, and marksmanship.
Credit Recommendation: In the lower-division baccalaureate/associate degree category, 3 semester hours in marksmanship/military science (9/03).

AR-2201-0443

TACTICS CERTIFICATION

Course Number: 010-F20; 2E-F163.
Location: Infantry School, Fort Benning, GA.
Length: 2 weeks (50 hours).
Exhibit Dates: 4/94–Present.
Learning Outcomes: Upon completion of the course, the student will be able to lead troops, apply tactics, and deploy and use weapons.
Instruction: Practical application and field exercises cover planning missions and tactical operations.
Credit Recommendation: Credit is not recommended because of the military-specific nature of the course (12/95).

AR-2201-0444

PATHFINDER

Course Number: 2E-SI5Q/011-ASIF7.
Location: Infantry School, Fort Benning, GA.
Length: 2–3 weeks (145–252 hours).
Exhibit Dates: 8/90–Present.
Learning Outcomes: Upon completion of the course, the student will be able to provide navigational assistance to drop-zone aircraft and deploy troops.
Instruction: Lectures and practical exercises cover the organization, mission, and training of pathfinder platoons; electronics; communications; and map reading.
Credit Recommendation: Credit is not recommended because of the military-specific nature of the course (8/02).

AR-2201-0445

FIGHTING VEHICLE INFANTRYMAN

Course Number: 11M10-OSUT.
Location: Infantry School, Fort Benning, GA.
Length: 2 weeks (80 hours).
Exhibit Dates: 11/94–9/98.
Learning Outcomes: Upon completion of the course, the student will be able to use and maintain the Bradley Fighting Vehicle.
Instruction: Lectures and field experiences cover the safe and effective operation of the Bradley Fighting Vehicle. This course includes an 8-9 week Basic Training component.
Credit Recommendation: Credit is not recommended because of the military-specific nature of the course (12/95).
Related Occupations: 11M.

AR-2201-0446

1. OSUT-HEAVY WEAPONS ANTIARMOR INFANTRYMAN
2. BASIC ANTIARMOR TRAINING

Course Number: ***Version 1:*** 11H10-OSUT. *Version 2:* 7-11H10.
Location: Infantry School, Fort Benning, GA.
Length: ***Version 1:*** 1 week (40 hours). *Version 2:* 2 weeks (97 hours).
Exhibit Dates: ***Version 1:*** 11/94–1/98. *Version 2:* 2/98–Present.
Learning Outcomes: ***Version 1:*** Upon completion of the course, the student will be able to function as a heavy antiarmor crewmember. *Version 2:* Upon completion of the course, the student will be able to perform all skills necessary to operate the TOW missile weapons launcher.
Instruction: ***Version 1:*** Lectures and field exercises cover the TOW-2 weapons system. *Version 2:* Practical exercise, laboratory, and lecture. Topics include the capabilities and operation of the TOW missile launcher.
Credit Recommendation: ***Version 1:*** Credit is not recommended because of the military-specific nature of the course (12/95). *Version 2:* Credit is not recommended because of the military-specific nature of the course (10/02).
Related Occupations: 11H.

AR-2201-0447

1. OSUT-INDIRECT FIRE INFANTRYMAN
2. BASIC MORTAR

Course Number: ***Version 1:*** 010-11C10. *Version 2:* 7-11C10.
Location: Infantry School, Fort Benning, GA.
Length: ***Version 1:*** 2 weeks (80 hours). *Version 2:* 2 weeks (98 hours).
Exhibit Dates: ***Version 1:*** 11/94–1/98. *Version 2:* 2/98–Present.
Learning Outcomes: ***Version 1:*** Upon completion of the course, the student will be proficient as a mortar crewmember. *Version 2:* Pending evaluation.
Instruction: ***Version 1:*** Practical field exercises cover the maintenance and operation of the 81mm mortar. *Version 2:* Pending evaluation.
Credit Recommendation: ***Version 1:*** Credit is not recommended because of the military-specific nature of the course (12/95). *Version 2:* Pending evaluation.
Related Occupations: 11C.

AR-2201-0449

JUMPMASTER

Course Number: ***Version 1:*** 2E-F60/011-F16; 011-F16; 2E-F60. *Version 2:* 2E-F60/011-F16.
Location: Infantry School, Fort Benning, GA.
Length: ***Version 1:*** 1–2 weeks (77–93 hours). *Version 2:* 2 weeks (98 hours).
Exhibit Dates: ***Version 1:*** 10/90–1/02. *Version 2:* 2/02–Present.
Learning Outcomes: ***Version 1:*** Before 10/90 see AR-2201-0273. Upon completion of the course, the student will be able to train students (qualified parachutists) in jumpmaster techniques. *Version 2:* Upon completion of the course, the student will be able to safely perform the duties of the Jumpmaster including parachute rigging, drop zone safety, and conducting a jump exercise.
Instruction: ***Version 1:*** Lectures, demonstrations, and practical exercises cover parachuting techniques. *Version 2:* Lectures, demonstrations, and practical exercises. Topics cover parachuting techniques.
Credit Recommendation: ***Version 1:*** Credit is not recommended because of the military-specific nature of the course (9/00). *Version 2:* In the lower-division baccalaureate/associate degree category, 2 semester hours in physical education-parachuting (10/02).

AR-2201-0450

INFANTRY PRE-COMMAND

Course Number: 2G-F26.
Location: Infantry School, Fort Benning, GA.
Length: 2 weeks (87–88 hours).
Exhibit Dates: 10/92–Present.
Learning Outcomes: Upon completion of the course, the student will be able to describe infantry doctrine, use infantry equipment, and be prepared for command.
Instruction: Demonstrations and field exercises cover arms tactics, operations in battle/logistics, and unit readiness.
Credit Recommendation: Credit is not recommended because of the military-specific nature of the course (12/95).

AR-2201-0451

INFANTRY MORTAR LEADER

Course Number: 2E-SI2Z/010-F1; 010-F1; 2E-SI3Z.
Location: Infantry School, Fort Benning, GA.
Length: 5–6 weeks (215–298 hours).
Exhibit Dates: 11/94–Present.
Learning Outcomes: Upon completion of the course, the student will be able to train officers and noncommissioned officers to command infantry heavy mortar units.
Instruction: Lectures and practical exercise cover the supervision of a heavy mortar platoon in support of infantry combat operations, including tactical training and employment, mechanical training, crew drill, forward observation procedures, fire direction center procedures, and heavy mortar platoon communications.
Credit Recommendation: Credit is not recommended because of the military-specific nature of the course (9/00).

AR-2201-0452

INFANTRYMAN ADVANCED NONCOMMISSIONED OFFICER (NCO)

Course Number: ***Version 1:*** 011-11B40; 010-11B40. *Version 2:* 010-11B40, Phase 2.
Location: Infantry School, Fort Benning, GA.
Length: ***Version 1:*** 10–11 weeks (404–427 hours). *Version 2:* 8 weeks (318 hours).

Exhibit Dates: *Version 1:* 1/96–8/01. *Version 2:* 9/01–Present.

Learning Outcomes: *Version 1:* Upon completion of the course, the student will be able to identify the benefits of physical training; perform physical assessment; and describe environmental issues, key factors of threat, and HIV transmission and prevention. *Version 2:* Upon completion of the course, the student will be able to demonstrate leadership skills in directing small group activities engaged in training and tactical operations.

Instruction: *Version 1:* This course includes small-group instructions, lectures, live demonstrations, field training and experiences, video presentations, and group discussions. *Version 2:* Audio-visual materials, practical and classroom exercises, discussion, laboratory, lecture, and computer-based training. Topics include development of leadership competence as applied to execution of infantry combat skills.

Credit Recommendation: *Version 1:* In the lower-division baccalaureate/associate degree category, 1 semester hour in map reading and 1 in materiel management. See AR-1404-0035 for common core credit (9/00). *Version 2:* In the lower-division baccalaureate/associate degree category, 1 semester hour in land navigation, 1 in material management, and 2 in principles of supervision. See AR-1404-0035 for common core credit (10/02).

Related Occupations: 11B.

AR-2201-0453

Mechanized Leaders
(Bradley Leaders)
(Bradley Fighting Vehicle Leader)

Course Number: 2E-SI3X/010-F5; 010-F5; 2E-SI3X.

Location: Infantry School, Fort Benning, GA.

Length: 7 weeks (291–360 hours).

Exhibit Dates: 4/95–Present.

Learning Outcomes: Upon completion of the course, the student will be able to demonstrate proper hand and arm safety procedures, operate hull controls and intercom system on the Bradley Fighting Vehicle, and disassemble the machine gun to perform safety checks.

Instruction: This course includes live demonstrations, video demonstrations, lectures, and slide and video presentations.

Credit Recommendation: Credit is not recommended because of the military-specific nature of the course (8/02).

AR-2201-0454

Bradley Fighting Vehicle Commander

Course Number: 010-F5-M; 2E-SI3X.

Location: Infantry School, Fort Benning, GA.

Length: 4 weeks (242 hours).

Exhibit Dates: 11/90–Present.

Learning Outcomes: Upon completion of the course, the student will be able to perform as a commander of the Bradley Fighting Vehicle.

Instruction: Lectures and practical exercises cover vehicle maintenance and operation, weapon systems, tactics, and safety.

Credit Recommendation: Credit is not recommended because of the military-specific nature of the course (12/95).

AR-2201-0455

Airborne

Course Number: *Version 1:* 2E-SI5PSQI7/011-SQIP. *Version 2:* 2E-SI5P/SQI7/011-SQIP.

Location: Infantry School, Fort Benning, GA.

Length: *Version 1:* 3–4 weeks (113–119 hours). *Version 2:* 2 weeks (107 hours).

Exhibit Dates: *Version 1:* 10/92–12/01. *Version 2:* 1/02–Present.

Learning Outcomes: *Version 1:* Upon completion of the course, the student will be able to perform battlefield parachuting. *Version 2:* Upon completion of the course, the student will be able to use the parachute as a means of combat deployment.

Instruction: *Version 1:* Methods include jump tower training, landing falls, malfunctions, and actual jump experience. *Version 2:* Topics include jump tower training, landing falls, malfunctions, and experience in jumping.

Credit Recommendation: *Version 1:* Credit is not recommended because of the military-specific nature of the course (10/00). *Version 2:* In the lower-division baccalaureate/associate degree category, 3 semester hours in physical educational-parachuting (10/02).

AR-2201-0456

Air Assault

Course Number: 2F-A1.

Location: Infantry School, Fort Benning, GA; Training Centers, US.

Length: 2 weeks (61 hours).

Exhibit Dates: 7/94–Present.

Learning Outcomes: Upon completion of the course, the student will be able to perform combat air assault operations and use helicopters to perform air missions including evacuations.

Instruction: Topics include principles and techniques of combat assaults, aircraft safety, aircraft missions, field exercises, and assault practice.

Credit Recommendation: Credit is not recommended because of the military-specific nature of the course (4/00).

AR-2201-0457

Individual Ready Reserve Premobilization

Course Number: 010-F10 (IRR).

Location: Infantry School, Fort Benning, GA.

Length: 2 weeks (162 hours).

Exhibit Dates: 1/90–Present.

Learning Outcomes: Upon completion of the course, the student will be able to perform as a fire team/squad leader.

Instruction: Lectures and field exercises cover land navigation, weapons, tactics, and enemy engagement.

Credit Recommendation: Credit is not recommended because of the military-specific nature of the course (12/95).

AR-2201-0458

TOW Master Gunner

Course Number: 010-ASIS1; 071-010-ASIS1.

Location: Infantry School, Fort Benning, GA.

Length: 4 weeks (208–209 hours).

Exhibit Dates: 1/93–Present.

Learning Outcomes: Upon completion of the course, the student will perform vehicle troubleshooting, describe systems operations, use gunnery strategies, and conduct training.

Instruction: The student qualifies on weaponry, conducts maintenance checks, mounts machine guns, and assembles/disassembles weapons. This course includes demonstrations and exercises on range performance.

Credit Recommendation: Credit is not recommended because of the military-specific nature of the course (12/95).

AR-2201-0459

Special Forces Assistant Operations/ Intelligence Sergeant

Course Number: 011-18F40.

Location: John F. Kennedy Special Warfare Center and School, Fort Bragg, NC.

Length: 10 weeks (423 hours).

Exhibit Dates: 1/95–9/98.

Learning Outcomes: Before 1/95 see AR-1606-0210. Upon completion of the course, the student will perform intelligence functions in support of Special Forces missions.

Instruction: Lectures, demonstrations, and field exercises cover Special Forces operations, intelligence, and technical skills and functions that support warfare.

Credit Recommendation: In the lower-division baccalaureate/associate degree category, 1 semester hour in photography, 2 in map reading, and 3 in military science (6/96).

Related Occupations: 18F.

AR-2201-0460

Ammunition Technician Senior Warrant Officer Reserve, Phase 1

Course Number: 4-9-C32-910A-RC.

Location: Ordnance, Missile and Munitions School, Redstone Arsenal, AL.

Length: 2–3 weeks (93 hours).

Exhibit Dates: 10/90–Present.

Learning Outcomes: Upon completion of the course, the student will be able describe career development paths of a warrant officer.

Instruction: Lectures provide the student with instruction in common Army subject matters, including munitions management, standard Army ammunition system, Army specific career development, and military-specific law.

Credit Recommendation: In the lower-division baccalaureate/associate degree category, 2 semester hours in military science (5/01).

Related Occupations: 910A.

AR-2201-0461

Ammunition Technician Senior Warrant Officer Reserve, Phase 2

Course Number: 4-9-C32-910A-RC.

Location: Ordnance, Missile and Munitions School, Redstone Arsenal, AL.

Length: 2 weeks (80 hours).

Exhibit Dates: 10/90–Present.

Learning Outcomes: Upon completion of the course, the student will perform various senior warrant officer ammunition technician duties.

Instruction: Lectures and practical exercises provide instruction in the areas of standard Army ammunition system operation and maintenance.

Credit Recommendation: Credit is not recommended because of the military-specific nature of the course (5/01).

AR-2201-0462

Ammunition Specialist Advanced Noncommissioned Officer (NCO)

Course Number: 645-55B40.
Location: Ordnance, Missile and Munitions School, Redstone Arsenal, AL.
Length: 10–11 weeks (402 hours).
Exhibit Dates: 6/95–9/00.
Learning Outcomes: Upon completion of the course, the student will be able to manage ammunition units, including safety, training, and maintenance management.
Instruction: Lectures, practical exercises, and small-group instruction cover surveillance operations; ammunition inspection, shipping, storage, and safety; and manual and computer-based inventory. This course contains a common core of leadership subjects.
Credit Recommendation: In the lower-division baccalaureate/associate degree category, 2 semester hours in military science. See AR-1404-0035 for common core credit (7/96).
Related Occupations: 55B.

AR-2201-0463

Ammunition Technician Warrant Officer Basic

Course Number: 4E-910A.
Location: Ordnance, Missile and Munitions School, Redstone Arsenal, AL.
Length: 10–11 weeks (432 hours).
Exhibit Dates: 10/94–Present.
Learning Outcomes: Before 10/94 see AR-0802-0021. Upon completion of the course, the student will demonstrate the technical skills necessary to perform duties as on Ammunition Technician.
Instruction: Lectures and practical exercises cover conventional ammunition, identification of ammunition, magazine storage areas, ammunition material management, ammunition destruction and disposal, and manual and automated inventory operations.
Credit Recommendation: Credit is not recommended because of the military-specific nature of the course (5/01).
Related Occupations: 910A.

AR-2201-0464

Combat Service Support Missile and Munitions Pre-Command

Course Number: 2G-F45.
Location: Ordnance, Missile and Munitions School, Redstone Arsenal, AL.
Length: 3–4 weeks (133 hours).
Exhibit Dates: 10/91–Present.
Learning Outcomes: Upon completion of the course, the student will be able to perform material management and support functions for missiles, munitions, and special weapons.
Instruction: Instruction is lecture-based. Topics include missile material management and support, combat service support operations, ammunition service support, and operation of munitions production facilities and depots.
Credit Recommendation: Credit is not recommended because of the military-specific nature of the course (5/01).

AR-2201-0465

Ammunition Specialist, Phase 1

Course Number: 645-55B10.
Location: Ordnance, Missile and Munitions School, Redstone Arsenal, AL.
Length: 2 weeks (71 hours).
Exhibit Dates: 3/94–6/00.
Learning Outcomes: Before 3/94 see AR-2201-0389. Upon completion of the course, the student will be able to identify ammunition items; destroy unserviceable ammunition; perform inventory and issue receipts; store and load munitions; and operate a forklift.
Instruction: Conferences and practical exercises cover ammunition material, service procedures, storage, handling, movement, inspection, maintenance, and destruction.
Credit Recommendation: Credit is not recommended because of the military-specific nature of the course (7/96).
Related Occupations: 55B.

AR-2201-0466

Ammunition Specialist, Phase 2

Course Number: 645-55B10.
Location: Ordnance, Missile and Munitions School, Redstone Arsenal, AL.
Length: 8–9 weeks (321–322 hours).
Exhibit Dates: 10/94–9/00.
Learning Outcomes: Upon completion of the course, the student will be able to identify ammunition items; destroy unserviceable ammunition; perform manual and computer-based inventory; issue, receive, store and load munitions, and operate a forklift.
Instruction: Conferences and practical exercises cover ammunition material, service procedures, storage, handling, movement, inspection, maintenance, and destruction.
Credit Recommendation: Credit is not recommended because of the military-specific nature of the course (7/96).
Related Occupations: 55B.

AR-2201-0467

Ammunition Specialist Basic Noncommissioned Officer (NCO)

Course Number: 645-55B30.
Location: Ordnance, Missile and Munitions School, Redstone Arsenal, AL.
Length: 16 weeks (551–583 hours).
Exhibit Dates: 10/92–9/00.
Learning Outcomes: Before 10/92 see AR-2201-0392. Upon completion of the course, the student will be able to use a computerized inventory system to receive, identify, handle, store, inspect, issue, account for, and move munitions; apply standard operating procedures for safety and fire prevention and control; and destroy conventional ammunition.
Instruction: Lectures and practical exercises cover the identification of ammunition, ammunition storage principles, planning and handling procedures, inventory control, and directing fire fighting. The use of a computer to facilitate inventory control and to operate an ammunition supply point is also included. End of course evaluation is a comprehensive written examination. This course contains a common core of leadership subjects.
Credit Recommendation: Credit is recommended for the common core only. See AR-1406-0090 (7/96).
Related Occupations: 55B.

AR-2201-0468

Ammunition Specialist Advanced Noncommissioned Officer (NCO) Reserve Phase 2

Course Number: 645-55B40-RC.
Location: Ordnance, Missile and Munitions School, Redstone Arsenal, AL.
Length: 3 weeks (110 hours).
Exhibit Dates: 6/94–6/00.
Learning Outcomes: Upon completion of the course, the student will demonstrate the technical and leadership skills necessary to function as an ammunition operations supervisor and/or magazine platoon sergeant.
Instruction: Lectures, practical exercises, and small-group instruction cover ammunition support, storage, safety, destruction, inspection, shipping, security, and inventory.
Credit Recommendation: Credit is not recommended because of the military-specific nature of the course (7/96).
Related Occupations: 55B.

AR-2201-0469

Ordnance Munitions Materiel Management (Branch Qualification) Reserve Component

Course Number: 4-9-C20-91D (BQ)-RC.
Location: Ordnance, Missile and Munitions School, Redstone Arsenal, AL.
Length: 3 weeks (108 hours).
Exhibit Dates: 10/92–10/00.
Learning Outcomes: Upon completion of the course, the student will be able to identify ammunition, store ammunition in a magazine, manage ammunition material, and perform emergency destruction of ammunition.
Instruction: Lectures and limited practical exercises cover identification of ammunition, magazine storage areas, ammunition materiel management, shipping, accountability, service support systems, and emergency destruction.
Credit Recommendation: Credit is not recommended because of the military-specific nature of the course (7/96).
Related Occupations: 91D.

AR-2201-0470

Missile and Electronic Maintenance Advanced Noncommissioned Officer (NCO)

Course Number: 1-27-C42.
Location: Ordnance, Missile and Munitions School, Redstone Arsenal, AL.
Length: 4–5 weeks (178 hours).
Exhibit Dates: 12/95–6/00.
Learning Outcomes: Upon completion of the course, the student will be able to provide supervision and leadership to in munition fundamentals, organization, and management.
Instruction: Instruction includes lectures, demonstrations, and practical exercises in maintenance, management, and personnel development. The course also includes military logistics topics, administrative procedures, and ammunition management. This course contains a common core of leadership subjects.
Credit Recommendation: In the lower-division baccalaureate/associate degree category, 3 semester hours in military science. See AR-1404-0035 for common core credit (7/96).

AR-2201-0471

Cavalry Scout Basic Noncommissioned Officer (NCO) Reserve Component

Course Number: RC-171-19D30-001.
Location: Armor Center and School, Fort Knox, KY.
Length: 2–3 weeks (139 hours).
Exhibit Dates: 10/91–Present.
Learning Outcomes: Upon completion of the course, the student will be able to lead, train, and direct subordinates to maintain, operate, and employ weapons and equipment.
Instruction: Lectures and practical exercises cover combat tasks, maintenance, demolitions, land navigation, tactics, and field exercises.
Credit Recommendation: Credit is not recommended because of the military-specific nature of the course (3/97).
Related Occupations: 19D.

AR-2201-0472

Cavalry Scout Basic Noncommissioned Officer (NCO) Reserve Component

Course Number: RC-171-19D30-HMMWV-001.
Location: Armor Center and School, Fort Knox, KY.
Length: 2 weeks (106 hours).
Exhibit Dates: 7/92–Present.
Learning Outcomes: Upon completion of the course, the student will be able to lead, train, and direct subordinates in maintaining, operating, and employing weapons and equipment.
Instruction: Lectures and practical exercises cover maintenance, demolitions, land navigation, and field training.
Credit Recommendation: Credit is not recommended because of the military-specific nature of the course (3/97).

AR-2201-0473

Cavalry Scout

Course Number: RC-171-19D30-M113-001.
Location: Armor Center and School, Fort Knox, KY.
Length: 2 weeks (133 hours).
Exhibit Dates: 10/90–Present.
Learning Outcomes: Upon completion of the course, the student will be able to train, lead, and direct subordinates during tactical operations and to maintain, operate, and employ cavalry weapons systems.
Instruction: Lectures and practical exercises cover tactics, land navigation, maintenance engineering, and communications.
Credit Recommendation: Credit is not recommended because of the limited, specialized nature of the course (3/97).
Related Occupations: 19D.

AR-2201-0474

Cavalry Scout Advanced Noncommissioned Officer (NCO) Reserve Component

Course Number: RC-171-19D40.
Location: Armor Center and School, Fort Knox, KY.
Length: Phases 1 and 2, 3 weeks (101–105 hours).
Exhibit Dates: 10/92–Present.
Learning Outcomes: Upon completion of the course, the student will be able to lead armored cavalry units in specific combat situations.
Instruction: Lecture-demonstration and practical experience are used to cover call for artillery support, roadmatches, rearming, reconnaissance, application of law of war, platoon movement, haste river crossing, urban terrain, and platoon leadership.
Credit Recommendation: Credit is not recommended because of the military-specific nature of the course (3/97).

AR-2201-0475

1. M1/M1A1 Armor Crewman Basic Noncommissioned Officer (NCO)
2. Basic Noncommissioned Officer (NCO) M1A1 Abrams Armor Crewman
3. M1A1 Abrams Armor Crewman Basic Noncommissioned Officer (NCO)

Course Number: ***Version 1:*** 020-1-19K30. *Version 2:* 020-1-19K30 (M1A1). *Version 3:* 020-1-19K30 (M1A1); 020-19K30 (M1A1).
Location: ***Version 1:*** NCO Academy, Fort Stewart, GA; NCO Academy, Fort Polk, LA; NCO Academy, Fort Riley, KS; NCO Academy, Fort Hood, TX; NCO Academy, Fort Carson, CO; NCO Academy, 7th Army, Europe; NCO Academy, Fort Knox, KY; Armor Center and School, Fort Knox, KY. *Version 2:* Armor Center and School, Fort Knox, KY. *Version 3:* Armor School, Fort Knox, KY.
Length: ***Version 1:*** 8 weeks (408–409 hours). *Version 2:* 8 weeks (453 hours). *Version 3:* 8 weeks (434 hours).
Exhibit Dates: ***Version 1:*** 10/91–9/96. *Version 2:* 10/96–7/97. *Version 3:* 8/97–Present.
Learning Outcomes: ***Version 1:*** Upon completion of the course, the student will be able to train, direct, and lead personnel in the maintenance, operation, and deployment of tanks. *Version 2:* Upon completion of the course, the student will be able to fight; maintain, train, and sustain a tank crew; and perform duties of a tank commander. *Version 3:* Upon completion of the course, the student will be able to maintain, train, and sustain a tank crew; and perform duties of a tank commander.
Instruction: ***Version 1:*** Course includes lectures, demonstrations, and simulated exercises in the areas of supervision, leadership, decision making, and basic mechanical device theory. This course contains a common core of leadership subjects. *Version 2:* Lecture-discussion is used to cover the topics of armor tactics, secure communications, maintenance records, tank gunnery, mine warfare, tank weapons, troop-leading procedures, physical fitness training, and conduct of fire trainer (weaponry). This course contains a common core of leadership subjects. *Version 3:* Lecture-discussion is used to cover the topics of armor tactics, secure communications, maintenance records, tank gunnery, mine warfare, tank weapons, troop-leading procedures, physical fitness training, and conduct of fire trainer (weaponry). This course contains a common core of leadership subjects.
Credit Recommendation: ***Version 1:*** Credit is recommended for the common core only. See AR-1406-0090 (3/97). *Version 2:* Credit is recommended for the common core only. See AR-1406-0090 (3/97). *Version 3:* Credit is recommended for the common core only. See AR-1406-0090 (6/01).
Related Occupations: 19K.

AR-2201-0476

Master Gunner M60A3

Course Number: 020-ASID8 (M60A3).
Location: Armor Center and School, Fort Knox, KY.
Length: 11 weeks (456 hours).
Exhibit Dates: 1/90–Present.
Learning Outcomes: Upon completion of the course, the student will be able to practice advanced gunnery methods, weapons maintenance, and training management.
Instruction: Lectures and practical exercises cover weapons, turret maintenance, turret electronics, targeting, and communications.
Credit Recommendation: Credit is not recommended because of the military-specific nature of the course (3/97).

AR-2201-0477

Master Gunner (M60A3) Transition

Course Number: 020-ASID8 (M60A3)-T.
Location: Armor Center and School, Fort Knox, KY.
Length: 2 weeks (72 hours).
Exhibit Dates: 1/90–Present.
Learning Outcomes: Upon completion of the course, the student will be able to maintain and operate the turret, perform general gunnery tasks, and operate fire control systems and test equipment.
Instruction: Lectures and practical exercises cover weapons, turret maintenance, ammunition, fire control systems, night mission systems, and technical manuals.
Credit Recommendation: Credit is not recommended because of the military-specific nature of the course (3/97).

AR-2201-0478

1. M1/M1A1 Abrams Master Gunner
2. M1A1 Abrams Master Gunner

Course Number: ***Version 1:*** 020-ASIA8. *Version 2:* 020-ASIA8.
Location: Armor Center and School, Fort Knox, KY.
Length: ***Version 1:*** 11 weeks (387–446 hours). *Version 2:* 11 weeks (405 hours).
Exhibit Dates: ***Version 1:*** 1/90–11/97. *Version 2:* 12/97–Present.
Learning Outcomes: ***Version 1:*** Upon completion of the course, the student will be able to plan and supervise a gunnery program; manage a range; and supervise machine gun, turret electronics, targeting, communications systems, and turret maintenance. *Version 2:* Upon completion of the course, the student will be able to perform advanced gunnery methods, turret weapons maintenance, and training management.
Instruction: ***Version 1:*** Lectures and practical exercises cover weapons systems, turret maintenance, communications systems, targeting, and training management. *Version 2:* Lectures, demonstrations, and simulation in tank gunnery, turret maintenance, fire control systems, and training management.
Credit Recommendation: ***Version 1:*** Credit is not recommended because of the military-specific nature of the course (9/00). *Version 2:* In the lower-division baccalaureate/associate degree category, 3 semester hours in maintenance management (6/01).

AR-2201-0479

1. M1A1 Abrams Armor Crewman One Station Unit Training (OSUT)
2. M1A1 Abrams Armor Crewman

Course Number: ***Version 1:*** 19K10-OSUT. *Version 2:* 020-19K10; 19K10-OSUT; 19K10-OSUT, Phase 2; 19K10-OSUT, Phase 1.

Location: ***Version 1:*** Armor Center and School, Fort Knox, KY. *Version 2:* Armor Center, Fort Knox, KY.

Length: ***Version 1:*** 15–16 weeks (612–613 hours). *Version 2:* 15 weeks (700 hours).

Exhibit Dates: ***Version 1:*** 2/95–9/02. *Version 2:* 10/02–Present.

Learning Outcomes: ***Version 1:*** Before 2/95 see AR-2201-0330. Upon completion of the course, the student will be able to perform basic soldier tasks including driving and loading tanks. *Version 2:* Upon completion of the course, the student will be able to perform the duties of an armor crewman.

Instruction: ***Version 1:*** Instruction consists of lectures, demonstrations, and practices in military subjects, physical fitness, first aid, hazardous materials, military communications, land navigation, weapons, tactical training, specific vehicle operation, and situation training. This course includes an 8-9 week Basic Training component. *Version 2:* Classroom and practical exercises, discussion, and lecture. Topics include weapons training, land navigation, first aid, physical conditioning, safety, tank field exercises, and basic military subjects.

Credit Recommendation: ***Version 1:*** In the lower-division baccalaureate/associate degree category, 1 semester hour in physical education (3/97). *Version 2:* In the lower-division baccalaureate/associate degree category, 5 semester hours in military science and 1 in first aid (6/03).

Related Occupations: 19K.

AR-2201-0480

Armor Officer Advanced Reserve Component

Course Number: 2-17-C23.

Location: Armor Center and School, Fort Knox, KY.

Length: 2–3 weeks (95–99 hours).

Exhibit Dates: 10/94–Present.

Learning Outcomes: Upon completion of the course, the student will be able to command company-level armor forces and administer training at the company/team level.

Instruction: Instruction includes lectures and practical exercises in armor tactics, defensive positions, battlefield communications, combat orders, and offensive operations.

Credit Recommendation: Credit is not recommended because of the military-specific nature of the course (9/00).

AR-2201-0482

M1A2 Tank Commander Certification

Course Number: 2E-SI3J/020-ASIK4.

Location: Armor Center and School, Fort Knox, KY.

Length: 3–4 weeks (136 hours).

Exhibit Dates: 1/96–Present.

Learning Outcomes: Upon completion of the course, the student will be able to operate, maintain, and use weapons on the M1A2 battle tank.

Instruction: Instruction is provided using conventional classroom demonstrations and lectures, simulators, and actual training in various systems of the M1A2 tank. Topics covered include crew stations and duties, boresighting, plumb and synchronization, ammunition, gunnery, troubleshooting, crew maintenance, driving, and weapons.

Credit Recommendation: Credit is not recommended because of the military-specific nature of the course (9/00).

AR-2201-0483

M3 Scout Commander Certification

Course Number: 250-ASID3; 2E-SI3X.

Location: Armor Center and School, Fort Knox, KY.

Length: 3–4 weeks (123–124 hours).

Exhibit Dates: 1/90–Present.

Learning Outcomes: Upon completion of the course, the student will be able to operate and perform routine maintenance on armored calvary vehicles, install and disarm mines, calculate demolitions and explosions, use communications equipment, read and interpret maps, use intelligence reports, and conduct a dismounted patrol.

Instruction: Course uses lectures, demonstrations, and practical exercises to cover the topics of vehicle specifics, patrolling, hazardous materials, communications, map reading and land navigation, and intelligence.

Credit Recommendation: Credit is not recommended because of the military-specific nature of the course (3/97).

AR-2201-0484

M60A3 Tank Commander Certification

Course Number: 020-ASIB8; 2E-F85.

Location: Armor Center and School, Fort Knox, KY.

Length: 2 weeks (88–89 hours).

Exhibit Dates: 1/90–Present.

Learning Outcomes: Upon completion of the course, the student will be able to operate all systems on an M60A3 tank and perform all tasks by the various crewmembers.

Instruction: Includes lectures, demonstrations, conferences, and performance exercises in crew stations and duties, boresighting, turret troubleshooting, crew maintenance, driving, weapons, and tank gunnery.

Credit Recommendation: Credit is not recommended because of the military-specific nature of the course (3/97).

AR-2201-0485

M60A3 Tank Commander Certification

Course Number: 020-F6; 2E-F85.

Location: Armor Center and School, Fort Knox, KY.

Length: 2 weeks (76 hours).

Exhibit Dates: 1/90–Present.

Learning Outcomes: Upon completion of the course, the student will be able to operate all systems specific to an M60A3 tank and perform all tasks by the various crewmembers.

Instruction: The course includes lectures, demonstrations, conferences, and performance exercises in crew stations and duties, boresighting, turret troubleshooting, crew maintenance, driving, weapons, and tank gunnery.

Credit Recommendation: Credit is not recommended because of the military-specific nature of the course (3/97).

AR-2201-0486

M113/M901 Scout Commander Certification

Course Number: 250-F19; 2E-F89.

Location: Armor Center and School, Fort Knox, KY.

Length: 2 weeks (111–112 hours).

Exhibit Dates: 1/90–Present.

Learning Outcomes: Upon completion of the course, the student will be able to perform the duties of scout commander.

Instruction: This course includes lectures, conferences, demonstrations, and performance exercises in communications, weapons, NBC, intelligence, vehicle operations, maintenance and tactics.

Credit Recommendation: Credit is not recommended because of the military-specific nature of the course (3/97).

AR-2201-0487

M1A1 Tank Commander Certification

Course Number: 020-19K20/30/40; 2E-SI3M.

Location: Armor Center and School, Fort Knox, KY.

Length: 2 weeks (87–89 hours).

Exhibit Dates: 10/90–Present.

Learning Outcomes: Upon completion of the course, the student will be able to operate a tank and weaponry and function as a tank commander.

Instruction: Using lectures, demonstrations, and practical exercises, the topics covered are tank weapon systems use and maintenance, tank driving, tank safety, tank crew gunnery, machine operation, and maintenance.

Credit Recommendation: Credit is not recommended because of the military-specific nature of the course (3/97).

Related Occupations: 19K.

AR-2201-0488

Cavalry Scout Reclassification Reserve Component

Course Number: RC-171-19D20; RC-171-19D10.

Location: Armor Center and School, Fort Knox, KY.

Length: 2–3 weeks (95 hours).

Exhibit Dates: 6/90–9/99.

Learning Outcomes: After 9/99 see AR-2201-0535. Upon completion of the course, the student will be able to perform military communications, nuclear/biological/chemical operations, and land navigation; use weapons in field exercises; and apply tactical training to intelligence, combat engineering, and communications.

Instruction: Lectures and practical exercises cover nuclear/biological/chemical warfare, land navigation, communications, weapons systems, and mine removal.

Credit Recommendation: Credit is not recommended because of the military-specific nature of the course (3/97).

Related Occupations: 19D.

AR-2201-0489

M1A2 Tank Operations and Maintenance (Unit Level)

Course Number: 611-ASIK4 (45E/63E).
Location: Armor Center and School, Fort Knox, KY.
Length: 3 weeks (120 hours).
Exhibit Dates: 1/96–Present.
Learning Outcomes: Upon completion of the course, the student will be able to troubleshoot, remove, and install electrical/electronic units on the M1A2 Abrams tank.
Instruction: This course includes lectures, demonstrations, conferences, and performance exercises in inspecting, troubleshooting, and removing and replacing electrical/electronic units, including driver's control panel, driver's electronic unit, gunner's primary sight, laser range finder, gunner's auxiliary sight, thermal receiver unit, miniature electronic unit, commander's independent thermal sight viewer electronic unit, gunner's control display panel, fire control electronic unit, commander's integrated display, image control unit, power control unit, and thermal sight unit.
Credit Recommendation: Credit is not recommended because of the military-specific nature of the course (3/97).

AR-2201-0490

M1A1 Abrams Tank Turret Mechanics

Course Number: 643-45E10.
Location: Armor Center and School, Fort Knox, KY.
Length: 9 weeks (353 hours).
Exhibit Dates: 7/93–12/99.
Learning Outcomes: Before 7/93 see AR-1703-0050. Upon completion of the course, the student will be able to perform unit maintenance on the vehicle-mounted armament, associated fire control, and related systems on the M1A1 tank turret system.
Instruction: This course includes lectures, demonstrations, videos, and performance exercises in servicing, and troubleshooting the tank turret systems, including mechanical, hydraulic, and electrical systems.
Credit Recommendation: Credit is not recommended because of the military-specific nature of the course (3/97).
Related Occupations: 45E.

AR-2201-0491

M901A1 (ITV) Gunner/Crew Training

Course Number: 643-ASIE9.
Location: Armor Center and School, Fort Knox, KY.
Length: 1 week (50 hours).
Exhibit Dates: 1/90–Present.
Learning Outcomes: Upon completion of the course, the student will be able to perform the task of a gunner on the M901A1 launch vehicle.
Instruction: Methods include lectures, demonstrations, and hands-on experience with the vehicle as well as the launch system.
Credit Recommendation: Credit is not recommended because of the military-specific nature of the course (3/97).

AR-2201-0492

M1/M1A1 Tank Commander Certification (T)

Course Number: 020-19K40 (T); 020-19K30 (T); 020-19K20 (T); 2E-SI3M.
Location: Armor Center and School, Fort Knox, KY.
Length: 2 weeks (87–88 hours).
Exhibit Dates: 10/90–10/00.
Learning Outcomes: Upon completion of the course, the student will be able to prepare, calibrate, operate, secure, and troubleshoot the operational systems of the A1/A1A1 tank, including the fire control system, loader's platform, driver's station, commander station, and gunner's station.
Instruction: This course includes lectures, demonstrations, and performance exercises in the duties and operation of the gunner's station, driver's station, loader's station, and the commander's station on the A1/A1A1 tank.
Credit Recommendation: Credit is not recommended because of the military-specific nature of the course (3/97).
Related Occupations: 19K.

AR-2201-0493

Recruiting Pre-Command

Course Number: *Version 1:* 7C-F41. *Version 2:* 7C-F41.
Location: *Version 1:* Soldier Support Institute, Fort Jackson, SC. *Version 2:* Recruiting and Retention School, Fort Jackson, SC.
Length: *Version 1:* 2 weeks (76 hours). *Version 2:* 2 weeks (80 hours).
Exhibit Dates: *Version 1:* 7/97–7/03. *Version 2:* 8/03–Present.
Learning Outcomes: *Version 1:* Upon completion of the course, the student will be able to become an officer in the recruiting command. *Version 2:* Pending evaluation.
Instruction: *Version 1:* Methods of instruction include conferences, practical exercises, and guest speakers. Topics covered include leadership training, command responsibilities, recounting integrity, and personnel procurement. *Version 2:* Pending evaluation.
Credit Recommendation: *Version 1:* Credit is not recommended because of the military-specific nature of the course (9/97). *Version 2:* Pending evaluation.

AR-2201-0494

National Guard Officer Candidate

Course Number: OCS.
Location: National Guard Academies, US.
Length: 11 weeks (497 hours).
Exhibit Dates: 9/94–Present.
Learning Outcomes: Upon completion of the course, the student will demonstrate leadership and management skills in training, motivating, counseling, disciplining, and evaluating personnel; planning, organizing, and directing organization processes and procedures; reading and interpreting complex maps; and making navigational calculations.
Instruction: This course includes lectures, readings, discussions, exercises, and examination by tests and performance related to human resource management, leadership, writing, records, training, direction, and motivation.
Credit Recommendation: In the lower-division baccalaureate/associate degree category, 3 semester hours in leadership and 3 in management (9/97).

AR-2201-0495

Small Arms/Artillery Repairer

Course Number: 641-45B10.
Location: Ordnance Center and School, Aberdeen Proving Ground, MD.
Length: 11–12 weeks (468–480 hours).
Exhibit Dates: 10/96–Present.
Learning Outcomes: Upon completion of the course, the student will be able to maintain and repair small arms, machine guns, mortars, and artillery weapons.
Instruction: Lectures and practical exercises cover diagnosis, repair, and maintenance of small arms, machine guns, mortars, and artillery.
Credit Recommendation: In the lower-division baccalaureate/associate degree category, 1 semester hour in basic hydraulics, 3 in fundamentals of gunsmithing, and 3 in maintenance and repair of pump shotguns and automatic rifles (11/02).
Related Occupations: 45B.

AR-2201-0496

National Guard Scout Sniper

Course Number: MTC-02-01.
Location: Professional Education Center, Cp. Joseph Robinson, North Little Rock, AR.
Length: 3 weeks (300 hours).
Exhibit Dates: 10/95–Present.
Learning Outcomes: Upon completion of the course, the student will be able to deliver long-range, precision fire to a designated target. Knowledge and skills acquired may be used to provide pre-sniper training to other personnel.
Instruction: Performance based instruction includes lectures and practical exercises on the planning and conduct of antisniper missions stealth skills, collecting and reporting information and marksmanship. Certification is dependent upon student performance on marksmanship range.
Credit Recommendation: In the lower-division baccalaureate/associate degree category, 3 semester hours in marksmanship/military science (9/03).

AR-2201-0497

Total Army Training System (TATS) Multiple Launch Rocket System (MLRS) Platoon Sergeant Advanced Noncommissioned Officer (NCO)

Course Number: 061-13M40 (F).
Location: Reserve Schools, Multiple locations; Field Artillery School, Fort Sill, OK.
Length: 92–110 hours.
Exhibit Dates: 10/98–Present.
Learning Outcomes: Upon completion of the course, the student will be able to perform leadership and management tasks in a firing battery.
Instruction: The course consists of lectures and practical experiences on mission directives, mapping tactics, platoon defense, logistics, leadership, and supervision.
Credit Recommendation: Credit is not recommended because of the military-specific nature of the course (9/03).
Related Occupations: 13M.

AR-2201-0498

Total Army Training System (TATS) Multiple Launch Rocket System (MLRS) Platoon Sergeant Advanced Noncommissioned Officer (NCO)

Course Number: *Version 1:* 042-13M40 (F). *Version 2:* 042-13M40 (F).

Location: Field Artillery School, Fort Sill, OK.

Length: *Version 1:* 3–4 weeks (253 hours). *Version 2:* 1–2 weeks (104–110 hours).

Exhibit Dates: *Version 1:* 10/98–9/00. *Version 2:* 10/00–Present.

Learning Outcomes: *Version 1:* Upon completion of the course, the student will be able to perform leadership and management tasks in a firing battery. *Version 2:* Upon completion of the course, the student will be able to perform as a platoon sergeant of firing battery.

Instruction: *Version 1:* The course consists of lectures and practical experiences on mission directives, mapping tactics, platoon defense, logistics, leadership, and supervision. This course includes a common core of leadership subjects. *Version 2:* The course consists of lectures and practical experiences on mission directives, mapping tactics, platoon defense, logistics, leadership, and supervision.

Credit Recommendation: *Version 1:* Credit is recommended for the common core only. See AR-1404-0035 (6/00). *Version 2:* Credit is not recommended because of the military-specific nature of the course (9/03).

Related Occupations: 13M.

AR-2201-0499

Total Army Training System (TATS) Multiple Launch Rocket System (MLRS) Operations/Fire Direction Section Chief Basic Noncommissioned Officer (NCO)

Course Number: *Version 1:* 043-13P30 (F). *Version 2:* 043-13P30 (F), Phase 2.

Location: Field Artillery School, Fort Sill, OK.

Length: *Version 1:* 4 weeks (191 hours). *Version 2:* 1–2 weeks (111 hours).

Exhibit Dates: *Version 1:* 10/97–9/00. *Version 2:* 10/00–Present.

Learning Outcomes: *Version 1:* Upon completion of the course, the student will be able to compute firing data, select communications equipment, write target/mission reports, and lead a section. *Version 2:* Upon completion of the course, the student will be able to perform firing data computation, equipment maintenance, and communications system operations.

Instruction: *Version 1:* Lecture and practical exercise cover the supervision, operation, and maintenance of the launch fire direction system. This course includes a common core of leadership subjects. *Version 2:* Lectures and field exercises cover computation of firing data, ammunition, equipment maintenance, fire missions, communications, and leadership.

Credit Recommendation: *Version 1:* In the lower-division baccalaureate/associate degree category, 1 semester hour in maintenance management and 1 in safety. See AR-1406-0090 for common core credit (6/00). *Version 2:* In the lower-division baccalaureate/associate degree category, 1 semester hour in maintenance management and 1 in safety (1/01).

Related Occupations: 13P.

AR-2201-0500

Total Army Training System (TATS) Multiple Launch Rocket System (MLRS) Operations/Fire Direction Section Chief Basic Noncommissioned Officer (NCO)

Course Number: *Version 1:* 061-13P30 (F), Phase 2. *Version 2:* 061-13P30 (F), Phase 2.

Location: *Version 1:* Reserve Component Training Centers, US; Field Artillery School, Fort Sill, OK. *Version 2:* Field Artillery School, Fort Sill, OK.

Length: *Version 1:* 106 hours. *Version 2:* 1–2 weeks (111 hours).

Exhibit Dates: *Version 1:* 10/97–9/00. *Version 2:* 10/00–Present.

Learning Outcomes: *Version 1:* Upon completion of the course, the student will be able to compute firing data, select communications equipment, write target/mission reports, and lead a section. *Version 2:* Upon completion of the course, the student will be able to perform firing data computation, equipment maintenance, and communications systems operation.

Instruction: *Version 1:* Lecture and practical exercise cover the supervision, operation, and maintenance of the fire direction system. *Version 2:* Lectures and field exercises cover computation of firing data, ammunition, equipment maintenance, fire missions, and communications and leadership.

Credit Recommendation: *Version 1:* In the lower-division baccalaureate/associate degree category, 1 semester hour in maintenance management and 1 in safety (6/00). *Version 2:* In the lower-division baccalaureate/associate degree category, 1 semester hour in maintenance management and 1 in safety (1/01).

Related Occupations: 13P.

AR-2201-0501

1. Total Army Training System (TATS) Field Artillery Senior Radar/Targeting Sergeant Advanced Noncommissioned Officer (NCO)
2. Total Army Training System (TATS) Field Artillery Senior Radar/Targeting Sergeant Advanced Noncommissioned Officer (NCO), Phase 2

Course Number: *Version 1:* 221-13R40 (F). *Version 2:* 221-13R40 (F), Phase 2.

Location: Field Artillery School, Fort Sill, OK.

Length: *Version 1:* 10 weeks (407 hours). *Version 2:* 5 weeks (245 hours).

Exhibit Dates: *Version 1:* 6/96–12/99. *Version 2:* 1/00–Present.

Learning Outcomes: *Version 1:* Upon completion of the course, the student will use tactical and technical knowledge and supervisory skills to perform targeting in field artillery tactical missions, recognize the limitations of field equipment, decide targeting processes for counterfire operations, and deploy field equipment. The graduate will be able to use digital communications devices and computers to support field operations. *Version 2:* Upon completion of the course, the student will be able to perform both technical and supervisory skills in a field artillery environment.

Instruction: *Version 1:* The courses consists of classroom lectures and discussions, home study using CDs, field exercises, and laboratory experiments. this course includes a common core of leadership subjects. *Version 2:* The courses consists of classroom lectures and field experiences in the supervisory skills needed to perform targeting. The skills taught include fire support structure, command estimate process, target value analysis, tactical operations, and employment of equipment. This course includes a common core of leadership subjects.

Credit Recommendation: *Version 1:* Credit is recommended for the common core only. See AR-1404-0035 (6/00). *Version 2:* Credit is recommended for the common core only. See AR-1404-0035 (9/03).

Related Occupations: 13R.

AR-2201-0502

Total Army Training System (TATS) Field Artillery Senior Radar/Targeting Sergeant Advanced Noncommissioned Officer (NCO)

Course Number: 061-13R40 (F), Phase 3.

Location: Field Artillery School, Fort Sill, OK.

Length: 130 hours.

Exhibit Dates: 6/96–12/99.

Learning Outcomes: After 12/99 see AR-2201-0511. Upon completion of the course, the student will use tactical and supervisory skills to perform targeting in field artillery tactical missions, recognize the limitations of field equipment, decide targeting processes for counterfire operations, and deploy field equipment. The graduate will be able to use digital communications devices and computers to support field operations.

Instruction: The courses consists of classroom lectures and discussions, home study using CDs, field exercises, and laboratory experiments.

Credit Recommendation: Credit is not recommended because of the military-specific nature of the course (6/00).

Related Occupations: 13R.

AR-2201-0503

Total Army Training System (TATS) Cannon Fire Direction Section Chief Basic Noncommissioned Officer (NCO)

Course Number: 250-13E30 (F).

Location: Field Artillery School, Fort Sill, OK.

Length: 5–6 weeks (232 hours).

Exhibit Dates: 10/97–Present.

Learning Outcomes: Upon completion of the course, the student will be able to supervise a battery fire direction center.

Instruction: Lectures and practical experiences cover leadership, tactics, communications, land navigation, weapons, fire data, and maintenance. This course contains a common core of leadership subjects.

Credit Recommendation: Credit is recommended for the common core only. See AR-1406-0090 (6/00).

Related Occupations: 13E.

AR-2201-0504

Total Army Training System (TATS) Cannon Fire Direction Section Chief Basic Noncommissioned Officer (NCO)

Course Number: 061-13E30 (F), Phase 2.

Location: Field Artillery School, Fort Sill, OK.

Length: 147 hours.

Exhibit Dates: 10/97–Present.
Learning Outcomes: Upon completion of the course, the student will be able to supervise a battery fire direction center.
Instruction: Lectures and practical experiences cover leadership, tactics, communications, land navigation, weapons, fire data, and maintenance.
Credit Recommendation: Credit is not recommended because of the military-specific nature of the course (6/00).
Related Occupations: 13E.

AR-2201-0505

Total Army Training System (TATS) Fire Support Sergeant Advanced Noncommissioned Officer (NCO)

Course Number: 250-13F40 (F).
Location: Field Artillery School, Fort Sill, OK.
Length: 8 weeks (334 hours).
Exhibit Dates: 9/98–Present.
Learning Outcomes: Upon completion of the course, the student will be able to train others in modern battlefield tactics, techniques, and procedures.
Instruction: The course consists of classroom instruction on the capability and organization of the fire support team, including the duties, responsibilities, and mission of fire support team personnel and techniques necessary to perform, lead, train, and evaluate personnel. This course contains a common core of leadership subjects.
Credit Recommendation: Credit is recommended for the common core only. See AR-1404-0035 (6/00).
Related Occupations: 13F.

AR-2201-0506

Total Army Training System (TATS) Fire Support Sergeant Advanced Noncommissioned Officer (NCO), Phase 2

Course Number: 061-13F40, Phase 2.
Location: Reserve Component Training Centers, US; Field Artillery School, Fort Sill, OK.
Length: 160 hours.
Exhibit Dates: 9/98–Present.
Learning Outcomes: Upon completion of the course, the student will be able to train others in modern battlefield tactics, techniques, and procedures.
Instruction: The course consists of classroom instruction on the capability and organization of the fire support team, including the duties, responsibilities, and mission of fire support team personnel and techniques necessary to perform, lead, train, and evaluate personnel.
Credit Recommendation: Credit is not recommended because of the military-specific nature of the course (6/00).
Related Occupations: 13F.

AR-2201-0507

Total Army Training System (TATS) Field Artillery Survey Senior Sergeant Advanced Noncommissioned Officer (NCO)

Course Number: 412-82C40 (F).
Location: Field Artillery School, Fort Sill, OK.
Length: 4 weeks (111–179 hours).
Exhibit Dates: 10/98–Present.
Learning Outcomes: Upon completion of the course, the student will be able to plan operations and train others to plan and perform field artillery survey operations and tactical communications.
Instruction: Lectures and exercises cover artillery surveying, communications, equipment, systems, and organization. This course contains a common core of leadership subjects.
Credit Recommendation: In the lower-division baccalaureate/associate degree category, 1 semester hour in land surveying and 1 in communications. See AR-1404-0035 for common core credit (9/03).
Related Occupations: 82C.

AR-2201-0508

Total Army Training System (TATS) Field Artillery Survey Senior Sergeant Advanced Noncommissioned Officer (NCO)

Course Number: 061-82C40 (F).
Location: Reserve Component Training Centers, US; Field Artillery School, Fort Sill, OK.
Length: 96–111 hours.
Exhibit Dates: 10/98–Present.
Learning Outcomes: Upon completion of the course, the student will be able to and train others to plan and perform field artillery survey operations and tactical communications.
Instruction: Lectures and exercises cover artillery surveying, communications, equipment, systems, and organization.
Credit Recommendation: In the lower-division baccalaureate/associate degree category, 1 semester hour in land surveying and 1 in communications (9/03).
Related Occupations: 82C.

AR-2201-0509

Total Army Training System (TATS) Multiple Launch Rocket System (MLRS) Section Chief Basic Noncommissioned Officer (NCO)

Course Number: 042-13M30 (F).
Location: Field Artillery School, Fort Sill, OK.
Length: 4 weeks (194 hours).
Exhibit Dates: 10/97–9/00.
Learning Outcomes: After 9/00 see AR-1408-0271. Upon completion of the course, the student will be able lead a rocket system firing section.
Instruction: Lectures and exercises cover leadership, fire planning, safety, training, preventive maintenance, and supervision. This course includes a common core of leadership subjects.
Credit Recommendation: Credit is recommended for the common core only. See AR-1406-0090 (6/00).
Related Occupations: 13M.

AR-2201-0510

Total Army Training System (TATS) Multiple Launcher Rocket System (MLRS) Section Chief Basic Noncommissioned Officer (NCO)

Course Number: 061-13M30 (F).
Location: Reserve Component Training Centers, US; Field Artillery School, Fort Sill, OK.
Length: 109 hours.
Exhibit Dates: 10/97–9/00.
Learning Outcomes: After 9/00 see AR-1408-0272. Upon completion of the course, the student will be able lead a rocket system firing section.
Instruction: Lectures and practical exercises cover leadership, fire planning, safety, training, preventive maintenance, and supervision.
Credit Recommendation: Credit is not recommended because of the military-specific nature of the course (6/00).
Related Occupations: 13M.

AR-2201-0511

Total Army Training System (TATS) Field Artillery Senior Radar/Targeting Sergeant Advanced Noncommissioned Officer (NCO), Phase 2

Course Number: 061-13R40 (F), Phase 2.
Location: Field Artillery School, Fort Sill, OK.
Length: 143–177 hours.
Exhibit Dates: 1/00–Present.
Learning Outcomes: Before 1/00 see AR-2201-0502. Upon completion of the course, the student will be able to apply both technical and supervisory skills in a field artillery environment.
Instruction: This course consists of classroom lectures and field experience in the supervisory skills needed to perform targeting. The skills taught include fire support structure, command estimate process, target value analysis, tactical operations, and employment of equipment.
Credit Recommendation: Credit is not recommended because of the military-specific nature of the course (9/03).
Related Occupations: 13R.

AR-2201-0512

Total Army Training System (TATS) Automated Fire Support Systems Senior Sergeant Advanced Noncommissioned Officer (NCO)

Course Number: 250-13C40 (F).
Location: Field Artillery School, Fort Sill, OK.
Length: 4–5 weeks (200 hours).
Exhibit Dates: 10/98–Present.
Learning Outcomes: Upon completion of the course, the student will be able to lead field artillery tactical data systems unit.
Instruction: Lectures and practical exercises cover leadership, training, management, and fire planning. This course contains a common core of leadership subjects.
Credit Recommendation: Credit is recommended for the common core only. See AR-1404-0035 (6/00).
Related Occupations: 13C.

AR-2201-0513

Total Army Training System (TATS) Automated Fire Support Systems Senior Sergeant Advanced Noncommissioned Officer (NCO)

Course Number: 061-13C40.
Location: Reserve Component Training Centers, US; Field Artillery School, Fort Sill, OK.
Length: 117 hours.

Exhibit Dates: 10/98–Present.

Learning Outcomes: Upon completion of the course, the student will be able to lead a field artillery tactical data systems unit.

Instruction: Lectures and practical exercises cover leadership, training, management, and fire planning.

Credit Recommendation: Credit is not recommended because of the military-specific nature of the course (6/00).

Related Occupations: 13C.

AR-2201-0516

Total Army Training System (TATS) Field Artillery Tactical Data Systems Specialist

Course Number: *Version 1:* 250-13D10 (F). *Version 2:* 250-13D10 (F).

Location: Field Artillery School, Fort Sill, OK.

Length: *Version 1:* 7 weeks (263 hours). *Version 2:* 7–8 weeks (287 hours).

Exhibit Dates: *Version 1:* 10/00–12/02. *Version 2:* 1/03–Present.

Learning Outcomes: *Version 1:* Upon completion of the course, the student will be able to operate a field artillery tactical data system and communications systems. *Version 2:* Upon completion of the course, the student will be able to operate a field artillery tactical data system and communications system.

Instruction: *Version 1:* Lectures and practical exercises cover field artillery communications, tactics, fire direction, and operation of the tactical data systems. *Version 2:* Audiovisual materials, practical and classroom exercises, lectures, and computer-based training. Topics include military tactics, data communications, Advanced Field Artillery Tactical Data System (AFATADS) operation, and manual fire direction.

Credit Recommendation: *Version 1:* Credit is not recommended because of the military-specific nature of the course (1/01). *Version 2:* In the lower-division baccalaureate/associate degree category, 1 semester hour in military science and 3 in computer fundamentals (10/03).

Related Occupations: 13D.

AR-2201-0517

Total Army Training System (TATS) Field Artillery (FA) Tactical Data Systems Specialist

Course Number: *Version 1:* 061-13D10 (F). *Version 2:* 061-13D10 (F).

Location: *Version 1:* Field Artillery School, Fort Sill, OK. *Version 2:* Reserve Schools, Multiple locations.

Length: *Version 1:* 128 hours. *Version 2:* 7 weeks (287 hours).

Exhibit Dates: *Version 1:* 10/00–12/02. *Version 2:* 1/03–Present.

Learning Outcomes: *Version 1:* Upon completion of the course, the student will be able to operate a field artillery tactical data system and communications systems. *Version 2:* Upon completion of the course, the student will be able to operate a field artillery tactical data system and communications system.

Instruction: *Version 1:* Lectures and practical exercises in field artillery communications, tactics, fire direction, and operation of the tactical data systems. *Version 2:* Practical and classroom exercises, lectures, and computer-based training. Topics include field artillery communications, tactics, fire direction, and operation of tactical data systems.

Credit Recommendation: *Version 1:* Credit is not recommended because of the military-specific nature of the course (1/01). *Version 2:* In the lower-division baccalaureate/associate degree category, 2 semester hours in computer fundamentals (10/03).

Related Occupations: 13D.

AR-2201-0518

Total Army Training System (TATS) Cannon Fire Direction Section Chief Basic Noncommissioned Officer (NCO)

Course Number: 061-13E30 (F), Phase 2.

Location: Reserve Schools, Multiple locations.

Length: 168 hours.

Exhibit Dates: 10/00–Present.

Learning Outcomes: Before 10/00 see AR-1402-0160. Upon completion of the course, the student will be able to lead, train, and direct the personnel of a fire direction section.

Instruction: Lectures and practical exercises cover leadership, maintenance, tactics, communications, navigation, weapons, training, management, and fire control.

Credit Recommendation: Credit is not recommended because of the military-specific nature of the course (1/01).

Related Occupations: 13E.

AR-2201-0519

Total Army Training System (TATS) Cannon Fire Direction Section Chief Basic Noncommissioned Officer (NCO)

Course Number: 250-13E30 (F), Phase 2.

Location: Field Artillery School, Fort Sill, OK.

Length: 4 weeks (192 hours).

Exhibit Dates: 10/00–Present.

Learning Outcomes: Before 10/00 see AR-1402-0160. Upon completion of the course, the student will be able to lead, train, and direct the personnel of a fire direction section.

Instruction: Lectures and practical exercises cover leadership, maintenance, tactics, communications, navigation, weapons, training, management, and fire control.

Credit Recommendation: Credit is not recommended because of the military-specific nature of the course (1/01).

Related Occupations: 13E; 82C.

AR-2201-0520

Total Army Training System (TATS) Field Artillery Survey Senior Sergeant Advanced Noncommissioned Officer (NCO)

Course Number: 412-82C40 (F), Phase 2.

Location: Field Artillery School, Fort Sill, OK.

Length: 1–2 weeks (108 hours).

Exhibit Dates: 10/00–Present.

Learning Outcomes: Before 10/00 see AR-2201-0293. Upon completion of the course, the student will apply leadership skills, plan and train personnel in planning and performing field artillery survey operations, use tactical communications equipment, and perform new surveying techniques.

Instruction: Classroom and field operations cover the use of survey equipment and artillery equipment.

Credit Recommendation: In the lower-division baccalaureate/associate degree category, 3 semester hours for field experience in management (1/01).

Related Occupations: 82C.

AR-2201-0521

Total Army Training System (TATS) Field Artillery Survey Senior Sergeant Advanced Noncommissioned Officer (NCO)

Course Number: 061-82C40 (F), Phase 2.

Location: Reserve Schools, Multiple locations.

Length: 108 hours.

Exhibit Dates: 10/00–Present.

Learning Outcomes: Before 10/00 see AR-2201-0293. Upon completion of the course, the student will apply common leadership skills, plan and train personnel in planning and performing field artillery survey operations, use tactical communications equipment, and perform new surveying techniques.

Instruction: Classroom and field operations cover the use of survey equipment and artillery equipment.

Credit Recommendation: In the lower-division baccalaureate/associate degree category, 3 semester hours for field experience in management (1/01).

Related Occupations: 82C.

AR-2201-0522

Total Army Training System (TATS) Cannon Section Chief Basic Noncommissioned Officer (NCO)

Course Number: 061-13B30 (F), Phase 3.

Location: Reserve Schools, Multiple locations.

Length: 243 hours.

Exhibit Dates: 1/97–9/00.

Learning Outcomes: Upon completion of the course the student will be able to lead, train, and direct subordinates in the maintenance, safe operation, and deployment of weapons and equipment.

Instruction: Lectures and practical exercises are used to prepare students to supervise the positioning, arming, maintaining, and firing a howitzer. Course covers safety issues in using ammunition aiming techniques, and positioning of the howitzer using navigational aids. This course contains a common core of leadership subjects.

Credit Recommendation: Credit is recommended for the common core only. See AR-2201-0337 (1/01).

Related Occupations: 13B.

AR-2201-0523

Total Army Training System (TATS) Cannon Section Chief Basic Noncommissioned Officer (NCO)

Course Number: 041-13B30 (F); 041-13B30.

Location: Field Artillery School, Fort Sill, OK.

Length: 6 weeks (370 hours).

Exhibit Dates: 1/97–9/00.

Learning Outcomes: Upon completion of the course, the student will be able to lead, train, and direct subordinates in the maintenance, safe operation, and deployment of weapons and equipment.

Instruction: Lectures and practical exercises prepare students to supervise the positioning, arming, maintaining, and firing a howitzer. Course covers safety issues in using ammunition, aiming techniques, and positioning of the howitzer using navigational aids and maintenance of a communications radio. This course contains a common core of leadership subjects.

Credit Recommendation: Credit is recommended for the common core only. See AR-1406-0090 (1/01).

Related Occupations: 13B.

AR-2201-0524

Total Army Training System (TATS) Field Artillery Cannon Section Chief Basic Noncommissioned Officer (NCO)

Course Number: 041-13B30 (F), Phase 2.

Location: Field Artillery School, Fort Sill, OK.

Length: 2–3 weeks (137 hours).

Exhibit Dates: 10/00–Present.

Learning Outcomes: Before 10/00 see AR-2201-0362. Upon completion of the course, the student will be able to supervise a crew to safely transport, load, and aim five types of field artillery.

Instruction: Laboratory and practical exercises cover safely transporting equipment; preparing a howitzer for firing; using a map, compass, and dead reckoning to aim the artillery piece; and supervising a crew to maintain and perform fire control alignment tests.

Credit Recommendation: Credit is not recommended because of the military-specific nature of the course (1/01).

Related Occupations: 13B.

AR-2201-0525

Total Army Training System (TATS) Field Artillery Cannon Section Chief Basic Noncommissioned Officer (NCO)

Course Number: 061-13B30 (F), Phase 3.

Location: Reserve Schools, Multiple locations.

Length: 137 hours.

Exhibit Dates: 10/00–Present.

Learning Outcomes: Upon completion of the course, the student will be able to supervise a crew to safely transport, load, and aim five types of field artillery.

Instruction: Laboratory and practical exercises cover safely transporting equipment; preparing a howitzer for firing; using a map, compass, and dead reckoning to aim the artillery piece; and supervising a crew to maintain and perform fire control alignment tests.

Credit Recommendation: Credit is not recommended because of the military-specific nature of the course (1/01).

Related Occupations: 13B.

AR-2201-0526

Total Army Training System (TATS) Cannon Crewmember

Course Number: *Version 1:* 13B10-OSUT (F). *Version 2:* 13B10-OSUT (F).

Location: Field Artillery School, Fort Sill, OK.

Length: *Version 1:* 14 weeks (804 hours). *Version 2:* 14 weeks (210 hours).

Exhibit Dates: *Version 1:* 10/97–3/03. *Version 2:* 4/03–Present.

Learning Outcomes: *Version 1:* Upon completion of the course, the student will be able to perform basic combat survival skills, load and fire Howitzers, and perform preventive and other maintenance on artillery equipment. *Version 2:* Upon completion of the course, the student will be able to perform basic combat survival skills, load and fire Howitzers, and perform preventive and other maintenance on artillery equipment.

Instruction: *Version 1:* The course consists of classroom lectures, computer-based training, and field exercises covering the maintenance and uses of Howitzers, including loading and firing, and basic training. *Version 2:* Computer-based training, field exercises, and lectures. Topics include maintenance and use of Howitzers, including loading and firing, and basic training.

Credit Recommendation: *Version 1:* In the lower-division baccalaureate/associate degree category, 1 semester hour in personal community health, 1 in physical education, and 2 in basic military science (1/01). *Version 2:* In the lower-division baccalaureate/associate degree category, 1 semester hour in personal community health, 1 in physical education, and 2 in basic military science (10/03).

Related Occupations: 13B.

AR-2201-0527

Total Army Training System (TATS) Cannon Crewmember

Course Number: 061-13B10-M119 (F); 061-13B10 M10945 (F); 061-13B10-M198 (F); 061-13B10-M102 (F).

Location: Reserve Schools, Multiple locations.

Length: 6–7 weeks (109–127 hours).

Exhibit Dates: 10/97–Present.

Learning Outcomes: Upon completion of the course, the student will be able to perform basic combat survival skills, load and fire Howitzers, and perform preventive and other maintenance on artillery equipment.

Instruction: The course consists of classroom lectures and field exercises covering the maintenance and uses of Howitzers, including loading, firing, and basic training.

Credit Recommendation: In the lower-division baccalaureate/associate degree category, 1 semester hour in personal community health and 1 in physical education (10/03).

Related Occupations: 13B.

AR-2201-0528

Special Forces Communications Sergeant, Phase 2

Course Number: 011-18E30, Phase 2.

Location: John F. Kennedy Special Warfare Center and School, Fort Bragg, NC.

Length: 6 weeks (590 hours).

Exhibit Dates: 4/00–Present.

Learning Outcomes: Upon completion of the course, the student will be able to perform mission planning, field training, special operations, and air operations.

Instruction: Lectures and practical exercises cover airborne operations, mission planning, weapons tactics, and operational techniques.

Credit Recommendation: In the lower-division baccalaureate/associate degree category, 1 semester hour in map reading, 1 in land navigation, and 1 in safety and survival techniques (4/01).

Related Occupations: 18E.

AR-2201-0529

Advanced Military Free-Fall Parachutist

Course Number: 011-F66.

Location: John F. Kennedy Special Warfare Center and School, Fort Bragg, NC.

Length: 10 weeks (364 hours).

Exhibit Dates: 10/95–Present.

Learning Outcomes: Upon completion of the course, the student will be able to execute advanced military free-fall parachute jumps.

Instruction: Lectures and practical exercises cover parachute rigging and packing, airborne operations, weather data analysis, jump techniques, aerial maneuvers, and jump safety.

Credit Recommendation: In the lower-division baccalaureate/associate degree category, 3 semester hours in advanced parachuting/skydiving (4/01).

AR-2201-0530

Air Defense Artillery Basic Noncommissioned Officer (NCO)

Course Number: 0-14-C40 (TATS); 0-14-C40, Phase 2.

Location: Air Defense Artillery School, Fort Bliss, TX.

Length: 5 weeks (230–234 hours).

Exhibit Dates: 1/01–Present.

Learning Outcomes: Before 1/01 see AR-2201-0422. Upon completion of the course, the student will be able to lead, train, and direct subordinates in combat operations.

Instruction: Lectures and practical exercises are conducted in a 24-hour a day environment. Topics include leadership, tactics, communications, and weapons. This course contains a common core of leadership subjects.

Credit Recommendation: Credit is not recommended because of the military-specific nature of the course (4/04).

AR-2201-0531

U.S. Military Academy Mounted Maneuver

Course Number: 000-MMT.

Location: Armor Center and School, Fort Knox, KY.

Length: 1 week (50 hours).

Exhibit Dates: 10/97–Present.

Learning Outcomes: Upon completion of the course, the student will be trained on combat combined arms equipment and operations.

Instruction: Lecture, observation, and exercises on heavy combat arms, operations, tactics, battle drills, and command/control.

Credit Recommendation: In the lower-division baccalaureate/associate degree category, 3 semester hours in military tactics (6/01).

AR-2201-0532

Total Army Training System (TATS) Armor Crewman (M1/M1A1) (Reclassification), Phase 2

Course Number: 171-19K10/20 (M1/M1A1) (F), Phase 2.

Location: Reserve Schools, Multiple locations.
Length: 120 hours.
Exhibit Dates: 10/00–Present.
Learning Outcomes: Upon completion of the course, the student will be able to perform all communications, map reading, intelligence, maintenance, and operations tasks associated with an M1 tank.
Instruction: Instruction is provided in classrooms with traditional facilities and in the field with the necessary equipment and vehicles. Hands-on vehicle training is required.
Credit Recommendation: In the lower-division baccalaureate/associate degree category, 3 semester hours in military tactics (6/01).
Related Occupations: 19K.

AR-2201-0533

Cavalry Scout Advanced Noncommissioned Officer (NCO), Phase 2

Course Number: 250-19D40, Phase 2.
Location: Armor Center and School, Fort Knox, KY.
Length: 11–12 weeks (606 hours).
Exhibit Dates: 10/00–Present.
Learning Outcomes: Upon completion of the course, the student will be able to apply leadership, supervision, and tactics skills in an armor environment.
Instruction: Learning outcomes will be accomplished by demonstrations, classroom activities, small-group instruction, and practical experience. Topics include tactics, training, maintenance, weapons, deployment, and peacekeeping operations.
Credit Recommendation: In the lower-division baccalaureate/associate degree category, 3 semester hours in tactics and 3 in principles of supervision (6/01).
Related Occupations: 19D; 19K.

AR-2201-0534

Total Army Training System (TATS) M1/M1A1 Tank Commander Certification

Course Number: 171-SI3M/19K2/3/4/ (F).
Location: Armor Center and School, Fort Knox, KY.
Length: 2 weeks (116 hours).
Exhibit Dates: 10/98–Present.
Learning Outcomes: Upon completion of the course, the student will be able to describe the operation and performance of the M1/M1A1 Abrams tank.
Instruction: Lecture, discussion, and practical experience is used to cover topics of maintenance, gunnery, and specific crew duties.
Credit Recommendation: Credit is not recommended because of military-specific nature of the course (6/01).
Related Occupations: 19K.

AR-2201-0535

Total Army Training System (TATS) Cavalry Scout (HMMWV)

Course Number: 171-19D10 (HMMWV).
Location: Armor Center and School, Fort Knox, KY.
Length: 100 hours.
Exhibit Dates: 10/99–Present.
Learning Outcomes: Before 10/99 see AR-2201-0488. Upon completion of the course, the student will be able to perform military communications, nuclear/biological/chemical operations, and land navigation; use weapons in field exercises; and apply tactical training to intelligence, combat engineering, and communications.
Instruction: Lectures and practical exercises cover nuclear/biological/chemical warfare, land navigation, communications, weapons systems, and mine removal.
Credit Recommendation: Credit is not recommended because of the military-specific nature of the course (6/01).
Related Occupations: 19D.

AR-2201-0536

Tactical Army Training System (TATS) Cavalry Scout (M113) Reclassification, Phase 2

Course Number: 171-19D10 (M113), Phase 2.
Location: Reserve Schools, Multiple locations.
Length: 110 hours.
Exhibit Dates: 10/99–Present.
Learning Outcomes: Upon completion of the course, the student will be prepared to perform tasks in the areas of demolitions, mines, communications, map reading, and vehicle maintenance.
Instruction: Lectures and practical exercises cover antitank mines, NBC, demolition, communications, map reading, weapon systems, and intelligence.
Credit Recommendation: In the lower-division baccalaureate/associate degree category, 1 semester hour in map reading and 1 in military tactics (6/01).
Related Occupations: 19D.

AR-2201-0537

Total Army Training System (TATS) Cavalry Scout (M3) Reclassification, Phase 2

Course Number: 171-19D10 (M3), Phase 2.
Location: Reserve Schools, Multiple locations.
Length: 120 hours.
Exhibit Dates: 10/99–Present.
Learning Outcomes: Upon completion of the course, the student will be prepared to perform tasks in the areas of demolition, navigation, and vehicle maintenance.
Instruction: Lectures and practical exercises cover antitank mines, tactics, signaling, map reading, night vision, weapon systems, and vehicle maintenance.
Credit Recommendation: In the lower-division baccalaureate/associate degree category, 1 semester hour in map reading and 1 in military tactics (6/01).
Related Occupations: 19D.

AR-2201-0538

Master Fox Scout

Course Number: 494-F13.
Location: Chemical School, Fort Leonard Wood, MO.
Length: 2–3 weeks (93 hours).
Exhibit Dates: 8/00–Present.
Learning Outcomes: Upon completion of the course, the student will be able to use a land/water vehicle in battlefield operations with emphasis on nuclear, biological, and chemical reconnaissance survey (NBCRS) scenarios.
Instruction: Methods of instruction include lectures, oral and written examinations, and extensive military battlefield operational scenarios emphasizing the NBCRS capability of the master fox scout vehicle.
Credit Recommendation: In the lower-division baccalaureate/associate degree category, 3 semester hours in military tactics (1/02).

AR-2201-0539

Drill Sergeant

Course Number: 615-SQIX.
Location: Drill Sergeant School, Fort Jackson, SC; Drill Sergeant School, Vancouver, WA; Drill Sergeant School, Fort Knox, KY; Drill Sergeant School, Mattydale, NY; Drill Sergeant School, Oklahoma City, OK; Drill Sergeant School, Milwaukee, WI; Drill Sergeant School, Fort A.P.Hill, VA.
Length: Maximum, 52 weeks (321 hours).
Exhibit Dates: 8/00–Present.
Learning Outcomes: Upon completion of the course, the student will be able to supervise, instruct, motivate, and evaluate trainees.
Instruction: Methods of instruction include classroom exercises, practical exercises, lectures, discussions, audiovisual materials, and computer laboratories. Topics include leadership, physical fitness, marksmanship, and methods of instruction.
Credit Recommendation: In the lower-division baccalaureate/associate degree category, 2 in leadership, 1 in methods of instruction, and 1 in marksmanship (3/02).

AR-2201-0540

Civil Military Operations

Course Number: CMS-1, Phase 2.
Location: Western Hemisphere Institute for Security Cooperation (WHINSEC), Fort Benning, GA.
Length: 6 weeks (184 hours).
Exhibit Dates: 2/01–Present.
Learning Outcomes: Upon completion of the course, the student will be familiar with military civic action, role of the military in civilian operations, civil defense, disaster relief, counter-drug operations, human rights, and the rule of law, with particular emphasis on the role of the military in a democratic society.
Instruction: Audio-visual materials, case studies, classroom and practical exercises. Topics include training, military support of civil authority, civil defense, disaster relief, and counter-drug operations.
Credit Recommendation: In the lower-division baccalaureate/associate degree category, 3 semester hours in political science (8/02).

AR-2201-0541

Cadet Leadership Development-Infantry

Course Number: LDR-1, Phase 1.
Location: Western Hemisphere Institute for Security Cooperation (WHINSEC), Fort Benning, GA.
Length: 4 weeks (308 hours).
Exhibit Dates: 2/01–Present.
Learning Outcomes: Upon completion of the course, the student will be able to demonstrate basic light infantry tactics, techniques, and skills.

Instruction: Audio visual materials, practical and classroom exercises, discussion, lecture, learner presentations, and computer-based training. Topics include physical fitness, land navigation, weapons operations, infantry tactics and principles, air mobile operations planning, risk assessment, and human rights awareness.

Credit Recommendation: In the lower-division baccalaureate/associate degree category, 3 semester hours in military science (10/02).

AR-2201-0542

Cadet Troop Leader

Course Number: LDR-4, Phase 2.

Location: Western Hemisphere Institute for Security Cooperation (WHINSEC), Fort Benning, GA.

Length: 2 weeks (87 hours).

Exhibit Dates: 5/01–Present.

Learning Outcomes: Upon completion of the course, the student will be able to demonstrate a knowledge of U.S. Army combined arms doctrine, U.S. Army technology, leadership, military navigation, weapons, and operational techniques.

Instruction: Audio visual materials, practical and classroom exercises, discussion, lecture, and computer-based training. Topics include physical fitness, technology, night operations techniques, integrated weapons systems, combined arms warfighting systems, leadership, land navigation, weapons qualification and operational techniques, human rights awareness, and risk management.

Credit Recommendation: In the lower-division baccalaureate/associate degree category, 2 semester hours in military science (10/02).

AR-2201-0543

Command and General Staff Officer

Course Number: OPME-4, Phase 1.

Location: Western Hemisphere Institute for Security Cooperation (WHINSEC), Fort Benning, GA.

Length: 49 weeks (2045 hours).

Exhibit Dates: 2/01–Present.

Learning Outcomes: Upon completion of the course, the student will be able to demonstrate the skills and abilities to command, lead and train battalions, brigades and equivalent-sized units, to accomplish their assigned missions; employ and sustain weapon and equipment systems in combined arms operations; serve as principal staff officers from brigade to corps and as staff officers of major Army, joint, unified, or combined headquarters and efficiently manage manpower, equipment, money, and time.

Instruction: Audio visual materials, practical and classroom exercises, discussion, case studies, lecture and computer-based training. Topics include tactics, leadership, strategy, resource management, logistics, decision making, military history, U.S. foreign policy, democracy, force protection, anti-terrorism, military writing, language training, and human rights awareness.

Credit Recommendation: In the upper-division baccalaureate category, 15 semester hours in military science, 6 in military history, 3 in principles of management, 2 in decision making and problem solving, 2 in materiels management, and 2 in management practicum (10/02).

AR-2201-0544

Engineer Operations

Course Number: TAC-8, Phase 1.

Location: Western Hemisphere Institute for Security Cooperation (WHINSEC), Fort Benning, GA.

Length: 10 weeks (395 hours).

Exhibit Dates: 2/01–Present.

Learning Outcomes: Upon completion of the course, the student will be able to conduct basic engineer operations through the use of light engineer and light infantry tactics, techniques and procedures; demonstrate basic medical, communication, and leadership skills; recognize and detect areas of mines and booby traps; detect, probe, and use mine detectors to locate mines and booby traps; complete road marches while carrying heavy loads; and perform continuous operations in a field environment with minimum amounts of food and sleep.

Instruction: Audio-visual materials, practical and classroom exercises, discussion, computer-based training, and lecture. Topics include basic military skills including physical fitness, field medical procedures, and transportation; dealing with combat fatigue; exercising field sanitation, map reading, and communication skills; special engineering skills including demolition detonation mines, and mine fields; fundamentals of tactical operations including preparing for combat; engineer operations including sapper battle techniques; counter narcotics operations; and human rights awareness.

Credit Recommendation: In the lower-division baccalaureate/associate degree category, 1 semester hour in first aid and safety. In the upper-division baccalaureate category, 3 semester hours in military science (10/02).

AR-2201-0545

NCO Career

Course Number: NPME-7(E), Phase 1.

Location: Western Hemisphere Institute for Security Cooperation (WHINSEC), Fort Benning, GA.

Length: 5 weeks (370 hours).

Exhibit Dates: 2/01–Present.

Learning Outcomes: Upon completion of the course, the student will be able to describe leadership responsibilities, explain plan of action and after-action review, identify and prepare troop readiness, and supervise troops under specific conditions.

Instruction: Audio visual materials, practical and classroom exercises, discussion, learner presentations, and lecture. Topics include leadership responsibilities, troop readiness, and plan of action.

Credit Recommendation: In the lower-division baccalaureate/associate degree category, 3 semester hours in principles of supervision (10/02).

AR-2201-0546

NCO Professional Development

Course Number: NPME-8, Phase 1.

Location: Western Hemisphere Institute for Security Cooperation (WHINSEC), Fort Benning, GA.

Length: 6 weeks (234 hours).

Exhibit Dates: 2/01–Present.

Learning Outcomes: Upon completion of the course, the student will be able to lead, train, and direct subordinates at squad and platoon level.

Instruction: Audio visual materials, practical and classroom exercises, discussion, case studies, learner presentations, laboratory, and lecture. Topics include physical fitness, management training, leadership, communications, land navigation, weapons, fire support, infantry tactics and or principles, risk assessment, and human rights awareness.

Credit Recommendation: In the lower-division baccalaureate/associate degree category, 1 semester hour in physical education, 2 in basic military science, and 1 in land navigation/tactical operations (10/02).

AR-2201-0547

Antiarmor Leaders

Course Number: 2E/F206/010-F27.

Location: Infantry School, Fort Benning, GA.

Length: 3 weeks (108 hours).

Exhibit Dates: 4/02–Present.

Learning Outcomes: Upon completion of the course the student will be able to maintain, operate, and employ the TOW missile weapons systems, and secondary weapons.

Instruction: Audio-visual materials, practical and classroom exercises, discussion, laboratory, and lecture. Topics include usage and maintenance of the TOW weapon system to include machine guns and armor.

Credit Recommendation: Credit is not recommended because of the military-specific nature of the course (10/02).

AR-2201-0548

Bradley Transition

Course Number: 010-F26.

Location: Infantry School, Fort Benning, GA.

Length: 3 weeks (139 hours).

Exhibit Dates: 4/02–Present.

Learning Outcomes: Upon completion of the course, the student will be able to operate and maintain the Bradley Fighting Vehicle.

Instruction: Practical and classroom exercises, discussion, laboratory, and lecture. Topics include vehicle operations, maintenance, safety, weapons systems, and gunnery tactics.

Credit Recommendation: Credit is not recommended because of the military-specific nature of the course (10/02).

AR-2201-0549

Light Leaders

Course Number: 010-F32.

Location: Infantry School, Fort Benning, GA.

Length: 2 weeks (91 hours).

Exhibit Dates: 4/02–Present.

Learning Outcomes: Upon completion of the course, the student will be able to demonstrate competence in marksmanship, night vision, weapons employment and training, map reading, land navigation, and personnel deployment.

Instruction: Audio-visual materials, practical and classroom exercises, discussion, laboratory, and lecture. Topics include marksmanship, land navigation, weapons training, and small group combat tactics.

Credit Recommendation: In the lower-division baccalaureate/associate degree category, 1 semester hour in marksmanship (10/02).

AR-2201-0551

Javelin Gunnery

Course Number: 010-F22.
Location: Infantry School, Fort Benning, GA.
Length: 2 weeks (95 hours).
Exhibit Dates: 10/00–Present.
Learning Outcomes: Upon completion of the course, the student will be able to operate and maintain the javelin weapon system.
Instruction: Audio-visual materials, practical and classroom exercises, discussion, and laboratory. Topics include javelin weapon system, target engagement, field maintenance, and gunnery skills.
Credit Recommendation: Credit is not recommended because of the military-specific nature of the course (10/02).

AR-2201-0552

Infantryman

Course Number: 11B10-OSUT.
Location: Infantry School, Fort Benning, GA.
Length: 13–14 weeks (751 hours).
Exhibit Dates: 7/01–Present.
Learning Outcomes: Upon completion of the course, the student will be able to demonstrate skills and knowledge for combat survival in the areas of tactics, physical fitness, and weapons systems.
Instruction: Topics include marksmanship, physical conditioning, first aid, combat tactics, land navigation, and weapons systems.
Credit Recommendation: In the lower-division baccalaureate/associate degree category, 3 semester hours in marksmanship, 1 in first aid and safety, 1 in land navigation, and 3 in physical conditioning (10/02).
Related Occupations: 11B.

AR-2201-0555

One Station Unit Training

Course Number: 7-11-B10.
Location: Infantry School, Fort Benning, GA.
Length: 13–14 weeks (690 hours).
Exhibit Dates: 10/98–6/01.
Learning Outcomes: Upon completion of the course, the student will be able to perform in a combat environment using infantry skills in marksmanship, physical conditioning, land navigations, and weapons systems.
Instruction: Topics include marksmanship, physical conditioning, weapons systems, navigation, and field exercises.
Credit Recommendation: In the lower-division baccalaureate/associate degree category, 2 semester hours in marksmanship, 1 in land navigation, and 2 in physical conditioning (10/02).

AR-2201-0556

One Station Unit Training, Fighting Vehicle Infantryman

Course Number: 11B/C-OSUT; 7-11M10.
Location: Infantry School, Fort Benning, GA.
Length: 15–16 weeks (765 hours).
Exhibit Dates: 10/98–6/01.
Learning Outcomes: Upon completion of the course, the student will be able to perform in combat environment using skills in marksmanship, physical conditioning, and various weapons systems.
Instruction: Topics include marksmanship, physical conditioning, machine gun operations, map reading communications, combat survival skills, tactics, and testing.
Credit Recommendation: In the lower-division baccalaureate/associate degree category, 2 semester hours in marksmanship, 2 in physical conditioning, and 1 in land navigation (10/02).

AR-2201-0557

Infantry Officer Advanced-Reserve

Course Number: 2-7-C23 (DT), Phase 2.
Location: Infantry School, Fort Benning, GA.
Length: 2–3 weeks (98 hours).
Exhibit Dates: 2/93–Present.
Learning Outcomes: Upon completion of the course, the student will be able to provide leadership and service support in a battlefield environment.
Instruction: Topics include tactical doctrine; company operations; battalion, artillery, and engineer operations.
Credit Recommendation: Credit is not recommended because of the limited, specialized nature of the course (10/02).

AR-2201-0558

Bradley Fighting Vehicle Master Gunner (M2A3)

Course Number: 010-F24.
Location: Infantry School, Fort Benning, GA.
Length: 5 weeks (260 hours).
Exhibit Dates: 4/02–Present.
Learning Outcomes: Upon completion of the course, the student will be able to operate and perform basic maintenance and program diagnostics on a computerized fire control system and instruct other personnel on the operation of this system.
Instruction: Computer-based training and practical and classroom exercises. Topics include classroom instruction on the operation, fault handling, and software management of a computerized fire control system; instruction on conducting and managing a class presentation; and practical exercises operating the fire control system under field conditions.
Credit Recommendation: In the lower-division baccalaureate/associate degree category, 1 semester hour in computer operations (10/02).

AR-2201-0559

Short Range (SR) Unmanned Aerial Vehicle-External Pilot

Course Number: 243-ASIE1.
Location: Intelligence Center and School, Fort Huachuca, AZ.
Length: 16 weeks (576 hours).
Exhibit Dates: 10/02–Present.
Learning Outcomes: Upon completion of the course, the student will be able to launch, fly, and recover short-range Unmanned Aerial Vehicle (UAV).
Instruction: Practical exercises. Topics include preparation for flight, take-off's, and flight and recovery of the Unmanned Aerial Vehicle (UAV).
Credit Recommendation: Credit is not recommended because of the military specific nature of the course (12/02).

AR-2201-0560

Hunter UAV Electronic Maintenance Specialist

Course Number: 102-ASUI2.
Location: Intelligence Center and School, Fort Huachuca, AZ.
Length: 11 weeks (411 hours).
Exhibit Dates: 1/00–Present.
Learning Outcomes: Upon completion of the course, the student will be able to troubleshoot and repair an Unmanned Aerial Vehicle (UAV) and associated payloads, electronics, and ground support equipment.
Instruction: Practical exercises, discussion, laboratory, and lecture. Topics include troubleshooting and maintenance of various systems associated with Unmanned Aerial Vehicles (UAV).
Credit Recommendation: Credit is not recommended because of the military-specific nature of the course (12/02).

AR-2201-0561

Trojan Spirit II Maintenance

Course Number: 4C-F60/102-F120.
Location: Intelligence Center and School, Fort Huachuca, AZ.
Length: 1–2 weeks (60 hours).
Exhibit Dates: 10/01–Present.
Learning Outcomes: Upon completion of the course, the student will be able to perform technical maintenance on the communication central legacy (Trojan Spirit II), and the P31 communication system.
Instruction: Practical exercises, laboratory, and computer-based training. Topics include safety rules, signal flow, major components of the Trojan Spirit II and P31 system, and repair and replacement of components of the Trojan Spirit II and P31 system.
Credit Recommendation: Credit is not recommended because of the military-specific nature of the course (12/02).

AR-2201-0562

Trojan Spirit II Operator

Course Number: 3A-F76/243-F26.
Location: Intelligence Center and School, Fort Huachuca, AZ.
Length: 2–3 weeks (100 hours).
Exhibit Dates: 10/02–Present.
Learning Outcomes: Upon completion of the course, the student will be able to operate, troubleshoot, setup, and configure the Trojan Spirit II satellite communication system.
Instruction: Practical exercises and computer-based training. Topics include troubleshooting the satellite communication system, and configuration of collateral LAN for internal and external users.
Credit Recommendation: Credit is not recommended because of the military-specific nature of the course (12/02).

AR-2201-0563

Emanations Analysis Technician WOTTC

Course Number: 3B-352J, Phase 2.

Location: Intelligence Center and School, Fort Huachuca, AZ.
Length: 4–5 weeks (170 hours).
Exhibit Dates: 7/01–Present.
Learning Outcomes: Upon completion of the course, the student will be able to analyze emanation collection systems.
Instruction: Case studies, learner presentations, classroom exercises, and computer-based training. Topics include abilities to analyze Intelligence Operating Systems (IBOS), Signal Intelligence (SIGINT), Intelligence and Electronic Warfare Systems (IEW), Electronic Intercept Systems (ELINT), Tactical Electronic (TACELINT), and Tactical Exploitation of National Capabilities (TENCAP) systems.
Credit Recommendation: Credit is not recommended because of the military-specific nature of the course (12/02).

AR-2201-0564

Force XXI Battle Command Brigade and Below (FBCB2) Systems Maintainer

Course Number: 101-F40.
Location: Signal School, Fort Gordon, GA.
Length: 1–2 weeks (52 hours).
Exhibit Dates: 10/01–Present.
Learning Outcomes: Upon completion of the course, the student will be able to install, troubleshoot, and perform unit level maintenance on automated signal support systems.
Instruction: Audio visual materials, practical exercises, and laboratory. Topics include battlefield logistics, tactical operations, troubleshooting and maintenance of the Force Battle Command Systems.
Credit Recommendation: Credit is not recommended because of the military-specific nature of the course (2/03).

AR-2201-0566

Ground Mobile Forces SATCOM Operator

Course Number: 202-F7 (OS), Phase 4.
Location: Signal School, Fort Gordon, GA.
Length: 4–5 weeks (166 hours).
Exhibit Dates: 1/95–Present.
Learning Outcomes: Upon completion of the course, the student will be able to operate, troubleshoot, and perform unit level maintenance on satellite communication equipment.
Instruction: Practical and classroom exercises. Topics include tactical satellite terminal operation installation and system troubleshooting.
Credit Recommendation: Credit is not recommended because of the military-specific nature of the course (2/03).

AR-2201-0567

Enhanced Position Location Reporting System Planners

Course Number: 4C-F57/201-F9.
Location: Signal School, Fort Gordon, GA.
Length: 2 weeks (78 hours).
Exhibit Dates: 4/96–Present.
Learning Outcomes: Upon completion of the course, the student will be able to perform Enhanced Position Location Reporting System (EPLRS) planning, and management.
Instruction: Practical and classroom exercises. Topics include network operational planning, network analysis, database development, and system overview.
Credit Recommendation: Credit is not recommended because of the limited, specialized nature of the course (2/03).

AR-2201-0568

Local COMSEC Workstation Operator

Course Number: 4C-F59/160-F39.
Location: Signal School, Fort Gordon, GA.
Length: 2 weeks (79 hours).
Exhibit Dates: 1/02–Present.
Learning Outcomes: Upon completion of the course, the student will be able to operate and manage the local COMSEC management software workstation and associated platforms.
Instruction: Audio visual materials, practical and classroom exercises, and computer-based training. Topics include Army Key management system, local COMSEC management software, workstation hardware and related systems, accounting for and ordering COMSEC material, conducting inventories, and preparing and submitting reports.
Credit Recommendation: Credit is not recommended because of the limited, specialized nature of the course (2/03).

AR-2201-0569

Warrant Officer Senior Staff

Course Number: 1-250-C9.
Location: Aviation Center, Fort Rucker, AL.
Length: 2 weeks (70 hours).
Exhibit Dates: 11/96–Present.
Learning Outcomes: Upon completion of the course, the student will be able to provide update and refresher training in Senior Army Warrant Officer subjects.
Instruction: Audiovisual materials, practical and classroom exercises, discussion, lecture and computer-based training. Topics include the total Army concept and how the Senior Warrant Officer is integrated, force integration, update on threat, Army doctrine, equipment and policy.
Credit Recommendation: Credit is not recommended because of the limited, specialized nature of the course (4/03).

AR-2201-0572

Bradley Fire Support Vehicle (BFIST) Operator

Course Number: 250-ASID3.
Location: Field Artillery School, Fort Sill, OK.
Length: 4 weeks (144 hours).
Exhibit Dates: 10/01–Present.
Learning Outcomes: Upon completion of the course, the student will be able to operate and perform routine maintenance on armored cavalry vehicles, install and disarm mines, calculate demolitions and explosions, use communications equipment, read and interpret maps, use intelligence reports, and conduct a dismounted patrol.
Instruction: Course uses lectures, demonstrations, and practical exercises. Topics include vehicle specifics, patrolling, hazardous materials, communications, map reading and land navigation, and intelligence.
Credit Recommendation: Credit is not recommended because of the military-specific nature of the course (10/03).

AR-2201-0573

Ammunition Specialist

Course Number: 645-55B10.
Location: Ordnance, Missile and Munitions School, Redstone Arsenal, AL.
Length: 10 weeks (373 hours).
Exhibit Dates: 10/00–Present.
Learning Outcomes: Upon completion of the course, the student will be able to identify ammunition items; destroy unserviceable ammunition; perform inventory and issue receipt; store and load munitions; operate a forklift.
Instruction: Computer-based training, and practical exercises. Topics include ammunition material, service procedures, storage, handling, movement, inspection, maintenance, and destruction.
Credit Recommendation: Credit is not recommended because of the military nature of the course (2/04).

AR-2201-0574

Ammunition Specialist, Advanced Noncommissioned Officer

Course Number: 645-55B40 (TATS), Phase 2.
Location: Ordnance, Missile and Munitions School, Redstone Arsenal, AL.
Length: 8 weeks (308 hours).
Exhibit Dates: 10/00–Present.
Learning Outcomes: Upon completion of the course, the student will be able to manage ammunition units, including operations and maintenance safety and training.
Instruction: Computer-based training, practical exercises, and lecture. Topics include surveillance procedures, storage and transportation, and inventory control.
Credit Recommendation: In the lower-division baccalaureate/associate degree category, 2 semester hours in military science (2/04).

AR-2201-0575

Multiple Launch Rocket System Repairer Transition

Course Number: 121-27M10 (T).
Location: Ordnance, Missile and Munitions School, Redstone Arsenal, AL.
Length: 5 weeks (197 hours).
Exhibit Dates: 4/02–Present.
Learning Outcomes: Upon completion of the course, the student will be able to inspect, test, and adjust system components to specified tolerance for a rocket launch system.
Instruction: Practical and classroom exercises, and lecture. Topics include system maintenance of rocket launch system including the electrical, mechanical, and hydraulic subsystems.
Credit Recommendation: Credit is not recommended because of the limited specialized nature of the course (2/04).

AR-2201-0576

Ammunition Specialist, Basic Noncommissioned Officer

Course Number: 645-55B30 (DL), Phase 2.
Location: Ordnance, Missile and Munitions School, Redstone Arsenal, AL.
Length: 8 weeks (79 hours).
Exhibit Dates: 10/03–Present.

Learning Outcomes: Upon completion of the course, the student will be able to identify, detect, decontaminate, report, and inspect chemical munitions; select appropriate levels of protective equipment and clothing; conduct inspections of munitions, small missiles; and use/maintain precision measurement instruments.

Instruction: Computer-based training. Topics include training through distance learning to cover explosive safety standards; munitions inspection procedures; chemical munitions identification; storage, leak detection, and decontamination; and inventory control.

Credit Recommendation: Credit is not recommended because of the military-specific nature of the course (2/04).

AR-2201-0577

Ammunition Specialist, Basic Noncommissioned Officer

Course Number: 645-55B30 (TATS), Phase 2.

Location: Ordnance, Missile and Munitions School, Redstone Arsenal, AL.

Length: 14 weeks (504 hours).

Exhibit Dates: 10/00–Present.

Learning Outcomes: Upon completion of the course, the student will be able to identify, detect, decontaminate, report, and inspect chemical munitions; select appropriate levels of protective equipment and clothing; conduct inspections of munitions, small missiles; and use/maintain precision measurement instruments.

Instruction: Computer-based training, practical exercises, and lecture. Topics include explosive safety standards; munitions inspection procedures; chemical munitions identification, storage, leak detection, and decontamination; and inventory control.

Credit Recommendation: Credit is not recommended because of the military-specific nature of the course (2/04).

AR-2201-0578

Indirect Fire Infantryman

Course Number: 11C10-OSUT.

Location: Infantry School, Fort Benning, GA.

Length: 13–14 weeks (768 hours).

Exhibit Dates: 7/01–Present.

Learning Outcomes: Upon completion of the course, the student will be able to demonstrate skills and knowledge for combat survival in the areas of tactics, physical fitness, and weapons systems.

Instruction: Topics include marksmanship, physical conditioning, first aid, combat tactics, land navigation, and weapons systems.

Credit Recommendation: In the lower-division baccalaureate/associate degree category, 3 semester hours in marksmanship, 1 in first aid and safety, 1 in land navigation, and 3 in physical conditioning (10/02).

Related Occupations: 11C.

AR-2202-0001

Democratic Sustainment

Course Number: CMS-2, Phase 3.

Location: Western Hemisphere Institute for Security Cooperation (WHINSEC), Fort Benning, GA.

Length: 6 weeks (189 hours).

Exhibit Dates: 6/01–Present.

Learning Outcomes: Upon completion of the course, the student will be familiar with governmental operations, civil and military law, development of regional democracy, and religious influences on cultures.

Instruction: Audio-visual materials, classroom and practical exercises, lecture, and learner presentations. Participants must be military officers (captains through lieutenants) or civilian equivalent and possess a postsecondary degree.

Credit Recommendation: In the graduate degree category, 3 semester hours in political science or international relations (8/02).

AR-2202-0002

Laundry Advanced System Maintenance

Course Number: 690-63J10 (LADS).

Location: Ordnance Center and School, Aberdeen Proving Ground, MD.

Length: 1–2 weeks (64 hours).

Exhibit Dates: 10/02–Present.

Learning Outcomes: Upon completion of the course, the student will be able to follow the wiring and maintain the electrical and mechanical components of laundry advanced system.

Instruction: Practical exercises, classroom exercises, discussion, laboratory, and lecture. Topics include study of wiring and maintenance of the laundry advanced system.

Credit Recommendation: In the lower-division baccalaureate/associate degree category, 1 semester hour in electromechanical systems maintenance (8/03).

AR-2204-0085

Ammunition Warrant Officer Entry (Ammunition Manager)

Course Number: GPL.

Location: Missile and Munitions School, Redstone Arsenal, AL.

Length: 4–5 weeks (167–174 hours).

Exhibit Dates: 5/89–Present.

Learning Outcomes: Upon completion of the course, the student will be able to prepare hazardous material for transportation, establish tactical ammunition supply points, configure magazine storage, and train and direct ammunition platoons in ammunition operations.

Instruction: Lectures and practical exercises cover ammunition storage, fitness reports, preparing ammunition messages, preparing operational milestones, ammunition disposal, weapon type determination, transportation planning, and determination of sources of supply.

Credit Recommendation: Credit is not recommended because of the military-specific nature of the course (5/01).

AR-2204-0086

Ammunition Noncommissioned Officer (NCO)

Course Number: GPK.

Location: Missile and Munitions School, Redstone Arsenal, AL.

Length: 4 weeks (152 hours).

Exhibit Dates: 5/89–Present.

Learning Outcomes: Upon completion of the course, the student will be able to supervise the care, handling, preservation, security, and storage of ammunition, explosives, and their components.

Instruction: Lectures and practical exercises cover ammunition inspection, allowance management, stock control, physical security, and magazine and field storage operations.

Credit Recommendation: Credit is not recommended because of the military-specific nature of the course (5/01).

AR-2204-0087

1. Reconnaissance and Surveillance Leaders (Long Range Surveillance Leader)
2. Long Range Surveillance Leader

Course Number: *Version 1:* 2E-F173/011-F67. *Version 2:* 2E-F173/011-F67.

Location: Infantry School, Fort Benning, GA.

Length: *Version 1:* 4–5 weeks (517 hours). *Version 2:* 4–5 weeks (460 hours).

Exhibit Dates: *Version 1:* 3/01–Present. *Version 2:* 7/96–2/01.

Learning Outcomes: *Version 1:* Upon completion of the course, the student will be able to use survival skills to operate behind enemy lines; identify military vehicles, equipment, and radio target acquisition to the support team; conduct reconnaissance and surveillance behind enemy lines; perform insertion/extraction techniques. *Version 2:* Upon completion of the course, the student will be able to identify military vehicles, equipment, and radio target acquisition to the support team; conduct reconnaissance and surveillance behind enemy lines; perform insertion/extraction techniques; and set-up a field communication system.

Instruction: *Version 1:* Audio-visual materials, practical and classroom exercises. Topics include learning how to live off the land, build fires, and find or make shelter; practical helicopter insertion/extraction; and reconnaissance behind enemy lines, and capture/evading techniques. *Version 2:* Audio-visual materials, practical and classroom exercises. Topics include helicopter insertion and extraction, reconnaissance behind enemy lines and capture/evading techniques, and some survival training topics.

Credit Recommendation: *Version 1:* In the lower-division baccalaureate/associate degree category, 1 semester hour in survival skills, and 1 in personal conditioning (10/02). *Version 2:* In the lower-division baccalaureate/associate degree category, 1 semester hour in personal conditioning (10/02).

AR-2204-0088

Long Range Surveillance Leaders

Course Number: 2E-F173/011-F67.

Location: Infantry School, Fort Benning, GA.

Length: 4–5 weeks (517 hours).

Exhibit Dates: 7/96–3/01.

Learning Outcomes: Upon completion of the course, the student will be able to identify military vehicles and equipment and radio target acquisition to the support team, conduct reconnaissance and surveillance behind enemy lines, perform insertion/extraction techniques, and set-up a field communication systems.

Instruction: Audio visual materials, practical, and classroom exercises. Topics include practical exercises on helicopter insertion and extraction, reconnaissance behind enemy lines and capture evading techniques, and same survival training topics.

Credit Recommendation: In the lower-division baccalaureate/associate degree category, 1 semester hour in survival skills (10/02).

AR-2204-0089

LONG RANGE SURVEILLANCE LEADER-
RESERVE COMPONENT

Course Number: 2E-F173/011-F67-RC.

Location: Infantry School, Fort Benning, GA.

Length: 9–10 weeks (366 hours).

Exhibit Dates: 6/96–3/01.

Learning Outcomes: Upon completion of the course, the student will be able to conduct reconnaissance and surveillance exercises, set-up a field communication system, read maps, and navigate on land.

Instruction: Audio-visual materials, lecture, practical and classroom exercises. Topics include surveillance operations, physical fitness training, map reading, and land navigation.

Credit Recommendation: In the lower-division baccalaureate/associate degree category, 1 semester hour in land navigation (10/02).

AR-2205-0002

LIGHTER AIR-CUSHION VEHICLE NAVIGATOR

Course Number: 062-F4 (88K).

Location: Transportation School, Fort Story, VA; Transportation and Aviation Logistics School, Fort Eustis, VA.

Length: 5 weeks (180–190 hours).

Exhibit Dates: 10/87–1/98.

Learning Outcomes: Upon completion of the course, the student will be able to carry out the duties and responsibilities of navigator of a lighter air-cushion vehicle with due regard for the operational limitations imposed by weight, speed, loading and trim angles, sea state, surf, turning circles and crab angles, and obstacle clearance; pilot a lighter air-cushion vehicle and apply the nautical rules of the road; perform emergency procedures and corrective action in the event of fire, collision, plow in, towing, damage control, man overboard, and abandon ship; apply the theory of radar to maritime navigation and collision avoidance; and supervise the vehicle in all phases of its water and overland operation. Student will hold a current unlimited marine radar observer's certificate.

Instruction: Lectures, demonstrations, and practical work cover vehicle performance, operational safety, basic seamanship, radar, rules of the road, and navigator's duties and responsibilities.

Credit Recommendation: In the lower-division baccalaureate/associate degree category, 3 semester hours in seamanship and navigation (3/94).

Related Occupations: 88K.

Army Enlisted Occupation Exhibits

MOS-00B-002

Diver

00B10
00B20
00B30
00B40

Exhibit Dates: 1/90–1/94.

Career Management Field: 51 (General Engineering), subfield 511 (Construction Engineering).

Description

Summary: Supervises or performs underwater reconnaissance, demolition, repair, and salvage. *Skill Level 10:* Performs as assistant to senior divers, operating power-driven air compressor and electric and engine-driven winches; makes simple repairs to diving gear and equipment; applies principles of water rescue and first aid; may perform as diver, using self-contained underwater breathing apparatus and surface-supplied air diving equipment. *Skill Level 20:* Able to perform the duties required for Skill Level 10; using self-contained underwater breathing apparatus, performs reconnaissance, salvage, repair, and demolition operations; serves as senior diver; collects and reports on hydrographic conditions; makes charts and maps of underwater conditions; takes underwater photographs; prepares beach or river intelligence reports; keeps diving log; uses demolition techniques to clear underwater obstacles; employs sophisticated first aid treatment, particularly as related to diving operations. *Skill Level 30:* Able to perform the duties required for Skill Level 20; using surface-supplied air and diving equipment, performs underwater repair by caulking seams, patching holes, and clearing fouled equipment; directs or assists in underwater construction projects; rigs submerged objects for surfacing; employs underwater welding techniques and other special construction tools; operates decompression chamber; performs underwater inspection; estimates weight of underwater objects; and calculates method of lifting underwater material and objects. *Skill Level 40:* Able to perform the duties required for Skill Level 30; supervises underwater reconnaissance, demolition, repair, and salvage in addition to scuba and standard diving activities; supervises diver's bell (or appropriate mechanical diving unit) and directs maintenance of diving equipment; interprets hydrographic charts, maps, and sketches; instructs in diving procedures and safety practices; prepares reports on diving operations.

Recommendation, Skill Level 10

In the vocational certificate category, 3 semester hours in the use, operation, and maintenance of mechanical equipment; if duties included diving, 3 semester hours in scuba diving. In the lower-division baccalaureate/associate degree category, 2 semester hours in first aid. (NOTE: This recommendation for skill level 10 is valid for the dates 1/90-10/91 only) (9/79).

Recommendation, Skill Level 20

In the vocational certificate category, 9 semester hours in the use, operation, and maintenance of mechanical equipment and 6 in scuba diving. In the lower-division baccalaureate/associate degree category, 3 semester hours in mapping, 3 in technical report writing, 3 in marine and oceanographic technologies, 4 in first aid, and 3 in scuba diving, and additional credit for detonation/demolition, photography, and scuba diving on the basis of institutional evaluation (9/79).

Recommendation, Skill Level 30

In the vocational certificate category, 12 semester hours in the use, operation, and maintenance of mechanical equipment, 12 in marine underwater repair and construction, and 6 in scuba diving. In the lower-division baccalaureate/associate degree category, 3 semester hours in physics, 3 in mapping, 3 in technical report writing, 6 in marine and oceanographic technologies, 6 in first aid, 3 in scuba diving, and 3 in human relations, and additional credit for detonation/demolition, photography, and scuba diving on the basis of institutional evaluation (9/79).

Recommendation, Skill Level 40

In the vocational certificate category, 18 semester hours in the use, operation, and maintenance of mechanical equipment, 18 in marine underwater repair and construction, and 6 in scuba diving. In the lower-division baccalaureate/associate degree category, 3 semester hours in mapping, 3 in marine geography, 6 in physics, 6 in technical report writing, 9 in marine and oceanographic technologies, 6 in first aid and safety, 3 in scuba diving, 3 in human relations, 3 in personnel supervision, and 6 for field experience in management, and additional credit for detonation/demolition/salvage, photography, and scuba diving on the basis of institutional evaluation (9/79).

MOS-00B-003

Diver

00B10
00B20
00B30
00B40
00B50

Exhibit Dates: 2/94–9/04. (Effective 9/04 MOS 00B was discontinued, and its duties were incorporated into MOS 21D)

Career Management Field: 51 (General Engineering).

Description

Summary: Supervises or performs underwater reconnaissance, demolition, repair, and salvage. *Skill Level 10:* Qualifies as a first class diver; swims 1,000 yards on the surface in open water with fins, face mask, and buoyancy compensator within designated time; performs a qualification dive using scuba and surface-supplied diving equipment; conducts day and night general underwater searches; conducts detailed ship-bottom survey; demonstrates cardiopulmonary resuscitation (CPR) according to American Red Cross or American Heart Association standards; dives and works using self-contained and surface-supplied diving equipment; uses basic rigging, including care and selection of ropes and knots commonly used during diving operations; applies safety precautions in use of underwater oxygen, gas, and electric welding and cutting equipment. *Skill Level 20:* Able to perform the duties required for Skill Level 10; meets all requirements for second class diver; has knowledge of diving physiology, standard decompression tables, recognizing symptoms of decompression sickness, and treatment required for all common diving injuries and illnesses; demonstrates the operation of the hyperbaric chamber for treatment of diving injuries, illnesses, and surface decompression; performs hydrographic surveys; performs underwater work, including taking measurements, connecting pipe flanges, placing patches, pouring concrete, using excavating nozzles, and removing propellers; demonstrates the operation and maintenance of apparatus required for underwater cutting and welding; under supervision, performs maintenance on all diving and life support equipment. *Skill Level 30:* Able to perform the duties required for Skill Level 20; demonstrates a working knowledge of diving physics including computing pressures and volumes of breathing gases necessary to support divers working at depth; knows the cause, symptoms, treatment, and prevention of arterial gas embolism and decompression sickness; applies the theory of inert gas saturation and desaturation of body fluids and tissues; applies the knowledge of anatomy and physiology of the neurological system; conducts neurological examinations, evaluates the data to determine location and extent of injuries associated with pulmonary over-inflation syndromes and decompression sickness, and provides treatment; knows the causes, symptoms, treatment, and preventive measures for all types of diving injuries and illnesses; knows how and when to use a hyperbaric chamber; performs as inside and outside chamber operator and supervisor during treatment of diving injuries and surface decompression; supervises two or more divers in their tasks working underwater; sets up a diving station for self-contained and surface-supplied diving operations; perform and supervises independent diving operations using self-contained and surface-supplied air breathing apparatus; applies the general principles of vessel salvage to structural strength and grounding; supervises demolition, training, and operations, including calculating, placing, and employing explosives; supervises the maintenance and repair of all diving equipment. *Skill Level 40:* Able to perform the duties required for Skill

Level 30; meets all requirements for first class diver; demonstrates a working knowledge of principles of diving physics including pressure and general gas laws; conduct duplicates neurological examinations, evaluates data to determine location and extent of injuries associated with pulmonary over-inflation syndromes and decompression sickness and provides treatment; plans and supervises all types of air diving operations; chooses different classes of divers to accomplish a variety of underwater missions; demonstrates knowledge of all types of air diving equipment, related life support systems, and diving support sets including their advantages and limitations; demonstrates a thorough knowledge of the types of air compressors ordinarily used in diving operations including the various filtration methods; uses and interprets hydrographic charts, maps, and sketches; orders equipment, tools, parts, materials, and supplies; demonstrates a knowledge of the administrative control when handling and using explosives; has comprehensive knowledge of the underwater construction of military port facilities and fixed and tactical bridges. *Skill Level 50:* Able to perform the duties required for Skill Level 40; performs as a qualified master diver for control and support diving detachments; responsible for the command and control of assigned lightweight diving teams; ensures that all diving operations are conducted safely; supervises deep sea diving missions and dives conducted deeper than 100 feet; formulates demolition plans for operational and training missions; assists the commander in planning, scheduling, and executing training and operational missions; provides expertise to staff planners and diving teams; writes and develops doctrinal, regulatory, training, and safety material related to the accomplishment of diving missions.

Recommendation, Skill Level 10

Credit may be granted on the basis of an individualized assessment of the student (8/97).

Recommendation, Skill Level 20

In the lower-division baccalaureate/associate degree category, 3 semester hours in the operation and maintenance of underwater diving equipment, 3 in scuba diving, 3 in operation of hyperbaric chambers, 3 in repair of underwater diving equipment, 3 in hydrographic survey techniques, and 3 in underwater salvage and repair work (including cutting and welding). (NOTE: This recommendation for skill level 20 is valid for the dates 2/94-2/95 only) (8/97).

Recommendation, Skill Level 30

In the lower-division baccalaureate/associate degree category, 3 semester hours in the operation and maintenance of underwater diving equipment, 3 in scuba diving, 3 in operation of hyperbaric chambers, 3 in repair of underwater diving equipment, 3 in hydrographic survey techniques, 3 in underwater salvage and repair work (including cutting and welding), and 1 in supervision of divers (8/97).

Recommendation, Skill Level 40

In the lower-division baccalaureate/associate degree category, 3 semester hours in the operation and maintenance of underwater diving equipment; 3 in scuba diving; 3 in operation of hyperbaric chambers; 3 in repair of underwater diving equipment; 3 in hydrographic survey techniques; 3 in underwater salvage and repair work (including cutting and welding); 3 in advanced use, operation, maintenance, and repair of underwater diving equipment; 3 in advanced marine underwater repair and construction; 2 in advanced diving; 1 in report writing; and 2 in supervision of divers. In the upper-division baccalaureate category, 3 semester hours in diving physics and medicine, 2 in complex diving operations planning and supervision, and 2 for field experience in management (8/97).

Recommendation, Skill Level 50

In the lower-division baccalaureate/associate degree category, 3 semester hours in the operation and maintenance of underwater diving equipment; 3 in scuba diving; 3 in operation of hyperbaric chambers; 3 in repair of underwater diving equipment; 3 in hydrographic survey techniques; 3 in underwater salvage and repair work (including cutting and welding); 3 in advanced use, operation, maintenance, and repair of underwater diving equipment; 3 in advanced marine underwater repair and construction; 4 in advanced diving; 1 in report writing; and 2 in supervision of divers. In the upper-division baccalaureate category, 6 semester hours in diving physics and medicine, 3 in complex diving operations planning and supervision, 3 in doctrinal diving material development and writing, and 4 for field experience in management (8/97).

MOS-00D-001

Special Duty Assignment

00D20
00D40

Exhibit Dates: 1/90–Present.

Career Management Field: 00 (Exceptional Management Specialties), subfield 000 (Special Assignment).

Description

Summary: This MOS code is used to identify duty positions for special assignments; its use is approved on an individual case basis by the Headquarters of the Department of the Army. Before this code is assigned, a position must encompass duties that are highly specialized, that cannot be classified elsewhere, that do not correspond with formal military courses, and that occur so rarely and in such small numbers that establishment of a separate MOS is not practical. *Skill Level 20:* Used to identify all journeymen and technician levels of skill. *Skill Level 40:* Used to identify all leader levels of skill.

Recommendation, Skill Level 20

Credit on the basis of institutional evaluation of the individual student seeking competency recognition (5/76).

Recommendation, Skill Level 40

Credit on the basis of institutional evaluation of the individual student seeking competency recognition (5/76).

MOS-00E-002

Recruiter

00E20
00E30
00E40
00E50

Exhibit Dates: 1/90–10/91.

Career Management Field: 79 (Recruitment and Retention).

Description

Summary: Recruits or supervises the recruitment of personnel for the Army. *Skill Level 20:* Contacts and conducts interviews of individuals who are prospective enlistees into the Army; contacts representatives of schools, corporations, civic groups and other agencies to present career and employment opportunities of the Army; presents formal and informal talks to organizations, groups, or individuals; writes, edits, and presents recruiting material for use by local communications media; interviews and counsels individuals on enlistment incentives, Army benefits, and educational opportunities; evaluates applicants, using screening tests; prepares forms and documents as part of enlistment processing; counsels disqualified applicants; assists in market research and analysis of recruiting area. *Skill Level 30:* Able to perform the duties required for Skill Level 20; provides technical guidance to Skill Level 20 personnel. *Skill Level 40:* Able to perform the duties required for Skill Level 30; may serve as Recruiting Station Commander, Professional Development Sergeant, or Reception Station Liaison Sergeant, supervising five to eight persons; plans and organizes recruitment programs; maintains statistics on recruiting; prepares enlistment reports and official correspondence; prepares station plans, and conducts professional development programs; supervises subordinate personnel, provides guidance for improvement, and evaluates their performance. *Skill Level 50:* Able to perform the duties required for Skill Level 40; may serve as principal noncommissioned officer of a recruiting region, Senior Professional Development Sergeant, Staff Operations Sergeant, or Management Noncommissioned Officer for recruiting; may supervise up to 30 persons; assists area commander in the areas of training, production, and personnel status; plans and conducts recruiting seminars and conferences; develops presentations reflecting requirements, sales techniques, and enlistment contract options; identifies, investigates, and takes corrective action in problem areas; trains, assigns duties, and evaluates performance of subordinates; develops recruiter training programs.

Recommendation, Skill Level 20

In the lower-division baccalaureate/associate degree category, 2 semester hours in principles of advertising, 2 in social psychology, 3 in sales techniques, 3 in public speaking, 3 in records management, and 2 in audiovisual techniques. In the upper-division baccalaureate category, 3 semester hours in advertising media, 3 for field experience in recruiting, and 2 in vocational counseling (7/79).

Recommendation, Skill Level 30

In the lower-division baccalaureate/associate degree category, 2 semester hours in principles of advertising, 2 in social psychology, 3 in sales techniques, 3 in public speaking, 3 in records management, and 2 in audiovisual techniques. In the upper-division baccalaureate category, 3 semester hours in advertising media, 3 for field experience in recruiting, 2 in vocational counseling, and 2 in publicity release writing (7/79).

Recommendation, Skill Level 40

In the lower-division baccalaureate/associate degree category, 3 semester hours in principles of advertising, 3 in social psychology, 3 in sales techniques, 3 in public speaking, 3 in records management, 3 in audiovisual techniques, and 3 in personnel supervision. In the upper-division baccalaureate category, 3 semester hours in advertising media, 3 for field experience in recruiting, 2 in vocational counseling, 2 in publicity release writing, and 3 in personnel management (7/79).

Recommendation, Skill Level 50

In the lower-division baccalaureate/associate degree category, 3 semester hours in principles of advertising, 3 in social psychology, 3 in sales techniques, 3 in public speaking, 3 in records management, 3 in audiovisual techniques, 3 in personnel supervision, and 3 in office practice. In the upper-division baccalaureate category, 3 semester hours in advertising media, 3 for field experience in recruiting, 2 in vocational counseling, 2 in publicity release writing, 3 in personnel management, and 3 in office management (7/79).

MOS-00E-003

Recruiter (Reserve Component)
00E20
00E30
00E40
00E50

Exhibit Dates: 11/91–7/94.

Career Management Field: 79 (Recruitment and Reenlistment).

Description

Summary: Recruits or supervises the recruitment of personnel for the Army. *Skill Level 20:* Contacts individuals and conducts interviews with those who are prospective enlistees into the Army; contacts representatives of schools, corporations, civic groups, and other agencies to present the career and employment opportunities of the Army; presents formal and informal talks to organizations, groups, or individuals; writes, edits, and presents recruiting material for use by local communications media; interviews and counsels individuals on enlistment incentives, Army benefits, and educational opportunities; evaluates applicants, using screening tests; prepares forms and documents as part of enlistment processing; counsels disqualified applicants; assists in market research and analysis of recruiting area. *Skill Level 30:* Able to perform the duties required for Skill Level 20; provides technical guidance to Skill Level 20 personnel. *Skill Level 40:* Able to perform the duties required for Skill Level 30; may serve as Recruiting Station Commander, Professional Development Sergeant, or Reception Station Liaison Sergeant, supervising five to eight persons; plans and organizes recruitment programs; maintains statistics on recruiting; prepares enlistment reports and official correspondence; prepares station plans, and conducts professional development programs; supervises subordinate personnel; provides guidance for improvement, and evaluates their performance. *Skill Level 50:* Able to perform the duties required for Skill Level 40; may serve as principal noncommissioned officer of a recruiting region, Senior Professional Development Sergeant, Staff Operations Sergeant, or Management Noncommissioned Officer for recruiting; may supervise up to 30 persons; assists area commander in the areas of training, production, and personnel status; plans and conducts recruiting seminars and conferences; develops presentations reflecting requirements, marketing techniques, and enlistment contract options; identifies, investigates, and takes corrective action in problem areas; trains, assigns duties, and evaluates performance of subordinates; develops recruiter training programs.

Recommendation, Skill Level 20

In the lower-division baccalaureate/associate degree category, 2 semester hours in principles of advertising, 2 in social psychology, 3 in marketing techniques, 3 in public speaking, 2 in record keeping, and 2 in audiovisual techniques. In the upper-division baccalaureate category, 3 semester hours in advertising media, 3 for field experience in marketing, and 2 in vocational counseling (11/91).

Recommendation, Skill Level 30

In the lower-division baccalaureate/associate degree category, 2 semester hours in principles of advertising, 2 in social psychology, 3 in marketing techniques, 3 in public speaking, 2 in record keeping, and 2 in audiovisual techniques. In the upper-division baccalaureate category, 3 semester hours in advertising media, 3 for field experience in marketing, 2 in vocational counseling, and 2 in publicity release writing (11/91).

Recommendation, Skill Level 40

In the lower-division baccalaureate/associate degree category, 3 semester hours in principles of advertising, 3 in social psychology, 3 in marketing techniques, 3 in public speaking, 3 in record keeping, 3 in audiovisual techniques, and 3 in personnel supervision. In the upper-division baccalaureate category, 3 semester hours in advertising media, 3 for field experience in marketing, 2 in vocational counseling, 2 in publicity release writing, and 3 in personnel management (11/91).

Recommendation, Skill Level 50

In the lower-division baccalaureate/associate degree category, 3 semester hours in principles of advertising, 3 in social psychology, 3 in marketing techniques, 3 in public speaking, 3 in record keeping, 3 in audiovisual techniques, 3 in personnel supervision, and 3 in clerical procedures. In the upper-division baccalaureate category, 3 semester hours in advertising media, 3 for field experience in marketing, 2 in vocational counseling, 2 in publicity release writing, 3 in personnel management, and 3 in human resource management (11/91).

MOS-00E-004

Recruiter (Reserve Component)
00E40
00E50

Exhibit Dates: 8/94–12/95. Effective 12/95, MOS 00E was discontinued and its duties were incorporated into MOS 79R, Recruiter.

Career Management Field: 79 (Recruitment and Reenlistment).

Description

Summary: Recruits or supervises the recruitment of personnel for the Army. *Skill Level 40:* May serve as Recruiting Station Commander, Professional Development Sergeant, or Reception Station Liaison Sergeant, supervising five to eight persons; plans and organizes recruitment programs; maintains statistics on recruiting; prepares enlistment reports and official correspondence; prepares station plans, and conducts professional development programs; supervises subordinate personnel; provides guidance for improvement, and evaluates their performance. *Skill Level 50:* Able to perform the duties required for Skill Level 40; may serve as principal noncommissioned officer of a recruiting region, Senior Professional Development Sergeant, Staff Operations Sergeant, or Management Noncommissioned Officer for recruiting; may supervise up to 30 persons; assists area commander in the areas of training, production, and personnel status; plans and conducts recruiting seminars and conferences; develops presentations reflecting requirements, marketing techniques, and enlistment contract options; identifies, investigates, and takes corrective action in problem areas; trains, assigns duties, and evaluates performance of subordinates; develops recruiter training programs.

Recommendation, Skill Level 40

In the lower-division baccalaureate/associate degree category, 3 semester hours in principles of advertising, 3 in social psychology, 3 in marketing techniques, 3 in public speaking, 3 in record keeping, 3 in audiovisual techniques, and 3 in personnel supervision. In the upper-division baccalaureate category, 3 semester hours in advertising media, 3 for field experience in marketing, 2 in vocational counseling, 2 in publicity release writing, and 3 in personnel management (11/91).

Recommendation, Skill Level 50

In the lower-division baccalaureate/associate degree category, 3 semester hours in principles of advertising, 3 in social psychology, 3 in marketing techniques, 3 in public speaking, 3 in record keeping, 3 in audiovisual techniques, 3 in personnel supervision, and 3 in clerical procedures. In the upper-division baccalaureate category, 3 semester hours in advertising media, 3 for field experience in marketing, 2 in vocational counseling, 2 in publicity release writing, 3 in personnel management, and 3 in human resource management (11/91).

MOS-00R-001

Recruiter/Retention NCO
00R20
00R30
00R40
00R50

Exhibit Dates: 1/90–10/91.

Career Management Field: 79 (Recruitment and Retention).

Description

Summary: Recruits or reenlists personnel for military service in the Army. NOTE: MOS 00E (Recruiter) and MOS 79D (Reenlistment NCO) have been incorporated into MOS 00R. The combining of the two former MOS's results in a career track in which an individual has principal assignment either in recruiting or in retention, with special duties available in counseling, recruiting, training management, instructing, professional development advising, or supervising as the principal noncommissioned officer of a recruiting facility or a recruiting area/region. The recruiting track includes skill levels 20, 30, 40, and 50; the retention track, skill levels 30, 40, and 50. Because of the principal assignment option after skill level 20, descriptions and recommendations below are cited in two groups at skill levels 30, 40, and 50. *Skill Level 20:* (Recruiter): Contacts and interviews individuals as prospective enlistees; contacts representatives of schools, corporations, civic groups, and other agencies to present employment and career opportunities of the Army; presents formal and informal talks to organizations and groups; distributes recruiting publicity materials; establishes liaison with communications media and prepares local news releases; evaluates applicants, using screening tests and vocational aptitude tests; prepares forms and documents as part of the enlistment process; arranges for transportation, meals, and lodging of enlistees; assists in market research and analysis of recruiting area. *Skill Level 30:* Able to perform the duties required for Skill Level 20; (Recruiter): Provides technical guidance to lower skill level personnel. (Retention): Knows Army occupations, career fields and skill levels; explains reenlistment options, benefits, and obligations; counsels persons who will be separating from the Army about reserve component benefits; explains training options, vocational opportunities, and reenlistment bonuses; counsels respective reenlistees and assists in designing career plan; processes forms, documents, and records for reenlistment; prepares status charts and statistics; conducts training and inspections; advises unit commanders on status of reenlistment programs; recommends corrective actions in deficient areas. *Skill Level 40:* Able to perform the duties required for Skill Level 30; (Recruiter): Commands a recruiting station and maintains records, statistics, administrative files, and publications library; prepares enlistment reports; supervises and evaluates subordinate personnel; conducts professional development programs for lower skill level personnel; prepares official correspondence and reports; assigns duties to subordinates; and supervises personnel. (Retention): Advises commanders on status of reenlistment programs; conducts training for commanders, staff persons, noncommissioned officers and others, when assigned as principal noncommissioned officer in a large facility or region. *Skill Level 50:* Able to perform the duties required for Skill Level 40; (Recruiter): Supervises recruiting programs and personnel at several recruiting stations within an area; assists area commanders with training, operations, administration, and personnel matters; plans and conducts recruiting conferences and seminars; develops presentations which reflect changing sales techniques, enlistment contracts, and enlistment requirements; supervises subordinates; identifies and takes corrective actions in problem areas; evaluates subordinates; develops training programs; and inspects recruiting stations to ensure efficient operation and management. (Retention): Supervises, assigns duties, and evaluates performance of subordinate personnel; organizes and coordinates retention/career counseling activities; conducts reenlistment seminars and serves as policy advisor in reenlistment matters; coordinates the use and management of reenlistment funds; presents information talks as needed on reenlistment programs.

Recommendation, Skill Level 20

In the lower-division baccalaureate/associate degree category, 2 semester hours in social psychology, 2 in audiovisual techniques, 3 in sales techniques, 3 in public speaking, and 2 in record keeping. In the upper-division baccalaureate category, 3 semester hours for field experience in recruiting and 3 in records management (9/83).

Recommendation, Skill Level 30

In the lower-division baccalaureate/associate degree category, (Recruiter/Retention): 2 semester hours in social psychology, 3 in sales techniques, 3 in public speaking, 2 in audiovisual techniques, and 3 in record keeping. In the upper-division baccalaureate category, (Recruiter): 3 semester hours for field experience in recruiting and 3 in records management. (Retention): 3 semester hour in records management and 3 in vocational guidance (9/83).

Recommendation, Skill Level 40

In the lower-division baccalaureate/associate degree category, (Recruiter/Retention): 3 semester hours in social psychology, 3 in record keeping, 3 in sales techniques, 3 in public speaking, and 3 in audiovisual techniques. (Recruiter only): 3 semester hours in personnel supervision. In the upper-division baccalaureate category, (Recruiter): 3 semester hours for field experience in recruiting, 3 in records management, and 3 in personnel management. (Retention): 3 semester hours in records management and 3 in vocational guidance (9/83).

Recommendation, Skill Level 50

In the lower-division baccalaureate/associate degree category, (Recruiter/Retention) 3 semester hours in social psychology, 3 in record keeping, 3 in sales techniques, 3 in public speaking, 3 in audiovisual techniques, and 3 in personnel supervision. (Recruiter only): 3 semester hours in office procedures. In the upper-division baccalaureate category, (Recruiter): 3 semester hours for field experience in recruiting, 3 in records management, 3 in personnel management, and 3 in office management. (Retention): 3 semester hours in records management, 3 in vocational guidance, and 3 in personnel management (9/83).

MOS-00R-002

Recruiter/Retention NCO
00R20
00R30
00R40
00R50

Exhibit Dates: 11/91–7/94.

Career Management Field: 79 (Recruitment and Reenlistment).

Description

Summary: Recruits or reenlists personnel for military service in the Army. NOTE: Individuals have principal assignment either in recruiting or in retention, with special duties available in counseling, recruiting, training management, instructing, professional development advising, or serving in a supervisory role as the principal noncommissioned officer of a recruiting facility or a recruiting area/region. The recruiting track includes skill levels 20, 30, 40, and 50; the retention track includes skill levels 30, 40, and 50. Because of the principal assignment option after skill level 20, descriptions and recommendations below are cited in two groups at skill levels 30, 40, and 50. Individuals in this MOS do not necessarily start at skill level 20 and work up through the skill levels; they may transfer from any MOS at a comparable skill level. *Skill Level 20:* (Recruiter): Contacts and interviews individuals as prospective enlistees; contacts representatives of schools, corporations, civic groups, and other agencies to present employment and career opportunities of the Army; presents formal and informal talks to organizations and groups; distributes recruiting publicity materials; establishes liaison with communications media and prepares local news releases; evaluates applicants, using screening tests and vocational aptitude tests; prepares forms and documents as part of the enlistment process; arranges for transportation, meals, and lodging of enlistees; assists in market research and analysis of recruiting area. *Skill Level 30:* Able to perform the duties required for Skill Level 20; (Recruiter): Provides technical guidance to five to ten subordinates. (Retention): Knows Army occupations, career fields, and skill levels; explains reenlistment options, benefits and obligations; counsels persons who will be separating from the Army about reserve component benefits; explains training options, vocational opportunities, and reenlistment bonuses; counsels respective reenlistees and assists in designing career plans; processes forms, documents, and records for reenlistment; prepares status charts and statistics; conducts training and inspections; advises unit commanders on status of reenlistment programs; uses computer-based system to contact Central Personnel Command for legal determinations of reenlistment contracts; recommends corrective actions in areas that are deficient. *Skill Level 40:* Able to perform the duties required for Skill Level 30; (Recruiter): Commands a recruiting station and maintains records, statistics, administrative files, and publications library; prepares enlistment reports; supervises and evaluates subordinate personnel; conducts professional development programs for 10 to 12 subordinates; prepares official correspondence and reports; assigns duties to subordinates and supervises resources. (Retention): Advises commanders on status of reenlistment programs; conducts training for commanders, staff persons, noncommissioned officers, and others when assigned as principal noncommissioned officer in a large facility or region. *Skill Level 50:* Able to perform the duties required for Skill Level 40; (Recruiter): Supervises recruiting programs and personnel at several recruiting stations within an area; assists area commanders with training, operations, administration, and personnel matters; plans and conducts recruiting conferences and seminars; develops presentations which reflect sales techniques, enlistment contracts, and enlistment requirements; supervises and evaluates subordinates; identifies and takes corrective actions in problem areas; develops training programs and inspects recruiting stations to ensure efficient operation and management. (Retention): Supervises, assigns duties, and evaluates the performance of 12-34 subordinates; organizes and coordinates retention/career counseling activities; conducts reenlistment seminars and serves as policy advisor in reenlistment matters; coordinates the use and management of reenlistment funds; presents informational talks as needed on reenlistment programs.

Recommendation, Skill Level 20

In the lower-division baccalaureate/associate degree category, (Recruiter): 2 semester hours in social psychology, 2 in audiovisual techniques, 3 in marketing techniques, 3 in public speaking, and 2 in record keeping. In the upper-division baccalaureate category, (Recruiter): 3 semester hours in records management and 3 for field experience in marketing (11/91).

Recommendation, Skill Level 30

In the lower-division baccalaureate/associate degree category, (Recruiter/Retention): 2 semester hours in social psychology, 2 in audiovisual techniques, 3 in marketing techniques, 3 in public speaking, and 3 in record keeping. (Retention only): 1 semester hour in computer applications. In the upper-division baccalaureate category, (Recruiter): 3 semester hours in records management and 3 for field experience in marketing. (Retention): 3 semester hours in records management and 3 in vocational counseling; if progressed from OOR20, add 3 semester hours for field experience in marketing (11/91).

Recommendation, Skill Level 40

In the lower-division baccalaureate/associate degree category, (Recruiter/Retention): 3 semester hours in social psychology, 3 in audiovisual techniques, 3 in marketing techniques, 3 in public speaking, and 3 in record keeping. (Recruiter only): 3 semester hours in personnel supervision. (Retention only): 1 semester hour in computer applications. In the upper-division baccalaureate category, (Recruiter): 3 semester hours in records management, 3 for field experience in marketing, and 3 in personnel management. (Retention): 3 semester hours in records management and 3 in vocational counseling; if progressed from 00R20, add 3 semester hours for field experience in marketing (11/91).

Recommendation, Skill Level 50

In the lower-division baccalaureate/associate degree category, (Recruiter/Retention): 3 semester hours in social psychology, 3 in audiovisual techniques, 3 in marketing techniques, 3 in public speaking, 3 in record keeping, and 3 in personnel supervision. (Recruiter only): 3 semester hours in office procedures. (Retention only): 1 semester hour in computer applications. In the upper-division baccalaureate category, (Recruiter): 3 semester hours in records management, 3 for field experience in marketing, 3 in personnel management, and 3 in office procedures. (Retention): 3 semester hours in records management, 3 in vocational counseling, and 3 in personnel management; if progressed from 00R20, add 3 semester hours for field experience in marketing (11/91).

MOS-00R-003

RECRUITER/RETENTION NCO
00R30
00R40
00R50

Exhibit Dates: 8/94–12/95. Effective 12/95, MOS 00R was discontinued and the recruiter duties were incorporated into MOS 79R, Recruiter; the retention duties were incorporated into MOS 79S, Career Counselor.

Career Management Field: 79 (Recruitment and Reenlistment).

Description

Summary: Recruits or reenlists personnel for military service in the Army. NOTE: Individuals have principal assignment either in recruiting or in retention, with special duties available in counseling, recruiting, training management, instructing, professional development advising, or serving in a supervisory role as the principal noncommissioned officer of a recruiting facility or a recruiting area/region. Because of the principal assignment option, descriptions and recommendations below are cited in two groups at skill levels 30, 40, and 50. Individuals in this MOS do not necessarily start at skill level 30 and work up through the skill levels; they may transfer from any MOS at a comparable skill level. *Skill Level 30:* (Recruiter): Provides technical guidance to five to ten subordinates. (Retention): Knows Army occupations, career fields, and skill levels; explains reenlistment options, benefits and obligations; counsels persons who will be separating from the Army about reserve component benefits; explains training options, vocational opportunities, and reenlistment bonuses; counsels respective reenlistees and assists in designing career plans; processes forms, documents, and records for reenlistment; prepares status charts and statistics; conducts training and inspections; advises unit commanders on status of reenlistment programs; uses computer-based system to contact Central Personnel Command for legal determinations of reenlistment contracts; recommends corrective actions in areas that are deficient. *Skill Level 40:* Able to perform the duties required for Skill Level 30; (Recruiter): Commands a recruiting station and maintains records, statistics, administrative files, and publications library; prepares enlistment reports; supervises and evaluates subordinate personnel; conducts professional development programs for 10 to 12 subordinates; prepares official correspondence and reports; assigns duties to subordinates and supervises resources. (Retention): Advises commanders on status of reenlistment programs; conducts training for commanders, staff persons, noncommissioned officers, and others when assigned as principal noncommissioned officer in a large facility or region. *Skill Level 50:* Able to perform the duties required for Skill Level 40; (Recruiter): Supervises recruiting programs and personnel at several recruiting stations within an area; assists area commanders with training, operations, administration, and personnel matters; plans and conducts recruiting conferences and seminars; develops presentations which reflect sales techniques, enlistment contracts, and enlistment requirements; supervises and evaluates subordinates; identifies and takes corrective actions in problem areas; develops training programs and inspects recruiting stations to ensure efficient operation and management. (Retention): Supervises, assigns duties, and evaluates the performance of 12-34 subordinates; organizes and coordinates retention/career counseling activities; conducts reenlistment seminars and serves as policy advisor in reenlistment matters; coordinates the use and management of reenlistment funds; presents informational talks as needed on reenlistment programs.

Recommendation, Skill Level 30

In the lower-division baccalaureate/associate degree category, (Recruiter/Retention): 2 semester hours in social psychology, 2 in audiovisual techniques, 3 in marketing techniques, 3 in public speaking, and 3 in record keeping. (Recruiter only): 3 semester hours in personnel supervision. (Retention only): 1 semester hour in computer applications. In the upper-division baccalaureate category, (Recruiter): 3 semester hours in records management, and 3 for field experience in marketing. (Retention): 3 semester hours in records management and 3 in vocational counseling; if progressed from 00R20, add 3 semester hours for field experience in marketing (11/91).

Recommendation, Skill Level 40

In the lower-division baccalaureate/associate degree category, (Recruiter/Retention): 3 semester hours in social psychology, 3 in audiovisual techniques, 3 in marketing techniques, 3 in public speaking, and 3 in record keeping. (Recruiter only): 3 semester hours in personnel supervision. (Retention only): 1 semester hour in computer applications. In the upper-division baccalaureate category, (Recruiter): 3 semester hours in records management, 3 for field experience in marketing, and 3 in personnel management. (Retention): 3 semester hours in records management and 3 in vocational counseling; if progressed from 00R20, add 3 semester hours for field experience in marketing (11/91).

Recommendation, Skill Level 50

In the lower-division baccalaureate/associate degree category, (Recruiter/Retention): 3 semester hours in social psychology, 3 in audiovisual techniques, 3 in marketing techniques, 3 in public speaking, 3 in record keeping, and 3 in personnel supervision. (Recruiter only): 3 semester hours in office procedures. (Retention only): 1 semester hour in computer applications. In the upper-division baccalaureate category, (Recruiter): 3 semester hours in records management, 3 for field experience in marketing, 3 in personnel management, and 3 in office procedures. (Retention): 3 semester hours in records management, 3 in vocational counseling, and 3 in personnel management; if progressed from 00R20, add 3 semester hours for field experience in marketing (11/91).

MOS-00Z-001

COMMAND SERGEANT MAJOR
00Z50

Exhibit Dates: 1/90–10/91.

Career Management Field: 00 (Exceptional Management Specialties), subfield 000 (Special Assignment).

Description

Is an experienced mid-level manager responsible for advising high-level administrators on matters of personnel assignment, utilization, promotion, and training and on operations and logistics; supervises subordinate staff members and evaluates their performance; evaluates the operational effectiveness of the organization. NOTE: Required to have attained Skill Level 50 in at least one other MOS.

Recommendation

In the upper-division baccalaureate category, 18 semester hours in personnel management, 3 in staff principles, procedures, and organization, and 3 in office administration (2/75).

MOS-00Z-002

COMMAND SERGEANT MAJOR
00Z50

Exhibit Dates: 11/91–Present.

Career Management Field: 00 (Exceptional Management Specialties).

Description

Serves as principal noncommissioned officer of a military organization at battalion or higher level; provides advice to commander on all matters pertaining to enlisted personnel, personnel assignment, utilization, and training and on operations and logistics; conducts meetings with Senior Sergeants and First Sergeants of component units; has span of influence over 600 or more subordinates of component units; supervises subordinate staff members and evaluates their performance; evaluates the operational effectiveness of the organization; provides advice and counsel to commander and staff on selection of unit First Sergeant; serves on promotion boards and provides counsel in matters of military law (UCMJ). NOTE: Required to have attained Skill Level 50 in at least one other MOS and completed the Command Sergeant's Major Academy.

Recommendation

In the upper-division baccalaureate category, 6 semester hours in personnel supervision, 6 for field experience in management, 3 in organizational management, 3 in principles of management, 3 in public speaking, and 3 in office management (11/91).

MOS-01H-002

Biological Sciences Assistant

01H10
01H20
01H30

Exhibit Dates: 1/90–12/94. (Effective 12/94, MOS 01H was discontinued.)

Career Management Field: 91 (Medical).

Description

Summary: Performs professional-level laboratory and research duties in the field of biological science. *Skill Level 10:* Conducts studies in biology, bacteriology, biochemistry, entomology, or pharmacology; performs culture work on animal diseases; keeps records of culturing and of purity, density, and viability tests of cultures prepared for use as vaccines; inoculates and performs autopsies on laboratory animals used in the preparation of cultures; identifies, prepares, and ships cultures; makes bio-assays to determine toxicity of drugs and other substances; plans and executes experiments; calculates and administers doses and observes toxic effects produced; prepares detailed reports on experiments and tests. NOTE: The prerequisite for the assignment of this MOS is a bachelor's degree and a minimum or six months of related experience, or a master's degree in biology, bacteriology, zoology, parisitology, botany, pharmacology, or entomology. *Skill Level 20:* Able to perform the duties required for Skill Level 10; analyzes scientific data for publications; maintains inventories and equipment. *Skill Level 30:* Able to perform the duties required for Skill Level 20; provides overall research division supervision including quality control and training.

Recommendation, Skill Level 10

In the graduate degree category, 3 semester hours in biological research. (NOTE: This recommendation for skill level 10 is valid for the dates 1/90-10/91 only) (3/90).

Recommendation, Skill Level 20

In the graduate degree category, 6 semester hours in biological research (3/90).

Recommendation, Skill Level 30

In the graduate degree category, 9 semester hours in biological research (3/90).

MOS-02A-001

Army Bandsperson

02A10
02A20
02A30
02A40

Exhibit Dates: Pending Evaluation. 4/04–Present.

MOS-02B-003

Trumpet Player

02B10
02B20
02B30
02B40

Exhibit Dates: 1/90–1/93.

Career Management Field: 97 (Band).

Description

Summary: Plays trumpet in appropriate musical organizations; performs in military marching and concert musical organizations and, as qualified, in associated choral, jazz, and other small ensembles. *Skill Level 10:* Plays trumpet; marches in military band in response to drum major's signals and oral commands; reads, interprets, and plays instrumental music; plays instrument in marching, concert, or dance/jazz band as appropriate; plays, at sight, secondary parts to ceremonial music and marches of medium difficulty; plays with minimum preparation secondary parts to all styles of standard band literature; doubles on another musical instrument as assigned; applies knowledge of minimum basic music theory; discriminates and matches pitches; performs preventive maintenance on trumpet. *Skill Level 20:* Able to perform the duties required for Skill Level 10; provides technical guidance to Skill Level 10 personnel; plays trumpet in small instrumental groups as appropriate; plays, at sight, principal parts to ceremonial music and marches of medium difficulty; plays, with minimum preparation, principal parts to all styles of standard band literature; performs music library duties as assigned. *Skill Level 30:* Able to perform the duties required for Skill Level 20; supervises section as instrumental section leader; serves in various intermediate staff/administrative positions within the band as assigned; organizes, tunes, instructs, and rehearses section or consolidated section of related musical instruments; organizes and leads small instrumental groups; plays, at sight, intermediate music of any style; balances chords and volumes, and interprets complex rhythms; supervises preventive maintenance of musical instruments in section. *Skill Level 40:* Able to perform the duties required for Skill Level 30; provides technical guidance to subordinate personnel; supervises section leaders; rehearses the stage band; assists in rehearsing the concert band; conducts the stage band in a performance; organizes, rehearses, and leads instrumental ensembles; organizes, tunes, and rehearses elements of the band in rehearsal; arranges and transcribes concert and dance/stage band music; tunes and prepares the concert band for rehearsal.

Recommendation, Skill Level 10

In the lower-division baccalaureate/associate degree category, 1-3 semester hours (equivalent of one semester) in applied study of major instrument, 2 in music theory, 1 in concert band, 2 in marching band, and credit in small ensemble on the basis of institutional evaluation; because of the wide range of individual abilities, additional credit may be awarded on the basis of institutional evaluation in music theory, applied music (individual instruction in performance), and small performing ensembles. (NOTE: This recommendation for skill level 10 is valid for the dates 1/90-10/91 only) (6/78).

Recommendation, Skill Level 20

In the lower-division baccalaureate/associate degree category, 1-3 semester hours (equivalent of one semester) in applied study of major instrument, 3 in music theory, 2 in concert band, 2 in marching band, and credit in small ensemble on the basis of institutional evaluation; because of the wide range of individual abilities, additional credit may be awarded on the basis of institutional evaluation in music theory, applied music (individual instruction in performance), and small performing ensembles (6/78).

Recommendation, Skill Level 30

In the lower-division baccalaureate/associate degree category, 2-6 semester hours (equivalent of two semesters) in applied study of major instrument, 4 in music theory (including ear training), 3 in concert band, 3 in marching band, and credit in small ensemble on the basis of institutional evaluation; because of the wide range of individual abilities, additional credit may be awarded on the basis of institutional evaluation in music theory, applied music (individual instruction in performance), and small performing ensembles (6/78).

Recommendation, Skill Level 40

In the lower-division baccalaureate/associate degree category, 2-6 semester hours (equivalent of two semesters) in applied study of major instrument, 4 in music theory (including ear training), 3 in concert band, 3 in marching band, 3 in personnel supervision, and credit in small ensemble on the basis of institutional evaluation; because of the wide range of individual abilities, additional credit may be awarded on the basis of institutional evaluation in music theory, applied music (individual instruction in performance), and small performing ensembles. In the upper-division baccalaureate category, 3 semester hours in conducting and 3 in music administration and supervision (3/86).

MOS-02B-004

Cornet or Trumpet Player

02B10
02B20
02B30
02B40

Exhibit Dates: 2/93–2/95.

Career Management Field: 97 (Band).

Description

Summary: Plays cornet or trumpet in appropriate musical organizations; performs in military marching and concert musical organizations and, as qualified, in associated choral, jazz, and other small ensembles. *Skill Level 10:* Plays cornet or trumpet; marches in military band in response to drum major's signals and oral commands; reads, interprets, and plays instrumental music; plays in marching, concert, or dance/jazz band as appropriate; plays, at sight, secondary parts to ceremonial music and marches of medium difficulty; plays, with minimum preparation, secondary parts to all styles of standard band literature; applies knowledge of minimum basic music theory; discriminates and matches pitches; performs preventive maintenance on cornet or trumpet. *Skill Level 20:* Able to perform the duties required for Skill Level 10; provides technical guidance to Skill Level 10 personnel; plays cornet or trumpet in small instrumental groups as appropriate; plays, at sight, principal parts to ceremonial music and marches of medium difficulty; plays, with minimum preparation, principal parts to all styles of standard band literature; performs music library duties as assigned; assists in instrument repair. *Skill Level 30:* Able to perform the duties required for Skill Level 20; supervises section as instrumental section leader; serves in various intermediate staff/administrative positions within the band as assigned; organizes, tunes, instructs, and rehearses section or consolidated section of related musical instruments; organizes and leads small instrument groups; plays, at sight, intermediate music of any style; balances chords and dynamic levels and interprets complex rhythms; supervises preventive maintenance of musical instruments in section. *Skill Level 40:* Able to perform the duties required for Skill Level 30; provides technical guidance to subordinate personnel; supervises section leaders; rehearses the stage band; assists in rehearsing the concert band; conducts the stage band in a performance; organizes, rehearses, and leads instrumental ensembles; organizes, tunes, and rehearses elements of the band in rehearsal; arranges and transcribes concert and dance/stage band music; tunes and prepares the concert band for rehearsal.

Recommendation, Skill Level 10

Credit may be granted on the basis of an individualized assessment of the student (3/94).

Recommendation, Skill Level 20

In the lower-division baccalaureate/associate degree category, 4 semester hours in music theory (harmony, ear training, and sight singing), 2 in jazz theory/improvisation, 4-6 in applied performance (individual instruction), and 4 in performing ensembles (concert, jazz, and marching bands) (2/93).

Recommendation, Skill Level 30

In the lower-division baccalaureate/associate degree category, 8 semester hours in music theory (harmony, ear training, and sight singing), 2 in jazz theory/improvisation, 4-6 in applied performance (individual instruction), 6 in performing ensembles (concert, jazz, and marching bands), and 3 in personnel supervision. In the upper-division baccalaureate category, 1 semester hour in conducting (2/93).

Recommendation, Skill Level 40

In the lower-division baccalaureate/associate degree category, 12 semester hours in music theory (harmony, ear training, and sight singing), 2 in jazz theory/improvisation, 4-6 in applied performance (individual instruction), 6-8 in performing ensembles (concert, jazz, and marching bands), and 3 in personnel supervision. In the upper-division baccalaureate category, 2 semester hours in conducting, 2 in rehearsal techniques of conducting (concert and jazz), 2 in arranging, and 3 in logistics management (2/93).

MOS-02B-005

Cornet or Trumpet Player

02B10
02B20
02B30
02B40

Exhibit Dates: 3/95–Present.

Career Management Field: 97 (Band).

Description

Summary: Plays cornet or trumpet in appropriate musical organizations; performs in military marching and concert musical organizations and, as qualified, in associated choral, jazz, and other small ensembles. *Skill Level 10:* Plays cornet or trumpet; marches in military band in response to drum major's signals and oral commands; reads, interprets, and plays instrumental music; plays in marching, concert, or dance/jazz band as appropriate; plays, at sight, secondary parts to ceremonial music and marches of medium difficulty; plays, with minimum preparation, secondary parts to all styles of standard band literature; applies knowledge of minimum basic music theory; discriminates and matches pitches; performs preventive maintenance on cornet or trumpet. *Skill Level 20:* Able to perform the duties required for Skill Level 10; provides technical guidance to Skill Level 10 personnel; plays cornet or trumpet in small instrumental groups as appropriate; plays, at sight, principal parts to ceremonial music and marches of medium difficulty; plays, with minimum preparation, principal parts to all styles of standard band literature; performs music library duties as assigned; assists in instrument repair. *Skill Level 30:* Able to perform the duties required for Skill Level 20; supervises section as instrumental section leader; serves in various intermediate staff/administrative positions within the band as assigned; organizes, tunes, instructs, and rehearses section or consolidated section of related musical instruments; organizes and leads small instrument groups; plays, at sight, intermediate music of any style; balances chords and dynamic levels and interprets complex rhythms; supervises preventive maintenance of musical instruments in section. *Skill Level 40:* Able to perform the duties required for Skill Level 30; provides technical guidance to subordinate personnel; supervises section leaders; rehearses the stage band; assists in rehearsing the concert band; conducts the stage band in a performance; organizes, rehearses, and leads instrumental ensembles; organizes, tunes, and rehearses elements of the band in rehearsal; arranges and transcribes concert and dance/stage band music; tunes and prepares the concert band for rehearsal.

Recommendation, Skill Level 10

Credit may be granted on the basis of an individualized assessment of the student (3/94).

Recommendation, Skill Level 20

Credit may be granted on the basis of an individualized assessment of the student (2/93).

Recommendation, Skill Level 30

In the lower-division baccalaureate/associate degree category, 4 semester hours in music theory (harmony, ear training, and sight singing), 2 in jazz theory/improvisation, 4-6 in applied performance (individual instruction), 4 in performing ensembles (concert, jazz, and marching bands), and 3 in personnel supervision. In the upper-division baccalaureate category, 1 semester hour in conducting (2/93).

Recommendation, Skill Level 40

In the lower-division baccalaureate/associate degree category, 8 semester hours in music theory (harmony, ear training, and sight singing), 2 in jazz theory/improvisation, 4-6 in applied performance (individual instruction), 6 in performing ensembles (concert, jazz, and marching bands), and 3 in personnel supervision. In the upper-division baccalaureate category, 2 semester hours in conducting, 2 in rehearsal techniques of conducting (concert and jazz), 2 in arranging, and 3 in logistics management (2/93).

MOS-02C-003

Baritone or Euphonium Player

02C10
02C20
02C30
02C40

Exhibit Dates: 1/90–1/93.

Career Management Field: 97 (Band).

Description

Summary: Plays baritone or euphonium in appropriate musical organizations; performs in military marching and concert musical organizations and, as qualified, in associated choral, jazz, and other small ensembles. *Skill Level 10:* Plays baritone or euphonium; marches in military band in response to drum major's signals and oral commands; reads, interprets, and plays instrumental music; plays instrument in marching, concert, or dance/jazz band as appropriate; plays, at sight, secondary parts to ceremonial music and marches of medium difficulty; plays with minimum preparation secondary parts to all styles of standard band literature; doubles on another musical instrument as assigned; applies knowledge of minimum basic music theory; discriminates and matches pitches; performs preventive maintenance on baritone or euphonium. *Skill Level 20:* Able to perform the duties required for Skill Level 10; provides technical guidance to Skill Level 10 personnel; plays baritone or euphonium in small instrumental groups as appropriate; plays, at sight, principal parts to ceremonial music and marches of medium difficulty; plays, with minimum preparation, principal parts to all styles of standard band literature; performs music library duties as assigned. *Skill Level 30:* Able to perform the duties required for Skill Level 20; supervises section as instrumental section leader; serves in various intermediate staff/administrative positions within the band as assigned; organizes, tunes, instructs, and rehearses section or consolidated section of related musical instruments; organizes and leads small instrumental groups; plays, at sight, intermediate music of any style; balances chords and volumes, and interprets complex rhythms; supervises preventive maintenance of musical instruments in section. *Skill Level 40:* Able to perform the duties required for Skill Level 30; provides technical guidance to subordinate personnel; supervises section leaders; rehearses the stage band; assists in rehearsing the concert band; conducts the stage band in a performance; organizes, rehearses, and leads instrumental ensembles; organizes, tunes, and rehearses elements of the band in rehearsal; arranges and transcribes concert and dance/stage band music; tunes and prepares the concert band for rehearsal.

Recommendation, Skill Level 10

In the lower-division baccalaureate/associate degree category, 1-3 semester hours (equivalent of one semester) in applied study of major instrument, 2 in music theory, 1 in concert band, 2 in marching band, and credit in small ensemble on the basis of institutional evaluation; because of the wide range of individual abilities, additional credit may be awarded on the basis of institutional evaluation in music theory, applied music (individual instruction in performance), and small performing ensembles. (NOTE: This recommendation for skill level 10 is valid for the dates 1/90-10/91 only) (6/78).

Recommendation, Skill Level 20

In the lower-division baccalaureate/associate degree category, 1-3 semester hours (equivalent of one semester) in applied study of major instrument, 3 in music theory, 2 in concert band, 2 in marching band, and credit in small ensemble on the basis of institutional evaluation; because of the wide range of individual abilities, additional credit may be awarded on the basis of institutional evaluation in music theory, applied music (individual instruction in performance), and small performing ensembles (6/78).

Recommendation, Skill Level 30

In the lower-division baccalaureate/associate degree category, 2-6 semester hours (equivalent of two semesters) in applied study of major instrument, 4 in music theory (including ear training), 3 in concert band, 3 in marching band, and credit in small ensemble on the basis of institutional evaluation; because of the wide range of individual abilities, additional credit may be awarded on the basis of institutional evaluation in music theory, applied music (individual instruction in performance), and small performing ensembles (6/78).

Recommendation, Skill Level 40

In the lower-division baccalaureate/associate degree category, 2-6 semester hours (equivalent of two semesters) in applied study of major instrument, 4 in music theory, (including ear training), 3 in concert band, 3 in marching band, 3 in personnel supervision, and credit in small ensemble on the basis of institutional evaluation; because of the wide range of individual abilities, additional credit may be awarded on the basis of institutional evaluation in music theory, applied music (individual instruction in performance), and small performing ensembles. In the upper-division baccalaureate category, 3 semester hours in conducting and 3 in music administration and supervision (3/86).

MOS-02C-004

Baritone or Euphonium Player
02C10
02C20
02C30
02C40

Exhibit Dates: 2/93–2/95.

Career Management Field: 97 (Band).

Description

Summary: Plays baritone or euphonium in appropriate musical organizations; performs in military marching and concert musical organizations and, as qualified, in associated choral, jazz, and other small ensembles. *Skill Level 10:* Plays baritone or euphonium; marches in military band in response to drum major's signals and oral commands; reads, interprets, and plays instrumental music; plays in marching, concert, or dance/jazz band as appropriate; plays, at sight, secondary parts to ceremonial music and marches of medium difficulty; plays, with minimum preparation, secondary parts to all styles of standard band literature; applies knowledge of minimum basic music theory; discriminates and matches pitches; performs preventive maintenance on baritone or euphonium. *Skill Level 20:* Able to perform the duties required for Skill Level 10; provides technical guidance to Skill Level 10 personnel; plays baritone or euphonium in small instrumental groups as appropriate; plays, at sight, principal parts to ceremonial music and marches of medium difficulty; plays, with minimum preparation, principal parts to all styles of standard band literature; performs music library duties as assigned; assists in instrument repair. *Skill Level 30:* Able to perform the duties required for Skill Level 20; supervises section as instrumental section leader; serves in various intermediate staff/administrative positions within the band as assigned; organizes, tunes, instructs, and rehearses section or consolidated section of related musical instruments; organizes and leads small instrument groups; plays, at sight, intermediate music of any style; balances chords and dynamic levels and interprets complex rhythms; supervises preventive maintenance of musical instruments in section. *Skill Level 40:* Able to perform the duties required for Skill Level 30; provides technical guidance to subordinate personnel; supervises section leaders; rehearses the stage band; assists in rehearsing the concert band; conducts the stage band in a performance; organizes, rehearses, and leads instrumental ensembles; organizes, tunes, and rehearses elements of the band in rehearsal; arranges and transcribes concert and dance/stage band music; tunes and prepares the concert band for rehearsal.

Recommendation, Skill Level 10

Credit may be granted on the basis of an individualized assessment of the student (3/94).

Recommendation, Skill Level 20

In the lower-division baccalaureate/associate degree category, 4 semester hours in music theory (harmony, ear training, and sight singing), 2 in jazz theory/improvisation, 4-6 in applied performance (individual instruction), and 4 in performing ensembles (concert, jazz, and marching bands) (2/93).

Recommendation, Skill Level 30

In the lower-division baccalaureate/associate degree category, 8 semester hours in music theory (harmony, ear training, and sight singing), 2 in jazz theory/improvisation, 4-6 in applied performance (individual instruction), 6 in performing ensembles (concert, jazz, and marching bands), and 3 in personnel supervision. In the upper-division baccalaureate category, 1 semester hour in conducting (2/93).

Recommendation, Skill Level 40

In the lower-division baccalaureate/associate degree category, 12 semester hours in music theory (harmony, ear training, and sight singing), 2 in jazz theory/improvisation, 4-6 in applied performance (individual instruction), 6-8 in performing ensembles (concert, jazz, and marching bands), and 3 in personnel supervision. In the upper-division baccalaureate category, 2 semester hours in conducting, 2 in rehearsal techniques of conducting (concert and jazz), 2 in arranging, and 3 in logistics management (2/93).

MOS-02C-005

Baritone or Euphonium Player
02C10
02C20
02C30
02C40

Exhibit Dates: 3/95–Present.

Career Management Field: 97 (Band).

Description

Summary: Plays baritone or euphonium in appropriate musical organizations; performs in military marching and concert musical organizations and, as qualified, in associated choral, jazz, and other small ensembles. *Skill Level 10:* Plays baritone or euphonium; marches in military band in response to drum major's signals and oral commands; reads, interprets, and plays instrumental music; plays in marching, concert, or dance/jazz band as appropriate; plays, at sight, secondary parts to ceremonial music and marches of medium difficulty; plays, with minimum preparation, secondary parts to all styles of standard band literature; applies knowledge of minimum basic music theory; discriminates and matches pitches; performs preventive maintenance on baritone or euphonium. *Skill Level 20:* Able to perform the duties required for Skill Level 10; provides technical guidance to Skill Level 10 personnel; plays baritone or euphonium in small instrumental groups as appropriate; plays, at sight, principal parts to ceremonial music and marches of medium difficulty; plays, with minimum preparation, principal parts to all styles of standard band literature; performs music library duties as assigned; assists in instrument repair. *Skill Level 30:* Able to perform the duties required for Skill Level 20; supervises section as instrumental section leader; serves in various intermediate staff/administrative positions within the band as assigned; organizes, tunes, instructs, and rehearses section or consolidated section of related musical instruments; organizes and leads small instrument groups; plays, at sight, intermediate music of any style; balances chords and dynamic levels and interprets complex rhythms; supervises preventive maintenance of musical instruments in section. *Skill Level 40:* Able to perform the duties required for Skill Level 30; provides technical guidance to subordinate personnel; supervises section leaders; rehearses the stage band; assists in rehearsing the concert band; conducts the stage band in a performance; organizes, rehearses, and leads instrumental ensembles; organizes, tunes, and rehearses elements of the band in rehearsal; arranges and transcribes concert and dance/stage band music; tunes and prepares the concert band for rehearsal.

Recommendation, Skill Level 10

Credit may be granted on the basis of an individualized assessment of the student (3/94).

Recommendation, Skill Level 20

Credit may be granted on the basis of an individualized assessment of the student (2/93).

Recommendation, Skill Level 30

In the lower-division baccalaureate/associate degree category, 4 semester hours in music theory (harmony, ear training, and sight singing), 2 in jazz theory/improvisation, 4-6 in applied performance (individual instruction), 4 in performing ensembles (concert, jazz, and marching bands), and 3 in personnel supervision. In the upper-division baccalaureate category, 1 semester hour in conducting (2/93).

Recommendation, Skill Level 40

In the lower-division baccalaureate/associate degree category, 8 semester hours in music theory (harmony, ear training, and sight singing), 2 in jazz theory/improvisation, 4-6 in applied performance (individual instruction), 6 in performing ensembles (concert, jazz, and marching bands), and 3 in personnel supervision. In the upper-division baccalaureate category, 2 semester hours in conducting, 2 in rehearsal techniques of conducting (concert and jazz), 2 in arranging, and 3 in logistics management (2/93).

MOS-02D-003

French Horn Player
02D10
02D20
02D30
02D40
02D50

Exhibit Dates: 1/90–1/93.

Career Management Field: 97 (Band).

Description

Summary: Plays French horn in appropriate musical organizations; performs in military marching and concert musical organizations and, as qualified, in associated choral, jazz, and other small ensembles. *Skill Level 10:* Plays French horn; marches in military band in response to drum major's signals and oral commands; reads, interprets, and plays instrumental music; plays instrument in marching, concert, or dance/jazz band as appropriate; plays, at sight, secondary parts to ceremonial music and marches of medium difficulty; plays with minimum preparation secondary parts to all styles of standard band literature; doubles on another musical instrument as assigned; applies knowledge of minimum basic music theory; discriminates and matches pitches; performs preventive maintenance on French horn. *Skill Level 20:* Able to perform the duties required for Skill Level 10; provides technical guidance to Skill Level 10 personnel; plays French horn in small instrumental groups as appropriate; plays, at sight, principal parts to ceremonial music and marches of medium difficulty; plays, with minimum preparation, principal parts to all styles of standard band literature; performs music library duties as assigned. *Skill Level 30:* Able to perform the duties required for Skill Level 20; supervises section as instrumental section leader; serves in various intermediate staff/administrative positions within the band as assigned; organizes, tunes, instructs, and rehearses section or consolidated section of related musical instruments; organizes and leads small instrumental groups; plays, at sight, intermediate music of any style; balances chords and volumes, and interprets complex rhythms; supervises preventive maintenance of musical instruments in section. *Skill Level 40:* Able to perform the duties required for Skill Level 30; provides technical guidance to subordinate personnel; supervises section leaders; rehearses the stage band; assists in rehearsing the concert band; conducts the stage band in a performance; organizes, rehearses, and leads instrumental ensembles; organizes, tunes, and rehearses elements of the band in rehearsal; arranges and transcribes concert and dance/stage band music; tunes and prepares the concert band for rehearsal. *Skill Level 50:* Able to perform the duties required for Skill Level 40; able to perform the duties required for Skill Level 30; provides technical guidance to subordinate personnel; supervises section leaders; rehearses the stage band; assists in rehearsing the concert band; conducts the stage band in a performance; organizes, rehearses, and leads instrumental ensembles; organizes, tunes, and rehearses elements of the band in rehearsal; arranges and transcribes concerts and dance/stage band music; tunes and prepares the concert band for rehearsal.

Recommendation, Skill Level 10

In the lower-division baccalaureate/associate degree category, 1-3 semester hours (equivalent of one semester) in applied study of major instrument, 2 in music theory, 1 in concert band, 2 in marching band, and credit in small ensemble on the basis of institutional evaluation; because of the wide range of individual abilities, additional credit may be awarded on the basis of institutional evaluation in music theory, applied music (individual instruction in performance), and small performing ensembles. (NOTE: This recommendation for skill level 10 is valid for the dates 1/90-9/91 only) (6/78).

Recommendation, Skill Level 20

In the lower-division baccalaureate/associate degree category, 1-3 semester hours (equivalent of one semester) in applied study of major instrument, 3 in music theory, 2 in concert band, 2 in marching band, and credit in small ensemble on the basis of institutional evaluation; because of the wide range of individual abilities, additional credit may be awarded on the basis of institutional evaluation in music theory, applied music (individual instruction in performance), and small performing ensembles (6/78).

Recommendation, Skill Level 30

In the lower-division baccalaureate/associate degree category, 2-6 semester hours (equivalent of two semesters) in applied study of major instrument, 4 in music theory (including ear training), 3 in concert band, 3 in marching band, and credit in small ensemble on the basis of institutional evaluation; because of the wide range of individual abilities, additional credit may be awarded on the basis of institutional evaluation in music theory, applied music (individual instruction in performance), and small performing ensembles (6/78).

Recommendation, Skill Level 40

In the lower-division baccalaureate/associate degree category, 2-6 semester hours (equivalent of two semesters) in applied study of major instrument, 4 in music theory (including ear training), 3 in concert band, 3 in marching band, 3 in personnel supervision, and credit in small ensemble on the basis of institutional evaluation; because of the wide range of individual abilities, additional credit may be awarded on the basis of institutional evaluation in music theory, applied music (individual instruction in performance), and small performing ensembles. In the upper-division baccalaureate category, 3 semester hours in conducting and 3 in music administration and supervision (3/86).

Recommendation, Skill Level 50

In the lower-division baccalaureate/associate degree category, 2-6 semester hours (equivalent of two semesters) in applied study of major instrument, 4 in music theory (including ear training), 3 in concert band, 3 in marching band, 3 in personnel supervision, and credit in small ensemble on the basis of institutional evaluation; because of the wide range of individual abilities, additional credit may be awarded on the basis of institutional evaluation in music theory, applied music (individual instruction in performance), and small performing ensembles. In the upper-division baccalaureate category, 3 semester hours in conducting and 3 in music administration and supervision (6/78).

MOS-02D-004

French Horn Player
02D10
02D20
02D30
02D40

Exhibit Dates: 2/93–2/95.

Career Management Field: 97 (Band).

Description

Summary: Plays French horn in appropriate musical organizations; performs in military marching and concert musical organizations and, as qualified, in associated choral, jazz, and other small ensembles. *Skill Level 10:* Plays French horn; marches in military band in response to drum major's signals and oral commands; reads, interprets, and plays instrumental music; plays in marching, concert, or dance/jazz band as appropriate; plays, at sight, secondary parts to ceremonial music and marches of medium difficulty; plays, with minimum preparation, secondary parts to all styles of standard band literature; applies knowledge of minimum basic music theory; discriminates and matches pitches; performs preventive maintenance on French horn. *Skill Level 20:* Able to perform the duties required for Skill Level 10; provides technical guidance to Skill Level 10 personnel; plays French horns in small instrumental groups as appropriate; plays, at sight, principal parts to ceremonial music and marches of medium difficulty; plays, with minimum preparation, principal parts to all styles of standard band literature; performs music library duties as assigned; assists in instrument repair. *Skill Level 30:* Able to perform the duties required for Skill Level 20; supervises section as instrumental section leader; serves in various intermediate staff/administrative positions within the band as assigned; organizes, tunes, instructs, and rehearses section or consolidated section of related musical instruments; organizes and leads small instrument groups; plays, at sight, intermediate music of any style; balances chords and dynamic levels and interprets complex rhythms; supervises preventive maintenance of musical instruments in section. *Skill Level 40:* Able to perform the duties required for Skill Level 30; provides technical guidance to subordinate personnel; supervises section leaders; rehearses the stage band; assists in rehearsing the concert band; conducts the stage band in a performance; organizes, rehearses, and leads instrumental ensembles; organizes, tunes, and rehearses elements of the band in rehearsal; arranges and transcribes concert and dance/stage band music; tunes and prepares the concert band for rehearsal.

Recommendation, Skill Level 10

Credit may be granted on the basis of an individualized assessment of the student (3/94).

Recommendation, Skill Level 20

In the lower-division baccalaureate/associate degree category, 4 semester hours in music theory (harmony, ear training, and sight singing), 2 in jazz theory/improvisation, 4-6 in applied performance (individual instruction), and 4 in performing ensembles (concert, jazz, and marching bands) (2/93).

Recommendation, Skill Level 30

In the lower-division baccalaureate/associate degree category, 8 semester hours in music theory (harmony, ear training, and sight singing), 2 in jazz theory/improvisation, 4-6 in applied performance (individual instruction), 6 in performing ensembles (concert, jazz, and marching bands), and 3 in personnel supervision. In the upper-division baccalaureate category, 1 semester hour in conducting (2/93).

Recommendation, Skill Level 40

In the lower-division baccalaureate/associate degree category, 12 semester hours in music theory (harmony, ear training, and sight singing), 2 in jazz theory/improvisation, 4-6 in applied performance (individual instruction), 6-8 in performing ensembles (concert, jazz, and marching bands), and 3 in personnel supervision. In the upper-division baccalaureate category, 2 semester hours in conducting, 2 in rehearsal techniques of conducting (concert and jazz), 2 in arranging, and 3 in logistics management (2/93).

MOS-02D-005

French Horn Player

02D10
02D20
02D30
02D40

Exhibit Dates: 3/95–Present.

Career Management Field: 97 (Band).

Description

Summary: Plays French horn in appropriate musical organizations; performs in military marching and concert musical organizations and, as qualified, in associated choral, jazz, and other small ensembles. *Skill Level 10:* Plays French horn; marches in military band in response to drum major's signals and oral commands; reads, interprets, and plays instrumental music; plays in marching, concert, or dance/jazz band as appropriate; plays, at sight, secondary parts to ceremonial music and marches of medium difficulty; plays, with minimum preparation, secondary parts to all styles of standard band literature; applies knowledge of minimum basic music theory; discriminates and matches pitches; performs preventive maintenance on French horn. *Skill Level 20:* Able to perform the duties required for Skill Level 10; provides technical guidance to Skill Level 10 personnel; plays French horns in small instrumental groups as appropriate; plays, at sight, principal parts to ceremonial music and marches of medium difficulty; plays, with minimum preparation, principal parts to all styles of standard band literature; performs music library duties as assigned; assists in instrument repair. *Skill Level 30:* Able to perform the duties required for Skill Level 20; supervises section as instrumental section leader; serves in various intermediate staff/administrative positions within the band as assigned; organizes, tunes, instructs, and rehearses section or consolidated section of related musical instruments; organizes and leads small instrument groups; plays, at sight, intermediate music of any style; balances chords and dynamic levels and interprets complex rhythms; supervises preventive maintenance of musical instruments in section. *Skill Level 40:* Able to perform the duties required for Skill Level 30; provides technical guidance to subordinate personnel; supervises section leaders; rehearses the stage band; assists in rehearsing the concert band; conducts the stage band in a performance; organizes, rehearses, and leads instrumental ensembles; organizes, tunes; and rehearses elements of the band in rehearsal; arranges and transcribes concert and dance/stage band music; tunes and prepares the concert band for rehearsal.

Recommendation, Skill Level 10

Credit may be granted on the basis of an individualized assessment of the student (3/94).

Recommendation, Skill Level 20

Credit may be granted on the basis of an individualized assessment of the student (2/93).

Recommendation, Skill Level 30

In the lower-division baccalaureate/associate degree category, 4 semester hours in music theory (harmony, ear training, and sight singing), 2 in jazz theory/improvisation, 4-6 in applied performance (individual instruction), 4 in performing ensembles (concert, jazz, and marching bands), and 3 in personnel supervision. In the upper-division baccalaureate category, 1 semester hour in conducting (2/93).

Recommendation, Skill Level 40

In the lower-division baccalaureate/associate degree category, 8 semester hours in music theory (harmony, ear training, and sight singing), 2 in jazz theory/improvisation, 4-6 in applied performance (individual instruction), 6 in performing ensembles (concert, jazz, and marching bands), and 3 in personnel supervision. In the upper-division baccalaureate category, 2 semester hours in conducting, 2 in rehearsal techniques of conducting (concert and jazz), 2 in arranging, and 3 in logistics management (2/93).

MOS-02E-003

Trombone Player

02E10
02E20
02E30
02E40

Exhibit Dates: 1/90–1/93.

Career Management Field: 97 (Band).

Description

Summary: Plays trombone in appropriate musical organizations; performs in military marching and concert musical organizations and, as qualified, in associated choral, jazz, and other small ensembles. *Skill Level 10:* Plays trombone; marches in military band in response to drum major's signals and oral commands; reads, interprets, and plays instrumental music; plays instrument in marching, concert, or dance/jazz band as appropriate; plays, at sight, secondary parts to ceremonial music and marches of medium difficulty; plays with minimum preparation secondary parts to all styles of standard band literature; doubles on another musical instrument as assigned; applies knowledge of minimum basic music theory; discriminates and matches pitches; performs preventive maintenance on trombone. *Skill Level 20:* Able to perform the duties required for Skill Level 10; provides technical guidance to Skill Level 10 personnel; plays trombone in small instrumental groups as appropriate; plays, at sight, principal parts to ceremonial music and marches of medium difficulty; plays, with minimum preparation, principal parts to all styles of standard band literature; performs music library duties as assigned. *Skill Level 30:* Able to perform the duties required for Skill Level 20; supervises section as instrumental section leader; serves in various intermediate staff/administrative positions within the band as assigned; organizes, tunes, instructs, and rehearses section or consolidated section of related musical instruments; organizes and leads small instrumental groups; plays, at sight, intermediate music of any style; balances chords and volumes, and interprets complex rhythms; supervises preventive maintenance of musical instruments in section. *Skill Level 40:* Able to perform the duties required for Skill Level 30; provides technical guidance to subordinate personnel; supervises section leaders; rehearses the stage band; assists in rehearsing the concert band; conducts the stage band in a performance; organizes, rehearses, and leads instrumental ensembles; organizes, tunes, and rehearses elements of the band in rehearsal; arranges and transcribes concert and dance/stage band music; tunes and prepares the concert band for rehearsal.

Recommendation, Skill Level 10

In the lower-division baccalaureate/associate degree category, 1-3 semester hours in applied study of major instrument, 2 in music theory, 1 in concert band, 2 in marching band, and credit in small ensemble on the basis of institutional evaluation; because of the wide range of individual abilities, additional credit may be awarded on the basis of institutional evaluation in music theory, applied music (individual instruction in performance), and small performing ensembles. (NOTE: This recommendation for skill level 10 is valid for the dates 1/90-9/91 only) (6/78).

Recommendation, Skill Level 20

In the lower-division baccalaureate/associate degree category, 1-3 semester hours (equivalent of one semester) in applied study of major instrument, 3 in music theory, 2 in concert band, 2 in marching band, and credit in small ensemble on the basis of institutional evaluation; because of the wide range of individual abilities, additional credit may be awarded on the basis of institutional evaluation in music theory, applied music (individual instruction in performance), and small performing ensembles (6/78).

Recommendation, Skill Level 30

In the lower-division baccalaureate/associate degree category, 2-6 semester hours (equivalent of two semesters) in applied study of major instrument, 4 in music theory (including ear training), 3 in concert band, 3 in marching band, and credit in small ensemble on the basis of institutional evaluation; because of the wide range of individual abilities, additional credit may be awarded on the basis of institutional evaluation in music theory, applied music (individual instruction in performance), and small performing ensembles (6/78).

Recommendation, Skill Level 40

In the lower-division baccalaureate/associate degree category, 2-6 semester hours (equivalent of two semesters) in applied study of major instrument, 4 in music theory (including ear training), 3 in concert band, 3 in marching band, 3 in personnel supervision, and credit in small ensemble on the basis of institutional evaluation; because of the wide range of individual abilities, additional credit may be awarded on the basis of institutional evaluation in music theory, applied music (individual instruction in performance), and small performing ensembles. In the upper-division baccalaureate category, 3 semester hours in conducting and 3 in music administration and supervision (3/86).

MOS-02E-004

TROMBONE PLAYER

02E10
02E20
02E30
02E40

Exhibit Dates: 2/93–2/95.

Career Management Field: 97 (Band).

Description

Summary: Plays trombone in appropriate musical organizations; performs in military marching and concert musical organizations and, as qualified, in associated choral, jazz, and other small ensembles. *Skill Level 10:* Plays trombone; marches in military band in response to drum major's signals and oral commands; reads, interprets, and plays instrumental music; plays in marching, concert, or dance/jazz band as appropriate; plays, at sight, secondary parts to ceremonial music and marches of medium difficulty; plays, with minimum preparation, secondary parts to all styles of standard band literature; applies knowledge of minimum basic music theory; discriminates and matches pitches; performs preventive maintenance on trombone. *Skill Level 20:* Able to perform the duties required for Skill Level 10; provides technical guidance to Skill Level 10 personnel; plays trombone in small instrumental groups as appropriate; plays, at sight, principal parts to ceremonial music and marches of medium difficulty; plays, with minimum preparation, principal parts to all styles of standard band literature; performs music library duties as assigned; assists in instrument repair. *Skill Level 30:* Able to perform the duties required for Skill Level 20; supervises section as instrumental section leader; serves in various intermediate staff/administrative positions within the band as assigned; organizes, tunes, instructs, and rehearses section or consolidated section of related musical instruments; organizes and leads small instrument groups; plays, at sight, intermediate music of any style; balances chords and dynamic levels and interprets complex rhythms; supervises preventive maintenance of musical instruments in section. *Skill Level 40:* Able to perform the duties required for Skill Level 30; provides technical guidance to subordinate personnel; supervises section leaders; rehearses the stage band; assists in rehearsing the concert band; conducts the stage band in a performance; organizes, rehearses, and leads instrumental ensembles; organizes, tunes, and rehearses elements of the band in rehearsal; arranges and transcribes concert and dance/stage band music; tunes and prepares the concert band for rehearsal.

Recommendation, Skill Level 10

Credit may be granted on the basis of an individualized assessment of the student (3/94).

Recommendation, Skill Level 20

In the lower-division baccalaureate/associate degree category, 4 semester hours in music theory (harmony, ear training, and sight singing), 2 in jazz theory/improvisation, 4-6 in applied performance (individual instruction), and 4 in performing ensembles (concert, jazz, and marching bands) (2/93).

Recommendation, Skill Level 30

In the lower-division baccalaureate/associate degree category, 8 semester hours in music theory (harmony, ear training, and sight singing), 2 in jazz theory/improvisation, 4-6 in applied performance (individual instruction), 6 in performing ensembles (concert, jazz, and marching bands), and 3 in personnel supervision. In the upper-division baccalaureate category, 1 semester hour in conducting (2/93).

Recommendation, Skill Level 40

In the lower-division baccalaureate/associate degree category, 12 semester hours in music theory (harmony, ear training, and sight singing), 2 in jazz theory/improvisation, 4-6 in applied performance (individual instruction), 6-8 in performing ensembles (concert, jazz, and marching bands), and 3 in personnel supervision. In the upper-division baccalaureate category, 2 semester hours in conducting, 2 in rehearsal techniques of conducting (concert and jazz), 2 in arranging, and 3 in logistics management (2/93).

MOS-02E-005

TROMBONE PLAYER

02E10
02E20
02E30
02E40

Exhibit Dates: 3/95–Present.

Career Management Field: 97 (Band).

Description

Summary: Plays trombone in appropriate musical organizations; performs in military marching and concert musical organizations and, as qualified, in associated choral, jazz, and other small ensembles. *Skill Level 10:* Plays trombone; marches in military band in response to drum major's signals and oral commands; reads, interprets, and plays instrumental music; plays in marching, concert, or dance/jazz band as appropriate; plays, at sight, secondary parts to ceremonial music and marches of medium difficulty; plays, with minimum preparation, secondary parts to all styles of standard band literature; applies knowledge of minimum basic music theory; discriminates and matches pitches; performs preventive maintenance on trombone. *Skill Level 20:* Able to perform the duties required for Skill Level 10; provides technical guidance to Skill Level 10 personnel; plays trombone in small instrumental groups as appropriate; plays, at sight, principal parts to ceremonial music and marches of medium difficulty; plays, with minimum preparation, principal parts to all styles of standard band literature; performs music library duties as assigned; assists in instrument repair. *Skill Level 30:* Able to perform the duties required for Skill Level 20; supervises section as instrumental section leader; serves in various intermediate staff/administrative positions within the band as assigned; organizes, tunes, instructs, and rehearses section or consolidated section of related musical instruments; organizes and leads small instrument groups; plays, at sight, intermediate music of any style; balances chords and dynamic levels and interprets complex rhythms; supervises preventive maintenance of musical instruments in section. *Skill Level 40:* Able to perform the duties required for Skill Level 30; provides technical guidance to subordinate personnel; supervises section leaders; rehearses the stage band; assists in rehearsing the concert band; conducts the stage band in a performance; organizes, rehearses, and leads instrumental ensembles; organizes, tunes, and rehearses elements of the band in rehearsal; arranges and transcribes concert and dance/stage band music; tunes and prepares the concert band for rehearsal.

Recommendation, Skill Level 10

Credit may be granted on the basis of an individualized assessment of the student (3/94).

Recommendation, Skill Level 20

Credit may be granted on the basis of an individualized assessment of the student (2/93).

Recommendation, Skill Level 30

In the lower-division baccalaureate/associate degree category, 4 semester hours in music theory (harmony, ear training, and sight singing), 2 in jazz theory/improvisation, 4-6 in applied performance (individual instruction), 4 in performing ensembles (concert, jazz, and marching bands), and 3 in personnel supervision. In the upper-division baccalaureate category, 1 semester hour in conducting (2/93).

Recommendation, Skill Level 40

In the lower-division baccalaureate/associate degree category, 8 semester hours in music theory (harmony, ear training, and sight singing), 2 in jazz theory/improvisation, 4-6 in applied performance (individual instruction), 6 in performing ensembles (concert, jazz, and marching bands), and 3 in personnel supervision. In the upper-division baccalaureate category, 2 semester hours in conducting, 2 in rehearsal techniques of conducting (concert and jazz), 2 in arranging, and 3 in logistics management (2/93).

MOS-02F-003

TUBA PLAYER

02F10
02F20
02F30
02F40

Exhibit Dates: 1/90–1/93.

Career Management Field: 97 (Band).

Description

Summary: Plays tuba in appropriate musical organizations; performs in military marching and concert musical organizations and, as qualified, in associated choral, jazz, and other small ensembles. *Skill Level 10:* Plays tuba; marches in military band in response to drum major's signals and oral commands; reads, interprets, and plays instrumental music; plays instrument in marching, concert, or dance/jazz band as appropriate; plays, at sight, secondary parts to ceremonial music and marches of medium difficulty; plays with minimum preparation secondary parts to all styles of standard band literature; doubles on another musical instrument as assigned; applies knowledge of minimum basic music theory; discriminates and matches pitches; performs preventive maintenance on tuba. *Skill Level 20:* Able to perform the duties required for Skill Level 10; provides technical guidance to Skill Level 10 personnel; plays tuba in small instrumental groups as appropriate; plays, at sight, principal parts to ceremonial music and marches of medium difficulty; plays, with minimum preparation, principal parts to all styles of standard band literature; performs music library duties as assigned. *Skill Level 30:* Able to perform the duties required for Skill Level 20; supervises section as instrumental section leader; serves in various intermediate staff/administrative positions within the band as assigned; organizes, tunes, instructs, and rehearses section or consolidated section of related musical instruments; organizes and leads small instrumental groups; plays, at sight, intermediate music of any style; balances chords and volumes, and interprets complex rhythms; supervises preventive maintenance of musical instruments in section. *Skill Level 40:* Able to perform the duties required for Skill Level 30; provides technical guidance to subordinate personnel; supervises section leaders; rehearses the stage band; assists in rehearsing the concert band; conducts the stage band in a performance; organizes, rehearses, and leads instrumental ensembles; organizes, tunes, and rehearses elements of the band in rehearsal; arranges and transcribes concert and dance/stage band music; tunes and prepares the concert band for rehearsal.

Recommendation, Skill Level 10

In the lower-division baccalaureate/associate degree category, 1-3 semester hours (equivalent of one semester) in applied study of major instrument, 2 in music theory, 1 in concert band, 2 in marching band, and credit in small ensemble on the basis of institutional evaluation; because of the wide range of individual abilities, additional credit may be awarded on the basis of institutional evaluation in music theory, applied music (individual instruction in performance), and small performing ensembles. (NOTE: This recommendation for skill level 10 is valid for the dates 1/90-9/91 only) (6/78).

Recommendation, Skill Level 20

In the lower-division baccalaureate/associate degree category, 1-3 semester hours (equivalent of one semester) in applied study of major instrument, 3 in music theory, 2 in concert band, 2 in marching band, and credit in small ensemble on the basis of institutional evaluation; because of the wide range of individual abilities, additional credit may be awarded on the basis of institutional evaluation in music theory, applied music (individual instruction in performance), and small performing ensembles (6/78).

Recommendation, Skill Level 30

In the lower-division baccalaureate/associate degree category, 2-6 semester hours (equivalent of two semesters) in applied study of major instrument, 4 in music theory (including ear training), 3 in concert band, 3 in marching band, and credit in small ensemble on the basis of institutional evaluation; because of the wide range of individual abilities, additional credit may be awarded on the basis of institutional evaluation in music theory, applied music (individual instruction in performance), and small performing ensembles (6/78).

Recommendation, Skill Level 40

In the lower-division baccalaureate/associate degree category, 2-6 semester hours (equivalent of two semesters) in applied study of major instrument, 4 in music theory (including ear training), 3 in concert band, 3 in marching band, 3 in personnel supervision, and credit in small ensemble on the basis of institutional evaluation; because of the wide range of individual abilities, additional credit may be awarded on the basis of institutional evaluation in music theory, applied music (individual instruction in performance), and small performing ensembles. In the upper-division baccalaureate category, 3 semester hours in conducting and 3 in music administration and supervision (3/86).

MOS-02F-004

TUBA PLAYER
02F10
02F20
02F30
02F40

Exhibit Dates: 2/93–2/95.

Career Management Field: 97 (Band).

Description

Summary: Plays tuba in appropriate musical organizations; performs in military marching and concert musical organizations and, as qualified, in associated choral, jazz, and other small ensembles. *Skill Level 10:* Plays tuba; marches in military band in response to drum major's signals and oral commands; reads, interprets, and plays instrumental music; plays in marching, concert, or dance/jazz band as appropriate; plays, at sight, secondary parts to ceremonial music and marches of medium difficulty; plays, with minimum preparation, secondary parts to all styles of standard band literature; applies knowledge of minimum basic music theory; discriminates and matches pitches; performs preventive maintenance on tuba. *Skill Level 20:* Able to perform the duties required for Skill Level 10; provides technical guidance to Skill Level 10 personnel; plays tuba in small instrumental groups as appropriate; plays, at sight, principal parts to ceremonial music and marches of medium difficulty; plays, with minimum preparation, principal parts to all styles of standard band literature; performs music library duties as assigned; assists in instrument repair. *Skill Level 30:* Able to perform the duties required for Skill Level 20; supervises section as instrumental section leader; serves in various intermediate staff/administrative positions within the band as assigned; organizes, tunes, instructs, and rehearses section or consolidated section of related musical instruments; organizes and leads small instrument groups; plays, at sight, intermediate music of any style; balances chords and dynamic levels and interprets complex rhythms; supervises preventive maintenance of musical instruments in section. *Skill Level 40:* Able to perform the duties required for Skill Level 30; provides technical guidance to subordinate personnel; supervises section leaders; rehearses the stage band; assists in rehearsing the concert band; conducts the stage band in a performance; organizes, rehearses, and leads instrumental ensembles; organizes, tunes, and rehearses elements of the band in rehearsal; arranges and transcribes concert and dance/stage band music; tunes and prepares the concert band for rehearsal.

Recommendation, Skill Level 10

Credit may be granted on the basis of an individualized assessment of the student (3/94).

Recommendation, Skill Level 20

In the lower-division baccalaureate/associate degree category, 4 semester hours in music theory (harmony, ear training, and sight singing), 2 in jazz theory/improvisation, 4-6 in applied performance (individual instruction), and 4 in performing ensembles (concert, jazz, and marching bands) (2/93).

Recommendation, Skill Level 30

In the lower-division baccalaureate/associate degree category, 8 semester hours in music theory (harmony, ear training, and sight singing), 2 in jazz theory/improvisation, 4-6 in applied performance (individual instruction), 6 in performing ensembles (concert, jazz, and marching bands), and 3 in personnel supervision. In the upper-division baccalaureate category, 1 semester hour in conducting (2/93).

Recommendation, Skill Level 40

In the lower-division baccalaureate/associate degree category, 12 semester hours in music theory (harmony, ear training, and sight singing), 2 in jazz theory/improvisation, 4-6 in applied performance (individual instruction), 6-8 in performing ensembles (concert, jazz, and marching bands), and 3 in personnel supervision. In the upper-division baccalaureate category, 2 semester hours in conducting, 2 in rehearsal techniques of conducting (concert and jazz), 2 in arranging, and 3 in logistics management (2/93).

MOS-02F-005

TUBA PLAYER
02F10
02F20
02F30
02F40

Exhibit Dates: 3/95–Present.

Career Management Field: 97 (Band).

Description

Summary: Plays tuba in appropriate musical organizations; performs in military marching and concert musical organizations and, as qualified, in associated choral, jazz, and other small ensembles. *Skill Level 10:* Plays tuba; marches in military band in response to drum major's signals and oral commands; reads, interprets, and plays instrumental music; plays in marching, concert, or dance/jazz band as appropriate; plays, at sight, secondary parts to ceremonial music and marches of medium difficulty; plays, with minimum preparation, secondary parts to all styles of standard band literature; applies knowledge of minimum basic music theory; discriminates and matches pitches; performs preventive maintenance on tuba. *Skill Level 20:* Able to perform the duties required for Skill Level 10; provides technical guidance to Skill Level 10 personnel; plays tuba in small instrumental groups as appropriate; plays, at sight, principal parts to ceremonial music and marches of medium difficulty; plays, with minimum preparation, principal parts to all styles of standard band literature; performs music library duties as assigned; assists in instrument repair. *Skill Level 30:* Able to perform the duties required for Skill Level 20; supervises section as instrumental section leader; serves in various intermediate staff/administrative positions within the band as assigned; organizes, tunes, instructs, and rehearses section or consolidated section of related musical instruments; organizes and leads small instrument groups; plays, at sight, intermediate music of any style; balances chords and dynamic levels and interprets complex rhythms; supervises preventive maintenance of musical instruments in section. *Skill*

Level 40: Able to perform the duties required for Skill Level 30; provides technical guidance to subordinate personnel; supervises section leaders; rehearses the stage band; assists in rehearsing the concert band; conducts the stage band in a performance; organizes, rehearses, and leads instrumental ensembles; organizes, tunes, and rehearses elements of the band in rehearsal; arranges and transcribes concert and dance/stage band music; tunes and prepares the concert band for rehearsal.

Recommendation, Skill Level 10

Credit may be granted on the basis of an individualized assessment of the student (3/94).

Recommendation, Skill Level 20

Credit may be granted on the basis of an individualized assessment of the student (2/93).

Recommendation, Skill Level 30

In the lower-division baccalaureate/associate degree category, 4 semester hours in music theory (harmony, ear training, and sight singing), 2 in jazz theory/improvisation, 4-6 in applied performance (individual instruction), 4 in performing ensembles (concert, jazz, and marching bands), and 3 in personnel supervision. In the upper-division baccalaureate category, 1 semester hour in conducting (2/93).

Recommendation, Skill Level 40

In the lower-division baccalaureate/associate degree category, 8 semester hours in music theory (harmony, ear training, and sight singing), 2 in jazz theory/improvisation, 4-6 in applied performance (individual instruction), 6 in performing ensembles (concert, jazz, and marching bands), and 3 in personnel supervision. In the upper-division baccalaureate category, 2 semester hours in conducting, 2 in rehearsal techniques of conducting (concert and jazz), 2 in arranging, and 3 in logistics management (2/93).

MOS-02G-003

Flute or Piccolo Player

02G10
02G20
02G30
02G40

Exhibit Dates: 1/90–1/93.

Career Management Field: 97 (Band).

Description

Summary: Plays flute or piccolo in appropriate musical organizations; performs in military marching and concert musical organizations and, as qualified, in associated choral, jazz, and other small ensembles. *Skill Level 10:* Plays flute or piccolo; marches in military band in response to drum major's signals and oral commands; reads, interprets, and plays instrumental music; plays instrument in marching, concert, or dance/jazz band as appropriate; plays, at sight, secondary parts to ceremonial music and marches of medium difficulty; plays with minimum preparation secondary parts to all styles of standard band literature; doubles on another musical instrument as assigned; applies knowledge of minimum basic music theory; discriminates and matches pitches; performs preventive maintenance on flute or piccolo. *Skill Level 20:* Able to perform the duties required for Skill Level 10; provides technical guidance to Skill Level 10 personnel; plays flute or piccolo in small instrumental groups as appropriate; plays, at sight, principal parts to ceremonial music and marches of medium difficulty; plays, with minimum preparation, principal parts to all styles of standard band literature; performs music library duties as assigned. *Skill Level 30:* Able to perform the duties required for Skill Level 20; supervises section as instrumental section leader; serves in various intermediate staff/administrative positions within the band as assigned; organizes, tunes, instructs, and rehearses section or consolidated section of related musical instruments; organizes and leads small instrumental groups; plays, at sight, intermediate music of any style; balances chords and volumes, and interprets complex rhythms; supervises preventive maintenance of musical instruments in section. *Skill Level 40:* Able to perform the duties required for Skill Level 30; provides technical guidance to subordinate personnel; supervises section leaders; rehearses the stage band; assists in rehearsing the concert band; conducts the stage band in a performance; organizes, rehearses, and leads instrumental ensembles; organizes, tunes, and rehearses elements of the band in rehearsal; arranges and transcribes concert and dance/stage band music; tunes and prepares the concert band for rehearsal.

Recommendation, Skill Level 10

In the lower-division baccalaureate/associate degree category, 1-3 semester hours (equivalent of one semester) in applied study of major instrument, 2 in music theory, 1 in concert band, 2 in marching band, and credit in small ensemble on the basis of institutional evaluation; because of the wide range of individual abilities, additional credit may be awarded on the basis of institutional evaluation in music theory, applied music (individual instruction in performance), and small performing ensembles. (NOTE: This recommendation for skill level 10 is valid for the dates 1/90-9/91 only) (6/78).

Recommendation, Skill Level 20

In the lower-division baccalaureate/associate degree category, 1-3 semester hours (equivalent of one semester) in applied study of major instrument, 3 in music theory, 2 in concert band, 2 in marching band, and credit in small ensemble on the basis of institutional evaluation; because of the wide range of individual abilities, additional credit may be awarded on the basis of institutional evaluation in music theory, applied music (individual instruction in performance), and small performing ensembles (6/78).

Recommendation, Skill Level 30

In the lower-division baccalaureate/associate degree category, 2-6 semester hours (equivalent of two semesters) in applied study of major instrument, 4 in music theory (including ear training), 3 in concert band, 3 in marching band, and credit in small ensemble on the basis of institutional evaluation; because of the wide range of individual abilities, additional credit may be awarded on the basis of institutional evaluation in music theory, applied music (individual instruction in performance), and small performing ensembles (6/78).

Recommendation, Skill Level 40

In the lower-division baccalaureate/associate degree category, 2-6 semester hours (equivalent of two semesters) in applied study of major instrument, 4 in music theory (including ear training), 3 in concert band, 3 in marching band, 3 in personnel supervision, and credit in small ensemble on the basis of institutional evaluation; because of the wide range of individual abilities, additional credit may be awarded on the basis of institutional evaluation in music theory, applied music (individual instruction in performance), and small performing ensembles. In the upper-division baccalaureate category, 3 semester hours in conducting and 3 in music administration and supervision (3/86).

MOS-02G-004

Flute or Piccolo Player

02G10
02G20
02G30
02G40

Exhibit Dates: 2/93–2/95.

Career Management Field: 97 (Band).

Description

Summary: Plays flute or piccolo in appropriate musical organizations; performs in military marching and concert musical organizations and, as qualified, in associated choral, jazz, and other small ensembles. *Skill Level 10:* Plays flute or piccolo; marches in military band in response to drum major's signals and oral commands; reads, interprets, and plays instrumental music; plays in marching, concert, or dance/jazz band as appropriate; plays, at sight, secondary parts to ceremonial music and marches of medium difficulty; plays, with minimum preparation, secondary parts to all styles of standard band literature; applies knowledge of minimum basic music theory; discriminates and matches pitches; performs preventive maintenance on flute or piccolo. *Skill Level 20:* Able to perform the duties required for Skill Level 10; provides technical guidance to Skill Level 10 personnel; plays flute or piccolo in small instrumental groups as appropriate; plays, at sight, principal parts to ceremonial music and marches of medium difficulty; plays, with minimum preparation, principal parts to all styles of standard band literature; performs music library duties as assigned; assists in instrument repair. *Skill Level 30:* Able to perform the duties required for Skill Level 20; supervises section as instrumental section leader; serves in various intermediate staff/administrative positions within the band as assigned; organizes, tunes, instructs, and rehearses section or consolidated section of related musical instruments; organizes and leads small instrument groups; plays, at sight, intermediate music of any style; balances chords and dynamic levels and interprets complex rhythms; supervises preventive maintenance of musical instruments in section. *Skill Level 40:* Able to perform the duties required for Skill Level 30; provides technical guidance to subordinate personnel; supervises section leaders; rehearses the stage band; assists in rehearsing the concert band; conducts the stage band in performance; organizes, rehearses, and leads instrumental ensembles; organizes, tunes, and rehearses elements of the band in rehearsal; arranges and transcribes concert and dance/stage band music; tunes and prepares the concert band for rehearsal.

Recommendation, Skill Level 10

Credit may be granted on the basis of an individualized assessment of the student (3/94).

Recommendation, Skill Level 20

In the lower-division baccalaureate/associate degree category, 4 semester hours in music theory (harmony, ear training, and sight singing), 2 in jazz theory/improvisation, 4-6 in applied performance (individual instruction), and 4 in performing ensembles (concert, jazz, and marching bands) (2/93).

Recommendation, Skill Level 30

In the lower-division baccalaureate/associate degree category, 8 semester hours in music theory (harmony, ear training, and sight singing), 2 in jazz theory/improvisation, 4-6 in applied performance (individual instruction), 6 in performing ensembles (concert, jazz, and marching bands), and 3 in personnel supervision. In the upper-division baccalaureate category, 1 semester hour in conducting (2/93).

Recommendation, Skill Level 40

In the lower-division baccalaureate/associate degree category, 12 semester hours in music theory (harmony, ear training, and sight singing), 2 in jazz theory/improvisation, 4-6 in applied performance (individual instruction), 6-8 in performing ensembles (concert, jazz, and marching bands), and 3 in personnel supervision. In the upper-division baccalaureate category, 2 semester hours in conducting, 2 in rehearsal techniques of conducting (concert and jazz), 2 in arranging, and 3 in logistics management (2/93).

MOS-02G-005

Flute or Piccolo Player

02G10
02G20
02G30
02G40

Exhibit Dates: 3/95–Present.

Career Management Field: 97 (Band).

Description

Summary: Plays flute or piccolo in appropriate musical organizations; performs in military marching and concert musical organizations and, as qualified, in associated choral, jazz, and other small ensembles. *Skill Level 10:* Plays flute or piccolo; marches in military band in response to drum major's signals and oral commands; reads, interprets, and plays instrumental music; plays in marching, concert, or dance/jazz band as appropriate; plays, at sight, secondary parts to ceremonial music and marches of medium difficulty; plays, with minimum preparation, secondary parts to all styles of standard band literature; applies knowledge of minimum basic music theory; discriminates and matches pitches; performs preventive maintenance on flute or piccolo. *Skill Level 20:* Able to perform the duties required for Skill Level 10; provides technical guidance to Skill Level 10 personnel; plays flute or piccolo in small instrumental groups as appropriate; plays, at sight, principal parts to ceremonial music and marches of medium difficulty; plays, with minimum preparation, principal parts to all styles of standard band literature; performs music library duties as assigned; assists in instrument repair. *Skill Level 30:* Able to perform the duties required for Skill Level 20; supervises section as instrumental section leader; serves in various intermediate staff/administrative positions within the band as assigned; organizes, tunes, instructs, and rehearses section or consolidated section of related musical instruments; organizes and leads small instrument groups; plays, at sight, intermediate music of any style; balances chords and dynamic levels and interprets complex rhythms; supervises preventive maintenance of musical instruments in section. *Skill Level 40:* Able to perform the duties required for Skill Level 30; provides technical guidance to subordinate personnel; supervises section leaders; rehearses the stage band; assists in rehearsing the concert band; conducts the stage band in performance; organizes, rehearses, and leads instrumental ensembles; organizes, tunes, and rehearses elements of the band in rehearsal; arranges and transcribes concert and dance/stage band music; tunes and prepares the concert band for rehearsal.

Recommendation, Skill Level 10

Credit may be granted on the basis of an individualized assessment of the student (3/94).

Recommendation, Skill Level 20

Credit may be granted on the basis of an individualized assessment of the student (2/93).

Recommendation, Skill Level 30

In the lower-division baccalaureate/associate degree category, 4 semester hours in music theory (harmony, ear training, and sight singing), 2 in jazz theory/improvisation, 4-6 in applied performance (individual instruction), 4 in performing ensembles (concert, jazz, and marching bands), and 3 in personnel supervision. In the upper-division baccalaureate category, 1 semester hour in conducting (2/93).

Recommendation, Skill Level 40

In the lower-division baccalaureate/associate degree category, 8 semester hours in music theory (harmony, ear training, and sight singing), 2 in jazz theory/improvisation, 4-6 in applied performance (individual instruction), 6 in performing ensembles (concert, jazz, and marching bands), and 3 in personnel supervision. In the upper-division baccalaureate category, 2 semester hours in conducting, 2 in rehearsal techniques of conducting (concert and jazz), 2 in arranging, and 3 in logistics management (2/93).

MOS-02H-003

Oboe Player

02H10
02H20
02H30
02H40

Exhibit Dates: 1/90–1/93.

Career Management Field: 97 (Band).

Description

Summary: Plays oboe in appropriate musical organizations; performs in military marching and concert musical organizations and, as qualified, in associated choral, jazz, and other small ensembles. *Skill Level 10:* Plays oboe; marches in military band in response to drum major's signals and oral commands; reads, interprets, and plays instrumental music; plays instrument in marching, concert, or dance/jazz band as appropriate; plays, at sight, secondary parts to ceremonial music and marches of medium difficulty; plays with minimum preparation secondary parts to all styles of standard band literature; doubles on another musical instrument as assigned; applies knowledge of minimum basic music theory; discriminates and matches pitches; performs preventive maintenance on oboe. *Skill Level 20:* Able to perform the duties required for Skill Level 10; provides technical guidance to Skill Level 10 personnel; plays oboe in small instrumental groups as appropriate; plays, at sight, principal parts to ceremonial music and marches of medium difficulty; plays, with minimum preparation, principal parts to all styles of standard band literature; performs music library duties as assigned. *Skill Level 30:* Able to perform the duties required for Skill Level 20; supervises section as instrumental section leader; serves in various intermediate staff/administrative positions within the band as assigned; organizes, tunes, instructs, and rehearses section or consolidated section of related musical instruments; organizes and leads small instrumental groups; plays, at sight, intermediate music of any style; balances chords and volumes, and interprets complex rhythms; supervises preventive maintenance of musical instruments in section. *Skill Level 40:* Able to perform the duties required for Skill Level 30; provides technical guidance to subordinate personnel; supervises section leaders; rehearses the stage band; assists in rehearsing the concert band; conducts the stage band in a performance; organizes, rehearses, and leads instrumental ensembles; organizes, tunes, and rehearses elements of the band in rehearsal; arranges and transcribes concert and dance/stage band music; tunes and prepares the concert band for rehearsal.

Recommendation, Skill Level 10

In the lower-division baccalaureate/associate degree category, 1-3 semester hours (equivalent of one semester) in applied study of major instrument, 2 in music theory, 1 in concert band, 2 in marching band, and credit in small ensemble on the basis of institutional evaluation; because of the wide range of individual abilities, additional credit may be awarded on the basis of institutional evaluation in music theory, applied music (individual instruction in performance), and small performing ensembles. (NOTE: This recommendation for skill level 10 is valid for the dates 1/90-9/91 only) (6/78).

Recommendation, Skill Level 20

In the lower-division baccalaureate/associate degree category, 1-3 semester hours (equivalent of one semester) in applied study of major instrument, 3 in music theory, 2 in concert band, 2 in marching band, and credit in small ensemble on the basis of institutional evaluation; because of the wide range of individual abilities, additional credit may be awarded on the basis of institutional evaluation in music theory, applied music (individual instruction in performance), and small performing ensembles (6/78).

Recommendation, Skill Level 30

In the lower-division baccalaureate/associate degree category, 2-6 semester hours (equivalent of two semesters) in applied study of major instrument, 4 in music theory (including ear training), 3 in concert band, 3 in marching band, and credit in small ensemble on the basis of institutional evaluation; because of the wide range of individual abilities, additional credit may be awarded on the basis of institutional evaluation in music theory, applied music (individual instruction in performance), and small performing ensembles (6/78).

Recommendation, Skill Level 40

In the lower-division baccalaureate/associate degree category, 2-6 semester hours (equivalent of two semesters) in applied study of major instrument, 4 in music theory (including ear training), 3 in concert band, 3 in marching band, 3 in personnel supervision, and credit in small ensemble on the basis of institutional evaluation; because of the wide range of individual abilities, additional credit may be awarded on the basis of institutional evaluation in music theory, applied music (individual instruction in performance), and small performing ensembles. In the upper-division baccalaureate category, 3 semester hours in conducting and 3 in music administration and supervision (3/86).

MOS-02H-004

Oboe Player
02H10
02H20
02H30
02H40

Exhibit Dates: 2/93–2/95.

Career Management Field: 97 (Band).

Description

Summary: Plays oboe in appropriate musical organizations; performs in military marching and concert musical organizations and, as qualified, in associated choral, jazz, and other small ensembles. *Skill Level 10:* Plays oboe; marches in military band in response to drum major's signals and oral commands; reads, interprets, and plays instrumental music; plays in marching, concert, or dance/jazz band as appropriate; plays, at sight, secondary parts to ceremonial music and marches of medium difficulty; plays, with minimum preparation, secondary parts to all styles of standard band literature; applies knowledge of minimum basic music theory; discriminates and matches pitches; performs preventive maintenance on oboe. *Skill Level 20:* Able to perform the duties required for Skill Level 10; provides technical guidance to Skill Level 10 personnel; plays oboe in small instrumental groups as appropriate; plays, at sight, principal parts to ceremonial music and marches of medium difficulty; plays, with minimum preparation, principal parts to all styles of standard band literature; performs music library duties as assigned; assists in instrument repair. *Skill Level 30:* Able to perform the duties required for Skill Level 20; supervises section as instrumental section leader; serves in various intermediate staff/administrative positions within the band as assigned; organizes, tunes, instructs, and rehearses section or consolidated section of related musical instruments; organizes and leads small instrument groups; plays, at sight, intermediate music of any style; balances chords and dynamic levels and interprets complex rhythms; supervises preventive maintenance of musical instruments in section. *Skill Level 40:* Able to perform the duties required for Skill Level 30; provides technical guidance to subordinate personnel; supervises section leaders; rehearses the stage band; assists in rehearsing the concert band; conducts the stage band in a performance; organizes, rehearses, and leads instrumental ensembles; organizes, tunes, and rehearses elements of the band in rehearsal; arranges and transcribes concert and dance/stage band music; tunes and prepares the concert band for rehearsal.

Recommendation, Skill Level 10

Credit may be granted on the basis of an individualized assessment of the student (3/94).

Recommendation, Skill Level 20

In the lower-division baccalaureate/associate degree category, 4 semester hours in music theory (harmony, ear training, and sight singing), 2 in jazz theory/improvisation, 4-6 in applied performance (individual instruction), and 4 in performing ensembles (concert, jazz, and marching bands) (2/93).

Recommendation, Skill Level 30

In the lower-division baccalaureate/associate degree category, 8 semester hours in music theory (harmony, ear training, and sight singing), 2 in jazz theory/improvisation, 4-6 in applied performance (individual instruction), 6 in performing ensembles (concert, jazz, and marching bands), and 3 in personnel supervision. In the upper-division baccalaureate category, 1 semester hour in conducting (2/93).

Recommendation, Skill Level 40

In the lower-division baccalaureate/associate degree category, 12 semester hours in music theory (harmony, ear training, and sight singing), 2 in jazz theory/improvisation, 4-6 in applied performance (individual instruction), 6-8 in performing ensembles (concert, jazz, and marching bands), and 3 in personnel supervision. In the upper-division baccalaureate category, 2 semester hours in conducting, 2 in rehearsal techniques of conducting (concert and jazz), 2 in arranging, and 3 in logistics management (2/93).

MOS-02H-005

Oboe Player
02H10
02H20
02H30
02H40

Exhibit Dates: 3/95–Present.

Career Management Field: 97 (Band).

Description

Summary: Plays oboe in appropriate musical organizations; performs in military marching and concert musical organizations and, as qualified, in associated choral, jazz, and other small ensembles. *Skill Level 10:* Plays oboe; marches in military band in response to drum major's signals and oral commands; reads, interprets, and plays instrumentally music; plays in marching, concert, or dance/jazz band as appropriate; plays, at sight, secondary parts to ceremonial music and marches of medium difficulty; plays, with minimum preparation, secondary parts to all styles of standard band literature; applies knowledge of minimum basic music theory; discriminates and matches pitches; performs preventive maintenance on oboe. *Skill Level 20:* Able to perform the duties required for Skill Level 10; provides technical guidance to Skill Level 10 personnel; plays oboe in small instrumental groups as appropriate; plays, at sight, principal parts to ceremonial music and marches of medium difficulty; plays, with minimum preparation, principal parts to all styles of standard band literature; performs music library duties as assigned; assists in instrument repair. *Skill Level 30:* Able to perform the duties required for Skill Level 20; supervises section as instrumental section leader; serves in various intermediate staff/administrative positions within the band as assigned; organizes, tunes, instructs, and rehearses section or consolidated section of related musical instruments; organizes and leads small instrument groups; plays, at sight, intermediate music of any style; balances chords and dynamic levels and interprets complex rhythms; supervises preventive maintenance of musical instruments in section. *Skill Level 40:* Able to perform the duties required for Skill Level 30; provides technical guidance to subordinate personnel; supervises section leaders; rehearses the stage band; assists in rehearsing the concert band; conducts the stage band in a performance; organizes, rehearses, and leads instrumental ensembles; organizes, tunes, and rehearses elements of the band in rehearsal; arranges and transcribes concert and dance/stage band music; tunes and prepares the concert band for rehearsal.

Recommendation, Skill Level 10

Credit may be granted on the basis of an individualized assessment of the student (3/94).

Recommendation, Skill Level 20

Credit may be granted on the basis of an individualized assessment of the student (2/93).

Recommendation, Skill Level 30

In the lower-division baccalaureate/associate degree category, 4 semester hours in music theory (harmony, ear training, and sight singing), 2 in jazz theory/improvisation, 4-6 in applied performance (individual instruction), 4 in performing ensembles (concert, jazz, and marching bands), and 3 in personnel supervision. In the upper-division baccalaureate category, 1 semester hour in conducting (2/93).

Recommendation, Skill Level 40

In the lower-division baccalaureate/associate degree category, 8 semester hours in music theory (harmony, ear training, and sight singing), 2 in jazz theory/improvisation, 4-6 in applied performance (individual instruction), 6 in performing ensembles (concert, jazz, and marching bands), and 3 in personnel supervision. In the upper-division baccalaureate category, 2 semester hours in conducting, 2 in rehearsal techniques of conducting (concert and jazz), 2 in arranging, and 3 in logistics management (2/93).

MOS-02J-003

Clarinet Player
02J10
02J20
02J30
02J40

Exhibit Dates: 1/90–1/93.

Career Management Field: 97 (Band).

Description

Summary: Plays clarinet in appropriate musical organizations; performs in military marching and concert musical organizations and, as qualified, in associated choral, jazz, and other small ensembles. *Skill Level 10:* Plays clarinet; marches in military band in response to drum major's signals and oral commands; reads, interprets, and plays instrumental music; plays instrument in marching, concert, or dance/jazz band as appropriate; plays, at sight, secondary parts to ceremonial music and marches of medium difficulty; plays with minimum preparation secondary parts to all styles of standard band literature; doubles on another musical instrument as assigned; applies knowledge of minimum basic music theory; discriminates and matches pitches; performs preventive maintenance on clarinet. *Skill Level 20:* Able to perform the duties required for Skill Level 10; provides technical guidance to Skill Level 10 personnel; plays clarinet in small instrumental groups as appropriate; plays, at sight, principal parts to ceremonial music and marches of medium difficulty; plays, with minimum preparation, principal parts to all styles of standard band literature; performs music library duties as assigned. *Skill Level 30:* Able to perform the duties required for Skill Level 20; supervises section as instrumental section leader; serves in various intermediate staff/administrative positions within the band as assigned; organizes, tunes, instructs, and rehearses section or consolidated section of related musical instruments; organizes and leads small instrumental groups; plays, at sight, intermediate music of any style; balances chords and volumes, and interprets complex rhythms; supervises preventive maintenance of musical instruments in section. *Skill Level 40:* Able to perform the duties required for Skill Level 30; provides technical guidance to subordinate personnel; supervises section leaders; rehearses the stage band; assists in rehearsing the concert band; conducts the stage band in a performance; organizes, rehearses, and leads instrumental ensembles; organizes, tunes, and rehearses elements of the band in rehearsal; arranges and transcribes concert and dance/stage band music; tunes and prepares the concert band for rehearsal.

Recommendation, Skill Level 10

In the lower-division baccalaureate/associate degree category, 1-3 semester hours (equivalent of one semester) in applied study of major instrument, 2 in music theory, 1 in concert band, 2 in marching band, and credit in small ensemble on the basis of institutional evaluation; because of the wide range of individual abilities, additional credit may be awarded on the basis of institutional evaluation in music theory, applied music (individual instruction in performance), and small performing ensembles. (NOTE: This recommendation for skill level 10 is valid for the dates 1/90-9/91 only) (6/78).

Recommendation, Skill Level 20

In the lower-division baccalaureate/associate degree category, 1-3 semester hours (equivalent of one semester) in applied study of major instrument, 3 in music theory, 2 in concert band, 2 in marching band, and credit in small ensemble on the basis of institutional evaluation; because of the wide range of individual abilities, additional credit may be awarded on the basis of institutional evaluation in music theory, applied music (individual instruction in performance), and small performing ensembles (6/78).

Recommendation, Skill Level 30

In the lower-division baccalaureate/associate degree category, 2-6 semester hours (equivalent of two semesters) in applied study of major instrument, 4 in music theory (including ear training), 3 in concert band, 3 in marching band, and credit in small ensemble on the basis of institutional evaluation; because of the wide range of individual abilities, additional credit may be awarded on the basis of institutional evaluation in music theory, applied music (individual instruction in performance), and small performing ensembles (6/78).

Recommendation, Skill Level 40

In the lower-division baccalaureate/associate degree category, 2-6 semester hours (equivalent of two semesters) in applied study of major instrument, 4 in music theory (including ear training), 3 in concert band, 3 in marching band, 3 in personnel supervision, and credit in small ensemble on the basis of institutional evaluation; because of the wide range of individual abilities, additional credit may be awarded on the basis of institutional evaluation in music theory, applied music (individual instruction in performance), and small performing ensembles. In the upper-division baccalaureate category, 3 semester hours in conducting and 3 in music administration and supervision (3/86).

MOS-02J-004

Clarinet Player

02J10
02J20
02J30
02J40

Exhibit Dates: 2/93–2/95.

Career Management Field: 97 (Band).

Description

Summary: Plays clarinet in appropriate musical organizations; performs in military marching and concert musical organizations and, as qualified, in associated choral, jazz, and other small ensembles. *Skill Level 10:* Plays clarinet; marches in military band in response to drum major's signals and oral commands; reads, interprets, and plays instrumental music; plays in marching, concert, or dance/jazz band as appropriate; plays, at sight, secondary parts to ceremonial music and marches of medium difficulty; plays, with minimum preparation, secondary parts to all styles of standard band literature; applies knowledge of minimum basic music theory; discriminates and matches pitches; performs preventive maintenance on clarinet. *Skill Level 20:* Able to perform the duties required for Skill Level 10; provides technical guidance to Skill Level 10 personnel; plays clarinet in small instrumental groups as appropriate; plays, at sight, principal parts to ceremonial music and marches of medium difficulty; plays, with minimum preparation, principal parts to all styles of standard band literature; performs music library duties as assigned; assists in instrument repair. *Skill Level 30:* Able to perform the duties required for Skill Level 20; supervises section as instrumental section leader; serves in various intermediate staff/administrative positions within the band as assigned; organizes, tunes, instructs, and rehearses section or consolidated section of related musical instruments; organizes and leads small instrument groups; plays, at sight, intermediate music of any style; balances chords and dynamic levels and interprets complex rhythms; supervises preventive maintenance of musical instruments in section. *Skill Level 40:* Able to perform the duties required for Skill Level 30; provides technical guidance to subordinate personnel; supervises section leaders; rehearses the stage band; assists in rehearsing the concert band; conducts the stage band in a performance; organizes, rehearses, and leads instrumental ensembles; organizes, tunes, and rehearses elements of the band in rehearsal; arranges and transcribes concert and dance/stage band music; tunes and prepares the concert band for rehearsal.

Recommendation, Skill Level 10

Credit may be granted on the basis of an individualized assessment of the student (3/94).

Recommendation, Skill Level 20

In the lower-division baccalaureate/associate degree category, 4 semester hours in music theory (harmony, ear training, and sight singing), 2 in jazz theory/improvisation, 4-6 in applied performance (individual instruction), and 4 in performing ensembles (concert, jazz, and marching bands) (2/93).

Recommendation, Skill Level 30

In the lower-division baccalaureate/associate degree category, 8 semester hours in music theory (harmony, ear training, and sight singing), 2 in jazz theory/improvisation, 4-6 in applied performance (individual instruction), 6 in performing ensembles (concert, jazz, and marching bands), and 3 in personnel supervision. In the upper-division baccalaureate category, 1 semester hour in conducting (2/93).

Recommendation, Skill Level 40

In the lower-division baccalaureate/associate degree category, 12 semester hours in music theory (harmony, ear training, and sight singing), 2 in jazz theory/improvisation, 4-6 in applied performance (individual instruction), 6-8 in performing ensembles (concert, jazz, and marching bands), and 3 in personnel supervision. In the upper-division baccalaureate category, 2 semester hours in conducting, 2 in rehearsal techniques of conducting (concert and jazz), 2 in arranging, and 3 in logistics management (2/93).

MOS-02J-005

Clarinet Player

02J10
02J20
02J30
02J40

Exhibit Dates: 3/95–Present.

Career Management Field: 97 (Band).

Description

Summary: Plays clarinet in appropriate musical organizations; performs in military marching and concert musical organizations and, as qualified, in associated choral, jazz, and other small ensembles. *Skill Level 10:* Plays clarinet; marches in military band in response to drum major's signals and oral commands; reads, interprets, and plays instrumental music; plays in marching, concert, or dance/jazz band as appropriate; plays, at sight, secondary parts to ceremonial music and marches of medium difficulty; plays, with minimum preparation, secondary parts to all styles of standard band literature; applies knowledge of minimum basic music theory; discriminates and matches pitches; performs preventive maintenance on clarinet. *Skill Level 20:* Able to perform the duties required for Skill Level 10; provides technical guidance to Skill Level 10 personnel; plays clarinet in small instrumental groups as appropriate; plays, at sight, principal parts to ceremonial music and marches of medium difficulty; plays, with minimum preparation, principal parts to all styles of standard band literature; performs music library duties as assigned; assists in instrument repair. *Skill Level 30:* Able to perform the duties required for Skill Level 20; supervises section as instrumental section leader; serves in various intermediate staff/administrative positions within the band as assigned; organizes, tunes, instructs, and rehearses section or consolidated section of related musical instruments; organizes and leads small instrument groups; plays, at sight, intermediate music of any style; balances chords and dynamic levels and interprets complex rhythms; supervises preventive maintenance of musical instruments in section. *Skill Level 40:* Able to perform the duties required for Skill Level 30; provides technical guidance to subordinate personnel; supervises section leaders; rehearses the stage band; assists in rehearsing the concert band; conducts the stage band in a performance; organizes, rehearses, and leads instrumental ensembles; organizes, tunes, and rehearses elements of the band in rehearsal; arranges and transcribes concert and dance/stage band music; tunes and prepares the concert band for rehearsal.

Recommendation, Skill Level 10

Credit may be granted on the basis of an individualized assessment of the student (3/94).

Recommendation, Skill Level 20

Credit may be granted on the basis of an individualized assessment of the student (2/93).

Recommendation, Skill Level 30

In the lower-division baccalaureate/associate degree category, 4 semester hours in music theory (harmony, ear training, and sight singing), 2 in jazz theory/improvisation, 4-6 in applied performance (individual instruction), 4 in performing ensembles (concert, jazz, and marching bands), and 3 in personnel supervision. In the upper-division baccalaureate category, 1 semester hour in conducting (2/93).

Recommendation, Skill Level 40

In the lower-division baccalaureate/associate degree category, 8 semester hours in music theory (harmony, ear training, and sight singing), 2 in jazz theory/improvisation, 4-6 in applied performance (individual instruction), 6 in performing ensembles (concert, jazz, and marching bands), and 3 in personnel supervision. In the upper-division baccalaureate category, 2 semester hours in conducting, 2 in rehearsal techniques of conducting (concert and jazz), 2 in arranging, and 3 in logistics management (2/93).

MOS-02K-003

Bassoon Player

02K10

02K20

02K30

02K40

Exhibit Dates: 1/90–1/93.

Career Management Field: 97 (Band).

Description

Summary: Plays bassoon in appropriate musical organizations; performs in military marching and concert musical organizations and, as qualified, in associated choral, jazz, and other small ensembles. *Skill Level 10:* Plays bassoon; marches in military band in response to drum major's signals and oral commands; reads, interprets, and plays instrumental music; plays instrument in marching, concert, or dance/jazz band as appropriate; plays, at sight, secondary parts to ceremonial music and marches of medium difficulty; plays with minimum preparation secondary parts to all styles of standard band literature; doubles on another musical instrument as assigned; applies knowledge of minimum basic music theory; discriminates and matches pitches; performs preventive maintenance on bassoon. *Skill Level 20:* Able to perform the duties required for Skill Level 10; provides technical guidance to Skill Level 10 personnel; plays bassoon in small instrumental groups as appropriate; plays, at sight, principal parts to ceremonial music and marches of medium difficulty; plays, with minimum preparation, principal parts to all styles of standard band literature; performs music library duties as assigned. *Skill Level 30:* Able to perform the duties required for Skill Level 20; supervises section as instrumental section leader; serves in various intermediate staff/administrative positions within the band as assigned; organizes, tunes, instructs, and rehearses section or consolidated section of related musical instruments; organizes and leads small instrumental groups; plays, at sight, intermediate music of any style; balances chords and volumes, and interprets complex rhythms; supervises preventive maintenance of musical instruments in section. *Skill Level 40:* Able to perform the duties required for Skill Level 30; provides technical guidance to subordinate personnel; supervises section leaders; rehearses the stage band; assists in rehearsing the concert band; conducts the stage band in a performance; organizes, rehearses, and leads instrumental ensembles; organizes, tunes, and rehearses elements of the band in rehearsal; arranges and transcribes concert and dance/stage band music; tunes and prepares the concert band for rehearsal.

Recommendation, Skill Level 10

In the lower-division baccalaureate/associate degree category, 1-3 semester hours (equivalent of one semester) in applied study of major instrument, 2 in music theory, 1 in concert band, 2 in marching band, and credit in small ensemble on the basis of institutional evaluation; because of the wide range of individual abilities, additional credit may be awarded on the basis of institutional evaluation in music theory, applied music (individual instruction in performance), and small performing ensembles. (NOTE: This recommendation for skill level 10 is valid for the dates 1/90-9/91 only) (6/78).

Recommendation, Skill Level 20

In the lower-division baccalaureate/associate degree category, 1-3 semester hours (equivalent of one semester) in applied study of major instrument, 3 in music theory, 2 in concert band, 2 in marching band, and credit in small ensemble on the basis of institutional evaluation; because of the wide range of individual abilities, additional credit may be awarded on the basis of institutional evaluation in music theory, applied music (individual instruction in performance), and small performing ensembles (6/78).

Recommendation, Skill Level 30

In the lower-division baccalaureate/associate degree category, 2-6 semester hours (equivalent of two semesters) in applied study of major instrument, 4 in music theory (including ear training), 3 in concert band, 3 in marching band, and credit in small ensemble on the basis of institutional evaluation; because of the wide range of individual abilities, additional credit may be awarded on the basis of institutional evaluation in music theory, applied music (individual instruction in performance), and small performing ensembles (6/78).

Recommendation, Skill Level 40

In the lower-division baccalaureate/associate degree category, 2-6 semester hours (equivalent of two semesters) in applied study of major instrument, 4 in music theory (including ear training), 3 in concert band, 3 in marching band, 3 in personnel supervision, and credit in small ensemble on the basis of institutional evaluation; because of the wide range of individual abilities, additional credit may be awarded on the basis of institutional evaluation in music theory, applied music (individual instruction in performance), and small performing ensembles. In the upper-division baccalaureate category, 3 semester hours in conducting and 3 in music administration and supervision (3/86).

MOS-02K-004

Bassoon Player

02K10

02K20

02K30

02K40

Exhibit Dates: 2/93–2/95.

Career Management Field: 97 (Band).

Description

Summary: Plays bassoon in appropriate musical organizations; performs in military marching and concert musical organizations and, as qualified, in associated choral, jazz, and other small ensembles. *Skill Level 10:* Plays bassoon; marches in military band in response to drum major's signals and oral commands; reads, interprets, and plays instrumental music; plays in marching, concert, or dance/jazz band as appropriate; plays, at sight, secondary parts to ceremonial music and marches of medium difficulty; plays, with minimum preparation, secondary parts to all styles of standard band literature; applies knowledge of minimum basic music theory; discriminates and matches pitches; performs preventive maintenance on bassoon. *Skill Level 20:* Able to perform the duties required for Skill Level 10; provides technical guidance to Skill Level 10 personnel; plays bassoon in small instrumental groups as appropriate; plays, at sight, principal parts to ceremonial music and marches of medium difficulty; plays, with minimum preparation, principal parts to all styles of standard band literature; performs music library duties as assigned; assists in instrument repair. *Skill Level 30:* Able to perform the duties required for Skill Level 20; supervises section as instrumental section leader; serves in various intermediate staff/administrative positions within the band as assigned; organizes, tunes, instructs, and rehearses section or consolidated section of related musical instruments; organizes and leads small instrument groups; plays, at sight, intermediate music of any style; balances chords and dynamic levels and interprets complex rhythms; supervises preventive maintenance of musical instruments in section. *Skill Level 40:* Able to perform the duties required for Skill Level 30; provides technical guidance to subordinate personnel; supervises section leaders; rehearses the stage band; assists in rehearsing the concert band; conducts the stage band in a performance; organizes, rehearses, and leads instrumental ensembles; organizes, tunes, and rehearses elements of the band in rehearsal; arranges and transcribes concert and dance/stage band music; tunes and prepares the concert band for rehearsal.

Recommendation, Skill Level 10

Credit may be granted on the basis of an individualized assessment of the student (3/94).

Recommendation, Skill Level 20

In the lower-division baccalaureate/associate degree category, 4 semester hours in music theory (harmony, ear training, and sight singing), 2 in jazz theory/improvisation, 4-6 in applied performance (individual instruction), and 4 in performing ensembles (concert, jazz, and marching bands) (2/93).

Recommendation, Skill Level 30

In the lower-division baccalaureate/associate degree category, 8 semester hours in music theory (harmony, ear training, and sight singing), 2 in jazz theory/improvisation, 4-6 in applied performance (individual instruction), 6 in performing ensembles (concert, jazz, and marching bands), and 3 in personnel supervision. In the upper-division baccalaureate category, 1 semester hour in conducting (2/93).

Recommendation, Skill Level 40

In the lower-division baccalaureate/associate degree category, 12 semester hours in music theory (harmony, ear training, and sight singing), 2 in jazz theory/improvisation, 4-6 in applied performance (individual instruction), 6-8 in performing ensembles (concert, jazz, and marching bands), and 3 in personnel supervision. In the upper-division baccalaureate category, 2 semester hours in conducting, 2 in rehearsal techniques of conducting (concert and jazz), 2 in arranging, and 3 in logistics management (2/93).

MOS-02K-005

Bassoon Player

02K10
02K20
02K30
02K40

Exhibit Dates: 3/95–Present.

Career Management Field: 97 (Band).

Description

Summary: Plays bassoon in appropriate musical organizations; performs in military marching and concert musical organizations and, as qualified, in associated choral, jazz, and other small ensembles. *Skill Level 10:* Plays bassoon; marches in military band in response to drum major's signals and oral commands; reads, interprets, and plays instrumental music; plays in marching, concert, or dance/jazz band as appropriate; plays, at sight, secondary parts to ceremonial music and marches of medium difficulty; plays, with minimum preparation, secondary parts to all styles of standard band literature; applies knowledge of minimum basic music theory; discriminates and matches pitches; performs preventive maintenance on bassoon. *Skill Level 20:* Able to perform the duties required for Skill Level 10; provides technical guidance to Skill Level 10 personnel; plays bassoon in small instrumental groups as appropriate; plays, at sight, principal parts to ceremonial music and marches of medium difficulty; plays, with minimum preparation, principal parts to all styles of standard band literature; performs music library duties as assigned; assists in instrument repair. *Skill Level 30:* Able to perform the duties required for Skill Level 20; supervises section as instrumental section leader; serves in various intermediate staff/administrative positions within the band as assigned; organizes, tunes, instructs, and rehearses section or consolidated section of related musical instruments; organizes and leads small instrument groups; plays, at sight, intermediate music of any style; balances chords and dynamic levels and interprets complex rhythms; supervises preventive maintenance of musical instruments in section. *Skill Level 40:* Able to perform the duties required for Skill Level 30; provides technical guidance to subordinate personnel; supervises section leaders; rehearses the stage band; assists in rehearsing the concert band; conducts the stage band in a performance; organizes, rehearses, and leads instrumental ensembles; organizes, tunes, and rehearses elements of the band in rehearsal; arranges and transcribes concert and dance/stage band music; tunes and prepares the concert band for rehearsal.

Recommendation, Skill Level 10

Credit may be granted on the basis of an individualized assessment of the student (3/94).

Recommendation, Skill Level 20

Credit may be granted on the basis of an individualized assessment of the student (2/93).

Recommendation, Skill Level 30

In the lower-division baccalaureate/associate degree category, 4 semester hours in music theory (harmony, ear training, and sight singing), 2 in jazz theory/improvisation, 4-6 in applied performance (individual instruction), 4 in performing ensembles (concert, jazz, and marching bands), and 3 in personnel supervision. In the upper-division baccalaureate category, 1 semester hour in conducting (2/93).

Recommendation, Skill Level 40

In the lower-division baccalaureate/associate degree category, 8 semester hours in music theory (harmony, ear training, and sight singing), 2 in jazz theory/improvisation, 4-6 in applied performance (individual instruction), 6 in performing ensembles (concert, jazz, and marching bands), and 3 in personnel supervision. In the upper-division baccalaureate category, 2 semester hours in conducting, 2 in rehearsal techniques of conducting (concert and jazz), 2 in arranging, and 3 in logistics management (2/93).

MOS-02L-003

Saxophone Player

02L10
02L20
02L30
02L40

Exhibit Dates: 1/90–1/93.

Career Management Field: 97 (Band).

Description

Summary: Plays saxophone in appropriate musical organizations; performs in military marching and concert musical organizations and, as qualified, in associated choral, jazz, and other small ensembles. *Skill Level 10:* Plays saxophone; marches in military band in response to drum major's signals and oral commands; reads, interprets, and plays instrumental music; plays instrument in marching, concert, or dance/jazz band as appropriate; plays, at sight, secondary parts to ceremonial music and marches of medium difficulty; plays with minimum preparation secondary parts to all styles of standard band literature; doubles on another musical instrument as assigned; applies knowledge of minimum basic music theory; discriminates and matches pitches; performs preventive maintenance on saxophone. *Skill Level 20:* Able to perform the duties required for Skill Level 10; provides technical guidance to Skill Level 10 personnel; plays saxophone in small instrumental groups as appropriate; plays, at sight, principal parts to ceremonial music and marches of medium difficulty; plays, with minimum preparation, principal parts to all styles of standard band literature; performs music library duties as assigned. *Skill Level 30:* Able to perform the duties required for Skill Level 20; supervises section as instrumental section leader; serves in various intermediate staff/administrative positions within the band as assigned; organizes, tunes, instructs, and rehearses section or consolidated section of related musical instruments; organizes and leads small instrumental groups; plays, at sight, intermediate music of any style; balances chords and volumes, and interprets complex rhythms; supervises preventive maintenance of musical instruments in section. *Skill Level 40:* Able to perform the duties required for Skill Level 30; provides technical guidance to subordinate personnel; supervises section leaders; rehearses the stage band; assists in rehearsing the concert band; conducts the stage band in a performance; organizes, rehearses, and leads instrumental ensembles; organizes, tunes, and rehearses elements of the band in rehearsal; arranges and transcribes concert and dance/stage band music; tunes and prepares the concert band for rehearsal.

Recommendation, Skill Level 10

In the lower-division baccalaureate/associate degree category, 1-3 semester hours (equivalent of one semester) in applied study of major instrument, 2 in music theory, 1 in concert band, 2 in marching band, and credit in small ensemble on the basis of institutional evaluation; because of the wide range of individual abilities, additional credit may be awarded on the basis of institutional evaluation in music theory, applied music (individual instruction in performance), and small performing ensembles. (NOTE: This recommendation for skill level 10 is valid for the dates 1/90-9/91 only) (6/78).

Recommendation, Skill Level 20

In the lower-division baccalaureate/associate degree category, 1-3 semester hours (equivalent of one semester) in applied study of major instrument, 3 in music theory, 2 in concert band, 2 in marching band, and credit in small ensemble on the basis of institutional evaluation; because of the wide range of individual abilities, additional credit may be awarded on the basis of institutional evaluation in music theory, applied music (individual instruction in performance), and small performing ensembles (6/78).

Recommendation, Skill Level 30

In the lower-division baccalaureate/associate degree category, 2-6 semester hours (equivalent of two semesters) in applied study of major instrument, 4 in music theory (including ear training), 3 in concert band, 3 in marching band, and credit in small ensemble on the basis of institutional evaluation; because of the wide range of individual abilities, additional credit may be awarded on the basis of institutional evaluation in music theory, applied music (individual instruction in performance), and small performing ensembles (6/78).

Recommendation, Skill Level 40

In the lower-division baccalaureate/associate degree category, 2-6 semester hours (equivalent of two semesters) in applied study of major instrument, 4 in music theory (including ear training), 3 in concert band, 3 in marching band, 3 in personnel supervision, and credit in small ensemble on the basis of institutional evaluation; because of the wide range of individual abilities, additional credit may be awarded on the basis of institutional evaluation in music theory, applied music (individual instruction in performance), and small performing ensembles. In the upper-division baccalaureate category, 3 semester hours in conducting and 3 in music administration and supervision (3/86).

MOS-02L-004

Saxophone Player

02L10

02L20

02L30

02L40

Exhibit Dates: 2/93–2/95.

Career Management Field: 97 (Band).

Description

Summary: Plays saxophone in appropriate musical organizations; performs in military marching and concert musical organizations and, as qualified, in associated choral, jazz, and other small ensembles. *Skill Level 10:* Plays saxophone; marches in military band in response to drum major's signals and oral commands; reads, interprets, and plays instrumental music; plays in marching, concert, or dance/jazz band as appropriate; plays, at sight, secondary parts to ceremonial music and marches of medium difficulty; plays, with minimum preparation, secondary parts to all styles of standard band literature; applies knowledge of minimum basic music theory; discriminates and matches pitches; performs preventive maintenance on saxophone. *Skill Level 20:* Able to perform the duties required for Skill Level 10; provides technical guidance to Skill Level 10 personnel; plays saxophone in small instrumental groups as appropriate; plays, at sight, principal parts to ceremonial music and marches of medium difficulty; plays, with minimum preparation, principal parts to all styles of standard band literature; performs music library duties as assigned; assists in instrument repair. *Skill Level 30:* Able to perform the duties required for Skill Level 20; supervises section as instrumental section leader; serves in various intermediate staff/administrative positions within the band as assigned; organizes, tunes, instructs, and rehearses section or consolidated section of related musical instruments; organizes and leads small instrument groups; plays, at sight, intermediate music of any style; balances chords and dynamic levels and interprets complex rhythms; supervises preventive maintenance of musical instruments in section. *Skill Level 40:* Able to perform the duties required for Skill Level 30; provides technical guidance to subordinate personnel; supervises section leaders; rehearses the stage band; assists in rehearsing the concert band; conducts the stage band in a performance; organizes, rehearses, and leads instrumental ensembles; organizes, tunes, and rehearses elements of the band in rehearsal; arranges and transcribes concert and dance/stage band music; tunes and prepares the concert band for rehearsal.

Recommendation, Skill Level 10

Credit may be granted on the basis of an individualized assessment of the student (3/94).

Recommendation, Skill Level 20

In the lower-division baccalaureate/associate degree category, 4 semester hours in music theory (harmony, ear training, and sight singing), 2 in jazz theory/improvisation, 4-6 in applied performance (individual instruction), and 4 in performing ensembles (concert, jazz, and marching bands) (2/93).

Recommendation, Skill Level 30

In the lower-division baccalaureate/associate degree category, 8 semester hours in music theory (harmony, ear training, and sight singing), 2 in jazz theory/improvisation, 4-6 in applied performance (individual instruction), 6 in performing ensembles (concert, jazz, and marching bands), and 3 in personnel supervision. In the upper-division baccalaureate category, 1 semester hour in conducting (2/93).

Recommendation, Skill Level 40

In the lower-division baccalaureate/associate degree category, 12 semester hours in music theory (harmony, ear training, and sight singing), 2 in jazz theory/improvisation, 4-6 in applied performance (individual instruction), 6-8 in performing ensembles (concert, jazz, and marching bands), and 3 in personnel supervision. In the upper-division baccalaureate category, 2 semester hours in conducting, 2 in rehearsal techniques of conducting (concert and jazz), 2 in arranging, and 3 in logistics management (2/93).

MOS-02L-005

Saxophone Player

02L10

02L20

02L30

02L40

Exhibit Dates: 3/95–Present.

Career Management Field: 97 (Band).

Description

Summary: Plays saxophone in appropriate musical organizations; performs in military marching and concert musical organizations and, as qualified, in associated choral, jazz, and other small ensembles. *Skill Level 10:* Plays saxophone; marches in military band in response to drum major's signals and oral commands; reads, interprets, and plays instrumental music; plays in marching, concert, or dance/jazz band as appropriate; plays, at sight, secondary parts to ceremonial music and marches of medium difficulty; plays, with minimum preparation, secondary parts to all styles of standard band literature; applies knowledge of minimum basic music theory; discriminates and matches pitches; performs preventive maintenance on saxophone. *Skill Level 20:* Able to perform the duties required for Skill Level 10; provides technical guidance to Skill Level 10 personnel; plays saxophone in small instrumental groups as appropriate; plays, at sight, principal parts to ceremonial music and marches of medium difficulty; plays, with minimum preparation, principal parts to all styles of standard band literature; performs music library duties as assigned; assists in instrument repair. *Skill Level 30:* Able to perform the duties required for Skill Level 20; supervises section as instrumental section leader; serves in various intermediate staff/administrative positions within the band as assigned; organizes, tunes, instructs, and rehearses section or consolidated section of related musical instruments; organizes and leads small instrument groups; plays, at sight, intermediate music of any style; balances chords and dynamic levels and interprets complex rhythms; supervises preventive maintenance of musical instruments in section. *Skill Level 40:* Able to perform the duties required for Skill Level 30; provides technical guidance to subordinate personnel; supervises section leaders; rehearses the stage band; assists in rehearsing the concert band; conducts the stage band in a performance; organizes, rehearses, and leads instrumental ensembles; organizes, tunes, and rehearses elements of the band in rehearsal; arranges and transcribes concert and dance/stage band music; tunes and prepares the concert band for rehearsal.

Recommendation, Skill Level 10

Credit may be granted on the basis of an individualized assessment of the student (3/94).

Recommendation, Skill Level 20

Credit may be granted on the basis of an individualized assessment of the student (2/93).

Recommendation, Skill Level 30

In the lower-division baccalaureate/associate degree category, 4 semester hours in music theory (harmony, ear training, and sight singing), 2 in jazz theory/improvisation, 4-6 in applied performance (individual instruction), 4 in performing ensembles (concert, jazz, and marching bands), and 3 in personnel supervision. In the upper-division baccalaureate category, 1 semester hour in conducting (2/93).

Recommendation, Skill Level 40

In the lower-division baccalaureate/associate degree category, 8 semester hours in music theory (harmony, ear training, and sight singing), 2 in jazz theory/improvisation, 4-6 in applied performance (individual instruction), 6 in performing ensembles (concert, jazz, and marching bands), and 3 in personnel supervision. In the upper-division baccalaureate category, 2 semester hours in conducting, 2 in rehearsal techniques of conducting (concert and jazz), 2 in arranging, and 3 in logistics management (2/93).

MOS-02M-003

Percussion Player

02M10

02M20

02M30

02M40

Exhibit Dates: 1/90–1/93.

Career Management Field: 97 (Band).

Description

Summary: Performs on specified percussion instrument(s) in appropriate musical organizations; performs in military marching and concert musical organizations and, as qualified, in associated choral, jazz, and other small ensembles. *Skill Level 10:* Plays percussion instruments; marches in military band in response to drum major's signals and oral commands; reads, interprets, and plays instrumental music; plays instrument in marching, concert, or dance/jazz band as appropriate; plays, at sight, indefinite pitch percussion parts of standard band literature; plays, with minimum preparation, definite and indefinite pitch percussion parts of standard band literature; performs preventive maintenance on all percussion instruments. *Skill Level 20:* Able to perform the duties required for Skill Level 10; provides technical guidance to Skill Level 10 personnel; plays instrument in small instrumental groups as appropriate; plays, with minimum preparation, any percussion part to medium and difficult concert band literature; performs music library duties as assigned. *Skill Level 30:* Able to perform the duties required for Skill Level 20; supervises percussion section; serves in various intermediate staff/administrative positions within the band as assigned; organizes, instructs, and rehearses the percussion section; organizes and leads small instrumental groups; plays at sight, intermediate definite and indefinite pitch percussion parts of all styles of music; plays, with minimum preparation, any moderately difficult to difficult percussion music; balances chords and volumes and interprets complex rhythms; supervises preventive maintenance of instruments in the percussion section. *Skill Level 40:* Able to perform the duties required for Skill Level 30; provides technical guidance to subordinate personnel; supervises section leaders; rehearses the stage band; assists in rehearsing the concert band; conducts the stage band in a performance; organizes, rehearses, and leads instrumental ensembles; organizes, tunes, and rehearses elements of the band in rehearsal; arranges and transcribes concert and dance/stage band music; tunes and prepares the concert band for rehearsal.

Recommendation, Skill Level 10

In the lower-division baccalaureate/associate degree category, 1-3 semester hours (equivalent of one semester) in applied study of major instrument, 2 in music theory, 1 in concert band, 2 in marching band, and credit in small ensemble on the basis of institutional evaluation; because of the wide range of individual abilities, additional credit may be awarded on the basis of institutional evaluation in music theory, applied music (individual instruction in performance), and small performing ensembles. (NOTE: This recommendation for skill level 10 is valid for the dates 1/90-9/91 only) (6/78).

Recommendation, Skill Level 20

In the lower-division baccalaureate/associate degree category, 1-3 semester hours (equivalent of one semester) in applied study of major instrument, 3 in music theory, 2 in concert band, 2 in marching band, and credit in small ensemble on the basis of institutional evaluation; because of the wide range of individual abilities, additional credit may be awarded on the basis of institutional evaluation in music theory, applied music (individual instruction in performance), and small performing ensembles (6/78).

Recommendation, Skill Level 30

In the lower-division baccalaureate/associate degree category, 2-6 semester hours (equivalent of two semesters) in applied study of major instrument, 4 in music theory (including ear training), 3 in concert band, 3 in marching band, and credit in small ensemble on the basis of institutional evaluation; because of the wide range of individual abilities, additional credit may be awarded on the basis of institutional evaluation in music theory, applied music (individual instruction in performance), and small performing ensembles (6/78).

Recommendation, Skill Level 40

In the lower-division baccalaureate/associate degree category, 2-6 semester hours (equivalent of two semesters) in applied study of major instrument, 4 in music theory (including ear training), 3 in concert band, 3 in marching band, 3 in personnel supervision, and credit in small ensemble on the basis of institutional evaluation; because of the wide range of individual abilities, additional credit may be awarded on the basis of institutional evaluation in music theory, applied music (individual instruction in performance), and small performing ensembles. In the upper-division baccalaureate category, 3 semester hours in conducting and 3 in music administration and supervision (3/86).

MOS-02M-004

Percussion Player
02M10
02M20
02M30
02M40

Exhibit Dates: 2/93–2/95.

Career Management Field: 97 (Band).

Description

Summary: Plays percussion instrument(s) in appropriate musical organizations; performs in military marching and concert musical organizations and, as qualified, in associated choral, jazz, and other small ensembles. *Skill Level 10:* Plays percussion instruments; marches in military band in response to drum major's signals and oral commands; reads, interprets, and plays instrumental music; plays in marching, concert, or dance/jazz band as appropriate; plays, at sight, secondary parts to ceremonial music and marches of medium difficulty; plays, with minimum preparation, secondary parts to all styles of standard band literature; applies knowledge of minimum basic music theory; discriminates and matches pitches; performs preventive maintenance on percussion instruments. *Skill Level 20:* Able to perform the duties required for Skill Level 10; provides technical guidance to Skill Level 10 personnel; plays percussion instruments in small instrumental groups as appropriate; plays, at sight, principal parts to ceremonial music and marches of medium difficulty; plays, with minimum preparation, principal parts to all styles of standard band literature; performs music library duties as assigned; assists in instrument repair. *Skill Level 30:* Able to perform the duties required for Skill Level 20; supervises section as instrumental section leader; serves in various intermediate staff/administrative positions within the band as assigned; organizes, tunes, instructs, and rehearses section or consolidated section of related musical instruments; organizes and leads small instrument groups; plays, at sight, intermediate music of any style; balances chords and dynamic levels and interprets complex rhythms; supervises preventive maintenance of musical instruments in section. *Skill Level 40:* Able to perform the duties required for Skill Level 30; provides technical guidance to subordinate personnel; supervises section leaders; rehearses the stage band; assists in rehearsing the concert band; conducts the stage band in a performance; organizes, rehearses, and leads instrumental ensembles; organizes, tunes, and rehearses elements of the band in rehearsal; arranges and transcribes concert and dance/stage band music; tunes and prepares the concert band for rehearsal.

Recommendation, Skill Level 10

Credit may be granted on the basis of an individualized assessment of the student (3/94).

Recommendation, Skill Level 20

In the lower-division baccalaureate/associate degree category, 4 semester hours in music theory (harmony, ear training, and sight singing), 2 in jazz theory/improvisation, 4-6 in applied performance (individual instruction), and 4 in performing ensembles (concert, jazz, and marching bands) (2/93).

Recommendation, Skill Level 30

In the lower-division baccalaureate/associate degree category, 8 semester hours in music theory (harmony, ear training, and sight singing), 2 in jazz theory/improvisation, 4-6 in applied performance (individual instruction), 6 in performing ensembles (concert, jazz, and marching bands), and 3 in personnel supervision. In the upper-division baccalaureate category, 1 semester hour in conducting (2/93).

Recommendation, Skill Level 40

In the lower-division baccalaureate/associate degree category, 12 semester hours in music theory (harmony, ear training, and sight singing), 2 in jazz theory/improvisation, 4-6 in applied performance (individual instruction), 6-8 in performing ensembles (concert, jazz, and marching bands), and 3 in personnel supervision. In the upper-division baccalaureate category, 2 semester hours in conducting, 2 in rehearsal techniques of conducting (concert and jazz), 2 in arranging, and 3 in logistics management (2/93).

MOS-02M-005

Percussion Player
02M10
02M20
02M30
02M40

Exhibit Dates: 3/95–Present.

Career Management Field: 97 (Band).

Description

Summary: Plays percussion instrument(s) in appropriate musical organizations; performs in military marching and concert musical organizations and, as qualified, in associated choral, jazz, and other small ensembles. *Skill Level 10:* Plays percussion instruments; marches in military band in response to drum major's signals and oral commands; reads, interprets, and plays instrumental music; plays in marching, concert, or dance/jazz band as appropriate; plays, at sight, secondary parts to ceremonial music and marches of medium difficulty; plays, with minimum preparation, secondary parts to all styles of standard band literature; applies knowledge of minimum basic music theory; discriminates and matches pitches; performs preventive maintenance on percussion instruments. *Skill Level 20:* Able to perform the duties required for Skill Level 10; provides technical guidance to Skill Level 10 personnel; plays percussion instruments in small instrumental groups as appropriate; plays, at sight, principal parts to ceremonial music and marches of medium difficulty; plays, with minimum preparation, principal parts to all styles of standard band literature; performs music library duties as assigned; assists in instrument repair. *Skill Level 30:* Able to perform the duties required for Skill Level 20; supervises section as instrumental section leader; serves in various intermediate staff/administrative positions within the band as assigned; organizes, tunes, instructs, and rehearses section or consolidated section of related musical instruments; organizes and leads small instrument groups; plays, at sight, intermediate music of any style; balances chords and dynamic levels and interprets complex rhythms; supervises preventive maintenance of musical instruments in section. *Skill Level 40:* Able to perform the duties required for Skill Level 30; provides technical guidance to subordinate personnel; supervises section leaders; rehearses the stage band; assists in rehearsing the concert band; conducts the stage band in a performance; organizes, rehearses, and leads instrumental ensembles; organizes, tunes, and rehearses elements of the band in rehearsal; arranges and transcribes concert and dance/stage band music; tunes and prepares the concert band for rehearsal.

Recommendation, Skill Level 10

Credit may be granted on the basis of an individualized assessment of the student (3/94).

Recommendation, Skill Level 20

Credit may be granted on the basis of an individualized assessment of the student (2/93).

Recommendation, Skill Level 30

In the lower-division baccalaureate/associate degree category, 4 semester hours in music theory (harmony, ear training, and sight singing), 2 in jazz theory/improvisation, 4-6 in applied performance (individual instruction), 4 in performing ensembles (concert, jazz, and marching bands), and 3 in personnel supervision. In the upper-division baccalaureate category, 1 semester hour in conducting (2/93).

Recommendation, Skill Level 40

In the lower-division baccalaureate/associate degree category, 8 semester hours in music theory (harmony, ear training, and sight singing), 2 in jazz theory/improvisation, 4-6 in applied performance (individual instruction), 6 in performing ensembles (concert, jazz, and marching bands), and 3 in personnel supervision. In the upper-division baccalaureate category, 2 semester hours in conducting, 2 in rehearsal techniques of conducting (concert and jazz), 2 in arranging, and 3 in logistics management (2/93).

MOS-02N-003

Piano Player
02N10
02N20
02N30
02N40

Exhibit Dates: 1/90–1/93.

Career Management Field: 97 (Band).

Description

Summary: Plays keyboard or percussion instrument(s) in appropriate musical organizations; performs in military marching and concert musical organizations, and, as qualified, in associated choral, jazz, and other small ensembles. *Skill Level 10:* Plays keyboard or percussion instruments; marches in military band in response to drum major's signals and oral commands; reads, interprets, and plays instrumental music; plays piano in dance/stage bands, ensembles, and combos; plays percussion instrument in marching band; plays, with minimum preparation, easy percussion parts to standard concert band music on keyboard type mallet instruments; applies knowledge of minimum basic music theory; discriminates and matches pitches; performs preventive maintenance on musical instruments. *Skill Level 20:* Able to perform the duties required for Skill Level 10; provides technical guidance to Skill Level 10 personnel; plays instrument in small instrumental group as appropriate; improvises on the piano utilizing chord progressions; performs music library duties as assigned. *Skill Level 30:* Able to perform the duties required for Skill Level 20; supervises consolidated piano/guitar section; serves in various intermediate staff/administrative positions within band as assigned; organizes, tunes, instructs, and rehearses section; organizes and leads small instrumental groups; plays, at sight, intermediate music of any style; balances chords and volumes and interprets complex rhythms; supervises preventive maintenance of instruments in section. *Skill Level 40:* Able to perform the duties required for Skill Level 30; provides technical guidance to subordinate personnel; supervises section leaders; rehearses the stage band; assists in rehearsing the concert band; conducts the stage band in a performance; organizes, rehearses, and leads instrumental ensembles; organizes, tunes, and rehearses elements of the band in rehearsal; arranges and transcribes concert and dance/stage band music; tunes and prepares the concert band for rehearsal.

Recommendation, Skill Level 10

In the lower-division baccalaureate/associate degree category, 1-3 semester hours (equivalent of one semester) in applied study of major instrument, 2 in music theory, 1 in concert band, 2 in marching band, and credit in small ensemble on the basis of institutional evaluation; because of the wide range of individual abilities, additional credit may be awarded on the basis of institutional evaluation in music theory, applied music (individual instruction in performance), and small performing ensembles. (NOTE: This recommendation for skill level 10 is valid for the dates 1/90-9/91 only) (6/78).

Recommendation, Skill Level 20

In the lower-division baccalaureate/associate degree category, 1-3 semester hours (equivalent of one semester) in applied study of major instrument, 3 in music theory, 2 in concert band, 2 in marching band, and credit in small ensemble on the basis of institutional evaluation; because of the wide range of individual abilities, additional credit may be awarded on the basis of institutional evaluation in music theory, applied music (individual instruction in performance), and small performing ensembles (6/78).

Recommendation, Skill Level 30

In the lower-division baccalaureate/associate degree category, 2-6 semester hours (equivalent of two semesters) in applied study of major instrument, 4 in music theory (including ear training), 3 in concert band, 3 in marching band, and credit in small ensemble on the basis of institutional evaluation; because of the wide range of individual abilities, additional credit may be awarded on the basis of institutional evaluation in music theory, applied music (individual instruction in performance), and small performing ensembles (6/78).

Recommendation, Skill Level 40

In the lower-division baccalaureate/associate degree category, 2-6 semester hours (equivalent of two semesters) in applied study of major instrument, 4 in music theory (including ear training), 3 in concert band, 3 in marching band, and credit in small ensemble on the basis of institutional evaluation; because of the wide range of individual abilities, additional credit may be awarded on the basis of institutional evaluation in music theory, applied music (individual instruction in performance), and small performing ensembles. In the upper-division baccalaureate category, 3 semester hours in conducting and 3 in music administration and supervision (6/78).

MOS-02N-004

Piano Player
02N10
02N20
02N30
02N40

Exhibit Dates: 2/93–2/95.

Career Management Field: 97 (Band).

Description

Summary: Plays piano in appropriate musical organizations; performs in military marching and concert musical organizations and, as qualified, in associated choral, jazz, and other small ensembles. *Skill Level 10:* Plays piano; marches in military band in response to drum major's signals and oral commands; reads, interprets, and plays instrumental music; plays in marching, concert, or dance/jazz band as appropriate; plays, at sight, secondary parts to ceremonial music and marches of medium difficulty; plays, with minimum preparation, secondary parts to all styles of standard band literature; applies knowledge of minimum basic music theory; discriminates and matches pitches; performs preventive maintenance on piano. *Skill Level 20:* Able to perform the duties required for Skill Level 10; provides technical guidance to Skill Level 10 personnel; plays piano in small instrumental groups as appropriate; plays, at sight, principal parts to ceremonial music and marches of medium difficulty; plays, with minimum preparation, principal parts to all styles of standard band literature; performs music library duties as assigned; assists in instrument repair. *Skill Level 30:* Able to perform the duties required for Skill Level 20; supervises section as instrumental section leader; serves in various intermediate staff/administrative positions within the band as assigned; organizes, tunes, instructs, and rehearses section or consolidated section of related musical instruments; organizes and leads small instrument groups; plays, at sight, intermediate music of any style; balances chords and dynamic levels and interprets complex rhythms; supervises preventive maintenance of musical instruments in section. *Skill Level 40:* Able to perform the duties required for Skill Level 30; provides technical guidance to subordinate personnel; supervises section leaders; rehearses the stage band; assists in rehearsing the concert band; conducts the stage band in a performance; organizes, rehearses, and leads instrumental ensembles; organizes, tunes, and rehearses elements of the band in rehearsal; arranges and transcribes concert and dance/stage band music; tunes and prepares the concert band for rehearsal.

Recommendation, Skill Level 10

Credit may be granted on the basis of an individualized assessment of the student (3/94).

Recommendation, Skill Level 20

In the lower-division baccalaureate/associate degree category, 4 semester hours in music theory (harmony, ear training, and sight singing), 2 in jazz theory/improvisation, 4-6 in applied performance (individual instruction), and 4 in performing ensembles (concert, jazz, and marching bands) (2/93).

Recommendation, Skill Level 30

In the lower-division baccalaureate/associate degree category, 8 semester hours in music theory (harmony, ear training, and sight singing), 2 in jazz theory/improvisation, 4-6 in applied performance (individual instruction), 6 in performing ensembles (concert, jazz, and marching bands), and 3 in personnel supervision. In the upper-division baccalaureate category, 1 semester hour in conducting (2/93).

Recommendation, Skill Level 40

In the lower-division baccalaureate/associate degree category, 12 semester hours in music theory (harmony, ear training, and sight singing), 2 in jazz theory/improvisation, 4-6 in applied performance (individual instruction), 6-8 in performing ensembles (concert, jazz, and marching bands), and 3 in personnel supervision. In the upper-division baccalaureate category, 2 semester hours in conducting, 2 in rehearsal techniques of conducting (concert and jazz), 2 in arranging, and 3 in logistics management (2/93).

MOS-02N-005

Piano Player
02N10
02N20
02N30
02N40

Exhibit Dates: 3/95–Present.

Career Management Field: 97 (Band).

Description

Summary: Plays piano in appropriate musical organizations; performs in military marching and concert musical organizations and, as qualified, in associated choral, jazz, and other small ensembles. *Skill Level 10:* Plays piano; marches in military band in response to drum major's signals and oral commands; reads, interprets, and plays instrumental music; plays in marching, concert, or dance/jazz band as appropriate; plays, at sight, secondary parts to ceremonial music and marches of medium difficulty; plays, with minimum preparation, secondary parts to all styles of standard band literature; applies knowledge of minimum basic music theory; discriminates and matches pitches; performs preventive maintenance on piano. *Skill Level 20:* Able to perform the duties required for Skill Level 10; provides technical guidance to Skill Level 10 personnel; plays piano in small instrumental groups as appropriate; plays, at sight, principal parts to ceremonial music and marches of medium difficulty; plays, with minimum preparation, principal parts to all styles of standard band literature; performs music library duties as assigned; assists in instrument repair. *Skill Level 30:* Able to perform the duties required for Skill Level 20; supervises section as instrumental section leader; serves in various intermediate staff/administrative positions within the band as assigned; organizes, tunes, instructs, and rehearses section or consolidated section of related musical instruments; organizes and leads small instrument groups; plays, at sight, intermediate music of any style; balances chords and dynamic levels and interprets complex rhythms; supervises preventive maintenance of musical instruments in section. *Skill Level 40:* Able to perform the duties required for Skill Level 30; provides technical guidance to subordinate personnel; supervises section leaders; rehearses the stage band; assists in rehearsing the concert band; conducts the stage band in a performance; organizes, rehearses, and leads instrumental ensembles; organizes, tunes, and rehearses elements of the band in rehearsal; arranges and transcribes concert and dance/stage band music; tunes and prepares the concert band for rehearsal.

Recommendation, Skill Level 10

Credit may be granted on the basis of an individualized assessment of the student (3/94).

Recommendation, Skill Level 20

Credit may be granted on the basis of an individualized assessment of the student (2/93).

Recommendation, Skill Level 30

In the lower-division baccalaureate/associate degree category, 4 semester hours in music theory (harmony, ear training, and sight singing), 2 in jazz theory/improvisation, 4-6 in applied performance (individual instruction), 4 in performing ensembles (concert, jazz, and marching bands), and 3 in personnel supervision. In the upper-division baccalaureate category, 1 semester hour in conducting (2/93).

Recommendation, Skill Level 40

In the lower-division baccalaureate/associate degree category, 8 semester hours in music theory (harmony, ear training, and sight singing), 2 in jazz theory/improvisation, 4-6 in applied performance (individual instruction), 6 in performing ensembles (concert, jazz, and marching bands), and 3 in personnel supervision. In the upper-division baccalaureate category, 2 semester hours in conducting, 2 in rehearsal techniques of conducting (concert and jazz), 2 in arranging, and 3 in logistics management (2/93).

MOS-02S-003

Special Bandsperson
02S20
02S30
02S40
02S50

Exhibit Dates: 1/90–1/93.

Career Management Field: 97 (Band).

Description

Summary: Performs as musician or in allied field as member of a special band. NOTE: Persons holding MOS 02S are also eligible for credit for their secondary MOS's, which identify their particular musical instruments. Consult the appropriate exhibit for the credit recommendation. *Skill Level 20:* Plays brass, woodwind, percussion, keyboard, or stringed instrument and executes marching movements in military formations; performs as instrumentalist in field music detachment or special band, or as technician in special band support activities; doubles on related instruments; provides technical guidance to musicians in field music detachment; performs as vocalist when appropriate; classifies music or performs as copyist, librarian, and commentator/annotator; performs basic instrument maintenance. *Skill Level 30:* Able to perform the duties required for Skill Level 20; assists with music procurement; provides technical guidance to subordinate personnel. *Skill Level 40:* Able to perform the duties required for Skill Level 30; performs as arranger or solo instrumental specialist; provides technical guidance to subordinate instrumental specialists. *Skill Level 50:* Able to perform the duties required for Skill Level 40; serves as principal noncommissioned officer in instrumental, choral or music support section, instrumental group, or special band; instructs, trains, rehearses, and conducts instrumental groups to achieve recognized standards; organizes separate musical groups, as needed; rehearses and trains section members to achieve correct musical rendition, coordination, and styles; conducts band as assistant bandleader in formations and ceremonies; supervises band segment in separate support activities, including designing, constructing, and assembling necessary facilities for band presentation; using public relations techniques, arranges performances.

Recommendation, Skill Level 20

In the lower-division baccalaureate/associate degree category, 3 semester hours in marching band, 3 in concert band, 2 in ensemble electives, and 4 in music theory (basic harmony and ear training), and additional credit in band, music theory, electronics, and small performing ensembles (jazz, brass, woodwind, percussion, keyboard, string, and vocal) on the basis of institutional evaluation (9/81).

Recommendation, Skill Level 30

In the lower-division baccalaureate/associate degree category, 3 semester hours in marching band, 3 in concert band, 2 in ensemble electives, and 4 in music theory (basic harmony and ear training), and additional credit in band, music theory, electronics, and small performing ensembles (jazz, brass, woodwind, percussion, keyboard, string, and vocal) on the basis of institutional evaluation (9/81).

Recommendation, Skill Level 40

In the lower-division baccalaureate/associate degree category, 3 semester hours in marching band, 3 in concert band, 2 in ensemble electives, and 8 in music theory (basic harmony and ear training), and additional credit in band, music theory, electronics, and small performing ensembles (jazz, brass, woodwind, percussion, keyboard, string, and vocal) on the basis of institutional evaluation (9/81).

Recommendation, Skill Level 50

In the lower-division baccalaureate/associate degree category, 3 semester hours in marching band, 3 in concert band, 2 in ensemble electives, and 8 in music theory (basic harmony and ear training), and additional credit in band, music theory, electronics, and small performing ensembles (jazz, brass, woodwind, percussion, keyboard, string, and vocal) on the basis of institutional evaluation. In the upper-division baccalaureate category, 3 semester hours in conducting, 3 in music administration and supervision, and credit on the basis of institutional evaluation of the individual's experience and area of specialization in the following subject areas: applied music, advanced music theory, composition, arranging, education, human relations, and management (9/81).

MOS-02S-004

Special Bandmember
02S20
02S30
02S40
02S50

Exhibit Dates: 2/93–2/95.

Career Management Field: 97 (Band).

Description

Summary: The special bandmember performs as a musician or in the direct support of the mission of the U.S. Army Band, U.S. Army Field Band, USMA Band, or the 3rd Infantry (The Old Guard) Fife and Drum Corps. Use of MOS 02S is restricted to special bands only. *Skill Level 20:* Plays brass, woodwind, percussion, keyboard, or stringed instrument and executes marching movements in military formations; performs as instrumentalist in field music detachment or special band, or as technician in special band support activities; doubles on related instruments; provides technical guidance to musicians in field music detachment; performs as vocalist when appropriate; classifies music or performs as copyist, librarian, and commentator/annotator; performs basic instrument maintenance. *Skill Level 30:* Able to perform the duties required for Skill Level 20; assists with music procurement; provides technical guidance to subordinate personnel. *Skill Level 40:* Able to perform the duties required for Skill Level 30; performs as arranger or solo instrumental specialist. *Skill Level 50:* Able to perform the duties required for Skill Level 40; serves as principal noncommissioned officer in instrumental, choral or music support section, instrumental group, or special band; instructs, trains, rehearses, and conducts instrumental groups to achieve recognized standards; organizes separate musical groups, as needed; rehearses and trains section members to achieve correct musical styles; conducts band as assistant bandleader in formations and ceremonies; supervises band segment in separate support activities, including designing, constructing, and assembling necessary facilities for band presentation; using public relations techniques, arranges performances.

Recommendation, Skill Level 20

In the lower-division baccalaureate/associate degree category, 4 semester hours in music theory (harmony, ear training, and sight singing), 2 in jazz theory/improvisation, 4-6 in applied performance (individual instruction), and 4 in performing ensembles (concert, jazz, and marching bands) (2/93).

Recommendation, Skill Level 30

In the lower-division baccalaureate/associate degree category, 8 semester hours in music theory (harmony, ear training, and sight singing), 2 in jazz theory/improvisation, 4-6 in applied performance (individual instruction), 6 in performing ensembles (concert, jazz and marching bands), and 3 in personnel supervision. In the upper-division baccalaureate category, 1 semester hours in conducting (2/93).

Recommendation, Skill Level 40

In the lower-division baccalaureate/associate degree category, 12 semester hours in music theory (harmony, ear training, and sight singing), 2 in jazz theory/improvisation, 4-6 in applied performance (individual instruction), 6-8 in performing ensembles (concert, jazz, and marching bands), and 3 in personnel supervision. In the upper-division baccalaureate category, 2 semester hours in conducting, 2 in rehearsal techniques of conducting (concert and jazz), 2 in arranging, and 3 in logistics management (2/93).

Recommendation, Skill Level 50

In the lower-division baccalaureate/associate degree category, 16 semester hours in music theory (harmony, ear training, and sight singing), 2 in jazz theory/improvisation, 4-6 in applied performance (individual instruction), 6-8 in performing ensembles (concert, jazz, and marching bands), and 3 in personnel supervision. In the upper-division baccalaureate category, if paygrade is E-8: 2 semester hours in conducting, 2 in rehearsal techniques of conducting (concert and jazz), 4 in arranging, 2 in organizational management, and 3 in logistics management; if paygrade is E-9: 2 semester hours in conducting, 4 in rehearsal techniques of conducting (concert and jazz), 6 in arranging, 4 in organizational management, 3 in logistics management, and 3 in music business management (2/93).

MOS-02S-005

Special Bandmember
02S20
02S30
02S40
02S50

Exhibit Dates: 3/95–Present.

Career Management Field: 97 (Band).

Description

Summary: The special bandmember performs as a musician or in the direct support of the mission of the U.S. Army Band, U.S. Army Field Band, USMA Band, or the 3rd Infantry (The Old Guard) Fife and Drum Corps. Use of MOS 02S is restricted to special bands only. *Skill Level 20:* Plays brass, woodwind, percussion, keyboard, or stringed instrument and executes marching movements in military formations; performs as instrumentalist in field music detachment or special band, or as technician in special band support activities; doubles on related instruments; provides technical guidance to musicians in field music detachment; performs as vocalist when appropriate; classifies music or performs as copyist, librarian, and commentator/annotator; performs basic instrument maintenance. *Skill Level 30:* Able to perform the duties required for Skill Level 20; assists with music procurement; provides technical guidance to subordinate personnel. *Skill Level 40:* Able to perform the duties required for Skill Level 30; performs as arranger or solo instrumental specialist. *Skill Level 50:* Able to perform the duties required for Skill Level 40; serves as principal noncommissioned officer in instrumental, choral or music support section, instrumental group, or special band; instructs, trains, rehearses, and conducts instrumental groups to achieve recognized standards; organizes separate musical groups, as needed; rehearses and trains section members to achieve correct musical styles; conducts band as assistant bandleader in formations and ceremonies; supervises band segment in separate support activities, including designing, constructing, and assembling necessary facilities for band presentation; using public relations techniques, arranges performances.

Recommendation, Skill Level 20

Credit may be granted on the basis of an individualized assessment of the student (2/93).

Recommendation, Skill Level 30

In the lower-division baccalaureate/associate degree category, 4 semester hours in music theory (harmony, ear training, and sight singing), 2 in jazz theory/improvisation, 4-6 in applied performance (individual instruction), 4 in performing ensembles (concert, jazz, and marching bands), and 3 in personnel supervision. In the upper-division baccalaureate category, 1 semester hours in conducting (2/93).

Recommendation, Skill Level 40

In the lower-division baccalaureate/associate degree category, 8 semester hours in music theory (harmony, ear training, and sight singing), 2 in jazz theory/improvisation, 4-6 in applied performance (individual instruction), 6 in performing ensembles (concert, jazz and marching bands), and 3 in personnel supervision. In the upper-division baccalaureate category, 2 semester hours in conducting, 2 in rehearsal techniques of conducting (concert and jazz), 2 in arranging, and 3 in logistics management (2/93).

Recommendation, Skill Level 50

In the lower-division baccalaureate/associate degree category, 12 semester hours in music theory (harmony, ear training, and sight singing), 2 in jazz theory/improvisation, 4-6 in applied performance (individual instruction), 6-8 in performing ensembles (concert, jazz, and marching bands), and 3 in personnel supervision. In the upper-division baccalaureate category, if paygrade is E-8: 2 semester hours in conducting, 2 in rehearsal techniques of conducting (concert and jazz), 4 in arranging, 2 in organizational management, and 3 in logistics management; if paygrade is E-9: 2 semester hours in conducting, 4 in rehearsal techniques of conducting (concert and jazz), 6 in arranging, 4 in organizational management, 3 in logistics management, and 3 in music business management (2/93).

MOS-02T-003

Guitar Player
02T10
02T20
02T30
02T40

Exhibit Dates: 1/90–1/93.

Career Management Field: 97 (Band).

Description

Summary: Plays guitar or percussion instruments in appropriate musical organizations; performs in military marching and concert musical organizations and, as qualified, in associated choral, jazz, and other small ensembles. *Skill Level 10:* Plays guitar or percussion instruments; marches in military band in response to drum major's signals and oral commands; reads, interprets, and plays instrumental music; plays guitar in dance/jazz bands, ensembles, and combos; plays percussion instrument in marching band; plays miscellaneous percussion instruments in concert band; plays, at sight, moderately easy melodic lines with chord symbols; plays, with minimum preparation, standard guitar music of any style; improvises on the guitar using chord progressions; applies knowledge of minimum basic music theory; discriminates and matches pitches; performs preventive maintenance on musical instrument. *Skill Level 20:* Able to perform the duties required for Skill Level 10; provides technical guidance to Skill Level 10 personnel; plays instrument in small instrumental groups as appropriate; plays, at sight, moderately easy guitar music of any style; performs music library duties as assigned. *Skill Level 30:* Able to perform the duties required for Skill Level 20; supervises consolidated piano/guitar section; serves in various intermediate staff/administrative positions within band as assigned; organizes, tunes, instructs, and rehearses section; organizes and leads small instrumental groups; plays, at sight, intermediate music of any style; balances chords and volumes, and interprets complex rhythms; supervises preventive maintenance of musical instruments in section. *Skill Level 40:* Able to perform the duties required for Skill Level 30; provides technical guidance to subordinate personnel; supervises section leaders; rehearses the stage band; assists in rehearsing the concert band; conducts the stage band in a performance; organizes, rehearses, and leads instrumental ensembles; organizes, tunes, and rehearses elements of the band in rehearsal; arranges and transcribes concert and dance/stage band music; tunes and prepares the concert band for rehearsal.

Recommendation, Skill Level 10

In the lower-division baccalaureate/associate degree category, 1-3 semester hours (equivalent of one semester) in applied study of major instrument, 2 in music theory, 1 in concert band, 2 in marching band, and credit in small ensemble on the basis of institutional evaluation; because of the wide range of individual abilities, additional credit may be awarded on the basis of institutional evaluation in music theory, applied music (individual instruction in performance), and small performing ensembles. (NOTE: This recommendation for skill level 10 is valid for the dates 1/90-9/91 only) (6/78).

Recommendation, Skill Level 20

In the lower-division baccalaureate/associate degree category, 1-3 semester hours (equivalent of one semester) in applied study of major instrument, 3 in music theory, 2 in concert band, 2 in marching band, and credit in small ensemble on the basis of institutional evaluation; because of the wide range of individual abilities, additional credit may be awarded on the basis of institutional evaluation in music theory, applied music (individual instruction in performance), and small performing ensembles (6/78).

Recommendation, Skill Level 30

In the lower-division baccalaureate/associate degree category, 2-6 semester hours (equivalent of two semesters) in applied study of major instrument, 4 in music theory, 3 in concert band, 3 in marching band, and credit in small ensemble on the basis of institutional evaluation; because of the wide range of individual abilities, additional credit may be awarded on the basis of institutional evaluation in music theory, applied music (individual instruction in performance), and small performing ensembles (6/78).

Recommendation, Skill Level 40

In the lower-division baccalaureate/associate degree category, 2-6 semester hours (equivalent of two semesters) in applied study of major instrument, 4 in music theory, 3 in concert band, 3 in marching band, and credit in small ensemble on the basis of institutional evaluation; because of the wide range of individual abilities, additional credit may be awarded on the basis of institutional evaluation in music theory, applied music (individual instruction in performance), and small performing ensembles. In the upper-division baccalaureate category, 3 semester hours in conducting and 3 in music administration and supervision (6/78).

MOS-02T-004

Guitar Player
02T10
02T20
02T30
02T40

Exhibit Dates: 2/93–2/95.

Career Management Field: 97 (Band).

Description

Summary: Plays guitar in appropriate musical organizations; performs in military marching and concert musical organizations and, as qualified, in associated choral, jazz, and other small ensembles. *Skill Level 10:* Plays guitar; marches in military band in response to drum major's signals and oral commands; reads, interprets, and plays instrumental music; plays in marching, concert, or dance/jazz band as appropriate; plays, at sight, secondary parts to ceremonial music and marches of medium difficulty; plays, with minimum preparation, secondary parts to all styles of standard band literature; applies knowledge of minimum basic music theory; discriminates and matches pitches; performs preventive maintenance on guitar. *Skill Level 20:* Able to perform the duties required for Skill Level 10; provides technical guidance to Skill Level 10 personnel; plays guitar in small instrumental groups as appropriate; plays, at sight, principal parts to ceremonial music and marches of medium difficulty; plays, with minimum preparation, principal parts to all styles of standard band literature; performs music library duties as assigned; assists in instrument repair. *Skill Level 30:* Able to perform the duties required for Skill Level 20; supervises section as instrumental section leader; serves in various intermediate staff/administrative positions within the band as assigned; organizes, tunes, instructs, and rehearses section or consolidated section of related musical instruments; organizes and leads small instrument groups; plays, at sight, intermediate music of any style; balances chords and dynamic levels and interprets complex rhythms; supervises preventive maintenance of musical instruments in section. *Skill Level 40:* Able to perform the duties required for Skill Level 30; provides technical guidance to subordinate personnel; supervises section leaders; rehearses the stage band; assists in rehearsing the concert band; conducts the stage band in a performance; organizes, rehearses, and leads instrumental ensembles; organizes, tunes, and rehearses elements of the band in rehearsal; arranges and transcribes concert and dance/stage band music; tunes and prepares the concert band for rehearsal.

Recommendation, Skill Level 10

Credit may be granted on the basis of an individualized assessment of the student (3/94).

Recommendation, Skill Level 20

In the lower-division baccalaureate/associate degree category, 4 semester hours in music theory (harmony, ear training, and sight singing), 2 in jazz theory/improvisation, 4-6 in applied performance (individual instruction), and 4 in performing ensembles (concert, jazz, and marching bands) (2/93).

Recommendation, Skill Level 30

In the lower-division baccalaureate/associate degree category, 8 semester hours in music theory (harmony, ear training, and sight singing), 2 in jazz theory/improvisation, 4-6 in applied performance (individual instruction), 6 in performing ensembles (concert, jazz, and marching bands), and 3 in personnel supervision. In the upper-division baccalaureate category, 1 semester hour in conducting (2/93).

Recommendation, Skill Level 40

In the lower-division baccalaureate/associate degree category, 12 semester hours in music theory (harmony, ear training, and sight singing), 2 in jazz theory/improvisation, 4-6 in applied performance (individual instruction), 6-8 in performing ensembles (concert, jazz, and marching bands), and 3 in personnel supervision. In the upper-division baccalaureate category, 2 semester hours in conducting, 2 in rehearsal techniques of conducting (concert and jazz), 2 in arranging, and 3 in logistics management (2/93).

MOS-02T-005

Guitar Player
02T10
02T20
02T30
02T40

Exhibit Dates: 3/95–Present.

Career Management Field: 97 (Band).

Description

Summary: Plays guitar in appropriate musical organizations; performs in military marching and concert musical organizations and, as qualified, in associated choral, jazz, and other small ensembles. *Skill Level 10:* Plays guitar; marches in military band in response to drum major's signals and oral commands; reads, interprets, and plays instrumental music; plays in marching, concert, or dance/jazz band as appropriate; plays, at sight, secondary parts to ceremonial music and marches of medium difficulty; plays, with minimum preparation, secondary parts to all styles of standard band literature; applies knowledge of minimum basic music theory; discriminates and matches pitches; performs preventive maintenance on guitar. *Skill Level 20:* Able to perform the duties required for Skill Level 10; provides technical guidance to Skill Level 10 personnel; plays guitar in small instrumental groups as appropriate; plays, at sight, principal parts to ceremonial music and marches of medium difficulty; plays, with minimum preparation, principal parts to all styles of standard band literature; performs music library duties as assigned; assists in instrument repair. *Skill Level 30:* Able to perform the duties required for Skill Level 20; supervises section as instrumental section leader; serves in various intermediate staff/administrative positions within the band as assigned; organizes, tunes, instructs, and rehearses section or consolidated section of related musical instruments; organizes and leads small instrument groups; plays, at sight, intermediate music of any style; balances chords and dynamic levels and interprets complex rhythms; supervises preventive maintenance of musical instruments in section. *Skill Level 40:* Able to perform the duties required for Skill Level 30; provides technical guidance to subordinate personnel; supervises section leaders; rehearses the stage band; assists in rehearsing the concert band; conducts the stage band in a performance; organizes, rehearses, and leads instrumental ensembles; organizes, tunes, and rehearses elements of the band in rehearsal; arranges and transcribes concert and dance/stage band music; tunes and prepares the concert band for rehearsal.

Recommendation, Skill Level 10

Credit may be granted on the basis of an individualized assessment of the student (3/94).

Recommendation, Skill Level 20

Credit may be granted on the basis of an individualized assessment of the student (2/93).

Recommendation, Skill Level 30

In the lower-division baccalaureate/associate degree category, 4 semester hours in music theory (harmony, ear training, and sight singing), 2 in jazz theory/improvisation, 4-6 in applied performance (individual instruction), 4 in performing ensembles (concert, jazz, and marching bands), and 3 in personnel supervision. In the upper-division baccalaureate category, 1 semester hour in conducting (2/93).

Recommendation, Skill Level 40

In the lower-division baccalaureate/associate degree category, 8 semester hours in music theory (harmony, ear training, and sight singing), 2 in jazz theory/improvisation, 4-6 in applied performance (individual instruction), 6 in performing ensembles (concert, jazz, and marching bands), and 3 in personnel supervision. In the upper-division baccalaureate category, 2 semester hours in conducting, 2 in rehearsal techniques of conducting (concert and jazz), 2 in arranging, and 3 in logistics management (2/93).

MOS-02U-001

Electric Bass Guitar Player

02U10
02U20
02U30
02U40

Exhibit Dates: 1/90–1/93.

Career Management Field: 97 (Band).

Description

Summary: Plays electric bass guitar in appropriate musical organizations; performs in military marching and concert musical organizations and, as qualified, in associated choral, jazz, and other small ensembles. *Skill Level 10:* Plays electric bass guitar; marches in military band in response to drum major's signals and oral commands; reads, interprets, and plays instrumental music; plays instrument in marching, concert, or dance/jazz band as appropriate; plays, at sight, secondary parts to ceremonial music and marches of medium difficulty; plays with minimum preparation secondary parts to all styles of standard band literature; doubles on another musical instrument as assigned; applies knowledge of minimum basic music theory; discriminates and matches pitches; performs preventive maintenance on electric bass guitar. *Skill Level 20:* Able to perform the duties required for Skill Level 10; provides technical guidance to Skill Level 10 personnel; plays electric bass guitar in small instrumental groups as appropriate; plays, at sight, principal parts to ceremonial music and marches of medium difficulty; plays, with minimum preparation, principal parts to all styles of standard band literature; performs music library duties as assigned. *Skill Level 30:* Able to perform the duties required for Skill Level 20; supervises section as instrumental section leader; serves in various intermediate staff/administrative positions within the band as assigned; organizes, tunes, instructs, and rehearses section or consolidated section of related musical instruments; organizes and leads small instrumental groups; plays, at sight, intermediate music of any style; balances chords and volumes, and interprets complex rhythms; supervises preventive maintenance of musical instruments in section. *Skill Level 40:* Able to perform the duties required for Skill Level 30; provides technical guidance to subordinate personnel; supervises section leaders; rehearses the stage band; assists in rehearsing the concert band; conducts the stage band in a performance; organizes, rehearses, and leads instrumental ensembles; organizes, tunes, and rehearses elements of the band in rehearsal; arranges and transcribes concert and dance/stage band music; tunes and prepares the concert band for rehearsal.

Recommendation, Skill Level 10

In the lower-division baccalaureate/associate degree category, 1-3 semester hours (equivalent of one semester) in applied study of major instrument, 2 in music theory, 1 in concert band, 2 in marching band, and credit in small ensemble on the basis of institutional evaluation; because of the wide range of individual abilities, additional credit may be awarded on the basis of institutional evaluation in music theory, applied music (individual instruction in performance), and small performing ensembles. (NOTE: This recommendation for skill level 10 is valid for the dates 1/90-9/91 only) (6/78).

Recommendation, Skill Level 20

In the lower-division baccalaureate/associate degree category, 1-3 semester hours (equivalent of one semester) in applied study of major instrument, 3 in music theory, 2 in concert band, 2 in marching band, and credit in small ensemble on the basis of institutional evaluation; because of the wide range of individual abilities, additional credit may be awarded on the basis of institutional evaluation in music theory, applied music (individual instruction in performance), and small performing ensembles (6/78).

Recommendation, Skill Level 30

In the lower-division baccalaureate/associate degree category, 2-6 semester hours (equivalent of two semesters) in applied study of major instrument, 4 in music theory (including ear training), 3 in concert band, 3 in marching band, and credit in small ensemble on the basis of institutional evaluation; because of the wide range of individual abilities, additional credit may be awarded on the basis of institutional evaluation in music theory, applied music (individual instruction in performance), and small performing ensembles (6/78).

Recommendation, Skill Level 40

In the lower-division baccalaureate/associate degree category, 2-6 semester hours (equivalent of two semesters) in applied study of major instrument, 4 in music theory (including ear training), 3 in concert band, 3 in marching band, 3 in personnel supervision, and credit in small ensemble on the basis of institutional evaluation; because of the wide range of individual abilities, additional credit may be awarded on the basis of institutional evaluation in music theory, applied music (individual instruction in performance), and small performing ensembles. In the upper-division baccalaureate category, 3 semester hours in conducting and 3 in music administration and supervision (3/86).

MOS-02U-002

Electric Bass Player

02U10
02U20
02U30
02U40

Exhibit Dates: 2/93–2/95.

Career Management Field: 97 (Band).

Description

Summary: Plays electric bass in appropriate musical organizations; performs in military marching and concert musical organizations and, as qualified, in associated choral, jazz, and other small ensembles. *Skill Level 10:* Plays electric bass; marches in military band in response to drum major's signals and oral commands; reads, interprets, and plays instrumental music; plays in marching, concert, or dance/jazz band as appropriate; plays, at sight, secondary parts to ceremonial music and marches of medium difficulty; plays, with minimum preparation, secondary parts to all styles of standard band literature; applies knowledge of minimum basic music theory; discriminates and matches pitches; performs preventive maintenance on electric bass. *Skill Level 20:* Able to perform the duties required for Skill Level 10; provides technical guidance to Skill Level 10 personnel; plays electric bass in small instrumental groups as appropriate; plays, at sight, principal parts to ceremonial music and marches of medium difficulty; plays, with minimum preparation, principal parts to all styles of standard band literature; performs music library duties as assigned; assists in instrument repair. *Skill Level 30:* Able to perform the duties required for Skill Level 20; supervises section as instrumental section leader; serves in various intermediate staff/administrative positions within the band as assigned; organizes, tunes, instructs, and rehearses section or consolidated section of related musical instruments; organizes and leads small instrument groups; plays, at sight, intermediate music of any style; balances chords and dynamic levels and interprets complex rhythms; supervises preventive maintenance of musical instruments in section. *Skill Level 40:* Able to perform the duties required for Skill Level 30; provides technical guidance to subordinate personnel; supervises section leaders; rehearses the stage band; assists in rehearsing the concert band; conducts the stage band in a performance; organizes, rehearses, and leads instrumental ensembles; organizes, tunes, and rehearses elements of the band in rehearsal; arranges and transcribes concert and dance/stage band music; tunes and prepares the concert band for rehearsal.

Recommendation, Skill Level 10

Credit may be granted on the basis of an individualized assessment of the student (3/94).

Recommendation, Skill Level 20

In the lower-division baccalaureate/associate degree category, 4 semester hours in music theory (harmony, ear training, and sight singing), 2 in jazz theory/improvisation, 4-6 in applied performance (individual instruction), and 4 in performing ensembles (concert, jazz, and marching bands) (2/93).

Recommendation, Skill Level 30

In the lower-division baccalaureate/associate degree category, 8 semester hours in music theory (harmony, ear training, and sight singing), 2 in jazz theory/improvisation, 4-6 in applied performance (individual instruction), 6 in performing ensembles (concert, jazz, and marching bands), and 3 in personnel supervision. In the upper-division baccalaureate category, 1 semester hour in conducting (2/93).

Recommendation, Skill Level 40

In the lower-division baccalaureate/associate degree category, 12 semester hours in music theory (harmony, ear training, and sight singing), 2 in jazz theory/improvisation, 4-6 in applied performance (individual instruction), 6-8 in performing ensembles (concert, jazz, and marching bands), and 3 in personnel supervision. In the upper-division baccalaureate category, 2 semester hours in conducting, 2 in rehearsal techniques of conducting (concert and jazz), 2 in arranging, and 3 in logistics management (2/93).

MOS-02U-003

Electric Bass Player
02U10
02U20
02U30
02U40

Exhibit Dates: 3/95–Present.

Career Management Field: 97 (Band).

Description

Summary: Plays electric bass in appropriate musical organizations; performs in military marching and concert musical organizations and, as qualified, in associated choral, jazz, and other small ensembles. *Skill Level 10:* Plays electric bass; marches in military band in response to drum major's signals and oral commands; reads, interprets, and plays instrumental music; plays in marching, concert, or dance/jazz band as appropriate; plays, at sight, secondary parts to ceremonial music and marches of medium difficulty; plays, with minimum preparation, secondary parts to all styles of standard band literature; applies knowledge of minimum basic music theory; discriminates and matches pitches; performs preventive maintenance on electric bass. *Skill Level 20:* Able to perform the duties required for Skill Level 10; provides technical guidance to Skill Level 10 personnel; plays electric bass in small instrumental groups as appropriate; plays, at sight, principal parts to ceremonial music and marches of medium difficulty; plays, with minimum preparation, principal parts to all styles of standard band literature; performs music library duties as assigned; assists in instrument repair. *Skill Level 30:* Able to perform the duties required for Skill Level 20; supervises section as instrumental section leader; serves in various intermediate staff/administrative positions within the band as assigned; organizes, tunes, instructs, and rehearses section or consolidated section of related musical instruments; organizes and leads small instrument groups; plays, at sight, intermediate music of any style; balances chords and dynamic levels and interprets complex rhythms; supervises preventive maintenance of musical instruments in section. *Skill Level 40:* Able to perform the duties required for Skill Level 30; provides technical guidance to subordinate personnel; supervises section leaders; rehearses the stage band; assists in rehearsing the concert band; conducts the stage band in a performance; organizes, rehearses, and leads instrumental ensembles; organizes, tunes, and rehearses elements of the band in rehearsal; arranges and transcribes concert and dance/stage band music; tunes and prepares the concert band for rehearsal.

Recommendation, Skill Level 10

Credit may be granted on the basis of an individualized assessment of the student (3/94).

Recommendation, Skill Level 20

Credit may be granted on the basis of an individualized assessment of the student (2/93).

Recommendation, Skill Level 30

In the lower-division baccalaureate/associate degree category, 4 semester hours in music theory (harmony, ear training, and sight singing), 2 in jazz theory/improvisation, 4-6 in applied performance (individual instruction), 4 in performing ensembles (concert, jazz, and marching bands), and 3 in personnel supervision. In the upper-division baccalaureate category, 1 semester hour in conducting (2/93).

Recommendation, Skill Level 40

In the lower-division baccalaureate/associate degree category, 8 semester hours in music theory (harmony, ear training, and sight singing), 2 in jazz theory/improvisation, 4-6 in applied performance (individual instruction), 6 in performing ensembles (concert, jazz, and marching bands), and 3 in personnel supervision. In the upper-division baccalaureate category, 2 semester hours in conducting, 2 in rehearsal techniques of conducting (concert and jazz), 2 in arranging, and 3 in logistics management (2/93).

MOS-02Z-001

Bands Senior Sergeant
02Z50

Exhibit Dates: 1/90–1/93.

Career Management Field: 97 (Band).

Description

Serves as the principal noncommissioned officer of an Army band; able to perform the duties required for 02P40 (Brass Group Leader), 02Q40 (Woodwind Group Leader), or 02R40 (Percussion Group Leader); selects, prepares, and presents concert, marching, dance, stage, and show band music and assumes any other musical or military functions of the bandmaster as delegated; assists in planning, coordinating, and supervising all activities in support of the band mission; applies public relations techniques to band employment; supervises on-the-job training; advises bandmaster on all personnel matters.

Recommendation

In the lower-division baccalaureate/associate degree category, 3 semester hours in marching band, 3 in concert band, 2 in ensemble electives, and 4 in music theory (basic harmony and ear training), and additional credit in music theory, applied music (individual instruction in performance), and small performing ensembles (jazz, brass, woodwind, percussion, string, and vocal) on the basis of institutional evaluation. In the upper-division baccalaureate category, 3 semester hours in conducting, 3 in music administration and supervision, 3 in introduction to management, 3 in personnel management, 3 in human relations or public relations, 6 for field experience in management, and additional credit in arranging and for field experience in education on the basis of institutional evaluation (6/76).

MOS-02Z-002

Bands Senior Sergeant
02Z50

Exhibit Dates: 2/93–Present.

Career Management Field: 97 (Bands).

Description

Serves as the principal noncommissioned officer of an Army band; able to perform the duties required for Skill Level 40 of any MOS in Career Management Field 97; selects, prepares, and presents concert, marching, dance, stage, and show band music and assumes any other musical or military functions of the bandmaster as delegated; assists in planning, coordinating, and supervising all activities in support of the band mission; serves as publications liaison for the band; supervises on-the-job training; advises bandmaster on all personnel matters.

Recommendation

In the lower-division baccalaureate/associate degree category, 16 semester hours in music theory (harmony, ear training, and sight singing), 2 in jazz theory/improvisation, 4-6 in applied performance (individual instruction), 6-8 in performing ensembles (concert, jazz and marching bands), and 3 in personnel supervision. In the upper-division baccalaureate category, if paygrade is E-8: 2 semester hours in conducting, 2 in rehearsal techniques of conducting (concert and jazz), 4 in arranging, 2 in organizational management, and 3 in logistics management; if paygrade is E-9: 2 semester hours in conducting, 4 in rehearsal techniques of conducting (concert and jazz), 6 in arranging, 4 in organizational management, 3 in logistics management, and 3 in music business management (2/93).

MOS-05D-003

Electronic Warfare/Signal Intelligence Emitter Identifier/Locator
05D10
05D20
05D30

Exhibit Dates: 1/90–3/90. Effective 3/90, MOS 05D was discontinued and its duties were incorporated into MOS 98D, Emitter Locator/Identifier.

Career Management Field: 98 (Electronic Warfare/Cryptologic).

Description

Summary: Operates or supervises the operation of radio direction-finding systems and other systems using advanced identification techniques; intercepts and acquires bearings on target transmitters and performs analysis on maps or charts to establish probable locations of target transmitters. NOTE: Many of the duties required for this MOS involve highly classified materials, equipment, and activities; therefore, not all the competencies and knowledge associated with the MOS were evaluated. *Skill Level 10:* Employs special transmitters; forwards bearing and identification information to a control center; selects, erects, and orients tactical antennas; obtains desired visual display on specialized monitor oscilloscopes; records electrical characteristics of signals displayed, using light sensitive recorders; operates direction-finding and related cryptological, communication, and automatic data processing equipment; prepares and maintains operation logs and card files; types at a minimum rate of 25 words per minute; copies international Morse code at a minimum rate of 20 groups per minute. *Skill Level 20:* Able to perform the duties required for Skill Level 10; maintains section management files; classifies, analyzes, and evaluates observed bearings and waveform oscillograms; establishes, plots, and evaluates bearings to determine probable geographical location of foreign transmitters; measures bands on oscillograms to determine ripple frequency, modulation percentages, and duration of other effects; maintains calibration and accuracy studies and computes statistical data, including standard deviation and systemic errors; monitors quality control of input and data for automatic data processing support; may serve as a first-line supervisor, assigning work loads and completing personnel evaluations; presents oral and written reports. *Skill Level 30:* Able to perform the duties required for Skill Level 20; inspects equipment to ensure proper alignment and orientation; provides guidance and assistance in site selection and equipment installation; prepares written and oral reports; establishes and maintains facilities and support for site personnel; implements emergency action plans; may have experience as the enlisted commander of a detachment.

Recommendation, Skill Level 10

In the vocational certificate category, 3 semester hours in electronic systems operations. In the lower-division baccalaureate/associate degree category, 3 semester hours in keyboarding and 1 in computer literacy (9/88).

Recommendation, Skill Level 20

In the vocational certificate category, 3 semester hours in electronic systems operations. In the lower-division baccalaureate/associate degree category, 3 semester hours in keyboarding, 3 in written communication, 1 in office practices, and 2 in computer literacy (9/88).

Recommendation, Skill Level 30

In the vocational certificate category, 3 semester hours in electronic systems operations. In the lower-division baccalaureate/associate degree category, 3 semester hours in keyboarding, 3 in written communication, 3 in office practices, 3 in computer literacy, and 3 for field experience in personnel supervision (9/88).

MOS-05H-003

Electronic Warfare/Signal Intelligence Morse Interceptor
05H10
05H20
05H30
05H40

Exhibit Dates: 1/90–3/90. Effective 3/90, MOS 05H was discontinued and its duties were incorporated into MOS 98H, Morse Interceptor.

Career Management Field: 98 (Electronic Warfare/Cryptologic).

Description

Summary: Operates international Morse code message interception and simple printer equipment and supervises the operation of such equipment in mobile or fixed installations for the purpose of detecting, identifying, and exploiting foreign communications. NOTE: Many of the duties required for this MOS involve highly classified materials, equipment, and activities; therefore, not all the competencies and knowledge associated with the MOS were evaluated. *Skill Level 10:* Operates Morse Code interception equipment, including radio receivers, special typewriters, teletypewriters, antenna selection devices, internal communications equipment, and magnetic tape recorders; searches for, identifies, and manually records foreign international Morse Code communications at a minimum rate of 20 groups per minute; performs first-level analysis of messages to detect anomalies and suspect items which may be of intelligence interest; maintains operator's log of messages and related data and delivers messages to analysts for interpretation; performs operator maintenance on equipment; types at a minimum speed of 25 words per minute. *Skill Level 20:* Able to perform the duties required for Skill Level 10; performs more detailed message analysis and evaluation prior to forwarding messages to other analysts; writes detailed reports regarding intercepted messages; conducts on-the-job training; presents oral reports to high-level command staff. *Skill Level 30:* Able to perform the duties required for Skill Level 20; supervises Morse intercept activities; establishes and maintains extensive intercept files for messages and related data; assists in formulating unit deployment plans. *Skill Level 40:* Able to perform the duties required for Skill Level 30; allocates personnel and equipment resources; assists in designing collection strategies; may have experience as the enlisted commander of a detachment; analyzes automatic data processing results and confers with computer programmers and analysts; uses counseling techniques to alleviate stress among subordinates. NOTE: May have progressed to 05H40 from 05H30 or 05D30 (Electronic Warfare/Signal Intelligence Emitter Identifier/Locator).

Recommendation, Skill Level 10

In the vocational certificate category, 3 semester hours in Morse code and 3 in electronic systems operations. In the lower-division baccalaureate/associate degree category, 3 semester hours in keyboarding (9/88).

Recommendation, Skill Level 20

In the vocational certificate category, 3 semester hours in Morse code and 3 in electronic systems operations. In the lower-division baccalaureate/associate degree category, 3 semester hours in keyboarding and 1 in office practices (9/88).

Recommendation, Skill Level 30

In the vocational certificate category, 3 semester hours in Morse code and 3 in electronic systems operations. In the lower-division baccalaureate/associate degree category, 3 semester hours in keyboarding, 3 in office practices, 3 in human relations, 3 for field experience in personnel supervision, and 3 in written communication (9/88).

Recommendation, Skill Level 40

In the vocational certificate category, 3 semester hours in Morse code and 3 in electronic systems operations. In the lower-division baccalaureate/associate degree category, 3 semester hours in keyboarding, 3 in office practices, 3 in human relations, 3 for field experience in personnel supervision, 3 in written communication, 3 for a counseling practicum, and 3 in social science and humanities. In the upper-division baccalaureate category, 3 semester hours for field experience in management (9/88).

MOS-05K-003

Electronic Warfare/Signal Intelligence Non-Morse Interceptor
05K10
05K20
05K30
05K40

Exhibit Dates: 1/90–3/90. Effective 3/90, MOS 05K was discontinued and its duties were incorporated into MOS 98K, Non-Morse Interceptor/Analyst.

Career Management Field: 98 (Electronic Warfare/Cryptologic).

Description

Summary: Operates non-Morse communications intercept and recording equipment and supervises the operation of such equipment in mobile or fixed environments for the purpose of identifying and recording foreign radioteletype, facsimile, and data communications transmissions. NOTE: Many of the duties for this MOS involve highly classified material, equipment, and activities; therefore, not all the competencies and knowledge associated with the MOS were evaluated. *Skill Level 10:* Operates radioteletype, facsimile, and data intercept and recording equipment; knows basic AC and DC theory, circuit electronic theory, basic frequency analysis, spectrum analysis, and functional algebra; searches for, identifies, and records foreign transmissions; maintains a log of interceptions; prepares technical reports; types at a minimum rate of 25 words per minute; performs operator maintenance on non-Morse intercept equipment; selects, erects, and orients tactical antennas. *Skill Level 20:* Able to perform the duties required for Skill Level 10; assists in the establishment of operational sites; maintains the technical data base to support collection operations; employs special electronic equipment for complex signal analysis; analyzes intercepted communications for items of intelligence interest; prepares detailed reports; provides technical guidance to Skill Level 10 personnel. *Skill Level 30:* Able to perform the duties required for Skill Level 20; supervises non-Morse intercept activities; allocates equipment and personnel resources; writes extensive reports to accurately provide intelligence information; analyzes long-term trends, using statistical analysis techniques; coordinates interaction with other collection and processing activities; conducts on-the-job training. *Skill Level 40:* Able to perform the duties required for Skill Level 30; supervises non-Morse intercept activities; interprets signal intelligence collection priorities; ensures proper handling of intelligence information; assesses procedures and operations for adequacy in meeting intelligence requirements and recommends changes; may have experience as the enlisted commander of a detachment; uses counseling techniques to alleviate stress among subordinates.

Recommendation, Skill Level 10

In the vocational certificate category, 3 semester hours in electronic systems operations. In the lower-division baccalaureate/associate degree category, 3 semester hours in keyboarding, 1 in computer literacy, and 2 in office practices (9/88).

Recommendation, Skill Level 20

In the vocational certificate category, 3 semester hours in electronic systems operations. In the lower-division baccalaureate/associate degree category, 3 semester hours in keyboarding, 2 in computer literacy, 3 in office practices, and 3 in written communication (9/88).

Recommendation, Skill Level 30

In the vocational certificate category, 3 semester hours in electronic systems operations. In the lower-division baccalaureate/associate degree category, 3 semester hours in keyboarding, 3 in computer literacy, 3 in office practices, 3 in written communication, 3 in human relations, and 3 in social sciences and humanities (9/88).

Recommendation, Skill Level 40

In the vocational certificate category, 3 semester hours in electronic systems operations. In the lower-division baccalaureate/associate degree category, 3 semester hours in keyboarding, 3 in computer literacy, 3 in office practices, 3 in written communication, 3 in human relations, 3 in social sciences and humanities, 3 for a counseling practicum, and 3 for field experience in personnel supervision. In the upper-division baccalaureate category, 3 semester hours for field experience in management (9/88).

MOS-11B-004

Infantryman

11B10
11B20
11B30
11B40
11B50

Exhibit Dates: 1/90–2/95.

Career Management Field: 11 (Infantry).

Description

Summary: Leads, supervises, and serves as a member of an infantry unit of 10-20 persons, employing individual weapons, machine guns, and antiarmor weapons in offensive and defensive ground combat. *Skill Level 10:* Uses individual infantry weapons; lays field wire; performs basic communications functions and operates communication equipment; utilizes camouflage to conceal weapons and personnel; constructs minor fortifications; performs land navigation; performs preventive maintenance on weapons, equipment, and some vehicles; makes verbal reports; administers first aid; operates wheeled vehicles to transport personnel, supplies, and equipment. *Skill Level 20:* Able to perform the duties required for Skill Level 10; serves as a team leader, directing deployment and employment of personnel; supervises maintenance and construction activities; reads, interprets, and collects intelligence information; distributes administrative and training documents; trains subordinate personnel. *Skill Level 30:* Able to perform the duties required for Skill Level 20; as a first-line supervisor, directs the utilization of personnel and equipment; coordinates unit actions with adjacent and supporting elements; insures proper collection and reporting of intelligence data. *Skill Level 40:* Able to perform the duties required for Skill Level 30; supervises and trains personnel in infantry operations and intelligence activities; assists in planning, organizing, directing, supervising, training, coordinating, and reporting activities of subordinate units; supervises receipt, storage, and distribution of supplies, equipment, and food to subordinate units; provides oral and written reports; assists in production and administration of staff journals, files, records, and reports. *Skill Level 50:* Able to perform the duties required for Skill Level 40; serves as the principal noncommissioned officer in an infantry company; assists in planning, coordinating, and supervising all company activities; advises superiors on all matters concerning subordinate personnel. NOTE: May have progressed to 11B50 from 11B40, 11C40 (Indirect Fire Infantryman), 11H40 (Heavy Antiarmor Weapons Crewman), or 11M40 (Fighting Vehicle Infantryman).

Recommendation, Skill Level 10

In the vocational certificate category, 3 semester hours in mechanical maintenance. In the lower-division baccalaureate/associate degree category, 1 semester hour in map reading, 1 in first aid, and credit in surveying on the basis of institutional evaluation. (NOTE: This recommendation for skill level 10 is valid for the dates 1/90-9/91 only) (10/83).

Recommendation, Skill Level 20

In the vocational certificate category, 3 semester hours in mechanical maintenance. In the lower-division baccalaureate/associate degree category, 1 semester hour in map reading, 1 in first aid, and credit in surveying on the basis of institutional evaluation (10/83).

Recommendation, Skill Level 30

In the vocational certificate category, 3 semester hours in mechanical maintenance. In the lower-division baccalaureate/associate degree category, 2 semester hours in map reading, 1 in first aid, 2 in record keeping, 3 in personnel supervision, 3 in human relations, and credit in surveying on the basis of institutional evaluation (10/83).

Recommendation, Skill Level 40

In the vocational certificate category, 3 semester hours in mechanical maintenance. In the lower-division baccalaureate/associate degree category, 3 semester hours in map reading, 1 in first aid, 3 in record keeping, 3 in personnel supervision, 3 in human relations, 3 in principles of instruction, and credit in surveying on the basis of institutional evaluation. In the upper-division baccalaureate category, 3 semester hours for field experience in management (10/83).

Recommendation, Skill Level 50

In the vocational certificate category, 3 semester hours in mechanical maintenance. In the lower-division baccalaureate/associate degree category, 3 semester hours in map reading, 1 in first aid, 3 in record keeping, 3 in personnel supervision, 3 in human relations, 3 in principles of instruction, 3 in office administration, 3 in introduction to management, and credit in surveying on the basis on institutional evaluation. In the upper-division baccalaureate category, 3 semester hours for field experience in management and 3 in management problems (10/83).

MOS-11B-005

Infantryman

11B10
11B20
11B30
11B40
11B50

Exhibit Dates: 3/95–Present.

Career Management Field: 11 (Infantry).

Description

Summary: Leads, supervises, and serves as a member of an infantry unit of 10-20 persons, employing individual weapons, machine guns, and antiarmor weapons in offensive and defensive ground combat. *Skill Level 10:* Uses individual infantry weapons; lays field wire; performs basic communications functions and operates communication equipment; utilizes camouflage to conceal weapons and personnel; constructs minor fortifications; performs land navigation; performs preventive maintenance on weapons, equipment, and some vehicles; makes verbal reports; administers first aid; operates wheeled vehicles to transport personnel, supplies, and equipment. *Skill Level 20:* Able to perform the duties required for Skill Level 10; serves as a team leader, directing deployment and employment of personnel; supervises maintenance and construction activities; reads, interprets, and collects intelligence information; distributes administrative and training documents; trains subordinate personnel. *Skill Level 30:* Able to perform the duties required for Skill Level 20; as a first-line supervisor, directs the utilization of personnel and equipment; coordinates unit actions with adjacent and supporting elements; insures proper collection and reporting of intelligence data. *Skill Level 40:* Able to perform the duties required for Skill Level 30; supervises and trains personnel in infantry operations and intelligence activities; assists in planning, organizing, directing, supervising, training, coordinating, and reporting activities of subordinate units; supervises receipt, storage, and distribution of supplies, equipment, and food to subordinate units; provides oral and written reports; assists in production and administration of staff journals, files, records, and reports. *Skill Level 50:* Able to perform the duties required for Skill Level 40; serves as the principal noncommissioned officer in an infantry company; assists in planning, coordinating, and supervising all company activities; advises superiors on all matters concerning subordinate personnel. NOTE: May have progressed to 11B50 from 11B40, 11C40 (Indirect Fire Infantryman), 11H40 (Heavy Antiarmor Weapons Crewman), or 11M40 (Fighting Vehicle Infantryman).

Recommendation, Skill Level 10

Credit may be granted on the basis of an individualized assessment of the student (10/83).

Recommendation, Skill Level 20

Credit may be granted on the basis of an individualized assessment of the student (10/83).

Recommendation, Skill Level 30

In the vocational certificate category, 3 semester hours in mechanical maintenance. In the lower-division baccalaureate/associate degree category, 1 semester hour in map reading, 1 in first aid, 2 in record keeping, 3 in personnel supervision, 3 in human relations, and credit in surveying on the basis of institutional evaluation (10/83).

Recommendation, Skill Level 40

In the vocational certificate category, 3 semester hours in mechanical maintenance. In the lower-division baccalaureate/associate degree category, 2 semester hours in map reading, 1 in first aid, 3 in record keeping, 3 in personnel supervision, 3 in human relations, 3 in principles of instruction, and credit in surveying on the basis of institutional evaluation. In the upper-division baccalaureate category, 3 semester hours for field experience in management (10/83).

Recommendation, Skill Level 50

In the vocational certificate category, 3 semester hours in mechanical maintenance. In the lower-division baccalaureate/associate degree category, 3 semester hours in map reading, 1 in first aid, 3 in record keeping, 3 in personnel supervision, 3 in human relations, 3 in principles of instruction, 3 in office administration, 3 in introduction to management, and credit in surveying on the basis on institutional evaluation. In the upper-division baccalaureate category, 3 semester hours for field experience in management and 3 in management problems (10/83).

MOS-11C-003

Indirect Fire Infantryman

11C10
11C20
11C30
11C40
11C50

Exhibit Dates: 1/90–1/92.

Career Management Field: 11 (Infantry).

Description

Summary: Leads or serves as a member of a mortar squad, section, or platoon employing crew-served and individual weapons in offensive and defensive operations. *Skill Level 10:* Employs individual weapons; assists in construction of minor fortifications; performs minor maintenance on weapons, equipment, and wheeled and tracked vehicles; performs land navigation; measures horizontal and vertical angles; estimates ranges; computes firing data; administers first aid; collects and verbally reports tactical information using basic communications equipment. *Skill Level 20:* Able to perform the duties required for Skill Level 10; leads mortar squad; supervises crew readiness; trains crew; supervises work details. *Skill Level 30:* Able to perform the duties required for Skill Level 20; leads mortar section; supervises employment and deployment of personnel and equipment; conducts surveys to determine weapon location. *Skill Level 40:* Able to perform the duties required for Skill Level 30; supervises mortar platoon; assists in planning, organizing, training, reporting, and coordinating activities of subordinate sections; supervises receipt, storage, and distribution of ammunition, supplies, equipment, and food. *Skill Level 50:* Able to perform the duties required for Skill Level 40; plans, coordinates, supervises, and participates in activities pertaining to organization, training, and combat operations and intelligence; serves as principal NCO in a mechanized infantry headquarters company; supervises the processing of operations and intelligence information at a battalion or higher level.

Recommendation, Skill Level 10

In the vocational certificate category, 3 semester hours in mechanical maintenance. In the lower-division baccalaureate/associate degree category, 3 semester hours in surveying and 1 in first aid. (NOTE: This recommendation for skill level 10 is valid for the dates 1/90-9/91 only) (6/79).

Recommendation, Skill Level 20

In the vocational certificate category, 3 semester hours in mechanical maintenance. In the lower-division baccalaureate/associate degree category, 3 semester hours in surveying, 3 for field experience in management, and 1 in first aid (6/79).

Recommendation, Skill Level 30

In the vocational certificate category, 3 semester hours in mechanical maintenance. In the lower-division baccalaureate/associate degree category, 3 semester hours in surveying, 3 in human relations, 3 in record keeping, 3 for field experience in management, and 1 in first aid (6/79).

Recommendation, Skill Level 40

In the vocational certificate category, 3 semester hours in mechanical maintenance. In the lower-division baccalaureate/associate degree category, 6 semester hours in human relations, 3 in office administration, 3 in record keeping, 3 in surveying, and 1 in first aid. In the upper-division baccalaureate category, 3 semester hours for field experience in management and 3 in personnel management (5/88).

Recommendation, Skill Level 50

In the vocational certificate category, 3 semester hours in mechanical maintenance. In the lower-division baccalaureate/associate degree category, 6 semester hours in human relations, 3 in office administration, 3 in record keeping, 3 in surveying, and 1 in first aid. In the upper-division baccalaureate category, 6 semester hours for field experience in management and 3 in personnel management (5/88).

MOS-11C-004

Indirect Fire Infantryman

11C10
11C20
11C30
11C40
11C50

Exhibit Dates: 2/92–Present.

Career Management Field: 11 (Infantry).

Description

Summary: Leads or serves as a member of a mortar squad, section, or platoon employing crew-served and individual weapons in offensive and defensive operations. *Skill Level 10:* Employs individual weapons; assists in construction of minor fortifications; performs minor maintenance on weapons, equipment, and wheeled and tracked vehicles; performs land navigation; measures horizontal and vertical angles; estimates ranges; computes firing data; administers first aid; collects and orally reports tactical information using basic communications equipment. *Skill Level 20:* Able to perform the duties required for Skill Level 10; leads mortar squad; supervises crew readiness; trains crew; supervises work details. *Skill Level 30:* Able to perform the duties required for Skill Level 20; leads mortar section; supervises employment and deployment of personnel and equipment; conducts surveys to determine weapons location; prepares efficiency reports on subordinates. *Skill Level 40:* Able to perform the duties required for Skill Level 30; supervises mortar platoon; assists in planning, organizing, training, reporting, and coordinating activities of subordinate sections; supervises receipt, storage, and distribution of ammunition, supplies, equipment, and food. *Skill Level 50:* Able to perform the duties required for Skill Level 40; plans, coordinates, supervises, and participates in activities pertaining to organization, training, and combat operations and intelligence; serves as principal noncommissioned officer in a mechanized infantry headquarters company; supervises the processing of operations and intelligence information at a battalion or higher level; counsels subordinates and reviews efficiency reports.

Recommendation, Skill Level 10

Credit may be granted on the basis of an individualized assessment of the student (3/94).

Recommendation, Skill Level 20

In the lower-division baccalaureate/associate degree category, 3 semester hours in military science and 3 in personnel supervision. (NOTE: This recommendation for skill level 20 is valid for the dates 2/92-2/95 only) (2/92).

Recommendation, Skill Level 30

In the lower-division baccalaureate/associate degree category, 3 semester hours in military science, 3 in personnel supervision, and 2 in records and information management (2/92).

Recommendation, Skill Level 40

In the lower-division baccalaureate/associate degree category, 3 semester hours in military science, 3 in personnel supervision, and 2 in records and information management. In the upper-division baccalaureate category, 3 semester hours for field experience in management and 3 in organizational management (2/92).

Recommendation, Skill Level 50

In the lower-division baccalaureate/associate degree category, 3 semester hours in military science, 3 in personnel supervision, and 2 in records and information management. In the upper-division baccalaureate category, 6 semester hours for field experience in management and 3 in organizational management; if paygrade E-9 has been attained, additional credit as follows: 3 semester hours in management problems, 3 in operations management, and 3 in communication techniques for managers (2/92).

MOS-11H-002

Heavy Antiarmor Weapons Crewman

11H10
11H20
11H30
11H40
11H50

Exhibit Dates: 1/90–1/92.

Career Management Field: 11 (Infantry).

Description

Summary: Leads or serves as a member of heavy antiarmor crew-served vehicle squad, section, or platoon. *Skill Level 10:* Serves as a loader of a special ammunition weapon; fires weapon or drives the vehicle; prepares ammunition; exercises safety precautions and knows first-aid procedures; lays field wire, performs basic communications function and operates communications equipment; performs preventive maintenance on vehicle, armament, and weapon fire control system; assists in clearing or breaching minefields or fortifications. *Skill Level 20:* Able to perform the duties required for Skill Level 10; serves as a vehicle crew commander, directing employment of personnel in field operations; evaluates terrain; selects emplacement locations and assigns target locations; supervises construction of fortifications; interprets maps and prepares field sketches; supervises crew training and various work details. *Skill Level 30:* Able to perform the duties required for Skill Level 20; supervises and leads an antiarmor section consisting of several vehicles; coordinates and supervises the employment of the section; supervises the receipt, storage, and distribution of food and supplies; coordinates firepower and advises on tactical situations; supervises maintenance program of all section vehicles and equipment; trains and instructs replacement personnel. *Skill Level 40:* Able to perform the duties required for Skill Level 30; assists in planning, organizing, and coordinating activities of subordinates; assists in coordinating administrative matters and implementing training programs and communications activities; advises commander on employment of weapons systems. *Skill Level 50:* Able to perform the duties required for Skill Level 40; plans, coordinates, supervises, and participates in activities pertaining to organization, training, combat operations, and intelligence; serves as a principal noncommissioned officer in a heavy antiarmor weapons company; supervises the processing of operations and intelligence information at battalion or higher level.

Recommendation, Skill Level 10

In the vocational certificate category, 3 semester hours in mechanical maintenance. In the lower-division baccalaureate/associate degree category, 1 semester hour in first aid. (NOTE: This recommendation for skill level 10 is valid for the dates 1/90-9/91 only) (11/79).

Recommendation, Skill Level 20

In the vocational certificate category, 3 semester hours in mechanical maintenance. In the lower-division baccalaureate/associate degree category, 1 semester hour in first aid and 3 in map reading (11/79).

Recommendation, Skill Level 30

In the vocational certificate category, 3 semester hours in mechanical maintenance. In the lower-division baccalaureate/associate degree category, 1 semester hour in first aid, 3 in map reading, and 3 in personnel supervision. In the upper-division baccalaureate category, 3 semester hours for field experience in management (11/79).

Recommendation, Skill Level 40

In the vocational certificate category, 3 semester hours in mechanical maintenance. In the lower-division baccalaureate/associate degree category, 1 semester hour in first aid, 3 in map reading, 3 in personnel supervision, and 3 in introduction to management. In the upper-division baccalaureate category, 3 semester hours for field experience in management and 3 in personnel management (11/79).

Recommendation, Skill Level 50

In the vocational certificate category, 3 semester hours in mechanical maintenance. In the lower-division baccalaureate/associate degree category, 1 semester hour in first aid, 3 in map reading, 3 in personnel supervision, and 3 in introduction to management. In the upper-division baccalaureate category, 6 semester hours for field experience in management and 3 in personnel management (5/88).

MOS-11H-003

HEAVY ANTIARMOR WEAPONS CREWMAN

11H10
11H20
11H30
11H40
11H50

Exhibit Dates: 2/92–Present.

Career Management Field: 11 (Infantry).

Description

Summary: Leads or serves as a member of heavy antiarmor crew-served vehicles squad, section, or platoon. *Skill Level 10:* Serves as a loader of a special ammunition weapon; fires weapon or drives the vehicle; prepares ammunition; exercises safety precautions and knows first aid procedures; lays field wire, performs basic communications function and operates communications equipment; performs preventive maintenance on vehicle, armament, and weapon fire control system; assists in clearing or breaching minefields or fortifications. *Skill Level 20:* Able to perform the duties required for Skill Level 10; serves as a vehicle crew commander, directing employment of personnel in field operations; evaluates terrain; selects emplacement location and assigns target locations; supervises construction of fortifications; interprets maps and prepares field sketches; supervises crew training and various work details. *Skill Level 30:* Able to perform the duties required for Skill Level 20; supervises and leads an antiarmor section consisting of several vehicles; coordinates and supervises squad of approximately six persons; supervises the receipt, storage, and distribution of food and supplies; coordinates firepower and advises on tactical situations; supervises maintenance program of all section vehicles and equipment; trains and instructs replacement personnel; prepares efficiency reports on subordinates. *Skill Level 40:* Able to perform the duties required for Skill Level 30; assists in planning, organizing, and coordinating activities of subordinates; assists in coordinating administrative matters and implementing training programs and communications activities; advises commander on employment of weapons systems. *Skill Level 50:* Able to perform the duties required for Skill Level 40; plans, coordinates, supervises, and participates in activities pertaining to organization, training, combat operations, and intelligence; serves as a principal noncommissioned officer in a heavy antiarmor weapons company; supervises the processing of operations and intelligence information at battalion or higher level; counsels subordinates and reviews efficiency reports.

Recommendation, Skill Level 10

Credit may be granted on the basis of an individualized assessment of the student (3/94).

Recommendation, Skill Level 20

In the lower-division baccalaureate/associate degree category, 3 semester hours in military science, 3 in introduction to heavy equipment maintenance, and 3 in personnel supervision. (NOTE: This recommendation for skill level 20 is valid for the dates 2/92-2/95 only) (2/92).

Recommendation, Skill Level 30

In the lower-division baccalaureate/associate degree category, 3 semester hours in military science, 3 in introduction to heavy equipment maintenance, 3 in personnel supervision, and 2 in records and information management (2/92).

Recommendation, Skill Level 40

In the lower-division baccalaureate/associate degree category, 3 semester hours in military science, 3 in introduction to heavy equipment maintenance, 3 in personnel supervision, and 2 in records and information management. In the upper-division baccalaureate category, 3 semester hours for field experience in management and 3 in organizational management (2/92).

Recommendation, Skill Level 50

In the lower-division baccalaureate/associate degree category, 3 semester hours in military science, 3 in introduction to heavy equipment maintenance, 3 in personnel supervision, and 2 in records and information management. In the upper-division baccalaureate category, 6 semester hours for field experience in management and 3 in organizational management; if pay-grade E-9 has been attained, additional credit as follows: 3 semester hours in management problems, 3 in operations management, and 3 in communication techniques for managers (2/92).

MOS-11M-002

FIGHTING VEHICLE INFANTRYMAN

11M10
11M20
11M30
11M40
11M50

Exhibit Dates: 1/90–Present.

Career Management Field: 11 (Infantry).

Description

Summary: Leads, supervises, manages, and serves as a member of a fighting vehicle unit or activity employing vehicular and dismounted weapons in combat operations. *Skill Level 10:* Operates and conducts preventive maintenance checks and services on the Infantry Fighting Vehicle (IFV) and components; performs operator maintenance on automotive components; selects routes and firing positions; provides a steady platform for stabilized weapons fire; assists in target detection, identification, and round sensing; assists in refueling and vehicle recovery operation; is familiar with turret recovery operation, loading, and controls; presents oral reports and operates communications equipment; collects and reports technical information; assists in construction of fortifications; carries and prepares ammunition; administers first aid and applies field sanitation methods; maintains assigned weapons and individual equipment. *Skill Level 20:* Able to perform the duties required for Skill Level 10; conducts preventive maintenance checks and services on IFV turret and weapon systems; performs operator maintenance on turret; assists in supervising automotive maintenance; loads and fires weapons; detects, acquires, identifies, and engages targets; responds to fire commands;

determines range to targets; employs battle sight gunnery techniques and adjusts direct fire; prepares range cards; sights and zeroes weapons; leads infantry dismount team in combat operations and processes intelligence and operations data; receives and implements combat orders; directs deployment of personnel in offensive, defensive, and retrograde combat operations; evaluates terrain, selects weapon placement sites, and assigns target engagement areas and fields of fire; utilizes IFV capabilities to maintain night surveillance; requests, observes, and adjusts direct supporting fires; supervises construction of hasty fortifications; supervises receipt, storage, issue, and loading of ammunition; records operational information on maps; reads and interprets maps and aerial photos; trains subordinates. *Skill Level 30:* Able to perform the duties required for Skill Level 20; receives and issues orders; controls vehicular and dismounted weapons fire in mounted and dismounted operations; engages targets from commander's position; coordinates action of squad with adjacent and supporting elements; coordinates organic and supporting firepower; leads and participates in patrols; insures collection and proper reporting of intelligence data to unit and responsible staff sections; supervises and manages all squad preventive and operator maintenance activities. *Skill Level 40:* Able to perform the duties required for Skill Level 30; performs duties as vehicle element or dismount element leader; assists platoon leader in controlling infantry fighting platoon in mounted or dismounted operations; acts as platoon leader in his absence; processes operations and intelligence information; assists in planning, organizing, directing, supervising, training, coordinating, and reporting activities of subordinate squads; supervises receipt, storage, and distribution of ammunition, supplies, equipment, and food to subordinate elements; supervises platoon preventive and operation maintenance activities of IFV; collects intelligence information to support combat operations; supervises and trains personnel in fighting vehicle operations, maintenance, and intelligence activities; assists in dissemination of intelligence information to unit and staff sections; assists in coordination and implementation of combat operations, training programs, and administrative and communication procedures; assists in production and administration of staff journals, files, records, and reports; assists in organization and operation of the tactical operations center. *Skill Level 50:* Able to perform the duties required for Skill Level 40; plans, coordinates, supervises, and participates in activities pertaining to organization, training, and combat operations and intelligence; serves as principal noncommissioned officer in a mechanized infantry company; supervises the processing of operations and intelligence information at battalion or higher level.

Recommendation, Skill Level 10

In the lower-division baccalaureate/associate degree category, 1 semester hour in first aid, 1 in vehicle maintenance, and 1 in map reading. (NOTE: This recommendation for skill level 10 is valid for the dates 1/90-9/91 only) (9/85).

Recommendation, Skill Level 20

In the lower-division baccalaureate/associate degree category, 1 semester hour in first aid, 2 in vehicle maintenance, 2 in map reading, and 2 in personnel supervision. (NOTE: This recommendation for skill level 20 is valid for the dates 1/90-2/95 only) (9/85).

Recommendation, Skill Level 30

In the lower-division baccalaureate/associate degree category, 1 semester hour in first aid, 2 in vehicle maintenance, 2 in map reading, and 3 in personnel supervision. In the upper-division baccalaureate category, 3 semester hours for field experience in management (9/85).

Recommendation, Skill Level 40

In the lower-division baccalaureate/associate degree category, 1 semester hour in first aid, 2 in vehicle maintenance, 2 in map reading, 3 in personnel supervision, and 3 in maintenance management. In the upper-division baccalaureate category, 6 semester hours for field experience in management (9/85).

Recommendation, Skill Level 50

In the lower-division baccalaureate/associate degree category, 1 semester hour in first aid, 2 in vehicle maintenance, 2 in map reading, 3 in personnel supervision, and 3 in maintenance management. In the upper-division baccalaureate category, 6 semester hours for field experience in management and 3 in personnel management (5/88).

MOS-11Z-002

Infantry Senior Sergeant

11Z50

Exhibit Dates: 1/90–Present.

Career Management Field: 11 (Infantry).

Description

Serves as the principal operations noncommissioned officer in an infantry brigade or comparable position in higher headquarters; may serve in a rear tactical operations center; plans, coordinates, and supervises organizational, training, and combat operations. NOTE: May have progressed to 11Z50 from 11B50, Infantryman, 11C50, Indirect Fire Infantryman, 11H50, Heavy Antiarmor Weapons Infantryman, or 11M50, Fighting Vehicle Infantryman.

Recommendation

In the lower-division baccalaureate/associate degree category, 3 semester hours in military science, 3 in personnel supervision, and 2 in records and information management. In the upper-division baccalaureate category, 3 semester hours for field experience in management and 3 in organizational management; if paygrade E-9 has been attained, additional credit as follows: 3 semester hours in management problems, 3 in operations management, and 3 in communication techniques for managers (2/92).

MOS-12B-003

Combat Engineer

12B10
12B20
12B30
12B40

Exhibit Dates: 1/90–5/91.

Career Management Field: 12 (Combat Engineering).

Description

Summary: Engages in vertical, road, and airfield construction and rigging, bridging, and demolition activities. *Skill Level 10:* Assists combat engineers and bridge and powered-bridge specialists in performance of duties; uses hand tools and engineering tools. *Skill Level 20:* Able to perform the duties required for Skill Level 10; assists with excavation, earth moving, rigging, and concrete work. *Skill Level 30:* Able to perform the duties required for Skill Level 20; acts as demolition expert. *Skill Level 40:* Able to perform the duties required for Skill Level 30; supervises teams of construction and demolition personnel.

Recommendation, Skill Level 10

In the vocational certificate category, 3 semester hours in hand tool operation (11/77).

Recommendation, Skill Level 20

In the vocational certificate category, 6 semester hours in construction equipment operation and 3 in hand tool operation (11/77).

Recommendation, Skill Level 30

In the vocational certificate category, 9 semester hours in demolition operations, 6 in construction equipment operation, and 3 in hand tool operation (11/77).

Recommendation, Skill Level 40

In the vocational certificate category, 15 semester hours in construction equipment operation, 9 in demolition operations, 3 in hand tool operation, 2 in geography (map interpretation), 3 in construction methods, 1 in blueprint reading, 2 in communication skills, and 3 in construction supervision. In the lower-division baccalaureate/associate degree category, 3 semester hours for field experience in management and additional credit in administration and in construction on the basis of institutional evaluation (11/77).

MOS-12B-004

Combat Engineer

12B10
12B20
12B30
12B40

Exhibit Dates: 6/91–9/04. (Effective 9/04, MOS 12B was discontinued, and its duties were incorporated into MOS 21B.)

Career Management Field: 12 (Combat Engineering).

Description

Summary: Engages in vertical, road, bridge, and airfield construction and rigging, bridging, and demolition activities. *Skill Level 10:* Assists combat engineers and bridge and powered-bridge specialists in performance of duties; reads, interprets, and plots maps, overlays, and photos; assists with tactical operations; uses hand tools and engineering tools. *Skill Level 20:* Able to perform the duties required for Skill Level 10; assists with excavation, earth moving, rigging, and concrete work; supervises and instructs subordinates; operates excavation and earth-moving heavy equipment; performs rigging, concrete and masonry work. *Skill Level 30:* Able to perform the duties required for Skill Level 20; acts as a demolition expert; with the knowledge of location and level of charge, coordinates work teams; writes orders; cross-checks material requirements. *Skill Level 40:* Able to perform the duties required for Skill Level 30; supervises teams of construction and demolition personnel; prepares construction procedure schedule; serves as liaison with supported units; plans construction operations.

Recommendation, Skill Level 10

In the vocational certificate category, 3 semester hours in construction site preparation.

In the lower-division baccalaureate/associate degree category, 3 semester hours in military science and 3 in construction materials and methods. (NOTE: This recommendation for skill level 10 is valid for the dates 6/91-9/91 only) (7/91).

Recommendation, Skill Level 20

In the vocational certificate category, 3 semester hours in construction site preparation. In the lower-division baccalaureate/associate degree category, 3 semester hours in military science, 3 in construction materials and methods, and 3 in personnel supervision. (NOTE: This recommendation for skill level 20 is valid for the dates 6/91-2/95 only) (7/91).

Recommendation, Skill Level 30

In the vocational certificate category, 3 semester hours in construction site preparation. In the lower-division baccalaureate/associate degree category, 3 semester hours in military science, 3 in construction materials and methods, 3 in personnel supervision, 3 in maintenance management, and 2 in records and information management (7/91).

Recommendation, Skill Level 40

In the vocational certificate category, 3 semester hours in construction site preparation. In the lower-division baccalaureate/associate degree category, 3 semester hours in military science, 3 in construction materials and methods, 3 in personnel supervision, 3 in maintenance management, and 2 in records and information management. In the upper-division baccalaureate category, 3 semester hours in organizational management and 3 for field experience in management (7/91).

MOS-12C-002

Bridge Crewman
12C10
12C20
12C30
12C40

Exhibit Dates: 1/90–5/91.

Career Management Field: 12 (Combat Engineering).

Description

Summary: Provides engineer bridging for streams and dry gaps. *Skill Level 10:* Assists in the assembly, erection, and disassembly of floating and fixed prefabricated bridges; uses and maintains hand tools; assists in rigging; prepares and operates assault boats. *Skill Level 20:* Able to perform the duties required for Skill Level 10; supervises rigging; performs as demolition expert. *Skill Level 30:* Able to perform the duties required for Skill Level 20; operates power boats; interprets maps; uses and maintains construction hand tools. *Skill Level 40:* Able to perform the duties required for Skill Level 30; supervises bridge construction and repair; determines bridge construction sites; provides work and material estimates; supervises use of heavy construction equipment; prepares technical reports; supervises a minimum of ten persons.

Recommendation, Skill Level 10

In the vocational certificate category, 3 semester hours in construction hand tool operation and maintenance (11/77).

Recommendation, Skill Level 20

In the vocational certificate category, 9 semester hours in bridge construction and 3 in construction hand tool operation and maintenance (11/77).

Recommendation, Skill Level 30

In the vocational certificate category, 12 semester hours in bridge construction, 4 in construction hand tool operation and maintenance, and 4 in power boat operation. In the lower-division baccalaureate/associate degree category, 6 semester hours in construction methods and 1 in blueprint reading (11/77).

Recommendation, Skill Level 40

In the vocational certificate category, 12 semester hours in bridge construction, 4 in construction hand tool operation and maintenance, 4 in power boat operation, 4 in construction methods, 3 in bridge maintenance, and 3 in construction supervision. In the lower-division baccalaureate/associate degree category, 6 semester hours in construction methods, 3 in construction supervision, 1 in blueprint reading, 2 in geography (map interpretation), 2 in technical writing, 2 in communication skills, and 1 in applied physics (11/77).

MOS-12C-003

Bridge Crewman
12C10
12C20
12C30
12C40

Exhibit Dates: 6/91–9/04. (Effective 9/04, MOS 12C was discontinued, and its duties were incorporated into MOS 21C.)

Career Management Field: 12 (Combat Engineering).

Description

Summary: Provides engineer bridging for streams and dry gaps. *Skill Level 10:* Assists in the assembly, erection, and disassembly of floating and fixed prefabricated bridges; uses and maintains hand tools; assists in rigging; prepares and operates assault boats; reads, interprets, and plots maps, overlays, and photos; assists with tactical operations. *Skill Level 20:* Able to perform the duties required for Skill Level 10; supervises and instructs subordinates; supervises rigging; performs as demolition expert; performs maintenance on specialized equipment needed to install bridges. *Skill Level 30:* Able to perform the duties required for Skill Level 20; operates power boats; interprets maps; evaluates existing bridges for load tolerances; calculates equipment requirements; inspects personnel, construction, and utilization of bridging and rafting equipment. *Skill Level 40:* Able to perform the duties required for Skill Level 30; supervises bridge construction and repair; determines bridge construction sites; provides work and material estimates; supervises the use of heavy construction equipment; prepares technical reports; supervises a minimum of ten persons.

Recommendation, Skill Level 10

In the lower-division baccalaureate/associate degree category, 3 semester hours in military science. (NOTE: This recommendation for skill level 10 is valid for the dates 6/91-9/91 only) (7/91).

Recommendation, Skill Level 20

In the vocational certificate category, 3 semester hours in construction site preparation. In the lower-division baccalaureate/associate degree category, 3 semester hours in military science, 3 in personnel supervision, and 2 in records and information management. (NOTE: This recommendation for skill level 20 is valid for the dates 6/91-2/95 only) (7/91).

Recommendation, Skill Level 30

In the vocational certificate category, 3 semester hours in construction site preparation. In the lower-division baccalaureate/associate degree category, 3 semester hours in military science, 3 in personnel supervision, 2 in records and information management, and 3 in bridge statics (civil engineering) (7/91).

Recommendation, Skill Level 40

In the vocational certificate category, 3 semester hours in construction site preparation. In the lower-division baccalaureate/associate degree category, 3 semester hours in military science, 3 in personnel supervision, 2 in records and information management, and 3 in bridge statics (civil engineering). In the upper-division baccalaureate category, 3 semester hours in organizational management and 3 for field experience in management (7/91).

MOS-12F-002

Engineer Tracked Vehicle Crewman
12F10
12F20
12F30
12F40

Exhibit Dates: 1/90–12/96. Effective 12/96, MOS 12F was discontinued and its duties were transferred to MOS 12B.

Career Management Field: 12 (Combat Engineer).

Description

Summary: Leads, supervises, or serves as member of combat engineer vehicle-launched bridge crew or operates engineer armored personnel carrier. *Skill Level 10:* Operates combat engineer vehicles over varied terrain and roads; reacts to commands and navigates over land; operates special equipment such as bridge launchers, lifting devices, power winches, and bulldozers; loads, fires, and unloads guns; installs and dismantles radios and antennas; services and adjusts vehicle components; participates in bridging and demolition operations. *Skill Level 20:* Able to perform the duties required for Skill Level 10; serves as squad leader; supervises, instructs, and provides technical guidance to subordinates. *Skill Level 30:* Able to perform the duties required for Skill Level 20; supervises crew and section personnel; directs overall operations and training exercises; conducts battle drills; identifies targets, and selects placement for demolitions; coordinates maintenance requirements; manages equipment and personnel; conducts reconnaissance and reports intelligence information. *Skill Level 40:* Able to perform the duties required for Skill Level 30; supervises platoon and section personnel; assists in planning, organizing, directing, supervising, and managing at the platoon, section, or team level; directs assignment of vehicles and personnel.

Recommendation, Skill Level 10

In the vocational certificate category, 2 semester hours in heavy equipment operation. In the lower-division baccalaureate/associate degree category, 1 semester hour in first aid, 1 in vehicle maintenance, and 1 in map reading. (NOTE: This recommendation for skill level 10 is valid for the dates 1/90-9/91 only) (9/85).

Recommendation, Skill Level 20

In the vocational certificate category, 3 semester hours in heavy equipment operation. In the lower-division baccalaureate/associate degree category, 1 semester hour in first aid, 2 in vehicle maintenance, 2 in map reading, and 2 in personnel supervision. (NOTE: This recommendation for skill level 20 is valid for the dates 1/90-2/95 only) (9/85).

Recommendation, Skill Level 30

In the vocational certificate category, 3 semester hours in heavy equipment operation. In the lower-division baccalaureate/associate degree category, 1 semester hour in first aid, 2 in vehicle maintenance, 2 in map reading, and 3 in personnel supervision. In the upper-division baccalaureate category, 3 semester hours for field experience in management (9/85).

Recommendation, Skill Level 40

In the vocational certificate category, 3 semester hours in heavy equipment operation. In the lower-division baccalaureate/associate degree category, 1 semester hour in first aid, 2 in vehicle maintenance, 2 in map reading, 3 in personnel supervision, and 3 in maintenance management. In the upper-division baccalaureate category, 6 semester hours for field experience in management (9/85).

MOS-12Z-002

COMBAT ENGINEERING SENIOR SERGEANT
12Z50

Exhibit Dates: 1/90–5/91.

Career Management Field: 12 (Combat Engineer).

Description

Serves as principal noncommissioned officer of a combat engineer company, bridge company, or engineer staff section; serves as a bridge inspector, operations, or intelligence noncommissioned officer; may advise higher headquarters, supported units, allied forces, Army Reserve units, and Army National Guard units; may serve as an instructor in a formal military school; capable of managing personnel and activities in a civilian engineering/construction firm; assists in planning, supervising, and coordinating all unit activities; advises commander on personnel equipment and technical matters; provides subordinate supervisors with instructions; plans and implements training programs; inspects work progress and enforces job specifications and safety standards; supervises maintenance of functional and project files; prepares correspondence, plans, orders, construction schedules, and reports; maintains equipment records; maintains technical publications. NOTE: May have progressed to 12Z50 from Skill Level 40 of any MOS in the combat engineering career management field (12).

Recommendation

In the upper-division baccalaureate category, 3 semester hours in elements of construction management, 3 in management problems, 3 in records management, 3 in quality control, 3 in personnel supervision, 3 in supervision and leadership, and 6 for field experience in management. NOTE: Has a prerequisite knowledge of engineering equivalent to an associate degree (30-40 semester hours depending on general education requirements) in engineering technology (11/77).

MOS-12Z-003

COMBAT ENGINEERING SENIOR SERGEANT
12Z50

Exhibit Dates: 6/91–9/04. (Effective 9/04, MOS12Z was discontinued, and its duties were incorporated into MOS 21Z.)

Career Management Field: 12 (Combat Engineer).

Description

Serves as principal noncommissioned officer of a combat engineer company, bridge company, or engineer staff section; serves as a bridge inspector, operations, or intelligence noncommissioned officer; may advise headquarters, supported units, allied forces, Army Reserve units, and Army National Guard units; may serve as an instructor in a formal military school; capable of managing personnel and activities in a civilian engineering/construction firm; assists in planning, supervising, and coordinating all unit activities; advises commander on personnel equipment and technical matters; provides subordinate supervisors with instructions; plans and implements training programs; inspects work progress and enforces job specifications and safety standards; supervises maintenance of functional and project files; prepares correspondence, plans, orders, construction schedules, and reports; maintains equipment records; maintains technical publications. NOTE: May have progressed to 12Z50 from Skill Level 40 of MOS 12B, Combat Engineer, 12C, Bridge Crewmember, or 12F, Engineer Tracked Vehicle Crewman.

Recommendation

In the lower-division baccalaureate/associate degree category, 3 semester hours in military science, 3 in personnel supervision, 3 in maintenance management, 3 in technical writing, and 3 in records and information management. In the upper-division baccalaureate category, 6 semester hours for field experience in management; if individual has attained paygrade E-9, additional credit as follows: 3 semester hours in management problems, 3 in operations management, and 3 in communication techniques for managers (7/91).

MOS-13B-003

CANNON CREWMEMBER
13B10
13B20
13B30
13B40

Exhibit Dates: 1/90–4/99.

Career Management Field: 13 (Field Artillery).

Description

Summary: Supervises or serves as a member of a field artillery cannon unit; places and fires cannon, maintains transport vehicles and weapons, and operates communications equipment. *Skill Level 10:* Serves as a vehicle driver, ammunition specialist, cannoneer, maintenance technician, and communications equipment operator; determines azimuth; reads maps; navigates; operates radio set; services hydraulic, pneumatic, and electromechanical systems. *Skill Level 20:* Able to perform the duties required for Skill Level 10; assists section chief; uses aiming devices; performs boresight; supervises preventive maintenance, ammunition supply, and equipment transport. *Skill Level 30:* Able to perform the duties required for Skill Level 20; serves as section chief; directs defense of section; instructs and supervises seven to twelve persons; supervises equipment maintenance, emplacement construction, safety procedures, and ammunition handling and distribution. *Skill Level 40:* Able to perform the duties required for Skill Level 30; supervises 12 to 60 persons; selects sites for weapon emplacement; supervises firing; supervises and conducts training of section personnel; lays fire direction; conducts weapon checks; supervises ammunition train and movement of weapons; prepares technical reports.

Recommendation, Skill Level 10

In the lower-division baccalaureate/associate degree category, 1 semester hour in communications system operation, 1 in hydraulic and electromechanical systems troubleshooting and maintenance, 1 in map reading, and 1 in applied mathematics. (NOTE: This recommendation for skill level 10 is valid for the dates 1/90-9/91 only) (6/89).

Recommendation, Skill Level 20

In the lower-division baccalaureate/associate degree category, 2 semester hours in communications system operation, 2 in hydraulic and electromechanical systems troubleshooting and maintenance, 2 in map reading, and 2 in applied mathematics. (NOTE: This recommendation for skill level 20 is valid for the dates 1/90-2/95 only) (6/89).

Recommendation, Skill Level 30

In the lower-division baccalaureate/associate degree category, 2 semester hours in communications system operation, 2 in hydraulic and electromechanical systems troubleshooting and maintenance, 2 in map reading, 2 in applied mathematics, 3 in principles of supervision, 3 in communication skills, and 1 in technical report writing (6/89).

Recommendation, Skill Level 40

In the lower-division baccalaureate/associate degree category, 2 semester hours in communications system operation, 2 in hydraulic and electromechanical systems troubleshooting and maintenance, 2 in map reading, 2 in applied mathematics, 3 in principles of supervision, 3 in communication skills, and 3 in technical report writing. In the upper-division baccalaureate category, 3 semester hours for field experience in management and 3 in personnel management (6/89).

MOS-13B-004

CANNON CREWMEMBER
13B10
13B20
13B30
13B40

Exhibit Dates: 5/99–Present.

Career Management Field: 13 (Field Artillery).

Description

Summary: Supervises or serves as a member of a field artillery cannon unit; places and fires cannon; maintains transport vehicles and weapons; uses and generates computer data; operates voice and digital communications equipment. *Skill Level 10:* Serves as a vehicle driver, ammunition specialist, cannoneer, maintenance technician, and communications equipment operator; determines azimuth; reads maps; navigates; operates radio set; services hydraulic, pneumatic, and electromechanical systems. *Skill Level 20:* Able to perform the duties required for Skill Level 10; assists section chief; uses aiming devices; performs boresight; supervises preventive maintenance, ammunition supply, and equipment transport. *Skill Level 30:* Able to perform the duties required for Skill Level 20; serves as section chief; directs defense of section; instructs and supervises seven to twelve persons; supervises equipment maintenance, emplacement construction, safety procedures, and ammunition handling and distribution. *Skill Level 40:* Able to perform the duties required for Skill Level 30; supervises 12 to 60 persons; selects sites for weapon emplacement; supervises firing; supervises and conducts training of section personnel; lays fire direction; conducts weapon checks; supervises ammunition train and movement of weapons; prepares technical reports.

Recommendation, Skill Level 10

Credit may be granted on the basis of an individualized assessment of the student (5/99).

Recommendation, Skill Level 20

Credit may be granted on the basis of an individualized assessment of the student (5/99).

Recommendation, Skill Level 30

In the lower-division baccalaureate/associate degree category, 2 semester hours in communications systems operation, 1 in computer familiarization, 2 in hydraulic and electromechanical systems troubleshooting and maintenance, 2 in map reading and land navigation, 2 in applied mathematics, 3 in principles of supervision, 3 in oral communication skills, and 1 in technical report writing (5/99).

Recommendation, Skill Level 40

In the lower-division baccalaureate/associate degree category, 2 semester hours in communications systems operation, 1 in computer familiarization, 2 in hydraulic and electromechanical systems troubleshooting and maintenance, 2 in map reading and land navigation, 2 in applied mathematics, 3 in principles of supervision, 3 in oral communication skills, and 3 in technical report writing. In the upper-division baccalaureate category, 3 semester hours for field experience in management and 3 in personnel management (5/99).

MOS-13C-002

TACFIRE Operations Specialist
13C10
13C20
13C30
13C40

Exhibit Dates: 1/90–4/99.

Career Management Field: 13 (Field Artillery).

Description

Summary: Supervises or serves as a member of an activity operating tactical fire direction (TACFIRE) equipment in a field artillery cannon battalion or higher unit. *Skill Level 10:* Operates equipment in conjunction with or in support of TACFIRE computer; assists in preparation and operation of the computer; assists in preparing the computer center for operation and shutdown; performs cabling and equipment installation and removal; operates computer center equipment during normal, alternate mode, and emergency operations; recognizes partial/complete computer center failure and performs alternate mode reconfiguration procedures; distributes printed data generated by TACFIRE computer; performs computer center preventive maintenance and performance checks; operates and performs operator maintenance on power generation equipment; prepares TACFIRE Remote Terminal (TRT) for operation; performs TRT cabling and installation/removal procedures; operates TRT to input and retrieve computer control information, tactical data base, artillery target intelligence, fire planning, survey, tactical, and technical fire control operations data; recognizes partial/complete TRT failure and performs alternate mode reconfiguration procedures; performs TRT troubleshooting procedures and operator maintenance; performs TRT shutdown procedures; lays field wire; installs field telephones and vehicle radio remote control equipment; erects/dismantles antennas; performs operator checks and maintenance on communications equipment; transmits and receives using communications equipment; monitors radio broadcasts. *Skill Level 20:* Able to perform the duties required for Skill Level 10; provides technical guidance to subordinates; inputs and retrieves computer center control information necessary to establish communications with subscribers and initiates computer center operations; performs maintenance on TACFIRE computer and remote devices; maintains required tools, test equipment, and appropriate spare parts to accomplish organizational maintenance duties; supervision and management experience may vary. *Skill Level 30:* Able to perform the duties required for Skill Level 20; operates TACFIRE computer in field artillery cannon battalion fire direction center; operates TACFIRE computer to input and retrieve computer initialization control and tactical data base information; processes tactical and technical fire control data, artillery target intelligence, non-nuclear fire planning data survey information and routines, and meteorological data; performs operator maintenance and checks on computer; supervises subordinates. *Skill Level 40:* Able to perform the duties required for Skill Level 30; assists in supervision of battalion or higher operations section; operates TACFIRE computer in field artillery brigade or higher unit; supervises TRT operators; prepares and distributes maps, operational information, and training materials; present briefings on current operations and situations; compiles information and prepares reports; ensures compliance with security precautions and regulations; supervises march order and emplacement of fire direction center equipment; operates TACFIRE computer to input and retrieve computer initialization, control, and tactical data base information. NOTE: May have progressed to 13C40 from 13C30 or 13E30 (Cannon Fire Direction Specialist).

Recommendation, Skill Level 10

Credit is not recommended because of the limited and specialized nature of the skills, competencies, and knowledge (6/89).

Recommendation, Skill Level 20

In the lower-division baccalaureate/associate degree category, if served in a supervisory position, 3 semester hours in principles of supervision and 3 in maintenance management. (NOTE: This recommendation for skill level 20 is valid for the dates 1/90-2/95 only) (6/89).

Recommendation, Skill Level 30

In the lower-division baccalaureate/associate degree category, 3 semester hours in principles of supervision, 3 in maintenance management, and 3 in communication skills (oral and written) (6/89).

Recommendation, Skill Level 40

In the lower-division baccalaureate/associate degree category, 3 semester hours in principles of supervision, 3 in maintenance management, and 3 in communication skills (oral and written). In the upper-division baccalaureate category, 3 semester hours for field experience in management and 3 in personnel management (6/89).

MOS-13C-003

TACFIRE Operations Specialist
13C10
13C20
13C30
13C40

Exhibit Dates: 5/99–Present.

Career Management Field: 13 (Field Artillery).

Description

Summary: Supervises or serves as a member of an activity operating tactical fire direction (TACFIRE) equipment in a field artillery cannon battalion or higher unit. *Skill Level 10:* Operates equipment in conjunction with or in support of TACFIRE computer; assists in preparation and operation of the computer; assists in preparing the computer center for operation and shutdown; performs cabling and equipment installation and removal; operates computer center equipment during normal, alternate mode, and emergency operations; recognizes partial/complete computer center failure and performs alternate mode reconfiguration procedures; distributes printed data generated by TACFIRE computer; performs computer center preventive maintenance and performance checks; operates and performs operator maintenance on power generation equipment; prepares TACFIRE Remote Terminal (TRT) for operation; performs TRT cabling and installation/removal procedures; operates TRT to input and retrieve computer control information, tactical data base, artillery target intelligence, fire planning, survey, tactical, and technical fire control operations data; recognizes partial/complete TRT failure and performs alternate mode reconfiguration procedures; performs TRT troubleshooting procedures and operator maintenance; performs TRT shutdown procedures; lays field wire; installs field telephones and vehicle radio remote control equipment; erects/dismantles antennas; performs operator checks and maintenance on communications equipment; transmits and receives using communications equipment; monitors radio broadcasts; uses voice and digital communication; uses and generates computer data. *Skill Level 20:* Able to perform the duties required for Skill Level 10; provides technical guidance to subordinates; inputs and retrieves computer center control information necessary to establish communications with subscribers and initiates computer center operations; performs maintenance on TACFIRE computer and remote devices; maintains required tools, test equipment, and appropriate spare parts to accomplish organizational maintenance duties. Supervision and management experience may vary. *Skill Level 30:* Able to perform the duties required for Skill Level 20; operates TACFIRE computer in field artillery cannon battalion fire direction center; operates TACFIRE computer to input and retrieve computer initialization control and tactical data base information; processes tactical and technical fire control data, artillery target intelligence, non-nuclear fire planning data survey information and routines, and meteorological data; performs operator maintenance and checks on computer; supervises subordinates. *Skill Level 40:* Able to perform the duties required for Skill Level 30; assists in supervision of battalion or higher operations section; operates TACFIRE computer in field artillery brigade or higher unit; supervises TRT operators; prepares and distributes maps, operational information, and training materials; presents briefings on current operations and situations; compiles information and prepares reports; ensures compliance with security precautions and regulations; supervises march order and emplacement of fire direction center equipment. NOTE: May have progressed to 13C40 from 13C30 or 13E30 (Cannon Fire Direction Specialist).

Recommendation, Skill Level 10

Credit may be granted on the basis of an individualized assessment of the student (5/99).

Recommendation, Skill Level 20

Credit may be granted on the basis of an individualized assessment of the student (5/99).

Recommendation, Skill Level 30

In the lower-division baccalaureate/associate degree category, 2 semester hours in applied mathematics, 2 in computer familiarization, 2 in communications systems operation, 2 in map reading and land navigation, 2 in technical report writing, 3 in principles of supervision, 3 in maintenance management, 3 in oral communication skills, and 1 in safety and hazardous material training (5/99).

Recommendation, Skill Level 40

In the lower-division baccalaureate/associate degree category, 2 semester hours in applied mathematics, 2 in computer familiarization, 2 in communications systems operation, 2 in map reading and land navigation, 2 in technical report writing, 3 in principles of supervision, 3 in maintenance management, 3 in oral communication skills, and 1 in safety and hazardous material training. In the upper-division baccalaureate category, 3 semester hours for field experience in management and 3 in personnel management (5/99).

MOS-13D-001

Field Artillery Automated Tactical Data Systems Specialist

((Field Artillery Tactical Data Systems Specialist))

13D10

13D20

13D30

13D40

Exhibit Dates: Pending Evaluation. 2/98–Present.

MOS-13E-004

Cannon Fire Direction Specialist

13E10

13E20

13E30

13E40

Exhibit Dates: 1/90–2/95.

Career Management Field: 13 (Field Artillery).

Description

Summary: Leads, supervises, or serves as a member of a field artillery cannon fire direction center (FDC) or of an operations section. *Skill Level 10:* Uses graphs, tables, charts, and maps for manual computation of firing locations, angles, altitude corrections, and displacements; constructs firing charts; computes corrections for meteorological, registration, and muzzle velocity; uses computer to make firing data computations; knows and can use either the Battery Computer System (BCS) or the Backup Computer System (BUCS) (military version of Hewlitt-Packard 71B (16K) computer); installs and operates field telephones, digital message system, and FM radio transmitter/receiver; knows and uses mathematics concepts (including some geometry and trigonometric functions) for manual computations; operates and maintains vehicles, section equipment, and generators; maintains situation map, fire support records, reports, and overlays; uses data from standardized tables to calculate data for nonstandard conditions. *Skill Level 20:* Able to perform the duties required for Skill Level 10; computes firing data and transmits to firing units; prepares status charts, target lists, and situation maps; enters database into computer unit; computes and enters meteorological and muzzle velocity corrections; prepares BUCS for operation and computes firing data. *Skill Level 30:* Able to perform the duties required for Skill Level 20; instructs personnel in fire direction techniques and operations; supervises computation of data; insures accuracy and completeness of firing data; evaluates and prepares annual efficiency reports on subordinates. *Skill Level 40:* Able to perform the duties required for Skill Level 30; supervises fire direction operations; assists operations sergeant in supervision of operations section; presents briefings on current operations; prepares training materials; insures compliance with security requirements; supervises and evaluates eight to twelve persons.

Recommendation, Skill Level 10

In the lower-division baccalaureate/associate degree category, 1 semester hour in applied mathematics, 1 in communications system operations, 1 in map reading, and 1 in computer operations. (NOTE: This recommendation for skill level 10 is valid for the dates 1/90-9/91 only. (6/89).

Recommendation, Skill Level 20

In the lower-division baccalaureate/associate degree category, 1 semester hour in applied mathematics, 1 in communications system operations, 1 in map reading, and 2 in computer operations (6/89).

Recommendation, Skill Level 30

In the lower-division baccalaureate/associate degree category, 2 semester hours in applied mathematics, 1 in communications system operations, 2 in map reading, 2 in computer operations, 3 in principles of supervision, and 2 in communication skills (6/89).

Recommendation, Skill Level 40

In the lower-division baccalaureate/associate degree category, 2 semester hours in applied mathematics, 1 in communications system operations, 2 in map reading, 2 in computer operations, 3 in principles of supervision, and 3 in communication skills. In the upper-division baccalaureate category, 3 semester hours for field experience in management and 3 in personnel management (6/89).

MOS-13E-005

Cannon Fire Direction Specialist

13E10

13E20

13E30

13E40

Exhibit Dates: 3/95–4/99.

Career Management Field: 13 (Field Artillery).

Description

Summary: Leads, supervises, or serves as a member of a field artillery cannon fire direction center (FDC) or of an operations section. *Skill Level 10:* Uses graphs, tables, charts, and maps for manual computation of firing locations, angles, altitude corrections, and displacements; constructs firing charts; computes corrections for meteorological, registration, and muzzle velocity; uses computer to make firing data computations; knows and can use either the Battery Computer System (BCS) or the Backup Computer System (BUCS) (military version of Hewlitt-Packard 71B (16K) computer); installs and operates field telephones, digital message system, and FM radio transmitter/receiver; knows and uses mathematics concepts (including some geometry and trigonometric functions) for manual computations; operates and maintains vehicles, section equipment, and generators; maintains situation map, fire support records, reports, and overlays; uses data from standardized tables to calculate data for nonstandard conditions. *Skill Level 20:* Able to perform the duties required for Skill Level 10; computes firing data and transmits to firing units; prepares status charts, target lists, and situation maps; enters database into computer unit; computes and enters meteorological and muzzle velocity corrections; prepares BUCS for operation and computes firing data. *Skill Level 30:* Able to perform the duties required for Skill Level 20; instructs personnel in fire direction techniques and operations; supervises computation of data; insures accuracy and completeness of firing data; evaluates and prepares annual efficiency reports on subordinates. *Skill Level 40:* Able to perform the duties required for Skill Level 30; supervises fire direction operations; assists operations sergeant in supervision of operations section; presents briefings on current operations; prepares training materials; insures compliance with security requirements; supervises and evaluates eight to twelve persons.

Recommendation, Skill Level 10

Credit may be granted on the basis of an individualized assessment of the student (6/89).

Recommendation, Skill Level 20

Credit may be granted on the basis of an individualized assessment of the student (6/89).

Recommendation, Skill Level 30

In the lower-division baccalaureate/associate degree category, 1 semester hour in applied mathematics, 1 in communications system operations, 1 in map reading, and 2 in computer operations, 3 in principles of supervision, and 2 in communication skills (6/89).

Recommendation, Skill Level 40

In the lower-division baccalaureate/associate degree category, 2 semester hours in applied mathematics, 1 in communications system operations, 2 in map reading, 2 in computer operations, 3 in principles of supervision, and 3 in communication skills. In the upper-division baccalaureate category, 3 semester hours for field experience in management and 3 in personnel management (6/89).

MOS-13E-006

Cannon Fire Direction Specialist

13E10
13E20
13E30
13E40

Exhibit Dates: 5/99–Present.

Career Management Field: 13 (Field Artillery).

Description

Summary: Leads, supervises, or serves as a member of a field artillery cannon fire direction center (FDC) or of an operations section. *Skill Level 10:* Uses graphs, tables, charts, and maps for manual computation of firing locations, angles, altitude corrections, and displacements; constructs firing charts; computes corrections for meteorological, registration, and muzzle velocity; uses computer to make firing data computations; can use either the Battery Computer System (BCS) or the Back-up Computer System (BUCS) (military version of Hewlitt-Packard 71B (16K) computer); installs and operates field telephones, digital message system, and FM radio transmitter/receiver; knows and uses mathematics concepts (including some geometry and trigonometric functions) for manual computations; operates and maintains vehicles, section equipment, and generators; maintains situation map, fire support records, reports, and overlays; uses data from standardized tables to calculate data for nonstandard conditions; uses and generates computer data; uses voice and digital communications. *Skill Level 20:* Able to perform the duties required for Skill Level 10; computes firing data and transmits to firing units; prepares status charts, target lists, and situation maps; enters data base into computer unit; computes and enters meteorological and muzzle velocity corrections; prepares BUCS for operation and computes firing data. *Skill Level 30:* Able to perform the duties required for Skill Level 20; instructs personnel in fire direction techniques and operations; supervise computation of data; ensure accuracy and completeness of firing data; evaluates and prepares annual efficiency reports on subordinates. *Skill Level 40:* Able to perform the duties required for Skill Level 30; supervises fire direction operations; assists operations sergeant in supervision of operations section; presents briefings on current operations; prepares training materials; ensures compliance with security requirements; supervises and evaluates eight to twelve persons.

Recommendation, Skill Level 10

Credit may be granted on the basis of an individualized assessment of the student (5/99).

Recommendation, Skill Level 20

Credit may be granted on the basis of an individualized assessment of the student (5/99).

Recommendation, Skill Level 30

In the lower-division baccalaureate/associate degree category, 2 semester hour in applied mathematics, 2 in communications systems operation, 2 in map reading and land navigation, 2 in computer familiarization, 3 in principles of supervision, 2 in oral communication skills, 2 in technical report writing, and 1 in safety and hazardous material training (5/99).

Recommendation, Skill Level 40

In the lower-division baccalaureate/associate degree category, 2 semester hours in applied mathematics, 2 in communications systems operation, 2 in map reading and land navigation, 2 in computer familiarization, 3 in principles of supervision, 3 in oral communication skills, 3 in technical report writing, and 1 in safety and hazardous material training. In the upper-division baccalaureate category, 3 semester hours for field experience in management and 3 in personnel management (5/99).

MOS-13F-002

Fire Support Specialist

13F10
13F20
13F30
13F40

Exhibit Dates: 1/90–4/99.

Career Management Field: 13 (Field Artillery).

Description

Summary: Serves as a member of a field artillery unit; leads or supervises fire support activities including intelligence, target processing, and observation. *Skill Level 10:* Reads maps and makes map entries; assists in fire support planning; prepares staff journals; prepares target records; encodes and decodes messages; requests munitions; emplaces and uses laser range finders; determines target location; erects antennas and installs field telephones; operates radios and telephones; transports personnel and weapons; maintains vehicles; administers first aid; performs patrol duty; fires defense weapons. *Skill Level 20:* Able to perform the duties required for Skill Level 10; leads and instructs forward observer and combat observation lasing team; assists in training subordinates in fire support; assists in developing fire support plan and tactics; prepares target lists and assists in compiling target information; sets up and maintains observation system; performs map reconnaissance. *Skill Level 30:* Able to perform the duties required for Skill Level 20; supervises five to ten persons; instructs, advises, and evaluates personnel; supervises operator maintenance on equipment; assists targeting noncommissioned officer; processes fire requests; prepares technical reports; integrates fire support plan; performs duties of the Fire Support Officer in his absence. *Skill Level 40:* Able to perform the duties required for Skill Level 30; leads and trains targeting and fire support elements in combat operations; instructs, advises, and evaluates subordinate unit fire support sergeants in fire planning and coordination techniques; writes and coordinates fire support plans; recommends fire support employment; recommends target selection; and assists in tactics and target analysis at the brigade, division, and corps levels.

Recommendation, Skill Level 10

In the lower-division baccalaureate/associate degree category, 2 semester hours in map reading, 1 in communications system operation, and 1 in applied mathematics. (NOTE: This recommendation for skill level 10 is valid for the dates 1/90-9/91 only) (6/89).

Recommendation, Skill Level 20

In the lower-division baccalaureate/associate degree category, 3 semester hours in map reading, 2 in communications system operation, and 2 in applied mathematics. (NOTE: This recommendation for skill level 20 is valid for the dates 1/90-2/95 only) (6/89).

Recommendation, Skill Level 30

In the lower-division baccalaureate/associate degree category, 3 semester hours in map reading, 2 in communications system operation, 2 in applied mathematics, 2 in technical writing, 3 in principles of supervision, and 2 in communication skills (6/89).

Recommendation, Skill Level 40

In the lower-division baccalaureate/associate degree category, 3 semester hours in map reading, 2 in communications system operation, 2 in applied mathematics, 3 in technical writing, 3 in principles of supervision, and 3 in communication skills. In the upper-division baccalaureate category, 3 semester hours for field experience in management and 3 in personnel management (6/89).

MOS-13F-003

Fire Support Specialist

13F10
13F20
13F30
13F40

Exhibit Dates: 5/99–Present.

Career Management Field: 13 (Field Artillery).

Description

Summary: Serves as a member of a field artillery unit; leads or supervises fire support activities, including intelligence, target processing, and observation; uses voice and digital communications systems; uses and generates computer data. *Skill Level 10:* Reads maps and makes map entries; assists in fire support planning; prepares staff journals; prepares target records; encodes and decodes messages; requests munitions; emplaces and uses laser range finders; determines target location; erects antennas and installs field telephones; operates radios and telephones; transports personnel and weapons; maintains vehicles; administers first aid; performs patrol duty; fires defense weapons. *Skill Level 20:* Able to perform the duties required for Skill Level 10; leads and instructs forward observer and combat observation lasing team; assists in training subordinates in fire support; assists in developing fire support plans and tactics; prepares target lists and assists in compiling target information; sets up and maintains observation system; performs map reconnaissance. *Skill Level 30:* Able to perform the duties required for Skill Level 20; supervises five to ten persons; instructs, advises, and evaluates personnel; supervises operator maintenance on equipment; assists targeting noncommissioned officer; processes fire requests; prepares technical reports; integrates fire support plan; in the absence of the Fire Support Officer, duties of that position. *Skill Level*

40: Able to perform the duties required for Skill Level 30; leads and trains targeting and fire support elements in combat operations; instructs, advises, and evaluates subordinate unit fire support sergeants in fire planning and coordination techniques; writes and coordinates fire support plans; recommends fire support employment; recommends target selection; assists in tactics and target analysis at the brigade, division, and corps levels.

Recommendation, Skill Level 10

Credit may be granted on the basis of an individualized assessment of the student (5/99).

Recommendation, Skill Level 20

Credit may be granted on the basis of an individualized assessment of the student (5/99).

Recommendation, Skill Level 30

In the lower-division baccalaureate/associate degree category, 2 semester hours in map reading and land navigation, 2 in communications systems operation, 2 in applied mathematics, 2 in computer familiarization, 1 in safety and hazardous materials training, 3 in principles of supervision, 2 in technical writing, and 2 in oral communication skills (5/99).

Recommendation, Skill Level 40

In the lower-division baccalaureate/associate degree category, 2 semester hours in map reading and land navigation, 2 in communications systems operation, 2 in applied mathematics, 3 in computer familiarization, 1 in safety and hazardous material training, 3 in technical writing, 3 in principles of supervision, and 3 in oral communication skills. In the upper-division baccalaureate category, 3 semester hours for field experience in management and 3 in personnel management (5/99).

MOS-13M-002

Multiple Launch Rocket System Crewmember
13M10
13M20
13M30
13M40

Exhibit Dates: 1/90–2/95.

Career Management Field: 13 (Field Artillery).

Description

Summary: Supervises or serves in a field artillery unit on the multiple launch rocket system; operates and maintains the launcher, transport vehicle, trailer, and other support equipment. *Skill Level 10:* Participates in launch maintenance and ammunition resupply activities; drives transport vehicles; participates in launcher operations; operates radio; maintains computer and performs system checks. *Skill Level 20:* Able to perform the duties required for Skill Level 10; assists the section chief; trains subordinate personnel; operates fire control panel; operates launcher; performs organizational maintenance; serves as firing platoon reconnaissance sergeant. *Skill Level 30:* Able to perform the duties required for Skill Level 20; serves as section chief supervising seven to twelve persons; instructs in launch procedures; supervises preventive maintenance on launch vehicles; supervises fire control panel maintenance; writes reports, and assures safety procedure compliance. *Skill Level 40:* Able to perform the duties required for Skill Level 30; supervises 12 to 60 persons; supervises all section maintenance; writes technical and personnel reports; supervises and performs training of personnel, and assures safety procedure compliance.

Recommendation, Skill Level 10

In the lower-division baccalaureate/associate degree category, 1 semester hour in communications system operations, 1 in electromechanical systems troubleshooting and maintenance, and 1 in electronic and computer systems troubleshooting and maintenance. (NOTE: This recommendation for skill level 10 is valid for the dates 1/90-9/91 only) (6/89).

Recommendation, Skill Level 20

In the lower-division baccalaureate/associate degree category, 2 semester hours in communications system operations, 2 in electromechanical systems troubleshooting and maintenance, 2 in electronic and computer systems troubleshooting and maintenance, and 1 in map reading (6/89).

Recommendation, Skill Level 30

In the lower-division baccalaureate/associate degree category, 2 semester hours in communications system operations, 2 in electromechanical systems troubleshooting and maintenance, 2 in electronic and computer systems troubleshooting and maintenance, 1 in map reading, 1 in applied mathematics, 1 in technical report writing, 3 in principles of supervision, and 3 in communication skills (6/89).

Recommendation, Skill Level 40

In the lower-division baccalaureate/associate degree category, 2 semester hours in communications system operations, 2 in electromechanical systems troubleshooting and maintenance, 2 in electronic and computer systems troubleshooting and maintenance, 1 in map reading, 1 in applied mathematics, 3 in technical report writing, 3 in principles of supervision, and 3 in communication skills. In the upper-division baccalaureate category, 3 semester hours for field experience in management and 3 in personnel management (6/89).

MOS-13M-003

Multiple Launch Rocket System Crewmember
13M10
13M20
13M30
13M40

Exhibit Dates: 3/95–4/99.

Career Management Field: 13 (Field Artillery).

Description

Summary: Supervises or serves in a field artillery unit on the multiple launch rocket system; operates and maintains the launcher, transport vehicle, trailer, and other support equipment. *Skill Level 10:* Participates in launch maintenance and ammunition resupply activities; drives transport vehicles; participates in launcher operations; operates radio; maintains computer and performs system checks. *Skill Level 20:* Able to perform the duties required for Skill Level 10; assists the section chief; trains subordinate personnel; operates fire control panel; operates launcher; performs organizational maintenance; serves as firing platoon reconnaissance sergeant. *Skill Level 30:* Able to perform the duties required for Skill Level 20; serves as section chief supervising seven to twelve persons; instructs in launch procedures; supervises preventive maintenance on launch vehicles; supervises fire control panel maintenance; writes reports, and assures safety procedure compliance. *Skill Level 40:* Able to perform the duties required for Skill Level 30; supervises 12 to 60 persons; supervises all section maintenance; writes technical and personnel reports; supervises and performs training of personnel, and assures safety procedure compliance.

Recommendation, Skill Level 10

Credit may be granted on the basis of an individualized assessment of the student (6/89).

Recommendation, Skill Level 20

Credit may be granted on the basis of an individualized assessment of the student (6/89).

Recommendation, Skill Level 30

In the lower-division baccalaureate/associate degree category, 2 semester hours in communications system operations, 2 in electromechanical systems troubleshooting and maintenance, 2 in electronic and computer systems troubleshooting and maintenance, 1 in map reading, 1 in technical report writing, 3 in principles of supervision, and 3 in communication skills (6/89).

Recommendation, Skill Level 40

In the lower-division baccalaureate/associate degree category, 2 semester hours in communications system operations, 2 in electromechanical systems troubleshooting and maintenance, 2 in electronic and computer systems troubleshooting and maintenance, 1 in map reading, 1 in applied mathematics, 3 in technical report writing, 3 in principles of supervision, and 3 in communication skills. In the upper-division baccalaureate category, 3 semester hours for field experience in management and 3 in personnel management (6/89).

MOS-13M-004

Multiple Launch Rocket System Crewmember
13M10
13M20
13M30
13M40

Exhibit Dates: 5/99–Present.

Career Management Field: 13 (Field Artillery).

Description

Summary: Supervises or serves in a field artillery unit on the multiple launch rocket system; operates voice and digital communications equipment; uses and generates computer data; operates and maintains the launcher, transport vehicle, trailer, and other support equipment. *Skill Level 10:* Participates in launch maintenance and ammunition resupply activities; drives transport vehicles; participates in launcher operations; operates radio; maintains computer fire control systems and performs system checks. *Skill Level 20:* Able to perform the duties required for Skill Level 10; assists the section chief; trains subordinate personnel; operates fire control panel; operates launcher; performs organizational maintenance; serves as firing platoon reconnaissance sergeant. *Skill Level 30:* Able to perform the duties required for Skill Level 20; serves as section chief supervising seven to twelve persons; instructs in launch procedures; supervises preventive maintenance on launch vehicles; supervises fire control panel maintenance; writes reports, and ensures safety procedure compliance. *Skill Level 40:* Able to perform the duties required for Skill Level 30; supervises 12 to 60 persons; supervises all section maintenance; writes technical and personnel reports; supervises and performs training of personnel, and ensures safety procedure compliance.

Recommendation, Skill Level 10

Credit may be granted on the basis of an individualized assessment of the student (5/99).

Recommendation, Skill Level 20

Credit may be granted on the basis of an individualized assessment of the student (5/99).

Recommendation, Skill Level 30

In the lower-division baccalaureate/associate degree category, 1 semester hour in electromechanical systems troubleshooting, 1 in computer familiarization, 2 in map reading and land navigation, 1 in technical writing, 1 in safety and hazardous material training, 2 in communications systems operations, 2 in principles of supervision, 3 in oral communication skills, and 1 in electronic and computer systems troubleshooting and maintenance (5/99).

Recommendation, Skill Level 40

In the lower-division baccalaureate/associate degree category, 2 semester hours in electromechanical systems troubleshooting, 1 in computer familiarization, 2 in electronic and computer systems troubleshooting and maintenance, 3 in principles of supervision, 3 in oral communication skills, 2 in communications systems operations, 1 in safety and hazardous material training, 2 in map reading and land navigation, 3 in technical writing, and 1 in applied mathematics. In the upper-division baccalaureate category, 3 semester hours for field experience in management and 3 in personnel management (5/99).

MOS-13N-001

LANCE CREWMEMBER

13N10
13N20
13N30
13N40

Exhibit Dates: 1/90–6/93. Effective 6/93, MOS 13N was discontinued.

Career Management Field: 13 (Field Artillery).

Description

Summary: Serves as a Lance missile crewmember or supervises Lance crewmembers. *Skill Level 10:* Under supervision, performs crewmember duties during placement, assembly, disassembly, or firing of missile; drives vehicles; emplaces missile launcher; operates power generator; performs maintenance; prepares missile handling equipment; provides proper storage locations for ammunition; performs guard duties and assists in defense of position; renders first aid and provides sanitation facilities. *Skill Level 20:* Able to perform the duties required for Skill Level 10; provides leadership and applicable training to subordinates; assists in instruction of security teams; assists or supervises orientation of launcher; monitors computer operation and programming; performs warhead settings; supervises or assists in maintenance and adjustment of equipment; establishes communication with proper facility; counsels subordinates. *Skill Level 30:* Able to perform the duties required for Skill Level 20; leads and supervises preparation of launcher and missile firing; provides leadership and instruction to personnel in receipt, storage, handling, and firing of missile; directs assembly of missile components; directs operation and maintenance of section equipment; supervises and coordinates ammunition supply; prepares maintenance and technical records; verifies sighting and emplacement operations. *Skill Level 40:* Able to perform the duties required for Skill Level 30; leads and directs missile placement and resupply operations; leads and directs maintenance; plans and organizes crew work schedules; ensures proper safety procedures; verifies final missile tests; originates and/or reviews technical personnel, administrative, readiness, and efficiency reports.

Recommendation, Skill Level 10

In the lower-division baccalaureate/associate degree category, 1 semester hour in communications system operation, 2 in electronic and computer systems troubleshooting and maintenance, and 2 in electrohydraulic systems troubleshooting and maintenance. (NOTE: This recommendation for skill level 10 is valid for the dates 1/90-9/91 only) (6/89).

Recommendation, Skill Level 20

In the lower-division baccalaureate/associate degree category, 2 semester hours in communications system operation, 4 in electronic and computer systems troubleshooting and maintenance, 2 in electrohydraulic systems troubleshooting and maintenance, and 2 in map reading (6/89).

Recommendation, Skill Level 30

In the lower-division baccalaureate/associate degree category, 2 semester hours in communications system operation, 4 in electronic and computer systems troubleshooting and maintenance, 2 in electrohydraulic systems troubleshooting and maintenance, 2 in map reading, 1 in applied mathematics, 2 in technical writing, 3 in principles of supervision, and 3 in communication skills (6/89).

Recommendation, Skill Level 40

In the lower-division baccalaureate/associate degree category, 2 semester hours in communications system operation, 4 in electronic and computer systems troubleshooting and maintenance, 2 in electrohydraulic systems troubleshooting and maintenance, 2 in map reading, 1 in applied mathematics, 3 in technical writing, 3 in principles of supervision, and 3 in communication skills. In the upper-division baccalaureate category, 3 semester hours for field experience in management and 3 in personnel management (6/89).

MOS-13P-001

MULTIPLE LAUNCH ROCKET SYSTEM/LANCE OPERATIONS/FIRE DIRECTION SPECIALIST

13P10
13P20
13P30
13P40

Exhibit Dates: 1/90–2/95.

Career Management Field: 13 (Field Artillery).

Description

Summary: Leads, supervises, or serves as a member of a Multiple Launch Rocket System (MLRS) or a Lance missile fire direction center (FDC), or an operations section for either nuclear or conventional missions. *Skill Level 10:* Uses graphs, tables, charts, and maps for manual computation of firing locations, angles, altitudes, corrections, and displacements; constructs firing charts; knows and uses mathematical concepts (including some plane geometry and trigonometric functions) for manual computations; plots information on situation maps; operates the following specialized computers: Fire Direction System (FDS), Platoon Leaders Digital Message Device (PLDMD), or the Back-up Computer System (BUCS) (military version of Hewlett-Packard 71B (16K) computer); installs and operates digital message systems, FM radio transmitter/receiver and secures voice equipment; operates and maintains vehicles, equipment, and generators; uses data from standardized tables to calculate data for nonstandard conditions. *Skill Level 20:* Able to perform the duties required for Skill Level 10; computes data for Lance targets; operates BUCS in Lance fire direction centers; assists in supervising and instructing fire direction center personnel; uses map reading skills in maintaining situation maps. *Skill Level 30:* Able to perform the duties required for Skill Level 20; instructs personnel in techniques and procedures for manual computation of firing data; maintains target and mission reports; may serve in a high level Fire Support Element to provide guidance and advice on matters of employment of Lance missiles; instructs in the use and operation of FDS, PLDMD, and BUCS communications equipment. *Skill Level 40:* Able to perform the duties required for Skill Level 30; supervises FDC to insure accurate computations, plotting, and issuing of fire commands; organizes work schedules and assigns duties to FDC personnel; instructs and supervises personnel; prepares technical, personnel, and administrative reports of section activities; provides emergency action plans and information in a Lance unit.

Recommendation, Skill Level 10

In the lower-division baccalaureate/associate degree category, 1 semester hour in applied mathematics, 1 in communications system operations, 1 in map reading, and 1 in computer operations. (NOTE: This recommendation for skill level 10 is valid for the dates 1/90-9/91 only) (6/89).

Recommendation, Skill Level 20

In the lower-division baccalaureate/associate degree category, 1 semester hour in applied mathematics, 1 in communications system operations, 1 in map reading, and 2 in computer operations (6/89).

Recommendation, Skill Level 30

In the lower-division baccalaureate/associate degree category, 2 semester hours in applied mathematics, 2 in communications system operations, 2 in map reading, 2 in computer operations, 2 in principles of supervision, and 2 in communication skills (6/89).

Recommendation, Skill Level 40

In the lower-division baccalaureate/associate degree category, 2 semester hours in applied mathematics, 2 in communications system operations, 2 in map reading, 2 in computer operations, 3 in principles of supervision, 3 in communication skills, and 3 in personnel management (6/89).

MOS-13P-002

Multiple Launch Rocket System/Lance Operations/Fire Direction Specialist
13P10
13P20
13P30
13P40

Exhibit Dates: 3/95–4/99.

Career Management Field: 13 (Field Artillery).

Description

Summary: Leads, supervises, or serves as a member of a Multiple Launch Rocket System (MLRS) or a Lance missile fire direction center (FDC), or an operations section for either nuclear or conventional missions. *Skill Level 10:* Uses graphs, tables, charts, and maps for manual computation of firing locations, angles, altitudes, corrections, and displacements; constructs firing charts; knows and uses mathematical concepts (including some plane geometry and trigonometric functions) for manual computations; plots information on situation maps; operates the following specialized computers: Fire Direction System (FDS), Platoon Leaders Digital Message Device (PLDMD), or the Back-up Computer System (BUCS) (military version of Hewlett-Packard 71B (16K) computer); installs and operates digital message systems, FM radio transmitter/receiver and secures voice equipment; operates and maintains vehicles, equipment, and generators; uses data from standardized tables to calculate data for nonstandard conditions. *Skill Level 20:* Able to perform the duties required for Skill Level 10; computes data for Lance targets; operates BUCS in Lance fire direction centers; assists in supervising and instructing fire direction center personnel; uses map reading skills in maintaining situation maps. *Skill Level 30:* Able to perform the duties required for Skill Level 20; instructs personnel in techniques and procedures for manual computation of firing data; maintains target and mission reports; may serve in a high level Fire Support Element to provide guidance and advice on matters of employment of Lance missiles; instructs in the use and operation of FDS, PLDMD, and BUCS communications equipment. *Skill Level 40:* Able to perform the duties required for Skill Level 30; supervises FDC to insure accurate computations, plotting, and issuing of fire commands; organizes work schedules and assigns duties to FDC personnel; instructs and supervises personnel; prepares technical, personnel, and administrative reports of section activities; provides emergency action plans and information in a Lance unit.

Recommendation, Skill Level 10

Credit may be granted on the basis of an individualized assessment of the student (6/89).

Recommendation, Skill Level 20

Credit may be granted on the basis of an individualized assessment of the student (6/89).

Recommendation, Skill Level 30

In the lower-division baccalaureate/associate degree category, 1 semester hour in applied mathematics, 1 in communications system operations, 1 in map reading, 2 in computer operations, 2 in principles of supervision, and 2 in communication skills (6/89).

Recommendation, Skill Level 40

In the lower-division baccalaureate/associate degree category, 2 semester hours in applied mathematics, 2 in communications system operations, 2 in map reading, 2 in computer operations, 3 in principles of supervision, 3 in communication skills, and 3 in personnel management (6/89).

MOS-13P-003

Multiple Launch Rocket System (MLRS Automated Tactical Data Systems Specialist
((Multiple Launch Rocket System/Lance Operations))
13P10
13P20
13P30
13P40

Exhibit Dates: 5/99–Present.

Career Management Field: 13 (Field Artillery).

Description

Summary: Leads, supervises, or serves as a member of a Multiple Launch Rocket System (MLRS) or an operations section for conventional missions. *Skill Level 10:* Uses graphs, tables, charts, and maps for manual computation of firing locations, angles, altitudes, corrections, and displacements; constructs firing charts; knows and uses mathematical concepts (including some plane geometry and trigonometric functions) for manual computations; plots information on situation maps; operates the following specialized computers: Fire Direction System (FDS), Platoon Leaders Digital Message Device (PLDMD), or the Back-up Computer System (BUCS) (military version of Hewlett-Packard 71B (16K) computer); installs and operates digital message systems, FM radio transmitter/receiver, and secure voice equipment; operates and maintains vehicles, equipment, and generators; uses data from standardized tables to calculate data for nonstandard conditions. *Skill Level 20:* Able to perform the duties required for Skill Level 10; computes data for targets; operates fire direction centers; assists in supervising and instructing fire direction center personnel; uses map reading skills in maintaining situation maps. *Skill Level 30:* Able to perform the duties required for Skill Level 20; instructs personnel in techniques and procedures for manual computation of firing data; maintains target and mission reports; may serve in a high level fire support element to provide guidance and advice on matters of employment of missiles; instructs in the use and operation of FDS, PLDMD, and BUCS communications equipment. *Skill Level 40:* Able to perform the duties required for Skill Level 30; supervises FDC to insure accurate computations, plotting, and issuing of fire commands; organizes work schedules and assigns duties to FDC personnel; instructs and supervises personnel; prepares technical, personnel, and administrative reports of section activities; provides emergency action plans and information in a the unit.

Recommendation, Skill Level 10

Credit may be granted on the basis of an individualized assessment of the student (5/99).

Recommendation, Skill Level 20

Credit may be granted on the basis of an individualized assessment of the student (5/99).

Recommendation, Skill Level 30

In the lower-division baccalaureate/associate degree category, 2 semester hours in applied mathematics, 1 in safety and hazardous material training, 2 in computer familiarization, 2 in communications systems operations, 2 in map reading and land navigation, 1 in technical writing, 2 in principles of supervision, and 2 in oral communication skills (5/99).

Recommendation, Skill Level 40

In the lower-division baccalaureate/associate degree category, 2 semester hours in applied mathematics, 1 in safety and hazardous material training, 2 in computer familiarization, 2 in communications systems operations, 2 in map reading and land navigation, 3 in technical writing, 3 in principles of supervision, and 3 in oral communication skills. In the upper-division baccalaureate category, 3 semester hours for field experiences in management and 3 in personnel management (5/99).

MOS-13R-002

Field Artillery Firefinder Radar Operator
13R10
13R20
13R30
13R40

Exhibit Dates: 1/90–4/99.

Career Management Field: 13 (Field Artillery).

Description

Summary: Operates or supervises operation of artillery and mortar firefinder radars to locate hostile firing weapons and adjusts friendly artillery. *Skill Level 10:* Prepares radar set for operation; operates radar set; directs and reports system malfunctions; administers first aid. *Skill Level 20:* Able to perform the duties required for Skill Level 10; provides technical guidance to subordinates; assists in radar emplacement; assists section chief; in the absence of the section chief, assumes duties of that position. *Skill Level 30:* Able to perform the duties required for Skill Level 20; supervises operation of firefinder radar section; prepares for radar operation and site selection; instructs personnel in radar operation; supervises organizational maintenance. *Skill Level 40:* Able to perform the duties required for Skill Level 30; supervises, trains, and evaluates subordinates; coordinates survey data and assists in site selection; prepares reports summarizing data obtained from radar sections on troop strengths, logistics, surveillance techniques, and tactical operations.

Recommendation, Skill Level 10

In the vocational certificate category, 2 semester hours in radar operations. (NOTE: This recommendation for skill level 10 is valid for the dates 1/90-9/91 only) (6/89).

Recommendation, Skill Level 20

In the vocational certificate category, 2 semester hours in radar operations. In the lower-division baccalaureate/associate degree category, if served as section chief, 3 semester hours in principles of supervision and 3 in maintenance management. (NOTE: This recommendation for skill level 20 is valid for the dates 1/90-2/95 only) (6/89).

Recommendation, Skill Level 30

In the vocational certificate category, 2 semester hours in radar operations. In the lower-division baccalaureate/associate degree category, 3 semester hours in principles of supervision, 3 in maintenance management, and 3 in communication skills (oral and written) (6/89).

Recommendation, Skill Level 40

In the vocational certificate category, 2 semester hours in radar operations. In the lower-division baccalaureate/associate degree category, 3 semester hours in principles of supervision, 3 in maintenance management, and 3 in communication skills (oral and written). In the upper-division baccalaureate category, 3 semester hours for field experience in management and 3 in personnel management (6/89).

MOS-13R-003

Field Artillery Firefinder Radar Operator
13R10
13R20
13R30
13R40

Exhibit Dates: 5/99–Present.

Career Management Field: 13 (Field Artillery).

Description

Summary: Operates or supervises operation of artillery and mortar firefinder radars to locate hostile firing weapons and adjusts friendly artillery. *Skill Level 10:* Prepares radar set for operation; operates radar set; directs and reports system malfunctions; administers first aid. *Skill Level 20:* Able to perform the duties required for Skill Level 10; provides technical guidance to subordinates; assists in radar emplacement; assists section chief; in the absence of the section chief, assumes duties of that position. *Skill Level 30:* Able to perform the duties required for Skill Level 20; supervises operation of firefinder radar section; prepares for radar operation and site selection; instructs personnel in radar operation; supervises organizational maintenance. *Skill Level 40:* Able to perform the duties required for Skill Level 30; supervises, trains, and evaluates subordinates; coordinates survey data and assists in site selection; prepares reports summarizing data obtained from radar sections on troop strengths, logistics, surveillance techniques, and tactical operations.

Recommendation, Skill Level 10

Credit may be granted on the basis of an individualized assessment of the student (5/99).

Recommendation, Skill Level 20

Credit may be granted on the basis of an individualized assessment of the student (5/99).

Recommendation, Skill Level 30

In the lower-division baccalaureate/associate degree category, 2 semester hours in map reading and land navigation, 1 in computer familiarization, 2 in electronic troubleshooting and maintenance, 3 in principles of supervision, 3 in oral communication skills, and 3 in maintenance management (5/99).

Recommendation, Skill Level 40

In the lower-division baccalaureate/associate degree category, 2 semester hours in map reading and land navigation, 1 in computer familiarization, 2 in electronic troubleshooting and maintenance, 3 in principles of supervision, 3 in oral communication skills, and 3 in maintenance management. In the upper-division baccalaureate category, 3 semester hours for field experience in management and 3 in personnel management (5/99).

MOS-13S-001

Field Artillery Surveyor
13S10
13S20
13S30
13S40

Exhibit Dates: 9/04–Present.

Career Management Field: 13 (Field Artillery).

Description

Summary: Engages in surveying activities. *Skill Level 10:* Determines distance between survey stations; marks survey stations; performs astronomical observations for determining locations; operates and maintains survey equipment, including theodolite, laser-activated infrared distance measuring instrument, PADS inertial system gyroscope unit for locations, and the Backup Computer Systems (Hewlett-Packard Computer) for computations; knows manual computation formulas, equations, procedures, and forms, as well as ephemeris. *Skill Level 20:* Able to perform the duties required for Skill Level 10; applies knowledge of basic algebra, basic geometry, and trigonometric functions; computes azimuth and distance coordinates, altimetric and trigonometric heights, and astronomical azimuths from field data. *Skill Level 30:* Able to perform the duties required for Skill Level 20; supervises a survey party; selects starting station and method of survey; reviews survey data for accuracy; supervises maintenance of survey party equipment; instructs in survey procedures and techniques. *Skill Level 40:* Able to perform the duties required for Skill Level 30; supervises and coordinates operations of several survey parties; prepares technical, personnel, and administrative reports.

Recommendation, Skill Level 10

Credit may be granted on the basis of an individualized assessment of the student (4/04).

Recommendation, Skill Level 20

Credit may be granted on the basis of an individualized assessment of the student (4/04).

Recommendation, Skill Level 30

In the lower-division baccalaureate/associate degree category, 2 semester hours in introduction to maps and aerial photos, 6 in surveying, 3 in applied algebra and trigonometry, 3 in computer operations, 1 in introduction to astronomy, 3 in oral communication skills, 3 in technical report writing, and 3 in principles of supervision (4/04).

Recommendation, Skill Level 40

In the lower-division baccalaureate/associate degree category, 3 semester hours in introduction to maps and aerial photos, 6 in surveying, 3 in applied algebra and trigonometry, 3 in computer operations, 2 in introduction to astronomy, 3 in oral and written communication skills, 3 in technical report writing, and 3 in personnel supervision. In the upper-division baccalaureate category, 3 semester hours for field experience in management and 3 in personnel management (4/04).

MOS-13W-001

Field Artillery Meteorological Crewmember
13W10
13W20
13W30
13W40

Exhibit Dates: 6/03–Present.

Career Management Field: 13 (Field Artillery).

Description

Summary: Supervises or participates in operation of field artillery meteorological observation station. *Skill Level 10:* Operates meteorological equipment in meteorological observation station; participates in assembly and emplacement of meteorological equipment and associated generators; operates recorders; tabulates pressure and wind data; assembles, operates, and disassembles balloon inflation and launching equipment; installs, tests, and adjusts radio weather instrument (Rawin) sets and radiosonde transmitters; operates computer that converts raw temperature, pressure, and wind velocities to digital data that presents temperature, density, and wind profiles; performs operator maintenance on equipment; has experience in use of rifle, grenade launcher, and machine gun; installs and operates digital and voice radio communications equipment; drives vehicles under adverse conditions; provides routine automotive service and limited automotive repair. *Skill Level 20:* Able to perform the duties required for Skill Level 10; employs applicable navigation aids to derive correct location; uses maps to best advantage; performs operational checks on Rawin set; is proficient in basic combat techniques; instructs subordinates. *Skill Level 30:* Able to perform the duties required for Skill Level 20; serves as team chief during periods of extended operation; writes reports; checks data for quality control; makes operational checks on equipment; operates communications equipment; encodes and decodes meteorological data; serves as instructor and counselor to subordinates. *Skill Level 40:* Able to perform the duties required for Skill Level 30; supervises operation of meteorological section or station; selects station sites; supervises emplacement, calibration, maintenance, and operation of equipment; supervises preparation and distribution of ballistic meteorological messages; assigns, instructs, and supervises crew members; reviews, consolidates, and prepares technical, personnel, and administrative reports; counsels subordinates.

Recommendation, Skill Level 10

Credit may be granted on the basis of an individualized assessment of the student (4/04).

Recommendation, Skill Level 20

Credit may be granted on the basis of an individualized assessment of the student (4/04).

Recommendation, Skill Level 30

In the lower-division baccalaureate/associate degree category, 3 semester hours in general meteorology, 3 in meteorological instrumentation, 1 in technical writing, 3 in map reading and land navigation, 2 in meteorological observation, 1 in applied mathematics, 3 in computer familiarization, 3 in oral communication skills, 1 in automotive service, 1 in safety and hazardous material, and 3 in principles of supervision (4/04).

Recommendation, Skill Level 40

In the lower-division baccalaureate/associate degree category, 3 semester hours in map reading and land navigation, 3 in general meteorology, 1 in applied mathematics, 3 in computer familiarization, 3 in principles of supervision, 3 in meteorological instrumentation, 3 in meteorological observation, 3 in technical writing, 3 in oral communication skills, 1 in automotive service, and 1 in safety and hazardous material. In the upper-division baccalaureate category, 3 semester hours for field experience in management and 3 in personnel management (4/04).

MOS-13Z-001

Field Artillery Senior Sergeant
13Z50

Exhibit Dates: 1/90–1/92.

Career Management Field: 13 (Field Artillery).

Description

Serves as the principal noncommissioned officer in a major field artillery organization; supervises operations, intelligence activities, and liaison activities; assists superiors in planning, coordinating, and supervising all operations and personnel matters; holds meetings with subordinate noncommissioned officers to provide guidance and direction; directs on-the-job training; is a mid-level manager overseeing 45-150 persons. NOTE: May have progressed to 13Z50 from 13B40 (Field Artillery Crewman), 13W50 (Field Artillery Target Acquisition Senior Sergeant), or 13Y50 (Cannon/Missile Senior Sergeant).

Recommendation

In the upper-division baccalaureate category, 3 semester hours in introduction to management, 3 in personnel management, 3 in human relations, 6 for field experience in management, and additional credit in public speaking and for an internship in education on the basis of institutional evaluation (4/77).

MOS-13Z-002

Field Artillery Senior Sergeant
13Z50

Exhibit Dates: 2/92–4/99.

Career Management Field: 13 (Field Artillery).

Description

Supervises operations, intelligence, fire support, and target acquisition activities in a field artillery battalion, brigade, division, or corps artillery unit; leads, supervises, and participates in all field artillery operations. NOTE: May progress to 13Z50 from Skill Level 40 of any of the following MOS's: 13B (Cannon Crewmember), 13C (TACFIRE Operations Specialist), 13E (Cannon Fire Direction Specialist), 13F (Fire Support Specialist), 13M (Multiple Launch Rocket System Crewmember), 13N (Lance Crewmember), 13P (Multiple Launch Rocket System/Lance Operations/Fire Direction Specialist), 13R (Field Artillery Firefinder Radar Operator), 13T (Remotely Piloted Vehicle Crewmember), 15E (Pershing Missile Crewmember), 17B (Field Artillery Radar Crewmember), 21G (Pershing Electronics Materiel Specialist), 82C (Field Artillery Surveyor), or 93F (Field Artillery Meteorological Crewmember).

Recommendation

In the lower-division baccalaureate/associate degree category, 3 semester hours in personnel supervision, 2 in records and information management, and 3 in military science. In the upper-division baccalaureate category, 3 semester hours for field experience in management and 3 in organizational management; if paygrade E-9 has been attained, additional credit as follows: 3 semester hours in management problems, 3 in operations management, and 3 in communication techniques for managers (2/92).

MOS-13Z-003

Field Artillery Senior Sergeant
13Z50

Exhibit Dates: 5/99–Present.

Career Management Field: 13 (Field Artillery).

Description

Supervises operations, intelligence, fire support, and target acquisition activities in a field artillery battalion, brigade, division, or corps artillery unit; leads, supervises, and participates in all field artillery operations. NOTE: May progress to 13Z50 from Skill Level 40 of any of the following MOS's: 13B (Cannon Crewmember), 13C (TACFIRE Operations Specialist), 13E (Cannon Fire Direction Specialist), 13F (Fire Support Specialist), 13M (Multiple Launch Rocket System Crewmember), 13P (Multiple Launch Rocket System/Lance Operations/Fire Direction Specialist), 13R (Field Artillery Firefinder Radar Operator), 82C (Field Artillery Surveyor), or 93F (Field Artillery Meteorological Crewmember).

Recommendation

In the lower-division baccalaureate/associate degree category, 3 semester hours in personnel supervision, 2 in records and information management, 3 in military science (drills, ceremonies, and regulations compliance), 3 in computer literacy, and 3 in oral communication skills. In the upper-division baccalaureate category, 3 semester hours for field experience in management and 3 in organizational management; if paygrade E-9 has been attained, additional credit as follows: 3 semester hours in management problems, 3 in operations management, and 3 in communication techniques for managers (5/99).

MOS-14D-001

Hawk Missile System Crewmember
14D10
14D20
14D30
14D40

Exhibit Dates: 11/90–Present.

Career Management Field: 14 (Air Defense Artillery).

Description

Summary: Leads, supervises or serves as member of a command and acquisition, missile resupply, radar firing, and fire control section; maintains the Hawk missile launching and storage area and radar equipment and performs or supervises operations and intelligence functions. *Skill Level 10:* Serves as crewmember in preparing launching area equipment; assembles, operates, and maintains Hawk fire control equipment and engagement simulators; performs intelligence duties as required; operates vehicles; participates in emplacement of generators, trailers, pallets, loaders, launchers, and associated equipment; performs missile loading and unloading; performs arming and disarming functions; operates and performs preventive maintenance on generators, launchers, and associated equipment; observes proper safety precautions; operates and performs operator maintenance; checks and makes adjustments on Hawk fire control and radar system; operates antijamming devices; prepares and posts daily journal, map overlays, and situation displays; operates radio and telephone equipment; assists organizational maintenance technicians. *Skill Level 20:* Able to perform the duties required for Skill Level 10; provides technical guidance to subordinates; assists platoon leader in tactical procedures; performs preventive maintenance; consolidates information from other radar sections; processes intelligence reports; prepares and processes operations and training material; coordinates training activities; transmits intelligence and locations of incoming targets to concerned units; alerts firing batteries and advises of change orders; prepares efficiency reports on subordinates. *Skill Level 30:* Able to perform the duties required for Skill Level 20; provides guidance to subordinates; enforces safety practices and standards of readiness; conducts training; coordinates maintenance procedures; assists in establishment of command and operation centers; prepares efficiency reports. *Skill Level 40:* Able to perform the duties required for Skill Level 30; enforces standard of readiness; conducts training to support these standards; supervises operations of battle management centers; assists in planning, organizing, directing, supervising, training, coordinating, and reporting activities of subordinate sections.

Recommendation, Skill Level 10

In the lower-division baccalaureate/associate degree category, 3 semester hours in hydraulic systems and 3 in military science. (NOTE: This recommendation for skill level 10 is valid for the dates 1/90-9/91 only) (3/92).

Recommendation, Skill Level 20

In the lower-division baccalaureate/associate degree category, 3 semester hours in hydraulic systems, 3 in military science, and 3 in personnel supervision. (NOTE: This recommendation for skill level 20 is valid for the dates 11/90-2/95 only) (3/92).

Recommendation, Skill Level 30

In the lower-division baccalaureate/associate degree category, 3 semester hours in hydraulic systems, 3 in military science, 3 in personnel supervision, and 2 in records and information management (3/92).

Recommendation, Skill Level 40

In the lower-division baccalaureate/associate degree category, 3 semester hours in hydraulic systems, 3 in military science, 3 in personnel supervision, and 2 in records and information management. In the upper-division baccalaureate category, 3 semester hours for field experience in management and 3 in organizational management (3/92).

MOS-14E-001

PATRIOT FIRE CONTROL ENHANCED OPERATOR/MAINTAINER
14E10
14E20
14E30
14E40
14E50

Exhibit Dates: 2/97–Present.

Career Management Field: 23 (Air Defense System Maintenance).

Description

Summary: Supervises, sets up, operates, or performs maintenance on the Patriot air defense missile system. *Skill Level 10:* Drives heavy equipment to move missile system to site; sets up and performs operator maintenance on system; operates truck, electrical generator, and launcher; performs repairs on the Patriot system; installs equipment modifications; prepares appropriate supplies, communications equipment, computer and computer peripherals, tools, test equipment, and shelters; removes and replaces faulty circuit cards; operates maintenance vehicles. *Skill Level 20:* Able to perform the duties required for Skill Level 10; provides technical guidance to subordinates; performs checks, adjustments, and repairs on major item interface; performs unit maintenance on Patriot information and coordination control, engagement control station, radar set, identification-friend-or-foe, launching station, antiradiation missile decoy, and antenna mast group; supervises maintenance procedures; assists in training operators in maintenance techniques; troubleshoots problems with external test equipment when isolation is not accomplished by built-in or other machine-aided equipment. *Skill Level 30:* Able to perform the duties required for Skill Level 20; serves as shift supervisor while performing as chief engagement controller; supervises emplacement of launching and fire control equipment; provides supervision in training programs for operator and maintenance personnel; supervises preparation of technical forms and reports; requisitions and stocks supplies and repair parts; determines faulty work practices and demonstrates proper maintenance and troubleshooting techniques. *Skill Level 40:* Able to perform the duties required for Skill Level 30; coordinates and supervises activities of subordinates, sections, and squadrons; supervises inspection of maintenance performance; supervises inspection teams; supervises repair parts stocking; establishes maintenance priorities. *Skill Level 50:* Able to perform the duties required for Skill Level 40; serves as maintenance chief; assists in planning, organization, and training operations; provides technical expertise; may serve as a member of a multinational evaluation team; applies knowledge of capabilities and limitations of the Patriot weapon, command and control, and radar systems.

Recommendation, Skill Level 10

Credit may be granted on the basis of an individualized assessment of the student (5/98).

Recommendation, Skill Level 20

Credit may be granted on the basis of an individualized assessment of the student (5/98).

Recommendation, Skill Level 30

In the lower-division baccalaureate/associate degree category, 3 semester hours in basic electronics, 3 in basic electronics laboratory, 3 in digital principles, 3 in personnel supervision, 3 in maintenance management, 2 in records and information management, 3 in electronic systems troubleshooting, and 3 in computer systems troubleshooting (5/98).

Recommendation, Skill Level 40

In the lower-division baccalaureate/associate degree category, 3 semester hours in basic electronics, 3 in basic electronics laboratory, 3 in digital principles, 3 in personnel supervision, 3 in maintenance management, 2 in records and information management, 3 in electronic systems troubleshooting, and 3 in computer systems troubleshooting. In the upper-division baccalaureate category, 3 semester hours in organizational management and 3 for field experience in management (5/98).

Recommendation, Skill Level 50

In the lower-division baccalaureate/associate degree category, 3 semester hours in basic electronics, 3 in basic electronics laboratory, 3 in digital principles, 3 in personnel supervision, 3 in maintenance management, 2 in records and information management, 3 in electronic systems troubleshooting, and 3 in computer system troubleshooting. In the upper-division baccalaureate category, 3 semester hours in organizational management and 6 for field experience in management (5/98).

MOS-14J-001

AIR DEFENSE COMMAND, CONTROL, COMMUNICATION, COMPUTERS AND INTELLIGENCE TACTICAL OPERATIONS CENTER ENHANCED OPERATOR/MAINTAINER
((Early Warning System Operator))
14J10
14J20
14J30
14J40

Exhibit Dates: 1/90–Present.

Career Management Field: 14 (Air Defense Artillery).

Description

Summary: Operates, supervises and performs preventive and corrective maintenance on a manual early warning network sensor system for operational and intelligence functions. *Skill Level 10:* Operates and performs preventive maintenance checks on early warning team vehicles; detects, correlates, and broadcasts early warning information and identifies aircraft; operates over radio nets; uses power generation equipment. *Skill Level 20:* Able to perform the duties required for Skill Level 10; provides technical guidance to subordinates; inventories and destroys classified materials; performs security functions. *Skill Level 30:* Able to perform the duties required for Skill Level 20; provides technical guidance to subordinates; supervises operations and maintenance of the MEWN/sensor section, team, or platoon; assists and supervises operations and intelligence functions in air defense units; observes, incorrect procedures and demonstrates correct procedures and techniques; supervises operation of tactical communications. *Skill Level 40:* Able to perform the duties required for Skill Level 30; prepares operation plans, orders, and standing operating procedures; identifies support maintenance requirements.

Recommendation, Skill Level 10

In the lower-division baccalaureate/associate degree category, 3 semester hours in electronic systems maintenance. (NOTE: This recommendation for skill level 10 is valid for the dates 1/90-9/91 only. (7/97).

Recommendation, Skill Level 20

In the lower-division baccalaureate/associate degree category, 3 semester hours in electronic systems maintenance and 2 in personnel supervision. (NOTE: This recommendation for skill level 20 is valid for the dates 1/90-2/95 only. (7/97).

Recommendation, Skill Level 30

In the lower-division baccalaureate/associate degree category, 3 semester hours in electronic systems maintenance, 3 in personal supervision, 3 in maintenance management, and 2 in records management (7/97).

Recommendation, Skill Level 40

In the lower-division baccalaureate/associate degree category, 3 semester hours in electronics systems maintenance, 3 in personnel supervision, 3 in maintenance management, and 2 in records management. In the upper-division baccalaureate category, 3 semester hours for field experience in management (7/97).

MOS-14L-001

AIR DEFENSE ARTILLERY (ADA) COMMAND AND CONTROL SYSTEM OPERATOR/MAINTAINER
14L10
14L20
14L30
14L40

Exhibit Dates: Pending Evaluation. 2/97–Present.

MOS-14M-001

MAN PORTABLE AIR DEFENSE SYSTEM CREWMEMBER
14M10
14M20
14M30
14M40

Exhibit Dates: Pending Evaluation. 2/97–Present.

MOS-14R-001

BRADLEY LINEBACKER CREWMEMBER
((Line of Sight-Forward-Heavy Crewmember))
14R10
14R20
14R30
14R40

Exhibit Dates: 8/91–Present.

Career Management Field: 14 (Air Defense Artillery).

Description

Summary: Supervises and operates the Line of Sight-Forward-Heavy (LOS-F-H) or the Bradley Linebacker vehicle and associated equipment. *Skill Level 10:* Operates and performs preventive maintenance checks and services; operates communications equipment; visually identifies aircraft and armored vehicles; operates weapons systems. *Skill Level 20:* Able to perform the duties required for Skill Level 10; provides technical guidance to subordinates; provides assistance in crew training; collects and consolidates intelligence information. *Skill Level 30:* Able to perform the duties required for Skill Level 20; coordinates and supervises preventive maintenance; processes special and periodic reports; commands the LOS-F-H vehicle. *Skill Level 40:* Able to perform the duties required for Skill Level 30; plans, organizes, directs, supervises, trains, coordinates, and reports activities of staff; supervises maintenance activities.

Recommendation, Skill Level 10
Credit is not recommended because the skills, competencies, and knowledge are uniquely military in nature (7/97).

Recommendation, Skill Level 20
Credit is not recommended because the skills, competencies, and knowledge are uniquely military in nature (7/97).

Recommendation, Skill Level 30
In the lower-division baccalaureate/associate degree category, 3 semester hours in personnel supervision (7/97).

Recommendation, Skill Level 40
In the lower-division baccalaureate/associate degree category, 3 semester hours in personnel supervision. In the upper-division baccalaureate category, 3 semester hours for field experience in management (7/97).

MOS-14S-001

Avenger Crewmember
14S10
14S20
14S30
14S40

Exhibit Dates: 4/91–Present.

Career Management Field: 14 (Air Defense Artillery).

Description
Summary: Supervises or serves as a member of an Avenger weapon unit. *Skill Level 10:* Prepares and fires Avenger weapon; identifies friendly and hostile aircraft; operates and assists in programming interrogator equipment; operates assigned vehicles, simulators, and interrogator and cryptological equipment; operates and maintains radio and/or wire communication with firing units and higher headquarters. *Skill Level 20:* Able to perform the duties required for Skill Level 10; provides technical guidance and supervision to subordinates. *Skill Level 30:* Able to perform the duties required for Skill Level 20; supervises operation of personnel in unit; assists in planning and conducting air defense; assigns personnel as required to provide air defense coverage in other units; supervises programming of interrogator equipment; supervises radio and wire communications and the issuance of ammunition. *Skill Level 40:* Able to perform the duties required for Skill Level 30; supervises operation of Avenger platoon of at least 40 persons; assists in planning, organizing, directing, supervising, training, coordinating, and reporting activities of subordinate sections; assists in coordination and implementation of training programs, administrative matters, and communication activities.

Recommendation, Skill Level 10
In the lower-division baccalaureate/associate degree category, 3 semester hours in military science. (NOTE: This recommendation for skill level 10 is valid for the dates 4/91-9/91 only) (3/92).

Recommendation, Skill Level 20
In the lower-division baccalaureate/associate degree category, 3 semester hours in military science and 3 in personnel supervision. (NOTE: This recommendation for skill level 20 is valid for the dates 4/91-2/95 only) (3/92).

Recommendation, Skill Level 30
In the lower-division baccalaureate/associate degree category, 3 semester hours in military science, 3 in personnel supervision, and 2 in records and information management (3/92).

Recommendation, Skill Level 40
In the lower-division baccalaureate/associate degree category, 3 semester hours in military science, 3 in personnel supervision, and 2 in records and information management. In the upper-division baccalaureate category, 3 semester hours for field experience in management and 3 in organizational management (3/92).

MOS-14T-001

Patriot Launching Station Enhanced Operator/Maintainer
14T10
14T20
14T30
14T40

Exhibit Dates: 2/97–Present.

Career Management Field: 23 (Air Defense System Maintenance).

Description
Summary: Supervises, sets up, operates, or performs maintenance on the Patriot air defense missile system. *Skill Level 10:* Drives heavy equipment to move missile system to site; sets up and performs operator maintenance on system; operates truck, electrical generator, and launcher; performs repairs on the Patriot system; installs equipment modifications; prepares appropriate supplies, communications equipment, computer and computer peripherals, tools, test equipment, and shelters; removes and replaces faulty circuit cards; operates maintenance vehicles. *Skill Level 20:* Able to perform the duties required for Skill Level 10; provides technical guidance to subordinates; performs checks, adjustments, and repairs on major item interface; performs unit maintenance on Patriot information and coordination control, engagement control station, radar set, identification-friend-or foe, launching station, antiradiation missile decoy, and antenna mast group; supervises maintenance procedures; assists in training operators in maintenance techniques; troubleshoots problems with external test equipment when isolation is not accomplished by built-in or other machine-aided equipment. *Skill Level 30:* Able to perform the duties required for Skill Level 20; serves as shift supervisor while performing as chief engagement controller; supervises emplacement of launching and fire control equipment; provides supervision in training programs for operator and maintenance personnel; supervises preparation of technical forms and reports; requisitions and stocks supplies and repair parts; determines faulty work practices and demonstrates proper maintenance and troubleshooting techniques. *Skill Level 40:* Able to perform the duties required for Skill Level 30; coordinates and supervises activities of subordinate sections and squads; supervises inspection of maintenance performance; supervises inspection teams; supervises repair parts stocking; establishes maintenance priorities.

Recommendation, Skill Level 10
Credit may be granted on the basis of an individualized assessment of the student (5/98).

Recommendation, Skill Level 20
Credit may be granted on the basis of an individualized assessment of the student (5/98).

Recommendation, Skill Level 30
In the lower-division baccalaureate/associate degree category, 3 semester hours in basic electronics, 3 in basic electronics laboratory, 3 in digital principles, 3 in personnel supervision, 3 in maintenance management, 2 in records and information management, 3 in electronic systems troubleshooting, and 3 in computer systems troubleshooting (5/98).

Recommendation, Skill Level 40
In the lower-division baccalaureate/associate degree category, 3 semester hours in basic electronics, 3 in basic electronics laboratory, 3 in digital principles, 3 in personnel supervision, 3 in maintenance management, 2 in records and information management, 3 in electronic systems troubleshooting, and 3 in computer systems troubleshooting. In the upper-division baccalaureate category, 3 semester hours in organizational management and 3 for field experience in management (5/98).

MOS-14Z-001

Air Defense Artillery (ADA) Senior Sergeant
14Z50

Exhibit Dates: 2/97–Present.

Career Management Field: 14 (Air Defense Artillery).

Description
Serves as the principal noncommissioned officer in the supervision of an air defense artillery unit or headquarters; assists superiors in the appraisal of air defense artillery operations, training, and intelligence information; prepares status board and situation map; collects, interprets, evaluates, and disseminates intelligence data; prepares and edits training and intelligence material and the operations manual; supervises security; advises superiors on all personnel matters; addresses personnel issues of subordinates; identifies counseling resources; assists superiors in planning, coordinating, and directing all air defense artillery activities. NOTE: May have progressed to 14Z50 from Skill Level 40 of MOS 14D (Hawk Missile System Crewmember), 14R (Line of Sight-Forward-Heavy Crewmember), 14S (Avenger Crewmember), 14J (Early Warning System Operator), 16P (Chaparral Crewmember), 16R (Vulcan Crewmember), 16S (Man Portable Air Defense System Crewmember), 16T (Patriot Missile Crewmember), or 25L (AN/TSQ-73 Air Defense Artillery Command and Control System Operator/Maintainer).

Recommendation
In the lower-division baccalaureate/associate degree category, 3 semester hours in military science, 3 in personnel supervision, and 2 in records and information management. In the upper-division baccalaureate category, 6 semester hours for field experience in management and 3 in organizational management. If individual has attained paygrade of E-9, additional credit as follows: 3 semester hours in manage-

ment problems, 3 in operations management, and 3 in communication techniques for managers (5/98).

MOS-15B-001

Aircraft Powerplant Repairer
15B10
15B20
15B30

Exhibit Dates: 10/03–Present.

Career Management Field: 15 (Aircraft Maintenance).

Description

Summary: Performs or supervises maintenance on turbine engines and their components. *Skill Level 10:* Performs maintenance on turbine power plants and assists in power plant removal; removes, replaces, services, prepares, preserves, and cleans engine assemblies or components; maintains ground support equipment; uses and performs maintenance on common and special tools; uses certain aircraft forms and records; prepares requests for turn-ins and repair parts. *Skill Level 20:* Able to perform the duties required for Skill Level 10; provides technical guidance to subordinates; applies troubleshooting techniques to diagnose malfunctions in specific engine components; disassembles, repairs, reassembles, adjusts, and tests turbine engine systems, subsystems and components; maintains selected ground support equipment; maintains diagnostic equipment; prepares forms and records; performs on-the-job training. *Skill Level 30:* Able to perform the duties required for Skill Level 20; plans work flow; applies production control, quality assurance, and other maintenance management principles and procedures to shop operations; prepares forms and records; conducts on-the-job training; supervises and evaluates work performance in power plant maintenance; may serve as technical inspector; monitors shop and flight-line safety.

Recommendation, Skill Level 10

Credit may be granted on the basis of an individualized assessment of the student (4/04).

Recommendation, Skill Level 20

Credit may be granted on the basis of an individualized assessment of the student (4/04).

Recommendation, Skill Level 30

In the lower-division baccalaureate/associate degree category, 3 semester hours in turbine engine principles, 3 in turbine engine maintenance, 3 in turbine engine troubleshooting, 3 in principles of supervision, 1 in introduction to computers, 1 in computer applications, 3 in maintenance management, 2 in records management, and 2 in aviation safety (4/04).

MOS-15D-003

Aircraft Powertrain Repairer
15D10
15D20
15D30

Exhibit Dates: 10/03–Present.

Career Management Field: 15 (Aircraft Maintenance).

Description

Summary: Supervises or performs maintenance and repair on helicopter power train subsystems and components. *Skill Level 10:* Performs maintenance and tests; assists with inspections on helicopter power train subsystems and components; removes and replaces power train quills, transmission-adapting parts, and rotary wing hub oil tanks; applies corrosion preventive procedures; cleans, preserves, and stores components; performs maintenance on selected ground support equipment; uses and performs maintenance on common and special tools; uses certain aircraft forms and records; prepares request for turn-in and repair parts. *Skill Level 20:* Able to perform the duties required for Skill Level 10; provides technical guidance to subordinates; applies troubleshooting techniques to diagnose malfunctions; performs nondestructive inspections on aircraft components; disassembles, repairs, reassembles, adjusts, and tests power train components, systems, and subsystems; balances main and tail rotor hub assembly and aligns main rotor and hub assembly; performs maintenance on diagnostic equipment; prepares forms and records; supervises on-the-job training. *Skill Level 30:* Able to perform the duties required for Skill Level 20; may serve as technical inspector; plans work; applies production control, quality assurance, and other maintenance management principles; supervises and evaluates work performance; conducts on-the-job training; monitors shop and flight-line safety.

Recommendation, Skill Level 10

Credit may be granted on the basis of an individualized assessment of the student (4/04).

Recommendation, Skill Level 20

Credit may be granted on the basis of an individualized assessment of the student (4/04).

Recommendation, Skill Level 30

In the lower-division baccalaureate/associate degree category, 3 semester hours in aircraft maintenance fundamentals, 3 in helicopter powertrain maintenance, 3 in helicopter powertrain repairs, 3 in nondestructive inspections, 3 in principles of supervision, 1 in introduction to computers, 3 in maintenance management, 3 in records management, 3 in industrial management, 1 in computer applications, and 3 in industrial maintenance (4/04).

MOS-15E-003

Pershing Missile Crew Member
15E10
15E20
15E30
15E40

Exhibit Dates: 1/90–4/92. Effective 4/92, MOS 15E was discontinued.

Career Management Field: 13 (Field Artillery).

Description

Summary: Supervises or serves as a member of a Pershing missile crew. *Skill Level 10:* Under supervision, assists in launcher emplacement, assembly, disassembly, transporting, and storing of missile and missile components; uses radio and telephone communications equipment; operates gas turbine generator set; operates wrecker boom and tracked vehicles; performs maintenance on vehicles and field communications equipment; participates in emplacement and operation of azimuth reference unit; runs system diagnostics. *Skill Level 20:* Able to perform the duties required for Skill Level 10; operates equipment necessary to position launch vehicles and to assemble missiles; installs and monitors special equipment; operates erector launcher control panel during mating, recovery, and firing operations; is familiar with the mechanical technology of guided missiles, including thrust propulsion, energy conversion, hydraulics, and pneumatic systems; supervises operations performed by missile handlers while transporting and storing missiles and components; assists in supervising crew during missile prefire, firing, postfiring, assembly, and disassembly operations; supervision experience may vary. *Skill Level 30:* Able to perform the duties required for Skill Level 20; serves as section chief and first-line supervisor for 9-11 subordinates; supervises emplacement and preparation of missile for firing; supervises work techniques, work distribution, and maintenance; supervises positioning and operation of azimuth-laying equipment or azimuth reference unit; supervises missile storage, maintenance, assembly, disassembly, and firing procedures. *Skill Level 40:* Able to perform the duties required for Skill Level 30; is a first-line supervisor for 11 or more persons; supervises operation of firing, ammunition, or security platoon or battery control central section; supervises procurement, distribution, and testing of missile and missile components; serves as principal noncommissioned officer; prepares operation reports; participates in coordination and implementation of Pershing missile operation, training programs, administrative matters, and communication activities.

Recommendation, Skill Level 10

In the lower-division baccalaureate/associate degree category, 2 semester hours in electromechanical and hydraulic systems troubleshooting and maintenance. (NOTE: This recommendation for skill level 10 is valid for the dates 1/90-9/91 only) (6/89).

Recommendation, Skill Level 20

In the lower-division baccalaureate/associate degree category, 3 semester hours in electromechanical and hydraulic systems troubleshooting and maintenance; if served as section chief, team leader, or squad/shift leader, 3 semester hours in principles of supervision and 3 in maintenance management (6/89).

Recommendation, Skill Level 30

In the lower-division baccalaureate/associate degree category, 3 semester hours in electromechanical and hydraulic systems troubleshooting and maintenance, 3 in principles of supervision, 3 in maintenance management, and 3 in communication skills (oral and written) (6/89).

Recommendation, Skill Level 40

In the lower-division baccalaureate/associate degree category, 3 semester hours in electromechanical and hydraulic systems troubleshooting and maintenance, 3 in principles of supervision, 3 in maintenance management, and 3 in communication skills (oral and written). In the upper-division baccalaureate category, 3 semester hours for field experience in management and 3 in personnel management (6/89).

MOS-15F-003

Aircraft Electrician
15F10
15F20
15F30

Exhibit Dates: 10/03–Present.

Career Management Field: 15 (Aircraft Maintenance).

Description

Summary: Supervises or performs maintenance on aircraft electrical/electronic systems and aircraft crew station instruments. *Skill Level 10:* Removes and replaces electrical/electronic elements of assemblies and components, including electrical wiring, aircraft battery systems, and instruments; applies proper soldering techniques during circuitry repair; cleans, preserves, and stores electrical/electronic components and aircraft instruments; uses and maintains common and special tools; maintains and repairs ground support equipment; uses aircraft forms and records; prepares requests for turn-in and repair parts. *Skill Level 20:* Able to perform the duties required for Skill Level 10; provides technical guidance to subordinates; applies troubleshooting techniques to diagnose and localize malfunctions to specific electrical/electronic components including solid state and transistorized subsystems; disassembles, repairs, reassembles, adjusts, and tests electrical/electronic elements of assemblies and components; applies principles of pneudraulics to troubleshooting of interfacing electropneudraulic systems and components; repairs and replaces printed circuits; repairs nickel cadmium batteries; performs maintenance on diagnostic equipment; prepares forms and records. *Skill Level 30:* Able to perform the duties required for Skill Level 20; may serve as technical inspector; plans work flow; applies production control, quality assurance and other maintenance management principles; supervises and evaluates work performance; conducts on-the-job training; monitors shop and flight-line safety.

Recommendation, Skill Level 10

Credit may be granted on the basis of an individualized assessment of the student (4/04).

Recommendation, Skill Level 20

Credit may be granted on the basis of an individualized assessment of the student (4/04).

Recommendation, Skill Level 30

In the lower-division baccalaureate/associate degree category, 3 semester hours in aircraft maintenance fundamentals, 3 in AC/DC circuit analysis, 3 in electronic devices and circuits, 3 in electronic systems troubleshooting and repair, 3 in principles of supervision, 1 in introduction to computers, 1 in computer applications, 3 in maintenance management, 2 in records management, and 2 in aviation safety (4/04).

MOS-15G-001

Aircraft Structural Repairer
15G10
15G20
15G30

Exhibit Dates: 10/03–Present.

Career Management Field: 15 (Aircraft Maintenance).

Description

Summary: Supervises, inspects, or performs maintenance and repair of aircraft structure. *Skill Level 10:* Makes airframe structural repairs by fabrication of new parts using approved fasteners, rivets, special-purpose fasteners, and fiberglass materials; uses and maintains shop tools and equipment; uses field and technical manuals to make specific repairs. *Skill Level 20:* Able to perform the duties required for Skill Level 10; inspects airframes and components before, during, and after repair or modification to ensure specifications are met; uses precision measuring gauges and instruments; determines shop and bench stock requirements; prepares or assists in the preparation of requests for parts, tools, and supplies; interprets blueprints; coordinates work load; applies maintenance management principles and procedures to shop operations; supervises on-the-job training. *Skill Level 30:* Able to perform the duties required for Skill Level 20; supervises maintenance; may perform technical inspections on aircraft structures and control surfaces; conducts and supervises on-the-job training; ensures quality and production control; applies management principles and procedures to shop operations; may supervise from 6 to 19 persons; monitors shop and flight-line safety.

Recommendation, Skill Level 10

Credit may be granted on the basis of an individualized assessment of the student (4/04).

Recommendation, Skill Level 20

Credit may be granted on the basis of an individualized assessment of the student (4/04).

Recommendation, Skill Level 30

In the lower-division baccalaureate/associate degree category, 3 semester hours in aviation safety, 3 in aircraft and helicopter structures, 6 in aircraft maintenance fundamentals, 3 in principles of supervision, 1 in introduction to computers, 2 in supply management, 3 in maintenance management, 1 in computer applications, and 2 in records management (4/04).

MOS-15H-001

Aircraft Pneudraulics Repairer
15H10
15H20
15H30

Exhibit Dates: 10/03–Present.

Career Management Field: 15 (Aircraft Maintenance).

Description

Summary: Supervises, inspects, or performs maintenance on aircraft hydraulic systems. *Skill Level 10:* Removes and installs tubing assemblies, hoses, valves, pressure transmitters, and switches; services filters and shock absorbers/struts; repairs tubing assemblies and hydraulic and pneumatic valves; flushes and bleeds hydraulic systems; uses and maintains common and special tools; prepares requests for turn-in and repair parts; uses aircraft forms and records. *Skill Level 20:* Able to perform the duties required for Skill Level 10; provides technical guidance to subordinates; applies troubleshooting techniques to diagnose and localize malfunctions to hydraulic systems, subsystems, and components; maintains selected items of ground support equipment; prepares forms and records; supervises on-the-job training. *Skill Level 30:* Able to perform the duties required for Skill Level 20; applies production/quality assurance and maintenance management principles and procedures to shop operations; plans work flow; conducts and supervises on-the-job training; may perform technical inspections of electrical/electronic and subsystems components before, during, and after repair or modification; may supervise from 6 to 19 persons.

Recommendation, Skill Level 10

Credit may be granted on the basis of an individualized assessment of the student (4/04).

Recommendation, Skill Level 20

Credit may be granted on the basis of an individualized assessment of the student (4/04).

Recommendation, Skill Level 30

In the lower-division baccalaureate/associate degree category, 3 semester hours in fluid power, 3 in aviation safety, 6 in aircraft maintenance fundamentals, 3 in principles of supervision, 2 in supply management, 3 in maintenance management, and 2 in records management (4/04).

MOS-15J-004

OH-58D Armament/Electrical/Avionics Systems Repairer
15J10
15J20
15J30
15J40

Exhibit Dates: 10/03–Present.

Career Management Field: 67 (Aircraft Maintenance).

Description

Summary: Maintains, repairs, and aligns armament, electrical, and avionics systems installed on the OH-58D aircraft. *Skill Level 10:* Performs troubleshooting and repair procedures; removes and replaces components of security, communications, navigation, transponder, and flight control systems; maintains, repairs, and installs armament systems and associated weapons on the OH-58D aircraft. Prepares requisitions and maintains stock inventories. *Skill Level 20:* Able to perform the duties required for Skill Level 10; coordinates team work schedules; trains and assists subordinates. *Skill Level 30:* Able to perform the duties required for Skill Level 20; coordinates and supervises unit and intermediate avionics/electrical/armament maintenance; implements quality assurance programs; supervises and schedules training; coordinates tasks with superiors; maintains appropriate forms and records. *Skill Level 40:* Able to perform the duties required for Skill Level 30; supervises aircraft armament, fire control, electrical, and avionics systems maintenance; supervises technical inspection activities; determines personnel, facility, and parts requirements; supervises training programs; prepares evaluations and reports.

Recommendation, Skill Level 10

Credit may be granted on the basis of an individualized assessment of the student (6/02).

Recommendation, Skill Level 20

Credit may be granted on the basis of an individualized assessment of the student (6/02).

Recommendation, Skill Level 30

In the lower-division baccalaureate/associate degree category, 3 semester hours in aviation safety, 3 in aircraft maintenance fundamentals, 3 in AC/DC circuit analysis, 3 in electronic devices and circuits, 2 in electronics communication, 3 in electronic systems troubleshooting and repair, 1 in digital electronics, 3 in principles of supervision, 1 in introduction to computers, 1 in computer applications, 3 in maintenance management, and 2 in records management (6/02).

Recommendation, Skill Level 40

In the lower-division baccalaureate/associate degree category, 3 semester hours in aviation safety, 3 in aircraft maintenance fundamentals, 3 in AC/DC circuit analysis, 3 in electronic devices and circuits, 2 in electronics communication, 3 in electronic systems troubleshooting and repair, 1 in digital electronics, 3 in principles of supervision, 1 in introduction to computers, 1 in computer applications, 3 in maintenance management, and 2 in records management. In the upper-division baccalaureate category, 3 semester hours for a management practicum, 3 in principles of management, and 3 in operations management (6/02).

MOS-15K-001

Aircraft Components Repair Supervisor
15K40

Exhibit Dates: 10/03–Present.

Career Management Field: 15 (Aircraft Maintenance).

Description

Able to supervise the duties of the following MOS's: 15B (Aircraft Powerplant Repairer), 15D (Aircraft Powertrain Repairer), 15F (Aircraft Electrician), 15G (Aircraft Structural Repairer), 15H (Aircraft Pneudraulics Repairer), and 15N (Avionic Mechanic); supervises maintenance operations of aircraft component repairs; plans work load in terms of resources and facilities; applies production control, quality assurance, and other maintenance management principles to aircraft component repair; supervises and evaluates the work performance of 20 or more persons; orients and instructs subordinates; conducts on-the-job training. NOTE: May have progressed from skill level 30 of MOS 15B, 15D, 15F, 15G, 15H,or 15N.

Recommendation

In the upper-division baccalaureate category, 3 semester hours for a management practicum, 3 in principles of management, and 3 in operations management (4/04).

MOS-15M-001

UH-1 Helicopter Repairer
15M10
15M20
15M30

Exhibit Dates: 10/03–Present.

Description

Summary: Performs or supervises maintenance on helicopters, excluding repairs of system components, or conducts maintenance inspections. *Skill Level 10:* Performs maintenance on helicopters; removes and installs such subsystem assemblies as engines, transmissions, gear boxes, rotor hubs, and rotor blades; performs extensive inspections and maintenance on helicopters; uses maintenance forms and records; performs operational checks; maintains ground support equipment; may perform crew chief duties. *Skill Level 20:* Able to perform the duties required for Skill Level 10; provides technical guidance to subordinates; troubleshoots helicopter systems; prepares maintenance forms and records; applies production control; instructs subordinates; conducts on-the-job training in helicopter maintenance, logistics, and safety practices; assists in performing weight and balance checks; performs section chief duties; uses computer management information systems. *Skill Level 30:* Able to perform the duties required for Skill Level 20; supervises all helicopter maintenance activities; plans work flow; applies quality control and other maintenance management principles to shop, flight line, and logistics; supervises on-the-job training; evaluates work performance; monitors operational checks and test flights; certifies airworthiness of the aircraft; performs management functions; analyzes computer management information systems products.

Recommendation, Skill Level 10

Credit may be granted on the basis of an individualized assessment of the student (1/01).

Recommendation, Skill Level 20

Credit may be granted on the basis of an individualized assessment of the student (1/01).

Recommendation, Skill Level 30

In the lower-division baccalaureate/associate degree category, 9 semester hours in aircraft maintenance fundamentals, 3 in aircraft systems fundamentals, 3 in principles of supervision, 3 in computer management information systems, and 3 in logistics management. In the upper-division baccalaureate category, 3 semester hours in organizational management, 3 in human resource management, and 3 in advanced computer management information systems (1/01).

MOS-15N-001

Avionic Mechanic
15N10
15N20
15N30

Exhibit Dates: 10/03–Present.

Career Management Field: 15 (Aircraft Maintenance).

Description

Summary: Performs unit maintenance on tactical security systems (COMSEC), communications, navigation, transponder, and flight control equipment installed in aircraft; maintains aircraft antennas and associated aircraft wiring; aligns radar antennas; maintains stock and bench components for repair of aircraft avionics equipment. *Skill Level 10:* Performs troubleshooting procedures to the level of individual units; removes and replaces modules of security systems, communications, navigation, transponder, and flight control equipment and their associated wiring and antennas; checks performance of individual systems; prepares forms and records related to aircraft maintenance. *Skill Level 20:* Able to perform the duties required for Skill Level 10; coordinates maintenance schedules; trains and assists subordinates. *Skill Level 30:* Able to perform the duties required for Skill Level 20; coordinates and supervises unit and intermediate avionic maintenance on all aircraft; implements quality assurance programs; supervises training; coordinates tasks with superiors; maintains equipment forms and records.

Recommendation, Skill Level 10

Credit may be granted on the basis of an individualized assessment of the student (4/04).

Recommendation, Skill Level 20

Credit may be granted on the basis of an individualized assessment of the student (4/04).

Recommendation, Skill Level 30

In the lower-division baccalaureate/associate degree category, 3 semester hours in aircraft maintenance fundamentals, 3 in AC/DC circuit analysis, 3 in electronic devices and circuits, 3 in electronic systems troubleshooting and repair, 2 in electronic communications, 1 in digital electronics, 3 in principles of supervision, 1 in introduction to computers, 1 in computer applications, 3 in maintenance management, 2 in records management, and 2 in aviation safety (4/04).

MOS-15P-001

Aviation Operations Specialist
15P10
15P20
15P30
15P40
15P50

Exhibit Dates: 10/03–Present.

Career Management Field: 93 (Aviation Operation).

Description

Summary: Schedules, clears, and dispatches aircraft. *Skill Level 10:* Processes cross-country and local flight plans with other agencies including the FAA; maintains flight information; prepares, types, and maintains records and reports on flight operations and activities; maintains current files on flying regulations and navigational aid information; arranges ground services for transient aircraft; interprets and posts teletype weather reports; understands terminology used in air navigation; knows air traffic control advisory procedures. *Skill Level 20:* Able to perform the duties required for Skill Level 10; supervises a small flight operations activity consisting of 5 to 12 subordinates; schedules aircraft missions, dispatches aircraft, and performs associated administrative duties; plans and schedules work assignments; checks work of subordinates and instructs them in proper work techniques and procedures; reviews, consolidates, and prepares technical, personnel, and administrative reports; assists in preparing pre-accident plans. *Skill Level 30:* Able to perform the duties required for Skill Level 20; supervises a medium-sized flight operations activity consisting of 13 or more subordinates; provides technical guidance to subordinates; supervises the preparation of situation map; prepares operations letters; assists administratively in aircraft accident investigations; assists in the preparation of letters of agreement, operations estimates, and operations orders. *Skill Level 40:* Able to perform the duties required for Skill Level 30; supervises a large flight operations activity; plans, coordinates, and supervises activities pertaining to organization, training, combat operations, and combat intelligence; coordinates implementation of operations, training programs, and communication activities. NOTE: May have progressed to 93P40 from 93P30 or 93B30, (Aeroscout Observer). *Skill Level 50:* Able to perform the duties required for Skill Level 40; serves as a mid-level manager; supervises the processing of operations and intelligence information at the brigade level.

Recommendation, Skill Level 10

Credit may be granted on the basis of an individualized assessment of the student (11/99).

Recommendation, Skill Level 20

Credit may be granted on the basis of an individualized assessment of the student (11/99).

Recommendation, Skill Level 30

In the lower-division baccalaureate/associate degree category, 3 semester hours in introduction to aircraft dispatching, 3 in aircraft operations management, 3 in FAA regulations, 3 in computer literacy, 2 in computer applications, 2 in office procedures, 3 in principles of supervision, 3 in airfield management, and 3 in advanced aircraft dispatching. In the upper-division baccalaureate category, 3 semester hours in aviation management and 2 in organizational management (11/99).

Recommendation, Skill Level 40

In the lower-division baccalaureate/associate degree category, 3 semester hours in introduction to aircraft dispatching, 3 in aircraft operations management, 3 in FAA regulations, 3 in computer literacy, 3 in computer applications, 2 in office procedures, 3 in principles of supervision, 3 in airfield management, and 3 in advanced aircraft dispatching. In the upper-division baccalaureate category, 3 semester hours in aviation management, 3 in organizational management, 3 in air traffic operations, and 3 for field experience in management (11/99).

Recommendation, Skill Level 50

In the lower-division baccalaureate/associate degree category, 3 semester hours in introduction to aircraft dispatching, 3 in aircraft operations management, 3 in FAA regulations, 3 in computer literacy, 3 in computer applications, 2 in office procedures, 3 in principles of supervision, 3 in airfield management, and 3 in advanced aircraft dispatching. In the upper-division baccalaureate category, 3 semester hours in aviation management, 3 in organizational management, 3 in air traffic operations, and 6 for field experience in management (11/99).

MOS-15Q-001

Air Traffic Control Operator

15Q10
15Q20
15Q30
15Q40

Exhibit Dates: 10/03–Present.

Career Management Field: 93 (Aviation Operation).

Description

Summary: Provides both radar and nonradar air traffic control services for military and civilian air traffic during takeoff, flight, and landing under Visual Flight Rules (VFR), Special Visual Flight Rules (SVFR), and Instrument Flight Rules (IFR). *Skill Level 10:* Controls takeoffs, flight, and landing of IFR/VFR/SVFR air traffic; provides flight following, enroute routing, terminal approach control, and ground-controlled approach services; operates radar; installs and moves ATC radar and associated equipment; applies FAA and Army air traffic rules and regulations; may hold an FAA radar controller certificate; issues special air traffic control instructions to aviators concerning airfield facilities, emergency landing areas, obstructions, landmarks, flying areas, restrictions, local regulations, weather advisories, and observed hazards that affect the safe operation of aircraft; keeps records and statistics, including tape recordings of voice radio communications, on daily air traffic operations; processes incoming and outgoing flight data information and analyzes air traffic; operates airfield lighting systems, light signals, and nonradar approach control boards; employs aeronautical charts, maps, radio, and ground communications; applies FAA and Army air traffic rules and regulations and holds an FAA control tower operator certificate; controls vehicular traffic on airport movement area. *Skill Level 20:* Able to perform the duties required for Skill Level 10; provides technical guidance to subordinates. *Skill Level 30:* Able to perform the duties required for Skill Level 20; serves as shift supervisor or facility chief; supervises inspection procedures, site operation, and record keeping; prepares shift duty rosters; develops and administers facility training program and on-the-job training. *Skill Level 40:* Able to perform the duties required for Skill Level 30; as ATC tower chief, plans, organizes, and supervises air traffic control tower activities; assigns duties; spot-checks work performed; instructs subordinates in work techniques and procedures; coordinates work activities; establishes and supervises on-the-job training programs; prepares technical, personnel, and administrative reports; monitors the handling and storage of tape-recorded conversations between air traffic control personnel and aircraft pilots; maintains records and statistics on daily air traffic operations.

Recommendation, Skill Level 10

Credit may be granted on the basis of an individualized assessment of the student (11/99).

Recommendation, Skill Level 20

Credit is not recommended Credit may be granted on the basis of an individualized assessment of the student (11/99).

Recommendation, Skill Level 30

In the lower-division baccalaureate/associate degree category, 6 semester hours in air traffic control, 6 in air traffic management, 3 in federal regulations, 3 in principles of supervision, 3 in computer literacy, and 3 in computer applications. In the upper-division baccalaureate category, 3 semester hours in principles of instruction (11/99).

Recommendation, Skill Level 40

In the lower-division baccalaureate/associate degree category, 6 semester hours in air traffic control, 6 in air traffic management, 3 in federal regulations, 3 in principles of supervision, 3 in computer literacy, and 3 in computer applications. In the upper-division baccalaureate category, 3 semester hours in organizational management, 6 for field experience in management, 6 in air traffic operations, and 3 in principles of instruction (11/99).

MOS-15R-001

AH-64 Attack Helicopter Repairer

15R10
15R20
15R30
15R40

Exhibit Dates: 10/03–Present.

Career Management Field: 67 (Aircraft Maintenance).

Description

Summary: Performs or supervises maintenance on attack helicopters, excluding repairs of system components. *Skill Level 10:* Assists in the removal and installation of such subsystem assemblies as engines, transmissions, gear boxes, rotor hubs, and rotor blades; prepares helicopter for extensive inspections and maintenance; assists in operational checks; performs maintenance on common and special tools and ground support equipment; uses maintenance forms and records. *Skill Level 20:* Serves as attack helicopter crew chief; provides technical guidance to subordinates; maintains ground support equipment; prepares maintenance forms and records. *Skill Level 30:* Able to perform the duties required for Skill Level 20; supervises all maintenance on attack helicopters; plans work flow; applies production control, quality control, and other maintenance management principles to shop, flight-line, and supply operations; instructs personnel and conducts on-the-job training in helicopter maintenance, supply, and safety techniques; evaluates work performance; monitors operational checks and test flights; certifies airworthiness of the aircraft; if duty position is section chief, serves as the principal noncommissioned officer in a section. *Skill Level 40:* Able to perform the duties required for Skill Level 30; supervises all maintenance activities; diagnoses complex malfunctions; determines personnel and parts requirements; coordinates work schedules and assigns duties; applies production, quality control, and other maintenance management principles and procedures; prepares evaluations, special reports, and maintenance records; supervises technical training programs; assists in the preparation of plans and policies; controls the flow of work orders, requisitions, and correspondence; plans the layout of helicopter maintenance areas, shops, and facilities.

Recommendation, Skill Level 10

Credit may be granted on the basis of an individualized assessment of the student (6/99).

Recommendation, Skill Level 20

Credit may be granted on the basis of an individualized assessment of the student (6/99).

Recommendation, Skill Level 30

In the lower-division baccalaureate/associate degree category, 12 semester hours in basic helicopter maintenance, 2 in aircraft forms and records, 2 in aircraft weight and balance, 5 in aviation maintenance management, 2 in quality control, and 3 in principles of supervision (6/99).

Recommendation, Skill Level 40

In the lower-division baccalaureate/associate degree category, 12 semester hours in basic helicopter maintenance, 3 in aircraft forms and records, 2 in aircraft weight and balance, 8 in aviation maintenance management, 4 in quality control, and 3 in principles of supervision. In the upper-division baccalaureate category, 3 semester hours in management practicum and 3 in principles of management (6/99).

MOS-15S-001

OH-58D Helicopter Repairer

15S10
15S20
15S30
15S40

Exhibit Dates: 10/03–Present.

Career Management Field: 67 (Aircraft Maintenance).

Description

Summary: Performs or supervises maintenance on OH-58D helicopters, excluding repairs of system components. *Skill Level 10:* Assists in the removal and installation of various subsystem assemblies, including engines, weapons systems, transmissions, gear boxes, rotor hubs, and rotor blades and associated equipment; prepares helicopters for extensive inspections and maintenance; assists in operational checks; performs maintenance on common and special tools and ground support equipment; uses maintenance forms and records; performs aircrew duties as required. *Skill Level 20:* Able to perform the duties required for Skill Level 10; provides technical guidance to subordinates; maintains ground support equipment; prepares equipment and aircraft maintenance forms and records. *Skill Level 30:* Able to perform the duties required for Skill Level 20; supervises all maintenance on utility helicopters; plans work flow; applies production control, quality control, and other maintenance management principles to shop, flight-line, and supply operations; instructs subordinates and conducts on-the-job training in helicopter maintenance, supply, and safety techniques; evaluates work performance; monitors operational checks and test flights; certifies airworthiness of the aircraft. *Skill Level 40:* Able to perform the duties required for Skill Level 30; supervises, counsels, and trains all subordinates; maintains flow charts; coordinates flight schedules; accounts shop equipment, calibration records, and files; uses problem-solving techniques; writes reports and personnel evaluations; conducts train-the-trainer programs.

Recommendation, Skill Level 10

Credit may be granted on the basis of an individualized assessment of the student (6/99).

Recommendation, Skill Level 20

Credit may be granted on the basis of an individualized assessment of the student (6/99).

Recommendation, Skill Level 30

In the lower-division baccalaureate/associate degree category, 12 semester hours in basic helicopter maintenance, 2 in aircraft forms and records, 2 in aircraft weight and balance, 5 in aviation maintenance management, 2 in quality control, and 3 in principles of supervision (6/99).

Recommendation, Skill Level 40

In the lower-division baccalaureate/associate degree category, 12 semester hours in basic helicopter maintenance, 3 in aircraft forms and records, 2 in aircraft weight and balance, 8 in aviation maintenance management, 4 in quality control, and 3 in principles of supervision. In the upper-division baccalaureate category, 3 semester hours in management practicum and 3 in principles of management (6/99).

MOS-15T-001

UH-60 Helicopter Repairer

15T10

15T20

15T30

15T40

Exhibit Dates: 1/03–Present.

Career Management Field: 67 (Aircraft Maintenance).

Description

Summary: Performs or supervises maintenance on tactical transport helicopters, excluding repairs of system components. *Skill Level 10:* Assists in the removal and installation of such subsystem assemblies as engines, transmissions, gear boxes, rotor hubs, and rotor blades; prepares helicopter for extensive inspections and maintenance; assists in operational checks; performs maintenance on common and special tools and ground support equipment; uses maintenance forms and records. *Skill Level 20:* Able to perform the duties required for Skill Level 10; serves as tactical transport helicopter crew chief; provides technical guidance to subordinates; maintains ground support equipment; prepares maintenance forms and records. *Skill Level 30:* Able to perform the duties required for Skill Level 20; supervises all maintenance on tactical transport helicopters; plans work flow; applies production control, quality control, and other maintenance management principles to shop, flightline, and supply operations; instructs personnel and conducts on-the-job training in helicopter maintenance, supply, and safety techniques; evaluates work performance; monitors operational checks and test flights; certifies airworthiness of the aircraft; if duty position is section chief, serves as the principal noncommissioned officer in a section. *Skill Level 40:* Able to perform the duties required for Skill Level 30; supervises all maintenance activities; diagnoses complex malfunctions; determines personnel and parts requirements; coordinates work schedules and assigns duties; applies production, quality control, and other maintenance management principles and procedures; prepares evaluations, special reports, and maintenance records; supervises technical training programs; assists in the preparation of plans and policies; controls the flow of work orders, requisitions, and correspondence; plans the layout of helicopter maintenance areas, shops, and facilities.

Recommendation, Skill Level 10

Credit may be granted on the basis of an individualized assessment of the student (6/99).

Recommendation, Skill Level 20

Credit may be granted on the basis of an individualized assessment of the student (6/99).

Recommendation, Skill Level 30

In the lower-division baccalaureate/associate degree category, 12 semester hours in basic helicopter maintenance, 2 in aircraft forms and records, 2 in aircraft weight and balance, 5 in aviation maintenance management, 2 in quality control, and 3 in principles of supervision (6/99).

Recommendation, Skill Level 40

In the lower-division baccalaureate/associate degree category, 12 semester hours in basic helicopter maintenance, 3 in aircraft forms and records, 2 in aircraft weight and balance, 8 in aviation maintenance management, 4 in quality control, and 3 in principles of supervision. In the upper-division baccalaureate category, 3 semester hours in management practicum and 3 in principles of management (6/99).

MOS-15U-001

CH-47 Helicopter Repairer

15U10

15U20

15U30

15U40

Exhibit Dates: Pending Evaluation. 4/03–Present.

MOS-15V-001

Observation/Scout Helicopter Repairer

15V10

15V20

15V30

Exhibit Dates: 10/03–Present.

Description

Summary: Performs or supervises maintenance on observation/scout helicopters, excluding repairs of system components, or conducts maintenance inspections. *Skill Level 10:* Performs maintenance on observation/scout helicopters; removes and installs such subsystem assemblies as engines, transmissions, gear boxes, rotor hubs, and rotor blades; performs extensive inspections and maintenance on helicopters; uses maintenance forms and records; performs operational checks; maintains ground support equipment; may perform crew chief duties. *Skill Level 20:* Able to perform the duties required for Skill Level 10; provides technical guidance to subordinates; troubleshoots helicopter systems; prepares maintenance forms and records; applies production control; instructs subordinates; conducts on-the-job training in helicopter maintenance, logistics, and safety practices; assists in performing weight and balance checks; performs section chief duties; uses computer management information systems. *Skill Level 30:* Able to perform the duties required for Skill Level 20; supervises all helicopter maintenance; plans work flow; applies quality control and other maintenance management principles to shop, flight line, and logistics; supervises on-the-job training; evaluates work performance; monitors operational checks and test flights; certifies airworthiness of the aircraft; performs management functions; analyzes computer management information system products.

Recommendation, Skill Level 10

Credit may be granted on the basis of an individualized assessment of the student (1/01).

Recommendation, Skill Level 20

Credit may be granted on the basis of an individualized assessment of the student (1/01).

Recommendation, Skill Level 30

In the lower-division baccalaureate/associate degree category, 9 semester hours in aircraft maintenance fundamentals, 3 in aircraft systems fundamentals, 3 in principles of supervision, 3 in computer management information systems, and 3 in logistics management. In the upper-division baccalaureate category, 3 semester hours in organizational management, 3 in human resource management, and 3 in advanced computer management information systems (1/01).

MOS-15X-001

AH-64A Armament/Electrical Systems Repairer
15X10
15X20
15X30
15X40

Exhibit Dates: Pending Evaluation. 4/03–Present.

MOS-15Y-001

AH-64D Armament/Electrical/Avionics Systems Repairer
15Y10
15Y20
15Y30
15Y40

Exhibit Dates: 10/03–Present.

Career Management Field: 67 (Aircraft Maintenance).

Description

Summary: Maintains, repairs, and aligns armament, electrical, and avionics systems installed on the AH-64D aircraft. *Skill Level 10:* Performs troubleshooting and repair procedures; removes and replaces components of security, communications, navigation, transponder, and flight control systems; maintains, repairs, and installs armament systems and associated weapons on the AH-64D aircraft; prepares requisitions and maintains stock inventories. *Skill Level 20:* Able to perform the duties required for Skill Level 10; coordinates team work schedules; trains and assists subordinates. *Skill Level 30:* Able to perform the duties required for Skill Level 20; coordinates and supervises unit and intermediate avionics/electrical/armament maintenance; implements quality assurance programs; supervises and schedules training; coordinates tasks with superiors; maintains appropriate forms and records. *Skill Level 40:* Able to perform the duties required for Skill Level 30; supervises aircraft armament, fire control, electrical, and avionics systems maintenance; supervises technical inspection activities; determines personnel, facility, and parts requirements; supervises training programs; prepares evaluations and reports.

Recommendation, Skill Level 10

Credit may be granted on the basis of an individualized assessment of the student (6/02).

Recommendation, Skill Level 20

Credit may be granted on the basis of an individualized assessment of the student (6/02).

Recommendation, Skill Level 30

In the lower-division baccalaureate/associate degree category, 3 semester hours in aviation safety, 3 in aircraft maintenance fundamentals, 3 in AC/DC circuit analysis, 3 in electronic devices and circuits, 2 in electronics communication, 3 in electronic systems troubleshooting and repair, 1 in digital electronics, 3 in principles of supervision, 1 in introduction to computers, 1 in computer applications, 3 in maintenance management, and 2 in records management (6/02).

Recommendation, Skill Level 40

In the lower-division baccalaureate/associate degree category, 3 semester hours in aviation safety, 3 in aircraft maintenance fundamentals, 3 in AC/DC circuit analysis, 3 in electronic devices and circuits, 2 in electronics communication, 3 in electronic systems troubleshooting and repair, 1 in digital electronics, 3 in principles of supervision, 1 in introduction to computers, 1 in computer applications, 3 in maintenance management, and 2 in records management. In the upper-division baccalaureate category, 3 semester hours for a management practicum, 3 in principles of management, and 3 in operations management (6/02).

MOS-15Z-001

Aircraft Maintenance Senior Sergeant
15Z50

Exhibit Dates: 10/03–Present.

Career Management Field: 67 (Aircraft Maintenance).

Description

Supervises aircraft maintenance at either unit, intermediate, or depot levels on activities having a mix of aircraft maintenance and/or component repair; serves as the principal noncommissioned officer in aircraft maintenance activities and may become first sergeant; plans and lays out aircraft maintenance areas and component repair shops and facilities; ensures that proper maintenance techniques have been employed by use of quality assurance inspections; supervises and participates in preparation of studies, evaluations, special reports, and records pertaining to aircraft maintenance and component repair operations and training and related activities; applies production quality control and other maintenance management principles and procedures; processes work orders and allocates maintenance responsibilities.

Recommendation

In the lower-division baccalaureate/associate degree category, 3 semester hours in technical writing, 3 in principles of supervision, and 3 in maintenance management. In the upper-division baccalaureate category, 3 semester hours in a management practicum and 3 in principles of management (6/99).

MOS-16D-001

Hawk Missile Crew Member
((Hawk Missile Crewman))
16D10
16D20
16D30
16D40

Exhibit Dates: 1/90–2/92.

Career Management Field: 16 (Air Defense Artillery), subfield 161 (Air Defense Artillery Missile and Gun Operations).

Description

Summary: Serves as a member of or supervises the operation of a firing platoon, an assembly section, or a service and maintenance section of a Hawk missile battery. *Skill Level 10:* Maintains and prepares Hawk missile and associated equipment for launching; participates in the unloading and emplacement of equipment; assists in assembling, arming, and disarming the missile; performs routine preventive maintenance on launch and handling equipment; assists in performing checks and adjustments; observes safety precautions. *Skill Level 20:* Able to perform the duties required for Skill Level 10; provides guidance to subordinates. *Skill Level 30:* Able to perform the duties required for Skill Level 20; supervises the assembly, testing, arming, disarming, loading, and unloading of Hawk missiles; supervises the transportation and emplacement of equipment; supervises preventive maintenance and coordinates organizational maintenance of launch and loading equipment. *Skill Level 40:* Able to perform the duties required for 16D30 or 16E30 (Hawk Fire Control Crewman); serves as platoon sergeant, supervising over 20 persons; plans and organizes security arrangements; interprets intelligence information; supervises maintenance of equipment.

Recommendation, Skill Level 10

Credit is not recommended because of the limited technical nature of the skills, competencies, and knowledge (2/76).

Recommendation, Skill Level 20

In the vocational certificate category, 3 semester hours in hydraulic and electrohydraulic systems (2/76).

Recommendation, Skill Level 30

In the vocational certificate category, 3 semester hours in hydraulic and electrohydraulic systems. In the lower-division baccalaureate/associate degree category, 3 semester hours in human relations (2/76).

Recommendation, Skill Level 40

In the vocational certificate category, 3 semester hours in hydraulic and electrohydraulic systems. In the lower-division baccalaureate/associate degree category, 3 semester hours in human relations, 3 in personnel supervision, and 3 for field experience in management (2/76).

MOS-16D-002

Hawk Missile Crewmember
16D10
16D20
16D30
16D40

Exhibit Dates: 3/92–6/95. Effective 6/95, MOS 16D was discontinued.

Career Management Field: 14 (Air Defense Artillery).

Description

Summary: Serves as a member of or supervises the operation of a firing platoon, an assembly section, or a service and maintenance section of a Hawk missile battery. *Skill Level 10:* Maintains and prepares Hawk missile and associated launching/storage area equipment; participates in the unloading and emplacement of equipment; assists in assembling, arming, and disarming the missile; performs routine preventive maintenance on launch and handling equipment; assists in performing checks and adjustments; observes safety precautions. *Skill Level 20:* Able to perform the duties required for Skill Level 10; provides guidance to subordinates. *Skill Level 30:* Able to perform the duties required for Skill Level 20; supervises the assembly, testing, arming, disarming, loading, and unloading of Hawk missiles; supervises the transportation and emplacement of equipment; supervises preventive maintenance and coordinates organizational maintenance of launch and loading equipment. *Skill Level 40:* Able to perform the duties required for Skill Level 30 of MOS 16D or 16E (Hawk Fire Control Crewmember); serves as platoon sergeant, supervising over 20 persons; plans and organizes security arrangements; interprets intelligence information; supervises maintenance of equipment. NOTE: May have progressed to 16D40 from 16D30 or 16E30 (Hawk Fire Control Crewmember).

Recommendation, Skill Level 10

Credit may be granted on the basis of an individualized assessment of the student (3/94).

Recommendation, Skill Level 20

In the lower-division baccalaureate/associate degree category, 3 semester hours in military science, 3 in hydraulic systems, and 3 in personnel supervision. (NOTE: This recommendation for skill level 20 is valid for the dates 3/92-2/95 only) (3/92).

Recommendation, Skill Level 30

In the lower-division baccalaureate/associate degree category, 3 semester hours in military science, 3 in hydraulic systems, 3 in personnel supervision, and 2 in records and information management (3/92).

Recommendation, Skill Level 40

In the lower-division baccalaureate/associate degree category, 3 semester hours in military science, 3 in hydraulic systems, 3 in personnel supervision, and 2 in records and information management. In the upper-division baccalaureate category, 3 semester hours in organizational management and 3 for field experience in management (3/92).

MOS-16E-001

Hawk Fire Control Crew Member
16E10
16E20
16E30

Exhibit Dates: 1/90–2/92.

Career Management Field: 16 (Air Defense Artillery), subfield 161 (Air Defense Artillery Missile and Gun Operations).

Description

Summary: Serves as a member of or supervises a command and acquisition section or firing section of a Hawk firing platoon. *Skill Level 10:* Operates fire control equipment or engagement simulator; operates and performs operator maintenance checks and adjustments on Hawk missile fire control equipment; observes safety precautions. *Skill Level 20:* Able to perform the duties required for Skill Level 10; provides technical guidance to subordinates. *Skill Level 30:* Able to perform the duties required for Skill Level 20; supervises approximately 10 persons operating Hawk fire control equipment; prepares radar coverage and clutter diagrams; determines known reference point; supervises moving and placement of equipment.

Recommendation, Skill Level 10

Credit is not recommended because of the limited technical nature of the skills, competencies, and knowledge (2/76).

Recommendation, Skill Level 20

Credit is not recommended because of the limited technical nature of the skills, competencies, and knowledge (2/76).

Recommendation, Skill Level 30

In the lower-division baccalaureate/associate degree category, 3 semester hours in human relations (2/76).

MOS-16E-002

Hawk Fire Control Crewmember
16E10
16E20
16E30

Exhibit Dates: 3/92–6/95. Effective 6/95, MOS 16E was discontinued.

Career Management Field: 14 (Air Defense Artillery).

Description

Summary: Serves as a member of or supervises a command and acquisition section or firing section of a Hawk firing platoon. *Skill Level 10:* Operates fire control equipment or engagement simulator; operates and performs operator maintenance checks and adjustments on Hawk missile fire control equipment; observes safety precautions. *Skill Level 20:* Able to perform the duties required for Skill Level 10; supervises and provides technical guidance to subordinates. *Skill Level 30:* Able to perform the duties required for Skill Level 20; supervises persons operating Hawk fire control equipment; prepares radar coverage and clutter diagrams; determines known reference point; supervises moving and placement of equipment.

Recommendation, Skill Level 10

Credit may be granted on the basis of an individualized assessment of the student (3/94).

Recommendation, Skill Level 20

In the lower-division baccalaureate/associate degree category, 3 semester hours in military science and 3 in personnel supervision. (NOTE: This recommendation for skill level 20 is valid for the dates 3/92-2/95 only) (3/92).

Recommendation, Skill Level 30

In the lower-division baccalaureate/associate degree category, 3 semester hours in military science, 3 in personnel supervision, and 2 in records and information management (3/92).

MOS-16F-001

Light Air Defense Artillery Crewman (Reserve Forces)
16F10
16F20
16F30
16F40

Exhibit Dates: 1/90–9/90.

Career Management Field: 16 (Air Defense Artillery), subfield 161 (Air Defense Artillery Missile and Gun Operations).

Description

Summary: Supervises or serves as a member of a light air defense artillery automatic weapons, machine gun, or ammunition platoon, section, or squad. *Skill Level 10:* Loads ammunition and fires automatic weapons; operates computing sight; operates communications equipment; performs operator maintenance on automatic weapons; requisitions, receives, stores, and distributes ammunition. *Skill Level 20:* Able to perform the duties required for Skill Level 10; leads air defense artillery automatic weapons squad of approximately 10 persons; serves as assistant ammunition sergeant. *Skill Level 30:* Able to perform the duties required for Skill Level 20; supervises and coordinates the activities of subordinate squads; assures that operator maintenance standards are attained. *Skill Level 40:* Able to perform the duties required for Skill Level 30; supervises the operations of an automatic weapons platoon of approximately 30 persons.

Recommendation, Skill Level 10

Credit is not recommended because the skills, competencies, and knowledge are uniquely military in nature (2/76).

Recommendation, Skill Level 20

Credit is not recommended because the skills, competencies, and knowledge are uniquely military in nature (2/76).

Recommendation, Skill Level 30

In the lower-division baccalaureate/associate degree category, 3 semester hours in human relations (2/76).

Recommendation, Skill Level 40

In the lower-division baccalaureate/associate degree category, 3 semester hours in personnel supervision, 3 in human relations, and 3 for field experience in management (2/76).

MOS-16F-002

Light Air Defense Artillery Crewmember
16F10
16F20
16F30
16F40

Exhibit Dates: 10/90–4/92. Effective 4/92, MOS 16F was discontinued.

Career Management Field: 14 (Air Defense Artillery).

Description

Summary: Supervises or serves as a member of a light air defense artillery automatic weapons, machine gun, or ammunition platoon, section, or squad. *Skill Level 10:* Loads ammunition and fires automatic weapons; operates computing sight; operates communications equipment; performs operator maintenance on automatic weapons; requisitions, receives, stores, and distributes ammunition. *Skill Level 20:* Able to perform the duties required for Skill Level 10; leads air defense artillery automatic weapons squad; supervises battery ammunition section. *Skill Level 30:* Able to perform the duties required for Skill Level 20; leads air defense artillery automatic weapons squad; supervises battalion ammunition section and the operation of Army Air Defense command post. *Skill Level 40:* Able to perform the duties required for Skill Level 30; supervises operations of an automatic weapons platoon.

Recommendation, Skill Level 10

In the lower-division baccalaureate/associate degree category, 3 semester hours in military science. (NOTE: This recommendation for skill level 10 is valid for the dates 9/90-9/91 only) (3/92).

Recommendation, Skill Level 20

In the lower-division baccalaureate/associate degree category, 3 semester hours in military science and 3 in personnel supervision (3/92).

Recommendation, Skill Level 30

In the lower-division baccalaureate/associate degree category, 3 semester hours in military science, 3 in personnel supervision, and 2 in records and information management (3/92).

Recommendation, Skill Level 40

In the lower-division baccalaureate/associate degree category, 3 semester hours in military science, 3 in personnel supervision, and 2 in records and information management. In the upper-division baccalaureate category, 3 semester hours in organizational management and 3 for field experience in management (3/92).

MOS-16H-001

Air Defense Artillery Operations and Intelligence Assistant
16H10
16H20
16H30
16H40

Exhibit Dates: 1/90–9/90. Effective 9/90, MOS 16H was discontinued and its duties were incorporated into MOS's 16F, Light Air Defense Artillery Crewmember; 16J, Forward Area Alerting Radar Operator; 16P, Chaparral Crewmember; 16R, Vulcan Crewmember; and 16S, Man Portable Air Defense System or Pedestal Mounted Stinger Crewmember.

Career Management Field: 16 (Air Defense Artillery), subfield 162 (Air Defense Operations and Intelligence).

Description

Summary: Supervises or serves as a member of an air defense artillery activity engaged in operations or intelligence in order to prepare and revise operations data and situation maps. *Skill Level 10:* Performs plotting and operations or intelligence duties in an air defense artillery unit; plots, reports, and records operations or intelligence information by preparing charts, overlays, strip maps, training aids, logs, reports, and correspondence; installs and operates field telephone equipment. *Skill Level 20:* Able to perform the duties required for Skill Level 10; provides technical guidance to subordinates; collects and consolidates information from related units; prepares training materials and coordinates training activities; transmits intelligence and grid locations to other units. *Skill Level 30:* Able to perform the duties required for Skill Level 20; supervises operations; demonstrates correct procedures and techniques; supervises the preparation of situation maps and overlays; coordinates operations and intelligence data; supervises operation of communications equipment. *Skill Level 40:* Able to perform the duties required for 16H30 or 16J30 (Defense Acquisition Radar Operator); as a first sergeant, supervises 20 or more persons in an airspace control element; supervises and evaluates training; prepares operations plans; assists in reconnaissance.

Recommendation, Skill Level 10

In the lower-division baccalaureate/associate degree category, 2 semester hours in map reading (2/76).

Recommendation, Skill Level 20

In the lower-division baccalaureate/associate degree category, 2 semester hours in map reading (2/76).

Recommendation, Skill Level 30

In the lower-division baccalaureate/associate degree category, 3 semester hours in map reading and 3 in human relations (2/76).

Recommendation, Skill Level 40

In the lower-division baccalaureate/associate degree category, 3 semester hours in map reading, 3 in human relations, 3 in personnel supervision, and 3 for field experience in management (2/76).

MOS-16J-001

Defense Acquisition Radar Crewman
16J10
16J20
16J30

Exhibit Dates: 1/90–9/90.

Career Management Field: 16 (Air Defense Artillery), subfield 162 (Air Defense Artillery Operations and Intelligence).

Description

Summary: Supervises or serves as a member of a defense acquisition radar section or a forward area alerting radar unit. *Skill Level 10:* Operates radars and IFF (identification-friend-or-foe) equipment to obtain early warning and target identification data; manipulates radar adjustments to achieve optimum operating performance; observes, tracks, and interprets targets; reads topographic maps to locate points by means of coordinates; prepares grid overlays to plot positions; employs electronic countermeasures to avoid signal jamming; prepares records; performs operator maintenance; follows safety procedures; uses special test equipment to insure proper radar operation and maintenance; installs and operates communications equipment; operates power generating equipment. *Skill Level 20:* Able to perform the duties required for Skill Level 10; provides technical guidance to subordinates. *Skill Level 30:* Able to perform the duties required for Skill Level 20; supervises over 20 persons; supervises operation and maintenance of radar and related equipment; directs on-the-job training; maintains equipment records; interprets intelligence information.

Recommendation, Skill Level 10

In the vocational certificate category, credit in surveying on the basis of institutional evaluation. In the lower-division baccalaureate/associate degree category, 2 semester hours in map reading (2/76).

Recommendation, Skill Level 20

In the vocational certificate category, credit in surveying on the basis of institutional evaluation. In the lower-division baccalaureate/associate degree category, 2 semester hours in map reading (2/76).

Recommendation, Skill Level 30

In the vocational certificate category, credit in surveying on the basis of institutional evaluation. In the lower-division baccalaureate/associate degree category, 2 semester hours in map reading and 3 in human relations (2/76).

MOS-16J-002

Forward Area Alerting Radar Crewman
16J10
16J20
16J30
16J40

Exhibit Dates: 10/90–4/92. Effective 4/92, MOS 16J was discontinued and its duties were incorporated into MOS 14J, Early Warning System Operator.

Career Management Field: 14 (Air Defense Artillery).

Description

Summary: Supervises or serves as a member of a forward area alerting radar section or unit. *Skill Level 10:* Operates radars and IFF (identification-friend-or foe) equipment to obtain early warning and target identification data; manipulates radar adjustments to achieve optimum operating performance; observes, tracks, and interprets targets; reads topographic maps to locate points by means of coordinates; prepares grid overlays to plot positions; employs electronic countermeasures to avoid signal jamming; prepares records; performs operator maintenance; follows safety procedures; uses special test equipment to insure proper radar operation and maintenance; installs and operates power generating equipment. *Skill Level 20:* Able to perform the duties required for Skill Level 10; provides technical guidance to subordinates. *Skill Level 30:* Able to perform the duties required for Skill Level 20; supervises the operation and maintenance of radar and related equipment; directs on-the-job training; maintains equipment records; interprets intelligence information. *Skill Level 40:* Able to perform the duties required for Skill Level 30; assists in design of air defense artillery defenses and in conduct of reconnaissance, selection, and occupation of position.

Recommendation, Skill Level 10

In the lower-division baccalaureate/associate degree category, 3 semester hours in military science. (NOTE: This recommendation for skill level 10 is valid for the dates 9/90-9/91 only) (3/92).

Recommendation, Skill Level 20

In the lower-division baccalaureate/associate degree category, 3 semester hours in military science and 3 in personnel supervision (3/92).

Recommendation, Skill Level 30

In the lower-division baccalaureate/associate degree category, 3 semester hours in military science, 3 in personnel supervision, and 2 in records and information management (3/92).

Recommendation, Skill Level 40

In the lower-division baccalaureate/associate degree category, 3 semester hours in military science, 3 in personnel supervision, and 2 in records and information management. In the upper-division baccalaureate category, 3 semester hours for field experience in management and 3 in organizational management (3/92).

MOS-16P-002

Chaparral Crewmember
16P10
16P20
16P30
16P40

Exhibit Dates: 1/90–2/92.

Career Management Field: 16 (Air Defense Artillery).

Description

Summary: Supervises or serves as a crewman in a Chaparral system unit or as a gunner in a Redeye section. *Skill Level 10:* Prepares Chaparral fire unit for firing; prepares and fires Redeye missiles; stores ammunition; operates Chaparral unit equipment, including tracked vehicles, power generating equipment, target alert area display set, and fire control devices; assists in equipment maintenance. *Skill Level 20:* Able to perform the duties required for Skill Level 10; provides technical guidance to subordinates; prepares and fires Chaparral missiles; leads Redeye team. *Skill Level 30:* Able to perform the duties required for Skill Level 20; leads Chaparral squad; is chief of Redeye section; supervises persons engaged in storing, loading, and transporting ammunition; supervises operation and operator maintenance on the equipment. *Skill Level 40:* Able to perform the duties required for Skill Level 30; serves as platoon sergeant; plans, organizes, directs, supervises, trains, coordinates, and reports activities of subordinate squads and sections; supervises platoon maintenance activities; supervises 20-30 persons; prepares and issues platoon operation orders.

Recommendation, Skill Level 10
Credit is not recommended because the skills, competencies, and knowledge are uniquely military (11/88).

Recommendation, Skill Level 20
Credit is not recommended because the skills, competencies, and knowledge are uniquely military (11/88).

Recommendation, Skill Level 30
In the lower-division baccalaureate/associate degree category, 3 semester hours in human relations (11/88).

Recommendation, Skill Level 40
In the lower-division baccalaureate/associate degree category, 3 semester hours in human relations, 3 in personnel supervision, and 3 for field experience in management (11/88).

MOS-16P-003

CHAPARRAL CREWMEMBER
16P10
16P20
16P30
16P40

Exhibit Dates: 3/92–6/98. Effective 6/98, MOS 16P was discontinued.

Career Management Field: 14 (Air Defense Artillery).

Description
Summary: Supervises or serves as a crewman in a Chaparral fire unit or platoon. *Skill Level 10:* Prepares Chaparral fire unit for firing; stores ammunition; operates Chaparral unit equipment, including tracked vehicles, power generating equipment, target alert area display set, and fire control devices; assists in equipment maintenance. *Skill Level 20:* Able to perform the duties required for Skill Level 10; provides technical guidance to subordinates. *Skill Level 30:* Able to perform the duties required for Skill Level 20; leads Chaparral squad; supervises persons engaged in storing, loading, and transporting ammunition; supervises operation and operator maintenance on the equipment. *Skill Level 40:* Able to perform the duties required for Skill Level 30; plans, organizes, directs, supervises, trains, coordinates, and reports activities of subordinate squads and sections; supervises platoon maintenance activities; prepares and issues platoon operation orders.

Recommendation, Skill Level 10
Credit may be granted on the basis of an individualized assessment of the student (3/94).

Recommendation, Skill Level 20
In the lower-division baccalaureate/associate degree category, 3 semester hours in military science and 3 in personnel supervision. (NOTE: This recommendation for skill level 20 is valid for the dates 3/92-2/95 only) (3/92).

Recommendation, Skill Level 30
In the lower-division baccalaureate/associate degree category, 3 semester hours in military science, 3 in personnel supervision, and 2 in records and information management (3/92).

Recommendation, Skill Level 40
In the lower-division baccalaureate/associate degree category, 3 semester hours in military science, 3 in personnel supervision, and 2 in records and information management. In the upper-division baccalaureate category, 3 semester hours in organizational management and 3 for field experience in management (3/92).

MOS-16R-002

AIR DEFENSE ARTILLERY SHORT RANGE GUNNERY CREWMAN
((Short Range Air Defense Artillery Crewman))
((Area Air Defense Artillery Crewman))
16R10
16R20
16R30
16R40

Exhibit Dates: 1/90–9/90.

Career Management Field: 16 (Air Defense Artillery), subfield 161 (Air Defense Artillery Missile and Gun Operations).

Description
Summary: Supervises or serves as a crewman in a Vulcan system section or serves as a platoon sergeant in a Chaparral or Vulcan platoon. *Skill Level 10:* Prepares and assists in firing Vulcan antiaircraft gun; operates and performs operator maintenance on wheeled and tracked vehicles, power generating equipment, communications equipment, and fire control devices; handles ammunition; assists in the emplacement and displacement of the weapons system; assists in performance of routine maintenance. *Skill Level 20:* Able to perform the duties required for Skill Level 10; provides technical guidance to subordinates. *Skill Level 30:* Able to perform the duties required for Skill Level 20; supervises equipment operation and operator maintenance; supervises the emplacement and displacement of weapons system; supervises squad training and coordinates squad activities. *Skill Level 40:* Able to perform the duties required for 16R30 or 16P30 (Air Defense Artillery Short Range Missile Crewman or Chaparral Crewman); serves as platoon sergeant in a Chaparral or Vulcan platoon of more than 20 persons; assists superiors in planning, directing, training, coordinating, and reporting the activities of the platoon.

Recommendation, Skill Level 10
Credit is not recommended because the skills, competencies, and knowledge are uniquely military (2/76).

Recommendation, Skill Level 20
Credit is not recommended because the skills, competencies, and knowledge are uniquely military (2/76).

Recommendation, Skill Level 30
In the lower-division baccalaureate/associate degree category, 3 semester hours in human relations (2/76).

Recommendation, Skill Level 40
In the lower-division baccalaureate/associate degree category, 3 semester hours in personnel supervision, 3 in human relations, and 3 for field experience in management (2/76).

MOS-16R-003

VULCAN CREWMEMBER
16R10
16R20
16R30
16R40

Exhibit Dates: 10/90–6/97. Effective 6/97, MOS 16R was discontinued and its duties were transferred to MOS 14S.

Career Management Field: 14 (Air Defense Artillery).

Description
Summary: Supervises or serves as a crewman in a Vulcan fire unit or platoon. *Skill Level 10:* Prepares and assists in firing Vulcan antiaircraft gun; operates and performs operator maintenance on wheeled and tracked vehicles and fire control units; handles ammunition; assists in the emplacement and displacement of the weapons system; assists in performance of routine maintenance. *Skill Level 20:* Able to perform the duties required for Skill Level 10; provides technical guidance to subordinates; supervises/instructs in loading/unloading Vulcan ammunition. *Skill Level 30:* Able to perform the duties required for Skill Level 20; supervises equipment operation and operator maintenance; supervises the emplacement and displacement of weapons system; supervises squad training and coordinates squad activities. *Skill Level 40:* Able to perform the duties required for Skill Level 30; supervises the operations and maintenance of airspace control element; assists in design of air defense artillery defense and in conduct of reconnaissance, selection, and occupation of position.

Recommendation, Skill Level 10
In the lower-division baccalaureate/associate degree category, 3 semester hours in military science. (NOTE: This recommendation for skill level 10 is valid for the dates 9/90-9/91 only) (3/92).

Recommendation, Skill Level 20
In the lower-division baccalaureate/associate degree category, 3 semester hours in military science and 3 in personnel supervision. (NOTE: This recommendation for skill level 20 is valid for the dates 9/90-2/95 only) (3/92).

Recommendation, Skill Level 30
In the lower-division baccalaureate/associate degree category, 3 semester hours in military science, 3 in personnel supervision, and 2 in records and information management (3/92).

Recommendation, Skill Level 40
In the lower-division baccalaureate/associate degree category, 3 semester hours in military science, 3 in personnel supervision, and 2 in records and information management. In the upper-division baccalaureate category, 3 semester hours in organizational management and 3 for field experience in management (3/92).

MOS-16S-001

MAN PORTABLE AIR DEFENSE SYSTEM (MANPADS) CREWMAN
16S10
16S20
16S30
16S40

Exhibit Dates: 1/90–9/90.

Career Management Field: 16 (Air Defense Artillery), subfield 161 (Air Defense Artillery Missile and Gun Operations).

Description

Summary: Supervises or serves as a member of MANPADS missile unit. *Skill Level 10:* Prepares and fires MANPADS missiles; identifies friendly and hostile aircraft; operates and assists in programming interrogator equipment; operates assigned vehicles, simulators, and interrogator and cryptological equipment; establishes and maintains radio and/or wire communication with firing units and higher headquarters. *Skill Level 20:* Able to perform the duties required for Skill Level 10; provides technical guidance to subordinates. *Skill Level 30:* Able to perform the duties required for Skill Level 20; supervises operation of MANPADS section; assists in planning and conducting air defense; assigns personnel as required to provide air defense coverage to other units; supervises programming of interrogator equipment; supervises radio and wire communications and the issuance of ammunition. *Skill Level 40:* Able to perform the duties required for Skill Level 30; supervises operation of MANPADS platoon; assists in planning, organizing, directing, supervising, training, coordinating, and reporting activities of subordinate sections; assists in coordination and implementation of training programs, administrative matters, and communication activities.

Recommendation, Skill Level 10

Credit is not recommended because the skills, competencies, and knowledge are uniquely military (7/82).

Recommendation, Skill Level 20

Credit is not recommended because the skills, competencies, and knowledge are uniquely military (7/82).

Recommendation, Skill Level 30

In the lower-division baccalaureate/associate degree category, 3 semester hours in personnel supervision (7/82).

Recommendation, Skill Level 40

In the lower-division baccalaureate/associate degree category, 3 semester hours in personnel supervision and 3 for field experience in management. In the upper-division baccalaureate category, 3 semester hours in organizational management and 3 for field experience in management (7/82).

MOS-16S-002

Man Portable Air Defense System (MANPADS) Crewmember or Pedestal Mounted Stinger Crewmember
16S10
16S20
16S30
16S40

Exhibit Dates: 10/90–6/97. Effective 6/97, MOS 16S was discontinued and its duties were transferred to MOS 14S.

Career Management Field: 14 (Air Defense Artillery).

Description

Summary: Supervises or serves as a member of MANPADS missile unit. *Skill Level 10:* Prepares and fires MANPADS missiles; identifies friendly and hostile aircraft; establishes and maintains proper radio and wire communications with firing units and higher headquarters. *Skill Level 20:* Able to perform the duties required for Skill Level 10; supervises and provides technical guidance to subordinates. *Skill Level 30:* Able to perform the duties required for Skill Level 20; assists in planning and conducting air defense; assigns personnel as required to provide air defense coverage to other units; supervises programming of interrogator equipment; supervises radio and wire communications and the issuance of ammunition. *Skill Level 40:* Able to perform the duties required for Skill Level 30; supervises operation of platoon; assists in planning, organizing, directing, supervising, training, coordinating, and reporting activities of subordinate sections; assists in coordination and implementation of training programs, administrative matters, and communication activities.

Recommendation, Skill Level 10

In the lower-division baccalaureate/associate degree category, 3 semester hours in military science. (NOTE: This recommendation for skill level 10 is valid for the dates 9/90-9/91 only) (3/92).

Recommendation, Skill Level 20

In the lower-division baccalaureate/associate degree category, 3 semester hours in military science and 3 in personnel supervision. (NOTE: This recommendation for skill level 20 is valid for the dates 9/90-2/95 only) (3/92).

Recommendation, Skill Level 30

In the lower-division baccalaureate/associate degree category, 3 semester hours in military science, 3 in personnel supervision, and 2 in records and information management (3/92).

Recommendation, Skill Level 40

In the lower-division baccalaureate/associate degree category, 3 semester hours in military science, 3 in personnel supervision, and 2 in records and information management. In the upper-division baccalaureate category, 3 semester hours in organizational management and 3 for field experience in management (3/92).

MOS-16T-002

Patriot Missile Crewmember
16T10
16T20
16T30
16T40

Exhibit Dates: 1/90–2/92.

Career Management Field: 16 (Air Defense Artillery).

Description

Summary: Supervises or serves as a crewmember in a Patriot air defense unit. *Skill Level 10:* Drives heavy equipment to move missile system to site; sets up and performs operator maintenance on system; operates truck, electrical generator, and launcher; may include familiarization with general surveying techniques. *Skill Level 20:* Able to perform the duties required for Skill Level 10; provides technical guidance to subordinates; assists in supervising on-the-job training. *Skill Level 30:* Able to perform the duties required for Skill Level 20; supervises emplacement of launching and fire control equipment; supervises crew maintenance of Patriot and related equipment. *Skill Level 40:* Able to perform the duties required for Skill Level 30; assists in planning, organizing, directing, supervising, training, coordinating, and reporting activities of subordinate sections and squads.

Recommendation, Skill Level 10

In the vocational certificate category, 3 semester hours in heavy truck operation. (NOTE: This recommendation for skill level 10 is valid for the dates 1/90-9/91 only) (11/86).

Recommendation, Skill Level 20

In the vocational certificate category, 3 semester hours in heavy truck operation (11/86).

Recommendation, Skill Level 30

In the vocational certificate category, 3 semester hours in heavy truck operation. In the lower-division baccalaureate/associate degree category, 3 semester hours in personnel supervision (11/86).

Recommendation, Skill Level 40

In the vocational certificate category, 3 semester hours in heavy truck operation. In the lower-division baccalaureate/associate degree category, 3 semester hours in personnel supervision and 3 in introduction to management. In the upper-division baccalaureate category, 3 semester hours in personnel management and 3 for field experience in management (11/86).

MOS-16T-003

Patriot Missile Crewmember
16T10
16T20
16T30
16T40

Exhibit Dates: 3/92–6/97. Effective 6/97, MOS 16T was discontinued and its duties were transferred to MOS14E.

Career Management Field: 14 (Air Defense Artillery).

Description

Summary: Supervises or serves as a crewmember in a Patriot air defense unit. *Skill Level 10:* Drives heavy equipment to move missile system to site; sets up and performs operator maintenance on system; operates truck, electrical generator, and launcher. *Skill Level 20:* Able to perform the duties required for Skill Level 10; supervises and provides technical guidance to subordinates. *Skill Level 30:* Able to perform the duties required for Skill Level 20; supervises emplacement of launching and fire control equipment; supervises crew maintenance of Patriot and related equipment. *Skill Level 40:* Able to perform the duties required for Skill Level 30; assists in planning, organizing, directing, supervising, training, coordinating, and reporting activities of subordinate sections and squads.

Recommendation, Skill Level 10

Credit may be granted on the basis of an individualized assessment of the student (3/94).

Recommendation, Skill Level 20

In the vocational certificate category, 3 semester hours in heavy equipment operation. In the lower-division baccalaureate/associate degree category, 3 semester hours in military science and 3 in personnel supervision. (NOTE: This recommendation for skill level 20 is valid for the dates 3/92-2/95 only) (3/92).

Recommendation, Skill Level 30

In the vocational certificate category, 3 semester hours in heavy equipment operation. In the lower-division baccalaureate/associate degree category, 3 semester hours in military science, 3 in personnel supervision, and 2 in records and information management (3/92).

Recommendation, Skill Level 40

In the vocational certificate category, 3 semester hours in heavy equipment operation. In the lower-division baccalaureate/associate degree category, 3 semester hours in military science, 3 in personnel supervision, and 2 in records and information management. In the upper-division baccalaureate category, 3 semester hours in organizational management and 3 for field experience in management (3/92).

MOS-16Z-001

Air Defense Artillery Senior Sergeant
16Z50

Exhibit Dates: 1/90–2/92.

Career Management Field: 16 (Air Defense Artillery), subfield 160 (Air Defense Artillery General).

Description

Serves as the principal noncommissioned officer in an air defense artillery unit or headquarters and supervises 44 or more persons; assists superiors in the appraisal of air defense artillery operations, training, and intelligence information; prepares status board and situation map; collects, interprets, evaluates, and disseminates intelligence data; prepares and edits training and intelligence material and the operations manual; supervises security; advises superiors on all personnel matters; coordinates operation of battery food service and supply activities; assists superiors in planning, coordinating, and directing all air defense artillery activities. NOTE: May have progressed to 16Z50 from Skill Level 40 of any MOS in the air defense artillery career management field (16).

Recommendation

In the lower-division baccalaureate/associate degree category, 3 semester hours in personnel supervision, 3 in technical writing, 3 in public speaking, 3 in map reading, and additional credit in electronics and mechanical maintenance on the basis of institutional evaluation. In the upper-division baccalaureate category, 3 semester hours in introduction to management, 3 in personnel management, 3 in human relations, 6 for field experience in management, and additional credit in industrial arts education on the basis of institutional evaluation (2/76).

MOS-16Z-002

Air Defense Artillery Senior Sergeant
16Z50

Exhibit Dates: 3/92–6/97. Effective 6/97, MOS 16Z was discontinued and its duties were transferred to MOS 14Z.

Career Management Field: 14 (Air Defense Artillery).

Description

Serves as the principal noncommissioned officer in the supervision of an air defense artillery unit or headquarters; assists superiors in the appraisal of air defense artillery operations, training, and intelligence information; prepares status board and situation map; collects, interprets, evaluates, and disseminates intelligence data; prepares and edits training and intelligence material and the operations manual; supervises security; advises superiors on all personnel matters; assists superiors in planning, coordinating, and directing all air defense artillery activities. NOTE: May have progressed to 16Z50 from Skill Level 40 of MOS 14D (Hawk Missile System Crewmember), 16D (Hawk Missile Crewmember), 16F (Light Air Defense Artillery Crewmember), 16J (Forward Area Alerting Radar Operator), 16P (Chaparral Crewmember), 16R (Vulcan Crewmember), 16S (Man Portable Air Defense System Crewmember), or 16T (Patriot Missile Crewmember).

Recommendation

In the lower-division baccalaureate/associate degree category, 3 semester hours in military science, 3 in personnel supervision, and 2 in records and information management. In the upper-division baccalaureate category, 3 semester hours for field experience in management and 3 in organizational management; if individual has attained pay grade E-9, additional credit as follows: 3 semester hours in management problems, 3 in operations management, and 3 in communication techniques for managers (3/92).

MOS-17B-002

Field Artillery Radar Crewmember
17B10
17B20
17B30
17B40

Exhibit Dates: 1/90–2/92.

Career Management Field: 13 (Field Artillery), subfield 132 (Field Artillery Target Acquisition Operations).

Description

Summary: Engages in operation of counterfire and moving target locating radars and associated equipment. *Skill Level 10:* Emplaces and conceals equipment; operates power unit; operates and performs minor maintenance on radars and associated equipment. *Skill Level 20:* Able to perform the duties required for Skill Level 10; provides technical guidance to subordinates; prepares site evaluation chart and surveillance cards; performs traverse to locate radar and establish directional control; assists in scheduling maintenance on radar and associated equipment; directs radar operations during periods of interference and jamming; employs electronic counter-countermeasures. *Skill Level 30:* Able to perform the duties required for Skill Level 20; supervises a radar section of approximately seven persons; selects and evaluates emplacement sites; instructs radar operators in radar techniques and procedures; plans and organizes work schedules; assures adherence to safety procedures involving radar operation and maintenance. *Skill Level 40:* Able to perform the duties required for Skill Level 30; supervises operation of radar platoon of approximately 20 persons; supervises and directs operation of radar sections; assists platoon commander in supervision of platoon operations, training, resupply, and maintenance; conducts reconnaissance and selection of tactical area for radar; prepares technical, personnel, and administrative reports.

Recommendation, Skill Level 10

In the lower-division baccalaureate/associate degree category, 1 semester hour for field experience in electromechanical maintenance, 1 in basic mathematics, and 1 in map reading. (NOTE: This recommendation for skill level 10 is valid for the dates 1/90-9/91 only) (4/77).

Recommendation, Skill Level 20

In the lower-division baccalaureate/associate degree category, 1 semester hour for field experience in electromechanical maintenance, 1 in basic mathematics, and 1 in map reading (4/77).

Recommendation, Skill Level 30

In the lower-division baccalaureate/associate degree category, 3 semester hours for field experience in management, 2 in communication skills, 2 in human relations, 2 in personnel supervision, 1 for field experience in electromechanical maintenance, 1 in basic mathematics, and 1 in map reading (4/77).

Recommendation, Skill Level 40

In the lower-division baccalaureate/associate degree category, 3 semester hours for field experience in management, 3 in communication skills, 3 in human relations, 3 in personnel supervision, 1 for field experience in electromechanical maintenance, 1 in basic mathematics, and 1 in map reading (4/77).

MOS-17B-003

Field Artillery Radar Crewmember
17B10
17B20
17B30
17B40

Exhibit Dates: 3/92–6/94. Effective 6/94, MOS 17B was discontinued.

Career Management Field: 13 (Field Artillery).

Description

Summary: Supervises or participates in operation of counterfire and moving target radars and associated equipment. *Skill Level 10:* Emplaces and conceals equipment; operates power unit; operates and performs minor maintenance on radars and associated equipment. *Skill Level 20:* Able to perform the duties required for Skill Level 10; provides technical guidance to subordinates; prepares site evaluation chart and surveillance cards; assists in scheduling maintenance on radar and associated equipment; directs radar operations during periods of interference and jamming; employs electronic counter-countermeasures. *Skill Level 30:* Able to perform the duties required for Skill Level 20; supervises a radar section; selects and evaluates emplacement sites; instructs radar operators in radar techniques and procedures; plans and organizes work schedules; assures adherence to safety procedures involving radar operation and maintenance. *Skill Level 40:* Able to perform the duties required for Skill Level 30; supervises operation of radar platoon; supervises and directs operation of radar sections; assists platoon commander in supervision of platoon operations, training, resupply, and maintenance; conducts reconnaissance and selection of tactical area for radars; prepares technical, personnel, and administrative reports.

Recommendation, Skill Level 10

Credit may be granted on the basis of an individualized assessment of the student (3/94).

Recommendation, Skill Level 20

In the lower-division baccalaureate/associate degree category, 3 semester hours in military science and 3 in personnel supervision (3/92).

Recommendation, Skill Level 30

In the lower-division baccalaureate/associate degree category, 3 semester hours in military science, 3 in personnel supervision, and 2 in records and information management (3/92).

Recommendation, Skill Level 40

In the lower-division baccalaureate/associate degree category, 3 semester hours in military science, 3 in personnel supervision, and 2 in records and information management. In the upper-division baccalaureate category, 3 semester hours in organizational management and 3 for field experience in management (3/92).

MOS-18B-001

Special Forces Weapons Sergeant
((Special Operations Weapons Sergeant))
18B30
18B40

Exhibit Dates: 1/90–5/96.

Career Management Field: 18 (Special Operations).

Description

Summary: Duties primarily involve participation in a special operations team or detachment involving unconventional warfare, foreign internal defense, strike operations, strategic reconnaissance, and counterterrorism; the operational detachment works unilaterally or with foreign military forces; duties frequently require regional orientation, including foreign language proficiency and in-country experience; duties include participation in waterborne, jungle, desert, mountain, and winter operations; many of the duties are highly classified. Serves as a detachment member in military operations, foreign internal defense, unconventional warfare, and counterterrorist operations; employs warfare tactics and techniques in small arms, antiaircraft, and antiarmor weapons; recruits, organizes, and trains forces. *Skill Level 30:* Conducts training in light infantry tactics and combat operations; prepares and teaches weapons skills to detachment members, indigenous, and nonindigenous personnel; reads and interprets maps and aerial photos; writes reports. *Skill Level 40:* Able to perform the duties required for Skill Level 30; provides tactical and technical guidance to detachment members; leads and trains small light infantry tactical units in highly specialized combat operations.

Recommendation, Skill Level 30

In the lower-division baccalaureate/associate degree category, 3 semester hours in technical report writing. In the upper-division baccalaureate category, 3 semester hours for a practicum in instructional techniques (5/87).

Recommendation, Skill Level 40

In the lower-division baccalaureate/associate degree category, 3 semester hours in technical report writing. In the upper-division baccalaureate category, 3 semester hours for a practicum in instructional techniques (5/87).

MOS-18B-002

Special Forces Weapons Sergeant
18B30
18B40

Exhibit Dates: 6/96–Present.

Career Management Field: 18 (Special Operations).

Description

Summary: Duties involve participation in a special operations team or detachment involving unconventional warfare, foreign internal defense, strike operations, strategic reconnaissance, and counterterrorism. The operational detachment works unilaterally or with foreign military forces. Duties frequently require regional orientation, including foreign language proficiency and in-country experience. Duties also include employing warfare tactics and techniques, recruiting, organizing, and training forces. *Skill Level 30:* Conducts training in light infantry tactics and combat operations; prepares and teaches weapons skills to detachment members and indigenous and nonindigenous personnel; reads and interprets maps and aerial photos; writes reports. *Skill Level 40:* Able to perform the duties required for Skill Level 30; provides tactical and technical guidance to detachment members; leads and trains small light infantry tactical units in highly specialized combat operations.

Recommendation, Skill Level 30

In the lower-division baccalaureate/associate degree category, 2 semester hours in technical report writing, 3 in leadership, and 3 in instructor training techniques. In the upper-division baccalaureate category, 3 semester hours for field experience in management (6/96).

Recommendation, Skill Level 40

In the lower-division baccalaureate/associate degree category, 3 semester hours in technical report writing, 3 in leadership, and 3 in instructor training techniques. In the upper-division baccalaureate category, 6 semester hours for field experience in management (6/96).

MOS-18C-001

Special Forces Engineer Sergeant
((Special Operations Engineer Sergeant))
18C30
18C40

Exhibit Dates: 1/90–5/96.

Career Management Field: 18 (Special Operations).

Description

Summary: Duties primarily involve participation in a special operations team or detachment involving unconventional warfare, foreign internal defense, strike operations, strategic reconnaissance, and counterterrorism; the operational detachment works unilaterally or with foreign military forces; duties frequently require regional orientation, including foreign language proficiency and in-country experience; duties include participation in waterborne, jungle, desert, mountain, and winter operations; many of the duties are highly classified. Performs, supervises, and instructs in engineering techniques and operations employed in warfare; advises, trains, and assists foreign nationals in their own environment. *Skill Level 30:* Performs the duties of a combat engineer and demolition specialist; leads, supervises, plans, and conducts training in all special operations engineering duties; writes reports. *Skill Level 40:* Able to perform the duties required for Skill Level 30; leads, supervises, plans, and instructs in all phases of combat engineering and special operations engineering; provides technical guidance to subordinates and other special operations detachment members.

Recommendation, Skill Level 30

In the lower-division baccalaureate/associate degree category, 3 semester hours in construction technology and 3 in technical report writing. In the upper-division baccalaureate category, 3 semester hours in construction project management, 3 for a practicum in instructional techniques, 6 for field experience in construction, and 3 for field experience in management (5/87).

Recommendation, Skill Level 40

In the lower-division baccalaureate/associate degree category, 6 semester hours in construction technology and 3 in technical report writing. In the upper-division baccalaureate category, 3 semester hours in construction project management, 3 for a practicum in instructional techniques, 6 for field experience in construction, and 3 for field experience in management (5/87).

MOS-18C-002

Special Forces Engineer Sergeant
18C30
18C40

Exhibit Dates: 6/96–Present.

Career Management Field: 18 (Special Operations).

Description

Summary: Duties primarily involve participation in a special operations team or detachment involving unconventional warfare, foreign internal defense, strike operations, strategic reconnaissance, and counterterrorism. The operational detachment works unilaterally or with foreign military forces. Duties frequently require regional orientation, including foreign language proficiency and in-country experience. Duties include participation in waterborne, jungle, desert, mountain, and winter operations. Many of the duties are highly classified. Duties also include performing, supervising, and instructing in construction and/or demolition techniques and operations employed in warfare; and advising, training, and assisting foreign nationals in their own environment. *Skill Level 30:* Performs the duties of a combat construction and demolition specialist; leads, supervises, plans, and conducts training in all special operations duties; writes reports. *Skill Level 40:* Able to perform the duties required for Skill Level 30; leads, supervises, plans, and instructs in all phases of combat construction; provides technical guidance to subordinates and other special operations detachment members.

Recommendation, Skill Level 30

In the lower-division baccalaureate/associate degree category, 3 semester hours for field experience in construction technology, 3 for field experience in explosive demolitions, 3 in technical report writing, and 3 in instructor training techniques. In the upper-division baccalaureate category, 3 semester hours for field experience in management (6/96).

Recommendation, Skill Level 40

In the lower-division baccalaureate/associate degree category, 6 semester hours for field experience in construction technology, 3 for field experience in explosive demolitions, 3 in technical report writing, and 3 in instructor training techniques. In the upper-division baccalaureate category, 6 semester hours for field experience in management (6/96).

MOS-18D-001

Special Forces Medical Sergeant
((Special Operations Medical Sergeant))
18D30
18D40

Exhibit Dates: 1/90–5/96.

Career Management Field: 18 (Special Operations).

Description

Summary: Duties primarily involve participation in a special operations team or detachment involving unconventional warfare, foreign internal defense, strike operations, strategic reconnaissance, and counterterrorism; the operational detachment works unilaterally or with foreign military forces; duties frequently require regional orientation, including foreign language proficiency and in-country experience; duties include participation in waterborne, jungle, desert, mountain, and winter operations; many of the duties are highly classified. Establishes field and unconventional warfare medical facilities to support operations; ensures medical preparedness. *Skill Level 30:* Provides emergency, routine, and long term medical care for special operations members, associated allied, and indigenous personnel; trains personnel in basic emergency and preventive medical care; operates with minimal or no medical supervision or assistance; examines and provides medical care for special operations detachment members; maintains records. *Skill Level 40:* Able to perform the duties required for Skill Level 30; supervises routine and emergency field medical activities, patient care, and medical service activities.

Recommendation, Skill Level 30

In the lower-division baccalaureate/associate degree category, 3 semester hours in personnel supervision and 3 in technical report writing. In the upper-division baccalaureate category, 9 semester hours for a practicum in medical care, 3 for field experience in management, 3 in case studies in medicine, and 3 for a practicum in instructional techniques (5/87).

Recommendation, Skill Level 40

In the lower-division baccalaureate/associate degree category, 3 semester hours in personnel supervision and 3 in technical report writing. In the upper-division baccalaureate category, 9 semester hours for a practicum in medical care, 3 for field experience in management, 3 in case studies in medicine, and 3 for a practicum in instructional techniques (5/87).

MOS-18D-002

Special Forces Medical Sergeant
18D30
18D40

Exhibit Dates: 6/96–Present.

Career Management Field: 18 (Special Operations).

Description

Summary: Duties primarily involve participation as a paramedic in a special operations team or detachment involving unconventional warfare, foreign internal defense, strike operations, strategic reconnaissance, and counterterrorism. The operational detachment works unilaterally or with foreign military forces. Duties frequently require regional orientation, including foreign language proficiency and in-country experience. Duties include participation in waterborne, jungle, desert, mountain, and winter operations. Duties also include establishing field and unconventional warfare medical facilities to support operations and ensuring medical preparedness of the team. Many of the duties are highly classified. *Skill Level 30:* Provides emergency and routine medical care as a paramedic and long-term medical care for special operations members, associated allied and indigenous personnel; conducts training in basic emergency and preventive medical care; operates with minimal or no medical supervision or assistance; maintains records; consults with other health team members; arranges for medical evacuation. *Skill Level 40:* Able to perform the duties required for Skill Level 30; supervises routine and emergency field medical activities, patient care, and medical service activities.

Recommendation, Skill Level 30

In the lower-division baccalaureate/associate degree category, 3 semester hours in leadership, 3 in technical report writing, 3 in records and information management, and 3 in instructor training techniques; if certified as a paramedic, 24 semester hours in emergency medical technology/paramedic. In the upper-division baccalaureate category, 3 semester hours for field experience in management (6/96).

Recommendation, Skill Level 40

In the lower-division baccalaureate/associate degree category, 3 semester hours in leadership, 3 in technical report writing, 3 in records and information management, and 3 in instructor training techniques; if certified as a paramedic, 24 semester hours in emergency medical technology/paramedic. In the upper-division baccalaureate category, 6 semester hours for field experience in management (6/96).

MOS-18E-001

Special Forces Communications Sergeant
((Special Operations Communications Sergeant))
18E30
18E40

Exhibit Dates: 1/90–5/96.

Career Management Field: 18 (Special Operations).

Description

Summary: Duties primarily involve participation in a special operations team or detachment involving unconventional warfare, foreign internal defense, strike operations, strategic reconnaissance, and counterterrorism; the operational detachment works unilaterally or with foreign military forces; duties frequently require regional orientation, including foreign language proficiency and in-country experience; duties include participation in waterborne, jungle, desert, mountain, and winter operations; many of the duties are highly classified. Installs, operates, supervises, and trains indigenous forces in the installation and operations of FM, AM, HF, VHF, and UHF/SHF radio communications equipment. *Skill Level 30:* Develops lesson plans and instructs military and nonmilitary personnel in the installation and use of radio communications equipment; writes reports; performs tests and maintains equipment. *Skill Level 40:* Able to perform the duties required for Skill Level 30; prepares communication plans and annexes for commander of special operations detachment; supervises preparation of records and reports.

Recommendation, Skill Level 30

In the lower-division baccalaureate/associate degree category, 3 semester hours in technical report writing. In the upper-division baccalaureate category, 3 semester hours for a practicum in instructional techniques (5/87).

Recommendation, Skill Level 40

In the lower-division baccalaureate/associate degree category, 3 semester hours in technical report writing. In the upper-division baccalaureate category, 3 semester hours for a practicum in instructional techniques (5/87).

MOS-18E-002

Special Forces Communications Sergeant
18E30
18E40

Exhibit Dates: 6/96–Present.

Career Management Field: 18 (Special Operations).

Description

Summary: Duties primarily involve participation in a special operations team or detachment involving unconventional warfare, foreign internal defense, strike operations, strategic reconnaissance, and counterterrorism. The operational detachment works unilaterally or with foreign military forces. Duties frequently require regional orientation, including foreign language proficiency and in-country experience. Duties include participation in waterborne, jungle, desert, mountain, and winter operations. Many of the duties are highly classified. Duties also include installing, operating, supervising, and training indigenous forces in the installation and operation of FM, AM, HF, VHF, and UHF/SHF radio communications equipment. *Skill Level 30:* Develops lesson plans and instructs military and nonmilitary personnel in the installation and use of radio communications equipment; writes reports; performs tests and maintains equipment. *Skill Level 40:* Able to perform the duties required for Skill Level 30; prepares communications plans and annexes for commander of special operations detachment; supervises preparation of records and reports.

Recommendation, Skill Level 30

In the lower-division baccalaureate/associate degree category, 3 semester hours in technical report writing, 3 in radio broadcasting/telecommunications, and 3 in instructional training techniques. In the upper-division baccalaureate category, 3 semester hours for field experience in management (6/96).

Recommendation, Skill Level 40

In the lower-division baccalaureate/associate degree category, 3 semester hours in technical report writing, 3 in radio broadcasting/telecommunications, and 3 in instructional training techniques. In the upper-division baccalaureate category, 6 semester hours for field experience in management (6/96).

MOS-18F-001

Special Forces Intelligence Sergeant
((Special Operations Intelligence Sergeant))
18F30
18F40

Exhibit Dates: 1/90–5/96.

Career Management Field: 18 (Special Operations).

Description

Summary: Duties primarily involve participation in a special operations team or detachment involving unconventional warfare, foreign internal defense, strike operations, strategic reconnaissance, and counterterrorism; the operational detachment works unilaterally or with foreign military forces; duties frequently require regional orientation, including foreign language proficiency and in-country experience; duties include participation in waterborne, jungle, desert, mountain, and winter operations; many of the duties are highly classified. Plans and executes small unit strike, foreign internal defense, unconventional warfare, and counterterrorist operations. *Skill Level 30:* Recruits, organizes, trains, and supervises indigenous and nonindigenous forces; writes reports; coordinates and conducts training in intelligence subjects; performs security duties; able to use fingerprint identification, operate photographic equipment, develop and print film, prepare area studies, and conduct area assessments. *Skill Level 40:* Able to perform the duties required for Skill Level 30; prepares tactical exercise plans and scenarios; establishes intelligence nets; prepares agent reports; conducts agent handling and training; prepares physical security plans; provides technical and tactical guidance to special operations members in intelligence-related subjects; advises special operations commander on all intelligence-related matters.

Recommendation, Skill Level 30

In the lower-division baccalaureate/associate degree category, 3 semester hours in technical report writing. In the upper-division baccalaureate category, 3 semester hours for a practicum in instructional techniques and 3 for field experience in management (5/87).

Recommendation, Skill Level 40

In the lower-division baccalaureate/associate degree category, 3 semester hours in technical report writing. In the upper-division baccalaureate category, 3 semester hours for a practicum in instructional techniques and 3 for field experience in management (5/87).

MOS-18F-002

Special Forces Assistant Operations and Intelligence Sergeant
18F40

Exhibit Dates: 6/96–Present.

Career Management Field: 18 (Special Operations).

Description

Duties involve participation in a special operations team or detachment involving unconventional warfare, foreign internal defense, strike operations, strategic reconnaissance, and counterterrorism. The operational detachment works unilaterally or with foreign military forces. Duties frequently require regional orientation, including foreign language proficiency and in-country experience.

Plans and executes small unit strike, foreign internal defense, unconventional warfare, and counterterrorist operations; recruits, organizes, trains, and supervises indigenous and nonindigenous forces; writes reports; coordinates and conducts training in intelligence subjects; performs security duties; able to use fingerprint identification, operate photographic equipment, develop and print film, prepare area studies, and conduct area assessments; prepares tactical exercise plans and scenarios; establishes intelligence nets; prepares agent reports; conducts agent handling and training; prepares physical security plans; provides technical and tactical guidance to special operations members in intelligence-related subjects; advises special operations commander on all intelligence-related matters. May have progressed to 18F40 from skill level 30 of MOS18B, Special Forced Weapons Sergeant; MOS 18C, Special Forces Engineer Sergeant; MOS18D, Special Forces Medical Sergeant; MOS18E, Special Forces Communications Sergeant.

Recommendation

In the lower-division baccalaureate/associate degree category, 3 semester hours in technical report writing, 3 in instructor training techniques, 3 in leadership, and 3 in photographic principles. In the upper-division baccalaureate category, 6 semester hours for field experience in management (6/96).

MOS-18Z-001

Special Forces Senior Sergeant
((Special Operations Senior Sergeant))
18Z50

Exhibit Dates: 1/90–5/96.

Career Management Field: 18 (Special Operations).

Description

Duties primarily involve participation in a special operations team or detachment involving unconventional warfare, foreign internal defense, strike operations, strategic reconnaissance, and counterterrorism; the operational detachment works unilaterally or with foreign military forces; duties frequently require regional orientation, including foreign language proficiency and in-country experience; duties include participation in waterborne, jungle, desert, mountain, and winter operations; many of the duties are highly classified. Heads, supervises, and serves as senior enlisted member for special operations activities; performs senior leadership, staff, and training functions within special operations; provides tactical and technical guidance and professional support to superiors and subordinates; recruits, organizes, trains, and supervises indigenous and nonindigenous personnel and organizations.

Recommendation

In the lower-division baccalaureate/associate degree category, 3 semester hours in personnel supervision and 3 in technical report writing. In the upper-division baccalaureate category, 3 semester hours for a practicum in instructional techniques and 3 for field experience in management (5/87).

MOS-18Z-002

Special Forces Senior Sergeant
18Z50

Exhibit Dates: 6/96–Present.

Career Management Field: 18 (Special Operations).

Description

Duties primarily involve participation in a special operations team or detachment involving unconventional warfare in a foreign country. This includes internal defense, strike operations, strategic reconnaissance, and counterterrorism. The operational detachment works unilaterally or with foreign military forces. Duties frequently require regional orientation, including foreign language proficiency and in-country experience. Duties include participation in waterborne, jungle, desert, mountain, and winter operations. Many of the duties are highly classified.

Heads, supervises, and serves as senior enlisted member for special operations activities; performs senior leadership, staff, and training functions; provides tactical and technical guidance and professional support to superiors and subordinates; recruits, organizes, trains, and supervises indigenous and nonindigenous personnel and organizations. May have progressed to 18Z50 from skill level 40 of MOS 18B, Special Forces Weapons Sergeant; MOS 18C, Special Forces Engineer Sergeant; MOS 18D, Special Forces Medical Sergeant; MOS 18E, Special Forces Communications Sergeant; or MOS 18F, Special Forces Assistant Operations and Intelligence Sergeant.

Recommendation

In the lower-division baccalaureate/associate degree category, 3 semester hours in personnel supervision and 3 in technical report writing. In the upper-division baccalaureate category, 6 semester hours for field experience in management, 6 in management problems, and 3 in training methods (6/96).

MOS-19D-001

Cavalry Scout
19D10
19D20
19D30
19D40

Exhibit Dates: 1/90–12/90.

Career Management Field: 19 (Armor).

Description

Summary: Serves as a member or leads members of a scout crew, squad, section, or platoon in reconnaissance security and other combat operations. *Skill Level 10:* Serves as a member of observation and listening team; gathers and reports information on terrain features, enemy strength, disposition, and equipment; identifies targets; requests and adjusts indirect and aerial fire; operates wheeled and tracked vehicles and operates communications equipment; uses maps and map symbols; navigates on ground; uses compass and distinguishes topographic features; performs operator maintenance on scout vehicles, weapons, and communications equipment; assists in firing of crew-served weapons; performs first aid and applies field sanitation methods; uses radiac instruments and chemical detection kits. *Skill Level 20:* Able to perform the duties required for Skill Level 10; leads scout crew or assists squad leader; selects, organizes, and supervises operation of observation and listening posts; prepares, distributes, and files maps and overlays; reproduces, distributes, and files operations and intelligence information; coordinates with adjacent and supporting elements. *Skill Level 30:* Able to perform the duties required for Skill Level 20; leads scout squad or section; directs tactical deployment; leads in combat; supervises maintenance of assigned vehicles and equipment; evaluates routes, assembly areas, and positioning for mounted combat operations. *Skill Level 40:* Able to perform the duties required for Skill Level 30; supervises armored cavalry or reconnaissance platoon and processes operations and intelligence information; assists in planning, organizing, directing, supervising, training, coordinating, and reporting activities of the scout or armored cavalry platoon and staff sections; directs distribution of fire; supervises platoon maintenance activities; collects, evaluates, and assists in interpretation and dissemination of combat information. NOTE: May have progressed to 19D40 from 19D30 or 19G30 (Armor Reconnaissance Vehicle Crewman).

Recommendation, Skill Level 10

In the vocational certificate category, 3 semester hours in mechanical maintenance. In the lower-division baccalaureate/associate degree category, 1 semester hour in first aid and 3 in map reading (11/79).

Recommendation, Skill Level 20

In the vocational certificate category, 6 semester hours in mechanical maintenance. In the lower-division baccalaureate/associate degree category, 1 semester hour in first aid, 3 in map reading, and 3 in record keeping (11/79).

Recommendation, Skill Level 30

In the vocational certificate category, 6 semester hours in mechanical maintenance. In the lower-division baccalaureate/associate degree category, 1 semester hour in first aid, 3 in map reading, 3 in record keeping, and 3 in personnel supervision. In the upper-division baccalaureate category, 3 semester hours for field experience in management (11/79).

Recommendation, Skill Level 40

In the vocational certificate category, 6 semester hours in mechanical maintenance. In the lower-division baccalaureate/associate degree category, 1 semester hour in first aid, 3 in map reading, 3 in record keeping, 3 in personnel supervision, and 3 in human relations. In the upper-division baccalaureate category, 3 semester hours for field experience in management and 3 in personnel management (11/79).

MOS-19D-002

Cavalry Scout
19D10
19D20
19D30
19D40

Exhibit Dates: 1/91–5/01.

Career Management Field: 19 (Armor).

Description

Summary: Leads, serves, or assists as a member of a scout crew, squad, section, or platoon in reconnaissance security and other combat operations. *Skill Level 10:* Serves as a member of observation and listening team; gathers and reports information on terrain features, enemy strength, disposition, and equipment; identifies targets; requests and adjusts indirect and aerial fire; operates wheeled and tracked vehicles and operates communications equipment; uses maps and map symbols; navigates on ground; uses compass and distinguishes topographic features; assists in firing of crew-served weapons; assists mechanics and makes minor repairs on scout vehicles and tanks; performs first aid and applies field sanitation methods; uses radiac instruments and chemical detection kits. *Skill Level 20:* Able to perform the duties required for Skill Level 10; leads scout crew or assists squad leader; selects, organizes, and supervises operation of observation and listening posts; prepares, distributes, and files operations and intelligence information; coordinates with adjacent and supporting elements. *Skill Level 30:* Able to perform the duties required for Skill Level 20; leads scout squad or section; directs tactical deployment; leads in combat; supervises maintenance of assigned vehicles and equipment; evaluates routes, assembly areas, and positioning for mounted combat operations. *Skill Level 40:* Able to perform the duties required for Skill Level 30; supervises armored cavalry or reconnaissance platoon; processes operations and intelligence information; assists in planning, organizing, directing, supervising, training, coordinating, and reporting activities of the scout or armored cavalry platoon and staff sections; directs distribution of fire; supervises platoon maintenance activities; collects, evaluates, and assists in interpretation and dissemination of combat information. NOTE: May have progressed to 19D40 from 19D30 or 19G30 (Armor Reconnaissance Vehicle Crewman).

Recommendation, Skill Level 10

In the lower-division baccalaureate/associate degree category, 3 semester hours in military science, 3 in automotive fundamentals, and 1 in first aid. (NOTE: This recommendation for skill level 10 is valid for the dates 1/91-9/91 only) (4/91).

Recommendation, Skill Level 20

In the lower-division baccalaureate/associate degree category, 3 semester hours in military science, 3 in automotive fundamentals, 1 in first aid, and 3 in personnel supervision. (NOTE: This recommendation for skill level 20 is valid for the dates 1/91-2/95 only) (4/91).

Recommendation, Skill Level 30

In the lower-division baccalaureate/associate degree category, 3 semester hours in military science, 3 in automotive fundamentals, 1 in first aid, 3 in personnel supervision, 3 in maintenance management, and 2 in records and information management (4/91).

Recommendation, Skill Level 40

In the lower-division baccalaureate/associate degree category, 3 semester hours in military science, 3 in automotive fundamentals, 1 in first aid, 3 in personnel supervision, 3 in maintenance management, and 2 in records and information management. In the upper-division baccalaureate category, 3 semester hours for field experience in management and 3 in organizational management (4/91).

MOS-19D-003

Cavalry Scout
19D10
19D20
19D30
19D40

Exhibit Dates: 6/01–Present.

Career Management Field: 19 (Armor).

Description

Summary: Leads, serves, or assists as a member of a scout crew, squad, section, or platoon in reconnaissance security and other combat operations. *Skill Level 10:* Serves as a member of observation and listening team; gathers and reports information on terrain features, enemy strength, disposition, and equipment; loads, cleans, and fires weapons; identifies targets; requests and adjusts indirect and aerial fire; operates wheeled and tracked vehicles and communications equipment; uses maps and map symbols; navigates on ground; uses compass and distinguishes topographic features; assists in firing of crew-served weapons; assists mechanics and makes minor repairs on scout vehicles and tanks; performs first aid and applies field sanitation methods; uses radiac instruments and chemical detection kits. *Skill Level 20:* Able to perform the duties required for Skill Level 10; supervises scout vehicle crew or assists squad leader; selects, organizes, and supervises operation of observation and listening posts; prepares, distributes, and files operations and intelligence information; coordinates actions of vehicles with platoon and supporting elements; supervises request, receipt, storage, and issue of ammunition. *Skill Level 30:* Able to perform the duties required for Skill Level 20; supervises scout squad or section; directs tactical deployment; leads in combat; supervises maintenance of assigned vehicles and equipment; evaluates routes, assembly areas, and positioning for mounted combat operations. *Skill Level 40:* Able to perform the duties required for Skill Level 30; supervises armored cavalry or reconnaissance platoon; processes operations and intelligence information; assists commander in planning, organizing, directing, supervising, training, coordinating, and reporting activities of the scout or armored cavalry platoon and staff sections; directs distribution of fire; supervises platoon maintenance activities; collects, evaluates, and assists in interpretation and dissemination of combat information.

Recommendation, Skill Level 10

Credit may be granted on the basis of an individualized assessment of the student (6/01).

Recommendation, Skill Level 20

Credit may be granted on the basis of an individualized assessment of the student (6/01).

Recommendation, Skill Level 30

In the lower-division baccalaureate/associate degree category, 2 semester hours in military science, 3 in introduction to automobile service, 1 in first aid, 1 in map reading, 3 in principles of supervision, 2 in maintenance management, and 2 in records and information management. In the upper-division baccalaureate category, 3 semester hours for a management practicum (6/01).

Recommendation, Skill Level 40

In the lower-division baccalaureate/associate degree category, 3 semester hours in military science, 3 in introduction to automobile service, 1 in first aid, 1 in map reading, 3 in principles of supervision, 3 in maintenance management, and 3 in records and information management. In the upper-division baccalaureate category, 6 semester hours for a management practicum (6/01).

MOS-19E-001

M48-M60A1/A3 Armor Crewman
19E10
19E20
19E30
19E40

Exhibit Dates: 1/90–12/90.

Career Management Field: 19 (Armor).

Description

Summary: Leads, supervises, or serves as a member of an armor tank unit. *Skill Level 10:* Secures, prepares, and stows communications equipment aboard vehicle; loads and unloads main gun; exercises safety precautions in handling of ammunition; installs and dismantles antennas; prepares radio equipment for operation and operates communications equipment; assists in target detection and identification; determines range to target; prepares range cards, reads maps, places turret in operation, and fires main gun; performs operator maintenance on turret, weapons, controls, and communications equipment; protects self, weapons, and equipment from chemical and other contaminants; employs principles of escape and evasion; administers first aid and applies field sanitation methods; breaches and clears minefields and obstacles. *Skill Level 20:* Able to perform the duties required for Skill Level 10; loads and fires main gun; leads ammunition supply section; helps process intelligence and operation data; provides technical guidance to subordinates; prepares, files, and distributes operation maps, situation maps, and overlays; reads and interprets maps and aerial photographs; reproduces, distributes, and files operation intelligence, administrative, and unit training documents, orders, publications, and evaluation results. *Skill Level 30:* Able to perform the duties required for Skill Level 20; supervises subordinates; trains crews; requests, observes, and adjusts supporting fire. NOTE: May have progressed to 19E30 from 19E20 or 19F20 (Tank Driver). *Skill Level 40:* Able to perform the duties required for Skill Level 30; supervises tank platoon and processing of operations and intelligence information; assists in planning, organizing, directing, training, supervising, coordinating, and reporting activities of tank or staff sections; collects, evaluates, and assists in interpretation and dissemination of combat information.

Recommendation, Skill Level 10

In the vocational certificate category, 3 semester hours in mechanical maintenance. In the lower-division baccalaureate/associate degree category, 1 semester hour in first aid (11/79).

Recommendation, Skill Level 20

In the vocational certificate category, 3 semester hours in mechanical maintenance. In the lower-division baccalaureate/associate degree category, 3 semester hours in map reading, 1 in first aid, and 2 in record keeping (11/79).

Recommendation, Skill Level 30

In the vocational certificate category, 3 semester hours in mechanical maintenance. In the lower-division baccalaureate/associate degree category, 3 semester hours in map reading, 1 in first aid, 3 in personnel supervision, and 3 in record keeping. In the upper-division baccalaureate category, 3 semester hours for field experience in management (11/79).

Recommendation, Skill Level 40

In the vocational certificate category, 3 semester hours in mechanical maintenance. In the lower-division baccalaureate/associate degree category, 3 semester hours in map reading, 1 in first aid, 3 in record keeping, 3 in personnel supervision, and 3 in human relations. In the upper-division baccalaureate category, 3 semester hours for field experience in management and 3 in personnel management (11/79).

MOS-19E-002

M48-M60 Armor Crewman
19E10
19E20
19E30
19E40

Exhibit Dates: 1/91–5/01.

Career Management Field: 19 (Armor).

Description

Summary: Leads, supervises, or serves as a member of an armor tank unit. *Skill Level 10:* Secures, prepares, and stows communications equipment aboard vehicle; loads and unloads main gun; exercises safety precautions in handling of ammunition; installs and dismantles antennas; prepares radio equipment for operation and operates communications equipment; assists in target detection and identification; determines range to target; prepares range cards, reads maps, places turret in operation, and fires main gun; performs operator maintenance on components; protects self, weapons, and equipment from chemical and other contaminants; employs principles of escape and evasion; administers first aid and applies field sanitation methods; breaches and clears minefields and obstacles. *Skill Level 20:* Able to perform the duties required for Skill Level 10; loads and fires main gun; leads ammunition supply section; helps process intelligence and operation data; provides technical guidance to subordinates; prepares, files, and distributes operation maps, situation maps, and aerial photographs; reproduces, distributes, and files operation intelligence, administrative, and unit training documents, orders, publications, and evaluation results. *Skill Level 30:* Able to perform the duties required for Skill Level 20; supervises subordinates; trains crews; requests, observes, and adjusts supporting fire. *Skill Level 40:* Able to perform the duties required for Skill Level 30; supervises tank platoon and processing of operations and intelligence information; assists in planning, organizing, directing, training, supervising, coordinating, and reporting activities of tank or staff sections; collects, evaluates, and assists in interpretation and dissemination of combat information; provides counseling; maintains personnel records for performance ratings.

Recommendation, Skill Level 10

In the lower-division baccalaureate/associate degree category, 3 semester hours in military science, 3 in automotive fundamentals, and 1 in first aid. (NOTE: This recommendation for skill level 10 is valid for the dates 1/91-9/91 only) (4/91).

Recommendation, Skill Level 20

In the lower-division baccalaureate/associate degree category, 3 semester hours in military science, 3 in automotive fundamentals, 1 in first aid, and 3 in personnel supervision. (NOTE: This recommendation for skill level 20 is valid for the dates 1/91-2/95 only) (4/91).

Recommendation, Skill Level 30

In the lower-division baccalaureate/associate degree category, 3 semester hours in military science, 3 in automotive fundamentals, 1 in first aid, 3 in personnel supervision, 3 in maintenance management, and 2 in records and information management (4/91).

Recommendation, Skill Level 40

In the lower-division baccalaureate/associate degree category, 3 semester hours in military science, 3 in automotive fundamentals, 1 in first aid, 3 in personnel supervision, 3 in maintenance management, and 2 in records and information management. In the upper-division baccalaureate category, 3 semester hours for field experience in management and 3 in organizational management (4/91).

MOS-19E-003

M48-M60 Armor Crewman
19E10
19E20
19E30
19E40

Exhibit Dates: 6/01–Present.

Description

Summary: Leads, supervises, or serves as a member of an armor tank unit. *Skill Level 10:* Secures, prepares, and stows communications equipment on armor vehicle; loads and unloads main gun; exercises safety precautions in handling ammunition; installs and dismantles antennas; prepares radio equipment for operation and operates communications equipment; assists in target detection and identification; determines the range of targets; prepares range cards; reads maps; places turret in operation; fires the main gun; performs operator maintenance on components; protects self, weapons, and equipment from chemical and other contaminants; employs principles of escape and evasion; administers first aid and applies field sanitation methods; breaches and clears minefields and obstacles. *Skill Level 20:* Able to perform the duties required for Skill Level 10; loads and fires main gun; leads ammunition supply section; helps process intelligence and operation data; provides technical guidance to subordinates; prepares, distributes, and files operation maps, situation maps, and aerial photographs; reproduces, distribute, and files operation intelligence, administrative and unit training documents, orders, and publications. *Skill Level 30:* Able to perform the duties required for Skill Level 20; supervises subordinates; trains crews; requests, observes, and adjusts supporting fire. *Skill Level 40:* Able to perform the duties required for Skill Level 30; supervises tank platoon and processing of operations and intelligence information; assists in planning, organizing, directing, training, supervising, coordinating, and reporting activities of tank or staff sections; collects, evaluates, and assists in interpretation and dissemination of combat information; provides counseling; maintains personnel records for performance ratings.

Recommendation, Skill Level 10

Credit may be granted on the basis of an individualized assessment of the student (6/01).

Recommendation, Skill Level 20

Credit may be granted on the basis of an individualized assessment of the student (6/01).

Recommendation, Skill Level 30

In the lower-division baccalaureate/associate degree category, 3 semester hours in military science, 3 in automotive fundamentals, 1 in first aid, 3 in principles of supervision, 2 in records and information management, and 2 in maintenance management. In the upper-division baccalaureate category, 3 semester hours for a management practicum (6/01).

Recommendation, Skill Level 40

In the lower-division baccalaureate/associate degree category, 3 semester hours in military science, 3 in automotive fundamentals, 1 in first aid, 3 in principles of supervision, 2 in records and information management, and 3 in maintenance management. In the upper-division baccalaureate category, 3 semester hours in operations management and 6 for a management practicum (6/01).

MOS-19K-001

XM1 Armor Crewman
19K10
19K20
19K30
19K40

Exhibit Dates: 1/90–12/90.

Career Management Field: 19 (Armor).

Description

Summary: Leads, supervises, or serves as a member of XM1 armor unit in offensive and defensive combat operations. *Skill Level 10:* Drives tank, loads and fires main gun; exercises techniques of land navigation; reads and interprets maps; secures, prepares, and stows ammunition; installs and dismantles antennas; prepares and operates communications equipment; assists in target detection and identification; prepares range finder for operation; performs operator maintenance on turret, weapons, controls, and communication equipment; conducts operational checks; services and maintains tank chassis, turret, and automotive components. *Skill Level 20:* Able to perform the duties required for Skill Level 10; inspects automotive and turret components of platoon vehicles for malfunctions; assists in the training readiness of platoon drivers and the proficiency of gunners and loaders; processes intelligence and operational data. *Skill Level 30:* Able to perform the duties required for Skill Level 20; leads tank crew; coordinates action of tank with platoon and supporting elements; supervises crew operator maintenance; coordinates maintenance requirements; evaluates work of subordinates; ensures collection and proper reporting of intelligence data to units and responsible staff sections. *Skill Level 40:* Able to perform the duties required for Skill Level 30; supervises tank platoon and processing of operations and intelligence data; assists in planning, organizing, directing, training, supervising, coordinating, and reporting of activities of tank or staff sections; supervises platoon maintenance activities; collects, evaluates, and assists in the interpretation and dissemination of combat information.

Recommendation, Skill Level 10

In the lower-division baccalaureate/associate degree category, 3 semester hours in automotive maintenance and 3 in mechanical systems (7/82).

Recommendation, Skill Level 20

In the lower-division baccalaureate/associate degree category, 3 semester hours in automotive maintenance and 3 in mechanical systems (7/82).

Recommendation, Skill Level 30

In the lower-division baccalaureate/associate degree category, 3 semester hours in automotive maintenance, 3 in mechanical systems, and 3 in personnel supervision (7/82).

Recommendation, Skill Level 40

In the lower-division baccalaureate/associate degree category, 3 semester hours in automotive maintenance, 3 in mechanical systems, and 3 in personnel supervision. In the upper-division baccalaureate category, 3 semester hours in organizational management and 3 for field experience in management (7/82).

MOS-19K-002

M1 Armor Crewman
19K10
19K20
19K30
19K40

Exhibit Dates: 1/91–12/98.

Career Management Field: 19 (Armor).

Description

Summary: Serves as a member or leads members of M1 armor unit in offensive and defensive combat operations. *Skill Level 10:* Drives tank; loads and fires main gun; exercises techniques of land navigation; reads and interprets maps; secures, prepares, and stows ammunition; installs and dismantles antennas; prepares and operates communication equipment; assists in target detection and identification; prepares range finder for operation; performs operator maintenance on turret, weapons, controls, and communication equipment; conducts operational checks; assists mechanics and makes minor repairs on tank turret and automotive components. *Skill Level 20:* Able to perform the duties required for Skill Level 10; inspects automotive and turret components of platoon vehicles for malfunctions; assists in the training readiness of platoon drivers and the proficiency of gunners and loaders; processes intelligence and operational data. *Skill Level 30:* Able to perform the duties required for Skill Level 20; leads tank crew; coordinates action of tank with platoon and supporting elements; supervises crew operator maintenance; coordinates maintenance requirements; evaluates work of subordinates; ensures collection and proper reporting of intelligence data to units and responsible staff sections. *Skill Level 40:* Able to perform the duties required for Skill Level 30; supervises tank platoon and processing of operations and intelligence data; assists in planning, organizing, directing, training, supervising, coordinating, and reporting activities of tank or staff sections; supervises platoon maintenance activities; collects, evaluates, and assists in the interpretation and dissemination of combat information.

Recommendation, Skill Level 10

In the lower-division baccalaureate/associate degree category, 3 semester hours in military science, 3 in automotive fundamentals, and 1 in first aid. (NOTE: This recommendation for skill level 10 is valid for the dates 1/91-9/91 only) (4/91).

Recommendation, Skill Level 20

In the lower-division baccalaureate/associate degree category, 3 semester hours in military science, 3 in automotive fundamentals, 1 in first aid, and 3 in personnel supervision. (NOTE: This recommendation for skill level 20 is valid for the dates 1/91-2/95 only) (4/91).

Recommendation, Skill Level 30

In the lower-division baccalaureate/associate degree category, 3 semester hours in military science, 3 in automotive fundamentals, 1 in first aid, 3 in personnel supervision, 3 in maintenance management, and 2 in records and information management (4/91).

Recommendation, Skill Level 40

In the lower-division baccalaureate/associate degree category, 3 semester hours in military science, 3 in automotive fundamentals, 1 in first aid, 3 in personnel supervision, 3 in maintenance management, and 2 in records and information management. In the upper-division baccalaureate category, 3 semester hours for field experience in management and 3 in organizational management (4/91).

MOS-19K-003

M1 Armor Crewman
19K10
19K20
19K30
19K40

Exhibit Dates: 1/99–5/01.

Career Management Field: 19 (Armor).

Description

Summary: Serves as a member, supervises or leads members of an M1 armor unit in offensive and defensive combat operations. *Skill Level 10:* Drives tank; loads and fires main gun; exercises techniques of land navigation; reads and interprets maps; secures, prepares, and stows ammunition; installs and dismantles antennas; prepares and operates communications equipment; assists in target detection and identification; prepares range finder for operation; performs operator maintenance on turret, weapons, controls, and communications equipment; conducts operational checks; assists mechanics and makes minor repairs on tank turrets and components. *Skill Level 20:* Able to perform the duties required for Skill Level 10; inspects tank and turret components of platoon vehicles for malfunctions; assists in the training readiness of platoon drivers and the proficiency of gunners and loaders; reads and processes operational data. *Skill Level 30:* Able to perform the duties required for Skill Level 20; leads tank crew; coordinates action of tank with platoon and supporting elements; supervises crew operator maintenance; coordinates maintenance requirements; evaluates work of subordinates; ensures collection and proper reporting of intelligence data to units and responsible staff sections. *Skill Level 40:* Able to perform the duties required for Skill Level 30; supervises tank platoon and processing of operations and intelligence data; assists in planning, organizing, directing, training, supervising, coordinating, and reporting of activities of tank or staff sections; supervises platoon maintenance activities; collects, evaluates, and assists in the interpretation and dissemination of combat information.

Recommendation, Skill Level 10

Credit may be granted on the basis of an individualized assessment of the student (6/99).

Recommendation, Skill Level 20

Credit may be granted on the basis of an individualized assessment of the student (6/99).

Recommendation, Skill Level 30

In the lower-division baccalaureate/associate degree category, 6 semester hours in heavy equipment fundamentals, 1 in first aid, 3 in heavy equipment repair and maintenance, 3 in maintenance management, 3 in principles of supervision, and 2 in records management (6/99).

Recommendation, Skill Level 40

In the lower-division baccalaureate/associate degree category, 6 semester hours in heavy equipment fundamentals, 1 in first aid, 3 in heavy equipment repair and maintenance, 6 in maintenance management, 3 in principles of supervision, and 4 in records management. In the upper-division baccalaureate category, 3 semester hours principles of management and 3 in a management practicum (6/99).

MOS-19K-004

M1 Armor Crewman

19K10

19K20

19K30

19K40

Exhibit Dates: 6/01–Present.

Career Management Field: 19 (Armor).

Description

Summary: Serves as a member, and supervises or leads members of an M1 armor unit in offensive and defensive combat operations. *Skill Level 10:* Drives tank; loads and fires main gun; exercises techniques of land navigation; reads and interprets maps; secures, prepares, and stows ammunition; installs and dismantles antennas; prepares and operates communications equipment; assists in target detection and identification; prepares range finder for operation; performs operator maintenance on turret, weapons, controls, and communications equipment; conducts operational checks; assists mechanics and makes minor repairs on tank turrets and components. *Skill Level 20:* Able to perform the duties required for Skill Level 10; inspects tank and turret components of platoon vehicles for malfunctions; assists in the training readiness of platoon drivers and the proficiency of gunners and loaders; reads and processes operational data. *Skill Level 30:* Able to perform the duties required for Skill Level 20; leads tank crew; coordinates action of tank with platoon and supporting elements; supervises crew operator maintenance; coordinates maintenance requirements; evaluates work of subordinates; ensures collection and proper reporting of intelligence data to units and responsible staff sections. *Skill Level 40:* Able to perform the duties required for Skill Level 30; supervises tank platoon and processing of operations and intelligence data; assists in planning, organizing, directing, training, supervising, coordinating, and reporting of activities of tank or staff sections; supervises platoon maintenance activities; collects, evaluates, and assists in the interpretation and dissemination of combat information.

Recommendation, Skill Level 10

Credit may be granted on the basis of an individualized assessment of the student (6/01).

Recommendation, Skill Level 20

Credit may be granted on the basis of an individualized assessment of the student (6/01).

Recommendation, Skill Level 30

In the lower-division baccalaureate/associate degree category, 3 semester hours in heavy equipment fundamentals, 3 in heavy equipment repair and maintenance, 1 in first aid, 2 in maintenance management, 3 in principles of supervision, 2 in report writing, and 2 in records and information management. In the upper-division baccalaureate category, 3 semester hours for a management practicum (6/01).

Recommendation, Skill Level 40

In the lower-division baccalaureate/associate degree category, 3 semester hours in heavy equipment fundamentals, 3 in heavy equipment repair and maintenance, 1 in first aid, 3 in maintenance management, 3 in principles of supervision, 3 in report writing, and 3 in records and information management. In the upper-division baccalaureate category, 3 semester hours in principles of management and 3 for a management practicum (6/01).

MOS-19Z-001

Armor Senior Sergeant

19Z50

Exhibit Dates: 1/90–12/90.

Career Management Field: 19 (Armor).

Description

Plans, coordinates, and supervises activities of a unit; supervises operations, intelligence, or liaison activities; as intelligence sergeant, collects, interprets, analyzes, and distributes intelligence information; writes routine and special reports, commendations, and operations plans; edits and prepares tactical and training plans; coordinates and implements training programs and directs on-the-job training; is a mid-level manager overseeing 90-150 persons; assists in making formal briefing presentations and in presenting instruction to larger groups (30-70); provides formal and informal counseling to individuals and groups. NOTE: May have progressed to 19Z50 from 19D40 (Cavalry Scout), 19E40 (M48-M60A1/A3 Armor Crewman), or 19J (M60A2 Armor Crewman).

Recommendation

In the lower-division baccalaureate/associate degree category, 6 semester hours in personnel supervision, 3 in office management, 3 in report writing, and 3 in public speaking. In the upper-division baccalaureate category, 6 semester hours in personnel management, 6 for field experience in management, 3 in human relations, and 3 in report writing (11/79).

MOS-19Z-002

Armor Senior Sergeant

19Z50

Exhibit Dates: 1/91–5/01.

Career Management Field: 19 (Armor).

Description

Plans, coordinates, and supervises activities of a unit; supervises operations, intelligence, or liaison activities; as intelligence sergeant, collects, interprets, analyzes, and distributes intelligence information; writes routine and special reports, commendations, and operations plans; edits and prepares tactical and training plans; coordinates and implements training programs and directs on-the-job training; is a mid-level manager overseeing 90-150 persons; assists in making formal briefing presentations and in presenting instruction to larger groups (30-70); provides formal and informal counseling to individuals and groups; works on budget with superiors; maintains computerized data file on past and future planned activity; provides daily briefing to staff sergeants; maintains computerized personnel data files, job evaluations, and awards. NOTE: May have progressed to 19Z50 from 19D40 (Cavalry Scout), 19E40 (M48-M60 Armor Crewman), or 19K (M1 Armor Crewman).

Recommendation

In the lower-division baccalaureate/associate degree category, 3 semester hours in military science, 3 in personnel supervision, 3 in maintenance management, 2 in records and information management, 3 in office management, and 3 in report writing. In the upper-division baccalaureate category, 3 semester hours in organizational management and 3 for field experience in management; if individual has attained paygrade E-9, additional credit as follows: 3 semester hours in management problems, 3 in operations management, and 3 in communication techniques for managers (4/91).

MOS-19Z-003

Armor Senior Sergeant

19Z50

Exhibit Dates: 6/01–Present.

Description

Supervises operations center at battalion and higher levels; plans, coordinates, and supervises activities of a unit; supervises operations, intelligence, or liaison activities; as intelligence sergeant, collects, interprets, analyzes, and distributes intelligence information; writes routine and special reports, commendations, and operations plans; prepares and edits tactical and training plans; coordinates and implements training programs; directs on-the-job training; assists in making formal briefing presentations and in presenting instruction to large groups; provides formal and informal counseling to individuals and groups; works on budget with superiors; maintains computerized data file on past and future planned activities; provides daily briefing to staff sergeants; maintains computerized personnel data files, job evaluations, and awards. NOTE: May have progressed to 19Z50 from 19D40, Cavalry Scout; 19E40, M48-M60 Armor Crewman; or 19K40, M1 Armor Crewman.

Recommendation

In the lower-division baccalaureate/associate degree category, 3 semester hours in principles of supervision, 3 in records and information management, 3 in report writing, and 3 in speech. In the upper-division baccalaureate category, 3 semester hours in operations management and 6 for a management practicum (6/01).

MOS-21B-001

Combat Engineer

21B10

21B20

21B30

21B40

Exhibit Dates: Pending Evaluation. 4/03–Present.

MOS-21C-001

Bridge Crew Member
21C10
21C20
21C30
21C40

Exhibit Dates: Pending Evaluation. 4/03–Present.

MOS-21D-001

Diver
21D10
21D20
21D30
21D40

Exhibit Dates: Pending Evaluation. 4/03–Present.

MOS-21E-001

Heavy Construction Equipment Operator
21E10
21E20

Exhibit Dates: Pending Evaluation. 4/03–Present.

MOS-21F-001

Crane Operator
21F10
21F20

Exhibit Dates: Pending Evaluation. 4/03–Present.

MOS-21G-004

Pershing Electronics Materiel Specialist
21G10
21G20
21G30
21G40

Exhibit Dates: 1/90–4/92. Effective 4/92, MOS 21G was discontinued.

Career Management Field: 13 (Field Artillery).

Description

Summary: Supervises or performs inspection, checkout, and troubleshooting on electronic, mechanical, hydraulic, and pneumatic equipment associated with the Pershing missile system. *Skill Level 10:* Operates and performs maintenance on the Pershing missile programmer-test station, missile, missile trainer, erector-launcher, power station, and associated equipment; performs limited isolation of malfunctioning equipment and components through the use of computer and adapter diagnostic program tapes. *Skill Level 20:* Able to perform the duties required for Skill Level 10; performs detailed organizational maintenance of the system and provides technical guidance to operators; isolates malfunctioning equipment and components through manual troubleshooting techniques, using volt ohm meters and functional schematics; adjusts or replaces malfunctioning components; prepares maintenance and supply forms and reports; in the absence of a supervisor or section chief, assumes duties of those positions. *Skill Level 30:* Able to perform the duties required for Skill Level 20; supervises maintenance activities; supervises the preparation of technical forms and reports; coordinates activities with supporting maintenance units; determines faulty work practices and demonstrates proper maintenance and troubleshooting techniques; supervises the requisitioning and storage of supplies and repair parts. *Skill Level 40:* Able to perform the duties required for Skill Level 30; supervises battery level organizational maintenance of Pershing missile system electronic equipment; supervises preparation of technical forms and reports and battery requisitioning of missile supplies; serves as senior enlisted maintenance advisor for the battery and conducts instruction on equipment operation and maintenance.

Recommendation, Skill Level 10

In the lower-division baccalaureate/associate degree category, 3 semester hours in electrical/electronic systems troubleshooting and maintenance. (NOTE: This recommendation for skill level 10 is valid for the dates 1/90-9/91 only) (6/89).

Recommendation, Skill Level 20

In the lower-division baccalaureate/associate degree category, 3 semester hours in electrical/electronic systems troubleshooting and maintenance; if served in a supervisory position, 3 semester hours in principles of supervision and 3 in maintenance management (6/89).

Recommendation, Skill Level 30

In the lower-division baccalaureate/associate degree category, 3 semester hours in electrical/electronic systems troubleshooting and maintenance, 3 in principles of supervision, 3 in maintenance management, and 3 in communication skills (written and oral) (6/89).

Recommendation, Skill Level 40

In the lower-division baccalaureate/associate degree category, 3 semester hours in electrical/electronic systems troubleshooting and maintenance, 3 in principles of supervision, 3 in maintenance management, and 3 in communication skills (written and oral). In the upper-division baccalaureate category, 3 semester hours for field experience in management and 3 in personnel management (6/89).

MOS-21G-005

Quarrying Specialist
21G10
21G20
21G30

Exhibit Dates: Pending Evaluation. 4/03–Present.

MOS-21H-001

Construction Engineering Supervision
21H30
21H40

Exhibit Dates: Pending Evaluation. 4/03–Present.

MOS-21J-001

General Construction Equipment Operator
21J10
21J20

Exhibit Dates: Pending Evaluation. 4/03–Present.

MOS-21K-001

Plumber
21K10
21K20

Exhibit Dates: Pending Evaluation. 4/03–Present.

MOS-21L-002

Pershing Electronics Repairer
21L10
21L20
21L30
21L40
21L50

Exhibit Dates: 1/90–9/91.

Career Management Field: 27 (Ballistic/Land Combat Missile and Light Air Defense Weapons Systems Maintenance).

Description

Summary: Supervises or performs maintenance and repair of electronic, mechanical, hydraulic, and pneumatic systems and components associated with the Pershing missile system, using a theoretical and working knowledge of electronic, digital, and logic circuits. *Skill Level 10:* Performs maintenance on the programmer test station, items for the guidance and control section, training set and associated training equipment, and on-system test equipment; reads electronic schematics and uses a wide range of electronic meters, including scope, to inspect, test, and adjust components to specific tolerances; determines deficiencies and causes of malfunctions in electronic, electrical, electromechanical, and pneumatic assemblies, subassemblies, modules, and circuit elements; determines serviceability and disposition of defective assemblies, subassemblies, and parts; serves on inspection teams. *Skill Level 20:* Able to perform the duties required for Skill Level 10; provides technical guidance to subordinates; installs equipment modifications; provides technical assistance to support units. *Skill Level 30:* Able to perform the duties required for Skill Level 20; establishes work load and repair priorities; supervises receipt, storage, inspection, testing, and repair of items; implements quality control measures; organizes and conducts on-the-job training programs; determines faulty work practices and demonstrates proper maintenance techniques; establishes and maintains maintenance records; serves as maintenance inspector; may supervise six or more persons. *Skill Level 40:* Able to perform the duties required for Skill Level 30 of MOS 21L, 21G (Pershing Electronic Materiel Specialist), or 46N (Pershing Electrical-Mechanical Repairer); supervises maintenance on electronic and electromechanical systems and components of the missile system; serves as section chief or platoon sergeant of a detachment, platoon company, or comparable organization; supervises and assists in the development of quality control programs; may supervise seven or more persons. *Skill Level 50:* Able to perform the duties required for Skill Level 40; supervises Pershing missile system maintenance activities; serves as maintenance chief; may supervise 30 or more persons; serves as principal noncommissioned officer in a missile maintenance unit.

Recommendation, Skill Level 10

In the vocational certificate category, 6 semester hours in basic electronics and 9 in computer fundamentals. In the lower-division baccalaureate/associate degree category, 3 semester hours in AC/DC theory, 3 in basic electronics, 3 in computer fundamentals, and 3 in technical mathematics (6/79).

Recommendation, Skill Level 20

In the vocational certificate category, 6 semester hours in basic electronics and 9 in computer fundamentals. In the lower-division baccalaureate/associate degree category, 3 semester hours in AC/DC theory, 3 in basic electronics, 3 in computer fundamentals, and 3 in technical mathematics (6/79).

Recommendation, Skill Level 30

In the vocational certificate category, 6 semester hours in basic electronics and 9 in computer fundamentals. In the lower-division baccalaureate/associate degree category, 3 semester hours in AC/DC theory, 3 in basic electronics, 3 in computer fundamentals, 3 in technical mathematics, 3 in maintenance management, 3 in personnel supervision, and 3 in human relations (6/79).

Recommendation, Skill Level 40

In the vocational certificate category, 6 semester hours in basic electronics and 9 in computer fundamentals. In the lower-division baccalaureate/associate degree category, 3 semester hours in AC/DC theory, 3 in basic electronics, 3 in computer fundamentals, 3 in technical mathematics, 3 in maintenance management, 3 in personnel supervision, 3 in human relations, and 2 in communication skills. In the upper-division baccalaureate category, 3 semester hours in introduction to management, 3 for field experience in management, and 3 in industrial arts education (electronics) (6/79).

Recommendation, Skill Level 50

In the vocational certificate category, 6 semester hours in basic electronics and 9 in computer fundamentals. In the lower-division baccalaureate/associate degree category, 3 semester hours in AC/DC theory, 3 in basic electronics, 3 in computer fundamentals, 3 in technical mathematics, 3 in maintenance management, 3 in personnel supervision, 3 in human relations, 2 in communication skills, and 3 for field experience in management. In the upper-division baccalaureate category, 3 semester hours in introduction to management, 6 for field experience in management, 3 in industrial arts education (electronics), and 3 in personnel management (6/79).

MOS-21L-003

Pershing Electronics Repairer
21L10
21L20
21L30
21L40
21L50

Exhibit Dates: 10/91–12/91. Effective 12/91, MOS 21L was discontinued.

Career Management Field: 27 (Land Combat and Air Defense System Direct and General Support Maintenance).

Description

Summary: Supervises or performs maintenance and repair of electronic, mechanical, hydraulic, and pneumatic systems and components associated with the Pershing missile system, using a theoretical and working knowledge of electronic, digital, and logic circuits. *Skill Level 10:* Performs maintenance on the programmer test station, items for the guidance and control section, training set and associated training equipment, and on-system test equipment; reads electronic schematics and uses a wide range of electronic meters, including scope, to inspect, test, and calibrate circuits to specific tolerances using computerized test set; identifies and corrects faults in electronic, electrical, electromechanical, and pneumatic assemblies, subassemblies, modules, and circuit elements; determines serviceability and disposition of defective assemblies, subassemblies, and parts; serves on inspection teams. *Skill Level 20:* Able to perform the duties required for Skill Level 10; provides technical guidance to subordinates; installs minor equipment modifications; provides technical assistance to support units. *Skill Level 30:* Able to perform the duties required for Skill Level 20; establishes work load and repair priorities; supervises receipt, storage, inspection, testing, and repair of items; implements quality control measures; organizes and conducts on-the-job training program; demonstrates proper work practices and maintenance techniques; establishes and maintains maintenance records; serves as maintenance inspector; may supervise six or more persons. *Skill Level 40:* Able to perform the duties required for Skill Level 30 of MOS 21L or 46N (Pershing Electrical-Mechanical Repairer); supervises maintenance on electronic and electromechanical systems and components of the missile system; serves as section chief or platoon sergeant of a detachment, platoon company, or comparable organization; supervises and assists in the development of quality control programs; may supervise seven or more persons. *Skill Level 50:* Able to perform the duties required for Skill Level 40; supervises Pershing missile system maintenance activities; serves as maintenance chief; may supervise 30 or more persons; serves as principal noncommissioned officer in a missile maintenance unit. NOTE: May progress to 21L50 from 21L40 or 46N40 (Pershing Electrical Mechanical Repairer).

Recommendation, Skill Level 10

Credit may be granted on the basis of an individualized assessment of the student (3/94).

Recommendation, Skill Level 20

In the lower-division baccalaureate/associate degree category, 3 semester hours in basic electronics, 4 in computer systems troubleshooting and maintenance, 3 in electronic systems troubleshooting and maintenance, 3 in computer literacy, and 3 in personnel supervision (9/91).

Recommendation, Skill Level 30

In the lower-division baccalaureate/associate degree category, 3 semester hours in basic electronics, 4 in computer systems troubleshooting and maintenance, 3 in electronic systems troubleshooting and maintenance, 3 in computer literacy, 3 in personnel supervision, 3 in maintenance management, and 2 in records and information management (9/91).

Recommendation, Skill Level 40

In the lower-division baccalaureate/associate degree category, 3 semester hours in basic electronics, 4 in computer systems troubleshooting and maintenance, 3 in electronic systems troubleshooting and maintenance, 3 in computer literacy, 3 in personnel supervision, 3 in maintenance management, and 2 in records and information management. In the upper-division baccalaureate category, 3 semester hours in organizational management and 3 for field experience in management (9/91).

Recommendation, Skill Level 50

In the lower-division baccalaureate/associate degree category, 3 semester hours in basic electronics, 4 in computer systems troubleshooting and maintenance, 3 in electronic systems troubleshooting and maintenance, 3 in computer literacy, 3 in personnel supervision, 3 in maintenance management, and 2 in records and information management. In the upper-division baccalaureate category, 3 semester hours in organizational management and 6 for field experience in management; if individual has attained paygrade E-9, additional credit as follows: 3 semester hours in management problems, 3 in operations management, and 3 in communication techniques for managers (9/91).

MOS-21L-004

Lithographer
21L10
21L20
21L30
21L40

Exhibit Dates: Pending Evaluation. 4/03–Present.

MOS-21M-001

Firefighter
21M10
21M20
21M30
21M40

Exhibit Dates: Pending Evaluation. 4/03–Present.

MOS-21N-001

Construction Equipment Supervisor
21N30
21N40

Exhibit Dates: Pending Evaluation. 4/03–Present.

MOS-21P-001

Prime Power Production
21P20
21P30
21P40

Exhibit Dates: Pending Evaluation. 4/03–Present.

MOS-21Q-001

Transmission and Distribution Specialist (RC)
21Q10
21Q20
21Q30
21Q40

Exhibit Dates: Pending Evaluation. 4/03–Present.

MOS-21R-002

Interior Electrician
21R10
21R20

Exhibit Dates: Pending Evaluation. 4/03–Present.

MOS-21S-002

Topographic Surveyor
21S10
21S20
21S30
21S40

Exhibit Dates: Pending Evaluation. 4/03–Present.

MOS-21T-002

TECHNICAL ENGINEER
21T10
21T20
21T30
21T40

Exhibit Dates: Pending Evaluation. 4/03–Present.

MOS-21U-002

TOPOGRAPHIC ANALYST
21U10
21U20
21U30
21U40

Exhibit Dates: Pending Evaluation. 4/03–Present.

MOS-21V-001

CONCRETE AND ASPHALT EQUIPMENT OPERATOR
21V10
21V20
21V30

Exhibit Dates: Pending Evaluation. 4/03–Present.

MOS-21W-001

CARPENTRY AND MASONRY SPECIALIST
21W10
21W20

Exhibit Dates: Pending Evaluation. 4/03–Present.

MOS-21X-001

GENERAL ENGINEERING SUPERVISOR
21X50

Exhibit Dates: 9/04–Present.

Career Management Field: 21 (General Engineering).

Description

Provides staff supervision to units engaged in structural building and repair and provision of water and utility services; serves as principal noncommissioned officer in an engineering combat (heavy) company or in an engineer combat battalion; may serve in an engineer section at engineer brigade or higher headquarters; directs and coordinates activities of staff personnel to ensure integration of all construction and utility operations; assists in appraisal of training status, construction planning, and material requests; inspects structural facilities and water and utility production and distribution systems; supervises conformance to OSHA standards of construction; supervises performance of skill level 40 of MOS 00B (Diver), 51H (Construction Engineering Supervisor), 51M (Firefighter), 51T (Technical Engineering Supervisor), 52E (Prime Power Production Specialist), 52G (Transmission and Distribution Specialist), or 62N (Construction Equipment Supervisor).

Recommendation

In the lower-division baccalaureate/associate degree category, 3 semester hours in personnel supervision and 3 in communication skills. In the upper-division baccalaureate category, 3 semester hours in human resource management, 3 for a practicum in management, 3 in operations/logistics management, 3 in organizational management, 3 in contract management, and 3 in code and regulations (4/04).

MOS-21Y-001

TOPOGRAPHIC ENGINEERING SUPERVISOR
21Y50

Exhibit Dates: Pending Evaluation. 4/03–Present.

MOS-21Z-001

COMBAT ENGINEERING SENIOR SERGEANT
21Z50

Exhibit Dates: Pending Evaluation. 4/03–Present.

MOS-23R-001

HAWK MISSILE SYSTEM MECHANIC
23R10
23R20
23R30
23R40
23R50

Exhibit Dates: 3/90–Present.

Career Management Field: 23 (Air Defense System Maintenance).

Description

Summary: Supervises or performs maintenance on Hawk PIP III missile system and associated equipment. *Skill Level 10:* Performs equipment checks and adjustments; troubleshoots and repairs to the block diagram level; assists in the training of operators and provides technical assistance in conduct of operational and maintenance inspections involving electrical, mechanical, and hydraulic systems. *Skill Level 20:* Able to perform the duties required for Skill Level 10; provides technical guidance to subordinates; supervises maintenance; completes maintenance forms and records; installs equipment modifications; initiates requests for spare parts, test equipment, tools, and supplies. *Skill Level 30:* Able to perform the duties required for Skill Level 20; supervises and performs technical inspections; supervises preparation of maintenance forms and records; supervises and conducts training programs for operator and maintenance personnel. *Skill Level 40:* Able to perform the duties required for Skill Level 30; supervises and coordinates equipment organizational maintenance; assists in establishing maintenance procedures; supervises the requisition of supplies and repair parts. *Skill Level 50:* Able to perform the duties required for Skill Level 40; evaluates, inspects, and coordinates maintenance activities; provides technical guidance to subordinates; advises battalion commander regarding unit maintenance program.

Recommendation, Skill Level 10

In the lower-division baccalaureate/associate degree category, 3 semester hours in basic electronics, 3 in radar systems, and 3 in digital electronics laboratory. In the upper-division baccalaureate category, 3 semester hours in hydraulic systems troubleshooting and maintenance and 3 in computer systems troubleshooting and maintenance. (NOTE: This recommendation for skill level 10 is valid for the dates 3/90-9/91 only) (6/91).

Recommendation, Skill Level 20

In the lower-division baccalaureate/associate degree category, 3 semester hours in basic electronics, 3 in radar systems, 3 in digital electronics laboratory, 2 in records and information management, 3 in personnel supervision, and 2 in maintenance management. In the upper-division baccalaureate category, 3 semester hours in hydraulic systems troubleshooting and maintenance, 3 in computer systems troubleshooting and maintenance, and 3 in electronic systems troubleshooting. (NOTE: This recommendation for skill level 20 is valid for the dates 3/90-2/95 only) (6/91).

Recommendation, Skill Level 30

In the lower-division baccalaureate/associate degree category, 3 semester hours in basic electronics, 3 in radar systems, 3 in digital electronics laboratory, 2 in records and information management, 3 in personnel supervision, and 3 in maintenance management. In the upper-division baccalaureate category, 3 semester hours in hydraulic systems troubleshooting and maintenance, 3 in computer systems troubleshooting and maintenance, and 3 in electronic systems troubleshooting (6/91).

Recommendation, Skill Level 40

In the lower-division baccalaureate/associate degree category, 3 semester hours in basic electronics, 3 in radar systems, 3 in digital electronics laboratory, 2 in records and information management, 3 in personnel supervision, and 3 in maintenance management. In the upper-division baccalaureate category, 3 semester hours in hydraulic systems troubleshooting and maintenance, 3 in computer systems troubleshooting and maintenance, 3 in electronic systems troubleshooting, 3 in organizational management, and 3 for field experience in management (6/91).

Recommendation, Skill Level 50

In the lower-division baccalaureate/associate degree category, 3 semester hours in basic electronics, 3 in radar systems, 3 in digital electronics laboratory, 2 in records and information management, 3 in personnel supervision, and 3 in maintenance management. In the upper-division baccalaureate category, 3 semester hours in hydraulic systems troubleshooting and maintenance, 3 in computer systems troubleshooting and maintenance, 3 in electronic systems troubleshooting, 3 in organizational management, and 6 for field experience in management; if individual has attained paygrade E-9, additional credit as follows: 3 semester hours in management problems, 3 in operations management, and 3 in communications techniques for managers (6/91).

MOS-24C-002

HAWK FIRING SECTION MECHANIC
((Improved Hawk Firing Section Mechanic))
24C10
24C20
24C30
24C40

Exhibit Dates: 1/90–5/91.

Career Management Field: 23 (Air Defense Missile Maintenance), subfield 235 (Improved Hawk Missile System Mechanics).

Description

Summary: Supervises or performs maintenance on improved Hawk firing section equipment. *Skill Level 10:* Performs maintenance on Hawk or improved Hawk radar, missile launching control box, electrical, mechanical, and hydraulic systems of the launcher, transporter, and associated equipment; determines malfunction; performs repairs by replacing subassembly units; assists in orientation, alignment, and synchronization of equipment; assembles external hardware of missile; makes prescribed tests to determine operational and maintenance status. *Skill Level 20:* Able to perform the duties required for Skill Level 10; installs equipment modifications; supervises transfer and handling of missile; provides technical training to missile crew supervisory and operational personnel in arming and disarming missile; initiates requests for repair parts, test equipment, tools, and supplies. *Skill Level 30:* Able to perform the duties required for Skill Level 20; provides technical supervision in training programs for operator and maintenance personnel; demonstrates proper maintenance and troubleshooting techniques; supervises maintenance and inspection teams; supervises the preparation of technical forms and reports. *Skill Level 40:* Able to perform the duties required for Skill Level 30; supervises and coordinates organizational maintenance of Hawk or Improved Hawk firing section equipment.

Recommendation, Skill Level 10

In the vocational certificate category, 3 semester hours in basic electricity and 3 in applied mathematics. In the lower-division baccalaureate/associate degree category, 3 semester hours in basic electronics, 3 in AC/DC circuits, 1 in radar/microwave, 3 in technical mathematics, and additional credit in electronics, mechanics, and hydraulics on the basis of institutional evaluation (1/80).

Recommendation, Skill Level 20

In the vocational certificate category, 3 semester hours in basic electricity and 3 in applied mathematics. In the lower-division baccalaureate/associate degree category, 3 semester hours in basic electronics, 3 in AC/DC circuits, 1 in radar/microwave, 3 in technical mathematics, and additional credit in electronics, mechanics, and hydraulics on the basis of institutional evaluation (1/80).

Recommendation, Skill Level 30

In the vocational certificate category, 3 semester hours in basic electricity and 3 in applied mathematics. In the lower-division baccalaureate/associate degree category, 3 semester hours in basic electronics, 3 in AC/DC circuits, 1 in radar/microwave, 3 in technical mathematics, 3 in electronics shop practices, 3 in personnel supervision, and 2 in records and report preparation, and additional credit in electronics, mechanics, and hydraulics on the basis of institutional evaluation. In the upper-division baccalaureate category, 3 semester hours for field experience in personnel management and 3 in industrial education (electronics) (1/80).

Recommendation, Skill Level 40

In the vocational certificate category, 3 semester hours in basic electricity and 3 in applied mathematics. In the lower-division baccalaureate/associate degree category, 3 semester hours in basic electronics, 3 in AC/DC circuits, 1 in radar/microwave, 3 in technical mathematics, 3 in electronics shop practices, 3 in personnel supervision, 2 in records and reports preparation, and 3 in maintenance management, and additional credit in electronics, mechanics, and hydraulics on the basis of institutional evaluation. In the upper-division baccalaureate category, 3 semester hours for field experience in personnel management, 3 in introduction to management, 3 in human relations, and 3 in industrial education (electronics) (1/80).

MOS-24C-003

Hawk Firing Section Mechanic
24C10
24C20
24C30

Exhibit Dates: 6/91–6/95. Effective 6/95, MOS 24C was discontinued.

Career Management Field: 23 (Air Defense System Maintenance).

Description

Summary: Supervises or performs maintenance on improved Hawk firing section equipment. *Skill Level 10:* Performs maintenance on Hawk radar, missile launching control box, and electrical, mechanical, and hydraulic systems of the launcher, transporter, and associated equipment; identifies malfunctions; performs repairs by replacing subassembly units; assists in orientation, alignment, and synchronization of equipment; assembles external hardware of missile; makes prescribed tests to determine operational and maintenance status. *Skill Level 20:* Able to perform the duties required for Skill Level 10; installs equipment modifications; supervises transfer and handling of missile; provides technical training to missile crew supervisory and operational personnel in arming and disarming missile; initiates requests for repair parts, test equipment, tools, and supplies. *Skill Level 30:* Able to perform the duties required for Skill Level 20; provides technical supervision in training programs for operator and maintenance personnel; serves as maintenance team leader; demonstrates proper maintenance and troubleshooting techniques; supervises maintenance and inspection teams; supervises the preparation of technical forms and reports.

Recommendation, Skill Level 10

In the lower-division baccalaureate/associate degree category, 3 semester hours in basic electronics and 3 in radar systems. In the upper-division baccalaureate category, 3 semester hours in hydraulic systems troubleshooting. (NOTE: This recommendation for skill level 10 is valid for the dates 6/91-9/91 only) (6/91).

Recommendation, Skill Level 20

In the lower-division baccalaureate/associate degree category, 3 semester hours in basic electronics, 3 in radar systems, 3 in personnel supervision, 2 in records and information management, and 2 in maintenance management. In the upper-division baccalaureate category, 3 semester hours in hydraulic systems troubleshooting and 3 in electronic systems troubleshooting. (NOTE: This recommendation for skill level 20 is valid for the dates 6/91-2/95 only) (6/91).

Recommendation, Skill Level 30

In the lower-division baccalaureate/associate degree category, 3 semester hours in basic electronics, 3 in radar systems, 3 in personnel supervision, 2 in records and information management, and 3 in maintenance management. In the upper-division baccalaureate category, 3 semester hours in hydraulic systems troubleshooting and maintenance and 3 in electronic systems troubleshooting (6/91).

MOS-24G-003

Hawk Information Coordination Central Mechanic
((Improved Hawk Information Coordination Central Mechanic))
24G10
24G20
24G30

Exhibit Dates: 1/90–5/91.

Career Management Field: 23 (Air Defense Missile Maintenance).

Description

Summary: Supervises or performs organizational maintenance on Hawk or improved Hawk information and coordination equipment. *Skill Level 10:* Performs equipment checks and adjustments; localizes malfunctions and repairs at the block diagram level; assists in the training of operators in operator maintenance; provides technical assistance in the conduct of operational and maintenance inspections. *Skill Level 20:* Able to perform the duties required for Skill Level 10; initiates requests for repair parts, tools, and supplies; completes maintenance and supply forms. *Skill Level 30:* Able to perform the duties required for Skill Level 20; supervises training programs for operator and maintenance personnel; prepares forms and reports; demonstrates proper maintenance and troubleshooting techniques; supervises requisition and stocking of supplies and repair parts; organizes and supervises maintenance and inspection teams.

Recommendation, Skill Level 10

In the lower-division baccalaureate/associate degree category, 6 semester hours in basic electronics, 3 in technical mathematics, 3 in electronic systems troubleshooting, 1 in microwave techniques, 3 in digital techniques, and credit in computer operations on the basis of institutional evaluation (12/84).

Recommendation, Skill Level 20

In the lower-division baccalaureate/associate degree category, 6 semester hours in basic electronics, 3 in technical mathematics, 3 in electronic systems troubleshooting, 1 in microwave techniques, 3 in digital techniques, and credit in computer operations on the basis of institutional evaluation (12/84).

Recommendation, Skill Level 30

In the vocational certificate category, 3 semester hours in electronics shop practices. In the lower-division baccalaureate/associate degree category, 6 semester hours in basic electronics, 3 in technical mathematics, 3 in electronic systems troubleshooting, 1 in microwave techniques, 3 in digital techniques, 3 in personnel supervision, and credit in computer operations on the basis of institutional evaluation (12/84).

MOS-24G-004

Hawk Information Coordination Central Mechanic
24G10
24G20
24G30

Exhibit Dates: 6/91–6/95. Effective 6/95, MOS 24G was discontinued.

Career Management Field: 23 (Air Defense Missile Maintenance).

Description

Summary: Supervises or performs organizational maintenance on Hawk information and coordination equipment. *Skill Level 10:* Performs equipment checks and adjustments; localizes malfunctions and repairs at the block diagram level; assists in the training of operators in operator maintenance and provides technical assistance in the conduct of operational and maintenance inspections. *Skill Level 20:* Able to perform the duties required for Skill Level 10; initiates requests for repair parts, tools, and supplies; serves as maintenance section team leader; provides technical supervision for training programs for operators; completes maintenance and supply forms. *Skill Level 30:* Able to perform the duties required for Skill Level 20; supervises training programs for operator and maintenance personnel; prepares forms and reports; demonstrates proper maintenance and troubleshooting techniques; supervises requisition and stocking of supplies and repair parts; organizes and supervises maintenance and inspection teams.

Recommendation, Skill Level 10

In the lower-division baccalaureate/associate degree category, 3 semester hours in basic electronics, 3 in radar systems, and 3 in digital electronics laboratory. In the upper-division baccalaureate category, 3 semester hours in computer systems troubleshooting. (NOTE: This recommendation for skill level 10 is valid for the dates 6/91-9/91 only) (6/91).

Recommendation, Skill Level 20

In the lower-division baccalaureate/associate degree category, 3 semester hours in basic electronics, 3 in radar systems, 3 in digital electronics laboratory, 3 in personnel supervision, 2 in records and information management, and 2 in maintenance management. In the upper-division baccalaureate category, 3 semester hours in computer systems troubleshooting and maintenance, and 3 in electronic systems troubleshooting. (NOTE: This recommendation for skill level 20 is valid for the dates 6/91-2/95 only) (6/91).

Recommendation, Skill Level 30

In the lower-division baccalaureate/associate degree category, 3 semester hours in basic electronics, 3 in radar systems, 3 in digital electronics laboratory, 3 in personnel supervision, 2 in records and information management, and 3 in maintenance management. In the upper-division baccalaureate category, 3 semester hours in computer systems troubleshooting and maintenance and 3 in electronic systems troubleshooting (6/91).

MOS-24H-003

Hawk Fire Control Repairer

24H10
24H20
24H30

Exhibit Dates: 1/90–8/90.

Career Management Field: 23 (Air Defense Missile Maintenance), subfield 236 (Improved Hawk Missile System Repair).

Description

Summary: Supervises or performs maintenance and inspections of Hawk or improved Hawk battery control centers, including radar consoles, electronic systems, communications systems, computers, and various types of advanced electronic test equipment. *Skill Level 10:* Inspects, tests, and adjusts components to specific tolerances; determines shortcomings and malfunctions in electronic assemblies, subassemblies, modules, and circuit elements with common and system-specific test equipment; repairs unserviceable items by removing and replacing defective components and parts; determines serviceability and disposition of defective assemblies, subassemblies, modules, and circuit elements; performs quality control measures; serves on inspection teams; performs maintenance and repair on test equipment. *Skill Level 20:* Able to perform the duties required for Skill Level 10; provides technical guidance to Skill Level 10 personnel; installs electrical and electronic equipment modifications; provides technical assistance to supported units; completes maintenance and supply forms. *Skill Level 30:* Able to perform the duties required for Skill Level 20; supervises subordinates; establishes work loads and repair priorities; recommends procedures for receipt, storage, inspection, testing, and repair of items; implements quality control measures; performs initial and final checkout and inspection of designated system items and their assemblies and subassemblies; supervises inspection and maintenance teams; organizes and conducts on-the-job training programs; determines faulty work practices and demonstrates proper maintenance and troubleshooting techniques; establishes and keeps maintenance records; prepares maintenance reports.

Recommendation, Skill Level 10

In the vocational certificate category, 3 semester hours in applied mathematics. In the lower-division baccalaureate/associate degree category, 3 semester hours in basic electricity, 3 in basic electronics, 3 in electronic instruments, 3 in technical mathematics, and additional credit in electronics and technical mathematics on the basis of institutional evaluation (1/80).

Recommendation, Skill Level 20

In the vocational certificate category, 3 semester hours in applied mathematics. In the lower-division baccalaureate/associate degree category, 3 semester hours in basic electricity, 3 in basic electronics, 3 in electronic instruments, 3 in technical mathematics, and additional credit in electronics and technical mathematics on the basis of institutional evaluation (1/80).

Recommendation, Skill Level 30

In the vocational certificate category, 3 semester hours in applied mathematics. In the lower-division baccalaureate/associate degree category, 3 semester hours in basic electricity, 3 in basic electronics, 3 in electronic instruments, 3 in technical mathematics, 3 in electronics shop practices, 3 in personnel supervision, 2 in records and reports preparation, and additional credit in electronics and technical mathematics on the basis of institutional evaluation. In the upper-division baccalaureate category, 3 semester hours for field experience in personnel management (1/80).

MOS-24H-004

Hawk Fire Control Repairer

24H10
24H20
24H30
24H40

Exhibit Dates: 9/90–9/99. Effective 9/99, MOS 24H was discontinued.

Career Management Field: 27 (Land Combat and Air Defense System Direct and General Support Maintenance).

Description

Summary: Supervises or performs maintenance and inspection of Hawk or improved Hawk battery control centers including radar consoles, electronic systems, communications systems, computers, and various types of advanced electronic test equipment. *Skill Level 10:* Inspects, tests, and adjusts components to specific tolerances; determines shortcomings and malfunctions in electronic assemblies, subassemblies, modules, and circuit elements with common and system-specific test equipment; repairs unserviceable items by removing and replacing defective components and parts; determines serviceability and disposition of defective assemblies, subassemblies, modules, and circuit elements; performs quality control measures; serves on inspection teams; performs maintenance and repair on test equipment. *Skill Level 20:* Able to perform the duties required for Skill Level 10; provides technical guidance to subordinates; installs electrical and electronic equipment modifications; provides technical assistance to supported units; performs initial and final equipment checkout and inspection of designated systems and their assemblies; supervises maintenance procedures; completes maintenance and supply forms. *Skill Level 30:* Able to perform the duties required for Skill Level 20; supervises subordinates; establishes work loads and repair priorities; recommends procedures for receipt, storage, inspection, testing, and repair of items; implements quality control measures; performs initial and final checkout and inspection of designated system items and their assemblies and subassemblies; supervises inspection and maintenance teams; organizes and conducts on-the-job training programs; determines faulty work practices and demonstrates proper maintenance and troubleshooting techniques; establishes and keeps maintenance records; prepares maintenance reports. *Skill Level 40:* Able to perform the duties required for Skill Level 30; supervises and coordinates organizational maintenance; supervises the preparation of maintenance forms and records; serves as platoon sergeant.

Recommendation, Skill Level 10

In the lower-division baccalaureate/associate degree category, 3 semester hours in basic electronics, 3 in solid state electronics laboratory, and 3 in digital circuit laboratory. (NOTE: This recommendation for skill level 10 is valid for the dates 9/90-9/91 only) (6/91).

Recommendation, Skill Level 20

In the lower-division baccalaureate/associate degree category, 3 semester hours in basic electronics, 3 in solid state electronics laboratory, 3 in digital circuit laboratory, 3 in personnel supervision, 2 in records and information management, and 2 in maintenance management. In the upper-division baccalaureate category, 3 semester hours in electronic systems troubleshooting. (NOTE: This recommendation for skill level 20 is valid for the dates 9/90-2/95 only) (6/91).

Recommendation, Skill Level 30

In the lower-division baccalaureate/associate degree category, 3 semester hours in basic electronics, 3 in solid state electronics laboratory, 3 in digital circuit laboratory, 3 in personnel supervision, 2 in records and information management, and 3 in maintenance management. In the upper-division baccalaureate category, 3 semester hours in electronic systems troubleshooting (6/91).

Recommendation, Skill Level 40

In the lower-division baccalaureate/associate degree category, 3 semester hours in basic electronics, 3 in solid state electronics laboratory, 3 in digital circuit laboratory, 3 in personnel supervision, 2 in records and information management, and 3 in maintenance management. In the upper-division baccalaureate category, 3 semester hours in electronic systems troubleshooting, 3 in organizational management, and 3 for field experience in management (6/91).

MOS-24K-003

HAWK CONTINUOUS WAVE RADAR REPAIRER

24K10
24K20
24K30

Exhibit Dates: 1/90–8/90.

Career Management Field: 23 (Air Defense Missile Maintenance), subfield 236 (Improved Hawk Missile System Repair).

Description

Summary: Supervises or performs maintenance on Hawk or improved Hawk continuous wave radars and associated equipment. *Skill Level 10:* Performs support maintenance on electronic systems of continuous wave radars and special high-power illuminator radars associated with the Hawk missile system; adjusts, inspects, and tests electronic assemblies, subassemblies, modules, and circuit elements, using common and specialized test equipment; performs initial and final checkout and inspections for the electronic portions of the radar systems; performs modifications and serves on inspection teams. *Skill Level 20:* Able to perform the duties required for Skill Level 10; provides technical guidance to subordinates; installs equipment modifications and provides technical assistance to supported units; prepares maintenance records. *Skill Level 30:* Able to perform the duties required for Skill Level 20; establishes work loads and repair priorities; implements quality control measures; performs initial and final checkout and inspection of the radar systems and their subassemblies; supervises inspection and maintenance teams; establishes and keeps maintenance records; prepares maintenance reports; organizes and conducts on-the-job training programs.

Recommendation, Skill Level 10

In the vocational certificate category, 3 semester hours in applied mathematics, 3 in DC and AC circuits, 3 in electronic test procedures, 3 in tube and transistor circuits, and 1 in microwave techniques. In the lower-division baccalaureate/associate degree category, 3 semester hours in technical mathematics, 3 in elementary DC and AC circuits, 3 in introduction to tube and transistor circuits, and 1 in microwave laboratory (1/80).

Recommendation, Skill Level 20

In the vocational certificate category, 3 semester hours in applied mathematics, 3 in DC and AC circuits, 3 in tube and transistor circuits, and 2 in microwave techniques. In the lower-division baccalaureate/associate degree category, 3 semester hours in technical mathematics, 3 in elementary DC and AC circuits, 3 in introduction to tube and transistor circuits, 3 in electronic test procedures, and 2 in microwave laboratory (1/80).

Recommendation, Skill Level 30

In the vocational certificate category, 3 semester hours in applied mathematics, 3 in DC and AC circuits, 3 in tube and transistor circuits, and 2 in microwave techniques. In the lower-division baccalaureate/associate degree category, 3 semester hours in technical mathematics, 3 in elementary DC and AC circuits, 3 in introduction to tube and transistor circuits, 3 in electronic test procedures, 3 in microwave laboratory, 3 in personnel supervision, and 2 in records and reports preparation. In the upper-division baccalaureate category, 3 semester hours for field experience in personnel management (1/80).

MOS-24K-004

HAWK CONTINUOUS WAVE RADAR REPAIRER

24K10
24K20
24K30
24K40

Exhibit Dates: 9/90–9/99. Effective 9/99 MOS 24K was discontinued.

Career Management Field: 27 (Land Combat and Air Defense System Direct and General Support Maintenance).

Description

Summary: Supervises or performs maintenance on Hawk continuous wave radars and associated equipment. *Skill Level 10:* Performs support maintenance on electronic systems of continuous wave radars associated with the Hawk missile system; tests and adjusts components to specific tolerances and determines shortcomings and malfunctions in electronic assemblies, subassemblies, modules, and circuit elements with common and specialized test equipment; replaces unserviceable defective parts; determines serviceability and disposition of defective elements; performs quality control measures; serves on inspection teams. *Skill Level 20:* Able to perform the duties required for Skill Level 10; provides technical guidance to subordinates; supervises maintenance procedures; installs equipment modifications and provides technical assistance to supported units; prepares maintenance records; performs initial and final checkout and inspection of designated system items and their assemblies. *Skill Level 30:* Able to perform the duties required for Skill Level 20; supervises inspection and maintenance teams; establishes and keeps maintenance records; prepares maintenance reports; organizes and conducts on-the-job training programs; recommends procedures for receipt, storage, inspection, testing, and repair of items; determines faulty work practices and demonstrates proper maintenance and troubleshooting techniques. *Skill Level 40:* Able to perform the duties required for Skill Level 30; provides technical guidance to subordinates; serves as platoon sergeant; establishes work loads and repair priorities.

Recommendation, Skill Level 10

In the lower-division baccalaureate/associate degree category, 3 semester hours in DC circuits, 2 in DC circuits laboratory, 3 in basic electronics, and 2 in basic electronics laboratory. In the upper-division baccalaureate category, 3 semester hours in electronic systems troubleshooting. (NOTE: This recommendation for skill level 10 is valid for the dates 9/90-9/91 only) (6/91).

Recommendation, Skill Level 20

In the lower-division baccalaureate/associate degree category, 3 semester hours in DC circuits, 2 in DC circuits laboratory, 3 in basic electronics, 2 in basic electronics laboratory, 3 in personnel supervision, 2 in records and information management, and 2 in maintenance management. In the upper-division baccalaureate category, 6 semester hours in electronic systems troubleshooting. (NOTE: This recommendation for skill level 20 is valid for the dates 9/90-2/95 only) (6/91).

Recommendation, Skill Level 30

In the lower-division baccalaureate/associate degree category, 3 semester hours in DC circuits, 2 in DC circuits laboratory, 3 in basic electronics, 2 in basic electronics laboratory, 3 in personnel supervision, 2 in records and information management, and 3 in maintenance management. In the upper-division baccalaureate category, 6 semester hours in electronic systems troubleshooting (6/91).

Recommendation, Skill Level 40

In the lower-division baccalaureate/associate degree category, 3 semester hours in DC circuits, 2 in DC circuits laboratory, 3 in basic electronics, 2 in basic electronics laboratory, 3 in personnel supervision, 2 in records and information management, and 3 in maintenance management. In the upper-division baccalaureate category, 6 semester hours in electronic systems troubleshooting, 3 in organizational management, and 3 for field experience in management (6/91).

MOS-24M-002

VULCAN SYSTEM MECHANIC

24M10
24M20
24M30
24M40

Exhibit Dates: 1/90–5/91.

Career Management Field: 27 (Ballistic/Land Combat Missile and Light Air Defense Weapons Systems Maintenance), subfield 272 (Light Air Defense Systems Maintenance).

Description

Summary: Supervises or performs maintenance on the Vulcan weapons system and the Forward Area Alerting Radar and associated equipment. *Skill Level 10:* Performs routine maintenance on range-only radar, the target display system, and test, diagnostic, and measuring equipment; maintains electromechanical and mechanical components associated with turret assemblies on Vulcan launch equipment; uses some basic test equipment including meters and oscilloscopes, as well as specialized test equipment; isolates malfunctions to the black box level. *Skill Level 20:* Able to perform the duties required for Skill Level 10; provides technical guidance to subordinates. *Skill Level 30:* Able to perform the duties required for Skill Level 20; supervises subordinates; assigns duties; maintains forms and records to insure proper maintenance procedures for all Chaparral systems. *Skill Level 40:* Able to perform the duties required for Skill Level 30; establishes maintenance procedures; conducts technical inspections; coordinates maintenance activities.

Recommendation, Skill Level 10

In the vocational certificate category, 2 semester hours in mechanical laboratory. In the lower-division baccalaureate/associate degree category, 2 semester hours in electronics laboratory (1/80).

Recommendation, Skill Level 20

In the vocational certificate category, 2 semester hours in mechanical laboratory. In the lower-division baccalaureate/associate degree category, 2 semester hours in electronics laboratory (1/80).

Recommendation, Skill Level 30

In the vocational certificate category, 2 semester hours in mechanical laboratory. In the lower-division baccalaureate/associate degree category, 2 semester hours in personnel supervision, 3 in electronics and mechanical shop practices, 2 in maintenance management, and 2 in electronics laboratory (1/80).

Recommendation, Skill Level 40

In the vocational certificate category, 2 semester hours in mechanical laboratory. In the lower-division baccalaureate/associate degree category, 3 semester hours in personnel supervision, 3 in electronics and mechanical shop practices, 4 in maintenance management, and 2 in electronics laboratory (1/80).

MOS-24M-003

Vulcan System Mechanic

24M10
24M20
24M30
24M40

Exhibit Dates: 6/91–6/97. Effective 6/97, MOS 24M was discontinued and its duties were transferred to MOS 14R.

Career Management Field: 23 (Air Defense Systems Mechanic).

Description

Summary: Supervises or performs maintenance on the Vulcan weapons system and associated equipment. *Skill Level 10:* Isolates malfunctions in analog and digital radar, computer, and servo systems to the module and component level; performs maintenance and repair on RF mechanical components; identifies and recommends changes to operations manuals and procedures; maintains, troubleshoots, and repairs servomechanism systems. *Skill Level 20:* Able to perform the duties required for Skill Level 10; provides technical guidance in operator maintenance training. *Skill Level 30:* Able to perform the duties required for Skill Level 20; supervises subordinates; assigns duties; maintains forms and records to insure proper maintenance procedures. *Skill Level 40:* Able to perform the duties required for Skill Level 30; establishes maintenance procedures; conducts technical inspections; coordinates maintenance activities; supervises the requisitioning and stocking of supplies and repair parts.

Recommendation, Skill Level 10

In the lower-division baccalaureate/associate degree category, 3 semester hours in basic electronics laboratory. In the upper-division baccalaureate category, 3 semester hours in mechanical systems troubleshooting. (NOTE: This recommendation for skill level 10 is valid for the dates 6/91-9/91 only) (6/91).

Recommendation, Skill Level 20

In the lower-division baccalaureate/associate degree category, 3 semester hours in basic electronics laboratory, 3 in personnel supervision, 2 in records and information management, and 2 in maintenance management. In the upper-division baccalaureate category, 3 semester hours in mechanical systems troubleshooting and 3 in electronic systems troubleshooting. (NOTE: This recommendation for skill level 20 is valid for the dates 6/91-2/95 only) (6/91).

Recommendation, Skill Level 30

In the lower-division baccalaureate/associate degree category, 3 semester hours in basic electronics laboratory, 3 in personnel supervision, 2 in records and information management, and 3 in maintenance management. In the upper-division baccalaureate category, 3 semester hours in mechanical systems troubleshooting and 3 in electronic systems troubleshooting (6/91).

Recommendation, Skill Level 40

In the lower-division baccalaureate/associate degree category, 3 semester hours in basic electronics laboratory, 3 in personnel supervision, 2 in records and information management, and 3 in maintenance management. In the upper-division baccalaureate category, 3 semester hours in mechanical systems troubleshooting, 3 in electronic systems troubleshooting, 3 in organizational management, and 3 for field experience in management (6/91).

MOS-24N-002

Chaparral System Mechanic

24N10
24N20
24N30
24N40

Exhibit Dates: 1/90–5/91.

Career Management Field: 27 (Ballistic/Land Combat Missile and Light Air Defense Weapons Systems Maintenance), subfield 272 (Light Air Defense Systems Maintenance).

Description

Summary: Supervises or performs maintenance on Chaparral weapon system and associated equipment. *Skill Level 10:* Performs routine maintenance on Chaparral weapon system and target display sets; maintains mechanical, electrical, hydraulic, pneumatic, and some electronic systems associated with the Chaparral launch equipment; uses meters and specialized test sets to localize malfunctions to the black box level. *Skill Level 20:* Able to perform the duties required for Skill Level 10; provides technical assistance to subordinates. *Skill Level 30:* Able to perform the duties required for Skill Level 20; supervises subordinates; assigns duties; maintains forms and records to insure proper maintenance procedures for all Chaparral systems. *Skill Level 40:* Able to perform the duties required for Skill Level 30; establishes maintenance procedures; conducts technical inspections of Chaparral systems; coordinates maintenance activities.

Recommendation, Skill Level 10

In the vocational certificate category, 2 semester hours in mechanical laboratory (1/80).

Recommendation, Skill Level 20

In the vocational certificate category, 2 semester hours in mechanical laboratory (1/80).

Recommendation, Skill Level 30

In the vocational certificate category, 2 semester hours in mechanical laboratory. In the lower-division baccalaureate/associate degree category, 2 semester hours in personnel supervision, 3 in mechanical shop practices, and 2 in maintenance management (1/80).

Recommendation, Skill Level 40

In the vocational certificate category, 2 semester hours in mechanical laboratory. In the lower-division baccalaureate/associate degree category, 3 semester hours in personnel supervision, 3 in mechanical shop practices, and 4 in maintenance management (1/80).

MOS-24N-003

Chaparral System Mechanic

24N10
24N20
24N30
24N40

Exhibit Dates: 6/91–6/98. Effective 6/98, MOS 24N was discontinued.

Career Management Field: 23 (Air Defense System Maintenance).

Description

Summary: Supervises or performs maintenance on Chaparral weapon system and associated equipment. *Skill Level 10:* Performs routine maintenance on Chaparral weapon system and target display sets; maintains mechanical, electrical, hydraulic, pneumatic, and some electronic systems associated with Chaparral launch equipment; uses meters and specialized test sets to localize malfunctions to the black box level. *Skill Level 20:* Able to perform the duties required for Skill Level 10; provides technical assistance to subordinates; assists in maintaining equipment records and repair lists; performs advanced troubleshooting, using systems theory and schematic interpretation; supervises maintenance procedures. *Skill Level 30:* Able to perform the duties required for Skill Level 20; supervises subordinate personnel; assigns duties; maintains forms and records to insure proper maintenance procedures for Chaparral systems. *Skill Level 40:* Able to perform the duties required for Skill Level 30; establishes maintenance procedures; conducts technical inspections of Chaparral systems and coordinates maintenance activities.

Recommendation, Skill Level 10

In the lower-division baccalaureate/associate degree category, 3 semester hours in mechanical shop practices. (NOTE: This recommendation for skill level 10 is valid for the dates 6/91-9/91 only) (6/91).

Recommendation, Skill Level 20

In the lower-division baccalaureate/associate degree category, 3 semester hours in mechanical shop practices, 3 in basic electronics laboratory, 3 in personnel supervision, 2 in records and information management, and 2 in maintenance management. (NOTE: This recommendation for skill level 20 is valid for the dates 6/91-2/95 only) (6/91).

Recommendation, Skill Level 30

In the lower-division baccalaureate/associate degree category, 3 semester hours in mechanical shop practices, 3 in basic electronics laboratory, 3 in personnel supervision, 2 in records and information management, and 3 in maintenance management (6/91).

Recommendation, Skill Level 40

In the lower-division baccalaureate/associate degree category, 3 semester hours in mechanical shop practices, 3 in basic electronics laboratory, 3 in personnel supervision, 2 in records and information management, and 3 in maintenance management. In the upper-division baccalaureate category, 3 semester hours in organizational management and 3 for field experience in management (6/91).

MOS-24R-001

Hawk Master Mechanic
24R40
24R50

Exhibit Dates: 1/90–5/91.

Career Management Field: 23 (Air Defense Missile Maintenance).

Description

Summary: Supervises or performs maintenance on improved Hawk equipment. *Skill Level 40:* Able to perform the duties required for Skill Level 30 of MOS 24E, Hawk (Improved Hawk) Fire Control Mechanic; MOS 24G, Hawk (Improved Hawk) Information Coordination Central Mechanic; and MOS 24C, Hawk Firing Section Mechanic; supervises and coordinates the maintenance of improved Hawk information coordination central, automatic data equipment, improved continuous wave acquisition radar, identification-friend-or-foe, battery terminal equipment, battery control central, pulse acquisition radar, range-only radar, and associated tools and test equipment; prepares maintenance forms and supply records; diagnoses and determines causes of malfunctions; reads, interprets, and explains complex circuit diagrams and schematics; provides technical advice and assistance. NOTE: May have progressed to 24R40 from 24E30 (Hawk (Improved Hawk) Fire Control Mechanic) or 24G30 (Hawk (Improved Hawk) Information Coordination Central Mechanic). *Skill Level 50:* Able to perform the duties required for Skill Level 40; supervises and coordinates maintenance on improved Hawk high-powered illuminator radar, missile launching section control box, launcher, mechanical, hydraulic, and electrical systems, and loader-transporter; serves as senior mechanic; establishes, maintains, and monitors on-the-job training programs; provides technical guidance to other units.

Recommendation, Skill Level 40

In the lower-division baccalaureate/associate degree category, 3 semester hours in introduction to management. In the upper-division baccalaureate category, 3 semester hours in personnel management and 3 for field experience in management (12/84).

Recommendation, Skill Level 50

In the lower-division baccalaureate/associate degree category, 3 semester hours in introduction to management. In the upper-division baccalaureate category, 3 semester hours in personnel management, 6 for field experience in management, and 3 in management problems (12/84).

MOS-24R-002

Hawk Master Mechanic
24R40
24R50

Exhibit Dates: 6/91–6/95. Effective 6/95, MOS 24R was discontinued.

Career Management Field: 23 (Air Defense System Maintenance).

Description

Summary: Supervises or performs maintenance on Hawk equipment. *Skill Level 40:* Able to perform the duties required for Skill Level 30 of MOS 24C, Hawk Firing Section Mechanic or MOS 24G, Hawk Information Coordination Central Mechanic; supervises and coordinates battery unit maintenance on Hawk high-powered illuminator radar, missile launching section control box, launcher, mechanical, hydraulic, and electrical systems, loader transporter, information and coordination control, continuous wave acquisition radar, and other tools and test equipment; diagnoses and determines cause of malfunctions; reads, interprets, and explains complex circuit diagrams and schematics; performs digital and computer fault isolation; performs fault isolation and maintenance procedures on continuous wave acquisition radar and the Hawk integrated system; enforces safety policies, procedures, and standards, as well as equipment standards of readiness; conducts training to support these standards. NOTE: May have progressed to 24R40 from 24C30, Hawk Firing Section Mechanic or 24G30, Hawk Information Coordination Central Mechanic. *Skill Level 50:* Able to perform the duties required for Skill Level 40; supervises and performs unit level operational checks, fault isolation, and maintenance; serves as senior mechanic at battalion and brigade levels; advises battalion commander on unit maintenance program.

Recommendation, Skill Level 40

In the lower-division baccalaureate/associate degree category, 3 semester hours in personnel supervision, 3 in maintenance management, and 2 in records and information management. In the upper-division baccalaureate category, 3 semester hours for field experience in management and 3 in organizational management (6/91).

Recommendation, Skill Level 50

In the lower-division baccalaureate/associate degree category, 3 semester hours in personnel supervision, 3 in maintenance management, and 2 in records and information management. In the upper-division baccalaureate category, 6 semester hours for field experience in management and 3 in organizational management; if individual has attained paygrade E-9, additional credit as follows: 3 semester hours in management problems, 3 in operations management, and 3 in communication techniques for managers (6/91).

MOS-24T-002

Patriot Operator and System Mechanic
24T10
24T20
24T30
24T40
24T50

Exhibit Dates: 1/90–5/91.

Career Management Field: 23 (Air Defense Missile Maintenance).

Description

Summary: Supervises, operates, or performs maintenance on the Patriot air defense missile system. *Skill Level 10:* Performs repairs on the Patriot radar set; installs authorized equipment modifications; prepares appropriate supplies, communication equipment, computer and computer peripherals, tools, and test equipment; removes and replaces faulty circuit cards; operates maintenance vehicle. *Skill Level 20:* Able to perform the duties required for Skill Level 10; provides technical guidance to subordinates; performs organizational maintenance on the Patriot radar set and assists in training operators in maintenance techniques; troubleshoots problems with external test equipment when isolation is not accomplished by built-in or other machine-aided equipment. *Skill Level 30:* Able to perform the duties required for Skill Level 20; serves as shift supervisor; provides supervision in training programs for operator and maintenance personnel; supervises preparation of technical forms and reports; requisitions and stocks supplies and repair parts. *Skill Level 40:* Able to perform the duties required for Skill Level 30; coordinates activities of maintenance personnel; supervises inspection of maintenance performance; supervises inspection teams; supervises repair parts stocking; establishes maintenance priorities. *Skill Level 50:* Able to perform the duties required for Skill Level 40; serves as maintenance chief; assists in planning, organizing, and training operations; provides technical expertise; may serve as a member of a multinational evaluation team.

Recommendation, Skill Level 10

In the lower-division baccalaureate/associate degree category, 6 semester hours in basic electronics, 3 in electronic systems troubleshooting, and 3 in digital circuits (12/84).

Recommendation, Skill Level 20

In the lower-division baccalaureate/associate degree category, 6 semester hours in basic electronics, 3 in electronic systems troubleshooting, and 3 in digital circuits (12/84).

Recommendation, Skill Level 30

In the lower-division baccalaureate/associate degree category, 6 semester hours in basic electronics, 3 in electronic systems troubleshooting, 3 in digital circuits, and 3 in personnel supervision (12/84).

Recommendation, Skill Level 40

In the lower-division baccalaureate/associate degree category, 6 semester hours in basic electronics, 3 in electronic systems troubleshooting, 3 in digital circuits, 3 in personnel supervision, and 3 in introduction to management. In the upper-division baccalaureate category, 3 semester hours in personnel management and 3 for field experience in management (12/84).

Recommendation, Skill Level 50

In the lower-division baccalaureate/associate degree category, 6 semester hours in basic electronics, 3 in electronic systems troubleshooting, 3 in digital circuits, 3 in personnel supervision, and 3 in introduction to management. In the upper-division baccalaureate category, 3 semester hours in personnel management and 6 for field experience in management (12/84).

MOS-24T-003

Patriot Operator and System Mechanic
24T10
24T20
24T30
24T40
24T50

Exhibit Dates: 6/91–6/97. Effective 6/97, MOS 24T was discontinued and its duties were transferred to the following MOS's: 14E, 14J, 14T, and 14Z.

Career Management Field: 23 (Air Defense System Maintenance).

Description

Summary: Supervises, operates, or performs maintenance on the Patriot air defense missile system. *Skill Level 10:* Performs repairs on the Patriot system; installs equipment modifications; prepares appropriate supplies, communication equipment, computer and computer peripherals, tools, test equipment, and shelters; removes and replaces faulty circuit cards; operates the maintenance vehicle. *Skill Level 20:* Able to perform the duties required for Skill Level 10; provides technical guidance to subordinates; performs checks, adjustments, and repairs on major item interface; performs unit maintenance on Patriot information and coordination control, engagement control station, radar set, identification-friend-or-foe, launching station, antiradiation missile decoy, and antenna mast group; supervises maintenance procedures; assists in training operators in maintenance techniques; troubleshoots problems with external test equipment when isolation is not accomplished by built-in or other machine-aided equipment. *Skill Level 30:* Able to perform the duties required for Skill Level 20; serves as shift supervisor while performing as chief engagement controller; provides supervision in training programs for operator and maintenance personnel; supervises preparation of technical forms and reports; requisitions and stocks supplies and repair parts; determines faulty work practices and demonstrates proper maintenance and troubleshooting techniques. *Skill Level 40:* Able to perform the duties required for Skill Level 30; coordinates activities of maintenance personnel; supervises inspection of maintenance performance; supervises inspection teams; supervises repair parts stocking; establishes maintenance priorities. *Skill Level 50:* Able to perform the duties required for Skill Level 40; serves as maintenance chief; assists in planning, organization, and training operations; provides technical expertise; may serve as a member of a multinational evaluation team; applies knowledge of capabilities and limitations of the Patriot weapon, command and control, and radar systems.

Recommendation, Skill Level 10

In the lower-division baccalaureate/associate degree category, 3 semester hours in basic electronics, 3 in basic electronics laboratory, and 3 in digital principles. (NOTE: This recommendation for skill level 10 is valid for the dates 6/91-9/91 only) (6/91).

Recommendation, Skill Level 20

In the lower-division baccalaureate/associate degree category, 3 semester hours in basic electronics, 3 in basic electronics laboratory, 3 in digital principles, 3 in personnel supervision, 2 in maintenance management, and 2 in records and information management. In the upper-division baccalaureate category, 3 semester hours in electronic systems troubleshooting and 3 in computer systems troubleshooting. (NOTE: This recommendation for skill level 20 is valid for the dates 6/91-2/95 only) (6/91).

Recommendation, Skill Level 30

In the lower-division baccalaureate/associate degree category, 3 semester hours in basic electronics, 3 in basic electronics laboratory, 3 in digital principles, 3 in personnel supervision, 3 in maintenance management, and 2 in records and information management. In the upper-division baccalaureate category, 3 semester hours in electronic systems troubleshooting and 3 in computer systems troubleshooting (6/91).

Recommendation, Skill Level 40

In the lower-division baccalaureate/associate degree category, 3 semester hours in basic electronics, 3 in basic electronics laboratory, 3 in digital principles, 3 in personnel supervision, 3 in maintenance management, and 2 in records and information management. In the upper-division baccalaureate category, 3 semester hours in electronic systems troubleshooting, 3 in computer systems troubleshooting, 3 in organizational management, and 3 for field experience in management (6/91).

Recommendation, Skill Level 50

In the lower-division baccalaureate/associate degree category, 3 semester hours in basic electronics, 3 in basic electronics laboratory, 3 in digital principles, 3 in personnel supervision, 3 in maintenance management, and 2 in records and information management. In the upper-division baccalaureate category, 3 semester hours in electronic systems troubleshooting, 3 in computer system troubleshooting, 3 in organizational management, and 6 for field experience in management; if individual has attained paygrade E-9, additional credit as follows: 3 semester hours in management problems, 3 in operations management, and 3 in communications techniques for managers (6/91).

MOS-24U-002

NIKE-HERCULES CUSTODIAL MECHANIC

24U10
24U20
24U30
24U40

Exhibit Dates: 1/90–9/90. Effective 9/90, MOS 24U was discontinued.

Career Management Field: 23 (Air Defense Missile Maintenance), subfield 232 (Nike Missile System Mechanics).

Description

Summary: Supervises or performs maintenance on Hercules missiles, airframes, and launch devices. *Skill Level 10:* Assembles, installs, maintains, and adjusts certain mechanical assemblies relating to on-missile guidance control; assembles and interconnects electronic missile components; localizes malfunctions and repairs authorized components using specialized test sets; initiates requisitions for repairs of mechanical and some electronic missile subsystems; checks calibration of test sets; makes adjustments; isolates malfunctions in test sets to component level and makes repairs. *Skill Level 20:* Able to perform the duties required for Skill Level 10; provides technical guidance to subordinates. *Skill Level 30:* Able to perform the duties required for Skill Level 20; supervises activities of subordinates; supervises the preparation of maintenance forms and reports; supervises training programs and requisitioning of parts for repairs; organizes and supervises maintenance and inspection teams. *Skill Level 40:* Able to perform the duties required for Skill Level 30; supervises and coordinates all maintenance on the Hercules missiles, airframe, and launch devices; may serve as chief electronics mechanic at battalion or higher level.

Recommendation, Skill Level 10

In the vocational certificate category, 3 semester hours in electric/electronic fundamentals, 3 in electronic maintenance, and 3 in electronic troubleshooting. In the lower-division baccalaureate/associate degree category, 3 semester hours in electronics fundamentals (1/80).

Recommendation, Skill Level 20

In the vocational certificate category, 3 semester hours in electric/electronic fundamentals, 3 in electronic maintenance, and 3 in electronic troubleshooting. In the lower-division baccalaureate/associate degree category, 3 semester hours in electronics fundamentals (1/80).

Recommendation, Skill Level 30

In the vocational certificate category, 3 semester hours in electric/electronic fundamentals, 3 in electronic maintenance, and 3 in electronic troubleshooting. In the lower-division baccalaureate/associate degree category, 3 semester hours in electronics fundamentals, 2 in maintenance management, 3 in personnel supervision, 3 in electronics shop practices, and 2 in records and reports preparation. In the upper-division baccalaureate category, 3 semester hours for field experience in personnel management (1/80).

Recommendation, Skill Level 40

In the vocational certificate category, 3 semester hours in electrical/electronic fundamentals, 3 in electronic maintenance, and 3 in electronic troubleshooting. In the lower-division baccalaureate/associate degree category, 3 semester hours in electronics fundamentals, 3 in maintenance management, 3 in personnel supervision, 3 in electronics shop practices, and 2 in records and reports preparation. In the upper-division baccalaureate category, 3 semester hours for field experience in personnel management, 3 in introduction to management, and 3 in human relations (1/80).

MOS-25B-001

INFORMATION SYSTEMS OPERATOR-ANALYST

25B10
25B20
25B30
25B40

Exhibit Dates: Pending Evaluation. 4/04–Present.

MOS-25C-001

RADIO OPERATOR-MAINTAINER

25C10
25C20
25C30

Exhibit Dates: Pending Evaluation. 4/04–Present.

MOS-25D-001

TELECOMMUNICATION OPERATOR-MAINTAINER

25D10
25D20
25D30
25D40

Exhibit Dates: Pending Evaluation. 4/04–Present.

MOS-25F-001

NETWORK SWITCHING SYSTEMS OPERATOR-MAINTAINER

25F10
25F20
25F30

Exhibit Dates: Pending Evaluation. 4/04–Present.

MOS-25L-001

AN/TSQ-73 Air Defense Artillery Command and Control System Operator/Repairer
25L10
25L20
25L30
25L40

Exhibit Dates: 1/90–5/91.

Career Management Field: 23 (Air Defense Missile Maintenance), subfield 237 (Fire Distribution Systems Repair).

Description

Summary: Supervises, operates, and performs maintenance on Air Defense Command and Control System AN/TSQ-73. *Skill Level 10:* Operates the command and control system which collects, evaluates, and distributes target data to air defense artillery fire units; operates general purpose target display systems and communication stations; monitors display status readouts to ensure correct target identification; although this skill level indicates primarily operator duties, maintenance skills are being developed; diagnostic procedures are, in general, limited to programmed diagnostic checks. *Skill Level 20:* Able to perform the duties required for Skill Level 10; performs full range of maintenance on display consoles, power supplies, communications equipment, general purpose computers, and peripheral devices; provides technical and operational assistance to subordinates; removes and replaces faulty assemblies and circuit cards. *Skill Level 30:* Able to perform the duties required for Skill Level 20; provides technical guidance to subordinates; able to detect and replace defective individual components; supervises on-the-job training and all operation/repair activities of subordinates. *Skill Level 40:* Able to perform the duties required for Skill Level 30; supervises and inspects overall organizational maintenance of the air defense command and control system.

Recommendation, Skill Level 10

In the vocational certificate category, 3 semester hours in applied mathematics. In the lower-division baccalaureate/associate degree category, 3 semester hours in introduction to electronics, 3 in basic electricity, 3 in technical mathematics, and additional credit in electronics and digital computer operations on the basis of institutional evaluation (1/80).

Recommendation, Skill Level 20

In the vocational certificate category, 3 semester hours in applied mathematics. In the lower-division baccalaureate/associate degree category, 3 semester hours in introduction to electronics, 3 in basic electricity, 3 in technical mathematics, 3 in electronic test procedures, and additional credit in electronics and digital computer operations on the basis of institutional evaluation (1/80).

Recommendation, Skill Level 30

In the vocational certificate category, 3 semester hours in applied mathematics and 3 in introduction to computers. In the lower-division baccalaureate/associate degree category, 3 semester hours in introduction to electronics, 3 in basic electricity, 3 in technical mathematics, 3 in electronic test procedures, 3 in introduction to computer science, 3 in electronics shop practices, 3 in personnel supervision, 2 in maintenance management, and additional credit in electronics and digital computer operations on the basis of institutional evaluation. In the upper-division baccalaureate category, 3 semester hours for field experience in personnel supervision (1/80).

Recommendation, Skill Level 40

In the vocational certificate category, 3 semester hours in applied mathematics and 3 in introduction to computers. In the lower-division baccalaureate/associate degree category, 3 semester hours in introduction to electronics, 3 in basic electricity, 3 in technical mathematics, 3 in electronic test procedures, 3 in introduction to computer science, 3 in electronics shop practices, 3 in personnel supervision, 3 in maintenance management, and additional credit in electronics and digital computer operations on the basis of institutional evaluation. In the upper-division baccalaureate category, 3 semester hours for field experience in personnel supervision, 3 in introduction to management, and 3 in human relations (1/80).

MOS-25L-002

AN/TSQ 73 Air Defense Artillery Command and Control System Operator/Maintainer
((AN/TSQ 73 Air Defense Artillery Command and Control System Operator/Repairer))
25L10
25L20
25L30
25L40

Exhibit Dates: 6/91–6/97. Effective 6/97, MOS 25L was discontinued and its duties were transferred to MOS 14L and MOS 14J.

Career Management Field: 23 (Air Defense System Maintenance).

Description

Summary: Supervises, operates, and performs maintenance on Air Defense Command and Control System AN/TSQ-73 for the Hawk Missile. *Skill Level 10:* Operates the command and control system which collects, evaluates, and distributes target data to air defense artillery fire units; operates general purpose target display systems and communication stations; monitors display status readouts to ensure correct target identification; although this skill level indicates primarily operator duties, maintenance skills are being developed; diagnostic procedures are, in general, limited to programmed diagnostic checks. *Skill Level 20:* Able to perform the duties required for Skill Level 10; performs full range of maintenance on display consoles, power supplies, communications equipment, general purpose computers, and peripheral devices; provides technical and operational assistance to subordinates; supervises maintenance; removes and replaces faulty assemblies and circuit cards. *Skill Level 30:* Able to perform the duties required for Skill Level 20; provides technical guidance to subordinates; able to detect and replace defective individual components; supervises on-the-job training and all operation/repair activities of subordinates. *Skill Level 40:* Able to perform the duties required for Skill Level 30; supervises and inspects overall organizational maintenance of the air defense command and control system.

Recommendation, Skill Level 10

In the lower-division baccalaureate/associate degree category, 3 semester hours in basic electronics laboratory. (NOTE: This recommendation for skill level 10 is valid for the dates 6/91-9/91 only) (6/91).

Recommendation, Skill Level 20

In the lower-division baccalaureate/associate degree category, 3 semester hours in basic electronics laboratory, 3 in personnel supervision, 2 in maintenance management, and 2 in records and information management. In the upper-division baccalaureate category, 3 semester hours in electronic systems troubleshooting and 3 in computer systems troubleshooting. (NOTE: This recommendation for skill level 20 is valid for the dates 6/91-2/95 only) (6/91).

Recommendation, Skill Level 30

In the lower-division baccalaureate/associate degree category, 3 semester hours in basic electronics laboratory, 3 in personnel supervision, 3 in maintenance management, and 3 in records and information management. In the upper-division baccalaureate category, 3 semester hours in electronic systems troubleshooting and 3 in computer systems troubleshooting (6/91).

Recommendation, Skill Level 40

In the lower-division baccalaureate/associate degree category, 3 semester hours in basic electronics laboratory, 3 in personnel supervision, 3 in maintenance management, and 3 in records and information management. In the upper-division baccalaureate category, 3 semester hours in electronic systems troubleshooting, 3 in computer systems troubleshooting, 3 in organizational management, and 3 for field experience in management (6/91).

MOS-25L-003

Cable System Installer/Maintainer
25L10
25L20
25L30

Exhibit Dates: Pending Evaluation. 4/04–Present.

MOS-25M-001

Multimedia Illustrator
25M10
25M20
25M30

Exhibit Dates: 2/94–3/04.

Career Management Field: 25 (Visual Information).

Description

Summary: Performs or supervises the design and production of graphic communication for briefings, posters, and training materials used in a variety of media formats including prints, slides, overhead transparencies, video, and computer CRT display; demonstrate creative problem solving skills in visualizing concrete and abstract concepts for effective communication. *Skill Level 10:* Serves as designer; solves visual communication problems with traditional drawing and graphic art techniques and computer graphics software; selects and uses oil paints, watercolors, inks, charcoal pencils, and crayons; prepares graphs, charts, overlays, lettering, and illustrations for print, audiovisual, and electronic presentations used for command briefings, training lessons, signs, information posters, manuals, operation plans, and overlays;

prepares mechanical drawing equipment and mounts photographs for static display and copy illustration; design layouts, color photo retouching, and the visualization process; executes freehand drawing of figures, faces, scenes, vehicles, and weapons; maintains illustration equipment and graphics workstation. *Skill Level 20:* Able to perform the duties required for Skill Level 10; provides guidance, critique, and technical assistance to beginning graphics specialist; has greater proficiency in visual problem solving, work planning and assignment, and scheduling. *Skill Level 30:* Able to perform the duties required for Skill Level 20; supervises other graphics personnel; conducts evaluation of graphics production; analyzes requirements in terms of illustration needs and scheduling; makes oral and written recommendations for use of graphics; maintains supplies; conducts maintenance inspections; recommends equipment and software purchases.

Recommendation, Skill Level 10

Credit may be granted on the basis of an individualized assessment of the student (9/95).

Recommendation, Skill Level 20

In the lower-division baccalaureate/associate degree category, 2 semester hours in graphic design applications, 2 in lettering and calligraphy, 2 in mechanical drawing, and 3 in audiovisual presentations. In the upper-division baccalaureate category, 3 semester hours in principles of typography, 2 in computer graphics, and 2 in visual communication problem solving. NOTE: This recommendation for skill level 20 is valid for the dates 2/94-2/95 only (9/95).

Recommendation, Skill Level 30

In the lower-division baccalaureate/associate degree category, 2 semester hours in graphic design applications, 2 in lettering and calligraphy, 2 in mechanical drawing, and 3 in audiovisual presentations. In the upper-division baccalaureate category, 3 semester hours in principles of typography, 2 in computer graphics, and 2 in visual communication problem solving (9/95).

MOS-25M-002

Multimedia Illustrator
25M10
25M20
25M30

Exhibit Dates: Pending Evaluation. 4/04–Present.

MOS-25P-001

Visual Information/Audio Documentation Systems Specialist
25P10
25P20
25P30

Exhibit Dates: 1/90–6/94. Effective 6/94, MOS 25P was discontinued and its duties were incorporated into MOS 25V, Combat Documentation/Production Specialist.

Career Management Field: 25 (Visual Information).

Description

Summary: Serves as producer in the preparation and production of live and recorded television or motion pictures; operates and performs operator maintenance on television, video, audio, and motion picture equipment. *Skill Level 10:* Is responsible for preproduction planning, preparation, and collection of production material, including site selection, lighting, camera placement, camera operation, and sound/audio setup and operation; maintains equipment. *Skill Level 20:* Able to perform the duties required for Skill Level 10; provides technical guidance to subordinates; assists in writing and producing productions; is responsible for postproduction, including reviewing and editing footage and use of special effect devices; assists in set construction; performs maintenance on equipment. *Skill Level 30:* Able to perform the duties required for Skill Level 20; supervises production team in all aspects; reviews collected material to evaluate quality; prepares scripts and supervises preproduction preparation, including set, scenery, graphics, and special effects; produces television, motion picture, and sound/audio productions.

Recommendation, Skill Level 10

In the lower-division baccalaureate/associate degree category, 1 semester hour in set design/lighting, 2 in basic television studio techniques (multicamera), 2 in basic video remote techniques (single camera), and 2 in basic audio techniques; if experience included working with motion pictures, 2 in basic cinematography. (NOTE: This recommendation for skill level 10 is valid for the dates 1/90-9/91 only) (11/91).

Recommendation, Skill Level 20

In the lower-division baccalaureate/associate degree category, 2 semester hours in set design/lighting, 3 in basic television studio techniques (multicamera), 3 in basic video remote techniques (single camera), 3 in basic audio techniques, 3 in basic postproduction processes, and 3 in personnel supervision; if experience included working with motion pictures, 3 in basic cinematography (11/91).

Recommendation, Skill Level 30

In the lower-division baccalaureate/associate degree category, 3 semester hours in set design/lighting, 3 in basic television studio techniques (multicamera), 3 in basic video remote techniques (single camera), 3 in basic audio techniques, 4 in basic postproduction processes, and 3 in personnel supervision; if experience included working with motion pictures, 3 in basic cinematography. In the upper-division baccalaureate category, 3 semester hours for field experience in media production, 2 in television/video production, and 1 in television/video writing (11/91).

MOS-25P-002

Microwave Systems Operator-Maintainer
25P10
25P20
25P30
25P40

Exhibit Dates: Pending Evaluation. 4/04–Present.

MOS-25Q-001

Graphics Documentation Specialist
25Q10
25Q20
25Q30

Exhibit Dates: 1/90–6/94. Effective 6/94, MOS 25Q was discontinued and its duties were incorporated into MOS 25M, Multimedia Illustrator.

Career Management Field: 25 (Visual Information).

Description

Summary: Performs or supervises the design and production of graphic communication for briefings, posters, and training materials; illustrations may be used in a variety of media formats including prints, slides, overhead transparencies, video, and computer CRT display; includes creative problem solving in visualizing concrete and abstract concepts for effective communication. *Skill Level 10:* Serves as designer; solves visual communication problems with traditional drawing and graphic art techniques and computer graphics software; proficient in the selection and use of oil paints, watercolors, inks, charcoal pencils, and crayons; prepares graphs, charts, overlays, lettering, and illustrations for print, audiovisual, and electronic presentations used for command briefings, training lessons, signs, information posters, manuals, operation plans, and overlays; prepares mechanical drawing equipment and mounts photographs for static display and copy illustration; knowledgeable in layout design, color, selection, photo retouching, and the visualization process; executes freehand drawing of figures, faces, scenes, vehicles, and weapons; maintains illustration equipment and graphics workstation. *Skill Level 20:* Able to perform the duties required for Skill Level 10; provides guidance, critique, and technical assistance to beginning graphics specialist; has greater proficiency in visual problem solving, work planning and assignment, and scheduling. *Skill Level 30:* Able to perform the duties required for Skill Level 20; supervises other graphics personnel; conducts evaluation of graphics production; analyzes requirements in terms of illustration needs and scheduling; makes oral and written recommendations for use of graphics; maintains supplies; conducts maintenance inspections; recommends equipment and software purchases.

Recommendation, Skill Level 10

In the lower-division baccalaureate/associate degree category, 2 semester hours in graphic design applications, 2 in lettering and calligraphy, 2 in mechanical drawing, and 2 in audiovisual presentations. In the upper-division baccalaureate category, 2 semester hours in principles of typography, 2 in computer graphics, and 1 in visual communication problem solving. (NOTE: This recommendation for skill level 10 is valid for the dates 1/90-9/91 only) (11/91).

Recommendation, Skill Level 20

In the lower-division baccalaureate/associate degree category, 2 semester hours in graphic design applications, 2 in lettering and calligraphy, 2 in mechanical drawing, and 3 in audiovisual presentations. In the upper-division baccalaureate category, 3 semester hours in principles of typography, 2 in computer graphics, and 2 in visual communication problem solving (11/91).

Recommendation, Skill Level 30

In the lower-division baccalaureate/associate degree category, 2 semester hours in graphic design applications, 2 in lettering and calligraphy, 2 in mechanical drawing, and 3 in audiovisual presentations. In the upper-division baccalaureate category, 3 semester hours in principles of typography, 3 in computer graphics, and 3 in visual communication problem solving (11/91).

MOS-25Q-002

Multichannel Transmission Systems Operator-Maintainer
25Q10
25Q20
25Q30

Exhibit Dates: Pending Evaluation. 4/04–Present.

MOS-25R-001

Visual Information/Audio Equipment Repairer
25R10
25R20
25R30

Exhibit Dates: 1/90–3/04.

Career Management Field: 25 (Visual Information).

Description

Summary: Installs and maintains audio and visual equipment and assists in the use and operation of video cameras, recorders, television receivers and transmitters, and audio transmitters. *Skill Level 10:* Installs, repairs, and provides necessary assistance in operation of audiovisual cameras, recorders, mixer equipment, and television and audio transmitters of various power levels; repairs and maintains all equipment utilized in an audio and television installation. *Skill Level 20:* Able to perform the duties required for Skill Level 10; supervises and assists subordinates in difficult repair and installation situations; counsels subordinates; prepares requisitions for supplies; keeps repair logs. *Skill Level 30:* Able to perform the duties required for Skill Level 20; supervises all subordinates; establishes training sessions; plans and coordinates installation, operation, and repair with superiors.

Recommendation, Skill Level 10

In the lower-division baccalaureate/associate degree category, 3 semester hours in electronic troubleshooting procedures, 2 in television camera repair, 2 in television transmitter repair, 2 in audio transmitter repair, and 2 in video recording procedures. (NOTE: This recommendation for skill level 10 is valid for the dates 1/90-9/91 only) (11/91).

Recommendation, Skill Level 20

In the lower-division baccalaureate/associate degree category, 3 semester hours in electronic troubleshooting procedures, 2 in television camera repair, 2 in television transmitter repair, 2 in audio transmitter repair, 2 in video recording procedures, 3 in personnel supervision, 1 in maintenance management, and 1 in records and information management. (NOTE: This recommendation for skill level 20 is valid for the dates 1/90-2/95 only) (11/91).

Recommendation, Skill Level 30

In the lower-division baccalaureate/associate degree category, 3 semester hours in electronic troubleshooting procedures, 2 in television camera repair, 2 in television transmitter repair, 2 in audio transmitter repair, 2 in video recording procedures, 3 in personnel supervision, 2 in maintenance management, and 1 in records and information management (11/91).

MOS-25S-001

Still Documentation Specialist
25S10
25S20
25S30

Exhibit Dates: 1/90–6/94. Effective 6/94, MOS 25S was discontinued and its duties were incorporated into MOS 25V, Combat Documentation/Production Specialist.

Career Management Field: 25 (Visual Information).

Description

Summary: Participates in still visual information activities including ground and aerial photographs; performs photographic lab processing services in black and white, color prints and color slides; performs some preventive maintenance on equipment. *Skill Level 10:* Uses photo flood lamps, electronic flash, light meters, and larger format cameras; shoots portraits, picture stories, and other photo assignments. *Skill Level 20:* Able to perform the duties required for Skill Level 10; evaluates photos; maintains photo library; provides guidance for photo shoots and laboratory procedures; monitors and inspects laboratory procedures. *Skill Level 30:* Able to perform the duties required for Skill Level 20; coordinates photo team coverage; establishes quality control; maintains photo supplies and inventory; ensures that all equipment is in operating order.

Recommendation, Skill Level 10

In the lower-division baccalaureate/associate degree category, 3 semester hours in still photography, 3 in black and white darkroom processing, and 3 in color darkroom processing. (NOTE: This recommendation for skill level 10 is valid for the dates 1/90-9/91 only) (11/91).

Recommendation, Skill Level 20

In the lower-division baccalaureate/associate degree category, 4 semester hours in still photography, 4 in black and white darkroom processing, 4 in color darkroom processing, and 3 in personnel supervision (11/91).

Recommendation, Skill Level 30

In the lower-division baccalaureate/associate degree category, 5 semester hours in still photography, 4 in black and white darkroom processing, 4 in color darkroom processing, and 3 in personnel supervision. In the upper-division baccalaureate category, 3 semester hours for field experience in media production (11/91).

MOS-25S-002

Satellite Communication Systems Operator-Maintainer
25S10
25S20
25S30
25S40

Exhibit Dates: Pending Evaluation. 4/04–Present.

MOS-25T-001

Satellite/Microwave Systems Chief
25T50

Exhibit Dates: Pending Evaluation. 4/04–Present.

MOS-25U-001

Signal Support Systems Specialist
25U10
25U20
25U30
25U40
25U50

Exhibit Dates: Pending Evaluation. 4/04–Present.

MOS-25V-001

Combat Documentation/Production Specialist
25V10
25V20
25V30

Exhibit Dates: 2/94–Present.

Career Management Field: 25 (Visual Information).

Description

Summary: Supervises, and operates electronic and film-based still, video, and audio equipment to document combat and noncombat operations; operates broadcast, collection, television production, and distribution equipment; creates visual information products in support of combat documentation, psychological operations, military intelligence, medical, public affairs, training, and other functions. *Skill Level 10:* Operates and performs maintenance on motion, still, and studio television cameras and electronic and film-based processing, editing, audio, and printing darkroom equipment; prepares captions for documentation images; operates and performs preventive maintenance on vehicles and power generators. *Skill Level 20:* Able to perform the duties required for Skill Level 10; provides technical guidance to subordinates; monitors, operates, and performs maintenance on master control systems; aligns and adjusts video cameras; prepares video reports; operates and performs maintenance on still video transmission systems. *Skill Level 30:* Able to perform the duties required for Skill Level 20; performs as team leader of combat camera documentation teams; selects documentation/production equipment and determines systems mission support requirements; coordinates and directs personnel and operational requirements to produce audiovisual, audio, and television productions in both fixed and tactical environments.

Recommendation, Skill Level 10

Credit may be granted on the basis of an individualized assessment of the student (9/95).

Recommendation, Skill Level 20

In the lower-division baccalaureate/associate degree category, 3 semester hours in personnel supervision; if experience included television production, 2 semester hours in set design/lighting, 3 in basic television studio techniques (multicamera), 3 in basic video remote techniques (single camera), 3 in basic audio techniques, and 3 in basic postproduction processes; if experience included working with motion pictures, 3 in basic cinematography; if experience included still photography, 4 semester hours in still photography, 4 in black-and-white darkroom processing, and 4 in color darkroom processing. (NOTE: This recommendation for skill level 20 is valid for the dates 2/94-2/95 only (9/95).

Recommendation, Skill Level 30

In the lower-division baccalaureate/associate degree category, 3 semester hours in personnel supervision; if experience included television production, 2 semester hours in set design/lighting, 3 in basic television studio techniques (multicamera), 3 in basic video remote techniques (single camera), 3 in basic audio techniques, and 3 in basic postproduction processes; if experience included working with motion pictures, 3 in basic cinematography; if experience included still photography, 4 semester hours in still photography, 4 in black-and-white darkroom processing, and 4 in color darkroom processing (9/95).

MOS-25W-001

TELECOMMUNICATIONS OPERATIONS CHIEF
25W40
25W50

Exhibit Dates: Pending Evaluation. 4/04–Present.

MOS-25X-001

SENIOR SIGNAL SERGEANT
25X50

Exhibit Dates: Pending Evaluation. 4/04–Present.

MOS-25Y-001

INFORMATION SYSTEMS CHIEF
25Y50

Exhibit Dates: Pending Evaluation. 4/04–Present.

MOS-25Z-001

VISUAL INFORMATION OPERATIONS CHIEF ((Audio-Visual Chief))
25Z40
25Z50

Exhibit Dates: 1/90–11/01.

Career Management Field: 25 (Visual Information).

Description

Summary: Supervises subordinates performing visual information duties and manages all documentation operations of an Army post or major unit; analyzes and defines; requirements, plan, execute, and evaluates visual information support, personnel supervision, and training; maintains equipment and facilities; and requisitions materials and supplies. *Skill Level 40:* Manages photographic facility operations, radio and television operations, and visual information documentation teams; provides personnel supervision, training, and guidance; oversees equipment repair and maintenance; schedules studio operations and mobile van and remote site operations; manages radio and television microwave, broadcast, and cable transmission system operations; plans and coordinates visual information operations, including determination of present and future personnel and equipment requirements for documentation teams and television and radio production; schedules and directs operations and evaluates performance in coordination with unit or agency being supported. NOTE: May have progressed to 25Z40 from 25M30 (Multimedia Illustrator); 25P30 (Visual Information/Audio Documentation Systems Specialist); 25Q30 (Graphics Documentation Specialist); 25R30 (Visual Information/Audio Equipment Repairer); 25S30 (Still Documentation Specialist); or 25V30 (Combat Documentary/Production Specialist). *Skill Level 50:* Able to perform the duties required for Skill Level 40; performs administrative tasks including personnel assignment, discipline, and career counseling; assists command and staff offices in determining appropriate visual information uses and documentation needs; maintains quality assurance of all visual information and documentation production.

Recommendation, Skill Level 40

In the upper-division baccalaureate category, 3 semester hours in personnel administration, 3 in organizational management, 3 in organizational planning, 3 in human resource management, and 3 for field experience in management (11/91).

Recommendation, Skill Level 50

In the upper-division baccalaureate category, 3 semester hours in personnel administration, 3 in organizational management, 3 in organizational planning, 3 in human resource management, and 6 for field experience in management (11/91).

MOS-25Z-002

VISUAL INFORMATION OPERATIONS CHIEF
25Z40
25Z50

Exhibit Dates: 12/01–Present.

Career Management Field: 25 (Visual Information).

Description

Summary: Supervises subordinates performing visual information duties and manages all documentation/production operations of an Army post or major unit; analyzes and defines requirements; plans, executes, and evaluates visual information support, personnel supervision, and training; maintains equipment and facilities; and requisitions materials and supplies. *Skill Level 40:* Manages photographic facility operations, radio and television operations, and visual information documentation teams; provides personnel supervision, training, and guidance; oversees equipment repair and maintenance; schedules studio operations and mobile van and remote site operations; manages radio and television microwave, broadcast, and cable transmission system operations; plans and coordinates visual information operations, including determination of present and future personnel and equipment requirements for documentation teams and television and radio production; supports web-based design initiatives using current industry-standard software; schedules and directs operations and evaluates performance in coordination with unit or agency being supported. NOTE: May have progressed to 25Z40 from 25M30 (Multimedia Illustrator), 25P30 (Visual Information/Audio Documentation Systems Specialist), 25Q30 (Graphics Documentation Specialist), 25R30 (Visual Information/Audio Equipment Repairer), 25S30 (Still Documentation Specialist), or 25V30 (Combat Documentary/Production Specialist). *Skill Level 50:* Able to perform the duties required for Skill Level 40; performs administrative tasks including personnel assignment, discipline, and career counseling; assists command and staff offices in determining appropriate visual information uses and documentation needs; maintains quality assurance of all visual information and documentation production; assists with budget development.

Recommendation, Skill Level 40

In the upper-division baccalaureate category, 3 semester hours in personnel management, 3 in audiovisual production management, 3 in organizational management, 3 in organizational planning, 3 in human resource management, and 3 in management problems (12/01).

Recommendation, Skill Level 50

In the upper-division baccalaureate category, 3 semester hours in personnel management, 3 in organizational management, 3 in organizational planning, 3 in audiovisual production management, 3 in human resource management, and 6 in management problems (12/01).

MOS-26H-002

AIR DEFENSE RADAR REPAIRER
26H10
26H20
26H30

Exhibit Dates: 1/90–3/90. Effective 3/90, MOS 26H was discontinued.

Career Management Field: 23 (Air Defense Missile Maintenance), subfield 231 (Nike Missile System Repair).

Description

Summary: Supervises or performs maintenance and repairs on air defense radars and associated equipment, including height, azimuth, and distance-finding radars, servo, data transmission systems, and computers. *Skill Level 10:* Performs highly complex repairs on solid state, integrated circuit, vacuum tube radar, and associated equipment; provides repair support and advice for various radar maintenance groups; uses advanced as well as basic electronics and schematic tracing to troubleshoot systems and subassemblies down to the component level; performs certain mathematical calculations during routine calibration or alignment procedures; uses common and advanced industrial test equipment as well as specialized test sets to diagnose system and subassembly failures; uses common and specialized test sets to diagnose system and subassembly failures; uses common and specialized hand tools to perform repair, adjustment, or modification on radar or associated equipment; may train maintenance and operator personnel on the proper handling or maintenance of specified radar equipment. *Skill Level 20:* Able to perform the duties required for Skill Level 10; provides technical guidance to subordinates; updates knowledge of advanced modifications being performed on existing radar equipment through on-the-job training or by attending workshops. *Skill Level 30:* Able to perform the duties required for Skill Level 20; supervises subordinates; assigns duties; manages maintenance schedules, inspections, and quality control; supervises inspection and maintenance teams; conducts on-the-job training; keeps maintenance records.

Recommendation, Skill Level 10

In the vocational certificate category, 3 semester hours in basic electronics, 3 in electronic control devices, 2 in technical mathematics, and 1 in electronic instruments. In the lower-division baccalaureate/associate degree category, 3 semester hours in basic electricity and electronics, 3 in radar/microwave technology, and 1 in digital circuitry. In the upper-division baccalaureate category, 3 semester hours in an electronics technical elective (non—engineering) and 3 in a radar/microwave technical elective (1/80).

Recommendation, Skill Level 20

In the vocational certificate category, 3 semester hours in basic electronics, 3 in electronic control devices, 3 in digital circuitry, 2 in technical mathematics, and 3 in electronic instruments. In the lower-division baccalaureate/associate degree category, 3 semester hours in basic electricity and electronics, 3 in radar/ microwave technology, 1 in digital circuitry, and 3 in personnel supervision. In the upper-division baccalaureate category, 3 semester hours in an electronics technical elective (non—engineering), and 3 in a radar/microwave technical elective (1/80).

Recommendation, Skill Level 30

In the vocational certificate category, 3 semester hours in basic electronics, 3 in electronic control devices, 3 in digital circuitry, 2 in technical mathematics, and 3 in electronic instruments. In the lower-division baccalaureate/associate degree category, 3 semester hours in basic electricity and electronics, 3 in radar/ microwave technology, 1 in digital circuitry, 3 in personnel supervision, and 3 in maintenance management. In the upper-division baccalaureate category, 3 semester hours in an electronics technical elective (non—engineering), 3 in a radar/microwave technical elective, and 3 for field experience in personnel supervision (1/ 80).

MOS-27B-002

Land Combat Support System Test Specialist/Lance Repairer
27B10
27B20
27B30
27B40

Exhibit Dates: 1/90–9/91.

Career Management Field: 27 (Ballistic/Land Combat Missile and Light Air Defense Weapons Systems Maintenance), subfield 271 (Ballistic/Land Combat Systems Maintenance).

Description

Summary: Maintains Land Combat Systems (LCSS) and the Lance missile system; inspects and tests equipment; locates card-level faults with special test equipment; replaces faulty assemblies and subassemblies; selects program tapes, patchboards and cable assemblies; adjusts and aligns mechanical, electrical, and optical assemblies; calibrates and repairs support equipment; performs quality control measures; prepares maintenance and supply forms and reports. *Skill Level 10:* Works under close supervision in performing maintenance functions; uses technical manuals; follows schematics and organized maintenance procedures; uses multimeters, oscilloscopes, signal generators, and power supplies; tests digital multimeters, waveform connectors, signal generators, pulse generators, power supplies, printers, and detector adapters; performs operator maintenance on AC power generators and replaces faulty components. *Skill Level 20:* Able to perform the duties required for Skill Level 10; supervises and provides technical assistance to subordinates; repairs multimeters, waveform connectors, pulse generators, power supplies, signal generators, and patchboards; performs modifications on support equipment. *Skill Level 30:* Able to perform the duties required for Skill Level 20; provides technical guidance to subordinates; schedules work; sets repair priorities; inspects, tests, and repairs missile system components; demonstrates proper work practices and maintenance techniques; implements quality control measures; performs system inspections; supervises maintenance and inspection teams; may supervise a minimum of five persons. *Skill Level 40:* Able to perform the duties required for Skill Level 30; able to supervise personnel performing the duties of MOS 27B, MOS 27E (TOW/Dragon Repairer), and MOS 27H (Shillelagh Repairer); supervises and coordinates support maintenance activities of system-associated electronic equipment; assists in developing quality control measures; serves as section chief performing missile system support maintenance; serves as senior instructor in a training facility. NOTE: May have progressed to 27B40 from 27B30, 27E30 (TOW/Dragon Repairer), or 27H30 (Shillelagh Repairer).

Recommendation, Skill Level 10

In the vocational certificate category, 3 semester hours in an electronics technician program. In the lower-division baccalaureate/associate degree category, 3 semester hours in introduction to electricity and electronics (5/ 78).

Recommendation, Skill Level 20

In the vocational certificate category, 3 semester hours in an electronics technician program. In the lower-division baccalaureate/associate degree category, 3 semester hours in introduction to electricity and electronics, 2 in electronic instrumentation, and 2 in digital computer principles (5/78).

Recommendation, Skill Level 30

In the vocational certificate category, 3 semester hours in an electronics technician program. In the lower-division baccalaureate/associate degree category, 3 semester hours in introduction to electricity and electronics, 2 in electronic instrumentation, 2 in digital computer principles, and 1 in personnel supervision; if the duty assignment was Lance Repair Supervisor or LCSS Repair Supervisor, 1 additional semester hour in personnel supervision and credit in human relations on the basis of institutional evaluation (5/78).

Recommendation, Skill Level 40

In the vocational certificate category, 3 semester hours in an electronics technician program. In the lower-division baccalaureate/associate degree category, 3 semester hours in introduction to electricity and electronics, 3 in personnel supervision, 3 in human relations, 2 in electronic instrumentation, and 2 in digital computer principles (5/78).

MOS-27B-003

Land Combat Support System Test Specialist
27B10
27B20
27B30
27B40

Exhibit Dates: 10/91–6/95. Effective 6/95, MOS 27B was discontinued, and its duties were incorporated into MOS 35B, 27E, 27M, and 35Y.

Career Management Field: 27 (Land Combat and Air Defense System Direct and General Support Maintenance).

Description

Summary: Maintains Land Combat Systems (LCSS) and the Lance Missile System; inspects and tests equipment; locates card-level faults with special test equipment; replaces faulty assemblies and subassemblies; selects program tapes, patchboards and cable assemblies; adjusts and aligns mechanical, electrical, and optical assemblies; calibrates and repairs support equipment; performs quality control measures; prepares maintenance and supply forms and reports. *Skill Level 10:* Works under close supervision in performing maintenance functions; uses technical manuals and troubleshooting trees; follows schematics and organized maintenance procedures; uses digital voltmeters, oscilloscopes, signal generators, and power supplies; tests waveform converter, computerized test set, computerized signal generators, pulse generators, power supplies, printers, and detector adapters; performs maintenance on AC power generators and replaces faulty components. *Skill Level 20:* Able to perform the duties required for Skill Level 10; supervises and provides technical assistance to subordinates; repairs multimeters, waveform generator, pulse generators, power supplies, signal generators, and patchboards; performs minor modifications on support equipment. *Skill Level 30:* Able to perform the duties required for Skill Level 20; provides technical guidance to subordinates; schedules work; sets repair priorities; inspects, tests, and repairs missile system assemblies; demonstrates proper work practices and maintenance techniques; implements quality control measures; performs system inspections; supervises maintenance and inspection teams; may supervise a minimum of five persons. *Skill Level 40:* Able to perform the duties required for Skill Level 30; supervises persons performing the duties of MOS 27B, MOS 27E (TOW/Dragon Repairer), MOS 27L (Lance System Repairer), and MOS 27M (Multiple Launcher Rocket System Repairer); supervises and coordinates support maintenance activities of system-associated electronic equipment; assists in developing quality control measures; serves as section chief performing missile system support maintenance; may serve as senior instructor in a training facility. NOTE: May have progressed to 27B40 from 27B30, 27E30 (TOW/Dragon Repairer), 27L30 (Lance System Repairer), or 27M30 (Multiple Launcher Rocket System Repairer).

Recommendation, Skill Level 10

Credit may be granted on the basis of an individualized assessment of the student (3/94).

Recommendation, Skill Level 20

In the lower-division baccalaureate/associate degree category, 3 semester hours in basic electronics, 4 in electronic systems troubleshooting and maintenance, 3 in computer systems troubleshooting and maintenance, 3 in computer literacy, and 3 in personnel supervision. (NOTE: This recommendation for skill level 20 is valid for the dates 10/91-2/95 only) (9/91).

Recommendation, Skill Level 30

In the lower-division baccalaureate/associate degree category, 3 semester hours in basic electronics, 4 in electronic systems troubleshooting and maintenance, 3 in computer systems troubleshooting and maintenance, 3 in computer literacy, 3 in personnel supervision, 2 in records and information management, and 3 in maintenance management (9/91).

Recommendation, Skill Level 40

In the lower-division baccalaureate/associate degree category, 3 semester hours in basic electronics, 4 in electronic systems troubleshooting and maintenance, 3 in computer systems troubleshooting and maintenance, 3 in computer literacy, 3 in personnel supervision, 2 in records and information management, and 3 in maintenance management. In the upper-division baccalaureate category, 3 semester hours in organizational management and 3 for field experience in management (9/91).

MOS-27D-001

Paralegal Specialist

27D10
27D20
27D30
27D40
27D50

Exhibit Dates: 10/00–Present.

Career Management Field: 71 (Administration).

Description

Summary: Prepares or supervises the preparation of court-martial records, board proceedings, preliminary hearings, and investigations. *Skill Level 10:* Under close supervision, assists in preparing legal papers and forms, conducting investigations and researching activities, and maintaining files; types at a minimum of 30 words per minute; assists in maintaining law library. *Skill Level 20:* Able to perform the duties required for Skill Level 10; prepares correspondence; interview clients to determine nature of problem; compiles records into quarterly reports; provides legal assistance (military and civil); refers legal problems to attorney; makes referrals as needed. *Skill Level 30:* Able to perform the duties required for Skill Level 20; supervises a legal office or maintains a law library; provides guidance to subordinates; maintains files; reviews claims; conducts claim investigations. *Skill Level 40:* Able to perform the duties required for Skill Level 30; plans and organizes legal office operation; previews legal documentation for proper and prompt disposition; researches decisions, statutes, and regulations under supervision of a judge advocate or civilian attorney. *Skill Level 50:* Able to perform the duties required for Skill Level 40; supervises the administration of a large legal office; prepares budgets; plans programs; secures special proceedings, highly sensitive, or specially-handled documents or evidence; assigns duties; evaluates performance of subordinates.

Recommendation, Skill Level 10

Credit may be granted on the basis of an individualized assessment of the student (6/97).

Recommendation, Skill Level 20

Credit may be granted on the basis of an individualized assessment of the student (6/97).

Recommendation, Skill Level 30

In the lower-division baccalaureate/associate degree category, 3 semester hours in legal research, 3 in keyboarding, 3 in office practices, 3 in legal investigations and research, 3 in legal writing, 3 in legal military practices and procedures, 3 in law office management, 3 in litigation, and 3 in personnel supervision. In the upper-division baccalaureate category, 3 semester hours in management problems (6/97).

Recommendation, Skill Level 40

In the lower-division baccalaureate/associate degree category, 3 semester hours in legal research, 3 in keyboarding, 3 in office practices, 3 in legal investigation and research, 3 in legal writing, 3 in legal military practices and procedures, 3 in law office management, 3 in litigation, and 3 in personnel supervision. In the upper-division baccalaureate category, 3 semester hours in management problems and 3 in human resource management (6/97).

Recommendation, Skill Level 50

In the lower-division baccalaureate/associate degree category, 3 semester hours in legal research, 3 in keyboarding, 3 in office practices, 3 in legal investigation and research, 3 in legal writing, 3 in legal military practices and procedures, 3 in law office management, 3 in litigation, and 3 in personnel supervision. In the upper-division baccalaureate category, 3 semester hours in management problems and 3 in human resource management (6/97).

MOS-27E-002

TOW/Dragon Repairer

27E10
27E20
27E30

Exhibit Dates: 1/90–9/91.

Career Management Field: 27 (Ballistic/Land Combat Missile and Light Air Defense Weapons Systems Maintenance), subfield 271 (Ballistics/Land Combat Systems Maintenance).

Description

Summary: Performs maintenance on TOW and Dragon missile system trainers, night sights, battery chargers, and system test equipment. *Skill Level 10:* Performs maintenance tasks under close supervision; follows technical manuals in servicing and maintaining systems; inspects, tests, and adjusts components to specified tolerances using special test equipment; locates and replaces malfunctioning electronic, electrical, mechanical, pneumatic and electromechanical subassemblies, modules, and circuit elements; prepares maintenance and supply forms and reports. *Skill Level 20:* Able to perform the duties required for Skill Level 10; provides technical guidance to subordinates; provides technical assistance to support units; installs equipment modifications; troubleshoots weapon system, launch effects trainer, TOW training set, and breakout box; performs operational checks on tracker test set, tracker and launch effects trainer, and monitoring set. *Skill Level 30:* Able to perform the duties required for Skill Level 20; supervises subordinates; supervises inspection and maintenance teams; establishes work loads and priorities; conducts in-service staff training; demonstrates proper maintenance and troubleshooting techniques; recommends maintenance procedures; implements quality control measures; establishes and maintains maintenance records; performs inspections; may supervise up to five persons.

Recommendation, Skill Level 10

In the lower-division baccalaureate/associate degree category, 3 semester hours in use and care of tools and test equipment (5/78).

Recommendation, Skill Level 20

In the lower-division baccalaureate/associate degree category, 3 semester hours in use and care of tools and test equipment, and 3 in introduction to electricity and electronics (5/78).

Recommendation, Skill Level 30

In the lower-division baccalaureate/associate degree category, 3 semester hours in use and care of tools and test equipment, 3 in introduction to electricity and electronics, and 1 in personnel supervision; if the duty position was TOW/Dragon Repair foreman,1 additional semester hour in personnel supervision and credit in human relations on the basis of institutional evaluation (5/78).

MOS-27E-003

Land Combat Electronic Missile System Repairer

((TOW/Dragon Missile Electronics Repairer))

((TOW/Dragon Repairer))

27E10
27E20
27E30

Exhibit Dates: 10/91–12/94.

Career Management Field: 27 (Land Combat and Air Defense System Direct and General Support Maintenance).

Description

Summary: Performs direct and general support maintenance on TOW and Dragon missile systems, trainers, night sights, battery chargers, and system test equipment. *Skill Level 10:* Performs maintenance tasks under close supervision; follows technical manuals in servicing and maintaining systems; inspects, tests, and adjusts components to specified tolerances using special test equipment and computerized test consoles; locates and replaces malfunctioning electronic, electrical, mechanical, optical, pneumatic and electromechanical subassemblies, modules, and circuit elements; determines serviceability and disposition of defective assemblies, subassemblies, and parts; prepares maintenance and supply forms and reports. *Skill Level 20:* Able to perform the duties required for Skill Level 10; provides technical guidance to subordinates; provides technical assistance to support units; installs equipment modifications; troubleshoots weapon system, launch effects trainer, TOW training set, tracker and launcher effect trainer and monitoring set. *Skill Level 30:* Able to perform the duties required for Skill Level 20; supervises subordinates; supervises inspection and maintenance teams; establishes work loads and priorities; conducts in-service staff training; demonstrates proper maintenance and troubleshooting techniques; recommends maintenance procedures; implements quality control measures; establishes and maintains maintenance records; performs inspections; may supervise up to eight persons.

Recommendation, Skill Level 10

Credit may be granted on the basis of an individualized assessment of the student (3/94).

Recommendation, Skill Level 20

In the lower-division baccalaureate/associate degree category, 3 semester hours in basic electronics, 4 in electronic systems troubleshooting and maintenance, 3 in computer literacy, 3 in digital principles, 3 in solid state electronics, and 3 in personnel supervision (9/91).

Recommendation, Skill Level 30

In the lower-division baccalaureate/associate degree category, 3 semester hours in basic electronics, 4 in electronic systems troubleshooting and maintenance, 3 in computer literacy, 3 in digital principles, 3 in solid state electronics, 3 in personnel supervision, 3 in maintenance management, and 2 in records and information management (9/91).

MOS-27E-004

LAND COMBAT ELECTRONIC MISSILE SYSTEM REPAIRER
27E10
27E20
27E30
27E40

Exhibit Dates: 1/95–9/03. Effective 10/03, MOS 27E was discontinued and its duties were transferred to MOS 35A, Land Combat Electronic Missile System Repairer.

Career Management Field: 27 (Land Combat and Air Defense Systems Direct and General Support Maintenance).

Description

Summary: Performs direct and general support maintenance on TOW and DRAGON missile systems, Bradley Fighting Vehicle System, JAVELIN system, trainers, night sights, battery chargers, and system test equipment. *Skill Level 10:* Performs maintenance tasks under close supervision; follows technical manuals in servicing and maintaining systems; inspects, tests, and adjusts components to specified tolerances using special test equipment and computerized test consoles; locates and replaces malfunctioning electronic, electrical, mechanical, optical, pneumatic, and electromechanical subassemblies, modules, and circuit elements; determines serviceability and disposition of defective assemblies, subassemblies, and parts; prepares maintenance and supply forms and reports. *Skill Level 20:* Able to perform the duties required for Skill Level 10; provides technical guidance to subordinates; provides technical assistance to support units; installs equipment modifications; troubleshoots weapon system, launch effects trainer, TOW training set, tracker, and launcher effect trainer and monitoring set. *Skill Level 30:* Able to perform the duties required for Skill Level 20; supervises subordinates; supervises inspection and maintenance teams; establishes work loads and priorities; conducts in-service staff training; demonstrates proper maintenance and troubleshooting techniques; recommends maintenance procedures; implements quality control measures; establishes and maintains maintenance records; performs inspections; may supervise up to eight soldiers. *Skill Level 40:* Able to perform the duties required for Skill Level 30; supervises and coordinates support maintenance; serves as section chief or platoon sergeant of detachment, platoon, company, or comparable unit.

Recommendation, Skill Level 10

Credit may be granted on the basis of an individualized assessment of the student (7/97).

Recommendation, Skill Level 20

Credit may be granted on the basis of an individualized assessment of the student (7/97).

Recommendation, Skill Level 30

In the lower-division baccalaureate/associate degree category, 3 semester hours in basic electronics, 3 in electronic systems troubleshooting and maintenance, 3 in personnel supervision, 3 in maintenance management, and 2 in records and information management (7/97).

Recommendation, Skill Level 40

In the lower-division baccalaureate/associate degree category, 3 semester hours in basic electronics, 3 in electronic systems troubleshooting and maintenance, 3 in personnel supervision, 3 in maintenance management, and 2 in records and information management. In the upper-division baccalaureate category, 3 semester hours for field experience in management (7/97).

MOS-27F-002

VULCAN REPAIRER
27F10
27F20
27F30

Exhibit Dates: 1/90–9/91.

Career Management Field: 27 (Ballistic/Land Combat Missile and Light Air Defense Weapons Systems Maintenance), subfield 272 (Light Air Defense Systems Maintenance).

Description

Summary: Performs maintenance on Vulcan electronic assemblies, range-only radar, and associated equipment. *Skill Level 10:* Works under close supervision, inspects, tests, and adjusts components to specific tolerances; determines malfunctions in assemblies, subassemblies, modules, and circuit elements with common and specialized test equipment; removes and replaces defective components and parts; tests transmitters/receivers, antennas, range computers, power supplies, amplifiers, control and electronic field test sets; calibrates power supplies, modular testers, receiver test sets, and other components unique to Vulcan electronic assemblies. *Skill Level 20:* Able to perform the duties required for Skill Level 10; provides technical assistance to subordinates; installs system modifications; provides technical assistance to support units. *Skill Level 30:* Able to perform the duties required for Skill Level 20; supervises and performs maintenance tasks; establishes work loads and priorities; establishes and conducts staff training; implements quality control measures; supervises inspection and maintenance teams; maintains records; recommends inspection, testing, and maintenance techniques; may supervise up to five persons.

Recommendation, Skill Level 10

In the lower-division baccalaureate/associate degree category, 3 semester hours in use and care of tools and test equipment (5/78).

Recommendation, Skill Level 20

In the lower-division baccalaureate/associate degree category, 3 semester hours in use and care of tools and test equipment and 3 in basic electronics (5/78).

Recommendation, Skill Level 30

In the lower-division baccalaureate/associate degree category, 3 semester hours in use and care of tools and test equipment, 3 in basic electronics, and 1 in personnel supervision; if the duty position was Vulcan Repair Supervisor, 1 additional semester hour in personnel supervision, and credit in human relations on the basis of institutional evaluation (5/78).

MOS-27F-003

VULCAN REPAIRER
27F10
27F20
27F30

Exhibit Dates: 10/91–12/95. Effective 12/95, MOS 27F was discontinued.

Career Management Field: 27 (Land Combat and Air Defense System Direct and General Support Maintenance).

Description

Summary: Performs direct and general support maintenance on Vulcan electronic assemblies, Vulcan cannon and feed system, towed Vulcan carriage, range-only radar, and associated equipment. *Skill Level 10:* Works under close supervision; inspects, tests, and adjusts components to specified tolerances; determines malfunctions in assemblies, subassemblies, modules, and circuit elements with common and specialized test equipment; removes and replaces defective components and parts; inspects, tests, and adjusts Vulcan system components and parts using automatic test equipment; performs initial, work-in-process, on-site technical, and quality control inspections. *Skill Level 20:* Able to perform the duties required for Skill Level 10; provides technical assistance to subordinates; installs system modifications; provides technical assistance to support units; completes maintenance and supply forms; calibrates radar and antenna test sets; installs equipment modifications. *Skill Level 30:* Able to perform the duties required for Skill Level 20; supervises and performs maintenance tasks; establishes work loads and priorities; establishes and conducts staff training; implements quality control measures; supervises inspection, testing, and maintenance techniques; may supervise up to ten persons.

Recommendation, Skill Level 10

Credit may be granted on the basis of an individualized assessment of the student (3/94).

Recommendation, Skill Level 20

In the lower-division baccalaureate/associate degree category, 3 semester hours in basic electronics, 4 in electronic systems troubleshooting and maintenance, 3 in computer literacy, 3 in radar systems, and 3 in personnel supervision. (NOTE: This recommendation for skill level 20 is valid for the dates 10/91-2/95 only) (9/91).

Recommendation, Skill Level 30

In the lower-division baccalaureate/associate degree category, 3 semester hours in basic electronics, 4 in electronic systems troubleshooting and maintenance, 3 in computer literacy, 3 in radar systems, 3 in personnel supervision, 3 in maintenance management, and 2 in records and information management (9/91).

MOS-27G-002

CHAPARRAL/REDEYE REPAIRER
27G10
27G20
27G30
27G40

Exhibit Dates: 1/90–9/91.

Career Management Field: 27 (Ballistic/Land Combat Missile and Light Air Defense Weapon Systems Maintenance), subfield 272 (Light Air Defense Systems Maintenance).

Description

Summary: Performs maintenance on the Chaparral and Redeye missiles systems, associated equipment, and trainers. *Skill Level 10:* Inspects, tests, and adjusts components to specific tolerances; determines malfunctions; uses system-associated test equipment; removes and replaces defective components and parts; serves on maintenance and inspection teams; prepares maintenance and supply forms and reports. *Skill Level 20:* Able to perform the duties required for Skill Level 10; provides technical guidance to subordinates; installs equipment modifications; provides technical assistance to supported units. *Skill Level 30:* Able to perform the duties required for Skill Level 20; establishes work load and repair priorities; supervises inspection and maintenance teams; organizes and conducts on-the-job training programs; demonstrates proper maintenance and troubleshooting techniques; implements quality control measures; establishes and maintains maintenance records, may supervise up to five persons. *Skill Level 40:* Able to perform the duties required for Skill Level 30 of MOS 27G or 27F (Vulcan Repairer); supervises and coordinates support maintenance of the Chaparral and Redeye missile systems, Vulcan weapon system electronics, forward area alerting radar, system-associated trainers and test equipment, and power generation equipment; calibrates system-associated electronic equipment; supervises the quality control program and assists in its development. NOTE: May have progressed to 27G40 from 27G30 or 27F30 (Vulcan Repairer).

Recommendation, Skill Level 10

In the lower-division baccalaureate/associate degree category, 3 semester hours in the use and care of tools and test equipment and credit in basic electronics on the basis of institutional evaluation (5/78).

Recommendation, Skill Level 20

In the lower-division baccalaureate/associate degree category, 3 semester hours in the use and care of tools and test equipment and credit in basic electronics on the basis of institutional evaluation (5/78).

Recommendation, Skill Level 30

In the lower-division baccalaureate/associate degree category, 3 semester hours in use and care of tools and test equipment, 1 semester hour in personnel supervision, and credit in basic electronics on the basis of institutional evaluation; if the duty position was Chaparral/Redeye Missile Systems Repair Supervisor, 1 additional semester hour in personnel supervision and credit in human relations on the basis of institutional evaluation (5/78).

Recommendation, Skill Level 40

In the lower-division baccalaureate/associate degree category, 3 semester hours in use and care of tools and test equipment, 3 in personnel supervision, 3 in human relations, and credit in basic electronics on the basis of institutional evaluation (5/78).

MOS-27G-003

CHAPARRAL/REDEYE REPAIRER

27G10
27G20
27G30
27G40

Exhibit Dates: 10/91–Present.

Career Management Field: 27 (Land Combat and Air Defense System Direct and General Support Maintenance).

Description

Summary: Performs maintenance on the Chaparral and Redeye missile systems, associated equipment, and trainers. *Skill Level 10:* Inspects, tests, and adjusts components to specific tolerances; determines malfunctions in electronic, electrical, and cryogenic assemblies using specialized test equipment; removes and replaces defective components and parts; performs repair and maintenance on hydraulic, pneumatic, and high pressure systems; prepares maintenance and supply forms and reports. *Skill Level 20:* Able to perform the duties required for Skill Level 10; provides technical guidance to subordinates; installs equipment modifications; provides technical assistance to support units. *Skill Level 30:* Able to perform the duties required for Skill Level 20; establishes work load and repair priorities; supervises inspection and maintenance teams; organizes and conducts on-the-job training programs; demonstrates proper maintenance and troubleshooting techniques; implements quality control measures; establishes and maintains maintenance records; may supervise up to ten persons. *Skill Level 40:* Able to perform the duties required for Skill Level 30 of MOS 27G, 27F (Vulcan Repairer), or 27N (Forward Area Alerting Radar Repairer); supervises and coordinates support maintenance of the Chaparral and Redeye missile systems, Vulcan weapon system electronics, forward area alerting radar, system-associated trainers and test equipment, and power generation equipment; calibrates system-associated electronic equipment; supervises the quality control program and assists in its development. NOTE: May have progressed to 27G40 from 27G30, 27F30 (Vulcan Repairer), or 27N30 (Forward Area Alerting Radar Repairer).

Recommendation, Skill Level 10

Credit may be granted on the basis of an individualized assessment of the student (3/94).

Recommendation, Skill Level 20

In the lower-division baccalaureate/associate degree category, 3 semester hours in basic electronics, 4 in electronic systems troubleshooting and maintenance, 3 in computer literacy, 3 in solid state electronics, 3 in hydraulics and pneumatics, and 3 in personnel supervision. (NOTE: This recommendation for skill level 20 is valid for the dates 10/91-2/95 only) (9/91).

Recommendation, Skill Level 30

In the lower-division baccalaureate/associate degree category, 3 semester hours in basic electronics, 4 in electronic systems troubleshooting and maintenance, 3 in computer literacy, 3 in solid state electronics, 3 in hydraulics and pneumatics, 3 in personnel supervision, 3 in maintenance management, and 2 in records and information management (9/91).

Recommendation, Skill Level 40

In the lower-division baccalaureate/associate degree category, 3 semester hours in basic electronics, 4 in electronic systems troubleshooting and maintenance, 3 in computer literacy, 3 in solid state electronics, 3 in hydraulics and pneumatics, 3 in personnel supervision, 3 in maintenance management, and 2 in records and information management. In the upper-division baccalaureate category, 3 semester hours in organizational management and 3 in management problems (9/91).

MOS-27H-003

HAWK FIELD MAINTENANCE EQUIPMENT AND FIRING REPAIRER
((Hawk Firing Section Repairer))

27H10
27H20
27H30
27H40

Exhibit Dates: 1/90–Present.

Career Management Field: 27 (Land Combat and Air Defense System Direct and General Support Maintenance).

Description

Summary: Supervises or performs maintenance on the electrical, hydraulic, hydropneumatic, mechanical, and electromechanical portions of the improved Hawk system and the electronic systems of Hawk or improved Hawk launcher. *Skill Level 10:* Inspects, tests, and adjusts equipment to specific tolerances; determines deficiencies in electronic, electrical, hydraulic, hydropneumatic, mechanical, and electromechanical assemblies, subassemblies, and circuits with common and specialized test equipment including computer-based automatic test equipment; performs required maintenance; repairs unserviceable items by removing and replacing defective components and parts and determines their disposition; performs quality control measures; performs load testing, calibration, and repair of tools and equipment. *Skill Level 20:* Able to perform the duties required for Skill Level 10; provides technical guidance to subordinates; installs electrical, hydraulic, hydropneumatic, mechanical, and electromechanical equipment modifications; provides technical assistance to supported units; completes maintenance and supply forms. *Skill Level 30:* Able to perform the duties required for Skill Level 20; supervises subordinates in the maintenance of the improved Hawk missile and launching systems; establishes work loads and repair priorities; recommends procedures for receiving, storing, inspecting, testing, and repairing items; supervises inspection and maintenance teams; organizes and conducts on-the-job training programs; determines faulty work practices and demonstrates proper maintenance and troubleshooting techniques; establishes and keeps records; prepares maintenance reports. *Skill Level 40:* Able to perform the duties required for Skill Level 30; supervises and coordinates maintenance of Hawk or improved Hawk system, including launcher, radar, battery control, information coordination, simulator station, test equipment, and associated training equipment; serves as a section chief or principal noncommissioned officer of a platoon performing missile system support maintenance; supervises and inspects activities of subordinates engaged in missile system maintenance; supervises quality assurance and assists in its development; advises superiors on technical and personnel matters; assists in planning and performing on-the-job training.

Recommendation, Skill Level 10

In the lower-division baccalaureate/associate degree category, 3 semester hours in hydraulic systems, 3 in electromechanical systems maintenance, 3 in basic electronics, 3 in servomechanism systems, and 3 in computer literacy.

(NOTE: This recommendation for skill level 10 is valid for the dates 1/90-9/91 only) (9/91).

Recommendation, Skill Level 20

In the lower-division baccalaureate/associate degree category, 3 semester hours in hydraulic systems, 3 in electromechanical systems maintenance, 3 in basic electronics, 3 in servomechanism systems, 3 in computer literacy, and 3 in personnel supervision. (NOTE: This recommendation for skill level 20 is valid for the dates 1/90-2/95 only) (9/91).

Recommendation, Skill Level 30

In the lower-division baccalaureate/associate degree category, 3 semester hours in hydraulic systems, 3 in electromechanical systems maintenance, 3 in basic electronics, 3 in servomechanism systems, 3 in computer literacy, 3 in personnel supervision, 3 in maintenance management, and 2 in records and information management (9/91).

Recommendation, Skill Level 40

In the lower-division baccalaureate/associate degree category, 3 semester hours in hydraulic systems, 3 in electromechanical systems maintenance, 3 in basic electronics, 3 in servomechanism systems, 3 in computer literacy, 3 in personnel supervision, 3 in maintenance management, and 2 in records and information management. In the upper-division baccalaureate category, 3 semester hours in organizational management and 3 for field experience in management (9/91).

MOS-27J-001

Hawk Field Maintenance Equipment/
Pulse Acquisition Radar Repairer
27J10
27J20
27J30
27J40

Exhibit Dates: 1/90–6/95. Effective 6/95, MOS 27J was discontinued and its duties were incorporated into MOS 27H.

Career Management Field: 27 (Land Combat and Air Defense System Direct and General Support Maintenance).

Description

Summary: Supervises or performs maintenance on Hawk or improved Hawk missile pulse radars and related equipment. *Skill Level 10:* Performs maintenance, calibration, and minor modifications on pulse-acquisition and pulse-ranging radars under direct supervision; inspects and calibrates electronic systems and subsystems using digital voltmeters, oscilloscopes, signal generators, and automatic computer-based test equipment; performs initial and final checkout and inspection for the electronic subsystems of the radar system. *Skill Level 20:* Able to perform the duties required for Skill Level 10; provides technical guidance to subordinates; performs unsupervised maintenance on Hawk missile radar. *Skill Level 30:* Able to perform the duties required for Skill Level 20; supervises and provides technical assistance to subordinates; completes maintenance and supply forms. *Skill Level 40:* Able to perform the duties required for Skill Level 30; establishes and controls work loads; establishes and keeps records; prepares maintenance reports; performs supervisory duties as section chief.

Recommendation, Skill Level 10

In the lower-division baccalaureate/associate degree category, 3 semester hours in basic electronics, 3 in electronic systems troubleshooting and maintenance, 3 in computer systems troubleshooting and maintenance, and 3 in computer literacy. (NOTE: This recommendation for skill level 10 is valid for the dates 1/90-9/91 only) (9/91).

Recommendation, Skill Level 20

In the lower-division baccalaureate/associate degree category, 3 semester hours in basic electronics, 4 in electronic systems troubleshooting and maintenance, 3 in computer systems troubleshooting and maintenance, 3 in computer literacy, and 3 in personnel supervision. (NOTE: This recommendation for skill level 20 is valid for the dates 1/90-2/95 only) (9/91).

Recommendation, Skill Level 30

In the lower-division baccalaureate/associate degree category, 3 semester hours in basic electronics, 4 in electronic systems troubleshooting and maintenance, 3 in computer systems troubleshooting and maintenance, 3 in computer literacy, 3 in personnel supervision, 3 in maintenance management, and 2 in records and information management (9/91).

Recommendation, Skill Level 40

In the lower-division baccalaureate/associate degree category, 3 semester hours in basic electronics, 4 in electronic systems troubleshooting and maintenance, 3 in computer systems troubleshooting and maintenance, 3 in computer literacy, 3 in personnel supervision, 3 in maintenance management, and 2 in records and information management. In the upper-division baccalaureate category, 3 semester hours in organizational management and 3 for field experience in management (9/91).

MOS-27K-001

Hawk Fire Control/Continuous Wave
Radar Repairer
27K10
27K20
27K30
27K40

Exhibit Dates: 1/90–9/99. Effective 9/99 MOS 27K was discontinued.

Career Management Field: 27 (Land Combat and Air Defense System Direct and General Support Maintenance).

Description

Summary: Supervises or performs maintenance on continuous wave radars, high-power illumination radars, radar signal processing and information display systems, and associated digital data processing computers and test equipment. *Skill Level 10:* Performs maintenance on continuous wave and pulse radar systems and associated power supplies; inspects, tests, and adjusts radar system circuits to the module level; performs maintenance on video display terminals; inspects, tests, adjusts video data monitors to the module and component level; performs maintenance on signal processing systems composed of linear and digital circuits and microprocessor-based data processing circuits; inspects, tests, adjusts, and repairs signal and data processing systems to the module level; diagnoses digital computer circuits using test programs and subroutines. *Skill Level 20:* Able to perform the duties required for Skill Level 10; provides technical assistance, guidance, and training to subordinates; installs modifications to radar, signal processing, and data processing systems; initiates requests for repair parts, tools, test equipment, and other supplies. *Skill Level 30:* Able to perform the duties required for Skill Level 20; supervises subordinates; establishes work loads and repair priorities; recommends procedures for receipt, storage, inspection, testing, and repair of items; implements quality control measures; performs initial and final checkout and inspection of designated items and their assemblies and subassemblies; supervises inspection and maintenance teams; organizes and conducts on-the-job training programs; determines faulty work practices and demonstrates proper maintenance and troubleshooting techniques; establishes and keeps maintenance records; prepares maintenance reports. *Skill Level 40:* Able to perform the duties required for Skill Level 30; plans, organizes, and directs the activities of a maintenance section, including overall responsibility for manpower allocation, inventory control, training personnel, and records maintenance.

Recommendation, Skill Level 10

In the lower-division baccalaureate/associate degree category, 3 semester hours in computer systems and organization, 3 in computer systems troubleshooting and maintenance, 3 in radar systems, 3 in electronic systems troubleshooting and maintenance, and 3 in basic electronics. (NOTE: This recommendation for skill level 10 is valid for the dates 1/90-9/91 only) (9/91).

Recommendation, Skill Level 20

In the lower-division baccalaureate/associate degree category, 3 semester hours in computer systems and organization, 3 in computer systems troubleshooting and maintenance, 3 in radar systems, 3 in electronic systems troubleshooting and maintenance, 3 in basic electronics, and 3 in personnel supervision. (NOTE: This recommendation for skill level 20 is valid for the dates 1/90-2/95 only) (9/91).

Recommendation, Skill Level 30

In the lower-division baccalaureate/associate degree category, 3 semester hours in computer systems and organization, 3 in computer systems troubleshooting and maintenance, 3 in radar systems, 3 in electronic systems troubleshooting and maintenance, 3 in basic electronics, 3 in personnel supervision, 3 in maintenance management, and 2 in records and information management (9/91).

Recommendation, Skill Level 40

In the lower-division baccalaureate/associate degree category, 3 semester hours in computer systems and organization, 3 in computer systems troubleshooting and maintenance, 3 in radar systems, 3 in electronic systems troubleshooting and maintenance, 3 in basic electronics, 3 in personnel supervision, 3 in maintenance management, and 2 in records and information management. In the upper-division baccalaureate category, 3 semester hours in organizational management and 3 for field experience in management (9/91).

MOS-27L-001

Lance System Repairer
27L10
27L20
27L30

Exhibit Dates: 1/90–12/93. Effective 12/93, MOS 27L was discontinued.

Career Management Field: 27 (Land Combat and Air Defense System Intermediate Maintenance).

Description

Summary: Performs or supervises intermediate maintenance on the Lance missile system trainer, loader, transporter, and launcher. *Skill Level 10:* Inspects, tests, troubleshoots, and repairs electronic, electrical, hydraulic, mechanical, and electromechanical assemblies of the Lance system trainer, loader, transporter, and launcher and system test equipment. *Skill Level 20:* Able to perform the duties required for Skill Level 10; supervises field maintenance of Lance missile system. *Skill Level 30:* Able to perform the duties required for Skill Level 20; schedules work loads and establishes priorities; recommends procedures for receipt and storage of items; supervises inspection of items and on-the-job training; establishes and maintains maintenance records.

Recommendation, Skill Level 10

In the vocational certificate category, 3 semester hours in care and use of hand tools and test equipment. (NOTE: This recommendation for skill level 10 is valid for the dates 1/90-9/91 only) (4/86).

Recommendation, Skill Level 20

In the vocational certificate category, 3 semester hours in care and use of hand tools and test equipment. In the lower-division baccalaureate/associate degree category, 3 semester hours in troubleshooting and diagnostic procedures in system analysis and maintenance (4/86).

Recommendation, Skill Level 30

In the vocational certificate category, 3 semester hours in care and use of hand tools and test equipment. In the lower-division baccalaureate/associate degree category, 3 semester hours in troubleshooting and diagnostic procedures in system analysis and maintenance, and 2 in personnel supervision (4/86).

MOS-27M-001

Multiple Launch Rocket System (MLRS) Repairer
27M10
27M20
27M30

Exhibit Dates: 1/90–4/99.

Career Management Field: 27 (Land Combat and Air Defense System Intermediate Maintenance).

Description

Summary: Supervises or performs intermediate maintenance on Multiple Launch Rocket System self-propelled launcher loader, launcher pad/container trainer, and test support group. *Skill Level 10:* Troubleshoots electrical, electronic, and mechanical assemblies, modules, and interconnecting cables of the self-propelled launcher loader module; isolates malfunctions; repairs and replaces chassis-mounted components; prepares and maintains equipment logs, modification and utilization records, and calibration charts. *Skill Level 20:* Able to perform the duties required for Skill Level 10; provides technical assistance to support units; supervises field maintenance of MLRS system. *Skill Level 30:* Able to perform the duties required for Skill Level 20; schedules work loads and establishes priorities; recommends procedures for receipt and storage of items; supervises inspection of items; conducts on-the-job training; establishes and maintains maintenance records.

Recommendation, Skill Level 10

In the vocational certificate category, 3 semester hours in care and use of hand tools and test equipment. (NOTE: This recommendation for skill level 10 is valid for the dates 1/90-9/91 only) (4/86).

Recommendation, Skill Level 20

In the vocational certificate category, 3 semester hours in care and use of hand tools and test equipment. In the lower-division baccalaureate/associate degree category, 3 semester hours in troubleshooting and diagnostic procedures in system analysis and maintenance. (NOTE: This recommendation for skill level 20 is valid for the dates 1/90-2/95 only) (4/86).

Recommendation, Skill Level 30

In the vocational certificate category, 3 semester hours in care and use of hand tools and test equipment. In the lower-division baccalaureate/associate degree category, 3 semester hours in troubleshooting and diagnostic procedures in system analysis and maintenance and 2 in personnel supervision (4/86).

MOS-27M-002

Multiple Launch Rocket System (MLRS) Repairer
27M10
27M20
27M30
27M40

Exhibit Dates: 5/99–3/04. (Effective 4/04, this MOS was discontinued, and its duties were incorporated into MOS 35P)

Career Management Field: 27 (Land Combat and Air Defense System Intermediate Maintenance).

Description

Summary: Supervises or performs intermediate maintenance on Multiple Launch Rocket system, self-propelled launcher loader, launcher pad/container trainer, and test support group. *Skill Level 10:* Troubleshoots electrical, electronic, and mechanical assemblies, modules, and interconnecting cables for the launcher loader module; replaces and repairs electrical, hydraulic, and mechanical assemblies and cables; uses breakout boxes and test equipment to make proper repairs and calibrations; performs unit maintenance on system-specific test equipment. *Skill Level 20:* Able to perform the duties required for Skill Level 10; supervises subordinates and provides technical guidance; provides technical assistance to supported units. *Skill Level 30:* Able to perform the duties required for Skill Level 20; supervises inspection and maintenance teams; serves as unit quality control/quality assurance inspector; monitors activities associated with MLRS maintenance; identifies faulty work practices; organizes and conducts on-the-job training programs. *Skill Level 40:* Able to perform the duties required for Skill Level 30; supervises and coordinates support maintenance of MLRS systems.

Recommendation, Skill Level 10

Credit may be granted on the basis of an individualized assessment of the student (5/99).

Recommendation, Skill Level 20

Credit may be granted on the basis of an individualized assessment of the student (5/99).

Recommendation, Skill Level 30

In the lower-division baccalaureate/associate degree category, 1 semester hour in hand tools and test equipment, 3 in AC/DC circuits, 1 in hydraulic troubleshooting, 2 in map reading and land navigation, 1 in schematic reading, 3 in computer literacy, 3 in personnel management, and 2 in technical writing (5/99).

Recommendation, Skill Level 40

In the lower-division baccalaureate/associate degree category, 1 semester hour in hand tools and test equipment, 3 in AC/DC circuits, 1 in hydraulic troubleshooting, 2 in map reading and land navigation, 1 in schematic reading, 3 in computer literacy, 3 in personnel management, 3 in technical writing, 2 in automotive service, and 3 in records and information management. In the upper-division baccalaureate category, 3 semester hours for field experience in management and 3 in organizational management (5/99).

MOS-27N-001

Forward Area Alerting Radar (FAAR) Repairer
27N10
27N20
27N30

Exhibit Dates: 1/90–4/92. Effective 4/92, MOS 27N was discontinued.

Career Management Field: 27 (Land Combat and Air Defense System Intermediate Maintenance).

Description

Summary: Performs or supervises direct support and general support maintenance on the Forward Area Alerting Radar system, including radar, Target Alert Data Display Set (TADDS) and Interrogator Set (IFF). *Skill Level 10:* Performs support level maintenance on radar, TADDS, and IFF electronic assemblies and associated specialized test equipment, including organizational maintenance test set and support maintenance test set; troubleshoots assemblies, subassemblies, and modular and circuit elements; repairs, removes and/or replaces defective components and parts; inspects, tests, and adjusts FAAR system components; performs inspections; prepares and maintains equipment logs, modification and utilization records, exchange tags, and calibration data cards. *Skill Level 20:* Able to perform the duties required for Skill Level 10; provides technical guidance to subordinates; provides technical assistance to supported units. *Skill Level 30:* Able to perform the duties required for Skill Level 20; supervises subordinates; establishes work load and repair priorities; recommends procedures for receipt, storage, inspection, testing, and repair of FAAR system items; organizes and conducts on-the-job training programs; implements quality control measures.

Recommendation, Skill Level 10

In the vocational certificate category, 3 semester hours in use and care of hand tools and test equipment. (NOTE: This recommendation for skill level 10 is valid for the dates 1/90-9/91 only) (4/86).

Recommendation, Skill Level 20

In the vocational certificate category, 3 semester hours in use and care of hand tools and test equipment. In the lower-division baccalaureate/associate degree category, 3 semester hours in troubleshooting and diagnostic techniques in system analysis and maintenance and 3 in electronic circuits (4/86).

Recommendation, Skill Level 30

In the vocational certificate category, 3 semester hours in use and care of hand tools and test equipment. In the lower-division baccalaureate/associate degree category, 3 semester hours in troubleshooting and diagnostic techniques in system analysis and maintenance, 3 in electronic circuits, and 2 in personnel supervision (4/86).

MOS-27T-001

AVENGER SYSTEM REPAIRER
((Pedestal Mounted Stinger, Line of Sight-Rear Air Defense Systems Repairer))
27T10
27T20
27T30
27T40

Exhibit Dates: 3/90–3/04. Effective 4/04, MOS 27T was discontinued and its duties were transferred to MOS 35T, Avenger System Repairer

Career Management Field: 27 (Land Combat and Air Defense System Intermediate Maintenance).

Description

Summary: Supervises or performs maintenance on a Pedestal Mounted Stinger/Avenger and associated components (less carrier and communications). *Skill Level 10:* Troubleshoots electrical, electronic, mechanical assemblies, and interconnecting cables; isolates malfunctions; repairs and replaces chassis-mounted components. *Skill Level 20:* Able to perform the duties required for Skill Level 10; supervises subordinates; provides technical assistance to support units. *Skill Level 30:* Able to perform the duties required for Skill Level 20; establishes work loads and repair priorities; supervises inspection and maintenance teams; conducts on-the-job training; demonstrates proper maintenance and troubleshooting techniques; establishes and manages maintenance records; prepares maintenance reports. *Skill Level 40:* Able to perform the duties required for Skill Level 30; supervises and coordinates maintenance activities; supervises calibration of system electronic equipment; performs inspections and ensures quality control; assists in planning, organizing, directing, supervising, training, coordinating, and reporting activities of subordinate sections.

Recommendation, Skill Level 10

In the lower-division baccalaureate/associate degree category, 3 semester hours in basic electricity and 3 in military science. (NOTE: This recommendation for skill level 10 is valid for the dates 3/90-9/91 only) (3/92).

Recommendation, Skill Level 20

In the lower-division baccalaureate/associate degree category, 3 semester hours in basic electricity, 3 in military science, and 3 in personnel supervision. (NOTE: This recommendation for skill level 20 is valid for the dates 3/90-2/95 only) (3/92).

Recommendation, Skill Level 30

In the lower-division baccalaureate/associate degree category, 3 semester hours in basic electricity, 3 in military science, 3 in personnel supervision, and 2 in records and information management (3/92).

Recommendation, Skill Level 40

In the lower-division baccalaureate/associate degree category, 3 semester hours in basic electricity, 3 in military science, 3 in personnel supervision, and 2 in records and information management. In the upper-division baccalaureate category, 3 semester hours for field experience in management and 3 in organizational management (3/92).

MOS-27V-001

HAWK MAINTENANCE CHIEF
27V50

Exhibit Dates: 1/90–6/94. Effective 6/94, MOS 27V was discontinued and its duties were incorporated into MOS 27Z, Land Combat/Air Defense Systems Maintenance Chief.

Career Management Field: 27 (Land Combat and Air Defense System Direct and General Support Maintenance).

Description

Provides technical management for Hawk missile unit; serves as maintenance chief or first sergeant in support of platoon at battalion or higher level; supervises and coordinates maintenance activities; serves as senior noncommissioned officer in logistical support activities; establishes maintenance schedules, training programs, and production and control schedules; communicates with supervisors and subordinates using verbal, visual, and graphics presentations. NOTE: May have progressed to 27V50 from 23R50 (Hawk Missile Systems Mechanic), 24R50 (Hawk Master Mechanic), 24H40 (Hawk Fire Control Repairer), 24K40 (Hawk Continuous Wave Radar Repairer), 27H40 (Hawk Firing Section Repairer), 27J40 (Hawk Field Maintenance Equipment/Pulse Acquisition Radar), 27K40 (Hawk Fire Control/Continuous Wave Radar Repairer), or 27X50 (Patriot System Mechanic).

Recommendation

In the lower-division baccalaureate/associate degree category, 3 semester hours in personnel supervision, 3 in records and information management, 3 in technical writing, and 3 in computer literacy. In the upper-division baccalaureate category, 6 semester hours for field experience in management and 3 in organizational management; if individual has attained paygrade E-9, additional credit as follows: 3 semester hours in management problems, 3 in operations management, and 3 in communication techniques for managers (9/91).

MOS-27X-001

PATRIOT SYSTEM REPAIRER
27X20
27X30
27X40
27X50

Exhibit Dates: 3/90–2/95.

Career Management Field: 27 (Land Combat and Air Defense System Direct and General Support Maintenance).

Description

Summary: Supervises and performs direct, general, and depot maintenance support on Patriot missile system, associated equipment, and trainers. *Skill Level 20:* Performs direct, general, and depot support maintenance on engagement control station, phased-array radar, identification friend or foe equipment, antenna, and the communications relay group; installs software modifications and develops specialized computer software test procedures; analyzes and interprets Patriot diagnostic test results; maintains VHF data links; performs some fiber optics repair. *Skill Level 30:* Able to perform the duties required for Skill Level 20; performs initial and final checkout and inspection of designated systems and their assemblies and subassemblies; serves as technical inspector of quality assurance section. *Skill Level 40:* Able to perform the duties required for Skill Level 30; supervises preparation of maintenance forms and reports; supervises organizational maintenance in an operational readiness float section. *Skill Level 50:* Able to perform the duties required for Skill Level 40; supervises direct, general, and depot-level maintenance performed on the missile system.

Recommendation, Skill Level 20

In the lower-division baccalaureate/associate degree category, 3 semester hours in radar systems, 3 in computer systems troubleshooting and maintenance, 3 in digital principles, 3 in basic electronics, 3 in electronic systems troubleshooting and maintenance, and 3 in personnel supervision (9/91).

Recommendation, Skill Level 30

In the lower-division baccalaureate/associate degree category, 3 semester hours in radar systems, 4 in computer systems troubleshooting and maintenance, 3 in digital principles, 3 in basic electronics, 3 in electronic systems troubleshooting and maintenance, 3 in electronic communications, 3 in personnel supervision, 3 in maintenance management, and 2 in records and information management (9/91).

Recommendation, Skill Level 40

In the lower-division baccalaureate/associate degree category, 3 semester hours in radar systems, 4 in computer systems troubleshooting and maintenance, 3 in digital principles, 3 in basic electronics, 3 in electronic systems troubleshooting and maintenance, 3 in electronic communications, 3 in personnel supervision, 3 in maintenance management, and 2 in records and information management. In the upper-division baccalaureate category, 3 semester hours in organizational management and 3 for field experience in management (9/91).

Recommendation, Skill Level 50

In the lower-division baccalaureate/associate degree category, 3 semester hours in radar systems, 4 in computer systems troubleshooting and maintenance, 3 in digital principles, 3 in basic electronics, 3 in electronic systems troubleshooting and maintenance, 3 in electronic communications, 3 in personnel supervision, 3 in maintenance management, and 2 in records and information management. In the upper-division baccalaureate category, 3 semester hours in organizational management and 6 for field experience in management; if individual has attained paygrade E-9, additional credit as follows: 3 semester hours in management problems, 3 in operations management, and 3 in communication techniques for managers (9/91).

MOS-27X-002

Patriot System Repairer
27X20
27X30
27X40
27X50

Exhibit Dates: 3/95–5/01.

Career Management Field: 27 (Land Combat and Air Defense System Direct and General Support Maintenance).

Description

Summary: Supervises and performs direct, general, and depot maintenance support on Patriot missile system, associated equipment, and trainers. *Skill Level 20:* Performs direct, general, and depot support maintenance on engagement control station, phased-array radar, identification friend or foe equipment, antenna, and the communications relay group; installs software modifications and develops specialized computer software test procedures; analyzes and interprets Patriot diagnostic test results; maintains VHF data links; performs some fiber optics repair. *Skill Level 30:* Able to perform the duties required for Skill Level 20; performs initial and final checkout and inspection of designated systems and their assemblies and subassemblies; serves as technical inspector of quality assurance section. *Skill Level 40:* Able to perform the duties required for Skill Level 30; supervises preparation of maintenance forms and reports; supervises organizational maintenance in an operational readiness float section. *Skill Level 50:* Able to perform the duties required for Skill Level 40; supervises direct, general, and depot-level maintenance performed on the missile system.

Recommendation, Skill Level 20

Credit may be granted on the basis of an individualized assessment of the student (9/91).

Recommendation, Skill Level 30

In the lower-division baccalaureate/associate degree category, 3 semester hours in radar systems, 3 in computer systems troubleshooting and maintenance, 3 in digital principles, 3 in basic electronics, 3 in electronic systems troubleshooting and maintenance, 3 in electronic communications, 3 in personnel supervision, 3 in maintenance management, and 2 in records and information management (9/91).

Recommendation, Skill Level 40

In the lower-division baccalaureate/associate degree category, 3 semester hours in radar systems, 4 in computer systems troubleshooting and maintenance, 3 in digital principles, 3 in basic electronics, 3 in electronic systems troubleshooting and maintenance, 3 in electronic communications, 3 in personnel supervision, 3 in maintenance management, and 2 in records and information management. In the upper-division baccalaureate category, 3 semester hours in organizational management and 3 for field experience in management (9/91).

Recommendation, Skill Level 50

In the lower-division baccalaureate/associate degree category, 3 semester hours in radar systems, 4 in computer systems troubleshooting and maintenance, 3 in digital principles, 3 in basic electronics, 3 in electronic systems troubleshooting and maintenance, 3 in electronic communications, 3 in personnel supervision, 3 in maintenance management, and 2 in records and information management. In the upper-division baccalaureate category, 3 semester hours in organizational management and 6 for field experience in management; if individual has attained paygrade E-9, additional credit as follows: 3 semester hours in management problems, 3 in operations management, and 3 in communication techniques for managers (9/91).

MOS-27X-003

Patriot System Repairer
27X20
27X30
27X40
27X50

Exhibit Dates: 6/01–9/03. Effective 10/03, MOS 27X was discontinued and its duties were transferred to MOS 35S, Patriot System Repairer.

Career Management Field: 27 (Land Combat and Air Defense System Direct and General Support Maintenance).

Description

Summary: Supervises and performs direct, general, and depot maintenance support on Patriot missile system, associated equipment, and trainers. *Skill Level 20:* Performs direct, general, and depot support maintenance on engagement control station, phased-array radar, identification friend or foe equipment, antenna, and the communications relay group; installs software modifications and develops specialized computer software test procedures; analyzes and interprets Patriot diagnostic test results; maintains VHF data links; performs some fiber-optics repair. *Skill Level 30:* Able to perform the duties required for Skill Level 20; performs initial and final checkout and inspection of designated systems and their assemblies and subassemblies; serves as technical inspector of quality assurance section. *Skill Level 40:* Able to perform the duties required for Skill Level 30; supervises preparation of maintenance forms and reports; supervises organizational maintenance in an operational readiness float section. *Skill Level 50:* Able to perform the duties required for Skill Level 40; supervises direct, general, and depot-level maintenance performed on the missile system.

Recommendation, Skill Level 20

Credit may be granted on the basis of an individualized assessment of the student (6/01).

Recommendation, Skill Level 30

In the lower-division baccalaureate/associate degree category, 2 semester hours in radar systems, 3 in electronic systems troubleshooting and maintenance, 3 in digital electronics, 3 in basic electronics, 3 in basic electricity, 2 in electronic communications, 3 in principles of supervision, and 3 in maintenance management (6/01).

Recommendation, Skill Level 40

In the lower-division baccalaureate/associate degree category, 2 semester hours in radar systems, 3 in electronic systems troubleshooting and maintenance, 3 in digital electronics, 3 in basic electronics, 3 in basic electricity, 2 in electronic communications, 4 in principles of supervision, and 4 in maintenance management. In the upper-division baccalaureate category, 3 semester hours in human resource management and 3 for field experience in management (6/01).

Recommendation, Skill Level 50

In the lower-division baccalaureate/associate degree category, 2 semester hours in radar systems, 3 in electronic systems troubleshooting and maintenance, 3 in digital electronics, 3 in basic electronics, 3 in basic electricity, 2 in electronic communications, 4 in principles of supervision, and 4 in maintenance management. In the upper-division baccalaureate category, 3 semester hours in human resource management, 6 for field experience in management, and 3 in operations management (6/01).

MOS-27Z-001

Land Combat Support System (LCCS)
Missile Maintenance Chief
27Z50

Exhibit Dates: 1/90–2/91.

Career Management Field: 27 (Combat Missile Maintenance), subfield 272 (Combat Missile Repair).

Description

Coordinates support maintenance of the LCCS, including the Lance, TOW, Dragon, Chaparral, Shillelagh, and Redeye missile systems, Vulcan weapon system electronics, Forward Area Alerting Radar, and system-associated trainers and test equipment; helps determine and administers policy; assists superiors in planning and conducting training; advises superiors on all matters concerning enlisted personnel; may serve as a chief instructor in a training facility. NOTE: May have progressed to 27Z50 from 27D40 (Land Combat Support System Test Specialist), 27D40 (Lance Missile System Repairman), 27E40 (Wire-Guided Missile System Repairman), 27F40 (Chaparral/Vulcan Defense System Repairman), 27G40 (Redeye Missile System Repairman), 27H40 (Shillelagh Missile System Repairman), 24M40 (Vulcan System Mechanic), or 24N40 (Chaparral System Mechanic).

Recommendation

In the lower-division baccalaureate/associate degree category, credit in electronics on the basis of institutional evaluation. In the upper-division baccalaureate category, 3 semester hours in introduction to management, 3 in personnel management, 3 in human relations, 6 for field experience in management, and additional credit in industrial arts education on the basis of institutional evaluation; if the duty assignment was chief instructor in a formal training facility, 3 additional semester hours for an internship in education (2/76).

MOS-27Z-002

Missile Systems Maintenance Chief
((Land Combat/Air Defense Systems Maintenance Chief))
27Z50

Exhibit Dates: 3/91–5/01.

Career Management Field: 27 (Land Combat and Air Defense System Direct and General Support Maintenance).

Description

Coordinates support maintenance of the Land Combat Support System (LCSS), including the Lance, TOW, Dragon, Chaparral, Shillelagh, and Redeye missile systems, Multiple Launch Rocket System (MLRS), Vulcan weapon system electronics, Forward Area

Alerting Radar (FAAR), and system-associated trainers and test equipment; uses military-specific software for inventory control, production control and integrated automation system; continually updates and validates maintenance records; helps determine and administers policy; assists superiors in planning and accomplishing training; advises superiors on all matters concerning enlisted personnel; may serve as a chief instructor in a training facility. NOTE: May have progressed to 27Z50 from 21L50 (Pershing Electronics Repairer), 24M40 (Vulcan System Mechanic), 24N40 (Chaparral System Mechanic), 27B40 (Land Combat Electrical-Mechanical Repairer).

Recommendation

In the lower-division baccalaureate/associate degree category, 3 semester hours in personnel supervision, 3 in records and information management, 3 in computer literacy, and 3 in maintenance management. In the upper-division baccalaureate category, 3 semester hours in organizational management and 6 for field experience in management; if individual has attained paygrade E-9, additional credit as follows: 3 semester hours in management problems, 3 in operations management, and 3 in communication techniques for managers (9/91).

MOS-27Z-003

Missile Systems Maintenance Chief

27Z50

Exhibit Dates: 6/01–3/04. (Effective 4/04, this MOS was discontinued, and its duties were incorporated into MOS 35V.)

Career Management Field: 35 (Electronic Maintenance and Calibration).

Description

Coordinates support maintenance of the Land Combat Support System (LCSS), including the AVENGER, ATACMS, Bradley Fighting Vehicle System, JAVELIN, TOW, and DRAGON missile systems, Multiple Launch Rocket System (MLRS), and system-associated trainers and test equipment; uses military-specific software for inventory control, production control and integrated automation system; continually updates and validates maintenance records; helps determine and administers policy; assists superiors in planning and accomplishing training; advises superiors on all matters concerning enlisted personnel; may serve as a chief instructor in a training facility. NOTE: May have progressed to 27Z50 from 27G40 (Chaparral and Redeye Repairer); 27E40 (Land Combat Electronic Missile System Repairer); 27M40 (Multiple Launch Rocket System (MLRS) Repairer); or 27T40 (Avenger Systems Repairer).

Recommendation

In the lower-division baccalaureate/associate degree category, 3 semester hours in principles of supervision, 3 in records and information management, 3 in computer applications, and 3 in maintenance management. In the upper-division baccalaureate category, 3 semester hours in human resource management, 6 for field experience in management, and 3 in operations management (6/01).

MOS-29E-001

Radio Repairer

29E10

29E20

29E30

Exhibit Dates: 1/90–1/94.

Career Management Field: 29 (Communications-Electronics Maintenance).

Description

Summary: Installs and maintains radio receiver s, transmitters, and associated equipment. *Skill Level 10:* Uses oscilloscopes, multimeters, radio frequency generators, and spectrum analyzers to troubleshoot equipment; troubleshoots and repairs equipment to the component level; uses logical troubleshooting methods to locate faults. *Skill Level 20:* Able to perform the duties required for Skill Level 10; assists and instructs subordinates; keeps detailed records of maintenance history on equipment; follows formal maintenance procedures. *Skill Level 30:* Able to perform the duties required for Skill Level 20; assigns duties to subordinates; establishes work load, work schedules, and priorities; responsible for quality control inspections; evaluates and counsels subordinates; maintains technical library; initiates and maintains forms and records; responsible for records pertaining to repair parts; prepares and reviews administrative reports.

Recommendation, Skill Level 10

In the lower-division baccalaureate/associate degree category, 3 semester hours in introduction to electronics, 2 in electronic troubleshooting and repair, and 2 in radio communications. (NOTE: This recommendation for skill level 10 is valid for the dates 1/90-9/91 only) (7/87).

Recommendation, Skill Level 20

In the lower-division baccalaureate/associate degree category, 3 semester hours in introduction to electronics, 4 in electronic troubleshooting and repair, 2 in radio communications, and 2 in record keeping (7/87).

Recommendation, Skill Level 30

In the lower-division baccalaureate/associate degree category, 3 semester hours in introduction to electronics, 4 in electronic troubleshooting and repair, 2 in radio communications, 2 in record keeping, 3 in maintenance management, and 3 in principles of supervision (7/87).

MOS-29E-002

Radio Repairer

29E10

29E20

29E30

Exhibit Dates: 2/94–6/95. Effective 6/95, MOS 29E was discontinued and its duties were incorporated into MOS 35E.

Career Management Field: 29 (Signal Maintenance).

Description

Summary: Installs and maintains radio receivers, transmitters, and associated equipment. *Skill Level 10:* Uses oscilloscopes, multimeters, radio frequency generators, and spectrum analyzers to troubleshoot equipment; troubleshoots and repairs equipment to the component level; uses logical troubleshooting methods to locate faults; performs tests to verify operability of repaired equipment. *Skill Level 20:* Able to perform the duties required for Skill Level 10; assists and instructs subordinates; keeps detailed records of equipment maintenance; performs final quality control inspection of repaired equipment. *Skill Level 30:* Able to perform the duties required for Skill Level 20; assigns duties; establishes work load, work schedules, and priorities; responsible for quality control inspections; evaluates and counsels subordinates; maintains technical library; initiates and maintains forms and records; responsible for records pertaining to repair parts; prepares and reviews administrative reports; develops operational procedures for service shop.

Recommendation, Skill Level 10

Credit may be granted on the basis of an individualized assessment of the student (3/95).

Recommendation, Skill Level 20

In the lower-division baccalaureate/associate degree category, 3 semester hours in basic electronics, 4 in electronic systems troubleshooting and repair, 2 in radio communications, and 2 in record keeping. (NOTE: This recommendation for skill level 20 is valid for the dates 2/94-2/95 only) (3/95).

Recommendation, Skill Level 30

In the lower-division baccalaureate/associate degree category, 3 semester hours in basic electronics, 4 in electronic systems troubleshooting and repair, 2 in radio communications, 2 in record keeping, 3 in maintenance management, and 3 in personnel supervision (3/95).

MOS-29F-001

Fixed Communications Security Equipment Repairer

29F10

29F20

29F30

Exhibit Dates: 1/90–3/91. Effective 3/91, MOS 29F was discontinued and its duties were incorporated into MOS 29S, Field Communications Security Equipment Repairer.

Career Management Field: 29 (Communications-Electronics Maintenance).

Description

Summary: Inspects, tests, and performs maintenance on fixed communications security (COMSEC) equipment. *Skill Level 10:* Operates conventional and specialized test equipment; performs repairs on equipment or system by removing and replacing defective equipment; installs and tests equipment; interprets system diagrams, troubleshooting charts, and technical publications; keeps maintenance records; individuals holding the following Additional Skill Identifiers (ASIs) perform maintenance at the component level: ASI G7, COMSEC Equipment Full Maintenance (Field); ASI G8, COMSEC Equipment Full Maintenance (Fixed); ASI X8, COMSEC System Maintenance Specialized Repair Activity (SRA); and ASI X9, General COMSEC Maintenance Specialized Repair Activity (SRA). *Skill Level 20:* Able to perform the duties required for Skill Level 10; provides technical guidance and procedural assistance to subordinates; prepares and maintains repair parts stock records. *Skill Level 30:* Able to perform the duties required for Skill Level 20; assigns duties to subordinates; establishes work load, work schedules, and maintenance priorities; responsible for quality control inspections; evaluates and counsels subordinate personnel;

provides technical assistance to subordinates; maintains technical library; maintains forms and records; responsible for records pertaining to repair parts; prepares and reviews technical and administrative reports.

Recommendation, Skill Level 10

In the lower-division baccalaureate/associate degree category, 3 semester hours in introduction to electronics and 1 in electronic troubleshooting and repair. If ASI G7, G8, X8, or X9 is held, 2 additional semester hours in electronic troubleshooting and repair (7/87).

Recommendation, Skill Level 20

In the lower-division baccalaureate/associate degree category, 3 semester hours in introduction to electronics, 2 in electronic troubleshooting and repair, and 2 in record keeping. If ASI G7, G8, X8, or X9 is held, 2 additional semester hours in electronic troubleshooting and repair (7/87).

Recommendation, Skill Level 30

In the lower-division baccalaureate/associate degree category, 3 semester hours in introduction to electronics, 2 in electronic troubleshooting and repair, 2 in record keeping, 3 in maintenance management, and 3 in principles of supervision. If ASI G7, G8, X8, or X9 is held, 2 additional semester hours in electronic troubleshooting and repair (7/87).

MOS-29J-001

Teletypewriter Equipment Repairer
29J10
29J20
29J30

Exhibit Dates: 1/90–1/94.

Career Management Field: 29 (Communications-Electronics Maintenance).

Description

Summary: Performs maintenance on electromechanical and electronic teletypewriter and facsimile equipment; individuals holding Additional Skill Identifier (ASI) T2, Teletypewriter Maintenance (MOD 40 TTY) also repair microcomputer systems. *Skill Level 10:* Installs, maintains, and repairs electrical and mechanical office machines at the module level; uses basic hand tools; reads elementary electrical and electronic diagrams; performs detailed testing to discover and diagnose causes of equipment malfunction. *Skill Level 20:* Able to perform the duties required for Skill Level 10; aligns electrical, electronic, and mechanical components as prescribed in applicable technical publications; reassembles and operates repaired equipment to ensure conformity with technical specifications; uses proper techniques to maintain, localize, and isolate electrical, electronic, and mechanical causes of equipment malfunction; prepares and keeps maintenance forms and records. *Skill Level 30:* Able to perform the duties required for Skill Level 20; supervises four or more subordinates; able to diagnose equipment malfunctions and repair requirements; conducts on-the-job training; organizes and supervises maintenance and inspection teams; requisitions and stocks supplies and repair parts; maintains equipment forms and records; plans and schedules equipment maintenance.

Recommendation, Skill Level 10

In the lower-division baccalaureate/associate degree category, 2 semester hours in electronic troubleshooting and repair and 3 in introduction to electronics. If ASI T2 is held, 3 semester hours in microcomputer systems repair. (NOTE: This recommendation for skill level 10 is valid for the dates 1/90-9/91 only) (7/87).

Recommendation, Skill Level 20

In the lower-division baccalaureate/associate degree category, 4 semester hours in electronic troubleshooting and repair, 3 in introduction to electronics, and 2 in record keeping. If ASI T2 is held, 3 semester hours in microcomputer systems repair (7/87).

Recommendation, Skill Level 30

In the lower-division baccalaureate/associate degree category, 4 semester hours in electronic troubleshooting and repair, 3 in introduction to electronics, 2 in record keeping, 3 in personnel supervision, and 3 in maintenance management. IF ASI T2 is held, 3 semester hours in microcomputer systems repair (7/87).

MOS-29J-002

Telecommunications Terminal Device Repairer
29J10
29J20
29J30
29J40

Exhibit Dates: 2/94–6/95. MOS 29J was discontinued from the Army MOS system before ACE was able to conduct an evaluation. It will not be evaluated for this time period.

Description ***Skill Level 10:*** *Skill Level 20: Skill Level 30: Skill Level 40:*

MOS-29M-001

Tactical Satellite/Microwave Repairer
29M10
29M20
29M30

Exhibit Dates: 1/90–12/91. Effective 12/91, MOS 29M was discontinued, and its duties were incorporated into MOS 29V, Strategic Microwave Systems Repairer.

Career Management Field: 29 (Communications-Electronics Maintenance).

Description

Summary: Supervises and performs maintenance on tactical satellite, microwave, multichannel radio, multiplexer and associated equipment. *Skill Level 10:* Operates vehicles and power generation equipment; inspects and repairs equipment using test equipment and procedures found in technical manuals to determine cause of malfunction and corrective action required; tests repaired item to insure proper operation; performs preventive maintenance checks; records maintenance activities on appropriate forms. *Skill Level 20:* Able to perform the duties required for Skill Level 10; performs complex maintenance duties; provides technical and procedural assistance to subordinates; requests and maintains authorized repair parts stock, supplies, and technical manuals. *Skill Level 30:* Able to perform the duties required for Skill Level 20; assigns duties to subordinates; establishes work load, work schedules, and priorities; responsible for quality control inspections; evaluates, counsels, and provides technical assistance to subordinates; maintains technical library; initiates and maintains forms and records; responsible for records pertaining to repair parts; prepares and reviews administrative reports; performs more complex troubleshooting.

Recommendation, Skill Level 10

In the lower-division baccalaureate/associate degree category, 3 semester hours in introduction to electronics, 3 in microwave communications systems, and 2 in electronic troubleshooting and repair. (NOTE: This recommendation for skill level 10 is valid for the dates 1/90-9/91 only) (7/87).

Recommendation, Skill Level 20

In the lower-division baccalaureate/associate degree category, 3 semester hours in introduction to electronics, 3 in microwave communications systems, 4 in electronic troubleshooting and repair, and 2 in record keeping (7/87).

Recommendation, Skill Level 30

In the lower-division baccalaureate/associate degree category, 3 semester hours in introduction to electronics, 3 in microwave communications systems, 6 in electronic troubleshooting and repair, 2 in record keeping, 3 in principles of supervision, and 3 in maintenance management (7/87).

MOS-29N-001

Switching Central Repairer
29N10
29N20
29N30

Exhibit Dates: 1/90–1/94.

Career Management Field: 29 (Communications-Electronics Maintenance).

Description

Summary: Performs and supervises maintenance on manual and semiautomatic and selected transportable automatic electronic telephone central office systems. *Skill Level 10:* Maintains relay-operated central office systems and electronic switching systems; uses test equipment including oscilloscopes and multimeters. *Skill Level 20:* Able to perform the duties required for Skill Level 10; serves as a crew chief; schedules tasks and maintains maintenance forms and records. *Skill Level 30:* Able to perform the duties required for Skill Level 20; assigns duties to subordinates; establishes work load, work schedules, and priorities; provides technical assistance to subordinates; maintains technical library; initiates and maintains forms and records; responsible for records pertaining to repair parts; prepares and reviews administrative reports.

Recommendation, Skill Level 10

In the vocational certificate category, 6 semester hours in telephone equipment and repair. In the lower-division baccalaureate/associate degree category, 2 semester hours in introduction to electronics and 2 in electronic troubleshooting and repair. (NOTE: This recommendation for skill level 10 is valid for the dates 1/90-9/91 only) (7/87).

Recommendation, Skill Level 20

In the vocational certificate category, 6 semester hours in telephone equipment and repair. In the lower-division baccalaureate/associate degree category, 3 semester hours in introduction to electronics, 4 in electronic troubleshooting and repair, and 2 in record keeping (7/87).

Recommendation, Skill Level 30

In the vocational certificate category, 6 semester hours in telephone equipment and repair. In the lower-division baccalaureate/associate degree category, 3 semester hours in introduction to electronics, 4 in electronic troubleshooting and repair, 2 in record keeping, 3 in principles of supervision, and 3 in maintenance management (7/87).

MOS-29N-002

SWITCHING CENTRAL REPAIRER
29N10
29N20
29N30

Exhibit Dates: 2/94–2/95.

Career Management Field: 29 (Signal Maintenance).

Description

Summary: Performs and supervises maintenance on manual, semiautomatic, and selected transportable automatic electronic telephone central office systems using schematics and test equipment. *Skill Level 10:* Maintains relay-operated central office systems and electronic switching systems; uses test equipment including oscilloscopes and multimeters; uses circuit and wiring diagrams and schematics. *Skill Level 20:* Able to perform the duties required for Skill Level 10; serves as a crew chief; schedules tasks and maintains maintenance forms and records; conducts quality control inspections on repaired equipment. *Skill Level 30:* Able to perform the duties required for Skill Level 20; assigns specific duties to subordinates; establishes work load, work schedules, and priorities; provides technical assistance to subordinates; maintains technical library; initiates and maintains forms and records; responsible for records pertaining to repair parts; prepares and reviews administrative reports; supervises quality control inspections.

Recommendation, Skill Level 10

Credit may be granted on the basis of an individualized assessment of the student (3/95).

Recommendation, Skill Level 20

In the lower-division baccalaureate/associate degree category, 3 semester hours in introduction to electricity and electronics, 2 in electronic equipment repair, and 1 in maintenance record keeping (3/95).

Recommendation, Skill Level 30

In the lower-division baccalaureate/associate degree category, 3 semester hours in introduction to electricity and electronics, 3 in electronic equipment repair, 2 in maintenance record keeping, 2 in personnel supervision, and 2 in maintenance management (3/95).

MOS-29N-003

SWITCHING CENTRAL REPAIRER
29N10
29N20
29N30

Exhibit Dates: 3/95–6/95. Effective 6/95, MOS 29N was discontinued, and its duties were incorporated into MOS 35N.

Career Management Field: 29 (Signal Maintenance).

Description

Summary: Performs and supervises maintenance on manual, semiautomatic, and selected transportable automatic electronic telephone central office systems using schematics and test equipment. *Skill Level 10:* Maintains relay-operated central office systems and electronic switching systems; uses test equipment including oscilloscopes and multimeters; uses circuit and wiring diagrams and schematics. *Skill Level 20:* Able to perform the duties required for Skill Level 10; serves as a crew chief; schedules tasks and maintains maintenance forms and records; conducts quality control inspections on repaired equipment. *Skill Level 30:* Able to perform the duties required for Skill Level 20; assigns specific duties to subordinates; establishes work load, work schedules, and priorities; provides technical assistance to subordinates; maintains technical library; initiates and maintains forms and records; responsible for records pertaining to repair parts; prepares and reviews administrative reports; supervises quality control inspections.

Recommendation, Skill Level 10

Credit may be granted on the basis of an individualized assessment of the student (3/95).

Recommendation, Skill Level 20

Credit may be granted on the basis of an individualized assessment of the student (3/95).

Recommendation, Skill Level 30

In the lower-division baccalaureate/associate degree category, 3 semester hours in introduction to electricity and electronics, 2 in electronic equipment repair, 1 in maintenance record keeping, 2 in personnel supervision, and 2 in maintenance management (3/95).

MOS-29P-001

COMMUNICATIONS SECURITY MAINTENANCE CHIEF
29P40

Exhibit Dates: 1/90–3/91. Effective 3/91, MOS 29P was discontinued, and its duties were incorporated into MOS 29S, Field Communications Security Equipment Repairer.

Career Management Field: 29 (Communications-Electronics Maintenance).

Description

Supervises the maintenance and repair of communications security (COMSEC) equipment.

May have progressed to 29P40 from 29F30, Fixed Communications Security Equipment Repairer, or 29S30, Field Communications Security Equipment Repairer; advises personnel, supervisors, and commanders on the operational matters dealing with COMSEC maintenance and supply; performs inspection of COMSEC accounting procedures; plans and supervises on-the-job training programs; evaluates and counsels subordinate personnel; supervises and coordinates maintenance and/or logistics support for COMSEC material.

Recommendation

In the upper-division baccalaureate category, 3 semester hours for field experience in management (7/87).

MOS-29S-001

FIELD COMMUNICATIONS SECURITY EQUIPMENT REPAIRER
29S10
29S20
29S30
29S40

Exhibit Dates: 1/90–2/91.

Career Management Field: 29 (Communications-Electronics Maintenance).

Description

Summary: Performs maintenance on field cryptographic equipment. *Skill Level 10:* Uses oscilloscopes and multimeters to troubleshoot cryptographic equipment; troubleshoots and repairs equipment to the board level; uses logical troubleshooting methods to locate faults; individuals holding the following Additional Skill Identifiers (ASIs) perform maintenance at the component level: ASI G7, COMSEC Equipment Full Maintenance (Field); ASI G8, COMSEC Equipment Full Maintenance (Fixed); ASI X8, COMSEC System Maintenance Specialized Repair Activity; and ASI X9, General COMSEC Maintenance Specialized Repair Activity. *Skill Level 20:* Able to perform the duties required for Skill Level 10; assists and instructs subordinates in equipment maintenance; keeps detailed records of maintenance history on equipment; follows formal maintenance system. *Skill Level 30:* Able to perform the duties required for Skill Level 20; assigns duties to subordinates; establishes work load, work schedules, and priorities; responsible for quality control inspections; evaluates, counsels, and provides technical assistance to subordinates; maintains technical library; initiates and maintains forms and records; responsible for records pertaining to repair parts; prepares and reviews administrative reports. *Skill Level 40:* Able to perform the duties required for Skill Level 30; supervises and provides technical guidance to subordinate personnel; supervises and organizes mobile maintenance inspection teams; advises commander on all matters pertaining to repair operations; plans and supervises on-the-job training programs.

Recommendation, Skill Level 10

In the lower-division baccalaureate/associate degree category, 3 semester hours in introduction to electronics and 1 in electronic troubleshooting and repair. If ASI G7, G8, X8, or X9 is held, 2 additional semester hours in electronic troubleshooting and repair (7/87).

Recommendation, Skill Level 20

In the lower-division baccalaureate/associate degree category, 3 semester hours in introduction to electronics, 2 in electronic troubleshooting and repair, and 2 in record keeping. If ASI G7, G8, X8, or X9 is held, 2 additional semester hours in electronic troubleshooting and repair (7/87).

Recommendation, Skill Level 30

In the lower-division baccalaureate/associate degree category, 3 semester hours in introduction to electronics, 2 in electronic troubleshooting and repair, 2 in record keeping, 3 in maintenance management, and 3 in principles of supervision. If ASI G7, G8, X8, or X9 is held, 2 additional semester hours in electronic troubleshooting and repair (7/87).

Recommendation, Skill Level 40

In the lower-division baccalaureate/associate degree category, 3 semester hours in introduction to electronics, 2 in electronic troubleshooting and repair, 2 in record keeping, 3 in maintenance management, and 3 in principles of supervision. If ASI G7, G8, X8, or X9 is held, 2 additional semester hours in electronic troubleshooting and repair. In the upper-division baccalaureate category, 3 semester hours for field experience in management (7/87).

MOS-29S-002

Field Communications Security
Equipment Repairer
29S10
29S20
29S30
29S40

Exhibit Dates: 3/91–1/94.

Career Management Field: 29 (Communications-Electronics Maintenance).

Description

Summary: Performs maintenance on field cryptographic equipment. *Skill Level 10:* Uses oscilloscopes and multimeters to troubleshoot cryptographic equipment; troubleshoots and repairs equipment to the board level; uses logical troubleshooting methods to locate faults; individuals holding the following Additional Skill Identifiers (ASIs) perform maintenance at the component level: ASI G7, COMSEC Equipment Full Maintenance (Field); ASI G8, COMSEC Equipment Full Maintenance (Fixed); ASI X8, COMSEC System Maintenance Specialized Repair Activity; and ASI X9, General COMSEC Maintenance Specialized Repair Activity. *Skill Level 20:* Able to perform the duties required for Skill Level 10; assists and instructs subordinates in equipment maintenance; keeps detailed records of maintenance history on equipment; follows formal maintenance system. *Skill Level 30:* Able to perform the duties required for Skill Level 20; assigns duties to subordinates; establishes work load, work schedules, and priorities; responsible for quality control inspections; evaluates, counsels, and provides technical assistance to subordinates; maintains technical library; initiates and maintains forms and records; responsible for records pertaining to repair parts; prepares and reviews administrative reports. *Skill Level 40:* Able to perform the duties required for Skill Level 30; supervises and provides technical guidance to subordinates; supervises and organizes mobile maintenance and inspection teams; advises commander on matters pertaining to repair operations; plans and supervises on-the-job training.

Recommendation, Skill Level 10

In the lower-division baccalaureate/associate degree category, 3 semester hours in introduction to electronics and 1 in electronic troubleshooting and repair. If ASI G7, G8, X8, or X9 is held, 2 additional semester hours in electronic troubleshooting and repair. (NOTE: This recommendation for skill level 10 is valid for the dates 3/91-9/91 only) (7/87).

Recommendation, Skill Level 20

In the lower-division baccalaureate/associate degree category, 3 semester hours in introduction to electronics, 2 in electronic troubleshooting and repair, and 2 in record keeping. If ASI G7, G8, X8, or X9 is held, 2 additional semester hours in electronic troubleshooting and repair (7/87).

Recommendation, Skill Level 30

In the lower-division baccalaureate/associate degree category, 3 semester hours in introduction to electronics, 2 in electronic troubleshooting and repair, 2 in record keeping, 3 in maintenance management, and 3 in principles of supervision. If ASI G7, G8, X8, or X9 is held, 2 additional semester hours in electronic troubleshooting and repair (7/87).

Recommendation, Skill Level 40

In the lower-division baccalaureate/associate degree category, 3 semester hours in introduction to electronics, 2 in electronic troubleshooting and repair, 2 in record keeping, 3 in maintenance management, and 3 in principles of supervision. If ASI G7, G8, X8, or X9 is held, 2 additional semester hours in electronic troubleshooting and repair. In the upper-division baccalaureate category, 3 semester hours for field experience in management (7/87).

MOS-29S-003

Communications Security Equipment
Repairer
29S10
29S20
29S30
29S40

Exhibit Dates: 2/94–2/95.

Career Management Field: 29 (Signal Maintenance).

Description

Summary: Performs maintenance on field cryptographic equipment. *Skill Level 10:* Operates, tests, troubleshoots, and calibrates all communications security (COMSEC) equipment; uses oscilloscopes and multimeters to determine board-level faults; records maintenance actions on appropriate forms; performs tests on repaired equipment and ensures compliance with security and calibration standards; individuals holding Additional Skill Identifier (ASI) G7 (COMSEC Equipment Full Maintenance) troubleshoot and repair equipment to the component level. *Skill Level 20:* Able to perform the duties required for Skill Level 10; provides technical assistance to subordinates; keeps detailed records of equipment maintenance history; follows formal maintenance system; prepares and maintains maintenance management and equipment control reports; controls and accounts for equipment repair items. *Skill Level 30:* Able to perform the duties required for Skill Level 20; assigns duties to subordinates; establishes work load, work schedules, and priorities; responsible for quality control inspections; evaluates, counsels, and provides technical assistance to subordinates; maintains technical library; initiates and maintains forms and records; responsible for records pertaining to repair parts; prepares and reviews administrative reports; supervises quality control programs; conducts technical and on-the-job training. *Skill Level 40:* Able to perform the duties required for Skill Level 30; supervises and provides technical guidance to subordinates; supervises and organizes mobile maintenance and inspection teams; advises commander on matters pertaining to repair operations; plans and supervises on-the-job training; writes and maintains shop operating procedures; acquires repair parts and ensures material accountability.

Recommendation, Skill Level 10

Credit may be granted on the basis of an individualized assessment of the student (3/95).

Recommendation, Skill Level 20

In the lower-division baccalaureate/associate degree category, 3 semester hours in basic electronics, 3 in electronic systems troubleshooting and repair, and 2 in introduction to computers (3/95).

Recommendation, Skill Level 30

In the lower-division baccalaureate/associate degree category, 3 semester hours in basic electronics, 3 in electronic systems troubleshooting and repair, 2 in record keeping, 2 in maintenance management, 2 in personnel supervision, and 3 in introduction to computers. In the upper-division baccalaureate category, 2 semester hours for field experience in management (3/95).

Recommendation, Skill Level 40

In the lower-division baccalaureate/associate degree category, 3 semester hours in basic electronics, 3 in electronic systems troubleshooting and repair, 2 in record keeping, 3 in maintenance management, 3 in personnel supervision, and 3 in introduction to computers. In the upper-division baccalaureate category, 3 semester hours for field experience in management (3/95).

MOS-29S-004

Communications Security Equipment
Repairer
29S10
29S20
29S30
29S40

Exhibit Dates: 3/95–6/95. Effective 6/95, MOS 29S was discontinued, and its duties were incorporated into MOS's 35E and 35W.

Career Management Field: 29 (Signal Maintenance).

Description

Summary: Performs maintenance on field cryptographic equipment. *Skill Level 10:* Operates, tests, troubleshoots, and calibrates all communications security (COMSEC) equipment; uses oscilloscopes and multimeters to determine board-level faults; records maintenance actions on appropriate forms; performs tests on repaired equipment and ensures compliance with security and calibration standards; individuals holding Additional Skill Identifier (ASI) G7 (COMSEC Equipment Full Maintenance) troubleshoot and repair equipment to the component level. *Skill Level 20:* Able to perform the duties required for Skill Level 10; provides technical assistance to subordinates; keeps detailed records of equipment maintenance history; follows formal maintenance system; prepares and maintains maintenance management and equipment control reports; controls and accounts for equipment repair items. *Skill Level 30:* Able to perform the duties required for Skill Level 20; assigns duties to subordinates; establishes work load, work schedules, and priorities; responsible for quality control inspections; evaluates, counsels, and provides technical assistance to subordinates; maintains technical library; initiates and maintains forms and records; responsible for records pertaining to repair parts; prepares and reviews administrative reports; supervises quality control programs; conducts technical and on-the-job training. *Skill Level 40:* Able to perform the duties required for Skill Level 30; supervises and provides technical guidance to subordinates; supervises and organizes mobile maintenance and inspection teams; advises commander on matters pertaining to repair operations; plans and supervises on-the-job training; writes and maintains shop operating procedures; acquires repair parts and ensures material accountability.

Recommendation, Skill Level 10

Credit may be granted on the basis of an individualized assessment of the student (3/95).

Recommendation, Skill Level 20

Credit may be granted on the basis of an individualized assessment of the student (5/96).

Recommendation, Skill Level 30

In the lower-division baccalaureate/associate degree category, 3 semester hours in basic electronics, 3 in electronic systems troubleshooting and repair, 2 in record keeping, 2 in maintenance management, 2 in personnel supervision, and 2 in introduction to computers. In the upper-division baccalaureate category, 2 semester hours for field experience in management (3/95).

Recommendation, Skill Level 40

In the lower-division baccalaureate/associate degree category, 3 semester hours in basic electronics, 3 in electronic systems troubleshooting and repair, 2 in record keeping, 3 in maintenance management, 3 in personnel supervision, and 3 in introduction to computers. In the upper-division baccalaureate category, 3 semester hours for field experience in management (3/95).

MOS-29T-001

SATELLITE/MICROWAVE COMMUNICATIONS CHIEF

29T40

Exhibit Dates: 1/90–12/91. Effective 12/91, MOS 29T was discontinued, and its duties were incorporated into MOS 29V, Strategic Microwave Systems Repairer and MOS 29Y, Satellite Communications Systems Repairer.

Career Management Field: 29 (Communications Electronics Maintenance).

Description

Supervises unit and intermediate level maintenance of microwave communications systems including satellite and ground station installations; serves as principal noncommissioned officer; provides direction to personnel performing maintenance at ground station microwave and satellite installations; supervises maintenance personnel; coordinates and assigns duties; evaluates and counsels subordinates; advises superiors on maintenance activities; prepares and coordinates technical reports and correspondence; initiates and conducts on-the-job training; instructs subordinates in the utilization of special test equipment; performs liaison between staff, operation, and maintenance personnel. NOTE: May have progressed to 29T40 from 29V30 (Strategic Microwave Systems Repairer) or 29Y30 (Satellite Communications Systems Repairer).

Recommendation

In the lower-division baccalaureate/associate degree category, 3 semester hours in principles of supervision. In the upper-division baccalaureate category, 3 semester hours for field experience in management and 3 in management problems (6/88).

MOS-29V-001

STRATEGIC MICROWAVE SYSTEMS REPAIRER

29V10
29V20
29V30

Exhibit Dates: 1/90–12/91.

Career Management Field: 29 (Communications Electronics Maintenance).

Description

Summary: Supervises, operates, and performs unit and intermediate maintenance on microwave communications equipment, terminals, and systems. *Skill Level 10:* Uses test equipment to troubleshoot microwave communications equipment, terminals, and systems; troubleshoots and repairs to the component level; operates equipment and completes preventive maintenance procedures. *Skill Level 20:* Able to perform the duties required for Skill Level 10; assists and instructs subordinates in equipment maintenance and provides technical assistance when needed; keeps detailed records of equipment maintenance. *Skill Level 30:* Able to perform the duties required for Skill Level 20; assigns duties; establishes work load, work schedules, and priorities; conducts quality control inspections; evaluates and counsels subordinates; maintains technical library; initiates and maintains forms and records; responsible for records pertaining to repair parts.

Recommendation, Skill Level 10

In the lower-division baccalaureate/associate degree category, 3 semester hours in basic electronics laboratory, 2 in digital principles, and 3 in electronic systems troubleshooting and maintenance. (NOTE: This recommendation for skill level 10 is valid for the dates 1/90-9/91 only) (6/88).

Recommendation, Skill Level 20

In the lower-division baccalaureate/associate degree category, 3 semester hours in basic electronics laboratory, 2 in digital principles, 6 in electronic systems troubleshooting and maintenance, 2 in maintenance management, and 2 in record keeping (6/88).

Recommendation, Skill Level 30

In the lower-division baccalaureate/associate degree category, 3 semester hours in basic electronics laboratory, 2 in digital principles, 6 in electronic systems troubleshooting and maintenance, 2 in maintenance management, 2 in record keeping, and 3 in principles of supervision (6/88).

MOS-29V-002

MICROWAVE SYSTEMS OPERATOR/REPAIRER

29V10
29V20
29V30
29V40

Exhibit Dates: 1/92–6/94. Effective 6/94, MOS 29V was discontinued, and its duties were incorporated into MOS 31P, Microwave Systems Operator-Maintainer.

Career Management Field: 29 (Communications-Electronics Maintenance).

Description

Summary: Supervises, operates, and performs unit and intermediate maintenance on microwave communications equipment, terminals, and systems in ground station installations. *Skill Level 10:* Uses test equipment to troubleshoot microwave communications equipment, terminals, and systems; troubleshoots and repairs to the component level; operates equipment and completes preventive maintenance procedures. *Skill Level 20:* Able to perform the duties required for Skill Level 10; assists and instructs subordinates in equipment maintenance and provides technical assistance when needed; keeps detailed records of equipment maintenance history. *Skill Level 30:* Able to perform the duties required for Skill Level 20; assigns duties; establishes work load, work schedules, and priorities; conducts quality control inspections; evaluates and counsels subordinates; maintains technical library; initiates and maintains forms and records; responsible for records pertaining to repair parts; prepares and reviews administrative reports. *Skill Level 40:* Able to perform the duties required for Skill Level 30; supervises unit and intermediate maintenance of microwave communications systems (ground station installations); serves as principal noncommissioned officer; provides direction to personnel performing maintenance at ground station microwave installations; supervises maintenance personnel; coordinates and assigns duties; evaluates and counsels subordinates; advises superiors on maintenance activities; prepares and coordinates technical reports and correspondence; initiates and conducts on-the-job training; instructs subordinates in the utilization of special test equipment; performs liaison among staff, operation, and maintenance personnel.

Recommendation, Skill Level 10

Credit may be granted on the basis of an individualized assessment of the student (3/94).

Recommendation, Skill Level 20

In the lower-division baccalaureate/associate degree category, 3 semester hours in basic electronics laboratory, 2 in digital principles, 6 in electronic systems troubleshooting and maintenance, 2 in maintenance management, and 2 in record keeping (6/88).

Recommendation, Skill Level 30

In the lower-division baccalaureate/associate degree category, 3 semester hours in basic electronics laboratory, 2 in digital principles, 6 in electronic systems troubleshooting and maintenance, 2 in maintenance management, 2 in record keeping, and 3 in principles of supervision (6/88).

Recommendation, Skill Level 40

In the lower-division baccalaureate/associate degree category, 3 semester hours in basic electronics laboratory, 2 in digital principles, 6 in electronic systems troubleshooting and maintenance, 2 in maintenance management, 2 in record keeping, and 3 in principles of supervision. In the upper-division baccalaureate category, 3 semester hours for field experience in management and 3 in management problems (6/88).

MOS-29W-001

COMMUNICATIONS MAINTENANCE SUPPORT CHIEF

29W40

Exhibit Dates: 1/90–1/94.

Career Management Field: 29 (Communications Electronics Maintenance).

Description

Supervises the performance of unit, intermediate, and depot maintenance of communications equipment; coordinates duties; evaluates and counsels subordinates; advises superiors on maintenance activities; reviews, prepares, and coordinates technical reports and correspondence; initiates and conducts on-the-job and apprentice training; instructs subordinates in the use of special test equipment; performs liaison among staff, operation, and maintenance personnel; makes plans for the defense of the unit. NOTE: May have progressed to 29W40 from 29E30 (Radio Repairer), 29J30 (Teletypewriter Equipment Repairer), 29M30 (Tactical Satellite/Microwave Repairer), or 29N30 (Telephone Central Office Repairer).

Recommendation

In the upper-division baccalaureate category, 3 semester hours for field experience in management and 3 in management problems (6/88).

MOS-29W-002

Electronics Maintenance Supervisor
((Communications Electronics Maintenance Chief))
29W40

Exhibit Dates: 2/94–6/95. Effective 6/95, MOS 29W was discontinued, and its duties were incorporated into MOS 35W.

Career Management Field: 29 (Signal Maintenance).

Description

Supervises the performance of unit, intermediate, and depot maintenance of communications equipment; coordinates duties; evaluates and counsels subordinates; advises superiors on maintenance activities; reviews, prepares, and coordinates technical reports and correspondence; initiates and conducts on-the-job and apprentice training programs; instructs subordinates in the utilization of special test equipment; serves as liaison among staff, operation, and maintenance facilities; makes plans for the defense of the unit. NOTE: May have progressed to 29W40 from 29E30 (Radio Repairer), 29J30 (Telecommunications Terminal Device Repairer), 29N30 (Switching Control Repairer), 39C30 (Target Acquisition/Surveillance Radar Repairer), or 39E30 (Special Electronic Repairer).

Recommendation

In the upper-division baccalaureate category, 3 semester hours for field experience in management and 3 in management problems (3/95).

MOS-29X-001

Communications Equipment Maintenance Chief
29X50

Exhibit Dates: 1/90–3/91. Effective 3/91, MOS 29X was discontinued, and its duties were incorporated into MOS 29Z, Electronics Maintenance Chief.

Career Management Field: 29 (Communications Electronics Maintenance).

Description

Supervises the Army's communications equipment maintenance mission; determines requirements, assigns overall duties, and coordinates activities of subordinates; develops and applies policies and procedures for a maintenance facility; selects communications sites and advises maintenance officer or unit commander on maintenance activities; conducts training and prepares technical and administrative reports. NOTE: May have progressed to 29X50 from 29P40 (Communications Security Maintenance Chief), 29U40 (Digital Equipment Maintenance Chief), 29W40 (Communications Maintenance Support Chief), or 29T40 (Satellite Microwave Communications Chief).

Recommendation

In the upper-division baccalaureate category, 3 semester hours in organizational management and 3 for field experience in management (6/88).

MOS-29Y-001

Satellite Communications (SATCOM) Systems Repairer
29Y10
29Y20
29Y30

Exhibit Dates: 1/90–12/91.

Career Management Field: 29 (Communications Electronics Maintenance).

Description

Summary: Operates and maintains satellite ground station equipment. *Skill Level 10:* Operates and performs preventive maintenance on satellite ground station equipment; uses common electronic test equipment to test and repair digital circuitry; performs a limited amount of troubleshooting. *Skill Level 20:* Able to perform the duties required for Skill Level 10; performs extensive troubleshooting and repair of equipment; keeps maintenance records; supervises and provides guidance to subordinates. *Skill Level 30:* Able to perform the duties required for Skill Level 20; prepares technical and administrative reports; maintains files and records; assigns duties; performs and supervises quality control inspections; coordinates logistical support; establishes on-the-job training; advises superiors.

Recommendation, Skill Level 10

In the lower-division baccalaureate/associate degree category, 3 semester hours in basic electronics laboratory, 2 in digital principles, and 1 in electronic systems troubleshooting and maintenance. (NOTE: This recommendation for skill level 10 is valid for the dates 1/90-9/91 only) (6/88).

Recommendation, Skill Level 20

In the lower-division baccalaureate/associate degree category, 3 semester hours in basic electronics laboratory, 2 in digital principles, 6 in electronic systems troubleshooting and maintenance, and 2 in record keeping (6/88).

Recommendation, Skill Level 30

In the lower-division baccalaureate/associate degree category, 3 semester hours in basic electronics laboratory, 2 in digital principles, 6 in electronic systems troubleshooting and maintenance, 2 in record keeping, 2 in maintenance management, and 3 in principles of supervision (6/88).

MOS-29Y-002

Satellite Communications (SATCOM) Systems Repairer
29Y10
29Y20
29Y30
29Y40

Exhibit Dates: 1/92–6/94. Effective 6/94, MOS 29Y was discontinued, and its duties were incorporated into MOS 31S, Satellite Communications Systems Operator-Maintainer.

Career Management Field: 29 (Communications Electronics Maintenance).

Description

Summary: Operates and maintains satellite ground station equipment. *Skill Level 10:* Operates and performs preventive maintenance on satellite ground station equipment; uses common electronic test equipment to test and repair digital circuitry; performs a limited amount of troubleshooting. *Skill Level 20:* Able to perform the duties required for Skill Level 10; performs extensive troubleshooting and repair of equipment; keeps maintenance records; supervises and provides guidance to subordinates. *Skill Level 30:* Able to perform the duties required for Skill Level 20; prepares technical and administrative reports; maintains files and records; assigns duties; performs and supervises quality control inspections; coordinates logistical support; establishes on-the-job training; advises superiors. *Skill Level 40:* Able to perform the duties required for Skill Level 30; supervises unit and intermediate maintenance of microwave communications systems (satellite installations); serves as principal noncommissioned officer; provides direction to personnel performing maintenance at satellite installations; supervises maintenance personnel; coordinates and assigns duties; evaluates and counsels subordinates; advises superiors on maintenance activities; prepares and coordinates technical reports and correspondence; initiates and conducts on-the-job training; instructs subordinates in the utilization of special test equipment; performs liaison among staff, operations, and maintenance personnel.

Recommendation, Skill Level 10

Credit may be granted on the basis of an individualized assessment of the student (3/94).

Recommendation, Skill Level 20

In the lower-division baccalaureate/associate degree category, 3 semester hours in basic electronics laboratory, 2 in digital principles, 6 in electronic systems troubleshooting and maintenance, and 2 in record keeping (6/88).

Recommendation, Skill Level 30

In the lower-division baccalaureate/associate degree category, 3 semester hours in basic electronics laboratory, 2 in digital principles, 6 in electronic systems troubleshooting and maintenance, 2 in record keeping, 2 in maintenance management, and 3 in principles of supervision (6/88).

Recommendation, Skill Level 40

In the lower-division baccalaureate/associate degree category, 3 semester hours in basic electronics laboratory, 2 in digital principles, 6 in electronic systems troubleshooting and maintenance, 2 in record keeping, 2 in maintenance management, and 3 in principles of supervision. In the upper-division baccalaureate category, 3 semester hours for field experience in management and 3 in management problems (6/88).

MOS-29Z-001

ELECTRONICS MAINTENANCE CHIEF
29Z50

Exhibit Dates: 1/90–6/95. Effective 6/95, MOS 29Z was discontinued, and its duties were incorporated into MOS's 35W, 35Z, and 31Z.

Career Management Field: 29 (Communications Electronics Maintenance).

Description

Supervises, monitors, and directs the Army's communications and electronics maintenance mission; determines requirements, assigns duties, and coordinates activities of subordinate units; develops and applies policies and procedures for maintenance facilities; briefs commander and staff on maintenance status of organization; prepares technical correspondence and administrative reports; supervises training in communications, operations, procedures, and maintenance practices. NOTE: May have progressed to 29Z50 from 29X50 (Communications Equipment Maintenance Chief), 39X50 (Electronics Equipment Maintenance Chief), or 35H50 (Calibration Specialist).

Recommendation

In the upper-division baccalaureate category, 3 semester hours in organizational management and 3 for field experience in management (6/88).

MOS-31B-002

MILITARY POLICE
31B10
31B20
31B30
31B40
31B50

Exhibit Dates: Pending Evaluation. 4/03–Present.

MOS-31C-001

RADIO OPERATOR-MAINTAINER
((Single-Channel Radio Operator))
31C10
31C20
31C30

Exhibit Dates: 1/90–4/99.

Career Management Field: 31 (Communications-Electronics Operations).

Description

Summary: Installs and operates high frequency radio, radio telephone, teletype, single-channel satellite equipment, field generator power supplies, communications security devices, and associated equipment. *Skill Level 10:* Installs and operates equipment. *Skill Level 20:* Able to perform the duties required for Skill Level 10; supervises the operation of equipment; supervises 8-15 subordinates; performs combat communications-electronics planning and maintenance administration. *Skill Level 30:* Able to perform the duties required for Skill Level 20; supervises 15-20 subordinates in the installation, operation, and maintenance of equipment; has additional skills in the areas of signal security, tactical satellite communications, and personnel guidance and supervision.

Recommendation, Skill Level 10

In the lower-division baccalaureate/associate degree category, if ASI was J7, WHCA Console Control Operations, credit as follows: 1 semester hour in computer literacy, 2 in files management, and 3 in introduction to voice communications networks. (NOTE: This recommendation for skill level 10 is valid for the dates 1/90-9/91 only) (9/85).

Recommendation, Skill Level 20

In the lower-division baccalaureate/associate degree category, if ASI was J7, WHCA Console Control Operations, credit as follows: 1 semester hour in computer literacy, 2 in files management, and 3 in introduction to voice communications networks. (NOTE: This recommendation for skill level 20 is valid for the dates 1/90-2/95 only) (9/85).

Recommendation, Skill Level 30

In the lower-division baccalaureate/associate degree category, 3 semester hours in personnel supervision. If ASI was J7, WHCA Console Control Operations, additional credit as follows 1 in computer literacy, 2 in files management, and 3 in introduction to voice communications networks (9/85).

MOS-31C-002

RADIO OPERATOR-MAINTAINER
31C10
31C20
31C30

Exhibit Dates: 5/99–Present.

Career Management Field: 31 (Communications-Electronics Operations).

Description

Summary: Installs, operates, and performs unit level maintenance on high-frequency radio, radio telephone, teletype, single-channel satellite equipment, field generator power supplies, communications security devices, and associated equipment. *Skill Level 10:* Installs and operates equipment; performs unit level maintenance. *Skill Level 20:* Able to perform the duties required for Skill Level 10; supervises the operation of equipment; supervises 8-15 subordinates; performs combat communications-electronics planning and maintenance administration. *Skill Level 30:* Able to perform the duties required for Skill Level 20; supervises 15-20 subordinates in the installation, operation, and maintenance of equipment; has additional skills in the areas of signal security, tactical satellite communications, and personnel guidance and supervision.

Recommendation, Skill Level 10

Credit may be granted on the basis of an individualized assessment of the student (5/99).

Recommendation, Skill Level 20

Credit may be granted on the basis of an individualized assessment of the student (5/99).

Recommendation, Skill Level 30

In the lower-division baccalaureate/associate degree category, 2 semester hours in record keeping, 2 in introduction to voice and data communications, 3 in personnel supervision, and 3 in maintenance management (5/99).

MOS-31D-002

MOBILE SUBSCRIBER EQUIPMENT
TRANSMISSION SYSTEM OPERATOR
31D10
31D20
31D30

Exhibit Dates: 1/90–6/95. Effective 6/95, MOS 31D was discontinued, and its duties were incorporated into MOS 31R.

Career Management Field: 31 (Signal Operations).

Description

Summary: Installs, operates, supervises, and performs unit maintenance on Mobile Subscriber Equipment (MSE) radio transmission equipment, radio access units, and interoperable devices. *Skill Level 10:* Installs, operates, and performs unit maintenance on line-of-sight multichannel radio terminals, radio access units, communications security (COMSEC) devices, multiplexing equipment, and antenna systems; interconnects equipment components and configures systems for operation; maintains records and reports pertaining to equipment operation; performs unit maintenance on MSE radio transmission and associated equipment. *Skill Level 20:* Able to perform the duties required for Skill Level 10; supervises and trains subordinate personnel; prepares work schedules and checks operation records and station files for completeness and accuracy; writes technical reports and ensures availability of spare parts and supplies. *Skill Level 30:* Able to perform the duties required for Skill Level 20; determines requirements, assigns duties, and coordinates activities of communications personnel; plans, supervises, and implements tactical training programs; prepares technical and administrative reports; supervises march orders and emplacement of section/team equipment; oversees and performs periodic and emergency adjustment on MSE transmission equipment, radio access units, and interoperable devices.

Recommendation, Skill Level 10

In the lower-division baccalaureate/associate degree category, 2 semester hours in data communications system troubleshooting and maintenance. (NOTE: This recommendation for skill level 10 is valid for the dates 1/90-9/91 only) (6/90).

Recommendation, Skill Level 20

In the lower-division baccalaureate/associate degree category, 3 semester hours in data communications system troubleshooting and maintenance. (NOTE: This recommendation for skill level 20 is valid for the dates 1/90-2/95) (6/90).

Recommendation, Skill Level 30

In the lower-division baccalaureate/associate degree category, 3 semester hours in data communications system troubleshooting and maintenance, 3 in principles of supervision, and 3 in maintenance management (6/90).

MOS-31D-003

CID SPECIAL AGENT
31D20
31D30
31D40
31D50

Exhibit Dates: Pending Evaluation. 4/03–Present.

MOS-31E-003

CORRECTIONS SPECIALIST
31E10
31E20
31E30
31E40
31E50

Exhibit Dates: 10/04–Present.

Career Management Field: 31 (Law Enforcement).

Description

Summary: Functions as a correctional officer, providing rehabilitative, health, welfare, and security services to prisoners within a correctional facility; conducts inspections; prepares written reports; coordinates activities of inmates and staff personnel. *Skill Level 10:* Assists with supervision and management of military prisoners; provides external security to confinement or corrections facility. *Skill Level 20:* Able to perform the duties required for Skill Level 10; functions as a correctional officer; provides individual counseling and guidance to prisoners within a rehabilitative program. *Skill Level 30:* Able to perform the duties required for Skill Level 20; functions as a first-line supervisor of up to ten correctional officers; prepares or reviews reports and records of individuals and programs; provides formal counseling. *Skill Level 40:* Able to perform the duties required for Skill Level 30; as a mid-level manager, supervises and coordinates all correctional, custodial, treatment, and rehabilitative activities of inmates and up to 100 staff members in correctional facility containing as many as 250 prisoners; responsible for administrative inspection and review of all records, reports, and activities; furnishes recommendations to superiors regarding clemency, restoration to duty, and further treatment of inmates; advises superior on matters concerning subordinates. *Skill Level 50:* Able to perform the duties required for Skill Level 40; supervises and coordinates inmate and staff activities in a large consolidated correctional facility; as an administrator, plans, coordinates, and implements all counseling, rehabilitative, and correctional programs in a permanent facility containing 250 or more prisoners; equivalent to a federal maximum-security prison warden.

Recommendation, Skill Level 10

Credit may be granted on the basis of an individualized assessment of the student (4/04).

Recommendation, Skill Level 20

In the lower-division baccalaureate/associate degree category, 3 semester hours in correctional operations. (NOTE: This recommendation for skill level 20 is valid for the dates 2/93-2/95 only) (4/04).

Recommendation, Skill Level 30

In the lower-division baccalaureate/associate degree category, 3 semester hours in correctional operations, 3 in correctional supervision, and 3 in a criminal justice elective (4/04).

Recommendation, Skill Level 40

In the lower-division baccalaureate/associate degree category, 3 semester hours in correctional operations, 3 in correctional supervision, 3 in a criminal justice elective, and 3 in correctional administration. In the upper-division baccalaureate category, 3 semester hours in principles of management (4/04).

Recommendation, Skill Level 50

In the lower-division baccalaureate/associate degree category, 3 semester hours in correctional operations, 3 in correctional supervision, 3 in a criminal justice elective, and 3 in correctional administration. In the upper-division baccalaureate category, 3 semester hours in principles of management and 3 in introduction to public administration (4/04).

MOS-31F-001

NETWORK SWITCHING SYSTEMS OPERATOR/MAINTAINER
((Electronic Switching Systems Operator))
((Mobile Subscriber Equipment Network Switching System Operator))
31F10
31F20
31F30

Exhibit Dates: 1/90–4/99.

Career Management Field: 31 (Signal Operations).

Description

Summary: Installs, operates, supervises, and performs unit maintenance on systems control center, node center switch, extension node switches, associated multiplexing equipment, net radio interface equipment, communications security devices, and remote transmission (down-the-hill) radios; troubleshoots to the board level using digital multimeters, computer terminals with diagnostic readouts, and board level LEDs and configuration switches. *Skill Level 10:* Initializes and loads data base; positions, assembles, and interconnects equipment components; operates motor vehicles and power generation equipment; adjusts and aligns equipment to prescribed tolerances; interprets and uses operating instructions, data base information, and system diagrams; establishes and maintains net connectivity; inputs and receives information from system control center for transmitting and receiving operating directives and data base input/output; provides subscriber assistance; interprets computer printouts to troubleshoot, repair, or replace faulty line replaceable units; coordinates with system operators and managers to resolve difficulties; places spare equipment into operation upon failure of on-line unit. *Skill Level 20:* Able to perform the duties required for Skill Level 10; supervises installation, operation, and maintenance; trains personnel in the employment and operation of equipment; assists in selection of sites; corrects faulty work practices and operating procedures; performs various supervisory duties pertaining to work schedules, records, supplies, and reports. *Skill Level 30:* Able to perform the duties required for Skill Level 20; determines requirements, assigns duties, and coordinates activities of communications personnel; plans, implements, and supervises training programs; supervises maintenance programs; coordinates logistical support; prepares and oversees technical and administrative reports; oversees/performs periodic and emergency adjustments.

Recommendation, Skill Level 10

In the lower-division baccalaureate/associate degree category, 2 semester hours in data communications system troubleshooting and maintenance. (NOTE: This recommendation for skill level 10 is valid for the dates 1/90-9/91 only) (6/90).

Recommendation, Skill Level 20

In the lower-division baccalaureate/associate degree category, 3 semester hours in data communications system troubleshooting and maintenance. (NOTE: This recommendation for skill level 20 is valid for the dates 1/90-2/95 only) (6/90).

Recommendation, Skill Level 30

In the lower-division baccalaureate/associate degree category, 3 semester hours in data communications system troubleshooting and maintenance, 3 in principles of supervision, and 3 in maintenance management (6/90).

MOS-31F-002

NETWORK SWITCHING SYSTEMS OPERATOR/MAINTAINER
31F10
31F20
31F30

Exhibit Dates: 5/99–Present.

Career Management Field: 31 (Signal Operations).

Description

Summary: Installs, operates, supervises, and performs unit maintenance on systems control center, node center switch, extension node switches, associated multiplexing equipment, net radio interface equipment, communications security devices, and remote transmission radios; troubleshoots using digital multimeters, computer terminals with diagnostic readouts and board level LEDs and configuration switches. *Skill Level 10:* Initializes and loads data base; positions, assembles, and interconnects equipment components; operates motor vehicles and power generation equipment; adjusts and aligns equipment to prescribed tolerances; interprets and uses operating instructions, data base information, and system diagrams; establishes and maintains net connectivity; inputs and receives information from system control center for transmitting and receiving operating directives and data base input/output; provides subscriber assistance; interprets computer printouts to troubleshoot and replace faulty line-replaceable units; coordinates with system operators and managers to resolve difficulties. *Skill Level 20:* Able to perform the duties required for Skill Level 10; supervises installation, operation, and maintenance; trains personnel in the employment and operation of equipment; assists in selection of sites; performs various supervisory duties pertaining to work schedules, records, supplies, and reports. *Skill Level 30:* Able to perform the duties required for Skill Level 20; determines requirements, assigns duties, and coordinates activities of communications personnel; plans, implements, and supervises training programs; supervises maintenance programs; coordinates logistical support; prepares and oversees technical and administrative reports; oversees/performs periodic and emergency adjustments.

Recommendation, Skill Level 10

Credit may be granted on the basis of an individualized assessment of the student (5/99).

Recommendation, Skill Level 20

Credit may be granted on the basis of an individualized assessment of the student (5/99).

Recommendation, Skill Level 30

In the lower-division baccalaureate/associate degree category, 3 semester hours in data communications troubleshooting and maintenance, 3 in principles of supervision, and 3 in maintenance management (5/99).

MOS-31G-002

TACTICAL COMMUNICATION CHIEF
31G30
31G40

Exhibit Dates: 1/90–6/93. Effective 6/93, MOS 31G was discontinued, and its duties were incorporated into MOS 31U, Signal Support Systems Specialist.

Career Management Field: 31 (Communications Electronics Operations).

Description

Summary: Supervises the installation, operation, and maintenance of communications systems in infantry, armor, artillery, or other units employing similar methods of communication. *Skill Level 30:* Plans, coordinates, and supervises the installation, operation, and maintenance of ground FM radio and field wire communications systems, air ground radio sets, radioteletypewriter sets, and switchboard systems; coordinates activities and assigns duties; conducts training programs; advises the commander on communications matters. NOTE: May have progressed to 31G30 from 31K20 (Combat Signaler) or 31V20 (Unit Level Communications Maintainer). *Skill Level 40:* Able to perform the duties required for Skill Level 30; monitors and coordinates the operation of the various elements of integrated communications systems; prepares, supervises, and conducts training programs; prepares technical and administrative reports; assists in selecting locations for communications facilities; performs liaison among staff, operations, and maintenance personnel.

Recommendation, Skill Level 30

In the lower-division baccalaureate/associate degree category, 3 semester hours in principles of supervision and 3 in maintenance management. If ASI was J7, WHCA Console Control Operations, additional credit as follows: 1 semester hour in computer literacy, 2 in files management, and 3 in introduction to voice communications networks (6/88).

Recommendation, Skill Level 40

In the lower-division baccalaureate/associate degree category, 3 semester hours in principles of supervision and 3 in maintenance management. If ASI was J7, WHCA Console Control Operations, additional credit as follows: 1 semester hour in computer literacy, 2 in files management, and 3 in introduction to voice communications networks. In the upper-division baccalaureate category, 3 semester hours in management problems and 3 for field experience in management (6/88).

MOS-31K-001

COMBAT SIGNALER
31K10
31K20

Exhibit Dates: 1/90–6/93. Effective 6/93, MOS 31K was discontinued and its duties were incorporated into MOS 31L, Wire Systems Installer, or 31U, Signal Support Systems Specialist.

Career Management Field: 31 (Communications-Electronics Operations).

Description

Summary: Supervises, installs, operates, and maintains basic wire, switchboard, and frequency modulation (FM) radio communications equipment and systems in combat, combat support, and combat service support units. *Skill Level 10:* Performs field wire system troubleshooting; splices wires; installs field telephones and FM radios; processes messages. *Skill Level 20:* Able to perform the duties required for Skill Level 10; supervises and trains subordinates; estimates supply requirements; interprets maps and overlays.

Recommendation, Skill Level 10

Credit is not recommended because the skills, competencies, and knowledge are uniquely military in nature (9/85).

Recommendation, Skill Level 20

Credit is not recommended because the skills, competencies, and knowledge are uniquely military in nature (9/85).

MOS-31L-002

CABLE SYSTEMS INSTALLER-MAINTAINER
((Wire Systems Installer))
31L10
31L20
31L30

Exhibit Dates: 1/90–4/99.

Career Management Field: 31 (Communications Electronics Operations).

Description

Summary: Installs, operates, supervises, and performs operator maintenance on wire and cable communications systems; operates pole-setting and reeling equipment; works on wood structures using tools and techniques compatible with industry practice; installs field phones and cabling to telephone switchboards; identifies and corrects faults in telephone sets; tests, repairs, and splices wire/cable; knows CPR procedures, first aid, and safety practices relating to line construction. *Skill Level 10:* Installs, operates, and performs operator maintenance on wire and cable systems, including repeaters, restorers, voltage protective devices, telephones, test stations, and telephone substation equipment; climbs poles; clears and maintains rights of way; assists in the construction of tactical wire/cable and pole lines; operates construction equipment; maintains and tests cable communication systems using a variety of test equipment. *Skill Level 20:* Able to perform the duties required for Skill Level 10; coordinates and supervises team member activities in construction, installation, and rehabilitation of wire and cable communication systems and auxiliary equipment; coordinates logistic requirements. *Skill Level 30:* Able to perform the duties required for Skill Level 20; supervises and coordinates all phases of construction, installation, and maintenance of wire/cable systems; schedules work load; estimates time, supplies, personnel, and equipment requirements; checks maintenance performed by subordinates; plans and conducts training and prepares required reports.

Recommendation, Skill Level 10

In the vocational certificate category, 6 semester hours in pole and line construction (communications systems), 3 in telephone installation and repair, and 1 in first aid/CPR. In the lower-division baccalaureate/associate degree category, if ASI was J7, WHCA Console Control Operations, additional credit as follows: 1 semester hour in computer literacy, 2 in files management, and 3 in introduction to voice communications networks. (NOTE: This recommendation for skill level 10 is valid for the dates 1/90-9/91 only) (6/88).

Recommendation, Skill Level 20

In the vocational certificate category, 6 semester hours in pole and line construction (communications systems), 3 in telephone installation and repair, 1 in first aid/CPR, and 3 in construction technology/heavy equipment. In the lower-division baccalaureate/associate degree category, 2 semester hours in principles of supervision. If ASI was J7, WHCA Console Control Operations, additional credit as follows: 1 semester hour in computer literacy, 2 in files management, and 3 in introduction to voice communications networks. (NOTE: This recommendation for skill level 20 is valid for the dates 1/90-2/95 only) (6/88).

Recommendation, Skill Level 30

In the vocational certificate category, 6 semester hours in pole and line construction (communications systems), 3 in telephone installation and repair, 1 in first aid/CPR, and 3 in construction technology/heavy equipment. In the lower-division baccalaureate/associate degree category, 3 semester hours in principles of supervision and 3 in maintenance management. If ASI was J7, WHCA Console Control Operations, additional credit as follows: 1 semester hour in computer literacy, 2 in files management, and 3 in introduction to voice communications networks (6/88).

MOS-31L-003

CABLE SYSTEMS INSTALLER-MAINTAINER
31L10
31L20
31L30

Exhibit Dates: 5/99–Present.

Career Management Field: 31 (Communications Electronics Operations).

Description

Summary: Installs, operates, supervises, and maintains cable and wire communications systems including digital group multiplexers, remote multiplexing combiners, repeaters, telephones, test stations, and related equipment; tests circuits/groups to detect and locate line faults; climbs poles, splices wire/cable and optical fibers; knows safety practices relating to line construction; assigns work crews and coordinates activities of personnel to meet workload demands; prepares and conducts MOS and unit training programs. *Skill Level 10:* Installs, operates, and performs operator maintenance on wire and cable systems, including repeaters, restorers, voltage protective devices, telephones, test stations, and telephone substation equipment; climbs poles; clears and maintains rights of way; assists in the construction of tactical wire/cable and pole lines; operates construction equipment; maintains and tests cable communications systems using a variety of test equipment. *Skill Level 20:* Able to perform the duties required for Skill Level 10; coordinates and supervises team member activities in construction, installation, and rehabilitation of wire and cable communications systems and auxiliary equipment; coordinates logistic requirements. *Skill Level 30:* Able to perform the duties required for Skill Level 20; supervises and coordinates all phases of construction, installation, and maintenance of wire/cable systems; schedules work load; estimates time, supplies, personnel, and equipment requirements; checks maintenance performed by subordinates; plans and conducts training; prepares required reports.

Recommendation, Skill Level 10

Credit may be granted on the basis of an individualized assessment of the student (5/99).

Recommendation, Skill Level 20

Credit may be granted on the basis of an individualized assessment of the student (5/99).

Recommendation, Skill Level 30

In the vocational certificate category, 3 semester hours in pole climbing and line construction and 3 in telephone installation and maintenance. In the lower-division baccalaureate/associate degree category, 3 semester hours in record keeping, 2 in introduction to voice and data communications networks, 3 in principles of supervision, and 3 in maintenance management (5/99).

MOS-31M-003

Multichannel Communications Equipment Operator
31M10
31M20
31M30

Exhibit Dates: 1/90–5/91.

Career Management Field: 31 (Communications-Electronics Operations).

Description

Summary: Installs, operates, and supervises multichannel communications equipment. *Skill Level 10:* Installs and operates multichannel communications equipment, including radio, communications security devices, and multiplexer equipment; performs elementary operating checks and replaces modules as needed. *Skill Level 20:* Able to perform the duties required for Skill Level 10; operates and supervises the operation of multichannel communications equipment; applies fundamentals of radio procedures, communications security procedures, and precautions; oversees periodic and emergency maintenance; may supervise up to six persons. *Skill Level 30:* Able to perform the duties required for Skill Level 20; supervises the installation and operation of multichannel communications systems; prepares and disseminates work schedules, operational procedures, and instructions; prepares traffic diagrams and administrative orders relative to system design; may supervise from five to nineteen subordinates.

Recommendation, Skill Level 10

In the vocational certificate category, 1 semester hour in laboratory practices. In the lower-division baccalaureate/associate degree category, if ASI was J7, WHCA Console Control Operations, additional credit as follows:1 semester hour in computer literacy, 2 in files management, and 3 in introduction to voice communications networks (10/79).

Recommendation, Skill Level 20

In the vocational certificate category, 2 semester hours in laboratory practices. In the lower-division baccalaureate/associate degree category, if ASI was J7, WHCA Console Control Operations, additional credit as follows: 1 semester hour in computer literacy, 2 in files management, and 3 in introduction to voice communications networks (10/79).

Recommendation, Skill Level 30

In the vocational certificate category, 2 semester hours in laboratory practices. In the lower-division baccalaureate/associate degree category, 3 semester hours in personnel supervision. If ASI was J7, WHCA Console Control Operations, additional credit as follows: 1 semester hour in computer literacy, 2 in files management, and 3 in introduction to voice communications networks (10/79).

MOS-31M-004

Multichannel Transmission Systems Operator
((Multichannel Communications Systems Operator))
31M10
31M20
31M30

Exhibit Dates: 6/91–6/95. Effective 6/95, MOS 31M was discontinued, and its duties were incorporated into MOS's 31R and 31S.

Career Management Field: 31 (Signal Operations).

Description

Summary: Installs, operates, and supervises multichannel communications equipment. *Skill Level 10:* Installs and operates multichannel communications equipment (both analog and digital), including radio, communications security devices, and multiplexer equipment; performs elementary operating checks and replaces modules as needed; may also install and operate fiber optic communications links. *Skill Level 20:* Able to perform the duties required for Skill Level 10; operates and supervises the operation of multichannel communications equipment; applies fundamentals of radio procedures, communications security procedures, and precautions; oversees periodic and emergency maintenance; may supervise up to six persons. *Skill Level 30:* Able to perform the duties required for Skill Level 20; supervises the installation and operation of multichannel communications systems; prepares and disseminates work schedules, operational procedures, and instructions; prepares traffic diagrams and administrative orders relative to system design; may supervise from five to nineteen subordinates.

Recommendation, Skill Level 10

In the lower-division baccalaureate/associate degree category, 2 semester hours in basic electronics laboratory and 3 in introduction to voice communications systems. If ASI was J7, WHCA Console Control Operations, additional credit as follows: 1 semester hour in computer literacy, 2 in files management, and 3 in introduction to voice communications networks. (NOTE: This recommendation for skill level 10 is valid for the dates 6/91-9/91 only) (6/91).

Recommendation, Skill Level 20

In the lower-division baccalaureate/associate degree category, 2 semester hours in basic electronics laboratory, 3 in introduction to voice communications systems, 2 in records and information management, 2 in maintenance management, and 3 in personnel supervision. If ASI was J7, WHCA Console Control Operations, additional credit as follows: 1 semester hour in computer literacy, 2 in files management, and 3 in introduction to voice communications networks. (NOTE: This recommendation for skill level 20 is valid for the dates 6/91-2/95 only) (6/91).

Recommendation, Skill Level 30

In the lower-division baccalaureate/associate degree category, 2 semester hours in basic electronics laboratory, 3 in introduction to voice communications systems, 2 in records and information management, 3 in maintenance management, and 3 in personnel supervision. If ASI was J7, WHCA Console Control Operations, additional credit as follows: 1 semester hour in computer literacy, 2 in files management, and 3 in introduction to voice communications networks (6/91).

MOS-31N-003

Communications Systems/Circuit Controller
31N10
31N20
31N30
31N40

Exhibit Dates: 1/90–6/94. Effective 6/94, MOS 31N was discontinued, and its duties were incorporated into MOS 31P, Microwave Systems Operator-Maintainer.

Career Management Field: 31 (Signal Operations).

Description

Summary: Supervises or operates communications patching panels to establish interconnections between communications facilities. *Skill Level 10:* Terminates and patches circuits in accordance with patching orders; designates cable pairs to be used between loop and terminal; assists in testing local wire, radio, data, and teletypewriter circuits; operates test equipment, including oscilloscope, signal generator, and multimeter, to identify circuit and equipment faults; operates telephone switchboard and teletypewriter components of patching panels; interprets system maps, overlays, and circuit diagrams. *Skill Level 20:* Able to perform the duties required for Skill Level 10; may supervise up to three subordinates; troubleshoots communications equipment, including VHF, microwave, troposcatter, satellite, and other narrow band or wideband systems; prepares circuit connection record and interprets routing charts and communications diagrams; maintains circuit control records; identifies sources of communications interference and circuit trouble. *Skill Level 30:* Able to perform the duties required for Skill Level 20; supervises circuit activities in a switching center; supervises the installation and operation of system control facilities; diagnoses and determines causes of malfunctions of circuits and equipment; interprets and prepares operating instructions, circuit routing charts, communications diagrams, route bulletins, circuit registers, and other pertinent diagrams, charts, and records; evaluates performance of subordinates; may supervise four or more persons. *Skill Level 40:* Able to perform the duties required for Skill Level 30; establishes maintenance procedures; coordinates and supervises organizational maintenance of circuit activities in a switching center of a system control facility.

Recommendation, Skill Level 10

Credit is not recommended because of the limited or specialized nature of the skills, competencies, and knowledge (6/91).

Recommendation, Skill Level 20

In the lower-division baccalaureate/associate degree category, 2 semester hours in records and information management, 2 in maintenance management, and 3 in personnel supervision (6/91).

Recommendation, Skill Level 30

In the lower-division baccalaureate/associate degree category, 2 semester hours in records and information management, 3 in maintenance management, and 3 in personnel supervision (6/91).

Recommendation, Skill Level 40

In the lower-division baccalaureate/associate degree category, 2 semester hours in records and information management, 3 in maintenance management, and 3 in personnel supervision. In the upper-division baccalaureate category, 3 semester hours in organizational management and 3 for field experience in management (6/91).

MOS-31P-001

Microwave Systems Operator-Maintainer
31P10
31P20
31P30
31P40

Exhibit Dates: 2/94–2/95.

Career Management Field: 31 (Signal Operations).

Description

Summary: Supervises, operates, and maintains microwave communications systems, antennas, and support equipment; performs quality control and testing of circuits, trunks, links, systems, and facilities. *Skill Level 10:* Configures, aligns, and performs maintenance on microwave communications and support equipment; monitors, diagnoses, and restores circuits, systems, and interface equipment. *Skill Level 20:* Able to perform the duties required for Skill Level 10; performs complex microwave system troubleshooting and maintenance tasks; provides technical assistance to personnel engaged in operation and maintenance of microwave systems; supervises and assists subordinates in performing preventive maintenance on assigned communications equipment. *Skill Level 30:* Able to perform the duties required for Skill Level 20; implements service orders; supervises the operation and maintenance of communications systems; performs limited maintenance on the equipment; establishes work load and repair priorities; performs or supervises quality control and technical inspections. *Skill Level 40:* Able to perform the duties required for Skill Level 30; supervises, plans, and directs the installation, operation, and maintenance of microwave communications systems and facilities; prepares equipment service requests; provides technical advice and assistance; writes operating policies and procedures; determines unit capability; develops and conducts training for subordinates.

Recommendation, Skill Level 10

Credit may be granted on the basis of an individualized assessment of the student (3/95).

Recommendation, Skill Level 20

In the lower-division baccalaureate/associate degree category, 3 semester hours in electronic systems troubleshooting and repair, 2 in maintenance management, and 2 in personnel supervision (3/95).

Recommendation, Skill Level 30

In the lower-division baccalaureate/associate degree category, 4 semester hours in electronic systems troubleshooting and repair, 3 in maintenance management, and 3 in personnel supervision. In the upper-division baccalaureate category, 3 semester hours for field supervision in management (3/95).

Recommendation, Skill Level 40

In the lower-division baccalaureate/associate degree category, 4 semester hours in electronic systems troubleshooting and repair, 4 in maintenance management, and 4 in personnel supervision. In the upper-division baccalaureate category, 4 semester hours for field experience in management (3/95).

MOS-31P-002

Microwave Systems Operator-Maintainer
31P10
31P20
31P30
31P40

Exhibit Dates: 3/95–Present.

Career Management Field: 31 (Signal Operations).

Description

Summary: Supervises, operates, and maintains microwave communications systems, antennas, and support equipment; performs quality control and testing of circuits, trunks, links, systems, and facilities. *Skill Level 10:* Configures, aligns, and performs maintenance on microwave communications and support equipment; monitors, diagnoses, and restores circuits, systems, and interface equipment. *Skill Level 20:* Able to perform the duties required for Skill Level 10; performs complex microwave system troubleshooting and maintenance tasks; provides technical assistance to personnel engaged in operation and maintenance of microwave systems; supervises and assists subordinates in performing preventive maintenance on assigned communications equipment. *Skill Level 30:* Able to perform the duties required for Skill Level 20; implements service orders; supervises the operation and maintenance of communications systems; performs limited maintenance on the equipment; establishes work load and repair priorities; performs or supervises quality control and technical inspections. *Skill Level 40:* Able to perform the duties required for Skill Level 30; supervises, plans, and directs the installation, operation, and maintenance of microwave communications systems and facilities; prepares equipment service requests; provides technical advice and assistance; writes operating policies and procedures; determines unit capability; develops and conducts training for subordinates.

Recommendation, Skill Level 10

Credit may be granted on the basis of an individualized assessment of the student (3/95).

Recommendation, Skill Level 20

Credit may be granted on the basis of an individualized assessment of the student (5/96).

Recommendation, Skill Level 30

In the lower-division baccalaureate/associate degree category, 3 semester hours in electronic systems troubleshooting and repair, 3 in maintenance management, and 3 in personnel supervision. In the upper-division baccalaureate category, 3 semester hours for field supervision in management (3/95).

Recommendation, Skill Level 40

In the lower-division baccalaureate/associate degree category, 4 semester hours in electronic systems troubleshooting and repair, 4 in maintenance management, and 4 in personnel supervision. In the upper-division baccalaureate category, 4 semester hours for field experience in management (3/95).

MOS-31Q-001

Tactical Satellite/Microwave Systems Operator
31Q10
31Q20
31Q30

Exhibit Dates: 1/90–12/91. Effective 12/91, MOS 31Q was discontinued, and its duties were incorporated into MOS 31M, Multichannel Communications Systems Operator, and ASI 7A, Satellite Communications Terminal AN/TSC-85 Operator/Maintainer.

Career Management Field: 31 (Communications-Electronics Operations).

Description

Summary: Supervises or performs installation, operation, and maintenance of microwave, troposcatter radio, satellite, and multiplexing equipment. *Skill Level 10:* Installs, operates, and maintains microwave, troposcatter radio satellite, multiplexing equipment, and associated power equipment and communications security devices; assembles, erects, and orients antennas; tunes, adjusts, and aligns equipment for maximum performance; adjusts multiplexing equipment for maximum efficiency; operates voice frequency and teletype equipment; performs satellite balancing procedures; employs test equipment and tools such as spectrum analyzers and voltmeters; corrects operational troubles; detects electronic countermeasures; interprets military maps and traffic diagrams; applies safety precautions around high voltages and performs emergency actions in event of injury. *Skill Level 20:* Able to perform the duties required for Skill Level 10; assists in selection of site for equipment installation; makes alignment adjustments for maximum efficiency; monitors operations of system equipment; recognizes electronic countermeasures and employs corrective action; maintains station logs; provides technical guidance to subordinates. *Skill Level 30:* Able to perform the duties required for Skill Level 20; supervises installation of station equipment and performance of personnel; instructs in proper operating procedures; maintains logs and records; prepares administrative and technical reports; requisitions supplies; provides assistance in complex or unusual operational problems; supervises and evaluates five or more subordinates.

Recommendation, Skill Level 10

In the lower-division baccalaureate/associate degree category, 2 semester hours in electronic troubleshooting and repair and 2 in satellite communication systems. (NOTE: This recom-

mendation for skill level 10 is valid for the dates 1/90-9/91 only) (7/87).

Recommendation, Skill Level 20

In the lower-division baccalaureate/associate degree category, 4 semester hours in electronic troubleshooting and repair, 2 in satellite communication systems, and 2 in record keeping (7/87).

Recommendation, Skill Level 30

In the lower-division baccalaureate/associate degree category, 4 semester hours in electronic troubleshooting and repair, 2 in satellite communication systems, 2 in record keeping, 3 in principles of supervision, and 3 in maintenance management (7/87).

MOS-31R-001

Multichannel Transmission Systems Operator/Maintainer

31R10
31R20
31R30

Exhibit Dates: 2/95–Present.

Career Management Field: 31 (Signal Operations).

Description

Summary: Supervises or performs installation, operation, and unit-level maintenance on multichannel line of site and tropospheric scatter communications systems, communications security devices, and associated equipment. *Skill Level 10:* Under supervision, installs, operates, and performs unit-level maintenance on multichannel line of site and tropospheric scatter communications systems, antennas, and associated equipment. *Skill Level 20:* Able to perform the duties required for Skill Level 10; provides technical assistance to subordinates. *Skill Level 30:* Able to perform the duties required for Skill Level 20; supervises and prepares work schedules; assigns duties; provides technical assistance to subordinates in resolving complex maintenance problems; prepares, reviews, and consolidates technical and administrative reports and requests; conducts on-the-job training.

Recommendation, Skill Level 10

Credit may be granted on the basis of an individualized assessment of the student (7/97).

Recommendation, Skill Level 20

Credit may be granted on the basis of an individualized assessment of the student (7/97).

Recommendation, Skill Level 30

In the vocational certificate category, 2 semester hours in radio operation and maintenance. In the lower-division baccalaureate/associate degree category, 3 semester hours in personnel supervision and 3 in maintenance management (7/97).

MOS-31S-004

Satellite Communication Systems Operator-Maintainer

31S10
31S20
31S30
31S40

Exhibit Dates: 2/94–Present.

Career Management Field: 31 (Signal Operations).

Description

Summary: Installs, operates, maintains, and supervises satellite communications ground terminals, systems, networks, and associated equipment; operates and performs preventive maintenance checks and services on communications equipment, vehicles, and power generators. *Skill Level 10:* Configures, aligns, tests, and maintains satellite communications equipment and associated devices; conducts network operations; performs quality control tests; prepares system and equipment forms and reports; operates and performs preventive maintenance on assigned equipment, vehicles, and power generators. *Skill Level 20:* Able to perform the duties required for Skill Level 10; provides technical assistance to subordinates; identifies and reports electronic jamming and applies appropriate electronic countermeasures; supervises and performs alignment of communications systems; ensures adequate inventory of spare parts; compiles system and network statistics for reports. *Skill Level 30:* Able to perform the duties required for Skill Level 20; plans, supervises, and integrates the installation, operation, and maintenance of satellite communication equipment; establishes work load, work schedules, and maintenance priorities; trains subordinates; consolidates system and network statistics and reports; implements service orders; performs and supervises quality control inspections. *Skill Level 40:* Develops and ensures compliance with policies and procedures related to facility management; determines capabilities and limitations of assigned equipment; handles service requests from joint defense or commercial activities; provides training to subordinates; manages security programs for satellite facilities.

Recommendation, Skill Level 10

Credit may be granted on the basis of an individualized assessment of the student (3/95).

Recommendation, Skill Level 20

In the lower-division baccalaureate/associate degree category, 3 semester hours in electronic systems troubleshooting and repair and 3 in basic electronics. (NOTE: This recommendation for skill level 20 is valid for the dates 2/94-2/95 only) (3/95).

Recommendation, Skill Level 30

In the lower-division baccalaureate/associate degree category, 3 semester hours in electronic systems troubleshooting and repair, 3 in basic electronics, 2 in personnel supervision, and 3 in maintenance management. In the upper-division baccalaureate category, 3 semester hours for field experience in management (3/95).

Recommendation, Skill Level 40

In the lower-division baccalaureate/associate degree category, 3 semester hours in electronic systems troubleshooting and repair, 3 in basic electronics, 2 in record keeping, 3 in personnel supervision, and 3 in maintenance management. In the upper-division baccalaureate category, 4 semester hours for field experience in management (3/95).

MOS-31T-004

Satellite/Microwave Systems Chief

31T50

Exhibit Dates: 2/94–Present.

Career Management Field: 31 (Signal Operations).

Description

Supervises and directs duties of subordinates in the installation and operation of satellite/microwave communications control system; keeps records and writes reports on operation, utilization, and maintenance of satellite microwave communications control system plans; directs personnel assignments; performs liaison among command, staff, operations, and maintenance personnel; delegates the preparation of technical studies, evaluation reports, and associated records. NOTE: May have progressed to 31T50 from Skill Level 40 of MOS 31S (Satellite Communication Systems Operator/Maintainer) or MOS 31P (Microwave Systems Operator/Maintainer).

Recommendation

In the lower-division baccalaureate/associate degree category, 3 semester hours in basic electronics, 3 in electronic systems troubleshooting and repair, 2 in record keeping, 3 in personnel supervision, and 3 in maintenance management. In the upper-division baccalaureate category, 4 semester hours for field experience in management (3/95).

MOS-31U-002

Signal Support Systems Specialist

31U10
31U20
31U30
31U40
31U50

Exhibit Dates: 2/93–2/95.

Career Management Field: 31 (Signal Operations).

Description

Summary: Supervises, installs, maintains, troubleshoots, and employs battlefield signal support systems; performs unit maintenance on signal equipment and trains users of equipment. *Skill Level 10:* Installs, maintains, and troubleshoots signal support equipment, radio systems, and data distribution systems; provides technical support and training for users. *Skill Level 20:* Able to perform the duties required for Skill Level 10; supervises, installs, maintains, and troubleshoots signal support systems, radio systems, and battlefield automated systems; provides training and unit technical assistance; prepares maintenance and supply requests. *Skill Level 30:* Able to perform the duties required for Skill Level 20; plans, supervises, and integrates the installation, operation, and maintenance of signal support systems; plans, supervises, and executes unit maintenance programs, operations orders, and reports; plans and provides unit training. *Skill Level 40:* Able to perform the duties required for Skill Level 30; plans and supervises the installation, operation, and maintenance of signal support systems and network integration; develops and implements unit maintenance programs; directs unit training; provides technical advice and logistics support for unit operations and maintenance. *Skill Level 50:* Able to perform the duties required for Skill Level 40; supervises, plans, and integrates the installation, employment, and maintenance of signal support systems; develops staff policies and procedures; provides technical advice and assistance to commanders and subordinate units; develops signal battlefield integration plans; coordinates all aspects of signal activities with higher, lower, and adjacent headquarters.

Recommendation, Skill Level 10

Credit may be granted on the basis of an individualized assessment of the student (3/95).

Recommendation, Skill Level 20

In the lower-division baccalaureate/associate degree category, 3 semester hours in basic electronics and 3 in electronic systems troubleshooting and repair (3/95).

Recommendation, Skill Level 30

In the lower-division baccalaureate/associate degree category, 3 semester hours in basic electronics, 4 in electronic systems troubleshooting and repair, 3 in maintenance management, and 3 in personnel supervision. In the upper-division baccalaureate category, 3 semester hours for field experience in management (3/95).

Recommendation, Skill Level 40

In the lower-division baccalaureate/associate degree category, 3 semester hours in basic electronics, 4 in electronic systems troubleshooting and repair, 4 in maintenance management, and 3 in personnel supervision. In the upper-division baccalaureate category, 4 semester hours for field experience in management (3/95).

Recommendation, Skill Level 50

In the lower-division baccalaureate/associate degree category, 3 semester hours in basic electronics, 4 in electronic systems troubleshooting and repair, 4 in maintenance management, and 3 in personnel supervision. In the upper-division baccalaureate category, 5 semester hours for field experience in management (3/95).

MOS-31U-003

SIGNAL SUPPORT SYSTEMS SPECIALIST

31U10
31U20
31U30
31U40
31U50

Exhibit Dates: 3/95–Present.

Career Management Field: 31 (Signal Operations).

Description

Summary: Supervises, installs, maintains, troubleshoots, and employs battlefield signal support systems; performs unit maintenance on signal equipment and trains users of equipment. *Skill Level 10:* Installs, maintains, and troubleshoots signal support equipment, radio systems, and data distribution systems; provides technical support and training for users. *Skill Level 20:* Able to perform the duties required for Skill Level 10; supervises, installs, maintains, and troubleshoots signal support systems, radio systems, and battlefield automated systems; provides training and unit technical assistance; prepares maintenance and supply requests. *Skill Level 30:* Able to perform the duties required for Skill Level 20; plans, supervises, and integrates the installation, operation, and maintenance of signal support systems; plans, supervises, and executes unit maintenance programs, operations orders, and reports; plans and provides unit training. *Skill Level 40:* Able to perform the duties required for Skill Level 30; plans and supervises the installation, operation, and maintenance of signal support systems and network integration; develops and implements unit maintenance programs; directs unit training; provides technical advice and logistics support for unit operations and maintenance. *Skill Level 50:* Able to perform the duties required for Skill Level 40; supervises, plans, and integrates the installation, employment, and maintenance of signal support systems; develops staff policies and procedures; provides technical advice and assistance to commanders and subordinate units; develops signal battlefield integration plans; coordinates all aspects of signal activities with higher, lower, and adjacent headquarters.

Recommendation, Skill Level 10

Credit may be granted on the basis of an individualized assessment of the student (3/95).

Recommendation, Skill Level 20

Credit may be granted on the basis of an individualized assessment of the student (3/95).

Recommendation, Skill Level 30

In the lower-division baccalaureate/associate degree category, 3 semester hours in basic electronics, 3 in electronic systems troubleshooting and repair, 3 in maintenance management, and 3 in personnel supervision. In the upper-division baccalaureate category, 3 semester hours for field experience in management (3/95).

Recommendation, Skill Level 40

In the lower-division baccalaureate/associate degree category, 3 semester hours in basic electronics, 4 in electronic systems troubleshooting and repair, 4 in maintenance management, and 3 in personnel supervision. In the upper-division baccalaureate category, 4 semester hours for field experience in management (3/95).

Recommendation, Skill Level 50

In the lower-division baccalaureate/associate degree category, 3 semester hours in basic electronics, 4 in electronic systems troubleshooting and repair, 4 in maintenance management, and 3 in personnel supervision. In the upper-division baccalaureate category, 5 semester hours for field experience in management (3/95).

MOS-31V-002

UNIT LEVEL COMMUNICATIONS MAINTAINER

31V10
31V20

Exhibit Dates: 1/90–6/93. Effective 6/93, MOS 31V was discontinued, and its duties were incorporated into MOS 31U, Signal Support Systems Specialist.

Career Management Field: 31 (Communications Electronics Operations).

Description

Summary: Installs, operates, supervises, and performs preventive maintenance on tactical communications and selected electronic equipment in tactical units. *Skill Level 10:* Installs and operates tactical radio equipment; performs limited maintenance, such as module replacement and antenna repair on HF and VHF ground and air-to-ground tactical equipment. *Skill Level 20:* Able to perform the duties required for Skill Level 10; supervises the preparation of equipment maintenance records; conducts on-the-job training programs.

Recommendation, Skill Level 10

In the lower-division baccalaureate/associate degree category, 3 semester hours in electronic communications and 3 in basic electronics laboratory. (NOTE: This recommendation for skill level 10 is valid for the dates 1/90-9/91 only) (6/88).

Recommendation, Skill Level 20

In the lower-division baccalaureate/associate degree category, 3 semester hours in electronic communications, 3 in basic electronics laboratory, 2 in personnel supervision, and 3 in record keeping (6/88).

MOS-31W-002

TELECOMMUNICATIONS OPERATIONS CHIEF

((Mobile Subscriber Equipment Communications Chief))

31W40
31W50

Exhibit Dates: 1/90–4/99.

Career Management Field: 31 (Signal Operations).

Description

Summary: Plans, coordinates, directs, provides technical assistance, and supervises the installation, operation and maintenance of Mobile Subscriber Equipment (MSE). *Skill Level 40:* Assigns, supervises and coordinates duties of subordinate personnel concerning setup and operation of MSE equipment, supervises preparation of records and reports; supervises and conducts training in operating techniques, equipment utilization, and maintenance; manages, develops, and supervises maintenance system; monitors adequacy and effectiveness of maintenance programs and supply support. NOTE: May have progressed to 31W40 from 31D30 (Mobile Subscriber Equipment Transmission System Operator) or 31F30 (Mobile Subscriber Equipment Network Switching System Operator). *Skill Level 50:* Able to perform the duties required for Skill Level 40; plans, coordinates, and directs deployments; performs liaison among command, staff, operations, and maintenance personnel; prepares and supervises the preparation of technical studies, evaluation reports, and associated records.

Recommendation, Skill Level 40

In the lower-division baccalaureate/associate degree category, if ASI was J7, WHCA Console Control Operations, the following credit applies: 3 semester hours in computer literacy, 2 in files management, and 3 in introduction to voice communications networks. In the upper-division baccalaureate category, for all individuals: 3 semester hours for field experience in management and 3 in management problems (6/90).

Recommendation, Skill Level 50

In the lower-division baccalaureate/associate degree category, if ASI was J7, WHCA Console Control Operations, the following credit applies: 3 semester hours in computer literacy, 2 in files management, and 3 in introduction to voice communications networks. In the upper-division baccalaureate category, for all individuals: 6 semester hours for field experience in management and 3 in management problems (6/90).

MOS-31W-003

TELECOMMUNICATIONS OPERATIONS CHIEF

31W40
31W50

Exhibit Dates: 5/99–Present.

Career Management Field: 31 (Signal Operations).

Description

Summary: Plans, coordinates, directs, provides technical assistance, and supervises the installation, operation and maintenance of Mobile Subscriber Equipment (MSE). *Skill Level 40:* Assigns, supervises, and coordinates duties of subordinates concerning setup and operation of MSE equipment; supervises preparation of records and reports; supervises and conducts training in operating techniques, equipment utilization, and maintenance; manages, develops, and supervises maintenance system; monitors adequacy and effectiveness of maintenance programs and supply support. NOTE: May have progressed to 31W40 from 31D30 (Mobile Subscriber Equipment Transmission System Operator) or 31F30 (Mobile Subscriber Equipment Network Switching System Operator). *Skill Level 50:* Able to perform the duties required for Skill Level 40; performs liaison among command, staff, operations, and maintenance personnel; prepares and supervises the preparation of technical studies, evaluation reports, and associated records.

Recommendation, Skill Level 40

In the upper-division baccalaureate category, 3 semester hours for field experience in management and 3 in maintenance management (5/99).

Recommendation, Skill Level 50

In the upper-division baccalaureate category, 3 semester hours for field experience in management and 6 in maintenance management (5/99).

MOS-31Y-001

Telecommunications Systems Supervisor
((Communications Systems Supervisor))
31Y40

Exhibit Dates: 1/90–6/95. Effective 6/95, MOS 31Y was discontinued, and its duties were incorporated into MOS's 31W and 31S.

Career Management Field: 31 (Communications Electronics Operations).

Description

Directs, coordinates, and manages the installation, operation, and maintenance of manual and automated communications equipment, systems, and facilities; provides technical guidance concerning all aspects of communications systems employment, maintenance, and logistical support; prepares technical reports; maintains system documentation; conducts briefings for staff and maintenance personnel; performs liaison with staff, operations, and maintenance personnel; manages human resource functions for personnel engaged in the maintenance and operations of wire and cable systems, switchboards, radioteletype, relay, satellite, and microwave communications; conducts training programs. NOTE: May have progressed to 31Y40 from 13C30 (Single Channel Radio Operator), 31L30 (Wire Systems Installer), 31M30 (Multichannel Communications Systems Operator), 31Q30 (Tactical Satellite/Microwave Systems Operator), 36L30 (Transportable Automatic Switching Systems Operator/Maintainer), or 36M30 (Switching Systems Operator).

Recommendation

In the lower-division baccalaureate/associate degree category, if ASI was J7, WHCA Console Control Operations, the following credit applies: 1 semester hour in computer literacy, 2 in files management, and 3 in introduction to voice communications networks. In the upper-division baccalaureate category, for all individuals: 3 semester hours in management problems and 3 for field experience in management (6/88).

MOS-31Z-003

Senior Signal Sergeant
((Communications-Operations Chief))
31Z50

Exhibit Dates: 1/90–4/99.

Career Management Field: 31 (Communications Electronics Operations).

Description

Supervises, coordinates, and directs the installation, operation, and maintenance of integrated electronic communications systems; manages communications facility; performs liaison with civilian, command staff, and operations and maintenance personnel; plans, organizes, directs, and supervises the deployment of integrated communications systems, including site selection, personnel assignments, and budget development. NOTE: May have progressed to 31Z50 from 31G40 (Tactical Communications Chief), 31Y40 (Communications Systems Supervisor), 31X40 (Communications Systems/Circuit Control Supervisor), 72E40 (Tactical Telecommunications Center Operator), or 72G40 (Automatic Data Telecommunications Center Operator).

Recommendation

In the lower-division baccalaureate/associate degree category, if ASI was J7, WHCA Console Control Operations, the following credit applies: 1 semester hour in computer literacy, 2 in files management, and 3 in introduction to voice communications networks. In the upper-division baccalaureate category, for all individuals: 3 semester hours in organizational management and 3 for field experience in management (6/88).

MOS-31Z-004

Senior Signal Sergeant
((Communications-Operations Chief))
31Z50

Exhibit Dates: 5/99–Present.

Career Management Field: 31 (Communications Electronics Operations).

Description

Supervises, coordinates, and directs the installation, operation, and maintenance of integrated electronic communications systems and information management networks; manages communications facility; uses computers to control and monitor network status in an operational environment; performs liaison with civilian, command staff, and operations and maintenance personnel; plans, organizes, directs, and supervises the deployment of integrated communications systems, including site selection, personnel assignments, and budget development. NOTE: May have progressed to 31Z50 from 31G40 (Tactical Communications Chief), 31Y40 (Communications Systems Supervisor), 31X40 (Communications Systems/Circuit Control Supervisor), 72E40 (Tactical Telecommunications Center Operator), or 72G40 (Automatic Data Telecommunications Center Operator).

Recommendation

In the lower-division baccalaureate/associate degree category, 3 semester hours in computer applications. In the upper-division baccalaureate category, 3 semester hours in organizational management, 3 for field experience in management, 3 network administration, and 3 in information management (5/99).

MOS-33M-001

Electronic Warfare/Intercept Strategic Systems Analyst and Command and Control Subsystems Repairer
33M20
33M30
33M40

Exhibit Dates: 1/90–9/91. Effective 9/91 MOS 33M was discontinued, and its duties were incorporated into MOS 33Y, Strategic Systems Repairer.

Career Management Field: 33 (Electronic Warfare/Intercept Systems Maintenance).

Description

Summary: Supervises or performs unit, intermediate, and depot maintenance on bus controllers, time code generators, microcomputers, and peripheral equipment. *Skill Level 20:* Replaces, repairs, and installs electronic warfare/intercept command and control equipment according to manufacturers' instructions; conducts electrical and mechanical tests using wiring diagrams and technical manuals; performs preventive maintenance on computers and computer peripherals. *Skill Level 30:* Able to perform the duties required for Skill Level 20; serves as maintenance team leader or assistant team leader; serves as quality assurance inspector and equipment calibration coordinator; manages, counsels, and evaluates subordinates; develops maintenance procedures; prepares and reviews technical reports; maintains a technical library; requisitions spare parts; maintains records; conducts on-the-job training. *Skill Level 40:* Able to perform the duties required for Skill Level 30; serves as team leader or shop foreman; assigns personnel to maintenance teams and supervises team activities; conducts and monitors on-the-job training programs; counsels and evaluates subordinates and determines their training requirements; prepares status reports and maintains administrative and maintenance files. NOTE: May have progressed to 33M40 from 33M30, 33P30, Electronic Warfare/Intercept Strategic Receiving Subsystems Repairer, or 33Q30, Electronic Warfare/Intercept Strategic Processing/Storage Subsystems Repairer.

Recommendation, Skill Level 20

In the vocational certificate category, 6 semester hours in troubleshooting techniques. In the lower-division baccalaureate/associate degree category, 3 semester hours in electronics laboratory and 3 in electronic shop practices (5/87).

Recommendation, Skill Level 30

In the vocational certificate category, 6 semester hours in troubleshooting techniques. In the lower-division baccalaureate/associate degree category, 3 semester hours in electronics laboratory, 3 in electronic shop practices, 3 in office procedures, and 3 in human relations (5/87).

Recommendation, Skill Level 40

In the vocational certificate category, 6 semester hours in troubleshooting techniques. In the lower-division baccalaureate/associate degree category, 3 semester hours in electronics laboratory, 3 in electronic shop practices, 3 in office procedures, and 3 in human relations. In the upper-division baccalaureate category, 3 semester hours in office management and 3 in personnel management (5/87).

MOS-33P-001

Electronic Warfare/Intercept Strategic Receiving Subsystems Repairer

33P10
33P20
33P30

Exhibit Dates: 1/90–9/91. Effective 9/91, MOS 33P was discontinued, and its duties were incorporated into MOS 33Y, Strategic Systems Repairer.

Career Management Field: 33 (Electronic Warfare/Intercept Systems Maintenance).

Description

Summary: Supervises or performs preventive maintenance on receivers, direction finders, antenna control, multiplexers, demultiplexers, and other equipment associated with RF and receiving subsystems. *Skill Level 10:* Troubleshoots electronic warfare/intercept receiving equipment and support equipment down to the component level; installs receiving equipment and support equipment according to manufacturers' specifications; performs tests using volt ohm meters, oscilloscopes, signal generators, spectrum analyzers, and other test equipment. *Skill Level 20:* Able to perform the duties required for Skill Level 10; assists in conducting on-the-job training and quality assurance inspections; performs advanced electrical and mechanical tests using wiring diagrams, signal flow charts, technical manuals, troubleshooting charts, and other diagnostic materials; supervises work performed by subordinates. *Skill Level 30:* Able to perform the duties required for Skill Level 20; serves as maintenance team leader or assistant team leader; serves as quality assurance inspector and equipment calibration coordinator; manages, counsels, and evaluates subordinates; develops maintenance procedures; prepares and reviews technical reports; maintains a technical library; requisitions spare parts; maintains records; conducts on-the-job training.

Recommendation, Skill Level 10

In the vocational certificate category, 6 semester hours in troubleshooting techniques (5/87).

Recommendation, Skill Level 20

In the vocational certificate category, 6 semester hours in troubleshooting techniques. In the lower-division baccalaureate/associate degree category, 3 semester hours in electronics laboratory and 3 in electronic shop practices (5/87).

Recommendation, Skill Level 30

In the vocational certificate category, 6 semester hours in troubleshooting techniques. In the lower-division baccalaureate/associate degree category, 3 semester hours in electronics laboratory, 3 in electronics shop practices, 3 in office procedures, and 3 in human relations (5/87).

MOS-33Q-001

Electronic Warfare/Intercept Strategic Processing/Storage Subsystems Repairer

33Q10
33Q20
33Q30

Exhibit Dates: 1/90–9/91. Effective 9/91, MOS 33Q was discontinued, and its duties were incorporated into MOS 33Y, Strategic Systems Repairer.

Career Management Field: 33 (Electronic Warfare/Intercept Systems Maintenance).

Description

Summary: Supervises or performs maintenance on electronic warfare/intercept processing/storage subsystems, multiplexers, recorder-reproducers, oscillographs, control units, video amplifiers, power supplies, and other associated equipment. *Skill Level 10:* Uses voltmeters, ohmmeters, oscilloscopes, signal generators, power meters, spectrum analyzers, and other testing devices; conducts electrical and mechanical tests using wiring diagrams, signal flow charts, technical manuals, troubleshooting charts, and job performance aids; uses high-reliability soldering techniques; bench tests to ensure alignment; troubleshoots and repairs to the component level. *Skill Level 20:* Able to perform the duties required for Skill Level 10; performs harmonic, wow and flutter, and other distortion tests and takes corrective action; tests analog and digital circuits to determine course of pulse, spectra, time phase and amplitude deterioration, frequency shifting and bandpass reduction/expansion; assists in conducting formal/informal on-the-job training; serves as quality assurance inspector and as assistant service school instructor. *Skill Level 30:* Able to perform the duties required for Skill Level 20; supervises the performance of unit, intermediate, and depot maintenance on electronic warfare/intercept equipment; manages, counsels, and evaluates subordinate personnel; maintains records; requisitions spare parts; conducts on-the-job training.

Recommendation, Skill Level 10

In the vocational certificate category, 6 semester hours in troubleshooting techniques (5/87).

Recommendation, Skill Level 20

In the vocational certificate category, 6 semester hours in troubleshooting techniques. In the lower-division baccalaureate/associate degree category, 3 semester hours in electronics laboratory and 3 in electronic shop practices (5/87).

Recommendation, Skill Level 30

In the vocational certificate category, 6 semester hours in troubleshooting techniques. In the lower-division baccalaureate/associate degree category, 3 semester hours in electronics laboratory, 3 in electronics shop practices, 3 in office procedures, and 3 in human relations (5/87).

MOS-33R-001

Aviation Systems Repairer

((Electronic Warfare/Intercept Aviation Systems Repairer))

33R10
33R20
33R30
33R40

Exhibit Dates: 1/90–6/98. Effective 6/98, MOS 33R was discontinued and its duties were incorporated into MOS 33W (Electronic Warfare/Intercept Systems Repairer).

Career Management Field: 33 (Electronic Warfare/Intercept Systems Maintenance).

Description

Summary: Repairs, supervises, or performs unit, intermediate, or depot maintenance on electronic warfare intercept, airborne systems, equipment, and assemblies. *Skill Level 10:* Performs maintenance on aviation equipment including wideband receivers, recorders, modulators, multiplexers, demodulators, antennas, monitors, and direction finding and positioning equipment. *Skill Level 20:* Able to perform the duties required for Skill Level 10; performs maintenance on additional equipment, including various airborne and air-to-ground intercept, recording, countermeasure, and computer peripheral devices. *Skill Level 30:* Able to perform the duties required for Skill Level 20; supervises the maintenance on electronic warfare intercept airborne equipment; manages, counsels, and evaluates subordinates; requisitions spare parts; prepares and reviews technical reports; maintains technical library; conducts on-the-job training. *Skill Level 40:* Able to perform the duties required for Skill Level 30; serves as maintenance team leader or foreman; prepares status reports and maintains administrative and maintenance files. NOTE: May have progressed to 33R40 from 33R30 or 33V30, Electronic Warfare/Intercept Aerial Sensor Repairer.

Recommendation, Skill Level 10

In the vocational certificate category, 6 semester hours in troubleshooting techniques. (NOTE: This recommendation for skill level 10 is valid for the dates 1/90-9/91 only) (5/87).

Recommendation, Skill Level 20

In the vocational certificate category, 6 semester hours in troubleshooting techniques. In the lower-division baccalaureate/associate degree category, 3 semester hours in electronics laboratory and 3 in electronics shop practices. (NOTE: This recommendation for skill level 20 is valid for the dates 1/90-2/95 only (5/87).

Recommendation, Skill Level 30

In the vocational certificate category, 6 semester hours in troubleshooting techniques. In the lower-division baccalaureate/associate degree category, 3 semester hours in electronics laboratory, 3 in electronics shop practices, 3 in office procedures, and 3 in human relations (5/87).

Recommendation, Skill Level 40

In the vocational certificate category, 6 semester hours in troubleshooting techniques. In the lower-division baccalaureate/associate degree category, 3 semester hours in electronics laboratory, 3 in electronic shop practices, 3 in office procedures, and 3 in human relations. In the upper-division baccalaureate category, 3 semester hours in office management and 3 in personnel management (5/87).

MOS-33T-001

Tactical Systems Repairer
((Electronic Warfare/Intercept Tactical Systems Repairer))
33T10
33T20
33T30
33T40

Exhibit Dates: 1/90–6/98. Effective 6/98, MOS 33T was discontinued and its duties were incorporated into MOS 33W (Electronic Warfare/Intercept Systems Repairer).

Career Management Field: 33 (Electronic Warfare/Intercept Systems Maintenance).

Description

Summary: Repairs, supervises, or performs unit, intermediate, or depot maintenance on electronic warfare/intercept equipment and assemblies. *Skill Level 10:* Performs maintenance on equipment including HF, VHF, UHF, and M F receivers, narrowband recorder-reproducers, tactical antennas and controls, modulators and demodulators, multiplexers and demultiplexers, signal monitors, direction finding and electronic countermeasure equipment. *Skill Level 20:* Able to perform the duties required for Skill Level 10; performs harmonic distortion and wow and flutter tests on tactical narrowband recorders; tests analog and digital circuits to determine the cause of pulse, spectrum, time phase and amplitude deterioration, frequency shifting, and bandpass reduction/expansion; assists in conducting formal and informal on-the-job training; provides technical guidance to subordinates. *Skill Level 30:* Able to perform the duties required for Skill Level 20; supervises the performance of maintenance activities; manages, counsels, and evaluates subordinates; maintains records; requisitions spare parts; conducts on-the-job training. *Skill Level 40:* Able to perform the duties required for Skill Level 30; serves as maintenance team leader or foreman; organizes and monitors performance of maintenance activities; prepares status reports; maintains administrative and maintenance files.

Recommendation, Skill Level 10

In the vocational certificate category, 6 semester hours in troubleshooting techniques. (NOTE: This recommendation for skill level 10 is valid for the dates 1/90-9/91 only) (5/87).

Recommendation, Skill Level 20

In the vocational certificate category, 6 semester hours in troubleshooting techniques. In the lower-division baccalaureate/associate degree category, 3 semester hours in electronics laboratory and 3 in electronics shop practices. (NOTE: This recommendation for skill level 20 is valid for the dates 1/90-2/95 only (5/87).

Recommendation, Skill Level 30

In the vocational certificate category, 6 semester hours in troubleshooting techniques. In the lower-division baccalaureate/associate degree category, 3 semester hours in electronics laboratory, 3 in electronics shop practices, 3 in office procedures, and 3 in human relations (5/87).

Recommendation, Skill Level 40

In the vocational certificate category, 6 semester hours in troubleshooting techniques. In the lower-division baccalaureate/associate degree category, 3 semester hours in electronics laboratory, 3 in electronics shop practices, 3 in office procedures, and 3 in human relations. In the upper-division baccalaureate category, 3 semester hours in office management and 3 in personnel management (5/87).

MOS-33V-001

Electronic Warfare/Intercept Aerial Sensor Repairer
33V10
33V20
33V30

Exhibit Dates: 1/90–12/93. Effective 12/93, MOS 33V was discontinued, and its duties were incorporated into MOS 33R, Aviation Systems Repairer.

Career Management Field: 33 (Electronic Warfare/Intercept Systems Maintenance).

Description

Summary: Repairs, supervises, or performs unit, intermediate, or depot maintenance on electronic warfare intercept airborne sensor systems, associated data links, and assemblies. *Skill Level 10:* Performs maintenance on aviation equipment including side-looking radars, signal processors, monitors, receiving-transmitting equipment, encoding and decoding devices, and aircraft annotation equipment. *Skill Level 20:* Able to perform the duties required for Skill Level 10; supervises and assists subordinates; repairs selected specialized equipment; performs technical inspections of sensor modules and assemblies. *Skill Level 30:* Able to perform the duties required for Skill Level 20; counsels subordinates; assigns, monitors, and inspects work; develops maintenance procedures; supervises installation of certain aerial sensor systems; supervises acquisition of parts; diagnoses unusual and complex malfunctions of aerial sensor systems and recommends modifications of sensor systems; maintains technical library; writes reports.

Recommendation, Skill Level 10

In the vocational certificate category, 6 semester hours in troubleshooting techniques. (NOTE: This recommendation for skill level 10 is valid for the dates 1/90-9/91 only) (5/87).

Recommendation, Skill Level 20

In the vocational certificate category, 6 semester hours in troubleshooting techniques. In the lower-division baccalaureate/associate degree category, 3 semester hours in electronics laboratory and 3 in electronics shop practices (5/87).

Recommendation, Skill Level 30

In the vocational certificate category, 6 semester hours in troubleshooting techniques. In the lower-division baccalaureate/associate degree category, 3 semester hours in electronics laboratory, 3 in electronics shop practices, 3 in office procedures, and 3 in human relations (5/87).

MOS-33W-001

Electronic Warfare/Intercept Systems Repairer
((Intelligence and Electronic Warfare/Intercept Systems Repairer))
33W10
33W20
33W30
33W40
33W50

Exhibit Dates: 2/98–Present.

Description

Summary: Performs or supervises maintenance on tactical and strategic Intelligence and Electronic Warfare (IEW) systems, equipment, and assemblies. *Skill Level 10:* Maintains, tests, troubleshoots, and repairs electronic HF through SHF transmitters and receivers, analog and digital recorders, antennas and transmission lines, radar and computer equipment, electronic countermeasures, and automatic direction finding equipment (ADF) associated with tactical, strategic and aerial sensor systems using specialized test and measuring equipment. *Skill Level 20:* Able to perform the duties required for Skill Level 10; supervises and leads teams of IEW systems repairers; troubleshoots and repairs complex malfunctions in IEW systems; performs distortion tests on recorders; performs critical alignment and calibration of IEW systems; assists in conducting on-the-job training on IEW systems. *Skill Level 30:* Able to perform the duties required for Skill Level 20; supervises and leads squad or sections of IEW systems repairers; tests and troubleshoots complex, computer-controlled IEW systems; installs and maintains IEW systems computer networks; serves as a quality control inspector for IEW systems maintains technical library and equipment repair maintenance schedules; recommends damages to work load schedules; conducts formal and informal training sessions on equipment and military practices. *Skill Level 40:* Able to perform the duties required for Skill Level 30; supervises and leads platoons or shops of IEW systems repairers at direct support/general support maintenance activities; supervises unit test, measurement, and diagnostic equipment calibration program; monitors maintenance files in accordance with Army regulations; inspects the Army maintenance management system and repair parts records; verifies condition codes assigned to equipment during technical inspections. *Skill Level 50:* Able to perform the duties required for Skill Level 40; supervises and leads companies and major IEW system maintenance facilities, training programs and logistic activities; responsible for all aspects of operation of IEW repair facility; develops plans to use resources and determines priorities and training requirements; prepares technical studies, special reports, correspondence and records pertaining to maintenance operations.

Recommendation, Skill Level 10

Credit may be granted on the basis of an individualized assessment of the student (5/99).

Recommendation, Skill Level 20

Credit may be granted on the basis of an individualized assessment of the student (5/99).

Recommendation, Skill Level 30

In the lower-division baccalaureate/associate degree category, 3 semester hours in computer systems and organization, 3 in computer systems troubleshooting and maintenance, 3 in electronic communications, 3 in electronic systems troubleshooting and maintenance, 3 in personnel supervision, 3 in recordkeeping, and 3 in maintenance management (5/99).

Recommendation, Skill Level 40

In the lower-division baccalaureate/associate degree category, 3 semester hours in computer systems and organization, 3 in computer systems troubleshooting and maintenance, 3 in electronic communications, 3 in electronic systems troubleshooting and maintenance, 3 in personnel supervision, 3 in recordkeeping, and 3 in maintenance management. In the upper-division baccalaureate category, 3 semester hours for field experience in management and 3 in organizational management (5/99).

Recommendation, Skill Level 50

In the lower-division baccalaureate/associate degree category, 3 semester hours in computer systems and organization, 3 in computer systems troubleshooting and maintenance, 3 in electronic communications, 3 in electronic systems troubleshooting and maintenance, 3 in personnel supervision, 3 in recordkeeping, and 3 in maintenance management. In the upper-division baccalaureate category, 3 semester hours in management problems, 3 for field experience in management and 3 in organizational management (5/99).

MOS-33Y-001

Strategic Systems Repairer
33Y10
33Y20
33Y30
33Y40

Exhibit Dates: 9/90–6/98. Effective 6/98, MOS 33Y was discontinued and its duties were incorporated into MOS 33W (Electronic Warfare/Intercept Systems Repairer).

Career Management Field: 33 (Electronics Warfare/Intercept Systems Maintenance).

Description

Summary: Performs unit, intermediate, and depot maintenance on electronic communications systems consisting of high-frequency receivers, digital computer, digital and analog recorders, and computer peripherals. *Skill Level 10:* Troubleshoots and isolates equipment malfunctions to the module and component level using digital volt meters, oscilloscopes, signal generators, spectrum analyzers, and diagnostic software; installs and tests system components; performs preventive maintenance on system; solders to high-reliability standards; generates work orders. *Skill Level 20:* Able to perform the duties required for Skill Level 10; performs intermediate testing and alignment of RF receivers including wow, flutter, and harmonic distortion; performs advanced testing of computer systems including resolution of hardware-software conflicts; maintains and calibrates test equipment; serves as quality assurance inspector; supervises and conducts on-the-job training. *Skill Level 30:* Able to perform the duties required for Skill Level 20; performs advanced testing and troubleshooting of RF and computer equipment; analyzes recurring malfunctions and recommends equipment modifications; performs critical alignment and calibration of subsystems; serves as maintenance assistant team leader or team leader and quality assurance inspector; coordinates equipment calibration and testing; manages, counsels, and evaluates subordinates; develops maintenance procedures; prepares and reviews technical reports; maintains a technical library; requisitions spare parts; conducts on-the-job training. *Skill Level 40:* Able to perform the duties required for Skill Level 30; serves as maintenance team leader; assigns personnel to maintenance teams and supervises team activities; conducts and monitors on-the-job training programs; counsels and evaluates subordinates and determines their training requirements; prepares status reports and maintains administrative and maintenance files; reviews and generates technical reports; determines maintenance budget requirements; controls preventive maintenance program; prepares and delivers maintenance briefings; performs acceptance inspections; supervises the installation of new equipment.

Recommendation, Skill Level 10

In the lower-division baccalaureate/associate degree category, 3 semester hours in computer systems and organization, 3 in computer systems troubleshooting and maintenance, 3 in electronic communications, and 3 in electronic systems troubleshooting and maintenance. (NOTE: This recommendation for skill level 10 is valid for the dates 9/90-9/91 only) (10/93).

Recommendation, Skill Level 20

In the lower-division baccalaureate/associate degree category, 3 semester hours in computer systems and organization, 6 in computer systems troubleshooting and maintenance, 3 in electronic communications, and 6 in electronic systems troubleshooting and maintenance. (NOTE: This recommendation for skill level 20 is valid for the dates 9/90-2/95 only) (10/93).

Recommendation, Skill Level 30

In the lower-division baccalaureate/associate degree category, 3 semester hours in computer systems and organization, 6 in computer systems troubleshooting and maintenance, 3 in electronic communications, 6 in electronic systems troubleshooting and maintenance, 3 in personnel supervision, and 3 in records and information management (10/93).

Recommendation, Skill Level 40

In the lower-division baccalaureate/associate degree category, 3 semester hours in computer systems and organization, 6 in computer systems troubleshooting and maintenance, 3 in electronic communications, 6 in electronic systems troubleshooting and maintenance, 3 in personnel supervision, and 3 in records and information management. In the upper-division baccalaureate category, 3 semester hours for field experience in management and 3 in organizational management (10/93).

MOS-33Z-002

Electronic Warfare/Intercept Systems Maintenance Supervisor
33Z50

Exhibit Dates: 1/90–6/98. Effective 6/98, MOS 33Z was discontinued and its duties were incorporated into MOS 33W (Electronic Warfare/Intercept Systems Repairer) and MOS 96D (Imagery Analyst).

Career Management Field: 33 (Electronic Warfare/Intercept Systems Maintenance).

Description

Supervises unit, intermediate, and depot maintenance on all types of electronic warfare/intercept equipment; plans layout and design of maintenance and facilities; identifies training requirements and develops training plans; serves as liaison with staff command, site, and unit personnel; prepares technical evaluations, reports, and correspondence pertaining to maintenance operations. NOTE: May have progressed to 33Z50 from 33R40, Electronic Warfare/Intercept Aviation Systems Repairer; 33T40, Electronic Warfare/Intercept Tactical Systems Repairer; or 33M40, Electronic Warfare/Intercept Strategic Systems Analyst and Command and Control Subsystems Repairer.

Recommendation

In the lower-division baccalaureate/associate degree category, 3 semester hours in applied psychology. In the upper-division baccalaureate category, 3 semester hours in technical writing, 3 in a management elective, 3 in industrial education, 3 in counseling, 3 for an education internship, and 3 for field experience in management (5/87).

MOS-35A-001

Land Combat Electronic Missile System Repairer
35A10
35A20
35A30
35A40

Exhibit Dates: 10/93–4/04. (Effective 5/04 this MOS was discontinued, and its duties were incorporated into MOS 94A)

Career Management Field: 27 (Land Combat and Air Defense Systems Direct and General Support Maintenance).

Description

Summary: Performs direct and general support maintenance on TOW and DRAGON missile systems, Bradley Fighting Vehicle System, JAVELIN system, trainers, night sights, battery chargers, and system test equipment. *Skill Level 10:* Performs maintenance tasks under close supervision; follows technical manuals in servicing and maintaining systems; inspects, tests, and adjusts components to specified tolerances using special test equipment and computerized test consoles; locates and replaces malfunctioning electronic, electrical, mechanical, optical, pneumatic, and electromechanical subassemblies, modules, and circuit elements; determines serviceability and disposition of defective assemblies, subassemblies, and parts; prepares maintenance and supply forms and reports. *Skill Level 20:* Able to perform the duties required for Skill Level 10; provides technical guidance to subordinates; provides technical assistance to support units; installs equipment modifications; troubleshoots weapon system, launch effects trainer, TOW training set, tracker, and launcher effect trainer and monitoring set. *Skill Level 30:* Able to perform the duties required for Skill Level 20; supervises subordinates; supervises inspection and maintenance teams; establishes work loads and priorities; conducts in-service staff training; demonstrates proper maintenance and troubleshooting techniques; recommends maintenance procedures; implements quality control measures; establishes and maintains maintenance records; performs inspections; may supervise up to eight soldiers. *Skill Level 40:* Able to perform the duties required for Skill Level 30; supervises and coordinates support maintenance; serves as section chief or platoon sergeant of detachment, platoon, company, or comparable unit.

Recommendation, Skill Level 10

Credit may be granted on the basis of an individualized assessment of the student (7/97).

Recommendation, Skill Level 20

Credit may be granted on the basis of an individualized assessment of the student (7/97).

Recommendation, Skill Level 30

In the lower-division baccalaureate/associate degree category, 3 semester hours in basic electronics, 3 in electronic systems troubleshooting and maintenance, 3 in personnel supervision, 3 in maintenance management, and 2 in records and information management (7/97).

Recommendation, Skill Level 40

In the lower-division baccalaureate/associate degree category, 3 semester hours in basic electronics, 3 in electronic systems troubleshooting and maintenance, 3 in personnel supervision, 3 in maintenance management, and 2 in records and information management. In the upper-division baccalaureate category, 3 semester hours for field experience in management (7/97).

MOS-35B-004

Land Combat Support System Test Specialist
35B10
35B20
35B30

Exhibit Dates: 2/95–12/00. Effective 12/00, MOS 35B was discontinued and its duties were incorporated into MOS 35Y, Integrated Family of Test Equipment Operator/Maintainer.

Career Management Field: 35 (Electronic Maintenance and Calibration).

Description

Summary: Maintains land combat systems; inspects and tests equipment; locates card-level faults using special test equipment; replaces faulty assemblies and subassemblies; adjusts and aligns mechanical, electrical, and optical assemblies; calibrates and repairs support equipment; performs quality control measures; prepares maintenance and supply reports. *Skill Level 10:* Works under close supervision in performing maintenance functions; uses technical manuals and troubleshooting trees; follows schematics and maintenance procedures; uses digital multimeters, oscilloscopes, signal generators, and power supplies; tests computer diagnostic equipment, including signal and pulse generators, power supplies, printers, and detectors; performs maintenance on AC power generators; troubleshoots and replaces faulty components. *Skill Level 20:* Able to perform the duties required for Skill Level 10; supervises and provides technical assistance to subordinates; repairs multimeters, waveform and pulse generators, and power supplies; performs minor modifications on support equipment. *Skill Level 30:* Able to perform the duties required for Skill Level 20; supervises maintenance and inspection teams; provides technical guidance to subordinates; schedules work; sets repair priorities; inspects, tests, and repairs missile system assemblies; demonstrates proper work practices and maintenance techniques; implements quality control measures; performs system inspections.

Recommendation, Skill Level 10

Credit may be granted on the basis of an individual assessment of the student (11/97).

Recommendation, Skill Level 20

Credit may be granted on the basis of an individual assessment of the student (11/97).

Recommendation, Skill Level 30

In the lower-division baccalaureate/associate degree category, 6 semester hours in DC/AC circuits, 3 in electronic systems troubleshooting and maintenance, 3 in computer systems troubleshooting and maintenance, 3 in computer literacy, 3 in personnel supervision, and 2 in records and information management (11/97).

MOS-35D-002

Air Traffic Control Equipment Repairer
35D10
35D20
35D30
35D40
35D50

Exhibit Dates: Pending Evaluation. 2/96–Present.

MOS-35E-003

Radio and Communications Security (COMSEC) Repairer
35E10
35E20
35E30

Exhibit Dates: 2/95–4/04. (Effective 5/04, this MOS was discontinued, and its duties were incorporated into MOS 94E.)

Career Management Field: 35 (Electronic Calibration and Maintenance).

Description

Summary: Performs repairs or supervises direct and general support level maintenance in radio receivers, transmitters, COMSEC equipment, and associated equipment. *Skill Level 10:* Uses multimeters, oscilloscopes, waveform generators, and spectrum analyzers to troubleshoot and repair radio communications systems equipment and subassemblies; adjusts, aligns, and replaces defective components and boards; performs bench equipment tests to verify operability of repaired equipment; uses computer-automated diagnostic tests to isolate faults. *Skill Level 20:* Able to perform the duties required for Skill Level 10; performs complex and multi-failure tasks; supervises, and provides technical and procedural assistance to subordinates; performs final or quality control inspection of repaired equipment and keeps detailed records of equipment maintenance; controls the accounts for cryptographic items with in the facility. *Skill Level 30:* Able to perform the duties required for Skill Level 20; assigns duties; establishes work load, work schedules, and priorities; provides technical information on repairs and teaches complex or special tasks to subordinates and supported units; ensures that repair functions comply with Army and National Security Agency communications security specifications and policies; develops standard operating procedures relative to direct and general support shop operations; conducts technical and on-the-job training programs in units.

Recommendation, Skill Level 10

Credit may be granted on the basis of an individualized assessment of the student (7/97).

Recommendation, Skill Level 20

Credit may be granted on the basis of an individualized assessment of the student (7/97).

Recommendation, Skill Level 30

In the lower-division baccalaureate/associate degree category, 3 semester hours in electronic systems troubleshooting and repair, 3 in personnel supervision, and 3 in maintenance management (7/97).

MOS-35F-002

Special Electronic Devices Repairer
35F10
35F20
35F30

Exhibit Dates: 2/95–Present.

Career Management Field: 35 (Electronic Maintenance and Calibration).

Description

Summary: Supervises or performs maintenance on specialized military combat equipment such as night vision, mine detection, and illumination device equipment and nuclear, biological, and chemical warning and measuring devices; uses schematic diagrams and basic electronic troubleshooting skills in identifying and correcting malfunction to the circuit board level; requests and maintains bench stock, parts, supplies, and technical publications. *Skill Level 10:* Working under direct supervision, inspects, adjusts, and repairs electronic surveillance, measurement, vision, and detection equipment. *Skill Level 20:* Able to perform the duties required for Skill Level 10; provides technical assistance to subordinates; assists in solving unique problems found in a wide range of equipment; requests and maintains stock, parts, supplies, and technical publications. *Skill Level 30:* Able to perform the duties required for Skill Level 20; prepares, reviews, and consolidates technical and administrative reports and requests; maintains files and records; assigns duties; establishes work load, work schedules, and repair priorities; conducts on-the-job training.

Recommendation, Skill Level 10

Credit may be granted on the basis of an individualized assessment of the student (7/97).

Recommendation, Skill Level 20

Credit may be granted on the basis of an individualized assessment of the student (7/97).

Recommendation, Skill Level 30

In the lower-division baccalaureate/associate degree category, 3 semester hours in electronic systems troubleshooting and maintenance, 3 in record keeping, 3 in personnel supervision, and 2 in maintenance management (7/97).

MOS-35G-003

Biomedical Equipment Specialist, Basic
35G10
35G20

Exhibit Dates: 1/90–5/92.

Career Management Field: 91 (Medical), subfield 915 (Biomedical Equipment Repair).

Description

Summary: Performs routine maintenance on medical equipment. *Skill Level 10:* Performs routine maintenance and repair on electrical, analog, and digital electronic, mechanical, and optical equipment; repairs and installs mechanical medical equipment, including sterilizers, stills, anesthesia apparatus, resuscitators, dental operating units, operating tables and lamps, food carts, hospital beds, etc.; dismantles and cleans equipment; installs water, air, and steam lines; performs routine tests on medical equipment to determine leakage currents and calibration. *Skill Level 20:* Able to perform the duties required for Skill Level 10; provides technical guidance to subordinates.

Recommendation, Skill Level 10

In the vocational certificate category, 3 semester hours in DC and AC circuit fundamentals and 6 in theory of operation and troubleshooting medical equipment. In the lower-division baccalaureate/associate degree category, 3 semester hours in medical equipment maintenance and repair and 2 in shop practices. (NOTE: This recommendation for skill level 10 is valid for the dates 1/90-9/91 only) (10/82).

Recommendation, Skill Level 20

In the vocational certificate category, 3 semester hours in DC and AC circuit fundamentals and 6 in theory of operation and troubleshooting medical equipment. In the lower-division baccalaureate/associate degree category, 3 semester hours in medical equipment maintenance and repair and 2 in shop practices (10/82).

MOS-35G-004

Medical Equipment Repairer, Unit Level
35G10
35G20

Exhibit Dates: 6/92–12/94. Effective 12/94, MOS 35G was discontinued, and its duties were transferred to MOS 91A.

Career Management Field: 91 (Medical).

Description

Summary: Performs routine maintenance, service, repair, and calibration on medical equipment; applies mechanical, electrical, electromechanical, pneumatic, hydraulic, solid state, linear and analog, digital, radiological, and optical principles. *Skill Level 10:* Must be able to unpack, inspect, inventory, assemble, clean, lubricate, calibrate, repair, and test all types of biomedical equipment, including defibrillators, patient monitoring systems, pulmonary equipment, electroencephalonary equipment, diathermy systems, diathermy apparatus, spectrophotometers, ultrasonic therapy and treatment apparatus, dental operating units, anesthesia apparatus, resuscitators, operating tables and lamps, hospital beds, etc.; dismantles and cleans equipment; installs water, air, and steam lines; performs routine tests on medical equipment to determine leakage currents and calibrations. *Skill Level 20:* Able to perform the duties required for Skill Level 10; provides technical guidance to subordinates; advises and assists equipment operators in the assembly and disassembly of field medical equipment; troubleshoots, performs, and supervises unit maintenance activities.

Recommendation, Skill Level 10

Credit may be granted on the basis of an individualized assessment of the student (3/94).

Recommendation, Skill Level 20

In the lower-division baccalaureate/associate degree category, 3 semester hours in DC circuits, 3 in solid state electronics, 3 in AC circuits, 3 in electronic instrumentation, 3 in mechanical and electromechanical controls, 3 in pneumatic and hydraulic controls, 3 in applications of sensors, 3 in digital principles, 3 in analog and linear circuit principles, 3 in electronic equipment diagnostics and repair, and 3 in personnel supervision (6/92).

MOS-35H-003

Calibration Specialist
35H10
35H20
35H30
35H40
35H50

Exhibit Dates: 1/90–9/91.

Career Management Field: 29 (Communications-Electronics Maintenance).

Description

Summary: Calibrates and repairs all common test, measuring, and diagnostic equipment; maintains all calibration standards and accessories; maintains electrical, electronic, pressure, vacuum, hydraulic, and mechanical measuring instruments. *Skill Level 10:* Calibrates and repairs test, measuring, and diagnostic equipment, including meters and other low-level systems; maintains calibration and maintenance records; assists with more advanced repair and calibration procedures. *Skill Level 20:* Able to perform the duties required for Skill Level 10; provides technical guidance to subordinates; performs cross-check of low-level calibration standards; maintains records applicable to these and other calibration and maintenance functions. *Skill Level 30:* Able to perform the duties required for Skill Level 20; provides direct supervision, quality control, technical support, and coordination of unit activities; supervises on-the-job training; supervises a calibration team and prepares its work schedule; calibrates using more sophisticated and accurate standards, including flow rate, viscosity, specific gravity, phase, sound-pressure level, irradiance, photometric, brightness, magnetic field strength, etc.; performs cross-checks of reference standards and maintains standard levels. *Skill Level 40:* Able to perform the duties required for Skill Level 30; supervises and provides on-the-job training to subordinates; maintains calibration files to provide historical, production, and scheduling information for control of operations; conducts liaison with supported units to schedule production advice on test, measuring, and diagnostic equipment; provides quality control by sample inspection against standards; previews correspondence and prepares interservice agreements relative to calibration support; establishes schedules; supervises the development and presentation of new equipment and technology training; may supervise four or more persons, or if duty position is Calibration Operations NCO, may supervise five or more calibration teams. *Skill Level 50:* Able to perform the duties required for Skill Level 40; as principal noncommissioned officer of a calibration facility, performs technical and administrative duties; plans, coordinates, and supervises training and operation activities; prepares correspondence; advises superiors on technical calibration matters; may supervise five or more calibration teams.

Recommendation, Skill Level 10

In the lower-division baccalaureate/associate degree category, 3 semester hours in introduction to electronics, 2 in AC/DC circuits, 1 in digital circuits, 1 in solid state circuitry, 1 in communications principles, 2 in instrumentation, 3 in troubleshooting techniques, 2 in physics, and 2 in mathematics (algebra and trigonometry) (5/79).

Recommendation, Skill Level 20

In the lower-division baccalaureate/associate degree category, 3 semester hours in introduction to electronics, 3 in AC/DC circuits, 3 in digital circuits, 3 in solid-state circuitry, 2 in communications principles, 4 in instrumentation, 6 in troubleshooting techniques, 2 in physics, 2 in mathematics (algebra and trigonometry), and 3 in microwave circuits (5/79).

Recommendation, Skill Level 30

In the lower-division baccalaureate/associate degree category, 3 semester hours in introduction to electronics, 3 in AC/DC circuits, 3 in digital circuits, 3 in solid state circuitry, 2 in communications principles, 5 in instrumentation, 6 in troubleshooting techniques, 2 in physics, 2 in mathematics (algebra and trigonometry), 3 in microwave circuits, 2 in human relations, 2 in personnel supervision, and 2 in maintenance management (5/79).

Recommendation, Skill Level 40

In the lower-division baccalaureate/associate degree category, 3 semester hours in introduction to electronics, 3 in AC/DC circuits, 3 in digital circuits, 3 in solid state circuitry, 2 in communications principles, 5 in instrumentation, 6 in troubleshooting techniques, 2 in physics, 2 in mathematics (algebra and trigonometry), 3 in microwave circuits, 3 in human relations, 3 in personnel supervision, and 2 in maintenance management. In the upper-division baccalaureate category, 3 semester hours for field experience in management; if the duty position was calibration operations NCO, 3 additional semester hours in management problems (5/79).

Recommendation, Skill Level 50

In the lower-division baccalaureate/associate degree category, 3 semester hours in introduction to electronics, 3 in AC/DC circuits, 3 in digital circuits, 3 in solid state circuitry, 2 in communications principles, 5 in instrumentation, 6 in troubleshooting techniques, 2 in physics, 2 in mathematics (algebra and trigonometry), 3 in microwave circuits, 3 in human relations, 3 in personnel supervision, and 3 in maintenance management. In the upper-division baccalaureate category, 6 semester hours for field experience in management and 3 in management problems; if the duty position was calibration operations NCO, 3 additional semester hours in management problems (5/79).

MOS-35H-004

Test, Measurement, and Diagnostic Equipment Maintenance Support Specialist
35H10
35H20
35H30
35H40
35H50

Exhibit Dates: 10/91–1/95.

Career Management Field: 35 (Electronic Maintenance and Calibration).

Description

Summary: Performs electronic and mechanical calibration and repair duties on all common test, measuring, and diagnostic equipment; maintains all calibration standards and accessories; maintains electrical, electronic, pressure, vacuum, hydraulic, fiber optic, and mechanical measuring instruments. *Skill Level 10:* Calibrates and repairs test, measuring, and diagnostic equipment including meters and other low-level systems; maintains calibration and maintenance records; assists with more advanced repair and calibration procedures. *Skill Level 20:* Able to perform the duties required for Skill Level 10; provides individual technical guidance to subordinates; performs cross-check of low-level calibration standards; maintains records applicable to these and other calibration and maintenance functions. *Skill Level 30:* Able to perform the duties required for Skill Level 20; provides direct supervision, quality control, technical support, and coordination of unit activities; supervises on-the-job training; supervises a calibration team and prepares its work schedule; performs high-precision calibration using more sophisticated and accurate standards, including flow rate, viscosity, specific gravity, phase-sound pressure level, voltage, current, frequency, irradiance, photometric, brightness, magnetic field strength, etc.; performs cross-checks of reference standards and maintains standard level. *Skill Level 40:* Able to perform the duties required for Skill Level 30; supervises and provides on-the-job training to subordinates; maintains calibration files to provide historical, production, and scheduling information for control of operations; conducts liaison with supported units to schedule production advice on test, measuring, and diagnostic equipment; provides quality control by sample inspection against standards; previews correspondence and prepares interservice agreements relative to calibration support; establishes schedules; supervises the development and presentation of new equipment and technology training; may supervise four or more persons, or if duty position is Calibration Operations NCO, may supervise five or more calibration teams. NOTE: May progress to 35H40 from 35H30 or 39B30, Automatic Test Equipment Operator/Maintainer. *Skill Level 50:* Able to perform the duties required for Skill Level 40; as principal noncommissioned officer of a calibration facility, performs technical and administrative duties; plans, coordinates, and supervises training and operation activities; prepares correspondence; advises superiors on technical calibration matters; may supervise five or more calibration teams. NOTE: May progress to 35H50 from 35H40 or 35Y40, Integrated Family or Test Equipment Operator/ Maintainer.

Recommendation, Skill Level 10

Credit may be granted on the basis of an individualized assessment of the student (3/94).

Recommendation, Skill Level 20

In the lower-division baccalaureate/associate degree category, 3 semester hours in basic electronics, 3 in electronic systems test and maintenance, 3 in electronic systems test and calibration, 3 in basic mechanical measurements laboratory, 3 in advanced electronics laboratory, 3 in solid state electronics, and 3 in personnel supervision (9/91).

Recommendation, Skill Level 30

In the lower-division baccalaureate/associate degree category, 3 semester hours in basic electronics, 4 in electronic systems test and maintenance, 4 in electronic systems test and calibration, 3 in basic mechanical measurements laboratory, 3 in advanced electronics laboratory, 3 in solid state electronics, 3 in advanced mechanical measurements laboratory, 3 in microwave laboratory, 3 in digital principles, 3 in personnel supervision, 2 in records and information management, and 3 in maintenance management (9/91).

Recommendation, Skill Level 40

In the lower-division baccalaureate/associate degree category, 3 semester hours in basic electronics, 4 in electronic systems test and maintenance, 4 in electronic systems test and calibration, 3 in basic mechanical measurements laboratory, 3 in advanced electronics laboratory, 3 in solid state electronics, 3 in advanced mechanical measurements laboratory, 3 in microwave laboratory, 3 in digital principles, 3 in personnel supervision, 2 in records and information management, and 3 in maintenance management. In the upper-division baccalaureate category, 3 semester hours in organizational management and 3 for field experience in management (9/91).

Recommendation, Skill Level 50

In the lower-division baccalaureate/associate degree category, 3 semester hours in basic electronics, 4 in electronic systems test and maintenance, 4 in electronic systems test and calibration, 3 in basic mechanical measurements laboratory, 3 in advanced electronics laboratory, 3 in solid state electronics, 3 in advanced mechanical measurements laboratory, 3 in microwave laboratory, 3 in digital principles, 3 in personnel supervision, 2 in records and information management, and 3 in maintenance management. In the upper-division baccalaureate category, 3 semester hours in organizational management and 6 for field experience in management; if individual has attained paygrade E-9, additional credit as follows: 3 semester hours in management problems, 3 in operations management, and 3 in communication techniques for managers (9/91).

MOS-35H-005

Test, Measurement, and Diagnostic Equipment Maintenance Support Specialist
35H10
35H20
35H30
35H40

Exhibit Dates: 2/95–Present.

Career Management Field: 35 (Electronic Maintenance and Calibration).

Description

Summary: Performs electronic and mechanical calibration and repair duties on all common test, measuring, and diagnostic equipment; maintains electrical, electronic, pressure, vacuum, hydraulic, fiber optic, and mechanical measuring instruments and calibration standards. *Skill Level 10:* Calibrates and repairs test, measuring, and diagnostic equipment including and other low-level systems; maintains calibration and maintenance records; assists with more advanced repair and calibration procedures. *Skill Level 20:* Able to perform the duties required for Skill Level 10; provides individual technical guidance to subordinates; performs cross-check of low-level calibration standards; maintains records applicable to these and other calibration and maintenance functions. *Skill Level 30:* Able to perform the duties required for Skill Level 20; provides direct supervision, quality control, technical support, and coordination of unit activities; supervises on-the-job training; supervises a calibration team and prepares its work schedule; performs high-precision calibration using more sophisticated and accurate standards, including flow rate, viscosity, specific gravity, phase-sound pressure level, voltage, current, frequency, irradiance, photometric, brightness, magnetic field strength, etc.; performs cross-checks of reference standards and maintains standard level. *Skill Level 40:* Able to perform the duties required for Skill Level 30; as principal noncommissioned officer of a calibration facility, performs technical and administrative duties; plans, coordinates, and supervises training and operation activities; prepares correspondence; advises superiors on technical calibration matters; supervises and provides on-the-job training to subordinates; maintains calibration files to provide historical, production, and scheduling information for control of operations; conducts liaison with supported units to schedule production advice on test, measuring, and diagnostic equipment; provides quality control by sample inspection against standards; previews correspondence and prepares interservice agreements relative to calibration support; establishes schedules; supervises the development and presentation of new equipment and technology training; may supervise four or more persons.

Recommendation, Skill Level 10

Credit may be granted on the basis of an individual assessment of the student (11/97).

Recommendation, Skill Level 20

Credit may be granted on the basis of an individual assessment of the student (11/97).

Recommendation, Skill Level 30

In the lower-division baccalaureate/associate degree category, 6 semester hours in DC and AC circuits; 6 in applied electronic system test, calibration, troubleshooting, and maintenance laboratory; 3 in basic mechanical measurements laboratory; 3 in solid state electronics; 3 in advanced mechanical measurements laboratory; 3 in personnel supervision; 3 in computerized records management; and 3 in maintenance management. In the upper-division baccalaureate category, 3 in organizational management, 3 for field experience in management, and 3 in management problems (11/97).

Recommendation, Skill Level 40

In the lower-division baccalaureate/associate degree category, 6 semester hours in DC and AC circuits; 6 in applied electronic system test, calibration, troubleshooting, and maintenance laboratory; 3 in basic mechanical measurements laboratory; 3 in solid state electronics; 3 in advanced mechanical measurements laboratory; 3 in personnel supervision; 3 in computerized records management; and 6 in maintenance management. In the upper-division baccalaureate category, 3 semester hours in organizational management, 3 for field experience in management, 3 in management problems, and 3 in communication techniques for managers (11/ 97).

MOS-35J-002

Computer/Automation Systems Repair
((Telecommunications Terminal Device Repairer))
35J10
35J20
35J30

Exhibit Dates: 2/95–Present.

Career Management Field: 35 (Electronic Maintenance and Calibration).

Description

Summary: Performs maintenance on telecommunications terminal equipment and microcomputer systems. *Skill Level 10:* Troubleshoots microcomputers and telecommunications equipment using built-in test equipment (BITE), test, measurement and diagnostic equipment (TMDE), schematics, flow charts and technical publications; repairs assemblies and subassemblies by disassembling, adjusting, aligning, and replacing faulty and shop-replaceable units; tests repaired equipment to ensure operability and compliance with technical specifications; maintains selected commercial off-the-shelf computers. *Skill Level 20:* Able to perform the duties required for Skill Level 10; prepares and processes maintenance supply requests and work orders; performs initial and final inspection of repaired equipment; provides technical assistance to subordinates; performs complex maintenance tasks. *Skill Level 30:* Able to perform the duties required for Skill Level 20; establishes work load, schedules, and repair priorities; provides technical assistance and teaches subordinates technical skills; assigns duties; coordinates logistical support; advises supervisors on shop equipment maintenance status; prepares maintenance reports and writes technical portion of maintenance shop operating procedures.

Recommendation, Skill Level 10

Credit may be granted on the basis of an individualized assessment of the student (7/97).

Recommendation, Skill Level 20

Credit may be granted on the basis of an individualized assessment of the student (7/97).

Recommendation, Skill Level 30

In the lower-division baccalaureate/associate degree category, 3 semester hours in electronic troubleshooting and repair, 3 in introduction to electronics, 2 in record keeping, 3 in personnel supervision, and 3 in maintenance management (7/97).

MOS-35K-001

Apache Attack Helicopter Systems Repairer
35K10
35K20
35K30

Exhibit Dates: Pending Evaluation. 4/04–Present.

MOS-35L-003

Avionic Communications Equipment Repairer
35L10
35L20
35L30

Exhibit Dates: 2/96–Present.

Career Management Field: 35 (Electronic Maintenance).

Description

Summary: Performs intermediate maintenance or repairs (to the component level) on avionic communications equipment. *Skill Level 10:* Within a shop environment, performs intermediate maintenance on aircraft communications equipment including troubleshooting, repairing, and testing of very high frequency, ultra high, and high frequency AM and FM receivers and transmitters and intercom and electronic interface devices. Prepares cables and sire harnesses; repairs and solders circuit boards containing surface mount devices. *Skill Level 20:* Able to perform the duties required for Skill Level 10; supervises, trains, and counsels subordinates; requisitions parts; keeps records; writes annual performance reports. *Skill Level 30:* Able to perform the duties required for Skill Level 20; initiates quality assurance programs; supervises subordinates; handles requisitions and communication with superiors; develops training programs; prepares and reviews administrative reports.

Recommendation, Skill Level 10

Credit may be granted on the basis of an individualized assessment of the student (4/98).

Recommendation, Skill Level 20

Credit may be granted on the basis of an individualized assessment of the student (4/98).

Recommendation, Skill Level 30

In the lower-division baccalaureate/associate degree category, 3 semester hours in electronic systems troubleshooting and maintenance, 3 in avionic communications systems, 3 in basic electronics laboratory, 3 in basic electronics, 3 in personnel supervision, 3 in maintenance management, 2 in high reliability soldering, and 2 in records and information management (4/98).

MOS-35M-003

Radar Repairer
35M10
35M20
35M30

Exhibit Dates: 2/95–Present.

Career Management Field: 35 (Electronic Maintenance and Calibration).

Description

Summary: Performs or supervises repair and maintenance of the Firefinder radar system. *Skill Level 10:* Troubleshoots and repairs the radar system to the component level; inspects, tests, and adjusts system components to specific tolerances; prepares and maintains equipment logs. *Skill Level 20:* Able to perform the duties required for Skill Level 10; provides technical guidance to subordinates; performs initial and final checkout and inspection of assemblies and subassemblies; calibrates radar, organizational maintenance, and antenna alignment test sets. *Skill Level 30:* Able to perform the duties required for Skill Level 20; establishes work load and repair priorities; demonstrates proper maintenance and troubleshooting techniques; organizes and conducts on-the-job training programs; coordinates and prepares maintenance reports and records.

Recommendation, Skill Level 10

Credit may be granted on the basis of an individualized assessment of the student (7/97).

Recommendation, Skill Level 20

Credit may be granted on the basis of an individualized assessment of the student (7/97).

Recommendation, Skill Level 30

In the lower-division baccalaureate/associate degree category, 3 semester hours in electronic systems troubleshooting and repair, 3 in personnel supervision, and 3 in maintenance management (7/97).

MOS-35N-002

Wire Systems Equipment Repairer
35N10
35N20
35N30

Exhibit Dates: 2/95–Present.

Career Management Field: 35 (Electronic Maintenance).

Description

Summary: Performs and supervises maintenance on manual, semiautomatic, and selected transportable automatic electronic telephone central office systems using schematics and test equipment. *Skill Level 10:* Maintains relay-operated central office systems and electronic switching systems; uses test equipment including multimeters; uses circuit and wiring diagrams and schematics. *Skill Level 20:* Able to perform the duties required for Skill Level 10; serves as a crew chief; schedules tasks and maintains maintenance forms and records; conducts quality control inspections on repaired equipment. *Skill Level 30:* Able to perform the duties required for Skill Level 20; assigns specific duties to subordinates; establishes work load, work schedules, and priorities; provides technical assistance to subordinates; maintains technical library; initiates and maintains forms and records; responsible for records pertaining to repair parts; prepares and reviews administrative reports; supervises quality control inspections.

Recommendation, Skill Level 10

Credit may be granted on the basis of an individualized assessment of the student (4/98).

Recommendation, Skill Level 20

Credit may be granted on the basis of an individualized assessment of the student (4/98).

Recommendation, Skill Level 30

In the lower-division baccalaureate/associate degree category, 3 semester hours in introduction to electricity, 2 in telephone system repair, 1 in maintenance record keeping, 3 in personnel supervision, and 3 in maintenance management (4/98).

MOS-35P-001

Multiple Launch Rocket System Repairer
35P10

Exhibit Dates: Pending Evaluation. 4/04–Present.

MOS-35Q-001

Avionic Flight Systems Repairer
35Q10
35Q20
35Q30

Exhibit Dates: 2/96–9/99. Effective 9/99, MOS 35Q was discontinued and its duties were incorporated into MOS 35R, Avionic Radar Repairer.

Career Management Field: 35 (Electronic Maintenance).

Description

Summary: Performs intermediate maintenance or repairs on avionic navigation, flight control, and stabilization equipment. *Skill Level 10:* Within a shop environment, performs intermediate and depot maintenance on avionic navigation flight control and stabilization systems that include: marker beacons; radio direction finders; VOR and glide slope receivers; automatic flight controls; global positioning systems; and altitude/heading navigation systems; troubleshoots system using test equipment; replaces faulty components. *Skill Level 20:* Able to perform the duties required for Skill Level 10; supervises maintenance activities; provides unit level maintenance on special test equipment used to repair flight systems; repairs faulty printed circuit boards; counsels subordinates. *Skill Level 30:* Able to perform the duties required for Skill Level 20; supervises and provides technical guidance to subordinates; evaluates maintenance operations for compliance with directives, technical manuals, work standards and safety; applies production and quality control principles; conducts technical training; monitors requisition of parts, tools, and supplies.

Recommendation, Skill Level 10

Credit may be granted on the basis of an individualized assessment of the student (4/98).

Recommendation, Skill Level 20

Credit may be granted on the basis of an individualized assessment of the student (4/98).

Recommendation, Skill Level 30

In the lower-division baccalaureate/associate degree category, 3 semester hours in electronic systems troubleshooting and maintenance, 3 in airborne electronic navigation systems, 3 in basic electronics laboratory, 3 in basic electronics, 3 in personnel supervision, 3 in maintenance management, and 2 in records and information management (4/98).

MOS-35R-003

Avionic Radar Repairer

35R10
35R20
35R30

Exhibit Dates: 2/96–Present.

Career Management Field: 35 (Electronic Maintenance).

Description

Summary: Performs intermediate and depot maintenance on avionic radar equipment. *Skill Level 10:* Troubleshoots, repairs, tests, and performs maintenance on test, diagnostic, and measurement equipment used to diagnose problems in avionic radar equipment; assists in intermediate maintenance of avionic airborne radar equipment, including terrain-following avoidance radars, Doppler navigation radars, weather radars, radar altimeters, radar transponders, navigational distance and bearing equipment, and inertial navigation instrument; replaces faulty components and individual parts down to flexible and plated-through hole circuit boards. *Skill Level 20:* Able to perform the duties required for Skill Level 10; performs maintenance on special test equipment and avionic equipment; trains subordinates; prepares maintenance records and requisitions; repairs circuit boards; tests support equipment. *Skill Level 30:* Able to perform the duties required for Skill Level 20; assists establishing production and quality assurance programs; teaches maintenance practices and techniques; supervises and inspects the maintenance of avionic radar equipment; develops training programs; provides technical assistance to subordinates; assists in the development of new types of electronic equipment; maintains equipment records, technical literature, and spare parts; inspects maintenance procedures for quality.

Recommendation, Skill Level 10

Credit may be granted on the basis of an individualized assessment of the student (4/98).

Recommendation, Skill Level 20

Credit may be granted on the basis of an individualized assessment of the student (4/98).

Recommendation, Skill Level 30

In the lower-division baccalaureate/associate degree category, 3 semester hours in electronic systems troubleshooting repair and maintenance, 3 in radar systems, 3 in basic electronics laboratory, 3 in basic electronics, 3 in personnel supervision, 3 in maintenance management, and 2 in records and information management (4/98).

MOS-35S-001

Patriot System Repairer

35S20
35S30
35S40
35S50

Exhibit Dates: 10/03–Present.

Career Management Field: 27 (Land Combat and Air Defense System Direct and General Support Maintenance).

Description

Summary: Supervises and performs direct, general, and depot maintenance support on Patriot missile system, associated equipment, and trainers. *Skill Level 20:* Performs direct, general, and depot support maintenance on engagement control station, phased-array radar, identification friend or foe equipment, antenna, and the communications relay group; installs software modifications and develops specialized computer software test procedures; analyzes and interprets Patriot diagnostic test results; maintains VHF data links; performs some fiber-optics repair. *Skill Level 30:* Able to perform the duties required for Skill Level 20; performs initial and final checkout and inspection of designated systems and their assemblies and subassemblies; serves as technical inspector of quality assurance section. *Skill Level 40:* Able to perform the duties required for Skill Level 30; supervises preparation of maintenance forms and reports; supervises organizational maintenance in an operational readiness float section. *Skill Level 50:* Able to perform the duties required for Skill Level 40; supervises direct, general, and depot-level maintenance performed on the missile system.

Recommendation, Skill Level 20

Credit may be granted on the basis of an individualized assessment of the student (6/01).

Recommendation, Skill Level 30

In the lower-division baccalaureate/associate degree category, 2 semester hours in radar systems, 3 in electronic systems troubleshooting and maintenance, 3 in digital electronics, 3 in basic electronics, 3 in basic electricity, 2 in electronic communications, 3 in principles of supervision, and 3 in maintenance management (6/01).

Recommendation, Skill Level 40

In the lower-division baccalaureate/associate degree category, 2 semester hours in radar systems, 3 in electronic systems troubleshooting and maintenance, 3 in digital electronics, 3 in basic electronics, 3 in basic electricity, 2 in electronic communications, 4 in principles of supervision, and 4 in maintenance management. In the upper-division baccalaureate category, 3 semester hours in human resource management and 3 for field experience in management (6/01).

Recommendation, Skill Level 50

In the lower-division baccalaureate/associate degree category, 2 semester hours in radar systems, 3 in electronic systems troubleshooting and maintenance, 3 in digital electronics, 3 in basic electronics, 3 in basic electricity, 2 in electronic communications, 4 in principles of supervision, and 4 in maintenance management. In the upper-division baccalaureate category, 3 semester hours in human resource management, 6 for field experience in management, and 3 in operations management (6/01).

MOS-35T-003

Avenger System Repairer

35T10
35T20
35T30
35T40

Exhibit Dates: Pending Evaluation. 4/04–Present.

MOS-35U-003

Biomedical Equipment Specialist, Advanced

35U20
35U30
35U40
35U50

Exhibit Dates: 1/90–5/92.

Career Management Field: 91 (Medical), subfield 915 (Biomedical Equipment Repair).

Description

Summary: Provides periodic scheduled maintenance and repairs all types of medical equipment; applies mechanical, hydraulic, high- and low-pressure gas and steam, electrical, electronic, solid state, digital, radiological, and optical principles or supervises biomedical equipment repair functions. *Skill Level 20:* Able to perform duties of Biomedical Equipment Specialist, Basic (35G10); provides periodic scheduled maintenance and repairs medical equipment; applies mechanical, hydraulic, high- and low-pressure gas and steam, electrical, electronic, solid state, digital, logic, radiological, and optical principles; develops sequence of operating events from schematic diagrams; unpacks, inspects, inventories, assembles, calibrates, installs, adjusts, cleans, lubricates, repairs, and tests all types of biomedical equipment, including electrocardiographs, electroencephalographs, patient monitoring systems, pulmonary equipment, diathermy apparatus, blood cell counters, spectrophotometers, ultrasonic therapy and treatment apparatus, heart-lung machines, radioisotope monitoring and measuring equipment, and X-ray systems with ancillary equipment, including tables, rapid film changers, dye contrast injection apparatus, automatic film processors, cinerography, closed circuit television, and image intensifiers; utilizes knowledge of electrical/electronic theory, including function of solid state devices, integrated circuits, servo systems, photoelectric cells, instrumentation and automation, digital principles, readout systems, and vacuum, cathode-ray, and X-ray tubes. *Skill Level 30:* Able to perform the duties required for Skill Level 20; provides technical guidance and supervision to subordinates; develops and establishes procedures for operation of biomedical equipment maintenance activity; establishes and maintains procedures to ensure effective preventive maintenance program; ensures that diagnostic and measuring equipment is calibrated in accordance with Army calibration program; determines personnel requirements; organizes work schedules; assigns duties, monitors personnel performance, and counsels subordinates; prepares evaluation reports; ensures adherence to standards of conduct, cleanliness, technical accuracy, and safety regulations in all areas of activity; prepares and conducts on-the-job training programs; advises and assists in administrative, fiscal, personnel, and supply matters; advises commander on specific medical systems including requirements for utilities and advantages and disadvantages of contract versus in-house maintenance; advises procurement personnel of functional and safety aspects of medical equipment and systems. *Skill Level 40:* Able to perform the duties required for Skill Level 30; supervises organizational, direct, and general support maintenance of biomedical equipment; organizes and supervises inspection and maintenance teams. *Skill Level 50:* Able to perform the duties required for Skill Level 40; supervises medical equipment maintenance operations; writes, develops, and coordinates command-wide regulations and policies relating to logistical material maintenance programs; provides technical information and guidance to subordinate units; participates in review and approval of subordinate unit submissions; evaluates unit maintenance programs and institutes methods for upgrading such programs; monitors utilization of medical equipment maintenance personnel and associated material assets of subordinate units; determines need for and schedules personnel for attendance at specialized manufacturers' courses; performs as team chief for medical maintenance assistance and instruction teams.

Recommendation, Skill Level 20

In the vocational certificate category, 3 semester hours in technical mathematics, 3 in DC circuits and laboratory, 3 in AC circuits and laboratory, 3 in physics, 3 in transistor laboratory and theory, 3 in electronic circuit theory, 3 in instrumentation, 1 in medical equipment safety practices, and 3 in electronics shop practices. In the lower-division baccalaureate/associate degree category, 1 semester hour in medical terminology, 3 in AC/DC circuits theory and laboratory, 1 in electronics laboratory, 3 in electrical/electronic construction methods, 1 in electrical safety, 3 in instrumentation, 2 in digital theory, 3 in X-ray equipment repair, and 3 in medical equipment repair (10/82).

Recommendation, Skill Level 30

In the vocational certificate category, 3 semester hours in technical mathematics, 3 in DC circuits and laboratory, 3 in AC circuits and laboratory, 3 in physics, 3 in transistor laboratory and theory, 3 in electronic circuit theory, 3 in instrumentation, 1 in medical equipment safety practices, 3 in electronics shop practices, 6 in X-ray equipment repair, and 6 in medical equipment repair. In the lower-division baccalaureate/associate degree category, 1 semester hour in medical terminology, 3 in AC/DC circuit theory and laboratory, 1 in electronics laboratory, 3 in electrical/electronic construction methods, 1 in electrical safety, 3 in instrumentation, 2 in digital theory, 3 in X-ray equipment repair, 3 in medical equipment repair, 3 in personnel supervision, 2 in human relations, and 2 in technical writing (10/82).

Recommendation, Skill Level 40

In the vocational certificate category, 3 semester hours in technical mathematics, 3 in DC circuits and laboratory, 3 in AC circuits and laboratory, 3 in physics, 3 in transistor laboratory and theory, 3 in electronic circuit theory, 3 in instrumentation, 1 in medical equipment safety practices, 3 in electronics shop practices, 6 in X-ray equipment repair, and 6 in medical equipment repair. In the lower-division baccalaureate/associate degree category, 1 semester hour in medical terminology, 3 in AC/DC circuit theory and laboratory, 1 in electronics laboratory, 3 in electrical/electronic construction methods, 1 in electrical safety, 3 in instrumentation, 2 in digital theory, 3 in X-ray equipment repair, 3 in medical equipment repair, 3 in personnel supervision, 2 in human relations, 2 in technical writing, 3 for field experience in management, and 3 in record keeping. In the upper-division baccalaureate category, 3 semester hours in industrial psychology, 3 in introduction to management, and 3 in personnel management (10/82).

Recommendation, Skill Level 50

In the vocational certificate category, 3 semester hours in technical mathematics, 3 in DC circuits and laboratory, 3 in AC circuits and laboratory, 3 in physics, 3 in transistor laboratory and theory, 3 in electronic circuit theory, 3 in instrumentation, 1 in medical equipment safety practices, 3 in electronics shop practices, 6 in X-ray equipment repair, and 6 in medical equipment repair. In the lower-division baccalaureate/associate degree category, 1 semester hour in medical terminology, 3 in AC/DC circuit theory and laboratory, 1 in electronics laboratory, 3 in electrical/electronic construction methods, 1 in electrical safety, 3 in instrumentation, 2 in digital theory, 3 in X-ray equipment repair, 3 in medical equipment repair, 3 in personnel supervision, 2 in human relations, 3 in technical writing, 3 for field experience in management, 3 in record keeping, and 3 in public speaking. In the upper-division baccalaureate category, 3 semester hours in industrial psychology, 3 in introduction to management, 6 in personnel management, and 6 in industrial arts education (electronics) (10/82).

MOS-35U-004

Medical Equipment Repairer, Advanced

35U30
35U40
35U50

Exhibit Dates: 6/92–12/94. (Effective 12/94, MOS 35U was discontinued, and its duties were transferred to MOS 91A.)

Career Management Field: 91 (Medical).

Description

Summary: Provides periodic scheduled maintenance and repairs all types of medical equipment; applies mechanical, hydraulic, high- and low-pressure gas and steam, electrical, electronic, solid state, digital, logic, radiological, and optical principles; supervises biomedical equipment repair functions. *Skill Level 30:* Able to perform duties of Medical Equipment Repairer, Unit Level (35G20); applies basic knowledge of microprocessors, analog-to-digital conversion, digital-to-analog conversion, and diagnostics of microprocessor circuits; develops and establishes procedures for operation of biomedical equipment maintenance activity; establishes and maintains procedures to ensure effective preventive maintenance program; ensures that diagnostic and measuring equipment is calibrated in accordance with Army calibration program; determines personnel requirements, organizes work schedules, assigns duties, monitors personnel performance, and counsels subordinates; prepares evaluation reports; ensures adherence to standards of conduct, cleanliness, technical accuracy, and safety regulations in all areas of activity; advises and assists in administrative, fiscal, personnel, and supply matters; advises commander on specific medical systems including requirements for utilities and advantages and disadvantages of contract versus in-house maintenance; advises procurement personnel of functional and safety aspects of medical equipment and systems. *Skill Level 40:* Able to perform the duties required for Skill Level 30; supervises organizational, direct, and general support maintenance of biomedical equipment; organizes and supervises inspection and maintenance teams. *Skill Level 50:* Able to perform the duties required for Skill Level 40; supervises medical equipment maintenance operations; writes, develops, and coordinates command-wide regulations and policies related to logistical material maintenance programs; provides technical information and guidance to subordinate units; participates in review and approval of subordinate unit submissions; evaluates unit maintenance programs and institutes methods for upgrading such programs; monitors utilization of medical equipment maintenance personnel and associated material assets of subordinate units; determines need for and schedules personnel for attendance at specialized manufacturers' courses; performs as team chief for medical maintenance assistance and instruction teams.

Recommendation, Skill Level 30

In the lower-division baccalaureate/associate degree category, 3 semester hours in DC circuits, 3 in solid state electronics, 3 in AC circuits, 3 in electronic instrumentation, 3 in mechanical and electromechanical controls, 3 in pneumatic and hydraulic controls, 3 in applications of sensors, 3 in digital principles, 3 in analog and linear circuit principles, 3 in microprocessors, 3 in personnel supervision, and 2 in records and information management (6/92).

Recommendation, Skill Level 40

In the lower-division baccalaureate/associate degree category, 3 semester hours in DC circuits, 3 in solid state electronics, 3 in AC circuits, 3 in electronic instrumentation, 3 in mechanical and electromechanical controls, 3 in pneumatic and hydraulic controls, 3 in applications of sensors, 3 in digital principles, 3 in analog and linear circuit principles, 3 in microprocessors, 3 in personnel supervision, and 2 in records and information management. In the upper-division baccalaureate category, 3 semester hours in organizational management and 3 for field experience in management (6/92).

Recommendation, Skill Level 50

In the lower-division baccalaureate/associate degree category, 3 semester hours in DC circuits, 3 in solid state electronics, 3 in AC circuits, 3 in electronic instrumentation, 3 in mechanical and electromechanical controls, 3 in pneumatic and hydraulic controls, 3 in applications of sensors, 3 in digital principles, 3 in analog and linear circuit principles, 3 in microprocessors, 3 in personnel supervision, and 2 in records and information management. In the upper-division baccalaureate category, 3 semester hours in organizational management and 3 for field experience in management; if individual has attained paygrade E-9, additional credit as follows: 3 semester hours in management problems, 3 in operations management, and 3 in communication techniques for managers (6/92).

MOS-35V-001

Electronic and Missile Systems Maintenance Chief
35V50

Exhibit Dates: Pending Evaluation. 4/04–Present.

MOS-35W-001

Electronic Maintenance Chief
35W40
35W50

Exhibit Dates: 2/95–Present.

Description ***Skill Level 40:*** Supervises and provides technical guidance to subordinates on the maintenance of all Army standard electronic equipment systems including communications security devices; develops and implements unit level electronic maintenance programs; develops and enforces policies and procedures for facility management; organizes work schedules; provides personal counseling and ensures compliance with standards. NOTE: May have progressed from Skill Level 30 of MOS 35C, (Surveillance Radar Repairer); 35E (Radio and Communications Security (COMSEC) Repairer); 35F (Special Electronic Devices Repairer); 35J (Telecommunications Terminal Device Repairer); or 35N (Wire Systems Equipment Repairer). *Skill Level 50:* Able to perform the duties required for Skill Level 40; interprets Army maintenance policies; writes technical and operational procedures on maintenance policies; inspects maintenance activities for compliance with repair priorities; advises commander and command staff on all aspects of electronic maintenance operations.

Recommendation, Skill Level 40

In the lower-division baccalaureate/associate degree category, 3 semester hours in personnel supervision, 3 in records and information management, 3 in computer literacy, and 3 in maintenance management. In the upper-division baccalaureate category, 3 semester hours for field experience in management and 3 in management problems (11/97).

Recommendation, Skill Level 50

In the lower-division baccalaureate/associate degree category, 3 semester hours in personnel supervision, 3 in records and information management, 3 in computer literacy, and 3 in maintenance management. In the upper-division baccalaureate category, 6 semester hours for field experience in management, 3 in management problems, and 3 in communication techniques for managers (11/97).

MOS-35Y-001

Integrated Family of Test Equipment Operator/Maintainer
35Y10
35Y20
35Y30
35Y40
35Y50

Exhibit Dates: 9/90–Present.

Description

Summary: Operates, performs, and supervises maintenance support units. *Skill Level 10:* Uses automated test equipment to troubleshoot equipment to the component level; repairs by adjusting, aligning, and replacing defective components. *Skill Level 20:* Able to perform the duties required for Skill Level 10; provides technical assistance to subordinates; keeps maintenance and supply records and forms; installs equipment modifications. *Skill Level 30:* Able to perform the duties required for Skill Level 20; establishes work load, work schedules, and priorities; assigns duties to subordinates; is responsible for quality control inspections; evaluates and counsels subordinates; initiates and maintains forms and records; prepares and reviews administrative reports; organizes and conducts on-the-job training programs; prepares maintenance reports. *Skill Level 40:* Able to perform the duties required for Skill Level 30; supervises calibration of system-associated electronic equipment; implements quality control measures; serves as maintenance quality assurance and quality control inspector; establishes and maintains maintenance records. *Skill Level 50:* Able to perform the duties required for Skill Level 40; plans, coordinates, and supervises activities pertaining to training and technical operation of unit; interprets and supervises execution of policies; provides liaison to support staff and command.

Recommendation, Skill Level 10

In the lower-division baccalaureate/associate degree category, 2 semester hours in DC circuits, 2 in AC circuits, and 3 in electronic systems troubleshooting and maintenance. (NOTE: This recommendation for skill level 10 is valid for the dates 1/90-9/91 only) (11/95).

Recommendation, Skill Level 20

In the lower-division baccalaureate/associate degree category, 2 semester hours in DC circuits, 2 in AC circuits, 3 in electronic systems troubleshooting and maintenance, and 2 in record keeping. (NOTE: This recommendation for skill level 20 is valid for the dates 9/90-2/95 only) (11/95).

Recommendation, Skill Level 30

In the lower-division baccalaureate/associate degree category, 2 semester hours in DC circuits, 2 in AC circuits, 3 in electronic systems troubleshooting and maintenance, 2 in record keeping, and 3 in personnel supervision (11/95).

Recommendation, Skill Level 40

In the lower-division baccalaureate/associate degree category, 2 semester hours in DC circuits, 2 in AC circuits, 3 in electronic systems troubleshooting and maintenance, 2 in record keeping, 3 in maintenance management, and 3 in personnel supervision. In the upper-division baccalaureate category, 3 semester hours for field experience in management (11/95).

Recommendation, Skill Level 50

In the lower-division baccalaureate/associate degree category, 2 semester hours in DC circuits, 2 in AC circuits, 3 in electronic systems troubleshooting and maintenance, 2 in record keeping, 3 in maintenance management, and 3 in personnel supervision. In the upper-division baccalaureate category, 3 semester hours for field experience in management, 3 in human resource management, and 3 in management problems (11/95).

MOS-35Z-001

Senior Electronics Maintenance Chief
35Z50

Exhibit Dates: 2/95–Present.

Career Management Field: 35 (Senior Electronic Maintenance Chief).

Description

Directs electronic maintenance functions, provides technical advice to commanders and command staff concerning Army electronic maintenance; writes or coordinates publication of Army directives concerning electronic maintenance. Individual assignment may be in any number of tasks (ranging from base facility support to human resources). NOTE: May have progressed from Skill Level 50 of MOS 35Y (Integrated Family of Test Equipment (IFTE) operator/maintainer); 23R (Hawk Missile System Mechanic Support); 24R (Hawk Missile Systems Mechanic); 27X (Patriot System Repairer); 35W (Electronic Maintenance Chief); or 27Z (Missile System Maintenance Chief).

Recommendation

In the lower-division baccalaureate/associate degree category, 3 semester hours in personal supervision, 3 in records/inventory management, 3 in computer literacy, and 3 in maintenance management. In the upper-division baccalaureate category, 3 semester hours in organizational management and 6 for field experience in management. If the individual has attained paygrade E-9, 3 semester hours in management problems and 3 in communication techniques for managers (11/97).

MOS-36L-003

Transportable Automatic Switching Systems Operator/Maintainer
36L10
36L20
36L30

Exhibit Dates: 1/90–6/95. Effective 6/95, MOS 36L was discontinued, and its duties were incorporated into MOS's 31F, Network Switching Systems Operator/Maintainer and 74G, Telecommunications Computer Operator-Maintainer.

Career Management Field: 31 (Communications-Electronics Operations).

Description

Summary: Supervises, installs, operates, and performs maintenance on transportable automatic circuit switches and central office equipment and installed communication security devices; performs preventive maintenance on vehicles and power generators. *Skill Level 10:* Installs, initializes, and performs maintenance on automatic switching systems, automatic message switching central systems, and ancillary equipment; uses specialized test equipment. *Skill Level 20:* Able to perform the duties required for Skill Level 10; provides technical assistance and guidance regarding installation and operation; performs off-line adjustments of the equipment; maintains maintenance and repair records. *Skill Level 30:* Able to perform the duties required for Skill Level 20; assigns specific duties to subordinates; establishes work load, work schedules, and priorities; responsible for quality control inspections; evaluates and counsels subordinates; provides technical assistance to subordinates; maintains technical library; initiates and maintains forms and records; prepares and reviews administrative reports.

Recommendation, Skill Level 10

In the lower-division baccalaureate/associate degree category, 3 semester hours in introduction to electronics and 3 in electronic troubleshooting and repair. If ASI was J7, WHCA Console Control Operations, additional credit as follows: 1 semester hour in computer literacy, 2 in files management, and 3 in introduction to voice communications networks. (NOTE: This recommendation for skill level 10 is valid for the dates 1/90-9/91 only) (7/87).

Recommendation, Skill Level 20

In the lower-division baccalaureate/associate degree category, 3 semester hours in introduction to electronics, 5 in electronic troubleshooting and repair, and 2 in record keeping. If ASI was J7, WHCA Console Control Operations, additional credit as follows: 1 semester hour in computer literacy, 2 in files management, and 3 in introduction to voice communications networks. (NOTE: This recommendation for skill level 20 is valid for the dates 1/90-2/95 only) (7/87).

Recommendation, Skill Level 30

In the lower-division baccalaureate/associate degree category, 3 semester hours in introduction to electronics, 5 in electronic troubleshooting and repair, 2 in record keeping, 3 in principles of supervision, and 3 in maintenance management. If ASI was J7, WHCA Console Control Operations, additional credit as follows: 1 semester hour in computer literacy, 2 in files management, and 3 in introduction to voice communications networks (7/87).

MOS-36M-001

Wire Systems Operator
36M10
36M20
36M30

Exhibit Dates: 1/90–4/92. Effective 4/92, MOS 36M was discontinued, and its duties were incorporated into MOS 31F, Electronic Switching Systems Operator.

Career Management Field: 31 (Communications-Electronics Operation).

Description

Summary: Installs, operates, maintains, and supervises manual, semiautomatic and automatic switchboards, systems, telephone central offices, and auxiliary equipment. *Skill Level 10:* Installs, programs, initiates, operates, and performs operator and organizational maintenance on switchboards, telephone central offices, and assigned equipment; makes appropriate message log entries and uses route diagrams to handle proper switching procedures. *Skill Level 20:* Able to perform the duties required for Skill Level 10; supervises the installation, operation, and maintenance of switchboards, telephone central offices, assigned equipment, vehicles, and power generation equipment; trains and advises subordinates on assigned duties; monitors operator proficiency, responsiveness, cooperation, tact, and voice discipline; maintains station logs. *Skill Level 30:* Able to perform the duties required for Skill Level 20; supervises, performs administrative functions, and coordinates the installation, operation, and maintenance performed by subordinates; prepares work schedules and coordinates overall activities; conducts training and coordinates all phases of switchboard communications activities within the unit.

Recommendation, Skill Level 10

In the lower-division baccalaureate/associate degree category, if ASI was J7, WHCA Console Control Operations, credit as follows: 1 semester hour in computer literacy, 2 in files management, and 3 in introduction to voice communications networks. (NOTE: This recommendation for skill level 10 is valid for the dates 1/90-9/91 only) (2/84).

Recommendation, Skill Level 20

In the lower-division baccalaureate/associate degree category, 3 semester hours in personnel supervision. If ASI was J7, WHCA Console Control Operations, additional credit as follows: 1 semester hour in computer literacy, 2 in files management, and 3 in introduction to voice communications networks (2/84).

Recommendation, Skill Level 30

In the lower-division baccalaureate/associate degree category, 3 semester hours in personnel supervision and 3 in principles of management. If ASI was J7, WHCA Console Control Operations, additional credit as follows: 1 semester hour in computer literacy, 2 in files management, and 3 in introduction to voice communications networks (2/84).

MOS-37F-001

Psychological Operations Specialist
37F10
37F20
37F30
37F40
37F50

Exhibit Dates: 9/90–5/96.

Career Management Field: 37 (Psychological Operations).

Description

Summary: Supervises, coordinates, and participates in analysis, planning, production, and dissemination of tactical, strategic, and consolidated psychological operations (PSYOP). *Skill Level 10:* Assists in collecting and reporting PSYOP-related information; assists in processing information into intelligence to support PSYOP; assists in identifying intelligence collection requirements for PSYOP support; assists in evaluating and analyzing current intelligence and PSYOP studies and estimates to determine PSYOP targets; assists in establishment and maintenance of situation maps to provide current intelligence/PSYOP information and identification, disposition, and movement of enemy forces; assists in the design of PSYOP products; assists in the development and administration of surveys to evaluate the effects of planned and executed products; operates PSYOP dissemination equipment and assists in the delivery of PSYOP products; assists in the packaging of PSYOP products for delivery by various means; maintains journals, status boards, visual displays, charts, and graphs required to manage PSYOP; either manually or using automatic data processing equipment, prepares, stores, and retrieves information on PSYOP-related intelligence, plans, campaigns, and products; assists and performs intelligence functions as required; maintains and operates organizational communications equipment, generators, and organic PSYOP production and dissemination team equipment; assists in establishing and maintaining systematic cross-referenced PSYOP records and files; safeguards classified information. *Skill Level 20:* Able to perform the duties required for Skill Level 10; provides technical guidance to subordinates; coordinates resource requirements for the development, production, and dissemination of PSYOP products; assists in the integration of PSYOP planning in support of conventional, special operations, and deception planning; develops and administers surveys to evaluate the effectiveness of PSYOP products; assists in determining the appropriate mix of media, relative to available assets, to disseminate PSYOP products; analyzes current intelligence holdings to identify intelligence gaps and subsequent intelligence collection requirements to support PSYOP; assists in identifying psychological vulnerability and susceptibility of PSYOP targets; assists in evaluating translations of captured enemy documents and reports of interrogations; assists in the analysis of enemy propaganda and other foreign media; informs superiors of information of immediate tactical PSYOP value; packages PSYOP products for delivery by various means; identifies and maintains information on the availability of products and delivery means; establishes and maintains reference files of translated material. *Skill Level 30:* Able to perform the duties required for Skill Level 20; supervises receipt, analysis, and storage of PSYOP-related intelligence information; coordinates PSYOP intelligence collection requirements with supported command and higher headquarters; spot-checks analyses performed by subordinates; plans and advises on PSYOP in direct and general support of operational forces; plans and implements PSYOP campaigns; assists in planning, identification, mobilization, and deployment of PSYOP resources; supervises propaganda writers, broadcast specialists, journalists, and illustrators to develop and produce PSYOP products; supervises PSYOP dissemination and delivery sections and teams; determines dissemination requirements for PSYOP products; assists in the conduct of liaison with the supported command; advises supported commands on psychological considerations for planning operations; provides guidance and training for subordinates; assists in preparing and conducting PSYOP training programs; analyzes enemy propaganda. *Skill Level 40:* Able to perform the duties required for Skill Level 30; plans and organizes work schedule and assigns tasks in support of tactical or strategic PSYOP missions; assists in supervision of the propaganda development and tactical operations centers; conducts general PSYOP training programs for the command; supervises planning and dissemination of PSYOP products; controls the execution of PSYOP campaigns; monitors preparation and production of PSYOP products and acquisition of resources necessary to support implementation of PSYOP campaigns;

conducts liaison with supported commands; supervises intelligence analysts, interpreters, interrogators, and translators attached to PSYOP units. *Skill Level 50:* Able to perform the duties required for Skill Level 40; supervises and provides technical guidance to subordinates; provides liaison to supported staff and commands; assists PSYOP commanders and staff officers in planning, organizing, directing, supervising, training, and coordinating activities pertaining to PSYOP at all levels of command.

Recommendation, Skill Level 10

In the lower-division baccalaureate/associate degree category, 1 semester hour in technical writing, 1 in record keeping, and 1 in audiovisual technology. (NOTE: This recommendation for skill level 10 is valid for the dates 9/90-9/91 only) (12/88).

Recommendation, Skill Level 20

In the lower-division baccalaureate/associate degree category, 2 semester hours in technical writing, 1 in record keeping, 2 in audiovisual technology, 2 in records management, 1 in oral communication, and 2 in personnel supervision. (NOTE: This recommendation for skill level 20 is valid for the dates 9/90-2/95 only) (12/88).

Recommendation, Skill Level 30

In the lower-division baccalaureate/associate degree category, 2 semester hours in technical writing, 1 in record keeping, 2 in audiovisual technology, 2 in records management, 2 in oral communication, 3 in personnel supervision, and 3 in business organization and management (12/88).

Recommendation, Skill Level 40

In the lower-division baccalaureate/associate degree category, 2 semester hours in technical writing, 1 in record keeping, 2 in audiovisual technology, 2 in records management, 2 in oral communication, 3 in personnel supervision, and 3 in business organization and management. In the upper-division baccalaureate category, 3 semester hours for field experience in management and 3 in principles of management (12/88).

Recommendation, Skill Level 50

In the lower-division baccalaureate/associate degree category, 2 semester hours in technical writing, 1 in record keeping, 2 in audiovisual technology, 2 in records management, 2 in oral communication, 3 in personnel supervision, and 3 in business organization and management. In the upper-division baccalaureate category, 3 semester hours for field experience in management, 3 in principles of management, and 3 in management problems (12/88).

MOS-37F-002

Psychological Operations Specialist

37F10
37F20
37F30
37F40
37F50

Exhibit Dates: 6/96–Present.

Career Management Field: 37 (Psychological Operations).

Description

Summary: Supervises, coordinates, and participates in analysis, planning, production, and dissemination of tactical, strategic, and consolidated psychological operations (PSYOP). *Skill Level 10:* Assists in identifying, collecting, and reporting PSYOP-related information; assists in processing information into intelligence to support PSYOP; assists in evaluating and analyzing current intelligence and PSYOP studies and estimates to determine PSYOP targets; assists in establishment and maintenance of situation maps to provide current intelligence/PSYOP information and identification, disposition, and movement of enemy forces; assists in the design of PSYOP products; assists in the development and administration of surveys to evaluate the effects of planned and executed products; assists in the packaging of PSYOP products for delivery; maintains journals, status boards, visual displays, charts, and graphs required to manage PSYOP; either manually or using automatic data processing equipment, prepares, stores, and retrieves information on PSYOP-related intelligence, plans, campaigns, and products; maintains and operates equipment, generators, and organic PSYOP production and dissemination team equipment; assists in establishing and maintaining systematic cross-referenced PSYOP records and files; safeguards classified information. *Skill Level 20:* Able to perform the duties required for Skill Level 10; provides technical guidance to subordinates; coordinates resource requirements for the development, production, and dissemination of PSYOP products; assists in the integration of PSYOP planning in support of conventional, special operations, and deception planning; develops and administers surveys to evaluate the effectiveness of PSYOP products; assists in determining the appropriate mix of media; analyzes current intelligence holdings to identify intelligence gaps and subsequent intelligence collection requirements to support PSYOP; assists in identifying psychological vulnerability and susceptibility of PSYOP targets; assists in evaluating translations of captured enemy documents and reports of interrogations; assists in the analysis of enemy propaganda and other foreign media; informs superiors of information of immediate tactical PSYOP value; packages PSYOP products for delivery by various means; identifies and maintains information on the availability of products and delivery means; establishes and maintains reference files of translated material. *Skill Level 30:* Able to perform the duties required for Skill Level 20; supervises receipt, analysis, and storage of PSYOP-related intelligence information; coordinates PSYOP intelligence collection requirements with supported command and higher headquarters; spot-checks analyses performed by subordinates; plans and advises on PSYOP in direct and general support of operational forces; plans and implements PSYOP campaigns; assists in planning, identification, mobilization, and deployment of PSYOP resources; supervises propaganda writers, broadcast specialists, journalists, and illustrators to develop and produce PSYOP products; supervises PSYOP dissemination and delivery sections and teams; determines dissemination requirements for PSYOP products; assists in the conduct of liaison with the supported command; provides guidance and training for subordinates; assists in preparing and conducting PSYOP training programs; analyzes enemy propaganda. *Skill Level 40:* Able to perform the duties required for Skill Level 30; plans and organizes work schedules and assigns tasks in support of tactical or strategic PSYOP missions; assists in supervision of the propaganda development and tactical operations centers; conducts general PSYOP training programs for the command; supervises planning and dissemination of PSYOP products; controls the execution of PSYOP campaigns; monitors preparation and production of PSYOP products and acquisition of resources necessary to support implementation of PSYOP campaigns; conducts liaison with supported commands; supervises intelligence analysts, interpreters, interrogators, and translators attached to PSYOP units. *Skill Level 50:* Able to perform the duties required for Skill Level 40; supervises and provides technical guidance to subordinates; provides liaison to supported staff and commands; assists PSYOP commanders and staff officers in planning, organizing, directing, supervising, training, and coordinating activities pertaining to PSYOP at all levels of command.

Recommendation, Skill Level 10

Credit may be granted on the basis of an individualized assessment of the student (6/96).

Recommendation, Skill Level 20

Credit may be granted on the basis of an individualized assessment of the student (6/96).

Recommendation, Skill Level 30

In the lower-division baccalaureate/associate degree category, 1 semester hour in technical report writing, 1 in record keeping, and 3 in leadership. In the upper-division baccalaureate category, 3 semester hours for field experience in management (6/96).

Recommendation, Skill Level 40

In the lower-division baccalaureate/associate degree category, 2 semester hours in technical report writing, 3 in record keeping, 3 in leadership, and 3 in instructor training techniques. In the upper-division baccalaureate category, 3 semester hours for field experience in management (6/96).

Recommendation, Skill Level 50

In the lower-division baccalaureate/associate degree category, 2 semester hours in technical report writing, 3 in record keeping, 3 in leadership, and 3 in instructor training techniques. In the upper-division baccalaureate category, 6 semester hours for field experience in management (6/96).

MOS-38A-001

Civil Affairs Specialist (Reserve Components)

38A10
38A20
38A30
38A40
38A50

Exhibit Dates: 8/91–3/01.

Career Management Field: 38 (Public Affairs).

Description

Summary: Supervises, researches, coordinates, conducts, and participates in analysis, planning, and production of civil affairs documents and actions. *Skill Level 10:* Maintains journals, status boards, and visual display charts and graphs; assists in the following activities: preparing required surveys; restoring government operations and services normally provided by the host nation; establishing programs to benefit the local population; locating and acquiring local resources needed for US military operations; coordinating civilian and military use of public communications; coordinating public works, utilities, and transportation operations in support of the civil affairs mission; coordinating the resettlement of dislocated civilians; updating area studies; researching the commander's legal and moral obligations concerning the treatment of the indigenous population; coordinating the consolidation of activities with psychological operations; establishing and maintaining systematic cross-referenced civil affairs records and files; safeguarding classified information. *Skill Level 20:* Able to perform the duties required for Skill Level 10; provides technical training and guidance to subordinates; assists in coordinating the integration of civil affairs planning in support of conventional and special operations forces and in the analysis of a foreign nation's organizational lines of authority and political power structure; coordinates treatment of noncombatants and establishment of emergency aid station; prepares facts for dissemination by radio and television and conducts research for troop information programs; identifies agencies available to support public communication mission; prepares communication reports. *Skill Level 30:* Able to perform the duties required for Skill Level 20; coordinates support of foreign national agencies or host nation for US military operations; provides guidance in assessing government agencies' resources and assists in coordinating the restoration of government functions; assists in analysis of existing treaties and agreements with foreign nations; assists in preparing plans for the security of sensitive foreign nation facilities; supervises the gathering of data on existing health problems in areas of operation and analyzes basic needs of the civilian population; conducts economic surveys; determines local civil defense requirements; coordinates the distribution of international relief supplies and identifies natural resources available within the civilian community; supervises the preparation of reports for the arts, monuments, and archives team; assists in the restoration of cultural structures; identifies civil information resources and supervises the preparation of reports; identifies legal obligations concerning public and civilian property and facilities. *Skill Level 40:* Able to perform the duties required for Skill Level 30; supervises, organizes, and conducts civil affairs training; supervises preparation and maintenance of surveys, journals, status boards, charts, reports, and files; supervises the Civil Military Operations Center, coordinating work schedules and assigning tasks; conducts liaison with support commands; supervises civil affairs message center and provides communication security. *Skill Level 50:* Able to perform the duties required for Skill Level 40; provides liaison with supported staff and commands and foreign or host nation support; assists civil affairs commander and staff officer in planning, organizing, directing, supervising, training, and coordinating activities related to civil affairs at all levels of command.

Recommendation, Skill Level 10

In the lower-division baccalaureate/associate degree category, 3 semester hours in cross-cultural communication. (NOTE: This recommendation for skill level 10 is valid for the dates 8/91-9/91 only) (12/92).

Recommendation, Skill Level 20

In the lower-division baccalaureate/associate degree category, 3 semester hours in cross-cultural communications, 3 in area studies, and 3 in political science. (NOTE: This recommendation for skill level 20 is valid for the dates 8/91-2/95 only) (12/92).

Recommendation, Skill Level 30

In the lower-division baccalaureate/associate degree category, 3 semester hours in cross-cultural communications, 3 in area studies, 3 in political science, and 3 in personnel supervision (12/92).

Recommendation, Skill Level 40

In the lower-division baccalaureate/associate degree category, 3 semester hours in cross-cultural communications, 3 in area studies, 3 in political science, and 3 in personnel supervision. In the upper-division baccalaureate category, 3 semester hours in organizational management (12/92).

Recommendation, Skill Level 50

In the lower-division baccalaureate/associate degree category, 3 semester hours in cross-cultural communications, 3 in area studies, 3 in political science, and 3 in personnel supervision. In the upper-division baccalaureate category, 3 semester hours in organizational management and 3 for field experience in management (12/92).

MOS-38A-002

Civil Affairs Specialist (Reserve Components)
38A10
38A20
38A30
38A40
38A50

Exhibit Dates: 4/01–Present.

Career Management Field: 38 (Public Affairs).

Description

Summary: Supervises, researches, coordinates, conducts, and participates in analysis, planning, and production of civil affairs documents and actions. *Skill Level 10:* Maintains journals, status boards, and visual display charts and graphs; assists in the following activities: preparing required surveys; restoring government operations and services normally provided by the host nation; establishing programs to benefit the local population; locating and acquiring local resources needed for US military operations; coordinating civilian and military use of public communications; coordinating public works, utilities, and transportation operations in support of the civil affairs mission; coordinating the resettlement of dislocated civilians; updating area studies; researching the commander's legal and moral obligations concerning the treatment of the indigenous population; coordinating the consolidation of activities with psychological operations; establishing and maintaining systematic cross-referenced civil affairs records and files; safeguarding classified information. *Skill Level 20:* Able to perform the duties required for Skill Level 10; provides technical training and guidance to subordinates; assists in coordinating the integration of civil affairs planning in support of conventional and special operations forces and in the analysis of a foreign nation's organizational lines of authority and political power structure; coordinates treatment of noncombatants and establishment of emergency aid station; prepares facts for dissemination by radio and television and conducts research for troop information programs; identifies agencies available to support public communication mission; prepares communication reports. *Skill Level 30:* Able to perform the duties required for Skill Level 20; coordinates support of foreign national agencies or host nation for US military operations; provides guidance in assessing government agencies' resources and assists in coordinating the restoration of government functions; assists in analysis of existing treaties and agreements with foreign nations; assists in preparing plans for the security of sensitive foreign nation facilities; supervises the gathering of data on existing health problems in areas of operation and analyzes basic needs of the civilian population; conducts economic surveys; determines local civil defense requirements; coordinates the distribution of international relief supplies and identifies natural resources available within the civilian community; supervises the preparation of reports for the arts, monuments, and archives team; assists in the restoration of cultural structures; identifies civil information resources and supervises the preparation of reports; identifies legal obligations concerning public and civilian property and facilities. *Skill Level 40:* Able to perform the duties required for Skill Level 30; supervises, organizes, and conducts civil affairs training; supervises preparation and maintenance of surveys, journals, status boards, charts, reports, and files; supervises the Civil Military Operations Center, coordinating work schedules and assigning tasks; conducts liaison with support commands; supervises civil affairs message center and provides communications security. *Skill Level 50:* Able to perform the duties required for Skill Level 40; provides liaison with supported staff and commands and foreign or host nation support; assists civil affairs commander and staff officer in planning, organizing, directing, supervising, training, and coordinating activities related to civil affairs at all levels of command.

Recommendation, Skill Level 10

Credit may be granted on the basis of an individualized assessment of the student (4/01).

Recommendation, Skill Level 20

Credit may be granted on the basis of an individualized assessment of the student (4/01).

Recommendation, Skill Level 30

In the lower-division baccalaureate/associate degree category, 3 semester hours in cross-cultural communications, 3 in area studies, 3 in political science, and 3 in personnel supervision (4/01).

Recommendation, Skill Level 40

In the lower-division baccalaureate/associate degree category, 3 semester hours in cross-cultural communications, 3 in area studies, 3 in political science, and 3 in personnel supervision. In the upper-division baccalaureate category, 3 semester hours in organizational management (4/01).

Recommendation, Skill Level 50

In the lower-division baccalaureate/associate degree category, 3 semester hours in cross-cultural communications, 3 in area studies, 3 in political science, and 3 in personnel supervision. In the upper-division baccalaureate category, 3 semester hours in organizational management and 3 for field experience in management (4/01).

MOS-39B-001

Automatic Test Equipment Operator/
Maintainer
39B10
39B20
39B30

Exhibit Dates: 1/90–4/99.

Career Management Field: 29 (Communications-Electronics Maintenance).

Description

Summary: Operates, maintains, and supervises maintenance on automatic test equipment. *Skill Level 10:* Operates automatic test equipment. *Skill Level 20:* Able to perform the duties required for Skill Level 10; maintains automatic test equipment; uses knowledge of digital electronics, microprocessors, and computer systems to troubleshoot equipment; uses oscilloscopes, meters, signal generators, and logic probes to locate faults; keeps records on maintenance procedures. *Skill Level 30:* Able to perform the duties required for Skill Level 20; assigns specific duties to subordinates; responsible for quality control inspections; evaluates and counsels subordinates; provides technical assistance to subordinates; maintains technical library; initiates and maintains forms and records; responsible for records pertaining to repair parts; prepares and reviews administrative reports.

Recommendation, Skill Level 10

Credit is not recommended because the skills, competencies, and knowledge are uniquely military in nature (7/87).

Recommendation, Skill Level 20

In the lower-division baccalaureate/associate degree category, 3 semester hours in introduction to electronics, 5 in electronic troubleshooting and repair, 3 in digital electronics, 3 in microprocessors, 3 in computer systems repair, and 2 in record keeping. (NOTE: This recommendation for skill level 20 is valid for the dates 1/90-2/95 only) (7/87).

Recommendation, Skill Level 30

In the lower-division baccalaureate/associate degree category, 3 semester hours in introduction to electronics, 5 in electronic troubleshooting and repair, 3 in digital electronics, 3 in microprocessors, 3 in computer systems repair, 2 in record keeping, 3 in principles of supervision, and 3 in maintenance management (7/87).

MOS-39B-002

Automatic Test Equipment Operator/
Maintainer
39B10
39B20
39B30

Exhibit Dates: 5/99–3/04. (Effective 4/04 this MOS was discontinued, and its duties were incorporated into MOS 35K.)

Career Management Field: 29 (Communications-Electronics Maintenance).

Description

Summary: Operates, automatic test equipment; maintains and supervises unit level and direct maintenance, troubleshooting, and repair of automatic test equipment; completes adjustments and calibration procedures to ensure compliance with technical specifications; prepares appropriate maintenance forms and records. *Skill Level 10:* Maintains automatic test equipment; uses knowledge of digital electronics, microprocessors, and computer systems to troubleshoot equipment; uses oscilloscopes, meters, signal generators, and logic probes to locate faults. *Skill Level 20:* Able to perform the duties required for Skill Level 10; assigns specific duties to subordinates; performs quality control inspections; provides technical assistance to subordinates; initiates and maintains forms and records. *Skill Level 30:* Able to perform the duties required for Skill Level 20; responsible for records pertaining to repair parts; prepares and reviews administrative reviews; evaluates and counsels subordinates; maintains technical library.

Recommendation, Skill Level 10

Credit may be granted on the basis of an individualized assessment of the student (5/99).

Recommendation, Skill Level 20

Credit may be granted on the basis of an individualized assessment of the student (5/99).

Recommendation, Skill Level 30

In the lower-division baccalaureate/associate degree category, 3 semester hours in electronic troubleshooting and repair, 3 in computer systems repair, 3 in record keeping, 3 in principles of supervision, and 3 in maintenance management (5/99).

MOS-39C-001

Target Acquisition/Surveillance Radar
Repairer
39C10
39C20
39C30

Exhibit Dates: 1/90–6/95. Effective 6/95, MOS 39C was discontinued, and its duties were incorporated into MOS's 35C, Surveillance Radar Repairer and 35M, Radar Repairer.

Career Management Field: 29 (Communications Electronics Maintenance).

Description

Summary: Performs or supervises maintenance on target acquisition and ground level surveillance radars and associated equipment. *Skill Level 10:* Uses oscilloscopes and multimeters to troubleshoot equipment to the component level; repairs equipment by adjusting, aligning, and replacing defective components. *Skill Level 20:* Able to perform the duties required for Skill Level 10; performs complex maintenance tasks; provides technical assistance to subordinates; keeps maintenance records and forms. *Skill Level 30:* Able to perform the duties required for Skill Level 20; assigns duties to subordinates; establishes work load, work schedules, and priorities; responsible for quality control inspections; evaluates and counsels subordinates; maintains technical library; initiates and maintains forms and records; responsible for records pertaining to repair parts; prepares and reviews administrative reports.

Recommendation, Skill Level 10

In the lower-division baccalaureate/associate degree category, 3 semester hours in introduction to electronics and 2 in electronic troubleshooting and repair. (NOTE: This recommendation for skill level 10 is valid for the dates 1/90-9/91 only) (7/87).

Recommendation, Skill Level 20

In the lower-division baccalaureate/associate degree category, 3 semester hours in introduction to electronics, 4 in electronic troubleshooting and repair, and 2 in record keeping. (NOTE: This recommendation for skill level 20 is valid for the dates 1/90-2/95 only) (7/87).

Recommendation, Skill Level 30

In the lower-division baccalaureate/associate degree category, 3 semester hours in introduction to electronics, 4 in electronic troubleshooting and repair, 2 in record keeping, 3 in principles of supervision, and 3 in maintenance management (7/87).

MOS-39D-001

Decentralized Automated Service
Support System (DAS3) Computer
Systems Repairer
39D10
39D20
39D30

Exhibit Dates: 1/90–6/94. Effective 6/94, MOS 39D was discontinued, and its duties were incorporated into MOS 39G, Automated Communications Computer Systems Repairer.

Career Management Field: 29 (Communications Electronics Maintenance).

Description

Summary: Supervises or performs maintenance on a military version of a commercial data processing computer. *Skill Level 10:* Working under direct supervision, inspects, adjusts, and repairs computer system components including central processing unit, disk drives, tape drives, video terminals, printers, modems, and card readers/punches; uses vendor-supplied diagnostic software to localize system malfunctions; uses vendor-supplied documentation and test equipment to repair to the board and component level; uses standard electronic test and measurement equipment to troubleshoot and repair equipment. *Skill Level 20:* Able to perform the duties required for Skill Level 10; supervises and provides on-the-job training to subordinates; manages inventory control. *Skill Level 30:* Able to perform the duties required for Skill Level 20; prepares, reviews and consolidates technical and administrative reports; maintains files, records, and technical library; assigns duties; establishes work load, work schedules, and repair priorities.

Recommendation, Skill Level 10

In the lower-division baccalaureate/associate degree category, 1 semester hour in DC circuits, 1 in AC circuits, 2 in digital principles, and 3 in computer systems troubleshooting and maintenance. (NOTE: This recommendation for skill level 10 is valid for the dates 1/90-9/91 only) (6/88).

Recommendation, Skill Level 20

In the lower-division baccalaureate/associate degree category, 1 semester hour in DC circuits, 1 in AC circuits, 2 in digital principles, and 6 in computer systems troubleshooting and maintenance (6/88).

Recommendation, Skill Level 30

In the lower-division baccalaureate/associate degree category, 1 semester hour in DC circuits, 1 in AC circuits, 2 in digital principles, 6 in computer systems troubleshooting and maintenance, 3 in record keeping, 3 in principles of supervision, and 2 in maintenance management (6/88).

MOS-39E-001

Special Electronic Devices Repairer
39E10
39E20
39E30

Exhibit Dates: 1/90–6/95. Effective 6/95, MOS 39E was discontinued and its duties were incorporated into MOS 35F, Special Electronic Devices Repairer.

Career Management Field: 29 (Communications Electronics Maintenance).

Description

Summary: Supervises or performs maintenance on specialized military combat equipment such as night vision, threat detection, and mine detection equipment; uses schematic diagrams and basic electronic troubleshooting skills in identifying and correcting malfunctions to the component level; updates maintenance manuals, records, and supplies. *Skill Level 10:* Working under direct supervision, inspects, adjusts, and repairs electronic surveillance, measurement, vision, and detection equipment. *Skill Level 20:* Able to perform the duties required for Skill Level 10; provides technical guidance to subordinates; assists in solving unique problems found in a wide range of equipment. *Skill Level 30:* Able to perform the duties required for Skill Level 20; prepares, reviews, and consolidates technical and administrative reports and requests; maintains files and records; assigns duties; establishes work load, work schedules, and repair priorities; conducts on-the-job training.

Recommendation, Skill Level 10

In the lower-division baccalaureate/associate degree category, 1 semester hour in AC circuits, 1 in DC circuits, 3 in basic electronics laboratory, and 3 in electronic systems troubleshooting and maintenance. (NOTE: This recommendation for skill level 10 is valid for the dates 1/90-9/91 only) (6/88).

Recommendation, Skill Level 20

In the lower-division baccalaureate/associate degree category, 1 semester hour in AC circuits, 1 in DC circuits, 3 in basic electronics laboratory, and 6 in electronic systems troubleshooting and maintenance. (NOTE: This recommendation for skill level 20 is valid for the dates 1/90-2/95 only) (6/88).

Recommendation, Skill Level 30

In the lower-division baccalaureate/associate degree category, 1 semester hour in AC circuits, 1 in DC circuits, 3 in basic electronics laboratory, 6 in electronic systems troubleshooting and maintenance, 3 in record keeping, 3 in principles of supervision, and 2 in maintenance management (6/88).

MOS-39G-001

Automated Communications Computer Systems Repairer
39G10
39G20
39G30

Exhibit Dates: 1/90–1/94.

Career Management Field: 29 (Signal Maintenance).

Description

Summary: Performs maintenance and repair on computer systems (mainframes, minicomputers, and microcomputers) and related peripheral equipment, including tape and disk drives, printers, monitors, data/voice I/O equipment and measurement and test equipment; performs troubleshooting techniques, including visual inspections, diagnostics, the use of built-in test equipment, and standard laboratory equipment, such as oscilloscopes, logic probes, multimeters, logic analyzers, and time domain reflectometers. *Skill Level 10:* Performs repairs to the board level; may also perform component-level repair; follows written procedures, schematics, and assembly language listings in troubleshooting procedures and diagnostics; performs modifications; repairs cables; requests parts; maintains service records. *Skill Level 20:* Able to perform the duties required for Skill Level 10; provides technical guidance and supervision to subordinates; tests repaired or modified equipment for compliance with specifications; requests logistical support; prepares and maintains technical and maintenance management records. *Skill Level 30:* Able to perform the duties required for Skill Level 20; establishes work loads, work schedules, and repair priorities; performs and supervises quality inspections; establishes and maintains on-the-job training programs; reviews technical and maintenance management reports; provides technical assistance to lower level maintenance and repair activities.

Recommendation, Skill Level 10

In the lower-division baccalaureate/associate degree category, 3 semester hours in digital fundamentals, 3 in electronic systems troubleshooting and maintenance, and 3 in solid state electronics. (NOTE: This recommendation for skill level 10 is valid for the dates 1/90-9/91 only) (6/90).

Recommendation, Skill Level 20

In the lower-division baccalaureate/associate degree category, 3 semester hours in digital fundamentals, 3 in electronic systems troubleshooting and maintenance, 3 in solid state electronics, 3 in computer systems and organization, and 3 in computer systems troubleshooting and maintenance (6/90).

Recommendation, Skill Level 30

In the lower-division baccalaureate/associate degree category, 3 semester hours in digital fundamentals, 3 in electronic systems troubleshooting and maintenance, 3 in solid state electronics, 3 in computer systems and organization, 3 in computer systems troubleshooting and maintenance, 3 in principles of supervision, and 3 in maintenance management (6/90).

MOS-39G-002

Automated Communications Computer System Repairer
39G10
39G20
39G30
39G40

Exhibit Dates: Pending Evaluation. 2/94–6/95.

MOS-39L-001

Field Artillery Digital Systems Repairer
39L10
39L20
39L30

Exhibit Dates: 1/90–4/92. Effective 4/92, MOS 39L was discontinued, and its duties were incorporated into MOS 29J, Telecommunications Terminal Device Repairer.

Career Management Field: 29 (Signal Maintenance).

Description

Summary: Performs and supervises intermediate maintenance on gun, rocket and missile fire control, directional, and digital computer systems; meteorological computers; motor fire control calculators; remotely piloted vehicle control stations; and peripheral equipment. *Skill Level 10:* Operates and performs intermediate maintenance on digital fire control systems and peripheral support equipment; fault detection and correction is at the board level. *Skill Level 20:* Able to perform the duties required for Skill Level 10; uses knowledge of digital electronics and computer system concepts to isolate and correct complex problems; keeps records on maintenance procedures; supervises subordinates; conducts on-the-job training; oversees installation of modifications on equipment; plans and recommends establishment of procedures for receipt, storage, inspection, testing, and repair of components. *Skill Level 30:* Able to perform the duties required for Skill Level 20; supervises the operation and maintenance performed by subordinates; prepares work schedules and coordinates overall activities; conducts training and counseling sessions.

Recommendation, Skill Level 10

In the lower-division baccalaureate/associate degree category, 3 semester hours in electronic systems troubleshooting and maintenance. (NOTE: This recommendation for skill level 10 is valid for the dates 1/90-9/91 only) (6/89).

Recommendation, Skill Level 20

In the lower-division baccalaureate/associate degree category, 3 semester hours in electronic systems troubleshooting and maintenance, 3 in personnel supervision, and 3 in maintenance management (6/89).

Recommendation, Skill Level 30

In the lower-division baccalaureate/associate degree category, 3 semester hours in electronic systems troubleshooting and maintenance, 3 in principles of supervision, 3 in maintenance management, and 3 in communication skills (oral and written) (6/89).

MOS-39T-001

Tactical Computer Systems Repairer
39T10
39T20
39T30

Exhibit Dates: 1/90–3/90. Effective 3/90, MOS 39T was discontinued.

Career Management Field: 29 (Communications Electronics Maintenance).

Description

Summary: Supervises or performs maintenance on a military version of a commercial desktop data processing microcomputer. *Skill Level 10:* Working under direct supervision, inspects, adjusts, and repairs computer system components including CPU, hard disk drives, tape drives, video terminals, printers, and modems; uses vendor-supplied diagnostic software to localize system malfunctions. *Skill Level 20:* Able to perform the duties required for Skill Level 10; has a basic knowledge of systems software and related utilities. *Skill Level 30:* Able to perform the duties required for Skill Level 20; supervises maintenance activities; prepares, reviews, and consolidates technical and administrative reports; maintains files, records, and technical library; assigns duties; establishes work load, schedules, and repair priorities.

Recommendation, Skill Level 10

In the lower-division baccalaureate/associate degree category, 1 semester hour in DC circuits, 1 in AC circuits, 2 in digital principles, and 3 in computer systems troubleshooting and maintenance (6/88).

Recommendation, Skill Level 20

In the lower-division baccalaureate/associate degree category, 1 semester hour in DC circuits, 1 in AC circuits, 2 in digital principles, 6 in computer systems troubleshooting and maintenance, and 3 in computer systems and organization (6/88).

Recommendation, Skill Level 30

In the lower-division baccalaureate/associate degree category, 1 semester hour in DC circuits, 1 in AC circuits, 2 in digital principles, 6 in computer systems troubleshooting and maintenance, 3 in computer systems and organization, 2 in maintenance management, 3 in record keeping, and 3 in principles of supervision (6/88).

MOS-39V-001

COMPUTERIZED SYSTEMS MAINTENANCE CHIEF
39V40

Exhibit Dates: 1/90–6/94. Effective 6/94, MOS 39V was discontinued and its duties were incorporated into MOS 39G, Automated Communications Computer Systems Repairer.

Career Management Field: 29 (Signal Maintenance).

Description

Supervises and manages direct and general support maintenance of computerized systems, such as field artillery digital and tactical fire direction systems and automated communications computer systems; determines requirements and assigns duties; coordinates activities involving installation and maintenance of minicomputers, peripherals, and test and measuring equipment; prepares, reviews, and consolidates technical reports; implements standard operating procedures; organizes and conducts on-the-job training; instructs and counsels subordinates; performs liaison with staff, operations, and maintenance personnel; determines effectiveness of performance to meet work load standards; reads, interprets, and explains complex concepts, specifications circuits, schematic and wiring diagrams, and minicomputer and mainframe organization. NOTE: May have progressed to 39V40 from 39D30 (Decentralized Automated Service Support System Computer Systems Repairer), 39G30 (Automated Communications Computer Systems Repairer), 39B30 (Automatic Test Equipment Operator/Maintainer), or 39Y30 (Field Artillery Tactical Fire Direction Systems Repairer).

Recommendation

In the lower-division baccalaureate/associate degree category, 3 semester hours in computer systems and organization, 3 in principles of supervision, and 3 in maintenance management. In the upper-division baccalaureate category, 3 semester hours for field experience in management and 3 in management problems (6/90).

MOS-39W-001

RADAR/SPECIAL ELECTRONIC DEVICES MAINTENANCE CHIEF
39W40

Exhibit Dates: 1/90–3/91. Effective 3/91, MOS 39W was discontinued, and its duties were incorporated into MOS 29W, Communications Maintenance Support Chief.

Career Management Field: 29 (Signal Maintenance).

Description

Supervises electronic maintenance shop operations by interpreting daily production reports and visual display loads of equipment in the shop; determines work priorities, parts needs, test equipment status, and personnel qualifications; assists subordinates in troubleshooting equipment; coordinates installation of new equipment; prepares and maintains records associated with planned maintenance programs; prepares and maintains support stock lists; determines causes of and reports unusual maintenance failures; organizes and conducts formal and on-the-job training for shop personnel. NOTE: May have progressed to 39W40 from 39C30 (Target Acquisition/Surveillance Radar Repairer) or 39E30 (Special Electronic Devices Repairer).

Recommendation

In the lower-division baccalaureate/associate degree category, 3 semester hours in principles of supervision and 3 in maintenance management. In the upper-division baccalaureate category, 3 semester hours for field experience in management and 3 in management problems (6/90).

MOS-39X-001

ELECTRONICS EQUIPMENT MAINTENANCE CHIEF
39X50

Exhibit Dates: 1/90–3/91. Effective 3/91, MOS 39X was discontinued, and its duties were incorporated into MOS 29Z, Electronics Maintenance Chief.

Career Management Field: 29 (Signal Maintenance).

Description

Supervises the electronic maintenance mission of radar/special electronic devices, automatic data processing, and automatic test equipment facility; determines work priorities and requirements; assigns and coordinates activities of shop personnel; applies policies and procedures for facility/unit/station management; prepares, maintains, and uses maintenance records and reports associated with routine and planned maintenance programs; plans and coordinates training in communication, operations procedures, and maintenance programs; assists staff in continuous appraisal of electronic equipment. NOTE: May have progressed to 39X50 from 39V40 (Computerized Systems Maintenance Chief) or 39W40 (Radar/Special Electronic Devices Maintenance Chief).

Recommendation

In the lower-division baccalaureate/associate degree category, 3 semester hours in principles of supervision and 3 in maintenance management. In the upper-division baccalaureate category, 6 semester hours for field experience in management and 3 in management problems (6/90).

MOS-39Y-001

FIELD ARTILLERY TACTICAL FIRE DIRECTION SYSTEMS REPAIRER
39Y10
39Y20
39Y30

Exhibit Dates: 1/90–4/92. Effective 4/92, MOS 39Y was discontinued, and its duties were incorporated into MOS 29J, Telecommunications Terminal Device Repairer.

Career Management Field: 29 (Communications Electronics Maintenance).

Description

Summary: Performs direct and general support maintenance on field artillery tactical fire direction systems, variable format entry devices, and related test equipment. *Skill Level 10:* Makes detailed tests using voltmeters, ohmmeters, oscilloscopes, logic probes, and other test equipment to localize and isolate malfunctions; replaces faulty components at the integrated circuit board level; may replace resistors, capacitors, integrated circuits, and other discrete components in emergency situations; maintains tools and test equipment; test operates computer-controlled system; checks and maintains electromechanical subsystems, including motors, generators, servos, and mechanical actuators; interprets complex block diagrams, schematics, and technical manuals. *Skill Level 20:* Able to perform the duties required for Skill Level 10; checks maintenance performed by subordinates; inspects systems for wear and possible future problem areas; supervises system repair contact team; interfaces with system users to identify cause of malfunctions. *Skill Level 30:* Able to perform the duties required for Skill Level 20; has complete responsibility for supervising a multimillion dollar maintenance van; directs and coordinates the activities of five to ten subordinates; performs performance appraisals; conducts on-the-job training; establishes work priorities.

Recommendation, Skill Level 10

In the lower-division baccalaureate/associate degree category, 3 semester hours in circuit diagnosis and 3 in computer system maintenance and repair. (NOTE: This recommendation for skill level 10 is valid for the dates 1/90-9/91 only) (11/85).

Recommendation, Skill Level 20

In the lower-division baccalaureate/associate degree category, 6 semester hours in circuit diagnosis, 6 in computer system maintenance and repair, 3 in record keeping, and 1 in personnel supervision (11/85).

Recommendation, Skill Level 30

In the lower-division baccalaureate/associate degree category, 6 semester hours in circuit diagnosis, 6 in computer system maintenance and repair, 3 in record keeping, and 3 in personnel supervision. In the upper-division baccalaureate category, 3 semester hours for field experience in management (11/85).

MOS-41B-002

TOPOGRAPHIC INSTRUMENT REPAIR SPECIALIST
41B10
41B20

Exhibit Dates: 1/90–9/90. Effective 9/90, MOS 41B was discontinued, and its duties were incorporated into MOS 82B, Construction Surveyor, and MOS 82D, Topographic Surveyor.

Career Management Field: 81 (Topographic Engineering), subfield 812 (Surveying).

Description

Summary: Performs maintenance on topographic survey and cartographic equipment. *Skill Level 10:* Inspects equipment for damaged, defective, or malfunctioning parts or assemblies; adjusts and performs operational tests on equipment; checks optics for such conditions as lens aberrations, parallax, tilt of reticule, and scratched lens; disassembles equipment and replaces defective parts with standard items; interprets blueprints and military or manufacturers' specifications; isolates and corrects faults; lays and fabricates unobtainable metal parts or coordinates their fabrication with machinists, welders, or other specialists; makes sketches or diagrams used in disassembling and reassembling equipment or fabricating parts; applies principles of basic electricity when making minor electrical repairs; makes precision bits and clearances using precision measuring instruments; requisitions and inventories spare parts; prepares maintenance records. *Skill Level 20:* Able to perform the duties required for Skill Level 10; provides technical guidance to subordinates; controls quality of topographic instrument maintenance by inspecting repairs performed; estimates cost of repairs, parts, and labor; prepares and reviews maintenance records and reports; schedules and assigns duties; fabricates special tools; overhauls and rebuilds equipment.

Recommendation, Skill Level 10

In the vocational certificate category, 6 semester hours in survey and optical instrument repair. In the lower-division baccalaureate/associate degree category, 3 semester hours in survey and optical instrument repair, 3 in basic surveying, 2 in care and use of tools and equipment, and credit in optics on the basis of institutional evaluation (5/78).

Recommendation, Skill Level 20

In the vocational certificate category, 6 semester hours in survey and optical instrument repair. In the lower-division baccalaureate/associate degree category, 3 semester hours in survey and optical instrument repair, 3 in basic surveying, 2 in use and care of tools and equipment, 2 in record keeping, and credit in optics on the basis of institutional evaluation (5/78).

MOS-41C-003

FIRE CONTROL INSTRUMENT REPAIRER
41C10
41C20
41C30

Exhibit Dates: 1/90–9/91.

Career Management Field: 63 (Mechanical Maintenance), subfield 634 (Armament Maintenance).

Description

Summary: Performs maintenance on fire control instruments and related equipment. *Skill Level 10:* Performs maintenance on fire control instruments and related equipment, including binoculars, aiming circles, telescopes, range finders, and ballistic computers; uses hand and power tools and testing devices to remove, repair, replace, and synchronize fire control systems; traces circuits using schematic diagrams; replaces defective parts; performs operating tests to ensure proper performance. *Skill Level 20:* Able to perform the duties required for Skill Level 10; troubleshoots and diagnoses fire control instruments; supervises and provides technical assistance to subordinates; conducts on-the-job training; prepares maintenance records. *Skill Level 30:* Able to perform the duties required for Skill Level 20; instructs and supervises subordinates; plans and organizes work schedules; applies management principles and techniques; performs quality control checks and technical inspections; requisitions parts and materials; establishes and supervises a shop safety program.

Recommendation, Skill Level 10

In the lower-division baccalaureate/associate degree category, 3 semester hours in applied physics, 3 in optical equipment repair laboratory, and 1 in use and care of hand and power tools (7/82).

Recommendation, Skill Level 20

In the lower-division baccalaureate/associate degree category, 3 semester hours in applied physics, 3 in optical equipment repair laboratory, 1 in use and care of hand and power tools, 1 in record keeping, 3 in shop practices, and 2 in personnel supervision (7/82).

Recommendation, Skill Level 30

In the lower-division baccalaureate/associate degree category, 3 semester hours in applied physics, 3 in optical equipment repair laboratory, 1 in use and care of hand and power tools, 1 in record keeping, 3 in shop practices, and 3 in personnel supervision. In the upper-division baccalaureate category, 3 semester hours for field experience in management and additional credit in management on the basis of institutional evaluation (7/82).

MOS-41C-004

FIRE CONTROL INSTRUMENT REPAIRER
41C10
41C20
41C30

Exhibit Dates: 10/91–4/92. Effective 4/92, MOS 41C was discontinued, and its duties were incorporated into MOS 45G, Fire Control Repairer and MOS 45K, Armament Repairer.

Career Management Field: 63 (Mechanical Maintenance).

Description

Summary: Performs maintenance on fire control instruments and related equipment. *Skill Level 10:* Performs maintenance on fire control instruments and related equipment, including binoculars, aiming circles, telescopes, range finders, and ballistic computers; uses hand and power tools and testing devices to remove, repair, replace, and synchronize fire control systems; traces circuits using schematic diagrams; uses oscilloscopes, digital volt meters, and associated test equipment; replaces defective parts; performs operating tests to ensure proper performance. *Skill Level 20:* Able to perform the duties required for Skill Level 10; troubleshoots and diagnoses fire control instruments; supervises and provides technical assistance to subordinates; conducts on-the-job training; prepares maintenance records. *Skill Level 30:* Able to perform the duties required for Skill Level 20; instructs and supervises subordinates; plans and organizes work schedules; applies management principles and techniques; performs quality control checks and technical inspections; requisitions parts and materials; establishes and supervises a shop safety program.

Recommendation, Skill Level 10

Credit may be granted on the basis of an individualized assessment of the student (3/94).

Recommendation, Skill Level 20

In the lower-division baccalaureate/associate degree category, 1 semester hour in applied physics, 3 in optical instrument repair, and 3 in personnel supervision (11/91).

Recommendation, Skill Level 30

In the lower-division baccalaureate/associate degree category, 1 semester hour in applied physics, 3 in optical instrument repair, and 3 in personnel supervision (11/91).

MOS-42A-001

HUMAN RESOURCE SPECIALIST
42A10
42A20
42A30
42A40
42A50

Exhibit Dates: Pending Evaluation. 4/03–Present.

MOS-42C-002

ORTHOTIC SPECIALIST
42C10
42C20
42C30
42C40

Exhibit Dates: 1/90–5/92.

Career Management Field: 91 (Medical), subfield 912 (Patient Care).

Description

Summary: Designs, fabricates, fits, adjusts, and repairs orthotic devices. *Skill Level 10:* Able to perform the duties required for Skill Level 10 of MOS 91B (Medical Specialist); assists in designing, fabricating, fitting, adjusting, and repairing orthotic appliances; makes impressions of injured body members and fabricates orthotic appliances of metal, plastic, synthetics, resins, plaster, leather, rubber, textiles, and adhesives, using welding, soldering, forging, riveting, sewing, laminating, brazing, woodworking, pattern and template machines and equipment, and all types of power and hand tools; adjusts constructed appliances and fits to patient; repairs, readjusts, reworks, and refits worn and defective appliances. *Skill Level 20:* Able to perform the duties required for Skill Level 10; measures patient in operating room and molds temporary plastic socket; designs, fabricates, and tests experimental devices for which there are no models or standard pattern; adapts techniques to fit local requirements and available equipment. *Skill Level 30:* Able to perform the duties required for Skill Level 20; provides technical guidance to subordinates;

instructs subordinates in work techniques and procedures; inspects completed appliances. *Skill Level 40:* Able to perform the duties required for Skill Level 30; establishes work priorities; organizes work schedules and assigns duties; supervises the equipment maintenance program; makes sure the shop is safe, clean, and orderly; determines personnel requirements and prepares and conducts training programs; evaluates personnel performance; prepares administrative, technical, and personnel reports; coordinates orthopedic activities with other elements of the medical treatment facility; advises and assists physicians in professional, fiscal, technical, and administrative matters; establishes stock level for supplies and equipment and supervises the requisitioning, storing, and issuing of supplies.

Recommendation, Skill Level 10

In the lower-division baccalaureate/associate degree category, 24 semester hours in orthotic techniques (including spinal, lower-extremity, and upper-extremity orthotics), 6 in applied design, 2 in anatomy, 2 in shop practices, 1 in physiology, and 1 in kinesiology. (NOTE: This recommendation for skill level 10 is valid for the dates 1/90-9/91 only) (6/77).

Recommendation, Skill Level 20

In the lower-division baccalaureate/associate degree category, 24 semester hours in orthotic techniques (including spinal, lower-extremity, and upper-extremity orthotics), 6 in applied design, 3 in clinical orthotics, 3 in human relations, 2 in anatomy, 2 in shop practices, 1 in physiology, and 1 in kinesiology (6/77).

Recommendation, Skill Level 30

In the lower-division baccalaureate/associate degree category, 24 semester hours in orthotic techniques (includes spinal, lower-extremity, and upper-extremity orthotics), 6 in applied design, 3 in clinical orthotics, 3 in human relations, 2 in anatomy, 2 in shop practices, 1 in physiology, and 1 in kinesiology (6/77).

Recommendation, Skill Level 40

In the lower-division baccalaureate/associate degree category, 24 semester hours in orthotic techniques (including spinal, lower-extremity, and upper-extremity orthotics), 6 in applied design, 3 in clinical orthotics, 3 in human relations, 3 in personnel supervision, 2 in anatomy, 2 in shop practices, 1 in physiology, and 1 in kinesiology. In the upper-division baccalaureate category, 2 semester hours in medical facility management (6/77).

MOS-42C-003

Orthotic Specialist

42C10
42C20
42C30
42C40

Exhibit Dates: 6/92–12/94. (Effective 12/94, MOS 42C was discontinued.)

Career Management Field: 91 (Medical).

Description

Summary: Designs, fabricates, fits, adjusts, and repairs orthotic devices. *Skill Level 10:* Assists in designing, fabricating, fitting, adjusting, and repairing orthotic appliances; makes impressions of injured body members and fabricates orthotic appliances of metal, plastic, synthetics, resins, plaster, leather, rubber, textiles, and adhesives, using welding, soldering, forging, riveting, sewing, laminating, brazing, woodworking, pattern and template machines and equipment, and all types of power and hand tools; adjusts constructed appliances and fits to patient; repairs, readjusts, reworks, and refits worn and defective appliances. *Skill Level 20:* Able to perform the duties required for Skill Level 10; measures patient in operating room and molds temporary plastic socket; designs, fabricates, and tests experimental devices for which there are no models or standard pattern; adapts techniques to fit local requirements and available equipment. *Skill Level 30:* Able to perform the duties required for Skill Level 20; provides technical guidance to subordinates; instructs subordinates in work techniques and procedures; inspects completed appliances. *Skill Level 40:* Able to perform the duties required for Skill Level 30; establishes work priorities; organizes work schedules and assigns duties; supervises the equipment maintenance program; makes sure the shop is safe, clean, and orderly; determines personnel requirements; prepares and conducts training programs; evaluates personnel performance; prepares administrative, technical, and personnel reports; coordinates orthopedic activities with other elements of the medical treatment facility; advises and assists physician in professional, fiscal, technical, and administrative matters; establishes stock level for supplies and equipment and supervises the requisitioning, storing, and issuing of supplies.

Recommendation, Skill Level 10

Credit may be granted on the basis of an individualized assessment of the student (3/94).

Recommendation, Skill Level 20

In the lower-division baccalaureate/associate degree category, 8 semester hours in orthotic fabrication, 6 in orthotic design, 3 in patient measurement and assessment, and 3 in personnel supervision (6/92).

Recommendation, Skill Level 30

In the lower-division baccalaureate/associate degree category, 8 semester hours in orthotic fabrication, 6 in orthotic design, 3 in patient measurement and assessment, 3 in personnel supervision, and 2 in records and information management (6/92).

Recommendation, Skill Level 40

In the lower-division baccalaureate/associate degree category, 8 semester hours in orthotic fabrication, 6 in orthotic design, 3 in patient measurement and assessment, 3 in personnel supervision, and 2 in records and information management. In the upper-division baccalaureate category, 3 semester hours for field experience in management and 3 in organizational management (6/92).

MOS-42D-002

Dental Laboratory Specialist

42D10
42D20
42D30
42D40

Exhibit Dates: 1/90–5/92.

Career Management Field: 91 (Medical), subfield 911 (Dental).

Description

Summary: Supervises or performs basic and advanced procedures in fabrication and repair of dental prosthodontic appliances. *Skill Level 10:* Able to perform the duties required for Skill Level 10 of MOS 91E (Dental Specialist); performs primary procedures in fabrication and repair of removable and fixed dental prosthodontic appliances; reviews and processes prosthodontic prescriptions and consultation requests; prepares and uses basic dental laboratory materials; fabricates casts from preliminary impressions; fabricates final impression trays for complete, immediate-complete, removable, and fixed partial dentures; prepares impressions and pours final casts for complete, immediate-complete, removable, and fixed partial dentures; fabricates stabilized and nonstabilized occlusion rims for complete, immediate-complete, removable, and fixed partial dentures; replaces missing or broken teeth and repairs fractured bases on complete and removable partial dentures. *Skill Level 20:* Able to perform the duties required for Skill Level 10; performs basic dental laboratory procedures; uses face-brow and measurements to mount casts on adjustable articulator; employs mold guide, occlusion rims, registrations, markings, and interarch measurements to select denture teeth; fabricates maxillary and mandibular complete dentures using conventional and liquid resin techniques; fabricates immediate-complete dentures with surgical templates; fabricates removable partial denture prostheses; modifies dental laboratory techniques to fabricate removable complete and partial dentures for cases with variant interarch and interridge relationships; repairs, relines, rebases, and duplicates removable complete and partial dentures; fabricates and repairs basic orthodontic, pedodontic, periodontic, and surgical appliances; fabricates and repairs conventional fixed partial dentures, including crowns, inlays, and performed pontics; fabricates transitional fixed partial dentures. *Skill Level 30:* Able to perform the duties required for Skill Level 20; performs advanced dental laboratory procedures or assists in managing a small dental laboratory; fabricates ceramic and porcelain-fused-to-metal crowns and fixed partial dentures; fabricates fixed and removable precision appliances; fabricates complex orthodontic, pedodontic, periodontic, and surgical appliances; assists in fabricating maxillofacial appliances; performs quality control procedures; establishes work priorities and assigns duties; organizes work schedules; demonstrates and explains work techniques and procedures to subordinates; evaluates personnel performance; supervises operational maintenance program of laboratory equipment. *Skill Level 40:* Able to perform the duties required for Skill Level 30; assists in managing a large dental laboratory or a section of a regional dental activity; advises and assists dentists in personnel matters, supply economy procedures, and fiscal, technical, and administrative matters; prepares administrative and technical reports.

Recommendation, Skill Level 10

In the lower-division baccalaureate/associate degree category, 2 semester hours in anatomy and tooth morphology, 1 in dental materials, and 1 in complete dentures. (NOTE: This recommendation for skill level 10 is valid for the dates 1/90-9/91 only) (6/77).

Recommendation, Skill Level 20

In the lower-division baccalaureate/associate degree category, 4 semester hours in complete dentures, 3 in partial dentures, 3 in fixed restorative, 2 in anatomy and tooth morphology, 1 in dental anatomy, and 1 in dental materials (6/77).

Recommendation, Skill Level 30

In the lower-division baccalaureate/associate degree category, 4 semester hours in complete dentures, 4 in dental ceramics, 3 in partial dentures, 3 in fixed restorative, 2 in anatomy and tooth morphology, 2 in orthodontic procedures, 2 in dental anatomy, 1 in dental materials, and 1 in dental anatomy (occlusion), and additional credit in personnel supervision, field experience in management, and communication skills on the basis of institutional evaluation (6/77).

Recommendation, Skill Level 40

In the lower-division baccalaureate/associate degree category, 4 semester hours in complete dentures, 4 in dental ceramics, 3 in partial dentures, 3 in fixed restorative, 2 in anatomy and tooth morphology, 2 in orthodontic procedures, 2 in dental anatomy, 2 in personnel supervision, 2 for field experience in management, 2 in oral communication skills, 1 in health care management, 1 in dental materials, and 1 in dental anatomy (occlusion) (6/77).

MOS-42D-003

Dental Laboratory Specialist
42D10
42D20
42D30
42D40

Exhibit Dates: 6/92–12/94. Effective 12/94, MOS 42D was discontinued, and its duties were transferred to MOS 91E.

Career Management Field: 91 (Medical).

Description

Summary: Supervises or performs basic and advanced procedures in fabrication and repair of dental prosthodontic appliances. *Skill Level 10:* Able to perform the duties required for Skill Level 10 of MOS 91E (Dental Specialist); performs primary procedures in fabrication and repair of removable and fixed dental prosthodontic appliances; reviews and processes prosthodontic prescriptions and consultation requests; prepares and uses basic dental laboratory materials; fabricates final impression trays for complete, immediate-complete, removable, and fixed partial dentures; prepares impressions and pours final casts for complete, immediate-complete, removable, and fixed partial dentures; fabricates stabilized and nonstabilized occlusion rims for complete, immediate-complete, removable, and fixed partial dentures; replaces missing or broken teeth and repairs fractured bases on complete and removable partial dentures; performs basic dental laboratory procedures; uses facebrow and measurements to mount casts on adjustable articulator; employs mold guide, occlusion rims, registrations, markings, and interarch measurements to select denture teeth; fabricates maxillary and mandibular complete dentures; fabricates removable partial denture prostheses; repairs, relines, rebases, and duplicates removable complete and partial dentures; fabricates and repairs basic orthodontic, pedodontic, periodontic, and surgical appliances; fabricates and repairs conventional fixed partial dentures, including crowns, inlays, and preformed pontics; fabricates transitional fixed partial dentures. *Skill Level 20:* Able to perform the duties required for Skill Level 10; reviews prescriptions and determines method of accomplishing task assignment; orders, receives, and issues supplies and equipment; assists subordinates; assists with training and administrative management; prepares and maintains laboratory, feeder requisitions, and reports. *Skill Level 30:* Able to perform the duties required for Skill Level 20; performs advanced dental laboratory procedures or assists in managing a small dental laboratory; fabricates ceramic and porcelain-fused-to-metal crowns and fixed partial dentures; fabricates fixed and removable precision appliances; fabricates complex orthodontic, pedodontic, periodontic, and surgical appliances; assists in fabricating maxillofacial appliances; performs quality control procedures; establishes work priorities and assigns duties; organizes work schedules; demonstrates and explains work techniques and procedures to subordinates; evaluates personnel performance; supervises operational maintenance program of laboratory equipment. *Skill Level 40:* Able to perform the duties required for Skill Level 30; assists in managing a large dental laboratory or a section of a regional dental activity; advises and assists dentists in personnel matters, supply economy procedures, and fiscal, technical, and administrative matters; prepares administrative and technical reports.

Recommendation, Skill Level 10

Credit may be granted on the basis of an individualized assessment of the student (3/94).

Recommendation, Skill Level 20

In the lower-division baccalaureate/associate degree category, 3 semester hours in basic denture preparation, 3 in dental laboratory procedures, 4 in directed practice in the dental laboratory, and 3 in personnel supervision (6/92).

Recommendation, Skill Level 30

In the lower-division baccalaureate/associate degree category, 3 semester hours in basic denture preparation, 3 in dental laboratory procedures, 4 in directed practice in the dental laboratory, 3 in personnel supervision, and 2 in records and information management (6/92).

Recommendation, Skill Level 40

In the lower-division baccalaureate/associate degree category, 3 semester hours in basic denture preparation, 3 in dental laboratory procedures, 4 in directed practice in the dental laboratory, 3 in personnel supervision, and 2 in records and information management. In the upper-division baccalaureate category, 3 semester hours in organizational management and 3 for field experience in management (6/92).

MOS-42E-002

Optical Laboratory Specialist
42E10
42E20
42E30
42E40
42E50

Exhibit Dates: 1/90–5/92.

Career Management Field: 91 (Medical), subfield 913 (Health Services).

Description

Summary: Makes prescription lenses or supervises optical laboratories. *Skill Level 10:* Makes, duplicates, and inserts prescription lenses; selects proper stock to fill requirements of spectacle prescription; computes and records curvature and thickness and marks lenses for surfacing; prepares lenses for blocking and surfacing; grinds and polishes lenses to prescribed specifications; locates and identifies optical center, optical axis, and cutting line on lenses; chucks lenses for edging; edges lenses to correct size and shape; hardens glass lenses to required impact resistance; selects and assembles frame components; mounts lenses and aligns frames; prepares completed spectacles for shipment. *Skill Level 20:* Able to perform the duties required for Skill Level 10; edits prescriptions and maintains records of prescriptions received by optical unit; calculates and records amount of prism required to permit necessary decentration; calculates and records distance required between top of bifocal or trifocal segments and distance-optical-center to produce prescribed segment height; dyes plastic lenses to specified tints; maintains files of completed spectacle prescriptions; prepares surface work sheets for plastic lens coating; calculates and records positional effective lens power; maintains inventory of supplies and equipment. *Skill Level 30:* Able to perform the duties required for Skill Level 20; supervises a small optical laboratory; inspects completed spectacles for compliance with prescription requirements and quality standards; assigns duties and trains subordinates in work techniques and procedures; determines stock levels; maintains supply records; requisitions, stores, and issues supplies; supervises preventive maintenance of equipment. *Skill Level 40:* Able to perform the duties required for Skill Level 30; supervises a medium-sized optical laboratory; maintains records on cost data and expenditures; ensures that the laboratory is safe, clean, and orderly; supervises quality control procedures; establishes work priorities; organizes work schedules; evaluates personnel performance; plans layout areas; prepares technical, personnel, and administrative reports; coordinates activities of the optical laboratory with medical treatment facilities; advises and assists professionals in fiscal, technical, and administrative matters. *Skill Level 50:* Able to perform the duties required for Skill Level 40; supervises a large optical laboratory; prepares periodic laboratory reports; determines manpower requirements; drafts budget estimates; prepares operating procedures for optical unit.

Recommendation, Skill Level 10

In the lower-division baccalaureate/associate degree category, 10 semester hours for an optical laboratory practicum, 8 in optical laboratory procedures, and 5 in physics (optics). (NOTE: This recommendation for skill level 10 is valid for the dates 1/90-9/91 only) (6/77).

Recommendation, Skill Level 20

In the lower-division baccalaureate/associate degree category, 10 semester hours for an optical laboratory practicum, 8 in optical laboratory procedures, 5 in physics (optics), 1 in anatomy and physiology, and 1 in office procedures (6/77).

Recommendation, Skill Level 30

In the lower-division baccalaureate/associate degree category, 10 semester hours for an optical laboratory practicum, 8 in optical laboratory procedures, 5 in physics (optics), 3 in office procedures, 1 in anatomy and physiology, and 1 in personnel supervision (6/77).

Recommendation, Skill Level 40

In the lower-division baccalaureate/associate degree category, 10 semester hours for an optical laboratory practicum, 8 in optical laboratory procedures, 5 in physics (optics), 3 in office procedures, 3 in personnel supervision, 3 for field experience in management, and 1 in anatomy and physiology (6/77).

Recommendation, Skill Level 50

In the lower-division baccalaureate/associate degree category, 10 semester hours for an optical laboratory practicum, 8 in optical laboratory procedures, 5 in physics (optics), 3 in office procedures, 3 in personnel supervision, 3 for field experience in management, 3 in introduction to management, and 1 in anatomy and physiology. In the upper-division baccalaureate category, 3 semester hours in medical facility management and 3 in management problems (6/77).

MOS-42E-003

Optical Laboratory Specialist
42E10
42E20
42E30
42E40
42E50

Exhibit Dates: 6/92–7/94.

Career Management Field: 91 (Medical).

Description

Summary: Makes prescription lenses or supervises optical laboratories. *Skill Level 10:* Makes, duplicates, and inserts prescription lenses; selects proper stock to fill requirements of spectacle prescription; computes and records curvature and thickness and marks lenses for surfacing; grinds and polishes lenses to prescribed specifications; locates and identifies optical center, optical axis, and cutting line on lenses; chucks lenses for edging; edges lenses to correct size and shape; hardens glass lenses to required impact resistance; selects and assembles frame components; mounts lenses and aligns frames; prepares completed spectacles for shipment. *Skill Level 20:* Able to perform the duties required for Skill Level 10; edits prescriptions and maintains records of prescriptions received by optical unit; calculates and records amount of prism required to permit necessary decentration; calculates and records distance required between top of bifocal or trifocal segments and distance-optical-center to produce prescribed segment height; dyes plastic lenses to specified tints; maintains files of completed spectacle prescriptions; prepares surface work sheets for plastic lens coating; calculates and records positional effective lens power; maintains inventory of supplies and equipment. *Skill Level 30:* Able to perform the duties required for Skill Level 20; supervises a small optical library; inspects completed spectacles for compliance with prescription requirements and quality standards; assigns duties and trains subordinates in work techniques and procedures; determines stock levels; maintains supply records; requisitions, stores, and issues supplies; supervises preventive maintenance of equipment. *Skill Level 40:* Able to perform the duties required for Skill Level 30; supervises a medium-sized optical laboratory; maintains records of cost data and expenditures; ensures that the laboratory is safe, clean, and orderly; supervises quality control procedures; establishes work priorities; organizes work schedules; evaluates personnel performance; plans layout areas; prepares technical, personnel, and administrative reports; coordinates activities of the optical laboratory with medical treatment facilities; advises and assists professionals in fiscal, technical, and administrative matters. *Skill Level 50:* Able to perform the duties required for Skill Level 40; supervises a large optical laboratory; prepares periodic laboratory reports; determines manpower requirements; drafts budget estimates; prepares operating procedures for optical unit.

Recommendation, Skill Level 10

Credit may be granted on the basis of an individualized assessment of the student (3/94).

Recommendation, Skill Level 20

In the lower-division baccalaureate/associate degree category, 7 semester hours in optical laboratory procedures and 3 in personnel supervision (6/92).

Recommendation, Skill Level 30

In the lower-division baccalaureate/associate degree category, 7 semester hours in optical laboratory procedures, 3 in personnel supervision, and 2 in records and information management (6/92).

Recommendation, Skill Level 40

In the lower-division baccalaureate/associate degree category, 7 semester hours in optical laboratory procedures, 3 in personnel supervision, and 2 in records and information management. In the upper-division baccalaureate category, 3 semester hours in organizational management and 3 for field experience in management (6/92).

Recommendation, Skill Level 50

In the lower-division baccalaureate/associate degree category, 7 semester hours in optical laboratory procedures, 3 in personnel supervision, and 2 in records and information management. In the upper-division baccalaureate category, 3 semester hours in organizational management and 3 for field experience in management; if paygrade E-9 was attained, additional credit as follows: 3 semester hours in management problems, 3 in operations management, and 3 in communication techniques for managers (6/92).

MOS-42E-004

Optical Laboratory Specialist
42E10
42E20
42E30
42E40

Exhibit Dates: 8/94–11/01. Effective 11/01, MOS 42E was discontinued and its duties were transferred into MOS 91H, Optical Laboratory Specialist.

Career Management Field: 91 (Medical).

Description

Summary: Makes prescription lenses or supervises optical laboratories. *Skill Level 10:* Makes, duplicates, and inserts prescription lenses; selects proper stock to fill requirements of spectacle prescription; computes and records curvature and thickness and marks lenses for surfacing; grinds and polishes lenses to prescribed specifications; locates and identifies optical center, optical axis, and cutting line on lenses; chucks lenses for edging; edges lenses to correct size and shape; hardens glass lenses to required impact resistance; selects and assembles frame components; mounts lenses and aligns frames; prepares completed spectacles for shipment. *Skill Level 20:* Able to perform the duties required for Skill Level 10; edits prescriptions and maintains records of prescriptions received by optical unit; calculates and records amount of prism required to permit necessary decentration; calculates and records distance required between top of bifocal or trifocal segments and distance-optical-center to produce prescribed segment height; dyes plastic lenses to specified tints; maintains files of completed spectacle prescriptions; prepares surface work sheets for plastic lens coating; calculates and records positional effective lens power; maintains inventory of supplies and equipment. *Skill Level 30:* Able to perform the duties required for Skill Level 20; supervises a small optical library; inspects completed spectacles for compliance with prescription requirements and quality standards; assigns duties and trains subordinates in work techniques and procedures; determines stock levels; maintains supply records; requisitions, stores, and issues supplies; supervises preventive maintenance of equipment. *Skill Level 40:* Able to perform the duties required for Skill Level 30; supervises a medium-sized optical laboratory; maintains records of cost data and expenditures; ensures that the laboratory is safe, clean, and orderly; supervises quality control procedures; establishes work priorities; organizes work schedules; evaluates personnel performance; plans layout areas; prepares technical, personnel, and administrative reports; coordinates activities of the optical laboratory with medical treatment facilities; advises and assists professionals in fiscal, technical, and administrative matters.

Recommendation, Skill Level 10

Credit may be granted on the basis of an individualized assessment of the student (3/94).

Recommendation, Skill Level 20

In the lower-division baccalaureate/associate degree category, 7 semester hours in optical laboratory procedures and 3 in personnel supervision. Effective 11/01, MOS 42E was discontinued and its duties were transferred to MOS 91H, Optical Laboratory Specialist (6/92).

Recommendation, Skill Level 30

In the lower-division baccalaureate/associate degree category, 7 semester hours in optical laboratory procedures, 3 in personnel supervision, and 2 in records and information management (6/92).

Recommendation, Skill Level 40

In the lower-division baccalaureate/associate degree category, 7 semester hours in optical laboratory procedures, 3 in personnel supervision, and 2 in records and information management. In the upper-division baccalaureate category, 3 semester hours in organizational management and 3 for field experience in management (6/92).

MOS-42F-002

Human Resource Information System Management Specialist
42F10
42F20
42F30

Exhibit Dates: Pending Evaluation. 4/03–Present.

MOS-42L-001

Administrative Specialist
42L10
42L20
42L30
42L40
42L50

Exhibit Dates: Pending Evaluation. 4/03–Present.

MOS-42R-001

Army Bandsperson
42R10
42R20
42R30
42R40

Exhibit Dates: Pending Evaluation. 4/03–Present.

MOS-42S-001

Special Band Member
42S20
42S30
42S40
42S50

Exhibit Dates: Pending Evaluation. 4/03–Present.

MOS-43E-002

Parachute Rigger
43E10
43E20
43E30
43E40
43E50

Exhibit Dates: 1/90–1/92.

Career Management Field: 76 (Supply and Service), subfield 762 (Service).

Description

Summary: Is a qualified parachutist; packs aircraft cargo and personnel parachutes; fabricates, assembles, repairs, and rigs airdrop equipment. *Skill Level 10:* Packs and repairs cargo and personnel parachutes; rigs, fabricates, and assembles airdrop equipment; loads, positions, and secures cargo for airdrops; tests hooks and releases; utilizes repair equipment for parachutists; reads maps; loads, unloads, receives, inspects, inventories, and stores air equipment. *Skill Level 20:* Able to perform the duties required for Skill Level 10; provides technical guidance to subordinates; conducts routine and in-storage inspections of airdrop equipment; tests rip cord and canopy release assemblies. *Skill Level 30:* Able to perform the duties required for Skill Level 20; supervises small and medium activities engaged in packing and repairing parachutes and rigging and loading supplies and equipment for airdrop; assigns packers, repairers, riggers, and inspectors; instructs personnel on job requirements and techniques and inspects work in progress; inspects working area to insure a clean and safe environment; plans and supervises operational equipment maintenance programs; establishes supply requirements; requisitions, stores, and issues supplies and equipment; establishes, develops, and maintains operational and maintenance records and prepares periodic reports; may supervises up to 35 persons. *Skill Level 40:* Able to perform the duties required for Skill Level 30; supervises large activities engaged in packing and repairing parachutes and rigging and loading of supplies and equipment for airdrop; may supervise up to 50 persons. *Skill Level 50:* Able to perform the duties required for Skill Level 40; serves as principal noncommissioned officer in airdrop equipment support company; may supervise more than 50 persons or serve in staff position.

Recommendation, Skill Level 10

In the lower-division baccalaureate/associate degree category, 5 semester hours in parachute inspection and rigging and 1 in map reading. (NOTE: This recommendation for skill level 10 is valid for the dates 1/90-9/91 only) (5/78).

Recommendation, Skill Level 20

In the lower-division baccalaureate/associate degree category, 5 semester hours in parachute inspection and rigging, 5 in air freight (on board) management, and 1 in map reading (5/78).

Recommendation, Skill Level 30

In the lower-division baccalaureate/associate degree category, 6 semester hours in parachute inspection and rigging, 6 in air freight (on board) management, 1 in map reading, 2 in technical report writing, 3 in human relations, and 3 in personnel supervision. In the upper-division baccalaureate category, 3 semester hours in personnel management, 3 for field experience in management, 3 in inventory control, 3 in safety procedures, and 2 in records administration (5/78).

Recommendation, Skill Level 40

In the lower-division baccalaureate/associate degree category, 6 semester hours in parachute inspection and rigging, 6 in air freight (on board) management, 1 in map reading, 3 in personnel supervision, 3 in human relations, and 2 in technical report writing. In the upper-division baccalaureate category, 3 semester hours in personnel management, 3 for field experience in management, 3 in inventory control, 3 in safety procedures, 3 in records administration, and 3 in introduction to management (5/78).

Recommendation, Skill Level 50

In the lower-division baccalaureate/associate degree category, 6 semester hours in parachute inspection and rigging, 6 in air freight (on board) management, 1 in map reading, 3 in personnel supervision, 3 in human relations, 2 in technical report writing, and 3 for field experience in management. In the upper-division baccalaureate category, 3 semester hours in personnel management, 6 for field experience in management, 3 in records administration, 3 in inventory control, 3 in safety procedures, and 3 in introduction to management (5/78).

MOS-43E-003

Parachute Rigger
43E10
43E20
43E30
43E40
43E50

Exhibit Dates: 2/92–12/95. Effective 12/95, MOS 43E was discontinued and its duties were incorporated into MOS 92R, Parachute Rigger.

Career Management Field: 76 (Supply and Service).

Description

Summary: Is a qualified parachutist; packs aircraft cargo and personnel parachutes; fabricates, assembles, repairs, and rigs airdrop equipment. *Skill Level 10:* Packs and repairs cargo and personnel parachutes; rigs, fabricates, and assembles airdrop equipment; loads, positions, and secures cargo for airdrops; tests hooks and releases; utilizes repair equipment for parachutists; reads maps; loads, unloads, receives, inspects, inventories, and stores airdrop equipment. *Skill Level 20:* Able to perform the duties required for Skill Level 10; provides technical guidance to subordinates; conducts routine and in-storage inspection of airdrop equipment; tests rip cord and canopy release assemblies; supervises squad of approximately six persons. *Skill Level 30:* Able to perform the duties required for Skill Level 20; supervises small and medium activities engaged in packing and repairing parachutes and rigging and loading supplies and equipment for airdrop; assigns packers, repairers, riggers, and inspectors; instructs personnel on job requirements and techniques and inspects work in progress; inspects working area to insure a clean and safe working environment; plans and supervises operational equipment maintenance programs; establishes supply requirements; requisitions, stores, and issues supplies and equipment; establishes, develops, and maintains operational and maintenance records and prepares periodic reports; may supervise up to 35 persons. *Skill Level 40:* Able to perform the duties required for Skill Level 30; supervises large activities engaged in packing and repairing parachutes and the rigging and loading of supplies and equipment for airdrop; may supervise up to 50 persons. *Skill Level 50:* Able to perform the duties required for Skill Level 40; serves as principal noncommissioned officer in airdrop equipment support company; may supervise more than 50 persons or serve in staff positions.

Recommendation, Skill Level 10

Credit may be granted on the basis of an individualized assessment of the student (3/94).

Recommendation, Skill Level 20

In the lower-division baccalaureate/associate degree category, 3 semester hours in military science, 3 in aircraft weight and balance, and 3 in personnel supervision. (NOTE: This recommendation for skill level 20 is valid for the dates 2/92-2/95 only) (2/92).

Recommendation, Skill Level 30

In the lower-division baccalaureate/associate degree category, 3 semester hours in military science, 3 in aircraft weight and balance, 3 in personnel supervision, 3 in technical report writing, and 2 in records and information management. In the upper-division baccalaureate category, 2 semester hours in inventory control (2/92).

Recommendation, Skill Level 40

In the lower-division baccalaureate/associate degree category, 3 semester hours in military science, 3 in aircraft weight and balance, 3 in personnel supervision, 3 in technical report writing, and 2 in records and information management. In the upper-division baccalaureate category, 3 semester hours in inventory control, 3 in organizational management, and 3 for field experience in management (2/92).

Recommendation, Skill Level 50

In the lower-division baccalaureate/associate degree category, 3 semester hours in military science, 3 in aircraft weight and balance, 3 in personnel supervision, 3 in technical report writing, and 2 in records and information management. In the upper-division baccalaureate category, 3 semester hours in inventory control, 3 in organizational management, and 6 for field experience in management; if paygrade E-9 has been attained, additional credit as follows: 3 semester hour in management problems, 3 in operations management, and 3 in communication techniques for managers (2/92).

MOS-43M-003

Fabric Repair Specialist

43M10

43M20

43M30

43M40

Exhibit Dates: 1/90–1/92.

Career Management Field: 76 (Supply and Service), subfield 762 (Service).

Description

Summary: Repairs or supervises the repair of textile and canvas items, webbed equipment, and clothing. *Skill Level 10:* Repairs textile and canvas items; determines repair requirement; operates sewing, stitching, and patching equipment; performs minor repairs and preventive maintenance on canvas repair and textile equipment. *Skill Level 20:* Able to perform the duties required for Skill Level 10; performs quality assurance inspections; inspects, classifies, and determines reparability of textile, canvas, and webbed items; salvages repairable parts. *Skill Level 30:* Able to perform the duties required for Skill Level 20; supervises and coordinates small- and medium-sized operations and preventive maintenance activities in mobile and fixed textile and canvas repair shop; establishes quality control standards; establishes work schedules and assignments; supervises from 8 to 30 persons; evaluates personnel performance; counsels personnel and prepares evaluation reports; maintains records. *Skill Level 40:* Able to perform the duties required for Skill Level 30; supervises large operation and preventive maintenance activities in textile and canvas repair shop; supervises more than 30 persons.

Recommendation, Skill Level 10

In the vocational certificate category, 5 semester hours in power sewing machine operations and 4 for field experience in sewing operations. (NOTE: This recommendation for skill level 10 is valid for the dates 1/90-9/91 only) (5/78).

Recommendation, Skill Level 20

In the vocational certificate category, 5 semester hours in power sewing machine operations and 6 for field experience in sewing operations. In the lower-division baccalaureate/associate degree category, 6 semester hours in sewing and tailoring operations (5/78).

Recommendation, Skill Level 30

In the vocational certificate category, 5 semester hours in power sewing machine operations and 8 for field experience in sewing operations. In the lower-division baccalaureate/associate degree category, 6 semester hours in sewing and tailoring operations, 3 in human relations, and 2 in personnel supervision. In the upper-division baccalaureate category, 3 semester hours for field experience in management (5/78).

Recommendation, Skill Level 40

In the vocational certificate category, 5 semester hours in power sewing machine operations and 8 for field experience in sewing operations. In the lower-division baccalaureate/associate degree category, 6 semester hours in sewing and tailoring operations, 3 in human relations, 3 in introductory organization management, and 2 in personnel supervision. In the upper-division baccalaureate category, 3 semester hours in introduction to management and 3 for field experience in management (5/78).

MOS-43M-004

Fabric Repair Specialist

43M10

43M20

43M30

43M40

Exhibit Dates: 2/92–2/95.

Career Management Field: 76 (Supply and Service).

Description

Summary: Repairs or supervises the repair of textile and canvas items, webbed equipment, and clothing. *Skill Level 10:* Repairs textile and canvas items; determines repair requirements; operates sewing, stitching, and patching equipment; performs minor repairs and preventive maintenance on canvas repair and textile equipment. *Skill Level 20:* Able to perform the duties required for Skill Level 10; performs quality assurance inspections; inspects, classifies and determines reparability of textile, canvas, and webbed items; salvages repairable parts; supervises squad of approximately six persons. *Skill Level 30:* Able to perform the duties required for Skill Level 20; supervises and coordinates small or medium operation and preventive maintenance activities in mobile and fixed textile and canvas repair shop; establishes quality control standards; establishes work schedules and assignments; supervises from 8 to 30 persons; evaluates personnel performance; counsels personnel; prepares evaluation reports; maintains records. *Skill Level 40:* Able to perform the duties required for Skill Level 30; supervises large operation and preventive maintenance activities in textile and canvas repair shop; supervises more than 30 persons.

Recommendation, Skill Level 10

Credit may be granted on the basis of an individualized assessment of the student (3/94).

Recommendation, Skill Level 20

In the vocational certificate category, 6 semester hours in power machine operations and 5 for field experience in power sewing. In the lower-division baccalaureate/associate degree category, 3 semester hours in personnel supervision (2/92).

Recommendation, Skill Level 30

In the vocational certificate category, 7 semester hours in power machine operations and 6 for field experience in power sewing. In the lower-division baccalaureate/associate degree category, 3 semester hours in personnel supervision and 2 in records and information management (2/92).

Recommendation, Skill Level 40

In the vocational certificate category, 8 semester hours in power machine operations and 6 for field experience in power sewing. In the lower-division baccalaureate/associate degree category, 3 semester hours in personnel supervision and 2 in records and information management. In the upper-division baccalaureate category, 3 semester hours for field experience in management and 3 in organizational management (2/92).

MOS-43M-005

Fabric Repair Specialist

43M10

43M20

43M30

43M40

Exhibit Dates: 3/95–9/01. Effective 10/01, MOS 43M was discontinued and its duties were incorporated into MOS 92S, Laundry and Textile Specialist.

Career Management Field: 76 (Supply and Service).

Description

Summary: Repairs or supervises the repair of textile and canvas items, webbed equipment, and clothing. *Skill Level 10:* Repairs textile and canvas items; determines repair requirements; operates sewing, stitching, and patching equipment; performs minor repairs and preventive maintenance on canvas repair and textile equipment. *Skill Level 20:* Able to perform the duties required for Skill Level 10; performs quality assurance inspections; inspects, classifies and determines reparability of textile, canvas, and webbed items; salvages repairable parts; supervises squad of approximately six persons. *Skill Level 30:* Able to perform the duties required for Skill Level 20; supervises and coordinates small or medium operation and preventive maintenance activities in mobile and fixed textile and canvas repair shop; establishes quality control standards; establishes work schedules and assignments; supervises from 8 to 30 persons; evaluates personnel performance; counsels personnel; prepares evaluation reports; maintains records. *Skill Level 40:* Able to perform the duties required for Skill Level 30; supervises large operation and preventive maintenance activities in textile and canvas repair shop; supervises more than 30 persons.

Recommendation, Skill Level 10

Credit may be granted on the basis of an individualized assessment of the student (3/94).

Recommendation, Skill Level 20

Credit may be granted on the basis of an individualized assessment of the student (2/92).

Recommendation, Skill Level 30

In the vocational certificate category, 6 semester hours in power machine operations and 5 for field experience in power sewing. In the lower-division baccalaureate/associate degree category, 3 semester hours in personnel supervision and 2 in records and information management (2/92).

Recommendation, Skill Level 40

In the vocational certificate category, 7 semester hours in power machine operations and 6 for field experience in power sewing. In the lower-division baccalaureate/associate degree category, 3 semester hours in personnel supervision and 2 in records and information management. In the upper-division baccalaureate category, 3 semester hours for field experience in management and 3 in organizational management (2/92).

MOS-44B-003

Metal Worker
44B10
44B20

Exhibit Dates: 1/90–9/91.

Career Management Field: 63 (Mechanical Maintenance), subfield 632 (Metalworking).

Description

Summary: Repairs and straightens metal body panels, fenders, and sheet metal parts; repairs and installs radiators and fuel tanks; welds ferrous and nonferrous metals. *Skill Level 10:* Repairs damaged metal body components with hammer, dolly blocks, spoon, and hydraulic jacks; removes and replaces body trim; cuts, grinds, and installs safety glass; cleans, repairs, and installs radiators and fuel tanks; prepares surface for painting by filling, sanding, masking, and cleaning; welds ferrous and nonferrous metals, using oxyacetylene, electric arc, and inert gas welding equipment. *Skill Level 20:* Able to perform the duties required for Skill Level 10; provides technical guidance and supervision to subordinates; demonstrates and performs more difficult phases of working metal by forming, welding, forging, and brazing; fabricates complex parts.

Recommendation, Skill Level 10

In the lower-division baccalaureate/associate degree category, 6 semester hours in automotive body repair, 3 in welding, and 3 in radiator repair (7/82).

Recommendation, Skill Level 20

In the lower-division baccalaureate/associate degree category, 6 semester hours in automotive body repair, 6 in welding, and 3 in radiator repair (7/82).

MOS-44B-004

Metal Worker
44B10
44B20

Exhibit Dates: 10/91–1/03.

Career Management Field: 63 (Mechanical Maintenance).

Description

Summary: Supervises, inspects, installs, modifies, and performs maintenance on metal body components, radiators, fuel tanks, hulls, and accessories of army land vehicles, watercraft, and amphibians; welds ferrous and nonferrous metals. *Skill Level 10:* Repairs damaged metal body components with hammer, dolly blocks, spoon, and hydraulic jacks; removes and replaces body trim; cuts, grinds, and installs safety glass; cleans, repairs, and installs radiators and fuel tanks; prepares surface for painting by filling, sanding, masking, and cleaning; welds ferrous and nonferrous metals, using oxyacetylene, electric arc, and inert gas welding equipment. *Skill Level 20:* Able to perform the duties required for Skill Level 10; conducts on-the-job training; completes work records and forms; demonstrates and performs more difficult phases of working metal by forming, welding, forging, and brazing; fabricates complex parts.

Recommendation, Skill Level 10

Credit may be granted on the basis of an individualized assessment of the student (3/94).

Recommendation, Skill Level 20

In the lower-division baccalaureate/associate degree category, 6 semester hours in automotive body repair, 3 in oxyacetylene welding, 3 in arc and inert gas welding, 2 in maintenance management, and 3 in personnel supervision. (NOTE: This recommendation for skill level 20 is valid for the dates 10/91-2/95 only) (11/91).

MOS-44B-005

Metal Worker
44B10
44B20

Exhibit Dates: 2/03–Present.

Career Management Field: 63 (Mechanical Maintenance).

Description

Summary: Supervises, inspects, installs, modifies, and performs maintenance on metal body components, radiators, fuel tanks, hulls, and accessories of army vehicles. *Skill Level 10:* Welds ferrous and nonferrous metals using oxyacetylene, electric arc, and inert gas welding equipment; repairs, repaints, and installs metal body components, radiators, and fuel tanks; and modifies other related items. *Skill Level 20:* Able to perform the duties required for Skill Level 10; supervises and provides technical guidance to subordinates; performs direct support and general support maintenance metal worker tasks.

Recommendation, Skill Level 10

Credit may be granted on the basis of an individualized assessment of the student (2/03).

Recommendation, Skill Level 20

Credit may be granted on the basis of an individualized assessment of the student (2/03).

MOS-44C-002

Financial Management Technician
44C10
44C20
44C30
44C40
44C50

Exhibit Dates: Pending Evaluation. 4/04–Present.

MOS-44E-003

Machinist
44E10
44E20
44E30
44E40

Exhibit Dates: 1/90–9/91.

Career Management Field: 63 (Mechanical Maintenance), subfield 632 (Metalworking).

Description

Summary: Performs and supervises the fabrication, repair, and modification of metallic and nonmetallic parts; supervises metalwork shop. *Skill Level 10:* Fabricates, repairs, and modifies metallic and nonmetallic parts using utility grinders, power cutoff saws, armature undercutters, arbor and hydraulic presses, drill presses and attachments, accessories, and tools; interprets job orders, blueprints, sketches, and specifications; lays out and sets up work; establishes and marks necessary reference points, centerlines, and matching guidelines; computes and sets up proper machine speeds to attain correct cutting speeds, and depth of cut and feed; checks dimensions of finished work using measuring devices, such as squares, rules, scales, calipers, micrometers, and thread gauges; operates hand tools such as files, taps, dies, punches, hammers, screwdrivers, and wrenches. *Skill Level 20:* Able to perform the duties required for Skill Level 10; provides technical guidance and supervision to subordinates; fabricates, repairs, and modifies metallic and nonmetallic parts, using milling machines, shapers, vertical and horizontal band saws, lathes, and tool and cutter grinders; applies heat treating procedures; conducts on-the-job training. *Skill Level 30:* Able to perform the duties required for Skill Level 20 of MOS 44E or 44B (Metal Worker); supervises support maintenance/metalworking shop; performs final inspection and approves disposition of completed work; supervises preparation of shop records and reports; assists in planning and setting up maintenance shops to obtain maximum utilization; schedules, assigns, and coordinates work; may supervise up to 19 persons. NOTE: May have progressed to 44E30 from 44E20 or 44B20 (Metal Worker). *Skill Level 40:* Able to perform the duties required for Skill Level 30; performs as support maintenance machine shop supervisor or metalworker supervisor; performs support maintenance administrative functions and supervises the preparation of records and training of subordinates; may supervise 20 or more persons.

Recommendation, Skill Level 10

In the lower-division baccalaureate/associate degree category, 6 semester hours in machine shop laboratory, and 3 in machine shop theory (7/82).

Recommendation, Skill Level 20

In the lower-division baccalaureate/associate degree category, 8 semester hours in machine shop laboratory and 4 in machine shop theory (7/82).

Recommendation, Skill Level 30

In the lower-division baccalaureate/associate degree category, 8 semester hours in machine shop laboratory, 4 in machine shop theory, and 3 in personnel supervision (7/82).

Recommendation, Skill Level 40

In the lower-division baccalaureate/associate degree category, 8 semester hours in machine shop laboratory, 4 in machine shop theory, and 3 in personnel supervision. In the upper-division baccalaureate category, 3 semester hours in personnel management and 3 for field experience in management (7/82).

MOS-44E-004

Machinist
44E10
44E20
44E30
44E40

Exhibit Dates: 10/91–1/03.

Career Management Field: 63 (Mechanical Maintenance).

Description

Summary: Performs and supervises the fabrication, repair, and modification of metallic and nonmetallic parts; supervises metalwork shop. *Skill Level 10:* Fabricates, repairs, and modifies metallic and nonmetallic parts using utility grinders, power cutoff saws, armature undercutters, lathes, milling machines, arbor and hydraulic presses, drill presses and attachments, accessories, and tools; interprets job orders, blueprints, sketches, and specifications; lays out and sets up work; establishes and marks necessary reference points, centerlines, and matching guidelines; computes and sets up proper machine speeds to attain correct cutting speeds and depth of cut and feed; checks dimensions of finished work using such measuring devices as squares, rules, scales, calipers, micrometers, and thread gauges; operates hand tools such as files, taps, dies, punches, hammers, screwdrivers, and wrenches. *Skill Level 20:* Able to perform the duties required for Skill Level 10; provides technical guidance and supervision to subordinates; fabricates, repairs, and modifies metallic and nonmetallic parts using milling machines, shapers, vertical and horizontal band saws, lathes, and tool and cutter grinders; conducts on-the-job training; completes work records and forms. *Skill Level 30:* Able to perform the duties required for Skill Level 20 of MOS 44E or 44B (Metal Worker); supervises support maintenance/metalworking shop; performs final inspection and approves disposition of completed work; supervises preparation of shop records and reports; assists in planning and setting up of maintenance shops to obtain maximum utilization; schedules, assigns, and coordinates work; supervises up to 19 persons. NOTE: May have progressed to 44E30 from 44E20 or 44B20 (Metal Worker). *Skill Level 40:* Able to perform the duties required for Skill Level 30; performs as support maintenance machine shop supervisor or metalworker supervisor; performs support maintenance administrative functions; supervises the preparation of personnel records and training; supervises 20 or more persons.

Recommendation, Skill Level 10

Credit may be granted on the basis of an individualized assessment of the student (3/94).

Recommendation, Skill Level 20

In the lower-division baccalaureate/associate degree category, 4 semester hours in machine shop theory, 8 in machine shop laboratory, 2 in maintenance management, and 3 in personnel supervision. (NOTE: This recommendation for skill level 20 is valid for the dates 10/91-2/95 only) (11/91).

Recommendation, Skill Level 30

In the lower-division baccalaureate/associate degree category, 4 semester hours in machine shop theory, 8 in machine shop laboratory, 3 in maintenance management, and 3 in personnel supervision. If progressed from 44B20, grant the following credit: 3 semester hours in automotive body repair, 3 in oxyacetylene welding, 3 in arc and inert gas welding, 3 in maintenance management, and 3 in personnel supervision (11/91).

Recommendation, Skill Level 40

In the lower-division baccalaureate/associate degree category, 4 semester hours in machine shop theory, 8 in machine shop laboratory, 3 in maintenance management, and 3 in personnel supervision. If progressed from 44B20, grant the following credit: 3 semester hours in automotive body repair, 3 in oxyacetylene welding, 3 in arc and inert gas welding, 3 in maintenance management, and 3 in personnel supervision. In the upper-division baccalaureate category, 3 semester hours for field experience in management (11/91).

MOS-44E-005

MACHINIST
44E10
44E20
44E30
44E40

Exhibit Dates: 2/03–Present.

Career Management Field: 63 (Mechanical Maintenance).

Description

Summary: Performs and supervises the fabrication, repair, and modification of metallic and nonmetallic parts; supervises metalwork shop activities. *Skill Level 10:* Fabricates, repairs, and modifies metallic and nonmetallic parts using utility grinders, power cutoff saws, armature undercutters, engine lathes, milling machines, arbor and hydraulic presses, drill presses and attachments, accessories, and tools; interprets job orders, blueprints, sketches, and specifications; lays out and sets up work; establishes and marks necessary reference point and centerlines; computes and sets up proper machine speeds to attain correct cutting speeds and depth of cut and feed; checks dimensions of finished work using such measuring devices as squares, rules, scales, calipers, micrometers, and thread gauges; uses hand tools such as files, taps, dies, punches, hammers, screwdrivers, and wrenches. *Skill Level 20:* Able to perform the duties required for Skill Level 10; provides technical guidance and supervision to subordinates; fabricates, repairs, and modifies metallic and nonmetallic parts; conducts on-the-job training; completes work records and reports. *Skill Level 30:* Able to perform the duties required for Skill Level 20 of MOS 44E or 44B (Metal Worker); supervises support maintenance/metalworking shop; performs final inspection and approves disposition of completed work; supervises preparation of shop records and reports; assists in planning and setting up of maintenance shops to obtain maximum utilization; schedules, assigns, and coordinates work; supervises subordinates. NOTE: May have progressed to 44E30 from 44E20 or 44B20 (Metal Worker). *Skill Level 40:* Able to perform the duties required for Skill Level 30; performs as support maintenance machine shop supervisor or metalworker supervisor; performs support maintenance administrative functions; supervises the preparation of personnel records and training; supervises subordinates.

Recommendation, Skill Level 10

Credit may be granted on the basis of an individualized assessment of the student (2/03).

Recommendation, Skill Level 20

Credit may be granted on the basis of an individualized assessment of the student (2/03).

Recommendation, Skill Level 30

In the lower-division baccalaureate/associate degree category, if progressed from 44E20, grant the following credit: 3 semester hours in machine shop theory, 9 in machine shop laboratory, 3 in introduction to welding, 3 in maintenance management, and 3 in principles of supervision. If progressed from 44B20, grant the following credit: 3 semester hours in arc welding, 3 in oxyacetylene fuel welding and cutting, 6 in MIG/TIG welding, 3 in metal fabrication, 3 in maintenance management, 3 in materiel management, and 3 in principles of supervision (2/03).

Recommendation, Skill Level 40

In the lower-division baccalaureate/associate degree category, if progressed from 44E20, grant the following credit: 3 semester hours in machine shop theory, 9 in machine shop laboratory, 3 in introduction to welding, 3 in maintenance management, and 3 in principles of supervision. If progressed from 44B20, grant the following credit: 3 semester hours in arc welding, 3 in oxyacetylene fuel welding and cutting, 6 in MIG/TIG welding, 3 in metal fabrication, 3 in maintenance management, 3 in materiel management, and 3 in principles of supervision. In the upper-division baccalaureate category, 3 semester hours for a practicum in management (2/03).

MOS-45B-002

SMALL ARMS REPAIRER
45B10
45B20

Exhibit Dates: 1/90–2/92.

Career Management Field: 63 (Mechanical Maintenance), subfield 634 (Armament Maintenance).

Description

Summary: Performs or supervises the performance of maintenance, repair, and overhaul of small arms. *Skill Level 10:* Disassembles weapons and cleans dirt, rust, and corrosion from components; inspects components for defects; performs repairs by grinding, sandblasting, soldering, and fitting component parts; replaces defective parts; lubricates and reassembles weapons; fits and adjusts parts to ensure smooth operation; uses common and specialized hand tools. *Skill Level 20:* Able to perform the duties required for Skill Level 10; troubleshoots and diagnoses small arms malfunctions; provides technical guidance and assistance to subordinates; supervises technical service assistance teams.

Recommendation, Skill Level 10

In the lower-division baccalaureate/associate degree category, 2 semester hours in use and care of hand tools, 2 in use and care of precision test equipment, and 3 in basic firearms repair. (NOTE: This recommendation for skill level 10 is valid for the dates 1/90-9/91 only) (7/82).

Recommendation, Skill Level 20

In the lower-division baccalaureate/associate degree category, 2 semester hours in use and care of hand tools, 2 in use and care of precision test equipment, 3 in basic firearms repair, and additional credit in basic firearms repair on the basis of institutional evaluation (7/82).

MOS-45B-003

SMALL ARMS/ARTILLERY REPAIRER
((Small Arms Repairer))
45B10
45B20

Exhibit Dates: 3/92–Present.

Career Management Field: 63 (Mechanical Maintenance).

Description

Summary: Performs or supervises the performance of maintenance, repair, and overhaul of small arms. *Skill Level 10:* Disassembles weapons and cleans dirt, rust, and corrosion from components; inspects components for defects; performs repairs by grinding, sandblasting, soldering, and fitting component parts; replaces defective parts; lubricates and reassembles weapons; fits and adjusts parts to ensure smooth operation; uses common and specialized hand tools. *Skill Level 20:* Able to perform the duties required for Skill Level 10; troubleshoots and diagnoses small arms malfunctions; provides technical guidance and assistance to subordinates; supervises technical service assistance teams; prepares efficiency reports on subordinates; keeps proper records and inventory; instructs and counsels subordinates.

Recommendation, Skill Level 10

Credit may be granted on the basis of an individualized assessment of the student (3/94).

Recommendation, Skill Level 20

In the vocational certificate category, 3 semester hours in basic small arms repair. In the lower-division baccalaureate/associate degree category, 3 semester hours in military science and 3 in personnel supervision. (NOTE: This recommendation for skill level 20 is valid for the dates 3/92-2/95 only) (3/92).

MOS-45D-001

Self-Propelled Field Artillery Turret Mechanic
45D10
45D20

Exhibit Dates: 1/90–2/92.

Career Management Field: 63 (Mechanical Maintenance), subfield 635 (Weapon System Maintenance).

Description

Summary: Troubleshoots and performs maintenance on armament, fire control, and related components of field artillery systems. *Skill Level 10:* Removes, repairs, installs, and tests armament sighting, fire control, electrical, and mechanical systems; repairs and adjusts previously diagnosed malfunctions of mechanical, electrical, and hydraulic systems; troubleshoots mechanical, electrical, and hydraulic systems using hand tools and test equipment. *Skill Level 20:* Able to perform the duties required for Skill Level 10; provides technical guidance and supervision to subordinates; diagnoses and troubleshoots to the component level; isolates malfunctions using schematic diagrams; tests and inspects components; conducts on-the-job training and completes maintenance records.

Recommendation, Skill Level 10

In the lower-division baccalaureate/associate degree category, 3 semester hours in hydraulics, 3 in basic electricity, and 1 in use and care of hand tools. (NOTE: This recommendation for skill level 10 is valid for the dates 1/90-9/91 only) (7/82).

Recommendation, Skill Level 20

In the lower-division baccalaureate/associate degree category, 3 semester hours in hydraulics, 3 in basic electricity, 1 in use and care of hand tools, 1 in records administration, and 1 in industrial safety (7/82).

MOS-45D-002

Self-Propelled Field Artillery Turret Mechanic
45D10
45D20

Exhibit Dates: 3/92–9/03. Effective 10/03, MOS 45D was discontinued and its duties were incorporated into MOS 63D, Artillery Mechanic.

Career Management Field: 63 (Mechanical Maintenance).

Description

Summary: Troubleshoots and performs maintenance on armament, fire control, and related components of field artillery systems. *Skill Level 10:* Removes, repairs, installs, and tests armament sighting, fire control, electrical, and mechanical systems; repairs and adjusts previously diagnosed malfunctions of mechanical, electrical, and hydraulic systems; troubleshoots mechanical, electrical, and hydraulic systems using hand tools and test equipment. *Skill Level 20:* Able to perform the duties required for Skill Level 10; provides technical guidance and supervision to subordinates; diagnoses and troubleshoots to the system level; isolates malfunctions using schematic diagrams; tests and inspects components; conducts on-the-job training and completes maintenance records.

Recommendation, Skill Level 10

Credit may be granted on the basis of an individualized assessment of the student (3/94).

Recommendation, Skill Level 20

In the lower-division baccalaureate/associate degree category, 3 semester hours in military science, 3 in hydraulics, 3 in basic electricity, and 3 in personnel supervision. (NOTE: This recommendation for skill level 20 is valid for the dates 3/92-2/95) (3/92).

MOS-45E-001

XM1 Tank Turret Mechanic
45E10
45E20

Exhibit Dates: 1/90–2/92.

Career Management Field: 63 (Mechanical Maintenance), subfield 635 (Weapon System Maintenance).

Description

Summary: Performs maintenance on armament, fire control, and related components of tanks. *Skill Level 10:* Services, repairs, and adjusts armament and fire control systems; performs corrective maintenance on mechanical, electrical, and hydraulic components; performs site adjustments and test-operates equipment; troubleshoots and repairs mechanical, electrical, and hydraulic systems using schematic diagrams; uses power tools, hand tools, and test equipment. *Skill Level 20:* Able to perform the duties required for Skill Level 10; provides technical guidance and supervision to subordinates; diagnoses and troubleshoots to the component level; conducts on-the-job training; completes maintenance records.

Recommendation, Skill Level 10

In the lower-division baccalaureate/associate degree category, 3 semester hours in hydraulics, 3 in basic electricity, and 1 in use and care of hand tools. (NOTE: This recommendation for skill level 10 is valid for the dates 1/90-9/91 only) (7/82).

Recommendation, Skill Level 20

In the lower-division baccalaureate/associate degree category, 3 semester hours in hydraulics, 3 in basic electricity, 1 in use and care of hand tools, 1 in records administration, and 1 in industrial safety (7/82).

MOS-45E-002

M1 Abrams Tank Turret Mechanic
45E10
45E20

Exhibit Dates: 3/92–5/01.

Career Management Field: 63 (Mechanical Maintenance).

Description

Summary: Performs maintenance on vehicular-mounted armament (including machine guns) and associated fire control and related systems on M1 turrets. *Skill Level 10:* Services, repairs, and adjusts armament and fire control systems; performs corrective maintenance on mechanical, electrical, and hydraulic components; performs site adjustments and test-operates equipment; troubleshoots and repairs mechanical, electrical, and hydraulic systems using schematic diagrams; uses power tools, hand tools and test equipment. *Skill Level 20:* Able to perform the duties required for Skill Level 10; provides technical guidance and supervision to subordinates; troubleshoots to the system level; conducts on-the-job training; completes maintenance records.

Recommendation, Skill Level 10

Credit may be granted on the basis of an individualized assessment of the student (3/94).

Recommendation, Skill Level 20

In the lower-division baccalaureate/associate degree category, 3 semester hours in military science, 3 in hydraulics, 3 in basic electricity, and 3 in personnel supervision. (NOTE: This recommendation for skill level 20 is valid for the dates 3/92-2/95 only) (3/92).

MOS-45E-003

M1 Abrams Tank Turret Mechanic
45E10
45E20

Exhibit Dates: 6/01–Present.

Career Management Field: 63 (Mechanical Maintenance).

Description

Summary: Performs maintenance on vehicular-mounted armament (including machine guns) and associated fire control and related systems on M1 turrets. *Skill Level 10:* Services, troubleshoots, adjusts and replaces components relating to the turret functions of the M1 Abrams tank turret, including mechanical, electrical, electronic, and hydraulic components and systems; uses schematics, troubleshooting guides, hand tools, power tools, and specialized test equipment. *Skill Level 20:* Able to perform the duties required for Skill Level 10; provides technical guidance to subordinates; troubleshoots to the system level; develops and conducts on-the-job training; completes maintenance records and reports.

Recommendation, Skill Level 10

Credit may be granted on the basis of an individualized assessment of the student (6/01).

Recommendation, Skill Level 20

Credit may be granted on the basis of an individualized assessment of the student (6/01).

MOS-45G-001

Fire Control Repairer
((Fire Control System Repairer))
45G10
45G20
45G30

Exhibit Dates: 1/90–5/99.

Career Management Field: 63 (Mechanical Maintenance, subfield 634 (Armament Maintenance).

Description

Summary: Performs and supervises maintenance of laser, infrared laser devises, and electronic ballistic computers. *Skill Level 10:* Uses electronic and mechanical test equipment to perform troubleshooting and diagnostic procedures; traces circuits, uses schematic diagrams; uses hand and power tools; repairs circuit boards and practices laser safety precautions; has a working knowledge of multimeter, oscilloscope, and other special meters; visually inspects circuits and removes and replaces defective electrical and electronic components; has a thorough knowledge and understanding of electrical and electronic theory. *Skill Level 20:* Able to perform the duties required for Skill Level 10; provides technical guidance to subordinates; inspects and tests repaired equipment; conducts on-the-job training; prepares reports and records. *Skill Level 30:* Able to perform the duties required for Skill Level 20; supervises and instructs subordinates in troubleshooting and proper maintenance procedures and practices; prepares reports and records; plans work schedules and maintenance priorities; administers shop safety program.

Recommendation, Skill Level 10

In the vocational certificate category, 5 semester hours in electrical fundamentals and 5 in electronic fundamentals. In the lower-division baccalaureate/associate degree category, 1 semester hour in electrical shop, 5 in electrical/electronics theory, and 2 in computer fundamentals. (NOTE: This recommendation for skill level 10 is valid for the dates 1/90-9/91 only) (2/84).

Recommendation, Skill Level 20

In the vocational certificate category, 5 in electrical fundamentals and 5 in electronic fundamentals. In the lower-division baccalaureate/associate degree category, 1 semester hour in electrical shop, 5 in electrical/electronics theory, and 2 in computer fundamentals. (NOTE: This recommendation for skill level 20 is valid for the dates 1/90-2/95 only) (2/84).

Recommendation, Skill Level 30

In the vocational certificate category, 5 in electrical fundamentals and 5 in electronic fundamentals. In the lower-division baccalaureate/associate degree category, 1 semester hour in electrical shop, 5 in electrical/electronics theory, 2 in computer fundamentals, and 3 in personnel supervision (2/84).

MOS-45G-002

Fire Control Repairer
45G10
45G20

Exhibit Dates: 6/99–Present.

Career Management Field: 63 (Mechanical Maintenance).

Description

Summary: Performs and supervises maintenance of laser, infrared laser devises, and electronic ballistic computers in fire control systems. *Skill Level 10:* Uses electronic test equipment to perform troubleshooting and diagnostic procedures; traces circuits; uses schematic diagrams; uses hand and power tools; replaces circuit boards and practices laser safety precautions; uses multimeter, oscilloscope, and other special test equipment; visually inspects circuits; uses electrical and electronic theory to systems. *Skill Level 20:* Able to perform the duties required for Skill Level 10; provides technical guidance to subordinates; inspects and tests repaired equipment; conducts on-the-job training; prepares reports and records.

Recommendation, Skill Level 10

Credit may be granted on the basis of an individualized assessment of the student (6/99).

Recommendation, Skill Level 20

Credit may be granted on the basis of an individualized assessment of the student (6/99).

MOS-45K-003

Tank Turret Repairer
45K10
45K20
45K30

Exhibit Dates: 1/90–2/92.

Career Management Field: 63 (Mechanical Maintenance), subfield 634 (Armament Maintenance).

Description

Summary: Performs maintenance and repair on turret mechanisms and weapons. *Skill Level 10:* Performs maintenance and repair on mechanical and hydraulic turret mechanisms and weapons; replaces bearings; cleans and repairs parts; removes and disassembles components for repair; uses precision instruments and hand tools; traces circuits and analyzes electrical problems using schematic diagrams. *Skill Level 20:* Able to perform the duties required for Skill Level 10; provides technical guidance and supervision to subordinates; troubleshoots and diagnoses turret mechanisms and weapons; isolates causes of malfunctions; determines necessary repairs, and performs final tests and inspections to ensure proper performance; conducts on-the-job training; completes maintenance records. *Skill Level 30:* Able to perform the duties required for skill level 20 of MOS 45K, 45B (Small Arms Repairer), or 45L (Artillery Repairer); performs, supervises, and inspects maintenance; may supervise up to 19 persons in armament maintenance section; plans and organizes work schedules; instructs personnel; establishes shop safety program. NOTE: May have progressed to 45K30 from 45K20, 45B20 (Small Arms Repairer) or 45L20 (Artillery Repairer).

Recommendation, Skill Level 10

In the lower-division baccalaureate/associate degree category, 1 semester hour in hydraulics, 1 in basic electrical wiring, and 1 in use and care of hand tools. (NOTE: This recommendation for skill level 10 is valid for the dates 1/90-9/91 only) (7/82).

Recommendation, Skill Level 20

In the lower-division baccalaureate/associate degree category, 2 semester hours in hydraulics, 2 in basic electrical wiring, 2 in use and care of hand tools, 3 in troubleshooting techniques (hydraulic and electrical systems), 1 in record keeping, and 2 in personnel supervision (7/82).

Recommendation, Skill Level 30

In the vocational certificate category, 2 semester hours in personnel supervision. In the lower-division baccalaureate/associate degree category, 2 semester hours in hydraulics, 2 in basic electrical wiring, 2 in use and care of hand tools, 3 in troubleshooting techniques (hydraulic and electrical systems), 1 in record keeping, 3 in personnel supervision, and additional credit in troubleshooting techniques (hydraulic and electrical systems) on the basis of institutional evaluation. In the upper-division baccalaureate category, 3 semester hours for field experience in management and additional credit in management on the basis of institutional evaluation (7/82).

MOS-45K-004

Armament Repairer
((Tank Turret Repairer))
45K10
45K20
45K30
45K40

Exhibit Dates: 3/92–1/03.

Career Management Field: 63 (Mechanical Maintenance).

Description

Summary: Performs direct support and general support maintenance and repair on turret mechanisms and weapons of tanks, cupolas, and similar material. *Skill Level 10:* Maintains and repairs mechanical and hydraulic turret mechanisms and weapons; replaces bearings; cleans and repairs parts; removes and disassembles components for repair; uses precision instruments and hand tools; traces circuits and analyzes electrical problems using schematic diagrams. *Skill Level 20:* Able to perform the duties required for Skill Level 10; provides technical guidance and supervision to subordinates; troubleshoots turret mechanisms and weapons; isolates causes of malfunctions; determines necessary repairs, and performs final tests and inspections to ensure proper performance; conducts on-the-job training; completes maintenance records. *Skill Level 30:* Able to perform the duties required for skill level 20 of MOS 45K, 45B (Small Arms Repairer), or 45L (Artillery Repairer); performs, supervises, and inspects maintenance; may supervise up to 19 persons in armament maintenance section; plans and organizes work schedules; instructs personnel; establishes a shop safety program. NOTE: May have progressed to 45K30 from 45K20, 45B20 (Small Arms Repairer) or 45L20 (Artillery Repairer). *Skill Level 40:* Able to perform the duties required for Skill Level 30; supervises repair crews; supervises the preparation of records and reports; supervises training of subordinates; applies production and quality control principles and procedures to maintenance operations; determines appropriate method for repair or fabrication; plans and organizes work area layout; schedules, assigns, and coordinates work according to the availability and capability of personnel and equipment.

Recommendation, Skill Level 10

Credit may be granted on the basis of an individualized assessment of the student (3/94).

Recommendation, Skill Level 20

In the lower-division baccalaureate/associate degree category, 3 semester hours in military science, 3 in hydraulics, 3 in basic electricity, and 3 in personnel supervision. (NOTE: This recommendation for skill level 20 is valid for the dates 3/92-2/95 only) (3/92).

Recommendation, Skill Level 30

In the lower-division baccalaureate/associate degree category, 3 semester hours in military science, 3 in hydraulics, 3 in basic electricity, 3 in personnel supervision, and 2 in records and information management (3/92).

Recommendation, Skill Level 40

In the lower-division baccalaureate/associate degree category, 3 semester hours in military science, 3 in hydraulics, 3 in basic electricity, 3 in personnel supervision, and 2 in records and information management. In the upper-division baccalaureate category, 3 semester hours for field experience in management and 3 in organizational management (3/92).

MOS-45K-005

Armament Repairer

45K10
45K20
45K30
45K40

Exhibit Dates: 2/03–Present.

Career Management Field: 63 (Mechanical Maintenance).

Description

Summary: Supervises and performs direct and general support and depot-level maintenance and repair on tank turret mechanisms and weapons of tanks, fighting vehicles, artillery, small arms, and infantry weapons. *Skill Level 10:* Performs ordnance armament maintenance and repair; maintains and repairs mechanical and hydraulic turret mechanisms and weapons; replaces bearings; cleans and repairs parts; removes and disassembles components for repair; uses precision instruments and hand tools; traces circuits and analyzes electrical and electronic problems using schematic diagrams and test equipment. *Skill Level 20:* Able to perform the duties required for Skill Level 10; provides technical guidance and supervision to subordinates; troubleshoots turret mechanisms, weapons and artillery; isolates causes of malfunctions; determines necessary repairs, and performs final tests and inspections to ensure proper performance; conducts on-the-job training; completes maintenance records. *Skill Level 30:* Able to perform the duties required for skill level 20 of MOS 45K, 45B (Small Arms Artillery Repairer), or 45G (Fire Control Repairer); performs, supervises, and inspects maintenance; supervises subordinates in armament maintenance section; plans and organizes work schedules; instructs personnel; establishes a shop safety program. NOTE: May have progressed to 45K30 from 45K20, 45B20 (Small Arms Artillery Repairer) or 45G20 (Fire Control Repairer). *Skill Level 40:* Able to perform the duties required for Skill Level 30; supervises subordinates; supervises the preparation of records and reports; supervises the training of subordinates; applies production and quality control principles and procedures to maintenance operations; determines appropriate method for repair or fabrication; serves as senior advisor of ordnance maintenance on turrets, artillery, and small arms.

Recommendation, Skill Level 10

Credit may be granted on the basis of an individualized assessment of the student (2/03).

Recommendation, Skill Level 20

Credit may be granted on the basis of an individualized assessment of the student (2/03).

Recommendation, Skill Level 30

In the lower-division baccalaureate/associate degree category, if progressed from 45K20, use this recommendation: 3 semester hours in fluid power, 3 in introduction to electricity/electronics, 4 in electronic devices and circuits, 3 in principles of supervision, 3 in maintenance management, and 3 in materiel management. If progressed from 45B20, use the following recommendation: 3 semester hours in military science, 3 in principles of supervision, 3 in maintenance management, and 3 in materiel management. If progressed from 45G20, use the following recommendation: 1 semester hour in electrical shop, 5 in electrical/electronics theory, 3 in principles of supervision, 3 in maintenance management, and 3 in materiel management (2/03).

Recommendation, Skill Level 40

In the lower-division baccalaureate/associate degree category, if progressed from 45K20, use this recommendation: 3 semester hours in fluid power, 3 in introduction to electricity/electronics, 4 in electronic devices and circuits, 3 in principles of supervision, 6 in maintenance management, and 3 in materiel management. If progressed from 45B20, use the following recommendation: 3 semester hours in military science, 3 in principles of supervision, 3 in maintenance management, and 3 in materiel management. If progressed from 45G20, use the following recommendation: 1 semester hour in electrical shop, 5 in electrical/electronics theory, 3 in principles of supervision, 3 in maintenance management, and 3 in materiel management. In the upper-division baccalaureate category, 3 semester hours for a practicum in management and 3 in organizational management (2/03).

MOS-45L-002

Artillery Repairer

45L10
45L20

Exhibit Dates: 1/90–2/92.

Career Management Field: 63 (Mechanical Maintenance), subfield 634 (Armament Maintenance).

Description

Summary: Maintains artillery turrets, weapons, and similar equipment. *Skill Level 10:* Maintains light, medium, and heavy self-propelled and towed artillery, turret mechanisms, and similar equipment; lubricates moving parts; uses common and specialized hand tools; measures tolerance with precision instruments and special tools; applies proper rigging procedures for hoisting heavy equipment and material; traces, tests, then repairs or replaces circuits. *Skill Level 20:* Able to perform the duties required for Skill Level 10; provides technical guidance to subordinates.

Recommendation, Skill Level 10

In the lower-division baccalaureate/associate degree category, 9 semester hours in shop practices and 2 in use and care of tools and precision test equipment. (NOTE: This recommendation for skill level 10 is valid for the dates 1/90-9/91 only) (7/82).

Recommendation, Skill Level 20

In the lower-division baccalaureate/associate degree category, 9 semester hours in shop practices and 2 in use and care of tools and precision test equipment (7/82).

MOS-45L-003

Artillery Repairer

45L10
45L20

Exhibit Dates: 3/92–4/92. Effective 4/92, MOS 45L was discontinued, and its duties were incorporated into MOS 45B, Small Arms Repair, and MOS 45K, Armament Repair.

Career Management Field: 63 (Mechanical Maintenance).

Description

Summary: Performs direct support and general support maintenance on artillery turrets, weapons and similar equipment. *Skill Level 10:* Maintains light, medium, and heavy self-propelled and towed artillery, turret mechanisms and similar equipment; lubricates moving parts; uses common and specialized hand tools; measures tolerance with precision instruments and special tools; applies proper rigging procedures for hoisting heavy equipment and material; traces and tests then repairs or replaces heavy equipment and material; traces and tests then repairs or replaces circuits. *Skill Level 20:* Able to perform the duties required for Skill Level 10; provides technical guidance and supervision to subordinates; employs additional test, measuring, and diagnostic equipment to identify malfunctions; authorizes final performance tests to assure equipment is operating within prescribed limits; conducts on-the-job training.

Recommendation, Skill Level 10

Credit may be granted on the basis of an individualized assessment of the student (3/92).

Recommendation, Skill Level 20

In the lower-division baccalaureate/associate degree category, 3 semester hours in military science, 3 in hydraulics, 3 in basic electricity, and 3 in personnel supervision (3/92).

MOS-45N-001

Tank Turret Mechanic

45N10
45N20

Exhibit Dates: 1/90–12/90.

Career Management Field: 63 (Mechanical Maintenance), subfield 634 (Armament Maintenance).

Description

Summary: Performs maintenance on conventional tank turret/cupola systems. *Skill Level 10:* Inspects, tests, troubleshoots, repairs, and adjusts electrical, hydraulic, fire control, mechanical, and firing systems; replaces components; cleans and lubricates turret/cupola systems; uses schematic diagrams; uses repair parts publications and orders parts. *Skill Level 20:* Able to perform the duties required for Skill Level 10; provides technical guidance to subordinates.

Recommendation, Skill Level 10

In the lower-division baccalaureate/associate degree category, 6 semester hours in hydraulics, 4 in basic electricity, and additional credit in hydraulics, electricity, and mechanics on the basis of institutional evaluation (12/75).

Recommendation, Skill Level 20

In the lower-division baccalaureate/associate degree category, 6 semester hours in hydraulics, 4 in basic electricity, and additional credit in hydraulics, electricity, and mechanics on the basis of institutional evaluation (12/75).

MOS-45N-002

M60A1/A3 Tank Turret Mechanic
45N10
45N20

Exhibit Dates: 1/91–9/01. Effective 10/01, MOS 45N was discontinued and its duties were incorporated into MOS 45E, M1 Abrams Tank Turret Mechanic.

Career Management Field: 63 (Mechanical Maintenance).

Description

Summary: Performs maintenance on conventional tank turret/cupola systems. *Skill Level 10:* Inspects, tests, troubleshoots, repairs, and adjusts electrical, hydraulic, fire control, mechanical, and firing systems; replaces components; cleans and lubricates turret/cupola systems; uses schematic diagrams; uses repair parts publications and orders parts. *Skill Level 20:* Able to perform the duties required for Skill Level 10; provides technical guidance to subordinates; employs applicable test, measuring, and diagnostic equipment in conjunction with technical publications and prior knowledge to diagnose and troubleshoot equipment malfunctions; troubleshoots, isolates, and identifies causes of system or component malfunctions; interprets complex schematic diagrams to trace and isolate malfunctions caused by cracked components; conducts on-the-job training; completes maintenance forms and records.

Recommendation, Skill Level 10

In the lower-division baccalaureate/associate degree category, 3 semester hours in hydraulics, 3 in basic electricity, 3 in drawing interpretation, 3 in automotive fundamentals, and 3 in military science. (NOTE: This recommendation for skill level 10 is valid for the dates 1/90-9/91 only) (4/91).

Recommendation, Skill Level 20

In the lower-division baccalaureate/associate degree category, 3 semester hours in hydraulics, 6 in basic electricity, 3 in drawing interpretation, 3 in automotive fundamentals, 3 in military science, and 3 in personnel supervision. (NOTE: This recommendation for skill level 20 is valid for the dates 1/91-2/95 only) (4/91).

MOS-45T-001

Bradley Fighting Vehicle System Turret Mechanic
((Bradley Fighting Vehicle System Mechanic))
45T10
45T20

Exhibit Dates: 1/90–9/99. Effective 10/99, MOS 45T was discontinued and its duties were incorporated into MOS 63M, Bradley Fighting Vehicle System Maintainer.

Career Management Field: 63 (Mechanical Maintenance).

Description

Summary: Performs maintenance of vehicular-mounted turret armaments, including turret weapons systems, fire control systems, and related electrical, mechanical, and hydraulic systems. *Skill Level 10:* Inspects, services, removes, replaces, adjusts, tests, and corrects previously diagnosed malfunctions of sighting and fire control systems and the vehicular-mounted turret electrical, mechanical, and hydraulic systems; cleans and maintains spare parts and hand tools; performs repair techniques as a member of a maintenance team; applies information from technical manuals and common schematic diagrams; completes applicable maintenance forms and records; practices safety precautions. *Skill Level 20:* Able to perform the duties required for Skill Level 10; provides technical supervision to subordinates; troubleshoots malfunctions using measuring and diagnostic equipment; interprets complex schematic diagrams; knows and applies map reading, traffic management, radio communication, and convoy operations skills.

Recommendation, Skill Level 10

Credit is not recommended because the skills, competencies, and knowledge are uniquely military in nature (7/87).

Recommendation, Skill Level 20

Credit is not recommended because the skills, competencies, and knowledge are uniquely military in nature (7/87).

MOS-45Z-002

Armament/Fire Control Maintenance Supervisor
45Z40

Exhibit Dates: 1/90–2/92.

Career Management Field: 63 (Mechanical Maintenance), subfield 634 (Armament Maintenance).

Description

Supervises repair crews engaged in maintenance of armament/fire control material; understands test and inspection procedure applicable to armament/fire control maintenance; plans and organizes work area layout to gain maximum efficiency and utilization of time, tools, equipment, and personnel; schedules, designs, and coordinates work according to availability and capabilities of personnel and equipment; estimates time required for specific maintenance effort; applies production control and quality control principles and procedures to maintenance operations; performs maintenance-related administrative functions; supervises the preparation of records and reports and the training of subordinates; may supervise 20 or more persons engaged in mechanical maintenance. NOTE: May have progressed to 45Z40 from 45G30 (Fire Control Systems Repairer), 41C30 (Fire Control Instrument Repairer), or 45K30 (Tank Turret Repairer).

Recommendation

In the lower-division baccalaureate/associate degree category, 3 semester hours in personnel supervision and 3 for field experience in management. In the upper-division baccalaureate category, 3 semester hours in introduction to management and 6 for field experience in management (7/82).

MOS-45Z-003

Armament/Fire Control Maintenance Supervisor
45Z40

Exhibit Dates: 3/92–4/92. Effective 4/92, MOS 45Z was discontinued, and its duties were incorporated into MOS 45K, Armament Repairer.

Career Management Field: 63 (Mechanical Maintenance).

Description

Supervises repair crews engaged in maintenance of armament/fire control material; understands test and inspection procedure applicable to armament/fire control maintenance; plans and organizes work area layout to gain maximum efficiency and utilization of time, tools, equipment, and personnel; schedules, designs, and coordinates work according to availability and capabilities of personnel and equipment; estimates time required for specific maintenance effort; applies production control and quality control principles and procedures to maintenance operation; performs maintenance-related administrative functions; supervises the preparation of records and reports and the training of subordinates; may supervise 20 or more persons engaged in mechanical maintenance; establishes and controls a file of technical publications. NOTE: May have progressed to 45Z40 from 45G30 (Fire Control Systems Repairer, 41C30 (Fire Control Instrument Repairer), or 45K30 (Tank Turret Repairer).

Recommendation

In the lower-division baccalaureate/associate degree category, 3 semester hours in military science, 3 in personnel supervision, and 2 in records and information management. In the upper-division baccalaureate category, 3 semester hours for field experience in management and 3 in organizational management (3/92).

MOS-46N-003

Pershing Electrical-Mechanical Repairer
46N10
46N20
46N30
46N40

Exhibit Dates: 1/90–10/92. Effective 10/92, MOS 46N was discontinued.

Career Management Field: 27 (Land Combat and Air Defense System Direct and General Support Maintenance).

Description

Summary: Supervises or performs direct and general support maintenance on the electrical, mechanical, and hydraulic portions of the Pershing missile system. *Skill Level 10:* Performs maintenance on the Pershing missile system erector-launcher, missile sections, system structures, missile containers, ground handling equipment, power distribution sets, and related training devices; inspects, tests, and adjusts components to specific tolerances; determines malfunctions in electrical, mechanical, electromechanical, hydraulic, pneumatic and hydropneumatic assemblies, subassemblies, and circuits with common and specialized test equipment; removes and replaces defective components and parts; prepares maintenance and supply forms and reports. *Skill Level 20:* Able to perform the duties required for Skill

Level 10; provides supervision and technical assistance to subordinates; prepares maintenance and supply forms and reports. *Skill Level 30:* Able to perform the duties required for Skill Level 20; serves as a first-line supervisor; organizes and conducts on-the-job training; supervises maintenance and inspection teams; establishes and keeps maintenance records, supervises quality control; sets work priorities. *Skill Level 40:* Able to perform the duties required for Skill Level 30; provides technical guidance and supervision to subordinates; assists in developing and executing quality control program.

Recommendation, Skill Level 10

In the lower-division baccalaureate/associate degree category, 3 semester hours in fluid power and 3 in automotive electrical systems troubleshooting. (NOTE: This recommendation for skill level 10 is valid for the dates 1/90-9/91 only) (8/92).

Recommendation, Skill Level 20

In the lower-division baccalaureate/associate degree category, 3 semester hours in fluid power, 3 in automotive electrical systems troubleshooting, and 3 in personnel supervision (8/92).

Recommendation, Skill Level 30

In the lower-division baccalaureate/associate degree category, 3 semester hours in fluid power, 3 in automotive electrical systems troubleshooting, 3 in personnel supervision, and 2 in records and information management. In the upper-division baccalaureate category, 3 semester hours in operations management (8/92).

Recommendation, Skill Level 40

In the lower-division baccalaureate/associate degree category, 3 semester hours in fluid power, 3 in automotive electrical systems troubleshooting, 3 in personnel supervision, and 2 in records and information management. In the upper-division baccalaureate category, 3 semester hours in operations management, 3 in organizational management, and 3 for field experience in management (8/92).

MOS-46Q-001

JOURNALIST

46Q10
46Q20
46Q30
46Q40

Exhibit Dates: 1/90–11/01.

Career Management Field: 46 (Public Affairs).

Description

Summary: Participates in or supervises the administration of information programs. *Skill Level 10:* Serving as a reporter, gathers information, prepares and edits news items and feature articles for newspapers, prepares publicity releases and information releases, and takes, develops, and selects photographs. *Skill Level 20:* Able to perform the duties required for Skill Level 10; provides guidance to subordinates; serving as an editor, edits news stories, prepares headlines and captions, designs format and layout, corrects and revises galley sheets, considers libel and copyright laws, coordinates newsgathering activities, and makes reporter assignments. *Skill Level 30:* Able to perform the duties required for Skill Level 20; supervises the administration of an information program; trains and supervises subordinates in all aspects of journalism including circulation methods and business policies. *Skill Level 40:* Able to perform the duties required for Skill Level 30; supervises and coordinates the management of an information program involving several media; implements new programs.

Recommendation, Skill Level 10

In the lower-division baccalaureate/associate degree category, 3 semester hours in basic news reporting, 3 in basic photography, and 3 in public relations writing. (NOTE: This recommendation for skill level 10 is valid for the dates 1/90-9/91 only) (6/91).

Recommendation, Skill Level 20

In the lower-division baccalaureate/associate degree category, 3 semester hours in basic news reporting, 3 in basic photography, 3 in public relations writing, 3 in copy editing, 3 in layout and design, and 3 in feature writing. (NOTE: This recommendation for skill level 20 is valid for the dates 1/90-2/95 only) (6/91).

Recommendation, Skill Level 30

In the lower-division baccalaureate/associate degree category, 3 semester hours in basic news reporting, 3 in basic photography, 3 in public relations writing, 3 in copy editing, 3 in layout and design, and 3 in feature writing. In the upper-division baccalaureate category, 3 semester hours in advanced news reporting, 3 in photojournalism, 3 in newspaper production, 3 in media relations, and 3 in personnel management (6/91).

Recommendation, Skill Level 40

In the lower-division baccalaureate/associate degree category, 3 semester hours in basic news reporting, 3 in basic photography, 3 in public relations writing, 3 in copy editing, 3 in layout and design, and 3 in feature writing. In the upper-division baccalaureate category, 3 semester hours in advanced news reporting, 3 in photojournalism, 3 in newspaper production, 3 in media relations, 3 in personnel management, 3 in public relations campaigns, 3 in survey methods, 3 in advanced media relations, 3 in publications management, and 3 in business administration (6/91).

MOS-46Q-002

JOURNALIST

46Q10
46Q20
46Q30
46Q40

Exhibit Dates: 12/01–Present.

Career Management Field: 46 (Public Affairs).

Description

Summary: Participates in or supervises the administration of information programs. *Skill Level 10:* Serving as a reporter, gathers information, prepares and edits news items and feature articles for newspapers, prepares publicity releases and information releases; takes, develops, and selects photographs; uses computer applications for writing articles, editing, researching information on the internet, and desktop publishing functions. *Skill Level 20:* Able to perform the duties required for Skill Level 10; provides guidance to subordinates; serving as an editor, edits news stories, prepares headlines and captions, designs format and layout, corrects and revises galley sheets, considers libel and copyright laws, coordinates newsgathering activities; makes reporter assignments; assists in public relations functions for internal communication, media relations, and community relations. *Skill Level 30:* Able to perform the duties required for Skill Level 20; supervises the administration of an information program; trains and supervises subordinates in all aspects of journalism and public affairs office activities, including circulation methods and business policies; provides counselling and training to subordinates; conducts performance appraisals; manages resource allocation and budgets. *Skill Level 40:* Able to perform the duties required for Skill Level 30; supervises and coordinates the management of an information program involving several media; implements new programs; develops plans; oversees personnel management functions; assists in budget preparation, resource allocation, and personnel assignments; responsible for mission and project accomplishment.

Recommendation, Skill Level 10

Credit may be granted on the basis of an individualized assessment of the student (12/01).

Recommendation, Skill Level 20

Credit may be granted on the basis of an individualized assessment of the student (12/01).

Recommendation, Skill Level 30

In the lower-division baccalaureate/associate degree category, 3 semester hours in basic news reporting, 3 in basic photography, 3 in public relations writing, 3 copy editing, 3 in layout and design, 3 in feature writing, and 3 in principles of supervision. In the upper-division baccalaureate category, 3 semester hours in project management (12/01).

Recommendation, Skill Level 40

In the lower-division baccalaureate/associate degree category, 3 semester hours in basic news reporting, 3 in basic photography, 3 in public relations writing, 3 in copy editing, 3 in layout and design, 3 in feature writing, 3 in desktop publishing, and 3 in principles of supervision. In the upper-division baccalaureate category, 3 semester hours in advanced news reporting, 3 in newspaper production, 3 in public relations management, 3 in project management, and 3 in media relations (12/01).

MOS-46R-001

BROADCAST JOURNALIST

46R10
46R20
46R30
46R40

Exhibit Dates: 1/90–11/01.

Career Management Field: 46 (Public Affairs).

Description

Summary: Participates in or supervises the operations of a radio and television broadcast center. *Skill Level 10:* Prepares scripts; announces, performs, and gives play-by-play coverage of events; operates broadcast station equipment for newscasts, radio music programs, and broadcast releases; conducts interviews; edits audio programs. *Skill Level 20:* Able to perform the duties required for Skill Level 10; provides guidance to subordinates; produces radio/television features, public service announcements, and newscasts; maintains program files and materials; processes traffic and continuity *Skill Level 30:* Able to perform the duties required for Skill Level 20; serves as script supervisor, producer, and director of radio and television programs; manages radio and television station and audio/video production and post production facility including field production and remote broadcasting; coordinates audio/video activities for public relations programs and campaigns. *Skill Level 40:* Able to perform the duties required for Skill Level 30; serves as station manager; coordinates programs and broadcast activities within the Armed Forces Network including feature and documentary productions and network news assignments; responsible for compliance with FCC and international broadcast regulations, audience measurements, and all administrative activities.

Recommendation, Skill Level 10

In the lower-division baccalaureate/associate degree category, 1 semester hour in audio production, 2 in broadcast newswriting, 3 in broadcast production, and 3 in radio announcing. (NOTE: This recommendation for skill level 10 is valid for the dates 1/90-9/91 only) (6/91).

Recommendation, Skill Level 20

In the lower-division baccalaureate/associate degree category, 1 semester hour in audio production, 2 in broadcast newswriting, 3 in broadcast production, and 3 in radio announcing. In the upper-division baccalaureate category, 3 semester hours in broadcast operations and 3 in field production. (NOTE: This recommendation for skill level 20 is valid for the dates 1/90-2/95 only) (6/91).

Recommendation, Skill Level 30

In the lower-division baccalaureate/associate degree category, 1 semester hour in audio production, 2 in broadcast newswriting, 3 in broadcast production, 3 in radio announcing, and 3 in public relations writing. In the upper-division baccalaureate category, 3 semester hours in broadcast operations, 3 in field production, 3 in broadcast station management, 3 in personnel management, 3 in advanced broadcast production, and 3 in broadcast programming (6/91).

Recommendation, Skill Level 40

In the lower-division baccalaureate/associate degree category, 1 semester hour in audio production, 2 in broadcast newswriting, 3 in broadcast production, and 3 in radio announcing. In the upper-division baccalaureate category, 3 semester hours in broadcast operations, 3 in field production, 3 in broadcast station management, 3 in personnel management, 3 in advanced broadcast production, 3 in broadcast programming, 3 in public relations writing, 6 in advanced field production, 3 in audience measurement, 3 in advanced station management, 3 in broadcast regulations, and 3 in business administration (6/91).

MOS-46R-002

Broadcast Journalist
46R10
46R20
46R30
46R40

Exhibit Dates: 12/01–Present.

Career Management Field: 46 (Public Affairs).

Description

Summary: Participates in or supervises the operations of a radio and television broadcast center. *Skill Level 10:* Prepares scripts; announces, performs, and gives play-by-play coverage of events; operates broadcast station equipment for newscasts, radio music programs, and broadcast releases; conducts interviews; digitally edits programs; produces audio and video field productions. *Skill Level 20:* Able to perform the duties required for Skill Level 10; provides guidance to subordinates; produces radio/television features, public affairs announcements, and newscasts; maintains program files and materials; processes traffic and continuity. *Skill Level 30:* Able to perform the duties required for Skill Level 20; serves as script supervisor, producer, and director of radio and television programs; manages radio and television station and audio/video production and post production facility, including field production and remote broadcasting; coordinates audio/video activities for public relations programs and campaigns. *Skill Level 40:* Able to perform the duties required for Skill Level 30; serves as a station manager; coordinates programs and broadcast activities within the Armed Forces Network, including feature and documentary productions and network news assignments; responsible for compliance with FCC and international broadcast regulations; performs audience measurement; prepares budget, policy, and procedures recommendations.

Recommendation, Skill Level 10

Credit may be granted on the basis of an individualized assessment of the student (12/01).

Recommendation, Skill Level 20

Credit may be granted on the basis of an individualized assessment of the student (12/01).

Recommendation, Skill Level 30

In the lower-division baccalaureate/associate degree category, 3 semester hours in audio production, 3 in broadcast newswriting, 3 in broadcast production, and 3 in radio announcing. In the upper-division baccalaureate category, 3 semester hours in broadcast operations, 3 in field production, and 3 in broadcast station management (12/01).

Recommendation, Skill Level 40

In the lower-division baccalaureate/associate degree category, 3 semester hour in audio production, 3 in broadcast newswriting, 3 in broadcast production, 3 in radio announcing, and 3 in public relations writing. In the upper-division baccalaureate category, 3 semester hours in broadcast operations, 3 in field production, 3 in broadcast station management, 3 in human resource management, 3 in advanced broadcast production, 3 in broadcast programming, 3 in advanced station management, and 3 in business administration (12/01).

MOS-46Z-001

Public Affairs Chief
46Z50

Exhibit Dates: 1/90–11/01.

Career Management Field: 46 (Public Affairs).

Description

Supervises, manages, and coordinates photographic operations, radio broadcasting, Army command and public affairs programs, and television production; supervises, coordinates, and provides technical assistance in the operation of all media activities; determines equipment and personnel requirements for an activity; serves as liaison with staff, maintenance, and operating personnel; prepares administrative and technical reports; assigns duties; evaluates and supervises training development and on-the-job training; conducts audience research and survey analysis; recommends establishment and modification of operations; ensures that public affairs releases comply with command policy, libel, slander, and copyright laws; presents and evaluates public affairs briefings. NOTE: May have progressed to 46Z50 from 46Q40, Journalist, or 46R40, Broadcast Journalist.

Recommendation

In the upper-division baccalaureate category, 3 semester hours in interpersonal communications, 6 in personnel management, 3 in public relations policy and management, 3 in public affairs planning, 3 in technical writing, and 3 in organizational management (6/91).

MOS-46Z-002

Public Affairs Chief
46Z50

Exhibit Dates: 12/01–Present.

Career Management Field: 46 (Public Affairs).

Description

Supervises, manages, and coordinates photographic operations, radio broadcasting, Army command and public affairs programs, and television production; supervises, coordinates, and provides technical assistance in the operation of all media activities; determines equipment and personnel requirements for an activity; serves as liaison with staff, maintenance, and operating personnel; prepares administrative and technical reports; assigns duties; evaluates and supervises training development and on-the-job training; conducts audience research and survey analysis; recommends establishment and modification of policy and operations; ensures that public affairs releases comply with command policy, libel, slander, and copyright laws; presents and evaluates public affairs briefings. NOTE: May have progressed to 46Z50 from 46Q40, (Journalist), or 46R40, (Broadcast Journalist).

Recommendation

In the upper-division baccalaureate category, 3 semester hours in interpersonal communication, 3 in strategic planning and forecasting, 3 in public relations policy and management, 3 in public affairs planning, 3 in operations management, and 3 in organizational management (12/01).

MOS-51B-002

Carpentry and Masonry Specialist
51B10
51B20

Exhibit Dates: 1/90–5/91.

Career Management Field: 51 (General Engineering), subfield 514 (Construction Engineering).

Description

Summary: Performs general and heavy carpentry or masonry duties in the fabrication, erection, maintenance, and repair of wooden and masonry structures and a variety of wooden articles. *Skill Level 10:* Uses basic carpentry tools and materials in building layout, framing, sheathing, fabricating, and roofing wooden structures; assists in erection of rough timber structures such as trestles, bridges, and wharves; assists in repair and renovation of all types of structures; uses basic masonry tools in construction of concrete walls, piers, and columns; places anchor bolts, rebar, and welded wire mesh; mixes mortar to specifications; lays brick and concrete blocks; maintains concrete mixes; cleans, lubricates, and sharpens woodworking and masonry tools. *Skill Level 20:* Able to perform the duties required for Skill Level 10; provides technical guidance to subordinates; interprets construction drawings and blueprints; computes material requirements and prepares building layout; constructs wooden or concrete foundation and erects building framework; installs flooring, walls, partitions, exterior siding, roofing, and stairs; fabricates and installs doors, windows, and screens; installs interior finishing such as cabinets, hardware, and wallboard; mixes, pours, finishes, and cures concrete; constructs concrete wall, pier, and column forms; selects mix specifications for mortar and grout; selects brick and concrete block pattern and joint type; installs lintels and finishes joints; cuts, trims, and faces stone, brick, and concrete blocks; installs and finishes tile; maintains carpentry and masonry tools and publications.

Recommendation, Skill Level 10

In the vocational certificate category, 9 semester hours in carpentry, 9 in masonry, and 3 in tool use and care (2/80).

Recommendation, Skill Level 20

In the vocational certificate category, 3 semester hours in tool use and care, 9 in carpentry, 9 in masonry, and additional credit in carpentry and masonry on the basis of institutional evaluation. In the lower-division baccalaureate/associate degree category, 3 semester hours in blueprint reading, 3 in applied mathematics, and 3 in industrial arts (construction methods) (2/80).

MOS-51B-003

Carpentry and Masonry Specialist

51B10

51B20

Exhibit Dates: 6/91–12/01.

Career Management Field: 51 (General Engineering).

Description

Summary: Performs general and heavy carpentry and masonry duties in the fabrication, erection, maintenance, and repair of wooden and masonry structures. *Skill Level 10:* Uses basic carpentry tools in framing, sheathing, fabricating, and roofing wooden structures; uses basic masonry tools in construction of foundations, piers, walls, and columns; uses basic concrete tools for foundations, site work, and slabs; places anchor bolts and steel reinforcement for concrete and masonry; erects buildings, including floor systems, partitions, roofing systems and stairs; installs doors and windows; maintains carpentry, masonry, and concrete tools; assists in erection of rough timber structures; repairs and renovates existing structures; assists in the performance of combat engineer missions. *Skill Level 20:* Able to perform the duties required for Skill Level 10; provides technical guidance and supervision to subordinates; interprets blueprints; estimates materials; performs building layout; schedules activities; installs finished carpentry products; performs specialized masonry activities.

Recommendation, Skill Level 10

In the vocational certificate category, 3 semester hours in construction tool use and care. In the lower-division baccalaureate/associate degree category, 3 semester hours in wood frame construction, 3 in concrete/masonry construction, and 3 in military science. (NOTE: This recommendation for skill level 10 is valid for the dates 6/91-9/91 only) (7/91).

Recommendation, Skill Level 20

In the vocational certificate category, 3 semester hours in construction tool use and care. In the lower-division baccalaureate/associate degree category, 3 semester hours in wood frame construction, 3 in concrete/masonry construction, 3 in military science, 3 in estimating and scheduling, 3 in personnel supervision, and 2 in records and information management. (NOTE: This recommendation for skill level 20 is valid for the dates 6/91-2/95 only) (7/91).

MOS-51B-004

Carpentry and Masonry Specialist

51B10

51B20

Exhibit Dates: 1/02–9/04. (Effective 9/04, MOS 51B was discontinued, and its duties were incorporated into MOS 21W.)

Career Management Field: 51 (General Engineering).

Description

Summary: Performs general and heavy carpentry and masonry duties in the fabrication, erection, maintenance, and repair of wooden and masonry structures. *Skill Level 10:* Uses basic carpentry tools in framing, sheathing, fabricating, and roofing wooden structures; uses basic masonry tools in construction of foundations, piers, walls, and columns; uses basic concrete tools for foundations, site work, and slabs; places anchor bolts and steel reinforcement for concrete and masonry; erects buildings, including floor systems, partitions, roofing systems and stairs; installs doors and windows; maintains carpentry, masonry, and concrete tools; assists in erection of rough timber structures; repairs and renovates existing structures; assists in the performance of combat engineer missions. *Skill Level 20:* Able to perform the duties required for Skill Level 10; provides technical guidance and supervision to subordinates; interprets blueprints; estimates materials; performs building layout; schedules activities; installs finished carpentry products; performs specialized masonry activities.

Recommendation, Skill Level 10

Credit may be granted on the basis of an individualized assessment of the student (1/02).

Recommendation, Skill Level 20

Credit may be granted on the basis of an individualized assessment of the student (1/02).

MOS-51G-002

Materials Quality Specialist

51G10

51G20

Exhibit Dates: 1/90–6/94. Effective 6/94, MOS 52G was discontinued, and its duties were incorporated into MOS 51T, Technical Engineering Specialist.

Career Management Field: 51 (General Engineering), subfield 511 (Technical Engineering).

Description

Summary: Surveys, samples, tests, evaluates, and classifies soils and aggregates; investigates and recommends solutions for soil stabilization; designs pavement mixes and inspects construction. *Skill Level 10:* Surveys, samples, tests, analyzes, evaluates, and classifies soils and aggregates; studies soils with respect to origin, distribution, and composition; performs physical testing and experimental work in the field and laboratory to determine engineering characteristics of soils and construction materials, including concrete, aggregates, bituminous, and other base materials; classifies soils according to standard engineering methods. *Skill Level 20:* Able to perform the duties required for Skill Level 10; designs pavement mixes and inspects construction; inspects and investigates problems of stabilization and drainage of soils used as foundations for roads, dams, buildings, and other structures; advises on soil use in construction.

Recommendation, Skill Level 10

In the vocational certificate category, 3 semester hours in soil classification, 3 in concrete mix design, and 3 in bituminous mix design. In the lower-division baccalaureate/associate degree category, 3 semester hours in soil and concrete testing laboratory. (NOTE: This recommendation for skill level 10 is valid for the dates 1/90-9/91 only) (2/80).

Recommendation, Skill Level 20

In the vocational certificate category, 3 semester hours in soil classification, 3 in concrete mix design, 3 in soil stabilization, and 3 in bituminous mix design. In the lower-division baccalaureate/associate degree category, 3 semester hours in soil and concrete testing laboratory, 3 in concrete and bituminous mix design, and 3 in soil engineering (2/80).

MOS-51H-002

Construction Engineering Supervisor

51H30

51H40

Exhibit Dates: 1/90–5/91.

Career Management Field: 51 (General Engineering), subfield 514 (Construction Engineering).

Description

Summary: Supervises construction and repair of buildings, warehouses, fixed bridges, port facilities, petroleum pipelines, tanks, and related equipment. *Skill Level 30:* Supervises construction and repair operations involving single structures or designated elements of military complexes, port facilities, or pipeline systems; provides direct supervision to personnel performing duties of carpentry and masonry specialists and of structural specialists; reads and interprets construction drawings; estimates material, time, utility, and labor requirements;

computes material stress factors under varying conditions; directs earth-moving machines during minor clearing and leveling operations; inspects structures and facilities to ensure compliance with specifications and quality of workmanship; reinforces existing structures to correct faulty construction or increase load capacity; supervises construction rigging operations; assists in planning pipeline systems and supervises installation and repair of pipelines, pumping stations, and storage tanks; conducts on-the-job training; directs operator maintenance on assigned vehicles; supervises performance of electrical and plumbing services on construction and repair projects. NOTE: May have progressed to 51H30 from 51B20 (Carpentry and Masonry Specialist) or 51C20 (Structures Specialist). *Skill Level 40:* Supervises construction and repair projects involving simultaneous operation of several small elements; interprets construction drawings and specifications; devises network flow diagrams, such as critical path method for company-sized or smaller projects; schedules work and equipment; prepares bills of material or material take-off lists; requisitions construction material; records and reports construction progress; coordinates work activities of supporting units; ensures compliance with directives, construction drawings, specifications, and verbal orders; develops and supervises vehicle and equipment operator maintenance program.

Recommendation, Skill Level 30

In the lower-division baccalaureate/associate degree category, 3 semester hours in technical mathematics, 3 in blueprint reading, 3 in construction methods, and 3 in personnel supervision. In the upper-division baccalaureate category, 3 semester hours in personnel management and 3 for field experience in management (2/80).

Recommendation, Skill Level 40

In the lower-division baccalaureate/associate degree category, 3 semester hours in technical mathematics, 3 in blueprint reading, 3 in construction methods, 3 in personnel supervision, 3 in records administration, 3 in construction project planning, and 3 in technical report writing. In the upper-division baccalaureate category, 3 semester hours in personnel management, 3 in construction project management, 3 in introduction to management, and 6 for field experience in management (2/80).

MOS-51H-003

Construction Engineering Supervisor
51H30
51H40

Exhibit Dates: 6/91–12/01.

Career Management Field: 51 (General Engineering).

Description

Summary: Supervises construction and repair of buildings, warehouses, fixed bridges, port facilities, petroleum pipelines, tanks, and related equipment. *Skill Level 30:* Supervises construction and repair operations involving single structures or designated elements of military complexes, port facilities, or pipeline systems; provides direct supervision to persons performing duties of carpentry and masonry specialists and of structural specialists; reads and interprets construction drawings; estimates material, time, utility, and labor requirements; computes material stress factors under varying conditions; directs earth-moving machines during minor clearing and leveling operations; inspects structures and facilities to ensure compliance with specifications and quality of workmanship; reinforces existing structures to correct faulty construction or increase load capacity; supervises construction rigging operations; assists in planning pipeline systems and supervises installation and repair of pipeline, pumping stations, and storage tanks; conducts on-the-job training; directs operator maintenance on assigned vehicles; supervises performance of electrical and plumbing services on construction and repair projects. NOTE: May have progressed to 51H30 from 51B20 (Carpentry and Masonry Specialist), 51K20 (Plumber), or 51R20 (Interior Electrician). *Skill Level 40:* Able to perform the duties required for Skill Level 30; supervises construction and repair projects involving simultaneous operation of several small elements; interprets construction drawings and specifications; devises network flow diagrams, such as critical path method for company-sized or smaller projects; schedules work and equipment; prepares bills of material or material take-off lists; requisitions construction material; records and reports construction progress; coordinates work activities of supporting units; ensures compliance with directives, construction drawings, specifications, and verbal orders; develops and supervises vehicle and equipment operator maintenance program.

Recommendation, Skill Level 30

In the lower-division baccalaureate/associate degree category, 3 semester hours in military science, 3 in construction methods, 3 in construction management, 3 in personnel supervision, and 2 in records and information management; if progressed from MOS 51K (Plumber) or MOS 51R (Interior Electrician), add 3 semester hours in construction estimating and scheduling (7/91).

Recommendation, Skill Level 40

In the lower-division baccalaureate/associate degree category, 3 semester hours in military science, 3 in construction methods, 3 in construction management, 3 in personnel supervision, and 2 in records and information management; if progressed from MOS 51K (Plumber) or MOS 51R (Interior Electrician), add 3 semester hours in construction estimating and scheduling. In the upper-division baccalaureate category, 3 semester hours for field experience in management and 3 in organizational management (7/91).

MOS-51H-004

Construction Engineering Supervisor
51H30
51H40

Exhibit Dates: 1/02–9/04. (Effective 9/04, MOS 51H was discontinued, and its duties were incorporated into MOS 21H.)

Career Management Field: 51 (General Engineering).

Description

Summary: Supervises construction and repair of buildings, warehouses, fixed bridges, port facilities, petroleum pipelines, tanks, and related equipment. *Skill Level 30:* Supervises construction and repair operations involving single structures or designated elements of military complexes, port facilities, or pipeline systems; provides direct supervision to persons performing duties of carpentry and masonry specialists and of structural specialists; reads and interprets construction drawings; estimates material, time, utility, and labor requirements; computes material stress factors under varying conditions; directs earth-moving machines during minor clearing and leveling operations; inspects structures and facilities to ensure compliance with specifications and quality of workmanship; reinforces existing structures to correct faulty construction or increase load capacity; supervises construction rigging operations; assists in planning pipeline systems and supervises installation and repair of pipeline, pumping stations, and storage tanks; conducts on-the-job training; directs operator maintenance on assigned vehicles; supervises performance of electrical and plumbing services on construction and repair projects. NOTE: May have progressed to 51H30 from 51B20 (Carpentry and Masonry Specialist), 51K20 (Plumber), or 51R20 (Interior Electrician). *Skill Level 40:* Able to perform the duties required for Skill Level 30; supervises construction and repair projects involving simultaneous operation of several small elements; interprets construction drawings and specifications; devises network flow diagrams, such as critical path method for company-sized or smaller projects; schedules work and equipment; prepares bills of material or material take-off lists; requisitions construction material; records and reports construction progress; coordinates work activities of supporting units; ensures compliance with directives, construction drawings, specifications, and verbal orders; develops and supervises vehicle and equipment operator maintenance program; supervises quality assurance and control of projects.

Recommendation, Skill Level 30

In the lower-division baccalaureate/associate degree category, 3 semester hours in construction methods, 3 in construction management, 3 in principles of supervision, 2 in records and information management, and 3 in cost estimating and scheduling. In the upper-division baccalaureate category, 3 semester hours for a practicum in management (1/02).

Recommendation, Skill Level 40

In the lower-division baccalaureate/associate degree category, 3 semester hours in construction methods, 3 in construction management, 3 in principles of supervision, 3 in records and information management, and 3 in cost estimating and scheduling. In the upper-division baccalaureate category, 3 semester hours for a practicum in management and 3 in advanced quality assurance/control (1/02).

MOS-51K-002

Plumber
51K10
51K20

Exhibit Dates: 1/90–1/92.

Career Management Field: 51 (General Engineering).

Description

Summary: Installs and repairs plumbing pipe systems and fixtures. *Skill Level 10:* Assists in installation and repair of pipe systems, plumbing fixtures, and equipment; reads and interprets drawings, plans, and specifications;

installs, tests, troubleshoots, and repairs pipe systems for water, air, oil, gas, and sewage; services plumbing tools and equipment; assists in laying and clearing minefields; primes and emplaces demolitions; constructs and removes wire, obstacles, emplacements, and bunkers. *Skill Level 20:* Able to perform the duties required for Skill Level 10; installs, maintains, and repairs pipe systems, fixtures, and equipment; explains and demonstrates techniques and procedures used in plumbing and pipefitting; conducts inspections of plumbing facilities.

Recommendation, Skill Level 10

In the vocational certificate category, 9 semester hours in general plumbing. (NOTE: This recommendation for skill level 10 is valid for the dates 1/90-9/91 only) (9/85).

Recommendation, Skill Level 20

In the vocational certificate category, 12 semester hours in general plumbing (9/85).

MOS-51K-003

PLUMBER

51K10
51K20

Exhibit Dates: 2/92–Present. (Effective 9/04, MOS 51K was discontinued, and its duties were incorporated into MOS 21K.)

Career Management Field: 51 (General Engineering).

Description

Summary: Installs and repairs plumbing pipe systems and fixtures. *Skill Level 10:* Assists in installation and repair of pipe systems, plumbing fixtures, and equipment; reads and interprets drawings, plans, and specifications; installs, tests, troubleshoots, and repairs pipe systems for water, air, oil, gas, and sewage; services plumbing tools and equipment; assists in laying and clearing minefields; primes and emplaces demolitions; constructs and removes wire, obstacles, emplacements, and bunkers. *Skill Level 20:* Able to perform the duties required for Skill Level 10; installs, maintains, and repairs pipe systems, fixtures, and equipment; explains and demonstrates techniques and procedures used in plumbing and pipefitting; conducts inspections of plumbing facilities; draws plumbing plans; coordinates and schedules plumbing activities, and prepares materials take-off lists and estimates; supervises squad of approximately six persons.

Recommendation, Skill Level 10

Credit may be granted on the basis of an individualized assessment of the student (2/92).

Recommendation, Skill Level 20

In the lower-division baccalaureate/associate degree category, 3 semester hours in military science, 3 in blueprint reading and drawing, 3 in plumbing and piping systems, 3 in personnel supervision, and 3 in estimating and scheduling. (NOTE: This recommendation for skill level 20 is valid for the dates 2/92-2/95 only) (2/92).

MOS-51M-002

FIREFIGHTER

51M10
51M20
51M30
51M40

Exhibit Dates: 1/90–5/91.

Career Management Field: 51 (General Engineering), subfield 513 (Fire Protection)

Description

Summary: Supervises or performs fire fighting, rescue, salvage, and fire protection operations. *Skill Level 10:* Operates all types of fire extinguishers for different types of fires; administers first aid; performs basic fire fighting duties including selection of nozzles and hoses for applying water or fog stream on various types of fires; maintains fire fighting equipment; operates fire fighting vehicles. *Skill Level 20:* Able to perform the duties required for Skill Level 10; conducts performance-oriented training sessions; performs as general crew chief in fire fighting and investigation of fires. *Skill Level 30:* Able to perform the duties required for Skill Level 20; performs the duties of fire chief/fire inspector; supervises fire prevention; supervises equipment maintenance; monitors and evaluates training procedures; inspects for fire regulation violations; serves as principal noncommissioned officer of fire company or team chief of crash rescue team of four or more persons. *Skill Level 40:* Able to perform the duties required for Skill Level 30; supervises and directs fire fighting and rescue units; identifies type of fires; recommends and enforces fire prevention regulations; inspects building facilities and installations after fires to determine damage and cause of fires; maintains records and reports; supervises two or more fire companies.

Recommendation, Skill Level 10

In the lower-division baccalaureate/associate degree category, 3 semester hours in introduction to fire science and 1 in first aid (2/80).

Recommendation, Skill Level 20

In the lower-division baccalaureate/associate degree category, 3 semester hours in introduction to fire science, 3 in fire prevention, and 2 in first aid (2/80).

Recommendation, Skill Level 30

In the lower-division baccalaureate/associate degree category, 3 semester hours in introduction to fire science, 6 in fire prevention, 2 in first aid, 3 in fire investigation, and 3 in fire administration. In the upper-division baccalaureate category, 3 semester hours in personnel management (2/80).

Recommendation, Skill Level 40

In the lower-division baccalaureate/associate degree category, 3 semester hours in introduction to fire science, 6 in fire prevention, 2 in first aid, 6 in fire investigation, 6 in fire administration, and 3 in technical report writing or communication. In the upper-division baccalaureate category, 3 semester hours in personnel management, and 3 for field experience in management (2/80).

MOS-51M-003

FIREFIGHTER

51M10
51M20
51M30
51M40

Exhibit Dates: 6/91–2/95.

Career Management Field: 51 (General Engineering).

Description

Summary: Supervises or performs fire fighting, rescue, salvage, and fire protection operations. *Skill Level 10:* Operates all types of fire extinguishers for different types of fires; administers first aid; performs basic fire fighting duties including selection of nozzles and hoses for applying water, foam, or fog stream on various types of fires; maintains fire fighting equipment; operates fire fighting vehicles. *Skill Level 20:* Able to perform the duties required for Skill Level 10; supervises and instructs subordinates; conducts performance-oriented training sessions; performs as general crew chief in fire fighting and investigation of fires; handles hazardous material and chemical spills; performs first aid procedures; performs maintenance on pneumatic equipment. *Skill Level 30:* Able to perform the duties required for Skill Level 20; performs the duties of fire chief/fire inspector; supervises fire prevention; supervises equipment maintenance; monitors and evaluates training procedures; inspects for fire regulation violations; serves as principal noncommissioned officer of fire company or team chief of aircraft and vehicle crash rescue team of four or more persons. *Skill Level 40:* Able to perform the duties required for Skill Level 30; supervises and directs fire fighting and rescue units; identifies type of fires; recommends and enforces fire prevention regulations; inspects building facilities and installations after fires to determine damage and cause of fires; maintains records and reports; supervises two or more fire companies.

Recommendation, Skill Level 10

In the lower-division baccalaureate/associate degree category, 3 semester hours in introduction to fire science, 3 in fire/crash rescue, and 2 in first aid. (NOTE: This recommendation for skill level 10 is valid for the dates 6/91-9/91 only) (7/91).

Recommendation, Skill Level 20

In the lower-division baccalaureate/associate degree category, 3 semester hours in introduction to fire science, 3 in fire/crash rescue, 3 in fire prevention, 2 in first aid, and 3 in personnel supervision (7/91).

Recommendation, Skill Level 30

In the lower-division baccalaureate/associate degree category, 3 semester hours in introduction to fire science, 3 in fire/crash rescue, 3 in fire prevention, 3 in first aid, 3 in personnel supervision, 3 in fire investigation, 3 in fire administration, and 3 in technical communication. In the upper-division baccalaureate category, 3 semester hours for field experience in management (7/91).

Recommendation, Skill Level 40

In the lower-division baccalaureate/associate degree category, 3 semester hours in introduction to fire science, 3 in fire/crash rescue, 6 in fire prevention, 3 in first aid, 3 in personnel supervision, 6 in fire investigation, 3 in fire administration, and 3 in technical communication. In the upper-division baccalaureate category, 3 semester hours for field experience in management and 3 in organizational management (7/91).

MOS-51M-004

FIREFIGHTER

51M10
51M20
51M30
51M40

Exhibit Dates: 3/95–12/01.

Career Management Field: 51 (General Engineering).

Description

Summary: Supervises or performs fire fighting, rescue, salvage, and fire protection operations. *Skill Level 10:* Operates all types of fire extinguishers for different types of fires; administers first aid; performs basic fire fighting duties including selection of nozzles and hoses for applying water, foam, or fog stream on various types of fires; maintains fire fighting equipment; operates fire fighting vehicles. *Skill Level 20:* Able to perform the duties required for Skill Level 10; supervises and instructs subordinates; conducts performance-oriented training sessions; performs as general crew chief in fire fighting and investigation of fires; handles hazardous material and chemical spills; performs first aid procedures; performs maintenance on pneumatic equipment. *Skill Level 30:* Able to perform the duties required for Skill Level 20; performs the duties of fire chief/fire inspector; supervises fire prevention; supervises equipment maintenance; monitors and evaluates training procedures; inspects for fire regulation violations; serves as principal noncommissioned officer of fire company or team chief of aircraft and vehicle crash rescue team of four or more persons. *Skill Level 40:* Able to perform the duties required for Skill Level 30; supervises and directs fire fighting and rescue units; identifies type of fires; recommends and enforces fire prevention regulations; inspects building facilities and installations after fires to determine damage and cause of fires; maintains records and reports; supervises two or more fire companies.

Recommendation, Skill Level 10

Credit may be granted on the basis of an individualized assessment of the student (7/91).

Recommendation, Skill Level 20

Credit may be granted on the basis of an individualized assessment of the student (7/91).

Recommendation, Skill Level 30

In the lower-division baccalaureate/associate degree category, 3 semester hours in introduction to fire science, 3 in fire/crash rescue, 3 in fire prevention, 2 in first aid, 3 in personnel supervision, 3 in fire administration, and 3 in technical communication. In the upper-division baccalaureate category, 3 semester hours for field experience in management (7/91).

Recommendation, Skill Level 40

In the lower-division baccalaureate/associate degree category, 3 semester hours in introduction to fire science, 3 in fire/crash rescue, 6 in fire prevention, 3 in first aid, 3 in personnel supervision, 3 in fire investigation, 3 in fire administration, and 3 in technical communication. In the upper-division baccalaureate category, 3 semester hours for field experience in management and 3 in organizational management (7/91).

MOS-51M-005

Firefighter

51M10
51M20
51M30
51M40

Exhibit Dates: 1/02–9/04. (Effective 9/04, MOS 51M was discontinued, and its duties were incorporated into MOS 21M.)

Career Management Field: 51 (General Engineering).

Description

Summary: Supervises or performs fire fighting, rescue, salvage, and fire protection operations. *Skill Level 10:* Operates all types of fire extinguishers for different types of fires; administers first aid; performs basic fire fighting duties, including selection of nozzles and hoses for applying water, foam, or fog stream on various types of fires; maintains fire fighting equipment; operates fire fighting vehicles; participates in continual update training and review; HAZMAT awareness and operations; and CPR certification; performs fire fighting duties in aircraft crash incidents. *Skill Level 20:* Able to perform the duties required for Skill Level 10; supervises and instructs subordinates; conducts performance-oriented training sessions; performs as general crew chief in fire fighting and investigation of fires; handles hazardous material and chemical spills; performs first aid procedures; performs maintenance on pneumatic equipment. *Skill Level 30:* Able to perform the duties required for Skill Level 20; performs the duties of fire chief/fire inspector; supervises fire prevention; supervises equipment maintenance; monitors and evaluates training procedures; inspects for fire regulation violations; serves as principal noncommissioned officer of fire company or team chief of aircraft and vehicle crash rescue team of four or more persons. *Skill Level 40:* Able to perform the duties required for Skill Level 30; supervises and directs fire fighting and rescue units; identifies type of fires; recommends and enforces fire prevention regulations; inspects building facilities and installations after fires to determine damage and cause of fires; maintains records and reports; supervises two or more fire companies.

Recommendation, Skill Level 10

Credit may be granted on the basis of an individualized assessment of the student (1/02).

Recommendation, Skill Level 20

Credit may be granted on the basis of an individualized assessment of the student (1/02).

Recommendation, Skill Level 30

In the lower-division baccalaureate/associate degree category, 3 semester hours in introduction to fire science, 3 in fire/crash rescue, 3 in fire prevention, 2 in first aid, 3 in principles of supervision, 3 in fire administration, and 3 in oral communication skills. In the upper-division baccalaureate category, 3 semester hours for a practicum in management (2/01).

Recommendation, Skill Level 40

In the lower-division baccalaureate/associate degree category, 3 semester hours in introduction to fire science, 3 in fire/crash rescue, 6 in fire prevention, 3 in first aid, 3 in principles of supervision, 3 in fire investigation, 3 in fire administration, and 3 in oral communication skills. In the upper-division baccalaureate category, 3 semester hours for a practicum in management and 3 in organizational management (2/01).

MOS-51R-003

Interior Electrician

51R10
51R20

Exhibit Dates: 1/90–1/92.

Career Management Field: 51 (General Engineering).

Description

Summary: Supervises or installs and maintains interior electrical systems and equipment; installs conduit service drops, fuse panels, switches, outlet, and lighting according to National Electrical Code. *Skill Level 10:* Installs service entrances from weatherhead to service panel and branch circuits; service protective panels; function and outlet boxes; thin wall, flexible, rigid, metallic, and nonmetallic conduit; metallic and nonmetallic sheathed cable; and switches, outlets, and special electrical equipment; uses test equipment, including multimeter, voltmeter, ammeter, and megger to test circuits. *Skill Level 20:* Able to perform the duties required for Skill Level 10; supervises installation and maintenance of interior electrical systems; provides technical guidance to subordinates; plans system layout using blueprints, sketches, specifications, and wiring diagrams; inspects systems for proper installation in accordance with the National Electrical Code.

Recommendation, Skill Level 10

In the vocational certificate category, 4 semester hours in residential wiring. (NOTE: This recommendation for skill level 10 is valid for the dates 1/90-9/91 only) (9/85).

Recommendation, Skill Level 20

In the vocational certificate category, 6 semester hours in residential wiring (9/85).

MOS-51R-004

Interior Electrician

51R10
51R20

Exhibit Dates: 2/92–9/04. (Effective 9/04, MOS 51R was discontinued, and its duties were incorporated into MOS 21R)

Career Management Field: 51 (General Engineering).

Description

Summary: Supervises or installs and maintains interior electrical systems and equipment; installs conduit service drops, fuse panels, switches, outlets, and lighting according to National Electrical Code. *Skill Level 10:* Reads and interprets drawings, plans, and specifications; installs service entrances from weatherhead to service panel and branch circuits; service protective panels; function and outlet boxes; thin wall, flexible, rigid, metallic, and nonmetallic conduit; metallic and nonmetallic sheathed cable; and switches, outlets, and special electrical equipment; uses test equipment including multimeter, voltmeter, ammeter, and megger to test circuits. *Skill Level 20:* Able to perform the duties required for Skill Level 10; supervises installation and maintenance of interior electrical systems; provides technical guidance to subordinates; plans system layout using blueprints and wiring diagrams; prepares sketches, plans, and specifications; prepares material take-off lists and estimates; coordinates and schedules electrical wiring activities; inspects systems for proper installation in accordance with the National Electrical Code; supervises squad of approximately six persons.

Recommendation, Skill Level 10

Credit may be granted on the basis of individualized assessment of the student (3/94).

Recommendation, Skill Level 20

In the lower-division baccalaureate/associate degree category, 3 semester hours in military science, 3 in blueprint reading and drawing, 3 in basic electrical wiring, 3 in personnel supervision, and 3 in estimating and scheduling. (NOTE: This recommendation for skill level 20 is valid for the dates 2/92-2/95 only) (2/92).

MOS-51T-001

TECHNICAL ENGINEERING SUPERVISOR
51T30
51T40

Exhibit Dates: 1/90–5/91.

Career Management Field: 51 (General Engineering), subfield 511 (Technical Engineering).

Description

Summary: Supervises or participates in construction site development, including technical investigation surveying, developing construction plans and specifications, and performing quality control inspections. *Skill Level 30:* Supervises construction surveying, drafting, and testing of construction materials; conducts surveys and preliminary engineering inspections; prepares layouts and detail drawings of pavements, structures, buildings, and utility systems from sketches and specified design criteria. Assists in inspection of all phases of construction and repair. NOTE: May have progressed to 51T30 from 51G20 (Materials Quality Specialist), 82B (Construction Surveyor), or 81B (Technical Drafting Specialist). *Skill Level 40:* Able to perform the duties required for Skill Level 30; provides technical guidance to subordinates; plans and supervises preparation and layout of drawings; performs simple load calculations to ensure construction of safe facilities; prepares network flow diagrams, such as critical path method.

Recommendation, Skill Level 30

In the lower-division baccalaureate/associate degree category, 3 semester hours in civil engineering technology, 3 in materials testing, 3 in introduction to surveying, 3 in personnel supervision, and 6 in engineering drafting (2/80).

Recommendation, Skill Level 40

In the lower-division baccalaureate/associate degree category, 3 semester hours in civil engineering technology, 3 in materials testing, 3 in introduction to surveying, 3 in personnel supervision, and 6 in engineering drafting. In the upper-division baccalaureate category, 3 semester hours in engineering management and 6 in construction project planning (2/80).

MOS-51T-002

TECHNICAL ENGINEERING SUPERVISOR
51T30
51T40

Exhibit Dates: 6/91–1/94.

Career Management Field: 51 (General Engineering).

Description

Summary: Supervises construction site development, including technical investigation surveying, developing construction plans and specifications, and performing quality control inspections. *Skill Level 30:* Supervises construction materials; conducts surveys and preliminary engineering inspections; prepares layouts and detail drawings of pavements, structures, buildings, and utility systems from sketches and specified design criteria; assists in inspection of all phases of construction and repair. NOTE: May have progressed to 51T30 from 51G20 (Materials Quality Specialist), 81B20 (Technical Drafting Specialist), or 82B20 (Construction Surveyor). *Skill Level 40:* Able to perform the duties required for Skill Level 30; provides technical guidance to subordinates; plans and supervises preparation and layout of drawings; performs simple load calculations to ensure construction of safe facilities; prepares network flow diagrams such as critical path method; has knowledge of asphalt and concrete quality testing procedures.

Recommendation, Skill Level 30

In the lower-division baccalaureate/associate degree category, 3 semester hours in military science, 3 in personnel supervision, 3 in maintenance management, 2 in records and information management, 3 in construction plans and specifications, 3 in surveying, and 3 in soil mechanics (7/91).

Recommendation, Skill Level 40

In the lower-division baccalaureate/associate degree category, 3 semester hours in military science, 3 in personnel supervision, 3 in maintenance management, 2 in records and information management, 3 in construction plans and specifications, 3 in surveying, and 3 in soil mechanics. In the upper-division baccalaureate category, 3 semester hours for field experience in management and 3 in organizational management (7/91).

MOS-51T-003

TECHNICAL ENGINEERING SPECIALIST
51T10
51T20
51T30
51T40

Exhibit Dates: 2/94–9/04. (Effective 9/04 MOS 51T was discontinued, and its duties were incorporated into MOS 21t.)

Career Management Field: 51 (General Engineering).

Description

Summary: Supervises or participates in construction site development, including technical investigation surveying, developing construction plans and specifications, and performing quality control inspections. *Skill Level 10:* Assists in perform field and laboratory tests on construction materials; surveys; draft and designs military construction, using Automated Computer-Assisted Drafting/Design (AUTO CAD). *Skill Level 20:* Able to perform the duties required for Skill Level 10; performs field and laboratory testing on construction materials; surveys; completes drafting and design of military construction. *Skill Level 30:* Able to perform the duties required for Skill Level 20; supervises the selection of construction materials; conducts surveys and preliminary engineering inspections; prepares layouts and detail drawings of pavements, structures, buildings, and utility systems from sketches and specified design criteria; assists in inspection of all phases of construction and repair. *Skill Level 40:* Able to perform the duties required for Skill Level 30; provides technical guidance to subordinates; plans and supervises preparation and layout of drawings; performs simple load calculations to ensure construction of safe facilities; prepares network flow diagrams such as critical path method; applies knowledge of asphalt and concrete quality testing procedures.

Recommendation, Skill Level 10

Credit may be granted on the basis of an individualized assessment of the student (6/97).

Recommendation, Skill Level 20

In the lower-division baccalaureate/associate degree category, 2 semester hours in construction surveying, 2 in drafting, and 2 in AUTO CAD. (NOTE: This recommendation for skill level 20 is valid for the dates 2/94-2/95 only) (6/97).

Recommendation, Skill Level 30

In the lower-division baccalaureate/associate degree category, 2 semester hours in construction surveying, 2 in drafting, 2 AUTO CAD, 3 in construction management, and 3 in personnel supervision (6/97).

Recommendation, Skill Level 40

In the lower-division baccalaureate/associate degree category, 2 semester hours in construction surveying, 2 in drafting, 2 AUTO CAD, 3 in construction management, 3 in personnel supervision, 3 in construction plans and specifications, 3 in construction surveying, and 3 in soil mechanics and testing. In the upper-division baccalaureate category, 3 semester hours for field experience in management and 3 in organizational management (6/97).

MOS-51Z-001

GENERAL ENGINEERING SUPERVISOR
51Z50

Exhibit Dates: 1/90–5/91.

Career Management Field: 51 (General Engineering), subfield 510 (General Engineering).

Description

Provides staff supervision to units engaged in structural building and repair and provision of water and utilities services; serves as principal noncommissioned officer in an engineering combat (heavy) company or in an engineer combat battalion; may serve in an engineer section at engineer brigade or higher headquarters; directs and coordinates activities of staff personnel to ensure proper integration of all construction and utility operations; assists in appraisal of training status, construction planning, and material requests; inspects structural facilities and water and utility production and distribution systems; supervises performance of combat engineer missions. NOTE: May have progressed to 51Z50 from 51T40 (Technical Engineering Supervisor), 52E40 (Prime Power Production Specialist), 51P40 (Utilities Engineering Supervisor), 51M40 (Firefighter), 51H40 (Construction Engineering Supervisor), 00B40 (Diver), 62N40 (Construction Equipment Supervisor), or 53B40 (Industrial Gas Production Specialist).

Recommendation

In the lower-division baccalaureate/associate degree category, 3 semester hours in records administration, 3 in technical mathematics, 3 in technical writing, 3 in industrial technology, 3 in human relations, and 3 in personnel supervision. In the upper-division baccalaureate category, 6 semester hours in construction management, 3 in personnel management, 6 for field experience in management, and additional credit in engineering and management on the basis of institutional evaluation (2/80).

MOS-51Z-002

General Engineering Supervisor
51Z50

Exhibit Dates: 6/91–12/01.

Career Management Field: 51 (General Engineering).

Description

Provides staff supervision to units engaged in structural building and repair and provision of water and utility services; serves as principal noncommissioned officer in an engineering combat (heavy) company or in an engineer combat battalion; may serve in an engineer section at engineer brigade or higher headquarters; directs and coordinates activities of staff personnel to insure integration of all construction and utility operations; assists in appraisal of training status, construction planning, and material requests; inspects structural facilities and water and utility production and distribution systems; supervises performance of skill level 40 of MOS 00B (Diver), 51H (Construction Engineering Supervisor), 51M (Firefighter), 51T (Technical Engineering Supervisor), 52E (Prime Power Production Specialist), 52G (Transmission and Distribution Specialist), or 62N (Construction Equipment Supervisor).

Recommendation

In the lower-division baccalaureate/associate degree category, 3 semester hours in personnel supervision, 3 in technical communication, and 3 in records and information management. In the upper-division baccalaureate category, 6 semester hours for field experience in management and 3 in organizational management; if individual has attained paygrade E-9, additional credit as follows: 3 semester hours in management problems, 3 in operations management, and 3 in communications techniques for managers (7/91).

MOS-51Z-003

General Engineering Supervisor
51Z50

Exhibit Dates: 1/02–8/04. (Effective 9/04, MOS 51Z was discontinued, and its duties were incorporated into MOS 21X.)

Career Management Field: 51 (General Engineering).

Description

Provides staff supervision to units engaged in structural building and repair and provision of water and utility services; serves as principal noncommissioned officer in an engineering combat (heavy) company or in an engineer combat battalion; may serve in an engineer section at engineer brigade or higher headquarters; directs and coordinates activities of staff personnel to ensure integration of all construction and utility operations; assists in appraisal of training status, construction planning, and material requests; inspects structural facilities and water and utility production and distribution systems; supervises conformance to OSHA standards of construction; supervises performance of skill level 40 of MOS 00B (Diver), 51H (Construction Engineering Supervisor), 51M (Firefighter), 51T (Technical Engineering Supervisor), 52E (Prime Power Production Specialist), 52G (Transmission and Distribution Specialist), or 62N (Construction Equipment Supervisor).

Recommendation

In the lower-division baccalaureate/associate degree category, 3 semester hours in personnel supervision and 3 in communication skills. In the upper-division baccalaureate category, 3 semester hours in human resource management, 3 for a practicum in management, 3 in operations/logistics management, 3 in organizational management, 3 in contract management, and 3 in code and regulations (1/02).

MOS-52C-004

Utilities Equipment Repairer
52C10
52C20
52C30

Exhibit Dates: 1/90–2/95.

Career Management Field: 63 (Mechanical Maintenance).

Description

Summary: Supervises or performs maintenance on refrigeration equipment, air conditioning units, multifuel forced air heaters, high-pressure air compressors, gasoline and diesel engines, and electrical motors. *Skill Level 10:* Maintains primarily equipment that does not require disassembly for repair; maintenance duties may include some system or equipment disassembly. *Skill Level 20:* Able to perform the duties required for Skill Level 10; supervises four or five subordinates; supervises maintenance and performs repairs requiring system disassembly, replacing of parts, and rebuilding. *Skill Level 30:* Able to perform the duties required for Skill Level 20; supervises 25-50 persons; performs depot maintenance, tests, and approves repairs; schedules training; prepares maintenance schedules and inspection reports. May progress to 52C30 from 52C20 or 63J20 (Quartermaster and Chemical Equipment Repairer).

Recommendation, Skill Level 10

In the vocational certificate category, 3 semester hours in electrical laboratory. In the lower-division baccalaureate/associate degree category, 3 semester hours in air conditioning and refrigeration, 3 in small engine repair, and 3 in power mechanics. (NOTE: This recommendation for skill level 10 is valid for the dates 1/90-9/91 only) (7/88).

Recommendation, Skill Level 20

In the vocational certificate category, 3 semester hours in electrical laboratory. In the lower-division baccalaureate/associate degree category, 5 semester hours in air conditioning and refrigeration, 3 in small engine repair, and 3 in power mechanics (7/88).

Recommendation, Skill Level 30

In the vocational certificate category, if progressed to 52C30 from 52C20, use the following credit recommendation: 3 semester hours in electrical laboratory. If progressed to 52C30 from 63J20, use the following credit recommendation: 1 semester hour in basic electricity laboratory. In the lower-division baccalaureate/associate degree category, if progressed to 52C30 from 52C20, use the following credit recommendation: 6 semester hours in air conditioning and refrigeration, 3 in small engine repair, 3 in power mechanics, 3 in personnel supervision, and 3 in maintenance management. If progressed to 52C30 from 63J20, use the following credit recommendation: 3 semester hours in small appliance repair, 1 in use and care of hand tools, 1 in use of testing equipment, 3 in personnel supervision, and 3 in maintenance management (7/88).

MOS-52C-005

Utilities Equipment Repairer
52C10
52C20
52C30

Exhibit Dates: 3/95–5/99.

Career Management Field: 63 (Mechanical Maintenance).

Description

Summary: Supervises or performs maintenance on refrigeration equipment, air conditioning units, multifuel forced air heaters, high-pressure air compressors, gasoline and diesel engines, and electrical motors. *Skill Level 10:* Maintains primarily equipment that does not require disassembly for repair; maintenance duties may include some system or equipment disassembly. *Skill Level 20:* Able to perform the duties required for Skill Level 10; supervises four or five subordinates; supervises maintenance and performs repairs requiring system disassembly, replacing of parts, and rebuilding. *Skill Level 30:* Able to perform the duties required for Skill Level 20; supervises 25-50 persons; performs depot maintenance, tests, and approves repairs; schedules training; prepares maintenance schedules and inspection reports. May progress to 52C30 from 52C20 or 63J20 (Quartermaster and Chemical Equipment Repairer).

Recommendation, Skill Level 10

Credit may be granted on the basis of an individualized assessment of the student (7/88).

Recommendation, Skill Level 20

Credit may be granted on the basis of an individualized assessment of the student (7/88).

Recommendation, Skill Level 30

In the vocational certificate category, if progressed to 52C30 from 52C20, use the following credit recommendation: 3 semester hours in electrical laboratory. If progressed to 52C30 from 63J20, use the following credit recommendation: 1 semester hour in basic electricity laboratory. In the lower-division baccalaureate/associate degree category, if progressed to 52C30 from 52C20, use the following credit recommendation: 5 semester hours in air conditioning and refrigeration, 3 in small engine repair, 3 in power mechanics, 3 in personnel supervision, and 3 in maintenance management. If progressed to 52C30 from 63J20, use the following credit recommendation: 2 semester hours in small appliance repair, 1 in use and care of hand tools, 1 in use of testing equipment, 3 in personnel supervision, and 3 in maintenance management (7/88).

MOS-52C-006

Utilities Equipment Repairer
52C10
52C20
52C30

Exhibit Dates: 6/99–Present.

Career Management Field: 63 (Mechanical Maintenance).

Description

Summary: Supervises or performs maintenance on refrigeration equipment, air conditioning units, multifuel forced air heaters, high-pressure air compressors, gasoline and diesel engines, and electrical motors. *Skill Level 10:* Maintains primarily equipment that does not require disassembly for repair; maintenance duties may include some system or equipment disassembly. *Skill Level 20:* Able to perform the duties required for Skill Level 10; supervises four or five subordinates; supervises maintenance and performs repairs requiring system disassembly, replacing of parts, and rebuilding; performs basic record keeping tasks; writes proficiency reports. *Skill Level 30:* Able to perform the duties required for Skill Level 20; supervises 25-50 persons; performs depot maintenance, tests, and approves repairs; schedules training; prepares maintenance schedules and inspection reports. May progress to 52C30 from 52C20 or 63J20 (Quartermaster and Chemical Equipment Repairer).

Recommendation, Skill Level 10

Credit may be granted on the basis of an individualized assessment of the student (6/99).

Recommendation, Skill Level 20

Credit may be granted on the basis of an individualized assessment of the student (6/99).

Recommendation, Skill Level 30

In the lower-division baccalaureate/associate degree category, if progressed to 52C30 from 52C20, use the following credit recommendation: 6 semester hours in air conditioning/refrigeration repair and maintenance, 3 in basic electricity laboratory, 3 in small engine repair, 3 in principles of supervision, and 3 in maintenance management. If progressed to 52C30 from 63J20, use the following credit recommendation: 3 semester hours in small gas engine repair, 3 in water treatment technology, 3 in principles of supervision, and 3 in maintenance management (6/99).

MOS-52D-003

Power Generation Equipment Repairer

52D10
52D20
52D30

Exhibit Dates: 1/90–1/98.

Career Management Field: 63 (Mechanical Maintenance).

Description

Summary: Supervises or maintains and overhauls power generation equipment. *Skill Level 10:* Maintains engines by adjusting valves, carburetors, ignition points, alternators, regulators, and solenoids; performs maintenance services such as lubrication, oil change, and radiator flushes; repairs small motor generators. *Skill Level 20:* Able to perform the duties required for Skill Level 10; repairs or overhauls starters, alternators, generators, voltage regulators, and control circuits; maintains and repairs diesel engines. *Skill Level 30:* Able to perform the duties required for Skill Level 20; supervises subordinates; instructs in the use of tools and test equipment; conducts maintenance training and assigns duties.

Recommendation, Skill Level 10

In the vocational certificate category, 3 semester hours in electrical laboratory. In the lower-division baccalaureate/associate degree category, 3 semester hours in small engine repair. (NOTE: This recommendation for skill level 10 is valid for the dates 1/90-9/91 only) (9/85).

Recommendation, Skill Level 20

In the vocational certificate category, 3 semester hours in electrical laboratory. In the lower-division baccalaureate/associate degree category, 3 semester hours in small engine repair, 3 in power generation equipment repair, and 3 in diesel engine maintenance. (NOTE: This recommendation for skill level 20 is valid for the dates 1/90-2/95 only) (9/85).

Recommendation, Skill Level 30

In the vocational certificate category, 3 semester hours in electrical laboratory. In the lower-division baccalaureate/associate degree category, 3 semester hours in small engine repair, 3 in power generation equipment repair, 3 in diesel engine maintenance, 3 in personnel supervision, and 3 in maintenance management (9/85).

MOS-52D-004

Power Generation Equipment Repairer

52D10
52D20
52D30

Exhibit Dates: 2/98–Present.

Career Management Field: 63 (Mechanical Maintenance).

Description

Summary: Maintains and repairs power generation equipment, small internal combustion engines, and air compressors. *Skill Level 10:* Troubleshoots, repairs, and maintains power generation equipment and engines by adjusting valves, carburetors, ignition points, alternators, regulators, and solenoids; performs maintenance services such as lubrication, oil change, and radiator flushes; repairs small motor generators. *Skill Level 20:* Able to perform the duties required for Skill Level 10; troubleshoots, repairs, or overhauls starters, alternators, generators, voltage regulators, and control circuits; maintains and repairs diesel engines. *Skill Level 30:* Able to perform the duties required for Skill Level 20; supervises subordinates; instructs in the use of tools and test equipment; conducts maintenance training assigns duties.

Recommendation, Skill Level 10

Credit may be granted on the basis of an individualized assessment of the student (6/99).

Recommendation, Skill Level 20

Credit may be granted on the basis of an individualized assessment of the student (6/99).

Recommendation, Skill Level 30

In the lower-division baccalaureate/associate degree category, 3 semester hours in electrical systems, 3 in small engine troubleshooting and repair, 6 in power generation equipment troubleshooting and repair, 3 in diesel engine maintenance, 3 in principles of supervision, and 3 in maintenance management (6/99).

MOS-52E-002

Prime Power Production Specialist

52E20
52E30
52E40

Exhibit Dates: 1/90–9/90.

Career Management Field: 51 (General Engineering), subfield 512 (Utilities Engineering).

Description

Summary: NOTE: MOS 52E contains four specialties. Each specialty is identified by an Additional Skill Identifier (ASI), a letter and a number appended to the MOS designation. For example, appending ASI S2 to the MOS designation (52E20S2) designates the Nuclear Power Plant Mechanic specialty. Addition of ASI S3 designates Nuclear Power Plant Electrician; ASI S4, Nuclear Power Plant Instrument Repairer; and ASI S5, Nuclear Power Plant Health Physicist. General duties for all skill levels of MOS 52E are described below, followed by a specific description for each specialty. Individuals holding this MOS supervise, operate, install, and maintain electric power plant and associated auxiliary systems and equipment; inspect, adjust, repair, and replace equipment/systems. *Skill Level 20:* Assists in the operation and maintenance of prime power plants and auxiliary systems and equipment; assists in starting, stopping, and regulating diesel engine, gas engine, and turbine engine driven generators to obtain power output or perform emergency procedures; assists in controlling instrumentation, power conversion equipment, station service electrical system, associated pumps, and chemical feed systems; observes and interprets instrument readings; maintains records of routine and unusual indications; adjusts temperatures, pressures, flow rates, and process liquid levels by regulating appropriate controls; operates controls to minimize or eliminate hazards associated with radiation exposure, high pressure steam, and high voltages; inspects, tests, and performs operator preventive maintenance on nuclear and conventional power plant equipment and auxiliary systems and equipment. *Skill Level 30:* Able to perform the duties required for Skill Level 20; operates and maintains prime power plant; establishes start-up condition in power plants, auxiliary systems, and equipment; starts, stops, and regulates diesel engine, gas engine, and turbine engine driven generators to obtain power output or perform emergency procedures; introduces control instrumentation, power conversion equipment, station service electrical systems, associated pumps, and chemical feed systems; brings plant to full power; manipulates nuclear power controls to adjust temperatures, pressures, flow rates, and process liquid levels; performs operator preventive maintenance on nuclear and conventional power plant equipment, auxiliary systems, and equipment; conducts on-the-job training; assists in establishing priorities and planning and scheduling work assignments; assists in establishing shop practices and inspection procedures; performs as shift supervisor in conventional power plant; provides technical guidance to subordinates. *Skill Level 40:* Able to perform the duties required for Skill Level 30; supervises operation and maintenance of prime power plants and equipment; operates control room in power plant; supervises preventive maintenance on nuclear/conventional power plant equipment, auxiliary systems, and equipment; supervises on-the-job training; plans, coordinates, and schedules work assignments, priorities, shop practices, and inspection procedures; spot checks work performance and informs subordinates of corrective action; prepares technical reports. For all skill levels: If ASI is S2, Nuclear Power Plant Mechanic, performs the following duties: installs, adjusts, repairs, replaces, and tests the operation of power plant mechanical equipment such as diesel engines, gas turbines, steam turbines, oil-fired high- and low-pressure boilers, oil purification systems, fuel and oil storage and transfer systems, air compressors, pumps, valves, heat exchangers, governing systems, piping systems, and hoisting equipment; performs nondestructive tests on equipment, components, and systems; fabricates parts by machining and welding metals and alloys; employs hand tools, precision-measuring devices, and electrical and pneumatic power tools; interprets schematics and mechanical drawings of equipment and systems to solve maintenance problems; performs as a member of refueling team and assists in preparing radioactive material for shipment and storage; performs scheduled and preventive maintenance on equipment and ensures historical and mechanical records are kept current. If ASI is S3, Nuclear Power Plant Electrician, performs the following duties: applies National Electric Code provisions that govern installation, modification, and repair of electrical wiring and equipment in power plants; employs safety practices while performing electrical maintenance; to solve maintenance problems, traces wiring and equipment diagrams related to plant electrical equipment and power distribution systems; troubleshoots electrical equipment by interpreting visual, mechanical, and audible symptoms; interprets mechanical and electrical problems that involve algebraic formulas; wires, repairs, and tests operation of switch gear, circuit breakers, motors, generators, transformers, high- and low-voltage distributors, transmission systems, and protective and synchronizing devices of power plant equipment; solders connections, splices cables, and makes cable and wire runs; installs, maintains, and troubleshoots batteries, inverters, battery chargers and motor generators; calibrates and maintains protective relays and associated auxiliary current transformers, potential transformers control wiring for generators, transformers, and motors; employs equipment such as multimeters, clamp-on ammeters, meggers, high pot tester, phase rotation indicators, oscilloscopes, and protective relay tests sets; performs scheduled preventive maintenance on electrical equipment/components and ensures that historical and maintenance records are kept current; assists in refueling operations and preparing radioactive materials for storage and shipment. If ASI is S4, Nuclear Power Plant Instrument Repairer, performs the following duties: troubleshoots all types of nuclear power plant instrumentation systems using schematic and wiring diagrams; installs, calibrates, repairs, adjusts, and test instrumentation and control equipment; installs electrical, electronic, and pneumatic instruments; calibrates control, regulatory, and electronic instruments; repairs or replaces circuit wiring, coaxial cables, vacuum tubes, semiconductors, and saturable core devices; employs equipment such as multimeters, oscilloscopes, transistor test sets, signal generators, and pressure and vacuum comparators; adjusts and aligns conventional and solid state electrical and electronic devices and pneumatic instruments used for motivation and control in other components of systems; interprets mechanical and electrical problems that involve algebraic formulas; interprets radiation survey instrument readings; performs scheduled preventive maintenance on equipment and maintains historical and maintenance records; assists in refueling operations and preparing radioactive materials for storage and shipment. If ASI is S5, Nuclear Power Plant Health Physicist, performs the following duties: monitors and treats process fluids for radioactivity and other impurities; prepares reagents to chemical standards; applies techniques for volumetric, gravimetric, colorimetric, and radiochemical analysis of water; operates testing equipment and keeps it in operating condition; determines alpha, beta, gamma, and neutron activities in air, water, and solid samples; converts radiation levels into millicuries; interprets results in relation to maximum permissible concentrations prescribed by regulations; computes radiation levels, exposure time limits, accumulated dosages, and radiation counter efficiencies; employs slide rule, logarithm tables, and chemical and algebraic formulas; performs decontamination procedures; establishes and monitors plant radiological safety program; ensures that scheduled preventive maintenance on radiation monitoring equipment is performed; maintains historical and maintenance records; assists in refueling operations and preparing radioactive materials for storage and shipment.

Recommendation, Skill Level 20

In the vocational certificate category, 6 semester hours in the use and care of hand and power tools. In the lower-division baccalaureate/associate degree category, 3 semester hours in industrial safety and 6 in troubleshooting techniques (mechanical and electrical). If ASI is S2 (Mechanic), see exhibit AR-1720-0004 in the Army courses volume of the GUIDE. If ASI is S3 (Electrician), see exhibit AR-1720-0003. If ASI is S4 (Instrument Repairer), see exhibit AR-1720-0005. If ASI is S5 (Health Physicist), see exhibit AR-1720-0002 (8/81).

Recommendation, Skill Level 30

In the vocational certificate category, 6 semester hours in the use and care of hand and power tools. In the lower-division baccalaureate/associate degree category, 3 semester hours in industrial safety, 6 in troubleshooting techniques (mechanical and electrical), 3 in human relations, and 3 in personnel supervision. In the upper-division baccalaureate category, 3 semester hours in personnel management and 3 in industrial safety. If ASI is S2 (Mechanic), see exhibit AR-1720-0004 in the Army courses volume of the GUIDE. If ASI is S3 (Electrician), see exhibit AR-1720-0003. If ASI is S4 (Instrument Repairer), see exhibit AR-1720-0005. If ASI is S5 (Health Physicist), see exhibit AR-1720-0002 (8/81).

Recommendation, Skill Level 40

In the vocational certificate category, 6 semester hours in use and care of hand and power tools and 2 in technical writing. In the lower-division baccalaureate/associate degree category, 3 semester hours in industrial safety, 6 in troubleshooting techniques (mechanical and electrical), 3 in human relations, 3 in personnel supervision, 3 in introduction to management, and 2 in technical report writing. In the upper-division baccalaureate category, 3 semester hours in personnel management, 3 in industrial safety, 3 for field experience in management, and 3 in management problems. If ASI is S2 (Mechanic), see exhibit AR-1720-0004 in the Army courses volume of the GUIDE. If ASI is S3 (Electrician), see exhibit AR-1720-0003. If ASI is S4 (Instrument Repairer), see exhibit AR-1720-0005. If ASI is S5 (Health Physicist), see exhibit AR-1720-0002 (8/81).

MOS-52E-003

Prime Power Production Specialist
52E20
52E30
52E40

Exhibit Dates: 10/90–9/04. (Effective 9/04 MOS 52E was discontinued, and its duties were incorporated into MOS 21P.)

Career Management Field: 51 (General Engineering).

Description

Summary: NOTE: MOS 52E contains three specialties. Each specialty is identified by an Additional Skill Identifier (ASI), a letter and a number appended to the MOS designation (e.g., 52E20E5). ASI E5 designates Instrument Maintenance (Power Station), ASI S2 designates Mechanical Equipment Maintenance (Power Station), and ASI S3 designates Electrical Equipment Maintenance (Power Station). General duties for all skill levels are described below, followed by a specific description for each specialty. Individuals holding this MOS supervise, operate, install, and maintain electric power plants and associated auxiliary systems and equipment. *Skill Level 20:* Operates and maintains electric power plants; performs start-up in power plants and auxiliary systems and equipment; starts, stops, and regulates diesel, gas, and turbine driven generators to obtain power output or perform emergency procedures; controls power conversion equipment, station service electrical system, associated pumps, and chemical feed systems; observes equipment operating parameters; interprets instrumentation data; adjusts temperatures, pressures, flow rates, and processed liquid levels by regulating appropriate controls; operates regulatory controls to minimize or eliminate hazards to high pressure steam and high voltages; diagnoses plant equipment/system abnormalities by interpreting fault indicators, applying troubleshooting techniques, and identifying faulty assemblies and components; determines malfunctions by using diagnostic test and troubleshooting procedures; performs scheduled maintenance; repairs or overhauls generating equipment and associated components; prepares maintenance data work sheets; selects, uses, and cares for common hand and shop tools; applies work and fire safety practices, precautions, and procedures; maintains CPR qualification; uses appropriate word processing, data base management, and spread sheet software. *Skill Level 30:* Able to perform the duties required for Skill Level 20; provides technical guidance to subordinates; coordinates changes in plant equipment/power systems; inventories equipment and supplies; prepares reports; enforces power plant and maintenance shop safety program; serves as technical inspector; determines and isolates complex malfunctions; performs quality assurance inspections. *Skill Level 40:* Able to perform the duties required for Skill Level 30; inspects plant site for proper preparations and placement of generating units; coordinates, plans, and schedules team activities; evaluates team safety program; maintains accountability of team's supplies and equipment; prepares power station team standard operating procedures; ensures licensing of equipment operators; evaluates management of plant operation and maintenance personnel; estimates manpower, equipment, and material necessary to accomplish installation of power station and construction of associated systems. If ASI is E5, Instrument Maintenance, performs the following duties: troubleshoots all types of power plant instrumentation systems, using schematic and wiring diagrams; installs, calibrates, repairs, adjusts, and tests instrumentation and control equipment; installs electrical, electronic, and pneumatic instruments; calibrates control, regulatory, and electronic instruments; repairs or replaces circuit wiring, coaxial cables, semiconductors, and saturable core devices; uses multimeters, oscilloscopes, transistor test sets, signal generators, and pressure and vacuum comparators; adjusts and aligns conventional and solid state electrical and electronic devices and pneumatic actuators used for motivation and control in other components of system; interprets and solves mechanical and electrical problems using algebraic formulas; performs scheduled preventive maintenance on equipment; maintains historical and maintenance records; assists in refueling operations and preparing materials for storage and shipment. If ASI is S2, Mechanical Equipment Maintenance, performs the following duties; installs, adjusts, repairs, replaces, and tests the operation of power plant mechanical equipment such as diesel engines, gas turbines, oil-fired high- and low-pressure boilers, oil purification systems, fuel and oil storage and transfer systems, air compressors, air conditioners and heat pumps, valves, heat exchangers, governing systems, piping systems, and hoisting equipment; performs nondestructive tests on equipment, components, and systems; fabricates parts by machining and welding (arc, gas, TIG) metals and alloys; uses hand tools, precision measuring devices, and electrical and pneumatic power tools; interprets schematics and mechanical drawings of equipment and systems to solve maintenance problems; performs as a member of refueling team and assists in preparing material for shipment and storage; performs scheduled and preventive maintenance on equipment; maintains historical and mechanical records. If ASI is S3, Electrical Equipment Maintenance, performs the following duties: applies National Electric Code provisions that govern installation, modification, and repair of electrical wiring and equipment in power plants; employs work safety practices while performing electrical maintenance; to solve maintenance problems, traces wiring and equipment diagrams related to plant electrical equipment by interpreting visual, mechanical, and audible symptoms; interprets and solves mechanical and electrical problems using algebraic formulas; wires, repairs, and tests operations of switch gear, circuit breakers, motors, generators, transformers, high- and low-voltage distributors, transmission systems, and protective and synchronizing devices of power plant equipment; solders connections, splices cables, and makes cable and wire runs; installs, maintains, and troubleshoots batteries, inverters, battery chargers, and motor generators; calibrates and maintains protective relays and associated auxiliary current transformers, potential transformers, control wiring for generators, transformers, and motors; uses multimeters, clamp-on ammeters, meggers, high pot tester, phase rotation indicators, oscilloscopes, and protective relay test sets; performs scheduled preventive maintenance on electrical equipment/components; maintains historical and maintenance records; assists in refueling operations and preparing materials for storage and shipment.

Recommendation, Skill Level 20

In the vocational certificate category, 3 semester hours in use and care of hand and power tools. In the lower-division baccalaureate/associate degree category, 3 semester hours in computer software applications, 2 in emergency medical training, 3 in power plant troubleshooting and maintenance, 3 in industrial safety, 2 in records and information management, 2 in personnel supervision, and 2 in report writing. If ASI is E5 (Instrument Maintenance), see exhibit AR-1720-0010 in the Army courses volume of the GUIDE. If ASI is S2 (Mechanical Equipment Maintenance), see exhibit AR-1720-0011. If ASI is S3 (Electrical Equipment Maintenance), see exhibit AR-1720-0009 (7/93).

Recommendation, Skill Level 30

In the vocational certificate category, 3 semester hours in use and care of hand and power tools. In the lower-division baccalaureate/associate degree category, 3 semester hours in computer software applications, 2 in emergency medical training, 6 in power plant troubleshooting and maintenance, 3 in industrial safety, 2 in records and information management, 3 in personnel supervision, and 2 in report writing. If ASI is E5 (Instrument Maintenance), see exhibit AR-1720-0010 in the Army courses volume of the GUIDE. If ASI is S2 (Mechanical Equipment Maintenance), see exhibit AR-1720-0011. If ASI is S3 (Electrical Equipment Maintenance), see exhibit AR-1720-0009. (NOTE: This recommendation for skill level 20 is valid for the dates 10/90-2/95 only) (7/93).

Recommendation, Skill Level 40

In the vocational certificate category, 3 semester hours in the use of hand and power tools. In the lower-division baccalaureate/associate degree category, 3 semester hours in computer software applications, 2 in emergency medical training, 6 in power plant troubleshooting and maintenance, 3 in industrial safety, 2 in records and information management, 3 in personnel supervision, and 2 in report writing. In the upper-division baccalaureate category, 3 semester hours for field experience in management and 3 in logistics management. If ASI is E5 (Instrument Maintenance), see exhibit AR-1720-0010 in the Army courses volume of the GUIDE. If ASI is S2 (Mechanical Equipment Maintenance), see exhibit AR-1720-0011. If ASI is S3 (Electrical Equipment Maintenance), see exhibit AR-1720-0009 (7/93).

MOS-52F-001

Turbine Engine Driven Generator Repairer
52F10
52F20
52F30

Exhibit Dates: 1/90–6/98. Effective 6/98, MOS 52F was discontinued and its duties were incorporated into MOS 52D (Power Generation Equipment Repairer).

Career Management Field: 63 (Mechanical Maintenance).

Description

Summary: Performs or supervises maintenance and overhaul of turbine engine driven generator sets and turbine engine prime movers of tactical generator sets. *Skill Level 10:* Performs maintenance on turbine engine driven generator and associated equipment including turbine engines used as prime movers on other units; troubleshoots mechanical and electrical systems; interprets schematic diagrams. *Skill Level 20:* Able to perform the duties required for Skill Level 10; repairs or overhauls components and assemblies including starters, generators, fuel pumps, exciters, AC and DC voltage regulators, electric and electronic load-sensing governors, contactors, switches, and control circuits; diagnoses faults in systems and components. *Skill Level 30:* Able to perform the duties required for Skill Level 20; provides technical assistance, guidance, and supervision to subordinates; plans and organizes work schedules and assigns duties; establishes maintenance priorities and allocates work loads; performs final inspection and testing of repaired equipment; conducts on-the-job training.

Recommendation, Skill Level 10

In the vocational certificate category, 3 semester hours in electrical laboratory. In the lower-division baccalaureate/associate degree category, 3 semester hours in turbine engine maintenance. (NOTE: This recommendation for skill level 10 is valid for the dates 1/90-9/91 only) (9/85).

Recommendation, Skill Level 20

In the vocational certificate category, 3 semester hours in electrical laboratory. In the lower-division baccalaureate/associate degree category, 6 semester hours in turbine engine maintenance. (NOTE: This recommendation for skill level 20 is valid for the dates 1/90-2/95 only) (9/85).

Recommendation, Skill Level 30

In the vocational certificate category, 3 semester hours in electrical laboratory. In the lower-division baccalaureate/associate degree category, 6 semester hours in turbine engine maintenance, 3 in personnel supervision, and 3 in maintenance management (9/85).

MOS-52G-001

Transmission and Distribution Specialist
52G10
52G20
52G30
52G40

Exhibit Dates: 1/90–9/04. (Effective 9/04, MOS 52G was discontinued, and its duties were incorporated into MOS 21Q)

Career Management Field: 51 (General Engineering).

Description

Summary: Supervises or installs and maintains electrical distribution systems. *Skill Level 10:* Installs and maintains electrical power distribution systems including poles, transformers, guy wires, and service drops. *Skill Level 20:* Able to perform the duties required for Skill Level 10; inspects, maintains, and repairs hot or power-on transmission systems; repairs, calibrates, and tests substation equipment. *Skill Level 30:* Able to perform the duties required for Skill Level 20; supervises subordinates; supervises power line and security/airport lighting installation; supervises operator maintenance on equipment and tools; insures adherence to proper safety and clearance procedures. *Skill Level 40:* Able to perform the duties required for Skill Level 30; estimates project manpower, equipment, and material requirements; coordinates work, maintenance, and planning schedules; advises on construction and maintenance procedures and safety training.

Recommendation, Skill Level 10

In the vocational certificate category, 3 semester hours in power line installation and maintenance. (NOTE: This recommendation for skill level 10 is valid for the dates 1/90-9/91 only) (9/85).

Recommendation, Skill Level 20

In the vocational certificate category, 4 semester hours in power line installation and maintenance. (NOTE: This recommendation for skill level 20 is valid for the dates 1/90-2/95 only) (9/85).

Recommendation, Skill Level 30

In the vocational certificate category, 4 semester hours in power line installation and maintenance. In the lower-division baccalaureate/associate degree category, 3 semester hours in personnel supervision and 3 in maintenance management (9/85).

Recommendation, Skill Level 40

In the vocational certificate category, 4 semester hours in power line installation and maintenance. In the lower-division baccalaureate/associate degree category, 3 semester hours in personnel supervision and 3 in maintenance management. In the upper-division baccalaureate category, 3 semester hours for field experience in management (9/85).

MOS-52X-001

Special Purpose Equipment Repairer
52X40

Exhibit Dates: 1/90–5/99.

Career Management Field: 63 (Mechanical Maintenance).

Description

Manages special purpose equipment repair facilities; special purpose equipment includes utilities equipment (air conditioning, heating, refrigeration), power generation equipment, turbine engine driven generators, and quartermaster and chemical equipment; supervises 25-60 persons; plans and organizes work area layout; applies production and quality control principles and procedures; trains and advises subordinates; performs administrative duties including control of maintenance publication file and supervision of the preparation of records and reports.

Recommendation

In the lower-division baccalaureate/associate degree category, 3 semester hours in personnel supervision and 3 in maintenance management. In the upper-division baccalaureate category, 3 semester hours for field experience in management and 3 in management problems (7/88).

MOS-52X-002

Special Purpose Equipment Repairer
52X40

Exhibit Dates: 6/99–Present.

Career Management Field: 63 (Mechanical Maintenance).

Description

Manages special purpose equipment repair facilities; special purpose equipment includes utilities equipment (air conditioning, heating, refrigeration), power generation equipment, turbine-engine-driven generators, and quartermaster and chemical equipment; supervises 15—40 persons; plans and organizes work area layout; applies production and quality control principles and procedures; trains and advises subordinates; performs administrative duties, including control of maintenance publication file and supervision of the preparation of records and reports.

Recommendation

In the lower-division baccalaureate/associate degree category, 3 semester hours in principles of supervision and 3 in maintenance management. In the upper-division baccalaureate category, 3 semester hours in a management practicum, 3 in principles of management, and 3 in human resource management (6/99).

MOS-54B-002

Chemical Operations Specialist
54B10
54B20
54B30
54B40
54B50

Exhibit Dates: 1/90–5/99.

Career Management Field: 54 (Chemical).

Description

Summary: Supervises or operates and maintains both smoke generating equipment and nuclear, biological, and chemical (NBC) detection and decontamination equipment; assists in the establishment, administration, and application of NBC defense measures. *Skill Level 10:* Participates in preparation and operation of smoke generating equipment; performs NBC reconnaissance; operates and maintains identification/detection and decontamination equipment. *Skill Level 20:* Able to perform the duties required for Skill Level 10; plans and organizes smoke generating fuel supply team work schedules; assigns duties; instructs and supervises subordinates in work techniques and procedures; supervises reconnaissance/decontamination operations and serves as company NBC operations noncommissioned officer; prepares and evaluates reports; conducts training; assists in computation of radiation factors; observes operating efficiency and preparedness of unit; prepares, processes, and distributes NBC intelligence reports. *Skill Level 30:* Able to perform the duties required for Skill Level 20; controls and coordinates supply and resupply efforts; supervises and coordinates maintenance of assigned equipment; responsible for reports concerning fuel and fog oil supplies; computes and plots NBC data; supervises reconnaissance and decontamination squads, and serves as an advisor. *Skill Level 40:* Able to perform the duties required for Skill Level 30; plans and organizes smoke platoon operations; assigns work to subordinates according to their capabilities; establishes priorities and inspects smoke production for compliance with operational procedures; reviews, consolidates, and prepares technical, personnel, and administrative reports covering unit activities; assists command and staff officers in continuous appraisal of chemical operations and training situations; supervises NBC decontamination platoon; monitors NBC company technical operations and serves as NBC advisor; reviews, consolidates, and prepares technical, personnel, and administrative reports; coordinates logistical needs. *Skill Level 50:* Able to perform the duties required for Skill Level 40; conducts operations and intelligence functions; prepares and maintains maps and operations information for NBC units and activities; prepares operations and inspection reports; prepares tactical NBC plans and training material; collects, interprets, analyzes, and evaluates NBC intelligence data; provides technical advice in handling, transporting, and storing chemical munitions; maintains NBC staff journals, files, records, and reports.

Recommendation, Skill Level 10

Credit is not recommended because the skills, competencies, and knowledge are uniquely military in nature (9/88).

Recommendation, Skill Level 20

In the lower-division baccalaureate/associate degree category, 2 semester hours in personnel supervision and 1 in technical writing. (NOTE: This recommendation for skill level 20 is valid for the dates 1/90-3/95 only) (9/88).

Recommendation, Skill Level 30

In the lower-division baccalaureate/associate degree category, 3 semester hours in personnel supervision, 2 in technical writing, 1 in records management, and 2 in human relations (9/88).

Recommendation, Skill Level 40

In the lower-division baccalaureate/associate degree category, 3 semester hours in personnel supervision, 3 in technical writing, 1 in records management, and 3 in human relations. In the upper-division baccalaureate category, 3 semester hours for field experience in management (9/88).

Recommendation, Skill Level 50

In the lower-division baccalaureate/associate degree category, 3 semester hours in personnel supervision, 3 in technical writing, 1 in records management, and 3 in human relations. In the upper-division baccalaureate category, 6 semester hours for field experience in management and 3 in management problems (9/88).

MOS-54B-003

Chemical Operations Specialist
54B10
54B20
54B30
54B40
54B50

Exhibit Dates: 6/99–9/04. (Effective 9/04, MOS 54B was discontinued, and its duties were incorporated into MOS 74D.)

Career Management Field: 54 (Chemical).

Description

Summary: Supervises or operates and maintains both smoke generating equipment and nuclear, biological, and chemical (NBC) detection and decontamination equipment; assists in the establishment, administration, and application of NBC defense measures. *Skill Level 10:* Participates in preparation and operation of smoke generating equipment; performs NBC reconnaissance; operates and maintains identification/detection and decontamination equipment. *Skill Level 20:* Able to perform the duties required for Skill Level 10; plans and organizes smoke generating fuel supply team work schedules; assigns duties; instructs and supervises subordinates in work techniques and procedures; supervises reconnaissance/decontamination operations and serves as company NBC operations noncommissioned officer; prepares and evaluates reports; conducts training; assists in computation of radiation factors; observes operating efficiency and preparedness of unit; prepares, processes, and distributes NBC intelligence reports. *Skill Level 30:* Able to perform the duties required for Skill Level 20; controls and coordinates supply and resupply efforts; plans and conducts training; supervises and coordinates maintenance of assigned equipment; responsible for reports concerning fuel and fog oil supplies; computes and plots NBC data; supervises reconnaissance and decontamination squads; serves as an advisor. *Skill Level 40:* Able to perform the duties required for Skill Level 30; plans and organizes smoke platoon operations; assigns work to subordinates according to their capabilities; establishes priorities; inspects smoke production for compliance with operational procedures; reviews, consolidates, and prepares technical, personnel, and administrative reports covering unit activities; assists command and staff officers in continuous appraisal of chemical operations and training situations; supervises NBC decontamination platoon; monitors NBC company technical operations and serves as NBC advisor; reviews, consolidates, and prepares technical, personnel, and administrative reports; coordinates logistical needs. *Skill Level 50:* Able to perform the duties required for Skill Level 40; conducts operations and intelligence functions; prepares and maintains maps and operations information for NBC units and activities; prepares operations and inspection reports; prepares tactical NBC plans and training material; collects, interprets, analyzes, and evaluates NBC intelligence data; provides technical advice in handling, transporting, and storing chemical munitions; maintains NBC staff journals, files, records, and reports.

Recommendation, Skill Level 10

Credit may be granted on the basis of an individualized assessment of the student (6/99).

Recommendation, Skill Level 20

Credit may be granted on the basis of an individualized assessment of the student (6/99).

Recommendation, Skill Level 30

In the lower-division baccalaureate/associate degree category, 3 semester hours in principles of supervision, 2 in technical writing, and 2 in records management (6/99).

Recommendation, Skill Level 40

In the lower-division baccalaureate/associate degree category, 3 semester hours in principles of supervision, 3 in technical writing, and 3 in records management. In the upper-division baccalaureate category, 3 semester hours in a management practicum (6/99).

Recommendation, Skill Level 50

In the lower-division baccalaureate/associate degree category, 3 semester hours in principles of supervision, 4 in technical writing, and 4 in records management. In the upper-division baccalaureate category, 6 semester hours in a management practicum (6/99).

MOS-55B-002

AMMUNITION SPECIALIST

((Ammunition Storage and Operations Specialist))

55B10

55B20

55B30

55B40

Exhibit Dates: 1/90–12/90.

Career Management Field: 55 (Ammunition).

Description

Summary: Receives, inspects, classifies, stores, and issues ammunition; keeps records of warehouse transactions. *Skill Level 10:* Serves as an ammunition storage assistant or records clerk; assists in the receipt, storage, issue, and maintenance of all types of conventional and chemical ammunition, ammunition components, and explosives; loads, unloads, stacks, and stores ammunition supplies and explosives including guided missiles using material handling equipment; prepares ammunition for shipment by bracing and staying loads; inventories ammunition in storage; assists in upkeep of operations area and facilities; employs safety precautions regarding ammunition storage, handling, and maintenance; assists in the routine destruction of unserviceable and irreparable conventional and chemical ammunition and explosives. *Skill Level 20:* Able to perform the duties required for Skill Level 10; provides technical guidance to subordinates; performs emergency destruction of ammunition; prepares periodic statistical reports on issue, receipt, and storage functions. *Skill Level 30:* Able to perform the duties required for Skill Level 20; supervises subordinates; plans and organizes ammunition storage facilities and maintenance operations; estimates requirements for personnel, tools, equipment, and supplies; supervises the establishment and maintenance of ammunition stock control records. *Skill Level 40:* Able to perform the duties required for Skill Level 30; supervises ammunition supply operations including maintenance of non-nuclear ammunition; performs as technical advisor; prepares activity reports; assigns duties to subordinates; supervises the setting up and camouflaging of ammunition supply and maintenance facilities.

Recommendation, Skill Level 10

In the vocational certificate category, 6 semester hours in explosives (6/76).

Recommendation, Skill Level 20

In the vocational certificate category, 6 semester hours in explosives and 3 in supply procedures (6/76).

Recommendation, Skill Level 30

In the vocational certificate category, 6 semester hours in explosives and 3 in supply procedures. In the lower-division baccalaureate/associate degree category, 3 semester hours in personnel supervision and 3 in human relations (6/76).

Recommendation, Skill Level 40

In the vocational certificate category, 6 semester hours in explosives and 3 in supply procedures. In the lower-division baccalaureate/associate degree category, 3 semester hours in personnel supervision, 3 in human relations, and 3 for field experience in management (6/76).

MOS-55B-003

AMMUNITION SPECIALIST

55B10
55B20
55B30
55B40

Exhibit Dates: 1/91–1/93.

Career Management Field: 55 (Ammunition).

Description

Summary: Supervises, performs, or assists in ammunition storage, receipt, issue, maintenance, stock control and accounting, and inspection and destruction procedures. *Skill Level 10:* Serves as an ammunition storage assistant or records clerk; assists in the receipt, storage, issue, and maintenance of all types of conventional and chemical ammunition, ammunition components, and explosives; loads, unloads, stacks, and stores ammunition supplies and explosives including guided missiles using material handling equipment; prepares ammunition for shipment by bracing and staying loads; inventories ammunition in storage; assists in upkeep of operations area and facilities; employs safety precautions regarding ammunition storage, handling, and maintenance; assists in the routine destruction of unserviceable and irreparable conventional and chemical ammunition and explosives. *Skill Level 20:* Able to perform the duties required for Skill Level 10; provides technical guidance to subordinates; performs emergency destruction of ammunition; prepares periodic statistical reports on issue, receipt, and storage functions. *Skill Level 30:* Able to perform the duties required for Skill Level 20; supervises subordinates; plans and organizes ammunition storage facilities and maintenance operations; estimates requirements for personnel, tools, equipment, and supplies; supervises the establishment and maintenance of ammunition stock control records. *Skill Level 40:* Able to perform the duties required for Skill Level 30; supervises ammunition supply operations including maintenance of non-nuclear ammunition; performs as technical advisor; prepares activity reports; assigns duties to subordinates; supervises the setting up and camouflaging of ammunition supply and maintenance facilities.

Recommendation, Skill Level 10

In the lower-division baccalaureate/associate degree category, 3 semester hours in military science and 3 in inventory control. (NOTE: This recommendation for skill level 10 is valid for the dates 1/91-9/91 only) (4/91).

Recommendation, Skill Level 20

In the lower-division baccalaureate/associate degree category, 3 semester hours in military science, 3 in inventory control, and 3 in personnel supervision (4/91).

Recommendation, Skill Level 30

In the lower-division baccalaureate/associate degree category, 3 semester hours in military science, 3 in inventory control, 3 in personnel supervision, 3 in maintenance management, and 2 in records and information management (4/91).

Recommendation, Skill Level 40

In the lower-division baccalaureate/associate degree category, 3 semester hours in military science, 3 in inventory control, 3 in personnel supervision, 3 in maintenance management, and 2 in records and information management. In the upper-division baccalaureate category, 3 semester hours for field experience in management and 3 in organizational management (4/91).

MOS-55B-004

Ammunition Specialist

55B10
55B20
55B30
55B40

Exhibit Dates: 2/93–6/96.

Career Management Field: 55 (Ammunition).

Description

Summary: Supervises, performs, or assists in ammunition storage, receipt, issue, maintenance, stock control and accounting, and inspection and destruction procedures. *Skill Level 10:* Serves as an ammunition storage assistant or records clerk; assists in the receipt, storage, issue, and maintenance of all types of conventional and chemical ammunition, ammunition components, and explosives; loads, unloads, stacks, and stores ammunition supplies and explosives, including guided missiles, using material handling equipment; prepares ammunition for shipment by bracing and staying loads; inventories ammunition in storage using both automated and manual procedures; employs safety precautions regarding ammunition storage, handling, maintenance, and hazardous materials; assists in the routine destruction of unserviceable and irreparable conventional and chemical ammunition and explosives. *Skill Level 20:* Able to perform the duties required for Skill Level 10; supervises and provides technical guidance to subordinates; performs emergency destruction of ammunition; prepares periodic statistical reports on issue, receipt, and storage functions; prepares ammunition, components, and explosives for transport. *Skill Level 30:* Able to perform the duties required for Skill Level 20; supervises subordinates; plans and organizes ammunition storage facilities and maintenance operations; estimates requirements for personnel, tools, equipment, and supplies; supervises the establishment and maintenance of ammunition stock control records; conducts inspections and tests. *Skill Level 40:* Able to perform the duties required for Skill Level 30; supervises ammunition supply operations; serves as technical advisor; prepares activity reports; assigns duties to subordinates; advises on surveillance and security matters.

Recommendation, Skill Level 10

Credit may be granted on the basis of an individualized assessment of the student (6/97).

Recommendation, Skill Level 20

In the lower-division baccalaureate/associate degree category, 2 semester hours in introduction to hazardous materials management and 3 in inventory control. (NOTE: This recommendation for skill level 20 is valid for the dates 2/93-2/95 only) (6/97).

Recommendation, Skill Level 30

In the lower-division baccalaureate/associate degree category, 2 semester hours in introduction to hazardous materials management, 3 in inventory control, 3 in personnel supervision, 3 in maintenance management, and 2 in records and information management (6/97).

Recommendation, Skill Level 40

In the lower-division baccalaureate/associate degree category, 3 semester hours in introduction to hazardous materials management, 3 in inventory control, 3 in personnel supervision, 3 in maintenance management, and 2 in records and information management. In the upper-division baccalaureate category, 3 semester hours for field experience in management (6/97).

MOS-55B-005

Ammunition Specialist

55B10
55B20
55B30
55B40
55B50

Exhibit Dates: 7/96–Present.

Career Management Field: 55 (Ammunition).

Description

Summary: Supervises, performs, or assists in ammunition storage, receipt, issue, maintenance, stock control and accounting, and inspection and destruction procedures. *Skill Level 10:* Serves as an ammunition storage assistant or records clerk; assists in the receipt, storage, issue, and maintenance of all types of conventional and chemical ammunition, ammunition components, and explosives; loads, unloads, stacks, and stores ammunition supplies and explosives, including guided missiles, using material handling equipment; prepares ammunition for shipment by bracing and staying loads; inventories ammunition in storage using both automated and manual procedures; employs safety precautions regarding ammunition storage, handling, maintenance, and hazardous materials; assists in the routine destruction of unserviceable and irreparable conventional and chemical ammunition and explosives. *Skill Level 20:* Able to perform the duties required for Skill Level 10; supervises and provides technical guidance to subordinates; performs emergency destruction of ammunition; prepares periodic statistical reports on issue, receipt, and storage functions; prepares ammunition, components, and explosives for transport. *Skill Level 30:* Able to perform the duties required for Skill Level 20; supervises subordinates; plans and organizes ammunition storage facilities and maintenance operations; estimates requirements for personnel, tools, equipment, and supplies; supervises the establishment and maintenance of ammunition stock control records; conducts inspections and tests. *Skill Level 40:* Able to perform the duties required for Skill Level 30; supervises ammunition supply operations; serves as technical advisor; prepares activity reports; assigns duties to subordinates; advises on surveillance and security matters. *Skill Level 50:* Able to perform the duties required for Skill Level 40; supervises and manages the Standard Army Ammunition System (SAAS); assists in the survey and layout of areas allocated for the establishment of ammunition storage areas; plans requirements for labor, equipment, and supplies in preparation and application of plans, work policies, and procedures; advises on packaging and rigging loads for movement by all types of transportation; prepares and reviews evaluations and reports on operations and training activities; advises commander on enlisted personnel matters.

Recommendation, Skill Level 10

Credit may be granted on the basis of an individualized assessment of the student (6/97).

Recommendation, Skill Level 20

In the lower-division baccalaureate/associate degree category, 2 semester hours in introduction to hazardous materials management and 3 in inventory control. (NOTE: This recommendation for skill level 20 is valid for the dates 2/93-2/95 only) (6/97).

Recommendation, Skill Level 30

In the lower-division baccalaureate/associate degree category, 2 semester hours in introduction to hazardous materials management, 3 in inventory control, 3 in personnel supervision, 3 in maintenance management, and 2 in records and information management (6/97).

Recommendation, Skill Level 40

In the lower-division baccalaureate/associate degree category, 3 semester hours in introduction to hazardous materials management, 3 in inventory control, 3 in personnel supervision, 3 in maintenance management, and 2 in records and information management. In the upper-division baccalaureate category, 3 semester hours for field experience in management (6/97).

Recommendation, Skill Level 50

In the lower-division baccalaureate/associate degree category, 3 semester hours in introduction to hazardous materials management, 3 in inventory control, 3 in personnel supervision, 3 in maintenance management, and 2 in records and information management. In the upper-division baccalaureate category, 3 semester hours for field experience in management, 3 in organizational management, 3 in personnel management, 3 in public administration, and 3 in technical writing (1/00).

MOS-55D-001

Explosive Ordnance Disposal Specialist

55D10
55D20
55D30
55D40
55D50

Exhibit Dates: 1/90–12/90.

Career Management Field: 55 (Ammunition).

Description

Summary: Supervises or performs disposal activities related to explosive devices. *Skill Level 10:* Assists in the location, identification, rendering safe, removal, and destruction of explosive ordnance other than nuclear fission or fusion materials; uses detection instruments to determine hazards and to delineate exclusion areas; decontaminates explosive devices as required; assists in isolating explosive and/or contaminated areas; removes obstructions surrounding explosive devices; makes photographs and radiographs; operates and performs maintenance on explosive disposal equipment, tools, and vehicles, including radiac equipment, radios, and chemical detection equipment; cooperates with law enforcement agencies and protection services. *Skill Level 20:* Able to perform the duties required for Skill Level 10; performs render-safe and disposal procedures for nuclear devices; destroys or neutralizes unserviceable and irreparable conventional explosive devices; reads and interprets radiographs, diagrams, drawings, and technical literature; identifies explosives by appearance and markings; performs radiation monitoring and evaluates existing and potential hazards associated with nuclear devices. *Skill Level 30:* Able to perform the duties required for Skill Level 20; supervises four or more persons; prepares technical intelligence and incident reports; develops and/or modifies render-safe procedures for conventional explosive devices; conducts classes; assists in unit administration, supply, security, and records management; assumes command of an explosives disposal team. *Skill Level 40:* Able to perform the duties required for Skill Level 30; supervises five or more persons; performs administrative and supervisory activities for the disposal unit including personnel, supply, maintenance, security, training, and management; consolidates, edits, and reviews disposal reports; maintains liaison with supporting security units; coordinates operations and movements; prepares orders and operating procedures. *Skill Level 50:* Able to perform the duties required for Skill Level 40; supervises disposal operation unit under direction of unit commander; coordinates operations and movements; supervises activities of several disposal teams over a wide geographical area; oversees and supervises coordination and liaison with supporting units; assumes command of unit in absence of commander.

Recommendation, Skill Level 10

In the vocational certificate category, 3 semester hours in explosives or mechanical technology and 1 in photography (6/76).

Recommendation, Skill Level 20

In the vocational certificate category, 6 semester hours in explosives or mechanical technology, 2 in blueprint interpretation, and 1 in photography. If the duty assignment was nuclear explosive ordnance specialist, 3 additional semester hours in nuclear technology (6/76).

Recommendation, Skill Level 30

In the vocational certificate category, 6 semester hours in explosives or mechanical technology, 2 in blueprint interpretation, and 1 in photography. If the duty assignment was nuclear explosive ordnance specialist, 3 additional semester hours in nuclear technology. In the lower-division baccalaureate/associate degree category, 2 semester hours in technical writing, 2 in maintenance management, 3 in human relations, and 3 in public speaking (6/76).

Recommendation, Skill Level 40

In the vocational certificate category, 6 semester hours in explosives or mechanical technology, 2 in blueprint interpretation, and 1 in photography. If the duty assignment was nuclear explosive ordnance specialist, 3 additional semester hours in nuclear technology. In the lower-division baccalaureate/associate degree category, 2 semester hours in technical writing, 2 in maintenance management, 3 in human relations, and 3 in public speaking. In the upper-division baccalaureate category, 3 semester hours in radiation physics, 3 in introduction to management, and 3 for field experience in management (6/76).

Recommendation, Skill Level 50

In the vocational certificate category, 6 semester hours in explosives or mechanical technology, 2 in blueprint interpretation, and 1 in photography; if the duty assignment was nuclear explosive ordnance specialist, 3 additional semester hours in nuclear technology. In the lower-division baccalaureate/associate degree category, 3 semester hours in technical writing, 3 in maintenance management, 3 in human relations, 3 in public speaking, and 6 for innovative explosive devices. In the upper-division baccalaureate category, 3 semester hours in radiation physics, 3 in introduction to management, and 6 for field experience in management (6/76).

MOS-55D-002

Explosive Ordnance Disposal Specialist

55D10
55D20
55D30
55D40
55D50

Exhibit Dates: 1/91–5/01.

Career Management Field: 55 (Ammunition).

Description

Summary: Supervises or performs disposal of explosive devices. *Skill Level 10:* Assists in the location, identification, rendering safe, removal, and destruction of explosive ordnance other than nuclear fission or fusion materials; uses detection instruments to determine hazards and delineate exclusion areas; decontaminates explosive devices as required; assists in isolating explosive and/or contaminated areas; removes obstructions surrounding explosive devices; makes photographs and radiographs; operates and performs maintenance on explosive disposal equipment, tools, and vehicles, including radiac equipment, radios, robots, and chemical detection equipment; cooperates with law enforcement agencies and protection services. *Skill Level 20:* Able to perform the duties required for Skill Level 10; performs render-safe and disposal procedures for nuclear devices; destroys or neutralizes unserviceable and irreparable conventional explosive devices; reads and interprets radiographs, diagrams, drawings, and technical literature; identifies explosives by appearance and markings; performs radiation monitoring and evaluates existing and potential hazards associated with nuclear devices. *Skill Level 30:* Able to perform the duties required for Skill Level 20; supervises four or more persons; prepares technical intelligence and incident reports using personal and mainframe computers; develops and/or modifies render-safe procedures for conventional explosive devices; conducts classes; assists in unit administration, supply, security, and records management; assumes command of an explosive disposal team. *Skill Level 40:* Able to perform the duties required for Skill Level 30; supervises five or more persons; performs administrative and supervisory activities for the disposal unit including personnel, supply, maintenance, security, training, and management; consolidates, edits, and reviews disposal reports; maintains liaison with supporting security units; coordinates operations and movements; establishes ordering and operating procedures. *Skill Level 50:* Able to perform the duties required for Skill Level 40; supervises disposal teams over a wide geographical area; oversees and supervises coordination and liaison with supporting units; assumes command of unit in absence of commander.

Recommendation, Skill Level 10

In the lower-division baccalaureate/associate degree category, 3 semester hours in military science, 3 in drawing interpretation, and 3 in instrumentation. (NOTE: This recommendation for skill level 10 is valid for the dates 1/91-9/91 only) (4/91).

Recommendation, Skill Level 20

In the lower-division baccalaureate/associate degree category, 3 semester hours in military science, 3 in drawing interpretation, 3 in instrumentation, 3 in computer applications, and 3 in personnel supervision. (NOTE: This recommendation for skill level 20 is valid for the dates 1/91-2/95 only) (4/91).

Recommendation, Skill Level 30

In the lower-division baccalaureate/associate degree category, 3 semester hours in military science, 3 in drawing interpretation, 3 in instrumentation, 3 in computer applications, 3 in personnel supervision, 3 in maintenance management, and 2 in records and information management (4/91).

Recommendation, Skill Level 40

In the lower-division baccalaureate/associate degree category, 3 semester hours in military science, 3 in drawing interpretation, 3 in instrumentation, 3 in computer applications, 3 in personnel supervision, 3 in maintenance management, 2 in records and information management, and 3 in public speaking. In the upper-division baccalaureate category, 3 semester hours for field experience in management and 3 in organizational management (4/91).

Recommendation, Skill Level 50

In the lower-division baccalaureate/associate degree category, 3 semester hours in military science, 3 in drawing interpretation, 3 in instrumentation, 3 in computer applications, 3 in personnel supervision, 3 in maintenance management, 2 in records and information management, and 3 in public speaking. In the upper-division baccalaureate category, 6 semester hours for field experience in management and 3 in organizational management; if individual has attained paygrade E-9, additional credit as follows: 3 semester hours in management problems, 3 in operations management, and 3 in communication techniques for managers (4/91).

MOS-55D-003

Explosive Ordnance Disposal Specialist

55D10
55D20
55D30
55D40
55D50

Exhibit Dates: 6/01–3/04. (Effective 4/04 this MOS was discontinued, and its duties were incorporated into MOS 89D.)

Description

Summary: Supervises or performs disposal of explosive devices. *Skill Level 10:* Determines the location, identification, rendering safe, removal, and destruction of explosive ordnance other than nuclear fission or fusion materials; uses detection instruments appropriately; destroys or neutralizes unserviceable and irreparable conventional explosive devices; reads and interprets radiographs, diagrams, drawings, and technical literature; identifies explosives by appearance and markings; cooperates with law enforcement agencies and protection services. *Skill Level 20:* Able to perform the duties required for Skill Level 10; provides technical guidance to subordinates; assists with the administrative duties necessary to support unit operations; assists in preparing technical intelligence and incident reports. *Skill Level 30:* Able to perform the duties required for Skill Level 20; supervises four or more persons; prepares technical intelligence and incident reports using personal and mainframe computers; develops and modifies render-safe procedures for conventional explosive devices; performs radiation monitoring and evaluates existing and potential hazards associated with nuclear devices; conducts classes; assists in unit administration, supply, security, and records management; assumes command of an explosive disposal team. *Skill Level 40:* Able to perform the duties required for Skill Level 30; supervises five or more persons; performs administrative and supervisory activities for the disposal unit including personnel, supply, maintenance, security, training, and management; consolidates, edits, and reviews disposal reports; maintains liaison with supporting security units; coordinates operations and movements; establishes ordering and operating procedures. *Skill Level 50:* Able to perform the duties required for Skill Level 40; supervises disposal teams over a wide geographical area; oversees and supervises coordination and liaison with supporting units; assumes command of unit in absence of commander.

Recommendation, Skill Level 10

Credit may be granted on the basis of an individualized assessment of the student (6/01).

Recommendation, Skill Level 20

Credit may be granted on the basis of an individualized assessment of the student (6/01).

Recommendation, Skill Level 30

In the lower-division baccalaureate/associate degree category, 3 semester hours in military science, 3 in drawing interpretation, 3 in instrumentation, 3 in computer applications, and 3 in hazardous materials (6/01).

Recommendation, Skill Level 40

In the lower-division baccalaureate/associate degree category, 3 semester hours in military science, 3 in drawing interpretation, 3 in instrumentation, 3 in computer applications, 3 in hazardous materials, 3 in oral communications, 3 in technical communications, 3 in principles of supervision, and 3 in information management. In the upper-division baccalaureate category, 3 semester hours in hazardous materials management, 3 in human resource management, and 3 for a management practicum (6/01).

Recommendation, Skill Level 50

In the lower-division baccalaureate/associate degree category, 3 semester hours in military science, 3 in drawing interpretation, 3 in instrumentation, 3 in computer applications, 3 in hazardous materials, 3 in oral communications, 3 in technical communications, 3 in principles of supervision, and 3 in information management. In the upper-division baccalaureate category, 3 semester hours in hazardous materials management, 3 in human resource management, and 6 for a management practicum (6/01).

MOS-55G-001

Nuclear Weapons Maintenance Specialist
55G10
55G20
55G30
55G40

Exhibit Dates: 1/90–12/90.

Career Management Field: 55 (Ammunition).

Description

Summary: Supervises or performs maintenance and surveillance of nuclear weapons, nuclear weapons trainers, and associated electrical, mechanical, and nuclear components. *Skill Level 10:* Assembles, tests, and adjusts electrical and mechanical components to ensure proper operation; uses common and specialized hand tools, power tools, and test equipment; performs surveillance and monitoring operations for radioactive matter; applies appropriate safety procedures; packages, unpacks, paints, and stencils weapons, components, and containers. *Skill Level 20:* Able to perform the duties required for Skill Level 10; test-operates weapons components; receives, stores, and inspects (preissue and preload) nuclear components; uses radioactive survey equipment; provides technical guidance to subordinates. *Skill Level 30:* Able to perform the duties required for Skill Level 20; supervises and instructs subordinates in MOS 55G and 35F (Nuclear Weapons Electronics Specialist); calculates work loads, coordinates work activities, and makes assignments; prepares training charts, documents, and materials; adheres strictly to regulations and inspection techniques and procedures in the maintenance, modification, repair, transportation, and maintenance calibration of nuclear weapons and associated trainers, testers, and handling equipment. *Skill Level 40:* Able to perform the duties required for Skill Level 30; supervises, coordinates, and inspects maintenance facilities, activities, and operations; helps plan maintenance activities; serves as a technical advisor in electronics; reviews safety and fire regulations, standard operating procedures, and the utilization of equipment and recommends changes; has office management responsibilities, including records and forms management, correspondence, security standards, and personnel processing.

Recommendation, Skill Level 10

In the vocational certificate category, 1 semester hour in mechanical maintenance, 2 in electrical maintenance, 2 in use and care of hand tools, power tools, and test equipment, 1 in safety practices and procedures, and 1 in shop operations. Advanced standing in an electrical maintenance apprentice training program on the basis of employer or trade association performance examination (6/76).

Recommendation, Skill Level 20

In the vocational certificate category, 1 semester hour in mechanical maintenance, 2 in electrical maintenance, 2 in use and care of hand tools, power tools, and test equipment, 1 in safety practices and procedures, and 1 in shop operations. Advanced standing in an electrical maintenance apprentice training program on the basis of employer or trade association performance examination. In the lower-division baccalaureate/associate degree category, 2 semester hours in toxic material handling and 1 in nuclear technology (6/76).

Recommendation, Skill Level 30

In the vocational certificate category, 3 semester hours in safety practices and procedures, 2 in electrical maintenance, 2 in use and care of hand tools, power tools, and test equipment, 2 in shop operations, and 1 in mechanical maintenance. Advanced standing in an electrical maintenance apprentice training program on the basis of employer or trade association performance examination. In the lower-division baccalaureate/associate degree category, 3 semester hours for field experience in maintenance management, 3 in human relations, 2 in toxic material handling, and 1 in nuclear technology, and additional credit in nuclear technology on the basis of institutional evaluation (6/76).

Recommendation, Skill Level 40

In the vocational certificate category, 3 semester hours in safety practices and procedures, 2 in electrical maintenance, 2 in use and care of hand tools, power tools, and test equipment, 2 in shop operations, and 1 in mechanical maintenance. Advanced standing in an electrical maintenance apprentice training program on the basis of employer of trade association performance examination. In the lower-division baccalaureate/associate degree category, 3 semester hours for field experience in maintenance management, 3 in human relations, 3 in office management, 3 in personnel supervision, 2 in toxic material handling, and 1 in nuclear technology, and additional credit in nuclear technology on the basis of institutional evaluation. In the upper-division baccalaureate category, 3 semester hours in introduction to management and 3 for field experience in management (6/76).

MOS-55G-002

Nuclear Weapons Specialist
55G10
55G20
55G30
55G40

Exhibit Dates: 1/91–12/94. Effective 12/94, MOS 55G was discontinued.

Career Management Field: 55 (Ammunition).

Description

Summary: Supervises or performs maintenance and surveillance of nuclear weapons, nuclear weapons trainers, and associated electronic test equipment. *Skill Level 10:* Assembles, tests, and adjusts electrical and mechanical components to ensure proper operation; uses common and specialized hand tools, power tools, and test equipment; performs surveillance and monitoring operations for radioactive matter; applies appropriate safety procedures; packages, unpacks, paints, and stencils weapons, components, and containers. *Skill Level 20:* Able to perform the duties required for Skill Level 10; test-operates weapons components; receives, stores, and inspects (preissue and preload) nuclear components; uses radioactive survey equipment; provides technical guidance to subordinates. *Skill Level 30:* Able to perform the duties required for Skill Level 20; supervises and instructs subordinates; calculates work loads, coordinates work activities, and makes assignments; prepares training charts, documents, and materials; adheres strictly to regulations and appropriate inspection techniques and procedures in the maintenance, modification, repair, transportation, and maintenance calibration of nuclear weapons and associated trainers, testers, and handling equipment. *Skill Level 40:* Able to perform the duties required for Skill Level 30; supervises, coordinates, and inspects maintenance facilities, activities, and operations; helps plan maintenance activities; serves as a technical advisor in electronics; reviews safety and fire regulations, standard operating procedures, and the utilization of equipment and recommends changes; has office management responsibilities, including records and forms management, correspondence, security standards, and personnel processing.

Recommendation, Skill Level 10

In the lower-division baccalaureate/associate degree category, 3 semester hours in military science, 3 in basic electronics, 3 in mechanical maintenance, and 3 in industrial safety. (NOTE: This recommendation for skill level 10 is valid for the dates 1/91-9/91 only) (4/91).

Recommendation, Skill Level 20

In the lower-division baccalaureate/associate degree category, 3 semester hours in military science, 3 in basic electronics, 3 in mechanical maintenance, 3 in industrial safety, 3 in inventory control, and 3 in personnel supervision (4/91).

Recommendation, Skill Level 30

In the lower-division baccalaureate/associate degree category, 3 semester hours in military science, 3 in basic electronics, 3 in mechanical maintenance, 3 in industrial safety, 3 in inventory control, 3 in personnel supervision, 3 in maintenance management, 2 in records and information management, and 3 in drawing interpretation (4/91).

Recommendation, Skill Level 40

In the lower-division baccalaureate/associate degree category, 3 semester hours in military science, 3 in basic electronics, 3 in mechanical maintenance, 3 in industrial safety, 3 in inventory control, 3 in personnel supervision, 3 in maintenance management, 2 in records and information management, and 3 in drawing interpretation. In the upper-division baccalaureate category, 3 semester hours for field experience in management and 3 in organizational management (4/91).

MOS-55R-001

Ammunition Stock Control and Accounting Specialist
55R10
55R20
55R30

Exhibit Dates: 1/90–6/93. Effective 6/93, MOS 55R was discontinued, and its duties were incorporated into MOS 55B, Ammunition Specialist.

Career Management Field: 55 (Ammunition).

Description

Summary: Supervises, performs, or assists in ammunition stock control and accounting duties. Duties are related to receipt, storage, issue, and distribution of munitions, guided missiles and large rockets, munition components, packaging materials, and explosives. *Skill Level 10:* Performs ammunition supply stock control and accounting duties using both automated and manual procedures; prepares and posts accountable stock records; prepares and processes issue, receipt, adjustment, and rewarehousing documents, as well as shipping documents and transaction and stock status reports. *Skill Level 20:* Able to perform the duties required for Skill Level 10; provides technical guidance to subordinates; prepares periodic reports by compiling statistics on issue, receipt, adjustment, and storage functions. *Skill Level 30:* Able to perform the duties required for Skill Level 20; supervises manual and automated stock control operations; assigns duties and instructs subordinates in proper work techniques and procedures; plans and organizes work schedules and performs as a technical advisor.

Recommendation, Skill Level 10

In the vocational certificate category, 3 semester hours in supply procedures. (NOTE: This recommendation for skill level 10 is valid for the dates 1/90-9/91 only) (4/86).

Recommendation, Skill Level 20

In the vocational certificate category, 3 semester hours in supply procedures. In the lower-division baccalaureate/associate degree category, 3 semester hours in introduction to inventory control (4/86).

Recommendation, Skill Level 30

In the vocational certificate category, 3 semester hours in supply procedures. In the lower-division baccalaureate/associate degree category, 3 semester hours in introduction to inventory control, 2 in records management, and 2 in personnel supervision (4/86).

MOS-55X-001

Ammunition Inspector
55X30
55X40

Exhibit Dates: 1/90–12/90.

Career Management Field: 55 (Ammunition).

Description

Summary: Conducts inspections and tests to determine the serviceability of conventional and non-nuclear special ammunition, components, and related package materials. *Skill Level 30:* Inspects ammunition to determine deterioration; evaluates safety procedures; inspects incoming and outgoing ammunition shipments; inspects magazines, storage sites, and surrounding areas for storage safety; inspects ammunition destruction sites for safety condition; selects ammunition samples for tests, evaluations, or investigations; teaches surveillance and safety techniques; performs function tests of selected ammunition items; maintains files. NOTE: May have progressed to 55X30 from 55B30 (Ammunition Specialist), 55D30 (Explosive Ordnance Disposal Specialist), or 55G30 (Nuclear Weapons Maintenance Specialist). *Skill Level 40:* Able to perform the duties required for Skill Level 30; serves as the principal advisor on ammunition surveillance and safety matters and provides technical guidance to subordinates; reviews safety requirements; coordinates surveillance activities with civilian authorities and inspectors.

Recommendation, Skill Level 30

In the vocational certificate category, 6 semester hours in explosives, 3 in supply procedures, 3 in quality control, and 3 in safety inspection procedures. In the lower-division baccalaureate/associate degree category, 3 semester hours in human relations and 1 in office procedures (6/76).

Recommendation, Skill Level 40

In the vocational certificate category, 6 semester hours in explosives, 3 in supply procedures, 3 in quality control, and 3 in safety inspection procedures. In the lower-division baccalaureate/associate degree category, 3 semester hours in human relations, 3 in personnel supervision, 3 for field experience in management, and 3 in office procedures (6/76).

MOS-55X-002

Ammunition Inspector
55X30
55X40

Exhibit Dates: 1/91–6/93. Effective 6/93, MOS 55X was discontinued, and its duties were incorporated into MOS 55B, Ammunition Specialist.

Career Management Field: 55 (Ammunition).

Description

Summary: Conducts inspections and tests to determine the serviceability of conventional and non-nuclear special ammunition, components, and related package materials. *Skill Level 30:* Supervises the inspection of ammunition to determine level of deterioration; evaluates safety procedures; inspects incoming and outgoing ammunition shipments, and vehicle suitability; inspects magazines, storage sites, and surrounding areas for storage safety; inspects ammunition destruction sites for safety condition; selects ammunition samples for tests of selected ammunition items; maintains files. NOTE: Must be able to perform the duties of skill level 30 in the following MOS's: 55B (Ammunition Specialist), 55D (Explosive Ordnance Disposal Specialist), 55G (Nuclear Weapons Specialist), or 55R (Ammunition Stock Control and Accounting Specialist); has four years of experience in one of the above MOS's. *Skill Level 40:* Able to perform the duties required for Skill Level 30; serves as the principal advisor on ammunition surveillance and safety matters and provides technical guidance to subordinates; reviews safety requirements; coordinates surveillance activities with civilian authorities and inspectors.

Recommendation, Skill Level 30

In the lower-division baccalaureate/associate degree category, 3 semester hours in military science, 3 in industrial safety, and 3 in personnel supervision. In the upper-division baccalaureate category, 3 semester hours in logistics management and 3 in quality assurance (4/91).

Recommendation, Skill Level 40

In the lower-division baccalaureate/associate degree category, 3 semester hours in military science, 3 in industrial safety, and 3 in personnel supervision. In the upper-division baccalaureate category, 3 semester hours in logistics management, 3 in quality assurance, 3 in organizational management, and 3 for field experience in management (4/91).

MOS-55Z-001

Ammunition Supervisor
55Z50

Exhibit Dates: 1/90–12/90.

Career Management Field: 55 (Ammunition).

Description

Serves as the principal supervisor and advisor for the receipt, issue, classification, storage, surveillance, maintenance, disposition, and decontamination of conventional ammunition and nuclear and non-nuclear special ammunition; advises on packaging and rigging loads for movement by all types of transportation; assists in the selection and layout of ammunition storage facilities; recommends employment of and requirements for technical assistance teams, labor, equipment, and supplies; assists in the preparation of plans, policies, and procedures; prepares and reviews evaluations and reports on operations and training activities; analyzes ammunition stock with automatic data processing data; advises commander on enlisted personnel matters. NOTE: May have progressed to 55Z50 from 55B40 (Ammunition Specialist), 55G40 (Nuclear Weapons Maintenance Specialist), or 55X40 (Ammunition Inspector).

Recommendation

In the upper-division baccalaureate category, 3 semester hours in introduction to management, 3 in personnel management, 3 in business administration, 3 in human relations, and 6 for field experience in management (6/76).

MOS-55Z-002

AMMUNITION SUPERVISOR

55Z50

Exhibit Dates: 1/91–6/96. Effective 6/96, MOS 55Z was discontinued and its duties were incorporated into MOS 55B.

Career Management Field: 55 (Ammunition).

Description

Serves as the principal supervisor and advisor for the receipt, issue, classification, storage, surveillance, maintenance, disposition, and decontamination of conventional ammunition and nuclear and non-nuclear special ammunition; advises on packaging and rigging loads for movement by all types of transportation; assists in the selection and layout of ammunition storage facilities; recommends employment of and requirements for technical assistance teams, labor, equipment, and supplies; assists in the preparation of plans, policies, and procedures; prepares and reviews evaluations and reports on operations and training activities; analyzes ammunition stock with automatic data processing data; advises commander on enlisted personnel matters. NOTE: May have progressed to 55Z50 from 55B40 (Ammunition Specialist), 55G40 (Nuclear Weapons Specialist), or 55X40 (Ammunition Inspector).

Recommendation

In the lower-division baccalaureate/associate degree category, 3 semester hours in inventory control and 3 in maintenance management. In the upper-division baccalaureate category, 3 semester hours in organizational management, 3 in public administration, 3 for field experience in management, 3 in personnel management, and 3 in technical writing; if individual has attained pay grade E-9, additional credit as follows: 3 semester hours in management problems, 3 in operations management, and 3 in communication techniques for managers (4/91).

MOS-56M-002

CHAPLAIN ASSISTANT

56M10
56M20
56M30
56M40
56M50

Exhibit Dates: 10/01–Present.

Career Management Field: 71 (Administration).

Description

Summary: Acts as administrative assistant to a military chaplain; performs or supervises office activities such as preparing correspondence and maintaining records; assists the chaplain in preparing for chapel and religious programs; comparable to a civilian hospital chaplain's assistant or administrative assistant in a church or synagogue. *Skill Level 10:* Prepares facility for services; prepares schedules and religious materials; operates and maintains audiovisual equipment; acts as receptionist; answers routine inquiries; requisitions, receives, and maintains equipment and supplies; types letters, messages, forms, and records; maintains files. *Skill Level 20:* Able to perform the duties required for Skill Level 10; provides technical guidance to subordinates; assists in planning and programming religious services and education. *Skill Level 30:* Able to perform the duties required for Skill Level 20; supervises five to eight persons; participates in preparation of budget; maintains fund records, including disbursements, receipts, and petty cash. *Skill Level 40:* Able to perform the duties required for Skill Level 30; supervises subordinates; coordinates volunteer, part-time, and other personnel; reviews prepared correspondence and reports; participates in planning and programming religious activities. *Skill Level 50:* Able to perform the duties required for Skill Level 40; performs supervisory and management duties in large installation.

Recommendation, Skill Level 10

Credit may be granted on the basis of an individualized assessment of the student (6/01).

Recommendation, Skill Level 20

Credit may be granted on the basis of an individualized assessment of the student (6/01).

Recommendation, Skill Level 30

In the lower-division baccalaureate/associate degree category, 3 semester hours in business communication, 3 in office procedures, 3 in human relations, 3 in principles of supervision, and 3 in budget administration. In the upper-division baccalaureate category, 3 semester hours in crisis intervention (6/01).

Recommendation, Skill Level 40

In the lower-division baccalaureate/associate degree category, 3 semester hours in business communication, 3 in interpersonal communication, 3 in office procedures, 3 in human relations, 3 in principles of supervision, and 3 in budget administration. In the upper-division baccalaureate category, 3 semester hours in crisis intervention (6/01).

Recommendation, Skill Level 50

In the lower-division baccalaureate/associate degree category, 3 semester hours in business communication, 3 in interpersonal communication, 3 in office procedures, 3 in human relations, 3 in principles of supervision, and 3 in budget administration. In the upper-division baccalaureate category, 3 semester hours in crisis intervention and 3 in applied psychology (6/01).

MOS-57E-002

LAUNDRY AND BATH SPECIALIST

57E10
57E20
57E30
57E40
57E50

Exhibit Dates: 1/90–1/92.

Career Management Field: 76 (Supply and Service), subfield 762 (Service).

Description

Summary: Performs or supervises laundry, bath, decontamination, and reimpregnation functions. *Skill Level 10:* Operates laundry, bath, fumigation, and delousing equipment; installs, inspects, and operates generators, compressors, water pumps and water heaters; cleans, lubricates, and performs preventive maintenance on laundry, bath, and related equipment. *Skill Level 20:* Able to perform the duties required for Skill Level 10; determines laundry sites, suitable water supplies, drainage, parking sites, and concealment; prepares facility layout; supervises marking, classifying, and washing operations; directs use of washing formulas, soaps, detergents, and other materials; establishes work priorities; prepares work schedules. *Skill Level 30:* Able to perform the duties required for Skill Level 20; supervises operations of medium-sized unit; supervises and coordinates with medical authorities; determines personnel requirements and plans and supervises training. *Skill Level 40:* Able to perform the duties required for Skill Level 30; supervises large units; establishes, evaluates, and monitors unit training and preventive maintenance; prepares reports, operation plans, and orders; monitors mobile laundry and renovation operations. *Skill Level 50:* Able to perform the duties required for Skill Level 40; serves as principal noncommissioned officer in field service company; advises superiors on personnel matters; assists in conducting unit training.

Recommendation, Skill Level 10

In the vocational certificate category, 5 semester hours in institutional laundry. (NOTE: This recommendation for skill level 10 is valid for the dates 1/90-9/91 only) (5/78).

Recommendation, Skill Level 20

In the vocational certificate category, 5 semester hours in institutional laundry (5/78).

Recommendation, Skill Level 30

In the vocational certificate category, 5 semester hours in institutional laundry. In the lower-division baccalaureate/associate degree category, 3 semester hours in personnel supervision and 3 in human relations. In the upper-division baccalaureate category, 3 semester hours for field experience in management (5/78).

Recommendation, Skill Level 40

In the vocational certificate category, 5 semester hours in institutional laundry. In the lower-division baccalaureate/associate degree category, 3 semester hours in personnel supervision, 3 in human relations, and 2 in record keeping. In the upper-division baccalaureate category, 3 semester hours in introduction to management and 3 for field experience in management (5/78).

Recommendation, Skill Level 50

In the vocational certificate category, 5 semester hours in institutional laundry. In the lower-division baccalaureate/associate degree category, 3 semester hours in personnel supervision, 3 in human relations, and 2 in record keeping. In the upper-division baccalaureate category, 3 semester hours in introduction to management and 6 for field experience in management (5/78).

MOS-57E-003

LAUNDRY AND BATH SPECIALIST
57E10
57E20
57E30
57E40
57E50

Exhibit Dates: 2/92–9/01. Effective 10/01, MOS 57E was discontinued and its duties were incorporated into MOS 92S, Laundry and Textile Specialist.

Career Management Field: 76 (Supply and Services).

Description

Summary: Performs or supervises laundry, bath, decontamination, and reimpregnation functions. *Skill Level 10:* Operates laundry, bath, fumigation, and delousing equipment; installs, inspects, and operates generators, compressors, water pumps, and water heaters; cleans, lubricates, and performs preventive maintenance on laundry, bath, and related equipment. *Skill Level 20:* Able to perform the duties required for Skill Level 10; determines laundry sites, suitable water supplies, drainage, parking sites, and concealment; prepares facility layout; supervises marking, classifying, and washing operations; directs use of washing formulas, soaps, detergents, and other materials; establishes work priorities; prepares work schedules. *Skill Level 30:* Able to perform the duties required for Skill Level 20; supervises operations of medium-sized unit; supervises and coordinates with medical authorities; determines personnel requirements and plans and supervises training. *Skill Level 40:* Able to perform the duties required for Skill Level 30; supervises large units; establishes, evaluates, and monitors unit training and preventive maintenance; prepares reports, operation plans, and orders; monitors mobile laundry and renovation operations. *Skill Level 50:* Able to perform the duties required for Skill Level 40; serves as principal noncommissioned officer in field service company; advises superiors on personnel matters; assists in conducting unit training. NOTE: May progress to 57E50 from 57E40 or 43M40 (Fabric Repair Specialist).

Recommendation, Skill Level 10

Credit is not recommended because of the limited, specialized nature of the skills, competencies, and knowledge (2/92).

Recommendation, Skill Level 20

In the lower-division baccalaureate/associate degree category, 3 semester hours in personnel supervision (2/92).

Recommendation, Skill Level 30

In the lower-division baccalaureate/associate degree category, 3 semester hours in personnel supervision and 2 in records and information management (2/92).

Recommendation, Skill Level 40

In the lower-division baccalaureate/associate degree category, 3 semester hours in personnel supervision and 2 in records and information management. In the upper-division baccalaureate category, 3 semester hours in organizational management and 3 for field experience in management (2/92).

Recommendation, Skill Level 50

In the lower-division baccalaureate/associate degree category, 3 semester hours in personnel supervision and 2 in records and information management. In the upper-division baccalaureate category, 3 semester hours in organizational management and 6 for field experience in management; if paygrade E-9 was attained, additional credit as follows: 3 semester hours in management problems, 3 in operations management, and 3 in communication techniques for managers (2/92).

MOS-57F-003

GRAVES REGISTRATION SPECIALIST
57F10
57F20
57F30
57F40
57F50

Exhibit Dates: 1/90–1/92.

Career Management Field: 76 (Supply and Service), subfield 762 (Service).

Description

Summary: Recovers, identifies, processes, and inters human remains; performs duties at disaster sites as well as within the community; provides technical data to validate the personal identification process; designs cemetery facilities and directs funeral activities. *Skill Level 10:* Serves as a mortuary assistant in recovery, transportation, identification, and interment of remains; reads or indicates the location of graves on maps, sketches, or overlays; determines recovery locations; records identification characteristics including fingerprints and skeletal and dental charts; operates cameras and assists X-ray operations; assists in burials, excavating of graves, and cemetery maintenance. *Skill Level 20:* Able to perform the duties required for Skill Level 10; supervises search and recovery operations for remains; teaches special handling, marking, and shipping of contagious disease cases and contaminated remains; prepares location sketches and overlays; prepares work schedules and assigns personnel to work positions. *Skill Level 30:* Able to perform the duties required for Skill Level 20; supervises medium-sized graves registration activities; serves as graves registration noncommissioned officer in staff position; assists in postmortem examinations; assists in special criminal investigations. *Skill Level 40:* Able to perform the duties required for Skill Level 30; supervises large graves registration activities; advises subordinates on health requirements; coordinates and advises on military burials; assists in review of mortuary services contracts. *Skill Level 50:* Able to perform the duties required for Skill Level 40; supervises memorial affairs and graves registration activities; provides technical and administrative advice and assistance on graves registration matters including acquisition of land for temporary cemeteries and facility, manpower, and equipment requirements; establishes and maintains official records; serves on special boards and councils; prepares reports on memorial affairs and graves registration activities.

Recommendation, Skill Level 10

In the lower-division baccalaureate/associate degree category, 4 semester hours in mortuary sciences and 1 map reading. (NOTE: This recommendation for skill level 10 is valid for the dates 1/90-9/91 only) (5/78).

Recommendation, Skill Level 20

In the lower-division baccalaureate/associate degree category, 6 semester hours in mortuary sciences, 1 in map reading, 2 in physical anthropology, and 2 in anatomy and physiology. In the upper-division baccalaureate category, 3 semester hours in mortuary sciences (5/78).

Recommendation, Skill Level 30

In the lower-division baccalaureate/associate degree category, 9 semester hours in mortuary sciences, 1 in map reading, 2 in physical anthropology, 3 in human relations, 2 in personnel supervision, and 2 in anatomy and physiology. In the upper-division baccalaureate category, 3 semester hours in mortuary sciences and 3 for field experience in management (5/78).

Recommendation, Skill Level 40

In the lower-division baccalaureate/associate degree category, 12 semester hours in mortuary sciences, 2 in physical anthropology, 3 in human relations, 3 in personnel supervision, 2 in anatomy and physiology, and 1 in map reading. In the upper-division baccalaureate category, 3 semester hours in mortuary sciences, 3 in introduction to management, 3 for field experience in management, and 3 in personnel management (5/78).

Recommendation, Skill Level 50

In the lower-division baccalaureate/associate degree category, 12 semester hours in mortuary sciences, 2 in physical anthropology, 3 in human relations, 3 in personnel supervision, 2 in anatomy and physiology, 1 in map reading, and 2 in technical report writing. In the upper-division baccalaureate category, 3 semester hours in mortuary sciences, 3 in introduction to management, 6 for field experience in management, 3 in management, 3 in personnel management, and 3 in records administration (5/78).

MOS-57F-004

MORTUARY AFFAIRS SPECIALIST
((Graves Registration Specialist))
57F10
57F20
57F30
57F40
57F50

Exhibit Dates: 2/92–12/95. Effective 12/95, MOS 57F was discontinued and its duties were incorporated into MOS 92M, Mortuary Affairs Specialist.

Career Management Field: 76 (Supply and Services).

Description

Summary: Recovers, identifies, processes, and inters human remains at disaster sites as well as within the community; provides technical data to validate the personal identification process; designs cemetery facilities; directs funeral activities. *Skill Level 10:* Serves as a mortuary assistant in recovery, transportation, identification, and interment of remains; reads or indicates the location of graves on maps, sketches, or overlays; determines recovery locations; records identification characteristics including fingerprints and skeletal and dental charts; operates cameras and assists X-ray operations; assists in burials, excavating of graves, and cemetery maintenance. *Skill Level 20:* Able to perform the duties required for Skill Level 10; supervises search and recovery operations for remains; teaches special handling, marking, and shipping of contagious disease cases and contaminated remains; prepares location sketches and overlays; prepares work schedules and assigns personnel to work positions; supervises a squad of approximately six persons. *Skill Level 30:* Able to perform the duties required for Skill Level 20; supervises medium-sized graves registration activities; serves as graves registration noncommissioned officer in staff position; assists in postmortem examinations; assists in special criminal investigations; prepares efficiency reports on subordinates. *Skill Level 40:* Able to perform the duties required for Skill Level 30; advises subordinates on health requirements; coordinates and advises on military burials; assists in review of mortuary services contracts; counsels subordinates. *Skill Level 50:* Able to perform the duties required for Skill Level 40; supervises memorial affairs and graves registration activities; provides technical and administrative advice and assistance on graves registration matters including acquisition of land for temporary cemeteries, and facility, manpower, and equipment requirements; establishes and maintains official records; serves on special boards and councils; prepares reports on memorial affairs and graves registration activities; reviews reports of subordinates.

Recommendation, Skill Level 10

Credit may be granted on the basis of an individualized assessment of the student (3/94).

Recommendation, Skill Level 20

In the lower-division baccalaureate/associate degree category, 3 semester hours in military science, 4 in mortuary sciences, 2 in forensic identification, 2 in anatomy and physiology, and 3 in personnel supervision. (NOTE: This recommendation for skill level 20 is valid for the dates 2/92-2/95 only) (2/92).

Recommendation, Skill Level 30

In the lower-division baccalaureate/associate degree category, 3 semester hours in military science, 4 in mortuary sciences, 2 in forensic identification, 2 in anatomy and physiology, 3 in personnel supervision, and 2 in records and information management (2/92).

Recommendation, Skill Level 40

In the lower-division baccalaureate/associate degree category, 3 semester hours in military science, 4 in mortuary sciences, 2 in forensic identification, 2 in anatomy and physiology, 3 in personnel supervision, and 2 in records and information management. In the upper-division baccalaureate category, 3 semester hours in organizational management and 3 for field experience in management (2/92).

Recommendation, Skill Level 50

In the lower-division baccalaureate/associate degree category, 3 semester hours in military science, 4 in mortuary sciences, 2 in forensic identification, 2 in anatomy and physiology, 3 in personnel supervision, and 2 in records and information management. In the upper-division baccalaureate category, 3 semester hours in organizational management and 6 for field experience in management; if paygrade E-9 has been attained, additional credit as follows: 3 semester hours in management problems, 3 in operations management, and 3 in communication techniques for managers (2/92).

MOS-62B-003

CONSTRUCTION EQUIPMENT REPAIRER

62B10

62B20

62B30

62B40

Exhibit Dates: 1/90–5/91.

Career Management Field: 63 (Mechanical Maintenance), subfield 633 (Machinery Maintenance).

Description

Summary: Supervises or performs maintenance on engineering construction equipment including powered bridging equipment and power generation equipment. *Skill Level 10:* Performs maintenance on gasoline, diesel, and electrically powered construction and associated equipment; performs preventive maintenance including all servicing; isolates malfunctions by using appropriate electric or hydraulic diagnostic equipment and technical publications; inspects, disassembles, repairs, and reassembles carburetors, fuel systems, cooling systems, electrical systems, transmission assemblies, and brake systems; reconditions and assembles clutches, steering clutches, final drives, hydraulic systems, and crawler assemblies; removes and overhauls engines; uses hand tools, test equipment, powered shop tools, and oxyacetylene and electrical welding equipment. *Skill Level 20:* Able to perform the duties required for Skill Level 10; provides technical guidance to subordinates; performs and records results of technical inspections; uses and maintains specialized shop equipment, test sets, and technical publications. *Skill Level 30:* Able to perform the duties required for Skill Level 20; supervises and inspects maintenance of engineering construction equipment; determines extent of repair or overhaul needed for equipment; controls requisitioning, storage, and inventory of tools and supplies; plans and schedules work assignments and priorities; performs complex diagnosis and repair of equipment; supervises record keeping; prepares activity reports. *Skill Level 40:* Able to perform the duties required for Skill Level 30; provides general management and supervision for maintenance on engineering construction equipment, power generation equipment, and associated items; monitors performance of subordinates; inspects work to ensure compliance with standard operating procedures, safety standards, and technical publications; analyzes productivity and work quality; participates in research and development activities.

Recommendation, Skill Level 10

In the vocational certificate category, 1 semester hour in use and care of tools. In the lower-division baccalaureate/associate degree category, 6 semester hours in heavy equipment mechanics and 1 in basic welding (7/79).

Recommendation, Skill Level 20

In the vocational certificate category, 1 semester hour in use and care of tools. In the lower-division baccalaureate/associate degree category, 6 semester hours in heavy equipment mechanics and 1 in basic welding (7/79).

Recommendation, Skill Level 30

In the vocational certificate category, 1 semester hour in use and care of tools. In the lower-division baccalaureate/associate degree category, 9 semester hours in heavy equipment mechanics, 2 in basic welding, 1 in records maintenance, and 2 in personnel supervision (7/79).

Recommendation, Skill Level 40

In the vocational certificate category, 1 semester hour in use and care of tools. In the lower-division baccalaureate/associate degree category, 9 semester hours in heavy equipment mechanics, 2 in basic welding, 1 in records maintenance, and 3 in personnel supervision. In the upper-division baccalaureate category, 3 semester hours for field experience in management and 3 in personnel management (7/79).

MOS-62B-004

CONSTRUCTION EQUIPMENT REPAIRER

62B10

62B20

62B30

62B40

Exhibit Dates: 6/91–Present.

Career Management Field: 63 (Mechanical Maintenance).

Description

Summary: Supervises or performs maintenance on engineering construction equipment including powered bridging equipment and power generation equipment. *Skill Level 10:* Performs maintenance on gasoline, diesel, and electrically powered construction and associated equipment; performs preventive maintenance including all servicing; isolates malfunctions by using appropriate electric or hydraulic diagnostic equipment and technical publications; inspects, disassembles, repairs, and reassembles carburetors, fuel systems, cooling systems, electrical systems, transmission assemblies, and brake systems; reconditions and assembles clutches, steering clutches, final drives, hydraulic systems, and crawler assemblies; removes and overhauls engines; uses hand tools, test equipment, powered shop tools, and oxyacetylene and electrical welding equipment; reads maps including topographic maps; recognizes military map symbols. *Skill Level 20:* Able to perform the duties required for Skill Level 10; provides technical guidance to subordinates; supervises repair and maintenance activities; performs and records results of technical inspections; uses and maintains specialized shop equipment, test sets, and technical publications. *Skill Level 30:* Able to perform the duties required for Skill Level 20; supervises and inspects maintenance of engineering construction equipment; determines extent of repair or overhaul needed for equipment; controls requisitioning, storage, and inventory of tools and supplies; plans and schedules work assignments and priorities; performs complex diagnosis and repair of all equipment; supervises record keeping; prepares activity reports. *Skill Level 40:* Able to perform the duties required for Skill Level 30; provides general management and supervision for maintenance on engineering construction equipment, power generation equipment, and associated items; monitors performance of subordinates; inspects work to ensure compliance with standard operating procedures, safety standards, and technical publications; analyzes productivity and work quality; participates in research and development activities.

Recommendation, Skill Level 10

In the lower-division baccalaureate/associate degree category, 3 semester hours in oxyacetylene and arc welding, 3 in heavy equipment mechanics, 3 in diesel and gasoline engine troubleshooting, and 3 in military science. (NOTE: This recommendation for skill level 10 is valid for the dates 6/91-9/91 only) (7/91).

Recommendation, Skill Level 20

In the lower-division baccalaureate/associate degree category, 3 semester hours in oxyacetylene and arc welding, 3 in heavy equipment mechanics, 3 in diesel and gasoline engine troubleshooting, 3 in military science, 3 in hydraulic and pneumatic systems, 3 in electrical systems, 2 in maintenance management, 3 in personnel supervision, and 3 in diesel and gasoline engine rebuilding. (NOTE: This recommendation for skill level 20 is valid for the dates 6/91-2/95 only) (7/91).

Recommendation, Skill Level 30

In the lower-division baccalaureate/associate degree category, 3 semester hours in oxyacetylene and arc welding, 3 in heavy equipment mechanics, 3 in diesel and gasoline engine troubleshooting, 3 in military science, 3 in hydraulic and pneumatic systems, 3 in electrical systems, 3 in maintenance management, 3 in personnel supervision, and 3 in diesel and gasoline engine rebuilding (7/91).

Recommendation, Skill Level 40

In the lower-division baccalaureate/associate degree category, 3 semester hours in oxyacetylene and arc welding, 3 in heavy equipment mechanics, 3 in diesel and gasoline engine troubleshooting, 3 in military science, 3 in hydraulic and pneumatic systems, 3 in electrical systems, 3 in maintenance management, 3 in personnel supervision, and 3 in diesel and gasoline engine rebuilding. In the upper-division baccalaureate category, 3 semester hours for field experience in management (7/91).

MOS-62E-002

Heavy Construction Equipment Operator
62E10
62E20

Exhibit Dates: 1/90–5/91.

Career Management Field: 51 (General Engineering), subfield 515 (Construction Equipment Operation).

Description

Summary: Operates crawler and wheeled tractors with dozer attachments, scoop loader, motorized grader, and towed or self-propelled scraper. *Skill Level 10:* Clears, grubs, strips, rough-grades, excavates, backfills, levels, slopes, cuts, ditches, and stockpiles with crawler, wheeled tractor with dozer attachment, and scoop loader; under supervision, operates motorized grader and towed or self-propelled scooper; loads equipment for movement and drives tractor-trailer for hauling; performs operator maintenance. *Skill Level 20:* Able to perform the duties required for Skill Level 10; strips, scarifies, cuts, fills, spreads, backfills, backblades, ditches, and slopes with motorized grader; removes overburden; loads and spreads with tractor-scraper; windrows, blends, and spreads asphalt road mix; mixes soil stabilization materials; performs surface and drainage maintenance; performs operator maintenance.

Recommendation, Skill Level 10

In the vocational certificate category, 5 semester hours in heavy equipment operation and 5 in mechanical maintenance (7/79).

Recommendation, Skill Level 20

In the vocational certificate category, 5 semester hours in heavy equipment operation and 5 in mechanical maintenance (7/79).

MOS-62E-003

Heavy Construction Equipment Operator
62E10
62E20

Exhibit Dates: 6/91–9/04. (Effective 9/04 MOS 62E was discontinued, and its duties were incorporated into MOS 21E.)

Career Management Field: 51 (General Engineering).

Description

Summary: Operates crawler and wheeled tractors with dozer attachments, scoop loader, motorized grader, and towed or self-propelled scraper. *Skill Level 10:* Clears, grubs, strips, rough-grades, excavates, backfills, levels, slopes, cuts, ditches and stockpiles with crawler, wheeled tractor with dozer attachment, and scoop loader; under supervision, operates motorized grader and towed or self-propelled scooper; loads equipment for movement and drives tractor-trailer for hauling; performs operator maintenance and assists mechanics; assists in combat engineer missions. *Skill Level 20:* Able to perform the duties required for Skill Level 10; supervises and instructs subordinates; strips, scarifies, cuts, fills, spreads, backfills, backblades, ditches, and slopes with motorized grader; removes overburden; loads and spreads with tractor-scraper; windrows, blends, and spreads asphalt road mix; performs surface and drainage maintenance; performs operator maintenance.

Recommendation, Skill Level 10

In the vocational certificate category, 5 semester hours in heavy construction equipment operation and maintenance. In the lower-division baccalaureate/associate degree category, 3 semester hours in military science. (NOTE: This recommendation for skill level 10 is valid for the dates 6/91-9/91 only) (7/91).

Recommendation, Skill Level 20

In the vocational certificate category, 10 semester hours in heavy construction equipment operation and maintenance. In the lower-division baccalaureate/associate degree category, 3 semester hours in military science and 3 in personnel supervision. (NOTE: This recommendation for skill level 20 is valid for the dates 6/91-2/95 only) (7/91).

MOS-62F-002

Lifting and Loading Equipment Operator
62F10
62F20

Exhibit Dates: 1/90–5/91.

Career Management Field: 51 (General Engineering), subfield 515 (Construction Equipment Operation).

Description

Summary: Operates cranes and crane shovels of stationary, crawler, and truck-mounted types; operates rough terrain forklifts. *Skill Level 10:* Drives wheel-mounted cranes; loads crane shovel on vehicle for transport; maneuvers boom with spreader, hook, sling, or other device to attach, lift, swing, or lower loads in response to signals; directs rigging load and estimates safe lifting capacity; loads and unloads containerized, palletized, and unpalletized items on various types of surfaces; assists in operation of crane shovel with attachments; performs operator maintenance. *Skill Level 20:* Able to perform the duties required for Skill Level 10; performs ditching, sloping, excavating, and stockpiling with dragline; digs, loads, and stockpiles with clamshell; ditches and excavates with backhoe; performs pile setting and driving operations; levels, face cuts, and excavates with shovel front; operates special-purpose heavy cranes at industrial facilities; performs operator maintenance.

Recommendation, Skill Level 10

In the vocational certificate category, 5 semester hours in crane operation and 5 in equipment maintenance (7/79).

Recommendation, Skill Level 20

In the vocational certificate category, 8 semester hours in crane operation and 5 in equipment maintenance (7/79).

MOS-62F-003

Crane Operator
62F10
62F20

Exhibit Dates: 6/91–9/04. (Effective 9/04, MOS 62F was discontinued, and its duties were incorporated into MOS 21F.)

Career Management Field: 51 (General Engineering).

Description

Summary: Operates stationary, crawler, and truck-mounted cranes and crane shovels. *Skill Level 10:* Drives wheel-mounted cranes; loads crane on vehicle for transports; maneuvers boom with spreader, hook, sling, or other devices to attach, lift, swing, or lower loads in response to signals; directs rigging load and estimates safe lifting capacity; loads and unloads containerized, palletized, and unpalletized items on various types of surfaces; performs operator maintenance and assists mechanics; dig, loads, and stockpiles with clamshell; performs pile setting and driving operations; assists in combat engineer missions. *Skill Level 20:* Able to perform the duties required for Skill Level 10; supervises and instructs subordinates; operates special purpose heavy crane.

Recommendation, Skill Level 10

In the vocational certificate category, 10 semester hours in crane operation and maintenance. In the lower-division baccalaureate/associate degree category, 3 semester hours in military science. (NOTE: This recommendation for skill level 10 is valid for the dates 6/91-9/91 only) (7/91).

Recommendation, Skill Level 20

In the vocational certificate category, 15 semester hours in crane operation and maintenance. In the lower-division baccalaureate/associate degree category, 3 semester hours in military science and 3 in personnel supervision. (NOTE: This recommendation for skill level 20 is valid for the dates 6/91-2/95 only) (7/91).

MOS-62G-002

Quarrying Specialist
62G10
62G20
62G30

Exhibit Dates: 1/90–5/91.

Career Management Field: 51 (General Engineering), subfield 515 (Construction Equipment).

Description

Summary: Operates or supervises the operation of electric, pneumatic, and internal combustion machines used in drilling, crushing, grading, and cleaning gravel and rock; detonates explosives to blast rock in quarries and at construction sites. *Skill Level 10:* Under supervision, assembles, erects, adjusts, and operates the crushing, screening, conveying, and washing units; operates rock drills; maintains tools and equipment; assists powder man. *Skill Level 20:* Able to perform the duties required for Skill Level 10; provides technical guidance to subordinates; determines screen sizes and crusher setting to produce aggregate to specifications; identifies the type and quality of aggregate by simple field tests; notes and corrects or assists in correcting machine malfunctions and deficiencies; performs routine maintenance; performing as blaster, is able to transport and use explosives; determines drilling pattern, depth, and spacing of bore holes; operates various rock drills. *Skill Level 30:* Able to perform the duties required for Skill Level 20; supervises all phases of quarry operation.

Recommendation, Skill Level 10

In the vocational certificate category, 5 semester hours in machine operation (8/79).

Recommendation, Skill Level 20

In the vocational certificate category, 5 semester hours in machine operation and 5 in testing and sampling. In the lower-division baccalaureate/associate degree category, 2 semester hours in machine operation, 1 in testing and sampling, and 2 in personnel supervision and training (8/79).

Recommendation, Skill Level 30

In the vocational certificate category, 5 semester hours in machine operation, 5 in testing and sampling, and 6 in plant management. In the lower-division baccalaureate/associate degree category, 2 semester hours in machine operation, 1 in testing and sampling, 3 in personnel supervision and training, 3 in plant management, and 3 in construction management. In the upper-division baccalaureate category, 3 semester hours for field experience in management (8/79).

MOS-62G-003

Quarrying Specialist
62G10
62G20
62G30

Exhibit Dates: 6/91–Present.

Career Management Field: 51 (General Engineering).

Description

Summary: Operates or supervises the operation of electric, pneumatic, and internal combustion machines used in drilling, crushing, grading, and cleaning gravel and rock; detonates explosives to blast rock in quarries and at construction sites. *Skill Level 10:* Under supervision, assembles, erects, adjusts, and operates the crushing, screening, conveying, and washing units; operates rock drills; maintains tools and equipment; assists powder man; assists in performance of combat engineer missions. *Skill Level 20:* Able to perform the duties required for Skill Level 10; supervises and instructs subordinates; determines screen sizes and crusher settings to produce aggregate to specifications; identifies type and quality of aggregate by simple field tests; notes and corrects or assists in correcting machine malfunctions and deficiencies; performs routine maintenance; performing as a blaster, is able to transport and use explosives; determines drilling pattern, depth, and spacing of bore holes; operates various rock drills. *Skill Level 30:* Able to perform the duties required for Skill Level 20; supervises all phases of quarry operations; has knowledge of rock formations and overlay in order to determine safe and efficient blasting patterns, depth of holes, and charge.

Recommendation, Skill Level 10

In the vocational certificate category, 3 semester hours in quarry equipment operations and maintenance. In the lower-division baccalaureate/associate degree category, 3 semester hours in military science. (NOTE: This recommendation for skill level 10 is valid for the dates 6/91-9/91 only) (7/91).

Recommendation, Skill Level 20

In the vocational certificate category, 3 semester hours in quarry equipment operations and maintenance. In the lower-division baccalaureate/associate degree category, 3 semester hours in military science and 3 in personnel supervision (7/91).

Recommendation, Skill Level 30

In the vocational certificate category, 3 semester hours in quarry equipment operations and maintenance. In the lower-division baccalaureate/associate degree category, 3 semester hours in military science, 3 in personnel supervision, 2 in records and information management, and 3 in applied geology (7/91).

MOS-62H-002

Concrete and Asphalt Equipment
Operator
62H10
62H20
62H30

Exhibit Dates: 1/90–1/92.

Career Management Field: 51 (General Engineering), subfield 515 (Construction Equipment Operator).

Description

Summary: Supervises or operates all equipment used in concrete and asphalt production and paving. *Skill Level 10:* Operates concrete mobile mixer truck; assists plant equipment operator in operation of equipment and machines used in asphalt production and paving; loads unmixed materials in concrete mobile mixer; adjusts controls to obtain required mix proportion and slump at the job site; assists in operation of spreading and finishing equipment; assists in erection, operation, and maintenance of asphalt producing plant; operates asphalt distributor and aggregate spreader; assists in operation of asphalt paving and surfacing equipment. *Skill Level 20:* Able to perform the duties required for Skill Level 10; operates all equipment and machines used in asphalt production and paving and concrete paving operations; erects, operates, and maintains asphalt plant; directs preparation and placement of concrete with mobile mixer truck. *Skill Level 30:* Able to perform the duties required for Skill Level 20; supervises all aspects of asphalt and concrete production and paving operations; plans and directs layout and erection of asphalt plant; supervises production of hot plant mix asphalt; directs asphalt placing and rolling operations; supervises concrete mobile mixer truck use and employment; supervises concrete and asphalt equipment maintenance and training programs.

Recommendation, Skill Level 10

In the vocational certificate category, if the individual has experience in the operation of asphalt paving equipment, 5 semester hours in operation of asphalt paving equipment; if the individual has experience in the operation of concrete equipment, 5 semester hours in operation of concrete equipment. (NOTE: This recommendation for skill level 10 is valid for the dates 1/90-9/91 only) (2/80).

Recommendation, Skill Level 20

In the vocational certificate category, 5 semester hours in handling and storage of materials and additional credit as follows: If the individual has experience in the operation of asphalt paving equipment, 5 semester hours in operation of asphalt paving equipment; if the individual has experience in the operation of concrete equipment, 5 semester hours in the operation of concrete equipment. In the lower-division baccalaureate/associate degree category, 2 semester hours in construction material processing and application (2/80).

Recommendation, Skill Level 30

In the vocational certificate category, 5 semester hours in handling and storage of materials and additional credit as follows: If the individual has experience in the operation of asphalt paving equipment, 5 semester hours in operation of asphalt paving equipment; if the individual has experience in the operation of concrete equipment, 5 semester hours in the operation of concrete equipment. In the lower-division baccalaureate/associate degree category, 2 semester hours in construction material processing and application, 3 in construction management, and 2 personnel supervision. In the upper-division baccalaureate category, 3 semester hours for field experience in management (2/80).

MOS-62H-003

Concrete and Asphalt Equipment
Operator
62H10
62H20
62H30

Exhibit Dates: 2/92–12/01.

Career Management Field: 51 (General Engineering).

Description

Summary: Supervises or operates all equipment used in concrete and asphalt production and paving. *Skill Level 10:* Operates concrete mobile mixer truck; troubleshoots, maintains, and performs light repair of mobile mixer truck; assists plant equipment operator in operation of equipment and machines used in asphalt production and paving; estimates coverage requirements of asphalt and concrete projects; loads unmixed materials in concrete mobile mixer; adjusts controls to obtain required mix proportion and slump at the job site; assists in operation of spreading and finishing equipment; performs quality control tests on plastic concrete; assists in erection, operation, and maintenance of asphalt producing plant; operates asphalt distributor and aggregate spreader; assists in operation of asphalt paving and surfacing equipment. *Skill Level 20:* Able to perform the duties required for Skill Level 10; operates all equipment and machines used in asphalt production and paving and concrete paving operations; erects, operates, and maintains asphalt plant; directs preparation and placement of concrete with mobile mixer truck; supervises squad of approximately six persons. *Skill Level 30:* Able to perform the duties required for Skill Level 20; designs concrete and asphalt mixes according to specifications; supervises all aspects of concrete and asphalt production and paving operations; plans and directs layout and erection of asphalt plant; supervises production of hot plant mix asphalt; directs asphalt placing and rolling operations; supervises concrete mobile mixer truck use and employment; supervises concrete and asphalt equipment maintenance and training programs.

Recommendation, Skill Level 10

Credit may be granted on the basis of an individualized assessment of the student (3/94).

Recommendation, Skill Level 20

In the lower-division baccalaureate/associate degree category, 3 semester hours in military science, 3 in maintenance and light repair of paving equipment, 3 in paving techniques and operations, and 3 in personnel supervision. (NOTE: This recommendation for skill level 20 is valid for the dates 2/92-2/95 only) (2/92).

Recommendation, Skill Level 30

In the lower-division baccalaureate/associate degree category, 3 semester hours in military science, 3 in maintenance and light repair of paving equipment, 3 in paving techniques and operations, 3 in personnel supervision, and 2 in records and information management (2/92).

MOS-62H-004

Concrete and Asphalt Equipment Operator
62H10
62H20
62H30

Exhibit Dates: 1/02–9/04. (Effective 9/04, MOS 62H was discontinued, and its duties were incorporated into MOS 21V.)

Career Management Field: 51 (General Engineering).

Description

Summary: Supervises or operates all equipment used in concrete and asphalt production and paving. *Skill Level 10:* Operates concrete mobile mixer truck; troubleshoots, maintains, and performs repair of paving vehicles; assists plant equipment operator in operation of equipment and machines used in asphalt production and paving; estimates coverage requirements of asphalt and concrete projects; loads unmixed materials in concrete mobile mixer; adjusts controls to obtain required mix proportion and slump at the job site; assists in operation of spreading and finishing equipment; performs quality control tests on plastic concrete; assists in erection, operation, and maintenance of asphalt producing plant; operates asphalt distributor and aggregate spreader; assists in operation of asphalt paving and surfacing equipment. *Skill Level 20:* Able to perform the duties required for Skill Level 10; operates all equipment and machines used in asphalt production and paving and concrete paving operations; erects, operates, and maintains asphalt plant; directs preparation and placement of concrete with mobile mixer truck; supervises squad of approximately six persons. *Skill Level 30:* Able to perform the duties required for Skill Level 20; designs concrete and asphalt mixes according to specifications; supervises all aspects of concrete and asphalt production and paving operations; plans and directs layout and erection of asphalt plant; supervises production of hot plant mix asphalt; directs asphalt placing and rolling operations; supervises concrete mobile mixer truck use and employment; supervises concrete and asphalt equipment maintenance and training programs.

Recommendation, Skill Level 10

Credit may be granted on the basis of an individualized assessment of the student (1/02).

Recommendation, Skill Level 20

Credit may be granted on the basis of an individualized assessment of the student (1/02).

Recommendation, Skill Level 30

In the lower-division baccalaureate/associate degree category, 3 semester hours in repair and maintenance of paving equipment, 3 in paving techniques and operation, 3 in principles of supervision, and 2 in records and information management (1/02).

MOS-62J-002

General Construction Equipment Operator
62J10
62J20

Exhibit Dates: 1/90–1/92.

Career Management Field: 51 (Engineering), subfield 515 (Construction Equipment Operation).

Description

Summary: Operates a variety of equipment in support of construction projects including air compressors and special-purpose construction machines engaged in compaction, ditching, pumping, augering, and well-drilling operations. *Skill Level 10:* Operates pneumatic wood borer, rock drill, air blower gun attachment, nail driver, sump pump, paving breaker, chain and circular saw, grinder, concrete vibrator, and backfill temper; operates power roller over foundations, filler stones, cover stone courses, and surface treatments, such as asphalt macadam, road mix, and hot-mix pavement; operates water distributor in compaction and dust control operations; performs operator maintenance; assists in operation of ditching machines. *Skill Level 20:* Able to perform the duties required for Skill Level 10; operates utility tractor with loader and backhoe, general purpose excavators and attachments, earth auger, and ditching machines; performs leveling, backfilling, rough grading, stockpiling, ditch digging, and loading with utility tractor-loader and backhoe; drills vertical and angled holes of appropriate diameter to proper depth with earth auger for demolition support, fence posts, power poles, and anchor posts.

Recommendation, Skill Level 10

In the vocational certificate category, 10 semester hours in machine operation and 10 in equipment maintenance. (NOTE: This recommendation for skill level 10 is valid for the dates 1/90-9/91 only) (7/79).

Recommendation, Skill Level 20

In the vocational certificate category, 15 semester hours in machine operation and 15 in equipment maintenance (7/79).

MOS-62J-003

General Construction Equipment Operator
62J10
62J20

Exhibit Dates: 2/92–9/04. (Effective 9/04, MOS 62J was discontinued, and its duties were incorporated into MOS 21J.)

Career Management Field: 51 (General Engineering).

Description

Summary: Operates a variety of equipment in support of construction projects including air compressors and special-purpose construction machines engaged in compaction, ditching, pumping, augering, and well-drilling operations. *Skill Level 10:* Operates pneumatic wood borer, rock drill, air blower gun attachment, nail driver, sump pump, paving breaker, chain and circular saw, grinder, concrete vibrato, and backfill temper; operates power roller over foundations, filler stones, cover stone courses, and such surface treatments as, asphalt macadam and road mix and hot-mix pavement; operates water distributor in compaction and dust control operations; performs operator maintenance and light repair; assists in operation of ditching machines. *Skill Level 20:* Able to perform the duties required for Skill Level 10; performs line and grade operations using surveyor's transit and level; performs analysis of various rock and soil structures; operates utility tractor with loader and backhoe, general purpose excavators and attachments, earth auger, and ditching machines; performs leveling, backfilling, rough grading, stockpiling, ditch digging, and loading with utility tractor loader and backhoe; drills vertical and angled holes of appropriate diameter to proper depth with earth auger for demolition support, fence posts, power poles, and anchor posts; supervises squad of approximately six persons.

Recommendation, Skill Level 10

Credit may be granted on the basis of an individualized assessment of the student (2/92).

Recommendation, Skill Level 20

In the lower-division baccalaureate/associate degree category, 3 semester hours in military science, 3 in maintenance and light repair of construction equipment, 3 in heavy equipment operation, 3 in personnel supervision, 2 in surveying operations, and 2 in basic soil mechanics. (NOTE: This recommendation for skill level 20 is valid for the dates 2/92-2/95 only) (2/92).

MOS-62N-002

Construction Equipment Supervisor
62N30
62N40

Exhibit Dates: 1/90–1/92.

Career Management Field: 51 (General Engineering) subfield 515 (Construction Equipment Operation).

Description

Summary: Supervises construction equipment and quarry, paving, and plant equipment operations. *Skill Level 30:* Provides general supervision to personnel operating loading, lifting, heavy, and general construction equipment; supervises and coordinates activities of above equipment including drilling operations; determines most efficient type of equipment to be used in construction, maintenance, and repair projects; estimates equipment and operator requirements for specific jobs; conducts on-the-job training and assists in supervision of unit engaged in construction equipment operations; maintains records; assists job superintendent at civilian construction site. NOTE: May have progressed to 62N30 from 62J20 (Heavy Construction Equipment Operator), or 62F20 (Lifting and Loading Equipment Operator). *Skill Level 40:* Able to perform the duties required for Skill Level 30; provides general supervision over all phases of construction operations; coordinates the activities of personnel and equipment; prepares operational reports; keeps production records; equivalent to job superintendent on civilian construction site.

Recommendation, Skill Level 30

In the lower-division baccalaureate/associate degree category, 3 semester hours in personnel supervision, 3 in construction management, and 3 in record keeping. In the upper-division baccalaureate category, 3 semester hours for field experience in management and 3 in maintenance management (8/79).

Recommendation, Skill Level 40

In the lower-division baccalaureate/associate degree category, 3 semester hours in personnel supervision, 3 in construction management, 3 in record keeping, and 3 for field experience in management. In the upper-division baccalaureate category, 6 semester hours for field experience in management, 3 in personnel management, 3 in maintenance management, and 3 in report writing (8/79).

MOS-62N-003

Construction Equipment Supervisor

62N30

62N40

Exhibit Dates: 2/92–9/04. (Effective 9/04 MOS 62N was discontinued, and its duties were incorporated into MOS 21N.)

Career Management Field: 51 (General Engineering).

Description

Summary: Supervises construction equipment and quarry, paving, and plant equipment operations. *Skill Level 30:* Provides general supervision to personnel operating loading, lifting, heavy, and general construction equipment; reads and interprets drawings, plans, and specifications; supervises and coordinates activities of above equipment including drilling operations; determines most efficient type of equipment to be used in construction, maintenance, and repair projects; estimates equipment and operator requirements for specific jobs; conducts on-the-job training and assists in supervision of unit engaged in construction equipment operations; maintains records; assists job superintendent at civilian construction site. NOTE: May have progressed to 62N30 from 62E20 (Heavy Construction Equipment Operator), 62F20 (Crane Operator), or 62J20 (General Construction Equipment Operator). *Skill Level 40:* Able to perform the duties required for Skill Level 30; performs line and grade operations using surveyor's transit and level; performs various soil tests; estimates and schedules operations; provides general supervision over all phases of construction operations; coordinates the activities of personnel and equipment machinery; prepares operational reports; keeps production records; equivalent to job superintendent on civilian construction site. NOTE: May have progressed to 62N40 from 62N30, 62G30 (Quarrying Specialist), or 62H30 (Concrete and Asphalt Equipment Operator).

Recommendation, Skill Level 30

In the lower-division baccalaureate/associate degree category, 2 semester hours in blueprint reading and drawing, 2 in records and information management, 3 in military science, and 3 in personnel supervision (2/92).

Recommendation, Skill Level 40

In the lower-division baccalaureate/associate degree category, 2 semester hours in blueprint reading and drawing, 2 in records and information management, 3 in military science, 3 in personnel supervision, 3 in basic soil mechanics, 3 in surveying operations, and 3 in estimating and scheduling. In the upper-division baccalaureate category, 3 semester hours in organizational management and 3 for field experience in management (2/92).

MOS-63A-001

M1 Abrams Tank System Maintainer

63A10

63A20

63A30

63A40

Exhibit Dates: Pending Evaluation. 10/99–Present.

MOS-63B-003

Light Wheel Vehicle Mechanic

63B10

63B20

63B30

63B40

63B50

Exhibit Dates: 1/90–7/92.

Career Management Field: 63 (Mechanical Maintenance), subfield 633 (Machinery Maintenance).

Description

Summary: Diagnoses malfunctions and performs and supervises corrective maintenance on internal combustion engines and power generating units, including accessories, power trains, and chassis components of wheel vehicles; adjusts operating mechanisms such as valve tappets, carburetors, governors, ignition system points, control linkage, clutches, and brakes; performs tune-ups; recommends appropriate power generating equipment to accommodate specific electrical loads; operates wreckers. *Skill Level 10:* Replaces engine components such as fuel pumps, generators, starters, regulators, radiators, universal joints, brake shoes, engine mounts, and lines and fittings; adjusts operating mechanisms including power generating units; prepares maintenance forms and records. *Skill Level 20:* Able to perform the duties required for Skill Level 10; able to diagnose problems and provides technical guidance to subordinates. *Skill Level 30:* Able to perform the duties required for skill level 20 of MOS 63B, 63S (Heavy Wheel Vehicle Mechanic), or 63Y (Track Vehicle Mechanic); plans and organizes work schedules; assigns duties; performs operational and administrative duties; instructs and supervises subordinates in appropriate practices, procedures, and safety. *Skill Level 40:* Able to perform the duties required for Skill Level 30; performs administrative functions and supervises the preparation of records, reports, and training of maintenance personnel; initiates action necessary to requisition repair parts and supplies; responsible for control of technical publications. *Skill Level 50:* Able to perform the duties required for Skill Level 40; evaluates subordinates; determines training requirements and quality control principles; develops operation plans, policies, and procedures.

Recommendation, Skill Level 10

In the lower-division baccalaureate/associate degree category, 9 semester hours in automotive, diesel, or truck mechanics (including engines and related systems, power train systems, and chassis systems). (NOTE: This recommendation for skill level 10 is valid for the dates 1/90-9/91 only) (9/82).

Recommendation, Skill Level 20

In the lower-division baccalaureate/associate degree category, 9 semester hours in automotive, diesel, or truck mechanics (including engines and related systems, power train systems, and chassis systems) and additional credit on the basis of institutional evaluation (9/82).

Recommendation, Skill Level 30

In the lower-division baccalaureate/associate degree category, 9 semester hours in automotive, diesel, or truck mechanics (including engines and related systems, power train systems, and chassis systems), 2 in personnel supervision, and additional credit in automotive, diesel, or truck mechanics on the basis of institutional evaluation (9/82).

Recommendation, Skill Level 40

In the lower-division baccalaureate/associate degree category, 9 semester hours in automotive, diesel, or truck mechanics (including engines and related systems, power train systems, and chassis systems), 3 in personnel supervision, 2 in records administration, and additional credit in automotive, diesel, or truck mechanics on the basis of institutional evaluation. In the upper-division baccalaureate category, 3 semester hours in personnel management and 3 for field experience in management (9/82).

Recommendation, Skill Level 50

In the lower-division baccalaureate/associate degree category, 9 semester hours in automotive, diesel, or truck mechanics (including engines and related systems, power train systems, and chassis systems), 3 in personnel supervision, 3 in records administration, and additional credit in automotive, diesel, or truck mechanics on the basis of institutional evaluation. In the upper-division baccalaureate category, 3 semester hours in personnel management, 3 for field experience in management, and 3 in applied psychology (9/82).

MOS-63B-004

Light Wheel Vehicle Mechanic
63B10
63B20
63B30
63B40
63B50

Exhibit Dates: 8/92–1/03.

Career Management Field: 63 (Mechanical Maintenance).

Description

Summary: Diagnoses malfunctions and performs and supervises corrective maintenance on diesel engines and power generating units, including accessories, power trains, and chassis components of wheel vehicles; adjusts operating mechanisms such as valve tappets, governors, ignition system points, control linkage, clutches, brakes, suspension, and steering; performs tune-up; recommends appropriate power generating equipment to accommodate specific electrical loads; operates wreckers. *Skill Level 10:* Replaces engine components such as fuel pumps, generators, starters, regulators, radiators, universal joints, brake shoes, engine mounts, and lines and fittings; adjusts operating mechanisms including power generating unit; prepares maintenance forms and records. *Skill Level 20:* Able to perform the duties required for Skill Level 10; provides supervision and technical guidance to subordinates; conducts on-the-job training; completes maintenance forms and records; troubleshoots to component level; isolates and identifies causes of malfunctions; interprets complex schematic diagrams. *Skill Level 30:* Able to perform the duties required for Skill Level 20 of MOS 63B, 63S (Heavy Wheel Vehicle Mechanic), or 63Y (Track Vehicle Mechanic); plans and organizes work schedules; assigns duties; performs operational and administrative duties; instructs and supervises subordinates in appropriate practices, procedures, and safety. *Skill Level 40:* Able to perform the duties required for Skill Level 30; performs administrative functions and supervises the preparation of records, reports, and training of maintenance personnel; initiates action necessary to requisition repair parts and supplies; maintains technical publications. *Skill Level 50:* Able to perform the duties required for Skill Level 40; evaluates subordinates; determines training requirements and quality control principles; develops operation plans, policies, and procedures; plans and organizes work area layout.

Recommendation, Skill Level 10

Credit may be granted on the basis of an individualized assessment of the student (3/94).

Recommendation, Skill Level 20

In the lower-division baccalaureate/associate degree category, 3 semester hours in automotive electrical systems, 3 in diesel engine fundamentals, 3 in diesel engine performance, 3 in heavy duty brake systems, 3 in suspension and steering systems, 3 in military science, and 3 in personnel supervision. (NOTE: This recommendation for skill level 20 is valid for the dates 8/92-2/95 only) (8/92).

Recommendation, Skill Level 30

In the lower-division baccalaureate/associate degree category, 3 semester hours in automotive electrical systems, 3 in diesel engine fundamentals, 3 in diesel engine performance, 3 in heavy duty brake systems, 3 in suspension and steering systems, 3 in military science, 3 in personnel supervision, 2 in records and information management, 3 in computer applications, and 3 in safety (8/92).

Recommendation, Skill Level 40

In the lower-division baccalaureate/associate degree category, 3 semester hours in automotive electrical systems, 3 in diesel engine fundamentals, 3 in diesel engine performance, 3 in heavy duty brake systems, 3 in suspension and steering systems, 3 in military science, 3 in personnel supervision, 2 in records and information management, 3 in computer applications, and 3 in safety. In the upper-division baccalaureate category, 3 semester hours in organizational management and 3 for field experience in management (8/92).

Recommendation, Skill Level 50

In the lower-division baccalaureate/associate degree category, 3 semester hours in automotive electrical systems, 3 in diesel engine fundamentals, 3 in diesel engine performance, 3 in heavy duty brake systems, 3 in suspension and steering systems, 3 in military science, 3 in personnel supervision, 2 in records and information management, 3 in computer applications, and 3 in safety. In the upper-division baccalaureate category, 3 semester hours in organizational management and 6 for field experience in management; if paygrade E-9 has been attained, additional credit as follows: 3 semester hours in management problems, 3 in operations management, and 3 in communication techniques for managers (8/92).

MOS-63B-005

Light Wheel Vehicle Mechanic
63B10
63B20
63B30
63B40

Exhibit Dates: 2/03–3/04. (Effective 4/04, this MOS was discontinued, and its duties were incorporated into MOS 63X.)

Career Management Field: 63 (Mechanical Maintenance).

Description

Summary: Diagnoses malfunctions and performs and supervises corrective maintenance on diesel engines and power generating units, including accessories, power trains, and chassis components of wheel vehicles; adjusts operating mechanisms such as governors, control linkage, clutches, brakes, suspension, and steering; performs tune-up; recommends appropriate power generating equipment to accommodate specific electrical loads; operates wreckers. *Skill Level 10:* Replaces engine components such as fuel pumps, generators, starters, voltage regulators, relays, radiators, universal joints, brake shoes, engine mounts, and lines and fittings; adjusts operating mechanisms including power generating unit; prepares maintenance forms and records. *Skill Level 20:* Able to perform the duties required for Skill Level 10; provides supervision and technical guidance to subordinates; conducts on-the-job training; completes maintenance forms and records; troubleshoots to component level; isolates and identifies causes of malfunctions; uses electronic scanners; interprets complex schematic diagrams. *Skill Level 30:* Able to perform the duties required for Skill Level 20 of MOS 63B, 63S (Heavy Wheel Vehicle Mechanic), or 63Y (Track Vehicle Mechanic); plans and organizes work schedules; assigns duties; performs operational and administrative duties; instructs and supervises subordinates in appropriate practices, procedures, and safety. *Skill Level 40:* Able to perform the duties required for Skill Level 30; performs administrative functions and supervises the preparation of records, reports, and the training of maintenance personnel; initiates action necessary to requisition repair parts and supplies; maintains technical publications.

Recommendation, Skill Level 10

Credit may be granted on the basis of an individualized assessment of the student (2/03).

Recommendation, Skill Level 20

Credit may be granted on the basis of an individualized assessment of the student (2/03).

Recommendation, Skill Level 30

In the lower-division baccalaureate/associate degree category, if progressed to 63B30 from 63B20, grant the following credit: 3 semester hours in automotive electrical systems, 3 in diesel engine fundamentals, 3 in diesel engine performance, 3 in heavy duty brake systems, 3 in suspension and steering systems, 3 in basic shop skills, 2 in mechanical blueprint interpretation, 3 in technical mathematics, 3 in principles of supervision, and 3 in computer applications. If progressed from 63S20, grant the following credit: 3 semester hours in diesel fuel systems, 3 in electrical systems, 3 in brake systems, 3 in heavy duty drive trains, 3 in diesel engine fundamentals, 3 in basic shop skills, 2 in mechanical blueprint interpretation, 3 in technical mathematics, 3 in principles of supervision, and 3 in computer applications. If progressed from 63Y20, grant the following credit: 3 semester hours in electrical systems, 3 in track vehicle drive trains, 3 in diesel engine performance, 3 in hydraulic systems, 3 in basic shop skills, 2 in mechanical blueprint interpretation, 3 in technical mathematics, 3 in principles of supervision, and 3 in computer applications (2/03).

Recommendation, Skill Level 40

In the lower-division baccalaureate/associate degree category, if progressed to 63B30 from 63B20, grant the following credit: 3 semester hours in automotive electrical systems, 3 in diesel engine fundamentals, 3 in diesel engine performance, 3 in heavy duty brake systems, 3 in suspension and steering systems, 3 in basic shop skills, 2 in mechanical blueprint interpretation, 3 in technical mathematics, 3 in principles of supervision, and 3 in computer applications. If progressed from 63S20, grant the following credit: 3 semester hours in diesel fuel systems, 3 in electrical systems, 3 in brake systems, 3 in heavy duty drive trains, 3 in diesel engine fundamentals, 3 in basic shop skills, 2 in mechanical blueprint interpretation, 3 in technical mathematics, 3 in principles of supervision, and 3 in computer applications. If progressed from 63Y20, grant the following credit: 3 semester hours in electrical systems, 3 in track vehicle drive trains, 3 in diesel engine performance, 3 in hydraulic systems, 3 in basic shop skills, 2 in mechanical blueprint interpretation, 3 in technical mathematics, 3 in principles of supervision, and 3 in computer applications. In the upper-division baccalaureate category, 3 semester hours in organizational management and 3 for a practicum in management (2/03).

MOS-63D-001

Self-Propelled Field Artillery System Mechanic
63D10
63D20
63D30
63D40
63D50

Exhibit Dates: 1/90–7/92.

Career Management Field: 63 (Mechanical Maintenance), subfield 635 (Weapon System Maintenance).

Description

Summary: Performs and supervises maintenance on tracked and wheel vehicles. *Skill Level 10:* Services, lubricates, removes, and replaces components and assemblies such as engine, fuel, cooling, final drive, track, and suspension; performs corrective maintenance on hydraulic, mechanical, and electrical systems and/or components; employs test, measuring, and diagnostic equipment in troubleshooting; operates wrecker. *Skill Level 20:* Able to perform the duties required for Skill Level 10; provides technical guidance to subordinates. *Skill Level 30:* Able to perform the duties required for Skill Level 20 of MOS 63D and 45D (Self-Propelled Field Artillery Turret Mechanic); plans and organizes work schedules; assigns duties; performs operational and administrative duties; instructs and supervises subordinates in appropriate practices, procedures, and safety; utilizes salvage techniques. *Skill Level 40:* Able to perform the duties required for Skill Level 30; performs administrative functions and supervises the preparation of records, reports, and training of maintenance personnel; initiates action necessary to requisition repair parts and supplies; maintains technical publications. *Skill Level 50:* Able to perform the duties required for Skill Level 40; evaluates subordinates; determines training requirements; develops operation plans, policies, and procedures.

Recommendation, Skill Level 10

In the lower-division baccalaureate/associate degree category, 9 semester hours in automotive, diesel, or truck mechanics (including engines and related systems, power train systems, and chassis systems). (NOTE: This recommendation for skill level 10 is valid for the dates 1/90-9/91 only) (9/82).

Recommendation, Skill Level 20

In the lower-division baccalaureate/associate degree category, 9 semester hours in automotive, diesel, or truck mechanics (including engines and related systems, power train systems, and chassis systems) and additional credit on the basis of institutional evaluation (9/82).

Recommendation, Skill Level 30

In the lower-division baccalaureate/associate degree category, 9 semester hours in automotive, diesel, or truck mechanics (including engines and related systems, power train systems, and chassis systems), 2 in personnel supervision, and additional credit in automotive, diesel, or truck mechanics on the basis of institutional evaluation (9/82).

Recommendation, Skill Level 40

In the lower-division baccalaureate/associate degree category, 9 semester hours in automotive, diesel, or truck mechanics (including engines and related systems, power train systems, and chassis systems), 3 in personnel supervision, 2 in records administration, and additional credit in automotive, diesel, or truck mechanics on the basis of institutional evaluation. In the upper-division baccalaureate category, 3 semester hours in personnel management and 3 for field experience in management (9/82).

Recommendation, Skill Level 50

In the lower-division baccalaureate/associate degree category, 9 semester hours in automotive, diesel, or truck mechanics (including engines and related systems, power train systems, and chassis systems), 3 in personnel supervision, 3 in records administration, and additional credit in automotive, diesel, or truck mechanics on the basis of institutional evaluation. In the upper-division baccalaureate category, 3 semester hours in personnel management, 3 for field experience in management, and 3 in applied psychology (9/82).

MOS-63D-002

Self-Propelled Field Artillery System Mechanic
63D10
63D20
63D30
63D40
63D50

Exhibit Dates: 8/92–9/03.

Career Management Field: 63 (Mechanical Maintenance).

Description

Summary: Performs and supervises maintenance on tracked and wheel vehicles. *Skill Level 10:* Services, lubricates, removes, and replaces components and assemblies such as engine, fuel, cooling, final drive, track, and suspension; performs corrective maintenance on hydraulic, mechanical, and electrical systems and/or components; employs test, measuring, and diagnostic equipment in troubleshooting; operates wrecker. *Skill Level 20:* Able to perform the duties required for Skill Level 10; provides supervision and technical guidance to subordinates; conducts on-the-job training; interprets complex schematic diagrams; completes maintenance forms and records. *Skill Level 30:* Able to perform the duties required for Skill Level 20 of MOS 63D and 45D (Self-Propelled Field Artillery Turret Mechanic); plans and organizes work schedules; assigns duties; performs operational and administration duties; instructs and supervises subordinates in appropriate practices, procedures, and safety; utilizes salvage techniques. *Skill Level 40:* Able to perform the duties required for Skill Level 30; performs administrative functions and supervises the preparation of records, reports, and training of maintenance personnel; initiates action necessary to requisition repair parts and supplies; maintains technical publications. *Skill Level 50:* Able to perform the duties required for Skill Level 40; evaluates subordinates; determines training requirements; develops operation plans, policies, and procedures; plans and organizes work area layout.

Recommendation, Skill Level 10

Credit may be granted on the basis of an individualized assessment of the student (3/94).

Recommendation, Skill Level 20

In the lower-division baccalaureate/associate degree category, 3 semester hours in hydraulics, 3 in diesel engine fundamentals, 3 in automotive electrical systems, 3 in track vehicle drive train, 3 in military science, and 3 in personnel supervision. (NOTE: This recommendation for skill level 20 is valid for the dates 8/92-2/95 only) (8/92).

Recommendation, Skill Level 30

In the lower-division baccalaureate/associate degree category, 3 semester hours in hydraulics, 3 in diesel engine fundamentals, 3 in automotive electrical systems, 3 in track vehicle drive train, 3 in military science, 3 in personnel supervision, 2 in records and information management, 3 in computer applications, and 3 in safety (8/92).

Recommendation, Skill Level 40

In the lower-division baccalaureate/associate degree category, 3 semester hours in hydraulics, 3 in diesel engine fundamentals, 3 in automotive electrical systems, 3 in track vehicle drive train, 3 in military science, 3 in personnel supervision, 2 in records and information management, 3 in computer applications, and 3 in safety. In the upper-division baccalaureate category, 3 semester hours in organizational management and 3 for field experience in management (8/92).

Recommendation, Skill Level 50

In the lower-division baccalaureate/associate degree category, 3 semester hours in hydraulics, 3 in diesel engine fundamentals, 3 in automotive electrical systems, 3 in track vehicle drive train, 3 in military science, 3 in personnel supervision, 2 in records and information management, 3 in computer applications, and 3 in safety. In the upper-division baccalaureate category, 3 semester hours in organizational management and 6 for field experience in management; if paygrade E-9 has been attained, additional credit as follows: 3 semester hours in management problems, 3 in operations management, and 3 in communication techniques for managers (8/92).

MOS-63D-003

Artillery Mechanic
63D10
63D20
63D30
63D40

Exhibit Dates: Pending Evaluation. 10/03–Present.

MOS-63E-001

M1 Abrams Tank System Mechanic
((XM1 Tank System Mechanic))
63E10
63E20
63E30
63E40
63E50

Exhibit Dates: 1/90–5/99.

Career Management Field: 63 (Mechanical Maintenance), subfield 635 (Weapon System Maintenance).

Description

Summary: Performs and supervises maintenance and recovery operations on M1 tanks; maintenance includes automotive, turret, fire control, and chemical protection systems. *Skill Level 10:* Troubleshoots and performs organizational maintenance on automotive systems and components of M1 Tank, Recovery Vehicle (M88/A1), M113 family vehicles, and 1/4 ton and 2 1/2 ton tactical trucks; services, lubricates, replaces, removes, installs, repairs and adjusts, purges, and tests following systems, components, and assemblies: the engine fuel exhaust, cooling, electrical, transmission, final drive, track and suspension, steering controls, hull, chemical protection, miscellaneous accessories, and special purpose kits; prepares vehicles for operation under abnormal conditions by sealing, waterproofing, and servicing with special fuel and lubricants; performs corrective organizational maintenance on previously diagnosed malfunctions of mechanical, electrical, and hydraulic systems and/or components; test-operates equipment as required and monitors for evidence of abnormal operations; employs applicable test, measuring, and diagnostic equipment in conjunction with technical publications to troubleshoot and test continuity of electrical circuits; troubleshoots mechanical, electrical, and hydraulic systems; interprets schematic diagrams; operates hand and power tools and equipment; performs or assists in vehicle recovery and evacuation operations. *Skill Level 20:* Able to perform the duties required for Skill Level 10; provides technical guidance to subordinates; diagnoses automotive system faults at the organizational maintenance level; troubleshoots to component level; isolates and identifies causes of malfunctions; operates vehicle for diagnoses and inspection purposes; interprets complex schematic diagrams; conducts on-the-job training; completes applicable maintenance forms and records. *Skill Level 30:* Able to perform the duties required for Skill Level 20; performs as M1 tank system mechanic, fulfilling duties of both automotive (MOS 63E20) and turret (MOS 45E20) mechanics; plans and organizes work schedules and assigns duties; instructs and supervises subordinates in troubleshooting and organizational maintenance practices; supervises maintenance training including preventive maintenance training of equipment operators; performs or supervises on-site maintenance as a member or leader of a maintenance team; performs inspections and tests repaired equipment; evaluates work performance of subordinates; determines training requirements; assists in preparation of maintenance plans, policies, and procedures; supervises or prepares technical studies, evaluations, special reports, correspondence, and records pertaining to maintenance operations and training. *Skill Level 40:* Able to perform the duties required for Skill Level 30; performs as M1 tank maintenance supervisor (master mechanic) for company maintenance personnel; plans and organizes work area layout to gain maximum efficiency and utilization of time, equipment, and personnel; supervises and instructs subordinates in proper maintenance practices, procedures, and techniques; supervises maintenance administration personnel; initiates necessary maintenance management and repair parts supply actions; employs rapid and accurate techniques to determine proper maintenance and repair parts requirements in order to return damaged system to serviceable condition; maintains technical publications. *Skill Level 50:* Able to perform the duties required for Skill Level 40; performs as M1 senior tank maintenance supervisor (senior master mechanic) for battalion maintenance section and personnel; supervises organizational maintenance on track and wheeled vehicles and upkeep of hand and power tools; plans work flow, assigns duties, and instructs in maintenance techniques; supervises processing of work orders; evaluates work performance of subordinates; determines training requirements; assists in preparation of maintenance operation plans, policies, and procedures; supervises or prepares technical studies, evaluations, special reports, correspondence, and records pertaining to maintenance and training operations.

Recommendation, Skill Level 10

In the lower-division baccalaureate/associate degree category, 2 semester hours in automotive fundamentals. (NOTE: This recommendation for skill level 10 is valid for the dates 1/90-9/91 only) (2/84).

Recommendation, Skill Level 20

In the lower-division baccalaureate/associate degree category, 3 semester hours in automotive fundamentals. (NOTE: This recommendation for skill level 20 is valid for the dates 1/90-2/95 only) (2/84).

Recommendation, Skill Level 30

In the lower-division baccalaureate/associate degree category, 3 semester hours in automotive fundamentals and 3 in personnel supervision (2/84).

Recommendation, Skill Level 40

In the lower-division baccalaureate/associate degree category, 3 semester hours in automotive fundamentals, 3 in personnel supervision, and 3 in maintenance management. In the upper-division baccalaureate category, 3 semester hours in personnel management and 3 for field experience in management (2/84).

Recommendation, Skill Level 50

In the lower-division baccalaureate/associate degree category, 3 semester hours in automotive fundamentals, 3 in personnel supervision, and 3 in maintenance management. In the upper-division baccalaureate category, 3 semester hours in personnel management, 6 for field experience in management, and 3 in management problems (2/84).

MOS-63E-002

M1 Abrams Tank System Mechanic
63E10
63E20
63E30
63E40
63E50

Exhibit Dates: 6/99–Present.

Career Management Field: 63 (Mechanical Maintenance).

Description

Summary: Performs and supervises maintenance and recovery operations on M1 tanks, including automotive, turret, fire control, and chemical protection systems. *Skill Level 10:* Troubleshoots and performs organizational maintenance on automotive systems and components of M1 Tank, recovery vehicle (M88/A1), M113 family vehicles, and 1/4 ton and 2 1/2 ton tactical trucks; services, lubricates, replaces, removes, installs, repairs and adjusts, purges, and tests the following systems, components, and assemblies: the engine fuel exhaust, cooling, electrical, transmission, final drive, track and suspension, steering controls, hull, chemical protection, miscellaneous accessories, and special purpose kits; prepares vehicles for operation under abnormal conditions by sealing, waterproofing, and servicing with special fuel and lubricants; performs corrective organizational maintenance on previously diagnosed malfunctions of mechanical, electrical, and hydraulic systems and/or components; test-operates equipment as required and monitors for evidence of abnormal operations; employs applicable test, measuring, and diagnostic equipment in conjunction with technical publications to troubleshoot and test continuity of electrical circuits; troubleshoots mechanical, electrical, and hydraulic systems; interprets schematic diagrams; operates hand and power tools and equipment; performs or assists in vehicle recovery and evacuation operations. *Skill Level 20:* Able to perform the duties required for Skill Level 10; provides technical guidance to subordinates; diagnoses automotive system faults at the organizational maintenance level; troubleshoots to component level; isolates and identifies causes of malfunctions; operates vehicle for diagnoses and inspection purposes; interprets complex schematic diagrams; conducts on-the-job training; completes applicable maintenance forms and records. *Skill Level 30:* Able to perform the duties required for Skill Level 20; performs as M1 tank system mechanic, fulfilling duties of both automotive (MOS 63E20) and turret (MOS 45E20) mechanics; plans and organizes work schedules and assigns duties; instructs and supervises subordinates in troubleshooting and organizational maintenance practices; supervises maintenance training including preventive maintenance training of equipment operators; performs or supervises on-site maintenance as a member or leader of a maintenance team; performs inspections and tests repaired equipment; evaluates work performance of subordinates; determines training requirements; assists in preparation of maintenance plans, policies, and procedures; supervises or prepares technical studies, evaluations, special reports, correspondence, and records pertaining to maintenance operations and training. *Skill Level 40:* Able to perform the duties required for Skill Level 30; performs as M1 tank maintenance supervisor (master mechanic) for company maintenance personnel; plans and organizes work area layout to gain maximum efficiency and utilization of time, equipment, and personnel; supervises and instructs subordinates in proper maintenance practices, procedures, and techniques; supervises maintenance administration personnel; initiates necessary maintenance management and repair parts supply actions; employs rapid and accurate techniques to determine proper maintenance and repair parts requirements in order to return damaged system to serviceable condition; maintains technical publications. *Skill Level 50:* Able to perform the duties required for Skill Level 40; performs as M1 senior tank maintenance supervisor (senior master mechanic) for battalion maintenance section and personnel; supervises organizational maintenance on track and wheeled vehicles and upkeep of hand and power tools; plans work flow, assigns duties, and instructs in maintenance techniques; supervises processing of work orders; evaluates work performance of subordinates; determines training requirements; assists in preparation of maintenance operation plans, policies, and procedures; supervises or prepares technical studies, evaluations, special reports, correspondence, and records pertaining to maintenance and training operations.

Recommendation, Skill Level 10

Credit may be granted on the basis of an individualized assessment of the student (6/99).

Recommendation, Skill Level 20

Credit may be granted on the basis of an individualized assessment of the student (6/99).

Recommendation, Skill Level 30

In the lower-division baccalaureate/associate degree category, 6 semester hours in heavy equipment maintenance and repair and 3 in principles of supervision (6/99).

Recommendation, Skill Level 40

In the lower-division baccalaureate/associate degree category, 6 semester hours in heavy equipment maintenance and repair, 3 in principles of supervision, and 3 in maintenance management. In the upper-division baccalaureate category, 3 semester hours in principles of management and 3 in a management practicum (6/99).

Recommendation, Skill Level 50

In the lower-division baccalaureate/associate degree category, 6 semester hours in heavy equipment maintenance and repair, 3 in principles of supervision, 6 in maintenance management, and 3 in records management. In the upper-division baccalaureate category, 3 semester hours in principles of management, 3 in a management practicum, and 3 in human resource management (6/99).

MOS-63G-002

Fuel and Electrical Systems Repairer

63G10
63G20

Exhibit Dates: 1/90–7/92.

Career Management Field: 63 (Mechanical Maintenance), subfield 633 (Machinery Maintenance).

Description

Summary: Performs maintenance on fuel and electrical systems, brake components, vehicles, engines of power generating equipment, and material-handing equipment. *Skill Level 10:* Tests and repairs fuel and electrical systems and components such as fuel pumps, fuel injection pumps, batteries, generators, distributors, and ignition systems; rebuilds components such as carburetors, fuel injection pumps, generators, distributors, fuel injectors, master cylinders, brake chambers, and wheel cylinders; repairs disk brake calipers; turns brake drums; relines brake shoes. *Skill Level 20:* Able to perform the duties required for Skill Level 10; provides technical guidance and supervision to subordinates; troubleshoots and uses diagnostic equipment to identify malfunctions in fuel and electrical systems; modifies equipment and recommends repair procedures.

Recommendation, Skill Level 10

In the lower-division baccalaureate/associate degree category, 9 semester hours in automotive, diesel, or truck mechanics. (NOTE: This recommendation for skill level 10 is valid for the dates 1/90-9/91 only) (9/82).

Recommendation, Skill Level 20

In the lower-division baccalaureate/associate degree category, 9 semester hours in automotive, diesel, or truck mechanics and additional credit on the basis of institutional evaluation (9/82).

MOS-63G-003

Fuel and Electrical Systems Repair

63G10
63G20

Exhibit Dates: 8/92–Present.

Career Management Field: 63 (Mechanical Maintenance).

Description

Summary: Performs maintenance on diesel fuel and electrical systems, brake components, vehicles, engines of power generating equipment, and material-handling equipment. *Skill Level 10:* Tests and repairs diesel fuel and electrical systems and components such as diesel fuel pumps, diesel fuel injection pumps, batteries, and generators; rebuilds components such as diesel fuel injection pumps, generators, fuel injectors, master cylinders, brake chambers, and wheel cylinders; repairs disk brake calipers; turns brake drums; relines brake shoes. *Skill Level 20:* Able to perform the duties required for Skill Level 10; provides technical guidance and supervision to subordinates; troubleshoots and uses diagnostic equipment to identify malfunctions in diesel fuel and electrical systems; modifies equipment and recommends repair procedures; interprets complex schematic diagrams; conducts on-the-job training; completes maintenance forms and records.

Recommendation, Skill Level 10

Credit may be granted on the basis of an individualized assessment of the student (3/94).

Recommendation, Skill Level 20

In the lower-division baccalaureate/associate degree category, 3 semester hours in automotive electrical systems, 3 in diesel engine fundamentals, 3 in diesel fuel systems, 3 in heavy duty brakes, 3 in military science, and 3 in personnel supervision. (NOTE: This recommendation for skill level 20 is valid for the dates 8/92-2/95 only) (8/92).

MOS-63H-004

Track Vehicle Repairer

63H10
63H20
63H30
63H40

Exhibit Dates: 1/90–7/92.

Career Management Field: 63 (Mechanical Maintenance), subfield 633 (Machinery Maintenance).

Description

Summary: Performs and supervises maintenance on wheel and track vehicles and material-handling equipment. *Skill Level 10:* Diagnoses and analyzes engine, power train, and chassis component malfunctions; tests, repairs, overhauls, adjusts, and replaces assemblies, subassemblies, and components such as engines (gasoline, diesel, and multifuel), clutches, transmissions, differentials, steering assemblies, transfer case and hydraulic cylinders, and components; replaces valves, shafts, gears, bearings, rings and seals, using jacks, pullers, gauges and jigs; operates wreckers. *Skill Level 20:* Able to perform the duties required for Skill Level 10; provides technical guidance to subordinates; determines serviceability of components by using measuring devices. *Skill Level 30:* Able to perform the duties required for skill level 20 of MOS 63H, 63G (Fuel and Electrical Systems Repairer), or 63W (Wheel Vehicle Repairer); performs and supervises maintenance and completes technical inspections; repairs or overhauls engine power trains and chassis components; applies information contained in technical publications to maintenance requirements; uses measuring and testing equipment for alignment and adjustment of components and assemblies; plans and organizes work schedules; assigns duties; supervises recovery operations; administers safety programs. NOTE: May have progressed to 63H30 from 63H20, 63G20 (Fuel and Electrical Systems Repairer), or 63W20 (Wheel Vehicle Repairer). *Skill Level 40:* Able to perform the duties required for Skill Level 20 of MOS 63H or 63J (Quartermaster and Chemical Equipment Repairer); supervises maintenance and repair of equipment and vehicles (automotive, chemical, and quartermaster); applies production and quality control procedures. NOTE: May have progressed to 63H40 from 63J30 (Quartermaster and Chemical Equipment Repairer).

Recommendation, Skill Level 10

In the lower-division baccalaureate/associate degree category, 9 semester hours in automotive mechanics (including engines and related systems, power train systems, and chassis systems). (NOTE: This recommendation for skill level 10 is valid for the dates 1/90-9/91 only) (9/82).

Recommendation, Skill Level 20

In the lower-division baccalaureate/associate degree category, 9 semester hours in automotive mechanics (including engines and related systems, power train systems, and chassis systems) and additional credit in automotive mechanics on the basis of institutional evaluation (9/82).

Recommendation, Skill Level 30

In the lower-division baccalaureate/associate degree category, 15 semester hours in automotive mechanics (including engines and related systems, power train systems, and chassis systems), 2 in personnel supervision, and additional credit in automotive mechanics on the basis of institutional evaluation. In the upper-division baccalaureate category, credit in management or administration on the basis of institutional evaluation (9/82).

Recommendation, Skill Level 40

In the lower-division baccalaureate/associate degree category, 15 semester hours in automotive mechanics (including engines and related systems, power train systems, and chassis systems), 3 in personnel supervision, 2 in records administration, and additional credit in automotive mechanics on the basis of institutional evaluation. In the upper-division baccalaureate category, 3 semester hours in personnel management and 3 for field experience in management (9/82).

MOS-63H-005

Track Vehicle Repairer

63H10
63H20
63H30
63H40

Exhibit Dates: 8/92–1/03.

Career Management Field: 63 (Mechanical Maintenance).

Description

Summary: Performs and supervises maintenance on wheel and track vehicles and material-handling equipment. *Skill Level 10:* Diagnoses and analyzes engine, power train, and chassis component malfunctions; tests, repairs, overhauls, adjusts, and replaces assemblies, subassemblies, and components such as diesel engines clutches, transmissions, differentials, steering assemblies, transfer case and hydraulic cylinders, and components; replaces valves, shafts, gears, bearings, rings and seals, using jacks, pullers, gauges, and jigs; operates wreckers. *Skill Level 20:* Able to perform the duties required for Skill Level 10; provides supervision and technical guidance to subordinates; determines serviceability of components using measuring devices; interprets complex schematic diagrams; conducts on-the-job training; completes maintenance forms and records. *Skill Level 30:* Able to perform the duties required for skill level 20 of MOS 63H, 63G (Fuel and Electrical Systems Repairer), or 63W (Wheel Vehicle Repairer); performs and supervises maintenance and completes technical inspections; repairs or overhauls diesel engines, power trains, and chassis components; applies information contained in technical publications to maintenance requirements; uses types of measuring and testing equipment for alignment and adjustment of components and assemblies; plans and organizes work schedules; assigns duties; supervises recovery operations; administers safety programs. NOTE: May have progressed to 63H30 from 63H20, 63G20 (Fuel and Electrical Systems Repairer), or 63W20 (Wheel Vehicle Repairer). *Skill Level 40:* Able to perform the duties required for Skill Level 30 of MOS 63H or 63J (Quartermaster and Chemical Equipment Repairer); supervises maintenance and repair of equipment and vehicles (automotive, chemical, and quartermaster); plans and organizes work area layout; supervises the preparation of records and reports; supervises training; applies production and quality control procedures.

Recommendation, Skill Level 10

Credit may be granted on the basis of an individualized assessment of the student (3/94).

Recommendation, Skill Level 20

In the lower-division baccalaureate/associate degree category, 3 semester hours in hydraulic systems, 3 in diesel fuel systems, 3 in diesel engine overhaul, 3 in automotive electrical systems, 3 in track vehicle drive trains, 3 in military science, and 3 in personnel supervision. (NOTE: This recommendation for skill level 20 is valid for the dates 8/92-2/95 only) (8/92).

Recommendation, Skill Level 30

In the lower-division baccalaureate/associate degree category, 3 semester hours in hydraulic systems, 3 in diesel fuel systems, 3 in diesel engine overhaul, 3 in automotive electrical systems, 3 in track vehicle drive trains, 3 in military science, 3 in personnel supervision, 2 in records and information management, 3 in computer applications, and 3 in safety (8/92).

Recommendation, Skill Level 40

In the lower-division baccalaureate/associate degree category, 3 semester hours in hydraulic systems, 3 in diesel fuel systems, 3 in diesel engine overhaul, 3 in automotive electrical systems, 3 in track vehicle drive trains, 3 in military science, 3 in personnel supervision, 2 in records and information management, 3 in computer applications, and 3 in safety. In the upper-division baccalaureate category, 3 semester hours in organizational management and 3 for field experience in management (8/92).

MOS-63H-006

TRACK VEHICLE REPAIRER
63H10
63H20
63H30
63H40

Exhibit Dates: 2/03–3/04. (Effective 4/04 this MOS was discontinued, and its duties were incorporated into MOS 63X.)

Career Management Field: 63 (Mechanical Maintenance).

Description

Summary: Performs and supervises maintenance on wheel and track vehicles and material-handling equipment. *Skill Level 10:* Develops understanding of preventative maintenance procedures; diagnoses and analyzes engine, power train, and chassis component malfunctions; tests, repairs, overhauls, adjusts, and replaces assemblies, subassemblies, and components such as diesel engines, gas turbine power plants, clutches, transmissions, differentials, steering assemblies, transfer case and hydraulic cylinders, and components; replaces valves, shafts, gears, bearings, rings and seals; develops proficiency in using jacks, pullers, gauges, and jigs; develops skill in the use of recovery vehicles such as wreckers and hoisting vehicles. *Skill Level 20:* Able to perform the duties required for Skill Level 10; provides supervision and technical guidance to subordinates; determines serviceability of components using measuring devices; interprets complex schematic diagrams; conducts on-the-job training; completes maintenance forms and records. *Skill Level 30:* Able to perform the duties required for skill level 20 of MOS 63H, 63G (Fuel and Electrical Systems Repairer), or 63W (Wheel Vehicle Repairer); performs and supervises maintenance and completes technical inspections; repairs or overhauls diesel engines, power trains, and chassis components; applies information contained in technical publications to maintenance requirements; uses types of measuring and testing equipment for alignment and adjustment of components and assemblies; plans and organizes work schedules; assigns duties; supervises recovery operations; administers safety programs; obtains competence in the use of appropriate computer service and instructions software. NOTE: May have progressed to 63H30 from 63H20, 63G20 (Fuel and Electrical Systems Repairer), or 63W20 (Wheel Vehicle Repairer). *Skill Level 40:* Able to perform the duties required for Skill Level 30 of MOS 63H or 63J (Quartermaster and Chemical Equipment Repairer); supervises maintenance and repair of equipment and vehicles (automotive, chemical, and quartermaster); plans and organizes work area layout; supervises the preparation of records and reports; supervises training; applies production and quality control procedures; develops use of computer software programs such as Word, Excel, PowerPoint, and Access.

Recommendation, Skill Level 10

Credit may be granted on the basis of an individualized assessment of the student (2/03).

Recommendation, Skill Level 20

Credit may be granted on the basis of an individualized assessment of the student (2/03).

Recommendation, Skill Level 30

In the lower-division baccalaureate/associate degree category, if progressed from 63H20, grant the following credit: 3 semester hours in diesel fuel systems, 3 in diesel engine overhaul, 3 in automotive electrical systems, 3 in track vehicle drive trains, 3 in fundamentals of fluid power, 3 in basic shop skills, 3 in technical mathematics, 3 in mechanical blueprint interpretation, 3 in principles of supervision, and 3 in computer applications. If progressed from 63G20, grant the following credit: 3 semester hours in automotive electrical systems, 3 in diesel engine fundamentals, 3 in diesel fuel systems, 3 in heavy duty brakes, 3 in basic shop skills, 3 in technical mathematics, 3 in mechanical blueprint interpretation, 3 in principles of supervision, and 3 in computer applications. If progressed from 63W20, grant the following credit: 3 semester hours in introduction to internal combustion engines, 3 in transmission repair, 3 in suspension and steering system repair, 3 in axle assembly repair, 3 in brake system repair, 3 in basic shop skills, 3 in technical mathematics, 3 in mechanical blueprint interpretation, 3 in principles of supervision, and 3 in computer applications (2/03).

Recommendation, Skill Level 40

In the lower-division baccalaureate/associate degree category, if progressed from 63H20, grant the following credit: 3 semester hours in diesel fuel systems, 3 in diesel engine overhaul, 3 in automotive electrical systems, 3 in track vehicle drive trains, 3 in fundamentals of fluid power, 3 in basic shop skills, 3 in technical mathematics, 3 in mechanical blueprint interpretation, 3 in principles of supervision, and 3 in computer applications. If progressed from 63G20, grant the following credit: 3 semester hours in automotive electrical systems, 3 in diesel engine fundamentals, 3 in diesel fuel systems, 3 in heavy duty brakes, 3 in basic shop skills, 3 in technical mathematics, 3 in mechanical blueprint interpretation, 3 in principles of supervision, and 3 in computer applications. If progressed from 63W20, grant the following credit: 3 semester hours in introduction to internal combustion engines, 3 in transmission repair, 3 in suspension and steering system repair, 3 in axle assembly repair, 3 in brake system repair, 3 in basic shop skills, 3 in technical mathematics, 3 in mechanical blueprint interpretation, 3 in principles of supervision, and 3 in computer applications. In the upper-division baccalaureate category, 3 semester hours in organizational management and 3 for a practicum in management (2/03).

MOS-63J-003

QUARTERMASTER AND CHEMICAL EQUIPMENT REPAIRER
63J10
63J20

Exhibit Dates: 1/90–5/99.

Career Management Field: 63 (Mechanical Maintenance).

Description

Summary: Supervises or performs maintenance on chemical, quartermaster, and special-purpose equipment. *Skill Level 10:* Disassembles, repairs, reassembles, adjusts, and tests various types of quartermaster machinery and equipment, including sewing machines, gasoline lanterns, laundry units, bath units, bakery units, and such special-purpose equipment as petroleum, oil, and lubricant equipment; performs maintenance on electrical circuits; performs welding, plumbing, and pipefitting tasks; uses supply and technical literature, regulations, and manufacturers' catalogs in performing maintenance and in keeping records. *Skill Level 20:* Able to perform the duties required for Skill Level 10; provides technical guidance to subordinates; if serves as foreman, supervises from six to ten persons.

Recommendation, Skill Level 10

In the vocational certificate category, 1 semester hour in basic electricity laboratory. In the lower-division baccalaureate/associate degree category, 2 semester hours in small appliance repair, 1 in use and care of hand tools, and 1 in use of testing equipment. (NOTE: This recommendation for skill level 10 is valid for the dates 1/90-9/91 only) (7/87).

Recommendation, Skill Level 20

In the vocational certificate category, 1 semester hour in basic electricity laboratory. In the lower-division baccalaureate/associate degree category, 2 semester hours in small appliance repair, 1 in use and care of hand tools, and 1 in use of testing equipment. (NOTE: This recommendation for skill level 20 is valid for the dates 1/90-2/95 only) (7/87).

MOS-63J-004

QUARTERMASTER AND CHEMICAL EQUIPMENT REPAIRER
63J10
63J20

Exhibit Dates: 6/99–Present.

Career Management Field: 63 (Mechanical Maintenance).

Description

Summary: Supervises or performs maintenance on chemical, quartermaster, special purpose equipment and heaters. *Skill Level 10:* Maintains and repairs various types of quartermaster machinery and equipment, including decontamination systems, protective filter systems, smoke generator systems, laundry asher/extractor systems, dryer/tumbler systems, bath units, bakery units, and such special purpose equipment as electrical/fuel heater systems, pumps, and reverse osmosis water purification equipment. *Skill Level 20:* Able to perform the duties required for Skill Level 10; supervises and provides technical guidance to subordinates.

Recommendation, Skill Level 10

Credit may be granted on the basis of an individualized assessment of the student (6/99).

Recommendation, Skill Level 20

Credit may be granted on the basis of an individualized assessment of the student (6/99).

MOS-63M-001

BRADLEY FIGHTING VEHICLE SYSTEM MAINTAINER
63M10
63M20
63M30
63M40

Exhibit Dates: Pending Evaluation. 10/99–Present.

MOS-63N-001

M60A1/A3 TANK SYSTEM MECHANIC
63N10
63N20
63N30
63N40
63N50

Exhibit Dates: 1/90–7/92.

Career Management Field: 63 (Mechanical Maintenance), subfield 635 (Weapon System Maintenance).

Description

Summary: Performs and supervises maintenance on tracked and wheel vehicles. *Skill Level 10:* Services, lubricates, removes, and replaces components and assemblies such as engine, fuel, cooling, final drive, and track and suspension; performs corrective maintenance on hydraulic, mechanical, and electrical systems and/or components; employs tests, measuring, and diagnostic equipment in troubleshooting; operates wrecker. *Skill Level 20:* Able to perform the duties required for Skill Level 10; provides technical guidance to subordinates. *Skill Level 30:* Able to perform the duties required for skill level 10 of MOS 63N, 63R (M60A2 Tank System Mechanic), and 45N (M60A1/A3 Tank Turret Mechanic); plans and organizes work schedules; assigns duties; performs operational and administrative duties; instructs and supervises subordinates in practices, procedures, and safety; utilizes salvage techniques. *Skill Level 40:* Able to perform the duties required for Skill Level 30; performs related administrative functions and supervises the preparation of records and reports and the training of maintenance personnel; initiates action necessary to requisition repair parts and supplies; maintains technical publications. *Skill Level 50:* Able to perform the duties required for Skill Level 40; evaluates subordinates; determines training requirements; develops operation plans, policies, and procedures.

Recommendation, Skill Level 10

In the lower-division baccalaureate/associate degree category, 9 semester hours in automotive, diesel, or truck mechanics. (NOTE: This recommendation for skill level 10 is valid for the dates 1/90-9/91 only) (9/82).

Recommendation, Skill Level 20

In the lower-division baccalaureate/associate degree category, 9 semester hours in automotive, diesel, or truck mechanics and additional credit on the basis of institutional evaluation (9/82).

Recommendation, Skill Level 30

In the lower-division baccalaureate/associate degree category, 9 semester hours in automotive, diesel, or truck mechanics, 2 in personnel supervision, and additional credit in automotive, diesel, or truck mechanics on the basis of institutional evaluation (9/82).

Recommendation, Skill Level 40

In the lower-division baccalaureate/associate degree category, 9 semester hours in automotive, diesel, or truck mechanics, 3 in personnel supervision, and 2 in records administration, and additional credit in automotive, diesel, or truck mechanics on the basis of institutional evaluation. In the upper-division baccalaureate category, 3 semester hours in personnel management and 3 for field experience in management (9/82).

Recommendation, Skill Level 50

In the lower-division baccalaureate/associate degree category, 9 semester hours in automotive, diesel, or truck mechanics, 3 in personnel supervision, and 3 in records administration, and additional credit in automotive, diesel, or truck mechanics on the basis of institutional evaluation. In the upper-division baccalaureate category, 3 semester hours in personnel management, 3 for field experience in management, and 3 in applied psychology (9/82).

MOS-63N-002

M60A1/A3 TANK SYSTEM MECHANIC
63N10
63N20
63N30
63N40
63N50

Exhibit Dates: 8/92–9/01. Effective 10/01, MOS 63N was discontinued and its duties were incorporated into MOS 63E.

Career Management Field: 63 (Mechanical Maintenance).

Description

Summary: Performs and supervises maintenance on M60A1 and A3 tanks. *Skill Level 10:* Services, lubricates, removes, and replaces components and assemblies such as engine, fuel, cooling, final drive, track, suspension turret assembly, and hull; performs corrective maintenance on hydraulic, mechanical, and electrical systems and/or components; employs measuring and diagnostic equipment in troubleshooting; operates wrecker. *Skill Level 20:* Able to perform the duties required for Skill Level 10; provides supervision and technical guidance to subordinates; conducts on-the-job training; isolates and identifies causes of malfunctions; interprets complex schematic diagrams. *Skill Level 30:* Able to perform the duties required for skill level 20 of MOS 63N and 45N (M60A1/A3 Tank Turret Mechanic); plans and organizes work schedules; assigns duties; performs operational and administrative duties; instructs and supervises subordinates in practices, procedures, and safety; utilizes salvage techniques. *Skill Level 40:* Able to perform the duties required for Skill Level 30; performs administration functions and supervises the preparation of records and reports and the training of maintenance personnel; initiates action necessary to requisition repair parts and supplies; plans and organizes work area layout; maintains technical publications. *Skill Level 50:* Able to perform the duties required for Skill Level 40; evaluates subordinates; determines training requirements; develops operation plans, policies, and procedures; supervises the preparation of records and reports.

Recommendation, Skill Level 10

Credit may be granted on the basis of an individualized assessment of the student (3/94).

Recommendation, Skill Level 20

In the lower-division baccalaureate/associate degree category, 3 semester hours in automotive electrical systems, 3 in hydraulic systems, 3 in track vehicle drive trains, 3 in diesel engine fundamentals, 3 in military science, and 3 in personnel supervision. (NOTE: This recom-

mendation for skill level 20 is valid for the dates 8/92-2/95 only) (8/92).

Recommendation, Skill Level 30

In the lower-division baccalaureate/associate degree category, 3 semester hours in automotive electrical systems, 3 in hydraulic systems, 3 in track vehicle drive trains, 3 in diesel engine fundamentals, 3 in military science, 3 in personnel supervision, 2 in records and information management, 3 in computer applications, and 3 in safety (8/92).

Recommendation, Skill Level 40

In the lower-division baccalaureate/associate degree category, 3 semester hours in automotive electrical systems, 3 in hydraulic systems, 3 in track vehicle drive trains, 3 in diesel engine fundamentals, 3 in military science, 3 in personnel supervision, 2 in records and information management, 3 in computer applications, and 3 in safety. In the upper-division baccalaureate category, 3 semester hours in organizational management and 3 for field experience in management (8/92).

Recommendation, Skill Level 50

In the lower-division baccalaureate/associate degree category, 3 semester hours in automotive electrical systems, 3 in hydraulic systems, 3 in track vehicle drive trains, 3 in diesel engine fundamentals, 3 in military science, 3 in personnel supervision, 2 in records and information management, 3 in computer applications, and 3 in safety. In the upper-division baccalaureate category, 3 semester hours in organizational management and 6 for field experience in management; if paygrade E-9 has been attained, additional credit as follows: 3 semester hours in management problems, 3 in operations management, and 3 in communication techniques for managers (8/92).

MOS-63S-001

Heavy Wheel Vehicle Mechanic
63S10
63S20

Exhibit Dates: 1/90–7/92.

Career Management Field: 63 (Mechanical Maintenance), subfield 633 (Machinery Maintenance).

Description

Summary: Performs maintenance on heavy wheel vehicles and material-handling equipment. *Skill Level 10:* Replaces engine components such as fuel pumps, generators, starters, regulators, radiators, universal joints, brake shoes, engine mounts, and lines and fittings; adjusts operating mechanisms; prepares maintenance forms and records. *Skill Level 20:* Able to perform the duties required for Skill Level 10; provides technical guidance and supervision to subordinates; troubleshoots and uses diagnostic equipment to identify malfunctions.

Recommendation, Skill Level 10

In the lower-division baccalaureate/associate degree category, 9 semester hours in automotive, diesel, or truck mechanics. (NOTE: This recommendation for skill level 10 is valid for the dates 1/90-9/91 only) (9/82).

Recommendation, Skill Level 20

In the lower-division baccalaureate/associate degree category, 9 semester hours in automotive, diesel, or truck mechanics, and additional credit on the basis of institutional evaluation (9/82).

MOS-63S-002

Heavy Wheel Vehicle Mechanic
63S10
63S20

Exhibit Dates: 8/92–Present.

Career Management Field: 63 (Mechanical Maintenance).

Description

Summary: Performs maintenance on heavy wheel vehicles and material-handling equipment. *Skill Level 10:* Replaces engine components such as fuel pumps, generators, starters, regulators, radiators, universal joints, brake shoes, engine mounts, and lines and fittings; adjusts operating mechanisms; prepares maintenance forms and records. *Skill Level 20:* Able to perform the duties required for Skill Level 10; provides technical guidance and supervision to subordinates; troubleshoots and uses diagnostic equipment to identify malfunctions; interprets complex schematic diagrams; completes maintenance forms and records; conducts on-the-job training.

Recommendation, Skill Level 10

Credit may be granted on the basis of individualized assessment of the student (3/94).

Recommendation, Skill Level 20

In the lower-division baccalaureate/associate degree category, 3 semester hours in diesel fuel systems, 3 in electrical systems, 3 in brake systems, 3 in heavy duty drive trains, 3 in diesel engine fundamentals, 3 in military science, and 3 in personnel supervision. (NOTE: This recommendation for skill level 20 is valid for the dates 8/92-2/95 only) (8/92).

MOS-63T-001

ITV/IFV/CFV System Mechanic
63T10
63T20
63T30
63T40
63T50

Exhibit Dates: 1/90–7/92.

Career Management Field: 63 (Mechanical Maintenance), subfield 635 (Weapon System Maintenance).

Description

Summary: Performs and supervises maintenance on tracked and wheel vehicles. *Skill Level 10:* Services, lubricates, removes, and replaces components and assemblies such as engine, fuel, cooling, final drive, track, and suspension; performs corrective maintenance on hydraulic, mechanical, and electrical systems and/or components; employs measuring and diagnostic equipment in troubleshooting; operates wrecker. *Skill Level 20:* Able to perform the duties required for Skill Level 10; provides technical guidance to subordinates. *Skill Level 30:* Able to perform the duties required for skill level 20 of MOS 63T or 45T (ITV/IFV/CFV Turret Mechanic); plans and organizes work schedules; assigns duties; performs operational and administrative duties; instructs and supervises subordinates in practices, procedures, and safety; utilizes salvage techniques. *Skill Level 40:* Able to perform the duties required for Skill Level 30; performs administrative functions and supervises the preparation of records and reports and the training of maintenance personnel; initiates action necessary to requisition repair parts and supplies; maintains technical publications. *Skill Level 50:* Able to perform the duties required for Skill Level 40; evaluates subordinates; determines training requirements; develops operation plans, policies, and procedures.

Recommendation, Skill Level 10

In the lower-division baccalaureate/associate degree category, 9 semester hours in automotive, diesel, or truck mechanics. (NOTE: This recommendation for skill level 10 is valid for the dates 1/90-9/91 only) (9/82).

Recommendation, Skill Level 20

In the lower-division baccalaureate/associate degree category, 9 semester hours in automotive, diesel, or truck mechanics and additional credit in this area on the basis of institutional evaluation (9/82).

Recommendation, Skill Level 30

In the lower-division baccalaureate/associate degree category, 9 semester hours in automotive, diesel, or truck mechanics, 2 in personnel supervision, and additional credit in automotive, diesel, or truck mechanics on the basis of institutional evaluation (9/82).

Recommendation, Skill Level 40

In the lower-division baccalaureate/associate degree category, 9 semester hours in automotive, diesel, or truck mechanics, 3 in personnel supervision, 2 in records administration, and additional credit in automotive, diesel, or truck mechanics on the basis of institutional evaluation. In the upper-division baccalaureate category, 3 semester hours in personnel management and 3 for field experience in management (9/82).

Recommendation, Skill Level 50

In the lower-division baccalaureate/associate degree category, 9 semester hours in automotive, diesel, or truck mechanics, 3 in personnel supervision, 3 in records administration, and additional credit in automotive, diesel, or truck mechanics on the basis of institutional evaluation. In the upper-division baccalaureate category, 3 semester hours in personnel management, 3 for field experience in management, and 3 in applied psychology (9/82).

MOS-63T-002

Bradley Fighting Vehicle System Mechanic
63T10
63T20
63T30
63T40
63T50

Exhibit Dates: 8/92–5/01.

Career Management Field: 63 (Mechanical Maintenance).

Description

Summary: Performs and supervises maintenance on tracked and wheel vehicles. *Skill Level 10:* Services, lubricates, removes, and replaces components and assemblies such as engine, fuel, cooling, engine performance, final drive, track, and suspension; performs corrective maintenance on hydraulic, mechanical, and electrical systems and/or components; employs measuring and diagnostic equipment in troubleshooting; operates wrecker. *Skill Level 20:* Able to perform the duties required for Skill Level 10; provides supervision and technical guidance to subordinates; conducts on-the-job training;

interprets complex schematic diagrams; completes maintenance forms and records. *Skill Level 30:* Able to perform the duties required for skill level 20 of MOS 63T or 45T (Bradley Fighting Vehicles System Turret Mechanic); plans and organizes work schedules; assigns duties; performs operational and administrative duties; instructs and supervises subordinates in practices, procedures, and safety; utilizes salvage techniques. *Skill Level 40:* Able to perform the duties required for Skill Level 30; performs administrative functions and supervises the preparation of records and reports; supervises the training of maintenance personnel; initiates action necessary to requisition repair parts and supplies; maintains technical publications. *Skill Level 50:* Able to perform the duties required for Skill Level 40; evaluates subordinates; determines training requirements; develops operation plans, policies, and procedures; plans work flow and assigns duties; prepares technical studies and special reports.

Recommendation, Skill Level 10

Credit may be granted on the basis of individualized assessment of the student (3/94).

Recommendation, Skill Level 20

In the lower-division baccalaureate/associate degree category, 3 semester hours in engine performance (gasoline and diesel), 3 in track and suspension systems, 3 in wheel and track vehicle drive trains, 3 in automotive electrical systems, 3 in hydraulic systems, 3 in military science, and 3 in personnel supervision. (NOTE: This recommendation for skill level 20 is valid for the dates 8/92-2/95 only) (8/92).

Recommendation, Skill Level 30

In the lower-division baccalaureate/associate degree category, 3 semester hours in engine performance (gasoline and diesel), 3 in track and suspension systems, 3 in wheel and track vehicle drive trains, 3 in automotive electrical systems, 3 in hydraulic systems, 3 in military science, 3 in personnel supervision, 2 in records and information management, 3 in computer applications, and 3 in safety (8/92).

Recommendation, Skill Level 40

In the lower-division baccalaureate/associate degree category, 3 semester hours in engine performance (gasoline and diesel), 3 in track and suspension systems, 3 in wheel and track vehicle drive trains, 3 in automotive electrical systems, 3 in hydraulic systems, 3 in military science, 3 in personnel supervision, 2 in records and information management, 3 in computer applications, and 3 in safety. In the upper-division baccalaureate category, 3 semester hours in organizational management and 3 for field experience in management (8/92).

Recommendation, Skill Level 50

In the lower-division baccalaureate/associate degree category, 3 semester hours in engine performance (gasoline and diesel), 3 in track and suspension systems, 3 in wheel and track vehicle drive trains, 3 in automotive electrical systems, 3 in hydraulic systems, 3 in military science, 3 in personnel supervision, 2 in records and information management, 3 in computer applications, and 3 in safety. In the upper-division baccalaureate category, 3 semester hours in organizational management and 6 for field experience in management; if paygrade E-9 has been attained, additional credit as follows: 3 semester hours in management problems, 3 in operations management, and 3 in communication techniques for managers (8/92).

MOS-63T-003

Bradley Fighting Vehicle System Mechanic
63T10
63T20
63T30
63T40

Exhibit Dates: 6/01–Present.

Career Management Field: 63 (Mechanical Maintenance).

Description

Summary: Performs and supervises maintenance on Bradley Fighting Vehicle. *Skill Level 10:* Services, troubleshoots, adjusts, and replaces components and assemblies of the Bradley Fighting Vehicle such as engine, fuel, cooling, final drive, track, suspension, and brakes; uses schematics, troubleshooting guides, hand tools, power tools, and special test equipment; operates recovery vehicle (wrecker). *Skill Level 20:* Able to perform the duties required for Skill Level 10; provides technical guidance to subordinates; troubleshoots to the system level; develops and conducts on-the-job training; completes maintenance records and reports. *Skill Level 30:* Able to perform the duties required for Skill Level 20; plans and organizes work schedules; assigns duties; performs operational and administrative duties; instructs and supervises subordinates in practices, procedures, and safety; utilizes salvage techniques. *Skill Level 40:* Able to perform the duties required for Skill Level 30; performs administrative and supervisory duties; manages personnel, resources, and time in order to accomplish the unit mission; counsels and trains subordinates; manages financial, personnel, and training records; supervises the development of training programs; initiates actions to requisition repair parts and supplies; maintains technical publications.

Recommendation, Skill Level 10

Credit may be granted on the basis of an individualized assessment of the student (6/01).

Recommendation, Skill Level 20

Credit may be granted on the basis of an individualized assessment of the student (6/01).

Recommendation, Skill Level 30

In the lower-division baccalaureate/associate degree category, 3 semester hours basic engine systems, 3 in suspension systems, 3 in drivetrain systems, 3 in brake systems, 3 in basic electrical systems, 3 in basic hydraulics, and 3 in principles of supervision. In the upper-division baccalaureate category, 3 semester hours for a practicum in management (6/01).

Recommendation, Skill Level 40

In the lower-division baccalaureate/associate degree category, 3 semester hours basic engine systems, 3 in suspension systems, 3 in drivetrain systems, 3 in brake systems, 3 in basic electrical systems, 3 in basic hydraulics, and 3 in principles of supervision. In the upper-division baccalaureate category, 3 semester hours in organization development and 3 for a practicum in management (6/01).

MOS-63W-001

Wheel Vehicle Repairer
63W10
63W20

Exhibit Dates: 1/90–7/92.

Career Management Field: 63 (Mechanical Maintenance), subfield 633 (Machinery Maintenance).

Description

Summary: Performs maintenance on wheel vehicles, material-handling equipment (except the motor on electrical material-handling equipment). *Skill Level 10:* Performs maintenance on engines, clutches, transmissions, differentials, steering assemblies, transfer cases, and hydraulic cylinders; repairs components by replacing valves, shafts, gears, bearings, rings, and seals; operates wrecker. *Skill Level 20:* Able to perform the duties required for Skill Level 10; provides technical guidance and supervision to subordinates; troubleshoots and uses diagnostic equipment to identify malfunctions; modifies equipment and recommends repair procedures.

Recommendation, Skill Level 10

In the lower-division baccalaureate/associate degree category, 9 semester hours in automotive, diesel, or truck mechanics. (NOTE: This recommendation for skill level 10 is valid for the dates 1/90-9/91 only) (9/82).

Recommendation, Skill Level 20

In the lower-division baccalaureate/associate degree category, 9 semester hours in automotive, diesel, or truck mechanics, and additional credit in this area on the basis of institutional evaluation (9/82).

MOS-63W-002

Wheel Vehicle Repairer
63W10
63W20

Exhibit Dates: 8/92–Present.

Career Management Field: 63 (Mechanical Maintenance).

Description

Summary: Performs maintenance on wheel vehicles, material-handling equipment (except the motor on electrical material-handling equipment). *Skill Level 10:* Performs maintenance on engines, clutches, transmissions, differentials, steering assemblies, transfer cases, and hydraulic cylinders; repairs components by replacing valves, shafts, gears, bearings, rings, and seals; operates wrecker. *Skill Level 20:* Able to perform the duties required for Skill Level 10; provides technical guidance and supervision to subordinates; troubleshoots and uses diagnostic equipment to identify malfunctions; modifies equipment and recommends repair procedures.

Recommendation, Skill Level 10

Credit may be granted on the basis of individualized assessment of the student (3/94).

Recommendation, Skill Level 20

In the lower-division baccalaureate/associate degree category, 3 semester hours in introduction to internal combustion engines, 3 in transmission repair, 3 in suspension and steering system repair, 3 in axle assembly repair, 3 in brake system repair, 3 in military science, and 3 in personnel supervision. (NOTE: This recommendation for skill level 20 is valid for the dates 8/92-2/95 only) (8/92).

MOS-63X-001

VEHICLE MAINTENANCE SUPERVISOR
63X10
63X20
63X30
63X40

Exhibit Dates: Pending Evaluation. 4/04–Present.

MOS-63Y-001

TRACK VEHICLE MECHANIC
63Y10
63Y20

Exhibit Dates: 1/90–7/92.

Career Management Field: 63 (Mechanical Maintenance), subfield 633 (Machinery Maintenance).

Description

Summary: Performs maintenance on track vehicles. *Skill Level 10:* Replaces engine components such as fuel pumps, generators, starters, regulators, radiators, universal joints, brake shoes, engine mounts, and lines and fittings; adjusts operating mechanisms; prepares maintenance forms and records. *Skill Level 20:* Able to perform the duties required for Skill Level 10; provides technical guidance and supervision to subordinates; troubleshoots and uses diagnostic equipment to identify malfunctions.

Recommendation, Skill Level 10

In the lower-division baccalaureate/associate degree category, 9 semester hours in automotive, diesel, or truck mechanics. (NOTE: This recommendation for skill level 10 is valid for the dates 1/90-9/91 only) (9/82).

Recommendation, Skill Level 20

In the lower-division baccalaureate/associate degree category, 9 semester hours in automotive, diesel, or truck mechanics, and additional credit in this area on the basis of institutional evaluation (9/82).

MOS-63Y-002

TRACK VEHICLE MECHANIC
63Y10
63Y20

Exhibit Dates: 8/92–Present.

Career Management Field: 63 (Mechanical Maintenance).

Description

Summary: Performs maintenance on track vehicles. *Skill Level 10:* Replaces engine components, including fuel pumps, generators, starters, regulators, radiators, universal joints, brake shoes, engine mounts, and lines and fittings; adjusts operating mechanisms; prepares maintenance forms and records. *Skill Level 20:* Able to perform the duties required for Skill Level 10; provides technical guidance and supervision to subordinates; troubleshoots and uses diagnostic equipment to identify malfunctions; interprets complex schematic diagrams.

Recommendation, Skill Level 10

Credit may be granted on the basis of an individualized assessment of the student (3/94).

Recommendation, Skill Level 20

In the lower-division baccalaureate/associate degree category, 3 semester hours in electrical systems, 3 in track vehicle drive trains, 3 in diesel engine performance, 3 in hydraulic systems, 3 in military science, and 3 in personnel supervision. (NOTE: This recommendation for skill level 20 is valid for the dates 8/92-2/95 only) (8/92).

MOS-63Z-002

MECHANICAL MAINTENANCE SUPERVISOR
63Z50

Exhibit Dates: 1/90–7/92.

Career Management Field: 63 (Mechanical Maintenance), subfield 633 (Machinery Maintenance).

Description

Manages and supervises all phases of maintenance on all types of mechanical equipment; plans layout of maintenance shops and facilities; supervises production and quality control; evaluates work performance and training requirements; assists in preparing maintenance policies and special reports. NOTE: May have progressed to 63Z50 from skill level 40 of any MOS in the mechanical maintenance career management field (63).

Recommendation

In the lower-division baccalaureate/associate degree category, 3 semester hours in personnel supervision, 3 in records administration, and additional credit on the basis of institutional evaluation. In the upper-division baccalaureate category, 3 semester hours in introduction to management, 6 for field experience in management, and additional credit on the basis of institutional evaluation (9/82).

MOS-63Z-003

MECHANICAL MAINTENANCE SUPERVISOR
63Z50

Exhibit Dates: 8/92–1/03.

Career Management Field: 63 (Mechanical Maintenance).

Description

Supervises unit, direct, and general support maintenance on all types of mechanical equipment; plans layout of maintenance shops and facilities; supervises production and quality control; evaluates work performance and training requirements of subordinates; assists in preparing maintenance policies, technical studies, correspondence, records, and planning and special reports. NOTE: If rank is Master Sergeant, may have progressed from skill level 40 of MOS 44E, Machinist; 45K, Armament Repairer; 52X, Special Purpose Equipment Repairer; 62B, Construction Equipment Repairer; or 63H, Track Vehicle Repairer; if rank is Sergeant Major, may have progressed from skill level 50 of MOS 63B, Light Wheel Vehicle Mechanic; 63D, Self-Propelled Field Artillery System Mechanic; 63E, M1 Abrams System Mechanic; 63N, M60A1/A3 Tank System Mechanic; or 63T, Bradley Fighting Vehicle System Mechanic.

Recommendation

In the lower-division baccalaureate/associate degree category, 3 semester hours in personnel supervision, 2 in records and information management, 3 in military science, and 3 in computer applications. In the upper-division baccalaureate category, 3 semester hours in organizational management and 3 for field experience in management; if paygrade E-9 has been attained, additional credit as follows: 3 semester hours in management problems, 3 in operations management, and 3 in communication techniques for managers; if served as First Sergeant, add 3 semester hours in fleet maintenance management (8/92).

MOS-63Z-004

MECHANICAL MAINTENANCE SUPERVISOR
63Z50

Exhibit Dates: 2/03–Present.

Career Management Field: 63 (Mechanical Maintenance).

Description

Supervises unit, direct, and general support maintenance on all types of mechanical equipment; plans layout of maintenance shops and facilities; supervises production and quality control; evaluates work performance and training requirements of subordinates; assists in preparing maintenance policies, technical studies, correspondence, records, planning reports, and special reports. NOTE: May have progressed from skill level 40 of MOS 44E, Machinist; 45K, Armament Repairer; 52X, Special Purpose Equipment Repairer; 62B, Construction Equipment Repairer; 63B, Light Wheel Vehicle Mechanic; 63D, Self-Propelled Field Artillery System Mechanic; 63E, M1 Abrams System Mechanic; 63H, Track Vehicle Repairer; 63N, M60A1/A3 Tank System Mechanic; or 63T, Bradley Fighting Vehicle System Mechanic.

Recommendation

In the lower-division baccalaureate/associate degree category, 3 semester hours in principles of supervision, 3 in maintenance management, 3 in technical report writing, and 3 in speech communications. In the upper-division baccalaureate category, 3 semester hours in organizational management and 3 for a practicum in management; if paygrade E-9 has been attained, additional credit as follows: 3 semester hours in management problems, 3 in operations management, and 3 in business communications; if served as First Sergeant, add 3 semester hours in fleet maintenance management (2/03).

MOS-66G-001

UTILITY AIRPLANE TECHNICAL INSPECTOR
UTILITY/CARGO AIRPLANE TECHNICAL INSPECTOR
66G20
66G30

Exhibit Dates: 1/90–3/90. Effective 3/90, MOS 66G was discontinued, and its duties were incorporated into MOS 67G, Utility Airplane Repairer.

Career Management Field: 67 (Aircraft Maintenance).

Description

Summary: Performs technical inspections on utility or cargo airplanes to determine maintenance requirements and to insure that appropriate maintenance has been performed. *Skill Level 20:* Trains and provides technical guidance to repairers on proper maintenance techniques and safety procedures; checks airplane flight and maintenance records to determine adherence to prescribed maintenance standards; maintains technical library; inspects systems, subsystems, and components according to applicable technical publications and checklists before, during, and after airplane maintenance and modifications; computes basic aircraft weight and balance; checks and troubleshoots systems; participates in maintenance test flights; annotates records of required maintenance; evaluates operational readiness of airplanes and recommends corrective action; inspects crash-damaged airplanes and estimates time, parts, and costs to repair; ensures compliance with aircraft configuration control, Army Oil Analysis Program, and test measuring and diagnostic calibration. NOTE: Progressed to 66G20 from 67G20, Utility Airplane Repairer. *Skill Level 30:* Able to perform the duties required for Skill Level 20; plans, directs, and supervises technical inspections; provides technical guidance to subordinates; evaluates technical training programs; prepares evaluation reports on subordinates; performs maintenance trend analysis and checks maintenance supply documents to ensure compliance with product quality control standards.

Recommendation, Skill Level 20

In the lower-division baccalaureate/associate degree category, 20 semester hours in aircraft maintenance, 3 in technical electives, and 3 in aviation maintenance management (7/87).

Recommendation, Skill Level 30

In the lower-division baccalaureate/associate degree category, 20 semester hours in aircraft maintenance, 3 in technical electives, 3 in aviation maintenance management, 3 in quality control, and 3 in personnel supervision. In the upper-division baccalaureate category, 3 semester hours in principles of management and 3 for field experience in management (7/87).

MOS-66H-001

Observation Airplane Technical Inspector
66H20
66H30

Exhibit Dates: 1/90–9/90. Effective 9/90, MOS 66H was discontinued, and its duties were incorporated into MOS 67H, Observation Airplane Repairer.

Career Management Field: 67 (Aircraft Maintenance).

Description

Summary: Performs technical inspections on observation airplanes to determine maintenance requirements and to ensure that appropriate maintenance has been performed. *Skill Level 20:* Trains and provides technical guidance to repairers on proper maintenance techniques and safety procedures; checks airplane flight and maintenance records to determine adherence to prescribed maintenance standards; maintains technical library; inspects systems, subsystems, and components according to applicable technical publications and checklists before, during, and after airplane maintenance and modifications; computes basic aircraft weight and balance; checks and troubleshoots systems; participates in maintenance test flights; annotates records of required maintenance; evaluates operational readiness of airplanes and recommends corrective action; inspects crash-damaged airplanes and estimates time, parts, and costs to repair; ensures compliance with aircraft configuration control, Army Oil Analysis Program, and test measuring and diagnostic calibration. NOTE: Progressed to 66H20 from 67H20, Observation Airplane Repairer. *Skill Level 30:* Able to perform the duties required for Skill Level 20; plans, directs, and supervises technical inspections; provides technical guidance to subordinates; evaluates technical training programs; prepares evaluation reports on subordinates; performs maintenance trend analysis and checks maintenance supply documents to ensure compliance with product quality control standards.

Recommendation, Skill Level 20

In the lower-division baccalaureate/associate degree category, 20 semester hours in aircraft maintenance, 3 in technical electives, and 3 in aviation maintenance management (7/87).

Recommendation, Skill Level 30

In the lower-division baccalaureate/associate degree category, 20 semester hours in aircraft maintenance, 3 in technical electives, 3 in aviation maintenance management, 3 in quality control, and 3 in personnel supervision. In the upper-division baccalaureate category, 3 semester hours in principles of management and 3 for field experience in management (7/87).

MOS-66J-001

Aircraft Armament Technical Inspector
66J30

Exhibit Dates: 1/90–9/90. Effective 9/90, MOS 66J was discontinued, and its duties were incorporated into MOS 68J, Aircraft Armament/Missile Systems Repairer.

Career Management Field: 67 (Aircraft Maintenance).

Description

Performs duties as an aircraft armament technical inspector; inspects and makes operational checks on armament systems and subsystems to isolate malfunctions using special test sets and applicable manuals and publications; plans and directs inspection activities; performs maintenance trend analysis; provides technical guidance and training to armament repairers on maintenance techniques and procedures; participates in armament system test flights. NOTE: Progressed to 66J30 from 68J30, Aircraft Fire Control Repairer.

Recommendation

In the lower-division baccalaureate/associate degree category, 3 semester hours in quality control, 3 in maintenance management, and 3 in personnel supervision. In the upper-division baccalaureate category, 3 semester hours in principles of management and 3 for field experience in management (7/87).

MOS-66N-001

Utility Helicopter Technical Inspector
66N20
66N30

Exhibit Dates: 1/90–9/90. Effective 9/90, MOS 66N was discontinued, and its duties were incorporated into MOS 67N, UH-1 Helicopter Repairer.

Career Management Field: 67 (Aircraft Maintenance).

Description

Summary: Performs technical inspections on UH-1 utility helicopters to determine maintenance requirements and to insure that appropriate maintenance has been performed. *Skill Level 20:* Trains and provides technical guidance to repairers on maintenance techniques and safety procedures; checks airplane flight and maintenance records to determine adherence to prescribed maintenance standards; maintains technical library; inspects systems, subsystems, and components according to applicable technical publications and checklists before, during, and after airplane maintenance and modifications; computes basic aircraft weight and balance; checks and troubleshoots systems; participates in maintenance test flights; annotates records of required maintenance; evaluates operational readiness of airplanes and recommends corrective action; inspects crash-damaged airplanes and estimates time, parts, and costs to repair; ensures compliance with aircraft configuration control, Army Oil Analysis Program, and test measuring and diagnostic calibration. NOTE: Progressed to 66N20 from 67N20, Utility Helicopter Repairer. *Skill Level 30:* Able to perform the duties required for Skill Level 20; plans, directs, and supervises technical inspections; provides technical guidance to subordinates; evaluates technical training programs; prepares evaluation reports on subordinates; performs maintenance trend analysis and checks maintenance supply documents to ensure compliance with product quality control standards.

Recommendation, Skill Level 20

In the lower-division baccalaureate/associate degree category, 20 semester hours in aircraft maintenance, 3 in technical electives, and 3 in aviation maintenance management (7/87).

Recommendation, Skill Level 30

In the lower-division baccalaureate/associate degree category, 20 semester hours in aircraft maintenance, 3 in technical electives, 3 in aviation maintenance management, 3 in quality control, and 3 in personnel supervision. In the upper-division baccalaureate category, 3 semester hours in principles of management and 3 for field experience in management (7/87).

MOS-66R-001

AH-64 Attack Helicopter Technical Inspector
66R20
66R30

Exhibit Dates: 1/90–9/90. Effective 9/90, MOS 66R was discontinued, and its duties were incorporated into MOS 67R, AH-64 Attack Helicopter Repairer.

Career Management Field: 67 (Aircraft Maintenance).

Description

Summary: Performs technical inspections on AH-64 attack helicopters to determine maintenance requirements and to ensure that appropriate maintenance has been performed. *Skill Level 20:* Trains and provides technical guidance to repairers on maintenance techniques and safety procedures; checks airplane flight and maintenance records to determine adherence to prescribed maintenance standards; maintains technical library; inspects systems, subsystems, and components according to applicable technical publications and checklists before, during, and after airplane maintenance and modifications; computes basic aircraft weight and balance; checks and troubleshoots systems; participates in maintenance test flights; annotates records of required maintenance; evaluates operational readiness of airplanes and recommends corrective action; inspects crash-damaged airplanes and estimates time, parts, and costs to repair; ensures compliance with aircraft configuration control, Army

Oil Analysis Program, and test measuring and diagnostic calibration. NOTE: Progressed to 66R20 from 67R20, AH-64 Attack Helicopter Repairer. *Skill Level 30:* Able to perform the duties required for Skill Level 20; plans, directs, and supervises technical inspections; provides technical guidance to subordinates; evaluates technical training programs; prepares evaluation reports on subordinates; performs maintenance trend analysis and checks maintenance supply documents to ensure compliance with product quality control standards.

Recommendation, Skill Level 20

In the lower-division baccalaureate/associate degree category, 20 semester hours in aircraft maintenance, 3 in technical electives, and 3 in aviation maintenance management (7/87).

Recommendation, Skill Level 30

In the lower-division baccalaureate/associate degree category, 20 semester hours in aircraft maintenance, 3 in technical electives, 3 in aviation maintenance management, 3 in quality control, and 3 in personnel supervision. In the upper-division baccalaureate category, 3 semester hours in principles of management and 3 for field experience in management (7/87).

MOS-66S-001

Scout Helicopter Technical Inspector

66S20

66S30

Exhibit Dates: 1/90–9/90. Effective 9/90, MOS 66S was discontinued, and its duties were incorporated into MOS 67S, OH-58D Helicopter Repairer.

Career Management Field: 67 (Aircraft Maintenance).

Description

Summary: Performs technical inspections on OH-58D scout helicopters to determine maintenance requirements and to ensure that appropriate maintenance has been performed. *Skill Level 20:* Trains and provides technical guidance to repairers on maintenance techniques and safety procedures; checks airplane flight and maintenance records to determine adherence to prescribed maintenance standards; maintains technical library; inspects systems, subsystems, and components according to applicable technical publications and checklists before, during, and after airplane maintenance and modifications; computes basic aircraft weight and balance; checks and troubleshoots systems; participates in maintenance test flights; annotates records of required maintenance; evaluates operational readiness of airplanes and recommends corrective action; inspects crash-damaged airplanes and estimates time, parts, and costs to repair; ensures compliance with aircraft configuration control, Army Oil Analysis Program, and test measuring and diagnostic calibration. NOTE: Progressed to 66S20 from 67S20, Scout Helicopter Repairer. *Skill Level 30:* Able to perform the duties required for Skill Level 20; plans, directs, and supervises technical inspections; provides technical guidance to subordinates; evaluates technical training programs; prepares evaluation reports on subordinates; performs maintenance trend analysis and checks maintenance supply documents to ensure compliance with product quality control standards.

Recommendation, Skill Level 20

In the lower-division baccalaureate/associate degree category, 20 semester hours in aircraft maintenance, 3 in technical electives, and 3 in aviation maintenance management (7/87).

Recommendation, Skill Level 30

In the lower-division baccalaureate/associate degree category, 20 semester hours in aircraft maintenance, 3 in technical electives, 3 in aviation maintenance management, 3 in quality control, and 3 in personnel supervision. In the upper-division baccalaureate category, 3 semester hours in principles of management and 3 for field experience in management (7/87).

MOS-66T-001

Tactical Transport Helicopter Technical Inspector

66T20

66T30

Exhibit Dates: 1/90–9/90. Effective 9/90, MOS 66T was discontinued, and its duties were incorporated into MOS 67T, UH-60 Helicopter Repairer.

Career Management Field: 67 (Aircraft Maintenance).

Description

Summary: Performs technical inspections on UH-60 tactical transport helicopters to determine maintenance requirements and to ensure that appropriate maintenance has been performed. *Skill Level 20:* Trains and provides technical guidance to repairers on maintenance techniques and safety procedures; checks airplane flight and maintenance records to determine adherence to prescribed maintenance standards; maintains technical library; inspects systems, subsystems, and components according to applicable technical publications and checklists before, during, and after airplane maintenance and modifications; computes basic aircraft weight and balance; checks and troubleshoots systems; participates in maintenance test flights; annotates records of required maintenance; evaluates operational readiness of airplanes and recommends corrective action; inspects crash-damaged airplanes and estimates time, parts, and costs to repair; ensures compliance with aircraft configuration control, Army Oil Analysis Program, and test measuring and diagnostic calibration. NOTE: Progressed to 66T20 from 67T20, Tactical Transport Helicopter Repairer. *Skill Level 30:* Able to perform the duties required for Skill Level 20; plans, directs, and supervises technical inspections; provides technical guidance to subordinates; evaluates technical training programs; prepares evaluation reports on subordinates; performs maintenance trend analysis and checks maintenance supply documents to insure compliance with product quality control standards.

Recommendation, Skill Level 20

In the lower-division baccalaureate/associate degree category, 20 semester hours in aircraft maintenance, 3 in technical electives, and 3 in aviation maintenance management (7/87).

Recommendation, Skill Level 30

In the lower-division baccalaureate/associate degree category, 20 semester hours in aircraft maintenance, 3 in technical electives, 3 in aviation maintenance management, 3 in quality control, and 3 in personnel supervision. In the upper-division baccalaureate category, 3 semester hours in principles of management and 3 for field experience in management (7/87).

MOS-66U-001

Medium Helicopter Technical Inspector

66U20

66U30

Exhibit Dates: 1/90–9/90. Effective 9/90, MOS 66U was discontinued, and its duties were incorporated into MOS 67U, CH-47 Helicopter Repairer.

Career Management Field: 67 (Aircraft Maintenance).

Description

Summary: Performs technical inspections on CH-47 medium helicopters to determine maintenance requirements and to ensure that appropriate maintenance has been performed. *Skill Level 20:* Trains and provides technical guidance to repairers on maintenance techniques and safety procedures; checks airplane flight and maintenance records to determine adherence to prescribed maintenance standards; maintains technical library; inspects systems, subsystems, and components according to applicable technical publications and checklists before, during, and after airplane maintenance and modifications; computes basic aircraft weight and balance; checks and troubleshoots systems; participates in maintenance test flights; annotates records of required maintenance; evaluates operational readiness of airplanes and recommends corrective action; inspects crash-damaged airplanes and estimates time, parts, and costs to repair; ensures compliance with aircraft configuration control, Army Oil Analysis Program, and test measuring and diagnostic calibration. NOTE: Progressed to 66U20 from 67U20, Medium Helicopter Repairer. *Skill Level 30:* Able to perform the duties required for Skill Level 20; plans, directs, and supervises technical inspections; provides technical guidance to subordinates; evaluates technical training programs; prepares evaluation reports on subordinates; performs maintenance trend analysis and checks maintenance supply documents to ensure compliance with product quality control standards.

Recommendation, Skill Level 20

In the lower-division baccalaureate/associate degree category, 20 semester hours in aircraft maintenance, 3 in technical electives, and 3 in aviation maintenance management (7/87).

Recommendation, Skill Level 30

In the lower-division baccalaureate/associate degree category, 20 semester hours in aircraft maintenance, 3 in technical electives, 3 in aviation maintenance management, 3 in quality control, and 3 in personnel supervision. In the upper-division baccalaureate category, 3 semester hours in principles of management and 3 for field experience in management (7/87).

MOS-66V-001

Observation/Scout Helicopter Technical Inspector

66V20

66V30

Exhibit Dates: 1/90–9/90. Effective 9/90, MOS 66V was discontinued, and its duties were incorporated into MOS 67V, Observation/Scout Helicopter Repairer.

Career Management Field: 67 (Aircraft Maintenance).

Description

Summary: Performs technical inspections on OH-58A/C Observation/Scout helicopters to determine maintenance requirements and to ensure that appropriate maintenance has been performed. *Skill Level 20:* Trains and provides technical guidance to repairers on maintenance techniques and safety procedures; checks airplane flight and maintenance records to determine adherence to prescribed maintenance standards; maintains technical library; inspects systems, subsystems, and components according to applicable technical publications and checklists before, during, and after airplane maintenance and modifications; computes basic aircraft weight and balance records; checks and troubleshoots systems; participates in maintenance test flights; annotates records of required maintenance; evaluates operational readiness of airplanes and recommends corrective action; inspects crash-damaged airplanes and estimates time, parts, and costs to repair; ensures compliance with aircraft configuration control, Army Oil Analysis Program, and test measuring and diagnostic calibration. NOTE: Progressed to 66V20 from 67V20, Observation/Scout Helicopter Repairer. *Skill Level 30:* Able to perform the duties required for Skill Level 20; plans, directs, and supervises technical inspections; provides technical guidance to subordinates; evaluates technical training programs; prepares evaluation reports on subordinates; performs maintenance trend analysis and checks maintenance supply documents to ensure compliance with product quality control standards.

Recommendation, Skill Level 20

In the lower-division baccalaureate/associate degree category, 20 semester hours in aircraft maintenance, 3 in technical electives, and 3 in aviation maintenance management (7/87).

Recommendation, Skill Level 30

In the lower-division baccalaureate/associate degree category, 20 semester hours in aircraft maintenance, 3 in technical electives, 3 in aviation maintenance management, 3 in quality control, and 3 in personnel supervision. In the upper-division baccalaureate category, 3 semester hours in principles of management and 3 for field experience in management (7/87).

MOS-66X-001

Heavy Lift Helicopter Technical Inspector

66X20

66X30

Exhibit Dates: 1/90–3/90. Effective 3/90, MOS 66S was discontinued, and its duties were incorporated into MOS 67X, Heavy Lift Helicopter Repairer.

Career Management Field: 67 (Aircraft Maintenance).

Description

Summary: Performs technical inspections on CH-54 heavy lift helicopters to determine maintenance requirements and to ensure that appropriate maintenance has been performed. *Skill Level 20:* Trains and provides technical guidance to repairers on maintenance techniques and safety procedures; checks airplane flight and maintenance records to determine adherence to prescribed maintenance standards; maintains technical library; inspects systems, subsystems, and components according to applicable technical publications and checklists before, during, and after airplane maintenance and modifications; computes basic aircraft weight and balance records; checks and troubleshoots systems; participates in maintenance test flights; annotates records of required maintenance; evaluates operational readiness of airplanes and recommends corrective action; inspects crash-damaged airplanes and estimates time, parts, and costs to repair; ensures compliance with aircraft configuration control, Army Oil Analysis Program, and test measuring and diagnostic calibration. NOTE: Progressed to 66X20 from 67X20, Heavy Lift Helicopter Repairer. *Skill Level 30:* Able to perform the duties required for Skill Level 20; plans, directs, and supervises technical inspections; provides technical guidance to subordinates; evaluates technical training programs; prepares evaluation reports on subordinates; performs maintenance trend analysis and checks maintenance supply documents to ensure compliance with product quality control standards.

Recommendation, Skill Level 20

In the lower-division baccalaureate/associate degree category, 20 semester hours in aircraft maintenance, 3 in technical electives, and 3 in aviation maintenance management (7/87).

Recommendation, Skill Level 30

In the lower-division baccalaureate/associate degree category, 20 semester hours in aircraft maintenance, 3 in technical electives, 3 in aviation maintenance management, 3 in quality control, and 3 in personnel supervision. In the upper-division baccalaureate category, 3 semester hours in principles of management and 3 for field experience in management (7/87).

MOS-66Y-001

AH-1 Attack Helicopter Technical Inspector

66Y20

66Y30

Exhibit Dates: 1/90–9/90. Effective 9/90, MOS 66Y was discontinued, and its duties were incorporated into MOS 67Y, AH-1 Attack Helicopter Repairer.

Career Management Field: 67 (Aircraft Maintenance).

Description

Summary: Performs technical inspections on AH-1 attack helicopters to determine maintenance requirements and to ensure that appropriate maintenance has been performed. *Skill Level 20:* Trains and provides technical guidance to repairers on maintenance techniques and safety procedures; checks airplane flight and maintenance records to determine adherence to prescribed maintenance standards; maintains technical library; inspects systems, subsystems, and components according to applicable technical publications and checklists before, during, and after airplane maintenance and modifications; computes basic aircraft weight and balance records; checks and troubleshoots systems; participates in maintenance test flights; annotates records of required maintenance; evaluates operational readiness of airplanes and recommends corrective action; inspects crash-damaged airplanes and estimates time, parts, and costs to repair; ensures compliance with aircraft configuration control, Army Oil Analysis Program, and test measuring and diagnostic calibration. NOTE: Progressed to 66Y20 from 67Y20, AH-1 Attack Helicopter Repairer. *Skill Level 30:* Able to perform the duties required for Skill Level 20; plans, directs, and supervises technical inspections; provides technical guidance to subordinates; evaluates technical training programs; prepares evaluation reports on subordinates; performs maintenance trend analysis and checks maintenance supply documents to ensure compliance with product quality control standards.

Recommendation, Skill Level 20

In the lower-division baccalaureate/associate degree category, 20 semester hours in aircraft maintenance, 3 in technical electives, and 3 in aviation maintenance management (7/87).

Recommendation, Skill Level 30

In the lower-division baccalaureate/associate degree category, 20 semester hours in aircraft maintenance, 3 in technical electives, 3 in aviation maintenance management, 3 in quality control, and 3 in personnel supervision. In the upper-division baccalaureate category, 3 semester hours in principles of management and 3 for field experience in management (7/87).

MOS-67A-001

General Aircraft Repairer

67A10

67A20

Exhibit Dates: 11/90–8/94. Effective 8/94, MOS 67A was discontinued.

Career Management Field: 67 (Aircraft Maintenance).

Description

Summary: Performs maintenance on both rotary wing (helicopter) and fixed wing aircraft at the intermediate level; maintenance responsibilities include the airframe systems, subsystems, and components. *Skill Level 10:* Performs maintenance checks on rotary and fixed wing aircraft; assists in performing aircraft inspections; assists in diagnosing and troubleshooting malfunctions of subsystems using special tools and equipment; prepares aircraft for shipments by surface and by air; removes and replaces subsystem assemblies and components; services and lubricates aircraft and related subsystems; uses and performs maintenance on ground support equipment and special tools; prepares maintenance forms and records. *Skill Level 20:* Able to perform the duties required for Skill Level 10; performs maintenance, operational checks, and scheduled inspections; accomplishes diagnostic and troubleshooting duties on aircraft subsystems; performs corrosion prevention processes.

Recommendation, Skill Level 10

In the lower-division baccalaureate/associate degree category, 3 semester hours in aircraft maintenance technology. (NOTE: This recommendation for skill level 10 is valid for the dates 11/90—9/91 only) (3/96).

Recommendation, Skill Level 20

In the lower-division baccalaureate/associate degree category, 6 semester hours in aircraft maintenance technology. (NOTE: This recommendation for skill level 20 is valid for the dates 11/90-2/95 only) (3/96).

MOS-67B-002

Certified General Airplane Repairer
67B10
67B20

Exhibit Dates: 4/92–8/94. Effective 8/94, MOS 67B was discontinued.

Career Management Field: 67 (Aircraft Maintenance).

Description

Summary: Supervises and performs maintenance on rotary-wing (helicopter) and fixed-wing aircraft. NOTE: Must have held an MOS in the 67 Career Management Field for a minimum of 12 months prior to being awarded this MOS. *Skill Level 10:* Performs maintenance on aircraft including the airframe, systems, subsystems, and components; assists in performing special aircraft inspections; assists in diagnosing and troubleshooting malfunctions of subsystems; uses and performs maintenance on ground support equipment and on common and special tools; prepares maintenance forms and records. *Skill Level 20:* Able to perform the duties required for Skill Level 10; supervises and provides technical guidance to subordinates.

Recommendation, Skill Level 10

In the lower-division baccalaureate/associate degree category, 3 semester hours in aircraft maintenance technology. (NOTE: This recommendation for skill level 10 is valid for the dates 11/90-9/91 only) (3/96).

Recommendation, Skill Level 20

In the lower-division baccalaureate/associate degree category, 6 semester hours in aircraft maintenance technology. (NOTE: This recommendation for skill level 20 is valid for the dates 11/90-2/95 only) (3/96).

MOS-67G-003

Utility Airplane Repairer
67G10
67G20
67G30
67G40

Exhibit Dates: 1/90–5/99.

Career Management Field: 67 (Aircraft Maintenance).

Description

Summary: Supervises and performs maintenance on utility airplanes at the unit, intermediate, and component levels, excluding repair of system components. *Skill Level 10:* Performs air crewmember duties, removes and installs subsystems such as engines, propellers, and landing gear; removes and installs components, including starters, generators, inverters, batteries, pumps, and hydraulic units; services and lubricates aircraft and aircraft subsystems; prepares aircraft for maintenance inspections; performs scheduled and special inspections and maintenance checks; assists in troubleshooting; prepares aircraft for air or surface shipment; taxis aircraft; uses and performs maintenance on ground support equipment and special tools; prepares forms, records, and reports. *Skill Level 20:* Able to perform the duties required for Skill Level 10; provides technical guidance to subordinates; performs maintenance and operational checks and scheduled inspections; troubleshoots aircraft subsystems. *Skill Level 30:* Able to perform the duties required for Skill Level 20; applies production, quality control, and other management principles to shop and flight line operation; plans work flow; instructs personnel and conducts training programs in maintenance, supply, and safety techniques; supervises the preparation of forms, records, and reports on aircraft maintenance and supply procedures. *Skill Level 40:* Able to perform the duties required for Skill Level 30; supervises the training of personnel; prepares evaluations, special reports, and maintenance records; plans and lays out airplane maintenance areas, repair shops, and facilities requirements; assists in the preparation of plans and policies; controls the flow of work orders, requisitions, and correspondence.

Recommendation, Skill Level 10

In the vocational certificate category, 3 semester hours in aircraft maintenance. In the lower-division baccalaureate/associate degree category, 3 semester hours in aircraft maintenance technology. (NOTE: This recommendation for skill level 10 is valid for the dates 1/90-9/91 only) (10/84).

Recommendation, Skill Level 20

In the vocational certificate category, 6 semester hours in aircraft maintenance. In the lower-division baccalaureate/associate degree category, 6 semester hours in aircraft maintenance technology. (NOTE: This recommendation for skill level 20 is valid for the dates 1/90-2/95 only) (10/84).

Recommendation, Skill Level 30

In the vocational certificate category, 6 semester hours in aircraft maintenance. In the lower-division baccalaureate/associate degree category, 6 semester hours in aircraft maintenance technology and 3 in personnel supervision (10/84).

Recommendation, Skill Level 40

In the vocational certificate category, 6 semester hours in aircraft maintenance. In the lower-division baccalaureate/associate degree category, 6 semester hours in aircraft maintenance technology, 3 in personnel supervision, and 3 in aircraft maintenance management (10/84).

MOS-67G-004

Utility Airplane Repairer
67G10
67G20
67G30
67G40

Exhibit Dates: 6/99–Present.

Career Management Field: 67 (Aircraft Maintenance).

Description

Summary: Supervises and performs maintenance on utility airplanes at the unit, intermediate, and component levels, excluding repair of system components. *Skill Level 10:* Performs air crewmember duties; removes and installs subsystems such as engines, propellers, and landing gear; removes and installs components, including starters, generators, inverters, batteries, pumps, and hydraulic units; services and lubricates aircraft and aircraft subsystems; prepares aircraft for maintenance inspections; performs scheduled and special inspections and maintenance checks; assists in troubleshooting; prepares aircraft for air or surface shipment; taxis aircraft; uses and performs maintenance on ground support equipment and special tools; prepares forms, records, and reports. *Skill Level 20:* Able to perform the duties required for Skill Level 10; provides technical guidance to subordinates; performs maintenance and operational checks and scheduled inspections; troubleshoots aircraft subsystems. *Skill Level 30:* Able to perform the duties required for Skill Level 20; applies production, quality control, and other management principles to shop and flight-line operation; plans work flow; instructs personnel and conducts training programs in maintenance, supply, and safety techniques; supervises the preparation of forms, records, and reports on aircraft maintenance and supply procedures. *Skill Level 40:* Able to perform the duties required for Skill Level 30; supervises the training of personnel; prepares evaluations, special reports, and maintenance records; plans and lays out airplane maintenance areas, repair shops, and facilities requirements; assists in the preparation of plans and policies; controls the flow of work orders, requisitions, and correspondence.

Recommendation, Skill Level 10

Credit may be granted on the basis of an individualized assessment of the student (6/99).

Recommendation, Skill Level 20

Credit may be granted on the basis of an individualized assessment of the student (6/99).

Recommendation, Skill Level 30

In the lower-division baccalaureate/associate degree category, 12 semester hours in basic helicopter maintenance, 2 in aircraft forms and records, 2 in aircraft weight and balance, 5 in aviation maintenance management, 2 in quality control, and 3 in principles of supervision (6/99).

Recommendation, Skill Level 40

In the lower-division baccalaureate/associate degree category, 12 semester hours in basic helicopter maintenance, 3 in aircraft forms and records, 2 in aircraft weight and balance, 8 in aviation maintenance management, 4 in quality control, and 3 in principles of supervision. In the upper-division baccalaureate category, 3 semester hours in management practicum and 3 in principles of management (6/99).

MOS-67H-003

Observation Airplane Repairer
67H10
67H20
67H30
67H40

Exhibit Dates: 1/90–12/96. Effective 12/96, MOS 67H was discontinued.

Career Management Field: 67 (Aircraft Maintenance).

Description

Summary: Supervises and performs maintenance on observation airplanes at the unit, intermediate, and component levels, excluding repair of system components. *Skill Level 10:* Removes and installs subsystem assemblies such as engine, propellers, and landing gear; removes and installs components, including starters, generators, inverters, batteries, pumps, and hydraulic units; services and lubricates aircraft and the aircraft subsystems; prepares aircraft for maintenance inspections; performs scheduled and special inspections and maintenance checks; assists in troubleshooting; prepares aircraft for air or surface shipment; taxis aircraft; uses and performs maintenance on ground support equipment and special tools; prepares forms, records, and reports. *Skill Level 20:* Able to perform the duties required for Skill Level 10; provides technical guidance to subordinates; performs maintenance and operational checks and scheduled inspections; accomplishes diagnostic and troubleshooting duties on aircraft subsystems. *Skill Level 30:* Able to perform the duties required for Skill Level 20; applies production, quality control, and other management principles to shop and flight line operation; plans work flow; instructs personnel and conducts training programs in maintenance, supply, and safety techniques; supervises the preparation of forms, records, and reports on aircraft maintenance and supply procedures. *Skill Level 40:* Able to perform the duties required for Skill Level 30; supervises the training of personnel; prepares evaluations, special reports, and maintenance records; plans and lays out airplane maintenance areas, repair shops, and facility requirements; assists in the preparation of plans and policies; controls the flow of work orders, requisitions, and correspondence.

Recommendation, Skill Level 10

In the vocational certificate category, 3 semester hours in aircraft maintenance. In the lower-division baccalaureate/associate degree category, 3 semester hours in aircraft maintenance technology. (NOTE: This recommendation for skill level 10 is valid for the dates 1/90-9/91 only) (10/84).

Recommendation, Skill Level 20

In the vocational certificate category, 6 semester hours in aircraft maintenance. In the lower-division baccalaureate/associate degree category, 6 semester hours in aircraft maintenance technology. (NOTE: This recommendation for skill level 20 is valid for the dates 1/90-2/95 only) (10/84).

Recommendation, Skill Level 30

In the vocational certificate category, 6 semester hours in aircraft maintenance. In the lower-division baccalaureate/associate degree category, 6 semester hours in aircraft maintenance technology and 3 in personnel supervision (10/84).

Recommendation, Skill Level 40

In the vocational certificate category, 6 semester hours in aircraft maintenance. In the lower-division baccalaureate/associate degree category, 6 semester hours in aircraft maintenance technology, 3 in personnel supervision, and 3 in aircraft maintenance management (10/84).

MOS-67N-002

Utility Helicopter Repairer
67N10
67N20
67N30

Exhibit Dates: 1/90–12/90.

Career Management Field: 67 (Aviation Maintenance), subfield 671 (Aircraft Maintenance).

Description

Summary: Performs or supervises maintenance on utility helicopters, excluding repairs of systems components, or conducts maintenance inspections. *Skill Level 10:* Assists in the removal and installation of subsystem assemblies such as engines, transmissions, gear boxes, rotor hubs, and rotor blades; prepares helicopters for extensive inspections and maintenance; assists in operational checks; performs maintenance on common and special tools and ground support equipment; uses maintenance forms and records; may perform duties of utility helicopter crew chief. *Skill Level 20:* Able to perform the duties required for Skill Level 10; serves as utility helicopter crew chief; provides technical guidance to subordinates; maintains ground support equipment; prepares maintenance forms and records. *Skill Level 30:* Able to perform the duties required for Skill Level 20; supervises all maintenance on utility helicopters; plans work flow; applies production control, quality control, and other maintenance management principles to shop, flight line, and supply operations; instructs personnel, and conducts on-the-job training in helicopter maintenance, supply, and safety techniques; evaluates work performance; monitors operational checks and test flights; certifies airworthiness of the aircraft.

Recommendation, Skill Level 10

In the lower-division baccalaureate/associate degree category, 6 semester hours in basic helicopter maintenance (5/78).

Recommendation, Skill Level 20

In the lower-division baccalaureate/associate degree category, 6 semester hours in basic helicopter maintenance. In the upper-division baccalaureate category, 2 semester hours in aviation maintenance management (5/78).

Recommendation, Skill Level 30

In the lower-division baccalaureate/associate degree category, 6 semester hours in basic helicopter maintenance, 2 in aviation maintenance management, 3 in human relations (or a management elective), and 2 in quality control; if the duty position was maintenance supervisor section chief, 3 semester hours in personnel supervision. In the upper-division baccalaureate category, 3 semester hours in aviation maintenance management, 3 in personnel management, and 3 for field experience in management; or 9 semester hours in management electives (5/78).

MOS-67N-003

UH-1 Helicopter Repairer
67N10
67N20
67N30

Exhibit Dates: 1/91–12/00.

Career Management Field: 67 (Aircraft Maintenance).

Description

Summary: Performs or supervises maintenance on utility helicopters, excluding repairs of system components, or conducts maintenance inspections. *Skill Level 10:* Assists in the removal and installation of such subsystem assemblies as engines, transmissions, gear boxes, rotor hubs, and rotor blades; prepares helicopters for extensive inspections and maintenance; assists in operational checks; performs maintenance on common and special tools and ground support equipment; uses maintenance forms and records; may perform duties of utility helicopter crew chief. *Skill Level 20:* Able to perform the duties required for Skill Level 10; maintains ground support equipment; prepares maintenance forms and records. *Skill Level 30:* Able to perform the duties required for Skill Level 20; supervises all maintenance on utility helicopters; plans work flow; applies production control, quality assurance, and other maintenance management principles to shop, flight line, and supply operations; instructs personnel, and conducts on-the-job training in helicopter maintenance, supply, and safety techniques; evaluates work performance; monitors operational checks and test flights; certifies airworthiness of the aircraft.

Recommendation, Skill Level 10

In the lower-division baccalaureate/associate degree category, 6 semester hours in basic aircraft maintenance. (NOTE: This recommendation for skill level 10 is valid for the dates 1/91-9/91 only) (3/91).

Recommendation, Skill Level 20

In the lower-division baccalaureate/associate degree category, 6 semester hours in basic aircraft maintenance, 3 in personnel supervision, and 3 in helicopter maintenance. (NOTE: This recommendation for skill level 20 is valid for the dates 1/91-2/95 only) (3/91).

Recommendation, Skill Level 30

In the lower-division baccalaureate/associate degree category, 6 semester hours in basic aircraft maintenance, 3 in personnel supervision, 3 in helicopter maintenance, 2 in occupational safety, 3 in maintenance management, and 2 in records and information management (3/91).

MOS-67N-004

UH-1 Helicopter Repairer
67N10
67N20
67N30

Exhibit Dates: 1/01–9/03. Effective 10/03, MOS 67N was discontinued and its duties were transferred to MOS 15M, UH-1 Helicopter Repairer.

Description

Summary: Performs or supervises maintenance on helicopters, excluding repairs of system components, or conducts maintenance inspections. *Skill Level 10:* Performs maintenance on helicopters; removes and installs such subsystem assemblies as engines, transmissions, gear boxes, rotor hubs, and rotor blades; performs extensive inspections and maintenance on helicopters; uses maintenance forms and records; performs operational checks; maintains ground support equipment; may perform crew chief duties. *Skill Level 20:* Able to perform the duties required for Skill Level 10; provides technical guidance to subordinates; troubleshoots helicopter systems; prepares maintenance forms and records; applies production control; instructs subordinates; conducts on-the-job training in helicopter maintenance, logistics, and safety practices; assists in performing weight and balance checks; performs section chief duties; uses computer management information systems. *Skill Level 30:* Able to perform the duties required for Skill Level 20; supervises all helicopter maintenance activities; plans work flow; applies quality control and other maintenance management principles to shop, flight line, and logistics; supervises on-the-job training; evaluates work performance; monitors operational checks and test flights; certifies airworthiness of the aircraft; performs management functions; analyzes computer management information systems products.

Recommendation, Skill Level 10

Credit may be granted on the basis of an individualized assessment of the student (1/01).

Recommendation, Skill Level 20

Credit may be granted on the basis of an individualized assessment of the student (1/01).

Recommendation, Skill Level 30

In the lower-division baccalaureate/associate degree category, 9 semester hours in aircraft maintenance fundamentals, 3 in aircraft systems fundamentals, 3 in principles of supervision, 3 in computer management information systems, and 3 in logistics management. In the upper-division baccalaureate category, 3 semester hours in organizational management, 3 in human resource management, and 3 in advanced computer management information systems (1/01).

MOS-67R-001

AH-64 Attack Helicopter Repairer
67R10
67R20
67R30
67R40

Exhibit Dates: 1/90–5/99.

Career Management Field: 67 (Aircraft Maintenance).

Description

Summary: Performs or supervises maintenance on attack helicopters, excluding repairs of system components. *Skill Level 10:* Assists in the removal and installation of such subsystem assemblies as engines, transmissions, gear boxes, rotor hubs, and rotor blades; prepares helicopter for extensive inspections and maintenance; assists in operational checks; performs maintenance on common and special tools and ground support equipment; uses maintenance forms and records. *Skill Level 20:* Able to perform the duties required for Skill Level 10; serves as attack helicopter crew chief; provides technical guidance to subordinates; maintains ground support equipment; prepares maintenance forms and records. *Skill Level 30:* Able to perform the duties required for Skill Level 20; supervises all maintenance on attack helicopters; plans work flow; applies production control, quality control, and other maintenance management principles to shop, flight line, and supply operations; instructs personnel, and conducts on-the-job training in helicopter maintenance, supply, and safety techniques; evaluates work performance; monitors operational checks and test flights; certifies airworthiness of the aircraft; if duty position is section chief, serves as the principal noncommissioned officer in a section. *Skill Level 40:* Able to perform the duties required for Skill Level 30; supervises all maintenance activities; diagnoses complex malfunctions; determines personnel and parts requirements; coordinates work schedules and assigns duties; applies production, quality control, and other maintenance management principles and procedures; prepares evaluations, special reports, and maintenance records; supervises technical training programs; assists in the preparation of plans and policies; controls the flow of work orders, requisitions, and correspondence; plans the layout of helicopter maintenance areas, shops, and facilities. NOTE: May have progressed to 67R40 from 67R30; 66R30, AH-64 Attack Helicopter Technical Inspector; 67S30, Scout Helicopter Repairer; or 66S30, Scout Helicopter Technical Inspector.

Recommendation, Skill Level 10

In the lower-division baccalaureate/associate degree category, 8 semester hours in basic helicopter maintenance. (NOTE: This recommendation for skill level 10 is valid for the dates 1/90-9/91 only) (7/87).

Recommendation, Skill Level 20

In the lower-division baccalaureate/associate degree category, 8 semester hours in basic helicopter maintenance and 2 in aviation maintenance management. (NOTE: This recommendation for skill level 20 is valid for the dates 1/90-2/95 only) (7/87).

Recommendation, Skill Level 30

In the lower-division baccalaureate/associate degree category, 8 semester hours in basic helicopter maintenance, 3 in aviation maintenance management, 2 in quality control, and 3 in personnel supervision; if duty position was section chief, 3 additional semester hours in personnel supervision. In the upper-division baccalaureate category, 3 semester hours for field experience in management and 3 in principles of management (7/87).

Recommendation, Skill Level 40

In the lower-division baccalaureate/associate degree category, 9 semester hours in basic helicopter maintenance, 3 in aviation maintenance management, 2 in quality control, 6 in personnel supervision, and 1 in records administration. In the upper-division baccalaureate category, 6 semester hours for field experience in management and 3 in principles of management (7/87).

MOS-67R-002

AH-64 Attack Helicopter Repairer
67R10
67R20
67R30
67R40

Exhibit Dates: 6/99–9/03. Effective 10/03, MOS 67R was discontinued and its duties were transferred to MOS 15R, AH-64 Attack Helicopter Repairer.

Career Management Field: 67 (Aircraft Maintenance).

Description

Summary: Performs or supervises maintenance on attack helicopters, excluding repairs of system components. *Skill Level 10:* Assists in the removal and installation of such subsystem assemblies as engines, transmissions, gear boxes, rotor hubs, and rotor blades; prepares helicopter for extensive inspections and maintenance; assists in operational checks; performs maintenance on common and special tools and ground support equipment; uses maintenance forms and records. *Skill Level 20:* Serves as attack helicopter crew chief; provides technical guidance to subordinates; maintains ground support equipment; prepares maintenance forms and records. *Skill Level 30:* Able to perform the duties required for Skill Level 20; supervises all maintenance on attack helicopters; plans work flow; applies production control, quality control, and other maintenance management principles to shop, flight-line, and supply operations; instructs personnel and conducts on-the-job training in helicopter maintenance, supply, and safety techniques; evaluates work performance; monitors operational checks and test flights; certifies airworthiness of the aircraft; if duty position is section chief, serves as the principal noncommissioned officer in a section. *Skill Level 40:* Able to perform the duties required for Skill Level 30; supervises all maintenance activities; diagnoses complex malfunctions; determines personnel and parts requirements; coordinates work schedules and assigns duties; applies production, quality control, and other maintenance management principles and procedures; prepares evaluations, special reports, and maintenance records; supervises technical training programs; assists in the preparation of plans and policies; controls the flow of work orders, requisitions, and correspondence; plans the layout of helicopter maintenance areas, shops, and facilities.

Recommendation, Skill Level 10

Credit may be granted on the basis of an individualized assessment of the student (6/99).

Recommendation, Skill Level 20

Credit may be granted on the basis of an individualized assessment of the student (6/99).

Recommendation, Skill Level 30

In the lower-division baccalaureate/associate degree category, 12 semester hours in basic helicopter maintenance, 2 in aircraft forms and records, 2 in aircraft weight and balance, 5 in aviation maintenance management, 2 in quality control, and 3 in principles of supervision (6/99).

Recommendation, Skill Level 40

In the lower-division baccalaureate/associate degree category, 12 semester hours in basic helicopter maintenance, 3 in aircraft forms and records, 2 in aircraft weight and balance, 8 in aviation maintenance management, 4 in quality control, and 3 in principles of supervision. In the upper-division baccalaureate category, 3 semester hours in management practicum and 3 in principles of management (6/99).

MOS-67S-001

OH-58D Helicopter Repairer
67S10
67S20
67S30

Exhibit Dates: 1/90–7/95.

Career Management Field: 67 (Aircraft Maintenance).

Description

Summary: Performs or supervises maintenance on Scout helicopters, excluding repairs of system components. *Skill Level 10:* Assists in the removal and installation of such subsystem assemblies as engines, transmissions, gear boxes, rotor hubs, and rotor blades; prepares helicopters for extensive inspections and maintenance; assists in operational checks; performs maintenance on common and special tools and ground support equipment; uses maintenance forms and records; may perform duties of Scout helicopter crew chief. *Skill Level 20:* Able to perform the duties required for Skill Level 10; serves as scout helicopter crew chief; provides technical guidance to subordinates; maintains ground support equipment; prepares maintenance forms and records. *Skill Level 30:* Able to perform the duties required for Skill Level 20; supervises all maintenance on utility helicopters; plans work flow; applies production control, quality control, and other maintenance management principles to shop, flight line, and supply operations; instructs personnel and conducts on-the-job training in helicopter maintenance, supply, and safety techniques; evaluates work performance; monitors operational checks and test flights; certifies airworthiness of the aircraft; if duty position is section chief, serves as the principal noncommissioned officer in a section.

Recommendation, Skill Level 10

In the lower-division baccalaureate/associate degree category, 8 semester hours in basic helicopter maintenance. (NOTE: This recommendation for skill level 10 is valid for the dates 1/90-9/91 only) (7/87).

Recommendation, Skill Level 20

In the lower-division baccalaureate/associate degree category, 8 semester hours in basic helicopter maintenance and 2 in aviation maintenance management. (NOTE: This recommendation for skill level 20 is valid for the dates 1/90-2/95 only) (7/87).

Recommendation, Skill Level 30

In the lower-division baccalaureate/associate degree category, 8 semester hours in basic helicopter maintenance, 3 in aviation maintenance management, 2 in quality control, and 3 in personnel supervision; if duty position was section chief, 3 additional semester hours in personnel supervision. In the upper-division baccalaureate category, 3 semester hours for field experience in management and 3 in principles of management (7/87).

MOS-67S-002

OH-58D Helicopter Repairer
67S10
67S20
67S30
67S40

Exhibit Dates: 8/95–5/99.

Career Management Field: 67 (Aircraft Maintenance).

Description

Summary: Performs or supervises maintenance on OH-58D helicopters, excluding repairs of system components. *Skill Level 10:* Assists in the removal and installation of various subsystem assemblies, including engines, weapons systems, transmissions, gear boxes, rotor hubs, and rotor blades and associated equipment; prepares helicopters for extensive inspections and maintenance; assists in operational checks; performs maintenance on common and special tools and ground support equipment; uses maintenance forms and records; performs air crew duties as required. *Skill Level 20:* Able to perform the duties required for Skill Level 10; provides technical guidance to subordinates; maintains ground support equipment; prepares equipment and aircraft maintenance forms and records. *Skill Level 30:* Able to perform the duties required for Skill Level 20; supervises all maintenance on utility helicopters; plans work flow; applies production control, quality control, and other maintenance management principles to shop, flight line, and supply operations; instructs subordinates and conducts on-the-job training in helicopter maintenance, supply, and safety techniques; evaluates work performance; monitors operational checks and test flights; certifies airworthiness of the aircraft. *Skill Level 40:* Able to perform the duties required for Skill Level 30; supervises, counsels, and trains all subordinates; maintains flow charts; coordinates flight schedules; for accounts of shop equipment, calibration, and files; uses problem-solving techniques; writes reports and personnel evaluations; conducts train-the-trainer programs.

Recommendation, Skill Level 10

Credit may be granted on the basis of an individualized assessment of the student (7/97).

Recommendation, Skill Level 20

Credit may be granted on the basis of an individualized assessment of the student (7/97).

Recommendation, Skill Level 30

In the lower-division baccalaureate/associate degree category, 8 semester hours in basic helicopter maintenance, 3 in aviation maintenance management, 2 in quality control, and 3 in personnel supervision (7/97).

Recommendation, Skill Level 40

In the lower-division baccalaureate/associate degree category, 8 semester hours in basic helicopter maintenance, 3 in aviation maintenance management, 2 in quality control, and 3 in personnel supervision. In the upper-division baccalaureate category, 3 semester hours for field experience in management and 3 in management problems (7/97).

MOS-67S-003

OH-58D Helicopter Repairer
67S10
67S20
67S30
67S40

Exhibit Dates: 6/99–9/03. Effective 10/03, MOS 67S was discontinued and its duties were transferred to MOS 15S, OH-58D Helicopter Repairer.

Career Management Field: 67 (Aircraft Maintenance).

Description

Summary: Performs or supervises maintenance on OH-58D helicopters, excluding repairs of system components. *Skill Level 10:* Assists in the removal and installation of various subsystem assemblies, including engines, weapons systems, transmissions, gear boxes, rotor hubs, and rotor blades and associated equipment; prepares helicopters for extensive inspections and maintenance; assists in operational checks; performs maintenance on common and special tools and ground support equipment; uses maintenance forms and records; performs aircrew duties as required. *Skill Level 20:* Able to perform the duties required for Skill Level 10; provides technical guidance to subordinates; maintains ground support equipment; prepares equipment and aircraft maintenance forms and records. *Skill Level 30:* Able to perform the duties required for Skill Level 20; supervises all maintenance on utility helicopters; plans work flow; applies production control, quality control, and other maintenance management principles to shop, flight-line, and supply operations; instructs subordinates and conducts on-the-job training in helicopter maintenance, supply, and safety techniques; evaluates work performance; monitors operational checks and test flights; certifies airworthiness of the aircraft. *Skill Level 40:* Able to perform the duties required for Skill Level 30; supervises, counsels, and trains all subordinates; maintains flow charts; coordinates flight schedules; accounts shop equipment, calibration records, and files; uses problem-solving techniques; writes reports and personnel evaluations; conducts train-the-trainer programs.

Recommendation, Skill Level 10

Credit may be granted on the basis of an individualized assessment of the student (6/99).

Recommendation, Skill Level 20

Credit may be granted on the basis of an individualized assessment of the student (6/99).

Recommendation, Skill Level 30

In the lower-division baccalaureate/associate degree category, 12 semester hours in basic helicopter maintenance, 2 in aircraft forms and records, 2 in aircraft weight and balance, 5 in aviation maintenance management, 2 in quality control, and 3 in principles of supervision (6/99).

Recommendation, Skill Level 40

In the lower-division baccalaureate/associate degree category, 12 semester hours in basic helicopter maintenance, 3 in aircraft forms and records, 2 in aircraft weight and balance, 8 in aviation maintenance management, 4 in quality control, and 3 in principles of supervision. In the upper-division baccalaureate category, 3 semester hours in management practicum and 3 in principles of management (6/99).

MOS-67T-002

UH-60 Helicopter Repairer
67T10
67T20
67T30
67T40

Exhibit Dates: 1/90–5/99.

Career Management Field: 67 (Aircraft Maintenance).

Description

Summary: Performs or supervises maintenance on tactical transport helicopters, excluding repairs of system components. *Skill Level 10:* Assists in the removal and installation of such subsystem assemblies as engines, transmissions, gear boxes, rotor hubs, and rotor blades; prepares helicopter for extensive inspections and maintenance; assists in operational checks; performs maintenance on common and special tools and ground support equipment; uses maintenance forms and records. *Skill Level 20:* Able to perform the duties required for Skill Level 10; serves as tactical transport helicopter crew chief; provides technical guidance to subordinates; maintains ground support equipment prepares maintenance forms and records. *Skill Level 30:* Able to perform the duties required for Skill Level 20; supervises all maintenance on tactical transport helicopters; plans work flow; applies production control, quality control, and other maintenance management principles to shop, flight line, and supply operations; instructs personnel, and conducts on-the-job training in helicopter maintenance, supply, and safety techniques; evaluates work performance; monitors operational checks and test flights; certifies airworthiness of the aircraft; if duty position is section chief, serves as the principal noncommissioned officer in a section. *Skill Level 40:* Able to perform the duties required for Skill Level 30; supervises all maintenance activities; diagnoses complex malfunctions; determines personnel and parts requirements; coordinates work schedules and assigns duties; applies production, quality control, and other maintenance management principles and procedures; prepares evaluations, special reports, and maintenance records; supervises technical training programs; assists in the preparation of plans and policies; controls the flow of work orders, requisitions, and correspondence; plans the layout of helicopter maintenance areas, shops, and facilities. NOTE: May have progressed to 67T40 from 67T30; 66T30, Tactical Transport Helicopter Technical Inspector; 67N30, Utility Helicopter Repairer; or 66N30, Utility Helicopter Technical Inspector.

Recommendation, Skill Level 10

In the lower-division baccalaureate/associate degree category, 3 semester hours in basic helicopter maintenance. (NOTE: This recommendation for skill level 10 is valid for the dates 1/90-9/91 only) (7/87).

Recommendation, Skill Level 20

In the lower-division baccalaureate/associate degree category, 3 semester hours in basic helicopter maintenance and 2 in aviation maintenance management. (NOTE: This recommendation for skill level 20 is valid for the dates 1/90-2/95 only) (7/87).

Recommendation, Skill Level 30

In the lower-division baccalaureate/associate degree category, 3 semester hours in basic helicopter maintenance, 3 in aviation maintenance management, 2 in quality control, and 3 in personnel supervision; if duty position was section chief, 3 additional semester hours in personnel supervision. In the upper-division baccalaureate category, 3 semester hours for field experience in management and 3 in principles of management (7/87).

Recommendation, Skill Level 40

In the lower-division baccalaureate/associate degree category, 4 semester hours in basic helicopter maintenance, 3 in aviation maintenance management, 2 in quality control, 6 in personnel supervision, and 1 in records administration. In the upper-division baccalaureate category, 6 semester hours for field experience in management and 3 in principles of management (7/87).

MOS-67T-003

UH-60 Helicopter Repairer

67T10
67T20
67T30
67T40

Exhibit Dates: 6/99–9/03. Effective 10/03, MOS 67T was discontinued and its duties were transferred to MOS 15T, UH-60 Helicopter Repairer.

Career Management Field: 67 (Aircraft Maintenance).

Description

Summary: Performs or supervises maintenance on tactical transport helicopters, excluding repairs of system components. *Skill Level 10:* Assists in the removal and installation of such subsystem assemblies as engines, transmissions, gear boxes, rotor hubs, and rotor blades; prepares helicopter for extensive inspections and maintenance; assists in operational checks; performs maintenance on common and special tools and ground support equipment; uses maintenance forms and records. *Skill Level 20:* Able to perform the duties required for Skill Level 10; serves as tactical transport helicopter crew chief; provides technical guidance to subordinates; maintains ground support equipment; prepares maintenance forms and records. *Skill Level 30:* Able to perform the duties required for Skill Level 20; supervises all maintenance on tactical transport helicopters; plans work flow; applies production control, quality control, and other maintenance management principles to shop, flightline, and supply operations; instructs personnel and conducts on-the-job training in helicopter maintenance, supply, and safety techniques; evaluates work performance; monitors operational checks and test flights; certifies airworthiness of the aircraft; if duty position is section chief, serves as the principal noncommissioned officer in a section. *Skill Level 40:* Able to perform the duties required for Skill Level 30; supervises all maintenance activities; diagnoses complex malfunctions; determines personnel and parts requirements; coordinates work schedules and assigns duties; applies production, quality control, and other maintenance management principles and procedures; prepares evaluations, special reports, and maintenance records; supervises technical training programs; assists in the preparation of plans and policies; controls the flow of work orders, requisitions, and correspondence; plans the layout of helicopter maintenance areas, shops, and facilities.

Recommendation, Skill Level 10

Credit may be granted on the basis of an individualized assessment of the student (6/99).

Recommendation, Skill Level 20

Credit may be granted on the basis of an individualized assessment of the student (6/99).

Recommendation, Skill Level 30

In the lower-division baccalaureate/associate degree category, 12 semester hours in basic helicopter maintenance, 2 in aircraft forms and records, 2 in aircraft weight and balance, 5 in aviation maintenance management, 2 in quality control, and 3 in principles of supervision (6/99).

Recommendation, Skill Level 40

In the lower-division baccalaureate/associate degree category, 12 semester hours in basic helicopter maintenance, 3 in aircraft forms and records, 2 in aircraft weight and balance, 8 in aviation maintenance management, 4 in quality control, and 3 in principles of supervision. In the upper-division baccalaureate category, 3 semester hours in management practicum and 3 in principles of management (6/99).

MOS-67U-003

CH-47 Helicopter Repairer

67U10
67U20
67U30
67U40

Exhibit Dates: 1/90–12/00.

Career Management Field: 67 (Aircraft Maintenance).

Description

Summary: Performs or supervises maintenance on medium helicopters, excluding repairs of system components. *Skill Level 10:* Assists in the removal and installation of subsystem assemblies such as engines, transmissions, gear boxes, rotor hubs, and rotor blades; prepares helicopter for extensive inspections and maintenance; assists in operational checks; performs maintenance on common and special tools and ground support equipment; uses maintenance forms and records. *Skill Level 20:* Able to perform the duties required for Skill Level 10; serves as medium helicopter crew chief; provides technical guidance to subordinates; maintains ground support equipment; prepares maintenance forms and records. *Skill Level 30:* Able to perform the duties required for Skill Level 20; supervises all maintenance on medium helicopters; plans work flow; applies production control, quality control, and other maintenance management principles to shop, flight line, and supply operations; instructs personnel, and conducts on-the-job training in helicopter maintenance, supply, and safety techniques; evaluates work performance; monitors operational checks and test flights; certifies airworthiness of the aircraft; if duty position is section chief, serves as the principal noncommissioned officer in a section. *Skill Level 40:* Able to perform the duties required for Skill Level 30; supervises all maintenance activities; diagnoses complex malfunctions; determines personnel and parts requirements; coordinates work schedules and assigns duties; applies production, quality control, and other maintenance management principles and procedures; prepares evaluations, special reports, and maintenance records; supervises technical training; assists in the preparation of plans and policies; controls the flow of work orders, requisitions, and correspondence; plans the layout of helicopter maintenance areas, shops, and facilities. NOTE: May have progressed to 67U40 from 67U30; 66U30, Medium Helicopter Technical Inspector; 67X30, Heavy Lift Helicopter Repairer; or 66X30, Heavy Lift Helicopter Technical Inspector.

Recommendation, Skill Level 10

In the lower-division baccalaureate/associate degree category, 8 semester hours in basic helicopter maintenance. (NOTE: This recommendation for skill level 10 is valid for the dates 1/90-9/91 only) (7/87).

Recommendation, Skill Level 20

In the lower-division baccalaureate/associate degree category, 8 semester hours in basic helicopter maintenance and 2 in aviation maintenance management. (NOTE: This recommendation for skill level 20 is valid for the dates 1/90-2/95 only) (7/87).

Recommendation, Skill Level 30

In the lower-division baccalaureate/associate degree category, 8 semester hours in helicopter maintenance, 3 in aviation maintenance management, 2 in quality control, and 3 in personnel supervision; if duty position was section chief, 3 additional semester hours in personnel supervision. In the upper-division baccalaureate category, 3 semester hours for field experience in management and 3 in principles of management (7/87).

Recommendation, Skill Level 40

In the lower-division baccalaureate/associate degree category, 9 semester hours in basic helicopter maintenance, 3 in aviation maintenance management, 2 in quality control, 6 in personnel supervision, and 1 in records administration. In the upper-division baccalaureate category, 6 semester hours for field experience in management and 3 in principles of management (7/87).

MOS-67U-004

CH-47 Helicopter Repairer

67U10
67U20
67U30
67U40

Exhibit Dates: 1/01–9/03. Effective 10/03, MOS 67U was discontinued and its duties were transferred to MOS 15U, CH-47 Helicopter Repairer.

Description

Summary: Performs or supervises maintenance on helicopters, excluding repairs of system components, or conducts maintenance inspections. *Skill Level 10:* Performs maintenance on helicopters; removes and installs such subsystem assemblies as engines, transmissions, gear boxes, rotor hubs, and rotor blades; performs extensive inspections and maintenance on helicopters; uses maintenance forms and records; performs operational checks; maintains ground support equipment; may perform crew chief duties. *Skill Level 20:* Able to perform the duties required for Skill Level 10; provides technical guidance to subordinates; troubleshoots helicopter systems; prepares maintenance forms and records; applies production control; instructs subordinates; conducts on-the-job training in helicopter maintenance, logistics, and safety practices; assists in performing weight and balance checks; performs section chief duties; uses computer management information systems. *Skill Level 30:* Able to perform the duties required for Skill Level 20; supervises all helicopter maintenance; plans work flow; applies quality control and other maintenance management principles to shop, flight line, and logistics; supervises on-the-job training; evaluates work performance; monitors operational checks and test flights; certifies airworthiness of the aircraft; performs management functions; analyzes computer management information system products. *Skill Level 40:* Able to perform the duties required for Skill Level 30; supervises all maintenance activities; diagnoses complex malfunctions; determines personnel and logistics requirements; coordinates work schedules and assigns duties; prepares special reports and maintenance records; supervises technical training; assists in the preparation of plans and policies; manages the preparation of work orders, requisitions, and correspondence; plans the layout of helicopter maintenance areas, shops, and facilities.

Recommendation, Skill Level 10

Credit may be granted on the basis of an individualized assessment of the student (1/01).

Recommendation, Skill Level 20

Credit may be granted on the basis of an individualized assessment of the student (1/01).

Recommendation, Skill Level 30

In the lower-division baccalaureate/associate degree category, 9 semester hours in aircraft maintenance fundamentals, 3 in aircraft systems fundamentals, 3 in principles of supervision, 3 in computer management information systems, and 3 in logistics management. In the upper-division baccalaureate category, 3 semester hours in organizational management, 3 in human resource management, and 3 in advanced computer management information systems (1/01).

Recommendation, Skill Level 40

In the lower-division baccalaureate/associate degree category, 9 semester hours in aircraft maintenance fundamentals, 3 in aircraft systems fundamentals, 3 in principles of supervision, 3 in computer management information systems, and 3 in logistics management. In the upper-division baccalaureate category, 3 semester hours in organizational management, 3 in human resource management, 3 in principles of management, and 3 in advanced computer management information systems (1/01).

MOS-67V-003

Observation/Scout Helicopter Repairer

67V10
67V20
67V30

Exhibit Dates: 1/90–1/98.

Career Management Field: 67 (Aircraft Maintenance).

Description

Summary: Supervises or performs maintenance on Observation/Scout helicopters, excluding the repair of system components. *Skill Level 10:* Performs maintenance on Observation/Scout helicopters; assists in the removal and installation of such subsystem assemblies as engines, transmissions, gear boxes, rotor hubs, and rotor blades; prepares helicopters for major inspections; assists in operational checks; uses maintenance forms and records. *Skill Level 20:* Able to perform the duties required for Skill Level 10; provides technical guidance to subordinates; performs operational inspection checks; troubleshoots rotor system malfunctions; performs all scheduled and special aircraft inspection maintenance tasks; prepares maintenance forms and records; maintains ground support equipment. *Skill Level 30:* Able to perform the duties required for Skill Level 20; applies quality control, production control, and other maintenance management principles to shop, flight line, and supply operations; instructs personnel and conducts on-the-job training in helicopter maintenance, supply, and safety techniques; assists in performing weight and balance checks; monitors operational checks and test flights; certifies airworthiness of aircraft; if duty position is section chief, serves as the principal noncommissioned officer in a section.

Recommendation, Skill Level 10

In the lower-division baccalaureate/associate degree category, 8 semester hours in basic helicopter maintenance. (NOTE: This recommendation for skill level 10 is valid for the dates 1/90-9/91 only) (7/87).

Recommendation, Skill Level 20

In the lower-division baccalaureate/associate degree category, 8 semester hours in basic helicopter maintenance and 2 in aviation maintenance management. (NOTE: This recommendation for skill level 20 is valid for the dates 1/90-2/95 only) (7/87).

Recommendation, Skill Level 30

In the lower-division baccalaureate/associate degree category, 8 semester hours in basic helicopter maintenance, 2 in quality control, 3 in aviation maintenance management, and 3 in personnel supervision; if the duty position was section chief, 3 additional semester hours in personnel supervision. In the upper-division baccalaureate category, 3 semester hours for field experience in management and 3 in principles of management (7/87).

MOS-67V-004

Observation/Scout Helicopter Repairer

67V10
67V20
67V30

Exhibit Dates: 2/98–9/03. Effective 10/03, MOS 67V was discontinued and its duties were transferred to MOS 15V, Observation/Scout Helicopter Repairer.

Description

Summary: Performs or supervises maintenance on observation/scout helicopters, excluding repairs of system components, or conducts maintenance inspections. *Skill Level 10:* Performs maintenance on observation/scout helicopters; removes and installs such subsystem assemblies as engines, transmissions, gear boxes, rotor hubs, and rotor blades; performs extensive inspections and maintenance on helicopters; uses maintenance forms and records; performs operational checks; maintains ground support equipment; may perform crew chief duties. *Skill Level 20:* Able to perform the duties required for Skill Level 10; provides technical guidance to subordinates; troubleshoots helicopter systems; prepares maintenance forms and records; applies production control; instructs subordinates; conducts on-the-job training in helicopter maintenance, logistics, and safety practices; assists in performing weight and balance checks; performs section chief duties; uses computer management information systems. *Skill Level 30:* Able to perform the duties required for Skill Level 20; supervises all helicopter maintenance; plans work flow; applies quality control and other maintenance management principles to shop, flight line, and logistics; supervises on-the-job training; evaluates work performance; monitors operational checks and test flights; certifies airworthiness of the aircraft; performs management functions; analyzes computer management information system products.

Recommendation, Skill Level 10

Credit may be granted on the basis of an individualized assessment of the student (1/01).

Recommendation, Skill Level 20

Credit may be granted on the basis of an individualized assessment of the student (1/01).

Recommendation, Skill Level 30

In the lower-division baccalaureate/associate degree category, 9 semester hours in aircraft maintenance fundamentals, 3 in aircraft systems fundamentals, 3 in principles of supervision, 3 in computer management information systems, and 3 in logistics management. In the upper-division baccalaureate category, 3 semester hours in organizational management, 3 in human resource management, and 3 in advanced computer management information systems (1/01).

MOS-67X-003

Heavy Lift Helicopter Repairer

67X10
67X20
67X30

Exhibit Dates: 1/90–6/94. MOS 67X was discontinued.

Career Management Field: 67 (Aircraft Maintenance).

Description

Summary: Performs or supervises helicopter maintenance, excluding repair of system components. *Skill Level 10:* Maintains and repairs heavy lift helicopters and prepares them for inspections; removes and replaces such subsystem components as wheels, tires, starters, generators, inverters, voltage regulators, valves, hydraulic cylinders, and hoses; assists in operational checks; uses maintenance forms and records. *Skill Level 20:* Able to perform the duties required for Skill Level 10; may serve as heavy lift helicopter crew chief; provides technical guidance to subordinates; removes and replaces aircraft power plant, power train components, flight controls, and flight control hydraulic components; troubleshoots rotor system malfunctions; performs operational checks; performs all scheduled and special maintenance tasks; maintains ground support equipment. *Skill Level 30:* Able to perform the duties required for Skill Level 20; plans work flow; applies production control, quality control, and other maintenance management principles to shop and flight-line operations; prepares forms and records; certifies airworthiness of aircraft.

Recommendation, Skill Level 10

In the lower-division baccalaureate/associate degree category, 8 semester hours in basic helicopter maintenance (heavy). (NOTE: This recommendation for skill level 10 is valid for the dates 1/90-9/91 only) (7/87).

Recommendation, Skill Level 20

In the lower-division baccalaureate/associate degree category, 8 semester hours in basic helicopter maintenance (heavy) and 2 in aviation maintenance management (7/87).

Recommendation, Skill Level 30

In the lower-division baccalaureate/associate degree category, 8 semester hours in basic helicopter maintenance (heavy), 2 in quality control, 3 in aviation maintenance management, and 3 in personnel supervision. In the upper-division baccalaureate category, 3 semester hours for field experience in management and 3 in principles of management (7/87).

MOS-67Y-003

AH-1 Attack Helicopter Repairer
67Y10
67Y20
67Y30
67Y40

Exhibit Dates: 1/90–5/99.

Career Management Field: 67 (Aircraft Maintenance).

Description

Summary: Performs or supervises maintenance on attack helicopters, excluding repairs of system components. *Skill Level 10:* Assists in the removal and installation of such subsystem assemblies as engines, transmissions, gear boxes, rotor hubs, and rotor blades; prepares helicopter for extensive inspections and maintenance; assists in maintenance checks; performs maintenance on common and special tools and ground support equipment; uses maintenance forms and records. *Skill Level 20:* Able to perform the duties required for Skill Level 10; serves as attack helicopter crew chief; provides technical guidance to subordinates; maintains ground support equipment; prepares maintenance forms and records. *Skill Level 30:* Able to perform the duties required for Skill Level 20; supervises all maintenance on attack helicopters; plans work flow; applies production control, quality control, and other maintenance management principles to shop, flight-line, and supply operations; instructs personnel, and conducts on-the-job training in helicopter maintenance, supply, and safety techniques; evaluates work performance; monitors operational checks and test flights; certifies airworthiness of the aircraft; if duty position is section chief, serves as the principal noncommissioned officer in a section. *Skill Level 40:* Able to perform the duties required for Skill Level 30; supervises all maintenance activities; diagnoses complex malfunctions; determines personnel and parts requirements; coordinates work schedules and assigns duties; applies production, quality control, and other maintenance management principles and procedures; prepares evaluations, special reports, and maintenance records; supervises technical training programs; assists in the preparation of plans and policies; controls the flow of work orders, requisitions, and correspondence; plans the layout of helicopter maintenance areas, shops, and facilities. NOTE: May have progressed to 67Y40 from 67Y30; 66Y30, AH-1 Attack Helicopter Technical Inspector; 67V30, Observation/Scout Helicopter Repairer; or 66V30, Observation/Scout Helicopter Technical Inspector.

Recommendation, Skill Level 10

In the lower-division baccalaureate/associate degree category, 6 semester hours in basic helicopter maintenance. (NOTE: This recommendation for skill level 10 is valid for the dates 1/90-9/91 only) (7/87).

Recommendation, Skill Level 20

In the lower-division baccalaureate/associate degree category, 6 semester hours in basic helicopter maintenance and 2 in aviation maintenance management. (NOTE: This recommendation for skill level 20 is valid for the dates 1/90-2/95 only) (7/87).

Recommendation, Skill Level 30

In the lower-division baccalaureate/associate degree category, 6 semester hours in basic helicopter maintenance, 3 in aviation maintenance management, 2 in quality control, and 3 in personnel supervision; if duty position was section chief, 3 additional semester hours in personnel supervision. In the upper-division baccalaureate category, 3 semester hours for field experience in management and 3 in principles of management (7/87).

Recommendation, Skill Level 40

In the lower-division baccalaureate/associate degree category, 7 semester hours in basic helicopter maintenance, 3 in aviation maintenance management, 2 in quality control, 6 in personnel supervision, and 1 in records administration. In the upper-division baccalaureate category, 6 semester hours for field experience in management and 3 in principles of management (7/87).

MOS-67Y-004

AH-1 Attack Helicopter Repairer
67Y10
67Y20
67Y30
67Y40

Exhibit Dates: 6/99–9/04. (Effective 9/04, MOS 67Y was discontinued, and its duties were incorporated into MOS 15R.)

Career Management Field: 67 (Aircraft Maintenance).

Description

Summary: Performs or supervises maintenance on attack helicopters, excluding repairs of system components. *Skill Level 10:* Assists in the removal and installation of such subsystem assemblies as engines, transmissions, gear boxes, rotor hubs, and rotor blades; prepares helicopter for extensive inspections and maintenance; assists in maintenance checks; performs maintenance on common and special tools and ground support equipment; uses maintenance forms and records. *Skill Level 20:* Able to perform the duties required for Skill Level 10; serves as attack helicopter crew chief; provides technical guidance to subordinates; maintains ground support equipment; prepares maintenance forms and records. *Skill Level 30:* Able to perform the duties required for Skill Level 20; supervises all maintenance on attack helicopters; plans work flow; applies production control, quality control, and other maintenance management principles to shop, flight-line, and supply operations; instructs personnel and conducts on-the-job training in helicopter maintenance, supply, and safety techniques; evaluates work performance; monitors operational checks and test flights; certifies airworthiness of the aircraft; if duty position is section chief, serves as the principal noncommissioned officer in a section. *Skill Level 40:* Able to perform the duties required for Skill Level 30; supervises all maintenance activities; diagnoses complex malfunctions; determines personnel and parts requirements; coordinates work schedules and assigns duties; applies production, quality control, and other maintenance management principles and procedures; prepares evaluations, special reports, and maintenance records; supervises technical training programs; assists in the preparation of plans and policies; controls the flow of work orders, requisitions, and correspondence; plans the layout of helicopter maintenance areas, shops, and facilities.

Recommendation, Skill Level 10

Credit may be granted on the basis of an individualized assessment of the student (6/99).

Recommendation, Skill Level 20

Credit may be granted on the basis of an individualized assessment of the student (6/99).

Recommendation, Skill Level 30

In the lower-division baccalaureate/associate degree category, 12 semester hours in basic helicopter maintenance, 2 in aircraft forms and records, 2 in aircraft weight and balance, 5 in aviation maintenance management, 2 in quality control, and 3 in principles of supervision (6/99).

Recommendation, Skill Level 40

In the lower-division baccalaureate/associate degree category, 12 semester hours in basic helicopter maintenance, 3 in aircraft forms and records, 2 in aircraft weight and balance, 8 in aviation maintenance management, 4 in quality control, and 3 in principles of supervision. In the upper-division baccalaureate category, 3 semester hours in management practicum and 3 in principles of management (6/99).

MOS-67Z-003

Aircraft Maintenance Senior Sergeant
67Z50

Exhibit Dates: 1/90–5/99.

Career Management Field: 67 (Aircraft Maintenance).

Description

Supervises aircraft maintenance at either unit, intermediate, or depot levels on activities having a mix of aircraft maintenance and/or component repair; serves as the principal noncommissioned officer in aircraft maintenance activities and may become first sergeant; plans and lays out aircraft maintenance areas and component repair shops and facilities; ensures that proper maintenance techniques have been employed by use of quality assurance inspections; supervises and participates in preparation of studies, evaluations, special reports, and records pertaining to aircraft maintenance and component repair operations and training and related activities; applies production quality control and other maintenance management principles and procedures; processes work orders and allocates maintenance responsibilities. NOTE: May have progressed to 67Z50 from 67G40, Utility Airplane Repairer; 67T40, Tactical Transport Helicopter Repairer; 67R40, AH-64 Attack Helicopter Repairer; 67H40, AH-1 Attack Helicopter Repairer; 67U40, Medium Helicopter Repairer; 68K40, Aircraft Components Repair Supervisor; or 68J40, Aircraft Fire Control Repairer.

Recommendation

In the lower-division baccalaureate/associate degree category, 3 semester hours in technical writing, 3 in applied psychology, and 3 in interpersonal communications. In the upper-division baccalaureate category, 3 semester hours for field experience in management and 6 in management problems (7/87).

MOS-67Z-004

Aircraft Maintenance Senior Sergeant
67Z50

Exhibit Dates: 6/99–9/03. Effective 10/03, MOS 67Z will be discontinued and its duties were transferred to MOS 15Z.

Career Management Field: 67 (Aircraft Maintenance).

Description

Supervises aircraft maintenance at either unit, intermediate, or depot levels on activities having a mix of aircraft maintenance and/or component repair; serves as the principal noncommissioned officer in aircraft maintenance activities and may become first sergeant; plans and lays out aircraft maintenance areas and component repair shops and facilities; ensures that proper maintenance techniques have been employed by use of quality assurance inspections; supervises and participates in preparation of studies, evaluations, special reports, and records pertaining to aircraft maintenance and component repair operations and training and related activities; applies production quality control and other maintenance management principles and procedures; processes work orders and allocates maintenance responsibilities.

Recommendation

In the lower-division baccalaureate/associate degree category, 3 semester hours in technical writing, 3 in principles of supervision, and 3 in maintenance management. In the upper-division baccalaureate category, 3 semester hours in a management practicum and 3 in principles of management (6/99).

MOS-68B-002

Aircraft Powerplant Repairer
68B10
68B20
68B30

Exhibit Dates: 1/90–12/90.

Career Management Field: 67 (Aviation Maintenance), subfield 672 (Aircraft Component Repair).

Description

Summary: Performs or supervises maintenance on turbine engines and their components. *Skill Level 10:* Performs maintenance on turbine power plants and assists in power plant removal; removes, replaces, services, prepares, preserves, and cleans engine assemblies or components; performs maintenance on ground support equipment; uses and maintains common and special tools; uses certain aircraft forms and records; prepares requests for turn-ins and repair parts. *Skill Level 20:* Able to perform the duties required for Skill Level 10; provides technical guidance to subordinates; applies troubleshooting techniques to diagnose malfunctions in specific engine components; disassembles, repairs, reassembles, adjusts, and tests turbine engine systems, subsystems, and components; maintains selected ground support equipment; performs maintenance on diagnostic equipment; prepares forms and records. *Skill Level 30:* Able to perform the duties required for Skill Level 20; plans work flow; applies production control, quality control, and other maintenance management principles and procedures to shop operations; prepares forms and records; conducts on-the-job training; supervises and evaluates work performance in power plant maintenance; may serve as technical inspector; monitors shop and flight-line safety.

Recommendation, Skill Level 10

In the vocational certificate category, 12 semester hours in turbine engine maintenance and 3 for technical electives. In the lower-division baccalaureate/associate degree category, 15 semester hours in aviation maintenance technology (or 12 in turbine engine maintenance and 3 for technical electives) (5/78).

Recommendation, Skill Level 20

In the vocational certificate category, 12 semester hours in turbine engine maintenance and 3 for technical electives. In the lower-division baccalaureate/associate degree category, 15 semester hours in aviation maintenance technology (or 12 in turbine engine maintenance and 3 for technical electives) and 3 in aviation maintenance management. In the upper-division baccalaureate category, 6 semester hours in aviation maintenance management (5/78).

Recommendation, Skill Level 30

In the vocational certificate category, 12 semester hours in turbine engine maintenance and 3 for technical electives. In the lower-division baccalaureate/associate degree category, 15 semester hours in aviation maintenance technology (or 12 in turbine engine maintenance and 3 for technical electives), 3 in aviation maintenance management, 3 in quality control, 3 in flight safety, and 3 in human relations (or a management elective); if the duty assignment was aircraft power plant repair supervisor section chief, 3 semester hours in personnel supervision. In the upper-division baccalaureate category, 6 semester hours in aviation maintenance management and 3 in personnel management (5/78).

MOS-68B-003

Aircraft Powerplant Repairer
68B10
68B20
68B30

Exhibit Dates: 1/91–5/02.

Career Management Field: 67 (Aircraft Maintenance).

Description

Summary: Performs or supervises maintenance on turbine engines and their components. *Skill Level 10:* Performs maintenance on turbine power plants and assists in power plant removal; removes, replaces, services, prepares, preserves, and cleans engine assemblies or components; maintains ground support equipment; uses and performs maintenance on common and special tools; uses certain aircraft forms and records; prepares requests for turn-ins and repair parts. *Skill Level 20:* Able to perform the duties required for Skill Level 10; provides technical guidance to subordinates; applies troubleshooting techniques to diagnose malfunctions in specific engine components; disassembles, repairs, reassembles, adjusts, and tests turbine engine systems, subsystems and components; maintains selected ground support equipment; maintains diagnostic equipment; prepares forms and records; performs on-the-job training. *Skill Level 30:* Able to perform the duties required for Skill Level 20; plans work flow; applies production control, quality assurance, and other maintenance management principles and procedures to shop operations; prepares forms and records; conducts on-the-job training; supervises and evaluates work performance in power plant maintenance; may serve as technical inspector; monitors shop and flight-line safety.

Recommendation, Skill Level 10

In the lower-division baccalaureate/associate degree category, 3 semester hours in basic aircraft maintenance and 3 in principles of turbine engines. (NOTE: This recommendation for skill level 10 is valid for the dates 1/91-9/91 only) (3/91).

Recommendation, Skill Level 20

In the lower-division baccalaureate/associate degree category, 6 semester hours in basic aircraft maintenance, 3 in principles of turbine engines, and 3 in personnel supervision. (NOTE: This recommendation for skill level 20 is valid for the dates 1/91-2/95 only) (3/91).

Recommendation, Skill Level 30

In the lower-division baccalaureate/associate degree category, 6 semester hours in basic aircraft maintenance, 3 in principles of turbine engines, 3 in personnel supervision, 3 in maintenance management, 2 in records and information management, and 2 in occupational safety (3/91).

MOS-68B-004

AIRCRAFT POWERPLANT REPAIRER

68B10
68B20
68B30

Exhibit Dates: 6/02–9/03. Effective 10/03, MOS 68B will be discontinued and its duties will be transferred to MOS 15B, Aircraft Powerplant Repairer.

Career Management Field: 67 (Aircraft Maintenance).

Description

Summary: Performs or supervises maintenance on turbine engines and their components. *Skill Level 10:* Performs maintenance on turbine power plants and assists in power plant removal; removes, replaces, services, prepares, preserves, and cleans engine assemblies or components; maintains ground support equipment; uses and performs maintenance on common and special tools; uses certain aircraft forms and records; prepares requests for turn-ins and repair parts. *Skill Level 20:* Able to perform the duties required for Skill Level 10; provides technical guidance to subordinates; applies troubleshooting techniques to diagnose malfunctions in specific engine components; disassembles, repairs, reassembles, adjusts, and tests turbine engine systems, subsystems and components; maintains selected ground support equipment; maintains diagnostic equipment; prepares forms and records; performs on-the-job training. *Skill Level 30:* Able to perform the duties required for Skill Level 20; plans work flow; applies production control, quality assurance, and other maintenance management principles and procedures to shop operations; prepares forms and records; conducts on-the-job training; supervises and evaluates work performance in power plant maintenance; may serve as technical inspector; monitors shop and flight-line safety.

Recommendation, Skill Level 10

Credit may be granted on the basis of an individualized assessment of the student (6/02).

Recommendation, Skill Level 20

Credit may be granted on the basis of an individualized assessment of the student (6/02).

Recommendation, Skill Level 30

In the lower-division baccalaureate/associate degree category, 3 semester hours in turbine engine principles, 3 in turbine engine maintenance, 3 in turbine engine troubleshooting, 3 in principles of supervision, 1 in introduction to computers, 1 in computer applications, 3 in maintenance management, 2 in records management, and 2 in aviation safety (6/02).

MOS-68D-002

AIRCRAFT POWERTRAIN REPAIRER

68D10
68D20
68D30

Exhibit Dates: 1/90–12/90.

Career Management Field: 67 (Aviation Maintenance), subfield 672 (Aircraft Component Repair)

Description

Summary: Performs maintenance and repair or supervises maintenance on helicopter power train subsystems and components. *Skill Level 10:* Performs maintenance and tests; assists with inspections on helicopter power train subsystems and components; removes and replaces power train quills, transmissions-adapting parts, and rotary wing hub oil tanks; obtains and prepares oil samples for analysis; applies corrosion preventive procedures; cleans, preserves, and stores components; performs maintenance on selected ground support equipment; uses and performs maintenance on common and special tools; uses aircraft forms and records; prepares request for turn-in and repair parts. *Skill Level 20:* Able to perform the duties required for Skill Level 10; provides technical guidance to subordinates; applies troubleshooting techniques to diagnose malfunctions; performs nondestructive inspections on aircraft components; disassembles, repairs, reassembles, adjusts, and tests power train components, systems, and subsystems; balances main and tail rotor hub assembly and aligns main rotor and hub assembly; maintains diagnostic equipment; prepares forms and records. *Skill Level 30:* Able to perform the duties required for Skill Level 20; may serve as technical inspector; plans work; applies production control, quality control, and other maintenance management principles; supervises and evaluates work performance; conducts on-the-job training; monitors shop and flight-line safety.

Recommendation, Skill Level 10

In the vocational certificate category, 12 semester hours in helicopter component maintenance and 3 for technical electives. In the lower-division baccalaureate/associate degree category, 15 semester hours in aviation maintenance technology (or 12 in helicopter component maintenance and 3 for technical electives) (5/78).

Recommendation, Skill Level 20

In the vocational certificate category, 12 semester hours in helicopter component maintenance and 3 for technical electives. In the lower-division baccalaureate/associate degree category, 15 semester hours in aviation maintenance technology (or 12 in helicopter component maintenance and 3 for technical electives) and 3 in aviation maintenance management. In the upper-division baccalaureate category, 6 semester hours in aviation maintenance management (5/78).

Recommendation, Skill Level 30

In the vocational certificate category, 12 semester hours in helicopter component maintenance and 3 for technical electives. In the lower-division baccalaureate/associate degree category, 15 semester hours in aviation maintenance technology (or 12 in helicopter component maintenance and 3 for technical electives), 3 in aviation maintenance management, 3 in quality control, 3 in flight safety, and 3 in human relations (or a management elective); if the duty position was aircraft power train repair supervisor, 3 semester hours in personnel supervision. In the upper-division baccalaureate category, 6 semester hours in aviation maintenance management and 3 in personnel management (5/78).

MOS-68D-003

AIRCRAFT POWERTRAIN REPAIRER

68D10
68D20
68D30

Exhibit Dates: 1/91–5/02.

Career Management Field: 67 (Aircraft Maintenance).

Description

Summary: Performs maintenance and repair or supervises maintenance on helicopter power train subsystems and components. *Skill Level 10:* Performs maintenance and tests; assists with inspections on helicopter power train subsystems and components; removes and replaces power train quills, transmission-adapting parts, and rotary wing hub oil tanks; obtains and prepares oil samples for analysis; applies corrosion preventive procedures; cleans, preserves, and stores components; performs maintenance on selected ground support equipment; uses and performs maintenance on common and special tools; uses certain aircraft forms and records; prepares request for turn-in and repair parts. *Skill Level 20:* Able to perform the duties required for Skill Level 10; provides technical guidance to subordinates; applies troubleshooting techniques to diagnose malfunctions; performs nondestructive inspections on aircraft components; disassembles, repairs, reassembles, adjusts, and tests power train components, systems, and subsystems; balances main and tail rotor hub assembly and aligns main rotor and hub assembly; performs maintenance on diagnostic equipment; prepares forms and records; supervises on-the-job training. *Skill Level 30:* Able to perform the duties required for Skill Level 20; may serve as technical inspector; plans work; applies production control, quality assurance, and other maintenance management principles; supervises and evaluates work performance; conducts on-the-job training; monitors shop and flight-line safety.

Recommendation, Skill Level 10

In the lower-division baccalaureate/associate degree category, 3 semester hours in basic aircraft maintenance and 3 in power trains. (NOTE: This recommendation for skill level 10 is valid for the dates 1/91-9/91 only) (3/91).

Recommendation, Skill Level 20

In the lower-division baccalaureate/associate degree category, 6 semester hours in basic aircraft maintenance, 3 in power trains, and 3 in personnel supervision. (NOTE: This recommendation for skill level 20 is valid for the dates 1/91-2/95 only) (3/91).

Recommendation, Skill Level 30

In the lower-division baccalaureate/associate degree category, 6 semester hours in basic aircraft maintenance, 3 in power trains, 3 in personnel supervision, 3 in maintenance management, 2 in records and information management, 2 in occupational safety, and 3 in nondestructive inspections (3/91).

MOS-68D-004

AIRCRAFT POWERTRAIN REPAIRER

68D10
68D20
68D30

Exhibit Dates: 6/02–9/03. Effective 10/03, MOS 68D will be discontinued and its duties will be transferred to MOS 15D, Aircraft Powertrain Repairer.

Career Management Field: 67 (Aircraft Maintenance).

Description

Summary: Supervises or performs maintenance and repair on helicopter power train subsystems and components. *Skill Level 10:* Performs maintenance and tests; assists with inspections on helicopter power train subsystems and components; removes and replaces power train quills, transmission-adapting parts, and rotary wing hub oil tanks; applies corrosion preventive procedures; cleans, preserves, and stores components; performs maintenance on selected ground support equipment; uses and performs maintenance on common and special tools; uses certain aircraft forms and records; prepares request for turn-in and repair parts. *Skill Level 20:* Able to perform the duties required for Skill Level 10; provides technical guidance to subordinates; applies troubleshooting techniques to diagnose malfunctions; performs nondestructive inspections on aircraft components; disassembles, repairs, reassembles, adjusts, and tests power train components, systems, and subsystems; balances main and tail rotor hub assembly and aligns main rotor and hub assembly; performs maintenance on diagnostic equipment; prepares forms and records; supervises on-the-job training. *Skill Level 30:* Able to perform the duties required for Skill Level 20; may serve as technical inspector; plans work; applies production control, quality assurance, and other maintenance management principles; supervises and evaluates work performance; conducts on-the-job training; monitors shop and flight-line safety.

Recommendation, Skill Level 10

Credit may be granted on the basis of an individualized assessment of the student (6/02).

Recommendation, Skill Level 20

Credit may be granted on the basis of an individualized assessment of the student (6/02).

Recommendation, Skill Level 30

In the lower-division baccalaureate/associate degree category, 3 semester hours in aircraft maintenance fundamentals, 3 in helicopter powertrain maintenance, 3 in helicopter powertrain repairs, 3 in nondestructive inspections, 3 in principles of supervision, 1 in introduction to computers, 3 in maintenance management, 3 in records management, 3 in industrial management, 1 in computer applications, and 3 in industrial maintenance (6/02).

MOS-68F-002

Aircraft Electrician

68F10
68F20
68F30

Exhibit Dates: 1/90–12/90.

Career Management Field: 67 (Aviation Maintenance), subfield 672 (Aircraft Component Repair).

Description

Summary: Supervises or performs maintenance on aircraft electrical/electronic systems and aircraft crew station instruments. *Skill Level 10:* Removes and replaces electrical/electronic elements of assemblies and components, including electrical wiring, aircraft battery systems, and instruments; applies proper soldering techniques during circuitry repair, cleans, preserves, and stores electrical/electronic components and aircraft instruments; uses and performs maintenance on common and special tools; maintains and repairs ground support equipment; uses aircraft forms and records; prepares requests for turn-in and repair parts. *Skill Level 20:* Able to perform the duties required for Skill Level 10; provides technical guidance to subordinates; applies troubleshooting techniques to diagnose and localize malfunctions to specific electrical/electronic components including solid state and transistorized subsystems; disassembles, repairs, reassembles, adjusts, and tests electrical/electronic elements of assemblies and components; applies principles of pneudraulics to troubleshooting of interfacing electropneudraulic systems and components; repairs and replaces printed circuits; repairs nickel cadmium batteries; applies principles of electricity/electronics, gyroscopic motion, pneumatics, and hydraulics to repair aircraft instrument subsystems; performs maintenance on diagnostic equipment; prepares forms and records. *Skill Level 30:* Able to perform the duties required for Skill Level 20; may serve as technical inspector; plans work flow; applies production control, quality control, and other maintenance management principles; supervises and evaluates work performance; conducts on-the-job training; monitors shop and flight-line safety.

Recommendation, Skill Level 10

In the vocational certificate category, 12 semester hours in aircraft electrical systems maintenance and repair and 3 for technical electives. In the lower-division baccalaureate/associate degree category, 15 semester hours in aviation maintenance technology (5/78).

Recommendation, Skill Level 20

In the vocational certificate category, 12 semester hours in aircraft electrical system maintenance and repair and 3 for technical electives. In the lower-division baccalaureate/associate degree category, 15 semester hours in aviation maintenance technology and 3 in aviation maintenance management. In the upper-division baccalaureate category, 6 semester hours in aviation maintenance management (5/78).

Recommendation, Skill Level 30

In the vocational certificate category, 12 semester hours in aircraft electrical systems maintenance and repair and 3 for technical electives. In the lower-division baccalaureate/associate degree category, 15 semester hours in aviation maintenance technology, 3 in aviation maintenance management, 3 in quality control, 3 in flight safety, and 3 in human relations (or a management elective); if the duty position was aircraft electrician supervisor section chief, 3 semester hours in personnel supervision. In the upper-division baccalaureate category, 6 semester hours in aviation maintenance management and 3 in personnel management (5/78).

MOS-68F-003

Aircraft Electrician

68F10
68F20
68F30

Exhibit Dates: 1/91–5/02.

Career Management Field: 67 (Aircraft Maintenance).

Description

Summary: Supervises or performs maintenance on aircraft electrical/electronic systems and aircraft crew station instruments. *Skill Level 10:* Removes and replaces electrical/electronic elements of assemblies and components, including electrical wiring, aircraft battery systems, and instruments; applies proper soldering techniques during circuitry repair; cleans, preserves, and stores electrical/electronic components and aircraft instruments; uses and maintains common and special tools; maintains and repairs ground support equipment; uses aircraft forms and records; prepares requests for turn-in and repair parts. *Skill Level 20:* Able to perform the duties required for Skill Level 10; provides technical guidance to subordinates; applies troubleshooting techniques to diagnose and localize malfunctions to specific electrical/electronic components including solid state and transistorized subsystems; disassembles, repairs, reassembles, adjusts, and tests electrical/electronic elements of assemblies and components; applies principles of pneudraulics to troubleshooting of interfacing electropneudraulic systems and components; repairs and replaces printed circuits; repairs nickel cadmium batteries; applies principles of electricity/electronics, gyroscopic motion, pneumatics, and hydraulics to repair aircraft instrument subsystems; performs maintenance on diagnostic equipment; prepares forms and records. *Skill Level 30:* Able to perform the duties required for Skill Level 20; may serve as technical inspector; plans work flow; applies production control, quality assurance and other maintenance management principles; supervises and evaluates work performance; conducts on-the-job training; monitors shop and flight-line safety.

Recommendation, Skill Level 10

In the lower-division baccalaureate/associate degree category, 3 semester hours in basic aircraft maintenance and 3 in basic electronics. (NOTE: This recommendation for skill level 10 is valid for the dates 1/91-9/91 only) (3/91).

Recommendation, Skill Level 20

In the lower-division baccalaureate/associate degree category, 3 semester hours in basic aircraft maintenance, 3 in basic electronics, 3 in basic electricity and electronic circuits, 3 in electronic systems troubleshooting and repair, and 3 in personnel supervision. (NOTE: This recommendation for skill level 20 is valid for the dates 1/91-2/95 only) (3/91).

Recommendation, Skill Level 30

In the lower-division baccalaureate/associate degree category, 3 semester hours in basic aircraft maintenance, 3 in basic electronics, 3 in basic electricity and electronic circuits, 3 in electronic systems troubleshooting and repair, 3 in personnel supervision, 3 in maintenance management, 2 in records and information management, and 2 in occupational safety (3/91).

MOS-68F-004

Aircraft Electrician

68F10
68F20
68F30

Exhibit Dates: 6/02–9/03. Effective 10/03, MOS 68F will be discontinued and its duties will be transferred to MOS 15F, Aircraft Electrician.

Career Management Field: 67 (Aircraft Maintenance).

Description

Summary: Supervises or performs maintenance on aircraft electrical/electronic systems and aircraft crew station instruments. *Skill Level 10:* Removes and replaces electrical/electronic elements of assemblies and components, including electrical wiring, aircraft battery systems, and instruments; applies proper soldering techniques during circuitry repair; cleans, preserves, and stores electrical/electronic components and aircraft instruments; uses and maintains common and special tools; maintains and repairs ground support equipment; uses aircraft forms and records; prepares requests for turn-in and repair parts. *Skill Level 20:* Able to perform the duties required for Skill Level 10; provides technical guidance to subordinates; applies troubleshooting techniques to diagnose and localize malfunctions to specific electrical/electronic components including solid state and transistorized subsystems; disassembles, repairs, reassembles, adjusts, and tests electrical/electronic elements of assemblies and components; applies principles of pneudraulics to troubleshooting of interfacing electropneudraulic systems and components; repairs and replaces printed circuits; repairs nickel cadmium batteries; performs maintenance on diagnostic equipment; prepares forms and records. *Skill Level 30:* Able to perform the duties required for Skill Level 20; may serve as technical inspector; plans work flow; applies production control, quality assurance and other maintenance management principles; supervises and evaluates work performance; conducts on-the-job training; monitors shop and flight-line safety.

Recommendation, Skill Level 10

Credit may be granted on the basis of an individualized assessment of the student (6/02).

Recommendation, Skill Level 20

Credit may be granted on the basis of an individualized assessment of the student (6/02).

Recommendation, Skill Level 30

In the lower-division baccalaureate/associate degree category, 3 semester hours in aircraft maintenance fundamentals, 3 in AC/DC circuit analysis, 3 in electronic devices and circuits, 3 in electronic systems troubleshooting and repair, 3 in principles of supervision, 1 in introduction to computers, 1 in computer applications, 3 in maintenance management, 2 in records management, and 2 in aviation safety (6/02).

MOS-68G-002

Aircraft Structural Repairer

68G10
68G20
68G30

Exhibit Dates: 1/90–12/90.

Career Management Field: 67 (Aviation Maintenance), subfield 672 (Aircraft Component Repair).

Description

Summary: Supervises, inspects, or performs maintenance and repair of aircraft structure. *Skill Level 10:* Makes airframe structural repairs by fabrication of new parts using approved fasteners, rivets, special-purpose fasteners, and fiberglass materials; prepares airframe surfaces for painting; mixes and applies primers and paints to aircraft; uses and maintains shop tools and equipment; uses field and technical manuals to make specific repairs. *Skill Level 20:* Able to perform the duties required for Skill Level 10; inspects airframes and components before, during, and after repair or modification to ensure that specifications are met; performs welding maintenance; uses precision measuring gauges and instruments; determines shop and bench stock requirements; prepares or assists in the preparation of requests for parts, tools, and supplies; interprets blueprints; coordinates work load; applies maintenance management principles and procedures to shop operations. *Skill Level 30:* Able to perform the duties required for Skill Level 20; supervises maintenance; may perform technical inspections on aircraft structures and control surfaces; conducts and supervises on-the-job training; ensures quality and production control; applies management principles and procedures to shop operations; may supervise from 6 to 19 persons; monitors shop and flight-line safety.

Recommendation, Skill Level 10

In the vocational certificate category, 6 semester hours in airframe structures, 3 for technical electives, and additional credit in airframe structures on the basis of institutional evaluation. In the lower-division baccalaureate/associate degree category, 9 semester hours in aviation maintenance technology (or 6 in airframe structures and 3 for technical electives) and additional credit in airframe structures or aviation maintenance technology on the basis of institutional evaluation (5/78).

Recommendation, Skill Level 20

In the vocational certificate category, 6 semester hours in airframe structures, 3 for technical electives, and additional credit in airframe structures on the basis of institutional evaluation. In the lower-division baccalaureate/associate degree category, 9 semester hours in aviation maintenance technology (or 6 in airframe structures and 3 for technical electives), 3 in aviation maintenance management, and additional credit in airframe structures or aviation maintenance technology on the basis of institutional evaluation. In the upper-division baccalaureate category, 6 semester hours in aviation maintenance management (5/78).

Recommendation, Skill Level 30

In the vocational certificate category, 9 semester hours in airframe structures, 3 for technical electives, and additional credit in airframe structures on the basis of institutional evaluation. In the lower-division baccalaureate/associate degree category, 12 semester hours in aviation maintenance technology (or 9 in airframe structures and 3 for technical electives), 3 in aviation maintenance management, 3 in quality control, 3 in flight safety, 3 in human relations (or a management elective), and additional credit in aviation maintenance technology or airframe structures on the basis of institutional evaluation; if duty position was aircraft structural repair supervisor section chief, 3 semester hours in personnel supervision. In the upper-division baccalaureate category, 6 semester hours in aviation maintenance management and 3 in personnel management (5/78).

MOS-68G-003

Aircraft Structural Repairer

68G10
68G20
68G30

Exhibit Dates: 1/91–5/02.

Career Management Field: 67 (Aircraft Maintenance).

Description

Summary: Supervises, inspects, or performs maintenance and repair of aircraft structure. *Skill Level 10:* Makes airframe structural repairs by fabrication of new parts using approved fasteners, rivets, special-purpose fasteners, and fiberglass materials; uses and maintains shop tools and equipment; uses field and technical manuals to make specific repairs. *Skill Level 20:* Able to perform the duties required for Skill Level 10; inspects airframes and components before, during, and after repair or modification to ensure specifications are met; uses precision measuring gauges and instruments; determines shop and bench stock requirements; prepares or assists in the preparation of requests for parts, tools, and supplies; interprets blueprints; coordinates work load; applies maintenance management principles and procedures to shop operations; supervises on-the-job training. *Skill Level 30:* Able to perform the duties required for Skill Level 20; supervises maintenance; may perform technical inspections on aircraft structures and control surfaces; conducts and supervises on-the-job training; ensures quality and production control; applies management principles and procedures to shop operations; may supervise from 6 to 19 persons; monitors shop and flight-line safety.

Recommendation, Skill Level 10

In the lower-division baccalaureate/associate degree category, 3 semester hours in basic aircraft maintenance and 3 in airframe structures. (NOTE: This recommendation for skill level 10 is valid for the dates 1/91-9/91 only) (3/91).

Recommendation, Skill Level 20

In the lower-division baccalaureate/associate degree category, 6 semester hours in basic aircraft maintenance, 3 in airframe structures, and 3 in personnel supervision. (NOTE: This recommendation for skill level 20 is valid for the dates 1/91-2/95 only) (3/91).

Recommendation, Skill Level 30

In the lower-division baccalaureate/associate degree category, 6 semester hours in basic aircraft maintenance, 3 in airframe structures, 3 in personnel supervision, 3 in maintenance management, 2 in occupational safety, and 2 in records and information management (3/91).

MOS-68G-004

Aircraft Structural Repairer

68G10
68G20
68G30

Exhibit Dates: 6/02–9/03. Effective 10/03, MOS 68G will be discontinued and its duties will be transferred to MOS 15G, Aircraft Structural Repairer.

Career Management Field: 67 (Aircraft Maintenance).

Description

Summary: Supervises, inspects, or performs maintenance and repair of aircraft structure.

Skill Level 10: Makes airframe structural repairs by fabrication of new parts using approved fasteners, rivets, special-purpose fasteners, and fiberglass materials; uses and maintains shop tools and equipment; uses field and technical manuals to make specific repairs. *Skill Level 20:* Able to perform the duties required for Skill Level 10; inspects airframes and components before, during, and after repair or modification to ensure specifications are met; uses precision measuring gauges and instruments; determines shop and bench stock requirements; prepares or assists in the preparation of requests for parts, tools, and supplies; interprets blueprints; coordinates work load; applies maintenance management principles and procedures to shop operations; supervises on-the-job training. *Skill Level 30:* Able to perform the duties required for Skill Level 20; supervises maintenance; may perform technical inspections on aircraft structures and control surfaces; conducts and supervises on-the-job training; ensures quality and production control; applies management principles and procedures to shop operations; may supervise from 6 to 19 persons; monitors shop and flight-line safety.

Recommendation, Skill Level 10

Credit may be granted on the basis of an individualized assessment of the student (6/02).

Recommendation, Skill Level 20

Credit may be granted on the basis of an individualized assessment of the student (6/02).

Recommendation, Skill Level 30

In the lower-division baccalaureate/associate degree category, 3 semester hours in aviation safety, 3 in aircraft and helicopter structures, 6 in aircraft maintenance fundamentals, 3 in principles of supervision,1 in introduction to computers, 2 in supply management, 3 in maintenance management, 1 in computer applications, and 2 in records management (6/02).

MOS-68H-002

Aircraft Pneudraulics Repairer
68H10
68H20
68H30

Exhibit Dates: 1/90–12/90.

Career Management Field: 67 (Aviation Maintenance), subfield 672 (Aircraft Component Repair).

Description

Summary: Supervises, inspects, or performs maintenance on aircraft hydraulic systems. *Skill Level 10:* Removes and installs tubing assemblies, hoses, valves, pressure transmitters, and switches; services filters and shock absorbers/struts; repairs tubing assemblies and hydraulic and pneumatic valves; flushes and bleeds hydraulic systems; uses and maintains common and special tools; prepares requests for turn-in and repair parts; uses aircraft forms and records. *Skill Level 20:* Able to perform the duties required for Skill Level 10; provides technical guidance to subordinates; applies troubleshooting techniques to diagnose and localize malfunctions to hydraulic systems, subsystems, and components; disassembles, repairs, reassembles, adjusts, and tests hydraulic systems, subsystems, and components; maintains selected items of ground support equipment; prepares forms and records. *Skill Level 30:* Able to perform the duties required for Skill Level 20; applies production/quality control and maintenance management principles and procedures to shop operations; plans work flow; conducts and supervises on-the-job training; may perform technical inspections on electrical/electronic and subsystems components before, during, and after repair or modification; may supervise from 6 to 19 persons.

Recommendation, Skill Level 10

In the vocational certificate category, 6 semester hours in aircraft hydraulic system maintenance, 3 for technical electives, and additional credit in aircraft hydraulic system maintenance on the basis of institutional evaluation. In the lower-division baccalaureate/associate degree category, 9 semester hours in aviation maintenance technology (or 6 in aircraft hydraulic system maintenance and 3 for technical electives), 3 in aviation maintenance management, and additional credit in aviation maintenance technology or aircraft hydraulic system maintenance on the basis of institutional evaluation (5/78).

Recommendation, Skill Level 20

In the vocational certificate category, 6 semester hours in aircraft hydraulic system maintenance, 3 for technical electives, and additional credit in aircraft hydraulic system maintenance on the basis of institutional evaluation. In the lower-division baccalaureate/associate degree category, 9 semester hours in aviation maintenance technology (or 6 in aircraft hydraulic system maintenance and 3 for technical electives), 3 in aviation maintenance management, and additional credit in aviation maintenance technology or aircraft hydraulic system maintenance on the basis of institutional evaluation. In the upper-division baccalaureate category, 6 semester hours in aviation maintenance management (5/78).

Recommendation, Skill Level 30

In the vocational certificate category, 9 semester hours in aircraft hydraulic system maintenance, 3 for technical electives, and additional credit in aircraft hydraulic system maintenance on the basis of institutional evaluation. In the lower-division baccalaureate/associate degree category, 12 semester hours in aviation maintenance technology (or 9 in aircraft hydraulic system maintenance and 3 in technical electives), 3 in aviation maintenance management, 3 in quality control, 3 in flight safety, 3 in human relations (or a management elective), and additional credit in aviation maintenance management or aircraft hydraulic system maintenance on the basis of institutional evaluation; if the duty position was aircraft pneudraulics repair supervisor section chief, 3 semester hours in personnel supervision. In the upper-division baccalaureate category, 6 semester hours in aviation maintenance management and 3 in personnel management (5/78).

MOS-68H-003

Aircraft Pneudraulics Repairer
68H10
68H20
68H30

Exhibit Dates: 1/91–5/02.

Career Management Field: 67 (Aircraft Maintenance).

Description

Summary: Supervises, inspects, or performs maintenance on aircraft hydraulic systems. *Skill Level 10:* Removes and installs tubing assemblies, hoses, valves, pressure transmitters, and switches; services filters and shock absorbers/struts; repairs tubing assemblies and hydraulic and pneumatic valves; flushes and bleeds hydraulic systems; uses and maintains common and special tools; prepares requests for turn-in and repair parts; uses aircraft forms and records. *Skill Level 20:* Able to perform the duties required for Skill Level 10; provides technical guidance to subordinates; applies troubleshooting techniques to diagnose and localize malfunctions to hydraulic systems, subsystems, and components; maintains selected items of ground support equipment; prepares forms and records; supervises on-the-job training. *Skill Level 30:* Able to perform the duties required for Skill Level 20; applies production/quality assurance and maintenance management principles and procedures to shop operations; plans work flow; conducts and supervises on-the-job training; may perform technical inspections of electrical/electronic and subsystems components before, during, and after repair or modification; may supervise from 6 to 19 persons.

Recommendation, Skill Level 10

In the lower-division baccalaureate/associate degree category, 3 semester hours in aircraft fluid power. (NOTE: This recommendation for skill level 10 is valid for the dates 1/91-9/91 only) (3/91).

Recommendation, Skill Level 20

In the lower-division baccalaureate/associate degree category, 6 semester hours in basic aircraft maintenance, 3 in aircraft fluid power, and 3 in personnel supervision. (NOTE: This recommendation for skill level 20 is valid for the dates 1/91-2/95 only) (3/91).

Recommendation, Skill Level 30

In the lower-division baccalaureate/associate degree category, 6 semester hours in basic aircraft maintenance, 3 in aircraft fluid power, 3 in personnel supervision, 3 in maintenance management, 2 in records and information management, and 2 in occupational safety (3/91).

MOS-68H-004

Aircraft Pneudraulics Repairer
68H10
68H20
68H30

Exhibit Dates: 6/02–9/03. Effective 10/03, MOS 68H will be discontinued and its duties will be transferred to MOS 15H, Aircraft Pneudraulics Repairer.

Career Management Field: 67 (Aircraft Maintenance).

Description

Summary: Supervises, inspects, or performs maintenance on aircraft hydraulic systems. *Skill Level 10:* Removes and installs tubing assemblies, hoses, valves, pressure transmitters, and switches; services filters and shock absorbers/struts; repairs tubing assemblies and hydraulic and pneumatic valves; flushes and bleeds hydraulic systems; uses and maintains common and special tools; prepares requests for turn-in and repair parts; uses aircraft forms and records. *Skill Level 20:* Able to perform the duties required for Skill Level 10; provides technical guidance to subordinates; applies troubleshooting techniques to diagnose and localize malfunctions to hydraulic systems, subsystems, and components; maintains selected items of ground support equipment; prepares forms and records; supervises on-the-job training. *Skill Level 30:* Able to perform the duties required for Skill Level 20; applies production/quality assurance and maintenance management principles and procedures to shop operations; plans work flow; conducts and supervises on-the-job training; may perform technical inspections of electrical/electronic and subsystems components before, during, and after repair or modification; may supervise from 6 to 19 persons.

Recommendation, Skill Level 10

Credit may be granted on the basis of an individualized assessment of the student (6/02).

Recommendation, Skill Level 20

Credit may be granted on the basis of an individualized assessment of the student (6/02).

Recommendation, Skill Level 30

In the lower-division baccalaureate/associate degree category, 3 semester hours in fluid power, 3 in aviation safety, 6 in aircraft maintenance fundamentals, 3 in principles of supervision, 2 in supply management, 3 in maintenance management, and 2 in records management (6/02).

MOS-68J-004

AIRCRAFT ARMAMENT/MISSILE SYSTEMS REPAIRER
68J10
68J20
68J30
68J40

Exhibit Dates: 1/90–9/04. (Effective 9/04, MOS 68J was discontinued, and its duties were incorporated into MOS 15J.)

Career Management Field: 67 (Aircraft Maintenance).

Description

Summary: Performs electrical and electronic maintenance on aircraft fire control systems, armament systems, and ground support equipment. *Skill Level 10:* Removes, installs, assembles, disassembles, and performs maintenance on aircraft fire control systems; repairs and/or replaces electrical and electronic components. *Skill Level 20:* Able to perform the duties required for Skill Level 10; tests and troubleshoots electrical and electronic components of fire control systems using schematics and test equipment. *Skill Level 30:* Able to perform the duties required for Skill Level 20; supervises maintenance and performs technical inspections on fire control systems; plans and supervises on-the-job training; supervises eight or more persons. *Skill Level 40:* Able to perform the duties required for Skill Level 30; supervises all maintenance activities; determines personnel and parts requirements; coordinates work schedules and assigns duties; applies production, quality control, and other maintenance management principles and procedures; prepares evaluations, special reports, and maintenance records; supervises technical training programs; assists in the preparation of plans and policies; controls the flow of work orders, requisitions, and correspondence; plans the layout of aircraft weapons and fire control system maintenance areas and facilities; advises personnel in diagnosing complex malfunctions.

Recommendation, Skill Level 10

In the lower-division baccalaureate/associate degree category, 3 semester hours in basic electronics. (NOTE: This recommendation for skill level 10 is valid for the dates 1/90-9/91 only) (3/91).

Recommendation, Skill Level 20

In the lower-division baccalaureate/associate degree category, 3 semester hours in basic electronics, 3 in electronic systems troubleshooting and repair, and 3 in personnel supervision. (NOTE: This recommendation for skill level 20 is valid for the dates 1/90-2/95 only) (3/91).

Recommendation, Skill Level 30

In the lower-division baccalaureate/associate degree category, 3 semester hours in basic electronics, 3 in electronic systems troubleshooting and repair, 3 in personnel supervision, 1 in occupational safety, 3 in maintenance management, and 2 in records and information management (3/91).

Recommendation, Skill Level 40

In the lower-division baccalaureate/associate degree category, 3 semester hours in basic electronics, 3 in electronic systems troubleshooting and repair, 3 in personnel supervision, 2 in occupational safety, 3 in maintenance management, and 2 in records and information management. In the upper-division baccalaureate category, 3 semester hours for field experience in management (3/91).

MOS-68K-001

AIRCRAFT COMPONENTS REPAIR SUPERVISOR
68K40

Exhibit Dates: 1/90–12/90.

Career Management Field: 67 (Aviation Maintenance), subfield 672 (Aircraft Component Repair).

Description

Able to perform the duties required for skill level 30 in at least one of the following MOS's: 68B (Aircraft Power Plant Repairer), 68D (Aircraft Powertrain Repairer), 68F (Aircraft Electrician), 68G (Aircraft Structural Repairer), 68H (Aircraft Pneudraulics Repairer), 68J (Helicopter Missile System Repairer), or 68M (Helicopter Weapon Systems Repairer); supervises maintenance operations of aircraft component repairs; plans work loads in terms of resources and facilities; applies production control, quality control, and other maintenance management principles to aircraft component repair; supervises and evaluates the work performance of a minimum of 20 persons; orients and instructs subordinates; conducts on-the-job training programs.

Recommendation

In the upper-division baccalaureate category, 3 semester hours in maintenance management and 3 for field experience in personnel management (5/78).

MOS-68K-002

AIRCRAFT COMPONENTS REPAIR SUPERVISOR
68K40

Exhibit Dates: 1/91–5/02.

Career Management Field: 67 (Aircraft Maintenance).

Description

Able to supervises the duties of the following MOS's: 68B (Aircraft Powerplant Repairer), 68D (Aircraft Powertrain Repairer), 68F (Aircraft Electrician), 68G (Aircraft Structural Repairer), and 68H (Aircraft Pneudraulics Repairer); supervises maintenance operations of aircraft component repairs; plans work load in terms of resources and facilities; applies production control, quality assurance, and other maintenance management principles to aircraft component repair; supervises and evaluates the work performance of 20 or more persons; orients and instructs subordinates; conducts on-the-job training. NOTE: May have progressed from skill level 30 of MOS 68B, 68D, 68F, 68G, or 68H.

Recommendation

In the upper-division baccalaureate category, 3 semester hours for field experience in management. NOTE: Add credit for skill level 30 of previously-held MOS (3/91).

MOS-68K-003

AIRCRAFT COMPONENTS REPAIR SUPERVISOR
68K40

Exhibit Dates: 6/02–9/03. Effective 10/03, MOS 68K will be discontinued and its duties will be transferred to MOS 15K, Aircraft Components Repair Supervisor.

Career Management Field: 67 (Aircraft Maintenance).

Description

Able to supervise the duties of the following MOS's: 68B (Aircraft Powerplant Repairer), 68D (Aircraft Powertrain Repairer), 68F (Aircraft Electrician), 68G (Aircraft Structural Repairer), 68H (Aircraft Pneudraulics Repairer), and 68N (Avionic Mechanic); supervises maintenance operations of aircraft component repairs; plans work load in terms of resources and facilities; applies production control, quality assurance, and other maintenance management principles to aircraft component repair; supervises and evaluates the work performance of 20 or more persons; orients and instructs subordinates; conducts on-the-job training. NOTE: May have progressed from skill level 30 of MOS 68B, 68D, 68F, 68G, 68H,or 68N.

Recommendation

In the upper-division baccalaureate category, 3 semester hours for a management practicum, 3 in principles of management, and 3 in operations management (6/02).

MOS-68L-001

AVIONIC COMMUNICATIONS EQUIPMENT REPAIRER
68L10
68L20
68L30

Exhibit Dates: 1/90–6/96. Effective 6/96, MOS 68L was discontinued and its duties were incorporated into MOS 35L, Avionic Communications Equipment Repairer.

Career Management Field: 67 (Aircraft Maintenance).

Description

Summary: Performs intermediate maintenance or repairs (to the component level) on avionic communications equipment. *Skill Level 10:* Performs intermediate maintenance, within a shop environment, on aircraft communications equipment including very high frequency, ultra high, and high frequency AM and FM receivers and transmitters and intercom and electronic interface devices. *Skill Level 20:* Able to perform the duties required for Skill Level 10; supervises, trains, and counsels subordinates; requisitions parts; keeps records. *Skill Level 30:* Able to perform the duties required for Skill Level 20; initiates quality assurance programs; supervises subordinates; handles requisitions and communication with superiors; develops training programs.

Recommendation, Skill Level 10

In the lower-division baccalaureate/associate degree category, 3 semester hours in electronic systems troubleshooting and maintenance, 2 in avionic communications systems, 3 in basic electronics laboratory, and 2 in basic electronics. (NOTE: This recommendation for skill level 10 is valid for the dates 1/90-9/91 only) (3/91).

Recommendation, Skill Level 20

In the lower-division baccalaureate/associate degree category, 3 semester hours in electronic systems troubleshooting and maintenance, 3 in avionic communications systems, 3 in basic electronics laboratory, 3 in basic electronics, and 3 in personnel supervision. (NOTE: This recommendation for skill level 20 is valid for the dates 1/90-2/95 only) (3/91).

Recommendation, Skill Level 30

In the lower-division baccalaureate/associate degree category, 3 semester hours in electronic systems troubleshooting and maintenance, 3 in avionic communications systems, 3 in basic electronics laboratory, 3 in basic electronics, 3 in personnel supervision, 3 in maintenance management, and 2 in records and information management (3/91).

MOS-68N-001

Avionic Mechanic
68N10
68N20
68N30

Exhibit Dates: 1/90–5/02.

Career Management Field: 67 (Aircraft Maintenance).

Description

Summary: Performs unit maintenance on tactical security systems (COMSEC), communications, navigation, transponder, and flight control equipment installed in aircraft; maintains aircraft antennas and associated aircraft wiring; aligns radar antennas. *Skill Level 10:* Performs troubleshooting procedures to the level of individual units; removes and replaces modules of security systems, communications, navigation, transponder, and flight control equipment and their associated wiring and antennas; checks performance of individual systems. *Skill Level 20:* Able to perform the duties required for Skill Level 10; coordinates maintenance schedules; trains and assists subordinates. *Skill Level 30:* Able to perform the duties required for Skill Level 20; coordinates and supervises unit and intermediate avionic maintenance on all aircraft; implements quality assurance programs; supervises training; coordinates tasks with superiors; maintains equipment forms and records.

Recommendation, Skill Level 10

In the lower-division baccalaureate/associate degree category, 3 semester hours in electronic systems troubleshooting and maintenance, 2 in basic electronics, and 2 in avionics systems. (NOTE: This recommendation for skill level 10 is valid for the dates 1/90-9/91 only) (3/91).

Recommendation, Skill Level 20

In the lower-division baccalaureate/associate degree category, 3 semester hours in electronic systems troubleshooting and maintenance, 3 in basic electronics, 3 in avionics systems, and 3 in personnel supervision. (NOTE: This recommendation for skill level 20 is valid for the dates 1/90-2/95 only) (3/91).

Recommendation, Skill Level 30

In the lower-division baccalaureate/associate degree category, 3 semester hours in electronic systems troubleshooting and maintenance, 3 in basic electronics, 3 in avionics systems, 3 in personnel supervision, 3 in maintenance management, 2 in records and information management, and 2 in occupational safety (3/91).

MOS-68N-002

Avionic Mechanic
68N10
68N20
68N30

Exhibit Dates: 6/02–9/03. Effective 10/03, MOS 68N will be discontinued and its duties will be transferred to MOS 15N, Avionic Mechanic.

Career Management Field: 67 (Aircraft Maintenance).

Description

Summary: Performs unit maintenance on tactical security systems (COMSEC), communications, navigation, transponder, and flight control equipment installed in aircraft; maintains aircraft antennas and associated aircraft wiring; aligns radar antennas; maintains stock and bench components for repair of aircraft avionics equipment. *Skill Level 10:* Performs troubleshooting procedures to the level of individual units; removes and replaces modules of security systems, communications, navigation, transponder, and flight control equipment and their associated wiring and antennas; checks performance of individual systems; prepares forms and records related to aircraft maintenance. *Skill Level 20:* Able to perform the duties required for Skill Level 10; coordinates maintenance schedules; trains and assists subordinates. *Skill Level 30:* Able to perform the duties required for Skill Level 20; coordinates and supervises unit and intermediate avionic maintenance on all aircraft; implements quality assurance programs; supervises training; coordinates tasks with superiors; maintains equipment forms and records.

Recommendation, Skill Level 10

Credit may be granted on the basis of an individualized assessment of the student (6/02).

Recommendation, Skill Level 20

Credit may be granted on the basis of an individualized assessment of the student (6/02).

Recommendation, Skill Level 30

In the lower-division baccalaureate/associate degree category, 3 semester hours in aircraft maintenance fundamentals, 3 in AC/DC circuit analysis, 3 in electronic devices and circuits, 3 in electronic systems troubleshooting and repair, 2 in electronic communications, 1 in digital electronics, 3 in principles of supervision, 1 in introduction to computers, 1 in computer applications, 3 in maintenance management, 2 in records management, and 2 in aviation safety (6/02).

MOS-68P-001

Avionic Maintenance Supervisor
68P40

Exhibit Dates: 1/90–12/99. Effective 12/99, MOS 68P was discontinued and its duties were incorporated into MOS 68K, Aircraft Components Repair Supervisor.

Career Management Field: 67 (Aircraft Maintenance).

Description

Able to supervise the duties of the following MOS's: 68L (Avionic Communications Equipment Repairer), 68N (Avionic Mechanic), 68Q (Avionic Flight Systems Repairer), and 68R (Avionic Radar Repairer); supervises intermediate and depot maintenance of aviation communications, navigation, radar, flight control and other electronic/electrical systems associated with Army aircraft; supervises production and quality assurance procedures; inspects work areas and maintenance equipment; assists in installation and modification of complex equipment; plans and schedules overall maintenance; provides technical advice; enforces shop safety procedures; supervises the maintenance of equipment records, spare parts, and supplies; reviews and directs appropriate action on technical directives; recommends changes to maintenance publications and drawings; supervises and participates in studies, evaluations, special reports, and records pertaining to radar, communications, flight controls, automatic test equipment, maintenance operations, and training issues; develops and supervises training programs for avionic maintenance personnel. NOTE: May have progressed to 68P40 from skill level 30 of 68L, 68N, 68Q, or 68R.

Recommendation

In the upper-division baccalaureate category, 3 semester hours for field experience in management. NOTE: Add credit for skill level 30 of previously-held MOS (3/91).

MOS-68Q-001

Avionic Flight Systems Repairer
68Q10
68Q20
68Q30

Exhibit Dates: 1/90–6/96. Effective 6/96, MOS 68Q was discontinued and it duties were incorporated into MOS 35Q, Avionic Flight Systems Repairer.

Career Management Field: 67 (Aircraft Maintenance).

Description

Summary: Performs intermediate maintenance or repairs on avionic navigation, flight control, and stabilization equipment. *Skill Level 10:* Performs, within a shop environment, intermediate repair of such flight control equipment as marker beacons, radio direction finders, tactical bearing and range equipment, instrument landing systems, and auto pilot and stabilization equipment; replaces parts to the component level. *Skill Level 20:* Able to perform the duties required for Skill Level 10; calibrates, repairs, and documents test instruments; trains and counsels subordinates; assists in maintenance duties. *Skill Level 30:* Able to perform the duties required for Skill Level 20; initiates quality assurance programs; supervises subordinates; handles requisitions, records, and communication with superiors; develops training programs.

Recommendation, Skill Level 10

In the lower-division baccalaureate/associate degree category, 3 semester hours in electronic systems troubleshooting and maintenance, 2 in airborne electronic navigation systems, 3 in basic electronics laboratory, and 2 in basic electronics. (NOTE: This recommendation for skill level 10 is valid for the dates 1/90-9/91 only) (3/91).

Recommendation, Skill Level 20

In the lower-division baccalaureate/associate degree category, 3 semester hours in electronic systems troubleshooting and maintenance, 3 in airborne electronic navigation systems, 3 in basic electronics laboratory, 3 in basic electronics, and 3 in personnel supervision. (NOTE: This recommendation for skill level 20 is valid for the dates 1/90-2/95 only) (3/91).

Recommendation, Skill Level 30

In the lower-division baccalaureate/associate degree category, 3 semester hours in electronic systems troubleshooting and maintenance, 3 in airborne electronic navigation systems, 3 in basic electronics laboratory, 3 in basic electronics, 3 in personnel supervision, 3 in maintenance management, and 2 in records and information management (3/91).

MOS-68R-001

Avionic Radar Repairer
68R10
68R20
68R30

Exhibit Dates: 1/90–6/96. Effective 6/96, MOS 68R was discontinued and its duties were incorporated into MOS 35R, Avionic Radar Repairer.

Career Management Field: 67 (Aircraft Maintenance).

Description

Summary: Performs intermediate and depot maintenance on avionic radar equipment. *Skill Level 10:* Performs organizational maintenance on test, diagnostic, and measurement equipment used to diagnose problems in avionic radar equipment; assists in intermediate maintenance of avionic airborne radar equipment, including terrain-following avoidance radars, Doppler navigation radars, weather radars, radar altimeters, radar transponders, navigational distance and bearing equipment, and inertial navigation instruments. *Skill Level 20:* Able to perform the duties required for Skill Level 10; performs maintenance on special test equipment and avionic equipment; trains subordinates; prepares maintenance records and requisitions; repairs circuit boards; tests support equipment for the OH-58D helicopter. *Skill Level 30:* Able to perform the duties required for Skill Level 20; assists establishing production and quality assurance programs; teaches maintenance practices and techniques; supervises and inspects the maintenance of avionic radar equipment; develops training programs; provides technical assistance to subordinates; assists in the development of new types of electronic equipment; maintains equipment records, technical literature, and spare parts; inspects maintenance procedures for quality.

Recommendation, Skill Level 10

In the lower-division baccalaureate/associate degree category, 2 semester hours in electronic systems troubleshooting and maintenance, 2 in radar systems, 3 in basic electronics laboratory, and 2 in basic electronics. (NOTE: This recommendation for skill level 10 is valid for the dates 1/90-9/91 only) (3/91).

Recommendation, Skill Level 20

In the lower-division baccalaureate/associate degree category, 3 semester hours in electronic systems troubleshooting and maintenance, 3 in radar systems, 3 in basic electronics laboratory, 3 in basic electronics, and 3 in personnel supervision. (NOTE: This recommendation for skill level 20 is valid for the dates 1/90-2/95 only) (3/91).

Recommendation, Skill Level 30

In the lower-division baccalaureate/associate degree category, 3 semester hours in electronic systems troubleshooting and maintenance, 3 in radar systems, 3 in basic electronics laboratory, 3 in basic electronics, 3 in personnel supervision, 3 in maintenance management, and 2 in records and information management (3/91).

MOS-68S-001

OH-58D Armament/Electrical/Avionics Systems Repairer
68S10
68S20
68S30
68S40

Exhibit Dates: 1/00–9/03. Effective 10/03, MOS 68S will be discontinued and its duties will be transferred to MOS 15J, OH-58D Armament/Electrical/Avionics Systems Repairer.

Career Management Field: 67 (Aircraft Maintenance).

Description

Summary: Maintains, repairs, and aligns armament, electrical, and avionics systems installed on the OH-58D aircraft. *Skill Level 10:* Performs troubleshooting and repair procedures; removes and replaces components of security, communications, navigation, transponder, and flight control systems; maintains, repairs, and installs armament systems and associated weapons on the OH-58D aircraft. Prepares requisitions and maintains stock inventories. *Skill Level 20:* Able to perform the duties required for Skill Level 10; coordinates team work schedules; trains and assists subordinates. *Skill Level 30:* Able to perform the duties required for Skill Level 20; coordinates and supervises unit and intermediate avionics/electrical/armament maintenance; implements quality assurance programs; supervises and schedules training; coordinates tasks with superiors; maintains appropriate forms and records. *Skill Level 40:* Able to perform the duties required for Skill Level 30; supervises aircraft armament, fire control, electrical, and avionics systems maintenance; supervises technical inspection activities; determines personnel, facility, and parts requirements; supervises training programs; prepares evaluations and reports.

Recommendation, Skill Level 10

Credit may be granted on the basis of an individualized assessment of the student (6/02).

Recommendation, Skill Level 20

Credit may be granted on the basis of an individualized assessment of the student (6/02).

Recommendation, Skill Level 30

In the lower-division baccalaureate/associate degree category, 3 semester hours in aviation safety, 3 in aircraft maintenance fundamentals, 3 in AC/DC circuit analysis, 3 in electronic devices and circuits, 2 in electronics communication, 3 in electronic systems troubleshooting and repair, 1 in digital electronics, 3 in principles of supervision, 1 in introduction to computers, 1 in computer applications, 3 in maintenance management, and 2 in records management (6/02).

Recommendation, Skill Level 40

In the lower-division baccalaureate/associate degree category, 3 semester hours in aviation safety, 3 in aircraft maintenance fundamentals, 3 in AC/DC circuit analysis, 3 in electronic devices and circuits, 2 in electronics communication, 3 in electronic systems troubleshooting and repair, 1 in digital electronics, 3 in principles of supervision, 1 in introduction to computers, 1 in computer applications, 3 in maintenance management, and 2 in records management. In the upper-division baccalaureate category, 3 semester hours for a management practicum, 3 in principles of management, and 3 in operations management (6/02).

MOS-68X-001

AH-64 Armament/Electrical Systems Repairer
((Armament/Electrical Systems Repairer))
68X10
68X20
68X30
68X40

Exhibit Dates: 4/92–9/03. Effective 10/03, MOS 68X was discontinued and its duties were transferred to MOS 15X, Armament/Electrical Systems Repairer.

Description

Summary: Supervises and performs maintenance on AH-64 armament/electrical systems at the unit, intermediate, and depot level; maintains and repairs electrical, electronics, mechanical, and pneudraulic systems associate with AH-64 armament, missile, fire control, electrical, and instrument systems. *Skill Level 10:* Performs aviation unit, intermediate and depot maintenance on AH-64 armament electrical systems, integrated electronic/instrument systems, fire control/missile systems, and auxiliary ground support equipment; isolates faults, diagnoses, troubleshoots, and repairs malfunctions to specific armament, electrical, instrument and fire control systems and components including solid state and transistorized subsystems; repairs, disassembles, and assembles equipment according to technical manuals, directives, and safety procedures; uses test sets and diagnostic equipment; maintains records on weapons and subsystems. *Skill Level 20:* Able to perform the duties required for Skill Level 10; provides guidance to subordinate personnel on technical aspects of duties. *Skill Level 30:* Able to perform the duties required for Skill Level 20; supervises and performs extensive diagnostic checks and services; plans work flow; conducts technical training and instruction; applies production control and quality control principles and procedures; evaluates subordinates; inspects maintenance to ensure that repairs are performed within prescribed specifications; monitors shop and flight line safety procedures. *Skill Level 40:* Able to perform the duties required for Skill Level 30; plans and manages AH-64 armament/electrical system maintenance areas and facilities; coordinates work, assigns duties, and instructs subordinates; advises in diagnosing complex malfunctions; ensures that shop safety principles and procedures are observed; prepares evaluations and special reports.

Recommendation, Skill Level 10

Credit may be granted on the basis of an individualized assessment of the student (1/96).

Recommendation, Skill Level 20

In the vocational certificate category, 3 semester hours in aircraft/avionics maintenance. In the lower-division baccalaureate/associate degree category, 6 semester hours in aircraft/avionics maintenance technology. (NOTE: This recommendation for skill level 20 is valid for the dates 4/92-2/95 only) (1/96).

Recommendation, Skill Level 30

In the vocational certificate category, 3 semester hours in aircraft/avionics maintenance. In the lower-division baccalaureate/associate degree category, 6 semester hours in aircraft/avionics maintenance technology and 2 in records and information management (1/96).

Recommendation, Skill Level 40

In the vocational certificate category, 3 semester hours in aircraft/avionics maintenance. In the lower-division baccalaureate/associate degree category, 6 semester hours in aircraft/avionics maintenance technology, 3 in personnel supervision, 3 in maintenance management, and 2 in records and information management. In the upper-division baccalaureate category, 3 semester hours for field experience in management (1/96).

MOS-68Y-001

AH-64D Armament/Electrical/Avionic Systems Repairer
68Y10
68Y20
68Y30
68Y40

Exhibit Dates: 4/92–9/03. Effective 10/03, MOS 68Y will be discontinued and its duties will be transferred to MOS 15Y, AH-64D Armament/Electrical/Avionic Systems Repairer.

Description

Summary: Supervises and performs maintenance on AH-64D armament, electrical, and avionic systems at the unit, intermediate, and depot level; maintains and repairs electrical, electronics, mechanical, and pneudraulic systems associate with AH-64 armament, missile, fire control, electrical, electronic, and instrument systems. *Skill Level 10:* Performs aviation unit, intermediate and depot maintenance on AH-64D armament electrical systems, integrated electronic/instrument systems, fire control/missile systems, and auxiliary ground support equipment; isolates faults, diagnoses, troubleshoots, and repairs malfunctions to specific armament, electrical, instrument and fire control systems and components including solid state and transistorized subsystems; repairs, disassembles, and assembles equipment according to technical manuals, directives, and safety procedures; uses test sets and diagnostic equipment; maintains records on weapons and subsystems. *Skill Level 20:* Able to perform the duties required for Skill Level 10; provides guidance to subordinate personnel on technical aspects of duties. *Skill Level 30:* Able to perform the duties required for Skill Level 20; supervises and performs extensive diagnostic checks and services; plans work flow; conducts technical training and instruction; applies production control and quality control principles and procedures; evaluates subordinates; inspects maintenance to ensure that repairs are performed within prescribed specifications; monitors shop and flight line safety procedures. *Skill Level 40:* Able to perform the duties required for Skill Level 30; plans and manages AH-64 armament/electrical system maintenance areas and facilities; coordinates work, assigns duties, and instructs subordinates; advises in diagnosing complex malfunctions; ensures that shop safety principles and procedures are observed; prepares evaluations and special reports.

Recommendation, Skill Level 10

Credit may be granted on the basis of an individualized assessment of the student (1/96).

Recommendation, Skill Level 20

In the vocational certificate category, 3 semester hours in aircraft/avionics maintenance. In the lower-division baccalaureate/associate degree category, 6 semester hours in aircraft/avionics maintenance technology. (NOTE: This recommendation for skill level 20 is valid for the dates 4/92-2/95 only) (1/96).

Recommendation, Skill Level 30

In the vocational certificate category, 3 semester hours in aircraft/avionics maintenance. In the lower-division baccalaureate/associate degree category, 6 semester hours in aircraft/avionics maintenance technology and 2 in records and information management (1/96).

Recommendation, Skill Level 40

In the vocational certificate category, 3 semester hours in aircraft/avionics maintenance. In the lower-division baccalaureate/associate degree category, 6 semester hours in aircraft/avionics maintenance technology, 3 in personnel supervision, 3 in maintenance management, and 2 in records and information management. In the upper-division baccalaureate category, 3 semester hours for field experience in management (1/96).

MOS-71C-003

Executive Administrative Assistant
71C10
71C20
71C30

Exhibit Dates: 1/90–3/94.

Career Management Field: 71 (Administration), subfield 711 (General Administration).

Description

Summary: Types correspondence, forms, and reports; transcribes recorded dictation; performs general office duties; maintains files; uses word processing equipment; drafts routine correspondence; may supervise office operations. *Skill Level 10:* Types at a minimum speed of 45 net words per minute; handles incoming and outgoing correspondence; types correspondence, forms, and reports; transcribes recorded dictation; files information and maintains files; proofreads and corrects work; serves as receptionist; prepares schedules and handles protocol matters; requisitions publications and forms; drafts routine correspondence. *Skill Level 20:* Able to perform the duties required for Skill Level 10; assumes some responsibility for assigning and verifying work of others; types at a minimum of 50-55 net words per minute. *Skill Level 30:* Able to perform the duties required for Skill Level 20; serves as secretary to general offices; plans and organizes office operations; prepares correspondence and writes procedures and directives.

Recommendation, Skill Level 10

In the lower-division baccalaureate/associate degree category, 3 semester hours in typing, 3 in machine transcription, 3 in business communication, 2 in filing and records management, 2 in clerical record keeping, 2 in office procedures, and 2 in word processing. (NOTE: This recommendation for skill level 10 is valid for the dates 1/90-9/91 only) (3/84).

Recommendation, Skill Level 20

In the lower-division baccalaureate/associate degree category, 6 semester hours in typing, 3 in machine transcription, 3 in business communication, 2 in filing and records management, 2 in clerical record keeping, 2 in office procedures, 2 in word processing, and 2 for a practicum in office procedures (3/84).

Recommendation, Skill Level 30

In the lower-division baccalaureate/associate degree category, 6 semester hours in typing, 3 in machine transcription, 3 in business communication, 2 in filing and records management, 2 in clerical record keeping, 2 in office procedures, 2 in word processing, 2 for a practicum in office procedures, and 2 in office management (3/84).

MOS-71C-004

Executive Administrative Assistant
71C10
71C20
71C30

Exhibit Dates: 4/94–6/96. Effective 6/96, MOS 71C was discontinued and its duties were incorporated into MOS 71L and ASI E3.

Career Management Field: 71 (Administration).

Description

Summary: Types correspondence, forms, and reports; transcribes recorded dictation; performs general office duties; maintains files; uses word processing equipment; drafts routine correspondence; may supervise office operations. *Skill Level 10:* Types at a minimum speed of 45 net words per minute; handles incoming and outgoing correspondence; types correspondence, forms, and reports; transcribes recorded dictation; files information and maintains files; proofreads and corrects work; serves as receptionist; prepares schedules and handles protocol matters; requisitions publications and forms; drafts routine correspondence. *Skill Level 20:* Able to perform the duties required for Skill Level 10; assumes some responsibility for assigning and verifying work of others; types at a minimum of 50-55 net words per minute. *Skill Level 30:* Able to perform the duties required for Skill Level 20; serves as secretary to general offices; plans and organizes office operations; prepares correspondence and writes procedures and directives.

Recommendation, Skill Level 10

Credit may be granted on the basis of an individualized assessment of the student (4/94).

Recommendation, Skill Level 20

In the lower-division baccalaureate/associate degree category, 3 semester hours in machine transcription, 3 in business communication, 3 in filing and records management, 3 in office procedures, 6 in word processing, and 3 for field experience in personnel procedures. (NOTE: This recommendation for skill level 20 is valid for the dates 4/94-2/95 only) (4/94).

Recommendation, Skill Level 30

In the lower-division baccalaureate/associate degree category, 3 semester hours in machine transcription, 3 in business communication, 3 in filing and records management, 3 in office procedures, 6 in word processing, 3 in office management, and 3 for field experience in personnel procedures (4/94).

MOS-71D-002

LEGAL SPECIALIST
71D10
71D20
71D30
71D40
71D50

Exhibit Dates: 1/90–12/90.

Career Management Field: 71 (Administration), subfield 714 (Legal).

Description

Summary: Performs or supervises the preparation of court martial records, board proceedings, preliminary hearings, and investigations. *Skill Level 10:* Under close supervision, assists in preparing legal papers and forms, conducting investigations and researching activities, and in maintaining files; types at a minimum of 30 words per minute; assists in maintaining law library. *Skill Level 20:* Able to perform the duties required for Skill Level 10; prepares correspondence; compiles records into quarterly reports; provides legal assistance (military and civil); makes referrals as needed. *Skill Level 30:* Able to perform the duties required for Skill Level 20; supervises a legal office or maintains a law library; provides guidance to subordinates; maintains files. *Skill Level 40:* Able to perform the duties required for Skill Level 30; plans and organizes legal office operation; previews legal documentation for proper and prompt disposition; researches decisions, statutes, and regulations under supervision of a judge advocate or civilian attorney. *Skill Level 50:* Able to perform the duties required for Skill Level 40; supervises the administrative functions of a large legal office; prepares budgets; plans programs.

Recommendation, Skill Level 10

In the lower-division baccalaureate/associate degree category, 3 semester hours in typing, 3 in office practices, 3 in legal terminology, and 3 in English composition (6/78).

Recommendation, Skill Level 20

In the lower-division baccalaureate/associate degree category, 3 semester hours in typing, 3 in office practices, 3 in business communication or technical writing, 3 in legal terminology, 2 in military legal practices and procedures, and 3 in English composition (6/78).

Recommendation, Skill Level 30

In the lower-division baccalaureate/associate degree category, 3 semester hours in typing, 3 in office practices, 3 in business communication or technical writing, 3 in legal terminology, 3 in military legal practices and procedures, 3 in English composition, 3 in human relations, and 3 in office management (6/78).

Recommendation, Skill Level 40

In the lower-division baccalaureate/associate degree category, 3 semester hours in typing, 3 in office practices, 3 in business communication or technical writing, 3 in legal terminology, 3 in military legal practices and procedures, 3 in English composition, 3 in human relations, 3 in office management, and 3 in introduction to management. In the upper-division baccalaureate category, 3 semester hours for field experience in office management (6/78).

Recommendation, Skill Level 50

In the lower-division baccalaureate/associate degree category, 3 semester hours in typing, 3 in office practices, 3 in business communication or technical writing, 3 in legal terminology, 3 in military legal practices and procedures, 3 in English composition, 3 in human relations, 3 in office management, and 3 in introduction to management. In the upper-division baccalaureate category, 6 semester hours for field experience in office management (6/78).

MOS-71D-003

LEGAL SPECIALIST
71D10
71D20
71D30
71D40
71D50

Exhibit Dates: 1/91–7/94.

Career Management Field: 71 (Administration).

Description

Summary: prepares or supervises the preparation of court martial records, board proceedings, preliminary hearings, and investigations. *Skill Level 10:* Under close supervision, assists in preparing legal papers and forms, conducting investigations and researching activities, and maintaining files; types at a minimum of 30 words per minute; assists in maintaining law library. *Skill Level 20:* Able to perform the duties required for Skill Level 10; prepares correspondence; interview clients to determine nature of problem; compiles records into quarterly reports; provides legal assistance (military and civil); refers legal problems to attorney; makes referrals as needed. *Skill Level 30:* Able to perform the duties required for Skill Level 20; supervises a legal office or maintains a law library; provides guidance to subordinates; responsible for maintaining files; reviews claims and conducts claim investigations. *Skill Level 40:* Able to perform the duties required for Skill Level 30; plans and organizes legal office operation; previews legal documentation for proper and prompt disposition; researches decisions, statutes, and regulations under supervision of a judge advocate or civilian attorney. *Skill Level 50:* Able to perform the duties required for Skill Level 40; supervises the administration of a large legal office; prepares budgets; plans programs; secures special proceedings, highly sensitive or specially-handled documents, or evidence; assigns duties; evaluates performance of subordinates.

Recommendation, Skill Level 10

In the lower-division baccalaureate/associate degree category, 3 semester hours in legal research, 3 in typewriting, 3 in office practices, and 3 in English composition. (NOTE: This recommendation for skill level 10 is valid for the dates 1/91-9/91 only) (4/91).

Recommendation, Skill Level 20

In the lower-division baccalaureate/associate degree category, 3 semester hours in legal research, 3 in typewriting, 3 in office practices, 3 in English composition, 3 in legal writing, 3 in military legal practices and procedures, 3 in office management, and 3 in litigation (civil) (4/91).

Recommendation, Skill Level 30

In the lower-division baccalaureate/associate degree category, 3 semester hours in legal research, 3 in typewriting, 3 in office practices, 3 in English composition, 3 in legal writing, 3 in military legal practices and procedures, 3 in office management, 3 in litigation (civil), and 3 in personnel supervision (4/91).

Recommendation, Skill Level 40

In the lower-division baccalaureate/associate degree category, 3 semester hours in legal research, 3 in typewriting, 3 in office practices, 3 in English composition, 3 in legal writing, 3 in military legal practices and procedures, 3 in office management, 3 in litigation (civil), and 3 in personnel supervision. In the upper-division baccalaureate category, 3 semester hours in organizational management (4/91).

Recommendation, Skill Level 50

In the lower-division baccalaureate/associate degree category, 3 semester hours in legal research, 3 in typewriting, 3 in office practices, 3 in English composition, 3 in legal writing, 3 in military legal practices and procedures, 3 in office management, 3 in litigation (civil), and 3 in personnel supervision. In the upper-division baccalaureate category, 3 semester hours in organizational management and 3 for field experience in management (4/91).

MOS-71D-004

LEGAL SPECIALIST
71D10
71D20
71D30
71D40
71D50

Exhibit Dates: 8/94–9/01. Effective 10/01, MOS 71D was discontinued and its duties were transferred to MOS 27D, Paralegal Specialist.

Career Management Field: 71 (Administration).

Description

Summary: Prepares or supervises the preparation of court martial records, board proceedings, preliminary hearings, and investigations. *Skill Level 10:* Under close supervision, assists in preparing legal papers and forms, conducting investigations and researching activities, and maintaining files; types at a minimum of 30 words per minute; assists in maintaining law library. *Skill Level 20:* Able to perform the duties required for Skill Level 10; prepares correspondence; interview clients to determine nature of problem; compiles records into quarterly reports; provides legal assistance (military and civil); refers legal problems to attorney; makes referrals as needed. *Skill Level 30:* Able to perform the duties required for Skill Level 20; supervises a legal office or maintains a law library; provides guidance to subordinates; maintains files; reviews claims conducts claim investigations. *Skill Level 40:* Able to perform the duties required for Skill Level 30; plans and organizes legal office operation; previews legal documentation for proper and prompt disposition; researches decisions, statutes, and regulations under supervision of a judge advocate or civilian attorney. *Skill Level 50:* Able to perform the duties required for Skill Level 40; supervises the administration of a large legal office; prepares budgets; plans programs; secures special proceedings, highly sensitive or specially-handled documents or evidence; assigns duties; evaluates performance of subordinates.

Recommendation, Skill Level 10

Credit may be granted on the basis of an individualized assessment of the student (6/97).

Recommendation, Skill Level 20

In the lower-division baccalaureate/associate degree category, 3 semester hours in legal research, 3 in keyboarding, 3 in office practices, 3 in legal investigations and research, 3 in legal writing, 3 in legal military practices and procedures, 3 in law office management, and 3 in litigation. (NOTE: This recommendation for skill level 20 is valid for the dates 8/94-2/95 only) (6/97).

Recommendation, Skill Level 30

In the lower-division baccalaureate/associate degree category, 3 semester hours in legal research, 3 in keyboarding, 3 in office practices, 3 in legal investigations and research, 3 in legal writing, 3 in legal military practices and procedures, 3 in law office management, 3 in litigation, and 3 in personnel supervision. In the upper-division baccalaureate category, 3 semester hours in management problems (6/97).

Recommendation, Skill Level 40

In the lower-division baccalaureate/associate degree category, 3 semester hours in legal research, 3 in keyboarding, 3 in office practices, 3 in legal investigation and research, 3 in legal writing, 3 in legal military practices and procedures, 3 in law office management, 3 in litigation, and 3 in personnel supervision. In the upper-division baccalaureate category, 3 semester hours in management problems and 3 in human resource management (6/97).

Recommendation, Skill Level 50

In the lower-division baccalaureate/associate degree category, 3 semester hours in legal research, 3 in keyboarding, 3 in office practices, 3 in legal investigation and research, 3 in legal writing, 3 in legal military practices and procedures, 3 in law office management, 3 in litigation, and 3 in personnel supervision. In the upper-division baccalaureate category, 3 semester hours in management problems and 3 in human resource management (6/97).

MOS-71E-004

COURT REPORTER

71E20
71E30
71E40

Exhibit Dates: 1/90–12/90.

Career Management Field: 71 (Administration), subfield 714 (Legal).

Description

Summary: Takes notes of activities and statements in legal proceedings and prepares them for inclusion in official legal documents. *Skill Level 20:* Able to perform the duties required for skill level 10 of MOS 71D (Legal Clerk); takes notes at legal proceedings at a minimum of 175 words per minute, usually by means of a stenomask (an oral dictating machine), although stenotype or shorthand may be used; transcribes testimony at a minimum speed of 40 words per minute; maintains the records of military legal proceedings; uses English grammar, punctuation, and composition skills. NOTE: Required to have held skill level 10 of MOS 71D (Legal Clerk). *Skill Level 30:* Able to perform the duties required for Skill Level 20; may provide technical guidance to subordinates. *Skill Level 40:* Able to perform the duties required for Skill Level 30; supervises and assists subordinates; coordinates work assignments.

Recommendation, Skill Level 20

In the lower-division baccalaureate/associate degree category, 6 semester hours in typing, 9 in dictation and transcription, 2 in business communication or technical writing, 3 in English grammar and composition, and 3 in office procedures (6/78).

Recommendation, Skill Level 30

In the lower-division baccalaureate/associate degree category, 6 semester hours in typing, 9 in dictation and transcription, 2 in business communication or technical writing, 3 in English grammar and composition, and 3 in office procedures (6/78).

Recommendation, Skill Level 40

In the lower-division baccalaureate/associate degree category, 6 semester hours in typing, 9 in dictation and transcription, 2 in business communication or technical writing, 3 in English grammar and composition, 3 in office procedures, 3 in office management, and 3 in legal practices and procedures (6/78).

MOS-71E-005

COURT REPORTER

71E20
71E30
71E40

Exhibit Dates: 1/91–12/94. (Effective 12/94, MOS 71E was discontinued, and its duties were incorporated into MOS 71D).

Career Management Field: 71 (Administration).

Description

Summary: Takes notes of activities and statements in legal proceedings and prepares them for inclusion in official legal documents. *Skill Level 20:* Able to perform the duties required for skill level 10 of MOS 71D (Legal Clerk); takes notes at legal proceedings at a minimum of 175 words per minute, usually by means of a stenomask (an oral dictating machine), although stenotype or shorthand may be used; transcribes testimony at a minimum speed of 40 words per minute; maintains the records of military legal proceedings; uses English grammar, punctuation, and composition skills; adheres to American Bar Association code of professional responsibilities and standards of criminal justice. NOTE: Required to have held skill level 10 of MOS 71D (Legal Specialist). *Skill Level 30:* Able to perform the duties required for Skill Level 20; may provide technical guidance to subordinates; under the supervision of an attorney, interviews witnesses and evaluates potential testimony; performs legal and administrative research. *Skill Level 40:* Able to perform the duties required for Skill Level 30; supervises and assists subordinates; coordinates work assignments; conducts legal research using military appellate court decisions, federal statutes, regulations, and local law and court decisions.

Recommendation, Skill Level 20

In the lower-division baccalaureate/associate degree category, 3 semester hours in English composition, 6 in typewriting, 9 in dictation and transcription, and 3 in personnel supervision. In the upper-division baccalaureate category, 3 semester hours in technical writing (4/91).

Recommendation, Skill Level 30

In the lower-division baccalaureate/associate degree category, 3 semester hours in English composition, 6 in typewriting, 9 in dictation and transcription, 3 in personnel supervision, and 2 in records and information management. In the upper-division baccalaureate category, 3 semester hours in technical writing (4/91).

Recommendation, Skill Level 40

In the lower-division baccalaureate/associate degree category, 3 semester hours in English composition, 6 in typewriting, 9 in dictation and transcription, 3 in personnel supervision, and 2 in records and information management. In the upper-division baccalaureate category, 3 semester hours in technical writing, 3 in organizational management, and 3 for field experience in management (4/91).

MOS-71G-002

PATIENT ADMINISTRATION SPECIALIST

71G10
71G20
71G30
71G40
71G50

Exhibit Dates: 1/90–3/92.

Career Management Field: 91 (Medical), subfield 914 (Medical Support).

Description

Summary: Supervises or performs administrative duties in patient administration division of hospital or other medical activity. *Skill Level 10:* Performs patient administrative duties; applies knowledge of basic medical terminology including construction or composition of medical terms; applies knowledge of provisions and limitations of Freedom of Information and Privacy Acts; receives patients being admitted to medical facility; interviews patients and reviews records and available data to obtain necessary information for preparation of admission records including patient data cards and appropriate records and forms, depending upon type of patient, illness or injury, and urgency of required treatment; receives valuables and monies from patients and deposits into patient's trust fund; initiates required notification of patient admission to unit commanders or other headquarters; compiles data for preparation and prepares daily admission and disposition reports; performs administrative duties and prepares appropriate forms to process and report very seriously ill patients including notification of next of kin and progress reports; prepares and dispatches casualty reports; prepares appropriate forms or letters and performs necessary administrative duties to evacuate or transfer patients to other medical facilities; completes forms and reports and performs tasks regarding disposition of personal effects; computes various charges for pay patients; performs administrative action involving third party liability; applies knowledge of administrative terminology to medical records; keeps repository for military health records and dependents' medical records of hospital, clinic, outpatient service, or dispensary; classifies, indexes, and files record folders in repository; reviews medical record for adequacy and completeness based upon nature of case and treatment given; answers inquiries and provides information on medical records; compiles statistical data pertaining to admission, diagnosis, treatment, and disposition of patients; prepares statistical tables, charts, and graphs on medical data; prepares reports and data on births, deaths, and reportable conditions for submission to military and civilian health authorities and to appropriate legal and governmental agencies; initiates correspondence pertaining to medical records, medical board proceedings, and lines of duty investigation; answers inquiries on results of board actions, eligibility for medical care under Uniform Service Health Benefit Program; prepares necessary medical forms to convene the medical board to determine medical fitness for duty; types military and nonmilitary letters, endorsements, disposition forms, messages, and special medical forms and documents in draft and final form; maintains nonmedical files. *Skill Level 20:* Able to perform the duties required for Skill Level 10; provides technical guidance to subordinates; performs administrative duties in medical, surgical, or similar department of the hospital, including obtaining records from repository, posting entries onto medical record, returning records, and coordinating administrative responsibilities with other departments. *Skill Level 30:* Able to perform the duties required for Skill Level 20; supervises subordinates; assigns tasks to subordinates, ensuring efficient work flow and timely accomplishment of tasks regarding patient record keeping and the movement of patients; instructs personnel in performance of assigned tasks and instills understanding of need for timely and precise completion and posting of records and an appreciation of the sensitivity and confidentiality of the actions performed. *Skill Level 40:* Able to perform the duties required for Skill Level 30; supervises a medium-sized patient administration activity or hospital department. *Skill Level 50:* Able to perform the duties required for Skill Level 40; supervises a large patient administration activity or hospital department.

Recommendation, Skill Level 10

In the lower-division baccalaureate/associate degree category, 2 semester hours in medical administrative techniques, 2 in medical record keeping, 2 in medical records administration, 2 in medical terminology, and 1 typing. (NOTE: This recommendation for skill level 10 is valid for the dates 1/90-9/91 only) (1/82).

Recommendation, Skill Level 20

In the lower-division baccalaureate/associate degree category, 2 semester hours in medical administrative techniques, 2 in medical record keeping, 2 in medical records administration, 2 in medical terminology, 1 in typing, 2 in personnel supervision, and 1 in medical statistical reporting (1/82).

Recommendation, Skill Level 30

In the lower-division baccalaureate/associate degree category, 2 semester hours in medical administrative techniques, 2 in medical record keeping, 2 in medical records administration, 2 in medical terminology, 1 in typing, 2 in personnel supervision, 1 in medical statistical reporting, 2 in business communication, and 1 in human relations (1/82).

Recommendation, Skill Level 40

In the lower-division baccalaureate/associate degree category, 2 semester hours in medical administrative techniques, 2 in medical record keeping, 2 in medical records administration, 2 in medical terminology, 1 in typing, 3 in personnel supervision, 1 in medical statistical reporting, 2 in business communication, and 3 in human relations. In the upper-division baccalaureate category, 3 semester hours for field experience in office management, 2 for field experience in hospital administration, and 2 in introduction to management (1/82).

Recommendation, Skill Level 50

In the lower-division baccalaureate/associate degree category, 2 semester hours in medical administrative techniques, 2 in medical record keeping, 2 in medical records administration, 2 in medical terminology, 1 in typing, 3 in personnel supervision, 1 in medical statistical reporting, 2 in business communication, and 3 in human relations. In the upper-division baccalaureate category, 3 semester hours for field experience in office management, 3 for field experience in hospital administration, and 3 in introduction to management (1/82).

MOS-71G-003

Patient Administration Specialist
71G10
71G20
71G30
71G40
71G50

Exhibit Dates: 4/92–5/01. Effective 6/01, MOS 71G was discontinued and its duties were transferred to MOS 91G, Patient Administration Specialist.

Career Management Field: 91 (Medical).

Description

Summary: Supervises or performs administrative duties in patient administration division of hospital or other medical activity. *Skill Level 10:* Performs patient administrative duties; applies knowledge of basic medical terminology, including construction or composition of medical terms; applies knowledge of provisions and limitations of Freedom of Information and Privacy Acts; receives patients; interviews patients and reviews records and available data to obtain necessary information for preparation of admission records including patient data cards and appropriate records and forms, depending upon type of patient, illness or injury, and urgency of required treatment; receives valuables and monies from patients and deposits into patient's trust fund; initiates required notification of patient admission to unit commanders or other headquarters; compiles data for preparation and prepares daily admission and disposition reports; performs administrative duties and prepares appropriate forms to process and report very seriously ill patients including notification of next of kin and progress reports; prepares and dispatches casualty reports; prepares appropriate forms or letters and performs necessary administrative duties to evacuate or transfer patients to other medical facilities; completes forms and reports and performs tasks regarding disposition of personal effects; computes various charges for pay patients; performs administrative action involving third party liability; applies knowledge of administrative terminology to medical records; keeps repository for military health records and dependents' medical records for hospital, clinic, outpatient service, or dispensary; classifies, indexes, and files record folders in repository; reviews medical record for adequacy and completeness based upon nature of case and treatment given; answers inquiries and provides information on medical records; compiles statistical data pertaining to admission, diagnosis, treatment, and disposition of patients; prepares statistical tables, charts, and graphs on medical data; prepares reports and data on births, deaths, and reportable conditions for submission to military and civilian health authorities and to appropriate legal and governmental agencies; initiates correspondence pertaining to medical records, medical board proceedings, and lines of duty investigation; answers inquiries on results of board actions, eligibility for medical care under Uniform Service Health Benefit Program; prepares necessary medical forms to convene the medical board to determine medical fitness for duty; types military and nonmilitary letters, endorsements, disposition forms, messages, and special medical forms and documents in draft and final form; maintains nonmedical files. *Skill Level 20:* Able to perform the duties required for Skill Level 10; provides technical guidance to subordinates; performs administrative duties in medical, surgical, or similar department of the hospital, including obtaining records from repository, posting entries onto medical records, returning records, and coordinating administrative responsibilities with other departments. *Skill Level 30:* Able to perform the duties required for Skill Level 20; supervises subordinates; assigns tasks to subordinates, ensuring efficient work flow and timely accomplishment of tasks regarding patient record keeping and the movement of patients; instructs personnel in performance of assigned tasks and instills understanding of need for timely and precise completion and posting of records and an appreciation of the sensitivity and confidentiality of the actions performed. *Skill Level 40:* Able to perform the duties required for Skill Level 30; supervises a medium-sized patient administration activity or hospital department. *Skill Level 50:* Able to perform the duties required for Skill Level 40; supervises a large patient administration activity or hospital department.

Recommendation, Skill Level 10

Credit may be granted on the basis of an individualized assessment of the student (3/94).

Recommendation, Skill Level 20

In the lower-division baccalaureate/associate degree category, 4 semester hours in introduction to medical record technology, 4 in directed practice in medical record technology, and 3 in personnel supervision. (NOTE: This recom-

mendation for skill level 20 is valid for the dates 4/92-2/95 only) (4/92).

Recommendation, Skill Level 30

In the lower-division baccalaureate/associate degree category, 4 semester hours in introduction to medical record technology, 5 in directed practice in medical record technology, 3 in personnel supervision, and 2 in records and information management (4/92).

Recommendation, Skill Level 40

In the lower-division baccalaureate/associate degree category, 4 semester hours in introduction to medical record technology, 6 in directed practice in medical record technology, 3 in personnel supervision, and 2 in records and information management. In the upper-division baccalaureate category, 3 semester hours in organizational management and 3 for field experience in management (4/92).

Recommendation, Skill Level 50

In the lower-division baccalaureate/associate degree category, 4 semester hours in introduction to medical record technology, 6 in directed practice in medical record technology, 3 in personnel supervision, and 2 in records and information management. In the upper-division baccalaureate category, 3 semester hours in organizational management and 6 for field experience in management; if individual has attained paygrade E-9, additional credit as follows: 3 semester hours in management problems, 3 in operations management, and 3 in communication techniques for managers (4/92).

MOS-71L-002

ADMINISTRATIVE SPECIALIST

71L10
71L20
71L30
71L40
71L50

Exhibit Dates: 1/90–10/92.

Career Management Field: 71 (Administration), subfield 711 (General Administration).

Description

Summary: Supervises or performs administrative, clerical, and typing duties; types a minimum of 25 net words a minute. *Skill Level 10:* Types routine correspondence, messages, reports, and forms; serves as a receptionist; answers the telephone; handles incoming mail and messages; maintains files; uses copying equipment; receives and maintains publications. *Skill Level 20:* Able to perform the duties required for Skill Level 10; prepares special correspondence; composes routine correspondence; proofreads correspondence for proper format and accuracy; provides technical guidance to typists and clerical personnel; prepares charts, graphs, and rosters. *Skill Level 30:* Able to perform the duties required for Skill Level 20; supervises personnel performing typing, general clerical, and administrative duties; distributes work load; reviews and edits correspondence; sets up and reviews files; controls duplicating facilities; plans and organizes office operations; determines requirements for office equipment, supplies, and space. *Skill Level 40:* Able to perform the duties required for Skill Level 30; performs supervisory duties at a higher level of command. NOTE: May have progressed to 71L40 from 71L30 or 71C30 (Stenographer). *Skill Level 50:* Able to perform the duties required for Skill Level 40; supervises performance of administrative functions at higher level headquarters; provides guidance and engages in public relations activities.

Recommendation, Skill Level 10

In the lower-division baccalaureate/associate degree category, 2 semester hours in typing, 2 in record keeping, 1 in business communication, and 3 in office procedures. (NOTE: This recommendation for skill level 10 is valid for the dates 1/90-9/91 only) (6/78).

Recommendation, Skill Level 20

In the lower-division baccalaureate/associate degree category, 2 semester hours in typing, 2 in business communication, 2 in record keeping, and 4 in office procedures (6/78).

Recommendation, Skill Level 30

In the lower-division baccalaureate/associate degree category, 2 semester hours in typing, 3 in business communication, 2 in record keeping, 4 in office procedures, 3 for field experience in office practices, 3 in office management, and 3 in human relations (6/78).

Recommendation, Skill Level 40

In the lower-division baccalaureate/associate degree category, 2 semester hours in typing, 3 in business communication, 2 in record keeping, 4 in office procedures, 3 for field experience in office practices, 3 in office management, 3 in human relations, and 3 in personnel supervision. In the upper-division baccalaureate category, 3 semester hours in personnel management (6/78).

Recommendation, Skill Level 50

In the lower-division baccalaureate/associate degree category, 2 semester hours in typing, 3 in business communication, 2 in record keeping, 3 in records administration, 4 in office procedures, 3 in office management, 3 for field experience in office practices, 4 in human relations, 2 in public relations, 3 in personnel supervision, and 3 for field experience in office management. In the upper-division baccalaureate category, 3 semester hours in personnel management and 3 for field experience in management (6/78).

MOS-71L-003

ADMINISTRATIVE SPECIALIST

71L10
71L20
71L30
71L40
71L50

Exhibit Dates: 11/92–5/01.

Career Management Field: 71 (Administration).

Description

Summary: Supervises or performs administrative, clerical, and keyboarding duties (at a minimum of 35 words a minute). *Skill Level 10:* Types routine correspondence, messages, reports, and forms; serves as a receptionist; answers the telephone; handles incoming mail and messages; maintains files; uses copying equipment; receives and maintains publications; performs data entry. *Skill Level 20:* Able to perform the duties required for Skill Level 10; prepares special correspondence; composes routine correspondence; proofreads correspondence for proper format and accuracy; provides technical guidance to typists and clerical personnel; prepares charts, graphs, and rosters. *Skill Level 30:* Able to perform the duties required for Skill Level 20; supervises personnel performing keyboarding, general clerical, and administrative duties; distributes work load; reviews and edits correspondence; sets up and reviews files; controls duplicating facilities; plans and organizes office operations; determines requirements for office equipment, supplies, and space. *Skill Level 40:* Able to perform the duties required for Skill Level 30; performs supervisory duties at a higher level of command. NOTE: May have progressed to 71L40 from 71L30 or 71C30 (Executive Administrative Assistant). *Skill Level 50:* Able to perform the duties required for Skill Level 40; supervises performance of administrative functions at higher level headquarters; provides guidance; engages in public relations activities.

Recommendation, Skill Level 10

Credit may be granted on the basis of an individualized assessment of the student (3/94).

Recommendation, Skill Level 20

In the lower-division baccalaureate/associate degree category, 2 semester hours in keyboarding, 2 in record keeping, 2 in business communication, 4 in office procedures, and 3 in computer applications. (NOTE: This recommendation for skill level 20 is valid for the dates 11/92-2/95 only) (11/92).

Recommendation, Skill Level 30

In the lower-division baccalaureate/associate degree category, 2 semester hours in keyboarding, 2 in record keeping, 3 in business communication, 4 in office procedures, 3 in computer applications, 3 for field experience in office practices, 3 in office management, and 3 in personnel supervision (11/92).

Recommendation, Skill Level 40

In the lower-division baccalaureate/associate degree category, 2 semester hours in keyboarding, 2 in record keeping, 3 in business communication, 4 in office procedures, 3 in computer applications, 3 for field experience in office practices, 3 in office management, and 3 in personnel supervision. In the upper-division baccalaureate category, 2 semester hours in human resource management and 3 for field experience in management (11/92).

Recommendation, Skill Level 50

In the lower-division baccalaureate/associate degree category, 2 semester hours in keyboarding, 2 in record keeping, 3 in business communication, 4 in office procedures, 3 in computer applications, 3 for field experience in office practices, 3 in office management, and 3 in personnel supervision. In the upper-division baccalaureate category, 3 semester hours in human resource management and 6 for field experience in management; if paygrade E-9 has been attained, additional credit may be granted as follows: 3 semester hours in management problems and 3 in communication techniques for managers (11/92).

MOS-71L-004

ADMINISTRATIVE SPECIALIST

71L10
71L20
71L30
71L40
71L50

Exhibit Dates: 6/01–3/03. (Effective 4/03, this MOS was discontinued, and its duties were incorporated into MOS 42L.)

Career Management Field: 71 (Administration).

Description

Summary: Supervises or performs administrative support duties. *Skill Level 10:* Prepares routine correspondence, messages, reports, and forms; serves as a receptionist; answers the telephone; handles incoming mail and messages; maintains files; uses copying equipment; receives and maintains publications; utilizes word processing, presentation, and spreadsheet software. *Skill Level 20:* Able to perform the duties required for Skill Level 10; prepares special correspondence; composes routine correspondence; proofreads correspondence for proper format and accuracy; provides technical guidance to clerical personnel. *Skill Level 30:* Able to perform the duties required for Skill Level 20; supervises personnel performing word processing, clerical, and administrative functions; distributes work load; reviews and edits correspondence; sets up and reviews files; plans and organizes office operations; determines requirements for office equipment, supplies, and space. *Skill Level 40:* Able to perform the duties required for Skill Level 30; performs supervisory duties at a higher level of command. NOTE: May have progressed to 71L40 from 71L30 or 71C30 (Executive Administrative Assistant). *Skill Level 50:* Able to perform the duties required for Skill Level 40; supervises performance of administrative functions at higher-level headquarters; provides guidance; engages in public relations activities.

Recommendation, Skill Level 10

Credit may be granted on the basis of an individualized assessment of the student (6/01).

Recommendation, Skill Level 20

Credit may be granted on the basis of an individualized assessment of the student (6/01).

Recommendation, Skill Level 30

In the lower-division baccalaureate/associate degree category, 3 semester hours in introduction to computers, 2 in record keeping, 3 in business communication, 3 in office procedures, 3 for an internship in office practices, and 3 in office administration (6/01).

Recommendation, Skill Level 40

In the lower-division baccalaureate/associate degree category, 3 semester hours in introduction to computers, 2 in record keeping, 3 in business communication, 3 in office procedures, 3 for an internship in office practices, 3 in office administration, and 3 in principles of supervision. In the upper-division baccalaureate category, 3 semester hours for a management practicum (6/01).

Recommendation, Skill Level 50

In the lower-division baccalaureate/associate degree category, 3 semester hours in introduction to computers, 2 in record keeping, 3 in business communication, 3 in office procedures, 3 for an internship in office practices, 3 in office administration, and 3 in principles of supervision. In the upper-division baccalaureate category, 3 semester hours in human resource management and 6 for a management practicum (6/01).

MOS-71M-003

Chaplain Assistant
71M10
71M20
71M30
71M40
71M50

Exhibit Dates: 1/90–5/90.

Career Management Field: 71 (Administration).

Description

Summary: Acts as administrative assistant to a military chaplain; performs or supervises office activities such as preparing correspondence and maintaining records; assists the chaplain in preparing for chapel and religious programs; comparable to a civilian hospital chaplain's assistant or administrative assistant in a church or synagogue. *Skill Level 10:* Prepares facility for services; prepares programs and bulletins; prepares schedules and religious materials; operates and maintains audiovisual equipment; acts as receptionist; answers routine inquiries; requisitions, receives, and maintains equipment and supplies; types letters, messages, forms, and records; maintains files. *Skill Level 20:* Able to perform the duties required for Skill Level 10; provides technical guidance to subordinates; assists in planning and programming religious services and education requirements. *Skill Level 30:* Able to perform the duties required for Skill Level 20; provides technical guidance to personnel; participates in preparation of budget; maintains fund records including disbursements, receipts, and petty cash. *Skill Level 40:* Able to perform the duties required for Skill Level 30; supervises subordinates; coordinates volunteer, part time, and other personnel; reviews prepared correspondence and reports; participates in planning and programming religious activities. *Skill Level 50:* Able to perform the duties required for Skill Level 40; performs supervisory and management duties in large installation.

Recommendation, Skill Level 10

In the lower-division baccalaureate/associate degree category, 2 semester hours in business communication, 2 in communication skills, 1 in filing, 1 in office management, and 2 in introduction to audiovisual equipment (6/78).

Recommendation, Skill Level 20

In the lower-division baccalaureate/associate degree category, 2 semester hours in office management, 3 in communication skills, 1 in filing, 3 in business communication, and 2 in introduction to audiovisual equipment (6/78).

Recommendation, Skill Level 30

In the lower-division baccalaureate/associate degree category, 2 semester hours in office management, 3 in communication skills, 1 in filing, 3 in business communication, 2 in introduction to audiovisual equipment, 2 in budget administration, and 3 in human relations (6/78).

Recommendation, Skill Level 40

In the lower-division baccalaureate/associate degree category, 2 semester hours in office management, 1 in filing, 3 in business communication, 6 in communication skills, 2 in introduction to audiovisual equipment, 2 in budget administration, and 3 in human relations. In the upper-division baccalaureate category, 3 semester hours in public relations (6/78).

Recommendation, Skill Level 50

In the lower-division baccalaureate/associate degree category, 3 semester hours in office management, 1 in filing, 3 in business communication, 6 in communication skills, 2 in introduction to audiovisual equipment, 2 in budget administration, 3 in human relations, and 3 in applied psychology. In the upper-division baccalaureate category, 3 semester hours in public relations (6/78).

MOS-71M-004

Chaplain Assistant
71M10
71M20
71M30
71M40
71M50

Exhibit Dates: 6/90–5/01.

Career Management Field: 71 (Administration).

Description

Summary: Acts as administrative assistant to a military chaplain; performs or supervises office activities such as preparing correspondence and maintaining records; assists the chaplain in preparing for chapel and religious programs; comparable to a civilian hospital chaplain's assistant or administrative assistant in a church or synagogue. *Skill Level 10:* Prepares facility for services; prepares schedules and religious materials; operates and maintains audiovisual equipment; acts as receptionist; answers routine inquiries; requisitions, receives, and maintains equipment and supplies; types letters, messages, forms, and records; maintains files. *Skill Level 20:* Able to perform the duties required for Skill Level 10; provides technical guidance to subordinates; assists in planning and programming religious services and education. *Skill Level 30:* Able to perform the duties required for Skill Level 20; may provide supervision of five to eight persons; participates in preparation of budget; maintains fund records, including disbursements, receipts, and petty cash. *Skill Level 40:* Able to perform the duties required for Skill Level 30; supervises subordinates; coordinates volunteer, part time, and other personnel; reviews prepared correspondence and reports; participates in planning and programming religious activities. *Skill Level 50:* Able to perform the duties required for Skill Level 40; performs supervisory and management duties in large installation.

Recommendation, Skill Level 10

In the lower-division baccalaureate/associate degree category, 2 semester hours in business communication, 2 in communication skills, and 1 in office administration. (NOTE: This recommendation for skill level 10 is valid for the dates 6/90-9/91 only) (6/90).

Recommendation, Skill Level 20

In the lower-division baccalaureate/associate degree category, 3 semester hours in business communication, 3 in communication skills, 2 in office administration, and 2 in human relations. (NOTE: This recommendation for skill level 20 is valid for the dates 6/90-2/95 only) (6/90).

Recommendation, Skill Level 30

In the lower-division baccalaureate/associate degree category, 3 semester hours in business communication, 3 in communication skills, 2 in office administration, 3 in human relations, 1 in technical writing, 2 in principles of supervision, and 2 in budget administration. In the upper-division baccalaureate category, 1 semester hour in counseling services and 2 in crisis intervention (6/90).

Recommendation, Skill Level 40

In the lower-division baccalaureate/associate degree category, 3 semester hours in business communication, 3 in communication skills, 2 in office administration, 3 in human relations, 1 in technical writing, 3 in principles of supervision, and 2 in budget administration. In the upper-division baccalaureate category, 1 semester hour in counseling services and 2 in crisis intervention (6/90).

Recommendation, Skill Level 50

In the lower-division baccalaureate/associate degree category, 3 semester hours in business communication, 3 in communication skills, 3 in office administration, 3 in human relations, 1 in technical writing, 3 in principles of supervision, and 2 in budget administration. In the upper-division baccalaureate category, 1 semester hour in counseling services, 2 in crisis intervention, and 3 in applied psychology (6/90).

MOS-71M-005

Chaplain Assistant
71M10
71M20
71M30
71M40
71M50

Exhibit Dates: 6/01–9/01. Effective 10/01, MOS 71M was discontinued and its duties were transferred to MOS 56M, Chaplain Assistant.

Career Management Field: 71 (Administration).

Description

Summary: Acts as administrative assistant to a military chaplain; performs or supervises office activities such as preparing correspondence and maintaining records; assists the chaplain in preparing for chapel and religious programs; comparable to a civilian hospital chaplain's assistant or administrative assistant in a church or synagogue. *Skill Level 10:* Prepares facility for services; prepares schedules and religious materials; operates and maintains audiovisual equipment; acts as receptionist; answers routine inquiries; requisitions, receives, and maintains equipment and supplies; types letters, messages, forms, and records; maintains files. *Skill Level 20:* Able to perform the duties required for Skill Level 10; provides technical guidance to subordinates; assists in planning and programming religious services and education. *Skill Level 30:* Able to perform the duties required for Skill Level 20; supervises five to eight persons; participates in preparation of budget; maintains fund records, including disbursements, receipts, and petty cash. *Skill Level 40:* Able to perform the duties required for Skill Level 30; supervises subordinates; coordinates volunteer, part-time, and other personnel; reviews prepared correspondence and reports; participates in planning and programming religious activities. *Skill Level 50:* Able to perform the duties required for Skill Level 40; performs supervisory and management duties in large installation.

Recommendation, Skill Level 10

Credit may be granted on the basis of an individualized assessment of the student (6/01).

Recommendation, Skill Level 20

Credit may be granted on the basis of an individualized assessment of the student (6/01).

Recommendation, Skill Level 30

In the lower-division baccalaureate/associate degree category, 3 semester hours in business communication, 3 in office procedures, 3 in human relations, 3 in principles of supervision, and 3 in budget administration. In the upper-division baccalaureate category, 3 semester hours in crisis intervention (6/01).

Recommendation, Skill Level 40

In the lower-division baccalaureate/associate degree category, 3 semester hours in business communication, 3 in interpersonal communication, 3 in office procedures, 3 in human relations, 3 in principles of supervision, and 3 in budget administration. In the upper-division baccalaureate category, 3 semester hours in crisis intervention (6/01).

Recommendation, Skill Level 50

In the lower-division baccalaureate/associate degree category, 3 semester hours in business communication, 3 in interpersonal communication, 3 in office procedures, 3 in human relations, 3 in principles of supervision, and 3 in budget administration. In the upper-division baccalaureate category, 3 semester hours in crisis intervention and 3 in applied psychology (6/01).

MOS-72E-003

Combat Telecommunications Center Operator
72E10
72E20
72E30
72E40

Exhibit Dates: 1/90–3/91. Effective 3/91, MOS 72E was discontinued, and its duties were incorporated into MOS 74C, Record Telecommunications Center Operator.

Career Management Field: 31 (Communications-Electronics Operations).

Description

Summary: Operates, monitors, and supervises record telecommunications equipment or assemblies. *Skill Level 10:* Processes and delivers messages; performs related clerical and administrative functions within combat telecommunications centers; prepares messages in proper format; reproduces and distributes messages; operates on-line and off-line cryptographic equipment using appropriate systems and materials; installs, operates, and performs routine maintenance on telecommunications equipment. *Skill Level 20:* Able to perform the duties required for Skill Level 10; processes messages; supervises and performs administrative functions in combat telecommunications centers; maintains files necessary for adequate accounting and control of cryptographic material; coordinates handling of messages in combat telecommunications centers. *Skill Level 30:* Able to perform the duties required for Skill Level 20; supervises the operation of combat telecommunications centers and remote terminals; establishes and supervises on-the-job training; evaluates capabilities of subordinates; reviews control and accounting procedures for cryptographic material and recommends improvements. *Skill Level 40:* Able to perform the duties required for Skill Level 30; supervises the overall operation of combat telecommunications centers; coordinates training and technical operations of center; evaluates capabilities of subordinates; makes work assignments; coordinates efforts with other units.

Recommendation, Skill Level 10

In the lower-division baccalaureate/associate degree category, 3 semester hours in typing and 3 in office procedures (2/84).

Recommendation, Skill Level 20

In the lower-division baccalaureate/associate degree category, 3 semester hours in typing, 3 in office procedures, and 3 in office management (2/84).

Recommendation, Skill Level 30

In the lower-division baccalaureate/associate degree category, 3 semester hours in typing, 3 in office procedures, 3 in office management, and 3 in personnel supervision (2/84).

Recommendation, Skill Level 40

In the lower-division baccalaureate/associate degree category, 3 semester hours in typing, 3 in office procedures, 3 in office management, 3 in personnel supervision, and 3 in principles of management. In the upper-division baccalaureate category, 3 semester hours for field experience in management and 3 in management problems (2/84).

MOS-72G-003

Automatic Data Telecommunications Center Operator
72G10
72G20
72G30
72G40

Exhibit Dates: 1/90–3/91. Effective 3/91, MOS 72G was discontinued, and its duties were incorporated into MOS 74C, Record Telecommunications Center Operator.

Career Management Field: 31 (Communications-Electronics Operations).

Description

Summary: Supervises, operates, and monitors data communications equipment and peripheral devices in automatic switching center. *Skill Level 10:* Operates data communications equipment; processes and transmits messages through automatic digital networks; prepares messages for transmission in paper tape, magnetic tape, and disk form; operates card readers, card punches, magnetic tape transports, magnetic tape stations, high speed printers, and teletypewriters. *Skill Level 20:* Able to perform the duties required for Skill Level 10; operates digital computers to receive, process, store, and forward data communications; assigns computer and peripheral equipment for on-line processing; prepares messages for encoding; maintains files; operates off-line message processor and teletypewriters; prepares station message-handling efficiency statistics. *Skill Level 30:* Able to perform the duties required for Skill Level 20; serves as shift supervisor; ensures message protection and efficiency of high speed traffic channels; employs automatic digital message switching start-up and restoring procedures; takes appropriate action to correct equipment malfunction, data discrepancies, or circuitry failure; prepares logs of equipment status and significant events. *Skill Level 40:* Able to perform the duties required for Skill Level 30; serves as data communications supervisor; supervises switching center operations, including console, traffic service operations, programming, and technical control; facilitates routing traffic by coordination with various organizations and other stations; corrects discrepancies and faulty work practices; prepares, conducts, and supervises on-the-job training and evaluates personnel; prepares and submits technical reports including technical information on cryptographic operations.

Recommendation, Skill Level 10

In the lower-division baccalaureate/associate degree category, 3 semester hours in data processing procedures (2/84).

Recommendation, Skill Level 20

In the lower-division baccalaureate/associate degree category, 3 semester hours in data processing procedures and 3 in introduction to telecommunications networks (2/84).

Recommendation, Skill Level 30

In the lower-division baccalaureate/associate degree category, 3 semester hours in data processing procedures, 3 in introduction to telecommunication networks, 3 in personnel supervision, and 3 in technical report writing (2/84).

Recommendation, Skill Level 40

In the lower-division baccalaureate/associate degree category, 3 semester hours in data processing procedures, 3 in introduction to telecommunications networks, 3 in personnel supervision, 3 in technical report writing, and 3 in principles of management. In the upper-division baccalaureate category, 3 semester hours for field experience in management and 3 in management problems (2/84).

MOS-73C-002

FINANCE SPECIALIST
73C10
73C20
73C30
73C40

Exhibit Dates: 1/90–9/91.

Career Management Field: 71 (Administration), subfield 713 (Finance).

Description

Summary: Completes forms, verifies data, computes amounts, and prepares reports of activity pertaining to pay, leave, travel, and personal finance records of military personnel. *Skill Level 10:* Determines pay entitlements, prepares input, and verifies leave and earnings; prepares pay vouchers; verifies, corrects, and maintains files of personnel financial records; operates calculator and other office equipment; answers inquiries from service members. *Skill Level 20:* Able to perform the duties required for Skill Level 10; provides technical guidance to subordinates; computes, reviews, verifies, and corrects items related to travel reports and allowances; disburses, collects, and prepares reconciliation reports; performs quality assurance audits and special reviews and reports findings. *Skill Level 30:* Able to perform the duties required for Skill Level 20; may supervise from 4-18 persons; prepares reports. *Skill Level 40:* Able to perform the duties required for Skill Level 30; may supervise from 3-24 persons in a large finance organization.

Recommendation, Skill Level 10

In the lower-division baccalaureate/associate degree category, 1 semester hour in business mathematics, 2 in office procedures, and 2 in office machines (6/78).

Recommendation, Skill Level 20

In the lower-division baccalaureate/associate degree category, 1 semester hour in business mathematics, 2 in office machines, and 3 in office procedures (6/78).

Recommendation, Skill Level 30

In the lower-division baccalaureate/associate degree category, 1 semester hour in business mathematics, 2 in office machines, 3 in office procedures, 2 in office management, and credit in human relations on the basis of institutional evaluation (6/78).

Recommendation, Skill Level 40

In the lower-division baccalaureate/associate degree category, 1 semester hour in business mathematics, 2 in office machines, 2 in business communications, 3 in office procedures, 3 in office management, and credit in human relations and personnel supervision on the basis of institutional evaluation (6/78).

MOS-73C-003

FINANCE SPECIALIST
73C10
73C20
73C30
73C40

Exhibit Dates: 10/91–5/01.

Career Management Field: 71 (Administration).

Description

Summary: Completes forms, verifies data, computes amounts, and prepares reports of activity pertaining to pay, leave, travel, and finance records of military personnel. *Skill Level 10:* Determines pay entitlements, prepares input, and verifies leave and earnings; prepares pay vouchers; verifies, corrects, and maintains files of personnel financial records; operates calculator, computer terminal, and other office equipment; answers inquiries from service members. *Skill Level 20:* Able to perform the duties required for Skill Level 10; provides technical guidance to subordinates; computes, reviews, verifies, and corrects items related to travel reports and allowances; disburses, collects, and prepares reconciliation reports; performs quality assurance audits and special reviews and reports findings. *Skill Level 30:* Able to perform the duties required for Skill Level 20; may supervise from 4-18 subordinates; prepares reports; maintains cash book. *Skill Level 40:* Able to perform the duties required for Skill Level 30; safeguards public funds and related documents; compiles daily activity summary reports; may supervise from 3-24 persons in a large finance organization.

Recommendation, Skill Level 10

Credit may be granted on the basis of an individualized assessment of the student (3/94).

Recommendation, Skill Level 20

In the lower-division baccalaureate/associate degree category, 2 semester hours in business mathematics, 3 in office procedures, 2 in office machines, 1 in computer applications, and 1 in personnel supervision. (NOTE: This recommendation for skill level 20 is valid for the dates 10/91-2/95 only) (11/91).

Recommendation, Skill Level 30

In the lower-division baccalaureate/associate degree category, 2 semester hours in business mathematics, 3 in office procedures, 2 in office machines, 1 in computer applications, 2 in office administration, and 3 in personnel supervision (11/91).

Recommendation, Skill Level 40

In the lower-division baccalaureate/associate degree category, 2 semester hours in business mathematics, 3 in office procedures, 2 in office machines, 1 in computer applications, 3 in office administration, 3 in personnel supervision, and 2 in business communications. In the upper-division baccalaureate category, 3 semester hours in human relations in business (11/91).

MOS-73C-004

FINANCE SPECIALIST
73C10
73C20
73C30
73C40

Exhibit Dates: 6/01–9/02. Effective 10/02, MOS 73C was discontinued and its duties were incorporated into MOS 44C, Financial Management Technician.

Career Management Field: 71 (Administration).

Description

Summary: Determines pay entitlements, prepares input, and verifies leave and earnings; prepares pay vouchers; verifies, corrects, and maintains files of personnel financial records; operates personal computers and other office equipment; answers inquiries from service members. *Skill Level 10:* Determines pay entitlements, prepares input, and verifies leave and earnings; prepares pay vouchers; verifies, corrects, and maintains files of personnel financial records; operates calculator, computer terminal, and other office equipment; answers inquiries from service members. *Skill Level 20:* Able to perform the duties required for Skill Level 10; provides technical guidance to subordinates; computes, reviews, verifies, and corrects items related to travel reports and allowances; disburses, collects, and prepares reconciliation reports; performs quality assurance audits and special reviews and reports findings. *Skill Level 30:* Able to perform the duties required for Skill Level 20; may supervise from 4-18 subordinates; prepares reports; maintains cash book. *Skill Level 40:* Able to perform the duties required for Skill Level 30; safeguards public funds and related documents; compiles daily activity summary reports; may supervise from 3-24 persons in a large finance organization.

Recommendation, Skill Level 10

Credit may be granted on the basis of an individualized assessment of the student (6/01).

Recommendation, Skill Level 20

Credit may be granted on the basis of an individualized assessment of the student (6/01).

Recommendation, Skill Level 30

In the lower-division baccalaureate/associate degree category, 3 semester hours in introduction to computers, 2 in business mathematics, 3 in office procedures, 2 in office administration, and 3 in principles of supervision (6/01).

Recommendation, Skill Level 40

In the lower-division baccalaureate/associate degree category, 3 semester hours in introduction to computers, 2 in business mathematics, 3 in office procedures, 3 in office administration, 3 in principles of supervision, and 2 in business communication. In the upper-division baccalaureate category, 3 semester hours for a management practicum (6/01).

MOS-73D-002

Accounting Specialist
73D10
73D20
73D30
73D40

Exhibit Dates: 1/90–9/91.

Career Management Field: 71 (Administration), subfield 713 (Finance).

Description

Summary: Records, verifies, and reconciles data in accordance with Army accounting procedures. *Skill Level 10:* Records and verifies data; prepares error correction documents and reports; maintains files; operates calculator and other office equipment. *Skill Level 20:* Able to perform the duties required for Skill Level 10; provides technical guidance to subordinates; prepares expense reconciliations; interprets budget guidelines from higher headquarters and develops cost factors; develops activity budget submissions, reports, estimates, and analyses; conducts audit of accounting records. *Skill Level 30:* Able to perform the duties required for Skill Level 20; supervises and instructs subordinates; may supervise 4-14 persons; plans and coordinates accounting operations; reviews reports for completeness and accuracy. *Skill Level 40:* Able to perform the duties required for Skill Level 30; supervises more than 15 persons; supervises operations in a large accounting activity.

Recommendation, Skill Level 10

In the lower-division baccalaureate/associate degree category, 2 semester hours in office procedures, 1 in business mathematics, and 2 in office machines (6/78).

Recommendation, Skill Level 20

In the lower-division baccalaureate/associate degree category, 3 semester hours in office procedures, 2 in office machines, 2 in business mathematics, 2 in record keeping, and 1 in accounting (6/78).

Recommendation, Skill Level 30

In the lower-division baccalaureate/associate degree category, 3 semester hours in office procedures, 2 in office machines, 2 in business mathematics, 3 in record keeping, 1 in accounting, 2 in office management, and credit in human relations on the basis of institutional evaluation (6/78).

Recommendation, Skill Level 40

In the lower-division baccalaureate/associate degree category, 3 semester hours in office procedures, 2 in office machines, 2 in business mathematics, 3 in record keeping, 2 in accounting, 3 in office management, 3 in management principles, and credit in human relations and personnel supervision on the basis of institutional evaluation (6/78).

MOS-73D-003

Accounting Specialist
73D10
73D20
73D30
73D40

Exhibit Dates: 10/91–5/01.

Career Management Field: 71 (Administration).

Description

Summary: Records, verifies, and reconciles data in accordance with Army accounting procedures. *Skill Level 10:* Records and verifies data; prepares error correction documents and reports; maintains files; operates computer terminal and other office equipment. *Skill Level 20:* Able to perform the duties required for Skill Level 10; provides technical guidance to subordinates; prepares expense reconciliations; interprets budget guidelines from higher headquarters and develops cost factors; develops activity budget submissions, reports, estimates, and analyses; conducts audit of accounting records; supervises one to five persons. *Skill Level 30:* Able to perform the duties required for Skill Level 20; supervises and instructs subordinates; may supervise 4-14 persons; plans and coordinates accounting operations; reviews reports for completeness and accuracy; prepares adjustments to the general ledger. *Skill Level 40:* Able to perform the duties required for Skill Level 30; supervises more than 15 persons in a large accounting activity.

Recommendation, Skill Level 10

Credit may be granted on the basis of an individualized assessment of the student (3/94).

Recommendation, Skill Level 20

In the lower-division baccalaureate/associate degree category, 3 semester hours in office procedures, 2 in business mathematics, 2 in office machines, 1 in computer applications, 2 in government accounting, and 2 in record keeping. (NOTE: This recommendation for skill level 20 is valid for the dates 10/91-2/95 only) (11/91).

Recommendation, Skill Level 30

In the lower-division baccalaureate/associate degree category, 3 semester hours in office procedures, 2 in business mathematics, 2 in office machines, 1 in computer applications, 2 in government accounting, 2 in record keeping, 2 in office administration, and 3 in personnel supervision (11/91).

Recommendation, Skill Level 40

In the lower-division baccalaureate/associate degree category, 3 semester hours in office procedures, 2 in business mathematics, 2 in office machines, 1 in computer applications, 2 in government accounting, 3 in record keeping, 3 in office administration, and 3 in personnel supervision. In the upper-division baccalaureate category, 3 semester hour in human relations in business (11/91).

MOS-73D-004

Accounting Specialist
73D10
73D20
73D30
73D40

Exhibit Dates: 6/01–9/02. Effective 10/02, MOS 73D was discontinued and its duties were incorporated into MOS 44C, Financial Management Technician.

Career Management Field: 71 (Administration).

Description

Summary: Records, verifies, and reconciles data in accordance with Army accounting procedures. *Skill Level 10:* Records and verifies data; prepares error correction documents and reports; maintains files; operates computer terminal and other office equipment. *Skill Level 20:* Able to perform the duties required for Skill Level 10; provides technical guidance to subordinates; prepares expense reconciliations; interprets budget guidelines from higher headquarters and develops cost factors; develops activity budget submissions, reports, estimates, and analyses; conducts audit of accounting records; supervises one to five persons. *Skill Level 30:* Able to perform the duties required for Skill Level 20; supervises and instructs subordinates; may supervise 4-14 persons; plans and coordinates accounting operations; reviews reports for completeness and accuracy; prepares adjustments to the general ledger. *Skill Level 40:* Able to perform the duties required for Skill Level 30; supervises more than 15 persons in a large accounting activity.

Recommendation, Skill Level 10

Credit may be granted on the basis of an individualized assessment of the student (6/01).

Recommendation, Skill Level 20

Credit may be granted on the basis of an individualized assessment of the student (6/01).

Recommendation, Skill Level 30

In the lower-division baccalaureate/associate degree category, 3 semester hours in introduction to computers, 3 in office procedures, 2 in business mathematics, 2 in fund accounting, 2 in record keeping, 2 in office administration, and 3 in principles of supervision (6/01).

Recommendation, Skill Level 40

In the lower-division baccalaureate/associate degree category, 3 semester hours in introduction to computers, 3 in office procedures, 2 in business mathematics, 2 in fund accounting, 2 in record keeping, 2 in office administration, and 3 in principles of supervision. In the upper-division baccalaureate category, 3 semester hour for a management practicum (6/01).

MOS-73Z-001

Finance Senior Sergeant
73Z50

Exhibit Dates: 1/90–9/91.

Career Management Field: 71 (Administration), subfield 713 (Finance).

Description

Able to perform the duties required for skill level 30 of MOS 73C (Finance Specialist) and 73D (Accounting Specialist); supervises or performs finance operations, accounting operations, or comptroller functions at military installations, agencies, or commands; supervises and reviews payroll accounting procedures and records, travel disbursements, and internal review and budget functions; performs staff functions such as research and preparation of financial accounting statements and statistical reports; as a finance operations chief at pay grade E-8, supervises a minimum of 25 persons; as a finance operations chief at pay grade E-9, supervises a minimum of 60 persons. NOTE: May have previously held MOS 71B (Clerk-Typist), 71C (Stenographer), or 71S (Attache Specialist).

Recommendation

In the vocational certificate category, 3 semester hours in payroll accounting, 1 in filing, and additional credit as follows: if the duty assignment was division chief at paygrade E-8, 2 semester hours in budgeting and systems; if the duty assignment was finance operations chief at paygrade E-8, 3 semester hours in budgeting and systems; if the duty assignment was finance operations chief at paygrade E-9, 4 semester hours in budgeting and systems. In the lower-division baccalaureate/associate degree category, 6 semester hours for field experience in accounting, 3 in business communications, and additional credit as follows: if the duty assignment was division chief at paygrade E-8, 4 semester hours in personnel supervision and 4 in office management; if the duty assignment was finance operations chief at paygrade E-8, 5 semester hours in personnel supervision, 5 in office management, and 1 in accounting systems; if the duty assignment was finance operations chief at paygrade E-9, 6 semester hours in personnel supervision, 6 in office management, and 2 in accounting systems. In the upper-division baccalaureate category, 3 semester hours in introduction to management, and additional credit as follows: if the duty assignment was either division chief at paygrade E-8 or finance operations chief at paygrade E-8, 3 semester hours for field experience in management; if the duty assignment was finance operation chief at paygrade E-9, 6 semester hours for field experience in management (11/75).

MOS-73Z-002

Finance Senior Sergeant
73Z50

Exhibit Dates: 10/91–5/01.

Career Management Field: 71 (Administration).

Description

Able to perform the duties required for skill level 40 of MOS 73C (Finance Specialist) and 73D (Accounting Specialist); supervises or performs finance operations, accounting operations, or comptroller functions at military installations, agencies, or commands; supervises and reviews payroll accounting procedures and records, travel disbursements, and internal review and budget functions; performs staff functions such as research and preparation of financial accounting statements and statistical reports; may supervise from 25 to over 60 persons. NOTE: May have progressed from 73C40 (Finance Specialist) or 73D40 (Accounting Specialist).

Recommendation

In the lower-division baccalaureate/associate degree category, 3 semester hours in government accounting, 3 for field experience in accounting, 3 in business communication, 3 in personnel supervision, and 3 in office management. In the upper-division baccalaureate category, 3 semester hours in principles of management, 3 for field experience in management, 3 in management problems, and 3 in organizational management (11/91).

MOS-73Z-003

Finance Senior Sergeant
73Z50

Exhibit Dates: 6/01–Present.

Career Management Field: 71 (Administration).

Description

Able to perform the duties required for skill level 40 of MOS 73C (Finance Specialist) and 73D (Accounting Specialist); supervises or performs finance operations, accounting operations, or comptroller functions at military installations, agencies, or commands; supervises and reviews payroll accounting procedures and records, travel disbursements, and internal review and budget functions; performs staff functions such as research and preparation of financial accounting statements and statistical reports; may supervise from 25 to over 60 persons. NOTE: May have progressed from 73C40 (Finance Specialist) or 73D40 (Accounting Specialist).

Recommendation

In the lower-division baccalaureate/associate degree category, 3 semester hours in fund accounting, 3 in computer applications, 3 in business communication, 3 in principles of supervision, and 3 in office management. In the upper-division baccalaureate category, 3 semester hours in principles of management, 6 for a management practicum, and 3 in organization development (6/01).

MOS-74B-002

Information Systems Operator-Analyst
74B10
74B20
74B30
74B40

Exhibit Dates: 2/95–Present.

Description

Summary: Supervises, installs, operates, and performs maintenance on multifunction/multiuser information processing systems, peripheral equipment, and associated devices in mobile and fixed facilities; writes and tests computer retrieval programs; prepares documentation for assigned duties. *Skill Level 10:* Operates and performs unit maintenance on multi-functional/multi-user information processing systems and peripheral equipment; transfers data between information processing equipment and systems; isolates malfunctions; assists in design and testing of computer programs; drafts technical documentation. *Skill Level 20:* Able to perform the duties required for Skill Level 10; configures equipment to meet operational needs; performs senior operator and systems administrator duties; compiles production statistics; conducts training; writes, tests, and modifies computer programs; drafts operating manuals; troubleshoots software. *Skill Level 30:* Able to perform the duties required for Skill Level 20; supervises the deployment, installation, operation, and maintenance of multi-functional/multi-user information processing systems; determines requirements, assigns duties, and coordinates activities; develops and administers on-site training programs; writes and/or approves computer programs, manuals, and procedures; analyzes and implements telecommunications and connectivity needs for local and wide area networks. *Skill Level 40:* Able to perform the duties required for Skill Level 30; plans, supervises, coordinates, and provides technical assistance for equipment installation; conducts quality assurance checks; develops and enforces policies and procedures; develops and supervises training programs; organizes various levels of system security; arranges contractor support maintenance for selected equipment; directs high-level programming projects.

Recommendation, Skill Level 10

Credit may be granted on the basis of an individualized assessment of the student (6/97).

Recommendation, Skill Level 20

Credit may be granted on the basis of an individualized assessment of the student (6/97).

Recommendation, Skill Level 30

In the lower-division baccalaureate/associate degree category, 3 semester hours in introduction to computers and computing, 3 in introduction to computer operations, 3 in micro computer operating systems, 3 in micro computer applications, 3 in local area network operating systems, and 3 in personnel supervision (6/97).

Recommendation, Skill Level 40

In the lower-division baccalaureate/associate degree category, 3 semester hours in introduction to computers and computing, 3 introduction to computer operations, 3 in microcomputer operating systems, 3 in microcomputer applications, 3 in personnel supervision, 3 in local area network operating systems, 3 in advanced computer operations, and 3 in advanced local area network operating systems. In the upper-division baccalaureate category, 3 semester hours in organizational management and 3 for field experience in management (6/97).

MOS-74C-001

Record Telecommunications Operator/Maintainer
((Record Telecommunications Operator))
((Record Telecommunications Center Operator))
74C10
74C20
74C30
74C40

Exhibit Dates: 11/90–5/01.

Career Management Field: 74 (Record Information Operations).

Description

Summary: Supervises, installs, operates, and performs unit maintenance on manual and automated telecommunications equipment that may link mainframes, minicomputers and microcomputers to networks; various protocols and topologies, including local area networks and wide area networks are used. *Skill Level 10:* Installs, operates, and maintains telecommunications and automated message switching equipment; performs network troubleshooting and problem diagnosis; performs tests to check signal flow and linkage to other installations; maintains records of message activity. *Skill Level 20:* Able to perform the duties required for Skill Level 10; supervises assigned personnel in performing above operations; provides technical leadership to subordinates and conducts ongoing training; serves as shift leader of a facility; performs system problem diagnosis at a higher level of difficulty; utilizes data analyzers and other devices to determine system faults. *Skill Level 30:* Able to perform the duties required for Skill Level 20; supervises a telecommunications center; coordinates with other sites to determine system faults and plan corrective action. *Skill Level 40:* Able to perform the duties required for Skill Level 30; plans, directs, and coordinates the installation of tactical telecommunications switching centers; prepares reports, studies, and evaluation of unit activities; directs and supervises training programs to ensure efficiency, successful implementation of changes, and career growth for subordinates.

Recommendation, Skill Level 10

In the lower-division baccalaureate/associate degree category, 3 semester hours in office procedures, 3 in principles of data communications, and 3 in introduction to computers and computing. (NOTE: This recommendation for skill level 10 is valid for the dates 11/90-9/91 only) (10/92).

Recommendation, Skill Level 20

In the lower-division baccalaureate/associate degree category, 3 semester hours in office procedures, 3 in principles of data communications, 3 in introduction to computers and computing, 3 in advanced data communications, and 3 in personnel supervision. (NOTE: This recommendation for skill level 20 is valid for the dates 11/90-2/95 only) (10/92).

Recommendation, Skill Level 30

In the lower-division baccalaureate/associate degree category, 3 semester hours in office procedures, 3 in principles of data communications, 3 in introduction to computers and computing, 3 in advanced data communications, 3 in personnel supervision, and 2 in records and information management (10/92).

Recommendation, Skill Level 40

In the lower-division baccalaureate/associate degree category, 3 semester hours in office procedures, 3 in principles of data communications, 3 in introduction to computers and computing, 3 in advanced data communications, 3 in personnel supervision, and 2 in records and information management. In the upper-division baccalaureate category, 3 semester hours in organizational management and 3 for field experience in management (10/92).

MOS-74C-002

Telecommunications Operator/
Maintainer
((Record Telecommunications Operator/
Maintainer))
74C10
74C20
74C30
74C40

Exhibit Dates: 6/01–Present.

Career Management Field: 74 (Record Information Operations).

Description

Summary: Supervises, installs, operates, and performs unit maintenance on manual and automated telecommunications equipment that may link mainframes, minicomputers and microcomputers to networks; uses various protocols and topologies, including local area networks and wide area networks. *Skill Level 10:* Installs, operates, and maintains telecommunications and automated message switching equipment; performs network troubleshooting and problem diagnosis; performs tests to check signal flow and linkage to other installations; maintains records of message activity. *Skill Level 20:* Able to perform the duties required for Skill Level 10; supervises assigned personnel in performing above operations; provides technical leadership to subordinates and conducts ongoing training; serves as shift leader of a facility; performs system problem diagnosis at a higher level of difficulty; utilizes data analyzers and other devices to determine system faults. *Skill Level 30:* Able to perform the duties required for Skill Level 20; supervises a telecommunications center; coordinates with other sites to determine system faults and plan corrective action. *Skill Level 40:* Able to perform the duties required for Skill Level 30; plans, directs, and coordinates the installation of tactical telecommunications switching centers; prepares reports, studies, and evaluation of unit activities; directs and supervises training programs to ensure efficiency, successful implementation of changes, and career growth for subordinates.

Recommendation, Skill Level 10

Credit may be granted on the basis of an individualized assessment of the student (6/01).

Recommendation, Skill Level 20

Credit may be granted on the basis of an individualized assessment of the student (6/01).

Recommendation, Skill Level 30

In the lower-division baccalaureate/associate degree category, 3 semester hours in office procedures, 3 in principles of data communications, 3 in introduction to computers and computing, 3 in advanced data communications, 3 in principles of supervision, and 3 in analysis and design of data communications systems (6/01).

Recommendation, Skill Level 40

In the lower-division baccalaureate/associate degree category, 3 semester hours in office procedures, 3 in principles of data communications, 3 in introduction to computers and computing, 3 in advanced data communications, 3 in principles of supervision, and 3 in analysis and design of data communications systems. In the upper-division baccalaureate category, 3 semester hours for a management practicum and 3 for a computer information systems internship (6/01).

MOS-74D-003

Computer/Machine Operator
74D10
74D20
74D30
74D40

Exhibit Dates: 1/90–9/92.

Career Management Field: 74 (Automatic Data Processing), subfield 741 (Data Processing Equipment Operations).

Description

Summary: Operates electronic computer console and auxiliary equipment. *Skill Level 10:* Performs computer console support activities; mounts magnetic disk and tape; operates line printers; keeps a magnetic tape library. *Skill Level 20:* Able to perform the duties required for Skill Level 10; operates a computer system consisting of a central processing unit, magnetic tape units and/or disk storage units in a non-multiprogramming or non-multiprocessing environment. *Skill Level 30:* Able to perform the duties required for Skill Level 20; operates a computer system in multiprogramming and multiprocessing environment; operates a computer system which has remote-inquiry stations and a program-interrupt capability. *Skill Level 40:* Able to perform the duties required for Skill Level 30; supervises computer system and auxiliary equipment operator personnel; prepares production schedules.

Recommendation, Skill Level 10

In the lower-division baccalaureate/associate degree category, 3 semester hours in introduction to data processing, 3 in file organization and processing, and 2 in computer system operations. (NOTE: This recommendation for skill level 10 is valid for the dates 1/90-9/91 only) (9/81).

Recommendation, Skill Level 20

In the lower-division baccalaureate/associate degree category, 3 semester hours in introduction to data processing, 3 in file organization and processing, and 3 in computer system operations (9/81).

Recommendation, Skill Level 30

In the lower-division baccalaureate/associate degree category, 5 semester hours in computer system operations, 3 in introduction to data processing, and 3 in file organization and processing. In the upper-division baccalaureate category, 2 semester hours in computer operating systems and 1 in introduction to systems analysis and design (9/81).

Recommendation, Skill Level 40

In the lower-division baccalaureate/associate degree category, 5 semester hours in computer system operations, 3 in introduction to data processing, and 3 in file organization and processing. In the upper-division baccalaureate category, 3 semester hours in operations management, 3 in personnel supervision, 2 in computer operating systems, and 1 in introduction to systems analysis and design (9/81).

MOS-74D-004

Information Systems Operator
74D10
74D20
74D30
74D40

Exhibit Dates: 10/92–6/95. Effective 6/95, MOS 74D was discontinued, and its duties were incorporated into MOS 74B.

Career Management Field: 74 (Record Information Operations).

Description

Summary: Supervises, installs, operates, and performs unit level maintenance on multifunction/multiuser information processing systems, peripheral equipment, and associated devices in mobile and fixed facilities. *Skill Level 10:* Performs computer console supporting activities; mounts tapes; operates line printers; maintains magnetic tape library; resolves problems encountered in running computer jobs. *Skill Level 20:* Able to perform the duties required for Skill Level 10; employs Job Control Language to ensure the successful completion of computer job and the conversion of storage from disk to tape or vice versa. *Skill Level 30:* Able to perform the duties required for Skill

Level 20; supervises the deployment, installation, operation, and unit maintenance of information processing systems, including comparative operating systems such as VSE, MVS, and VM; works with users to ensure successful completion of computer jobs. *Skill Level 40:* Able to perform the duties required for Skill Level 30; responsible for computer operations across all shifts at a site; incorporates planning for local area network installation and operation and the integration of microcomputers into the computer network; interacts with other areas of the facility outside computer operations as necessary to plan installation and resolve problems.

Recommendation, Skill Level 10
Credit may be granted on the basis of an individualized assessment of the student (3/94).

Recommendation, Skill Level 20
In the lower-division baccalaureate/associate degree category, 3 semester hours in introduction to computers and computing, 3 in introduction to computer operations, 3 in advanced computer operations, 3 in Job Control Language and utilities, and 3 in personnel supervision. (NOTE: This recommendation for skill level 20 is valid for the dates 11/90-2/95 only) (10/92).

Recommendation, Skill Level 30
In the lower-division baccalaureate/associate degree category, 3 semester hours in introduction to computers and computing, 3 in introduction to computer operations, 3 in advanced computer operations, 3 in Job Control Language and utilities, 3 in personnel supervision, and 2 in records and information management (10/92).

Recommendation, Skill Level 40
In the lower-division baccalaureate/associate degree category, 3 semester hours in introduction to computers and computing, 3 in introduction to computer operations, 3 in advanced computer operations, 3 in Job Control Language and utilities, 3 in personnel supervision, and 2 in records and information management. In the upper-division baccalaureate category, 3 semester hours in organizational management and 3 for field experience in management (10/92).

MOS-74D-005

Chemical Operations Specialist
74D10
74D20
74D30
74D40

Exhibit Dates: Pending Evaluation. 4/03–Present.

MOS-74F-003

Programmer/Analyst
74F10
74F20
74F30
74F40

Exhibit Dates: 1/90–9/92.

Career Management Field: 74 (Automatic Data Processing), subfield 741 (Data Processing Equipment Operations).

Description
Summary: Analyzes, writes, tests, and implements computer programs and/or conducts data system studies involving investigations, evaluation, development, and implementation of new or modified data processing systems for an application area. *Skill Level 10:* Assists in the preparation, editing, testing, and implementation of computer programs; prepares and analyzes program and system flow charts; prepares program documentation; reviews and revises computer programs; writes simple Assembler and COBOL programs, as well as programs in other universal languages, such as FORTRAN, RPG, and BASIC. *Skill Level 20:* Able to perform the duties required for Skill Level 10; analyzes, writes, edits, tests, and implements computer programs; sorts, merges, and processes runs; writes detailed program specifications for minor problems; may supervise computer programming activities. NOTE: May have progressed to 74F20 from 74B10 (Card and Tape Writer) or 74D10 (Computer/Machine Operator). *Skill Level 30:* Able to perform the duties required for Skill Level 20; applies advanced programming techniques; produces computer operations documentation; performs or assists in supervising programming and system studies in personnel, intelligence, transportation, supply, maintenance, medical, finance, and data communications areas; develops procedures to produce flow charts, block diagrams, and detailed instructions, routines, and codes for data processing application. NOTE: May have progressed to 74F30 from 74F20, 74B20 (Card and Tape Writer), or 74D20 (Computer/Machine Operator). *Skill Level 40:* Able to perform the duties required for Skill Level 30; supervises systems analysis and programming activities; supervises, advises, guides, and evaluates programmers and systems analysts.

Recommendation, Skill Level 10
In the lower-division baccalaureate/associate degree category, 3 semester hours in introduction to data processing, 3 in Assembler language programming, 3 in ANSI COBOL programming, and additional credit in other programming languages on the basis of institutional evaluation. In the upper-division baccalaureate category, 3 semester hours in systems analysis and design. (NOTE: This recommendation for skill level 10 is valid for the dates 1/90-9/91 only) (9/81).

Recommendation, Skill Level 20
In the lower-division baccalaureate/associate degree category, 3 semester hours in introduction to data processing, 3 in computer file organization, 3 in Assembler language programming, 3 in ANSI COBOL programming, and additional credit in other programming languages on the basis of institutional evaluation. In the upper-division baccalaureate category, 4 semester hours in systems analysis and design (9/81).

Recommendation, Skill Level 30
In the lower-division baccalaureate/associate degree category, 3 semester hours in introduction to data processing, 3 in computer file organization, 3 in computer operating systems, 3 in Assembler language programming, 3 in ANSI COBOL programming, and additional credit in other programming languages on the basis of institutional evaluation. In the upper-division baccalaureate category, 8 semester hours in systems analysis and design and 3 in advanced programming techniques (9/81).

Recommendation, Skill Level 40
In the lower-division baccalaureate/associate degree category, 3 semester hours in introduction to data processing, 3 in computer file organization, 3 in computer operating systems, 3 in Assembler language programming, 3 in ANSI COBOL programming, and additional credit in personnel supervision and other programming languages on the basis of institutional evaluation. In the upper-division baccalaureate category, 8 semester hours in systems analysis and design, 3 in system and programming management, 3 in advanced programming techniques, 3 in human relations, and 3 for field experience in management (9/81).

MOS-74F-004

Software/Analyst
74F10
74F20
74F30
74F40

Exhibit Dates: 10/92–6/95. Effective 6/95, MOS 74F was discontinued, and its duties were incorporated into MOS 74B.

Career Management Field: 74 (Record Information Operations).

Description
Summary: Develops charts, diagrams and specifications; prepares and processes test data; writes computer programs; tests and integrates programs into systems that meet functional needs of users. *Skill Level 10:* Converts program specifications into executable code; tests code and integrates modules into programs; programming languages may include Ada, COBOL and CICS; Job Control Language is prepared for computer operations production of the system. *Skill Level 20:* Able to perform the duties required for Skill Level 10; provides technical guidance and assistance to subordinates; assists in on-the-job training; establishes utility routines to be integrated into production job streams; may serve as project leader for small projects; plans and supervises conversion of programs from one language into another. *Skill Level 30:* Able to perform the duties required for Skill Level 20; develops skills in using data base management systems and various query processes including SQL; prepares and plans the turnover of production jobs to computer operations. *Skill Level 40:* Able to perform the duties required for Skill Level 30; supervises a site; works with users requesting services; prepares reports on the status of projects; ensures compliance with directives, policies, and organizational standards of design and programming.

Recommendation, Skill Level 10
Credit may be granted on the basis of an individualized assessment of the student (3/94).

Recommendation, Skill Level 20
In the lower-division baccalaureate/associate degree category, 3 semester hours in introduction to computers and computing, 3 in Ada programming language, 3 in COBOL programming language I, 3 in COBOL programming language II, 3 in systems analysis and design, 3 in Job Control Language and utilities, 3 in documentation/technical writing, and 3 in personnel supervision. (NOTE: This recommendation for skill level 20 is valid for the dates 10/92-2/95 only) (10/92).

Recommendation, Skill Level 30
In the lower-division baccalaureate/associate degree category, 3 semester hours in introduction to computers and computing, 3 in Ada pro-

gramming language, 3 in COBOL programming language I, 3 in COBOL programming language II, 3 in systems analysis and design, 3 in Job Control Language and utilities, 3 in documentation/technical writing, 3 in data base management systems, 3 in personnel supervision, and 2 in records and information management (10/92).

Recommendation, Skill Level 40

In the lower-division baccalaureate/associate degree category, 3 semester hours in introduction to computers and computing, 3 in Ada programming language, 3 in COBOL programming language I, 3 in COBOL programming language II, 3 in systems analysis and design, 3 in Job Control Language and utilities, 3 in documentation/technical writing, 3 in data base management systems, 3 in personnel supervision, 2 in records and information management, and 3 in local networking and connectivity. In the upper-division baccalaureate category, 3 semester hours in organizational management and 3 for field experience in management (10/92).

MOS-74G-002

Telecommunications Computer Operator-Maintainer
74G10
74G20
74G30
74G40

Exhibit Dates: 2/95–12/00. Effective 12/00, MOS 74G was discontinued and its duties were incorporated into the following MOSs: 31C, 31F, 31P, 31W, 74B, and 74C.

Description

Summary: Supervises, installs, operates, and performs maintenance on telecommunications computer systems, automatic message switches, and associated peripheral equipment. *Skill Level 10:* Performs fault isolation, diagnoses malfunctions, and restores and repairs telecommunications computer systems, automatic message switches and peripheral equipment; provides technical advice and assistance to other equipment operators regarding system initialization, capabilities, limitations, interfaces, and troubleshooting. *Skill Level 20:* Able to perform the duties required for Skill Level 10; provides technical assistance to subordinates on complex maintenance or computer operations; ensures repair parts availability; coordinates communications interface requirements to ensure network access; supervises and assists in maintenance of communications equipment. *Skill Level 30:* Able to perform the duties required for Skill Level 20; supervises and prepares work schedules for subordinates engaged in operations and maintenance of telecommunications computer systems, automatic message switches and peripheral equipment; plans and implements communications interfaces with military, host nation, and commercial networks; develops and administers on-site training programs; supervises and performs maintenance management duties on assigned equipment. *Skill Level 40:* Able to perform the duties required for Skill Level 30; supervises and manages subordinates; develops and enforces policies and procedures; determines capabilities and limitations of assigned equipment; organizes work schedules; arranges higher-level or contractor support maintenance; assists and trains operators.

Recommendation, Skill Level 10

Credit may be granted on the basis of an individualized assessment of the student (6/97).

Recommendation, Skill Level 20

Credit may be granted on the basis of an individualized assessment of the student (6/97).

Recommendation, Skill Level 30

In the lower-division baccalaureate/associate degree category, 3 semester hours in introduction to computers and computing, 3 in introduction to computer operations, 3 in introduction to telecommunications systems, 3 in personnel supervision, and 3 in digital fundamentals (6/97).

Recommendation, Skill Level 40

In the lower-division baccalaureate/associate degree category, 3 semester hours in introduction to computers and computing, 3 in introduction to computer operations, 3 in introduction to telecommunications systems, 3 in personnel supervision, 3 in digital fundamentals, 3 in electronic system troubleshooting and maintenance, 3 in network hardware and software, and 3 in network systems analysis and design. In the upper-division baccalaureate category, 3 semester hours in organizational management, 3 for field experience in management, and 3 in advanced telecommunications (6/97).

MOS-74Z-001

Data Processing NCO
74Z50

Exhibit Dates: 1/90–9/92.

Career Management Field: 74 (Automatic Data Processing), subfield 741 (Data Processing Equipment Operations).

Description

Supervises a data processing installation, including systems analysts, computer programmers, computer operators, and related support groups; delegates tasks to subordinates and manages their activities; maintains a current knowledge of accepted systems and programming standards, personnel procedures, and office operation techniques; knows the basic principles of systems analysis, computer hardware, and unit-record equipment; possesses an overall understanding of computer programming. NOTE: May have detailed knowledge in one or more of the following areas: computer operations, computer equipment maintenance, computer programming, and systems analysis, through previous MOS experience in the automatic data processing career management field (74).

Recommendation

In the vocational certificate category, 3 semester hours in unit record data processing and 2 in computer operations. In the lower-division baccalaureate/associate degree category, 3 semester hours in data processing principles and 1 in computer programming. In the upper-division baccalaureate category, 6 semester hours for field experience in management, 3 in introduction to management, 3 in office management, 3 in personnel management, and 2 in business systems analysis (12/75).

MOS-74Z-002

Information Systems Chief
((Record Information Systems Chief))
74Z50

Exhibit Dates: 10/92–5/01.

Career Management Field: 74 (Record Information Operations).

Description

Supervises a data processing installation, including systems analysts, computer programmers, computer operators, and related support groups; delegates tasks to subordinates and manages their activities; maintains a current knowledge of accepted systems and programming standards, personnel procedures, and office operation techniques; knows the principles of systems analysis, computer hardware, telecommunications, and message switching; possesses an overall understanding of computer programming. NOTE: May have detailed knowledge in one or more of the above areas based on previous MOS experience in the record information operations career management field (74). May have progressed to 74Z50 from 74C40, Record Telecommunications Center Operator; 74D40, Information Systems Operator; or 74F40, Software Analyst.

Recommendation

In the lower-division baccalaureate/associate degree category, 3 semester hours in personnel supervision and 2 in records and information management. In the upper-division baccalaureate category, 3 semester hours in organizational management, 3 for field experience in management, and 3 in data processing management; if individual has attained paygrade E-9, additional credit as follows: 3 semester hours in management problems, 3 in operations management, and 3 in communication techniques for managers (10/92).

MOS-74Z-003

Information Systems Chief
74Z50

Exhibit Dates: 6/01–Present.

Career Management Field: 74 (Record Information Operations).

Description

Supervises a data processing installation, including systems analysis, computer programming, computer operations, data communications and networking, and related support groups; delegates tasks to subordinates and manages their activities; maintains a current knowledge of accepted systems and programming standards, personnel procedures, and office operating techniques; understands the principles of systems analysis, computer hardware, data communications, network planning, and message switching; possesses an overall understanding of computer languages and operating systems. May supervise 60 - 100 staff. NOTE: May have progressed to 74Z50 from Skill Level 40 of: MOS 74B, Information Systems Operator - Analyst; MOS 74C, Telecommunications Operator - Maintainer; or MOS 74G, Telecommunications Computer Operator - Maintainer.

Recommendation

In the lower-division baccalaureate/associate degree category, 3 semester hours in introduction to computers and computing, 3 in introduction to computer applications, 3 in introduction to data communications, 3 in computer operations management, 3 in principles of supervision, 3 in business communication, and 3 in distributed computing systems. In the upper-division baccalaureate category, 3 semester hours in principles of management, 6 for a management practicum, and 3 in human resource management (6/01).

MOS-75B-003

Personnel Administration Specialist
75B10
75B20
75B30

Exhibit Dates: 1/90–3/94.

Career Management Field: 71 (Administration), subfield 712 (Personnel).

Description

Summary: Performs clerical and administrative functions regarding general personnel matters. *Skill Level 10:* Types and completes forms and records; prepares, reviews, and processes specific personnel reports; maintains files; answers telephone; serves as receptionist; receives and dispatches mail; provides input data for computer system. *Skill Level 20:* Able to perform the duties required for Skill Level 10; provides technical assistance to subordinates; composes correspondence. *Skill Level 30:* Able to perform the duties required for Skill Level 20; assigns work to subordinates; counsels staff and clients on personnel matters; reviews and compiles data for reports and personnel actions; evaluates work of assigned staff.

Recommendation, Skill Level 10

In the lower-division baccalaureate/associate degree category, 2 semester hours in typing and 1 in office procedures. (NOTE: This recommendation for skill level 10 is valid for the dates 1/90-9/91 only) (3/84).

Recommendation, Skill Level 20

In the lower-division baccalaureate/associate degree category, 3 semester hours in typing, 3 in office procedures, and 1 in business communications (3/84).

Recommendation, Skill Level 30

In the lower-division baccalaureate/associate degree category, 3 semester hours in typing, 3 in office procedures, 3 in business communications, 3 for a practicum in office procedures, 3 in personnel supervision, and 2 in office management (3/84).

MOS-75B-004

Personnel Administrative Specialist
75B10
75B20
75B30

Exhibit Dates: 4/94–6/96.

Career Management Field: 71 (Administration).

Description

Summary: Performs clerical and administrative support for general personnel matters. *Skill Level 10:* Prepares correspondence and orders and completes forms and records using of word processing and other software; maintains files; answers telephone; interviews personnel; provides input for computer system. *Skill Level 20:* Able to perform the duties required for Skill Level 10; provides technical guidance and training to subordinates. *Skill Level 30:* Able to perform the duties required for Skill Level 20; prepares correspondence and reports; evaluates and counsels subordinates; supervises maintenance of files; prioritizes, organizes, and assigns work; prepares training programs.

Recommendation, Skill Level 10

Credit may be granted on the basis of an individualized assessment of the student (4/94).

Recommendation, Skill Level 20

In the lower-division baccalaureate/associate degree category, 3 semester hours in word processing, 3 in office procedures, 2 in record keeping, and 3 for field experience in personnel procedures. (NOTE: This recommendation for skill level 20 is valid for the dates 4/94-2/95 only) (4/94).

Recommendation, Skill Level 30

In the lower-division baccalaureate/associate degree category, 3 semester hours in word processing, 3 in business communications, 3 in office procedures, 3 in record keeping, 3 for field experience in personnel procedures, 3 in personnel supervision, and 3 in office administration (4/94).

MOS-75B-005

Personnel Administration Specialist
75B10
75B20

Exhibit Dates: 7/96–Present.

Career Management Field: 71 (Administration).

Description

Summary: Performs clerical and administrative support for general personnel matters. *Skill Level 10:* Prepares correspondence and orders and completes forms and records using word processing and other software; maintains files; answers telephone; interviews personnel; provides input for computer system. *Skill Level 20:* Able to perform the duties required for Skill Level 10; supervises, counsels, evaluates, and provides technical guidance and training to subordinates.

Recommendation, Skill Level 10

Credit may be granted on the basis of an individualized assessment of the student (6/97).

Recommendation, Skill Level 20

Credit may be granted on the basis of an individualized assessment of the student (6/97).

MOS-75C-003

Personnel Management Specialist
75C10
75C20
75C30

Exhibit Dates: 1/90–3/94.

Career Management Field: 71 (Administration), subfield 712 (Personnel).

Description

Summary: Participates in the occupational classification of personnel. *Skill Level 10:* Compares personnel records with classification standards to recommend assignment of individuals; types forms and correspondence; may interview personnel; provides input for computer system; maintains files. *Skill Level 20:* Able to perform the duties required for Skill Level 10; provides technical assistance to subordinates. *Skill Level 30:* Able to perform the duties required for Skill Level 20; coordinates and makes work assignments; supervises and evaluates performance of staff; organizes training programs; provides counseling on personnel matters.

Recommendation, Skill Level 10

In the lower-division baccalaureate/associate degree category, 1 semester hour in office procedures and 2 in typing. (NOTE: This recommendation for skill level 10 is valid for the dates 1/90-9/91 only) (3/84).

Recommendation, Skill Level 20

In the lower-division baccalaureate/associate degree category, 3 semester hours in office procedures, 3 in typing, and 1 in business communications (3/84).

Recommendation, Skill Level 30

In the lower-division baccalaureate/associate degree category, 3 semester hours in office procedures, 3 in typing, 1 in business communications, 3 for a practicum in office procedures, 2 in office management, 2 in human relations, and 3 in personnel supervision (3/84).

MOS-75C-004

Personnel Management Specialist
75C10
75C20
75C30

Exhibit Dates: 4/94–6/96. Effective 6/96, MOS 75C was discontinued and its duties were incorporated into MOS 75H, Personnel Services Specialist.

Career Management Field: 71 (Administration).

Description

Summary: Participates in occupational classification and management of manpower resources or supervises personnel management activities. *Skill Level 10:* Prepares correspondence and orders and completes forms and records using word processing and other software; maintains files; answers telephone; interviews personnel; provides input for computer system. *Skill Level 20:* Able to perform the duties required for Skill Level 10; provides technical guidance and training to subordinates. *Skill Level 30:* Able to perform the duties required for Skill Level 20; prepares correspondence and reports; evaluates and counsels subordinates; supervises maintenance of files; prioritizes, organizes, and assigns work; prepares training programs.

Recommendation, Skill Level 10

Credit may be granted on the basis of an individualized assessment of the student (4/94).

Recommendation, Skill Level 20

In the lower-division baccalaureate/associate degree category, 3 semester hours in word processing, 3 in office procedures, 2 in record keeping, and 3 for field experience in personnel procedures. (NOTE: This recommendation for skill level 20 is valid for the dates 4/94-2/95 only) (4/94).

Recommendation, Skill Level 30

In the lower-division baccalaureate/associate degree category, 3 semester hours in word processing, 3 in business communications, 3 in office procedures, 3 in record keeping, 3 for field experience in personnel procedures, 3 in personnel supervision, and 3 in office administration (4/94).

MOS-75D-003

Personnel Records Specialist
75D10
75D20
75D30

Exhibit Dates: 1/90–3/94.

Career Management Field: 71 (Administration), subfield 712 (Personnel).

Description

Summary: Maintains military personnel records or supervises records maintenance. *Skill Level 10:* Types correspondence, memorandums, and forms; prepares computer input data for personnel records; maintains files; posts regulation changes; prepares and verifies a variety of personnel records. *Skill Level 20:* Able to perform the duties required for Skill Level 10; provides technical assistance to subordinates; drafts correspondence and reports. *Skill Level 30:* Able to perform the duties required for Skill Level 20; determines personnel requirements; assigns duties and organizes work schedules; prepares and conducts training; evaluates and counsels personnel; monitors and audits personnel records; reviews computer system input and reconciles with output.

Recommendation, Skill Level 10

In the lower-division baccalaureate/associate degree category, 1 semester hour in office procedures, 2 in typing, and 2 in filing or records management. (NOTE: This recommendation for skill level 10 is valid for the dates 1/90-9/91 only) (3/84).

Recommendation, Skill Level 20

In the lower-division baccalaureate/associate degree category, 3 semester hours in office procedures, 3 in typing, 2 in filing or records management, and 2 in business communication or technical writing (3/84).

Recommendation, Skill Level 30

In the lower-division baccalaureate/associate degree category, 3 semester hours in office procedures, 3 in typing, 2 in filing or records management, 3 in business communication or technical writing, 2 in office management, 3 in personnel supervision, and 3 for a practicum in office procedures (3/84).

MOS-75D-004

Personnel Records Specialist
75D10
75D20
75D30

Exhibit Dates: 4/94–6/96. Effective 6/96, MOS 75D was discontinued and its duties were incorporated into MOS 75H, Personnel Services Specialist.

Career Management Field: 71 (Administration).

Description

Summary: Prepares and maintains personnel records or supervises records preparation and maintenance. *Skill Level 10:* Prepares correspondence forms and records using word processing software; maintains files; prepares and verifies a variety of personnel records; and provides computer data input. *Skill Level 20:* Able to perform the duties required for Skill Level 10; provides technical guidance to subordinates. *Skill Level 30:* Able to perform the duties required for Skill Level 20; prepares reports, correspondence, and recommendations; evaluates and counsels subordinates; prepares and conducts training; establishes work priorities.

Recommendation, Skill Level 10

Credit may be granted on the basis of an individualized assessment of the student (4/94).

Recommendation, Skill Level 20

In the lower-division baccalaureate/associate degree category, 3 semester hours in word processing, 3 in office procedures, 3 in record keeping, and 3 for field experience in personnel procedures. (NOTE: This recommendation for skill level 20 is valid for the dates 4/94-2/95 only) (4/94).

Recommendation, Skill Level 30

In the lower-division baccalaureate/associate degree category, 3 semester hours in word processing, 3 in office procedures, 3 in technical writing, 3 in record keeping, 3 for field experience in personnel procedures, 3 in personnel supervision, and 3 in office administration (4/94).

MOS-75E-003

Personnel Actions Specialist
75E10
75E20
75E30

Exhibit Dates: 1/90–3/94.

Career Management Field: 71 (Administration), subfield 712 (Personnel).

Description

Summary: Processes personnel actions concerning service members and their dependents. *Skill Level 10:* Completes forms required for training schools, awards and decorations, release, transfer, reenlistment, retirement, and casualties; provides information on veterans', social security, and retirement benefits; types forms, correspondence, and messages; verifies information; prepares computer input data; handles telephone calls; maintains and updates files. *Skill Level 20:* Able to perform the duties required for Skill Level 10; provides technical assistance to subordinates. *Skill Level 30:* Able to perform the duties required for Skill Level 20; determines personnel requirements, work priorities, duty assignments, and work schedules; prepares and conducts training programs; prepares administrative, technical, and personnel reports; advises and counsels individuals on personnel matters; supervises personnel action activities.

Recommendation, Skill Level 10

In the lower-division baccalaureate/associate degree category, 2 semester hours in typing and 2 in office procedures. (NOTE: This recommendation for skill level 10 is valid for the dates 1/90-9/91 only) (3/84).

Recommendation, Skill Level 20

In the lower-division baccalaureate/associate degree category, 3 semester hours in typing, 3 in office procedures, and 2 in business communication (3/84).

Recommendation, Skill Level 30

In the lower-division baccalaureate/associate degree category, 3 semester hours in typing, 3 in office procedures, 3 in business communications, 3 for a practicum in office procedures, and 3 in personnel supervision (3/84).

MOS-75E-004

Personnel Actions Specialist
75E10
75E20
75E30

Exhibit Dates: 4/94–6/96. Effective 6/96, MOS 75E was discontinued and its duties were incorporated into MOS 75H, Personnel Services Specialist.

Career Management Field: 71 (Administration).

Description

Summary: Performs personnel actions concerning service members and their dependents or supervises personnel actions activities. *Skill Level 10:* Prepares correspondence and completes forms and records using word processing software; maintains and updates files; answers telephone; verifies information; and provides computer data input. *Skill Level 20:* Able to perform the duties required for Skill Level 10; provides technical guidance and training to subordinates. *Skill Level 30:* Able to perform the duties required for Skill Level 20; prepares correspondence and recommendations; evaluates and counsels subordinates.

Recommendation, Skill Level 10

Credit may be granted on the basis of an individualized assessment of the student (4/94).

Recommendation, Skill Level 20

In the lower-division baccalaureate/associate degree category, 3 semester hours in word processing, 3 in office procedures, 2 in record keeping, 2 in business communication, and 3 for field experience in office procedures. (NOTE: This recommendation for skill level 20 is valid for the dates 4/94-2/95 only) (4/94).

Recommendation, Skill Level 30

In the lower-division baccalaureate/associate degree category, 3 semester hours in word processing, 3 in office procedures, 3 in business communication, 3 in record keeping, 3 for field experience in office procedures, 3 in office administration, and 3 in personnel supervision (4/94).

MOS-75F-001

Personnel Information System
Management Specialist
75F10
75F20
75F30

Exhibit Dates: 1/90–3/94.

Career Management Field: 71 (Administration), subfield 712 (Personnel).

Description

Summary: Operates and manages personnel information systems and assists or supervises users of systems. *Skill Level 10:* Analyzes, processes, distributes, and maintains personnel files; reviews input data for accuracy and prepares system documents; operates keypunch machines, optical scan equipment, and other data reduction devices, such as IBM 129, DATA 200, and computer components; monitors accuracy of personnel files and makes corrections as necessary; performs automated interface procedures with other automated systems; types entries onto forms and records; maintains files within the Army Functional Files System. *Skill Level 20:* Able to perform the duties required for Skill Level 10; operates personnel information systems and trains system users; provides technical assistance to subordinates; controls incoming/outgoing automatic data information transmissions; prepares output distribution plan; visits system users, and gives assistance or formal instruction as needed. *Skill Level 30:* Able to perform the duties required for Skill Level 20; supervises personnel information systems, assigns duties, organizes work schedules, and ensures work completion; determines training needs of subordinates and conducts training programs; reviews reports to assess system performance; develops production schedules; maintains liaison with managers of interfaced systems; establishes and maintains office procedures; prepares administrative and technical reports; counsels subordinates and prepares evaluation reports; may supervise four to nine personnel; establishes stock level of needed supplies and blank forms.

Recommendation, Skill Level 10

In the lower-division baccalaureate/associate degree category, 2 semester hours in computer systems operations and 2 in data processing. (NOTE: This recommendation for skill level 10 is valid for the dates 1/90-9/91 only) (3/84).

Recommendation, Skill Level 20

In the lower-division baccalaureate/associate degree category, 2 semester hours in computer system operations, 3 in data processing, 2 in computer file organization, and 1 in office procedures (3/84).

Recommendation, Skill Level 30

In the lower-division baccalaureate/associate degree category, 3 semester hours in computer system operations, 3 in data processing, 2 in office procedures, 2 in personnel supervision, and 2 in computer file organization (3/84).

MOS-75F-002

Personnel Information System Management Specialist
75F10
75F20
75F30

Exhibit Dates: 4/94–2/95.

Career Management Field: 71 (Administration).

Description

Summary: Operates and manages personnel information systems, trains and assists system users, or supervises systems activities. *Skill Level 10:* Prepares correspondence and completes forms and records using word processing software; maintains and updates files; answers telephone; verifies information; provides computer data input. *Skill Level 20:* Able to perform the duties required for Skill Level 10; provides technical guidance and training to subordinates; identifies training needs of system users; maintains system; prepares and maintains output distribution scheme. *Skill Level 30:* Able to perform the duties required for Skill Level 20; prepares correspondence and recommendations; supervises, evaluates, and counsels subordinates; establishes work priorities; assigns and instructs subordinates in duties and procedures; maintains liaison with servicing facility and interface systems; prepares and reviews reports.

Recommendation, Skill Level 10

Credit may be granted on the basis of an individualized assessment of the student (4/94).

Recommendation, Skill Level 20

In the lower-division baccalaureate/associate degree category, 3 semester hours in keyboarding, 2 in office procedures, 2 in computer operations, and 3 for field experience in personnel procedures (4/94).

Recommendation, Skill Level 30

In the lower-division baccalaureate/associate degree category, 3 semester hours in keyboarding, 3 in office procedures, 3 in business communication, 3 in computer operations, 3 for field experience in personnel procedures, 3 in office administration, and 3 in personnel supervision (4/94).

MOS-75F-003

Personnel Information System Management Specialist
75F10
75F20
75F30

Exhibit Dates: 3/95–1/96.

Career Management Field: 71 (Administration).

Description

Summary: Operates and manages personnel information systems, trains and assists system users, or supervises systems activities. *Skill Level 10:* Prepares correspondence and completes forms and records using word processing software; maintains and updates files; answers telephone; verifies information; provides computer data input. *Skill Level 20:* Able to perform the duties required for Skill Level 10; provides technical guidance and training to subordinates; identifies training needs of system users; maintains system; prepares and maintains output distribution scheme. *Skill Level 30:* Able to perform the duties required for Skill Level 20; prepares correspondence and recommendations; supervises, evaluates, and counsels subordinates; establishes work priorities; assigns and instructs subordinates in duties and procedures; maintains liaison with servicing facility and interface systems; prepares and reviews reports.

Recommendation, Skill Level 10

Credit may be granted on the basis of an individualized assessment of the student (4/94).

Recommendation, Skill Level 20

Credit may be granted on the basis of an individualized assessment of the student (4/94).

Recommendation, Skill Level 30

In the lower-division baccalaureate/associate degree category, 3 semester hours in keyboarding, 2 in office procedures, 3 in business communication, 2 in computer operations, 3 for field experience in personnel procedures, 3 in office administration, and 3 in personnel supervision (4/94).

MOS-75F-004

Personnel Information System Management Specialist
75F10
75F20
75F30

Exhibit Dates: 2/96–Present.

Career Management Field: 71 (Administration).

Description

Summary: Operates and manages field personnel information systems; trains and assists system users; or supervises systems activities. *Skill Level 10:* Prepares correspondence and completes forms and records using word processing software; maintains and updates files; answers telephone; verifies information; provides computer data input. *Skill Level 20:* Able to perform the duties required for Skill Level 10; provides technical guidance and training to subordinates; identifies training needs of system users; maintains system; prepares and maintains output distribution scheme. *Skill Level 30:* Able to perform the duties required for Skill Level 20; prepares correspondence and recommendations; supervises, evaluates, and counsels subordinates; establishes work priorities; assigns and trains subordinates in duties and procedures; maintains liaison with servicing facility and interface systems; prepares and reviews reports.

Recommendation, Skill Level 10

Credit may be granted on the basis of an individualized assessment of the student (6/97).

Recommendation, Skill Level 20

Credit may be granted on the basis of an individualized assessment of the student (6/97).

Recommendation, Skill Level 30

In the lower-division baccalaureate/associate degree category, 3 semester hours in computer applications, 2 in office administration, 3 in business communication, 2 in office procedures, 3 in personnel supervision, and 3 in record keeping. In the upper-division baccalaureate category, 3 semester hours in management problems and 3 in human resource management (6/97).

MOS-75H-001

Personnel Services Specialist
75H10
75H20
75H30
75H40
75H50

Exhibit Dates: 2/96–3/03. (Effective 4/03 this MOS was discontinued, and its duties were incorporated into MOS 42A.)

Career Management Field: 71 (Administration).

Description

Summary: Participates in occupational classification and management of manpower resources; supervises personnel management of manpower resources or supervises personnel management activities including maintaining personnel records and processing personnel actions service members and their family members. *Skill Level 10:* Prepares reports on strength levels and status of personnel; evaluates personnel qualifications for special assignments; prepares and processes requests for transfer or reassignment; processes classification/reclassification actions; prepares orders and requests for orders; prepares and maintains officer and enlisted personnel records; prepares and reviews personnel casualty documents; monitors suspense actions; initiates, monitors, and processes personnel evaluations; transfers records; processes soldiers for separation and retirement: processes and executes personnel service center level procedures and actions; processes applications for officer candidate school, warrant officer flight training, and other training; processes recommendations for awards and decorations; processes bars to reenlistment and suspension of favorable personnel actions; initiates applications for passports and visas; monitors appointment of line of duty, survivor assistance, and summary court officers; processes line of duty investigations; prepares letters of sympathy to next of kin; types correspondence and forms in draft and final copy; posts changes to Army regulations and other publications; prepares and maintains files on an automated data processing system; applies knowledge of provisions and limitations of Freedom of Information and Privacy acts. *Skill Level 20:* Able to perform the duties required for Skill Level 10; provides technical guidance and training to subordinates. *Skill Level 30:* Able to perform the duties required for Skill Level 20; supervises specific personnel functions in a small personnel office, battalion, and personnel services support activity; advises commanders on soldiers and personnel readiness and strength levels of supported reporting units; reviews consolidated reports, statistics, applications, and prepares recommendations for personnel actions to higher headquarters; reviews and prepares reports and data on strength (gains and losses) of personnel and makes duty assignments of enlisted personnel; reviews cyclic and other reports to assess systems performance; maintains liaison with servicing data processing facility and field managers of interfaced systems. *Skill Level 40:* Able to perform the duties required for Skill Level 30; supervises small personnel office, specific personnel functions, battalion and personnel services; supervises quality assurance procedures; advises commander, adjutant, and other staff members on personnel administration activities. *Skill Level 50:* Able to perform the duties required for Skill Level 40; supervises a large personnel activity, while performing specialized or all-encompassing personnel functions; manages all personnel services specialists within the command, including division, corps, and Department of Army; provides planning information for short-and long-range personnel requirements for operations planning, strength reduction and augmentation; prepares recommendations for staff officers.

Recommendation, Skill Level 10

Credit may be granted on the basis of an individualized assessment of the student (6/97).

Recommendation, Skill Level 20

Credit may be granted on the basis of an individualized assessment of the student (6/97).

Recommendation, Skill Level 30

In the lower-division baccalaureate/associate degree category, 3 semester hours in word processing/computer applications, 3 in record keeping, 2 in clerical procedures, 2 in office administration, 3 in business communication, and 3 in personnel supervision. In the upper-division baccalaureate category, 3 semester hours in human resource management and 3 in management problems (6/97).

Recommendation, Skill Level 40

In the lower-division baccalaureate/associate degree category, 3 semester hours in word processing/computer applications, 3 in record keeping, 3 in clerical procedures, 3 in office administration, 3 in business communication, and 3 in personnel supervision. In the upper-division baccalaureate category, 3 semester hours in human resource management, 3 in management problems, and 3 for field experience in management (6/97).

Recommendation, Skill Level 50

In the lower-division baccalaureate/associate degree category, 3 semester hours in word processing/computer application, 3 in record keeping, 3 in clerical procedures, 3 in office administration, 3 in business communication, and 3 in personnel supervision. In the upper-division baccalaureate category, 3 semester hours in human resource management, 6 in management problems, and 3 for field experience in management (6/97).

MOS-75Z-003

Personnel Sergeant

75Z40

75Z50

Exhibit Dates: 1/90–3/94.

Career Management Field: 71 (Administration), subfield 712 (Personnel).

Description

Summary: Supervises the operation of a personnel office, including personnel administration, personnel management, personnel records, and information systems. *Skill Level 40:* Supervises performance of legal, reenlistment, and administrative matters as well as personnel actions; reviews, consolidates, and drafts reports and surveys; researches specific policies and procedures relating to officer and enlisted personnel administration; reviews data prepared for computer input and reconciles output; as a mid-level manager, advises superiors on personnel administrative activities; supervises up to 18 persons in a segment of a large personnel office or in a small or medium-sized personnel office. NOTE: May have progressed to 75Z40 from 75B30 (Personnel Administration Specialist), 75C30 (Personnel Management Specialist), 75D30 (Personnel Records Specialist), 75E30 (Personnel Actions Specialist), or 75F30 (Information System Management Specialist). *Skill Level 50:* Able to perform the duties required for Skill Level 40; supervises up to 39 persons in a large segment of a large personnel office or serves as the principal noncommissioned officer of a large personnel office; may also perform the duties of a First Sergeant.

Recommendation, Skill Level 40

In the vocational certificate category, 3 semester hours in computer concepts. In the lower-division baccalaureate/associate degree category, 3 semester hours in office procedures, 3 in office management, 3 in business communication or technical writing, 3 in introduction to computer concepts, 3 in human relations, 3 in personnel supervision, and 3 in records management on the basis of institutional evaluation. In the upper-division baccalaureate category, 3 semester hours in personnel management and 3 for field experience in management (3/84).

Recommendation, Skill Level 50

In the vocational certificate category, 3 semester hours in computer concepts. In the lower-division baccalaureate/associate degree category, 3 semester hours in office procedures, 3 in office management, 3 in business communication or technical writing, 3 in introduction to computer concepts, 3 in human relations, 3 in personnel supervision, 3 for field experience in management, and 3 in records management on the basis of institutional evaluation. In the upper-division baccalaureate category, 3 semester hours in personnel management and 6 for field experience in management (3/84).

MOS-75Z-004

Personnel Sergeant

75Z40

75Z50

Exhibit Dates: 4/94–6/96. Effective 6/96, MOS 75Z was discontinued and its duties were incorporated into MOS 75H, Personnel Services Specialist.

Career Management Field: 71 (Administration).

Description

Summary: Supervises the operation of a personnel office, including personnel administration, personnel management, personnel records, and information systems. *Skill Level 40:* Supervises performance of legal, reenlistment, and administrative matters as well as personnel actions; reviews, consolidates, and drafts reports and surveys; researches specific policies and procedures related to officer and enlisted personnel administration; reviews data prepared for computer input and reconciles output; as a mid-level manager, advises superiors on personnel administrative activities; supervises up to 18 persons in a segment of a large personnel office or in a small or medium-sized personnel office. NOTE: May have progressed to 75Z40 from 75B30 (Personnel Administration Specialist), 75C30 (Personnel Management Specialist), 75D30 (Personnel Records Specialist), 75E30 (Personnel Actions Specialist), or 75F30 (Information System Management Specialist). *Skill Level 50:* Able to perform the duties required for Skill Level 40; supervises up to 39 persons in a large segment of a large personnel office or serves as the principal noncommissioned officer of a large personnel office; may also perform the duties of a first sergeant.

Recommendation, Skill Level 40

In the lower-division baccalaureate/associate degree category, 3 semester hours in office management, 3 in records management, 3 in computer applications, 3 in human relations, and 3 for field experience in management. In the upper-division baccalaureate category, 3 semester hours in human resource management and 3 in management problem solving (4/94).

Recommendation, Skill Level 50

In the lower-division baccalaureate/associate degree category, 3 semester hours in office management, 3 in records management, 3 in computer applications, 3 in human relations, and 3 for field experience in management. In the upper-division baccalaureate category, 3 semester hours in human resource management and 3 in management problems solving (4/94).

MOS-76C-001

Equipment Records and Parts Specialist

76C10

76C20

Exhibit Dates: 1/90–3/92.

Career Management Field: 76 (Supply and Service), subfield 761 (Supply).

Description

Summary: Maintains records for repair parts inventory, equipment use, and equipment maintenance. *Skill Level 10:* Obtains or prepares repair parts lists; locates information using catalog data and technical manuals; prepares repair parts requests; receives, stores, and issues repair parts; keeps inventory records of repair parts; prepares and maintains records on equipment usage, operation, and maintenance; initiates, receives, and maintains manual control records; may initiate, receive, and maintain automated maintenance control records. *Skill Level 20:* Able to perform the duties required for Skill Level 10; provides technical guidance to subordinates; obtains repair parts to meet equipment maintenance needs; assists in planning maintenance needs for the unit.

Recommendation, Skill Level 10

In the lower-division baccalaureate/associate degree category, 2 semester hours in record keeping and 2 in inventory control; if person worked with automated maintenance control records, 2 semester hours in computer operations. (NOTE: This recommendation for skill level 10 is valid for the dates 1/90-9/91 only) (9/82).

Recommendation, Skill Level 20

In the lower-division baccalaureate/associate degree category, 2 semester hours in record keeping and 2 in inventory control; if person worked with automated maintenance control records, 2 semester hours in computer operations (9/82).

MOS-76C-002

Equipment Records and Parts Specialist
76C10
76C20

Exhibit Dates: 4/92–6/93. Effective 6/93, MOS 76C was discontinued, and its duties were incorporated into MOS 92A, Automated Logistical Specialist.

Career Management Field: 76 (Supply and Services).

Description

Summary: Maintains records for repair parts inventory, equipment use, and equipment maintenance including hazardous material records. *Skill Level 10:* Obtains or prepares repair parts lists; locates information using catalog data and technical manuals; prepares repair parts requests; receives, stores, exchanges, and issues repair parts; keeps inventory records of repair parts including hazardous materials records; prepares and maintains records on equipment usage, operation, maintenance, modification, and calibration; initiates, receives, and maintains manual control records; may initiate, receive, and maintain automated maintenance control records; records equipment readiness codes and maintenance data for computer applications. *Skill Level 20:* Able to perform the duties required for Skill Level 10; supervises and provides technical guidance to subordinates; obtains repair parts to meet equipment maintenance needs; assists in planning maintenance needs for the unit.

Recommendation, Skill Level 10

Credit may be granted on the basis of an individualized assessment of the student (3/94).

Recommendation, Skill Level 20

In the lower-division baccalaureate/associate degree category, 3 semester hours in computer applications, 3 in record keeping, and 3 in personnel supervision (4/92).

MOS-76J-002

Medical Supplyman
76J10
76J20
76J30
76J40
76J50

Exhibit Dates: 1/90–3/92.

Career Management Field: 76 (Supply and Services), subfield 761 (Supply).

Description

Summary: Supervises or performs requisitioning, receipt, inventory management, storage, preservation, stock control, and accounting of medical supplies and equipment. *Skill Level 10:* Transports and handles medical supplies and equipment; inventories and issues medical supplies; performs in-storage care and preservation; applies special procedures for handling, storage, packaging, and shipping pharmaceuticals, biologicals, blood fractions, and medicines; interprets medical terminology and vocabulary directly related to supply function; operates office machines; keeps administrative and medical supply files; assists in keeping records on equipment operation and maintenance. *Skill Level 20:* Able to perform the duties required for Skill Level 10; provides technical guidance to subordinates; responsible for local purchasing. *Skill Level 30:* Able to perform the duties required for Skill Level 20; supervises small medical supply, stock control, or storage activity; plans medical supply and equipment operations; assigns duties, plans work loads, and instructs personnel in work techniques and procedures; responsible for budgeting and managing funds; works with automated property book system. *Skill Level 40:* Able to perform the duties required for Skill Level 30; supervise medium-sized medical supply, stock control, or storage activity. *Skill Level 50:* Able to perform the duties required for Skill Level 40; supervises a large medical supply, stock control, or storage activity or serves as principal noncommissioned officer of medical depot; analyzes medical logistical management data.

Recommendation, Skill Level 10

In the lower-division baccalaureate/associate degree category, 3 semester hours in warehouse practices, 2 in office machines, 1 in medical terminology, 1 in filing, and credit in record keeping and office management on the basis of institutional evaluation. (NOTE: This recommendation for skill level 10 is valid for the dates 1/90-9/91 only) (6/78).

Recommendation, Skill Level 20

In the lower-division baccalaureate/associate degree category, 3 semester hours in warehouse practices, 3 in record keeping, 2 in office machines, 1 in medical terminology, 2 in office management, and 2 in filing (6/78).

Recommendation, Skill Level 30

In the lower-division baccalaureate/associate degree category, 3 semester hours in warehouse practices, 3 in record keeping, 2 in budgeting practices, 2 in inventory control, 2 in office machines, 2 in office procedures, 1 in medical terminology, 2 in filing, 3 in office management, 2 in computer concepts, and 3 in human relations (6/78).

Recommendation, Skill Level 40

In the lower-division baccalaureate/associate degree category, 3 semester hours in warehouse practices, 3 in record keeping, 2 in budgeting practices, 2 in inventory control, 2 in office machines, 2 in office procedures, 1 in medical terminology, 2 in filing, 3 in office management, 2 in computer concepts, and 3 in human relations. In the upper-division baccalaureate category, 3 semester hours for field experience in management (6/78).

Recommendation, Skill Level 50

In the vocational certificate category, 6 semester hours for field experience in management. In the lower-division baccalaureate/associate degree category, 3 semester hours in warehouse practices, 3 in record keeping, 2 in budgeting practices, 2 in inventory control, 2 in office machines, 2 in office procedures, 1 in medical terminology, 2 in filing, 3 in office management, 2 in computer concepts, and 3 in human relations. In the upper-division baccalaureate category, 3 semester hours for field experience in management and 3 in personnel supervision; if served as principal noncommissioned officer, 3 semester hours in personnel management (6/78).

MOS-76J-003

Medical Supply Specialist
76J10
76J20
76J30
76J40
76J50

Exhibit Dates: 4/92–2/95.

Career Management Field: 91 (Medical).

Description

Summary: Supervises or performs requisitioning, receipt, inventory management, storage, preservation, stock control and quality, parts repair, and accounting of medical supplies and equipment. *Skill Level 10:* Transports and handles medical supplies and equipment; inventories and issues medical supplies; performs in-storage care and preservation; applies special procedures for handling, storage, packaging, and shipping pharmaceuticals, biologicals, blood fractions, and medicines; interprets medical terminology and vocabulary directly related to supply function; operates office machines; keeps administrative and medical supply files; assists in keeping records on equipment operation and maintenance. *Skill Level 20:* Able to perform the duties required for Skill Level 10; supervises and provides technical guidance to subordinates; responsible for local purchasing. *Skill Level 30:* Able to perform the duties required for Skill Level 20; supervises small medical supply, stock control, or storage activity; plans medical supply and equipment operations; assigns duties, plans work loads, and instructs personnel in work techniques and procedures; responsible for budgeting and managing funds; works with automated property book system. *Skill Level 40:* Able to perform the duties required for Skill Level 30; supervises a medium-sized medical supply, stock control, or storage activity. *Skill Level 50:* Able to perform the duties required for Skill Level 40; supervises a large medical supply, stock control, or storage activity or serves as principal noncommissioned officer of medical depot; analyzes medical logistical management data.

Recommendation, Skill Level 10

Credit may be granted on the basis of an individualized assessment of the student (3/94).

Recommendation, Skill Level 20

In the lower-division baccalaureate/associate degree category, 3 semester hours in computer applications, 3 in supply management, 3 in warehouse operations, and 3 in personnel supervision (4/92).

Recommendation, Skill Level 30

In the lower-division baccalaureate/associate degree category, 3 semester hours in computer applications, 3 in supply management, 4 in warehouse operations, 3 in personnel supervision, and 2 in records and information management (4/92).

Recommendation, Skill Level 40

In the lower-division baccalaureate/associate degree category, 3 semester hours in computer applications, 3 in supply management, 5 in warehouse operations, 3 in personnel supervision, and 2 in records and information management. In the upper-division baccalaureate category, 3 semester hours in organizational management and 3 for field experience in management (4/92).

Recommendation, Skill Level 50

In the lower-division baccalaureate/associate degree category, 3 semester hours in computer applications, 3 in supply management, 5 in warehouse operations, 3 in personnel supervision, and 2 in records and information management. In the upper-division baccalaureate category, 3 semester hours in organizational management and 6 for field experience in management; if paygrade E-9 has been attained, additional credit as follows: 3 semester hours in management problems, 3 in operations management, and 3 in communication techniques for managers (4/92).

MOS-76J-004

Medical Supply Specialist
76J10
76J20
76J30
76J40
76J50

Exhibit Dates: 3/95–5/01.

Career Management Field: 91 (Medical).

Description

Summary: Supervises or performs requisitioning, receipt, inventory management, storage, preservation, stock control and quality, parts repair, and accounting of medical supplies and equipment. *Skill Level 10:* Transports and handles medical supplies and equipment; inventories and issues medical supplies; performs in-storage care and preservation; applies special procedures for handling, storage, packaging, and shipping pharmaceuticals, biologicals, blood fractions, and medicines; interprets medical terminology and vocabulary directly related to supply function; operates office machines; keeps administrative and medical supply files; assists in keeping records on equipment operation and maintenance. *Skill Level 20:* Able to perform the duties required for Skill Level 10; supervises and provides technical guidance to subordinates; responsible for local purchasing. *Skill Level 30:* Able to perform the duties required for Skill Level 20; supervises small medical supply, stock control, or storage activity; plans medical supply and equipment operations; assigns duties, plans work loads, and instructs personnel in work techniques and procedures; responsible for budgeting and managing funds; works with automated property book system. *Skill Level 40:* Able to perform the duties required for Skill Level 30; supervises a medium-sized medical supply, stock control, or storage activity. *Skill Level 50:* Able to perform the duties required for Skill Level 40; supervises a large medical supply, stock control, or storage activity or serves as principal noncommissioned officer of medical depot; analyzes medical logistical management data.

Recommendation, Skill Level 10

Credit may be granted on the basis of an individualized assessment of the student (3/94).

Recommendation, Skill Level 20

Credit may be granted on the basis of an individualized assessment of the student (4/92).

Recommendation, Skill Level 30

In the lower-division baccalaureate/associate degree category, 3 semester hours in computer applications, 3 in supply management, 3 in warehouse operations, 3 in personnel supervision, and 2 in records and information management (4/92).

Recommendation, Skill Level 40

In the lower-division baccalaureate/associate degree category, 3 semester hours in computer applications, 3 in supply management, 4 in warehouse operations, 3 in personnel supervision, and 2 in records and information management. In the upper-division baccalaureate category, 3 semester hours in organizational management and 3 for field experience in management (4/92).

Recommendation, Skill Level 50

In the lower-division baccalaureate/associate degree category, 3 semester hours in computer applications, 3 in supply management, 5 in warehouse operations, 3 in personnel supervision, and 2 in records and information management. In the upper-division baccalaureate category, 3 semester hours in organizational management and 6 for field experience in management; if paygrade E-9 has been attained, additional credit as follows: 3 semester hours in management problems, 3 in operations management, and 3 in communication techniques for managers (4/92).

MOS-76J-005

Medical Supply Specialist
76J10
76J20
76J30
76J40
76J50

Exhibit Dates: 6/01–9/01. Effective 10/01, MOS 76J was discontinued and its duties were transferred to MOS 91J, Medical Logistics Specialist.

Career Management Field: 91 (Medical).

Description

Summary: Supervises or performs requisitioning, receipt, inventory management, storage, preservation, stock control and quality, and accounting of medical supplies and equipment. *Skill Level 10:* Transports and handles medical supplies and equipment; inventories and issues medical supplies; performs in-storage care and preservation; applies special procedures for handling, storage, packaging, and shipping pharmaceuticals, biologicals, blood fractions, and medicines; interprets medical terminology and vocabulary directly related to supply function; operates office machines; keeps administrative and medical supply files; assists in keeping records on equipment operation and maintenance. *Skill Level 20:* Able to perform the duties required for Skill Level 10; supervises and provides technical guidance to subordinates; responsible for local purchasing. *Skill Level 30:* Able to perform the duties required for Skill Level 20; supervises small medical supply, stock control, or storage activity; plans medical supply and equipment operations; assigns duties, plans work loads, and instructs personnel in work techniques and procedures; responsible for budgeting and managing funds; works with automated property book system. *Skill Level 40:* Able to perform the duties required for Skill Level 30; supervises a medium-sized medical supply, stock control, or storage activity. *Skill Level 50:* Able to perform the duties required for Skill Level 40; supervises a large medical supply, stock control, or storage activity or serves as principal noncommissioned officer of medical depot; analyzes medical logistical management data.

Recommendation, Skill Level 10

Credit may be granted on the basis of an individualized assessment of the student (6/01).

Recommendation, Skill Level 20

Credit may be granted on the basis of an individualized assessment of the student (6/01).

Recommendation, Skill Level 30

In the lower-division baccalaureate/associate degree category, 3 semester hours in computer applications, 3 in supply management, 3 in warehouse operations, 3 in principles of supervision, and 3 in records management (6/01).

Recommendation, Skill Level 40

In the lower-division baccalaureate/associate degree category, 3 semester hours in computer applications, 3 in supply management, 3 in warehouse operations, 3 in principles of supervision, and 3 in records management. In the upper-division baccalaureate category, 3 semester hours in organization development and 3 supply chain management (6/01).

Recommendation, Skill Level 50

In the lower-division baccalaureate/associate degree category, 3 semester hours in computer applications, 3 in supply management, 3 in warehouse operations, 3 in principles of supervision, and 3 in records management. In the upper-division baccalaureate category, 3 semester hours in organization development, 3 in supply chain management, and 3 for a management practicum (6/01).

MOS-76P-002

Materiel Control and Accounting Specialist
76P10
76P20
76P30
76P40

Exhibit Dates: 1/90–3/92.

Career Management Field: 76 (Supply and Service), subfield 761 (Supply).

Description

Summary: Performs or supervises inventory management functions and stock record keeping pertaining to receipt, distribution, and issue of materiel. *Skill Level 10:* Prepares stock records and other documents such as inventory, stock control, and accounting reports; operates office machines and data processing equipment; maintains technical reference library; reviews and processes supply requests; edits catalog data; performs financial inventory accounting administration functions; interprets supply documents in an automated environment for input and output processing; may perform accounting and sales functions in self-service supply, repair and utilities, clothing sales, and military exchange activities; receives, stores, and issues direct exchange items. *Skill Level 20:* Able to perform the duties required for Skill Level 10; reviews stock items and recommends additions, deletions, or changes; reconciles supply transactions and processes inventory adjustment reports; performs financial management functions. *Skill Level 30:* Able to perform the duties required for Skill Level 20; supervises personnel in stock control and supply accounting, sales activity, or property disposal; plans and organizes work schedules and instructs subordinates; plans and organizes operations to conform with work standards, current regulations, and directives; reconciles problems in automated supply accounting system; assists in all aspects of purchasing and contracting activities and may act as contracting officer; serves as imprest fund custodian cashier; manages materials, determining and directing necessary maintenance or overhaul. *Skill Level 40:* Able to perform the duties required for Skill Level 30; supervises material managers, supply activities, and sales; determines methods and procedures to improve material operations; assists in development and preparation of operations, including plans, maps, sketches, overlays, and other data; analyzes data to determine trends, quality control, and efficiency of operations.

Recommendation, Skill Level 10

In the lower-division baccalaureate/associate degree category, 2 semester hours in record keeping, 1 in warehouse practices, 2 in office machines, 1 in retailing, and 2 in computer operations. (NOTE: This recommendation for skill level 10 is valid for the dates 1/90-9/91 only) (9/82).

Recommendation, Skill Level 20

In the lower-division baccalaureate/associate degree category, 3 semester hours in record keeping, 2 in warehouse practices, 2 in office machines, 2 in retailing, 2 in computer operations, and 2 in business mathematics (9/82).

Recommendation, Skill Level 30

In the lower-division baccalaureate/associate degree category, 3 semester hours in record keeping, 2 in warehouse practices, 2 in office machines, 2 in retailing, 2 in computer operations, 2 in business mathematics, 2 in office procedures, 2 in inventory control, and 3 in personnel supervision (9/82).

Recommendation, Skill Level 40

In the lower-division baccalaureate/associate degree category, 3 semester hours in record keeping, 2 in warehouse practices, 2 in office machines, 2 in retailing, 2 in computer operations, 2 in business mathematics, 2 in office procedures, 2 in inventory control, 3 in personnel supervision, and 3 for field experience in management. In the upper-division baccalaureate category, 3 semester hours for field experience in sales management (9/82).

MOS-76P-003

MATERIEL CONTROL AND ACCOUNTING SPECIALIST
76P10
76P20
76P30
76P40

Exhibit Dates: 4/92–6/93. Effective 6/93, MOS 76P was discontinued, and its duties were incorporated into MOS 92A, Automated Logistical Specialist.

Career Management Field: 76 (Supply and Services).

Description

Summary: Performs or supervises inventory management functions and stock record keeping pertaining to receipt, distribution, and issue of materiel. *Skill Level 10:* Prepares stock records and other documents such as inventory, stock control, and accounting reports; operates office machines and data processing equipment; maintains technical reference library; reviews and processes supply requests; edits catalog data; performs financial inventory accounting administration functions; interprets supply documents in an automated environment for input and output processing; may perform accounting and sales functions in self-service supply, repair and utilities, clothing sales, and military exchange activities; receives, stores, and issues direct exchange items. *Skill Level 20:* Able to perform the duties required for Skill Level 10; supervises and provides technical guidance to subordinates; reviews stock items and recommends additions, deletions, or changes; reconciles supply transactions and processes inventory adjustment reports; maintains recorded property disposal activities; performs financial management functions. *Skill Level 30:* Able to perform the duties required for Skill Level 20; supervises personnel in stock control and supply accounting, sales activity, or property disposal; plans and organizes work schedules and instructs subordinates; plans and organizes operations to conform with work standards, current regulations, and directives; reconciles problems in automated supply accounting system; assists in all aspects of purchasing and contracting activities and may act as contracting officer; serves as imprest fund custodian cashier; manages materials, determining and directing necessary maintenance or overhaul. *Skill Level 40:* Able to perform the duties required for Skill Level 30; supervises material managers, supply activities, and sales; determines methods and procedures to improve material operations; assists in development and preparation of operations, including plans, maps, sketches, overlays, and other data; analyzes data to determine trends, quality control, and efficiency of operations.

Recommendation, Skill Level 10

Credit may be granted on the basis of an individualized assessment of the student (3/94).

Recommendation, Skill Level 20

In the lower-division baccalaureate/associate degree category, 3 semester hours in office machines, 3 in computer applications, 3 in materiel management, and 3 in personnel supervision (4/92).

Recommendation, Skill Level 30

In the lower-division baccalaureate/associate degree category, 3 semester hours in office machines, 3 in computer applications, 3 in materiel management, 3 in personnel supervision, and 2 in records and information management (4/92).

Recommendation, Skill Level 40

In the lower-division baccalaureate/associate degree category, 3 semester hours in office machines, 3 in computer applications, 3 in materiel management, 3 in personnel supervision, and 2 in records and information management. In the upper-division baccalaureate category, 3 semester hours in organizational management and 3 for field experience in management (4/92).

MOS-76V-002

MATERIEL STORAGE AND HANDLING SPECIALIST
76V10
76V20
76V30
76V40

Exhibit Dates: 1/90–3/92.

Career Management Field: 76 (Supply and Service), subfield 761 (Supply).

Description

Summary: Receives or supervises receipt, storage, issue, inspection, packing, and shipment of materiel. *Skill Level 10:* Reviews and verifies shipping documents; inspects incoming equipment for serviceability and damage; operates materiel-handling equipment; packs, crates, stencils, weighs, and bands equipment for shipment or storage; assigns storage location for rotation of stock, maximum utilization of sensitive, classified, or hazardous storage, or warehouse space; maintains file of personnel authorized to receive supplies; performs inventory and location surveys; maintains centralized stock locater files; receives, inventories, and issues Quick Supply Store items. *Skill Level 20:* Able to perform the duties required for Skill Level 10; plans, organizes, controls, and supervises small materiel storage activity; provides guidance to subordinates; determines requirements and assigns use of warehouse equipment and personnel; supervises equipment storage including document control and location inventories. *Skill Level 30:* Able to perform the duties required for Skill Level 20; conducts inspections to assure compliance with standards and regulations; prepares reports regarding personnel, storage, and relocation of materiel; supervises a medium-sized warehouse activity. *Skill Level 40:* Able to perform the duties required for Skill Level 30; supervises a large warehousing activity; conducts surveillance inspection of materiel in storage; develops training programs for subordinates.

Recommendation, Skill Level 10

In the lower-division baccalaureate/associate degree category, 3 semester hours in warehouse practices. (NOTE: This recommendation for skill level 10 is valid for the dates 1/90-9/91 only) (9/82).

Recommendation, Skill Level 20

In the lower-division baccalaureate/associate degree category, 3 semester hours in warehouse practices, 3 in record keeping, and 2 in inventory control (9/82).

Recommendation, Skill Level 30

In the lower-division baccalaureate/associate degree category, 3 semester hours in personnel supervision, 3 in warehouse practices, 3 in record keeping, 2 in inventory control, and 2 in office procedures (9/82).

Recommendation, Skill Level 40

In the lower-division baccalaureate/associate degree category, 3 semester hours in personnel supervision, 3 for field experience in management, 3 in warehouse practices, 3 in record keeping, 2 in inventory control, 2 in office procedures, and 2 in material management (9/82).

MOS-76V-003

Materiel Storage and Handling Specialist
76V10
76V20
76V30
76V40

Exhibit Dates: 4/92–6/93. Effective 6/93, MOS 76V was discontinued, and its duties were incorporated into MOS 92A, Automated Logistical Specialist.

Career Management Field: 76 (Supply and Services).

Description

Summary: Performs or supervises receipt, storage, issue, inspection, packing, shipment, and accounting of materiel. *Skill Level 10:* Reviews and verifies shipping documents; inspects incoming equipment for serviceability and damage; operates materiel-handling equipment; packs, crates, stencils, weighs, and bands equipment for shipment or storage; operates office equipment including computers; assigns storage location for rotation of stock, maximum utilization of sensitive, classified, or hazardous storage, or warehouse space; maintains file of personnel authorized to receive supplies; performs inventory and location surveys; maintains centralized stock locater files; receives, inventories, and issues Quick Supply Store items. *Skill Level 20:* Able to perform the duties required for Skill Level 10; plans, organizes, controls, and supervises small materiel storage activity; provides guidance to subordinates; determines requirements and assigns use of warehouse equipment and personnel; supervises equipment storage including document control and location inventories. *Skill Level 30:* Able to perform the duties required for Skill Level 20; conducts inspections to ensure compliance with standards and regulations; prepares reports regarding personnel, storage, and relocation of materiel; prepares and updates warehouse planographs; selects materiel-handling equipment; supervises warehouse activity. *Skill Level 40:* Able to perform the duties required for Skill Level 30; supervises warehouse activity; conducts surveillance inspection of materiel in storage; develops training programs.

Recommendation, Skill Level 10

Credit may be granted on the basis of an individualized assessment of the student (3/94).

Recommendation, Skill Level 20

In the lower-division baccalaureate/associate degree category, 3 semester hours in office machines, 3 in materiel management, and 3 in personnel supervision (4/92).

Recommendation, Skill Level 30

In the lower-division baccalaureate/associate degree category, 3 semester hours in office machines, 3 in materiel management, 3 in personnel supervision, and 2 in records and information management (4/92).

Recommendation, Skill Level 40

In the lower-division baccalaureate/associate degree category, 3 semester hours in office machines, 3 in materiel management, 3 in personnel supervision, and 2 in records and information management. In the upper-division baccalaureate category, 3 semester hours in organizational management and 3 for field experience in management (4/92).

MOS-76X-002

Subsistence Supply Specialist
76X10
76X20
76X30
76X40

Exhibit Dates: 1/90–3/92.

Career Management Field: 76 (Supply and Service), subfield 761 (Supply).

Description

Summary: Receives, stores, issues, distributes, inventories, and ships subsistence (food) supplies; works or supervises foodstuff warehousing operations, including receipt and storage of commodities in freezers, refrigerators, or dry storage bins and distribution to organizations, commissaries, and dining facilities. *Skill Level 10:* Receives and checks accuracy of quantities received and their condition; segregates damaged supplies; stores supplies and maintains locater records; issues stock and performs inventories; performs stock control and accounting; requisitions supplies to maintain stock levels; operates office machines and stock-handling equipment; checks quality and quantities of outgoing supplies and checks documents; distributes rations to each separate organization or account; may perform commissary store functions, including stocking, pricing, inventory, security, and construction of displays; serves as cashier. *Skill Level 20:* Able to perform the duties required for Skill Level 10; supervises food distribution and small commissary operations; plans and coordinates supply activities; directs supply personnel; applies principles of automatic data processing to supply system; conducts on-the-job training; checks accuracy of work; inspects warehouse coolers and freezer units; recommends corrective action to minimize spoilage and improve supply procedures; administers stock control; performs sales functions in commissary stores; analyzes statistical data and reports; prices stock; computes resale item prices and reorders; maintains commissary consumption records. *Skill Level 30:* Able to perform the duties required for Skill Level 20; supervises a medium-sized commissary or food supply activity; maintains monthly and quarterly records for each individual account supplied by the warehouse; maintains inventory control for 50 to 60 separate accounts and for all items. *Skill Level 40:* Able to perform the duties required for Skill Level 30; supervises a large commissary or food supply activity; interprets and uses annual food plans; manages food reserve stock; performs advisory duties in planning and coordinating food reserve stocks; performs advisory duties in planning and coordinating food storage and distribution operations; ensures compliance with food supply policies.

Recommendation, Skill Level 10

In the lower-division baccalaureate/associate degree category, 3 semester hours in warehouse practices and 2 in office machines; if the duty assignment included commissary work, 3 semester hours in retail sales. (NOTE: This recommendation for skill level 10 is valid for the dates 1/90-9/91 only) (1/82).

Recommendation, Skill Level 20

In the lower-division baccalaureate/associate degree category, 3 semester hours in warehouse practices, 3 in record keeping, 2 in office machines, and 2 in business mathematics; if the duty assignment included commissary work, 3 semester hours in retail sales (1/82).

Recommendation, Skill Level 30

In the lower-division baccalaureate/associate degree category, 3 semester hours in warehouse practices, 3 in record keeping, 3 in personnel supervision, 2 in office machines, 2 in business mathematics, 2 in inventory control, and 2 in office procedures; if the duty assignment included commissary work, 3 semester hours in retail sales (1/82).

Recommendation, Skill Level 40

In the lower-division baccalaureate/associate degree category, 3 semester hours in warehouse practices, 3 in record keeping, 3 in personnel supervision, 2 in office machines, 2 in business mathematics, 2 in inventory control, and 2 in office procedures; if the duty assignment included commissary work, 3 semester hours in retail sales. In the upper-division baccalaureate category, 3 semester hours for field experience in management (1/82).

MOS-76X-003

Subsistence Supply Specialist
76X10
76X20
76X30
76X40

Exhibit Dates: 4/92–6/93. Effective 6/93, MOS 76X was discontinued, and its duties were incorporated into MOS 92A, Automated Logistical Specialist.

Career Management Field: 76 (Supply and Service).

Description

Summary: Receives, stores, issues, distributes, inventories, and ships subsistence (food) supplies; works or supervises foodstuff warehousing operations, including receipt and storage of commodities in freezers, refrigerators, or dry storage bins and distribution to organizations, commissaries, and dining facilities. *Skill Level 10:* Receives and checks accuracy of quantities received and their condition; segregates damaged supplies; stores supplies and maintains locater records; issues stock and performs inventories; performs stock control and accounting; requisitions supplies to maintain stock levels; operates office machines, stock-handling equipment, and computers; checks quality and quantities of outgoing supplies and checks documents; distributes rations to each separate organization or account; may perform commissary store functions, including stocking, pricing of goods, and inventory, security, and construction of displays; serves as cashier. *Skill*

Level 20: Able to perform the duties required for Skill Level 10; supervises food distribution and small commissary operations; plans and coordinates supply activities; directs supply personnel; applies principles of computer applications to supply system; conducts on-the-job training; checks accuracy of work; inspects warehouse coolers and freezer units; recommends corrective action to minimize spoilage and improve supply procedures; administers stock control; performs sales functions in commissary stores; analyzes statistical data and reports; prices stock; computes resale item prices and reorders; maintains commissary consumption records. *Skill Level 30:* Able to perform the duties required for Skill Level 20; supervises a medium-sized commissary or food supply activity; maintains monthly and quarterly records for each individual account supplied by the warehouse; maintains inventory control for multiple accounts and for all items. *Skill Level 40:* Able to perform the duties required for Skill Level 30; supervises a large commissary or food supply activity; interprets and uses annual food plans; manages food reserve stock; performs advisory duties in planning and coordinating food reserve stocks; performs advisory duties in planning and coordinating food storage and distribution operations; ensures compliance with food supply policies.

Recommendation, Skill Level 10

Credit may be granted on the basis of an individualized assessment of the student (3/94).

Recommendation, Skill Level 20

In the lower-division baccalaureate/associate degree category, 3 semester hours in computer applications, 3 in supply management, and 3 in personnel supervision; if additional skill identifier (ASI) was U5, Commissary Operations, 3 semester hours in retail management and 3 in sales management (4/92).

Recommendation, Skill Level 30

In the lower-division baccalaureate/associate degree category, 3 semester hours in computer applications, 3 in supply management, 3 in personnel supervision, and 2 in records and information management; if additional skill identifier (ASI) was U5, Commissary Operations, 3 semester hours in retail management and 3 in sales management (4/92).

Recommendation, Skill Level 40

In the lower-division baccalaureate/associate degree category, 3 semester hours in computer applications, 3 in supply management, 3 in personnel supervision, and 2 in records and information management; if additional skill identifier (ASI) was U5, Commissary Operations, 3 semester hours in retail management and 3 in sales management. In the upper-division baccalaureate category, 3 semester hours in organizational management and 3 for field experience in management (4/92).

MOS-76Y-002

Unit Supply Specialist
76Y10
76Y20
76Y30
76Y40

Exhibit Dates: 1/90–3/92.

Career Management Field: 76 (Supply and Services), subfield 761 (Supply).

Description

Summary: Supervises or performs duties connected with the requisition, receipt, storage, accountability, and issuance of individual, organizational, and installation supplies, clothing, and equipment. *Skill Level 10:* Serves as a unit supply clerk; prepares supply records and forms, inventory control listings and count cards, and inventory reports; uses supply catalogs in requisitioning and distributing supplies; uses typewriter, adding machines, calculators, copiers, and microfiche in performing duties; keeps administrative files; practices safety, security, and accountability in storage operations; prepares, reviews, and corrects property listings and annexes; posts transactions to records; prepares clothing for issue; processes laundry including laundry rosters and payroll deduction forms for laundry; maintains security and records for weapons; prepares weapons reports and ammunition reports for commander. *Skill Level 20:* Able to perform the duties required for Skill Level 10; provides technical guidance to subordinates; establishes priorities and assigns work; conducts on-the-job training; inspects completed work, records, and reports for accuracy and for compliance with directives; maintains automated supply accounting system; computes supply usage factors; applies principles of automatic data processing input filing, processing, and output techniques; supervises the issue, handling, and security of sensitive items and small arms. *Skill Level 30:* Able to perform the duties required for Skill Level 20; supervises supply personnel in a large supply activity (serving over 450 military personnel); trains and supervises personnel in external load rigging for rotary wing aircraft; provides technical assistance to Records Specialists and Parts Specialists; ensures professional development of subordinates. *Skill Level 40:* Able to perform the duties required for Skill Level 30; serves on the staff at a battalion headquarters; may supervise other supply sergeants in the Supply Career Management Field (76); analyzes statistical data to determine trends; coordinates logistical activities; reviews policies; improves supply methods and procedures; conducts assistance visits and develops training programs; coordinates supply activities with other units.

Recommendation, Skill Level 10

In the lower-division baccalaureate/associate degree category, 3 semester hours in record keeping, 3 in office procedures, 2 in office machines, and 1 in typing. (NOTE: This recommendation for skill level 10 is valid for the dates 1/90-9/91 only) (1/82).

Recommendation, Skill Level 20

In the lower-division baccalaureate/associate degree category, 3 semester hours in record keeping, 3 in office procedures, 3 in inventory control, 2 in office machines, and 1 in typing (1/82).

Recommendation, Skill Level 30

In the lower-division baccalaureate/associate degree category, 3 semester hours in personnel supervision, 3 in human relations, 3 in record keeping, 3 in office procedures, 3 in inventory control, 2 in office machines, and 1 in typing (1/82).

Recommendation, Skill Level 40

In the lower-division baccalaureate/associate degree category, 3 semester hours in personnel supervision, 3 in human relations, 3 in record keeping, 3 in office procedures, 3 in office administration, 3 in inventory control, 2 in office machines, and 1 in typing. In the upper-division baccalaureate category, 3 semester hours for field experience in management (1/82).

MOS-76Y-003

Unit Supply Specialist
76Y10
76Y20
76Y30
76Y40

Exhibit Dates: 4/92–6/93. Effective 6/93, MOS 76Y was discontinued, and its duties were incorporated into MOS 92A, Automated Logistical Specialist.

Career Management Field: 76 (Supply and Services).

Description

Summary: Supervises or performs duties connected with the requisition, receipt, storage, accountability, and issuance of individual, organizational, and installation supplies, clothing, and equipment. *Skill Level 10:* Serves as a unit supply clerk; prepares supply records and forms, inventory control listings and count cards, and inventory reports using Army computer software; uses supply catalogs in requisitioning and distributing supplies; uses microcomputer, typewriter, adding machines, calculators, copiers, and microfiche files; practices safety, security, and accountability in storage operations; prepares, reviews, and corrects property listings and annexes; posts transactions to records; prepares cash collection vouchers for lost, damaged, or destroyed items; prepares clothing for issue; processes laundry including laundry rosters and payroll deduction forms for laundry; prepares documents for turn-in of supplies and equipment; maintains security and records for weapons; prepares weapons and ammunition reports for commander. *Skill Level 20:* Able to perform the duties required for Skill Level 10; provides technical guidance to subordinates; establishes priorities and assigns work; conducts on-the-job training; inspects completed work, records, and reports for accuracy and for compliance with directives; maintains automated supply accounting system; computes supply usage factors; determines method of relief from responsibility for lost, damaged, and destroyed items; applies principles of automatic data processing input filing, processing, and output techniques; supervises the issue, handling, and security of sensitive items and small arms. *Skill Level 30:* Able to perform the duties required for Skill Level 20; supervises supply personnel in a large supply activity (serving over 450 military personnel); trains and supervises personnel in external load rigging for rotary wing aircraft; provides technical assistance to Records Specialists and Parts Specialists; reviews property book and adjustment document for correctness; ensures professional development of subordinates. NOTE: May have progressed to 76Y30 from 76Y20 or 76C20, Equipment Records and Parts Specialist. *Skill Level 40:* Able to perform the duties required for Skill Level 30; serves on the staff at a battalion headquarters; may supervise other supply sergeants in the Supply Career Management Field (76); analyzes statistical data to determine trends; coordinates logistical activities; reviews policies; improves supply methods and procedures; conducts assistance visits; develops training programs; coordinates supply activities with other units.

Recommendation, Skill Level 10

Credit may be granted on the basis of an individualized assessment of the student (3/94).

Recommendation, Skill Level 20

In the lower-division baccalaureate/associate degree category, 3 semester hours in computer applications, 3 in supply and logistics management, and 3 in personnel supervision (4/92).

Recommendation, Skill Level 30

In the lower-division baccalaureate/associate degree category, 3 semester hours in computer applications, 3 in supply and logistics management, 3 in personnel supervision, and 2 in records and information management (4/92).

Recommendation, Skill Level 40

In the lower-division baccalaureate/associate degree category, 3 semester hours in computer applications, 3 in supply and logistics management, 3 in personnel supervision, and 2 in records and information management. In the upper-division baccalaureate category, 3 semester hours in organizational management and 3 for field experience in management (4/92).

MOS-76Z-002

Senior Supply Sergeant
76Z50

Exhibit Dates: 1/90–3/92.

Career Management Field: 76 (Supply and Service), subfield 760 (Supply General).

Description

Able to perform the duties required for skill level 40 of any MOS in the supply career management field (76); serves as a mid-level manager, supervising personnel engaged in large supply and service operations such as laundry, bath, graves, registration, decontamination, transportation, property disposal, medical supply, commissary operations, and resupply by airdrops; analyzes reports and evaluates supply operations and training; manages stock control and accounting, procurement, and inventory control; supervises storage, receiving, issue, materiel handling, and supply locater systems and facility safety and supply security operations; coordinates supply support data systems, salvage operations, and repair parts supply procedures; supervises the preparation of operating instructions, reports, and related technical material; recommends units for assignment to support special operations. May be assigned as first sergeant with the following duties: assists the commander in planning, supervising, inspecting, developing, and executing unit policy; advises commander on enlisted personnel matters; provides counsel and guidance to subordinates; coordinates administration of unit and support services; may serve as the principal noncommissioned officer in a unit of approximately 150 persons.

Recommendation

In the lower-division baccalaureate/associate degree category, 3 semester hours in record keeping, 3 in office administration, 3 in personnel supervision, 3 in inventory control, 3 in office procedures, 3 in warehouse practices, 3 in business mathematics, 2 in office machines, and 1 in typing. In the upper-division baccalaureate category, 3 semester hours in introduction to management, 3 in personnel management, 3 in records administration, 3 in human relations, and 6 for field experience in management (1/82).

MOS-76Z-003

Senior Supply/Service Sergeant
76Z50

Exhibit Dates: 4/92–6/93. Effective 6/93, MOS 76Z was discontinued, and its duties were incorporated into MOS 92A, Automated Logistical Specialist.

Career Management Field: 76 (Supply and Services).

Description

Able to perform the duties required for skill level 40 of any MOS in the supply career management field (76); serves as a mid-level manager, supervising personnel engaged in large supply and service operations such as laundry, bath, graves, registration, decontamination, transportation, property disposal, medical supply, commissary operations, and resupply by airdrops; analyzes reports and evaluates supply operations and training; manages stock control and accounting, procurement, and inventory control; supervises storage, receiving, issue, materiel handling, and supply locater systems and facility safety and supply security operations; coordinates supply support data systems, salvage operations, and repair parts supply procedures; supervises the preparation of operating instructions, reports, and related technical material; recommends units for assignment to support special operations. May be assigned as first sergeant with the following duties: assists the commander in planning, supervising, inspecting, developing and executing unit policy; advises commander on enlisted personnel matters; provides counsel and guidance to subordinates; coordinates administration of unit and support services; may serve as the principal noncommissioned officer in a unit of approximately 150 persons. NOTE: May have progressed to 76Z50 from 76P40, Materiel Control and Accounting Specialist; 76V40, Materiel Storage and Handling Specialist; 76X40, Subsistence Supply Specialist; 76Y40, Unit Supply Specialist; 43M40, Fabric Repair Specialist; 43E50, Parachute Rigger; 57E50, Laundry and Bath Specialist; or 57F50, Graves Registration Specialist.

Recommendation

In the lower-division baccalaureate/associate degree category, 3 semester hours in personnel supervision and 2 in records and information management. In the upper-division baccalaureate category, 3 semester hours in organizational management and 3 for field experience in management; if individual has attained paygrade E-9, additional credit as follows: 3 semester hours in management problems, 3 in operations management, and 3 in communication techniques for managers (4/92).

MOS-77F-001

Petroleum Supply Specialist
77F10
77F20
77F30
77F40
77F50

Exhibit Dates: 1/90–11/01.

Career Management Field: 77 (Petroleum and Water).

Description

Summary: Operates and maintains petroleum storage, dispensing, and distribution facilities and pipeline systems. *Skill Level 10:* Receives and dispenses bulk and packaged petroleum products; operates forklift trucks, conveyors, and cranes in loading, unloading, moving, and storing petroleum supplies; marks petroleum containers for proper identification; conducts safety inspections of storage facilities; performs basic preventive maintenance on petroleum storage and handling equipment. *Skill Level 20:* Able to perform the duties required for Skill Level 10; supervises aircraft refueling; assures adherence to safety procedures; maintains inventory records of petroleum products; inspects petroleum storage sites; may supervise small petroleum storage or dispensing activity; maintains pipeline systems and pressure reducing stations; applies fire fighting and fire prevention techniques. *Skill Level 30:* Able to perform the duties required for Skill Level 20; schedules, orders, dispatches, and otherwise recommends movement of petroleum products; supervises pipeline or pump station operation and petroleum supply storage facilities; uses product cycles to minimize product contamination in storage facilities or pipelines; supervises dispersion and camouflage of supplies and equipment. *Skill Level 40:* Able to perform the duties required for Skill Level 30; ensures compliance with fire and safety regulations; coordinates petroleum operations at staff level; furnishes required reports on equipment to higher headquarters; performs quality surveillance; may supervise up to 16 persons. *Skill Level 50:* Able to perform the duties required for Skill Level 40; as the principal noncommissioned officer in a large petroleum operation, assists commissioned officers in the planning, coordination, and supervision of petroleum storage and distribution and in the evaluation of operations and training; supervises and inspects work performed by subordinate noncommissioned officers; collects and prepares material pertaining to petroleum operations and training; may serve as a first sergeant of a company, a mid-level managerial position involving personnel, supply, training, and inspection responsibilities. NOTE: May have progressed to 77F50 from 77F40; 77L40 (Petroleum Laboratory Specialist); or 77W40 (Water Treatment Specialist).

Recommendation, Skill Level 10

In the vocational certificate category, 2 semester hours in fire safety. (NOTE: This recommendation for skill level 10 is valid for the dates 1/90-9/91 only) (10/88).

Recommendation, Skill Level 20

In the vocational certificate category, 2 semester hours in fire safety, 2 in basic mathematics, and 2 in pump maintenance. (NOTE: This recommendation for skill level 20 is valid for the dates 1/90-2/95 only) (10/88).

Recommendation, Skill Level 30

In the vocational certificate category, 2 semester hours in fire safety, 2 in basic mathematics, and 2 in pump maintenance. In the lower-division baccalaureate/associate degree category, 3 semester hours for field experience in personnel supervision (10/88).

Recommendation, Skill Level 40

In the vocational certificate category, 2 semester hours in fire safety, 2 in basic mathematics, and 2 in pump maintenance. In the lower-division baccalaureate/associate degree category, 3 semester hours in industrial safety, 3 in personnel supervision, 3 for field experience in personnel supervision, 3 in human relations, and 2 in technical writing (10/88).

Recommendation, Skill Level 50

In the vocational certificate category, 2 semester hours in fire safety, 2 in basic mathematics, and 2 in pump maintenance. In the lower-division baccalaureate/associate degree category, 3 semester hours in industrial safety, 3 in personnel supervision, 3 for field experience in personnel supervision, 3 in human relations, 3 in technical writing, 3 in record keeping, 3 in office administration, and 2 in material management. In the upper-division baccalaureate category, 3 semester hours for field experience in management (10/88).

MOS-77F-002

Petroleum Supply Specialist

77F10
77F20
77F30
77F40
77F50

Exhibit Dates: 12/01–9/04. (Effective 9/04, MOS 77F was discontinued, and its duties were incorporated into MOS 92F.)

Career Management Field: 77 (Petroleum and Water).

Description

Summary: Operates and maintains storage, dispensing, and distribution facilities and pipeline systems for petroleum, oils, and lubricants (POL). *Skill Level 10:* Receives and dispenses bulk and packaged petroleum products; operates forklift trucks, conveyors, and cranes in loading, unloading, moving, and storing petroleum supplies; marks petroleum containers for proper identification; conducts safety inspections of storage facilities; performs basic preventive maintenance on POL storage and handling equipment. *Skill Level 20:* Able to perform the duties required for Skill Level 10; supervises aircraft refueling; ensures adherence with safety procedures; maintains inventory records of petroleum products; inspects petroleum storage sites; maintains pipeline systems; applies fire fighting and fire prevention techniques. *Skill Level 30:* Able to perform the duties required for Skill Level 20; schedules, orders, dispatches, and otherwise recommends movement of petroleum products; supervises pipeline or pump station operation and petroleum supply storage facilities; uses product cycles to minimize product contamination in storage facilities or pipelines; supervises dispersion and camouflage of supplies and equipment. *Skill Level 40:* Able to perform the duties required for Skill Level 30; ensures compliance with fire and safety regulations; coordinates POL operations at staff level; furnishes required reports on equipment to higher headquarters; performs quality surveillance. *Skill Level 50:* Able to perform the duties required for Skill Level 40; as the principal noncommissioned officer in a large petroleum operation, assists commissioned officers in the planning, coordination, and supervision of petroleum storage and distribution and in the evaluation of operations and training; supervises and inspects work performed by subordinate noncommissioned officers; collects and prepares material pertaining to petroleum operations and training; may serve as a first sergeant of a company, a mid-level managerial position involving personnel, supply, training, and inspection responsibilities. NOTE: May have progressed to 77F50 from 77F40; 77L40 (Petroleum Laboratory Specialist); or 77W40 (Water Treatment Specialist).

Recommendation, Skill Level 10

Credit may be granted on the basis of an individualized assessment of the student (12/01).

Recommendation, Skill Level 20

Credit may be granted on the basis of an individualized assessment of the student (12/01).

Recommendation, Skill Level 30

In the lower-division baccalaureate/associate degree category, 2 semester hours in fire safety, 2 in pump maintenance, 3 in principles of supervision, and 3 in industrial safety (12/01).

Recommendation, Skill Level 40

In the lower-division baccalaureate/associate degree category, 2 semester hours in fire safety, 2 in pump maintenance, 3 in principles of supervision, 3 in industrial safety, and 2 in technical writing. In the upper-division baccalaureate category, 3 semester hours for a management practicum (12/01).

Recommendation, Skill Level 50

In the lower-division baccalaureate/associate degree category, 2 semester hours in fire safety, 2 in pump maintenance, 6 in principles of supervision, 3 in industrial safety, 3 in technical writing, 3 in office administration, and 2 in material management. In the upper-division baccalaureate category, 6 semester hours for a management practicum and 3 in human resource management (12/01).

MOS-77L-001

Petroleum Laboratory Specialist

77L10
77L20
77L30
77L40

Exhibit Dates: 1/90–11/01.

Career Management Field: 77 (Petroleum and Water).

Description

Summary: Performs or supervises physical and chemical tests on petroleum products to determine suitability for intended use. *Skill Level 10:* Prepares laboratory equipment for operation; prepares reagents; calibrates instruments; obtains petroleum test samples; performs preventive maintenance on technical laboratory equipment; performs specification tests, including water content, sediment, color, carbon residue, penetration, and oxidation stability; performs quality surveillance tests; uses fire prevention and safety control measures. *Skill Level 20:* Able to perform the duties required for Skill Level 10; provides technical guidance to subordinates; employs cooperative fuel research engine to test fuels; prepares laboratory reports; evaluates test results with specification requirements and makes product recommendations. *Skill Level 30:* Able to perform the duties required for Skill Level 20; establishes calibration program; assists in establishing quality surveillance programs; supervises medium-sized petroleum laboratory activities; plans work schedules and assigns duties; inspects laboratory to ensure proper procedures are being followed, and work is being done efficiently; supervises inspection of anti-icing additives and corrosion inhibitors during pipeline transfers; supervises placement, setup, and maintenance of laboratory equipment; coordinates petroleum product storage and distribution activities; reviews work reports and evaluates results; maintains inventory control and performs all related supply activity; ensures adherence to laboratory fire and safety procedures. *Skill Level 40:* Able to perform the duties required for Skill Level 30; supervises large petroleum laboratory; assists and plans quality surveillance operations; applies OSHA and EPA regulations; determines methods of reclaiming or downgrading contaminated and deteriorated petroleum products; performs quality surveillance of loading and unloading bulk shipments to ensure safety and to prevent contamination; ensures proper sampling and laboratory analysis of petroleum products; monitors sampling, testing, use, and blending of additives; reports on analysis results, environmental considerations, contracts for petroleum products, transporting, and storage; generally supervises from four to ten persons.

Recommendation, Skill Level 10

In the lower-division baccalaureate/associate degree category, 3 semester hours in petroleum product analysis, 2 in applied chemistry, 1 in chemical safety, and 1 in first aid. (NOTE: This recommendation for skill level 10 is valid for the dates 1/90-9/91 only) (10/88).

Recommendation, Skill Level 20

In the lower-division baccalaureate/associate degree category, 6 semester hours in petroleum product analysis, 3 in basic analytical chemistry laboratory, 2 in applied chemistry, 1 in chemical safety, and 1 in first aid. (NOTE: This recommendation for skill level 20 is valid for the dates 1/90-2/95 only) (10/88).

Recommendation, Skill Level 30

In the lower-division baccalaureate/associate degree category, 6 semester hours in petroleum product analysis, 3 in basic analytical chemistry laboratory, 3 in personnel supervision, 2 in applied chemistry, 1 in chemical safety, and 1 in first aid (10/88).

Recommendation, Skill Level 40

In the lower-division baccalaureate/associate degree category, 6 semester hours in petroleum product analysis, 3 in basic analytical chemistry laboratory, 3 in personnel supervision, 3 in technical writing, 3 in human relations, 2 in applied chemistry, 2 in industrial safety, 1 in chemical safety, and 1 in first aid. In the upper-division baccalaureate category, 3 semester hours for field experience in management (10/88).

MOS-77L-002

Petroleum Laboratory Specialist

77L10
77L20
77L30
77L40

Exhibit Dates: 12/01–9/04. (Effective 9/04, MOS 77L was discontinued, and its duties were incorporated into MOS 92L.)

Career Management Field: 77 (Petroleum and Water).

Description

Summary: Performs or supervises physical and chemical tests on petroleum, oils, and lubricants (POL) to determine suitability for intended use. *Skill Level 10:* Prepares laboratory equipment for operation; prepares reagents; calibrates instruments; obtains petroleum test samples; performs preventive maintenance on technical laboratory equipment; performs ASTD (American Society for Testing Materials) specification tests, including water content, sediment, color, carbon residue, penetration, and oxidation stability; performs quality surveillance tests; uses fire prevention and safety control measures. *Skill Level 20:* Able to perform the duties required for Skill Level 10; provides technical guidance to subordinates; prepares laboratory reports; evaluates test results with specification requirements; makes product recommendations; performs fire and safety inspections. *Skill Level 30:* Able to perform the duties required for Skill Level 20; establishes calibration program; assists in establishing quality surveillance programs; supervises medium-sized petroleum laboratory activities; plans work schedules and assigns duties; inspects laboratory to ensure proper procedures are being followed, and work is being done efficiently; supervises placement, setup, and maintenance of laboratory equipment; coordinates petroleum product storage and distribution activities; reviews work reports and evaluates results; maintains inventory control and performs all related supply activity; ensures adherence to laboratory fire and safety procedures; applies OSHA and EPA regulations. *Skill Level 40:* Able to perform the duties required for Skill Level 30; supervises large petroleum laboratory; assists and plans quality surveillance operations; performs quality surveillance of loading and unloading bulk shipments to ensure safety and to prevent contamination; ensures proper sampling and laboratory analysis of petroleum products; ensures contractor compliance with POL contracts.

Recommendation, Skill Level 10

Credit may be granted on the basis of an individualized assessment of the student (12/01).

Recommendation, Skill Level 20

Credit may be granted on the basis of an individualized assessment of the student (12/01).

Recommendation, Skill Level 30

In the lower-division baccalaureate/associate degree category, 3 semester hours in college algebra, 3 in analytical chemistry, 1 in chemical safety, 3 in analytical chemistry laboratory, 3 in principles of supervision, 2 in computer applications, and 2 in technical writing (12/01).

Recommendation, Skill Level 40

In the lower-division baccalaureate/associate degree category, 3 semester hours in college algebra, 3 in analytical chemistry, 1 in chemical safety, 3 in analytical chemistry laboratory, 3 in principles of supervision, 2 in computer applications, and 2 in technical writing. In the upper-division baccalaureate category, 3 semester hours for a management practicum (12/01).

MOS-77W-001

Water Treatment Specialist
77W10
77W20
77W30
77W40

Exhibit Dates: 1/90–11/01.

Career Management Field: 77 (Petroleum and Water).

Description

Summary: Supervises or installs, operates, and maintains water supply and treatment equipment and systems. *Skill Level 10:* Assists in water reconnaissance and setup, operation, and maintenance of water points; assists in site preparations, setup, operation, troubleshooting, and maintenance of water purification equipment and storage facilities; operates and maintains equipment used in production of potable water in field locations; performs water quality analysis tests. *Skill Level 20:* Able to perform the duties required for Skill Level 10; installs, operates, and maintains water supply equipment and systems; provides technical guidance to subordinates; conducts water reconnaissance; develops water sources and water points and analyzes both raw and treated water; supervises operation of water purification equipment. *Skill Level 30:* Able to perform the duties required for Skill Level 20; supervises the installation, operation, and maintenance of water supply and treatment systems; provides direct supervision to individual water team functions; inspects operational condition of equipment and systems; prepares reports on the operation and maintenance of water supply systems. *Skill Level 40:* Able to perform the duties required for Skill Level 30; develops and supervises operation of area water supply and treatment plan; directs water reconnaissance and water point development for divisional areas; manages operation and ensures quality control in water supply and purification activities.

Recommendation, Skill Level 10

In the vocational certificate category, 3 semester hours in waste treatment. In the lower-division baccalaureate/associate degree category, 3 semester hours in water purification. (NOTE: This recommendation for skill level 10 is valid for the dates 1/90-9/91 only) (10/88).

Recommendation, Skill Level 20

In the vocational certificate category, 3 semester hours in waste treatment, 3 in water and sewage facility operation, and 3 in utility plumbing. In the lower-division baccalaureate/associate degree category, 3 semester hours in water purification and 3 in water supply. (NOTE: This recommendation for skill level 20 is valid for the dates 1/90-2/95 only) (10/88).

Recommendation, Skill Level 30

In the vocational certificate category, 3 semester hours in waste treatment, 3 in utility plumbing, and 6 in water and sewage facility operation. In the lower-division baccalaureate/associate degree category, 3 semester hours in water purification, 3 in water supply, and 3 in personnel supervision (10/88).

Recommendation, Skill Level 40

In the vocational certificate category, 3 semester hours in waste treatment, 3 in utility plumbing, and 12 in water and sewage facility operation. In the lower-division baccalaureate/associate degree category, 3 semester hours in water purification, 3 in water supply, and 3 in personnel supervision. In the upper-division baccalaureate category, 3 semester hours for field experience in management (10/88).

MOS-77W-002

Water Treatment Specialist
77W10
77W20
77W30
77W40

Exhibit Dates: 12/01–9/04. (Effective 9/04, MOS 77W was discontinued, and its duties were incorporated into MOS 92W.)

Career Management Field: 77 (Petroleum and Water).

Description

Summary: Supervises or installs, operates, and maintains water supply and treatment equipment and systems. *Skill Level 10:* Assists in water reconnaissance and setup, operation, and maintenance of water points; assists in site preparations, setup, operation, troubleshooting, and maintenance of water purification equipment and storage facilities; operates and maintains equipment used in production of potable water in field locations; performs water quality analysis tests. *Skill Level 20:* Able to perform the duties required for Skill Level 10; installs, operates, and maintains water supply equipment and systems; provides technical guidance to subordinates; conducts water reconnaissance; develops water sources and water points and analyzes both raw and treated water; supervises operation and maintenance of water purification equipment; inspects operational condition of equipment and systems; prepares reports on the operation and maintenance of water supply systems. *Skill Level 30:* Able to perform the duties required for Skill Level 20; supervises the installation, operation, and maintenance of water supply and treatment systems; provides direct supervision to individual water team functions. *Skill Level 40:* Able to perform the duties required for Skill Level 30; develops and supervises operation of area water supply and treatment plan; directs water reconnaissance and water point development for divisional areas; manages operation and ensures quality control in water supply and purification activities.

Recommendation, Skill Level 10

Credit may be granted on the basis of an individualized assessment of the student (12/01).

Recommendation, Skill Level 20

Credit may be granted on the basis of an individualized assessment of the student (12/01).

Recommendation, Skill Level 30

In the lower-division baccalaureate/associate degree category, 3 semester hours in college algebra, 6 in water treatment, 3 in principles of supervision, and 2 in technical writing (12/01).

Recommendation, Skill Level 40

In the lower-division baccalaureate/associate degree category, 6 semester hours in water treatment, 3 in principles of supervision, and 2 in technical writing. In the upper-division baccalaureate category, 3 semester hours for management practicum (12/01).

MOS-79D-001

Reenlistment NCO
79D30
79D40
79D50

Exhibit Dates: 1/90–10/91.

Career Management Field: 79 (Recruitment and Retention).

Description

Summary: Counsels military personnel on reenlistment and performs related reenlistment duties. *Skill Level 30:* Contacts, interviews, and advises enlisted personnel on reenlistment prerequisites, options, obligations, opportunities, and benefits; assists in designing individual career plans, using personnel records and data from interview and counseling sessions; relates physical, mental, and moral qualifications of individuals to prescribed standards and determines if service schooling is needed; processes reenlistment documents; maintains publications; compiles data; prepares reports; plans reenlistment ceremonies; and performs follow-ups. *Skill Level 40:* Able to perform the duties required for Skill Level 30; provides technical guidance to subordinates. *Skill Level 50:* Able to perform the duties required for Skill Level 40; supervises and evaluates subordinates; assigns duties; recommends reenlistment policy changes; develops and assesses reenlistment programs.

Recommendation, Skill Level 30

In the vocational certificate category, 2 semester hours in salesmanship. In the lower-division baccalaureate/associate degree category, 2 semester hours in public speaking, 2 in interview techniques, 1 in educational techniques, 1 in record keeping, and 1 in technical report writing. In the upper-division baccalaureate category, 1 semester hour for field experience in counseling (7/79).

Recommendation, Skill Level 40

In the vocational certificate category, 2 semester hours in salesmanship. In the lower-division baccalaureate/associate degree category, 2 semester hours in public speaking, 3 in interview techniques, 2 in educational techniques, 1 in record keeping, and 1 in technical report writing. In the upper-division baccalaureate category, 2 semester hours for field experience in counseling and 2 in vocational counseling (7/79).

Recommendation, Skill Level 50

In the vocational certificate category, 2 semester hours in salesmanship. In the lower-division baccalaureate/associate degree category, 3 semester hours in public speaking, 3 in interview techniques, 2 in educational techniques, 1 in record keeping, 1 in technical report writing, and 2 in personnel supervision. In the upper-division baccalaureate category, 3 semester hours for field experience in counseling, 2 in vocational counseling, and 1 in personnel management (7/79).

MOS-79D-002

Reenlistment NCO (Reserve Components)
79D30
79D40
79D50

Exhibit Dates: 11/91–12/95. Effective 12/95, MOS 79D was discontinued and its duties were incorporated into MOS 79S, Career Counselor.

Career Management Field: 79 (Recruitment and Reenlistment).

Description

Summary: Counsels military personnel on reenlistment and performs related reenlistment duties. *Skill Level 30:* Contracts, interviews, and advises enlisted personnel on reenlistment prerequisites, options, obligations, opportunities, and benefits; assists in designing individual career plans, using personnel records and data from interview and counseling sessions; relates physical, mental, and moral qualifications of individuals to prescribed standards and determines if service schooling is needed; processes reenlistment documents; maintains publications; compiles data; prepares reports; plans reenlistment ceremonies; performs follow-ups. *Skill Level 40:* Able to perform the duties required for Skill Level 30; provides technical guidance to subordinates. *Skill Level 50:* Able to perform the duties required for Skill Level 40; supervises and evaluates subordinates; assigns duties; recommends reenlistment policy changes; develops and assesses reenlistment programs.

Recommendation, Skill Level 30

In the lower-division baccalaureate/associate degree category, 3 semester hours in public speaking, 2 in interview techniques, 3 in record keeping, and 1 in technical report writing. In the upper-division baccalaureate category, 3 semester hours in vocational counseling (11/91).

Recommendation, Skill Level 40

In the lower-division baccalaureate/associate degree category, 3 semester hours in public speaking, 3 in interview techniques, 3 in record keeping, and 1 in technical report writing. In the upper-division baccalaureate category, 3 semester hours in vocational counseling and 2 for field experience in counseling (11/91).

Recommendation, Skill Level 50

In the lower-division baccalaureate/associate degree category, 3 semester hours in public speaking, 3 in interview techniques, 3 in record keeping, 1 in technical report writing, and 2 in personnel supervision. In the upper-division baccalaureate category, 3 semester hours in vocational counseling, 3 for field experience in counseling, and 2 in personnel management (11/91).

MOS-79R-001

Recruiter
79R30
79R40
79R50

Exhibit Dates: 8/95–Present.

Career Management Field: 79 (Recruitment and Reenlistment).

Description

Summary: Recruits personnel for service in the Army. NOTE: May have progressed to Recruiter from any Army MOS. *Skill Level 30:* Able to perform the duties required for Skill Level 20 of previously-held MOS. Contacts individuals and conducts interviews with those who are prospective enlistees into the Army; contacts representatives of schools, corporations, civic groups, and other agencies to present the career and employment opportunities of the Army; presents formal and informal talks to organizations, groups, or individuals; writes, edits, and presents recruiting material for use by local communications media; interviews and counsels individuals on enlistment incentives, Army benefits, and educational opportunities; evaluates applicants, using screening tests; prepares forms and documents as part of enlistment processing; counsels disqualified applicants; assists in market research and analysis of recruiting area. *Skill Level 40:* Able to perform the duties required for Skill Level 30; commands a recruiting station and maintains records, statistics, administrative files, and publications library; supervises and evaluates subordinates; conducts professional development programs for 10 to 12 subordinates; prepares official correspondence and enlistment reports; assigns duties to subordinates and supervises resources; plans and organizes recruitment programs; provides guidance to soldiers and evaluates their progress. *Skill Level 50:* Able to perform the duties required for Skill Level 40; supervises recruiting programs and personnel at several recruiting stations within an area; assists area commanders with training, operations, administration, and personnel matters; plans and conducts recruiting conferences and seminars; develops presentations that reflect sales techniques, enlistment contracts, and enlistment requirements; supervises and evaluates subordinates; identifies and takes corrective actions in problem areas; develops training programs and inspects recruiting stations to ensure efficient operation and management; may supervise up to 30 soldiers.

Recommendation, Skill Level 30

In the lower-division baccalaureate/associate degree category, 2 semester hours in social psychology, 2 in audiovisual techniques, 3 in marketing techniques, 3 in public speaking, 2 in record keeping, 2 in principles of advertising, and 3 in office procedures. In the upper-division baccalaureate category, 3 semester hours for field experience in marketing, 2 in vocational counseling, 3 in advertising media, and 2 in publicity release writing (11/96).

Recommendation, Skill Level 40

In the lower-division baccalaureate/associate degree category, 3 semester hours in social psychology, 3 in audiovisual techniques, 3 in marketing techniques, 3 in public speaking, 3 in record keeping, 3 in principles of advertising, 3 in office procedures, and 3 in personnel supervision. In the upper-division baccalaureate category, 3 semester hours in records management, 3 for field experience in marketing, 3 in personnel management, 3 in advertising media, 2 in vocational counseling, and 2 in publicity release writing (11/91).

Recommendation, Skill Level 50

In the lower-division baccalaureate/associate degree category, 3 semester hours in social psychology, 3 in audiovisual techniques, 3 in marketing techniques, 3 in public speaking, 3 in record keeping, 3 in principles of advertising, 3 in personnel supervision, and 3 in office procedures. In the upper-division baccalaureate category, 3 semester hours in records management, 3 for field experience in marketing, 3 in vocational counseling, 3 in advertising media, 2 in publicity release writing, 3 in human resource management, and 3 in personnel management (11/91).

MOS-79S-001

Career Counselor
79S30
79S40
79S50

Exhibit Dates: 8/95–Present.

Career Management Field: 79 (Recruitment and Reenlistment).

Description

Summary: Counsels military personnel on reenlistment and performs related reenlistment duties. NOTE: May have progressed to Career Counselor from any MOS. *Skill Level 30:* Able to perform the duties required for Skill Level 20 of any MOS. Conducts interviews; advises enlisted personnel on reenlistment prerequisites, options, obligations, opportunities, and benefits; assists in designing individual career plans, using personnel records and data from interview and counseling sessions; relates physical, mental, and moral qualifications of individuals to prescribed standards and determines if service schooling is needed; processes reenlistment documents; maintains publications; compiles data; prepares reports; plans reenlistment ceremonies; performs follow-ups. *Skill Level 40:* Able to perform the duties required for Skill Level 30; advises commanders on status of reenlistment programs; conducts training for commanders, staff persons, noncommissioned officers, and others when assigned as principal noncommissioned officer in a large facility or region. *Skill Level 50:* Able to perform the duties required for Skill Level 40; supervises, assigns duties, and evaluates the performance of 12-34 subordinates; organizes and coordinates retention/career counseling activities; conducts reenlistment seminars and serves as policy advisor in reenlistment matters; coordinates the use and management of reenlistment funds; presents informational talks as needed on reenlistment programs; recommends policy changes; develops and assesses reenlistment programs.

Recommendation, Skill Level 30

In the lower-division baccalaureate/associate degree category, 3 semester hours in social psychology, 3 in audiovisual techniques, 3 in marketing techniques, 3 in public speaking, 3 in record keeping, 2 in interview techniques, 1 in technical report writing, and 1 in computer applications. In the upper-division baccalaureate category, 3 semester hours in vocational counseling (11/96).

Recommendation, Skill Level 40

In the lower-division baccalaureate/associate degree category, 3 semester hours in social psychology, 3 in audiovisual techniques, 3 in marketing techniques, 3 in public speaking, 3 in record keeping, 3 in interview techniques, 1 in technical report writing, and 1 in computer applications. In the upper-division baccalaureate category, 3 semester hours in records management, 3 in vocational counseling, and 2 for field experience in counseling (11/96).

Recommendation, Skill Level 50

In the lower-division baccalaureate/associate degree category, 3 semester hours in social psychology, 3 in audiovisual techniques, 3 in marketing techniques, 3 in public speaking, 3 in record keeping, 3 in interview techniques, 1 in technical report writing, 3 in personnel supervision, and 1 in computer applications. In the upper-division baccalaureate category, 3 semester hours in records management, 3 in vocational counseling, 3 for field experience in counseling, and 3 in personnel management (11/96).

MOS-79T-001

Recruiter/Retention NCO (National Guard)
79T40
79T50

Exhibit Dates: 2/97–Present.

Career Management Field: 79 (Recruitment and Reenlistment).

Description

Summary: Recruits or reenlists personnel for military service in the Army. NOTE: Individuals have principal assignment either in recruiting or in retention, with special duties available in counseling, recruiting, training management, instructing, professional development advising, or serving in a supervisory role as the principal noncommissioned officer of a recruiting facility or a recruiting area/region. Because of the principal assignment option, descriptions and recommendations below are cited in two groups (recruiting and retention) at skill levels 40 and 50. Individuals may transfer to MOS 79T from any MOS at a comparable skill level. *Skill Level 40:* (Recruiter): Commands a recruiting station and maintains records, statistics, administrative files, and publications library; prepares enlistment reports; supervises and evaluates subordinate personnel; conducts professional development programs for 10 to 12 subordinates; prepares official correspondence and reports; assigns duties to subordinates and supervises resources. (Retention): Advises commanders on status of reenlistment programs; conducts training for commanders, staff persons, noncommissioned officers, and others when assigned as principal noncommissioned officer in a large facility or region. *Skill Level 50:* Able to perform the duties required for Skill Level 40; (Recruiter): Supervises recruiting programs and personnel at several recruiting stations within an area; assists area commanders with training, operations, administration, and personnel matters; plans and conducts recruiting conferences and seminars; develops presentations which reflect sales techniques, enlistment contracts, and enlistment requirements; supervises and evaluates subordinates; identifies and takes corrective actions in problem areas; develops training programs and inspects recruiting stations to ensure efficient operation and management. (Retention): Supervises, assigns duties, and evaluates the performance of 12-34 subordinates; organizes and coordinates retention/career counseling activities; conducts reenlistment seminars and serves as policy advisor in reenlistment matters; coordinates the use and management of reenlistment funds; presents informational talks as needed on reenlistment programs.

Recommendation, Skill Level 40

In the lower-division baccalaureate/associate degree category, (Recruiter/Retention): 3 semester hours in social psychology, 3 in audiovisual techniques, 3 in marketing techniques, 3 in public speaking, and 3 in record keeping. (Recruiter only): 3 semester hours in personnel supervision. (Retention only): 1 semester hour in computer applications. In the upper-division baccalaureate category, (Recruiter): 3 semester hours in records management, 3 for field experience in marketing, and 3 in personnel management. (Retention): 3 semester hours in records management and 3 in vocational counseling (11/91).

Recommendation, Skill Level 50

In the lower-division baccalaureate/associate degree category, (Recruiter/Retention): 3 semester hours in social psychology, 3 in audiovisual techniques, 3 in marketing techniques, 3 in public speaking, 3 in record keeping, and 3 in personnel supervision. (Recruiter only): 3 semester hours in office procedures. (Retention only): 1 semester hour in computer applications. In the upper-division baccalaureate category, (Recruiter): 3 semester hours in records management, 3 for field experience in marketing, 3 in personnel management, and 3 in office procedures. (Retention): 3 semester hours in records management, 3 in vocational counseling, and 3 in personnel management (11/91).

MOS-79V-001

Retention and Transition NCO (Reserves)
79V30
79V40
79V50

Exhibit Dates: 3/00–Present.

Career Management Field: 79 (Recruitment and Reenlistment).

Description

Summary: Counsels Army Reserve personnel on retention and transfers. NOTE: May have progressed to MOS 79V from any MOS. *Skill Level 30:* Able to perform the duties required for Skill Level 20 of any MOS. Conducts interviews; advises enlisted personnel on reenlistment prerequisites, options, obligations, opportunities, and benefits; assists in designing individual career plans, using personnel records and data from interview and counseling sessions; relates physical, mental, and moral qualifications of individuals to prescribed standards and determines if service schooling is needed; processes reenlistment documents; conducts professional development programs; maintains publications; compiles data; prepares reports; plans reenlistment ceremonies; performs follow-ups. *Skill Level 40:* Able to perform the duties required for Skill Level 30; advises commanders on status of reenlistment programs; conducts training for commanders, staff persons, noncommissioned officers, and others; identifies soldiers to fill vacant warrant officer positions; coordinates reenlistment ceremonies; manages retention and transition support activities. *Skill Level 50:* Able to perform the duties required for Skill Level 40; supervises, assigns duties, and evaluates the performance of subordinates; organizes and coordinates retention/career counseling activities; conducts reenlistment seminars and conferences; serves as policy advisor in reenlistment matters; coordinates the use and management of reenlistment funds; presents informational talks as needed on reenlistment programs; recommends policy changes; develops and assesses reenlistment programs.

Recommendation, Skill Level 30

In the lower-division baccalaureate/associate degree category, 3 semester hours in social psychology, 3 in audiovisual techniques, 3 in marketing techniques, 3 in public speaking, 3 in record keeping, 2 in interview techniques, 1 in technical report writing, and 1 in computer applications. In the upper-division baccalaureate category, 3 semester hours in vocational counseling (3/00).

Recommendation, Skill Level 40

In the lower-division baccalaureate/associate degree category, 3 semester hours in social psychology, 3 in audiovisual techniques, 3 in marketing techniques, 3 in public speaking, 3 in record keeping, 3 in interview techniques, 1 in technical report writing, and 1 in computer applications. In the upper-division baccalaureate category, 3 semester hours in records management, 3 in vocational counseling, and 2 for field experience in counseling (3/00).

Recommendation, Skill Level 50

In the lower-division baccalaureate/associate degree category, 3 semester hours in social psychology, 3 in audiovisual techniques, 3 in marketing techniques, 3 in public speaking, 3 in record keeping, 3 in interview techniques, 1 in technical report writing, 3 in personnel supervision, and 1 in computer applications. In the upper-division baccalaureate category, 3 semester hours in records management, 3 in vocational counseling, 3 for field experience in counseling, and 3 in personnel management (3/00).

MOS-81B-003

TECHNICAL DRAFTING SPECIALIST
81B10
81B20

Exhibit Dates: 1/90–9/91.

Career Management Field: 51 (General Engineering), subfield 511 (Technical Engineering).

Description

Summary: Drafts detailed working plans for construction of bridges, roads, railroads, piers, buildings, and utility installation. *Skill Level 10:* Assists in construction or design drafting; prepares drawings and plans of miscellaneous structures and mechanical and electrical devices using standard drafting equipment; plots profiles for construction of roads, railroads, and other construction projects from survey notes or contour maps; prepares charts and graphs to portray various data. *Skill Level 20:* Able to perform the duties required for Skill Level 10; provides technical guidance to subordinates; reviews, modifies, and corrects finished drawings to conform to specifications and assists design officer in modifying designs of steel, concrete, and wood structures.

Recommendation, Skill Level 10

In the vocational certificate category, 10 semester hours in construction drafting. In the lower-division baccalaureate/associate degree category, 3 semester hours in construction drafting (7/79).

Recommendation, Skill Level 20

In the vocational certificate category, 30 semester hours in construction drafting. In the lower-division baccalaureate/associate degree category, 8 semester hours in construction drafting, 3 in algebra, 3 in trigonometry, 2 in communication skills, and 2 in personnel supervision (7/79).

MOS-81B-004

TECHNICAL DRAFTING SPECIALIST
81B10
81B20

Exhibit Dates: 10/91–6/94. Effective 6/94, MOS 81B was discontinued, and its duties were incorporated into MOS 51T, Technical Engineering Specialist.

Career Management Field: 51 (General Engineering).

Description

Summary: Drafts detailed working plans for construction of bridges, roads, railroads, piers, buildings, and utility installation. *Skill Level 10:* Assists in performance of construction or design drafting; prepares drawings and plans of miscellaneous structures and mechanical and electrical devices, using standard drafting equipment; plots profiles for construction using standard drafting equipment; plots profiles for construction of roads, railroads, and other construction projects from survey notes or contour maps; prepares charts and graphs to portray various data; may have worked with computer-aided design software (CAD). *Skill Level 20:* Able to perform the duties required for Skill Level 10; supervises and provides technical guidance to subordinates; conducts on-the-job training; completes work records and forms; reviews, modifies, and corrects finished drawings to conform to specifications; checks construction schedules and bills of material; assists design officer in modifying designs of steel, concrete, and wood structures.

Recommendation, Skill Level 10

Credit may be granted on the basis of an individualized assessment of the student (3/94).

Recommendation, Skill Level 20

In the lower-division baccalaureate/associate degree category, 3 semester hours in construction drafting and 3 in personnel supervision (11/91).

MOS-81C-002

CARTOGRAPHER
81C10
81C20
81C30
81C40

Exhibit Dates: 1/90–9/92.

Career Management Field: 81 (Topographic Engineering), subfield 811 (Cartography).

Description

Summary: Supervises or performs cartographic drafting activities or compiles or revises planimetric and topographic maps. *Skill Level 10:* Draws or scribes cultural, topographic, hydrographic, or other features on drawings, transparent overlays, and scribing surfaces for reproduction of maps; compiles and revises planimetric and topographic maps; constructs mosaics from aerial photographs; constructs map projections and grids; uses photo interpretation methods for delineating aerial photographs; performs field classification of aerial photographs. *Skill Level 20:* Able to perform the duties required for Skill Level 10; provides technical guidance to subordinates; keeps records and requests supplies; makes recommendations to supervisors. *Skill Level 30:* Able to perform the duties required for Skill Level 20; supervises cartographic laboratory; edits computations, map compilation base, manuscripts, overlays, color proofs, and color separation drawings; makes production recommendations; maintains files; may supervise more than 12 persons. *Skill Level 40:* Able to perform the duties required for Skill Level 30; determines work priorities; supervises final inspection and editing of maps; plans, coordinates, and administers activities of cartographic laboratory with other units; may supervise more than 30 persons.

Recommendation, Skill Level 10

In the lower-division baccalaureate/associate degree category, 3 semester hours in mechanical drawing, 6 in cartography, 3 in photogrammetry, and 3 in aerial photographic interpretation. (NOTE: This recommendation for skill level 10 is valid for the dates 1/90-9/91 only) (6/78).

Recommendation, Skill Level 20

In the lower-division baccalaureate/associate degree category, 3 semester hours in mechanical drawing, 6 in cartography, 3 in photogrammetry, 3 for field experience in cartography, 3 in aerial photographic interpretation, and credit in human relations on the basis of institutional evaluation (6/78).

Recommendation, Skill Level 30

In the lower-division baccalaureate/associate degree category, 3 semester hours in mechanical drawing, 6 in cartography, 3 in photogrammetry, 3 in aerial photographic interpretation, 3 for field experience in cartography, 3 for field experience in management, and 3 in human relations (6/78).

Recommendation, Skill Level 40

In the lower-division baccalaureate/associate degree category, 3 semester hours in mechanical drawing, 6 in cartography, 3 in photogrammetry, 3 in aerial photographic interpretation, 3 for field experience in cartography, 3 in introduction to management, 3 in human relations, 3 in personnel supervision, and 3 for field experience in management (6/78).

MOS-81C-003

CARTOGRAPHER
81C10
81C20
81C30
81C40

Exhibit Dates: 10/92–6/97. Effective 6/97, MOS 81C was discontinued and its duties were transferred to MOS 81T.

Career Management Field: 81 (Topographic Engineering).

Description

Summary: Supervises or performs cartographic drafting activities or revises planimetric and topographic maps. *Skill Level 10:* Draws or scribes cultural, topographic, hydrographic, or other features on drawings, transparent overlays, and scribing surfaces for reproduction of maps; compiles and revises planimetric and topographic maps; constructs mosaics from aerial photographs; constructs map projections and grids; employs computer-assisted cartography; uses photo interpretation methods for delineating aerial photographs; performs field classification of aerial photographs. *Skill Level 20:* Able to perform the duties required for Skill Level 10; provides technical guidance to subordinates; keeps records; requests supplies; may supervise more than eight persons; makes recommendations to supervisors. *Skill Level 30:* Able to perform the duties required for Skill Level 20; supervises cartographic laboratory; edits computations, map compilation base, manuscripts, overlays, color proofs, and color separation drawings; makes production recommendations; maintains files; may supervise more than 25 persons. *Skill Level 40:* Able to perform the duties required for Skill Level 30; determines work priorities; supervises final inspection and editing of maps; plans, coordinates, and administers activities of cartographic laboratory with other units; may supervise more than 50 persons.

Recommendation, Skill Level 10

Credit may be granted on the basis of an individualized assessment of the student (3/94).

Recommendation, Skill Level 20

In the lower-division baccalaureate/associate degree category, 3 semester hours in personnel supervision and 2 in records and information management. In the upper-division baccalaureate category, 3 semester hours in elements of cartography, 3 in computer-assisted cartography, 3 in photogrammetry, and 3 in air photo interpretation. (NOTE: This recommendation for skill level 20 is valid for the dates 10/92-2/95 only) (10/92).

Recommendation, Skill Level 30

In the lower-division baccalaureate/associate degree category, 3 semester hours in personnel supervision and 2 in records and information management. In the upper-division baccalaureate category, 3 semester hours in elements of cartography, 3 in computer-assisted cartography, 3 in photogrammetry, 3 in air photo interpretation, and 3 for field experience in cartography (10/92).

Recommendation, Skill Level 40

In the lower-division baccalaureate/associate degree category, 3 semester hours in personnel supervision and 2 in records and information management. In the upper-division baccalaureate category, 3 semester hours in elements of cartography, 3 in computer-assisted cartography, 3 in photogrammetry, 3 in air photo interpretation, 3 for field experience in cartography, 3 in organizational management, and 3 for field experience in management (10/92).

MOS-81L-001

LITHOGRAPHER
81L10
81L20
81L30
81L40

Exhibit Dates: 8/95–9/04. (Effective 9/04, MOS 81L was discontinued, and its duties were incorporated into MOS 21L.)

Description

Summary: Operates and performs operator maintenance on offset duplicators/presses, copy cameras, plate making, and various types of bindery and film processing equipment; supervises and performs all printing and binding, camera operations, layout, and plate making activities. *Skill Level 10:* Produces lithographic film products such as contact negative and positive film, contact paper prints and line negatives and positives; produces color proofs, deep etches, peel coat images, scribes, presensitized plates and color key images; produces flats for press operations; performs bindery operations and operates offset duplicators and presses to reproduce printed materials; performs preventive maintenance on all photolithographic equipment. *Skill Level 20:* Able to perform the duties required for Skill Level 10; provides technical guidance to subordinates; produces film products, including continuous-tone and half-tone negatives; produces multicolor printed matter using a two-color offset duplicator; performs and supervises operator and preventive maintenance on all photolithographic equipment. *Skill Level 30:* Able to perform the duties required for Skill Level 20; directs and prepares mobile and modular print system sections; directs photolithographic section operations; directs preventive maintenance on photolithographic equipment. *Skill Level 40:* Able to perform the duties required for Skill Level 30; supervises preparation of mobile and modular print system organization for operations and movement; prepares field operations layout for mobile organization; supervises preventive maintenance schedules for photolithographic equipment.

Recommendation, Skill Level 10

Credit may be granted on the basis of an individualized assessment of the student (6/97).

Recommendation, Skill Level 20

Credit may be granted on the basis of an individualized assessment of the student (6/97).

Recommendation, Skill Level 30

In the lower-division baccalaureate/associate degree category, 3 semester hours in process photography, 3 in printing process and equipment, 3 in plate making and stripping, 3 in personnel supervision, and 3 in layout, finishing, and bindery. In the upper-division baccalaureate category, 3 semester hours in press operations management and 3 in estimation and supplies (6/97).

Recommendation, Skill Level 40

In the lower-division baccalaureate/associate degree category, 3 semester hours in process photography, 3 in printing process and equipment, 3 in plate making and stripping, 3 for field experience in press operations, 3 in personnel supervision, and 3 in layout, finishing, and bindery. In the upper-division baccalaureate category, 3 semester hours in press operations management, 3 in estimation and supplies, 3 in operations management, and 3 in graphic arts management and quality control (6/97).

MOS-81Q-002

TERRAIN ANALYST
81Q10
81Q20
81Q30
81Q40

Exhibit Dates: 1/90–2/95.

Career Management Field: 81 (Topographic Engineering).

Description

Summary: Performs or supervises terrain analysis activities, including reconnaissance, collection and maintenance of source materials, data extraction, analysis and synthesis, and prediction of terrain and weather effects. *Skill Level 10:* Participates in the collection of data or physical and cultural terrain features; uses stereoscopic and monocular instruments to extract terrain data from remotely sensed imagery and existing topographic products; records analysis results on overlays; adds to geographic data base; employs basic drafting techniques; assists in preparation of manuscripts that accompany maps and charts; reviews terrain descriptions and records findings; assists senior personnel. *Skill Level 20:* Able to perform the duties required for Skill Level 10; provides technical guidance to subordinates; uses various techniques to measure terrain data from remotely sensed imagery and existing topographic products; estimates conditions not available from existing sources; manipulates data and transforms results to product specifications; edits finished manuscripts for completeness and accuracy; uses data base information to predict the effects of weather on terrain; compiles analysis, overlays, and narrative for publication as special maps or terrain studies; inspects and edits completed work. *Skill Level 30:* Able to perform the duties required for Skill Level 20; provides technical assistance to subordinates and superiors; supervises and conducts on-the-job training; supervises and maintains administrative, intelligence, and reference files; supervises division terrain analysis team; advises commanding officer. *Skill Level 40:* Able to perform the duties required for Skill Level 30; supervises terrain analysis activities in corps or theater Army terrain teams or higher unit; coordinates collection requirements; plans and coordinates organization, training, and technical operations; coordinates with numerous other staff elements in the Army.

Recommendation, Skill Level 10

In the lower-division baccalaureate/associate degree category, 3 semester hours in physical geography, 3 in aerial photographic interpretation, and 3 in technical communications. In the upper-division baccalaureate category, 3 semester hours in photogrammetry and 3 in geographic field methods. (NOTE: This recommendation for skill level 10 is valid for the dates 1/90-9/91 only) (7/88).

Recommendation, Skill Level 20

In the lower-division baccalaureate/associate degree category, 3 semester hours in physical geography, 3 in aerial photographic interpretation, and 3 in technical communications. In the upper-division baccalaureate category, 3 semester hours in photogrammetry, 3 in geographic field methods, 3 in terrain analysis, and 3 in cartography (7/88).

Recommendation, Skill Level 30

In the lower-division baccalaureate/associate degree category, 3 semester hours in physical geography, 3 in aerial photographic interpretation, 3 in technical communications, 3 in project management, and 3 in personnel supervision. In the upper-division baccalaureate category, 3 semester hours in photogrammetry, 3 in geographic field methods, 3 in terrain analysis, 3 in cartography, and 3 in advanced photogrammetry (7/88).

Recommendation, Skill Level 40

In the lower-division baccalaureate/associate degree category, 3 semester hours in physical geography, 3 in aerial photographic interpretation, 3 in technical communications, 3 in project management, and 3 in personnel supervision. In the upper-division baccalaureate category, 3 semester hours in photogrammetry, 3 in geographic field methods, 3 in terrain analysis, 3 in cartography, 3 in advanced photogrammetry, 3 for field experience in management, and 3 in management problems (7/88).

MOS-81Q-003

TERRAIN ANALYST

81Q10
81Q20
81Q30
81Q40

Exhibit Dates: 3/95–6/97. Effective 6/97, MOS 81Q was discontinued and its duties were transferred to MOS 81T.

Career Management Field: 81 (Topographic Engineering).

Description

Summary: Performs or supervises terrain analysis activities, including reconnaissance, collection and maintenance of source materials, data extraction, analysis and synthesis, and prediction of terrain and weather effects. *Skill Level 10:* Participates in the collection of data or physical and cultural terrain features; uses stereoscopic and monocular instruments to extract terrain data from remotely sensed imagery and existing topographic products; records analysis results on overlays; adds to geographic data base; employs basic drafting techniques; assists in preparation of manuscripts that accompany maps and charts; reviews terrain descriptions and records findings; assists senior personnel. *Skill Level 20:* Able to perform the duties required for Skill Level 10; provides technical guidance to subordinates; uses various techniques to measure terrain data from remotely sensed imagery and existing topographic products; estimates conditions not available from existing sources; manipulates data and transforms results to product specifications; edits finished manuscripts for completeness and accuracy; uses data base information to predict the effects of weather on terrain; compiles analysis, overlays, and narrative for publication as special maps or terrain studies; inspects and edits completed work. *Skill Level 30:* Able to perform the duties required for Skill Level 20; provides technical assistance to subordinates and superiors; supervises and conducts on-the-job training; supervises and maintains administrative, intelligence, and reference files; supervises division terrain analysis team; advises commanding officer. *Skill Level 40:* Able to perform the duties required for Skill Level 30; supervises terrain analysis activities in corps or theater Army terrain teams or higher unit; coordinates collection requirements; plans and coordinates organization, training, and technical operations; coordinates with numerous other staff elements in the Army.

Recommendation, Skill Level 10

Credit may be granted on the basis of an individualized assessment of the student (7/88).

Recommendation, Skill Level 20

Credit may be granted on the basis of an individualized assessment of the student (7/88).

Recommendation, Skill Level 30

In the lower-division baccalaureate/associate degree category, 3 semester hours in physical geography, 3 in aerial photographic interpretation, 3 in technical communications, 3 in project management, and 3 in personnel supervision. In the upper-division baccalaureate category, 3 semester hours in photogrammetry, 3 in geographic field methods, 3 in terrain analysis, and 3 in cartography (7/88).

Recommendation, Skill Level 40

In the lower-division baccalaureate/associate degree category, 3 semester hours in physical geography, 3 in aerial photographic interpretation, 3 in technical communications, 3 in project management, and 3 in personnel supervision. In the upper-division baccalaureate category, 3 semester hours in photogrammetry, 3 in geographic field methods, 3 in terrain analysis, 3 in cartography, 3 in advanced photogrammetry, 3 for field experience in management, and 3 in management problems (7/88).

MOS-81T-001

TOPOGRAPHIC ANALYST

81T10
81T20
81T30
81T40

Exhibit Dates: 2/97–9/04. (Effective 9/04, MOS 81T was discontinued, and its duties were incorporated into MOS 21U.)

Career Management Field: 81 (Topographic Engineering).

Description

Summary: Supervises and/or performs cartographic and terrain analysis duties; collects and processes military geographic information from remote-sensed imagery, digital data, intelligence data, existing topographic products, and other collateral data sources; edits cartographic and terrain analysis products; advises command and staff officers on topographic operations and special map product planning. *Skill Level 10:* Extracts terrain data from remote-sensed imagery, field reconnaissance, digital data, existing topographic products, and other collateral data sources; records analysis results to a military graphic data base; draws, scribes, digitizes, and scans cultural, topographic, hydrographic, and/or other features on overlay/scribing surfaces or in digital formats; utilizes basic drafting techniques to tailor terrain products; revises planimetric and topographic maps; performs digital manipulation of topographic information by querying, viewing, evaluating, and downloading digital data. *Skill Level 20:* Able to perform the duties required for Skill Level 10; supervises and provides technical guidance to subordinates; analyzes collected military geographic information to produce tactical decision aids; predicts terrain and weather effects as applied to command, control, communications, computer, and intelligence systems; performs data base management for the storage of aerial photographs, maps, digital data bases, and collateral source materials; performs quality assurance during all stages of topographic operations. *Skill Level 30:* Able to perform the duties required for Skill Level 20; supervises topographic analysts; supervises equipment maintenance; evaluates source materials for military geographic information analysis; supervises quality assurance during all stages of topographic operations; ensures required administrative, intelligence, source data, and reference files are maintained; advises command and staff officers on all aspects of topographic operations and doctrine. *Skill Level 40:* Able to perform the duties required for Skill Level 30; supervises quality assurance program for all topographic operations; plans and coordinates activities pertaining to the organization, training, and tactical and technical operations; directs preparation for movement and operation of mobile/transportable work sections; advises command and staff officers on all aspects of topographic operations and doctrine; coordinates the planning and production of military geographic information; supervises preparation of transportable/mobile work sections; advises command and staff officers on all aspects of topographic operations and doctrine; coordinates the planning and production of military geographic information.

Recommendation, Skill Level 10

Credit may be granted on the basis of an individualized assessment of the student (5/99).

Recommendation, Skill Level 20

Credit may be granted on the basis of an individualized assessment of the student (5/99).

Recommendation, Skill Level 30

In the lower-division baccalaureate/associate degree category, 3 semester hours in applied algebra and trigonometry, 3 in computer operating systems, 2 in physical geography, 2 in geographic field methods, 2 in aerial photographic interpretation, 2 in photogrammetry, 2 in remote sensing interpretation, 2 in terrain analysis, 3 in cartography, 2 in technical report writing, 3 in geographic information systems, 3 in principles of supervision, and 3 in oral communication skills (5/99).

Recommendation, Skill Level 40

In the lower-division baccalaureate/associate degree category, 3 semester hours in applied algebra and trigonometry, 3 in computer operating systems, 3 in physical geography, 3 in geographical field methods, 3 in aerial photographic interpretation, 3 in photogrammetry, 2 in remote sensing interpretation, 3 in terrain analysis, 3 in cartography, 3 in technical report writing, 3 geographic information systems, 3 in principles of supervision, and 3 in oral communication skills. In the upper-division baccalaureate category, 3 semester hours for field experience in management and 3 in personnel management (5/99).

MOS-81Z-001

TOPOGRAPHIC ENGINEERING SUPERVISOR

81Z50

Exhibit Dates: 1/90–9/92.

Career Management Field: 81 (Topographic Engineering), subfield 810 (Topographic Engineering, General).

Description

Supervises topographic surveying, cartography, and photolithography activities; serves as first sergeant of a company or as a topographic operations sergeant; assists in topographic planning and control activities; assists in determining requirements and providing program and technical supervision of topographic mapping and other military intelligence programs including geodetic and topographic surveying activities; assists in command supervision and coordination of map reproduction and topographic map supply programs. NOTE: May have progressed to 81Z50 from 81C40, Cartographer; 82D40, Topographic Surveyor; or 83F40, Photolithographer.

Recommendation

In the lower-division baccalaureate/associate degree category, 6 semester hours for field experience in graphic arts, 3 in record keeping, 3 in human relations, 3 in personnel supervision, 3 in introduction to management, and 6 in printing management. In the upper-division baccalaureate category, 3 semester hours in industrial arts education (graphic arts), 3 in industrial management, 3 in plant organization, and 6 for field experience in management (6/78).

MOS-81Z-002

Topographic Engineering Supervisor
81Z50

Exhibit Dates: 10/92–9/04. (Effective 9/04, MOS 81Z was discontinued, and its duties were incorporated into MOS 21Y.)

Career Management Field: 81 (Topographic Engineering).

Description

Supervises topographic surveying, cartography, and photolithography activities; serves as first sergeant of a company or as topographic operations sergeant; assists in topographic planning and control activities; assists in determining requirements and providing program and technical supervision of topographic mapping and other military intelligence programs including geodetic and topographic surveying activities; assists in command supervision and coordination of map reproduction and topographic map supply programs. NOTE: May have progressed to 81Z50 from 81C40, Cartographer; 81Q40, Terrain Analyst; 82D40, Topographic Surveyor; or 83F40, Printing and Bindery Specialist.

Recommendation

In the lower-division baccalaureate/associate degree category, 3 semester hours in personnel supervision and 2 in records and information management. In the upper-division baccalaureate category, 3 semester hours in organizational management and 3 for field experience in management; if individual has attained pay grade E-9, additional credit as follows: 3 semester hours in operations management, 3 in management problems, and 3 in communication techniques for managers (10/92).

MOS-82B-002

Construction Surveyor
82B10
82B20

Exhibit Dates: 1/90–9/91.

Career Management Field: 51 (General Engineering), subfield 511 (Technical Engineering).

Description

Summary: Performs construction surveys to determine relative positions of points on the ground to provide data for construction layout and earthwork. *Skill Level 10:* Serves as construction surveyor; reads maps and places appropriate reference points for construction layout and earthwork; operates surveying instruments and equipment such as tape, alidade, plane table, level, transit and theodolite; employs such instruments as aneroid barometer and compass; records observations and makes minor computations; maintains and adjusts survey instruments and equipment. *Skill Level 20:* Able to perform the duties required for Skill Level 10; directs survey party and works independently while conducting construction surveys; conducts initial reconnaissance of construction area; interprets construction drawings and briefs survey party; directs use of stadia; establishes centerlines for horizontal curves and spirals, and grade lines for vertical curves; responsibilities include surveys for all features of bridges, buildings, sewer lines, roads, airfields, and drainage; maintains records of survey party work.

Recommendation, Skill Level 10

In the lower-division baccalaureate/associate degree category, 3 semester hours in basic land surveying (7/79).

Recommendation, Skill Level 20

In the lower-division baccalaureate/associate degree category, 6 semester hours in basic and advanced land surveying. In the upper-division baccalaureate category, 3 semester hours for field experience in management (7/79).

MOS-82B-003

Construction Surveyor
82B10
82B20

Exhibit Dates: 10/91–6/94. Effective 6/94, MOS 82B was discontinued, and its duties were incorporated into MOS 51T, Technical Engineering Specialist.

Career Management Field: 51 (General Engineering).

Description

Summary: Performs construction surveys to determine relative positions of points on the ground to provide data for construction layout and earthwork. *Skill Level 10:* Serves as construction surveyor; reads maps and places appropriate reference points for construction layout and earthwork; operates surveying instruments and equipment such as tape, alidade, plane table, level, transit and theodolite; employs such instruments as aneroid barometer, compass and optical range finders; records observations and makes minor computations; maintains and adjusts survey instruments and equipment. *Skill Level 20:* Able to perform the duties required for Skill Level 10; directs survey party and works independently while conducting construction surveys; conducts initial reconnaissance of construction area; interprets construction drawings and briefs survey party; directs use of stadia; establishes centerlines for horizontal curves and spirals, and grade lines for vertical curves; responsibilities include surveys for all features of bridges, buildings, sewer lines, roads, airfields, and drainage; maintains records of survey party work; conducts on-the-job training; completes work records and forms.

Recommendation, Skill Level 10

Credit may be granted on the basis of an individualized assessment of the student (3/94).

Recommendation, Skill Level 20

In the lower-division baccalaureate/associate degree category, 4 semester hours in land surveying and 3 in personnel supervision (11/91).

MOS-82C-003

Field Artillery Surveyor
82C10
82C20
82C30
82C40

Exhibit Dates: 1/90–4/99.

Career Management Field: 13 (Field Artillery).

Description

Summary: Engages in surveying activities. *Skill Level 10:* Determines distance between survey stations; marks survey stations; performs astronomical observations for determining locations; operates and maintains survey equipment, including theodolite, laser-activated infrared distance measuring instrument, PADS inertial system gyroscope unit for locations, and the Backup Computer Systems (Hewlett-Packard Computer) for computations; knows manual computation formulas, equations, procedures, and forms, as well as ephemeris and logarithmic tables. *Skill Level 20:* Able to perform the duties required for Skill Level 10; applies knowledge of basic algebra, basic geometry, and trigonometric functions and logarithms; computes azimuth and distance, coordinates, altimetric, and trigonometric heights, and astronomical azimuths from field data. *Skill Level 30:* Able to perform the duties required for Skill Level 20; supervises a survey party; selects starting station and method of survey; reviews survey data for accuracy; supervises maintenance of survey party equipment; instructs in survey procedures and techniques. *Skill Level 40:* Able to perform the duties required for Skill Level 30; supervises and coordinates operations of several survey parties; prepares technical, personnel, and administrative reports.

Recommendation, Skill Level 10

In the lower-division baccalaureate/associate degree category, 2 semester hours in introduction to maps and aerial photos, 2 in surveying, 1 in trigonometry, 1 in algebra, 1 in computer operations, and 1 in introduction to astronomy. (NOTE: This recommendation for skill level 10 is valid for the dates 1/90-9/91 only) (6/89).

Recommendation, Skill Level 20

In the lower-division baccalaureate/associate degree category, 2 semester hours in introduction to maps and aerial photos, 8 in surveying, 2 in algebra, 1 in trigonometry, 2 in computer operations, and 1 in introduction to astronomy. (NOTE: This recommendation for skill level 20 is valid for the dates 1/90-2/95 only) (6/89).

Recommendation, Skill Level 30

In the lower-division baccalaureate/associate degree category, 2 semester hours in introduction to maps and aerial photos, 8 in surveying, 2 in algebra, 1 in trigonometry, 2 in computer operations, 1 in introduction to astronomy, 2 in communication skills, and 3 in principles of supervision (6/89).

Recommendation, Skill Level 40

In the lower-division baccalaureate/associate degree category, 2 semester hours in introduction to maps and aerial photos, 8 in surveying, 2 in algebra, 1 in trigonometry, 2 in computer operations, 1 in introduction to astronomy, 3 in communication skills, and 3 in personnel supervision. In the upper-division baccalaureate category, 3 semester hours for field experience in management and 3 in personnel management (6/89).

MOS-82C-004

Field Artillery Surveyor
82C10
82C20
82C30
82C40

Exhibit Dates: 5/99–8/04. (Effective 9/04, MOS 82C was discontinued, and its duties were incorporated into MOS 13S.)

Career Management Field: 13 (Field Artillery).

Description

Summary: Engages in surveying activities. *Skill Level 10:* Determines distance between survey stations; marks survey stations; performs astronomical observations for determining locations; operates and maintains survey equipment, including theodolite, laser-activated infrared distance measuring instrument, PADS inertial system gyroscope unit for locations, and the Backup Computer Systems (Hewlett-Packard Computer) for computations; knows manual computation formulas, equations, procedures, and forms, as well as ephemeris. *Skill Level 20:* Able to perform the duties required for Skill Level 10; applies knowledge of basic algebra, basic geometry, and trigonometric functions; computes azimuth and distance coordinates, altimetric and trigonometric heights, and astronomical azimuths from field data. *Skill Level 30:* Able to perform the duties required for Skill Level 20; supervises a survey party; selects starting station and method of survey; reviews survey data for accuracy; supervises maintenance of survey party equipment; instructs in survey procedures and techniques. *Skill Level 40:* Able to perform the duties required for Skill Level 30; supervises and coordinates operations of several survey parties; prepares technical, personnel, and administrative reports.

Recommendation, Skill Level 10

Credit may be granted on the basis of an individualized assessment of the student (5/99).

Recommendation, Skill Level 20

Credit may be granted on the basis of an individualized assessment of the student (5/99).

Recommendation, Skill Level 30

In the lower-division baccalaureate/associate degree category, 2 semester hours in introduction to maps and aerial photos, 6 in surveying, 3 in applied algebra and trigonometry, 3 in computer operations, 1 in introduction to astronomy, 3 in oral communication skills, 3 in technical report writing, and 3 in principles of supervision (5/99).

Recommendation, Skill Level 40

In the lower-division baccalaureate/associate degree category, 3 semester hours in introduction to maps and aerial photos, 6 in surveying, 3 in applied algebra and trigonometry, 3 in computer operations, 2 in introduction to astronomy, 3 in oral and written communication skills, 3 in technical report writing, and 3 in personnel supervision. In the upper-division baccalaureate category, 3 semester hours for field experience in management and 3 in personnel management (5/99).

MOS-82D-002

Topographic Surveyor
82D10
82D20
82D30
82D40

Exhibit Dates: 1/90–9/92.

Career Management Field: 81 (Topographic Engineering), subfield 812 (Surveying).

Description

Summary: Supervises or conducts surveys to provide control data for map making and supervises or performs topographic or geodetic computations. *Skill Level 10:* Records topographic survey data; operates survey instruments; performs topographic computations; conducts topographic surveys; interprets maps and aerial photographs; records measurements and other field data; computes horizontal distances from taped, electronic, or stadia measurement; computes angular closures and triangulations; computes differences in elevation from trigonometric, altimetric, and differential leveling; computes azimuths from stellar observations and from grid coordinates; employs planetable with alidade, theodolite, level, microwave, infrared, and base line taping distance measuring equipment; operates programmable electronic calculator. *Skill Level 20:* Able to perform the duties required for Skill Level 10; performs geodetic computation and adjustments and topographic survey observations to extend horizontal and vertical control; employs theodolite, precise levels, altimeters, and lasers; makes field checks; obtains data to calibrate instruments; computes and adjusts second-order surveys; computes elevation of tidal bench marks from tidal observation data for secondary stations; determines and writes required numbers and types of condition equation used in least squares adjustments; operates and writes programs for programmable electronic calculation; directs and controls personnel when acting as survey party chief. *Skill Level 30:* Able to perform the duties required for Skill Level 20; must be able to supervise personnel performing the duties of 82D20 or 41B20 (Topographic Instrument Repair Specialist); performs astronomical observations; computes geographic azimuths to verify topographic surveys; supervises survey section or computing activities in a survey party; calibrates equipment; performs survey reconnaissance; evaluates field data and results obtained; computes geographic positions to include elevation from observation derived from triangulations, trilateration, and traverse and spirit leveling; computes astronomical and la-place azimuth, latitude, and longitude from astronomical observations; computes gravity survey field data and earth tide gravity correction; evaluates and verifies results of all computations; supervises programming of electronic calculators. *Skill Level 40:* Able to perform the duties required for Skill Level 30; supervises geodetic or topographic surveying party and computing activities of survey operations; plans work schedule; collects relevant data; supervises reconnaissance; plans and arranges for field party support in remote areas; coordinates surveying and computing activities; assigns duties; plans training and instructs personnel; supervises field survey activities in support of photogrammetric requirements; prepares, reviews, and consolidates technical, personnel, and administrative reports.

Recommendation, Skill Level 10

In the lower-division baccalaureate/associate degree category, 3 semester hours in technical mathematics (including algebra, trigonometry, geometry, and spherical geometry), 3 in physics (optics), 3 in surveying, 3 in geodetic surveying, and 3 in engineering computations. (NOTE: This recommendation for skill level 10 is valid for the dates 1/90-9/91 only) (6/78).

Recommendation, Skill Level 20

In the lower-division baccalaureate/associate degree category, 3 semester hours in technical mathematics (including algebra, trigonometry, geometry, and spherical geometry), 3 in physics (optics), 3 in surveying, 3 in geodetic surveying, 3 in engineering computations, 3 in data processing, and 6 for field experience in surveying (6/78).

Recommendation, Skill Level 30

In the vocational certificate category, 3 semester hours in technical writing and 3 in engineering computations. In the lower-division baccalaureate/associate degree category, 3 semester hours in technical mathematics (including algebra, trigonometry, geometry, and spherical geometry), 3 in physics (optics), 3 in surveying, 3 in geodetic surveying, 3 in personnel supervision, 6 in engineering computations, 9 for field experience in surveying, 3 in data processing, and 3 in human relations (6/78).

Recommendation, Skill Level 40

In the vocational certificate category, 3 semester hours in engineering computations and 3 in record keeping. In the lower-division baccalaureate/associate degree category, 3 semester hours in technical mathematics (including algebra, trigonometry, geometry, and spherical geometry), 3 in physics (optics), 3 in surveying, 3 in geodetic surveying, 3 in technical writing, 3 in personnel supervision, 9 in engineering computations, 12 for field experience in surveying, 3 in human relations, 3 in introduction to management, and 3 in data processing. In the upper-division baccalaureate category, 3 semester hours in personnel management and 5 in project layout (6/78).

MOS-82D-003

Topographic Surveyor
82D10
82D20
82D30
82D40

Exhibit Dates: 10/92–9/04. (Effective 9/04, MOS 82D was discontinued, and its duties were incorporated into MOS 21S.)

Career Management Field: 81 (Topographic Engineering).

Description

Summary: Supervises or conducts surveys to provide control data for map making and supervises or performs topographic or geodetic computations. *Skill Level 10:* Records topographic survey data; operates survey instruments; performs topographic computations; conducts topographic surveys; interprets maps and aerial photographs; records measurements and other field data; computes horizontal distances from taped, electronic, or stadia measurement; computes angular closures and triangulations; computes differences in elevation from trigonometric, altimetric, and differential leveling; computes azimuths from stellar observations and from grid coordinates; employs planetable with alidade, theodolite, level, microwave, infrared, and base line taping distance measuring equipment; operates programmable electronic calculator. *Skill Level 20:* Able to perform the duties required for Skill Level 10; performs geodetic computation and adjustments and topographic survey observations to extend horizontal and vertical control; employs theodolite, precise levels, altimeters, and lasers;

makes field checks; obtains data to calibrate instruments; computes and adjusts second-order surveys; computes elevation of tidal bench marks from tidal observations data for secondary stations; utilizes electronic calculators and microcomputers for electronic calculations; directs and controls personnel when acting as survey party chief. *Skill Level 30:* Able to perform the duties required for Skill Level 20; must be able to supervise personnel performing the duties of 82D20 or 41B20 (Topographic Instrument Repair Specialist); performs astronomical observations; computes geographic azimuths to verify topographic surveys; supervises survey section or computing activities in a survey party; calibrates equipment; performs survey reconnaissance; evaluates field data and results obtained; computes geographic positions including elevation from observation derived from triangulations, trilateration, and traverse and spirit leveling; computes azimuth, latitude, and longitude from astronomical observations; computes gravity survey field data and earth tide gravity correction; evaluates and verifies results of all computations; supervises the use of electronic calculators and microcomputers for electronic calculations. *Skill Level 40:* Able to perform the duties required for Skill Level 30; supervises geodetic or topographic surveying party and computing activities of survey operations; plans work schedule; collects relevant data; supervises reconnaissance; plans and arranges for field party support in remote areas; coordinates surveying and computing activities; assigns duties; plans training and instructs personnel; supervises field survey activities in support of photogrammetric requirements; prepares, reviews, and consolidates technical, personnel, and administrative reports.

Recommendation, Skill Level 10

Credit may be granted on the basis of an individualized assessment of the student (3/94).

Recommendation, Skill Level 20

In the lower-division baccalaureate/associate degree category, 3 semester hours in surveying, 3 in geodetic surveying, 4 in technical mathematics, 3 in introduction to microcomputers, 6 for field experience in surveying, and 3 in personnel supervision. (NOTE: This recommendation for skill level 20 is valid for the dates 10/92-2/95 only) (10/92).

Recommendation, Skill Level 30

In the lower-division baccalaureate/associate degree category, 3 semester hours in surveying, 3 in geodetic surveying, 4 in technical mathematics, 3 in introduction to microcomputers, 9 for field experience in surveying, 3 in personnel supervision, and 2 in records and information management (10/92).

Recommendation, Skill Level 40

In the lower-division baccalaureate/associate degree category, 3 semester hours in surveying, 3 in geodetic surveying, 4 in technical mathematics, 3 in introduction to microcomputers, 9 for field experience in surveying, 3 in personnel supervision, and 2 in records and information management. In the upper-division baccalaureate category, 3 semester hours in organizational management and 3 for field experience in management (10/92).

MOS-83E-003

Photo and Layout Specialist

83E10

83E20

83E30

Exhibit Dates: 1/90–12/95. Effective 12/95, MOS 83E was discontinued and its duties were incorporated into MOS 81L, Lithographer.

Career Management Field: 81 (Topographic Engineering).

Description

Summary: Operates lithographic copy cameras and prepares lithographic negatives and positives, flats, contact prints, color proofs, scribe and peels coat products, and offset press plates. *Skill Level 10:* Operates copy cameras, film processors, and densitometers to produce negatives from camera-ready copy; selects correct materials, exposures, and chemicals; uses developer and fixer solutions; washes and dries negatives; performs stripping and/or platemaking activities; operates platemaking equipment; produces rectified aerial prints; performs operator maintenance on photographic and platemaking equipment. *Skill Level 20:* Able to perform the duties required for Skill Level 10; provides technical guidance to subordinates; makes quality checks; performs detailed topographic process photography and lithographic stripping and platemaking; performs preventive maintenance on automatic film processor; calibrates digital densitometers. *Skill Level 30:* Able to perform the duties required for Skill Level 20; supervises camera, photomechanical, layout, and platemaking activities and personnel; ensures quality in all operations; trains personnel in photolithographic operations; maintains supply inventory.

Recommendation, Skill Level 10

In the lower-division baccalaureate/associate degree category, 3 semester hours in process photography and 3 in platemaking and/or stripping for offset printing (based on institutional evaluation of the individual's experience). (NOTE: This recommendation for skill level 10 is valid for the dates 1/90-9/91 only) (7/88).

Recommendation, Skill Level 20

In the lower-division baccalaureate/associate degree category, 3 semester hours in process photography, 3 in platemaking and/or stripping for offset printing (based on institutional evaluation of the individual's experience), and 3 for field experience in process photography. (NOTE: This recommendation for skill level 20 is valid for the dates 1/90-2/95 only) (7/88).

Recommendation, Skill Level 30

In the lower-division baccalaureate/associate degree category, 3 semester hours in process photography, 3 in platemaking and/or stripping for offset printing (based on institutional evaluation of the individual's experience), 6 for field experience in process photography, and 3 in personnel supervision (7/88).

MOS-83F-002

Printing and Bindery Specialist

83F10

83F20

83F30

83F40

Exhibit Dates: 1/90–9/92.

Career Management Field: 81 (Topographic Engineering), subfield 813 (Photolithographer).

Description

Summary: Operates bindery equipment, offset presses, and duplicating equipment and supervises photolithographic activities. *Skill Level 10:* Performs bindery operations and operates offset presses and duplicators; services and adjusts equipment; mixes etch solutions; stores supplies; cleans work area and equipment. *Skill Level 20:* Able to perform the duties required for Skill Level 10; provides technical guidance to subordinates; makes quality checks; detects equipment malfunctions; demonstrates operation and maintenance procedures. *Skill Level 30:* Able to perform the duties required for Skill Level 20; must be able to supervise personnel performing the duties of 83F20; 41K20 (Reproduction Equipment Repair Specialist; and 83E20, (Photo and Layout Specialist); supervises lithographic photography and platemaking, offset press and duplicating, and bindery operations in a small lithographic activity; may supervise more than 19 persons. NOTE: May have progressed to 83F30 from 83F20; 83E20, Photo and Layout Specialist; or 41K20, Reproduction Equipment Repair Specialist. *Skill Level 40:* Able to perform the duties required for Skill Level 30; supervises a large photolithographic facility; organizes and controls work flow; prepares administrative and operational reports; determines work and supply requirements; may supervise more than 30 persons.

Recommendation, Skill Level 10

In the lower-division baccalaureate/associate degree category, 3 semester hours in operation and maintenance of duplication and reproduction machines and 3 for field experience in bindery operations. (NOTE: This recommendation for skill level 10 is valid for the dates 1/90-9/91 only) (6/78).

Recommendation, Skill Level 20

In the lower-division baccalaureate/associate degree category, 3 semester hours in operation and maintenance of duplication and reproduction machines, 3 for field experience in bindery operations, and 3 for field experience in reproduction operations (6/78).

Recommendation, Skill Level 30

In the lower-division baccalaureate/associate degree category, 3 semester hours in operation and maintenance of duplication and reproduction machines, 3 for field experience in bindery operations, 3 for field experience in reproduction operations, 3 in basic photography, 3 in platemaking, 3 in stripping for offset printing, 3 in human relations, and 3 in personnel supervision (6/78).

Recommendation, Skill Level 40

In the lower-division baccalaureate/associate degree category, 3 semester hours in operation and maintenance of duplication and reproduction machines, 3 for field experience in bindery operations, 3 for field experience in reproduction operations, 3 in basic photography, 3 in platemaking, 3 in stripping for offset printing, 3 in human relations, 3 in personnel supervision, 3 in technical report writing, and 3 for field experience in management (6/78).

MOS-83F-003

Printing And Bindery Specialist

83F10

83F20

83F30

83F40

Exhibit Dates: 10/92–2/95.

Career Management Field: 81 (Topographic Engineering).

Description

Summary: Operates and performs operator maintenance on offset presses and duplicating and bindery equipment; supervises various photolithographic activities. *Skill Level 10:* Operates offset press and duplicating equipment, including single color printing, mounting plates and blankets, and preparing and adjusting inking assembly; performs bindery operations, including cutting paper, stitching, drilling, and collecting printed matter. *Skill Level 20:* Able to perform the duties required for Skill Level 10; utilizes a second color head on offset presses; supervises and provides technical guidance to subordinates; conducts on-the-job training. *Skill Level 30:* Able to perform the duties required for Skill Level 20; leads offset press team; supervises small or medium-sized offset press/offset duplicator and bindery activities; ensures quality of printed material and bindery operations; performs cost estimating for basic printing operations. *Skill Level 40:* Able to perform the duties required for Skill Level 30; coordinates activities and sets priorities for the entire unit; coordinates with prepress operations (camera, stripping, and platemaking) to ensure that schedules are met; may supervise up to 100 persons; schedules reproduction projects; maintains photolithographic operations. NOTE: May progress to 83F40 from 83F30 or 83E30, Photo and Layout Specialist.

Recommendation, Skill Level 10

Credit may be granted on the basis of an individualized assessment of the student (3/94).

Recommendation, Skill Level 20

In the lower-division baccalaureate/associate degree category, 3 semester hours in printing processes and equipment, 3 in press procedures, 3 in finishing and bindery operations, and 3 in personnel supervision (10/92).

Recommendation, Skill Level 30

In the lower-division baccalaureate/associate degree category, 3 semester hours in printing processes and equipment, 3 in press procedures, 3 in finishing and bindery operations, 3 in printing planning and control, 3 in estimating and supplies, 3 in personnel supervision, and 2 in records and information management (10/92).

Recommendation, Skill Level 40

In the lower-division baccalaureate/associate degree category, 3 semester hours in printing processes and equipment, 3 in press procedures, 3 in finishing and bindery operations, 3 in printing planning and control, 3 in estimating and supplies, 3 in personnel supervision, and 2 in records and information management. In the upper-division baccalaureate category, 3 semester hours in organizational management, 3 for field experience in management, and 3 in printing/graphic arts management (10/92).

MOS-83F-004

Printing And Bindery Specialist

83F10

83F20

83F30

83F40

Exhibit Dates: 3/95–12/95. Effective 12/95, MOS 83F was discontinued and its duties were incorporated into MOS 81L, Lithographer.

Career Management Field: 81 (Topographic Engineering).

Description

Summary: Operates and performs operator maintenance on offset presses and duplicating and bindery equipment; supervises various photolithographic activities. *Skill Level 10:* Operates offset press and duplicating equipment, including single color printing, mounting plates and blankets, and preparing and adjusting inking assembly; performs bindery operations, including cutting paper, stitching, drilling, and collecting printed matter. *Skill Level 20:* Able to perform the duties required for Skill Level 10; utilizes a second color head on offset presses; supervises and provides technical guidance to subordinates; conducts on-the-job training. *Skill Level 30:* Able to perform the duties required for Skill Level 20; leads offset press team; supervises small or medium-sized offset press/offset duplicator and bindery activities; ensures quality of printed material and bindery operations; performs cost estimating for basic printing operations. *Skill Level 40:* Able to perform the duties required for Skill Level 30; coordinates activities and sets priorities for the entire unit; coordinates with prepress operations (camera, stripping, and platemaking) to ensure that schedules are met; may supervise up to 100 persons; schedules reproduction projects; maintains photolithographic operations. NOTE: May progress to 83F40 from 83F30 or 83E30, Photo and Layout Specialist.

Recommendation, Skill Level 10

Credit may be granted on the basis of an individualized assessment of the student (3/94).

Recommendation, Skill Level 20

Credit may be granted on the basis of an individualized assessment of the student (10/92).

Recommendation, Skill Level 30

In the lower-division baccalaureate/associate degree category, 3 semester hours in printing processes and equipment, 3 in press procedures, 3 in finishing and bindery operations, 3 in personnel supervision, and 2 in records and information management (10/92).

Recommendation, Skill Level 40

In the lower-division baccalaureate/associate degree category, 3 semester hours in printing processes and equipment, 3 in press procedures, 3 in finishing and bindery operations, 3 in printing planning and control, 3 in estimating and supplies, 3 in personnel supervision, and 2 in records and information management. In the upper-division baccalaureate category, 3 semester hours in organizational management, 3 for field experience in management, and 3 in printing/graphic arts management (10/92).

MOS-88H-001

Cargo Specialist

88H10

88H20

88H30

88H40

Exhibit Dates: 1/90–10/92.

Career Management Field: 88 (Transportation).

Description

Summary: Transfers or supervises the transfer of cargo to and from water, land, and air transports by manual and mechanical methods. *Skill Level 10:* Serves as cargo handler and checker; loads, unloads, checks, and tallies cargo; uses slings, nets, hooks, pallets, spreaders, lifting bars, winches, hoists, and other cargo-handling equipment; employs safety procedures; signals winchman, hoistman, or cargo equipment operator where to move cargo. *Skill Level 20:* Able to perform the duties required for Skill Level 10; serves as hatch foreman or senior cargo checker and enforces safety practices. *Skill Level 30:* Able to perform the duties required for Skill Level 20; supervises a medium-sized terminal operation work force by planning and organizing work schedules, assigning duties, and designating work groups (hatch gangs) as required for each operation; advises on handling of all types of cargo and demonstrates techniques and use of equipment; coordinates heavy lift equipment and special terminal operations gear; prepares technical, personnel, and administrative reports. *Skill Level 40:* Able to perform the duties required for Skill Level 30; supervises a large terminal work force normally consisting of 56 or more persons.

Recommendation, Skill Level 10

In the vocational certificate category, 4 semester hours in cargo handling and rigging. (NOTE: This recommendation for skill level 10 is valid for the dates 1/90-9/91 only) (5/76).

Recommendation, Skill Level 20

In the vocational certificate category, 7 semester hours in cargo handling and rigging (5/76).

Recommendation, Skill Level 30

In the vocational certificate category, 13 semester hours in cargo handling, stowage, and rigging. In the lower-division baccalaureate/associate degree category, 3 semester hours in human relations, 3 in personnel supervision, 2 in report writing, 2 in cargo planning, 2 in cargo processing and documentation, and 1 in transportation of dangerous materials (5/76).

Recommendation, Skill Level 40

In the vocational certificate category, 13 semester hours in cargo handling, stowage, and rigging. In the lower-division baccalaureate/associate degree category, 3 semester hours in human relations, 3 in personnel supervision, 3 in records administration, 3 in report writing, 3 for field experience in management, 2 in cargo processing and documentation, 2 in cargo planning, and 1 in transportation of dangerous materials (5/76).

MOS-88H-002

Cargo Specialist
88H10
88H20
88H30
88H40

Exhibit Dates: 11/92–2/95.

Career Management Field: 88 (Transportation).

Description

Summary: Transfers or supervises the transfer of cargo to and from water, land, and air transports by manual and mechanical methods. *Skill Level 10:* Serves as cargo handler and checker; loads, unloads, checks, and tallies cargo; uses slings, nets, hooks, pallets, spreaders, lifting bars, winches, hoists, and other cargo-handling equipment; employs safety procedures; signals winchman, hoistman, or cargo equipment operator where to move cargo. *Skill Level 20:* Able to perform the duties required for Skill Level 10; serves as hatch foreman or senior cargo checker; enforces safety practices; operates cranes; supervises rough terrain equipment maintenance. *Skill Level 30:* Able to perform the duties required for Skill Level 20; inspects and organizes storage operations; requisitions and issues supplies and equipment; supervises a medium-sized terminal operations work force by planning and organizing work schedules, assigning duties, and designating work groups (hatch gangs) as required for each operation; advises on handling of all types of cargo and demonstrates techniques and use of equipment; coordinates heavy lift equipment and special terminal operations gear; prepares technical, personnel, and administrative reports. *Skill Level 40:* Able to perform the duties required for Skill Level 30; supervises a large terminal work force normally consisting of 50 or more persons; operates and supervises storage areas employing foreign nations.

Recommendation, Skill Level 10

Credit may be granted on the basis of an individualized assessment of the student (11/92).

Recommendation, Skill Level 20

In the vocational certificate category, 6 semester hours in cargo handling and rigging. In the lower-division baccalaureate/associate degree category, 3 semester hours in cargo handling equipment operations and maintenance and 3 in personnel supervision (11/92).

Recommendation, Skill Level 30

In the vocational certificate category, 6 semester hours in cargo handling and rigging, 3 in cargo stowage, and 1 in transportation of hazardous materials. In the lower-division baccalaureate/associate degree category, 3 semester hours in cargo handling equipment operations and maintenance and 3 in personnel supervision (11/92).

Recommendation, Skill Level 40

In the vocational certificate category, 9 semester hours in cargo handling and rigging, 3 in cargo stowage, and 1 in transportation of hazardous materials. In the lower-division baccalaureate/associate degree category, 3 semester hours in cargo handling equipment operations and maintenance and 3 in personnel supervision. In the upper-division baccalaureate category, 3 semester hours for field experience in management and 3 in organizational management (11/92).

MOS-88H-003

Cargo Specialist
88H10
88H20
88H30
88H40

Exhibit Dates: 3/95–11/01.

Career Management Field: 88 (Transportation).

Description

Summary: Transfers or supervises the transfer of cargo to and from water, land, and air transports by manual and mechanical methods. *Skill Level 10:* Serves as cargo handler and checker; loads, unloads, checks, and tallies cargo; uses slings, nets, hooks, pallets, spreaders, lifting bars, winches, hoists, and other cargo-handling equipment; employs safety procedures; signals winchman, hoistman, or cargo equipment operator where to move cargo. *Skill Level 20:* Able to perform the duties required for Skill Level 10; serves as hatch foreman or senior cargo checker; enforces safety practices; operates cranes; supervises rough terrain equipment maintenance. *Skill Level 30:* Able to perform the duties required for Skill Level 20; inspects and organizes storage operations; requisitions and issues supplies and equipment; supervises a medium-sized terminal operations work force by planning and organizing work schedules, assigning duties, and designating work groups (hatch gangs) as required for each operation; advises on handling of all types of cargo and demonstrates techniques and use of equipment; coordinates heavy lift equipment and special terminal operations gear; prepares technical, personnel, and administrative reports. *Skill Level 40:* Able to perform the duties required for Skill Level 30; supervises a large terminal work force normally consisting of 50 or more persons; operates and supervises storage areas employing foreign nations.

Recommendation, Skill Level 10

Credit may be granted on the basis of an individualized assessment of the student (11/92).

Recommendation, Skill Level 20

Credit may be granted on the basis of an individualized assessment of the student (11/92).

Recommendation, Skill Level 30

In the vocational certificate category, 6 semester hours in cargo handling and rigging. In the lower-division baccalaureate/associate degree category, 3 semester hours in cargo handling equipment operations and maintenance and 3 in personnel supervision (11/92).

Recommendation, Skill Level 40

In the vocational certificate category, 6 semester hours in cargo handling and rigging, 3 in cargo stowage, and 1 in transportation of hazardous materials. In the lower-division baccalaureate/associate degree category, 3 semester hours in cargo handling equipment operations and maintenance and 3 in personnel supervision. In the upper-division baccalaureate category, 3 semester hours for field experience in management and 3 in organizational management (11/92).

MOS-88H-004

Cargo Specialist
88H10
88H20
88H30
88H40

Exhibit Dates: 12/01–Present.

Career Management Field: 88 (Transportation).

Description

Summary: Transfers or supervises the transfer of cargo and passengers to and from water, land, and air transports by manual and mechanical methods. *Skill Level 10:* Serves as cargo handler and checker; loads, unloads, checks, and tallies cargo; uses and maintains slings, nets, hooks, pallets, spreaders, lifting bars, winches, hoists, and other cargo-handling equipment; employs safety procedures; signals winchman, hoistman, or cargo equipment operator where to move cargo. *Skill Level 20:* Able to perform the duties required for Skill Level 10; serves as hatch foreman or senior cargo checker; enforces safety practices; operates cranes; supervises cargo handling equipment maintenance. *Skill Level 30:* Able to perform the duties required for Skill Level 20; inspects and organizes storage operations; requisitions and issues supplies and equipment; supervises a medium-sized terminal operations work force by planning and organizing work schedules, assigning duties, and designating work groups as required for each operation; advises on handling of all types of cargo and demonstrates techniques and use of equipment; coordinates heavy lift equipment and special terminal operations gear; prepares technical, personnel, and administrative reports. *Skill Level 40:* Able to perform the duties required for Skill Level 30; supervises a large terminal work force; operates and supervises storage areas employing foreign nationals.

Recommendation, Skill Level 10

Credit may be granted on the basis of an individualized assessment of the student (12/01).

Recommendation, Skill Level 20

Credit may be granted on the basis of an individualized assessment of the student (12/01).

Recommendation, Skill Level 30

In the lower-division baccalaureate/associate degree category, 3 semester hours in cargo handling equipment operations and maintenance, 3 in computer applications, and 3 in principles of supervision (12/01).

Recommendation, Skill Level 40

In the lower-division baccalaureate/associate degree category, 3 semester hours in cargo handling equipment operations and maintenance, 3 in computer applications, and 3 in principles of supervision. In the upper-division baccalaureate category, 3 semester hours for a management practicum and 3 in organizational management (12/01).

MOS-88K-001

Watercraft Operator
88K10
88K20
88K30
88K40

Exhibit Dates: 1/90–10/92.

Career Management Field: 88 (Transportation).

Description

Summary: Operates and performs deck duties on Army watercraft and, in some instances, amphibians. *Skill Level 10:* Performs general seaman duties; under close supervision, assists in docking, unloading, and anchoring watercraft; assists in loading and unloading cargo; stands lookout watch; assists in inspecting, servicing, and operating fire equipment, lifeboats, and rafts; performs preventive maintenance and inspections on Army watercraft; maneuvers Army watercraft; uses simple sounding devices, aids to navigation, steering compass courses, and running lights; communicates with stations ashore and afloat, using accepted communications equipment and techniques; applies basic water safety and rescue procedures. *Skill Level 20:* Able to perform the duties required for Skill Level 10; operates and dispatches Army watercraft and, under close supervision, operates Army watercraft for specific missions in accordance with navigational and communications rules and principles; keeps records of watercraft status and movements. *Skill Level 30:* Able to perform the duties required for Skill Level 20; supervises the operation of watercraft and small boat crew; knows capabilities and limitations of Army watercraft and assigns and dispatches them accordingly; provides berthing and unberthing services to large or oceangoing vessels; computes fuel requirements and plans maintenance; knows procedures for abandoning ship including sea rescue and lifeboat operations; supervises embarking and debarking of troops; instructs subordinates in watercraft operational practices, procedures, and techniques. NOTE: Some Skill Level 30 personnel are trained as operators of large amphibian vehicles and can supervise their operation. *Skill Level 40:* Able to perform the duties required for Skill Level 30; supervises the operations of a large boat crew or assists in controlling Army watercraft operations; coordinates work activities; reviews, consolidates, and prepares technical, personnel, and administrative reports; supervises up to 20 persons.

Recommendation, Skill Level 10

In the vocational certificate category, 4 semester hours in seamanship, 1 in piloting and navigation, 1 in communications, and 1 in safety. (NOTE: This recommendation for skill level 10 is valid for the dates 1/90-9/91 only) (6/76).

Recommendation, Skill Level 20

In the vocational certificate category, 8 semester hours in seamanship, 6 in piloting and navigation, 3 in communications, and 1 in safety (6/76).

Recommendation, Skill Level 30

In the vocational certificate category, 8 semester hours in seamanship, 6 in piloting and navigation, 3 in communications, 3 in basic navigation, 3 in advanced navigation, 1 in safety, and additional credit in amphibian operation on the basis of institutional evaluation. In the lower-division baccalaureate/associate degree category, 3 semester hours in navigation and 3 in human relations (6/76).

Recommendation, Skill Level 40

In the vocational certificate category, 8 semester hours in seamanship, 6 in piloting and navigation, 3 in communications, 3 in basic navigation, 3 in advanced navigation, 1 in safety, and additional credit in amphibian operation on the basis of institutional evaluation. In the lower-division baccalaureate/associate degree category, 3 semester hours in navigation, 3 in human relations, 3 in marine management, 3 for field experience in management, and 3 in personnel supervision (6/76).

MOS-88K-002

Watercraft Operator

88K10

88K20

88K30

88K40

Exhibit Dates: 11/92–11/01.

Career Management Field: 88 (Transportation).

Description

Summary: Operates and performs deck duties on Army watercraft and, in some instances, amphibians. *Skill Level 10:* Performs general seaman duties; under close supervision, assists in docking, unloading, and anchoring watercraft; assists in loading and unloading cargo; stands lookout watch; assists in inspecting, servicing, and operating fire equipment, lifeboats, and rafts; performs preventive maintenance and inspections on Army watercraft; maneuvers Army watercraft; uses simple sounding devices, aids to navigation, and steering compass courses; communicates with stations ashore and afloat using accepted communications equipment and techniques; applies basic water safety and rescue procedures. *Skill Level 20:* Able to perform the duties required for Skill Level 10; operates and dispatches Army watercraft and, under close supervision, operates Army watercraft for specific missions in accordance with navigation and communications rules and principles; keeps records of watercraft status and movements. *Skill Level 30:* Able to perform the duties required for Skill Level 20; supervises the operation of watercraft and small boat crew; knows capabilities and limitations of Army watercraft and assigns and dispatches them accordingly; provides berthing and unberthing services to large or oceangoing vessels; computes fuel requirements and plans maintenance; knows procedures for abandoning ship including sea rescue and lifeboat operations; supervises embarking and debarking of troops; instructs subordinates in watercraft operation practices, procedures, and techniques. NOTE: Some Skill Level 30 personnel are trained as operators of large amphibian vehicles and can supervise their operation. *Skill Level 40:* Able to perform the duties required for Skill Level 30; supervises the operations of a large boat crew or assists in controlling Army watercraft operations; coordinates work activities; reviews, consolidates, and prepares technical, personnel, and administrative reports; supervises up to 20 persons.

Recommendation, Skill Level 10

Credit may be granted on the basis of an individualized assessment of the student (3/94).

Recommendation, Skill Level 20

In the vocational certificate category, 3 semester hours in communications and 1 in safety. In the lower-division baccalaureate/associate degree category, 3 semester hours in seamanship, 3 in basic navigation, and 3 in personnel supervision. (NOTE: This recommendation for skill level 20 is valid for the dates 11/92-2/95 only) (11/92).

Recommendation, Skill Level 30

In the vocational certificate category, 3 semester hours in communications and 1 in safety. In the lower-division baccalaureate/associate degree category, 3 semester hours in seamanship, 3 in basic navigation, 3 in personnel supervision, and 2 in records management (11/92).

Recommendation, Skill Level 40

In the vocational certificate category, 3 semester hours in communications and 1 in safety. In the lower-division baccalaureate/associate degree category, 3 semester hours in seamanship, 3 in basic navigation, 3 in personnel supervision, and 2 in records management. In the upper-division baccalaureate category, 3 semester hours in organizational management and 3 for field experience in management (11/92).

MOS-88K-003

Watercraft Operator

88K10

88K20

88K30

88K40

Exhibit Dates: 12/01–Present.

Career Management Field: 88 (Transportation).

Description

Summary: Operates and performs deck duties on Army watercraft and, in some instances, amphibians. *Skill Level 10:* Performs general seaman duties; under close supervision, assists in docking, unloading, and anchoring watercraft; assists in loading and unloading cargo; stands lookout watch; assists in inspecting, servicing, and operating fire equipment, lifeboats, and rafts; performs preventive maintenance and inspections on Army watercraft; maneuvers Army watercraft; uses simple sounding devices, aids to navigation and steer compass courses; communicates with stations ashore and afloat using accepted communications equipment and techniques; applies basic water safety and rescue procedures. *Skill Level 20:* Able to perform the duties required for Skill Level 10; operates and dispatches Army watercraft and, under close supervision, operates Army watercraft for specific missions in accordance with navigation and communications rules and principles; keeps records of watercraft status and movements. *Skill Level 30:* Able to perform the duties required for Skill Level 20; supervises the operation of watercraft and small boat crew; knows capabilities and limitations of Army watercraft and assigns and dispatches them accordingly; provides berthing and unberthing services to large or oceangoing vessels; computes fuel requirements and plans maintenance; knows procedures for abandoning ship, including sea rescue and lifeboat operations; supervises embarking and debarking of troops; instructs subordinates in watercraft operation practices, procedures, and techniques. NOTE:

Some Skill Level 30 personnel are trained as operators of large amphibian vehicles and can supervise their operation. *Skill Level 40:* Able to perform the duties required for Skill Level 30; supervises the operations of a large boat crew or assists in controlling Army watercraft operations; coordinates work activities; reviews, consolidates, and prepares technical, personnel, and administrative reports; supervises up to 20 persons.

Recommendation, Skill Level 10

Credit may be granted on the basis of an individualized assessment of the student (12/01).

Recommendation, Skill Level 20

Credit may be granted on the basis of an individualized assessment of the student (12/01).

Recommendation, Skill Level 30

In the lower-division baccalaureate/associate degree category, 3 semester hours in business communication, 1 in safety, 3 in seamanship, 3 in basic navigation, 3 in principles of supervision, 2 in records management, and 3 in computer applications (12/01).

Recommendation, Skill Level 40

In the lower-division baccalaureate/associate degree category, 3 semester hours in business communication, 1 in safety, 3 in seamanship, 3 in basic navigation, 3 in principles of supervision, 2 in records management, and 3 in computer applications. In the upper-division baccalaureate category, 3 semester hours in organizational management and 3 for a management practicum (12/01).

MOS-88L-001

WATERCRAFT ENGINEER

88L10
88L20
88L30
88L40

Exhibit Dates: 1/90–10/92.

Career Management Field: 88 (Transportation).

Description

Summary: Supervises or performs maintenance on propulsion systems and auxiliary equipment of marine vessels, and in some instances, on amphibians. *Skill Level 10:* Under close supervision, performs maintenance on Army watercraft; applies safety precautions pertinent to starting, stopping, and operating main diesel engines, electrical machinery, and other auxiliary components installed on Army watercraft; reads and records pressures and temperatures to ensure safe and efficient operation of machinery; uses hand and power tools; assists in performing preventive maintenance inspections and correcting maintenance shortcomings and deficiencies on watercraft. *Skill Level 20:* Able to perform the duties required for Skill Level 10; provides technical guidance to subordinates; computes fuel consumption, tank capacities, and pumping rates for loading and discharging; maintains a working knowledge of damage control, fire prevention, and federal, state, and local environmental pollution control regulations; adjusts or troubleshoots malfunctioning diesel engines, electrical machinery, and auxiliary equipment; makes emergency repairs or takes emergency damage control measures at sea; performs soldering and pipefitting; uses precision tools; interprets technical publications; replaces engine and auxiliary equipment subassemblies; performs preventive maintenance inspections on watercraft; in some instances, may also perform these duties on amphibians. *Skill Level 30:* Able to perform the duties required for Skill Level 20; evaluates job performance of subordinates; maintains a reference library of technical manuals and service bulletins; interprets technical blueprints; troubleshoots, repairs, purges, recharges, and tests watercraft refrigeration systems; uses special test equipment and instruments in repairing diesel engines and accessory items; assists supervisor in engineering maintenance functions; plans and organizes work schedules and assigns duties to subordinates; instructs subordinates in work production and quality control procedures; is capable of supervising a medium-sized watercraft maintenance facility; establishes physical layout of work site and ensures an efficient, clean, and safe work environment; requisitions, issues, stores, and maintains records of supplies and equipment; coordinates work activities with other maintenance elements; recommends changes to technical publications as required; in some instances, may also perform these duties on amphibians. *Skill Level 40:* Able to perform the duties required for Skill Level 30; supervises a large marine maintenance facility; estimates and prepares preliminary budget forecasts of supply and maintenance activities; allocates funds and resources for the efficient operation of supply and maintenance activities; establishes operating procedures on the basis of directives and information received from superiors.

Recommendation, Skill Level 10

In the vocational certificate category, 4 semester hours in internal-combustion engine principles, 1 in internal-combustion engine operation, and 1 in shipboard operations. (NOTE: This recommendation for skill level 10 is valid for the dates 1/90-9/91 only) (6/76).

Recommendation, Skill Level 20

In the vocational certificate category, 7 semester hours in internal-combustion engine principles, 4 in internal-combustion engine operation, 5 in shipboard operations, and additional credit in amphibian maintenance on the basis of institutional evaluation (6/76).

Recommendation, Skill Level 30

In the vocational certificate category, 7 semester hours in internal-combustion engine principles, 5 in shipboard operations, 4 in internal-combustion engine operation, 3 in basic refrigeration, and 3 in use and care of tools and test equipment, and additional credit in amphibian maintenance on the basis of institutional evaluation. In the lower-division baccalaureate/associate degree category, 3 semester hours in watercraft maintenance, 3 in marine refrigeration, and 3 in human relations, and additional credit in amphibian maintenance on the basis of institutional evaluation (6/76).

Recommendation, Skill Level 40

In the vocational certificate category, 7 semester hours in internal-combustion engine principles, 5 in shipboard operations, 4 in internal-combustion engine operation, 3 in basic refrigeration, and 3 in use and care of tools and test equipment, and additional credit in amphibian maintenance on the basis of institutional evaluation. In the lower-division baccalaureate/associate degree category, 3 semester hours in watercraft maintenance, 3 in marine refrigeration, 3 in human relations, 3 in introduction to management, 3 in personnel supervision, 3 for field experience in management, and additional credit in amphibian maintenance on the basis of institutional evaluation (6/76).

MOS-88L-002

WATERCRAFT ENGINEER

88L10
88L20
88L30
88L40

Exhibit Dates: 11/92–11/01.

Career Management Field: 88 (Transportation).

Description

Summary: Supervises or performs maintenance on propulsion systems and auxiliary equipment of marine vessels, and in some instances, on amphibians. *Skill Level 10:* Under close supervision, performs maintenance on Army watercraft; applies safety precautions pertinent to starting, stopping, and operating main diesel engines, electrical machinery, and other auxiliary components installed on Army watercraft; reads and records pressures and temperatures to ensure safe and efficient operation of machinery; uses hand and power tools; assists in performing preventive maintenance inspections and correcting maintenance shortcomings and deficiencies on watercraft. *Skill Level 20:* Able to perform the duties required for Skill Level 10; provides technical guidance to subordinates; computes fuel consumption, tank capacities, and pumping rates for loading and discharging; troubleshoots, repairs, purges, recharges, and tests watercraft refrigeration systems; maintains a working knowledge of damage control, fire prevention, and federal, state, and local environmental pollution control regulations; adjusts or troubleshoots malfunctioning diesel engines, hydraulic and pneumatic systems, marine drive gear, electrical machinery, and auxiliary equipment; makes emergency repairs or takes emergency damage control measures at sea; performs soldering, oxyacetylene and arc welding, and pipefitting; uses precision tools; interprets technical publications; replaces engine and auxiliary equipment subassemblies; performs preventive maintenance inspections on watercraft; in some instances, may also perform these duties on amphibians. *Skill Level 30:* Able to perform the duties required for Skill Level 20; evaluates job performance of subordinates; maintains a reference library of technical manuals and service bulletins; interprets technical blueprints; uses special test equipment and instruments in repairing diesel engines and accessory items; assists supervisor in engineering maintenance functions; plans and organizes work schedules and assigns duties to subordinates; instructs subordinates in work production and quality control procedures; is capable of supervising a medium-sized watercraft maintenance facility; establishes physical layout of work site and ensures an efficient, clean, and safe work environment; requisitions, issues, stores, and maintains records of supplies and equipment; coordinates work activities with other maintenance elements; recommends changes to technical publications as required; in some instances, may also perform these duties on amphibians. *Skill Level 40:* Able to perform the duties required for Skill Level 30; supervises a large marine maintenance facility; performs managerial duties; estimates and prepares preliminary budget forecasts of supply and maintenance activities; allocates funds and resources for the efficient operation of supply and maintenance activities; establishes operating procedures on the basis of directives and information received from superiors.

Recommendation, Skill Level 10

Credit may be granted on the basis of an individualized assessment of the student (3/94).

Recommendation, Skill Level 20

In the lower-division baccalaureate/associate degree category, 2 semester hours in hydraulics and pneumatics, 2 in automotive electricity, 2 in marine gear transmissions, 2 in diesel engine maintenance, 3 in diesel engine overhaul, 2 in air conditioning and refrigeration, 1 in oxyacetylene welding, 1 in arc welding, and 3 in personnel supervision. (NOTE: This recommendation for skill level 20 is valid for the dates 11/92-2/95 only) (11/92).

Recommendation, Skill Level 30

In the lower-division baccalaureate/associate degree category, 2 semester hours in hydraulics and pneumatics, 2 in automotive electricity, 2 in marine gear transmissions, 2 in diesel engine maintenance, 3 in diesel engine overhaul, 2 in air conditioning and refrigeration, 1 in oxyacetylene welding, 1 in arc welding, 3 in personnel supervision, and 3 in maintenance management (11/92).

Recommendation, Skill Level 40

In the lower-division baccalaureate/associate degree category, 2 semester hours in hydraulics and pneumatics, 2 in automotive electricity, 2 in marine gear transmissions, 2 in diesel engine maintenance, 3 in diesel engine overhaul, 2 in air conditioning and refrigeration, 1 in oxyacetylene welding, 1 in arc welding, 3 in personnel supervision, and 3 in maintenance management. In the upper-division baccalaureate category, 3 semester hours in organizational management and 3 for field experience in management (11/92).

MOS-88L-003

Watercraft Engineer

88L10
88L20
88L30
88L40

Exhibit Dates: 12/01–Present.

Career Management Field: 88 (Transportation).

Description

Summary: Supervises or performs maintenance on propulsion systems and auxiliary equipment of marine vessels, and in some instances, on amphibians. *Skill Level 10:* Under close supervision, performs maintenance on Army watercraft; applies safety precautions pertinent to starting, stopping, and operating main diesel engines, electrical machinery, and other auxiliary components installed on Army watercraft; reads and records pressures and temperatures to ensure safe and efficient operation of machinery; uses hand and power tools; assists in performing preventive maintenance inspections and correcting maintenance shortcomings and deficiencies on watercraft. *Skill Level 20:* Able to perform the duties required for Skill Level 10; provides technical guidance to subordinates; computes fuel consumption, tank capacities, and pumping rates for loading and discharging; troubleshoots, repairs, purges, recharges, and tests watercraft refrigeration systems; maintains a working knowledge of damage control, fire prevention, and federal, state, and local environmental pollution control regulations; adjusts or troubleshoots malfunctioning diesel engines, hydraulic and pneumatic systems, marine drive gear, electrical machinery, and auxiliary equipment; makes emergency repairs or takes emergency damage control measures at sea; performs soldering, oxyacetylene and arc welding, and pipefitting; uses precision tools; interprets technical publications; replaces engine and auxiliary equipment subassemblies; performs preventive maintenance inspections on watercraft; in some instances, may also perform these duties on amphibians. *Skill Level 30:* Able to perform the duties required for Skill Level 20; evaluates job performance of subordinates; maintains a reference library of technical manuals and service bulletins; interprets technical blueprints; uses special test equipment and instruments in repairing diesel engines and accessory items; assists supervisor in engineering maintenance functions; plans and organizes work schedules and assigns duties to subordinates; instructs subordinates in work production and quality control procedures; is capable of supervising a medium-sized watercraft maintenance operation; establishes physical layout of work site and ensures an efficient, clean, and safe work environment; requisitions, issues, stores, and maintains records of supplies and equipment; coordinates work activities with other maintenance elements; recommends changes to technical publications as required; in some instances, may also perform these duties on amphibians. *Skill Level 40:* Able to perform the duties required for Skill Level 30; supervises a large marine maintenance operation; performs managerial duties; estimates and prepares preliminary budget forecasts of supply and maintenance activities; allocates funds and resources for the efficient operation of supply and maintenance activities; establishes operating procedures on the basis of directives and information received from superiors.

Recommendation, Skill Level 10

Credit may be granted on the basis of an individualized assessment of the student (12/01).

Recommendation, Skill Level 20

Credit may be granted on the basis of an individualized assessment of the student (12/01).

Recommendation, Skill Level 30

In the lower-division baccalaureate/associate degree category, 2 semester hours in hydraulics/pneumatics, 2 in marine electricity, 2 in marine gear transmissions, 2 in diesel engine maintenance, 3 in diesel engine overhaul, 2 in air conditioning and refrigeration, 1 in oxyacetylene welding, 1 in arc welding, 3 in principles of supervision, 3 in computer applications, and 3 in maintenance management (12/01).

Recommendation, Skill Level 40

In the lower-division baccalaureate/associate degree category, 2 semester hours in hydraulics/pneumatics, 2 in marine electricity, 2 in marine gear transmissions, 2 in diesel engine maintenance, 3 in diesel engine overhaul, 2 in air conditioning and refrigeration, 1 in oxyacetylene welding, 1 in arc welding, 3 in principles of supervision, 3 in computer applications, and 3 in maintenance management. In the upper-division baccalaureate category, 3 semester hours in organizational management and 3 for management practicum (12/01).

MOS-88M-001

Motor Transport Operator

88M10
88M20
88M30
88M40

Exhibit Dates: 1/90–10/92.

Career Management Field: 88 (Transportation).

Description

Summary: Operates wheel vehicles to transport personnel and cargo. *Skill Level 10:* Operates single-unit wheel vehicles with a capacity of five tons or less in all conditions of light, weather, and terrain; applies safety rules and practices; ensures proper loading of vehicle; performs simple vehicular maintenance; keeps records of operation and performance; may serve as chauffeur or dispatcher; depending on the type of unit to which assigned, may also perform the same duties for tractors, semitrailers, tank transporters, and other vehicles with a capacity of more than five tons. *Skill Level 20:* Able to perform the duties required for Skill Level 10; provides technical guidance to subordinates. *Skill Level 30:* Able to perform the duties required for Skill Level 20; supervises small motor transport activity; participates in convoy operations and organization, including planning, the establishment of control measures, and the observance of civil laws and military regulations; reviews and prepares operation reports; determines the number of vehicles required for a given operation; schedules route and controls movements of motor transport equipment; may perform or direct driver testing functions. *Skill Level 40:* Able to perform the duties required for Skill Level 30; supervises a large motor transport activity.

Recommendation, Skill Level 10

In the vocational certificate category, 3 semester hours in motor vehicle operation (single-unit truck); if the duty assignment was heavy vehicle driver (tractor-trailer), 3 additional semester hours in motor vehicle operation. (NOTE: This recommendation for skill level 10 is valid for the dates 1/90-9/91 only) (5/76).

Recommendation, Skill Level 20

In the vocational certificate category, 4 semester hours in motor vehicle operation (single-unit truck); if the duty assignment was heavy vehicle driver (tractor-trailer), 4 additional semester hours in motor vehicle operation (5/76).

Recommendation, Skill Level 30

In the vocational certificate category, 4 semester hours in motor vehicle operation (single-unit truck); if the duty assignment was heavy vehicle driver (tractor-trailer), 4 additional semester hours in motor vehicle operation. In the lower-division baccalaureate/associate degree category, 6 semester hours in motor vehicle transportation management and 3 in motor maintenance management; if the duty assignment was driver test sergeant, 3 additional semester hours for an internship in education (driver training) (5/76).

Recommendation, Skill Level 40

In the vocational certificate category, 8 semester hours in motor vehicle operation (all road vehicles). In the lower-division baccalaureate/associate degree category, 6 semester hours in motor transportation management, 3 in motor maintenance management, 3 in personnel supervision, 3 in human relations, and 3 for field experience in management; if the duty assignment was driver test sergeant, 3 additional semester hours for an internship in education (driver training) (5/76).

MOS-88M-002

Motor Transport Operator

88M10
88M20
88M30
88M40

Exhibit Dates: 11/92–Present.

Career Management Field: 88 (Transportation).

Description

Summary: Operates wheel vehicles to transport personnel and cargo. *Skill Level 10:* Operates single-unit wheel vehicles with a capacity of five tons or less in all conditions of light, weather, and terrain; applies safety rules and practices; ensures proper loading of vehicle; performs simple vehicular maintenance; keeps records of operation and performance; may serve as chauffeur or dispatcher; depending on the type of unit to which assigned, may also perform the same duties for tractors, semitrailers, tank transporters, and other vehicles with a capacity of more than five tons. *Skill Level 20:* Able to perform the duties required for Skill Level 10; provides technical guidance to subordinates; may serve as dispatcher; performs recovery of vehicles. *Skill Level 30:* Able to perform the duties required for Skill Level 20; supervises small motor transportation activity; participates in convoy operations and organization, including planning, the establishment of control measures, and the observance of civil laws and military regulations; may perform as convoy march unit commander; reviews and prepares reports; determines the number of vehicles required for a given operation; schedules routes and controls movements of motor transport equipment; may perform or direct driver training functions. *Skill Level 40:* Able to perform the duties required for Skill Level 30; supervises a large motor transport activity; prepares reconnaissance data for transport movement; prepares map overlays; establishes motor park; establishes engineer requirements to support a truck battalion.

Recommendation, Skill Level 10

Credit may be granted on the basis of an individualized assessment of the student (11/92).

Recommendation, Skill Level 20

In the vocational certificate category, 3 semester hours in motor vehicle operation. In the lower-division baccalaureate/associate degree category, 3 semester hours in personnel supervision. (NOTE: This recommendation for skill level 20 is valid for the dates 11/92-2/95 only) (11/92).

Recommendation, Skill Level 30

In the vocational certificate category, 3 semester hours in motor vehicle operation. In the lower-division baccalaureate/associate degree category, 3 semester hours in personnel supervision, 3 in motor vehicle transportation management, and 3 in motor vehicle maintenance management. In the upper-division baccalaureate category, 3 semester hours for field experience in management (11/92).

Recommendation, Skill Level 40

In the vocational certificate category, 3 semester hours in motor vehicle operation. In the lower-division baccalaureate/associate degree category, 3 semester hours in personnel supervision, 6 in motor vehicle transportation management, and 3 in motor vehicle maintenance management. In the upper-division baccalaureate category, 3 semester hours for field experience in management and 3 in organizational management (11/92).

MOS-88N-001

Traffic Management Coordinator

88N10
88N20
88N30
88N40

Exhibit Dates: 1/90–10/92.

Career Management Field: 88 (Transportation).

Description

Summary: Coordinates the departure and arrival of freight and personnel by air, rail, highway, and water. *Skill Level 10:* Performs clerical duties associated with the logistics of freight and personnel movements; issues government and civilian shipping documents, including freight bills, bills of lading, and freight manifest sheets; assists military members in the preparation of travel itineraries, personal property shipping documents, and passenger movement forms. *Skill Level 20:* Able to perform the duties required for Skill Level 10; provides guidance to subordinates. *Skill Level 30:* Able to perform the duties required for Skill Level 20; supervises medium-scale freight movement office; plans, directs, and controls subordinates; assists in the development of future transportation plans; supervises freight reconsignment, maintains liaison with military and civilian transportation facilities; reviews, consolidates and prepares reports covering transportation movement operations. *Skill Level 40:* Able to perform the duties required for Skill Level 30; supervises personnel at a large-scale freight movement office; when requested, reviews recommendations concerning facilities and site selection for depots, terminals, and railheads; advises superiors on present and future transportation capabilities and technical problems.

Recommendation, Skill Level 10

In the lower-division baccalaureate/associate degree category, 1 semester hour in traffic and transportation or in physical distribution management. (NOTE: This recommendation for skill level 10 is valid for the dates 1/90-9/91 only) (5/76).

Recommendation, Skill Level 20

In the lower-division baccalaureate/associate degree category, 2 semester hours in traffic and transportation or physical distribution management (5/76).

Recommendation, Skill Level 30

In the lower-division baccalaureate/associate degree category, 4 semester hours in traffic and transportation or in physical distribution management, 3 in human relations, 3 in personnel supervision, and additional credit in traffic and transportation or in physical distribution management on the basis of institutional evaluation (5/76).

Recommendation, Skill Level 40

In the lower-division baccalaureate/associate degree category, 6 semester hours in traffic and transportation or in physical distribution management, 3 in human relations, 3 in personnel supervision, 3 for field experience in management, and additional credit in traffic and transportation or in physical distribution management on the basis of institutional evaluation. In the upper-division baccalaureate category, 3 semester hours in introduction to management (5/76).

MOS-88N-002

Traffic Management Coordinator

88N10
88N20
88N30
88N40

Exhibit Dates: 11/92–2/95.

Career Management Field: 88 (Transportation).

Description

Summary: Coordinates the departure and arrival of freight and personnel by air, rail, highway, and water. *Skill Level 10:* Performs clerical duties associated with the logistics of freight and personnel movements; issues government and civilian shipping documents, including freight bills, bills of lading, and freight manifest sheets; assists military members in the preparation of travel itineraries, personal property shipping documents, and passenger movement forms; has a working knowledge of Lotus 1,2,3, Multimate, and Harvard Graphics. *Skill Level 20:* Able to perform the duties required for Skill Level 10; provides guidance to subordinates; assists in providing on-the-job training; advises passengers on the preparation and use of transportation documentation. *Skill Level 30:* Able to perform the duties required for Skill Level 20; supervises medium-scale freight movement office; plans, directs, and controls personnel; assists in the development of future transportation plans; supervises freight reconsignment, maintains liaison with military and civilian transportation facilities; reviews, consolidates, and prepares reports covering transportation movement operations. *Skill Level 40:* Able to perform the duties required for Skill Level 30; supervises personnel at a large freight movement office; when requested, reviews recommendations concerning facilities and site selection for depots, terminals, and railheads; advises superiors on present and future transportation capabilities and technical problems.

Recommendation, Skill Level 10

Credit may be granted on the basis of an individualized assessment of the student (3/94).

Recommendation, Skill Level 20

In the lower-division baccalaureate/associate degree category, 2 semester hours in traffic and transportation or physical distribution management and 3 in computer applications (2/93).

Recommendation, Skill Level 30

In the lower-division baccalaureate/associate degree category, 4 semester hours in traffic and transportation or in physical distribution management, 3 in personnel supervision, 3 in business communications, and 3 in computer applications (2/93).

Recommendation, Skill Level 40

In the lower-division baccalaureate/associate degree category, 6 semester hours in traffic and transportation or in physical distribution management, 3 in personnel supervision, 3 in business communications, and 3 in computer applications. In the upper-division baccalaureate category, 3 semester hours for field experience in management and 3 in organizational management (2/93).

MOS-88N-003

Traffic Management Coordinator

88N10
88N20
88N30
88N40

Exhibit Dates: 3/95–11/01.

Career Management Field: 88 (Transportation).

Description

Summary: Coordinates the departure and arrival of freight and personnel by air, rail, highway, and water. *Skill Level 10:* Performs clerical duties associated with the logistics of freight and personnel movements; issues government and civilian shipping documents, including freight bills, bills of lading, and freight manifest sheets; assists military members in the preparation of travel itineraries, personal property shipping documents, and passenger movement forms; has a working knowledge of Lotus 1,2,3, Multimate, and Harvard Graphics. *Skill Level 20:* Able to perform the duties required for Skill Level 10; provides guidance to subordinates; assists in providing on-the-job training; advises passengers on the preparation and use of transportation documentation. *Skill Level 30:* Able to perform the duties required for Skill Level 20; supervises medium-scale freight movement office; plans, directs, and controls personnel; assists in the development of future transportation plans; supervises freight reconsignment, maintains liaison with military and civilian transportation facilities; reviews, consolidates, and prepares reports covering transportation movement operations. *Skill Level 40:* Able to perform the duties required for Skill Level 30; supervises personnel at a large freight movement office; when requested, reviews recommendations concerning facilities and site selection for depots, terminals, and railheads; advises superiors on present and future transportation capabilities and technical problems.

Recommendation, Skill Level 10

Credit may be granted on the basis of an individualized assessment of the student (3/94).

Recommendation, Skill Level 20

Credit may be granted on the basis of an individualized assessment of the student (2/93).

Recommendation, Skill Level 30

In the lower-division baccalaureate/associate degree category, 2 semester hours in traffic and transportation or in physical distribution management, 3 in personnel supervision, 3 in business communications, and 3 in computer applications (2/93).

Recommendation, Skill Level 40

In the lower-division baccalaureate/associate degree category, 4 semester hours in traffic and transportation or in physical distribution management, 3 in personnel supervision, 3 in business communications, and 3 in computer applications. In the upper-division baccalaureate category, 3 semester hours for field experience in management and 3 in organizational management (2/93).

MOS-88N-004

Traffic Management Coordinator

88N10
88N20
88N30
88N40

Exhibit Dates: 12/01–Present.

Career Management Field: 88 (Transportation).

Description

Summary: Coordinates the departure and arrival of freight and personnel by air, rail, highway, and water. *Skill Level 10:* Performs clerical duties associated with the logistics of freight and personnel movements; issues government and civilian shipping documents, including freight bills, bills of lading, and freight manifest sheets; assists military members in the preparation of travel itineraries, personal property shipping documents, and passenger movement forms; has a working knowledge of electronic software programs. *Skill Level 20:* Able to perform the duties required for Skill Level 10; provides guidance to subordinates; assists in providing on-the-job training; advises passengers on the preparation and use of transportation documentation. *Skill Level 30:* Able to perform the duties required for Skill Level 20; supervises medium-scale freight movement office; plans, directs, and controls personnel; assists in the development of future transportation plans; supervises freight reconsignment; maintains liaison with military and civilian transportation facilities; reviews, consolidates, and prepares reports covering transportation movement operations. *Skill Level 40:* Able to perform the duties required for Skill Level 30; supervises personnel at a large freight movement office; when requested, reviews recommendations concerning facilities and site selection for depots, terminals, and railheads; advises superiors on present and future transportation capabilities and technical problems.

Recommendation, Skill Level 10

Credit may be granted on the basis of an individualized assessment of the student (12/01).

Recommendation, Skill Level 20

Credit may be granted on the basis of an individualized assessment of the student (12/01).

Recommendation, Skill Level 30

In the lower-division baccalaureate/associate degree category, 2 semester hours in traffic and transportation or 2 in physical distribution management, 3 in principles of supervision, 3 in business communication, and 3 in computer applications (12/01).

Recommendation, Skill Level 40

In the lower-division baccalaureate/associate degree category, 4 semester hours in traffic and transportation or in physical distribution management, 3 in principles of supervision, 3 in business communication, and 3 in computer applications. In the upper-division baccalaureate category, 3 semester hours for a management practicum and 3 in organizational management (12/01).

MOS-88P-001

Locomotive Repairer

88P10
88P20
88P30
88P40

Exhibit Dates: 1/90–1/94.

Career Management Field: 88 (Transportation).

Description

Summary: Supervises or maintains nonelectric sections of locomotives. *Skill Level 10:* Inspects locomotives to diagnose malfunctioning parts; accompanies locomotive operator on test runs and observes functions of locomotives; cleans and lubricates components as required; performs technical inspections on locomotives and auxiliary equipment to ensure proper maintenance has been performed; interprets technical drawings and sketches. *Skill Level 20:* Able to perform the duties required for Skill Level 10; provides technical guidance to subordinates. *Skill Level 30:* Able to perform the duties required for Skill Level 20; performs initial and final inspections on locomotives scheduled for maintenance; supervises and instructs repairers in methods and techniques of repairs; enforces safety practices; prepares technical reports on locomotives and estimates personnel, supplies, parts, and equipment necessary to restore locomotives, tanks, and tank cars to efficient operating condition; plans, organizes, and coordinates work activity of subordinates engaged in maintaining, inspecting, and repairing locomotives and auxiliary equipment. *Skill Level 40:* Able to perform the duties required for skill level 30 of MOS 88P or 88S (Locomotive Electrician); supervises or controls repairs of mechanical and electrical portions of locomotives.

Recommendation, Skill Level 10

In the vocational certificate category, 6 semester hours in locomotive maintenance. (NOTE: This recommendation for skill level 10 is valid for the dates 1/90-9/91 only) (5/76).

Recommendation, Skill Level 20

In the vocational certificate category, 8 semester hours in locomotive maintenance (5/76).

Recommendation, Skill Level 30

In the vocational certificate category, 8 semester hours in locomotive maintenance. In the lower-division baccalaureate/associate degree category, 9 semester hours in management and supervision of locomotive maintenance (5/76).

Recommendation, Skill Level 40

In the vocational certificate category, 8 semester hours in locomotive maintenance. In the lower-division baccalaureate/associate degree category, 9 semester hours in management and supervision of locomotive maintenance, 3 in personnel supervision, and 3 in industrial management (5/76).

MOS-88P-002

Railway Equipment Repairer (Reserve Components)

88P10
88P20
88P30
88P40

Exhibit Dates: Pending Evaluation. 2/94–Present.

MOS-88Q-001

Railway Car Repairer

88Q10
88Q20
88Q30
88Q40

Exhibit Dates: 1/90–6/94. Effective 6/94, MOS 88Q was discontinued, and its duties were incorporated into MOS 88P, Locomotive Repairer, Reserve Components.

Career Management Field: 88 (Transportation).

Description

Summary: Supervises or performs maintenance on railway passenger, freight, and hospital cars. *Skill Level 10:* Dismantles and reassembles major components and auxiliary equipment; examines wheels, axles, journals, journal bearings, and draft gear for structural and safety defects; inspects cars for defects which interfere with proper loading and makes necessary repairs; repairs hand brake mechanisms, gear ratchets, linkage, center sills, and body bolsters; straightens bent sections of bodies; paints and stencils railway cars; performs technical inspection of railway cars to ensure proper maintenance; interprets blueprints and sketches; assists in salvage and rehabilitation of cars. *Skill Level 20:* Able to perform the duties required for Skill Level 10; provides technical guidance to subordinates; inspects railway cars prior and subsequent to maintenance. *Skill Level 30:* Able to perform the duties required for skill level 20 of MOS 88Q or 88R (Airbrake Repairer); plans, organizes, coordinates, and supervises work activity of a railway car section team; instructs repairers in proper maintenance techniques and safety procedures; calculates personnel and material requirements for specific repair work; maintains records and prepares detailed technical reports. *Skill Level 40:* Able to perform the duties required for Skill Level 30; supervises a large facility (with a minimum of 39 subordinates) involved in repairing railway cars and airbrake systems of locomotives and cars.

Recommendation, Skill Level 10

In the vocational certificate category, 6 semester hours in practices and procedures of railroad car repair. (NOTE: This recommendation for skill level 10 is valid for the dates 1/90-9/91 only) (5/76).

Recommendation, Skill Level 20

In the vocational certificate category, 6 semester hours in practices and procedures of railroad car repair. In the lower-division baccalaureate/associate degree category, 1 semester hour in personnel supervision (5/76).

Recommendation, Skill Level 30

In the vocational certificate category, 6 semester hours in practices and procedures of railroad car repair. In the lower-division baccalaureate/associate degree category, 3 semester hours in human relations, 3 in personnel supervision, and 1 in technical writing (5/76).

Recommendation, Skill Level 40

In the vocational certificate category, 6 semester hours in practices and procedures of railroad car repair. In the lower-division baccalaureate/associate degree category, 3 semester hours in human relations, 3 in personnel supervision, 3 for field experience in management, and 1 in technical writing (5/76).

MOS-88R-001

Air brake Repairer

88R10
88R20

Exhibit Dates: 1/90–6/94. Effective 6/94, MOS 88R was discontinued, and its duties were incorporated into MOS 88P, Locomotive Repairer, Reserve Components.

Career Management Field: 88 (Transportation).

Description

Summary: Inspects, services, and repairs air brake systems of locomotives and cars. *Skill Level 10:* Inspects air brake components and system for defects; dismantles, cleans, and lubricates components; removes and replaces defective parts; tests and adjusts components for maximum efficiency of air brake system; makes appropriate entries in parts and log books; prepares written reports. *Skill Level 20:* Able to perform the duties required for Skill Level 10; provides technical guidance to subordinates.

Recommendation, Skill Level 10

In the vocational certificate category, 6 semester hours in inspecting, servicing, and repairing air brake systems. (NOTE: This recommendation for skill level 10 is valid for the dates 1/90-9/91 only) (5/76).

Recommendation, Skill Level 20

In the vocational certificate category, 8 semester hours in inspecting, servicing, and repairing air brake systems (5/76).

MOS-88S-001

Locomotive Electrician

88S10
88S20
88S30

Exhibit Dates: 1/90–6/94. Effective 6/94, MOS 88S was discontinued, and its duties were incorporated into MOS 88P, Locomotive Repairer, Reserve Components.

Career Management Field: 88 (Transportation).

Description

Summary: Supervises or repairs electrical portions of diesel-electric systems of diesel-electric locomotives. *Skill Level 10:* Inspects, dismounts, dismantles, cleans, repairs, and replaces defective components of electrical systems of diesel-electric locomotives, including generators, traction motors, auxiliary motors, reversers, contactors, relays, and regulators; load-tests repaired electrical components; restores commutator surfaces, rebands armatures, and repairs armature bearings; calibrates electric measuring instruments; reads schematic wiring diagrams. *Skill Level 20:* Able to perform the duties required for Skill Level 10; provides technical guidance to subordinates. *Skill Level 30:* Able to perform the duties required for Skill Level 20; supervises medium-sized activity involved in repairing electrical portions of electrical systems of diesel-electric locomotives; interprets complex wiring diagrams; prepares shop layouts; develops plans for utilization of personnel, materials, and equipment; conducts on-the-job training; requisitions tools, parts, and supplies.

Recommendation, Skill Level 10

In the vocational certificate category, 6 semester hours in repair of electrical systems. (NOTE: This recommendation for skill level 10 is valid for the dates 1/90-9/91 only) (5/76).

Recommendation, Skill Level 20

In the vocational certificate category, 8 semester hours in repair of electrical systems (5/76).

Recommendation, Skill Level 30

In the vocational certificate category, 8 semester hours in repair of electrical systems. In the lower-division baccalaureate/associate degree category, 3 semester hours in personnel supervision, 3 in shop management, and 3 in human relations (5/76).

MOS-88T-001

Railway Section Repairer

88T10
88T20
88T30
88T40

Exhibit Dates: 1/90–1/94.

Career Management Field: 88 (Transportation).

Description

Summary: Performs and/or supervises maintenance, repair, and cleanup of tracks, roadbeds, switches, and other railway facilities. *Skill Level 10:* Inspects, repairs, and maintains railway right of way and adjoining structures, including switches, signals, degree of elevation, track curvature, and tie spacing; controls the issuing of repair and maintenance equipment. *Skill Level 20:* Able to perform the duties required for Skill Level 10; determines need for repair and construction equipment and arranges for movement to job site; conducts on-the-job training in minor repairs, emphasizing safety to both train and work repair crews; supervises limited size maintenance, repair, and clean up crews. *Skill Level 30:* Able to perform the duties required for Skill Level 20; supervises medium-sized crews (normally 29-38 subordinates or 2 section gangs); adjusts work assignments to best reflect capabilities of subordinates; prepares reports relating to the maintenance, repairs, and cleanup of railway right of way facilities. *Skill Level 40:* Able to perform the duties required for Skill Level 30; supervises the maintenance, repair, and cleanup activities of a large railway facility, normally consisting of more than 39 subordinates.

Recommendation, Skill Level 10

In the vocational certificate category, 5 semester hours in railroad operations. (NOTE: This recommendation for skill level 10 is valid for the dates 1/90-9/91 only) (5/76).

Recommendation, Skill Level 20

In the vocational certificate category, 7 semester hours in railroad operations (5/76).

Recommendation, Skill Level 30

In the vocational certificate category, 7 semester hours in railroad operations. In the lower-division baccalaureate/associate degree category, 3 semester hours in human relations, 3 in personnel supervision, and 3 for field experience in management (5/76).

Recommendation, Skill Level 40

In the vocational certificate category, 7 semester hours in railroad operations. In the lower-division baccalaureate/associate degree category, 3 semester hours in human relations, 3 in personnel supervision, and 6 for field experience in management (5/76).

MOS-88T-002

Railway Section Repairer (Reserve Components)
88T10
88T20
88T30
88T40

Exhibit Dates: Pending Evaluation. 2/94–Present.

MOS-88U-001

Locomotive Operator
88U10
88U20
88U30
88U40

Exhibit Dates: 1/90–1/94.

Career Management Field: 88 (Transportation).

Description

Summary: Operates steam, electric, and diesel-electric locomotives and related equipment. *Skill Level 10:* Operates steam, electric, and diesel-electric locomotives and related equipment; fires and sustains steam pressure by either hand or stoker firing of coal-burning locomotives; assists in performing the duties required for skill level 20. *Skill Level 20:* Able to perform the duties required for Skill Level 10; interprets specific operating instructions; efficiently operates locomotives and trains, observing appropriate gauges and meters; lubricates moving parts of locomotive; interprets and executes operating instructions; inspects equipment for proper operating condition; interprets train orders; compiles performance and delay reports for each trip; submits locomotive inspection reports noting corrective action needed for defects. *Skill Level 30:* Able to perform the duties required for Skill Level 20; reviews operation and inspection reports; investigates schedule delays, accidents, and unusual operating incidents and recommends corrective action; assigns duties to locomotive operators and other crew and monitors their performance; provides on-the-job training; enforces safety procedures; coordinates maintenance of locomotives; prepares technical, personnel, and administrative reports. *Skill Level 40:* Able to perform the duties required for Skill Level 30; supervises operation of steam, electric, and diesel-electric locomotives and related equipment.

Recommendation, Skill Level 10

In the vocational certificate category, 3 semester hours in steam, electric, and diesel-electric mechanics, 3 in operation and maintenance of locomotive engines, and 3 in railway safety. (NOTE: This recommendation for skill level 10 is valid for the dates 1/90-9/91 only) (5/76).

Recommendation, Skill Level 20

In the vocational certificate category, 3 semester hours in steam, electric, and diesel-electric mechanics, 3 in operation and maintenance of locomotive engines, and 3 in railway safety (5/76).

Recommendation, Skill Level 30

In the vocational certificate category, 9 semester hours in operation and maintenance of locomotive engines, 3 in steam, electric, and diesel mechanics, and 3 in railway safety. In the lower-division baccalaureate/associate degree category, 3 semester hours in human relations, 3 in personnel supervision, 3 in report writing, and 1 in safety (5/76).

Recommendation, Skill Level 40

In the vocational certificate category, 9 semester hours in operation and maintenance of locomotive engines, 3 in steam, electric, and diesel mechanics, and 3 in railway safety. In the lower-division baccalaureate/associate degree category, 3 semester hours in human relations, 3 in personnel supervision, 3 for field experience in management, 3 in report writing, and 1 in safety (5/76).

MOS-88U-002

Railway Operations Crewmember
88U10
88U20
88U30
88U40

Exhibit Dates: Pending Evaluation. 2/94–Present.

MOS-88V-001

Train Crew Member
88V10
88V20
88V30
88V40

Exhibit Dates: 1/90–6/94. Effective 6/94, MOS 88V was discontinued, and its duties were incorporated into MOS 88U, Railway Operations Crewmember.

Career Management Field: 88 (Transportation).

Description

Summary: Serves as yardmaster, conductor, or brakeman in the make-up and movement of railway cars and trains. *Skill Level 10:* Performs as a brakeman; receives orders for switching and performs car coupling and uncoupling operations; applies safety rules; gives and interprets railway signals for train control and operation; places placarded cars in trains and ensures that cars are properly loaded and lashed for safe handling; prepares and interprets train orders; prepares reports and forms for yard and railway operations; inspects rolling stock to determine operating condition. *Skill Level 20:* Able to perform the duties required for Skill Level 10; serves as a conductor; supervises the operation of train and coordinates train movements; reviews and interprets train orders and timetables; instructs and supervises enginemen and brakemen in connecting and disconnecting trains and cars and switching cars to sidings or spurs. *Skill Level 30:* Able to perform the duties required for Skill Level 20; serves as an assistant yardmaster; supervises and trains personnel in railway yard operation, including the handling, classification, and switching of trains and cars; maintains record of daily movement of railway cars; assures adherence to railway safety procedures; reviews, consolidates, and prepares technical, personnel, and administrative reports covering railway yard activities. *Skill Level 40:* Able to perform the duties required for Skill Level 30; supervises railway yard operations; normally supervises two skill level 30 yardmasters.

Recommendation, Skill Level 10

In the vocational certificate category, 7 semester hours in the make-up and movement of railway cars and trains. (NOTE: This recommendation for skill level 10 is valid for the dates 1/90-9/91 only) (5/76).

Recommendation, Skill Level 20

In the vocational certificate category, 10 semester hours in the make-up and movement of railway cars and trains. In the lower-division baccalaureate/associate degree category, 1 semester hour in personnel supervision (5/76).

Recommendation, Skill Level 30

In the vocational certificate category, 13 semester hours in the make-up and movement of railway cars and trains. In the lower-division baccalaureate/associate degree category, 3 semester hours in personnel supervision and 3 in human relations (5/76).

Recommendation, Skill Level 40

In the vocational certificate category, 13 semester hours in the make-up and movement of railway cars and trains. In the lower-division baccalaureate/associate degree category, 3 semester hours in personnel supervision, 3 in human relations, and 6 for field experience in management (5/76).

MOS-88W-001

Railway Movement Coordinator
88W10
88W20
88W30
88W40

Exhibit Dates: 1/90–6/94. Effective 6/94, MOS 88W was discontinued, and its duties were incorporated into MOS 88U, Railway Operations Crewmember.

Career Management Field: 88 (Transportation).

Description

Summary: Supervises train dispatching or operates railway stations, railway signals, and switches; assists in control and coordination of train movements. *Skill Level 10:* Accounts for car and train movement by compiling records reflecting locations of all rolling stock; keeps records of arriving and departing trains; computes operational and statistical reports on traffic volume, demurrage, expediting, and train movements; requisitions cars; supervises loading; compiles and checks bills of lading (waybills); seals cars; receives requisitions for empty cars and directs movement to loading areas; establishes liaison with other operating personnel and civilian carrier operating personnel to ensure safe and prompt handling of freight. *Skill Level 20:* Able to perform the duties required for Skill Level 10; provides guidance to subordinates. *Skill Level 30:* Able to perform the duties required for Skill Level 20; supervises receipt and transmission of messages relating to train movements; reviews transit time; coordinates and arranges for train crews; establishes close liaison with representatives of civilian and military railroads; supervises on-the-job training. *Skill Level 40:* Able to perform the duties required for Skill Level 30; provides staff supervision to subordinates in railway operating activities.

Recommendation, Skill Level 10

In the lower-division baccalaureate/associate degree category, 1 semester hour in traffic and transportation or in physical distribution management. (NOTE: This recommendation for skill level 10 is valid for the dates 1/90-9/91 only) (5/76).

Recommendation, Skill Level 20

In the lower-division baccalaureate/associate degree category, 2 semester hours in traffic and transportation or in physical distribution management (5/76).

Recommendation, Skill Level 30

In the lower-division baccalaureate/associate degree category, 6 semester hours in traffic and transportation or in physical distribution management (5/76).

Recommendation, Skill Level 40

In the lower-division baccalaureate/associate degree category, 6 semester hours in traffic and transportation or in physical distribution management, 3 in personnel supervision, and 3 in human relations (5/76).

MOS-88X-001

Railway Senior Sergeant
88X50

Exhibit Dates: 1/90–Present.

Career Management Field: 88 (Transportation).

Description

Supervises and coordinates maintenance of way and maintenance of equipment and railway operations, including the overall supervision of 88P40 (Locomotive Repairer), 88Q40 (Railway Car Repairer), 88T40 (Railway Section Repairer), 88U40 (Locomotive Operator), 88V40 (Train Crew Member), and 88W40 (Railway Movement Coordinator) personnel; advises superiors on all matters relating to operations, maintenance, and repair; monitors and controls railway passenger and freight traffic; prepares operational reports and training material.

Recommendation

In the upper-division baccalaureate category, 3 semester hours in records administration, 3 in personnel management, 3 in human relations, 3 for field experience in management, 3 in maintenance management, 3 in introduction to management, and 6 in transportation management (5/76).

MOS-88Y-001

Marine Senior Sergeant
88Y50

Exhibit Dates: 1/90–10/92.

Career Management Field: 88 (Transportation).

Description

Is a mid-level manager responsible for approximately 150 subordinates engaged in watercraft operation and maintenance; assists in planning, implementing, and coordinating administrative activities, operations, and training programs; prepares, edits, and consolidates reports, records, plans, and training materials. NOTE: May have progressed to 88Y50 from 88K40 (Watercraft Operator) or 88L40 (Watercraft Engineer).

Recommendation

In the upper-division baccalaureate category, 6 semester hours for field experience in management, 3 in introduction to management, 3 in personnel management, and 3 in records administration; if the duty assignment was chief instructor, 3 additional semester hours for an internship in education (6/76).

MOS-88Y-002

Marine Senior Sergeant
88Y50

Exhibit Dates: 11/92–12/95. Effective 12/95, MOS 88Y was discontinued and its duties were incorporated into MOS 88K, Watercraft Operator, and MOS 88Z, Transportation Senior Sergeant.

Career Management Field: 88 (Transportation).

Description

Manages 80-150 subordinates engaged in watercraft operation and maintenance; assists in planning, implementing, and coordinating administrative activities, operations, and training programs; prepares, edits, and consolidates reports, records, plans, and training materials. NOTE: May have progressed to 88Y50 from 88K40 (watercraft Operator) or 88L40 (Watercraft Engineer).

Recommendation

In the lower-division baccalaureate/associate degree category, 3 semester hours in computer applications, 3 in records and information management, 3 in personnel supervision, and 3 in report writing. In the upper-division baccalaureate category, 3 semester hours for field experience in management, 3 in organizational management, and 3 in human resource management; if paygrade E-9 has been attained, additional credit may be granted as follows: 3 semester hours in management problems and 3 in communication techniques for managers (11/92).

MOS-88Z-001

Transportation Senior Sergeant
88Z50

Exhibit Dates: 1/90–10/92.

Career Management Field: 88 (Transportation).

Description

Supervises operation and control of movement of personnel and cargo by air, rail, motor transport, and water; prepares tactical plans and training material; evaluates transportation operations; assists in planning, coordinating, and supervising all supporting activities; prepares studies, routine and special reports, and records on transportation operations; supervises personnel and technical operations. NOTE: May have progressed to 88Z50 from 88M40 (Motor Transport Operator), 88N40 (Traffic Management Coordinator), or 88H40 (Cargo Specialist).

Recommendation

In the lower-division baccalaureate/associate degree category, 9 semester hours in transportation operations, 3 in records administration, and 3 in report writing. In the upper-division baccalaureate category, 6 semester hours for field experience in management, 3 in introduction to management, and 3 for field experience in education (5/76).

MOS-88Z-002

Transportation Senior Sergeant
88Z50

Exhibit Dates: 11/92–11/01.

Career Management Field: 88 (Transportation).

Description

Supervises operation and control of movement of personnel and cargo by air, rail, motor transport, and water; prepares and distributes reports of operations; evaluates transportation operations; manages technical operations and supervises management coordinators or terminal operations coordinators; assists in providing staff supervision, policy, and guidance relating to movement of both personnel and cargo; assists in coordinating operations, training, administration, and communications activity; may also serve as operations sergeant, transportation supervisor, first sergeant, or sergeant major; serves more as a manager than as a supervisor in human resources management. NOTE: May have progressed to 88Z50 from 88M40 (Motor Transport Operator), 88N40 (Traffic Management Coordinator), or 88H40 (Cargo Specialist).

Recommendation

In the lower-division baccalaureate/associate degree category, 6 semester hours in transportation operations, 3 in records and information management, 3 in personnel supervision, and 3 in report writing. In the upper-division baccalaureate category, 3 semester hours for field experience in management, 3 in organizational management, 3 in human resource management, and 3 in transportation management; if paygrade E-9 has been attained, additional credit may be granted as follows: 3 semester hours in management problems and 3 in communication techniques for managers (11/92).

MOS-88Z-003

Transportation Senior Sergeant
88Z50

Exhibit Dates: 12/01–Present.

Career Management Field: 88 (Transportation).

Description

Supervises operation and control of movement of personnel and cargo by air, rail, motor transport, and water; prepares and distributes reports of operations; evaluates transportation operations; manages technical operations and supervises management coordinators or terminal operations coordinators; assists in providing staff supervision, policy, and guidance relating to movement of both personnel and cargo; assists in coordinating operations, training, administration, and communications activity;

may also serve as operations sergeant, transportation supervisor, first sergeant, or sergeant major; serves more as a manager than as a supervisor in human resources management. NOTE: May have progressed to 88Z50 from 88M40 (Motor Transport Operator), 88N40 (Traffic Management Coordinator), or 88H40 (Cargo Specialist).

Recommendation

In the lower-division baccalaureate/associate degree category, 6 semester hours in transportation operations, 3 in records and information management, 3 in principles of supervision, 3 in computer applications, and 3 in business communications. In the upper-division baccalaureate category, 3 semester hours for a management practicum, 3 in organizational management, 3 in human resource management, and 3 in transportation management (12/01).

MOS-89B-001

Ammunition Specialist

89B10
89B20
89B30
89B40

Exhibit Dates: Pending Evaluation. 4/04–Present.

MOS-89D-001

Explosive Ordnance Disposal Specialist

89D10
89D20
89D30
89D40

Exhibit Dates: Pending Evaluation. 4/04–Present.

MOS-91A-002

Medical Specialist

91A10
91A20

Exhibit Dates: 1/90–4/92. Effective 4/92, MOS 91A was discontinued, and its duties were transferred into MOS 91B, Medical Specialist.

Career Management Field: 91 (Medical).

Description

Summary: Administers emergency medical treatment to battlefield casualties; assists with outpatient care and treatment; assists with inpatient care and treatment under supervision of physician, nurse, physician's assistant, or noncommissioned officer. *Skill Level 10:* Administers emergency treatment in the field; assists with outpatient and inpatient care; maintains health records and clinical files; assists with nursing care of patients including medical examinations; takes and records temperature, pulse, respiration, and blood pressure; applies and removes surgical, wound, or skin dressings; collects and prepares specimens for analysis; assists in clinic and dispensary; performs routine admission tests; administers immunizations; assists with treatment of patients with common diseases; provides emergency medical care; surveys and sorts casualties; determines requirements for and administers emergency treatment; assists with triage of mass casualties; performs duties related to emergency care. *Skill Level 20:* Able to perform the duties required for Skill Level 10; provides technical guidance to subordinates; manages casualties until admission to medical facility; performs sorting and screening procedures on battlefield or in dispensaries or clinics; reads and interprets medical tags or records.

Recommendation, Skill Level 10

In the lower-division baccalaureate/associate degree category, 3 semester hours in fundamentals of nursing, 3 in emergency nursing procedures, 3 in first aid, and 3 for an internship in emergency nursing procedures. (NOTE: This recommendation for skill level 10 is valid for the dates 1/90-9/91 only) (9/85).

Recommendation, Skill Level 20

In the lower-division baccalaureate/associate degree category, 3 semester hours in fundamentals of nursing, 3 in emergency nursing procedures, 3 in first aid, 6 for an internship in emergency nursing procedures, and 1 in principles of administration (9/85).

MOS-91A-003

Medical Equipment Repairer

91A10
91A20
91A30
91A40
91A50

Exhibit Dates: 8/94–Present.

Description

Summary: Provides routine and periodic scheduled maintenance and repairs on all types of medical equipment; applies mechanical, hydraulic, high and low-pressure gas and steam, electrical, electronic, solid-state, digital, logic, radiological, and optical principles; supervises biomedical equipment repair functions. *Skill Level 10:* Must be able to perform preventive maintenance checks, clean, lubricate, calibrate, repair, and test all types of biomedical equipment, including defibrillators, patient monitoring systems, pulmonary equipment, electroencephalonary equipment, diathermy systems, diathermy apparatus, spectrophotometers, ultrasonic therapy and treatment apparatus, dental operating units, anesthesia apparatus, resuscitators, operating tables and lamps, hospital beds, etc.; dismantles and cleans equipment; installs water, air, and steam lines; performs routine tests on medical equipment to determine leakage currents and calibrations. *Skill Level 20:* Able to perform the duties required for Skill Level 10; provides technical guidance to subordinates; advises and assists equipment operators in the assembly and disassembly of field medical equipment; troubleshoots, performs, and supervises unit maintenance activities; prepares medical equipment reports and maintenance schedules. *Skill Level 30:* Able to perform the duties required for Skill Level 20; applies basic knowledge of microprocessors, analog-to-digital conversion, digital-to-analog conversion, and diagnostics of microprocessor circuits; develops and establishes procedures for operation of biomedical equipment maintenance activity; establishes and maintains procedures to ensure effective preventive maintenance program; ensures that diagnostic and measuring equipment is calibrated in accordance with Army calibration program; determines personnel requirements, organizes work schedules, assigns duties, monitors personnel performance, and counsels subordinates; prepares evaluation reports; ensures adherence to standards of conduct, cleanliness, technical accuracy, and safety regulations in all areas of activity; advises and assists in administrative, fiscal, personnel, and supply matters; advises commander on specific medical systems including requirements for utilities and advantages and disadvantages of contract versus in-house maintenance; advises procurement personnel of functional and safety aspects of medical equipment and systems. *Skill Level 40:* Able to perform the duties required for Skill Level 30; supervises organizational, direct, and general support maintenance of biomedical equipment; organizes and supervises inspection and maintenance teams. *Skill Level 50:* Able to perform the duties required for Skill Level 40; supervises medical equipment maintenance operations; writes, develops, and coordinates command-wide regulations and policies related to logistical material maintenance programs; provides technical information and guidance to subordinate units; determines need for and schedules personnel for attendance at specialized manufacturers' courses; performs as team chief for medical maintenance assistance and instruction teams.

Recommendation, Skill Level 10

Credit may be granted on the basis of an individualized assessment of the student (4/98).

Recommendation, Skill Level 20

In the lower-division baccalaureate/associate degree category, 3 semester hours in DC circuits, 3 in solid state electronics, 3 in AC circuits, 3 in electronic instrumentation, 3 in mechanical and electromechanical controls, 3 in pneumatic and hydraulic controls, 3 in applications of sensors, 3 in digital principles, 3 in analog and linear circuit principles, 3 in electronic equipment diagnostics and repair, and 2 in personnel supervision. (NOTE: This recommendation for skill level 20 is valid for the dates 9/94-2/95 only) (4/98).

Recommendation, Skill Level 30

In the lower-division baccalaureate/associate degree category, 3 semester hours in DC circuits, 3 in solid state electronics, 3 in AC circuits, 3 in electronic instrumentation, 3 in mechanical and electromechanical controls, 3 in pneumatic and hydraulic controls, 3 in applications of sensors, 3 in digital principles, 3 in analog and linear circuit principles, 3 in electronic equipment diagnostics and repair, 3 in personnel supervision, and 2 in records information and management (4/98).

Recommendation, Skill Level 40

In the lower-division baccalaureate/associate degree category, 3 semester hours in DC circuits, 3 in solid state electronics, 3 in AC circuits, 3 in electronic instrumentation, 3 in mechanical and electromechanical controls, 3 in pneumatic and hydraulic controls, 3 in applications of sensors, 3 in digital principles, 3 in analog and linear circuit principles, 3 in electronic equipment diagnostics and repair, 3 in personnel supervision, and 2 in records information and management. In the upper-division baccalaureate category, 3 semester hours in organizational management and 3 for field experience in management (4/98).

Recommendation, Skill Level 50

In the lower-division baccalaureate/associate degree category, 3 semester hours in DC circuits, 3 in solid state electronics, 3 in AC circuits, 3 in electronic instrumentation, 3 in mechanical and electromechanical controls, 3 in pneumatic and hydraulic controls, 3 in applications of sensors, 3 in digital principles, 3 in analog and linear circuit principles, 3 in electronic equipment diagnostics and repair, 3 in personnel supervision, and 2 in records information and management. In the upper-division baccalaureate category, 3 semester hours in organizational management and 3 for field experience in management; if individual has attained paygrade E-9, additional credit as follows: 3 semester hours in management problems, 3 in operations management, and 3 in communication techniques for managers (4/98).

MOS-91B-004

MEDICAL NCO
91B20
91B30
91B40
91B50

Exhibit Dates: 1/90–7/91.

Career Management Field: 91 (Medical).

Description

Summary: Supervises field and clinical medical facilities; assists with technical and administrative management of medical treatment facilities under supervision of physician, nurse, or physician's assistants; administers emergency and routine outpatient medical treatment to battle and nonbattle casualties; assists with outpatient care and treatment. *Skill Level 20:* Able to perform the duties of 91A10 or 91A20, Medical Specialist; administers emergency and routine medical treatment to casualties including minor surgery and IV fluid administration; assists with outpatient care and treatment and supervises outpatient facilities under the supervision of a physician, nurse, or physician's assistant; establishes priorities for medical emergency care; stabilizes patients and accompanies them to medical facilities; provides guidance and supervision to subordinates. NOTE: May have progressed to 91B20 from 91A10 or 91A20, Medical Specialist. *Skill Level 30:* Able to perform the duties required for skill level 20 of MOS 91B or MOS 91A, Medical Specialist; supervises activities in dispensaries, large clinics, and field medical services; coordinates activities of clinic with medical treatment facility; supervises ordering of supplies; determines personnel requirements and conducts training programs. NOTE: May have progressed to 91B30 from 91B20 or 91A20, Medical Specialist. *Skill Level 40:* Able to perform the duties required for Skill Level 30; supervises paraprofessional medical service activities in large fixed and mobile treatment facilities; coordinates activities of wards, clinics, and combined medical care facilities; maintains intelligence information and records. *Skill Level 50:* Able to perform the duties required for skill level 40 of MOS 91B and MOS 42C, Orthotic Specialist, and skill level 50 of MOS 91C, Practical Nurse; serves as the principal noncommissioned officer of staff sections, hospitals, other medical facilities, or service teams; serves on special boards and councils incident to medical service activities; supervises general administrative functions in a medical or treatment facility; supervises activities within the Army medical department or multifunctional medical staff activities; assists command surgeon in technical supervision of subordinate medical activities; evaluates personnel and the operational effectiveness of medical facilities; advises superiors on medical administrative matters; keeps official records and patient files; establishes report control systems; prepares periodic and special reports concerning assigned personnel, patients, and medical care and treatment; makes recommendations for improving working procedures and conditions; evaluates training programs and requirements; determines requirements and adequacy of medical service supporting elements such as ambulance, supply, and transportation. NOTE: May have progressed to 91B50 from 91B40, 42C40 (Orthotic Specialist), or 91C50 (Practical Nurse).

Recommendation, Skill Level 20

In the lower-division baccalaureate/associate degree category, 3 semester hours in fundamentals of nursing, 6 in emergency nursing procedures, 3 in first aid, 3 for an internship in emergency nursing procedures, and 3 in personnel supervision (9/85).

Recommendation, Skill Level 30

In the lower-division baccalaureate/associate degree category, 3 semester hours in fundamentals of nursing, 6 in emergency nursing procedures, 3 in first aid, 3 for an internship in emergency nursing procedures, 3 in personnel supervision, 3 in supply management, and 1 in methods of instruction (9/85).

Recommendation, Skill Level 40

In the lower-division baccalaureate/associate degree category, 3 semester hours in fundamentals of nursing, 6 in emergency nursing procedures, 3 in first aid, 3 for an internship in emergency nursing procedures, 3 in personnel supervision, 3 in supply management, and 1 in methods of instruction. In the upper-division baccalaureate category, 3 semester hours for field experience in management and 3 in patient care administration (9/85).

Recommendation, Skill Level 50

In the lower-division baccalaureate/associate degree category, 3 semester hours in fundamentals of nursing, 6 in emergency nursing procedures, 3 in first aid, 3 for an internship in emergency nursing procedures, 3 in personnel supervision, 3 in supply management, and 1 in methods of instruction. In the upper-division baccalaureate category, 3 semester hours for field experience in management, 3 in patient care administration, 3 in personnel management, and 3 in hospital administration (9/85).

MOS-91B-005

MEDICAL SPECIALIST
91B10
91B20
91B30
91B40
91B50

Exhibit Dates: 8/91–7/94.

Career Management Field: 91 (Medical).

Description

Summary: Supervises field and clinical medical facilities; assists with technical and administrative management of medical treatment facilities under supervision of physician, nurse, or physician's assistant; administers emergency and routine outpatient medical treatment to battle and nonbattle casualties; assists with outpatient care and treatment. *Skill Level 10:* Administers emergency treatment in the field; assists with outpatient and inpatient care; maintains health records and clinical files; assists with nursing care of patients including medical examinations; takes and records temperature, pulse, respiration, and blood pressure; applies and removes surgical, wound, or skin dressings; collects and prepares specimens for analysis; assists in clinic and dispensary; performs routine admission tests; administers immunizations; assists with treatment of patients with common diseases; provides emergency medical care; surveys and sorts casualties; determines requirements for and administers emergency treatment; assists with triage of mass casualties; performs duties related to emergency care. *Skill Level 20:* Able to perform the duties required for Skill Level 10; administers emergency and routine medical treatment to casualties including minor surgery and IV fluid administration; assists with outpatient care and supervises outpatient facilities under the supervision of a physician, nurse, or physician's assistant; establishes priorities for medical emergency care; stabilizes patients and accompanies them to medical facilities; provides guidance and supervision to subordinates. *Skill Level 30:* Able to perform the duties required for Skill Level 20; supervises activities in dispensaries, large clinics, and field medical services; coordinates activities of clinic with medical treatment facility; supervises ordering of supplies; determines personnel requirements; conducts training programs. *Skill Level 40:* Able to perform the duties required for Skill Level 30; supervises paraprofessional medical service activities in large fixed and mobile treatment facilities; coordinates activities of wards, clinics, and combined medical care facilities; maintains intelligence information and records. *Skill Level 50:* Able to perform the duties required for Skill Level 40; serves as the principal noncommissioned officer of staff sections, hospitals, other medical facilities or service teams; serves on special boards and councils incident to medical service activities; supervises general administrative functions in a medical or treatment facility; supervises activities within the Army medical department or multifunctional medical staff activities; assists command surgeon in technical supervision of subordinate activities; evaluates personnel and the operational effectiveness of medical facilities; advises superiors on medical administrative matters; keeps official records and patient files; establishes report control systems; prepares periodic and special reports concerning assigned personnel, patients, and medical care and treatment; makes recommendations for improving working procedures and conditions; evaluates training programs and requirements; determines requirements and adequacy of medical service supporting elements such as ambulance, supply, and transportation.

Recommendation, Skill Level 10

In the lower-division baccalaureate/associate degree category, 3 semester hours in fundamentals of nursing, 3 in basic medical emergency procedures, and 3 for field experience in medical emergency procedures. (NOTE: This recommendation for skill level 10 is valid for the dates 8/91-9/91 only) (6/92).

Recommendation, Skill Level 20

In the lower-division baccalaureate/associate degree category, 3 semester hours in fundamentals of nursing, 3 in basic medical emergency procedures, 4 for field experience in medical emergency procedures, and 3 in personnel supervision (6/92).

Recommendation, Skill Level 30

In the lower-division baccalaureate/associate degree category, 3 semester hours in fundamentals of nursing, 3 in basic medical emergency procedures, 5 for field experience in medical emergency procedures, 3 in personnel supervision, and 2 in records and information management (6/92).

Recommendation, Skill Level 40

In the lower-division baccalaureate/associate degree category, 3 semester hours in fundamentals of nursing, 3 in basic medical emergency procedures, 5 for field experience in medical emergency procedures, 3 in personnel supervision, and 2 in records and information management. In the upper-division baccalaureate category, 3 semester hours in organizational management and 3 for field experience in management (6/92).

Recommendation, Skill Level 50

In the lower-division baccalaureate/associate degree category, 3 semester hours in fundamentals of nursing, 3 in basic medical emergency procedures, 5 for field experience in medical emergency procedures, 3 in personnel supervision, and 2 in records and information management. In the upper-division baccalaureate category, 3 semester hours in organizational management and 6 for field experience in management; if paygrade E-9 has been attained, additional credit as follows: 3 semester hours in management problems, 3 in operations management, and 3 in communication techniques for managers (6/92).

MOS-91B-006

MEDICAL SPECIALIST

91B10
91B20
91B30
91B40
91B50

Exhibit Dates: 8/94–9/01. Effective 10/01, MOS 91B was discontinued and its duties were incorporated into MOS 91W, Health Care Specialist.

Career Management Field: 91 (Medical).

Description

Summary: Supervises field and clinical medical facilities; assists with technical and administrative management of medical treatment facilities under supervision of physician, nurse, or physician's assistant; administers emergency and routine outpatient medical treatment to battle and nonbattle casualties; assists with outpatient care and treatment. *Skill Level 10:* Administers emergency treatment in the field; assists with outpatient and inpatient care; maintains health records and clinical files; assists with nursing care of patients including medical examinations; takes and records temperature, pulse, respiration, and blood pressure; applies and removes surgical, wound, or skin dressings; collects and prepares specimens for analysis; assists in clinic and dispensary; performs routine admission tests; administers immunizations; assists with treatment of patients with common diseases; provides emergency medical care; surveys and sorts casualties; determines requirements for and administers emergency treatment; assists with triage of mass casualties; performs duties related to emergency care. *Skill Level 20:* Able to perform the duties required for Skill Level 10; administers emergency and routine medical treatment to casualties, including minor surgery and IV fluid administration; assists with outpatient care and supervises outpatient facilities under the supervision of a physician, nurse, or physician's assistant; establishes priorities for medical emergency care; stabilizes patients and accompanies them to medical facilities; provides guidance and supervision to subordinates. *Skill Level 30:* Able to perform the duties required for Skill Level 20; supervises activities in dispensaries, large clinics, and field medical services; coordinates activities of clinic with medical treatment facility; supervises ordering of supplies; determines personnel requirements; conducts training programs. *Skill Level 40:* Able to perform the duties required for Skill Level 30; supervises paraprofessional medical service activities in large fixed and mobile treatment facilities; coordinates activities of wards, clinics, and combined medical care facilities; maintains intelligence information and records. *Skill Level 50:* Able to perform the duties required for Skill Level 40; serves as the principal noncommissioned officer of staff sections, hospitals, other medical facilities, or service teams; serves on special boards and councils incident to medical service activities; supervises general administrative functions in a medical or treatment facility; supervises activities within the Army medical department or multifunctional medical staff activities; assists command surgeon in technical supervision of subordinate activities; evaluates personnel and the operational effectiveness of medical facilities; advises superiors on medical administrative matters; keeps official records and patient files; establishes report control systems; prepares periodic and special reports concerning assigned personnel, patients, and medical care and treatment; makes recommendations for improving working procedures and conditions; evaluates training programs and requirements; determines requirements and adequacy of medical service supporting elements such as ambulance, supply, and transportation.

Recommendation, Skill Level 10

Credit may be granted on the basis of an individualized assessment of the student (11/96).

Recommendation, Skill Level 20

In the lower-division baccalaureate/associate degree category, 3 semester hours in fundamentals of nursing, 3 in basic medical emergency procedures, 3 for field experience in medical emergency procedures, 5 in clinical experience, and 3 in personnel supervision. (NOTE: This recommendation for skill level 20 is valid for the dates 8/94-2/95 only) (11/96).

Recommendation, Skill Level 30

In the lower-division baccalaureate/associate degree category, 3 semester hours in fundamentals of nursing, 3 in basic medical emergency procedures, 3 for field experience in medical emergency procedures, 10 in clinical experience, 3 in personnel supervision, and 2 in records and information management (11/96).

Recommendation, Skill Level 40

In the lower-division baccalaureate/associate degree category, 3 semester hours in fundamentals of nursing, 3 in basic medical emergency procedures, 3 for field experience in medical emergency procedures, 10 in clinical experience, 3 in personnel supervision, and 2 in records and information management. In the upper-division baccalaureate category, 3 semester hours in organizational management and 3 for field experience in management (11/96).

Recommendation, Skill Level 50

In the lower-division baccalaureate/associate degree category, 3 semester hours in fundamentals of nursing, 3 in basic medical emergency procedures, 3 for field experience in medical emergency procedures, 10 in clinical experience, 3 in personnel supervision, and 2 in records and information management. In the upper-division baccalaureate category, 3 semester hours in organizational management and 6 for field experience in management (11/96).

MOS-91C-004

PRACTICAL NURSE

91C20
91C30
91C40
91C50

Exhibit Dates: 1/90–6/99.

Career Management Field: 91 (Medical).

Description

Summary: Supervises or performs preventive therapeutic and emergency nursing care under the supervision of a physician or registered nurse. *Skill Level 20:* Able to perform the duties required for skill level 10 or skill level 20 of MOS 91A (Medical Specialist); assists with all levels of patient administration; prepares and maintains clinical records; reads patient charts and diagnostic reports and makes entries in medical records; requisitions, receives, packs, unpacks, stores, and safeguards clinical equipment and supplies; operates clinical equipment and maintains records; ensures maximum patient hygiene and safety; assists with providing emergency medical treatment and management; interprets and carries out nursing care plans; cares for patients of all ages with common diseases and minor injuries and those who are seriously, chronically, and acutely ill; assists during labor and childbirth; assesses condition of newborns; assists and teaches mothers the principles of infant care; prepares and administers prescribed medication; obtains and prepares samples and specimens for laboratory cultures and analysis; administers intravenous fluids; performs traction care; prepares patients for surgery and assists with transportation; assists with maintenance of nosograstic tubes; cares for patients with asthma, pneumonia, or bronchitis with oxygen therapy, bladder catheterization, colostomy irrigation, and intravenous therapy procedures; carefully observes patients for signs of problems such as dehydration, mental stress, and life-threatening symptoms; obtains appropriate admission data and assists with patient transfer and discharge. This specialty has been approved by the state of Texas as a practical nursing program and graduates must pass the state practical nursing exam to practice. NOTE: May have progressed to 91C20 from 91A10 or 91A20 (Medical Specialist). *Skill Level 30:* Able to perform the duties required for Skill Level 20; performs advanced nursing procedures and provides technical guidance and advice to subordinates; manages small wards and provides nursing care leadership. *Skill Level 40:* Able to perform the duties required for Skill Level 30; manages large or multiward units; prepares nursing care plans and supervises implementation of these plans; conducts in-service programs. *Skill Level 50:*

Able to perform the duties required for Skill Level 40; advises and assists professional staff in personnel matters; develops and evaluates training programs; manages supplies and equipment stores; establishes priorities, distributes work load and assigns personnel; compiles management data; inspects organizational activities, observes discrepancies, and initiates corrective action.

Recommendation, Skill Level 20

In the lower-division baccalaureate/associate degree category, 12 semester hours for clinical nursing experience, 8 in medical-surgical nursing, 4 in anatomy and physiology, 3 in maternal and child nursing, 3 in environmental health, 2 in theory and practice of patient care, 2 in pharmacology, 1 in nutrition, 1 in preventive medicine, and 1 in psychiatric nursing. (NOTE: This recommendation for skill level 20 is valid for the dates 1/90-2/95 only) (9/85).

Recommendation, Skill Level 30

In the lower-division baccalaureate/associate degree category, 12 semester hours for clinical nursing experience, 8 in medical-surgical nursing, 4 in anatomy and physiology, 3 in maternal and child nursing, 3 in environmental health, 2 in theory and practice of patient care, 2 in pharmacology, 1 in nutrition, 1 in preventive medicine, 1 in psychiatric nursing, 3 in patient care, and 3 in introduction to management (9/85).

Recommendation, Skill Level 40

In the lower-division baccalaureate/associate degree category, 12 semester hours for clinical nursing experience, 8 in medical-surgical nursing, 4 in anatomy and physiology, 3 in maternal and child nursing, 3 in environmental health, 2 in theory and practice of patient care, 2 in pharmacology, 1 in nutrition, 1 in preventive medicine, 1 in psychiatric nursing, 3 in patient care, and 3 in introduction to management. In the upper-division baccalaureate category, 3 semester hours in nursing care management (9/85).

Recommendation, Skill Level 50

In the lower-division baccalaureate/associate degree category, 12 semester hours for clinical nursing experience, 8 in medical-surgical nursing, 4 in anatomy and physiology, 3 in maternal and child nursing, 3 in environmental health, 2 in theory and practice of patient care, 2 in pharmacology, 1 in nutrition, 1 in preventive medicine, 1 in psychiatric nursing, 3 in patient care, and 3 in introduction to management. In the upper-division baccalaureate category, 3 semester hours in nursing care management and 3 in administration (9/85).

MOS-91C-005

Practical Nurse

91C10
91C20
91C30
91C40
91C50

Exhibit Dates: 7/99–9/01. Effective 10/01, MOS 91C was discontinued and its duties were incorporated into MOS 91W, Health Care Specialist.

Career Management Field: 91 (Medical).

Description

Summary: Supervises or performs preventive therapeutic and emergency nursing care under the supervision of a physician or registered nurse. *Skill Level 10:* Performs preventive, therapeutic, and emergency nursing care procedures under supervision of physician, nurse, or medical NCO. *Skill Level 20:* Able to perform the duties required for Skill Level 10; assists with all levels of patient administration; prepares and maintains clinical record; supervises clinical records preparation and maintenance; reads patient charts and diagnostic reports and makes entries in medical records; requisitions, receives, packs, unpacks, stores, and safeguards clinical equipment and supplies; operates clinical equipment and maintains records; ensures maximum patient hygiene and safety; assists with providing emergency medical treatment and management; interprets and carries out nursing care plans; cares for patients of all ages with common diseases and minor injuries and those who are seriously, chronically, and acutely ill; assists during labor and childbirth; assesses condition of newborns; assists and teaches mothers the principles of infant care; prepares and administers prescribed medication; obtains and prepares samples and specimens for laboratory cultures and analysis; administers intravenous fluids; performs traction care; prepares patients for surgery and assists with transportation; assists with maintenance of nosograstic tubes; cares for patients with asthma, pneumonia, or bronchitis with oxygen therapy, bladder catheterization, colostomy irrigation, and intravenous therapy procedures; carefully observes patients for signs of problems such as dehydration, mental stress, and life-threatening symptoms; obtains appropriate admission data and assists with patient transfer and discharge. This specialty has been approved by the state of Texas as a practical nursing program and graduates must pass the state practical nursing exam to practice. *Skill Level 30:* Able to perform the duties required for Skill Level 20; performs advanced nursing procedures and provides technical guidance and advice to subordinates; manages small wards and provides nursing care leadership. *Skill Level 40:* Able to perform the duties required for Skill Level 30; manages large or multiward units; prepares nursing care plans and supervises implementation of these plans; conducts in-service programs. *Skill Level 50:* Able to perform the duties required for Skill Level 40; advises and assists professional staff in personnel matters; develops and evaluates training programs; manages supplies and equipment stores; establishes priorities, distributes work load, and assigns personnel; compiles management data; inspects organizational activities, observes discrepancies, and initiates corrective action.

Recommendation, Skill Level 10

Credit may be granted on the basis of an individualized assessment of the student (7/99).

Recommendation, Skill Level 20

Credit may be granted on the basis of an individualized assessment of the student (7/99).

Recommendation, Skill Level 30

In the lower-division baccalaureate/associate degree category, 12 semester hours for clinical nursing experience, 8 in medical-surgical nursing, 4 in anatomy and physiology, 3 in maternal and child nursing, 3 in environmental health, 2 in theory and practice of patient care, 2 in pharmacology, 1 in nutrition, 1 in preventive medicine, 1 in psychiatric nursing, 3 in patient care, 3 in introduction to management, and 3 in computer information technology (7/99).

Recommendation, Skill Level 40

In the lower-division baccalaureate/associate degree category, 12 semester hours for clinical nursing experience, 8 in medical-surgical nursing, 4 in anatomy and physiology, 3 in maternal and child nursing, 3 in environmental health, 2 in theory and practice of patient care, 2 in pharmacology, 1 in nutrition, 1 in preventive medicine, 1 in psychiatric nursing, 3 in patient care, 3 in introduction to management, and 3 in computer information technology. In the upper-division baccalaureate category, 3 semester hours in nursing care management (7/99).

Recommendation, Skill Level 50

In the lower-division baccalaureate/associate degree category, 12 semester hours for clinical nursing experience, 8 in medical-surgical nursing, 4 in anatomy and physiology, 3 in maternal and child nursing, 3 in environmental health, 2 in theory and practice of patient care, 2 in pharmacology, 1 in nutrition, 1 in preventive medicine, 1 in psychiatric nursing, 3 in patient care, 3 in introduction to management, and 3 in computer information technology. In the upper-division baccalaureate category, 3 semester hours in nursing care management and 3 in administration (7/99).

MOS-91D-003

Operating Room Specialist

91D10
91D20
91D30
91D40
91D50

Exhibit Dates: 1/90–7/94.

Career Management Field: 91 (Medical).

Description

Summary: Under the supervision of a registered nurse, supervises or assists in preparation and issue of sterile medical supplies and special equipment or prepares patient and operating suite for surgical procedures and assists during surgery. *Skill Level 10:* Under close supervision, prepares and issues central material supplies and equipment, including sterile and nonsterile solutions, surgical packs, and instruments; performs operating room duties, including transporting, positioning, draping, and skin preparation of patient; circulates and/or scrubs to assist in minor operative procedures; collects and maintains records of specimens; maintains a record of items and inventories items after surgery is performed; applies dressings. *Skill Level 20:* Able to perform the duties required for Skill Level 10; assists in supervision, guidance, and training of subordinates; maintains and operates highly technical surgical equipment; assists in surgical procedures; prepares tissue exam and culture slips; maintains emergency equipment; assists with cardiopulmonary resuscitation; inspects operating room to ensure clean, safe, and organized environment. *Skill Level 30:* Able to perform the duties required for Skill Level 20; manages central material service; supervises maintenance of operating room equipment; supervises the requisitioning, storing, and issuing of supplies; manages the operating room suite including the assignment of personnel; supervises, trains, and evaluates subordinates; establishes work priorities; organizes work schedules; assigns duties; determines stock level of supplies and equipment. *Skill Level 40:* Able to perform the duties required for Skill Level 30; administers central material service in large hospitals; supervises subordinates and activities of central material service and operating rooms; advises and assists professional staff in clinical, fiscal, technical, and administrative matters; coordinates the activities of the operating room and central material service with other elements of the medical treatment facility; compiles and evaluates management data and personnel requirements. *Skill Level 50:* Able to perform the duties required for Skill Level 40; assists with management of operating room suite in large hospital.

Recommendation, Skill Level 10

In the lower-division baccalaureate/associate degree category, 3 semester hours in basic principles of operating room procedures (includes operating room orientation and technique, microbiology, and surgical terminology), 3 in surgical basic sciences (anatomy and physiology in relation to surgical approach to disease and injury), and 8 for a practicum in operating room techniques and procedures. (NOTE: This recommendation for skill level 10 is valid for the dates 1/90-9/91 only) (9/85).

Recommendation, Skill Level 20

In the lower-division baccalaureate/associate degree category, 3 semester hours in basic principles of operating room procedures (includes operating room orientation and technique, microbiology, and surgical terminology), 3 in surgical basic sciences (anatomy and physiology in relation to surgical approach to disease and injury), and 8 for a practicum in operating room techniques and procedures (9/85).

Recommendation, Skill Level 30

In the lower-division baccalaureate/associate degree category, 3 semester hours in basic principles of operating room procedures (includes operating room orientation and technique, microbiology, and surgical terminology), 3 in surgical basic sciences (anatomy and physiology in relation to the surgical approach to disease and injury), 3 in personnel supervision, and 8 for a practicum in operating room techniques and procedures (9/85).

Recommendation, Skill Level 40

In the lower-division baccalaureate/associate degree category, 3 semester hours in basic principles of operating room procedures (includes operating room orientation and technique, microbiology, and surgical terminology), 3 in surgical basic sciences (anatomy and physiology in relation to the surgical approach to disease and injury), 3 in personnel supervision, 3 for field experience in management, and 8 for a practicum in operating room techniques and procedures. In the upper-division baccalaureate category, credit in patient care administration and health facility management on the basis of institutional evaluation (9/85).

Recommendation, Skill Level 50

In the lower-division baccalaureate/associate degree category, 3 semester hours in basic principles of operating room procedures (includes operating room orientation techniques, microbiology, and surgical terminology), 3 in surgical basic sciences (anatomy and physiology in relation to the surgical approach to disease and injury), 3 in personnel supervision, 3 for field experience in management, and 8 for a practicum in operating room techniques and procedures. In the upper-division baccalaureate category, 3 semester hours in personnel management and credit in patient care administration and health facility management on the basis of institutional evaluation (9/85).

MOS-91D-004

Operating Room Specialist

91D10
91D20
91D30
91D40
91D50

Exhibit Dates: 8/94–Present.

Career Management Field: 91 (Medical).

Description

Summary: Under the supervision of a registered nurse, supervises or assists in preparation and issue of sterile medical supplies and special equipment or prepares patient and operating suite for surgical procedures and assists during surgery. *Skill Level 10:* Under close supervision, prepares and issues central material supplies and equipment, including sterile and nonsterile solutions, surgical packs, and instruments; performs operating room duties, including transporting, positioning, draping, and skin preparation of patient; circulates and/or scrubs to assist in minor operative procedures; collects and maintains records of specimens; maintains a record of items and inventories items after surgery is performed; applies dressings. *Skill Level 20:* Able to perform the duties required for Skill Level 10; assists in supervision, guidance, and training of subordinates; assists in surgical procedures; prepares tissue exam and culture slips; maintains emergency equipment; assists with cardiopulmonary resuscitation; inspects operating room to ensure clean, safe, and organized environment. *Skill Level 30:* Able to perform the duties required for Skill Level 20; manages central material service; supervises maintenance of operating room equipment; maintains and operates highly technical surgical equipment; supervises the requisitioning, storing, and issuing of supplies; manages the operating room suite including the assignment of personnel; supervises, trains, and evaluates subordinates; establishes work priorities; organizes work schedules; assigns duties; determines stock level of supplies and equipment. *Skill Level 40:* Able to perform the duties required for Skill Level 30; administers central material service in large hospitals; supervises subordinates and activities of central material service and operating rooms; advises and assists professional staff in clinical, fiscal, technical, and administrative matters; coordinates the activities of the operating room and central material service with other elements of the medical treatment facility; compiles and evaluates management data and personnel requirements. *Skill Level 50:* Able to perform the duties required for Skill Level 40; assists with management of operating room suite in large hospital.

Recommendation, Skill Level 10

Credit may be granted on the basis of an individualized assessment of the student (11/96).

Recommendation, Skill Level 20

In the lower-division baccalaureate/associate degree category, 3 semester hours in basic principles of operating room procedures (includes operating room orientation and technique, microbiology, and surgical terminology), 3 in surgical basic sciences (anatomy and physiology in relation to surgical approach to disease and injury), and 8 for clinical experience in operating room techniques and procedures. (NOTE: This recommendation for skill level 20 is valid for the dates 8/94-2/95 only) (11/96).

Recommendation, Skill Level 30

In the lower-division baccalaureate/associate degree category, 3 semester hours in basic principles of operating room procedures (includes operating room orientation and technique, microbiology, and surgical terminology), 3 in surgical basic sciences (anatomy and physiology in relation to surgical approach to disease and injury), and 8 for clinical experience in operating room techniques and procedures (11/96).

Recommendation, Skill Level 40

In the lower-division baccalaureate/associate degree category, 3 semester hours in basic principles of operating room procedures (includes operating room orientation and technique, microbiology, and surgical terminology), 3 in surgical basic sciences (anatomy and physiology in relation to the surgical approach to disease and injury), 3 in specialized surgical methods, 3 in personnel supervision, 3 in supply management, and 15 for clinical experience in operating room techniques and procedures. In the upper-division baccalaureate category, credit in patient care administration and health facility management on the basis of institutional evaluation (11/96).

Recommendation, Skill Level 50

In the lower-division baccalaureate/associate degree category, 3 semester hours in basic principles of operating room procedures (includes operating room orientation techniques, microbiology, and surgical terminology), 3 in surgical basic sciences (anatomy and physiology in relation to the surgical approach to disease and injury), 3 in specialized surgical methods, 3 in personnel supervision, 3 in supply management, and 15 for clinical experience in operating room techniques and procedures. In the upper-division baccalaureate category, 3 semester hours in personnel management and credit in patient care administration and health facility management on the basis of institutional evaluation (11/96).

MOS-91E-002

Dental Specialist

91E10
91E20
91E30
91E40
91E50

Exhibit Dates: 1/90–3/92.

Career Management Field: 91 (Medical), subfield 911 (Dental).

Description

Summary: Assists in managing a dental facility or assists dentist in preventing, examining, and treating diseases of the teeth and oral region. *Skill Level 10:* Assists dentist in dental care and treatment; protects patient from excessive exposure to ionizing radiation; exposes and mounts bitewing, periapical, and occlusal radiographs using bisecting angle and long-cone paralleling techniques; exposes extraoral panoramic radiographs; prepares developer and fixer solutions; prepares equipment and materials to measure and mix such commonly used permanent and nonpermanent restorative materials as dental amalgam, composite and unfilled acrylic resins, zinc oxide and eugenol materials, cavity liners and bases, intermediate restorative materials, and zinc phosphate cement; prepares equipment and materials to measure and mix commonly-used impression materials and gypsum products, including irreversible hydrocolloid, polysulfide impression material, plaster of Paris, and artificial stone; fabricates casts from preliminary impressions; prepares and maintains dental chair, unit, and light; sharpens dental instruments; prepares instrument setups for commonly performed dental procedures; identifies, selects, and passes to dentist appropriate instruments for dental treatment procedures in the oral surgical, oral diagnostic, endodontic, restorative, periodontic, pedodontic, orthodontic, and prosthodontic specialties; performs resuscitative procedures and operates resuscitative equipment; schedules dental appointments; maintains patient records and records information and data using authorized symbols and abbreviations; applies aseptic techniques in cleaning, packaging, sterilizing, and storing dental instruments, equipment, and supplies. *Skill Level 20:* Able to perform the duties required for Skill Level 10; serves as a dental hygienist; selects and prepares dental hand pieces, contra-angles, ultrasonic equipment, solutions, and other materials to perform oral prophylaxis; applies disclosing solutions; assists dentists in carrying out oral disease control programs including individual and group patient education and motivation; removes anatomical and acquired stains, calculus, and bacterial plaque from teeth, using hand instruments, ultrasonic equipment, dental floss, and other adjuncts; polishes teeth with abrasives; applies topical fluoride solutions for prevention of dental caries; applies pit and fissure sealants; records information on abnormal appearance and characteristics of the tissues of teeth and oral region; refers patient to dentist for consultation and treatment; performs user maintenance on hand instruments, contra-angles, and ultrasonic equipment. *Skill Level 30:* Able to perform the duties required for Skill Level 20; performs reversible dental procedures under the direct supervision of a dentist; assists in managing a small dental facility; performs diagnostic procedures such as vitality tests, diet analysis, and nutritional consultation; exposes and develops special intraoral and extraoral radiographs; isolates operative site by placement of rubber dam; select and applies matrix bands and wedges; inserts, condenses, carves, polishes, and finishes various restorations; assists dentist in treating pericoronitis and local osteritis; removes sutures; manages surgical patient preoperatively and postoperatively; makes preliminary impressions; isolates teeth for endodontic treatment; establishes stock level for supplies and equipment; orders, stores, and issues supplies. *Skill Level 40:* Able to perform the duties required for Skill Level 30; assists in managing medium-sized dental facility; assists dentist in performing complex dental procedures. *Skill Level 50:* Able to perform the duties required for skill level 40 of MOS 91E or 42D (Dental Laboratory Specialist); assists in managing a dental clinic, dental laboratory, or dental organization; normally supervises from 20 to 120 subordinates; inspects dental facilities to assure cleanliness, safety, and comfort; assists dentists in analyzing clinical and laboratory operations to ensure efficient utilization of personnel and equipment; supervises maintenance of administrative and professional dental files and technical library; reviews and consolidates technical, administrative, and personnel reports; coordinates the activities of dental units; provides technical assistance in planning and staffing new facilities and in modifying existing facilities; plans, schedules, coordinates, and supervises dental specialty training.

Recommendation, Skill Level 10

In the lower-division baccalaureate/associate degree category, 8 semester hours in clinical training (dental assistant), 3 in chairside assisting, 3 in dental radiology, 2 in basic dental science, and 1 in emergencies. (NOTE: This recommendation for skill level 10 is valid for the dates 1/90-9/91 only) (6/77).

Recommendation, Skill Level 20

In the lower-division baccalaureate/associate degree category, 8 semester hours in clinical training (dental assistant), 8 in clinical training (dental hygienist), 3 in chairside assisting, 3 in dental radiology, 3 in associated dental science, 3 in dental materials, 2 in basic dental science, 2 in dental health, and 1 in emergencies (6/77).

Recommendation, Skill Level 30

In the lower-division baccalaureate/associate degree category, 8 semester hours in clinical training (dental assistant), 8 in clinical training (dental hygienist), 8 in clinical training (dental therapy assistant), 6 in dental materials, 4 in dental health, 3 in chairside assisting, 3 in dental radiology, 3 in associated dental science, 2 in basic dental science, 2 in related dental science, 1 in emergencies, and 1 in four-handed dentistry (6/77).

Recommendation, Skill Level 40

In the lower-division baccalaureate/associate degree category, 8 semester hours in clinical training (dental assistant), 8 in clinical training (dental hygienist), 8 in clinical training (dental therapy assistant), 6 in dental materials, 4 in dental health, 3 in chairside assisting, 3 in dental radiology, 3 in associated dental science, 2 in basic dental science, 2 in related dental science, 1 in emergencies, and 1 in four-handed dentistry (6/77).

Recommendation, Skill Level 50

In the lower-division baccalaureate/associate degree category, 8 semester hours in clinical training (dental assistant), 8 in clinical training (dental hygienist), 8 in clinical training (dental therapy assistant), 6 in dental materials, 4 in dental health, 3 in chairside assisting, 3 in dental radiology, 3 in associated dental science, 2 in basic dental science, 2 in related dental science, 1 in emergencies, 1 in four-handed dentistry, 3 in personnel supervision, 2 in health care management, and 2 for field experience in management (6/77).

MOS-91E-003

DENTAL SPECIALIST

91E10
91E20
91E30
91E40
91E50

Exhibit Dates: 4/92–11/01.

Career Management Field: 91 (Medical).

Description

Summary: Assists in managing a dental facility or assists dentist in preventing, examining, and treating diseases of the teeth and oral region. *Skill Level 10:* Assists dentist in dental care and treatment; protects patient from excessive exposure to ionizing radiation; exposes and mounts bitewing, periapical, and occlusal radiographs using bisecting angle and long-cone paralleling techniques; exposes extraoral panoramic radiographs; prepares developer and fixer solutions; prepares equipment and materials to measure and mix such commonly-used permanent and nonpermanent restorative materials as dental amalgam, composite and unfilled acrylic resins, zinc oxide and eugenol materials, cavity liners and bases, intermediate restorative materials, and zinc phosphate cement; prepares equipment and materials to measure and mix commonly used impression materials and gypsum products, including irreversible hydrocolloid, polysulfide impression material, plaster of Paris, and artificial stone; fabricates casts from preliminary impressions; prepares and maintains dental chair, unit, and light; sharpens dental instruments; prepares instrument setups for commonly performed dental procedures; identifies, selects, and passes to dentist appropriate instruments for dental treatment procedures in the oral surgical, oral diagnostic endodontic, restorative, periodontic, pedodontic, orthodontic, and prosthodontic specialties; performs resuscitative procedures and operates resuscitative equipment; schedules dental appointments; maintains patient records and records information and data using authorized symbols and abbreviations; applies aseptic techniques in cleaning, packaging, sterilizing, and storing dental instruments, equipment, and supplies. *Skill Level 20:* Able to perform the duties required for Skill Level 10; serves as a dental hygienist; selects and prepares dental hand pieces, contra-angles, ultrasonic equipment, solutions, and other materials to perform oral prophylaxis; applies disclosing solutions; assists dentists in carrying out oral disease control programs including individual and group patient education and motivation; removes anatomical and acquired stains, calculus, and bacterial plaque from teeth, using hand instruments, ultrasonic equipment, dental floss, and other adjuncts; polishes teeth with abrasives; applies topical fluoride solutions for prevention of dental caries; applies pit and fissure sealants; records information on abnormal appearance and characteristics of the tissues of teeth and oral region; refers patient to dentist for consultation and treatment; performs user maintenance on hand instruments, contra-angles, and ultrasonic equipment. *Skill Level 30:* Able to perform the duties required for Skill Level 20; performs reversible dental procedures under the direct supervision of a dentist; assists in managing a small dental facility; performs diagnostic procedures such as vitality tests, diet analysis, and nutritional consultation; exposes and develops special intraoral and extraoral radiographs; isolates operative site by placement of rubber dam; selects and applies matrix bands and wedges; inserts, condenses, carves, polishes and finishes various restorations; assists dentist in treating pericoronitis and local osteritis; removes sutures; manages surgical patient preoperatively and postoperatively; makes preliminary impressions; isolates teeth for endodontic treatment; establishes stock level for supplies and equipment; orders, stores, and issues supplies. *Skill Level 40:* Able to perform the duties required for Skill Level 30; assists in managing medium-sized dental facility; assists dentist in performing complex dental procedures. *Skill Level 50:* Able to perform the duties required for skill level 40 of MOS 91E or 42D (Dental Laboratory Specialist); assists in managing a dental clinic, dental laboratory, or dental organization; normally supervises from 20 to 120 subordinates; inspects dental facilities to assure cleanliness, safety, and comfort; assists dentists in analyzing clinical and laboratory operations to ensure efficient utilization of personnel and equipment; supervises maintenance of administrative and professional dental files and technical library; reviews and consolidates technical, administrative, and personnel reports; coordinates the activities of dental units; provides technical assistance in planning and staffing new facilities and in modifying existing facilities; plans, schedules, coordinates, and supervises dental specialty training.

Recommendation, Skill Level 10

Credit may be granted on the basis of an individualized assessment of the student (3/94).

Recommendation, Skill Level 20

In the lower-division baccalaureate/associate degree category, 2 semester hours in dental radiography, 3 in chairside dental assisting, 4 in dental assisting clinical practice, and 3 in personnel supervision; if additional skill identifier was X2, Preventive Dentistry Specialty, 4 semester hours in dental hygiene clinical practice. (NOTE: This recommendation for skill level 20 is valid for the dates 4/92-2/95 only) (4/92).

Recommendation, Skill Level 30

In the lower-division baccalaureate/associate degree category, 2 semester hours in dental radiography, 3 in chairside dental assisting, 5 in dental assisting clinical practice, 3 in personnel supervision, and 2 in records and information management; if additional skill identifier was X2, Preventive Dentistry Specialty, 5 semester hours in dental hygiene clinical practice (4/92).

Recommendation, Skill Level 40

In the lower-division baccalaureate/associate degree category, 2 semester hours in dental radiography, 3 in chairside dental assisting, 6 in dental assisting clinical practice, 3 in personnel supervision, and 2 in records and information management; if additional skill identifier was X2, Preventive Dentistry Specialty, 6 semester hours in dental hygiene clinical practice. In the upper-division baccalaureate category, 3 semester hours in organizational management and 3 for field experience in management (4/92).

Recommendation, Skill Level 50

In the lower-division baccalaureate/associate degree category, 2 semester hours in dental radiography, 3 in chairside dental assisting, 6 in dental assisting clinical practice, 3 in personnel supervision, and 2 in records and information management; if additional skill identifier was X2, Preventive Dentistry Specialty, 6 semester hours in dental hygiene clinical practice. In the upper-division baccalaureate category, 3 semester hours in organizational management and 6 for field experience in management; if individual has attained paygrade E-9, additional credit as follows: 3 semester hours in management problems, 3 in operations management, and 3 in communication techniques for managers (4/92).

MOS-91E-004

DENTAL SPECIALIST
91E10
91E20
91E30
91E40
91E50

Exhibit Dates: 12/01–Present.

Career Management Field: 91 (Medical).

Description

Summary: Assists in managing a dental facility or assists dentist in preventing, examining, and treating diseases of the teeth and oral region. *Skill Level 10:* Assists dentist in dental care and treatment; protects patient from excessive exposure to ionizing radiation; exposes and mounts bitewing, periapical, and occlusal radiographs using bisecting angle and long-cone paralleling techniques; exposes extraoral panoramic radiographs; prepares developer and fixer solutions; prepares equipment and materials to measure and mix such commonly-used permanent and nonpermanent restorative materials as dental amalgam, composite and unfilled acrylic resins, zinc oxide, and eugenol materials, cavity liners and bases, intermediate restorative materials, and zinc phosphate cement; prepares equipment and materials to measure and mix commonly used impression materials and gypsum products, including irreversible hydrocolloid, polysulfide impression material, plaster of Paris, and artificial stone; fabricates casts from preliminary impressions; prepares and maintains dental chair, unit, and light; sharpens dental instruments; prepares instrument setups for commonly performed dental procedures; identifies, selects, and passes to dentist appropriate instruments for dental treatment procedures in the oral surgical, oral diagnostic endodontic, restorative, periodontic, pedodontic, orthodontic, and prosthodontic specialties; performs resuscitative procedures and operates resuscitative equipment; schedules dental appointments; maintains patient records and records information and data using authorized symbols and abbreviations; applies aseptic techniques in cleaning, packaging, sterilizing, and storing dental instruments, equipment, and supplies. Individuals holding the Additional Skill Identifier (ASI) X2 Preventive Dentistry Specialist, also perform the following duties: selects and prepares dental hand pieces, contra-angles, ultrasonic equipment, solutions, and other materials to perform oral prophylaxis; applies disclosing solutions; assists dentists in carrying out oral disease control programs including individual and group patient education and motivation; removes anatomical and acquired stains, calculus, and bacterial plaque from teeth, using hand instruments, ultrasonic equipment, dental floss, and other adjuncts; polishes teeth with abrasives; applies topical fluoride solutions for prevention of dental caries; applies pit and fissure sealants; records information on abnormal appearance and characteristics of the tissues of teeth and oral region; refers patient to dentist for consultation and treatment; performs user maintenance on hand instruments, contra-angles, and ultrasonic equipment. *Skill Level 20:* Able to perform the duties required for Skill Level 10; manages staff; prepares reports and maintains records; supervises data entry and control; maintains supplies and equipment. *Skill Level 30:* Able to perform the duties required for Skill Level 20; performs reversible dental procedures under direct supervision of a dentist; assists in managing a small dental facility; performs diagnostic procedures such as vitality tests, diet analysis, and nutritional consultation; exposes and develops special intraoral and extraoral radiographs; isolates operative site by placement of rubber dam; selects and applies matrix bands and wedges; inserts, condenses, carves, polishes, and finishes various restorations; assists dentist in treating pericoronitis and local osteritis; removes sutures; manages surgical patient preoperatively and postoperatively; makes preliminary impressions; isolates teeth for endodontic treatment; established stock level for supplies and equipment; orders, stores, and issues supplies; provides education in dental health to patients, families, and the community. *Skill Level 40:* Able to perform the duties required for Skill Level 30; assist in managing medium-sized dental facility; assists dentist in performing complex dental procedures. *Skill Level 50:* Able to perform the duties required for Skill Level 40; of MOS 91E or 42D (Dental Laboratory Specialist); assists in managing a dental clinic, dental laboratory, or dental organization; normally supervises from 20 to 120 subordinates; inspects dental facilities to assure cleanliness, safety, and comfort; assist dentists in analyzing clinical and laboratory operations to ensure efficient utilization of personnel and equipment; supervises maintenance of administrative and professional dental files and technical library; reviews and consolidates technical, administrative, and personnel reports; coordinates the activities of dental units; provides technical assistance in planning and staffing new facilities and in modifying existing facilities; plans, schedules, coordinates, and supervises dental specialty training.

Recommendation, Skill Level 10

Credit may be granted on the basis of an individualized assessment of the student (12/01).

Recommendation, Skill Level 20

Credit may be granted on the basis of an individualized assessment of the student (12/01).

Recommendation, Skill Level 30

In the lower-division baccalaureate/associate degree category, 2 semester hours in dental radiography, 3 in chairside dental assisting, 5 in dental assisting clinical practice, 3 in principles of supervision, 3 in dental health promotion, and 2 in records and information management; if additional skill identifier was X2, Preventive Dentistry Specialty, 5 in dental hygiene clinical practice (12/01).

Recommendation, Skill Level 40

In the lower-division baccalaureate/associate degree category, 2 semester hours in dental radiography, 3 in chairside dental assisting, 6 in dental assisting clinical practice, 3 in principles of supervision, 3 in dental health promotion, and 2 in records and information management; if Additional Skill Identifier was X2, Preventive Dentistry Specialty, 6 in dental hygiene clinical practice. In the upper-division baccalaureate category, 3 semester hours in organizational management and 3 for a management practicum (12/01).

Recommendation, Skill Level 50

In the lower-division baccalaureate/associate degree category, 2 semester hours in dental radiography, 3 in chairside dental assisting, 6 in dental assisting clinical practice, 3 in principles of supervision, 3 in dental health promotion, and 2 in records and information management; if Additional Skill Identifier was X2, Preventive Dentistry Specialty, 6 hours in dental hygiene clinical practice. In the upper-division baccalaureate category, 3 semester hours in organizational management and 6 for a management practicum (12/01).

MOS-91F-002

Psychiatric Specialist
91F10
91F20
91F30
91F40

Exhibit Dates: 1/90–3/92.

Career Management Field: 91 (Medical), subfield 912 (Patient Care).

Description

Summary: Assists in the care and treatment of psychiatric, drug, and alcohol patients and in the management of a psychiatric facility under the supervision of a psychiatric nurse. *Skill Level 10:* Able to perform the duties of 91B10 (Medical Specialist); assists with diagnostic procedures; observes, records, and reports significant patient behavior; identifies and responds to psychopathological behaviors; provides intensive care for patients who are receiving special treatment; participates in individual and group therapies and somatic therapy; safeguards confidentiality of patients' clinical records and reports; orients patients to the services and routines of the treatment facility; supervises personal hygiene of patients and assists them in meeting their physical needs; accompanies patients to appointments and therapeutic activities within treatment facility and local community; functions as a member of electroconvulsive therapy team; assists with general nursing care activities, measuring and recording vital signs, administering prescribed medication, and collecting and labeling specimens. *Skill Level 20:* Able to perform the duties required for Skill Level 10; dispenses medication; supervises such patient activities, as physical reconditioning, occupational therapy, recreational therapy, and group therapy. *Skill Level 30:* Able to perform the duties required for Skill Level 20; assists with management of psychiatric ward; organizes work schedules and assigns duties; evaluates personnel; requisitions, stores, and issues supplies and equipment; assists with in-service training programs; serves as a group facilitator under professional supervision. *Skill Level 40:* Able to perform the duties required for Skill Level 30; assists with management of psychiatric nursing activity; advises and assists professional staff in technical, administrative, personnel, and fiscal matters; establishes work priorities; compiles management data; inspects facility to ensure a comfortable, safe, and sanitary environment; coordinates activities with other elements of medical treatment facility; reviews and consolidates technical and administrative reports.

Recommendation, Skill Level 10

In the lower-division baccalaureate/associate degree category, 3 semester hours in psychiatric nursing, 3 in psychiatric therapies, 3 for clinical experience in mental health, 2 in abnormal psychology, and 2 in human development. (NOTE: This recommendation for skill level 10 is valid for the dates 1/90-9/91 only) (6/77).

Recommendation, Skill Level 20

In the lower-division baccalaureate/associate degree category, 3 semester hours in psychiatric nursing, 3 in psychiatric therapies, 5 for clinical experience in mental health, 2 in abnormal psychology, and 2 in human development (6/77).

Recommendation, Skill Level 30

In the lower-division baccalaureate/associate degree category, 5 semester hours for clinical experience in mental health, 3 in psychiatric nursing, 3 in psychiatric therapies, 3 in psychiatric ward management, 3 in personnel supervision, 2 in abnormal psychology, and 2 in human development (6/77).

Recommendation, Skill Level 40

In the lower-division baccalaureate/associate degree category, 5 semester hours for clinical experience in mental health, 3 in psychiatric nursing, 3 in psychiatric therapies, 3 in psychiatric ward management, 3 in personnel supervision, 2 in abnormal psychology, and 2 in human development. In the upper-division baccalaureate category, credit in patient care administration and medical facility management on the basis of institutional evaluation (6/77).

MOS-91F-003

Psychiatric Specialist
91F10
91F20
91F30
91F40

Exhibit Dates: 4/92–12/96. Effective 12/96, MOS 91F was discontinued and its duties were transferred to MOS 91X.

Career Management Field: 91 (Medical).

Description

Summary: Assists in the care and treatment of psychiatric, drug, and alcohol patients and in the management of a psychiatric facility under the supervision of a physician or psychiatric nurse. *Skill Level 10:* Assists with diagnostic procedures; observes, records, and reports significant patient behavior; identifies and responds to psychopathological behaviors; provides intensive care for patients who are receiving special treatment; participates in individual and group therapies and somatic therapy; safeguards confidentiality of patients' clinical records and reports; orients patients to the services and routines of the treatment facility; supervises personal hygiene of patients and assists them in meeting their physical needs; accompanies patients to appointments and therapeutic activities within treatment facility and local community; functions as a member of electroconvulsive therapy team; assists with general nursing care activities, measuring and recording vital signs, and collecting and labeling specimens. *Skill Level 20:* Able to perform the duties required for Skill Level 10; supervises such patient activities as physical reconditioning, occupational therapy, recreational therapy, and group therapy. *Skill Level 30:* Able to perform the duties required for Skill Level 20; assists with management of psychiatric ward; organizes work schedules and assigns duties; evaluates personnel; requisitions, stores, and issues supplies and equipment; assists with in-service training programs; serves as a group facilitator under professional supervision. *Skill Level 40:* Able to perform the duties required for Skill Level 30; assists with management of psychiatric nursing activity; advises and assists professional staff in technical, administrative, personnel, and fiscal matters; establishes work priorities; compiles management data; inspects facility to ensure a comfortable, safe, and sanitary environment; coordinates activities with other elements of medical treatment facility; reviews and consolidates technical and administrative reports.

Recommendation, Skill Level 10

Credit may be granted on the basis of an individualized assessment of the student (3/94).

Recommendation, Skill Level 20

In the lower-division baccalaureate/associate degree category, 3 semester hours in introduction to human services, 4 in human services clinical practice, and 3 in personnel supervision. (NOTE: This recommendation for skill level 20 is valid for the dates 4/92-2/95 only) (4/92).

Recommendation, Skill Level 30

In the lower-division baccalaureate/associate degree category, 3 semester hours in introduction to human services, 5 in human services clinical practice, 3 in personnel supervision, and 2 in records and information management (4/92).

Recommendation, Skill Level 40

In the lower-division baccalaureate/associate degree category, 3 semester hours in introduction to human services, 6 in human services clinical practice, 3 in personnel supervision, and 2 in records and information management. In the upper-division baccalaureate category, 3 semester hours in organizational management and 3 for field experience in management (4/92).

MOS-91G-002

Behavioral Science Specialist
91G10
91G20
91G30
91G40

Exhibit Dates: 1/90–3/92.

Career Management Field: 91 (Medical), subfield 912 (Patient Care).

Description

Summary: Under professional supervision, assists with management of mental health activity or obtains social case histories; collects basic psychological data; performs counseling functions. *Skill Level 10:* Able to perform the duties required for skill level 10 of MOS 91B (Medical Specialist); interviews clients; conducts collateral interviews and screens records; administers and scores intelligence tests under supervision; assists in determining client need for referral to a professional; records psychosocial data and behavior observations; orients and refers clients to military, civilian, and agency resources; prepares and presents cases for staffings and conferences; requests and transmits client records. *Skill Level 20:* Able to perform the duties required for Skill Level 10; under professional supervision, provides supportive counseling and follow-up service to individuals experiencing a wide range of social or emotional problems; assists with group counseling and therapy sessions; leads discussion groups; assists in determination of need for hospitalization; under supervision, administers and scores achievement tests, objective personality tests, and tests of organic impairment and records the results; assists patients with discharge planning; consults on management of individual behavioral problems; assists professional staff in setting up home health care programs and makes follow-up home visits; gathers statistical data; establishes and maintains clinical and general office files. *Skill Level 30:* Able to perform the duties required for Skill Level 20; provides advice and limited technical guidance to subordinates; provides supportive counseling and crisis intervention services to clients; assists in staffing and coordination of highly complex cases; collects and organizes research data and makes statistical computations; assists in analysis of research data and report writing; compiles case load data and assigns clients to specific staff members; evaluates manpower requirements; prepares administrative reports; requisitions, stores, and issues supplies, equipment, publications, and training aids; develops work procedures and practices; prepares and conducts in-service training programs. *Skill Level 40:* Able to perform the duties required for Skill Level 30; under supervision, administers and records thematic projective tests; assists in determining program and management objectives and procedures; establishes work priorities and work schedules; assigns duties and instructs subordinates in work techniques and procedures; prepares and counsels personnel and prepares evaluation reports; coordinates activity of facility with related elements of the medical treatment facility; advises and assists professionals with fiscal, technical, and administrative matters; establishes stock level for supplies and equipment.

Recommendation, Skill Level 10

In the lower-division baccalaureate/associate degree category, 6 semester hours in clinical psychology and counseling, 3 in psychology, 3 in human relations, 1 in sociology, and 1 in psychological testing. (NOTE: This recommendation for skill level 10 is valid for the dates 1/90-9/91 only) (6/77).

Recommendation, Skill Level 20

In the lower-division baccalaureate/associate degree category, 6 semester hours in clinical psychology and counseling, 3 in psychology, 3 in human relations, 3 in psychological testing, and 1 in sociology (6/77).

Recommendation, Skill Level 30

In the lower-division baccalaureate/associate degree category, 8 semester hours in clinical psychology and counseling, 3 in psychology, 3 in human relations, 3 in psychological testing, 2 in personnel supervision, and 1 in sociology. In the upper-division baccalaureate category, credit in patient care administration on the basis of institutional evaluation (6/77).

Recommendation, Skill Level 40

In the lower-division baccalaureate/associate degree category, 8 semester hours in clinical psychology and counseling, 4 in psychological testing, 3 in psychology, 3 in human relations, 3 in personnel supervision, 3 in principles of administration, and 1 in sociology. In the upper-division baccalaureate category, credit in patient care administration and medical facility management on the basis of institutional evaluation (6/77).

MOS-91G-003

Behavioral Science Specialist

91G10

91G20

91G30

91G40

Exhibit Dates: 4/92–12/96. Effective 12/96, MOS 91G was discontinued and its duties were transferred to MOS 91X.

Career Management Field: 91 (Medical).

Description

Summary: Under professional supervision, assists with management of outpatient mental health clients; obtains social case histories; collects basic psychological data; counsels. *Skill Level 10:* Interviews clients; conducts collateral interviews; administers and scores intelligence tests under supervision; assists in determining client need for referral to a professional; records psychosocial data and behavior observations; orients and refers clients to military, civilian, and agency resources; prepares and presents cases for staffings and conferences; requests and transmits client records. *Skill Level 20:* Able to perform the duties required for Skill Level 10; under professional supervision, provides supportive counseling and follow-up service to individuals experiencing a wide range of social or emotional problems; assists with group counseling and therapy sessions; leads discussion groups; assists in determining need for hospitalization; under supervision, administers and scores achievement tests, objective personality tests, and tests of organic impairment and records the results; assists patients with discharge planning; consults on management of individual behavioral problems; assists professional staff in setting up home health care programs and makes follow-up home visits; gathers statistical data; establishes and maintains clinical and general office files. *Skill Level 30:* Able to perform the duties required for Skill Level 20; provides advice and limited technical guidance to subordinates; provides supportive counseling and crisis intervention services to clients; collects and organizes research data and makes statistical computations; assists in analysis of research data and report writing; compiles case load data and assigns clients to specific staff members; evaluates manpower requirements; prepares administrative reports; requisitions, stores, and issues supplies, equipment, publications, and training aids; develops work procedures and practices; prepares and conducts in-service training programs. *Skill Level 40:* Able to perform the duties required for Skill Level 30; under supervision, administers and records thematic projective tests; assists in determining program and management objectives and procedures; establishes work priorities and work schedules; assigns duties and instructs subordinates in work techniques and procedures; prepares and counsels personnel and prepares evaluation reports; coordinates activity of facility with related elements of the medical treatment facility; advises and assists professionals with fiscal, technical, and administrative matters; establishes stock level for supplies and equipment.

Recommendation, Skill Level 10

Credit may be granted on the basis of an individualized assessment of the student (3/94).

Recommendation, Skill Level 20

In the lower-division baccalaureate/associate degree category, 3 semester hours in introduction to human services, 4 in human services clinical practice, 3 in behavior modification techniques, and 3 in personnel supervision. (NOTE: This recommendation for skill level 20 is valid for the dates 4/92-2/95 only) (4/92).

Recommendation, Skill Level 30

In the lower-division baccalaureate/associate degree category, 3 semester hours in introduction to human services, 5 in human services clinical practice, 3 in behavior modification techniques, 3 in personnel supervision, and 2 in records and information management (4/92).

Recommendation, Skill Level 40

In the lower-division baccalaureate/associate degree category, 3 semester hours in introduction to human services, 6 in human services clinical practice, 3 in behavior modification techniques, 3 in personnel supervision, and 2 in records and information management. In the upper-division baccalaureate category, 3 semester hours in organizational management and 3 for field experience in management (4/92).

MOS-91G-004

Patient Administration Specialist

91G10

91G20

91G30

91G40

91G50

Exhibit Dates: 6/01–Present.

Career Management Field: 91 (Medical).

Description

Summary: Supervises or performs administrative duties in patient administration division of hospital or other medical activity. *Skill Level 10:* Performs patient administrative duties; applies knowledge of basic medical terminology, including construction or composition of medical terms; applies knowledge of provisions and limitations of Freedom of Information and Privacy Acts; receives patients; interviews patients and reviews records and available data to obtain necessary information for preparation of admission records including patient data cards and appropriate records and forms, depending upon type of patient, illness or injury, and urgency of required treatment; receives valuables and monies from patients and deposits into patient's trust fund; initiates required notification of patient admission to unit commanders or other headquarters; compiles data for preparation and prepares daily admission and disposition reports; performs administrative duties and prepares appropriate forms to process and report very seriously ill patients including notification of next of kin and progress reports; prepares and dispatches casualty reports; prepares appropriate forms or letters and performs necessary administrative duties to evacuate or transfer patients to other medical facilities; completes forms and reports and performs tasks regarding disposition of personal effects; computes various charges for pay patients; performs administrative action involving third party liability; applies knowledge of administrative terminology to medical records; keeps repository for military health records and dependents' medical records for hospital, clinic, outpatient service, or dispensary; classifies, indexes, and files record folders in repository;

reviews medical record for adequacy and completeness based upon nature of case and treatment given; answers inquiries and provides information on medical records; compiles statistical data pertaining to admission, diagnosis, treatment, and disposition of patients; prepares statistical tables, charts, and graphs on medical data; prepares reports and data on births, deaths, and reportable conditions for submission to military and civilian health authorities and to appropriate legal and governmental agencies; initiates correspondence pertaining to medical records, medical board proceedings, and lines of duty investigation; answers inquiries on results of board actions, eligibility for medical care under Uniform Service Health Benefit Program; prepares necessary medical forms to convene the medical board to determine medical fitness for duty; types military and nonmilitary letters, endorsements, disposition forms, messages, and special medical forms and documents in draft and final form; maintains nonmedical files. *Skill Level 20:* Able to perform the duties required for Skill Level 10; provides technical guidance to subordinates; performs administrative duties in medical, surgical, or similar department of the hospital, including obtaining records from repository, posting entries onto medical records, returning records, and coordinating administrative responsibilities with other departments. *Skill Level 30:* Able to perform the duties required for Skill Level 20; supervises subordinates; assigns tasks to subordinates, ensuring efficient work flow and timely accomplishment of tasks regarding patient record keeping and the movement of patients; instructs personnel in performance of assigned tasks and instills understanding of need for timely and precise completion and posting of records and an appreciation of the sensitivity and confidentiality of the actions performed. *Skill Level 40:* Able to perform the duties required for Skill Level 30; supervises a medium-sized patient administration activity or hospital department. *Skill Level 50:* Able to perform the duties required for Skill Level 40; supervises a large patient administration activity or hospital department.

Recommendation, Skill Level 10

Credit may be granted on the basis of an individualized assessment of the student (3/94).

Recommendation, Skill Level 20

Credit may be granted on the basis of an individualized assessment of the student (6/01).

Recommendation, Skill Level 30

In the lower-division baccalaureate/associate degree category, 3 semester hours in introduction to computers, 3 in medical record terminology, 3 in medical records coding, 3 in principles of supervision, 3 in data management, and 3 in medical records management (6/01).

Recommendation, Skill Level 40

In the lower-division baccalaureate/associate degree category, 3 semester hours in introduction to computers, 3 in medical record terminology, 3 in medical records coding, 3 in principles of supervision, 3 in data management, and 3 in medical records management. In the upper-division baccalaureate category, 3 semester hours in organization development (6/01).

Recommendation, Skill Level 50

In the lower-division baccalaureate/associate degree category, 3 semester hours in introduction to computers, 3 in medical record terminology, 3 in medical records coding, 3 in principles of supervision, 3 in data management, and 3 in medical records management. In the upper-division baccalaureate category, 3 semester hours in organization development and 3 for a management practicum (6/01).

MOS-91H-002

Orthopedic Specialist

91H10
91H20
91H30
91H40

Exhibit Dates: 1/90–3/92.

Career Management Field: 91 (Medical), subfield 912 (Patient Care).

Description

Summary: Assists physician in treating patients with orthopedic conditions and injuries or assists in managing an orthopedic clinic. *Skill Level 10:* Able to perform the duties required for skill level 10 of MOS 91B (Medical Specialist); under close supervision of a physician, treats orthopedic patients; may perform the duties required for skill level 20. *Skill Level 20:* Able to perform the duties required for Skill Level 10; assists with minor surgery in orthopedic clinic; prepares treatment area and patient for surgery or treatment; transfers patients with unstable fractures to fracture table or other work surface; operates fracture table and assists in positioning patient; changes dressings and bandages; removes sutures; cleans and performs minor debridement of open wounds; fabricates, modifies, and removes short leg, long leg, arm casts, and hip and shoulder spica; rolls or manually shapes plaster around area to be immobilized; constructs and applies plaster and mechanized splints; makes support bandages and slings; instructs patients regarding care and danger of casts; assists in applying and adjusting skeletal traction. *Skill Level 30:* Able to perform the duties required for Skill Level 20; provides technical guidance to subordinates. *Skill Level 40:* Able to perform the duties required for Skill Level 30; assists in managing an orthopedic clinic; organizes work schedules; assigns duties; instructs subordinates in work techniques and procedures; evaluates personnel performance and prepares evaluation reports; establishes safety procedures; inspects work area to ensure a safe working environment; supervises operator maintenance on assigned equipment; requisitions, stores, and issues supplies and equipment; reviews, consolidates, and prepares technical, personnel, and administrative reports; coordinates activities of orthopedic clinic with other elements of treatment facility; assists physician with technical, personnel, and fiscal matters.

Recommendation, Skill Level 10

In the lower-division baccalaureate/associate degree category, credit in anatomy, physiology, principles of orthopedics, orthopedic methods and materials, and clinical application of orthopedic principles on the basis of institutional evaluation. (NOTE: This recommendation for skill level 10 is valid for the dates 1/90-9/91 only) (6/77).

Recommendation, Skill Level 20

In the lower-division baccalaureate/associate degree category, 5 semester hours in clinical application of orthopedic principles, 4 in principles of orthopedics, 4 in orthopedic methods and materials, 3 in clinical orthopedic assisting, 3 in anatomy, and 1 in physiology (6/77).

Recommendation, Skill Level 30

In the lower-division baccalaureate/associate degree category, 5 semester hours in clinical application of orthopedic principles, 4 in principles of orthopedics, 4 in orthopedic methods and materials, 3 in clinical orthopedic assisting, 3 in anatomy, and 1 in physiology (6/77).

Recommendation, Skill Level 40

In the lower-division baccalaureate/associate degree category, 5 semester hours in clinical application of orthopedic principles, 4 in principles of orthopedics, 4 in orthopedic methods and materials, 3 in clinical orthopedic assisting, 3 in anatomy, 3 in personnel supervision, 2 in communication skills, 2 in principles of management, 1 in health care management, 1 in physiology, and credit in medical facility management on the basis of institutional evaluation (6/77).

MOS-91H-003

Orthopedic Specialist

91H10
91H20
91H30
91H40

Exhibit Dates: 4/92–12/94. (Effective 12/94, MOS 91H was discontinued, and its duties were incorporated into MOS 91B).

Career Management Field: 91 (Medical).

Description

Summary: Assists physician in treating patients with orthopedic conditions and injuries or assists in managing an orthopedic clinic. *Skill Level 10:* Assists with minor surgery in orthopedic clinic; prepares treatment area and patient for surgery or treatment; transfers patients with unstable fractures to fracture table or other work surface; operates fracture table and assists in positioning patient; changes dressings and bandages; removes sutures; cleans and performs minor debridement of open wounds; fabricates, modifies, and removes casts; applies plaster and mechanized splints; makes support bandages and slings; instructs patients regarding care and danger of casts; assists in applying and adjusting skeletal traction. *Skill Level 20:* Able to perform the duties required for Skill Level 10; provides supervision and technical guidance to subordinates. *Skill Level 30:* Able to perform the duties required for Skill Level 20; assists in managing an orthopedic clinic; organizes work schedules; assigns duties; instructs subordinates in work techniques and procedures; evaluates personnel performance and prepares evaluation reports; establishes safety procedures; inspects work area to ensure a safe working environment; supervises operator maintenance on assigned equipment; requisitions, stores, and issues supplies and equipment; reviews, consolidates, and prepares technical, personnel, and administrative reports; coordinates activities of orthopedic clinic with other elements of treatment facility; assists physician with technical, personnel, and fiscal matters. *Skill Level 40:* Able to perform the duties required for Skill Level 30; prepares, conducts, and supervises continuing education and training; assists with operations to ensure compliance with Joint Commission Accreditation Hospital standards.

Recommendation, Skill Level 10
Credit may be granted on the basis of an individualized assessment of the student (3/94).

Recommendation, Skill Level 20
In the lower-division baccalaureate/associate degree category, 4 semester hours in orthopedic methods and materials, 4 in orthopedics clinical experience, and 3 in personnel supervision (4/92).

Recommendation, Skill Level 30
In the lower-division baccalaureate/associate degree category, 4 semester hours in orthopedic methods and materials, 5 in orthopedics clinical experience, 3 in personnel supervision, and 2 in records and information management (4/92).

Recommendation, Skill Level 40
In the lower-division baccalaureate/associate degree category, 4 semester hours in orthopedic methods and materials, 6 in orthopedics clinical experience, 3 in personnel supervision, and 2 in records and information management. In the upper-division baccalaureate category, 3 semester hours in organizational management and 3 for field experience in management (4/92).

MOS-91H-004

Optical Laboratory Specialist
91H10
91H20
91H30
91H40

Exhibit Dates: 12/01–Present.

Career Management Field: 91 (Medical).

Description
Summary: Makes prescription lenses or supervises optical laboratories. *Skill Level 10:* Makes, duplicates, and inserts prescription lenses; selects proper stock to fill requirements of spectacle prescription; computes and records curvature and thickness and marks lenses for surfacing; grinds and polishes lenses to prescribed specifications; locates and identifies optical center, optical axis, and cutting line on lenses; chucks lenses for edging; edges lenses to correct size and shape; hardens glass lenses to required impact resistance; selects and assembles frame components; mounts lenses and aligns frames; prepares completed spectacles for shipment. *Skill Level 20:* Able to perform the duties required for Skill Level 10; edits prescriptions and maintains records of prescriptions received by optical unit; calculates and records amount of prism required to permit necessary decentration; calculates and records distance required between top of bifocal or trifocal segments and distance-optical-center to produce prescribed segment height; dyes plastic lenses to specified tints; maintains files of completed spectacle prescriptions; prepares surface work sheets for plastic lens coating; calculates and records positional effective lens power; maintains inventory of supplies and equipment. *Skill Level 30:* Able to perform the duties required for Skill Level 20; supervises an optical laboratory; establishes work priorities; organizes work schedules; inspects completed spectacles for compliance with prescription requirements and quality standards; assigns duties and trains subordinates in work techniques and procedures; evaluates personnel performance; determines stock levels; maintains supply records; requisitions, stores, and issues supplies; supervises preventive maintenance of equipment. *Skill Level 40:* Able to perform the duties required for Skill Level 30; supervises an optical laboratory; maintains records of cost data and expenditures; ensures that the laboratory is safe, clean, and orderly; supervises quality control procedures; plans layout areas; prepares technical, personnel, and administrative reports; coordinates activities of the optical laboratory with medical treatment facilities; advises and assists professionals in fiscal, technical, and administrative matters.

Recommendation, Skill Level 10
Credit may be granted on the basis of an individualized assessment of the student (12/01).

Recommendation, Skill Level 20
Credit may be granted on the basis of an individualized assessment of the student (12/01).

Recommendation, Skill Level 30
In the lower-division baccalaureate/associate degree category, 4 semester hours in optical laboratory procedures, 3 in optical laboratory technology, 3 in principles of supervision, and 3 in records and information management. In the upper-division baccalaureate category, 3 semester hours for a management practicum (12/01).

Recommendation, Skill Level 40
In the lower-division baccalaureate/associate degree category, 4 semester hours in optical laboratory procedures, 3 in optical laboratory technology, 3 in principles of supervision, and 3 in records and information management. In the upper-division baccalaureate category, 3 in organizational management and 3 for a management practicum (12/01).

MOS-91J-002

Physical Therapy Specialist
91J10
91J20
91J30
91J40

Exhibit Dates: 1/90–5/92.

Career Management Field: 91 (Medical), subfield 912 (Patient Care).

Description
Summary: Under the direction of a physical therapist and following medical prescription, supervises or administers physical therapy treatment and exercises to decrease physical disabilities and to promote physical fitness of patients. *Skill Level 10:* Able to perform the duties required for skill level 10 of MOS 91B (Medical Specialist); under close supervision, administers physical therapy treatments to patients. *Skill Level 20:* Able to perform the duties required for Skill Level 10; transports, positions, and drapes patients for treatment; administers treatment involving use of heat, ice, cervical and pelvic traction, ultraviolet light, ultrasound treatment, whirlpool, hydrotherapy, therapeutic massage, electrical stimulation, and manual exercise procedures; assists in performing postural drainage; instructs and assists patients with casts in ambulation and gait training; administers progressive resistance and physical reconditioning exercises; stores and issues supplies and equipment. *Skill Level 30:* Able to perform the duties required for Skill Level 20; assists with management of medium-sized physical therapy clinic; instructs and assists patients in exercise activities; performs exercise procedures for amputees and patients with central nervous system disorders; supervises operational management of unit, including personnel, supplies, equipment, and scheduling; prepares patient and administrative reports; coordinates activities of physical therapy clinic with other elements of medical treatment facility. *Skill Level 40:* Able to perform the duties required for Skill Level 30; assists with management of large physical therapy clinic; supervises integration of segmented and diversified physical therapy programs conducted in several clinics.

Recommendation, Skill Level 10
In the lower-division baccalaureate/associate degree category, 6 semester hours in health science. (NOTE: This recommendation for skill level 10 is valid for the dates 1/90-9/91 only) (6/77).

Recommendation, Skill Level 20
In the lower-division baccalaureate/associate degree category, 6 semester hours in health science and 6 in physical therapy (6/77).

Recommendation, Skill Level 30
In the lower-division baccalaureate/associate degree category, 6 semester hours in health science, 6 in physical therapy, 2 in human relations, and 1 for field experience in management. In the upper-division baccalaureate category, credit in patient care administration on the basis of institutional evaluation (6/77).

Recommendation, Skill Level 40
In the lower-division baccalaureate/associate degree category, 6 semester hours in health science, 6 in physical therapy, 3 in human relations, and 3 for field experience in management. In the upper-division baccalaureate category, credit in patient care administration and medical facility management on the basis of institutional evaluation (6/77).

MOS-91J-003

Physical Therapy Specialist
91J10
91J20
91J30
91J40

Exhibit Dates: 6/92–12/94. (Effective 12/94, MOS 91J was discontinued, and its duties were incorporated into MOS 91B.)

Career Management Field: 91 (Medical).

Description
Summary: Under the direction of a physical therapist and following medical prescription, supervises or administers physical therapy treatment and exercises to decrease physical disabilities and to promote physical fitness of patients. *Skill Level 10:* Transports, positions, and drapes patients for treatment; administers treatment involving use of heat, ice, cervical and pelvic traction, ultraviolet light, ultrasound treatment, whirlpool, hydrotherapy, therapeutic massage, electrical stimulation, and manual exercise procedures; assists in performing postural drainage; instructs and assists patients with casts in ambulation and gait training; administers progressive resistance and physical reconditioning exercises; stores and issues supplies and equipment. *Skill Level 20:* Able to perform the duties required for Skill Level 10; assists with management of medium-sized physical therapy clinic; instructs and assists patients in exercise activities; performs exercise

procedures for amputees and patients with central nervous system disorders; supervises operational management of unit, including personnel, supplies, equipment, and scheduling; prepares patient and administrative reports; coordinates activities of physical therapy clinic with other elements of medical treatment facility. *Skill Level 30:* Able to perform the duties required for Skill Level 20; controls patient flow through clinic; when appropriate, administers treatment independently; prepares budget, technical, administrative, and patient case reports; maintains supplies and equipment; recommends new equipment; assists in personnel matters. *Skill Level 40:* Able to perform the duties required for Skill Level 30; assists with management of large physical therapy clinic; supervises integration of segmented and diversified physical therapy programs conducted in several clinics.

Recommendation, Skill Level 10

Credit may be granted on the basis on an individualized assessment of the student (3/94).

Recommendation, Skill Level 20

In the lower-division baccalaureate/associate degree category, 5 semester hours in physical therapy clinical procedures and 3 in personnel supervision (6/92).

Recommendation, Skill Level 30

In the lower-division baccalaureate/associate degree category, 8 semester hours in physical therapy clinical procedures, 3 in personnel supervision, 3 in patient assessment procedures, and 2 in records and information management (6/92).

Recommendation, Skill Level 40

In the lower-division baccalaureate/associate degree category, 8 semester hours in physical therapy clinical procedures, 3 in personnel supervision, 3 in patient assessment procedures, and 2 in records and information management. In the upper-division baccalaureate category, 3 semester hours in organizational management and 3 for field experience in management (6/92).

MOS-91J-004

Medical Logistics Specialist

91J10
91J20
91J30
91J40
91J50

Exhibit Dates: 10/01–Present.

Career Management Field: 91 (Medical).

Description

Summary: Supervises or performs requisitioning, receipt, inventory management, storage, preservation, stock control and quality, and accounting of medical supplies and equipment. *Skill Level 10:* Transports and handles medical supplies and equipment; inventories and issues medical supplies; performs in-storage care and preservation; applies special procedures for handling, storage, packaging, and shipping pharmaceuticals, biologicals, blood fractions, and medicines; interprets medical terminology and vocabulary directly related to supply function; operates office machines; keeps administrative and medical supply files; assists in keeping records on equipment operation and maintenance. *Skill Level 20:* Able to perform the duties required for Skill Level 10; supervises and provides technical guidance to subordinates; responsible for local purchasing. *Skill Level 30:* Able to perform the duties required for Skill Level 20; supervises small medical supply, stock control, or storage activity; plans medical supply and equipment operations; assigns duties, plans work loads, and instructs personnel in work techniques and procedures; responsible for budgeting and managing funds; works with automated property book system. *Skill Level 40:* Able to perform the duties required for Skill Level 30; supervises a medium-sized medical supply, stock control, or storage activity. *Skill Level 50:* Able to perform the duties required for Skill Level 40; supervises a large medical supply, stock control, or storage activity or serves as principal noncommissioned officer of medical depot; analyzes medical logistical management data.

Recommendation, Skill Level 10

Credit may be granted on the basis of an individualized assessment of the student (6/01).

Recommendation, Skill Level 20

Credit may be granted on the basis of an individualized assessment of the student (6/01).

Recommendation, Skill Level 30

In the lower-division baccalaureate/associate degree category, 3 semester hours in computer applications, 3 in supply management, 3 in warehouse operations, 3 in principles of supervision, and 3 in records management (6/01).

Recommendation, Skill Level 40

In the lower-division baccalaureate/associate degree category, 3 semester hours in computer applications, 3 in supply management, 3 in warehouse operations, 3 in principles of supervision, and 3 in records management. In the upper-division baccalaureate category, 3 semester hours in organization development and 3 supply chain management (6/01).

Recommendation, Skill Level 50

In the lower-division baccalaureate/associate degree category, 3 semester hours in computer applications, 3 in supply management, 3 in warehouse operations, 3 in principles of supervision, and 3 in records management. In the upper-division baccalaureate category, 3 semester hours in organization development, 3 in supply chain management, and 3 for a management practicum (6/01).

MOS-91K-001

Medical Laboratory Specialist

91K10
91K20
91K30
91K40
91K50

Exhibit Dates: 8/94–Present.

Career Management Field: 91 (Medical).

Description

Summary: NOTE: MOS 91K is a consolidation of three MOS's: 92B, Medical Laboratory Specialist; 92E, Cytology Specialist; and 01H, Biological Sciences Assistant. Duties formerly associated with MOS 92B are incorporated in MOS 91K; duties formerly associated with MOS 92E have been transferred to MOS 91K with the Additional Skill Identifier (ASI) M2 (Cytology Specialist); duties formerly associated with MOS 01H have been transferred to MOS 91K with the ASI P9 (Biological Sciences).

For MOS 91K without ASI M2 or P9: Performs laboratory procedures and elementary and advanced examinations of biological and environmental specimens to aid in diagnosis, treatment, and prevention of disease.

For MOS 91K with ASI M2: Provides professional support and service to pathologists by supervising or performing diagnostic cytology services and assisting in rendering diagnoses.

For MOS 91K with ASI P9: Performs professional-level laboratory and research duties in the field of biological science. NOTE: The prerequisite for the assignment of ASI P9 is a bachelor's degree and a minimum of six months of related experience; or a master's degree in biology, bacteriology, zoology, parisitology, botany, pharmacology, or entomology. *Skill Level 10:* For 91K10: Receives and logs requests for laboratory tests; collects labels and accessions specimens; prepares equipment, reagents, and supplies for routine laboratory testing; performs routine laboratory tests in urinalysis, hematology, clinical chemistry, microbiology, blood bank and serology; runs appropriate quality control; calculates and reports results of analyses under supervision; performs preventative maintenance on laboratory equipment.

For 91K10P9: Conducts studies in biology, bacteriology, biochemistry, entomology, or pharmacology; performs culture work on animal diseases; keeps records of culturing and of purity, density, and viability tests of cultures prepared for use as vaccines; inoculates and performs autopsies on laboratory animals used in the preparation of cultures; identifies, prepares, and ships cultures; makes bioassays to determine toxicity of drugs and other substances; plans and executes experiments; calculates and administers doses and observes toxic effects produced; prepares detailed reports on experiments and tests. *Skill Level 20:* Able to perform the duties required for Skill Level 10; for 91K20: Maintains supply and equipment inventory in assigned laboratory section; performs first echelon maintenance on instrumentation; prepares specimens for shipment to reference laboratories; collects and processes donor blood; prepares blood components; identifies microorganisms; performs antibody identification; prepares histological sections; participates in technical problem solving and quality control monitoring.

For 91K20M2: Logs, processes, and examines a variety of cellular samples of tissues and body fluids and renders diagnoses; examines microscope slides containing cells and processes cytological specimens from vagina and cervix, pleural fluid, gastric juice, bronchial washings, ascitic fluid, spinal fluid, prostatic fluid or needle aspiration, breast secretions or aspirates, urine, sputum, buccal smears, other body secretions, excretions, or scrapings, bones, and soft tissue tumors; identifies a variety of cell types from various parts of body and interprets morphological abnormalities; interprets variations that fall within normal limits and subtle variations in gross and microscopic appearance of cytological specimens and correlates clinical history with microscopic findings; evaluates and interprets cellular morphological changes associated with infectious processes, including viral, bacterial, fungal, and parasitic diseases; identifies organisms where possible on morphological basis; interprets endocrinological states, including both normal variations and abnormal conditions as shown by cellular reactions; evaluates cellular changes brought about by radiation, drugs, and other forms of therapy; applies knowledge of carcinogenesis to interpretation of premalignant and malignant cellular changes; recognizes and reports a wide variety of neoplasms, including both epithelial and mesodermal; differentiates between carcinoma-in-situ and invasive squamous carcinoma; classifies and grades neoplasm according to current classification schemes; determines, where possible, whether neoplasm is primary or metastatic; prepares written diagnostic reports on all cases based on correlation of clinical data with results for cytological study, including interpretation of benign versus malignant, classification or grade of neoplasm, and recommendations for the need of further diagnostic procedures; assists in, selects, and advises on proper mode of collection of cytological specimens; recognizes and evaluates departures from expected performance norms and determines and performs necessary actions for correcting procedural deficiencies; examines specimens describing color, viscosity, amount, and specific gravity; determines, selects, and smears portions of the specimen that may have particular pathological significance; performs cytocentrifuge preparations; selects and performs quality control procedures; performs routine histology procedures; inspects, cleans, adjusts, and calibrates laboratory equipment.

For 91K20P9: Analyzes scientific data publications; maintains inventories and equipment. *Skill Level 30:* Able to perform the duties required for Skill Level 20; for 91K30: Performs complex procedures in clinical laboratory sections and research laboratories; supervises pother laboratory personnel in the performance of routine laboratory procedures, instrument operations, and quality control; participates in the acquisition and setup of new instrumentation and procedures; provides technical assistance to subordinates.

For 91K30M2: Provides technical guidance to subordinates; supervises small cytology section; inspects laboratory to ensure orderly, clean, and safe environment; organizes work schedules and assigns duties; instructs subordinates in work techniques and procedures; evaluates personnel performances and counsels personnel; prepares evaluation reports; establishes and maintains teaching file of interesting or unusual cases; implements and conducts inservice and performance-oriented training programs; evaluates and initiates new procedures; performs advanced cytological procedures including chromosomal analysis by karyotyping; establishes and administers program for correlation of cytological diagnosis with histological findings; reports findings to pathologist; reviews findings with pathologist; advises and assists in administrative, fiscal, personnel, and supply matters; prepares administrative, technical, and personnel reports; requisitions, safeguards, issues, and monitors supplies and equipment; establishes report control system; prepares and reviews correspondence; supervises maintenance of office files and technical library; supervises operational and preventive maintenance program for assigned equipment.

For 91K30P9: Provides overall research division supervision, including quality control and training. *Skill Level 40:* Able to perform the duties required for Skill Level 30; for 91K40: develops and supervises quality control measures; establishes work priorities; organizes work schedules and assigns duties; instructs subordinates in technical procedures; insure that the laboratory complies with OSHA standards; evaluates personnel performance and counsels subordinates; prepares administrative, technical, and personnel reports; coordinates activities of the laboratory with other elements of the medical facility; advises and assists the laboratory officer with all aspects of the laboratory management; supervises ordering and storage of supplies and equipment.

For 91K40M2: supervises medium-sized and large cytology laboratories; develops and supervises quality control program; participates in training of residents physicians in pathology; reviews and consolidates reports. *Skill Level 50:* Able to perform the duties required for Skill Level 40; supervises all activities in a large medical laboratory; prepares reports; determines personnel requirements and conducts inservice presentations.

Recommendation, Skill Level 10

Credit may be granted on the basis of an individualized assessment of the student (11/96).

Recommendation, Skill Level 20

In the lower-division baccalaureate/associate degree category, for 91K20 without ASI M2 or P9: 15 semester hours in clinical laboratory experience. In the upper-division baccalaureate category, for 91K20M2, 4 semester hours in introduction to cytology, 3 in female genital tract (benign), 4 in female genital tract (malignant), 2 in respiratory tract cytology, 2 in gastrointestinal tract cytology, 3 in central nervous system and effusion, 2 in urinary tract cytology, 3 in fine needle aspiration cytology, 2 in research, 18 for clinical experience, and 1 in organizational management. In the graduate degree category, for 91K20P9, 6 semester hours in biological research. (NOTE: This recommendation for skill level 20 is valid for the dates 8/94-2/95 only) (11/96).

Recommendation, Skill Level 30

In the lower-division baccalaureate/associate degree category, for 91K30 without ASI M2 or P9, 15 semester hours for clinical laboratory experience, 5 in laboratory instrumentation, 5 in advanced clinical laboratory procedures, 3 in personnel supervision, 3 in laboratory information management, 3 in inventory control, and 3 in special clinical methods. In the upper-division baccalaureate category, for 91K30M2, 4 semester hours in introduction to cytology, 3 in female genital tract (benign), 4 in female genital tract (malignant), 2 in respiratory tract cytology, 2 in gastrointestinal tract cytology, 3 in central nervous system and effusion, 2 in urinary tract cytology, 3 in fine needle aspiration cytology, 2 in research, 24 for clinical experience, 3 in organizational management, and 3 in laboratory administration. In the graduate degree category, for 91K30P9, 9 semester hours in biological research (11/96).

Recommendation, Skill Level 40

In the lower-division baccalaureate/associate degree category, for MOS 91K40 without ASI M2 or P9: 15 semester hours for clinical laboratory experience, 5 in laboratory instrumentation, 5 in advanced clinical laboratory procedures, 3 in personnel supervision, 3 in laboratory information management, 3 in inventory control, and 3 in special clinical methods. In the upper-division baccalaureate category, for MOS 91K40: 3 semester hours in organizational management and 3 for field experience in management. For MOS 91K40M2: 4 semester hours in introduction to cytology, 3 in female genital tract (benign), 4 in female genital tract (malignant), 2 in respiratory tract cytology, 2 in gastrointestinal tract cytology, 3 in central nervous system and effusion, 2 in urinary tract cytology, 3 in fine needle aspiration cytology, 2 in research, 24 for clinical experience, 6 in organizational management, 6 in laboratory administration, and 3 in educational methodology. In the graduate degree category, for MOS 91K40P2: 9 semester hours in biological research (11/96).

Recommendation, Skill Level 50

In the lower-division baccalaureate/associate degree category, 15 semester hours for clinical laboratory experience, 5 in laboratory instrumentation, 5 in advanced clinical laboratory procedures, 3 in personnel supervision, 3 in laboratory information management, 3 in inventory control, and 3 in special clinical methods. In the upper-division baccalaureate category, for 91K50 without ASI M2 or P9: 3 semester hours in organizational management and 3 for field experience in management. For 91K50M2: 4 semester hours in introduction to cytology, 3 in female genital tract (benign), 4 in female genital tract (malignant), 2 in respiratory tract cytology, 2 in gastrointestinal tract cytology, 3 in central nervous system and effusion, 2 in urinary tract cytology, 3 in fine needle aspiration cytology, 2 in research, 24 for clinical experience, 6 in organizational management, 6 in laboratory administration, and 3 in educational methodology. In the graduate degree category, for 91K50P9: 12 semester hours in biological research (11/96).

MOS-91L-003

Occupational Therapy Specialist

91L10
91L20
91L30
91L40

Exhibit Dates: 1/90–12/94. (Effective 12/94, MOS 91L was discontinued, and its duties were incorporated into MOS 91B.)

Career Management Field: 9 (Medical).

Description

Summary: Under the direction of a registered occupational therapist, assists patients in adapting skills for daily living and in work therapy activities; conducts interviews, administers tests, and assists patients with daily living activities to promote physical and mental health. *Skill Level 10:* Interviews, tests, teaches, and assists clients in occupational therapy activities; assists in evaluating test results; interprets physician's referrals to determine therapeutic strategies; arranges for and reports physiological and psychological patient responses; maintains progress records; measures, fabricates, fits, adjusts, and instructs in the use of adaptive devices; maintains equipment and supplies; maintains safe and comfortable work area. *Skill Level 20:* Able to perform the duties required for Skill Level 10; assists in administering occupational therapy programs; plans ward, clinic, and recreational activity programs and assists with planning for home activity programs; teaches activities of daily living. *Skill Level 30:* Able to perform the duties required for Skill Level 20; assists with management of medium-sized occupational therapy clinic; prepares work schedules; instructs subordinates;

evaluates personnel performance; prepares administrative and patient reports; coordinates activities of occupational therapy program; advises and assists therapist in personnel and administrative matters; maintains supply levels as needed; coordinates activities of occupational therapy clinic with other elements of the medical treatment facility. *Skill Level 40:* Able to perform the duties required for Skill Level 30; assists with management of a large occupational therapy clinic; coordinates integration of segmented and diversified occupational therapy programs conducted in several clinics; develops, supervises, and evaluates work methods and training programs; plans clinic layout; requests supporting services.

Recommendation, Skill Level 10

In the lower-division baccalaureate/associate degree category, 3 semester hours in occupational therapy skills, 3 for field experience in occupational therapy, 3 in creative arts, 3 in communication skills, and 3 in human relations. (NOTE: This recommendation for skill level 10 is valid for the dates 1/90-9/91 only) (3/90).

Recommendation, Skill Level 20

In the lower-division baccalaureate/associate degree category, 6 semester hours in occupational therapy skills, 6 for field experience in occupational therapy, 6 in creative arts, 3 in communication skills, and 3 in human relations (3/90).

Recommendation, Skill Level 30

In the lower-division baccalaureate/associate degree category, 6 semester hours in occupational therapy skills, 6 for field experience in occupational therapy, 6 in creative arts, 3 in communication skills, 3 in human relations, 3 in principles of supervision, and 3 in patient care administration (3/90).

Recommendation, Skill Level 40

In the lower-division baccalaureate/associate degree category, 6 semester hours in occupational therapy skills, 6 for field experience in occupational therapy, 6 in creative arts, 3 in communication skills, 3 in human relations, 3 in principles of supervision, 3 in patient care administration, and 3 in educational techniques (3/90).

MOS-91M-001

Hospital Food Service Specialist

91M10
91M20
91M30
91M40
91M50

Exhibit Dates: 3/90–2/95.

Career Management Field: 91 (Medical).

Description

Summary: Assists in the planning and implementation of the food service program in hospitals; prepares, cooks, and serves food to patients and staff. *Skill Level 10:* Interviews patients and family members to determine dietary status; completes dietary history and individual menu pattern; evaluates, records, and reviews patient's diet; monitors patient's food consumption and maintains dietary records; measures, weighs, blends, and mixes various foodstuffs; roasts, fries, broils, boils, and stews meats; prepares fruits, vegetables, salads, desserts, beverages, and dairy products; bakes yeast breads, quick breads, pies, cakes, and cookies; sets up and serves regular and modified diets for patients; assembles and checks patient trays; instructs patients on modified diets to aid them in food selection; maintains records as to patient food preferences and diet changes and adjusts diet menus; prepares food in accordance with instruction from dietician or physician; performs preventive maintenance on kitchen equipment. *Skill Level 20:* Able to perform the duties required for Skill Level 10; provides technical guidance to subordinates; modifies dietary treatments in compliance with nutrition care plan; assists in evaluating nutrition problems; completes clinical dietetic reports; performs quality assurance checks in nutrition care operations; ensures adherence to procedures, temperatures, and time periods in food preparation; adjusts individual standard diet menus; participates in planning of food service activities; conducts food acceptability studies; applies storage standards; completes food production forms and records; monitors tasks in support of patient tray service operation, including preparation, assembly, and delivery to wards. *Skill Level 30:* Able to perform the duties required for Skill Level 20; supervises food production or diet therapy activities in a hospital; plans and prepares hospital menus for regular and modified diets and estimates the number of persons to be fed; schedules and supervises food preparation for compliance with dietetic requirements and for quality and quantity standards; inspects prepared food for palatability and appearance and to determine compliance with menus and recipes; inspects hospital food service facilities and corrects deficiencies; supervises procurement, storage, distribution, accounting for, and servicing of food in a hospital food service; conducts classroom and on-the-job training for subordinates. *Skill Level 40:* Able to perform the duties required for Skill Level 30; performs supervisory duties; develops, coordinates, implements, and advises on food programs; evaluates and designs improvements in food programs and training programs; assists in preparation of hospital food service budget. *Skill Level 50:* Able to perform the duties required for Skill Level 40; prepares budget and determines financial requirements; instructs military and civilian personnel; utilizes office management techniques; advises superiors on food service equipment layout and space design.

Recommendation, Skill Level 10

In the lower-division baccalaureate/associate degree category, 4 semester hours in quantity food preparation and 4 in nutrition and diet therapy. (NOTE: This recommendation for skill level 10 is valid for the dates 3/90-9/91 only) (6/92).

Recommendation, Skill Level 20

In the lower-division baccalaureate/associate degree category, 4 semester hours in quantity food preparation, 6 in nutrition and diet therapy, 2 in food service operations, and 3 in personnel supervision (6/92).

Recommendation, Skill Level 30

In the lower-division baccalaureate/associate degree category, 4 semester hours in quantity food preparation, 6 in nutrition and diet therapy, 4 in food service operations, 3 in personnel supervision, and 2 in records and information management (6/92).

Recommendation, Skill Level 40

In the lower-division baccalaureate/associate degree category, 4 semester hours in quantity food preparation, 6 in nutrition and diet therapy, 4 in food service operations, 3 in personnel supervision, and 2 in records and information management. In the upper-division baccalaureate category, 3 semester hours in organizational management and 3 for field experience in management (6/92).

Recommendation, Skill Level 50

In the lower-division baccalaureate/associate degree category, 4 semester hours in quantity food preparation, 6 in nutrition and diet therapy, 4 in food service operations, 3 in personnel supervision, and 2 in records and information management. In the upper-division baccalaureate category, 3 semester hours in organizational management and 6 for field experience in management; if paygrade E-9 has been attained, additional credit as follows: 3 semester hours in management problems, 3 in operations management, and 3 in communications techniques for managers (6/92).

MOS-91M-002

Hospital Food Service Specialist

91M10
91M20
91M30
91M40
91M50

Exhibit Dates: 3/95–11/01.

Career Management Field: 91 (Medical).

Description

Summary: Assists in the planning and implementation of the food service program in hospitals; prepares, cooks, and serves food to patients and staff. *Skill Level 10:* Interviews patients and family members to determine dietary status; completes dietary history and individual menu pattern; evaluates, records, and reviews patient's diet; monitors patient's food consumption and maintains dietary records; measures, weighs, blends, and mixes various foodstuffs; roasts, fries, broils, boils, and stews meats; prepares fruits, vegetables, salads, desserts, beverages, and dairy products; bakes yeast breads, quick breads, pies, cakes, and cookies; sets up and serves regular and modified diets for patients; assembles and checks patient trays; instructs patients on modified diets to aid them in food selection; maintains records as to patient food preferences and diet changes and adjusts diet menus; prepares food in accordance with instruction from dietician or physician; performs preventive maintenance on kitchen equipment. *Skill Level 20:* Able to perform the duties required for Skill Level 10; provides technical guidance to subordinates; modifies dietary treatments in compliance with nutrition care plan; assists in evaluating nutrition problems; completes clinical dietetic reports; performs quality assurance checks in nutrition care operations; ensures adherence to procedures, temperatures, and time periods in food preparation; adjusts individual standard diet menus; participates in planning of food service activities; conducts food acceptability studies; applies storage standards; completes food production forms and records; monitors tasks in support of patient tray service operation, including preparation, assembly, and delivery to wards. *Skill Level 30:* Able to perform the duties required for Skill Level 20; supervises food production or diet therapy activities in a hospital; plans and prepares hospital menus for regular and modified diets and estimates the number of persons to be fed; schedules and supervises food preparation for compliance with dietetic requirements and for quality and quantity standards; inspects prepared food for palatability and appearance and to determine compliance with menus and recipes; inspects hospital food service facilities and corrects deficiencies; supervises procurement, storage, distribution, accounting for, and servicing of food in a hospital food service; conducts classroom and on-the-job training for subordinates. *Skill Level 40:* Able to perform the duties required for Skill Level 30; performs supervisory duties; develops, coordinates, implements, and advises on food programs; evaluates and designs improvements in food programs and training programs; assists in preparation of hospital food service budget. *Skill Level 50:* Able to perform the duties required for Skill Level 40; prepares budget and determines financial requirements; instructs military and civilian personnel; utilizes office management techniques; advises superiors on food service equipment layout and space design.

Recommendation, Skill Level 10

Credit may be granted on the basis of an individualized assessment of the student (6/92).

Recommendation, Skill Level 20

Credit may be granted on the basis of an individualized assessment of the student (6/92).

Recommendation, Skill Level 30

In the lower-division baccalaureate/associate degree category, 4 semester hours in quantity food preparation, 6 in nutrition and diet therapy, 2 in food service operations, and 3 in personnel supervision (6/92).

Recommendation, Skill Level 40

In the lower-division baccalaureate/associate degree category, 4 semester hours in quantity food preparation, 6 in nutrition and diet therapy, 4 in food service operations, 3 in personnel supervision, and 2 in records and information management. In the upper-division baccalaureate category, 3 semester hours in organizational management and 3 for field experience in management (6/92).

Recommendation, Skill Level 50

In the lower-division baccalaureate/associate degree category, 4 semester hours in quantity food preparation, 6 in nutrition and diet therapy, 4 in food service operations, 3 in personnel supervision, and 2 in records and information management. In the upper-division baccalaureate category, 3 semester hours in organizational management and 6 for field experience in management; if paygrade E-9 has been attained, additional credit as follows: 3 semester hours in management problems, 3 in operations management, and 3 in communications techniques for managers (6/92).

MOS-91M-003

Hospital Food Service Specialist

91M10
91M20
91M30
91M40
91M50

Exhibit Dates: 12/01–Present.

Career Management Field: 91 (Medical).

Description

Summary: Assists in the planning and implementation of the food service program in hospitals; prepares, cooks, and serves food to patients and staff. *Skill Level 10:* Interviews patients and family members to determine dietary status; completes dietary history and individual menu pattern; evaluates, records, and reviews patient's diet; monitors patient's food consumption and maintains dietary records; measures, weighs, blends, and mixes various foodstuffs; roasts, fries, broils, boils, and stews meats; prepares fruits, vegetables, salads, desserts, beverages, and dairy products; bakes yeast breads, quick breads, pies, cakes, and cookies; sets up and serves regular and modified diets for patients; assembles and checks patient trays; instructs patients on modified diets to aid them in food selection; maintains records as to patient food preferences and diet changes and adjusts diet menus; prepares food in accordance with instruction from dietician or physician; performs preventive maintenance on kitchen equipment. *Skill Level 20:* Able to perform the duties required for Skill Level 10; supervises subordinates; modifies dietary treatments in compliance with nutrition care plan; assists in evaluating nutrition problems; completes clinical dietetic reports; performs quality assurance checks in nutrition care operations; ensures adherence to procedures, temperatures, and time periods in food preparation; adjusts individual standard diet menus; participates in planning of food service activities; conducts food acceptability studies; applies storage standards; completes food production forms and records; monitors tasks in support of patient tray service operation, including preparation, assembly, and delivery to wards; provides education to the staff and the community. *Skill Level 30:* Able to perform the duties required for Skill Level 20; supervises food production or diet therapy activities in a hospital; plans and prepares hospital menus for regular and modified diets and estimates the number of persons to be fed; schedules and supervises food preparation for compliance with dietetic requirements and for quality and quantity standards; inspects prepared food for palatability and appearance and to determine compliance with menus and recipes; inspects hospital food service facilities and corrects deficiencies; supervises procurement, storage, distribution, accounting for, and servicing of food in a hospital food service; conducts classroom and on-the-job training for subordinates; coordinates work schedules; evaluates personnel performance; prepares reports; maintains records; performs data management. *Skill Level 40:* Able to perform the duties required for Skill Level 30; performs supervisory duties; develops, coordinates, implements, and advises on food programs; evaluates and designs improvements in food programs and training programs; assists in preparation of hospital food service budget. *Skill Level 50:* Able to perform the duties required for Skill Level 40; supervises the nutrition care divisions or appropriate headquarters staff position.

Recommendation, Skill Level 10

Credit may be granted on the basis of an individualized assessment of the student (12/01).

Recommendation, Skill Level 20

Credit may be granted on the basis of an individualized assessment of the student (12/01).

Recommendation, Skill Level 30

In the lower-division baccalaureate/associate degree category, 4 semester hours in quantity food preparation, 6 in nutrition and diet therapy, 2 in food service operations, 3 in records and information management, and 3 in personnel supervision. In the upper-division baccalaureate category, 3 semester hours in organizational management, 3 for management practicum, and 3 in nutrition education (12/01).

Recommendation, Skill Level 40

In the lower-division baccalaureate/associate degree category, 4 semester hours in quantity food preparation, 6 in nutrition and diet therapy, 4 in food service operations, 3 in personnel supervision, and 3 in records and information management. In the upper-division baccalaureate category, 3 semester hours in organizational management, 3 in nutrition education, and 3 for a management practicum (12/01).

Recommendation, Skill Level 50

In the lower-division baccalaureate/associate degree category, 4 semester hours in quantity food preparation, 6 in nutrition and diet therapy, 4 in food service operations, 3 in personnel supervision, and 3 in records and information management. In the upper-division baccalaureate category, 3 semester hours in organizational management, 3 in nutrition education, and 6 for a management practicum (12/01).

MOS-91N-002

Cardiac Specialist

91N10
91N20
91N30
91N40

Exhibit Dates: 1/90–3/92.

Career Management Field: 91 (Medical), subfield 912 (Patient Care).

Description

Summary: Assists with management of cardiac clinic or performs specialized noninvasive cardiac tests and examinations. *Skill Level 10:* Able to perform the duties required for skill level 10 of MOS 91B (Medical Specialist); under close supervision, assists skill level 20 personnel. *Skill Level 20:* Able to perform the duties required for Skill Level 10; sets up, calibrates, operates, and maintains electrocardiograph, vectorcardiograph, and phonocardiograph; attaches electrodes and conducts tests according to established procedures; transmits record to physician; cuts and mounts tracing for inclusion in medical records; administers endurance tests; under supervision, performs technical examination during operation of artificial pacemaker; performs cardiopulmonary resuscitation and uses defibrillator; performs maintenance on equipment; receives, stores, and maintains supplies. *Skill Level 30:* Able to perform the duties required for Skill Level 20; provides technical guidance to subordinates. *Skill Level 40:* Able to perform the duties required for Skill Level 30; establishes priorities and work schedules; assigns duties and instructs subordinates in procedures; inspects facilities; determines personnel requirements; conducts training programs; supervises operational maintenance program of laboratory and clinical equipment; evaluates and counsels personnel; establishes stock level for supplies and equipment and supervises requisition, storage, and issue of supplies.

Recommendation, Skill Level 10

In the lower-division baccalaureate/associate degree category, 3 semester hours in cardiovascular anatomy and physiology, 3 in cardiovascular patient care, 2 in cardiovascular technology, 2 in cardiovascular electronics, 2 in physiological chemistry, and 1 in technical mathematics. (NOTE: This recommendation for skill level 10 is valid for the dates 1/90-9/91 only) (6/77).

Recommendation, Skill Level 20

In the lower-division baccalaureate/associate degree category, 10 semester hours for a cardiovascular internship, 3 in cardiovascular anatomy and physiology, 3 in cardiovascular patient care, 2 in cardiovascular technology, 2 in cardiovascular electronics, 2 in physiological chemistry, and 1 in technical mathematics (6/77).

Recommendation, Skill Level 30

In the lower-division baccalaureate/associate degree category, 10 semester hours for a cardiovascular internship, 3 in cardiovascular anatomy and physiology, 3 in cardiovascular patient care, 2 in cardiovascular technology, 2 in cardiovascular electronics, 2 in physiological chemistry, 1 in technical mathematics, and 2 in personnel supervision (6/77).

Recommendation, Skill Level 40

In the lower-division baccalaureate/associate degree category, 10 semester hours for a cardiovascular internship, 3 in cardiovascular anatomy and physiology, 3 in cardiovascular patient care, 2 in cardiovascular technology, 2 in cardiovascular electronics, 2 in physiological chemistry, 1 in technical mathematics, 3 in personnel supervision, 3 in human relations, and 2 for field experience in management (6/77).

MOS-91N-003

Cardiac Specialist
91N10
91N20
91N30
91N40

Exhibit Dates: 4/92–6/95. Effective 6/95, MOS 91N was discontinued, and its duties were incorporated into MOS 91B.

Career Management Field: 91 (Medical).

Description

Summary: Assists with management of cardiac clinic or performs specialized noninvasive cardiac tests and examinations. *Skill Level 10:* Sets up, calibrates, operates, and maintains electrocardiograph, vectorcardiograph, and phonocardiograph; attaches electrodes and conducts tests according to established procedures; transmits record to physician; cuts and mounts tracing for inclusion in medical records; administers endurance tests; under supervision, performs technical examination during operation of artificial pacemaker; performs cardiopulmonary resuscitation and uses defibrillator; performs maintenance on equipment; receives, stores, and maintains supplies. *Skill Level 20:* Able to perform the duties required for Skill Level 10; prepares hand receipts; prepares maintenance requests; requisitions equipment and supplies. *Skill Level 30:* Able to perform the duties required for Skill Level 20; provides technical guidance to subordinates; assists with supervisory duties in a cardiac clinic; organizes work schedules. *Skill Level 40:* Able to perform the duties required for Skill Level 30; establishes work priorities and work schedules; assigns duties and instructs subordinates in procedures; inspects facilities; determines personnel requirements and conducts training programs; supervises operational maintenance program of laboratory and clinical equipment; evaluates and counsels personnel; establishes stock level for supplies and equipment and supervises requisition, storage, and issue of supplies.

Recommendation, Skill Level 10

Credit may be granted on the basis of an individualized assessment of the student (3/94).

Recommendation, Skill Level 20

In the lower-division baccalaureate/associate degree category, 4 semester hours in principles of cardiovascular technology, 4 in cardiovascular technology clinical practice, and 3 in personnel supervision; if additional skill identifier is Y6, Cardiac Catheterization Specialty, 4 semester hours in cardiac catheterization technology clinical practice. (NOTE: This recommendation for skill level 20 is valid for the dates 4/92-2/95 only) (4/92).

Recommendation, Skill Level 30

In the lower-division baccalaureate/associate degree category, 4 semester hours in principles of cardiovascular technology, 5 in cardiovascular technology clinical practice, 3 in personnel supervision, and 2 in records and information management; if additional skill identifier is Y6, Cardiac Catheterization Specialty, 5 semester hours in cardiac catheterization technology clinical practice (4/92).

Recommendation, Skill Level 40

In the lower-division baccalaureate/associate degree category, 4 semester hours in principles of cardiovascular technology, 6 in cardiovascular technology clinical practice, 3 in personnel supervision, and 2 in records and information management; if additional skill identifier is Y6, Cardiac Catheterization Specialty, 6 semester hours in cardiac catheterization technology clinical practice. In the upper-division baccalaureate category, 3 semester hours in organizational management and 3 for field experience in management (4/92).

MOS-91P-003

X-Ray Specialist
91P10
91P20
91P30
91P40
91P50

Exhibit Dates: 1/90–7/94.

Career Management Field: 91 (Medical).

Description

Summary: Operates or supervises the operation of fixed and portable X-ray equipment to take radiographs and to assist with treatment procedures. *Skill Level 10:* Able to perform the duties required for skill level 10 of MOS 91B (Medical Specialist); performs routine radiographic procedures; operates X-ray machines, utilizing proper exposure factors; applies radiation safety principles to protect self and patients from ionizing radiation; assists radiologists with fluoroscopic examination and simple special radiographic procedures; processes radiographic films by manual and automatic methods; performs routine patient administration including the maintenance of radiographic records. *Skill Level 20:* Able to perform the duties required for Skill Level 10; assembles radiographs for reading; performs body section, stereoscopic foreign body localization, prenatal, and pediatric radiographic procedures; performs follow-up radiographic examinations of the digestive, urogenital, respiratory, vascular, and nervous systems; takes radiographs of the extremities, trunk, and skull using portable equipment; assists with special radiographic and fluoroscopic procedures. *Skill Level 30:* Able to perform the duties required for Skill Level 20; performs complex or specialized radiograph procedures or supervises a medium-sized X-ray activity; performs routine examinations or assists with special examinations of the urogenital, respiratory, vascular, and nervous systems; inspects X-ray activities for compliance with radiation safety procedures and initiates corrective action if necessary; organizes work schedules; assigns duties; instructs personnel in technical procedures; supervises preventive maintenance of equipment; supervises procedures to enable unit to function in a toxic environment; evaluates personnel performance and prepares evaluation reports. *Skill Level 40:* Able to perform the duties required for Skill Level 30; supervises a large X-ray activity; inspects clinic to ensure a safe, clean, orderly, comfortable environment for patients; establishes work priorities; reviews, consolidates, and prepares technical, administrative, and personnel reports; coordinates activities of clinic with other elements of the medical treatment facility. *Skill Level 50:* Able to perform the duties required for Skill Level 40; supervises X-ray activity in large medical center; assumes departmental management duties including budget, personnel, and supply.

Recommendation, Skill Level 10

In the lower-division baccalaureate/associate degree category, 9 semester hours for a clinical internship, 6 in radiographic techniques, and 3 in radiation protection. (NOTE: This recommendation for skill level 10 is valid for the dates 1/90-9/91 only) (9/85).

Recommendation, Skill Level 20

In the lower-division baccalaureate/associate degree category, 15 semester hours for a clinical internship, 6 in radiographic techniques, 3 in radiation protection, and 3 in special radiographic procedures (9/85).

Recommendation, Skill Level 30

In the lower-division baccalaureate/associate degree category, 15 semester hours for a clinical internship, 6 in radiographic techniques, 3 in radiation protection, 3 in special radiographic procedures, and 3 in personnel supervision. In the upper-division baccalaureate category, credit in patient care administration on the basis of institutional evaluation (9/85).

Recommendation, Skill Level 40

In the lower-division baccalaureate/associate degree category, 15 semester hours for a clinical internship, 6 in radiographic techniques, 3 in radiation protection, 3 in special radiographic procedures, 3 in personnel supervision, and 3 for field experience in management. In the upper-division baccalaureate category, 3 semester hours in radiology department administration and credit in patient care administration on the basis of institutional evaluation (9/85).

Recommendation, Skill Level 50

In the lower-division baccalaureate/associate degree category, 15 semester hours for a clinical internship, 6 in radiographic techniques, 3 in radiation protection, 3 in special radiographic procedures, 3 in personnel supervision, and 3 for field experience in management. In the upper-division baccalaureate category, 6 semester hours in radiology department administration and credit in patient care administration on the basis of institutional evaluation (9/85).

MOS-91P-004

Radiology Specialist
((X-Ray Specialist))
91P10
91P20
91P30
91P40
91P50

Exhibit Dates: 8/94–Present.

Career Management Field: 91 (Medical).

Description

Summary: Operates or supervises the operation of fixed and portable X-ray equipment to take radiographs and to assist with treatment procedures. *Skill Level 10:* Able to perform the duties required for skill level 10 of MOS 91B (Medical Specialist); performs routine radiographic procedures; operates X-ray machines, using proper exposure factors; applies radiation safety principles to protect self and patients from ionizing radiation; assists radiologists with fluoroscopic examination and simple radiographic procedures; processes radiographic films by manual and automatic methods; performs routine patient administration including the maintenance of radiographic records. *Skill Level 20:* Able to perform the duties required for Skill Level 10; assembles radiographs for reading; performs body section, stereoscopic foreign body localization, prenatal, and pediatric radiographic procedures; performs follow-up radiographic examinations of the digestive, urogenital, respiratory, vascular, and nervous systems; takes radiographs of the extremities, trunk, and skull using portable equipment; assists with special radiographic and fluoroscopic procedures. *Skill Level 30:* Able to perform the duties required for Skill Level 20; performs complex diagnostic imaging procedures such as ultrasound, CT, MRI, and mammography; supervises a medium-sized X-ray activity; performs routine examinations or assists with special examinations of the urogenital, respiratory, vascular, and nervous systems; inspects X-ray activities for compliance with radiation safety procedures and initiates corrective action if necessary; organizes work schedules; assigns duties; instructs personnel in technical procedures; supervises preventive maintenance of equipment; supervises procedures to enable unit to function in a toxic environment; evaluates personnel performance and prepares evaluation reports. *Skill Level 40:* Able to perform the duties required for Skill Level 30; supervises a large X-ray activity; inspects clinic to ensure a safe, clean, orderly, comfortable environment for patients; establishes work priorities; reviews, consolidates, and prepares technical, administrative, and personnel reports; coordinates activities of clinic with other elements of the medical treatment facility. *Skill Level 50:* Able to perform the duties required for Skill Level 40; supervises X-ray activity in large medical center; assumes departmental management duties including budget, personnel, and supply.

Recommendation, Skill Level 10

Credit may be granted on the basis of an individualized assessment of the student (11/96).

Recommendation, Skill Level 20

In the lower-division baccalaureate/associate degree category, 20 semester hours for a clinical internship, 6 in radiographic techniques, and 3 in radiation protection. (NOTE: This recommendation for skill level 20 is valid for the dates 8/94-2/95 only) (11/96).

Recommendation, Skill Level 30

In the lower-division baccalaureate/associate degree category, 20 semester hours for a clinical internship, 6 in radiographic techniques, 3 in radiation protection, 3 in special radiographic procedures, 3 in special diagnostic imaging procedures, and 3 in personnel supervision. In the upper-division baccalaureate category, 3 semester hours in radiology department administration (11/96).

Recommendation, Skill Level 40

In the lower-division baccalaureate/associate degree category, 20 semester hours for a clinical internship, 6 in radiographic techniques, 3 in radiation protection, 3 in special radiographic procedures, 4 in special diagnostic imaging procedures, and 3 in personnel supervision. In the upper-division baccalaureate category, 3 semester hours in radiology department administration, 3 in personnel management, and 3 for field experience in management (11/96).

Recommendation, Skill Level 50

In the lower-division baccalaureate/associate degree category, 20 semester hours for a clinical internship, 6 in radiographic techniques, 3 in radiation protection, 3 in special radiographic procedures, 4 in personal diagnostic imaging procedures, and 3 in personnel supervision. In the upper-division baccalaureate category, 6 semester hours in radiology department administration, 3 in personnel management, and 3 for field experience in management (11/96).

MOS-91Q-003

Pharmacy Specialist
91Q10
91Q20
91Q30
91Q40
91Q50

Exhibit Dates: 1/90–7/94.

Career Management Field: 91 (Medical).

Description

Summary: Supervises or prepares, controls, and issues pharmaceutical products under supervision of pharmacist or physician. *Skill Level 10:* Able to perform the duties required for skill level 10 of MOS 91B (Medical Specialist); under supervision, interprets, compounds, manufactures, and files prescription orders, bulk drug orders, or unit dose orders; performs storage, accounting, inventory, and control procedures for pharmaceuticals; operates and maintains manufacturing and packaging equipment; assists in setting up unit equipment and shelters. *Skill Level 20:* Able to perform the duties required for Skill Level 10; issues and requisitions medication; evaluates prescription orders, bulk drug orders, or unit dose orders; checks drug orders for interactions, incompatibilities, and availability of dosage forms; issues medication to patients, wards, clinics, and other agencies. *Skill Level 30:* Able to perform the duties required for Skill Level 20; provides guidance to subordinates; assists in pharmacy inspections; orders, standard and nonstandard supplies; maintains stock levels; supervises sections within pharmacy service. *Skill Level 40:* Able to perform the duties required for Skill Level 30; assists in compiling information for meetings of Therapeutic Agent Board; assists in revision and update of hospital formulary or drug list; organizes work schedules; instructs personnel in procedures and conducts in-service training; evaluates personnel and prepares evaluation reports; sets stock levels and prepares requisitions; prepares technical, personnel, and administrative reports; coordinates pharmacy activities with other elements of medical treatment facility. *Skill Level 50:* Able to perform the duties required for Skill Level 40; supervises pharmacy activities in large hospital, medical center, or general hospital; assists in establishment and operation of formal training programs; ensures compliance with quality control standards for doctrinal material in formal training programs.

Recommendation, Skill Level 10

In the lower-division baccalaureate/associate degree category, 5 semester hours in introduction to pharmacy practice, 3 in pharmacy assisting, and additional credit in general chemistry, pharmaceutical chemistry, pharmacology, health sciences, and physiology on the basis of institutional evaluation. In the upper-division baccalaureate category, 3 semester hours in pharmaceutical laboratory and 2 in pharmaceutical mathematics. (NOTE: This recommendation for skill level 10 is valid for the dates 1/90-9/91 only) (9/85).

Recommendation, Skill Level 20

In the lower-division baccalaureate/associate degree category, 10 semester hours in introduction to pharmacy practice, 3 in pharmacy assisting, and additional credit in general chemistry, pharmaceutical chemistry, pharmacology, health sciences, and physiology on the basis of institutional evaluation. In the upper-division baccalaureate category, 3 semester hours in pharmaceutical laboratory and 2 in pharmaceutical mathematics (9/85).

Recommendation, Skill Level 30

In the lower-division baccalaureate/associate degree category, 10 semester hours in introduction to pharmacy practice, 3 in pharmacy assisting, 1 in inventory control, and additional credit in general chemistry, pharmaceutical chemistry, pharmacology, pharmaceutical preparations, health sciences, physiology, and laboratory techniques on the basis of institutional evaluation. In the upper-division baccalaureate category, 3 semester hours in pharmaceutical laboratory and 2 in pharmaceutical mathematics (9/85).

Recommendation, Skill Level 40

In the lower-division baccalaureate/associate degree category, 10 semester hours in introduction to pharmacy practice, 3 in pharmacy assisting, 3 in personnel supervision, 2 in principles of administration, 1 in inventory control, and additional credit in general chemistry, pharmaceutical chemistry, pharmacology, pharmaceutical preparations, health sciences, physiology, and laboratory techniques on the basis of institutional evaluation. In the upper-division baccalaureate category, 3 semester hours in pharmaceutical laboratory, 2 in pharmaceutical mathematics, and additional credit in medical facility management on the basis of institutional evaluation (9/85).

Recommendation, Skill Level 50

In the lower-division baccalaureate/associate degree category, 10 semester hours in introduction to pharmacy practice, 3 in pharmacy assisting, 3 in personnel supervision, 2 in principles of administration, 1 in inventory control, and additional credit in general chemistry, pharmaceutical chemistry, pharmacology, pharmaceutical preparations, health sciences, physiology, and laboratory techniques on the basis of institutional evaluation. In the upper-division baccalaureate category, 3 semester hours in pharmaceutical laboratory, 2 in pharmaceutical mathematics, 3 in methods of instruction, 3 for field experience in management, and additional credit in medical facility management on the basis of institutional evaluation (9/85).

MOS-91Q-004

Pharmacy Specialist
91Q10
91Q20
91Q30
91Q40
91Q50

Exhibit Dates: 8/94–Present.

Career Management Field: 91 (Medical).

Description

Summary: Supervises or prepares, controls, and issues pharmaceutical products under supervision of pharmacist or physician. *Skill Level 10:* Able to perform the duties required for skill level 10 of MOS 91B (Medical Specialist); under supervision, interprets, compounds, manufactures, and files prescription orders, bulk drug orders, or unit dose orders; performs storage, accounting, inventory, and control procedures for pharmaceuticals; operates and maintains manufacturing and packaging equipment; assists in setting up unit equipment and shelters. *Skill Level 20:* Able to perform the duties required for Skill Level 10; issues and requisitions medication; evaluates and prepares prescription orders (including oral and intravenous solutions), bulk drug orders, or unit dose orders; checks drug orders for interactions, incompatibilities, and availability of dosage forms; issues medication to patients, wards, clinics, and other agencies. *Skill Level 30:* Able to perform the duties required for Skill Level 20; provides guidance to subordinates; assists in pharmacy inspections; orders standard and nonstandard supplies; maintains stock levels; supervises sections within pharmacy service; manages computer information systems; provides appropriate counseling. *Skill Level 40:* Able to perform the duties required for Skill Level 30; assists in compiling information for meetings of Therapeutic Agent Board; assists in revision and update of hospital formulary or drug list; organizes work schedules; instructs personnel in procedures and conducts in-service training; evaluates personnel and prepares evaluation reports; sets stock levels and prepares requisitions; prepares technical, personnel, and administrative reports; coordinates pharmacy activities with other elements of medical treatment facility. *Skill Level 50:* Able to perform the duties required for Skill Level 40; supervises pharmacy activities in large hospital, medical center, or general hospital; assists in establishment and operation of formal training programs; ensures compliance with quality control standards for doctrinal material in formal training programs.

Recommendation, Skill Level 10

Credit may be granted on the basis of an individualized assessment of the student (11/96).

Recommendation, Skill Level 20

In the lower-division baccalaureate/associate degree category, 5 semester hours in clinical pharmacy practice; 3 in pharmacy assisting; and additional credit in general chemistry, pharmaceutical chemistry, pharmacology, health sciences, and physiology on the basis of institutional evaluation. In the upper-division baccalaureate category, 3 semester hours in pharmaceutical laboratory and 2 in pharmaceutical mathematics. (NOTE: This recommendation for skill level 20 is valid for the dates 8/94-2/95 only) (11/96).

Recommendation, Skill Level 30

In the lower-division baccalaureate/associate degree category, 10 semester hours in clinical pharmacy practice; 3 in pharmacy assisting; and additional credit in general chemistry, pharmaceutical chemistry, pharmacology, health sciences, and physiology on the basis of institutional evaluation. In the upper-division baccalaureate category, 3 semester hours in pharmaceutical laboratory, 2 in pharmaceutical mathematics, and 2 in pharmacy facility management (11/96).

Recommendation, Skill Level 40

In the lower-division baccalaureate/associate degree category, 15 semester hours in clinical pharmacy practice; 3 in pharmacy assisting; 2 in pharmaceutical preparations; 3 in personnel supervision; 2 in principles of administration; 3 in inventory control; 3 in information management; and additional credit in general chemistry, pharmaceutical chemistry, pharmacology, pharmaceutical preparations, health sciences, physiology, and laboratory techniques on the basis of institutional evaluation. In the upper-division baccalaureate category, 3 semester hours in pharmaceutical laboratory, 2 in pharmaceutical mathematics, and 4 in pharmacy facility management (11/96).

Recommendation, Skill Level 50

In the lower-division baccalaureate/associate degree category, 15 semester hours in clinical pharmacy practice; 2 in pharmaceutical preparations; 3 in pharmacy assisting; 3 in personnel supervision; 2 in principles of administration; 3 in inventory control; 3 in information management; and additional credit in general chemistry, pharmaceutical chemistry, pharmacology, pharmaceutical preparations, health sciences, physiology, and laboratory techniques on the basis of institutional evaluation. In the upper-division baccalaureate category, 3 semester hours in pharmaceutical laboratory, 2 in pharmaceutical mathematics, 3 in methods of instruction, 3 for field experience in management, and 4 in pharmacy facility management (11/96).

MOS-91R-002

Veterinary Specialist
91R10
91R20
91R30
91R40
91R50

Exhibit Dates: 1/90–5/92.

Career Management Field: 91 (Medical), subfield 913 (Health Services).

Description

Summary: Inspects food and food products for quality, safety, and adherence to contractual specifications. *Skill Level 10:* Assists with food hygiene and quality assurance inspections, using principles of statistical sampling, food microbiology, laboratory analysis, and knowledge of food processing and preservation. *Skill Level 20:* Able to perform the duties required for Skill Level 10; conducts subsistence procurement inspections; inspects processing, handling, packaging, packing, and quality control procedures in canneries, slaughterhouses, and dairies, and in water, food, meat packing, and dehydration plants; examines poultry, eggs, fruit, vegetables, seafood, meats, and dairy products; selects, prepares, and transmits samples for laboratory testing; reviews test results for contractual requirements; records contractual deficiencies and violations and initiates recommendations for corrective action; develops and applies double and multiple statistical sampling procedures and switching techniques; reviews and updates publications relative to quality assurance procedures; assists veterinarian in antemortem and postmortem examination of food animals. *Skill Level 30:* Able to perform the duties required for Skill Level 20; supervises procurement quality assurance and surveillance inspections; establishes work priorities; organizes work schedules; assigns duties; instructs subordinates; supervises equipment maintenance programs; prepares and conducts training programs; evaluates personnel performance; maintains technical records; requisitions, stores, and issues supplies. *Skill Level 40:* Able to perform the duties required for Skill Level 30; trains subordinates in food inspection and sanitary control programs; analyzes reports and surveys and recommends remedial action; coordinates transportation, administration, and logistical support of food inspectors; reviews, consolidates, and prepares team inspection records and other technical, personnel, and administrative reports. *Skill Level 50:* Able to perform the duties required for skill level 40 of MOS 91R, 91S (Environmental Health Specialist), or 91T (Animal Specialist); develops work methods and procedures; develops, directs, and evaluates training programs; coordinates activities of veterinary and environmental health activities; assists professional staff with technical, administrative, and fiscal matters.

Recommendation, Skill Level 10

In the lower-division baccalaureate/associate degree category, 7 semester hours in food science, 6 in inspection procedures and regulations, and 1 in veterinary preventive medicine. (NOTE: This recommendation for skill level 10 is valid for the dates 1/90-9/91 only) (6/77).

Recommendation, Skill Level 20

In the lower-division baccalaureate/associate degree category, 16 semester hours in food science, 7 in inspection procedures and regulations, and 2 in veterinary preventive medicine (6/77).

Recommendation, Skill Level 30

In the lower-division baccalaureate/associate degree category, 16 semester hours in food science, 7 in inspection procedures and regulations, 2 in veterinary preventive medicine, and 2 in personnel supervision (6/77).

Recommendation, Skill Level 40

In the lower-division baccalaureate/associate degree category, 16 semester hours in food science, 7 in inspection procedures and regulations, 3 in personnel supervision, 3 in human relations, and 2 in veterinary preventive medicine (6/77).

Recommendation, Skill Level 50

In the lower-division baccalaureate/associate degree category, 16 semester hours in food science, 7 in inspection procedures and regulations, 3 in personnel supervision, 3 in human relations, 3 in principles of management, and 2 in veterinary preventive medicine. In the upper-division baccalaureate category, 3 semester hours for field experience in management and 3 in management problems (6/77).

MOS-91R-003

Veterinary Food Inspection Specialist

91R10
91R20
91R30
91R40
91R50

Exhibit Dates: 6/92–11/01.

Career Management Field: 91 (Medical).

Description

Summary: Under supervision, inspects food and food products designated for consumption for quality, safety, and adherence to contractual specifications. *Skill Level 10:* Assists in inspection of meat, poultry, water, foods, eggs, dairy products, fruits, and vegetables in depots, warehouses, distribution points, installations, and conveyances; determines class of inspection to be performed, lot size, inspection levels, proper inspection tables, acceptable quality levels, identity, condition, and quantity of subsistence; selects sampling plan and samples, and applies sampling procedures to subsistence items on individual or lot basis and determines and verifies age, weight, and physical conformance of samples; makes sensory evaluations; inspects for packaging, packing, and marketing requirements; stamps inspected subsistence and related documents then classifies defects; advises supervisor of nonconformance and assists in evaluation of abnormalities; determines disposition of sample units; collects, prepares, and transmits samples to laboratory for subsistence examination or testing; identifies unsanitary conditions in food storage facilities and commissary stores; prepares subsistence inspection reports; prepares, utilizes, cleans and makes minor repairs and adjustments to food inspection equipment; receives, unpacks, stores, and safeguards inspection supplies and equipment; uses approved security measures to safeguard samples and inspection stamps. *Skill Level 20:* Able to perform the duties required for Skill Level 10; provides technical guidance to subordinates; conducts subsistence quality assurance inspections in food handling and processing establishments on food items procured, shipped, or stored; develops and applies statistical sampling procedures; reviews laboratory test results for product requirements; performs inspection of subsistence to determine compliance with contractual requirements; performs surveillance examinations of subsistence items to determine suitability for issue and continued serviceability; determines deteriorative condition of meat and meat products, fresh fruit and vegetables, dairy products, water, foods, and semiperishable subsistence; records contractual deficiencies and initiates recommendations for corrective actions; inspects commissary stores and military food storage facilities for unsanitary conditions and recommends corrective action where necessary; prepares and conducts in-service training for the food inspection activity; prepares and maintains quality assurance procedures on fresh dairy products; prepares and maintains quality history record and product verification record and distributes as required; reviews and updates publications that direct quality assurance procedures. *Skill Level 30:* Able to perform the duties required for Skill Level 20; develops and directs procurement quality assurance activities in contractor's establishments and receipt and surveillance inspections at military installations, subsistence distribution points, and depots; trains subordinates in technical aspects of food inspection; reviews, consolidates, and analyzes reports and surveys and recommends remedial action to assure effective inspection procedures; records violations of state and other federal requirements concerning food items; evaluates sanitary compliance of commercial and government-controlled food facilities; utilizes proper channels and procedures to report violations of food wholesomeness, animal disease, and food establishment sanitary requirements; develops and enforces security measures; supervises equipment maintenance program; monitors and evaluates personnel performance; counsels subordinates and prepares evaluation reports; ensures adherence to standards of conduct, cleanliness, technical accuracy, and safety regulations; prepares and conducts training programs; supervises procedures to enable unit to function in nuclear, biological, and chemical environment; assists veterinarian in control of zoonotic diseases; supervises emergency field slaughter procedures of food animals; advises and assists in administrative, fiscal, personnel, and supply matters; monitors requisition, storage, issue, and utilization of supplies and equipment; establishes report control system; prepares and reviews correspondence; maintains food inspection files. *Skill Level 40:* Able to perform the duties required for Skill Level 30; develops, evaluates, and directs food inspection training programs; coordinates transportation, administration, and logistical support of food inspectors and animal care specialists assigned to activities other than the parent activity. *Skill Level 50:* Able to perform the duties required for Skill Level 40; evaluates veterinary service training programs and requirements and provides recommendations for improvement; develops work methods and procedures; develops, evaluates, and directs training programs; coordinates activities of veterinary element with other elements of medical facility; advises and assists professional staff in supply economy procedures and fiscal, technical, and administrative matters.

Recommendation, Skill Level 10

Credit may be granted on the basis of an individualized assessment of the student (3/94).

Recommendation, Skill Level 20

In the lower-division baccalaureate/associate degree category, 3 semester hours in food science technology, 3 in food inspection, 3 in quality assurance, and 3 in personnel supervision. (NOTE: This recommendation for skill level 20 is valid for the dates 6/92-2/95 only) (6/92).

Recommendation, Skill Level 30

In the lower-division baccalaureate/associate degree category, 3 semester hours in food science technology, 3 in food inspection, 3 in quality assurance, 3 in personnel supervision, and 2 in records and information management (6/92).

Recommendation, Skill Level 40

In the lower-division baccalaureate/associate degree category, 3 semester hours in food science technology, 3 in food inspection, 3 in quality assurance, 3 in personnel supervision, and 2 in records and information management. In the upper-division baccalaureate category, 3 semester hours in operations management, 3 for field experience in management, and 3 in organizational management (6/92).

Recommendation, Skill Level 50

In the lower-division baccalaureate/associate degree category, 3 semester hours in food science technology, 3 in food inspection, 3 in quality assurance, 3 in personnel supervision, and 2 in records and information management. In the upper-division baccalaureate category, 3 semester hours in operations management, 3 for field experience in management, and 3 in organizational management; if paygrade E-9 has been attained, additional credit as follows: 3 semester hours in management problems and 3 in communication techniques for managers (6/92).

MOS-91R-004

Veterinary Food Inspection Specialist

91R10
91R20
91R30
91R40
91R50

Exhibit Dates: 12/01–Present.

Career Management Field: 91 (Medical).

Description

Summary: Under supervision, inspects food and food products designated for consumption for quality, safety, and adherence to contractual specifications. *Skill Level 10:* Assists in inspection of meat, poultry, water, foods, eggs, dairy products, fruits, and vegetables in depots, warehouses, distribution points, installations, and conveyances; determines class of inspection to be performed, lot size, inspection levels, proper inspection tables, acceptable quality levels, identity, condition, and quantity of subsistence; selects sampling plan and samples and applies sampling procedures to subsistence items on individual or lot basis and determines and verifies age, weight, and physical conformance of samples; makes sensory evaluations; inspects for packaging, packing, and marketing requirements; stamps inspected subsistence and related documents then classifies defects; advises supervisor of nonconformance and assists in evaluation of abnormalities; determines disposition of sample units; collects, prepares, and transmits samples to laboratory for subsistence examination or testing; identifies unsanitary conditions in food storage facilities and commissary stores; prepares subsistence inspection reports; prepares, utilizes, cleans and makes minor repairs and adjustments to food inspection equipment; receives, unpacks, stores, and safeguards inspection supplies and equipment; uses approved security measures to safeguard samples and inspection stamps. *Skill Level 20:* Able to perform the duties required for Skill Level 10; provides technical guidance to subordinates; conducts subsistence quality assurance inspections in food handling and processing establishments on food items procured, shipped, or stored; develops and applies statistical sampling procedures; reviews laboratory test results for product requirements; performs inspection of subsistence to determine compliance with contractual requirements; performs surveillance examinations of subsistence items to determine suitability for issue and continued serviceability; determines deteriorative condition of meat and meat products, fresh fruit and vegetables, dairy products, water, foods, and semiperishable subsistence; records contractual deficiencies and initiates recommendations for corrective actions; inspects commissary stores and military food storage facilities for unsanitary conditions and recommends corrective action where necessary; prepares and conducts in-service training for the food inspection activity; prepares and maintains quality assurance procedures on fresh dairy products; prepares and maintains quality history record and product verification record and distributes as required; reviews and updates publications that direct quality assurance procedures. *Skill Level 30:* Able to perform the duties required for Skill Level 20; develops and directs procurement quality assurance activities in contractor's establishments and receipt and surveillance inspections at military installations, subsistence distribution points, and depots; supervises and trains subordinates in technical aspects of food inspection; reviews, consolidates, and analyzes reports and surveys and recommends remedial action to assure effective inspection procedures; records violations of state and other federal requirements concerning food items; evaluates sanitary compliance of commercial and government-controlled food facilities; utilizes proper channels and procedures to report violations of food wholesomeness, animal disease, and food establishment sanitary requirements; develops and enforces security measures; supervises equipment maintenance program; monitors and evaluates personnel performance; counsels subordinates and prepares evaluation reports; ensures adherence to standards of conduct, cleanliness, technical accuracy, and safety regulations; prepares and conducts training programs; supervises procedures to enable unit to function in nuclear, biological, and chemical environment; assists veterinarian in control of zoonotic diseases; supervises emergency field slaughter procedures of food animals; advises and assists in administrative, fiscal, personnel, and supply matters; monitors requisition, storage, issue, and utilization of supplies and equipment; establishes report control system; prepares and reviews correspondence; maintains food inspection files. *Skill Level 40:* Able to perform the duties required for Skill Level 30; develops, evaluates, and directs food inspection training programs; coordinates transportation, administration, and logistical support of food inspectors and animal care specialists assigned to activities other than the parent activity. *Skill Level 50:* Able to perform the duties required for Skill Level 40; evaluates veterinary service training programs and requirements and provides recommendations for improvement; develops work methods and procedures; develops, evaluates, and directs training programs; coordinates activities of veterinary element with other elements of medical facility; advises and assists professional staff in supply economy procedures and fiscal, technical, and administrative matters.

Recommendation, Skill Level 10

Credit may be granted on the basis of an individualized assessment of the student (12/01).

Recommendation, Skill Level 20

Credit may be granted on the basis of an individualized assessment of the student (12/01).

Recommendation, Skill Level 30

In the lower-division baccalaureate/associate degree category, 3 semester hours in food science technology, 3 in food inspection, 3 in quality assurance, 3 in principles of supervision, and 3 in records and information management (12/01).

Recommendation, Skill Level 40

In the lower-division baccalaureate/associate degree category, 3 semester hours in food science technology, 3 in food inspection, 3 in quality assurance, 3 in principles of supervision, and 3 in records and information management. In the upper-division baccalaureate category, 3 semester hours in operations management, 3 for a practicum in management, and 3 in organizational management (12/01).

Recommendation, Skill Level 50

In the lower-division baccalaureate/associate degree category, 3 semester hours in food science technology, 3 in food inspection, 3 in quality assurance, 3 in principles of supervision, and 3 in records and information management. In the upper-division baccalaureate category, 3 semester hours in operations management, 6 in management practicum, and 3 in organizational management (12/01).

MOS-91S-004

Preventive Medicine Specialist
91S10
91S20
91S30
91S40
91S50

Exhibit Dates: 1/90–7/94.

Career Management Field: 91 (Medical).

Description

Summary: Supervises, conducts, or assists in surveys and inspections and establishes control measures of environmental factors affecting health of persons in areas of concern. *Skill Level 10:* Conducts environmental health surveys and inspections and assists in preventive medicine by inspecting living quarters, food handling establishments, water systems, liquid and solid waste disposal systems, barber and beauty shops, swimming facilities, nurseries, and industrial areas; inspects personal hygiene practices; reports deviations from prescribed health standards and recommends corrective action; collects specimens and samples and submits them to the laboratory; mixes and applies insecticides, rodenticides, and repellants; performs bacteriological examination and chemical analysis of water and inspects for quarantine standards; performs operator maintenance on preventive medicine equipment; collects data for environmental health reports. *Skill Level 20:* Able to perform the duties required for Skill Level 10; conducts environmental health surveys, inspections, and laboratory procedures; performs stream surveys; inspects hospital infectious waste control and disposal procedures; interviews food poisoning patients and tuberculosis, VD, and other communicable disease patients and contacts; conducts related communicable disease investigations; completes case and contact reports; assists with monitoring electromagnetic radiation; identifies medically important parasites, disease vectors, and their animal hosts and recommends insect control procedures; inspects and provides technical assistance for water supply systems, liquid and solid waste collection, treatment, and disposal; inspects and provides technical assistance for collection, treatment, and disposal of hazardous waste. *Skill Level 30:* Able to perform the duties required for Skill Level 20; supervises preventive medicine sections; collects and analyzes data on occupational illness, injuries, and environmental health; organizes work schedules; assigns duties; instructs in work techniques and procedures; plans and conducts training programs; supervises operational maintenance program of assigned equipment; evaluates personnel performance and prepares evaluation reports as well as administrative, technical, and personnel reports; prepares supply and equipment requisitions. *Skill Level 40:* Able to perform the duties required for Skill Level 30; supervises preventive medicine activities; organizes environmental stress surveillance programs; coordinates toxicology data with poison control center; plans vector control program; analyzes and evaluates data pertaining to preventive medicine activities; determines personnel requirements and establishes work priorities; advises and assists professionals in supply, fiscal, technical, and administrative matters. *Skill Level 50:* Able to perform the duties required for Skill Level 40; writes, develops, and coordinates command-wide regulations and policies relating to environmental health services; evaluates training programs and recommends improvements in such programs; participates in studies and reviews and makes recommendations regarding proposals; prepares and presents briefings; assists command environmental health officer in technical supervision of subordinate unit environmental health activities.

Recommendation, Skill Level 10

In the lower-division baccalaureate/associate degree category, 15 semester hours in sanitation and disease control (6 semester hours in sanitation and disease control, 3 in laboratory exercises in preventive medicine, 3 for field exercises in prevention and control of communicable diseases, and 3 in entomological and pest management). (NOTE: This recommendation for skill level 10 is valid for the dates 1/90-9/91 only) (9/88).

Recommendation, Skill Level 20

In the lower-division baccalaureate/associate degree category, 18 semester hours in sanitation and disease control (6 semester hours in sanitation and disease control, 3 in laboratory exercises in preventive medicine, 3 for field exercises in prevention and control of communicable diseases, 3 in entomological and pest management, and 3 in preventive medicine administration) and 9 in environmental science (9/88).

Recommendation, Skill Level 30

In the lower-division baccalaureate/associate degree category, 18 semester hours in sanitation and disease control (6 semester hours in sanitation and disease control, 3 in laboratory exercises in preventive medicine, 3 for field exercises in prevention and control of communicable diseases, 3 in entomological and pest management, and 3 in preventive medicine administration), 9 in environmental science, 3 in human relations, and 3 in personnel administration (9/88).

Recommendation, Skill Level 40

In the lower-division baccalaureate/associate degree category, 18 semester hours in sanitation and disease control (6 semester hours in sanitation and disease control, 3 in laboratory exercises in preventive medicine, 3 for field exercises in prevention and control of communicable diseases, 3 in entomological and pest management, and 3 in preventive medicine administration), 9 in environmental science, 3 in human relations, and 3 in personnel administration. In the upper-division baccalaureate category, 3 semester hours in management problems and 3 for field experience in management (9/88).

Recommendation, Skill Level 50

In the lower-division baccalaureate/associate degree category, 18 semester hours in sanitation and disease control (6 semester hours in sanitation and disease control, 3 in laboratory exercises in preventive medicine, 3 for field exercises in prevention and control of communicable diseases, 3 in entomological and pest management, and 3 in preventive medicine administration), 9 in environmental science, 3 in human relations, and 3 in personnel administration. In the upper-division baccalaureate category, 3 semester hours in management problems and 6 for field experience in management (9/88).

MOS-91S-005

Preventive Medicine Specialist

91S10
91S20
91S30
91S40
91S50

Exhibit Dates: 8/94–Present.

Career Management Field: 91 (Medical).

Description

Summary: Supervises, conducts, or assists in surveys and inspections and establishes control measures of environmental factors affecting health of persons in areas of concern. *Skill Level 10:* Conducts environmental health surveys and inspections and assists in preventive medicine by inspecting living quarters, food handling establishments, water systems, liquid and solid waste disposal systems, barber and beauty shops, swimming facilities, nurseries, and industrial areas; inspects personal hygiene practices; reports deviations from prescribed health standards and recommends corrective action; collects specimens and samples and submits them to the laboratory; mixes and applies insecticides, rodenticides, and repellants; performs bacteriological examination and chemical analysis of water and inspects for quarantine standards; performs operator maintenance on preventive medicine equipment; collects data for environmental health reports. *Skill Level 20:* Able to perform the duties required for Skill Level 10; conducts environmental health surveys, inspections, and laboratory procedures; performs stream surveys; inspects hospital infectious waste control and disposal procedures; interviews food poisoning patients and tuberculosis, STD, and other communicable disease patients and contacts; conducts related communicable disease investigations; completes case and contact reports; assists with monitoring electromagnetic radiation; identifies medically important parasites, disease vectors, and their animal hosts and recommends insect control procedures; inspects and provides technical assistance for water supply systems, liquid and solid waste collection, treatment, and disposal; inspects and provides technical assistance and supervision in collection, treatment, and disposal of hazardous waste. *Skill Level 30:* Able to perform the duties required for Skill Level 20; supervises preventive medicine sections; collects and analyzes data on occupational illness, injuries, and environmental health; organizes work schedules; assigns duties; instructs in work techniques and procedures; plans and conducts training programs; supervises operational maintenance program on assigned equipment; evaluates personnel performance and prepares evaluation reports, as well as administrative, technical, and personnel reports; prepares supply and equipment requisitions. *Skill Level 40:* Able to perform the duties required for Skill Level 30; supervises preventive medicine activities; organizes environmental stress surveillance programs; coordinates toxicology data with poison control center; plans vector control program; analyzes and evaluates data pertaining to preventive medicine activities; determines personnel requirements and establishes work priorities; advises and assists professionals in supply, fiscal, technical, and administrative matters. *Skill Level 50:* Able to perform the duties required for Skill Level 40; writes, develops, and coordinates command-wide regulations and policies relating to environmental health services; evaluates training programs and recommends improvements; participates in studies and reviews and makes recommendations regarding proposals; prepares and presents briefings; assists command environmental health officer in technical supervision of subordinate unit environmental health activities.

Recommendation, Skill Level 10

Credit may be granted on the basis of an individualized assessment of the student (11/96).

Recommendation, Skill Level 20

In the lower-division baccalaureate/associate degree category, 12 semester hours in sanitation and disease control (6 semester hours in sanitation and disease control, 3 in entomological and pest management, and 3 in preventive medicine administration), 9 in environmental science, and 5 for clinical experience in preventive medicine. (NOTE: This recommendation for skill level 20 is valid for the dates 8/94-2/95 only) (11/96).

Recommendation, Skill Level 30

In the lower-division baccalaureate/associate degree category, 12 semester hours in sanitation and disease control (6 semester hours in sanitation and disease control, 3 in entomological and pest management, and 3 in preventive medicine administration), 9 in environmental science,10 for clinical experience in preventive medicine, and 3 in personnel supervision (11/96).

Recommendation, Skill Level 40

In the lower-division baccalaureate/associate degree category, 12 semester hours in sanitation and disease control (6 semester hours in sanitation and disease control, 3 in entomological and pest management, and 3 in preventive medicine administration), 9 in environmental science,10 for clinical experience in preventive medicine, 3 in human relations, and 3 in personnel supervision. In the upper-division baccalaureate category, 3 semester hours in management problems and 3 for field experience in management (11/96).

Recommendation, Skill Level 50

In the lower-division baccalaureate/associate degree category, 12 semester hours in sanitation and disease control (6 semester hours in sanitation and disease control, 3 in entomological and pest management, and 3 in preventive medicine administration), 9 in environmental science,10 for clinical experience in preventive medicine, 3 in human relations, and 3 in personnel supervision. In the upper-division baccalaureate category, 3 semester hours in management problems and 6 for field experience in management (11/96).

MOS-91T-003

Animal Specialist

91T10
91T20
91T30
91T40

Exhibit Dates: 1/90–5/92.

Career Management Field: 91 (Medical), subfield 913 (Health Services).

Description

Summary: Under the direction of a veterinarian, supervises or provides care and treatment for small, medium-sized, and large animals. *Skill Level 10:* Provides animal care, management, and treatment; feeds and waters small and medium-sized animals; cleans and sanitizes animal facility, equipment, and utensils; receives and identifies incoming animals; quarantines incoming and sick animals; administers parenteral and oral medication and immunizations; collects and prepares blood and urine samples and fecal specimens; performs common laboratory tests including examining for microfilaria; assists in taking radiographs; prepares animals and equipment for surgery and anesthesia; administers postoperative treatment and monitors postoperative recovery; provides emergency treatment when appropriate; records laboratory results in health records; prepares health and vaccination certificates for veterinarian's signature; requisitions, receives, stores, and inventories pharmaceuticals, supplies, and equipment; prepares reagents, emergency drugs, and special animal rations; maintains animal files, including registration, health data, and immunization record; identifies common animal diseases and administers standard treatment; debrides and sutures superficial wounds;

calculates and administers drugs; induces general anesthesia and operates anesthetic equipment; assists in surgical procedures; maintains operating room supplies and cleanliness; performs appropriate laboratory analyses; informs owners of proper treatment instructions and medication prescribed by veterinarian; supervises animal registry and animal pound. *Skill Level 20:* Able to perform the duties required for Skill Level 10; conducts on-the-job training and provides technical guidance to subordinates; may supervise up to nine persons. *Skill Level 30:* Able to perform the duties required for Skill Level 20; supervises small and medium-sized veterinary activities; organizes work schedules and assigns duties; instructs personnel in proper work techniques and procedures; evaluates personnel performance and prepares evaluation reports; establishes safety procedures; supervises operator maintenance on assigned equipment; conducts in-service training programs; supervises requisition of supplies and equipment; prepares, reviews, and consolidates technical, personnel, and administrative reports; inspects facility; assists veterinarian in personnel, supply, and fiscal matters. *Skill Level 40:* Able to perform the duties required for Skill Level 30; supervises large veterinary activities.

Recommendation, Skill Level 10

In the vocational certificate category, 15 semester hours in veterinary laboratory techniques. In the lower-division baccalaureate/associate degree category, 2 semester hours for a practicum in veterinary science, 4 in veterinary laboratory techniques, 3 in microbiology, 3 in anatomy and physiology of animals, 3 in administration of drugs, 2 in animal hygiene, 2 in animal diseases, and 1 in surgical procedures for animals. In the upper-division baccalaureate category, 6 semester hours in research on the basis of institutional review of scientific papers or projects. (NOTE: This recommendation for skill level 10 is valid for the dates 1/90-9/91 only) (9/80).

Recommendation, Skill Level 20

In the vocational certificate category, 15 semester hours in veterinary laboratory techniques and 3 in office practices and procedures. In the lower-division baccalaureate/associate degree category, 2 semester hours for a practicum in veterinary science, 4 in veterinary laboratory techniques, 3 in microbiology, 3 in anatomy and physiology of animals, 3 in administration of drugs, 2 in animal hygiene, 2 in animal diseases, 1 in surgical procedures for animals, 3 in personnel supervision, and 3 in advanced veterinary laboratory techniques. In the upper-division baccalaureate category, 6 semester hours in research on the basis of institutional review of scientific papers or projects (9/80).

Recommendation, Skill Level 30

In the vocational certificate category, 15 semester hours in veterinary laboratory techniques and 3 in office practices and procedures. In the lower-division baccalaureate/associate degree category, 2 semester hours for a practicum in veterinary science, 4 in veterinary laboratory techniques, 3 in microbiology, 3 in anatomy and physiology of animals, 3 in administration of drugs, 3 in personnel supervision, 3 in technical report writing, 3 in office management, 2 in animal hygiene, 2 in animal diseases, 1 in surgical procedures for animals, and 3 in advanced veterinary laboratory techniques. In the upper-division baccalaureate category, 6 semester hours in research on the basis of institutional review of scientific papers or projects (9/80).

Recommendation, Skill Level 40

In the vocational certificate category, 15 semester hours in veterinary laboratory techniques and 3 in office practices and procedures. In the lower-division baccalaureate/associate degree category, 2 semester hours for a practicum in veterinary science, 4 in veterinary laboratory techniques, 3 in microbiology, 3 in anatomy and physiology of animals, 3 in administration of drugs, 3 in personnel supervision, 3 in technical report writing, 3 in office management, 2 in animal hygiene, 2 in animal diseases, 1 in surgical procedures for animals, and 3 in advanced veterinary laboratory techniques. In the upper-division baccalaureate category, credit in veterinary facility management and for field experience in management on the basis of institutional evaluation and 6 semester hours in research on the basis of institutional review of scientific papers or projects (9/80).

MOS-91T-004

ANIMAL CARE SPECIALIST
91T10
91T20
91T30
91T40

Exhibit Dates: 6/92–11/01.

Career Management Field: 91 (Medical).

Description

Summary: Under the direction of a veterinarian, supervises or provides care and treatment for small, medium-sized, and large animals. *Skill Level 10:* Provides animal care, management, and treatment; feeds and waters small and medium-sized animals; cleans and sanitizes animal facility, equipment, and utensils; receives and identifies incoming animals; quarantines incoming and sick animals; administers parenteral and oral medication and immunizations; collects and prepares blood and urine samples and fecal specimens; performs common laboratory tests including examining for microfilaria; assists in taking radiographs; prepares animals and equipment for surgery and anesthesia; administers postoperative treatment and monitors postoperative recovery; provides emergency treatment when appropriate; records laboratory results in health records; prepares health and vaccination certificates for veterinarian's signature; requisitions, receives, stores and inventories pharmaceuticals, supplies, and equipment; prepares reagents, emergency drugs, and special animal rations; maintains animal files, including registration, health data, and immunization record; identifies common animal diseases and performs standard treatment techniques; debrides and sutures superficial wounds; calculates and administers drugs; induces general anesthesia and operates anesthetic equipment; assists in surgical procedures; maintains operating room supplies and cleanliness; performs appropriate laboratory analyses; informs owners of proper treatment instructions and medication prescribed by veterinarian; supervises animal registry and animal pound. *Skill Level 20:* Able to perform the duties required for Skill Level 10; conducts on-the-job training and provides technical guidance to subordinates; observes animals for common diseases; inspects animal facilities; performs advanced emergency medical treatment; operates mechanical apparatus during surgery; performs diagnostic laboratory procedures; supervises small and medium-sized veterinary activities; organizes work schedules and assigns duties; instructs personnel in proper work techniques and procedures; evaluates personnel performance and prepares evaluation reports; establishes safety procedures; supervises operator maintenance on assigned equipment; conducts in-service training programs; supervises requisition of supplies and equipment; prepares, reviews, and consolidates technical, personnel, and administrative reports; inspects facility; assists veterinarian in personnel, supply, and fiscal matters. *Skill Level 30:* Able to perform the duties required for Skill Level 20; develops procedures for veterinary facility; performs as liaison for veterinary officer in various capacities; reviews and prepares technical, personnel, and administrative reports; supervises preventive maintenance and supply program. *Skill Level 40:* Able to perform the duties required for Skill Level 30; supervises large veterinary activities.

Recommendation, Skill Level 10

Credit may be granted on the basis of an individualized assessment of the student (3/94).

Recommendation, Skill Level 20

In the lower-division baccalaureate/associate degree category, 4 semester hours in veterinary laboratory techniques, 5 in clinical practice in veterinary science, 3 in animal hygiene, and 3 in personnel supervision. (NOTE: This recommendation for skill level 20 is valid for the dates 6/92-2/95 only) (6/92).

Recommendation, Skill Level 30

In the lower-division baccalaureate/associate degree category, 4 semester hours in veterinary laboratory techniques, 5 in clinical practice in veterinary science, 3 in animal hygiene, 3 in personnel supervision, and 2 in records and information management (6/92).

Recommendation, Skill Level 40

In the lower-division baccalaureate/associate degree category, 4 semester hours in veterinary laboratory techniques, 5 in clinical practice in veterinary science, 3 in animal hygiene, 3 in personnel supervision, and 2 in records and information management. In the upper-division baccalaureate category, 3 semester hours in organizational management and 3 for field experience in management (6/92).

MOS-91T-005

ANIMAL CARE SPECIALIST
91T10
91T20
91T30
91T40

Exhibit Dates: 12/01–Present.

Career Management Field: 91 (Medical).

Description

Summary: Under the direction of a veterinarian, supervises or provides care and treatment for small, medium-sized, and large animals. *Skill Level 10:* Provides animal care, management, and treatment; feeds and waters small and medium-sized animals; cleans and sanitizes animal facility, equipment, and utensils; receives and identifies incoming animals; quarantines incoming and sick animals; administers parenteral and oral medication and immunizations; collects and prepares blood and urine samples and fecal specimens; performs common laboratory tests including examining for microfilaria; assists in taking radiographs; prepares animals and equipment for surgery and anesthesia; administers postoperative treatment and monitors postoperative recovery; provides emergency treatment when appropriate; records laboratory results in health records; prepares health and vaccination certificates for veterinarian's signature; requisitions, receives, stores, and inventories pharmaceuticals, supplies, and equipment; prepares reagents, emergency drugs, and special animal rations; maintains animal files, including registration, health data, and immunization record; identifies common animal diseases and performs standard treatment techniques; debrides and sutures superficial wounds; calculates and administers drugs; induces general anesthesia and operates anesthetic equipment; assists in surgical procedures; maintains operating room supplies and cleanliness; performs appropriate laboratory analyses; informs owners of proper treatment instructions and medication prescribed by veterinarian; supervises animal registry and animal pound; enters and maintains all records electronically; complies with regulatory requirements. *Skill Level 20:* Able to perform the duties required for Skill Level 10; conducts on-the-job training and provides technical guidance to subordinates; observes animals for common diseases; inspects animal facilities; performs advanced emergency medical treatment; operates mechanical apparatus during surgery; performs diagnostic laboratory procedures; supervises small and medium-sized veterinary activities; organizes work schedules and assigns duties; instructs personnel in proper work techniques and procedures; evaluates personnel performance and prepares evaluation reports; establishes safety procedures; supervises operator maintenance on assigned equipment; conducts in-service training programs; supervises requisition of supplies and equipment; prepares, reviews, and consolidates technical, personnel, and administrative reports; inspects facility; assists veterinarian in personnel, supply, fiscal matters; and personnel supervision. *Skill Level 30:* Able to perform the duties required for Skill Level 20; trains and supervises subordinates in a veterinary facility; develops procedures for a veterinary facility; performs as liaison for veterinary officer in various capacities; reviews and prepares technical, personnel, and administrative reports; supervises preventive maintenance and supply program. *Skill Level 40:* Able to perform the duties required for Skill Level 30; supervises large veterinary facility.

Recommendation, Skill Level 10

Credit may be granted on the basis of an individualized assessment of the student (12/01).

Recommendation, Skill Level 20

Credit may be granted on the basis of an individualized assessment of the student (12/01).

Recommendation, Skill Level 30

In the lower-division baccalaureate/associate degree category, 4 semester hours in veterinary laboratory techniques, 5 in clinical practice in veterinary science, 3 in animal hygiene, 3 in principles of supervision, and 3 in records and information management (12/01).

Recommendation, Skill Level 40

In the lower-division baccalaureate/associate degree category, 4 semester hours in veterinary laboratory techniques, 5 in clinical practice in veterinary science, 3 in animal hygiene, 3 in principles of supervision, 3 in records and information management. In the upper-division baccalaureate category, 3 semester hours in organizational management and 3 for a practicum in management (12/01).

MOS-91U-002

Ear, Nose, and Throat (ENT) Specialist
91U10
91U20
91U30
91U40

Exhibit Dates: 1/90–3/92.

Career Management Field: 91 (Medical), subfield 912 (Patient Care).

Description

Summary: Conducts routine diagnostic tests and assists physician in care and treatment of ear, nose, and throat patients. *Skill Level 10:* Able to perform the duties required for skill level 10 of MOS 91B (Medical Specialist); administers and records patient responses to various diagnostic and labyrinthine tests; measures auditory acuity; irrigates patients' ears; sets up equipment and assists physician with clinical procedures; obtains specimens for culture and prepares laboratory requests; prepares equipment trays and basins; assists the physician in the performance of minor surgery; sterilizes and cares for instruments and equipment; operates resuscitation and oxygenation equipment; performs and records the results of diagnostic tests; assists audiologist in hearing aid evaluation. *Skill Level 20:* Able to perform the duties required for Skill Level 10; provides technical guidance to subordinates; administers prescribed medication. *Skill Level 30:* Able to perform the duties required for Skill Level 20; assists in managing a small ENT clinic; organizes work schedules; assigns duties; instructs subordinates in work techniques and procedures; supervises operational maintenance program of clinical equipment; prepares training programs; evaluates personnel performance and prepares evaluation reports; coordinates activities of ENT clinic with other elements of medical treatment facility; assists physician with personnel, fiscal, and administrative matters; prepares, reviews, and consolidates technical and administrative reports. *Skill Level 40:* Able to perform the duties required for Skill Level 30; assists in managing a large ear, nose, and throat clinic.

Recommendation, Skill Level 10

In the lower-division baccalaureate/associate degree category, 4 semester hours for a practicum in otolaryngology, 3 in principles of otolaryngology, and 1 in anatomy and physiology. (NOTE: This recommendation for skill level 10 is valid for the dates 1/90-9/91 only) (6/77).

Recommendation, Skill Level 20

In the lower-division baccalaureate/associate degree category, 7 semester hours for a practicum in otolaryngology, 3 in principles of otolaryngology, 2 in principles of audiometry, and 1 in anatomy and physiology (6/77).

Recommendation, Skill Level 30

In the lower-division baccalaureate/associate degree category, 7 semester hours for a practicum in otolaryngology, 3 in principles of otolaryngology, 3 in personnel supervision, 2 in principles of audiometry, 2 in communication skills, 2 in principles of management, 1 in anatomy and physiology, and 1 in health care management. In the upper-division baccalaureate category, credit in patient care administration on the basis of institutional evaluation (6/77).

Recommendation, Skill Level 40

In the lower-division baccalaureate/associate degree category, 7 semester hours for a practicum in otolaryngology, 3 in principles of otolaryngology, 3 in personnel supervision, 2 in principles of audiometry, 2 in communication skills, 2 in principles of management, 1 in anatomy and physiology, and 1 in health care management. In the upper-division baccalaureate category, credit in patient care administration and medical facility management on the basis of institutional evaluation (6/77).

MOS-91U-003

Ear, Nose, and Throat (ENT) Specialist
91U10
91U20
91U30
91U40

Exhibit Dates: 4/92–12/94. (Effective 12/94, MOS 91U was discontinued, and its duties were incorporated into MOS 91B.)

Career Management Field: 91 (Medical).

Description

Summary: Conducts routine diagnostic tests and assists physician in care and treatment of ear, nose, and throat patients. *Skill Level 10:* Administers and records patient responses to various diagnostic and labyrinthine tests; measures auditory acuity; irrigates patients' ears; sets up equipment and assists physician with clinical procedures; obtains specimens for culture and prepares laboratory requests; prepares equipment trays and basins; assists the physician in the performance of minor surgery; sterilizes and cares for instruments and equipment; operates resuscitation and oxygenation equipment; performs and records the results of diagnostic tests; assists audiologist in hearing aid evaluation. *Skill Level 20:* Able to perform the duties required for Skill Level 10; provides technical guidance and supervision to subordinates; administers prescribed medication. *Skill Level 30:* Able to perform the duties required for Skill Level 20; assists in managing a small ENT clinic; organizes work schedules; assigns duties; instructs subordinates in work techniques and procedures; supervises operational maintenance program of clinical equipment; prepares training programs; evaluates personnel performance and prepares evaluation reports; coordinates activities of ENT clinic with other elements of medical treatment facility; assists physician with personnel, fiscal, and administrative matters; prepares, reviews, and consolidates technical and administrative reports. *Skill Level 40:* Able to perform the duties required for Skill Level 30; assists in managing a large ear, nose, and throat clinic.

Recommendation, Skill Level 10

Credit may be granted on the basis of an individualized assessment of the student (3/94).

Recommendation, Skill Level 20

In the lower-division baccalaureate/associate degree category, 2 semester hours in audiology assisting, 4 in audiology clinical practice, and 3 in personnel supervision (4/92).

Recommendation, Skill Level 30

In the lower-division baccalaureate/associate degree category, 2 semester hours in audiology assisting, 5 in audiology clinical practice, 3 in personnel supervision, and 2 in records and information management (4/92).

Recommendation, Skill Level 40

In the lower-division baccalaureate/associate degree category, 2 semester hours in audiology assisting, 6 in audiology clinical practice, 3 in personnel supervision, and 2 in records and information management. In the upper-division baccalaureate category, 3 semester hours in organizational management and 3 for field experience in management (4/92).

MOS-91V-002

Respiratory Specialist
91V10
91V20
91V30
91V40

Exhibit Dates: 1/90–5/92.

Career Management Field: 91 (Medical), subfield 912 (Patient Care).

Description

Summary: Assists with the care and treatment of patients requiring respiratory assistance; performs pulmonary function tests under the supervision of a physician or anesthetist. *Skill Level 10:* Able to perform the duties required for skill level 10 of MOS 91B (Medical Specialist); administers respiratory therapy and performs pulmonary function tests; receives and interprets treatment request; prepares and tests equipment; mixes prescribed cardiopulmonary medication; verifies identity and instructs, orients, and positions patient for therapy; adjusts controls of intermittent positive pressure breathing apparatus and administers therapy; assesses tidal volume and adjusts apparatus to meet ventilatory changes; observes, corrects, and records patient response; administers aerosol and gas therapy; adjusts and maintains equipment during operation and observes safety regulations; administers pulmonary drainage procedures; percusses and vibrates patient; collects and visually analyzes sputum specimens for foreign substances; assists with cardiopulmonary resuscitation, evaluates effectiveness, and records results; prepares and tests nebulizer for administration of medication and adjusts for drug concentration; follows isolation procedures; disassembles, cleans, sterilizes, assembles, and tests respiratory therapy equipment; adjusts volume ventilator controls for initial operation and during therapy; observes and corrects patient response and evaluates ventilator effectiveness; weans patient from ventilator dependence; prepares, tests, and administers special volume ventilator equipment, such as chronic positive pressure breathing, intermittent mandatory ventilation, chronic positive airway pressure, and positive and expiratory pressure; applies electromechanical monitors; evaluates airway potency; instills solution prior to tracheobronchial aspiration; hyperinflates and hyperoxygenates patient; aspirates tracheobronchial passage; inflates and deflates tracheostomy cuff; provides tracheostomy tube and stoma care; evaluates patient cardiorespiratory status; utilizes emergency ventilators; evaluates resuscitator's effectiveness; implements culturing procedures; applies corrective actions for contaminated equipment; reads and evaluates patient charts; instructs patient and family in breathing exercises and postural drainage procedures and evaluates ability to operate home therapy equipment; maintains safe, clean, and orderly work area; stores and maintains supplies and equipment. *Skill Level 20:* Able to perform the duties required for Skill Level 10; calibrates and slopes arterial blood gas analyzer; performs arterial blood gas procedures and interprets and records results; applies safety precautions relevant to pulmonary function testing; calculates data and compares results; performs auscultation of lungs; receives and issues supplies and equipment. *Skill Level 30:* Able to perform the duties required for Skill Level 20; assists in defibrillation of patient; plans and monitors culturing procedures; determines corrective action for contaminated equipment; inspects working areas to ensure an orderly, clean, and safe environment for patients; supervises operational maintenance program of clinical equipment; prepares and conducts training programs; evaluates personnel performance and prepares evaluation reports; prepares administrative technical and patient reports; supervises packing, unpacking, loading, unloading, and setting up of unit equipment and shelters. *Skill Level 40:* Able to perform the duties required for Skill Level 30; supervises respiratory unit activities; applies and monitors safety procedures relevant to modular pulmonary function testing; administers tests, calculates test data, records results, and compares results with normal predicted values; develops, implements, and evaluates training programs; compiles management data, evaluates and determines personnel requirements, and assists in the planning and operation of respiratory unit; coordinates activities of unit with other elements of the medical treatment facility; advises and assists professional staff in supply economy procedures and fiscal, technical, and administrative matters; reviews, consolidates, and prepares technical, personnel, and administrative reports.

Recommendation, Skill Level 10

In the vocational certificate category, 30 semester hours in a respiratory therapy program. In the lower-division baccalaureate/associate degree category, 8 semester hours for clinical experience in respiratory therapy, 4 in respiratory therapy in clinical medicine, 4 in anatomy and physiology, 3 in principles of pulmonary therapy, 3 in principles of ventilation therapy, 3 in cardiopulmonary physiology, 2 in microbiology and immunology, 2 in basic sciences (chemistry and physics), and 1 in pharmacology. (NOTE: This recommendation for skill level 10 is valid for the dates 1/90-9/91 only) (6/77).

Recommendation, Skill Level 20

In the vocational certificate category, 33 semester hours in a respiratory therapy program. In the lower-division baccalaureate/associate degree category, 11 semester hours for clinical experience in respiratory therapy, 4 in respiratory therapy in clinical medicine, 4 in anatomy and physiology, 3 in principles of pulmonary therapy, 3 in principles of ventilation therapy, 3 in cardiopulmonary physiology, 2 in microbiology and immunology, 2 in basic sciences (chemistry and physics), and 1 in pharmacology (6/77).

Recommendation, Skill Level 30

In the vocational certificate category, 33 semester hours in a respiratory therapy program. In the lower-division baccalaureate/associate degree category, 11 semester hours for clinical experience in respiratory therapy, 4 in respiratory therapy in clinical medicine, 4 in anatomy and physiology, 3 in principles of pulmonary therapy, 3 in principles of ventilation therapy, 3 in cardiopulmonary physiology, 2 in microbiology and immunology, 2 in basic sciences (chemistry and physics), and 1 in pharmacology (6/77).

Recommendation, Skill Level 40

In the vocational certificate category, 33 semester hours in a respiratory therapy program. In the lower-division baccalaureate/associate degree category, 11 semester hours for clinical experience in respiratory therapy, 4 in respiratory therapy in clinical medicine, 4 in anatomy and physiology, 3 in principles of pulmonary therapy, 3 in principles of ventilation therapy, 3 in cardiopulmonary physiology, 2 in microbiology and immunology, 2 in basic sciences (chemistry and physics), 1 in pharmacology, 3 in personnel supervision, and 3 in clinic management. In the upper-division baccalaureate category, credit in patient care administration, introduction to management, and for field experience in management on the basis of institutional evaluation (6/77).

MOS-91V-003

Respiratory Specialist
91V10
91V20
91V30
91V40

Exhibit Dates: 6/92–11/01.

Career Management Field: 91 (Medical).

Description

Summary: Assists with the care and treatment of patients requiring respiratory therapy; performs pulmonary function tests under the supervision of a physician or anesthetist. *Skill Level 10:* Administers respiratory therapy and performs pulmonary function tests; receives and interprets treatment request; prepares and tests equipment; mixes prescribed cardiopulmonary medications; verifies identity and instructs, orients, and positions patient for therapy; adjusts controls of intermittent positive pressure breathing apparatus and administers therapy; assesses tidal volume and adjusts apparatus to meet ventilatory changes; observes, corrects, and records patient response; administers aerosol and gas therapy; adjusts and maintains equipment during operation and observes safety regulations; administers pulmonary drainage procedures; percusses and vibrates patient; collects and visually analyzes sputum specimens; assists with cardiopulmonary resuscitation, evaluates effectiveness, and records results; prepares and tests nebulizer for administration of medication; follows isolation procedures; disassembles, cleans, sterilizes; assembles, and tests respiratory therapy equipment; adjusts volume ventilator controls for initial operation and during therapy; observes patient response and evaluates ventilator effectiveness; weans patient from ventilator dependence; prepares, tests, and administers all modes of special volume ventilator equipment; applies electromechanical monitors; evaluates airway potency; instills solutions; hyperinflates and hyperoxygenates patient; analyzes arterial blood samples; records and interprets results; auscultates lungs; aspirates tracheobronchial passage; inflates and deflates tracheostomy cuff; provides tracheostomy tube and stoma care; evaluates patient cardiorespiratory status; utilizes emergency ventilators; evaluates patient cardiorespiratory status; utilizes emergency ventilators; evaluates resuscitator's effectiveness; implements culturing procedures; applies corrective actions for contaminated equipment; reads and evaluates patient charts; instructs patient and family in breathing exercises and postural drainage procedures; evaluates ability to operate home therapy equipment; maintains safe, clean, and orderly work area; stores and maintains supplies and equipment. *Skill Level 20:* Able to perform the duties required for Skill Level 10; calibrates and slopes arterial blood gas analyzer; applies safety precautions relevant to pulmonary function testing; calculates data and compares results; receives and issues supplies and equipment. *Skill Level 30:* Able to perform the duties required for Skill Level 20; assists in defibrillation of patient; plans and monitors culturing procedures; determines corrective action for contaminated equipment; inspects working areas to ensure an orderly, clean, and safe environment for patients; supervises operational maintenance program of clinical equipment; prepares and conducts training programs; evaluates personnel performance and prepares evaluation reports; prepares administrative technical and patient reports; supervises packing, unpacking, loading, unloading, and setting up of unit equipment and shelter. *Skill Level 40:* Able to perform the duties required for Skill Level 30; supervises respiratory unit activities; applies and monitors safety procedures relevant to modular pulmonary function testing; administers tests, calculates test data, records results, and compares results with normal predicted values; develops, implements, and evaluates training programs; compiles management data, evaluates and determines personnel requirements, and assists in the planning and operation of respiratory unit; coordinates activities of unit with other elements of the medical treatment facility; advises and assists professional staff in supply economy procedures and fiscal, technical, and administrative matters; reviews, consolidates, and prepares technical, personnel, and administrative reports.

Recommendation, Skill Level 10

Credit may be granted on the basis of an individualized assessment of the student (3/94).

Recommendation, Skill Level 20

In the lower-division baccalaureate/associate degree category, 4 semester hours in mechanical ventilation, 4 in respiratory care clinical practice, 4 in arterial blood gases, and 3 in personnel supervision. (NOTE: This recommendation for skill level 20 is valid for the dates 6/92-2/95 only) (6/92).

Recommendation, Skill Level 30

In the lower-division baccalaureate/associate degree category, 4 semester hours in mechanical ventilation, 4 in respiratory care clinical practice, 4 in arterial blood gases, 3 in personnel supervision, and 2 in records and information management (6/92).

Recommendation, Skill Level 40

In the lower-division baccalaureate/associate degree category, 4 semester hours in mechanical ventilation, 4 in respiratory care clinical practice, 4 in arterial blood gases, 3 in personnel supervision, and 2 in records and information management. In the upper-division baccalaureate category, 3 semester hours in organizational management and 3 for field experience in management (6/92).

MOS-91V-004

Respiratory Specialist
91V20
91V30
91V40

Exhibit Dates: 12/01–Present.

Career Management Field: 91 (Medical).

Description

Summary: Assists with the care and treatment of patients requiring respiratory therapy; performs pulmonary function tests under the supervision of a physician or anesthetist. *Skill Level 20:* Administers respiratory therapy and performs pulmonary function tests; receives and interprets treatment requests; prepares and tests equipment; mixes prescribed cardiopulmonary medications; verifies identity and instructs, orients, and positions patient for therapy; adjusts controls of intermittent positive pressure breathing apparatus and administers therapy; assesses tidal volume and adjusts apparatus to meet ventilatory changes; observes, corrects, and records patient response; administers aerosol and gas therapy; adjusts and maintains equipment during operation and observes safety regulations; administers pulmonary drainage procedures; percusses and vibrates patient; collects and visually analyzes sputum specimens; assists with cardiopulmonary resuscitation, evaluates effectiveness, and records results; prepares and tests nebulizer for administration of medication; follows isolation procedures; disassembles, cleans, sterilizes; assembles, and tests respiratory therapy equipment; adjusts volume ventilator controls for initial operation and during therapy; observes patient response and evaluates ventilator effectiveness; weans patient from ventilator dependence; prepares, tests, and administers all modes of special volume ventilator equipment; applies electromechanical monitors; evaluates airway potency; instills solutions; hyperinflates and hyperoxygenates patient; analyzes arterial blood samples; records and interprets results; auscultates lungs; aspirates tracheobronchial passage; inflates and deflates tracheostomy cuff; provides tracheostomy tube and stoma care; evaluates patient cardiorespiratory status; utilizes emergency ventilators; evaluates patient cardiorespiratory status; utilizes emergency ventilators; evaluates resuscitator's effectiveness; implements culturing procedures; applies corrective actions for contaminated equipment; reads and evaluates patient charts; instructs patient and family in breathing exercises and postural drainage procedures; evaluates ability to operate home therapy equipment; maintains safe, clean, and orderly work area; stores and maintains supplies and equipment; calibrates and slopes arterial blood gas analyzer; applies safety precautions relevant to pulmonary function testing; calculates data and compares results; receives and issues supplies and equipment. *Skill Level 30:* Able to perform the duties required for Skill Level 20; supervises a respiratory unit; assists in defibrillation of patient; plans and monitors culturing procedures; determines corrective action for contaminated equipment; inspects working areas to ensure an orderly, clean, and safe environment for patients; supervises operational maintenance program of clinical equipment; prepares and conducts training programs for the staff and the community; manages budget; procures supplies; evaluates personnel performance and prepares evaluation reports; prepares administrative, technical, and patient reports; maintains computerized data bases; supervises packing, unpacking, loading, unloading, and setting up of unit equipment and shelter. *Skill Level 40:* Able to perform the duties required for Skill Level 30; applies and monitors safety procedures relevant to modular pulmonary function testing; administers tests, calculates test data, records results, and compares results with normal predicted values; develops, implements, and evaluates training programs; compiles management data, evaluates and determines personnel requirements, and assists in the planning and operation of respiratory unit; coordinates activities of unit with other elements of the medical treatment facility; advises and assists professional staff in supply economy procedures and fiscal, technical, and administrative matters; reviews, consolidates, and prepares technical, personnel, and administrative reports.

Recommendation, Skill Level 20

Credit may be granted on the basis of an individualized assessment of the student (12/01).

Recommendation, Skill Level 30

In the lower-division baccalaureate/associate degree category, 4 semester hours in mechanical ventilation, 4 in respiratory care clinical practice, 4 in arterial blood gases, 3 in principles of supervision, and 3 in records and information management. In the upper-division baccalaureate category, 3 semester hours in organizational management and 2 in respiratory care education (12/01).

Recommendation, Skill Level 40

In the lower-division baccalaureate/associate degree category, 4 semester hours in mechanical ventilation, 4 in respiratory care clinical practice, 4 in arterial blood gases, 3 in principles of supervision, and 3 in records and information management. In the upper-division baccalaureate category, 3 semester hours in organizational management, 2 in respiratory care education, and 3 for a management practicum (12/01).

MOS-91W-002

Nuclear Medicine Specialist
91W10
91W20
91W30
91W40

Exhibit Dates: 1/90–5/92.

Career Management Field: 91 (Medical), subfield 912 (Patient Care).

Description

Summary: Supervises or performs medical diagnostic and therapeutic procedures with radioactive isotopes. *Skill Level 10:* Able to perform the duties required for skill level 10 of MOS 91P (X-Ray Specialist), 92B (Medical Laboratory Specialist), or 91B (Medical Specialist); prepares and administers radioisotopes to patients and operates radiological imaging devices and counters. *Skill Level 20:* Able to perform the duties required for Skill Level 10; performs operational checks, standardizing procedures, and preventive maintenance on equipment; calculates dosage, prepares dilution, and administers tests under direct supervision of physician; operates imaging device to make scans of organs, including brain, liver, thyroid, pancreas, spleen, and lungs; makes photographs of images; collects appropriate specimens for such procedures as blood volume determination, Schilling test, thyroid uptake, red cell survival time, and fat metabolism studies; performs assay procedures using beta and gamma radioassay techniques; determines activity of radioisotope stocks and waste; determines presence of health hazards from therapeutic, diagnostic, and experimental radioactive isotopes; measures and records degree of contamination; controls spread of contamination; and institutes decontamination procedures. *Skill Level 30:* Able to perform the duties required for Skill Level 20; provides technical guidance to subordinates. *Skill Level 40:* Able to perform the duties required for Skill Level 30; assists with the management of a nuclear medicine facility; requisitions, accounts for, and stores radioisotopes used in the laboratory; is responsible for radiation safety program in laboratory area; supervises radiation survey, monitoring, and contamination control procedures; disposes of stocks when radioactivity decays to nonusable level; establishes work priorities and procedures; organizes work schedules; assigns duties; establishes safety procedures; instructs subordinates in work techniques and procedures; supervises operational maintenance program of clinical equipment; evaluates personnel performance and prepares evaluation reports; prepares patient, technical, and administrative reports; coordinates activities of nuclear medicine facility with other elements of the medical treatment facility; advises and assists professional staff with technical, administrative, and fiscal matters.

Recommendation, Skill Level 10

In the lower-division baccalaureate/associate degree category, 6 semester hours for a clinical internship, 6 in algebra, 6 in physics, and additional credit in nuclear medicine technology, radiology, and occupational health and safety on the basis of institutional evaluation. (NOTE: This recommendation for skill level 10 is valid for the dates 1/90-9/91 only) (6/77).

Recommendation, Skill Level 20

In the lower-division baccalaureate/associate degree category, 6 semester hours for a clinical internship, 6 in nuclear medicine procedures, 6 in algebra, 6 in physics, and additional credit in nuclear medicine technology, radiology, and occupational health and safety on the basis of institutional evaluation (6/77).

Recommendation, Skill Level 30

In the lower-division baccalaureate/associate degree category, 6 semester hours for a clinical internship, 6 in nuclear medicine procedures, 6 in algebra, 6 in physics, and 3 in advanced nuclear medicine procedures, and additional credit in nuclear medicine technology, radiology, and occupational health and safety on the basis of institutional evaluation (6/77).

Recommendation, Skill Level 40

In the lower-division baccalaureate/associate degree category, 6 semester hours for a clinical internship, 6 in nuclear medicine procedures, 6 in algebra, 6 in physics, 3 in advanced nuclear medicine procedures, 3 in personnel supervision, and 3 in human relations, and additional credit in nuclear medicine technology, radiology, and occupational health and safety on the basis of institutional evaluation. In the upper-division baccalaureate category, 3 semester hours in administration of nuclear medicine facility (6/77).

MOS-91W-003

Nuclear Medicine Specialist

91W10

91W20

91W30

91W40

Exhibit Dates: 6/92–12/94. (Effective 12/94, MOS 91W was discontinued, and its duties were incorporated into MOS 91P.)

Career Management Field: 91 (Medical).

Description

Summary: Supervises or performs medical diagnostic and therapeutic procedures with radioactive isotopes. *Skill Level 10:* Prepares and administers radioisotopes to patients and operates radiological imaging devices and counters. NOTE: Progresses to 91W10 from 91P10, X-Ray Specialist; 92B10, Medical Laboratory Specialist; or 91A10, Medical Specialist. *Skill Level 20:* Able to perform the duties required for Skill Level 10; performs operational checks, standardizing procedures, and preventive maintenance on equipment; calculates dosage, prepares dilution, and administers tests under direct supervision of physician; operates imaging device to make scans of organs, including brain, liver, thyroid, pancreas, spleen, and lungs; makes photographs of images; collects appropriate specimens for such procedures as blood volume determination, Schilling test, thyroid uptake, red cell survival time, and fat metabolism studies; performs assay procedures using beta and gamma radioassay techniques; determines activity of radioisotope stocks and waste; determines presence of health hazards from therapeutic, diagnostic, and experimental radioactive isotopes; measures and records degree of contamination; controls spread of contamination; and institutes decontamination procedures. *Skill Level 30:* Able to perform the duties required for Skill Level 20; provides supervision and technical guidance to subordinates. *Skill Level 40:* Able to perform the duties required for Skill Level 30; assists with the management of a nuclear medicine facility; requisitions, accounts for, and stores radioisotopes used in the laboratory; is responsible for radiation safety program in laboratory area; supervises radiation survey, monitoring, and contamination control procedures; disposes of stocks when radioactivity decays to nonusable level; establishes work priorities and procedures; organizes work schedules; assigns duties; establishes safety procedures; instructs subordinates in work techniques and procedures; supervises operational maintenance program of clinical equipment; evaluates personnel performance and prepares evaluation reports; prepares patient, technical, and administrative reports; coordinates activities of nuclear medicine facility with other elements of the medical treatment facility; advises and assists professional staff with technical, administrative, and fiscal matters.

Recommendation, Skill Level 10

Credit may be granted on the basis of an individualized assessment of the student (3/94).

Recommendation, Skill Level 20

In the lower-division baccalaureate/associate degree category, 4 semester hours in directed practice in nuclear medicine, 4 in clinical applications in nuclear medicine, and 3 in personnel supervision (6/92).

Recommendation, Skill Level 30

In the lower-division baccalaureate/associate degree category, 4 semester hours in directed practice in nuclear medicine, 4 in clinical applications in nuclear medicine, 3 in personnel supervision, and 2 in records and information management (6/92).

Recommendation, Skill Level 40

In the lower-division baccalaureate/associate degree category, 4 semester hours in directed practice in nuclear medicine, 4 in clinical applications in nuclear medicine, 3 in personnel supervision, 2 in records and information management, and 3 in quality assurance and safety. In the upper-division baccalaureate category, 3 semester hours in organizational management and 3 for field experience in management (6/92).

MOS-91W-004

Health Care Specialist

91W10

91W20

91W30

91W40

91W50

Exhibit Dates: Pending Evaluation. 10/01–Present.

MOS-91X-001

Health Physics Specialist

91X20

91X30

91X40

Exhibit Dates: 1/90–12/94. (Effective 12/94, MOS 91X was discontinued, and its duties were incorporated into MOS 91S.)

Career Management Field: 91 (Medical).

Description

Summary: Supervises or performs technician-level health physics functions in support of radiation protection programs either in medical treatment facilities or other agencies operating under Army control. *Skill Level 20:* Conducts ionizing and non-ionizing radiation protection surveys, including diagnostic X-ray, nuclear medicine inpatient, sealed and unsealed source radiation therapy, and teletherapy in research and vitro clinical laboratories, industrial facilities, gamma irradiators, waste storage and disposal areas, electronic microscope, laser, and microwave equipment, ultraviolet and high intensity visible light source facilities, and environmental radiation and occupational air sampling; evaluates outpatient potential hazards and develops techniques to minimize risk of radiation exposure to workers; coordinates collection and disposal of radioactive waste with material users; calibrates radiation survey meters and other health physics detectors, monitors, and measuring equipment; arranges outside calibration support; conducts preliminary investigation of radioactive contamination incidents and apparent overexposure to radiation;

prepares results of surveys; performs quality control procedures; develops and presents special radiation safety instruction for technician users. *Skill Level 30:* Able to perform the duties required for Skill Level 20; assumes basic supervisory activities; supervises personnel dosimetry program; establishes work schedules and priorities; conducts nonroutine radiation protection surveys; performs nonroutine analysis; reports incidence of noncompliance; evaluates complex radiation activities for potential hazards; evaluates working conditions for workers; conducts studies to define radiation hazards; repairs and calibrates complex equipment; develops requirements for new equipment; prepares data for budget planning; conducts training of emergency team members. *Skill Level 40:* Able to perform the duties required for Skill Level 30; designs and conducts sophisticated radiation protection surveys; reviews plans for proposed construction of radiation facilities for design adequacy; conducts thorough investigations of unusual or complex radiation incidents.

Recommendation, Skill Level 20

In the lower-division baccalaureate/associate degree category, 4 semester hours in principles of physical sciences, 4 in introduction to health physical sciences, 2 in environmental and legal aspects of radiation, 5 in medical X-ray survey procedures, and 5 for clinical experience in applied health physics (9/85).

Recommendation, Skill Level 30

In the lower-division baccalaureate/associate degree category, 4 semester hours in principles of physical sciences, 4 in introduction to health physical sciences, 2 in environmental and legal aspects of radiation, 5 in medical X-ray procedures, 5 for clinical experience in applied health physics, 2 in oral communications, and 3 for field experience in radiation protection (9/85).

Recommendation, Skill Level 40

In the lower-division baccalaureate/associate degree category, 4 semester hours in principles of physical sciences, 4 in introduction to health physical sciences, 2 in environmental and legal aspects of radiation, 5 in medical X-ray procedures, 5 for clinical experience in applied health physics, 3 in oral communications, and 6 for field experience in radiation protection (9/85).

MOS-91X-002

Mental Health Specialist
91X10
91X20
91X30
91X40

Exhibit Dates: 8/96–Present.

Career Management Field: 91 (Medical).

Description

Summary: Under professional supervision, assists with management of outpatient mental health, drug, and alcohol patients; obtains social case histories; collects basic psychological data; counsels patients with personal, behavioral, or mental health problems. *Skill Level 10:* Interviews clients; conducts collateral interviews; administers and scores intelligence tests under supervision; assists in determining client need for referral to a professional; records psychosocial data and behavior observations; orients and refers clients to military, civilian, and agency resources; prepares and presents cases for staffings and conferences; requests and transmits client records. *Skill Level 20:* Able to perform the duties required for Skill Level 10; under professional supervision, provides supportive counseling and follow-up service to individuals experiencing a wide range of social or emotional problems; assists with group counseling and therapy sessions; leads discussion groups; assists in determining need for hospitalization; under supervision, administers and scores achievement tests, objective personality tests, and tests of organic impairment and records the results; assists patients with discharge planning; consults on management of individual behavioral problems; assists professional staff in setting up home health care programs and makes follow-up home visits; gathers statistical data; establishes and maintains clinical and general office files. *Skill Level 30:* Able to perform the duties required for Skill Level 20; provides advice and limited technical guidance to subordinates; provides supportive counseling and crisis intervention services to clients; collects and organizes research data and makes statistical computations; assists in analysis of research data and report writing; compiles caseload data and assigns clients to specific staff members; evaluates manpower requirements; prepares administrative reports; requisitions, stores, and issues supplies, equipment, publications, and training aids; develops work procedures and practices; prepares and conducts in-service training programs. *Skill Level 40:* Able to perform the duties required for Skill Level 30; under supervision, administers and records thematic projective tests; assists in determining program and management objectives and procedures; establishes work priorities and work schedules; assigns duties and instructs subordinates in work techniques and procedures; prepares and counsels personnel and prepares evaluation reports; coordinates activity of facility with related elements of the medical treatment facility; advises and assists professionals with fiscal, technical, and administrative matters; establishes stock level for supplies and equipment.

Recommendation, Skill Level 10

Credit may be granted on the basis of an individualized assessment of the student (7/99).

Recommendation, Skill Level 20

Credit may be granted on the basis of an individualized assessment of the student (7/99).

Recommendation, Skill Level 30

In the lower-division baccalaureate/associate degree category, 3 semester hours in introduction to human services, 5 in human services clinical practice, 3 in behavior modification techniques, 3 in personnel supervision, 2 in quality assurance, and 2 in records and information management (7/99).

Recommendation, Skill Level 40

In the lower-division baccalaureate/associate degree category, 3 semester hours in introduction to human services, 6 in human services clinical practice, 3 in behavior modification techniques, 3 in personnel supervision, 2 in quality assurance, and 2 in records and information management. In the upper-division baccalaureate category, 3 semester hours in organizational management and 3 for field experience in management (7/99).

MOS-91Y-002

Eye Specialist
91Y10
91Y20
91Y30
91Y40

Exhibit Dates: 1/90–3/92.

Career Management Field: 91 (Medical), subfield 912 (Patient Care).

Description

Summary: Conducts routine diagnostic tests and assists in care and treatment of ophthalmology or optometry patients. *Skill Level 10:* Able to perform the duties required for skill level 10 of MOS 91B (Medical Specialist); conducts routine diagnostic tests, administers prescribed medications, and assists in caring for eye patients; determines purpose of patient visit; administers various visual tests and records results; measures accommodative ability; measures muscle balance with and without correction, near and distant; measures limits of central and peripheral field, intraocular pressure, interpupillary distance, bridge size, and temple length; performs minor repairs and adjustments to spectacles; instills drops and ointments in eye as prescribed; removes nonembedded foreign objects from surface of eye or underside of lid; assists in obtaining specimens; assists the physician in performing minor surgery; maintains ophthalmology and optometry equipment and instruments; performs prescribed sterilization procedures; performs ophthalmic photography; measures prismatic power of lenses; operates resuscitation and oxygenation equipment; instructs patients in insertion, removal, and care of contact lenses; prepares slides for weights and grams staining; applies and removes dressings as directed; performs emergency procedures as required; maintains inventory of supplies and equipment. *Skill Level 20:* Able to perform the duties required for Skill Level 10; provides technical guidance to subordinates. *Skill Level 30:* Able to perform the duties required for Skill Level 20; assists in managing a small ophthalmology or optometry clinic; organizes work schedules and determines personnel requirements; assigns duties and establishes work priorities; instructs subordinates in work techniques and procedures; supervises operational maintenance program of clinical equipment; evaluates personnel performance and prepares evaluation reports; coordinates activities of ophthalmology and optometry clinics with other elements of the medical treatment facility; assists physician with personnel, fiscal, and administrative matters; prepares patient and administrative reports. *Skill Level 40:* Able to perform the duties required for Skill Level 30; assists in managing a large ophthalmology or optometry clinic.

Recommendation, Skill Level 10

In the lower-division baccalaureate/associate degree category, 4 semester hours in principles of ophthalmology, 4 for a practicum in ophthalmology, 4 for a practicum in optometry, 2 in principles of optometry, and 1 in anatomy and physiology. (NOTE: This recommendation for skill level 10 is valid for the dates 1/90-9/91 only) (6/77).

Recommendation, Skill Level 20

In the lower-division baccalaureate/associate degree category, 4 semester hours in principles of ophthalmology, 6 for a practicum in ophthalmology, 6 for a practicum in optometry, 2 in principles of optometry, and 1 in anatomy and physiology (6/77)

Recommendation, Skill Level 30

In the lower-division baccalaureate/associate degree category, 4 semester hours in principles of ophthalmology, 6 for a practicum in ophthalmology, 6 for a practicum in optometry, 2 in principles of optometry, 1 in anatomy and physiology, 3 in personnel supervision, 2 in communication skills, 2 in principles of management, and 1 in health care administration. In the upper-division baccalaureate category, credit in patient care administration on the basis of institutional evaluation (6/77).

Recommendation, Skill Level 40

In the lower-division baccalaureate/associate degree category, 4 semester hours in principles of ophthalmology, 6 for a practicum in ophthalmology, 6 for a practicum in optometry, 2 in principles of optometry, 1 in anatomy and physiology, 3 in personnel supervision, 2 in communication skills, 2 in principles of management, and 1 in health care administration. In the upper-division baccalaureate category, credit in patient care administration and medical facility management on the basis of institutional evaluation (6/77).

MOS-91Y-003

Eye Specialist

91Y10
91Y20
91Y30
91Y40

Exhibit Dates: 4/92–12/94. (Effective 12/94, MOS 91Y was discontinued, and its duties were incorporated into MOS 91B.)

Career Management Field: 91 (Medical).

Description

Summary: Conducts routine diagnostic tests and assists in care and treatment of ophthalmology or optometry patients. *Skill Level 10:* Conducts routine diagnostic tests; administers prescribed medications; assists in caring for eye patients; determines purpose of patient visit; administers various visual tests and records results; measures accommodative ability; measures muscle balance with and without correction, near and distant; measures limits of central and peripheral field, intraocular pressure, interpupillary distance, bridge size, and temple length; performs minor repairs and adjustments to spectacles; instills drops and ointments in eye as prescribed; removes nonembedded foreign objects from surface of eye or underside of lid; assists in obtaining specimens; assists the physician in performing minor surgery; maintains ophthalmology and optometry equipment and instruments; performs prescribed sterilization procedures; performs ophthalmic photography; measures prismatic power of lenses; operates resuscitation and oxygenation equipment; instructs patients in insertion, removal, and care of contact lenses; prepares slides for weights and grams staining; applies and removes dressings as directed; performs emergency procedures as required; maintains inventory of supplies and equipment. *Skill Level 20:* Able to perform the duties required for Skill Level 10; provides technical guidance to subordinates. *Skill Level 30:* Able to perform the duties required for Skill Level 20; assists in managing a small ophthalmology or optometry clinic; organizes work schedules and determines personnel requirements; assigns duties and establishes work priorities; instructs subordinates in work techniques and procedures; supervises operational maintenance program of clinical equipment; evaluates personnel performance and prepares evaluation reports; coordinates activities of ophthalmology and optometry clinics with other elements of the medical treatment facility; assists physician with personnel, fiscal, and administrative reports. *Skill Level 40:* Able to perform the duties required for Skill Level 30; assists in managing a large ophthalmology or optometry clinic.

Recommendation, Skill Level 10

Credit may be granted on the basis of an individualized assessment of the student (3/94).

Recommendation, Skill Level 20

In the lower-division baccalaureate/associate degree category, 4 semester hours in principles of optometric assisting, 4 in optometric assisting clinical practice, and 3 in personnel supervision (4/92).

Recommendation, Skill Level 30

In the lower-division baccalaureate/associate degree category, 4 semester hours in principles of optometric assisting, 5 in optometric assisting clinical practice, 3 in personnel supervision, and 2 in records and information management (4/92).

Recommendation, Skill Level 40

In the lower-division baccalaureate/associate degree category, 4 semester hours in principles of optometric assisting, 6 in optometric assisting clinical practice, 3 in personnel supervision, and 2 in records and information management. In the upper-division baccalaureate category, 3 semester hours in organizational management and 3 for field experience in management (4/92).

MOS-92A-001

Automated Logistical Specialist

92A10
92A20
92A30
92A40
92A50

Exhibit Dates: 2/93–2/95.

Career Management Field: 92 (Supply and Service).

Description

Summary: Supervises or performs stock record and warehouse functions, including stock receipt, storage, distribution, and issue; maintains equipment records and parts. *Skill Level 10:* Establishes and maintains stock records and other records; establishes and maintains automated and manual accounting records, posts receipts and turn-ins; reviews and verifies quantities received against shipping documents; prepares and maintains records on equipment usage, operation, maintenance, modification, and calibration; processes inventories, surveys, and warehousing documents; prepares, annotates, and distributes shipping documents; performs accounting and sales functions in self-service supply. *Skill Level 20:* Able to perform the duties required for Skill Level 10; supervises and provides technical guidance to subordinates; reviews stock items and recommends additions, deletions, or changes; obtains repair parts to meet equipment maintenance needs; assists in planning maintenance needs for the unit; performs financial management functions. *Skill Level 30:* Able to perform the duties required for Skill Level 20; inspects and evaluates inventory management activities; analyzes statistical data to determine effectiveness of technical edit; plans and organizes receipt, issue, salvage, and maintenance of records for all classes of supply; prepares reports regarding personnel, storage, and relocation of material; supervises warehouse activity. *Skill Level 40:* Able to perform the duties required for Skill Level 30; conducts surveillance inspection of material in storage; develops training programs; assists in development and preparation of operations information including plans, maps, sketches, and other data related to supply organization employment. *Skill Level 50:* Able to perform the duties required for Skill Level 40; serves as mid-level manager supervising supply and related service operations; assists commander in planning, supervising, inspecting, developing, and executing unit policy; advises commander on enlisted personnel matters; provides counsel and guidance to subordinates; analyzes reports on supply and service support operations.

Recommendation, Skill Level 10

Credit may be granted on the basis of an individualized assessment of the student (3/95).

Recommendation, Skill Level 20

In the lower-division baccalaureate/associate degree category, 3 semester hours in supply management, 3 in records and information management, 2 in computer applications, and 1 in personnel supervision (3/95).

Recommendation, Skill Level 30

In the lower-division baccalaureate/associate degree category, 3 semester hours in supply management, 3 in records and information management, 3 in computer applications, and 3 in personnel supervision. In the upper-division baccalaureate category, 3 semester hours for field experience in management (3/95).

Recommendation, Skill Level 40

In the lower-division baccalaureate/associate degree category, 3 semester hours in supply management, 3 in records and information management, 3 in personnel supervision; and 3 in computer applications based on individual assessment of the student. In the upper-division baccalaureate category, 6 semester hours for field experience in management (3/95).

Recommendation, Skill Level 50

In the lower-division baccalaureate/associate degree category, 3 semester hours in supply management, 3 in records and information management, 3 in personnel supervision, and 3 in computer applications based on individual assessment of the student. In the upper-division baccalaureate category, 6 semester hours for field experience in management and 3 in organizational management (3/95).

MOS-92A-002

Automated Logistical Specialist

92A10
92A20
92A30
92A40
92A50

Exhibit Dates: 3/95–Present.

Career Management Field: 92 (Supply and Service).

Description

Summary: Supervises or performs stock record and warehouse functions, including stock receipt, storage, distribution, and issue; maintains equipment records and parts. *Skill Level 10:* Establishes and maintains stock records and other records; establishes and maintains automated and manual accounting records, posts receipts and turn-ins; reviews and verifies quantities received against shipping documents; prepares and maintains records on equipment usage, operation, maintenance, modification, and calibration; processes inventories, surveys, and warehousing documents; prepares, annotates, and distributes shipping documents; performs accounting and sales functions in self-service supply. *Skill Level 20:* Able to perform the duties required for Skill Level 10; supervises and provides technical guidance to subordinates; reviews stock items and recommends additions, deletions, or changes; obtains repair parts to meet equipment maintenance needs; assists in planning maintenance needs for the unit; performs financial management functions. *Skill Level 30:* Able to perform the duties required for Skill Level 20; inspects and evaluates inventory management activities; analyzes statistical data to determine effectiveness of technical edit; plans and organizes receipt, issue, salvage, and maintenance of records for all classes of supply; prepares reports regarding personnel, storage, and relocation of material; supervises warehouse activity. *Skill Level 40:* Able to perform the duties required for Skill Level 30; conducts surveillance inspection of material in storage; develops training programs; assists in development and preparation of operations information including plans, maps, sketches, and other data related to supply organization employment. *Skill Level 50:* Able to perform the duties required for Skill Level 40; serves as mid-level manager supervising supply and related service operations; assists commander in planning, supervising, inspecting, developing, and executing unit policy; advises commander on enlisted personnel matters; provides counsel and guidance to subordinates; analyzes reports on supply and service support operations.

Recommendation, Skill Level 10

Credit may be granted on the basis of an individualized assessment of the student (3/95).

Recommendation, Skill Level 20

Credit may be granted on the basis of an individualized assessment of the student (3/95).

Recommendation, Skill Level 30

In the lower-division baccalaureate/associate degree category, 3 semester hours in supply management, 3 in records and information management, 2 in computer applications, and 3 in personnel supervision. In the upper-division baccalaureate category, 3 semester hours for field experience in management (3/95).

Recommendation, Skill Level 40

In the lower-division baccalaureate/associate degree category, 3 semester hours in supply management, 3 in records and information management, 3 in personnel supervision; and 3 in computer applications based on individual assessment of the student. In the upper-division baccalaureate category, 6 semester hours for field experience in management (3/95).

Recommendation, Skill Level 50

In the lower-division baccalaureate/associate degree category, 3 semester hours in supply management, 3 in records and information management, 3 in personnel supervision, and 3 in computer applications based on individual assessment of the student. In the upper-division baccalaureate category, 6 semester hours for field experience in management and 3 in organizational management (3/95).

MOS-92B-002

Medical Laboratory Specialist

92B10
92B20
92B30
92B40
92B50

Exhibit Dates: 1/90–5/92.

Career Management Field: 91 (Medical), subfield 913 (Health Services).

Description

Summary: Performs blood banking procedures and elementary and advanced examinations of biological and environmental specimens to aid in diagnosis, treatment, and prevention of disease. *Skill Level 10:* Receives, collects, and labels specimens; calculates and records results of analyses; prepares laboratory specimens for the determination of parasites, plasmodium species, ABO group and Rh type blood; performs antihuman globulin, microhematocrit, and manual hemoglobulin procedures; performs manual white blood cell counts; prepares and examines thin-blood film slides and stain procedures; performs erythrocyte sedimentation rate procedures; determines specific gravity and performs chemical tests on urine to detect glucose, protein, blood, ketones, mucin, or melanogen; prepares urine samples for determination of pH, microscopic examination, cells, and mucous threads; inoculates bacteriological specimens and stains bacteriological smears and examines for stain reaction; prepares and sterilizes culture media; performs manual biochemical determinations on blood; examines semiautomatically for chlorides, sodium, and potassium; performs basic serological screening for syphilis and rheumatoid arthritis; performs basic preventive maintenance on laboratory equipment. *Skill Level 20:* Able to perform the duties required for Skill Level 10; performs blood banking and elementary laboratory procedures related to immunohematology, hematology, biochemistry, serology, bacteriology, and parasitology; selects, pretests, and bleeds donors and stores and issues blood; determines ABO group and Rh type of donor and recipient; performs immunohematology procedures such as antibody titers and compatibility testing; prepares and stores blood components; performs ABO subgrouping and DU testing to detect antibodies; performs complete blood count and cell counts; performs coagulation tests; uses automated, semiautomated, or manual procedures to determine cholesterol, total protein and A/G ratio, blood pH, pO2, and pCO2; performs basic serology procedures for syphilis, pregnancy, infectious mononucleosis, and arthritis; prepares red cell suspensions and dilutions; examines body fluids and feces for presence of protozoa, helminths, and extraintestinal parasites; prepares and sterilizes culture media; inoculates specimens for microscopic examination; isolates, identifies, and studies such microorganisms as streptococcus, staphylococcus, pneumococcus, and common gram negative organisms. *Skill Level 30:* Able to perform the duties required for Skill Level 20; supervises small medical laboratory or performs advanced procedures in all phases of blood banking and clinical laboratory testing, including virology, mycology, histology, and toxicology; performs complex laboratory tests, examinations, and analyses under general supervision and provides technical assistance to subordinates; performs cultures and antimicrobial sensitivity determinations including those taken from autopsies; performs clinical parasitology procedures; performs fluorescent antibody agglutination, complement fixation, and other serological tests; processes tissue; performs biochemical analysis of blood, urine, sweat, and spinal fluid for carbohydrates, proteins, hemoglobins, lipids, hormones, minerals, electrolytes, vitamins, enzymes, pigments, and drugs; performs metabolic, liver function, and renal function tests; screens exfoliative cytology specimens; performs complete blood counts and identifies maturation stages in normal and abnormal cells; performs special procedures in staining and preparing bone marrow smears, erythrocyte osmotic fragility studies, lupus erythematosus determinations, detection of coagulation disorders, complex electrophoresis studies, and abnormal cell morphology in anemias and leukemias; analyzes air, food, water, and biological fluids for contaminants; supervises operator maintenance procedures. *Skill Level 40:* Able to perform the duties required for Skill Level 30; supervises medium-sized medical laboratory; develops and supervises quality control measures and procedures; assigns duties; organizes work schedules; instructs subordinates; inspects laboratory; supervises operational maintenance; determines personnel requirements and prepares and conducts training; evaluates personnel performance; prepares administrative, technical, and personnel reports; coordinates activities of medical laboratories with other units of medical facility; assists laboratory officer in administrative matters. *Skill Level 50:* Able to perform the duties required for Skill Level 40; supervises large medical laboratory activities.

Recommendation, Skill Level 10

In the vocational certificate category, 10 semester hours in a certified laboratory assistant program. In the lower-division baccalaureate/associate degree category, 15 semester hours in a medical laboratory technician program (basic procedures in hematology, clinical chemistry, and urinalysis, immunohematology and serology, medical bacteriology, and parasitology). (NOTE: This recommendation for skill level 10 is valid for the dates 1/90-9/91 only) (6/77).

Recommendation, Skill Level 20

In the vocational certificate category, 10 semester hours in a certified laboratory assistant program. In the lower-division baccalaureate/associate degree category, 24 semester hours in a medical laboratory technician program (hematology, immunology, immunohematology and blood banking; and basic procedures in hematology, clinical chemistry and urinalysis, immunohematology and serology, medical bacteriology, and parasitology) (6/77).

Recommendation, Skill Level 30

In the vocational certificate category, 10 semester hours in a certified laboratory assistant program. In the lower-division baccalaureate/associate degree category, 54 semester hours in a medical laboratory technician program (hematology, immunology, immunohematology and blood banking, histopathology, mycology, virology, anatomy and physiology, cytology, parasitology, advanced clinical chemistry, advanced medical bacteriology, and advanced hematology; and basic procedures in hematology, clinical chemistry and urinalysis, immunohematology and serology, medical bacteriology, and parasitology). In the upper-division baccalaureate category, credit in medical technology, chemistry, and biology on the basis of institutional evaluation (6/77).

Recommendation, Skill Level 40

In the vocational certificate category, 10 semester hours in a certified laboratory assistant program. In the lower-division baccalaureate/associate degree category, 54 semester hours in a medical laboratory technician program (hematology, immunology, immunohematology and blood banking, histopathology, mycology, virology, anatomy and physiology, cytology, parasitology, advanced clinical chemistry, advanced medical bacteriology, advanced hematology; and basic procedures in hematology, clinical chemistry and urinalysis, immunohematology and serology, medical bacteriology, and parasitology), 3 in personnel supervision, and 3 in principles of administration. In the upper-division baccalaureate category, credit in medical technology, chemistry, biology, and medical facility management on the basis of institutional evaluation (6/77).

Recommendation, Skill Level 50

In the vocational certificate category, 10 semester hours in a certified laboratory assistant program. In the lower-division baccalaureate/associate degree category, 54 semester hours in a medical laboratory technician program (hematology, immunology, immunohematology and blood banking, histopathology, mycology, virology, anatomy and physiology, cytology, parasitology, advanced clinical chemistry, advanced medical bacteriology, advanced hematology; and basic procedures in hematology, clinical chemistry and urinalysis, immunohematology and serology, medical bacteriology, and parasitology), 3 in personnel supervision, and 3 in principles of administration. In the upper-division baccalaureate category, credit in medical technology, chemistry, biology, and medical facility management on the basis of institutional evaluation (6/77).

MOS-92B-003

Medical Laboratory Specialist

92B10
92B20
92B30
92B40
92B50

Exhibit Dates: 6/92–12/94. (Effective 12/94, MOS 92B was discontinued, and its duties were incorporated into MOS 91K.)

Career Management Field: 91 (Medical).

Description

Summary: Performs laboratory procedures and elementary and advanced examinations of biological and environmental specimens to aid in diagnosis, treatment, and prevention of disease. *Skill Level 10:* Receives, logs, and files requests for laboratory examination and analysis: receives, collects, and labels specimens; assembles and prepares supplies, equipment, reagents, and various types of glassware, and calculates and records results of analyses; performs formalin ether fecal concentrations and examines concentrate for presence of parasites; prepares thick and thin blood smears, performs giemsa stain techniques, and examines smear for presence of plasmodium species; determines ABO group and Rh type of blood specimens; performs direct and indirect antihuman globulin procedures; performs microhematocrit, manual hemoglobin procedures, and manual white blood cell counts; prepares thin blood film slides and performs Wright's stain procedure; examines blood films to enumerate normal leucocytes and confirms normal erythrocyte morphology; performs erythrocyte sedimentation rate procedures; determines urine-specific gravity and performs chemical tests on urine to detect glucose, Bence-Jones protein, blood, ketones, mucin, or melanogen; determines urine pH; concentrates urine for microscopic examination and identifies squamous epithelial cells, white blood cells, red blood cells, cast, and mucous threads in urine; inoculates bacteriological smears by gram stain technique; examines smears for stain reaction limited to gram, positive/negative rods/cocci and prepares and sterilizes culture media; performs manual biochemical determinations on blood, limited to glucose, area nitrogen, CO2 bilrubin, and titrimetric fluoride; performs semiautomated determinations on chlorides, sodium, and potassium; performs basic serological screening tests for syphilis and rheumatoid arthritis; performs preventive maintenance on laboratory equipment. *Skill Level 20:* Able to perform the duties required for Skill Level 10; collects and processes specimens for shipments to consultant laboratories; screens, selects, pretests, determines ABO group and pH type, and bleeds donors; stores and issues blood; performs immunohematology procedures such as antibody titers and compatibility testing; prepares and stores blood and blood components such as packed cells and plasma packs; performs ABO subgrouping and DU testing and detects presence of antibodies; performs hematology tests and determinations such as complete blood count, differential and cell counts of body fluids, and thrombocyte counts; performs coagulation tests such as prothrombin and partial thromboplastin time; uses automated, semiautomated, or manual procedures to conduct basic biochemical procedures such as tests for enzyme determination, cholesterol, total protein and A/G ratio, blood pH, pO2, and pCO2; performs basic serology procedures such as tests for syphilis, pregnancy, infectious mononucleosis, and arthritis; prepares red cell suspensions and dilutions; examines body fluids and feces for presence of protozoa, helminths, and extraintestinal parasites; prepares and sterilizes culture media and inoculates specimens into media; prepares and stains smears for microscopic examination; isolates, identifies, and studies microorganisms such as streptococcus, staphylococcus, pneumococcus, and common gram negative organisms. *Skill Level 30:* Able to perform the duties required for Skill Level 20; supervises operator maintenance on assigned equipment; supervises procedures to enable the unit to function in toxic environment; supervises packing, unpacking, loading, unloading, and setting up unit equipment and shelters; performs complex laboratory tests, examinations, and analyses under general supervision and provides technical assistance to subordinates; performs culture and antimicrobial sensitivity determinations including those taken from autopsies to isolate and identify microorganisms; performs clinical parasitology procedures related to life cycles, epidemiology, identification, and concentration of parasites in blood, feces, and body fluids; performs fluorescent antibody agglutination, complement fixation, and other serological tests for febrile, viral, rickettsial, syphilitic, parasitic, and mycotic infections; processes tissue removed by biopsy, from autopsy, or discharged from the body by fixing the tissue in preservative, dehydrating, embedding, sectioning, staining, and mounting on slides; performs biochemical analyses of blood, urine, sweat, spinal fluid, and tissue for carbohydrates, proteins, hemoglobin, lipids, hormones, minerals, electrolytes, vitamins, enzymes, pigments, and drugs; performs metabolic, liver, and renal function tests; determines total bases, acidbase balance, and blood gases in blood; identifies blood cell maturation stages in normal and abnormal cells; performs special procedures in staining and preparing bone marrow smears, erythrocyte osmotic fragility studies, lupus erythematous determinations, detection of coagulation disorders, complex eletrophoresis studies, and abnormal cell morphology in anemias and leukemias; analyzes air, water, food, and biological fluids for contaminants. *Skill Level 40:* Able to perform the duties required for Skill Level 30; develops and supervises quality control measures and procedures; establishes work priorities; organizes work schedules and assigns duties; instructs subordinates in work techniques and procedures; inspects laboratory to ensure orderly, clean, and safe working environment; supervises operation maintenance program of laboratory equipment; determines personnel requirements and prepares and conducts training programs; evaluates personnel requirements and prepares and conducts training programs; evaluates personnel performance and counsels personnel; prepares evaluation reports; prepares administrative, technical, and personnel reports; coordinates activities of medical laboratory with other elements of the medical treatment facility; advises and assists laboratory officer in professional matters, supply economy procedures, and fiscal, technical, and administrative matters; establishes stock levels and supervises the requisitioning, storage, and issue of supplies and equipment. NOTE: May progress to 92B40 from 92B30 or 01H30, Biological Sciences Assistant. *Skill Level 50:* Able to perform the duties required for Skill Level 40; supervises large medical laboratory activities; prepares reports and determines personnel requirements. NOTE: May progress to 92B50 from 92B40 or 92E40, Cytology Specialist.

Recommendation, Skill Level 10

Credit may be granted on the basis of an individualized assessment of the student (3/94).

Recommendation, Skill Level 20

In the lower-division baccalaureate/associate degree category, 5 semester hours in clinical laboratory procedures, 5 in laboratory instrumentation, and 3 in personnel supervision (6/92).

Recommendation, Skill Level 30

In the lower-division baccalaureate/associate degree category, 5 semester hours in clinical laboratory procedures, 5 in laboratory instrumentation, 5 in advanced clinical laboratory procedures, 3 in personnel supervision, and 2 in records and information management (6/92).

Recommendation, Skill Level 40

In the lower-division baccalaureate/associate degree category, 5 semester hours in clinical laboratory procedures, 5 in laboratory instrumentation, 5 in advanced clinical laboratory procedures, 3 in personnel supervision, and 2 in records and information management. In the upper-division baccalaureate category, 3 semester hours in organizational management and 3 for field experience in management (6/92).

Recommendation, Skill Level 50

In the lower-division baccalaureate/associate degree category, 5 semester hours in clinical laboratory procedures, 5 in laboratory instrumentation, 5 in advanced clinical laboratory procedures, 3 in personnel supervision, and 2 in records and information management. In the upper-division baccalaureate category, 3 semester hours in organizational management and 3 for field experience in management; if paygrade E-9 has been attained, additional credit as follows: 3 semester hours in management problems, 3 in operations management, and 3 in communication techniques for managers (6/92).

MOS-92E-001

Cytology Specialist

92E20
92E30
92E40

Exhibit Dates: 1/90–12/94. Effective 12/94, MOS 92E was discontinued, and its duties were incorporated into MOS 91K.

Career Management Field: 91 (Medical).

Description

Summary: Provides professional support and service to pathologists by supervising or performing diagnostic cytology services and rendering diagnoses. *Skill Level 20:* Logs, processes, and examines a variety of cellular samples of tissues and body fluids and renders diagnoses; examines microscope slides containing cells and processes cytological specimens from vagina and cervix, pleural fluid, gastric juice, bronchial washings, ascitic fluid, spinal fluid, prostatic fluid or needle aspiration, breast secretions or aspirates, urine, sputum, buccal smears, other body secretions, excretions, or scrapings, bones, and soft tissue tumors; identifies a variety of cell types from various parts of body and interprets morphological abnormalities; interprets variations that fall within normal limits and subtle variations in gross and microscopic appearance of cytological specimens and correlates clinical history with microscopic findings; evaluates and interprets cellular morphological changes associated with infectious processes, including viral, bacterial, fungal, and parasitic diseases; identifies organisms where possible on morphological basis; interprets endocrinological states including both normal variations and abnormal conditions as shown by cellular reactions; evaluates cellular changes brought about by radiation, drugs, and other forms of therapy; applies knowledge of carcinogenesis to interpretation of premalignant and malignant cellular changes; recognizes and reports a wide variety of neoplasms including both epithelial and mesodermal; differentiates between carcinoma-in-situ and invasive squamous carcinoma; classifies and grades neoplasm according to current classification schemes; determines, where possible, whether neoplasm is primary or metastatic; prepares written diagnostic reports on all cases based on correlation of clinical data with results for cytological study; includes interpretation of benign versus malignant, classification or grade of neoplasm, and recommendations for the need of further diagnostic procedures in these reports; assists in, selects, and advises on proper mode of collection of cytological specimens; recognizes and evaluates departures from expected performance norms and determines and accomplishes necessary actions for correcting procedural deficiencies; examines specimens describing color, viscosity, amount, and specific gravity; determines, selects, and smears portions of the specimen that may have particular pathological significance; prepares cell blocks for transmittal to histology; prepares cytocentrifuge preparations; stains and coverslips specimens; selects and performs quality control procedures; performs routine histology procedures; inspects, cleans, adjusts, and calibrates laboratory equipment; receives, unpacks, inspects, and stores supplies and equipment; maintains statistics and compiles monthly reports. *Skill Level 30:* Able to perform the duties required for Skill Level 20; provides technical guidance to subordinates; supervises small cytology section; inspects laboratory to ensure orderly, clean, and safe environment; organizes work schedules and assigns duties; instructs subordinates in work techniques and procedures; evaluates personnel performances and counsels personnel; prepares evaluation reports; establishes and maintains teaching file of interesting or unusual cases; implements and conducts in-service and performance-oriented training programs; evaluates and initiates new procedures; performs advanced cytological procedures including chromosomal analysis by karyotyping; establishes and administers program for correlation of cytological diagnosis with histological findings; reports findings to pathologist; reviews findings with pathologist; advises and assists in administrative, fiscal, personnel, and supply matters; prepares administrative, technical, and personnel reports; requisitions, safeguards, issues, and monitors effective utilization of supplies and equipment; establishes report control system; prepares and reviews correspondence; supervises maintenance of office files and technical library; supervises operational and preventive maintenance program for assigned equipment. *Skill Level 40:* Able to perform the duties required for Skill Level 30; supervises medium-sized and large cytology laboratories; develops and supervises quality control program; participates in training of resident physicians in pathology; reviews and consolidates reports.

Recommendation, Skill Level 20

In the upper-division baccalaureate category, 4 semester hours in introduction to cytology, 3 in female genital tract (benign), 4 in female genital tract (malignant), 2 in respiratory tract cytology, 2 in gastrointestinal tract cytology, 3 in central nervous system and effusion, 2 in urinary tract cytology, 3 in fine needle aspiration cytology, 2 in research, 18 for clinical experience, and 1 in organizational management (3/90).

Recommendation, Skill Level 30

In the upper-division baccalaureate category, 4 semester hours in introduction to cytology, 3 in female genital tract (benign), 4 in female genital tract (malignant), 2 in respiratory tract cytology, 2 in gastrointestinal tract cytology, 3 in central nervous system and effusion, 2 in urinary tract cytology, 3 in fine needle aspiration cytology, 2 in research, 24 for clinical experience, 3 in organizational management, and 3 in laboratory administration (3/90).

Recommendation, Skill Level 40

In the upper-division baccalaureate category, 4 semester hours in introduction to cytology, 3 in female genital tract (benign), 4 in female genital tract (malignant), 2 in respiratory tract cytology, 2 in gastrointestinal tract cytology, 3 in central nervous system and effusion, 2 in urinary tract cytology, 3 in fine needle aspiration cytology, 2 in research, 24 for clinical experience, 6 in organizational management, 6 in laboratory administration, and 3 in educational methodology (3/90).

MOS-92F-001

Petroleum Supply Specialist

92F10
92F20
92F30
92F40

Exhibit Dates: Pending Evaluation. 4/03–Present.

MOS-92G-001

Food Service Operations

92G10
92G20
92G30
92G40
92G50

Exhibit Dates: 8/95–Present.

Career Management Field: 94 (Food Service).

Description

Summary: Prepares, cooks, and serves food in field or garrison food service operations; supervises food service operations. *Skill Level 10:* Performs preliminary food preparation procedures; prepares and/or cooks menu items listed on the production schedule; bakes, fries, steams, braises, boils, simmers, steams, and sautes as prescribed by Army recipes; sets up serving lines; garnishes food items; applies food protection and sanitation measures in field and garrison environments; receives and stores subsistence items; performs general housekeeping duties; operates, maintains, and cleans field kitchen equipment; erects, strikes, and stores all types of field kitchens, and performs preventive maintenance on garrison and field kitchen equipment. *Skill Level 20:* Able to perform the duties required for Skill Level 10; ensures that proper procedures, temperatures, and time periods are adhered to during food preparation; performs limited supervisory and inspection functions including shift supervision. *Skill Level 30:* Able to perform the duties required for Skill Level 20; prepares more complex menu items; supervises shift, unit, or consolidated food service operations in field or garrison environments; establishes operating and work procedures; inspects food preparation/storage areas; supervises dining facility staff; determines subsistence requirements; requests, receives, and accounts for subsistence items; applies food service accounting procedures; prepares production schedule; makes necessary menu adjustments; establishes and maintains on-the-job and apprentice training programs; prepares technical, personnel, and administrative reports concerning food service operations; implements emergency, disaster, and combat feeding plans; coordinates logistical support. *Skill Level 40:* Able to perform the duties required for Skill Level 30; assigns staff to duty positions; administers on-the-job training program; coordinates with food service officer, food advisor, assistant food service sergeants, and first cooks; provides guidance to subordinates; coordinates support requirements with facility engineers and veterinary activity; plans and implements menus to insure nutritionally-balanced meals insures accuracy of accounting and equipment records; develops and initiates standard operating procedures on safety, energy, security, and fire prevention programs. *Skill Level 50:* Able to perform the duties required for Skill Level 40; develops, coordinates, implements, advises, and evaluates food service programs; monitors requests for food items and equipment; develops and analyzes troop menus and coordinates menu substitutions; evaluates operation of garrison and field kitchens, field bakeries, and food service training facilities; surveys individual preferences, food preparation, and food conservation; prepares reports, studies, and briefings on food service activities.

Recommendation, Skill Level 10

Credit may be granted on the basis of an individualized assessment of the student (6/97).

Recommendation, Skill Level 20

Credit may be granted on the basis of an individualized assessment of the student (6/97).

Recommendation, Skill Level 30

In the lower-division baccalaureate/associate degree category, 3 semester hours in basic food preparation, 3 in sanitation, 3 for field experience in food service, 3 in baking, and 3 in personnel supervision (6/97).

Recommendation, Skill Level 40

In the lower-division baccalaureate/associate degree category, 3 semester hours in basic food preparation, 3 in sanitation, 3 in baking, 3 for field experience in food service, 3 in kitchen food service operations, 3 in menu planning, 3 in food service management, and 3 in personnel supervision. In the upper-division baccalaureate category, 3 semester hours in human resource management (6/97).

Recommendation, Skill Level 50

In the lower-division baccalaureate/associate degree category, 3 semester hours in basic food preparation, 3 in sanitation, 3 in baking, 3 for field experience in food service, 3 in kitchen food service operations, 3 in menu planning, 3 in food service management, and 3 in personnel supervision. In the upper-division baccalaureate category, 3 semester hours in human resource management (6/97).

MOS-92L-001

Petroleum Laboratory Specialist

92L10
92L20
92L30
92L40

Exhibit Dates: Pending Evaluation. 4/03–Present.

MOS-92M-001

Mortuary Affairs Specialist

92M10
92M20
92M30
92M40
92M50

Exhibit Dates: 8/95–Present.

Career Management Field: 76 (Supply and Services).

Description

Summary: Recovers, identifies, processes, and inters human remains at disaster sites as well as within the community; provides technical data to validate the personal identification process; designs cemetery facilities; directs funeral activities. *Skill Level 10:* Serves as a mortuary assistant in recovery, transportation, identification, and interment of remains; reads or indicates the location of graves on maps, sketches, or overlays; determines recovery locations; records identification characteristics, including fingerprints and skeletal and dental charts; operates cameras and assists X-ray operations; assists in burials, excavating of graves, and cemetery maintenance. *Skill Level 20:* Able to perform the duties required for Skill Level 10; serves as mortuary assistant in recovery, transportation, identification, and interment of remains; reads or indicates the location of graves on maps, sketches, or overlays; determines recovery locations; records identification characteristics, including fingerprints and skeletal and dental charts; operates cameras and assists X-ray operations; assists in burials, excavating of graves, and cemetery maintenance. *Skill Level 30:* Able to perform the duties required for Skill Level 20; supervises graves registration activities; serves as graves registration noncommissioned officer in staff position; assists in postmortem examinations; assists in special criminal investigations; prepares efficiency reports on subordinates. *Skill Level 40:* Able to perform the duties required for Skill Level 30; advises subordinates on health requirements; coordinates and advises on military burials; assists in review of mortuary services contracts; counsels subordinates. *Skill Level 50:* Able to perform the duties required for Skill Level 40; supervises memorial affairs and graves registration activities; provides technical and administrative advice and assistance on graves registration matters, including acquisition of land for temporary cemeteries and facility, manpower, and equipment requirements; establishes and maintains official records; serves on special boards and councils; prepares reports on memorial affairs and graves registration activities; reviews reports of subordinates.

Recommendation, Skill Level 10

Credit may be granted on the basis of an individualized assessment of the student (6/97).

Recommendation, Skill Level 20

Credit may be granted on the basis of an individualized assessment of the student (6/97).

Recommendation, Skill Level 30

In the lower-division baccalaureate/associate degree category, 3 semester hours in military science, 4 in mortuary sciences, 2 in forensic identification, 2 in anatomy and physiology, 3 in personnel supervision, and 2 in records and information management (6/97).

Recommendation, Skill Level 40

In the lower-division baccalaureate/associate degree category, 3 semester hours military science, 4 in mortuary sciences, 2 in forensic identification, 2 in anatomy and physiology, 3 in personnel supervision, 3 in technical writing, and 2 in records and information management. In the upper-division baccalaureate category, 3 semester hours in organizational management and 3 for field experience in management (6/97).

Recommendation, Skill Level 50

In the lower-division baccalaureate/associate degree category, 3 semester hours in military science, 4 in mortuary sciences, 2 in forensic identification, 2 in anatomy and physiology, 3 in personnel supervision, 3 in technical writing, and 2 in records and information management. In the upper-division baccalaureate category, 3 semester hours in organizational management and 3 for field experience in management (6/97).

MOS-92R-001

Parachute Rigger

92R10
92R20
92R30
92R40
92R50

Exhibit Dates: 8/95–Present.

Career Management Field: 92 (Supply and Service).

Description

Summary: The parachute rigger is a qualified parachutist who supervises or packs and repairs cargo and personnel parachutes, and rigs equipment and supply containers for airdrop. *Skill Level 10:* Inventories, cleans, inspects, packs, and stores personnel and cargo parachutes and associated equipment; rigs supplies, equipment, and vehicles for airdrop; fabricates and assembles airdrop platforms, cushioning materials, and rigging components; loads, positions, and secures supplies and equipment within aircraft; tests and installs extraction and release systems; inspects cargo and personnel parachutes and other airdrop equipment before and after each use, during packing and rigging, and at prescribed intervals to determine serviceability; repairs or replaces airdrop equipment; uses and maintains machines and tools for fabrication, modification, and repair of parachutes and other airdrop equipment. *Skill Level 20:* Able to perform the duties required for Skill Level 10; performs duties in packing and air delivery and airdrop equipment repair; provides technical guidance to subordinates; inspects airdrop equipment; conducts sampling and quality assurance technical inspections; disposes of unserviceable airdrop equipment. *Skill Level 30:* Able to perform the duties required for Skill Level 20; supervises operations, instructs personnel on job requirements and techniques, and inspects work in progress; supervises, inspects, and certifies rigging and loading of cargo or vehicles for airdrop operations; performs as safety noncommissioned officer in aircraft and parachute issue point and as malfunction noncommissioned officer at drop zone; diagnoses malfunctions occurring in airdrop equipment; inspects air items to ensure manufacturer quality control; controls and expedites airborne support activities; assists in planning and coordinating training for standard and nonstandard rigging and sling loading and airdrop procedures. *Skill Level 40:* Able to perform the duties required for Skill Level 30; provides supervision and technical guidance to subordinates; examines, classifies, and disposes of equipment used in critical or fatal injury and other unserviceable equipment. *Skill Level 50:* Able to perform the duties required for Skill Level 40; supervises operations of organizations performing airborne and resupply by airdrop missions; supervises development and preparation of operations information, plans, maps, sketches, overlays, and related data to facilitate airborne operations; advises on airborne operational matters and performs liaison between staff and supported personnel.

Recommendation, Skill Level 10

Credit may be granted on the basis of an individualized assessment of the student (6/97).

Recommendation, Skill Level 20

Credit may be granted on the basis of an individualized assessment of the student (6/97).

Recommendation, Skill Level 30

In the vocational certificate category, 3 semester hours in sewing machine operation. In the lower-division baccalaureate/associate degree category, 3 semester hours in military science, 3 in shop operation and quality control, 3 in personnel supervision, 3 in technical report writing, and 2 in records and information management. In the upper-division baccalaureate category, 3 semester hours in inventory control, 3 in operations management, and 3 for field experience in management (6/97).

Recommendation, Skill Level 40

In the vocational certificate category, 3 semester hours in sewing machine operation. In the lower-division baccalaureate/associate degree category, 3 semester hours in military science, 3 in shop operation and quality control, 3 in personnel supervision, 3 in technical report writing, and 2 in records and information management. In the upper-division baccalaureate category, 3 semester hours in inventory control, 3 in operations management, and 3 for field experience in management (6/97).

Recommendation, Skill Level 50

In the vocational certificate category, 3 semester hours in sewing machine operation. In the lower-division baccalaureate/associate degree category, 3 semester hours in military science, 3 in shop operation and quality control, 3 in personnel supervision, 3 in technical report writing, and 2 in records and information management. In the upper-division baccalaureate category, 3 semester hours in inventory control, 3 in operations management, 3 in organizational management, 3 for field experience in management, and 3 in management problems (6/97).

MOS-92S-001

Laundry and Textile Specialist

92S10
92S20
92S30
92S40
92S50

Exhibit Dates: Pending Evaluation. 10/01–Present.

MOS-92W-001

Water Treatment Specialist

92W10
92W20
92W30
92W40

Exhibit Dates: Pending Evaluation. 4/03–Present.

MOS-92Y-001

Unit Supply Specialist

92Y10
92Y20
92Y30
92Y40
92Y50

Exhibit Dates: 2/93–2/95.

Career Management Field: 92 (Supply and Service).

Description

Summary: Supervises or performs duties involving request, receipt, storage, issue, accountability, and preservation of expendable supplies and equipment. *Skill Level 10:* Prepares and maintains organizational supply records and forms; receives and inspects inventories for future deliveries; maintains accounting system associated with supply management; secures and controls supplies; uses computer applications in work assignments; practices general clerical procedures. *Skill Level 20:* Able to perform the duties required for Skill Level 10; provides technical guidance in areas of supply management; reviews records and information documents for accuracy and completeness; coordinates supply activities; posts transactions to organization and installation property books and supporting transaction files. *Skill Level 30:* Able to perform the duties required for Skill Level 20; uses general supervisory skills to assist others in performing supply and inventory control management duties; reviews records and documents for accuracy; recommends professional development activities for subordinates; assists and advises supply officer and commander. *Skill Level 40:* Able to perform the duties required for Skill Level 30; uses basic management skills in coordinating employee activities; coordinates logistical activities with other supply and service units and motor transport units; develops and executes training programs. *Skill Level 50:* Able to perform the duties required for Skill Level 40; uses organizational management techniques to plan, analyze, direct, and coordinate activities; supervises development and preparation of operations information, plans, maps, sketches, overlays, and related data; contributes to staff development activities and operations of supply support data systems.

Recommendation, Skill Level 10

Credit may be granted on the basis of an individualized assessment of the student (3/95).

Recommendation, Skill Level 20

In the lower-division baccalaureate/associate degree category, 3 semester hours in supply management, 3 in records and information management, 3 in computer applications, 1 in personnel supervision, 2 in bookkeeping, and 3 in clerical procedures (3/95).

Recommendation, Skill Level 30

In the lower-division baccalaureate/associate degree category, 3 semester hours in supply management, 3 in records and information management, 3 in computer applications, 3 in personnel supervision, 3 in bookkeeping, and 3 in clerical procedures. In the upper-division baccalaureate category, 3 semester hours for field experience in management (3/95).

Recommendation, Skill Level 40

In the lower-division baccalaureate/associate degree category, 3 semester hours in supply management, 3 in records and information management, 3 in personnel supervision, 3 in bookkeeping, 3 in clerical procedures; and 3 in computer applications based on an individualized assessment of the student. In the upper-division baccalaureate category, 6 semester hours for field experience in management (3/95).

Recommendation, Skill Level 50

In the lower-division baccalaureate/associate degree category, 3 semester hours in supply management, 3 in records and information management, 3 in personnel supervision, 3 in bookkeeping, 3 in clerical procedures; and 3 in computer applications based on an individualized assessment of the student. In the upper-division baccalaureate category, 6 semester hours in field experience in management and 3 in organizational management (3/95).

MOS-92Y-002

Unit Supply Specialist

92Y10
92Y20
92Y30
92Y40
92Y50

Exhibit Dates: 3/95–Present.

Career Management Field: 92 (Supply and Service).

Description

Summary: Supervises or performs duties involving request, receipt, storage, issue, accountability, and preservation of expendable supplies and equipment. *Skill Level 10:* Prepares and maintains organizational supply records and forms; receives and inspects inventories for future deliveries; maintains accounting system associated with supply management; secures and controls supplies; uses computer applications in work assignments; practices general clerical procedures. *Skill Level 20:* Able to perform the duties required for Skill Level 10; provides technical guidance in areas of supply management; reviews records and information documents for accuracy and completeness; coordinates supply activities; posts transactions to organization and installation property books and supporting transaction files. *Skill Level 30:* Able to perform the duties required for Skill Level 20; uses general supervisory skills to assist others in performing supply and inventory control management duties; reviews records and documents for accuracy; recommends professional development activities for subordinates; assists and advises supply officer and commander. *Skill Level 40:* Able to perform the duties required for Skill Level 30; uses basic management skills in coordinating employee activities; coordinates logistical activities with other supply and service units and motor transport units; develops and executes training programs. *Skill Level 50:* Able to perform the duties required for Skill Level 40; uses organizational management techniques to plan, analyze, direct, and coordinate activities; supervises development and preparation of operations information, plans, maps, sketches, overlays, and related data; contributes to staff development activities and operations of supply support data systems.

Recommendation, Skill Level 10

Credit may be granted on the basis of an individualized assessment of the student (3/95).

Recommendation, Skill Level 20

Credit may be granted on the basis of an individualized assessment of the student (3/95).

Recommendation, Skill Level 30

In the lower-division baccalaureate/associate degree category, 3 semester hours in supply management, 3 in records and information management, 3 in computer applications, 3 in personnel supervision, 2 in bookkeeping, and 3 in clerical procedures. In the upper-division baccalaureate category, 3 semester hours for field experience in management (3/95).

Recommendation, Skill Level 40

In the lower-division baccalaureate/associate degree category, 3 semester hours in supply management, 3 in records and information management, 3 in personnel supervision, 3 in bookkeeping, 3 in clerical procedures; and 3 in computer applications based on an individualized assessment of the student. In the upper-division baccalaureate category, 6 semester hours for field experience in management (3/95).

Recommendation, Skill Level 50

In the lower-division baccalaureate/associate degree category, 3 semester hours in supply management, 3 in records and information management, 3 in personnel supervision, 3 in bookkeeping, 3 in clerical procedures; and 3 in computer applications based on an individualized assessment of the student. In the upper-division baccalaureate category, 6 semester hours in field experience in management and 3 in organizational management (3/95).

MOS-92Z-001

Senior Noncommissioned Logistician
92Z50

Exhibit Dates: 2/93–Present.

Career Management Field: 92 (Supply and Service).

Description

Serves as a mid-level manager, supervising, advising, counseling, evaluating, and directing subordinates in supply and related service MOS's; assists commander in planning, supervising, inspecting, developing, and executing policy; advises commander on personnel matters; provides counsel and guidance to subordinates; coordinates administration and support services of unit activities; performs as liaison between staff and supported soldiers to manage supply, material, and such related service operations as transportation, property disposal, and commissary operations. NOTE: May have progressed to 92Z50 from 92A50 (Automated Logistical Specialist) or 92Y50 (Unit Supply Specialist).

Recommendation

In the lower-division baccalaureate/associate degree category, 3 semester hours in supply management and 3 in personnel supervision. In the upper-division baccalaureate category, 3 semester hours in organizational management, 3 in logistics management, 3 in personnel management, 3 for field experience in management, and 3 communication techniques for managers (3/95).

MOS-93B-001

Aeroscout Observer
93B10
93B20
93B30

Exhibit Dates: 1/90–6/98. Effective 6/98, MOS 93B was discontinued and its duties were incorporated into MOS 67V (Observation/Scout Helicopter Repairer).

Career Management Field: 93 (Aviation Operations).

Description

Summary: Participates in aerial observation, communication, and navigation in the operation of the observation/scout helicopter (OH-58A/C). *Skill Level 10:* Develops flight plan; navigates by navigation aids or dead reckoning; locates target and determines coordinates or deviation from known point; uses map graphics to control security of operation by other units; communicates by open as well as secure means; performs as a crew member; recognizes hazards and executes evasive action; prepares situation reports relative to communication and navigation; maintains aircrew training manuals; assists pilot with visual and instrument navigation; performs aerial reconnaissance; directs close air support; relays requests for close support; formulates and transmits pilot reports relative to weather; assists in emergency procedures; is able to act as a helicopter pilot in emergencies. *Skill Level 20:* Able to perform the duties required for Skill Level 10; supervises and provides technical guidance to subordinates; prepares operational maps; indicates location and deployment of enemy and friendly units; reproduces and distributes required reports. *Skill Level 30:* Able to perform the duties required for Skill Level 20; provides information for tactical deployment to the commander; supervises maintenance equipment and ammunition issues; prepares weight and balance computations; collects, evaluates, and coordinates with adjacent units; prepares and updates enemy/friendly situation map; evaluates and counsels subordinates.

Recommendation, Skill Level 10

In the lower-division baccalaureate/associate degree category, 1 semester hour in basic helicopter maneuvers and 2 in aerial navigation. (NOTE: This recommendation for skill level 10 is valid for the dates 1/90-9/91 only) (4/90).

Recommendation, Skill Level 20

In the lower-division baccalaureate/associate degree category, 2 semester hours in basic helicopter maneuvers and 4 in aerial navigation. (NOTE: This recommendation for skill level 20 is valid for the dates 1/90-2/95 only) (4/90).

Recommendation, Skill Level 30

In the lower-division baccalaureate/associate degree category, 2 semester hours in basic helicopter maneuvers, 4 in aerial navigation, and 3 in principles of supervision (4/90).

MOS-93C-002

Air Traffic Control (ATC) Operator
93C10
93C20
93C30
93C40
93C50

Exhibit Dates: 1/90–7/95.

Career Management Field: 93 (Aviation Operation).

Description

Summary: Provides both radar and nonradar air traffic control services for military and civilian air traffic during takeoff, flight, and landing under Visual Flight Rules (VFR), Special Visual Flight Rules (SVFR), and Instrument Flight Rules (IFR). *Skill Level 10:* Controls takeoffs, flight, and landing of IFR/VFR/SVFR air traffic; provides flight following, enroute route, terminal approach control, and ground controlled approach services; operates radar; installs and moves ATC radar and associated equipment; applies FAA and Army air traffic rules and regulations; may hold an FAA radar controller certificate; issues special air traffic control instructions to aviators concerning airfield facilities, emergency landing areas, obstructions, landmarks, flying areas, restrictions, local regulations, weather advisories, and observed hazards that affect the safe operation of aircraft; keeps records and statistics, including tape recordings of voice radio communications, on daily air traffic operations; processes incoming and outgoing flight data information and analyzes air traffic; operates airfield lighting systems, light signals, and nonradar approach control boards; employs aeronautical charts, maps, radio, and ground communications; applies FAA and Army air traffic rules and regulations and holds an FAA control tower operator certificate; controls vehicular traffic on airport movement area. *Skill Level 20:* Able to perform the duties required for Skill Level 10; provides technical guidance to subordinates. *Skill Level 30:* Able to perform the duties required for Skill Level 20; serves as shift supervisor or facility chief; supervises inspection procedures, site operation, and record keeping; prepares shift duty rosters; administers facility training program and on-the-job training. *Skill Level 40:* Able to perform the duties required for Skill Level 30; as ATC tower chief, plans, organizes, and supervises air traffic control tower activities; assigns duties; spot-checks work performed; instructs subordinates in work techniques and procedures; coordinates work activities; establishes and supervises on-the-job training programs; prepares technical, personnel, and administrative reports; monitors the handling and storage of tape-recorded conversations between air traffic control personnel and aircraft pilots; maintains records and statistics on daily air traffic operations. *Skill Level 50:* Able to perform the duties required for Skill Level 40; supervises all air traffic control operations in a large facility; assigns personnel; establishes physical layout of work sites and ensures an efficient and safe environment; reviews, consolidates, and prepares technical, personnel, and administrative reports covering all phases of air traffic control operations.

Recommendation, Skill Level 10

In the lower-division baccalaureate/associate degree category, 6 semester hours in air traffic control, 6 in air traffic management, and 3 in federal regulations. (NOTE: This recommendation for skill level 10 is valid for the dates 1/90-9/91 only) (4/90).

Recommendation, Skill Level 20

In the lower-division baccalaureate/associate degree category, 6 semester hours in air traffic control, 6 in air traffic management, and 3 in federal regulations. (NOTE: This recommendation for skill level 20 is valid for the dates 1/90-2/95 only) (4/90).

Recommendation, Skill Level 30

In the lower-division baccalaureate/associate degree category, 6 semester hours in air traffic control, 6 in air traffic management, 3 in federal regulations, and 3 in principles of supervision (4/90).

Recommendation, Skill Level 40

In the lower-division baccalaureate/associate degree category, 6 semester hours in air traffic control, 6 in air traffic management, 3 in federal regulations, and 3 in principles of supervision. In the upper-division baccalaureate category, 3 semester hours in organizational management, 3 for field experience in management, and 6 in air traffic operations (4/90).

Recommendation, Skill Level 50

In the lower-division baccalaureate/associate degree category, 6 semester hours in air traffic control, 6 in air traffic management, 3 in federal regulations, and 3 in principles of supervision. In the upper-division baccalaureate category, 3 semester hours in organizational management, 6 for field experience in management, and 6 in air traffic operations (4/90).

MOS-93C-003

Air Traffic Control (ATC) Operator

93C10
93C20
93C30
93C40

Exhibit Dates: 8/95–10/99.

Career Management Field: 93 (Aviation Operation).

Description

Summary: Provides both radar and nonradar air traffic control services for military and civilian air traffic during takeoff, flight, and landing under Visual Flight Rules (VFR), Special Visual Flight Rules (SVFR), and Instrument Flight Rules (IFR). *Skill Level 10:* Controls takeoffs, flight, and landing of IFR/VFR/SVFR air traffic; provides flight following, enroute route, terminal approach control, and ground controlled approach services; operates radar; installs and moves ATC radar and associated equipment; applies FAA and Army air traffic rules and regulations; may hold an FAA radar controller certificate; issues special air traffic control instructions to aviators concerning airfield facilities, emergency landing areas, obstructions, landmarks, flying areas, restrictions, local regulations, weather advisories, and observed hazards that affect the safe operation of aircraft; keeps records and statistics, including tape recordings of voice radio communications, on daily air traffic operations; processes incoming and outgoing flight data information and analyzes air traffic; operates airfield lighting systems, light signals, and nonradar approach control boards; employs aeronautical charts, maps, radio, and ground communications; applies FAA and Army air traffic rules and regulations and holds an FAA control tower operator certificate; controls vehicular traffic on airport movement area. *Skill Level 20:* Able to perform the duties required for Skill Level 10; provides technical guidance to subordinates. *Skill Level 30:* Able to perform the duties required for Skill Level 20; serves as shift supervisor or facility chief; supervises inspection procedures, site operation, and record keeping; prepares shift duty rosters; administers facility training program and on-the-job training. *Skill Level 40:* Able to perform the duties required for Skill Level 30; as ATC tower chief, plans, organizes, and supervises air traffic control tower activities; assigns duties; spot-checks work performed; instructs subordinates in work techniques and procedures; coordinates work activities; establishes and supervises on-the-job training programs; prepares technical, personnel, and administrative reports; monitors the handling and storage of tape-recorded conversations between air traffic control personnel and aircraft pilots; maintains records and statistics on daily air traffic operations.

Recommendation, Skill Level 10

Credit may be granted on the basis of an individualized assessment of the student (4/90).

Recommendation, Skill Level 20

Credit may be granted on the basis of an individualized assessment of the student (4/90).

Recommendation, Skill Level 30

In the lower-division baccalaureate/associate degree category, 6 semester hours in air traffic control, 6 in air traffic management, 3 in federal regulations, and 3 in principles of supervision (4/90).

Recommendation, Skill Level 40

In the lower-division baccalaureate/associate degree category, 6 semester hours in air traffic control, 6 in air traffic management, 3 in federal regulations, and 3 in principles of supervision. In the upper-division baccalaureate category, 3 semester hours in organizational management, 3 for field experience in management, and 6 in air traffic operations (4/90).

MOS-93C-004

Air Traffic Control Operator

93C10
93C20
93C30
93C40

Exhibit Dates: 11/99–9/03. Effective 10/03, MOS 93C was discontinued and its duties were transferred to MOS 15Q, Air Traffic Controller.

Career Management Field: 93 (Aviation Operation).

Description

Summary: Provides both radar and nonradar air traffic control services for military and civilian air traffic during takeoff, flight, and landing under Visual Flight Rules (VFR), Special Visual Flight Rules (SVFR), and Instrument Flight Rules (IFR). *Skill Level 10:* Controls takeoffs, flight, and landing of IFR/VFR/SVFR air traffic; provides flight following, enroute routing, terminal approach control, and ground-controlled approach services; operates radar; installs and moves ATC radar and associated equipment; applies FAA and Army air traffic rules and regulations; may hold an FAA radar controller certificate; issues special air traffic control instructions to aviators concerning airfield facilities, emergency landing areas, obstructions, landmarks, flying areas, restrictions, local regulations, weather advisories, and observed hazards that affect the safe operation of aircraft; keeps records and statistics, including tape recordings of voice radio communications, on daily air traffic operations; processes incoming and outgoing flight data information and analyzes air traffic; operates airfield lighting systems, light signals, and nonradar approach control boards; employs aeronautical charts, maps, radio, and ground communications; applies FAA and Army air traffic rules and regulations and holds an FAA control tower operator certificate; controls vehicular traffic on airport movement area. *Skill Level 20:* Able to perform the duties required for Skill Level 10; provides technical guidance to subordinates. *Skill Level 30:* Able to perform the duties required for Skill Level 20; serves as shift supervisor or facility chief; supervises inspection procedures, site operation, and record keeping; prepares shift duty rosters; develops and administers facility training program and on-the-job training. *Skill Level 40:* Able to perform the duties required for Skill Level 30; as ATC tower chief, plans, organizes, and supervises air traffic control tower activities; assigns duties; spot-checks work performed; instructs subordinates in work techniques and procedures; coordinates work activities; establishes and supervises on-the-job training programs; prepares technical, personnel, and administrative reports; monitors the handling and storage of tape-recorded conversations between air traffic control personnel and aircraft pilots; maintains records and statistics on daily air traffic operations.

Recommendation, Skill Level 10

Credit may be granted on the basis of an individualized assessment of the student (11/99).

Recommendation, Skill Level 20

Credit is not recommended Credit may be granted on the basis of an individualized assessment of the student (11/99).

Recommendation, Skill Level 30

In the lower-division baccalaureate/associate degree category, 6 semester hours in air traffic control, 6 in air traffic management, 3 in federal regulations, 3 in principles of supervision, 3 in computer literacy, and 3 in computer applications. In the upper-division baccalaureate category, 3 semester hours in principles of instruction (11/99).

Recommendation, Skill Level 40

In the lower-division baccalaureate/associate degree category, 6 semester hours in air traffic control, 6 in air traffic management, 3 in federal regulations, 3 in principles of supervision, 3 in computer literacy, and 3 in computer applications. In the upper-division baccalaureate category, 3 semester hours in organizational management, 6 for field experience in management, 6 in air traffic operations, and 3 in principles of instruction (11/99).

MOS-93D-001

Air Traffic Control Equipment Repairer

((Air Traffic Control Systems, Subsystems and Equipment Repairer))
93D10
93D20
93D30
93D40
93D50

Exhibit Dates: 1/90–6/96. Effective 6/96, MOS 93D was discontinued and its Skill Level 10—40 duties were incorporated into MOS 35D; Skill Level 50 duties were incorporated into MOS 35Z.

Career Management Field: 93 (Aviation Operations).

Description

Summary: Installs and performs maintenance on air traffic control communications equipment, navigational aids, and landing systems. *Skill Level 10:* Installs air traffic control and navigation aids; troubleshoots to locate equipment malfunctions; tests, aligns, and adjusts equipment; completes maintenance records; maintains spare parts and tools; makes checks and adjustments to associated avionic components; updates and uses technical manuals and trouble analysis charts; replaces faulty components and individual parts. *Skill Level 20:* Able to perform the duties required for Skill Level 10; provides technical guidance to subordinates; performs comparison checks and modifications to equipment. *Skill Level 30:* Able to perform the duties required for Skill Level 20; provides technical guidance and instruction to subordinates; assists in the establishment of production and quality control procedures; inspects work areas and maintenance equipment; assists in the diagnosis of complex malfunctions and in component modifications. *Skill Level 40:* Able to perform the duties required for Skill Level 30; supervises maintenance procedures and practices; conducts training programs for unit personnel; interprets and explains complex specifications, circuits, and schematic and wiring diagrams; diagnoses and determines the cause of unusual equipment malfunctions; prepares technical reports; coordinates maintenance certification programs. *Skill Level 50:* Able to perform the duties required for Skill Level 40; supervises and evaluates maintenance and installation procedures and practices; assists in logistical appraisal of air traffic control systems; develops policies and procedures for unit management and determines the effectiveness of unit maintenance performance; supervises maintenance certification programs.

Recommendation, Skill Level 10

In the lower-division baccalaureate/associate degree category, 2 semester hours in advanced electronics laboratory and 2 in navigation/communications systems. (NOTE: This recommendation for skill level 10 is valid for the dates 1/90-9/91 only) (7/87).

Recommendation, Skill Level 20

In the lower-division baccalaureate/associate degree category, 3 semester hours in advanced electronics laboratory, 3 in navigation/communications systems, and 3 in instrument landing systems. (NOTE: This recommendation for skill level 20 is valid for the dates 1/90-2/95 only) (7/87).

Recommendation, Skill Level 30

In the lower-division baccalaureate/associate degree category, 3 semester hours in advanced electronics laboratory, 3 in navigation/communications systems, and 3 in instrument landing systems (7/87).

Recommendation, Skill Level 40

In the lower-division baccalaureate/associate degree category, 3 semester hours in advanced electronics laboratory, 3 in navigation/communications systems, 3 in instrument landing systems, 3 in personnel supervision, and 2 in records administration. In the upper-division baccalaureate category, 3 semester hours in principles of management and 3 for field experience in management (7/87).

Recommendation, Skill Level 50

In the lower-division baccalaureate/associate degree category, 3 semester hours in advanced electronics laboratory, 3 in navigation/communications systems, 3 in instrument landing systems, 3 in personnel supervision, and 3 in records administration. In the upper-division baccalaureate category, 3 semester hours in principles of management, 3 for field experience in management, and 3 in applied psychology (7/87).

MOS-93F-003

Field Artillery Meteorological Crewmember
93F10
93F20
93F30
93F40

Exhibit Dates: 1/90–2/95.

Career Management Field: 13 (Field Artillery).

Description

Summary: Supervises or participates in operation of field artillery meteorological observation station. *Skill Level 10:* Operates meteorological equipment in meteorological observation station; participates in assembly and emplacement of meteorological equipment and associated generators; operates recorders; tabulates pressure and wind data; assembles, operates, and disassembles balloon inflation and launching equipment; installs, tests, and adjusts radio weather instrument (Rawin) sets and radiosonde transmitters; operates computer that converts raw temperature, pressure, and wind velocities to digital data that presents temperature, density, and wind profiles; performs operator maintenance on equipment; has experience in use of rifle, grenade launcher, and machine gun; installs and operates digital and voice radio communications equipment; drives vehicles under adverse conditions. *Skill Level 20:* Able to perform the duties required for Skill Level 10; employs applicable navigation aids to derive correct location; uses maps to best advantage; performs operational checks on Rawin set; is proficient in basic combat techniques; instructs subordinates. *Skill Level 30:* Able to perform the duties required for Skill Level 20; serves as team chief during periods of extended operation; writes reports; checks data for quality control; makes operational checks on equipment; operates communications equipment; encodes and decodes meteorological data; serves as instructor and counselor to subordinates. *Skill Level 40:* Able to perform the duties required for Skill Level 30; supervises operation of meteorological section or station; selects station sites; supervises emplacement, calibration, maintenance, and operation of equipment; supervises preparation and distribution of ballistic meteorological messages: assigns, instructs, and supervises crewmembers; reviews, consolidates, and prepares technical, personnel, and administrative reports; counsels subordinates.

Recommendation, Skill Level 10

In the lower-division baccalaureate/associate degree category, 2 semester hours in general meteorology and 3 in meteorological instrumentation. (NOTE: This recommendation for skill level 10 is valid for the dates 1/90-9/91 only) (6/89).

Recommendation, Skill Level 20

In the lower-division baccalaureate/associate degree category, 2 semester hours in general meteorology, 3 in meteorological instrumentation, 1 in technical writing, 1 in map reading, and 2 in meteorological observation (6/89).

Recommendation, Skill Level 30

In the lower-division baccalaureate/associate degree category, 2 semester hours in general meteorology, 3 in meteorological instrumentation, 3 in technical writing, 1 in map reading, 2 in meteorological observation, and 3 in principles of supervision (6/89).

Recommendation, Skill Level 40

In the lower-division baccalaureate/associate degree category, 2 semester hours in general meteorology, 3 in meteorological instrumentation, 3 in technical writing, 1 in map reading, 2 in meteorological observation, and 3 in principles of supervision. In the upper-division baccalaureate category, 3 semester hours for field experience in management and 3 in personnel management (6/89).

MOS-93F-004

Field Artillery Meteorological Crewmember
93F10
93F20
93F30
93F40

Exhibit Dates: 3/95–4/99.

Career Management Field: 13 (Field Artillery).

Description

Summary: Supervises or participates in operation of field artillery meteorological observation station. *Skill Level 10:* Operates meteorological equipment in meteorological observation station; participates in assembly and emplacement of meteorological equipment and associated generators; operates recorders; tabulates pressure and wind data; assembles, operates, and disassembles balloon inflation and launching equipment; installs, tests, and adjusts radio weather instrument (Rawin) sets and radiosonde transmitters; operates computer that converts raw temperature, pressure, and wind velocities to digital data that presents temperature, density, and wind profiles; performs operator maintenance on equipment; has experience in use of rifle, grenade launcher, and machine gun; installs and operates digital and voice radio communications equipment; drives vehicles under adverse conditions. *Skill Level 20:* Able to perform the duties required for Skill Level 10; employs applicable navigation aids to derive correct location; uses maps to best advantage; performs operational checks on Rawin set; is proficient in basic combat techniques; instructs subordinates. *Skill Level 30:* Able to perform the duties required for Skill Level 20; serves as team chief during periods of extended operation; writes reports; checks data for quality control; makes operational checks on equipment; operates communications equipment; encodes and decodes meteorological data; serves as instructor and counselor to subordinates. *Skill Level 40:* Able to perform the duties required for Skill Level 30; supervises operation of meteorological section or station; selects station sites; supervises emplacement, calibration, maintenance, and operation of equipment; supervises preparation and distribution of ballistic meteorological messages; assigns, instructs, and supervises crew members; reviews, consolidates, and prepares technical, personnel, and administrative reports; counsels subordinates.

Recommendation, Skill Level 10

Credit may be granted on the basis of an individualized assessment of the student (6/89).

Recommendation, Skill Level 20

Credit may be granted on the basis of an individualized assessment of the student (6/89).

Recommendation, Skill Level 30

In the lower-division baccalaureate/associate degree category, 2 semester hours in general meteorology, 3 in meteorological instrumentation, 1 in technical writing, 1 in map reading, 2 in meteorological observation, and 3 in principles of supervision (6/89).

Recommendation, Skill Level 40

In the lower-division baccalaureate/associate degree category, 2 semester hours in general meteorology, 3 in meteorological instrumentation, 3 in technical writing, 1 in map reading, 2 in meteorological observation, and 3 in principles of supervision. In the upper-division baccalaureate category, 3 semester hours for field experience in management and 3 in personnel management (6/89).

MOS-93F-005

Field Artillery Meteorological Crewmember
93F10
93F20
93F30
93F40

Exhibit Dates: 5/99–5/03. Effective 6/03, MOS 93F was discontinued and it duties were incorporated into MOS 13B (Cannon Crewmember), 13M (Multiple Launch Rocket System Crewmember), and 13W (Field Artillery Meteorological Crewmember).

Career Management Field: 13 (Field Artillery).

Description

Summary: Supervises or participates in operation of field artillery meteorological observation station. *Skill Level 10:* Operates meteorological equipment in meteorological observation station; participates in assembly and emplacement of meteorological equipment and associated generators; operates recorders; tabulates pressure and wind data; assembles, operates, and disassembles balloon inflation and launching equipment; installs, tests, and adjusts radio weather instrument (Rawin) sets and radiosonde transmitters; operates computer that converts raw temperature, pressure, and wind velocities to digital data that presents temperature, density, and wind profiles; performs operator maintenance on equipment; has experience in use of rifle, grenade launcher, and machine gun; installs and operates digital and voice radio communications equipment; drives vehicles under adverse conditions; provides routine automotive service and limited automotive repair. *Skill Level 20:* Able to perform the duties required for Skill Level 10; employs applicable navigation aids to derive correct location; uses maps to best advantage; performs operational checks on Rawin set; is proficient in basic combat techniques; instructs subordinates. *Skill Level 30:* Able to perform the duties required for Skill Level 20; serves as team chief during periods of extended operation; writes reports; checks data for quality control; makes operational checks on equipment; operates communications equipment; encodes and decodes meteorological data; serves as instructor and counselor to subordinates. *Skill Level 40:* Able to perform the duties required for Skill Level 30; supervises operation of meteorological section or station; selects station sites; supervises emplacement, calibration, maintenance, and operation of equipment; supervises preparation and distribution of ballistic meteorological messages; assigns, instructs, and supervises crew members; reviews, consolidates, and prepares technical, personnel, and administrative reports; counsels subordinates.

Recommendation, Skill Level 10

Credit may be granted on the basis of an individualized assessment of the student (5/99).

Recommendation, Skill Level 20

Credit may be granted on the basis of an individualized assessment of the student (5/99).

Recommendation, Skill Level 30

In the lower-division baccalaureate/associate degree category, 3 semester hours in general meteorology, 3 in meteorological instrumentation, 1 in technical writing, 3 in map reading and land navigation, 2 in meteorological observation, 1 in applied mathematics, 3 in computer familiarization, 3 in oral communication skills, 1 in automotive service, 1 in safety and hazardous material, and 3 in principles of supervision (5/99).

Recommendation, Skill Level 40

In the lower-division baccalaureate/associate degree category, 3 semester hours in map reading and land navigation, 3 in general meteorology, 1 in applied mathematics, 3 in computer familiarization, 3 in principles of supervision, 3 in meteorological instrumentation, 3 in meteorological observation, 3 in technical writing, 3 in oral communication skills, 1 in automotive service, and 1 in safety and hazardous material. In the upper-division baccalaureate category, 3 semester hours for field experience in management and 3 in personnel management (5/99).

MOS-93P-002

Aviation Operations Specialist
93P10
93P20
93P30
93P40
93P50

Exhibit Dates: 1/90–10/99.

Career Management Field: 93 (Aviation Operation).

Description

Summary: Schedules, clears, and dispatches aircraft. *Skill Level 10:* Processes cross-country and local flight plans with other agencies including the FAA; maintains flight information; prepares, types, and maintains records and reports on flight operations and activities; maintains current files on flying regulations and navigational aid information; arranges ground services for transient aircraft; interprets and posts teletype weather reports; understands terminology used in air navigation; is aware of air traffic control advisory procedures. *Skill Level 20:* Able to perform the duties required for Skill Level 10; supervises a small flight operations activity consisting of 5 to 12 subordinates; schedules aircraft missions, dispatches aircraft, and performs associated administrative duties; plans and schedules work assignments; checks work of subordinates and instructs them in proper work techniques and procedures; reviews, consolidates, and prepares technical, personnel, and administrative reports; assists in preparing preaccident plans. *Skill Level 30:* Able to perform the duties required for Skill Level 20; supervises a medium-sized flight operations activity consisting of 13 or more subordinates; provides technical guidance to subordinates; supervises the preparation of situation map; prepares operations letters; assists administratively in aircraft accident investigations; assists in the preparation of letters of agreement, operations estimates, and operations orders. *Skill Level 40:* Able to perform the duties required for Skill Level 30; supervises a large flight operations activity; plans, coordinates, and supervises activities pertaining to organization, training, combat operations, and combat intelligence; coordinates implementation of operations, training programs, and communication activities. NOTE: May have progressed to 93P40 from 93P30 or 93B30, Aeroscout Observer. *Skill Level 50:* Able to perform the duties required for Skill Level 40; serves as a mid-level manager; supervises the processing of operations and intelligence information at the brigade level.

Recommendation, Skill Level 10

In the lower-division baccalaureate/associate degree category, 3 semester hours in introduction to aircraft dispatching, 3 in aircraft operations management, 3 in FAA regulations, and 2 in office procedures. (NOTE: This recommendation for skill level 10 is valid for the dates 1/90-9/91 only) (4/90).

Recommendation, Skill Level 20

In the lower-division baccalaureate/associate degree category, 3 semester hours in introduction to aircraft dispatching, 3 in aircraft operations management, 3 in FAA regulations, 2 in office procedures, and 3 in principles of supervision. In the upper-division baccalaureate category, 3 semester hours in aviation management. (NOTE: This recommendation for skill level 20 is valid for the dates 1/90-2/95 only) (4/90).

Recommendation, Skill Level 30

In the lower-division baccalaureate/associate degree category, 3 semester hours in introduction to aircraft dispatching, 3 in aircraft operations management, 3 in FAA regulations, 2 in office procedures, 3 in principles of supervision, 3 in airfield management, and 3 in advanced aircraft dispatching. In the upper-division baccalaureate category, 3 semester hours in aviation management and 2 in organizational management (4/90).

Recommendation, Skill Level 40

In the lower-division baccalaureate/associate degree category, 3 semester hours in introduction to aircraft dispatching, 3 in aircraft operations management, 3 in FAA regulations, 2 in office procedures, 3 in principles of supervision, 3 in airfield management, and 3 in advanced aircraft dispatching. In the upper-division baccalaureate category, 3 semester hours in aviation management, 3 in organizational management, 3 in air traffic operations, and 3 for field experience in management (4/90).

Recommendation, Skill Level 50

In the lower-division baccalaureate/associate degree category, 3 semester hours in introduction to aircraft dispatching, 3 in aircraft operations management, 3 in FAA regulations, 2 in office procedures, 3 in principles of supervision, 3 in airfield management, and 3 in advanced aircraft dispatching. In the upper-division baccalaureate category, 3 semester hours in aviation management, 3 in organizational management, 3 in air traffic operations, and 6 for field experience in management (4/90).

MOS-93P-003

AVIATION OPERATIONS SPECIALIST

93P10
93P20
93P30
93P40
93P50

Exhibit Dates: 11/99–9/03. Effective 10/03, MOS 93P was discontinued and its duties were transferred to MOS 15P, Aviation Operations Specialist.

Career Management Field: 93 (Aviation Operation).

Description

Summary: Schedules, clears, and dispatches aircraft. *Skill Level 10:* Processes cross-country and local flight plans with other agencies including the FAA; maintains flight information; prepares, types, and maintains records and reports on flight operations and activities; maintains current files on flying regulations and navigational aid information; arranges ground services for transient aircraft; interprets and posts teletype weather reports; understands terminology used in air navigation; knows air traffic control advisory procedures. *Skill Level 20:* Able to perform the duties required for Skill Level 10; supervises a small flight operations activity consisting of 5 to 12 subordinates; schedules aircraft missions, dispatches aircraft, and performs associated administrative duties; plans and schedules work assignments; checks work of subordinates and instructs them in proper work techniques and procedures; reviews, consolidates, and prepares technical, personnel, and administrative reports; assists in preparing pre-accident plans. *Skill Level 30:* Able to perform the duties required for Skill Level 20; supervises a medium-sized flight operations activity consisting of 13 or more subordinates; provides technical guidance to subordinates; supervises the preparation of situation map; prepares operations letters; assists administratively in aircraft accident investigations; assists in the preparation of letters of agreement, operations estimates, and operations orders. *Skill Level 40:* Able to perform the duties required for Skill Level 30; supervises a large flight operations activity; plans, coordinates, and supervises activities pertaining to organization, training, combat operations, and combat intelligence; coordinates implementation of operations, training programs, and communication activities. NOTE: May have progressed to 93P40 from 93P30 or 93B30, (Aeroscout Observer). *Skill Level 50:* Able to perform the duties required for Skill Level 40; serves as a mid-level manager; supervises the processing of operations and intelligence information at the brigade level.

Recommendation, Skill Level 10

Credit may be granted on the basis of an individualized assessment of the student (11/99).

Recommendation, Skill Level 20

Credit may be granted on the basis of an individualized assessment of the student (11/99).

Recommendation, Skill Level 30

In the lower-division baccalaureate/associate degree category, 3 semester hours in introduction to aircraft dispatching, 3 in aircraft operations management, 3 in FAA regulations, 3 in computer literacy, 2 in computer applications, 2 in office procedures, 3 in principles of supervision, 3 in airfield management, and 3 in advanced aircraft dispatching. In the upper-division baccalaureate category, 3 semester hours in aviation management and 2 in organizational management (11/99).

Recommendation, Skill Level 40

In the lower-division baccalaureate/associate degree category, 3 semester hours in introduction to aircraft dispatching, 3 in aircraft operations management, 3 in FAA regulations, 3 in computer literacy, 3 in computer applications, 2 in office procedures, 3 in principles of supervision, 3 in airfield management, and 3 in advanced aircraft dispatching. In the upper-division baccalaureate category, 3 semester hours in aviation management, 3 in organizational management, 3 in air traffic operations, and 3 for field experience in management (11/99).

Recommendation, Skill Level 50

In the lower-division baccalaureate/associate degree category, 3 semester hours in introduction to aircraft dispatching, 3 in aircraft operations management, 3 in FAA regulations, 3 in computer literacy, 3 in computer applications, 2 in office procedures, 3 in principles of supervision, 3 in airfield management, and 3 in advanced aircraft dispatching. In the upper-division baccalaureate category, 3 semester hours in aviation management, 3 in organizational management, 3 in air traffic operations, and 6 for field experience in management (11/99).

MOS-94A-001

LAND CONTROL ELECTRONIC MISSILE SYSTEM REPAIRER

94A10
94A20
94A30
94A40

Exhibit Dates: Pending Evaluation. 5/04–Present.

MOS-94B-003

FOOD SERVICE SPECIALIST

94B10
94B20
94B30
94B40
94B50

Exhibit Dates: 1/90–3/94.

Career Management Field: 94 (Food Service).

Description

Summary: Procures, prepares, and cooks food; identifies and uses appropriate equipment. *Skill Level 10:* Either has undergone on-the-job training program in food preparation and serving techniques or has recently completed a program of instruction in the basic principles of food preparation and service; under the supervision of an experienced cook, weighs, blends, mixes, and cooks food in accordance with prescribed procedures; washes, peels, cuts, and dices fruits, vegetables, meats, salads, and dairy products; prepares simple soups, sauces, and gravies; under the supervision of an experienced baker, prepares simple baked items, including breadstuffs and desserts; assists in receiving and storing food and supplies; operates and performs preventive maintenance on food service equipment; portions and serves food on serving lines; applies sanitation procedures in handling, storing, preparing, and serving food. *Skill Level 20:* Able to perform the duties required for Skill Level 10; provides guidance to subordinates; prepares meats, fruits, vegetables, salads, desserts, beverages, and dairy products for serving; performs small-scale baking and meat cutting; operates and performs preventive maintenance on kitchen equipment; applies hygiene and sanitation procedures. *Skill Level 30:* Able to perform the duties required for Skill Level 20; supervises, as first cook, the scheduling of personnel and facilities and the preparation of food; inspects food prior to serving; supervises procurement and storage of foods; alternative job paths are available with assignment either as cook (leading to dining facility manager) or as meat cutter (leading to chief meat cutter). *Skill Level 40:* Able to perform the duties required for Skill Level 30; develops work sheet according to master menu; maintains records and files; applies accounting procedures to operate within budget; prepares reports; serves as dining facilities manager. *Skill Level 50:* Able to perform the duties required for Skill Level 40; has management responsibilities at the staff level including developing plans and conducting evaluations; develops, coordinates, implements, and advises on food programs; evaluates operation of dining and serving areas, kitchens, field bakeries, and food service training facilities and maintenance of equipment; is a member of the menu board; assists in menu development; may be in complete charge of food service operations during partial unit deployment or at a large facility with multiple food service outlets; prepares reports and studies on food service activities.

Recommendation, Skill Level 10

In the lower-division baccalaureate/associate degree category, 3 semester hours in quantity food preparation, 1 in kitchen operations, and 1 in food service operations. (NOTE: This recom-

mendation for skill level 10 is valid for the dates 1/90-9/91 only) (11/77).

Recommendation, Skill Level 20

In the lower-division baccalaureate/associate degree category, 6 semester hours in quantity food preparation, 3 in kitchen operations, and 1 in food service operations (11/77).

Recommendation, Skill Level 30

In the lower-division baccalaureate/associate degree category, 6 semester hours in quantity food preparation, 3 in kitchen operations, 3 in personnel supervision, 3 for field experience in food service, and 1 in food service operations; if the duty assignment was cook, 2 semester hours in food service operations; if the duty assignment was meat cutter, 3 semester hours in meat cutting (11/77).

Recommendation, Skill Level 40

In the lower-division baccalaureate/associate degree category, 6 semester hours in quantity food preparation, 3 in kitchen operations, 3 in personnel supervision, 3 in human relations, 3 for field experience in food service, and 1 in food service operations; if the duty assignment was cook, 2 semester hours in food service operations; if the duty assignment was meat cutter, 3 semester hours in meat cutting (11/77).

Recommendation, Skill Level 50

In the lower-division baccalaureate/associate degree category, 6 semester hours in quantity food preparation, 3 in kitchen operations, 3 in personnel supervision, 3 in human relations, 3 in report writing, 3 in communication skills (oral), 3 for field experience in food service, and 1 in food service operations; if the duty assignment was cook, 2 semester hours in food service operations; if the duty assignment was meat cutter, 3 semester hours in meat cutting. In the upper-division baccalaureate category, 6 semester hours for a food service internship (11/77).

MOS-94B-004

Food Service Specialist
94B10
94B20
94B30
94B40
94B50

Exhibit Dates: 4/94–12/95. Effective 12/95, MOS 94B was discontinued and its duties were transferred to MOS 92G.

Career Management Field: 94 (Food Service).

Description

Summary: Prepares, cooks, and serves food in field or garrison food service operations; supervises food service operations. *Skill Level 10:* Performs preliminary food preparation procedures; prepares and/or cooks menu items listed on the production schedule; bakes, fries, steams, braises, boils, simmers, steams, and sautes as prescribed by Army recipes; sets up serving lines; garnishes food items; applies food protection and sanitation measures in field and garrison environments; receives and stores subsistence items; performs general housekeeping duties; operates, maintains, and cleans field kitchen equipment; erects, strikes, and stores all types of field kitchens, and performs preventive maintenance on garrison and field kitchen equipment. *Skill Level 20:* Able to perform the duties required for Skill Level 10; ensures that proper procedures, temperatures, and time periods are adhered to during food preparation; performs limited supervisory and inspection functions including shift supervision. *Skill Level 30:* Able to perform the duties required for Skill Level 20; prepares more complex menu items; supervises shift, unit, or consolidated food service operations in field or garrison environments; establishes operating and work procedures; inspects food preparation/storage areas; supervises dining facility staff; determines subsistence requirements; requests, receives, and accounts for subsistence items; applies food service accounting procedures; prepares production schedule; makes necessary menu adjustments; establishes and maintains on-the-job and apprentice training programs; prepares technical, personnel, and administrative reports concerning food service operations; implements emergency, disaster, and combat feeding plans; coordinates logistical support. *Skill Level 40:* Able to perform the duties required for Skill Level 30; assigns staff to duty positions; administers on-the-job training program; coordinates with food service officer, food advisor, assistant food service sergeants, and first cooks; provides guidance to subordinates; coordinates support requirements with TISA, facility engineers, and veterinary activity; plans and implements menus to ensure nutritionally balanced meals; ensures accuracy of accounting and equipment records; develops standard operating procedures on safety, energy, security, and fire prevention programs. *Skill Level 50:* Able to perform the duties required for Skill Level 40; develops, coordinates, implements, advises, and evaluates food service programs; monitors requests for food items and equipment; develops and analyzes troop menus and coordinates menu substitutions; evaluates operation of garrison and field kitchens, field bakeries, and food service training facilities; surveys individual preferences, food preparation, and food conservation; prepares reports, studies, and briefings on food service activities.

Recommendation, Skill Level 10

Credit may be granted on the basis of an individualized assessment of the student (4/94).

Recommendation, Skill Level 20

In the lower-division baccalaureate/associate degree category, 6 semester hours in quantity food preparation, 3 in kitchen operations, and 3 for field experience in food service. (NOTE: This recommendation for skill level 20 is valid for the dates 4/94-2/95 only) (4/94).

Recommendation, Skill Level 30

In the lower-division baccalaureate/associate degree category, 6 semester hours in quantity food preparation, 3 in kitchen operations, 3 in food service operations, 3 in personnel supervision, and 3 for field experience in food service (4/94).

Recommendation, Skill Level 40

In the lower-division baccalaureate/associate degree category, 6 semester hours in quantity food preparation, 3 in kitchen operations, 3 in food service operations, 3 in personnel supervision, 3 in menu planning, 3 in food service management, and 3 for field experience in food service. In the upper-division baccalaureate category, 3 semester hours in human resource management (4/94).

Recommendation, Skill Level 50

In the lower-division baccalaureate/associate degree category, 6 semester hours in quality food preparation, 3 in kitchen operations, 3 in food service operations, 3 in personnel supervision, 3 in menu planning, 3 in business communications, 3 in food service management, and 3 for field experience in food service. In the upper-division baccalaureate category, 3 semester hours in human resource management (4/94).

MOS-94E-001

Radio & Communications Security Repairer
94E10
94E20
94E30
94E40

Exhibit Dates: Pending Evaluation. 5/04–Present.

MOS-94F-002

Hospital Food Service Specialist
94F10
94F20
94F30
94F40
94F50

Exhibit Dates: 1/90–3/91. Effective 3/91, MOS 94F was discontinued, and its duties were incorporated into MOS 91M, Hospital Food Service Specialist.

Career Management Field: 94 (Food Service).

Description

Summary: Assists in the planning and implementation of the food service program in hospitals; prepares, cooks, and serves food to patients and staff. *Skill Level 10:* Measures, weighs, blends, and mixes various foodstuffs; roasts, fries, broils, boils, and stews meats; prepares fruits, vegetables, salads, desserts, beverages, and dairy products; bakes yeast breads, quick breads, pies, cakes, and cookies; sets up and serves regular and modified diets for patients; assembles and checks patient trays; instructs patients on modified diets to aid them in food selection; maintains records as to patient food preferences and diet changes and adjusts diet menus; prepares food in accordance with instructions from dietician or physician; performs preventive maintenance on kitchen equipment. *Skill Level 20:* Able to perform the duties required for Skill Level 10; provides technical guidance to subordinates; ensures adherence to procedures, temperatures, and time periods in food preparation; writes and adjusts individual standard diet menus; participates in planning of food service activities; conducts food acceptability studies; applies storage standards, completes food production forms and records; monitors tasks in support of patient tray service operation, including preparation, assembly, and delivery to wards. *Skill Level 30:* Able to perform the duties required for Skill Level 20; supervises food production or diet therapy activities in a hospital; plans and prepares hospital menus for regular and modified diets and estimates the number of persons to be fed; schedules and supervises food preparation for compliance with dietetic requirements and for quality and quantity standards; inspects prepared food for palatability and appearance and to determine compliance with menus and recipes; inspects hospital food service facilities and initiates appropriate corrective action to correct deficiencies; supervises procurement, storage, distribution, accounting for, and serving of food in a hospital food service; conducts classroom and on-the-job training for subordinates. *Skill Level 40:* Able to perform the duties required for Skill Level 30; performs supervisory duties; develops, coordinates, implements, and advises on food programs; evaluates and designs improvements in food programs and training programs; assists in preparation of hospital food service budget. *Skill Level 50:* Able to perform the duties required for Skill Level 40; prepares budget and determines financial requirements; instructs military and civilian personnel; applies office management techniques; advises superiors on food service equipment layout and space design.

Recommendation, Skill Level 10

In the lower-division baccalaureate/associate degree category, 6 semester hours in quantity food preparation, 3 in nutrition, 2 in diet therapy, and 1 in food service administration and sanitation (6/77).

Recommendation, Skill Level 20

In the lower-division baccalaureate/associate degree category, 6 semester hours in quantity food preparation, 3 in nutrition, 3 in kitchen operations, 2 in diet therapy, and 1 in food service administration and sanitation (6/77).

Recommendation, Skill Level 30

In the lower-division baccalaureate/associate degree category, 6 semester hours in quantity food preparation, 3 in nutrition, 3 in kitchen operations, 2 in diet therapy, 2 in personnel supervision, 1 in food service administration and sanitation, and additional credit in menu planning, food service operation, and for a food service internship on the basis of institutional evaluation (6/77).

Recommendation, Skill Level 40

In the vocational certificate category, 2 semester hours in bookkeeping. In the lower-division baccalaureate/associate degree category, 6 semester hours in quantity food preparation, 3 in nutrition, 3 in kitchen operations, 3 in personnel supervision, 3 in human relations, 2 in diet therapy, 1 in food service administration and sanitation, and additional credit in menu planning, food service operation, and for a food service internship on the basis of institutional evaluation. In the upper-division baccalaureate category, credit for a food service internship on the basis of institutional evaluation (6/77).

Recommendation, Skill Level 50

In the vocational certificate category, 2 semester hours in bookkeeping. In the lower-division baccalaureate/associate degree category, 6 semester hours in quantity food preparation, 3 in nutrition, 3 in kitchen operations, 3 in personnel supervision, 3 in human relations, 2 in diet therapy, 1 in accounting, 1 in food service administration and sanitation, and additional credit in menu planning, food service operations, and for a food service internship on the basis of institutional evaluation. In the upper-division baccalaureate category, 6 semester hours for field experience in management, 3 in food service equipment layout design, and additional credit in introduction to management, industrial arts (food service), and diet therapy on the basis of institutional evaluation (6/77).

MOS-94F-003

Special Electronic Device Repairer
94F10
94F20
94F30
94F40

Exhibit Dates: Pending Evaluation. 5/04–Present.

MOS-94H-001

Test Measurement and Diagnostic Equipment Repairer
94H10
94H20
94H30
94H40

Exhibit Dates: Pending Evaluation. 5/04–Present.

MOS-94K-001

Apache Attack Helicopter
94K10
94K20
94K30
94K40

Exhibit Dates: Pending Evaluation. 5/04–Present.

MOS-94L-001

Avionic Communications Equipment Repairer
94L10
94L20
94L30
94L40

Exhibit Dates: Pending Evaluation. 5/04–Present.

MOS-94M-001

Radar Repairer
94M10
94M20
94M30
94M40

Exhibit Dates: Pending Evaluation. 5/04–Present.

MOS-94P-001

Multiple Launch Rocket System Repairer
94P10
94P20
94P30
94P40

Exhibit Dates: Pending Evaluation. 5/04–Present.

MOS-94R-001

Avionic System Repairer
94R10
94R20
94R30
94R40

Exhibit Dates: Pending Evaluation. 5/04–Present.

MOS-94S-001

Patriot Systems Repairer
94S10
94S20
94S30
94S40

Exhibit Dates: Pending Evaluation. 5/04–Present.

MOS-94T-001

Avenger System Repairer
94T10
94T20
94T30
94T40

Exhibit Dates: Pending Evaluation. 5/04–Present.

MOS-94V-001

Electronic and Missile Systems Maintenance Chief
94V10
94V20
94V30
94V40

Exhibit Dates: Pending Evaluation. 5/04–Present.

MOS-94W-001

Electronic Maintenance
94W10
94W20
94W30
94W40

Exhibit Dates: Pending Evaluation. 5/04–Present.

MOS-94Y-001

Integrated Family of Test Equipment Operator/Maintainer
94Y10
94Y20
94Y30
94Y40

Exhibit Dates: Pending Evaluation. 5/04–Present.

MOS-94Z-001

Senior Electronic Maintenance Chief
94Z10
94Z20
94Z30
94Z40

Exhibit Dates: Pending Evaluation. 5/04–Present.

MOS-95B-004

Military Police
95B10
95B20
95B30
95B40
95B50

Exhibit Dates: 1/90–9/04. (Effective 9/04 MOS 95B was discontinued, and its duties were incorporated into MOS 31B.)

Career Management Field: 95 (Law Enforcement).

Description

Summary: Supervises or provides law enforcement; preserves military control; provides security; controls traffic; quells disturbances; protects property and personnel; handles prisoners of war, refugees, or evacuees; investigates incidents. *Skill Level 10:* Enforces traffic regulations and law and order; exercises military control and discipline and guards prisoners of war; responsible for traffic accident investigation; provides physical security for designated individuals, installations, facilities, and equipment; maintains traffic control and enforces traffic regulations and safety; participates in civil disturbances and riot control operations; responsible for law enforcement investigations; performs foot and motorized patrol and applies crime prevention measures; prepares military police reports including sworn statements and processes evidence. *Skill Level 20:* Able to perform the duties required for Skill Level 10; leads military police patrol, small squad, and small detachment; supervises compound or work project; coordinates MP activities with civil police organizations; directs MP activities to quell disturbances and cope with disasters; supervises traffic safety activity and riot and crowd control; prepares reports, forms, and records on MP operations and activities. *Skill Level 30:* Able to perform the duties required for Skill Level 20; leads military police squad, medium-sized section, detachment, or platoon; assists in planning, organizing, directing, supervising, training, coordinating, and reporting activities of subordinates; organizes work schedules, assigns duties, and instructs personnel in techniques and procedures; evaluates personnel performance. *Skill Level 40:* Able to perform the duties required for Skill Level 30; leads large military police detachment, section, or platoon; collects offensive and defensive intelligence information and trains personnel in police operations and intelligence activities; assists in coordinating and implementing military police operations, training programs, and communication activities; assists in production and administration of staff journals, files, records, and reports. *Skill Level 50:* Able to perform the duties required for Skill Level 40; provides staff supervision or principal noncommissioned officer (NCO) direction to units engaged in military police or criminal investigation operations and confinement facilities; may have served as the principal NCO of a military police unit; determines requirements for police support; evaluates training programs and requirements.

Recommendation, Skill Level 10

In the lower-division baccalaureate/associate degree category, 3 semester hours in patrol operations. (NOTE: This recommendation for skill level 10 is valid for the dates 1/90-9/91 only) (11/86).

Recommendation, Skill Level 20

In the lower-division baccalaureate/associate degree category, 3 semester hours in patrol operations. (NOTE: This recommendation for skill level 20 is valid for the dates 1/90-2/95 only) (11/86).

Recommendation, Skill Level 30

In the lower-division baccalaureate/associate degree category, 3 semester hours in patrol operations, 3 in police supervision, and 3 in a law enforcement elective (11/86).

Recommendation, Skill Level 40

In the lower-division baccalaureate/associate degree category, 3 semester hours in patrol operations, 3 in police supervision, 3 in a law enforcement elective, and 3 in office records management. In the upper-division baccalaureate category, 3 semester hours in principles of management (11/86).

Recommendation, Skill Level 50

In the lower-division baccalaureate/associate degree category, 3 semester hours in patrol operations, 3 in police supervision, 3 in a law enforcement elective, and 3 in office records management. In the upper-division baccalaureate category, 3 semester hours in principles of management and 3 in introduction to public administration (11/86).

MOS-95C-002

Corrections Specialist
((Corrections NCO))
95C20
95C30
95C40
95C50

Exhibit Dates: 1/90–1/93.

Career Management Field: 95 (Law Enforcement).

Description

Summary: Functions as a correctional officer, providing rehabilitative, health, welfare, and security services to prisoners within a correctional facility; conducts inspections; prepares written reports; coordinates activities of inmates and staff personnel. *Skill Level 20:* Functions as a correctional officer; provides individual counseling and guidance to prisoners within a rehabilitative program. *Skill Level 30:* Able to perform the duties required for Skill Level 20; functions as a first-line supervisor of up to ten correctional officers; prepares or reviews reports and records of individuals and programs; provides formal counseling. *Skill Level 40:* Able to perform the duties required for Skill Level 30; as a mid-level manager, supervises and coordinates all correctional, custodial, treatment, and rehabilitative activities of inmates and up to 100 staff members in correctional facility containing as many as 250 prisoners; responsible for administrative inspection and review of all records, reports, and activities; furnishes recommendations to superiors regarding clemency, restoration to duty, and further treatment of inmates; advises superior on matters concerning subordinates. *Skill Level 50:* Able to perform the duties required for Skill Level 40; supervises and coordinates inmate and staff activities in a large consolidated correctional facility; as an administrator, plans, coordinates, and implements all counseling, rehabilitative, and correctional programs in a permanent facility containing 250 or more prisoners; equivalent to a federal maximum-security prison warden.

Recommendation, Skill Level 20

In the lower-division baccalaureate/associate degree category, 3 semester hours in correctional operations. (NOTE: This recommendation for skill level 20 is valid for the dates 1/90-2/95 only) (11/86).

Recommendation, Skill Level 30

In the lower-division baccalaureate/associate degree category, 3 semester hours in correctional operations, 3 in correctional supervision, and 3 in a criminal justice elective (11/86).

Recommendation, Skill Level 40

In the lower-division baccalaureate/associate degree category, 3 semester hours in correctional operations, 3 in correctional supervision, 3 in a criminal justice elective, and 3 in correctional administration. In the upper-division baccalaureate category, 3 semester hours in principles of management (11/86).

Recommendation, Skill Level 50

In the lower-division baccalaureate/associate degree category, 3 semester hours in correctional operations, 3 in correctional supervision, 3 in a criminal justice elective, and 3 in correctional administration. In the upper-division baccalaureate category, 3 semester hours in principles of management and 3 in introduction to public administration (11/86).

MOS-95C-003

Corrections Specialist
((Corrections NCO))
95C10
95C20
95C30
95C40
95C50

Exhibit Dates: 2/93–9/04. (Effective 9/04, MOS 95C was discontinued, and its duties were incorporated into MOS 31E.)

Career Management Field: 95 (Law Enforcement).

Description

Summary: Functions as a correctional officer, providing rehabilitative, health, welfare, and security services to prisoners within a correctional facility; conducts inspections; prepares written reports; coordinates activities of inmates and staff personnel. *Skill Level 10:* Assists with supervision and management of military prisoners; provides external security to confinement or corrections facility. *Skill Level 20:* Able to perform the duties required for Skill Level 10; functions as a correctional officer; provides individual counseling and guidance to prisoners within a rehabilitative program. *Skill Level 30:* Able to perform the duties required for Skill Level 20; functions as a first-line supervisor of up to ten correctional officers; prepares or reviews reports and records of individuals and programs; provides formal counseling. *Skill Level 40:* Able to perform the duties required for Skill Level 30; as a mid-level manager, supervises and coordinates all correctional, custodial, treatment, and rehabilitative activities of inmates and up to 100 staff members in correctional facility containing as many as 250 prisoners; responsible for administrative inspection and review of all records, reports, and activities; furnishes recommendations to superiors regarding clemency, restoration to duty, and further treatment of inmates; advises superior on matters concerning subordinates. *Skill Level 50:* Able to perform the duties required for Skill Level 40; supervises and coordinates inmate and staff activities in a large consolidated correctional facility; as an administrator, plans, coordinates, and implements all counseling, rehabilitative, and correctional programs in a permanent facility containing 250 or more prisoners; equivalent to a federal maximum-security prison warden.

Recommendation, Skill Level 10

Credit may be granted on the basis of an individualized assessment of the student (2/96).

Recommendation, Skill Level 20

In the lower-division baccalaureate/associate degree category, 3 semester hours in correctional operations. (NOTE: This recommendation for skill level 20 is valid for the dates 2/93-2/95 only) (11/86).

Recommendation, Skill Level 30

In the lower-division baccalaureate/associate degree category, 3 semester hours in correctional operations, 3 in correctional supervision, and 3 in a criminal justice elective (11/86).

Recommendation, Skill Level 40

In the lower-division baccalaureate/associate degree category, 3 semester hours in correctional operations, 3 in correctional supervision, 3 in a criminal justice elective, and 3 in correctional administration. In the upper-division baccalaureate category, 3 semester hours in principles of management (11/86).

Recommendation, Skill Level 50

In the lower-division baccalaureate/associate degree category, 3 semester hours in correctional operations, 3 in correctional supervision, 3 in a criminal justice elective, and 3 in correctional administration. In the upper-division baccalaureate category, 3 semester hours in principles of management and 3 in introduction to public administration (11/86).

MOS-95D-003

CID Special Agent
95D20
95D30
95D40
95D50

Exhibit Dates: 1/90–10/94.

Career Management Field: 95 (Law Enforcement).

Description

Summary: Supervises or conducts investigations of incidents, offenses, and allegations of criminality. *Skill Level 20:* Assists in investigations of incidents, offenses, and allegations of criminality; under close supervision, performs duties shown for MOS 95D30. *Skill Level 30:* Able to perform the duties required for Skill Level 20; conducts investigations of incidents, offenses, or allegations of criminality; applies laws governing investigation, search, and apprehension; photographs and sketches crime scenes and takes investigative notes; interviews/interrogates witnesses, suspects, subjects, and victims of crimes, obtaining necessary statements/confessions and/or polygraph examinations; analyzes evidence, laboratory findings, and statements relating to investigations; testifies before courts-martial and other tribunals; examines crime scenes for evidence such as fingerprints, bloodstains, weapons, footprints, documents, and other trace evidence; collects, preserves, and tags evidence including suspected drugs and determines need for crime laboratory analysis and identification; conducts surveillance, searches of premises, and lineups associated with investigations of allegations of criminality; conducts crime prevention surveys on nonlogistic facilities and operations; prepares chronology of investigative activities, investigative reports, and action records relating to incidents, offenses, and allegations of criminality. *Skill Level 40:* Able to perform the duties required for Skill Level 30; provides technical guidance and assistance to subordinates; reviews completed cases and processes evidence; maintains chain of custody accountability for evidence gathered during investigations; preserves and safeguards evidence in order to meet court admissibility standards; assists in conduct of crime prevention surveys of logistic facilities and operations; supervises general administrative functions and personnel engaged in investigative support activities; reviews reports and records relating to investigative activity for administrative accuracy and sufficiency; maintains investigative complaint log, status board, and suspense system; prepares periodic and special reports concerning assigned personnel and investigative activity and develops statistical data; monitors criminalistics field tests of newly-acquired investigative equipment; coordinates implementation of new investigative techniques; determines equipment needs of operational elements; prepares unit training program; supervises word processing center; supervises administrative and supply personnel; evaluates personnel performance; counsels personnel; prepares evaluation reports. *Skill Level 50:* Able to perform the duties required for Skill Level 40; provides investigative support in central headquarters; manages general administrative functions and personnel engaged in investigative support activities; determines equipment and budgetary needs of operational elements; prepares and provides unit training programs and instruction.

Recommendation, Skill Level 20

In the lower-division baccalaureate/associate degree category, 3 semester hours in criminal investigation, 3 in introduction to criminal justice, 3 in police organization and administration, 3 in principles of criminal evidence, and 3 in technical report writing. (NOTE: This recommendation for skill level 20 is valid for the dates 1/90-2/95 only) (9/81).

Recommendation, Skill Level 30

In the lower-division baccalaureate/associate degree category, 3 semester hours in criminal investigation, 3 in introduction to criminal justice, 3 in police organization and administration, 3 in principles of criminal evidence, and 3 in technical report writing (9/81).

Recommendation, Skill Level 40

In the lower-division baccalaureate/associate degree category, 3 semester hours in criminal investigation, 3 in introduction to criminal justice, 3 in police organization and administration, 3 in principles of criminal evidence, 3 in technical report writing, 4 in introduction to criminalistics (including laboratory), 3 in crime and delinquency, and 3 in personnel supervision (9/81).

Recommendation, Skill Level 50

In the lower-division baccalaureate/associate degree category, 3 semester hours in criminal investigation, 3 in introduction to criminal justice, 3 in police organization administration, 3 in principles of criminal evidence, 3 in technical report writing, 4 in introduction to criminalistics (including laboratory), 3 in crime and delinquency, and 3 in personnel supervision. In the upper-division baccalaureate category, 3 semester hours in crime prevention and control, 3 in principles of management, and 3 in resource management (9/81).

MOS-95D-004

CID Special Agent
95D20
95D30
95D40
95D50

Exhibit Dates: 11/94–9/04. (Effective 9/04, MOS 95D was discontinued, and its duties were incorporated into MOS 31D.)

Career Management Field: 95 (Military Police).

Description

Summary: Supervises or conducts investigations of incidents, offenses, and allegations of criminality. *Skill Level 20:* Assists in investigations of incidents, offenses, and allegations of criminality; under close supervision, performs duties shown for MOS 95D30. *Skill Level 30:* Able to perform the duties required for Skill Level 20; conducts investigations of incidents, offenses, or allegations of criminality; applies laws governing investigation, search, and apprehension; photographs and sketches crime scenes and takes investigative notes; interviews/interrogates witnesses, suspects, subjects, and victims of crimes obtaining necessary statements/confessions; analyzes evidence, laboratory findings, and statements relating to investigations; testifies before courts-martial and other tribunals; examines crime scenes for evidence such as fingerprints, bloodstains, weapons, footprints, documents, and other trace evidence; collects, preserves, and tags evidence including suspected drugs and determines need for crime laboratory analysis and identification; conducts surveillance, searches of premises, and lineups associated with investigations of allegations of criminality; conducts crime prevention surveys on nonlogistic facilities and operations; prepares chronology of investigative activities, investigative reports, and action records relating to incidents, offenses, and allegations of criminality. *Skill Level 40:* Able to perform the duties required for Skill Level 30; provides technical guidance and assistance to subordinates; reviews completed cases and processes evidence; maintains chain of custody accountability for evidence gathered during investigations; preserves and safeguards evidence in order to meet court admissibility standards; assists in conduct of crime prevention surveys of logistic facilities and operations; supervises general administrative functions and soldiers engaged in investigative support activities; reviews reports and records relating to investigative activity for administrative accuracy and sufficiency; maintains investigative complaint log, status board, and suspense system; prepares periodic and special reports concerning assigned personnel and investigative activity and develops statistical data; monitors criminalistics field tests of newly acquired investigative equipment; coordinates implementation of new investigative techniques; determines equipment needs of operational elements; prepares unit training program; supervises word processing center; supervises administrative and supply personnel; evaluates personnel performance; counsels personnel; prepares evaluation reports. *Skill Level 50:* Able to perform the duties required for Skill Level 40; provides investigative support in central headquarters; manages general administrative functions and soldiers engaged in investigative support activities; determines equipment and budgetary needs of operational elements; prepares and provides unit training programs and instruction.

Recommendation, Skill Level 20

In the lower-division baccalaureate/associate degree category, 3 semester hours in criminal investigation, 3 in criminal justice, 3 in police organization and administration, 3 in criminal evidence, and 3 in technical report writing. (NOTE: This recommendation for skill level 20 is valid for the dates 1/90-2/95 only) (11/94).

Recommendation, Skill Level 30

In the lower-division baccalaureate/associate degree category, 3 semester hours in criminal investigation, 3 in introduction to criminal justice, 3 in police organization and administration, 3 in criminal evidence, and 3 in technical report writing (11/94).

Recommendation, Skill Level 40

In the lower-division baccalaureate/associate degree category, 3 semester hours in criminal investigation, 3 in introduction to criminal justice, 3 in police organization and administration, 3 in criminal evidence, 3 in technical report writing, 4 in criminalistics (laboratory), and 3 in personnel supervision. In the upper-division baccalaureate category, 3 semester hours for field experience in management (11/94).

Recommendation, Skill Level 50

In the lower-division baccalaureate/associate degree category, 3 semester hours in criminal investigation, 3 in introduction to criminal justice, 3 in police organization and administration, 3 in criminal evidence, 3 in technical report writing, 4 in criminalistics (including laboratory), and 3 in personnel supervision. In the upper-division baccalaureate category, 3 semester hours in crime prevention and control, 6 for field experience in management, and 3 in human resource management (11/94).

MOS-96B-002

Intelligence Analyst

96B10
96B20
96B30
96B40
96B50

Exhibit Dates: 1/90–9/91.

Career Management Field: 96 (Military Intelligence); subfield 962 (Technical Intelligence Production).

Description

Summary: Assembles, integrates, analyzes, and disseminates intelligence information collected from tactical, strategic, and technical sources; analyzes intelligence information and serves as a country or geographic area specialist; may function as censor, editor, mail examiner, or file clerk. *Skill Level 10:* Processes incoming reports and messages; assists in maintaining intelligence records, files, and situation maps; assists in preparation of consolidated reports, maps, overlays, and aerial photographs; proofreads and assembles individual intelligence reports; safeguards classified information; assists in integration of information. *Skill Level 20:* Able to perform the duties required for Skill Level 10; analyzes intelligence holdings and identifies intelligence collection requirements; assists in coordination with technical intelligence personnel and preparation of reports on captured enemy material; prepares drafts of intelligence reports; reads maps. *Skill Level 30:* Able to perform the duties required for Skill Level 20; provides technical guidance to subordinates; supervises receipt, analysis, and storage of intelligence information; completes, edits, and disseminates intelligence reports; assists in conducting intelligence training programs; assists in establishing personnel security; assists in the collection of military, economic, political, sociological, and geographic information; serves as senior intelligence editor or as intelligence team chief; supervises from four to ten individuals. *Skill Level 40:* Able to perform the duties required for Skill Level 30; serves as a chief strategic intelligence analyst or as a first-line supervisor for intelligence operations teams; assists in supervision of combined information/intelligence coordination centers; conducts general intelligence training programs; supervises and coordinates the preparation of intelligence reports and prepares consolidated reports containing maps, aerial photographs, and technical data; conducts current situation briefings; may also supervise personnel holding MOS 96C (Interrogator). *Skill Level 50:* Able to perform the duties required for Skill Level 40; serves as chief operations and intelligence sergeant of large military units; supervises intelligence activities; assists supervisors in the appraisal of intelligence procedures and operations; supervises the training of personnel assigned to tactical and technical operations of intelligence sections; supervises up to 200 persons. NOTE: May have progressed to 96B50 from 96B40 or 96C40 (Interrogator).

Recommendation, Skill Level 10

In the lower-division baccalaureate/associate degree category, 3 semester hours in clerical practices (11/77).

Recommendation, Skill Level 20

In the lower-division baccalaureate/associate degree category, 3 semester hours in clerical practices, 3 in records administration, and 1 in map reading (11/77).

Recommendation, Skill Level 30

In the lower-division baccalaureate/associate degree category, 3 semester hours in clerical practices, 3 in records administration, 3 in technical writing, and 1 in map reading (11/77).

Recommendation, Skill Level 40

In the lower-division baccalaureate/associate degree category, 3 semester hours in clerical practices, 3 in records administration, 3 in technical writing, 3 in oral communication, 3 in personnel supervision, and 1 in map reading (11/77).

Recommendation, Skill Level 50

In the lower-division baccalaureate/associate degree category, 3 semester hours in clerical practices, 3 in records administration, 3 in technical writing, 3 in oral communication, 3 in personnel supervision, and 1 in map reading. In the upper-division baccalaureate category, 3 semester hours in management problems, 3 for field experience in management, and 3 for a practicum in education (11/77).

MOS-96B-003

Intelligence Analyst

96B10
96B20
96B30
96B40
96B50

Exhibit Dates: 10/91–12/99.

Career Management Field: 96 (Military Intelligence).

Description

Summary: Assembles, integrates, analyzes, and disseminates intelligence information collected from tactical, strategic, and technical sources; analyzes intelligence information; serves as a country or geographic area specialist. *Skill Level 10:* Processes incoming reports and messages; assists in maintaining intelligence records, files, and situation maps; assists in preparation of consolidated reports, maps, overlays, and aerial photographs; proofreads and assembles individual intelligence reports; safeguards classified information. *Skill Level 20:* Able to perform the duties required for Skill Level 10; analyzes intelligence holdings and identifies intelligence collection requirements; assists in coordination with technical intelligence personnel and preparation of reports on captured enemy material; prepares drafts of intelligence reports; has a working knowledge of personal computers. *Skill Level 30:* Able to perform the duties required for Skill Level 20; provides technical guidance to subordinates; supervises receipt, analysis, and storage of intelligence information; completes, edits, and disseminates intelligence reports; assists in conducting intelligence training programs and briefings; writes technical materials; supervises and evaluates from four to ten individuals. *Skill Level 40:* Able to perform the duties required for Skill Level 30; assists in supervision of combined information/intelligence coordination centers; conducts general intelligence training programs; evaluates personnel and completes evaluation reports. *Skill Level 50:* Able to perform the duties required for Skill Level 40; serves as chief operations and intelligence sergeant of large military units; supervises intelligence activities; assists commander and staff officer in appraisal of intelligence operations and training procedures; coordinates operating requirements of subordinate units with major or supported units.

Recommendation, Skill Level 10

Credit may be granted on the basis of an individualized assessment of the student (3/94).

Recommendation, Skill Level 20

In the lower-division baccalaureate/associate degree category, 3 semester hours in computer literacy and 3 in personnel supervision. (NOTE: This recommendation for skill level 20 is valid for the dates 10/91-2/95 only) (9/91).

Recommendation, Skill Level 30

In the lower-division baccalaureate/associate degree category, 3 semester hours in computer literacy, 3 in personnel supervision, 3 in technical writing, and 2 in records and information management (9/91).

Recommendation, Skill Level 40

In the lower-division baccalaureate/associate degree category, 3 semester hours in computer literacy, 3 in personnel supervision, 3 in technical writing, and 2 in records and information management. In the upper-division baccalaureate category, 3 semester hours in organizational management and 3 for field experience in management (9/91).

Recommendation, Skill Level 50

In the lower-division baccalaureate/associate degree category, 3 semester hours in computer literacy, 3 in personnel supervision, 3 in technical writing, and 2 in records and information management. In the upper-division baccalaureate category, 3 semester hours in organizational management and 6 for field experience in management; if individual has attained paygrade E-9, additional credit as follows: 3 semester hours in management problems, 3 in operations management, and 3 in communications techniques for managers (9/91).

MOS-96B-004

Intelligence Analyst

96B10
96B20
96B30
96B40
96B50

Exhibit Dates: 1/00–Present.

Career Management Field: 96 (Military Intelligence).

Description

Summary: Assembles, integrates, analyzes, and disseminates intelligence information collected from tactical, strategic, and technical sources including multispectral, electro-optical data; analyzes intelligence information; serves as a country or geographic area specialist; uses computers for a variety of networking and data base operations. *Skill Level 10:* Processes incoming reports and messages; assists in maintaining intelligence records, files, and situation maps; assists in preparation of consolidated reports, maps, overlays, and aerial photographs; proofreads and assembles individual intelligence reports; safeguards classified information; uses computers for analytical tools, data base access, networking, and office applications. *Skill Level 20:* Able to perform the duties required for Skill Level 10; analyzes intelligence holdings and identifies intelligence collection requirements; assists in coordination with technical intelligence personnel and preparation of reports on captured enemy material; prepares standard intelligence reports; uses a variety of commercial and military computer programs; supervises two to five people. *Skill Level 30:* Able to perform the duties required for Skill Level 20; provides technical guidance to subordinates; supervises receipt, analysis, and storage of intelligence information; completes, edits, and disseminates standard and specialized intelligence reports; assists in conducting intelligence training programs; writes technical materials; supervises and evaluates from four to ten individuals; prepares and conducts briefings for commanders and other high-ranking officials. *Skill Level 40:* Able to perform the duties required for Skill Level 30; assists in supervision of combined information/intelligence coordination centers; conducts general intelligence training programs; evaluates personnel and completes evaluation reports. *Skill Level 50:* Able to perform the duties required for Skill Level 40; serves as chief operations and intelligence sergeant of large military units; supervises intelligence activities; assists commander and staff officer in appraisal of intelligence operations and training procedures; coordinates operating requirements of subordinate units with major or supported units; coordinates planning for contingency operations.

Recommendation, Skill Level 10

Credit may be granted on the basis of an individualized assessment of the student (1/00).

Recommendation, Skill Level 20

Credit may be granted on the basis of an individualized assessment of the student (1/00).

Recommendation, Skill Level 30

In the lower-division baccalaureate/associate degree category, 3 semester hours in computer software applications, 3 in principles of supervision, 3 in technical report writing, 3 in regional geography, 3 in information systems, 3 in office administration, and 3 in speech communication. In the upper-division baccalaureate category, 3 semester hours in professional writing (1/00).

Recommendation, Skill Level 40

In the lower-division baccalaureate/associate degree category, 3 semester hours in computer software applications, 3 in principles of supervision, 3 in technical report writing, 3 in regional geography, 3 in information systems, 3 in office administration, and 3 in speech communication. In the upper-division baccalaureate category, 3 semester hours for a practicum in management, 3 in professional writing, and 3 in information systems management (1/00).

Recommendation, Skill Level 50

In the lower-division baccalaureate/associate degree category, 3 semester hours in computer software applications, 3 in principles of supervision, 3 in technical report writing, 3 in regional geography, 3 in information systems, 3 in office administration, and 3 in speech communication. In the upper-division baccalaureate category, 3 semester hours for a practicum in management, 3 in professional writing, 3 in organizational development, 3 in human resource management, and 3 in information systems management (1/00).

MOS-96D-003

Imagery Analyst

96D10
96D20
96D30
96D40
96D50

Exhibit Dates: 1/90–2/95.

Career Management Field: 96 (Military Intelligence).

Description

Summary: Interprets aerial and ground imagery developed by photographic and electronic means. *Skill Level 10:* Utilizes principles and techniques of photogrammetry and employs electronic, mechanical, and optical devices to study and interpret photographs; assists in preparation of map overlays, plots, mosaics, charts, and other graphics; computes distances and field coordinates; assists in the preparation of written reports on image interpretation findings. *Skill Level 20:* Able to perform the duties required for Skill Level 10; provides technical guidance to subordinates; studies and interprets imagery produced by aerial sensor systems and hand-held camera; prepares maps; conducts briefings; computes areas and volumes; selects entry zones. *Skill Level 30:* Able to perform the duties required for Skill Level 20; supervises image interpretation activities; assists air reconnaissance liaison officer and tactical surveillance officer; performs quality control on imagery analysis and reports; coordinates planning for ground and air reconnaissance and surveillance plans and requests; receives, evaluates, and disseminates mission results; plans and organizes work schedules and assigns duties; instructs subordinates in work techniques. *Skill Level 40:* Able to perform the duties required for Skill Level 30; assists in planning activities related to image utilization; arranges priorities for conduct of operation; assists in planning; levies imagery reproduction requirements; supervises preparation of reconnaissance and surveillance plan; coordinates with collection agencies and requesters; instructs subordinates in proper image interpretation techniques; supervises preparation and maintenance of files. *Skill Level 50:* Able to perform the duties required for Skill Level 40; supervises the duties of 96D40 and 96H40 (Aerial Intelligence Specialist); prepares and coordinates surveillance plan; assists command and staff officers; coordinates operation requirements; prepares, coordinates, and supervises activities related to image interpretation; directs training; supervises staff; advises superiors. NOTE: May have progressed to 96D50 from 96D40 or 96H40 (Aerial Intelligence Specialist).

Recommendation, Skill Level 10

In the lower-division baccalaureate/associate degree category, 3 semester hours in map reading, 2 in mechanical drawing, 3 in aerial photographic interpretation, 3 in photogrammetry, 1 in technical writing, and 3 in technical mathematics. (NOTE: This recommendation for skill level 10 is valid for the dates 1/90-9/91 only) (6/90).

Recommendation, Skill Level 20

In the lower-division baccalaureate/associate degree category, 3 semester hours in map reading, 3 in photogrammetry, 3 in aerial photographic interpretation, 3 in technical mathematics, 2 in mechanical drawing, and 1 in technical writing (6/90).

Recommendation, Skill Level 30

In the lower-division baccalaureate/associate degree category, 3 semester hours in map reading, 3 in photogrammetry, 3 in aerial photographic interpretation, 3 in technical mathematics, 3 in technical writing, 3 in principles of supervision, and 2 in mechanical drawing (6/90).

Recommendation, Skill Level 40

In the lower-division baccalaureate/associate degree category, 3 semester hours in map reading, 3 in photogrammetry, 3 in aerial photographic interpretation, 3 in technical mathematics, 3 in technical writing, 3 in principles of supervision, and 2 in mechanical drawing. In the upper-division baccalaureate category, 3 semester hours in management problems and 3 for field experience in management (6/90).

Recommendation, Skill Level 50

In the lower-division baccalaureate/associate degree category, 3 semester hours in map reading, 3 in photogrammetry, 3 in aerial photographic interpretation, 3 in technical mathematics, 3 in technical writing, 3 in principles of supervision, and 2 in mechanical drawing. In the upper-division baccalaureate category, 3 semester hours in management problems and 6 for field experience in management (6/90).

MOS-96D-004

Imagery Analyst
96D10
96D20
96D30
96D40
96D50

Exhibit Dates: 3/95–12/99.

Career Management Field: 96 (Military Intelligence).

Description

Summary: Interprets aerial and ground imagery developed by photographic and electronic means. *Skill Level 10:* Utilizes principles and techniques of photogrammetry and employs electronic, mechanical, and optical devices to study and interpret photographs; assists in preparation of map overlays, plots, mosaics, charts, and other graphics; computes distances and field coordinates; assists in the preparation of written reports on image interpretation findings. *Skill Level 20:* Able to perform the duties required for Skill Level 10; provides technical guidance to subordinates; studies and interprets imagery produced by aerial sensor systems and hand-held camera; prepares maps; conducts briefings; computes areas and volumes; selects entry zones. *Skill Level 30:* Able to perform the duties required for Skill Level 20; supervises image interpretation activities; assists air reconnaissance liaison officer and tactical surveillance officer; performs quality control on imagery analysis and reports; coordinates planning for ground and air reconnaissance and surveillance plans and requests; receives, evaluates, and disseminates mission results; plans and organizes work schedules and assigns duties; instructs subordinates in work techniques. *Skill Level 40:* Able to perform the duties required for Skill Level 30; assists in planning activities related to image utilization; arranges priorities for conduct of operation; assists in planning; levies imagery reproduction requirements; supervises preparation of reconnaissance and surveillance plan; coordinates with collection agencies and requesters; instructs subordinates in proper image interpretation techniques; supervises preparation and maintenance of files. *Skill Level 50:* Able to perform the duties required for Skill Level 40; supervises the duties of 96D40 and 96H40 (Aerial Intelligence Specialist); prepares and coordinates surveillance plan; assists command and staff officers; coordinates operation requirements; prepares, coordinates, and supervises activities related to image interpretation; directs training; supervises staff; advises superiors. NOTE: May have progressed to 96D50 from 96D40 or 96H40 (Aerial Intelligence Specialist).

Recommendation, Skill Level 10

Credit may be granted on the basis of an individualized assessment of the student (6/90).

Recommendation, Skill Level 20

Credit may be granted on the basis of an individualized assessment of the student (6/90).

Recommendation, Skill Level 30

In the lower-division baccalaureate/associate degree category, 3 semester hours in map reading, 3 in photogrammetry, 3 in aerial photographic interpretation, 3 in technical mathematics, 1 in technical writing, 3 in principles of supervision, and 2 in mechanical drawing (6/90).

Recommendation, Skill Level 40

In the lower-division baccalaureate/associate degree category, 3 semester hours in map reading, 3 in photogrammetry, 3 in aerial photographic interpretation, 3 in technical mathematics, 3 in technical writing, 3 in principles of supervision, and 2 in mechanical drawing. In the upper-division baccalaureate category, 3 semester hours in management problems and 3 for field experience in management (6/90).

Recommendation, Skill Level 50

In the lower-division baccalaureate/associate degree category, 3 semester hours in map reading, 3 in photogrammetry, 3 in aerial photographic interpretation, 3 in technical mathematics, 3 in technical writing, 3 in principles of supervision, and 2 in mechanical drawing. In the upper-division baccalaureate category, 3 semester hours in management problems and 6 for field experience in management (6/90).

MOS-96D-005

Imagery Analyst
96D10
96D20
96D30
96D40
96D50

Exhibit Dates: 1/00–Present.

Career Management Field: 96 (Military Intelligence).

Description

Summary: Interprets aerial and ground imagery developed by photographic and electronic means. *Skill Level 10:* Uses principles and techniques of photogrammetry and employs computer-based electronic, mechanical, and optical devices to study and interpret photographs; assists in preparation of map overlays, plots, mosaics, charts, and other graphics; computes distances and field coordinates; assists in the preparation of written reports on image interpretation findings. *Skill Level 20:* Able to perform the duties required for Skill Level 10; provides technical guidance to as many as six subordinates; studies and interprets imagery produced by aerial sensor systems and hand-held camera; prepares maps; conducts briefings; computes areas and volumes; selects entry zones. *Skill Level 30:* Able to perform the duties required for Skill Level 20; supervises image interpretation activities; assists air reconnaissance liaison officer and tactical surveillance officer; performs quality control on imagery analysis and reports; coordinates planning for ground and air reconnaissance and surveillance plans and requests; receives, evaluates, and disseminates mission results; plans and organizes work schedules and assigns duties; instructs subordinates in work techniques. *Skill Level 40:* Able to perform the duties required for Skill Level 30; assists in planning activities related to image utilization; arranges priorities for conduct of operation; assists in planning; levies imagery reproduction requirements; supervises preparation of reconnaissance and surveillance plans; coordinates with collection agencies and requesters; instructs subordinates in proper image interpretation techniques; supervises preparation and maintenance of files. *Skill Level 50:* Able to perform the duties required for Skill Level 40; supervises the duties of 96D40 and 96H40 (Aerial Intelligence Specialist); prepares and coordinates surveillance plans; assists command and staff officers; coordinates operation requirements; prepares, coordinates, and supervises activities related to image interpretation; directs training; supervises staff; advises superiors.

Recommendation, Skill Level 10

Credit may be granted on the basis of an individualized assessment of the student (1/00).

Recommendation, Skill Level 20

Credit may be granted on the basis of an individualized assessment of the student (1/00).

Recommendation, Skill Level 30

In the lower-division baccalaureate/associate degree category, 3 semester hours in map reading, 3 in computer software applications, 2 in technical mathematics, 1 in technical writing, 3 in principles of supervision, 3 in speech communication, and 3 in office administration. In the upper-division baccalaureate category, 2 semester hours in photogrammetry and 2 in aerial photographic and satellite image interpretation (1/00).

Recommendation, Skill Level 40

In the lower-division baccalaureate/associate degree category, 3 semester hours in map reading, 3 in computer software applications, 2 in technical mathematics, 1 in technical writing, 3 in principles of supervision, 3 in speech communication, and 3 in office administration. In the upper-division baccalaureate category, 2 semester hours in photogrammetry, 2 in aerial photographic and satellite image interpretation, 3 for a practicum in management, and 3 in organizational development (1/00).

Recommendation, Skill Level 50

In the lower-division baccalaureate/associate degree category, 3 semester hours in map reading, 3 in computer software applications, 2 in technical mathematics, 1 in technical writing, 3 in principles of supervision, 3 in speech communication, and 3 in office administration. In the upper-division baccalaureate category, 2 semester hours in photogrammetry, 2 in aerial photographic and satellite image interpretation, 3 for a practicum in management, 3 in human resource management, and 3 in organizational development (1/00).

MOS-96F-001

Psychological Operations Specialist
96F10
96F20
96F30
96F40
96F50

Exhibit Dates: 1/90–9/91. Effective 9/91, MOS 96F was discontinued, and its duties were incorporated into MOS 37F, Psychological Operations Specialist.

Career Management Field: 96 (Military Intelligence).

Description

Summary: Supervises, coordinates, and participates in analysis, planning, production, and dissemination of tactical, strategic, and consolidated psychological operations (PSYOP). *Skill Level 10:* Assists in collecting and reporting PSYOP-related information; assists in processing information into intelligence to support PSYOP; assists in identifying intelligence collection requirements for PSYOP support; assists in evaluating and analyzing current intelligence, PSYOP studies, and estimates to determine PSYOP targets; assists in establishment and maintenance of situation maps to provide current intelligence/PSYOP information and identification, disposition, and movement of enemy forces; assists in the design of PSYOP products; assists in the development and administration of surveys to evaluate the effects of planned and executed products; operates PSYOP dissemination equipment and assists in the delivery of PSYOP products; assists in the packaging of PSYOP products for delivery by various means; maintains journals, status boards, visual displays, charts, and graphs required to manage PSYOP; either manually or using automatic data processing equipment, prepares, stores, and retrieves information on PSYOP-related intelligence, plans, campaigns, and products; assists and performs intelligence functions as required; maintains and operates organizational communications equipment, generators, and organic PSYOP production and dissemination team equipment; assists in establishing and maintaining systematic cross-referenced PSYOP records and files; safeguards classified information. *Skill Level 20:* Able to perform the duties required for Skill Level 10; provides technical guidance to subordinates; coordinates resource requirements for the development, production, and dissemination of PSYOP products; assists in the integration of PSYOP planning in support of conventional, special operations, and deception planning; develops and administers surveys to evaluate the effectiveness of PSYOP products; assists in determining the appropriate mix of media relative to available assets to disseminate PSYOP products; analyzes current intelligence holdings to identify intelligence gaps and subsequent intelligence collection requirements to support PSYOP; evaluates current intelligence, PSYOP studies, and estimates to determine PSYOP targets; assists in identifying psychological vulnerabilities and susceptibilities of PSYOP targets; assists in evaluating translations of captured enemy documents and intelligence information reports from interrogations; assists in the analysis of enemy foreign propaganda and other foreign media; informs superiors of information having immediate tactical PSYOP value; packages PSYOP products for delivery by various means; identifies and maintains information on the availability of products and delivery means; establishes and maintains reference files of translated material. *Skill Level 30:* Able to perform the duties required for Skill Level 20; supervises receipt, analysis, and storage of PSYOP-related intelligence information; coordinates PSYOP intelligence collection requirements with supported command and higher headquarters; spot-checks analyses performed by subordinates; plans and advises on PSYOP in direct and general support of operational forces; plans and implements PSYOP campaigns; assists in planning, identification, mobilization, and deployment of PSYOP resources; supervises propaganda writers, broadcast specialists, journalists, and illustrators to develop and produce PSYOP products; supervises PSYOP dissemination and delivery sections and teams; determines dissemination requirements for PSYOP products; assists in the conduct of liaison with the supported command; advises supported commands on psychological considerations of planning operations; provides guidance and training for subordinates; assists in preparing and conducting PSYOP training programs; analyzes enemy propaganda. *Skill Level 40:* Able to perform the duties required for Skill Level 30; plans and organizes work schedules and assigns specific tasks in support of tactical or strategic PSYOP missions; assists in supervision of propaganda development and tactical operations centers; conducts general PSYOP training programs for the command; supervises planning and dissemination of PSYOP products; controls the execution of PSYOP campaigns; monitors preparation and production of PSYOP products and acquisition of resources necessary to support implementation of PSYOP campaigns; conducts liaison with supported commands; supervises intelligence analysts, interpreters, interrogators, and translators assigned or attached to PSYOP units. *Skill Level 50:* Able to perform the duties required for Skill Level 40; supervises and provides technical guidance to subordinates; provides liaison with supported staff and commands; assists PSYOP commanders and staff officers in planning, organizing, directing, supervising, training, and coordinating activities pertaining to PSYOP at all levels of command.

Recommendation, Skill Level 10

In the lower-division baccalaureate/associate degree category, 1 semester hour in technical writing, 1 in record keeping, and 1 in audiovisual technology (12/88).

Recommendation, Skill Level 20

In the lower-division baccalaureate/associate degree category, 2 semester hours in technical writing, 1 in record keeping, 2 in audiovisual technology, 2 in records management, 1 in oral communication, and 2 in personnel supervision (12/88).

Recommendation, Skill Level 30

In the lower-division baccalaureate/associate degree category, 2 semester hours in technical writing, 1 in record keeping, 2 in audiovisual technology, 2 in records management, 2 in oral communication, 3 in personnel supervision, and 3 in business organization and management (12/88).

Recommendation, Skill Level 40

In the lower-division baccalaureate/associate degree category, 2 semester hours in technical writing, 1 in record keeping, 2 in audiovisual technology, 2 in records management, 2 in oral communication, 3 in personnel supervision, and 3 in business organization and management. In the upper-division baccalaureate category, 3 semester hours for field experience in management and 3 in principles of management (12/88).

Recommendation, Skill Level 50

In the lower-division baccalaureate/associate degree category, 2 semester hours in technical writing, 1 in record keeping, 2 in audiovisual technology, 2 in records management, 2 in oral communication, 3 in personnel supervision, and 3 in business organization and management. In the upper-division baccalaureate category, 3 semester hours for field experience in management, 3 in principles of management, and 3 in management problems (12/88).

MOS-96H-002

Aerial Intelligence Specialist
96H10
96H20
96H30
96H40

Exhibit Dates: 1/90–2/95.

Career Management Field: 96 (Military Intelligence).

Description

Summary: Supervises or participates in aerial surveillance and electronic intercept operations to provide information and imagery of intelligence value. *Skill Level 10:* Participates in mission planning for aerial surveillance, aerial visual reconnaissance, aerial search and rescue, aerial radiological survey and similar missions; operates data link terminal station and performs side-looking airborne radar imagery analysis; prepares and operates aerial surveillance/electronic intercept systems; performs aerial missions operating aerial infrared, radar, and photographic equipment including data transmission links and ground data terminal station; operates radio and aids in aerial navigation; recognizes enemy electronic countermeasures and performs countermeasures; acquires targets; participates in mission debriefing; assists in the analysis of images; performs preflight, preoperation, and operator and unit maintenance on surveillance and electronic intercept equipment; troubleshoots systems to determine nature and location of problems; records operation and maintenance data. *Skill Level 20:* Able to perform the duties required for Skill Level 10; provides technical guidance to subordinates; may serve as assistant instructor at service school; operates and supervises the operation of aerial surveillance and aerial surveillance/electronic intercept systems; assists in flight planning, weather analysis, navigational computation and aircraft preflight inspections; prepares aerial surveillance equipment for operation; assists imagery analyst in interpreting imagery recording using keys and reference material. *Skill Level 30:* Able to perform the duties required for Skill Level 20; supervises operation and activities of aerial surveillance, electronic intercept, and data terminal section; supervises inspection and operator maintenance of aerial surveillance and electronic intercept equipment; assists in planning, employment, and management of aerial surveillance, electronic intercept, and data terminal systems; serves as instructor at service school; assists in the analysis and interpretation of aerial sensor imagery to determine geographical features of terrain and physical features of enemy installation; conducts or participates in briefing of commander and staff of supported headquarters on capabilities and limitations of aerial surveillance and similar missions. *Skill Level 40:* Able to perform the duties required for Skill Level 30; supervises operations and activities of platoon-sized detachment; plans and organizes work schedules; assigns duties and instructs section sergeant in work techniques; reviews and critiques mission results; advises commander; coordinates personnel on team employment, deployment, operational supply, and maintenance requirements; serves as instructor at service school.

Recommendation, Skill Level 10

In the lower-division baccalaureate/associate degree category, 3 semester hours in technical mathematics and 1 in map/air photo reading. In the upper-division baccalaureate category, 1 semester hour in remote sensing. (NOTE: This recommendation for skill level 10 is valid for the dates 1/90-9/91 only) (6/90).

Recommendation, Skill Level 20

In the lower-division baccalaureate/associate degree category, 3 semester hours in technical mathematics, 1 in map/air photo reading, and 1 in basic electronic systems. In the upper-division baccalaureate category, 1 semester hour in remote sensing (6/90).

Recommendation, Skill Level 30

In the lower-division baccalaureate/associate degree category, 3 semester hours in technical mathematics, 3 in map/air photo reading, 1 in basic electronic systems, 3 in principles of supervision, and 3 in communication skills (oral and written). In the upper-division baccalaureate category, 1 semester hour in remote sensing and 3 for field experience in management (6/90).

Recommendation, Skill Level 40

In the lower-division baccalaureate/associate degree category, 3 semester hours in technical mathematics, 3 in map/air reading, 1 in basic electronic systems, 3 in principles of supervision, and 3 in communication skills (oral and written). In the upper-division baccalaureate category, 1 semester hour in remote sensing, 3 for field experience in management, and 3 in management problems (6/90).

MOS-96H-003

Aerial Intelligence Specialist

96H10
96H20
96H30
96H40

Exhibit Dates: 3/95–5/97.

Career Management Field: 96 (Military Intelligence).

Description

Summary: Supervises or participates in aerial surveillance and electronic intercept operations to provide information and imagery of intelligence value. *Skill Level 10:* Participates in mission planning for aerial surveillance, aerial visual reconnaissance, aerial search and rescue, aerial radiological survey and similar missions; operates data link terminal station and performs side-looking airborne radar imagery analysis; prepares and operates aerial surveillance/electronic intercept systems; performs aerial missions operating aerial infrared, radar, and photographic equipment including data transmission links and ground data terminal station; operates radio and aids in aerial navigation; recognizes enemy electronic countermeasures and performs countermeasures; acquires targets; participates in mission debriefing; assists in the analysis of images; performs preflight, preoperation, and operator and unit maintenance on surveillance and electronic intercept equipment; troubleshoots systems to determine nature and location of problems; records operation and maintenance data. *Skill Level 20:* Able to perform the duties required for Skill Level 10; provides technical guidance to subordinates; may serve as assistant instructor at service school; operates and supervises the operation of aerial surveillance and aerial surveillance/electronic intercept systems; assists in flight planning, weather analysis, navigational computation and aircraft preflight inspections; prepares aerial surveillance equipment for operation; assists imagery analyst in interpreting imagery recording using keys and reference material. *Skill Level 30:* Able to perform the duties required for Skill Level 20; supervises operation and activities of aerial surveillance, electronic intercept, and data terminal section; supervises inspection and operator maintenance of aerial surveillance and electronic intercept equipment; assists in planning, employment, and management of aerial surveillance, electronic intercept, and data terminal systems; serves as instructor at service school; assists in the analysis and interpretation of aerial sensor imagery to determine geographical features of terrain and physical features of enemy installation; conducts or participates in briefing of commander and staff of supported headquarters on capabilities and limitations of aerial surveillance and similar missions. *Skill Level 40:* Able to perform the duties required for Skill Level 30; supervises operations and activities of platoon-sized detachment; plans and organizes work schedules; assigns duties and instructs section sergeant in work techniques; reviews and critiques mission results; advises commander; coordinates personnel on team employment, deployment, operational supply, and maintenance requirements; serves as instructor at service school.

Recommendation, Skill Level 10

Credit may be granted on the basis of an individualized assessment of the student (6/90).

Recommendation, Skill Level 20

Credit may be granted on the basis of an individualized assessment of the student (6/90).

Recommendation, Skill Level 30

In the lower-division baccalaureate/associate degree category, 3 semester hours in technical mathematics, 1 in map/air photo reading, 1 in basic electronic systems, 3 in principles of supervision, and 3 in communication skills (oral and written). In the upper-division baccalaureate category, 1 semester hour in remote sensing and 3 for field experience in management (6/90).

Recommendation, Skill Level 40

In the lower-division baccalaureate/associate degree category, 3 semester hours in technical mathematics, 3 in map/air photo reading, 1 in basic electronic systems, 3 in principles of supervision, and 3 in communication skills (oral and written). In the upper-division baccalaureate category, 1 semester hour in remote sensing, 3 for field experience in management, and 3 in management problems (6/90).

MOS-96H-004

Imagery Ground Station Operator

96H10
96H20
96H30
96H40
96H50

Exhibit Dates: 6/97–Present.

Career Management Field: 96 (Military Intelligence).

Description

Summary: Supervises or participates in detecting, locating, and tracking ground targets and rotary wing and slow moving fixed wing aircraft. *Skill Level 10:* Participates in premission planning for aerial surveillance, reconnaissance missions; deploys and redeploys the Imagery Ground Station (IGS) in a tactical environment; operates and maintains data systems and communications equipment; analyzes near-real-time radar imagery and soft copy imagery, including infrared and electro-optics; prepares imagery analysis reports; performs maintenance on mission equipment. *Skill Level 20:* Able to perform the duties required for Skill Level 10; provides guidance to subordinates; supervises shift operations; performs reconnaissance and IGS site selection; prepares and disseminates near-real-time radar imagery data. *Skill Level 30:* Able to perform the duties required for Skill Level 20; supervises IGS operations and maintenance of related equipment; supervises the analysis of near-real-time imagery, aerial intelligence information and imagery; supervises development and dissemination of intelligence products; participates in and conducts mission briefings; performs liaison with outside agencies; manages security requirements. *Skill Level 40:* Able to perform the duties required for Skill Level 30; supervises Joint Surveillance Target Attack Radar system operations and activities; assists in planning and using IGS and other surveillance resources; incorporates IGS mission into overall collection plan; instructs subordinates in proper soft copy imagery analysis techniques and procedures. *Skill Level 50:* Able to perform the duties required for Skill Level 40; plans, coordinates, supervises, and participates in activities pertaining to organization, training, operations, and intelligence of Joint Surveillance Target Attack Radar Systems.

Recommendation, Skill Level 10

Credit may be granted on the basis of an individualized assessment of the student (1/00).

Recommendation, Skill Level 20

Credit may be granted on the basis of an individualized assessment of the student (1/00).

Recommendation, Skill Level 30

In the lower-division baccalaureate/associate degree category, 3 semester hours in map/air photo reading, 3 in network systems troubleshooting, 2 in maintenance technology, 3 in principles of supervision, 3 in technical report writing, and 3 in computer software applications. In the upper-division baccalaureate category, 3 semester hours for field experience in management (1/00).

Recommendation, Skill Level 40

In the lower-division baccalaureate/associate degree category, 3 semester hours in map/air photo reading, 3 in network systems troubleshooting, 2 in maintenance technology, 3 in principles of supervision, 3 in technical report writing, 3 in computer software applications, and 3 in speech communications. In the upper-division baccalaureate category, 3 semester hours for a practicum in management, and 3 in human resource management (1/00).

Recommendation, Skill Level 50

In the lower-division baccalaureate/associate degree category, 3 semester hours in map/air photo reading, 3 in network systems troubleshooting, 2 in maintenance technology, 3 in principles of supervision, 3 in technical report writing, 3 in computer software applications, and 3 in speech communications. In the upper-division baccalaureate category, 6 semester hours for a practicum in management, 3 in human resource management, and 3 in operations management (1/00).

MOS-96R-001

Ground Surveillance Systems Operator

96R10
96R20
96R30
96R40
96R50

Exhibit Dates: 1/90–4/99.

Career Management Field: 96 (Military Intelligence).

Description

Summary: Supervises or operates ground surveillance systems engaged in intelligence and information gathering. *Skill Level 10:* Operates ground surveillance systems; reads military maps, overlays, and aerial photographs; uses radar surveillance cards and plotters; assists in emplacement, camouflage, and recovery of equipment; detects, locates, and reports target data; operates organic communications equipment, light wheeled vehicles, and power sources; assists in performing preventive maintenance on ground surveillance systems and associated equipment. *Skill Level 20:* Able to perform the duties required for Skill Level 10; provides technical guidance to subordinates; selects specific equipment emplacement sites and supervises emplacement and operation of equipment; prepares overlays and surveillance cards. *Skill Level 30:* Able to perform the duties required for Skill Level 20; supervises organizational maintenance, preventive maintenance, emplacement and displacement of equipment, and preparation of surveillance cards; notifies supported unit of team locations; recommends methods of employment; conducts tactical and equipment-oriented training; maintains operational records. *Skill Level 40:* Able to perform the duties required for Skill Level 30; plans, recommends, and determines employment and operational techniques; plans and organizes work schedules and assigns duties; instructs squad and team leaders in work techniques and procedures; supervises operations, activities of platoon, and requisitioning of replacement parts; applies military intelligence collection and surveillance planning procedures; participates in electromagnetic intelligence surveillance and collection training of personnel. *Skill Level 50:* Able to perform the duties required for Skill Level 40; assists tactical surveillance officer in planning and coordinating employment of ground surveillance equipment and in preparing and implementing reconnaissance and surveillance plans; provides input to collection plans; prepares and updates ground reconnaissance and surveillance situation maps; may serve as principal noncommissioned officer (identified by the special qualifications identifier M - First Sergeant).

Recommendation, Skill Level 10

Credit is not recommended because the skills, competencies, and knowledge are uniquely military (7/87).

Recommendation, Skill Level 20

Credit is not recommended because the skills, competencies, and knowledge are uniquely military (7/87).

Recommendation, Skill Level 30

In the lower-division baccalaureate/associate degree category, 2 semester hours in personnel supervision (7/87).

Recommendation, Skill Level 40

In the lower-division baccalaureate/associate degree category, 3 semester hours in personnel supervision. In the upper-division baccalaureate category, 3 semester hours in principles of management and 3 for field experience in management (7/87).

Recommendation, Skill Level 50

In the lower-division baccalaureate/associate degree category, 3 semester hours in personnel supervision. In the upper-division baccalaureate category, 3 semester hours in principles of management, 3 for field experience in management, and 3 in applied psychology (7/87).

MOS-96R-002

Ground Surveillance Systems Operator

96R10
96R20
96R30
96R40
96R50

Exhibit Dates: 5/99–Present.

Career Management Field: 96 (Military Intelligence).

Description

Summary: Supervises or operates ground surveillance systems engaged in intelligence and information gathering and remote sensing. *Skill Level 10:* Operates ground surveillance systems; reads military maps, overlays, and aerial photographs; uses radar surveillance cards and plotters; assists in emplacement, camouflage, and recovery of equipment; detects, locates, and reports target data; operates communications equipment, light wheeled vehicles, and power sources; assists in performing preventive maintenance on ground surveillance systems and associated equipment. *Skill Level 20:* Able to perform the duties required for Skill Level 10; provides technical guidance to subordinates; selects specific equipment emplacement sites and supervises emplacement and operation of equipment; prepares overlays and surveillance cards. *Skill Level 30:* Able to perform the duties required for Skill Level 20; supervises organizational maintenance, preventive maintenance, emplacement and displacement of equipment, and preparation of surveillance cards; notifies supported unit of team locations; recommends methods of employment; conducts tactical and equipment-oriented training; maintains operational records. *Skill Level 40:* Able to perform the duties required for Skill Level 30; plans, recommends, and determines employment and operational techniques; plans and organizes work schedules and assigns duties; instructs squad and team leaders in work techniques and procedures; supervises operations, activities of platoon, and requisitioning of replacement parts; applies military intelligence collection and surveillance planning procedures; participates in electromagnetic intelligence surveillance and collection training of personnel. *Skill Level 50:* Able to perform the duties required for Skill Level 40; assists tactical surveillance officer in planning and coordinating employment of ground surveillance equipment and in preparing and implementing reconnaissance and surveillance plans; provides input to collection plans; prepares and updates ground reconnaissance and surveillance situation maps; may serve as principal noncommissioned officer.

Recommendation, Skill Level 10

Credit may be granted on the basis of an individualized assessment of the student (5/99).

Recommendation, Skill Level 20

Credit may be granted on the basis of an individualized assessment of the student (5/99).

Recommendation, Skill Level 30

In the lower-division baccalaureate/associate degree category, 3 semester hours in map reading and land navigation, 1 in automotive service, 1 in electro-optics, 1 in applied mathematics, 3 in radar, 2 in introduction to electronics, 2 in computer familiarization, 3 in personnel supervision, 3 in oral communication skills, and 2 in technical writing (5/99).

Recommendation, Skill Level 40

In the lower-division baccalaureate/associate degree category, 3 semester hours map reading and land navigation, 1 in automotive service, 1 in electro-optics, 1 in applied mathematics, 3 in radar, 2 in introduction to electronics, 3 in personnel supervision, 3 in computer familiarization, 3 in oral communication skills, and 3 in technical writing. In the upper-division baccalaureate category, 3 semester hours in personnel management and 3 for field experience in management (5/99).

Recommendation, Skill Level 50

In the lower-division baccalaureate/associate degree category, 3 semester hours map reading and land navigation, 1 in automotive service, 1 in electro-optics, 1 in applied mathematics, 3 in radar, 2 in introduction to electronics, 3 in personnel supervision, 3 in computer familiarization, 3 in oral communication skills, and 3 in technical writing. In the upper-division baccalaureate category, 3 semester hours in personnel management and 6 for field experience in management (5/99).

MOS-96U-001

Unmanned Aerial Vehicle Operator

96U10
96U20
96U30
96U40
96U50

Exhibit Dates: 4/92–Present.

Career Management Field: 96 (Military Intelligence).

Description

Summary: Assembles, launches, flies, and operates on-board surveillance equipment, and retrieves unmanned aerial vehicles; performs operator maintenance on RF and computer equipment. *Skill Level 10:* Assembles, launches, flies, and retrieves aircraft; performs minor airframe repair. *Skill Level 20:* Able to perform the duties required for Skill Level 10; supervises activities of subordinates; directs emplacement of equipment; supervises or performs airframe repair. *Skill Level 30:* Able to perform the duties required for Skill Level 20; provides technical guidance to subordinates; performs shift supervisor duties; coordinates mission with other units; maintains operational records. *Skill Level 40:* Able to perform the duties required for Skill Level 30; coordinates shift operations. *Skill Level 50:* Able to perform the duties required for Skill Level 40; assists commander in site selection; supervises area security; coordinates activities of subordinate aerial vehicle platoons.

Recommendation, Skill Level 10

Credit may be granted on the basis of an individualized assessment of the student (10/93).

Recommendation, Skill Level 20

In the lower-division baccalaureate/associate degree category, 3 semester hours in computer operations. (NOTE: This recommendation for skill level 20 is valid for the dates 4/92-2/95 only) (10/93).

Recommendation, Skill Level 30

In the lower-division baccalaureate/associate degree category, 3 semester hours in computer operations and 3 in personnel supervision (10/93).

Recommendation, Skill Level 40

In the lower-division baccalaureate/associate degree category, 3 semester hours in computer operations and 3 in personnel supervision. In the upper-division baccalaureate category, 3 semester hours for field experience in management (10/93).

Recommendation, Skill Level 50

In the lower-division baccalaureate/associate degree category, 3 semester hours in computer operations and 3 in personnel supervision. In the upper-division baccalaureate category, 6 semester hours for field experience in management (10/93).

MOS-96Z-001

Intelligence Senior Sergeant
96Z50

Exhibit Dates: 1/90–8/91.

Career Management Field: 96 (Military Intelligence), subfield 960 (General Tactical Intelligence).

Description

Supervises intelligence collection, analysis, processing, and surveillance activities at group, division, corps, or higher headquarters; supervises training activities; supervises tactical and technical operations; assists commissioned officers in appraisal of intelligence operations and training; instructs in specific phases of command intelligence procedures; supervises collection, preparation, and distribution of intelligence operations and training data; advises supervisors on matters concerning intelligence, operations, and administration. NOTE: May have progressed to 96Z50 from 96B50 (Intelligence Analyst), 96D50 (Image Interpreter), or 17K50 (Ground Surveillance Radar Crewman).

Recommendation

In the lower-division baccalaureate/associate degree category, 3 semester hours in clerical practices, 3 in records administration, 3 in technical writing, 3 in oral communication, 3 in personnel supervision, and 1 in map reading. In the upper-division baccalaureate category, 6 semester hours for a practicum in education, 3 for field experience in management, and 3 in management problems (11/77).

MOS-96Z-002

Intelligence Senior Sergeant
96Z50

Exhibit Dates: 9/91–12/99.

Career Management Field: 96 (Military Intelligence).

Description

Supervises intelligence collection, analysis, processing, and surveillance activities at group, division, corps, or comparable higher headquarters; supervises training activities; supervises tactical and technical operations; assists commissioned officers in appraisal of intelligence operations and training; instructs in specific phases of command intelligence procedures; supervises collection, preparation, and distribution of intelligence operations and training data; advises supervisors on matters concerning intelligence, operations, and administration. NOTE: May have progressed to 96Z50 from 96B50 (Intelligence Analyst), 96D50 (Imagery Analyst), or 96R50 (Ground Surveillance Systems Operator).

Recommendation

In the lower-division baccalaureate/associate degree category, 3 semester hours in personnel supervision, 3 in records and information management, and 3 in computer literacy. In the upper-division baccalaureate category, 3 semester hours in organizational management and 3 for field experience in management; if individual has attained pay grade E-9, additional credit as follows: 3 semester hours in management problems, 3 in operations management, and 3 in communication techniques for managers (9/91).

MOS-96Z-003

Intelligence Senior Sergeant
96Z50

Exhibit Dates: 1/00–Present.

Career Management Field: 96 (Military Intelligence).

Description

Supervises intelligence collection, analysis, processing, and surveillance activities at group, division, corps, or comparable higher headquarters; supervises training activities; supervises tactical and technical operations, including computer software and hardware maintenance and system management; assists commissioned officers in appraisal of intelligence operations and training; instructs in specific phases of command intelligence procedures; supervises collection, preparation, and distribution of intelligence operations and training data; advises supervisors on matters concerning intelligence, operations, and administration. NOTE: May have progressed to 96Z50 from 96B50 (Intelligence Analyst), 96D50 (Imagery Analyst), 96H50 (Imagery Ground Station Operator), 96R50 (Ground Surveillance Systems Operator), or 96U (Unmanned Aerial Vehicle Operator).

Recommendation

In the upper-division baccalaureate category, 6 semester hours for a practicum in management, 3 in organizational planning, 3 in personnel management, and 3 in principles of management (1/00).

MOS-97B-003

Counterintelligence Agent
97B10
97B20
97B30
97B40
97B50

Exhibit Dates: 1/90–10/94.

Career Management Field: 96 (Military Intelligence), subfield 961 (Controlled Intelligence).

Description

Summary: NOTE: The skill levels of MOS 97B are not necessarily indicative of progressively complex duties, increased skills, and greater responsibilities; the higher skill levels in this MOS usually indicate longer tenure. Individual in this MOS conducts security-oriented investigations when there is a threat to military operations, installations, and personnel; collects background information concerning foreign and domestic individuals, groups, and organizations posing such a threat; collects information through personal interviews, screening of public and private records made available to them, and surveillance techniques; is knowledgeable about the geography, political, and economic systems, culture, and customs of at least one foreign country; operates specialized photographic and sound recording equipment; conducts physical surveys of installations and facilities to minimize threats to security; maintains files; prepares written reports and gives oral reports of findings; conducts security-oriented classes and briefings; may be proficient in a foreign language. NOTE: Required to have held another MOS for at least two years prior to holding this MOS. *Skill Level 10:* Under close supervision, assists in tactical noninvestigative counterintelligence and operational security duties; assists in determining enemy intelligence collection assets, organizations, composition, personnel, methods of operation, capabilities, vulnerabilities, limitations, and missions; assists in evaluating sources of information; assists in formulating investigation plans; prepares recording and photographic equipment for operation; compiles collected counterintelligence information; prepares and types summaries and reports; prepares interrogation rooms; disseminates intelligence reports; stores, inventories, and controls classified documents. *Skill Level 20:* Able to perform the duties required for Skill Level 10; conducts counterintelligence investigations, often without close supervision; functions as a basic counterintelligence agent; applies fundamentals of military and civil law in conducting investigations; evaluates sources and information; prepares and types reports and summaries; conducts security surveys. *Skill Level 30:* Able to perform the duties required for Skill Level 20; may provide counterintelligence support at higher headquarters; may assist in on-the-job training of newer agents; functions as chief counterintelligence agent; plans and conducts counterintelligence operations to include analyzing, selecting, exploiting, and neutralizing targets of counterintelligence interest. *Skill Level 40:* Able to perform the duties required for Skill Level 30; may be used in a supervisory capacity; plans, organizes, and coordinates activities of counterintelligence support teams; functions as a senior counterintelligence agent. *Skill Level 50:* Able to perform the duties required for Skill Level 40; may conduct especially sensitive investigations; may be used as a staff specialist, planning and supervising intelligence operations. NOTE: May have progressed to 97B50 from 97B40 or 97C50 (Area Intelligence Specialist).

Recommendation, Skill Level 10

Credit is not recommended because of the limited or specialized nature of the skills, competencies, and knowledge (10/84).

Recommendation, Skill Level 20

In the lower-division baccalaureate/associate degree category, 6 semester hours in social science, 3 in police administration, 3 in industrial security, 3 in United States government, 3 in applied psychology, 3 in report writing, 1 in typing, and additional credit in personnel supervision and office management on the basis of institutional evaluation. NOTE: Add credit for the specific foreign language in accordance with the recommendation in the DoD volume of the Guide to the Evaluation of Educational Experiences in the Armed Services (10/84).

Recommendation, Skill Level 30

In the lower-division baccalaureate/associate degree category, 6 semester hours in social science, 3 in police administration, 3 in industrial security, 3 in United States government, 3 in applied psychology, 3 in report writing, 1 in typing, and additional credit in personnel supervision and office management on the basis of institutional evaluation. NOTE: Add credit for the specific foreign language in accordance with the recommendation in the DoD volume of the Guide to the Evaluation of Educational Experiences in the Armed Services (10/84).

Recommendation, Skill Level 40

In the lower-division baccalaureate/associate degree category, 6 semester hours in social science, 3 in police administration, 3 in industrial security, 3 in United States government, 3 in applied psychology, 3 in report writing, 1 in typing, and additional credit in personnel supervision and office management on the basis of institutional evaluation. In the upper-division baccalaureate category, 3 semester hours in political ideologies, 3 in constitutional law, and 3 in public relations. NOTE: Add credit for the specific foreign language in accordance with the recommendation in the DoD volume of the Guide to the Evaluation of Educational Experiences in the Armed Services (10/84).

Recommendation, Skill Level 50

In the lower-division baccalaureate/associate degree category, 6 semester hours in social science, 3 in police administration, 3 in industrial security, 3 in United States government, 3 in applied psychology, 3 in report writing, 1 in typing, and additional credit in personnel supervision and office management on the basis of institutional evaluation. In the upper-division baccalaureate category, 3 semester hours in political ideologies, 3 in constitutional law, and 3 in public relations. NOTE: Add credit for the specific foreign language in accordance with the recommendation in the DoD volume of the Guide to the Evaluation of Educational Experiences in the Armed Services (10/84).

MOS-97B-004

Counterintelligence Agent

97B10

97B20

97B30

97B40

97B50

Exhibit Dates: 11/94–12/99.

Career Management Field: 96 (Military Intelligence).

Description

Summary: NOTE: The skill levels of MOS 97B are not necessarily indicative of progressively complex duties, increased skills, and greater responsibilities; the higher skill levels in this MOS usually indicate longer tenure. Individual in this MOS conducts security-oriented investigations when there is a threat to military operations, installations, and personnel; collects background information concerning foreign and domestic individuals, groups, and organizations posing such a threat; collects information through personal interviews, screening of public and private records made available to them, and surveillance techniques; operates specialized photographic and sound-recording equipment; conducts physical surveys of installations and facilities to minimize threats to security; maintains files; prepares written reports and gives oral reports of findings; conducts security-oriented classes and briefings; may be proficient in a foreign language. *Skill Level 10:* Under close supervision, assists in tactical non-investigative counterintelligence and operational security duties; assists in determining enemy intelligence collection assets, organizations, composition, personnel, methods of operation, capabilities, vulnerabilities, limitations, and missions; assists in evaluating sources of information; assists in formulating investigation plans; prepares recording and photographic equipment for operation; compiles collected counterintelligence information; prepares and types summaries and reports; prepares interrogation rooms; disseminates intelligence reports; stores, inventories, and controls classified data bases. *Skill Level 20:* Able to perform the duties required for Skill Level 10; conducts counterintelligence investigations, often without close supervision; functions as a basic counterintelligence agent; applies fundamentals of military and civil law in conducting investigations; evaluates sources and information; prepares and types reports and summaries; conducts security surveys. *Skill Level 30:* Able to perform the duties required for Skill Level 20; may provide counterintelligence support at higher headquarters; may assist in on-the-job training of newer agents; functions as chief counterintelligence agent; plans and conducts counterintelligence operations to include analyzing, selecting, exploiting, and neutralizing targets of counterintelligence interest. *Skill Level 40:* Able to perform the duties required for Skill Level 30; may be used in a supervisory capacity; plans, organizes, and coordinates activities of counterintelligence support teams; functions as a senior counterintelligence agent. *Skill Level 50:* Able to perform the duties required for Skill Level 40; may conduct especially sensitive investigations; may be used as a staff specialist, planning and supervising intelligence operations.

Recommendation, Skill Level 10

Credit is not recommended because of the limited or specialized nature of the skills, competencies, and knowledge (11/94).

Recommendation, Skill Level 20

In the lower-division baccalaureate/associate degree category, 3 semester hours in social science, 3 in police administration, 3 in physical security, 3 in United States government, 3 in applied psychology, 3 in report writing, 1 in keyboarding, and 3 in interpersonal communications. NOTE: Add credit for the specific foreign language in accordance with the recommendation in the DoD volume of the Guide. (NOTE: This recommendation for skill level 20 is valid for the dates 11/94-2/95 only) (11/94).

Recommendation, Skill Level 30

In the lower-division baccalaureate/associate degree category, 3 semester hours in social science, 3 in police administration, 3 in physical security, 3 in United States government, 3 in applied psychology, 3 in report writing, 1 in keyboarding, 3 in interpersonal communications, 3 in personnel supervision, and 2 in records and information management. NOTE: Add credit for the specific foreign language in accordance with the recommendation in the DoD volume of the Guide (11/94).

Recommendation, Skill Level 40

In the lower-division baccalaureate/associate degree category, 3 semester hours in social science, 3 in police administration, 3 in physical security, 3 in United States government, 3 in applied psychology, 3 in report writing, 1 in keyboarding, 3 in personnel supervision, and 2 in records and information management. In the upper-division baccalaureate category, 3 semester hours in public relations, 3 in ethnic studies, and 3 in international relations. NOTE: Add credit for the specific foreign language in accordance with the recommendation in the DoD volume of the Guide (11/94).

Recommendation, Skill Level 50

In the lower-division baccalaureate/associate degree category, 3 semester hours in social science, 3 in police administration, 3 in physical security, 3 in United States government, 3 in applied psychology, 3 in report writing, 1 in keyboarding, 3 in personnel supervision, and 2 in records and information management. In the upper-division baccalaureate category, 3 semester hours in public relations, 3 in ethnic studies, 3 in international relations, and 3 for field experience in management. NOTE: Add credit for the specific foreign language in accordance with the recommendation in the ODD volume of the Guide (11/94).

MOS-97B-005

Counterintelligence Agent

97B10
97B20
97B30
97B40
97B50

Exhibit Dates: 1/00–Present.

Career Management Field: 96 (Military Intelligence).

Description

Summary: NOTE: The skill levels of MOS 97B are not necessarily indicative of progressively complex duties, increased skills, and greater responsibilities; the higher skill levels in this MOS usually indicate longer tenure. Individual in this MOS conducts security-oriented investigations when there is a threat to military operations, installations, and personnel; collects background information concerning foreign and domestic individuals, groups, and organizations posing such a threat; collects information through personal interviews, screening of public and private records made available to them, and surveillance techniques; operates specialized photographic and sound-recording equipment; conducts physical surveys of installations and facilities to minimize threats to security; maintains files; prepares written reports and gives oral reports of findings; conducts security-oriented classes and briefings; may be proficient in a foreign language. *Skill Level 10:* Under close supervision, assists in tactical non-investigative counterintelligence and operational security duties; assists in determining enemy intelligence collection assets, organizations, composition, personnel, methods of operation, capabilities, vulnerabilities, limitations, and missions; assists in evaluating sources of information; assists in formulating investigation plans; prepares recording and photographic equipment for operation; compiles collected counterintelligence information; prepares and types summaries and reports; prepares interrogation rooms; disseminates intelligence reports; stores, inventories, and controls classified data bases. *Skill Level 20:* Able to perform the duties required for Skill Level 10; conducts counterintelligence investigations, often without close supervision; functions as a basic counterintelligence agent; applies fundamentals of military and civil law in conducting investigations; evaluates sources and information; prepares and types reports and summaries; conducts security surveys. *Skill Level 30:* Able to perform the duties required for Skill Level 20; may provide counterintelligence support at higher headquarters; may assist in on-the-job training of newer agents; functions as chief counterintelligence agent; plans and conducts counterintelligence operations including analyzing, selecting, exploiting, and neutralizing targets of counterintelligence interest. *Skill Level 40:* Able to perform the duties required for Skill Level 30; may be used in a supervisory capacity; plans, organizes, and coordinates activities of counterintelligence support teams; functions as a senior counterintelligence agent. *Skill Level 50:* Able to perform the duties required for Skill Level 40; may conduct especially sensitive investigations; may be used as a staff specialist, planning and supervising intelligence operations.

Recommendation, Skill Level 10

Credit may be granted on the basis of an individual assessment of the student (1/00).

Recommendation, Skill Level 20

Credit may be granted on the basis of an individual assessment of the student (1/00).

Recommendation, Skill Level 30

In the lower-division baccalaureate/associate degree category, 3 semester hours in social science, 3 in physical security, 3 in applied psychology, 3 in technical writing, 3 in computer software applications, 3 in speech communication, and 3 in principles of supervision (1/00).

Recommendation, Skill Level 40

In the lower-division baccalaureate/associate degree category, 3 semester hours in social science, 3 in physical security, 3 in applied psychology, 3 in technical writing, 3 in computer software applications, 3 in speech communication, 3 in information systems, and 3 in principles of supervision. In the upper-division baccalaureate category, 3 semester hours in regional geography and 3 in international relations (1/00).

Recommendation, Skill Level 50

In the lower-division baccalaureate/associate degree category, 3 semester hours in social science, 3 in physical security, 3 in applied psychology, 3 in technical writing, 3 in computer software applications, 3 in speech communication, 3 in information systems, and 3 in principles of supervision. In the upper-division baccalaureate category, 3 semester hours in regional geography, 3 in international relations, 3 for a practicum in management, and 3 in organizational development (1/00).

MOS-97E-002

Interrogator
97E10
97E20
97E30
97E40
97E50

Exhibit Dates: 1/90–9/91.

Career Management Field: 96 (Military Intelligence).

Description

Summary: Conducts and supervises interrogations in foreign language; knows geography, political system, economic system, and customs of the countries in which the foreign language is spoken; prepares translation reports. *Skill Level 10:* Conducts interrogations in foreign language of prisoners of war, enemy deserters, civilians, and refugees to obtain information necessary for the development of military intelligence; compares and verifies information obtained with information contained in other interrogation reports, captured documents, and intelligence reports; prepares notes and keeps detailed records on all interrogations performed; translates and prepares summaries, extracts, and full translations of written foreign material (directives, records, messages, combat orders, technical publications) into English; translates speeches, announcements, radio scripts, and other materials into foreign language for use in non-English speaking countries; types translated materials; establishes reference files of translation materials. *Skill Level 20:* Able to perform the duties required for Skill Level 10; provides technical guidance to subordinates; reviews and edits translations. *Skill Level 30:* Able to perform the duties required for Skill Level 20; provides technical guidance to subordinates; performs more difficult interrogations and translations including scientific information; determines requirements for summaries, extracts, or complete translations of documents; performs as a team chief for interrogations and translator/interpretation functions; monitors interrogations and translations for accuracy, adequacy, and completeness; organizes and conducts on-the-job training; assists in preparation and presentation of information to superiors. *Skill Level 40:* Able to perform the duties required for Skill Level 30; supervises an interrogation center or interrogation, translation/interpreter sections; plans and organizes work schedules; assists in specific tasks; plans training activities; ensures establishment of required files and preparation of required reports. *Skill Level 50:* Able to perform the duties required for Skill Level 40; supervises strategic intelligence interrogation center; plans, coordinates, and supervises intelligence interrogation training and operation; provides expert interrogation capability for high-level, technical interrogations; advises commander and staff on distribution and assignment of interrogation personnel.

Recommendation, Skill Level 10

In the vocational certificate category, 3 semester hours in record keeping and 2 in filing. In the lower-division baccalaureate/associate degree category, 6 semester hours in social science, 3 in typing, 3 in report writing, 3 in oral communication, and additional credit in area studies on the basis of institutional evaluation. NOTE: Add credit for the specific foreign language in accordance with the recommendation in the DoD volume 4 of the Guide to the Evaluation of Educational Experiences in the Armed Services (11/77).

Recommendation, Skill Level 20

In the vocational certificate category, 3 semester hours in record keeping and 2 in filing. In the lower-division baccalaureate/associate degree category, 6 semester hours in social science, 3 in typing, 3 in report writing, 3 in oral communication, and additional credit in area studies on the basis of institutional evaluation. NOTE: Add credit for the specific foreign language in accordance with the recommendation in the DoD volume of the Guide to the Evaluation of Educational Experiences in the Armed Services (11/77).

Recommendation, Skill Level 30

In the vocational certificate category, 3 semester hours in record keeping and 2 in filing. In the lower-division baccalaureate/associate degree category, 6 semester hours in social science, 3 in typing, 3 in report writing, 3 in oral communication, 3 in personnel supervision, and additional credit in area studies on the basis of institutional evaluation. NOTE: Add credit for the specific foreign language in accordance with the recommendation in the DoD volume of the Guide to the Evaluation of Educational Experiences in the Armed Services (11/77).

Recommendation, Skill Level 40

In the vocational certificate category, 3 semester hours in record keeping and 2 in filing. In the lower-division baccalaureate/associate degree category, 6 semester hours in social science, 3 in typing, 3 in report writing, 3 in oral communication, 3 in personnel supervision, and additional credit in area studies on the basis of institutional evaluation. In the upper-division baccalaureate category, 3 semester hours in personnel management and additional credit in area studies, for field experience in management, and for a practicum in education on the basis of institutional evaluation. NOTE: Add credit for the specific foreign language in accordance with the recommendation in the DoD volume of the Guide to the Evaluation of Educational Experiences in the Armed Services (11/77).

Recommendation, Skill Level 50

In the vocational certificate category, 3 semester hours in record keeping and 2 in filing. In the lower-division baccalaureate/associate degree category, 6 semester hours in social science, 3 in typing, 3 in report writing, 3 in oral communication, 3 in personnel supervision, and additional credit in area studies on the basis of institutional evaluation. In the upper-division baccalaureate category, 3 semester hours in personnel management, 3 for field experience in management, and additional credit in area studies and for a practicum in education on the basis of institutional evaluation. NOTE: Add credit for the specific foreign language in accordance with the recommendation in the DoD volume of the Guide to the Evaluation of Educational Experiences in the Armed Services (5/88).

MOS-97E-003

INTERROGATOR
97E10
97E20
97E30
97E40
97E50

Exhibit Dates: 10/91–12/99.

Career Management Field: 96 (Military Intelligence).

Description

Summary: Conducts and supervises interrogations in foreign language; knows geography, political system, economic system, and customs of the countries in which the foreign language is spoken; prepares and edits translation reports. *Skill Level 10:* Conducts interrogations in foreign language of prisoners of war, enemy deserters, civilians, and refugees to obtain information necessary for the development of military intelligence reports; prepares notes and keeps detailed records on all interrogations performed; translates and prepares summaries, extracts, and full translations of written foreign material (directives, records, messages, combat orders, technical publications) into English; translates speeches, announcements, radio scripts, and other materials into foreign language for use in non-English speaking countries; establishes reference files of translation materials. *Skill Level 20:* Able to perform the duties required for Skill Level 10; supervises and provides technical guidance to subordinates; reviews and edits translations; performs difficult translations; ensures the accurate exchange of ideas, statements, and intent; has a basic understanding and working knowledge of personal computers. *Skill Level 30:* Able to perform the duties required for Skill Level 20; provides technical guidance to subordinates; performs as a team chief for interrogations and translator/interpretation functions; monitors interrogations and translations for accuracy, adequacy, and completeness; organizes and conducts on-the-job training; assists in preparation and presentation of information to superiors. *Skill Level 40:* Able to perform the duties required for Skill Level 30; supervises an interrogation center or interrogation, translation/interpreter sections; plans and organizes work schedules; assigns specific tasks; plans training activities; ensures establishment of required files and preparation of required reports; performs as platoon sergeant in a military intelligence company. *Skill Level 50:* Able to perform the duties required for Skill Level 40; supervises strategic intelligence interrogation center; plans, coordinates, and supervises interrogation capability for high-level, technical interrogations; advises commander and staff on distribution and assignment of interrogation personnel.

Recommendation, Skill Level 10

Credit may be granted on the basis of an individualized assessment of the student (3/94).

Recommendation, Skill Level 20

In the lower-division baccalaureate/associate degree category, 3 semester hours in report writing, 3 in oral communication, 3 in computer literacy, and 3 in personnel supervision. In the upper-division baccalaureate category, 3 semester hours in applied psychology, 3 in regional geography, and 3 in geopolitics. NOTE: Add credit for the specific foreign language in accordance with the recommendations in the DoD volume of the Guide to the Evaluation of Educational Experiences in the Armed Services. (NOTE: This recommendation for skill level 20 is valid for the dates 10/91-2/95 only) (9/91).

Recommendation, Skill Level 30

In the lower-division baccalaureate/associate degree category, 3 semester hours in report writing, 3 in oral communication, 3 in computer literacy, 3 in personnel supervision, and 2 in records and information management. In the upper-division baccalaureate category, 3 semester hours in applied psychology, 3 in regional geography, and 3 in geopolitics. NOTE: Add credit for the specific foreign language in accordance with the recommendations in the DoD volume of the Guide to the Evaluation of Educational Experiences in the Armed Services (9/91).

Recommendation, Skill Level 40

In the lower-division baccalaureate/associate degree category, 3 semester hours in report writing, 3 in oral communication, 3 in computer literacy, 3 in personnel supervision, and 2 in records and information management. In the upper-division baccalaureate category, 3 semester hours in applied psychology, 3 in regional geography, 3 in geopolitics, 3 for field experience in management, and 3 in organizational management. NOTE: Add credit for the specific foreign language in accordance with the recommendations in the DoD volume of the Guide to the Evaluation of Educational Experiences in the Armed Services (9/91).

Recommendation, Skill Level 50

In the lower-division baccalaureate/associate degree category, 3 semester hours in report writing, 3 in oral communication, 3 in computer literacy, 3 in personnel supervision, and 2 in records and information management. In the upper-division baccalaureate category, 3 semester hours in applied psychology, 3 in regional geography, 3 in geopolitics, 6 for field experience in management, and 3 in organizational management; if individual has attained pay-grade E-9, additional credit as follows: 3 semester hours in management problems, 3 in operations management, and 3 in communication techniques for managers. NOTE: Add credit for the specific foreign language in accordance with the recommendations in the DoD volume of the Guide to the Evaluation of Educational Experiences in the Armed Services (9/91).

MOS-97E-004

HUMAN INTELLIGENCE COLLECTOR
((Interrogator))
97E10
97E20
97E30
97E40
97E50

Exhibit Dates: 1/00–Present.

Career Management Field: 96 (Military Intelligence).

Description

Summary: Conducts and supervises interrogations in foreign language; knows geography, political system, economic system, and customs of the countries in which the foreign language is spoken; prepares and edits translation reports. *Skill Level 10:* Conducts interrogations in foreign language of prisoners of war, enemy deserters, civilians, and refugees to obtain information necessary for the development of military intelligence reports; prepares notes and keeps detailed records on all interrogations performed; translates and prepares summaries, extracts, and full translations of written foreign material (directives, records, messages, combat orders, technical publications) into English; translates speeches, announcements, radio scripts, and other materials into foreign language for use in non-English speaking countries; establishes reference files of translation materials. *Skill Level 20:* Able to perform the duties required for Skill Level 10; supervises and provides technical guidance to subordinates; reviews and edits translations; performs difficult translations; ensures the accurate exchange of ideas, statements, and intent; has a basic understanding and working knowledge of personal computers. *Skill Level 30:* Able to perform the duties required for Skill Level 20; provides technical guidance to subordinates; performs as a team chief for interrogations and translator/interpretation functions; monitors interrogations and translations for accuracy, adequacy, and completeness; organizes and conducts on-the-job training; assists in preparation and presentation of information to superiors. *Skill Level 40:* Able to perform the duties required for Skill Level 30; supervises an interrogation center or interrogation, translation/interpreter sections; plans and organizes work schedules; assigns specific tasks; plans training activities; ensures establishment of required files and preparation of required reports; performs as platoon sergeant in a military intelligence company. *Skill Level 50:* Able to perform the duties required for Skill Level 40; supervises strategic intelligence interrogation center; plans, coordinates, and supervises interrogation capability for high-level, technical interrogations; advises commander and staff on distribution and assignment of interrogation personnel.

Recommendation, Skill Level 10

Credit may be granted on the basis of an individualized assessment of the student (1/00).

Recommendation, Skill Level 20

Credit may be granted on the basis of an individualized assessment of the student (1/00).

Recommendation, Skill Level 30

In the lower-division baccalaureate/associate degree category, 3 semester hours in technical writing, 3 in speech communication, 3 in applied psychology, 3 in regional geography, 3 in computer software applications, 3 in information systems, and 3 in principles of supervision. In the upper-division baccalaureate category, 3 semester hours in international relations (1/00).

Recommendation, Skill Level 40

In the lower-division baccalaureate/associate degree category, 3 semester hours in technical writing, 3 in speech communication, 3 in applied psychology, 3 in regional geography, 3 in computer software applications, 3 in information systems, and 3 in principles of supervision. In the upper-division baccalaureate category, 3 semester hours in international relations and 3 for a practicum in management (1/00).

Recommendation, Skill Level 50

In the lower-division baccalaureate/associate degree category, 3 semester hours in technical writing, 3 in speech communication, 3 in applied psychology, 3 in regional geography, 3 in computer software applications, 3 in information systems, and 3 in principles of supervision. In the upper-division baccalaureate category, 3 semester hours in international relations, 3 for a practicum in management, 3 in human resource management, and 3 in organizational development (1/00).

MOS-97G-001

Signal Security Specialist
97G10
97G20
97G30
97G40
97G50

Exhibit Dates: 1/90–9/91.

Career Management Field: 96 (Military Intelligence).

Description

Summary: Supervises and conducts signal security operations throughout the Army; monitors and analyzes communications signals; performs electronic security inspections; reviews documents for adequacy of security; prepares comprehensive written reports and makes oral presentations. NOTE: Many of the duties required for this MOS involve highly classified materials, equipment, and activities; therefore, not all the competencies and knowledge associated with the MOS were evaluated. *Skill Level 10:* Performs transcription, monitoring, and basic analysis of radiotelephone and conventional telephone messages to detect communications security discrepancies and violations; reviews communications, electronic signal instructions, communications electronic operating instructions, and operations plans and orders to determine adequacy of communications security; reads and interprets maps and map overlays; prepares basic reports on signal security activities; operates and performs user maintenance on communications security and monitoring equipment; selects sites for, erects, and orients tactical antennas; selects and uses commercial battery and generator power; performs general office duties such as filing, typing (35 words per minute minimum), preparation of reports, and record keeping; presents oral reports concerning communications security activities. *Skill Level 20:* Able to perform the duties required for Skill Level 10; serves as team chief, supervising two to five persons; provides basic signal security advice and assistance to commanders; performs security analysis of security codes; establishes and supervises operating sites; writes reports including statistical analyses; presents oral reports on data collected; makes recommendations to eliminate disclosure of intelligence information; provides advice and assistance concerning cryptographic systems and the interpretation of basic documents and regulations; determines need for written communications security instructions; prepares special, periodic, and project reports on signal security activities. *Skill Level 30:* Able to perform the duties required for Skill Level 20; serves as first-line supervisor of 10-12 persons; provides advice and assistance on electronic security procedures and signal security training; prepares signal security monitoring cover plan; analyzes the signal output of communications and noncommunications equipment to establish methods of operations, cryptonetting, and other types of intelligence; reviews documents for signal security; plans and supervises electronic security support; schedules signal security operational activities. *Skill Level 40:* Able to perform the duties required for Skill Level 30; supervises 15 or more persons; determines needed signal intelligence support; presents oral reports to commanders on operational security principles; plans signal security actions to support the operational security plan; serves as advisor to commanders on signal security matters; plans and conducts classes on signal security activities. *Skill Level 50:* Able to perform the duties required for Skill Level 40; serves as signal security operations chief, first sergeant of a company, or signal security chief; supervises 50-300 persons; has experience as the enlisted commander of a detachment; develops and writes signal security doctrine; advises commander and manages signal security personnel; plans career progression training; allocates personnel; advises commander and manages signal security equipment resources; plans and carries out signal security missions; advises generals and high-level government officials on signal security activities.

Recommendation, Skill Level 10

In the lower-division baccalaureate/associate degree category, 3 semester hours in typing, 2 in written communication skills, 2 in oral communication skills, 2 in electronic system operations, and 1 in office practices (6/77).

Recommendation, Skill Level 20

In the lower-division baccalaureate/associate degree category, 3 semester hours in typing, 3 in written communication skills, 3 in oral communication skills, 3 for field experience in personnel supervision, 2 in electronic system operations, 2 in office practices, and 2 in introductory mathematics (6/77).

Recommendation, Skill Level 30

In the lower-division baccalaureate/associate degree category, 4 semester hours in written communication skills, 3 in typing, 3 in oral communication skills, 3 for field experience in personnel supervision, 3 in human relations, 2 in electronic system operations, 2 in office practices, and 2 in introductory mathematics. In the upper-division baccalaureate category, 2 semester hours for field experience in management (6/77).

Recommendation, Skill Level 40

In the lower-division baccalaureate/associate degree category, 6 semester hours in written communication skills, 3 in typing, 3 in oral communication skills, 3 for field experience in personnel supervision, 3 in human relations, 2 in electronic system operations, 2 in office practices, and 2 in introductory mathematics. In the upper-division baccalaureate category, 3 semester hours for field experience in management and 3 for an internship in education (6/77).

Recommendation, Skill Level 50

In the lower-division baccalaureate/associate degree category, 6 semester hours in written communication skills, 3 in typing, 3 in oral communication skills, 3 for field experience in personnel supervision, 3 in human relations, 2 in electronic system operations, 2 in office practices, and 2 in introductory mathematics. In the upper-division baccalaureate category, 6 semester hours for field experience in management, 3 in introduction to management, 3 in personnel management, and 3 for an internship in education (6/77).

MOS-97G-002

Multi-Discipline Counter-Intelligence Operator/Analyst
((Counter-Signals Intelligence Specialist))
97G10
97G20
97G30
97G40
97G50

Exhibit Dates: 10/91–6/98. Effective 6/98, MOS 97G was discontinued and its duties were incorporated into the following MOS's: 96B (Intelligence Analyst), 97B (Counterintelligence Agent), and 98K (Non-Morse Interceptor/Analyst).

Career Management Field: 96 (Military Intelligence).

Description

Summary: Supervises and conducts signal security operations throughout the Army; monitors and analyzes communications signals; performs electronic security; prepares comprehensive written reports and makes oral presentations. NOTE: Many of the duties required for this MOS involve highly classified materials, equipment, and activities; therefore, not all the competencies and knowledge associated with the MOS were evaluated. *Skill Level 10:* Performs transcription, monitoring, and basic analysis of radiotelephone and conventional telephone messages to detect communications security discrepancies and violations; reviews communications, electronic signal instructions, communications electronic operating instructions, and operations plans and orders to determine adequacy of communications security; reads and interprets maps and map overlays; prepares basic reports on signal security activities; operates and performs user maintenance on communications security and monitoring equipment; selects sites for, erects, and orients tactical antennas; selects and uses commercial battery and generator power; performs general office duties such as filing, typing (35 words per minute minimum), preparation of reports, and record keeping; presents oral reports concerning communications security activities; has a working knowledge of personal computers. *Skill Level 20:* Able to perform the duties required for Skill Level 10; serves as team chief, supervising two to five persons; provides basic signal security advice and assistance to commanders; analyzes security codes; establishes and supervises operating sites; writes reports including statistical analyses; presents oral reports on data collected; makes recommendations to eliminate disclosure of intelligence information; provides advice and assistance concerning cryptographic systems and the interpretation of basic documents and regulations; determines need for written communications security instructions; prepares special, periodic, and project reports on signal security activities. *Skill Level 30:* Able to perform the duties required for Skill Level 20; is a first-line supervisor of 10-20 persons; provides advice and assistance on electronic security procedures and signal security training; prepares signal security monitoring cover plan; analyzes the signal output of communications and noncommunications equipment to establish methods of operations, cryptonetting, and other types of intelligence; reviews documents for signal security; plans and supervises electronic security support; schedules signal security operational activities. *Skill Level 40:* Able to perform the duties required for Skill Level 30; supervises 15 or more persons; determines needed signal intelligence support; presents oral reports to commanders on operational security principles; plans signal security actions to support the operational security plan; serves as advisor to commanders on signal security matters; plans and conducts classes on signal security activities. *Skill Level 50:* Able to perform the duties required for Skill Level 40; serves as signal security operations chief, first sergeant of a company, or signal security chief; supervises 50-300 persons; has experience as the enlisted commander of a detachment; develops and writes signal security doctrine; advises commander and manages signal security personnel; plans career progression training; allocates personnel; advises commander and manages signal security equipment resources; plans and carries out signal security missions; advises generals and high-level government officials on signal security activities.

Recommendation, Skill Level 10

Credit may be granted on the basis of an individualized assessment of the student (3/94).

Recommendation, Skill Level 20

In the lower-division baccalaureate/associate degree category, 3 semester hours in computer literacy and 3 in personnel supervision. (NOTE: This recommendation for skill level 20 is valid for the dates 10/91-2/95 only) (9/91).

Recommendation, Skill Level 30

In the lower-division baccalaureate/associate degree category, 3 semester hours in computer literacy, 3 in personnel supervision, and 2 in records and information management (9/91).

Recommendation, Skill Level 40

In the lower-division baccalaureate/associate degree category, 3 semester hours in computer literacy, 3 in personnel supervision, and 2 in records and information management. In the upper-division baccalaureate category, 3 semester hours for field experience in management and 3 in organizational management (9/91).

Recommendation, Skill Level 50

In the lower-division baccalaureate/associate degree category, 3 semester hours in computer literacy, 3 in personnel supervision, and 2 in records and information management. In the upper-division baccalaureate category, 6 semester hours for field experience in management and 3 in organizational management; if individual has attained paygrade E-9, additional credit as follows: 3 semester hours in management problems, 3 in operations management, and 3 in communication techniques for managers (9/91).

MOS-97L-001

TRANSLATOR/INTERPRETER
97L10
97L20
97L30
97L40
97L50

Exhibit Dates: Pending Evaluation. 8/94–Present.

MOS-97Z-001

COUNTERINTELLIGENCE/HUMAN INTELLIGENCE SENIOR SERGEANT
97Z50

Exhibit Dates: 1/90–12/99.

Career Management Field: 96 (Military Intelligence).

Description

Able to perform the duties required for 97B, Counterintelligence Agent; 97E, Interrogator; or 97G, Counter Signal Intelligence Specialist; supervises and guides all counterintelligence activities; plans mission assignments; participates in long and short range planning for organizational operations including contingency operations and needs; responsible for coordinating and managing career and professional development of senior enlisted personnel in unit; responsible for supervising the maintenance records pertaining to materials, resources, personnel actions, and reports; oversees employment, substance abuse, and counseling; supervises daily schedules, ensuring that the unit is efficient and ready; responsible for development and maintenance of all unit standard operating procedures. NOTE: May have progressed to MOS 97Z50 from 97B40 (Counterintelligence Agent), 97E40 (Interrogator), or 97G40 (Signal Security Specialist).

Recommendation

In the upper-division baccalaureate category, 3 semester hours in management problems, 3 in human resource management, and 3 in organizational planning (6/90).

MOS-97Z-002

COUNTERINTELLIGENCE/HUMAN INTELLIGENCE SENIOR SERGEANT
97Z50

Exhibit Dates: 1/00–Present.

Career Management Field: 96 (Military Intelligence).

Description

Able to perform the duties required for 97B, Counterintelligence Agent; 97E, Interrogator; or 97L, Translator/Interpreter; supervises and guides all counterintelligence activities; plans mission assignments; participates in long- and short-range planning for organizational operations, including contingency operations and needs; responsible for coordinating and managing career and professional development of senior enlisted personnel in unit; responsible for supervising the maintenance records pertaining to materials, resources, personnel actions, and reports; oversees employment, substance abuse, and counseling; supervises daily schedules, ensuring that the unit is efficient and ready; responsible for development and maintenance of all unit standard operating procedures. NOTE: May have progressed to MOS 97Z50 from 97B50 (Counterintelligence Agent), 97E50 (Interrogator), or 97L50 (Translator/Interpreter-Reserve components).

Recommendation

In the upper-division baccalaureate category, 6 semester hours for a management practicum, 3 in personnel management, 3 in principles of management, and 3 in organizational planning (1/00).

MOS-98C-003

SIGNALS INTELLIGENCE ANALYST
98C10
98C20
98C30
98C40

Exhibit Dates: 1/90–2/95.

Career Management Field: 98 (Electronic Warfare/Cryptologic).

Description

Summary: Supervises, analyzes, and reports intercepted foreign communications at a mobile or fixed site. NOTE: Many of the duties required for this MOS involve highly classified materials, equipment, and activities; therefore, not all the competencies and knowledge associated with the MOS were evaluated. *Skill Level 10:* Gathers, sorts, and scans intercepted messages and signals and performs initial analysis to establish communications patterns; isolates valid message traffic; reduces communications data into automatic data processing format; operates communications equipment for reporting and coordination; types at a minimum rate of 25 words per minute; has knowledge of the geography and culture of the area from which intercepted communications originate; may acquire a technical vocabulary in one or more foreign language. *Skill Level 20:* Able to perform the duties required for Skill Level 10; may supervise five to eight persons; analyzes foreign communications including encrypted material; uses data processing techniques to analyze communications; presents written and oral technical and intelligence reports. *Skill Level 30:* Able to perform the duties required for Skill Level 20; supervises traffic analysis and provides guidance on the interpretation of collected information; devises methods for solving complex analytical problems; maintains analytical files; compiles, writes, edits, evaluates, and disseminates intelligence reports; analyzes automatic data processing results and confers with computer programmers and analysts; coordinates activities and schedules work; implements emergency action plans. *Skill Level 40:* Able to perform the duties required for Skill Level 30; supervises traffic and signal intelligence analysis activities and coordinates collection processing, analysis, and reports; supervises 18-40 persons; may serve as an operations sergeant, supervising up to 75 persons; may have experience as the enlisted commander of a detachment; uses counseling techniques to alleviate stress among subordinates.

Recommendation, Skill Level 10

In the vocational certificate category, 3 semester hours in electronic systems operations. In the lower-division baccalaureate/associate degree category, 3 semester hours in written communication, 3 in keyboarding, 1 in computer literacy, 2 in geography, and credit for foreign language proficiency on the basis of institutional evaluation. (NOTE: This recom-

mendation for skill level 10 is valid for the dates 1/90-9/91 only) (9/88).

Recommendation, Skill Level 20

In the vocational certificate category, 3 semester hours in electronic systems operations. In the lower-division baccalaureate/associate degree category, 6 semester hours in written communication, 3 in geography, 2 in oral communication, 2 in office practices, 3 in keyboarding, 2 in computer literacy, and credit in foreign language proficiency on the basis of institutional evaluation (9/88).

Recommendation, Skill Level 30

In the vocational certificate category, 3 semester hours in electronic systems operations. In the lower-division baccalaureate/associate degree category, 6 semester hours in written communication, 3 in geography, 3 in oral communication, 3 in office practices, 3 in keyboarding, 3 in social science and humanities, 3 in human relations, 3 in computer literacy, and credit in foreign language proficiency on the basis of institutional evaluation (9/88).

Recommendation, Skill Level 40

In the vocational certificate category, 3 semester hours in electronic systems operations. In the lower-division baccalaureate/associate degree category, 6 semester hours in written communication, 3 in geography, 3 in oral communication, 3 in office practices, 3 in keyboarding, 3 in social science and humanities, 3 in human relations, 3 for a counseling practicum, 3 in computer literacy, 3 for field experience in personnel supervision, and credit in foreign language proficiency on the basis of institutional evaluation. In the upper-division baccalaureate category, 3 semester hours for field experience in management (9/88).

MOS-98C-004

Signals Intelligence Analyst

98C10
98C20
98C30
98C40

Exhibit Dates: 3/95–1/98.

Career Management Field: 98 (Electronic Warfare/Cryptologic).

Description

Summary: Supervises, analyzes, and reports intercepted foreign communications at a mobile or fixed site. NOTE: Many of the duties required for this MOS involve highly classified materials, equipment, and activities; therefore, not all the competencies and knowledge associated with the MOS were evaluated. *Skill Level 10:* Gathers, sorts, and scans intercepted messages and signals and performs initial analysis to establish communications patterns; isolates valid message traffic; reduces communications data into automatic data processing format; operates communications equipment for reporting and coordination; types at a minimum rate of 25 words per minute; has knowledge of the geography and culture of the area from which intercepted communications originate; may acquire a technical vocabulary in one or more foreign language. *Skill Level 20:* Able to perform the duties required for Skill Level 10; may supervise five to eight persons; analyzes foreign communications including encrypted material; uses data processing techniques to analyze communications; presents written and oral technical and intelligence reports. *Skill Level 30:* Able to perform the duties required for Skill Level 20; supervises traffic analysis and provides guidance on the interpretation of collected information; devises methods for solving complex analytical problems; maintains analytical files; compiles, writes, edits, evaluates, and disseminates intelligence reports; analyzes automatic data processing results and confers with computer programmers and analysts; coordinates activities and schedules work; implements emergency action plans. *Skill Level 40:* Able to perform the duties required for Skill Level 30; supervises traffic and signal intelligence analysis activities and coordinates collection processing, analysis, and reports; supervises 18-40 persons; may serve as an operations sergeant, supervising up to 75 persons; may have experience as the enlisted commander of a detachment; uses counseling techniques to alleviate stress among subordinates.

Recommendation, Skill Level 10

Credit may be granted on the basis of an individualized assessment of the student (9/88).

Recommendation, Skill Level 20

Credit may be granted on the basis of an individualized assessment of the student (9/88).

Recommendation, Skill Level 30

In the vocational certificate category, 3 semester hours in electronic systems operations. In the lower-division baccalaureate/associate degree category, 6 semester hours in written communication, 3 in geography, 2 in oral communication, 2 in office practices, 3 in keyboarding, 3 in human relations, 2 in computer literacy, and credit in foreign language proficiency on the basis of institutional evaluation (9/88).

Recommendation, Skill Level 40

In the vocational certificate category, 3 semester hours in electronic systems operations. In the lower-division baccalaureate/associate degree category, 6 semester hours in written communication, 3 in geography, 3 in oral communication, 3 in office practices, 3 in keyboarding, 3 in social science and humanities, 3 in human relations, 3 for a counseling practicum, 3 in computer literacy, 3 for field experience in personnel supervision, and credit in foreign language proficiency on the basis of institutional evaluation. In the upper-division baccalaureate category, 3 semester hours for field experience in management (9/88).

MOS-98C-005

Signals Intelligence Analyst

98C10
98C20
98C30
98C40
98C50

Exhibit Dates: 2/98–Present.

Career Management Field: 98 (Signals Intelligence/Electronic Warfare Operations).

Description

Summary: Supervises, analyzes, and reports intercepted foreign communications at a mobile or fixed site; performs collection management; produces combat, strategic, and tactical intelligence reports. NOTE: Many of the duties required for this MOS involve highly classified materials, equipment, and activities; therefore, not all the competencies and knowledge associated with the MOS were evaluated. *Skill Level 10:* Gathers, sorts, and scans intercepted messages and signals and performs initial analysis to establish communication patterns; isolates valid message traffic; uses various informational data bases; operates communications equipment for reporting and coordination; types at a minimum rate of 25 words per minute; has knowledge of the geography and culture of the area from which intercepted communications originate; may acquire a technical vocabulary in one or more foreign language. *Skill Level 20:* Able to perform the duties required for Skill Level 10; may supervise five to eight persons; analyzes foreign communications including encrypted material; uses informational data bases to analyze communication; presents written and oral technical and intelligence reports. *Skill Level 30:* Able to perform the duties required for Skill Level 20; supervises traffic analysis and provides guidance on the interpretation of collected information; devises methods for solving complex analytical problems; maintains analytical files; compiles, writes, edits, evaluates, and disseminates intelligence reports; analyzes informational data base results and confers with computer programmers and analysts; coordinates activities and schedules work; implements emergency action plans. *Skill Level 40:* Able to perform the duties required for Skill Level 30; supervises traffic and signal intelligence analysis activities and coordinates collection processing, analysis, and reports; supervises 18-40 persons; may serve as an operations sergeant, supervising up to 75 persons; may have experience as the enlisted commander of a detachment; uses counseling techniques to alleviate stress among subordinates. *Skill Level 50:* Able to perform the duties required for Skill Level 40; provides guidance and supervision to subordinates in operation and command intelligence procedures; coordinates tactical and strategic analysis, reporting, and processing operation; provides appraisals of intelligence products, activities, and operations; develops unit intelligence training strategy.

Recommendation, Skill Level 10

Credit may be granted on the basis of individual assessment of the student (1/00).

Recommendation, Skill Level 20

Credit may be granted on the basis of individual assessment of the student (1/00).

Recommendation, Skill Level 30

In the lower-division baccalaureate/associate degree category, 3 semester hours in computer software applications, 3 in technical report writing, 3 in principles of supervision, 3 in speech communication, and 3 in applied mathematics (1/00).

Recommendation, Skill Level 40

In the lower-division baccalaureate/associate degree category, 3 semester hours in computer software applications, 3 in technical report writing, 3 in principles of supervision, 3 in speech communication, 3 in applied mathematics, and 3 in office administration. In the upper-division baccalaureate category, 3 semester hours for a practicum in management (1/00).

Recommendation, Skill Level 50

In the lower-division baccalaureate/associate degree category, 3 semester hours in computer software applications, 3 in technical report writing, 3 in principles of supervision, 3 in speech communication, 3 in applied mathematics, and

3 in office administration. In the upper-division baccalaureate category, 3 semester hours for a practicum in management, 3 in human resource management, 3 in professional writing, and 3 in organizational development (1/00).

MOS-98D-001

Emitter Locator/Identifier

98D10

98D20

98D30

98D40

Exhibit Dates: 3/90–2/95.

Career Management Field: 98 (Signals Intelligence (SIGINT)/Electronic Warfare (EW) Operations).

Description

Summary: Operates or establishes and supervises the operation of radio direction-finding systems and other systems using advanced identification and automated computer analysis techniques; intercepts and acquires bearings on target transmitters and performs analysis on maps or charts to establish probable locations of target transmitters; relays direction-finding information to other stations in the signals identification and electronic warfare network. *Skill Level 10:* Employs special transmitters; forwards bearings and identification information to a control center; selects, erects, and orients tactical antennas; obtains desired visual display on specialized monitor oscilloscopes; records electrical characteristics of signals displayed using light-sensitive recorders; operates direction-finding and related cryptological, communications, and automatic data processing equipment; prepares and maintains operation logs and card files; types at a minimum rate of 25 words per minute with no errors; copies international Morse code at a minimum rate of 20 groups per minute. *Skill Level 20:* Able to perform the duties required for Skill Level 10; employs surveying techniques for antenna site selection; maintains section management files; classifies, analyzes, and evaluates observed bearings and waveform oscillograms; establishes, plots, and evaluates bearings to determine probable geographical location of foreign transmitters; performs measurement of bands on oscillograms to determine ripple frequency, modulation percentages, and duration of other effects; maintains calibration and accuracy studies; conducts computer-assisted analysis of signals; relays direction-finding information to stations in network; monitors quality control of input and data for automatic data processing support; may serve as a first-line supervisor, assigning work loads and completing personnel evaluations; presents oral and written reports. *Skill Level 30:* Able to perform the duties required for Skill Level 20; inspects equipment to ensure proper alignment and orientation; provides guidance and assistance in site selection and equipment installation; prepares and supervises the preparation of written and oral reports; establishes and maintains facilities and support for site personnel; implements emergency action plans. *Skill Level 40:* Able to perform the duties required for Skill Level 30; may serve as detachment commander for about 20 persons; provides guidance to subordinates; allocates personnel resources; conducts emitter locator/identifier briefings.

Recommendation, Skill Level 10

In the lower-division baccalaureate/associate degree category, 2 semester hours in keyboarding and 2 in computer systems and organization. (NOTE: This recommendation for skill level 10 is valid for the dates 3/90-9/91 only) (10/93).

Recommendation, Skill Level 20

In the lower-division baccalaureate/associate degree category, 2 semester hours in keyboarding, 2 in computer systems and organization, 3 in report writing, and 2 in surveying techniques (10/93).

Recommendation, Skill Level 30

In the lower-division baccalaureate/associate degree category, 2 semester hours in keyboarding, 3 in computer systems and organization, 3 in report writing, 2 in surveying techniques, and 3 in personnel supervision (10/93).

Recommendation, Skill Level 40

In the lower-division baccalaureate/associate degree category, 2 semester hours in keyboarding, 3 in computer systems and organization, 3 in report writing, 2 in surveying techniques, and 3 in personnel supervision. In the upper-division baccalaureate category, 3 semester hours for field experience in management and 3 in organizational management (10/93).

MOS-98D-002

Emitter Locator/Identifier

98D10

98D20

98D30

98D40

Exhibit Dates: 3/95–6/98. Effective 6/98, MOS 98D was discontinued and its duties were incorporated into MOS 98H (Morse Interceptor.

Career Management Field: 98 (Signals Intelligence (SIGINT)/Electronic Warfare (EW) Operations).

Description

Summary: Operates or establishes and supervises the operation of radio direction-finding systems and other systems using advanced identification and automated computer analysis techniques; intercepts and acquires bearings on target transmitters and performs analysis on maps or charts to establish probable locations of target transmitters; relays direction-finding information to other stations in the signals identification and electronic warfare network. *Skill Level 10:* Employs special transmitters; forwards bearings and identification information to a control center; selects, erects, and orients tactical antennas; obtains desired visual display on specialized monitor oscilloscopes; records electrical characteristics of signals displayed using light-sensitive recorders; operates direction-finding and related cryptological, communications, and automatic data processing equipment; prepares and maintains operation logs and card files; types at a minimum rate of 25 words per minute with no errors; copies international Morse code at a minimum rate of 20 groups per minute. *Skill Level 20:* Able to perform the duties required for Skill Level 10; employs surveying techniques for antenna site selection; maintains section management files; classifies, analyzes, and evaluates observed bearings and waveform oscillograms; establishes, plots, and evaluates bearings to determine probable geographical location of foreign transmitters; performs measurement of bands on oscillograms to determine ripple frequency, modulation percentages, and duration of other effects; maintains calibration and accuracy studies; conducts computer-assisted analysis of signals; relays direction-finding information to stations in network; monitors quality control of input and data for automatic data processing support; may serve as a first-line supervisor, assigning work loads and completing personnel evaluations; presents oral and written reports. *Skill Level 30:* Able to perform the duties required for Skill Level 20; inspects equipment to ensure proper alignment and orientation; provides guidance and assistance in site selection and equipment installation; prepares and supervises the preparation of written and oral reports; establishes and maintains facilities and support for site personnel; implements emergency action plans. *Skill Level 40:* Able to perform the duties required for Skill Level 30; may serve as detachment commander for about 20 persons; provides guidance to subordinates; allocates personnel resources; conducts emitter locator/identifier briefings.

Recommendation, Skill Level 10

Credit may be granted on the basis of an individualized assessment of the student (10/93).

Recommendation, Skill Level 20

Credit may be granted on the basis of an individualized assessment of the student (10/93).

Recommendation, Skill Level 30

In the lower-division baccalaureate/associate degree category, 2 semester hours in keyboarding, 2 in computer systems and organization, 3 in report writing, 2 in surveying techniques, and 3 in personnel supervision (10/93).

Recommendation, Skill Level 40

In the lower-division baccalaureate/associate degree category, 2 semester hours in keyboarding, 3 in computer systems and organization, 3 in report writing, 2 in surveying techniques, and 3 in personnel supervision. In the upper-division baccalaureate category, 3 semester hours for field experience in management and 3 in organizational management (10/93).

MOS-98G-003

Electronic Warfare (EW)/Signal Intelligence Voice Interceptor

98G10

98G20

98G30

98G40

98G50

Exhibit Dates: 1/90–9/90.

Career Management Field: 98 (Electronic Warfare/Cryptologic).

Description

Summary: Conducts and supervises the interception, transcription, translation, and reporting of foreign voice transmissions in a mobile or fixed station environment. NOTE: Many of the duties required for this MOS involve highly classified materials, equipment, and activities; therefore, not all the competencies and knowledge associated with the MOS were evaluated. *Skill Level 10:* Operates intercept receivers including radiotelephone and multichannel systems and recording equipment; selects, erects, and orients tactical antennas; makes written records of foreign voice transmissions which are composed of limited terminology and simple syntactic structures; identifies languages spoken in the geographic area to which assigned; categorizes foreign voice signals by type of activity; prepares voice activity records; makes verbatim translation from foreign language to English; scans written foreign language materials for key words and indicators; provides translation assistance to non-language-qualified analysts; extracts specific intelligence information from voice radio transmissions; researches and develops special project reports; presents oral reports; types in English and in language of proficiency; performs operation maintenance on equipment. *Skill Level 20:* Able to perform the duties required for Skill Level 10; intercepts, identifies, and records foreign voice transmissions; supervises a small unit; assembles, integrates, analyzes, and disseminates intelligence information covering political, economic, sociological, historical, and psychological factors of a geographic area; prepares special studies and reports as required; must have the following minimum capabilities in at least one foreign language: vocabulary (aural recognition) of 6,000-8,000 words; 750-1,000 technical term items; knowledge of complex grammar and syntax; 85-90 percent of all existing word functions; 85-90 percent of all kinship terms; total comprehension of writing systems, except in languages with ideographic systems, where the ability to write 500 and read 1,500 characters is required. *Skill Level 30:* Able to perform the duties required for Skill Level 20; supervises voice communication intercept activities; operates sophisticated equipment designed to collect and simultaneously produce on-line activity records of complex foreign voice radio transmissions containing technical terminology, advanced grammar/syntax, and colloquial conversational forms; directs voice signal collection and processing priorities; identifies and performs limited analysis on nonclear voice and nonvoice signals; writes complex reports; makes oral presentations to general staff; must achieve a fluent and accurate proficiency in one or more foreign languages and have the following minimum foreign language capabilities: vocabulary (aural recognition) of 8,000-10,000 words; 1,000-1,500 technical term items; advanced knowledge of grammar and syntax; 90-100 percent of all existing kinship terms and forms of address; must be able to write 700 and read 2,000 characters in an ideographic writing system. *Skill Level 40:* Able to perform the duties required for Skill Level 30; supervises voice communication countermeasure activities; refines essential elements of information needed to support assigned mission; performs voice intercept and processing of highly complex foreign voice transmissions; prepares papers for use at high military and government levels; must achieve the following minimum capabilities in one or more foreign languages; vocabulary (aural recognition) of 10,000-15,000 words; 1,500-3,000 technical terms; total knowledge of word functions; total knowledge of kinship terms and forms of address; must be able to write 1,000 and read 2,500-3,000 characters in an ideographic writing system; uses counseling techniques to alleviate stress among subordinates. *Skill Level 50:* Able to perform the duties required for Skill Level 40; serves as electronic warfare/signal intelligence voice operations chief; evaluates and refines job requirements and system capabilities for communication intelligence linguists; must attain a foreign language proficiency equivalent to that of a well-educated native speaker and have the following minimum capabilities: vocabulary (aural recognition) of 15,000-plus words; 3,000-plus technical terms; advanced knowledge of grammar and syntax; must be able to write 1,500 and read 2,500-3,000 characters in ideographic writing systems.

Recommendation, Skill Level 10

In the vocational certificate category, 3 semester hours in electronic systems operations. In the lower-division baccalaureate/associate degree category, 3 semester hours in keyboarding, 1 in computer literacy, and 3 in library research techniques. In the upper-division baccalaureate category, 3 semester hours in advanced written communications. NOTE: Add credit for the specific foreign language(s), using the recommendation in the DoD volume of the Guide to the Evaluation of Educational Experiences in the Armed Services as a minimum and awarding additional credit for foreign language proficiency on the basis of institutional evaluation (9/88).

Recommendation, Skill Level 20

In the vocational certificate category, 3 semester hours in electronic systems operations. In the lower-division baccalaureate/associate degree category, 3 semester hours in keyboarding, 2 in computer literacy, 3 in library research techniques, 3 in records administration, and 3 in geography. In the upper-division baccalaureate category, 6 semester hours in advanced written communications. NOTE: Add credit for the specific foreign language(s), using the recommendation in the DoD volume of the Guide to the Evaluation of Educational Experiences in the Armed Services as a minimum and awarding additional credit for foreign language proficiency on the basis of institutional evaluation (9/88).

Recommendation, Skill Level 30

In the vocational certificate category, 3 semester hours in electronic systems operations. In the lower-division baccalaureate/associate degree category, 3 semester hours in keyboarding, 3 in computer literacy, 3 in library research techniques, 3 in records administration, 3 in geography, 3 in social science and humanities, 3 in human relations, and 2 in office practices. In the upper-division baccalaureate category, 6 semester hours in advanced written communications. NOTE: Add credit for the specific foreign language(s), using the recommendation in the DoD volume of the Guide to the Evaluation of Educational Experiences in the Armed Services as a minimum and awarding additional credit for foreign language proficiency on the basis of institutional evaluation (9/88).

Recommendation, Skill Level 40

In the vocational certificate category, 3 semester hours in electronic systems operations. In the lower-division baccalaureate/associate degree category, 3 semester hours in keyboarding, 3 in computer literacy, 3 in library research techniques, 3 in records administration, 3 in geography, 6 in social science and humanities, 3 in human relations, 2 in office practices, 3 for a counseling practicum, and 3 for field experience in personnel supervision. In the upper-division baccalaureate category, 6 semester hours in advanced written communications and 3 for field experience in management. NOTE: Add credit for the specific foreign language(s), using the recommendation in the DoD volume of the Guide to the Evaluation of Educational Experiences in the Armed Services as a minimum and awarding additional credit for foreign language proficiency on the basis of institutional evaluation (9/88).

Recommendation, Skill Level 50

In the vocational certificate category, 3 semester hours in electronic systems operations. In the lower-division baccalaureate/associate degree category, 3 semester hours in keyboarding, 3 in computer literacy, 3 in library research techniques, 3 in records administration, 3 in geography, 6 in social science and humanities, 3 in human relations, 2 in office practices, 3 for a counseling practicum, and 3 for field experience in personnel supervision. In the upper-division baccalaureate category, 6 semester hours in advanced written communications, 6 for field experience in management, 3 in principles of management, and 3 in personnel management. NOTE: Add credit for the specific foreign language(s), using the recommendation in the DoD volume of the Guide to the Evaluation of Educational Experiences in the Armed Services as a minimum and awarding additional credit for foreign language proficiency on the basis of institutional evaluation (9/88).

MOS-98G-004

VOICE INTERCEPTOR

98G10
98G20
98G30
98G40

Exhibit Dates: 10/90–2/95.

Career Management Field: 98 (Signals Intelligence/Electronic Warfare Operations).

Description

Summary: Conducts and supervises the interception, transcription, translation, and reporting of foreign voice transmissions in a mobile or fixed station environment. NOTE: Many of the duties required for this MOS involve highly classified materials, equipment, and activities; therefore, not all the competencies and knowledge associated with the MOS were evaluated. *Skill Level 10:* Operates intercept receivers including radio telephone and multichannel systems and recording equipment; selects, erects, and orients tactical antennas; makes written records of foreign voice transmissions which are composed of limited terminology and simple syntactic structures; identifies languages spoken in the geographic area to which assigned; categorizes foreign voice signals by type of activity; prepares voice activity records; makes verbatim translation from foreign language to English; scans written foreign language materials for keywords and indicators; provides translation assistance to non-language-qualified analysts; extracts specific intelligence information from voice radio transmissions; researches and develops special project reports; presents oral reports; types in English and in language of proficiency; performs operator maintenance on equipment; has a working knowledge of personal computers. *Skill Level 20:* Able to perform the duties required for Skill Level 10; supervises and provides technical guidance to subordinates; intercepts, identifies, and records foreign voice transmissions; supervises a small unit; assembles, integrates, analyzes, and disseminates intelligence information covering political, economic, sociological, historical, and psychological factors of a geographical area; prepares special studies and reports as required; must have the following minimum capabilities in at least one foreign language: vocabulary (aural recognition) of 6,000-8,000 words; 750-1,000 technical term items; knowledge of complex grammar and syntax; 85-90 percent of all kinship terms; total comprehension of functions; 85-90 percent of all kinship terms; total comprehension of writing systems, except in languages with ideographic systems where the ability to write 500 and read 1,500 characters is required. *Skill Level 30:* Able to perform the duties required for Skill Level 20; supervises voice communication intercept activities; evaluates subordinates; operates sophisticated equipment designed to collect and simultaneously produce on-line activity records of complex foreign voice radio transmissions containing technical terminology, advanced grammar/syntax, and colloquial conversational forms; directs voice signal collection and processing priorities; identifies and performs limited analysis on nonclear voice and nonvoice signals; writes complex reports; makes oral presentations to general staff; must achieve a fluent and accurate proficiency in one or more foreign language and have the following minimum foreign language capabilities: vocabulary (aural recognition) of 8,000-10,000 words; 1,000-1,500 technical term items; advanced knowledge of grammar and syntax; 90-100 percent of all existing kinship terms and forms of address; must be able to write 700 and read 2,000 characters in an ideographic writing system. *Skill Level 40:* Able to perform the duties required for Skill Level 30; supervises voice communication countermeasure activities; produces personnel evaluation reports; refines essential elements of information needed to support assigned mission; performs voice intercept and processing of highly complex foreign voice transmissions; prepares papers for use at high military and government levels; must achieve the following minimum capabilities in one or more foreign languages; vocabulary (aural recognition) of 10,000-15,000 words; 1,500-3,000 technical terms; total knowledge of word functions; total knowledge of kinship terms and forms of address; must be able to write 1,000 and read 2,500-3,000 characters in an ideographic writing system; uses counseling techniques to alleviate stress among subordinates.

Recommendation, Skill Level 10

In the lower-division baccalaureate/associate degree category, 3 semester hours in library research, 3 in computer literacy, and 3 in technical report writing. (NOTE: This recommendation for skill level 10 is valid for the dates 9/90-9/91 only). NOTE: Add credit for specific foreign language(s) using the recommendation in the DoD volume of the Guide to the Evaluation of Educational Experiences in the Armed Services as a minimum and awarding additional credit for foreign language proficiency on the basis of institutional evaluation (9/91).

Recommendation, Skill Level 20

In the lower-division baccalaureate/associate degree category, 3 semester hours in library research, 3 in computer literacy, 3 in technical report writing, 3 in cultural geography, and 3 in personnel supervision. NOTE: Add credit for the specific foreign language(s) using the recommendation in the DoD volume of the Guide to the Evaluation of Educational Experience in the Armed Services as a minimum and awarding additional credit for foreign language proficiency on the basis of institutional evaluation (9/91).

Recommendation, Skill Level 30

In the lower-division baccalaureate/associate degree category, 3 semester hours in library research, 3 in computer literacy, 3 in technical report writing, 3 in cultural geography, 3 in personnel supervision, 2 in records and information management, 3 in office practices, and 3 in social sciences. NOTE: Add credit for the specific foreign language(s) using the recommendation in the DoD volume of the Guide to the Evaluation of Educational Experience in the Armed Services as a minimum and awarding additional credit for foreign language proficiency on the basis of institutional evaluation (9/91).

Recommendation, Skill Level 40

In the lower-division baccalaureate/associate degree category, 3 semester hours in library research, 3 in computer literacy, 3 in technical report writing, 3 in cultural geography, 3 in personnel supervision, 2 in records and information management, 3 in office practices, and 3 in social sciences. In the upper-division baccalaureate category, 3 semester hours for field experience in management and 3 in organizational management. NOTE: Add credit for the specific foreign language(s) using the recommendation in the DoD volume of the Guide to the Evaluation of Educational Experiences in the Armed Services as a minimum and awarding additional credit for foreign language proficiency on the basis of institutional evaluation (9/91).

MOS-98G-005

Voice Interceptor
98G10
98G20
98G30
98G40

Exhibit Dates: 3/95–1/98.

Career Management Field: 98 (Signals Intelligence/Electronic Warfare Operations).

Description

Summary: Conducts and supervises the interception, transcription, translation, and reporting of foreign voice transmissions in a mobile or fixed station environment. NOTE: Many of the duties required for this MOS involve highly classified materials, equipment, and activities; therefore, not all the competencies and knowledge associated with the MOS were evaluated. *Skill Level 10:* Operates intercept receivers including radio telephone and multichannel systems and recording equipment; selects, erects, and orients tactical antennas; makes written records of foreign voice transmissions which are composed of limited terminology and simple syntactic structures; identifies languages spoken in the geographic area to which assigned; categorizes foreign voice signals by type of activity; prepares voice activity records; makes verbatim translation from foreign language to English; scans written foreign language materials for keywords and indicators; provides translation assistance to non-language-qualified analysts; extracts specific intelligence information from voice radio transmissions; researches and develops special project reports; presents oral reports; types in English and in language of proficiency; performs operator maintenance on equipment; has a working knowledge of personal computers. *Skill Level 20:* Able to perform the duties required for Skill Level 10; supervises and provides technical guidance to subordinates; intercepts, identifies, and records foreign voice transmissions; supervises a small unit; assembles, integrates, analyzes, and disseminates intelligence information covering political, economic, sociological, historical, and psychological factors of a geographical area; prepares special studies and reports as required; must have the following minimum capabilities in at least one foreign language: vocabulary (aural recognition) of 6,000-8,000 words; 750-1,000 technical term items; knowledge of complex grammar and syntax; 85-90 percent of all kinship terms; total comprehension of functions; 85-90 percent of all kinship terms; total comprehension of writing systems, except in languages with ideographic systems where the ability to write 500 and read 1,500 characters is required. *Skill Level 30:* Able to perform the duties required for Skill Level 20; supervises voice communication intercept activities; evaluates subordinates; operates sophisticated equipment designed to collect and simultaneously produce on-line activity records of complex foreign voice radio transmissions containing technical terminology, advanced grammar/syntax, and colloquial conversational forms; directs voice signal collection and processing priorities; identifies and performs limited analysis on nonclear voice and nonvoice signals; writes complex reports; makes oral presentations to general staff; must achieve a fluent and accurate proficiency in one or more foreign language and have the following minimum foreign language capabilities: vocabulary (aural recognition) of 8,000-10,000 words; 1,000-1,500 technical term items; advanced knowledge of grammar and syntax; 90-100 percent of all existing kinship terms and forms of address; must be able to write 700 and read 2,000 characters in an ideographic writing system. *Skill Level 40:* Able to perform the duties required for Skill Level 30; supervises voice communication countermeasure activities; produces personnel evaluation reports; refines essential elements of information needed to support assigned mission; performs voice intercept and processing of highly complex foreign voice transmissions; prepares papers for use at high military and government levels; must achieve the following minimum capabilities in one or more foreign languages; vocabulary (aural recognition) of 10,000-15,000 words; 1,500-3,000 technical terms; total knowledge of word functions; total knowledge of kinship terms and forms of address; must be able to write 1,000 and read 2,500-3,000 characters in an ideographic writing system; uses counseling techniques to alleviate stress among subordinates.

Recommendation, Skill Level 10

Credit may be granted on the basis of an individualized assessment of the student (9/91).

Recommendation, Skill Level 20

Credit may be granted on the basis of an individualized assessment of the student (9/91).

Recommendation, Skill Level 30

In the lower-division baccalaureate/associate degree category, 3 semester hours in library research, 3 in computer literacy, 3 in technical report writing, 3 in cultural geography, 3 in personnel supervision, 2 in records and information management, and 3 in office practices. NOTE: Add credit for the specific foreign language(s) using the recommendation in the DoD volume of the Guide to the Evaluation of Educational Experience in the Armed Services as a minimum and awarding additional credit for foreign language proficiency on the basis of institutional evaluation (9/91).

Recommendation, Skill Level 40

In the lower-division baccalaureate/associate degree category, 3 semester hours in library research, 3 in computer literacy, 3 in technical report writing, 3 in cultural geography, 3 in personnel supervision, 2 in records and information management, 3 in office practices, and 3 in social sciences. In the upper-division baccalaureate category, 3 semester hours for field experience in management and 3 in organizational management. NOTE: Add credit for the specific foreign language(s) using the recommendation in the DoD volume of the *Guide to the Evaluation of Educational Experiences in the Armed Services* as a minimum and awarding additional credit for foreign language proficiency on the basis of institutional evaluation (9/91).

MOS-98G-006

Voice Interceptor

98G10
98G20
98G30
98G40
98G50

Exhibit Dates: 2/98–Present.

Career Management Field: 98 (Signals Intelligence/Electronic Warfare Operations).

Description

Summary: Conducts and supervises the interception, transcription, translation, and reporting of foreign voice transmissions in a mobile or fixed station environment. NOTE: Many of the duties required for this MOS involve highly classified materials, equipment, and activities; therefore, not all the competencies and knowledge associated with the MOS were evaluated. *Skill Level 10:* Operates intercept receivers, including radio telephone and multichannel systems and recording equipment; selects, erects, and orients tactical antennas; makes written records of foreign voice transmissions which are composed of limited terminology and simple syntactic structures; identifies languages spoken in the geographic area to which assigned; categorizes foreign voice signals by type of activity; prepares voice activity records; makes verbatim translation from foreign language to English; scans written foreign language materials for keywords and indicators; provides translation assistance to non-language-qualified analysts; extracts specific intelligence information from voice radio transmissions; researches and develops special project reports; presents oral reports; types in English and in language of proficiency; performs operator maintenance on equipment; has a working knowledge of personal computers and software applications. *Skill Level 20:* Able to perform the duties required for Skill Level 10; supervises and provides technical guidance to subordinates; intercepts, identifies, and records foreign voice transmissions; supervises a small unit; assembles, integrates, analyzes, and disseminates intelligence information covering political, economic, sociological, historical, and psychological factors of a geographical area; prepares special studies and reports as required; must have the following minimum capabilities in at least one foreign language: vocabulary (aural recognition) of 6,000-8,000 words; 750-1,000 technical term items; knowledge of complex grammar and syntax; 85-90 percent of all kinship terms; total comprehension of functions; total comprehension of writing systems, except in languages with ideographic systems where the ability to write 500 and read 1,500 characters is required. *Skill Level 30:* Able to perform the duties required for Skill Level 20; supervises voice communication intercept activities; evaluates subordinates; operates sophisticated equipment designed to collect and simultaneously produce on-line activity records of complex foreign voice radio transmissions containing technical terminology, advanced grammar/syntax, and colloquial conversational forms; directs voice signal collection and processing priorities; identifies and performs limited analysis on nonclear voice and nonvoice signals; writes complex reports; makes oral presentations to general staff; must achieve a fluent and accurate proficiency in one or more foreign language and have the following minimum foreign language capabilities: vocabulary (aural recognition) of 8,000-10,000 words; 1,000-1,500 technical term items; advanced knowledge of grammar and syntax; 90-100 percent of all existing kinship terms and forms of address; must be able to write 700 and read 2,000 characters in an ideographic writing system. *Skill Level 40:* Able to perform the duties required for Skill Level 30; supervises voice communication countermeasure activities; produces personnel evaluation reports; refines essential elements of information needed to support assigned mission; performs voice intercept and processing of highly complex foreign voice transmissions; prepares papers for use at high military and government levels; must achieve the following minimum capabilities in one or more foreign languages; vocabulary (aural recognition) of 10,000-15,000 words; 1,500-3,000 technical terms; total knowledge of word functions; total knowledge of kinship terms and forms of address; must be able to write 1,000 and read 2,500-3,000 characters in an ideographic writing system; uses counseling techniques to alleviate stress among subordinates. *Skill Level 50:* Able to perform the duties required for Skill Level 40; provides guidance to subordinates; supervises immediate subordinate in operations and command intelligence procedures; coordinates operations; provides appraisals of intelligence products, activities and operations; develops and plans training.

Recommendation, Skill Level 10

Credit may be granted on the basis of an individualized assessment of the student (1/00).

Recommendation, Skill Level 20

Credit may be granted on the basis of an individualized assessment of the student (1/00).

Recommendation, Skill Level 30

In the lower-division baccalaureate/associate degree category, 3 semester hours in technical report writing, 3 in speech communication, 3 in regional geography, 3 in computer software applications, and 3 in principles of supervision (1/00).

Recommendation, Skill Level 40

In the lower-division baccalaureate/associate degree category, 3 semester hours in technical report writing, 3 in speech communication, 3 in regional geography, 3 in computer software applications, 3 in principles of supervision, and 3 in office administration. In the upper-division baccalaureate category, 3 semester hours in organizational development and 3 in professional writing (1/00).

Recommendation, Skill Level 50

In the lower-division baccalaureate/associate degree category, 3 semester hours in technical report writing, 3 in speech communication, 3 in regional geography, 3 in computer software applications, 3 in principles of supervision, and 3 in office administration. In the upper-division baccalaureate category, 3 semester hours in organizational development, 3 in human resource management, 3 for a practicum in management, and 3 in professional writing (1/00).

MOS-98H-001

Morse Interceptor

98H10
98H20
98H30
98H40

Exhibit Dates: 3/90–1/98.

Career Management Field: 98 (Signals Intelligence/Electronic Warfare Operations).

Description

Summary: Operates international Morse code message interception equipment, keyboard entry devices, and printer equipment; supervises the operation of such equipment in mobile or fixed installations for the purpose of detecting, identifying, and exploiting foreign communications. NOTE: Many of the duties required for this MOS involve highly classified materials, equipment, and activities; therefore, not all the competencies and knowledge associated with the MOS were evaluated. *Skill Level 10:* Operates Morse code interception equipment, including radio receivers, special typewriters, teletypewriters, computer input keyboards, antenna selection devices, internal communications equipment, and magnetic tape recorders; searches for, identifies, and manually records foreign international Morse code communications at a minimum rate of 20 groups per minute; performs first-level analysis of message to detect anomalies and suspect items which may be of intelligence interest; maintains operator's log of messages and related data and delivers messages to analysts for interpretation; performs operator maintenance on equipment; types at a minimum speed of 25 words per minute. *Skill Level 20:* Able to perform the duties required for Skill Level 10; supervises and provides technical guidance to subordinates; performs more detailed message analysis and evaluation prior to forwarding messages to other analysts; writes detailed reports regarding intercepted messages; conducts on-the-job training; presents oral reports to high-level command staff. *Skill Level 30:* Able to perform the duties required for Skill Level 20; supervises Morse intercept activities; establishes and maintains extensive intercept files for messages and related data; evaluates subordinates; assists in formulating unit deployment plans. *Skill Level 40:* Able to perform the duties required for Skill Level 30; allocates personnel and equipment resources; assists in designing collection strategies; may have experience as the enlisted commander of a detachment; analyzes automatic data processing results and confers with computer programmers and analysts; uses counseling techniques to alleviate stress among subordinates; produces evaluation reports of subordinates.

Recommendation, Skill Level 10

In the lower-division baccalaureate/associate degree category, 3 semester hours in computer literacy and 3 in international Morse code. (NOTE: This recommendation for skill level 10 is valid for the dates 3/90-9/91 only) (9/91).

Recommendation, Skill Level 20

In the lower-division baccalaureate/associate degree category, 3 semester hours in computer literacy, 3 in international Morse code, and 3 in personnel supervision. (NOTE: This recommendation for skill level 20 is valid for the dates 3/90-2/95 only) (9/91).

Recommendation, Skill Level 30

In the lower-division baccalaureate/associate degree category, 3 semester hours in computer literacy, 3 in international Morse code, 3 in personnel supervision, 3 in records and information management, and 3 in technical report writing (9/91).

Recommendation, Skill Level 40

In the lower-division baccalaureate/associate degree category, 3 semester hours in computer literacy, 3 in international Morse code, 3 in personnel supervision, 3 in records and information management, and 3 in technical report writing. In the upper-division baccalaureate category, 3 semester hours for field experience in management and 3 in organizational management (9/91).

MOS-98H-002

COMMUNICATIONS INTERCEPTOR/LOCATOR
((Communications Locator/Interceptor))
98H10
98H20
98H30
98H40
98H50

Exhibit Dates: 2/98–Present.

Career Management Field: 98 (Signals Intelligence/Electronic Warfare Operations).

Description

Summary: Operates international Morse code message interception equipment, keyboard entry devices, and printer equipment; supervises the operation of such equipment in mobile or fixed installations for the purpose of detecting, identifying, and exploiting foreign communications. NOTE: Many of the duties required for this MOS involve highly classified materials, equipment, and activities; therefore, not all the competencies and knowledge associated with the MOS were evaluated. *Skill Level 10:* Operates Morse code interception equipment, including radio receivers, special typewriters, teletypewriters, computer input keyboards, antenna selection devices, internal communications equipment, and magnetic tape recorders; searches for, identifies, and manually records foreign international Morse code communications at a minimum rate of 20 groups per minute; performs first-level analysis of message to detect anomalies and suspect items which may be of intelligence interest; maintains operator's log of messages and related data and delivers messages to analysts for interpretation; performs operator maintenance on equipment; types at a minimum speed of 25 words per minute. *Skill Level 20:* Able to perform the duties required for Skill Level 10; supervises and provides technical guidance to subordinates; performs more detailed message analysis and evaluation prior to forwarding messages to other analysts; writes detailed reports regarding intercepted messages; conducts on-the-job training; presents oral reports to high-level command staff. *Skill Level 30:* Able to perform the duties required for Skill Level 20; supervises Morse intercept activities; establishes and maintains extensive intercept files for messages and related data; evaluates subordinates; assists in formulating unit deployment plans. *Skill Level 40:* Able to perform the duties required for Skill Level 30; allocates personnel and equipment resources; assists in designing collection strategies; may have experience as the enlisted commander of a detachment; analyzes automatic data processing results and confers with computer programmers and analysts; uses counseling techniques to alleviate stress among subordinates; produces evaluation reports of subordinates. *Skill Level 50:* Able to perform the duties required for Skill Level 40; provides staff planning for contingency operations; determines personnel assignments; prepares individual and unit training schedules; uses operations management and human resource management techniques; coordinates tactical and strategic signal intelligence activities; provides appraisals of intelligence products and operations.

Recommendation, Skill Level 10

Credit may be granted on the basis of individual assessment of the student (1/00).

Recommendation, Skill Level 20

Credit may be granted on the basis of individual assessment of the student (1/00).

Recommendation, Skill Level 30

In the lower-division baccalaureate/associate degree category, 3 semester hours in computer software applications, 3 in applied mathematics, 3 in principles of supervision, 3 in technical writing, and 3 in speech communication (1/00).

Recommendation, Skill Level 40

In the lower-division baccalaureate/associate degree category, 3 semester hours in computer software applications, 3 in applied mathematics, 3 in principles of supervision, 3 in technical writing, 3 in information systems, and 3 in speech communication. In the upper-division baccalaureate category, 3 semester hours in human resource management and 3 in organizational development (1/00).

Recommendation, Skill Level 50

In the lower-division baccalaureate/associate degree category, 3 semester hours in computer software applications, 3 in applied mathematics, 3 in principles of supervision, 3 in technical writing, 3 in information systems, and 3 in speech communication. In the upper-division baccalaureate category, 3 semester hours in organizational development, 3 for a practicum in management, 3 in human resource management, and 3 in professional writing (1/00).

MOS-98J-003

NONCOMMUNICATIONS INTERCEPTOR/
ANALYST
98J10
98J20
98J30
98J40

Exhibit Dates: 1/90–2/95.

Career Management Field: 98 (Electronic Warfare/Cryptologic).

Description

Summary: Operates noncommunications intercept recording and analysis equipment; intercepts, identifies, interprets, and analyzes signals from noncommunications sources; operates electro-optical receiving and analysis equipment. NOTE: Many of the duties required for this MOS involve highly classified materials, equipment, and activities; therefore, not all the competencies and knowledge associated with the MOS were evaluated. *Skill Level 10:* Erects noncommunications collection systems; conducts search for selected categories or classes of noncommunications or electro-optical signals; uses technical references and equipment to record and perform preliminary analysis and identification of intercepted signals of interest; determines line bearings or intercepted signals within specified limits; reports acquired information; posts entries and keeps logs; operates communications equipment for intelligence reporting and coordination. *Skill Level 20:* Able to perform the duties required for Skill Level 10; searches for general categories or classes of noncommunications or electro-optical signals; exercises quality control over intercept effort; uses indexed data base and technical references to fix origin of intercept and analysis reports, maps, and overlays; assists with fusion product reporting; establishes and maintains files. *Skill Level 30:* Able to perform the duties required for Skill Level 20; serves as a first-line supervisor of an intercept/analysis (noncommunications signals) activity; assigns search missions; intercepts and analyzes more complex noncommunications or electro-optical signals; prepares technical and administrative reports; makes recommendations on the employment of noncommunications intercept units; selects and establishes operations sites; interprets and analyzes technical documents and reports; performs fusion analysis and intelligence reporting. *Skill Level 40:* Able to perform the duties required for Skill Level 30; supervises noncommunications intercept, countermeasures, and analysis activities of 20 or more persons; based on requirements, determines objectives and priorities; assigns work loads and allocates resources among subordinate supervisors; coordinates activities with other units; evaluates requirements and assists in developing plans to meet future operation, personnel, and equipment needs; uses counseling techniques to alleviate stress among subordinates.

Recommendation, Skill Level 10

In the vocational certificate category, 3 semester hours in electronic systems operations. In the lower-division baccalaureate/associate degree category, 2 semester hours in computer literacy and 3 in keyboarding. (NOTE: This recommendation for skill level 10 is valid for the dates 1/90-9/91 only) (9/88).

Recommendation, Skill Level 20

In the vocational certificate category, 3 semester hours in electronic systems operations. In the lower-division baccalaureate/associate degree category, 3 semester hours in computer literacy, 3 in keyboarding, 3 in written communications, and 2 in office practices (9/88).

Recommendation, Skill Level 30

In the vocational certificate category, 3 semester hours in electronic systems operations. In the lower-division baccalaureate/associate degree category, 3 semester hours in computer literacy, 3 in keyboarding, 3 in written communications, 3 in office practices, 3 in human relations, 3 in oral communications, and 3 in social science and humanities (9/88).

Recommendation, Skill Level 40

In the vocational certificate category, 3 semester hours in electronic systems operations. In the lower-division baccalaureate/associate degree category, 3 semester hours in computer literacy, 3 in keyboarding, 3 in written communications, 3 in office practices, 3 in human relations, 3 for a counseling practicum, 3 in oral communications, 3 in social science and humanities, and 3 for field experience in personnel supervision. In the upper-division baccalaureate category, 3 semester hours for field experience in management (9/88).

MOS-98J-004

Noncommunications Interceptor/
Analyst
98J10
98J20
98J30
98J40

Exhibit Dates: 3/95–1/98.

Career Management Field: 98 (Electronic Warfare/Cryptologic).

Description

Summary: Operates noncommunications intercept recording and analysis equipment; intercepts, identifies, interprets, and analyzes signals from noncommunications sources; operates electro-optical receiving and analysis equipment. NOTE: Many of the duties required for this MOS involve highly classified materials, equipment, and activities; therefore, not all the competencies and knowledge associated with the MOS were evaluated. *Skill Level 10:* Erects noncommunications collection systems; conducts search for selected categories or classes of noncommunications or electro-optical signals; uses technical references and equipment to record and perform preliminary analysis and identification of intercepted signals of interest; determines line bearings or intercepted signals within specified limits; reports acquired information; posts entries and keeps logs; operates communications equipment for intelligence reporting and coordination. *Skill Level 20:* Able to perform the duties required for Skill Level 10; searches for general categories or classes of noncommunications or electro-optical signals; exercises quality control over intercept effort; uses indexed data base and technical references to fix origin of intercept and analysis reports, maps, and overlays; assists with fusion product reporting; establishes and maintains files. *Skill Level 30:* Able to perform the duties required for Skill Level 20; serves as a first-line supervisor of an intercept/analysis (noncommunications signals) activity; assigns search missions; intercepts and analyzes more complex noncommunications or electro-optical signals; prepares technical and administrative reports; makes recommendations on the employment of noncommunications intercept units; selects and establishes operations sites; interprets and analyzes technical documents and reports; performs fusion analysis and intelligence reporting. *Skill Level 40:* Able to perform the duties required for Skill Level 30; supervises noncommunications intercept, countermeasures, and analysis activities of 20 or more persons; based on requirements, determines objectives and priorities; assigns work loads and allocates resources among subordinate supervisors; coordinates activities with other units; evaluates requirements and assists in developing plans to meet future operation, personnel, and equipment needs; uses counseling techniques to alleviate stress among subordinates.

Recommendation, Skill Level 10

Credit may be granted on the basis of an individualized assessment of the student (9/88).

Recommendation, Skill Level 20

Credit may be granted on the basis of an individualized assessment of the student (9/88).

Recommendation, Skill Level 30

In the vocational certificate category, 3 semester hours in electronic systems operations. In the lower-division baccalaureate/associate degree category, 3 semester hours in computer literacy, 3 in keyboarding, 3 in written communications, 2 in office practices, 3 in human relations, and 3 in oral communications (9/88).

Recommendation, Skill Level 40

In the vocational certificate category, 3 semester hours in electronic systems operations. In the lower-division baccalaureate/associate degree category, 3 semester hours in computer literacy, 3 in keyboarding, 3 in written communications, 3 in office practices, 3 in human relations, 3 for a counseling practicum, 3 in oral communications, 3 in social science and humanities, and 3 for field experience in personnel supervision. In the upper-division baccalaureate category, 3 semester hours for field experience in management (9/88).

MOS-98J-005

Electronic Intelligence Interceptor/
Analyst
98J10
98J20
98J30
98J40
98J50

Exhibit Dates: 2/98–Present.

Career Management Field: 98 (Electronic Warfare/Cryptologic).

Description

Summary: Operates noncommunications intercept recording and analysis equipment; intercepts, identifies, interprets, and analyzes signals from noncommunications sources; operates electro-optical receiving and analysis equipment. NOTE: Many of the duties required for this MOS involve highly classified materials, equipment, and activities; therefore, not all the competencies and knowledge associated with the MOS were evaluated. *Skill Level 10:* Erects noncommunications collection systems; conducts search for selected categories or classes of noncommunications or electro-optical signals; uses technical references and equipment to record and perform preliminary analysis and identification of intercepted signals of interest; determines line bearings or intercepted signals within specified limits; reports acquired information; posts entries and keeps logs; operates communications equipment for intelligence reporting and coordination. *Skill Level 20:* Able to perform the duties required for Skill Level 10; searches for general categories or classes of noncommunications or electro-optical signals; exercises quality control over intercept effort; uses indexed data base and technical references to fix origin of intercept and analysis reports, maps, and overlays; assists with fusion product reporting; establishes and maintains files. *Skill Level 30:* Able to perform the duties required for Skill Level 20; serves as a first-line supervisor of an intercept/analysis (noncommunications signals) activity; assigns search missions; intercepts and analyzes more complex noncommunications or electro-optical signals; prepares technical and administrative reports; makes recommendations on the employment of noncommunications intercept units; selects and establishes operations sites; interprets and analyzes technical documents and reports; performs fusion analysis and intelligence reporting. *Skill Level 40:* Able to perform the duties required for Skill Level 30; supervises noncommunications intercept, countermeasures, and analysis activities of 20 or more persons; based on requirements, determines objectives and priorities; assigns work loads and allocates resources among subordinate supervisors; coordinates activities with other units; evaluates requirements and assists in developing plans to meet future operation, personnel, and equipment needs; uses counseling techniques to alleviate stress among subordinates. *Skill Level 50:* Able to perform the duties required for Skill Level 40; provides staff planning for contingency operations, determines personnel assignments; prepares individual and unit training schedules; uses operations management and human resource management techniques; provides appraisals of intelligence products, activities, and operations.

Recommendation, Skill Level 10

Credit may be granted on the basis of an individualized assessment of the student (1/00).

Recommendation, Skill Level 20

Credit may be granted on the basis of an individualized assessment of the student (1/00).

Recommendation, Skill Level 30

In the lower-division baccalaureate/associate degree category, 3 semester hours in computer software applications, 3 in applied mathematics, 3 in principles of supervision, 3 in technical writing, and 3 in speech communication (1/00).

Recommendation, Skill Level 40

In the lower-division baccalaureate/associate degree category, 3 semester hours in computer software applications, 3 in applied mathematics, 3 in principles of supervision, 3 in technical writing, 3 in information systems, and 3 in speech communication. In the upper-division baccalaureate category, 3 semester hours in organizational development and 3 in professional writing (1/00).

Recommendation, Skill Level 50

In the lower-division baccalaureate/associate degree category, 3 semester hours in computer software applications, 3 in applied mathematics, 3 in principles of supervision, 3 in technical writing, 3 in information systems, and 3 in speech communication. In the upper-division baccalaureate category, 3 semester hours in organizational development, 3 for a practicum in management, 3 in professional writing, and 3 in human resource management (1/00).

MOS-98K-001

Non-Morse Interceptor/Analyst
98K10
98K20
98K30
98K40

Exhibit Dates: 3/90–1/98.

Career Management Field: 98 (Signals Intelligence/Electronic Warfare Operations).

Description

Summary: Operates non-Morse communications intercept and recording equipment and supervises the operation of such equipment in mobile or fixed environments for the purpose of identifying and recording foreign radioteletype, facsimile, and data communications transmissions. NOTE: Many of the duties for this MOS involve highly classified material, equipment, and activities; therefore, not all the competencies and knowledge associated with the MOS were evaluated. *Skill Level 10:* Operates radioteletype, facsimile, data intercept, and recording equipment; knows basic AC and DC theory, circuit electronic theory, basic frequency analysis, spectrum analysis, and functional algebra; searches for, identifies, and records foreign transmissions; maintains a log of interceptions; prepares technical reports; types at a minimum rate of 25 words per minute; has a working knowledge of personal computers; performs operator maintenance on non-Morse intercept equipment; selects, erects, and orients tactical antennas. *Skill Level 20:* Able to perform the duties required for Skill Level 10; assists in the establishment of operational sites; maintains the technical data base to support collection operations; employs special electronic equipment for complex signal analysis; analyzes intercepted communications for items of intelligence interest; prepares detailed reports; supervises and provides technical guidance to subordinates. *Skill Level 30:* Able to perform the duties required for Skill Level 20; supervises non-Morse intercept activities; evaluates subordinates; allocates equipment and personnel resources; writes extensive reports to provide intelligence information; analyzes long-term trends using statistical analysis techniques; coordinates interaction with other collection and processing activities; conducts on-the-job training. *Skill Level 40:* Able to perform the duties required for Skill Level 30; supervises non-Morse intercept activities; produces evaluation reports for subordinate personnel; interprets signal intelligence collection priorities; ensures proper handling of intelligence information; assesses procedures and operations for adequacy in meeting intelligence requirements and recommends changes; may have experience as the enlisted commander of detachment; uses counseling techniques to alleviate stress among subordinates.

Recommendation, Skill Level 10

In the lower-division baccalaureate/associate degree category, 3 semester hours in computer literacy and 3 in office practices. (NOTE: This recommendation for skill level 10 is valid for the dates 3/90-9/91 only) (9/91).

Recommendation, Skill Level 20

In the lower-division baccalaureate/associate degree category, 3 semester hours in computer literacy, 3 in office practices, 3 in technical report writing, and 3 in personnel supervision. (NOTE: This recommendation for skill level 20 is valid for the dates 3/90-2/95 only) (9/91).

Recommendation, Skill Level 30

In the lower-division baccalaureate/associate degree category, 3 semester hours in computer literacy, 3 in office practices, 3 in technical report writing, 3 in personnel supervision, 3 in social sciences, and 2 in records and information management (9/91).

Recommendation, Skill Level 40

In the lower-division baccalaureate/associate degree category, 3 semester hours in computer literacy, 3 in office practices, 3 in technical report writing, 3 in personnel supervision, 3 in social sciences, and 2 in records and information management. In the upper-division baccalaureate category, 3 semester hours for field experience in management and 3 in organizational management (9/91).

MOS-98K-002

Signals Collection/Identification Analyst
98K10
98K20
98K30
98K40
98K50

Exhibit Dates: Pending Evaluation. 2/98–Present.

MOS-98P-001

Multi-Sensor Operator
98P10
98P20
98P30
98P40

Exhibit Dates: Pending Evaluation. 4/03–Present.

MOS-98Y-001

Signals Collector/Analyst
98Y10
98Y20
98Y30
98Y40

Exhibit Dates: Pending Evaluation. 5/04–Present.

MOS-98Z-002

Signals Intelligence/Electronic Warfare Chief
98Z50

Exhibit Dates: 1/90–12/99.

Career Management Field: 98 (Electronic Warfare./Cryptologic).

Description

NOTE: Many of the duties required for this MOS involve highly classified materials, equipment, and activities; therefore, not all the competencies and knowledge associated with this MOS were evaluated.

Serves as the principal authority on signal intelligence matters in a high-level military organization; directs signal intelligence activities and provides technical, operational, and administrative guidance to subordinates; interprets all signal intelligence requirements and develops signal intelligence tasks; evaluates military posture of other nations and determines the capabilities and limitations of US signal intelligence activities to meet the determined threats; reviews, evaluates, prepares, and implements plans and orders; assists high-level commanders in appraising the effectiveness of signal intelligence functions, publications, and procedures and recommends changes; advises commanders on matters of personal welfare, morale, assignment, utilization, promotion, privileges, discipline, and training; provides written and oral evaluations of subordinate personnel; prepares and presents written and oral reports to high-level personnel in the Department of the Army and other national agencies involved in the preparation of national security and intelligence policy; manages equipment, material, and support service resources. NOTE: May have progressed to 98Z50 from 98C40 (EW/Signal Intelligence Analyst), 98G50 (EW/ Signal Intelligence Voice Interceptor), 98J40 (EW/Signal Intelligence Noncommunications Interceptor), 05H40 (EW/Signal Intelligence Morse Interceptor), 05K40 (EW/Signal Intelligence Non-Morse Interceptor); may have acquired proficiency in one or more foreign languages through the performance of a previously held MOS.

Recommendation

In the upper-division baccalaureate category, 6 semester hours for field experience in management, 3 in office management, 3 in personnel management, and 3 in principles of management. NOTE: Add credit for the specific foreign language(s) by referring to the recommendation in the DoD volume of the Guide to the Evaluation of Educational Experiences in the Armed Services as a minimum and awarding additional credit for foreign language proficiency on the basis of institutional evaluation (9/88).

MOS-98Z-003

Signals Intelligence Senior Sergeant
((Signals Intelligence/Electronic Warfare Chief))
98Z50

Exhibit Dates: 1/00–Present.

Career Management Field: 98 (Electronic Warfare./Cryptologic).

Description

NOTE: Many of the duties required for this MOS involve highly classified materials, equipment, and activities; therefore, not all the competencies and knowledge associated with this MOS were evaluated.

Serves as the principal authority on signal intelligence matters in a high-level military organization; directs signal intelligence activities and provides technical, operational, and administrative guidance to subordinates; interprets all signal intelligence requirements and develops signal intelligence tasks; evaluates military posture of other nations and determines the capabilities and limitations of US signal intelligence activities to meet the determined threats; reviews, evaluates, prepares, and implements plans and orders; assists high-level commanders in appraising the effectiveness of signal intelligence functions, publications, and procedures and recommends changes; advises commanders on matters of personnel welfare, morale, assignment, utilization, promotion, privileges, discipline, and training; provides written and oral evaluations of subordinate personnel; prepares and presents written and oral reports to high-level personnel in the Department of the Army and other national agencies involved in the preparation of national security and intelligence policy; manages equipment, material, and support service resources. NOTE: May have progressed to 98Z50 from 98C50 (Signals Intelligence Analyst), 98G50 (Voice Interceptor), 98H50 (Communications Locator/Interceptor), 98J50 (Electronic Intelligence Interceptor/Analyst), or 98K50 (Signals Collection/Identification Analyst); may have acquired proficiency in one or more foreign language through the performance of a previously-held MOS.

Recommendation

In the upper-division baccalaureate category, 6 semester hours for a management practicum, 3 in organizational planning, 3 in personnel management, and 3 in principles of management. NOTE: Add credit for the specific foreign language(s) (1/00).

Army Warrant Occupation Exhibits

MOS-130A-001

Pershing Missile System Technician
130A0

Exhibit Dates: 1/90–5/92. MOS 130A has been discontinued.

Career Pattern

May have progressed to Pershing Missile System Technician from MOS 15E (Pershing Missile Crewmember), MOS 21G (Pershing Electronics Materiel Specialist), MOS 21L (Pershing Electronics Repairer), or MOS 46N (Pershing Electrical-Mechanical Repairer).

Description

Supervises the maintenance of field artillery missiles, fire control systems, and missile launching and handling equipment; oversees assembly of missiles; supervises utilization, testing, and maintenance of tools, specialized test sets, cable sets, and warhead assemblies; inspects and directs testing, servicing, and repair of missile launching and guidance equipment; supervises maintenance of missile components, determines repair/replacement requirements, ensures that repairs and adjustments are made, and makes final inspections; advises superiors on technical and tactical considerations affecting the employment of missiles; serves as a technical advisor on matters of preventive maintenance and inspection of radar, missiles, fire control systems, missile launching and handling equipment, and on-missile materiel; observes and corrects improper technical procedures and repair techniques; coordinates maintenance techniques and standards with preventive maintenance personnel to ensure uniformity; implements changes in inspection, repair, and test procedures and instructs personnel on new techniques of missile maintenance and repair; ensures that equipment modifications are accomplished; examines and interprets procedures, directives, schematics, and technical publications for data pertinent to the employment of radar, missiles, and related equipment and transmits data to subordinates; reviews operating and maintenance records and tests equipment to ascertain adequacy of maintenance; monitors storage of missile guidance and motor sections; ensures compliance with safety and security regulations and procedures; writes and reviews technical reports; reviews and consolidates requests for tools, spare parts, technical supplies, publications, and equipment; may serve as an instructor at a formal training school or be employed in missile procurement activities.

Recommendation

In the vocational certificate category, 1 semester hour in care and use of hand tools. In the lower-division baccalaureate/associate degree category, 3 semester hours in electronics laboratory, 3 in personnel supervision, 3 in technical report writing, 2 in maintenance management, 2 in industrial/human relations, and additional credit in instructional methods on the basis of institutional evaluation. In the upper-division baccalaureate category, 3 semester hours for field experience in management; if rank was CW2, 2 additional semester hours for field experience in management; if rank was CW3, 4 additional semester hours for field experience in management; if rank was CW4, 6 additional semester hours for field experience in management (3/77).

MOS-130B-001

Lance Missile System Technician
130B0

Exhibit Dates: 1/90–6/93. Effective 6/93, MOS 130B was discontinued.

Career Pattern

May have progressed to Lance Missile System Technician from MOS 13N (Lance Crewmember).

Description

Supervises the maintenance of field artillery missiles, fire control systems, and missile launching and handling equipment; oversees assembly of missiles; supervises utilization, testing, and maintenance of tools, specialized test sets, cable sets, and warhead assemblies; inspects and directs testing, servicing, and repair of missile launching and guidance equipment; supervises maintenance of missile components; determines repair/replacement requirements; ensures that repairs and adjustments are made and makes final inspections; advises superiors on technical and tactical considerations affecting the employment of missiles; serves as a technical advisor on matters of preventive maintenance and inspection of radar, missiles, fire control systems, missile launching and handling equipment, and on-missile materiel; observes and corrects improper technical procedures and repair techniques; coordinates technical maintenance procedures and repair techniques; coordinates maintenance techniques and standards with preventive maintenance personnel to ensure uniformity; implements changes in inspection, repair, and test procedures and instructs personnel on new techniques of missile maintenance and repair; ensures that equipment modifications are accomplished; examines and interprets procedures, directives, schematics, and technical publications for data pertinent to the employment of radar, missiles, and related equipment and transmits data to subordinates; reviews operating and maintenance records and tests equipment to ascertain adequacy of maintenance; monitors storage of missile guidance and motor sections; ensures compliance with safety and security regulations and procedures; writes and reviews technical reports; reviews and consolidates requests for tools, spare parts, technical supplies, publications, and equipment; may serve as an instructor at a formal training school or be employed in missile procurement activities.

Recommendation

In the vocational certificate category, 1 semester hour in care and use of hand tools. In the lower-division baccalaureate/associate degree category, 3 semester hours in electronics laboratory, 3 in personnel supervision, 3 for field experience in management, 3 in technical report writing, 2 in maintenance management, 2 in industrial/human relations and additional credit in instructional methods on the basis of institutional evaluation (3/77).

MOS-131A-001

Target Acquisition Radar Technician
131A0

Exhibit Dates: 1/90–8/94.

Career Pattern

May have progressed to Target Acquisition Radar Technician from any MOS in Career Management Field 13 (except MOS's 13B and 13M) and Career Management Field 29.

Description

Normally serves as chief of a section or platoon engaged in maintaining and operating field artillery radars to provide target location; has a detailed knowledge of the operational aspects of field artillery radar, the technical principles of equipment construction, the scope and techniques of field artillery radar, and safety precautions relevant to operations and maintenance; advises on technical considerations involving field artillery radar; coordinates the activities of field artillery radar personnel; supervises the movement and emplacement of radar and associated equipment; interprets technical data; implements changes in inspection, repair, and test procedures; instructs personnel on new or revised techniques of radar maintenance and employment; inspects completed work and assists personnel in isolating and correcting malfunctions in equipment; must be able to transmit target and counterfire data to support artillery units; reviews maintenance records; supervises the requisitioning of tools, repair parts, technical supplies, publications, and equipment; may serve as an instructor at a formal training school, on a headquarters staff, or in research, development, test, and evaluation activities.

Recommendation

In the vocational certificate category, 2 semester hours in use of basic hand tools. In the lower-division baccalaureate/associate degree category, 6 semester hours in electronics theory and laboratory, 3 in personnel supervision, 3 for field experience in management, 2 in industrial/human relations, 2 in technical report writing, and 2 in maintenance management (3/77).

MOS-131A-002

Target Acquisition Radar Technician
131A0

Exhibit Dates: 9/94–Present.

Career Pattern

May have progressed to Target Acquisition Radar Technician from any MOS in Career Management Field 13 (except MOS's 13B and 13M) and Career Management Field 29.

Description

Normally serves as chief of a section or platoon engaged in maintaining and operating field artillery radars to provide target location; may supervise 6-11 subordinates; has a detailed knowledge of the operational aspects of field artillery Doppler radar, the technical principles of equipment construction, the scope of field artillery radar, and safety precautions relevant to operations and maintenance; advises on technical considerations involving field artillery radar; coordinates the activities of field artillery radar personnel; supervises the movement and emplacement of radar and associated equipment; interprets technical data; implements changes in inspection, repair, and test procedures; may prepare written revisions to technical manuals or revisions to Modification Work Orders when equipment is modified/updated; writes reports of evaluations and inspections of equipment problems; instructs personnel on new or revised techniques of radar electronics maintenance and employment; inspects completed work and assists personnel in isolating and correcting malfunctions in equipment; must be able to transmit target and counterfire data to support artillery units; reviews maintenance records; supervises the requisitioning of tools, repair parts, technical supplies, publications, and equipment; may serve as an instructor/writer at a formal service school; prepares, reviews, and presents instruction on radar systems; provides coordination between military and industry during the developing, testing, and fielding of new target acquisition systems; updates doctrine for target acquisition radars; may serve on a headquarters staff, or in research, development, test, and evaluation activities.

Recommendation

In the lower-division baccalaureate/associate degree category, 3 semester hours in maintenance management, 3 in technical writing, 3 in records and information management, 3 in computer and data processing management, and 3 in personnel supervision. In the upper-division baccalaureate category, 3 semester hours in personnel management, 3 for field experience in management, and 3 in an education elective (instruction). If rank was CW2, 2 additional semester hours for field experience in management; if rank was CW3, 4 additional semester hours for field experience in management; if rank was CW4, 6 additional semester hours for field experience in management (9/94).

MOS-131B-001

Remotely Piloted Vehicle Technician
131B0

Exhibit Dates: 1/90–4/92. NOTE: This MOS was deleted from the MOS system before an evaluation could be scheduled. Credit may be granted on the basis of institutional evaluation

Career Pattern

Description

MOS-132A-001

Meteorology Technician
132A0

Exhibit Dates: 1/90–12/93. Effective 12/93, MOS 132A was discontinued.

Career Pattern

May have progressed to Meteorology Technician from MOS 93F (Field Artillery Meteorological Crewman).

Description

Normally serves as the chief of a meteorology section that provides ballistic meteorological data, not involving weather forecasting, to artillery units; is responsible for the organization and operation of the section and its physical facilities, as well as the supervision and training of enlisted specialists in the section; manages and allocates resources to install, operate, and maintain mercurial, aneroid, electronic, optical, and mechanical meteorological instruments and equipment; determines layout of ballistic meteorological stations and facilities and plans equipment installation; knows methods of observing and plotting meteorological data; inspects and tests meteorological equipment; determines operating procedures and techniques; allocates work to subordinates; determines requirements for, procures, and manages supplies, repair parts, and equipment; supervises calibration of graphs and other recording devices; evaluates ballistic meteorological data and produces messages to be transmitted to artillery units; instructs subordinates on established maintenance techniques and operating procedures; may serve as an instructor at a formal service school; may be employed in research, development, and evaluation activities.

Recommendation

In the lower-division baccalaureate/associate degree category, 3 semester hours for field experience in management, 3 in industrial/human relations, 2 in electronics laboratory, 1 in basic mathematics, and 1 in meteorology. In the upper-division baccalaureate category, 3 semester hours in personnel management and 3 in management electives (3/77).

MOS-140A-001

Command and Control Systems Technician
140A0

Exhibit Dates: 1/90–5/91.

Career Pattern

May have progressed to Command and Control Maintenance Technician from MOS 24C (Hawk Firing Section Mechanic), MOS 24E (Hawk Fire Control Mechanic), MOS 24G (Hawk Information Coordination Central Mechanic), MOS 24H (Hawk Fire Control Repairer), MOS 24J (Pulse Radar Repairer), MOS 24K (Hawk Continuous Wave Radar Repairer), MOS 24L (Hawk Launcher and Mechanical Systems Repairer), MOS 24R (Hawk Master Mechanic), MOS 24T (Patriot Operator and Systems Mechanic), MOS 24U (Nike-Hercules Custodial Mechanic), MOS 24V (Hawk Maintenance Chief), or MOS 24L (AN/TSQ 73 Air Defense Artillery Command and Control System Operator/Repairer).

Description

Coordinates the activities of 25-30 maintenance technicians and manages equipment for the installation, repair, maintenance, and modification of Army air defense command and control systems, including data processing equipment, radar equipment, communications equipment, and power system equipment; develops maintenance procedures; performs hands-on troubleshooting when subordinates are unable to isolate problems; advises and instructs repair personnel; estimates repair priorities based on mission, type of work to be performed, and availability of parts and personnel; advises commissioned officers on command and control system capabilities and limitations.

Recommendation

In the lower-division baccalaureate/associate degree category, 3 semester hours in shop management, 3 in maintenance management, 3 in computer and data processing concepts, and 3 in training management. In the upper-division baccalaureate category, 3 semester hours in personnel management, 3 in human resource management, and 6 for field experience in management. Credit in the following areas on the basis of institutional evaluation: computer electronics, computer system repair, radar systems, electronic communications, and power systems (11/88).

MOS-140A-002

Command and Control Systems Technician
140A0

Exhibit Dates: 6/91–Present.

Career Pattern

May have progressed to Command and Control Maintenance Technician from MOS 24C (Hawk Firing Section Mechanic), 24E (Hawk Fire Control Mechanic), 24G (Hawk Information Coordination Central Mechanic), 24H (Hawk Fire Control Repairer), 24J (Hawk Pulse Radar Repairer), 24K (Hawk Continuous Wave Radar Repairer), 24L (Hawk Launcher and Mechanical System Repairer), 24R (Hawk Master Mechanic), 24T (Patriot Operator and Systems Mechanic), 24U (Nike-Hercules Custodial Mechanic), 24V (Hawk Maintenance Chief), or 25L (AN/TSQ 73 Air Defense Artillery Command and Control System Operator/Repairer).

Description

Coordinates the activities of 25-30 maintenance technicians and manages equipment for the installation, repair, maintenance, and modification of Army air defense command and control systems, including data processing equipment, radar equipment, communications equipment, and power system equipment; develops maintenance procedures; performs hands-on troubleshooting when subordinates are unable to isolate problems; advises and instructs repair personnel; estimates repair priorities based on mission, type of work to be performed, and availability of parts and personnel; advises commissioned officers on command and control system capabilities and limitations.

Recommendation

In the lower-division baccalaureate/associate degree category, 3 semester hours in shop management, 3 in maintenance management, 3 in computer systems, and 3 in training management. In the upper-division baccalaureate category, 3 semester hours in personnel management and 3 in human resource management; if rank was CW2, 2 semester hours for field experience in management; if rank was CW3, 4 semester hours for field experience in management; if rank was CW4, 6 semester hours for field experience in management; if served as a Master Warrant Officer, 8 semester hours for field experience in management (6/91).

MOS-140B-001

Chaparral/Vulcan Systems Technician
140B0

Exhibit Dates: 1/90–5/91.

Career Pattern

May have progressed to Chaparral/Vulcan Systems Technician from any of the following MOS's: 24C, 24E, 24G, 24H, 24J, 24K, 24L, 24M, 24N, 24R, 24S, 24T, 24U, 24V, 25L, 27F, or 27G.

Description

Supervises repair and maintenance of the Chaparral and Vulcan light air defense electronic systems and land combat support missile systems, including forward area alert radar and target alerting identification systems; supervises operation of maintenance shops, on-site maintenance activities, and shop quality control activities; knows theory, operation, and functioning of Chaparral/Vulcan systems and associated equipment including test and training equipment and motor support units; has knowledge of electricity, electronics, pneumatics, hydraulics, and mechanics; reads, uses, and interprets schematic diagrams and other technical publications; uses special tools and test equipment to maintain, adjust, troubleshoot, and repair system equipment; organizes, supervises, manages, and evaluates most phases of maintenance programs peculiar to the system and to its support equipment; acts in the capacity of maintenance officer in the absence of a line officer; coordinates activities with other maintenance organizations; supervises 20-50 subordinates; advises and assists in establishing training programs for operators and crewmembers; provides staff supervision for supply functions; supervises inventory and purchasing of maintenance parts, tools, and equipment; applies safety procedures pertaining to the missile systems and associated equipment; writes technical and administrative reports; advises superiors on the technical and tactical capabilities, limitations, and employment of the missile systems; may serve as an instructor at a formal training school or as a technical advisor on a headquarters staff.

Recommendation

In the vocational certificate category, 6 semester hours in electricity/electronics laboratory. In the lower-division baccalaureate/associate degree category, 5 semester hours in automotive mechanics, 5 in diesel mechanics, 5 in industrial truck mechanics, 3 in basic electronics, 3 in troubleshooting, 1 in shop practices and procedures, 1 in industrial safety, 3 in maintenance management, 3 in personnel supervision, 2 in inventory management, 2 in maintenance shop record keeping, 2 in report writing, and credit in public speaking/communication on the basis of institutional evaluation. In the upper-division baccalaureate category, 3 semester hours in personnel management and training and 3 in management electives; if rank was CW2, 2 additional semester hours for field experience in management; if rank was CW3, 4 additional semester hours for field experience in management; if rank was CW4, 6 additional semester hours for field experience in management; if duty assignment was platoon leader, credit in leadership and supervision on the basis of institutional evaluation. If duty assignment was platoon leader, credit in leadership and supervision on the basis of institutional evaluation (5/77).

MOS-140B-002

Chaparral/Vulcan Systems Technician
140B0

Exhibit Dates: 6/91–3/92.

Career Pattern

May have progressed to Chaparral and Vulcan Systems Technician from MOS 24C (Hawk Firing Section Mechanic), 24E (Hawk Fire Control Mechanic), 24G (Hawk Information Coordination Central Mechanic), 24H (Hawk Fire Control Repairer), 24J (Hawk Pulse Radar Repairer), 24K (Hawk Continuous Wave Radar Repairer), 24L (Hawk Launcher and Mechanical Systems Repairer), 24M (Vulcan System Mechanic), 24N (Chaparral System Mechanic), 24R (Hawk Master Mechanic), 24S (Roland System Mechanic), 24T (Patriot Operator and Systems Mechanic), 24U (Nike-Hercules Custodial Mechanic), 24V (Hawk Maintenance Chief), 25L (AN/TSQ-73 Air Defense Artillery Command and Control System Operator/Repairer), 27F (Vulcan Repairer), or 27G (Chaparral/Redeye Repairer).

Description

Supervises repair and maintenance of the Chaparral and Vulcan light air defense electronic systems and land combat support missile systems, including forward area alert radar and target alerting identification systems; supervises operation of maintenance shops, on-site maintenance activities, and shop quality control activities; knows theory, operation, and functioning of Chaparral/Vulcan systems and associated equipment including test and training equipment and motor support units; applies knowledge of electricity, electronics, pneumatics, hydraulics, and mechanics; reads, uses, and interprets schematic diagrams and other technical publications; uses special tools and test equipment to maintain, adjust, troubleshoot, and repair system equipment; organizes, supervises, manages, and evaluates most phases of maintenance programs peculiar to the system and to its support equipment; acts in the capacity of maintenance officer in the absence of a line officer; coordinates activities with the other maintenance organizations; supervises 20-50 subordinates; advises and assists in establishing training programs for operators and crewman; provides staff supervision for supply functions; supervises inventory and purchasing of maintenance parts, tools, and equipment; applies safety procedures pertaining to the missile systems and associated equipment; writes technical and administrative reports; advises superiors on the technical and tactical capabilities, limitations, and employment of the missile systems; may serve as an instructor at a formal training school or as a technical advisor on a headquarters staff.

Recommendation

In the lower-division baccalaureate/associate degree category, 3 semester hours in maintenance management, 3 in inventory management, 3 in personnel supervision, and 3 in report writing. In the upper-division baccalaureate category, 3 semester hours in personnel management and 3 in management electives; if rank was CW2, 2 semester hours for field experience in management; if rank was CW3, 4 semester hours for field experience in management; if rank was CW4, 6 semester hours for field experience in management; if served as Master Warrant Officer (MW), 8 semester hours for field experience in management (6/91).

MOS-140B-003

SHORAD Systems Technician
140B0

Exhibit Dates: 4/92–9/00. Effective 10/00, MOS 140B was discontinued and its duties were transferred to MOS 140A, Command and Control Systems Integrator.

Career Pattern

May have progressed to SHORAD Systems Technician from MOS 23R (Hawk Missile System), 24C (Hawk Firing Section Mechanic), 24G (Hawk Information Coordination Central Mechanic), 24H (Hawk Fire Control Repairer), 24K (Hawk Continuous Wave Radar Repairer), 24M (Vulcan System Mechanic), 24N (Chaparral System Mechanic), 24R (Hawk Master Mechanic), 24T (Patriot Operator and Systems Mechanic), or 25L (AN/TSQ-73 Air Defense Artillery Command and Control System Operator/Repairer).

Description

Supervises repair and maintenance of the SHORAD light air defense electronic systems and land combat support missile systems including forward area alert radar and target alerting identification systems; supervises operation of maintenance shops, on-site maintenance activities, and shop quality control activities; knows theory, operation, and functioning of SHORAD systems and associated equipment including test and training equipment and motor support units; applies knowledge of electricity, electronics, pneumatics, hydraulics, and mechanics; reads, uses and interprets schematic diagrams and other technical publications; uses special tools and test equipment to maintain, adjust, troubleshoot, and repair system equipment; organizes, supervises, manages, and evaluates most phases of maintenance programs peculiar to the system and to its support equipment; acts in the capacity of maintenance officer in the absence of a line officer; coordinates activities with the other maintenance organizations; supervises 20-50 subordinates; advises and assists in establishing training programs for operators and crewman; knows and provides staff supervision for supply functions; supervises inventory and purchasing of maintenance parts, tools, and equipment; knows safety procedures pertaining to the missile systems and associated equipment; writes technical and administrative reports; advises superiors on the technical and tactical capabilities, limitations, and employment of the missile systems; may serve as an instructor at a formal training school or as a technical advisor on a headquarters staff.

Recommendation

In the lower-division baccalaureate/associate degree category, 3 semester hours in maintenance management, 3 in inventory management, 3 in personnel supervision, and 3 in report writing. In the upper-division baccalaureate category, 3 semester hours in personnel management and 3 in management electives; if rank was CW2, 2 semester hours for field experience in management; if rank was CW3, 4 semester hours for field experience in management; if rank was CW4, 6 semester hours for field experience in management; if served as Master Warrant Officer, 8 semester hours for field experience in management (6/91).

MOS-140C-001

CUSTODIAL SYSTEMS TECHNICIAN
140C0

Exhibit Dates: 1/90–8/90. Effective 9/90, MOS 140C was discontinued.

Career Pattern

May have progressed to Custodial Systems Technician from MOS 24U (Nike-Hercules Custodial Mechanic).

Description

Supervises the assembly, testing, and maintenance of all components of Air Defense/Ballistic Missile Defense (AD/BMD) missiles; knows operations, theory, and on-line maintenance of missile testing and servicing equipment associated with warhead sections, fuses, electronic and mechanical on-missile guidance equipment, hydraulic systems, and propulsion units; knows theory and function of hydraulic systems, vacuum tubes, solid state devices, and associated circuits and circuit elements; performs checkouts of electronic, electrical, mechanical, electromechanical, and hydraulic systems; applies authorized procedure for modifying existing equipment; applies AD/BMD missile transport, storage, and safety program regulations and policies for nuclear and explosive components; applies theory, operation, and maintenance procedures for all launch area equipment; oversees uncrating and mating of missiles, rocket motor assemblies, and warhead sections; supervises emplacement of missiles and launch area equipment in mobile air defense units; applies maintenance record keeping and inventory control procedures; trains and evaluates missile maintenance personnel in troubleshooting and maintenance procedures; advises commander on the tactical aspects of missile employment, emplacement, readiness status, and firing.

Recommendation

In the lower-division baccalaureate/associate degree category, 6 semester hours in troubleshooting, 3 in technical mathematics, 3 in basic electronics, 3 in AC/DC circuit analysis, 3 in transistor theory, 3 in maintenance management, 2 in instrumentation, 2 in controls, 2 in pulse circuits, 2 in inventory management, 2 in personnel supervision, 2 in report writing, 1 in maintenance shop record keeping, and additional credit in controls, transistor theory, mathematics, hydraulics, and public speaking/communication on the basis of institutional evaluation. In the upper-division baccalaureate category, 3 semester hours in personnel management and training and 3 in management electives; if rank was CW2, 2 additional semester hours for field experience in management; if rank was CW3, 4 additional semester hours for field experience in management; if rank was CW4, 6 additional semester hours for field experience in management; if duty assignment was platoon leader, credit in leadership and supervision on the basis of institutional evaluation. If duty assignment was platoon leader, credit in leadership and supervision on the basis of institutional evaluation (5/77).

MOS-140D-001

HAWK MISSILE SYSTEMS TECHNICIAN
140D0

Exhibit Dates: 1/90–5/91.

Career Pattern

May have progressed to Hawk Missile Systems Technician from any of the following MOS's: 24C, 24E, 24G, 24H, 24J, 24K, 24L, 24R, 24S, 24T, 24U, 24V, 25L, 27F, or 27G.

Description

Supervises the assembly, inspection, and repair of the improved Hawk missile system, simulator equipment, and associated test equipment; understands theory, functions, and operation of on-missile guidance equipment, ranging radars, pulse radars, launchers, fire control equipment, simulator stations, computers, and automatic data transmission systems used with the improved Hawk air defense missile system and associated power generation and air conditioning equipment; applies theory and functions of solid state devices and associated circuits and circuit elements; assembles and tests components of missile, radar, and fire control equipment; supervises testing of system specialized test sets and procedures; supervises repairers in isolating malfunctions and in repairing, adjusting, and aligning systems and system components; advises personnel on equipment modifications and of changes to inspection, repair, test, and maintenance calibration procedures; applies safety procedures and regulations pertaining to high-energy electromagnetic radiation hazards, high-voltage hazards, and high-pressure hydraulic equipment; applies missile transport and storage methods; knows operation, function, and maintenance of battery terminal equipment; trains personnel in safety and security procedures and maintenance techniques; interprets technical publications; supervises parts inventory and supply system including procurement; reviews and prepares reports; evaluates supply and maintenance problems, advises superiors, and recommends actions to correct deficiencies; may serve as an instructor at a formal training school or in missile procurement activities.

Recommendation

In the lower-division baccalaureate/associate degree category, 6 semester hours in technical mathematics, 6 in electronics troubleshooting, 3 in basic electronics, 3 in AC/DC circuit analysis, 3 in transistor theory, 3 in instrumentation, 2 in pulse circuits, 2 in control systems, 2 in digital circuitry, 2 in computer fundamentals, 1 in radar/microwave theory, 1 in machine language, 3 in maintenance management, 2 in inventory management, 2 in personnel supervision, 2 in report writing, 1 in maintenance shop record keeping, and additional credit in electronics, mathematics, data processing, computer science, and public speaking/communication on the basis of institutional evaluation. In the upper-division baccalaureate category, 3 semester hours in personnel management and training and 3 in management electives; if rank was CW2, 2 additional semester hours for field experience in management; if rank was CW3, 4 additional semester hours for field experience in management; if rank was CW4 6 additional semester hours for field experience in management; if duty assignment was platoon leader, credit in leadership and supervision on the basis of institutional evaluation. If duty assignment was platoon leader, credit in leadership and supervision on the basis of institutional evaluation (5/77).

MOS-140D-002

HAWK MISSILE SYSTEM TECHNICIAN
140D0

Exhibit Dates: 6/91–9/00. Effective 10/00, MOS 140D was discontinued and its duties were transferred to MOS 140E, PATRIOT Systems Technician.

Career Pattern

May have progressed to Hawk Missile Systems Technician from MOS 24C (Hawk Firing Section Mechanic), 24E (Hawk Fire Control Mechanic), 24G (Hawk Information Coordination Central Mechanic), 24H (Hawk Fire Control Repairer), 24J (Hawk Pulse Radar Repairer), 24K (Hawk Continuous Wave Radar Repairer), 24L (Hawk Launcher and Mechanical Systems Repairer), 24R (Hawk Master Mechanic), 24S (Roland System Mechanic), 24T (Patriot Operator and Systems Mechanic), 24U (Nike-Hercules Custodial Mechanic), 24V (Hawk Maintenance Chief), 25L (AN/TSQ 73 Air Defense Artillery Command and Control System Operator/Repairer), 27F (Vulcan Repairer), or 27G (Chaparral/Redeye Repairer).

Description

Supervises the assembly, inspection, and repair of the Hawk and Hawk PIP III missile system, simulator equipment, and associated test equipment; understands theory, functions, and operation of on-missile guidance equipment, ranging radars, pulse radars, launchers, fire control equipment, simulator stations, computers, and automatic data transmission systems used with the Hawk and Hawk PIP III air defense missile systems and associated power generation and air conditioning equipment; knows theory and functions of solid state devices and associated circuits and circuit elements; assembles and tests components of missile, radar, and fire control equipment; supervises testing of system specialized test sets and procedures; supervises repairers in isolating malfunctions and in repairing, adjusting, and aligning systems and system components; advises personnel on equipment modifications and of changes to inspection, repair, test, and maintenance calibration procedures; applies safety procedures and regulations pertaining to high-energy electromagnetic radiation hazards, high-voltage hazards, and high-pressure hydraulic equipment; applies missile transport and storage methods; knows operation, function, and maintenance of battery terminal equipment; trains personnel in safety and security procedures and maintenance techniques; interprets technical publications; supervises parts inventory and supply system including procurement; reviews and prepares reports; evaluates supply and maintenance problems, advises superiors, and recommends actions to correct deficiencies; may serve as an instructor at a formal training school or in missile procurement activities.

Recommendation

In the lower-division baccalaureate/associate degree category, 3 semester hours in maintenance management, 3 in inventory management, 3 in personnel supervision, and 3 in report writing. In the upper-division baccalaureate category, 3 semester hours in personnel management and 3 in management electives; if rank was CW2, 2 semester hours for field experience in management; if rank was CW3, 4 semester hours for field experience in management; if rank was CW4, 6 semester hours for field experience in management; if served as Master Warrant Officer, 8 semester hours for field experience in management (6/91).

MOS-140E-001

Patriot System Technician
((Patriot Missile System Technician))
140E0

Exhibit Dates: 1/90–5/91.

Career Pattern

May have progressed to Patriot System Technician from MOS 24C (Hawk Firing Section Mechanic), 24E (Hawk Fire Control Mechanic), 24G (Hawk Information Coordination Central Mechanic), 24H (Hawk Fire Control Repairer), 24J (Pulse Radar Repairer), 24K (Hawk Continuous Wave Radar Repairer), 24L (Hawk Launcher and Mechanical Systems Repairer), 24R (Hawk Master Mechanic), 24T (Patriot Operator and Systems Mechanic), 24U (Nike-Hercules Custodial Mechanic), 24V (Hawk Maintenance Chief), 25L (AN/TSQ 73 Air Defense Artillery Command and Control System Operator/Repairer), 27F (Vulcan Repairer), or 27G (Chaparral/Redeye Repairer).

Description

Supervises the repair and maintenance of Patriot air defense missile control equipment, data processing and simulator equipment, and radars; supervises the movement, emplacement, and testing of fire control and auxiliary equipment; understands theory, function, and operation of solid state devices, associated circuits, tracking radars, acquisition radars, interrogator equipment, computer data transmission systems, tactical control circuits, counter-countermeasure systems, and power supply equipment for Patriot air defense missile fire control equipment; supervises the use, testing, and maintenance of tools and test equipment used in diagnosing complex malfunctions in missile electronic ground guidance and tactical data processing equipment; monitors equipment during testing, simulations, and missile firing exercises to detect operator error and/or system malfunctions and initiates corrective action, if required; applies regulations and procedure required for working with high voltage, electromagnetic radiation, and X-rays; supervises emplacement, testing, and maintenance of fire control equipment; advises commander on employment of counter-countermeasures; knows operation, function, and maintenance of battery terminal equipment; trains maintenance and operator personnel in Patriot air defense missile control equipment; supervises parts inventory and supply system including procurement for Patriot air defense missile; supervises maintenance and operation inventory; evaluates effectiveness of maintenance programs, operator training, and parts and equipment supply operations; prepares reports.

Recommendation

In the lower-division baccalaureate/associate degree category, 6 semester hours in electronics troubleshooting, 3 in technical mathematics, 3 in basic electronics, 3 in AC/DC circuit analysis, 3 in transistor theory, 3 in maintenance management, 2 in instrumentation, 2 in pulse circuits, 2 in digital circuitry, 2 in computer fundamentals, 2 in inventory management, 2 in personnel supervision, 2 in report writing, 1 in maintenance shop recordkeeping, 1 in control systems, 1 in radar/microwave theory, 1 in machine language, and additional credit in electronics, mathematics, data processing, computer science, and public speaking/communication on the basis of institutional evaluation. In the upper-division baccalaureate category, 3 semester hours in personnel management and training and 3 in management electives; if rank was CW2, 2 additional semester hours for field experience in management; if rank was CW3, 4 additional semester hours for field experience in management; if rank was CW4, 6 additional semester hours for field experience in management; if duty assignment was platoon leader, credit in leadership and supervision on the basis of institutional evaluation (5/77).

MOS-140E-002

Patriot Missile System Technician
140E0

Exhibit Dates: 6/91–Present.

Career Pattern

NOTE: May have progressed to Patriot Missile System Technician from MOS 24C (Hawk Firing Section Mechanic), 24E (Hawk Fire Control Mechanic), 24G (Hawk Information Coordination Central Mechanic), 24H (Hawk Fire Control Repairer), 24J (Hawk Pulse Radar Repairer), 24K (Hawk Continuous Wave Radar Repairer), 24L (Hawk Launcher and Mechanical Systems Repairer), 24R (Hawk Master Mechanic), 24T (Patriot Operator and Systems Mechanic), 24U (Nike-Hercules Custodial Mechanic), 24V (Hawk Maintenance Chief), 25L (AN/TSQ 73 Air Defense Artillery Command and Control System Operator/Repairer), 27F (Vulcan Repairer), or 27G (Chaparral/Redeye Repairer).

Description

Supervises the repair and maintenance of Patriot air defense missile control equipment, data processing and simulator equipment, and radars; supervises the movement, emplacement, and testing of fire control and auxiliary equipment; understands theory, function, and operation of solid state devices, associated circuits, tracking radars, acquisition radars, interrogator equipment, computer data transmission systems, tactical control circuits, counter-countermeasure systems, and power supply equipment for Patriot air defense missile fire control equipment; supervises the use, testing, and maintenance of tools and test equipment used in diagnosing complex malfunctions in missile electronic ground guidance and tactical data processing equipment; monitors equipment during testing, simulations, and missile firing exercises to detect operator error and/or systems malfunctions and initiates corrective action, if required; applies regulations and procedures required for working with high voltage, electromagnetic radiation, and X-rays; supervises emplacement, testing, and maintenance of fire control equipment; advises commander on employment of counter-countermeasures; knows operation, function, and maintenance of battery terminal equipment; trains maintenance and operator personnel in Patriot air defense missile control equipment; supervises parts inventory and supply system including procurement for Patriot air defense missile; supervises maintenance and operation inventory; evaluates effectiveness of maintenance programs, operator training, and parts and equipment supply operations; prepares reports.

Recommendation

In the lower-division baccalaureate/associate degree category, 3 semester hours in maintenance management, 3 in inventory management, 3 in personnel supervision, and 3 in report writing. In the upper-division baccalaureate category, 3 semester hours in personnel management and 3 in management electives; if rank was CW2, 2 semester hours for field experience in management; if rank was CW3, 4 semester hours for field experience in management; if rank was CW4, 6 semester hours for field experience in management; if served as Master Warrant Officer (MW), 8 semester hours for field experience in management (6/91).

MOS-150A-001

Air Traffic Control Technician
150A0

Exhibit Dates: 1/90–Present.

Career Pattern

May have progressed to MOS 150A from 93C, Air Traffic Control Operator.

Description

Supervises and manages personnel in an Air Traffic Control organization. Senior warrant officer: Develops, revises, and reviews instrument en route and terminal procedures; provides technical expertise pertaining to all aspects of en route and terminal aircraft operations; supervises air traffic control training requirements. Master warrant officer: Able to perform the duties required for Senior warrant officer; conducts liaison with collateral units and functions as senior advisor for air traffic control operations; coordinates and establishes special-use airspace; develops and monitors standards for instruction of personnel.

Recommendation

In the lower-division baccalaureate/associate degree category, 3 semester hours in principles of supervision. In the upper-division baccalaureate category, 3 semester hours in organizational management, 3 in management problems, and 3 for field experience in management; if served as master warrant officer, 3 additional semester hours for field experience in management (4/90).

MOS-151A-001

Aviation Maintenance Technician
151A0

Exhibit Dates: 1/90–12/90.

Career Pattern

May have progressed to Aviation Maintenance Technician from any aircraft maintenance MOS.

Description

Supervises aviation maintenance and repair shops, teams, sections, or platoons that maintain or repair army rotary- and fixed-wing aircraft; organizes and manages maintenance facilities, equipment, and personnel to inspect, service, test, disassemble, repair, reassemble, adjust, and retest aircraft or aircraft components; assigns work to subordinates and supervises work in progress, final inspection, and final testing; develops operating procedures and performs administrative duties related to supply and maintenance activities; provides technical assistance to pilots and flight crews concerning required crew maintenance of aircraft and aircraft components; interprets technical material related to aircraft maintenance.

Recommendation

In the vocational certificate category, 3 semester hours in essentials of supervision, 3 in technical writing, 2 in warehousing and storage, and 3 in maintenance management and inventory control. In the lower-division baccalaureate/associate degree category, 3 semester hours in maintenance management, 3 in shop practices and procedures, 2 in inventory management, 2 in material handling and safety, 3 in personnel supervision, 3 in survey of computer concepts, 2 in maintenance shop record keeping, and 3 in communication skills (speech/technical writing). In the upper-division baccalaureate category, 3 semester hours in personnel management and 3 in management electives; if rank was CW2, 2 additional semester hours for field experience in management; if rank was CW3, 4 additional semester hours for field experience in management; if rank was CW4, 6 additional semester hours for field experience in management; if duty assignment was platoon leader, section leader, or equivalent leadership position, credit in leadership and supervision on the basis of institutional evaluation (7/78).

MOS-151A-002

Aviation Maintenance Technician
151A0

Exhibit Dates: 1/91–12/00.

Career Pattern

May have progressed to Aviation Maintenance technician from any enlisted aircraft maintenance MOS.

Description

Supervises aviation maintenance and repair shops, teams, sections, or facilities that maintain or repair army rotary- and fixed-wing aircraft; organizes and manages maintenance facilities, equipment, and personnel to inspect, service, test, disassemble, repair, reassemble, adjust, and retest aircraft or aircraft components; assigns work to subordinates and supervises work in progress, final inspection, and final testing; develops operating procedures and performs administrative duties related to supply and maintenance activities; provides technical assistance to pilot and flight crews concerning required crew maintenance of aircraft and aircraft components; interprets technical material related to aircraft maintenance.

Recommendation

In the lower-division baccalaureate/associate degree category, 3 semester hours in maintenance management, 3 in personnel supervision, and 3 in records and information management. In the upper-division baccalaureate category, 3 semester hours in human resource management, 3 in management problems, 3 in operations management, 3 in organizational management, and 3 for field experience in management. If served as a senior warrant officer, 2 additional semester hours for field experience in management; if served as a master warrant officer, 3 additional semester hours for field experience in management (3/91).

MOS-151A-003

Aviation Maintenance Technician
151A0

Exhibit Dates: 1/01–Present.

Career Pattern

May have progressed to Aviation Maintenance Technician from any enlisted aircraft maintenance MOS.

Description

Supervises aviation maintenance and repair shops, teams, or facilities that maintain or repair aircraft; organizes and manages maintenance facilities, equipment, and personnel to inspect, service, test, disassemble, repair, reassemble, adjust, and retest aircraft or aircraft components; assigns work to subordinates and supervises work in progress, final inspection, and final testing; develops operating procedures and performs administrative duties related to logistics and maintenance activities; provides technical assistance to pilots and flight crews concerning required crew maintenance of aircraft and aircraft components; interprets technical material related to aircraft maintenance; uses computer management information systems to support aircraft maintenance activities.

Recommendation

In the lower-division baccalaureate/associate degree category, 3 semester hours in maintenance management, 3 in principles of supervision, and 3 in records/information management. In the upper-division baccalaureate category, 3 semester hours in human resource management, 3 in management problems, 3 in operations management, 3 in organizational management, and 3 for field experience in management (1/01).

MOS-152B-002

OH-58A/C Scout Pilot
152B0

Exhibit Dates: 1/90–10/99.

Career Pattern

May have progressed to MOS 152B from any enlisted MOS.

Description

Pilots and commands attack and scout helicopters under tactical and nontactical conditions; operates aircraft during all types of meteorological conditions during day, night, and while using night vision goggles; coordinates, conducts, and directs tactical helicopter operations; participates in reconnaissance and security missions; maintains current flight status in accordance with established training requirements; knowledge and skills are equivalent to FAA commercial/instrument ratings on rotorcraft category aircraft. NOTE: Knowledge and skills may be equivalent to other FAA certificates/ratings (flight instructor, instrument instructor, ground instructor, airline transport pilot, flight engineer, etc.) or the applicant may be trained on additional categories of aircraft.

Recommendation

In the lower-division baccalaureate/associate degree category, 18 semester hours in commercial/instrument rating, 3 in aircraft rules and procedures, 3 in navigation (VFR), 3 in flight physiology, 3 in aviation meteorology, 3 in instrument flight planning and procedures, 3 in aircraft engines and systems, 3 in aircraft performance, 3 in general mathematics, 3 in aviation safety, 3 in communication skills (oral), and 3 in principles of supervision. In the upper-division baccalaureate category, if served as instructor pilot (designated on Army records as 152BC), 3 semester hours in educational psychology, 3 in personnel management, and 3 in interpersonal communication; if served as an instrument flight examiner (designated on Army records as 152BF), 3 semester hours in airline transport pilot training; if served as section leader or aircraft repair maintenance technician (designated on Army records as 152BG), 3 semester hours in organizational management, 3 in accident investigation procedures, and 3 in aircraft systems management (4/90).

MOS-152B-003

OH-58A/C Scout Pilot
152B0

Exhibit Dates: 11/99–Present.

Career Pattern

May have progressed to MOS 152B from any enlisted MOS.

Description

Pilots and commands attack and scout helicopters under tactical and nontactical conditions; operates aircraft during all types of meteorological conditions during day, night, and while using night vision goggles; coordinates, conducts, and directs tactical helicopter operations; participates in reconnaissance and security missions; maintains current flight status in accordance with established training requirements; knowledge and skills are equivalent to FAA commercial/instrument ratings on rotorcraft category aircraft. NOTE: Knowledge and skills may be equivalent to other FAA certificates/ratings (flight instructor, instrument instructor, ground instructor, airline transport pilot, flight engineer, etc.) or the applicant may be trained on additional categories of aircraft.

Recommendation

In the lower-division baccalaureate/associate degree category, 3 semester hours in aircraft rules and procedures, 3 in navigation (VFR), 3 in flight physiology, 3 in aviation meteorology, 3 in instrument flight planning, 3 in aircraft engines and systems, 3 in aircraft performance, 3 in general mathematics, 3 in aviation safety, 3 in communication skills, 3 in principles of supervision, 3 in computer literacy, 3 in technical writing, 3 in turbine theory and operations, and 3 in risk management. In the upper-division baccalaureate category, 3 semester hours in crew resource management and 3 in flight guidance systems; if served as instructor pilot (designated on Army records as 152BC), 3 semester hours in principles of instruction, 3 in personnel management, 3 in interpersonal communication, 3 in computer applications, and 3 in curriculum development; if served as an instrument flight examiner (designated on Army records as 152BF), 3 semester hours in advanced instructional techniques (helicopters); if served as section leader or aircraft repair maintenance technician (designated on Army records as

152BG), 3 semester hours in organizational management, 3 in accident investigation procedures, 3 in aviation maintenance management, and 3 in logistics (11/99).

MOS-152C-002

OH-6 Scout Pilot
152C0

Exhibit Dates: 1/90–10/99.

Career Pattern

May have progressed to MOS 152C from any enlisted MOS.

Description

Pilots and commands attack and scout helicopters under tactical and nontactical conditions; operates aircraft during day, night, and while using night vision goggles; coordinates, conducts, and directs tactical helicopter operations; participates in reconnaissance and security missions; maintains current flight status in accordance with established training requirements; knowledge and skills are equivalent to FAA commercial/instrument ratings on rotorcraft category aircraft. NOTE: Knowledge and skills may be equivalent to other FAA certificates/ratings (flight instructor, instrument instructor, ground instructor, airline transport pilot, flight engineer, etc.) or the applicant may be trained on additional categories of aircraft.

Recommendation

In the lower-division baccalaureate/associate degree category, 18 semester hours in commercial/instrument rating, 3 in aircraft rules and procedures, 3 in navigation (VFR), 3 in flight physiology, 3 in aviation meteorology, 3 in instrument flight planning and procedures, 3 in aircraft engines and systems, 3 in aircraft performance, 3 in general mathematics, 3 in aviation safety, 3 in communication skills (oral), and 3 in principles of supervision. In the upper-division baccalaureate category, if served as instructor pilot (designated on Army records as 152CC), 3 semester hours in educational psychology, 3 in personnel management, and 3 in interpersonal communication; if served as an instrument flight examiner (designated on Army records as 152CF), 3 semester hours in airline transport pilot training; if served as section leader or aircraft repair maintenance technician (designated on Army records as 152CG), 3 semester hours in organizational management, 3 in accident investigation procedures, and 3 in aircraft systems management (4/90).

MOS-152C-003

OH-6 Scout Pilot
152C0

Exhibit Dates: 11/99–Present.

Career Pattern

May have progressed to MOS 152C from any enlisted MOS.

Description

Pilots and commands attack and scout helicopters under tactical and nontactical conditions; operates aircraft during day, night, and while using night vision goggles; coordinates, conducts, and directs tactical helicopter operations; participates in reconnaissance and security missions; maintains current flight status in accordance with established training requirements; knowledge and skills are equivalent to FAA commercial/instrument ratings on rotorcraft category aircraft. NOTE: Knowledge and skills may be equivalent to other FAA certificates/ratings (flight instructor, instrument instructor, ground instructor, airline transport pilot, flight engineer, etc.) or the applicant may be trained on additional categories of aircraft.

Recommendation

In the lower-division baccalaureate/associate degree category, 3 semester hours in aircraft rules and procedures, 3 in navigation (VFR), 3 in flight physiology, 3 in aviation meteorology, 3 in instrument flight planning and procedures, 3 in aircraft engines and systems, 3 in aircraft performance, 3 in general mathematics, 3 in aviation safety, 3 in communication skills, 3 in principles of supervision, 3 in computer literacy, 3 in technical writing, 3 in turbine theory and operations, and 3 in risk management. In the upper-division baccalaureate category, 3 semester hours in crew resource management and 3 in flight guidance systems; if served as instructor pilot (designated on Army records as 152CC), 3 semester hours in principles of instruction, 3 in personnel management, 3 in interpersonal communication, 3 in computer applications, and 3 in curriculum development; if served as an instrument flight examiner (designated on Army records as 152CF), 3 semester hours in advanced instructional techniques (helicopters); if served as section leader or aircraft repair maintenance technician (designated on Army records as 152CG), 3 semester hours in organizational management, 3 in accident investigation procedures, 3 in aviation maintenance management, and 3 in logistics (11/99).

MOS-152D-002

OH-58D Scout Pilot
152D0

Exhibit Dates: 1/90–10/99.

Career Pattern

May have progressed to MOS 152D from any enlisted MOS.

Description

Pilots and commands attack and scout helicopters under tactical and nontactical conditions; operates aircraft during all types of meteorological conditions during day, night, and while using night vision goggles; coordinates, conducts, and directs tactical helicopter operations; participates in reconnaissance and security missions; maintains current flight status in accordance with established training requirements; knowledge and skills are equivalent to FAA commercial/instrument ratings on rotorcraft category aircraft. NOTE: Knowledge and skills may be equivalent to other FAA certificates/ratings (flight instructor, instrument instructor, ground instructor, airline transport pilot, flight engineer, etc.) or the applicant may be trained on additional categories of aircraft.

Recommendation

In the lower-division baccalaureate/associate degree category, 18 semester hours in commercial/instrument rating, 3 in aircraft rules and procedures, 3 in navigation (VFR), 3 in flight physiology, 3 in aviation meteorology, 3 in instrument flight planning and procedures, 3 in aircraft engines and systems, 3 in aircraft performance, 3 in general mathematics, 3 in aviation safety, 3 in communication skills (oral), and 3 in principles of supervision. In the upper-division baccalaureate category, if served as instructor pilot (designated on Army records as 152DC), 3 semester hours in educational psychology, 3 in personnel management, and 3 in interpersonal communication; if served as an instrument flight examiner (designated on Army records as 152DF), 3 semester hours in airline transport pilot training; if served as section leader or aircraft repair maintenance technician (designated on Army records as 152DG), 3 semester hours in organizational management, 3 in accident investigation procedures, and 3 in aircraft systems management (4/90).

MOS-152D-003

OH-58D Scout Pilot
152D0

Exhibit Dates: 11/99–Present.

Career Pattern

May have progressed to MOS 152D from any enlisted MOS.

Description

Pilots and commands attack and scout helicopters under tactical and nontactical conditions; operates aircraft during all types of meteorological conditions during day, night, and while using night vision goggles; coordinates, conducts, and directs tactical helicopter operations; participates in reconnaissance and security missions; maintains current flight status in accordance with established training requirements; knowledge and skills are equivalent to FAA commercial/instrument ratings on rotorcraft category aircraft. NOTE: Knowledge and skills may be equivalent to other FAA certificates/ratings (flight instructor, instrument instructor, ground instructor, airline transport pilot, flight engineer, etc.) or the applicant may be trained on additional categories of aircraft.

Recommendation

In the lower-division baccalaureate/associate degree category, 3 semester hours in aircraft rules and procedures, 3 in navigation (VFR), 3 in flight physiology, 3 in aviation meteorology, 3 in instrument flight planning, 3 in aircraft engines and systems, 3 in aircraft performance, 3 in general mathematics, 3 in aviation safety, 3 in communication skills, 3 in principles of supervision, 3 in computer literacy, 3 in technical writing, 3 in turbine theory and operations, and 3 in risk management. In the upper-division baccalaureate category, 3 semester hours in crew resource management and 3 in flight guidance systems; if served as instructor pilot (designated on Army records as 152DC), 3 semester hours in principles of instruction, 3 in personnel management, 3 in interpersonal communication, 3 in computer applications, and 3 in curriculum development; if served as an instrument flight examiner (designated on Army records as 152DF), 3 semester hours in advanced instructional techniques (helicopters); if served as section leader or aircraft repair maintenance technician (designated on Army records as 152DG), 3 semester hours in organizational management, 3 in accident investigation procedures, 3 in aviation maintenance management, and 3 in logistics (11/99).

MOS-152E-001

RAH-66 Comanche Pilot
152E0

Exhibit Dates: 11/99–Present.

Career Pattern

May have progressed to MOS 152E from any enlisted MOS.

Description

Pilots and commands attack and scout helicopters under tactical and nontactical conditions; operates aircraft during all types of meteorological conditions during day, night, and while using night vision goggles; coordinates, conducts, and directs tactical helicopter operations; participates in reconnaissance and security missions; maintains current flight status in accordance with established training requirements; knowledge and skills are equivalent to FAA commercial/instrument ratings on rotorcraft category aircraft. NOTE: Knowledge and skills may be equivalent to other FAA certificates/ratings (flight instructor, instrument instructor, ground instructor, airline transport pilot, flight engineer, etc.) or the applicant may be trained on additional categories of aircraft.

Recommendation

In the lower-division baccalaureate/associate degree category, 3 semester hours in aircraft rules and procedures, 3 in navigation (VFR), 3 in flight physiology, 3 in aviation meteorology, 3 in instrument flight planning, 3 in aircraft engines and systems, 3 in aircraft performance, 3 in general mathematics, 3 in aviation safety, 3 in communication skills, 3 in principles of supervision, 3 in computer literacy, 3 in technical writing, 3 in turbine theory and operations, and 3 in risk management. In the upper-division baccalaureate category, 3 semester hours in crew resource management and 3 in flight guidance systems; if served as instructor pilot (designated on Army records as 152BC), 3 semester hours in principles of instruction, 3 in personnel management, 3 in interpersonal communication, 3 in computer applications, and 3 in curriculum development; if served as an instrument flight examiner (designated on Army records as 152BF), 3 semester hours in advanced instructional techniques (helicopters); if served as section leader or aircraft repair maintenance technician (designated on Army records as 152BG), 3 semester hours in organizational management, 3 in accident investigation procedures, 3 in aviation maintenance management, and 3 in logistics (11/99).

MOS-152F-002

AH-64A Attack Pilot
((AH-64 Attack Pilot))
152F0

Exhibit Dates: 1/90–10/99.

Career Pattern

May have progressed to MOS 152F from any enlisted MOS.

Description

Pilots and commands attack and scout helicopter under tactical and nontactical conditions; operates aircraft during all types of meteorological conditions during day, night, and while using night vision goggles; coordinates, conducts, and directs tactical helicopter operations; participates in reconnaissance and security missions; maintains current flight status in accordance with established training requirements; knowledge and skills are equivalent to FAA commercial/instrument ratings on rotorcraft category aircraft. NOTE: Knowledge and skills may be equivalent to other FAA certificates/ratings (flight instructor, instrument instructor, ground instructor, airline transport pilot, flight engineer, etc.) or the applicant may be trained on additional categories of aircraft.

Recommendation

In the lower-division baccalaureate/associate degree category, 18 semester hours in commercial/instrument rating, 3 in aircraft rules and procedures, 3 in navigation (VFR), 3 in flight physiology, 3 in aviation meteorology, 3 in instrument flight planning and procedures, 3 in aircraft engines and systems, 3 in aircraft performance, 3 in general mathematics, 3 in aviation safety, 3 in communication skills (oral), and 3 in principles of supervision. In the upper-division baccalaureate category, if served as instructor pilot (designated on Army records as 152FC), 3 semester hours in educational psychology, 3 in personnel management, and 3 in interpersonal communication; if served as an instrument flight examiner (designated on Army records as 152FF), 3 semester hours in airline transport pilot training; if served as section leader or aircraft repair maintenance technician (designated on Army records as 152FG), 3 semester hours in organizational management, 3 in accident investigation procedures, and 3 in aircraft systems management (4/90).

MOS-152F-003

AH-64A Attack Pilot
((AH-64 Attack Pilot))
152F0

Exhibit Dates: 11/99–Present.

Career Pattern

May have progressed to MOS 152F from any enlisted MOS.

Description

Pilots and commands attack and scout helicopter under tactical and nontactical conditions; operates aircraft during all types of meteorological conditions during day, night, and while using night vision goggles; coordinates, conducts, and directs tactical helicopter operations; participates in reconnaissance and security missions; maintains current flight status in accordance with established training requirements; knowledge and skills are equivalent to FAA commercial/instrument ratings on rotorcraft category aircraft. NOTE: Knowledge and skills may be equivalent to other FAA certificates/ratings (flight instructor, instrument instructor, ground instructor, airline transport pilot, flight engineer, etc.) or the applicant may be trained on additional categories of aircraft.

Recommendation

In the lower-division baccalaureate/associate degree category, 3 semester hours in aircraft rules and procedures, 3 in navigation (VFR), 3 in flight physiology, 3 in aviation meteorology, 3 in instrument flight planning, 3 in aircraft engines and systems, 3 in aircraft performance, 3 in general mathematics, 3 in aviation safety, 3 in communication skills, 3 in principles of supervision, 3 in computer literacy, 3 in technical writing, 3 in turbine theory and operations, and 3 in risk management. In the upper-division baccalaureate category, 3 semester hours in crew resource management and 3 in flight guidance systems; if served as instructor pilot (designated on Army records as 152FC), 3 semester hours in principles of instruction, 3 in personnel management, 3 in interpersonal communication, 3 in computer applications, and 3 in curriculum development; if served as an instrument flight examiner (designated on Army records as 152FF), 3 semester hours in advanced instructional techniques (helicopters); if served as section leader or aircraft repair maintenance technician (designated on Army records as 152FG), 3 semester hours in organizational management, 3 in accident investigation procedures, 3 in aviation maintenance management, and 3 in logistics (11/99).

MOS-152G-002

AH-1 Attack Pilot
152G0

Exhibit Dates: 1/90–10/99.

Career Pattern

May have progressed to MOS 152G from any enlisted MOS.

Description

Pilots and commands attack and scout helicopters under tactical and nontactical conditions; operates aircraft during all types of meteorological conditions during day, night, and while using night vision goggles; coordinates, conducts, and directs tactical helicopter operations; participates in reconnaissance and security missions; maintains current flight status in accordance with established training requirements; knowledge and skills are equivalent to FAA commercial/instrument ratings on rotorcraft category aircraft. NOTE: Knowledge and skills may be equivalent to other FAA certificates/ratings (flight instructor, instrument instructor, ground instructor, airline transport pilot, flight engineer, etc.) or the applicant may be trained on additional categories of aircraft.

Recommendation

In the lower-division baccalaureate/associate degree category, 18 semester hours in commercial/instrument rating, 3 in aircraft rules and procedures, 3 in navigation (VFR), 3 in flight physiology, 3 in aviation meteorology, 3 in instrument flight planning and procedures, 3 in aircraft engines and systems, 3 in aircraft performance, 3 in general mathematics, 3 in aviation safety, 3 in communication skills (oral), and 3 in principles of supervision. In the upper-division baccalaureate category, if served as instructor pilot (designated on Army records as 152GC), 3 semester hours in educational psychology, 3 in personnel management, and 3 in interpersonal communication; if served as an instrument flight examiner (designated on Army records as 152GF), 3 semester hours in airline transport pilot training; if served as section leader or aircraft repair maintenance technician (designated on Army records as 152GG), 3 semester hours in organizational management, 3 in accident investigation procedures, and 3 in aircraft systems management (4/90).

MOS-152G-003

AH-1 Attack Pilot
152G0

Exhibit Dates: 11/99–Present.

Career Pattern

May have progressed to MOS 152G from any enlisted MOS.

Description

Pilots and commands attack and scout helicopters under tactical and nontactical conditions; operates aircraft during all types of meteorological conditions during day, night, and while using night vision goggles; coordinates, conducts, and directs tactical helicopter operations; participates in reconnaissance and security missions; maintains current flight status in accordance with established training requirements; knowledge and skills are equivalent to FAA commercial/instrument ratings on rotorcraft category aircraft. NOTE: Knowledge and skills may be equivalent to other FAA certificates/ratings (flight instructor, instrument instructor, ground instructor, airline transport pilot, flight engineer, etc.) or the applicant may be trained on additional categories of aircraft.

Recommendation

In the lower-division baccalaureate/associate degree category, 3 semester hours in aircraft rules and procedures, 3 in navigation (VFR), 3 in flight physiology, 3 in aviation meteorology, 3 in instrument flight planning and procedures, 3 in aircraft engines and systems, 3 in aircraft performance, 3 in general mathematics, 3 in aviation safety, 3 in communication skills, 3 in principles of supervision, 3 in computer literacy, 3 in technical writing, 3 in turbine theory and operations, and 3 in risk management. In the upper-division baccalaureate category, 3 semester hours in crew resource management and 3 in flight guidance systems; if served as instructor pilot (designated on Army records as 152GC), 3 semester hours in principles of instruction, 3 in personnel management, 3 in interpersonal communication, 3 in computer applications, and 3 in curriculum development; if served as an instrument flight examiner (designated on Army records as 152GF), 3 semester hours in advanced instructional techniques (helicopters); if served as section leader or aircraft repair maintenance technician (designated on Army records as 152GG), 3 semester hours in organizational management, 3 in accident investigation procedures, 3 in aviation maintenance management, and 3 in logistics (11/99).

MOS-152H-001

AH-64D Attack Pilot
152H0

Exhibit Dates: 8/98–Present.

Career Pattern

May have progressed to MOS 152H from any enlisted MOS.

Description

Pilots and commands attack and scout helicopters under tactical and nontactical conditions; operates aircraft during all types of meteorological conditions during day, night, and while using night vision goggles; coordinates, conducts, and directs tactical helicopter operations; participates in reconnaissance and security missions; maintains current flight status in accordance with established training requirements; knowledge and skills are equivalent to FAA commercial/instrument ratings on rotorcraft category aircraft. NOTE: Knowledge and skills may be equivalent to other FAA certificates/ratings (flight instructor, instrument instructor, ground instructor, airline transport pilot, flight engineer, etc.) or the applicant may be trained on additional categories of aircraft.

Recommendation

In the lower-division baccalaureate/associate degree category, 3 semester hours in aircraft rules and procedures, 3 in navigation (VFR), 3 in flight physiology, 3 in aviation meteorology, 3 in instrument flight planning, 3 in aircraft engines and systems, 3 in aircraft performance, 3 in general mathematics, 3 in aviation safety, 3 in communication skills, 3 in principles of supervision, 3 in computer literacy, 3 in technical writing, 3 in turbine theory and operations, and 3 in risk management. In the upper-division baccalaureate category, 3 semester hours crew resource management and 3 in flight guidance systems; if served as an instructor pilot (designated on Army records as 152HC), 3 semester hours in principles of instruction, 3 in personnel management, 3 in interpersonal communication, 3 in computer applications, and 3 in curriculum development; if served as instrument flight examiner (designated Army records as 152HF) 3 semester hours in advanced instructional techniques (helicopters); if served as a section leader or aircraft repair maintenance technician (designated on Army records as 152HG), 3 semester hours in organizational management, 3 in accident investigation procedures, 3 in aviation maintenance management, and 3 in logistics (11/99).

MOS-153A-002

Rotary Wing Aviator (Aircraft Nonspecific)
153A0

Exhibit Dates: 1/90–10/99.

Career Pattern

May have progressed to MOS 153A from any enlisted MOS.

Description

Pilots and commands utility helicopters under tactical and nontactical conditions; operates aircraft during all types of meteorological conditions during day, night, and while using night vision goggles; performs aircraft operations, including rescue hoist, air assaults, aerial mine and flare deliveries, internal/external loads, and paradrop/rappelling procedures; may perform aerial reconnaissance and be used to transport passengers or cargo; routinely participates in combat and combat support training; maintains current flight status in accordance with established training requirements; knowledge and skills are equivalent to FAA commercial/instrument ratings on rotorcraft category aircraft. Knowledge and skills may be equivalent to other FAA certificates/ratings (flight instructor, instrument instructor, ground instructor, airline transport pilot, flight engineer, etc.) or the applicant may be trained on additional categories of aircraft.

Recommendation

In the lower-division baccalaureate/associate degree category, 18 semester hours in commercial/instrument rating, 3 in aircraft rules and procedures, 3 in navigation (VFR), 3 in flight physiology, 3 in aviation meteorology, 3 in instrument flight planning and procedures, 3 in aircraft engines and systems, 3 in aircraft performance, 3 in general mathematics, 3 in aviation safety, 3 in communication skills (oral), and 3 in principles of supervision. In the upper-division baccalaureate category, if served as instructor pilot (designated on Army records as 153AC), 3 semester hours in educational psychology, 3 in personnel management, and 3 in interpersonal communication; if served as an instrument flight examiner (designated on Army records as 153AF), 3 semester hours in airline transport pilot training; if served as section leader or aircraft repair maintenance technician (designated on Army records as 153AG), 3 semester hours in organizational management, 3 in accident investigation procedures, and 3 in aircraft systems management (4/90).

MOS-153A-003

Rotary Wing Aviator (Aircraft Nonspecific)
153A0

Exhibit Dates: 11/99–Present.

Career Pattern

May have progressed to MOS 153A from any pilot warrant officer MOS.

Description

Pilots and commands fixed-wing utility aircraft under tactical and nontactical conditions; operates aircraft during all types of meteorological conditions, during day, night, and while using night vision goggles; performs aircraft operations, including rescue hoist, air assaults, aerial mine and flare deliveries, internal/external loads, and paradrop/rappelling procedures; may perform aerial reconnaissance and be used to transport passengers or cargo; routinely participates in combat support training; maintains current flight status in accordance with established training requirements; knowledge and skills are equivalent to FAA commercial/instrument ratings on rotorcraft category aircraft. Knowledge and skills may be equivalent to other FAA certificates/ratings (flight instructor, instrument instructor, ground instructor, airline transport pilot, flight engineer, etc.) or the applicant may be trained on additional categories of aircraft.

Recommendation

In the lower-division baccalaureate/associate degree category, 3 semester hours in aircraft rules and procedures, 3 in navigation (VFR), 3 in flight physiology, 3 in aviation meteorology, 3 in instrument flight planning, 3 in aircraft engines and systems, 3 in aircraft performance, 3 in general mathematics, 3 in aviation safety, 3 in communication skills, 3 in technical writing, 3 in principles of supervision, 3 in computer literacy, 3 in turbine theory and operation, and 3 in risk management. In the upper-division baccalaureate category, 3 semester hours in crew resource management and 3 in flight guidance systems; if served as instructor pilot (designated on Army records as 153C), 3 semester hours in principles of instruction, 3 in personnel management, 3 in interpersonal communication, 3 in computer applications, and 3 in curriculum development; if served as an instrument flight examiner (designated on Army records as 153AF), 3 semester hours in advanced instructional techniques; if served as section leader or aircraft repair maintenance technician (designated on Army records as 153AG), 3 semester hours in organizational management, 3 in accident investigation procedures, 3 in aircraft aviation management, and 3 in logistics (11/99).

MOS-153B-002

UH-1 Pilot
153B0

Exhibit Dates: 1/90–10/99.

Career Pattern

May have progressed to MOS 153B from any enlisted MOS.

Description

Pilot and commands utility helicopters under tactical and nontactical conditions; operates aircraft during all types of meteorological conditions, during day, night, and while using night vision goggles; performs aircraft operations, including rescue hoist, air assaults, aerial mine and flare deliveries, internal/external loads, and paradrop/rappelling procedures; may perform aerial reconnaissance and be used to transport passengers or cargo; routinely participates in combat support training requirements; knowledge and skills are equivalent to FAA commercial/instrument ratings on rotorcraft category aircraft. Knowledge and skills may be equivalent to other FAA certificates/ratings (flight instructor, instrument instructor, ground instructor, airline transport pilot, flight engineer, etc.) or the applicant may be trained on additional categories of aircraft.

Recommendation

In the lower-division baccalaureate/associate degree category, 18 semester hours in commercial/instrument rating, 3 in aircraft rules and procedures, 3 in navigation (VFR), 3 in flight physiology, 3 in aviation meteorology, 3 in instrument flight planning and procedures, 3 in aircraft engines and systems, 3 in aircraft performance, 3 in general mathematics, 3 in aviation safety, 3 in communication skills (oral), and 3 in principles of supervision. In the upper-division baccalaureate category, if served as instructor pilot (designated on Army records as 153BC), 3 semester hours in educational psychology, 3 in personnel management, and 3 in interpersonal communication; if served as an instrument flight examiner (designated on Army records as 153BF), 3 semester hours in airline transport pilot training; if served as section leader or aircraft repair maintenance technician (designated on Army records as 153BG), 3 semester hours in organizational management, 3 in accident investigation procedures, and 3 in aircraft systems management (4/90).

MOS-153B-003

UH-1 Pilot
153B0

Exhibit Dates: 11/99–Present.

Career Pattern

May have progressed to MOS 153B from any pilot warrant officer MOS.

Description

Pilots and commands fixed-wing utility aircraft under tactical and nontactical conditions; operates aircraft during all types of meteorological conditions, during day, night, and while using night vision goggles; performs aircraft operations, including rescue hoist, air assaults, aerial mine and flare deliveries, internal/external loads, and paradrop/rappelling procedures; may perform aerial reconnaissance and be used to transport passengers or cargo; routinely participates in combat support training; maintains current flight status in accordance with established training requirements; knowledge and skills are equivalent to FAA commercial/instrument ratings on rotorcraft category aircraft. Knowledge and skills may be equivalent to other FAA certificates/ratings (flight instructor, instrument instructor, ground instructor, airline transport pilot, flight engineer, etc.) or the applicant may be trained on additional categories of aircraft.

Recommendation

In the lower-division baccalaureate/associate degree category, 3 semester hours in aircraft rules and procedures, 3 in navigation (VFR), 3 in flight physiology, 3 in aviation meteorology, 3 in instrument flight planning and procedures, 3 in aircraft engines and systems, 3 in aircraft performance, 3 in general mathematics, 3 in aviation safety, 3 in communication skills, 3 in technical writing, 3 in principles of supervision, 3 in computer literacy, 3 in turbine theory and operation, and 3 in risk management. In the upper-division baccalaureate category, 3 semester hours in crew resource management and 3 in flight guidance systems; if served as instructor pilot (designated on Army records as 153BC), 3 semester hours in principles of instruction, 3 in personnel management, 3 in interpersonal communication, 3 in computer applications, and 3 in curriculum development; if served as an instrument flight examiner (designated on Army records as 153BF), 3 semester hours in advanced instructional techniques; if served as section leader or aircraft repair maintenance technician (designated on Army records as 153BG), 3 semester hours in organizational management, 3 in accident investigation procedures, 3 in aircraft aviation management, and 3 in logistics (11/99).

MOS-153C-002

OH-58A/C Observation Pilot
153C0

Exhibit Dates: 1/90–6/95. Effective 6/95, MOS 153C was discontinued and its duties were incorporated into MOS 152B.

Career Pattern

May have progressed to MOS 153C from any enlisted MOS.

Description

Pilots and commands utility helicopters under tactical and nontactical conditions; operates aircraft during all types of meteorological conditions, during day, night, and while using night vision goggles; performs aircraft operations, including rescue hoist, air assaults, aerial mine and flare deliveries, internal/external loads, and paradrop/rappelling procedures; may perform aerial reconnaissance and be used to transport passengers or cargo; routinely participates in combat and combat support training; maintains current flight status in accordance with established training requirements; knowledge and skills are equivalent to FAA commercial/instrument ratings on rotorcraft category aircraft. Knowledge and skills may be equivalent to other FAA certificates/ratings (flight instructor, instrument instructor, ground instructor, airline transport pilot, flight engineer, etc.) or the applicant may be trained on additional categories of aircraft.

Recommendation

In the lower-division baccalaureate/associate degree category, 18 semester hours in commercial/instrument rating, 3 in aircraft rules and procedures, 3 in navigation (VFR), 3 in flight physiology, 3 in aviation meteorology, 3 in instrument flight planning and procedures, 3 in aircraft engines and systems, 3 in aircraft performance, 3 in general mathematics, 3 in aviation safety, 3 in communication skills (oral), and 3 in principles of supervision. In the upper-division baccalaureate category, if served as instructor pilot (designated on Army records as 153CC), 3 semester hours in educational psychology, 3 in personnel management, and 3 in interpersonal communication; if served as an instrument flight examiner (designated on Army records as 153CF), 3 semester hours in airline transport pilot training; if served as section leader or aircraft repair maintenance technician (designated on Army records as 153CG), 3 semester hours in organizational management, 3 in accident investigation procedures, and 3 in aircraft systems management (4/90).

MOS-153D-002

UH-60 Pilot
153D0

Exhibit Dates: 1/90–10/99.

Career Pattern

May have progressed to MOS 153D from any enlisted MOS.

Description

Pilots and commands utility helicopters under tactical and nontactical conditions; operates aircraft during all types of meteorological conditions, during day, night, and while using night vision goggles; performs aircraft operations, including rescue hoist, air assaults, aerial mine and flare deliveries, internal/external loads, and paradrop/rappelling procedures; may perform aerial reconnaissance and be used to transport passengers or cargo; routinely participates in combat and combat support training; maintains current flight status in accordance with established training requirements; knowledge and skills are equivalent to FAA commercial/instrument ratings on rotorcraft category aircraft. Knowledge and skills may be equivalent to other FAA certificates/ratings (flight instructor, instrument instructor, ground instructor, airline transport pilot, flight engineer, etc.) or the applicant may be trained on additional categories of aircraft.

Recommendation

In the lower-division baccalaureate/associate degree category, 18 semester hours in commercial/instrument rating, 3 in aircraft rules and procedures, 3 in navigation (VFR), 3 in flight physiology, 3 in aviation meteorology, 3 in instrument flight planning and procedures, 3 in aircraft engines and systems, 3 in aircraft performance, 3 in general mathematics, 3 in aviation safety, 3 in communication skills (oral), and 3 in principles of supervision. In the upper-division baccalaureate category, if served as instructor pilot (designated on Army records as 153DC), 3 semester hours in educational psychology, 3 in personnel management, and 3 in interpersonal communication; if served as an instrument flight examiner (designated on Army records as 153DF), 3 semester hours in airline transport pilot training; if served as section leader or aircraft repair maintenance technician (designated on Army records as 153DG), 3 semester hours in organizational management, 3 in accident investigation procedures, and 3 in aircraft systems management (4/90).

MOS-153D-003

UH-60 Pilot
153D0

Exhibit Dates: 11/99–Present.

Career Pattern

May have progressed to MOS 153D from any pilot warrant officer MOS.

Description

Pilots and commands fixed-wing utility aircraft under tactical and nontactical conditions; operates aircraft during all types of meteorological conditions, during day, night, and while using night vision goggles; performs aircraft operations, including rescue hoist, air assaults, aerial mine and flare deliveries, internal/external loads, and paradrop/rappelling procedures; may perform aerial reconnaissance and be used to transport passengers or cargo; routinely participates in combat and combat support training; maintains current flight status in accordance with established training requirements; knowledge and skills are equivalent to FAA commercial/instrument ratings on rotorcraft category aircraft. Knowledge and skills may be equivalent to other FAA certificates/ratings (flight instructor, instrument instructor, ground instructor, airline transport pilot, flight engineer, etc.) or the applicant may be trained on additional categories of aircraft.

Recommendation

In the lower-division baccalaureate/associate degree category, 3 semester hours in aircraft rules and procedures, 3 in navigation (VFR), 3 in flight physiology, 3 in aviation meteorology, 3 in instrument flight planning, 3 in aircraft engines and systems, 3 in aircraft performance, 3 in general mathematics, 3 in aviation safety, 3 in communication skills, 3 in technical writing, 3 in principles of supervision, 3 in computer literacy, 3 in turbine theory and operation, and 3 in risk management. In the upper-division baccalaureate category, 3 semester hours in crew resource management and 3 in flight guidance systems; if served as instructor pilot (designated on Army records as 153DC), 3 semester hours in principles of instruction, 3 in personnel management, 3 in interpersonal communication, 3 in computer applications, and 3 in curriculum development; if served as an instrument flight examiner (designated on Army records as 153DF), 3 semester hours in advanced instructional techniques; if served as section leader or aircraft repair maintenance technician (designated on Army records as 153DG), 3 semester hours in organizational management, 3 in accident investigation procedures, 3 in aircraft aviation management, and 3 in logistics (11/99).

MOS-154A-002

CH-54 Pilot (Heavy Lift RC only)
154A0

Exhibit Dates: 1/90–6/94. Effective 6/94, MOS 154A was discontinued, and its duties were incorporated into MOS 154C, CH-47D Pilot.

Career Pattern

May have progressed to MOS 154A from any enlisted MOS.

Description

Pilots and commands cargo helicopters under tactical and nontactical conditions; operates aircraft during all types of meteorological conditions during day, night, and while using night vision goggles; performs hoist, internal/external load, supply, paradrop/rappelling, and water operations; routinely participates in combat and combat support training; maintains current flight status in accordance with established training requirements; knowledge and skills are equivalent to FAA commercial/instrument ratings on rotorcraft category aircraft. Knowledge and skills may be equivalent to other FAA certificates/ratings (flight instructor, instrument instructor, ground instructor, airline transport pilot, flight engineer, etc.) or the applicant may be trained on additional categories of aircraft.

Recommendation

In the lower-division baccalaureate/associate degree category, 18 semester hours in commercial/instrument rating, 3 in aircraft rules and procedures, 3 in navigation (VFR), 3 in flight physiology, 3 in aviation meteorology, 3 in instrument flight planning and procedures, 3 in aircraft engines and systems, 3 in aircraft performance, 3 in general mathematics, 3 in aviation safety, 3 in communication skills (oral), and 3 in principles of supervision. In the upper-division baccalaureate category, if served as instructor pilot (designated on Army records as 154AC), 3 semester hours in educational psychology, 3 in personnel management, and 3 in interpersonal communication; if served as an instrument flight examiner (designated on Army records as 154AF), 3 semester hours in airline transport pilot training; if served as section leader or aircraft repair maintenance technician (designated on Army records as 154AG), 3 semester hours in organizational management, 3 in accident investigation procedures, and 3 in aircraft systems management (4/90).

MOS-154B-002

CH-47A/B/C Pilot
((CH-47 Pilot))
154B0

Exhibit Dates: 1/90–6/94. Effective 6/94, MOS 154B was discontinued and its duties were incorporated into MOS 154C, CH-47D Pilot.

Career Pattern

May have progressed to MOS 154B from any enlisted MOS.

Description

Pilots and commands cargo helicopters under tactical and nontactical conditions; operates aircraft during all types of meteorological conditions during day, night, and while using night vision goggles; performs hoist, internal/external load, supply, paradrop/rappelling, and water operations; routinely participates in combat and combat support training; maintains current flight status in accordance with established training requirements; knowledge and skills are equivalent to FAA commercial/instrument ratings on rotorcraft category aircraft. Knowledge and skills may be equivalent to other FAA certificates/ratings (flight instructor, instrument instructor, ground instructor, airline transport pilot, flight engineer, etc.) or the applicant may be trained on additional categories of aircraft.

Recommendation

In the lower-division baccalaureate/associate degree category, 18 semester hours in commercial/instrument rating, 3 in aircraft rules and procedures, 3 in navigation (VFR), 3 in flight physiology, 3 in aviation meteorology, 3 in instrument flight planning and procedures, 3 in aircraft engines and systems, 3 in aircraft performance, 3 in general mathematics, 3 in aviation safety, 3 in communication skills (oral), and 3 in principles of supervision. In the upper-division baccalaureate category, if served as instructor pilot (designated on Army records as 154BC), 3 semester hours in educational psychology, 3 in personnel management, and 3 in interpersonal communication; if served as an instrument flight examiner (designated on Army records as 154BF), 3 semester hours in airline transport pilot training; if served as section leader or aircraft repair maintenance technician (designated on Army records as 154BG), 3 semester hours in organizational management, 3 in accident investigation procedures, and 3 in aircraft systems management (4/90).

MOS-154C-002

CH-47D Pilot
154C0

Exhibit Dates: 1/90–10/99.

Career Pattern

May have progressed to MOS 154C from any enlisted MOS.

Description

Pilots and commands cargo helicopters under tactical and nontactical conditions; operates aircraft during all types of meteorological conditions during day, night, and while using night vision goggles; performs hoist, internal/external load, supply, paradrop/rappelling, and water operations; routinely participates in combat and combat support training; maintains current flight status in accordance with established training requirements; knowledge and skills are equivalent to FAA commercial/instrument ratings on rotorcraft category aircraft. Knowledge and skills may be equivalent to other FAA certificates/ratings (flight instructor, instrument instructor, ground instructor, airline transport pilot, flight engineer, etc.) or the applicant may be trained on additional categories of aircraft.

Recommendation

In the lower-division baccalaureate/associate degree category, 18 semester hours in commercial/instrument rating, 3 in aircraft rules and procedures, 3 in navigation (VFR), 3 in flight physiology, 3 in aviation meteorology, 3 in instrument flight planning and procedures, 3 in aircraft engines and systems, 3 in aircraft performance, 3 in general mathematics, 3 in aviation safety, 3 in communication skills (oral), and 3 in principles of supervision. In the upper-division baccalaureate category, if served as instructor pilot (designated on Army records as 154CC), 3 semester hours in educational psychology, 3 in personnel management, and 3 in interpersonal communication; if served as an instrument flight examiner (designated on Army records as 154CF), 3 semester hours in airline transport pilot training; if served as section leader or aircraft repair maintenance technician (designated on Army records as 154CG), 3 semester hours in organizational management, 3 in accident investigation procedures, and 3 in aircraft systems management (4/90).

MOS-154C-003

CH-47D Pilot
154C0

Exhibit Dates: 11/99–Present.

Career Pattern

May have progressed to MOS 154C from any enlisted MOS.

Description

Pilots and commands cargo helicopters under tactical and nontactical conditions; operates aircraft during all types of meteorological conditions during day, night, and while using night vision goggles; performs hoist, internal/external load, supply, paradrop/rappelling, and water operations; routinely participates in combat and combat support training; maintains current flight status in accordance with established training requirements; knowledge and skills are equivalent to FAA commercial/instrument ratings on rotorcraft category aircraft. Knowledge and skills may be equivalent to other FAA certificates/ratings (flight instructor, instrument instructor, ground instructor, airline transport pilot, flight engineer, etc.) or the applicant may be trained on additional categories of aircraft.

Recommendation

In the lower-division baccalaureate/associate degree category, 3 semester hours in aircraft rules and procedures, 3 in navigation (VFR), 3 in flight physiology, 3 in aviation meteorology, 3 in instrument flight planning, 3 in aircraft engines and systems, 3 in aircraft performance, 3 in general mathematics, 3 in aviation safety, 3 in communication skills, 3 in principles of supervision, 3 in computer literacy, 3 in technical writing, 3 in turbine theory and operations, and 3 in risk management. In the upper-division baccalaureate category, 3 semester hours in crew resource management and 3 in flight guidance systems; if served as instructor pilot (designated on Army records as 154CC), 3 semester hours in principles of instruction, 3 in personnel management, 3 in interpersonal communication, 3 in computer applications, and 3 in curriculum development; if served as an instrument flight examiner (designated on Army records as 154CF), 3 semester hours in advanced instructional techniques (helicopters). If served as a section leader or aircraft repair maintenance technician (designated on Army records as 154CG), 3 semester hours in organizational management, 3 in accident investigation procedures, 3 in aviation maintenance management, and 3 in logistics (11/99).

MOS-155A-002

Fixed Wing Aviator (Aircraft Nonspecific)
155A0

Exhibit Dates: 1/90–10/99.

Career Pattern

May have progressed to MOS 155A from any pilot warrant officer MOS in the 152, 153, or 154 MOS series.

Description

Pilots and commands fixed-wing utility aircraft under tactical and nontactical conditions; operates aircraft during day and night; transports passengers and cargo; routinely performs instrument flight procedures, navigation, and airborne weather radar interpretation; performs military intelligence and airborne radio relay missions; maintains current flight status in accordance with established training requirements; knowledge and skills are equivalent to FAA commercial/instrument ratings on rotorcraft category aircraft. Knowledge and skills may be equivalent to other FAA certificates/ratings (flight instructor, instrument instructor, ground instructor, airline transport pilot, flight engineer, etc.) or the applicant may be trained on additional categories of aircraft.

Recommendation

In the lower-division baccalaureate/associate degree category, 18 semester hours in commercial/instrument rating, 3 in aircraft rules and procedures, 3 in navigation (VFR), 3 in flight physiology, 3 in aviation meteorology, 3 in instrument flight planning and procedures, 3 in aircraft engines and systems, 3 in aircraft performance, 3 in general mathematics, 3 in aviation safety, 3 in communication skills (oral), and 3 in principles of supervision. In the upper-division baccalaureate category, if served as instructor pilot (designated on Army records as 155AC), 3 semester hours in educational psychology, 3 in personnel management, and 3 in interpersonal communication; if served as an instrument flight examiner (designated on Army records as 155AF), 3 semester hours in airline transport pilot training; if served as section leader or aircraft repair maintenance technician (designated on Army records as 154AG), 3 semester hours in organizational management, 3 in accident investigation procedures, and 3 in aircraft systems management (4/90).

MOS-155A-003

Fixed Wing Aviator (Aircraft Nonspecific)
155A0

Exhibit Dates: 11/99–Present.

Career Pattern

May have progressed to MOS 155A from any pilot warrant officer MOS in the 152, 153, or 154 MOS series.

Description

Pilots and commands fixed-wing utility aircraft under tactical and nontactical conditions; operates aircraft during day and night; transports passengers and cargo; routinely performs instrument flight procedures, navigation, and airborne weather radar interpretation; performs military intelligence and airborne radio relay missions; maintains current flight status in accordance with established training requirements; knowledge and skills are equivalent to FAA commercial/instrument ratings on fixed wing category aircraft. Knowledge and skills may be equivalent to other FAA certificates/ratings (flight instructor, instrument instructor, ground instructor, airline transport pilot, flight engineer, etc.) or the applicant may be trained on additional categories of aircraft.

Recommendation

In the lower-division baccalaureate/associate degree category, 3 semester hours in aircraft rules and procedures, 3 in navigation (VFR), 3 in flight physiology, 3 in aviation meteorology, 3 in instrument flight planning and procedures, 3 in aircraft engines and systems, 3 in aircraft performance, 3 in general mathematics, 3 in aviation safety, 3 in communication skills (oral), 3 in technical writing, 3 in principles of supervision, 3 in computer literacy, 3 in turbine theory, and 3 in risk management. In the upper-division baccalaureate category, 3 semester hours in crew resource management and 3 in flight guidance systems; if served as instructor pilot (designated on Army records as 155AC), 3 semester hours in principles of instruction, 3 in personnel management, 3 in computer applications, 3 in curriculum development, and 3 in interpersonal communication; if served as an instrument flight examiner (designated on Army records as 155AF), 3 semester hours in advanced instructional techniques; if served as section leader or aircraft repair maintenance technician (designated on Army records as 154AG), 3 semester hours in organizational management, 3 in accident investigation procedures, 3 in aircraft aviation management, and 3 in logistics (11/99).

MOS-155D-002

U-21 Pilot
155D0

Exhibit Dates: 1/90–Present.

Career Pattern

May have progressed to MOS 155D from any pilot warrant officer MOS in the 152, 153, or 154 MOS series.

Description

Pilots and commands fixed-wing utility aircraft under tactical and nontactical conditions; operates aircraft during all types of meteorological conditions during day and night; transports passengers and cargo; routinely performs instrument flight procedures, navigation, and airborne weather radar interpretation; performs military intelligence and airborne radio relay missions; maintains current flight status in accordance with established training requirements; knowledge and skills are equivalent to FAA commercial/instrument ratings on rotorcraft category aircraft. Knowledge and skills may be equivalent to other FAA certificates/ratings (flight instructor, instrument instructor, ground instructor, airline transport pilot, flight engineer, etc.) or the applicant may be trained on additional categories of aircraft.

Recommendation

In the lower-division baccalaureate/associate degree category, 18 semester hours in commercial/instrument rating, 3 in aircraft rules and procedures, 3 in navigation (VFR), 3 in flight physiology, 3 in aviation meteorology, 3 in instrument flight planning and procedures, 3 in aircraft engines and systems, 3 in aircraft performance, 3 in general mathematics, 3 in aviation safety, 3 in communication skills (oral), and 3 in principles of supervision. In the upper-division baccalaureate category, if served as instructor pilot (designated on Army records as 155DC), 3 semester hours in educational psychology, 3 in personnel management, and 3 in interpersonal communication; if served as an instrument flight examiner (designated on Army records as 155DF), 3 semester hours in airline transport pilot training; if served as section leader or aircraft repair maintenance technician (designated on Army records as 155DG), 3 semester hours in organizational management, 3 in accident investigation procedures, and 3 in aircraft systems management (4/90).

MOS-155E-002

C-12 Pilot
155E0

Exhibit Dates: 1/90–10/99.

Career Pattern

May have progressed to MOS 155E from any pilot warrant officer MOS in the 152, 153, or 154 MOS series.

Description

Pilots and commands fixed-wing utility aircraft under tactical and nontactical conditions; operates aircraft during all types of meteorological conditions during day and night; transports passengers and cargo; routinely performs instrument flight procedures, navigation, and airborne weather radar interpretation; performs military intelligence and airborne radio relay missions; maintains current flight status in accordance with established training requirements; knowledge and skills are equivalent to FAA commercial/instrument ratings on rotorcraft category aircraft. Knowledge and skills may be equivalent to other FAA certificates/ratings (flight instructor, instrument instructor, ground instructor, airline transport pilot, flight engineer, etc.) or the applicant may be trained on additional categories of aircraft.

Recommendation

In the lower-division baccalaureate/associate degree category, 18 semester hours in commercial/instrument rating, 3 in aircraft rules and procedures, 3 in navigation (VFR), 3 in flight physiology, 3 in aviation meteorology, 3 in instrument flight planning and procedures, 3 in aircraft engines and systems, 3 in aircraft performance, 3 in general mathematics, 3 in aviation safety, 3 in communication skills (oral), and 3 in principles of supervision. In the upper-division baccalaureate category, if served as instructor pilot (designated on Army records as 155EC), 3 semester hours in educational psychology, 3 in personnel management, and 3 in interpersonal communication; if served as an instrument flight examiner (designated on Army records as 155EF), 3 semester hours in airline transport pilot training; if served as section leader or aircraft repair maintenance technician (designated on Army records as 155EG), 3 semester hours in organizational management, 3 in accident investigation procedures, and 3 in aircraft systems management (4/90).

MOS-155E-003

C-12 Pilot
155E0

Exhibit Dates: 11/99–Present.

Career Pattern

May have progressed to MOS 155E from any pilot warrant officer MOS in the 152, 153, or 154 MOS series.

Description

Pilots and commands fixed-wing utility aircraft under tactical and nontactical conditions; operates aircraft during all types of meteorological conditions during day and night; transports passengers and cargo; routinely performs instrument flight procedures, navigation, and airborne weather radar interpretation; performs military intelligence and airborne radio relay missions; maintains current flight status in accordance with established training requirements; knowledge and skills are equivalent to FAA commercial/instrument ratings on fixed wing category aircraft. Knowledge and skills may be equivalent to other FAA certificates/ratings (flight instructor, instrument instructor, ground instructor, airline transport pilot, flight engineer, etc.) or the applicant may be trained on additional categories of aircraft.

Recommendation

In the lower-division baccalaureate/associate degree category, 3 semester hours in aircraft rules and procedures, 3 in navigation (VFR), 3 in flight physiology, 3 in aviation meteorology, 3 in instrument flight planning and procedures, 3 in aircraft engines and systems, 3 in aircraft performance, 3 in general mathematics, 3 in aviation safety, 3 in communication skills (oral), 3 in technical writing, 3 in principles of supervision, 3 in computer literacy, 3 in turbine theory, and 3 in risk management. In the upper-division baccalaureate category, 3 semester hours in crew resource management and 3 in flight guidance systems; if served as instructor pilot (designated on Army records as 155EC), 3 semester hours in principles of instruction, 3 in personnel management, 3 in computer applications, 3 in curriculum development, and 3 in interpersonal communication; if served as an instrument flight examiner (designated on Army records as 155EF), 3 semester hours in advanced instructional techniques; if served as section leader or aircraft repair maintenance technician (designated on Army records as 155EG), 3 semester hours in organizational management, 3 in accident investigation procedures, 3 in aircraft aviation management, and 3 in logistics (11/99).

MOS-155F-001

Jet Aircraft Pilot
155F0

Exhibit Dates: 8/98–Present.

Career Pattern

May have progressed to MOS 155F from any pilot warrant officer MOS in the 152, 153, or 154 MOS series.

Description

Pilots and commands fixed-wing utility aircraft under tactical and nontactical conditions; operates aircraft during all types of meteorological conditions during day and night; transports passengers and cargo; routinely performs instrument flight procedures, navigation, and airborne weather radar interpretation; performs military intelligence and airborne radio relay missions; maintains current flight status in accordance with established training requirements; knowledge and skills are equivalent to FAA commercial/instrument ratings on fixed wing category aircraft. Knowledge and skills may be equivalent to other FAA certificates/ratings (flight instructor, instrument instructor, ground instructor, airline transport pilot, flight engineer, etc.) or the applicant may be trained on additional categories of aircraft.

Recommendation

In the lower-division baccalaureate/associate degree category, 3 semester hours in aircraft rules and procedures, 3 in navigation (VFR), 3 in flight physiology, 3 in aviation meteorology, 3 in instrument flight planning and procedures, 3 in aircraft engines and systems, 3 in aircraft performance, 3 in general mathematics, 3 in aviation safety, 3 in communication skills (oral), 3 in technical writing, 3 in principles of supervision, 3 in computer literacy, 3 in turbine theory, and 3 in risk management. In the upper-division baccalaureate category, 3 semester hours in crew resource management, 3 in flight guidance systems, and 3 in long-range navigation; if served as instructor pilot (designated on Army records as 155FC), 3 semester hours in principles of instruction, 3 in personnel management, 3 in computer applications, 2 in curriculum development, and 3 in interpersonal communication; if served as an instrument flight examiner (designated on Army records as 155FF), 3 semester hours in advanced instructional techniques; if served as section leader or aircraft repair maintenance technician (designated on Army records as 155FG), 3 semester hours in organizational management, 3 in accident investigation procedures, 3 in aircraft aviation management, and 3 in logistics (11/99).

MOS-155G-001

05A/E0-5B/RC-7 Pilot
155G0

Exhibit Dates: 11/99–Present.

Career Pattern

May have progressed to MOS 155G from any pilot warrant officer MOS in the 152, 153, or 154 MOS series.

Description

Pilots and commands fixed-wing utility aircraft under tactical and nontactical conditions; operates aircraft during all types of meteorological conditions during day and night; transports passengers and cargo; routinely performs instrument flight procedures, navigation, and airborne weather radar interpretation; performs military intelligence and airborne radio relay missions; maintains current flight status in accordance with established training requirements; knowledge and skills are equivalent to FAA commercial/instrument ratings on fixed wing category aircraft. Knowledge and skills may be equivalent to other FAA certificates/ratings (flight instructor, instrument instructor, ground instructor, airline transport pilot, flight engineer, etc.) or the applicant may be trained on additional categories of aircraft.

Recommendation

In the lower-division baccalaureate/associate degree category, 3 semester hours in aircraft rules and procedures, 3 in navigation (VFR), 3 in flight physiology, 3 in aviation meteorology, 3 in instrument flight planning and procedures, 3 in aircraft engines and systems, 3 in aircraft performance, 3 in general mathematics, 3 in aviation safety, 3 in communication skills (oral), 3 in technical writing, 3 in principles of supervision, 3 in computer literacy, 3 in turbine theory, and 3 in risk management. In the upper-division baccalaureate category, 3 semester hours in crew resource management and 3 in flight guidance systems; if served as instructor pilot (designated on Army records as 155GC), 3 semester hours in principles of instruction, 3 in personnel management, and 3 in computer applications, 3 in curriculum development, and 3 in interpersonal communication; if served as an instrument flight examiner (designated on Army records as 155GF), 3 semester hours in advanced instructional techniques; if served as section leader or aircraft repair maintenance technician (designated on Army records as 155GG), 3 semester hours in organizational management, 3 in accident investigation procedures, 3 in aircraft aviation management, and 3 in logistics (11/99).

MOS-156A-002

OV-1/RV-1 Pilot
156A0

Exhibit Dates: 1/90–10/99.

Career Pattern

May have progressed to MOS 156A from warrant officer pilot MOS's 155A, 155D, or 155E.

Description

Pilots and commands fixed-wing surveillance aircraft under tactical and nontactical conditions; operates aircraft during all types of meteorological conditions during day and night; routinely performs instrument flight procedures, navigation, and airborne weather radar interpretation; performs military intelligence gathering, threat penetration, reconnaissance, imagery, message drop, and electronic countermeasure missions; maintains current flight status in accordance with established training requirements; knowledge and skills are equivalent to FAA commercial/instrument ratings on rotorcraft category aircraft. Knowledge and skills may be equivalent to other FAA certificates/ratings (flight instructor, instrument instructor, ground instructor, airline transport pilot, flight engineer, etc.) or the applicant may be trained on additional categories of aircraft.

Recommendation

In the lower-division baccalaureate/associate degree category, 18 semester hours in commercial/instrument rating, 3 in aircraft rules and procedures, 3 in navigation (VFR), 3 in flight physiology, 3 in aviation meteorology, 3 in instrument flight planning and procedures, 3 in aircraft engines and systems, 3 in aircraft performance, 3 in general mathematics, 3 in aviation safety, 3 in communication skills (oral), and 3 in principles of supervision. In the upper-division baccalaureate category, if served as instructor pilot (designated on Army records as 156AC), 3 semester hours in educational psychology, 3 in personnel management, and 3 in interpersonal communication; if served as an instrument flight examiner (designated on Army records as 156AF), 3 semester hours in airline transport pilot training; if served as section leader or aircraft repair maintenance technician (designated on Army records as 156AG), 3 semester hours in organizational management, 3 in accident investigation procedures, and 3 in aircraft systems management (4/90).

MOS-156A-003

OV-1/RV-1 Pilot
156A0

Exhibit Dates: 11/99–Present.

Career Pattern

May have progressed to MOS 156A from warrant officer pilot MOS's 155A, 155D, or 155E.

Description

Pilots and commands fixed-wing surveillance aircraft under tactical and nontactical conditions; operates aircraft during all types of meteorological conditions during day and night; routinely performs instrument flight procedures, navigation, and airborne weather radar interpretation; performs military intelligence gathering, threat penetration, reconnaissance, imagery, message drop, and electronic countermeasure missions; maintains current flight status in accordance with established training requirements; knowledge and skills are equivalent to FAA commercial/instrument ratings on fixed wing category aircraft. Knowledge and skills may be equivalent to other FAA certificates/ratings (flight instructor, instrument instructor, ground instructor, airline transport pilot, flight engineer, etc.) or the applicant may be trained on additional categories of aircraft.

Recommendation

In the lower-division baccalaureate/associate degree category, 3 semester hours in aircraft rules and procedures, 3 in navigation (VFR), 3 in flight physiology, 3 in aviation meteorology, 3 in instrument flight planning and procedures, 3 in aircraft engines and systems, 3 in aircraft performance, 3 in general mathematics, 3 in aviation safety, 3 in communication skills (oral), 3 in principles of supervision, 3 in computer literacy, 3 in turbine theory, and 3 in risk management. In the upper-division baccalaureate category, 3 semester hours in crew resource management and 3 in flight guidance systems; if served as instructor pilot (designated on Army records as 156AC), 3 semester hours in principles of instruction, 3 in personnel management, 3 in computer applications, 3 in curriculum development, and 3 in interpersonal communication; if served as an instrument flight examiner (designated on Army records as 156AF), 3 semester hours in advanced instructional techniques; if served as section leader or aircraft repair maintenance technician (designated on Army records as 154AG), 3 semester hours in organizational management, 3 in accident investigation procedures, 3 in aircraft aviation management, and 3 in logistics (11/99).

MOS-180A-001

Special Forces Warrant Officer
((Special Forces Technician))
((Special Operations Technician))
180A0

Exhibit Dates: 1/90–3/01.

Career Pattern

May have progressed to MOS 180A from MOS 18B (Special Operations Weapons Sergeant), 18C (Special Operations Engineer Sergeant), 18D (Special Operations Medical Sergeant), 18E (Special Operations Communications Sergeant), 18F (Special Operations Intelligence Sergeant), or 18Z (Special Operations Senior Sergeant).

Description

Duties primarily involve participation in a special operations team or detachment involving unconventional warfare, foreign internal defense, strike operations, strategic reconnaissance, and counterterrorism; the operational detachment works unilaterally or with foreign military forces; duties frequently require regional orientation including foreign language proficiency and in-country experience; duties include participation in waterborne, jungle, desert, mountain, and winter operations; many of the duties are highly classified. Manages and supervises special operations team members and technicians in communications, weapons, engineering, medicine, and operations techniques; conducts management assistance, planning, and training of special operations detachments; provides technical expertise in administration, logistics, intelligence, planning, and training; manages, supervises, and trains from 12 to 100 persons.

Recommendation

In the lower-division baccalaureate/associate degree category, 3 semester hours in personnel supervision and 3 in technical report writing. In the upper-division baccalaureate category, 3 semester hours in personnel management, 3 for a practicum in instructional techniques, and 6 for field experience in management (5/87).

MOS-180A-002

Special Forces Warrant Officer
180A0

Exhibit Dates: 4/01–Present.

Career Pattern

May have progressed to MOS 180A from MOS 18B (Special Forces Weapons Sergeant), 18C (Special Forces Engineer Sergeant), 18D (Special Forces Medical Sergeant), 18E (Special Forces Communications Sergeant), 18F (Special Forces Intelligence Sergeant), or 18Z (Special Forces Senior Sergeant).

Description

Duties primarily involve participation in a special operations team or detachment involving unconventional warfare, foreign internal defense, strike operations, strategic reconnaissance, and counterterrorism; the operational detachment works unilaterally or with foreign military forces; duties frequently require regional orientation, including foreign language proficiency and in-country experience; duties include participation in water, jungle, desert, mountain, and winter operations; many of the duties are highly classified. Manages and supervises special operations team members and technicians in communications, weapons, engineering, medicine, and operations techniques; conducts management assistance, planning, and training of special operations detachments; provides technical expertise in administration, logistics, intelligence, planning, and training; manages, supervises, and trains from 12 to 100 persons.

Recommendation

In the lower-division baccalaureate/associate degree category, 3 semester hours in principles of supervision, 3 in business communication, and 3 in computer applications. In the upper-division baccalaureate category, 3 semester hours in principles of management, 3 in instructional techniques, and 6 for field experience in management (4/01).

MOS-210A-001

Utilities Operation and Maintenance Technician
210A0

Exhibit Dates: 1/90–12/01.

Career Pattern

May have progressed to Utilities Operation and Maintenance Technician from MOS 52D (Power Generation Equipment Repairer), 52E (Prime Power Production Specialist), or 52F (Turbine Engine Driven Generator).

Description

Manages personnel engaged in utility operation and maintenance; performs administrative, supply management, and quality control functions that ensure operation, delivery, and maintenance of utilities for an installation, hospital, or associated activity; manages and applies the fundamentals of installation, operation, and maintenance of water supply, plumbing, heating, sewage, electrical, refrigeration, air conditioning, and fire fighting systems and power stations; estimates material and personnel requirements for maintenance and repair of utility plants and systems; reads and interprets blueprints, schematics, and electrical diagrams; applies electrical, electronic, and mechanical theory relevant to electrical power plants using steam-, gas-, or diesel-powered generating units; applies basic principles of military, civilian, and contractor procedures and personnel management; uses technical publications of operating principles of utilities plants and systems; diagnoses difficulties; plans maintenance and repair schedules and procedures; manages personnel and equipment engaged in the production of electrical power; determines maintenance and repair requirements and establishes work priorities; allocates maintenance and repair resources; establishes shop practices and policies; directs activities of plant and shift supervisors, electricians, mechanics, instrumentation and process control technicians, and health physics specialists; supervises installation and maintenance of high-voltage electrical power transmission lines and substations; establishes operating procedures for power plant operations, maintenance, equipment inspections, and industrial safety programs; supervises preparation of operating, malfunction, and supply reports; directs procurement and distribution of supplies, tools, and equipment, administers appropriate budgets; acts as technical advisor to commanders concerned with utilities operations; may serve as service school instructor; serves as station or power plant superintendent, power system technician, or section or team chief; may directly supervise up to 100 persons; may be able to apply occupational skill to specialized features of nuclear power plants or of ballistic missile (Safeguard System) support equipment.

Recommendation

In the lower-division baccalaureate/associate degree category, 3 semester hours in industrial safety, 3 in shop management, 3 in material management, 3 in personnel supervision, 3 in office administration, 3 in record keeping, 3 in technical writing, 3 in human relations, and credit in ballistic missile support equipment repair and nuclear power on the basis of institutional evaluation. In the upper-division baccalaureate category, 3 semester hours in management problems, 3 in management and training, and 2 for field experience in management; if rank was CW2, 2 additional semester hours for field experience in management; if rank was CW3, 4 additional semester hours for field experience in management; if rank was CW4, 6 additional semester hours for field experience in management; if duty assignment was service school instructor, 6 semester hours for a practicum in education (11/77).

MOS-210A-002

Utilities Operation and Maintenance Technician
210A0

Exhibit Dates: 1/02–Present.

Career Pattern

May have progressed to Utilities Operation and Maintenance Technician from MOS 52D (Power Generation Equipment Repairer), 52E (Prime Power Production Specialist), or 52F (Turbine and Fixed Engine Driven Generator).

Description

Manages personnel engaged in utility operation and maintenance; performs administrative, supply management, and quality control functions that ensure operation, delivery, and maintenance of utilities for an installation, hospital, or associated activity; manages and applies the fundamentals of installation, operation, and maintenance of water supply, plumbing, heating, sewage, electrical, refrigeration, air conditioning, and fire fighting systems and power stations for field hospitals; estimates material and personnel requirements for maintenance and repair of utility plants and systems; reads and interprets blueprints, schematics, and electrical diagrams; applies electrical, electronic, and mechanical theory relevant to electrical power plants using steam-, gas-, or diesel-powered generating units; applies basic principles of military, civilian, and contractor procedures for personnel management; uses technical publications of operating principles of utilities plants and systems; diagnoses difficulties; plans maintenance and repair schedules and procedures; manages personnel and equipment engaged in the production of electrical power; determines maintenance and repair requirements and establishes work priorities; allocates maintenance and repair resources; establishes shop practices and policies; directs activities of plant and shift supervisors, electricians, mechanics, and instrumentation and process control technicians; supervises installation and maintenance of high-voltage electrical power transmission lines and substations; establishes operating procedures for power plant operations, maintenance, equipment inspections, and industrial safety programs; supervises preparation of operating, malfunction, and supply reports; directs procurement and distribution of supplies, tools, and equipment; administers appropriate budgets; acts as technical advisor to commanders concerned with utilities operations; may serve as service school instructor; serves as station or power plant superintendent, power system technician, or section or team chief; may directly supervise up to 100 persons; applies and supervises safety management practices to conform to codes and regulations; manages contracts with other agencies; CW1/CW2: generally work with hospital and construction units; CW3/CW4: concentrate on primary power, instruction, and supervision; CW5: functions as a general manager, liaison officer, and makes job assignments to warrant officers in the career field.

Recommendation

In the lower-division baccalaureate/associate degree category, 3 semester hours in principles of supervision, 3 in oral/written communication skills, and 3 in installation and maintenance of prime power generating systems. In the upper-division baccalaureate category, 3 semester hours in codes and regulations, 3 in human resource management, 3 for a practicum in management, 3 in operations/logistics management, 3 in organizational management, 3 in contract management, 3 in prime power generation planning (CW3 and higher), and 3 in management problems (CW4 and higher) (1/02).

MOS-213A-001

Engineer Equipment Repair Technician
213A0

Exhibit Dates: 1/90–12/93. Effective 12/93, MOS 213A was discontinued, and its duties were incorporated into MOS 919A, Engineer Equipment Repair Technician.

Career Pattern

May have progressed to Engineer Equipment Repair Technician from MOS 52C (Utilities Equipment Repairer), MOS 52D (Power Generation Equipment Repairer), MOS 52F (Turbine Engine Driven Generator Repairer), MOS 62B (Construction Equipment Repairer), or MOS 62N (Construction Equipment Supervisor).

Description

Supervises personnel engaged in maintenance of engineer equipment; analyzes malfunctions and supervises minor repair and adjustment of engineer equipment utilized for power generation, earth moving, shaping and compacting, lifting and loading, quarrying and rock crushing, asphalt/concrete mixing and surfacing, water purification, refrigeration and air conditioning, missile system support, water gap crossing, transfer, and engineer electronic applications; supervises modification of equipment required by work orders; inspects incoming equipment to determine repair requirements; assigns work and inspects outgoing equipment; establishes maintenance and repair schedules; establishes internal administrative procedures for procurement, storage, and distribution of tools, parts, and publications.

Recommendation

In the lower-division baccalaureate/associate degree category, 3 semester hours in shop management, 3 in maintenance management, 3 in administrative office management, 3 in business communication, 3 in record keeping, 3 in computer and data processing concepts, and 3 in personnel supervision. In the upper-division baccalaureate category, 3 semester hours in personnel management, 3 in management problems, 3 for field experience in management, 3 in education electives (instruction and counseling), and additional credit in leadership and supervision on the basis of institutional evaluation; if rank was CW2, 2 additional semester hours for field experience in management; if rank was CW3, 4 additional semester hours for field experience in management; if rank was CW4, 6 additional semester hours for field experience in management (5/78).

MOS-215A-001

Photomapping Technician
215A0

Exhibit Dates: 1/90–3/91. Effective 4/91, MOS 215A was discontinued.

Career Pattern

May have progressed to Photomapping Technician from MOS 81C, Cartographer.

Description

Supervises activities and personnel engaged in preparation and revision of military maps, charts, and orthophotographs; supervises technical and specialized photogrammetric and drafting procedures used in the preparation of detailed military maps, mosaics, overlays, and prints; applies photogrammetric techniques and principles and performs topographic drafting operations and procedures; establishes work priorities; assigns duties; solves unusual problems in map making; prepares records and reports; writes technical reports.

Recommendation

In the lower-division baccalaureate/associate degree category, 3 semester hours in personnel supervision, 3 in plant or office management, 3 in technical report writing, 3 in photogrammetry, and 3 in aerial photographic interpretation. In the upper-division baccalaureate category, 3 semester hours in personnel management and training, 3 for management electives, and 3 for field experience in management; if rank was CW2, 2 additional semester hours for field experience in management; if rank was CW3, 4 additional semester hours for field experience in management; if rank was CW4, 6 additional semester hours for field experience in management (6/78).

MOS-215B-001

Survey Technician
215B0

Exhibit Dates: 1/90–3/91. Effective 4/91, MOS 215B was discontinued.

Career Pattern

NOTE: This MOS was formerly designated 821A. May have progressed to Survey Technician from MOS 82C (Field Artillery Surveyor) or MOS 82D (Topographic Surveyor).

Description

Manages field surveying activities and personnel in support of map making and artillery survey requirements; supervises technical and special procedures employed by survey parties; applies land, highway, and topographic surveying procedures; prepares and interprets aerial photographs and topographic maps; applies basic survey methods such as traverse, triangulation, and resection; uses standard survey computing forms; applies methods for determining azimuth by astronomical or other means; uses astronomical publications such as star lists and ephemeris; coordinates the activities of crews engaged in land, highway, and topographic surveys; adapts survey methods to terrain conditions; converts grid and geographical coordinates; uses trigonometry in various calculations; interprets ground maps, aerial photographs, and map substitutes; ensures exact location of points, distances, elevations, lines, areas, and contours in map making efforts; keeps accurate records of secured data; oversees proper use of surveying instruments; verifies accuracy of data obtained from surveys; makes appropriate calculations for map making, using reference tables, standard data, and mathematical formulas; establishes work priorities; designates survey parties and teams; solves complex and unusual problems encountered during survey operations; supervises accountability and maintenance of equipment; maintains records and reports of surveying activities; may serve as a service school instructor.

Recommendation

In the upper-division baccalaureate category, 3 semester hours in introduction to management, 3 in management problems, 3 in quality control, 3 in records management, 3 in personnel management, 3 in supervision and leadership, and 2 for field experience in management; if rank was CW2, 2 additional semester hours for field experience in management; if rank was CW3, 4 additional semester hours for field experience in management; if rank was CW4, 6 additional semester hours for field experience in management; and additional credit in geodetic or related sciences and cartography on the basis of institutional evaluation. Has knowledge of surveying or geodetic science equivalent to an associate degree (30-40 semester hours, depending on the general education requirements) in geometronic technology (11/77).

MOS-215C-001

Reproduction Technician
215C0

Exhibit Dates: 1/90–3/91. Effective 4/91, MOS 215C was discontinued.

Career Pattern

NOTE: This MOS was formerly designated 833A. May have progressed to Reproduction Technician from MOS 83E (Photo and Layout Specialist) or MOS 83F (Printing and Bindery Specialist).

Description

Supervises fixed and mobile platoons and teams engaged in the reproduction of maps, related material, or psychological operations material; organizes activities, schedules production, and establishes procedures; oversees maintenance of equipment; supervises reproduction of topographic maps, pictomaps, overlays, and overprints; applies lithographic reproduction quality standards and color separation processes; prepares records and reports; writes technical reports.

Recommendation

In the lower-division baccalaureate/associate degree category, 3 semester hours in personnel supervision, 3 in production control, 3 in technical report writing, and 6 for field experience in graphic arts. In the upper-division baccalaureate category, 3 semester hours in personnel management, 3 in management electives, and 2 for field experience in management; if rank was CW2, 2 additional semester hours for field experience in management; if rank was CW3, 4 additional semester hours for field experience in management; if rank was CW4, 6 additional semester hours for field experience in management (5/78).

MOS-215D-001

Terrain Analysis Technician
215D0

Exhibit Dates: 1/90–4/99.

Career Pattern

May have progressed to Terrain Analysis Technician from MOS 81Q (Terrain Analyst), 81C (Cartographer), 82D (Topographic Surveyor), 93E (Meteorological Observer), or 96B (Intelligence Analyst).

Description

Supervises activities and personnel in identification, analysis, synthesis, production, and dissemination of terrain information; collects data; designs and develops graphics; identifies project needs; analyzes impact of weather; briefs commanders; develops automated and manual filing systems; supervises use and operation of computer-aided equipment; interacts with other units; understands elements of soils, vegetation, geology, weather hydrology, and their relationship to each other; prepares lengthy reports; presents oral briefings.

Recommendation

In the lower-division baccalaureate/associate degree category, 3 semester hours in public speaking and 3 in expository writing. In the upper-division baccalaureate category, 3 semester hours in advanced physical geography, 3 in geodesy, 3 in geographic information systems, 3 in human resource management, and 3-6 semester hours for field experience in management (7/88).

MOS-215D-002

Terrain Analysis Technician
215D0

Exhibit Dates: 5/99–Present.

Career Pattern

May have progressed to Terrain Analysis Technician from MOS 81Q (Terrain Analyst), 81C (Cartographer).

Description

Supervises activities and personnel in identification, analysis, synthesis, production, and dissemination of terrain information; collects data; designs and develops graphics; identifies project needs; analyzes impact of weather; briefs commanders; develops automated and manual filing systems; supervises use and operation of computer-aided equipment; interacts with other units; understands elements of soils, vegetation, geology, weather hydrology, and their relationship to each other; prepares lengthy reports; presents oral briefings.

Recommendation

In the lower-division baccalaureate/associate degree category, 3 semester hours in oral communication skills, 3 in data base management, 3 in principles of supervision, 3 in computer operations, and 3 in technical writing. In the upper-division baccalaureate category, 3 semester hours in advanced physical geography, 3 in geodesy, 3 in geographic information systems, 3 in personnel management, 3 in operations management, and 3 for field experience in management (5/99).

MOS-250A-001

Communications Security Technician
((Telecommunications Technician))
250A0

Exhibit Dates: 1/90–9/93.

Career Pattern

May have progressed to MOS 250A from MOS 72E (Tactical Telecommunications Center Operator), 72G (Automatic Data Telecommunications Center Operator), 29F (Fixed Communications Security Equipment Repairer), 29S (Field Communications Security Equipment Repairer), and 29P (Communications Security Maintenance Chief).

Description

Manages and supervises personnel, equipment, and facilities for the operation of mobile and fixed communication center or activities; supervises personnel engaged in transmitting messages and operating communications or data processing equipment; performs systems analysis; assigns work to subordinates; plans, supervises and conducts training; assumes responsibility for accounts; is accountable for procurement, storage, and use of communications equipment; may supervise, inspect and evaluate operations associated with automated message processing centers and automatic digital switches or terminals; supervises testing, adjusting, and repairing of communications security equipment in either fixed locations or in logistic support facilities; may serve as a service school instructor; may serve as an advisor to members of allied armies; may serve as a commodity manager for communications sections, material, and in procurement, research, developmental, tests, and evaluation activities.

Recommendation

In the lower-division baccalaureate/associate degree category, 3 semester hours in shop management, 3 in maintenance management, 3 in administrative office management, 3 in business communications, 3 in record keeping, 3 in computer and data processing concepts, and 3 in personnel supervision. In the upper-division baccalaureate category, 3 semester hours in personnel management, 3 in management problems, 3 for field experience in management, 3 in education electives (instruction and counseling) and additional credit in leadership and supervision on the basis of institutional evaluation; if rank was CW2, 2 additional semester hours for field experience in management; if rank was CW3, 4 additional semester hours for field experience in management; if rank was CW4, 6 additional semester hours for field experience in management (5/78).

MOS-250A-002

Communications Security Technician

250A0

Exhibit Dates: 10/93–12/98. Effective 12/98, MOS 250A was discontinued and its duties were incorporated into MOS 250N, Network Management Technician.

Career Pattern

May have progressed to MOS 250A from MOS 29S (Communications Security Equipment Repairer) or 74C (Record Telecommunications Center Operator).

Description

Manages, plans, designs, and engineers automation, communications, and visual information systems at all command levels of army; installs, supervises, and evaluates systems; operates state of the art communications systems, including telephone, cable, switching circuit control, high frequency radio, microwave, and satellite; uses word processing equipment in preparation of technical reports and in writing revisions for technical training manuals; manages and supervises personnel, equipment, and facilities for the operation of mobile and fixed communications center or activities; supervises personnel engaged in transmitting messages and operating communications or data processing equipment; performs systems analysis; organizes communications center activities; assigns work to subordinates; plans, supervises and conducts training; is accountable for procurement, storage, and use of communications equipment; supervises testing, adjusting, and repairing of communications security equipment in either fixed locations or in logistic support facilities; may serve as a service school instructor; may serve as an advisor to member of allied armies; may serve as a commodity manager in procurement, research, development, tests, and evaluation activities; develops new equipment fielding plans; serves as communications security advisor to the commander and ensures compliance with security regulations; ensures proper employment and support of communications that are fully secure.

Recommendation

In the lower-division baccalaureate/associate degree category, 3 semester hours in service center management, 3 in maintenance management, 3 in office administration, 3 in business communications, 3 in records and information management, 3 in computer and data processing concepts, and 3 in technical writing. In the upper-division baccalaureate category, 3 semester hours in personnel management, 3 for field experience in management, 3 in education electives (instruction and counseling), 3 in strategic planning, 3 in systems management, 3 in organizational management, and 3 in material requirements planning; if rank was CW2, 2 additional semester hours for field experience in management; if rank was CW3, 4 additional semester hours for field experience in management; if rank was CW4, 6 additional semester hours for field experience in management (10/93).

MOS-250B-001

Tactical Automated Network Technician

250B0

Exhibit Dates: 1/90–12/98. Effective 12/98, MOS 250B was discontinued and its duties were incorporated into MOS 250N, Network Management Technician.

Career Pattern

May have progressed to Tactical Automated Network Technician from MOS 31F (Mobile Subscriber Equipment Network Switching System Operator), 31W (Mobile Subscriber Equipment Communications Chief), 31Y (Communications Systems Supervisor), 31Z (Communications-Operations Chief), 36L (Transportable Electronic Switching Systems Operator/Maintainer), or 72E (Telecommunications Center Operator).

Description

Manages and supervises the operation of tactical automated network message circuits and data switching systems; plans, conducts, and supervises training on automated switching networks; coordinates network troubleshooting and restoration of systems; serves as network manager for mobile subscriber equipment tactical communications systems; plans, directs, and manages the deployment, installation, activation, and movement of communications facilities comprising the networks; loads and runs system diagnostics.

Recommendation

In the lower-division baccalaureate/associate degree category, 3 semester hours in maintenance management. In the upper-division baccalaureate category, 3 semester hours for field experience in management and 3 in management problems (6/90).

MOS-250N-001

Network Management Technician

250N0

Exhibit Dates: 12/97–Present.

Career Pattern

May have progressed to 250N from MOS's 31F, 31P, 31R, 31S, 31W, and 74C.

Description

Plans and manages the deployment, installation, and activation of communications facilities comprised of local and wide area networks; manages assets of personnel, equipment, and facilities for the operation of communications and data networks; ensures compliance with operational procedures, and administrative and physical security of classified material; plans, constructs, tests, and administers the operation of local area networks; plans and conducts training programs and sessions on communications networks; maintains systems files and records.

Recommendation

In the lower-division baccalaureate/associate degree category, 3 semester hours in maintenance management, 3 in record keeping, 3 in computer operating systems, and 3 in public speaking. In the upper-division baccalaureate category, 3 semester hours in personnel management, 3 in local area networking, and 3 in network systems administration (5/99).

MOS-251A-001

Data Processing Technician

251A0

Exhibit Dates: 1/90–5/01.

Career Pattern

May have progressed to Data Processing Technician from MOS 74D (Computer/Machine Operator), MOS 74F (Programmer Analyst), or MOS 74Z (Data Processing NCO).

Description

Directs and coordinates production activities of electronic data processing element or unit, including functional or machine design applications, programming, or operation of automatic data processing (ADP) equipment; analyzes, plans, and manages the development, test, evaluation, and modification of automatic data processing systems; plans and coordinates activities of data processing personnel engaged in programming and debugging programs; establishes work standards; supervises coding and machine processing of data; develops training programs; interprets policies, purposes, and goals of ADP activity for subordinates; participates in decisions regarding personnel staffing, allocation of ADP resources, and equipment acquisition.

Recommendation

In the lower-division baccalaureate/associate degree category, 3 semester hours in data processing principles, 3 in computer fundamentals, 3 in technical report writing, 3 in business communications, 3 in computer operating systems, 3 in computer facility management, 3 in personnel supervision, and additional credit in the appropriate computer languages on the basis of institutional evaluation. In the upper-division baccalaureate category, 3 semester hours in computer systems and program design, 3 in personnel management, and 3 in management electives; if rank was CW2, 2 additional semester hours for field experience in management; if rank was CW3, 4 additional semester hours for field experience in management; if rank was CW4, 6 additional semester hours for field experience in management (6/78).

MOS-251A-002

Data Processing Technician

251A0

Exhibit Dates: 6/01–Present.

Career Pattern

May have progressed to 251A from MOS 74B (Information Systems Operator-Analyst).

Description

Directs and coordinates production activities of a data processing element or unit, including functional or machine design applications and the operation of computing equipment; designs, installs, and troubleshoots networks; analyzes, plans, and manages the development, test, evaluation, and modification of computing systems; establishes work standards; develops training programs; interprets policies, purposes, and goals of the activity for subordinates; participates in decisions regarding personnel staffing, allocation of resources, equipment acquisition, and deployment.

Recommendation

In the lower-division baccalaureate/associate degree category, 3 semester hours in introduction to computers and computing, 3 in introduction to computer operations, 3 in introduction to data communications, 3 in computer operations systems, 3 in business communication, 3 in analysis and design of data communications systems, and 3 in distributed computing systems. In the upper-division baccalaureate category, 3 semester hours in principles of management and 3 for a management practicum; if rank was CW2, add 2 semester hours for a computing systems internship; if rank was CW3, add 4 semester hours for a computing systems internship; if rank was CW4, add 6 semester hours for a computing systems internship (6/01).

MOS-252A-001

Test Measurement and Diagnostic Equipment (TMDE) Technician
((Calibration and Repair Technician))
((Calibration Technician))
252A0

Exhibit Dates: 1/90–8/91. Effective 9/91, MOS 252A was redesignated MOS 918A; see exhibit MOS-918A-001.

Career Pattern

May have progressed to Calibration Technician from MOS 35H (Calibration Specialist).

Description

Supervises or performs repair and calibration of precision equipment used for the measuring, testing, controlling, and indicating of temperature, pressure, vacuum, fluid flow, liquid level, mechanical motion, rotation, humidity, density, acidity, alkalinity, and combustion (equipment may include tools, dial pressure gauges, scales, balances, direction and sighting devices, oscilloscopes, meters, digital test equipment, signal generators, power supplies, and specialized mechanical and electronic testing equipment); maintains equipment within prescribed manufacturer standards and national or international bureau of standards tolerances and ensures that calibration surveillance standards are met; manages activities and personnel engaged in calibration of instruments or system components of all types of Army equipment and weapons; interprets and analyzes measurement data and specifications of equipment, instruments, or gauges to conform to specified standards; develops procedures for testing and diagnosing malfunctions or implements established guidance to adjust or repair item requiring calibration; interprets technical data, schematics, and transfer standards in the use and operation of material; supervises testing systems to isolate malfunctions and faulty components; instructs subordinates on the correctional adjustments to obtain special readings or characteristics such as frequency or inductance; certifies instruments for accuracy; may be assigned chief of section engaged in calibration activities and may supervise up to 50 calibration technicians; may be employed as a contracting officer's representative to monitor and evaluate the performance of contractor personnel performing maintenance calibration functions for the Army; may be assigned as a service school instructor at Army or joint service schools and typically is involved in writing of training materials and teaching of update courses to subordinates; may oversee budget of calibration unit, office, personnel, job placement, employee relations, shipping of equipment, and other essential officer-level duties.

Recommendation

In the vocational certificate category, 3 semester hours in technical electives. In the lower-division baccalaureate/associate degree category, 3 semester hours in electricity/electronics, 6 in circuits, 3 in troubleshooting techniques, 3 in electronic measurement laboratory, 3 in personnel supervision and human relations, and 3 in technical writing and report preparation. In the upper-division baccalaureate category, 6 semester hours in management electives and 3 in education electives (8/79).

MOS-254A-001

Signal Systems Support Technician
254A0

Exhibit Dates: Pending Evaluation. 10/00–Present.

MOS-255Z-001

Signal Systems Operations
255Z0

Exhibit Dates: Pending Evaluation. 10/00–Present.

MOS-256A-001

Communications-Electronics Repair Technician
256A0

Exhibit Dates: 1/90–9/93.

Career Pattern

May have progressed to Communications-Electronics Equipment Repair Technician from any of the following MOS's: 39C, 29E, 29J, 29M, 29T, 29U, 29W, 29Y, 35L, 35M, 35P, 35R, and 39E.

Description

Manages personnel and equipment in the installation, maintenance, and modification of radio, radar, television, microwave, navigation, avionic, wire communication equipment, and associated tools, tests sets and accessory equipment; receives and inspects equipment; prepares work assignments for subordinates; plans and implements maintenance schedules; develops repair and operating procedures and ensures quality control; instructs personnel on test and maintenance procedures; supervises general technical and complex repair of communications-electronic equipment at all command levels; estimates repair requirements and costs; advises commander or staff officers on communications equipment development, procurement, capabilities, limitations, and employment; may serve as service school instructor; may serve as advisor to members of allied armies; may serve with procurement; may serve in research, development, and evaluation activities related to communications equipment.

Recommendation

In the lower-division baccalaureate/associate degree category, 3 semester hours in shop management, 3 in maintenance management, 3 in administrative office management, 3 in business communications, 3 in record keeping, 3 in computer and data processing concepts, and 3 in personnel supervision. In the upper-division baccalaureate category, 3 semester hours in personnel management, 3 in management problems, 3 for field experience in management, 3 in education electives (instruction and counseling) and additional credit in leadership and supervision on the basis of institutional evaluation; if rank was CW2, 2 additional semester hours for field experience in management; if rank was CW3, 4 additional semester hours for field experience in management; if rank was CW4, 6 additional semester hours for field experience in management (5/78).

MOS-256A-002

Signal Systems Maintenance Technician
256A0

Exhibit Dates: 10/93–6/95. Effective 6/95, MOS 256A was discontinued and its duties were incorporated into MOS 918B and 250B.

Career Pattern

May have progressed to Signal Systems Maintenance Technician from any of the following MOS's: 29E, 29J, 29T, 29V, 29W, 29Y, 39C, 39E, 39G, 39L, 39V, or 68P.

Description

Manages personnel and equipment in the installation, maintenance, and modification of radio, radar, television, microwave, navigation, avionic, wire communication equipment, and associated tools, tests sets, and accessory equipment; receives and inspects equipment; prepares work assignments for subordinates; plans and implements maintenance schedules; develops repair and operating procedures and ensures quality control; instructs personnel on test and maintenance procedures; diagnoses and supervises complex repair of communications equipment development, procurement, capabilities, limitations, and employment; knows and uses operating systems and software, including DOS, OS2, UNIX, Harvard Graphics, Word Perfect, and Windows; writes extensive reports of evaluations and inspections of communications equipment problems; writes recommended changes to user advisor to members of allied armies; may serve as advisor to member of allied armies; may serve with procurement; may serve in research, development, and evaluation activities related to communications equipment.

Recommendation

In the lower-division baccalaureate/associate degree category, 3 semester hours in service center management, 3 in maintenance management, 3 in administrative office management, 3 in technical writing, 3 in records and information management, 3 in computer and data processing concepts, and 3 in personnel supervision. In the upper-division baccalaureate category, 3 semester hours in personnel management, 3 for field experience in management, 3 in education electives (instruction and counseling), 3 in systems management, and 3 in logistics management; if rank was CW2, 2 additional semester hours for field experience in management; if rank was CW3, 4 additional semester hours for field experience in management; if rank was CW4, 6 additional semester hours for field experience in management (10/93).

MOS-257A-001

Data Processing Systems Repair Technician
257A0

Exhibit Dates: 1/90–8/90. Effective 9/90, MOS 257A was discontinued.

Career Pattern

May have progressed to Data Processing Systems Repair Technician from any of the following MOS's: 25L, 29G, 29H, 39D, 39K, and 39Y.

Description

Manages personnel and equipment involved in the installation, operation, repair, maintenance, and modification of punch card, electronic data processing, command and control systems, and ancillary equipment and tools; develops operating procedures; interprets technical data; advises and instructs repair personnel; solves unusually complex problems of diagnosis, repair, and modification; establishes repair priorities; estimates repair requirements and costs; advises commander on use, capability, and limitations of data processing systems; may serve as an instructor in a service school; may serve in research, development, test, and evaluation activities.

Recommendation

In the lower-division baccalaureate/associate degree category, 3 semester hours in shop management, 3 in maintenance management, 3 in administrative office management, 3 in business communications, 3 in recordkeeping, 3 in computer and data processing concepts, and 3 in personnel supervision. In the upper-division baccalaureate category, 3 semester hours in personnel management, 3 in management problems, 3 for field experience in management, 3 in education electives (instruction and counseling) and additional credit in leadership and supervision on the basis of institutional evaluation; if rank was CW2, 2 additional semester hours for field experience in management; if rank was CW3, 4 additional semester hours for field experience in management; if rank was CW4, 6 additional semester hours for field experience in management (5/78).

MOS-270A-001

Legal Administrator
270A0

Exhibit Dates: 10/01–Present.

Career Pattern

May have progressed to Legal Administrator from MOS 27D, Paralegal Specialist.

Description

Performs the duties of a paralegal, as well the duties of a legal administrator; manages the office of the Staff Judge Advocate or of the headquarters exercising general court-martial jurisdiction; analyzes legal and administrative documents, claims, and records of court-martial and nonjudicial punishment; drafts documents dealing with civil matters, including wills, bills of sale, and separation and support agreements; interprets civil statutes and Army regulations governing the administration of military justice, contracts, and the procurement program; provides technical supervision to unit legal personnel and office staff; issues orders appointing members of court, boards, and committees; receives, processes, and authenticates official technical correspondence and makes referrals to attorneys or to other agencies when required; maintains law and administrative libraries; implements Law Library Services Policies, Procedures, and Systems; researches cases and other references to support legal, administrative, or military decisions; maintains statistical information; participates in manpower surveys; acts as office security manager; supervises military and civilian personnel; is directly responsible for office financial and budgetary matters; analyzes legal operations to determine where automated systems will enhance legal services; reviews internal automated legal research utilization reports; directs training of personnel in operation of computers, peripherals, and off-line equipment.

Recommendation

In the lower-division baccalaureate/associate degree category, 3 semester hours in office management, 3 in legal ethics, interviewing, and investigations, 3 in computer applications, 3 in principles of supervision, 3 in law office systems, 3 in litigation, and 3 in public administration. In the upper-division baccalaureate category, 3 semester hours in law office management, 3 in computer-assisted legal research, 3 in computerized litigation, 3 in human resource management, 3 in organization development, and 3 in systems design and analysis (6/01).

MOS-311A-001

CID Special Agent
311A0

Exhibit Dates: 1/90–10/94.

Career Pattern

May have progressed to CID Special Agent from MOS 95D (CID Special Agent).

Description

Conducts investigations and supervises technical and other personnel in the investigation of known or suspected crimes involving government property and individuals subject to military jurisdiction; examines scene of incident and collects and submits all relevant physical evidence (fingerprints, blood stains, suspected narcotics, castings, documents) to crime laboratory; studies and evaluates evidence to determine motives and responsible individuals; develops investigative plans; prepares reports of investigations; apprehends violators or suspects based on probable cause; testifies at courts-martial or other appropriate judiciary tribunals; engages in crime prevention efforts and physical security surveys; performs worldwide protective services for Department of Defense executives, visiting foreign officials, and other designated principals; applies techniques for effective investigation of any type of crime involving US government property and individuals subject to the Uniform Code of Military Justice; knows what constitutes an offense under applicable criminal codes and laws; applies techniques governing search and apprehension; applies rules of evidence and methods of collecting, preserving, and protecting evidence; applies complete procedures for identifying, protecting, and searching a crime scene; applies techniques of surveillance, covert operation, raids, and search and seizure and conducts such activities; applies evidence, records, and laboratory reports to criminal investigations; applies techniques of crime prevention, physical security, and industrial defense surveys; evaluates findings and develops recommendations resulting from such surveys; interviews and interrogates complainants, witnesses, informants, suspects, and other persons considered knowledgeable of or connected with crimes; develops information on suspects concerning habits, associates, aliases, characteristics, and other personal information and uses such information in the investigative process; prepares proper records of information obtained through interviews and interrogations; supervises criminal investigation teams; applies basic techniques for conducting laboratory examinations and the analysis of physical evidence; evaluates all information obtained in connection with a specific crime; performs administrative procedures and office management; writes reports and correspondence; testifies as expert witness on any standard technique or procedure used in criminal investigation; coordinates with appropriate civil and military agencies involving jurisdictional authority, exchange of information, and similar problems and considerations; may be an operations officer giving direction to other personnel involved in various investigations; may serve as an instructor or supervise instruction of personnel in various crime-related topics; may have foreign language proficiency; has extensive knowledge about narcotics and narcotics trafficking, including clandestine laboratories, drug traffic patterns, drug identification, and drug field testing; conducts drug abuse education programs.

Recommendation

In the lower-division baccalaureate/associate degree category, 6 semester hours in laws/rules of evidence, 4 in introduction to law enforcement, 3 in criminal law, 3 in criminal investigation, 3 in report writing, and 3 in oral communication, and additional credit in a foreign language on the basis of institutional evaluation. In the upper-division baccalaureate category, 6 semester hours for field experience in criminology (crime scene and laboratory) and 3 for a practicum in education, and additional credit in foreign language and education on the basis of institutional evaluation (11/77).

MOS-311A-002

CID Special Agent
311A0

Exhibit Dates: 11/94–Present.

Career Pattern

May have progressed to CID Special Agent from MOS 95D (CID Special Agent).

Description

Conducts investigations and supervises technical and other personnel in the investigation of known or suspected crimes involving government property and individuals subject to military jurisdiction; examines scene of incident and collects and submits all relevant physical evidence (fingerprints, blood stains, suspected narcotics, castings, documents) to crime laboratory; studies and evaluates evidence to determine motives and responsible individuals; develops investigative plans; prepares reports of investigations; apprehends violators or suspects based on probable cause; testifies at court-martial or other appropriate judiciary tribunals; engages in crime prevention efforts and physical security surveys; performs worldwide protective services for Department of Defense executives, visiting foreign officials, and other designated principals; applies techniques for effective investigation of any type of crime involving US government property and individuals subject to the Uniform Code of Military Justice; knows what constitutes an offense under applicable criminal codes and laws; applies techniques governing search and apprehension; applies rules of evidence and methods of collecting, preserving, and protecting evidence; applies complete procedures for identifying, protecting, and searching a crime scene; applies techniques of surveillance, covert operation, raids, and search and seizure and conducts such activities; applies evidence, records, and laboratory reports to criminal investigations; applies techniques of crime prevention, physical security, and industrial defense surveys; evaluates findings and develops recommendations resulting from such surveys; interviews and interrogates complainants, witnesses, informants, suspects, and other persons considered knowledgeable of or connected with crimes; develops information on suspects concerning habits, associates, aliases, characteristics, and other personnel information and uses such information in the investigative process; prepares proper records of information obtained through interviews and interrogations; supervises criminal investigation teams; applies basic techniques for conducting laboratory examinations and the analysis of physical evidence; evaluates all information obtained in connection with a specific crime; knows administrative procedures, report writing, correspondence formats, supply channels, and office management; testifies as expert witness concerning application of any standard technique or procedure used in criminal investigation; coordinates with appropriate civil and military agencies involving jurisdictional authority, exchange of information, and similar problems and considerations; may be an operations officer giving direction to other personnel involved in various investigations; may serve as an instructor or supervise instruction of personnel in various crime-related topics; may have foreign language proficiency; has extensive knowledge about narcotics and narcotics trafficking, including clandestine laboratories, drug traffic patterns, drug identification, and drug field testing; conducts drug abuse education programs.

Recommendation

In the lower-division baccalaureate/associate degree category, 6 semester hours in laws/rules of evidence, 3 in introduction to law enforcement, 3 in criminal law, 3 in criminal investigation, 3 in report writing, and 3 in oral communication, and additional credit in a foreign language on the basis of institutional evaluation. In the upper-division baccalaureate category, 6 semester hours for field experience in criminalistics (crime scene and laboratory) (11/94).

MOS-350B-001

All Source Intelligence Technician
((Order of Battle Technician))
350B0

Exhibit Dates: 1/90–8/91.

Career Pattern

May have progressed to MOS 350B from MOS 96B (Intelligence Analyst).

Description

Collects and evaluates tactical and strategic information related to organization, operations, capabilities, and limitations of armed forces; uses the accumulated information to develop order-of-battle data; makes reliability assessments of information through comparison with previously evaluated information; maintains close liaison with other intelligence activities, including counterintelligence, photo interpretation, interrogation, and language interpretation units; develops and maintains maps and overlays to provide complete and accurate intelligence information relating to friendly and enemy armed forces; interprets maps; prepares written reports; gives briefings and oral reports; maintains current information concerning both friendly and enemy forces, including identification, disposition, personalities, combat efficiency, and history; evaluates the significance of armed forces vulnerability studies for use in predicting probable courses of actions; has a working knowledge of automatic data processing procedures as applied to tactical and strategic intelligence; supervises, commands, or acts as chief of a section, detachment, or team engaged in the development of the data or information required in this area; may serve as a service school instructor.

Recommendation

In the lower-division baccalaureate/associate degree category, 3 semester hours in clerical practices, 3 in records administration, 3 in technical writing, 3 in oral communication, 3 in personnel supervision, and 1 in map reading. In the upper-division baccalaureate category, 3 semester hours in management problems; if rank was CW2, 2 additional semester hours for field experience in management; if rank was CW3, 4 additional semester hours for field experience in management; if rank was CW4, 6 additional semester hours for field experience in management; if duty assignment was service school instructor, 6 semester hours for a practicum in education (11/77).

MOS-350B-002

All Source Intelligence Technician
350B0

Exhibit Dates: 9/91–12/99.

Career Pattern

May progress to All Source Intelligence Technician from MOS 96B (Intelligence Analyst).

Description

Collects and evaluates tactical and strategic information related to organization, operations, capabilities, and limitations of armed forces; uses the accumulated information to develop order-of-battle data; makes reliability assessments of information through comparison with previously evaluated information; maintains close liaison with other intelligence, photo interpretation, interrogation, and language interpretation units; develops and maintains maps and overlays to provide complete and accurate intelligence information relating to friendly and enemy armed forces; interprets maps; prepares written reports; gives briefings and oral reports; maintains current information concerning both friendly and enemy forces, including identification, disposition, personalities, combat efficiency, and history; evaluates the significance of armed forces vulnerability studies for use in predicting probable courses of action; has a working knowledge of automatic data processing procedures as applied to tactical and strategic intelligence; performs basic computer functions; writes technical reports; supervises, commands, or acts as chief of a section, detachment, or team engaged in the development of the data or information required in this area.

Recommendation

In the lower-division baccalaureate/associate degree category, 3 semester hours in social sciences, 3 in technical report writing, 3 in oral communication, 3 in office management, and 3 in computer literacy. In the upper-division baccalaureate category, 3 semester hours in regional geography, 3 in applied psychology, 3 in organizational management, and 3 for field experience in management; if rank was CW2, add 3 semester hours in management problems, 3 in operations management, 3 in communication techniques for managers, and 3 additional semester hours for field experience in management; if rank was CW3, add 3 semester hours in management problems, 3 in operations management, 3 in communication techniques for managers, and 5 additional semester hours for field experience in management; if rank was CW4, add 3 semester hours in management problems, 3 in operations management, 3 in communication techniques for managers, and 7 additional semester hours for field experience in management (9/91).

MOS-350B-003

All Source Intelligence Technician
350B0

Exhibit Dates: 1/00–Present.

Career Pattern

May progress to All Source Intelligence Technician from MOS 96B (Intelligence Analyst).

Description

Collects and evaluates tactical and strategic information related to organization, operations, capabilities, and limitations of armed forces; uses the accumulated information to develop order-of-battle data; makes reliability assessments of information through comparison with previously-evaluated information; maintains close liaison with other intelligence, photo interpretation, interrogation, and language interpretation units; develops and maintains maps and overlays to provide complete and accurate intelligence information relating to friendly and enemy armed forces; interprets maps; prepares written reports; gives briefings and oral reports; maintains current information concerning both friendly and enemy forces, including identification, disposition, personalities, combat efficiency, and history; evaluates the significance of armed forces vulnerability studies for use in predicting probable courses of action; has a working knowledge of information systems and data processing as applied to tactical and strategic intelligence; uses advanced computer applications, including commercial and specialized software applications; performs software and hardware maintenance and replacement; writes technical reports; supervises, commands, or acts as chief of a section, detachment, or team engaged in the development of data or information required in this area; assists in developing long-range intelligence plans and managing projects related to intelligence technical improvements.

Recommendation

In the lower-division baccalaureate/associate degree category, 3 semester hours in regional geography, 3 in applied psychology, 3 in computer software applications, 3 in information systems maintenance, and 3 in information systems operations. In the upper-division baccalaureate category, 3 semester hours in professional writing, 3 in organizational management, 3 for a management practicum, 3 in human resource management, and 3 in office administration; if rank was CW2, add 3 semester hours in management problems, 3 in operations management, and 3 in communication techniques for managers; if rank was CW3, add 3 semester hours in management problems, 3 in operations management, 3 in communication techniques for managers, and 3 in long-range planning; if rank was CW4, add 3 semester hours in management problems, 3 in operations management, 3 in communication techniques for managers, 3 in long-range planning, and 3 in project management; if rank was CW5, add 3 semester hours in management problems, 3 in operations management, 3 in communication techniques for managers, 3 in long-range planning, 3 in project management, and 3 in managerial planning and forecasting (1/00).

MOS-350D-001

Imagery Intelligence Technician
350D0

Exhibit Dates: 1/90–8/91.

Career Pattern
May have progressed to Image Interpretation Technician from MOS 96D (Imagery Analyst).

Description
Manages activities or performs duties relative to image interpretations; applies techniques and principles of image interpretation, photogrammetry, and topographic drafting; knows and applies basic principles of geology and human and physical geography; understands basic cartography, surveying, geometry, trigonometry, and the metric system; prepares written summaries and gives oral reports; conducts briefings; establishes and maintains files; instructs subordinates in proper work techniques and procedures; directs training; supervises personnel; advises superiors; develops map overlays; may serve as a service school instructor.

Recommendation
In the lower-division baccalaureate/associate degree category, 3 semester hours in map reading, 3 in photogrammetry, 3 in aerial photographic interpretation, 3 in physical geography, 3 in technical writing, 3 in technical mathematics, 2 in mechanical drawing, and 3 in human relations. In the upper-division baccalaureate category, 3 semester hours in management problems, 3 for field experience in management, and 3 in personnel management; if duty assignment was instructor at a formal training facility, 6 additional semester hours for a practicum in education (11/77).

MOS-350D-002

Imagery Intelligence Technician
350D0

Exhibit Dates: 9/91–12/99.

Career Pattern
May have progressed to Imagery Intelligence Technician from MOS 96D, Imagery Analyst.

Description
Manages activities or performs duties relative to image interpretations; applies techniques and principles of image interpretation and photogrammetry; knows and applies basic principles of geology and human and physical geography; understands basic cartography, surveying, geometry, trigonometry, and the metric system; prepares written summaries and gives oral reports; conducts briefings; establishes and maintains files; instructs subordinates in proper work techniques and procedures; directs training; supervises personnel; advises superiors; develops map overlays; performs basic computer functions; writes technical reports.

Recommendation
In the lower-division baccalaureate/associate degree category, 3 semester hours in map and air photo interpretation, 3 in landform geography, 3 in technical report writing, and 3 in computer literacy. In the upper-division baccalaureate category, 3 semester hours in remote sensing, 3 in photogrammetry, 3 in aerial photographic interpretation, 3 in organizational management, and 3 for field experience in management; if rank was CW2, add 3 semester hours in management problems, 3 in operations management, 3 in communication techniques for managers, and 3 additional semester hours for field experience in management; if rank was CW3, add 3 semester hours in management problems, 3 in operations management, 3 in communication techniques for managers, and 5 additional semester hours for field experience in management; if rank was CW4, add 3 semester hours in management problems, 3 in operations management, 3 in communication techniques for managers, and 7 additional semester hours for field experience in management (9/91).

MOS-350D-003

Imagery Intelligence Technician
350D0

Exhibit Dates: 1/00–Present.

Career Pattern
May have progressed to Imagery Intelligence Technician from MOS 96D, Imagery Analyst.

Description
Manages activities or performs duties relative to image interpretations; applies techniques and principles of image interpretation and photogrammetry; knows and applies basic principles of geology and human and physical geography; understands basic cartography, surveying, geometry, trigonometry, and the metric system; prepares written summaries and gives oral reports; conducts briefings; establishes and maintains files; instructs subordinates in proper work techniques and procedures; directs training; supervises personnel; advises superiors; develops map overlays; performs basic computer functions; writes technical reports.

Recommendation
In the lower-division baccalaureate/associate degree category, 3 semester hours in map and aerial photographic reading, 3 in landform geography, 3 in technical writing, 3 in computer software applications, 3 in office management, and 3 in oral communication. In the upper-division baccalaureate category, 3 semester hours in remote sensing, 3 in photogrammetry, 3 in aerial photographic interpretation, 3 for a management practicum, 3 in human resource management, and 3 in office administration; if rank was CW2, add 3 semester hours in management problems, 3 in operations management, and 3 in communication techniques for managers; if rank was CW3, add 3 semester hours in management problems, 3 in operations management, 3 in communication techniques for managers, and 3 in long-range planning; if rank was CW4, add 3 semester hours in management problems, 3 in operations management, 3 in communication techniques for managers, 3 in long-range planning, and 3 in project management; if rank was CW5, add 3 semester hours in management problems, 3 in operations management, 3 in communication techniques for managers, 3 in long-range planning, 3 in project management, and 3 in managerial planning and forecasting (1/00).

MOS-350L-001

Attache Technician
350L0

Exhibit Dates: 1/90–Present.

Career Pattern
May have progressed to Attache Technician from MOS 71L (Administrative Specialist).

Description
Performs general administrative and logistics functions in support of Defense Army Attache office located in an embassy of the United States of America; must know regulations, directives, and procedures necessary for managing and operating administrative and logistics support functions; secures and manages housing accommodations for personnel assigned to the embassy; knows history, political and economic institutions, social customs, and, when possible, the language of the country to which assigned; advises other attache office personnel and visitors regarding matters of protocol, military courtesies, and public affairs; supervises enlisted and civilian support specialists; manages internal activities of the defense attache office; compiles and prepares reports and communications; receives, interviews, and schedules meetings with US and foreign military and civilian visitors; supervises internal communications; maintains files and administers the record keeping process; types correspondence, messages, forms, and manuscripts; authenticates vouchers; processes and reviews requisitions; may serve as service school instructor; may perform other officer-level duties as required.

Recommendation
In the lower-division baccalaureate/associate degree category, 4 semester hours in typing, 2 in filing, 3 in office practices, 3 in business communications, 3 in records administration, 1 in business machines, 3 in human relations, 6 for field experience in office management, 6 in social science, 2 in public relations, and 3 in personnel supervision. In the upper-division baccalaureate category, 3 semester hours in management problems and 3 in real property management; if rank was CW2, 2 additional semester hours for field experience in management; if rank was CW3, 4 additional semester hours for field experience in management; if rank was CW4, 6 additional semester hours for field experience in management. Add credit for the specific foreign language in accordance with the recommendation in the DoD volume of the Guide to the Evaluation of Educational Experiences in the Armed Services (11/77).

MOS-350U-001

Unmanned Aerial Vehicle Operations Technician
350U0

Exhibit Dates: Pending Evaluation. 6/01–Present.

MOS-351B-001

Counterintelligence Special Agent
351B0

Exhibit Dates: 1/90–8/91.

Career Pattern

May have progressed to Counterintelligence Technician from MOS 97B (Counterintelligence Agent).

Description

Manages activities or performs duties that provide intelligence information; investigates personnel under US Army jurisdiction to determine their suitability for assignment to sensitive duties; conducts interrogations, briefings, and debriefings; prepares intelligence and investigative reports; analyzes and interprets intelligence and counterintelligence data; works closely with civilian law enforcement officials and investigative agencies; supervises and trains subordinates; knows organization, mission, tactics, and operating methods of friendly and enemy intelligence units and personnel; applies principles and procedures involved in counterintelligence investigations; applies interrogation techniques; applies techniques for conducting effective liaison with local, state, and national investigative and other governmental agencies of friendly and occupied countries; knows operation, characteristics, and regulations governing the use of recording and monitoring equipment; applies techniques and regulations governing investigative surveillance; applies fundamentals of military and civil law including due process and the application of legal principles to counterintelligence operations; collects evidence admissible for legal action; obtains depositions; knows the culture, custom, history, social, economic, and political structures of the area of operation; conducts security investigations, surveys, and vulnerability studies of installations; applies principles for communications and signal security; processes photographic film; may be a service school instructor.

Recommendation

In the lower-division baccalaureate/associate degree category, 6 semester hours in social science, 3 in police administration, 3 in industrial security, 3 in US government, 3 in applied psychology, 3 in report writing, 3 in oral communication, 3 in office management, 2 in personnel supervision, and 1 in typing. In the upper-division baccalaureate category, 3 semester hours in political ideologies, 3 in constitutional law, and 3 in public relations; if the duty assignment was service school instructor, 6 additional semester hours for a practicum in education (11/77).

MOS-351B-002

Counterintelligence Technician
351B0

Exhibit Dates: 9/91–Present.

Career Pattern

May progress to Counterintelligence Technician from MOS 97B (Counterintelligence Agent).

Description

Manages activities or performs duties that provide intelligence information; investigates personnel under Army jurisdiction to determine their suitability for assignment to sensitive duties; conducts interrogations, briefings, and debriefings; prepares intelligence and investigative reports; analyzes and interprets intelligence and counterintelligence data; works closely with civilian law enforcement officials and investigative agencies; supervises and trains subordinates; knows organization, mission, tactics, and operating methods of friendly and enemy intelligence units and personnel; applies principles and procedures involved in counterintelligence investigations; applies interrogation techniques; applies techniques for conducting effective liaison with local, state, and national investigative and other governmental agencies of friendly and occupied countries; knows operation, characteristics, and regulations governing the use of recording and monitoring equipment; applies techniques and regulations governing investigative surveillance; applies fundamentals of military and civil law including due process and application of legal principles to counterintelligence operations; collects evidence admissible for legal action; obtains depositions; knows the culture, custom, history, social, economic, and political structures of the area of operation; conducts security investigations, surveys, and vulnerability studies of installations; applies principles of communications and signal security; processes photographic film; performs basic computer functions; writes technical reports.

Recommendation

In the lower-division baccalaureate/associate degree category, 6 semester hours in social science, 3 in police administration, 3 in industrial security, 3 in US government, 3 in technical report writing, 3 in oral communication, 3 in office management, and 3 in computer literacy. In the upper-division baccalaureate category, 3 semester hours in political ideologies, 3 in constitutional law, 3 in public relations, 3 in applied psychology, 3 in organizational management, and 3 for field experience in management; if rank was CW2, add 3 semester hours in management problems, 3 in operations management, 3 in communication techniques for managers, and 3 additional semester hours for field experience in management; if rank was CW3, add 3 semester hours in management problems, 3 in operations management, 3 in communication techniques for managers, and 5 additional semester hours for field experience in management; if rank was CW4, add 3 semester hours in management problems, 3 in operations management, 3 in communication techniques for managers, and 7 additional semester hours for field experience in management (9/91).

MOS-351C-001

Area Intelligence Technician
351C0

Exhibit Dates: 1/90–Present.

Career Pattern

This MOS was formerly designated 972A.

Description

Collects intelligence information on a specific geographic area through the use of human resources; works with a minimum of supervision; uses extremely demanding operational security procedures in a foreign environment; knows culture, customs, history, politics, geography, economics, and social structure of geographic area(s); applies procedures for preparing and forwarding reports; conducts interviews, interrogations, briefings, and debriefings; prepares informational and operational reports; analyzes and interprets information for intelligence implications; reads maps and prepares map overlays; supervises area intelligence units and teams. NOTE: Many of the required duties for this MOS involve highly classified materials, equipment, techniques, and activities; therefore, not all the competencies and knowledge associated with the MOS were evaluated.

Recommendation

In the lower-division baccalaureate/associate degree category, 6 semester hours in area studies, 3 in regional geography, 3 in applied psychology, 3 in report writing, 3 in speech, 3 in photography, 1 in electronic systems, 1 in typing, and 1 in human relations. In the upper-division baccalaureate category, 6 semester hours in history electives, 3 in economics electives, 3 in comparative cultures, 3 in applied research, 3 in personnel management, and 3 in international relations; if the duty assignment was service school instructor, 6 semester hours for a practicum in education. In the graduate degree category, 3 semester hours for a seminar in military intelligence strategy and policy and additional credit in international relations on the basis of institutional evaluation. Add credit for the specific foreign language in accordance with the recommendation in the DoD volume of the Guide to the Evaluation of Educational Experiences in the Armed Services (11/77).

MOS-351E-001

Interrogation Technician
351E0

Exhibit Dates: 1/90–8/91.

Career Pattern

May have progressed to Interrogation Technician from MOS 97E (Interrogator).

Description

Knows history, culture, geography, and current politics and economics of country or countries in area to which assigned; knows techniques and principles of interrogation and document exploitation; knows interrogation reporting procedures; conducts interrogations of informants, prisoners of war, and refugees in a foreign language; prepares reports of interrogations and identifies and resolves conflicting information provided by different sources; disseminates reports and makes assessments of the validity of information for dissemination to users; summarizes findings; functions as a supervisor, chief, or commander of a team or larger unit; translates technical publications related to recent scientific discoveries and inventions; usually is proficient in reading, writing, speaking, and translating two or more foreign languages.

Recommendation

In the lower-division baccalaureate/associate degree category, 6 semester hours in social science, 3 in report writing, 3 in oral communication, 3 in personnel supervision, and additional credit in area studies on the basis of institutional evaluation. In the upper-division baccalaureate category, 3 semester hours in personnel management, 3 in management problems, and additional credit in area studies on the basis of institutional evaluation. Add credit for the specific foreign language in accordance with the recommendation in the DoD volume of the Guide to the Evaluation of Educational Experiences in the Armed Services (11/77).

MOS-351E-002

Human Intelligence Collection
Technician
((Interrogation Technician))
351E0

Exhibit Dates: 9/91–12/99.

Career Pattern

May progress to Interrogation Technician from MOS 97E (Interrogator).

Description

Knows history, culture, geography, and current politics and economics of country or countries in area to which assigned; applies techniques and principles of interrogation and document exploitation; applies interrogation reporting procedures; conducts interrogations of informants, prisoners of war, and refugees in a foreign language; prepares reports of interrogations and identifies and resolves conflicting information provided by different sources; disseminates reports and makes assessments of the validity of information for dissemination to users; summarizes findings; functions as a supervisor, chief, or commander of a team or larger unit; translates technical publications related to recent scientific discoveries and inventions; usually is proficient in reading, writing, speaking, and translating foreign language; performs basic computer functions; writes technical reports.

Recommendation

In the lower-division baccalaureate/associate degree category, 6 semester hours in social science, 3 in technical report writing, 3 in oral communication, and 3 in computer literacy. In the upper-division baccalaureate category, 3 semester hours in applied psychology, 3 in regional geography, 3 in organizational management, and 3 for field experience in management; if rank was CW2, add 3 semester hours in management problems, 3 in operations management, 3 in communication techniques for managers, and 3 additional semester hours for field experience in management; if rank was CW3, add 3 semester hours in management problems, 3 in operations management, 3 in communication techniques for managers, and 5 additional semester hours for field experience in management; if rank was CW4, add 3 semester hours in management problems, 3 in operations management, 3 in communication techniques for managers, and 7 additional semester hours for field experience in management. Add credit for the specific foreign language in accordance with the recommendation in the DoD volume of the Guide to the Evaluation of Educational Experiences in the Armed Services (9/91).

MOS-351E-003

Human Intelligence Collection
Technician
((Interrogation Technician))
351E0

Exhibit Dates: 1/00–Present.

Career Pattern

May progress to Interrogation Technician from MOS 97E (Interrogator).

Description

Knows history, culture, geography, and current politics and economics of country or countries in area to which assigned; applies techniques and principles of interrogation and document exploitation; applies interrogation reporting procedures; conducts interrogations of informants, prisoners of war, and refugees in a foreign language; prepares reports of interrogations and identifies and resolves conflicting information provided by different sources; disseminates reports and makes assessments of the validity of information for dissemination to users; summarizes findings; functions as a supervisor, chief, or commander of a team or larger unit; translates technical publications related to recent scientific discoveries and inventions; usually is proficient in reading, writing, speaking, and translating foreign language; performs basic computer functions; writes technical reports; manages human intelligence assets for the command; assists in long-range planning and asset management.

Recommendation

In the lower-division baccalaureate/associate degree category, 6 semester hours in social science, 3 in technical report writing, 3 in speech communication, 3 in computer software application, 3 in regional geography, 3 in applied psychology, and 3 in principles of supervision. In the upper-division baccalaureate category, 3 semester hours in professional writing, 3 in organizational management, 3 in political ideologies, 3 for a management practicum, 3 in human resource management, and 3 in office administration; if rank was CW2, add 3 semester hours in management problems, 3 in operations management, and 3 in communication techniques for managers; if rank was CW3, add 3 semester hours in management problems, 3 in operations management, 3 in communication techniques for managers, and 3 in long-range planning; if rank was CW4, add 3 semester hours in management problems, 3 in operations management, 3 in communication techniques for managers, 3 in long-range planning, and 3 in project management; if rank was CW5, add 3 semester hours in management problems, 3 in operations management, 3 in communication techniques for managers, 3 in long-range planning, 3 in project management, and 3 in managerial planning and forecasting. Add credit for the specific foreign language (1/00).

MOS-352C-001

Traffic Analysis Technician
352C0

Exhibit Dates: 1/90–8/91.

Career Pattern

May have progressed to Traffic Analysis Technician from MOS 98C (Electronic Warfare/Signal Intelligence Analyst).

Description

Manages personnel and technical equipment engaged in intercepting, decoding, and analyzing communication traffic for content of possible intelligence value; establishes priorities of intercept missions for acquisition of desired traffic; plans, organizes, and coordinates activities; reviews data obtained in the traffic intercept process for accuracy and relevance; prepares and transmits information reports; advises commanders and staff officers on matters concerning traffic analysis and use of the data.

Recommendation

In the lower-division baccalaureate/associate degree category, 3 semester hours in personnel supervision. In the upper-division baccalaureate category, 3 semester hours in personnel management and training and 3 for management electives; if rank was CW2, 2 semester hours for field experience in management; if rank was CW3, 4 semester hours for field experience in management; if rank was CW4, 6 semester hours for field experience in management (8/78).

MOS-352C-002

Traffic Analysis Technician
352C0

Exhibit Dates: 9/91–Present.

Career Pattern

May progress to Traffic Analysis Technician from MOS 98C (Signals Intelligence Analyst).

Description

Manages personnel and technical equipment engaged in intercepting, decoding, and analyzing communication and electronics traffic for content of possible intelligence value; establishes priorities of intercept missions for acquisition of desired traffic; plans, organizes, and coordinates activities; reviews data obtained in the traffic intercept process for accuracy and relevance; prepares and transmits information reports; advises commanders and staff officers on matters concerning traffic analysis and use of the data; performs basic computer functions; if rank was CW2, also plans, organizes, and supervises teams, section and/or platoons, or other traffic analysis organization; as service school instructor develops, conducts, or supervises the instruction of student personnel and analytical procedures and techniques; assists research and development elements to develop collection, processing, location, identification, and analytical equipment; fuses all source information into all-source intelligence; if rank was CW4, may perform as the senior staff officer at national and headquarters level.

Recommendation

In the lower-division baccalaureate/associate degree category, 3 semester hours in computer literacy. In the upper-division baccalaureate category, 3 semester hours in computer software design, 3 in organizational management, and 3 for field experience in management; if rank was CW2, add 3 semester hours in management problems, 3 in operations management, 3 in communication techniques for managers, and 3 additional semester hours for field experience in management; if rank was CW3, add 3 semester hours in management problems, 3 in operations management, 3 in communication techniques for managers, and 5 additional semester hours for field experience in management; if rank was CW4, add 3 semester hours in management, 3 in communication techniques for managers, and 7 additional semester hours for field experience in management (9/91).

MOS-352D-001

Emitter Location/Identification
Technician
352D0

Exhibit Dates: 1/90–2/92.

Career Pattern

May have progressed to Emitter Location/Identification Technician from MOS 05D (Electronic Warfare/Signal Intelligence Emitter Identifier/Locator).

Description

Manages personnel and technical equipment assets engaged in establishment and employment of Emitter Location/Identification System (ELI) activities in a communications-electronic intelligence environment; supervises maintenance, calibration, adjustment, and testing of equipment; operates or supervises operation of radio (CW, voice, teletype), and cryptographic equipment; ensures that personnel follow established direction-finding communications nets; advises commanders or staff officers on the employment, deployment, and utilization of ELI assets; analyzes collected information and converts to usable intelligence for use of commanders.

Recommendation

In the lower-division baccalaureate/associate degree category, 3 semester hours in basic mathematics and 3 in personnel supervision; if served as platoon leader, 3 semester hours in principles of administration. In the upper-division baccalaureate category, 3 semester hours in personnel management and training, 3 in management problems, and 3 for management electives; if rank was CW2, 2 additional semester hours for field experience in management; if rank was CW3, 4 semester hours for field experience in management; if rank was CW4, 6 semester hours for field experience in management (8/78).

MOS-352D-002

Emitter Location/Identification Technician
352D0

Exhibit Dates: 3/92–12/00. Effective 12/00, MOS 352D was discontinued and its duties were incorporated into MOS 352H, Communications Interceptor/Locator Technician.

Career Pattern

May have progressed to Emitter Location/Identification Technician from MOS 98D (Emitter Identifier/Locator).

Description

Manages personnel and technical equipment assets engaged in establishment and employment of Emitter Location/Identification System (ELI) activities in a communications-electronic intelligence environment; supervises maintenance, calibration, adjustment, and testing of equipment; operates or supervises operation of radio (CW, voice, teletype), and cryptographic equipment; ensures that personnel follow established direction-finding communications nets; advises commanders or staff officers on the employment, deployment, and utilization of ELI assets; analyzes collected information and converts to usable intelligence for use of commanders.

Recommendation

In the lower-division baccalaureate/associate degree category, 3 semester hours in personnel supervision and 2 in records and information management. In the upper-division baccalaureate category, 3 semester hours in organizational management and 3 for field experience in management; if rank was CW2, add 3 semester hours in management problems, 3 in operations management, 3 in communication techniques for managers, and 3 additional semester hours for field experience in management; if rank was CW3, add 3 semester hours in management problems, 3 in operations management, 3 in communication techniques for managers, and 5 additional semester hours for field experience in management; if rank was CW4, add 3 semester hours in management problems, 3 in operations management, 3 in communication techniques for managers, and 7 additional semester hours for field experience in management (3/92).

MOS-352G-001

Voice Intercept Technician
352G0

Exhibit Dates: 1/90–8/91.

Career Pattern

May have progressed to Voice Intercept Technician from MOS 98G (Electronic Warfare/Signal Intelligence Voice Interceptor).

Description

Manages and plans personnel and equipment engaged in the intercept and analysis of voice communications; supervises installation of and provides operational direction to voice intercept and electronic warfare activities; supervises the calibration, adjustment, and testing of voice intercept and standard communications equipment; applies knowledge of wave propagation antenna and electronic theory to ensure proper antenna emplacement and orientation; reads, writes, comprehends, transcribes, and translates at least one foreign language; supervises intercept, transcription, and translation of voice communications in a foreign language; advises commanders and staff officers on the employment and utilization of voice intercept and electronic equipment and personnel; conducts training programs for voice intercept and electronic countermeasures personnel; advises the commander on language problem areas.

Recommendation

In the lower-division baccalaureate/associate degree category, 3 semester hours in personnel supervision; if served as platoon leader, 3 semester hours in principles of administration. In the upper-division baccalaureate category, 6 semester hours for humanities or social science electives, 3 in personnel management and training, 3 in management problems, and 3 for management electives; if rank was CW2, 2 semester hours for field experience in management; if rank was CW3, 4 semester hours for field experience in management; if rank was CW4, 6 semester hours for field experience in management. Add credit for the specific foreign language(s) by referring to the recommendation in the DoD volume of the Guide to the Evaluation of Educational Experiences in the Armed Services and modifying the recommendation on the basis of institutional evaluation (8/77).

MOS-352G-002

Voice Intercept Technician
352G0

Exhibit Dates: 9/91–Present.

Career Pattern

May have progressed to Voice Intercept Technician from MOS 98G (Voice Interceptor).

Description

Manages and plans personnel and equipment engaged in the intercept and analysis of voice communications; supervises installation of and provides operational direction to voice intercept and electronic warfare activities; supervises operators in established techniques and procedures; performs basic computer functions; reads, writes, comprehends, transcribes, and translates at least one foreign language; supervises intercept, transcription, and translation of voice communications in a foreign language; advises commanders and staff officers on the employment and utilization of voice intercept and electronic equipment and personnel; conducts training programs for voice intercept and electronic countermeasures personnel; advises the commander on language problem areas.

Recommendation

In the lower-division baccalaureate/associate degree category, 3 semester hours in computer literacy. In the upper-division baccalaureate category, 3 semester hours in regional geography, 3 in geopolitics, 3 in management problems, 3 in organizational management, and 3 for field experience in management; if rank was CW2, add 3 semester hours in management problems, 3 in operations management, 3 in communication techniques for managers, and 3 additional semester hours for field experience in management; if rank was CW3, add 3 semester hours in management problems, 3 in operations management, 3 in communication techniques for managers, and 5 additional semester hours for field experience in management; if rank was CW4, add 3 semester hours in management problems, 3 in operations management, 3 in communication techniques for managers, and 7 additional semester hours for field experience in management (9/91).

MOS-352H-001

Morse Intercept Technician
352H0

Exhibit Dates: 1/90–8/91.

Career Pattern

May have progressed to Morse Intercept Technician from MOS 05H (Electronic Warfare/Signal Intelligence Morse Interceptor).

Description

Manages personnel and technical equipment in establishment and employment of Morse intercept and electronic warfare activities; plans, coordinates, and supervises personnel engaged in these operations; may supervise several hundred persons; establishes work schedules and priorities; evaluates performance of subordinates; supervises calibration, adjustment, and testing of Morse intercept equipment; conducts training programs for equipment operators; applies antenna theory and wave propagation characteristics; supervises message analysis; supervises cryptanalysis principles and procedures.

Recommendation

In the lower-division baccalaureate/associate degree category, 3 semester hours in personnel supervision; if served as platoon leader, 3 semester hours in principles of administration. In the upper-division baccalaureate category, 3 semester hours in personnel management and training, 3 in management problems, and 3 for management electives; if rank was CW2, 2 semester hours for field experience in management; if rank was CW3, 4 semester hours for field experience in management; if rank was CW4, 6 semester hours for field experience in management (8/78).

MOS-352H-002

Morse Intercept Technician
352H0

Exhibit Dates: 9/91–12/00.

Career Pattern

May have progressed to Morse Intercept Technician from MOS 98H (Morse Interceptor).

Description

Manages personnel and technical equipment in establishment and employment of Morse intercept and electronic warfare activities; plans, coordinates, and supervises personnel engaged in these operations; may supervise several hundred persons; establishes work schedules and priorities; evaluates performance of subordinates; supervises calibration, adjustment, and testing of Morse intercept equipment; conducts training programs for equipment operators; applies antenna theory and wave propagation characteristics; supervises message analysis; supervises cryptanalysis principles and procedures; performs basic computer functions; writes technical reports.

Recommendation

In the lower-division baccalaureate/associate degree category, 3 semester hours in social sciences, 3 in technical report writing, 3 in oral communication, 3 in office management, and 3 in computer literacy. In the upper-division baccalaureate category, 3 semester hours in organizational management and 3 for field experience in management; if rank was CW2, add 3 semester hours in management problems, 3 in operations management, 3 in communication techniques for managers, and 3 additional semester hours for field experience in management; if rank was CW3, add 3 semester hours in management problems, 3 in operations management, 3 in communication techniques for managers, and 5 additional semester hours for field experience in management; if rank was CW4, add 3 semester hours in management problems, 3 in operations management, 3 in communication techniques for managers, and 7 additional semester hours for field experience in management (9/91).

MOS-352H-003

Communications Interceptor/Locator Technician
352H0

Exhibit Dates: Pending Evaluation. 1/01–Present.

MOS-352J-001

Emanations Analysis Technician
352J0

Exhibit Dates: 1/90–Present.

Career Pattern

May have progressed to Emanations Analysis Technician from MOS 98J (Electronic Warfare/Signal Intelligence Noncommunications Interceptor).

Description

Manages personnel and technical equipment assets engaged in intercept and analysis of noncommunications electromagnetic emissions; plans, organizes, and supervises establishment and operation of facilities and units engaged in electronic intelligence activities; selects sites for tactical noncommunications intercept/analysis equipment; ensures that electronic intelligence operators and analysts follow established techniques and analytical procedures in the intercept and interpretation of electromagnetic signals; advises operators and analysts; conducts training programs for noncommunications intercept operators/analysts; may serve as a service school instructor; may serve as platoon leader.

Recommendation

In the lower-division baccalaureate/associate degree category, 3 semester hours in personnel supervision; if served as platoon leader, 3 semester hours in principles of administration. In the upper-division baccalaureate category, 3 semester hours in personnel management and training, 3 in management problems, and 3 in a management elective; if rank was CW2, 2 semester hours for field experience in management; if rank was CW3, 4 semester hours for field experience in management; if rank was CW4, 6 semester hours for field experience in management (9/88).

MOS-352K-001

Non-Morse Intercept Technician
352K0

Exhibit Dates: 1/90–8/91.

Career Pattern

May have progressed to Non-Morse Intercept Technician from MOS 05K (Electronic Warfare/Signal Intelligence Non-Morse Interceptor).

Description

Manages personnel and technical equipment in establishment and employment of non-Morse intercept and electronic warfare activities; plans, coordinates, and supervises personnel engaged in these operations; may supervise several hundred persons; establishes work schedules and priorities; evaluates performance of subordinates; supervises calibration, adjustment, and testing of various electronic intercept equipment including radios and printers; conducts training programs for equipment operators; applies antenna theory and wave propagation characteristics; supervises message analysis; supervises cryptanalysis procedures.

Recommendation

In the lower-division baccalaureate/associate degree category, 3 semester hours in personnel supervision; if served as platoon leader, 3 semester hours in principles of administration. In the upper-division baccalaureate category, 3 semester hours in personnel management and training, 3 in management problems, and 3 for management electives; if rank was CW2, 2 additional semester hours for field experience in management; if rank was CW3, 4 additional semester hours for field experience in management; if rank was CW4, 6 semester hours for field experience in management (8/78).

MOS-352K-002

Non-Morse Intercept Technician
352K0

Exhibit Dates: 9/91–Present.

Career Pattern

May progress to Non-Morse Intercept Technician from MOS 98K (Non-Morse Interceptor Analyst).

Description

Manages personnel and technical equipment in establishment and employment of non-Morse intercept and electronic warfare activities; plans, coordinates, and supervises personnel engaged in these operations; may supervise several hundred persons; establishes work schedules and priorities; evaluates performance of subordinates; supervises calibration, adjustment, and testing of various electronic intercept equipment including radios and printers; conducts training programs for equipment operators; performs basic computer functions; writes technical reports; applies antenna theory and wave propagation characteristics; supervises message analysis; supervises cryptanalysis procedures.

Recommendation

In the lower-division baccalaureate/associate degree category, 3 semester hours in computer literacy and 3 in technical report writing. In the upper-division baccalaureate category, 3 semester hours in organizational management and 3 for field experience in management; if rank was CW2, add 3 semester hours in management problems, 3 in operations management, 3 in communication techniques for managers, and 3 additional semester hours for field experience in management; if rank was CW3, add 3 semester hours in management problems, 3 in operations management, 3 in communication techniques for managers, and 5 additional semester hours for field experience in management; if rank was CW4, add 3 semester hours in management problems, 3 in operations management, 3 in communication techniques for managers, and 7 additional semester hours for field experience in management (9/91).

MOS-353A-001

Intelligence and Electronic Warfare (IEW) Equipment Technician
353A0

Exhibit Dates: 1/90–8/91.

Career Pattern

May have progressed to Intelligence and Electronic Warfare Equipment Technician from any of the following MOS's: 33M, 33P, 33Q, 33R, 33T, and 33V.

Description

Manages personnel and equipment involved in installing, maintaining, repairing, and modifying complex electronic warfare, intercept, and related ancillary electronic equipment; applies theory and knowledge of vacuum tubes, transistors, and other solid state devices and digital techniques; applies techniques of circuit analysis and performance measurement; keeps maintenance records; supervises requisitioning of supplies and spare parts; recommends equipment modifications; instructs repair personnel on troubleshooting and repair and on specialized tests and procedures; estimates repair requirements and costs; prepares and implements maintenance schedules; advises commander on use of electronic systems equipment; may serve as a service school instructor.

Recommendation

In the lower-division baccalaureate/associate degree category, 3 semester hours in shop management, 3 in maintenance management, 3 in administrative office management, 3 in business communication, 3 in record keeping, 3 in computer and data processing concepts, and 3 in personnel supervision. In the upper-division baccalaureate category, 3 semester hours in personnel management, 3 in management problems, 3 for field experience in management, 3 in education electives (instruction and counseling), and additional credit in leadership and supervision on the basis of institutional evaluation; if rank was CW2, 2 additional semester hours for field experience in management; if rank was CW3, 4 additional semester hours for field experience in management; if rank was CW4, 6 additional semester hours for field experience in management (5/78).

MOS-353A-002

INTELLIGENCE AND ELECTRONIC WARFARE EQUIPMENT TECHNICIAN
353A0

Exhibit Dates: 9/91–Present.

Career Pattern

May have progressed to Intelligence and Electronic Warfare Equipment Technician from any of the following MOS's: 33M (EW/Intercept Strategic Systems Analyst and Command and Control Subsystems Repairer), 33P (EW/Intercept Strategic Receiving Subsystems Repairer), 33Q (EW/Intercept Strategic Processing and Storage Subsystems Repairer), 33R (EW/Intercept Aviation Systems Repairer), 33T (EW/Intercept Tactical Systems Repairer), or 33V (EW/Intercept Aerial Sensor Repairer).

Description

Manages personnel and equipment involved in installing, maintaining, repairing, and modifying complex electronic warfare, intercept, and related ancillary electronic equipment; applies theory and knowledge of vacuum tubes, transistors, and other solid state devices and digital techniques; applies techniques of circuit analysis and performance measurement; keeps maintenance records; supervises requisitioning of supplies and spare parts; recommends equipment modifications; instructs repair personnel on troubleshooting and repair and on specialized tests and procedures; estimates repair requirements and costs; prepares and implements maintenance schedules; advises commander on use of electronic systems equipment; performs basic computer and records management functions; writes technical reports; may serve as a service school instructor.

Recommendation

In the lower-division baccalaureate/associate degree category, 3 semester hours in computer literacy, 3 in technical report writing, and 3 in records management. In the upper-division baccalaureate category, 3 semester hours in organizational management and 3 for field experience in management; if rank was CW2, add 3 semester hours in management problems, 3 in operations management, 3 in communication techniques for managers, and 3 additional semester hours for field experience in management; if rank was CW3, add 3 semester hours in management problems, 3 in operations management, 3 in communication techniques for managers, and 5 additional semester hours for field experience in management; if rank was CW4, add 3 semester hours in management problems, 3 in operations management, 3 in communication techniques for managers, and 7 additional semester hours for field experience in management (9/91).

MOS-420A-001

MILITARY PERSONNEL TECHNICIAN
420A0

Exhibit Dates: 1/90–3/94.

Career Pattern

May have progressed from any enlisted MOS in Career Management Field 71.

Description

Supervises administrative activities related to personnel management, including office organization and operation, counseling of personnel, and personnel records; develops input for and interprets output from automated systems supporting the personnel function; oversees the selection and use of office equipment, including typewriters, calculators, and word processing equipment; supervises military and civilian personnel engaged in specialized administrative and personnel management duties; assists in hiring and evaluating civilian employees; participates in personnel selection, duty assignment, evaluation, and training to ensure full utilization of personnel; counsels personnel from a wide variety of backgrounds, assisting them with career and personal decisions; refers individuals with problems to the chaplain, Judge Advocate General, medical personnel, or appropriate agency, if necessary; initiates and prepares correspondence, messages, and reports; interprets regulations; makes decisions based on a variety of information sources and requirements; processes changes to manpower documents.

Recommendation

In the lower-division baccalaureate/associate degree category, 3 semester hours in administrative office management, 3 in business communication, 3 in records management, 3 in introduction to computer concepts, 3 in personnel supervision, and 6 for field experience in management. In the upper-division baccalaureate category, 3 semester hours in personnel management and 3 in counseling; if rank was CW2, 3 additional semester hours for field experience in personnel management and counseling; if rank was CW3, 6 additional semester hours for field experience in personnel management and counseling; if rank was CW4, 9 additional semester hours for field experience in personnel management and counseling; if duty assignment was platoon leader, additional credit in leadership and supervision on the basis of institutional evaluation. In the graduate degree category, credit for an internship in personnel management and counseling on the basis of institutional evaluation (10/77).

MOS-420A-002

MILITARY PERSONNEL TECHNICIAN
420A0

Exhibit Dates: 4/94–Present.

Career Pattern

May have progressed from any enlisted MOS in Career Management Field 71.

Description

Supervises administrative activities related to personnel management, including office organization and operation, counseling of personnel, and personnel records; develops input for and interprets output from automated systems supporting the personnel function; oversees the selection and use of office equipment, including typewriters, calculators, and word processing equipment; supervises military and civilian personnel engaged in specialized administrative and personnel management duties; assists in hiring and evaluating civilian employees; participates in personnel selection, duty assignment, evaluation, and training to ensure full utilization of personnel; counsels personnel from a wide variety of backgrounds, assisting them with career and personal decisions; refers individuals with problems to the chaplain, Judge Advocate General, medical personnel, or appropriate agency, if necessary; initiates and prepares correspondence, messages, and reports; interprets regulations; makes decisions based on a variety of information sources and requirements; processes changes to manpower documents.

Recommendation

In the lower-division baccalaureate/associate degree category, 3 semester hours in computer applications, 3 in office administration, and 6 for field experience in personnel management. In the upper-division baccalaureate category, 3 semester hours in human resource management and 3 in counseling (4/94).

MOS-420C-001

BANDMASTER
420C0

Exhibit Dates: 1/90–1/93.

Career Pattern

Normally progresses to Bandmaster from an Army enlisted bandsman (currently any MOS in Career Management Field 97, Band, which includes MOS 02B through MOS 02Z), completing from one to three preparatory programs of instruction.

Description

Possesses skill and understanding in applied music ensemble playing, basic music theory, planning concerts, conducting, etc.; acquires basic understanding of the techniques of all wind and percussion instruments; has extensive and practical experience in arranging music for bands and jazz ensembles and, in some cases, choral groups; conducts band literature equivalent to that included in upper-division and graduate courses; acquires a thorough grasp of military marching evolutions, commands, and procedures; is responsible for band administration and the supervision of the musical development of band units; directs personnel, administration, supply (uniforms, instruments, and publications), and repair and maintenance activities of band units; has fiscal control of band units; routinely serves as a company commander; may serve as an instructor in the Army element of the School of Music.

Recommendation

In the lower-division baccalaureate/associate degree category, 16 semester hours in music theory (includes harmony, sight singing, ear training, and keyboard), 8 in applied music, 4 in performing ensemble, 3 in woodwind, brass, and percussion techniques, 3 in personnel supervision, and additional credit in applied music on the basis of institutional evaluation. In the upper-division baccalaureate category, 6 semester hours in arranging, 4 in performing ensemble, 6 in conducting, 2 in marching band techniques, 3 for field experience in management, and additional credit in introduction to management on the basis of institutional evaluation. If the duty assignment was an instructor in the Army element of the School of Music, credit in teaching methods in music, education materials, evaluation and testing in education, and for a practicum in education on the basis of institutional evaluation; if rank was CW2, 2 additional semester hours for field experience in management; if rank was CW3, 4 additional semester hours for field experience in management; if rank was CW4, 6 additional semester hours for field experience in management. In the graduate degree category, 3 semester hours in conducting, 3 in band administration, 2 in performing ensemble, and additional credit in educational administration on the basis of institutional evaluation (4/77).

MOS-420C-002

BANDMASTER
420C0

Exhibit Dates: 2/93–Present.

Career Pattern

Normally progresses to Bandmaster from an Army enlisted bandsman (any MOS in Career Management Field 97, Band, which includes MOS 02B through MOS 02Z), completing from one to three preparatory programs of instruction.

Description

Possesses skill and understanding in applied music, ensemble playing, basic music theory, planning concerts, and conducting; acquires basic understanding of techniques of all wind and percussion instruments; has extensive and practical experience in arranging music for bands and jazz ensembles and, in some cases, choral groups; conducts band literature equivalent to that included in upper-division and graduate courses; acquires a thorough grasp of military marching evolutions, commands, and procedures; is responsible for band administration and the supervision of the musical development of band units; directs personnel, administration, supply (uniforms, instruments, and publications), and repair and maintenance activities of band units; has fiscal control of band units; serves routinely as company commander; may serve as an instructor in the Army element of the School of Music.

Recommendation

In the lower-division baccalaureate/associate degree category, 16 semester hours in music theory (harmony, ear training, and sight singing), 2 in jazz theory/improvisation, 8 in applied music, 6-8 in performing ensembles, 3 in woodwind, brass, and percussion techniques, and 3 in personnel supervision. In the upper-division baccalaureate category, 6 semester hours in arranging, 6 in rehearsal techniques of conducting, 6 in organizational management, 3 in logistics management, and 3 in music business management (2/93).

MOS-420D-001

CLUB MANAGER
420D0

Exhibit Dates: 1/90–Present.

Career Pattern

May have progressed to warrant officer Club Manager from enlisted MOS 00J (Club Manager).

Description

Manages the operation of Army or joint service clubs, club systems, hotels, or similar activities involving up to several thousand members; supervises all business operations of the activities, including financial operations, beverage and food services, membership activities, and social and recreational activities; knows regulations and policies governing financial management and accountability for property and funds, including in-depth knowledge of internal controls, budgeting, credit card operations, and financial statements and reports; applies fundamentals of business law and contractual relationships; applies military and civilian protocol and social customs; monitors operations to ensure control of costs, waste, pilferage, inventory rotation, and accountability; has knowledge of appropriate laws, codes, and regulations; supervises purchase, storage, preparation, and service of food and beverages; monitors club operations to ensure control of costs, inventory, accountability, and receipts; completes required reports; supervises catering of social functions, parties, and special events; selects and recommends contracting for entertainment groups; enforces safety requirements, club rules of conduct, and dress codes; supervises, directs, and trains subordinates and civilian employees.

Recommendation

In the lower-division baccalaureate/associate degree category, 3 semester hours in report writing, 3 in communication skills (oral), 3 in food and beverage management, 3 in food and labor cost control systems, 3 in personnel supervision, and 3 in office practices. In the upper-division baccalaureate category, 3 semester hours in hotel, motel, or club management and 3 in personnel management and training; if rank was CW2, 2 additional semester hours for field experience in management; if rank was CW3, 4 additional semester hours for field experience in management; if rank was CW4, 6 additional semester hours for field experience in management (11/77).

MOS-550A-001

LEGAL ADMINISTRATOR
550A0

Exhibit Dates: 1/90–12/90.

Career Pattern

May have progressed to Legal Administrator from MOS 71D (Legal Specialist) or MOS 71E (Court Reporter).

Description

Performs the duties of a paralegal, as well the duties of a legal administrator; manages the office of the Staff Judge Advocate or of the headquarters exercising general court-martial jurisdiction; analyses legal and administrative documents, claims, and records of court-martial and nonjudicial punishment; drafts documents dealing with civil matters, including wills, bills of sale, and separation and support agreements; interprets civil statutes and Army regulations governing the administration of military justice, contracts, and the procurement program; provides technical supervision to unit legal personnel and office staff; issues orders appointing members of courts, boards, and committees; receives, processes, and authenticates official technical correspondence and makes referrals to attorneys or to other agencies when required; maintains law and administrative libraries; researches cases and other references to support legal, administrative, or military decisions; maintains statistical information; participates in manpower surveys; acts as office security manager; supervises military and civilian personnel; is directly responsible for office financial and budgetary matters; if qualified as court reporter, records legal proceedings by dictation for later transcription.

Recommendation

In the lower-division baccalaureate/associate degree category, 3 semester hours in principles of supervision, 3 in principles of financial budgeting, 3 in legal terminology, 3 in law office management, 3 in trial preparation and procedures, 3 in legal research, 3 in paralegalism (duties and responsibilities of the legal assistant, code of ethics, principles of conduct, relationship to the attorney, etc.); if qualified as a court reporter, 8 additional semester hours in machine shorthand, 6 in typewriting, 3 in machine transcription, 3 in introduction to court reporting, and 3 in business communication. In the upper-division baccalaureate category, 3 semester hours in personnel management, 3 in management electives, and 3 in legal environment and processes; if rank was CW2, 2 additional semester hours for field experience in management; if rank was CW3, 4 additional semester hours for field experience in management; if rank was CW4, 6 additional semester hours for field experience in management (10/77).

MOS-550A-002

LEGAL ADMINISTRATOR
550A0

Exhibit Dates: 1/91–5/01.

Career Pattern

May have progressed to Legal Administrator from MOS 71D (Legal Specialist) or MOS 71E (Court Reporter).

Description

Performs the duties of a paralegal, as well the duties of a legal administrator; manages the office of the Staff Judge Advocate or of the headquarters exercising general court-martial jurisdiction; analyzes legal and administrative documents, claims, and records of court-martial and nonjudicial punishment; drafts documents dealing with civil matters, including wills, bills of sale, and separation and support agreements; interprets civil statutes and Army regulations governing the administration of military justice, contracts, and the procurement program; provides technical supervision to unit legal personnel and office staff; issues orders appointing members of court, boards, and committees; receives, processes, and authenticates official technical correspondence and makes referrals to attorneys or to other agencies when required; maintains law and administrative libraries; implements Law Library Services Policies, Procedures, and Systems; researches cases and other references to support legal, administrative, or military decisions; maintains statistical information; participates in manpower surveys; acts as office security manager; supervises military and civilian personnel; is directly responsible for office financial and budgetary matters; if qualified as court reporter, records legal proceedings by dictation for later transcription; analyzes legal operations to determine where automated systems will enhance legal services; reviews internal automated legal research utilization reports; directs training of personnel in operation of computers, peripherals, and off-line equipment.

Recommendation

In the lower-division baccalaureate/associate degree category, 3 semester hours in introduction to law, 3 in legal ethics, interviewing, and investigations, 3 in microcomputer applications, 3 in personnel supervision, 3 in law office systems, 3 in litigation, and 3 in public administration. In the upper-division baccalaureate category, 3 semester hours in law office management, 3 in computer-assisted legal research, 3 in computerized litigation, 3 in personnel management, 3 in organizational management, and 3 for field experience in management (4/91).

MOS-550A-003

Legal Administrator
550A0

Exhibit Dates: 6/01–9/01. Effective 10/01, MOS 550A was discontinued and its duties were transferred to MOS 270A, Legal Administrator.

Career Pattern

May have progressed to Legal Administrator from MOS 71D (Legal Specialist) or MOS 71E (Court Reporter).

Description

Performs the duties of a paralegal, as well the duties of a legal administrator; manages the office of the Staff Judge Advocate or of the headquarters exercising general court-martial jurisdiction; analyzes legal and administrative documents, claims, and records of court-martial and nonjudicial punishment; drafts documents dealing with civil matters, including wills, bills of sale, and separation and support agreements; interprets civil statutes and Army regulations governing the administration of military justice, contracts, and the procurement program; provides technical supervision to unit legal personnel and office staff; issues orders appointing members of court, boards, and committees; receives, processes, and authenticates official technical correspondence and makes referrals to attorneys or to other agencies when required; maintains law and administrative libraries; implements Law Library Services Policies, Procedures, and Systems; researches cases and other references to support legal, administrative, or military decisions; maintains statistical information; participates in manpower surveys; acts as office security manager; supervises military and civilian personnel; is directly responsible for office financial and budgetary matters; analyzes legal operations to determine where automated systems will enhance legal services; reviews internal automated legal research utilization reports; directs training of personnel in operation of computers, peripherals, and off-line equipment.

Recommendation

In the lower-division baccalaureate/associate degree category, 3 semester hours in office management; 3 in legal ethics; interviewing, and investigations; 3 in computer applications; 3 in principles of supervision; 3 in law office systems; 3 in litigation; and 3 in public administration. In the upper-division baccalaureate category, 3 semester hours in law office management, 3 in computer-assisted legal research, 3 in computerized litigation, 3 in human resource management, 3 in organization development, and 3 in systems design and analysis (6/01).

MOS-600A-001

Physician Assistant
600A0

Exhibit Dates: 1/90–5/92.

Career Pattern

May have progressed to Physician Assistant from any of the following MOS's: 91B, 91C, 91D, 91F, 91G, 91H, 91J, 91L, 91N, 91P, 91Q, 91S, 91U, 91V, 91W, 91Y, 92B, or 92E; has from three to ten years of clinical experience prior to appointment as a Physician Assistant.

Description

Working under the guidance and supervision of a physician, provides general medical care for the sick and injured; makes diagnoses of diseases, disorders, and injuries; performs preventive, diagnostic, and therapeutic medical and surgical procedures; obtains and records medical data and case histories on prescribed forms; recognizes potential zoonotic and food-borne disease problems. NOTE: Appointment to Warrant Officer as a Physician Assistant is based upon satisfactory completion of the two-year Physician Assistant program of instruction at the Army's Academy of Health Sciences, Fort Sam Houston, Texas. This program consists of 52 weeks of didactic instruction and 52 weeks of clinical practicum. Didactic instruction is provided in human anatomy and physiology, clinical physiology, inorganic chemistry, biological chemistry, community health, mental health, pharmacology, clinical medicine, clinical surgery, pediatrics, pathology and laboratory practices, preventive medicine methods and practices, and medical and surgical procedures. A practicum of either four or six weeks duration are taken in the following clinics: obstetrics/gynecology, pediatric, eye, ear, nose and throat, neuro-psychiatric, dermatology, orthopedic, surgical medicine; clinical practices are also taken in preventive medicine, ambulatory care, and laboratory service.

Recommendation

In the upper-division baccalaureate category, 60 semester hours in physician assisting; if rank was CW2, 2 additional semester hours for field experience in management; if rank was CW3, 4 additional semester hours for field experience in management; if rank was CW4, 6 additional semester hours for field experience in management (6/77).

MOS-600A-002

Physician Assistant
600A0

Exhibit Dates: 6/92–Present.

Career Pattern

May have progressed to Physician Assistant from any of the following MOS's: 91A, 91B, 91C, 91D, 91F, 91G, 91H, 91J, 91L, 91N, 91P, 91Q, 91S, 91U, 91V, 91W, 91Y, 92B, or 92E; has from three to ten years of clinical experience prior to appointment as a Physician Assistant.

Description

Working under the guidance and supervision of a physician, provides primary and specialized care of the sick and wounded; interviews and examines patients to collect history and physical data; orders diagnostics and laboratory procedures; diagnoses, treats, and prescribes courses of treatment and medication; consults and refers patients for specialized procedures or a physician's evaluation; organizes, instructs, and supervises enlisted personnel in technical aspects of patient care; prepares reports; supervises preventive medicine programs.

Recommendation

In the lower-division baccalaureate/associate degree category, 3 semester hours in personnel supervision and 2 in records and information management. In the upper-division baccalaureate category, 5 semester hours in primary care practice, 3 in organizational management, and 3 for field experience in management; if rank was CW2, add 3 semester hours in management problems, 3 in operations management, 3 in communication techniques for managers, and 3 additional semester hours for field experience in management; if rank was CW3, add 3 semester hours in management problems, 3 in operations management, 3 in communication techniques for managers, and 5 additional semester hours for field experience in management; if rank was CW4, add 3 semester hours in management problems, 3 in operations management, 3 in communication techniques for managers, and 7 additional semester hours for field experience in management (6/92).

MOS-640A-001

Veterinary Services Technician
640A0

Exhibit Dates: 1/90–11/01.

Career Pattern

Progresses to Veterinary Services Technician from MOS 91R (Veterinary Food Inspection Specialist).

Description

Supervises, directs, and manages food inspection personnel and equipment essential for maintenance of military food hygiene and food quality assurance under the guidance of a veterinarian; supervises subordinates in the conduct of tests used in veterinary medical procedures; applies specialized technical knowledge and skill regarding nonmedical aspects of animal care and control; supervises the preparation of veterinary activity reports; coordinates food inspection operations of veterinary detachments and activities; manages and directs personnel, facilities, and equipment required for military food hygiene, safety, and quality assurance under the guidance and supervision of the Deputy Command for Veterinary Services; organizes and conducts sanitary inspections, recognizes sanitary deficiencies, and completes sanitary compliance ratings of civilian and government facilities that produce, process, prepare, manufacture, store, or otherwise handle subsistence; supervises the slaughter and processing of food animals in emergency situations; directs the sampling of subsistence for testing/detection of nuclear, biological, and chemical contamination; interfaces with Department of Defense, federal inspection authorities and other agencies/offices concerning food inspection matters; identifies microbiological, chemical, and physical deterioration of subsistence; interprets toxicological, microbiological, chemical, and physical findings of subsistence; provides technical advice to installation subsistence supply activities; provides assistance to the veterinarian in programs to control animal disease and prevents zoonotic and foodborne illnesses; interviews patients and collects other data on foodborne disease and zoonotic disease to assist epidemiological investigations; observes, records, and reports on subsistence programs that may have public health significance; supervises preparation of reports pertaining to veterinary activities; performs emergency procedures for relief of poisoning, shock, heat/cold, or other life-threatening injuries to animals; provides assistance to the veterinarian in the administration and management of the veterinary nonappropriated fund, including accounting, inventory requirements, and ordering procedures; assists the veterinarian in developing and managing an effective stray animal program.

Recommendation

In the lower-division baccalaureate/associate degree category, 3 semester hours in food science technology, 3 in food inspection and regulation, 3 in personnel supervision, and 2 in records and information management. In the upper-division baccalaureate category, 3 semester hours in organizational management and 3 for field experience in management; if rank was CW2, add 3 semester hours in management problems, 3 in operations management, 3 in communication techniques for managers, and 3 additional semester hours for field experience in management; if rank was CW3, add 3 semester hours in management problems, 3 in operations management, 3 in communication techniques for managers, 3 in education and training, and 5 additional semester hours for field experience in management; if rank was CW4, add 3 semester hours in management problems, 3 in operations management, 3 in communication techniques for managers, 3 in education and training, and 7 additional semester hours for field experience in management (3/92).

MOS-640A-002

Veterinary Services Technician
640A0

Exhibit Dates: 12/01–Present.

Career Pattern

Progresses to Veterinary Services Technician from MOS 91R (Veterinary Food Inspection Specialist).

Description

Supervises, directs, and manages food inspection personnel and equipment essential for maintenance of military food hygiene and food quality assurance under the guidance of a veterinarian; supervises subordinates in the conduct of tests used in veterinary medical procedures; coordinates food inspection operations of veterinary detachments and activities; manages and directs personnel, facilities, and equipment required for military food hygiene, safety, and quality assurance under the guidance and supervision of a senior officer; organizes and conducts sanitary inspections, recognizes sanitary deficiencies, and completes sanitary compliance ratings of civilian and government facilities that produce, process, prepare, manufacture, store, or otherwise handle subsistence; supervises the slaughter and processing of food animals in emergency situations; supervises the procurement and acceptance of food stocks from foreign establishments; directs the sampling of subsistence for testing/detection of nuclear, biological, and chemical contamination; interfaces with Department of Defense, federal inspection authorities, and other agencies/offices concerning food inspection matters; identifies microbiological, chemical, and physical deterioration of subsistence; interprets toxicological, microbiological, chemical, and physical findings of subsistence; provides technical advice to installation subsistence supply activities; provides assistance to the veterinarian in programs to control animal disease and preventing zoonotic and foodborne illnesses; interviews patients and collects other data on foodborne disease and zoonotic disease to assist epidemiological investigations; observes, records, and reports on subsistence programs that may have public health significance; supervises preparation of reports pertaining to food safety; provides assistance to the veterinarian in the administration and management of accounting, inventory requirements, and ordering procedures.

Recommendation

In the lower-division baccalaureate/associate degree category, 3 semester hours in food science technology, 3 in food inspection and regulation, 3 in principles of supervision, and 2 in records and information management. In the upper-division baccalaureate category, 3 semester hours for a practicum in management; if rank was CW2, add 3 semester hours in management problems, 3 in operations management, 3 in communication techniques for managers, and 4 for a practicum in management; if rank was CW3, add 4 semester hours in management problems, 4 in operations management, 3 in communication techniques for managers, 3 in educational design and delivery, and 5 for a practicum in management; if rank was CW4, add 5 semester hours in management problems, 5 in operations management, 6 in communication techniques for managers, 3 in educational design and delivery, 2 in current problems in food safety, and 6 for a practicum in management; if rank was CW5, add 6 semester hours in management problems, 6 in operations management, 6 in communication techniques for managers, 4 in current problems in food safety, 3 in educational design and delivery and 7 for a practicum in management (12/01).

MOS-670A-001

Biomedical Equipment Repair Technician
670A0

Exhibit Dates: 1/90–5/92.

Career Pattern

May have progressed to Biomedical Equipment Repair Technician from MOS 35U (Biomedical Equipment Specialist, Advanced).

Description

Organizes, supervises, and manages personnel and organizations responsible for installing, operating, maintaining, and repairing optical, dental, and medical diagnostic and treatment equipment; plans and develops detailed procedures for installation of instruments or equipment; develops internal operating procedures and manages allocated resources; anticipates and determines repair requirements, parts, and replacement of equipment; instructs on established maintenance techniques and operating procedures; advises on and supervises use of schematics, tube testing equipment, oscilloscopes, voltmeters, and other test equipment to isolate equipment malfunctions; inspects inoperative equipment, develops cost estimates for repair, and makes appropriate recommendations regarding repair or replacement; maintains records; distributes work load; makes final inspections; develops plans for design and layout of medical laboratory or equipment facilities; prepares drawings, sketches, notes, and materials for installation of equipment; may serve as a service school instructor; may perform duties in support of the procurement, research, development, testing, and evaluation of medical equipment.

Recommendation

In the lower-division baccalaureate/associate degree category, 3 semester hours in personnel supervision, 3 in human relations, 3 in communication skills (oral and written), and 3 in record keeping. In the upper-division baccalaureate category, 6 semester hours in personnel management, 3 for field experience in management, 3 in management problems, 3 in introduction to management, 3 in public speaking, 3 in technical report writing, 3 in records administration, and 3 in office management. If rank was CW2, 2 additional semester hours for field experience in management; if rank was CW3, 4 additional semester hours for field experience in management; if rank was CW4, 6 additional semester hours for field experience in management (11/79).

MOS-670A-002

Health Services Maintenance Technician
670A0

Exhibit Dates: 6/92–4/99.

Career Pattern

Progresses to Biomedical Equipment Repair Technician from MOS 35U (Biomedical Equipment Specialist, Advanced).

Description

Organizes, supervises, and manages personnel and organizations responsible for installing, operating, maintaining, and repairing optical, dental, and medical diagnostic and treatment equipment; plans and develops detailed procedures for installation of instruments or equipment; develops internal operating procedures and manages allocated resources; anticipates and determines repair requirements, parts, and replacement of equipment; instructs on established maintenance techniques and operating procedures; advises on and supervises use of schematics, digital multimeters, functional generators, oscilloscopes, and other test equipment to isolate equipment malfunctions; inspects inoperative equipment, develops cost estimates for repair, and makes appropriate recommendations regarding repair or replacement; maintains records; distributes work load; makes final inspections; develops or plans for design and layout of medical laboratory or equipment facilities; prepares drawings, sketches, notes, and materials for installation of equipment; may serve as a service school instructor; may perform duties in support of the procurement, research, development, testing, and evaluation of medical equipment.

Recommendation

In the lower-division baccalaureate/associate degree category, 3 semester hours in computer applications, 3 in personnel supervision, and 2 in records and information management. In the upper-division baccalaureate category, 3 semester hours in organizational management and 3 for field experience in management; if rank was CW2, add 3 semester hours in management problems, 3 in operations management, 3 in communication techniques for managers, and 3 additional semester hours for field experience in management; if rank was CW3, add 3 semester hours in management problems, 3 in operations management, 3 in communication techniques for managers, and 5 additional semester hours for field experience in management; if rank was CW4, add 3 semester hours in management problems, 3 in operations management, 3 in communication techniques for managers, and 7 additional semester hours for field experience in management (6/92).

MOS-670A-003

Health Services Maintenance Technician
670A0

Exhibit Dates: 5/99–Present.

Career Pattern

Progresses to Health Services Maintenance Technician from MOS 91A (Medical Equipment Repairer).

Description

Organizes, supervises, and manages personnel and organizations responsible for installing, operating, maintaining, and repairing optical, dental, and medical diagnostic and therapeutic equipment; plans and develops detailed procedures for installation of instruments or equipment; develops internal operating procedures and manages allocated resources; anticipates and determines repair requirements, parts, and replacement of equipment; instructs on established maintenance techniques and operating procedures; advises on and supervises use of test, measurement, diagnostic, and special purpose test equipment to isolate equipment malfunctions; inspects inoperative equipment, develops cost estimates for repair, and makes appropriate recommendations regarding repair or replacement; maintains records; distributes work load; makes final inspections; develops or plans for design and layout of medical laboratory or equipment facilities; prepares drawings, sketches, notes, and materials for installation of equipment; may serve as a service school instructor; may perform duties in support of the procurement, research, development, testing, and evaluation of medical equipment.

Recommendation

In the lower-division baccalaureate/associate degree category, 3 semester hours in personnel supervision, 3 in communication skills (oral and written), and 2 in records and information management. In the upper-division baccalaureate category, 3 semester hours in management problems, 3 in personnel management, 3 for field experience in management, and 3 in records administration; if rank was CW2, add 2 semester hours for field experience in management; if rank was CW3, add 3 semester hours for field experience in management; if rank was CW4, add 3 semester hours for field experience in management (5/99).

MOS-880A-001

Marine Deck Officer
880A0

Exhibit Dates: 1/90–10/92.

Career Pattern

May have progressed to Marine Deck Officer from MOS 88K (Watercraft Operator).

Description

Commands Army watercraft on oceans, bays, sounds, lakes, and coastal waters; is responsible for proper operation and safety of the vessel including discipline of the crew and seaworthiness of the vessel; must know maritime law and have a working knowledge of international and inland rules of the road; is proficient in seamanship and ship handling; operates all deck machinery and equipment, including ground tackle, cargo gear, lifeboat, raft, fire fighting and lifesaving equipment, emergency steering, and damage control systems; qualified to control destination of vessel using all navigational aids including electronic equipment (radio direction finders, radar, and fathometers); conducts radio and visual communication with government and commercial vessels and shore stations, including flashing light, semaphore, and flag-hoist signalling using international procedures when required; knows cargo planning, stowage procedures, and techniques of loading and securing cargo to ensure seaworthiness; is proficient in celestial navigation; uses navigational aids, including charts, compasses, sextants, lighthouse, and light buoys both at sea and when approaching shore; holds qualifying license validated by the Marine Qualification Board of the Army for command position; if holding a masters license, must also have radar operation qualifications.

Recommendation

In the lower-division baccalaureate/associate degree category, 3 semester hours in personnel supervision, 3 for field experience in management, 3 in record keeping, and 3 in principles of management. In the upper-division baccalaureate category, 3 semester hours in management electives, 3 in management problems, 3 for field experience in management, and 2 in human relations; if rank was CW2, 2 additional semester hours for field experience in management; if rank was CW3; 4 additional semester hours for field experience in management; if rank was CW4, 6 additional semester hours for field experience in management (9/79).

MOS-880A-002

Marine Deck Officer
880A0

Exhibit Dates: 11/92–11/01.

Career Pattern

Progresses to Marine Deck Officer from MOS 88K (Watercraft Operator).

Description

Commands Army watercraft vessels on oceans, bays, sounds, lakes, and coastal waters; is responsible for proper operation and safety of the vessel including discipline of the crew and seaworthiness of the vessel; must know maritime law and have a working knowledge of international and inland rules of the road; is proficient in seamanship and ship handling; operates all deck machinery and equipment, including ground tackle, cargo gear, lifeboat, raft, fire fighting and lifesaving equipment, emergency steering, and damage control systems; qualified to control destination of vessel using all navigational aids, including electronic equipment (radio direction finders, radar, Loran-C, SATNAV, GPS, and fathometers); conducts radio and visual communication with government and commercial vessels and shore stations, including flashing light, semaphore, and flag-hoist signaling using international procedures when required; knows cargo planning, stowage procedures, and techniques of loading and securing cargo to ensure seaworthiness; is proficient in celestial navigation; uses navigational aids, including charts, compasses, sextants, lighthouse, and light buoys both at sea and when approaching shore; hold qualifying license for command position, validated by the Marine Qualification Board of the Army; if holding a masters license, must also have radar operation qualifications; may also serve as a nautical science instructor.

Recommendation

In the lower-division baccalaureate/associate degree category, 3 semester hours in personnel supervision and 2 in records and information management. In the upper-division baccalaureate category, 3 semester hours in organizational management and 3 for field experience in management; if rank was CW2, add 3 semester hours in management problems, 3 in operations management, 3 in communication techniques for managers, and 3 additional semester hours for field experience in management; if rank was CW3, add 3 semester hours in management problems, 3 in operations management, 3 in communication techniques for managers, and 5 additional semester hours for field experience in management; if rank was CW4, add 3 semester hours in management problems, 3 in operations management, 3 in communication techniques for managers, 3 in budget management, and 7 additional semester hours for field experience in management (11/92).

MOS-880A-003

Marine Deck Officer
880A0

Exhibit Dates: 12/01–Present.

Career Pattern

Progresses to Marine Deck Officer from MOS 88K (Watercraft Operator).

Description

Commands Army watercraft vessels on oceans, bays, sounds, lakes, and coastal waters; is responsible for proper operation and safety of the vessel including discipline of the crew and seaworthiness of the vessel; must know maritime law and have a working knowledge of international and inland rules of the road; is proficient in seamanship and ship handling; operates all deck machinery and equipment, including ground tackle, cargo gear, lifeboat, raft, fire fighting and lifesaving equipment, emergency steering, and damage control systems; qualified to control destination of vessel using all navigational aids; conducts radio and visual communication with government and commercial vessels and shore stations, including flashing light, semaphore, and flag-hoist signaling using international procedures when required; knows cargo planning, stowage procedures, and techniques of loading and securing cargo to ensure seaworthiness; is proficient in celestial navigation; uses navigational aids, including charts, compasses, sextants, lighthouse, and light buoys both at sea and when approaching shore; hold qualifying license for command position, validated by the Marine Qualification Board of the Army; if holding a masters license, must also have radar operation qualifications; may also serve as a nautical science instructor.

Recommendation

In the lower-division baccalaureate/associate degree category, 3 semester hours in principles of supervision, 3 in computer applications, and

2 in records and information management. In the upper-division baccalaureate category, 3 semester hours in organizational management and 3 for a management practicum; if rank was CW2, add 3 semester hours in management problems, 3 in operations management, 3 in communication techniques for managers, and 3 additional semester hours for a management practicum; if rank was CW3, add 3 semester hours in management problems, 3 in operations management, and 5 additional semester hours for a management practicum; if rank was CW4, add 3 semester hours in management problems, 3 in operations management, 3 in budget management, 3 in financial management, and 7 additional semester hours for a management practicum (12/01).

MOS-881A-001

Marine Engineering Officer
881A0

Exhibit Dates: 1/90–10/92.

Career Pattern
May have progressed to Marine Engineering Officer from MOS 88L (Watercraft Engineer).

Description
Supervises the operation, maintenance, repair, and overhaul of marine power plants, propulsion systems, heating and ventilating systems, and other mechanical, plumbing, and electrical equipment in ships, docks, and marine port facilities; inspects ship's machinery to determine compliance with maintenance standards or to determine extent, cost, and nature of repairs required; provides instruction in the proper maintenance and repair of machinery and shipboard equipment; monitors performance on contracts for marine repair and overhaul; supervises loading of fuel, water, and ballast; maintains and records work performed and prepares maintenance reports; when assigned as Marine Engineering Technician or Marine Inspector, inspects vessel and prepares and processes repair orders; may serve as a school instructor.

Recommendation
In the lower-division baccalaureate/associate degree category, 3 semester hours in personnel supervision, 2 for field experience in management, 3 in record keeping, and 3 in principles of management. In the upper-division baccalaureate category, 3 semester hours in management electives, 3 in management problems, and 2 for field experience in management; if rank was CW2, 2 additional semester hours for field experience in management; if rank was CW3, 4 additional semester hours for field experience in management; if rank was CW4, 6 additional semester hours for field experience in management (9/79).

MOS-881A-002

Marine Engineering Officer
881A0

Exhibit Dates: 11/92–11/01.

Career Pattern
Progresses to Marine Engineering Officer from MOS 88L (Watercraft Engineer).

Description
Supervises the operation, maintenance, repair, and overhaul of marine power plants, propulsion systems, heating and ventilating systems, and other mechanical, plumbing, and electrical equipment in ships, docks, and marine port facilities; inspects ship's machinery and shipboard equipment; monitors performance of contracts for marine repair and overhaul; supervises loading of fuel, water, and ballast; maintains and records work performed and prepares maintenance reports; when assigned as Marine Engineering Technician or Marine Inspector, inspects vessels and prepares and processes repair orders; may serve as a school instructor; holds qualifying license for command position validated by the Marine Qualification Board of the Army; may serve as a marine engineering instructor.

Recommendation
In the lower-division baccalaureate/associate degree category, 3 semester hours in personnel supervision and 2 in records and information management. In the upper-division baccalaureate category, 3 semester hours in organizational management and 3 for field experience in management; if rank was CW2, add 3 semester hours in management problems, 3 in operations management, 3 in communication techniques for managers, and 3 additional semester hours for field experience in management; if rank was CW3, add 3 semester hours in management problems, 3 in operations management, 3 in communication techniques for managers, and 5 additional semester hours for field experience in management; if rank was CW4, add 3 semester hours in management problems, 3 in operations management, 3 in communication techniques for managers, 3 in budget management, and 7 additional semester hours for field experience in management (3/92).

MOS-881A-003

Marine Engineering Officer
881A0

Exhibit Dates: 12/01–Present.

Career Pattern
Progresses to Marine Engineering Officer from MOS 88L (Watercraft Engineer).

Description
Supervises the operation, maintenance, repair, and overhaul of marine power plants, propulsion systems, heating and ventilating systems, and other mechanical, plumbing, and electrical equipment in ships, docks, and marine port facilities; inspects ship's machinery and shipboard equipment; monitors performance of contracts for marine repair and overhaul; supervises loading of fuel, water, and ballast; maintains and records work performed and prepares maintenance reports; when assigned as Marine Engineering Technician or Marine Inspector, inspects vessels and prepares and processes repair orders; may serve as a school instructor; holds qualifying license for command position validated by the Marine Qualification Board of the Army.

Recommendation
In the lower-division baccalaureate/associate degree category, 3 semester hours in principles of supervision, 3 in computer applications, and 2 in records and information management. In the upper-division baccalaureate category, 3 semester hours in organizational management and 3 for a management practicum; if rank was CW2, add 3 semester hours in management problems, 3 in operations management, 3 in communication techniques for managers, and 3 additional semester hours for a management practicum; if rank was CW3, add 3 semester hours in management problems, 3 in operations management, and 5 additional semester hours for a management practicum; if rank was CW4, add 3 semester hours in management problems, 3 in operations management, 3 in financial management, and 7 additional semester hours for a management practicum (12/01).

MOS-910A-001

Ammunition Technician
910A0

Exhibit Dates: 1/90–12/90.

Career Pattern
May have progressed to Ammunition Technician from MOS 55B (Ammunition Specialist), MOS 55D (Explosive Ordnance Disposal Specialist), MOS 55G (Nuclear Weapons Specialist), MOS 55R (Ammunition Stock Control and Accounting Specialist), MOS 55X (Ammunition Inspector), or MOS 55Z (Ammunition Supervisor).

Description
Manages personnel, equipment, supply, and facility assets for the receipt, storage, and issue of all classes of ammunition; plans, coordinates and supervises 50-150 persons engaged in the requisitioning, receipt, inspection, surveillance, testing, maintenance, repair, modification, storage, issue, crating, packaging, transportation, and destruction of ammunition, missiles, rockets, explosive components, and non-nuclear warheads; plans for safe and efficient employment of munition and high explosives; conducts surveillance inspections and determines serviceability of conventional ammunition; monitors handling and storage of special ammunition; applies policies and procedures for technical escort of chemical, biological, and nuclear materials and radioactive wastes; provides technical assistance regarding conventional ammunition to field activities; interprets technical publications and trains enlisted specialists; classifies, evacuates, or determines disposition of captured or damaged ammunition; supervises load testing, calibration, and use of ammunition test and handling equipment; participates in the investigation of accidents; prepares and reviews reports including investigative reports; is familiar with military and civilian tariffs and the logistics of supply and transportation; ensures compliance with safety and security regulations pertaining to ammunition and missiles; applies Occupational Safety and Health Act regulations and procedures; knows physiological effects of, protective measures against, and emergency and first aid procedures employed after exposure to toxic chemicals; knows fire fighting procedures for conventional and special ammunition; plans and conducts safety programs for personnel handling ammunition and missiles; reports on problem areas related to accidents and changes to procedures for storing and handling hazardous chemicals; establishes and maintains control over ammunition stocks and ensures accountability of munitions material; may serve as detachment commander, instructor at a formal service school, advisor to members of allied armies, or advisor to research, development, testing, and evaluation activities.

Recommendation
In the lower-division baccalaureate/associate degree category, 8 semester hours in physical science, 5 in industrial safety, 3 in basic elec-

tronics, 3 in personnel supervision, 3 in inventory management, 3 in report writing, 2 in maintenance shop record keeping, 2 in security practices, and additional credit in basic chemistry, chemistry of explosives, basic physics, nuclear physics, and public speaking/communication on the basis of institutional evaluation. In the upper-division baccalaureate category, 3 semester hours in personnel management and training, 3 in management electives, and 3 in logistics management; if rank was CW2, 2 additional semester hours for field experience in management; if rank was CW3, 4 additional semester hours for field experience in management; if rank was CW4, 6 additional semester hours for field experience in management; if duty assignment was platoon leader, credit in leadership and supervision on the basis of institutional evaluation (5/77).

MOS-910A-002

AMMUNITION TECHNICIAN
910A0

Exhibit Dates: 1/91–Present.

Career Pattern

May have progressed to Ammunition Technician from MOS 55B (Ammunition Specialist), MOS 55D (Explosive Ordnance Disposal Specialist), MOS 55G (Nuclear Weapons Specialist), MOS 55R (Ammunition Stock Control and Accounting Specialist), MOS 55X (Ammunition Inspector), or MOS 55Z (Ammunition Supervisor).

Description

Manages personnel, equipment, supply, and facility assets for the receipt, storage, and issue of all classes of ammunition; plans, coordinates and supervises 50-150 persons engaged in the requisitioning, receipt, inspection, surveillance, testing, maintenance, repair, modification, storage, issue, crating, packaging, transportation, and destruction of ammunition, missiles, rockets, explosive components, and non-nuclear warheads; plans for safe and efficient employment of munition and high explosives; conducts surveillance inspections and determines serviceability of conventional ammunition; monitors handling and storage of special ammunition; applies policies and procedures for technical escort of chemical, biological, and nuclear materials and radioactive wastes; provides technical assistance regarding conventional ammunition to field activities; interprets technical publications and trains enlisted specialists; classifies, evaluates, or determines disposition of captured or damaged ammunition; supervises load testing and calibration and use of ammunition test and handling equipment; participates in the investigation of accidents; prepares and reviews reports including investigative reports; is familiar with military and civilian tariffs and the logistics of supply and transportation; ensures compliance with safety and security regulations pertaining to ammunition and missiles; applies Occupational Safety and Health Act regulations and procedures; knows physiological effects of, protective measures against, and emergency and first aid procedures employed after exposure to toxic chemicals; knows fire fighting procedures for conventional and special ammunition; plans and conducts safety programs for personnel handling ammunition and missiles; reports on problem areas related to accidents and changes to procedures for storing and handling hazardous chemicals; establishes and maintains control of ammunition stocks and ensures accountability of munitions material; may serve as detachment commander, instructor at a formal service school, advisor to member of allied armies, or advisor to research, development, testing, and evaluation activities.

Recommendation

In the lower-division baccalaureate/associate degree category, 3 semester hours in physical science, 3 in industrial safety, and 3 in instrumentation. In the upper-division baccalaureate category, 3 semester hours in human relations in business, 3 in technical writing, 3 in security management, 3 in logistics management, 3 in maintenance management, 3 in operations management, 3 in communication techniques for managers, and 3 for field experience in management; if served as CW3 or CW4, 3 additional semester hours for field experience in management; if served as MW, 6 additional semester hours for field experience in management (4/91).

MOS-911A-002

NUCLEAR WEAPONS TECHNICIAN
911A0

Exhibit Dates: 1/90–12/90.

Career Pattern

May have progressed to Nuclear Weapons Technician from MOS 55D (Explosive Ordnance Disposal Specialist), MOS 55G (Nuclear Weapons Specialist), or 55Z (Ammunition Supervisor).

Description

Supervises the assembly, disassembly, inspection, test, repair, modification, maintenance, security, and safety of nuclear warhead subsections, projectiles, munitions, components, and associated trainers; supervises from 8 to 40 persons engaged in the maintenance calibration of the test and handling equipment associated with nuclear materials and weapons; determines the acceptability of nuclear warhead sections and subassemblies to ensure their reliability through close supervision and personal inspection of maintenance operations, including electronic component replacement, nuclear material replacement, and proper storage and handling; ensures that equipment modifications are accomplished as announced; examines, interprets, rewrites, and disseminates technical materials for revised operating procedures, orders, technical publications, new test equipment, and new safety procedures pertinent to nuclear materials and weapons; enforces the compliance with weapons safety rules, OSHA regulations, security regulations, and unique safety requirements; performs emergency disarming and destruction of nuclear weapons, which involves handling nuclear materials and conventional high explosives; supervises all administrative duties associated with repair and maintenance including requisitions for tools, parts, and supplies; manages personnel assignments; writes reports and maintains records; performs inspections; provides technical guidance and assistance to all support units, operating agencies, and company commanders regarding matters involving nuclear weapons.

Recommendation

In the vocational certificate category, 1 semester hour in electronic circuits and 3 in explosives handling. In the lower-division baccalaureate/associate degree category, 3 semester hours in electronic measurements laboratory, 6 in nuclear materials handling, 6 in nuclear security and safety procedures, 3 in personnel supervision and human relations, and 3 in technical writing and report preparation. In the upper-division baccalaureate category, 6 semester hours in management electives and 3 in education electives (8/79).

MOS-911A-003

NUCLEAR WEAPONS TECHNICIAN
911A0

Exhibit Dates: 1/91–6/94. Effective 6/94, MOS 911A was discontinued.

Career Pattern

May have progressed to Nuclear Weapons Technician from MOS 55D (Explosive Ordnance Disposal Specialist), MOS 55G (Nuclear Weapons Specialist), or 55Z (Ammunition Supervisor).

Description

Supervises the assembly, disassembly, inspection, test, repair, modification, maintenance, security and safety of nuclear warhead subsections, projectiles, munitions, components, and associated trainers; supervises from 8 to 40 persons engaged in the maintenance calibration of the test and handling equipment associated with nuclear materials and weapons; determines the acceptability of nuclear warhead sections and subassemblies to ensure their reliability through close supervision and inspection of maintenance operations, including electronic component replacement, nuclear material replacement, and proper storage and handling; ensures that equipment modifications are accomplished as announced; examines, interprets, rewrites, and disseminates technical materials for revised operating procedures, orders, technical publications, new test equipment, and new safety procedures pertinent to nuclear materials and weapons; enforces the compliance with weapons safety rules, OSHA regulations, security regulations, and unique safety requirements; performs emergency disarming and destruction of nuclear weapons, which involves handling nuclear materials and conventional high explosives; supervises all administrative duties associated with repair and maintenance including requisitions for tools, parts, and supplies; manages personnel assignments; writes reports and maintains records; performs inspections; provides technical guidance and assistance to all support units, operating agencies, and company commanders regarding matters involving nuclear weapons.

Recommendation

In the lower-division baccalaureate/associate degree category, 3 semester hours in physical science, 3 in industrial safety, and 3 in instrumentation. In the upper-division baccalaureate category, 3 semester hours in human relations in business, 3 in technical writing, 3 in security management, 3 in logistics management, 3 in management problems, 3 in operations management, 3 in communication techniques for managers, and 3 for field experience in management; if served as CW3 or CW4, 3 additional semester hours for field experience in management; if served as MW, 6 additional semester hours for field experience in management (4/91).

MOS-912A-001

Land Combat Support Missile Systems Technician
912A0

Exhibit Dates: 1/90–8/91.

Career Pattern

May have progressed to Land Combat Support Missile Systems Technician from any of the following MOS's: 27B, 27E, 27F, 27G, 27L, 27M, 27N, and 27Z.

Description

Manages supply, equipment, facility, and personnel in the maintenance and repair of land combat support missile systems (short or medium-distance guided missile weapon systems), associated system-designed electronic test systems, and training devices; directs the activities of personnel engaged in isolating malfunctions occurring in electronic firing and guiding systems and components; analyzes complex malfunctions; supervises repair of and modifications to system components, including guidance control systems, optical trackers, modulators, infrared sensors, conduct-of-fire trainers, missiles, and specialized test equipment; oversees supply management and preparation of work reports and maintenance records; forecasts supply requirements; ensures that final inspections and checkout procedures are conducted; examines, interprets, and disseminates technical material, including orders, bulletins, and manuals; instructs enlisted specialists in troubleshooting techniques, diagnostic procedures, and the interpretation of tests on system components, circuits, and optical elements; applies techniques and safety procedures for handling explosive components; may serve as instructor at a formal training school.

Recommendation

In the vocational certificate category, 2 semester hours in use of basic hand tools. In the lower-division baccalaureate/associate degree category, 6 semester hours in electronics theory and laboratory, 3 in personnel supervision, 3 for field experience in management, 3 in industrial/human relations, 2 in maintenance management, and 1 in technical report writing. In the upper-division baccalaureate category, 3 semester hours in personnel management and training and 3 in management electives (3/77).

MOS-912A-002

Land Combat Missile Systems Technician
912A0

Exhibit Dates: 9/91–5/01.

Career Pattern

May have progressed to Land Combat Support Missile System Technician from any of the following MOS's: 27B (Land Combat Support System Test Specialist), 27E (TOW/Dragon Repairer), 27F (Vulcan Repairer), 27G (Chaparral/Redeye Repairer), 27L (Lance System Repairer), 27M (Multiple Launch Rocket System Repairer), 27N (Forward Area Alerting Radar Repairer), or 27Z (Land Combat/Air Defense Systems Maintenance Chief).

Description

Serves as a technical manager for the land combat missile system; manages the logistics of supply, production and control, personnel management, equipment and facility in maintenance and repair of land combat support missile systems; directs the activities of personnel engaged in isolating malfunctions occurring in electronic firing and guiding systems and components; analyzes complex malfunctions; supervises repair of and modifications to system components, including guidance control systems, optical trackers, modulators, infrared sensors, conduct-of-fire trainers, missiles, and specialized test equipment; oversees supply management and preparation of work reports and maintenance records; forecasts supply requirements; ensures that final inspections and checkout procedures are conducted; examines, interprets, and disseminates technical material, including orders, bulletins, and manuals; instructs enlisted specialists in troubleshooting techniques, diagnostic procedures, and the interpretation of tests on system components, circuits, and optical elements; knows techniques and safety procedures for handling explosive components; may serve as instructor at a formal training school.

Recommendation

In the lower-division baccalaureate/associate degree category, 3 semester hours in computer literacy and 3 in technical writing. In the upper-division baccalaureate category, 3 semester hours in organizational management and 3 for field experience in management; if rank was CW2, add 3 semester hours in management problems, 3 in operations management, 3 in communication techniques for managers, and 3 additional semester hours for field experience in management; if rank was CW3, add 3 semester hours in management problems, 3 in operations management, 3 in communication techniques for managers, and 5 additional semester hours for field experience in management; if rank was CW4, add 3 semester hours in management problems, 3 in operations management, 3 in communication techniques for managers, and 7 additional semester hours for field experience in management (6/01).

MOS-912A-003

Land Combat Missile Systems Technician
912A0

Exhibit Dates: 6/01–Present.

Career Pattern

May have progressed to Land Combat Support Missile System Technician from any of the following MOS's: 27E (Land Combat Electronic Missile System Repairer), 27M (Multiple Launch Rocket System Repairer), 27Z (Missile Systems Maintenance Chief), 35Y (Integrated Family of Test Equipment Operator and Maintainer, or 39B (Automatic Test Equipment Operator and Maintainer).

Description

Serves as a technical manager for the land combat missile system; manages the logistics of supply, production and control, personnel management, equipment, and facility in maintenance and repair of land combat support missile systems; directs the activities of personnel engaged in isolating malfunctions occurring in electronic firing and guiding systems and components; analyzes complex malfunctions; supervises repair of and modifications to system components, including guidance control systems, optical trackers, modulators, infrared sensors, conduct-of-fire trainers, missiles, and specialized test equipment; oversees supply management and preparation of work reports and maintenance records; forecasts supply requirements; ensures that final inspections and checkout procedures are conducted; examines, interprets, and disseminates technical material, including orders, bulletins, and manuals; instructs enlisted specialists in troubleshooting techniques, diagnostic procedures, and the interpretation of tests on system components, circuits, and optical elements; knows techniques and safety procedures for handling explosive components; may serve as instructor at a formal training school.

Recommendation

In the lower-division baccalaureate/associate degree category, 3 semester hours in computer applications and 3 in technical writing. In the upper-division baccalaureate category, 3 semester hours in human resource management and 3 for field experience in management; if rank was CW2, add 3 semester hours in operations management and 3 additional semester hours for field experience in management; if rank was CW3, add 3 semester hours in operations management and 5 additional semester hours for field experience in management; if rank was CW4, add 3 semester hours in operations management and 7 additional semester hours for field experience in management (6/01).

MOS-913A-001

Armament Repair Technician
913A0

Exhibit Dates: 1/90–2/92.

Career Pattern

This MOS was formerly designated MOS 421A. May have progressed to Armament Repair Technician from any of the following MOS's: 41C, 45B, 45D, 45E, 45G, 45K, 45L, 45N, 45T, or 45Z.

Description

Manages activities and personnel engaged in the maintenance and repair of small arms, artillery and armor weapons, and crew-served weapons; oversees the inspection, repair, replacement, and modification of armament materiel; supervises inspections and check-out procedures; directs shop operations, including planning work flow, requisitioning parts and supplies, inspecting maintenance procedures, and providing technical assistance to maintenance personnel in artillery or armor units; must know nomenclature design, operation, and employment of the Army's current inventory of field artillery weapons, armor weapons, and small arms; trains enlisted specialists in the use of shop equipment and power tools, in inspection and test procedures, and in the disassembly, repair, and adjustment of armament material; uses technical publications and interprets specifications; uses automated and manual systems for supply of technical publications, tools, repair parts, and related maintenance supplies; applies regulations and procedures pertaining to physical security, accountability, and shipment of weapons; may serve as chief of a team, section, or platoon in field activities; may serve as instructor in a formal training school.

Recommendation

In the lower-division baccalaureate/associate degree category, 3 semester hours in personnel supervision, 3 for field experience in management, 3 in industrial/human relations, 3 in maintenance/shop management and record keeping, 3 in use and care of hand and power tools, and 1 in industrial safety. In the upper-division baccalaureate category, 3 semester hours in personnel management and training (3/77).

MOS-913A-002

ARMAMENT REPAIR TECHNICIAN
913A0

Exhibit Dates: 3/92–1/03.

Career Pattern

May have progressed to Armament Repair Technician from any of the following MOS's: 41C, 45B, 45D, 45E, 45G, 45K, 45L, 45N, 45T, or 45Z.

Description

Manages activities and personnel engaged in the maintenance and repair of small arms, artillery and armor weapons, and crew-served weapons; oversees the inspection, repair, replacement, and modification of armament materiel; supervises inspections and checkout procedures; directs shop operations, including planning work flow, requisitioning parts and supplies, inspecting maintenance procedures, and providing technical assistance to maintenance personnel in artillery or armor units; must know nomenclature design, operation, and employment of the Army's current inventory of field artillery weapons, armor weapons, and small arms; trains enlisted specialists in the use of shop equipment and power tools, in inspection and test procedures, and in the disassembly, repair, and adjustment of armament material; uses technical publications and interprets specifications; uses automated and manual systems for supply of technical publications, tools, repair parts, and related maintenance supplies; applies regulations and procedures pertaining to physical security, accountability, and shipment of weapons; may serve as chief of a team, section, or platoon in field activities; may serve as instructor in a formal training school.

Recommendation

In the lower-division baccalaureate/associate degree category, 3 semester hours in personnel supervision and 2 in records and information management. In the upper-division baccalaureate category, 3 semester hours in organizational management and 3 for field experience in management; if rank was CW2, add 3 semester hours in management problems, 3 in operations management, 3 in communication techniques for managers, and 3 additional semester hours for field experience in management; if rank was CW3, add 3 semester hours in management problems, 3 in operations management, 3 in communication techniques for managers, and 5 additional semester hours for field experience in management; if rank was CW4, add 3 semester hours in management problems, 3 in operations management, 3 in communication techniques for managers, and 7 additional semester hours for field experience in management (3/92).

MOS-913A-003

ARMAMENT REPAIR TECHNICIAN
913A0

Exhibit Dates: 2/03–Present.

Career Pattern

May have progressed to Armament Repair Technician from any of the following MOS's: 45B, 45D, 45E, 45G, 45K, 45N, or 45T.

Description

Manages activities and personnel engaged in the maintenance and repair of small arms, artillery, and armament systems; directs the inspection, repair, replacement, and modification of armament materiel; supervises inspections and checkout procedures; directs shop operations, including planning work flow, safety, requisitioning parts and supplies, inspecting maintenance procedures, and providing technical assistance to maintenance personnel in artillery or armor units; must know nomenclature design, operation, and employment of the Army's current inventory of field artillery weapons, armor weapons, and small arms; trains enlisted specialists in the use of shop equipment and power tools, in inspection and test procedures, and in the disassembly, repair, and adjustment of armament material; uses technical publications and interprets specifications; uses automated and manual systems for supply of technical publications, tools, repair parts, and related maintenance supplies; applies regulations and procedures pertaining to physical security, accountability, and shipment of weapons; may serve as chief of a team, section, or platoon in field activities; may serve as instructor in a formal training school. Since 1977, understands, adheres to, and instructs subordinates in OSHA and EPA policies on safety and hazardous materials, in compliance with the Army mandate to follow OSHA standards.

Recommendation

In the lower-division baccalaureate/associate degree category, 3 semester hours in principles of supervision, 3 in maintenance management, 3 in materiel management, 3 in technical report writing, 3 in ethics (military) and 2 in speech communications; after 1997, add 3 semester hours in industrial safety. In the upper-division baccalaureate category, 3 semester hours in human resource management, 3 in operations management, and 3 for a practicum in management; if rank was CW3, add 3 semester hours for a practicum in management; if rank was CW4, add 6 semester hours for a practicum in management; if rank was CW5, add 9 semester hours for a practicum in management (2/03).

MOS-914A-001

ALLIED TRADES TECHNICIAN
914A0

Exhibit Dates: 1/90–2/92.

Career Pattern

May have progressed to Allied Trades Technician from MOS 44B (Metal Worker) or MOS 44E (Machinist).

Description

Supervises setup, maintenance, and operation of machine tools used to make or repair metal parts, mechanisms, tools, or machinery; applies knowledge of mechanics, shop mathematics, metal properties, and layout machinery procedures; manages service section shop engaged in metalworking, automotive component repair, and the allied trades; supervises personnel who operate general or specialized metalworking machines such as lathes and turning machines; supervises automotive body repair and painting, glass, plastic, machine, canvas, leather, welding, and woodworking shop activities and the recovery of all types of wheeled or track vehicles or equipment; controls work quality by instruction of subordinates and inspection of work in progress and completed work; interprets regulations, orders, and specifications and demonstrates correct procedures; plans shop layout; conducts training and orientation of personnel in technical procedures and equipment innovations; directs shop operations, including supply, scheduling, work flow, and personnel management; uses automated and manual systems in managing supplies of repair parts, tools, fuel, and technical publications; prepares, implements, and maintains standard operating procedures for management of maintenance activities; may be employed as a service school instructor; performs other officer-level duties required by mission of the unit to which assigned.

Recommendation

In the lower-division baccalaureate/associate degree category, 3 semester hours in shop management, 3 in maintenance management, 3 in administrative office management, 3 in business communication, 3 in record keeping, 3 in computer and data processing concepts, and 3 personnel supervision. In the upper-division baccalaureate category, 3 semester hours in personnel management, 3 in management problems, 3 for field experience in management, 3 in education electives (instruction and counseling), and additional credit in leadership and supervision on the basis of institutional evaluation; if rank was CW2, 2 additional semester hours for field experience in management; if rank was CW3, 4 additional semester hours for field experience in management; if rank was CW4, 6 additional semester hours for field experience in management (5/78).

MOS-914A-002

ALLIED TRADES TECHNICIAN
914A0

Exhibit Dates: 3/92–1/03.

Career Pattern

May have progressed to Allied Trades Technician from MOS 44B (Metal Worker) or MOS 44E (Machinist).

Description

Supervises setup, maintenance, and operation of machine tools used to make or repair metal parts, mechanisms, tools or machinery; applies knowledge of mechanics, shop mathematics, metal properties, and layout machinery procedures; manages service section shop engaged in metalworking, automotive component repair, and the allied trades; supervises personnel who operate general or specialized metalworking machines such as lathes and turning machines; supervises automotive body repair and painting, glass, plastic, machine, canvas, leather, welding, and woodworking shop activities and the recovery of all types of wheeled or track vehicles or equipment; controls work quality by instruction of subordinates and inspection of work in progress and completed work; interprets regulations, orders, and specifications and demonstrates correct procedures; plans shop layout; conducts training and orientation of personnel in technical procedures and equipment innovations; directs shop operations, including supply, scheduling, work flow, and personnel management; uses automated and manual systems in managing supplies of repair parts, tools, fuel, and technical publications; prepares, implements, and maintains standard operating procedures for management of maintenance activities; may serve as a service school instructor; performs other officer-level duties required by mission of the unit to which assigned.

Recommendation

In the lower-division baccalaureate/associate degree category, 3 semester hours in personnel supervision and 2 in records and information management. In the upper-division baccalaureate category, 3 semester hours in organizational management and 3 for field experience in management; if rank was CW2, add 3 semester hours in management problems, 3 in operations management, 3 in communication techniques for managers, and 3 additional semester hours for field experience in management; if rank was CW3, add 3 semester hours in management problems, 3 in operations management, 3 in communication techniques for managers, and 5 additional semester hours for field experience in management; if rank was CW4, add 3 semester hours in management problems, 3 in operations management, 3 in communication techniques for managers, and 7 additional semester hours for field experience in management (3/92).

MOS-914A-003

ALLIED TRADES TECHNICIAN
914A0

Exhibit Dates: 2/03–Present.

Career Pattern

May have progressed to Allied Trades Technician from MOS 44B (Metal Worker) or MOS 44E (Machinist).

Description

Directs and supervises setup, maintenance, and operation of machine tools and welding equipment used to make or repair metal parts, mechanisms, tools or machinery; applies knowledge of mechanics, shop mathematics, metal properties, and layout machinery procedures; manages service section shop engaged in metalworking, automotive component repair, and the allied trades; supervises personnel who operate general or specialized metalworking or welding machines; supervises automotive body repair and painting, glass, plastic, machine, canvas, leather, welding, and woodworking shop activities and the recovery of all types of wheeled or track vehicles; controls work quality by instruction of subordinates and inspection of work in progress and completed work; interprets regulations, orders, and specifications and demonstrates correct procedures; plans shop layout; conducts training and orientation of personnel in technical procedures, safety, and equipment innovations; directs shop operations, including supply, scheduling, work flow, and personnel management; uses automated and manual systems in managing supplies of repair parts, tools, fuel, and technical publications; prepares, implements, and maintains standard operating procedures for management of maintenance activities; may serve as a service school instructor; performs other officer-level duties required by mission of the unit to which assigned; and since 1997, understands, adheres to, and instructs subordinates in OSHA and EPA policies on safety and hazardous materials, in compliance with the Army mandate to follow OSHA standards.

Recommendation

In the lower-division baccalaureate/associate degree category, 3 semester hours in principles of supervision, 3 in maintenance management, 3 in materiel management, 3 in technical report writing, 3 in ethics (military), and 3 in speech communications; after 1997, add 3 semester hours in industrial safety In the upper-division baccalaureate category, 3 semester hours in human resource management, 3 for a practicum in management, and 3 in operations management; if rank was CW3, add 3 semester hours for a practicum in management; if rank was CW4, add 6 semester hours for a practicum in management; if rank was CW5, add 9 semester hours for a practicum in management (2/03).

MOS-915A-001

WHEEL VEHICLE MAINTENANCE TECHNICIAN
915A0

Exhibit Dates: 1/90–7/92.

Career Pattern

May have progressed from any of the following MOS's: 63B, 63D, 63E, 63F, 63G, 63H, 63N, 63S, 63T, 63Y, 63W, or 63Z.

Description

Supervises the maintenance and repair of wheel and track vehicles and accessory equipment; inspects vehicles; prepares, implements, and maintains procedures for management of maintenance activities; interprets regulations, technical manuals, and orders pertaining to automotive and accessory equipment maintenance functions; supervises personnel assigned to maintenance sections, motor pools, or similar activities; knows the theory, function, and operation of internal combustion engines and the assembly, nomenclature, checkout, inspection procedures, diagnostic tests, tuning, and other adjustments of automotive and nonengineer mechanical equipment, including electrical, mechanical, hydraulic, pneumatic, and hydropneumatic subsystems; uses special tools and test equipment; performs emergency maintenance procedures, including use of salvaged parts, reconditioning of equipment with hand tools, and substitution of parts and minor assemblies; uses automated and manual systems for supply of technical publications, tools, and repair parts; uses unit readiness reporting system; knows the use of automotive, metal, and tire shop equipment; diagnoses complex malfunctions and instructs personnel on isolation and repair of unusual malfunctions; may serve as a service school instructor; writes technical memoranda and other correspondence.

Recommendation

In the lower-division baccalaureate/associate degree category, 3 semester hours in shop management, 3 in maintenance management, 3 in administrative office management, 3 in business communication, 3 in record keeping, 3 in computer and data processing concepts, and 3 in personnel supervision. In the upper-division baccalaureate category, 3 semester hours in personnel management, 3 in management problems, 3 for field experience in management, 3 in education electives (instruction and counseling), and additional credit in leadership and supervision on the basis of institutional evaluation; if rank was CW2, 2 additional semester hours for field experience in management; if rank was CW3, 4 additional semester hours for field experience in management; if rank was CW4, 6 additional semester hours for field experience in management (5/78).

MOS-915A-002

WHEEL VEHICLE MAINTENANCE TECHNICIAN (LIGHT)
915A0

Exhibit Dates: 8/92–1/03.

Career Pattern

May have progressed from any of the following MOS's: 62B, 63B, 63D, 63E, 63F, 63G, 63N, 63S, 63T, 63Y, 63W, or 63Z.

Description

Plans, organizes, and performs unit maintenance on wheel vehicles, light track vehicles (except Bradley), self-propelled artillery systems, fire control, armament, ground support, and power-driven chemical equipment; analyzes unit equipment malfunctions; directs unit work loads; enforces fire and safety programs; manages calibration, oil analysis, and readiness reports; using the Army maintenance management system, manages periodic maintenance, vehicle dispatcher, vehicle repair, and common task training; conducts technical inspections of units; establishes procedures; instructs subordinate personnel; supervises personnel assigned to maintenance sections, motor pools, or similar activities.

Recommendation

In the lower-division baccalaureate/associate degree category, 3 semester hours in personnel supervision, 2 in records and information management, and 2 in computer applications. If rank was CW2 or CW3: 1 additional semester hour in computer applications. If rank was CW4: 2 additional semester hours in computer applications. In the upper-division baccalaureate category, 3 semester hours in organizational management and 3 for field experience in management. If rank was CW2, 3 additional semester hours for field experience in management, 3 in management problems, 3 in operations management, 3 in communication techniques for managers, and 3 in fleet maintenance management. If rank was CW3, 5 additional semester hours for field experience in management, 3 in management problems, 3 in operations management, 3 in communication techniques for managers, and 3 in fleet maintenance management. If rank was CW4, 7 additional semester hours for field experience in management, 3 in management problems, 3 in operations management, 3 in communication techniques for managers, and 3 in fleet maintenance management (8/92).

MOS-915B-001

LIGHT TRACK SYSTEMS MAINTENANCE TECHNICIAN
915B0

Exhibit Dates: 1/90–3/91. The skills, competencies, and knowledge required to perform the duties of MOS 915B are comparable to those of MOS 915A. Use the recommendation in exhibit MOS-915A-001. In 4/91, MOS 915B was discontinued and its duties were transferred to MOS 915A.

Career Pattern

Description

MOS-915C-001

Field Artillery Systems Maintenance Technician
915C0

Exhibit Dates: 1/90–3/91. The skills, competencies, and knowledge required to perform the duties of MOS 915C are comparable to those of MOS 915A. Use the recommendation in exhibit MOS-915A-001. In 4/91 MOS 915C was discontinued and its duties were transferred to MOS 915A.

Career Pattern

Description

MOS-915D-001

Armor/Cavalry Systems Maintenance Technician
915D0

Exhibit Dates: 1/90–7/92.

Career Pattern

May have progressed from warrant officer MOS 915A (Wheel Vehicle Maintenance Technician), MOS 915B (Light Track Systems Maintenance Technician), or MOS 915C (Field Artillery Systems Maintenance Technician).

Description

Supervises the maintenance and repair of wheel and track vehicles and accessory equipment; inspects vehicles; prepares, implements, and maintains procedures for management of maintenance activities; interprets regulations, technical manuals, and orders pertaining to automotive and accessory equipment maintenance functions; supervises personnel assigned to maintenance sections, motor pools, or similar activities; knows the theory, function, and operation of internal combustion engines and the assembly, nomenclature, checkout, inspection procedures, diagnostic tests, tuning, and other adjustments of automotive and nonengineer mechanical equipment, including electrical, mechanical, hydraulic, pneumatic, and hydropneumatic subsystems; uses special tools and test equipment; performs emergency maintenance methods, including use of salvaged parts, reconditioning of equipment with hand tools, and substitution of parts and minor assemblies; uses automated and manual systems for supply of technical publications, tools, and repair parts; uses unit readiness reporting system; knows the use and functioning of automotive, metal, and tire shop equipment; diagnoses complex malfunctions and instructs personnel on isolation and repair of unusual malfunctions; may serve as a service school instructor; writes technical memoranda and other correspondence.

Recommendation

In the lower-division baccalaureate/associate degree category, 3 semester hours in shop management, 3 in maintenance management, 3 in administrative office management, 3 in business communication, 3 in record keeping, 3 in computer and data processing concepts, and 3 in personnel supervision. In the upper-division baccalaureate category, 3 semester hours in personnel management, 3 in management problems, 3 for field experience in management, 3 in education electives (instruction and counseling), and additional credit in leadership and supervision on the basis of institutional evaluation; if rank was CW2, 2 additional semester hours for field experience in management; if rank was CW3, 4 additional semester hours for field experience in management; if rank was CW4, 6 additional semester hours for field experience in management (5/78).

MOS-915D-002

Unit Maintenance Technician (Heavy)
915D0

Exhibit Dates: 8/92–Present.

Career Pattern

Progressed from MOS 915A (Unit Maintenance Technician, Light).

Description

Plans, organizes, and executes unit maintenance of wheel, light track vehicles and armament, fire control, ground support, and power-driven chemical equipment; analyzes equipment malfunctions; directs unit repair work loads; establishes and enforces shop fire and safety programs; manages unit calibration, oil analysis, and readiness programs; directs recovery of unit equipment; manages the Army maintenance management system at unit level; oversees maintenance schedules, vehicle dispatch, parts requisitioning, and common task training; conducts unit technical inspections and instructs subordinate personnel; supervises personnel assigned to maintenance sections, motor pools, or similar activities.

Recommendation

In the lower-division baccalaureate/associate degree category, 3 semester hours in personnel supervision, 2 in records and information management, and 2 in computer applications. If rank was CW2 or CW 3: 1 additional semester hour in computer applications. If rank was CW4: 2 additional semester hours in computer applications. In the upper-division baccalaureate category, 3 semester hours in organizational management and 3 for field experience in management. If rank was CW2: 3 additional semester hours for field experience in management, 3 in management problems, 3 in operations management, 3 in communication techniques for managers, and 3 in fleet maintenance management. If rank was CW3: 5 additional semester hours for field experience in management, 3 in management problems, 3 in operations management, 3 in communication techniques for managers, and 3 in fleet maintenance management. If rank was CW4: 7 additional semester hours for field experience in management, 3 in management problems, 3 in operations management, 3 in communication techniques for managers, and 3 in fleet maintenance management (8/92).

MOS-915E-001

Support/Staff Maintenance Technician
915E0

Exhibit Dates: 1/90–7/92.

Career Pattern

May have progressed from warrant officer MOS 915A (Wheel Vehicle Maintenance Technician), MOS 915B (Light Track Systems Maintenance Technician), MOS 915C (Field Artillery Systems Maintenance Technician), or MOS 915D (Armor/Cavalry Systems Maintenance Technician).

Description

Supervises the maintenance and repair of wheel and track vehicles and accessory equipment; inspects vehicles; prepares, implements, and maintains procedures for management of maintenance activities; interprets regulations, technical manuals, and orders pertaining to automotive and accessory equipment maintenance functions; supervises personnel assigned to maintenance sections, motor pools, or similar activities; applies the theory, function, and operation of internal combustion engines, and the assembly, nomenclature, checkout, inspection procedures, diagnostic tests, tuning, and other adjustments of automotive and nonengineer mechanical equipment, including electrical, mechanical, hydraulic, pneumatic, and hydropneumatic subsystems; uses special tools and test equipment; knows emergency maintenance methods, including use of salvaged parts, reconditioning of equipment with hand tools, and substitution of parts and minor assemblies; uses automated and manual systems for supply of technical publications, tools, and repair parts; uses unit readiness reporting system; knows the use and functioning of automotive, metal, and tire shop equipment; diagnoses complex malfunctions and instructs personnel on isolation and repair of unusual malfunctions; may serve as a service school instructor; writes technical memoranda and other correspondence.

Recommendation

In the lower-division baccalaureate/associate degree category, 3 semester hours in shop management, 3 in maintenance management, 3 in administrative office management, 3 in business communications, 3 in record keeping, 3 in computer and data processing concepts, and 3 in personnel supervision. In the upper-division baccalaureate category, 3 semester hours in personnel management, 3 in management problems, 3 for field experience in management, 3 in education electives (instruction and counseling), and additional credit in leadership and supervision on the basis of institutional evaluation; if rank was CW2, 2 additional semester hours for field experience in management; if rank was CW3, 4 additional semester hours for field experience in management; if rank was CW4, 6 additional semester hours for field experience in management (5/78).

MOS-915E-002

Support/Staff Maintenance Technician
915E0

Exhibit Dates: 8/92–1/03.

Career Pattern

May have progressed from warrant officer MOS 915A (Unit Maintenance Technician, Light) or MOS 91D (Unit Maintenance Technician, Heavy).

Description

Knows and has performed duties described for Unit Maintenance Technician (Light); manages major repair of automotive equipment and work flow and ensures work quality; establishes maintenance and repair schedules and internal administrative procedures; writes technical training materials; instructs students in school environment; diagnoses complex auto equipment malfunctions and instructs personnel on repair; establishes shop fire and safety programs; manages equipment calibration, used product disposal, and monitors common task training; advises higher command levels as a subject matter expert; supervises personnel assigned to maintenance sections, motor pools, or similar activities.

Recommendation

In the lower-division baccalaureate/associate degree category, 3 semester hours in personnel supervision, 2 in records and information management, and 2 in computer applications. If rank was CW2 or CW3: 1 additional semester hour in computer applications. If rank was CW4: 2 additional semester hours in computer applications. In the upper-division baccalaureate category, 3 semester hours in organizational management and 3 for field experience in management. If rank was CW2: 3 additional semester hours for field experience in management, 3 in management problems, 3 in operations management, 3 in communication techniques for managers, and 3 in fleet maintenance management. If rank was CW3: 5 additional semester hours for field experience in management, 3 in management problems, 3 in operations management, 3 in communication techniques for managers, and 3 in fleet maintenance management. If rank was CW4: 7 additional semester hours for field experience in management, 3 in management problems, 3 in operations management, 3 in communication techniques for managers, and 3 in fleet maintenance management (8/92).

MOS-915E-003

Support Maintenance Technician
915E0

Exhibit Dates: 2/03–Present.

Career Pattern

May have progressed from warrant officer MOS 915A (Unit Maintenance Technician, Light) or MOS 915D (Unit Maintenance Technician, Heavy).

Description

Manages major repair of automotive equipment; establishes work flow and ensures work quality; establishes maintenance and repair schedules and internal administrative procedures; develops technical training materials; diagnoses complex auto equipment malfunctions and instructs personnel on repair procedures; establishes shop fire and safety programs; manages equipment calibration; monitors common task training; advises higher command levels as a subject matter expert; supervises personnel assigned to maintenance sections, motor pools, or similar activities. Since 1997, understands, adheres to, and instructs subordinates in OSHA and EPA policies on safety and hazardous materials, in compliance with the Army mandate to follow OSHA standards.

Recommendation

In the lower-division baccalaureate/associate degree category, 3 semester hours in principles of supervision, 2 in ethics (military), 3 in information management, 3 in technical report writing, 3 in computer information systems, and 3 in speech communications; after 1997, 3 semester hours in industrial safety. In the upper-division baccalaureate category, 3 semester hours in organizational management, 3 for a practicum in management, 3 in operations management, and 3 in human resource management; if rank was CW3, add 3 semester hours for a practicum in management; if rank was CW4, add 6 semester hours for a practicum in management; if rank was CW5, add 9 semester hours for a practicum in management (2/03).

MOS-916A-001

High-to-Medium Altitude Air Defense Direct Support/General Support Maintenance Technician
916A0

Exhibit Dates: 4/92–Present.

Career Pattern

May have progressed to 916A from MOS's 24H (Hawk Fire Control Repairer), 24K (Hawk Continuous Wave Radar Repairer), 27K (Hawk Firing Control/Continuous Wave Radar Repairer), 27J (Hawk Field Maintenance Equipment/Pulse Acquisition Radar), 27H (Hawk Firing Section Repairer), or 27X (Patriot System Repairer).

Description

Directs test procedures, diagnostic system analysis, and troubleshooting techniques; employs digital switching theory, logic, and computer fundamentals applicable to the High-to-Mid Air Defense (HIMAD) missile systems; monitors and directs work flow and test procedures, diagnostic system analysis, and troubleshooting techniques for automated test equipment supporting HIMAD systems; ensures quality assurance/quality control procedures; manages the scheduling of periodic maintenance services and calibration of organic equipment; monitors, coordinates, and analyzes contracting, procurement, and material acquisition programs for new and existing HIMAD systems; presents written and oral briefings.

Recommendation

In the lower-division baccalaureate/associate degree category, 3 semester hours in introduction to computers and 3 in technical writing. In the upper-division baccalaureate category, 3 semester hours in organizational management and 3 for field experience in management. If rank was CW2; add 3 semester hours in management problems, 3 in operations management, 3 in communication techniques for managers, and 3 for field experience in management. If rank was CW3, add: 3 semester hours in management problems, 3 in operations management, 3 in communication techniques for managers, and 5 for field experience in management. If rank was CW4 or CW5; add 3 semester hours in management problems, 3 in operations management, 3 in communication techniques for managers, and 7 for field experience in management (5/98).

MOS-917A-001

Maneuver Forces Air Defense Systems Technician
917A0

Exhibit Dates: 4/92–Present.

Career Pattern

May have progressed to 917A from MOS 27F, Vulcan Repairer, 27G, Chaparral/Redeye Repairer, or 27T, Avenger System Repairer.

Description

Directs test procedures, diagnostic system analysis, and troubleshooting of mechanical, electrical, electronic, hydraulic, pneumatic, and optical malfunctions of Maneuver Forces Air Defense (MFAD) systems and their components; monitors and directs work flow and test procedures, diagnostic system analysis, and troubleshooting techniques for automated test equipment supporting MFAD systems; directs and assists personnel in solving unusable complex problems of diagnosis, and modification of automated test equipment used in support of MFAD systems; ensures quality assurance/quality control procedures; manages the scheduling of periodic maintenance, services, and calibration of organic equipment; monitors, coordinates, and analyzes contracting, procurement, and material acquisition programs for new and existing MFAD systems; present written and oral briefings.

Recommendation

In the lower-division baccalaureate/associate degree category, 3 semester hours in introduction to computers and 3 in technical writing. In the upper-division baccalaureate category, 3 semester hours in organizational management and 3 for field experience in management; if rank was CW2, add 3 semester hours in management problems, 3 in operations management, 3 in communication techniques for managers, and 3 additional semester hours for field experience in management; if rank was CW3, add 3 semester hours in management problems, 3 in operations management, 3 in communication techniques for managers, and 5 additional semester hours for field experience in management; if rank was CW4 or CW5, add 3 semester hours in management problems, 3 in operations management, 3 in communication techniques for managers, and 7 additional semester hours for field experience in management (4/98).

MOS-918A-001

Test, Measurement, and Diagnostic Equipment (TMDE) Maintenance Support Technician
918A0

Exhibit Dates: 9/91–Present.

Career Pattern

This MOS was formerly designated MOS 252A. May have progressed to Test, Measurement, and Diagnostic Equipment Maintenance Support Technician from MOS 35H (Test, Measurement, and Diagnostic Equipment Maintenance Support Specialist).

Description

Supervises or repairs and calibrates precision equipment used for the measuring, testing, controlling, and indicating temperature, pressure, vacuum, fluid flow, liquid level, mechanical motion, rotation, humidity, density, acidity, alkalinity, and combustion (equipment may include tools, dial pressure gauges, scales, balances, direction and sighting devices, oscilloscopes, meters, digital test equipment, signal generators, power supplies, and specialized mechanical and electronic testing equipment); maintains equipment within prescribed manufacturer standards and national or international bureau of standards tolerances and ensures that calibration surveillance standards are met; manages activities and personnel engaged in calibration of instruments or system components of all types of Army equipment and weapons; interprets and analyzes measurement data and specifications of equipment, instruments, or gauges to conform to specified standards; develops procedures for testing and diagnosing malfunctions or implements established guidance to adjust or repair item requiring calibration; interprets technical data, schematics, and transfer standards in the use and operation of this material; supervises testing systems to isolate malfunctions and faulty components;

instructs subordinates on the correctional adjustments to obtain special readings or characteristics including frequency or inductance; certifies instruments for accuracy; may be chief of section engaged in calibration activities and may supervise up to 50 calibration technicians; may be employed as a contracting officer's representative to monitor and evaluate the performance of contractor personnel performing maintenance calibration functions for the Army; may be assigned as a service school instructor at Army or joint service school and typically is involved in writing training materials and teaching update courses to subordinates; oversee budget of calibration unit; oversees office, personnel, job placement, employee relations, shipping of equipment, and other essential officer-level duties.

Recommendation

In the lower-division baccalaureate/associate degree category, 3 semester hours in computer literacy and 3 in technical writing. In the upper-division baccalaureate category, 3 semester hours in organizational management and 3 for field experience in management; if rank was CW2, add 3 semester hours in management problems, 3 in operations management, 3 in communication techniques for managers, and 3 additional semester hours for field experience in management; if rank was CW3, add 3 semester hours in management problems, 3 in operations management, 3 in communication techniques for managers, and 5 additional semester hours for field experience in management; if rank was CW4, add 3 semester hours in management problems, 3 in operations management, 3 in communication techniques for managers, and 7 additional semester hours for field experience in management (9/91).

MOS-918B-001

Electronic Systems Maintenance Technician
918B0

Exhibit Dates: 2/95–Present.

Career Pattern

May have progressed to MOS 918B from MOS 35E, Radio/COMSEC Repairer, 35J, Telecommunications Terminal Device Repairer, 35N, Wire Systems Equipment Repairer, 35W, Electronic Maintenance Chief; or 68P, Avionic Maintenance Supervisor.

Description

Manages, supervises, and coordinates the installation, operation, repair, maintenance, and modification of radio, radar, computer, electronic data processing, television, navigation, avionics, communications, and cryptographic equipment, and associated tools, test sets, and accessory equipment; inspects incoming faulty equipment; prepares work assignments; plans and implements maintenance schedules; develops repair and operating procedures and ensures quality control; manages training on test and maintenance procedures; diagnoses and supervises complex repair of communications equipment; writes extensive reports of evaluations and inspections of equipment repair and associated problems; manages depot level budget; writes evaluation reports of subordinates; advises commander or staff officers on electronics equipment development, procurement, capabilities, limitations, and employment.

Recommendation

In the lower-division baccalaureate/associate degree category, 3 semester hours in introduction to computers and 3 in technical writing. In the upper-division baccalaureate category, 3 semester hours in organizational management and 3 for field experience in management; if rank was CW2, add 3 semester hours in management problems, 3 in operations management, 3 in communication techniques for managers, and 3 additional semester hours for field experience in management; if rank was CW3, add 3 semester hours in management problems, 3 in operations management, 3 in communication techniques for managers, and 5 additional semester hours for field experience in management; if rank was CW4 or CW5, add 3 semester hours in management problems, 3 in operations management, 3 in communication techniques for managers, and 7 additional semester hours for field experience in management (4/98).

MOS-919A-001

Engineer Equipment Repair Technician
919A0

Exhibit Dates: 8/93–1/03.

Career Pattern

May have progressed to MOS 919A from MOS 52C, Utilities Equipment Repairer, 52D, Power Generation Equipment Repairer, 52F, Turbine Engine Driven Generator Repairer, 52X, Special Purpose Equipment Repairer, or 62B, Construction Equipment Repairer.

Description

.Supervises personnel engaged in maintenance of engineer equipment; analyzes malfunctions and supervises minor repair and adjustment of engineer equipment utilized for power generation, earth moving, shaping and compacting, lifting and loading, quarrying and rock crushing, asphalt/concrete mixing and surfacing, water purification, refrigeration and air conditioning, missile system support, water gap crossing, transfer, and engineer electronic applications; supervises modification of equipment required by work orders; inspects incoming equipment to determine repair requirements; assigns work and inspects outgoing equipment; establishes maintenance and repair schedules; establishes internal administrative procedures for procurement, storage, and distribution of tools, parts, and publications.

Recommendation

In the lower-division baccalaureate/associate degree category, 3 semester hours in introduction to computers and 3 in technical writing. In the upper-division baccalaureate category, 3 semester hours in organizational management and 3 for field experience in management; if rank was CW2, add 3 semester hours in management problems, 3 in operations management, 3 in communication techniques for managers, and 3 additional semester hours for field experience in management; if rank was CW3, add 3 semester hours in management problems, 3 in operations management, 3 in communication techniques for managers, and 5 additional semester hours for field experience in management; if rank was CW4 or CW5, add 3 semester hours in management problems, 3 in operations management, 3 in communication techniques for managers, and 7 additional semester hours for field experience in management (4/98).

MOS-919A-002

Engineer Equipment Repair Technician
919A0

Exhibit Dates: 2/03–Present.

Career Pattern

May have progressed to MOS 919A from MOS 52C, Utilities Equipment Repairer; 52D, Power Generation Equipment Repairer; 52F, Turbine Engine Driven Generator Repairer; 52X, Special Purpose Equipment Repairer; or 62B, Construction Equipment Repairer.

Description

Supervises the maintenance of engineer equipment; analyzes malfunctions and supervises minor repair and adjustment of engineer equipment utilized for power generation, earth moving, shaping and compacting, lifting and loading, quarrying and rock crushing, asphalt/concrete mixing and surfacing, water purification, refrigeration and air conditioning, missile system support, water gap crossing, transfer, and engineer electronic applications; supervises modification of equipment required by work orders; inspects incoming equipment to determine repair requirements; assigns work and inspects outgoing equipment; establishes maintenance and repair schedules; establishes internal administrative procedures for procurement, storage, and distribution of tools, parts, and publications. Since 1997, understands, adheres to, and instructs subordinates in OSHA and EPA policies on safety and hazardous materials, in compliance with the Army mandate to follow OSHA standards.

Recommendation

In the lower-division baccalaureate/associate degree category, 3 semester hours in principles of supervision, 3 in maintenance management, 3 in materiel management, 3 in technical report writing, 3 in ethics (military), and 2 in speech communications; after 1997, add 3 semester hours in industrial safety. In the upper-division baccalaureate category, 3 semester hours in organizational management, 3 for a practicum in management, 3 in human resource management, and 3 in operations management; if rank was CW3, add 3 semester hours for a practicum in management; if rank was CW4, add 6 semester hours for a practicum in management; if rank was CW5, add 9 semester hours for a practicum in management (2/03).

MOS-920A-001

Property Book Technician
920A0

Exhibit Dates: 1/90–3/92.

Career Pattern

May have progressed to Property Book Technician from MOS 76J (Medical Supply Specialist) or MOS 76Y (Unit Supply Specialist).

Description

Administers and manages the Army supply systems; maintains inventory control; supervises purchasing procedures, inventory handling and storage, record keeping, and stock control; is responsible for accounting and management of capital stock for units and installations; uses manual and automated property accounting procedures; develops and administers supply budgets for units or installations; forecasts and plans supply requirements; uses automatic data processing in the supply areas;

gives technical advice to subordinate unit and activity commanders; establishes procedures for requisitioning, receiving, storing, and issuing supplies; establishes and maintains formal and informal lines of communication with supply control, maintenance, and transportation units with supported activities; may monitor or evaluate supply contractor's performance; prepares reports and correspondence.

Recommendation

In the vocational certificate category, 3 semester hours in material handling and safety. In the lower-division baccalaureate/associate degree category, 3 semester hours in technical report writing, 3 in personnel supervision, 3 in inventory management, 3 in introduction to computer concepts, 3 in warehousing and storage, 3 in record keeping, 3 in purchasing, 3 in material handling, 3 in principles of financial budgeting, and 2 in keypunch. In the upper-division baccalaureate category, 3 semester hours in management electives and 3 in personnel management and training; if rank was CW2, 2 additional semester hours for field experience in management; if rank was CW3, 4 additional semester hours for field experience in management; if rank was CW4, 6 additional semester hours for field experience in management (10/77).

MOS-920A-002

Property Accounting Technician
920A0

Exhibit Dates: 4/92–5/01.

Career Pattern

May have progressed to Property Book Technician from MOS 76J (Medical Supply Specialist) or MOS 76Y (Unit Supply Specialist).

Description

Maintains an organization's property accountability through the management of property books, both automated and nonautomated; responsible for accounting and management of capital stock for units and installations; uses manual and automated property accounting procedures; performs financial inventory accounting as applied to the Army's budget system; forecasts and plans supply requirements; uses automatic data processing in the supply area; gives technical advice to subordinate unit and activity commanders; establishes procedures for requisitioning, receiving, storing, and issuing supplies; establishes and maintains formal and informal lines of communication with supply control, maintenance, and transportation units in supported activities; may monitor or evaluate supply contractor's performance; prepares reports and correspondence.

Recommendation

In the lower-division baccalaureate/associate degree category, 3 semester hours in computer applications, 3 in personnel supervision, and 2 in records and information management. In the upper-division baccalaureate category, 3 semester hours in organizational management and 3 for field experience in management; if rank was CW2, add 3 semester hours in management problems, 3 in operations management, 3 in communication techniques for managers, and 3 additional semester hours for field experience in management; if rank was CW3, add 3 semester hours in management problems, 3 in operations management, 3 in communication techniques for managers, and 5 additional semester hours for field experience in management; if rank was CW4, add 3 semester hours in management problems, 3 in operations management, 3 in communication techniques for managers, and 7 additional semester hours for field experience in management (4/92).

MOS-920A-003

Property Accounting Technician
920A0

Exhibit Dates: 6/01–Present.

Career Pattern

May have progressed to 920A from MOS 92Y, (Unit Supply Specialist) or 76J, (Medical Supply Specialist).

Description

Maintains an organization's property accountability through the management of automated property books; responsible for accounting and management of capital stock for units and installations; uses automated property accounting procedures; performs financial inventory accounting; forecasts and plans supply requirements; uses computerized systems in the supply area.

Recommendation

In the lower-division baccalaureate/associate degree category, 3 semester hours in computer applications, 3 in principles of supervision, and 3 in data management. In the upper-division baccalaureate category, 3 semester hours in organizational management, 3 for a management practicum, 3 in operations management, and 3 in technical communication; if rank was CW2, add 3 semester hours for a management practicum; if rank was CW3, add 3 semester hours for a management practicum and 3 in supply chain management; if rank was CW4, add 3 semester hours for a management practicum, 3 in supply chain management, and 3 in logistics management (6/01).

MOS-920B-001

Repair Parts Technician
920B0

Exhibit Dates: 1/90–3/92.

Career Pattern

May have progressed to Repair Parts Technician from MOS 76J (Medical Supply Technician), MOS 76P (Materiel Control and Accounting Specialist), or MOS 76V (Materiel Storage and Handling Specialist).

Description

Administers and manages the Army supply system as it applies to maintenance support units usually at overseas locations; supervises the requisitioning, storage, distribution, and accounting of repair parts and maintenance-related supply items; applies manual and automated property accounting procedures; knows fundamentals of data processing as they apply to inventory management; knows the interrelationship of supply, maintenance, and transportation; knows policies and procedures for and interprets regulations regarding crating, packaging, preserving, shipping, and storing materiel and equipment; uses supply catalogs, technical manuals, bulletins, modification work orders, and cross-reference lists in requisitioning, receiving, storing, issuing, and identifying equipment, supplies, and repair parts; forecasts the requirements of repair parts and maintenance-related supply items; develops and administers supply budgets; inspects supported units to ensure that stock levels are adequate and within prescribed limits; makes inventory recommendations; provides technical guidance to supported unit personnel; develops operating procedures and performs administrative duties related to the supply activity; evaluates efficiency and effectiveness of supply operations; prepares correspondence and reports.

Recommendation

In the lower-division baccalaureate/associate degree category, 2 semester hours in keypunch, 3 in record keeping, 3 in technical report writing, 3 in inventory management, 3 in warehousing and storage, 3 in introduction to computer concepts, 3 in purchasing, 3 in material handling and safety, 3 in principles of financial budgeting, 3 in personnel supervision, and additional credit in traffic management on the basis of institutional evaluation. In the upper-division baccalaureate category, 3 semester hours in management electives, 3 in personnel management and training, and additional credit in transportation and physical distribution on the basis of institutional evaluation. If rank was CW2, 2 additional semester hours for field experience in management; if rank was CW3, 4 additional semester hours for field experience in management; if rank was CW4, 6 additional semester hours for field experience in management (10/77).

MOS-920B-002

Supply Systems Technician
920B0

Exhibit Dates: 4/92–5/01.

Career Pattern

May have progressed to Repair Parts Technician from MOS 76J (Medical Supply Specialist), MOS 76P (Materiel Control and Accounting Specialist), or MOS 76V (Materiel Storage and Handling Specialist).

Description

Administers and manages the Army supply system as it applies to maintenance support units usually at overseas locations; supervises the requisitioning, storage, distribution, and accounting of repair parts and maintenance-related supply items; applies manual and automated property accounting procedures; applies fundamentals of data processing as they apply to inventory management; knows the interrelationship of supply, maintenance, and transportation; applies policies and procedures for and interprets regulations regarding crating, packaging, preserving, shipping, and storing materiel and equipment; uses supply catalogs, technical manuals, bulletins, modification work orders, and cross-reference lists in requisitioning, receiving, storing, issuing, and identifying equipment, supplies, and repair parts; forecasts the requirements of repair parts and maintenance-related supply items; develops and administers supply budgets; inspects supported units to ensure that stock levels are adequate and within prescribed limits; makes inventory recommendations; provides technical guidance to supported unit personnel; develops operating procedures and performs administrative duties related to the supply activity; evaluates efficiency and effectiveness of supply operations; prepares correspondence and reports; instructs, manages, and supervises supply personnel.

Recommendation

In the lower-division baccalaureate/associate degree category, 3 semester hours in computer applications, 3 in personnel supervision, and 2 in records and information management. In the upper-division baccalaureate category, 3 semester hours in organizational management and 3 for field experience in management; if rank was CW2, add 3 semester hours in management problems, 3 in operations management, 3 in communication techniques for managers, and 3 additional semester hours for field experience in management; if rank was CW3, add 3 semester hours in management problems, 3 in operations management, 3 in communication techniques for managers, and 5 additional semester hours for field experience in management; if rank was CW4, add 3 semester hours in management problems, 3 in operations management, 3 in communication techniques for managers, and 7 additional semester hours for field experience in management (4/92).

MOS-920B-003

Supply Systems Technician
920B0

Exhibit Dates: 6/01–Present.

Career Pattern

May have progressed to 920B from MOS 92A, (Automated Logistical Specialist).

Description

Administers and manages the Army supply system as it applies to maintenance support units; supervises the requisitioning, storage, distribution, and accounting of repair parts and maintenance-related supply items; applies automated property accounting procedures; applies fundamentals of computer application systems as they apply to inventory management; knows the interrelationship of supply, maintenance, and transportation; applies policies and procedures for and interprets regulations regarding crating, packaging, preserving, shipping, and storing materiel and equipment; uses supply catalogs, technical manuals, bulletins, modification work orders, and cross-reference lists in requisitioning, receiving, storing, issuing, and identifying equipment, supplies, and repair parts; forecasts the requirements of repair parts and maintenance-related supply items; develops and administers supply budgets; inspects supported units to ensure that stock levels are adequate and within prescribed limits; makes inventory recommendations; provides technical guidance to supported unit personnel; develops operating procedures and performs administrative duties related to the supply activity; evaluates efficiency and effectiveness of supply operations; prepares correspondence and reports; instructs, manages, and supervises supply personnel.

Recommendation

In the lower-division baccalaureate/associate degree category, 3 semester hours in computer applications, 3 in principles of supervision, and 3 in data management. In the upper-division baccalaureate category, 3 semester hours in organization development, 3 in management issues, 3 in operations management, 3 in technical communication, and 3 for a management practicum. If rank was CW2, add 3 semester hours for a management practicum; if rank was CW3, add 3 semester hours for a management practicum and 3 in supply chain management; if rank was CW4, add 3 semester hours for a management practicum, 3 in supply chain management, and 3 in logistics management (6/01).

MOS-921A-001

Airdrop Equipment Technician
921A0

Exhibit Dates: 1/90–1/92.

Career Pattern

May have progressed to Airdrop Equipment Technician from MOS 43E (Parachute Rigger).

Description

Supervises preparation of equipment for airdrop, airdrop of supplies and equipment, parachute packing, and repair of parachutes and associated equipment; plans and organizes activities for repair of airdrop equipment and establishes rigging sites to support airborne operations; instructs personnel in airdrop procedures; prescribes procedures used in adjustment, care, and preservation of parachutes and other support equipment tools and machinery; provides technical guidance and serves as liaison to commanders; provides technical assistance to the personnel responsible for life cycle management of items, including design, research and development, testing, supply management, and classification; responsible for providing serviceable and safe parachutes and equipment to units engaged in parachute jumping and related airdrop operations; supervises repacking of parachutes in use; coordinates loading of equipment and supplies into aircraft and participates in joint airdrop load inspections; may serve as a service school instructor; applies maintenance record keeping and inventory control procedures.

Recommendation

In the lower-division baccalaureate/associate degree category, 2 semester hours in communication skills (oral). In the upper-division baccalaureate category, 3 semester hours in personnel management and 3 for management electives. If rank was CW2, 2 semester hours for field experience in management; if rank was CW3, 4 semester hours for field experience in management; if rank was CW4, 6 semester hours for field experience in management; if duty assignment was platoon leader, additional credit in leadership and supervision on the basis of institutional evaluation (5/78).

MOS-921A-002

Airdrop Systems Technician
921A0

Exhibit Dates: 2/92–5/02.

Career Pattern

May have progressed to Airdrop Systems Technician from MOS 43E (Parachute Rigger).

Description

Supervises preparation of equipment for airdrop, airdrop of supplies and equipment, parachute packing, and repair of parachutes and associated equipment; plans and organizes activities for repair of airdrop equipment and establishes rigging sites to support airborne operations; instructs personnel in airdrop procedures; prescribes procedures used in adjustment, care, and preservation of parachutes and other support equipment tools and machinery; provides technical guidance and serves as liaison to commanders; provides technical assistance to the personnel responsible for life cycle management of items, including design, research and development, testing, supply management, and classification; responsible for providing serviceable and safe parachutes and equipment to units engaged in parachute jumping and related airdrop operations; supervises repacking of parachutes in use; coordinates loading of equipment and supplies into aircraft and participates in joint airdrop load instructions; may serve as a service school instructor; applies maintenance record keeping and inventory control procedures.

Recommendation

In the lower-division baccalaureate/associate degree category, 3 semester hours in personnel supervision and 2 in records and information management. In the upper-division baccalaureate category, 3 semester hours in organizational management and 3 for field experience in management. If rank was CW2, add 3 semester hours in management problems, 3 in operations management, 3 in communication techniques for managers, and 3 additional semester hours for field experience in management; if rank was CW3, add 3 semester hours in management problems, 3 in operations management, 3 in communication techniques for managers, and 5 additional semester hours for field experience in management; if rank was CW4, add 3 semester hours in management problems, 3 in operations management, 3 in communication techniques for managers, and 7 additional semester hours for field experience in management (2/92).

MOS-921A-003

Airdrop Systems Technician
921A0

Exhibit Dates: 6/02–Present.

Career Pattern

May have progressed to Airdrop Systems Technician from MOS 92R (Parachute Rigger).

Description

Supervises preparation of equipment for airdrop, airdrop of supplies and equipment, parachute packing, and repair of parachutes and associated equipment; plans and organizes activities for repair of airdrop equipment and establishes rigging sites to support airborne operations; instructs personnel in airdrop procedures; prescribes procedures used in adjustment, care, and preservation of parachutes and other support equipment tools and machinery; provides technical guidance and serves as liaison to commanders; provides technical assistance to the personnel responsible for life cycle management of items, including design, research and development, testing, supply management, and classification; responsible for providing serviceable and safe parachutes and equipment to units engaged in parachute jumping and related airdrop operations; supervises repacking of parachutes in use; coordinates loading of equipment and supplies into aircraft and participates in joint airdrop load instructions; may serve as a service school instructor; applies maintenance record keeping and inventory control procedures.

Recommendation

In the lower-division baccalaureate/associate degree category, 3 semester hours in principles of supervision, 2 in records management, 1 in introduction to computers, 1 in computer applications, 2 in material management, and 2 in aviation safety. In the upper-division baccalaureate category, 3 semester hours for a management practicum, 3 in managerial planning and forecasting, 3 in principles of management, 3 in operations management, and 3 in communication techniques for managers (6/02).

MOS-922A-001

Food Service Technician
922A0

Exhibit Dates: 1/90–3/94.

Career Pattern
May have progressed to Food Service Technician from MOS 94B (Food Service Specialist).

Description
Supervises and administers food service activities for installations, commands, or organizations; maintains complete operational control over facilities, personnel, and specialized equipment; supervises the procurement, storage, distribution, and preparation of foods; writes reports; develops procedures for implementing policies from higher authority; uses specialized accounting methods; plans use, layout, and maintenance of food service equipment; in carrying out specialized duties, is required to devote most energies to the technical and human resource areas of food service administration; interprets contracts and monitors contractor's performance; if assigned to hospital food service duties, may be required to perform additional specialized supervisory functions in nutritional and diet therapy areas; may serve as staff advisor to a general officer or provide technical assistance to nonappropriated fund food service activities; may serve as a service school instructor.

Recommendation
In the lower-division baccalaureate/associate degree category, 6 semester hours in food service management, 3 in principles of management, 3 in personnel supervision, 3 in food service equipment and layout, and 3 in business report writing. In the upper-division baccalaureate category, 3 semester hours in management problems and 2 for field experience in management; if rank was CW2, 2 additional semester hours for field experience in management; if rank was CW3, 4 additional semester hours for field experience in management; if rank was CW4, 6 additional semester hours for field experience in management; if duty assignment was service school instructor, 3 additional semester hours for a practicum in education; if assigned to a hospital as a food service technician, 3 additional semester hours in nutrition or diet therapy (11/77).

MOS-922A-002

Food Service Technician
922A0

Exhibit Dates: 4/94–Present.

Career Pattern
May have progressed to Food Service Technician from MOS 94B (Food Service Specialist).

Description
Advises commander and administers food service activities for installation, commands, or organizations; advises on the procurement, storage, distribution, and preparation of food; writes reports; develops procedures for implementing policies from higher authority; uses food management formulas in carrying out specialized duties; focuses on the technical aspects of food service administration; interprets contracts and monitors contractor's performance; may serve as a staff advisor to a general officer or provide technical assistance to nonappropriated fund food service activities; may serve as a service school instructor.

Recommendation
In the lower-division baccalaureate/associate degree category, 6 semester hours in food service management and 3 in business communications. In the upper-division baccalaureate category, 3 semester hours in management problems, 3 in food cost control, and 3 for field experience in food service management. If duty assignment was a service school instructor, 3 semester hours in training methods (4/94).

Appendix A

The Evaluation Systems

BACKGROUND

Early editions of the *Guide to the Evaluation of Educational Experiences in the Armed Services* were prepared in response to specific needs. Immediately after World War II, the consensus in the education community was that the practice of granting blanket credit to World War I veterans as a reward for length of service was educationally unsound. Educators concluded that military learning experiences applicable to civilian curricula should be assessed by faculty for potential credit. Therefore, in December 1945, at the request of civilian educational institutions and the regional accrediting associations, the American Council on Education (ACE) established the Commission on Accreditation of Service Experiences, renamed the Commission on Educational Credit and Credentials in 1979, to evaluate military educational programs and to assist institutions in granting credit for such experiences. The first edition of the *Guide* was published in 1946.

The extension of the World War II G.I. Bill to include veterans of the Korean conflict, and the subsequent enrollment of many veterans in colleges and universities, created a need for a second edition, published in 1954.

The 1968 edition was prepared in anticipation of the increased enrollment of veterans resulting from the educational assistance provided under the Veterans Readjustment Benefits Act of 1966, and with the expectation that many would apply for educational credit for their learning experiences in the armed services. In addition, technological advances had necessitated major changes in service training, with a resulting need for new or revised educational credit recommendations.

The 1974 edition was prepared primarily to respond to three emerging considerations. First, because of the growth in vocational and technical programs there was a need to evaluate courses for possible credit in the vocational and technical categories in addition to the baccalaureate and graduate categories of previous editions. Second, active-duty servicemembers were enrolling in increasing numbers in civilian educational programs and were seeking credit for military formal courses soon after completing their service school training. Third, credit recommendations were needed for the many courses initiated or revised by the military since 1968.

The 1974 edition of the *Guide* marked the beginning of a new approach to reporting evaluations of formal military training. At its fall 1973 meeting, the Commission approved the concept of an ongoing *Guide* system. Elements of that system included the publication of biennial editions of the *Guide* through computerized composition, continual staff review of courses, and the computerized storage of course information for a more rapid updating of credit recommendations. In 1994, the computerized *Guide* system came in-house, and all data are now managed by the Military Evaluations Program staff.

Over the years the recommendations contained in the *Guide* have assisted educational institutions in granting credit to hundreds of thousands of servicemembers. Surveys show that most of the nation's colleges and universities use the formal course recommendations in awarding credit to veterans and active-duty service personnel. The recommendations have been widely accepted because military formal courses share certain key elements with traditional postsecondary programs. They are formally approved and administered, are designed for the purpose of achieving learning outcomes, conducted by qualified persons with specific subject-matter expertise, and structured to provide for the reliable and valid assessment of student learning.

The recommendations reflect the Commission's belief that it is sound education practice to give recognition for learning, no matter how or where that learning has been attained, provided that the learning is at the appropriate level, is in the appropriate area, and is applicable to an individual's postsecondary program of study.

Until 1975, however, no mechanism existed for providing recognition for the learning a servicemember attained through such learning experiences as self-instruction, on-the-job training, and work experience. In 1975 the Commission implemented a program for the evaluation of learning represented by demonstrated proficiency in Army enlisted military occupational specialties (MOSs). The MOS evaluation procedures were developed, tested, and refined during a feasibility study conducted by ACE and sponsored by the Department of the Army. Evaluators made recommendations for educational credit and advanced standing in apprentice training programs. Subsequently, the occupational assessment program of the Commission was expanded to include Navy general rates, ratings, warrant officers and limited duty officers, Army warrant officer MOSs, Navy warrant officer and limited duty officer specialties, Coast Guard enlisted ratings and warrant officers, and selected Marine Corps MOSs. A small number of Naval Enlisted Classifications (NECs) have also been evaluated.

In 1994, ACE published the *1954–1989 Guide to the Evaluation of Educational Experiences in the Armed Services*. It contains all courses and occupations with exhibit dates of 1954 to 12/89. **Please retain the *1954–1989 Guide* as a permanent reference and use it with the current**

Guide to the Evaluation of Educational Experiences in the Armed Services.

Beginning with the 1994 edition, all subsequent *Guides* contain all course and occupation exhibits with start dates of 1/90 and later.

THE ACE COURSE EVALUATION SYSTEM

Courses listed in the *Guide* are service school courses conducted on a formal basis, i.e., approved by a central authority within each service and listed by the service in its catalog. These courses are conducted for a specified period of time with a prescribed course of instruction, in a structured learning situation, and with qualified instructors.

Most courses are given on a full-time basis. After 1981, ACE began evaluation of courses that are 45 academic hours in length. Before that time courses evaluated were at least two weeks long, with a minimum of 60 contact hours of instruction. Before 1973, the minimum length requirement was three weeks or 90 contact hours.

In fall 1973, the Commission approved the following procedures and guidelines for the evaluation of military formal courses.

Selection of Evaluators

Nominations for course evaluators are requested from postsecondary institutions, professional and disciplinary societies, education associations, other evaluators, and regional accrediting associations.

The criteria for the selection of formal course evaluators are as follows:

1. The area of an evaluator's competence will closely approximate the area of the training to be evaluated.
2. Preference will be given to candidates with five or more years of postsecondary teaching or administrative experience, including curriculum development.
3. Preference will be given to candidates who are generally receptive to the recognition of learning that occurs in a variety of settings.

An evaluator candidate is interviewed by a staff member to determine whether the individual meets the selection criteria.

An effort is also made to obtain a diverse geographic representation on the team. Subject-matter specialists represent a variety of postsecondary institutional types.

The Evaluation Process

Courses are evaluated by teams of at least three subject-matter specialists (college and university professors, deans, and other academicians). Through discussion and the application of evaluation procedures and guidelines, team members reach a consensus on the amount and category of credit to be recommended.

Evaluation materials include the course syllabus, training materials, tests, textbooks, technical manuals, and examinations. Additional information may be obtained from discussions with instructors and program administrators, classroom observations, and examination of instructional equipment and laboratory facilities.

Evaluators have two major tasks for each course: the formulation of a credit recommendation and the preparation of the course description. The credit recommendation consists of the category of credit, the number of semester hours recommended, and the appropriate subject area. Evaluators write the course description (which appears in the *Guide* exhibits under the headings Learning Outcomes or Objectives and Instruction) in terms meaningful to civilian educators. The course description supplements the credit recommendations by summarizing the nature of the course.

Throughout the evaluation process, evaluators exercise professional judgment in applying the evaluative criteria and procedures. This position reflects the belief of the Commission on Educational Credit and Credentials that sound educational evaluation depends on professional judgment and expertise rather than on rigid application of criteria.

ARMY ENLISTED OCCUPATION CLASSIFICATION SYSTEM

The following paragraphs include background information on Army enlisted MOSs, a description of the enlisted MOS classification system, and detailed information on the enlisted MOS evaluation score.

The Army Enlisted MOS Classification System

The Army Enlisted Military Occupational Specialty (MOS) Classification System is a comprehensive taxonomy of Army enlisted duty positions. Closely related positions that require similar qualifications and the performance of similar duties are grouped as an MOS under a generic title. The job title *Legal Specialist,* for example, encompasses duty assignments such as preparing correspondence, maintaining files, and researching.

Army MOSs are usually referred to with a designation containing at least five characters (e.g., 71D10, 71D20, 71D30). The first three characters (two numbers and a letter) identify the MOS (e.g., 71D is the designation for Legal Specialist). The fourth character, a number from one to five, indicates the level of skill within the MOS. Only the skills, competencies, and knowledge represented by *the first four characters* of the MOS designation are evaluated for comparability with civilian learning.

The fifth character is normally zero; in some cases, a letter is used as a Special Qualification Identifier (SQI) to indicate a soldier's specific duty assignment or special qualifications. In fact, an enlisted soldier's occupation may be expressed by as many as nine characters, but because the last five characters indicate special qualifications that are variable, the learning they represent is not evaluated by ACE.

The first three characters, then, represent the MOS, or occupational designation, and the fourth character represents the skill level within the MOS. In accordance with Army practice, *skill levels are referred to in this publication with a neutral fifth character: zero* (i.e., 10, 20, 30, 40, and 50). Each enlisted MOS has from one to five skill levels, depending on the types of duty positions encompassed by the MOS. The five skill levels may be broadly characterized as follows:

Skill Level 10 identifies entry-level positions requiring performance of tasks under direct supervision.

Skill Level 20 identifies positions requiring performance of more difficult tasks under general supervision and, in some instances, involving supervision of soldiers in Skill Level 10.

Skill Level 30 identifies positions requiring performance of still more difficult tasks and involving first-line supervision of soldiers in Skill Levels 10 and 20.

Skill Level 40 identifies positions requiring relatively detailed knowledge of all tasks specified for a given MOS, normally involving first-line supervision of soldiers in Skill Levels 10, 20, and 30, as well as managerial duties.

Skill Level 50 identifies managerial and supervisory positions requiring broad knowledge of the tasks performed at all subordinate levels in a given MOS and related MOSs in order to coordinate and give direction to work activities.

Thus, each skill level represents progressively more complex duties, increased skills, and greater responsibility; and proficiency in a higher skill level includes the ability to perform the tasks required for the lower skill level(s), as well as additional tasks.

Enlisted MOSs are grouped into career management fields within the enlisted MOS classification system. Each career management field provides opportunities for advancement and career progression among related MOSs.

The Enlisted MOS Evaluation System

The Army regularly evaluates each enlisted soldier's MOS proficiency through the Enlisted Evaluation System. Individuals are evaluated to determine whether they have acquired and maintained the necessary MOS skills, competencies, and knowledge, as codified and described in the enlisted MOS classification system. The means used to evaluate soldiers have undergone a series of changes over the years.

An individual is awarded an MOS skill level when the skills, competencies, and knowledge for a particular MOS have been acquired. The MOS may be awarded (1) following successful completion of a period of supervised on-the-job training (reflected on the individual's set of orders) or (2) following successful completion of an MOS-producing course (reflected on a Course Completion Certificate).

Subsequently, an individual is periodically evaluated to determine whether he or she has *maintained* the MOS skills, competencies, and knowledge.

The following list traces the evaluation process from October 1973 to the present:

- **October 1973–December 1976. The Enlisted MOS Evaluation Score.** Until January 1977, the Army's Enlisted Evaluation System comprised a written examination (and in some MOSs, a performance examination such as typing or musical instrument performance) and the supervisor's rating of the soldier's job performance. The written examination was a 125-item, multiple-choice test that covered all major areas of skill-level proficiency. On the supervisor's rating, or Enlisted Evaluation Report (EER) and Senior Enlisted Evaluation Report (SEER), the soldier was rated on a wide range of characteristics and traits.

  The score that resulted from the application of the Enlisted Evaluation System was normally a composite score, consisting of the score on the written MOS test, the score from the supervisor's rating of job performance on the Enlisted Evaluation Report (EER), and the score on the performance test, if one was required. Composite MOS evaluation scores ranged from a low of 40 to a high of 160, with the minimum qualification score set at 70.

- **January 1977–July 1983. The SQTs.** The Skill Qualification Tests assessed the soldier's occupational proficiency through a combination of a hands-on performance evaluation, a written test, and the supervisor's performance certification of specific tasks. Development of the SQTs was based on an analysis of critical skills required for proficiency in a given MOS skill level. The test was used for soldiers to requalify in their assigned skill level. They requalified in their skill level by achieving the verified minimum score for the particular MOS. SQTs were not developed for pay grades E–8 and E–9. During this period, the tests were not standardized, and ACE suggests using the scores (60 or above) in conjunction with EERs/SEERs.

  EER/SEER. Because of the suspension of the use of MOS Evaluation Tests in December 1976 and the gradual phasing-in of SQTs, some soldiers have been evaluated solely on the basis of EER/SEERs.

An "EER-only" enlisted MOS evaluation score was not a composite score and did not include a score from a standardized examination. See Step 1 in How to Find and Use MOS Exhibits for more information on EERs and SEERs.

- **August 1983–October 1991. An Improved SQT.** After August 1983, the procedures for selecting the skills to be measured on each SQT were standardized and were being applied uniformly to the new test edition. These new procedures produced tests that consistently sampled skills that were representative of the MOS and only of the MOS. This standardization overcame two of ACE's prior concerns: The original SQTs tested general military-specific items (e.g., maintenance of firearms) as well as MOS-specific items, and they used test items in both the skill level being tested as well as those in the next higher skill level. The revised SQT tested only MOS-related competencies and only those at the current skill level.

  The SQT was a major component in a soldier's initial MOS certification and in the annual recertification process. Those who failed had to retake the test the following year. Failing the test two years in a row resulted in either losing the MOS or having the skill level reduced.

  Passing the SQT was one of several criteria used in promotion decisions. Other factors considered were training and schooling records, supervisor's ratings, leadership activities, and awards and citations. The Army set a score of 60 to indicate that MOS proficiency standards were met.

  Supervisor's Ratings. If the SQT score is not available, refer to EERs, SEERs, and NCO Evaluation Reports. (See Step 2 in How to Find and Use Army Enlisted MOS Exhibits.)

- **October 1991–September 1993. Self-Development Tests.** The SQT was replaced by the Self-Development Tests (SDT) for Skill Levels 20–40. The SDT was phased in gradually during this time period and was not reviewed by ACE during its pilot stage. *NCO Evaluation Reports (DA Form 2166-7) should be used to document MOS proficiency during this time period.* In Part IV, check for a success rating (meets standards) or an excellence rating (exceeds standards) in Sections b, d, and f. In Part V, determine that an overall rating of "fully capable" or "among the best" has been achieved. There is no longer an Army formal assessment for Skill Level 10 soldiers.

- **October 1993–February 1995. Self-Development Tests.** The SDT received ACE's endorsement for use as an indicator of MOS proficiency. Credit may now be awarded for enlisted MOSs at Skill Levels 20–40 when an SDT score of 70 percent or greater is achieved. SDT scores will be reported on the AARTS transcript and on the Individual Soldiers Report (ISR).

  Soldiers in Skill Level 10 (paygrades E–l through E–4) are not included in the SDT system and do not receive evaluation reports. Promotion from paygrades E–l through E–4, approved by the commanding officer, is automatic. There are no official Army documents verifying MOS proficiency at this level.

- *March 1995-Present.* The Army discontinued offering the SDT as of February 1995. ACE looked into the possibility of using NCOERs as the sole indicator of MOS proficiency. A feasibility study was conducted. It was found that while the NCOER is an indicator of competency in management skills, it falls short as an indicator of technical competence.

Further study led to the following policy:

After 3/95, only soldiers in Skill levels 30, 40, and 50 will be eligible for management credit based on the NCOER. They will also be eligible to receive the technical credit recommended for the preceding skill level. The MOS exhibits in this *Guide* have been modified to reflect this.

ACE's criteria for evaluating an occupational system hold that the system must be codified, adequately described, and provide for the assessment of the individual. Since these criteria are no longer met by the Army, ACE is unable to provide credit recommendations for Skill Levels 10 and 20. However, ACE will continue to provide credit recommendations for Skill Level 10-related courses. Credit recommendations for courses leading to and related to Skill Level 10, such as basic training, advanced individual training (AIT), and other Army courses taken by Skill Level 10 soldiers, may be found in the *Guide*.

Although the Skill Level 10 MOS recommendations are no longer provided, descriptions will continue to be a part of the MOS exhibit, because MOS skill level progression begins with Skill Level 10. ACE recommends that an institution wishing to grant credit for Skill Level 10 do so on the basis of an individualized assessment of the student.

ARMY WARRANT OFFICER MOSs

The following paragraphs provide background information on Army warrant officer and describe the warrant officer MOS classification system and the procedures used by the Army to select warrant officers and evaluate their MOS skills and knowledge.

The Army Warrant Officer MOS Classification System

Army warrant officers are highly skilled technicians. They are normally assigned to middle management or administrative positions that require highly specialized or technical skills and knowledge and the supervision of enlisted technical specialists.

The warrant officer MOS classification system currently includes approximately 76 MOSs grouped into 16 career fields. Each warrant officer MOS represents a set of duties and qualifications that are highly consistent from one duty assignment to another.

Warrant officer MOSs are normally identified by *at least* four characters (three numbers and a letter, e.g., 214E). Warrant officer MOSs do not have skill levels.

A warrant officer's occupational qualifications may be expressed with as many as nine characters, but because the last five characters indicate special qualifications that are variable, the learning they represent is not evaluated. Only the skills, competencies, and knowledge represented by the *first four characters* are evaluated by ACE. The fifth character, a number or a letter, designates a Special Qualification Identifier (SQI); when a warrant officer has not been awarded an SQI, the fifth character is a zero. The sixth and seventh characters (a number and a letter) are an Additional Skill Identifier (ASI) that relates a specific occupational skill or item of equipment to an MOS. The eighth and ninth characters represent competency in a specific foreign language.

Users may wish to conduct an individual assessment to grant credit for the additional learning represented by the last five characters. Because these additional characters are frequently awarded on the basis of successful completion of formal courses, the user may refer to the appropriate formal course recommendations in the *Guide*.

There are four grades of warrant officers: Warrant Officer, W–1 (WO1); Chief Warrant Officer, W–2 (CW2); Chief Warrant Officer, W–3 (CW3); and Chief Warrant Officer, W–4 (CW4) The grades reflect rank and do not signify differences in job duties.

Warrant Officer Selection and Evaluation

The procedures used in selecting warrant officers and evaluating their MOS skills and knowledge are different from those used for enlisted soldiers.

Selection. Warrant officers are appointed by the Secretary of the Army. Warrant officer vacancies are routinely announced, and interested persons undergo a competitive application process. Most applicants have had Army enlisted service experience, but such experience is not required.

To qualify for appointment, applicants must meet several criteria. For ACE purposes, the relevant criteria are:

1. Achieve a standard score of 110 or higher on the Aptitude Area General Technical Test of the Army Classification Battery or the Army Qualification Battery (this score is also required for commissioned officer applicants).
2. Demonstrate understanding of and proficiency in the English language.
3. Have sufficient education, technical training, and practical experience to ensure outstanding technical ability in the MOS for which application is being made.
4. Be a high school graduate or the equivalent and, when the MOS requires more than a high-school-level education, meet the additional education requirement for the specific MOS. (Two years of college or the equivalent is the desired goal for Regular Army warrant officers.)

Selected candidates must successfully complete a warrant officer candidate course before being appointed.

Evaluation. Under the present evaluation system, the normal procedure is for each warrant officer to be evaluated at least once a year. The evaluation is conducted by officers who serve as rater and senior rater. Each performs independently. The rater, the warrant officer's immediate supervisor, completes the full evaluation report. The senior rater, the officer who supervises the rater, ensures that the evaluation report has been accurately and properly completed and adds comments and a rating of the rated warrant officer.

Warrant officer evaluations focus on technical competence in the MOS. Raters assign adjectival and numerical ratings for each of several important duty areas.

The ratings are recorded on the Officer Evaluation Report (OER) (DA Form 67-8), which also contains a description of the specific functions, duties, and tasks that the rated warrant officer is required to perform during the assignment covered by the OER. Each report provides an appraisal of the rated warrant officer's professional attributes, quality of performance, and potential demonstrated during a specific period while in a particular duty assignment. Complete files of each warrant officer's OERs are maintained at the U.S. Total Army Personnel Command (PERSCOM), where all OERs are reviewed for accuracy and completeness. (See Appendix B for an OER form.)

ACE OCCUPATION EVALUATION SYSTEM

The ACE evaluation system for MOSs has three major components: the selection of evaluators, the materials required for evaluation, and the procedures and guidelines evaluators use in reaching decisions and making recommendations.

Selection of Evaluators

Nominations for evaluators are requested from postsecondary institutions, professional and disciplinary societies, education associations, and regional accrediting associations.

The criteria for selection of MOS evaluators are as follows:

1. The area of an evaluator's competence will closely approximate the area of the training to be evaluated.

2. Preference will be given to candidates with five or more years of postsecondary teaching or administrative experience, including curriculum development.

3. Preference will be given to candidates who are generally receptive to the recognition of learning that occurs in a variety of settings.

An evaluator candidate is interviewed by a staff member to determine whether the individual meets the selection criteria.

An effort is also made to obtain a diverse geographic representation on the team. Subject-matter specialists represent a variety of postsecondary institutional types.

Materials Required for Evaluation

In order to make a recommendation, evaluators must first identify the skills, competencies, and knowledge associated with a given warrant officer or enlisted MOS. The materials relevant to the evaluation of each warrant officer MOS or enlisted MOS skill level are made available to staff members and evaluators by the Army. Materials include the official Army MOS manual, which describes the duties and qualifications for each MOS; technical manuals, field manuals, and other publications used by enlisted soldiers and warrant officers in the day-to-day performance of their duties and by enlisted soldiers to prepare for their MOS evaluation tests; enlisted MOS skill-level evaluation tests; and study guides that outline the proficiency requirements for each enlisted MOS skill level. Additional information is obtained by observing and interviewing enlisted soldiers and warrant officers on-the-job during site visits to Army installations.

The Evaluation Process

Evaluators identify the skills, competencies, and knowledge required of warrant officers who are qualified in a given MOS and enlisted soldiers who are qualified in a given MOS skill level and relate that demonstrated learning to the same attributes acquired by students who have completed a comparable postsecondary course or curriculum. Because the evaluations are based on a comparison of learning outcomes, the amount of time an enlisted soldier or warrant officer may have spent acquiring MOS proficiency is not taken into consideration. The emphasis is on translating the learning demonstrated through MOS proficiency into terms used in formal civilian postsecondary education systems to recognize the same learning. This reflects the belief of the Commission that the value of learning does not depend on where or how the learning occurs.

Evaluation teams are assigned three tasks in the evaluation process: to identify the learning represented by proficiency in the MOS by reviewing the written materials and by observing warrant officers or soldiers performing in the MOS and interviewing them and their supervisors; to prepare a description of the duties, skills, competencies, and knowledge required for each specialty; and to make recommendations for each specialty level based on discussion and consensus.

Throughout the evaluation process, evaluators exercise professional judgment in applying the evaluative criteria and procedures. This position reflects the Commission's belief that sound educational evaluation depends on professional judgment and expertise rather than on rigid application of criteria.

The Commission continually reviews its criteria and procedures. Evaluators are encouraged to provide feedback and recommendations for consideration by the Commission.

CREDIT CATEGORIES

Educational credit is used by postsecondary institutions to quantify and record a student's successful completion of a unit of study. Postsecondary education consists of courses and programs of instruction for persons who are high school graduates or the equivalent, or who are beyond compulsory school age. ACE evaluators utilize the following categories of educational credit when formulating credit recommendations:

Vocational Certificate. This category describes coursework of the type normally found in certificate or diploma (nondegree) programs that are usually a year or less in length and designed to provide students with occupational skills. Course content is specialized, and the accompanying shop, laboratory, or similar practical components emphasize procedural more than analytical skills.

Lower-Division Baccalaureate/Associate Degree. This category describes coursework of the type normally found in the first two years of a baccalaureate program and in programs leading to the associate degree. The instruction stresses development of analytical abilities at the introductory level. Verbal, mathematical, and scientific concepts associated with an academic discipline are introduced, as are basic principles. Occupationally oriented courses in this category are normally designed to prepare a student to function as a technician in a particular field.

Upper-Division Baccalaureate. This category describes courses of the type found in the last two years of a baccalaureate program. The courses involve specialization of a theoretical or analytical nature beyond the introductory level. Successful performance by students normally requires prior study in the area.

Graduate Degree. This category describes courses with content of the type found in graduate programs. These courses often require independent study, original research, critical analysis, and the scholarly and professional application of the specialized knowledge or discipline. Students enrolled in such courses normally have completed a baccalaureate program.

Semester Hours

Credit recommendations for courses are not derived by simple arithmetic conversion. Evaluators exercise professional judgment and consider only those competencies that can be equated with civilian postsecondary curricula. Intensive courses offered by the military do not necessarily require as much outside preparation as many regular college courses. Evaluators consider the factors of pre- and post-course assignments, prior work-related experience, the concentrated nature of the learning experience, and the reinforcement of the course material gained in the subsequent work setting.

The MOS recommendations are based on the skills, competencies, and knowledge gained, as demonstrated through proficiency in a given enlisted MOS skill level or in a given warrant officer MOS, without reference to how much time elapsed during the learning process. The semester hour is used as a standard to express how many semester hours of appropriate course work a student would normally complete to attain the same learning outcomes or attest to the same level of competency.

Credit recommendations are expressed in semester credit hours. In determining semester hour recommendations, evaluators will be guided by, but not restricted to, the following standard definitions:

1. One semester credit hour for the equivalent of 15 hours of classroom contact plus 30 hours of outside preparation; or
2. One semester credit hour for the equivalent of 30 hours of laboratory work plus necessary outside preparation, normally expected to be 15 hours; or
3. One semester credit hour for the equivalent of no fewer than 45 hours (contact hours) of shop instruction.

Other Resources

The Defense Activity for Non-Traditional Education Support (DANTES) maintains the educational records of the servicemembers who have completed DANTES Subject Standardized Tests (DSSTs), College Level Examination Program (CLEP) examinations, United States Armed Forces Institute (USAFI) (disestablished in 1974), and GED tests.

The results of courses and tests taken under the auspices of USAFI before July 1, 1974 are available from the DANTES Program:

DANTES Program
The Chauncey Group International
P.O. Box 6605
Princeton, NJ 08541-6605

There is a $10 fee for each transcript. There is no charge for transcripts sent to military Test Control Officers (TCOs) for counseling purposes.

For GED tests taken overseas after July 1, 1974, write to:
GED Testing Service
One Dupont Circle
Washington, DC 20036-1193

For GED tests taken within the United States after July 1, 1974, write to the Department of Education in the state where the test was taken.

The results of DANTES Subject Standardized Tests (DSSTs) and CLEP tests taken under the auspices of the DANTES Program after July 1, 1974, will appear on the AARTS transcript if the tests are taken while the servicemember was on active duty. For spouses of servicemembers and other civilian personnel, test results may be obtained from:

DANTES Program
The Chauncey Group International
P.O. Box 6604
Princeton, NJ 08541-6604
Telephone: 609-720-6740

A fee of $8 is charged for each transcript. A transcript may include any or all DSST and CLEP examinations taken while in the military. There is no charge for transcripts sent to military Test Control Officers (TCOs) for counseling purposes.

Students may call the DANTES Program at 609-720-6740 to order a transcript request form. Or they may request a transcript if they include all of the following information along with the appropriate fee:

- Name (include all names tests will be registered under)
- Current address and phone number
- Social Security Number
- The address to where you would like your transcript mailed
- Which test you would like included on the transcript; i.e., passing scores only, certain tests, etc.

A signature is required in order to authorize the DANTES Program to release the information

Transcript request fees may be paid by check or money order payable to the DANTES program. Payment may also be made with a Master Card or Visa (please include the expiration date). If paying by credit card, you may fax your request to 609-720-6800.

Records of individuals tested by the Veterans Administration after October 1, 1989, may be obtained from:

Manager
Military Testing
GED Testing Service
American Council on Education
One Dupont Circle, Suite 250
Washington, DC 20036-1193

Appendix B

Sample Military Records

APPLICATION FOR THE EVALUATION OF
LEARNING EXPERIENCES DURING MILITARY SERVICE

____
(Date) (YYYYMMDD)

TO: *(Name and address of educational institution, agency, or employer)*

EVALUATION REQUEST FOR:

____
(Name of Applicant)

____
(Social Security Number)

ATTENTION:

Dear Official:

The applicant named above has requested that the attached summary of educational achievements, accomplished while in the Armed Forces of the United States, be forwarded to you for review and evaluation.

The American Council on Education publishes the *Guide to the Evaluation of Educational Experiences in the Armed Services.* The Guide series contains postsecondary credit recommendations for selected military courses and occupations. The 1954-1989 Guide contains recommendations spanning the dates 1/1954 - 12/1989, and should be kept as a permanent resource. The current edition contains credit recommendations from 1/1990 to the present, and is published every two years. In addition, supplemental handbooks are issued at 6-month intervals between Guide publications. The handbook contains recommendations for all evaluations conducted after the publication of the current Guide.

This form contains a record of a Service member's military courses and occupations. It should be signed by a military official whose signature certifies that the information that is entered on the form is accurate and is taken directly from original records. ACE ID numbers are entered in column 18 by military education officers.

The American Council on Education maintains an advisory service to provide credit recommendations for courses, tests, and occupations that cannot be located in any of the Guide publications. If ACE ID numbers have been entered into column 18, it is not necessary to submit this form to ACE. If there are questions about any of the entries, the institutional official may contact ACE for additional information. Credit recommendations are not provided to institutions at the applicant's request.

Authorized persons may submit questions to ACE at the following address: American Council on Education, Center for Adult Learning and Educational Credentials, One Dupont Circle, Washington, DC 20036-1193, ATTN: Military Evaluations. Telephone: (202) 939-9470; Fax: (202) 775-8578; e-mail: mileval@ace.nche.edu.

The evaluation of this applicant's learning experiences, as well as any guidance you may provide, should be sent directly to the applicant at the address shown in Block 6 on page 3.

Sincerely,

____
(Education Officer)

PRIVACY ACT STATEMENT

AUTHORITY: 10 U.S.C. 2007; P.L. 104-106; and E.O. 9397.

PRINCIPAL PURPOSE(S): To facilitate an individual's request for evaluation of educational experiences while in the military services.

ROUTINE USE(S): None.

DISCLOSURE: Voluntary; however, you will not be evaluated for your educational experiences during military service if you fail to provide requested information.

INSTRUCTIONS TO APPLICANT

DD Form 295 is for your convenience in applying for evaluation of your educational experiences during military service. Give as much detailed information as possible. Include additional information on separate sheets, if necessary. Do not use abbreviations.

You are encouraged to write a preliminary letter to the school or agency concerned, explaining your interest in its evaluation of your records for the continuance of your education. Training, correspondence study, or special experiences not described on this form, which you believe would be of interest to those reviewing your case, should be included in this letter.

The applicant should:

a. Complete items 1 through 14.

b. If you have attended college or completed any college correspondence courses, ask that college to send a transcript to the Registrar of the evaluating agency that this form is addressed to. DO NOT LIST ANY COLLEGE OR UNIVERSITY COURSES ON THIS FORM.

c. If you have completed any college-level standardized examinations for credit, such as USAFI or DANTES Subject Standardized Tests, or CLEP, ask the appropriate agency to send a score report to the Registrar of the evaluating agency that this form is addressed to. DO NOT LIST ANY EXAMINATIONS ON THIS FORM.

d. After completion, submit this DD Form 295 to the Certifying Officer.

INSTRUCTIONS TO CERTIFYING OFFICER

(Custodian of Personnel Records)

DD Form 295 is intended to provide factual information that schools and other evaluating agencies require for evaluation of the applicant's educational achievement. By your signature, you verify that all information is accurate and taken directly from military records. CERTIFYING OFFICERS WILL NOT MAKE RECOMMENDATIONS REGARDING CREDIT TO BE AWARDED.

The certifying officer should:

a. Complete items 15 through 17, in ink (or type). Supplemental sheets may be used.

b. Insure that the information provided in Section II is documented in the applicant's Service Record. Names of schools or courses should not be abbreviated.

c. Send this DD Form 295 to the Education Officer.

INSTRUCTIONS TO EDUCATION OFFICER

The education officer should:

a. Complete item 18.

b. Counsel the service member.

c. Complete page 1. The name and address of the evaluating agency should be the same as that listed at the top of page 3 of this form.

PAGE 1 IS IN ADDITION TO, AND NOT A SUBSTITUTE FOR, THE LETTER TO BE WRITTEN TO THE EVALUATING AGENCY BY THE APPLICANT.

d. Mail DD Form 295 directly to the designated evaluating agency.

APPLICATION FOR THE EVALUATION OF LEARNING EXPERIENCES DURING MILITARY SERVICE

TO *(Name and address of educational institution, agency, or employer)*

SECTION I - TO BE COMPLETED BY APPLICANT

1. NAME *(Last, First, Middle Initial)*	2. GRADE/RANK OR RATING	3. SOCIAL SECURITY NO.	4. PREVIOUS SERVICE NUMBER(S)

5. PRESENT BRANCH OF SERVICE *(Includes National Guard and Reserve components)*

☐ ARMY ☐ NAVY ☐ AIR FORCE ☐ MARINE CORPS ☐ COAST GUARD

6. APPLICANT'S MAILING ADDRESS FOR REPLY FROM EDUCATIONAL INSTITUTION

7. DATE OF BIRTH *(YYYYMMDD)*	8. PERMANENT HOME ADDRESS

CIVILIAN EDUCATION

9. HIGHEST GRADE OF SCHOOL COMPLETED *(X one)*

☐ 6 ☐ 7 ☐ 8 ☐ 9 ☐ 10 ☐ 11 ☐ 12

10. HIGHEST YEAR OF COLLEGE COMPLETED *(X one)*

☐ NONE ☐ FRESHMAN *(1 - 29 S.H.)* ☐ SOPHOMORE *(33 - 59 S.H.)* ☐ JUNIOR *(60 - 89 S.H.)* ☐ SENIOR *(90 - 100 S.H.)*

11. COLLEGE DEGREE EARNED *(X if applicable)*

☐ ASSOCIATE ☐ BACHELOR

12. EDUCATIONAL INSTITUTION LAST ATTENDED

a. NAME	b. MAILING ADDRESS

13. MILITARY CORRESPONDENCE COURSES COMPLETED *(The applicant should attach a copy of the course completion letter or certificate.)*

a. COURSE NAME *(If no courses were taken, print NONE)*	b. ACE GUIDE COURSE OR OCCUPATION IDENTIFICATION NUMBER	c. COURSE SPONSOR *(AIPD, MCI, ECI, CGI)*	d. DATE COURSE COMPLETED *(YYYYMMDD)*
(1)			
(2)			
(3)			
(4)			
(5)			
(6)			
(7)			
(8)			
(9)			
(10)			
(11)			
(12)			
(13)			
(14)			
(15)			
(16)			
(17)			
(18)			
(19)			
(20)			

14. APPLICANT CERTIFICATION: I have read the Privacy Act Statement on Page 2.

a. SIGNATURE	b. DATE SIGNED *(YYYYMMDD)*

SECTION II - TO BE COMPLETED BY CERTIFYING OFFICER

(Read Instructions on Page 2 before completing this page)

15. FORMAL SERVICE SCHOOLS ATTENDED *(If 40 hours in 5 consecutive days, or if longer than 32 hours in 5 consecutive days.) (If none, print NONE.)*

a. COURSE TITLE *(Do Not Abbreviate)*	b. MILITARY COURSE NUMBER	c. NAME OF SCHOOL, CITY, STATE	d. DATE ENTERED *(YYYYMMDD)*	e. LENGTH *(In weeks)* *(Note 1)*	f. DATE COMPLETED *(YYYYMMDD)*	g. FINAL MARK AND/OR CLASS STANDING *(Note 2)*	18. ACE GUIDE COURSE OR OCCUPATION IDENTIFICATION NO. *(To be filled out in Education Center)*
(1)							
(2)							
(3)							
(4)							
(5)							
(6)							

16. MILITARY OCCUPATIONAL HISTORY

a. MILITARY SPEC. CODE *(MOS, AFSC, Rate, etc.)* *(Note 3)*	b. MILITARY OCCUPATIONAL TITLE *(Do Not Abbreviate)*	c. DATES HELD (1) FROM *(YYYYMMDD)*	c. DATES HELD (2) TO *(YYYYMMDD)*	d. MOS/SQT/SDT SCORE *(For Army Enlisted Personnel)* *(Note 4)*	
(1)					
(2)					
(3)					

NOTES:
1. Print SP if course length was self paced.
2. If information is available, give grade received. If class standing is shown, give number in class, e.g., 10 in 241.
3. List most recent skill levels or grade.
4. MOS/SQT/SDT Evaluation Score and date of evaluation.

THIS APPLICATION MUST BE SIGNED BY AN OFFICER OR A DULY AUTHORIZED NONCOMMISSIONED OFFICER.

I certify that the information contained herein has been compared with official records, and that this information is correct.

17. CERTIFYING OFFICER

a. NAME *(Print or Type)*	b. GRADE/RANK	c. MILITARY ADDRESS *(Include ZIP Code)*
d. SIGNATURE	e. DATE SIGNED *(YYYYMMDD)*	

DD FORM 295, APR 2000

CAUTION: NOT TO BE USED FOR IDENTIFICATION PURPOSES

THIS IS AN IMPORTANT RECORD SAFEGUARD IT

ANY ALTERATIONS IN SHADED AREAS RENDER FORM VOID

DD FORM 1 JUL 79 214 PREVIOUS EDITIONS OF THIS FORM ARE OBSOLETE.

CERTIFICATE OF RELEASE OR DISCHARGE FROM ACTIVE DUTY

1. NAME *(Last, first, middle)*	2. DEPARTMENT, COMPONENT AND BRANCH	3. SOCIAL SECURITY NO.
	NAVY - USN	123 | 45 | 6789

4a. GRADE, RATE OR RANK	4b. PAY GRADE	5. DATE OF BIRTH	6. PLACE OF ENTRY INTO ACTIVE DUTY
EWCM	E9	14 AUG 42	San Diego, California

7. LAST DUTY ASSIGNMENT AND MAJOR COMMAND	8. STATION WHERE SEPARATED
USS ALBANY (CG-10)	USS ALBANY (CG-10) at Gaeta, Italy

9. COMMAND TO WHICH TRANSFERRED	10. SGLI COVERAGE
Not applicable	AMOUNT $ 20,000 ☐ NONE

11. PRIMARY SPECIALTY NUMBER, TITLE AND YEARS AND MONTHS IN SPECIALTY *(Additional specialty numbers and titles involving periods of one or more years)*

EW-1774 Electronics Warfare Systems Technician (SLQ-22/24)

12. RECORD OF SERVICE	YEAR(s)	MON(s)	DAY(s)
a. Date Entered AD This Period	74	01	04
b. Separation Date This Period	80	05	13
c. Net Active Service This Period	06	04	10
d. Total Prior Active Service	13	03	22
e. Total Prior Inactive Service	00	00	00
f. Foreign Service	00	00	00
g. Sea Service	03	01	07
h. Effective Date of Pay Grade	79	12	16
i. Reserve Oblig. Term. Date	NA	NA	NA

13. DECORATIONS, MEDALS, BADGES, CITATIONS AND CAMPAIGN RIBBONS AWARDED OR AUTHORIZED *(All periods of service)*

Good Conduct Award (FIFTH)
Navy Expeditionary Medal
Armed Forces Expeditionary Medal
National Defense Service Medal
Republic of Viet-Nam Campaign Medal w/device
Viet-Nam Service Medal (7 awards)
Navy Unit Commendation Medal (2 awards)
Joint Service Commendation Medal

14. MILITARY EDUCATION *(Course Title, number weeks, and month and year completed)*

Instructor Basic - 4 weeks - Mar 74
OHS Advanced - 2 weeks - Apr 77
RD "B" School - 32 weeks - Apr 71
CET WLR-1 Series (7107) - 7 weeks - Jul 71
CET AN/ULQ-6 - 8 weeks - Sep 71
ASW Tactical Course - 1 week - Jan 63
Electronics Warfare Operator "C" - 6 weeks - May 71
RD "A" - 24 weeks - Jan 62

15. MEMBER CONTRIBUTED TO POST-VIETNAM ERA VETERANS' EDUCATIONAL ASSISTANCE PROGRAM	16. HIGH SCHOOL GRADUATE OR EQUIVALENT	17. DAYS ACCRUED LEAVE PAID
☐ YES ☒ NO	☒ YES ☐ NO	"NONE"

18. REMARKS

Immediate reenlistment, 14 May 1980. This form was administratively issued 14 May 1980
Block 13 continued: Combat Action Ribbon - Navy "E" Ribbon (2 awards). X X
X X X X X X X X X X
X X X X X X X X X X
X X X X X X X X X X
X X X X X X X X X X
X X X X X X X X X X
X X X X X X X X X
X X X X X X X X X
X X X X X X X X X
X X X X X X X X X

19. MAILING ADDRESS AFTER SEPARATION

404 Twin Oaks Drive, Havertown, PA 19083

20. MEMBER REQUESTS COPY 6 BE SENT TO DIR. OF VET AFFAIRS ☐ YES ☒ NO

21. SIGNATURE OF MEMBER BEING SEPARATED

22. TYPED NAME, GRADE, TITLE AND SIGNATURE OF OFFICIAL AUTHORIZED TO SIGN

Assistant Personnel Officer

S/N 0102-LF-000-2140

MEMBER - 1

CAUTION: NOT TO BE USED FOR IDENTIFICATION PURPOSES

THIS IS AN IMPORTANT RECORD. SAFEGUARD IT.

ANY ALTERATIONS IN SHADED AREAS RENDER FORM VOID

CERTIFICATE OF RELEASE OR DISCHARGE FROM ACTIVE DUTY

1. NAME *(Last, First, Middle)*

2. DEPARTMENT, COMPONENT AND BRANCH

3. SOCIAL SECURITY NO.

4.a. GRADE, RATE OR RANK

4.b. PAY GRADE

5. DATE OF BIRTH *(YYMMDD)*

6. RESERVE OBLIG. TERM. DATE — Year | Month | Day

7.a. PLACE OF ENTRY INTO ACTIVE DUTY

7.b. HOME OF RECORD AT TIME OF ENTRY *(City and state, or complete address if known)*

8.a. LAST DUTY ASSIGNMENT AND MAJOR COMMAND

8.b. STATION WHERE SEPARATED

9. COMMAND TO WHICH TRANSFERRED

10. SGLI COVERAGE — None — Amount: $

11. PRIMARY SPECIALTY *(List number, title and years and months in specialty. List additional specialty numbers and titles involving periods of one or more years.)*

12. RECORD OF SERVICE	Year(s)	Month(s)	Day(s)
a. Date Entered AD This Period			
b. Separation Date This Period			
c. Net Active Service This Period			
d. Total Prior Active Service			
e. Total Prior Inactive Service			
f. Foreign Service			
g. Sea Service			
h. Effective Date of Pay Grade			

13. DECORATIONS, MEDALS, BADGES, CITATIONS AND CAMPAIGN RIBBONS AWARDED OR AUTHORIZED *(All periods of service)*

14. MILITARY EDUCATION *(Course title, number of weeks, and month and year completed)*

15.a. MEMBER CONTRIBUTED TO POST-VIETNAM ERA VETERANS' EDUCATIONAL ASSISTANCE PROGRAM — Yes | No

15.b. HIGH SCHOOL GRADUATE OR EQUIVALENT — Yes | No

16. DAYS ACCRUED LEAVE PAID

17. MEMBER WAS PROVIDED COMPLETE DENTAL EXAMINATION AND ALL APPROPRIATE DENTAL SERVICES AND TREATMENT WITHIN 90 DAYS PRIOR TO SEPARATION — Yes | No

18. REMARKS

19.a. MAILING ADDRESS AFTER SEPARATION *(Include Zip Code)*

19.b. NEAREST RELATIVE *(Name and address - include Zip Code)*

20. MEMBER REQUESTS COPY 6 BE SENT TO ______ DIR. OF VET AFFAIRS — Yes | No

21. SIGNATURE OF MEMBER BEING SEPARATED

22. OFFICIAL AUTHORIZED TO SIGN *(Typed name, grade, title and signature)*

SAMPLE

DD Form 214, NOV 88 S/N 0102-LF-006-5500 *Previous editions are obsolete.* MEMBER - 1

****************************** SAMPLE COPY *************************

FOR IMMEDIATE DELIVERY TO:

DEPARTMENT OF THE ARMY
HEADQUARTERS US ARMY ARMOR CENTER AND FORT KNOX
Fort Knox, Kentucky 40121

ORDERS XX-X 4 May 1990

DOE, JOHN 000-00-000 CSM Unit of Assignment (UIC Code) Ft Knox KY 40121

The following MOS action is directed.

Awarded: P19E10B800; S54E100000
Withdrawn: P54E100000
Effective date: 4 May 1990
Reclassification control number: Not applicable
Additional instructions: Authority: Paragraph xx-x, AR xxx-xxx
Format: 310

FOR THE COMMANDER:

GI JOE
LTC, AG
Adjutant General

DISTIRBUTION:
Orders Unit (4)
Indiv (3)
Rec Sec (1)
Unit of Asg (3)

HQPERSCOM (Branch) (1 ea Indiv)

****************************** SAMPLE COPY **********************

DEPARTMENT OF THE ARMY

CERTIFICATE OF TRAINING

This is to certify that

____

has successfully completed

Given at _______________

DA FORM 87, 1 OCT 78

☆U.S. GOVERNMENT PRINTING OFFICE: 1988-206-376

PERSONNEL QUALIFICATION RECORD — PART II

SECTION I — IDENTIFICATION DATA

1. NAME	2. S.S.N.

SECTION II — CLASSIFICATION AND ASSIGNMENT DATA

(3) MOS EVALUATION SCORES ☐ CONT

MOSC	YR & MO	SCORE	YR & MO	SCORE	YR & MO	SCORE

4. ASSIGNMENT CONSIDERATIONS ☐ CONT

5. OVERSEA SERVICE ☐ CONT

FROM	THRU	AREA AND COUNTRY	MO	TYPE	NTC	DEPN ARR OS

SECTION II — CLASSIFICATION AND ASSIGNMENT DATA *(Continued)*

6. MILITARY OCCUPATIONAL SPECIALTIES ☐ CONT

MOSC	TITLE	DATE

7 AVIATION ASI & GUNNERY QUALIFICATION ☐ CONT

AIRCRAFT		INSTR PILOT		GUNNERY SYSTEM	
F/W	R/W	F/W	R/W	TNG	INSTR

(8) APTITUDE AREA SCORES ☐ CONT

AREA	SCORE	AREA	SCORE

9. AWARDS, DECORATIONS & CAMPAIGNS ☐ CONT

DATE	
PLACE	

10. OTHER TESTS ☐ CONT

TEST	DATE	SCORE
MDB-		
OCT		
DLAT		
OQI-1		
FAST-OB		
WOCB		

11. AMERICAN BOARD CERTIFICATION & LICENSES OR CERTIFICATES HELD ☐ CONT

12. LANGUAGE PROFICIENCY

DA FORM 330 SUBMITTED	DATE

DA FORM 2-1 1 JAN 73

SECTION II – CLASSIFICATION AND ASSIGNMENT DATA *(Continued)*

13 PILOT RATINGS

ORIGINAL	DATE	CURRENT	DATE

14 FLYING STATUS ☐ CONT

INSTRUMENT CERTIFICATION

15 INTERNSHIPS, RESIDENCIES AND FELLOWSHIPS ☐ CONT

HOSPITAL	TYPE OR SERVICE	MONTHS	YEAR

16 HOSPITAL/TEACHING APPOINTMENTS AND PRIVATE PRACTICE ☐ CONT

FROM	THRU	INSTITUTION/LOCATION	TYPE	DURAT

17. CIVILIAN EDUCATION AND MILITARY SCHOOLS ☐ CONT

SCHOOL	MAJOR/COURSE/MOSC	DURAT	COMP	YEAR

SECTION III – SERVICE, TRAINING AND OTHER DATES

18. APPOINTMENTS AND REDUCTIONS ☐ CONT

GRADE	COMP	EFFECTIVE DATE	DATE OF ELIG./RANK

19. SPECIALIZED TRAINING ☐ CONT

SUBJECT	DATE
ATP 21-114 *(BCT)*	
Geneva-Hague Conventions	
Military Justice	
Benefits of Honorable Discharge	

(20) BASIC ENLISTED SERVICE DATE *(BESD)*

(21) TIME LOST *Sec. 972, Title 10, USC)* ☐ CONT

FROM	THRU	DAYS	REASON

SECTION IV – PERSONAL AND FAMILY DATA

22. PHYSICAL STATUS

HEIGHT	WEIGHT	GLASSES
		☐ YES ☐ NO

DATE OF EXAM

23. PLACE OF BIRTH AND CITIZENSHIP

SELF

SPOUSE

CITIZENSHIP OF SPOUSE

24. NUMBER OF DEPENDENTS

ADULT	CHILDREN

25. HOME OF RECORD/ADDRESS

26. CIVILIAN OCCUPATION

JOB TITLE:

DOT CODE	CRITICAL OCCUPATION	NO. MONTHS EMPLOYED	MOSC
	☐ YES ☐ NO		

DUTIES PERFORMED

EMPLOYER

– Fold Here –

SECTION V – MISCELLANEOUS

27. REMARKS

28. ITEM CONTINUATION

ITEM NO.	DATA

(29) DATE DA FORM 20B PREPARED:

(30) DATE DUPLICATE DA FORM 2-1 SUBMITTED:

31 REPORT OF CHANGES

1	2	3	4	5	6	7	8	9	10	11	12	13	14	15	16	17	18	19	20	21	22	23
24	25	26	27	28	29	30	31	32	33	34	35	36	37	38	39	40	41	42	43	44	45	46
47	48	49	50	51	52	53	54	55	56	57	58	59	60	61	62	63	64	65	66	67	68	69
70	71	72	73	74	75	76	77	78	79	80	81	82	83	84	85	86	87	88	89	90	91	92

SECTION IX – RESERVE COMPONENT DATA

32a. READY RESERVE OBLIGATION EXPIRATION DATE:

b. DA FORM 3726 OR 3726-1 AGREEMENT EXPIRATION DATE:

c. SERVICE OBLIGATION EXPIRATION DATE:

d. MANDATORY REMOVAL FROM ACTIVE STATUS:

e. RETIREMENT YEAR ENDING DATE:

33. DATE

PREPARED	REVIEWED

34. SIGNATURE

SECTION VII — CURRENT AND PREVIOUS ASSIGNMENTS

35. RECORD OF ASSIGNMENTS ☐ CONT

EFFECTIVE DATE	DUTY MOSC	PRINCIPAL DUTY	ORGANIZATION AND STATION OR OVERSEA COUNTRY	NON-DUTY DAYS BP YR/MO	NON-RATED DAYS EP YR/MO	TYPE REPORT

— Fold Here —

ENLISTED EVALUATION REPORT (AR 600-200)

For preparation, see DA Pamphlet 623-1.

C. SSN

0 1 2 3 4 5 6 7 8 9

PART I PERSONAL DATA

A. GRADE (ABBR) NAME (LAST) (FIRST) (MI) SSN

B. TYPE OF REPORT: INIT ANL CR SP OTHER ___ SPECIFY

D. ORGANIZATION AND STATION

E. PMOSC | F. DMOSC | G. SMOSC

H. PERIOD OF REPORT

BEGIN — MO. JAN FEB MAR APR MAY JUN JUL AUG SEP OCT NOV DEC — YR 74 75 76 77 78 79 80 81 82 83 84 85

END — MO. JAN FEB MAR APR MAY JUN JUL AUG SEP OCT NOV DEC — YR 74 75 76 77 78 79 80 81 82 83 84 85

I. NONRATED PERIOD

NO. OF MONTHS ______

REASON CODES ______

J. DUTY POSITION TITLE

AUTH PAY GR

PART II RATINGS

A. BRIEF DESCRIPTION OF DUTIES

B. INDORSER HAS NOT OBSERVED AND CANNOT RATE SOLDIER

C. REPORT BASED ON: DAILY CONTACT | FREQ OBSN | INFREQ OBSN | REPT & REC (R / I)

D. SOLDIER SUPPORTS THE ARMY'S EQUAL OPPORTUNITY PROGRAM: YES | NO (R / I)

E. DUTY PERFORMANCE TRAITS

DUTY PERFORMANCE TRAITS		RANKS WITH VERY BEST 5	SUPERIOR TO MOST 4		NEEDS IMPROVEMENT Some 1	Much 0	SCORE R	SCORE I
1. Is well informed on all phases of assigned duties. (Scope of knowledge about duties)	R / I							
2. Carries out orders without constant supervision. (Dependability in performing without supervision)	R / I							
3. Shows interest and enthusiasm for duties. (Attitude toward duties)	R / I							
4. Demonstrates qualities of leadership. (Exerts positive influence on others)	R / I							
5. Seeks out opportunities for self-improvement. (Effort directed toward realization of potential)	R / I							
6. Displays ability to initiate action without direction from others. (Aggressive pursuit of methods to improve duty performance)	R / I							
7. Is successful in working with others. (Ability to work in harmony with others)	R / I							
8. Personal behavior sets a good example for others. (High standards of personal conduct)	R / I							
9. Takes pride in dress and appearance. (Neat and military in bearing)	R / I							
10. Is physically fit, as required, for MOS/grade during combat. (Physical condition)	R / I							
TOTALS								

F. DEMONSTRATED OVERALL PERFORMANCE OF ASSIGNED DUTIES

	Ranks With Very Best	Superior to Most		Demonstrates Shortcomings: Minor	Demonstrates Shortcomings: Major	SCORE
R / I	44 43	42 38 34		14 10 6	5 3 1	

G. ADVANCEMENT POTENTIAL

IF I HAD THE AUTHORITY AND RESPONSIBILITY TO DO SO, I WOULD: (DISREGARD TIME IN GRADE REQUIREMENTS)

	Promote Immediately	Promote Ahead of Peers		Not Promote	Deny Continued Active Duty	SCORE
R / I	31 30	28 26 24		7 5 3	0	

H. SCORE

BLOCKS	RATER	INDORSER
B		
F		
G		
SUM	+	= ÷ 2 = REPT SCORE

DA Form 2166-5 1 Jul 75 This form, together with DA Form 2166-5A, 1 Jul 75, replaces DA Form 2166-4, 1 Jul 70, which is obsolete.

RATED SOLDIER'S LAST NAME AND SSN

PART II CONTINUED

I. CAREER DEVELOPMENT (RECOMMENDATIONS ON SCHOOLING AND ASSIGNMENTS)

J. 1. COMMENTS ARE MANDATORY TO JUSTIFY RATINGS IN PART II AS FOLLOWS:
a. BLOCK E SCORE BELOW 10 OR OVER 40, BLOCK F SCORE BELOW 6 OR OVER 42, BLOCK G SCORE BELOW 10 OR OVER 22, OR BLOCK D IF SOLDIER DOES NOT SUPPORT ARMY'S EQUAL OPPORTUNITY PROGRAM.
b. INDORSER WHO CHECKS BLOCK II B.
2. REMARKS OTHERWISE OPTIONAL.

RATER

INDORSER

PART III RATER AUTHENTICATION

A. ORGANIZATION AND DUTY ASSIGNMENT	B. NAME AND GRADE	C. DATE
	D. SIGNATURE	

PART IV INDORSER AUTHENTICATION

A. ORGANIZATION AND DUTY ASSIGNMENT	B. NAME AND GRADE	C. DATE
	D. SIGNATURE	

PART V SOLDIER AUTHENTICATION

A. I HAVE SEEN A COPY OF THIS REPORT COMPLETE THROUGH ACTION BY THE INDORSER. I HAVE BEEN COUNSELED CONCERNING THE REPORT.	B. NAME AND GRADE	C. DATE
	D. SIGNATURE	

PART VI REVIEWER AUTHENTICATION

A. SOLDIER WAS RATED BY CORRECT RATER AND INDORSER. NO FURTHER ACTION REQUIRED. ☐
MY REVIEW RESULTS IN ACTION INDICATED BY INCLOSURES. ☐

B. ORGANIZATION AND DUTY ASSIGNMENT	C. NAME AND GRADE	D. DATE
	E. SIGNATURE	

PART VII MILPO CERTIFICATION

A. DATE REPORT ENTERED ON DA FM 2-1:	C.	D.
B. SOLDIER'S COPY: GIVEN TO SOLDIER ☐ FORWARDED TO SOLDIER ☐ MAILED TO SOLDIER ☐ CERTIFIED MAIL NO. ________	MILPO SIGNATURE	UIC

SENIOR ENLISTED EVALUATION REPORT (AR 600-200)

For preparation, see DA Pamphlet 623-1.

PART I PERSONAL DATA

A. GRADE (ABBR) NAME (LAST) (FIRST) (MI) SSN

B. TYPE OF REPORT: INIT ▯ ANL ▮ CR ▯ SP ▯ OTHER ▯ SPECIFY

C. SSN: 2 6 5 5 2 6 9 1 5

D. ORGANIZATION AND STATION: HHT, 2d Armd Cav Regt, APO NY 09093

E. PMOSC	F. DMOSC	G. SMOSC
11E50	11E50	11Z50

H. PERIOD OF REPORT

BEGIN: MO. JAN FEB MAR APR MAY JUN JUL AUG SEP OCT NOV DEC — YR 74 75 76 77 78 79 80 81 82 83 84 85

END: MO. JAN FEB MAR APR MAY JUN JUL AUG SEP OCT NOV DEC — YR 74 75 76 77 78 79 80 81 82 83 84 85

I. NONRATED PERIOD

NO. OF MONTHS ______

REASON CODES ______

J. DUTY POSITION TITLE: Intelligence Sergeant

AUTH PAY GR E8

PART II RATINGS

A. BRIEF DESCRIPTION OF DUTIES: NCOIC of the Regimental S-2 Section, responsible for assisting the S-2 in all regimental level intelligence functions to include combat intelligence, personnel document, and physical security.

B. INDORSER HAS NOT OBSERVED AND CANNOT RATE SOLDIER ▯

C. REPORT BASED ON: DAILY CONTACT (R ▮, I ▮) — FREQ OBSN — INFREQ OBSN — REPT & REC

D. SOLDIER SUPPORTS THE ARMY'S EQUAL OPPORTUNITY PROGRAM: YES (R ▮, I ▮) NO

E. PERFORMANCE QUALITIES

		Ranks with very best 5	Superior to most 4	Exceeds or meets duty requirements 3	2	Needs improvement Some 1	Much 0	Score R	Score I
1. Anticipates requirements and actively pursues methods of improving duty performance. (Initiative)	R / I	▮ / ▮						5	5
2. Is physically fit, as required, for MOS/grade during combat. (Physical Condition)	R / I	▮ / ▮						5	5
3. Takes pride in high standards of dress, grooming, and military manner. (Military Bearing)	R / I	▮ / ▮						5	5
4. Behavior on and off duty is in accordance with highest Army standards. (Personal Conduct)	R / I	▮ / ▮						5	5
5. Is dependable and conscientious in fulfilling obligations. (Responsibility)	R / I	▮ / ▮						5	5
6. Is well informed on the scope of knowledge required for assigned duties. (Technical Competence)	R / I	▮ / ▮						5	5
TOTALS								30	30

F. LEADERSHIP SKILLS

		Ranks with very best 5	Superior to most 4	Exceeds or meets duty requirements 3	2	Needs improvement Some 1	Much 0	Score R	Score I
1. Is clear and to the point in conveying information and in giving directions. (Communications)	R / I	▮ / ▮						5	5
2. Promotes personal and professional growth of subordinates through personal interest in their problems. (Counseling)	R / I	▮ / ▮						5	5
3. Provides effective instruction (formal or informal) to improve the professional competence of subordinates. (Training)	R / I	▮ / ▮						5	5
4. Selects best course of action after weighing the alternatives. (Decision Ability)	R / I	▮ / ▮						5	5
5. Is fair, inspires confidence, accepts guidance, and has earned respect. (Relationships with Others)	R / I	▮ / ▮						5	5
TOTALS								25	25

G. DEMONSTRATED OVERALL PERFORMANCE

	Ranks With Very Best	Superior to Most	Exceeds or Meets Duty Requirements	Demonstrates Shortcomings Minor	Major	SCORE
RATER	▮ ▯	▯ ▯ ▯ ▯	▯ ▯ ▯ ▯	▯ ▯ ▯	▯ ▯	40
	40 38	36 33 30 27	23 20 17 14	10 8 6	3 1	
INDORSER	▮ ▯	▯ ▯ ▯ ▯	▯ ▯ ▯ ▯	▯ ▯ ▯	▯ ▯	40

H. ADVANCEMENT POTENTIAL – If I had the authority and responsibility to do so, I would: (Disregard time in grade requirements.)

	Promote Immediately	Promote Ahead of Peers	Promote With Peers	Not Promote	Deny Continued Active Duty	SCORE
R	▮ ▯	▯ ▯ ▯	▯ ▯ ▯	▯	▯	30
	30 28	26 22 18	14 10 6	2	0	
I	▮ ▯	▯ ▯ ▯	▯ ▯ ▯	▯	▯	30

I.

	BLK E		BLK F		BLK G		BLK H		SCORES	TEST SCORE
R	30	+	25	+	40	+	30	=	125	
I	30		25		40		30		125	
								SUM OF SCORES	250	÷2 125

DA Form 2166-5A 1 Jul 75 — This form, together with DA Form 2166-5 1 Jul 75, replaces DA Form 2166-4, 1 Jul 70, which is obsolete.

RATED SOLDIER'S LAST NAME AND SSN **PRESTON**

PART II CONTINUED

J. CAREER DEVELOPMENT (RECOMMENDATIONS ON SCHOOLING AND ASSIGNMENTS) RECOMMEND FOR: CSM (E-8 & E-9) R ■ I ■ 1SG (E-7) R ☐ I ☐

SGM Academy at earliest practical date.

K. 1. COMMENTS ARE MANDATORY TO JUSTIFY RATINGS IN PART II AS FOLLOWS:
a. BLOCK E SCORE BELOW 6 OR OVER 24, BLOCK F SCORE BELOW 5 OR OVER 20, BLOCK G SCORE BELOW 6 OR OVER 36, BLOCK H SCORE BELOW 6 OR OVER 14, OR BLOCK D IF SOLDIER DOES NOT SUPPORT ARMY'S EQUAL OPPORTUNITY PROGRAM.
b. INDORSER WHO CHECKS BLOCK II B.
2. REMARKS OTHERWISE OPTIONAL.

RATER: MSG Preston is one of the most conscientious and dedicated soldiers whom I have ever met. He consistently accomplishes his duties in an exemplary manner. Indicative of this was his performance during field training exercises, specifically Reforger 75. He was responsible for receiving, compiling, and forwarding to higher, all battlefield intelligence. His efforts greatly assisted the Regiment in accomplishing its mission in an outstanding manner. MSG Preston possesses drive, initiative, and stamina. He is consistently striving to improve himself by enrolling in college courses during his off-duty time. His military leadership principles, traits, and personal conduct and integrity won him the respect and admiration of his subordinates and superiors. MSG Preston's ability to get the job done places him among those at the top of the NCO Corps.

INDORSER: MSG Preston clearly possesses those characteristics which place him in that select group of truly outstanding NCOs. Placing the needs of the organization first, he accepted reassignment to an area foreign to his previous experience and through personal initiative and enthusiasm rapidly gained the required technical knowledge. Having observed this NCO transition from operations to intelligence with complete ease and total competence, I highly recommend that he be flagged for special consideration in assignments and schooling. An obvious leader, he possesses talents and qualifications far in excess of his contemporaries.

PART III RATER AUTHENTICATION

A. ORGANIZATION AND DUTY ASSIGNMENT	B. NAME AND GRADE	C. DATE
HHT, 2d Armd Cav Regt Asst S2	D. SIGNATURE	12 MAY 76

PART IV INDORSER AUTHENTICATION

A. ORGANIZATION AND DUTY ASSIGNMENT	B. NAME AND GRADE	C. DATE
HHT, 2d Armd Cav Regt Regimental S2	D. SIGNATURE	12 MAY 76

PART V SOLDIER AUTHENTICATION

A. I HAVE SEEN A COPY OF THIS REPORT COMPLETE THROUGH ACTION BY THE INDORSER. I HAVE BEEN COUNSELED CONCERNING THE REPORT.	B. NAME AND GRADE	C. DATE
	D. SIGNATURE	12 May 76

PART VI REVIEWER AUTHENTICATION

A. SOLDIER WAS RATED BY CORRECT RATER AND INDORSER. NO FURTHER ACTION REQUIRED. ☒
MY REVIEW RESULTS IN ACTION INDICATED BY INCLOSURES. ☐

B. ORGANIZATION AND DUTY ASSIGNMENT	C. NAME AND GRADE	D. DATE
HHT, 2d Armd Cav Regt Regimental Executive Officer	E. SIGNATURE	12 May 76

PART VII MILPO CERTIFICATION

A. DATE REPORT ENTERED ON DA FM 2-1: 2 JUN 1976	C. TCO Supv, Nbg RPC MILPO SIGNATURE	D. FLOAAA UIC
B. SOLDIER'S COPY: GIVEN TO SOLDIER ☐ FORWARDED TO SOLDIER ■ MAILED TO SOLDIER ☐ CERTIFIED MAIL NO. ______		

See Privacy Act Statement in AR 623-205, APPENDIX E.

ENLISTED EVALUATION REPORT
(AR 623-205)

Proponent agency for this form is the US Army Military Personnel Center

PART I. ADMINISTRATIVE DATA

A. LAST NAME – FIRST NAME – MIDDLE INITIAL	B. SSN	C. RANK (ABBR)	D. DATE OF RANK
		CSM	801102

E. PRIMARY MOSC	F. SECONDARY MOSC	G. UNIT, ORGANIZATION, STATION, ZIP CODE/APO, MACOM
00Z50	63Z50	HQ, 1st Armor Training Brigade, Ft Knox, Ky 40121-5250 (TC)

H. CODE/TYPE OF REPORT	I. PERIOD OF REPORT: FROM YEAR	FROM MONTH	THRU YEAR	THRU MONTH	J. RATED MONTHS	K. NONRATED MONTHS	L. NONRATED CODES
3 Change of Rater	85	01	85	04	4		

PART II. DUTY DESCRIPTION

A. PRINCIPAL DUTY TITLE: Brigade Command Sergeant Major

B. DUTY MOSC: 00Z50

C. DESCRIPTION OF DUTIES: Responsible to keep the Commander informed on the morale, esprit and overall condition of the Brigade. Assigns all noncommissioned officers and enlisted personnel assigned to the Brigade. Insures that a viable NCO Development Program is on-going in the Brigade. Helps insure that all units within the Brigade are conducting SQT training. Counsels and provides guidance to enlisted personnel assigned to the Brigade. Acts as advisor to the Commander concerning enlisted matters or any other matters that affect the Brigade mission.

PART III. EVALUATION OF PROFESSIONALISM AND PERFORMANCE

RATER	INDORSER	A. PROFESSIONAL COMPETENCE
5	5	1. Demonstrates initiative.
5	5	2. Adapts to changes.
5	5	3. Seeks self-improvement.
5	5	4. Performs under pressure.
5	5	5. Attains results.
5	5	6. Displays sound judgment.
5	5	7. Communicates effectively.
5	5	8. Develops subordinates
5	5	9. Demonstrates technical skills.
5	5	10. Physical fitness.
50	50	SUBTOTALS

SCORING SCALE

(High)

5

4

3

2

1

0

(Low)

RATER	INDORSER	B. PROFESSIONAL STANDARDS
5	5	1. Integrity
5	5	2. Loyalty.
5	5	3. Moral courage
5	5	4. Self-discipline
5	5	5. Military appearance
5	5	6. Earns respect
5	5	7. Supports EO EEO
35	35	SUBTOTALS

(Add the Rater's SUBTOTALS (A&B) and enter sum in the appropriate box in PART VI, SCORE SUMMARY. Do the same for Indorser.)

C. DEMONSTRATED PERFORMANCE OF PRESENT DUTY PASS 8410 66/152 YES

1. Rater's Evaluation: performance during his first four months as Brigade Command Sergeant Major has been one of untiring effort and professional, outstanding performance. His initiatives and organizational abilities contributed significantly to the present high standards of appearance in the 1st Armor Training Brigade. He always provides sound and timely advice on a broad range of issues. His recommendations have been well thought out and appropriate. is in top physical condition and his highly visible participation in all aspects of physical training has set the example for the cadre, NCO and officer. His excellent performance is due mainly to his attention to detail in all aspects of his job and his commitment to perfection. is a community oriented soldier, he and his family are actively involved in numerous organizational and helping activities for the Fort Knox community.

2. Indorser's Evaluation: has vigorously stepped forward into the 1st Brigade Command Sergeant Major position and thus far has proven to be well suited for the demanding position. During his short tenure, he has become instrumental in the brigade's daily operation; of particular note is his involvement in the brigade's building refurbishing program. His ability to coordinate post agency support with the work effort of six battalions has been a major undertaking which has been superbly executed. He sets the example in all aspects of soldiering for the 1st Brigade cadre.

DA FORM OCT 81 2166-6 REPLACES DA FORM 2166-5A, OCT 79, WHICH IS OBSOLETE.

PART IV. EVALUATION OF POTENTIAL

1. Rater's Evaluation: (Place score in applicable box)

40-38		37-20		19-0		40-0	
☐	Promote ahead of peers.	☐	Promote with peers.	☐	Do not promote.	40	E9 Soldiers Only.

Comments: (potential for higher-level school, assignment, and supervisory responsibility)

is capable of performing as a Command Sergeant Major at any level. Select for division CSM now and the 35 year program as soon as possible.

2. Indorser's Evaluation: (Place score in applicable box)

40-38		37-20		19-0		40-0	
☐	Promote ahead of peers.	☐	Promote with peers.	☐	Do not promote.	40	E9 Soldiers Only.

Comments: (potential for higher-level school, assignment, and supervisory responsibility)

Division, Corps CSM potential. A top-notch performer with limitless potential.

PART V. AUTHENTICATION

A. NAME OF RATER (Last, First, MI) | SSN | SIGNATURE

RANK, ORGANIZATION, AND DUTY ASSIGNMENT: HQ, 1st Armor Tng Bde, Ft Knox, Ky 40121-5250 Brigade Commander | DATE: 29 Apr 85

Refer to AR 623-205 for requirements to discuss contents of report with the rated soldier.

B. NAME OF INDORSER (Last, First, MI) | SSN | SIGNATURE

RANK, ORGANIZATION, AND DUTY ASSIGNMENT: BG HQ, USAARMS, Ft Knox, Ky Assistant Commandant | DATE: 10 May 1985

C. NAME OF RATED SOLDIER (Last, First, MI)

I have verified Administrative Data, PART I, and Duty Description, PART II. I have seen this report as prepared by the Rater and Indorser. I understand that my signature does not constitute agreement nor disagreement with their evaluations.

SSN: 2__-52-6915 | DATE: 10 May 85 | Signature:

D. NAME OF REVIEWER (Last, First, MI) | SSN

I have reviewed this report in accordance with AR 623-205 on 10 May 1985 (date)

Signature:

RANK, ORGANIZATION, AND DUTY ASSIGNMENT: BG HQ, USAARMS, Ft Knox, Ky Asst. Commandant

PART VI. SCORE SUMMARY

PART	RATER SCORE	INDORSER SCORE
III.	85	85
IV.	40	40
Sum	125	125
REPORT SCORE (R-1 + 2)		125

PART VII. MILPO CERTIFICATION

A. SOLDIER'S COPY:

☐ Given to Soldier ______ (date)

☑ Forwarded to Soldier 17 MAY 1985 (date)

☐ Mailed to Soldier ______ (date)

B. FORWARDING ADDRESS:

C. NO. OF INC.	D. DATE ENTERED ON DA FORM 2-1	E. MILPO SIGNATURE	F. MILPO
0	17 MAY 1985		TD02

NCO EVALUATION REPORT

For use of this form, see AR 623-205, the proponent agency is DCSPER

SEE PRIVACY ACT STATEMENT IN AR 623-205, APPENDIX E.

PART I - ADMINISTRATIVE DATA

a. NAME *(Last, First, Middle Initial)*	b. SSN	c. RANK	d. DATE OF RANK	e. PMOSC

f. UNIT, ORG., STATION, ZIP CODE OR APO, MAJOR COMMAND	g. REASON FOR SUBMISSION

h. PERIOD COVERED				i. RATED MONTHS	j. NON-RATED CODES	k. NO OF ENCL	l. RATED NCO COPY *(Check one and Date)*	Date	m. PSC Initials	n. CMD CODE	o. PSC CODE
FROM		THRU					1. Given to NCO				
YY	MM	YY	MM				2. Forwarded to NCO				

PART II - AUTHENTICATION

a. NAME OF RATER *(Last, First, Middle Initial)*	SSN	SIGNATURE
RANK, PMOSC/BRANCH, ORGANIZATION, DUTY ASSIGNMENT		DATE
b. NAME OF SENIOR RATER *(Last, First, Middle Initial)*	SSN	SIGNATURE
RANK, PMOSC/BRANCH, ORGANIZATION, DUTY ASSIGNMENT		DATE
C. RATED NCO: I understand my signature does not constitute agreement or disagreement with the evaluations of the rater and senior rater. Part I, height/weight and APFT entries are verified. I have seen this report completed through Part V. I am aware of the appeals process (AR 623-205)	SIGNATURE	DATE
d. NAME OF REVIEWER *(Last, First, Middle Initial)*	SSN	SIGNATURE
RANK, PMOSC/BRANCH, ORGANIZATION, DUTY ASSIGNMENT		DATE

e. ☐ CONCUR WITH RATER AND SENIOR RATER EVALUATIONS ☐ NONCONCUR WITH RATER AND/OR SENIOR RATER EVAL *(See attached comments)*

PART III - DUTY DESCRIPTION (Rater)

a. PRINCIPAL DUTY TITLE	b. DUTY MOSC

c. DAILY DUTIES AND SCOPE *(To include, as appropriate, people, equipment, facilities and dollars)*

d. AREAS OF SPECIAL EMPHASIS

e. APPOINTED DUTIES

f. *Counseling dates from checklist/record*	INITIAL	LATER	LATER	LATER

PART IV - VALUES/NCO RESPONSIBILITIES *(Rater)*

a. Complete each question. *(Comments are mandatory for "No" entries; optional for "Yes" entries.)*

V A L U E S

PERSONAL: *Commitment*, *Competence*, *Candor*, *Courage*

ARMY ETHIC: *Loyalty*, *Duty*, *Selfless Service*, *Integrity*

Question		YES	NO
1. Places dedication and commitment to the goals and missions of the Army and nation above personal welfare.	1		
2. Is committed to and shows a sense of pride in the unit - works as a member of the team.	2		
3. Is disciplined and obedient to the spirit and letter of a lawful order	3		
4. Is honest and truthful in word and deed	4		
5. Maintains high standards of personal conduct on and off duty.	5		
6. Has the courage of convictions and the ability to overcome fear - stands up for and does, what's right.	6		
7. Supports EO/EEO	7		

Bullet comments

DA FORM 2166-7, SEP 87 REPLACES DA FORM 2166-6, OCT 81, WHICH IS OBSOLETE

RATED NCO'S NAME *(Last, First, Middle Initial)*	SSN	THRU DATE

PART IV (Rater) - **VALUES/NCO RESPONSIBILITIES**

Specific Bullet examples of "EXCELLENCE" or "NEEDS IMPROVEMENT" are mandatory.
Specific Bullet examples of "SUCCESS" are optional.

b. COMPETENCE
- o Duty proficiency; MOS competency
- o Technical & tactical; knowledge, skills, and abilities
- o Sound judgment
- o Seeking self-improvement; always learning
- o Accomplishing tasks to the fullest capacity; committed to excellence

EXCELLENCE *(Exceeds std)* ☐ **SUCCESS** *(Meets std)* ☐ **NEEDS IMPROVEMENT** *(Some)* ☐ *(Much)* ☐

c. PHYSICAL FITNESS & MILITARY BEARING
- o Mental and physical toughness
- o Endurance and stamina to go the distance
- o Displaying confidence and enthusiasm; looks like a soldier

APFT | HEIGHT/WEIGHT

EXCELLENCE *(Exceeds std)* ☐ **SUCCESS** *(Meets std)* ☐ **NEEDS IMPROVEMENT** *(Some)* ☐ *(Much)* ☐

d. LEADERSHIP
- o Mission first
- o Genuine concern for soldiers
- o Instilling the spirit to achieve and win
- o Setting the example; Be, Know, Do

EXCELLENCE *(Exceeds std)* ☐ **SUCCESS** *(Meets std)* ☐ **NEEDS IMPROVEMENT** *(Some)* ☐ *(Much)* ☐

e. TRAINING
- o Individual and team
- o Mission focused; performance oriented
- o Teaching soldiers how; common tasks, duty-related skills
- o Sharing knowledge and experience to fight, survive and win

EXCELLENCE *(Exceeds std)* ☐ **SUCCESS** *(Meets std)* ☐ **NEEDS IMPROVEMENT** *(Some)* ☐ *(Much)* ☐

f. RESPONSIBILITY & ACCOUNTABILITY
- o Care and maintenance of equip./facilities
- o Soldier and equipment safety
- o Conservation of supplies and funds
- o Encouraging soldiers to learn and grow
- o Responsible for good, bad, right & wrong

EXCELLENCE *(Exceeds std)* ☐ **SUCCESS** *(Meets std)* ☐ **NEEDS IMPROVEMENT** *(Some)* ☐ *(Much)* ☐

PART V - OVERALL PERFORMANCE AND POTENTIAL

a. RATER. Overall potential for promotion and/or service in positions of greater responsibility.

AMONG THE BEST ☐ FULLY CAPABLE ☐ MARGINAL ☐

b. RATER. List 3 positions in which the rated NCO could best serve the Army at his/her current or next higher grade.

____
____
____

e. SENIOR RATER BULLET COMMENTS

c. SENIOR RATER. Overall performance

1	2	3	4	5
Successful			Fair	Poor

d. SENIOR RATER. Overall potential for promotion and/or service in positions of greater responsibility.

1	2	3	4	5
Superior			Fair	Poor

NCO COUNSELING CHECKLIST/RECORD

For use of this form, see AR 623-205; the proponent agency MILPERCEN

NAME OF RATED NCO	RANK	DUTY POSITION	UNIT

PURPOSE: The primary purpose of counseling is to improve performance and to professionally develop the rated NCO. The best counseling is always looking forward. It does not dwell on the past and on what was done, rather on the future and what can be done better. Counseling at the end of the rating period is too late since there is no time to improve before evaluation.

RULES:

1. Face-to-face performance counseling is mandatory for all Noncommissioned Officers.
2. This form is for use along with a working copy of the NCO-ER for conducting NCO performance counseling and recording counseling content and dates. Its use is mandatory for counseling all NCOs, CPL thru SFC/PSG, and is optional for counseling other senior NCOs.
3. Active Component. Initial counseling must be conducted within the first 30 days of each rating period, and at least quarterly thereafter. Reserve Components. (ARNG, USAR). Counseling must be conducted at least semiannually. There is no mandatory counseling at the end of the rating period.

CHECKLIST – FIRST COUNSELING SESSION AT THE BEGINNING OF THE RATING PERIOD

PREPARATION

1. Schedule counseling session, notify rated NCO.
2. Get copy of last duty description used for rated NCO's duty position, a blank copy of the NCO-ER, and the names of the new rating chain.
3. Update duty description (see page 2).
4. Fill out rating chain and duty description on working copy of NCO-ER. Parts II and III.
5. Read each of the values/responsibilities in Part IV of NCO-ER and the expanded definitions and examples on page 3 and 4 of this form.
6. Think how each value and responsibility in Part IV of NCO-ER applies to the rated NCO and his/her duty position.

Note: Leadership and training may be more difficult to apply than the other values/responsibilities when the rated NCO has no subordinates. Leadership is simply influencing others in the accomplishment of the mission and that can include peers and superiors. It also can be applied directly to additional duties and other areas of Army community life. Individual training is the responsibility of all NCOs whether or not there are subordinates. Every NCO knows something that can be taught to others and should be involved in some way in a training program.

7. Decide what you consider necessary for success (a meets standards rating) for each value/responsibility. Use the examples listed on pages 3 and 4 of this form as a guide in developing your own standards for success. Some may apply exactly, but you may have to change them or develop new ones that apply to your situation. Be specific so the rated NCO will know what is expected.
8. Make notes in blank spaces in Part IV of NCO-ER to help when counseling.
9. Review counseling tips in FM 22-101.

COUNSELING

1. Make sure rated NCO knows rating chain.
2. Show rated NCO the draft duty description on your working copy of the NCO-ER. Explain all parts. If rated NCO performed in position before, ask for any ideas to make duty description better.
3. Discuss the meaning of each value/responsibility in Part IV of NCO-ER. Use the trigger words on the NCO-ER, and the expanded definitions on pages 3 and 4 of this form to help.
4. Explain how each value/responsibility applies to the specific duty position by showing or telling your standards for success (a meets standards rating). Use examples on pages 3 and 4 of this form as a start point. Be specific so the rated NCO really knows what's expected.
5. When possible, give specific examples of excellence that could apply. This gives the rated NCO something special to strive for, Remember that only a few achieve real excellence and that real excellence always includes specific results and often includes accomplishments of subordinates.
6. Give rated NCO opportunity to ask questions and make suggestions.

AFTER COUNSELING

1. Record rated NCO's name and counseling date on this form.
2. Write key points made in counseling session on this form.
3. Show key points to rated NCO and get his initials.
4. Save NCO-ER with this checklist for next counseling session.

CHECKLIST – LATER COUNSELING SESSIONS DURING THE RATING PERIOD

PREPARATION

1. Schedule counseling session, notify rated NCO, and tell him/her to come prepared to discuss what has been accomplished in each value/responsibility area.
2. Look at working copy of NCO-ER you used during last counseling session.
3. Read and update duty description. Especially note the area of special emphasis; the priorities may have changed.
4. Read again, each of the values/responsibilities in Part IV of NCO-ER and the expanded definitions and examples on pages 3 and 4 of this form; then think again, about your standards for success.
5. Look over the notes you wrote down on page 2 of this form about the last counseling session.
6. Think about what the rated NCO has done so far during this rating period (specifically, observed action, demonstrated behavior, and results).
7. For each value/responsibility area, answer three questions: First, what has happened in response to any discussion you had during the last counseling session? Second, what has been done well?; and Third, what could be done better?
8. Make notes in blank spaces in Part IV of NCO-ER to help focus when counseling. (Use new NCO-ER if old one is full from last counseling session).
9. Review counseling tips in FM 22-101.

COUNSELING

1. Go over each part of the duty description with rated NCO. Discuss any changes, especially to the area of special emphasis.

2. Tell rated NCO how he/she is doing. Use your success standards as a guide for the discussion (the examples on pages 3 and 4 may help). First, for each value/responsibility, talk about what has happened in response to any discussion you had during the last counseling session (remember, observed action, demonstrated behavior and results). Second, talk about what was done well. Third, talk about how to do better. The goal is to get all NCOs to be successful and meet standards.

3. When possible, give examples of excellence that could apply. This gives the rated NCO something to strive for, REMEMBER, EXCELLENCE IS SPECIAL, ONLY A FEW ACHIEVE IT! Excellence includes results and often involves subordinates.

4. Ask rated NCO for ideas, examples and opinions on what has been done so far and what can be done better. (This step can be done first or last).

AFTER COUNSELING

1. Record counseling date on this form.

2. Write key points made in counseling session on this form.

3. Show key points to rated NCO and get his initials.

4. Save NCO-ER with this checklist for next counseling session. (Notes should make record NCO-ER preparation easy at end of rating period).

COUNSELING RECORD

DATE OF COUNSELING	RATED NCO's INITIALS	KEY POINTS MADE
INITIAL		
LATER		
LATER		
LATER		

DUTY DESCRIPTION (PART III of NCO-ER)

The duty description is essential to performance counseling and evaluation. It is used during the first counseling session to tell rated NCO what the duties are and what needs to be emphasized. It may change somewhat during the rating period. It is used at the end of the rating period to record what was important about the duties.

The five elements of the duty description:

1 & 2. Principal Duty Title and Duty MOS Code. Enter principal duty title and DMOS that most accurately reflects actual duties performed.

3. Daily Duties and Scope. This portion should address the most important routine duties and responsibilities. Ideally, this should include number of people supervised, equipment, facilities, and dollars involved and any other routine duties and responsibilities critical to mission accomplishment.

4. Area of Special Emphasis. This portion is most likely to change somewhat during the rating period. For the first counseling session, it includes those items that require top priority effort at least for the first part of the upcoming rating period. At the end of the rating period, it should include the most important items that applied at any time during the rating period (examples are preparation for REFORGER deployment, combined arms drills training for FTX, preparation for NTC rotation, revision of battalion maintenance SOP, training for tank table qualification, ITEP and company AMTP readiness, related tasks cross-training, reserve components annual training support (AT) and SIDPERS acceptance rate).

5. Appointed Duties. This portion should include those duties that are appointed and are not normally associated with the duty description.

VALUES/NCO RESPONSIBILITIES (PART IV of NCO-ER)

VALUES: Values are what soldiers, as a profession, judge to be right. They are the moral, ethical, and professional attributes of character. They are the heart and soul of a great Army. Part IVa of the NCO-ER includes some of the most important values. These are: Putting the welfare of the nation, the assigned mission and teamwork before individual interests; Exhibiting absolute honesty and courage to stand up for what is right; Developing a sense of obligation and support between those who are led, those who lead, and those who serve alongside; Maintaining high standards of personal conduct on and off duty; And finally, demonstrating obedience, total adherence to the spirit and letter of a lawful order, discipline, and ability to overcome fear despite difficulty or danger.

Examples of standards for "YES" ratings:

- Put the Army, the mission and subordinates first before own personal interest.
- Meet challenges without compromising integrity.
- Personal conduct, both on and off duty, reflects favorably on NCO corps.
- Obey lawful orders and do what is right without orders.
- Choose the hard right over the easy wrong.
- Exhibit pride in unit, be a team player.
- Demonstrate respect for all soldiers regardless of race, creed, color, sex, or national origin.

COMPETENCE: The knowledge, skills and abilities necessary to be expert in the current duty assignment and to perform adequately in other assignments within the MOS when required. Competence is both technical and tactical and includes reading, writing, speaking and basic mathematics. It also includes sound judgment, ability to weigh alternatives, form objective opinions and make good decisions. Closely allied with competence is the constant desire to be better, to listen and learn more and to do each task completely to the best of one's ability. Learn, grow, set standards, and achieve them, create and innovate, take prudent risks, never settle for less than best. Committed to excellence.

Examples of standards for "Success/Meets Standards" rating:

- Master the knowledge, skills and abilities required for performance in your duty position.
- Meet PMOS SQT standards for your grade.
- Accomplish completely and promptly those tasks assigned or required by duty position.
- Constantly seek ways to learn, grow and improve.

Examples of "Excellence":

- Picked as SSG to be a platoon sergeant over twelve other SSGs.
- Maintained SIDPERS rating of 98% for six months.
- Scored 94% on last SQT.
- Selected best truck master in annual battalion competition.
- Designated Installation Drill Sergeant of Quarter.
- Exceeded recruiting objectives two consecutive quarters.
- Awarded Expert Infantryman Badge (EIB).

PHYSICAL FITNESS AND MILITARY BEARING: Physical fitness is the physical and mental ability to accomplish the mission – combat readiness. Total fitness includes weight control, diet and nutrition, smoking cessation, control of substance abuse, stress management, and physical training. It covers strength, endurance, stamina, flexibility, speed, agility, coordination and balance. NCOs are responsible for their own physical fitness and that of their subordinates. Military Bearing consists of posture, dress, overall appearance, and manner of physical movement. Bearing also includes an outward display of inner-feelings, fears, and overall confidence and enthusiasm. An inherent NCO responsibility is concern with the military bearing of the individual soldier, to include on-the-spot corrections.

Examples of standards for "Success/Meets Standards" rating:

- Maintain weight within Army limits for age and sex.
- Obtain passing score in APFT and participate in a regular exercise program.
- Maintain personal appearance and exhibit enthusiasm to the point of setting an example for junior enlisted soldiers.
- Monitor and encourage improvement in the physical and military bearing of subordinates.

Examples of "Excellence":

- Received Physical Fitness Badge for 292 score on APFT.
- Selected soldier of the month/quarter/year.
- Three of the last four soldiers of the month were from his/her platoon.
- As Master Fitness Trainer, established battalion physical fitness program.
- His entire squad was commended for scoring above 270 on APFT.

LEADERSHIP: Influencing others to accomplish the mission. It consists of applying leadership attributes (Beliefs, Values, Ethics, Character, Knowledge, and Skills). It includes setting tough, but achievable standards and demanding that they be met; Caring deeply and sincerely for subordinates and their families and welcoming the opportunity to serve them; Conducting counseling; Setting the example by word and act/deed; Can be summarized by BE (Committed to the professional Army ethic and professional traits); KNOW (The factors of leadership, yourself, human nature, your job, and your unit); DO (Provide direction, implement, and motivate). Instill the spirit to achieve and win: Inspire and develop excellence. A soldier cared for today, leads tomorrow.

Examples of standards for "Success/Meets Standards" rating:

- Motivate subordinates to perform to the best of their ability as individuals and together as a disciplined cohesive team to accomplish the mission.
- Demonstrate that you care deeply and sincerely for soldiers and welcome the opportunity to serve them.
- Instill the spirit to achieve and win; Inspire and develop excellence through counseling.
- Set the example: BE, KNOW, DO.

Examples of "Excellence":

- Motivated entire squad to qualify expert with M-16.
- Won last three platoon quad inspections.
- Selected for membership in Sergeant Morales Club.
- Inspired mechanics to maintain operational readiness rating of 95% for two consecutive quarters.
- Led his squad through map orienteering course to win the battalion competition.
- Counseled two marginal soldiers ultimately selected for promotion.

TRAINING: Preparing individuals, units and combined arms teams for duty performance; The teaching of skills and knowledge. NCOs contribute to team training, are often responsible for unit training (Squads, Crews, Sections), but individual training is the most important, exclusive responsibility of the NCO Corps. Quality training bonds units: Leads directly to good discipline; Concentrates on wartime missions; Is tough and demanding without being reckless; Is performance oriented; Sticks to Army doctrine to standardize what is taught to fight, survive, and win, as small units when AirLand battle actions dictate. "Good training means learning from mistakes and allowing plenty of room for professional growth. Sharing knowledge and experience is the greatest legacy one can leave subordinates."

Examples of standards for "Success/Meets Standards" rating:

- Make sure soldiers-
  a. Can do identified common tasks.
  b. Are prepared for SQT and Commander's Evaluation.
  c. Develop and practice skills for duty position.
  d. Train as a squad/crew/section.
- Identify and recommend subordinates for professional development courses.
- Participate in unit training program.
- Share knowledge and experience with subordinates.

Examples of "Excellence":

- Taught five common tasks resulting in 100% GO on Annual CTT for all soldiers in directorate.
- Trained best howitzer section of the year in battalion.
- Coached subordinates to win consecutive soldier of month competitions.
- Established company Expert Field Medical Badge program resulting in 85% of all eligible soldiers receiving EFMB.
- Distinguished 1 tank and qualified 3 tanks in platoon on first run of tank table VIII.
- Trained platoon to fire honor battery during annual service practice.

RESPONSIBILITY AND ACCOUNTABILITY: The proper care, maintenance, use, handling, and conservation of personnel, equipment, supplies, property, and funds. Maintenance of weapons, vehicles, equipment, conservation of supplies, and funds is a special NCO responsibility because of its links to the success of all missions, especially those on the battlefield. It includes inspecting soldier's equipment often, using manual or checklist; Holding soldiers responsible for repairs and losses; Learning how to use and maintain all the equipment soldiers use; Being among the first to operate new equipment; Keeping up-to-date component lists; Setting aside time for inventories; and Knowing the readiness status of weapons, vehicles, and other equipment. It includes knowing where each soldier is during duty hours; Why he is going on sick call, where he lives, and his family situation; It involves reducing accidental manpower and monetary losses by providing a safe and healthful environment; It includes creating a climate which encourages young soldiers to learn and grow, and, to report serious problems without fear of repercussions. Also, NCOs must accept responsibility for their own actions and for those of their subordinates.

Examples of standards for "Success/Meets Standards" rating:

- Make sure your weapons, equipment, and vehicles are serviceable, maintained and ready for accomplishing the mission.
- Stop waste of supplies and limited funds.
- Be aware of those things that impact on soldier readiness e.g., family affairs, SQT, CTT, PQR, special duty, medical conditions, etc.
- Be responsible for your actions and those of your subordinates.

Examples of "Excellence":

- His emphasis on safety resulted in four tractor trailer drivers logging 10,000 miles accident free.
- Received commendation from CG for organizing post special olympics program.
- Won the installation award for Quarters of the Month.
- His constant instruction on maintenance resulted in six of eight mechanics earning master mechanic badges.
- Commended for no APCs on deadline report for six months.
- His learn and grow climate resulted in best platoon ARTEP results in the battalion.

★U.S.GPO:1988-0-201-424/80287

NCO EVALUATION REPORT

For use of this form, see AR 623-205; the proponent agency is ODCSPER

SEE PRIVACY ACT STATEMENT IN AR 623-205, APPENDIX C.

PART I - ADMINISTRATIVE DATA

a. NAME *(Last, First, Middle Initial)*	b. SSN	c. RANK	d. DATE OF RANK	e. PMOSC

f. UNIT, ORG., STATION, ZIP CODE OR APO, MAJOR COMMAND	g. REASON FOR SUBMISSION

h. PERIOD COVERED		i. RATED MONTHS	j. NON-RATED CODES	k. NO. OF ENCL	l. RATED NCO COPY *(Check one and Date)*	*Date*	m. PSC Initials	n. CMD CODE	o. PSB CODE
FROM	THRU				1. Given to NCO				
YYYY MM	YYYY MM				2. Forwarded to NCO				

PART II - AUTHENTICATION

a. NAME OF RATER *(Last, First, Middle Initial)*	SSN	SIGNATURE
RANK, PMOSC/BRANCH, ORGANIZATION, DUTY ASSIGNMENT		DATE
b. NAME OF SENIOR RATER *(Last, First, Middle Initial)*	SSN	SIGNATURE
RANK, PMOSC/BRANCH, ORGANIZATION, DUTY ASSIGNMENT		DATE
c. RATED NCO: I understand my signature does not constitute agreement or disagreement with the evaluations of the rater and senior rater. I further understand my signature verifies that the administrative data in Part I, the rating officials in Part II, the duty description to include the counseling dates in Part III, and the APFT and height/weight entries in Part IVc are correct. I have seen the report completed through Part V, except Parts IId and IIe. I am aware of the appeals process of AR 623-205.	SIGNATURE	DATE
d. NAME OF REVIEWER *(Last, First, Middle Initial)*	SSN	SIGNATURE
RANK, PMOSC/BRANCH, ORGANIZATION, DUTY ASSIGNMENT		DATE

e. ☐ CONCUR WITH RATER AND SENIOR RATER EVALUATIONS ☐ NONCONCUR WITH RATER AND/OR SENIOR RATER EVAL *(See attached comments)*

PART III - DUTY DESCRIPTION *(Rater)*

a. PRINCIPAL DUTY TITLE	b. DUTY MOSC

c. DAILY DUTIES AND SCOPE *(To include, as appropriate, people, equipment, facilities and dollars)*

d. AREAS OF SPECIAL EMPHASIS

e. APPOINTED DUTIES

f. COUNSELING DATES	INITIAL	LATER	LATER	LATER

PART IV - ARMY VALUES/ATTRIBUTES/SKILLS/ACTIONS *(Rater)*

a. **ARMY VALUES. Check either "YES" or "NO".** *Comments are mandatory for "No" entries; optional for "Yes" entries.)*

V A L U E S

Loyalty
Duty
Respect
Selfless-Service
Honor
Integrity
Personal Courage

	YES	NO
1. LOYALTY: Bears true faith and allegiance to the U. S. Constitution, the Army, the unit, and other soldiers.		
2. DUTY: Fulfills their obligations		
3. RESPECT/EO/EEO: Treats people as they should be treated.		
4. SELFLESS-SERVICE: Puts the welfare of the nation, the Army, and subordinates before their own.		
5. HONOR: Lives up to all the Army values		
6. INTEGRITY: Does what is right - legally and morally		
7. PERSONAL COURAGE: Faces fear, danger, or adversity (physical and moral).		

Bullet comments

RATED NCO'S NAME *(Last, First, Middle Initial)*	SSN	THRU DATE

PART IV *(Rater)* **- VALUES/NCO RESPONSIBILITIES**

Specific Bullet examples of "EXCELLENCE" or "NEEDS IMPROVEMENT" are mandatory.
Specific Bullet examples of "SUCCESS" are optional

b. COMPETENCE
- o Duty proficiency; MOS competency
- o Technical & tactical; knowledge, skills, and abilities
- o Sound judgment
- o Seeking self-improvement; always learning
- o Accomplishing tasks to the fullest capacity; committed to excellence

EXCELLENCE *(Exceeds std)*	**SUCCESS** *(Meets std)*	**NEEDS IMPROVEMENT** *(Some)*	*(Much)*
☐	☐	☐	☐

c. PHYSICAL FITNESS & MILITARY BEARING
- o Mental and physical toughness
- o Endurance and stamina to go the distance
- o Displaying confidence and enthusiasm; looks like a soldier

APFT	HEIGHT/WEIGHT

EXCELLENCE *(Exceeds std)*	**SUCCESS** *(Meets std)*	**NEEDS IMPROVEMENT** *(Some)*	*(Much)*
☐	☐	☐	☐

LEADERSHIP
- o Mission first
- o Genuine concern for soldiers
- o Instilling the spirit to achieve and win
- o Setting the example; Be, Know, Do

EXCELLENCE *(Exceeds std)*	**SUCCESS** *(Meets std)*	**NEEDS IMPROVEMENT** *(Some)*	*(Much)*
☐	☐	☐	☐

TRAINING
- o Individual and team
- o Mission focused; performance oriented
- o Teaching soldiers how; common tasks, duty-related skills
- o Sharing knowledge and experience to fight, survive and win

EXCELLENCE *(Exceeds std)*	**SUCCESS** *(Meets std)*	**NEEDS IMPROVEMENT** *(Some)*	*(Much)*
☐	☐	☐	☐

RESPONSIBILITY & ACCOUNTABILITY
- o Care and maintenance of equipment/facilities
- o Soldier and equipment safety
- o Conservation of supplies and funds
- o Encouraging soldiers to learn and grow
- o Responsible for good, bad, right & wrong

XCELLENCE *xceeds std)*	**SUCCESS** *(Meets std)*	**NEEDS IMPROVEMENT** *(Some)*	*(Much)*
☐	☐	☐	☐

PART V - OVERALL PERFORMANCE AND POTENTIAL

ATER. Overall potential for promotion and/or service in ons of greater responsibility.

ONG THE BEST	FULLY CAPABLE	MARGINAL
☐	☐	☐

ATER. List 3 positions in which the rated CO could best serve the Army at his/her rrent or next higher grade.

____

____

____

e. SENIOR RATER BULLET COMMENTS

NIOR RATER. Overall performance

1	2	3	4	5
Successful			Fair	Poor

d. SENIOR RATER. Overall potential for promotion and/or service in positions of greater responsibility.

1	2	3	4	5
Superior			Fair	Poor

NCO COUNSELING CHECKLIST/RECORD

For use of this form, see AR 623-205; the proponent agency is ODCSPER

NAME OF RATED NCO	RANK	DUTY POSITION	UNIT

PURPOSE: The primary purpose of counseling is to improve performance and to professionally develop the rated NCO. The best counseling is always looking forward. It does not dwell on the past and on what was done, rather on the future and what can be done better. Counseling at the end of the rating period is too late since there is no time to improve before evaluation.

RULES:

1. Face-to-face performance counseling is mandatory for all Noncommissioned Officers.
2. This form is for use along with a working copy of the NCO-ER for conducting NCO performance counseling and recording counseling content and dates. Its use is mandatory for counseling all NCOs, CPL through CSM.
3. Active Component. Initial counseling must be conducted within the first 30 days of each rating period, and at least quarterly thereafter. Reserve Components. (ARNG, USAR). Counseling must be conducted at least semiannually. There is no mandatory counseling at the end of the rating period.

CHECKLIST - FIRST COUNSELING SESSION AT THE BEGINNING OF THE RATING PERIOD

PREPARATION

1. Schedule counseling session, notify rated NCO.
2. Get copy of last duty description used for rated NCO's duty position, a blank copy of the NCO-ER, and the names of the new rating chain.
3. Update duty description (see page 2).
4. Fill out rating chain and duty description on working copy of NCO-ER. Parts II and III.
5. Read each of the values/responsibilities in Part IV of NCO-ER and the expanded definitions and examples on page 3 and 4 of this form.
6. Think how each value and responsibility in Part IV of NCO-ER applies to the rated NCO and his/her duty position.

Note: Leadership and training may be more difficult to apply than the other values/responsibilities when the rated NCO has no subordinates. Leadership is simply influencing others in the accomplishment of the mission and that can include peers and superiors. It also can be applied directly to additional duties and other areas of Army community life. Individual training is the responsibility of all NCOs whether or not there are subordinates. Every NCO knows something that can be taught to others and should be involved in some way in a training program.

7. Decide what you consider necessary for success (a meets standards rating) for each value/responsibility. Use the examples listed on pages 3 and 4 of this form as a guide in developing your own standards for success. Some may apply exactly, but you may have to change them or develop new ones that apply to your situation. Be specific so the rated NCO will know what is expected.
8. Make notes in blank spaces in Part IV of NCO-ER to help when counseling.
9. Record rated NCO's name, rank, duty position, and unit date on this form.
10. Write key points to be made during the counseling session on this form.
11. Review Developmental Counseling in FM 22-100, appendix C.

COUNSELING

1. Make sure rated NCO knows rating chain.
2. Show rated NCO the draft duty description on your working copy of the NCO-ER. Explain all parts. If rated NCO performed in position before, ask for any ideas to make duty description better.
3. Discuss the meaning of each value/responsibility in Part IV of NCO-ER. Use the trigger words on the NCO-ER, and the expanded definitions on pages 3 and 4 of this form to help.
4. Explain how each value/responsibility applies to the specific duty position by showing or telling your standards for success (a meets standards rating). Use examples on pages 3 and 4 of this form as a start point. Be specific so the rated NCO really knows what's expected.
5. When possible, give specific examples of excellence that could apply. This gives the rated NCO something special to strive for. Remember that only a few achieve real excellence and that real excellence always includes specific results and often includes accomplishments of subordinates.
6. Give rated NCO opportunity to ask questions and make suggestions.

BEFORE THE NCO DEPARTS THE COUNSELING SESSION

1. Record counseling date on this form.
2. Write any additional key points that came up during the counseling session on this form.
3. Show key points to rated NCO and get his/her initials.
4. Save NCO-ER with this checklist for next counseling session.

CHECKLIST - LATER COUNSELING SESSIONS DURING THE RATING PERIOD

PREPARATION

1. Schedule counseling session, notify rated NCO, and tell him/her to come prepared to discuss what has been accomplished in each value/responsibility area.
2. Look at working copy of NCO-ER you used during last counseling session.
3. Read and update duty description. Especially note the area of special emphasis; the priorities may have changed.
4. Read again, each of the values/responsibilities in Part IV of NCO-ER and the expanded definitions and examples on pages 3 and 4 of this form; then think again, about your standards for success.
5. Look over the notes you wrote down on page 2 of this form about the last counseling session.
6. Think about what the rated NCO has done so far during this rating period (specifically, observed action, demonstrated behavior, and results).
7. For each value/responsibility area, answer three questions: First, what has happened in response to any discussion you had during the last counseling session? Second, what has been done well?; and Third, what could be done better?
8. Makes notes in blank spaces in Part IV of NCO-ER to help focus when counseling. (Use new NCO-ER if old one is full from last counseling session.)
9. Write key points to be made during the counseling session on this form.
10. Review Developmental Counseling in FM 22-100, appendix C.

COUNSELING

1. Go over each part of the duty description with rated NCO. Discuss any changes, especially to the area of special emphasis.

2. Tell rated NCO how he / she is doing. Use your success standards as a guide for the discussion (the examples on pages 3 and 4 may help). First, for each value/responsibility, talk about what has happened in response to any discussion you had during the last counseling session (remember, observed action, demonstrated behavior and results). Second, talk about what was done well. Third, talk about how to do better. The goal is to get all NCOs to be successful and meet standards.

3. When possible, give examples of excellence that could apply. This gives the rated NCO something to strive for, REMEMBER, EXCELLENCE IS SPECIAL, ONLY A FEW ACHIEVE IT! Excellence includes results and often involves subordinates.

4. Ask rated NCO for ideas, examples and opinions on what has been done so far and what can be done better. (This step can be done first or last).

BEFORE THE NCO DEPARTS THE COUNSELING SESSION

1. Record counseling date on this form.
2. Write any additional key points that came up during the counseling session on this form.
3. Show key points to rated NCO and get his/her initials.
4. Save NCO-ER with this checklist for next counseling session. (Notes should make record NCO-ER preparation easy at the end of the rating period.

COUNSELING RECORD/KEY POINTS MADE

INITIAL

DATE RATED NCO'S INITIALS

LATER

DATE RATED NCO'S INITIALS

LATER

DATE RATED NCO'S INITIALS

LATER

DATE RATED NCO'S INITIALS

DUTY DESCRIPTION (PART III of NCO-ER)

The duty description is essential to performance counseling and evaluation. It is used during the first counseling session to tell rated NCO what the duties are and what needs to be emphasized. It may change somewhat during the rating period. It is used at the end of the rating period to record what was important about the duties.

The five elements of the duty description:

1 & 2. Principal Duty Title and Duty MOS Code. Enter principal duty title and DMOS that most accurately reflects actual duties performed.

3. Daily Duties and Scope. This portion should address the most important routine duties and responsibilities. Ideally, this should include number of people supervised, equipment, facilities, and dollars involved and any other routine duties and responsibilities critical to mission accomplishment.

4. Area of Special Emphasis. This portion is most likely to change somewhat during the rating period. For the first counseling session, it includes those items that require top priority effort at least for the first part of the upcoming rating period. At the end of the rating period, it should include the most important items that applied at any time during the rating period (examples are preparation for deployment, combined arms drills training for FTX, preparation for NTC rotation, revision of battalion maintenance SOP, training for tank table qualification, ITEP and company AMTP readiness, related tasks cross-training, reserve components annual training support (AT) and SIDPERS acceptance rate).

5. Appointed Duties. This portion should include those duties that are appointed and are not normally associated with the duty description.

VALUES/NCO RESPONSIBILITIES (PART IV of NCO-ER)

VALUES: Values tell us what we need to be, every day, in every action we take. Army values form the identity of America's Army, the solid rock upon which everything else stands. They are the glue that binds us together as members of a noble profession. They make the whole much greater than the sum of the parts. They are nonnegotiable: they apply to everyone all the time and in every stituation. Part IVa of the NCO-ER includes these values. They are: Loyalty - Bears true faith and allegiance to the U.S. Constitution, the Army, the unit, and the soldier. Duty - Fulfills their obligations. Respect - Treats people as they should be trea Selfless-Service - Puts the welfare of the nation, the Army, and subordinates before own. Honor - Lives up to all the Army values. Integrity - Does what's right - legall morally. Personal Courage - Faces fear, danger, or adversity (physical and moral).

Examples of standards for "YES" ratings:

- Put the Army, the mission and subordinates first before own personal interest.
- Meet challenges without compromising integrity.
- Personal conduct, both on and off duty, reflects favorably on NCO corps.
- Obey lawful orders and do what is right without orders.
- Choose the hard right over the easy wrong.
- Exhibit pride in unit, be a team player.
- Demonstrate respect for all soldiers regardless of race, creed, color, sex, or national origin.

COMPETENCE: The knowledge, skills and abilities necessary to be expert in the current duty assignment and to perform adequately in other assignments within the MOS when required. Competence is both technical and tactical and includes reading, writing, speaking and basic mathematics. It also includes sound judgment, ability to weigh alternatives, form objective opinions and make good judgments. Closely allied with competence is the constant desire to be better, to listen and learn m and do each task completely to the best of one's ability. Learn, grow, set standards, achieve them, create and innovate, take prudent risks, never settle for less than b Committed to excellence.

Examples of standards for "Success/Meets Standards" rating:

- Master the knowledge, skills and abilities required for performance in your duty position.
- Accomplish completely and promptly those tasks assigned or required by duty position.
- Constantly seek ways to learn, grow and improve.
- Formulate and develop new ideas.

Examples of "Excellence":

- Picked as SSG to be a platoon sergeant over twelve other SSGs. Maintained SIDPERS rating of 98% for six months.
- Selected best truck master in annual battalion competition.
- Designated Installation Drill Sergeant of Quarter.
- Exceeded recruiting objectives two consecutive quarters.
- Awarded Expert Infantryman Badge (EIB).
-

PHYSICAL FITNESS AND MILITARY BEARING: Physical fitness is the physical and mental ability to accomplish the mission - combat readiness. Total fitness includes weight control, diet and nutrition, smoking cessation, control of substance abuse, stress management, and physical training. It covers strength, endurance, stamina, flexibility, speed, agility, coordination and balance. NCOs are responsible for their own physical fitness and that of their subordinates. Military Bearing consists of posture, dress, overall appearance, and manner o physical movement. Bearing also includes an outward display of inner-feelings, fears, an overall confidence and enthusiasm. An inherent NCO responsibility is concern with th military bearing of the individual soldier, to include on-the-spot corrections.

Examples of standards for "Success/Meets Standards" rating:

- Maintain weight within Army limits for age and sex.
- Obtain passing score in APFT and participate in a regular exercise program.
- Maintain personal appearance and exhibit enthusiasm to the point of setting an example for junior enlisted soldiers.
- Monitor and encourage improvement in the physical and military bearing of subordinates.

Examples of "Excellence":

- Received Physical Fitness Badge for 292 score on APFT.
- Selected soldier of the month/quarter/year.
- Three of the last four soldiers of the month were from his/her platoon. As Master Fitness Trainer, established battalion physical fitness program.
- His entire squad was commended for scoring above 270 on APFT.
-

LEADERSHIP: Influencing others to accomplish the mission. It consists of applying leadership attributes (Beliefs, Values, Ethics, Character, Knowledge, and Skills). It includes setting tough, but achievable standards and demanding that they be met; Caring deeply and sincerely for subordinates and their families and welcoming the opportunity to serve them; Conducting counseling; Setting the example by word and act/deed; Can be summarized by BE (Committed to the professional Army ethic and professional traits); KNOW (The factors of leadership, yourself, human nature, your job, and your unit); DO (Provide direction, implement, and motivate). Instill the spirit to achieve and win: Inspire and develop excellence. A soldier cared for today, leads tomorrow.

Examples of standards for "Success/Meets Standards" rating:

- Motivate subordinates to perform to the best of their ability as individuals and together as a disciplined cohesive team to accomplish the mission.
- Demonstrate that you care deeply and sincerely for soldiers and welcome the opportunity to serve them.
- Instill the spirit to achieve and win; Inspire and develop excellence through counseling.
- Set the example: BE, KNOW, DO.
-

Examples of "Excellence":

- Motivated entire squad to qualify expert with assigned weapon.
- Won last three platoon squad inspections.
- Selected for membership in Sergeant Morales Club.
- Inspired mechanics to maintain operational readiness rating of 95% for two consecutive quarters.
- Led his squad through map orienteering course to win the battalion competition.
- Counseled two marginal soldiers ultimately selected for promotion.
-

TRAINING: Preparing individuals, units and combined arms teams for duty performance; The teaching of skills and knowledge. NCOs contribute to team training, are often responsible for unit training (Squads, Crews, Sections), but individual training is the most important, exclusive responsibility of the NCO Corps. Quality training bonds units: Leads directly to good discipline; Concentrates on wartime missions; Is tough and demanding without being reckless; Is performance oriented; Sticks to Army doctrine to standardize what is taught to fight, survive, and win, as small units when wartime actions dictate. "Good training means learning from mistakes and allowing plenty of room for professional growth. Sharing knowledge and experience is the greatest legacy one can leave subordinates."

Examples of standards for "Success/Meets Standards" rating:

- Make sure soldiers:-
  a. Can do identified common tasks.
  b. Are prepared for Commander's Evaluation.
  c. Develop and practice skills for duty position.
  d. Train as a squad/crew/section.
- Identify and recommend subordinates for professional development courses.
- Participate in unit training program.
- Share knowledge and experience with subordinates.

Examples of "Excellence":

- Taught five common tasks resulting in 100% GO on Annual CTT for all soldiers in directorate.
- Trained best howitzer section of the year in battalion.
- Coached subordinates to win consecutive soldier of month competitions.
- Established company Expert Field Medical Badge program resulting in 85% of all eligible soldiers receiving EFMB.
- Distinguished 1 tank and qualified 3 tanks in platoon on first run of tank table VIII.
- Trained platoon to fire honor battery during annual service practice.
-

RESPONSIBILITY AND ACCOUNTABILITY: The proper care, maintenance, use, handling, and conservation of personnel, equipment, supplies, property, and funds. Maintenance of weapons, vehicles, equipment, conservation of supplies, and funds is a special NCO responsibility because of its links to the success of all missions, especially those on the battlefield. It includes inspecting soldier's equipment often, using manual or checklist; holding soldiers responsible for repairs and losses; learning how to use and maintain all the equipment soldiers use; being among the first to operate new equipment; keeping up-to-date component lists; setting aside time for inventories; and knowing the readiness status of weapons, vehicles, and other equipment. It includes knowing where each soldier is during duty hours; why he/she is going on sick call, where he/she lives, and his/her family situation; it involves reducing accidental manpower and monetary losses by providing a safe and healthful environment; it includes creating a climate which encourages young soldiers to learn and grow, and, to report serious problems without fear of repercussions. Also, NCOs must accept responsibility for their own actions and for those of their subordinates.

Examples of standards for "Success/Meets Standards" rating:

- Make sure your weapons, equipment, and vehicles are serviceable, maintained and ready for accomplishing the mission.
- Stop waste of supplies and limited funds.
- Be aware of those things that impact on soldier readiness e.g., family affairs, CTT, PQR, special duty, medical conditions, etc.
- Be responsible for your actions and those of your subordinates.
-

Examples of "Excellence":

- His/her emphasis on safety resulted in four tractor trailer drivers logging 10,000 miles accident free.
- Received commendation from CG for organizing post special olympics program.
- Won the installation award for Quarters of the Month.
- His/her constant instruction on maintenance resulted in six of eight mechanics earning master mechanic badges.
- Commended for no Bradley's on deadline report for six months.
- His/her learn and grow climate resulted in best platoon ARTEP results in the battalion.
-

ROUTINE

PT 00092 310/2106Z PAGE 01

* * * * * *

* TC:_____*

CONTROL NR:_________ PROPONENT:______________ INFO: AG DRF * OPR;____*

ARMAG ARMBD ASD CHAPLAIN CPO DOC DEH DOL DOIM DPCA DPTM DRCS * * * * * *

DRM DSEC EEO IG NCOA PAO PCF PMO PROTOCOL RDC-L SGS SJA TNGGP 194BDE

1BDE 4BDE AC C\&SDEPT DCD DOES DOTD MAINTDEPT OCOA TSMT WPNSDEPT 12CAV

CID COMSY DENTAC MEDDAC LAO RGKNOX SSG 2ROTC 430RD 902MI USAISC 1000IV

TCC FILE ____

RCTUADFN RUCLEJB6800 3090135-UUUU--RUCIBAA.
ZNR UUUUU
R 031758Z NOV 88
FM CDR TNGSPTCEN FT EUSTIS VA //ATIC-ITF //
TO CDRUSAARMC FT KNOX KY//ATZK-DPT-PO-SQT-TSO 085//
BT
UNCLAS
SUBJ: INDIVIDUAL SOLDIER'S REPORT

PERSONAL IN NATURE

*** INDIVIDUAL SOLDIER'S REPORT ***

COMMANDER TSO: 085 TESTED: 27 OCT 88
SCORED: 03 NOV 88
SQT: 11B418800

REPORT ON: SCORE: 97
SSN: 576-62-2945
MOSC: 11B4 INFANTRYMAN

****

THE S Q T INDICATES POTENTIAL TRAINING WEAKNESS ON TASK(S):

051-192-3032 HASTY MINEFIELD

THE SOLDIER ALSO MISSED QUESTIONS ON TASK(S):

071-326-0543 ORGN SQUAD FOR ATTCK
071-326-5509 CONSOLID/REORG SQUAD
071-326-5630 PLATOON MOVEMENT
071-331-0002 COND LOCAL SECURITY
BT

ROUTINE

SEE PRIVACY ACT STATEMENT ON DA FORM 67–8–1

For use of this form, see AR 623-105; proponent agency is US Army Military Personnel Center.

PART I – ADMINISTRATIVE DATA

a. LAST NAME FIRST NAME MIDDLE INITIAL	b. SSN	c. GRADE	d. DATE OF RANK (Year / Month / Day)	e. BR	f. DESIGNATED SPECIALTIES	g. PMOS (WO)	h. STA CODE

i. UNIT, ORGANIZATION, STATION, ZIP CODE OR APO, MAJOR COMMAND	j. REASON FOR SUBMISSION	k. CMD CODE

l. PERIOD COVERED FROM (Year / Month / Day)	THRU (Year / Month / Day)	m. NO. OF MONTHS	n. MILPO CODE	o. RATED OFFICER COPY (Check one and date)	p. FORWARDING ADDRESS
				☐ 1. GIVEN TO OFFICER ☐ 2. FORWARDED TO OFFICER	

q. EXPLANATION OF NONRATED PERIODS

PART II – AUTHENTICATION *(Rated officer signature verifies PART I data and RATING OFFICIALS ONLY)*

a. NAME OF RATER *(Last, First, MI)*	SSN	SIGNATURE	
GRADE, BRANCH, ORGANIZATION, DUTY ASSIGNMENT			DATE
b. NAME OF INTERMEDIATE RATER *(Last, First, MI)*	SSN	SIGNATURE	
GRADE, BRANCH, ORGANIZATION, DUTY ASSIGNMENT			DATE
c. NAME OF SENIOR RATER *(Last, First, MI)*	SSN	SIGNATURE	
GRADE, BRANCH, ORGANIZATION, DUTY ASSIGNMENT			DATE

d. SIGNATURE OF RATED OFFICER	DATE	e. DATE ENTERED ON DA FORM 2-1	f. RATED OFFICER MPO INITIALS	g. SR MPO INITIALS	h. NO. OF INCL

PART III – DUTY DESCRIPTION *(Rater)*

a. PRINCIPAL DUTY TITLE	b. SSI/MOS

c. REFER TO PART IIIa, DA FORM 67–8–1

PART IV – PERFORMANCE EVALUATION – PROFESSIONALISM *(Rater)*

a. PROFESSIONAL COMPETENCE *(In Items 1 through 14 below, indicate the degree of agreement with the following statements as being descriptive of the rated officer. Any comments will be reflected in b below.)*

HIGH DEGREE ← → LOW DEGREE
1 2 3 4 5

1. Possesses capacity to acquire knowledge/grasp concepts		8. Displays sound judgment	
2. Demonstrates appropriate knowledge and expertise in assigned tasks		9. Seeks self-improvement	
3. Maintains appropriate level of physical fitness		10. Is adaptable to changing situations	
4. Motivates, challenges and develops subordinates		11. Sets and enforces high standards	
5. Performs under physical and mental stress		12. Possesses military bearing and appearance	
6. Encourages candor and frankness in subordinates		13. Supports EO/EEO	
7. Clear and concise in written communication		14. Clear and concise in oral communication	

b. PROFESSIONAL ETHICS *(Comment on any area where the rated officer is particularly outstanding or needs improvement)*

1. DEDICATION
2. RESPONSIBILITY
3. LOYALTY
4. DISCIPLINE
5. INTEGRITY
6. MORAL COURAGE
7. SELFLESSNESS
8. MORAL STANDARDS

PERIOD COVERED

PART V – PERFORMANCE AND POTENTIAL EVALUATION *(Rater)*

a. RATED OFFICER'S NAME ______ SSN ______

RATED OFFICER IS ASSIGNED IN ONE OF HIS/HER DESIGNATED SPECIALTIES/MOS ☐ YES ☐ NO

b. PERFORMANCE DURING THIS RATING PERIOD. REFER TO PART III, DA FORM 67–8 AND PART III a, b, AND c, DA FORM 67–8–1

☐ ALWAYS EXCEEDED REQUIREMENTS ☐ USUALLY EXCEEDED REQUIREMENTS ☐ MET REQUIREMENTS ☐ OFTEN FAILED REQUIREMENTS ☐ USUALLY FAILED REQUIREMENTS

c. COMMENT ON SPECIFIC ASPECTS OF THE PERFORMANCE. REFER TO PART III, DA FORM 67–8 AND PART III a, b, AND c, DA FORM 67–8–1. DO NOT USE FOR COMMENTS ON POTENTIAL!

d. THIS OFFICER'S POTENTIAL FOR PROMOTION TO THE NEXT HIGHER GRADE IS

☐ PROMOTE AHEAD OF CONTEMPORARIES ☐ PROMOTE WITH CONTEMPORARIES ☐ DO NOT PROMOTE ☐ OTHER *(Explain below)*

e. COMMENT ON POTENTIAL

PART VI – INTERMEDIATE RATER

a. COMMENTS

PART VII – SENIOR RATER

a. POTENTIAL EVALUATION *(See Chapter 4, AR 623-105)*

SR

DA USE ONLY

HI

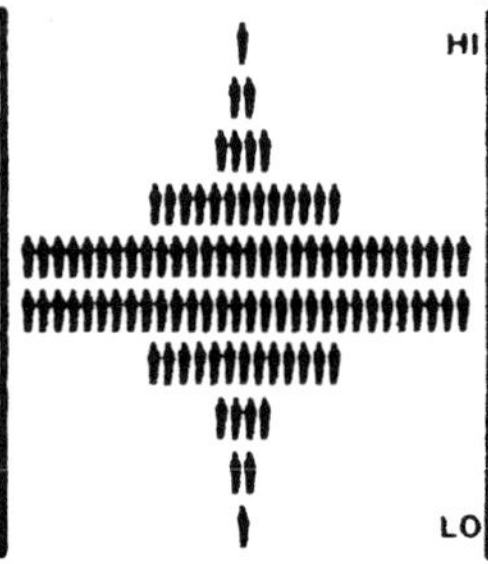

LO

A COMPLETED DA FORM 67–8–1 WAS RECEIVED WITH THIS REPORT AND CONSIDERED IN MY EVALUATION AND REVIEW

☐ YES ☐ NO *(Explain in b)*

b. COMMENTS

OFFICER EVALUATION REPORT SUPPORT FORM

For use of this form, see AR 623-105; the proponent agency is DCSPER.

Read Privacy Act Statement on Reverse before Completing this form

PART I – RATED OFFICER IDENTIFICATION

NAME OF RATED OFFICER *(Last, First, MI)*	GRADE	ORGANIZATION

PART II – RATING CHAIN – YOUR RATING CHAIN FOR THE EVALUATION PERIOD IS:

RATER	NAME	GRADE	POSITION
INTERMEDIATE RATER	NAME	GRADE	POSITION
SENIOR RATER	NAME	GRADE	POSITION

PART III – VERIFICATION OF INITIAL FACE-TO-FACE DISCUSSION

AN INITIAL FACE-TO-FACE DISCUSSION OF DUTIES, RESPONSIBILITIES, AND PERFORMANCE OBJECTIVES FOR THE CURRENT RATING PERIOD TOOK PLACE ON ______________________.

RATED OFFICER'S INITIALS ______________ RATER'S INITIALS ______________

PART IV – RATED OFFICER *(Complete a, b, and c below for this rating period)*

a. STATE YOUR SIGNIFICANT DUTIES AND RESPONSIBILITIES

DUTY TITLE IS ______________________, THE POSITION CODE IS ______________.

b. INDICATE YOUR MAJOR PERFORMANCE OBJECTIVES

c. LIST YOUR SIGNIFICANT CONTRIBUTIONS

SIGNATURE AND DATE

PART V – RATER AND/OR INTERMEDIATE RATER *(Review and comment on Part IVa, b, and c above. Insure remarks are consistent with your performance and potential evaluation on DA Form 67—8.)*

a. RATER COMMENTS *(Optional)*

SIGNATURE AND DATE *(Mandatory)*

b. INTERMEDIATE RATER COMMENTS *(Optional)*

SIGNATURE AND DATE *(Mandatory)*

DATA REQUIRED BY THE PRIVACY ACT OF 1974 *(5 U.S.C. 552a)*

1. **AUTHORITY:** Sec 301 Title 5 USC; Sec 3012 Title 10 USC.

2. **PURPOSE:** DA Form 67—8, Officer Evaluation Report, serves as the primary source of information for officer personnel management decisions. DA Form 67—8—1, Officer Evaluation Support Form, serves as a guide for the rated officer's performance, development of the rated officer, enhances the accomplishment of the organization mission, and provides additional performance information to the rating chain.

3. **ROUTINE USE:** DA Form 67—8 will be maintained in the rated officer's official military Personnel File (OMPF) and Career Management Individual File (CMIF). A copy will be provided to the rated officer either directly or sent to the forwarding address shown in Part I, DA Form 67—8. DA Form 67—8—1 is for organizational use only and will be returned to the rated officer after review by the rating chain.

4. **DISCLOSURE:** Disclosure of the rated officer's SSN (Part I, DA Form 67—8) is voluntary. However, failure to verify the SSN may result in a delayed or erroneous processing of the officer's OER. Disclosure of the information in Part IV, DA Form 67—8—1 is voluntary. However, failure to provide the information requested will result in an evaluation of the rated officer without the benefits of that officer's comments. Should the rated officer use the Privacy Act as a basis not to provide the information requested in Part IV, the Support Form will contain the rated officer's statement to that effect and be forwarded through the rating chain in accordance with AR 623—105.

★ U.S. Government Printing Office: 1985—491-003/43128

OFFICER EVALUATION REPORT

For use of this form, see AR 623-105; the proponent agency is ODCSPER

SEE PRIVACY ACT STATEMENT ON DA FORM 67-9-1

PART I - ADMINISTRATIVE DATA

a. NAME (Last, First, Middle Initial) | b. SSN | c. RANK | d. DATE OF RANK (Year, Month, Day) | e. BRANCH | f. DESIGNATED SPECIALTIES | PMOS (WO)

g. UNIT, ORG., STATION, ZIP CODE OR APO, MAJOR COMMAND | h. REASON FOR SUBMISSION

i. PERIOD COVERED: FROM (Year, Month, Day) THRU (Year, Month, Day) | j. RATED MONTHS | k. NONRATED CODES | l. NO. OF ENCL | m. RATED OFFICER COPY (Check one and date): 1. Given to Officer; 2. Forwarded to Officer; Date | n. PSB INITIAL | o. CMD CODE | p. PSB CODE

PART II - AUTHENTICATION (Rated officer's signature verifies officer has seen completed OER Parts I-VII and the admin data is correct)

a. NAME OF RATER (Last, First, MI) | SSN | RANK | POSITION | SIGNATURE | DATE

b. NAME OF INTERMEDIATE RATER (Last, First, MI) | SSN | RANK | POSITION | SIGNATURE | DATE

c. NAME OF SENIOR RATER (Last, First, MI) | SSN | RANK | POSITION | SIGNATURE | DATE

SENIOR RATER'S ORGANIZATION | BRANCH | SENIOR RATER TELEPHONE NUMBER | E-MAIL ADDRESS

d. This is a referred report, do you wish to make comments? ☐ ☐ Yes, comments are attached ☐ No | e. SIGNATURE OF RATED OFFICER | DATE

PART III - DUTY DESCRIPTION

a. PRINCIPAL DUTY TITLE | b. POSITION AOC/BR

c. SIGNIFICANT DUTIES AND RESPONSIBILITIES. REFER TO PART IVa, DA FORM 67-9-1

PART IV - PERFORMANCE EVALUATION - PROFESSIONALISM (Rater)

CHARACTER Disposition of the leader: combination of values, attributes, and skills affecting leader actions

a. ARMY VALUES (Comments mandatory for all "NO" entries. Use PART Vb.)

	Yes	No		Yes	No
1. HONOR: Adherence to the Army's publicly declared code of values			5. RESPECT: Promotes dignity, consideration, fairness, & EO		
2. INTEGRITY: Possesses high personal moral standards; honest in word and deed			6. SELFLESS-SERVICE: Places Army priorities before self		
3. COURAGE: Manifests physical and moral bravery			7. DUTY: Fulfills professional, legal, and moral obligations		
4. LOYALTY: Bears true faith and allegiance to the U.S. Constitution, the Army, the unit, and the soldier					

b. LEADER ATTRIBUTES / SKILLS / ACTIONS: First, mark "YES" or "NO" for each block. Second, choose a total of six that best describe the rated officer. Select one from ATTRIBUTES, two from SKILLS (Competence), and three from ACTIONS (LEADERSHIP). Place an "X" in the appropriate numbered box with optional comments in PART Vb. Comments are mandatory in Part Vb for all "No" entries.

b.1. ATTRIBUTES (Select 1) Fundamental qualities and characteristics	1. MENTAL (YES / NO) Possesses desire, will, initiative, and discipline	2. PHYSICAL (YES / NO) Maintains appropriate level of physical fitness and military bearing	3. EMOTIONAL (YES / NO) Displays self-control; calm under pressure
b.2 SKILLS (Competence) (Select 2) Skill development is part of self-development; prerequisite to action	1. CONCEPTUAL (YES / NO) Demonstrates sound judgment, critical/creative thinking, moral reasoning	2. INTERPERSONAL (YES / NO) Shows skill with people: coaching, teaching, counseling, motivating and empowering	3. TECHNICAL (YES / NO) Possesses the necessary expertise to accomplish all tasks and functions
	4. TACTICAL Demonstrates proficiency in required professional knowledge, judgment, and warfighting (YES / NO)		

b.3. ACTIONS (LEADERSHIP) (Select 3) Major activities leaders perform: influencing, operating, and improving

INFLUENCING Method of reaching goals while operating / improving	1. COMMUNICATING (YES / NO) Displays good oral, written, and listening skills for individuals / groups	2. DECISION-MAKING (YES / NO) Employs sound judgment, logical reasoning and uses resources wisely	3. MOTIVATING (YES / NO) Inspires, motivates, and guides others toward mission accomplishment
OPERATING Short-term mission accomplishment	4. PLANNING (YES / NO) Develops detailed, executable plans that are feasible, acceptable, and suitable	5. EXECUTING (YES / NO) Shows tactical proficiency, meets mission standards, and takes care of people/resources	6. ASSESSING (YES / NO) Uses after-action and evaluation tools to facilitate consistent improvement
IMPROVING Long-term improvement in the Army its people and organizations	7. DEVELOPING (YES / NO) Invests adequate time and effort to develop individual subordinates as leaders	8 BUILDING (YES / NO) Spends time and resources improving teams, groups and units; fosters ethical climate	9. LEARNING (YES / NO) Seeks self-improvement and organizational growth; envisioning, adapting and leading change

c. APFT: DATE: HEIGHT: WEIGHT:

d. JUNIOR OFFICER DEVELOPMENT - MANDATORY YES OR NO ENTRY FOR RATERS OF LTs AND WO1s. YES NO NA

WERE DEVELOPMENTAL TASKS RECORDED ON DA FORM 67-9-1a AND QUARTERLY FOLLOW-UP COUNSELINGS CONDUCTED?

DA FORM 67-9, OCT 97 REPLACES DA FORM 67-8, 1 SEP 79, WHICH IS OBSOLETE, 1 OCT 97 USAPA V2.00

NAME | SSN | PERIOD COVERED

PART V - PERFORMANCE AND POTENTIAL EVALUATION (Rater)

a. EVALUATE THE RATED OFFICER'S PERFORMANCE DURING THE RATING PERIOD AND HIS/HER POTENTIAL FOR PROMOTION

☐ OUTSTANDING PERFORMANCE, MUST PROMOTE ☐ SATISFACTORY PERFORMANCE, PROMOTE ☐ UNSATISFACTORY PERFORMANCE, DO NOT PROMOTE ☐ OTHER *(Explain)*

b. COMMENT ON SPECIFIC ASPECTS OF THE PERFORMANCE AND POTENTIAL FOR PROMOTION. REFER TO PART III, DA FORM 67-9 AND PART IVa, b, AND c DA FORM 67-9-1.

c. IDENTIFY ANY UNIQUE PROFESSIONAL SKILLS OR AREAS OF EXPERTISE OF VALUE TO THE ARMY THAT THIS OFFICER POSSESSES. FOR ARMY COMPETITIVE CATEGORY CPT THROUGH LTC, ALSO INDICATE A POTENTIAL CAREER FIELD FOR FUTURE SERVICE.

PART VI - INTERMEDIATE RATER

PART VII - SENIOR RATER

a. EVALUATE THE RATED OFFICER'S PROMOTION POTENTIAL TO THE NEXT HIGHER GRADE

☐ BEST QUALIFIED ☐ FULLY QUALIFIED ☐ DO NOT PROMOTE ☐ OTHER *(Explain below)*

I currently senior rate ______________ officer(s) in this grade

A completed DA Form 67-9-1 was received with this report and considered in my evaluation and review ☐ YES ☐ NO *(Explain in c)*

b. POTENTIAL COMPARED WITH OFFICERS SENIOR RATED IN SAME GRADE (OVERPRINTED BY DA)

☐ ABOVE CENTER OF MASS (Less than 50% in top box; Center of Mass if 50% or more in top box)

☐ CENTER OF MASS

☐ BELOW CENTER OF MASS RETAIN

☐ BELOW CENTER OF MASS DO NOT RETAIN

c. COMMENT ON PERFORMANCE/POTENTIAL

d. LIST 3 FUTURE ASSIGNMENTS FOR WHICH THIS OFFICER IS BEST SUITED. FOR ARMY COMPETITIVE CATEGORY CPT THROUGH LTC, ALSO INDICATE A POTENTIAL CAREER FIELD FOR FUTURE SERVICE.

c. LIST YOUR SIGNIFICANT CONTRIBUTIONS

SIGNATURE AND DATE

PART V - RATER AND/OR INTERMEDIATE RATER *(Review and comment on Part IVa, b, and c above).*

Insure remarks are consistent with your performance and potential evaluation on DA Form 67-9

a. RATER COMMENTS (Optional)

SIGNATURE AND DATE (Mandatory)

b. INTERMEDIATE RATER COMMENTS (Optional)

SIGNATURE AND DATE (Mandatory)

DATA REQUIRED BY THE PRIVACY ACT *(U.S.C. 552a)*

1. **AUTHORITY:** Sec 301 Title 5 USC; Sec 3012 Title 10

2. **PURPOSE:** DA Form 67-9, Officer Evaluation Report, serves as the primary source of information for officer personnel management decisions. DA Form 67-9-1, Officer Evaluation Support Form, serves as a guide for the rated officer's performance and development, enhances the accomplishment of the organization mission, and provides additional performance information to the rating chain. DA Form 67-9-1a, Junior Officer Developmental Support Form, serves as a common framework for Junior Officer Development and standardizes Junior Officer counseling.

3. **ROUTINE USE:** DA Form 67-9 will be maintained in the rated officer's Official Military Personnel File (OMPF) and Career Management Individual File (CMIF). A copy will be provided to the rated officer either directly or forwarded to the rated officer. DA Form 67-9-1 and DA Form 67-9-1a are for organizational use only and will be returned to the rated officer after review by the rating chain.

4. **DISCLOSURE:** Disclosure of the rated officer's SSN (Part I, DA Form 67-9) is voluntary. However, failure to verify the SSN may result in a delayed or erroneous processing of the officer's OER. Disclosure of the information in Part IV, DA Form 67-9-1 is voluntary. However, failure to provide the information requested will result in an evaluation of the rated officer without the benefits of that officer's comments. Should the rated officer use the Privacy Act as a basis not to provide the information requested in Part IV, the Support Form will contain the rated officer's statement to that effect and be forwarded through the rating chain in accordance with AR 623-105.

DA Form 67-9-1, page 1 of 2

OFFICER EVALUATION REPORT SUPPORT FORM

For use of this form, see AR 623-105; the proponent agency is ODCSPER

Read Privacy Act Statement on Reverse before Completing this form

PART I - RATED OFFICER IDENTIFICATION

NAME OF RATED OFFICER (Last, First, MI)	RANK	ORGANIZATION

PART II - RATING CHAIN - YOUR RATING CHAIN FOR THE EVALUATION PERIOD IS:

RATER	NAME	RANK	POSITION
INTERMEDIATE RATER	NAME	RANK	POSITION
SENIOR RATER	NAME	RANK	POSITION

PART III - VERIFICATION OF FACE-TO-FACE DISCUSSION

MANDATORY RATER / RATED OFFICER INITIAL FACE-TO-FACE COUNSELING ON DUTIES, RESPONSIBILITIES AND PERFORMANCE OBJECTIVES FOR THE CURRENT RATING PERIOD TOOK PLACE ON ____________ (Date) Rated Officer Initials ________ Rater Initials ________ Senior Rater Initials (Review) ________

PERIODIC RATER / RATED OFFICER FOLLOW-UP FACE-TO-FACE COUNSELINGS:

Dates ________ Rated Officer Initials ________ Rater Initials ________ Senior Rater Initials (Review) ________

PART IV - RATED OFFICER *(Complete a, b, and c below for this rating period)*

PRINCIPAL DUTY TITLE	POSITION AOC / BR

a. STATE YOUR SIGNIFICANT DUTIES AND RESPONSIBILITIES

b. INDICATE YOUR MAJOR PERFORMANCE OBJECTIVES

OFFICER RECORD BRIEF (DA Pam 600-8)

CON NO | BRIEF DATE | BASIC/CON BR | BR DTL EXPIRES | COMPONENT | SSN | NAME

SECTION I - ASSIGNMENT INFORMATION

OVERSEAS DUTY

YR MO RTN	COUNTRY	MONTHS	TCS	NUMBER OF OS TOURS	
				SHORT	LONG
				DROS	DEROS
				CONUS DEPARTURE DATE	

DATE DEPENDENTS ARRIVED OS

SPECIALTY/MOS DATA

ADDITIONAL SSI MOS

ASI DATA

SPECIALITIES

PREV DSGNATED SPEC1

PREV DSGNATED SPEC2

CONTROL SPECIALTY

PROJECTED SPECIALTY

FAO GEOG AREA

AVIATOR/GUNNERY QUALIFICATIONS

ASED

TOFDC AS OF

PILOT STATUS		AIRCRAFT	QUAL	AIRCRAFT	QUAL	AIRCRAFT	QUAL	AIRCRAFT	QUAL
INST CERT									
S/T COURSES									
RATING DATE		GUNNERY SYSTEMS							

SECTION II - SECURITY DATA

SCTY CLEARANCE

COMPL DATE OF SCTY INVES

TYPE/COMD OF SCTY INVES

SECTION V - FOREIGN LANGUAGE

LANGUAGE	READ	LISTEN

DLAT

SECTION VI - MILITARY EDUCATION

COURSE	MEL / YEAR

SECTION III - SERVICE DATA

PEBD	CURRENT PPN	EAD CURRENT TOUR
BASIC DATE OF APT	BASIC YR GP	SOURCE OF ORIG APT
MO DAYS AFCS	MO AFS	TYPE OF ORIG APT
CURR SVC AGRMT/EXPR DATE		DATE OF PROJ/MAND RET

	2LT-WO1	1LT-CW2	CPT-CW3	MAJ-CW4
TDOR				
PDOR				

	LTC	COL	BG	MG
TDOR				
PDOR				
TDOR	LTG		GEN	

SECTION VII - CIVILIAN EDUCATION

CEL

INSTITUTION

DISCIPLINE DEG YR

INSTITUTION

DISCIPLINE DEG YR

INSTITUTION

DISCIPLINE DEG YR

SECTION VIII - AWARDS AND DECORATIONS

SECTION IV - PERSONAL/FAMILY DATA

DATE OF BIRTH	BIRTHPLACE
COUNTRY OF CIT	SEX/RACE
NO DEPENDENT ADULTS/CHILDREN	RELIGION
MARITAL STATUS	SPOUSE BIRTHPLACE/CIT
PULHES/DATE	HEIGHT/WEIGHT

HOME OF RECORD AT EAD

MAILING ADDRESS

SECTION X - REMARKS

SECTION IX - ASSIGNMENT HISTORY

DATE OF AVAILABILITY | DATE OF LAST PCS | DATE OF LAST OER | ORG ZIP CODE

ASGT	FROM DATE	MO	UNIT NO	ORGANIZATION	STATION	LOC	COMD	DUTY TITLE	DMOS
PROJ									
CURRENT									
1ST PREV									
2ND PREV									
3RD PREV									
4TH PREV									
5TH PREV									
6TH PREV									
7TH PREV									
8TH PREV									
9TH PREV									
10TH PREV									
11TH PREV									
12TH PREV									
13TH PREV									
14TH PREV									
15TH PREV									
16TH PREV									
17TH PREV									
18TH PREV									
19TH PREV									

DA FORM 4037, JUL 82

PREVIOUS EDITIONS OF THIS FORM ARE OBSOLETE

For detailed explanations of data items, see Procedure 5-1 DA Pam 600-8.

REQUEST FOR ARMY/AMERICAN COUNCIL ON EDUCATION REGISTRY TRANSCRIPT

For use of this form, see AR 621-5; the proponent agency is ODCSPER.

DATA REQUIRED BY THE PRIVACY ACT OF 1974

AUTHORITY:	10 USC, Section 4302.
ROUTINE USES:	Upon initiation of individual.
PRINCIPAL PURPOSE(S):	To enable the Army/American Council on Education Registry Transcript System *(AARTS)* to access its computerized files, retrieve data, and produce a transcript for forwarding to individual or other addressee designated by the individual. Use of Social Security Number is necessary to make positive identification of individual and records.
DISCLOSURE:	Voluntary. Failure to provide required information will complicate, delay, and/or prevent administrative actions needed to produce the transcript and forward it to desired addressee.
ELIGIBLES:	(1) ONLY Regular Army *(RA)* enlisted soldiers and veterans whose Basic Active Service Dates *(BASD)* fall on or after 1 October 1981. (2) ONLY Army National Guard *(ARNG)* enlisted soldiers and veterans on active rolls as of 1 January 1993 whose Basic Pay Entry Dates *(BPED)* fall on or after 1 October 1981.

MAIL TO: AARTS Operation Center, Fort Leavenworth, KS 66027-5073
FAX TO: AARTS Operation Center, Commercial (913) 684-2011 or Autovon 552-2011

1. SSN
2. NAME *(Last First, MI) (List in parentheses other last names used)*
3. ENLISTED RANK
4. DATE OF BIRTH
5. SIGNATURE
6. ENLISTED STATUS
   - ☐ ACTIVE
   - ☐ VETERAN
   - ☐ ARMY NATIONAL GUARD

7.a. ENLISTMENT DATE *(BASD OR BPED)* *(Must be on or after 1 Oct 81)*

b. ETS DATE

8. FOR YOUR PERSONAL COPY/ARMY EDUCATION RECORD

a. SEND TO

b. MAILING ADDRESS *(Include ZIP Code + 4)*

9. FOR YOUR OFFICIAL COPY

a. SEND TO *(REGISTRAR, PERSONNEL OFFICER, OR EMPLOYER)*

b. NAME OF COLLEGE, EMPLOYER, OR OTHER RECIPIENT

c. MAILING ADDRESS *(Include ZIP Code + 4)*

10. ***FOR OFFICIAL USE ONLY***

a. RESEARCH RECORD

b. CROSS REF ID #

c. DATA ENTRY

	INITIATOR	ID #		FICE CODE	ID#
A			C		
S			O		

DA FORM 5454-R, JUL 93 PREVIOUS EDITIONS ARE OBSOLETE USAPPC V3.10

ARMY/AMERICAN COUNCIL ON EDUCATION
REGISTRY TRANSCRIPT

ISSUED: 09/25/2000 ** INSTITUTIONAL COPY **

S A M P L E

EVALUATIONS
UNIV OF MARYLAND
UNIT 29216
APO AE 09102

NAME: DOE JANE Q
SSN: 999-99-9991
RANK: SERGEANT
STATUS: INACTIVE
SERVICE: 3 YEARS, 8 MONTHS

AARTS ID: 2000-CN012516

MILITARY COURSE COMPLETIONS

COURSE: BASIC COMBAT TRAINING
(BASIC TRAINING)
US ARMY TRAINING CENTER FT JACKSON, SC

ACE GUIDE NUMBER AR-2201-0399

DATES TAKEN: 03/01/1997-04/30/1997 **ARMY COURSE NUMBER:** 750-BT

DESCRIPTION:

UPON COMPLETION OF THE COURSE, THE RECRUIT WILL BE ABLE TO DEMONSTRATE GENERAL KNOWLEDGE OF MILITARY ORGANIZATION AND CULTURE, MASTERY OF INDIVIDUAL AND GROUP COMBAT SKILLS INCLUDING MARKSMANSHIP AND FIRST AID, ACHIEVEMENT OF MINIMAL PHYSICAL CONDITIONING STANDARDS, AND APPLICATION OF BASIC SAFETY AND LIVING SKILLS IN AN OUTDOOR ENVIRONMENT. INSTRUCTION INCLUDES LECTURES, DEMONSTRATIONS, AND PERFORMANCE EXERCISES IN BASIC MILITARY CULTURE/SUBJECTS, INCLUDING MARKSMANSHIP, PHYSICAL CONDITIONING, FIRST AID, AND OUTDOOR ADAPTATION/LIVING SKILLS.

ACE CREDIT RECOMMENDATION:

IN THE LOWER-DIVISION BACCALAUREATE/ASSOCIATE DEGREE CATEGORY, 1 SEMESTER HOUR IN PERSONAL PHYSICAL CONDITIONING, 1 IN OUTDOOR SKILLS PRACTICUM, 1 IN MARKSMANSHIP, AND 1 IN FIRST AID (7/95).

COURSE: MEDICAL SPECIALIST
ACADEMY OF HEALTH SCIENCE FT SAM HOUSTON, TX

ACE GUIDE NUMBER AR-0709-0032

DATES TAKEN: 05/01/1997-07/15/1997 **ARMY COURSE NUMBER:** 300-91B10

DESCRIPTION:

UPON COMPLETION OF THE COURSE, THE STUDENT WILL BE ABLE TO MEASURE AND RECORD VITAL SIGNS, PRACTICE MEDICAL-SURGICAL ASEPSIS AND INFECTION CONTROL, PERFORM CPR AND FIRST AID, ADMINISTER OXYGEN THERAPY, PERFORM PATIENT ASSESSMENT, CONTROL BLEEDING, CARE FOR PATIENTS WITH SHOCK, IMMOBILIZE ORTHOPEDIC INJURIES, ADMINISTER IMMUNIZATIONS AND INJECTIONS, INITIATE AND MAINTAIN INTRAVENOUS THERAPY, AND MANAGE PATIENTS WITH ARTIFICIAL AIRWAYS. STUDENTS COMPLETE NATIONAL REGISTRY EXAM FOR EMERGENCY MEDICAL TECHNICIAN BASIC CERTIFICATION DURING THE COURSE. COURSE COVERS BLEEDING AND SHOCK, SPLINTING, MEDICAL EMERGENCIES, INVASIVE SKILLS, INJECTION, INTRAVENOUS INFUSION AND BLOOD DRAWING, NURSING SKILLS, ANATOMY AND PHYSIOLOGY, ACUTE MINOR ILLNESS, AND RECORD KEEPING. METHODOLOGY INCLUDES LECTURES, DEMONSTRATIONS, PRACTICAL EXERCISES, AND FIELD TRAINING IN CPR.

"THE FEDERAL FAMILY EDUCATIONAL RIGHTS AND PRIVACY ACT (FERPA) PROHIBITS FURTHER RELEASE OF THIS RECORD WITHOUT THE WRITTEN CONSENT OF THE INDIVIDUAL"

ACE CREDIT RECOMMENDATION:

IN THE LOWER-DIVISION BACCALAUREATE/ASSOCIATE DEGREE CATEGORY, 1 SEMESTER HOUR IN ANATOMY AND PHYSIOLOGY, 4 IN EMERGENCY MEDICAL TECHNOLOGY, 2 IN EMERGENCY MEDICAL TECHNICIAN CLINICAL EXPERIENCE, 2 IN BASIC NURSING SKILLS, AND 1 IN NURSING CLINICAL EXPERIENCE (12/99).

COURSE: PRIMARY LEADERSHIP DEVELOPMENT
NCO ACADEMY FT CARSON, CO

ACE GUIDE NUMBER
AR-2201-0253

DATES TAKEN: 01/02/2000-02/03/2000 **ARMY COURSE NUMBER:** 640-10-PLDC

DESCRIPTION:

UPON COMPLETION OF THE COURSE, THE STUDENT WILL BE ABLE TO PERFORM ALL BASIC TASKS RELATED TO NONCOMMISSIONED OFFICER LEADERSHIP RESPONSIBILITIES. THIS COURSE INCLUDES INTRODUCTORY MATERIAL ON PERSONAL LEADERSHIP, COMMUNICATIONS, RESOURCE MANAGEMENT, AND TRAINING MANAGEMENT. THE LEADERSHIP BLOCK INCLUDES MATERIAL ON LEADER CHARACTERISTICS, ETHICS, PROBLEM SOLVING, STYLES OF LEADERSHIP, HUMAN MOTIVATION, PERSONAL COUNSELING, AND THE EXERCISE OF AUTHORITY. EMPHASIS IS PLACED ON TEACHING TO TEACH AND ON LEADING PERSONNEL IN A SMALL-UNIT ENVIRONMENT. SUBSTANTIAL ATTENTION IS GIVEN TO PRACTICAL APPLICATION IN A FIELD-BASED ENVIRONMENT.

ACE CREDIT RECOMMENDATION:

IN THE LOWER-DIVISION BACCALAUREATE/ASSOCIATE DEGREE CATEGORY, 2 SEMESTER HOURS IN PRINCIPLES OF SUPERVISION AND 2 IN MILITARY SCIENCE (6/95).

TEST SCORES

COLLEGE LEVEL EXAMINATION PROGRAM (CLEP) - SUBJECT

- SE551 *MANAGEMENT INFORMATION SYSTEMS*

  DATE: 03/03/2000 SCORE: 54 Subscore A: Subscore B:

  ACE RECOMMENDED PASSING SCORE: 046 ACE RECOMMENDED CREDIT: AS FOLLOWS

  3 CREDITS UPPER LEVEL BACCALAUREATE

DANTES SUBJECT STANDARDIZED TESTS (DSST)

- *SH530* *HUMAN RESOURCE MANAGEMENT*

  DATE: 12/12/1999 SCORE: 62

  ACE RECOMMENDED PASSING SCORE: 048 ACE RECOMMENDED CREDIT: 03 SH

MILITARY EXPERIENCE

MILITARY OCCUPATIONAL SPECIALTIES HELD: 91B10 PRIMARY 07/1997-01/2000
91B20 PRIMARY 2/2000-PRESENT
91B10 DUTY

SQT (THRU OCT 1991)/SDT (NOV 1991 THRU FEB 1995) TAKEN: NONE

MILITARY OCCUPATIONAL SPECIALTY GROUP: 91B **ACE GUIDE NUMBER:**

TITLE: MEDICAL SPECIALIST MOS 91B-006

DESCRIPTION OF 91B10:

ADMINISTERS EMERGENCY TREATMENT IN THE FIELD; ASSISTS WITH OUTPATIENT AND INPATIENT CARE; MAINTAINS HEALTH RECORDS AND CLINICAL FILES; ASSISTS WITH NURSING CARE OF PATIENTS INCLUDING MEDICAL EXAMINATIONS; TAKES AND RECORDS TEMPERATURE, PULSE, RESPIRATION, AND BLOOD PRESSURE; APPLIES AND REMOVES SURGICAL, WOUND, OR SKIN DRESSINGS; COLLECTS AND PREPARES SPECIMENS FOR ANALYSIS; ASSISTS IN CLINIC AND DISPENSARY; PERFORMS ROUTINE ADMISSION TESTS; ADMINISTERS IMMUNIZATIONS; ASSISTS WITH TREATMENT OF PATIENTS WITH COMMON DISEASES; PROVIDES EMERGENCY MEDICAL CARE; SURVEYS AND SORTS CASUALTIES; DETERMINES REQUIREMENTS FOR AND ADMINISTERS EMERGENCY TREATMENT; ASSISTS WITH TRIAGE OF MASS CASUALTIES; PERFORMS DUTIES RELATED TO EMERGENCY CARE.

ACE CREDIT RECOMMENDATION FOR 91B10:

CREDIT MAY BE GRANTED ON THE BASIS OF AN INDIVIDUALIZED ASSESSMENT OF THE STUDENT (11/96).

"THE FEDERAL FAMILY EDUCATIONAL RIGHTS AND PRIVACY ACT (FERPA) PROHIBITS FURTHER RELEASE OF THIS RECORD WITHOUT THE WRITTEN CONSENT OF THE INDIVIDUAL"

MILITARY OCCUPATIONAL SPECIALTY GROUP: 91B **ACE GUIDE NUMBER:**

TITLE: MEDICAL SPECIALIST MOS 91B-006

DESCRIPTION OF 91B20:

ABLE TO PERFORM THE DUTIES REQUIRED FOR SKILL LEVEL 10; ADMINISTERS EMERGENCY AND ROUTINE MEDICAL TREATMENT TO CASUALTIES, INCLUDING MINOR SURGERY AND IV FLUID ADMINISTRATION; ASSISTS WITH OUTPATIENT CARE AND SUPERVISES OUTPATIENT FACILITIES UNDER THE SUPERVISION OF A PHYSICIAN, NURSE, OR PHYSICIAN'S ASSISTANT; ESTABLISHES PRIORITIES FOR MEDICAL EMERGENCY CARE; STABILIZES PATIENTS AND ACCOMPANIES THEM TO MEDICAL FACILITIES; PROVIDES GUIDANCE AND SUPERVISION TO SUBORDINATES.

ACE CREDIT RECOMMENDATION FOR 91B20:

IN THE LOWER-DIVISION BACCALAUREATE/ASSOCIATE DEGREE CATEGORY, 3 SEMESTER HOURS IN FUNDAMENTALS OF NURSING, 3 IN BASIC MEDICAL EMERGENCY PROCEDURES, 3 FOR FIELD EXPERIENCE IN MEDICAL EMERGENCY PROCEDURES, 5 IN CLINICAL EXPERIENCE, AND 3 IN PERSONNEL SUPERVISION. (NOTE: THIS RECOMMENDATION FOR SKILL LEVEL 20 IS VALID FOR THE DATES 8/94-2/95 ONLY) (11/96).

ADDITIONAL SKILL IDENTIFIERS: NONE

SPECIAL QUALIFICATION IDENTIFIERS: NONE

EXPLANATION OF RECORD

The Army/American Council on Education Registry Transcript System (AARTS) only contains information for the following:

a. Regular Army enlisted soldiers and veterans with basic active service dates falling on or after October 1, 1981.
b. Army National Guard enlisted soldiers and veterans on the active rolls January 1, 1993 with pay entry basic dates/basic active service dates falling on or after October 1, 1981.
c. Army Reserve enlisted soldiers and veterans on the active rolls April 1, 1997 with basic active service dates falling on or after October 1, 1981.

ALL information in the AARTS database comes directly from official sources: the Army Personnel Command, National Guard Bureau, US Army Reserve Command, Army Training Requirements and Resources System, American College Testing Program, Regents College Examination Program, and Educational Testing Service.

TRANSCRIPT INQUIRIES

Please address transcript inaccuracy reports, inquiries, and requests to:

MANAGER
AARTS OPERATIONS CENTER
415 MCPHERSON AVE
FT LEAVENWORTH KS 66027-1373

FIELD INQUIRY LINE: 1-866-297-4427 or (913) 684-3269
FAX: (913) 684-2011
E-MAIL: aarts@leavenworth.army.mil
WEB PAGE: http://www-leav.army.mil/AARTS

CREDIT RECOMMENDATIONS

Credit recommendations do not include grades and are always listed in semester hours.

CERTIFICATION

The American Council on Education (ACE) certifies all AARTS transcripts. The descriptions and college credit recommendations on the transcripts are extracted from the ACE "Guide to the Evaluation of Educational Experiences in the Armed Services". Please address questions concerning the ACE descriptions, credit recommendations, or pending evaluations to:

MILITARY REGISTRIES COORDINATOR
CENTER FOR ADULT LEARNING AND EDUCATIONAL CREDENTIALS
AMERICAN COUNCIL ON EDUCATION
ONE DUPONT CIRCLE NW SUITE 250
WASHINGTON DC 20036-1193
(202) 939-9475
WEB PAGE: http://www.acenet.edu

TERMINOLOGY

"INSTITUTIONAL COPY" designates an official transcript sent directly from the AARTS Operations Center to schools or employers.
"PERSONAL COPY" designates an unofficial transcript.

ACRONYMS AND DESCRIPTIONS

The Military Occupational Specialty (MOS) code encompasses both the specialty and the skill level held within an occupational group. For example: if the MOS code is 43E10, the specialty is contained in the 43E group (Parachute Rigger) and skill level (10) is the first level in that group.

The Special Qualification Identifier (SQI) designates a qualification such as Instructor (H) and Parachutist (P).

An Additional Skill Identifier (ASI) identifies specialized skills closely related to an MOS. For example, Master Fitness Trainer (P5) and Special Forces Military Free Fall Operations (W8).

The Skill Qualification Test (SQT) and Self-Development Test (SDT) measure proficiency in the primary MOS; the Army's passing score is 60 for the SQT and 70 for the SDT. These tests were not administered for secondary or duty MOS's. As of February 1995, MOS skill qualification tests were discontinued. Refer to the ACE Guide, Volume 2, for additional information.

REQUEST PERTAINING TO MILITARY RECORDS

Please read instructions on the reverse. If more space is needed, use plain paper.

PRIVACY ACT OF 1974 COMPLIANCE INFORMATION. The following information is provided in accordance with 5 U.S.C. 552a(e)(3) and applies to this form. Authority for collection of the information is 44 U.S.C. 2907, 3101, and 3103, and E.O. 9397 of November 22, 1943. Disclosure of the information is voluntary. The principal purpose of the information is to assist the facility servicing the records in locating and verifying the correctness of the requested records or information to answer your inquiry. Routine uses of the information as established and published in accordance with 5 U.S.C.a(e)(4)(D) include the transfer of relevant information to appropriate Federal, State, local, or foreign agencies for use in civil, criminal, or regulatory investigations or prosecution. In addition, this form will be filed with the appropriate military records and may be transferred along with the record to another agency in accordance with the routine uses established by the agency which maintains the record. If the requested information is not provided, it may not be possible to service your inquiry.

SECTION I—INFORMATION NEEDED TO LOCATE RECORDS (Furnish as much as possible)

1. NAME USED DURING SERVICE (*Last, first, and middle*)	2. SOCIAL SECURITY NO.	3. DATE OF BIRTH	4. PLACE OF BIRTH

5. ACTIVE SERVICE, PAST AND PRESENT (For an effective records search, it is important that ALL service be shown below)

BRANCH OF SERVICE (*Also, show last organization, if known*)	DATES OF ACTIVE SERVICE		*Check one*		SERVICE NUMBER DURING THIS PERIOD
	DATE ENTERED	DATE RELEASED	OFFI-CER	EN-LISTED	

6. RESERVE SERVICE, PAST OR PRESENT *If "none," check here* ▶ ☐

a. BRANCH OF SERVICE	b. DATES OF MEMBERSHIP		c. *Check one*		d. SERVICE NUMBER DURING THIS PERIOD
	FROM	TO	OFFI-CER ☐	EN-LISTED ☐	

7. NATIONAL GUARD MEMBERSHIP (*Check one*): ☐ a. ARMY ☐ b. AIR FORCE ☐ c. NONE

d. STATE	e. ORGANIZATION	f. DATES OF MEMBERSHIP		g. *Check one*		h. SERVICE NUMBER DURING THIS PERIOD
		FROM	TO	OFFI-CER ☐	EN-LISTED ☐	

8. IS SERVICE PERSON DECEASED ☐ YES ☐ NO *If "yes," enter date of death.*

9. IS (WAS) INDIVIDUAL A MILITARY RETIREE OR FLEET RESERVIST ☐ YES ☐ NO

SECTION II—REQUEST

1. EXPLAIN WHAT INFORMATION OR DOCUMENTS YOU NEED; OR, CHECK ITEM 2; OR, COMPLETE ITEM 3

2. IF YOU ONLY NEED A STATEMENT OF SERVICE *check here* ☐

3. *LOST SEPARATION DOCUMENT REPLACEMENT REQUEST (Complete a or b, and c.)*

☐	a. REPORT OF SEPARATION (*DD Form 214 or equivalent*)	YEAR ISSUED	*This contains information normally needed to determine eligibility for benefits. It may be furnished only to the veteran, the surviving next of kin, or to a representative with veteran's signed release (item 5 of this form).*
☐	b. DISCHARGE CERTIFICATE	YEAR ISSUED	*This shows only the date and character at discharge. It is of little value in determining eligibility for benefits. It may be issued only to veterans discharged honorably or under honorable conditions; or, if deceased, to the surviving spouse.*

c. EXPLAIN HOW SEPARATION DOCUMENT WAS LOST

4. EXPLAIN PURPOSE FOR WHICH INFORMATION OR DOCUMENTS ARE NEEDED

6. REQUESTER

a. IDENTIFICATION (*check appropriate box*)

☐ Same person identified in Section I ☐ Surviving spouse

☐ Next of kin (relationship) ______

☐ Other (specify)

b. SIGNATURE (*see instruction 3 on reverse side*) DATE OF REQUEST

5. RELEASE AUTHORIZATION, IF REQUIRED (*Read instruction 3 on reverse side*)

I hereby authorize release of the requested information/documents to the person indicated at right (item 7).

VETERAN SIGN HERE ▶ ______

(If signed by other than veteran show relationship to veteran.)

7. *Please type or print clearly* — COMPLETE RETURN ADDRESS

Name, number and street, city, State and ZIP code

TELEPHONE NO. (*include area code*) ▶

INSTRUCTIONS

1. Information needed to locate records. Certain identifying information is necessary to determine the location of an individual's record of military service. Please give careful consideration to and answer each item on this form. If you do not have and cannot obtain the information for an item, show "NA," meaning the information is "not available." Include as much of the requested information as you can. This will help us to give you the best possible service.

2. Charges for service. A nominal fee is charged for certain types of service. In most instances service fees cannot be determined in advance. If your request involves a service fee you will be notified as soon as that determination is made.

3. Restrictions on release of information. Information from records of military personnel is released subject to restrictions imposed by the military departments consistent with the provisions of the Freedom of Information Act of 1967 (as amended in 1974) and the Privacy Act of 1974. A service person has access to almost any information contained in his own record. The next of kin, if the veteran is deceased, and Federal officers for official purposes, are authorized to receive information from a military service or medical record only as specified in the above cited Acts. Other requesters must have the release authorization, in item 5 of the form, signed by the veteran or, if deceased, by the next of kin. Employers and others needing proof of military service are expected to accept the information shown on documents issued by the Armed Forces at the time a service person is separated.

4. Location of military personnel records. The various categories of military personnel records are described in the chart below. For each category there is a code number which indicates the address at the bottom of the page to which this request should be sent. For each military service there is a note explaining approximately how long the records are held by the military service before they are transferred to the National Personnel Records Center, St. Louis. Please read these notes carefully and make sure you send your inquiry to the right address. Please note especially that the record is not sent to the National Personnel Records Center as long as the person retains any sort of reserve obligation, whether drilling or non-drilling.

(If the person has two or more periods of service within the same branch, send your request to the office having the record for the last period of service.)

5. Definitions for abbreviations used below:
NPRC—National Personnel Records Center PERS—Personnel Records
TDRL—Temporary Disability Retirement List MED—Medical Records

SERVICE	NOTE: (See paragraph 4 above.)	CATEGORY OF RECORDS — WHERE TO WRITE ADDRESS CODE	▼
AIR FORCE (USAF)	*Except for TDRL and general officers retired with pay, Air Force records are transferred to NPRC from Code 1, 90 days after separation and from Code 2, 150 days after separation.*	Active members (includes National Guard on active duty in the Air Force), TDRL, and general officers retired with pay.	1
		Reserve, retired reservist in nonpay status, current National Guard officers not on active duty in Air Force, and National Guard released from active duty in Air Force.	2
		Current National Guard enlisted not on active duty in Air Force.	13
		Discharged, deceased, and retired with pay.	14
COAST GUARD (USCG)	*Coast Guard officer and enlisted records are transferred to NPRC 7 months after separation.*	Active, reserve, and TDRL members.	3
		Discharged, deceased, and retired members *(see next item).*	14
		Officers separated before 1/1/29 and enlisted personnel separated before 1/1/15.	6
MARINE CORPS (USMC)	*Marine Corps records are transferred to NPRC between 6 and 9 months after separation.*	Active, TDRL, and Selected Marine Corps Reserve members.	4
		Individual Ready Reserve and Fleet Marine Corps Reserve members.	5
		Discharged, deceased, and retired members *(see next item).*	14
		Members separated before 1/1/1905.	6
ARMY (USA)	*Army records are transferred to NPRC as follows: Active Army and Individual Ready Reserve Control Groups: About 60 days after separation. U.S. Army Reserve Troop Unit personnel: About 120 to 180 days after separation.*	Reserve, living retired members, retired general officers, and active duty records of current National Guard members who performed service in the U.S. Army before 7/1/72.*	7
		Active officers (including National Guard on active duty in the U.S. Army).	8
		Active enlisted (including National Guard on active duty in the U.S. Army) and enlisted TDRL.	9
		Current National Guard officers not on active duty in the U.S. Army.	12
		Current National Guard enlisted not on active duty in the U.S. Army.	13
		Discharged and deceased members *(see next item).*	14
		Officers separated before 7/1/17 and enlisted separated before 11/1/12.	6
		Officers and warrant officers TDRL.	8
NAVY (USN)	*Navy records are transferred to NPRC 6 months after retirement or complete separation.*	Active members (including reservists on duty)—PERS and MED	10
		Discharged, deceased, retired (with and without pay) less than six months, TDRL, drilling and nondrilling reservists — PERS ONLY	10
		Discharged, deceased, retired (with and without pay) less than six months, TDRL, drilling and nondrilling reservists — MED ONLY	11
		Discharged, deceased, retired (with and without pay) more than six months *(see next item)*—PERS & MED	14
		Officers separated before 1/1/03 and enlisted separated before 1/1/1886—PERS and MED	6

**Code 12 applies to active duty records of current National Guard officers who performed service in the U.S. Army after 6/30/72.*
Code 13 applies to active duty records of current National Guard enlisted members who performed service in the U.S. Army after 6/30/72.

ADDRESS LIST OF CUSTODIANS (BY CODE NUMBERS SHOWN ABOVE)—Where to write / send this form for each category of records

1	Air Force Manpower and Personnel Center Military Personnel Records Division Randolph AFB, TX 78150-6001	5	Marine Corps Reserve Support Center 10950 El Monte Overland Park, KS 66211-1408	8	USA MILPERCEN ATTN: DAPC-MSR 200 Stoval Street Alexandria, VA 22332-0400	12	Army National Guard Personnel Center Columbia Pike Office Building 5600 Columbia Pike Falls Church, VA 22041
2	Air Reserve Personnel Center Denver, CO 80280-5000	6	Military Archives Division National Archives and Records Administration Washington, DC 20408	9	Commander U.S. Army Enlisted Records and Evaluation Center Ft. Benjamin Harrison, IN 46249-5301	13	The Adjutant General *(of the appropriate State, DC, or Puerto Rico)*
3	Commandant U.S. Coast Guard Washington, DC 20593-0001	7	Commander U.S. Army Reserve Personnel Center ATTN: DARP-PAS 9700 Page Boulevard St. Louis, MO 63132-5200	10	Commander Naval Military Personnel Command ATTN: NMPC-036 Washington, DC 20370-5036	14	National Personnel Records Center (Military Personnel Records) 9700 Page Boulevard St. Louis, MO 63132
4	Commandant of the Marine Corps (Code MMRB-10) Headquarters, U.S. Marine Corps Washington, DC 20380-0001			11	Naval Reserve Personnel Center New Orleans, LA 70146-5000		

*U.S. Government Printing Office: 1992 — 342-196/50147

Occupational Title Index

This index is designed to provide access to the occupation exhibits in this volume. The titles are listed in alphabetical order. When the occupational title is found, note the exhibit ID number to its right. Locate that number in the proper occupation exhibit section.

Occupations are grouped by military service, using the following prefixes: **MOS:** Army; **CGR** and **CGW**: Coast Guard; **MCE:** Marine Corps; and **NER, LDO,** and **NWO:** Navy.

Keyword Index

This index is designed to provide access to the courses listed in the course exhibit sections of this volume. Course titles are arranged alphabetically under keywords extracted directly from the titles. For example, the keyword *Dental* is followed by all course titles containing that word.

To use this index:

- Identify a word (or group of words) that appears to be unique or descriptive.
- Locate the keyword(s) in this index.
- If the keyword or the course title cannot be found, identify another descriptive word in the title and try again.
- When the course title is found, note the exhibit ID number to the right of the title. Locate that number in the course exhibit section.

Course exhibits are grouped by military service, using the following prefixes: **AF:** Air Force; **AR:** Army; **CG:** Coast Guard; **DD:** Department of Defense; **MC:** Marine Corps; and **NV:** Navy.

Course Number Index

This index is designed to provide access to courses listed in the course exhibit section of this volume. Military course numbers are listed in alphanumeric order. When the official military course number is found, note the exhibit ID number to its right. Locate that number in the proper course exhibit section.

Course exhibits are grouped by military service, using the following prefixes: **AF:** Air Force; **AR:** Army; **CG:** Coast Guard; **DD:** Department of Defense; **MC:** Marine Corps; and **NV:** Navy.

Appendix F

Conversion of Army Enlisted and Warrant Officer Military Occupational Specialties to Department of Defense Occupational Codes

This appendix is designed to aid the user in determining the Army formal course number when only the MOS code is known.

The number that the Army assigns to a formal course for enlisted personnel often consists of a three-digit Department of Defense (DoD) occupational code followed by the five-digit code for the MOS and skill level associated with the course, e.g., 300-91C10. In order to use the Course Number Index to locate a specific course exhibit ID number, the user must know the entire official Army course number. However, on some soldiers' records the entire course number is not given; instead, the military record may list the course number as only the MOS and skill level, without the DoD code.

For example, the Army course number for the course that trains persons for MOS 13E (Cannon Fire Direction Specialist) is 250-13E10. When a soldier's record incompletely lists the course number as 13E10, thc user can refer to MOS 13E in this appendix to find the DoD occupational code, 250. The user can then turn to 250-13E10 in the Course Number Index to find the ID number of the course exhibit, which in this example is AR-1715-0844.

MOS	DOD CODE
00B	433
00C	830
00D	920
00E	501
00F	012
00G	600
00H	801
00J	800
00R	501
00U	500,762
00Z	521
01B	440
01C	440
01D	440
01E	440
01F	440
01G	440
01H	311,440,496
01K	500
01L	541
02B	450
02C	450
02D	450
02E	450
02F	450
02G	450
02H	450
02J	450
02K	450
02L	450
02M	450
02N	450
02P	450
02Q	450
02R	450
02S	450
02T	450
02U	450
02Z	450
03B	562
03C	562
03D	562
03Z	562
04B	241
04C	241
05B	201
05C	201
05D	231
05E	201
05F	201
05G	231
05H	231
05K	231
09C	920
09D	912
09R	912
09S	911
09T	911
09W	911
11B	010
11C	010
11D	250
11E	020
11F	250
11G	010
11H	010
11M	010
11Z	010
12A	030
12B	030
12C	030
12D	030
12E	030,431
12F	030
12Z	030
13A	041
13B	041
13C	250
13E	250
13F	250
13M	042
13N	042,043,221
13P	043
13R	221
13T	221,250
13W	250
13Y	041
13Z	041
14D	043
14J	043
14R	043
14S	043
15B	043
15D	043
15E	043
15F	042
15J	043,250
15Z	043
16B	043
16C	043
16D	043
16E	043
16F	043
16G	043
16H	250
16J	221
16K	221
16L	041
16P	043
16R	041,043
16S	043
16T	043
16Z	043

MOS	DOD CODE
17A	221
17B	221
17C	241,250,412
17D	412
17E	030
17K	221
17L	221,233
17M	221
17Z	221
18B	011
18C	011
18D	011
18E	011
18F	011
18Z	011
19D	250
19E	020
19F	020
19G	020
19H	020
19J	020
19K	020
19L	020
19Z	020
21G	121
21L	121
21M	121
21R	121
21S	121
21T	121
21U	121
22G	121
22K	121
22L	121
22M	121
22N	121
23N	104
23Q	121
23R	632
23S	104
23T	104
23U	104
23V	121
23W	104
24B	104
24C	121,632
24D	121
24E	121
24F	121
24G	104,121,632
24H	104,121,122
24J	104
24K	104
24L	121
24M	121,632
24N	121,632
24P	104
24Q	121
24R	121,632
24S	121
24T	632
24U	121
24V	121
24W	121
25D	150
25G	150
25H	150
25J	150
25K	150
25L	150
25P	400
25Q	414
25R	400,570
25S	400
25Z	150,400,570
26B	104
26C	104
26D	104
26E	104
26F	198
26H	104
26K	104
26L	101
26M	104
26N	104
26P	198
26Q	201
26R	201
26T	191,570
26V	101
26W	104
26Y	102
27B	121
27C	121
27D	121,122
27E	121
27F	101,121
27G	121
27H	121
27J	104
27K	104
27L	632
27M	121,611,632
27N	104
27P	121
27Q	122
27T	121
27V	121
27X	121
27Z	121
28M	121
29E	101
29F	101,160,820
29G	150
29H	150
29J	160,822
29M	101,720
29N	622,822
29P	160,820
29S	160,820
29T	101,201
29U	150
29V	101,102
29W	101
29X	101
29Y	102
29Z	101
31B	101
31C	201
31D	101,260
31E	101
31F	260
31G	101
31J	160
31K	201
31L	101,621
31M	202
31N	101,202,260
31Q	201
31R	198
31S	160
31T	160
31U	101,160
31V	101
31W	101,260
31X	101
31Y	260
31Z	260
32A	101
32C	101
32D	101
32E	101
32F	160
32G	160
32H	101
32Z	101
33B	102
33C	102
33D	102
33F	102
33G	102
33M	102
33P	102
33Q	102
33R	102
33S	102
331	102
33V	102
33Y	102
33Z	102
34B	150
34C	150
34D	150
34E	150
34F	150
34G	113
34H	150
34J	150
34K	150
34L	150
34M	150
34T	150
34Y	113
34Z	150
35B	198

MOS	DOD CODE
35C	198
35D	198
35E	198
35F	140
35G	198,326
35H	198
35J	112
35K	102
35L	102
35M	102
35N	102
35P	102
35R	102
35S	198
35T	198
35U	198,326
35Y	198
36C	621
36D	621
36E	621
36G	622
36H	622
36L	622
36M	621
37F	243
38A	570
39B	198,201,726
39C	104
39D	150
39E	198
39G	150
39K	150
39L	113,150
39T	150
39V	101,150
39W	101
39X	101
39Y	113
41B	670
41C	670
41E	198
41F	198
41G	198
41H	198
41J	670
41K	690
42C	304
42D	331
42E	311,323
42F	331
43A	760
43E	860
43J	760
43K	760
43L	760
43M	760
44A	700
44B	704
44C	701
44D	701
44E	702
44K	704
44Z	700
45A	640
45B	641
45D	642,643
45E	643
45G	113
45J	646
45K	643
45L	642
45M	646
45N	643
45P	643
45R	643
45T	611,643
45Z	640
46A	632
46D	121
46L	631
46N	631
46Q	570,631
46R	570
46Z	570
51A	710
51B	712
51C	710
51D	710
51E	790
51F	710
51G	491
51H	710
51J	720
51K	720
51L	720
51M	495,780
51N	720
51P	720
51Q	491
51R	721
51T	413
51Z	710
52A	721
52B	662
52C	662
52D	662
52E	662
52F	662
52G	721
52H	661
52J	661
52K	661
52L	661
52M	661
52X	610
53B	750
53C	750
54A	030
54B	030,494
54C	030
54D	690
54E	494
54F	030
54Z	494
55A	645
55B	645
55C	645
55D	431
55G	644
55R	551
55X	645
55Z	645
57A	850
57C	760
57D	710
57E	840
57F	492
57G	850
57H	822
61A	062
61B	062
61C	652
61D	062
61E	652
61F	704
61Z	062
62B	612
62C	633
62D	730
62E	713,730
62F	713,730
62G	713,730
62H	713,730
62J	713,730
62K	730
62L	730
62M	730
62N	713,730
63A	610
63B	610
63C	611
63D	611
63E	611
63F	610
63G	610
63H	610,611
63J	690
63K	690
63N	611
63R	611
63S	610
63T	611
63W	610
63Y	611
63Z	610
64C	811
64Z	811
65A	690
65B	690
65C	662
65D	690
65E	690

MOS	DOD CODE
65F	662
65G	850
65H	850
65J	812
65K	553,812
65Z	812
66G	600
66H	600
66J	600
66N	600
66R	600
66S	600
66T	600
66U	600
66V	600
66X	600
66Y	600
67A	600
67B	600
67C	600
67F	600
67G	600
67H	600
67M	600
67N	600
67P	600
67R	600
67S	600
67T	600
67U	600
67V	600
67W	600
67X	600
67Y	600
67Z	600
68A	600
68B	601
68C	601
68D	602
68E	602
68F	602
68G	603
68H	602
68J	102,602,646
68R	602
68L	102
68M	602,646
68N	102
68P	102,602
68Q	102
68R	102
68X	646
71B	510
71C	511
71D	512
71E	511,512
71F	515
71G	340,513
71H	500
71L	510
71M	561
71N	514,553
71P	517,556

MOS	DOD CODE
71Q	570
71R	570
71S	243
71T	510
71U	531
72B	580
72C	580
72D	580
72E	260,580
72F	580
72G	260,580
72H	260,580
73C	542
73D	541
73Z	542
74B	531
74C	260,531
74D	531
74E	531
74F	532
74G	531
74Z	531
75B	500
75C	500
75D	500
75E	500
75F	500
75Z	500
76A	551
76C	551
76D	551
76J	340,551
76L	400
76N	551
76P	551
76Q	551
76R	551
76S	551
76T	551
76U	551
76V	551
76W	821
76X	551,822
76Y	552
76Z	551
77F	821
77L	491
77W	720
79D	501
81A	413
81B	413
81C	411
81D	411
81E	414
81F	790
81Q	411
81Z	412
82A	412
82B	412

MOS	DOD CODE
82C	412
82D	412
82E	412
83A	740
83D	740
83E	740
83F	740
83Z	740
84H	400
84C	400
84D	191
84E	400
84F	400
84G	400
84T	570
84Z	570
88H	822
88K	062
88L	652
88M	811
88N	553
88P	690
88Q	690
88R	690
88S	662
88T	850
88U	812
88V	812
88W	553
88X	812
88Y	062
88Z	811
91A	300
91B	300
91C	300
91D	301
91E	330
91F	302
91G	302
91H	304
91J	303
91K	303
91L	303
91M	311,800
91N	300,311
91P	313
91Q	312
91R	321
91S	322
91T	321
91U	300,301
91V	300
91W	300,311,313
91X	322,324
91Y	300,323
91Z	300
92A	551
92B	311
92D	491
92E	311
92Y	551
92Z	552

MOS	DOD CODE
93B	250
93C	222
93D	102,191
93E	420
93F	420
93H	222
93J	222
93K	222
93T	222
93P	556
94A	800
94B	800
94C	800
94D	800
94F	325,800
94Z	800
95B	830
95C	831
95D	832
96B	243
96D	242
96F	243
96H	233
96R	243
96U	243
96Z	243
97B	244
97C	244
97D	244
97E	241,243
97F	120
97G	231
97Z	231,243,244
98B	232
98C	232
98D	231
98G	231
98H	231
98J	233
98K	231
98Z	230
001A	9E
002A	9A
003A	9B
004A	9E
005A	9E
006A	9E
007A	9E
008A	9E
011A	6H
021A	8E
031A	7N
041A	8E
051A	6G
100A	2C
100K	2C
100C	2C
100D	2C
100E	2C
100K	2C
100Q	2B
100R	2B
130A	4F
130B	4F
131A	4C
131B	4F
132A	5B
140A	4F
140B	4F
140C	4F
140D	4F
140E	4F
150A	2G
151A	4D
152B	2C
152C	2C
152D	2C
152F	2C
152G	2C
153A	2C
153B	2C
153C	2C
153D	2C
154A	2C
154B	2C
154C	2C
155A	2B
155D	2B
155E	2B
156A	2B
160A	4D
180A	2E
201A	5B
202A	5B,6I
210A	4A
211A	4C
212A	4F
213A	4L
214E	4F
214G	4F
215A	4M
215B	4M
215C	8G
215D	4N
221B	4F
222R	4B
222C	4B
223B	4F
224B	4F
224D	4F
225B	4F
225D	4F
250A	4C
250B	4C
251A	7E
252A	4B
256A	4C
257A	4B
260A	4E
271A	4F
285A	4C
286A	4C
287A	4B
290A	4C
310A	4A
311A	7H
350B	3A
350D	3A
350L	3A
351B	3C
351C	3A
351D	3A
351E	3A
352C	3B
352D	3B
352G	3B
352H	3B
352J	3B
352K	3B
353A	4C
401A	4N
411A	4E
420A	7C
420C	7N
420D	8E
421A	4E
441A	4L
500A	8C
510A	4H
550A	5F
600A	6H
621A	4L
630A	4L
630B	4L
630C	4L
630L	4L
630E	4L
640A	6G
670A	6H
711A	7C
712A	7A
713A	5F
741A	7E
761A	8B
762A	8B
811A	4M
821A	4M
833A	8G
841A	4N
880A	8C
881A	4H
910A	4E
911A	4E
912A	4F
913A	4E
914A	4L
915A	4L
915B	4L
915C	4L
915D	4L
915E	4L
916A	4F
917A	4F
918A	4B
919A	4L
920A	8B
920B	8B
921A	4N
922A	8E,8G
951A	7H
961A	3A
962A	3A
964A	3A

MOS	DOD CODE	MOS	DOD CODE	MOS	DOD CODE
971A	3C	982A	3B	985A	3B
972A	3A	983A	3B	986A	3B
973A	3A	984A	3B	988A	3B

REQUEST FOR COURSE RECOMMENDATION

The applicant for credit must fill out one form for *each* service school course completed. The institutional official is responsible for verifying from official military records that the student completed the entire course, and for submitting the form to the Center for Lifelong Learning, American Council on Education, One Dupont Circle, Washington, DC 20036-1193. ATTN: Military Evaluations. *Please Print.*

1. *Exact* course title *(do not abbreviate)* ______________________________

____

2. Service branch offering the course: ☐ Air Force ☐ Army ☐ Coast Guard ☐ Department of Defense ☐ Marine Corps ☐ Navy

3. Name of service school attended: ______________________________

4. Location (installation, state): ______________________________

5. Length of course *(in weeks):* ______________________________

6. Dates of attendance: From:____________ day/month/year To:____________ day/month/year

7. Official military course number: ______________________________

8. MOS/AFSC/Rating: ______________________________

9. Course was designed for: ☐ Warrant Officers ☐ Officer Candidates ☐ Commissioned Officers ☐ Enlisted Personnel ☐ Aviation Cadets ☐ Noncommissioned Officers

10. Rank or rating upon completion of the course: ______________________________

11. Please give some indication of subjects studied in course:

____

____

____
NAME OF STUDENT

____

DO NOT WRITE IN THIS SPACE
STAFF USE

____
SIGNATURE OF COLLEGE OFFICIAL

____
NAME OF COLLEGE OFFICIAL

____
TITLE

____
INSTITUTION

____
STREET

____
CITY STATE ZIP CODE

____
AREA CODE NUMBER EXT.

REQUEST FOR ARMY ENLISTED AND WARRANT OFFICER MOS EXHIBITS

This form should be used to request MOS exhibits from 1/90 to the present that are not found in the *Guide* or the *Handbook*. Clearly mark the MOS designation (e.g., 91B10), the MOS title, and the name of the applicant. Submit the form to The Center for Lifelong Learning, American Council on Education, One Dupont Circle, Washington, DC 20036-1193, Attention: Military Evaluations.

MOS	MOS Title Please print; do not abbreviate.	Name of Applicant

DO NOT WRITE IN THIS SPACE
STAFF USE

SIGNATURE OF OFFICIAL

NAME OF OFFICIAL

TITLE

INSTITUTION OR ORGANIZATION

STREET

CITY STATE ZIP CODE

AREA CODE NUMBER EXT.

Please retain file copies of any occupation recommendations received from the Advisory Service.